Weather of U.S. Cities

Other Gale Books About Weather

Climate Normals for the U.S.
Climates of the States
Storm Data, 1970-1979
Sun, Weather, and Climate
Sunrise and Sunset Tables
Understanding Climatic Change
The Weather Almanac
Weather and Climate Modification
Weather Atlas of the United States

Write for descriptive brochure

Weather of U.S. Cities

A compilation of weather records for 281 key cities and weather observation points in the United States and its island territories to provide insight into their diverse climates and normal weather tendencies. Supplies narrative statements about the various cities' climates, and complements the descriptions with statistical cumulations to quantify "normals, means, and extremes" for each.

THIRD EDITION

Volume 2

City Reports, Montana - Wyoming

James A. Ruffner
and
Frank E. Bair
Editors

GALE RESEARCH COMPANY
BOOK TOWER • DETROIT, MICHIGAN 48226

James A. Ruffner and Frank E. Bair, *Editors*

Mary Beth Trimper, *Production Manager*
Michael Vargas, *Production Associate*
Arthur Chartow, *Art Director*

Frederick G. Ruffner, *Chairman*
J. Kevin Reger, *President*
Dedria Bryfonski, *Publisher*
Ellen T. Crowley, *Associate Editorial Director*
Ann Evory, *Director, Indexes and Dictionaries Division*

Library of Congress Cataloging-in-Publication Data

Weather of U.S. cities.

    Subtitle: A compilation of weather records for 281
key cities and weather observation points in the United
States and its island territories to provide insight
into their diverse climates and normal weather tendencies,
supplies narrative statements about the various cities'
climates, and complements the descriptions with
statistical cumulations to quantify "normals, means,
and extremes" for each.
    Contents: v. 1. City reports, Alabama-Missouri —
v. 2. City reports, Montana-Wyoming.
    1. Urban climatology—United States—Handbooks,
manuals, etc. 2. United States—Climate—Handbooks,
manuals, etc. I. Ruffner, James A. II. Bair, Frank E.
III. Gale Research Company. IV. Title: Weather of US
cities.
QC983.W393   1987   551.6973   87-11869
ISBN 0-8103-2102-5 (set)

BIBLIOGRAPHIC NOTE: The narratives and statistical tables are based
on *Local Climatological Data, Annual Summary with Comparative Data,*
National Oceanic and Atmospheric Administration, 1985, or in the case
of inactive stations, the most recent revisions.

Contents

Volume 1: Alabama through Missouri

Introduction ix
How to read these reports 1

City Reports, state-by-state

Alabama
Birmingham (Airport) 8
Birmingham (City Office) 12
Huntsville 16
Mobile 20
Montgomery 24

Alaska
Anchorage 28
Annette 32
Barrow 36
Barter Island 40
Bethel 44
Bettles 48
Big Delta 52
Cold Bay 56
Fairbanks 60
Gulkana 64
Homer 68
Juneau 72
King Salmon 76
Kodiak 80
Kotzebue 84
McGrath 88
Nome 92
St. Paul Island .96
Talkeetna 100
Valdez 104
Yakutat 108

Arizona
Flagstaff 112
Phoenix 116
Tucson 120
Winslow 124
Yuma 128

Arkansas
Fort Smith 132
Little Rock 136
North Little Rock 140

California
Bakersfield 144

Bishop 148
Blue Canyon 152
Eureka 156
Fresno 160
Long Beach 164
Los Angeles (Airport) 168
Los Angeles (Civic Center) 172
Mt. Shasta 176
Red Bluff 180
Sacramento 184
San Diego 188
San Francisco (Downtown) 192
San Francisco (International Airport) . .196
Santa Maria 200
Stockton 204

Colorado
Alamosa 208
Colorado Springs 212
Denver 216
Grand Junction 220
Pueblo 224

Connecticut
Bridgeport 228
Hartford 232

Delaware
Wilmington 236

District of Columbia
Washington (Dulles Int. Airport) 240
Washington (National Airport) 244

Florida
Apalachicola 248
Daytona Beach 252
Fort Myers 256
Jacksonville 260
Key West 264
Miami 268
Orlando 272
Pensacola 276
Tallahassee 280
Tampa 284
West Palm Beach 288

Georgia
Athens 292
Atlanta 296
Augusta 300
Columbus. 304
Macon. 308
Savannah 312

Hawaii
Hilo. 316
Honolulu 320
Kahului. 324
Lihue. 328

Idaho
Boise 332
Lewiston. 336
Pocatello 340

Illinois
Cairo. 344
Chicago 348
Moline 352
Peoria 356
Rockford 360
Springfield 364

Indiana
Evansville. 368
Fort Wayne 372
Indianapolis 376
South Bend. 380

Iowa
Des Moines. 384
Dubuque. 388
Sioux City. 392
Waterloo 396

Kansas
Concordia 400
Dodge City 404
Goodland 408
Topeka 412
Wichita. 416

Kentucky
Jackson. 420
Lexington 424

Louisville 428
Paducah 432

Louisiana
Baton Rouge 436
Lake Charles 440
New Orleans. 444
Shreveport 448

Maine
Caribou 452
Portland 456

Maryland
Baltimore 460

Massachusetts
Blue Hill Observatory (Milton) 464
Boston 468
Worcester 472

Michigan
Alpena 476
Detroit 480
Flint 484
Grand Rapids. .488
Houghton Lake 492
Lansing. 496
Marquette. 500
Muskegon. 504
Sault Sainte Marie 508

Minnesota
Duluth 512
International Falls. 516
Minneapolis-St. Paul 520
Rochester 524
Saint Cloud 528

Mississippi
Jackson. 532
Meridian. 536
Tupelo 540

Missouri
Columbia 544
Kansas City (Downtown Airport) 548
Kansas City (Int. Airport) 552
St. Louis. 556
Springfield 560

Volume 2: Montana through Wyoming

How to read these reports ix

Montana
Billings 564
Glasgow 568
Great Falls 572
Havre 576
Helena 580
Kalispell 584
Miles City 588
Missoula 592

Nebraska
Grand Island 596
Lincoln 600
Norfolk............................604
North Platte 608
Omaha 612
Omaha (North) 616
Scottsbluff 620
Valentine 624

Nevada
Elko 628
Ely...............................632
Las Vegas 636
Reno 640
Winnemucca 644

New Hampshire
Concord 648
Mt. Washington Observatory (Gorham).652

New Jersey
Atlantic City (Airport) 656
Atlantic City (State Marina)..........660
Newark............................664

New Mexico
Albuquerque 668
Clayton...........................672
Roswell...........................676

New York
Albany 680
Binghamton 684
Buffalo 688
New York City (Central Park)........692
New York City (JFK Int. Airport)......696

New York City (La Guardia Field).....700
Rochester 704
Syracuse 708

North Carolina
Asheville..........................712
Cape Hatteras 716
Charlotte 720
Greensboro 724
Raleigh............................728
Wilmington 732

North Dakota
Bismarck 736
Fargo 740
Williston 744

Ohio
Akron-Canton 748
Cincinnati.........................752
Cleveland 756
Columbus.........................760
Dayton 764
Mansfield 768
Toledo 772
Youngstown.......................776

Oklahoma
Oklahoma City 780
Tulsa..............................784

Oregon
Astoria 788
Burns 792
Eugene 796
Medford 800
Pendleton 804
Portland 808
Salem 812
Sexton Summit 816

Pacific Islands
Guam 820
Johnston..........................824
Koror 828
Kwajalein (Marshall Is.) 832
Majuro (Marshall Is.)..............836
Pago Pago, American Samoa 840
Ponape 844

Truk (Moen; Caroline Is.) 848
Wake 852
Yap 856

Pennsylvania
Allentown 860
Erie 864
Harrisburg 868
Philadelphia 872
Pittsburgh 876
Scranton/Wilkes-Barre (Avoca) 880
Williamsport 884

Rhode Island
Block Island 888
Providence 892

South Carolina
Charleston 896
Columbia 900
Greenville-Spartanburg 904

South Dakota
Aberdeen 908
Huron 912
Rapid City 916
Sioux Falls 920

Tennessee
Bristol-Johnson City-Kingsport 924
Chattanooga 928
Knoxville 932
Memphis 936
Nashville 940
Oak Ridge 944

Texas
Abilene 948
Amarillo 952
Austin 956
Brownsville 960
Corpus Christi 964
Dallas-Fort Worth 968
Del Rio 972
El Paso 976
Galveston 980
Houston 984
Lubbock 988
Midland-Odessa 992

Port Arthur 996
San Angelo 1000
San Antonio 1004
Victoria 1008
Waco 1012
Wichita Falls 1016

Utah
Milford 1020
Salt Lake City 1024

Vermont
Burlington 1028

Virginia
Lynchburg 1032
Norfolk 1036
Richmond 1040
Roanoke 1044

Washington
Olympia 1048
Quillayute Airport 1052
Seattle-Tacoma (Airport) 1056
Seattle (Urban Site) 1060
Spokane 1064
Stampede Pass 1068
Walla Walla 1072
Yakima 1076

West Indies
San Juan, Puerto Rico 1080

West Virginia
Beckley 1084
Charleston 1088
Elkins 1092
Huntington 1096

Wisconsin
Green Bay 1100
La Crosse 1104
Madison 1108
Milwaukee 1112

Wyoming
Casper 1116
Cheyenne 1120
Lander 1124
Sheridan 1128

HOW TO READ THESE REPORTS

Weather in various cities

The information about the weather history of key cities is easy to find. The reports were planned to be informative, yet simple to read. The terminology used and the various standard formats are explained below.

The climate of each city is first presented as a narrative description, prepared by a local climatologist, whose job puts him in an unequalled position to know the weather of the particular area as no one else would.

The narrative report

Typically, the report begins with a description of the local area in terms of terrain, water bodies, and other topographical features because these features exercise key influences on the local weather. They are usually the cause if an area's weather differs sharply from weather in areas only a few miles away. For example, if a lake is near a city it will always influence the city's weather and, in fact, may even create climatic differences from one part of the city to another. Mountains, swamps, even plowed fields exercise their influences on air masses as these masses move toward a city. The climatologist generally will describe these features and their climatic effect as the opening statement of the narrative report.

The report will usually discuss the range of temperatures in the city, rainfall tendencies, snowfall history, and other points. It will typically close with notes about the area's agricultural adaptability. The history of first fall frost and last spring freeze is usually described, along with suggestions about the types of crops for which the area is climatologically suited. This narrative may answer all of your questions without requiring that you study the tables of statistics.

Statistics

Two kinds of statistics are presented. The first type distills many years of history to give you a profile of the city's weather (e.g., NORMALS, MEANS AND EXTREMES). The second group offers data for individual years, going back thirty years to allow the user to see what variances have tended to occur, as well as to see to what extent weather conditions have repeated themselves.

There are several meteorological terms used in the table, many of which have been reduced to symbols. The following thirty notes are designed to clarify the information included in the tables on an item-by-item basis:

TAMPA, FLORIDA City report (page 1)

Tampa is on west central coast of the Florida Peninsula. Very near the Gulf of Mexico at the upper end of Tampa Bay, land and sea breezes modify the subtropical climate. Major rivers flowing into the area are the Hillsborough, the Alafia, and the Little Manatee.

Winters are mild. Summers are long, rather warm and humid. Low temperatures are about 50 degrees in the winter and 70 degrees during the summer. Afternoon highs range from the low 70s in the winter to around 90 degrees from June through September. Invasions of cold northern air produce an occasional cool winter morning. Freezing temperatures occur on one or two mornings per year during December, January, and February. In some years no freezing temperatures occur. Temperatures rarely fail to recover to the 60s on the cooler winter days. Temperatures above the low 90s are uncommon because of the afternoon sea breezes and thunderstorms. An outstanding feature of the Tampa climate is the summer thunderstorm season. Most of the thunderstorms occur in the late afternoon hours from June through September. The resulting sudden drop in temperature from about 90 degrees to around 70 degrees makes for a pleasant change. Between a dry spring and a dry fall, some 30 inches of rain, about 60 percent of the annual total, falls during the summer months. Snowfall is very rare. Measurable snows under 1/2 inch have occurred only a few times in the last one hundred years.

A large part of the generally flat sandy land near the coast has an elevation of under 15 feet above sea level. This does make the area vulnerable to tidal surges. Tropical storms threaten the area on ...

HOW TO READ THESE REPORTS (continued)

1. **DATA** provides precise geographic location and elevation of weather stations.

2. **NORMAL** as applied to temperature, degree days, and precipitation refers to the value of that particular element averaged over the period from 1951-1980. When the station does not have continuous records from an instrument site with the same "exposure," a "difference factor" between the old site and the new site is used to adjust the observed values to a common series. The difference factor is determined from a period of simultaneous measurements. The base period is revised every ten years by adding the averages for the most recent decade and dropping them for the first decade of the former normals for 1951-1980. *Normal* does not refer to "normalcy" or "expectation," but only to the actual averages for a particular thirty-year period.

TAMPA, FLORIDA City report (page 2)

NORMALS, MEANS AN[...]

TABLE 1   TAM[...]

	JAN	FEB	MAR	APR
① LATITUDE: 27°58'N ③ LONGITUDE: 82°32' W ELEVATIO[...]				
② TEMPERATURE °F:				
Normals		70.0	71.0	76.2	81[...]
⑤ Daily Maximum		49.5	50.4	56.1	61[...]
⑥ Daily Minimum		59.8	60.7	66.2	7[...]
⑦ Monthly	④				
Extremes		84	88	91	
⑧ Record Highest	39	1975	1971	1949	
⑨ Year		21	24	29	
Record Lowest	39	1985	1958	1980	
Year					
NORMAL DEGREE DAYS:		228	186	87	
⑩ Heating (base 65°F)		66	68	124	
⑪ Cooling (base 65°F)				71	
⑫ % OF POSSIBLE SUNSHINE	38	64	66		
⑬ MEAN SKY COVER (tenths)					
Sunrise - Sunset	39	5.6	5.5	5.4	
⑭ MEAN NUMBER OF DAYS:					
Sunrise to Sunset	39	9.6	9.1	10.	
⑮ Clear	39	9.9	9.1	10.	
Partly Cloudy	39	11.5	10.1	10	
Cloudy					
Precipitation	39	6.4	6.8		
.01 inches or more					
Snow,Ice pellets	39	0.0	0.0		
1.0 inches or more					
Thunderstorms	39	1.0	1.6		
Heavy Fog Visibility	39	5.7	2.9		
1/4 mile or less					
Temperature °F	22	0.0	0.0		
⑯ Maximum	22	0.0	0.0		
90° and above					
⑰ 32° and below	22	2.3	0.8		
⑱ Minimum	22	0.0	0.0		
32° and below					
⑲ 0° and below					
⑳ AVG. STATION PRESS.(mb)	12	1019.9	1019.1		
㉑ RELATIVE HUMIDITY (%)		84	83		
Hour 01	22	86	85		
Hour 07 (Local Time)	22	59	5[...]		
Hour 13	22	73			
Hour 19	22				
PRECIPITATION (inches):		2.17	3.[...]		
Water Equivalent		8.02	7.[...]		
㉒ Normal	39	1948			
㉓ Maximum Monthly		T			
Year	39	1950			
㉔ Minimum Monthly		3.29			
Year	39	1953			
㉕ Maximum in 24 hrs					
Year					
Snow,Ice pellets	39	0.2			
Maximum Monthly		1977			
Year	39	0.2			
㉖ Maximum in 24 hrs		1977			
Year					
WIND	39	8.7			
㉗ Mean Speed (mph)		N			
Prevailing Direction					
through 1963					
㉘ Fastest Obs. 1 Min.	33	3[...]			
Direction	33	3[...]			
㉙ Speed (MPH)		195[...]			
Year					
Peak Gust	2	2			
Direction		19[...]			

HOW TO READ THESE REPORTS (continued)

3. (Note "a") **MEANS AND EXTREMES** are based on the period of years in which observations have been made under comparable conditions of instrument exposure. Data are included for dates through 1985 unless otherwise noted. The DATE OF AN EXTREME is the *most recent* one in cases of repeated occurences.

4. **LENGTH OF OBSERVATIONAL RECORD** for *Means and Extremes* is based on the length of January data for the present instrument site exposure (15 equal 15 years). The length of the record is through 1985. The table *does not* give the all time *high* or *low* value if it was recorded at a *different site* within the area. The *Mean* (or average) values for *Relative Humidity, Wind, Sunshine, Sky Condition,* and the *Mean Number of Days* with the various other weather conditions listed are also based on length of record noted in each instance, down through 1985. Check the first column for each of these rows to read this length of record for each item.

5. **AVERAGE of the HIGHEST TEMPERATURE (°F)** on each day of the month and year for the *period 1951-1980.* This value is obtained by taking the sum of the highest temperature for each day of the period (adjusted for site exposure if necessary) and dividing by the number of days included.

6. **AVERAGE of the LOWEST TEMPERATURE (°F)** on each day of the month and year for the *period 1951-1980.*

7. **AVERAGE of ALL DAILY TEMPERATURES (°F)** for the month and year for the *period 1951-1980*; computed as being the temperature one-half way between the average daily maximum and minimum values in items 5 and 6 above.

8. **EXTREMES—HIGHEST TEMPERATURE (°F)** ever recorded during any month at present site exposure.

9. **EXTREMES—LOWEST TEMPERATURE (°F)** ever recorded during any month at present site exposure.

10. **AVERAGE number of HEATING DEGREE DAYS** for each month and year for the *period 1951-1980.* The statistic is based on the amount that the *Daily Mean Temperature* falls below 65°F. Each degree of mean temperature below 65 is counted as one *Heating Degree Day.* If the *Daily Mean Temperature* is 65 degrees or higher, the *Heating Degree Day* value for that day is zero. Monthly and annual sums are calculated for each period and averaged over the appropriate thirty years of record to establish these "normal" values. Compare this with *Cooling Degree Days.*

11. **AVERAGE number of COOLING DEGREE DAYS** for each month and year for the *period 1951-1980.* The concept of this statistic is the mirror image of the concept of *Heating Degree Days* and is based on the amount that the *Daily Mean Temperature* exceeds 65°F. Each degree of mean temperature above 65 is counted as one *Cooling Degree Day.* If the *Daily Mean Temperature* is 65 degrees or below, the *Cooling Degree Day* is zero. PLEASE NOTE: *Heating and Cooling Degree Days* are calculated independently and do not cancel each other out. Both concepts are discussed at length in the paragraph headed *Energy Consumption Indices* which follows note 30.

12. **SUNSHINE**—The average percent of daytime hours subject to direct radiation from the sun at the present site. The percentage is given without regard for the intensity of sunshine. That is, thin clouds, light haze, or other minor obstructions to direct solar rays may be present but would not mitigate the full counting of an hour.

13. **VERTICAL OBSERVATION**—Average amount of daytime sky obscured by any type of cover expressed in tenths (e.g.,4.8 equal 4⁸/₁₀...or 48%).

14. **ACTIVITY LIMITING WEATHER**—Average number of days in month with specified weather conditions based on *present exposure.* An asterisk (*) indicates less than 1/2 day.

15. **CLOUDINESS**—Average number of days in month at the *present site* with various amounts of cloud cover. *Clear* indicates average daytime cloudiness of 0.3 or less; *partly cloudy* indicates average daytime cloudiness between 0.4 and 0.7; *cloudy* indicates average daytime cloudiness of 0.8 or more.

16. **VERY HOT DAYS**—Average number of days in month and year when the temperatures at the *present site* reached 90° or above (70°F or above at Alaskan stations).

17. **COLD DAYS**—Average number of days at the *present site* when the temperatures remained below 32°F at all times.

18. **FREEZING DAYS**—Average number of days at the *present site* when the temperature dropped to a minimum of 32°F or below.

19. **VERY COLD DAYS**—Average number of days at the *present site* when the minimum temperature was 0°F or below.

20. **AVERAGE STATION PRESSURE**—given in millibars.

		12												
(20)	AVG. STATION PRESS. (mb)	12	1019.9	1019.1	1017.6	1017.0	1015.5	1016.2	1017.6	1017.1	1015.4	1016.6	1018.5	1020.1	1017.6
(21)	RELATIVE HUMIDITY (%) Hour 01	22	84	83	82	82	81	84	85	87	86	85	85	84	84
	Hour 07	22	86	85	86	87	86	87	88	91	91	89	88	87	88
	Hour 13 (Local Time)	22	59	56	55	51	53	60	63	64	62	57	57	59	58
	Hour 19	22	73	69	67	62	62	68	73	76	75	71	74	74	70
	PRECIPITATION (inches): Water Equivalent														
(22)	-Normal		2.17	3.04	3.46	1.82	3.38	5.29	7.35	7.64	6.23	2.34	1.87	2.14	46.73
(23)	-Maximum Monthly	39	8.02	7.95	12.64	6.59	17.64	13.75	20.59	18.59	13.98	7.36	6.12	6.66	20.59
	-Year		1948	1963	1959	1957	1979	1974	1960	1949	1979	1952	1963	1950	JUL 1960
(24)	-Minimum Monthly	39	T	0.21	0.06	T	0.17	1.86	1.65	2.35	1.28	0.16	T	0.07	T
	-Year		1950	1950	1956	1981	1973	1951	1981	1952	1972	1979	1960	1984	APR 1981
(25)	-Maximum in 24 hrs	39	3.29	3.68	5.20	3.70	11.84	5.53	12.11	5.37	4.99	2.93	4.22	3.28	12.11
	-Year		1953	1981	1960	1951	1979	1974	1960	1949	1985	1985	1963	1969	JUL 1960
	Snow,Ice pellets														
	-Maximum Monthly	39	0.2	T	T										0.2
(26)	-Year		1977	1951	1980										JAN 1977
	-Maximum in 24 hrs	39	0.2	T	T										0.2
	-Year		1977	1951	1980										JAN 1977
	WIND:														
(27)	-Mean Speed (mph)	39	8.7	9.4	9.6	9.5	8.9	8.1	7.3	7.1	8.0	8.6	8.5	8.6	8.5
(28)	-Prevailing Direction through 1963		N	E	S	ENE	E	E	E	ENE	ENE	NNE	NNE	N	E
	Fastest Obs. 1 Min.														
(29)	-Direction	33	29	32	29	29	36	31	32	11	34	02	29	36	31
	-Speed (MPH)	33	35	50	43	37	46	67	58	38	56	38	40	45	67
	-Year		1959	1954	1956	1961	1958	1964	1963	1961	1960	1953	1963	1953	JUN 1964
	Peak Gust -Direction	2	N	NW	W	NW	E	NE	N	S	W	SW	N	NW	N
	-Speed (mph)	2	32	46	32	33		48	52	45	45	41	37	32	52
	-Date	2	1985	1984	1985			1985	1984	1985	1984	1985	1984	1984	1984

City report (page3)

TAMPA, FLORIDA

TABLE 2 — PRECIPITATION (inches)

TABLE 3 — AVERAGE TEMPERATURE (deg. F)

REFERENCE NOTES FOR TABLES 1, 2, 3 and 6

EXCEPTIONS

TABLES 2, 3, and 6

RECORD MEANS AND BEGINNING IN

GENERAL

T - TRACE AMOUNT
BLANK ENTRIES DENOTE MISSING/UNREPORTED DATA
# INDICATES A STATION OR INSTRUMENT RELOCATION

SPECIFIC

TABLE 1

(A) - LENGTH OF RECORD IN YEARS ALTHOUGH INDIVIDUAL MONTHS MAY BE MISSING
* LESS THAN .05

NORMALS — BASED ON THE 1951-1980 RECORD PERIOD
EXTREMES — DATES ARE THE MOST RECENT OCCURRENCE
WIND DIR. — NUMERALS SHOW TENS OF DEGREES CLOCKWISE FROM TRUE NORTH.
"00" INDICATES CALM
RESULTANT WIND DIRECTIONS ARE GIVEN TO WHOLE DEGREES.

City report (page 4)

TAMPA, FLORIDA

TABLE 4 — HEATING DEGREE DAYS Base 65 deg. F

TABLE 5 — COOLING DEGREE DAYS Base 65 deg. F

TABLE 6 — SNOWFALL (inches)

See Reference Notes, relative to all above tables, on preceding page.

HOW TO READ THESE REPORTS (continued)

21. AVERAGE RELATIVE HUMIDITY at various hours of the day. The time is expressed in terms of the 24 hour clock (00 is midnight, 06 is 6 a.m., 12 is noon, and 18 is 6 p.m.). Values are for present site only.

22. AVERAGE PRECIPITATION in inches of water equivalent for each month and year during the *period 1951-1980*. As in the other precipitation data, the values are expressed in inches of depth of the liquid water content of all forms of precipitation even if initially frozen. As in all "normal" values, when the station does not have continuous records from the same instrument site, a "ratio factor" between the new and old exposure is used to adjust the observed values to a common series.

(See also, *Precipitation*, under item 14, *Mean Number of Days*)

23. GREATEST PRECIPITATION in inches of water equivalent ever recorded during any month *at present site.*

24. LEAST PRECIPITATION in inches of water equivalent ever recorded during any month *at present site.*

25. GREATEST PRECIPITATION in inches of water equivalent ever recorded during any *day at present site.*

26. Summaries for **SNOW and ICE PELLETS** including *sleet* are similar to those for total precipitation. The values are expressed in inches of actual snow or ice fall. The water equivalent can be estimated roughly by using the rule-of-thumb that 10" of snow equals 1" of water.

27. WINDINESS—Average speed of wind is expressed in miles-per-hour without regard of direction.

28. PREVAILING WIND DIRECTION—The most common single wind direction without regard for wind speed or any minimum amount of persistence. The aggregate total of wind from the other directions may be very much greater than that from the "prevailing" direction. Direction is coded in two different ways; some reports use letters, some use numbers. When letters are used, they have the usual meaning, such as WSW indicating west-south-west. When numbers are used, they are given in tens of degrees clockwise from true north, so that 09 is 90° clockwise from north (*east*), 18 is 180° (*south*), 27 is 270° (*west*), and 36 is 360° (*north*). The statistic is based on *data through 1963 only.*

29. HIGH WIND—The greatest speed in miles per hour of any "mile" of wind passing the station. The accompanying direction and year of occurrence are also given. A mile of wind passing the station 1 minute has an average speed of 60 mph, 2 minutes—30 mph, 5 minutes—12 mph, etc. The wind cups of the particular instrument involved operate much like the wheel of a car in actuating the car's odometer. The instrument does not record the strength of individual wind gusts which usually last less than 20 seconds and which may be very much greater than the value given here. The fastest mile however does give some idea of the *extremes* of wind that can be encountered.

30. SPECIAL SYMBOLS that appear on many of the individual summaries:

   * **Less than one-half**
   T **Trace, an amount too small to measure.**
   - **Below zero temperatures are preceded by a minus sign.**

Weather averages year-by-year

PRECIPITATION refers to the inches of water equivalent in the total of all forms of liquid or frozen precipitation that fell during each month. *Snowfall* refers to the actual amount of snow in inches that fell during the month. T (trace) is a precipitation amount of less than 0.005 inches. (Note: in estimating the water equivalent of snow a ratio of 10" of snow equal 1" of water is customarily employed.)

AVERAGE TEMPERATURE equals the *average* of the *maximum* and *minimum temperatures* for each day of the month for the given year; afternoon temperatures were typically higher than these values and late night/early morning temperatures were typically *below* them.

ENERGY CONSUMPTION INDICES. See items 10 and 11 above. *Heating Degree Days* provide a well established index of *relative fuel consumption* for space heating in a given place—a month of 2000 *HDD* requires about twice the amount of space heating energy as one of 1000 *HDD*, while 100 *HDD* will require about the same fuel whether accumulated in 2 or 4 days. Regional differences in the *Heating Degree Day Index* are only partially useful in estimating comparative fuel requirements because the building construction and cultural expectations tend to be different in different parts of the country (e.g., subjective ideas of comfort in relation to temperature will vary). For example, the average standards of efficiency in heating equipment and insulation are generally lower in warmer climates so that fuel requirements tend not to decrease as rapidly as the *heating degree days* decrease.*

The complementary index of *Cooling Degree Days* provides only a rough guide to relative energy consumption in air conditioning. A proper air conditioning index will almost certainly require a factor for humidity variation and possibly factors for cloudiness or other weather variables. However, the *Cooling Degree Days Index* will have some usefulness in indicating relative outdoor comfort and relative indoor air conditioning requirements.

* See page 3 for discussion of Heating and Cooling Degree Days

City Reports
Montana—Wyoming

Billings, Montana, at an elevation of 3,100 to 3,600 feet above sea level, is situated in the borderline area between the Great Plains and the Rocky Mountains, and has a climate which takes on some of the characteristics of both regions. Its climate may be classified as semi—arid, but with irrigation and the favorable distribution of the precipitation, it is possible to raise a variety of crops in the area.

About a third of the annual precipitation falls during May and June, with June being the wettest month. The period of least precipitation is from November through February. These four months normally produce less than 20 percent of the annual precipitation. The heaviest snows occur during the spring and fall months when the temperature and moisture conditions are most favorable. Heavy snows of 6 inches or more also occur during November and December. The occurrence of thawing periods normally prevents the snow from accumulating to great depths on the ground. Thunderstorms are most frequent during the summer months. These storms are frequently accompanied by strong, gusty winds and occasionally by hail. Destructive hailstorms, however, are rather infrequent.

Winter is usually cold, though not extremely so, and generally affords several mild periods of a week to several weeks in length. The winter cold periods are ushered in by moderately strong north to northeast winds and snow. The coldest temperatures occur after the snow ends and the sky clears. True blizzard conditions are not observed very often in town, but in the surrounding rural areas, blizzard conditions may develop several times during the winter. Cold weather improves with the onset of moderate to strong southwest winds. This wind is sometimes a foehn condition (chinook), but is more often a drainage wind moving down the Yellowstone Valley which transports warmer air of Pacific origin to the area. Occasionally an open winter occurs when cold Arctic outbreaks pass far to the east and temperatures stay above zero degrees.

Spring brings a period of frequent and rapid fluctuations in the weather. It is usually cloudy and cool with frequent periods of rain and/or snow. As the season progresses, snows become less frequent until late May and June when rain is the rule. The last freezing temperatures in spring usually occur before mid—May though they have occurred as late as late June.

The summer season is characterized by warm days with abundant sunshine and low humidities. The nights are cool because of the altitude and the cool air drainage into the valley from the higher terrain. Seldom is there a protracted rainy spell during this season. Frequent thunderstorms bring threatening afternoon cloudiness but usually only small amounts of rain.

The first freezing temperatures of the fall season occur in late September, but they have been noted as early as late August. Over the years, the fall months have been about evenly distributed between cold, wet ones, and mild, dry, pleasant ones. The change to severe winter weather usually arrives after the middle of November. There have been years when the more severe type of winter weather have been delayed until late in December.

TABLE 1 NORMALS, MEANS AND EXTREMES

BILLINGS, MONTANA

LATITUDE: 45°48'N LONGITUDE: 108°32'W ELEVATION: FT. GRND 3567 BARO 03583 TIME ZONE: MOUNTAIN WBAN: 24033

	(a)	JAN	FEB	MAR	APR	MAY	JUNE	JULY	AUG	SEP	OCT	NOV	DEC	YEAR
TEMPERATURE °F:														
Normals														
-Daily Maximum		29.9	37.9	44.0	55.9	66.4	76.3	86.6	84.3	72.3	61.0	44.4	36.0	57.9
-Daily Minimum		11.8	18.8	23.6	33.2	43.3	51.6	58.0	56.2	46.5	37.5	25.5	18.2	35.4
-Monthly		20.9	28.4	33.8	44.5	54.9	64.0	72.3	70.3	59.4	49.3	35.0	27.1	46.6
Extremes														
-Record Highest	51	68	72	79	92	96	105	106	105	103	90	77	69	106
-Year		1953	1961	1978	1939	1936	1984	1937	1961	1983	1963	1983	1980	JUL 1937
-Record Lowest	51	-30	-38	-19	-5	14	32	41	40	22	2	-22	-32	-38
-Year		1937	1936	1951	1936	1954	1969	1972	1939	1984	1984	1959	1983	FEB 1936
NORMAL DEGREE DAYS:														
Heating (base 65°F)		1367	1025	967	612	318	111	9	27	214	487	900	1175	7212
Cooling (base 65°F)		0	0	0	0	0	81	235	191	46	0	0	0	553
% OF POSSIBLE SUNSHINE	46	48	54	61	60	60	64	76	75	68	61	46	44	60
MEAN SKY COVER (tenths)														
Sunrise - Sunset	46	7.1	7.2	7.2	7.1	6.6	5.9	4.2	4.3	5.2	5.8	6.8	6.8	6.2
MEAN NUMBER OF DAYS:														
Sunrise to Sunset														
-Clear	46	5.4	4.0	4.4	4.3	5.4	7.1	13.9	13.7	10.3	9.3	5.8	5.8	89.5
-Partly Cloudy	46	7.8	8.3	8.8	8.7	10.6	11.8	12.1	11.0	9.6	9.4	7.8	8.5	114.6
-Cloudy	46	17.7	16.0	17.8	17.0	14.9	11.1	5.0	6.2	10.1	12.3	16.4	16.6	161.2
Precipitation														
.01 inches or more	51	8.1	7.5	9.1	9.3	11.0	11.1	7.2	6.5	7.1	6.4	6.1	6.8	96.1
Snow, Ice pellets														
1.0 inches or more	43	3.5	2.5	3.5	2.2	0.4	0.*	0.0	0.0	0.5	1.2	2.2	2.8	18.8
Thunderstorms	46	0.*	0.*	0.1	1.1	4.1	7.2	7.6	5.7	1.8	0.2	0.0	0.*	28.1
Heavy Fog Visibility														
1/4 mile or less	38	1.4	2.3	1.9	2.5	1.3	0.7	0.3	0.3	1.2	1.9	2.3	1.7	17.8
Temperature °F														
-Maximum														
90° and above	26	0.0	0.0	0.0	0.*	0.3	3.3	12.9	11.1	1.9	0.*	0.0	0.0	29.7
32° and below	26	14.8	7.7	5.2	0.8	0.0	0.0	0.0	0.0	0.*	0.7	5.6	12.2	47.0
-Minimum														
32° and below	26	27.7	23.8	24.2	13.2	1.7	0.1	0.0	0.0	1.7	8.1	21.9	27.6	150.0
0° and below	26	9.2	2.9	1.3	0.0	0.0	0.0	0.0	0.0	0.0	0.0	1.3	4.9	19.6
AVG. STATION PRESS. (mb)	13	891.2	890.2	887.6	889.1	888.9	889.9	891.8	891.8	892.5	891.8	890.8	890.4	890.5
RELATIVE HUMIDITY (%)														
Hour 05	26	65	65	68	67	69	71	63	61	65	63	66	65	66
Hour 11	26	61	58	54	48	48	46	39	39	48	49	58	61	51
Hour 17 (Local Time)	26	57	51	46	40	42	40	31	30	38	42	54	58	44
Hour 23	26	64	62	62	58	59	59	49	46	54	56	62	63	58
PRECIPITATION (inches):														
Water Equivalent														
-Normal		0.97	0.71	1.05	1.93	2.39	2.07	0.85	1.05	1.26	1.16	0.85	0.80	15.09
-Maximum Monthly	51	2.35	1.77	2.70	4.42	7.71	7.64	3.12	3.50	4.99	3.80	2.34	2.00	7.71
-Year		1972	1978	1954	1955	1981	1944	1958	1965	1941	1971	1978	1973	MAY 1981
-Minimum Monthly	51	0.04	0.05	0.13	0.06	0.53	0.24	0.07	0.05	0.06	0.02	T	0.05	T
-Year		1941	1977	1936	1962	1937	1961	1976	1955	1964	1948	1954	1957	NOV 1954
-Maximum in 24 hrs	51	1.41	0.61	1.01	3.19	2.83	2.78	1.87	2.47	2.19	1.98	1.37	0.96	3.19
-Year		1972	1952	1973	1978	1952	1937	1958	1965	1966	1974	1959	1978	APR 1978
Snow, Ice pellets														
-Maximum Monthly	51	27.7	22.4	27.6	42.3	15.6	2.0			9.3	23.1	25.2	28.8	42.3
-Year		1963	1978	1935	1955	1981	1950			1984	1949	1978	1955	APR 1955
-Maximum in 24 hrs	47	16.6	9.0	10.5	23.7	15.3	2.0			7.5	11.2	15.3	13.7	23.7
-Year		1972	1944	1964	1955	1981	1950			1983	1980	1959	1978	APR 1955
WIND:														
Mean Speed (mph)	46	13.0	12.4	11.5	11.6	10.9	10.3	9.6	9.6	10.3	11.1	12.1	13.1	11.3
Prevailing Direction														
through 1963		SW	SW	SW	SW	NE	SW	SW	SW	SW	SW	SW	WSW	SW
Fastest Mile														
-Direction (!!!)	42	W	W	NW	NW	NN	NW	N	NW	NW	NW	NW	NW	NW
-Speed (MPH)	42	66	72	61	72	68	79	73	69	61	68	63	66	79
-Year		1953	1963	1956	1947	1939	1968	1947	1983	1949	1949	1948	1953	JUN 1968
Peak Gust														
-Direction (!!!)	2	W	W	W	NW	SE	S	NE	S	NW	NW	W	SW	NW
-Speed (mph)	2	48	43	45	49	53	51	58	53	49	61	45	48	61
-Date		1985	1985	1985	1985	1984	1984	1985	1985	1985	1985	1984	1984	OCT 1985

See Reference Notes to this table on the following pages.

TABLE 2 PRECIPITATION (inches) BILLINGS, MONTANA

YEAR	JAN	FEB	MAR	APR	MAY	JUNE	JULY	AUG	SEP	OCT	NOV	DEC	ANNUAL
1956	0.56	0.43	0.30	1.59	2.47	0.66	1.07	0.87	0.55	0.60	0.97	0.51	10.58
1957	0.82	0.55	1.52	3.55	3.39	4.49	0.25	1.47	1.24	2.58	1.31	0.05	21.22
1958	0.48	1.66	1.35	2.07	0.65	3.28	3.12	0.42	0.50	0.55	1.16	1.50	16.74
1959	1.02	1.13	0.23	2.36	1.07	0.97	0.31	0.19	1.38	0.69	1.92	0.33	11.60
1960	0.66	0.53	0.53	1.00	1.31	1.11	0.17	1.95	0.16	0.81	0.55	0.61	9.39
1961	0.15	0.20	0.72	2.33	1.57	0.24	0.85	0.23	3.99	1.47	1.55	0.22	13.52
1962	1.90	1.26	1.23	0.06	3.67	1.72	0.86	1.19	0.95	0.29	0.86	0.23	14.22
1963	2.23	0.31	0.39	2.38	2.49	3.16	0.49	0.39	1.29	0.31	0.05	1.43	14.92
1964	0.11	0.35	1.38	4.11	3.91	5.13	0.10	2.08	0.06	0.22	0.72	0.59	18.76
1965	0.66	1.18	0.90	1.18	1.89	2.30	1.28	3.50	2.17	0.09	1.07	0.74	16.55
1966	0.43	0.32	1.58	1.10	0.84	1.56	1.45	1.09	2.46	0.50	1.07	0.95	13.35
1967	0.36	0.39	1.55	1.63	1.84	5.18	0.37	0.54	0.66	1.04	0.50	0.79	14.85
1968	1.22	0.58	0.66	1.50	1.79	3.86	0.25	2.35	1.38	0.51	1.71	0.81	16.62
1969	0.99	0.17	0.57	1.48	0.78	5.74	1.69	0.42	0.36	1.56	0.66	0.31	14.73
1970	0.87	1.10	0.95	3.04	3.48	1.61	0.37	0.21	1.92	0.93	0.82	0.79	16.09
1971	1.30	0.56	0.83	1.60	2.07	0.70	0.40	0.43	1.80	3.80	0.17	1.13	14.79
1972	2.35	0.81	0.80	1.63	2.51	0.91	1.91	1.61	2.24	1.01	1.08	1.31	18.17
1973	1.30	0.43	1.59	3.00	0.73	0.81	0.29	1.23	2.21	1.36	1.21	2.00	16.16
1974	0.78	0.22	1.20	1.65	2.80	1.94	0.91	2.18	1.80	2.24	0.28	0.30	16.30
1975	2.05	0.75	1.30	1.91	3.62	1.62	2.64	0.37	0.24	2.66	1.39	1.96	20.51
1976	0.58	1.12	0.95	3.53	1.85	2.70	0.07	0.60	1.21	0.91	0.68	0.31	14.51
1977	1.44	0.05	1.37	0.64	1.35	0.63	0.80	1.05	0.65	1.20	1.42	1.65	12.25
1978	2.03	1.77	0.18	4.12	6.97	1.55	1.54	0.52	3.78	0.27	2.34	1.73	26.80
1979	0.72	0.56	1.11	1.20	0.92	1.06	0.46	0.87	0.16	2.45	0.53	0.09	8.41
1980	1.11	0.78	1.53	0.46	4.47	1.64	0.39	1.17	0.77	0.14	1.33	0.53	15.52
1981	0.21	0.24	1.75	0.35	7.71	1.58	1.65	0.55	0.14	1.15	0.42	1.07	16.45
1982	0.71	0.34	1.81	1.53	2.63	5.03	1.91	0.45	1.22	1.32	0.90	0.92	18.27
1983	0.11	0.31	0.73	0.56	2.23	0.88	1.52	1.12	2.26	0.37	0.95	0.84	12.86
1984	0.65	0.93	0.84	1.38	1.12	1.65	0.29	0.58	1.32	0.37	0.95	0.84	10.92
1985	0.31	0.39	2.05	0.31	1.27	1.07	1.40	1.66	1.89	0.69	1.43	0.20	12.67
Record Mean	0.76	0.62	1.07	1.52	2.27	2.40	0.95	0.93	1.33	1.13	0.77	0.71	14.46

TABLE 3 AVERAGE TEMPERATURE (deg. F) BILLINGS, MONTANA

YEAR	JAN	FEB	MAR	APR	MAY	JUNE	JULY	AUG	SEP	OCT	NOV	DEC	ANNUAL
1956	21.9	25.0	35.5	41.3	56.9	69.3	72.1	68.4	61.3	51.1	37.5	30.9	47.6
1957	11.2	27.2	34.6	42.3	56.2	62.7	74.5	70.4	61.4	44.7	34.3	37.3	46.2
1958	35.3	26.5	33.7	44.6	63.4	60.1	66.5	73.7	61.4	52.0	36.5	28.8	48.6
#1959	23.0	22.0	40.4	44.2	50.7	66.5	72.8	71.5	59.0	47.2	30.8	34.1	46.9
1960	23.2	25.3	35.0	44.8	56.4	66.2	77.9	69.8	62.2	51.5	36.9	30.3	48.3
1961	32.2	38.6	41.0	42.8	56.6	72.7	75.0	76.2	52.5	47.3	31.2	24.5	49.2
1962	19.6	23.9	29.5	50.7	55.0	65.0	69.6	69.4	59.0	53.7	41.4	32.7	47.5
1963	13.9	36.8	40.8	44.4	53.9	62.1	71.4	71.0	65.2	54.8	38.0	22.4	47.9
1964	29.1	31.4	31.8	45.2	56.7	62.7	75.6	68.1	56.4	51.5	31.2	28.5	46.5
1965	28.1	25.2	21.4	45.8	52.6	61.4	71.5	69.5	47.1	54.4	38.0	28.5	45.3
1966	15.1	25.7	37.1	40.1	55.4	61.6	76.2	68.3	65.5	50.4	34.5	29.2	46.6
1967	30.2	34.3	31.3	39.7	50.1	60.8	72.7	71.0	61.9	49.9	37.2	24.7	47.0
1968	23.4	34.5	43.5	45.3	54.0	62.6	72.1	67.6	61.6	52.1	38.3	19.9	47.9
1969	6.3	24.3	31.1	50.2	56.6	58.7	69.3	73.2	62.7	40.0	38.8	31.0	45.2
1970	20.2	33.0	29.5	38.6	54.5	66.0	74.6	74.6	55.8	45.1	33.6	24.7	45.7
1971	18.7	26.8	33.5	44.2	54.5	64.3	69.2	76.7	55.6	43.4	35.9	19.7	45.2
1972	13.2	26.2	40.1	43.7	53.8	67.0	66.6	70.1	54.8	43.6	18.1	14.3	44.3
1973	23.1	29.2	38.6	41.4	54.3	65.0	71.7	72.3	57.7	50.4	29.5	31.4	47.1
1974	22.6	35.8	35.8	47.7	55.0	66.9	75.0	64.8	56.0	50.5	37.4	29.9	47.7
1975	23.6	18.1	30.2	37.1	51.3	60.2	74.5	67.8	59.5	47.9	32.9	29.4	44.4
1976	26.5	32.9	32.0	46.1	56.2	60.9	73.7	71.1	62.0	46.0	31.5	31.5	47.9
1977	18.0	37.6	36.0	50.3	56.6	69.2	72.6	66.1	59.0	49.7	31.8	20.3	46.2
1978	11.0	16.0	34.8	46.6	53.6	64.0	67.4	71.5	67.4	52.6	33.5	36.3	42.6
1979	7.7	18.9	36.2	43.6	53.8	61.0	73.4	71.5	66.2	52.6	36.3	30.3	46.7
1980	17.0	29.2	34.4	54.6	61.0	66.5	75.7	67.4	67.0	50.7	40.6	30.3	49.1
1981	36.0	32.5	41.7	50.5	55.6	64.6	73.8	73.2	63.3	46.0	40.5	28.5	50.5
1982	13.1	28.0	38.4	42.6	52.4	61.4	70.8	75.8	60.4	50.2	33.8	8.7	45.9
1983	35.3	38.4	38.6	42.9	53.2	63.9	72.0	77.5	59.3	53.5	37.8	19.6	48.4
1984	29.3	38.3	38.1	45.0	55.5	64.5	74.5	74.7	54.0	49.6	37.6	25.9	47.8
1985	20.3	24.1	34.1	50.5	55.3	64.0	74.8	65.6	53.0	49.6	25.9	20.4	44.8
Record Mean	22.1	27.7	34.2	45.4	55.3	63.6	72.7	70.8	59.7	49.5	35.1	27.3	47.0
Max	31.5	37.5	44.4	57.0	67.2	75.9	87.0	85.0	72.6	61.1	44.5	36.3	58.3
Min	12.8	17.9	23.9	33.8	43.4	51.2	58.4	56.6	46.8	37.8	25.7	18.3	35.6

REFERENCE NOTES FOR TABLES 1, 2, 3 and 6   (BILLINGS, MT)

GENERAL

T - TRACE AMOUNT
BLANK ENTRIES DENOTE MISSING/UNREPORTED DATA.
# INDICATES A STATION OR INSTRUMENT RELOCATION.

SPECIFIC

TABLE 1

(a) - LENGTH OF RECORD IN YEARS. ALTHOUGH
      INDIVIDUAL MONTHS MAY BE MISSING.

* LESS THAN .05

NORMALS — BASED ON THE 1951-1980 RECORD PERIOD.
EXTREMES — DATES ARE THE MOST RECENT OCCURRENCE.
WIND DIR. — NUMERALS SHOW TENS OF DEGREES
            CLOCKWISE FROM TRUE NORTH.
            "00" INDICATES CALM.
RESULTANT WIND DIRECTIONS ARE GIVEN TO WHOLE DEGREES.

EXCEPTIONS

TABLES 2, 3, and 6

RECORD MEANS ARE THROUGH THE CURRENT YEAR,
BEGINNING IN   1935 FOR TEMPERATURE
                1935 FOR PRECIPITATION
                1935 FOR SNOWFALL

TABLE 4 HEATING DEGREE DAYS Base 65 deg. F   BILLINGS, MONTANA

SEASON	JULY	AUG	SEP	OCT	NOV	DEC	JAN	FEB	MAR	APR	MAY	JUNE	TOTAL
1956-57	7	50	140	425	820	1050	1663	1050	936	673	275	109	7198
1957-58	3	25	188	623	916	851	913	1073	960	601	105	158	6416
1958-59	44	11	162	403	848	1116	1297	1200	755	616	436	68	6956
#1959-60	23	3	251	543	1023	950	1291	1145	925	600	287	53	7094
1960-61	1	37	150	416	836	1067	1010	733	738	660	285	6	5939
1961-62			379	643	1008	1549	1640	1148	1094	498	409	89	7641
1962-63	4	46	187	342	701	990	1586	783	750	612	343	124	6468
1963-64	10	9	61	333	802	1316	1107	972	1025	587	275	111	6608
1964-65	0	67	253	415	1009	1454	1135	1110	1347	570	381	120	7861
1965-66	3	33	533	327	806	1125	1545	1094	860	742	319	155	7542
1966-67	0	54	83	446	910	1104	1069	852	1039	755	461	152	6925
1967-68	1	17	149	463	827	1240	1285	880	1044	589	334	125	6568
1968-69	16	42	147	394	793	1394	1818	1134	1044	438	283	218	7721
1969-70	17	4	116	766	780	1050	1382	891	1095	787	329	77	7294
1970-71	1	0	296	614	936	1244	1429	1063	972	617	322	86	7580
1971-72	25	7	305	659	869	1397	1601	1118	764	633	353	36	7767
1972-73	85	26	311	660	914	1451	1296	994	810	699	333	99	7678
1973-74	4	3	226	446	1058	1035	1313	811	896	511	443	94	6840
1974-75	7	76	269	444	821	1078	1280	1308	1071	829	418	152	7753
1975-76	1	31	185	536	955	1095	1186	926	1015	559	265	138	6892
1976-77	0	6	131	582	884	1030	1451	759	896	437	258	28	6462
1977-78	0	59	208	463	988	1380	1668	1366	933	545	357	97	8064
1978-79	41	45	204	487	1208	1509	1771	1286	887	636	350	71	8495
1979-80	2	7	44	383	937	884	1484	1033	939	324	159	46	6242
1980-81	0	25	127	462	724	1073	891	905	717	427	292	79	5722
1981-82	12	6	124	583	729	1124	1603	1028	987	666	386	142	7390
1982-83	12	0	215	453	926	1118	911	741	810	656	381	82	6305
1983-84	29	5	234	359	811	1741	1101	769	828	592	316	97	6882
1984-85	12	3	351	701	812	1404	1381	1140	950	428	184	103	7469
1985-86	13	65	358	471	1492	1207							

TABLE 5 COOLING DEGREE DAYS Base 65 deg. F   BILLINGS, MONTANA

YEAR	JAN	FEB	MAR	APR	MAY	JUNE	JULY	AUG	SEP	OCT	NOV	DEC	TOTAL
1969	0	0	0	0	29	34	156	266	56	0	0	0	541
1970	0	0	0	0	10	115	261	305	28	3	0	0	722
1971	0	0	0	0	4	69	162	377	29	0	0	0	641
1972	0	0	0	0	12	102	139	192	8	0	0	0	453
1973	0	0	0	0	13	107	219	235	16	2	0	0	590
1974	0	0	0	1	2	158	324	78	7	2	0	0	572
1975	0	0	0	0	2	15	299	126	26	10	0	0	478
1976	0	0	0	0	1	21	276	200	49	0	0	0	547
1977	0	0	0	0	7	160	242	102	34	0	0	0	545
1978	0	0	0	0	12	71	140	128	78	0	0	0	429
1979	0	0	0	0	12	126	270	216	86	6	0	0	716
1980	0	0	0	20	40	99	339	105	46	26	0	0	675
1981	0	0	0	0	6	74	291	268	76	3	0	0	718
1982	0	0	0	0	0	41	198	342	82	0	0	0	663
1983	0	0	0	0	24	54	256	400	69	7	2	0	812
1984	0	0	0	0	30	91	315	310	29	0	0	0	775
1985	0	0	0	0	42	83	325	92	6	0	0	0	548

TABLE 6 SNOWFALL (inches)   BILLINGS, MONTANA

SEASON	JULY	AUG	SEP	OCT	NOV	DEC	JAN	FEB	MAR	APR	MAY	JUNE	TOTAL
1956-57	0.0	0.0	0.0	3.6	8.7	5.8	9.0	6.8	16.7	6.5	0.0	0.0	57.1
1957-58	0.0	0.0	5.6	9.2	8.2	0.3	4.4	19.2	8.3	13.1	0.0	0.0	68.3
1958-59	0.0	0.0	0.0	8.2	14.5		10.1	13.2	3.1	11.3	0.0	0.0	60.4
1959-60	0.0	0.0	T	1.8	21.9	3.9	10.1	7.8	5.1	4.8	T	0.0	55.4
1960-61	0.0	0.0	0.0	T	3.6	7.5	1.6	2.1	2.4	14.0	0.0	0.0	31.2
1961-62	0.0	0.0	2.3	11.3	15.3	2.7	23.2	11.7	11.5	T	0.0	0.0	78.0
1962-63	0.0	0.0	6.3	0.0	8.2	1.3	27.7	2.3	2.7	15.4	0.0	0.0	63.9
1963-64	0.0	0.0	0.0	1.0	T	21.6	1.8	5.1	15.9	16.3	2.2	0.0	63.9
1964-65	0.0	0.0	0.0	T	7.2	7.8	8.0	13.9	11.4	3.4	3.5	0.0	55.2
1965-66	0.0	0.0	5.6	0.0	6.3	10.0	6.3	3.6	17.3	2.5	T	0.0	51.6
1966-67	0.0	0.0	0.0	T	12.0	10.3	6.1	3.7	16.2	11.4	7.7	0.0	67.4
1967-68	0.0	0.0	0.0	T	5.1	9.0	13.2	1.6	6.8	12.3	T	0.0	48.0
1968-69	0.0	0.0	T	T	4.6	8.6	12.2	2.0	5.5	3.0	T	T	35.9
1969-70	0.0	0.0	0.0	10.2	5.0	2.3	9.8	11.3	7.2	22.3	T	T	68.1
1970-71	0.0	0.0	4.9	T	4.9	8.4	11.5	5.7	8.0	2.5	T	0.0	45.9
1971-72	0.0	0.0	T	13.0	1.1	12.6	27.6	7.6	8.2	7.3	T	0.0	77.0
1972-73	0.0	0.0	4.7	3.2	T	12.9	13.3	5.4	7.4	14.3	1.8	0.0	63.0
1973-74	0.0	0.0	T	8.5	11.9	14.9	8.4	2.5	10.6	2.5	0.8	0.0	60.1
1974-75	0.0	0.0	T	0.6	1.1	3.5	20.8	6.8	14.0	12.9	1.1	0.0	60.8
1975-76	0.0	0.0	0.0	3.3	13.9	20.0	5.9	11.1	9.8	6.1	0.0	0.0	70.1
1976-77	0.0	0.0	0.0	2.1	7.0	6.2	16.1	0.3	13.7	5.7	T	0.0	51.1
1977-78	0.0	0.0	0.0	3.4	14.1	18.1	25.9	22.4	1.3	T	0.0	0.0	85.2
1978-79	0.0	0.0	T	0.0	25.2	22.2	10.1	8.7	5.9	3.7	0.3	0.0	76.1
1979-80	0.0	0.0	0.0	1.4	7.7	1.7	16.9	10.0	17.3	4.2	0.0	0.0	59.2
1980-81	0.0	0.0	T	17.8	4.5	5.5	2.1	3.4	16.3	0.7	15.6	0.0	65.9
1981-82	0.0	0.0	0.0	5.7	3.2	5.4	9.8	5.3	18.2	13.5	2.0	0.0	63.1
1982-83	0.0	0.0	5.7	1.5	5.6	11.2	0.1	1.0	6.4	5.8	11.9	0.0	49.2
1983-84	0.0	0.0	7.5	T	5.5	10.9	5.3	6.8	4.5	9.0	T	0.0	49.5
1984-85	0.0	0.0	9.3	6.5	9.9	16.1	4.8	3.8	21.7	1.9	0.0	0.0	74.0
1985-86	0.0	0.0	3.6	6.0	17.1	2.0							
Record Mean	0.0	0.0	1.3	3.5	7.0	8.9	9.4	8.0	10.3	7.4	1.5	T	57.3

See Reference Notes, relative to all above tables, on preceding page.

GLASGOW, MONTANA

Founded in the days of national expansion as a railroad shop town, Glasgow is situated in the valley of the Milk River, about 20 miles upstream from where the Milk River joins the Missouri. It lies on the natural route from the plains to Marias Pass in the northern Rockies. The city is located on the valley floor at an average elevation of about 2,100 feet above sea level. Hills rise sharply from the northern edge of the city to flat tableland about 200 feet higher than the valley. The Weather Service Office is located on this flat land about 1 mile north-northeast of the city. A gradual incline commences 3 to 4 miles to the south and southwest of the city and reaches to the rolling hills which separate the Milk River drainage from the Fort Peck Reservoir on the Missouri. The northern shore of Fort Peck Reservoir lies about 15 miles south of Glasgow. This is a body of water impounded by Fort Peck Dam which was completed in 1939. The dam, at full capacity, backs water up the Missouri Valley for over 180 miles. The shape of the reservoir is very irregular, but its average width south of Glasgow is about 10 miles.

The climate in the Glasgow area is continental with a large annual range in temperature and limited precipitation. Fort Peck Reservoir, to the south, seems to have little climatic effect as far north as Glasgow, except for brief periods of morning fog in the late fall which occasionally drift northward from the lake before it freezes. Seventy-eight percent of the annual precipitation falls from April through September, with May and June accounting for about 38 percent of the annual total. This distribution of precipitation helps to make the climate quite favorable for the growing of small grains. Winter precipitation nearly always falls as snow, but as a rule, although snow seldom accumulates to any great depth, it usually is formed into drifts in the open, unprotected areas. Blizzards during the winter months occur occasionally, but usually are of short duration. However, it is wise for travelers and stockmen to be on the alert for this danger during the winter months. Glasgow itself is well protected from most strong winds and blizzard conditions by hills to the north of the city, but occasionally the unprotected surrounding areas feel the full brunt of these winter storms.

Glasgow has a wide range of temperature. Winters are quite cold, but mild winter weather occasionally does occur, sometimes caused when the chinook or foehn wind, which descends the eastern slopes of the Rocky Mountains, reaches as far east as Glasgow. Very cold spells also occur, at least once each winter, but as a rule, these last only a few days. Summers are characterized by warm, sunny weather which can last for several weeks at a time. Sunny weather predominates during the warmer season, but interruptions in the form of clouds and showers do occur, usually in the afternoons and evenings. A few days of hot weather in July and August occur at times, but hot days are seldom oppressive because they are usually accompanied by low humidity.

As is usually the case with a continental climate in northern latitudes, the transitional fall and spring seasons at Glagow are quite rapid.

TABLE 1 — NORMALS, MEANS AND EXTREMES

GLASGOW, MONTANA

LATITUDE: 48°13'N   LONGITUDE: 106°37'W   ELEVATION: FT. GRND 2284 BARO 02279   TIME ZONE: MOUNTAIN   WBAN: 94008

	(a)	JAN	FEB	MAR	APR	MAY	JUNE	JULY	AUG	SEP	OCT	NOV	DEC	YEAR
TEMPERATURE °F:														
Normals														
-Daily Maximum		17.7	25.5	36.6	54.2	67.0	75.8	84.1	82.7	70.5	58.7	39.5	26.4	·53.2
-Daily Minimum		-1.3	6.3	15.7	30.7	42.1	51.0	56.7	54.9	44.2	33.6	18.8	7.3	30.0
-Monthly		8.2	15.9	26.1	42.5	54.5	63.4	70.4	68.8	57.4	46.2	29.1	16.9	41.6
Extremes														
-Record Highest	30	55	62	75	91	99	99	104	108	103	88	75	59	108
-Year		1968	1984	1966	1980	1980	1984	1983	1983	1983	1970	1975	1979	AUG 1983
-Record Lowest	30	-47	-37	-27	-3	20	33	41	37	20	-5	-26	-38	-47
-Year		1969	1982	1960	1975	1976	1979	1977	1956	1985	1984	1985	1977	JAN 1969
NORMAL DEGREE DAYS:														
Heating (base 65°F)		1761	1375	1203	675	332	113	23	50	260	583	1074	1491	8940
Cooling (base 65°F)		0	0	0	0	10	65	190	168	32	0	0	0	465
% OF POSSIBLE SUNSHINE														
MEAN SKY COVER (tenths)														
Sunrise - Sunset	30	7.1	7.2	6.9	6.9	6.6	5.9	4.5	4.7	5.6	6.2	7.0	7.2	6.3
MEAN NUMBER OF DAYS:														
Sunrise to Sunset														
-Clear	30	5.5	4.7	5.4	5.4	5.6	7.0	12.8	12.3	9.2	7.8	6.0	4.6	86.3
-Partly Cloudy	30	7.3	7.3	9.1	8.3	10.5	11.9	12.4	11.0	9.3	8.6	7.0	8.3	111.0
-Cloudy	30	18.2	16.2	16.5	16.4	14.9	11.1	5.9	7.7	11.5	14.6	17.0	18.1	168.0
Precipitation														
.01 inches or more	30	8.6	6.8	7.2	7.0	9.6	10.6	7.8	7.2	6.4	4.9	6.1	8.5	90.7
Snow, Ice pellets														
1.0 inches or more	30	2.2	1.4	1.2	0.8	0.1	0.0	0.0	0.0	0.1	0.5	1.2	1.6	9.2
Thunderstorms	26	0.0	0.0	0.1	0.5	3.6	7.2	7.7	5.7	1.5	0.2	0.*	0.0	26.4
Heavy Fog Visibility														
1/4 mile or less	26	2.1	2.3	1.9	0.7	0.3	0.3	0.*	0.1	0.2	0.6	1.8	2.7	12.8
Temperature °F														
-Maximum														
90° and above	21	0.0	0.0	0.0	0.*	0.6	2.1	9.4	9.0	1.6	0.0	0.0	0.0	22.8
32° and below	21	23.5	16.8	9.1	0.9	0.0	0.0	0.0	0.0	0.0	0.8	9.2	20.7	81.1
-Minimum														
32° and below	21	30.9	27.6	28.2	15.6	2.8	0.0	0.0	0.0	2.2	12.6	27.3	30.9	178.1
0° and below	21	16.4	9.8	3.3	0.1	0.0	0.0	0.0	0.0	0.0	0.1	2.4	10.7	42.7
AVG. STATION PRESS. (mb)	13	934.9	933.6	931.0	932.0	930.8	931.0	932.9	933.0	934.0	933.3	933.7	933.5	932.8
RELATIVE HUMIDITY (%)														
Hour 05	21	73	77	79	75	73	76	72	69	72	74	78	77	75
Hour 11	21	70	72	66	51	46	47	43	42	49	54	67	73	57
Hour 17 (Local Time)	21	70	71	59	43	40	40	33	32	39	47	63	73	51
Hour 23	21	73	77	75	65	62	65	58	54	61	66	74	76	67
PRECIPITATION (inches):														
Water Equivalent														
-Normal		0.46	0.38	0.40	0.87	1.76	2.47	1.65	1.43	0.89	0.56	0.31	0.37	11.55
-Maximum Monthly	30	1.24	0.74	0.93	1.99	3.74	5.36	5.17	5.74	4.14	1.77	1.26	1.03	5.74
-Year		1969	1979	1974	1969	1982	1963	1962	1985	1978	1975	1958	1982	AUG 1985
-Minimum Monthly	30	T	0.05	0.05	0.07	0.03	0.09	0.01	0.03	0.04	T	T	0.03	T
-Year		1973	1985	1957	1956	1958	1985	1984	1983	1960	1965	1969	1959	JAN 1973
-Maximum in 24 hrs	30	0.37	0.27	0.40	1.16	2.07	2.47	3.98	4.99	1.98	1.21	0.37	0.36	4.99
-Year		1969	1982	1964	1969	1974	1972	1962	1985	1978	1981	1981	1982	AUG 1985
Snow, Ice pellets														
-Maximum Monthly	30	24.2	15.9	14.8	13.7	10.7				2.2	7.0	17.2	13.9	24.2
-Year		1971	1979	1967	1970	1983				1983	1975	1958	1972	JAN 1971
-Maximum in 24 hrs	30	8.8	4.7	5.0	8.4	10.1				2.1	5.3	4.7	7.0	10.1
-Year		1971	1979	1967	1970	1983				1983	1975	1985	1972	MAY 1983
WIND:														
Mean Speed (mph)	16	10.2	10.4	11.4	12.6	11.9	11.0	10.5	10.9	11.0	10.6	9.4	9.8	10.8
Prevailing Direction														
Fastest Obs. 1 Min.														
-Direction (!!!)	17	36	28	30	32	12	24	29	27	30	32	27	32	27
-Speed (MPH)	17	39	44	41	46	41	39	46	44	46	44	48	39	48
-Year		1971	1974	1979	1978	1980	1972	1983	1975	1971	1968	1978	1979	NOV 1978
Peak Gust														
-Direction (!!!)	2	NW	W	E	N	NW	NW	N	SE	S	NW	NW	NW	SE
-Speed (mph)	2	44	53	43	51	58	49	43	61	52	60	41	46	61
-Date		1985	1984	1985	1984	1984	1985	1985	1985	1985	1985	1984	1985	AUG 1985

See Reference Notes to this table on the following page.

TABLE 2 PRECIPITATION (inches) GLASGOW, MONTANA

YEAR	JAN	FEB	MAR	APR	MAY	JUNE	JULY	AUG	SEP	OCT	NOV	DEC	ANNUAL
1956	0.11	0.49	0.13	0.07	2.29	1.68	1.62	2.03	0.10	0.03	0.38	0.26	9.19
1957	0.48	0.56	0.05	1.33	1.00	2.86	0.98	2.01	0.51	0.45	0.09	0.07	10.39
1958	0.13	0.33	0.29	0.58	0.03	2.06	1.15	0.52	0.31	0.13	1.26	0.28	7.07
1959	0.48	0.50	0.14	0.42	1.18	2.79	0.57	0.59	1.26	1.02	0.72	0.03	9.70
1960	0.43	0.32	0.27	0.72	1.43	1.61	0.55	0.95	0.04	0.34	0.30	0.34	7.30
1961	0.09	0.45	0.35	0.55	1.04	1.17	1.66	0.38	2.04	0.78	0.19	0.19	8.89
1962	0.34	0.45	0.55	0.21	2.50	4.43	5.17	0.48	0.15	1.15	0.43	0.19	17.77
1963	0.32	0.47	0.27	1.22	1.13	5.36	1.65	3.30	0.32	0.12	0.21	0.30	14.67
1964	0.21	0.19	0.61	0.41	2.52	2.78	1.20	0.85	0.52	0.20	0.26	0.78	10.53
1965	0.46	0.20	0.30	0.51	3.25	4.64	0.91	3.01	0.79	T	0.12	0.12	14.31
1966	0.51	0.14	0.10	0.59	1.91	0.89	3.13	3.65	0.41	0.68	0.34	0.23	12.58
1967	0.84	0.25	0.83	0.84	0.68	2.23	0.12	0.18	2.20	1.13	0.36	0.26	9.92
1968	0.13	0.15	0.10	0.67	0.23	2.24	0.82	2.00	0.75	0.10	0.07	0.39	7.65
1969	1.24	0.14	0.17	1.99	0.25	1.23	3.45	0.05	0.27	1.33	T	0.36	10.48
1970	0.49	0.05	0.27	1.51	1.88	2.43	1.07	0.04	1.32	0.44	0.49	0.26	10.25
1971	0.99	0.21	0.27	0.42	0.91	1.26	0.20	0.78	0.69	0.22	0.02	0.61	6.90
1972	0.55	0.59	0.50	0.84	2.73	3.77	2.30	2.40	0.69	0.24	0.46	0.78	15.22
1973	T	0.09	0.27	1.24	0.54	4.35	0.59	1.21	1.43	0.42	0.25	0.25	11.20
1974	0.18	0.25	0.93	0.49	3.27	2.00	2.31	2.69	0.07	0.42	0.26	0.38	13.12
1975	0.11	0.23	0.61	1.30	1.93	1.93	4.18	1.37	0.42	1.77	0.40	0.38	13.90
1976	0.27	0.13	0.30	0.60	0.64	4.27	5.00	1.16	0.41	0.21	0.26	0.28	13.53
1977	0.42	0.12	0.11	0.17	1.83	1.16	0.78	0.58	2.62	0.40	0.24	0.86	9.29
1978	0.19	0.40	0.25	0.51	3.65	2.72	2.63	0.16	4.14	0.28	0.48	0.45	15.86
1979	0.15	0.74	0.43	1.27	2.46	0.48	0.75	0.76	0.27	0.57	0.18	0.05	8.11
1980	0.50	0.14	0.36	0.22	0.46	2.55	0.44	1.36	0.87	1.35	0.20	0.50	8.95
1981	0.06	0.10	0.31	0.19	2.13	1.83	1.72	0.23	0.08	1.40	0.55	0.19	8.79
1982	0.75	0.43	0.72	0.22	3.74	1.03	0.97	1.19	0.97	0.98	0.16	1.03	12.21
1983	0.18	0.24	0.39	0.10	1.59	0.77	2.62	0.01	0.43	0.81	0.18	0.56	7.56
1984	0.43	0.20	0.68	0.10	0.78	1.99	0.81	0.01	0.43	0.57	0.18	0.56	6.74
1985	0.05	0.05	0.27	0.91	2.34	0.09	0.69	5.74	0.83	0.64	0.74	0.50	12.85
Record Mean	0.44	0.37	0.46	0.80	1.72	2.51	1.56	1.51	0.95	0.57	0.35	0.44	11.68

TABLE 3 AVERAGE TEMPERATURE (deg. F) GLASGOW, MONTANA

YEAR	JAN	FEB	MAR	APR	MAY	JUNE	JULY	AUG	SEP	OCT	NOV	DEC	ANNUAL
1956	5.6	9.9	26.2	38.7	55.1	67.6	68.7	67.1	57.7	47.3	32.9	20.9	41.5
1957	0.5	12.5	29.5	41.0	57.1	63.3	74.6	69.2	57.7	42.5	33.0	28.1	42.4
1958	25.1	13.7	27.4	44.9	61.5	60.8	67.0	72.7	58.6	47.4	25.9	15.2	43.4
1959	5.5	3.1	32.0	42.4	50.7	65.6	71.7	69.1	55.8	40.1	22.8	26.1	40.4
1960	9.1	12.9	22.8	42.5	55.2	62.3	75.5	68.6	60.2	48.0	28.4	17.4	41.9
1961	18.8	24.9	37.8	39.7	56.4	71.6	72.9	75.0	51.1	44.7	27.6	10.4	44.2
1962	11.9	12.8	19.3	47.2	53.6	64.8	66.6	68.5	56.9	49.1	38.9	24.5	42.8
1963	3.3	23.3	37.9	43.3	52.9	63.6	72.0	70.8	65.1	53.9	31.9	16.4	44.6
#1964	20.9	27.7	24.1	44.7	57.0	63.9	74.3	68.0	55.0	49.3	27.1	4.2	42.8
1965	6.4	11.1	10.5	41.8	53.2	61.9	71.0	69.4	48.1	52.2	26.8	22.7	39.6
1966	-1.2	7.8	32.0	37.8	55.3	62.3	71.6	65.4	62.1	45.9	23.9	18.9	40.1
1967	12.5	15.1	21.8	37.0	50.9	60.6	71.0	70.5	62.2	46.4	30.6	15.9	41.2
1968	11.2	18.5	38.1	42.0	52.1	60.0	69.0	65.0	57.6	45.8	32.9	9.9	41.9
1969	-7.0	8.2	18.3	47.6	53.8	58.3	66.1	73.4	61.9	38.7	34.5	22.8	39.7
1970	7.0	18.4	24.4	39.5	54.1	67.4	72.6	72.2	56.1	42.8	26.8	12.0	41.1
1971	5.1	14.3	24.6	44.9	55.8	65.0	68.8	77.2	56.5	44.3	32.3	10.8	41.6
1972	4.3	10.5	30.3	44.3	55.6	65.6	64.7	71.1	53.6	42.1	32.8	6.4	40.1
1973	18.1	23.6	39.0	41.6	56.2	65.6	69.8	72.9	56.6	49.3	21.3	15.7	44.2
1974	10.4	25.5	29.6	47.5	50.7	66.2	73.3	64.0	55.4	49.2	33.2	25.1	44.2
1975	18.5	12.6	24.7	35.9	52.8	62.4	73.1	64.9	56.3	45.0	30.1	18.5	41.2
1976	13.8	29.5	29.3	47.8	58.2	62.5	70.3	70.0	59.8	42.2	26.4	18.9	44.1
1977	3.1	26.5	33.6	48.2	58.3	65.7	69.1	61.2	55.2	46.0	24.5	8.3	41.6
1978	-2.1	7.4	24.5	43.8	54.8	62.7	67.1	65.9	58.5	45.4	21.3	9.2	38.2
1979	-4.3	1.5	23.4	36.3	50.4	64.3	72.2	69.7	62.5	49.0	26.9	26.5	39.9
1980	7.3	16.4	27.9	51.9	61.1	66.1	72.3	64.7	57.8	47.7	36.3	18.2	44.2
1981	25.3	25.8	38.9	48.2	56.8	62.0	71.7	73.1	60.8	44.1	37.1	17.7	46.7
1982	-5.1	9.8	24.6	40.6	51.2	62.3	69.7	69.3	56.5	45.2	25.7	19.0	39.1
1983	20.6	27.1	33.6	43.1	51.7	63.4	71.8	76.5	55.3	47.7	32.9	-3.2	43.4
1984	18.6	34.3	31.9	47.6	55.5	64.6	74.6	74.4	51.8	42.0	30.0	6.9	44.4
1985	8.5	14.9	33.6	48.6	58.9	60.6	72.3	64.6	50.5	43.1	9.7	12.8	39.8
Record Mean	9.6	16.4	27.4	43.7	55.0	65.4	71.2	69.5	57.4	46.4	28.9	15.9	42.0
Max	19.6	26.8	38.1	56.0	67.8	76.1	85.4	83.8	70.8	59.3	39.4	25.8	54.1
Min	-0.4	6.0	16.6	31.3	42.2	50.7	56.9	55.2	43.9	33.4	18.4	6.1	30.0

REFERENCE NOTES FOR TABLES 1, 2, 3 and 6 (GLASGOW, MT)

GENERAL

T - TRACE AMOUNT
BLANK ENTRIES DENOTE MISSING/UNREPORTED DATA.
# INDICATES A STATION OR INSTRUMENT RELOCATION.

SPECIFIC

TABLE 1

(a) - LENGTH OF RECORD IN YEARS. ALTHOUGH INDIVIDUAL MONTHS MAY BE MISSING.
* LESS THAN .05

NORMALS — BASED ON THE 1951-1980 RECORD PERIOD.
EXTREMES — DATES ARE THE MOST RECENT OCCURRENCE.
WIND DIR. — NUMERALS SHOW TENS OF DEGREES CLOCKWISE FROM TRUE NORTH. "00" INDICATES CALM.
RESULTANT WIND DIRECTIONS ARE GIVEN TO WHOLE DEGREES.

EXCEPTIONS

TABLES 2, 3, and 6

RECORD MEANS ARE THROUGH THE CURRENT YEAR, BEGINNING IN 1944 FOR TEMPERATURE
1944 FOR PRECIPITATION
1956 FOR SNOWFALL

TABLE 4 HEATING DEGREE DAYS Base 65 deg. F   GLASGOW, MONTANA

SEASON	JULY	AUG	SEP	OCT	NOV	DEC	JAN	FEB	MAR	APR	MAY	JUNE	TOTAL
1956-57	34	66	220	540	956	1363	2001	1465	1093	713	250	84	8785
1957-58	4	42	225	702	952	1135	1231	1433	1153	600	154	148	7779
1958-59	42	21	234	540	1167	1540	1842	1735	1018	670	439	78	9326
1959-60	20	19	295	765	1259	1198	1732	1507	1303	669	310	108	9185
1960-61	11	35	207	521	1092	1468	1429	1116	836	753	294	11	7773
1961-62	0	2	419	623	1116	1691	1643	1460	1410	530	347	66	9307
1962-63	23	43	243	486	780	1247	1917	1162	830	643	378	84	7836
#1963-64	4	13	72	341	984	1503	1359	1077	1261	602	267	92	7575
1964-65	0	52	298	482	1135	1974	1816	1507	1689	690	359	118	10120
1965-66	11	41	502	389	1137	1305	2053	1599	1016	812	316	132	9313
1966-67	5	78	136	588	1225	1422	1624	1396	1335	835	429	149	9222
1967-68	19	12	136	571	1024	1519	1664	1344	826	683	397	158	8353
1968-69	38	78	235	589	956	1706	2236	1584	1440	515	361	219	9957
1969-70	39	7	150	807	908	1301	1798	1297	1251	758	334	46	8696
1970-71	5	7	296	680	1142	1638	1416	1245	1245	595	285	59	9224
1971-72	27	0	267	635	973	1676	1883	1579	1067	614	305	55	9081
1972-73	71	11	338	703	958	1814	1452	1152	802	691	277	68	8337
1973-74	8	7	255	480	1306	1522	1692	1103	1089	526	436	70	8494
1974-75	3	88	283	481	1229	1437	1463	1244	945	867	373	102	8515
1975-76	5	68	260	613	1042	1437	1583	1022	1099	510	215	109	7963
1976-77	9	10	185	698	1151	1421	1917	1074	967	497	233	59	8221
1977-78	32	140	292	583	1208	1752	2081	1610	1252	628	319	111	10008
1978-79	42	70	233	601	1306	1725	2145	1780	1283	853	455	86	10579
1979-80	6	17	114	489	1139	1187	1788	1404	1145	398	175	49	7911
1980-81	2	63	232	536	854	1446	1225	1091	799	499	250	145	7142
1981-82	10	6	169	639	834	1459	2171	1546	1246	728	418	121	9347
1982-83	16	39	271	607	1171	1417	1369	1055	969	649	412	97	8072
1983-84	16	5	322	528	956	2113	1431	885	1018	515	318	85	8192
1984-85	0	20	406	705	1042	1797	1754	1400	967	485	218	156	8950
1985-86	11	84	429	673	1657	1617							

TABLE 5 COOLING DEGREE DAYS Base 65 deg. F   GLASGOW, MONTANA

YEAR	JAN	FEB	MAR	APR	MAY	JUNE	JULY	AUG	SEP	OCT	NOV	DEC	TOTAL
1969	0	0	0	0	20	26	79	276	62	0	0	0	463
1970	0	0	0	0	3	128	250	236	38	0	0	0	655
1971	0	0	0	0	9	67	153	383	22	0	0	0	634
1972	0	0	0	0	20	80	69	205	4	0	0	0	378
1973	0	0	0	0	13	91	165	259	10	0	0	0	538
1974	0	0	0	5	0	113	268	62	3	0	0	0	451
1975	0	0	0	0	1	33	263	75	6	0	0	0	378
1976	0	0	0	0	9	41	180	172	35	0	0	0	437
1977	0	0	0	1	34	85	166	29	6	0	0	0	321
1978	0	0	0	0	7	47	114	107	45	0	0	0	320
1979	0	0	0	0	0	70	236	170	48	1	0	0	533
1980	0	0	0	12	60	88	234	59	23	7	0	0	483
1981	0	0	0	0	2	22	227	265	48	0	0	0	564
1982	0	0	0	0	0	44	168	180	32	0	0	0	424
1983	0	0	0	0	7	54	234	369	40	0	0	0	704
1984	0	0	0	0	30	78	303	318	15	0	0	0	744
1985	0	0	0	0	33	29	242	78	3	0	0	0	385

TABLE 6 SNOWFALL (inches)   GLASGOW, MONTANA

SEASON	JULY	AUG	SEP	OCT	NOV	DEC	JAN	FEB	MAR	APR	MAY	JUNE	TOTAL
1956-57	0.0	0.0	0.0	T	0.7	4.1	7.6	8.4	0.5	T	0.0	0.0	21.3
1957-58	0.0	0.0	0.4	1.3	0.6	1.2	1.7	6.0	3.0	T	0.0	0.0	14.2
1958-59	0.0	0.0	T	T	17.2	4.5	7.6	6.7	0.7	1.8	T	0.0	38.5
1959-60	0.0	0.0	T	5.4	5.6	0.3	5.8	3.8	0.9	5.3	T	0.0	27.1
1960-61	0.0	0.0	0.0	0.8	1.7	4.2	1.5	1.8	5.0	1.3	0.0	0.0	16.3
1961-62	0.0	0.0	T	5.0	1.4	1.9	3.0	4.8	5.5	T	0.0	0.0	21.6
1962-63	0.0	0.0	T	2.0	T	1.4	3.2	3.9	1.4	5.3	T	0.0	17.2
1963-64	0.0	0.0	0.0	T	2.0	2.5	2.1	1.9	3.6	2.3	0.0	0.0	14.4
1964-65	0.0	0.0	T	1.6	2.4	10.8	10.7	6.6	4.9	3.1	T	0.0	40.1
1965-66	0.0	0.0	1.1	T	1.0	1.2	2.4	2.5	2.8	1.7	1.7	0.0	20.4
1966-67	0.0	0.0	0.0	1.6	4.6	3.9	14.2	4.3	14.8	12.1	3.1	0.0	58.6
1967-68	0.0	0.0	0.0	0.0	0.4	4.6	2.7	2.1	T	6.2	0.9	0.0	16.9
1968-69	0.0	0.0	0.0	T	0.7	5.7	24.1	2.7	3.3	0.6	0.0	0.0	37.1
1969-70	0.0	0.0	0.0	1.0	T	4.1	5.5	0.6	2.6	13.7	T	0.0	27.5
1970-71	0.0	0.0	0.1	1.1	6.7	24.2	4.3	1.1	1.1	T	0.4	0.0	41.3
1971-72	0.0	0.0	0.0	0.1	0.6	6.3	12.7	10.6	3.3	0.7	T	0.0	34.3
1972-73	0.0	0.0	0.2	2.2	0.3	13.9	T	1.2	0.1	4.0	T	0.0	21.9
1973-74	0.0	0.0	0.0	0.0	5.1	10.6	2.1	2.6	6.1	0.6	T	0.0	27.1
1974-75	0.0	0.0	0.0	0.2	1.0	1.5	3.5	7.4	4.0	T	0.0	0.0	19.9
1975-76	0.0	0.0	0.0	0.0	7.0	5.5	4.7	3.9	0.5	4.9	0.2	0.0	26.7
1976-77	0.0	0.0	0.0	1.6	2.3	4.0	7.7	0.3	0.7	0.4	0.0	0.0	17.0
1977-78	0.0	0.0	0.0	0.2	4.3	13.2	3.8	8.4	3.0	0.5	0.0	0.0	33.4
1978-79	0.0	0.0	T	T	7.6	6.5	2.4	15.9	7.1	7.8	0.5	0.0	47.8
1979-80	0.0	0.0	0.0	T	2.7	0.4	9.4	1.3	2.6	0.7	0.0	0.0	17.1
1980-81	0.0	0.0	0.0	3.1	1.0	11.0	1.1	0.8	0.1	T	0.0	0.0	17.1
1981-82	0.0	0.0	0.0	4.4	1.8	2.9	16.4	7.3	7.5	1.7	T	0.0	42.0
1982-83	0.0	0.0	1.5	0.6	0.4	11.2	1.1	1.3	3.5	0.1	10.7	0.0	30.4
1983-84	0.0	0.0	2.2	T	1.8	4.7	4.2	1.6	4.3	1.1	0.2	0.0	20.1
1984-85	0.0	0.0	2.0	4.2	1.7	12.7	1.0	0.7	4.6	1.7	0.0	0.0	28.6
1985-86	0.0	0.0	0.2	2.2	13.0	9.1							
Record Mean	0.0	0.0	0.3	1.5	3.1	5.6	6.4	4.1	3.6	2.6	0.8	0.0	27.8

See Reference Notes, relative to all above tables, on preceding page.

The city of Great Falls is located along the main stem of the Missouri River at its confluence with the Sun River. The Weather Service Office is located at the Municipal Airport on a plateau between the Sun and Missouri Rivers. This plateau is about 200 feet higher than most of the immediate valley area, and the airport is about two miles southwest of the Sun and Missouri River Junction. Except to the north and northeast, the valley is encircled by mountain ranges, which lie about 30 miles away from east to south, 40 miles to the southwest, and 60 to 100 miles distant from west to northwest. Topography plays an important part in the climate of Great Falls. The Continental Divide to the west, and Big and Little Belt Ranges to the south, are primary factors in producing the frequent wintertime chinook winds observed in this part of Montana. The combination of valleys and plateaus in the immediate area, contributes to marked temperature differences between the airport and the city proper, either on calm, clear mornings, or when chinook winds reach the airport before they are felt at the lower elevations in town.

Summertime in the area generally is quite pleasant, with cool nights, moderately warm and sunny days, and very little hot, humid weather. Most of the summer rainfall occurs in showers or thunderstorms, and steady rains may occur during late spring or early summer. At the airport, freezing temperatures do not occur in July or August and very rarely in June. Frost occurs frequently in April and October, but more often in the valleys than on the surrounding hills or plateaus. However, frost may occur on rare occasions in nearby low lying areas at any time of the year.

Winters are not as cold as is usually expected of a continental location at this latitude, largely as a result of the chinook winds for which this area is noted. While sub-zero weather is experienced normally several times during a winter, the coldest weather seldom lasts more than a few days at a time, and is usually terminated by southwest chinook winds which can produce sharp temperature rises of 40 degrees or more in 24 hours.

As a result of recurring chinooks throughout the winter season, snow seldom lies on the ground for more than a few days. In fact, the ground usually is bare, or nearly bare, of snow most of the winter, except in the surrounding mountains and higher foothills. On the other hand, invasions of cold air from the polar regions occur a few times each winter, and sharp temperature falls from above freezing to below zero within 24 hours are observed occasionally.

Precipitation generally falls as snow during late fall, winter, and early spring, although rain can occur in any month. Late spring, summer, and early fall precipitation is almost always rain, but some hail is observed occasionally during summer thunderstorms.

Although average annual precipitation at Great Falls would normally classify the area as semi-arid, it is important to note that about 70 percent of the annual total falls normally during the April to September growing season. The combination of ideal temperatures during the peak of the growing season, long hours of summer sunshine, and adequate precipitation during the six critical months, makes the climate very favorable for dryland farming. Heavy fog occurs about one day per month, but each case lasts only a small part of the day. Although the average windspeed is relatively high, strong winds over 70 mph are seldom observed. Visibility normally is excellent.

TABLE 1   NORMALS, MEANS AND EXTREMES

GREAT FALLS, MONTANA

LATITUDE: 47°29'N   LONGITUDE: 111°22'W   ELEVATION: FT. GRND 3663 BARO 03665   TIME ZONE: MOUNTAIN   WBAN: 24143

	(a)	JAN	FEB	MAR	APR	MAY	JUNE	JULY	AUG	SEP	OCT	NOV	DEC	YEAR
TEMPERATURE °F:														
Normals														
-Daily Maximum		28.2	36.5	41.7	54.0	65.3	74.3	84.2	82.0	70.5	59.5	43.5	34.7	56.2
-Daily Minimum		9.2	16.8	21.1	31.3	41.1	49.4	54.4	53.0	44.2	36.2	24.5	16.6	33.2
-Monthly		18.7	26.6	31.4	42.7	53.2	61.9	69.3	67.5	57.4	47.9	34.0	25.6	44.7
Extremes														
-Record Highest	48	62	67	78	89	93	99	105	106	98	91	76	69	106
-Year		1981	1981	1978	1980	1980	1976	1973	1969	1980	1943	1975	1939	AUG 1969
-Record Lowest	48	-37	-35	-29	-6	15	31	40	35	21	-9	-25	-43	-43
-Year		1969	1939	1951	1975	1954	1950	1980	1939	1985	1984	1985	1968	DEC 1968
NORMAL DEGREE DAYS:														
Heating (base 65°F)		1435	1072	1042	669	369	141	20	66	268	536	930	1218	7766
Cooling (base 65°F)		0	0	0	0	0	48	153	144	40	6	0	0	391
% OF POSSIBLE SUNSHINE	41	49	55	66	61	62	64	80	76	67	60	46	44	61
MEAN SKY COVER (tenths)														
Sunrise - Sunset	44	7.3	7.4	7.4	7.3	7.0	6.6	4.3	4.8	5.7	6.4	7.1	7.3	6.5
MEAN NUMBER OF DAYS:														
Sunrise to Sunset														
-Clear	48	5.4	4.0	4.1	4.0	4.7	5.4	13.6	12.5	9.5	7.0	4.9	4.8	79.9
-Partly Cloudy	48	6.1	6.5	8.8	8.0	9.7	10.6	11.5	10.6	9.2	8.9	7.5	7.7	105.1
-Cloudy	48	19.5	17.7	18.1	17.9	16.6	14.0	6.0	7.9	11.3	15.0	17.6	18.6	180.2
Precipitation														
.01 inches or more	48	9.2	7.8	9.3	8.9	11.4	12.0	7.3	7.6	7.3	5.8	7.1	7.7	101.5
Snow, Ice pellets														
1.0 inches or more	48	3.5	2.8	3.5	2.1	0.5	0.1	0.0	0.0	0.5	1.1	2.6	3.1	19.8
Thunderstorms	48	0.*	0.1	0.2	0.7	3.4	6.7	7.2	6.0	1.4	0.3	0.*	0.*	26.0
Heavy Fog Visibility														
1/4 mile or less	46	1.1	1.5	2.0	1.6	0.8	0.6	0.3	0.3	0.7	1.2	2.0	1.0	13.0
Temperature °F														
-Maximum														
90° and above	24	0.0	0.0	0.0	0.0	0.1	2.0	8.5	8.0	1.8	0.0	0.0	0.0	20.5
32° and below	24	14.6	8.1	6.6	1.6	0.0	0.0	0.0	0.0	0.*	1.0	5.4	12.8	50.1
-Minimum														
32° and below	24	27.4	23.9	25.8	16.0	3.6	0.1	0.0	0.0	2.9	10.8	21.5	26.6	158.4
0° and below	24	11.3	4.6	2.7	0.1	0.0	0.0	0.0	0.0	0.0	0.1	2.4	7.4	28.6
AVG. STATION PRESS.(mb)	13	888.1	887.0	884.8	886.5	886.4	887.3	889.3	889.2	889.8	888.7	887.6	887.3	887.7
RELATIVE HUMIDITY (%)														
Hour 05	24	67	67	67	66	68	69	64	63	66	63	64	66	66
Hour 11	24	63	59	55	47	46	44	37	39	46	47	55	61	50
Hour 17 (Local Time)	24	61	55	49	40	40	39	29	30	37	43	55	61	45
Hour 23	24	66	66	64	59	59	59	50	50	57	59	63	66	60
PRECIPITATION (inches):														
Water Equivalent														
-Normal		1.00	0.75	0.93	1.49	2.52	2.75	1.10	1.31	1.03	0.82	0.74	0.80	15.24
-Maximum Monthly	48	2.05	2.16	2.18	4.63	8.13	5.37	4.32	4.90	3.56	3.43	2.27	1.92	8.13
-Year		1969	1958	1967	1975	1953	1965	1955	1985	1941	1975	1955	1977	MAY 1953
-Minimum Monthly	48	T	0.01	0.15	0.05	0.67	0.52	0.04	0.03	0.10	T	0.02	T	T
-Year		1944	1950	1960	1981	1950	1960	1959	1969	1962	1965	1954	1954	OCT 1965
-Maximum in 24 hrs	48	0.74	0.88	1.14	2.43	3.42	2.74	2.40	2.00	1.82	1.15	0.97	0.82	3.42
-Year		1966	1951	1977	1951	1980	1964	1983	1985	1982	1954	1946	1972	MAY 1980
Snow, Ice pellets														
-Maximum Monthly	48	22.6	26.1	23.4	35.4	8.6	11.1	T	T	10.4	16.6	22.1	25.0	35.4
-Year		1969	1958	1982	1967	1983	1950	1966	1985	1984	1975	1955	1945	APR 1967
-Maximum in 24 hrs	48	10.2	11.0	11.0	16.8	8.6	11.0	T	T	6.1	8.3	10.8	9.8	16.8
-Year		1984	1951	1977	1973	1983	1950	1966	1985	1954	1957	1946	1945	APR 1973
WIND:														
Mean Speed (mph)	44	15.2	14.5	13.1	13.0	11.5	11.2	10.2	10.3	11.4	13.4	14.6	15.6	12.8
Prevailing Direction														
through 1963		SW	SW	SW	SW	SW	SW	SW	SW	SW	SW	SW	SW	SW
Fastest Mile														
-Direction (!!!)	41	SW	W	W	W	SW	NW	W	SW	NW	W	SW	SW	SW
-Speed (MPH)	41	65	72	73	70	65	70	73	71	73	73	73	82	82
-Year		1946	1954	1951	1946	1951	1980	1951	1954	1945	1949	1955	1956	DEC 1956
Peak Gust														
-Direction (!!!)	2	SW	SW	SW	SW	NW	W	W	W	SW	SW	SW	SW	W
-Speed (mph)	2	60	47	39	51	59	53	56	63	61	60	48	51	63
-Date		1984	1985	1985	1985	1984	1985	1985	1984	1984	1984	1984	1985	AUG 1984

See Reference Notes to this table on the following page

TABLE 2 PRECIPITATION (inches)   GREAT FALLS, MONTANA

YEAR	JAN	FEB	MAR	APR	MAY	JUNE	JULY	AUG	SEP	OCT	NOV	DEC	ANNUAL
1956	0.52	0.37	0.38	0.53	1.33	2.58	1.02	1.68	0.44	1.02	0.30	0.59	10.76
1957	1.80	0.73	0.77	0.95	2.82	2.94	0.75	1.68	1.89	0.61	0.11	16.16
1958	0.56	2.16	1.24	0.67	1.09	4.68	2.32	0.42	0.28	0.31	1.24	1.17	16.14
1959	1.57	1.00	0.58	0.36	2.94	1.88	0.04	0.37	1.55	1.20	1.71	0.43	13.63
1960	0.28	0.52	0.15	2.13	1.71	0.52	0.39	2.66	0.43	0.04	0.79	0.19	9.81
1961	0.22	0.19	0.88	0.96	1.80	0.73	1.01	0.63	1.95	0.32	1.49	0.30	10.48
1962	1.28	0.95	0.74	0.58	5.18	2.30	1.09	1.69	0.10	1.17	0.36	0.51	15.95
1963	1.71	0.32	0.35	1.24	1.27	2.88	0.96	0.49	0.87	0.63	0.21	1.02	11.95
1964	0.65	0.52	1.74	1.91	3.36	4.34	1.50	1.66	0.28	T	0.72	1.23	17.91
1965	0.84	1.18	0.79	2.51	1.47	5.37	1.03	1.58	1.90	T	1.13	0.59	18.39
1966	1.63	0.51	0.79	0.74	1.54	2.17	1.81	0.77	0.21	1.32	1.62	0.99	14.10
1967	1.12	0.28	2.18	3.69	2.17	3.65	0.91	0.23	1.59	1.13	0.26	1.47	18.68
1968	1.29	0.24	0.90	1.11	2.64	2.89	0.06	2.16	2.92	0.11	0.68	1.36	16.36
1969	2.05	0.40	0.44	0.38	1.14	5.33	1.11	0.03	0.13	0.89	0.11	0.40	12.41
1970	0.99	1.02	1.14	1.88	3.16	2.32	1.16	0.77	0.67	1.00	0.53	0.70	15.34
1971	1.22	0.65	1.12	0.66	3.03	0.62	0.27	1.16	0.61	0.30	0.36	1.48	11.48
1972	1.47	0.62	1.01	0.77	1.59	0.94	1.51	1.26	0.85	1.17	0.20	1.68	13.07
1973	0.33	0.26	0.30	2.89	0.95	1.43	0.13	0.88	1.29	0.97	1.36	1.37	12.16
1974	1.44	0.26	1.10	1.03	3.16	1.08	0.48	4.76	0.73	0.26	0.60	0.15	15.26
1975	1.14	0.71	1.34	4.63	3.89	4.47	1.20	2.13	0.74	3.43	1.01	0.55	25.24
1976	0.57	0.53	0.75	2.33	0.88	4.10	2.07	1.91	0.61	0.19	0.65	0.51	15.10
1977	1.04	0.19	1.90	0.26	2.11	0.54	1.87	1.94	2.22	0.51	0.43	1.92	14.93
1978	1.68	1.21	0.41	1.76	3.20	2.56	1.99	1.04	2.56	0.27	1.44	1.05	19.17
1979	0.71	0.57	1.00	2.05	0.69	2.61	0.27	0.29	0.33	0.84	0.29	0.26	9.91
1980	0.67	1.03	0.74	0.62	5.12	3.91	0.27	0.67	0.98	1.75	0.19	0.27	16.22
1981	0.34	0.44	2.09	0.05	5.20	1.32	1.04	1.21	0.39	1.06	0.29	0.43	13.86
1982	1.09	0.99	1.97	1.04	3.63	3.09	0.66	0.41	2.43	0.75	0.63	0.99	17.68
1983	0.10	0.33	1.61	0.26	1.34	3.03	3.78	1.10	1.89	0.77	1.28	0.70	16.19
1984	0.72	0.69	1.31	0.94	1.34	2.10	0.05	1.01	0.71	1.20	0.49	1.25	11.81
1985	0.35	0.22	1.02	0.41	3.28	0.58	0.47	4.90	3.23	1.10	1.16	0.47	17.19
Record Mean	0.84	0.68	1.00	1.21	2.45	2.78	1.18	1.24	1.23	0.81	0.75	0.73	14.91

TABLE 3 AVERAGE TEMPERATURE (deg. F)   GREAT FALLS, MONTANA

YEAR	JAN	FEB	MAR	APR	MAY	JUNE	JULY	AUG	SEP	OCT	NOV	DEC	ANNUAL
1956	20.8	23.5	33.8	41.4	54.3	63.8	69.4	65.1	58.8	47.7	38.3	29.2	45.5
1957	7.9	22.4	33.8	42.8	56.4	61.9	71.8	65.6	59.3	40.7	35.0	36.6	44.5
1958	36.6	23.6	28.1	44.0	61.9	58.9	63.8	71.2	58.5	52.2	33.3	29.9	46.8
1959	23.1	16.9	38.1	43.4	48.5	63.0	70.8	65.8	55.7	44.8	29.7	36.0	44.7
1960	22.8	23.9	31.3	42.8	52.3	62.2	74.5	65.5	60.2	50.2	35.5	30.2	46.0
#1961	32.8	36.0	37.7	40.3	54.8	69.4	70.2	72.4	50.2	46.6	30.0	20.3	46.7
1962	19.0	20.7	25.9	48.1	50.8	61.4	65.0	67.3	56.7	50.5	40.3	32.6	44.8
1963	12.8	36.8	39.4	42.9	52.9	61.4	68.9	68.8	64.0	54.1	38.6	23.5	47.0
1964	28.7	33.0	26.9	42.3	54.8	62.5	72.4	64.5	53.9	53.0	31.9	11.2	44.7
1965	23.9	27.0	20.9	44.7	51.5	60.3	69.6	68.0	45.0	53.4	34.9	29.1	44.0
1966	13.9	27.1	36.4	40.2	55.8	59.9	70.0	65.1	64.2	47.4	29.0	27.9	44.8
1967	26.3	32.1	27.2	35.5	52.5	60.2	73.0	71.0	62.9	50.2	36.2	22.1	45.8
1968	22.0	33.2	40.5	40.6	49.5	59.2	67.8	64.4	56.5	48.0	35.8	16.1	44.5
1969	-2.8	16.6	26.2	50.3	55.8	59.3	68.3	72.1	61.3	38.1	40.1	29.9	42.9
1970	14.7	32.3	29.0	38.5	53.8	66.5	71.4	70.9	54.6	44.7	30.2	23.0	44.1
1971	16.2	29.4	31.6	45.0	54.6	62.0	67.5	76.0	54.7	44.7	36.3	18.0	44.6
1972	12.8	22.5	38.3	42.4	53.5	65.2	64.7	69.6	53.9	43.2	36.3	17.9	43.4
1973	24.9	29.6	39.5	40.2	55.4	63.6	71.4	71.2	58.2	49.2	25.2	28.9	46.4
1974	19.8	33.8	33.3	47.1	49.5	66.9	72.9	62.9	54.8	51.4	38.7	32.5	47.0
1975	22.7	13.1	27.4	30.9	50.0	58.2	71.8	64.8	57.3	45.7	32.9	28.6	42.0
1976	26.4	30.5	31.6	46.1	56.9	61.0	70.3	67.9	61.2	46.5	36.6	31.6	47.2
1977	21.6	39.2	34.0	47.1	51.3	65.4	68.0	62.5	56.1	47.8	30.8	16.5	45.1
1978	7.7	14.5	33.6	44.1	51.3	62.5	67.1	66.5	58.8	48.8	17.4	41.4
1979	6.5	18.8	34.7	40.7	51.5	62.9	69.0	68.5	62.9	49.4	33.4	34.6	44.4
1980	15.2	28.1	32.4	52.9	57.1	60.9	69.4	62.4	58.0	49.0	39.4	23.8	45.7
1981	33.8	30.8	37.2	46.5	52.8	58.1	66.5	69.8	59.8	45.3	40.8	24.6	47.2
1982	6.3	19.5	27.9	37.9	48.4	60.2	66.8	65.0	53.9	46.6	32.1	26.8	41.0
1983	32.2	36.7	35.8	41.7	50.7	60.1	65.8	72.4	53.6	48.9	35.1	4.0	44.8
1984	29.5	36.9	35.3	44.4	51.7	59.8	69.8	71.6	51.8	40.2	35.5	13.0	45.0
1985	19.2	21.6	33.4	48.5	57.3	62.2	73.0	61.8	48.2	44.5	12.3	24.6	42.2
Record Mean	20.9	26.4	31.9	43.6	53.3	60.9	69.4	67.7	57.3	48.0	34.2	26.0	45.0
Max	30.2	36.2	42.0	54.9	65.0	72.8	83.7	81.7	69.8	59.1	43.3	34.8	56.1
Min	11.6	16.6	21.7	32.3	41.5	49.0	55.1	53.6	44.8	36.9	25.1	17.2	33.8

REFERENCE NOTES FOR TABLES 1, 2, 3 and 6   (GREAT FALLS, MT)

GENERAL

T - TRACE AMOUNT
BLANK ENTRIES DENOTE MISSING/UNREPORTED DATA.
# INDICATES A STATION OR INSTRUMENT RELOCATION.

SPECIFIC

TABLE 1

(a) - LENGTH OF RECORD IN YEARS. ALTHOUGH INDIVIDUAL MONTHS MAY BE MISSING.

* LESS THAN .05

NORMALS — BASED ON THE 1951-1980 RECORD PERIOD.
EXTREMES — DATES ARE THE MOST RECENT OCCURRENCE.
WIND DIR. — NUMERALS SHOW TENS OF DEGREES CLOCKWISE FROM TRUE NORTH. "00" INDICATES CALM.
RESULTANT WIND DIRECTIONS ARE GIVEN TO WHOLE DEGREES.

EXCEPTIONS

TABLES 2, 3, and 6

RECORD MEANS ARE THROUGH THE CURRENT YEAR, BEGINNING IN   1938 FOR TEMPERATURE
1938 FOR PRECIPITATION
1938 FOR SNOWFALL

TABLE 4 HEATING DEGREE DAYS Base 65 deg. F GREAT FALLS, MONTANA

SEASON	JULY	AUG	SEP	OCT	NOV	DEC	JAN	FEB	MAR	APR	MAY	JUNE	TOTAL
1956-57	40	77	203	532	790	1107	1770	1189	962	661	259	124	7714
1957-58	7	69	197	750	894	877	874	1153	1138	623	144	194	6920
1958-59	106	13	235	400	944	1081	1297	1347	828	637	505	109	7502
1959-60	26	67	310	621	1055	890	1301	1187	1041	660	409	114	7681
#1960-61	11	78	187	455	878	1073	989	807	839	732	326	14	6389
1961-62	4	9	449	566	1043	1378	1426	1236	1208	501	432	140	8392
1962-63	55	57	250	445	734	997	1615	786	768	655	381	144	6907
1963-64	23	23	103	338	782	1279	1118	923	1176	674	318	118	6875
1964-65	0	93	337	375	990	1666	1271	1055	1361	602	411	164	8325
1965-66	17	54	594	355	894	1106	1585	1054	880	736	304	182	7761
1966-67	7	85	101	537	1075	1143	1190	914	1166	878	396	168	7660
1967-68	2	8	136	453	858	1327	1329	914	753	724	472	198	7174
1968-69	38	93	261	520	867	1511	2104	1350	1198	432	292	199	8865
1969-70	27	8	172	828	743	1081	1559	908	1108	785	345	95	7659
1970-71	9	8	319	625	1036	1296	1506	993	1031	594	320	134	7874
1971-72	34	5	326	628	854	1455	1616	1226	820	667	368	77	8076
1972-73	109	23	331	668	856	1458	1240	983	785	739	477	85	7620
1973-74	6	27	226	483	1191	1111	1397	865	974	530	372	190	7372
1974-75	6	109	311	419	783	1000	1304	1450	1159	1015	460	190	8206
1975-76	12	60	235	592	961	1122	1192	994	1030	565	250	165	7178
1976-77	3	20	144	572	845	1031	1339	715	953	529	419	70	6640
1977-78	37	119	280	527	1021	1502	1776	1410	966	622	421	106	8787
1978-79	54	57	236	496	1228	1473	1808	1292	931	722	417	111	8825
1979-80	19	15	106	482	939	934	1538	1066	1004	370	267	148	6888
1980-81	16	110	225	504	763	1275	960	953	855	548	373	218	6800
1981-82	34	14	201	603	718	1244	1819	1271	1142	806	511	161	8524
1982-83	44	66	342	565	978	1181	1007	786	899	692	437	154	7151
1983-84	59	2	356	490	891	1888	1094	810	915	620	419	183	7727
1984-85	12	18	415	760	879	1611	1415	1212	971	489	249	134	8165
1985-86	4	147	498	629	1581	1246							

TABLE 5 COOLING DEGREE DAYS Base 65 deg. F GREAT FALLS, MONTANA

YEAR	JAN	FEB	MAR	APR	MAY	JUNE	JULY	AUG	SEP	OCT	NOV	DEC	TOTAL
1969	0	0	0	0	15	34	136	235	67	0	0	0	487
1970	0	0	0	0	4	144	214	197	15	6	0	0	580
1971	0	0	0	0	6	50	120	351	22	8	0	0	557
1972	0	0	0	0	19	87	108	175	5	0	0	0	394
1973	0	0	0	0	14	87	213	226	30	0	0	0	570
1974	0	0	0	0	0	148	253	54	11	7	0	0	473
1975	0	0	0	0	0	10	231	62	12	0	0	0	315
1976	0	0	0	0	6	51	174	116	37	5	0	0	389
1977	0	0	0	0	0	86	139	48	20	0	0	0	293
1978	0	0	0	0	0	36	125	111	60	0	0	0	332
1979	0	0	0	0	2	55	152	132	50	5	0	0	396
1980	0	0	0	12	30	31	156	37	21	18	0	0	305
1981	0	0	0	0	3	17	85	168	49	0	0	0	322
1982	0	0	0	0	0	24	104	73	15	0	0	0	216
1983	0	0	0	0	4	14	90	241	19	0	0	0	368
1984	0	0	0	5	15	33	169	229	26	0	0	0	477
1985	0	0	0	0	20	58	260	51	0	0	0	0	389

TABLE 6 SNOWFALL (inches) GREAT FALLS, MONTANA

SEASON	JULY	AUG	SEP	OCT	NOV	DEC	JAN	FEB	MAR	APR	MAY	JUNE	TOTAL
1956-57	0.0	0.0	T	2.4	2.4	10.3	20.7	9.5	6.9	4.0	0.0	0.0	56.2
1957-58	0.0	0.0	3.6	14.2	5.7	0.7	5.3	26.1	15.9	1.2	T	0.0	72.7
1958-59	0.0	0.0	T	1.0	13.8	17.6	19.3	14.7	5.0	2.7	T	0.0	74.1
1959-60	0.0	0.0	0.8	8.9	16.6	4.5	2.9	6.1	1.9	7.9	1.7	0.0	59.8
1960-61	0.0	0.0	0.0	0.0	T	7.8	2.0	2.9	0.6	7.9	6.1	T	27.3
1961-62	0.0	0.0	0.4	2.6	15.0	3.0	14.7	8.9	5.9	3.8	T	0.0	54.3
1962-63	0.0	0.0	0.4	T	2.1	3.7	17.7	3.2	3.9	11.1	T	0.0	42.1
1963-64	0.0	0.0	0.0	0.0	2.1	10.6	8.5	5.7	17.4	13.2	T	0.0	57.5
1964-65	0.0	0.0	T	T	7.9	13.8	6.9	12.5	7.7	11.5	T	0.7	61.0
1965-66	0.0	0.0	4.2	0.0	9.7	6.5	16.9	5.8	7.7	6.0	T	0.0	56.8
1966-67	T	0.0	0.0	3.4	15.2	9.8	10.9	2.8	21.3	35.4	8.0	0.0	106.8
1967-68	0.0	0.0	0.0	0.7	2.4	14.9	12.8	1.9	7.6	6.1	T	0.0	46.4
1968-69	0.0	0.0	3.9	0.2	2.6	13.6	5.2	5.5	0.9	22.6	0.1	5.3	59.9
1969-70	0.0	0.0	0.0	5.0	1.1	4.3	9.4	11.4	10.9	18.7	1.4	0.0	62.2
1970-71	0.0	0.0	3.0	6.4	7.3	16.2	8.9	9.7	4.7	T	0.0	0.0	61.1
1971-72	0.0	0.0	T	3.8	4.2	19.2	18.6	6.3	11.4	3.5	0.2	0.0	67.2
1972-73	0.0	0.0	0.5	9.5	2.0	16.2	3.3	1.8	1.8	24.8	0.3	0.0	60.2
1973-74	0.0	0.0	6.0	3.6	12.2	12.8	13.4	2.8	9.5	6.1	T	T	66.4
1974-75	0.0	0.0	1.3	1.1	1.3	5.9	13.2	7.2	12.4	29.2	5.6	0.0	77.2
1975-76	0.0	0.0	T	16.6	9.7	5.7	6.1	5.6	8.9	16.7	0.0	0.0	69.3
1976-77	0.0	0.0	0.0	0.4	8.8	8.3	13.9	1.8	21.5	1.0	2.1	T	57.8
1977-78	0.0	0.0	0.0	3.2	4.8	18.2	19.3	16.8	8.1	5.0	T	T	70.6
1978-79	0.0	0.0	0.0	T	16.5	11.5	12.0	8.1	14.8	8.6	2.6	T	74.1
1979-80	0.0	0.0	0.0	5.0	3.1	3.0	7.0	9.2	6.9	4.4	T	0.0	34.3
1980-81	0.0	0.0	0.0	7.7	3.3	5.4	4.1	7.1	11.5	0.1	T	0.0	39.2
1981-82	0.0	0.0	0.0	7.9	1.0	5.9	19.7	16.3	23.4	18.5	7.6	T	100.3
1982-83	0.0	0.0	0.7	1.5	8.8	13.0	0.9	4.1	6.6	1.4	8.6	0.0	45.6
1983-84	0.0	0.0	7.8	T	14.4	11.9	16.2	7.7	19.5	5.2	1.0	0.0	83.7
1984-85	0.0	0.0	10.4	10.9	5.5	16.6	5.4	3.8	11.9	2.4	0.0	0.0	66.9
1985-86	0.0	T	2.5	8.5	18.1	7.9							
Record Mean	T	T	1.5	3.3	7.6	8.9	10.1	8.4	10.2	7.2	1.6	0.4	59.0

See Reference Notes, relative to all above tables, on preceding page.

Havre, Montana, is located in a level valley formed by the Milk River, which courses through the city from west to east. Most of the city lies on the south side of the river. On the north side, hills rise abruptly to about 200 feet above the valley floor. The land mass north to the Canadian border is gently rolling and increases slightly in elevation. During winter months, frequent invasions of cold polar continental air move down across these rolling plains, bringing snow and sub-zero temperatures.

The Bearpaw Mountains extend from 15 to 30 miles south of Havre. Most of the peaks are from 4,000 to 5,000 feet above sea level, and several are above 6,000 feet. The highest is Old Baldy, 6,916 feet above sea level.

Winters are cold in the Havre area, but snow cover is seldom more than a few inches, and usually some ground is bare. Spells of mild weather do occur at least a few times each winter, arriving with sometimes fresh to strong southwest to west foehn winds. During winter months, rain rarely falls. Winter precipitation is almost always in the form of snow. The transition from winter to spring conditions is fairly rapid in the usual year, but cold snaps and snow can occur as late as early May or as early as September.

Summers are characterized by warm weather, seldom exceeding 95 degrees. Daytime warmest readings usually run from the 80s to the mid-90s during most of July and August, but summer relative humidities are seldom as high as 50 percent during afternoon hours. Summertime night temperatures are rarely oppressively warm. Most spring and summer precipitation falls as showers, but occasionally steady rains lasting several hours are observed in May and June, and again in September. Fall seasons are characterized by much clear weather, although cold snaps of a day or two, with some snow, can occur as early as mid-September.

TABLE 1   NORMALS, MEANS AND EXTREMES

HAVRE, MONTANA

LATITUDE: 48°33'N   LONGITUDE: 109°46'W   ELEVATION: FT. GRND 2584 BARO 02585   TIME ZONE: MOUNTAIN   WBAN: 94012

	(a)	JAN	FEB	MAR	APR	MAY	JUNE	JULY	AUG	SEP	OCT	NOV	DEC	YEAR
TEMPERATURE °F:														
Normals														
-Daily Maximum		21.5	30.3	39.8	55.3	67.8	76.4	85.4	83.6	71.6	60.0	41.5	29.6	55.3
-Daily Minimum		0.4	8.8	16.9	30.1	41.2	49.1	53.8	52.0	42.0	31.8	17.9	8.0	29.3
-Monthly		10.9	19.6	28.4	42.7	54.5	62.8	69.6	67.8	56.8	45.9	29.7	18.8	42.3
Extremes														
-Record Highest	25	63	68	75	91	98	105	106	111	101	90	78	65	111
-Year		1968	1981	1977	1980	1980	1984	1975	1961	1967	1980	1975	1962	AUG 1961
-Record Lowest	25	-52	-35	-28	-14	24	31	39	36	18	-16	-30	-50	-52
-Year		1969	1982	1978	1975	1967	1979	1981	1981	1984	1984	1985	1983	JAN 1969
NORMAL DEGREE DAYS:														
Heating (base 65°F)		1674	1271	1135	669	333	131	21	64	279	592	1059	1432	8660
Cooling (base 65°F)		0	0	0	0	8	65	163	151	33	0	0	0	420
% OF POSSIBLE SUNSHINE	23	50	61	69	70	73	76	85	82	74	65	50	46	67
MEAN SKY COVER (tenths)														
Sunrise - Sunset	20	7.4	7.2	7.2	7.1	6.6	6.2	4.2	4.7	5.7	6.5	7.1	7.4	6.4
MEAN NUMBER OF DAYS:														
Sunrise to Sunset														
-Clear	20	4.3	4.8	4.8	4.3	5.7	6.7	14.4	12.9	9.1	7.6	5.1	4.6	84.3
-Partly Cloudy	20	7.0	6.7	7.9	8.9	10.9	10.9	10.5	10.1	8.5	7.8	7.6	7.1	103.9
-Cloudy	20	19.5	16.6	18.3	16.8	14.4	12.4	6.2	7.9	12.4	15.7	17.4	19.4	176.9
Precipitation														
.01 inches or more	25	9.0	6.0	7.7	7.8	10.0	9.6	7.4	6.9	7.0	5.1	5.4	7.9	89.6
Snow, Ice pellets														
1.0 inches or more	24	2.7	2.1	2.1	2.0	0.2	0.0	0.0	0.0	0.1	0.8	1.8	3.0	15.0
Thunderstorms	17	0.0	0.0	0.0	0.6	2.8	5.1	5.6	5.3	1.2	0.2	0.0	0.0	20.9
Heavy Fog Visibility														
1/4 mile or less	17	0.4	1.1	1.3	1.1	0.5	0.1	0.1	0.1	0.4	0.8	1.6	0.9	8.3
Temperature °F														
-Maximum														
90° and above	25	0.0	0.0	0.0	0.*	0.3	3.6	10.5	10.5	2.1	0.*	0.0	0.0	27.1
32° and below	25	18.6	11.9	7.2	1.0	0.0	0.0	0.0	0.0	0.0	0.9	7.7	16.6	63.9
-Minimum														
32° and below	25	29.7	26.5	27.6	16.7	3.3	0.1	0.0	0.0	3.9	16.0	26.6	29.6	180.0
0° and below	25	15.4	8.0	4.1	0.3	0.0	0.0	0.0	0.0	0.0	0.2	3.7	11.0	42.8
AVG. STATION PRESS. (mb)	10	923.3	922.7	919.6	921.5	920.7	920.8	922.5	922.7	923.6	923.5	923.0	922.7	922.2
RELATIVE HUMIDITY (%)														
Hour 05	25	74	77	78	74	73	75	71	68	72	74	75	75	74
Hour 11 (Local Time)	25	70	68	61	49	44	44	37	37	45	50	62	70	53
Hour 17	19	70	64	54	40	37	35	28	28	35	45	61	70	47
Hour 23	11	73	74	73	60	58	60	53	50	63	65	74	75	65
PRECIPITATION (inches):														
Water Equivalent														
-Normal		0.59	0.41	0.51	1.09	1.61	2.12	1.25	1.15	0.92	0.57	0.43	0.52	11.17
-Maximum Monthly	25	2.33	1.04	2.03	2.59	4.99	4.72	4.03	3.57	3.11	2.06	1.23	2.04	4.99
-Year		1971	1979	1977	1967	1974	1965	1983	1974	1985	1980	1978	1977	MAY 1974
-Minimum Monthly	25	T	0.06	0.03	0.07	0.28	0.16	0.01	T	0.07	0.06	T	0.05	T
-Year		1973	1976	1973	1981	1967	1985	1967	1967	1969	1965	1969	1962	JAN 1973
-Maximum in 24 hrs	25	0.51	0.36	1.80	1.62	2.12	2.23	2.57	2.34	1.71	1.29	0.43	0.78	2.57
-Year		1971	1973	1977	1973	1983	1970	1983	1968	1978	1980	1978	1977	JUL 1983
Snow, Ice pellets														
-Maximum Monthly	24	41.4	18.6	30.1	32.8	30.6	T			6.2	10.5	18.9	25.1	41.4
-Year		1971	1978	1977	1975	1982	1969			1982	1984	1978	1977	JAN 1971
-Maximum in 24 hrs	25	11.7	6.2	26.9	16.2	23.5	T			6.2	8.5	6.6	6.1	26.9
-Year		1971	1973	1977	1967	1982	1969			1982	1981	1978	1977	MAR 1977
WIND:														
Mean Speed (mph)	16	10.3	10.0	9.9	10.9	10.2	9.6	9.1	8.9	9.6	9.8	9.6	10.2	9.9
Prevailing Direction														
through 1963		SW	SW	SW	SW	E	SW	SW	W	E	SW	SW	SW	SW
Fastest Mile														
-Direction (!!!)	25	SW	SW	SW	SW	W	W	NW	W	W	NW	SW	SW	NW
-Speed (MPH)	25	47	58	52	59	54	63	71	59	52	52	54	47	71
-Year		1964	1963	1964	1983	1965	1961	1961	1984	1971	1963	1978	1968	JUL 1961
Peak Gust														
-Direction (!!!)														
-Speed (mph)														
-Date														

See Reference Notes to this table on the following page.

TABLE 2 PRECIPITATION (inches) HAVRE, MONTANA

YEAR	JAN	FEB	MAR	APR	MAY	JUNE	JULY	AUG	SEP	OCT	NOV	DEC	ANNUAL
1956	0.16	0.13	0.14	0.30	1.57	1.75	1.41	1.63	1.19	0.41	0.18	0.46	9.33
1957	0.92	0.66	0.32	0.68	1.20	2.63	0.90	1.29	0.98	1.78	0.25	0.04	11.65
1958	0.25	1.06	0.39	0.23	0.52	1.81	1.71	0.43	0.54	1.22	2.42	0.57	11.15
1959	0.77	0.34	0.08	1.34	1.50	4.37	0.07	0.58	0.66	2.24	1.27	0.13	13.35
#1960	0.67	0.32	0.11	2.82	1.19	1.44	0.10	1.98	0.19	0.52	0.99	0.33	10.66
1961	0.18	0.16	0.18	1.32	1.31	0.70	0.89	0.19	1.26	0.73	0.22	0.29	7.43
1962	0.18	0.36	0.41	0.57	2.15	3.36	1.66	0.93	0.74	1.09	0.09	0.05	11.59
1963	0.54	0.78	0.10	0.80	1.01	1.88	2.13	0.70	0.36	0.64	0.07	0.52	9.53
1964	0.34	0.11	0.20	1.18	2.28	1.88	1.18	1.84	1.49	0.13	0.45	1.06	12.14
1965	0.56	0.51	0.31	1.12	1.01	4.72	3.77	2.77	2.45	0.06	0.64	0.19	18.11
1966	0.68	0.11	0.48	0.41	0.53	2.04	1.69	0.50	0.17	0.83	0.65	0.14	8.23
1967	0.35	0.46	1.28	2.59	0.28	0.69	0.01	T	1.77	0.37	0.18	0.29	8.27
1968	0.27	0.35	0.77	0.93	1.17	2.53	0.49	3.03	2.47	0.17	0.06	0.77	13.01
1969	1.66	0.23	0.19	0.33	0.82	2.17	2.13	0.01	0.07	0.48	T	0.24	8.33
1970	0.63	0.24	0.36	1.55	1.05	3.69	2.04	0.39	0.60	0.28	0.34	0.67	11.84
1971	2.33	0.33	0.07	0.39	0.56	1.75	0.37	0.70	0.15	0.33	0.23	0.84	8.05
1972	0.86	0.49	1.04	0.75	1.26	0.54	1.17	1.74	1.17	0.26	0.02	0.59	9.89
1973	T	0.77	0.03	2.35	1.01	2.97	0.32	0.65	0.88	1.00	1.04	0.71	11.73
1974	0.51	0.19	0.75	0.85	4.99	0.19	0.71	3.57	0.11	0.14	0.06	0.35	12.42
1975	0.13	0.32	1.40	2.32	1.74	3.30	2.99	3.11	0.52	1.02	0.80	0.59	18.24
1976	0.19	0.06	0.69	0.43	0.72	2.66	1.21	1.11	0.13	0.23	0.41	0.38	8.22
1977	0.78	0.07	2.03	0.12	3.27	0.58	0.43	0.82	1.74	0.10	0.61	2.04	12.59
1978	1.11	1.01	0.40	1.57	3.20	1.25	2.09	0.58	2.59	0.22	1.23	0.59	15.84
1979	0.49	1.04	0.55	1.73	0.87	1.25	1.18	0.22	0.65	0.55	0.12	0.20	8.85
1980	0.41	0.39	0.65	1.25	1.62	2.97	0.26	0.55	0.45	2.06	0.20	0.67	11.42
1981	0.08	0.08	1.09	0.07	2.60	1.19	1.73	0.99	0.69	1.15	0.11	0.27	10.05
1982	1.26	0.52	1.00	0.54	4.50	0.63	2.08	1.77	1.22	0.70	0.14	0.63	14.99
1983	0.26	0.11	1.15	0.60	2.62	1.60	4.03	0.21	0.83	0.25	0.40	0.54	12.60
1984	0.61	0.22	0.85	0.35	0.85	0.93	0.02	0.52	0.70	1.01	0.34	0.59	6.99
1985	0.13	0.23	0.98	0.89	0.96	0.16	0.40	2.37	3.11	0.71	1.08	0.12	11.14
Record Mean	0.63	0.44	0.58	0.96	1.74	2.57	1.56	1.17	1.15	0.67	0.53	0.55	12.55

TABLE 3 AVERAGE TEMPERATURE (deg. F) HAVRE, MONTANA

YEAR	JAN	FEB	MAR	APR	MAY	JUNE	JULY	AUG	SEP	OCT	NOV	DEC	ANNUAL
1956	10.7	15.3	32.3	40.9	56.0	66.0	70.1	66.5	58.1	47.4	35.3	25.5	43.7
1957	3.0	17.8	32.2	43.8	57.8	63.0	73.2	67.9	58.5	40.5	33.9	33.9	43.8
1958	30.6	15.4	25.2	45.5	62.0	61.5	66.5	72.4	58.6	48.4	30.9	22.5	45.0
1959	10.5	8.0	38.6	43.9	51.6	65.0	71.3	66.1	54.5	42.9	27.0	31.8	42.6
#1960	16.0	19.0	28.1	44.3	54.4	62.9	75.8	66.9	59.3	49.8	30.3	21.2	44.0
1961	27.5	29.8	37.3	39.4	56.2	69.9	70.9	72.9	50.8	45.4	26.2	11.0	44.8
1962	16.0	16.3	20.5	47.2	53.2	64.0	65.7	67.7	55.6	48.9	38.9	26.6	43.4
1963	6.7	28.7	38.3	43.8	54.0	63.3	70.4	69.9	63.8	52.6	31.9	16.2	45.0
1964	22.3	31.8	27.1	42.8	54.7	61.7	71.7	65.1	52.5	48.1	27.1	0.8	42.1
1965	8.8	16.2	14.9	41.2	54.7	60.0	68.4	67.4	46.2	50.5	24.8	21.4	39.3
1966	1.0	14.0	33.2	39.1	56.6	60.7	70.8	65.7	61.4	45.6	22.3	18.3	40.8
1967	17.1	23.2	24.3	33.6	51.4	59.7	71.5	71.1	62.6	48.1	32.2	18.8	42.8
1968	11.9	24.3	39.1	41.8	51.1	60.3	68.6	64.4	55.9	44.7	32.2	8.8	41.9
1969	-11.3	6.7	19.6	49.3	55.2	60.7	67.7	70.7	60.9	38.8	36.8	22.4	39.8
1970	7.1	20.5	25.5	39.6	55.7	66.8	70.8	69.9	55.4	42.0	23.6	10.4	40.6
1971	6.3	19.8	30.2	45.2	56.1	63.4	68.0	75.5	54.6	43.1	32.9	7.3	41.9
1972	3.5	12.5	33.6	43.6	56.3	67.3	66.0	72.5	52.9	40.9	33.2	12.2	41.2
1973	23.8	22.9	39.1	41.0	56.5	63.9	69.8	70.6	57.4	48.5	20.0	21.0	44.5
1974	13.9	29.8	31.3	48.3	50.8	67.3	73.1	63.6	55.0	47.7	34.3	29.1	45.4
1975	20.0	11.2	25.6	33.7	53.3	60.8	73.1	65.2	56.5	45.6	29.9	24.7	41.6
1976	22.5	33.1	30.8	48.5	57.9	62.0	71.7	70.6	61.3	43.6	30.2	26.7	46.6
1977	9.6	34.1	34.8	47.4	56.4	66.6	70.2	63.6	57.3	47.6	27.4	12.1	43.9
1978	2.5	12.2	29.1	46.6	55.6	63.9	67.9	66.7	59.3	47.4	21.1	12.4	40.4
1979	-2.7	3.3	31.8	40.4	52.3	65.1	70.0	70.3	62.7	49.0	31.2	28.9	41.8
1980	11.6	22.1	28.7	52.2	59.4	64.3	69.7	62.3	57.5	46.4	36.0	18.9	44.1
1981	28.5	27.3	38.1	46.4	55.3	57.9	67.4	70.7	59.2	43.9	38.1	17.1	45.8
1982	-3.1	12.2	27.3	39.3	49.6	62.3	67.0	67.4	55.1	45.4	25.7	23.9	39.3
1983	24.7	31.5	35.5	42.3	52.6	61.2	67.0	73.0	53.7	47.4	31.2	-5.0	42.9
1984	23.6	36.2	31.8	45.0	53.1	61.9	71.1	70.8	50.2	38.0	29.0	9.2	43.3
1985	12.8	15.1	33.7	47.6	58.2	60.2	72.6	63.8	48.9	43.6	9.1	19.9	40.5
Record Mean	13.4	17.6	28.9	44.0	54.5	62.4	69.6	67.4	56.3	45.7	30.1	19.9	42.5
Max	23.6	28.2	39.9	56.6	67.2	75.1	84.2	82.3	70.2	58.7	41.1	29.9	54.7
Min	3.2	7.1	17.9	31.5	41.7	49.7	54.9	52.5	42.5	32.6	19.2	9.9	30.2

REFERENCE NOTES FOR TABLES 1, 2, 3 and 6 (HAVRE, MT)

GENERAL

T - TRACE AMOUNT
BLANK ENTRIES DENOTE MISSING/UNREPORTED DATA.
# INDICATES A STATION OR INSTRUMENT RELOCATION.

SPECIFIC

TABLE 1

(a) - LENGTH OF RECORD IN YEARS. ALTHOUGH INDIVIDUAL MONTHS MAY BE MISSING.
* LESS THAN .05

NORMALS — BASED ON THE 1951-1980 RECORD PERIOD.
EXTREMES — DATES ARE THE MOST RECENT OCCURRENCE.
WIND DIR. — NUMERALS SHOW TENS OF DEGREES CLOCKWISE FROM TRUE NORTH.
     "00" INDICATES CALM.
RESULTANT WIND DIRECTIONS ARE GIVEN TO WHOLE DEGREES.

EXCEPTIONS

TABLE 1

1. THUNDERSTORMS AND HEAVY FOG ARE THROUGH 1977.

TABLES 2, 3, and 6

RECORD MEANS ARE THROUGH THE CURRENT YEAR, BEGINNING IN 1980 FOR TEMPERATURE
     1880 FOR PRECIPITATION
     1961 FOR SNOWFALL

TABLE 4 HEATING DEGREE DAYS Base 65 deg. F   HAVRE, MONTANA

SEASON	JULY	AUG	SEP	OCT	NOV	DEC	JAN	FEB	MAR	APR	MAY	JUNE	TOTAL
1956-57	20	56	210	538	885	1218	1924	1319	1006	628	220	86	8110
1957-58	3	35	204	753	925	960	1062	1385	1227	579	133	128	7394
1958-59	60	11	226	512	1017	1312	1686	1596	815	628	407	78	8348
1959-60	9	48	320	677	1136	1025	1512	1329	1139	612	334	98	8239
#1960-61	5	47	208	467	1036	1351	1157	982	857	759	297	7	7173
1961-62	4	7	432	602	1158	1672	1513	1360	1373	529	360	86	9096
1962-63	41	49	281	495	779	1185	1808	1010	819	632	345	116	7560
1963-64	16	19	110	383	988	1508	1318	958	1169	658	322	119	7568
1964-65	4	85	370	518	1129	1992	1740	1361	1551	706	405	179	10040
1965-66	17	57	553	443	1202	1348	1984	1425	978	769	295	161	9232
1966-67	3	72	140	594	1274	1444	1480	1164	1253	935	418	178	8955
1967-68	7	11	125	519	980	1428	1647	1171	797	688	421	166	7960
1968-69	42	69	273	620	975	1741	2370	1631	1403	465	315	170	10074
1969-70	26	15	159	808	842	1317	1793	1242	1216	753	282	54	8507
1970-71	7	22	304	708	1237	1689	1821	1261	1074	589	285	93	9090
1971-72	27	5	312	673	957	1785	1906	1520	969	635	281	32	9102
1972-73	74	9	360	741	947	1635	1273	1173	792	713	266	99	8082
1973-74	8	22	235	505	1344	1359	1585	979	1037	497	436	60	8067
1974-75	1	91	299	530	1109	1388	1506	1216	935	355	134		8479
1975-76	7	49	257	597	1046	1245	1313	915	1055	492	217	120	7313
1976-77	4	13	146	663	1037	1179	1714	860	929	521	263	50	7379
1977-78	20	95	236	531	1124	1640	1473	1108	986	546	286	79	9074
1978-79	40	48	218	544	1317	1625	2098	1728	1024	733	392	72	9839
1979-80	23	13	110	495	1005	1118	1654	1239	1120	380	206	81	7444
1980-81	15	102	236	580	865	1426	1125	1051	828	550	297	219	7294
1981-82	26	15	189	648	798	1477	2113	1475	1162	764	471	112	9250
1982-83	33	38	311	600	1174	1266	1241	930	906	674	384	135	7692
1983-84	46	5	347	540	1007	2173	1280	827	1024	593	377	145	8364
1984-85	10	29	453	832	1073	1727	1614	1395	965	516	234	162	9010
1985-86	5	110	475	654	1677	1393							

TABLE 5 COOLING DEGREE DAYS Base 65 deg. F   HAVRE, MONTANA

YEAR	JAN	FEB	MAR	APR	MAY	JUNE	JULY	AUG	SEP	OCT	NOV	DEC	TOTAL
1969	0	0	0	0	15	47	118	199	45	0	0	0	424
1970	0	0	0	0	0	114	193	179	23	3	0	0	512
1971	0	0	0	0	18	52	126	338	7	0	0	0	541
1972	0	0	0	0	19	108	110	249	3	0	0	0	489
1973	0	0	0	0	9	73	166	199	15	0	0	0	462
1974	0	0	0	4	0	138	260	55	8	0	0	0	465
1975	0	0	0	0	0	14	267	64	8	0	0	0	353
1976	0	0	0	0	2	36	220	193	42	3	0	0	496
1977	0	0	0	0	4	103	188	62	14	0	0	0	371
1978	0	0	0	0	2	52	136	108	54	0	0	0	352
1979	0	0	0	0	5	83	187	185	48	6	0	0	514
1980	0	0	0	2	38	64	169	27	18	11	0	0	329
1981	0	0	0	0	4	12	111	195	22	0	0	0	344
1982	0	0	0	0	0	36	103	115	20	0	0	0	274
1983	0	0	0	0	6	24	116	259	16	0	0	0	421
1984	0	0	0	0	17	58	207	217	15	0	0	0	514
1985	0	0	0	0	31	25	247	79	0	0	0	0	382

TABLE 6 SNOWFALL (inches)   HAVRE, MONTANA

SEASON	JULY	AUG	SEP	OCT	NOV	DEC	JAN	FEB	MAR	APR	MAY	JUNE	TOTAL
1956-57	0.0	0.0	0.0	0.4	1.8	9.2	14.8	9.7	1.8	4.1	0.0	0.0	41.8
1957-58	0.0	0.0	0.7	7.6	2.6	T	3.0	16.8	2.1	T	0.0	0.0	32.8
1958-59	0.0	0.0	0.0	T	26.4	8.3	10.9	4.0	0.1	T	T	0.0	49.7
1959-60	0.0	0.0	T	10.3	9.6	2.8	9.8	6.3	3.3	9.6	T	0.0	51.7
#1960-61	0.0	0.0	0.0	0.7	9.0	3.7	2.0	0.4	1.4	1.4	0.0	0.0	18.6
1961-62	0.0	0.0	0.0	1.7	4.0	4.5	3.1	5.3	3.1	T	0.0	0.0	21.7
1962-63	0.0	0.0	T	0.0	T	T	6.8	6.1	T	5.2	0.0	0.0	18.1
1963-64	0.0	0.0	0.0	0.0	1.4	8.4	4.2	1.4	3.5	5.3	0.0	0.0	24.2
1964-65	0.0	0.0	T	T	5.4	14.1	7.1	7.2	5.6	6.4	T	0.0	45.8
1965-66	0.0	0.0	2.8	0.0	9.5	3.7	11.6	2.4	5.8	5.9	0.0	0.0	41.7
1966-67	0.0	0.0	0.0	T	9.3	2.8	6.3	6.1	14.4	31.1	1.7	0.0	71.7
1967-68	0.0	0.0	0.0	T	1.8	5.1	4.8	5.8	4.6	7.0	T	0.0	29.1
1968-69	0.0	0.0	T	T	1.1	11.3	25.2	5.1	2.9	2.0	0.0	T	47.6
1969-70	0.0	0.0	0.0	1.3	T	3.9	8.7	7.1	6.2	13.2	T	0.0	40.4
1970-71	0.0	0.0	T	1.0	5.9	11.4	41.4	4.2	2.1	0.2	0.0	0.0	66.2
1971-72	0.0	0.0	T	3.7	T	12.0	15.0	8.5	10.6	7.0	T	0.0	56.8
1972-73	0.0	0.0	2.2	4.6	T	9.3	T	11.1	0.0	19.3	T	0.0	46.5
1973-74	0.0	0.0	0.0	0.4	10.4	10.8	8.7	1.0	8.3	3.1	1.5	0.0	44.2
1974-75	0.0	0.0	T	1.3	0.4	4.1	2.2	5.6	18.2	32.8	T	0.0	64.6
1975-76	0.0	0.0	0.0	1.5	10.8	7.4	3.0	0.7	8.3	0.2	0.0	0.0	31.9
1976-77	0.0	0.0	0.0	2.0	4.0	4.8	11.0	0.4	30.1	1.4	0.4	0.0	54.1
1977-78	0.0	0.0	0.0	T	5.1	25.1	17.9	18.6	3.7	3.1	0.0	0.0	73.5
1978-79	0.0	0.0	0.0	1.0	18.9	10.8	8.1	16.4	8.0	9.2	1.7	0.0	74.1
1979-80	0.0	0.0	0.0	0.0	1.2	0.7	5.7	5.2	4.1	1.9	0.0	0.0	19.8
1980-81	0.0	0.0	T	7.4	1.9	9.9	1.1	1.8	4.1	T	0.0	0.0	26.2
1981-82	0.0	0.0	0.0	10.1	0.2	4.6		0.0	0.0	0.0	0.0	0.0	
1982-83	0.0	0.0	0.0	0.0	3.7	9.4	3.9	2.8	6.5	3.0	T	0.0	29.3
1983-84	0.0	0.0	T	0.0	5.1	8.4	7.7	1.4	10.6	2.0	T	0.0	35.2
1984-85	0.0	0.0	T	10.5	4.9	11.7	2.3	5.9	11.8	10.9	T	0.0	58.0
1985-86	0.0	0.0	0.5	4.6	15.1	1.7							
Record Mean	0.0	0.0	0.5	2.1	4.8	7.9	8.7	5.6	7.5	7.2	1.4	T	45.7

See Reference Notes, relative to all above tables, on preceding page.

Helena is located on the south side of an intermountain valley bounded on the west and south by the main chain of the Continental Divide. The valley is approximately 25 miles in width from north to south and 35 miles long from east to west. The average height of the mountains above the valley floor is about 3,000 feet.

The climate of Helena may be described as modified continental. Several factors enter into modifying the continental climate characteristics. Some of these are invasion by Pacific Ocean air masses, drainage of cool air into the valley from the surrounding mountains, and the protecting mountain shield in all directions.

The mountains to the north and east sometimes deflect shallow masses of invading cold Arctic air to the east. Following periods of extreme cold, when the return circulation of maritime air has brought warming to most of the eastern part of the state, cold air may remain trapped in the valley for several days before being replaced by warmer air. During these periods of transition from cold-to-warm temperatures, inversions are often quite pronounced.

As may be expected in a northern latitude, cold waves may occur from November through February, with temperatures occasionally dropping to zero or lower.

Summertime temperatures are moderate, with maximum readings generally under 90 degrees and very seldom reaching 100 degrees. Like all mountain stations, there is usually a marked change in temperature from day to night. During the summer this tends to produce an agreeable combination of fairly warm days and cool nights.

Most of the precipitation falls from April through July from frequent showers or thunderstorms, but usually with some steady rains in June, the wettest month of the year. Like summer, fall and winter months are relatively dry. During the April to September growing season, precipitation varies considerably.

Thunderstorms are rather frequent from May through August. Snow can be expected from September through May, but amounts during the spring and fall are usually light, and snow on the ground ordinarily lasts only a day or two. During the winter months snow may remain on the ground for several weeks at a time. There is little drifting of snow in the valley, and blizzard conditions are very infrequent.

Severe ice, sleet, and hailstorms are very seldom observed. Since 1880, only a few hailstorms have caused extensive damage in the city of Helena.

In winter, hours of sunshine are more than would be expected at a mountain location.

Due to the sheltering influence of the mountains, Foehn (Chinook) winds are not as pronounced as might be expected for a location on the eastern slopes of the Rocky Mountains. Strong winds can occur at any time throughout the year, but generally do not last more than a few hours at a time.

Based on the 1951–1980 period, the average first occurrence of 32 degrees Fahrenheit in the fall is September 18 and the average last occurrence in the spring is May 18.

TABLE 1 NORMALS, MEANS AND EXTREMES

HELENA, MONTANA·

LATITUDE: 46°36'N LONGITUDE: 112°00' W ELEVATION: FT. GRND 3828 BARO 03898 TIME ZONE: MOUNTAIN WBAN: 24144

	(a)	JAN	FEB	MAR	APR	MAY	JUNE	JULY	AUG	SEP	OCT	NOV	DEC	YEAR
TEMPERATURE °F:														
Normals														
-Daily Maximum		28.1	36.2	42.5	54.7	64.9	73.1	83.6	81.3	70.3	58.6	42.3	33.3	55.7
-Daily Minimum		8.1	15.7	20.6	29.8	39.5	47.0	52.2	50.3	40.8	31.5	20.4	13.5	30.8
-Monthly		18.1	26.0	31.6	42.3	52.2	60.0	67.9	65.8	55.5	45.0	31.4	23.4	43.3
Extremes														
-Record Highest	45	62	68	77	85	90	98	102	105	99	85	70	64	105
-Year		1953	1950	1978	1980	1966	1941	1981	1969	1967	1963	1953	1980	AUG 1969
-Record Lowest	45	-42	-31	-30	1	17	30	36	32	18	-3	-39	-38	-42
-Year		1957	1956	1955	1954	1954	1969	1971	1956	1970	1972	1959	1964	JAN 1957
NORMAL DEGREE DAYS:														
Heating (base 65°F)		1454	1092	1035	681	397	179	41	77	308	617	1008	1287	8176
Cooling (base 65°F)		0	0	0	0	0	32	131	105	26	0	0	0	294
% OF POSSIBLE SUNSHINE	45	46	54	60	59	61	62	78	74	67	60	44	41	59
MEAN SKY COVER (tenths)														
Sunrise - Sunset	45	7.5	7.4	7.4	7.2	6.9	6.4	4.0	4.5	5.4	6.2	7.2	7.5	6.5
MEAN NUMBER OF DAYS:														
Sunrise to Sunset														
-Clear	45	4.7	4.1	3.8	3.8	4.8	5.6	14.8	13.2	10.0	8.1	4.7	4.0	81.6
-Partly Cloudy	45	5.9	6.5	8.3	8.7	9.8	11.0	10.8	10.8	8.8	8.6	7.4	6.9	103.6
-Cloudy	45	20.4	17.7	19.0	17.5	16.4	13.3	5.4	7.0	11.2	14.3	17.9	20.1	180.0
Precipitation														
.01 inches or more	45	8.2	6.4	8.6	8.0	11.0	11.5	7.3	7.5	6.8	5.8	7.0	8.0	96.2
Snow, Ice pellets														
1.0 inches or more	45	2.8	1.9	2.4	1.4	0.3	0.*	0.0	0.0	0.5	0.7	1.7	2.5	14.2
Thunderstorms	45	0.1	0.1	0.1	0.9	4.2	7.6	9.1	8.0	1.9	0.4	0.1	0.*	32.4
Heavy Fog Visibility 1/4 mile or less	45	1.6	1.3	0.8	0.2	0.2	0.*	0.*	0.1	0.2	0.4	1.4	1.9	8.2
Temperature °F														
-Maximum														
90° and above	22	0.0	0.0	0.0	0.0	0.*	1.4	7.5	6.9	1.1	0.0	0.0	0.0	17.0
32° and below	22	15.1	7.9	4.5	0.5	0.0	0.0	0.0	0.0	0.0	0.6	5.9	14.8	49.3
-Minimum														
32° and below	22	29.3	26.6	27.5	18.9	4.0	0.1	0.0	0.0	3.9	17.4	27.3	29.5	184.3
0° and below	22	9.2	4.1	1.6	0.0	0.0	0.0	0.0	0.0	0.0	0.1	1.9	6.4	23.5
AVG. STATION PRESS.(mb)	13	881.7	880.4	877.7	879.3	879.2	880.2	882.0	881.9	882.7	882.2	881.1	881.2	880.8
RELATIVE HUMIDITY (%)														
Hour 05	22	69	71	71	69	70	71	66	66	72	73	74	71	70
Hour 11	22	66	62	55	46	44	45	39	41	48	53	62	67	52
Hour 17 (Local Time)	22	62	54	46	38	37	38	29	31	36	42	58	65	45
Hour 23	22	68	68	65	59	59	59	50	52	59	64	70	70	62
PRECIPITATION (inches):														
Water Equivalent														
-Normal		0.66	0.44	0.69	1.01	1.72	2.01	1.04	1.18	0.83	0.65	0.54	0.60	11.37
-Maximum Monthly	45	2.78	1.13	1.62	3.00	6.09	4.74	3.89	4.23	3.37	2.68	1.50	1.48	6.09
-Year		1969	1959	1982	1975	1981	1944	1975	1974	1965	1975	1950	1977	MAY 1981
-Minimum Monthly	45	0.03	0.06	0.02	0.10	0.29	0.08	0.08	0.10	0.08	0.02	0.04	0.04	0.02
-Year		1944	1961	1959	1977	1979	1985	1973	1981	1972	1978	1969	1976	OCT 1978
-Maximum in 24 hrs	45	0.77	0.58	1.01	1.25	2.31	1.78	2.26	1.86	1.61	0.85	0.82	0.51	2.31
-Year		1969	1953	1957	1951	1981	1979	1983	1974	1980	1954	1959	1982	MAY 1981
Snow, Ice pellets														
-Maximum Monthly	45	35.6	19.7	21.6	20.6	12.7	2.7	T		13.7	11.0	32.9	22.8	35.6
-Year		1969	1959	1955	1967	1967	1969	1972		1965	1969	1959	1967	JAN 1969
-Maximum in 24 hrs	45	11.5	8.6	8.7	12.9	12.5	2.7	T		13.3	7.4	21.5	10.7	21.5
-Year		1969	1959	1955	1960	1967	1969	1972		1957	1969	1959	1941	NOV 1959
WIND:														
Mean Speed (mph)	45	6.8	7.5	8.5	9.3	8.9	8.6	7.9	7.5	7.5	7.2	7.1	6.9	7.8
Prevailing Direction through 1963		W	W	W	W	W	W	W	W	W	W	W	W	W
Fastest Mile														
-Direction	45	SW	W	SW	W	SW	W	SW	S	NW	W	SW	NW	W
-Speed (MPH)	45	73	73	61	52	56	56	65	65	54	62	56	59	73
-Year		1944	1949	1955	1950	1975	1970	1975	1947	1943	1948	1962	1946	FEB 1949
Peak Gust														
-Direction	2	W	W	NW	W	N	N	NW	W	W	W	W	W	W
-Speed (mph)	2	59	41	40	45	51	52	51	52	51	48	44	46	59
-Date		1984	1985	1985	1985	1985	1985	1985	1985	1984	1985	1985	1985	JAN 1984

See Reference Notes to this table on the following page.

TABLE 2 PRECIPITATION (inches)   HELENA, MONTANA

YEAR	JAN	FEB	MAR	APR	MAY	JUNE	JULY	AUG	SEP	OCT	NOV	DEC	ANNUAL
1956	0.54	0.09	0.42	0.67	1.28	1.80	1.04	1.43	0.16	1.02	0.12	0.46	9.13
1957	0.95	0.30	1.19	0.67	3.25	2.23	1.26	0.59	0.42	0.20	0.61	1.01	14.65
1958	0.16	0.95	1.13	0.83	1.34	4.28	0.11	0.36	0.46	0.95	1.45	0.28	12.91
1959	0.43	1.13	0.02	0.35	1.93	1.90	1.02	2.12	0.13	0.26	0.19	0.34	9.37
1960	0.23	0.25	0.22	1.56	0.94	0.25	1.02	2.12	0.13	0.26	0.19	0.34	7.51
1961	0.12	0.06	1.03	0.90	1.36	0.78	1.05	0.62	1.16	0.16	0.37	0.55	8.16
1962	0.67	0.51	0.69	0.90	3.77	2.50	1.27	1.80	0.31	0.95	0.57	0.14	14.08
1963	0.50	0.25	0.44	0.81	1.34	2.59	0.80	0.80	1.10	1.39	0.29	1.27	11.58
1964	0.31	0.27	0.51	1.56	3.52	2.98	0.83	1.91	0.16	0.04	0.53	0.99	13.61
1965	0.36	0.49	0.85	0.98	2.20	3.85	0.60	1.92	3.37	0.13	0.62	0.15	15.52
1966	0.46	0.33	0.28	0.51	0.43	0.96	0.32	0.42	0.34	0.75	1.04	0.62	6.46
1967	0.61	0.62	1.43	2.38	2.08	2.36	0.46	0.58	0.68	1.50	0.31	1.39	14.40
1968	0.59	0.16	0.53	1.21	1.62	2.68	0.26	2.00	2.22	0.23	0.92	0.75	13.17
1969	2.78	0.22	0.57	0.60	1.13	3.50	1.77	0.38	0.33	1.06	0.04	0.31	12.69
1970	0.51	0.67	0.96	0.81	1.20	2.11	0.93	0.63	0.36	0.58	0.44	0.54	9.74
1971	1.38	0.63	0.41	0.58	1.77	0.93	0.56	1.22	0.89	0.39	0.34	1.02	10.12
1972	1.12	0.54	0.63	0.41	0.77	1.12	0.56	1.63	0.08	0.57	0.33	0.46	8.22
1973	0.22	0.13	0.05	0.66	1.08	0.73	0.08	0.56	0.43	0.66	1.03	0.63	6.26
1974	0.66	0.23	0.38	0.76	2.07	0.34	0.49	4.23	0.22	0.51	0.30	0.26	10.45
1975	1.26	0.72	0.88	3.00	1.95	2.83	3.89	2.47	0.47	2.68	0.48	0.31	20.94
1976	0.26	0.38	0.41	1.34	0.87	2.74	0.29	1.58	1.82	0.04	0.30	0.04	10.07
1977	0.65	0.13	1.11	0.10	1.82	1.37	1.37	0.72	1.93	0.17	0.48	1.48	11.33
1978	0.96	0.61	0.31	0.94	1.20	0.44	0.44	2.83	1.11	0.02	1.19	0.76	10.96
1979	0.77	0.72	1.34	2.26	0.29	2.75	0.32	0.59	0.79	0.38	0.06	0.59	10.39
1980	0.62	0.74	0.88	0.63	4.32	3.16	1.92	0.28	2.57	1.21	0.32	0.40	17.05
1981	0.15	0.10	1.10	0.75	6.09	1.15	1.78	0.10	0.90	0.82	0.54	0.33	13.81
1982	0.80	0.58	1.62	0.54	1.77	2.99	0.49	0.74	2.74	0.35	0.31	1.05	13.98
1983	0.24	0.07	0.36	0.29	1.79	2.20	3.48	2.67	1.56	0.35	0.26	0.76	14.03
1984	0.17	0.15	0.49	1.45	1.03	2.14	0.11	1.11	0.73	0.74	0.47	0.41	9.00
1985	0.16	0.38	0.32	0.46	0.75	0.08	0.10	2.64	2.11	0.76	0.84	0.35	8.95
Record Mean	0.72	0.52	0.73	0.97	1.91	2.20	1.09	0.93	1.14	0.76	0.63	0.64	12.24

TABLE 3 AVERAGE TEMPERATURE (deg. F)   HELENA, MONTANA

YEAR	JAN	FEB	MAR	APR	MAY	JUNE	JULY	AUG	SEP	OCT	NOV	DEC	ANNUAL
1956	16.5	18.6	33.4	41.8	54.2	62.2	67.8	63.7	57.6	45.2	32.6	27.9	43.5
1957	1.7	21.9	32.1	42.1	55.0	60.5	69.0	66.2	56.2	41.0	32.3	32.6	42.6
1958	28.4	25.1	30.1	42.2	60.7	58.8	62.9	69.4	56.5	46.7	30.9	24.4	44.7
1959	20.7	14.7	36.3	43.1	47.4	61.9	68.1	64.2	53.7	43.7	24.3	28.9	42.3
1960	15.7	23.8	32.6	42.1	50.5	60.8	71.5	63.4	57.0	46.4	33.6	23.9	43.5
1961	27.4	36.6	37.0	40.6	53.7	68.5	70.9	71.3	48.9	43.7	27.8	22.6	45.8
1962	15.7	24.5	29.1	47.2	51.7	60.9	65.2	64.9	55.0	48.1	36.9	29.2	44.0
# 1963	10.5	34.8	37.6	42.6	53.4	60.0	68.2	68.0	61.2	50.9	34.8	16.9	44.9
1964	22.0	27.2	29.9	41.8	51.7	59.0	70.2	63.4	53.8	47.4	30.5	19.1	43.0
1965	29.6	26.6	24.6	44.4	50.1	59.5	67.7	65.8	45.6	49.3	34.7	27.5	43.8
1966	20.6	27.0	36.5	42.5	56.5	59.3	71.5	66.7	62.7	46.4	33.8	27.4	45.9
1967	28.7	32.7	29.0	37.8	52.1	59.9	70.5	71.4	61.9	47.4	33.0	19.0	45.3
1968	15.7	28.9	39.6	39.7	49.6	58.3	68.2	63.2	55.0	44.4	33.5	16.7	42.8
1969	7.2	14.5	21.3	46.7	55.2	57.8	67.9	70.1	58.0	38.1	33.5	25.9	41.4
1970	18.8	32.5	28.2	36.5	51.5	62.3	63.2	67.9	49.7	40.2	33.6	15.6	41.9
1971	19.0	28.5	31.9	41.5	51.4	57.0	63.2	69.0	49.7	39.0	31.1	13.6	41.1
1972	12.4	26.8	39.1	39.8	50.9	61.8	61.7	66.0	54.4	45.8	24.5	28.2	43.3
1973	17.7	22.8	36.0	40.3	52.9	61.2	68.9	70.4	61.5	45.8	34.7	27.7	44.8
1974	18.1	32.8	33.9	46.0	48.6	64.9	70.2	61.5	53.5	43.3	29.9	26.0	40.5
1975	21.1	13.7	28.8	32.9	48.6	56.9	69.2	61.2	54.1	43.3	29.9	26.0	40.5
1976	24.9	29.5	30.4	43.2	54.1	56.6	67.6	64.8	57.9	44.9	33.8	28.7	44.7
1977	18.0	34.1	32.4	46.9	50.4	63.6	66.3	63.6	56.3	45.6	31.2	20.7	44.1
1978	16.9	22.8	38.8	47.7	53.2	63.2	67.2	65.5	57.9	46.5	22.7	15.2	43.2
1979	1.1	20.1	34.7	42.7	53.2	62.0	69.2	67.7	61.3	47.8	29.0	28.2	43.1
1980	14.3	25.3	31.3	49.0	55.4	59.8	67.3	62.9	56.8	45.4	34.9	26.9	44.1
1981	28.4	29.7	38.0	46.4	52.8	58.9	67.7	69.5	58.8	43.8	36.4	25.3	46.3
1982	16.5	23.7	33.9	40.5	50.9	61.2	68.6	69.1	55.5	44.9	27.4	22.7	42.9
1983	30.6	35.3	38.3	43.1	51.5	60.2	66.0	70.8	55.5	46.0	34.7	5.5	44.6
1984	27.3	32.4	37.2	43.5	52.8	60.0	70.0	69.8	52.3	41.7	33.1	11.6	44.3
1985	12.3	18.8	33.4	46.9	56.2	63.3	75.0	63.1	49.6	42.3	12.6	15.0	40.7
Record Mean	19.6	24.5	32.4	43.4	52.1	59.8	67.8	66.3	55.7	45.4	33.4	24.0	43.7
Max	28.6	33.8	42.4	54.7	63.7	71.8	82.0	80.5	68.5	56.8	41.8	32.5	54.8
Min	10.7	15.1	22.4	32.1	40.5	47.8	53.6	52.1	42.8	33.9	22.9	15.5	32.5

REFERENCE NOTES FOR TABLES 1, 2, 3 and 6   (HELENA, MT)

GENERAL

T - TRACE AMOUNT
BLANK ENTRIES DENOTE MISSING/UNREPORTED DATA.
# INDICATES A STATION OR INSTRUMENT RELOCATION.

SPECIFIC

TABLE 1

(a) - LENGTH OF RECORD IN YEARS. ALTHOUGH INDIVIDUAL MONTHS MAY BE MISSING.

LESS THAN .05

NORMALS — BASED ON THE 1951-1980 RECORD PERIOD.
EXTREMES — DATES ARE THE MOST RECENT OCCURRENCE.
WIND DIR. — NUMERALS SHOW TENS OF DEGREES CLOCKWISE FROM TRUE NORTH. "00" INDICATES CALM.
RESULTANT WIND DIRECTIONS ARE GIVEN TO WHOLE DEGREES.

EXCEPTIONS

TABLES 2, 3, and 6

RECORD MEANS ARE THROUGH THE CURRENT YEAR, BEGINNING IN   1881 FOR TEMPERATURE
1881 FOR PRECIPITATION
1940 FOR SNOWFALL

TABLE 4 HEATING DEGREE DAYS Base 65 deg. F   HELENA, MONTANA

SEASON	JULY	AUG	SEP	OCT	NOV	DEC	JAN	FEB	MAR	APR	MAY	JUNE	TOTAL
1956-57	49	85	216	608	967	1145	1962	1202	1013	679	302	145	8373
1957-58	7	57	264	738	976	998	1129	1113	1073	679	167	189	7390
1958-59	99	14	263	561	1018	1253	1367	1405	886	651	540	126	8183
1959-60	41	69	351	652	1219	1112	1522	1192	996	682	446	134	8416
1960-61	7	114	242	572	934	1266	1158	789	863	726	354	21	7046
1961-62	1	7	476	652	1109	1311	1528	1133	1104	527	404	147	8399
1962-63	58	77	295	515	837	1106	1687	842	842	665	358	172	7454
#1963-64	32	32	131	427	897	1484	1328	1089	1080	690	411	192	7793
1964-65	1	101	332	542	1028	1420	1090	1070	1250	612	454	169	8069
1965-66	14	69	578	478	903	1153	1370	1058	874	670	267	188	7622
1966-67	8	51	117	570	930	1157	1120	898	1107	807	396	171	7332
1967-68	0	1	130	539	954	1422	1525	1042	780	751	472	206	7822
1968-69	23	102	294	633	937	1493	1788	1407	1348	543	301	226	9095
1969-70	35	14	219	826	938	1207	1427	904	1132	849	409	153	8113
1970-71	22	12	413	763	1063	1372	1422	1013	1020	699	414	251	8464
1971-72	91	22	454	707	936	1524	1628	1102	796	747	430	120	8557
1972-73	136	49	418	798	1010	1588	1465	1174	892	732	374	155	8791
1973-74	23	47	317	588	1208	1136	1452	897	956	564	500	99	7787
1974-75	16	130	338	588	905	1149	1355	1429	1114	954	501	235	8714
1975-76	14	119	322	666	1045	1199	1236	1023	1064	649	331	257	7925
1976-77	15	45	219	615	928	1120	1449	862	1005	535	443	90	7326
1977-78	52	92	270	593	1008	1367	1485	1175	806	512	361	87	7808
1978-79	32	60	244	564	1263	1540	1979	1250	930	665	359	128	9014
1979-80	11	15	127	528	1072	1138	1566	1148	1039	473	304	164	7585
1980-81	25	81	242	602	899	1175	1127	986	832	552	371	191	7083
1981-82	21	16	195	650	853	1227	1497	1153	959	726	428	136	7861
1982-83	30	16	304	618	1120	1306	1059	828	823	649	417	152	7322
1983-84	76	0	351	584	901	1842	1164	941	856	640	380	174	7909
1984-85	2	7	377	716	954	1654	1625	1291	973	538	266	97	8500
1985-86	3	105	455	696	1571	1545							

TABLE 5 COOLING DEGREE DAYS Base 65 deg. F   HELENA, MONTANA

YEAR	JAN	FEB	MAR	APR	MAY	JUNE	JULY	AUG	SEP	OCT	NOV	DEC	TOTAL
1969	0	0	0	0	1	18	128	178	20	0	0	0	345
1970	0	0	0	0	0	76	130	109	5	0	0	0	320
1971	0	0	0	0	0	14	42	154	0	0	0	0	210
1972	0	0	0	0	2	30	42	89	0	0	0	0	163
1973	0	0	0	0	4	45	151	108	3	0	0	0	311
1974	0	0	0	0	0	102	190	31	0	0	0	0	323
1975	0	0	0	0	0	1	154	12	0	0	0	0	167
1976	0	0	0	0	0	14	102	45	10	0	0	0	171
1977	0	0	0	0	0	57	101	54	13	0	0	0	225
1978	0	0	0	0	3	37	109	85	37	0	0	0	271
1979	0	0	0	0	1	45	152	103	21	0	0	0	322
1980	0	0	0	0	14	14	104	25	4	0	0	0	161
1981	0	0	0	0	0	15	109	165	16	0	0	0	305
1982	0	0	0	0	0	30	147	151	25	0	0	0	353
1983	0	0	0	0	4	16	115	186	12	0	0	0	333
1984	0	0	0	0	10	31	165	163	4	0	0	0	373
1985	0	0	0	0	2	55	318	54	0	0	0	0	429

TABLE 6 SNOWFALL (inches)   HELENA, MONTANA

SEASON	JULY	AUG	SEP	OCT	NOV	DEC	JAN	FEB	MAR	APR	MAY	JUNE	TOTAL
1956-57	0.0	0.0	0.0	1.9	0.7	5.4	18.7	5.8	12.9	6.3	0.0	0.0	51.7
1957-58	0.0	0.0	13.4	8.2	10.3	2.8	1.0	16.4	8.6	3.4	0.0	0.0	64.1
1958-59	0.0	0.0	0.0	0.3	7.4	15.3	7.4	19.7	0.4	1.5	T	0.0	52.0
1959-60	0.0	0.0	T	3.4	32.9	4.5	4.9	3.7	4.5	20.0	T	0.0	73.9
1960-61	0.0	0.0	0.0	4.1	2.0	2.9	1.8	0.4	9.1	8.3	T	0.0	28.6
1961-62	0.0	0.0	T	1.3	6.4	4.2	10.1	7.3	7.3	8.7	0.0	0.0	45.3
1962-63	0.0	0.0	0.4	0.0	1.2	1.5	9.8	2.0	5.9	2.3	T	0.0	23.1
1963-64	0.0	0.0	0.0	T	2.7	21.2	6.2	6.3	7.4	6.6	10.1	0.0	60.5
1964-65	0.0	0.0	0.0	0.0	6.3	13.2	2.3	9.4	12.9	6.3	5.0	0.0	55.4
1965-66	0.0	0.0	13.7	0.0	2.1	1.4	6.3	5.1	2.3	1.9	T	0.0	32.8
1966-67	0.0	0.0	0.0	T	9.5	5.7	7.3	10.5	14.9	20.6	12.7	0.0	81.2
1967-68	0.0	0.0	0.0	T	2.7	22.8	8.7	0.9	0.8	10.0	T	0.0	45.9
1968-69	0.0	0.0	3.0	0.4	7.4	11.0	35.6	3.6	7.0	2.6	0.7	2.7	74.0
1969-70	0.0	0.0	0.0	11.0	0.1	6.1	7.0	9.5	9.0	7.9	1.4	0.0	52.0
1970-71	0.0	0.0	T	1.4	5.4	7.6	16.2	3.5	2.5	0.7	T	0.0	37.2
1971-72	0.0	0.0	T	0.8	4.6	14.5	14.9	3.3	4.5	3.8	T	0.0	46.4
1972-73	T	0.0	0.3	4.7	0.7	7.8	3.2	1.8	0.1	6.5	T	0.0	25.1
1973-74	0.0	0.0	1.3	7.2	12.5	7.2	9.9	5.9	2.2	1.5	0.2	0.0	47.9
1974-75	0.0	0.0	T	1.5	0.8	2.7	15.2	10.7	12.3	15.4	0.2	0.0	58.8
1975-76	0.0	0.0	0.0	6.3	4.9	3.9	3.3	4.4	6.7	10.9	0.0	0.0	40.4
1976-77	0.0	0.0	0.0	T	2.9	0.9	13.8	0.8	14.1	0.5	1.0	0.0	34.0
1977-78	0.0	0.0	0.0	0.4	6.8	9.5	15.7	13.3	2.6	0.9	T	0.0	59.2
1978-79	0.0	0.0	T	T	22.1	13.8	11.7	7.2	12.4	9.3	T	T	76.5
1979-80	0.0	0.0	0.0	0.2	0.6	6.5	9.3	10.5	10.1	3.1	0.0	0.0	40.3
1980-81	0.0	0.0	0.0	3.7	1.2	3.8	2.7	2.1	3.3	0.1	T	0.0	16.9
1981-82	0.0	0.0	0.0	5.2	3.4	6.1	18.4	4.8	13.9	4.1	0.8	0.0	56.7
1982-83	0.0	0.0	6.5	0.5	4.1	11.3	3.2	0.2	2.2	1.1	9.9	0.0	39.0
1983-84	0.0	0.0	6.4	0.0	1.3	13.0	1.3	1.5	5.7	2.3	0.8	0.0	32.3
1984-85	0.0	0.0	6.3	9.0	5.7	7.5	3.9	6.2	4.0	0.8	0.0	0.0	43.4
1985-86	0.0	0.0	2.9	8.8	10.4	8.5							
Record Mean	T	0.0	1.7	2.2	6.6	8.7	9.0	6.0	7.3	4.8	1.5	0.1	47.8

See Reference Notes, relative to all above tables, on preceding page.

The climate of the Flathead Valley is influenced by the topography. The high mountains to the east form an effective barrier to many severe winter cold waves that move into areas east of the Rockies from Alberta. The mountains to the east rise abruptly 4,500 feet above the valley floor. The mountain snows and spring rains assure an adequate supply of water for the area.

In addition to Flathead Lake, the valley contains many smaller lakes, three rivers, and numerous streams and sloughs. Until late in the winter when a large portion of the lakes and sloughs become frozen, this water surface tends to limit temperature extremes. This effect is most noticeable in the southern end of the valley, because of the influence of Flathead Lake. Due to its size, Flathead Lake seldom freezes over.

The weather at the airport is considerably different in some respects from the weather in Kalispell. Generally there is more cloudiness at the airport since it is closer to the mountains to the east and north. Moist air moving in from the west and southwest, lifting and cooling as it moves over the mountains, is the major cause. On average there is more precipitation on the east side of the valley than on the west side. Average snowfall during the winter at the airport is 68 inches and in Kalispell it is 49 inches.

The annual prevailing wind direction at Kalispell is from the west. At the airport it is from the south. Wind speeds average considerably stronger at the airport than in Kalispell.

In the winter, when a cold wave moving down the east side of the Continental Divide does come over the mountains, the airport is in direct line of the pass the cold air comes through. During these cold waves the wind is from the northeast and will usually have speeds reaching 30 to 40 mph. The strongest gusts reported during these storms exceed 80 mph. As the cold air moves down the valley it spreads out, decreasing the wind velocity, and mixes with the warmer air of the valley. Unless these cold strong winds persist for 3 or 4 days, the wind in the lower part of the valley will be from the northwest, because of the influence of Flathead Lake and the mountains to the west. This wind is always much stronger in the northeast end of the valley where the airport is located than any other place in the valley. In the northwest corner where Whitefish is located, and in the southeast part of the valley, there is rarely much wind from this storm.

TABLE 1   NORMALS, MEANS AND EXTREMES

KALISPELL, MONTANA

LATITUDE: 48°18'N   LONGITUDE: 114°16' W   ELEVATION: FT. GRND 2965 BARO 02978   TIME ZONE: MOUNTAIN   WBAN: 24146

	(a)	JAN	FEB	MAR	APR	MAY	JUNE	JULY	AUG	SEP	OCT	NOV	DEC	YEAR
TEMPERATURE °F:														
Normals														
-Daily Maximum		27.4	35.0	42.1	54.6	64.8	72.1	82.1	80.3	69.2	55.3	39.0	31.5	54.4
-Daily Minimum		11.2	17.5	21.6	30.5	38.1	44.5	47.9	46.7	38.6	29.6	22.7	16.9	30.5
-Monthly		19.3	26.3	31.9	42.5	51.5	58.3	65.0	63.5	53.9	42.5	30.9	24.2	42.5
Extremes														
-Record Highest	36	53	56	71	84	89	96	104	105	99	82	65	57	105
-Year		1953	1981	1966	1977	1966	1955	1960	1961	1967	1979	1975	1979	AUG 1961
-Record Lowest	36	-38	-36	-29	10	19	28	31	31	16	-3	-28	-35	-38
-Year		1950	1950	1960	1951	1954	1973	1971	1969	1970	1984	1959	1968	JAN 1950
NORMAL DEGREE DAYS:														
Heating (base 65°F)		1417	1084	1026	672	419	218	66	128	345	698	1023	1265	8361
Cooling (base 65°F)		0	0	0	0	0	17	66	82	12	0	0	0	177
% OF POSSIBLE SUNSHINE														
MEAN SKY COVER (tenths)														
Sunrise - Sunset	36	8.7	8.3	7.8	7.5	6.9	6.5	4.1	4.7	5.6	7.0	8.4	8.9	7.0
MEAN NUMBER OF DAYS:														
Sunrise to Sunset														
-Clear	36	2.0	2.1	3.5	4.0	5.4	6.7	14.9	13.0	10.1	6.1	2.1	1.7	71.6
-Partly Cloudy	36	3.6	4.5	6.3	6.3	8.8	9.3	9.5	8.9	7.5	6.7	5.1	3.3	79.8
-Cloudy	36	25.3	21.6	21.3	19.7	16.8	14.1	6.6	9.1	12.3	18.2	22.8	26.0	213.8
Precipitation														
.01 inches or more	36	15.6	11.8	11.4	9.3	11.0	12.2	6.6	8.3	8.3	9.1	12.3	15.6	131.6
Snow, Ice pellets														
1.0 inches or more	36	6.1	3.9	2.1	0.8	0.3	0.0	0.0	0.0	0.*	0.6	2.6	5.4	22.0
Thunderstorms	35	0.0	0.1	0.3	0.8	2.7	5.2	5.4	5.3	2.0	0.4	0.1	0.0	22.1
Heavy Fog Visibility														
1/4 mile or less	25	4.6	4.4	2.3	0.5	1.0	0.8	0.7	0.9	2.3	4.7	4.6	4.7	31.6
Temperature °F														
-Maximum														
90° and above	26	0.0	0.0	0.0	0.0	0.0	1.0	6.6	6.3	0.3	0.0	0.0	0.0	14.3
32° and below	26	18.0	8.5	3.5	0.*	0.0	0.0	0.0	0.0	0.0	0.3	6.3	16.7	53.3
-Minimum														
32° and below	26	29.5	25.7	27.6	19.0	6.3	0.6	0.2	0.2	6.0	21.9	25.7	28.7	191.4
0° and below	26	7.6	2.8	1.2	0.0	0.0	0.0	0.0	0.0	0.0	0.*	0.8	4.5	17.0
AVG. STATION PRESS.(mb)	13	913.8	911.8	908.9	910.4	910.2	910.8	912.1	911.8	913.1	913.2	912.5	913.0	911.8
RELATIVE HUMIDITY (%)														
Hour 05	21	79	81	78	76	78	83	82	80	82	83	82	82	81
Hour 11 (Local Time)	26	78	76	64	52	51	53	45	46	55	66	77	81	62
Hour 17	21	74	68	54	43	43	46	35	35	43	53	71	77	54
Hour 23	20	78	79	75	68	69	73	69	68	75	79	81	81	75
PRECIPITATION (inches):														
Water Equivalent														
-Normal		1.62	1.06	0.84	1.06	1.76	2.24	0.94	1.44	1.11	0.98	1.29	1.59	15.93
-Maximum Monthly	36	3.11	1.99	1.87	2.37	3.90	4.72	2.74	3.78	3.97	2.96	4.44	4.23	4.72
-Year		1970	1981	1950	1978	1980	1966	1950	1976	1985	1951	1959	1964	JUN 1966
-Minimum Monthly	36	0.20	0.42	0.31	0.26	0.43	0.43	0.02	T	0.12	0.04	0.26	0.32	T
-Year		1985	1967	1965	1968	1950	1977	1953	1955	1952	1953	1969	1954	AUG 1955
-Maximum in 24 hrs	36	1.09	0.65	0.75	1.74	1.49	2.71	1.23	1.76	1.25	0.78	0.95	1.35	2.71
-Year		1982	1961	1981	1951	1980	1982	1977	1976	1959	1957	1959	1964	JUN 1982
Snow, Ice pellets														
-Maximum Monthly	36	34.8	21.2	14.8	8.1	8.9	0.3			3.1	11.1	39.0	49.7	49.7
-Year		1970	1975	1977	1961	1964	1962			1968	1984	1959	1951	DEC 1951
-Maximum in 24 hrs	36	11.8	7.2	7.6	10.0	7.5	0.3			3.0	6.2	10.1	15.4	15.4
-Year		1982	1979	1953	1951	1964	1962			1968	1951	1959	1951	DEC 1951
WIND:														
Mean Speed (mph)	23	6.0	6.1	7.3	8.2	7.7	7.3	6.7	6.6	6.5	5.3	5.6	5.7	6.6
Prevailing Direction														
Fastest Obs. 1 Min.														
-Direction (!!)	19	04	01	02	21	23	03	31	32	04	32	03	03	04
-Speed (MPH)	19	52	40	40	35	40	38	38	37	35	38	35	52	52
-Year		1982	1965	1965	1969	1965	1982	1985	1968	1980	1971	1978	1972	JAN 1982
Peak Gust														
-Direction (!!)	2	NE	N	NE	NE	SW	NW	NW	SW	NE	N	NE	N	SW
-Speed (mph)	2	36	49	37	37	39	43	45	58	47	43	44	46	58
-Date		1984	1985	1985	1985	1984	1985	1985	1984	1985	1984	1985	1984	AUG 1984

See Reference Notes to this table on the following pages.

KALISPELL, MONTANA

TABLE 2 PRECIPITATION (inches)   KALISPELL, MONTANA

YEAR	JAN	FEB	MAR	APR	MAY	JUNE	JULY	AUG	SEP	OCT	NOV	DEC	ANNUAL
1956	1.52	1.27	0.89	1.35	1.44	2.92	2.32	1.93	0.92	0.88	0.55	1.08	17.07
1957	1.12	1.28	0.49	1.39	1.25	1.87	0.85	0.45	0.20	2.00	0.56	0.98	12.44
1958	1.46	1.93	0.72	1.42	1.80	2.20	0.86	0.22	1.84	0.60	2.65	2.30	18.00
1959	2.03	1.08	0.61	1.22	3.08	1.58	0.04	0.68	3.84	1.97	4.44	0.40	20.97
1960	1.30	0.83	1.11	0.82	3.34	·0.62	0.03	2.31	0.33	1.00	1.59	0.96	14.24
1961	0.75	1.32	1.12	2.01	2.35	0.71	1.20	0.67	2.08	0.92	1.07	2.20	16.40
1962	0.97	0.92	1.09	1.05	2.04	0.82	0.16	0.95	0.42	1.25	1.00	0.81	11.48
1963	1.61	1.00	1.25	0.88	0.81	3.94	0.95	0.81	1.26	0.67	0.71	1.33	15.22
1964	1.24	0.50	1.26	0.60	2.56	3.56	1.69	1.77	1.87	0.95	2.13	4.23	22.36
1965	1.84	0.69	0.31	1.44	0.74	2.73	0.85	3.47	1.17	0.11	0.72	0.57	14.64
1966	1.61	1.02	0.79	0.65	1.63	4.72	0.37	1.05	0.29	0.88	2.52	1.58	17.11
1967	1.75	0.42	1.09	0.59	0.95	2.38	0.07	0.01	0.33	1.60	0.58	1.45	11.22
1968	1.01	1.00	0.43	0.26	2.59	2.16	0.46	3.10	3.33	1.50	1.55	2.56	19.95
1969	2.97	0.50	0.64	1.30	0.68	3.88	0.10	0.09	1.44	1.07	0.26	1.21	14.14
1970	3.11	1.27	0.84	0.68	2.18	2.58	1.59	0.34	0.90	1.20	1.29	1.39	17.37
1971	1.81	0.81	0.95	0.44	2.17	3.59	0.93	1.47	0.46	0.95	1.36	1.53	16.47
1972	1.58	1.57	1.14	0.86	1.50	1.69	1.51	1.03	0.76	0.84	0.55	1.60	14.63
1973	0.69	0.62	0.46	0.47	0.91	1.52	0.05	0.56	0.71	1.19	2.80	1.87	11.85
1974	1.94	1.02	1.39	1.92	1.06	1.75	0.64	0.70	1.18	0.12	1.04	1.21	13.97
1975	1.95	1.52	1.33	0.83	0.98	1.96	0.98	2.79	0.58	1.67	1.20	1.19	16.98
1976	1.36	1.40	0.35	0.97	1.79	1.69	1.69	3.78	0.36	0.38	0.47	0.65	14.84
1977	0.81	0.97	1.18	0.43	1.40	0.43	2.57	1.13	2.18	0.13	1.46	3.53	16.22
1978	2.30	0.71	0.67	2.37	2.47	0.91	1.50	2.64	0.80	0.07	1.43	0.87	16.74
1979	1.42	1.57	0.86	1.49	1.64	0.84	0.67	1.10	0.39	1.45	0.42	1.27	13.12
1980	2.15	1.92	0.86	1.52	3.90	2.96	0.81	1.60	0.74	0.78	0.49	2.54	20.27
1981	1.44	1.99	1.43	0.94	3.37	3.62	0.72	1.32	0.48	0.20	1.30	1.81	18.62
1982	2.66	1.60	0.92	1.32	0.78	4.05	1.59	0.82	1.92	0.52	1.43	1.88	19.49
1983	1.09	0.93	1.50	2.18	0.78	3.09	2.06	0.73	1.31	0.83	1.61	1.69	17.80
1984	0.78	0.66	1.34	1.53	1.56	1.77	0.51	0.88	1.88	1.85	1.77	1.22	15.75
1985	0.20	1.39	0.71	0.58	1.60	1.56	0.23	1.12	3.97	0.92	1.63	0.72	14.63
Record Mean	1.39	1.01	0.95	0.97	1.64	2.16	1.03	1.14	1.23	1.04	1.39	1.39	15.34

TABLE 3 AVERAGE TEMPERATURE (deg. F)   KALISPELL, MONTANA

YEAR	JAN	FEB	MAR	APR	MAY	JUNE	JULY	AUG	SEP	OCT	NOV	DEC	ANNUAL
1956	22.2	20.0	30.8	43.5	54.4	57.3	66.0	62.1	54.8	43.7	29.6	27.9	42.7
1957	9.6	21.6	32.6	43.2	56.4	59.0	64.5	63.0	55.4	40.7	31.1	31.0	42.4
1958	28.0	29.2	33.6	43.2	60.3	61.5	65.2	68.3	54.5	43.6	30.9	27.0	45.5
#1959	22.7	22.6	34.3	43.5	47.5	59.2	64.8	59.9	52.9	42.4	24.2	24.2	41.3
1960	14.3	19.3	27.3	43.0	50.2	58.0	69.8	61.0	54.7	43.4	31.3	21.5	41.2
1961	23.9	35.5	36.7	40.5	52.4	64.2	68.9	69.1	49.1	40.0	26.6	20.8	44.0
1962	17.0	25.7	29.8	45.8	50.7	57.7	62.8	62.0	53.9	43.0	36.1	31.2	42.9
1963	10.4	31.6	37.2	43.2	50.2	58.7	63.0	65.2	58.9	45.8	34.0	22.0	43.3
1964	24.6	23.9	27.9	40.7	49.7	56.8	64.0	58.4	50.1	42.3	31.0	18.0	40.6
1965	25.9	24.2	24.5	43.5	48.1	56.0	64.5	63.3	45.1	45.3	34.3	26.1	41.7
1966	24.3	23.9	32.5	41.7	53.0	54.8	64.4	61.6	59.3	41.2	30.7	27.3	42.9
1967	28.8	30.3	30.7	39.2	50.3	58.7	65.8	67.7	59.5	44.5	32.3	23.8	44.3
1968	20.4	31.9	39.1	40.2	49.8	58.0	65.5	61.8	53.0	41.0	31.4	17.8	42.5
1969	11.3	20.2	27.0	46.2	53.8	57.9	62.1	62.9	55.7	39.8	33.2	26.2	41.3
1970	19.1	27.2	30.6	40.1	52.1	62.4	66.1	64.3	48.9	39.1	30.6	23.6	42.0
1971	21.7	28.8	32.9	43.6	53.7	54.8	63.1	68.8	49.1	40.0	33.1	18.4	42.3
1972	13.9	26.2	38.4	40.0	52.8	59.6	62.5	57.6	51.1	39.3	32.8	18.3	41.9
1973	19.4	27.0	39.3	42.5	52.4	58.8	67.1	65.7	54.8	43.8	29.0	29.0	44.1
1974	20.2	31.7	33.7	44.8	48.3	64.9	65.9	63.2	53.8	42.8	34.3	29.0	44.2
1975	18.2	19.2	28.0	38.9	50.0	56.2	69.9	61.5	54.5	43.6	30.1	27.3	41.5
1976	24.1	29.0	31.6	44.8	53.4	56.3	64.6	62.8	56.8	41.4	31.3	27.2	43.6
1977	19.4	30.9	34.7	46.4	50.4	61.5	63.5	63.6	52.2	42.0	28.8	19.4	42.8
1978	19.7	25.5	34.6	44.0	48.8	59.5	64.5	61.6	54.0	42.7	25.9	14.2	41.3
1979	-0.2	23.1	33.7	42.8	51.7	60.3	67.7	67.3	58.5	45.6	29.0	32.2	42.7
1980	12.1	27.4	32.8	48.0	55.4	57.6	63.8	59.4	55.5	43.9	33.2	28.5	43.1
1981	29.7	30.5	38.8	45.0	52.6	53.8	63.0	67.9	55.3	40.2	32.6	21.5	44.2
1982	17.0	21.1	35.7	38.2	48.8	60.4	61.5	63.2	53.3	40.9	28.5	23.3	41.0
1983	29.6	32.9	38.0	42.3	51.5	57.3	61.2	66.3	50.6	42.8	33.9	7.9	42.9
1984	25.7	31.7	37.9	42.9	48.6	56.4	65.4	65.6	50.5	38.5	32.1	16.3	42.7
1985	17.6	15.4	28.0	44.7	54.3	57.1	68.5	59.4	47.9	40.4	16.8	18.2	39.0
Record Mean	20.9	25.4	33.3	43.6	52.0	58.4	65.3	63.5	53.8	43.5	31.8	24.4	43.0
Max	28.1	33.7	42.7	55.0	64.2	70.9	80.7	78.9	67.2	54.8	39.0	30.8	53.8
Min	13.6	17.1	23.8	32.2	39.7	45.8	49.8	48.0	40.3	32.1	24.6	18.0	32.1

REFERENCE NOTES FOR TABLES 1, 2, 3 and 6   (KALISPELL, MT)

GENERAL

T - TRACE AMOUNT
BLANK ENTRIES DENOTE MISSING/UNREPORTED DATA.
# INDICATES A STATION OR INSTRUMENT RELOCATION.

SPECIFIC

TABLE 1

(a) - LENGTH OF RECORD IN YEARS. ALTHOUGH INDIVIDUAL MONTHS MAY BE MISSING.

  * LESS THAN .05

NORMALS — BASED ON THE 1951-1980 RECORD PERIOD.
EXTREMES — DATES ARE THE MOST RECENT OCCURRENCE.
WIND DIR. — NUMERALS SHOW TENS OF DEGREES CLOCKWISE FROM TRUE NORTH.
    "00" INDICATES CALM.
RESULTANT WIND DIRECTIONS ARE GIVEN TO WHOLE DEGREES.

EXCEPTIONS

TABLES 2, 3, and 6

RECORD MEANS ARE THROUGH THE CURRENT YEAR, BEGINNING IN   1897 FOR TEMPERATURE
1897 FOR PRECIPITATION
1950 FOR SNOWFALL

TABLE 4 HEATING DEGREE DAYS Base 65 deg. F   KALISPELL, MONTANA

SEASON	JULY	AUG	SEP	OCT	NOV	DEC	JAN	FEB	MAR	APR	MAY	JUNE	TOTAL
1956-57	61	106	302	651	1056	1144	1716	1209	998	648	259	183	8333
1957-58	60	101	286	744	1007	1042	1139	996	964	646	171	119	7275
1958-59	59	31	330	656	1016	1171	1305	1180	945	636	537	185	8051
#1959-60	94	169	365	693	1283	1260	1568	1321	1164	654	452	212	9235
1960-61	15	149	307	664	1006	1342	1269	821	870	730	393	70	7636
1961-62	11	15	467	768	1146	1365	1488	1094	1089	570	439	222	8674
1962-63	119	120	326	674	858	1042	1690	928	855	649	449	200	7910
1963-64	106	73	186	591	926	1327	1246	1187	1143	726	470	239	8220
1964-65	69	199	439	700	1011	1454	1204	1138	1249	638	515	270	8886
1965-66	62	115	591	602	914	1201	1252	1144	1002	693	373	301	8250
1966-67	68	128	162	733	1023	1161	1118	964	1059	766	449	190	7821
1967-68	32	22	176	627	972	1272	1379	953	795	735	468	211	7642
1968-69	80	133	354	737	999	1458	1658	1246	1172	556	339	211	8943
1969-70	109	102	290	774	947	1197	1418	1054	1062	743	393	127	8216
1970-71	48	62	478	798	1026	1276	1338	1006	986	635	344	305	8302
1971-72	116	28	470	768	950	1440	1581	1118	817	742	381	181	8592
1972-73	113	26	413	792	959	1444	1410	1059	787	668	383	205	8259
1973-74	47	72	314	649	1076	1107	1385	926	965	598	510	125	7774
1974-75	54	91	328	681	914	1109	1445	1277	1140	777	459	258	8533
1975-76	15	133	306	654	1040	1160	1261	1037	1028	598	354	271	7857
1976-77	64	104	244	723	1002	1162	1405	949	935	553	443	114	7698
1977-78	85	97	375	707	1081	1405	1397	1098	934	624	497	169	8469
1978-79	60	139	331	685	1167	1567	2023	1168	964	662	404	152	9322
1979-80	50	17	191	595	1071	1008	1639	1084	990	503	299	218	7665
1980-81	74	178	279	647	948	1121	1087	962	806	593	376	329	7400
1981-82	84	36	287	760	966	1342	1484	1226	901	797	495	154	8532
1982-83	133	88	343	742	1088	1286	1090	893	831	676	417	227	7814
1983-84	130	32	427	680	927	1768	1209	958	835	657	501	254	8378
1984-85	60	50	433	816	982	1503	1463	1387	1140	601	333	240	9008
1985-86	5	180	507	755	1439	1444							

TABLE 5 COOLING DEGREE DAYS Base 65 deg. F   KALISPELL, MONTANA

YEAR	JAN	FEB	MAR	APR	MAY	JUNE	JULY	AUG	SEP	OCT	NOV	DEC	TOTAL
1969	0	0	0	0	0	6	27	44	17	0	0	0	94
1970	0	0	0	0	0	57	88	48	1	0	0	0	194
1971	0	0	0	0	0	6	66	155	0	0	0	0	227
1972	0	0	0	0	7	26	41	113	0	0	0	0	187
1973	0	0	0	0	1	26	118	100	13	0	0	0	258
1974	0	0	0	0	0	83	89	43	0	0	0	0	215
1975	0	0	0	0	0	0	177	30	0	0	0	0	207
1976	0	0	0	0	0	15	59	44	3	0	0	0	121
1977	0	0	0	2	0	17	48	62	0	0	0	0	129
1978	0	0	0	0	0	9	51	39	6	0	0	0	105
1979	0	0	0	0	0	18	142	96	2	0	0	0	258
1980	0	0	0	0	6	0	44	13	0	0	0	0	63
1981	0	0	0	0	0	0	32	135	2	0	0	0	169
1982	0	0	0	0	0	21	33	39	0	0	0	0	93
1983	0	0	0	0	4	4	17	83	4	0	0	0	112
1984	0	0	0	0	0	9	78	76	5	0	0	0	168
1985	0	0	0	0	6	9	120	12	0	0	0	0	147

TABLE 6 SNOWFALL (inches)   KALISPELL, MONTANA

SEASON	JULY	AUG	SEP	OCT	NOV	DEC	JAN	FEB	MAR	APR	MAY	JUNE	TOTAL
1956-57	0.0	0.0	0.0	1.8	3.9	5.8	24.3	16.8	6.5	0.4	0.0	0.0	59.5
1957-58	0.0	0.0	T	2.3	4.5	9.2	8.4	20.6	3.7	0.3	0.0	0.0	49.0
1958-59	0.0	0.0	0.0	1.6	8.0	22.3	18.9	17.1	3.4	0.6	T	0.0	71.9
1959-60	0.0	0.0	0.0	0.4	39.0	3.9	20.6	13.9	8.8	3.6	T	0.0	90.2
1960-61	0.0	0.0	0.0	0.5	10.1	17.5	5.5	3.5	7.2	8.1	4.1	0.0	56.5
1961-62	0.0	0.0	T	1.8	19.1	30.0	15.1	7.7	6.5	3.1	0.0	0.3	83.6
1962-63	0.0	0.0	T	0.0	1.5	2.4	27.2	6.5	2.3	3.0	0.2	0.0	43.1
1963-64	0.0	0.0	0.0	T	2.5	18.3	20.6	6.7	13.3	1.1	8.9	0.0	71.4
1964-65	0.0	0.0	0.0	1.6	10.5	32.0	20.7	10.7	4.9	4.0	0.2	0.0	84.6
1965-66	0.0	0.0	0.4	0.0	3.9	8.0	22.6	17.6	1.9	0.9	0.6	0.0	55.9
1966-67	0.0	0.0	0.0	T	11.9	15.7	8.9	5.8	10.0	2.1	2.8	0.0	57.2
1967-68	0.0	0.0	0.0	T	2.3	21.3	14.1	3.0	0.7	1.9	1.4	0.0	44.7
1968-69	0.0	0.0	3.1	T	11.1	30.3	34.2	7.2	4.9	0.1	0.0	0.0	90.9
1969-70	0.0	0.0	0.0	2.2	0.8	14.4	34.8	5.7	7.6	5.0	4.7	T	75.2
1970-71	0.0	0.0	T	1.0	6.6	21.7	25.7	8.4	10.2	T	T	0.0	73.6
1971-72	0.0	0.0	0.0	2.4	12.4	29.6	24.5	13.3	10.3	3.8	T	T	96.3
1972-73	0.0	0.0	0.2	10.6	4.0	9.4	6.5	8.3	0.9	1.8	T	T	41.7
1973-74	0.0	0.0	0.0	0.6	18.2	10.6	13.8	8.2	8.7	2.5	T	0.0	62.6
1974-75	0.0	0.0	0.0	0.2	2.8	11.9	20.6	21.2	10.7	0.1	2.1	0.0	69.6
1975-76	0.0	0.0	0.0	3.8	12.6	9.2	15.9	10.2	3.3	2.0	0.0	0.0	57.0
1976-77	0.0	0.0	0.0	T	2.3	7.2	8.8	6.3	14.8	2.3	T	0.0	41.7
1977-78	0.0	0.0	0.0	T	13.1	32.5	26.7	8.0	2.6	4.8	T	0.0	87.7
1978-79	0.0	0.0	T	0.1	11.8	14.6	19.4	18.6	7.1	4.3	T	0.0	75.9
1979-80	0.0	0.0	0.0	0.4	2.2	12.0	27.4	11.6	11.3	0.2	T	0.0	65.1
1980-81	0.0	0.0	0.0	0.3	0.7	18.1	9.3	12.6	4.3	4.9	0.0	0.0	50.2
1981-82	0.0	0.0	0.0	T	1.7	16.3	32.7	6.6	1.9	6.8	0.2	0.0	66.2
1982-83	0.0	0.0	T	10.4	17.6	5.9	5.2	1.0	4.6	T	0.0		44.7
1983-84	0.0	0.0	0.0	0.0	6.8	21.3	8.6	7.4	2.8	0.4	T	0.0	47.3
1984-85	0.0	0.0	0.0	11.1	9.6	14.0	2.5	19.1	6.7	1.0	T	0.0	64.0
1985-86	0.0	0.0	T	1.0	12.7	11.9							
Record Mean	0.0	0.0	0.1	1.5	8.3	16.5	18.5	10.8	6.2	2.5	0.9	T	65.4

See Reference Notes, relative to all above tables, on preceding page.

Miles City is located on the western edge of the northern great plains in a shallow part of the Yellowstone Valley. The Tongue River runs south from its confluence with the Yellowstone just west of the city. To the north the river bluffs are from 200 to 300 feet above the valley floor. There are no nearby mountain ranges to influence climatic conditions. Temperatures range from very cold in winter to quite warm in summer, which is characteristic of continental locations. The climate is classed as semi–arid with less than 10 inches of rainfall in about one year in seven.

The temperature has ranged from less than −65 degrees to more than 110 degrees. Cold waves with temperatures of zero or lower occur frequently during the winter. They are usually accompanied by northerly winds and snow, and last 2 to 4 days. Periods of several days with zero or lower can be expected during the winter months. Spring and fall are cool with temperatures of 90 degrees or above rarely occurring. High temperatures of 90 degrees or more do occur frequently in July and August, but humidities are low.

About 70 percent of the precipitation falls during the growing season, April through September, with greatest monthly amounts usually falling during May and June. Precipitation during the spring and summer often falls during periods of shower or thunderstorm activity, however, general rains also are frequent in late spring and early summer. Measurable snowfall can be expected as late as May and as early as September.

Sunny growing seasons, with May and June rainfall being the heaviest of the year, encourage rapid crop development. Crops grown in this area seldom have difficulty in reaching maturity, although hail sometimes causes local damage during the middle of the summer.

Based on the 1951–1980 period, the average first occurrence of 32 degrees Fahrenheit in the fall is September 29 and the average last occurrence in the spring is May 7.

TABLE 1 — NORMALS, MEANS AND EXTREMES

MILES CITY, MONTANA

LATITUDE: 46°26'N   LONGITUDE: 105°52'W   ELEVATION: FT. GRND 2629 BARO 02630   TIME ZONE: MOUNTAIN   WBAN: 24037

	(a)	JAN	FEB	MAR	APR	MAY	JUNE	JULY	AUG	SEP	OCT	NOV	DEC	YEAR
TEMPERATURE °F:														
Normals														
-Daily Maximum		24.3	32.6	42.6	57.1	69.2	79.2	88.9	86.6	73.8	61.3	42.7	31.5	57.5
-Daily Minimum		3.9	11.6	20.0	32.7	44.2	53.5	60.3	57.9	46.4	35.5	21.3	11.4	33.2
-Monthly		14.1	22.1	31.3	44.9	56.7	66.3	74.6	72.3	60.1	48.4	32.0	21.5	45.4
Extremes														
-Record Highest	48	62	67	83	92	99	104	109	110	106	93	75	69	110
-Year		1953	1980	1943	1980	1980	1961	1980	1949	1983	1963	1965	1939	AUG 1949
-Record Lowest	48	-37	-37	-27	6	15	32	41	35	20	0	-25	-38	-38
-Year		1969	1939	1947	1982	1954	1951	1945	1966	1942	1984	1985	1983	DEC 1983
NORMAL DEGREE DAYS:														
Heating (base 65°F)		1578	1201	1045	603	271	86	10	28	215	515	990	1349	7891
Cooling (base 65°F)		0	0	0	0	17	128	308	255	68	0	0	0	776
% OF POSSIBLE SUNSHINE														
MEAN SKY COVER (tenths)														
Sunrise - Sunset	13	6.8	7.0	6.9	6.8	6.5	5.5	4.0	4.0	5.1	5.4	6.6	6.5	5.9
MEAN NUMBER OF DAYS:														
Sunrise to Sunset														
-Clear	18	6.0	5.1	4.6	5.4	5.9	7.7	14.5	15.1	10.1	11.8	6.5	6.9	99.6
-Partly Cloudy	18	7.9	8.2	8.4	7.9	9.8	10.7	12.6	11.2	9.8	8.4	7.3	7.7	109.8
-Cloudy	18	17.1	15.0	18.1	16.7	15.3	11.6	3.9	4.8	10.1	10.8	16.2	16.4	155.8
Precipitation														
.01 inches or more	48	8.1	6.8	7.5	7.9	10.6	11.4	7.8	6.5	6.6	5.6	5.8	7.2	91.8
Snow, Ice pellets														
1.0 inches or more	36	2.1	1.8	2.1	1.3	0.1	0.*	0.0	0.0	0.1	0.3	1.8	2.4	12.2
Thunderstorms	35	0.0	0.*	0.1	0.7	4.0	7.8	7.3	6.1	1.8	0.2	0.*	0.*	28.1
Heavy Fog Visibility														
1/4 mile or less	33	1.4	1.9	1.2	0.7	0.7	0.3	0.1	0.1	0.6	0.7	1.2	1.9	10.7
Temperature °F														
-Maximum														
90° and above	41	0.0	0.0	0.0	0.1	1.1	4.2	15.6	13.5	3.1	0.1	0.0	0.0	37.7
32° and below	41	18.0	12.4	6.9	0.6	0.*	0.0	0.0	0.0	0.0	0.4	6.8	15.7	61.0
-Minimum														
32° and below	41	30.5	27.3	27.0	14.1	2.2	0.1	0.0	0.0	1.7	10.8	26.2	30.4	170.3
0° and below	41	12.4	7.1	3.0	0.0	0.0	0.0	0.0	0.0	0.0	0.*	2.1	7.2	31.8
AVG. STATION PRESS. (mb)	13	923.3	922.0	919.1	920.2	919.4	919.9	921.5	921.6	922.7	922.4	922.4	922.1	921.4
RELATIVE HUMIDITY (%)														
Hour 05	25	75	79	80	76	76	77	68	65	72	75	78	77	75
Hour 11	25	70	70	62	51	48	46	38	38	47	52	65	72	55
Hour 17 (Local Time)	25	70	67	56	44	43	40	31	29	38	47	63	72	50
Hour 23	25	74	77	74	66	65	64	52	50	59	66	74	76	66
PRECIPITATION (inches):														
Water Equivalent														
-Normal		0.57	0.56	0.59	1.37	2.31	2.75	1.52	1.26	1.08	0.90	0.60	0.60	14.11
-Maximum Monthly	48	1.78	1.30	1.83	4.22	6.81	9.78	4.58	4.00	4.67	6.31	2.17	1.78	9.78
-Year		1971	1959	1950	1973	1978	1944	1948	1951	1941	1971	1978	1968	JUN 1944
-Minimum Monthly	48	0.05	0.02	0.07	0.02	0.24	0.77	0.10	T	T	T	0.02	0.02	T
-Year		1981	1985	1959	1983	1958	1979	1971	1967	1960	1965	1953	1957	AUG 1967
-Maximum in 24 hrs	29	0.42	0.77	0.61	1.36	2.39	2.71	2.22	1.65	2.67	1.38	1.18	0.50	2.71
-Year		1944	1952	1944	1947	1955	1964	1985	1943	1941	1953	1957	1941	JUN 1964
Snow, Ice pellets														
-Maximum Monthly	45	17.2	19.0	17.8	16.6	12.0	2.0			7.1	12.6	19.4	18.0	19.4
-Year		1971	1949	1967	1967	1983	1950			1972	1949	1977	1968	NOV 1977
-Maximum in 24 hrs	19	7.5	6.5	5.6	10.5	2.0	2.0			2.2	3.8	8.0	7.0	10.5
-Year		1964	1952	1985	1947	1953	1950			1984	1946	1964	1958	APR 1947
WIND:														
Mean Speed (mph)	36	9.6	9.7	10.7	11.8	11.1	10.3	9.7	9.8	10.0	9.9	9.7	9.8	10.2
Prevailing Direction through 1963		NW	NW	NW	NW	SE	SE	SE	SE	NW	SSE	SSE	SSE	NW
Fastest Obs. 1 Min.														
-Direction (!!!)														
-Speed (MPH)														
-Year														
Peak Gust														
-Direction (!!!)														
-Speed (mph)														
-Date														

See Reference Notes to this table on the following page.

TABLE 2 PRECIPITATION (inches) MILES CITY, MONTANA

YEAR	JAN	FEB	MAR	APR	MAY	JUNE	JULY	AUG	SEP	OCT	NOV	DEC	ANNUAL
1956	0.52	0.13	0.08	0.43	2.43	0.98	1.83	1.52	0.26	0.54	0.61	0.22	9.55
1957	0.58	0.35	0.90	1.92	1.89	3.01	0.38	1.24	0.58	0.50	1.25	0.02	12.62
1958	0.32	0.47	0.64	1.61	0.24	2.85	3.58	0.45	0.14	1.16	1.26	0.50	13.22
1959	0.81	1.30	0.07	1.02	1.34	1.59	0.22	0.40	1.26	0.53	0.91	0.03	9.48
1960	0.51	0.37	0.68	0.72	1.32	1.25	0.33	0.78	T	0.22	0.12	0.69	6.99
1961	0.08	0.77	0.38	0.97	2.34	0.84	0.47	0.36	2.83	0.09	0.56	0.28	9.97
1962	0.35	0.41	0.65	0.57	4.97	4.59	4.51	0.37	1.07	0.61	0.43	0.47	19.00
1963	0.80	0.66	0.28	2.69	1.68	4.78	1.19	1.11	1.92	T	0.18	0.66	15.95
1964	0.72	0.57	0.93	1.25	1.42	4.72	0.54	2.33	0.23	0.25	1.33	0.62	14.91
1965	0.96	0.53	0.75	2.12	1.41	3.46	3.59	0.22	1.77	T	0.39	0.40	15.60
1966	0.67	0.53	0.77	0.85	1.07	3.26	1.61	0.62	0.52	0.34	0.80	0.54	11.58
1967	0.33	0.87	1.78	2.11	1.79	5.23	0.86	T	2.35	0.47	0.38	1.33	17.50
1968	0.84	0.77	0.29	0.48	1.64	5.18	0.93	3.88	0.48	0.63	0.72	1.78	17.62
1969	0.70	0.24	0.73	2.82	1.99	2.63	3.40	0.17	0.05	1.09	0.03	1.33	15.18
1970	0.69	1.06	0.83	2.59	3.18	2.01	1.06	0.03	1.43	0.99	0.88	0.43	15.18
1971	1.78	0.76	0.56	1.26	1.55	2.12	0.10	1.61	1.01	6.31	0.42	1.24	18.72
1972	1.48	0.73	0.60	0.45	2.83	3.62	1.97	3.34	0.68	0.66	0.15	1.03	17.54
1973	0.14	0.33	0.16	4.22	1.92	3.30	0.27	1.72	4.02	0.57	0.17	0.82	17.64
1974	0.22	0.19	0.72	1.17	4.13	1.29	1.35	1.44	0.32	2.24	0.23	0.34	13.64
1975	0.81	0.33	1.23	2.70	4.77	3.68	1.66	0.67	0.44	2.31	0.91	0.56	20.07
1976	0.23	0.18	0.36	1.43	1.00	3.41	1.11	0.78	1.06	0.75	0.19	0.27	10.77
1977	0.68	0.10	0.97	0.24	2.45	1.38	1.91	2.26	1.91	1.16	1.50	1.23	15.79
1978	0.51	1.19	0.12	0.47	6.81	1.39	2.51	0.81	3.40	0.27	2.17	0.63	20.28
1979	0.33	1.14	0.26	0.76	1.36	2.79	0.77	0.67	0.03	0.32	0.20	0.05	8.68
1980	0.33	0.33	0.28	0.66	0.28	3.04	0.53	2.04	0.73	1.61	0.29	0.33	10.45
1981	0.05	0.06	0.28	0.20	2.87	2.57	0.36	1.12	0.72	1.39	0.78	0.23	10.63
1982	0.96	0.20	0.73	0.53	2.61	5.10	0.69	0.61	2.23	1.61	0.10	1.01	16.38
1983	0.26	0.09	0.95	0.02	1.36	1.56	1.89	0.33	1.36	0.26	0.43	0.28	8.79
1984	0.13	0.13	0.27	1.34	0.90	3.55	0.18	0.89	0.58	0.11	0.20	0.53	8.81
1985	0.23	0.02	0.97	0.86	1.16	1.12	3.13	1.86	1.45	0.64	0.72	0.54	12.70
Record Mean	0.49	0.45	0.62	1.19	2.15	3.06	1.47	1.23	1.19	0.95	0.51	0.50	13.81

TABLE 3 AVERAGE TEMPERATURE (deg. F) MILES CITY, MONTANA

YEAR	JAN	FEB	MAR	APR	MAY	JUNE	JULY	AUG	SEP	OCT	NOV	DEC	ANNUAL
1956	10.6	17.7	34.9	41.6	58.0	73.0	73.1	70.8	61.8	51.1	34.3	28.4	46.3
1957	6.3	19.9	32.8	42.9	58.1	63.8	78.4	72.3	59.3	46.3	34.4	33.3	45.7
1958	30.5	19.4	32.0	46.4	64.8	62.6	68.5	75.9	63.0	50.9	32.7	21.2	47.3
1959	11.7	9.3	36.8	44.0	52.7	69.5	75.8	74.2	58.7	43.4	26.3	30.6	44.4
1960	17.5	18.2	29.1	45.6	58.7	66.7	80.2	71.7	62.9	51.4	32.8	23.1	46.5
1961	24.4	31.7	40.1	42.1	58.2	74.2	75.7	78.3	52.6	47.0	29.9	17.6	47.7
1962	16.7	20.7	26.1	49.3	56.4	67.5	70.8	71.3	58.7	51.0	39.1	27.6	46.3
1963	7.5	28.6	40.3	44.1	56.6	66.8	75.5	73.9	67.2	55.7	37.7	19.8	47.8
1964	24.6	27.5	28.3	46.5	57.7	64.4	78.2	69.0	56.5	50.3	25.6	7.7	44.8
1965	13.5	18.2	17.3	44.6	54.5	64.9	74.2	71.2	48.0	52.7	32.7	25.3	43.1
1966	6.1	13.6	35.0	41.1	59.1	66.2	78.3	69.0	65.0	48.0	25.7	21.2	44.1
1967	23.4	25.7	29.2	41.1	52.6	63.3	74.3	73.9	64.5	49.1	32.6	18.1	45.6
1968	14.3	24.1	40.1	43.6	53.2	62.2	73.6	68.9	61.3	48.7	34.5	12.3	44.8
1969	-0.7	16.5	25.5	51.0	58.1	60.7	71.8	77.5	66.0	40.3	37.5	25.7	44.1
1970	11.7	27.4	27.7	39.6	56.4	69.0	76.7	76.3	58.1	44.0	31.4	17.7	44.7
1971	11.2	20.4	31.6	46.4	56.9	68.0	72.1	79.2	57.9	44.7	34.6	12.2	44.6
1972	9.1	17.4	36.0	45.8	57.0	68.6	68.1	72.2	56.3	43.7	33.4	13.2	43.4
1973	20.3	26.8	40.5	42.4	56.2	66.6	72.8	74.8	58.4	50.3	29.6	22.8	46.8
1974	18.8	31.3	34.1	48.0	52.3	69.0	77.8	66.3	57.3	50.2	27.1	27.7	47.3
1975	22.2	16.4	27.7	38.8	54.3	64.0	76.8	69.4	58.4	47.3	31.0	22.6	44.1
1976	18.6	32.4	32.5	48.6	59.1	65.6	74.1	74.1	64.2	44.2	29.4	25.5	47.4
1977	7.7	31.7	36.2	51.2	61.4	71.1	74.4	65.8	59.2	49.1	28.5	14.2	45.9
1978	2.6	11.4	31.4	46.7	56.7	66.0	71.5	70.4	63.0	48.6	22.7	12.8	42.0
1979	0.2	8.0	29.8	41.4	53.7	67.3	74.6	72.0	65.7	50.7	29.8	29.1	43.6
1980	13.5	24.0	31.7	53.0	62.3	68.0	76.5	67.4	60.8	49.2	38.4	24.5	47.5
1981	31.0	27.9	40.7	51.3	57.1	63.9	75.1	74.7	63.8	45.0	37.0	21.9	49.1
1982	1.9	17.2	30.8	41.9	52.6	63.7	73.1	74.2	59.4	46.7	28.4	21.4	42.6
1983	27.7	34.9	36.5	44.9	53.7	65.7	76.1	80.4	58.3	49.9	34.0	0.9	46.9
1984	25.5	35.1	34.3	46.7	56.2	66.2	77.1	72.1	55.2	43.3	33.0	8.9	46.6
1985	12.1	17.1	33.8	51.2	62.4	64.1	76.5	68.3	55.0	46.9	14.4	17.9	43.3
Record Mean	15.5	21.8	31.7	45.7	56.5	65.4	74.6	72.5	60.2	48.5	31.8	21.3	45.5
Max	25.8	32.6	42.9	58.1	69.1	78.1	89.0	87.0	73.8	61.4	42.4	31.3	57.6
Min	5.2	11.0	20.4	33.2	44.0	52.8	60.1	58.0	46.5	35.6	21.2	11.2	33.3

REFERENCE NOTES FOR TABLES 1, 2, 3 and 6   (MILES CITY, MT)

GENERAL

T - TRACE AMOUNT
BLANK ENTRIES DENOTE MISSING/UNREPORTED DATA.
# INDICATES A STATION OR INSTRUMENT RELOCATION.

SPECIFIC

TABLE 1

(a) - LENGTH OF RECORD IN YEARS. ALTHOUGH INDIVIDUAL MONTHS MAY BE MISSING.

* LESS THAN .05

NORMALS — BASED ON THE 1951-1980 RECORD PERIOD.
EXTREMES — DATES ARE THE MOST RECENT OCCURRENCE.
WIND DIR. — NUMERALS SHOW TENS OF DEGREES CLOCKWISE FROM TRUE NORTH. "00" INDICATES CALM.
RESULTANT WIND DIRECTIONS ARE GIVEN TO WHOLE DEGREES.

EXCEPTIONS

TABLE 1

1. MEAN SKY COVER, AND DAYS CLEAR-PARTLY CLOUDY-CLOUDY ARE THROUGH 1962.
2. THROUGH 1964, MAXIMUM 24-HOUR PRECIPITATION AND SNOW ARE FOR CALENDAR DAY (MIDNIGHT TO MIDNIGHT) INSTEAD OF FOR ANY CONSECUTIVE 24-HOURS.

TABLES 2, 3, and 6

RECORD MEANS ARE THROUGH THE CURRENT YEAR, BEGINNING IN

1938 FOR TEMPERATURE
1938 FOR PRECIPITATION
1941 FOR SNOWFALL

TABLE 4 HEATING DEGREE DAYS Base 65 deg. F   MILES CITY, MONTANA

SEASON	JULY	AUG	SEP	OCT	NOV	DEC	JAN	FEB	MAR	APR	MAY	JUNE	TOTAL
1956-57	1	44	144	424	913	1129	1815	1258	991	662	231	91	7703
1957-58	0	13	185	586	912	976	1065	1274	1015	548	89	119	6782
1958-59	26	6	150	431	965	1355	1648	1558	866	625	377	43	8050
1959-60	2	3	259	664	1158	1059	1467	1353	1109	577	231	51	7933
1960-61	0	24	156	417	960	1291	1249	927	765	678	246	6	6719
1961-62	0	3	386	552	1046	1465	1497	1239	1199	471	259	53	8170
1962-63	10	39	205	427	769	1156	1784	1012	758	620	269	38	7087
1963-64	0	5	34	294	814	1396	1248	1081	1131	547	259	68	6877
1964-65	0	66	253	448	1176	1775	1591	1307	1472	606	329	66	9089
1965-66	1	28	510	375	963	1224	1826	1434	923	711	251	72	8318
1966-67	0	64	87	521	1169	1353	1284	1094	1103	711	402	91	7879
1967-68	5		116	489	969	1447	1570	1180	767	636	361	122	7667
1968-69	11	34	159	498	909	1628	2038	1352	1217	415	260	172	8693
1969-70	0	0	69	759	817	1214	1653	1045	1149	754	275	32	7767
1970-71	0	0	259	646	1003	1459	1664	1243	1032	550	257	39	8152
1971-72	8	3	245	621	905	1636	1732	1376	890	569	277	28	8290
1972-73	55	10	277	653	941	1603	1379	1063	752	671	281	74	7759
1973-74	3	0	209	446	1056	1301	1431	941	951	507	389	56	7290
1974-75	4	53	241	452	891	1168	1323	1355	1151	779	331	83	7831
1975-76	0	20	211	546	1013	1311	1437	941	999	486	195	81	7240
1976-77	1	3	140	642	1061	1219	1774	923	888	415	171	3	7240
1977-78	0	58	200	485	1091	1574	1934	1499	1035	543	272	68	8759
1978-79	14	24	167	502	1263	1614	2010	1594	1086	705	368	62	9409
1979-80	1	14	66	437	1048	1107	1595	1183	1026	377	156	32	7042
1980-81	1	38	168	495	792	1248	1047	1036	747	407	248	91	6318
1981-82	1	4	115	616	831	1331	1960	1337	1055	685	383	108	8426
1982-83	10	12	226	559	1092	1344	1149	838	875	597	357	66	7125
1983-84	9	1	261	461	924	1991	1220	862	946	542	295	72	7584
1984-85	0	2	332	665	952	1739	1636	1339	964	409	143	105	8286
1985-86	13	37	323	555	1516	1453							

TABLE 5 COOLING DEGREE DAYS Base 65 deg. F   MILES CITY, MONTANA

YEAR	JAN	FEB	MAR	APR	MAY	JUNE	JULY	AUG	SEP	OCT	NOV	DEC	TOTAL
1969	0	0	0	1	53	51	219	394	104	0	0	0	822
1970	0	0	0	0	16	176	367	358	60	3	0	0	980
1971	0	0	0	0	15	136	236	449	38	0	0	0	874
1972	0	0	0	0	34	143	157	238	20	0	0	0	592
1973	0	0	0	0	15	128	250	312	19	0	0	0	724
1974	0	0	0	5	2	182	407	101	14	0	0	0	711
1975	0	0	0	0	9	60	373	165	18	5	0	0	630
1976	0	0	0	0	17	106	365	292	84	4	0	0	868
1977	0	0	0	8	66	192	296	90	33	0	0	0	685
1978	0	0	0	0	22	105	222	198	114	0	0	0	661
1979	0	0	0	4	26	138	305	237	92	0	0	0	802
1980	0	0	0	22	78	133	364	119	48	12	0	0	776
1981	0	0	0	2	8	65	318	312	88	0	0	0	793
1982	0	0	0	0	5	79	267	304	67	0	0	0	722
1983	0	0	0	0	14	93	365	485	66	0	0	0	1023
1984	0	0	0	0	37	114	375	387	43	0	0	0	956
1985	0	0	0	3	65	84	373	145	28	0	0	0	698

TABLE 6 SNOWFALL (inches)   MILES CITY, MONTANA

SEASON	JULY	AUG	SEP	OCT	NOV	DEC	JAN	FEB	MAR	APR	MAY	JUNE	TOTAL
1956-57	0.0	0.0	0.0	T	4.0	1.5	5.6	3.8	8.6	2.0	0.0	0.0	25.5
1957-58	0.0	0.0	0.0	3.0	1.9	0.1	2.8	4.4	1.7	T	0.0	0.0	13.9
1958-59	0.0	0.0	0.0	8.9	10.8	10.3	11.8	0.6	3.5	0.0	0.0	0.0	45.9
1959-60	0.0	0.0	T	T	7.0	T	6.0	4.3	3.1	1.0	0.0	0.0	21.4
1960-61	0.0	0.0	0.0	T	2.0	6.0	1.4	6.0	3.0	T	0.0	0.0	18.4
1961-62	0.0	0.0	T	T	4.0	3.0	5.9	4.5	2.5	T	0.0	0.0	19.9
1962-63	0.0	0.0	T	T	1.4	3.1	9.6	6.0	0.4	9.0	0.0	0.0	29.5
1963-64	0.0	0.0	0.0	0.0	0.2	6.1	11.2	5.5	11.1	4.8	0.0	0.0	38.9
1964-65	0.0	0.0	0.0	0.0	14.4	7.3	13.4	3.5	6.9	4.0	T	0.0	49.5
1965-66	0.0	0.0	T	0.0	3.0	4.2	8.3	5.9	4.0	0.9	T	0.0	26.3
1966-67	0.0	0.0	0.0	T	7.9	3.7	2.4	5.8	17.8	16.6	8.6	0.0	62.8
1967-68	0.0	0.0	0.0	T	2.5	12.6	7.5	7.2	6.0	3.9	T	0.0	34.3
1968-69	0.0	0.0	0.0	T	2.2	18.0	6.9	2.4	6.0	1.9	0.0	0.1	37.5
1969-70	0.0	0.0	0.0	2.6	T	10.5	3.4	10.7	5.2	11.5	T	0.0	46.8
1970-71	0.0	0.0	0.6	0.3	3.3	3.4	17.2	6.6	5.1	T	0.1	0.0	36.6
1971-72	0.0	0.0	0.0	2.5	4.6	12.6	14.8	6.0	2.9	T	0.1	0.0	43.5
1972-73	0.0	0.0	7.1	3.3	T	10.0	1.7	3.3	0.4	5.1	0.0	T	30.9
1973-74	0.0	0.0	T	T	1.7	8.6	2.1	1.9	5.8	2.9	2.8	0.0	25.8
1974-75	0.0	0.0	T	0.1	0.7	3.4	5.6	3.3	12.3	10.6	T	0.0	36.0
1975-76	0.0	0.0	0.0	2.8	6.8	4.8	1.4	0.9	3.5	1.6	0.0	0.0	21.8
1976-77	0.0	0.0	0.0	4.4	1.7	2.6	6.4	0.8	5.4	T	0.0	0.0	21.3
1977-78	0.0	0.0	0.0	T	19.4	14.8	5.3	11.9	0.2	T	0.0	0.0	51.6
1978-79	0.0	0.0	T	0.0	16.8	6.7	11.8	2.2	2.2	3.5	T	0.0	44.3
1979-80	0.0	0.0	0.0	T	2.2	0.8	3.7	2.7	2.2	T	0.0	0.0	11.6
1980-81	0.0	0.0	0.0	3.0	0.9	2.0	T	1.0	T	T	0.0	0.0	6.9
1981-82	0.0	0.0	0.0	T	2.0	2.0	7.2	1.9	5.8	2.6	0.0	0.0	21.5
1982-83	0.0	0.0	T	T	1.0	10.4	1.5	0.4	4.3	T	12.0	0.0	29.6
1983-84	0.0	0.0	1.0	0.0	2.1	2.7	1.1	1.6	3.2	12.1	0.0	0.0	23.8
1984-85	0.0	0.0	3.3	0.6	0.8	8.8	4.4	2.0	12.1	1.1	0.0	0.0	33.1
1985-86	0.0	0.0	T	2.5	13.7	3.7							
Record Mean	0.0	0.0	0.3	1.4	4.2	5.5	5.8	5.0	5.0	3.1	0.7	T	31.1

See Reference Notes, relative to all above tables, on preceding page.

Missoula is situated in the heart of the Montana Rocky Mountains in the extreme north portion of the Bitterroot Valley, and about 5 miles east of the confluence of the Bitterroot and Clark Fork Rivers. The Clark Fork Valley begins at Missoula and extends about 20 miles west-northwestward. The Bitterroot Valley extends about 70 miles due southward from Missoula. The Continental Divide is 60 to 80 miles east of Missoula, and the Bitterroot Range is only about 20 miles away to the southwest. These two mountain ranges have a marked effect on the climate of Missoula.

The prevailing flow of air aloft over western Montana is from the west and southwest during spring and summer months, and from the west and northwest during the winter months. Since this air must pass over the Bitterroot Range, it loses much of its moisture on the western slopes of these mountains. As a result, Missoula receives only between 12 inches and 15 inches of precipitation annually. This small amount of precipitation makes for a semi-arid climate. There is sufficient irrigation water, however, from the nearby mountains. The heaviest precipitation, of about 2 inches, is received in each month of May and June.

Generally the spring months are cool and a little damp, with almost daily shower activity during May and June. There are about 137 growing days each year. The summer months are dry with moderate temperatures and cool nights. Seldom does the temperature reach 100 degrees. Oppressively warm nighttime temperatures are unknown.

In the winter, the Continental Divide shields the Missoula area from much of the severely cold air which moves down the continent from arctic regions. Because of this shielding effect, many of the cold waves which sweep down over eastern Montana miss the Missoula area entirely. Under certain conditions, however, the cold Arctic air does break over the Continental Divide, and moves with force into the Bitterroot and Clark Fork Valleys. When this happens, Missoula experiences severe blizzard conditions. The cold air is funnelled to the city through Hell Gate which is the mouth of the Clark Fork River canyon at Missoula. Locally these blizzards are referred to as Hell Gate Blizzards. After the valleys of western Montana are filled with the cold air, prolonged cold spells may occur. January is the coldest month, although periods of sub-zero weather occur occasionally in December and February. Rarely, there are brief periods of sub-zero weather in November and March. During the winter months the sunshine is limited to about 30 percent of the possible amount.

TABLE 1   NORMALS, MEANS AND EXTREMES

MISSOULA, MONTANA

LATITUDE: 46°55'N   LONGITUDE: 114°05'W   ELEVATION: FT. GRND 3197 BARO 3203   TIME ZONE: MOUNTAIN   WBAN: 24153

	(a)	JAN	FEB	MAR	APR	MAY	JUNE	JULY	AUG	SEP	OCT	NOV	DEC	YEAR
TEMPERATURE °F:														
Normals														
-Daily Maximum		28.8	36.4	44.4	56.5	65.8	74.0	84.8	82.7	71.3	57.0	40.4	31.8	56.2
-Daily Minimum		13.7	19.8	23.8	31.3	38.4	45.1	49.5	48.3	40.1	31.2	23.2	17.9	31.9
-Monthly		21.3	28.1	34.1	43.9	52.1	59.5	67.2	65.5	55.7	44.1	31.8	24.9	44.0
Extremes														
-Record Highest	41	59	60	75	86	92	98	105	105	99	85	66	60	105
-Year		1953	1967	1978	1977	1956	1955	1973	1961	1967	1980	1983	1956	JUL 1973
-Record Lowest	41	-33	-27	-13	14	21	31	31	32	20	0	-23	-30	-33
-Year		1957	1982	1955	1951	1985	1984	1971	1980	1985	1971	1955	1983	JAN 1957
NORMAL DEGREE DAYS:														
Heating (base 65°F)		1355	1033	958	633	400	180	29	75	289	648	996	1243	7839
Cooling (base 65°F)		0	0	0	0	0	18	97	91	10	0	0	0	216
% OF POSSIBLE SUNSHINE	40	31	41	51	54	57	59	80	75	67	53	34	28	53
MEAN SKY COVER (tenths)														
Sunrise - Sunset	41	8.4	8.2	7.9	7.5	7.0	6.4	3.8	4.5	5.5	6.8	8.2	8.6	6.9
MEAN NUMBER OF DAYS:														
Sunrise to Sunset														
-Clear	41	2.6	2.5	3.3	3.8	5.3	6.4	16.1	13.9	9.8	6.4	2.6	1.9	74.5
-Partly Cloudy	41	3.9	4.4	5.8	6.7	8.4	9.0	9.9	9.1	8.5	7.2	4.6	4.0	81.5
-Cloudy	41	24.5	21.3	21.8	19.5	17.2	14.6	5.1	8.0	11.8	17.4	22.8	25.2	209.2
Precipitation														
.01 inches or more	41	14.3	10.8	11.6	10.0	11.6	11.5	6.9	7.2	7.7	8.1	11.4	13.1	124.3
Snow, Ice pellets														
1.0 inches or more	41	4.4	2.1	2.0	0.7	0.3	0.0	0.0	0.0	0.0	0.3	2.1	4.0	15.9
Thunderstorms	41	0.*	0.*	0.*	0.5	3.4	5.2	6.3	6.1	2.0	0.3	0.*	0.0	23.9
Heavy Fog Visibility 1/4 mile or less	41	4.8	3.9	1.3	0.2	0.2	0.3	0.2	0.5	1.1	3.4	4.5	6.4	27.0
Temperature °F														
-Maximum														
90° and above	25	0.0	0.0	0.0	0.0	0.1	2.2	9.2	9.1	0.9	0.0	0.0	0.0	21.4
32° and below	25	16.8	6.9	2.5	0.0	0.0	0.0	0.0	0.0	0.0	0.2	6.0	17.4	49.8
-Minimum														
32° and below	25	29.4	26.1	26.3	18.0	5.8	0.2	0.*	0.*	4.4	18.3	26.2	29.4	184.2
0° and below	25	5.5	1.6	0.3	0.0	0.0	0.0	0.0	0.0	0.0	0.*	0.4	3.2	11.0
AVG. STATION PRESS.(mb)	13	906.9	905.0	901.9	903.5	903.4	903.9	905.3	904.9	906.2	906.6	905.7	906.4	905.0
RELATIVE HUMIDITY (%)														
Hour 05	25	85	85	84	80	81	83	77	75	84	86	87	87	83
Hour 11 (Local Time)	25	82	79	67	55	54	54	44	46	58	70	80	84	64
Hour 17	25	77	67	52	42	43	43	31	31	40	51	71	80	52
Hour 23	25	82	81	75	67	67	68	57	57	70	77	84	85	73
PRECIPITATION (inches):														
Water Equivalent														
-Normal		1.41	0.81	0.83	1.01	1.62	1.85	0.85	0.95	1.02	0.85	0.88	1.21	13.29
-Maximum Monthly	41	2.94	1.82	1.62	2.46	7.38	4.19	3.05	3.29	3.60	3.51	2.51	3.15	7.38
-Year		1969	1972	1972	1956	1980	1958	1946	1985	1985	1975	1973	1964	MAY 1980
-Minimum Monthly	41	0.16	0.17	0.20	0.08	0.25	0.35	0.09	T	0.05	0.01	0.22	0.25	T
-Year		1981	1973	1953	1977	1963	1961	1985	1967	1979	1978	1976	1976	AUG 1967
-Maximum in 24 hrs	41	0.88	1.03	0.63	1.65	1.92	1.61	1.47	1.43	1.34	1.49	0.81	0.94	1.92
-Year		1948	1975	1972	1951	1980	1964	1946	1947	1954	1946	1973	1964	MAY 1980
Snow, Ice pellets														
-Maximum Monthly	41	42.5	20.1	13.4	8.2	8.1	T			0.4	5.4	17.7	31.6	42.5
-Year		1963	1975	1955	1970	1978	1973			1983	1973	1947	1964	JAN 1963
-Maximum in 24 hrs	41	11.3	14.4	6.9	6.9	8.1	T			0.4	5.4	6.2	9.6	14.4
-Year		1980	1975	1977	1950	1978	1973			1983	1973	1961	1955	FEB 1975
WIND:														
Mean Speed (mph)	41	5.2	5.6	6.7	7.5	7.3	7.0	6.8	6.6	6.0	5.0	5.0	4.8	6.1
Prevailing Direction through 1963		ESE	NW	NW	NW	NW	NW	NW	NW	NW	NW	NW	E	NW
Fastest Mile														
-Direction (!!!)	40	S	NW	SW	NW	SW	S	SE	SW	N	SW	SW	W	SE
-Speed (MPH)	40	52	47	50	51	57	51	72	58	43	51	42	56	72
-Year		1953	1963	1972	1950	1954	1956	1957	1956	1974	1950	1978	1957	JUL 1957
Peak Gust														
-Direction (!!!)	2	SE	W	NW	NW	SW	NE	NW	W	SW	SW	SE	NW	SW
-Speed (mph)	2	38	38	45	45	56	56	48	51	53	51	40	35	56
-Date		1984	1985	1985	1984	1985	1985	1985	1984	1984	1985	1985	1984	MAY 1985

See Reference Notes to this table on the following page.

TABLE 2 PRECIPITATION (inches) MISSOULA, MONTANA

YEAR	JAN	FEB	MAR	APR	MAY	JUNE	JULY	AUG	SEP	OCT	NOV	DEC	ANNUAL
1956	0.80	0.95	1.17	2.46	0.66	2.87	1.96	1.00	0.49	1.35	0.46	0.99	15.16
1957	1.12	0.51	0.94	0.53	3.22	0.86	0.69	0.92	0.21	1.35	0.45	1.41	12.21
1958	0.52	1.14	0.41	1.42	1.27	4.19	1.67	0.67	1.15	1.35	1.78	1.41	16.98
1959	2.08	1.14	0.62	0.88	2.43	1.69	0.13	0.93	3.11	1.71	1.30	0.27	16.29
1960	1.04	0.53	0.72	0.72	1.41	0.53	0.13	1.69	0.37	0.67	1.09	0.97	9.87
1961	0.53	1.01	1.39	1.52	2.58	0.35	0.40	0.89	1.53	1.34	1.38	1.18	14.10
1962	1.21	1.32	0.66	0.72	1.88	1.49	0.40	0.59	1.54	0.62	0.61	1.06	11.99
1963	2.38	0.80	1.26	0.68	0.25	3.29	1.11	1.29	1.54	0.50	0.56	3.15	14.89
1964	0.70	0.51	0.67	1.49	1.22	3.13	0.94	1.84	0.51	0.17	1.05	0.43	15.22
1965	1.46	0.62	0.67	1.68	0.63	1.55	1.59	2.24	2.11				14.20
1966	1.44	0.96	0.86	0.66	0.50	1.81	0.71	1.01	0.96	0.81	0.59	0.79	11.10
1967	1.22	0.27	1.09	0.93	1.33	2.67	0.40	T	0.49	1.64	0.65	1.71	12.40
1968	0.87	0.73	0.60	0.94	1.05	1.89	0.21	2.38	1.92	0.54	0.55	1.17	12.85
1969	2.94	0.52	0.72	0.64	1.21	4.18	0.25	0.04	0.66	0.67	0.42	0.92	13.17
1970	2.87	0.34	1.06	1.07	1.75	2.84	1.68	0.08	0.48	1.00	0.97	0.94	15.08
1971	1.81	0.56	0.78	2.09	1.35	1.74	0.53	0.91	0.30	0.24	1.18	1.63	13.12
1972	2.04	1.82	1.62	0.96	0.69	1.37	0.64	0.24	1.66	0.78	0.41	1.46	13.69
1973	0.44	0.17	0.23	0.33	0.54	1.57	0.09	0.31	0.60	0.60	2.51	1.62	9.01
1974	2.07	0.68	1.26	0.61	0.44	1.36	1.03	1.18	0.70	0.25	0.50	0.68	10.76
1975	2.03	1.77	0.74	1.01	1.35	2.02	1.51	2.03	0.51	3.51	1.15	0.85	18.48
1976	0.90	1.04	0.40	0.94	0.79	1.52	1.20	0.88	0.58	0.33	0.22	0.25	9.05
1977	0.66	0.18	0.98	0.08	2.13	0.66	0.72	1.28	1.67	0.72	1.02	2.88	12.98
1978	1.15	0.66	0.67	1.08	1.98	0.77	0.57	1.11	1.78	0.01	1.00	0.99	11.77
1979	1.25	1.04	1.22	1.04	0.74	0.67	0.77	1.31	0.05	0.97	0.50	0.81	10.37
1980	1.80	0.60	0.88	0.96	7.38	2.04	1.58	0.62	0.77	0.75	0.63	1.34	19.35
1981	0.16	0.77	1.43	0.74	4.19	2.70	1.07	1.61	1.01	0.62	1.07	1.98	17.35
1982	2.07	1.31	1.52	1.34	2.03	1.83	0.94	0.38	2.09	0.43	0.37	1.07	15.38
1983	0.62	0.95	1.10	0.72	2.65	2.26	2.44	1.27	1.37	0.37	1.17	1.79	16.71
1984	0.86	0.44	1.32	2.04	2.02	1.47	0.38	1.47	0.79	0.96	0.89	0.66	13.30
1985	0.19	0.70	0.44	0.55	1.57	0.38	0.09	3.29	3.60	0.80	0.51	0.38	12.50
Record Mean	1.08	0.84	0.88	1.01	1.80	1.96	0.92	0.90	1.20	1.01	1.03	1.13	13.78

TABLE 3 AVERAGE TEMPERATURE (deg. F) MISSOULA, MONTANA

YEAR	JAN	FEB	MAR	APR	MAY	JUNE	JULY	AUG	SEP	OCT	NOV	DEC	ANNUAL
1956	22.8	20.4	31.9	46.0	55.7	59.1	67.8	63.1	57.4	44.6	30.7	30.0	44.1
1957	10.5	26.2	35.6	44.8	56.8	60.3	67.3	65.1	59.2	42.2	32.7	32.0	44.4
1958	28.8	33.3	36.2	43.7	60.5	61.4	65.8	68.9	55.6	45.2	32.6	29.3	46.8
1959	26.1	23.4	36.5	44.3	48.5	61.6	67.2	62.4	54.0	44.8	26.4	26.6	43.5
#1960	16.7	25.7	32.9	43.6	49.7	60.1	72.2	62.4	56.4	43.6	32.9	18.3	42.9
1961	21.6	34.2	37.0	42.2	51.1	64.7	69.3	70.3	49.9	41.1	27.1	21.5	44.2
1962	15.2	24.9	30.9	45.4	50.1	57.3	64.1	62.9	54.8	43.0	33.9	30.3	42.7
1963	8.7	31.2	38.7	44.0	53.6	59.1	64.1	66.6	59.4	47.5	34.8	18.0	43.8
1964	21.9	22.6	28.5	42.1	48.9	57.0	66.2	60.8	51.1	42.1	30.4	19.4	40.9
1965	24.6	24.6	25.6	44.7	48.9	57.2	65.0	64.1	47.8	47.0	36.4	27.2	42.7
1966	26.3	26.5	35.3	43.7	55.1	58.1	69.1	66.8	62.9	44.0	34.0	29.5	45.9
1967	31.0	34.5	34.0	40.9	51.4	60.3	70.0	71.2	62.2	44.6	31.9	22.9	46.2
1968	20.2	32.4	41.7	42.1	52.6	59.8	69.8	64.2	54.7	42.0	33.7	21.8	44.6
1969	17.9	20.7	29.0	46.5	54.3	63.0	64.3	67.0	56.8	40.2	31.8	26.0	42.8
1970	22.9	32.2	34.3	39.8	52.6	63.0	68.2	67.7	50.6	41.5	33.9	22.5	44.1
1971	26.3	32.2	35.5	44.8	54.5	58.3	67.1	71.5	51.9	42.2	33.7	21.1	44.9
1972	20.6	28.1	40.6	42.6	54.2	62.0	66.4	66.4	51.4	42.5	33.6	19.3	43.9
1973	21.5	30.9	39.0	43.1	53.2	60.0	69.6	67.4	56.5	45.3	30.2	30.3	45.6
1974	21.2	31.9	35.4	45.9	48.5	65.0	68.1	65.1	57.1	46.3	34.7	27.8	45.6
1975	22.7	22.4	33.0	39.3	49.2	55.4	71.8	62.1	56.1	43.6	30.4	25.7	42.7
1976	28.0	31.7	34.8	46.3	55.0	56.8	66.8	63.2	57.4	43.9	32.3	24.4	45.1
1977	18.6	31.9	34.2	46.9	50.3	63.4	65.8	67.6	54.6	44.2	31.3	24.5	44.4
1978	24.2	27.3	39.8	46.1	48.3	59.6	65.3	63.5	55.1	45.3	26.7	15.9	43.1
1979	5.6	25.5	35.9	43.9	52.2	62.3	69.2	69.1	61.4	48.1	27.2	31.2	44.3
1980	16.3	29.6	34.6	49.8	54.7	56.8	66.5	61.8	57.0	45.3	34.2	30.5	44.8
1981	29.5	31.4	40.0	46.7	53.4	57.5	65.0	69.4	56.7	41.7	32.7	24.0	45.7
1982	21.2	22.7	38.3	41.6	51.0	63.0	64.6	66.0	56.2	45.0	29.0	22.7	43.5
1983	30.0	34.5	39.9	44.6	51.8	58.6	62.1	68.9	51.1	43.8	34.3	11.8	44.3
1984	25.4	32.1	39.7	43.8	49.3	56.8	67.2	68.9	52.8	42.5	33.9	20.1	44.3
1985	19.2	23.7	36.2	47.7	55.4	62.0	74.8	62.2	50.5	40.3	21.7	14.7	42.4
Record Mean	21.8	27.6	35.5	44.8	52.6	59.8	67.7	66.0	56.0	45.0	32.7	24.7	44.5
Max	29.7	36.5	46.0	57.4	66.2	74.1	84.9	83.1	71.0	57.5	41.0	31.8	56.6
Min	13.9	18.8	25.1	32.1	39.1	45.5	50.4	48.9	41.0	32.4	24.4	17.6	32.5

REFERENCE NOTES FOR TABLES 1, 2, 3 and 6   (MISSOULA, MT)

GENERAL

T - TRACE AMOUNT
BLANK ENTRIES DENOTE MISSING/UNREPORTED DATA.
# INDICATES A STATION OR INSTRUMENT RELOCATION.

SPECIFIC

TABLE 1

(a) - LENGTH OF RECORD IN YEARS. ALTHOUGH
INDIVIDUAL MONTHS MAY BE MISSING.

* LESS THAN .05

NORMALS — BASED ON THE 1951-1980 RECORD PERIOD.
EXTREMES — DATES ARE THE MOST RECENT OCCURRENCE.
WIND DIR. — NUMERALS SHOW TENS OF DEGREES
CLOCKWISE FROM TRUE NORTH.
"00" INDICATES CALM.
RESULTANT WIND DIRECTIONS ARE GIVEN TO WHOLE DEGREES.

EXCEPTIONS

TABLES 2, 3, and 6

RECORD MEANS ARE THROUGH THE CURRENT YEAR,
BEGINNING IN 1892 FOR TEMPERATURE
1886 FOR PRECIPITATION
1945 FOR SNOWFALL

TABLE 4   HEATING DEGREE DAYS Base 65 deg. F   MISSOULA, MONTANA

SEASON	JULY	AUG	SEP	OCT	NOV	DEC	JAN	FEB	MAR	APR	MAY	JUNE	TOTAL
1956-57	54	90	226	626	1024	1078	1686	1078	906	599	256	161	7784
1957-58	26	58	178	699	964	1017	1116	879	889	634	173	132	6765
1958-59	47	30	295	608	964	1098	1201	1159	872	613	505	134	7526
#1959-60	67	101	337	617	1151	1185	1494	1135	988	636	466	148	8325
1960-61	8	136	256	655	957	1442	1340	856	862	679	425	67	7683
1961-62	6	10	448	734	1128	1343	1541	1117	1048	580	455	234	8644
1962-63	83	104	298	674	925	1065	1745	941	808	623	350	204	7820
1963-64	76	51	183	535	899	1449	1327	1222	1124	680	492	240	8278
1964-65	30	153	409	698	1031	1246	1246	1124	1218	602	491	230	8638
1965-66	44	88	509	555	849	1165	1194	1075	914	631	304	216	7544
1966-67	23	46	94	645	921	1097	1045	848	952	712	416	145	6944
1967-68	2	4	116	630	987	1299	1383	941	718	678	379	168	7305
1968-69	10	97	305	705	935	1333	1456	1234	1109	546	325	201	8256
1969-70	69	46	262	763	991	1200	1297	911	942	751	377	140	7749
1970-71	32	12	425	722	928	1313	1195	908	911	600	319	211	7576
1971-72	75	12	386	700	934	1352	1371	1062	749	666	338	133	7778
1972-73	81	39	400	688	936	1416	1343	946	800	651	353	186	7839
1973-74	18	47	262	606	1036	1070	1357	919	910	569	505	103	7402
1974-75	27	60	230	575	903	1147	1302	1184	985	765	483	281	7942
1975-76	9	116	263	657	1032	1213	1141	959	929	554	302	258	7433
1976-77	25	89	231	647	974	1253	1431	918	950	536	450	76	7580
1977-78	59	76	310	637	1001	1245	1259	1050	771	564	511	171	7654
1978-79	67	110	308	604	1141	1515	1841	1099	893	625	389	126	8718
1979-80	37	13	121	517	1130	1043	1506	1018	935	448	318	198	7284
1980-81	53	119	243	603	917	1063	1092	934	768	542	353	228	6915
1981-82	39	16	256	718	963	1266	1351	1182	824	698	425	104	7842
1982-83	71	48	267	611	1072	1304	1079	845	772	607	402	189	7267
1983-84	126	10	413	652	913	1647	1222	947	779	631	479	259	8078
1984-85	37	18	370	692	926	1383	1413	1154	885	513	304	130	7825
1985-86	0	121	429	755	1294	1554							

TABLE 5   COOLING DEGREE DAYS Base 65 deg. F   MISSOULA, MONTANA

YEAR	JAN	FEB	MAR	APR	MAY	JUNE	JULY	AUG	SEP	OCT	NOV	DEC	TOTAL
1969	0	0	0	0	0	25	56	117	22	0	0	0	220
1970	0	0	0	0	0	88	138	104	3	0	0	0	333
1971	0	0	0	0	1	14	146	220	0	0	0	0	381
1972	0	0	0	0	11	47	85	85	0	0	0	0	228
1973	0	0	0	0	0	43	165	124	15	0	0	0	347
1974	0	0	0	0	0	106	128	68	1	0	0	0	303
1975	0	0	0	0	0	0	227	31	0	0	0	0	258
1976	0	0	0	0	0	21	89	41	7	0	0	0	158
1977	0	0	0	0	1	36	92	163	3	0	0	0	295
1978	0	0	0	0	0	14	83	69	19	0	0	0	185
1979	0	0	0	0	0	50	177	146	17	0	0	0	390
1980	0	0	0	0	4	13	74	26	8	0	0	0	125
1981	0	0	0	0	0	10	48	158	10	0	0	0	226
1982	0	0	0	0	0	51	66	83	7	0	0	0	207
1983	0	0	0	0	0	4	43	138	5	0	0	0	190
1984	0	0	0	0	0	16	114	129	12	0	0	0	271
1985	0	0	0	0	11	47	311	39	0	0	0	0	408

TABLE 6   SNOWFALL (inches)   MISSOULA, MONTANA

SEASON	JULY	AUG	SEP	OCT	NOV	DEC	JAN	FEB	MAR	APR	MAY	JUNE	TOTAL
1956-57	0.0	0.0	0.0	2.5	1.6	1.8	19.9	6.5	9.6	T	0.0	0.0	41.9
1957-58	0.0	0.0	T	1.2	4.5	4.0	5.3	14.5	2.2	2.9	T	0.0	34.6
1958-59	0.0	0.0	0.0	4.3	9.0	13.8	18.5	7.4	7.2	2.3	T	0.0	52.3
1959-60	0.0	0.0	T	0.2	13.8	2.7	17.9	7.4	7.2	2.3	T	0.0	51.5
1960-61	0.0	0.0	0.0	0.2	5.1	11.7	7.9	3.3	5.2	4.7	6.5	0.0	44.6
1961-62	0.0	0.0	T	4.6	15.1	11.3	9.0	10.4	6.9	T	0.0	0.0	57.3
1962-63	0.0	0.0	T	T	0.8	1.1	42.5	3.5	6.7	1.5	0.1	0.0	56.2
1963-64	0.0	0.0	0.0	T	3.0	20.6	12.2	8.9	11.1	0.5	7.5	0.0	63.8
1964-65	0.0	0.0	0.0	1.7	3.7	31.6	21.5	11.2	8.4	4.0	T	0.0	82.1
1965-66	0.0	0.0	0.2	0.0	3.2	6.4	14.3	18.6	9.0	5.7	T	0.0	57.4
1966-67	0.0	0.0	0.0	0.2	3.4	11.2	7.4	2.2	8.9	5.7	2.6	0.0	41.6
1967-68	0.0	0.0	0.0	0.3	5.1	12.6	14.5	2.8	1.5	5.8	T	0.0	42.6
1968-69	0.0	0.0	T	T	1.5	19.5	27.5	9.2	9.0	T	0.0	0.0	66.7
1969-70	0.0	0.0	0.0	T	1.9	8.6	23.5	1.5	9.5	8.2	3.0	0.0	56.2
1970-71	0.0	0.0	T	1.1	4.4	12.7	9.9	3.4	6.3	3.5	T	T	41.3
1971-72	0.0	0.0	T	0.7	13.1	16.7	22.5	15.2	9.9	3.0	0.0	0.0	81.1
1972-73	0.0	0.0	0.2	1.2	4.5	8.9	3.0	2.1	0.5	0.2	T	T	20.6
1973-74	0.0	0.0	0.0	5.4	13.4	13.9	11.1	8.2	4.6	0.3	T	0.0	56.9
1974-75	0.0	0.0	0.0	0.2	2.7	7.5	16.9	20.1	7.1	3.0	T	0.0	57.5
1975-76	0.0	0.0	0.0	5.0	10.4	6.1	8.9	9.9	2.5	2.0	0.0	0.0	44.8
1976-77	0.0	0.0	0.0	0.4	1.2	3.0	10.3	1.8	12.5	1.0	T	0.0	30.2
1977-78	0.0	0.0	T	0.1	9.6	20.6	16.1	6.6	5.2	T	8.1	0.0	66.3
1978-79	0.0	0.0	0.0	0.0	12.3	14.1	20.0	7.1	7.3	1.9	0.0	0.0	62.7
1979-80	0.0	0.0	0.0	T	7.4	3.1	25.5	3.7	11.8	2.2	1.0	0.0	54.7
1980-81	0.0	0.0	0.0	T	0.9	2.1	1.4	8.0	2.0	T	0.0	0.0	14.4
1981-82	0.0	0.0	0.0	1.4	1.2	14.4	27.0	9.7	9.0	6.6	T	0.0	69.3
1982-83	0.0	0.0	0.0	0.5	2.8	10.4	4.8	5.2	0.5	T	T	0.0	24.2
1983-84	0.0	0.0	0.4	0.0	5.6	27.0	4.1	3.6	1.7	T	T	0.0	42.4
1984-85	0.0	0.0	T	3.9	3.3	10.4	2.5	11.8	3.3	0.1	T	0.0	34.9
1985-86	0.0	0.0	0.0	3.5	6.1	6.2							
Record Mean	0.0	0.0	T	0.9	6.0	11.1	13.1	7.8	6.0	2.3	0.9	T	48.2

See Reference Notes, relative to all above tables, on preceding page.

GRAND ISLAND, NEBRASKA

The Grand Island Weather Service Office is located at the Hall County Regional Airport, 3 miles northeast of downtown Grand Island. It is situated just west of the mid-point of the north-south runway. The site is less than 50 miles from the geographical center of the contiguous United States and in the shallow Platte River valley. The complex of the Loup River and its tributaries converge approximately 15 miles northwest of the station, then flows eastward across the state. The terrain immediately surrounding the station is flat, sandy, loam. Just to the north is the southern boundary of the Nebraska sandhills. The terrain slopes gently upward from the Missouri River valley in eastern Nebraska to the Rocky Mountains of Colorado and Wyoming.

The climate is primarily continental in nature with occasional incursions of maritime tropical air from the Gulf of Mexico and modified maritime polar air from the Pacific Ocean. Wintertime outbreaks of cold, dry, Arctic air from Canada are common, usually accompanied by strong biting winds.

The east-west upslope produces periods of fog and low stratus when winds have an easterly component, and the characteristics of a Chinook, with warm dry air, when the component is westerly. Dry-season dust storms occur occasionally with these Chinook winds. These have been reduced in recent years by increased farm irrigation. Growing season humidities have also been increased by the expanding irrigation projects. Summers are usually hot and dry with temperatures often reaching 100 degrees or more. Late spring and early summer is the peak season for severe thunderstorms with frequent hail and tornados occasionally occurring. Winters are punctuated by occasional severe blizzards and have wide variations in temperatures that range from mild to bitterly cold.

Based on the 1951-1980 period, the average first occurrence of 32 degrees Fahrenheit in the fall is October 9 and the average last occurrence in the spring is April 29.

TABLE 1   NORMALS, MEANS AND EXTREMES

GRAND ISLAND, NEBRASKA

LATITUDE: 40°58'N   LONGITUDE: 98°19'W   ELEVATION: FT. GRND 1841 BARO 01845   TIME ZONE: CENTRAL   WBAN: 14935

	(a)	JAN	FEB	MAR	APR	MAY	JUNE	JULY	AUG	SEP	OCT	NOV	DEC	YEAR
TEMPERATURE °F:														
Normals														
-Daily Maximum		31.2	38.1	47.1	62.3	73.0	83.6	88.8	86.9	77.1	66.7	49.4	37.3	61.8
-Daily Minimum		9.9	16.2	24.7	37.8	49.1	59.2	64.4	62.3	51.5	39.4	25.7	16.0	38.0
-Monthly		20.6	27.1	35.9	50.0	61.0	71.4	76.6	74.6	64.3	53.0	37.5	26.6	49.9
Extremes														
-Record Highest	40	72	77	90	94	101	107	109	110	104	96	82	76	110
-Year		1981	1972	1978	1964	1967	1970	1954	1983	1947	1947	1980	1964	AUG 1983
-Record Lowest	40	-28	-19	-21	7	23	38	42	40	23	16	-11	-23	-28
-Year		1963	1979	1960	1975	1967	1954	1971	1950	1984	1981	1976	1983	JAN 1963
NORMAL DEGREE DAYS:														
Heating (base 65°F)		1376	1058	902	447	169	27	7	6	104	377	822	1187	6482
Cooling (base 65°F)		0	0	0	0	49	219	366	303	83	8	0	0	1028
% OF POSSIBLE SUNSHINE														
MEAN SKY COVER (tenths)														
Sunrise - Sunset	36	6.1	6.3	6.4	6.1	6.0	5.1	4.4	4.5	4.6	4.8	6.0	6.0	5.5
MEAN NUMBER OF DAYS:														
Sunrise to Sunset														
-Clear	47	9.0	7.2	7.9	8.0	7.8	10.3	12.8	13.2	13.5	13.5	9.4	9.5	122.2
-Partly Cloudy	47	8.0	7.5	7.8	8.4	9.4	10.1	11.1	10.3	7.3	7.7	7.4	7.4	102.4
-Cloudy	47	14.0	13.6	15.4	13.5	13.6	9.6	7.1	7.5	9.0	9.8	13.2	14.1	140.5
Precipitation														
.01 inches or more	47	5.2	6.0	7.5	9.2	10.9	9.9	8.7	7.8	7.1	5.0	4.7	4.6	86.6
Snow, Ice pellets														
1.0 inches or more	47	2.0	2.2	2.0	0.6	0.*	0.0	0.0	0.0	0.*	0.1	1.0	1.7	9.6
Thunderstorms	47	0.*	0.2	1.2	3.8	7.7	10.0	9.0	8.0	5.3	2.0	0.5	0.1	47.9
Heavy Fog Visibility														
1/4 mile or less	47	1.9	2.3	1.8	1.3	0.8	0.6	0.7	1.0	1.2	1.6	2.1	2.3	17.8
Temperature °F														
-Maximum														
90° and above	24	0.0	0.0	0.*	0.5	1.8	7.6	14.4	11.9	3.6	0.3	0.0	0.0	40.2
32° and below	24	16.3	10.5	4.6	0.1	0.0	0.0	0.0	0.0	0.0	0.0	3.2	13.0	47.7
-Minimum														
32° and below	24	30.6	26.8	23.4	7.2	0.7	0.0	0.0	0.0	0.5	6.3	22.5	30.1	148.2
0° and below	24	8.7	3.7	0.5	0.0	0.0	0.0	0.0	0.0	0.0	0.0	0.4	3.8	17.0
AVG. STATION PRESS.(mb)	13	952.4	951.0	947.4	947.3	946.9	947.6	949.5	949.8	950.7	950.7	950.6	951.2	949.6
RELATIVE HUMIDITY (%)														
Hour 00	24	74	75	73	71	73	73	73	76	78	73	76	75	74
Hour 06 (Local Time)	24	76	77	78	78	81	81	82	84	84	79	80	78	80
Hour 12	24	64	62	57	51	52	51	52	52	54	49	57	63	55
Hour 18	24	67	63	54	47	49	47	49	50	54	53	63	68	55
PRECIPITATION (inches):														
Water Equivalent														
-Normal		0.52	0.81	1.55	2.64	3.70	3.72	2.71	2.59	2.51	1.09	0.80	0.67	23.31
-Maximum Monthly	47	1.65	3.39	5.57	7.34	8.88	13.96	9.60	8.73	9.00	4.42	3.77	2.17	13.96
-Year		1960	1971	1973	1984	1982	1967	1950	1977	1965	1946	1983	1968	JUN 1967
-Minimum Monthly	47	T	0.07	0.01	0.35	0.43	0.50	0.63	0.50	0.12	0.00	T	0.02	0.00
-Year		1961	1974	1967	1946	1964	1978	1970	1940	1939	1958	1939	1943	OCT 1958
-Maximum in 24 hrs	47	1.38	2.21	3.15	3.30	3.07	4.54	5.41	4.12	5.88	2.75	1.90	1.20	5.88
-Year		1947	1971	1979	1964	1985	1967	1950	1977	1977	1968	1973	1968	SEP 1977
Snow, Ice pellets														
-Maximum Monthly	47	16.1	21.5	20.5	9.0	4.5				3.8	6.6	17.1	26.0	26.0
-Year		1960	1969	1984	1984	1947				1985	1980	1983	1973	DEC 1973
-Maximum in 24 hrs	47	8.3	15.0	12.2	5.5	4.5				3.8	6.6	11.2	12.0	15.0
-Year		1960	1984	1984	1984	1947				1985	1980	1983	1968	FEB 1984
WIND:														
Mean Speed (mph)	36	11.8	12.0	13.6	14.2	12.7	12.0	10.7	10.6	11.1	11.4	11.9	11.8	12.0
Prevailing Direction														
through 1963		NNW	NNW	NNW	NNW	S	S	S	S	S	S	NNW	NNW	S
Fastest Obs. 1 Min.														
-Direction (!!!)	23	35	01	34	18	18	32	19	24	25	34	34	27	19
-Speed (MPH)	23	48	41	55	47	53	57	57	46	40	41	51	46	57
-Year		1969	1976	1971	1968	1982	1964	1972	1978	1970	1966	1975	1970	JUL 1972
Peak Gust														
-Direction (!!!)	2	NW	N	NW	26	S	S	N	SW	SW	N	N	NW	S
-Speed (mph)	2	51	52	49	62	76	55	39	53	48	52	46	53	76
-Date		1984	1984	1985	1984	1985	1985	1985	1985	1985	1985	1984	1985	MAY 1985

See reference Notes to this table on the following page.

TABLE 2 PRECIPITATION (inches)   GRAND ISLAND, NEBRASKA

YEAR	JAN	FEB	MAR	APR	MAY	JUNE	JULY	AUG	SEP	OCT	NOV	DEC	ANNUAL
1956	0.93	0.23	0.52	1.96	2.43	3.51	0.94	0.77	0.91	0.92	0.27	0.24	13.63
1957	0.30	0.24	2.17	3.49	5.64	4.31	1.68	4.08	3.36	1.08	0.70	0.38	27.43
1958	0.35	1.06	1.23	3.26	2.24	2.72	7.09	1.38	1.24	0.00	0.67	0.10	21.34
1959	0.66	0.41	4.35	3.09	6.69	4.09	0.95	3.11	2.36	1.48	0.56	0.62	28.37
1960	1.65	1.14	1.44	2.72	4.07	3.87	3.16	2.60	2.10	0.65	0.41	0.12	23.93
1961	T	0.49	1.53	1.79	7.49	4.01	4.76	1.33	2.55	0.20	0.86	1.03	26.04
1962	0.13	1.64	1.58	0.52	3.43	2.75	8.78	1.65	1.59	1.35	0.11	0.68	24.21
1963	0.70	0.21	1.17	1.39	3.28	3.22	2.18	3.06	3.45	0.06	0.25	0.13	19.10
1964	0.07	0.92	1.64	4.97	0.43	4.50	3.00	3.44	1.27	0.13	0.14	0.16	20.67
1965	0.67	1.23	1.26	2.18	5.97	5.21	2.36	3.35	9.00	0.37	0.50	0.53	32.63
1966	0.22	0.78	0.58	1.20	1.03	2.98	3.47	1.68	0.43	0.65	0.09	0.66	13.77
1967	0.59	0.07	0.01	0.99	3.40	13.96	0.98	1.30	1.02	1.37	0.22	0.46	24.37
1968	0.13	0.32	0.39	3.47	2.23	7.13	4.82	4.41	2.33	3.61	0.61	2.17	31.62
1969	0.91	2.48	0.19	2.55	4.13	3.46	3.10	3.75	2.01	3.43	0.19	0.82	27.02
1970	0.04	0.24	0.42	2.72	2.49	0.81	0.63	3.36	5.76	1.44	0.33	0.07	18.31
1971	0.82	3.39	1.12	0.95	5.32	5.62	2.27	0.66	1.35	1.75	1.82	0.50	25.57
1972	0.19	0.17	0.23	3.00	5.91	1.86	4.98	1.43	2.50	1.04	2.40	2.03	25.74
1973	0.77	0.45	5.57	1.63	3.85	0.84	2.90	1.38	8.39	1.51	2.37	2.07	31.73
1974	0.62	0.07	0.51	1.73	2.44	2.67	1.35	1.55	0.50	1.41	0.25	1.33	14.43
1975	0.86	0.46	1.02	2.76	1.50	6.86	2.35	1.18	1.01	0.11	3.26	0.16	21.53
1976	0.39	0.73	2.15	2.79	3.21	2.11	1.12	0.78	2.42	0.07	0.12	0.04	15.93
1977	0.42	0.18	3.30	4.89	6.85	1.37	1.73	8.73	7.77	1.42	1.13	0.43	38.22
1978	0.27	1.18	0.83	6.12	1.87	0.50	2.88	3.05	1.62	0.56	1.35	0.64	20.87
1979	0.82	0.43	5.56	3.27	3.99	2.65	2.66	1.39	2.54	3.02	1.78	0.48	28.59
1980	0.80	0.65	2.23	1.92	2.06	3.62	0.85	4.42	0.83	1.46	0.12	0.18	19.14
1981	0.16	0.19	3.14	1.14	4.28	0.60	3.45	4.38	0.93	1.08	3.21	0.63	23.19
1982	0.65	0.56	2.41	3.15	8.88	4.47	2.65	5.78	2.37	2.06	1.49	1.18	35.65
1983	0.66	0.35	3.40	1.30	4.59	6.29	1.71	2.04	2.67	1.08	3.77	0.76	28.62
1984	0.24	1.65	3.18	7.34	5.75	3.95	1.99	1.21	0.19	3.49	1.37	1.34	31.70
1985	0.27	0.27	1.15	4.37	4.62	3.98	3.20	2.25	5.80	1.83	0.61	0.27	28.62
Record Mean	0.54	0.77	1.45	2.47	3.97	3.84	3.03	2.90	2.62	1.49	0.97	0.68	24.75

TABLE 3 AVERAGE TEMPERATURE (deg. F)   GRAND ISLAND, NEBRASKA

YEAR	JAN	FEB	MAR	APR	MAY	JUNE	JULY	AUG	SEP	OCT	NOV	DEC	ANNUAL
1956	20.6	24.9	37.0	44.2	62.2	76.2	75.5	75.3	66.0	59.3	38.1	31.8	50.9
1957	17.1	32.5	36.6	47.6	57.9	69.4	79.3	75.9	61.6	51.6	36.8	34.5	50.1
1958	26.5	22.3	29.0	47.7	62.7	68.3	71.8	75.4	66.3	55.0	39.4	26.6	49.2
1959	20.1	23.6	38.2	49.0	61.0	73.1	73.2	78.8	63.7	48.7	32.8	33.9	49.7
1960	18.2	18.6	24.7	52.0	59.7	68.3	74.0	73.8	66.5	54.7	39.5	28.2	48.2
#1961	26.0	32.5	39.2	46.3	57.7	71.5	76.9	75.4	61.1	54.5	35.4	19.7	49.7
1962	20.0	25.6	28.5	50.9	66.9	69.4	71.9	73.2	60.3	53.3	38.3	26.0	48.7
1963	10.4	28.1	40.1	51.8	60.2	74.1	77.7	73.5	64.8	59.5	42.5	21.9	50.4
1964	29.2	27.3	34.6	52.0	66.6	70.8	79.4	69.5	62.2	52.1	37.7	23.1	50.4
1965	22.8	22.3	25.1	51.2	64.0	69.7	74.8	73.6	56.8	51.5	41.3	33.9	49.3
1966	16.7	25.8	42.1	45.5	61.1	71.0	81.0	70.7	63.9	53.8	38.1	26.3	49.7
1967	24.4	28.8	42.6	52.2	56.5	68.2	74.8	72.3	63.1	52.8	37.3	27.1	50.0
1968	23.9	27.0	44.4	51.1	58.2	73.2	75.2	74.4	64.2	54.2	36.8	20.4	50.3
1969	16.9	24.7	29.2	53.0	61.8	66.1	77.2	76.0	67.3	46.0	40.1	25.7	48.7
1970	17.9	31.9	33.5	49.5	65.8	73.6	78.5	77.7	64.4	48.8	37.2	29.3	50.7
1971	19.9	26.1	36.1	51.7	59.0	76.0	72.8	75.4	64.4	56.0	39.9	29.9	50.6
1972	21.8	27.2	41.7	49.6	61.2	72.5	73.9	72.8	64.7	48.9	35.0	20.0	49.1
1973	23.7	29.7	41.1	47.4	57.7	72.3	75.9	77.7	62.3	55.1	36.6	22.8	50.2
1974	17.2	32.4	42.4	52.6	64.3	69.7	82.3	69.0	61.2	54.8	38.3	25.2	50.8
1975	22.7	21.1	30.8	47.4	62.2	69.0	76.8	77.1	61.0	56.1	35.8	27.3	49.0
1976	24.4	37.5	38.7	53.3	59.3	71.0	77.3	76.0	65.4	47.4	33.2	28.1	51.0
1977	17.7	34.1	41.3	56.2	67.1	73.7	78.9	71.7	65.7	52.6	39.2	26.5	52.0
1978	12.2	15.0	35.5	51.3	59.8	73.7	76.3	73.8	68.3	52.6	35.2	18.4	47.6
1979	7.5	13.8	36.9	48.3	57.5	70.5	74.0	74.2	67.6	53.0	34.4	33.2	47.6
1980	21.6	23.7	34.2	51.5	62.0	73.1	80.7	76.9	66.9	51.5	41.7	29.8	51.0
1981	28.9	31.4	41.7	58.2	57.0	73.2	75.8	71.8	64.6	50.6	41.0	24.6	51.6
1982	12.9	23.8	35.0	47.0	60.8	65.7	76.5	72.4	63.2	52.5	35.0	28.6	47.8
1983	25.5	31.4	37.9	43.6	56.8	69.6	79.5	82.4	68.3	53.8	38.3	8.4	49.6
1984	23.6	34.6	33.4	47.4	60.4	73.5	76.9	77.2	62.0	51.9	39.3	26.6	50.6
1985	20.5	23.8	43.3	55.3	65.0	68.8	75.6	70.6	61.9	52.0	27.4	21.8	48.8
Record Mean	22.5	27.3	37.5	50.3	61.0	71.3	77.5	75.2	65.4	53.5	38.3	26.8	50.6
Max	33.1	38.0	48.8	62.6	72.9	83.5	90.1	87.9	78.2	66.6	49.8	36.9	62.4
Min	12.0	16.5	26.1	38.1	49.1	59.2	64.8	62.5	52.7	40.5	26.9	16.6	38.8

REFERENCE NOTES FOR TABLES 1, 2, 3 and 6   (GRAND ISLAND, NE)

GENERAL

T - TRACE AMOUNT
BLANK ENTRIES DENOTE MISSING/UNREPORTED DATA.
# INDICATES A STATION OR INSTRUMENT RELOCATION.

SPECIFIC

TABLE 1

(a) - LENGTH OF RECORD IN YEARS. ALTHOUGH
      INDIVIDUAL MONTHS MAY BE MISSING.
  * LESS THAN .05

NORMALS — BASED ON THE 1951-1980 RECORD PERIOD.
EXTREMES — DATES ARE THE MOST RECENT OCCURRENCE.
WIND DIR. — NUMERALS SHOW TENS OF DEGREES
            CLOCKWISE FROM TRUE NORTH.
            "00" INDICATES CALM.
RESULTANT WIND DIRECTIONS ARE GIVEN TO WHOLE DEGREES.

EXCEPTIONS

TABLES 2, 3, and 6

RECORD MEANS ARE THROUGH THE CURRENT YEAR,
BEGINNING IN   1901 FOR TEMPERATURE
                 1901 FOR PRECIPITATION
                 1939 FOR SNOWFALL

TABLE 4 HEATING DEGREE DAYS Base 65 deg. F   GRAND ISLAND, NEBRASKA

SEASON	JULY	AUG	SEP	OCT	NOV	DEC	JAN	FEB	MAR	APR	MAY	JUNE	TOTAL
1956-57	0	14	93	213	803	1022	1478	902	876	517	227	14	6159
1957-58	0	0	132	418	836	937	1189	1191	1109	516	115	44	6487
1958-59	0	9	74	331	760	1183	1384	1152	823	474	180	22	6392
1959-60	4	0	147	498	958	958	1445	1341	1241	406	185	23	7206
#1960-61	0	0	90	324'	759	1136	1205	904	791	555	260	18	6042
1961-62	0	0	203	327	883	1399	1391	1097	1122	431	48	37	6938
1962-63	10	0	168	372	795	1202	1689	1027	764	400	183	3	6613
1963-64	0	12	77	193	670	1328	1102	1084	934	391	85	39	5915
1964-65	0	44	159	399	810	1294	1301	1190	1229	422	97	8	6953
1965-66	0	11	271	284	705	956	1492	1091	703	577	188	27	6305
1966-67	0	20	108	353	799	1192	1252	1009	699	391	331	36	6190
1967-68	4	15	93	389	820	1168	1272	1095	634	414	227	13	6144
1968-69	1	9	84	346	840	1374	1483	1121	1102	357	154	82	6953
1969-70	0	0	25	586	740	1212	1455	919	973	471	93	14	6488
1970-71	0	0	146	513	828	1102	1392	1080	891	396	194	0	6542
1971-72	6	0	131	284	743	1083	1335	1089	714	463	181	19	6048
1972-73	6	9	116	494	894	1389	1273	980	734	520	240	11	6666
1973-74	2	0	123	306	847	1304	1479	904	691	369	91	40	6156
1974-75	0	27	165	315	793	1229	1302	1221	1052	524	127	31	6786
1975-76	0	0	178	296	870	1166	1251	791	808	354	198	11	5923
1976-77	0	0	95	551	944	1139	1461	861	729	280	15	0	6075
1977-78	0	1	47	380	768	1185	1632	1395	914	409	205	25	6961
1978-79	0	3	64	400	886	1438	1777	1431	866	492	258	38	7653
1979-80	9	10	49	366	915	978	1340	1191	946	420	176	11	6411
1980-81	0	2	71	420	694	1084	1113	934	716	221	255	7	5517
1981-82	9	0	80	439	713	1245	1612	1148	925	537	148	61	6917
1982-83	0	12	139	385	896	1120	1216	933	832	635	269	37	6474
1983-84	0	0	89	349	793	1751	1276	875	974	521	177	3	6808
1984-85	0	0	184	405	764	1185	1372	1147	664	319	76	39	6155
1985-86	0	13	217	399	1120	1334							

TABLE 5 COOLING DEGREE DAYS Base 65 deg. F   GRAND ISLAND, NEBRASKA

YEAR	JAN	FEB	MAR	APR	MAY	JUNE	JULY	AUG	SEP	OCT	NOV	DEC	TOTAL
1969	0	0	0	1	61	122	381	348	100	6	0	0	1019
1970	0	0	0	12	128	279	426	401	132	19	0	0	1397
1971	0	0	0	6	15	338	256	327	122	15	0	0	1079
1972	0	0	0	8	70	251	287	257	111	0	0	0	984
1973	0	0	0	0	19	238	345	401	53	9	0	0	1065
1974	0	0	0	6	77	189	541	159	56	6	0	0	1034
1975	0	0	0	2	49	158	373	384	66	28	0	0	1060
1976	0	0	0	9	29	196	387	344	111	12	0	0	1088
1977	0	0	0	24	85	268	437	214	77	0	0	0	1105
1978	0	0	5	5	47	293	355	283	169	2	0	0	1159
1979	0	0	0	0	36	211	294	303	132	2	0	0	978
1980	0	0	0	20	56	261	493	377	133	9	0	0	1349
1981	0	0	0	25	15	259	356	219	74	0	0	0	948
1982	0	0	0	1	24	90	364	248	92	2	0	0	821
1983	0	0	0	0	23	183	460	546	194	8	0	0	1414
1984	0	0	0	2	42	264	374	386	105	4	0	0	1177
1985	0	0	0	33	81	158	335	195	134	0	0	0	936

TABLE 6 SNOWFALL (inches)   GRAND ISLAND, NEBRASKA

SEASON	JULY	AUG	SEP	OCT	NOV	DEC	JAN	FEB	MAR	APR	MAY	JUNE	TOTAL
1956-57	0.0	0.0	0.0	0.0	1.8	2.7	3.7	1.7	3.3	7.2	0.0	0.0	20.4
1957-58	0.0	0.0	0.0	0.9	4.2	3.3	3.7	7.0	10.3	T	0.0	0.0	29.4
1958-59	0.0	0.0	0.0	0.0	3.0	1.9	6.3	5.0	12.2	1.2	0.0	0.0	29.6
1959-60	0.0	0.0	0.0	0.0	0.2	4.8	16.1	7.7	14.6	3.2	T	0.0	46.6
1960-61	0.0	0.0	0.0	T	T	1.1	T	3.5	6.6	4.7	0.0	0.0	15.9
1961-62	0.0	0.0	T	0.0	3.7	16.9	1.4	10.1	8.4	0.1	0.0	0.0	40.6
1962-63	0.0	0.0	0.0	0.0	0.8	6.3	7.9	1.5	4.0	0.0	0.0	0.0	20.5
1963-64	0.0	0.0	0.0	0.0	T	2.3	0.8	8.5	4.2	0.9	0.0	0.0	16.7
1964-65	0.0	0.0	0.0	0.0	T	0.8	7.9	15.3	6.3	0.0	0.0	0.0	30.3
1965-66	0.0	0.0	0.0	0.0	0.4	1.9	3.0	1.0	4.1	0.4	T	0.0	10.8
1966-67	0.0	0.0	0.0	0.5	0.2	8.5	6.4	1.0	T	4.1	4.3	0.0	25.0
1967-68	0.0	0.0	0.0	T	0.7	4.3	1.1	3.5	0.3	0.8	0.0	0.0	10.7
1968-69	0.0	0.0	0.0	0.0	1.7	21.8	7.6	21.5	1.8	T	0.0	0.0	54.4
1969-70	0.0	0.0	0.0	4.4	1.5	10.7	0.6	3.2	3.6	0.3	0.0	0.0	24.3
1970-71	0.0	0.0	0.0	0.1	0.1	0.4	10.3	12.7	11.7	0.2	0.0	0.0	35.5
1971-72	0.0	0.0	0.0	T	4.1	3.5	2.6	1.0	1.0	0.1	0.0	0.0	14.0
1972-73	0.0	0.0	0.0	0.2	11.5	17.5	2.6	4.7	2.1	3.8	T	0.0	42.4
1973-74	0.0	0.0	0.0	0.0	6.6	26.0	11.0	1.3	1.2	2.4	0.0	0.0	48.5
1974-75	0.0	0.0	0.0	0.0	0.3	16.0	8.3	7.4	3.1	2.6	0.0	0.0	37.7
1975-76	0.0	0.0	0.0	0.0	16.0	1.2	3.6	1.4	11.7	0.0	0.0	0.0	33.9
1976-77	0.0	0.0	0.0	T	1.1	0.4	5.4	1.1	12.5	4.8	0.0	0.0	25.3
1977-78	0.0	0.0	0.0	0.0	2.5	7.7	4.6	17.9	8.3	T	0.0	0.0	41.0
1978-79	0.0	0.0	0.0	0.0	10.4	8.1	11.0	3.9	10.4	2.2	0.0	0.0	46.0
1979-80	0.0	0.0	0.0	0.6	1.9	6.2	8.3	8.2	11.6	T	0.0	0.0	36.8
1980-81	0.0	0.0	0.0	6.6	0.8	1.6	1.6	2.0	2.8	5.4	T	0.0	19.2
1981-82	0.0	0.0	0.0	0.1	3.9	8.4	7.5	6.9	8.8	1.1	0.0	0.0	36.7
1982-83	0.0	0.0	0.0	0.2	2.7	10.7	5.0	5.9	11.8	3.9	0.0	0.0	40.2
1983-84	0.0	0.0	T	0.0	17.1	13.8	3.1	15.1	20.5	9.0	0.0	0.0	78.6
1984-85	0.0	0.0	0.0	T	1.6	7.5	5.9	3.4	4.5	0.0	0.0	0.0	22.9
1985-86	0.0	0.0	3.8	0.0	7.4	4.5							
Record Mean	0.0	0.0	0.1	0.3	3.4	6.7	5.8	6.2	6.5	1.9	0.2	0.0	31.1

See Reference Notes, relative to all above tables, on preceding page.

Lincoln is near the center of Lancaster County in southeastern Nebraska. The surrounding area is gently rolling prairie. The western edge of the city is in the flat valley of Salt Creek, which receives a number of tributaries in or near the city and flows northeastward to the lower Platte. The terrain slopes upward to the west and is sufficient to cause instability in moist easterly winds in the Lincoln area. Precipitation with westerly winds is infrequent since they are downslope. The upward slope to the west is a part of the general rise in elevation that begins at the Missouri River 45 miles east of Lincoln and culminates in the Continental Divide about 575 miles to the west. The chinook or foehn effect often produces rapid rises in temperature here during the winter with a shift of the wind to westerly.

The maximum temperature has exceeded 110 degrees. Hot winds, combining unusual wind force and high temperatures, occasionally cause serious injury to crops.

The majority of winter outbreaks of severely cold air from northwestern Canada move over the Lincoln area. The temperature has remained below zero degrees for more than 8 consecutive days. The center of some of the cold air masses move southward far enough to the east that their full effect is usually not felt here.

Normally the crop season, April through September, receives over three-fourths of the annual precipitation. Nighttime thunderstorms are predominant in the summer months, so that the needed moisture is received during much of the growing season at a time of least interference with outdoor work.

Annual snowfall is about 25 inches, although the annual snowfall has exceeded 59 inches. Much of the snow is light and melts rapidly. However, at times a considerable amount accumulates on the ground and has exceeded a depth of 21 inches.

In the summer the higher winds are associated with thunderstorms. Lincoln has been relatively free from tornadoes and more than slight hail damage seldom occurs. There is much sunshine, averaging 64 percent of the possible duration. Moderate to low humidities are at comfortable levels except for short periods during the summer when warm, moist, tropical air occasionally reaches this area.

TABLE 1 NORMALS, MEANS AND EXTREMES

LINCOLN, NEBRASKA

LATITUDE: 40°51'N   LONGITUDE: 96°45'W   ELEVATION: FT. GRND 1178 BARO 01197   TIME ZONE: CENTRAL   WBAN: 14939

	(a)	JAN	FEB	MAR	APR	MAY	JUNE	JULY	AUG	SEP	OCT	NOV	DEC	YEAR
TEMPERATURE °F:														
Normals														
–Daily Maximum		30.4	37.5	47.7	63.6	74.4	84.3	89.5	87.2	78.0	67.4	50.0	37.2	62.3
–Daily Minimum		8.9	15.4	25.1	38.6	49.9	60.2	65.6	63.4	52.9	40.6	26.9	16.0	38.6
–Monthly		19.6	26.5	36.4	51.1	62.2	72.3	77.6	75.3	65.5	54.0	38.5	26.6	50.4
Extremes														
–Record Highest	14	72	84	87	91	93	105	106	107	101	93	82	68	107
–Year		1981	1972	1978	1972	1983	1974	1983	1983	1975	1975	1980	1976	AUG 1983
–Record Lowest	14	-33	-24	-19	3	25	39	42	45	26	12	-5	-27	-33
–Year		1974	1979	1978	1975	1976	1978	1972	1978	1984	1972	1976	1983	JAN 1974
NORMAL DEGREE DAYS:														
Heating (base 65°F)		1404	1078	887	417	151	16	5	0	79	353	795	1190	6375
Cooling (base 65°F)		0	0	0	0	64	235	396	323	94	12	0	0	1124
% OF POSSIBLE SUNSHINE	29	57	57	56	59	62	70	73	72	66	63	54	52	62
MEAN SKY COVER (tenths)														
Sunrise – Sunset	21	6.2	6.3	6.5	6.3	6.1	5.4	4.8	4.9	5.0	5.4	6.3	6.5	5.8
MEAN NUMBER OF DAYS:														
Sunrise to Sunset														
–Clear	21	9.2	8.0	7.8	8.1	8.0	9.7	12.3	11.9	12.4	11.3	8.2	7.9	114.8
–Partly Cloudy	21	6.2	6.3	7.5	7.7	9.9	10.0	10.3	10.3	7.3	7.5	7.1	7.4	97.6
–Cloudy	21	15.6	13.9	15.7	14.1	13.0	10.4	8.3	8.9	10.3	12.2	14.6	15.8	152.9
Precipitation														
.01 inches or more	14	6.5	5.2	8.8	9.9	11.2	7.9	8.1	8.6	7.8	7.1	5.9	5.5	92.6
Snow,Ice pellets														
1.0 inches or more	14	2.2	2.0	1.9	0.5	0.0	0.0	0.0	0.0	0.0	0.1	1.0	1.9	9.6
Thunderstorms	14	0.1	0.2	1.7	4.9	7.0	7.9	8.1	7.5	5.1	2.6	0.7	0.4	46.4
Heavy Fog Visibility														
1/4 mile or less	14	0.9	1.7	2.1	0.4	0.4	0.4	0.1	0.6	0.4	1.2	1.7	1.7	11.6
Temperature °F														
–Maximum														
90° and above	14	0.0	0.0	0.0	0.1	0.6	7.8	17.1	12.3	4.5	0.2	0.0	0.0	42.6
32° and below	14	17.3	11.3	2.7	0.1	0.0	0.0	0.0	0.0	0.0	0.0	3.4	13.1	47.9
–Minimum														
32° and below	14	30.7	26.4	21.5	7.0	0.8	0.0	0.0	0.0	0.6	5.9	21.2	30.1	144.2
0° and below	14	9.1	4.7	0.6	0.0	0.0	0.0	0.0	0.0	0.0	0.0	0.2	4.4	19.1
AVG. STATION PRESS.(mb)	13	977.2	975.6	971.5	971.0	970.3	970.7	972.5	972.9	974.1	974.3	974.5	975.6	973.3
RELATIVE HUMIDITY (%)														
Hour 00	13	76	80	77	74	77	75	73	78	79	75	79	78	77
Hour 06 (Local Time)	13	77	81	82	81	84	83	81	85	85	81	82	80	82
Hour 12	13	66	67	62	56	57	54	52	57	56	54	62	66	59
Hour 18	13	70	67	59	54	55	51	49	54	57	57	66	71	59
PRECIPITATION (inches):														
Water Equivalent														
–Normal		0.64	1.01	1.94	2.81	3.84	3.84	3.20	3.42	2.93	1.68	0.96	0.65	26.92
–Maximum Monthly	14	1.59	1.26	6.65	7.21	7.97	7.67	6.93	8.57	7.52	5.29	3.81	3.42	8.57
–Year		1975	1984	1973	1978	1984	1983	1985	1982	1973	1979	1981	1984	AUG 1982
–Minimum Monthly	14	0.15	0.08	0.49	1.09	2.07	0.63	0.37	0.07	0.29	0.01	0.03	0.04	0.01
–Year		1981	1977	1972	1983	1980	1976	1983	1976	1974	1975	1976	1976	OCT 1975
–Maximum in 24 hrs	14	0.76	0.84	1.82	2.34	3.31	4.24	4.66	2.98	3.52	4.29	1.93	2.28	4.66
–Year		1975	1984	1979	1974	1984	1985	1985	1982	1977	1979	1977	1984	JUL 1985
Snow,Ice pellets														
–Maximum Monthly	14	14.6	13.8	17.0	7.2					0.8	3.3	8.6	19.8	19.8
–Year		1975	1978	1984	1983					1985	1980	1983	1973	DEC 1973
–Maximum in 24 hrs	14	8.0	7.7	8.3	4.2					0.8	3.3	7.7	10.4	10.4
–Year		1975	1978	1977	1979					1985	1980	1972	1973	DEC 1973
WIND:														
Mean Speed (mph)	13	10.1	10.5	11.9	12.7	10.5	10.1	9.8	9.6	9.8	10.1	10.2	10.3	10.5
Prevailing Direction														
Fastest Mile														
–Direction (!!!)	13	NW	NW	N	NW	W	NE	SW	N	SW	NW	NW	NW	NE
–Speed (MPH)	13	45	48	54	52	51	67	52	59	40	42	48	42	67
–Year		1983	1972	1976	1982	1972	1985	1981	1977	1978	1985	1975	1981	JUN 1985
Peak Gust														
–Direction (!!!)	2	NW	N	S	SW	SW	NE	NW	NW	SW	NW	N	NW	NE
–Speed (mph)	2	63	49	52	64	49	84	48	58	43	58	49	49	84
–Date		1984	1984	1985	1985	1985	1985	1984	1985	1984	1985	1984	1985	JUN 1985

See Reference Notes to this table on the following page

TABLE 2 PRECIPITATION (inches)   LINCOLN, NEBRASKA

YEAR	JAN	FEB	MAR	APR	MAY	JUNE	JULY	AUG	SEP	OCT	NOV	DEC	ANNUAL
1956	1.00	0.38	0.25	1.65	2.61	3.93	4.38	3.96	3.59	0.82	1.03	0.23	23.83
1957	0.44	0.28	3.33	2.20	3.42	6.07	6.05	4.35	1.55	4.36	1.91	0.67	34.63
1958	1.25	2.88	1.81	2.60	2.16	1.14	11.40	2.86	6.78	0.05	0.89	0.08	33.90
#1959	1.13	0.89	4.56	2.16	8.91	4.91	1.66	1.80	2.94	2.42	0.40	0.88	32.66
1960	1.48	2.10	2.03	2.02	4.35	4.96	3.42	6.29	2.97	1.49	0.33	31.52	31.52
1961	0.24	1.10	3.32	1.67	3.50	1.14	1.11	5.71	2.41	2.14	0.64	0.50	23.03
1962	0.52	1.28	1.09	0.79	2.91	3.03	6.50	5.25	3.54	0.72	0.20	0.33	23.52
1963	0.64	0.27	2.56	0.88	2.53	6.80	2.82	2.27	3.50	0.37	0.32	0.51	27.02
1964	0.21	0.76	1.29	2.54	2.72	8.56	3.18	3.61	2.95	0.41	0.84	1.36	41.33
1965	0.34	3.06	1.49	3.69	6.27	10.71	4.23	2.17	6.76	0.41			
1966	0.65	1.10	0.70	0.66	1.74	4.88	2.63	3.83	2.11	0.45	0.24	0.71	19.70
1967	0.47	0.14	0.88	1.63	4.47	12.93	1.91	1.91	2.91	1.91	0.41	0.68	32.33
1968	0.38	0.09	0.11	3.28	2.10	3.28	3.71	3.63	6.33	2.93	1.30	1.89	29.03
1969	0.68	0.89	1.45	4.62	4.46	2.78	3.93	2.09	0.77	2.88	0.12	1.23	25.90
1970	0.08	0.65	0.71	2.41	3.28	2.84	3.48	3.97	5.84	3.93	1.27	0.28	28.74
1971	1.30	2.79	0.70	0.78	6.30	1.75	3.71	0.95	0.96	4.29	3.41	0.83	27.77
#1972	0.22	0.23	0.49	4.23	4.24	2.58	3.18	3.63	2.78	3.22	3.58	1.37	29.75
1973	1.12	0.62	6.65	2.59	5.86	0.77	4.48	0.75	7.52	4.92	1.78	2.15	39.21
1974	0.56	0.08	0.73	3.88	5.22	0.91	0.46	4.52	0.29	0.01	2.53	0.66	21.38
1975	1.59	1.26	1.35	2.75	2.58	3.08	1.63	1.37	1.53	0.32	0.03	0.04	20.34
1976	0.36	1.15	2.59	3.60	3.03	0.63	2.99	0.07	3.09	1.86	2.03	0.35	17.90
1977	0.63	0.08	3.54	1.83	5.20	0.99	3.75	7.48	6.05	1.46	1.50	0.38	33.79
1978	0.34	1.19	1.11	7.21	3.68	2.37	5.05	1.90	4.27	5.29	1.28	0.38	30.46
1979	1.11	0.48	4.93	2.93	2.95	2.99	3.61	2.80	0.40	0.07	0.77	29.15	29.15
1980	1.12	0.55	1.80	1.93	2.07	3.00	1.82	6.21	0.33	1.96	0.08		21.64
1981	0.15	0.22	1.96	1.88	3.99	0.85	3.24	5.07	2.51	0.92	3.81	0.71	25.31
1982	0.89	0.30	2.68	3.30	5.48	5.57	4.05	8.57	3.22	1.17	1.24	1.97	38.44
1983	0.92	0.66	3.84	1.09	5.00	7.67	0.37	1.17	2.62	1.73	3.64	0.67	29.38
1984	0.27	1.26	3.04	6.52	7.97	5.94	1.35	1.40	1.43	3.89	0.18	3.42	36.67
1985	0.27	0.55	1.37	3.45	3.34	6.17	6.93	2.02	4.37	1.45	0.67	0.32	30.91
Record Mean	0.71	0.96	1.55	2.56	3.87	4.29	3.58	3.42	2.98	1.89	1.24	0.83	27.88

TABLE 3 AVERAGE TEMPERATURE (deg. F)   LINCOLN, NEBRASKA

YEAR	JAN	FEB	MAR	APR	MAY	JUNE	JULY	AUG	SEP	OCT	NOV	DEC	ANNUAL
1956	22.7	27.4	39.1	48.4	65.7	77.2	77.8	78.4	68.9	62.0	40.8	34.5	53.6
1957	19.7	34.2	39.1	50.4	61.0	71.5	82.3	77.5	68.4	52.5	37.5	36.9	52.3
1958	28.4	24.1	35.1	51.0	65.7	70.5	74.0	76.8	68.6	51.4	43.4	29.1	52.1
#1959	21.8	26.4	40.6	51.9	63.2	75.0	75.1	81.0	66.0	51.4	34.6	35.8	51.9
1960	21.4	20.2	24.9	54.2	62.0	70.3	76.5	76.5	68.6	56.9	41.8	29.7	50.3
1961	26.1	32.7	39.9	48.1	59.4	72.3	77.7	75.9	62.4	56.6	38.4	21.5	51.5
1962	20.5	26.7	32.9	51.4	70.3	72.1	75.4	75.8	63.7	57.1	42.3	29.2	51.5
1963	15.1	30.1	44.1	54.8	63.5	77.6	77.9	75.9	68.4	65.5	46.2	20.9	53.5
#1964	31.6	31.3	35.3	53.2	67.9	71.7	82.0	72.0	65.7	54.8	42.2	26.5	52.9
1965	24.4	22.6	26.7	52.5	66.7	70.8	75.8	74.8	60.1	57.8	42.7	36.2	51.0
1966	18.4	28.3	43.2	46.7	61.4	72.5	81.4	72.6	63.9	55.6	40.5	29.9	51.3
1967	25.2	29.3	43.6	54.1	59.0	70.1	74.9	72.8	63.5	53.7	39.4	24.1	52.1
1968	24.9	28.2	46.0	52.9	58.8	74.6	77.8	76.0	66.2	57.2	38.1	27.0	50.8
1969	19.6	27.1	31.6	54.1	64.3	68.5	79.1	76.0	66.7	50.4	42.7	31.2	52.5
1970	18.6	33.2	34.6	53.4	68.3	74.9	79.1	78.0	66.0	52.0	39.0		
1971	19.9	26.3	37.5	54.3	60.4	78.0	74.6	76.2	68.1	58.8	41.8	31.2	52.2
#1972	20.8	26.1	41.5	49.7	60.7	71.2	74.1	72.2	65.1	49.0	36.2	21.6	49.0
1973	23.4	28.4	42.6	49.6	59.0	73.4	75.0	77.2	63.2	56.1	39.0	21.8	50.7
1974	15.6	31.1	42.1	51.8	62.1	70.9	83.8	70.7	60.9	56.1	39.4	26.9	51.0
1975	21.7	20.1	30.6	48.4	63.4	70.9	78.5	79.5	61.7	57.3	39.5	29.9	50.1
1976	24.3	36.8	39.2	54.3	59.0	71.6	78.1	76.2	66.9	51.4	38.4	21.5	51.9
1977	13.1	31.7	43.5	56.7	68.4	74.4	80.9	72.3	66.0	51.4	38.3	25.1	51.9
1978	9.9	13.2	33.6	51.3	60.0	71.6	76.8	74.9	69.1	52.2	36.5	22.3	47.6
1979	7.2	13.0	38.4	49.6	59.7	72.1	75.0	75.2	68.7	54.5	37.3	32.3	48.6
1980	23.4	21.9	35.3	52.2	62.8	74.3	82.2	78.8	66.9	51.9	42.6	28.6	51.7
1981	28.2	31.5	42.8	58.8	59.8	75.0	78.7	72.4	65.9	52.1	42.1	24.9	52.7
1982	11.9	23.5	36.3	48.3	63.4	66.7	78.7	74.1	65.1	54.7	36.9	29.9	49.1
1983	26.9	31.5	39.4	45.3	58.1	71.7	81.1	83.5	69.3	54.8	39.4	8.2	50.8
1984	21.8	35.7	33.4	48.4	59.2	73.5	78.0	78.0	63.0	53.2	40.2	27.9	51.0
1985	20.1	22.5	42.6	54.9	64.5	69.0	76.0	76.0	71.3	53.8	29.0	21.4	49.1
Record Mean	23.1	27.4	38.3	52.0	62.2	72.3	77.9	75.7	67.0	55.1	39.8	28.2	51.5
Max	32.6	37.2	48.6	63.1	73.1	83.1	89.1	86.8	78.4	66.7	49.9	37.3	62.1
Min	13.7	17.6	27.9	40.8	51.3	61.4	66.7	64.6	55.5	43.6	29.6	19.1	41.0

REFERENCE NOTES FOR TABLES 1, 2, 3 and 6   (LINCOLN, NE)

GENERAL

T - TRACE AMOUNT
BLANK ENTRIES DENOTE MISSING/UNREPORTED DATA.
# INDICATES A STATION OR INSTRUMENT RELOCATION.

SPECIFIC

TABLE 1

(a) - LENGTH OF RECORD IN YEARS. ALTHOUGH
      INDIVIDUAL MONTHS MAY BE MISSING.
 * LESS THAN .05

NORMALS — BASED ON THE 1951-1980 RECORD PERIOD.
EXTREMES — DATES ARE THE MOST RECENT OCCURRENCE.
WIND DIR. — NUMERALS SHOW TENS OF DEGREES
            CLOCKWISE FROM TRUE NORTH.
            ''00'' INDICATES CALM.
RESULTANT WIND DIRECTIONS ARE GIVEN TO WHOLE DEGREES.

EXCEPTIONS

TABLES 2, 3, and 6

RECORD MEANS ARE THROUGH THE CURRENT YEAR,
BEGINNING IN   1887 FOR TEMPERATURE
                1878 FOR PRECIPITATION
                1956 FOR SNOWFALL

TABLE 4 HEATING DEGREE DAYS Base 65 deg. F   LINCOLN, NEBRASKA

SEASON	JULY	AUG	SEP	OCT	NOV	DEC	JAN	FEB	MAR	APR	MAY	JUNE	TOTAL
1956-57	0	5	53	145	719	938	1397	857	793	438	162	6	5513
1957-58	0	0	76	386	816	863	1128	1137	918	417	66	15	5822
1958-59	0	4	45	249	640	1106	1331	1071	746	393	141	14	5740
#1959-60	2	0	109	414	908	894	1344	1292	1237	355	131	11	6697
1960-61	0	0	55	266	690	1087	1199	899	769	509	211	7	5692
1961-62	0	4	183	268	'793	1337	1372	1068	989	414	22	23	6473
1962-63	0	0	109	278	675	1101	1541	970	642	310	115	0	5741
1963-64	0	6	35	90	558	1362	1028	971	913	354	73	34	5424
#1964-65	0	14	94	319	684	1187	1252	1182	1182	387	60	1	6362
1965-66	0	4	194	227	657	888	1439	1026	672	541	179	18	5845
1966-67	0	8	106	314	729	1133	1232	993	671	338	264	25	5813
1967-68	2	9	83	372	761	1082	1239	1060	587	371	211	11	5788
1968-69	0	4	48	275	799	1263	1400	1054	1026	321	129	53	6372
1969-70	0	0	12	469	664	1171	1433	883	891	379	61	7	5970
1970-71	1	0	108	424	774	1040	1393	1077	847	321	173	0	6158
#1971-72	4	0	89	223	693	1042	1361	1121	720	453	182	24	5912
1972-73	8	15	108	492	855	1340	1281	1018	688	458	199	1	6463
1973-74	1	0	99	281	772	1333	1533	941	704	403	139	32	6238
1974-75	0	14	161	278	762	1177	1338	1250	1059	495	117	20	6671
1975-76	0	0	168	269	759	1079	1255	810	792	328	206	11	5677
1976-77	0	0	83	553	979	1254	1604	927	659	256	11	0	6326
1977-78	0	0	42	413	793	1230	1703	1447	972	410	193	27	7230
1978-79	0	5	62	392	848	1315	1787	1454	816	463	197	16	7355
1979-80	3	7	42	324	825	1007	1281	1241	912	387	133	5	6167
1980-81	0	0	80	402	666	1120	1433	933	680	209	192	0	5415
1981-82	1	3	71	393	678	1237	1639	1159	880	501	88	52	6702
1982-83	0	3	118	324	840	1081	1173	931	785	584	238	25	6102
1983-84	0	0	75	326	763	1758	1334	841	970	496	196	3	6762
1984-85	0	0	167	366	737	1142	1385	1186	687	326	72	25	6093
1985-86	0	8	198	343	1071	1345							

TABLE 5 COOLING DEGREE DAYS Base 65 deg. F   LINCOLN, NEBRASKA

YEAR	JAN	FEB	MAR	APR	MAY	JUNE	JULY	AUG	SEP	OCT	NOV	DEC	TOTAL
1969	0	0	0	0	116	165	445	372	130	27	0	0	1255
1970	0	0	0	38	172	309	445	410	146	26	0	0	1546
1971	0	0	0	8	40	397	309	356	188	38	0	0	1336
#1972	0	0	0	0	55	217	296	247	119	2	0	0	936
1973	0	0	0	2	23	258	318	387	53	14	0	0	1055
1974	0	0	0	14	57	215	589	197	46	10	0	0	1128
1975	0	0	0	4	76	204	428	459	74	37	0	0	1282
1976	0	0	0	15	26	216	413	354	144	10	0	0	1178
1977	0	0	0	14	124	289	501	231	91	0	0	0	1250
1978	0	0	5	7	46	230	371	320	193	3	0	0	1175
1979	0	0	0	7	42	235	320	330	160	4	0	0	1098
1980	0	0	0	13	73	293	542	433	142	1	0	0	1497
1981	0	0	0	29	37	308	430	237	102	0	0	0	1143
1982	0	0	0	8	45	111	438	292	124	10	0	0	1028
1983	0	0	0	0	30	235	505	580	211	20	0	0	1581
1984	0	0	0	5	24	264	395	412	114	9	0	0	1223
1985	0	0	0	30	63	152	351	210	162	2	0	0	970

TABLE 6 SNOWFALL (inches)   LINCOLN, NEBRASKA

SEASON	JULY	AUG	SEP	OCT	NOV	DEC	JAN	FEB	MAR	APR	MAY	JUNE	TOTAL
1956-57	0.0	0.0	0.0	0.0	5.1	2.5	5.1	2.5	13.6	5.4	0.0	0.0	34.2
1957-58	0.0	0.0	0.0	T	12.6	5.6	7.4	3.2	10.0	T	0.0	0.0	38.8
1958-59	0.0	0.0	0.0	0.0	1.0	0.3	12.2	5.3	12.1	T	0.0	0.0	30.9
#1959-60	0.0	0.0	0.0	0.0	0.6	5.1	11.1	19.2	17.8	0.5	0.0	0.0	54.3
1960-61	0.0	0.0	0.0	0.0	T	0.7	2.0	4.5	5.1	0.8	0.0	0.0	13.1
1961-62	0.0	0.0	0.0	0.0	6.8*	12.2	3.3	4.3	2.5	0.1	0.0	0.0	29.2
1962-63	0.0	0.0	0.0	0.0	0.3	4.6	6.5	2.2	11.0	T	0.0	0.0	24.6
1963-64	0.0	0.0	0.0	0.0	0.0	4.2	2.5	2.2	4.1	T	0.0	0.0	13.0
1964-65	0.0	0.0	0.0	0.0	0.5	1.5	3.7	26.1	10.3	0.0	0.0	0.0	42.1
1965-66	0.0	0.0	0.0	0.0	T	2.0	1.9	1.5	3.6	T	0.0	0.0	9.0
1966-67	0.0	0.0	0.0	0.0	0.2	9.9	5.5	0.5	T	3.7	3.0	0.0	22.8
1967-68	0.0	0.0	0.0	T	T	1.9	4.3	1.0	T	T	0.0	0.0	7.2
1968-69	0.0	0.0	0.0	0.0	1.7	11.7	7.8	13.6	5.0	0.0	0.0	0.0	39.8
1969-70	0.0	0.0	0.0	T	0.2	15.8	1.7	3.2	4.7	0.6	0.0	0.0	26.2
1970-71	0.0	0.0	0.0	6.6	0.6	0.4	15.1	17.6	8.3	0.4	0.0	0.0	49.0
#1971-72	0.0	0.0	0.0	0.0	10.0	2.3	3.6	3.9	0.9	0.9	0.0	0.0	21.6
1972-73	0.0	0.0	0.0	T	T	8.6	6.5	5.1	0.1	2.2	0.0	0.0	29.2
1973-74	0.0	0.0	0.0	T	T	19.8	11.0	0.9	1.6	0.3	0.0	0.0	33.6
1974-75	0.0	0.0	0.0	0.0	1.9	8.3	14.6	10.9	4.4	2.0	0.0	0.0	42.1
1975-76	0.0	0.0	0.0	0.0	6.3	1.3	3.9	4.3	5.3	0.0	0.0	0.0	21.1
1976-77	0.0	0.0	0.0	0.4	0.4	0.6	8.7	T	8.3	3.4	0.0	0.0	21.8
1977-78	0.0	0.0	0.0	0.0	3.3	3.7	4.5	13.8	5.7	T	0.0	0.0	31.0
1978-79	0.0	0.0	0.0	0.0	7.3	3.0	11.5	4.4	3.7	4.5	0.0	0.0	34.4
1979-80	0.0	0.0	0.0	0.4	0.1	2.4	8.1	5.6	6.5	0.2	0.0	0.0	23.3
1980-81	0.0	0.0	0.0	3.3	T	2.3	1.7	1.9	3.8	0.0	0.0	0.0	13.0
1981-82	0.0	0.0	0.0	T	2.1	9.3	3.8	4.4	7.8	4.9	0.0	0.0	32.3
1982-83	0.0	0.0	0.0	T	T	5.2	5.1	7.6	12.9	7.2	0.0	0.0	38.0
1983-84	0.0	0.0	T	0.0	8.6	13.8	2.5	5.1	17.0	0.5	0.0	0.0	47.5
1984-85	0.0	0.0	0.0	0.0	0.5	6.2	4.7	3.1	7.0	0.0	0.0	0.0	21.5
1985-86	0.0	0.0	0.8	0.0	6.9	5.2							
Record Mean	0.0	0.0	T	0.4	2.9	5.6	6.4	6.1	6.5	1.3	0.1	0.0	29.1

See Reference Notes, relative to all above tables, on preceding page.

Norfolk is located in northeastern Nebraska, in the valley of the Elkhorn River. The city of Norfolk lies at an average elevation of 1,550 feet above sea level. The surrounding country is moderately rolling in all directions. The terrain becomes more level to the south and southwest. Norfolk is situated near the western limit of the Corn Belt. To the east the climate and soils are favorable for diversified farming and dairying. To the west precipitation becomes lighter, and the farming country gives way to the grazing lands of the Great Plains. There are no local topographic features of sufficient importance to affect the climate of the area.

Northeast Nebraska has a climate typical of the interior of large continents in middle latitudes. The rainfall is moderate. Summers are hot and winters cold, and there are great variations in temperature and precipitation from day to day and from season to season. Most of the moisture which falls over this area is brought in from the Gulf of Mexico. The rapid changes in temperature are caused by the interchange of warm air from the south and southwest with cold air from the north. The rapid day to day changes in weather conditions produce an invigorating and healthful climate in northeast Nebraska.

Daily temperature ranges of 30 to 40 degrees are not uncommon. Summertime precipitation is almost wholly in the form of showers and thunderstorms. Practically all precipitation in the colder months is in the form of snow. As a rule, nearly 85 percent of the snowfall occurs from December to March and the ground is covered by snow during this period.

Norfolk is subject to the strong and persistent winds which prevail over the Great Plains states. Winds of 40 to 50 mph are not uncommon in this area, and gusts up to 100 mph have been recorded at Norfolk. Prevailing winds are from the south and southwest from May through September, with prevailing northwesterly winds during the remainder of the year.

Based on the 1951–1980 period, the average first occurrence of 32 degrees Fahrenheit in the fall is October 5 and the average last occurrence in the spring is May 1.

TABLE 1 NORMALS, MEANS AND EXTREMES

NORFOLK, NEBRASKA

LATITUDE: 41°59'N   LONGITUDE: 97°26'W   ELEVATION: FT. GRND 1544 BARO 01547   TIME ZONE: CENTRAL   WBAN: 14941

	(a)	JAN	FEB	MAR	APR	MAY	JUNE	JULY	AUG	SEP	OCT	NOV	DEC	YEAR
TEMPERATURE °F:														
Normals														
-Daily Maximum		27.8	34.2	43.8	60.5	72.2	82.4	87.4	85.0	75.5	64.7	47.1	34.1	59.6
-Daily Minimum		6.9	13.3	23.2	36.9	48.8	58.9	64.2	61.9	51.0	38.8	24.8	13.7	36.9
-Monthly		17.4	23.8	33.5	48.7	60.5	70.7	75.8	73.5	63.3	51.8	36.0	23.9	48.2
Extremes														
-Record Highest	40	71	73	87	95	103	106	113	107	101	95	82	71	113
-Year		1981	1946	1968	1980	1967	1946	1954	1983	1971	1963	1945	1962	JUL 1954
-Record Lowest	40	-27	-26	-20	2	24	38	42	40	26	13	-15	-22	-27
-Year		1974	1981	1960	1975	1976	1964	1971	1967	1984	1972	1964	1983	JAN 1974
NORMAL DEGREE DAYS:														
Heating (base 65°F)		1476	1154	977	489	181	27	6	9	125	417	870	1274	7005
Cooling (base 65°F)		0	0	0	0	45	198	341	269	74	8	0	0	935
% OF POSSIBLE SUNSHINE														
MEAN SKY COVER (tenths)														
Sunrise - Sunset	40	6.2	6.3	6.7	6.3	6.1	5.4	4.7	4.7	4.8	5.0	6.2	6.3	5.7
MEAN NUMBER OF DAYS:														
Sunrise to Sunset														
-Clear	40	8.8	7.6	7.2	7.7	7.6	10.0	12.3	12.6	12.9	12.5	8.5	8.5	116.3
-Partly Cloudy	40	7.5	6.9	7.8	8.2	9.9	10.1	11.6	10.4	7.4	7.6	7.4	7.6	102.4
-Cloudy	40	14.8	13.7	16.1	14.1	13.4	9.9	7.1	7.9	9.6	10.9	14.1	14.9	146.6
Precipitation														
.01 inches or more	40	5.7	5.6	7.9	9.1	11.1	9.9	8.6	8.5	7.7	5.6	5.0	5.5	90.0
Snow, Ice pellets														
1.0 inches or more	40	1.9	2.0	2.2	0.6	0.*	0.0	0.0	0.0	0.*	0.2	1.2	1.9	10.0
Thunderstorms	28	0.*	0.2	0.9	3.9	8.0	10.0	9.5	9.7	3.9	2.6	0.6	0.1	49.3
Heavy Fog Visibility														
1/4 mile or less	28	1.2	2.1	1.6	0.7	0.6	0.7	0.7	1.2	0.9	1.1	1.6	1.5	13.7
Temperature °F														
-Maximum														
90° and above	40	0.0	0.0	0.0	0.4	1.2	6.6	11.9	10.0	3.5	0.3	0.0	0.0	34.0
32° and below	40	17.6	12.6	6.7	0.2	0.0	0.0	0.0	0.0	0.0	0.0	3.8	14.2	55.0
-Minimum														
32° and below	40	30.8	27.2	24.9	9.6	1.0	0.0	0.0	0.0	0.7	7.1	23.6	30.5	155.4
0° and below	40	10.1	5.0	0.9	0.0	0.0	0.0	0.0	0.0	0.0	0.0	0.6	4.5	21.1
AVG. STATION PRESS. (mb)	9	963.9	962.2	958.7	958.2	957.5	958.0	959.6	960.0	960.5	960.9	961.2	961.9	960.2
RELATIVE HUMIDITY (%)														
Hour 00	9	70	74	76	70	72	72	74	80	75	74	77	76	74
Hour 06 (Local Time)	40	75	78	80	79	80	82	82	85	83	79	79	78	80
Hour 12	40	64	64	61	50	51	52	53	54	52	50	58	65	56
Hour 18	40	68	67	61	48	49	49	49	52	53	53	63	70	57
PRECIPITATION (inches):														
Water Equivalent														
-Normal		0.52	0.80	1.54	2.21	3.71	4.35	3.21	2.65	2.09	1.36	0.72	0.63	23.79
-Maximum Monthly	40	2.33	3.18	5.14	7.47	8.61	12.22	9.11	5.53	8.13	4.57	3.97	2.25	12.22
-Year		1949	1971	1973	1984	1977	1967	1950	1951	1970	1968	1983	1982	JUN 1967
-Minimum Monthly	40	0.10	0.04	0.06	0.23	1.01	0.86	0.33	0.53	0.30	T	0.03	0.08	T
-Year		1970	1949	1967	1969	1948	1978	1954	1971	1956	1958	1980	1958	OCT 1958
-Maximum in 24 hrs	40	1.30	2.41	2.43	1.96	4.12	5.51	3.46	2.89	3.88	2.79	1.53	1.34	5.51
-Year		1982	1971	1981	1985	1973	1974	1952	1978	1970	1968	1975	1981	JUN 1974
Snow, Ice pellets														
-Maximum Monthly	40	16.4	22.8	20.8	13.2	2.7				1.1	3.1	22.6	19.1	22.8
-Year		1982	1984	1960	1984	1947				1985	1982	1983	1968	FEB 1984
-Maximum in 24 hrs	40	12.8	22.5	9.7	9.7	2.7				1.1	3.1	14.6	12.1	22.5
-Year		1982	1984	1966	1984	1947				1985	1982	1983	1978	FEB 1984
WIND:														
Mean Speed (mph)	9	12.6	12.1	13.4	13.7	11.8	10.8	10.0	9.9	10.9	11.2	11.9	12.4	11.7
Prevailing Direction														
Fastest Obs. 1 Min.														
-Direction (!!!)	4	33	34	30	32	09	05	35	36	28	33	35	35	32
-Speed (MPH)	4	46	43	44	51	41	39	35	40	46	35	45	46	51
-Year		1984	1984	1985	1982	1983	1984	1983	1982	1982	1985	1982	1982	APR 1982
Peak Gust														
-Direction (!!!)	2	NW	N	NW	NW	W	E	NE	S	S	NW	NW	NW	NW
-Speed (mph)	2	60	54	55	63	47	55	48	55	44	53	47	53	63
-Date		1984	1984	1984	1984	1985	1984	1984	1985	1984	1985	1984	1985	APR 1984

See Reference Notes to this table on the following pages.

TABLE 2   PRECIPITATION (inches)   NORFOLK, NEBRASKA

YEAR	JAN	FEB	MAR	APR	MAY	JUNE	JULY	AUG	SEP	OCT	NOV	DEC	ANNUAL
1956	0.70	0.40	0.14	1.18	1.84	3.89	3.48	2.11	0.30	1.16	0.95	0.21	16.36
1957	0.17	0.17	1.96	1.33	4.55	3.11	3.32	2.41	4.91	1.80	1.26	0.41	25.40
1958	0.65	1.20	0.78	3.45	1.48	2.27	8.43	2.39	0.50	T	0.36	0.08	21.59
1959	0.33	0.55	3.19	1.58	7.50	2.45	2.16	2.24	1.32	2.03	0.60	0.46	24.41
1960	1.74	1.02	2.68	2.53	5.93	6.30	3.17	3.78	1.61	0.62	0.93	0.70	31.01
1961	0.20	0.91	1.55	1.78	4.02	3.45	6.49	1.84	0.61	0.52		1.01	23.03
1962	0.30	1.71	2.48	1.04	3.77	4.86	4.24	1.29	1.54	0.65	0.09	0.67	22.64
1963	0.59	0.23	0.88	1.57	1.95	4.14	3.43	3.89	1.30	0.50	0.20	0.24	18.92
1964	0.22	0.73	1.50	3.30	2.95	7.46	2.52	4.19	2.35	0.22	0.19	0.73	26.36
1965	0.35	0.90	0.44	1.04	6.62	3.61	3.86	2.78	6.88	0.52	0.13	0.45	27.58
1966	0.52	1.52	1.17	0.94	1.38	5.80	2.17	3.78	2.08	1.07	0.04	0.56	21.03
1967	0.48	0.06	0.06	1.37	2.27	12.22	0.82	1.10	1.40	1.68	0.07	0.74	22.27
1968	0.19	0.06	0.72	3.54	1.38	2.49	2.42	3.12	2.38	4.57	0.38	1.75	23.00
1969	1.21	1.86	0.43	0.23	2.93	8.09	3.40	3.69	1.62	2.63	0.07	1.14	27.30
1970	0.10	0.35	1.38	2.83	3.47	2.17	1.54	1.44	8.13	2.55	0.97	0.37	25.30
1971	0.28	3.18	0.47	1.21	2.19	5.95	4.12	0.53	0.85	2.28	1.48	0.69	23.23
1972	0.42	0.14	0.74	3.98	4.64	3.70	4.85	2.67	1.74	0.77	1.02	1.20	25.87
1973	1.12	0.46	5.14	1.58	5.56	1.33	4.14	1.31	5.12	0.77	1.94	0.78	29.25
1974	0.36	0.12	0.46	1.70	3.38	7.09	0.84	2.72	0.55	1.71	0.11	0.67	19.71
1975	1.26	0.39	0.86	4.03	2.56	7.62	3.40	2.71	0.69	0.16	3.67	0.19	27.54
1976	0.29	0.71	2.68	1.79	3.53	2.48	1.23	0.66	2.56	0.38	0.08	0.21	16.60
1977	0.26	0.81	3.58	3.45	8.61	5.04	3.25	4.33	1.82	3.00	1.52	0.51	36.18
1978	0.21	0.92	0.82	4.35	2.84	0.86	4.48	3.86	0.82	0.62	0.79	0.92	21.49
1979	1.09	0.44	4.45	1.92	5.14	3.08	4.08	2.84	1.81	4.26	1.65	0.54	31.30
1980	0.25	0.54	0.94	1.11	1.77	3.79	1.06	4.66	0.86	1.91	0.03	0.24	17.16
1981	0.14	0.39	2.57	0.36	2.34	4.16	3.63	0.57	1.49	2.10		0.73	21.87
1982	1.63	0.35	1.68	1.47	7.70	1.62	2.48	3.30	2.72	3.53	2.66	2.25	31.39
1983	0.75	0.55	3.68	2.27	3.28	6.82	2.73	1.35	1.61	1.16	3.97	0.68	28.85
1984	0.34	2.34	2.33	7.47	4.14	5.48	2.31	1.50	0.72	4.30	1.62	1.13	33.68
1985	0.25	0.15	1.46	5.49	2.85	4.53	1.62	2.48	4.61	0.96	0.89	0.47	25.76
Record Mean	0.56	0.75	1.64	2.27	3.73	4.44	3.14	2.65	2.15	1.52	0.96	0.68	24.50

TABLE 3   AVERAGE TEMPERATURE (deg. F)   NORFOLK, NEBRASKA

YEAR	JAN	FEB	MAR	APR	MAY	JUNE	JULY	AUG	SEP	OCT	NOV	DEC	ANNUAL
1956	16.4	20.1	34.1	43.3	61.6	75.3	73.8	74.7	64.0	58.1	36.6	28.6	48.9
1957	14.0	28.7	34.5	46.4	58.1	68.8	79.1	74.5	61.1		36.0	32.5	48.6
1958	24.4	17.1	30.8	47.6	63.3	67.2	70.6	74.5	65.6	54.0	37.8	23.6	48.1
1959	16.3	20.8	36.3	47.8	59.9	73.3	73.0	77.1	62.3	47.4	29.7	31.7	48.0
1960	15.5	14.9	19.8	49.0	58.6	67.3	74.4	73.9	65.2	52.8	38.1	25.8	46.3
1961	20.3	28.3	36.6	44.4	57.3	70.2	74.5	74.4	59.3	52.9	35.2	17.8	47.6
1962	16.6	20.9	26.0	48.6	66.3	68.9	73.1	72.7	59.9	53.7	39.9	25.4	47.7
1963	10.5	26.5	41.1	51.2	60.9	74.6	76.7	72.8	66.6	60.7	40.1	17.9	50.0
1964	25.6	24.6	30.1	49.6	64.6	70.2	78.9	72.6	68.0	51.4	36.5	19.2	48.4
1965	20.0	18.7	23.7	49.3	63.6	69.3	74.1	72.0	54.6	54.3	37.4	32.0	47.4
1966	11.2	22.7	38.3	44.0	58.8	69.7	78.7	69.6	61.7	51.7	34.9	23.7	47.1
1967	20.9	23.3	40.3	49.6	55.4	67.5	73.4	70.3	62.1	49.6	35.8	24.1	47.7
1968	19.8	23.6	42.2	49.3	55.2	72.0	74.8	73.3	62.1	52.1	35.3	18.7	48.2
1969	13.4	23.1	25.5	51.2	61.5	65.4	75.4	74.2	65.4	45.2	38.8	23.1	46.9
1970	13.4	28.5	29.7	48.0	64.2	72.2	76.3	74.8	63.1	47.9	34.7	23.7	48.0
1971	16.0	24.4	33.8	50.3	57.7	74.2	71.3	73.7	63.2	54.4	37.3	24.2	48.4
1972	17.1	21.6	38.1	48.1	60.5	71.4	73.0	72.0	62.9	47.2	35.3	17.9	47.1
1973	21.2	26.1	44.1	48.2	58.5	71.6	75.2	77.5	61.0	54.9	36.2	22.1	49.5
1974	16.7	29.5	39.5	51.2	61.2	68.7	81.2	69.0	60.1	54.0	37.5	25.9	49.6
1975	20.7	19.3	29.2	44.6	63.5	69.6	77.5	76.3	60.0	55.0	35.5	24.7	48.0
1976	21.5	33.8	37.2	53.4	59.7	70.9	77.2	75.2	64.5	46.0	30.8	23.2	49.4
1977	12.5	30.9	41.1	56.2	67.8	72.8	78.1	70.2	65.7	51.2	35.5	22.3	50.3
1978	8.3	12.1	33.8	49.1	60.4	72.0	75.4	72.7	69.0	50.8	34.8	17.2	46.3
1979	5.7	11.2	32.9	47.1	58.0	70.7	74.2	72.9	67.2	51.0	32.9	31.0	46.2
1980	21.8	21.4	33.6	51.4	61.1	72.4	78.9	75.5	65.3	49.7	40.2	26.4	49.8
1981	26.7	29.3	40.8	57.8	57.8	72.2	76.0	71.5	63.8	49.8	40.5	23.5	50.8
1982	9.1	22.5	33.9	46.9	61.7	65.4	75.6	72.7	62.8	52.3	34.3	28.3	47.1
1983	24.5	29.5	37.1	43.6	57.2	69.8	78.2	81.0	66.6	52.2	37.0	6.9	48.6
1984	22.8	31.4	29.0	46.0	58.5	72.1	75.3	76.5	60.8	52.8	38.9	24.2	49.0
1985	18.1	23.4	41.3	54.5	64.6	67.4	73.9	70.0	60.7	51.1	25.0	17.4	47.3
Record Mean	17.9	23.9	34.0	49.0	60.2	70.3	75.6	73.6	63.3	52.1	35.9	23.4	48.3
Max	28.3	34.3	44.2	60.7	71.8	81.9	87.2	85.1	75.5	64.9	46.7	33.5	59.5
Min	7.4	13.5	23.8	37.2	48.6	58.7	64.0	62.1	51.0	39.3	25.0	13.3	37.0

REFERENCE NOTES FOR TABLES 1, 2, 3 and 6   (NORFOLK, NE)

GENERAL

T - TRACE AMOUNT
BLANK ENTRIES DENOTE MISSING/UNREPORTED DATA.
# INDICATES A STATION OR INSTRUMENT RELOCATION.

SPECIFIC

TABLE 1

(a) - LENGTH OF RECORD IN YEARS. ALTHOUGH INDIVIDUAL MONTHS MAY BE MISSING.

\#   LESS THAN .05

NORMALS — BASED ON THE 1951-1980 RECORD PERIOD.
EXTREMES — DATES ARE THE MOST RECENT OCCURRENCE.
WIND DIR. — NUMERALS SHOW TENS OF DEGREES
CLOCKWISE FROM TRUE NORTH.
"00" INDICATES CALM.
RESULTANT WIND DIRECTIONS ARE GIVEN TO WHOLE DEGREES.

EXCEPTIONS

TABLE 1

1. THUNDERSTORMS AND HEAVY FOG ARE THROUGH 1964 AND 1977 TO DATE AND MAY BE INCOMPLETE, DUE TO PART-TIME OPERATIONS.
2. FASTEST MILE WINDS ARE THROUGH 1981.

TABLES 2, 3, and 6

RECORD MEANS ARE THROUGH THE CURRENT YEAR, BEGINNING IN   1946 FOR TEMPERATURE
1946 FOR PRECIPITATION
1946 FOR SNOWFALL

TABLE 4   HEATING DEGREE DAYS Base 65 deg. F   NORFOLK, NEBRASKA

SEASON	JULY	AUG	SEP	OCT	NOV	DEC	JAN	FEB	MAR	APR	MAY	JUNE	TOTAL
1956-57	0	12	128	241	842	1122	1576	1008	939	550	234	20	6672
1957-58	0	0	151	461	865	1005	1248	1336	1051	514	112	50	6793
1958-59	0	6	84	361	809	1277	1503	1231	883	507	206	25	6892
1959-60	3	0	170	539	1051	1027	1526	1446	1397	488	210	23	7880
1960-61	0	0	118	381	801	1208	1379	1020	876	612	257	27	6679
1961-62	0	4	233	372	891	1455	1498	1228	1201	501	68	39	7490
1962-63	5	3	187	367	746	1222	1686	1069	734	412	171	1	6603
1963-64	0	10	53	168	741	1455	1212	1167	1075	460	110	38	6489
1964-65	0	48	146	416	848	1412	1389	1290	1271	473	113	7	7413
1965-66	0	14	318	330	825	1016	1667	1178	820	625	234	34	7061
1966-67	0	22	149	415	894	1275	1362	1160	764	463	352	34	6890
1967-68	15	32	118	483	874	1260	1395	1193	700	469	308	30	6877
1968-69	4	6	129	407	886	1432	1596	1165	1219	404	171	87	7506
1969-70	0	3	44	612	776	1293	1596	1015	1089	511	126	23	7088
1970-71	6	1	165	532	903	1272	1512	1126	962	435	222	5	7141
1971-72	14	0	148	328	824	1257	1481	1251	824	506	204	25	6862
1972-73	12	20	145	545	884	1457	1353	1086	735	497	221	12	6967
1973-74	4	0	141	314	858	1324	1493	987	783	414	147	45	6510
1974-75	0	18	190	340	817	1208	1369	1275	1107	604	105	25	7058
1975-76	2	0	207	326	877	1247	1343	900	854	352	191	15	6314
1976-77	0	1	108	587	1018	1287	1624	951	733	287	6	0	6602
1977-78	0	5	45	423	882	1318	1756	1477	965	471	187	28	7557
1978-79	1	9	49	434	900	1476	1836	1506	988	533	248	33	8013
1979-80	3	14	56	430	954	1046	1332	1258	966	423	174	8	6664
1980-81	0	1	100	473	738	1190	1180	995	741	238	235	5	5896
1981-82	9	1	98	464	727	1283	1730	1188	958	537	123	60	7178
1982-83	0	10	140	389	915	1132	1246	991	858	637	258	38	6614
1983-84	0	0	120	391	838	1798	1302	969	1109	563	220	6	7316
1984-85	0	1	196	375	775	1260	1451	1160	730	344	79	52	6423
1985-86	1	15	241	424	1195	1469							

TABLE 5   COOLING DEGREE DAYS Base 65 deg. F   NORFOLK, NEBRASKA

YEAR	JAN	FEB	MAR	APR	MAY	JUNE	JULY	AUG	SEP	OCT	NOV	DEC	TOTAL
1969	0	0	0	0	69	105	329	296	63	7	0	0	869
1970	0	0	0	9	107	244	364	311	115	10	0	0	1160
1971	0	0	0	3	3	288	214	276	103	7	0	0	894
1972	0	0	0	2	72	225	269	243	90	0	0	0	901
1973	0	0	0	0	24	218	326	395	31	9	0	0	1003
1974	0	0	0	4	34	164	506	149	50	4	0	0	911
1975	0	0	0	1	67	171	398	359	60	24	0	0	1080
1976	0	0	0	11	32	200	387	324	100	6	0	0	1060
1977	0	0	0	30	101	241	415	174	73	0	0	0	1034
1978	0	0	4	0	56	244	330	257	176	2	0	0	1069
1979	0	0	0	2	39	212	297	268	128	0	0	0	946
1980	0	0	0	23	61	236	440	334	115	5	0	0	1214
1981	0	0	0	29	19	229	357	206	70	0	0	0	910
1982	0	0	0	2	26	76	336	255	80	2	0	0	777
1983	0	0	0	2	23	188	416	501	175	3	0	0	1306
1984	0	0	0	2	23	226	326	365	76	6	0	0	1024
1985	0	0	0	34	73	135	283	177	119	0	0	0	821

TABLE 6   SNOWFALL (inches)   NORFOLK, NEBRASKA

SEASON	JULY	AUG	SEP	OCT	NOV	DEC	JAN	FEB	MAR	APR	MAY	JUNE	TOTAL
1956-57	0.0	0.0	0.0	0.0	2.6	3.3	2.3	1.6	5.9	4.3	0.0	0.0	20.0
1957-58	0.0	0.0	0.0	T	5.4	4.1	7.2	2.7	6.1	1.0	0.0	0.0	26.5
1958-59	0.0	0.0	0.0	0.0	0.4	1.4	4.8	6.3	14.9	3.0	0.0	0.0	30.8
1959-60	0.0	0.0	0.0	T	3.4	4.2	16.2	10.3	20.8	4.2	0.0	0.0	59.1
1960-61	0.0	0.0	0.0	0.0	0.2	2.2	2.7	8.7	12.1	1.9	0.0	0.0	27.8
1961-62	0.0	0.0	0.7	0.0	2.4	10.3	3.5	14.5	12.0	0.8	0.0	0.0	44.2
1962-63	0.0	0.0	0.0	T	0.2	7.9	8.6	2.9	3.4	T	0.0	0.0	23.0
1963-64	0.0	0.0	0.0	0.0	0.0	2.8	2.9	9.3	6.0	3.2	0.0	0.0	24.2
1964-65	0.0	0.0	0.0	0.0	1.4	8.4	5.9	13.0	4.1	0.2	0.0	0.0	33.0
1965-66	0.0	0.0	0.0	0.0	0.3	1.7	5.7	1.5	11.1	T	0.0	0.0	20.3
1966-67	0.0	0.0	0.0	1.6	T	6.7	5.5	0.6	0.5	T	T	0.0	14.9
1967-68	0.0	0.0	0.0	0.5	0.7	4.9	2.4	0.5	T	T	T	0.0	9.0
1968-69	0.0	0.0	0.0	0.0	1.3	19.1	9.8	19.1	1.4	0.0	0.0	0.0	50.7
1969-70	0.0	0.0	0.0	1.4	0.8	12.8	1.8	3.6	12.8	0.1	0.0	0.0	33.3
1970-71	0.0	0.0	0.0	1.3	0.9	2.5	5.1	2.6	5.5	0.3	0.0	0.0	18.2
1971-72	0.0	0.0	0.0	0.0	3.9	4.6	2.5	4.2	1.3	4.5	0.0	0.0	21.0
1972-73	0.0	0.0	0.0	0.0	1.1	14.8	7.2	5.9	0.2	0.5	0.0	0.0	29.7
1973-74	0.0	0.0	0.0	T	3.1	10.9	8.6	1.8	3.3	1.6	0.0	0.0	29.3
1974-75	0.0	0.0	0.0	0.0	1.0	8.6	15.5	6.2	3.9	4.1	0.0	0.0	39.3
1975-76	0.0	0.0	0.0	T	15.8	0.7	3.0	4.7	9.6	0.0	0.0	0.0	33.8
1976-77	0.0	0.0	0.0	1.6	1.0	3.2	4.2	4.0	8.9	2.3	0.0	0.0	25.2
1977-78	0.0	0.0	0.0	T	7.6	8.9	8.4	15.1	8.4	T	0.0	0.0	43.4
1978-79	0.0	0.0	0.0	0.0	7.3	15.1	12.2	6.1	10.0	1.8	0.0	0.0	52.5
1979-80	0.0	0.0	0.0	0.1	4.2	2.5	4.4	4.7	3.7	2.6	0.0	0.0	22.2
1980-81	0.0	0.0	0.0	2.3	0.6	2.0	1.4	3.8	T	0.0	0.0	0.0	10.1
1981-82	0.0	0.0	0.0	0.5	4.3	8.7	16.4	3.8	4.8	8.6	0.0	0.0	47.1
1982-83	0.0	0.0	0.0	3.1	4.0	11.1	7.1	4.6	15.4	6.3	0.0	0.0	51.6
1983-84	0.0	0.0	T	0.0	22.6	7.8	3.5	22.8	16.2	13.2	0.0	0.0	86.1
1984-85	0.0	0.0	T	0.4	1.1	5.6	3.1	0.9	6.3	0.5	0.0	0.0	17.9
1985-86	0.0	0.0	1.1	T	7.7	6.2							
Record Mean	0.0	0.0	T	0.3	3.6	6.4	5.9	6.0	7.0	1.8	0.1	0.0	31.3

See Reference Notes, relative to all above tables, on preceding page.

The climate of North Platte is characterized throughout the year by frequent rapid changes in the weather. During the winter, most North Pacific lows cross the country north of North Platte. The passage usually brings little or no snowfall, and only a moderate drop in temperature. Only when there is a major outbreak of cold air from Canada does the temperature fall to zero or below. The duration of below-zero temperature is hardly more than two mornings, and by the third or fourth day the temperature is ordinarily rising to the 40s or higher. Snowfall at the onset of a cold outbreak is usually less than 2 inches.

Only when a low moves from the middle Rockies through Nebraska, allowing easterly winds to draw moist air into the low circulation, does snowfall of appreciable amounts occur. Few of these storms move slowly enough, or are intense enough, to deposit much precipitation in the North Platte area. However, during some winters the cold outbreak and intense low from the mid-Rockies combine to produce severe cold and snow several inches in depth, with blizzard conditions following. During and after these snowfalls and blizzards, rail and highway traffic may be stalled until the snow is cleared. Widespread loss of unsheltered livestock and wild life results from such conditions.

The sudden and frequent weather changes of the winter continue through spring with decreasing intensity of temperature changes but increasing precipitation. The summer and fall months bring frequent changes from hot to cool weather. Most summer and fall precipitation is associated with thunderstorms, so the amounts are extremely variable. The surrounding area is occasionally damaged by locally severe winds and hailstorms.

Temperatures may reach into the upper 90s and lower 100s frequently during the summer months, but the elevation and clear skies bring rapid cooling after sunset to lows in the 60s or below by daybreak. Since the humidity is generally low, the extremely hot days of summer are not uncomfortable.

Based on the 1951–1980 period, the average first occurrence of 32 degrees Fahrenheit in the fall is September 24 and the average last occurrence in the spring is May 11.

TABLE 1 **NORMALS, MEANS AND EXTREMES**

NORTH PLATTE, NEBRASKA

LATITUDE: 41°08'N LONGITUDE: 100°41'W ELEVATION: FT. GRND 2775 BARO 02782 TIME ZONE: CENTRAL WBAN: 24023

	(a)	JAN	FEB	MAR	APR	MAY	JUNE	JULY	AUG	SEP	OCT	NOV	DEC	YEAR
TEMPERATURE °F:														
Normals														
-Daily Maximum		34.2	40.5	47.8	61.5	71.5	81.6	87.8	86.4	77.3	66.7	49.4	39.3	62.0
-Daily Minimum		8.3	14.1	21.5	33.6	44.8	55.0	60.6	58.3	46.5	33.7	20.5	12.5	34.1
-Monthly		21.3	27.3	34.7	47.5	58.2	68.3	74.2	72.3	61.9	50.2	35.0	25.9	48.1
Extremes														
-Record Highest	34	70	79	86	94	97	107	112	105	102	91	82	75	112
-Year		1981	1962	1968	1980	1953	1952	1954	1954	1979	1978	1980	1980	JUL 1954
-Record Lowest	34	-23	-22	-22	7	20	29	40	35	17	11	-13	-34	-34
-Year		1979	1981	1962	1975	1976	1969	1952	1976	1984	1969	1976	1983	DEC 1983
NORMAL DEGREE DAYS:														
Heating (base 65°F)		1355	1056	939	522	235	59	8	9	151	463	900	1212	6909
Cooling (base 65°F)		0	0	0	0	24	158	294	239	58	0	0	0	773
% OF POSSIBLE SUNSHINE	33	60	59	60	62	63	69	75	73	70	68	59	60	65
MEAN SKY COVER (tenths)														
Sunrise - Sunset	33	6.2	6.3	6.6	6.3	6.4	5.2	4.7	4.7	4.6	4.8	5.9	5.9	5.6
MEAN NUMBER OF DAYS:														
Sunrise to Sunset														
-Clear	33	8.3	6.9	7.3	7.2	6.7	10.4	12.1	12.0	13.1	12.9	9.2	9.4	115.4
-Partly Cloudy	33	8.5	7.8	7.6	9.1	10.4	10.9	12.2	11.7	8.5	8.5	8.1	8.2	111.4
-Cloudy	33	14.2	13.6	16.1	13.7	13.9	8.7	6.7	7.4	8.4	9.6	12.7	13.5	138.5
Precipitation														
.01 inches or more	33	5.2	5.3	7.0	8.3	10.7	9.5	9.8	7.6	6.6	4.9	4.7	4.5	84.0
Snow, Ice pellets														
1.0 inches or more	33	1.7	1.6	1.9	0.9	0.1	0.0	0.0	0.0	0.*	0.4	1.1	1.5	9.2
Thunderstorms	33	0.1	0.1	0.8	2.7	6.7	10.1	10.3	8.5	4.3	1.1	0.2	0.0	44.9
Heavy Fog Visibility 1/4 mile or less	33	1.3	1.9	2.0	0.9	0.7	0.8	1.0	1.7	1.9	2.0	2.4	1.5	18.1
Temperature °F														
-Maximum														
90° and above	21	0.0	0.0	0.0	0.2	0.8	5.2	13.1	11.2	3.9	0.2	0.0	0.0	34.7
32° and below	21	14.3	8.5	4.3	0.3	0.*	0.0	0.0	0.0	0.0	0.1	4.1	11.4	43.1
-Minimum														
32° and below	21	31.0	28.0	27.4	12.7	2.6	0.*	0.0	0.0	2.7	14.5	27.9	30.9	177.7
0° and below	21	9.6	3.6	0.7	0.0	0.0	0.0	0.0	0.0	0.0	0.0	0.7	5.0	19.6
AVG. STATION PRESS.(mb)	13	919.2	918.1	914.7	915.2	915.2	916.1	918.1	918.3	918.9	918.7	918.2	918.4	917.4
RELATIVE HUMIDITY (%)														
Hour 00	21	76	75	73	71	73	74	72	74	73	71	75	76	74
Hour 06 (Local Time)	21	77	77	78	79	81	83	82	83	81	79	80	78	80
Hour 12	21	63	58	53	48	50	52	50	50	48	46	55	61	53
Hour 18	21	65	55	48	43	46	48	47	46	45	47	57	64	51
PRECIPITATION (inches):														
Water Equivalent														
-Normal		0.40	0.55	1.12	1.85	3.36	3.72	2.98	1.92	1.67	0.91	0.56	0.43	19.47
-Maximum Monthly	34	1.12	1.98	2.89	5.01	8.01	6.81	7.05	5.36	6.03	2.91	2.89	1.22	8.01
-Year		1960	1978	1977	1984	1962	1965	1979	1957	1963	1969	1979	1977	MAY 1962
-Minimum Monthly	34	T	0.01	0.09	0.12	0.77	0.33	0.42	0.06	T	0.08	0.04	0.01	T
-Year		1964	1954	1967	1954	1966	1952	1955	1967	1953	1955	1974	1976	JAN 1964
-Maximum in 24 hrs	34	0.69	1.15	2.26	2.42	2.95	3.80	3.15	2.93	2.53	1.37	1.48	0.79	3.80
-Year		1960	1971	1959	1971	1962	1965	1964	1957	1963	1982	1979	1978	JUN 1965
Snow, Ice pellets														
-Maximum Monthly	34	17.1	20.6	21.9	14.5	3.6				3.1	15.7	17.5	14.1	21.9
-Year		1976	1978	1980	1984	1967				1985	1969	1979	1973	MAR 1980
-Maximum in 24 hrs	34	11.9	9.7	15.1	8.5	2.3				3.1	8.8	8.8	8.6	15.1
-Year		1976	1955	1980	1984	1967				1985	1969	1979	1968	MAR 1980
WIND:														
Mean Speed (mph)	33	9.3	10.0	11.7	12.8	11.7	10.5	9.6	9.4	9.8	9.6	9.6	9.2	10.3
Prevailing Direction through 1963		NW	NW	N	N	SE	SE	SE	SSE	SSE	SSE	NW	NW	NW
Fastest Obs. 1 Min.														
-Direction	6	31	35	17	32	31	32	31	18	36	34	03	02	02
-Speed (MPH)	6	40	39	44	45	44	52	46	41	35	37	41	52	52
-Year		1980	1981	1982	1984	1980	1981	1981	1985	1983	1985	1983	1982	DEC 1982
Peak Gust														
-Direction	2	NW	N	NW	S	NE	W	N	S	NW	W	NW	N	S
-Speed (mph)	2	49	48	53	76	72	51	52	72	47	55	43	45	76
-Date		1984	1984	1985	1985	1985	1984	1984	1985	1984	1985	1984	1985	APR 1985

See Reference Notes to this table on the following page.

TABLE 2 PRECIPITATION (inches)   NORTH PLATTE, NEBRASKA

YEAR	JAN	FEB	MAR	APR	MAY	JUNE	JULY	AUG	SEP	OCT	NOV	DEC	ANNUAL
1956	0.47	0.22	0.24	1.60	1.20	5.25	3.16	2.04	0.03	1.79	0.78	0.12	16.90
1957	0.14	0.29	1.27	3.07	7.21	3.59	3.20	5.36	2.68	1.45	0.96	0.50	29.72
1958	0.13	1.51	1.35	2.28	3.42	1.72	5.83	2.06	1.91	0.35	0.90	0.31	21.77
1959	0.31	0.59	2.61	1.35	3.51	3.91	1.27	2.43	1.89	2.12	0.15	0.23	20.37
1960	1.12	1.28	0.92	1.77	2.29	4.05	1.33	1.14	0.35	0.26	0.32	0.49	15.32
1961	T	0.07	2.24	2.49	5.15	2.58	1.23	1.08	2.78	0.15	0.50	0.44	18.71
1962	0.07	0.72	1.15	0.51	8.01	5.30	5.33	0.10	2.38	0.43	0.09	0.66	24.75
1963	0.48	0.15	0.71	1.09	3.98	1.82	1.60	2.18	6.03	0.67	0.39	0.24	19.34
1964	T	0.62	1.26	3.76	2.06	4.78	4.52	2.63	1.02	0.17	0.04	0.10	20.96
1965	0.61	0.35	0.25	1.42	4.18	6.81	6.68	1.82	5.69	0.73	0.05	1.02	29.61
1966	0.30	0.15	1.12	1.05	0.77	5.14	3.35	2.74	2.09	0.31	0.04	0.51	17.57
1967	0.47	0.03	0.09	1.03	3.94	6.05	3.74	0.06	1.12	0.58	0.19	0.15	17.45
1968	0.12	0.36	0.11	3.04	1.72	2.25	1.82	3.97	0.82	1.50	0.46	0.87	17.04
1969	0.94	0.26	0.17	0.15	1.35	4.20	2.52	1.61	0.80	2.91	0.18	0.10	15.19
1970	0.23	0.28	0.97	2.46	1.31	4.33	1.68	0.28	2.27	1.27	0.91	0.22	16.21
1971	0.47	1.28	0.97	3.94	2.54	5.84	3.48	0.91	2.01	1.75	0.90	0.16	24.25
1972	0.16	0.08	0.65	1.19	3.18	2.96	3.58	0.95	1.46	0.62	1.12	0.42	16.37
1973	0.40	0.10	2.45	1.45	3.85	0.88	2.87	2.82	3.98	1.26	0.54	1.13	21.73
1974	0.27	0.08	0.42	1.17	1.64	3.86	2.27	1.15	0.24	0.61	0.04	0.42	12.17
1975	0.26	0.17	0.92	1.77	2.12	6.12	2.51	0.25	0.56	0.14	1.15	0.22	16.19
1976	0.99	0.14	2.03	2.80	2.88	2.43	0.92	2.87	2.03	1.18	0.08	0.01	18.36
1977	0.18	0.27	2.89	4.85	5.90	1.31	4.41	1.79	1.48	0.18	0.39	1.22	24.87
1978	0.52	1.98	0.40	1.96	4.84	1.75	3.70	1.92	0.33	0.43	1.03	0.99	19.85
1979	0.86	0.09	2.78	1.46	2.96	3.37	7.05	1.49	0.45	1.32	2.89	0.28	25.00
1980	0.52	0.82	2.56	0.77	2.59	1.89	0.64	3.04	0.34	1.00	0.13	0.02	14.32
1981	0.07	0.05	2.72	2.47	5.37	2.32	5.09	2.48	0.25	0.60	1.94	0.43	23.79
1982	0.20	0.15	0.99	1.42	6.32	2.35	1.78	1.18	1.26	2.44	0.73	1.08	19.90
1983	0.33	0.25	1.54	2.12	3.20	3.32	3.74	1.98	0.14	0.56	1.56	0.46	19.20
1984	0.36	0.87	1.20	5.01	2.82	4.37	0.94	1.38	0.39	2.41	0.69	0.72	21.16
1985	0.55	0.14	0.44	1.84	4.01	0.87	3.98	1.16	3.23	1.24	1.09	0.79	19.34
Record Mean	0.40	0.48	0.96	2.08	3.00	3.24	2.74	2.17	1.55	1.04	0.54	0.51	18.72

TABLE 3 AVERAGE TEMPERATURE (deg. F)   NORTH PLATTE, NEBRASKA

YEAR	JAN	FEB	MAR	APR	MAY	JUNE	JULY	AUG	SEP	OCT	NOV	DEC	ANNUAL
1956	23.3	25.3	37.1	41.9	60.2	74.0	73.2	71.6	63.5	55.0	35.1	31.2	49.3
1957	16.1	32.5	34.7	44.3	55.6	66.2	76.5	74.2	59.2	49.4	35.3	33.7	48.1
1958	26.9	25.3	23.7	45.0	61.3	66.8	71.0	73.7	64.5	51.3	36.8	26.6	47.7
1959	23.2	24.2	36.6	46.4	58.0	72.5	71.8	75.3	60.6	45.7	30.3	32.6	48.1
1960	19.6	19.7	28.0	49.7	57.5	66.5	74.2	74.1	65.2	52.3	36.7	27.3	47.6
1961	26.6	32.2	37.9	42.8	55.2	69.0	73.4	74.8	57.5	49.3	33.9	22.5	47.9
1962	21.9	26.9	32.0	49.6	63.9	66.8	72.0	72.7	60.2	53.2	39.7	27.6	48.9
1963	13.1	32.3	40.8	50.8	60.5	72.9	77.5	73.4	65.9	58.1	38.8	21.9	50.5
#1964	28.1	24.9	32.3	47.0	60.9	67.2	77.5	69.8	63.1	50.3	35.8	25.2	48.5
1965	26.3	23.4	25.5	51.4	60.1	67.1	73.1	70.5	53.8	54.0	40.9	28.9	47.9
1966	15.6	21.4	38.5	43.8	60.0	68.8	77.6	67.8	62.1	50.5	36.0	24.5	47.2
1967	24.9	30.8	39.9	49.3	51.9	64.6	70.3	68.8	61.2	49.8	33.9	22.8	47.3
1968	22.2	27.1	39.9	46.2	53.3	69.5	72.6	71.6	61.1	51.1	34.4	19.0	47.4
1969	16.2	24.2	29.0	51.4	61.0	62.8	73.0	73.6	65.8	41.7	38.3	28.4	47.1
1970	22.3	31.8	31.3	45.6	61.9	68.1	74.3	75.6	60.8	45.4	34.8	26.8	48.2
1971	22.1	24.5	33.4	48.0	54.7	70.2	69.8	72.2	59.7	49.5	37.1	27.6	47.4
1972	20.7	29.9	40.7	46.6	58.1	68.4	70.4	70.8	61.2	46.4	29.8	17.2	46.7
1973	21.5	30.3	40.4	45.8	56.0	67.7	72.8	73.7	58.4	51.7	35.0	23.2	48.0
1974	16.0	33.0	40.1	50.0	59.5	66.5	77.5	67.1	57.5	52.9	34.8	24.2	48.3
1975	25.7	23.4	31.5	46.6	57.1	67.4	74.6	73.2	58.5	50.9	30.2	26.4	47.0
1976	17.8	32.7	34.0	47.4	53.1	64.6	72.8	70.6	59.8	45.2	30.5	28.2	46.4
1977	16.6	32.0	37.0	51.7	63.1	70.5	74.5	68.5	63.8	50.0	35.8	24.9	49.0
1978	11.2	14.7	35.1	48.4	56.6	68.2	74.5	71.0	65.4	49.2	32.1	14.6	45.1
1979	6.0	17.2	37.0	48.4	57.4	68.1	74.1	72.9	67.1	53.8	32.2	34.0	47.3
1980	25.1	26.8	35.5	50.5	60.2	72.5	78.0	74.0	63.7	48.9	38.5	32.0	50.5
1981	29.7	29.1	40.2	55.8	54.8	68.6	73.8	70.1	63.0	49.0	39.9	26.7	50.1
1982	17.1	28.1	36.3	44.8	57.3	63.4	75.0	73.1	61.7	49.2	33.3	28.1	47.3
1983	27.1	35.1	37.2	42.3	54.0	66.2	75.3	78.0	64.8	51.3	36.7	7.5	47.9
1984	20.4	33.2	34.1	43.3	58.1	68.0	73.7	75.5	58.4	48.0	37.2	22.5	47.7
1985	18.3	23.1	40.5	52.2	60.8	65.4	75.3	70.1	59.8	48.6	24.6	19.2	46.5
Record Mean	23.1	27.7	36.3	48.5	58.7	68.5	74.7	73.0	63.3	51.1	36.6	27.0	49.0
Max	35.4	40.2	49.1	61.6	71.1	80.9	87.6	86.2	77.5	66.0	50.1	39.2	62.1
Min	10.8	15.2	23.4	35.5	46.2	56.0	61.8	59.8	49.1	36.2	23.1	14.7	36.0

REFERENCE NOTES FOR TABLES 1, 2, 3 and 6   **(NORTH PLATTE, NE)**

GENERAL

T - TRACE AMOUNT
BLANK ENTRIES DENOTE MISSING/UNREPORTED DATA.
# INDICATES A STATION OR INSTRUMENT RELOCATION.

SPECIFIC

TABLE 1

(a) - LENGTH OF RECORD IN YEARS. ALTHOUGH INDIVIDUAL MONTHS MAY BE MISSING.

* LESS THAN .05

NORMALS — BASED ON THE 1951-1980 RECORD PERIOD.
EXTREMES — DATES ARE THE MOST RECENT OCCURRENCE.
WIND DIR. — NUMERALS SHOW TENS OF DEGREES CLOCKWISE FROM TRUE NORTH.
"00" INDICATES CALM.
RESULTANT WIND DIRECTIONS ARE GIVEN TO WHOLE DEGREES.

EXCEPTIONS

TABLES 2, 3, and 6

RECORD MEANS ARE THROUGH THE CURRENT YEAR, BEGINNING IN   1875 FOR TEMPERATURE
1875 FOR PRECIPITATION
1953 FOR SNOWFALL

TABLE 4 HEATING DEGREE DAYS Base 65 deg. F   NORTH PLATTE, NEBRASKA

SEASON	JULY	AUG	SEP	OCT	NOV	DEC	JAN	FEB	MAR	APR	MAY	JUNE	TOTAL
1956-57	0	31	120	315	891	1040	1512	903	935	615	290	50	6702
1957-58	0	0	182	482	882	968	1175	1106	1272	594	137	56	6854
1958-59	3	6	110	416	840	1183	1292	1137	871	550	224	22	6654
1959-60	6	1	198	594	1034	1000	1400	1310	1143	457	242	48	7433
1960-61	0	4	124	389	844	1162	1182	914	833	660	314	37	6463
1961-62	0	0	278	479	926	1312	1333	1062	1017	463	90	57	7017
1962-63	3	4	171	363	755	1152	1607	911	745	423	165	10	6309
1963-64	0	1	65	219	780	1329	1140	1157	1006	534	190	63	6484
#1964-65	0	43	143	450	870	1229	1189	1158	1217	405	173	26	6903
1965-66	0	11	345	335	718	1113	1530	1213	815	628	199	48	6955
1966-67	0	34	136	442	863	1250	1240	950	771	465	427	70	6648
1967-68	19	29	132	479	926	1305	1321	1096	770	558	363	26	7024
1968-69	17	18	150	428	911	1421	1509	1133	1106	399	178	122	7392
1969-70	1	0	45	716	795	1129	1315	923	1041	582	136	55	6738
1970-71	1	2	214	600	899	1177	1320	1127	974	501	311	13	7139
1971-72	12	0	210	472	829	1157	1368	1013	745	549	250	32	6637
1972-73	24	15	169	567	1047	1479	1343	967	756	567	283	34	7251
1973-74	10	0	219	407	892	1290	1518	889	765	445	190	62	6687
1974-75	0	40	256	373	900	1262	1216	1160	1032	558	247	62	7106
1975-76	6	0	228	437	1035	1191	1460	929	956	521	363	70	7196
1976-77	0	13	178	608	1028	1133	1493	920	858	395	81	2	6709
1977-78	2	34	96	458	869	1236	1662	1400	924	491	275	71	7518
1978-79	5	24	99	488	982	1560	1828	1335	862	491	259	56	7989
1979-80	4	11	52	341	975	957	1233	1102	909	439	166	10	6199
1980-81	0	6	107	491	790	1019	1089	1000	762	283	318	26	5891
1981-82	9	4	101	492	749	1179	1479	1030	885	601	239	111	6879
1982-83	0	18	160	484	946	1138	1167	833	854	672	343	90	6705
1983-84	2	0	128	419	840	1780	1379	915	953	647	236	33	7332
1984-85	0	0	247	519	829	1312	1440	1168	752	393	156	83	6899
1985-86	0	23	252	502	1205	1416							

TABLE 5 COOLING DEGREE DAYS Base 65 deg. F   NORTH PLATTE, NEBRASKA

YEAR	JAN	FEB	MAR	APR	MAY	JUNE	JULY	AUG	SEP	OCT	NOV	DEC	TOTAL
1969	0	0	0	0	58	63	257	276	76	0	0	0	730
1970	0	0	0	4	48	155	300	341	96	0	0	0	944
1971	0	0	0	0	1	177	167	229	59	1	0	0	634
1972	0	0	0	0	43	137	199	202	64	0	0	0	645
1973	0	0	0	0	11	121	260	278	29	3	0	0	702
1974	0	0	0	5	24	117	394	115	40	3	0	0	698
1975	0	0	0	10	11	89	311	260	39	7	0	0	727
1976	0	0	0	0	5	62	247	192	27	0	0	0	533
1977	0	0	0	2	26	174	305	149	64	0	0	0	720
1978	0	0	0	1	21	174	307	218	117	5	0	0	843
1979	0	0	0	0	27	156	294	262	123	0	0	0	862
1980	0	0	0	10	27	243	411	289	74	0	0	0	1054
1981	0	0	0	10	9	141	288	168	47	1	0	0	664
1982	0	0	0	0	8	70	314	276	68	0	0	0	736
1983	0	0	0	0	8	103	331	412	128	1	0	0	983
1984	0	0	0	0	27	129	281	331	55	2	0	0	825
1985	0	0	0	14	32	100	326	189	100	0	0	0	761

TABLE 6 SNOWFALL (inches)   NORTH PLATTE, NEBRASKA

SEASON	JULY	AUG	SEP	OCT	NOV	DEC	JAN	FEB	MAR	APR	MAY	JUNE	TOTAL
1956-57	0.0	0.0	0.0	0.5	5.2	1.3	2.8	2.0	7.2	13.1	0.0	0.0	32.1
1957-58	0.0	0.0	0.0	0.9	3.0	6.4	2.2	15.5	16.9	3.5	0.0	0.0	48.4
1958-59	0.0	0.0	0.0	0.7	5.8	4.7	5.2	9.3	10.2	1.4	0.0	0.0	37.3
1959-60	0.0	0.0	0.0	7.3	0.7	0.3	12.4	7.3	10.6	2.2	0.0	0.0	40.8
1960-61	0.0	0.0	0.0	T	2.9	4.4	T	0.7	20.0	13.0	0.0	0.0	41.0
1961-62	0.0	0.0	T	0.0	4.0	7.3	1.7	10.7	7.1	T	0.0	0.0	30.8
1962-63	0.0	0.0	0.0	0.0	1.5	6.1	8.7	0.9	9.5	0.2	0.0	0.0	26.9
1963-64	0.0	0.0	0.0	0.0	T	5.3	T	8.6	5.4	5.6	0.0	0.0	24.9
1964-65	0.0	0.0	0.0	0.0	T	1.8	9.2	7.2	5.0	T	0.0	0.0	23.2
1965-66	0.0	0.0	T	0.0	0.5	6.3	10.5	1.7	9.6	5.0	0.2	0.0	33.8
1966-67	0.0	0.0	0.0	1.4	0.5	8.5	9.3	0.8	0.7	4.5	3.6	0.0	29.3
1967-68	0.0	0.0	0.0	T	2.9	3.4	1.6	3.0	0.8	2.2	0.0	0.0	13.9
1968-69	0.0	0.0	0.0	0.0	0.3	10.3	12.4	3.2	1.8	T	T	0.0	28.0
1969-70	0.0	0.0	0.0	15.7	0.8	1.8	3.6	4.0	18.4	0.4	0.0	0.0	44.7
1970-71	0.0	0.0	0.0	9.0	3.6	1.8	5.6	2.7	19.0	0.8	T	0.0	42.5
1971-72	0.0	0.0	0.0	2.0	2.5	2.1	2.6	1.9	0.1	0.1	0.0	0.0	11.3
1972-73	0.0	0.0	0.0	0.8	8.0	8.7	3.8	0.4	1.6	1.7	0.0	0.0	25.0
1973-74	0.0	0.0	0.0	T	4.3	14.1	4.2	1.0	2.3	0.7	0.0	0.0	26.6
1974-75	0.0	0.0	0.0	0.0	T	4.5	2.2	2.6	3.9	3.0	0.0	0.0	16.2
1975-76	0.0	0.0	0.0	0.8	10.9	1.2	17.1	1.6	5.2	T	0.0	0.0	36.8
1976-77	0.0	0.0	0.0	1.0	0.6	T	2.5	2.2	9.9	8.1	0.0	0.0	24.3
1977-78	0.0	0.0	0.0	T	0.3	6.4	6.2	20.6	2.0	T	T	0.0	35.5
1978-79	0.0	0.0	0.0	T	7.6	10.3	6.3	0.3	5.9	2.3	T	0.0	32.7
1979-80	0.0	0.0	0.0	2.9	17.5	2.3	5.2	9.1	21.9	7.4	0.0	0.0	66.3
1980-81	0.0	0.0	0.0	0.6	1.2	T	0.7	0.5	0.5	0.4	0.0	0.0	3.9
1981-82	0.0	0.0	0.0	T	5.3	4.4	4.4	1.7	6.4	0.9	0.0	0.0	25.1
1982-83	0.0	0.0	0.0	1.0	2.0	9.7	1.6	0.1	5.9	5.4	0.0	0.0	25.7
1983-84	0.0	0.0	T	0.0	12.1	7.3	5.3	9.6	8.8	14.5	0.2	0.0	57.8
1984-85	0.0	0.0	T	0.3	0.9	8.7	8.5	0.8	2.1	0.0	0.0	0.0	21.3
1985-86	0.0	0.0	3.1	T	13.0	8.1							
Record Mean	0.0	0.0	0.1	1.3	4.0	5.2	5.3	4.9	7.3	3.2	0.2	0.0	31.4

See Reference Notes, relative to all above tables, on preceding page.

Omaha, Nebraska, is situated on the west bank of the Missouri River. The river level at Omaha is normally about 965 feet above sea level and the rolling hills in and around Omaha rise to about 1,300 feet above sea level. The climate is typically continental with relatively warm summers and cold, dry winters. It is situated midway between two distinctive climatic zones, the humid east and the dry west. Fluctuations between these two zones produce weather conditions for periods that are characteristic of either zone, or combinations of both. Omaha is also affected by most low pressure systems that cross the country. This causes periodic and rapid changes in weather, especially during the winter months.

Most of the precipitation in Omaha falls during sharp showers or thunderstorms, and these occur mostly during the growing season from April to September. Of the total precipitation, about 75 percent falls during this six-month period. The rain occurs mostly as evening or nighttime showers and thunderstorms. Although winters are relatively cold, precipitation is light, with only 10 percent of the total annual precipitation falling during the winter months.

Sunshine is fairly abundant, ranging around 50 percent of the possible in the winter to 75 percent of the possible in the summer.

TABLE 1   NORMALS, MEANS AND EXTREMES

OMAHA (EPPLEY AIRFIELD), NEBRASKA

LATITUDE: 41°18'N   LONGITUDE: 95°54' W   ELEVATION: FT. GRND   997 BARO   00985   TIME ZONE: CENTRAL   WBAN: 14942

	(a)	JAN	FEB	MAR	APR	MAY	JUNE	JULY	AUG	SEP	OCT	NOV	DEC	YEAR
TEMPERATURE °F:														
Normals														
-Daily Maximum		30.2	37.3	47.7	64.0	74.7	84.2	88.5	86.2	77.5	67.0	50.3	36.9	62.0
-Daily Minimum		10.2	17.1	26.9	40.3	51.8	61.7	66.8	64.2	54.0	42.0	28.6	17.4	40.1
-Monthly		20.2	27.2	37.3	52.2	63.3	73.0	77.7	75.2	65.8	54.5	39.5	27.1	51.1
Extremes														
-Record Highest	49	69	78	89	93	99	105	114	110	104	96	80	72	114
-Year		1944	1972	1968	1965	1939	1953	1936	1936	1939	1938	1980	1939	JUL 1936
-Record Lowest	49	-23	-21	-16	5	27	38	44	43	25	13	-9	-21	-23
-Year		1982	1981	1948	1975	1980	1983	1972	1967	1984	1972	1964	1983	JAN 1982
NORMAL DEGREE DAYS:														
Heating (base 65°F)		1389	1058	859	390	130	16	0	0	73	342	765	1172	6194
Cooling (base 65°F)		0	0	0	6	77	256	394	320	97	16	0	0	1166
% OF POSSIBLE SUNSHINE														
MEAN SKY COVER (tenths)														
Sunrise - Sunset	41	6.1	6.3	6.6	6.3	6.2	5.6	4.8	4.7	4.8	4.7	6.0	6.4	5.7
MEAN NUMBER OF DAYS:														
Sunrise to Sunset														
-Clear	41	8.9	7.4	7.0	7.3	7.4	8.5	11.7	12.8	12.7	13.3	8.9	8.1	114.0
-Partly Cloudy	41	7.9	7.5	8.1	8.7	9.8	11.2	12.4	10.1	7.4	8.1	7.6	7.4	106.1
-Cloudy	41	14.2	13.3	15.9	14.1	13.8	10.3	6.9	8.1	10.0	9.6	13.5	15.4	145.2
Precipitation														
.01 inches or more	49	6.5	6.6	8.7	9.5	11.5	10.6	8.9	9.1	8.4	6.4	5.3	6.2	97.7
Snow,Ice pellets														
1.0 inches or more	50	2.5	1.9	2.2	0.4	0.*	0.0	0.0	0.0	0.0	0.1	0.8	1.9	9.8
Thunderstorms	50	0.1	0.4	1.4	4.0	7.4	9.4	8.0	7.8	5.2	2.5	0.8	0.2	47.3
Heavy Fog Visibility														
1/4 mile or less	50	1.9	2.0	1.5	0.5	0.8	0.4	0.5	1.5	1.4	1.5	1.6	2.0	15.5
Temperature °F														
-Maximum														
90° and above	21	0.0	0.0	0.0	0.2	1.4	7.0	13.8	9.4	3.0	0.3	0.0	0.0	35.1
32° and below	21	16.0	10.2	3.8	0.1	0.0	0.0	0.0	0.0	0.0	0.0	2.6	12.5	45.2
-Minimum														
32° and below	21	30.1	26.1	21.6	6.3	0.5	0.0	0.0	0.0	0.5	5.5	19.8	29.2	139.7
0° and below	21	8.1	3.6	0.5	0.0	0.0	0.0	0.0	0.0	0.0	0.0	0.2	3.4	15.9
AVG. STATION PRESS.(mb)	4	982.7	981.6	977.7	978.2	977.0	977.0	979.8	980.5	982.2	982.4	982.0	982.0	980.3
RELATIVE HUMIDITY (%)														
Hour 00	21	76	77	73	69	73	76	77	79	81	76	77	77	76
Hour 06 (Local Time)	21	78	79	78	77	80	82	83	85	86	82	81	80	81
Hour 12	21	66	64	58	53	54	55	56	58	59	56	63	68	59
Hour 18	21	68	64	55	49	51	52	54	58	60	57	65	71	59
PRECIPITATION (inches):														
Water Equivalent														
-Normal		0.77	0.91	1.91	2.94	4.33	4.08	3.62	4.10	3.50	2.09	1.32	0.77	30.34
-Maximum Monthly	49	3.70	2.97	5.96	6.45	10.33	10.81	9.60	9.12	13.75	4.99	4.70	5.42	13.75
-Year		1949	1965	1973	1951	1959	1947	1958	1959	1965	1961	1983	1984	SEP 1965
-Minimum Monthly	49	0.05	0.09	0.12	0.23	0.56	1.03	0.39	0.61	0.41	T	0.03	T	T
-Year		1943	1981	1956	1936	1948	1972	1983	1984	1953	1952	1976	1943	OCT 1952
-Maximum in 24 hrs	43	1.52	2.24	1.45	2.56	3.58	3.48	3.37	3.40	6.47	3.13	2.53	3.03	6.47
-Year		1967	1954	1959	1938	1964	1942	1983	1958	1959	1965	1968	1948	1984	SEP 1965
Snow,Ice pellets														
-Maximum Monthly	50	25.7	25.4	27.2	8.6	2.0				T	7.2	12.0	19.9	27.2
-Year		1936	1965	1948	1945	1945				1985	1941	1957	1969	MAR 1948
-Maximum in 24 hrs	43	13.1	18.3	13.0	8.6	2.0	T			T	7.2	8.7	10.2	18.3
-Year		1949	1965	1948	1945	1945	1985			1985	1941	1957	1969	FEB 1965
WIND:														
Mean Speed (mph)	49	10.9	11.2	12.3	12.8	11.0	10.2	8.9	8.9	9.5	9.8	10.8	10.7	10.6
Prevailing Direction														
through 1963		NNW	NNW	NNW	NNW	SSE	SSE	SSE	SSE	SSE	SSE	SSE	SSE	SSE
Fastest Mile														
-Direction	41	NW	NW	NW	NW	NW	N	N	N	E	NW	NW	NW	N
-Speed (MPH)	41	57	57	73	65	73	72	109	66	47	62	56	52	109
-Year		1938	1947	1950	1937	1936	1942	1936	1944	1948	1966	1951	1938	JUL 1936
Peak Gust														
-Direction														
-Speed (mph)														
-Date														

See Reference Notes to this table on the following page.

TABLE 2 PRECIPITATION (inches) OMAHA (EPPLEY AIRFIELD), NEBRASKA

YEAR	JAN	FEB	MAR	APR	MAY	JUNE	JULY	AUG	SEP	OCT	NOV	DEC	ANNUAL
1956	0.71	0.32	0.12	1.19	1.31	2.67	7.44	3.52	2.07	1.93	1.81	0.18	23.27
1957	0.38	0.13	2.44	2.37	5.01	6.27	1.74	2.44	4.18	3.81	1.07	0.42	34.45
1958	1.18	0.85	0.83	4.18	1.87	1.29	9.60	3.07	2.20	0.08	2.04	1.07	28.26
1959	0.56	0.92	3.35	2.63	10.33	5.53	0.90	9.12	2.20	2.15	0.65	1.13	39.47
1960	1.17	1.17	2.07	1.99	6.53	5.69	1.85	7.12	4.04	1.05	0.11	0.40	33.19
1961	0.23	0.79	3.59	1.64	5.98	4.44	3.12	4.56	4.77	4.99	2.23	1.50	37.84
1962	0.41	1.94	1.53	0.64	5.26	2.56	7.55	3.80	3.79	1.74	0.68	0.80	30.70
1963	1.09	0.36	3.05	3.43	1.29	4.67	2.25	4.82	2.18	0.50	2.34	0.84	25.06
1964	0.54	0.36	1.58	4.40	7.46	6.39	4.61	4.13	2.60	0.76	1.24	1.40	35.75
1965	0.60	2.97	2.63	3.71	6.19	5.15	4.39	2.06	13.75				44.85
1966	0.81	0.41	0.88	0.83	3.67	5.93	4.70	1.85	2.16	0.64	0.22	0.62	22.72
1967	2.00	0.16	0.82	2.61	2.26	9.86	4.33	1.53	4.15	1.69	0.28	0.79	30.48
1968	0.49	0.10	0.57	3.93	4.35	4.44	2.61	3.30	5.74	4.62	1.50	2.04	33.69
1969	1.10	1.33	1.25	3.95	4.48	3.27	4.27	5.06	1.33	2.24	1.80	0.11	30.19
1970	0.20	0.14	0.85	2.84	2.64	2.48	1.98	4.66	4.93	4.88	1.23	0.42	27.25
1971	0.95	2.43	0.49	0.90	7.21	3.30	1.77	1.60	0.57	4.20	3.12	1.05	27.59
1972	0.38	0.36	1.14	4.63	5.13	1.03	7.28	2.60	4.93	3.21	3.25	1.62	35.56
1973	1.44	0.87	5.96	3.60	4.94	1.56	4.98	1.27	8.04	2.60	1.43	1.65	38.34
1974	0.63	0.16	0.80	1.72	2.65	1.79	0.79	4.17	2.54	2.74	1.41	0.81	20.21
1975	2.01	1.06	1.78	3.37	3.72	4.30	0.46	1.80	2.16	0.01	2.85	0.46	23.98
1976	0.11	1.39	2.13	2.98	3.55	2.84	1.52	0.62	1.64	1.35	2.90	0.35	18.37
1977	0.93	0.38	3.72	3.28	6.05	2.40	4.69	8.63	6.01	4.38	1.09	0.50	42.76
1978	0.15	0.76	1.04	4.61	5.05	2.19	5.89	3.87	6.01	0.96	1.23	0.14	32.12
1979	1.11	0.30	4.59	2.58	2.84		6.98		0.82	2.70	0.11	0.04	30.34
1980	0.93	0.52	1.40	1.72	2.50	8.99	3.63		6.98		2.70	0.86	
1981	0.20	0.09	0.88	1.33	4.13	2.14	1.87	4.80	1.51	1.92	2.60	0.86	22.33
1982	1.83	0.26	1.90	1.22	9.92	4.16	2.46	3.21	2.27	1.10	1.81	1.17	31.31
1983	0.86	0.68	3.65	1.00	2.81	6.52	0.39	1.24	2.45	2.16	4.70	0.63	27.09
1984	0.38	0.62	2.32	4.77	4.92	5.56	1.58	0.61	2.55	3.87	0.52	5.42	33.12
1985	0.56	1.88	1.36	3.16	2.46	1.73	3.27	1.50	2.71	1.36	0.85	0.37	21.21
Record Mean	0.74	0.89	1.48	2.64	3.85	4.50	3.60	3.39	3.14	2.10	1.26	0.90	28.50

TABLE 3 AVERAGE TEMPERATURE (deg. F) OMAHA (EPPLEY AIRFIELD), NEBRASKA

YEAR	JAN	FEB	MAR	APR	MAY	JUNE	JULY	AUG	SEP	OCT	NOV	DEC	ANNUAL
1956	21.1	26.1	37.6	48.4	65.5	77.6	76.2	76.8	66.8	61.1	39.7	32.4	52.5
1957	17.6	32.5	38.3	50.8	61.3	71.0	81.5	76.3	63.4	51.8	37.1	35.0	51.4
1958	26.7	20.3	35.7	51.0	65.7	69.7	73.1	75.8	67.5	57.1	42.2	26.6	51.0
1959	18.7	25.0	39.4	50.8	63.1	74.0	74.7	78.9	65.0	50.8	33.1	34.9	50.7
1960	20.5	19.2	23.1	53.2	62.2	70.0	70.0	75.6	67.7	56.2	41.3	28.9	49.5
1961	24.6	31.1	39.5	47.5	60.1	71.6	77.0	75.1	62.3	55.8	38.0	20.1	50.2
1962	18.2	25.0	31.9	50.9	60.1	72.3	75.0	75.0	63.2	57.0	42.0	28.2	50.8
#1963	13.7	27.1	42.7	54.7	63.1	76.3	79.0	74.3	67.8	64.4	44.1	17.8	52.1
1964	28.6	30.5	33.7	52.8	68.0	71.9	80.0	70.5	65.3	53.7	42.2	25.2	51.9
1965	23.0	22.2	27.1	53.2	68.1	71.8	75.7	74.1	61.0	57.2	42.4	36.9	51.1
1966	18.4	28.8	44.1	48.2	61.0	72.6	79.4	71.0	63.7	53.9	40.1	28.3	50.8
1967	25.5	28.4	45.5	55.6	59.1	70.8	74.4	71.7	62.5	52.1	38.4	28.4	51.1
1968	22.6	27.3	44.6	52.5	58.0	74.0	76.4	75.0	65.4	55.8	37.4	24.0	51.1
1969	18.7	26.7	31.5	53.1	63.2	67.9	78.2	75.6	67.7	50.1	41.0	24.8	49.9
1970	16.3	32.0	35.9	53.0	68.1	74.3	77.3	76.3	66.2	51.7	38.2	29.8	51.3
1971	17.8	25.3	37.2	54.4	60.4	77.8	74.2	74.7	67.8	58.8	41.2	28.9	51.5
1972	20.0	25.4	40.8	51.2	62.4	72.6	74.6	73.6	65.4	49.1	37.1	21.1	49.5
1973	22.6	28.6	44.3	50.4	59.4	73.3	75.5	77.6	64.2	57.3	39.8	22.7	51.3
1974	18.8	30.6	42.1	52.6	61.9	69.8	82.2	70.5	59.7	55.3	40.7	28.9	51.1
1975	22.5	22.4	31.7	49.4	66.2	72.5	78.9	79.7	62.6	58.2	41.6	30.9	51.4
1976	25.5	37.5	40.4	56.8	60.2	72.7	79.0	76.3	67.4	48.5	33.3	24.1	51.8
1977	13.3	33.2	41.6	59.0	70.2	75.1	80.6	72.5	67.4	53.1	39.7	26.2	53.0
1978	12.0	15.7	35.7	52.6	61.7	73.7	75.5	75.7	66.7	53.7	39.6	24.4	49.4
1979	10.7	17.3	39.8	50.0	61.8	72.7	76.0	76.6	64.7	49.0	35.8	30.3	49.8
1980	22.4	20.3	32.9	50.7	61.6	73.1	79.6	76.6	64.8	50.2	40.6	26.1	49.8
1981	24.1	28.4	40.8	57.4	58.6	72.6	76.6	71.0	64.8	50.2	37.1	22.9	50.6
1982	9.4	22.6	34.8	47.7	62.9	65.5	77.1	72.6	64.7	55.0	28.3	7.3	48.1
1983	24.9	30.1	37.3	43.5	56.5	69.6	79.4	81.5	66.7	52.2	38.5	27.1	48.9
1984	19.6	33.2	30.3	46.6	57.7	71.8	75.1	76.5	61.9	52.2	39.4	16.4	49.3
1985	19.1	23.6	43.4	54.9	63.4	67.2	74.4	69.4	62.1	52.7	28.5		47.9
Record Mean	21.6	26.5	37.5	51.7	62.5	72.2	77.4	75.1	66.3	54.7	39.1	27.0	51.0
Max	30.7	35.7	47.2	62.2	72.8	82.3	87.7	85.3	76.9	65.5	48.5	35.6	60.9
Min	12.5	17.2	27.8	41.2	52.2	62.0	67.1	64.9	55.6	43.9	29.7	18.5	41.1

REFERENCE NOTES FOR TABLES 1, 2, 3 and 6 (OMAHA, NE)

GENERAL

T - TRACE AMOUNT
BLANK ENTRIES DENOTE MISSING/UNREPORTED DATA.
# INDICATES A STATION OR INSTRUMENT RELOCATION.

SPECIFIC

TABLE 1

(a) - LENGTH OF RECORD IN YEARS. ALTHOUGH INDIVIDUAL MONTHS MAY BE MISSING.
* LESS THAN .05

NORMALS — BASED ON THE 1951-1980 RECORD PERIOD.
EXTREMES — DATES ARE THE MOST RECENT OCCURRENCE.
WIND DIR. — NUMERALS SHOW TENS OF DEGREES CLOCKWISE FROM TRUE NORTH.
"00" INDICATES CALM.
RESULTANT WIND DIRECTIONS ARE GIVEN TO WHOLE DEGREES.

EXCEPTIONS

TABLE 1

1. MEAN SKY COVER, AND DAYS CLEAR - PARTLY CLOUDY CLOUDY ARE THROUGH
2. MAXIMUM 24-HOUR PRECIPITATION AND SNOW, AND FASTEST MILE WINDS ARE THROUGH MAY 1977.

TABLES 2, 3, and 6

RECORD MEANS ARE THROUGH THE CURRENT YEAR,
BEGINNING IN   1873 FOR TEMPERATURE
1871 FOR PRECIPITATION
1936 FOR SNOWFALL

OMAHA, NEBRASKA

TABLE 4 HEATING DEGREE DAYS Base 65 deg. F   OMAHA (EPPLEY AIRFIELD), NEBRASKA

SEASON	JULY	AUG	SEP	OCT	NOV	DEC	JAN	FEB	MAR	APR	MAY	JUNE	TOTAL
1956-57	0	5	78	161	752	1004	1463	904	821	431	159	5	5783
1957-58	0	0	91	404	828	921	1181	1250	904	419	63	15	6076
1958-59	0	9	52	272	678	1184	1431	1114	790	426	135	14	6105
1959-60	3	0	115	433	948	923	1370	1319	1290	379	119	12	6911
1960-61	0	0	63	282	704	1110	1245	942	784	528	188	4	5850
1961-62	0	4	174	283	805	1384	1444	1116	1020	431	22	19	6702
1962-63	0	0	117	284	685	1134	1590	1058	681	314	120	0	5983
#1963-64	0	5	35	96	623	1458	1117	994	964	364	55	19	5730
1964-65	0	24	93	353	676	1227	1298	1191	1168	361	55	0	6446
1965-66	0	5	169	257	670	864	1440	1008	642	501	185	13	5754
1966-67	0	13	107	354	741	1132	1216	1020	615	295	249	17	5759
1967-68	8	15	109	417	791	1129	1307	1085	631	380	228	11	6111
1968-69	2	1	60	313	824	1268	1430	1066	1028	353	140	56	6541
1969-70	0	0	25	478	714	1242	1507	917	894	388	68	2	6235
1970-71	0	0	105	426	799	1085	1458	1106	853	319	172	1	6324
1971-72	5	1	95	228	707	1113	1392	1143	741	408	147	12	5992
1972-73	6	7	106	488	831	1357	1307	1014	636	437	191	0	6380
1973-74	0	0	90	254	750	1302	1427	955	702	379	140	31	6030
1974-75	0	15	191	300	726	1115	1311	1189	1024	469	72	11	6423
1975-76	0	0	141	251	695	1051	1219	791	757	261	177	4	5347
1976-77	0	0	61	522	947	1265	1598	883	579	219	10	1	6085
1977-78	0	1	28	361	754	1196	1637	1375	910	372	160	17	6811
1978-79	0	0	39	350	754	1255	1676	1333	775	451	156	12	6801
1979-80	1	6	65		867	1070	1318	1290	987	440	158	4	
1980-81	0	3	108	491	735	1198	1259	1018	743	241	221	0	6017
1981-82	7	0	85	452	723	1299	1721	1183	930	518	102	56	7079
1982-83	0	13	115	315	829	1131	1240	971	854	638	278	37	6421
1983-84	0	0	102	405	789	1786	1401	916	1071	552	243	7	7272
1984-85	0	3	184	391	766	1166	1416	1153	666	325	88	45	6203
1985-86	0	13	217	378	1089	1501							

TABLE 5 COOLING DEGREE DAYS Base 65 deg. F   OMAHA (EPPLEY AIRFIELD), NEBRASKA

YEAR	JAN	FEB	MAR	APR	MAY	JUNE	JULY	AUG	SEP	OCT	NOV	DEC	TOTAL
1969	0	0	0	2	95	149	416	338	113	25	0	0	1138
1970	0	0	0	32	172	286	386	361	148	20	0	0	1405
1971	0	0	0	12	34	393	295	308	188	42	0	0	1272
1972	0	0	0	3	74	249	314	280	124	1	0	0	1045
1973	0	0	0	3	23	257	332	395	74	22	0	0	1106
1974	0	0	0	11	52	182	540	193	39	4	0	0	1021
1975	0	0	0	7	115	242	441	464	76	44	0	0	1389
1976	0	0	0	21	34	240	440	358	139	17	0	0	1249
1977	0	0	0	45	179	310	489	236	105	0	0	0	1364
1978	0	0	7	5	67	287	386	333	231	5	0	0	1321
1979	0	0	0	8	64	249	344	345	122	0	0	0	
1980	0	0	0	15	61	254	459	368	107	0	0	0	1264
1981	0	0	0	24	29	235	372	196	85	0	0	0	941
1982	0	0	0	5	43	78	383	252	113	12	0	0	886
1983	0	0	0	0	20	183	453	519	167	17	0	0	1359
1984	0	0	0	6	22	220	320	366	96	4	0	0	1034
1985	0	0	0	30	44	116	290	156	137	1	0	0	774

TABLE 6 SNOWFALL (inches)   OMAHA (EPPLEY AIRFIELD), NEBRASKA

SEASON	JULY	AUG	SEP	OCT	NOV	DEC	JAN	FEB	MAR	APR	MAY	JUNE	TOTAL
1956-57	0.0	0.0	0.0	0.0	5.2	2.5	6.8	1.8	11.6	2.8	0.0	0.0	30.7
1957-58	0.0	0.0	0.0	T	12.0	5.5	8.8	3.7	7.8	T	0.0	0.0	37.8
1958-59	0.0	0.0	0.0	0.0	1.0	1.5	6.0	5.3	13.2	0.7	0.0	0.0	27.7
1959-60	0.0	0.0	0.0	0.0	1.5	5.1	14.0	13.4	22.7	T	0.0	0.0	56.7
1960-61	0.0	0.0	0.0	0.0	T	1.1	4.7	6.3	6.4	1.3	0.0	0.0	19.8
1961-62	0.0	0.0	T	0.0	4.0	19.6	5.0	13.2	8.9	0.9	0.0	0.0	51.6
1962-63	0.0	0.0	0.0	T	0.8	8.1	14.5	3.2	13.0	0.0	0.0	0.0	39.6
1963-64	0.0	0.0	0.0	0.0	0.0	6.6	4.9	3.5	10.5	T	0.0	0.0	25.5
1964-65	0.0	0.0	0.0	0.0	2.6	5.4	6.8	25.4	16.1	0.0	0.0	0.0	56.3
1965-66	0.0	0.0	0.0	0.0	0.1	1.7	2.8	1.3	8.3	T	T	0.0	14.2
1966-67	0.0	0.0	0.0	T	0.1	6.8	7.2	0.8	0.3	0.7	1.0	0.0	16.9
1967-68	0.0	0.0	0.0	1.5	T	5.7	4.6	1.1	T	T	T	0.0	12.9
1968-69	0.0	0.0	0.0	0.0	4.7	8.6	8.3	14.0	3.2	0.0	0.0	0.0	38.8
1969-70	0.0	0.0	0.0	T	0.5	19.9	3.2	1.1	5.8	T	0.0	0.0	30.5
1970-71	0.0	0.0	0.0	3.5	T	0.5	13.1	17.4	4.3	0.5	0.0	0.0	39.3
1971-72	0.0	0:0	0.0	0.0	9.0	4.4	1.9	5.0	3.1	1.5	0.0	0.0	24.9
1972-73	0.0	0.0	0.0	T	9.7	5.9	15.8	4.6	T	2.3	0.0	0.0	38.3
1973-74	0.0	0.0	0.0	0.0	4.1	10.7	11.5	2.4	4.2	0.8	0.0	0.0	33.7
1974-75	0.0	0.0	0.0	0.0	5.4	8.1	22.7	11.9	5.1	4.6	0.0	0.0	57.8
1975-76	0.0	0.0	0.0	0.0	0.3	0.6	1.8	7.1	4.2	0.0	0.0	0.0	20.0
1976-77	0.0	0.0	0.0	1.4	0.5	2.8	14.4	0.1	3.7	1.6	0.0	0.0	24.5
1977-78	0.0	0.0	0.0	0.0	T	4.0	1.4	17.0	6.5	T	0.0	0.0	28.9
1978-79	0.0	0.0	0.0	0.0	2.0	5.6	10.0	1.4	6.0	1.1	0.0	0.0	26.1
1979-80	0.0	0.0	0.0	2.0	T	1.3	5.0	3.0	9.2	T	0.0	0.0	20.5
1980-81	0.0	0.0	0.0	2.0	T	0.5	3.6	3.0	T	0.0	0.0	0.0	9.1
1981-82	0.0	0.0	0.0	T	1.0	7.0	2.4	3.0	8.0	2.9	0.0	0.0	24.3
1982-83	0.0	0.0	0.0	T	T	4.0	6.5	9.0	9.0	3.0	0.0	0.0	31.5
1983-84	0.0	0.0	0.0	0.0	6.9	13.5	3.3	2.0	14.2	T	0.0	0.0	39.9
1984-85	0.0	0.0	0.0	T	T	5.0	5.0	4.0	6.0	T	0.0	0.0	19.0
1985-86	0.0	0.0	T	0.0	5.0	4.5							
Record Mean	0.0	0.0	T	0.4	2.4	5.7	7.9	6.9	6.8	0.9	0.1	0.0	30.9

See Reference Notes, relative to all above tables, on preceding page.

Omaha, Nebraska, is situated on the west bank of the Missouri River. The river level at Omaha is normally about 965 feet above sea level and the rolling hills in and around Omaha rise to about 1,300 feet above sea level. The climate is typically continental with relatively warm summers and cold, dry winters. It is situated midway between two distinctive climatic zones, the humid east and the dry west. Fluctuations between these two zones produce weather conditions for periods that are characteristic of either zone, or combinations of both. Omaha is also affected by most low pressure systems that cross the country. This causes periodic and rapid changes in weather, especially during the winter months.

Most of the precipitation in Omaha falls during sharp showers or thunderstorms, and these occur mostly during the growing season from April to September. Of the total precipitation, about 75 percent falls during this six-month period. The rain occurs mostly as evening or nighttime showers and thunderstorms. Although winters are relatively cold, precipitation is light, with only 10 percent of the total annual precipitation falling during the winter months.

Sunshine is fairly abundant, ranging around 50 percent of the possible in the winter to 75 percent of the possible in the summer.

TABLE 1   NORMALS, MEANS AND EXTREMES

OMAHA (NORTH), NEBRASKA

LATITUDE: 41°22'N   LONGITUDE: 96°01'W   ELEVATION: FT. GRND 1309 BARO 01315   TIME ZONE: CENTRAL   WBAN: 94918

	(a)	JAN	FEB	MAR	APR	MAY	JUNE	JULY	AUG	SEP	OCT	NOV	DEC	YEAR
TEMPERATURE °F:														
Normals														
-Daily Maximum		27.8	34.5	44.8	61.4	72.4	81.6	85.7	83.7	75.0	64.6	47.6	34.5	59.5
-Daily Minimum		9.6	16.0	25.6	39.3	51.0	60.8	65.6	63.3	53.8	42.5	28.3	16.9	39.4
-Monthly		18.7	25.3	35.2	50.4	61.7	71.2	75.7	73.5	64.4	53.5	38.0	25.7	49.4
Extremes														
-Record Highest	31	66	76	87	91	100	101	107	106	103	93	79	66	107
-Year		1981	1972	1968	1981	1967	1969	1974	1983	1955	1975	1980	1976	JUL 1974
-Record Lowest	31	-22	-20	-16	7	25	41	44	45	28	16	-11	-24	-24
-Year		1982	1981	1960	1975	1967	1956	1971	1966	1984	1972	1964	1983	DEC 1983
NORMAL DEGREE DAYS:														
Heating (base 65°F)		1435	1112	924	438	158	25	6	6	94	366	810	1218	6592
Cooling (base 65°F)		0	0	0	0	56	211	338	270	76	13	0	0	964
% OF POSSIBLE SUNSHINE	49	54	53	54	57	61	67	74	70	67	65	51	47	60
MEAN SKY COVER (tenths)														
Sunrise - Sunset	10	6.2	6.4	6.4	6.3	6.3	5.3	5.1	5.1	4.7	5.7	6.2	6.4	5.8
MEAN NUMBER OF DAYS:														
Sunrise to Sunset														
-Clear	10	8.6	7.6	8.4	7.7	6.8	9.4	11.5	10.2	13.2	10.2	8.3	8.1	110.0
-Partly Cloudy	10	8.1	6.1	7.0	8.5	10.7	11.7	9.6	11.9	7.6	7.6	7.5	7.2	103.5
-Cloudy	10	14.3	14.6	15.6	13.8	13.5	8.9	9.9	8.9	9.2	13.2	14.2	15.7	151.8
Precipitation														
.01 inches or more	10	6.1	6.4	9.9	11.1	11.2	9.8	8.6	8.6	8.4	7.8	6.5	7.3	101.7
Snow, Ice pellets														
1.0 inches or more	10	2.5	1.9	2.1	0.9	0.0	0.0	0.0	0.0	0.0	0.3	1.1	2.5	11.3
Thunderstorms	10	0.1	0.6	1.6	3.7	7.5	9.6	8.4	8.5	6.0	2.6	0.8	0.3	49.7
Heavy Fog Visibility														
1/4 mile or less	10	1.1	2.6	2.8	0.5	0.8	0.7	0.1	0.9	0.7	1.2	2.6	2.2	16.2
Temperature °F														
-Maximum														
90° and above	31	0.0	0.0	0.0	0.2	0.8	4.3	9.3	7.0	2.2	0.1	0.0	0.0	23.9
32° and below	31	18.7	13.3	5.5	0.2	0.0	0.0	0.0	0.0	0.0	0.1	3.8	13.9	55.4
-Minimum														
32° and below	31	30.4	26.1	22.3	6.9	0.4	0.0	0.0	0.0	0.1	3.9	19.2	29.4	138.7
0° and below	31	8.6	3.9	0.4	0.0	0.0	0.0	0.0	0.0	0.0	0.0	0.3	3.6	16.8
AVG. STATION PRESS.(mb)	1	972.9	972.6	967.7	965.5	964.1	966.8	968.0	968.8	968.8	969.2	970.2	971.0	968.8
RELATIVE HUMIDITY (%)														
Hour 00	1	76	79	71	74	69	69	70	82	83	72	73	72	74
Hour 06 (Local Time)	1	78	81	79	81	79	79	79	89	86	79	76	76	80
Hour 12	1	68	68	57	60	55	57	57	65	68	63	64	68	63
Hour 18	1	69	68	51	58	52	54	55	67	71	61	66	69	62
PRECIPITATION (inches):														
Water Equivalent														
-Normal		0.70	0.95	2.00	2.74	4.26	4.21	3.50	4.19	3.36	2.11	1.16	0.76	29.94
-Maximum Monthly	31	1.85	2.86	5.27	7.12	9.09	8.16	9.77	11.77	14.10	4.87	5.11	4.45	14.10
-Year		1975	1965	1983	1984	1959	1984	1958	1960	1965	1971	1983	1984	SEP 1965
-Minimum Monthly	31	0.15	0.09	0.06	0.15	0.98	0.95	0.29	0.63	1.26	0.06	0.04	0.02	0.02
-Year		1976	1964	1956	1962	1955	1972	1975	1971	1971	1958	1976	1958	DEC 1958
-Maximum in 24 hrs	9	0.95	0.64	2.04	1.69	2.43	2.65	3.72	2.63	1.97	2.57	2.16	3.10	3.72
-Year		1982	1978	1982	1984	1979	1980	1977	1981	1977	1979	1983	1984	JUL 1977
Snow, Ice pellets														
-Maximum Monthly	31	21.5	23.2	23.3	10.3	0.7				0.3	5.2	13.9	19.3	23.3
-Year		1975	1965	1960	1983	1967				1985	1980	1957	1969	MAR 1960
-Maximum in 24 hrs	9	6.0	10.0	7.8	4.8					0.3	5.2	8.5	7.5	10.0
-Year		1979	1978	1983	1979	1977	1977	1977	1977	1985	1980	1983	1984	FEB 1978
WIND:														
Mean Speed (mph)	1	11.6	10.6	10.8	10.5	9.1	10.0	7.3	7.8	10.0	9.3	9.5	11.0	9.8
Prevailing Direction														
Fastest Mile														
-Direction (!!)	7	NW	NW	NW	NW	N	NW	NW	NW	NW	NW	NW	NW	NW
-Speed (MPH)	7	41	38	38	46	34	34	46	39	35	34	38	37	46
-Year		1978	1978	1982	1982	1983	1983	1980	1980	1980	1979	1982	1981	APR 1982
Peak Gust														
-Direction (!!)	1	NW	S	NW	S	SW	NW	N	NW	NW	NW	NW	NW	S
-Speed (mph)	1	49	37	47	55	43	40	39	35	41	48	45	46	55
-Date		1985	1985	1985	1985	1985	1985	1985	1985	1985	1985	1985	1985	APR 1985

See Reference Notes to this table on the following page.

TABLE 2 AVERAGE TEMPERATURE (deg. F)   OMAHA (NORTH), NEBRASKA

YEAR	JAN	FEB	MAR	APR	MAY	JUNE	JULY	AUG	SEP	OCT	NOV	DEC	ANNUAL
1956	19.0	23.6	35.6	46.0	63.2	76.0	73.7	74.6	64.9	60.2	37.3	30.3	50.4
1957	14.8	30.4	35.8	48.6	59.2	68.4	78.8	73.9	61.2	49.9	34.5	33.0	49.0
1958	24.2	18.3	33.1	49.2	63.5	67.5	70.7	73.4	65.9	55.6	40.1	24.3	48.8
1959	16.5	22.2	37.1	49.2	60.9	72.1	72.7	77.0	63.3	49.0	31.1	33.3	48.7
1960	18.2	16.9	21.0	51.6	60.4	68.0	73.7	73.5	66.1	54.8	39.8	26.7	47.6
1961	22.2	29.6	37.7	45.8	58.5	69.8	74.6	73.4	60.8	54.6	36.5	18.3	48.5
1962	16.2	22.9	29.7	49.6	68.5	70.1	73.4	73.1	61.6	55.7	40.8	26.3	49.0
1963	11.8	25.4	41.3	52.9	61.4	74.4	76.4	73.3	66.6	64.1	43.3	17.0	50.7
1964	28.0	29.2	32.3	51.1	66.1	69.8	77.7	68.9	63.4	52.7	39.7	22.3	50.1
1965	20.5	20.1	24.1	50.7	64.7	69.6	73.9	72.1	58.0	56.5	40.4	34.4	48.8
1966	15.5	25.8	41.4	45.4	59.8	70.3	77.6	70.1	61.8	53.2	37.7	25.6	48.7
1967	22.8	24.0	41.6	52.0	57.6	69.1	72.8	71.0	62.5	51.4	37.4	27.1	49.1
1968	21.3	25.4	43.8	51.2	56.9	72.5	74.8	73.5	63.7	54.4	36.1	21.7	49.6
1969	16.1	25.5	28.8	51.8	62.4	66.9	76.2	74.0	66.2	48.3	39.7	23.2	48.3
1970	13.7	29.8	33.2	51.1	66.1	72.2	74.7	74.6	64.5	50.4	36.2	27.0	49.5
1971	15.7	23.0	34.4	52.7	58.4	75.8	71.7	73.4	66.4	56.6	39.0	27.3	49.5
1972	18.7	23.8	39.1	49.6	61.6	71.4	73.0	72.2	64.4	49.1	35.7	20.2	48.2
1973	22.6	27.0	42.8	49.2	59.5	72.7	73.9	76.3	62.9	56.8	39.2	22.4	50.4
1974	18.4	29.7	41.2			69.6	81.8	69.5	59.9	53.9			
1975	21.7	20.5	29.8	46.9	64.9	70.7	77.4	77.5	61.3	57.7	40.0	28.3	49.7
1976	23.6	35.7	38.6	54.6	59.6	71.5	77.5	75.9	66.0	48.2	32.7	23.5	50.6
1977	11.7	31.4	43.3	57.2	68.0	72.9	78.7	70.7	65.9	52.0	37.9	23.9	51.1
1978	10.0	13.5	33.9	50.5	60.7	72.4	75.0	74.0	67.9	52.5	37.0	21.0	47.5
1979	7.9	13.6	35.0	47.6	59.8	70.8	74.3	74.1	67.9	53.3	36.9	31.9	47.8
1980	23.9	22.0	34.3	52.5	63.5	72.4	72.4	79.3	76.4	67.1	50.8	42.4	51.0
1981	27.3	30.5	42.9	58.4	60.0	73.7	76.9	71.5	65.7	50.9	42.2	24.5	52.1
1982	10.2	24.1	36.0	47.9	63.1	66.6	76.8	72.5	64.0	54.4	36.8	29.7	48.5
1983	25.8	31.0	37.6	43.9	57.5	70.6	79.5	81.8	66.0	53.2	38.7	7.8	49.6
1984	21.5	33.9	30.8	47.0	58.8	72.3	75.9	77.3	62.8	53.2	40.5	26.8	50.1
1985	18.6	24.1	43.6	55.6	65.2	68.4	75.8	70.8	62.9	53.8	28.9	17.5	48.8
Record Mean	18.8	24.9	35.9	48.9	59.7	70.8	75.8	73.8	64.3	53.5	37.8	25.0	49.1
Max	27.7	34.0	45.5	59.5	70.0	81.2	85.8	83.9	74.6	64.1	47.1	33.7	58.9
Min	9.8	15.7	26.3	38.2	49.4	60.4	65.8	63.6	54.0	42.9	28.5	16.2	39.2

TABLE 3 PRECIPITATION (inches)   OMAHA (NORTH), NEBRASKA

YEAR	JAN	FEB	MAR	APR	MAY	JUNE	JULY	AUG	SEP	OCT	NOV	DEC	ANNUAL
1956	0.61	0.33	0.06	1.23	2.11	2.74	5.92	3.72	1.90	0.96	1.62	0.19	21.39
1957	0.35	0.13	2.80	1.83	5.33	6.94	1.96	5.03	3.41	3.00	2.52	0.57	33.87
1958	1.02	1.02	0.91	3.75	2.21	2.86	9.77	3.88	4.05	0.06	0.66	0.02	30.21
1959	0.69	0.90	3.84	2.28	9.09	3.74	1.50	7.79	2.16	2.08	1.13	1.47	36.67
1960	1.57	1.59	2.26	2.02	6.42	2.81	2.81	11.77	3.81	1.30	0.25	0.54	41.81
1961	0.19	0.87	3.21	1.81	4.15	3.70	2.75	2.60	4.74	3.95	2.38	1.18	31.53
1962	0.36	1.89	1.39	0.51	3.98	3.02	7.10	5.59	4.08	1.77	0.71	0.64	31.04
1963	0.92	0.38	3.77	2.43	1.52	6.97	1.32	3.62	2.23	1.12	0.31	0.47	25.06
1964	0.30	0.34	1.50	5.42	5.43	6.30	3.90	3.71	3.25	0.36	0.84	0.79	32.14
1965	0.51	2.86	2.40	3.27	6.89	3.72	4.42	2.17	14.10	0.88	1.33	0.83	43.38
1966	0.78	0.42	0.87	0.89	5.04	6.12	3.41	3.94	1.82	0.79	0.13	0.53	24.74
1967	0.91	0.18	0.71	2.41	2.74	7.85	2.00	1.99	3.94	1.80	0.29	0.87	25.69
1968	0.42	0.09	0.82	4.43	4.64	3.55	3.19	3.62	4.33	4.40	1.35	1.78	32.62
1969	1.06	1.47	1.04	2.89	3.64	3.01	6.00	4.60	1.32	2.45	0.10	1.80	29.38
1970	0.25	0.15	0.89	2.43	3.31	2.41	4.25	3.43	3.52	4.09	1.03	0.13	25.89
1971	1.04	2.42	0.82	0.67	7.03	2.29	1.71	0.63	1.26	4.87	2.28	0.72	25.74
1972	0.40	0.29	1.28	4.65	5.37	0.95	5.63	2.78	4.54	3.61	3.22	1.63	34.35
1973	1.61	0.82	5.17	2.44	6.68	1.86	5.66	0.74	6.97	2.88	1.53	1.67	38.03
1974	0.64	0.20	0.70		2.57	5.68	0.10	4.98	1.89	3.17			
1975	1.85	1.06	1.88	3.15	3.98	3.36	0.29	1.96	2.30	0.08	2.73	0.94	23.58
1976	0.15	1.67	2.04	2.60	4.07	2.96	0.86	0.73	2.40	0.88	0.04	0.21	18.61
1977	0.75	0.17	3.84	2.26	5.59	2.19	7.28	7.13	5.33	4.66	1.70	0.62	41.52
1978	0.20	1.35	1.21	4.33	3.39	1.84	3.35	2.46	4.64	0.63	1.12	0.71	25.23
1979	1.22	0.38	4.13	2.70	2.99	3.55	2.22	2.54	2.31	4.46	1.58	0.22	28.30
1980	0.61	0.75	1.55	1.32	1.87	7.54	2.48	7.72	1.40	3.43	0.12	0.33	29.12
1981	0.30	0.17	0.93	1.69	3.90	1.99	4.45	7.92	1.49	1.74	3.61	0.60	28.79
1982	1.33	0.24	3.22	1.79	9.00	3.97	2.15	3.03	3.96	1.84	2.05	1.68	34.26
1983	1.14	1.03	5.27	2.02	4.82	4.93	1.15	1.03	2.84	1.91	5.11	0.69	31.94
1984	0.32	0.82	2.74	7.12	3.96	8.16	1.18	0.88	3.42	4.50	0.83	4.45	38.38
1985	0.35	0.73	1.62	2.44	4.16	2.52	2.71	1.55	2.84	1.94	0.65	0.34	21.85
Record Mean	0.72	0.82	2.06	2.57	4.41	4.16	3.44	3.71	3.53	2.26	1.33	0.87	29.90

REFERENCE NOTES FOR TABLES 1, 2, 3 and 6   (OMAHA [NORTH], NE)

GENERAL

T - TRACE AMOUNT
BLANK ENTRIES DENOTE MISSING/UNREPORTED DATA.
# INDICATES A STATION OR INSTRUMENT RELOCATION.

SPECIFIC

TABLE 1

(a) - LENGTH OF RECORD IN YEARS. ALTHOUGH INDIVIDUAL MONTHS MAY BE MISSING.
* LESS THAN .05

NORMALS — BASED ON THE 1951-1980 RECORD PERIOD.
EXTREMES — DATES ARE THE MOST RECENT OCCURRENCE.
WIND DIR. — NUMERALS SHOW TENS OF DEGREES CLOCKWISE FROM TRUE NORTH.
  "00" INDICATES CALM.
RESULTANT WIND DIRECTIONS ARE GIVEN TO WHOLE DEGREES.

EXCEPTIONS

TABLE 1

1. PERCENT OF POSSIBLE SUNSHINE IS FROM OMAHA (EPPLEY) THROUGH MAY 1977.

TABLES 2, 3, and 6

RECORD MEANS ARE THROUGH THE CURRENT YEAR,
BEGINNING IN   1954 FOR TEMPERATURE
  1954 FOR PRECIPITATION
  1954 FOR SNOWFALL

TABLE 4   HEATING DEGREE DAYS Base 65 deg. F   OMAHA (NORTH), NEBRASKA

SEASON	JULY	AUG	SEP	OCT	NOV	DEC	JAN	FEB	MAR	APR	MAY	JUNE	TOTAL
1956-57	0	11	105	184	824	1070	1553	963	898	491	209	19	6327
1957-58	0	0	136	460	909	984	1256	1303	983	473	94	39	6637
1958-59	2	15	75	304	740	1254	1494	1190	856	469	180	22	6601
1959-60	6	0	140	490	1011	978	1444	1388	1359	422	160	22	7420
1960-61	0	0	90	324	749	1179	1322	985	839	574	225	18	6305
1961-62	0	6	204	322	848	1442	1507	1173	1087	467	34	26	7116
1962-63	0	0	149	317	718	1194	1643	1103	726	364	158	4	6376
1963-64	0	11	54	98	644	1487	1145	1032	1010	416	84	28	6009
1964-65	0	31	132	379	753	1319	1373	1254	1262	433	95	5	7036
1965-66	0	9	230	270	731	944	1531	1092	723	582	222	25	6359
1966-67	0	18	141	375	813	1212	1303	1141	727	387	300	23	6440
1967-68	12	18	105	435	823	1170	1351	1144	657	414	264	24	6417
1968-69	2	3	90	345	858	1338	1514	1098	1114	388	162	63	6975
1969-70	0	3	38	532	750	1290	1585	982	979	440	92	13	6704
1970-71	4	0	131	464	860	1169	1519	1168	941	368	215	2	6841
1971-72	9	3	111	276	776	1162	1431	1188	795	461	171	21	6404
1972-73	7	12	116	491	871	1383	1308	1057	681	468	199	3	6596
1973-74	2	0	112	263	770	1315	1439	981	734				
1974-75	0	10	188	340			1338	1242	1084	541	90	19	
1975-76	1	0	167	261	744	1131	1275	841	811	314	190	12	5747
1976-77	0	0	87	538	962	1279	1646	935	664	262	16	1	6390
1977-78	0	3	41	394	803	1268	1702	1437	963	432	185	23	7251
1978-79	0	3	52	385	837	1358	1767	1435	921	514	199	24	7495
1979-80	4	15	48	354	837	1016	1267	1241	945	396	129	9	6261
1980-81	0	3	67	444	671	1157	1161	961	677	226	189	1	5557
1981-82	8	4	73	430	677	1251	1695	1143	894	512	94	50	6831
1982-83	0	12	123	333	844	1091	1206	946	844	625	244	26	6294
1983-84	0	0	94	376	782	1768	1344	896	1054	538	212	3	7067
1984-85	0	1	165	363	726	1177	1436	1139	656	315	65	35	6078
1985-86	0	7	209	342	1077	1467							

TABLE 5   COOLING DEGREE DAYS Base 65 deg. F   OMAHA (NORTH), NEBRASKA

YEAR	JAN	FEB	MAR	APR	MAY	JUNE	JULY	AUG	SEP	OCT	NOV	DEC	TOTAL
1977	0	0	0	35	115	246	436	186	76	0	2	0	1094
1978	0	0	5	2	58	252	317	290	202	2	0	0	1128
1979	0	0	0	1	47	206	301	303	142	2	0	0	1002
1980	0	0	0	25	87	237	451	362	137	11	0	0	1310
1981	0	0	0	34	38	269	384	209	101	0	0	0	1035
1982	0	0	0	4	43	104	375	250	99	12	0	0	887
1983	0	0	0	0	21	200	454	526	193	17	0	0	1411
1984	0	0	0	5	27	225	346	389	105	4	0	0	1101
1985	0	0	0	39	79	142	344	196	153	1	0	0	954

TABLE 6   SNOWFALL (inches)   OMAHA (NORTH), NEBRASKA

SEASON	JULY	AUG	SEP	OCT	NOV	DEC	JAN	FEB	MAR	APR	MAY	JUNE	TOTAL
1956-57	0.0	0.0	0.0	0.0	5.4	2.0	7.5	1.2	15.6	4.2	0.0	0.0	35.9
1957-58	0.0	0.0	0.0	T	13.9	4.6	7.5	2.7	9.0	0.0	0.0	0.0	37.7
1958-59	0.0	0.0	0.0	0.0	0.7	0.5	6.4	5.5	11.8	3.0	0.0	0.0	27.9
1959-60	0.0	0.0	0.0	0.0	1.4	4.2	15.8	16.9	23.3	0.3	0.0	0.0	61.9
1960-61	0.0	0.0	0.0	0.0	T	1.0	4.4	5.1	6.7	0.8	0.0	0.0	18.0
1961-62	0.0	0.0	T	0.0	4.2	15.6	4.5	12.6	8.0	0.6	0.0	0.0	45.5
1962-63	0.0	0.0	0.0	T	0.5	5.9	11.2	3.1	14.6	0.0	0.0	0.0	35.3
1963-64	0.0	0.0	0.0	0.0	0.0	5.9	4.7	2.4	9.0	T	0.0	0.0	22.0
1964-65	0.0	0.0	0.0	0.0	2.8	5.6	6.5	23.2	15.9	0.0	0.0	0.0	54.0
1965-66	0.0	0.0	0.0	0.0	0.2	1.2	3.5	0.4	8.4	T	T	0.0	13.7
1966-67	0.0	0.0	0.0	0.0	T	6.2	7.8	1.1	0.4	0.8	0.7	0.0	17.0
1967-68	0.0	0.0	0.0	0.0	T	6.2	4.9	0.7	T	T	T	0.0	11.8
1968-69	0.0	0.0	0.0	T	5.1	8.5	7.6	14.1	2.1	0.0	0.0	0.0	37.4
1969-70	0.0	0.0	0.0	T	0.2	19.3	2.9	1.6	5.1	T	T	0.0	29.1
1970-71	0.0	0.0	0.0	4.1	T	12.8	12.8	17.6	6.3	0.4	0.0	0.0	41.5
1971-72	0.0	0.0	0.0	T	7.8	3.3	3.6	4.3	3.0	1.2	0.0	0.0	23.2
1972-73	0.0	0.0	0.0	T	8.8	7.2	16.5	6.3	0.0	2.9	0.0	0.0	41.7
1973-74	0.0	0.0	0.0	0.0	3.6	9.0	10.6	2.9	4.6	0.0	0.0	0.0	
1974-75	0.0	0.0	0.0	0.0			21.5	12.2	5.0	3.7	0.0	0.0	
1975-76	0.0	0.0	0.0	0.0		8.0	0.6	2.6	7.6	5.0	0.0	0.0	23.8
1976-77	0.0	0.0	0.0	1.4	0.7	2.8	13.7	T	3.3	3.0	0.0	0.0	24.9
1977-78	0.0	0.0	0.0	0.0	3.0	7.4	4.0	22.0	8.6	T	0.0	0.0	45.0
1978-79	0.0	0.0	0.0	0.0	5.3	8.8	13.6	3.9	6.6	4.8	0.0	0.0	43.0
1979-80	0.0	0.0	0.0	1.9	0.2	1.5	7.2	9.8	8.0	0.2	0.0	0.0	28.8
1980-81	0.0	0.0	0.0	5.2	0.2	0.8	4.3	3.6	0.6	0.0	0.0	0.0	14.7
1981-82	0.0	0.0	0.0	0.9	4.7	5.9	5.9	3.5	5.2	3.9	0.0	0.0	31.9
1982-83	0.0	0.0	0.0	0.8	0.4	5.1	8.9	11.4	14.1	10.3	0.0	0.0	51.0
1983-84	0.0	0.0	0.0	0.0	13.6	15.9	3.1	3.4	18.3	1.5	0.0	0.0	55.8
1984-85	0.0	0.0	0.0	T	1.6	9.7	4.6	3.3	8.7	T	0.0	0.0	27.9
1985-86	0.0	0.0	0.3	0.0	6.9	5.5							
Record Mean	0.0	0.0	T	0.4	3.2	5.6	7.8	6.9	7.6	1.4	T	0.0	33.0

See Reference Notes, relative to all above tables, on preceding page.

Scottsbluff is located in the North Platte river valley that extends from central Wyoming southeast across western Nebraska. The valley is approximately 20 miles wide in the vicinity of Scottsbluff with a range of hills both to the north and south, parallel to the river. To the south the hills average 600 to 700 feet above the river with some projections upward to 1,000 feet. To the north, rolling hills range from 300 to 400 feet higher than the river.

Due to the protection of the higher hills to the south, southerly winds in the valley are rare. Prevailing winds are west to northwest during the winter months and east to southeast during the summer months. West to northwest winds are intensified by the funneling action of the valley and velocities of 30 to 50 mph are common during the winter and early spring. Quite often these winds are warmed by the downslope (chinook) effect from the higher elevations to the west and bring rapid warming and melting of the snow. Outbreaks of Arctic air bring cold wave conditions about five times each season. Snow with strong winds causing blowing and drifting snow occur several times each winter with a severe blizzard of extended duration occurring about once every thirty years. Easterly winds during the winter and early spring cause upslope conditions with low cloudiness and precipitation.

The average temperature is in the upper 40s. Summertime highs generally range from the 80s to the 90s with lows around 60. Summer temperatures of 100 degrees are reached or exceeded at least once each summer. In winter, highs average about 40 degrees with lows in the teens. Temperatures of zero or below occur about 15 times each winter.

Most of the precipitation occurs as thunderstorms during the spring and summer months. Severe thunderstorms with destructive hail are quite common during the late spring and summer. Tornadoes are infrequent and usually of short duration.

The Platte River in the vicinity of Scottsbluff is a wide shallow stream and has very little effect on the climate. Water stored in numerous upstream reservoirs is used for extensive irrigation in the valley. Lowland flooding occurs when heavy rains fall upstream and a greater than normal amount of water is being released from the upstream reservoirs.

Based on the 1951–1980 period, the average first occurrence of 32 degrees Fahrenheit in the fall is September 29 and the average last occurrence in the spring is May 7.

TABLE 1   NORMALS, MEANS AND EXTREMES

SCOTTSBLUFF, NEBRASKA

LATITUDE: 41°52'N   LONGITUDE: 103°36'W   ELEVATION: FT. GRND 3957 BARO 03950   TIME ZONE: MOUNTAIN   WBAN: 24028

	(a)	JAN	FEB	MAR	APR	MAY	JUNE	JULY	AUG	SEP	OCT	NOV	DEC	YEAR
TEMPERATURE °F:														
Normals														
-Daily Maximum		37.2	43.4	48.8	60.3	70.8	81.7	89.2	86.7	77.5	66.0	49.8	40.8	62.7
-Daily Minimum		11.2	16.7	22.3	32.5	43.6	53.3	59.2	56.5	45.6	34.3	22.0	14.8	34.3
-Monthly		24.2	30.1	35.5	46.4	57.2	67.5	74.2	71.6	61.5	50.2	35.9	27.8	48.5
Extremes														
-Record Highest	43	74	77	87	89	95	105	109	104	101	92	78	77	109
-Year		1982	1962	1943	1981	1969	1954	1973	1980	1948	1967	1980	1980	JUL 1973
-Record Lowest	43	-32	-28	-27	-8	15	30	40	39	19	9	-13	-27	-32
-Year		1963	1962	1948	1975	1983	1969	1959	1944	1985	1969	1947	1983	JAN 1963
NORMAL DEGREE DAYS:														
Heating (base 65°F)		1265	977	911	558	254	69	6	5	172	459	873	1153	6702
Cooling (base 65°F)		0	0	0	0	13	144	291	210	70	0	0	0	728
% OF POSSIBLE SUNSHINE														
MEAN SKY COVER (tenths)														
Sunrise – Sunset	35	6.2	6.2	6.6	6.4	6.4	5.2	4.3	4.4	4.3	4.8	6.0	6.0	5.6
MEAN NUMBER OF DAYS:														
Sunrise to Sunset														
-Clear	42	8.4	7.5	6.6	6.9	6.2	10.1	13.7	13.5	14.5	13.3	8.6	8.6	117.7
-Partly Cloudy	42	8.2	8.0	9.5	8.8	10.6	11.1	11.7	11.2	7.5	7.7	8.0	8.6	110.8
-Cloudy	42	14.4	12.8	14.9	14.4	14.2	8.8	5.6	6.4	8.0	10.0	13.4	13.8	136.8
Precipitation														
.01 inches or more	42	5.6	4.8	7.5	8.6	11.6	11.2	8.6	6.6	6.7	4.7	4.6	5.3	85.9
Snow, Ice pellets														
1.0 inches or more	35	2.1	1.7	2.8	1.6	0.4	0.0	0.0	0.0	0.1	0.7	1.7	2.0	13.0
Thunderstorms	41	0.0	0.0	0.3	1.7	7.2	11.3	10.7	8.0	4.0	0.8	0.1	0.*	44.0
Heavy Fog Visibility														
1/4 mile or less	41	0.9	0.8	1.5	0.6	0.7	0.6	0.4	0.5	0.8	0.9	1.2	1.1	10.0
Temperature °F														
-Maximum														
90° and above	21	0.0	0.0	0.0	0.0	1.1	7.2	16.8	13.4	4.4	0.1	0.0	0.0	43.0
32° and below	21	9.9	5.9	3.3	0.3	0.0	0.0	0.0	0.0	0.*	0.3	3.8	8.4	31.9
-Minimum														
32° and below	21	29.5	27.0	26.7	12.7	1.8	0.1	0.0	0.0	1.9	12.4	26.4	30.0	168.6
0° and below	21	7.0	3.2	0.8	0.1	0.0	0.0	0.0	0.0	0.0	0.0	1.0	4.3	16.4
AVG. STATION PRESS.(mb)	12	879.0	878.5	875.4	876.8	877.1	878.5	880.6	880.6	880.9	880.5	879.1	878.9	878.8
RELATIVE HUMIDITY (%)														
Hour 05	20	73	73	74	75	77	79	79	80	78	74	75	74	76
Hour 11	21	58	51	49	45	45	43	42	43	42	42	52	56	47
Hour 17 (Local Time)	21	58	47	43	40	41	39	37	38	36	39	52	58	44
Hour 23	20	71	69	67	65	67	67	66	67	66	65	69	71	68
PRECIPITATION (inches):														
Water Equivalent														
-Normal		0.44	0.37	0.97	1.43	2.66	2.93	1.96	0.97	1.08	0.75	0.52	0.51	14.59
-Maximum Monthly	43	1.26	1.17	2.62	3.89	6.28	8.33	4.82	2.51	4.22	3.02	1.75	1.54	8.33
-Year		1978	1978	1961	1984	1962	1947	1978	1979	1973	1969	1983	1978	JUN 1947
-Minimum Monthly	43	T	T	T	0.17	0.29	0.27	0.59	0.07	T	0.04	T	0.09	T
-Year		1961	1954	1966	1962	1966	1950	1946	1973	1953	1956	1943	1949	JAN 1961
-Maximum in 24 hrs	43	0.92	0.53	1.68	1.48	2.33	3.74	2.53	1.37	3.28	1.32	1.16	0.90	3.74
-Year		1976	1978	1974	1943	1955	1953	1948	1974	1951	1948	1979	1975	JUN 1953
Snow, Ice pellets														
-Maximum Monthly	43	23.7	15.3	23.5	18.0	7.5	0.1			5.1	21.6	18.5	18.1	23.7
-Year		1949	1978	1980	1957	1967	1951			1985	1969	1983	1985	JAN 1949
-Maximum in 24 hrs	43	11.2	6.6	15.0	9.9	3.7	0.1			4.8	8.5	8.4	10.9	15.0
-Year		1976	1978	1974	1970	1979	1951			1985	1969	1979	1975	MAR 1974
WIND:														
Mean Speed (mph)	34	10.6	11.2	12.3	12.7	11.7	10.4	9.2	9.0	9.4	9.6	10.1	10.5	10.6
Prevailing Direction through 1963		WNW	WNW	WNW	NW	ESE	ESE	ESE	ESE	ESE	NW	NW	WNW	ESE
Fastest Obs. 1 Min.														
-Direction (!!!)	35	34	29	29	32	32	29	05	35	32	29	29	32	29
-Speed (MPH)	35	53	60	62	54	80	80	52	52	46	44	56	47	80
-Year		1975	1953	1963	1964	1951	1954	1967	1980	1959	1966	1970	1954	JUN 1954
Peak Gust														
-Direction (!!!)														
-Speed (mph)														
-Date														

See Reference Notes to this table on the following page.

TABLE 2   PRECIPITATION (inches)   SCOTTSBLUFF, NEBRASKA

YEAR	JAN	FEB	MAR	APR	MAY	JUNE	JULY	AUG	SEP	OCT	NOV	DEC	ANNUAL
1956	0.27	0.25	0.29	1.54	2.75	0.29	2.08	0.29	0.05	0.04	1.02	0.14	10.02
1957	0.19	0.05	0.79	2.33	6.25	4.23	4.54	1.17	0.30	1.49	0.39	0.18	21.91
1958	0.01	0.72	1.18	1.30	1.69	5.07	2.56	0.27	0.38	0.07	0.39	1.09	14.73
1959	0.26	0.31	1.33	0.74	2.52	4.14	0.56	0.21	2.77	1.14	0.12	0.24	14.34
1960	0.26	0.50	0.35	0.96	2.04	1.21	0.55	1.20	0.65	0.63	0.53	0.52	9.40
1961	T	0.32	2.62	0.92	4.02	0.70	2.55	0.23	1.90	0.24	0.50	0.18	14.18
1962	0.37	0.70	0.34	0.29	6.28	4.23	3.65	0.28	0.05	1.12	0.20	0.40	17.91
1963	1.14	0.14	0.66	1.02	3.13	3.56	1.60	0.99	1.36	1.07	0.11	0.31	15.09
1964	0.04	0.25	0.38	2.40	1.45	1.47	0.69	0.19	0.09	0.08	0.01	0.65	7.70
1965	0.46	0.44	0.20	0.71	3.30	6.53	2.06	0.52	3.15	0.98	0.09	0.65	19.09
1966	0.46	0.21	0.17	1.11	0.27	2.09	3.34	1.86	1.59	0.81	0.20	0.17	12.28
1967	0.18	0.04	0.29	2.42	4.42	4.22	2.29	0.99	0.53	0.17	0.17	0.82	16.54
1968	0.15	0.23	0.64	1.76	2.69	2.29	1.17	2.08	0.25	0.86	0.19	0.55	12.86
1969	0.72	0.52	0.51	1.08	2.44	2.44	0.72	1.26	0.79	3.02	0.40	0.32	13.79
1970	0.57	0.18	0.93	2.26	1.38	2.49	1.10	0.40	0.34	1.57	0.38	0.39	11.99
1971	0.23	0.44	1.31	1.71	4.73	2.47	1.46	1.47	2.09	0.69	0.33	0.13	17.06
1972	0.35	0.25	0.69	2.91	1.54	4.43	3.77	1.71	1.64	0.79	1.75	0.77	20.60
1973	0.58	0.43	1.97	2.24	0.80	0.88	3.69	0.09	4.22	0.74	1.36	1.24	18.24
1974	0.57	0.10	1.99	0.35	0.79	0.98	0.66	1.58	0.94	0.48	0.42	0.18	9.04
1975	0.32	0.46	1.73	1.78	2.25	1.47	1.60	0.47	0.26	0.74	0.45	1.18	12.71
1976	0.96	0.29	0.36	2.48	2.27	1.31	0.25	0.78	0.22	0.35	0.35	0.10	9.72
1977	0.52	0.02	2.04	2.12	2.07	4.06	1.22	0.63	0.09	0.74	0.62	1.54	14.93
1978	1.26	1.17	0.68	1.24	4.37	2.41	4.82	1.25	0.74	1.66	1.60	0.49	20.19
1979	0.74	0.10	1.22	0.90	1.33	2.59	3.17	2.51	0.74	0.76	0.57	0.15	17.05
1980	1.21	0.99	2.16	0.57	2.82	0.79	1.07	0.47	0.47	0.47			12.03
1981	0.69	0.14	0.59	1.47	2.75	2.54	3.54	1.10	0.39	0.34	0.26	0.19	14.00
1982	0.32	0.20	0.46	0.50	2.93	6.63	4.78	1.66	1.78	1.22	0.80	0.57	21.85
1983	0.29	0.04	1.94	2.33	4.20	1.81	0.69	1.23	0.13	0.68	1.75	0.60	15.69
1984	0.44	0.50	1.47	3.89	1.23	1.23	1.80	0.57	0.45	0.88	0.28	0.50	13.24
1985	0.64	0.20	0.37	1.23	0.86	1.76	0.80	0.18	2.71	1.01	1.28	1.17	12.21
Record Mean	0.39	0.44	0.90	1.78	2.70	2.67	1.87	1.23	1.25	0.88	0.50	0.51	15.12

TABLE 3   AVERAGE TEMPERATURE (deg. F)   SCOTTSBLUFF, NEBRASKA

YEAR	JAN	FEB	MAR	APR	MAY	JUNE	JULY	AUG	SEP	OCT	NOV	DEC	ANNUAL
1956	26.8	25.6	36.5	41.6	59.3	73.4	72.4	69.4	63.2	52.7	32.9	33.3	48.9
1957	19.0	34.7	36.1	41.7	54.2	65.4	74.6	72.0	58.9	49.9	33.7	35.1	48.0
1958	29.4	30.0	29.8	43.3	62.7	66.6	69.6	73.4	63.3	51.9	37.5	25.2	48.6
1959	24.2	25.6	34.8	44.9	55.4	71.3	72.5	73.3	58.2	45.9	31.5	32.8	47.5
1960	24.0	24.2	33.9	48.3	57.2	67.0	74.5	71.1	62.7	50.6	37.0	22.6	47.8
1961	27.4	32.6	38.4	42.7	54.7	73.3	73.8	75.6	56.1	48.6	32.8	25.6	47.9
1962	21.0	28.3	33.1	50.2	60.1	65.4	71.0	70.4	61.4	52.4	40.8	30.5	48.7
1963	14.8	35.5	37.9	48.5	60.2	69.8	76.9	73.1	67.0	57.4	39.9	23.8	50.4
#1964	28.7	26.2	30.7	44.9	58.8	66.4	77.9	68.9	59.6	50.0	37.0	25.1	47.9
1965	32.2	26.9	26.1	51.1	57.4	64.4	73.5	69.6	53.1	43.4	41.5	30.9	48.3
1966	19.6	22.4	40.6	42.5	59.8	67.4	77.6	68.9	63.0	48.7	38.1	27.3	48.0
1967	31.0	32.0	41.3	48.7	51.6	63.1	72.7	70.7	61.9	51.3	34.7	19.7	48.2
1968	24.1	33.1	41.0	43.2	52.4	67.3	73.4	68.7	59.9	50.8	35.5	22.7	47.7
1969	23.6	32.4	30.8	50.8	59.2	61.4	74.9	75.1	66.7	41.1	38.3	30.8	48.8
1970	24.5	33.0	30.4	41.4	56.9	64.5	72.9	73.2	57.3	42.6	36.0	25.7	46.5
1971	26.8	28.6	34.7	48.2	54.5	69.3	70.2	72.9	58.4	47.5	37.9	29.4	48.2
1972	24.9	33.4	41.7	46.0	56.5	68.2	70.0	71.0	61.1	48.8	30.1	20.4	47.7
1973	24.6	29.8	38.1	41.9	54.9	67.5	71.6	72.9	57.2	50.7	35.3	25.6	47.5
1974	19.6	33.9	38.9	47.4	58.1	69.5	76.5	68.3	56.4	52.1	36.3	28.1	48.8
1975	26.6	25.3	33.1	44.5	56.5	65.6	77.2	72.2	59.7	50.9	35.6	31.9	48.3
1976	25.1	37.2	36.6	49.1	58.2	68.1	76.4	72.2	63.1	47.3	34.7	32.2	50.0
1977	20.8	36.6	38.1	53.0	63.5	73.1	76.3	70.2	65.4	51.6	38.1	28.3	51.2
1978	18.3	22.0	40.5	49.6	56.6	69.0	74.9	70.5	66.2	51.6	33.8	18.9	47.7
1979	10.4	28.1	41.0	50.1	56.1	68.6	75.0	71.7	67.0	53.1	32.0	34.3	49.0
1980	23.1	29.7	36.1	49.4	58.6	72.3	78.2	73.1	66.3	51.6	40.8	37.0	51.3
1981	32.5	32.8	42.7	56.4	57.4	71.3	76.2	72.8	66.9	50.9	42.7	31.2	52.8
1982	22.5	27.7	37.6	43.9	55.6	62.3	72.8	74.2	62.1	48.5	34.2	27.9	47.5
1983	32.6	36.5	36.8	40.7	51.1	63.9	74.2	76.7	63.9	51.5	34.2	12.4	47.9
1984	27.3	35.5	37.4	42.2	58.8	64.4	74.8	76.4	59.9	46.2	37.6	24.8	49.0
1985	20.8	25.6	39.5	50.7	61.7	66.3	75.2	71.6	57.4	47.5	22.0	20.2	46.5
Record Mean	25.4	29.0	36.2	46.8	56.8	66.8	73.7	71.7	61.6	49.8	36.5	27.7	48.5
Max	39.0	43.0	50.3	61.5	71.2	81.6	89.3	87.6	78.2	66.2	51.1	41.0	63.3
Min	11.7	15.0	22.1	32.1	42.3	52.0	58.0	55.8	45.1	33.4	21.8	14.4	33.6

REFERENCE NOTES FOR TABLES 1, 2, 3 and 6   (SCOTTSBLUFF, NE)

GENERAL

T - TRACE AMOUNT
BLANK ENTRIES DENOTE MISSING/UNREPORTED DATA.
# INDICATES A STATION OR INSTRUMENT RELOCATION.

SPECIFIC

TABLE 1

(a) - LENGTH OF RECORD IN YEARS. ALTHOUGH
     INDIVIDUAL MONTHS MAY BE MISSING.
*   LESS THAN .05

NORMALS — BASED ON THE 1951-1980 RECORD PERIOD.
EXTREMES — DATES ARE THE MOST RECENT OCCURRENCE.
WIND DIR. — NUMERALS SHOW TENS OF DEGREES
           CLOCKWISE FROM TRUE NORTH.
           "00" INDICATES CALM.
RESULTANT WIND DIRECTIONS ARE GIVEN TO WHOLE DEGREES.

EXCEPTIONS

TABLES 2, 3, and 6

RECORD MEANS ARE THROUGH THE CURRENT YEAR,
BEGINNING IN   1889 FOR TEMPERATURE
               1889 FOR PRECIPITATION
               1944 FOR SNOWFALL

TABLE 4 HEATING DEGREE DAYS Base 65 deg. F SCOTTSBLUFF, NEBRASKA

SEASON	JULY	AUG	SEP	OCT	NOV	DEC	JAN	FEB	MAR	APR	MAY	JUNE	TOTAL
1956-57	0	28	99	376	955	975	1421	843	890	690	332	72	6681
1957-58	0	2	191	469	935	920	1093	973	1083	647	124	44	6481
1958-59	5	1	126	398	821	1228	1258	1097	928	597	302	36	6797
1959-60	5	8	250	585	998	991	1265	1174	958	493	253	40	7020
1960-61	4	17	145	436	835	1305	1155	901	819	661	329	49	6656
1961-62	1	0	291	501	961	1211	1361	1021	981	439	177	73	7017
1962-63	3	32	133	383	719	1062	1552	819	833	487	164	13	6200
1963-64	0	0	46	236	746	1269	1116	1120	1055	594	234	61	6477
#1964-65	0	52	185	458	831	1232	1010	1060	1198	414	245	56	6741
1965-66	0	12	358	353	698	1050	1402	1187	748	667	215	74	6764
1966-67	0	31	106	500	801	1165	1049	917	729	484	428	92	6302
1967-68	0	24	125	431	904	1396	1265	917	738	648	382	39	6869
1968-69	11	35	171	434	878	1306	1278	906	1052	419	219	149	6858
1969-70	0	0	30	737	793	1054	1247	891	1065	703	252	109	6881
1970-71	0	0	261	689	861	1212	1177	1013	937	497	321	16	6984
1971-72	13	0	243	537	807	1096	1237	908	714	561	273	19	6408
1972-73	41	14	146	495	1041	1379	1245	980	825	684	312	57	7219
1973-74	18	0	249	438	883	1216	1407	863	801	523	230	53	6681
1974-75	0	28	265	394	854	1135	1183	1105	981	609	268	75	6897
1975-76	0	11	193	434	873	1019	1235	799	874	469	222	46	6175
1976-77	0	8	120	542	903	1011	1362	789	826	360	84	4	6009
1977-78	0	23	72	407	800	1131	1199	752	737	458	286	54	6626
1978-79	0	2	99	412	930	1425	1690	1029	737	447	288	50	7129
1979-80	0	0	55	363	984	946	1295	1019	889	463	215	8	6247
1980-81	0	0	61	411	721	859	998	897	684	261	253	16	5162
1981-82	6	0	42	432	662	1044	1307	1039	842	629	289	101	6393
1982-83	4	1	155	503	921	1142	997	792	864	722	429	112	6642
1983-84	7	0	126	412	919	1627	1165	851	850	677	219	60	6913
1984-85	0	0	223	574	812	1238	1367	1096	780	422	139	75	6726
1985-86	0	16	286	534	1284	1378							

TABLE 5 COOLING DEGREE DAYS Base 65 deg. F SCOTTSBLUFF, NEBRASKA

YEAR	JAN	FEB	MAR	APR	MAY	JUNE	JULY	AUG	SEP	OCT	NOV	DEC	TOTAL
1969	0	0	0	0	46	48	315	319	88	0	0	0	816
1970	0	0	0	0	7	99	254	263	37	0	0	0	660
1971	0	0	0	0	1	150	179	254	50	0	0	0	634
1972	0	0	0	0	16	121	203	207	36	0	0	0	583
1973	0	0	0	0	4	138	228	250	21	0	0	0	641
1974	0	0	0	0	24	192	363	135	15	0	0	0	729
1975	0	0	0	1	12	101	386	245	43	4	0	0	792
1976	0	0	0	0	18	146	359	240	66	1	0	0	830
1977	0	0	0	6	45	254	356	194	90	0	0	0	945
1978	0	0	0	0	34	180	316	198	141	0	0	0	869
1979	0	0	0	8	19	163	317	224	121	1	0	0	853
1980	0	0	0	0	22	234	417	263	104	3	0	0	1044
1981	0	0	0	10	23	213	361	250	107	3	0	0	967
1982	0	0	0	0	3	40	251	291	74	0	0	0	659
1983	0	0	0	0	6	82	297	369	99	0	0	0	853
1984	0	0	0	0	35	124	311	359	78	0	0	0	907
1985	0	0	0	3	42	121	324	225	66	0	0	0	781

TABLE 6 SNOWFALL (inches) SCOTTSBLUFF, NEBRASKA

SEASON	JULY	AUG	SEP	OCT	NOV	DEC	JAN	FEB	MAR	APR	MAY	JUNE	TOTAL
1956-57	0.0	0.0	0.0	T	15.2	0.9	4.9	1.3	8.6	18.0	T	0.0	48.9
1957-58	0.0	0.0	0.0	2.5	4.4	2.3	0.1	11.9	17.4	3.4	0.0	0.0	42.0
1958-59	0.0	0.0	0.0	T	3.9	11.1	6.1	6.8	17.5	6.7	1.3	0.0	53.4
1959-60	0.0	0.0	T	8.6	0.8	2.8	3.8	4.8	5.5	3.6	T	0.0	29.9
1960-61	0.0	0.0	0.0	0.6	6.8	7.0	0.1	3.4	13.4	7.0	1.0	0.0	39.3
1961-62	0.0	0.0	T	1.8	5.8	3.2	4.7	9.2	2.9	T	0.0	0.0	27.6
1962-63	0.0	0.0	T	0.0	0.6	5.2	14.7	1.8	7.2	1.2	0.0	0.0	30.7
1963-64	0.0	0.0	0.0	0.0	T	4.8	0.8	4.1	5.6	7.8	0.0	0.0	23.1
1964-65	0.0	0.0	0.0	0.0	0.1	8.3	4.8	6.6	3.6	2.0	1.0	0.0	26.4
1965-66	0.0	0.0	0.8	0.0	1.0	4.2	5.0	2.7	2.3	5.6	0.2	0.0	21.8
1966-67	0.0	0.0	0.0	T	2.3	3.4	3.0	0.6	2.1	3.3	7.5	0.0	22.2
1967-68	0.0	0.0	0.0	0.0	1.6	12.9	2.1	2.1	6.7	8.1	T	0.0	33.5
1968-69	0.0	0.0	0.0	0.0	1.1	6.1	8.5	5.4	6.1	0.3	0.0	T	27.5
1969-70	0.0	0.0	0.0	21.6	1.0	2.9	6.6	2.0	10.2	17.6	T	0.0	61.9
1970-71	0.0	0.0	T	11.2	0.4	4.4	2.2	4.4	14.2	3.0	T	0.0	39.8
1971-72	0.0	0.0	T	1.0	3.4	1.3	4.1	2.3	1.0	6.2	0.0	0.0	19.3
1972-73	0.0	0.0	0.0	5.3	14.0	10.2	6.9	5.9	4.8	5.2	T	0.0	52.3
1973-74	0.0	0.0	T	4.6	12.2	11.8	9.9	1.1	17.9	1.7	0.0	0.0	59.2
1974-75	0.0	0.0	3.8	T	T	2.7	5.9	5.8	17.7	10.7	T	0.0	46.6
1975-76	0.0	0.0	0.0	6.1	4.7	12.0	12.0	5.5	5.9	0.9	0.0	T	50.6
1976-77	0.0	0.0	0.0	0.1	5.4	1.3	8.1	0.1	10.9	1.0	0.0	0.0	26.9
1977-78	0.0	0.0	0.0	0.4	6.3	11.8	14.9	15.3	1.6	3.4	0.0	0.0	54.5
1978-79	0.0	0.0	0.0	2.9	7.2	17.5	10.5	1.6	7.7	4.3	6.4	0.0	58.1
1979-80	0.0	0.0	0.0	3.8	13.5	5.3	17.1	10.6	23.5	3.9	0.8	0.0	78.5
1980-81	0.0	0.0	0.0	1.2	3.7	1.7	8.3	2.1	3.2	1.3	0.0	0.0	21.5
1981-82	0.0	0.0	0.0	0.6	0.2	1.8	4.4	3.8	2.8	2.1	T	0.0	15.7
1982-83	0.0	0.0	0.0	5.3	4.3	4.9	2.8	T	16.4	8.9	2.6	0.0	45.2
1983-84	0.0	0.0	T	0.0	18.5	7.9	4.8	2.7	13.5	13.3	1.2	0.0	61.9
1984-85	0.0	0.0	1.6	2.2	2.5	7.1	7.2	2.3	3.2	1.7	0.0	0.0	27.8
1985-86	0.0	0.0	5.1	0.3	17.5	18.1							
Record Mean	0.0	0.0	0.4	2.3	5.2	6.4	6.6	4.6	8.8	4.7	1.0	T	39.9

See Reference Notes, relative to all above tables, on preceding page.

Valentine, located near the northern edge of the Sandhills and cattle country of Nebraska, is near the extreme northern border of the state. The city lies in the valley of the Niobrara River, a branch of the Missouri River, about 160 miles above the junction with the Missouri. It is the county seat of Cherry County and had its beginning in the fall of 1882. The name, Valentine, was selected for the new town in honor of Congressman E. K. Valentine, who represented the Third Congressional District of which Cherry County was a part.

The inland location offers a wide variety of weather. The high afternoon temperatures during the two warmest months, July and August, average nearly 90 degrees and the corresponding humidity averages about 40 percent. Uncomfortably warm nights are few with low morning temperatures averaging about 60 degrees. The temperature seldom reaches 100 degrees or more during the summer. The two coldest months are January and February. The minimum temperature generally reaches -20 degrees or colder once each winter.

Valentines location frequently places it in the path of cold Canadian air mass outbreaks during the cold season, alternating with mild, dry air moving across the Rockies from the Pacific. One or two bitterly cold days usually occur each winter when the temperature will stay below zero throughout the day. Blizzards are not frequent, but at least one is likely each winter season. An occasional severe blizzard occurs about once in every three or four winters. Blowing and drifting

snow reduce visibility to zero and bring outdoor activities and travel to a complete stop. Lives may be lost for anyone caught away from shelter, and there is usually loss of livestock which varies with the intensity and duration of the storm. Temperatures below 32 degrees have occurred as late as mid-June and as early as early September, and low temperature records in the 30s have occurred during the summer with light frost in low places. However, these are rare occurrences and temperatures below 50 degrees are not common in July and August.

About 65 percent of the annual precipitation falls during the growing season, May through September, and is predominantly the nighttime thunderstorm type with June being the wettest month.

The spring and fall seasons have mostly pleasant days, with the fall season having the most uniform character with lighter winds and gradually falling temperatures as the season progresses. In the spring the weather is windy and extremely variable, with summer-like days mixed with some of cold of winter. The widest extremes of temperature occur in March.

Some of the damaging weather elements other than blizzards, are high winds, which dig blow outs in the sand hills, and an occasional hailstorm. Damage from hail is not extensive and confined mostly to buildings and gardens since the area is mostly prairie. Floods are unknown. Tornadoes occasionally occur but seldom do much damage because ranches and towns are widely scattered.

TABLE 1 NORMALS, MEANS AND EXTREMES

VALENTINE, NEBRASKA

LATITUDE: 42°52'N   LONGITUDE: 100°33'W   ELEVATION: FT. GRND 2587 BARO 02590   TIME ZONE: CENTRAL   WBAN: 24032

	(a)	JAN	FEB	MAR	APR	MAY	JUNE	JULY	AUG	SEP	OCT	NOV	DEC	YEAR
TEMPERATURE °F:														
Normals														
—Daily Maximum		31.6	36.9	44.2	59.0	70.6	81.3	88.7	86.7	76.4	64.7	47.3	36.5	60.3
—Daily Minimum		5.8	11.4	19.4	32.6	44.0	54.4	60.3	57.8	46.5	34.3	20.7	11.4	33.2
—Monthly		18.7	24.1	31.8	45.8	57.3	67.8	74.5	72.3	61.5	49.5	34.0	24.0	46.8
Extremes														
—Record Highest	31	70	78	85	96	98	107	110	108	103	95	82	70	110
—Year		1974	1982	1978	1980	1969	1974	1980	1965	1983	1963	1965	1979	JUL 1980
—Record Lowest	31	-29	-28	-29	3	19	30	38	37	17	10	-22	-37	-37
—Year		1963	1975	1980	1975	1967	1969	1971	1965	1984	1984	1959	1983	DEC 1983
NORMAL DEGREE DAYS:														
Heating (base 65°F)		1435	1142	1029	576	253	62	7	10	161	486	930	1271	7362
Cooling (base 65°F)		0	0	0	0	14	149	301	236	56	5	0	0	761
% OF POSSIBLE SUNSHINE	29	63	62	59	59	62	69	76	76	71	68	61	60	66
MEAN SKY COVER (tenths)														
Sunrise - Sunset	30	6.3	6.4	6.6	6.4	6.1	5.2	4.4	4.3	4.5	4.9	5.9	6.0	5.6
MEAN NUMBER OF DAYS:														
Sunrise to Sunset														
—Clear	30	8.3	7.3	7.2	7.4	7.9	10.5	13.7	13.9	14.0	12.9	8.8	9.1	120.9
—Partly Cloudy	30	7.7	7.6	8.0	8.3	10.3	10.6	11.7	10.8	7.4	8.0	8.3	7.7	106.4
—Cloudy	30	15.0	13.3	15.8	14.3	12.9	8.9	5.6	6.3	8.6	10.1	12.9	14.2	137.9
Precipitation														
.01 inches or more	30	4.9	4.9	6.5	8.5	10.1	10.4	9.3	8.1	6.2	4.5	4.1	4.8	82.4
Snow, Ice pellets														
1.0 inches or more	30	1.6	1.7	2.0	1.1	0.*	0.0	0.0	0.0	0.*	0.4	1.5	1.5	10.0
Thunderstorms	10	0.0	0.0	0.2	1.5	7.6	10.6	11.8	9.3	3.7	0.8	0.2	0.0	45.7
Heavy Fog Visibility														
1/4 mile or less	10	0.4	0.7	0.8	0.6	0.5	0.1	0.5	0.4	0.4	0.4	0.1	0.3	5.4
Temperature °F														
—Maximum														
90° and above	30	0.0	0.0	0.0	0.3	1.4	6.1	14.4	13.4	4.9	0.5	0.0	0.0	41.0
32° and below	30	15.1	10.9	6.4	0.6	0.0	0.0	0.0	0.0	0.0	0.3	5.1	12.0	50.5
—Minimum														
32° and below	30	30.8	27.9	27.5	14.7	2.5	0.1	0.0	0.0	2.1	13.1	27.2	30.7	176.7
0° and below	30	11.3	6.3	1.8	0.0	0.0	0.0	0.0	0.0	0.0	0.0	1.4	6.6	27.4
AVG. STATION PRESS.(mb)	7	925.6	924.7	921.2	922.4	921.1	922.0	923.1	923.4	923.9	924.9	924.7	924.2	923.4
RELATIVE HUMIDITY (%)														
Hour 00	18	73	76	76	67	70	72	67	71	68	67	71	73	71
Hour 06	18	76	78	80	77	79	80	79	80	77	75	75	75	78
Hour 12 (Local Time)	18	64	62	58	49	49	48	47	45	44	46	53	60	52
Hour 18	18	64	61	55	45	45	43	41	41	40	44	56	64	50
PRECIPITATION (inches):														
Water Equivalent														
—Normal		0.28	0.52	0.83	1.82	2.86	2.97	2.42	2.42	1.42	0.83	0.41	0.33	17.11
—Maximum Monthly	30	0.82	1.43	4.23	4.01	8.96	7.09	8.96	6.71	5.91	1.95	2.62	1.00	8.96
—Year		1979	1962	1977	1968	1962	1983	1983	1966	1973	1981	1985	1977	JUL 1983
—Minimum Monthly	30	T	0.01	0.14	0.25	0.45	0.44	0.28	0.37	0.11	T	0.01	0.01	T
—Year		1961	1957	1956	1967	1966	1976	1980	1969	1958	1958	1962	1974	JAN 1961
—Maximum in 24 hrs	30	0.49	0.81	1.64	1.91	2.83	2.10	3.40	3.38	2.45	1.34	1.12	0.61	3.40
—Year		1973	1962	1977	1971	1962	1983	1983	1971	1973	1973	1985	1981	JUL 1983
Snow, Ice pellets														
—Maximum Monthly	30	15.2	17.5	51.0	16.1	3.3				18.4	4.7	34.5	14.8	51.0
—Year		1982	1962	1977	1970	1979				1985	1971	1985	1978	MAR 1977
—Maximum in 24 hrs	30	6.9	8.3	24.0	9.0	3.3				18.4	3.8	13.5	8.9	24.0
—Year		1971	1977	1977	1968	1979				1985	1971	1985	1983	MAR 1977
WIND:														
Mean Speed (mph)	17	9.5	9.4	10.7	11.5	11.1	10.0	9.3	9.5	9.9	9.7	9.8	9.5	10.0
Prevailing Direction														
through 1964		W	N	N	N	S	S	S	S	N	S	N	W	N
Fastest Mile														
—Direction (!!!)	25	NW	NW	NW	NW	SW	SE	SW	S	S	SE	NW	NW	SE
—Speed (MPH)	25	64	60	68	59	62	73	56	72	70	65	68	51	73
—Year		1972	1963	1976	1978	1974	1975	1972	1962	1973	1971	1975	1979	JUN 1975
Peak Gust														
—Direction (!!!)														
—Speed (mph)														
—Date														

See Reference Notes to this table on the following page.

TABLE 2 PRECIPITATION (inches)   VALENTINE, NEBRASKA

YEAR	JAN	FEB	MAR	APR	MAY	JUNE	JULY	AUG	SEP	OCT	NOV	DEC	ANNUAL
1956	0.30	0.21	0.14	0.98	1.18	1.72	1.49	3.38	0.46	1.20	0.64	0.04	11.74
1957	0.10	0.01	0.32	2.12	2.73	2.80	2.58	2.93	1.70	0.72	0.58	0.06	16.65
1958	0.20	0.43	0.38	1.96	2.97	3.82	5.39	0.50	0.11	T	0.26	0.10	16.12
1959	0.12	0.43	0.47	1.27	2.89	1.80	0.89	3.62	2.00	0.78	0.35	0.04	14.66
1960	0.04	1.19	0.64	2.56	4.08	2.75	1.27	4.08	1.86	0.23	0.41	0.26	19.37
1961	T	0.21	0.82	0.55	3.63	2.31	2.09	2.49	1.85	0.36	0.07	0.10	14.48
1962	0.03	1.43	0.82	0.55	8.96	6.84	2.64	1.69	0.71	1.00	0.01	0.10	24.78
1963	0.48	0.09	1.36	1.33	2.52	2.76	3.59	2.21	2.22	0.61	0.51	0.17	17.85
1964	0.14	0.23	0.30	2.30	1.49	2.79	3.38	0.68	1.51	0.02	0.04	0.45	13.33
1965	0.23	0.08	0.20	1.33	3.73	4.23	2.40	0.92	3.70	0.78	0.32	0.29	18.21
1966	0.28	0.39	0.39	1.06	0.45	3.65	1.71	6.71	2.62	1.01	0.11	0.44	18.82
1967	0.14	0.08	0.71	0.25	2.73	4.74	2.04	1.82	0.78	0.98	0.28	0.38	14.93
1968	0.20	0.33	0.22	4.01	2.98	5.73	1.70	3.10	0.60	0.56	0.41	0.62	20.46
1969	0.33	0.64	0.51	0.89	0.68	2.02	2.42	0.37	0.63	1.87	0.10	0.27	10.73
1970	0.02	0.11	0.64	2.80	2.26	2.40	2.64	0.59	0.71	1.11	0.26	0.28	13.82
1971	0.49	0.67	0.51	3.26	2.99	1.46	1.60	3.45	0.97	1.76	0.33	0.27	17.76
1972	0.41	0.13	0.34	2.32	3.64	2.55	4.63	1.00	0.46	0.63	1.37	0.32	17.80
1973	0.69	0.23	2.28	2.10	3.51	0.95	2.81	1.93	5.91	1.60	1.12	0.13	23.26
1974	0.13	0.21	0.55	1.30	2.57	1.26	1.26	1.51	0.74	0.84	0.19	0.01	10.57
1975	0.52	0.16	1.04	0.84	0.76	2.40	1.84	1.75	0.32	0.33	1.01	0.23	11.20
1976	0.27	0.32	0.57	2.03	1.56	0.44	2.12	0.99	2.22	0.38	0.09	0.22	11.21
1977	0.34	0.79	4.23	3.54	6.14	2.93	6.12	4.40	1.72	0.88	0.59	1.00	32.68
1978	0.25	1.22	0.19	3.25	3.38	2.52	4.17	3.19	0.56	0.12	0.55	0.68	20.08
1979	0.82	0.29	1.80	1.52	3.16	4.54	2.37	0.80	1.41	1.61	0.61	0.02	18.95
1980	0.38	0.68	1.52	0.46	2.67	2.79	0.28	3.03	0.43	1.37	0.09	0.42	14.12
1981	0.09	0.13	1.30	0.49	2.73	1.28	5.51	3.65	0.88	1.95	0.67	0.76	19.44
1982	0.50	0.07	1.05	0.74	6.70	2.11	3.20	3.37	2.70	1.80	0.88	0.53	23.65
1983	0.10	0.06	1.31	1.39	5.19	7.09	8.96	0.83	1.18	0.66	1.17	0.56	28.50
1984	0.11	0.34	0.90	3.40	2.73	2.09	6.53	1.40	0.57	0.55	0.57	0.12	19.31
1985	0.56	0.06	0.57	1.54	0.70	2.23	2.03	2.90	2.58	0.82	2.62	0.29	16.90
Record Mean	0.45	0.50	1.09	2.02	2.82	3.05	2.80	2.33	1.36	1.04	0.58	0.46	18.49

TABLE 3 AVERAGE TEMPERATURE (deg. F)   VALENTINE, NEBRASKA

YEAR	JAN	FEB	MAR	APR	MAY	JUNE	JULY	AUG	SEP	OCT	NOV	DEC	ANNUAL
1956	19.5	19.3	33.1	39.7	57.8	74.6	73.5	70.7	62.1	53.7	32.9	30.9	47.3
1957	11.0	27.8	31.9	42.4	54.9	64.9	77.8	73.2	59.2	48.6	34.7	32.8	46.6
1958	27.4	20.3	27.2	44.2	61.4	65.0	69.2	73.8	63.0	50.8	36.3	23.8	46.9
1959	19.4	19.7	35.7	44.9	55.7	72.2	73.5	74.8	59.6	43.7	26.6	32.0	46.5
1960	21.3	14.8	24.8	46.4	56.9	65.9	75.4	72.9	62.4	51.4	36.1	23.5	46.0
1961	24.1	29.8	36.5	40.9	54.8	68.9	72.7	73.9	57.2	45.7	34.3	21.6	47.0
1962	18.4	21.7	26.2	48.2	60.3	65.8	70.8	71.6	60.3	51.6	39.5	27.2	46.8
1963	10.1	29.7	39.8	47.6	58.6	71.4	77.1	73.5	65.6	58.4	38.7	19.8	49.2
1964	26.2	27.2	28.9	46.3	61.1	67.9	78.0	68.7	61.0	49.6	33.2	19.5	47.3
1965	24.0	23.7	21.1	48.2	58.1	67.8	73.7	71.1	51.5	53.3	36.1	30.3	46.6
1966	12.8	16.9	36.8	41.5	58.0	68.3	79.1	67.9	60.9	49.4	33.9	24.9	45.9
1967	25.5	26.2	37.7	47.7	51.7	64.5	71.8	70.2	62.1	50.1	33.6	20.0	46.8
1968	20.8	25.0	39.1	44.3	52.1	67.5	72.5	71.1	60.3	50.5	34.9	16.5	46.2
1969	12.7	22.4	23.5	51.3	60.1	62.4	74.5	75.6	66.7	42.3	38.5	26.1	46.3
1970	17.0	29.4	27.6	42.6	60.7	74.9	75.8	74.9	61.6	45.1	34.4	21.1	46.6
1971	16.6	20.3	32.4	47.8	55.5	71.3	70.9	75.4	60.2	49.0	36.1	24.4	46.7
1972	16.5	24.0	38.0	45.4	58.3	64.7	70.3	71.0	61.4	45.6	29.6	17.1	45.4
1973	21.2	28.6	39.6	45.0	55.8	68.4	73.8	76.8	57.7	52.7	34.8	23.8	48.2
1974	19.3	30.2	37.2	47.9	57.3	67.5	79.1	69.0	57.9	52.0	35.5	27.3	48.3
1975	24.3	19.0	28.4	44.0	57.9	66.8	77.6	73.4	58.7	51.3	31.2	25.6	46.5
1976	21.4	33.5	33.5	48.6	56.5	68.0	75.3	74.5	62.4	45.1	28.3	24.3	47.6
1977	11.1	30.7	32.8	50.6	63.0	71.2	74.7	67.4	62.6	48.7	34.4	19.8	47.2
1978	8.1	10.3	32.8	44.2	57.5	67.9	72.4	70.6	66.0	48.1	29.7	12.5	43.3
1979	4.1	13.3	33.0	45.7	54.9	67.1	73.4	70.7	66.5	49.3	30.1	32.2	45.0
1980	19.4	23.8	30.5	48.7	58.4	70.5	77.7	72.6	63.7	47.6	38.1	20.4	48.3
1981	28.5	26.4	38.7	54.1	54.7	68.3	73.9	71.2	63.3	48.2	39.8	23.6	49.3
1982	10.2	25.6	34.6	42.8	57.0	62.2	75.1	73.9	61.0	47.5	31.4	26.6	45.7
1983	29.5	34.4	34.7	40.5	52.7	65.1	75.4	78.3	63.3	50.0	34.7	3.9	46.9
1984	21.8	32.0	32.2	42.5	56.7	67.8	73.9	75.0	57.3	47.5	35.2	21.3	46.9
1985	17.7	21.6	37.2	51.3	62.5	64.1	75.2	69.5	58.3	47.6	18.0	15.9	44.9
Record Mean	20.4	24.0	33.0	46.3	56.9	67.2	74.2	72.2	62.0	49.7	34.6	24.6	47.0
Max	32.6	36.2	44.9	58.7	69.3	79.6	87.4	85.6	76.0	63.8	47.2	36.5	59.8
Min	8.3	11.8	21.0	33.8	44.5	54.7	60.9	58.7	47.9	35.5	22.0	12.8	34.3

REFERENCE NOTES FOR TABLES 1, 2, 3 and 6   (VALENTINE, NE)

GENERAL

T - TRACE AMOUNT
BLANK ENTRIES DENOTE MISSING/UNREPORTED DATA.
# INDICATES A STATION OR INSTRUMENT RELOCATION.

SPECIFIC

TABLE 1

(a) - LENGTH OF RECORD IN YEARS. ALTHOUGH INDIVIDUAL MONTHS MAY BE MISSING.

* LESS THAN .05

NORMALS — BASED ON THE 1951-1980 RECORD PERIOD.
EXTREMES — DATES ARE THE MOST RECENT OCCURRENCE.
WIND DIR. — NUMERALS SHOW TENS OF DEGREES CLOCKWISE FROM TRUE NORTH.
     "00" INDICATES CALM.
RESULTANT WIND DIRECTIONS ARE GIVEN TO WHOLE DEGREES.

EXCEPTIONS

TABLE 1

1. THUNDERSTORMS AND HEAVY FOG ARE THROUGH 1964 AND MAY BE INCOMPLETE, DUE TO PART-TIME OPERATIONS.

2. RELATIVE HUMIDITY AND MEAN WIND SPEED ARE THROUGH 1964 AND 1977 TO DATE.

3. FASTEST MILE WIND IS THROUGH AUGUST 1982.
   TABLES 2, 3, and 6

RECORD MEANS ARE THROUGH THE CURRENT YEAR,
BEGINNING IN   1889 FOR TEMPERATURE
                1889 FOR PRECIPITATION
                1956 FOR SNOWFALL

TABLE 4 HEATING DEGREE DAYS Base 65 deg. F   VALENTINE, NEBRASKA

SEASON	JULY	AUG	SEP	OCT	NOV	DEC	JAN	FEB	MAR	APR	MAY	JUNE	TOTAL
1956-57	0	31	139	350	957	1051	1673	1034	1018	670	320	70	7313
1957-58	0	7	186	511	902	993	1157	1245	1165	618	148	83	7015
1958-59	6	12	130	449	854	1273	1409	1262	901	597	315	33	7241
1959-60	5	9	235	653	1146	1017	1348	1451	1243	551	268	47	7973
1960-61	0	9	179	422	860	1278	1264	978	873	719	335	52	6969
1961-62	0	0	301	474	915	1338	1441	1209	1196	510	167	69	7620
1962-63	9	15	169	421	759	1163	1701	984	774	519	224	10	6748
1963-64	0	2	70	214	782	1397	1196	1088	1112	558	177	46	6642
1964-65	0	74	188	469	947	1266	1150	1150	1354	496	235	22	7604
1965-66	0	16	410	355	864	1068	1615	1341	865	699	247	67	7547
1966-67	0	51	179	476	928	1237	1221	1081	843	512	439	62	7029
1967-68	26	34	133	466	934	1389	1365	1154	798	614	393	56	7362
1968-69	15	24	179	451	896	1497	1618	1188	1278	401	223	132	7902
1969-70	0	8	46	696	787	1198	1482	991	1152	665	172	47	7244
1970-71	1	0	204	611	910	1353	1494	1244	1002	508	291	7	7625
1971-72	25	0	215	492	861	1251	1500	1180	827	584	257	43	7235
1972-73	30	38	165	596	1056	1483	1356	1013	779	592	297	28	7433
1973-74	7	0	236	376	899	1270	1413	970	855	510	243	65	6844
1974-75	0	39	233	400	876	1163	1256	1284	1126	642	236	67	7322
1975-76	5	5	231	435	1008	1215	1348	907	969	485	268	58	6933
1976-77	5	3	151	613	1094	1255	1672	952	990	431	101	8	7275
1977-78	0	35	122	499	914	1397	1760	1526	993	618	254	72	8190
1978-79	4	32	109	515	1053	1622	1888	1446	986	574	326	67	8622
1979-80	14	24	77	478	1040	1010	1408	1189	1063	495	231	26	7055
1980-81	0	14	112	536	804	1128	1124	1075	806	325	328	37	6289
1981-82	21	7	174	514	748	1274	1697	1100	936	659	250	108	7428
1982-83	1	16	189	536	1000	1183	1096	850	931	730	385	91	7008
1983-84	6	0	168	461	903	1892	1334	952	1008	667	288	43	7722
1984-85	0	0	280	542	886	1351	1461	1211	854	415	136	112	7248
1985-86	13	29	279	531	1404	1517							

TABLE 5 COOLING DEGREE DAYS Base 65 deg. F   VALENTINE, NEBRASKA

YEAR	JAN	FEB	MAR	APR	MAY	JUNE	JULY	AUG	SEP	OCT	NOV	DEC	TOTAL
1969	0	0	0	0	80	61	301	343	104	0	0	0	889
1970	0	0	0	0	49	181	341	315	108	0	0	0	994
1971	0	0	0	0	4	201	214	330	78	0	0	0	827
1972	0	0	0	0	56	132	202	231	63	0	0	0	684
1973	0	0	0	0	18	136	285	373	24	1	0	0	837
1974	0	0	0	4	13	144	446	171	25	3	0	0	806
1975	0	0	0	16	24	124	400	272	47	18	0	0	901
1976	0	0	0	1	11	157	330	306	81	2	0	0	888
1977	0	0	0	3	46	203	307	116	57	0	0	0	732
1978	0	0	0	0	27	163	239	213	148	0	0	0	790
1979	0	0	0	4	20	136	280	208	129	0	0	0	777
1980	0	0	0	15	34	198	404	255	78	4	0	0	988
1981	0	0	0	6	16	145	304	207	70	0	0	0	748
1982	0	0	0	0	10	39	321	299	73	0	0	0	742
1983	0	0	0	0	12	99	338	422	125	0	0	0	996
1984	0	0	0	0	38	134	280	318	55	5	0	0	830
1985	0	0	0	9	66	92	337	176	87	0	0	0	767

TABLE 6 SNOWFALL (inches)   VALENTINE, NEBRASKA

SEASON	JULY	AUG	SEP	OCT	NOV	DEC	JAN	FEB	MAR	APR	MAY	JUNE	TOTAL
1956-57	0.0	0.0	0.0	T	3.8	0.4	6.4	0.2	5.0	9.0	0.0	0.0	24.8
1957-58	0.0	0.0	0.0	T	1.0	0.8	3.2	5.6	3.2	2.3	0.0	0.0	16.1
1958-59	0.0	0.0	0.0	0.0	0.7	4.6	4.4	5.0	4.9	5.9	0.0	0.0	25.5
1959-60	0.0	0.0	0.0	3.9	9.2	0.6	3.2	5.2	6.8	6.1	0.0	0.0	35.0
1960-61	0.0	0.0	0.0	0.0	3.0	2.6	0.3	4.0	6.9	1.0	T	0.0	17.8
1961-62	0.0	0.0	T	T	2.2	1.8	0.7	17.5	6.4	0.4	0.0	0.0	29.0
1962-63	0.0	0.0	0.0	0.0	T	1.8	5.9	1.3	10.0	0.5	0.0	0.0	19.5
1963-64	0.0	0.0	0.0	0.0	3.1	2.5	2.4	3.0	5.0	3.0	0.0	0.0	19.0
1964-65	0.0	0.0	0.0	0.0	1.0	4.5	2.3	2.4	4.0	T	0.0	0.0	14.2
1965-66	0.0	0.0	0.0	0.0	2.2	1.3	4.4	4.5	2.9	6.2	T	0.0	21.5
1966-67	0.0	0.0	0.0	T	3.5	5.6	1.5	0.9	3.0	T	0.1	0.0	14.6
1967-68	0.0	0.0	0.0	0.0	1.1	5.9	5.9	3.2	0.6	9.6	T	0.0	26.3
1968-69	0.0	0.0	0.0	0.0	1.7	9.6	5.3	6.7	4.6	T	0.0	0.0	27.9
1969-70	0.0	0.0	0.0	4.7	0.9	3.3	0.9	2.2	10.1	16.1	0.0	0.0	38.2
1970-71	0.0	0.0	0.0	2.9	1.9	4.2	9.3	9.3	8.0	7.7	0.0	0.0	43.3
1971-72	0.0	0.0	0.0	4.7	1.0	3.7	5.7	2.0	0.4	0.4	T	0.0	17.9
1972-73	0.0	0.0	0.0	2.6	12.0	7.6	9.2	2.6	1.1	0.4	T	0.0	35.5
1973-74	0.0	0.0	T	T	13.3	3.0	2.5	3.0	4.1	0.5	0.0	0.0	26.4
1974-75	0.0	0.0	0.0	0.0	1.5	0.2	5.8	5.7	16.1	3.1	0.0	0.0	32.4
1975-76	0.0	0.0	0.0	1.1	11.5	1.4	5.8	5.2	8.3	T	0.0	0.0	33.3
1976-77	0.0	0.0	0.0	0.8	1.5	3.5	7.8	10.3	51.0	2.5	0.0	0.0	77.4
1977-78	0.0	0.0	0.0	1.1	3.0	8.3	3.8	15.9	2.5	9.7	0.0	0.0	44.3
1978-79	0.0	0.0	0.0	0.0	7.0	14.8	10.8	5.1	7.3	8.0	3.3	0.0	56.3
1979-80	0.0	0.0	0.0	1.9	6.2	0.5	7.6	12.6	21.6	2.9	T	0.0	53.3
1980-81	0.0	0.0	0.0	2.6	0.5	5.8	1.5	3.8	2.2	0.0	0.0	0.0	16.4
1981-82	0.0	0.0	0.0	1.4	8.0	11.2	15.2	1.5	6.0	4.6	0.0	0.0	47.9
1982-83	0.0	0.0	0.0	3.0	2.0	5.3	0.8	0.8	5.7	1.9	0.0	0.0	18.9
1983-84	0.0	0.0	T	0.0	12.0	12.4	1.5	4.0	7.2	8.5	0.1	0.0	45.7
1984-85	0.0	0.0	0.1	0.3	1.5	1.5	5.5	0.7	3.0	0.1	0.0	0.0	12.7
1985-86	0.0	0.0	18.4	T	34.5	4.9							
Record Mean	0.0	0.0	0.6	1.0	5.0	4.5	4.9	4.9	7.4	3.9	0.1	0.0	32.3

See Reference Notes, relative to all above tables, on preceding page.

Located in the Humbolt River Valley of northeastern Nevada, the Elko station is at the Municipal Airport, one mile west of town at an elevation just above 5,000 feet. Several mountain ranges with many peaks near or exceeding 10,000 feet in height dominate the landscape, but the immediate terrain consists of sagebrush-covered valleys and low foothills, the highest of which are approximately 2,500 feet above the station. A few areas, mostly in the higher mountains, are covered with sparse stands of juniper, aspen, pinion pine, and spruce. The only heavily forested area in northeastern Nevada is in the Jarbidge Mountains near the Idaho border.

Owing to the high elevation and proximity of the mountains, there is a wide temperature range. High nighttime radiation makes cool nights the rule even in mid-summer.

Rainfall is light with the heaviest amounts falling during the winter months as snow. Summer precipitation occurs mostly as showers and does not contribute much toward crop growth. Irrigation is necessary for the principal crops which are hay and some small grains. Vegetables requiring only a short growing season are cultivated on a small scale.

Cattle and sheep ranching is the most important industry of this section. The ranges ordinarily furnish excellent summer pasture for cattle with winter feeding being necessary from the hay crops. Sheep are moved into the foothills in early spring, to higher elevations during the summer, and then south to the east-central Nevada ranges for the winter months. Spring lambs born during April or early May are marketed in September.

Mining, another important local industry, is not particularly affected by the climate except where poor road conditions or excessively heavy mountain snows may halt winter operations.

Transportation by air, rail, or road is seldom affected by the weather for more than short periods.

Based on the 1951-1980 period, the average first occurrence of 32 degrees Fahrenheit in the fall is September 8 and the average last occurrence in the spring is June 5.

TABLE 1   NORMALS, MEANS AND EXTREMES

ELKO, NEVADA

LATITUDE: 40°50'N   LONGITUDE: 115°47'W   ELEVATION: FT. GRND 5050 BARO 05080   TIME ZONE: PACIFIC   WBAN: 24121

	(a)	JAN	FEB	MAR	APR	MAY	JUNE	JULY	AUG	SEP	OCT	NOV	DEC	YEAR
TEMPERATURE °F:														
Normals														
-Daily Maximum		36.6	42.6	48.9	58.2	68.5	79.2	90.4	87.8	78.8	66.3	49.4	38.3	62.1
-Daily Minimum		13.2	19.4	23.0	28.6	36.1	43.3	49.8	47.3	38.0	28.7	21.2	13.9	30.2
-Monthly		24.9	31.0	36.0	43.4	52.3	61.3	70.1	67.6	58.4	47.5	35.3	26.1	46.1
Extremes														
-Record Highest	55	63	67	77	86	92	104	107	107	99	88	78	64	107
-Year		1981	1981	1966	1981	1977	1981	1981	1978	1950	1980	1980	1940	JUL 1981
-Record Lowest	55	-43	-37	-9	-2	10	23	30	24	9	8	-12	-38	-43
-Year		1937	1933	1952	1936	1965	1976	1932	1932	1934	1958	1931	1932	JAN 1937
NORMAL DEGREE DAYS:														
Heating (base 65°F)		1240	952	899	648	396	166	19	58	230	543	891	1206	7248
Cooling (base 65°F)		0	0	0	0	6	52	178	138	32	0	0	0	406
% OF POSSIBLE SUNSHINE														
MEAN SKY COVER (tenths)														
Sunrise - Sunset	36	6.7	6.6	6.8	6.5	5.9	4.4	3.4	3.4	3.3	4.4	6.1	6.5	5.3
MEAN NUMBER OF DAYS:														
Sunrise to Sunset														
-Clear	49	6.7	6.2	6.3	6.4	8.2	13.0	17.4	17.9	18.5	14.0	8.6	7.5	130.8
-Partly Cloudy	49	7.8	7.3	8.2	9.2	10.3	9.5	9.7	8.9	6.6	8.0	6.9	6.9	99.3
-Cloudy	49	16.6	14.7	16.5	14.3	12.5	7.4	3.9	4.2	4.9	9.1	14.5	16.6	135.1
Precipitation														
.01 inches or more	55	9.2	8.7	8.9	7.4	7.9	5.8	3.5	3.6	3.8	4.9	7.0	8.7	79.2
Snow,Ice pellets														
1.0 inches or more	37	3.3	2.0	2.3	1.2	0.3	0.0	0.0	0.0	0.*	0.4	1.7	3.5	14.7
Thunderstorms	37	0.2	0.3	0.3	0.8	3.3	3.6	4.8	4.6	2.0	0.5	0.2	0.1	20.7
Heavy Fog Visibility														
1/4 mile or less	37	1.6	0.7	0.4	0.2	0.4	0.1	0.1	0.1	0.1	0.3	0.5	1.2	5.6
Temperature °F														
-Maximum														
90° and above	21	0.0	0.0	0.0	0.0	0.3	5.7	20.4	15.0	3.5	0.0	0.0	0.0	45.0
32° and below	21	9.1	3.7	0.8	0.1	0.0	0.0	0.0	0.0	0.0	0.1	1.5	8.8	24.0
-Minimum														
32° and below	21	28.6	25.4	25.6	20.3	6.9	0.8	0.0	0.*	5.5	19.6	24.0	28.8	185.5
0° and below	21	4.2	1.5	0.1	0.0	0.0	0.0	0.0	0.0	0.0	0.0	0.3	4.0	10.2
AVG. STATION PRESS.(mb)	11	846.8	845.5	841.9	842.6	843.1	844.2	845.7	845.9	845.8	846.7	845.3	847.0	845.0
RELATIVE HUMIDITY (%)														
Hour 04	20	77	78	76	72	70	67	56	56	62	68	75	76	69
Hour 10	20	72	67	55	44	38	33	26	28	33	43	61	70	48
Hour 16 (Local Time)	20	59	52	42	35	30	25	20	21	24	30	48	60	37
Hour 22	20	76	74	69	59	54	48	38	40	47	58	70	75	59
PRECIPITATION (inches):														
Water Equivalent														
-Normal		1.16	0.81	0.85	0.79	1.03	0.91	0.33	0.58	0.47	0.56	0.83	0.98	9.30
-Maximum Monthly	55	3.35	2.93	2.37	2.17	4.09	2.61	2.35	4.61	3.22	2.76	2.77	4.21	4.61
-Year		1956	1932	1975	1963	1971	1963	1950	1970	1978	1938	1942	1983	AUG 1970
-Minimum Monthly	55	0.04	0.08	0.13	0.10	T	T	0.00	T	T	T	T	T	0.00
-Year		1961	1967	1977	1949	1974	1974	1963	1969	1951	1958	1959	1976	JUL 1963
-Maximum in 24 hrs	55	1.27	0.89	1.02	1.10	1.73	1.85	1.04	4.13	2.32	1.31	1.31	1.62	4.13
-Year		1951	1936	1975	1943	1971	1968	1950	1970	1978	1939	1950	1950	AUG 1970
Snow,Ice pellets														
-Maximum Monthly	54	27.4	26.1	23.2	15.6	11.3	T	T	T	2.0	5.6	16.8	33.2	33.2
-Year		1950	1932	1967	1975	1971	1982	1950	1949	1982	1984	1985	1983	DEC 1983
-Maximum in 24 hrs	37	16.7	9.1	13.8	10.0	8.6	T	T	T	2.0	5.2	9.0	9.2	16.7
-Year		1951	1949	1967	1975	1971	1982	1950	1949	1982	1963	1965	1955	JAN 1951
WIND:														
Mean Speed (mph)	34	5.3	5.8	6.7	7.2	6.9	6.7	6.2	5.9	5.4	5.1	5.2	5.2	6.0
Prevailing Direction														
through 1963		SW	SW	SW	SW	SW	SW	SW	SW	SW	SW	SW	SW	SW
Fastest Obs. 1 Min.														
-Direction (!!)	31	23	27	29	25	34	27	16	16	27	29	20	27	27
-Speed (MPH)	31	40	39	41	48	55	61	36	35	58	35	40	50	61
-Year		1952	1963	1952	1956	1955	1984	1955	1966	1959	1953	1984	1952	JUN 1984
Peak Gust														
-Direction (!!)														
-Speed (mph)														
-Date														

See Reference Notes to this table on the following pages.

TABLE 2 PRECIPITATION (inches) ELKO, NEVADA

YEAR	JAN	FEB	MAR	APR	MAY	JUNE	JULY	AUG	SEP	OCT	NOV	DEC	ANNUAL
1956	3.35	0.48	0.18	1.51	1.74	0.58	0.05	T	0.02	0.97	0.08	1.08	10.04
1957	0.89	0.65	1.82	0.80	2.41	1.05	0.07	0.23	0.24	0.42	1.02	0.50	10.10
1958	0.98	1.43	0.92	0.19	0.24	0.52	0.14	0.68	0.17	T	0.51	0.62	6.40
1959	0.56	0.26	0.60	0.29	0.82	0.79	0.18	0.09	1.74	0.02	T	0.16	5.51
1960	0.62	1.13	1.63	0.94	0.50	0.04	0.49	0.32	0.04	0.45	1.28	0.40	7.84
1961	0.04	0.76	0.93	0.14	0.50	0.35	0.48	2.15	0.07	0.56	0.63	0.99	7.60
1962	0.81	1.67	0.63	0.26	2.34	1.12	0.46	0.26	0.09	0.12	0.36	0.12	8.24
1963	1.74	0.65	0.65	2.17	2.10	2.61	0.00	0.18	0.81	1.76	1.94	0.42	15.03
1964	1.27	0.10	0.97	1.14	1.15	2.24	0.18	0.10	0.14	0.48	3.30		12.14
1965	0.84	0.31	0.44	1.81	1.08	1.21	0.62	1.31	0.20	0.63	1.96	0.76	11.17
1966	0.24	0.37	0.26	0.64	0.55	0.36	0.69	0.52	0.44	0.02	0.58	1.83	6.50
1967	1.00	0.08	1.79	0.87	0.84	1.19	1.03	0.08	0.21	0.27	0.60	0.66	8.62
1968	1.16	1.45	1.12	0.53	1.15	2.60	0.04	1.94	0.36	0.79	1.56	1.93	14.63
1969	1.24	1.76	0.35	0.28	0.27	2.11	0.24	T	0.17	1.11	0.34	1.83	9.70
1970	2.36	0.40	0.50	0.37	0.37	1.29	0.48	4.61	0.52	0.64	1.32	1.70	14.56
1971	0.58	1.03	0.53	0.96	4.09	1.01	0.21	0.98	0.74	0.74	1.23	1.57	13.67
1972	0.39	0.57	0.56	0.36	0.16	1.46	T	0.29	1.01	1.92	0.98	0.77	8.47
1973	1.17	0.96	0.56	0.88	0.70	0.56	0.46	0.29	0.18	0.64	1.40	1.30	9.10
1974	0.61	0.38	0.86	0.58	T	T	0.19	0.12	0.00	1.16	0.34	0.53	4.77
1975	1.79	1.06	2.37	1.70	0.98	0.40	0.15	0.10	0.18	1.42	0.94	0.25	11.34
1976	0.35	0.68	0.25	0.65	0.50	0.64	0.44	0.91	1.84	0.58	0.26	T	7.10
1977	0.30	0.26	0.13	0.18	1.44	1.03	0.22	0.77	0.26	0.01	0.96	0.90	6.46
1978	0.68	0.97	1.88	1.98	0.25	0.18	0.58	0.02	3.22	0.25	0.61	0.52	11.14
1979	1.91	1.20	0.59	0.43	0.42	0.38	0.32	0.36	0.25	0.43	1.10	0.35	7.74
1980	3.11	1.89	0.77	1.22	3.15	0.80	0.33	0.10	0.42	0.19	0.62	0.21	12.81
1981	0.64	0.33	1.20	0.75	0.80	0.24	0.02	0.19	0.13	0.69	0.60	3.19	8.78
1982	0.82	0.65	1.94	0.50	1.04	0.54	0.69	1.24	2.55	1.11	1.78	0.86	13.72
1983	1.73	1.34	1.91	1.28	0.60	0.47	0.01	1.25	1.57	1.21	2.76	4.21	18.34
1984	0.57	0.80	1.25	1.00	0.24	1.29	1.04	0.46	0.11	1.75	1.40	0.45	10.36
1985	0.54	0.15	1.09	0.23	0.60	0.17	0.25	0.02	1.17	0.16	2.14	0.78	7.30
Record Mean	1.20	0.91	0.93	0.72	0.87	0.71	0.35	0.36	0.41	0.68	0.83	1.08	9.05

TABLE 3 AVERAGE TEMPERATURE (deg. F) ELKO, NEVADA

YEAR	JAN	FEB	MAR	APR	MAY	JUNE	JULY	AUG	SEP	OCT	NOV	DEC	ANNUAL
1956	30.6	20.5	36.6	44.6	53.4	60.1	68.6	63.8	58.5	44.6	30.5	26.1	44.8
1957	16.9	33.5	37.8	42.7	50.1	60.1	67.8	66.1	57.3	44.0	30.5	29.3	44.7
1958	25.4	36.1	32.3	41.0	56.5	60.8	67.2	70.8	57.6	48.5	35.6	32.4	47.0
1959	31.8	29.5	35.7	46.2	48.5	65.4	71.8	66.2	55.1	48.0	34.9	23.3	46.4
1960	20.5	26.9	40.0	44.5	51.1	64.3	71.8	65.2	61.0	46.2	36.1	26.0	46.1
1961	27.2	33.0	35.8	42.4	52.5	66.9	72.1	70.8	53.5	45.2	32.0	26.8	46.5
1962	18.9	28.5	33.3	48.0	50.2	60.1	66.4	65.1	59.2	49.9	35.4	28.5	45.3
1963	21.0	38.9	35.2	38.6	54.8	55.0	65.3	66.3	61.6	51.2	34.7	25.1	45.7
#1964	18.8	20.1	29.5	40.8	50.0	56.9	69.9	65.3	54.2	49.3	30.3	30.3	42.9
1965	30.3	31.4	33.5	44.9	48.5	57.7	67.7	66.1	52.3	50.2	38.3	19.2	45.0
1966	22.5	24.0	38.0	44.3	57.8	62.7	70.1	69.0	61.3	45.9	39.1	18.7	46.1
1967	25.2	31.5	37.3	40.1	51.4	62.7	74.9	75.0	64.9	49.9	41.6	23.1	48.2
1968	27.0	39.8	42.2	42.6	54.0	64.1	75.4	65.6	59.2	50.6	38.2	28.1	48.9
1969	28.8	25.4	30.9	45.4	57.6	61.0	71.1	70.8	61.6	41.1	34.3	27.1	46.2
1970	27.8	36.8	33.4	36.4	50.9	60.1	70.0	70.0	53.0	41.0	38.0	22.9	45.0
1971	30.3	34.0	36.1	41.8	48.2	58.1	68.7	70.8	53.2	43.7	33.9	18.8	44.8
1972	23.4	32.4	42.0	42.3	54.0	64.1	67.5	67.0	54.3	46.3	31.9	19.1	45.3
1973	20.0	31.4	36.8	43.0	56.1	62.8	70.8	69.0	57.0	47.9	36.8	30.3	46.8
1974	27.9	32.2	41.1	43.5	52.6	65.2	69.8	65.2	58.7	47.9	37.6	26.7	47.4
1975	18.9	29.2	35.8	37.6	51.0	59.7	71.3	64.1	58.8	46.4	32.4	29.4	44.6
1976	25.8	32.8	33.5	41.4	55.6	59.2	69.4	63.0	61.1	46.6	37.7	27.6	46.1
1977	24.1	35.6	34.8	49.3	50.6	69.3	71.7	70.9	60.9	51.0	38.9	34.7	49.3
1978	32.2	34.7	44.9	45.3	52.1	60.6	69.8	69.2	56.6	50.6	35.3	23.6	47.9
1979	22.7	33.0	39.8	46.4	56.5	65.8	73.9	71.2	66.1	54.6	33.6	31.5	49.6
1980	32.2	39.0	37.0	47.3	51.5	60.5	71.6	67.2	61.6	49.6	40.0	34.2	49.3
1981	34.6	35.5	40.8	48.2	54.2	67.3	73.9	72.8	61.6	44.8	40.5	34.1	50.7
1982	23.2	29.6	37.8	42.3	52.5	61.6	69.1	70.6	58.6	45.4	34.1	28.0	46.1
1983	30.4	31.8	41.8	42.7	52.6	62.1	68.7	72.1	62.7	50.8	36.5	28.7	48.4
1984	17.1	23.8	36.0	41.9	54.7	59.7	71.9	70.7	60.7	43.3	37.7	20.9	44.9
1985	21.8	26.8	35.1	48.0	54.0	64.7	75.9	65.9	53.7	46.2	30.1	23.1	45.4
Record Mean	23.8	29.9	36.8	44.3	52.8	61.2	70.1	67.4	57.9	47.0	35.3	26.0	46.0
Max	36.4	41.9	50.1	59.8	69.8	79.8	91.1	88.7	79.0	65.7	49.9	38.5	62.6
Min	11.2	17.9	23.4	28.7	35.7	42.5	49.0	46.2	36.8	28.4	20.7	13.5	29.5

REFERENCE NOTES FOR TABLES 1, 2, 3 and 6 (ELKO, NV)

GENERAL

T - TRACE AMOUNT
BLANK ENTRIES DENOTE MISSING/UNREPORTED DATA.
# INDICATES A STATION OR INSTRUMENT RELOCATION.

SPECIFIC

TABLE 1

(a) - LENGTH OF RECORD IN YEARS. ALTHOUGH INDIVIDUAL MONTHS MAY BE MISSING.

* LESS THAN .05

NORMALS — BASED ON THE 1951-1980 RECORD PERIOD.
EXTREMES — DATES ARE THE MOST RECENT OCCURRENCE.
WIND DIR. — NUMERALS SHOW TENS OF DEGREES CLOCKWISE FROM TRUE NORTH. "00" INDICATES CALM.
RESULTANT WIND DIRECTIONS ARE GIVEN TO WHOLE DEGREES.

EXCEPTIONS

TABLES 2, 3, and 6

RECORD MEANS ARE THROUGH THE CURRENT YEAR, BEGINNING IN 1910 FOR TEMPERATURE
1871 FOR PRECIPITATION
1932 FOR SNOWFALL

TABLE 4 HEATING DEGREE DAYS Base 65 deg. F   ELKO, NEVADA

SEASON	JULY	AUG	SEP	OCT	NOV	DEC	JAN	FEB	MAR	APR	MAY	JUNE	TOTAL
1956-57	23	77	195	625	1028	1199	1489	877	834	661	454	175	7637
1957-58	4	31	227	644	1028	1101	1217	801	1005	708	258	147	7171
1958-59	21	1	231	505	876	1004	1020	986	898	557	505	70	6674
1959-60	7	64	303	521	896	1288	1372	1098	769	608	425	57	7408
1960-61	0	102	130	575	858	1201	1166	889	896	673	381	60	6931
1961-62	0	2	341	606	985	1177	1424	1018	974	504	453	164	7648
1962-63	23	63	169	458	881	1124	1359	724	920	787	309	297	7114
1963-64	26	57	113	422	905	1228	1426	1298	1091	720	458	246	7990
#1964-65	7	85	317	481	1029	1067	1071	937	969	596	505	215	7279
1965-66	11	43	375	452	793	1416	1312	1142	830	615	226	128	7343
1966-67	8	27	126	584	769	1427	1227	932	852	737	413	123	7225
1967-68	0	0	56	459	695	1291	1168	725	700	664	334	100	6192
1968-69	0	94	195	438	799	1137	1119	1100	1051	581	225	138	6880
1969-70	24	8	120	736	916	1166	1148	782	974	849	431	190	7344
1970-71	5	1	357	736	804	1297	1068	863	891	690	515	205	7432
1971-72	0	6	351	651	928	1426	1284	940	704	673	339	59	7375
1972-73	17	28	317	573	985	1419	1386	931	867	652	273	135	7583
1973-74	5	19	236	522	844	1064	1145	910	734	637	382	87	6585
1974-75	28	55	187	523	815	1184	1422	994	899	815	427	168	7517
1975-76	8	77	180	570	970	1094	1209	927	967	701	286	177	7166
1976-77	12	80	145	563	810	1154	1262	816	927	464	441	16	6690
1977-78	0	27	164	430	773	933	1009	845	617	583	398	135	5914
1978-79	23	47	267	439	888	1277	1306	889	775	553	265	86	6815
1979-80	0	4	32	320	933	1032	1013	747	862	523	410	165	6041
1980-81	0	34	119	467	743	947	933	819	744	497	336	67	5706
1981-82	0	4	131	618	726	952	1289	987	836	676	381	130	6730
1982-83	32	0	225	598	925	1143	1066	924	713	661	390	114	6791
1983-84	29	0	105	434	847	1119	1480	1187	894	686	318	201	7300
1984-85	0	10	163	664	811	1360	1331	1060	921	505	335	69	7229
1985-86	0	42	338	573	1042	1294							

TABLE 5 COOLING DEGREE DAYS Base 65 deg. F   ELKO, NEVADA

YEAR	JAN	FEB	MAR	APR	MAY	JUNE	JULY	AUG	SEP	OCT	NOV	DEC	TOTAL
1969	0	0	0	0	1	26	220	194	25	0	0	0	466
1970	0	0	0	0	0	48	166	162	2	0	0	0	378
1971	0	0	0	0	0	7	140	191	4	0	0	0	342
1972	0	0	0	0	3	38	102	96	3	0	0	0	242
1973	0	0	0	0	3	74	192	152	2	0	0	0	423
1974	0	0	0	0	4	97	185	68	6	0	0	0	360
1975	0	0	0	0	0	15	212	56	0	0	0	0	283
1976	0	0	0	0	0	11	157	24	36	0	0	0	228
1977	0	0	0	0	3	152	217	216	48	0	0	0	636
1978	0	0	0	0	0	9	179	186	21	0	0	0	395
1979	0	0	0	0	9	120	284	207	70	4	0	0	694
1980	0	0	0	0	0	36	211	109	25	0	0	0	381
1981	0	0	0	2	9	145	282	254	36	0	0	0	728
1982	0	0	0	0	0	36	165	181	39	0	0	0	421
1983	0	0	0	0	11	33	151	228	40	0	0	0	463
1984	0	0	0	0	3	48	223	197	42	0	0	0	513
1985	0	0	0	0	0	66	342	77	7	0	0	0	492

TABLE 6 SNOWFALL (inches)   ELKO, NEVADA

SEASON	JULY	AUG	SEP	OCT	NOV	DEC	JAN	FEB	MAR	APR	MAY	JUNE	TOTAL
1956-57	0.0	0.0	0.0	1.2	1.4	3.7	9.3	1.0	2.9	0.1	T	T	19.6
1957-58	0.0	0.0	0.0	T	5.3	4.0	13.6	7.4	10.4	2.2	0.0	0.0	42.9
1958-59	0.0	0.0	0.0	T	T	6.3	3.4	4.8	1.7	2.2	0.4	0.0	18.8
1959-60	0.0	0.0	0.0	T	T	2.7	12.8	5.1	2.8	1.3	1.7	0.0	26.4
1960-61	0.0	0.0	0.0	T	2.2	5.2	0.3	5.9	11.0	0.7	T	0.0	25.3
1961-62	0.0	0.0	T	3.0	6.2	8.6	12.7	4.7	14.9	T	2.4	0.0	52.5
1962-63	0.0	0.0	0.0	T	1.9	T	11.4	1.0	5.6	14.3	0.6	0.0	34.8
1963-64	0.0	0.0	0.0	5.2	5.3	5.5	22.4	3.0	12.3	4.9	1.9	T	60.5
1964-65	0.0	0.0	0.0	0.0	12.2	11.0	9.3	4.9	4.5	7.7	0.8	0.0	50.4
1965-66	0.0	0.0	T	0.8	9.4	14.3	4.8	7.2	2.6	2.0	T	0.0	41.1
1966-67	0.0	0.0	0.0	0.1	0.9	17.4	12.5	1.2	23.2	11.1	1.2	0.0	67.6
1967-68	0.0	0.0	0.0	T	6.4	8.5	16.1	0.5	10.4	3.8	0.4	T	46.1
1968-69	0.0	0.0	T	0.0	3.3	24.9	10.8	17.4	4.9	2.4	0.0	0.0	63.7
1969-70	0.0	0.0	0.0	2.3	T	11.1	1.5	1.0	4.6	4.1	0.4	0.0	25.0
1970-71	0.0	0.0	0.0	4.0	3.1	17.8	9.0	11.8	2.0	10.1	11.3	T	69.1
1971-72	0.0	0.0	T	5.4	8.9	13.4	1.1	9.1	3.6	0.6	T	0.0	42.1
1972-73	0.0	0.0	0.0	1.0	7.3	8.6	10.5	5.5	4.2	1.0	0.0	0.0	38.1
1973-74	0.0	0.0	0.0	0.0	6.8	11.8	5.1	1.0	2.2	2.1	T	0.0	29.0
1974-75	0.0	0.0	0.0	0.0	T	6.7	14.5	5.9	6.3	15.6	8.0	0.0	57.0
1975-76	0.0	0.0	0.0	0.6	7.3	2.5	4.4	5.2	3.5	0.8	0.0	0.0	24.3
1976-77	0.0	0.0	0.0	0.0	0.5	T	4.3	1.8	1.5	2.0	2.2	0.0	12.3
1977-78	0.0	0.0	0.0	0.0	3.0	5.3	5.5	5.5	T	T	0.0		19.3
1978-79	0.0	0.0	T	T	1.0	2.0	8.3	13.0	2.4	2.1	0.6	T	29.4
1979-80	0.0	0.0	0.0	1.5	9.2	2.3	12.5	3.4	3.9	0.5	2.0	0.0	35.3
1980-81	0.0	0.0	0.0	1.2	0.5	0.5	4.4	1.4	2.6	2.9	0.2	T	13.7
1981-82	0.0	0.0	0.0	T	0.3	9.8	11.2	0.3	13.8	3.5	T	T	38.9
1982-83	0.0	0.0	2.0	T	8.1	6.6	16.5	10.6	9.1	1.9	0.2	0.0	55.5
1983-84	0.0	0.0	0.0	0.0	13.4	33.2	6.6	5.8	5.1	5.9	T	0.0	70.0
1984-85	0.0	0.0	0.0	5.6	5.4	4.9	5.6	2.0	7.7	0.4	0.0	0.0	31.6
1985-86	0.0	0.0	0.0	0.7	16.8	5.1							
Record Mean	0.0	0.0	0.1	0.8	4.6	8.2	9.8	6.2	5.6	2.6	0.8	T	38.8

See Reference Notes, relative to all above tables, on preceding page.

Ely, Nevada, is located within but near the southern rim of the Great Basin. The neighboring terrain consists of alternate mountain ranges and sagebrush covered valleys. Principal cover on the mountains is juniper, pinion, and, at higher elevations, white fir, and white pine. Valley floors in this region are near 6,000 feet above sea level. This high elevation is conducive to sharp nighttime radiation, which produces pleasant summer nights but also reduces the season that is free from freezing temperatures.

The Ely weather station is near the center of Steptoe Valley, which is 5 miles wide at this point. The mountains of the Egan Range to the west and the Schell Creek Range to the east range up to 4,000 feet above the station elevation and prevent strong surface winds from these directions. A very pronounced drainage wind sweeps down the valley during the morning hours. More precipitation is noted near the mountains than is measured in the center of the valley.

Because of low annual precipitation, farming is limited to areas that can be irrigated from mountain streams or wells. The livestock industry is predominant in agriculture. Cultivated crops consist almost entirely of grains and forage.

The mountain ranges provide fairly good summer pastures for cattle and the lowlands provide food for a good portion of the winter in dry or snow-softened desert plants. All stock, however, has to be finished for market in the feed yards. Sheep share the mountain pastures with cattle in the summer, and as winter approaches move out on the wide flat valleys. These browsers eat snow for water and consume a wide variety of desert plants, including the lowly sagebrush. It is not uncommon for bands of sheep to spend an entire winter without supplemental feed.

Based on the 1951–1980 period, the average first occurrence of 32 degrees Fahrenheit in the fall is September 6 and the average last occurrence in the spring is June 16.

TABLE 1 — NORMALS, MEANS AND EXTREMES

ELY, NEVADA

LATITUDE: 39°17'N   LONGITUDE: 114°51'W   ELEVATION: FT. GRND 6253 BARO 06257   TIME ZONE: PACIFIC   WBAN: 23154

	(a)	JAN	FEB	MAR	APR	MAY	JUNE	JULY	AUG	SEP	OCT	NOV	DEC	YEAR
TEMPERATURE °F:														
Normals														
-Daily Maximum		39.0	42.6	47.3	56.2	66.5	77.5	86.8	84.2	76.0	64.0	49.2	40.9	60.9
-Daily Minimum		9.7	15.0	19.4	25.7	33.6	40.4	48.1	46.6	37.3	28.0	18.5	11.0	27.8
-Monthly		24.4	28.8	33.4	41.0	50.0	59.0	67.5	65.4	56.7	46.0	33.9	26.0	44.3
Extremes														
-Record Highest	47	68	66	73	78	89	99	100	97	93	84	75	67	100
-Year		1951	1977	1966	1985	1984	1954	1985	1981	1950	1967	1975	1958	JUL 1985
-Record Lowest	47	-27	-25	-13	-5	7	18	30	24	15	-3	-15	-28	-28
-Year		1949	1949	1952	1982	1950	1976	1983	1960	1968	1971	1985	1972	DEC 1972
NORMAL DEGREE DAYS:														
Heating (base 65°F)		1259	1014	980	723	462	196	10	64	261	589	933	1209	7700
Cooling (base 65°F)		0	0	0	0	0	16	88	76	12	0	0	0	192
% OF POSSIBLE SUNSHINE	46	67	67	70	69	72	80	80	81	82	75	66	65	73
MEAN SKY COVER (tenths)														
Sunrise - Sunset	42	6.2	6.4	6.4	6.1	5.9	4.3	3.9	3.8	3.3	4.3	5.8	6.0	5.2
MEAN NUMBER OF DAYS:														
Sunrise to Sunset														
-Clear	47	8.6	6.9	7.6	7.7	7.7	13.2	14.9	14.8	17.3	14.4	9.7	8.8	131.6
-Partly Cloudy	47	7.6	6.9	8.3	9.0	11.4	10.3	11.3	11.7	7.7	8.2	8.1	7.9	108.4
-Cloudy	47	14.7	14.4	15.1	13.3	12.0	6.5	4.9	4.4	4.9	8.4	12.2	14.3	125.1
Precipitation														
.01 inches or more	47	6.9	7.0	8.5	7.4	7.2	4.8	5.5	5.3	4.5	5.0	5.1	6.4	73.7
Snow,Ice pellets														
1.0 inches or more	47	2.5	2.4	3.2	2.1	0.9	0.1	0.0	0.0	0.1	0.8	1.4	2.6	16.1
Thunderstorms	47	0.2	0.3	0.4	1.2	4.1	4.6	8.3	8.0	3.3	1.4	0.3	0.2	32.3
Heavy Fog Visibility 1/4 mile or less	47	0.3	0.2	0.3	0.3	0.1	0.*	0.0	0.*	0.1	0.2	0.3	0.4	2.4
Temperature °F														
-Maximum														
90° and above	47	0.0	0.0	0.0	0.0	0.0	2.2	9.8	5.3	0.5	0.0	0.0	0.0	17.7
32° and below	47	7.9	4.5	2.2	0.4	0.*	0.0	0.0	0.0	0.0	0.2	2.6	6.4	24.1
-Minimum														
32° and below	47	30.6	27.5	29.5	24.6	13.2	3.6	0.1	0.5	7.8	22.5	28.1	30.5	218.5
0° and below	47	7.4	3.9	1.3	0.1	0.0	0.0	0.0	0.0	0.0	0.*	1.3	4.7	18.8
AVG. STATION PRESS.(mb)	13	809.0	808.5	805.1	806.3	807.0	809.0	811.3	811.2	810.6	810.7	809.1	809.9	809.0
RELATIVE HUMIDITY (%)														
Hour 04	33	71	74	72	68	66	59	52	56	59	65	70	71	65
Hour 10 (Local Time)	33	60	58	50	40	35	28	24	27	31	39	51	58	42
Hour 16	33	55	50	43	34	31	23	22	23	24	32	46	55	37
Hour 22	33	70	71	66	57	53	43	39	42	46	56	66	69	57
PRECIPITATION (inches):														
Water Equivalent														
-Normal		0.72	0.68	0.91	0.92	1.08	0.80	0.65	0.62	0.70	0.59	0.60	0.75	9.02
-Maximum Monthly	47	1.92	2.19	2.40	3.41	3.26	3.53	2.18	2.51	4.99	3.67	1.82	2.11	4.99
-Year		1952	1969	1952	1978	1977	1963	1984	1983	1982	1981	1960	1966	SEP 1982
-Minimum Monthly	47	T	0.01	0.07	0.16	T	T	T	T	T	0.00	T	T	0.00
-Year		1948	1972	1972	1966	1948	1978	1948	1985	1953	1952	1959	1976	OCT 1952
-Maximum in 24 hrs	47	0.95	1.54	0.86	1.04	1.42	1.50	1.22	0.91	2.87	1.39	1.29	1.12	2.87
-Year		1952	1969	1954	1947	1955	1963	1952	1984	1982	1976	1960	1966	SEP 1982
Snow,Ice pellets														
-Maximum Monthly	47	24.8	20.0	24.8	24.5	12.1	5.6			6.3	12.1	17.3	22.3	24.8
-Year		1967	1976	1958	1963	1975	1939			1982	1981	1985	1968	JAN 1967
-Maximum in 24 hrs	47	13.1	10.4	10.6	10.7	8.0	5.6			2.9	7.3	12.9	12.7	13.1
-Year		1943	1956	1954	1970	1975	1939			1982	1954	1978	1970	JAN 1943
WIND:														
Mean Speed (mph)	47	10.3	10.4	10.8	11.0	10.7	10.6	10.3	10.5	10.4	10.1	10.0	10.0	10.4
Prevailing Direction through 1963		S	S	S	S	S	S	S	S	S	S	S	S	S
Fastest Mile														
-Direction (!!!)	42	SE	S	SE	S	S	SW	S	E	S	S	S	SE	S
-Speed (MPH)	42	66	56	59	59	74	63	50	57	57	65	51	61	74
-Year		1952	1954	1951	1951	1948	1952	1957	1954	1953	1950	1954	1952	MAY 1948
Peak Gust														
-Direction (!!!)	2	N	SE	SW	S	S	W	SE	SW	SW	SW	SE	S	S
-Speed (mph)	2	39	44	47	52	63	51	52	51	47	61	55	44	63
-Date		1984	1984	1985	1984	1984	1984	1985	1985	1985	1985	1985	1984	MAY 1984

See Reference Notes to this table on the following page.

TABLE 2   PRECIPITATION (inches)   ELY, NEVADA

YEAR	JAN	FEB	MAR	APR	MAY	JUNE	JULY	AUG	SEP	OCT	NOV	DEC	ANNUAL
1956	0.99	0.94	0.34	0.63	1.61	0.38	0.18	T	0.65	0.54	0.04	0.06	6.36
1957	1.01	0.17	1.14	0.53	2.68	0.41	0.66	0.71	0.02	0.77	0.54	0.50	9.14
1958	0.53	1.08	2.25	0.69	0.58	0.35	0.12	0.49	0.79	T	0.53	0.17	7.58
1959	0.17	1.43	0.31	0.46	1.07	0.17	0.16	0.44	0.99	0.15	T	0.62	5.97
1960	0.80	0.70	0.77	0.67	0.79	0.21	0.26	0.19	0.98	0.37	1.82	0.33	7.89
1961	0.15	0.36	1.21	0.80	0.64	0.56	0.64	1.14	0.41	0.52	0.36	0.48	7.27
1962	0.81	1.51	1.09	0.18	1.26	0.45	0.62	T	0.10	1.06	0.28	T	7.36
1963	0.11	0.49	0.84	2.12	0.40	3.53	0.01	0.29	2.18	0.37	0.60	0.20	11.14
1964	1.41	0.07	1.24	2.77	1.17	2.44	0.02	0.58	0.09	0.19	0.93	1.79	12.70
1965	0.46	0.64	0.46	0.74	0.54	1.25	1.12	1.52	1.56	0.27	0.93	1.28	10.77
1966	0.23	0.31	0.16	0.16	0.46	0.14	0.16	0.61	1.34	0.10	0.30	2.11	6.08
1967	1.86	0.10	0.37	1.38	3.05	2.83	0.84	0.41	2.23	0.13	0.84	0.69	14.73
1968	0.15	0.92	0.67	1.26	1.00	1.12	1.32	1.04	0.10	1.44	0.22	0.79	10.03
1969	1.24	2.19	0.41	0.98	0.28	2.80	0.55	0.34	0.37	0.91	0.79	0.59	11.45
1970	0.11	0.14	0.59	1.55	0.01	1.09	1.81	1.45	0.45	0.23	1.69	1.57	10.69
1971	0.63	0.57	0.20	1.31	2.89	0.09	0.17	0.25	0.39	1.08	0.59	1.25	9.42
1972	0.17	0.01	0.07	0.88	0.32	0.83	0.17	0.47	1.82	1.02	0.14	0.69	6.59
1973	1.34	0.71	2.17	0.20	0.38	1.14	0.43	2.06	0.07	0.88	1.10	0.75	11.23
1974	0.41	0.29	0.67	0.18	0.30	T	0.29	0.02	0.01	1.54	0.23	0.28	4.22
1975	0.74	0.76	1.59	1.20	1.48	0.31	1.04	0.51	0.55	0.91	0.29	0.39	9.77
1976	0.38	1.51	0.77	0.77	0.45	0.34	1.57	0.16	0.66	1.48	0.16	T	8.25
1977	0.39	0.09	0.74	0.17	3.26	0.49	0.49	1.59	0.50	0.33	0.24	0.90	9.19
1978	0.64	1.27	2.00	3.41	0.45	T	0.19	0.23	1.33	0.82	1.42	0.71	12.47
1979	0.89	0.59	1.07	0.22	1.44	0.15	1.27	0.58	0.07	0.76	0.28	0.07	7.39
1980	1.55	1.08	1.57	0.51	2.55	0.72	0.76	0.35	1.65	0.37	0.55	1.12	12.78
1981	0.77	0.16	1.32	1.10	2.02	0.15	0.24	0.07	0.36	3.67	0.17	0.26	10.29
1982	1.06	0.31	2.07	0.72	1.57	0.05	0.58	1.41	4.99	1.28	1.03	0.46	15.53
1983	1.41	1.33	1.18	1.87	0.38	2.28	0.09	2.51	0.88	0.50	0.96	1.45	14.84
1984	0.36	0.39	1.09	0.94	0.35	0.63	2.18	2.01	3.73	1.41	0.99	0.76	14.84
1985	0.49	0.42	1.07	0.17	1.33	0.43	0.58	T	1.82	1.44	1.55	0.59	9.89
Record Mean	0.69	0.62	0.95	1.00	1.02	0.79	0.63	0.63	0.87	0.81	0.63	0.67	9.30

TABLE 3   AVERAGE TEMPERATURE (deg. F)   ELY, NEVADA

YEAR	JAN	FEB	MAR	APR	MAY	JUNE	JULY	AUG	SEP	OCT	NOV	DEC	ANNUAL
1956	31.9	19.5	35.8	41.3	50.5	60.5	65.6	62.9	59.8	44.2	30.6	26.8	44.1
1957	18.7	34.3	35.9	41.0	47.5	60.0	67.0	66.5	56.4	43.5	29.1	30.6	44.2
1958	27.5	33.9	30.2	38.2	54.1	59.5	65.7	68.6	57.8	47.5	34.6	33.9	46.0
1959	28.8	26.0	35.6	44.8	47.4	62.6	69.4	65.5	53.5	46.1	34.4	25.9	45.0
1960	19.0	23.3	39.3	43.2	50.3	62.6	68.0	65.4	60.6	46.3	35.3	26.6	45.0
1961	27.0	31.7	34.0	41.4	50.2	62.4	68.4	66.1	52.6	43.4	30.5	25.1	44.4
1962	20.4	29.4	29.3	47.4	49.0	59.6	65.5	65.8	58.9	48.7	38.0	29.7	45.2
1963	22.9	36.8	32.3	36.1	54.3	57.7	66.5	65.3	59.8	50.7	34.3	27.6	45.0
1964	19.3	22.3	27.9	39.6	48.6	56.6	68.4	64.7	54.4	49.6	29.3	26.0	42.2
1965	29.4	27.8	32.4	41.9	45.9	55.2	65.5	63.6	50.5	49.2	37.6	24.5	43.6
1966	20.1	22.5	36.7	42.6	55.5	60.6	68.6	66.3	59.1	45.7	38.1	27.0	45.2
1967	26.2	28.4	38.4	34.7	49.2	54.6	68.1	68.1	57.9	47.7	37.6	17.6	44.1
1968	23.1	35.8	36.7	38.0	49.3	59.4	68.4	61.1	54.5	46.7	34.8	23.8	44.3
1969	31.2	25.9	26.2	43.2	56.9	57.4	68.5	69.6	60.7	40.4	35.1	29.8	45.4
1970	29.3	35.0	32.9	34.8	50.7	58.1	67.3	68.1	52.2	41.1	36.0	21.7	43.9
1971	24.2	29.9	34.7	41.3	47.0	59.1	68.2	68.4	52.5	40.3	32.0	18.9	43.1
1972	24.1	32.5	41.2	41.6	51.1	61.9	68.1	63.8	54.1	45.4	30.9	20.1	44.6
1973	20.3	26.9	30.3	39.1	52.7	59.6	66.7	66.1	53.9	45.7	32.6	27.0	43.4
1974	23.3	28.2	38.6	39.9	52.7	63.8	67.7	63.9	57.4	44.8	36.4	24.4	45.1
1975	24.1	28.4	31.8	34.1	47.5	57.2	68.0	63.6	57.5	44.8	32.1	28.5	43.1
1976	26.3	30.8	30.7	39.8	53.6	58.1	67.1	60.7	56.8	44.1	36.7	27.3	44.3
1977	22.1	30.5	28.4	45.0	44.9	62.5	67.2	65.8	57.4	47.6	36.4	30.8	44.9
1978	29.3	30.6	40.8	40.9	47.9	59.2	67.0	64.0	53.9	48.9	29.9	20.8	44.4
1979	17.1	26.6	34.7	42.3	51.8	60.0	67.4	63.6	61.1	49.3	31.3	31.4	44.8
1980	28.0	34.9	33.1	44.2	47.3	58.3	68.2	64.9	57.5	45.1	36.1	31.7	45.8
1981	31.5	31.3	35.6	46.5	50.0	63.3	69.4	68.1	59.7	42.9	38.8	33.8	47.6
1982	22.5	30.6	33.6	38.8	49.6	57.2	65.7	67.7	54.7	41.8	33.1	25.0	43.3
1983	28.9	30.2	36.7	37.6	47.7	57.5	65.8	65.5	59.2	48.1	33.2	22.1	44.8
1984	24.8	29.7	35.2	39.3	54.6	57.4	67.5	65.5	58.1	40.2	34.2	22.1	44.1
1985	19.5	24.4	32.5	46.4	52.9	63.4	68.5	65.5	52.0	44.3	27.5	25.0	43.5
Record Mean	23.9	28.3	33.5	41.4	50.2	58.5	67.3	65.5	56.6	45.7	33.8	26.5	44.3
Max	38.5	42.0	47.2	56.5	66.7	77.0	86.6	84.3	75.7	63.0	48.9	41.0	60.6
Min	9.3	14.5	19.7	26.2	33.7	40.0	48.0	46.6	37.5	28.3	18.7	12.0	27.9

REFERENCE NOTES FOR TABLES 1, 2, 3 and 6   (ELY, NV)

GENERAL

T - TRACE AMOUNT
BLANK ENTRIES DENOTE MISSING/UNREPORTED DATA.
# INDICATES A STATION OR INSTRUMENT RELOCATION.

SPECIFIC

TABLE 1

(a) - LENGTH OF RECORD IN YEARS. ALTHOUGH
INDIVIDUAL MONTHS MAY BE MISSING.
* LESS THAN .05

NORMALS — BASED ON THE 1951-1980 RECORD PERIOD.
EXTREMES — DATES ARE THE MOST RECENT OCCURRENCE.
WIND DIR. — NUMERALS SHOW TENS OF DEGREES
CLOCKWISE FROM TRUE NORTH.
"00" INDICATES CALM.
RESULTANT WIND DIRECTIONS ARE GIVEN TO WHOLE DEGREES.

EXCEPTIONS

TABLES 2, 3, and 6

RECORD MEANS ARE THROUGH THE CURRENT YEAR,
BEGINNING IN   1939 FOR TEMPERATURE
1939 FOR PRECIPITATION
1939 FOR SNOWFALL

TABLE 4 HEATING DEGREE DAYS Base 65 deg. F ELY, NEVADA

SEASON	JULY	AUG	SEP	OCT	NOV	DEC	JAN	FEB	MAR	APR	MAY	JUNE	TOTAL
1956-57	36	82	159	637	1025	1176	1432	855	893	712	535	176	7718
1957-58	12	33	252	659	1074	1058	1154	864	1068	796	332	173	7475
1958-59	34	3	227	534	903	958	1115	1085	901	598	537	98	6993
1959-60	11	51	342	580	909	1205	1418	1200	792	649	447	86	7690
1960-61	10	75	137	574	883	1183	1170	925	953	702	454	128	7194
1961-62	6	19	368	661	1029	1230	1381	987	1098	522	490	176	7967
1962-63	29	49	177	500	802	1091	1299	786	1007	861	322	335	7258
1963-64	20	35	152	435	913	1152	1411	1230	1143	755	500	251	7997
1964-65	17	74	314	470	1064	1203	1097	1037	1006	684	586	285	7837
1965-66	27	76	429	485	814	1248	1387	1188	869	664	291	154	7632
1966-67	9	35	177	592	801	1169	1193	1019	817	904	485	313	7514
1967-68	3	10	210	530	814	1462	1293	840	870	802	483	182	7499
1968-69	10	151	316	559	900	1268	1039	1087	1198	649	244	229	7650
1969-70	26	7	127	757	892	1084	1100	834	990	900	435	234	7386
1970-71	12	7	376	734	863	1334	1259	979	933	705	549	183	7934
1971-72	18	4	369	760	983	1422	1257	936	732	693	425	102	7701
1972-73	16	86	320	599	1019	1384	1379	1059	1067	769	376	195	8269
1973-74	18	42	326	591	965	1170	1285	1021	809	747	373	93	7440
1974-75	12	54	237	622	852	1250	1262	1021	1023	921	533	228	8015
1975-76	19	81	217	617	982	1123	1194	985	1058	749	346	220	7591
1976-77	31	130	245	641	842	1162	1324	959	1126	596	614	91	7761
1977-78	9	43	238	535	850	1056	1101	959	742	716	521	172	6942
1978-79	43	103	336	490	1047	1361	1479	1072	933	675	400	179	8118
1979-80	22	95	124	464	1005	1035	1138	866	983	617	539	208	7096
1980-81	11	76	221	608	859	1026	1033	937	905	546	456	107	6785
1981-82	1	3	159	680	781	960	1311	960	970	778	472	236	7311
1982-83	60	3	311	715	951	1236	1111	967	870	815	533	208	7780
1983-84	45	50	180	518	948	1172	1238	1018	917	764	318	240	7408
1984-85	19	21	209	759	917	1321	1404	1129	1000	548	371	98	7796
1985-86	8	42	386	635	1118	1232							

TABLE 5 COOLING DEGREE DAYS Base 65 deg. F ELY, NEVADA

YEAR	JAN	FEB	MAR	APR	MAY	JUNE	JULY	AUG	SEP	OCT	NOV	DEC	TOTAL
1969	0	0	0	0	1	6	144	154	7	0	0	0	312
1970	0	0	0	0	0	31	89	110	0	0	0	0	230
1971	0	0	0	0	0	8	122	117	2	0	0	0	249
1972	0	0	0	0	0	16	117	58	0	0	0	0	191
1973	0	0	0	0	0	38	77	85	0	0	0	0	200
1974	0	0	0	0	0	63	102	30	14	0	0	0	209
1975	0	0	0	0	0	2	120	46	0	0	0	0	168
1976	0	0	0	0	0	18	103	2	4	0	0	0	127
1977	0	0	0	0	0	24	86	74	14	0	0	0	198
1978	0	0	0	0	0	7	110	79	9	0	0	0	205
1979	0	0	0	0	0	35	103	56	14	0	0	0	208
1980	0	0	0	0	0	12	116	78	2	0	0	0	208
1981	0	0	0	0	0	64	143	108	9	0	0	0	324
1982	0	0	0	0	0	6	90	93	11	0	0	0	200
1983	0	0	0	0	0	0	78	74	11	0	0	0	163
1984	0	0	0	0	1	18	103	44	11	0	0	0	177
1985	0	0	0	0	0	57	124	65	2	0	0	0	248

TABLE 6 SNOWFALL (inches) ELY, NEVADA

SEASON	JULY	AUG	SEP	OCT	NOV	DEC	JAN	FEB	MAR	APR	MAY	JUNE	TOTAL
1956-57	0.0	0.0	0.0	1.1	1.6	0.4	18.5	T	12.2	5.9	3.8	0.0	43.5
1957-58	0.0	0.0	0.0	0.6	10.0	4.0	9.1	12.6	24.8	7.5	0.0	0.0	68.6
1958-59	0.0	0.0	0.8	T	5.3	1.8	2.0	19.9	3.3	0.3	3.9	0.0	37.3
1959-60	0.0	0.0	0.1	1.2	T	10.6	9.9	8.3	5.1	4.5	2.6	0.0	42.3
1960-61	0.0	0.0	0.0	3.9	2.4	4.9	0.5	5.0	18.5	7.3	4.1	0.0	46.6
1961-62	0.0	0.0	T	6.6	4.6	5.5	13.9	5.9	15.0	0.3	T	0.0	51.8
1962-63	0.0	0.0	0.0	T	1.1	T	T	2.1	7.7	24.5	0.7	T	36.1
1963-64	0.0	0.0	0.0	0.0	5.0	3.5	19.3	1.2	17.4	21.8	10.8	T	79.0
1964-65	0.0	0.0	T	T	12.9	17.9	6.6	9.0	4.6	6.8	1.0	0.0	58.8
1965-66	0.0	0.0	T	2.0	5.5	12.7	3.5	4.2	1.6	1.4	T	0.0	30.9
1966-67	0.0	0.0	0.0	T	1.0	7.1	24.8	2.1	4.4	12.8	3.6	0.8	56.6
1967-68	0.0	0.0	0.0	0.0	12.7	11.8	2.4	8.6	9.5	16.7	0.7	0.0	62.4
1968-69	0.0	0.0	T	3.5	1.9	22.3	8.3	19.1	6.3	3.9	0.0	T	65.3
1969-70	0.0	0.0	0.0	2.2	2.9	2.6	1.0	0.9	4.2	19.0	0.0	0.0	33.7
1970-71	0.0	0.0	0.0	T	8.0	17.5	8.8	7.7	4.1	13.1	6.3	0.0	65.5
1971-72	0.0	0.0	2.2	9.7	6.0	13.6	3.3	1.0	T	4.3	1.3	0.0	40.5
1972-73	0.0	0.0	0.0	T	0.7	10.5	14.6	10.0	24.0	1.7	0.4	0.0	61.9
1973-74	0.0	0.0	0.0	3.2	11.4	7.9	7.2	5.1	8.9	0.7	2.2	0.0	46.6
1974-75	0.0	0.0	0.0	1.6	0.9	4.4	10.0	8.8	16.4	15.6	12.1	T	69.8
1975-76	0.0	0.0	0.0	5.6	4.7	6.2	5.9	20.0	10.4	6.9	0.0	0.0	59.7
1976-77	0.0	0.0	0.0	T	1.4	T	5.9	0.6	11.2	2.6	10.9	0.0	32.6
1977-78	0.0	0.0	0.2	T	7.9	7.9	8.0	9.6	5.9	18.5	4.4	0.0	56.5
1978-79	0.0	0.0	1.3	6.8	17.0	9.1	13.6	7.1	13.7	2.8	2.5	1.5	75.4
1979-80	0.0	0.0	0.0	0.7	2.2	0.8	17.8	6.9	19.0	1.4	8.6	0.2	57.6
1980-81	0.0	0.0	0.0	1.6	4.0	11.5	8.0	2.2	15.6	6.3	0.7	0.0	49.9
1981-82	0.0	0.0	0.0	12.1	1.5	1.9	13.1	1.4	16.2	7.8	3.9	0.1	58.0
1982-83	0.0	0.0	6.3	1.0	9.3	3.6	13.1	10.3	11.1	10.4	3.8	0.0	71.1
1983-84	0.0	0.0	0.0	9.9	13.1	5.1	5.7	6.5	6.4	0.0	T	46.7	
1984-85	0.0	0.0	T	3.8	10.4	11.3	6.3	5.5	15.4	1.0	2.0	0.0	55.7
1985-86	0.0	0.0	T	8.7	17.3	4.5							
Record Mean	0.0	0.0	0.2	2.3	5.3	7.5	8.9	6.8	10.0	6.4	2.4	0.2	49.9

See Reference Notes, relative to all above tables, on preceding page.

Las Vegas is situated near the center of a broad desert valley, which is almost surrounded by mountains ranging from 2,000 to 10,000 feet higher than the floor of the valley. This Vegas Valley, comprising about 600 square miles, runs from northwest to southeast, and slopes gradually upward on each side toward the surrounding mountains. Weather observations are taken at McCarran Airport, 7 miles south of downtown Las Vegas, and about 5 miles southwest and 300 feet higher than the lower portions of the valley. Since mountains encircle the valley, drainage winds are usually downslope toward the center, or lowest portion of the valley. This condition also affects minimum temperatures, which in lower portions of the valley can be from 15 to 25 degrees colder than recorded at the airport on clear, calm nights.

The four seasons are well defined. Summers display desert conditions, with maximum temperatures usually in the 100 degree range. The proximity of the mountains contributes to the relatively cool summer nights, with the majority of minimum temperatures in the mid 70s. During about 2 weeks almost every summer warm, moist air predominates in this area, and causes scattered thunderstorms, occasionally quite severe, together with higher than average humidity. Soil erosion, especially near the mountains and foothills surrounding the valley, is evidence of the intensity of some of the thunderstorm activity. Winters, on the whole, are mild and pleasant. Daytime temperatures average near 60 degrees with mostly clear skies. The spring and fall seasons are generally considered most ideal, although rather sharp temperature changes can occur during these months. There are very few days during the spring and fall months when outdoor activities are affected in any degree by the weather.

The Sierra Nevada Mountains of California and the Spring Mountains immediately west of the Vegas Valley, the latter rising to elevations over 10,000 feet above the valley floor, act as effective barriers to moisture moving eastward from the Pacific Ocean. It is mainly these barriers that result in a minimum of dark overcast and rainy days. Rainy days average less than one in June to three per month in the winter months. Snow rarely falls in this valley and it usually melts as it falls, or shortly thereafter. Notable exceptions have occurred.

Strong winds, associated with major storms, usually reach this valley from the southwest or through the pass from the northwest. Winds over 50 mph are infrequent but, when they do occur, are probably the most provoking of the elements experienced in the Vegas Valley, because of the blowing dust and sand associated with them.

Based on the 1951–1980 period, the average first occurrence of 32 degrees Fahrenheit in the fall is November 21 and the average last occurrence in the spring is March 7.

TABLE 1 NORMALS, MEANS AND EXTREMES

LAS VEGAS, NEVADA

LATITUDE: 36°05'N   LONGITUDE: 115°10'W   ELEVATION: FT. GRND 2162 BARO 02179   TIME ZONE: PACIFIC   WBAN: 23169

	(a)	JAN	FEB	MAR	APR	MAY	JUNE	JULY	AUG	SEP	OCT	NOV	DEC	YEAR
TEMPERATURE °F:														
Normals														
-Daily Maximum		56.0	62.4	68.3	77.2	87.4	98.6	104.5	101.9	94.7	81.5	66.0	57.1	79.6
-Daily Minimum		33.0	37.7	42.3	49.8	59.0	68.6	75.9	73.9	65.6	53.5	41.2	33.6	52.8
-Monthly		44.5	50.0	55.3	63.5	73.2	83.6	90.2	87.9	80.2	67.5	53.6	45.4	66.2
Extremes														
-Record Highest	37	77	82	91	99	109	115	116	116	113	103	85	77	116
-Year		1975	1981	1966	1981	1951	1970	1985	1979	1950	1978	1980	1980	JUL 1985
-Record Lowest	37	8	18	23	31	40	49	61	56	46	26	21	15	8
-Year		1963	1985	1971	1975	1964	1955	1984	1968	1965	1971	1952	1968	JAN 1963
NORMAL DEGREE DAYS:														
Heating (base 65°F)		632	417	313	131	22	0	0	0	0	63	346	608	2532
Cooling (base 65°F)		0	0	12	86	279	558	784	713	453	144	0	0	3029
% OF POSSIBLE SUNSHINE	36	77	81	83	87	88	93	87	88	91	87	80	78	85
MEAN SKY COVER (tenths)														
Sunrise - Sunset	37	4.7	4.7	4.5	3.6	3.3	2.1	2.9	2.5	2.0	2.7	3.9	4.5	3.5
MEAN NUMBER OF DAYS:														
Sunrise to Sunset														
-Clear	37	14.0	12.5	13.8	16.6	18.5	22.4	19.5	21.5	22.6	20.6	15.6	14.6	212.1
-Partly Cloudy	37	6.2	7.1	8.9	7.5	7.9	5.1	8.1	6.6	5.1	6.3	7.4	6.8	83.0
-Cloudy	37	10.8	8.7	8.2	5.9	4.7	2.5	3.4	2.8	2.4	4.2	7.0	9.6	70.2
Precipitation														
.01 inches or more	37	3.0	2.6	2.9	1.8	1.3	0.7	2.7	3.0	1.8	1.7	2.1	2.5	26.1
Snow,Ice pellets														
1.0 inches or more	37	0.3	0.0	0.0	0.0	0.0	0.0	0.0	0.0	0.0	0.0	0.1	0.*	0.4
Thunderstorms	37	0.*	0.2	0.3	0.4	0.9	1.0	4.4	4.2	1.7	0.5	0.2	0.*	14.0
Heavy Fog Visibility														
1/4 mile or less	37	0.2	0.1	0.1	0.0	0.0	0.0	0.0	0.0	0.*	0.*	0.1	0.2	0.7
Temperature °F														
-Maximum														
90° and above	25	0.0	0.0	0.*	2.6	15.4	25.4	30.5	29.8	22.0	5.5	0.0	0.0	131.3
32° and below	25	0.2	0.0	0.0	0.0	0.0	0.0	0.0	0.0	0.0	0.0	0.0	0.*	0.2
-Minimum														
32° and below	25	13.4	4.6	1.5	0.1	0.0	0.0	0.0	0.0	0.0	0.2	2.3	11.5	33.5
0° and below	25	0.0	0.0	0.0	0.0	0.0	0.0	0.0	0.0	0.0	0.0	0.0	0.0	0.0
AVG. STATION PRESS.(mb)	13	942.1	940.9	936.9	935.8	933.9	933.6	935.1	935.6	935.9	938.6	940.3	942.4	937.6
RELATIVE HUMIDITY (%)														
Hour 04	25	55	50	44	35	31	24	29	35	34	38	46	55	40
Hour 10	25	41	36	30	22	19	15	19	24	23	26	33	41	27
Hour 16 (Local Time)	25	31	26	22	15	13	10	15	17	17	19	27	33	20
Hour 22	25	49	42	36	26	22	16	22	26	27	31	40	49	32
PRECIPITATION (inches):														
Water Equivalent														
-Normal		0.50	0.46	0.41	0.22	0.20	0.09	0.45	0.54	0.32	0.25	0.43	0.32	4.19
-Maximum Monthly	37	2.41	2.49	1.83	2.44	0.96	0.82	2.48	2.59	1.58	1.12	2.22	1.68	2.59
-Year		1949	1976	1973	1965	1969	1967	1984	1957	1963	1972	1965	1984	AUG 1957
-Minimum Monthly	37	T	0.00	0.00	0.00	0.00	0.00	0.00	0.00	0.00	0.00	0.00	0.00	0.00
-Year		1984	1977	1972	1962	1970	1982	1981	1980	1971	1979	1980	1981	JUN 1982
-Maximum in 24 hrs	37	1.01	1.19	1.14	0.97	0.80	0.75	1.36	2.59	1.07	0.70	1.78	0.95	2.59
-Year		1952	1976	1952	1965	1969	1967	1984	1957	1963	1976	1960	1977	AUG 1957
Snow,Ice pellets														
-Maximum Monthly	37	16.7	0.3	0.1	T						T	4.0	2.0	16.7
-Year		1949	1979	1976	1970						1956	1964	1967	JAN 1949
-Maximum in 24 hrs	37	9.0	6.9	0.1	T						T	4.0	2.0	9.0
-Year		1974	1979	1976	1970						1956	1964	1967	JAN 1974
WIND:														
Mean Speed (mph)	37	7.4	8.5	10.2	11.0	11.0	11.0	10.1	9.5	8.9	8.1	7.6	7.2	9.2
Prevailing Direction														
through 1963		W	SW	SW	SW	SW	SW	SW	SW	SW	WSW	W	W	SW
Fastest Mile														
-Direction	21	SW	NW	NW	SW	SW	SW	NE	NE	NW	NW	S	SW	NE
-Speed (MPH)	21	52	60	52	52	54	52	64	62	54	52	63	54	64
-Year		1965	1976	1977	1976	1977	1977	1976	1981	1971	1970	1983	1964	JUL 1976
Peak Gust														
-Direction	2	NW	NW	NW	NW	SE	NE	SW	S	SW	SW	SW	NW	NW
-Speed (mph)	2	43	67	82	55	62	59	53	55	41	52	51	52	82
-Date	2	1984	1984	1984	1984	1984	1984	1984	1984	1985	1984	1985	1984	MAR 1984

See Reference Notes to this table on the following page.

TABLE 2 PRECIPITATION (inches)   LAS VEGAS, NEVADA

YEAR	JAN	FEB	MAR	APR	MAY	JUNE	JULY	AUG	SEP	OCT	NOV	DEC	ANNUAL
1956	0.23	0.00	0.00	0.08	T	0.00	1.64	0.00	0.00	0.09	0.00	0.00	2.04
1957	0.26	0.16	0.10	0.55	0.15	T	0.41	2.59	T	0.49	0.26	0.01	4.98
1958	0.43	0.73	0.55	0.64	0.18	0.00	0.01	0.18	0.21	0.63	0.96	0.00	4.52
1959	0.11	0.72	T	T	T	T	0.02	0.33	0.01	0.51	1.09	1.38	4.17
1960	0.56	0.42	0.05	0.10	T	T	0.41	T	0.23	0.49	1.88	0.26	4.40
1961	0.22	0.01	0.51	0.02	T		0.53	0.80	0.26	0.26	0.10	0.46	3.17
1962	0.10	0.39	0.17	0.00	0.06	0.01	T	T	0.03	0.45	T	0.24	1.45
1963	0.12	0.33	0.23	0.10	T	0.15	0.00	0.42	1.58	0.61	0.33	0.00	3.87
1964	0.05	0.02	0.02	0.03	0.05	0.03	0.24	0.05	T	T	0.63	T	1.12
1965	0.05	0.45	0.74	2.44	0.40	T	0.28	0.38	T	T	2.22	1.00	7.96
1966	T	0.07	0.04	0.01	T	0.15	0.30	0.09	0.35	0.09	0.33	0.48	1.91
1967	0.47	0.00	T	0.09	0.21	0.82	0.20	0.38	1.03	0.00	1.52	0.82	5.54
1968	0.01	0.22	0.22	0.10	T	0.31	0.11	0.04	0.01	T	0.02	0.07	1.11
1969	1.57	0.96	0.57	T	0.96	0.23	0.06	0.33	0.08	0.27	0.06	T	5.09
1970	0.01	0.86	0.28	0.04	0.00	0.18	0.58	1.79	0.00	0.02	0.38	0.15	4.29
1971	T	0.03	T	T	0.84	T	0.08	0.90	0.00	0.06	0.12	0.51	2.54
1972	0.00	T	T	0.07	0.46	0.32	0.13	0.84	0.63	1.12	1.09	0.19	4.85
1973	0.49	1.64	1.83	0.35	0.09	0.03	T	0.08	T	0.02	0.14	0.01	4.68
1974	2.00	0.11	0.16	T	T	0.00	0.58	0.08	0.16	0.61	0.23	0.59	4.52
1975	0.01	0.05	1.07	0.42	0.35	T	0.26	0.06	1.17	0.03	T	0.05	3.47
1976	0.00	2.49	0.02	0.13	0.34	0.00	1.95	0.00	1.09	0.70	0.02	0.03	6.77
1977	0.21	0.00	0.28	0.01	0.72	0.05	T	1.38	0.19	0.06	0.01	1.06	3.97
1978	1.00	1.51	1.13	0.36	0.54	0.00	0.19	0.53	0.03	0.62	0.59	1.15	7.65
1979	2.18	0.07	0.96	0.06	0.35	0.00	0.78	2.12	T	0.00	0.03	0.24	6.79
1980	1.45	2.25	0.94	0.18	0.15		0.43	0.00	0.18	0.04	0.00	0.01	5.63
1981	0.09	0.20	1.44	0.02	0.50	T	0.00	0.20	0.25	0.15	0.29	0.00	3.14
1982	0.09	1.10	0.29	0.01	0.31	0.00	0.05	0.71	0.07	0.04	0.60	0.72	3.99
1983	0.43	0.32	0.90	0.45	0.16	0.00	0.06	1.25	0.50	0.26	0.10	0.43	4.86
1984	T	0.03	T	0.04	0.00	0.22	2.48	0.99	0.47	T	0.94	1.68	6.85
1985	0.19	0.02	0.06	0.31	T	0.02	0.13	0.00	0.08	0.07	0.37	0.02	1.27
Record Mean	0.49	0.44	0.45	0.23	0.17	0.07	0.45	0.50	0.35	0.24	0.37	0.39	4.15

TABLE 3 AVERAGE TEMPERATURE (deg. F)   LAS VEGAS, NEVADA

YEAR	JAN	FEB	MAR	APR	MAY	JUNE	JULY	AUG	SEP	OCT	NOV	DEC	ANNUAL
1956	48.8	45.6	56.5	63.4	74.3	85.5	88.2	86.2	83.4	65.3	50.6	45.8	66.2
1957	43.6	55.0	56.9	63.5	69.3	85.8	88.9	86.3	80.1	63.7	49.0	46.8	65.8
1958	46.0	52.2	52.1	62.4	78.0	84.0	88.8	90.5	81.0	69.9	53.5	48.8	67.3
1959	46.4	47.7	57.2	68.7	72.2	86.8	93.4	87.1	77.8	68.2	53.4	47.4	67.2
#1960	41.0	46.0	59.7	66.0	73.4	87.4	91.0	88.3	82.0	66.1	53.0	44.3	66.5
1961	45.1	51.0	56.2	65.1	72.9	87.0	91.0	87.7	75.6	64.1	50.3	42.5	65.7
1962	44.3	49.9	51.3	70.3	70.6	82.3	88.3	89.8	81.4	68.7	57.3	45.9	66.7
1963	41.1	55.8	54.0	58.5	75.9	78.6	90.4	87.9	80.5	70.1	55.0	44.6	66.0
1964	42.0	45.6	52.3	61.8	70.9	80.9	90.7	87.6	78.4	72.0	50.0	45.3	64.8
1965	47.1	49.5	53.2	61.2	69.6	78.0	88.7	87.9	74.8	69.8	55.9	45.0	65.1
1966	42.7	45.8	57.9	66.4	77.5	83.9	89.3	89.6	80.1	66.5	55.4	44.1	66.8
1967	45.3	50.6	59.3	56.2	72.5	79.6	91.7	90.3	80.0	69.1	56.7	41.6	66.1
1968	44.2	55.7	57.5	62.0	73.5	84.0	89.0	83.5	79.7	67.0	54.5	40.8	66.0
1969	47.5	46.3	53.0	64.4	76.8	81.4	89.7	92.2	82.5	62.8	53.5	45.8	66.3
1970	44.0	52.5	54.9	58.6	75.3	83.4	91.1	88.8	77.2	63.8	55.0	44.5	65.8
1971	44.4	49.7	55.8	63.0	68.0	83.3	92.8	89.0	77.6	61.7	50.9	41.4	64.8
1972	42.3	52.0	63.7	65.1	74.5	84.7	93.1	86.5	78.0	63.5	49.7	41.3	66.2
1973	40.9	49.6	50.7	62.2	76.7	85.2	91.7	87.6	78.9	67.7	53.4	46.2	65.9
1974	41.0	48.9	59.5	63.4	77.0	89.1	88.8	87.7	83.4	64.5	54.8	44.4	67.3
1975	45.3	48.8	53.9	56.6	72.5	83.8	90.3	87.5	81.7	66.1	53.0	48.2	65.6
1976	46.9	53.2	53.4	62.6	77.8	81.5	86.9	85.5	78.7	66.5	58.0	46.4	66.5
1977	45.7	54.2	52.6	68.6	67.7	88.0	92.4	90.1	80.6	71.4	57.2	51.9	68.3
1978	47.9	52.1	59.9	63.1	73.1	87.1	91.9	89.0	79.0	73.5	54.2	42.9	67.8
1979	41.1	48.4	56.0	66.1	75.4	85.5	91.1	85.9	85.3	70.7	51.6	47.2	67.1
1980	49.5	53.2	54.2	63.5	69.0	83.9	92.0	90.2	81.4	68.9	56.8	52.7	67.9
1981	51.1	52.5	56.4	70.6	74.3	88.8	92.7	90.0	82.5	64.4	58.0	48.8	69.2
1982	45.6	50.5	55.1	63.8	73.6	81.5	88.1	87.3	77.9	63.0	50.5	47.9	65.1
1983	46.6	51.7	56.4	58.5	72.8	82.8	88.5	83.8	82.5	67.8	55.3	47.9	66.2
1984	47.1	50.1	57.9	63.1	80.7	83.5	88.2	85.4	81.7	63.0	52.7	44.0	66.5
1985	44.4	47.4	54.9	68.2	76.9	87.4	92.0	89.1	75.4	67.3	51.7	48.3	67.0
Record Mean	44.4	49.6	55.3	64.0	73.7	83.3	89.7	87.6	79.8	66.9	53.3	45.8	66.1
Max	56.4	62.2	68.7	78.1	88.4	98.6	104.7	102.3	94.9	81.4	66.6	58.0	80.0
Min	32.3	36.9	42.0	49.9	58.9	67.9	74.7	72.8	64.7	52.4	40.0	33.5	52.2

REFERENCE NOTES FOR TABLES 1, 2, 3 and 6   (LAS VEGAS, NV)

GENERAL

T - TRACE AMOUNT
BLANK ENTRIES DENOTE MISSING/UNREPORTED DATA.
# INDICATES A STATION OR INSTRUMENT RELOCATION.

SPECIFIC

TABLE 1

(a) - LENGTH OF RECORD IN YEARS. ALTHOUGH
      INDIVIDUAL MONTHS MAY BE MISSING.
  * LESS THAN .05

NORMALS — BASED ON THE 1951-1980 RECORD PERIOD.
EXTREMES — DATES ARE THE MOST RECENT OCCURRENCE.
WIND DIR. — NUMERALS SHOW TENS OF DEGREES
            CLOCKWISE FROM TRUE NORTH.
            "00" INDICATES CALM.
RESULTANT WIND DIRECTIONS ARE GIVEN TO WHOLE DEGREES.

EXCEPTIONS

TABLES 2, 3, and 6

RECORD MEANS ARE THROUGH THE CURRENT YEAR,
BEGINNING IN   1937 FOR TEMPERATURE
                 1937 FOR PRECIPITATION
                 1949 FOR SNOWFALL

TABLE 4 HEATING DEGREE DAYS Base 65 deg. F   LAS VEGAS, NEVADA

SEASON	JULY	AUG	SEP	OCT	NOV	DEC	JAN	FEB	MAR	APR	MAY	JUNE	TOTAL
1956-57	0	0	0	85	427	590	657	276	243	93	32	0	2403
1957-58	0	0	0	89	475	556	580	350	394	154	3	0	2601
1958-59	0	0	0	42	338	497	571	477	239	21	19	0	2204
#1959-60	0	0	2	48	340	534	736	544	177	77	4	0	2462
1960-61	0	0	0	63	351	636	611	384	268	63	5	0	2381
1961-62	0	0	0	136	438	693	635	418	420	13	30	0	2783
1962-63	0	0	0	28	229	588	733	254	337	109	7	0	2285
1963-64	0	0	0	17	295	626	703	557	394	141	72	0	2805
1964-65	0	0	0	12	444	606	551	427	358	220	49	0	2667
1965-66	0	0	15	17	266	615	685	529	235	54	0	0	2416
1966-67	0	0	0	47	286	578	606	397	189	261	25	0	2389
1967-68	0	0	0	18	244	716	638	265	231	110	8	0	2230
1968-69	0	0	1	28	304	743	536	518	381	74	16	0	2601
1969-70	0	0	0	112	341	589	643	344	304	208	8	0	2549
1970-71	0	0	0	111	295	631	630	421	306	105	47	0	2546
1971-72	0	0	4	207	417	724	697	373	373	69	6	0	2596
1972-73	0	0	0	108	453	727	744	428	437	132	12	0	3041
1973-74	0	0	0	42	349	576	738	443	188	82	13	0	2431
1974-75	0	0	0	55	300	634	607	446	340	249	37	0	2668
1975-76	0	0	0	73	354	516	553	339	357	124	1	0	2317
1976-77	0	0	0	39	212	569	593	297	374	45	56	0	2185
1977-78	0	0	0	3	226	399	522	356	168	91	16	0	1781
1978-79	0	0	1	2	324	676	737	458	270	66	18	0	2552
1979-80	0	0	0	44	395	546	474	335	328	108	32	0	2262
1980-81	0	0	0	82	255	374	426	344	263	29	2	0	1775
1981-82	0	0	0	74	214	497	594	398	301	98	9	0	2185
1982-83	0	0	10	84	429	631	564	364	263	198	21	0	2564
1983-84	0	0	0	3	297	524	548	424	216	111	0	0	2123
1984-85	0	0	0	127	363	641	629	487	308	41	0	0	2596
1985-86	0	0	1	31	393	512							

TABLE 5 COOLING DEGREE DAYS Base 65 deg. F   LAS VEGAS, NEVADA

YEAR	JAN	FEB	MAR	APR	MAY	JUNE	JULY	AUG	SEP	OCT	NOV	DEC	TOTAL
1969	0	0	13	62	390	500	772	852	532	54	0	0	3175
1970	0	0	0	21	334	560	818	748	371	81	1	0	2934
1971	0	0	24	53	148	556	871	752	390	112	0	0	2906
1972	0	2	66	80	308	597	876	675	398	69	0	0	3071
1973	0	0	0	54	382	612	833	708	424	134	8	0	3155
1974	0	0	24	43	394	731	744	713	559	195	0	0	3403
1975	0	0	2	2	276	570	792	704	508	117	2	0	2973
1976	0	0	2	57	404	500	687	641	419	93	6	0	2809
1977	0	0	0	161	149	694	858	781	476	210	3	0	3332
1978	0	0	17	40	277	672	841	752	425	268	8	0	3300
1979	0	0	0	104	346	625	813	656	614	229	0	0	3387
1980	0	0	0	68	160	575	842	788	498	211	15	0	3157
1981	0	0	5	205	296	721	866	781	531	64	12	0	3481
1982	0	0	2	70	281	501	721	699	404	30	0	0	2708
1983	0	0	2	9	269	541	735	589	534	94	10	0	2783
1984	0	0	3	61	496	563	724	641	508	74	1	0	3071
1985	0	0	0	143	377	678	844	778	319	110	2	0	3251

TABLE 6 SNOWFALL (inches)   LAS VEGAS, NEVADA

SEASON	JULY	AUG	SEP	OCT	NOV	DEC	JAN	FEB	MAR	APR	MAY	JUNE	TOTAL
1970-71	0.0	0.0	0.0	0.0	0.0	T	T	0.0	0.0	0.0	0.0	0.0	T
1971-72	0.0	0.0	0.0	0.0	0.0	T	0.0	0.0	0.0	0.0	0.0	0.0	T
1972-73	0.0	0.0	0.0	0.0	0.0	0.0	0.3	0.0	T	0.0	0.0	0.0	0.3
1973-74	0.0	0.0	0.0	0.0	0.0	0.0	13.4	0.0	0.0	0.0	0.0	0.0	13.4
1974-75	0.0	0.0	0.0	0.0	0.0	0.0	T	0.0	T	0.0	0.0	0.0	T
1975-76	0.0	0.0	0.0	0.0	T	0.0	0.0	0.0	0.1	0.0	0.0	0.0	0.1
1976-77	0.0	0.0	0.0	0.0	0.0	0.0	0.0	0.0	0.0	0.0	0.0	0.0	0.0
1977-78	0.0	0.0	0.0	0.0	0.0	0.0	0.0	0.0	0.0	0.0	0.0	0.0	0.0
1978-79	0.0	0.0	0.0	0.0	0.0	T	9.9	0.3	0.0	0.0	0.0	0.0	10.2
1979-80	0.0	0.0	0.0	0.0	0.0	T	0.0	0.0	0.0	0.0	0.0	0.0	0.0
1980-81	0.0	0.0	0.0	0.0	0.0	0.0	0.0	0.0	0.0	0.0	0.0	0.0	0.0
1981-82	0.0	0.0	0.0	0.0	0.0	0.0	0.0	0.0	0.0	0.0	0.0	0.0	0.0
1982-83	0.0	0.0	0.0	0.0	0.0	0.0	0.0	0.0	0.0	0.0	0.0	0.0	0.0
1983-84	0.0	0.0	0.0	0.0	0.0	0.0	0.0	0.0	0.0	0.0	0.0	0.0	0.0
1984-85	0.0	0.0	0.0	0.0	0.0	T	0.0	T	0.0	0.0	0.0	0.0	T
1985-86	0.0	0.0	0.0	0.0	0.0	T							
Record Mean	0.0	0.0	0.0	T	0.1	0.1	1.2	T	T	T	0.0	0.0	1.4

See Reference Notes, relative to all above tables, on preceding page.

At an elevation of 4,400 feet above mean sea level, Reno is located at the west edge of Truckee Meadows in a semi-arid plateau lying in the lee of the Sierra Nevada Mountain Range. To the west, the Sierras rise to elevations of 9,000 to 11,000 feet. Hills to the east reach 6,000 to 7,000 feet. The Truckee River, flowing from the Sierras eastward through Reno, drains into Pyramid Lake to the northeast of the city.

The daily temperatures on the whole are mild, but the difference between the high and low often exceeds 45 degrees. While the afternoon high may exceed 90 degrees, a light wrap is often needed shortly after sunset. Nights with low temperatures over 60 degrees are rare. Afternoon temperatures in winter are moderate.

Based on the 1951–1980 period, the average first occurrence of 32 degrees Fahrenheit in the fall is September 16 and the average last occurrence in the spring is June 1.

More than half of the precipitation in Reno occurs mainly as mixed rain and snow, and falls from December to March. Although there is an average of about 25 inches of snow a year, it seldom remains on the ground for more than three or four days at a time. Summer rain comes mainly as brief thunderstorms in the middle and late afternoons. While precipitation is scarce, considerable water is available from the high altitude reservoirs in the Sierra Nevada, where precipitation is heavy.

Humidity is very low during the summer months, and moderately low during the winter. Fogs are rare, and are usually confined to the early morning hours of midwinter. Sunshine is abundant throughout the year.

TABLE 1   NORMALS, MEANS AND EXTREMES

RENO, NEVADA

LATITUDE: 39°30'N   LONGITUDE: 119°47'W   ELEVATION: FT. GRND 4404 BARO 4402   TIME ZONE: PACIFIC   WBAN: 23185

	(a)	JAN	FEB	MAR	APR	MAY	JUNE	JULY	AUG	SEP	OCT	NOV	DEC	YEAR
TEMPERATURE °F:														
Normals														
-Daily Maximum		44.8	51.1	55.8	63.3	72.3	81.8	91.3	88.7	81.4	70.0	55.6	46.2	66.9
-Daily Minimum		19.5	23.5	25.4	29.4	36.9	43.0	47.7	45.2	38.9	30.5	23.8	18.9	31.9
-Monthly		32.2	37.3	40.6	46.4	54.6	62.4	69.5	67.0	60.2	50.3	39.7	32.5	49.4
Extremes														
-Record Highest	44	70	74	83	89	95	101	104	105	101	91	77	70	105
-Year		1967	1967	1966	1981	1970	1985	1980	1983	1950	1980	1980	1969	AUG 1983
-Record Lowest	44	-16	-12	-2	13	18	25	33	24	20	8	1	-16	-16
-Year		1949	1962	1945	1956	1964	1954	1976	1962	1965	1971	1958	1972	DEC 1972
NORMAL DEGREE DAYS:														
Heating (base 65°F)		1017	773	756	558	333	124	16	59	171	456	759	1008	6030
Cooling (base 65°F)		0	0	0	0	11	46	156	117	27	0	0	0	357
% OF POSSIBLE SUNSHINE	38	66	68	75	80	82	85	92	93	92	83	71	64	79
MEAN SKY COVER (tenths)														
Sunrise - Sunset	43	6.3	6.3	6.0	5.6	4.9	3.6	2.3	2.3	2.6	4.0	5.8	6.2	4.7
MEAN NUMBER OF DAYS:														
Sunrise to Sunset														
-Clear	43	8.6	6.8	8.2	8.9	12.2	16.7	22.3	22.2	20.8	15.6	9.2	8.2	159.8
-Partly Cloudy	43	7.1	7.6	9.2	10.0	9.7	7.6	6.2	6.3	5.6	7.7	7.8	7.8	92.5
-Cloudy	43	15.3	13.8	13.5	11.2	9.1	5.7	2.5	2.6	3.6	7.7	13.0	15.1	113.0
Precipitation														
.01 inches or more	43	6.1	5.8	6.4	4.1	4.3	3.1	2.3	2.3	2.4	3.0	5.1	6.1	51.1
Snow,Ice pellets														
1.0 inches or more	43	2.1	1.6	1.6	0.6	0.3	0.0	0.0	0.0	0.*	0.2	0.8	1.4	8.4
Thunderstorms	43	0.0	0.*	0.1	0.3	2.0	2.6	3.5	3.0	1.2	0.6	0.0	0.0	13.2
Heavy Fog Visibility														
1/4 mile or less	43	2.2	0.8	0.3	0.1	0.1	0.0	0.0	0.*	0.1	0.1	0.2	0.7	2.9	7.5
Temperature °F														
-Maximum														
90° and above	22	0.0	0.0	0.0	0.0	1.2	7.0	21.1	16.5	4.8	0.1	0.0	0.0	50.8
32° and below	22	3.4	0.3	0.*	0.0	0.0	0.0	0.0	0.0	0.0	0.0	0.5	3.9	8.1
-Minimum														
32° and below	22	27.9	25.2	25.4	19.6	6.7	0.7	0.0	0.2	4.5	19.1	24.0	27.9	181.2
0° and below	22	1.0	0.*	0.*	0.0	0.0	0.0	0.0	0.0	0.0	0.0	0.0	1.0	2.1
AVG. STATION PRESS.(mb)	13	868.0	866.6	863.8	864.8	864.7	865.6	866.8	866.6	866.6	867.9	867.1	868.7	866.4
RELATIVE HUMIDITY (%)														
Hour 04	22	78	74	70	67	67	66	65	68	70	74	75	77	71
Hour 10 (Local Time)	22	70	58	49	39	34	31	28	30	35	43	58	67	45
Hour 16	22	51	40	34	28	25	22	19	20	22	28	43	52	32
Hour 22	22	73	63	56	49	45	41	36	39	47	57	67	72	54
PRECIPITATION (inches):														
Water Equivalent														
-Normal		1.24	0.95	0.74	0.46	0.74	0.34	0.30	0.27	0.30	0.34	0.60	1.21	7.49
-Maximum Monthly	44	4.13	3.69	2.02	2.04	2.89	1.31	1.06	1.65	2.31	2.14	3.08	5.25	5.25
-Year		1969	1962	1952	1958	1963	1965	1971	1965	1982	1945	1983	1955	DEC 1955
-Minimum Monthly	44	T	T	0.03	T	T	0.00	0.00	0.00	0.00	T	0.00	0.01	0.00
-Year		1966	1967	1972	1985	1985	1959	1951	1957	1974	1966	1959	1976	SEP 1974
-Maximum in 24 hrs	44	2.37	1.55	1.21	1.64	1.29	0.79	0.80	0.97	0.91	1.55	1.33	2.16	2.37
-Year		1943	1962	1943	1958	1963	1969	1949	1965	1982	1962	1981	1955	JAN 1943
Snow,Ice pellets														
-Maximum Monthly	44	20.0	23.5	29.0	7.5	14.1	0.2			1.5	5.1	16.5	25.6	29.0
-Year		1956	1969	1952	1958	1964	1970			1982	1971	1985	1971	MAR 1952
-Maximum in 24 hrs	44	12.0	13.9	16.9	7.3	9.0	0.2			1.5	3.7	15.4	14.9	16.9
-Year		1956	1959	1952	1958	1962	1970			1982	1971	1985	1971	MAR 1952
WIND:														
Mean Speed (mph)	43	5.6	6.1	7.7	8.2	7.8	7.5	6.9	6.4	5.7	5.3	5.4	5.2	6.5
Prevailing Direction														
through 1963		S	S	WNW	WNW	WNW	WNW	WNW	WNW	WNW	WNW	S	SW	WNW
Fastest Mile														
-Direction (!!!)	25	SW	SW	SW	SW	SW	N	SW	W	W	SW	S	SW	SW
-Speed (MPH)	25	80	66	80	48	48	50	44	47	42	74	61	68	80
-Year		1968	1975	1968	1982	1973	1985	1962	1984	1970	1985	1983	1968	JAN 1968
Peak Gust														
-Direction (!!!)	2	N	S	SE	S	SW	N	W	W	S	SW	SW	S	SW
-Speed (mph)	2	31	59	70	52	44	58	51	55	44	81	62	48	81
-Date		1985	1984	1984	1984	1984	1985	1985	1984	1985	1985	1984	1984	OCT 1985

See Reference Notes to this table on the following page.

TABLE 2 PRECIPITATION (inches)   RENO, NEVADA

YEAR	JAN	FEB	MAR	APR	MAY	JUNE	JULY	AUG	SEP	OCT	NOV	DEC	ANNUAL
1956	2.58	0.53	0.05	1.38	0.75	0.02	0.49	T	0.35	0.89	T	0.05	7.09
1957	1.20	0.80	0.26	0.21	0.83	0.06	0.03	0.00	0.61	1.27	0.58	1.52	7.37
1958	0.78	1.70	1.76	2.04	0.25	0.45	0.22	0.56	0.49	T	0.38	0.24	8.87
#1959	0.64	2.80	0.07	T	1.84	0.00	0.05	0.09	0.04	T	0.00	0.45	5.98
1960	1.21	1.19	0.59	0.05	T	T	0.87	T	T	0.18	1.45	0.24	5.78
1961	0.80	0.30	0.40	0.27	0.91	0.56	0.30	0.47	0.39	0.15	0.64	0.18	·5.37
1962	0.35	3.69	0.84	T	1.40	0.04	0.38	0.08	0.10	1.55	0.02	0.60	9.05
1963	2.51	1.09	0.41	0.82	2.89	1.10	T	0.17	0.18	0.24	1.44	0.08	10.93
1964	0.68	0.01	0.72	0.53	1.79	0.29	0.13	0.02	0.00	0.04	0.63	2.89	7.73
1965	1.64	0.01	0.98	0.27	0.18	1.31	0.35	1.65	0.50	0.02	1.62	1.19	9.72
1966	T	0.20	0.04	0.03	0.33	0.03	T	0.02	0.10	T	1.07	1.45	3.27
1967	2.11	T	1.93	0.95	0.47	0.59	0.57	1.22	0.82	0.04	0.22	0.55	9.47
1968	1.11	0.92	0.84	0.02	0.30	0.16	0.05	0.13	0.15	0.01	0.73	1.03	5.45
1969	4.13	1.74	0.07	0.10	0.20	1.29	0.17	T	0.01	0.40	0.04	2.07	10.22
1970	1.73	0.32	0.19	0.60	T	0.88	0.05	0.02	0.01	0.03	1.47	1.65	6.95
1971	0.75	0.33	1.54	0.59	2.38	0.09	1.06	0.09	0.10	0.44	0.24	2.97	10.58
1972	0.37	0.14	0.03	0.14	1.02	0.18	0.01	0.14	0.30	1.30	1.00	0.89	5.52
1973	1.54	1.66	0.73	0.13	0.75	0.07	0.27	0.48	0.01	0.56	1.74	1.27	9.21
1974	1.60	0.34	1.16	0.23	0.01	T	0.33	0.19	0.00	0.69	0.27	0.56	5.38
1975	0.32	1.74	1.59	0.62	0.21	0.21	0.03	1.03	0.92	0.15	0.12	0.01	6.95
1976	0.16	1.20	0.36	0.20	0.10	T	0.96	0.62	1.10	0.28	0.07	0.01	5.06
1977	0.67	0.71	0.19	T	1.24	1.03	0.07	0.01	0.01	0.14	0.23	2.54	6.84
1978	1.66	0.98	1.49	0.20	0.31	0.07	0.19	0.15	0.68	0.08	1.30	0.82	7.93
1979	0.66	0.82	0.52	0.41	0.16	T	0.58	0.38	T	0.31	0.17	2.02	6.03
1980	2.77	1.90	0.76	0.51	0.78	0.12	0.54	0.32	0.48	0.14	0.28	0.60	9.20
1981	0.85	0.21	0.58	0.21	0.57	T	0.01	0.36	0.07	0.64	2.13	1.05	6.68
1982	1.20	0.41	1.14	0.34	0.10	1.07	0.04	0.09	2.31	1.65	1.71	1.04	11.10
1983	1.72	1.58	1.31	1.35	0.21	0.53	T	0.78	0.84	0.36	3.08	1.47	13.23
1984	0.36	0.22	0.20	0.24	0.06	0.34	0.45	0.02	0.04	0.60	1.68	0.07	4.28
1985	0.24	0.68	1.07	T	T	0.12	T	0.01	0.63	0.46	1.23	0.55	4.99
Record Mean	1.34	1.04	0.77	0.46	0.62	0.36	0.27	0.26	0.31	0.42	0.70	1.06	7.64

TABLE 3 AVERAGE TEMPERATURE (deg. F)   RENO, NEVADA

YEAR	JAN	FEB	MAR	APR	MAY	JUNE	JULY	AUG	SEP	OCT	NOV	DEC	ANNUAL
1956	34.1	27.6	41.4	46.2	54.9	61.7	68.6	64.8	61.3	47.7	36.8	30.5	48.0
1957	25.8	38.9	42.4	45.7	53.5	64.9	68.6	65.0	61.5	46.8	36.6	35.2	48.7
1958	34.3	41.2	37.0	45.1	59.2	61.4	68.4	72.0	60.2	52.6	38.7	37.7	50.7
#1959	36.5	33.9	42.6	50.3	51.3	65.3	72.9	66.8	57.4	51.8	39.6	29.4	49.8
1960	27.5	36.1	44.3	48.2	51.7	65.2	70.4	65.5	61.0	48.9	40.1	32.1	49.3
1961	33.4	40.7	41.6	47.0	53.4	66.6	69.4	69.1	55.9	49.7	36.5	31.7	49.6
1962	26.7	34.1	39.6	50.1	51.5	61.7	67.0	61.5	60.8	51.2	40.8	33.9	48.3
#1963	28.2	44.4	40.5	41.8	57.2	59.0	65.1	64.7	63.3	52.4	40.2	35.1	49.3
1964	32.5	34.2	37.7	45.3	51.9	61.3	70.7	66.9	58.1	55.3	38.0	36.0	49.0
1965	33.6	40.1	43.2	49.6	54.0	59.2	66.1	65.8	54.9	52.4	41.6	29.9	49.2
1966	33.4	34.8	43.4	50.2	60.7	67.5	65.7	67.5	58.7	49.6	41.0	34.0	50.2
1967	37.0	40.3	40.6	40.3	54.9	59.9	69.6	71.0	62.6	51.0	42.7	28.3	49.8
1968	31.7	42.4	42.4	44.6	52.8	62.9	73.0	64.5	59.6	50.6	40.8	31.4	49.7
1969	37.0	34.2	40.7	48.4	58.1	62.0	70.8	68.8	63.4	45.3	39.0	36.7	50.4
1970	37.3	42.0	41.4	44.1	58.0	65.1	73.0	71.5	57.2	48.0	45.4	29.2	51.0
1971	33.0	36.9	40.9	45.9	50.9	59.7	71.5	70.7	56.5	45.8	37.5	26.6	48.0
1972	25.5	40.2	47.1	46.2	56.9	65.1	70.7	67.6	56.6	48.5	38.0	25.1	49.0
1973	28.4	37.7	37.9	43.8	58.8	64.3	70.1	67.4	58.5	49.2	42.8	38.3	49.8
1974	31.6	36.2	42.8	46.5	54.9	63.9	67.0	65.7	61.4	49.3	40.4	31.1	49.2
1975	32.3	34.1	38.4	39.9	54.2	62.4	70.4	64.8	64.2	49.5	35.9	33.8	48.3
1976	32.1	35.0	38.3	44.5	57.5	61.1	69.0	62.1	62.0	51.3	42.4	30.5	48.8
1977	32.3	40.0	38.1	51.3	48.5	68.6	69.5	69.8	60.7	52.2	41.9	38.3	51.0
1978	37.1	38.4	47.2	45.3	52.3	61.6	69.5	67.2	57.6	53.5	37.2	24.9	49.3
1979	28.9	36.9	41.9	46.9	57.1	63.9	69.9	67.7	64.2	54.0	38.0	35.5	50.4
1980	36.9	40.6	38.5	49.8	54.4	60.8	71.3	67.5	63.0	50.9	41.8	36.2	51.0
1981	36.1	38.8	41.7	50.7	57.5	68.3	67.9	69.4	64.7	46.9	42.5	39.0	52.0
1982	28.5	40.3	40.2	44.0	55.0	61.8	70.4	68.8	57.0	46.9	36.2	32.3	48.5
1983	34.2	38.6	40.5	43.8	53.3	62.5	67.2	69.9	62.9	54.1	41.2	38.8	50.6
1984	31.9	37.2	44.3	45.8	59.4	71.3	73.4	69.8	63.1	46.2	39.6	30.7	50.3
1985	30.6	37.0	38.7	52.7	56.3	68.1	73.4	68.5	57.4	50.4	34.8	31.2	50.0
Record Mean	31.9	36.7	41.2	47.4	54.9	62.7	70.2	68.4	60.8	50.9	40.8	33.3	50.0
Max	43.6	49.1	54.7	62.4	70.4	79.8	89.4	87.6	79.2	67.9	55.2	45.3	65.4
Min	20.2	24.3	27.7	32.4	39.3	45.5	51.0	49.2	42.3	33.8	26.4	21.4	34.5

REFERENCE NOTES FOR TABLES 1, 2, 3 and 6   (RENO, NV)

GENERAL

T - TRACE AMOUNT
BLANK ENTRIES DENOTE MISSING/UNREPORTED DATA.
# INDICATES A STATION OR INSTRUMENT RELOCATION.

SPECIFIC

TABLE 1

(a) - LENGTH OF RECORD IN YEARS. ALTHOUGH
INDIVIDUAL MONTHS MAY BE MISSING.
* LESS THAN .05

NORMALS — BASED ON THE 1951-1980 RECORD PERIOD.
EXTREMES — DATES ARE THE MOST RECENT OCCURRENCE.
WIND DIR. — NUMERALS SHOW TENS OF DEGREES
CLOCKWISE FROM TRUE NORTH.
"00" INDICATES CALM.
RESULTANT WIND DIRECTIONS ARE GIVEN TO WHOLE DEGREES.

EXCEPTIONS

TABLE 1

1. PERCENT OF POSSIBLE SUNSHINE IS THROUGH 1979 AND
1984 TO DATE.

TABLES 2, 3, and 6

RECORD MEANS ARE THROUGH THE CURRENT YEAR,
BEGINNING IN   1888 FOR TEMPERATURE
1870 FOR PRECIPITATION
1943 FOR SNOWFALL

TABLE 4   HEATING DEGREE DAYS Base 65 deg. F   RENO, NEVADA

SEASON	JULY	AUG	SEP	OCT	NOV	DEC	JAN	FEB	MAR	APR	MAY	JUNE	TOTAL
1956-57	20	43	116	528	838	1063	1211	724	693	572	348	78	6234
1957-58	1	45	116	556	845	916	943	658	861	589	176	139	5845
1958-59	20	0	155	377	782	839	875	867	686	435	415	58	5509
#1959-60	6	44	244	404	757	1096	1156	832	638	497	405	42	6121
1960-61	5	78	127	493	742	1015	970	674	721	531	352	55	5763
1961-62	9	7	267	470	848	1024	1178	858	780	440	409	118	6408
1962-63	7	110	127	422	718	957	1134	572	750	691	237	182	5907
#1963-64	38	62	82	389	738	921	1002	887	840	585	400	135	6079
1964-65	6	28	203	296	802	892	968	695	667	454	336	174	5521
1965-66	34	31	298	383	694	1079	974	838	664	439	135	105	5674
1966-67	45	45	184	473	715	951	859	685	750	737	317	177	5938
1967-68	1	3	87	429	663	1132	1025	650	694	606	369	113	5772
1968-69	0	91	180	441	721	1033	861	856	745	494	208	104	5734
1969-70	5	12	79	605	772	871	638	851	724	622	216	90	5485
1970-71	0	0	234	519	585	1102	985	780	739	567	429	164	6104
1971-72	9	13	264	588	819	1184	1216	709	548	558	259	54	6221
1972-73	8	26	249	503	803	1229	1127	760	834	630	199	87	6455
1973-74	7	36	193	483	659	819	1027	800	680	545	310	73	5632
1974-75	52	42	120	478	733	1046	1006	858	818	746	332	113	6344
1975-76	26	65	51	477	868	961	1012	863	821	610	224	138	6116
1976-77	15	116	116	419	670	1062	1002	695	828	405	509	30	5867
1977-78	10	12	154	392	687	821	858	739	544	582	387	113	5299
1978-79	21	62	234	347	826	1236	1113	781	709	536	247	91	6203
1979-80	9	28	56	339	805	908	865	701	813	451	319	152	5446
1980-81	13	35	79	430	688	885	890	727	715	424	228	48	5162
1981-82	12	7	83	554	669	800	1123	687	760	623	307	133	5758
1982-83	15	11	278	556	855	1006	947	732	752	630	371	77	6230
1983-84	40	8	104	332	708	805	1019	801	637	570	183	133	5340
1984-85	0	8	111	575	753	1056	1060	781	810	359	266	45	5824
1985-86	5	12	230	446	896	1039							

TABLE 5   COOLING DEGREE DAYS Base 65 deg. F   RENO, NEVADA

YEAR	JAN	FEB	MAR	APR	MAY	JUNE	JULY	AUG	SEP	OCT	NOV	DEC	TOTAL
1969	0	0	0	0	2	19	193	136	35	0	0	0	385
1970	0	0	0	0	6	100	256	210	7	0	0	0	579
1971	0	0	0	0	0	11	217	197	15	0	0	0	440
1972	0	0	0	0	12	65	191	113	4	0	0	0	385
1973	0	0	0	0	13	75	173	117	3	0	0	0	381
1974	0	0	0	0	2	47	118	71	20	0	0	0	258
1975	0	0	0	0	4	43	199	66	33	3	0	0	348
1976	0	0	0	0	0	25	144	34	33	0	0	0	236
1977	0	0	0	0	6	141	154	167	32	0	0	0	500
1978	0	0	0	0	0	16	166	139	19	0	0	0	340
1979	0	0	0	0	9	63	169	122	38	3	0	0	404
1980	0	0	0	0	2	32	218	119	25	1	0	0	397
1981	0	0	0	2	4	153	112	151	80	0	0	0	502
1982	0	0	0	0	2	45	188	135	47	0	0	0	417
1983	0	0	0	0	16	9	115	170	49	0	0	0	359
1984	0	0	0	0	16	42	264	162	61	0	0	0	545
1985	0	0	0	0	3	157	273	126	6	0	0	0	565

TABLE 6   SNOWFALL (inches)   RENO, NEVADA

SEASON	JULY	AUG	SEP	OCT	NOV	DEC	JAN	FEB	MAR	APR	MAY	JUNE	TOTAL
1956-57	0.0	0.0	0.0	1.9	T	0.4	4.4	T	0.4	0.1	0.0	0.0	7.2
1957-58	0.0	0.0	0.0	2.0	3.1	3.1	1.0	4.6	10.0	7.5	0.0	0.0	31.3
1958-59	0.0	0.0	0.0	0.0	4.1	T	0.5	22.1	0:1	0.0	0.1	0.0	26.9
#1959-60	0.0	0.0	T	0.0	0.0	1.8	15.4	3.0	T	0.6	T	0.0	20.8
1960-61	0.0	0.0	0.0	0.4	2.0	T	2.6	T	1.1	0.6	0.0	0.0	6.7
1961-62	0.0	0.0	0.0	T	8.7	0.7	3.5	13.3	12.3	T	9.0	0.0	47.5
1962-63	0.0	0.0	0.0	T	T	T	1.7	0.0	3.4	5.9	T	T	11.0
1963-64	0.0	0.0	0.0	0.0	3.1	1.0	4.7	1.0	9.2	6.6	14.1	0.0	39.7
1964-65	0.0	0.0	0.0	0.0	5.2	8.0	5.2	T	4.6	T	T	0.0	23.0
1965-66	0.0	0.0	0.0	T	2.2	9.8	T	1.4	0.4	T	0.0	0.0	13.8
1966-67	0.0	0.0	0.0	0.0	T	5.8	12.4	T	15.6	6.8	1.5	0.0	42.1
1967-68	0.0	0.0	0.0	0.0	T	1.9	12.4	2.2	3.5	T	T	0.0	20.0
1968-69	0.0	0.0	0.0	1.3	10.7	23.5	9.1	1.9	0.3	T		0.0	46.8
1969-70	0.0	0.0	0.0	0.0	T	2.8	T	0.6	1.8	2.3	T	0.2	7.7
1970-71	0.0	0.0	0.0	0.0	0.2	15.4	11.4	4.9	7.0	3.3	6.3	0.0	48.5
1971-72	0.0	0.0	T	5.1	0.2	25.6	5.7	0.7	0.2	1.0	0.0	0.0	38.5
1972-73	0.0	0.0	0.0	T	T	8.8	11.9	8.2	6.5	0.7	T	0.0	36.1
1973-74	0.0	0.0	0.0	0.0	2.5	6.8	8.7	1.0	6.5	2.6	T	0.0	28.1
1974-75	0.0	0.0	0.0	0.0	0.0	4.0	2.7	19.0	12.1	4.8	2.2	T	44.8
1975-76	0.0	0.0	0.0	T	1.7	0.3	15.8	7.8	2.7	T	0.0	0.0	28.3
1976-77	0.0	0.0	0.0	0.0	0.0	0.5	0.7	1.4	1.2	T	2.9	0.0	6.7
1977-78	0.0	0.0	0.0	0.0	0.5	5.6	0.3	5.1	0.2	0.4	0.0	0.0	12.2
1978-79	0.0	0.0	T	0.8	9.7	9.1	6.4	3.9	0.2	1.1	0.3	0.0	31.5
1979-80	0.0	0.0	0.0	0.3	T	2.1	4.7	6.1	7.6	1.2	T	0.0	22.0
1980-81	0.0	0.0	0.0	T	T	T	3.9	T	2.2	T	0.0	0.0	6.1
1981-82	0.0	0.0	0.0	1.1	2.0	0.1	12.5	2.3	6.7	0.8	0.5	0.0	26.0
1982-83	0.0	0.0	1.5	0.0	8.6	1.8	1.5	1.0	3.0	2.9	3.5	0.0	23.8
1983-84	0.0	0.0	0.0	0.0	5.7	0.5	6.7	1.5	0.1	T	0.0	0.0	14.5
1984-85	0.0	0.0	0.0	3.4	3.0	1.3	4.3	0.8	7.0	T	T	0.0	19.8
1985-86	0.0	0.0	T	1.2	16.5	1.4							
Record Mean	0.0	0.0	T	0.4	2.3	4.3	6.1	4.9	4.9	1.4	1.0	T	25.4

See Reference Notes, relative to all above tables, on preceding page.

Winnemucca lies at an elevation about 4300 feet above sea level and is effectively cut off by the Sierra Nevada Mountains from the moisture source of the Pacific Ocean. Winnemucca has a climate marked by warm days, cool nights, and light precipitation. Sixty-six percent of the annual rainfall occurs as rain and snow between December and May. The winter snow pack in the surrounding mountains is generally sufficient for essential summertime irrigation. Reservoirs along the streams hold surplus water for less favorable years. As a result of the characteristic dryness of the climate, the neighboring valleys and hills are covered with sagebrush, and trees are found only along streams and in other places where water is sufficient the year round. Though it is heavier in the mountains, snowfall at Winnemucca itself has had measurable amounts fall in every month except July, August and September. During the winter months, snow on the ground permits grazing in many desert regions where there is no other source of stock water. Grazing in the summer months is restricted to mountain range tracts where there is sufficient water. Streams in many areas have been stocked with fish and provide good fishing each year.

Temperatures in this plateau area tend to rise sharply right after sunrise and remain comparatively high during the daylight hours, then drop rapidly about sundown. Daily temperature variations of 50 degrees are not uncommon.

Based on the 1951–1980 period, the average first occurrence of 32 degrees Fahrenheit in the fall is September 10 and the average last occurrence in the spring is June 8.

TABLE 1　　NORMALS, MEANS AND EXTREMES

WINNEMUCCA, NEVADA

LATITUDE: 40°54'N　　LONGITUDE: 117°48'W　　ELEVATION: FT. GRND 4298 BARO 4301　　TIME ZONE: PACIFIC　　WBAN: 24128

	(a)	JAN	FEB	MAR	APR	MAY	JUNE	JULY	AUG	SEP	OCT	NOV	DEC	YEAR
TEMPERATURE °F:														
Normals														
-Daily Maximum		42.3	48.7	53.6	61.8	72.0	81.8	92.7	89.7	80.8	68.5	53.2	43.9	65.8
-Daily Minimum		17.2	22.6	23.8	28.8	37.4	45.1	51.2	47.6	38.3	28.9	22.2	17.0	31.7
-Monthly		29.8	35.7	38.7	45.3	54.7	63.5	72.0	68.7	59.5	48.7	37.7	30.5	48.7
Extremes														
-Record Highest	36	68	74	81	90	96	105	106	108	103	91	77	67	108
-Year		1971	1981	1972	1981	1977	1974	1979	1983	1950	1980	1980	1980	AUG 1983
-Record Lowest	36	-24	-28	-3	6	12	23	29	28	12	7	-8	-34	-34
-Year		1963	1985	1971	1972	1953	1954	1983	1960	1958	1970	1985	1972	DEC 1972
NORMAL DEGREE DAYS:														
Heating (base 65°F)		1091	820	815	591	334	126	0	42	193	505	819	1073	6409
Cooling (base 65°F)		0	0	0	0	14	81	222	157	31	0	0	0	505
% OF POSSIBLE SUNSHINE	36	49	54	57	64	70	76	85	84	81	73	53	49	66
MEAN SKY COVER (tenths)														
Sunrise - Sunset	36	7.0	6.9	6.7	6.4	5.7	4.5	2.8	3.0	3.2	4.5	6.3	6.6	5.3
MEAN NUMBER OF DAYS:														
Sunrise to Sunset														
-Clear	36	6.3	5.9	6.8	6.7	9.0	13.7	20.2	19.3	18.4	14.2	8.1	7.6	136.1
-Partly Cloudy	36	6.6	6.4	7.4	9.0	9.6	8.6	7.0	7.6	6.4	7.4	6.9	6.5	89.4
-Cloudy	36	18.1	16.0	16.9	14.3	12.4	7.7	3.8	4.1	5.2	9.4	15.0	17.0	139.8
Precipitation														
.01 inches or more	36	8.3	6.8	7.6	6.2	6.2	5.5	2.3	2.7	3.3	4.6	7.4	8.7	69.5
Snow, Ice pellets														
1.0 inches or more	36	1.9	1.3	1.8	0.8	0.3	0.0	0.0	0.0	0.0	0.1	1.0	1.8	9.0
Thunderstorms	35	0.*	0.2	0.2	0.7	2.5	2.9	3.3	3.5	1.8	0.5	0.1	0.1	15.8
Heavy Fog Visibility														
1/4 mile or less	35	1.9	0.4	0.1	0.1	0.*	0.1	0.0	0.*	0.1	0.2	0.3	1.1	4.4
Temperature °F														
-Maximum														
90° and above	36	0.0	0.0	0.0	0.*	1.1	8.1	22.6	18.4	5.4	0.1	0.0	0.0	55.7
32° and below	36	5.6	1.1	0.4	0.0	0.0	0.0	0.0	0.0	0.0	0.0	0.6	3.8	11.4
-Minimum														
32° and below	36	28.1	24.5	26.3	20.6	8.1	1.5	0.1	0.2	6.3	21.0	24.8	28.1	189.6
0° and below	36	3.7	0.6	0.1	0.0	0.0	0.0	0.0	0.0	0.0	0.0	0.5	2.2	7.1
AVG. STATION PRESS. (mb)	13	870.9	869.5	866.4	867.3	867.0	867.9	868.7	868.8	869.3	870.5	869.8	871.5	869.0
RELATIVE HUMIDITY (%)														
Hour 04	36	77	75	71	66	63	57	46	47	53	63	74	78	64
Hour 10　　(Local Time)	36	67	59	49	39	34	31	22	24	29	39	56	66	43
Hour 16	36	57	47	38	29	26	22	15	16	20	29	46	58	34
Hour 22	36	75	70	62	52	48	42	30	32	39	52	68	76	54
PRECIPITATION (inches):														
Water Equivalent														
-Normal		0.89	0.67	0.67	0.81	0.80	0.92	0.18	0.39	0.34	0.59	0.75	0.86	7.87
-Maximum Monthly	36	2.70	2.17	1.66	2.92	2.82	2.86	1.74	1.74	1.51	2.19	2.66	3.66	3.66
-Year		1956	1962	1952	1978	1957	1958	1984	1979	1976	1951	1950	1983	DEC 1983
-Minimum Monthly	36	0.04	0.08	0.06	0.06	T	T	0.00	0.00	0.00	T	T	0.03	0.00
-Year		1966	1967	1959	1959	1985	1981	1963	1969	1974	1978	1959	1976	SEP 1974
-Maximum in 24 hrs	36	0.74	0.72	0.67	1.01	1.34	1.79	0.86	0.75	0.83	1.64	1.58	0.95	1.79
-Year		1980	1960	1979	1958	1984	1958	1984	1958	1976	1951	1950	1969	JUN 1958
Snow, Ice pellets														
-Maximum Monthly	36	16.5	13.8	23.4	12.0	5.4	T			0.2	7.4	19.6	17.5	23.4
-Year		1950	1969	1952	1964	1965	1954			1971	1984	1985	1971	MAR 1952
-Maximum in 24 hrs	36	7.4	9.9	9.0	7.2	4.3	T			0.2	4.9	5.7	8.4	9.9
-Year		1983	1959	1982	1971	1971	1954			1971	1984	1985	1955	FEB 1959
WIND:														
Mean Speed (mph)	36	7.5	7.9	8.5	8.7	8.6	8.5	8.3	7.8	7.6	7.3	7.2	7.3	7.9
Prevailing Direction														
through 1963		NE	S	S	W	W	W	W	W	W	S	S	S	W
Fastest Mile														
-Direction (!!!)	36	W	W	W	W	N	W	W	W	NE	S	SW		W
-Speed (MPH)	36	56	59	66	52	61	57	56	51	54	40	61		66
-Year		1956	1961	1963	1963	1951	1963	1957	1960	1952	1960	1955		MAR 1963
Peak Gust														
-Direction (!!!)	2	N	SW	SW	NW	SE	W	SW	NW	NW	W	SW	W	W
-Speed (mph)	2	32	43	47	44	46	58	56	48	45	43	49	41	58
-Date		1984	1985	1985	1984	1985	1984	1985	1984	1985	1984	1984	1985	JUN 1984

See Reference Notes to this table on the following page.

TABLE 2   PRECIPITATION (inches)   WINNEMUCCA, NEVADA

YEAR	JAN	FEB	MAR	APR	MAY	JUNE	JULY	AUG	SEP	OCT	NOV	DEC	ANNUAL
1956	2.70	0.27	0.06	0.78	1.57	0.05	0.11	T	0.42	0.83	0.07	0.43	7.29
1957	0.91	0.71	1.14	1.08	2.82	0.31	T	T	0.26	1.77	0.58	0.44	10.02
1958	0.51	1.09	0.82	1.46	0.03	2.86	0.10	1.26	0.05	0.07	0.90	0.46	9.61
1959	0.34	0.86	0.06	0.06	0.61	0.39	T	0.38	0.93	0.02	T	0.42	4.07
1960	1.17	1.31	0.68	0.47	0.60	T	0.26	0.08	0.73	0.34	2.05	0.42	8.11
1961	0.04	0.60	1.10	0.12	0.61	0.49	0.30	1.08	0.88	0.69	0.80	1.08	7.79
1962	0.76	2.17	0.89	0.27	0.48	0.38	T	0.14	0.02	0.49	0.97	0.26	6.83
1963	0.79	0.76	0.34	1.09	1.89	2.58	0.00	0.71	0.25	0.62	1.07	0.76	10.86
1964	1.22	0.11	0.89	1.50	0.87	2.03	0.40	0.50	0.03	0.38	0.81	1.73	10.47
1965	0.98	0.38	0.28	0.94	1.08	0.95	0.36	0.72	0.10	0.17	1.20	0.59	7.75
1966	0.04	0.58	0.38	0.42	0.33	0.38	0.02	0.01	0.34	0.01	1.00	1.00	4.51
1967	0.93	0.08	1.32	1.86	0.25	1.55	0.37	T	0.20	0.09	0.73	0.32	7.70
1968	0.81	0.92	0.31	0.47	0.18	0.85	T	0.69	0.00	0.49	2.54	1.22	8.48
1969	1.30	1.48	0.28	1.29	T	1.65	0.01	0.00	0.01	1.43	2.07	2.07	9.67
1970	1.85	0.30	0.47	0.92	0.42	1.97	0.29	0.40	0.05	0.28	1.10	1.04	9.09
1971	0.19	0.37	0.75	1.47	1.33	1.60	0.03	0.23	0.37	0.26	1.10	1.47	9.17
1972	0.16	0.61	0.38	0.48	0.18	0.55	0.02	T	0.66	1.53	1.08	0.97	6.62
1973	1.35	0.33	0.89	0.73	1.33	0.06	0.10	0.05	0.48	0.64	0.78	1.18	7.92
1974	0.67	0.28	1.13	0.69	T	0.00	0.29	0.39	0.00	1.67	0.06	1.15	6.33
1975	1.02	0.85	1.29	0.53	0.68	0.70	0.06	0.79	0.79	1.74	0.47	0.27	8.59
1976	0.41	0.50	0.29	0.82	0.37	1.14	0.36	1.29	1.51	0.43	0.16	0.03	7.31
1977	0.34	0.14	0.37	0.25	1.56	2.36	0.26	0.64	0.22	T	0.75	1.19	8.08
1978	0.88	0.58	1.08	2.92	0.26	0.08	0.17	T	1.12	T	0.62	0.39	8.10
1979	0.86	1.22	0.96	1.04	0.25	0.39	0.74	1.74	T	0.61	0.61	0.31	8.73
1980	1.93	0.44	0.53	0.26	2.05	0.57	0.13	0.23	0.18	0.27	0.52	0.25	7.36
1981	0.44	0.51	0.97	0.19	1.94	T	0.01	0.03	0.23	1.08	1.40	1.64	8.44
1982	0.30	0.14	1.39	0.29	0.18	1.50	0.88	0.15	1.08	1.19	1.41	0.52	9.03
1983	1.18	0.97	1.46	0.82	0.78	0.85	0.73	1.53	0.59	0.65	1.25	3.66	14.47
1984	0.11	0.91	1.56	1.24	1.41	1.38	1.74	0.19	0.78	1.62	1.57	0.36	12.87
1985	0.83	0.68	0.73	0.20	T	0.04	0.45	0.02	0.52	0.19	2.51	0.84	7.01
Record Mean	0.98	0.84	0.88	0.83	0.86	0.77	0.25	0.25	0.38	0.68	0.79	1.00	8.52

TABLE 3   AVERAGE TEMPERATURE (deg. F)   WINNEMUCCA, NEVADA

YEAR	JAN	FEB	MAR	APR	MAY	JUNE	JULY	AUG	SEP	OCT	NOV	DEC	ANNUAL
1956	34.7	28.8	38.6	46.6	55.1	61.2	72.2	66.6	60.4	46.7	34.1	28.1	47.8
1957	21.7	38.3	40.3	44.8	52.9	64.0	69.6	65.7	60.1	45.3	32.9	33.1	47.4
1958	31.6	40.8	35.1	44.1	58.7	62.8	68.8	72.7	58.9	49.8	38.5	36.0	49.8
1959	34.6	31.2	38.8	47.0	49.4	66.9	74.1	68.0	56.6	49.6	35.8	25.3	48.1
1960	24.9	31.7	42.4	45.4	52.9	66.9	74.3	66.7	61.9	48.1	38.2	30.2	48.7
1961	31.0	36.5	39.5	44.7	52.9	68.9	71.5	71.5	54.9	48.2	34.3	30.1	48.7
1962	21.2	30.2	36.0	50.1	52.7	63.4	69.8	66.7	61.1	50.7	39.3	31.9	47.8
1963	22.6	42.8	37.3	40.6	58.1	58.9	67.5	67.9	63.6	53.4	40.0	31.0	48.6
1964	28.6	31.9	36.1	43.2	51.8	59.2	71.3	66.2	56.0	52.2	34.5	35.1	47.2
1965	32.7	35.2	37.1	47.9	50.3	61.2	70.2	66.5	54.0	51.4	40.9	28.8	48.0
1966	30.7	31.5	40.2	46.6	60.0	63.2	70.0	70.0	61.6	46.9	40.5	30.6	49.3
1967	32.6	37.4	39.0	39.4	53.6	60.6	72.5	73.7	63.5	48.6	40.5	26.9	49.0
1968	29.8	42.4	41.1	42.4	53.7	65.3	74.3	63.8	57.9	49.2	40.2	29.7	49.2
1969	33.2	31.3	37.7	47.3	58.6	62.3	71.5	69.7	63.0	42.6	36.0	31.8	48.8
1970	34.2	39.9	35.9	37.9	55.2	64.8	73.8	71.0	55.0	44.3	41.9	30.0	48.7
1971	33.7	34.9	37.9	44.5	52.7	62.4	74.8	75.0	54.0	44.6	36.1	24.2	47.9
1972	27.6	36.9	44.6	44.2	58.9	68.1	74.1	72.7	56.9	47.9	36.3	20.8	49.1
1973	26.6	37.0	37.2	44.2	58.4	65.7	72.8	69.7	58.6	47.7	40.8	36.3	49.6
1974	30.3	36.0	43.9	45.4	56.9	69.1	71.3	67.1	60.6	48.0	37.9	28.5	49.6
1975	27.4	34.6	40.7	40.4	53.8	61.8	74.0	67.1	63.3	49.6	36.3	34.8	48.6
1976	33.1	39.9	37.7	45.9	60.5	64.7	73.8	65.6	64.0	49.9	41.8	29.1	50.5
1977	26.9	38.5	37.2	50.4	50.4	70.6	72.4	72.4	58.9	50.4	38.9	26.0	50.7
1978	36.0	37.6	46.6	47.4	54.1	62.6	71.7	69.4	57.3	50.5	34.6	26.0	49.5
1979	25.4	35.9	41.8	47.1	59.1	66.5	72.9	69.5	63.2	53.1	35.3	32.9	50.2
1980	32.7	41.8	38.1	49.1	55.7	60.9	73.2	68.0	60.5	49.6	39.2	33.9	50.2
1981	34.6	36.9	38.6	49.2	53.2	65.9	71.4	71.5	63.0	45.5	41.5	37.4	50.8
1982	28.3	35.8	38.5	44.4	54.3	63.0	70.1	70.3	57.0	44.9	33.3	30.7	47.5
1983	34.3	39.0	41.7	42.7	53.3	62.0	67.2	70.6	59.8	50.2	31.5	31.5	49.3
1984	24.2	34.3	40.3	42.6	55.1	59.0	71.7	71.5	60.0	43.9	38.3	26.4	47.3
1985	19.1	27.2	36.9	48.9	54.7	68.0	74.7	66.9	55.1	44.9	28.0	26.7	45.9
Record Mean	28.3	34.3	39.7	46.7	54.8	63.3	71.8	69.5	59.6	48.7	37.9	30.4	48.7
Max	39.8	45.9	53.0	61.7	70.6	80.0	90.6	89.0	78.8	66.3	52.4	42.3	64.2
Min	16.8	22.6	26.3	31.6	39.0	46.6	53.0	49.9	40.4	31.0	23.3	18.4	33.2

REFERENCE NOTES FOR TABLES 1, 2, 3 and 6   (WINNEMUCA, NV)

GENERAL

T - TRACE AMOUNT
BLANK ENTRIES DENOTE MISSING/UNREPORTED DATA.
# INDICATES A STATION OR INSTRUMENT RELOCATION.

SPECIFIC

TABLE 1

(a) - LENGTH OF RECORD IN YEARS. ALTHOUGH
      INDIVIDUAL MONTHS MAY BE MISSING.
  * LESS THAN .05

NORMALS — BASED ON THE 1951-1980 RECORD PERIOD.
EXTREMES — DATES ARE THE MOST RECENT OCCURRENCE.
WIND DIR. — NUMERALS SHOW TENS OF DEGREES
            CLOCKWISE FROM TRUE NORTH.
            "00" INDICATES CALM.
RESULTANT WIND DIRECTIONS ARE GIVEN TO WHOLE DEGREES.

EXCEPTIONS

TABLES 2, 3, and 6

RECORD MEANS ARE THROUGH THE CURRENT YEAR,
BEGINNING IN   1878 FOR TEMPERATURE
                  1878 FOR PRECIPITATION
                  1950 FOR SNOWFALL

TABLE 4 HEATING DEGREE DAYS Base 65 deg. F WINNEMUCCA, NEVADA

SEASON	JULY	AUG	SEP	OCT	NOV	DEC	JAN	FEB	MAR	APR	MAY	JUNE	TOTAL
1956-57	9	38	150	562	922	1138	1336	741	760	601	364	104	6725
1957-58	1	39	153	604	957	983	1025	670	921	620	206	122	6301
1958-59	16	0	198	462	790	891	936	941	804	533	484	61	6116
1959-60	6	44	276	469	868	1225	1238	961	694	585	368	23	6757
1960-61	0	81	109	518	795	1071	1049	791	787	603	369	44	6217
1961-62	3	0	297	512	915	1072	1351	966	890	441	374	113	6934
1962-63	6	48	141	436	764	1022	1305	616	852	725	217	197	6329
1963-64	6	23	93	359	744	1048	1125	953	886	648	404	184	6473
1964-65	7	69	262	389	907	919	996	829	859	504	448	137	6326
1965-66	10	39	321	417	716	1116	1055	933	759	548	172	121	6207
1966-67	12	31	119	552	726	996	996	767	800	760	358	170	6350
1967-68	0	0	80	499	730	1174	1083	648	735	672	346	96	6063
1968-69	3	118	217	483	736	1087	979	938	837	521	194	110	6223
1969-70	16	9	85	685	862	1022	949	699	897	805	298	120	6447
1970-71	0	0	295	635	686	1080	962	837	834	604	374	117	6424
1971-72	3	3	306	625	860	1260	1151	808	626	617	218	36	6513
1972-73	1	6	255	525	855	1370	1184	780	859	616	218	97	6766
1973-74	1	26	200	531	719	881	1070	807	646	581	263	52	5777
1974-75	32	41	153	521	806	1127	1156	844	749	730	339	140	6638
1975-76	13	41	69	474	855	925	982	718	841	568	139	71	5696
1976-77	8	45	81	461	691	1104	1176	734	855	430	450	15	6050
1977-78	2	28	203	444	729	798	894	762	566	523	335	96	5380
1978-79	21	53	241	442	907	1202	1221	809	710	532	205	63	6406
1979-80	1	18	75	363	875	988	995	667	827	471	290	162	5732
1980-81	0	23	146	472	766	958	935	779	809	474	365	89	5816
1981-82	2	0	115	600	697	848	1131	811	812	610	327	124	6077
1982-83	29	8	258	618	943	1054	947	721	714	658	377	103	6430
1983-84	35	6	156	451	785	1032	1257	881	757	662	307	210	6539
1984-85	0	11	185	645	794	1188	1415	1049	863	479	314	49	6992
1985-86	3	39	293	617	1101	1179							

TABLE 5 COOLING DEGREE DAYS Base 65 deg. F WINNEMUCCA, NEVADA

YEAR	JAN	FEB	MAR	APR	MAY	JUNE	JULY	AUG	SEP	OCT	NOV	DEC	TOTAL
1969	0	0	0	0	2	37	222	162	33	0	0	0	456
1970	0	0	0	0	2	124	279	191	0	0	0	0	596
1971	0	0	0	0	0	45	295	313	13	0	0	0	666
1972	0	0	0	0	40	136	287	250	17	1	0	0	731
1973	0	0	0	0	18	124	248	180	15	0	0	0	585
1974	0	0	0	0	19	184	236	112	29	0	0	0	580
1975	0	0	0	0	0	50	297	113	25	3	0	0	488
1976	0	0	0	0	7	68	287	71	56	0	0	0	489
1977	0	0	0	0	4	188	257	265	27	0	0	0	741
1978	0	0	0	0	5	31	234	197	18	0	0	0	485
1979	0	0	0	0	29	114	256	165	26	3	0	0	593
1980	0	0	0	0	8	48	262	126	19	2	0	0	465
1981	0	0	0	6	5	124	208	208	61	0	0	0	612
1982	0	0	0	0	2	71	195	177	22	0	0	0	467
1983	0	0	0	0	21	21	108	189	10	0	0	0	349
1984	0	0	0	0	8	37	217	218	40	0	0	0	520
1985	0	0	0	0	3	148	311	105	3	0	0	0	570

TABLE 6 SNOWFALL (inches) WINNEMUCCA, NEVADA

SEASON	JULY	AUG	SEP	OCT	NOV	DEC	JAN	FEB	MAR	APR	MAY	JUNE	TOTAL
1956-57	0.0	0.0	0.0	1.5	0.1	3.3	12.2	T	3.9	T	0.0	0.0	21.0
1957-58	0.0	0.0	0.0	T	2.0	0.7	1.3	4.1	7.8	6.5	0.0	0.0	22.4
1958-59	0.0	0.0	0.0	T	1.0	2.5	2.5	12.5	T	T	0.0	0.0	18.5
1959-60	0.0	0.0	0.0	T	0.0	3.4	8.0	2.6	0.5	0.6	1.0	0.0	16.1
1960-61	0.0	0.0	0.0	T	0.9	T	T	0.2	2.1	1.1	T	0.0	4.3
1961-62	0.0	0.0	0.0	0.4	5.4	4.0	12.0	8.4	10.0	1.3	0.1	0.0	41.6
1962-63	0.0	0.0	0.0	T	1.3	0.4	4.4	T	2.7	6.7	T	0.0	15.5
1963-64	0.0	0.0	0.0	0.0	0.3	4.1	15.8	2.4	9.6	12.0	0.5	0.0	44.7
1964-65	0.0	0.0	0.0	0.0	6.3	4.5	3.2	1.8	1.5	3.3	5.4	0.0	26.0
1965-66	0.0	0.0	0.0	0.0	0.5	5.7	0.4	7.4	2.0	0.0	0.0	0.0	16.0
1966-67	0.0	0.0	0.0	T	T	10.8	4.1	0.4	11.1	6.3	T	0.0	32.7
1967-68	0.0	0.0	0.0	0.0	5.3	4.5	8.2	T	0.6	T	0.0	0.0	18.6
1968-69	0.0	0.0	0.0	0.0	6.1	11.3	6.9	13.8	3.5	2.9	0.0	0.0	44.5
1969-70	0.0	0.0	0.0	3.5	0.3	6.2	0.1	0.8	1.6	9.2	0.2	0.0	21.9
1970-71	0.0	0.0	0.0	0.4	0.1	4.9	1.8	2.4	2.0	8.8	4.3	0.0	24.7
1971-72	0.0	0.0	0.2	2.3	5.5	17.5	0.8	6.2	2.7	2.6	T	0.0	37.8
1972-73	0.0	0.0	0.0	0.1	1.2	13.3	9.9	1.1	7.4	3.4	0.0	0.0	36.4
1973-74	0.0	0.0	0.0	2.8	4.6	5.5	2.5	10.0	3.1	T	0.0	28.5
1974-75	0.0	0.0	0.0	T	T	5.7	7.0	7.2	4.9	2.9	4.3	0.0	32.0
1975-76	0.0	0.0	0.0	T	3.3	4.1	3.0	1.5	3.7	4.8	0.0	0.0	20.4
1976-77	0.0	0.0	0.0	T	T	0.4	4.5	0.9	4.8	0.7	1.6	0.0	12.9
1977-78	0.0	0.0	T	0.0	3.0	3.6	4.4	3.8	0.0	5.8	1.9	0.0	22.5
1978-79	0.0	0.0	T	0.0	1.8	4.2	5.2	12.1	4.6	1.9	0.1	0.0	29.9
1979-80	0.0	0.0	0.0	0.0	1.6	0.8	2.7	T	0.5	0.4	T	0.0	6.0
1980-81	0.0	0.0	0.0	T	T	0.2	3.7	1.2	1.1	T	T	0.0	6.2
1981-82	0.0	0.0	0.0	T	0.5	3.8	4.6	0.8	15.9	0.9	T	0.0	26.5
1982-83	0.0	0.0	0.0	T	5.8	2.3	9.2	2.1	0.9	1.3	0.3	0.0	21.9
1983-84	0.0	0.0	0.0	0.0	2.2	14.6	1.4	2.4	5.5	2.1	T	0.0	28.2
1984-85	0.0	0.0	0.0	7.4	2.6	3.9	8.8	7.2	5.8	0.4	T	0.0	36.1
1985-86	0.0	0.0	T	T	19.6	0.8							
Record Mean	0.0	0.0	T	0.4	2.5	4.9	5.6	3.7	4.8	2.6	0.6	T	25.1

See Reference Notes, relative to all above tables, on preceding page.

Concord, the Capital of New Hampshire, is situated near the geographical center of New England at an altitude of approximately 300 feet above sea level on the Merrimack River. Its surroundings are hilly with many lakes and ponds. The countryside is generously wooded, mostly on land reclaimed from fields which were formerly cleared for farming. From the coast about 50 miles to the southeast, the terrain slopes gently upward to the city. West of the city, the land rises some 2,000 feet higher in only half that distance. Mount Washington, at an elevation of 6,288 feet is in the White Mountains 75 miles north of town.

Northwesterly winds are prevalent. They bring cold, dry air during the winter and pleasantly cool, dry air in the summer. Stronger southerly winds occur during July and August, and easterly winds usually accompany summer and winter storms. Winter breezes are somewhat lighter, and winds are frequently calm during the night and early morning hours. Low temperatures, as a rule, do not interrupt normal out-of-doors activity because winds are calm or light, producing a low wind chill factor.

Very hot summer weather is infrequent. During any month, temperatures considerably above the average maxima and much below the normal minima are observed.

The average amount of precipitation for the warmer half of the year differs little from that for the colder half. Precipitation occurrences average approximately one day of three for the year, with a somewhat higher frequency for the April-May period, offsetting the lower frequency of August-October. The more significant rains and heavier snowfalls are associated with easterly winds, especially northeasterly winds. The first snowfall of an inch or more is likely to come between the middle of November and the middle of December. The snow cover normally lasts from mid-December until the last week of March, but bare ground is not rare in the winter, nor is a snowscape rare earlier or later in the season. Rain, sleet, or freezing rain may also occur.

Agriculture is neither intensive nor large-scale in the vicinity of the station. Potatoes and other frost-resistant vegetables, hardy fruits such as apples, forage for the dairy industry, and maple sugar are the principal crops.

Based on the 1951-1980 period, the average first occurrence of 32 degrees Fahrenheit in the fall is September 22 and the average last occurrence in the spring is May 23. Freezing temperatures have occurred as late as June and as early as August.

TABLE 1   # NORMALS, MEANS AND EXTREMES

CONCORD, NEW HAMPSHIRE

LATITUDE: 43°12'N   LONGITUDE: 71°30'W   ELEVATION: FT. GRND 342 BARO 00343   TIME ZONE: EASTERN   WBAN: 14745

	(a)	JAN	FEB	MAR	APR	MAY	JUNE	JULY	AUG	SEP	OCT	NOV	DEC	YEAR
TEMPERATURE °F:														
Normals														
-Daily Maximum		30.8	33.2	41.9	56.5	68.9	77.7	82.6	80.1	71.9	61.0	47.2	34.4	57.2
-Daily Minimum		9.0	11.0	22.2	31.6	41.4	51.6	56.4	54.5	46.2	35.5	27.3	14.5	33.4
-Monthly		19.9	22.1	32.0	44.0	55.2	64.7	69.5	67.3	59.0	48.3	37.3	24.5	45.3
Extremes														
-Record Highest	44	68	66	85	95	97	98	102	101	98	90	80	68	102
-Year		1950	1957	1977	1976	1962	1980	1966	1975	1953	1963	1950	1982	JUL 1966
-Record Lowest	44	-33	-37	-16	8	21	30	35	29	21	10	-1	-22	-37
-Year		1984	1943	1967	1969	1966	1972	1965	1965	1947	1972	1972	1951	FEB 1943
NORMAL DEGREE DAYS:														
Heating (base 65°F)		1398	1198	1020	627	314	67	20	39	191	521	831	1256	7482
Cooling (base 65°F)		0	0	0	0	10	58	160	111	14	0	0	0	353
% OF POSSIBLE SUNSHINE	44	52	55	52	53	54	58	63	60	56	54	42	47	54
MEAN SKY COVER (tenths)														
Sunrise - Sunset	44	6.1	6.1	6.4	6.5	6.6	6.3	6.1	5.9	5.9	5.9	6.8	6.4	6.2
MEAN NUMBER OF DAYS:														
Sunrise to Sunset														
-Clear	44	9.3	7.6	7.7	7.3	6.4	6.3	6.8	7.9	9.0	9.5	6.3	7.8	91.8
-Partly Cloudy	44	7.1	7.8	8.1	8.2	9.7	11.8	12.6	11.8	9.0	8.8	7.5	8.0	110.5
-Cloudy	44	14.5	12.8	15.2	14.6	15.0	11.9	11.5	11.3	12.0	12.7	16.1	15.2	162.9
Precipitation														
.01 inches or more	44	10.8	9.7	11.0	11.5	12.0	10.8	9.8	9.7	8.8	8.6	11.4	11.0	125.0
Snow,Ice pellets														
1.0 inches or more	44	4.5	4.0	3.1	0.6	0.*	0.0	0.0	0.0	0.0	0.*	1.3	4.1	17.7
Thunderstorms	44	0.*	0.1	0.3	0.9	2.5	4.5	5.4	3.8	1.9	0.6	0.1	0.*	20.0
Heavy Fog Visibility														
1/4 mile or less	44	2.1	1.8	2.9	1.9	3.2	3.8	5.4	7.0	9.2	6.6	3.6	2.7	50.0
Temperature °F														
-Maximum														
90° and above	20	0.0	0.0	0.0	0.2	0.9	2.3	4.8	2.5	0.6	0.0	0.0	0.0	11.4
32° and below	20	19.3	13.4	4.6	0.3	0.0	0.0	0.0	0.0	0.0	0.1	1.8	13.7	53.0
-Minimum														
32° and below	20	30.5	27.1	25.8	17.7	5.8	0.3	0.0	0.1	2.2	13.9	21.1	29.1	173.7
0° and below	20	10.8	7.4	1.4	0.0	0.0	0.0	0.0	0.0	0.0	0.0	0.2	5.6	25.4
AVG. STATION PRESS.(mb)	13	1002.2	1003.2	1001.8	1001.0	1001.7	1002.0	1002.1	1004.4	1005.4	1005.9	1004.1	1003.6	1003.1
RELATIVE HUMIDITY (%)														
Hour 01	19	74	72	73	74	82	88	89	90	90	85	81	78	81
Hour 07	20	74	76	76	74	77	83	84	88	90	87	84	80	81
Hour 13 (Local Time)	20	59	56	53	45	47	53	51	53	55	53	60	63	54
Hour 19	20	66	62	60	53	58	64	64	70	76	73	73	71	66
PRECIPITATION (inches):														
Water Equivalent														
-Normal		2.78	2.47	2.93	3.01	2.93	2.91	2.93	3.26	3.12	3.10	3.66	3.43	36.53
-Maximum Monthly	44	8.09	7.77	7.81	5.88	9.52	10.10	5.91	6.88	7.78	8.78	7.36	7.52	10.10
-Year		1979	1981	1953	1983	1984	1944	1967	1973	1960	1962	1983	1973	JUN 1944
-Minimum Monthly	44	0.40	0.67	0.86	1.02	0.60	0.64	0.96	0.95	0.41	0.59	0.75	0.58	0.40
-Year		1970	1978	1981	1985	1965	1979	1955	1944	1948	1947	1976	1943	JAN 1970
-Maximum in 24 hrs	44	2.12	2.26	2.27	2.10	2.59	4.47	2.54	3.71	4.12	4.24	2.89	3.31	4.47
-Year		1979	1981	1974	1945	1984	1944	1971	1973	1960	1962	1947	1969	JUN 1944
Snow,Ice pellets														
-Maximum Monthly	44	42.3	49.8	38.3	15.3	5.0					2.1	18.4	38.1	49.8
-Year		1979	1969	1956	1982	1945					1969	1971	1956	FEB 1969
-Maximum in 24 hrs	44	19.0	14.2	14.4	13.9	5.0					2.1	9.5	14.6	19.0
-Year		1944	1972	1984	1982	1945					1969	1961	1946	JAN 1944
WIND														
Mean Speed (mph)	43	7.3	7.9	8.3	7.9	7.0	6.4	5.7	5.3	5.5	5.9	6.6	7.0	6.7
Prevailing Direction														
through 1963		NW	NW	NW	NW	NW	NW	NW	NW	NW	NW	NW	NW	NW
Fastest Mile														
-Direction	42	NW	N	NE	NW	NW	W	SW	E	E	NW	NE	NW	NE
-Speed (MPH)	42	44	42	71	52	48	38	45	56	42	39	72	52	72
-Year		1972	1950	1950	1945	1945	1971	1971	1934	1960	1944	1950	1962	NOV 1950
Peak Gust														
-Direction	2	NW	NW	NE	NW	NW	NW	W	NW	SE	NW	W	NW	NW
-Speed (mph)	2	40	47	43	48	38	33	31	29	44	41	38	45	48
-Date		1985	1985	1984	1985	1985	1985	1984	1985	1985	1985	1984	1985	APR 1985

See reference Notes to this table on the following page.

TABLE 2   PRECIPITATION (inches)   CONCORD, NEW HAMPSHIRE

YEAR	JAN	FEB	MAR	APR	MAY	JUNE	JULY	AUG	SEP	OCT	NOV	DEC	ANNUAL
1956	4.40	2.70	3.67	3.34	1.75	1.69	3.27	2.08	5.57	1.60	2.12	4.56	36.75
1957	1.65	1.13	1.22	1.56	1.95	2.95	2.87	2.56	3.60	3.21	6.59	4.95	32.01
1958	5.83	1.51	1.43	3.95	1.92	2.50	3.58	2.61	3.60	1.96	4.40	1.40	34.69
1959	3.27	2.28	4.45	2.60	0.60	3.68	1.93	5.90	1.66	6.79	4.60	3.44	41.20
1960	2.46	2.82	2.48	4.59	4.00	3.47	3.62	2.57	7.78	2.09	2.58	2.47	40.93
1961	1.07	2.33	1.48	3.35	2.85	2.15	3.13	3.55	3.44	2.06	3.70	2.88	31.99
1962	2.66	2.59	1.64	2.13	2.58	3.30	2.99	1.76	8.78	2.65	3.15	3.68	36.82
1963	2.05	2.37	2.12	1.23	2.17	1.22	2.94	2.74	2.12	1.22	6.78	1.57	28.53
1964	3.64	1.93	3.01	2.05	1.15	0.81	3.37	2.74	0.42	2.24	3.47	1.52	27.90
1965	0.97	3.06	0.93	2.49	0.60	2.72	1.73	3.12	1.87	2.71	2.45	1.52	24.17
1966	2.69	2.13	2.19	1.16	2.16	1.78	2.30	4.05	5.40	3.31	3.01	2.42	32.60
1967	1.23	2.36	1.75	3.52	3.92	3.82	5.91	1.97	2.04	0.99	2.86	3.82	34.19
1968	1.79	0.93	3.80	2.85	5.20	5.90	1.62	3.60	2.34	2.23	5.24	5.82	41.32
1969	1.34	3.69	2.36	2.75	1.26	4.70	4.40	2.84	4.43	1.56	5.87	7.10	42.30
1970	0.40	4.27	2.78	3.38	3.04	2.26	2.33	3.06	3.40	3.64	3.03	3.08	34.67
1971	1.63	3.87	2.28	2.19	3.36	1.67	5.14	2.79	1.91	2.69	2.91	2.36	32.80
1972	1.44	2.60	4.16	2.71	4.20	3.54	5.40	2.12	2.55	2.23	6.57	4.55	42.07
1973	2.44	1.91	2.58	4.55	4.20	4.86	1.05	6.88	1.77	2.46	1.82	7.52	42.04
1974	2.80	2.32	3.98	2.58	3.74	1.82	1.41	2.20	4.74	1.64	3.20	4.02	34.45
1975	4.12	2.36	3.12	2.47	1.22	3.87	3.71	3.94	5.15	4.29	4.91	3.12	42.28
1976	3.40	2.36	2.02	2.43	3.90	2.74	3.20	2.66	2.73	4.05	0.75	2.27	32.51
1977	2.16	2.02	4.51	4.04	2.44	3.47	1.26	3.51	5.64	5.52	3.07	4.00	41.64
1978	6.32	0.67	2.16	2.06	2.67	3.18	1.08	2.87	0.46	2.72	1.77	2.91	28.87
1979	8.09	2.29	2.85	3.10	4.86	0.64	3.45	4.20	3.15	3.79	2.92	1.93	41.27
1980	0.43	0.78	3.37	3.72	0.86	2.83	2.35	3.99	2.19	2.63	3.12	0.79	27.06
1981	0.48	7.77	0.86	3.12	3.21	2.81	5.54	3.25	4.61	6.51	3.51	4.17	45.84
1982	3.98	2.88	2.47	3.08	1.91	7.84	2.83	2.54	1.85	1.52	2.93	0.91	34.74
1983	3.92	2.17	7.07	5.88	5.19	2.52	2.07	2.07	1.21	3.28	7.36	5.35	48.09
1984	1.89	5.06	2.92	3.74	9.52	2.83	4.44	0.97	1.08	4.42	2.67	2.70	42.24
1985	0.95	1.99	2.86	1.02	2.05	3.05	2.83	2.51	3.78	3.62	4.58	1.65	30.89
Record Mean	2.91	2.64	3.08	3.00	3.17	3.30	3.53	3.38	3.37	3.18	3.45	3.06	38.07

TABLE 3   AVERAGE TEMPERATURE (deg. F)   CONCORD, NEW HAMPSHIRE

YEAR	JAN	FEB	MAR	APR	MAY	JUNE	JULY	AUG	SEP	OCT	NOV	DEC	ANNUAL
1956	25.6	25.3	25.6	39.8	50.8	65.4	66.2	66.3	54.9	47.5	38.3	27.0	44.4
1957	13.5	26.8	34.4	47.6	56.2	67.4	68.2	64.1	61.4	49.3	40.9	32.9	46.9
1958	23.9	19.3	35.7	45.6	52.5	60.0	68.9	68.0	59.8	47.2	37.8	17.8	44.7
1959	21.2	19.1	31.1	45.5	59.4	63.4	72.1	70.2	62.6	49.2	38.3	29.3	46.8
1960	23.2	29.2	27.6	45.2	58.3	65.1	68.5	68.0	59.7	46.8	39.9	20.8	46.0
1961	14.5	24.8	32.4	42.6	53.2	65.4	69.5	68.5	67.1	52.4	39.5	27.0	46.4
1962	20.6	17.8	34.0	45.3	54.6	66.3	66.6	67.8	57.7	48.1	36.1	22.6	44.8
1963	20.8	18.8	33.5	43.4	55.5	66.4	72.1	65.6	56.5	53.8	42.3	17.3	45.5
1964	22.9	22.4	33.6	43.6	58.8	65.6	70.8	62.7	57.4	47.4	37.1	25.1	45.6
#1965	18.9	22.2	31.9	41.3	57.1	67.2	65.9	66.6	58.8	46.8	34.8	27.3	44.5
1966	20.4	22.5	34.6	42.2	53.4	66.8	72.2	69.6	57.0	48.0	41.5	26.2	46.2
1967	24.8	17.5	27.1	42.0	48.8	66.1	70.2	67.2	58.4	49.3	32.7	22.8	44.3
1968	15.8	18.0	34.8	44.8	51.1	62.0	69.3	64.6	61.0	51.2	34.8	22.5	44.2
1969	21.9	23.2	28.3	44.3	52.3	63.0	65.4	68.9	59.7	47.5	38.9	23.7	44.8
1970	11.0	22.6	30.7	45.8	57.3	63.1	70.9	68.7	60.6	51.0	39.3	20.3	45.1
1971	12.5	23.3	30.4	42.1	54.1	65.5	68.3	67.0	61.1	51.9	32.4	26.6	44.6
1972	22.0	21.2	27.9	40.2	56.8	62.8	69.3	64.1	57.5	42.3	31.2	23.4	43.2
1973	21.0	20.2	35.5	45.0	52.9	66.9	70.3	72.3	58.8	48.2	36.1	28.9	46.3
1974	21.4	21.1	31.8	45.8	51.1	62.8	67.5	67.0	58.9	42.5	35.9	26.7	44.4
1975	21.6	21.5	30.1	40.4	61.3	65.0	72.9	66.5	56.1	47.6	39.9	21.9	45.4
1976	10.9	24.7	31.9	46.7	53.6	68.9	66.8	65.3	57.2	45.0	31.7	16.1	43.3
1977	10.6	20.5	36.7	45.1	58.3	62.9	69.0	68.0	58.5	47.0	39.4	21.0	44.8
1978	17.5	13.5	28.0	40.6	57.5	65.2	69.7	69.6	55.9	46.7	35.3	22.7	43.5
1979	23.3	15.1	37.6	44.2	56.3	63.8	71.2	67.5	59.8	47.4	42.2	29.4	46.5
1980	22.4	19.1	31.8	44.4	55.6	62.8	70.6	68.4	58.2	45.1	34.8	19.2	44.4
1981	12.5	30.8	34.2	47.1	57.5	66.1	69.9	66.6	58.7	45.2	37.8	25.4	46.0
1982	10.9	20.8	30.2	41.6	57.3	60.9	69.5	65.4	60.2	47.6	41.7	32.4	44.9
1983	23.1	26.1	35.9	45.2	53.1	65.1	70.1	69.3	62.3	48.2	39.4	23.4	46.8
1984	16.0	30.5	28.2	44.6	52.9	65.9	68.7	69.3	57.8	50.4	38.3	30.3	46.1
1985	15.9	25.8	35.7	45.2	56.1	62.2	70.3	66.7	60.5	49.2	38.5	22.0	45.7
Record Mean	20.9	22.8	32.0	44.4	56.1	64.8	70.0	67.3	59.7	49.0	37.6	25.5	45.9
Max	31.2	33.5	41.6	55.9	68.7	77.1	82.1	79.0	71.4	60.6	46.9	34.7	56.9
Min	10.6	12.1	22.4	32.8	43.5	52.6	57.9	55.6	47.9	37.3	28.2	16.2	34.8

REFERENCE NOTES FOR TABLES 1, 2, 3 and 6   (CONCORD, NH)

GENERAL

T - TRACE AMOUNT
BLANK ENTRIES DENOTE MISSING/UNREPORTED DATA.
# INDICATES A STATION OR INSTRUMENT RELOCATION.

SPECIFIC

TABLE 1

(a) - LENGTH OF RECORD IN YEARS. ALTHOUGH
    INDIVIDUAL MONTHS MAY BE MISSING.
  * LESS THAN .05

NORMALS — BASED ON THE 1951-1980 RECORD PERIOD.
EXTREMES — DATES ARE THE MOST RECENT OCCURRENCE.
WIND DIR. — NUMERALS SHOW TENS OF DEGREES
    CLOCKWISE FROM TRUE NORTH.
    "00" INDICATES CALM.
RESULTANT WIND DIRECTIONS ARE GIVEN TO WHOLE DEGREES.

EXCEPTIONS

TABLES 2, 3, and 6

RECORD MEANS ARE THROUGH THE CURRENT YEAR,
BEGINNING IN   1871 FOR TEMPERATURE
    1855 FOR PRECIPITATION
    1942 FOR SNOWFALL

TABLE 4   HEATING DEGREE DAYS Base 65 deg. F   CONCORD, NEW HAMPSHIRE

SEASON	JULY	AUG	SEP	OCT	NOV	DEC	JAN	FEB	MAR	APR	MAY	JUNE	TOTAL
1956-57	50	50	323	532	799	1170	1592	1062	942	517	284	66	7387
1957-58	17	67	167	482	714	990	1269	1272	900	574	386	169	7007
1958-59	21	24	182	544	811	1459	1353	1279	1044	577	235	107	7636
1959-60	0	28	174	484	790	1100	1287	1032	1151	591	201	69	6907
1960-61	8	22	180	560	745	1363	1562	1121	1002	665	383	54	7665
1961-62	21	25	86	389	756	1172	1369	1316	956	585	341	33	7049
1962-63	31	30	234	519	860	1312	1363	966	639	291	70	70	7602
1963-64	20	57	267	342	675	1472	1295	1231	964	636	221	70	7250
#1964-65	8	106	246	539	831	1230	1418	1191	1015	704	257	133	7678
1965-66	50	78	227	556	899	1163	1376	1184	935	681	362	73	7584
1966-67	6	5	244	520	697	1196	1241	1323	1164	683	496	57	7632
1967-68	8	34	208	482	965	1144	1520	1358	928	598	424	118	7787
1968-69	18	92	133	424	899	1311	1330	1165	1128	613	389	119	7621
1969-70	64	40	193	534	777	1275	1668	1179	1055	572	256	108	7721
1970-71	4	25	181	431	760	1379	1622	1165	1064	682	332	73	7718
1971-72	26	49	165	396	970	1185	1327	1267	1142	736	262	112	7637
1972-73	27	82	223	695	1007	1284	1357	1250	905	596	370	78	7874
1973-74	15	9	244	518	860	1112	1345	1223	1025	573	432	99	7455
1974-75	34	26	213	694	865	1182	1339	1218	1075	730	152	98	7626
1975-76	10	78	260	532	747	1330	1672	1162	1019	565	356	60	7791
1976-77	37	84	234	615	992	1506	1683	1242	870	594	259	119	8235
1977-78	37	58	222	551	760	1360	1466	1435	1138	725	270	72	8094
1978-79	45	34	275	563	882	1304	1284	1392	841	617	280	99	7616
1979-80	33	64	199	546	675	1098	1317	1324	1022	610	290	123	7301
1980-81	13	33	245	611	899	1417	1626	953	951	530	267	40	7585
1981-82	12	43	192	608	810	1222	1674	1233	1072	695	246	136	7943
1982-83	25	66	169	535	692	1007	1291	1086	895	588	364	82	6800
1983-84	14	33	167	521	760	1283	1516	993	1135	607	382	85	7496
1984-85	19	27	238	446	792	1072	1514	1092	901	588	286	106	7081
1985-86	9	38	166	485	785	1326							

TABLE 5   COOLING DEGREE DAYS Base 65 deg. F   CONCORD, NEW HAMPSHIRE

YEAR	JAN	FEB	MAR	APR	MAY	JUNE	JULY	AUG	SEP	OCT	NOV	DEC	TOTAL
1969	0	0	0	0	4	65	82	168	44	0	0	0	363
1970	0	0	0	1	27	58	196	145	56	6	0	0	489
1971	0	0	0	0	0	95	135	118	55	0	0	0	403
1972	0	0	0	0	17	52	167	60	5	0	0	0	301
1973	0	0	0	3	2	145	184	242	67	2	0	0	645
1974	0	0	0	3	9	40	118	92	40	0	0	0	302
1975	0	0	0	0	48	108	263	134	0	0	0	0	553
1976	0	0	0	18	9	184	100	99	8	1	0	0	419
1977	0	0	0	5	57	64	168	156	38	0	0	0	488
1978	0	0	0	0	46	83	198	186	8	0	0	0	521
1979	0	0	0	1	15	69	232	150	46	6	0	0	519
1980	0	0	0	0	5	64	193	145	51	0	0	0	458
1981	0	0	0	0	38	80	172	101	11	0	0	0	402
1982	0	0	0	0	11	17	171	87	31	0	0	0	317
1983	0	0	0	0	2	93	179	172	93	5	0	0	544
1984	0	0	0	0	15	117	139	165	28	0	0	0	464
1985	0	0	0	0	16	27	184	93	36	1	0	0	357

TABLE 6   SNOWFALL (inches)   CONCORD, NEW HAMPSHIRE

SEASON	JULY	AUG	SEP	OCT	NOV	DEC	JAN	FEB	MAR	APR	MAY	JUNE	TOTAL
1956-57	0.0	0.0	0.0	0.0	10.3	38.1	20.5	2.8	2.7	2.8	0.0	0.0	77.2
1957-58	0.0	0.0	0.0	0.0	T	4.6	30.9	14.2	16.2	2.7	0.0	0.0	68.6
1958-59	0.0	0.0	0.0	T	1.6	11.1	9.8	16.1	30.3	T	0.0	0.0	68.9
1959-60	0.0	0.0	0.0	0.0	1.4	15.3	16.5	12.6	12.1	T	0.0	0.0	57.9
1960-61	0.0	0.0	0.0	T	T	23.1	11.4	10.8	14.7	3.9	0.0	0.0	63.9
1961-62	0.0	0.0	0.0	T	10.0	19.7	6.7	29.8	4.5	0.4	0.0	0.0	71.1
1962-63	0.0	0.0	0.0	T	0.4	12.7	15.3	14.8	13.9	0.5	T	0.0	57.6
1963-64	0.0	0.0	0.0	T	T	16.9	24.3	19.1	15.8	T	0.0	0.0	76.1
1964-65	0.0	0.0	0.0	T	T	12.4	13.2	6.5	5.2	1.3	0.0	0.0	38.6
1965-66	0.0	0.0	0.0	T	1.0	4.6	32.9	17.7	3.7	T	1.1	0.0	61.0
1966-67	0.0	0.0	0.0	0.0	T	14.3	8.3	32.6	19.6	6.0	T	0.0	80.8
1967-68	0.0	0.0	0.0	0.0	15.0	15.0	15.8	6.7	7.4	0.0	0.0	0.0	59.9
1968-69	0.0	0.0	0.0	0.0	11.6	13.4	4.7	49.8	5.7	T	0.0	0.0	85.2
1969-70	0.0	0.0	0.0	2.1	T	20.7	5.0	13.8	14.2	2.8	0.0	0.0	58.6
1970-71	0.0	0.0	0.0	0.0	T	30.1	15.6	19.8	24.0	6.3	0.0	0.0	95.8
1971-72	0.0	0.0	0.0	0.0	18.4	17.4	10.8	29.8	13.6	10.0	0.0	0.0	100.0
1972-73	0.0	0.0	0.0	0.0	12.9	23.2	13.1	5.6	0.1	3.4	0.0	0.0	58.3
1973-74	0.0	0.0	0.0	0.0	T	6.6	15.4	7.0	6.5	3.9	0.0	0.0	39.4
1974-75	0.0	0.0	0.0	0.0	2.9	18.3	19.5	17.1	4.7	5.5	0.0	0.0	68.0
1975-76	0.0	0.0	0.0	0.0	2.2	25.1	14.7	8.4	24.3	T	0.0	0.0	74.7
1976-77	0.0	0.0	0.0	T	3.1	11.6	37.1	11.7	22.3	0.5	T	0.0	86.3
1977-78	0.0	0.0	0.0	0.0	1.5	20.2	37.1	13.5	11.5	0.4	T	0.0	84.2
1978-79	0.0	0.0	0.0	0.0	10.5	16.2	42.3	4.5	0.2	5.1	0.0	0.0	78.8
1979-80	0.0	0.0	0.0	1.3	T	2.1	3.1	11.9	8.6	T	0.0	0.0	27.0
1980-81	0.0	0.0	0.0	T	9.4	9.8	9.2	20.9	5.4	T	0.0	0.0	54.7
1981-82	0.0	0.0	0.0	0.0	T	33.0	26.2	9.0	6.5	15.3	0.0	0.0	90.0
1982-83	0.0	0.0	0.0	0.0	1.3	3.6	9.0	20.8	4.0	T	T	0.0	38.7
1983-84	0.0	0.0	0.0	0.0	T	17.5	20.4	12.7	25.0	T	T	0.0	75.6
1984-85	0.0	0.0	0.0	0.0	T	16.5	11.6	11.0	12.4	1.0	0.0	0.0	52.5
1985-86	0.0	0.0	0.0	0.0	8.3	11.2							
Record Mean	0.0	0.0	0.0	0.1	3.9	14.2	17.7	14.9	11.2	2.3	0.1	0.0	64.4

See Reference Notes, relative to all above tables, on preceding page.

The Mount Washington Observatory is located at the summit of Mount Washington, New Hampshire, highest mountain of the Presidential range. The weather is very severe most of the year, conditions approximating those that would be encountered at a much higher latitude. The upper limits of timberline extend to 4,500 to 5,000 feet.

Prevailing winds are from the west and west-northwest, although the most severe storms are usually from the southeast. Winds are stronger at the summit than at the same elevation at a distance from the mountain, due to the Bernouilli effect. Mount Washington is near the mid-point of a 60-mile-long mountain front trending northeast to southwest. Wind speeds in excess of 100 mph are not uncommon, and the stations highest measured wind, 231 mph, still stands as a world record.

The station is in the clouds approximately 55 percent of the time. This is due partly to the effect of orographic uplift and partly due to the fact that the summit is often above the cloud base when there are low clouds in the area.

Minimum temperatures are not extreme compared to some U. S. valley stations. Annual temperature variations are not as great as they are in the surrounding lowlands, which may actually be colder than the summit when there is a strong inversion. Rime or glaze icing occurs often in winter, when the mountain is frequently in supercooled clouds.

Because of its severe climate, Mount Washington has for many years been used as a natural laboratory for cloud physics research and for the development and testing of instruments, aircraft components, and structures which are required to withstand high winds and icing conditions.

TABLE 1 NORMALS, MEANS AND EXTREMES

MOUNT WASHINGTON OBS. GORHAM, NEW HAMPSHIRE

LATITUDE: 44°16'N   LONGITUDE: 71°18'W   ELEVATION: FT. GRND 6262 BARO 6274   TIME ZONE: EASTERN   WBAN: 14755

	(a)	JAN	FEB	MAR	APR	MAY	JUNE	JULY	AUG	SEP	OCT	NOV	DEC	YEAR
TEMPERATURE °F:														
Normals														
-Daily Maximum		13.4	13.1	19.1	28.9	40.7	50.7	54.4	52.7	46.3	36.5	26.9	17.0	33.3
-Daily Minimum		-3.3	-3.4	4.7	15.9	28.1	38.4	42.9	41.5	34.9	24.5	13.8	1.2	19.9
-Monthly		5.1	4.8	11.9	22.4	34.4	44.5	48.7	47.1	40.6	30.5	20.4	9.1	26.6
Extremes														
-Record Highest	53	44	43	48	60	66	71	71	72	67	59	52	45	72
-Year		1950	1981	1946	1976	1977	1933	1953	1975	1960	1938	1982	1982	AUG 1975
-Record Lowest	53	-47	-46	-38	-20	-2	8	25	20	11	-5	-20	-46	-47
-Year		1934	1943	1950	1954	1966	1945	1982	1976	1942	1939	1958	1933	JAN 1934
NORMAL DEGREE DAYS:														
Heating (base 65°F)		1857	1686	1643	1278	949	612	505	555	732	1070	1341	1733	13961
Cooling (base 65°F)		0	0	0	0	0	0	0	0	0	0	0	0	0
% OF POSSIBLE SUNSHINE	47	32	34	34	35	36	32	31	31	36	39	29	29	33
MEAN SKY COVER (tenths)														
Sunrise - Sunset	47	7.7	7.6	7.8	7.7	7.7	8.1	8.2	7.9	7.4	7.0	7.9	8.0	7.8
MEAN NUMBER OF DAYS:														
Sunrise to Sunset														
-Clear	53	4.8	4.5	4.3	4.2	3.6	2.4	1.5	2.8	4.6	6.4	3.6	4.2	47.0
-Partly Cloudy	53	5.3	5.1	5.6	6.0	7.3	6.6	7.6	8.1	6.5	6.1	5.2	5.1	74.4
-Cloudy	53	20.9	18.6	21.1	19.9	20.2	21.0	21.9	20.1	18.9	18.5	21.2	21.7	243.9
Precipitation														
.01 inches or more	53	19.1	17.7	19.2	18.0	17.2	16.0	16.5	15.6	15.0	15.0	19.3	20.3	209.1
Snow,Ice pellets														
1.0 inches or more	42	11.3	10.2	10.8	7.8	3.0	0.5	0.*	0.*	0.6	3.8	8.5	11.5	68.0
Thunderstorms	53	0.2	0.1	0.3	0.9	1.7	3.6	4.4	2.8	1.0	0.5	0.3	0.1	16.0
Heavy Fog Visibility 1/4 mile or less	53	26.2	24.7	27.3	24.8	24.1	25.9	27.3	27.5	25.9	25.5	26.8	27.5	313.6
Temperature °F														
-Maximum														
90° and above	53	0.0	0.0	0.0	0.0	0.0	0.0	0.0	0.0	0.0	0.0	0.0	0.0	0.0
32° and below	53	29.0	26.5	26.4	18.6	6.4	0.8	0.*	0.2	2.1	10.2	20.3	27.1	167.6
-Minimum														
32° and below	53	31.0	28.0	30.7	28.3	20.1	6.5	1.5	3.1	12.0	23.1	28.2	30.8	243.2
0° and below	53	18.2	16.2	11.1	2.2	0.1	0.0	0.0	0.0	0.0	0.2	3.8	14.0	65.8
AVG. STATION PRESS.(mb)														
RELATIVE HUMIDITY (%)														
Hour 01	33	82	82	86	86	86	91	93	92	86	84	85	85	87
Hour 07	33	83	81	84	85	84	89	91	90	85	82	84	84	85
Hour 13 (Local Time)	33	83	83	84	84	81	84	87	87	85	81	84	84	84
Hour 19	33	83	82	85	86	84	87	91	91	88	83	85	84	86
PRECIPITATION (inches):														
Water Equivalent														
-Normal		7.31	8.01	8.19	7.03	6.46	7.06	6.90	7.60	7.15	6.73	8.54	8.94	89.92
-Maximum Monthly	53	18.23	25.56	15.98	14.62	18.82	16.00	15.53	13.14	14.07	13.49	19.56	17.95	25.56
-Year		1958	1969	1977	1974	1984	1973	1969	1955	1938	1959	1983	1973	FEB 1969
-Minimum Monthly	53	1.29	0.98	2.15	2.19	1.78	2.43	2.69	2.77	2.74	0.75	2.31	1.49	0.75
-Year		1981	1980	1946	1959	1951	1979	1955	1947	1948	1947	1939	1955	OCT 1947
-Maximum in 24 hrs	53	3.74	10.38	3.99	8.30	4.60	6.50	7.37	5.20	5.38	7.03	6.07	8.64	10.38
-Year		1975	1970	1962	1984	1967	1973	1969	1955	1985	1959	1968	1969	FEB 1970
Snow, Ice pellets														
-Maximum Monthly	53	94.6	172.8	98.0	89.3	52.2	8.1	1.1	2.5	7.8	34.4	86.6	103.7	172.8
-Year		1978	1969	1970	1975	1967	1959	1957	1965	1949	1969	1968	1968	FEB 1969
-Maximum in 24 hrs	53	24.0	49.3	27.4	25.4	22.2	4.4	1.1	2.5	5.9	17.0	25.0	37.5	49.3
-Year		1978	1969	1969	1974	1967	1965	1957	1965	1949	1969	1968	1968	FEB 1969
WIND:														
Mean Speed (mph)	51	45.7	44.2	41.9	36.7	29.6	27.3	25.3	24.9	28.6	33.5	38.9	44.5	35.1
Prevailing Direction through 1963		W	W	W	W	W	W	W	W	W	W	W	W	W
Fastest Obs. 1 Min.														
-Direction (!!)														
-Speed (MPH)														
-Year														
Peak Gust														
-Direction (!!)	53	NW	E	W	SE	W	NW	NW	ENE	SE	W	NW	NW	SE
-Speed (mph)	53	173	166	180	231	164	136	110	142	174	161	163	178	231
-Date		1985	1972	1942	1934	1945	1949	1933	1954	1979	1943	1983	1980	APR 1934

See reference Notes to this table on the following page.

TABLE 2 PRECIPITATION (inches) MOUNT WASHINGTON OBS. GORHAM, NEW HAMPSHIRE

YEAR	JAN	FEB	MAR	APR	MAY	JUNE	JULY	AUG	SEP	OCT	NOV	DEC	ANNUAL
1956	7.04	4.19	7.49	6.18	7.75	6.90	10.32	10.54	5.99	4.57	5.15	5.41	81.53
1957	3.46	3.32	3.58	4.44	6.06	6.69	10.70	2.82	5.58	4.65	8.50	13.35	73.15
1958	18.23	9.34	8.97	8.32	5.30	4.25	9.79	4.66	8.28	7.97	7.78	4.55	97.44
1959	5.47	4.04	5.21	2.19	3.63	7.46	3.67	7.58	5.46	13.49	9.21	5.70	73.11
1960	4.35	14.67	4.54	4.79	13.41	5.85	5.86	3.77	8.09	6.12	4.74	5.02	81.21
1961	3.00	5.88	3.85	10.69	3.00	5.75	5.17	4.11	5.09	2.52	6.38	4.64	60.08
1962	3.91	6.35	6.53	4.47	3.98	4.53	11.48	11.20	9.42	13.30	8.15	6.68	90.00
1963	4.66	5.60	7.66	6.97	4.31	4.99	6.93	12.77	3.10	3.89	17.57	7.47	85.92
1964	9.21	5.64	8.55	5.14	6.83	5.62	5.57	8.27	4.86	5.30	5.88	5.79	76.66
1965	4.55	7.52	2.71	3.97	2.40	8.15	4.42	8.19	10.01	7.88	6.90	4.65	71.35
1966	9.35	6.40	5.19	3.61	4.55	6.88	5.63	10.84	5.57	5.47	5.79	6.66	75.94
1967	4.92	7.28	2.82	6.23	12.68	6.92	6.80	5.36	7.61	6.67	7.45	7.26	82.00
1968	4.12	3.95	7.53	4.74	6.27	11.02	6.18	5.53	3.89	5.75	16.07	16.10	91.15
1969	8.60	25.56	12.22	6.59	6.01	5.06	15.53	7.50	4.17	7.26	14.41	17.23	130.14
1970	5.53	22.29	13.92	8.43	8.36	5.03	5.59	6.76	10.98	6.89	6.15	13.06	112.99
1971	8.18	8.32	10.19	7.57	10.96	4.89	9.25	9.29	6.24	5.71	10.18	9.51	100.29
1972	7.01	13.37	11.75	9.73	6.36	10.40	7.62	8.31	7.34	8.38	15.28	16.06	121.61
1973	11.74	5.72	9.01	13.33	11.59	16.00	4.38	5.16	6.93	7.71	11.87	16.24	121.39
1974	8.26	9.43	15.96	14.62	7.57	7.00	9.84	7.30	9.03	4.34	11.45	16.24	121.04
1975	13.61	8.92	15.95	13.45	4.32	6.75	8.31	6.50	12.32	6.59	6.37	10.48	113.57
1976	11.87	12.81	9.42	6.04	11.08	9.07	6.29	6.65	10.36	13.61	13.57		121.40
1977	11.11	12.09	15.98	8.90	3.73	11.12	3.99	7.21	10.64	8.89	11.37	12.08	117.11
1978	18.19	2.98	14.16	9.96	8.13	10.84	3.75	4.23	5.67	4.41	3.24	6.83	92.39
1979	12.62	4.34	5.30	4.98	7.62	2.43	3.80	7.53	6.62	6.75	5.97	3.38	71.34
1980	2.59	0.98	7.99	6.65	2.55	5.37	1.55	7.96	9.96	8.41	11.55	4.31	74.24
1981	1.29	19.81	3.72	5.41	8.59	8.90	11.02	9.38	10.82	9.59	5.72	7.09	101.34
1982	8.03	4.43	5.56	8.25	4.49	10.16	5.90	9.89	6.20	5.54	14.64	7.97	91.06
1983	7.93	5.67	10.26	10.73	11.43	3.60	6.96	10.32	5.57	7.29	19.56	17.38	116.70
1984	4.99	11.56	9.31	14.19	18.82	10.71	8.51	4.79	4.48	5.66	9.84	12.92	115.78
1985	5.48	10.88	12.72	6.77	6.97	9.69	7.93	6.51	9.60	6.15	8.07	7.02	97.79
Record Mean	6.41	7.11	7.37	6.93	6.54	7.01	6.66	7.18	7.03	6.43	8.29	8.18	85.13

TABLE 3 AVERAGE TEMPERATURE (deg. F) MOUNT WASHINGTON OBS. GORHAM, NEW HAMPSHIRE

YEAR	JAN	FEB	MAR	APR	MAY	JUNE	JULY	AUG	SEP	OCT	NOV	DEC	ANNUAL
1956	14.5	6.0	7.7	19.8	27.2	45.1	44.9	45.2	36.6	34.7	21.1	12.5	26.3
1957	-2.9	9.7	13.8	23.2	34.7	48.0	46.2	43.9	42.0	30.8	21.8	15.2	27.2
1958	13.0	0.6	16.1	24.8	31.8	39.6	49.3	47.1	40.1	28.2	19.3	1.0	25.9
1959	3.9	1.1	8.2	24.0	37.3	44.1	52.5	50.1	43.8	30.0	20.6	12.4	27.3
1960	7.0	9.5	5.4	26.3	41.2	44.6	46.1	48.5	44.3	27.4	20.6	4.4	27.1
1961	-2.3	9.5	12.3	20.4	31.5	44.1	48.6	48.7	49.2	34.9	22.7	11.4	27.6
1962	2.3	5.3	12.5	22.5	37.5	45.4	43.9	48.7	36.8	29.1	20.8	6.7	26.0
1963	5.4	-1.5	13.1	19.9	33.6	46.8	49.4	43.5	38.8	35.1	23.0	-0.3	25.6
1964	7.9	3.4	13.0	22.4	37.4	42.9	50.9	42.2	37.7	28.5	20.4	13.6	26.7
1965	1.3	3.0	7.7	18.2	36.4	42.5	45.2	46.2	43.1	27.2	15.1	11.6	24.8
1966	5.4	6.1	14.1	20.1	30.1	45.3	48.1	47.4	36.8	28.5	27.0	13.4	26.9
1967	8.9	-1.6	9.1	20.4	25.3	48.2	50.1	47.4	42.0	31.8	15.0	11.6	25.7
1968	4.5	-3.4	15.8	26.6	32.6	43.1	49.1	43.8	43.4	32.1	17.0	7.4	26.0
1969	8.0	8.2	9.1	23.7	31.8	45.8	47.3	47.7	43.0	30.8	22.5	10.7	27.4
1970	-3.6	4.8	11.0	22.5	36.9	44.9	51.1	48.3	41.0	34.2	24.2	6.4	26.9
1971	-2.1	8.2	8.6	20.3	34.1	45.0	47.9	46.9	45.2	39.7	18.0	9.6	26.8
1972	5.7	1.3	12.0	16.6	37.9	43.2	49.4	46.0	42.2	26.3	18.6	10.8	25.8
1973	6.0	5.2	22.6	23.5	33.1	46.0	50.4	51.2	38.4	32.6	17.8	16.7	28.6
1974	6.6	2.9	8.5	24.0	29.4	45.0	47.1	48.5	39.0	23.2	23.0	13.5	25.9
1975	7.5	4.3	9.6	16.4	40.3	44.2	51.4	48.3	38.8	32.2	24.8	9.0	27.2
1976	2.6	8.4	13.5	23.7	33.6	48.3	47.1	47.6	38.7	26.5	11.0	0.0	25.1
1977	-3.3	5.3	18.1	22.3	36.2	42.1	47.4	46.9	39.2	29.6	23.7	6.9	26.2
1978	2.9	-1.3	6.6	18.4	37.0	43.5	48.7	28.4	37.7	28.4	19.6	3.5	24.8
1979	7.5	-1.0	18.4	23.9	37.7	45.1	50.9	45.2	40.9	29.7	25.2	11.4	27.9
1980	4.7	1.1	12.2	12.2	35.1	44.8	48.7	48.6	39.0	25.2	16.5	3.5	25.3
1981	-2.5	15.5	11.5	23.2	36.2	44.8	48.9	47.1	38.7	29.8	21.3	11.2	27.2
1982	-2.2	5.1	12.4	18.5	38.3	40.9	47.9	42.1	43.1	32.3	24.1	16.9	26.6
1983	7.9	10.6	17.9	25.6	32.9	46.1	48.4	48.2	43.0	31.2	22.9	7.5	28.5
1984	4.0	16.7	8.0	25.4	32.1	44.5	47.8	50.6	36.5	34.9	21.8	14.1	28.0
1985	-1.8	6.9	10.9	22.1	35.9	39.3	47.9	46.1	43.2	31.0	23.2	3.7	25.7
Record Mean	4.8	5.4	11.9	22.4	34.7	44.4	48.2	47.3	40.8	31.1	20.5	9.2	26.8
Max	13.1	13.7	19.3	29.0	41.0	50.5	54.5	52.8	46.5	37.0	27.0	17.1	33.5
Min	-3.6	-2.8	4.6	15.7	28.4	38.3	41.9	41.7	35.0	25.1	14.0	1.3	20.0

REFERENCE NOTES FOR TABLES 1, 2, 3 and 6 (MT. WASH'N OBS'Y [GORHAM] MA)

GENERAL

T - TRACE AMOUNT
BLANK ENTRIES DENOTE MISSING/UNREPORTED DATA.
# INDICATES A STATION OR INSTRUMENT RELOCATION.

SPECIFIC

TABLE 1

(a) - LENGTH OF RECORD IN YEARS. ALTHOUGH INDIVIDUAL MONTHS MAY BE MISSING.
* LESS THAN .05

NORMALS — BASED ON THE 1951-1980 RECORD PERIOD.
EXTREMES — DATES ARE THE MOST RECENT OCCURRENCE.
WIND DIR. — NUMERALS SHOW TENS OF DEGREES CLOCKWISE FROM TRUE NORTH.
"00" INDICATES CALM.
RESULTANT WIND DIRECTIONS ARE GIVEN TO WHOLE DEGREES.

EXCEPTIONS

TABLE 1

1. SUNSHINE DATA IS BY VISUAL OBSERVATION

2. PRECIPITATION IS MEASURED BY GAGE WITH TILTED ORIFICE DURING SUMMER AND ESTIMATED FROM SNOWFALL/DEPTH MEASUREMENTS DURING WINTER.

3. TEMPERATURE AND PRECIPITATION MAY BE DOUBTFUL DURING PERIODS OF EXTREME WIND VELOCITY AND/OR ICING CONDITIONS

TABLES 2, 3, and 6

RECORD MEANS ARE THROUGH THE CURRENT YEAR, BEGINNING IN   1933 FOR TEMPERATURE
1933 FOR PRECIPITATION
1933 FOR SNOWFALL

TABLE 4 HEATING DEGREE DAYS Base 65 deg. F   MOUNT WASHINGTON OBS. GORHAM, NEW HAMPSHIRE

SEASON	JULY	AUG	SEP	OCT	NOV	DEC	JAN	FEB	MAR	APR	MAY	JUNE	TOTAL
1956-57	615	608	847	932	1313	1623	2106	1541	1582	1247	933	501	13848
1957-58	577	646	683	1054	1289	1539	1608	1806	1507	1199	1022	756	13686
1958-59	480	548	742	1133	1368	1985	1895	1789	1758	1222	850	622	14392
1959-60	382	455	632	1089	1326	1636	1796	1603	1848	1153	732	605	13257
1960-61	578	502	614	1158	1325	1880	2089	1552	1630	1334	1033	621	14316
1961-62	503	502	466	924	1261	1657	1942	1669	1622	1269	847	579	13241
1962-63	648	495	839	1107	1322	1808	1844	1863	1603	1345	966	537	14377
1963-64	478	659	779	924	1252	2024	1768	1786	1607	1271	847	657	14052
1964-65	431	699	812	1125	1330	1590	1973	1733	1771	1397	878	668	14407
1965-66	606	576	651	1164	1489	1649	1848	1645	1574	1339	1077	587	14205
1966-67	515	539	837	1124	1132	1595	1738	1862	1729	1328	1224	495	14118
1967-68	456	540	684	1026	1497	1650	1876	1986	1521	1147	994	650	14027
1968-69	484	651	640	1011	1434	1782	1766	1587	1733	1230	1022	569	13909
1969-70	540	527	650	1058	1269	1676	2127	1683	1667	1266	867	592	13922
1970-71	424	511	713	924	1217	1812	2079	1583	1746	1334	949	593	13885
1971-72	527	554	587	780	1404	1714	1838	1846	1639	1450	832	646	13817
1972-73	474	579	675	1196	1388	1677	1830	1672	1306	1239	983	567	13586
1973-74	445	422	793	997	1409	1495	1809	1738	1747	1225	1096	591	13767
1974-75	544	505	771	1291	1252	1589	1777	1699	1713	1453	758	616	13968
1975-76	415	513	780	1009	1197	1735	1935	1638	1591	1233	967	496	13509
1976-77	547	534	781	1184	1617	2017	2114	1670	1445	1275	887	686	14757
1977-78	538	551	768	1091	1233	1796	1924	1855	1805	1393	861	634	14449
1978-79	497	493	811	1128	1355	1792	1782	1849	1440	1229	841	593	13810
1979-80	430	607	715	1090	1185	1657	1867	1853	1633	1184	919	717	13857
1980-81	499	500	774	1227	1448	1907	2093	1381	1651	1246	890	598	14214
1981-82	492	545	781	1085	1305	1660	2085	1676	1623	1391	821	718	14182
1982-83	524	703	649	1009	1221	1488	1766	1520	1455	1176	988	559	13058
1983-84	508	513	654	1042	1256	1780	1885	1394	1766	1180	1012	609	13599
1984-85	527	437	849	926	1290	1574	2068	1622	1675	1280	899	766	13913
1985-86	525	579	646	1044	1249	1898							

TABLE 5 COOLING DEGREE DAYS Base 65 deg. F   MOUNT WASHINGTON OBS. GORHAM, NEW HAMPSHIRE

YEAR	JAN	FEB	MAR	APR	MAY	JUNE	JULY	AUG	SEP	OCT	NOV	DEC	TOTAL
1969	0	0	0	0	0	0	0	0	0	0	0	0	0
1970	0	0	0	0	0	0	0	0	0	0	0	0	0
1971	0	0	0	0	0	0	0	0	0	0	0	0	0
1972	0	0	0	0	0	0	0	0	0	0	0	0	0
1973	0	0	0	0	0	0	0	0	0	0	0	0	0
1974	0	0	0	0	0	0	0	0	0	0	0	0	0
1975	0	0	0	0	0	0	0	1	0	0	0	0	1
1976	0	0	0	0	0	0	0	0	0	0	0	0	0
1977	0	0	0	0	0	0	0	0	0	0	0	0	0
1978	0	0	0	0	0	0	0	0	0	0	0	0	0
1979	0	0	0	0	0	0	0	0	0	0	0	0	0
1980	0	0	0	0	0	0	0	0	0	0	0	0	0
1981	0	0	0	0	0	0	0	0	0	0	0	0	0
1982	0	0	0	0	0	0	0	0	0	0	0	0	0
1983	0	0	0	0	0	0	0	0	0	0	0	0	0
1984	0	0	0	0	0	0	0	0	0	0	0	0	0
1985	0	0	0	0	0	0	0	0	0	0	0	0	0

TABLE 6 SNOWFALL (inches)   MOUNT WASHINGTON OBS. GORHAM, NEW HAMPSHIRE

SEASON	JULY	AUG	SEP	OCT	NOV	DEC	JAN	FEB	MAR	APR	MAY	JUNE	TOTAL
1956-57	0.0	0.0	5.5	3.9	30.7	47.1	31.8	17.9	20.3	25.6	10.7	0.1	193.6
1957-58	1.1	T	T	1.3	12.1	59.0	92.5	65.2	59.7	43.9	7.8	1.5	344.1
1958-59	0.0	T	T	17.9	30.6	39.1	32.5	43.0	52.0	4.2	4.5	8.1	231.9
1959-60	0.0	0.0	4.4	4.9	16.9	39.5	41.6	102.4	27.1	11.4	5.7	0.0	253.9
1960-61	0.0	0.0	0.0	20.2	21.2	28.1	23.9	23.0	24.1	59.6	6.5	0.0	206.6
1961-62	T	T	0.0	5.7	40.6	41.1	22.9	42.1	28.1	11.8	1.1	T	193.4
1962-63	0.1	0.0	1.0	24.7	36.9	40.5	31.0	46.0	39.7	27.9	3.5	T	251.3
1963-64	0.0	T	1.2	12.7	29.6	48.6	43.2	38.5	40.0	23.6	3.6	5.9	246.9
1964-65	0.0	1.2	0.7	23.5	21.1	31.2	31.7	32.7	19.8	23.3	1.1	7.9	194.2
1965-66	0.0	2.5	0.1	28.2	37.8	29.1	63.9	51.5	29.4	22.5	23.6	0.6	289.2
1966-67	0.0	T	3.9	13.3	18.3	37.0	30.9	52.1	22.8	30.1	52.2	0.0	260.6
1967-68	0.0	0.0	6.9		41.6	40.0	36.8	26.7	34.8	6.6	1.5		211.1
1968-69	0.0	0.0	1.5	19.0	86.6	103.7	56.8	172.8	95.0	21.1	9.9	0.0	566.4
1969-70	0.0	0.0	T	34.4	20.3	84.9	30.2	95.7	98.0	34.9	9.2	T	407.6
1970-71	T	0.0	4.9	3.9	23.8	84.5	57.2	45.9	86.6	55.9	33.8	0.0	396.5
1971-72	0.0	0.2	0.0	0.4	52.8	60.3	38.9	79.1	63.1	74.4	7.0	2.0	378.2
1972-73	0.0	T	0.0	12.5	80.7	91.6	70.5	35.2	46.1	71.9	25.3	0.1	433.9
1973-74	0.0	0.0	0.6	10.5	46.3	55.4	49.5	53.7	77.3	64.0	10.6	T	367.9
1974-75	T	0.0	3.7	16.3	44.6	77.1	80.7	50.9	86.3	89.3	1.7	1.1	451.7
1975-76	0.0	0.0	4.3	4.7	11.5	61.5	66.2	65.2	52.0	23.0	19.5	0.2	308.1
1976-77	0.0	1.6	1.5	22.9	76.1	82.0	90.8	65.5	97.4	44.8	16.3	0.3	499.2
1977-78	T	T	0.4	14.9	43.0	73.4	94.6	21.4	71.4	44.3	7.4	3.5	374.3
1978-79	T	0.0		8.7	19.9	43.5	67.7	22.3	18.0	24.1	1.8	0.6	206.6
1979-80	0.8	T	0.4	11.5	8.7	23.5	15.6	14.2	43.3	18.1	4.8	1.4	142.3
1980-81	T	0.0		18.7	55.1	35.8	12.2	36.4	32.3	18.3	5.7	0.0	215.2
1981-82	0.0	0.0	5.6	10.3	22.4	54.3	54.0	34.7	40.6	35.2	0.4	T	257.5
1982-83	0.0	0.6	T	9.4	20.2	13.4	40.8	20.6	36.4	39.2	8.2	T	188.8
1983-84	T	0.0	0.3	11.3	69.6	78.1	40.3	62.9	49.5	41.4	26.3	T	379.7
1984-85	0.0	0.0	1.7	18.3	28.3	62.2	45.6	54.7	71.5	44.3	13.3	2.1	342.0
1985-86	0.0	0.0	0.2	5.1	18.5	51.9							
Record Mean	T	0.1	1.6	11.6	30.5	42.5	38.1	40.4	41.7	29.5	10.8	1.1	248.0

See Reference Notes, relative to all above tables, on preceding page.

The Atlantic City State Marina is located on Abescon Island on the southeast coast of New Jersey. Surrounding terrain, composed of tidal marshes and beach sand, is flat and lies slightly above sea level. The climate is principally continental in character. However, the moderating influence of the Atlantic Ocean is apparent throughout the year, being more marked in the city than at the airport. As a result, summers are relatively cooler and winters milder than elsewhere at the same latitude.

Land and sea breezes, local circulations resulting from the differential heating and cooling of the land and sea, often prevail. These winds occur when moderate or intense storms are not present in the area, thus enabling the local circulation to overcome the general wind pattern. During the warm season sea breezes in the late morning and afternoon hours prevent excessive heating. Frequently, the temperature at Atlantic City during the afternoon hours in the summer averages several degrees lower than at the airport and the airport averages several degrees lower than localities farther inland. On occasions, sea breezes have lowered the temperature as much as 15 to 20 degrees within a half hour. However, the major effect of the sea breeze at the airport is preventing the temperature from rising above the 80s. Because the change in ocean temperature lags behind the air temperature from season to season, the weather tends to remain comparatively mild late into the fall, but on the other hand, warming is retarded in the spring. Normal ocean temperatures range from an average near 37 degrees in January to near 72 degrees in August.

Precipitation is moderate and well distributed throughout the year, with June the driest month and August the wettest. Tropical storms or hurricanes occasionally bring excessive rainfall to the area. The bulk of winter precipitation results from storms which move northeastward along or near the east coast of the United States. Snowfall is considerably less than elsewhere at the same latitude and does not remain long on the ground. Precipitation, often beginning as snow, will frequently become mixed with or change to rain while continuing as snow over more interior sections. In addition, ice storms and resultant glaze are relatively infrequent.

TABLE 1 — NORMALS, MEANS AND EXTREMES
ATLANTIC CITY, NEW JERSEY N.A.F.E.C.

LATITUDE: 39°27'N LONGITUDE: 74°34'W ELEVATION: FT. GRND 64 BARO 00118 TIME ZONE: EASTERN WBAN: 93730

	(a)	JAN	FEB	MAR	APR	MAY	JUNE	JULY	AUG	SEP	OCT	NOV	DEC	YEAR
TEMPERATURE °F: Normals														
-Daily Maximum		40.6	42.4	50.3	61.6	71.0	79.6	84.0	82.5	76.7	66.1	55.4	45.0	62.9
-Daily Minimum		22.9	23.9	31.6	40.4	49.9	58.8	64.8	63.5	56.4	44.8	35.8	26.6	43.3
-Monthly		31.8	33.2	41.0	51.0	60.5	69.2	74.4	73.0	66.6	55.5	45.6	35.8	53.1
Extremes -Record Highest	42	78	75	87	94	99	106	104	102	99	90	84	75	106
-Year		1967	1985	1945	1969	1969	1969	1966	1948	1983	1959	1950	1984	JUN 1969
-Record Lowest	42	-10	-11	5	12	25	37	46	40	32	23	11	-7	-11
-Year		1977	1979	1984	1969	1966	1980	1965	1976	1969	1969	1964	1950	FEB 1979
NORMAL DEGREE DAYS: Heating (base 65°F)		1029	890	744	420	165	26	0	0	27	298	582	905	5086
Cooling (base 65°F)		0	0	0	0	26	152	291	248	75	0	0	0	792
% OF POSSIBLE SUNSHINE	25	49	52	53	55	54	58	60	63	60	56	48	44	54
MEAN SKY COVER (tenths) Sunrise - Sunset	27	6.3	6.3	6.3	6.2	6.4	6.1	6.2	6.0	5.7	5.5	6.2	6.4	6.1
MEAN NUMBER OF DAYS: Sunrise to Sunset -Clear	27	8.4	7.7	7.7	7.4	6.4	6.9	7.3	7.4	10.0	10.2	7.7	8.1	95.3
-Partly Cloudy	27	7.8	6.9	8.1	9.4	10.6	11.0	10.8	11.5	8.3	8.5	8.7	7.6	109.2
-Cloudy	27	14.7	13.6	15.2	13.2	14.0	12.1	13.0	12.1	11.7	12.3	13.6	15.4	160.7
Precipitation .01 inches or more	42	10.8	9.7	10.8	10.7	10.4	8.9	8.6	8.6	7.4	7.2	9.3	9.6	112.1
Snow, Ice pellets 1.0 inches or more	41	1.7	1.4	0.6	0.1	0.0	0.0	0.0	0.0	0.0	0.0	0.1	0.6	4.6
Thunderstorms	27	0.2	0.5	0.9	2.2	3.4	4.5	5.7	5.4	1.9	0.7	0.4	0.2	26.0
Heavy Fog Visibility 1/4 mile or less	27	3.1	3.4	3.6	3.4	4.6	4.4	4.1	3.6	3.4	4.8	3.5	2.8	44.6
Temperature °F -Maximum 90° and above	21	0.0	0.0	0.0	0.2	0.4	3.0	6.4	5.0	1.6	0.0	0.0	0.0	16.7
32° and below	21	8.6	5.5	0.6	0.0	0.0	0.0	0.0	0.0	0.0	0.0	0.1	3.0	17.8
-Minimum 32° and below	21	26.2	21.8	17.0	6.3	0.4	0.0	0.0	0.0	0.1	3.4	12.4	21.6	109.1
0° and below	21	1.0	0.6	0.0	0.0	0.0	0.0	0.0	0.0	0.0	0.0	0.0	0.3	1.9
AVG. STATION PRESS.(mb)	13	1015.0	1015.3	1013.3	1012.5	1012.3	1013.0	1013.4	1014.8	1015.7	1016.7	1016.0	1016.1	1014.5
RELATIVE HUMIDITY (%) Hour 01	21	75	76	75	77	83	87	86	88	87	86	81	76	81
Hour 07	21	77	78	77	76	78	81	83	85	87	87	83	78	81
Hour 13 (Local Time)	21	59	56	55	51	56	57	56	57	57	56	58	58	56
Hour 19	21	70	68	65	64	69	71	70	74	78	78	75	71	71
PRECIPITATION (inches): Water Equivalent -Normal		3.47	3.34	4.04	3.20	3.07	2.78	4.02	4.72	2.89	3.06	3.73	3.61	41.93
-Maximum Monthly	42	7.71	5.98	6.80	7.95	11.51	6.36	13.09	6.27	6.27	7.50	9.65	7.33	13.09
-Year		1948	1958	1953	1952	1948	1970	1959	1967	1966	1943	1972	1969	JUL 1959
-Minimum Monthly	42	0.26	0.82	0.62	0.84	0.40	0.10	0.51	0.34	0.41	0.15	0.68	0.62	0.10
-Year		1955	1980	1945	1976	1957	1954	1954	1983	1970	1963	1976	1955	JUN 1954
-Maximum in 24 hrs	42	2.86	2.59	2.66	3.37	4.15	2.91	6.46	6.40	3.98	2.95	3.93	2.75	6.46
-Year		1944	1966	1979	1952	1959	1952	1959	1966	1954	1958	1953	1951	JUL 1959
Snow, Ice pellets -Maximum Monthly	41	15.9	35.2	17.6	3.2	T					T	7.8	8.6	35.2
-Year		1961	1967	1969	1965	1977					1980	1967	1960	FEB 1967
-Maximum in 24 hrs	41	14.4	17.1	11.5	3.2	T					T	7.8	7.5	17.1
-Year		1964	1979	1969	1965	1977					1980	1967	1960	FEB 1979
WIND: Mean Speed (mph)	27	11.3	11.6	12.1	11.9	10.4	9.3	8.6	8.2	8.6	9.1	10.6	10.8	10.2
Prevailing Direction through 1963		WNW	W	WNW	S	S	S	S	S	ENE	W	W	WNW	S
Fastest Obs. 1 Min. -Direction (!!!)	26	29	27	24	07	31	29	26	12	32	29	27	36	32
-Speed (MPH)	26	47	43	46	46	35	37	37	35	60	41	40	55	60
-Year		1971	1960	1973	1961	1973	1964	1970	1971	1960	1961	1960	1960	SEP 1960
Peak Gust -Direction (!!!)	2	NW	W	NE	W	S	NW	W	SW	NW	NE	E	W	NW
-Speed (mph)	2	38	49	55	47	38	46	45	33	69	31	40	46	69
-Date		1985	1985	1984	1985	1984	1985	1985	1985	1985	1984	1985	1985	SEP 1985

ATLANTIC CITY (NAFEC Airport) NEW JERSEY

TABLE 2   PRECIPITATION (inches)   ATLANTIC CITY, NEW JERSEY N.A.F.E.C.

YEAR	JAN	FEB	MAR	APR	MAY	JUNE	JULY	AUG	SEP	OCT	NOV	DEC	ANNUAL
1956	2.87	3.58	4.40	2.71	2.36	4.12	4.79	3.25	3.03	5.88	2.44	4.15	43.58
1957	1.86	4.01	4.30	2.28	0.54	2.76	0.31	3.38	2.14	1.62	4.54	6.78	34.52
#1958	5.24	6.86	7.57	4.33	4.42	4.38	9.72	9.48	3.71	6.11	2.71	2.64	67.17
1959	1.59	1.86	3.84	3.33	5.03	1.29	13.09	3.57	1.48	4.37	5.15	3.15	47.75
1960	2.69	5.71	3.67	2.34	2.21	0.77	5.84	2.36	4.03	3.08	1.74	3.46	37.90
1961	4.06	4.51	6.36	3.12	3.17	3.00	3.40	1.73	3.36	4.73	2.83	3.38	43.65
1962	4.21	3.47	5.42	3.50	1.77	4.20	1.72	5.29	3.09	2.04	4.88	3.84	43.43
1963	2.94	2.50	5.21	1.39	2.95	3.07	2.60	2.93	4.35	0.15	6.46	2.35	36.90
1964	6.35	4.32		7.59	1.46	0.84	2.79	1.63	5.91	2.67		3.47	41.01
1965	3.58	2.44	3.75	2.00	2.59	1.24	2.61	2.40	1.60	1.18	0.79	1.09	25.27
1966	3.45	5.17	0.70	2.58	3.17	1.87	2.59	9.04	6.27	3.73	1.91	4.81	45.29
1967	1.16	3.24	3.78	2.76	3.68	1.37	4.19	11.98	1.50	2.86	1.72	5.57	43.81
1968	2.77	1.69	4.99	1.50	5.55	2.86	1.75	2.20	0.43	2.73	3.10	3.89	33.46
1969	1.68	2.38	3.19	3.55	1.68	1.42	12.64	2.56	1.65	2.09	4.28	7.33	44.45
1970	1.50	3.08	3.11	4.66	1.81	6.36	2.83	2.70	0.41	3.73	5.73	3.04	38.96
1971	2.67	5.26	1.64	1.29	1.88	0.69	3.65	10.40	4.39	4.20	5.02	2.08	43.17
1972	2.93	4.31	3.59	4.62	3.51	4.82	2.81	0.44	3.66	5.11	9.65	3.63	49.08
1973	3.26	3.63	3.08	4.39	3.08	4.32	3.28	2.05	4.73	2.74	1.43	5.48	41.47
1974	3.47	2.40	4.62	2.66	2.61	2.52	1.99	5.50	2.95	1.90	1.08	4.76	36.46
1975	5.94	3.08	3.84	3.90	5.44	3.86	6.02	5.01	5.16	1.76	3.76	2.53	50.30
1976	4.52	2.70	1.39	0.84	3.61	0.97	2.23	4.70	3.06	6.60	0.68	2.52	33.82
1977	3.45	1.41	3.42	2.13	0.64	1.53	2.79	6.49	2.97	4.19	4.75	4.69	38.46
1978	5.70	1.11	5.17	1.53	6.71	3.00	5.77	6.82	1.51	1.21	2.96	3.52	45.01
1979	7.13	5.76	3.82	2.98	2.80	3.15	7.63	4.11	3.32	2.19	3.23	2.19	48.11
1980	2.63	0.82	6.38	5.40	1.61	3.58	2.47	2.63	1.74	3.20	3.63	0.75	34.84
1981	0.56	3.72	1.41	6.20	3.18	4.91	1.28	3.25	1.96	2.96	1.12	3.94	34.49
1982	4.11	2.06	2.70	3.85	2.42	3.03	3.62	1.63	1.34	1.14	4.17	2.85	32.92
1983	2.46	3.32	5.85	7.45	5.21	3.01	0.51	2.90	2.22	3.48	6.70	5.06	48.17
1984	2.41	3.70	5.92	4.84	6.58	1.62	4.35	2.44	1.31	1.46	3.02	1.79	39.44
1985	2.07	1.71	2.38	1.02	5.04	1.55	4.36	3.94	2.26	0.90	3.81	0.93	29.97
Record Mean	3.31	3.16	3.77	3.39	3.31	2.62	4.23	4.29	2.87	2.95	3.48	3.38	40.77

TABLE 3   AVERAGE TEMPERATURE (deg. F)   ATLANTIC CITY, NEW JERSEY N.A.F.E.C.

YEAR	JAN	FEB	MAR	APR	MAY	JUNE	JULY	AUG	SEP	OCT	NOV	DEC	ANNUAL
1956	33.2	39.2	40.0	47.1	56.0	68.8	72.4	73.5	66.3	58.9	47.3	45.2	54.0
1957	31.9	38.9	42.1	51.7	60.0	69.9	73.8	70.6	69.6	56.5	50.6	42.3	54.8
#1958	33.8	29.8	40.3	51.2	57.8	65.8	76.9	72.8	66.0	55.5	47.8	28.3	52.2
1959	31.5	34.3	40.9	52.6	64.1	71.4	74.6	76.1	69.7	59.5	45.3	39.0	54.9
1960	35.3	36.7	32.3	55.2	61.0	70.7	73.7	74.8	67.9	56.5	47.7	30.1	53.5
1961	26.9	36.1	43.0	49.1	59.2	69.4	76.6	74.9	73.4	58.1	47.7	35.1	54.6
1962	33.2	34.1	40.8	52.2	63.2	70.6	72.9	73.0	64.6	57.1	43.7	32.3	53.1
1963	31.3	29.5	45.5	52.5	60.3	70.5	76.1	72.4	63.9	59.9	50.2	30.5	53.5
#1964	35.3	33.2	43.1	49.7	64.0	70.9	76.3	72.2	67.6	53.5	47.0	36.8	54.1
1965	27.6	33.0	37.8	48.7	64.1	67.6	72.8	72.1	68.0	53.1	44.9	36.4	52.2
1966	28.5	30.1	41.4	47.2	56.6	70.2	75.3	73.4	65.2	52.7	47.3	36.2	52.0
1967	38.9	29.2	38.6	51.4	54.1	69.0	73.4	71.5	61.2	52.4	39.5	35.1	51.2
1968	26.1	27.1	41.0	51.0	58.9	70.1	74.2	73.9	66.5	55.0	43.8	31.5	51.8
1969	30.3	31.8	37.4	53.6	62.0	69.5	73.2	73.3	65.5	53.9	43.9	34.5	52.4
1970	26.9	33.8	38.2	49.6	61.5	68.9	73.7	73.9	65.9	58.2	47.6	36.5	53.2
1971	28.8	35.7	39.0	47.6	57.6	69.7	72.0	70.4	67.5	60.7	44.5	41.1	52.9
1972	33.9	32.5	38.4	46.5	59.1	66.1	73.5	68.2	51.6	44.3	42.5	52.7	
1973	36.2	34.7	48.0	51.3	58.8	72.5	75.1	75.9	66.4	57.4	47.4	39.9	55.3
1974	39.3	33.2	44.4	54.5	60.3	68.0	75.0	74.2	65.9	52.2	46.3	39.0	54.3
1975	37.2	36.5	40.1	45.9	63.3	69.6	74.5	73.8	62.8	57.0	47.9	35.5	53.7
1976	28.4	39.5	43.2	51.5	57.4	69.6	72.5	71.4	64.9	51.2	39.6	30.1	51.6
1977	19.7	32.8	45.8	52.5	61.7	67.5	75.3	75.3	69.5	54.9	45.4	35.4	53.4
1978	30.4	23.8	38.1	49.3	57.2	68.7	72.1	76.5	63.4	53.1	47.8	37.9	51.5
1979	31.5	21.7	44.8	51.2	62.9	66.9	74.6	72.7	65.6	54.4	38.0	35.5	52.7
1980	30.8	28.6	38.7	51.6	61.6	64.3	72.4	73.1	67.9	52.3	41.3	31.2	51.2
1981	22.8	34.2	36.7	50.4	57.8	70.2	76.5	73.8	66.7	53.4	48.8	34.4	51.9
1982	26.4	36.8	43.1	49.8	64.1	69.9	76.9	71.8	64.8	55.2	48.5	41.0	54.0
1983	33.4	35.5	44.9	51.7	60.3	70.7	78.7	75.9	66.9	56.5	46.5	33.9	54.7
1984	28.0	40.1	36.6	49.7	60.7	73.6	75.3	77.1	65.9	62.3	44.6	44.0	54.8
1985	26.8	35.4	44.8	55.1	64.5	70.0	74.7	73.7	68.4	58.1	53.4	33.1	55.0
Record Mean	30.6	32.9	41.0	50.8	60.6	69.5	74.8	73.7	66.6	55.5	46.2	35.7	53.2
Max	39.5	42.2	50.6	61.5	71.1	79.8	84.3	83.2	76.8	66.1	55.8	44.9	63.0
Min	21.6	23.6	31.5	40.1	50.1	59.1	65.2	64.1	56.4	45.0	36.5	26.4	43.3

REFERENCE NOTES FOR TABLES 1, 2, 3 and 6   (ATLANTIC CITY [N.A.F.E.C.], NJ)

GENERAL

T - TRACE AMOUNT
BLANK ENTRIES DENOTE MISSING/UNREPORTED DATA.
# INDICATES A STATION OR INSTRUMENT RELOCATION.

SPECIFIC

TABLE 1

(a) - LENGTH OF RECORD IN YEARS. ALTHOUGH INDIVIDUAL MONTHS MAY BE MISSING.

* LESS THAN .05

NORMALS — BASED ON THE 1951-1980 RECORD PERIOD.
EXTREMES — DATES ARE THE MOST RECENT OCCURRENCE.
WIND DIR. — NUMERALS SHOW TENS OF DEGREES CLOCKWISE FROM TRUE NORTH.
"00" INDICATES CALM.
RESULTANT WIND DIRECTIONS ARE GIVEN TO WHOLE DEGREES.

EXCEPTIONS

TABLE 1

1. TEMPERATURE AND PRECIPITATION INCLUDE DATA FROM U.S. NAVAL AIR STATION RECORDS.

TABLES 2, 3, and 6

RECORD MEANS ARE THROUGH THE CURRENT YEAR, BEGINNING IN   1958 FOR TEMPERATURE
1958 FOR PRECIPITATION
1945 FOR SNOWFALL

TABLE 4   HEATING DEGREE DAYS Base 65 deg. F   ATLANTIC CITY, NEW JERSEY N.A.F.E.C.

SEASON	JULY	AUG	SEP	OCT	NOV	DEC	JAN	FEB	MAR	APR	MAY	JUNE	TOTAL
1956-57	1	0	70	183	524	605	1019	721	701	394	173	29	4420
#1957-58	0	1	40	258	427	699	962	981	757	408	221	41	4795
1958-59	0	1	62	306	510	1132	1029	852	740	367	131	23	5153
1959-60	0	0	48	254	586	800	914	816	740	311	140	12	4890
1960-61	0	0	20	267	505	1074	1174	804	677	478	198	14	5211
1961-62	0	0	29	226	525	920	977	860	743	392	134	12	4818
1962-63	0	2	83	250	630	1007	1040	986	596	375	178	13	5160
1963-64	0	1	102	170	438	1059	916	913	674	461	121	33	4888
#1964-65	0	1	42	351	532	869	1152	890	838	485	109	77	5346
1965-66	3	24	70	368	596	879	1125	971	726	529	275	43	5609
1966-67	0	0	85	373	513	888	801	996	813	411	341	23	5244
1967-68	0	1	149	399	757	922	1195	1094	674	417	196	13	5817
1968-69	0	18	45	314	629	1031	1068	926	847	354	150	29	5411
1969-70	4	16	108	356	626	939	1174	867	820	453	166	13	5542
1970-71	0	0	49	224	515	876	1117	814	798	514	228	29	5164
1971-72	4	15	47	143	620	736	958	936	816	550	187	44	5056
1972-73	5	8	28	410	612	691	885	842	523	414	207	3	4628
1973-74	0	0	64	245	519	773	788	885	632	329	184	27	4446
1974-75	0	0	68	393	557	800	857	790	763	566	121	17	4932
1975-76	0	2	108	251	505	904	1128	733	668	409	242	55	5005
1976-77	0	17	71	424	757	1075	1398	895	587	380	145	42	5791
1977-78	2	0	28	311	457	910	1069	1146	826	463	256	24	5492
1978-79	6	0	116	365	512	831	1032	1208	623	406	110	39	5248
1979-80	6	19	80	341	483	830	1053	1050	808	394	145	102	5311
1980-81	3	5	38	392	705	1042	1299	855	871	436	251	14	5911
1981-82	0	0	47	356	600	940	1187	787	670	448	79	13	5127
1982-83	0	14	64	318	498	740	973	820	617	405	166	16	4631
1983-84	0	6	94	288	547	958	1137	714	874	451	171	8	5248
1984-85	0	0	80	129	604	646	1179	821	621	319	106	14	4519
1985-86	0	0	53	225	346	981							

TABLE 5   COOLING DEGREE DAYS Base 65 deg. F   ATLANTIC CITY, NEW JERSEY N.A.F.E.C.

YEAR	JAN	FEB	MAR	APR	MAY	JUNE	JULY	AUG	SEP	OCT	NOV	DEC	TOTAL
1969	0	0	0	17	65	170	264	282	129	18	0	0	945
1970	0	0	0	0	63	136	279	285	187	20	0	0	970
1971	0	0	0	0	6	178	228	191	129	17	12	0	761
1972	0	0	0	0	11	83	341	282	132	4	0	0	853
1973	0	0	0	10	21	233	321	345	113	20	0	0	1063
1974	0	0	0	19	44	121	319	292	103	2	6	0	906
1975	0	0	0	0	74	164	305	282	48	10	0	0	883
1976	0	0	0	13	15	201	238	221	74	3	0	0	765
1977	0	0	2	15	51	120	328	325	168	7	2	0	1018
1978	0	0	0	0	19	143	233	364	74	4	0	0	837
1979	0	0	5	0	50	103	308	266	103	15	0	0	850
1980	0	0	0	0	47	88	238	262	132	6	0	0	773
1981	0	0	0	6	31	177	364	279	132	1	0	0	990
1982	0	0	0	0	56	165	375	231	64	23	9	0	923
1983	0	0	0	15	30	196	431	350	190	33	0	0	1245
1984	0	0	0	0	43	273	326	382	115	53	0	1	1193
1985	0	0	3	30	99	170	376	278	163	18	3	0	1140

TABLE 6   SNOWFALL (inches)   ATLANTIC CITY, NEW JERSEY N.A.F.E.C.

SEASON	JULY	AUG	SEP	OCT	NOV	DEC	JAN	FEB	MAR	APR	MAY	JUNE	TOTAL
1956-57	0.0	0.0	0.0	0.0	0.0	T	7.9	1.0	0.2	1.0	0.0	0.0	10.1
#1957-58	0.0	0.0	0.0	0.0	T	4.6	12.7	13.3	2.9	0.0	0.0	0.0	33.5
1958-59	0.0	0.0	0.0	0.0	0.0	5.1	2.7	T	T	2.0	0.0	0.0	9.8
1959-60	0.0	0.0	0.0	0.0	T	1.5	0.6	3.6	13.4	T	0.0	0.0	19.1
1960-61	0.0	0.0	0.0	T	0.0	8.6	15.9	7.6	0.2	0.0	0.0	0.0	32.3
1961-62	0.0	0.0	0.0	0.0	0.0	1.0	7.1	4.6	3.9	T	0.0	0.0	16.6
1962-63	0.0	0.0	0.0	T	T	5.1	4.4	0.3	T	0.0	0.0	0.0	9.8
1963-64	0.0	0.0	0.0	0.0	0.0	7.6	15.1	12.0	3.4	T	0.0	0.0	38.1
1964-65	0.0	0.0	0.0	0.0	0.0	1.0	8.2	3.3	2.8	3.2	0.0	0.0	18.5
1965-66	0.0	0.0	0.0	0.0	0.0	0.0	15.1	8.0	T	0.0	0.0	0.0	23.1
1966-67	0.0	0.0	0.0	0.0	T	8.5	1.1	35.2	2.1	T	0.0	0.0	46.9
1967-68	0.0	0.0	0.0	0.0	7.8	3.9	0.8	4.2	1.8	0.0	0.0	0.0	18.5
1968-69	0.0	0.0	0.0	0.0	T	4.3	7.0	17.6	0.0	0.0	0.0	0.0	29.4
1969-70	0.0	0.0	0.0	0.0	T	0.6	10.3	5.9	T	0.0	0.0	0.0	16.8
1970-71	0.0	0.0	0.0	0.0	0.0	1.4	7.2	1.9	0.9	T	0.0	0.0	11.4
1971-72	0.0	0.0	0.0	0.0	0.0	0.1	2.9	5.9	T	T	0.0	0.0	8.9
1972-73	0.0	0.0	0.0	T	0.0	T	0.0	0.4	T	T	0.0	0.0	0.4
1973-74	0.0	0.0	0.0	0.0	0.1	T	0.4	9.9	T	T	0.0	0.0	10.4
1974-75	0.0	0.0	0.0	0.0	0.2	T	3.3	1.6	2.0	T	0.0	0.0	7.1
1975-76	0.0	0.0	0.0	0.0	0.0	0.8	4.9	1.0	3.3	0.0	0.0	0.0	10.0
1976-77	0.0	0.0	0.0	0.0	T	4.0	7.8	0.5	0.0	0.0	T	0.0	12.3
1977-78	0.0	0.0	0.0	0.0	T	T	2.4	14.6	8.1	T	0.0	0.0	25.1
1978-79	0.0	0.0	0.0	0.0	1.2	T	14.2	27.7	T	0.0	0.0	0.0	43.1
1979-80	0.0	0.0	0.0	T	0.0	4.8	6.3	0.8	2.6	T	0.0	0.0	14.5
1980-81	0.0	0.0	0.0	T	T	0.1	3.2	0.0	T	0.0	0.0	0.0	3.3
1981-82	0.0	0.0	0.0	0.0	0.0	0.6	7.8	4.4	T	2.0	0.0	0.0	14.8
1982-83	0.0	0.0	0.0	0.0	0.0	6.7	T	14.9	T	0.7	0.0	0.0	22.3
1983-84	0.0	0.0	0.0	0.0	0.1	0.4	3.8	T	4.0	T	0.0	0.0	8.3
1984-85	0.0	0.0	0.0	0.0	T	T	15.1	0.0	T	1.3	0.0	0.0	16.4
1985-86	0.0	0.0	0.0	0.0	0.0	4.2							
Record Mean	0.0	0.0	0.0	T	0.4	2.3	5.2	5.4	2.8	0.3	T	0.0	16.4

See Reference Notes, relative to all above tables, on preceding page.

The Atlantic City State Marina is located on Abescon Island on the southeast coast of New Jersey. Surrounding terrain, composed of tidal marshes and beach sand, is flat and lies slightly above sea level. The climate is principally continental in character. However, the moderating influence of the Atlantic Ocean is apparent throughout the year, being more marked in the city than at the airport. As a result, summers are relatively cooler and winters milder than elsewhere at the same latitude.

Land and sea breezes, local circulations resulting from the differential heating and cooling of the land and sea, often prevail. These winds occur when moderate or intense storms are not present in the area, thus enabling the local circulation to overcome the general wind pattern. During the warm season sea breezes in the late morning and afternoon hours prevent excessive heating. Frequently, the temperature at Atlantic City during the afternoon hours in the summer averages several degrees lower than at the airport and the airport averages several degrees lower than localities farther inland. On occasions, sea breezes have lowered the temperature as much as 15 to 20 degrees within a half hour. However, the major effect of the sea breeze at the airport is preventing the temperature from rising above the 80s. Because the change in ocean temperature lags behind the air temperature from season to season, the weather tends to remain comparatively mild late into the fall, but on the other hand, warming is retarded in the spring. Normal ocean temperatures range from an average near 37 degrees in January to near 72 degrees in August.

Precipitation is moderate and well distributed throughout the year, with June the driest month and August the wettest. Tropical storms or hurricanes occasionally bring excessive rainfall to the area. The bulk of winter precipitation results from storms which move northeastward along or near the east coast of the United States. Snowfall is considerably less than elsewhere at the same latitude and does not remain long on the ground. Precipitation, often beginning as snow, will frequently become mixed with or change to rain while continuing as snow over more interior sections. In addition, ice storms and resultant glaze are relatively infrequent.

TABLE 1   NORMALS, MEANS AND EXTREMES

ATLANTIC CITY, NEW JERSEY STATE MARINA

LATITUDE: 39°23'N   LONGITUDE: 74°26' W   ELEVATION: FT. GRND   11 BARO   TIME ZONE: EASTERN   WBAN: 13724

	(a)	JAN	FEB	MAR	APR	MAY	JUNE	JULY	AUG	SEP	OCT	NOV	DEC	YEAR
TEMPERATURE °F:														
Normals														
-Daily Maximum		40.3	41.8	47.7	57.7	65.7	74.5	79.9	79.5	73.8	64.4	54.3	44.8	60.4
-Daily Minimum		27.9	28.9	35.6	44.2	53.3	62.3	68.2	68.0	61.8	50.7	41.1	31.9	47.8
-Monthly		34.1	35.4	41.7	51.0	59.5	68.4	74.1	73.8	67.8	57.5	47.7	38.4	54.1
Extremes														
-Record Highest	27	70	72	76	91	94	97	100	102	92	87	78	72	102
-Year		1967	1976	1985	1960	1969	1968	1983	1983	1985	1967	1982	1984	AUG 1983
-Record Lowest	27	-2	1	12	22	36	45	53	53	42	27	19	4	-2
-Year		1985	1961	1980	1982	1966	1972	1979	1983	1974	1974	1958	1983	JAN 1985
NORMAL DEGREE DAYS:														
Heating (base 65°F)		958	829	722	420	189	23	0	0	32	246	519	825	4763
Cooling (base 65°F)		0	0	0	0	19	125	282	273	116	17	0	0	832
% OF POSSIBLE SUNSHINE														
MEAN SKY COVER (tenths)														
Sunrise - Sunset														
MEAN NUMBER OF DAYS:														
Sunrise to Sunset														
-Clear														
-Partly Cloudy														
-Cloudy														
Precipitation														
.01 inches or more	26	10.5	9.5	10.5	9.7	10.2	9.0	8.3	8.4	7.4	7.5	9.4	10.0	110.4
Snow,Ice pellets														
1.0 inches or more														
Thunderstorms														
Heavy Fog Visibility														
1/4 mile or less														
Temperature °F														
-Maximum														
90° and above	26	0.0	0.0	0.0	0.1	0.3	0.8	1.9	1.3	0.4	0.0	0.0	0.0	4.8
32° and below	26	7.4	4.0	0.6	0.0	0.0	0.0	0.0	0.0	0.0	0.0	0.1	2.5	14.7
-Minimum														
32° and below	26	22.8	18.4	10.0	1.5	0.0	0.0	0.0	0.0	0.0	0.4	4.5	16.4	74.0
0° and below	26	0.*	0.0	0.0	0.0	0.0	0.0	0.0	0.0	0.0	0.0	0.0	0.0	*
AVG. STATION PRESS.(mb)														
RELATIVE HUMIDITY (%)														
Hour 01														
Hour 07														
Hour 13 (Local Time)														
Hour 19														
PRECIPITATION (inches):														
Water Equivalent														
-Normal		3.25	3.22	3.71	3.12	2.91	2.88	3.89	4.54	2.73	2.76	3.54	3.51	40.06
-Maximum Monthly	27	6.96	6.29	6.50	6.62	6.01	7.28	15.69	14.77	5.55	5.47	7.57	6.56	15.69
-Year		1979	1979	1984	1964	1984	1973	1959	1967	1964	1959	1972	1969	JUL 1959
-Minimum Monthly	27	0.35	0.79	0.68	0.82	0.90	0.61	0.57	0.93	0.39	0.02	0.82	0.71	0.02
-Year		1981	1980	1966	1976	1977	1971	1981	1964	1959	1963	1965	1985	OCT 1963
-Maximum in 24 hrs	27	3.40	2.40	2.60	1.80	3.13	4.10	6.62	8.60	2.54	2.67	3.90	2.39	8.60
-Year		1962	1966	1979	1970	1978	1968	1969	1967	1971	1972	1977	1969	AUG 1967
Snow,Ice pellets														
-Maximum Monthly														
-Year					1985		1985	1985	1985	1985				
-Maximum in 24 hrs														
-Year														
WIND:														
Mean Speed (mph)														
Prevailing Direction														
Fastest Mile														
-Direction	25	W		W	SSW	SW	NW	W	SE	NNW	SE	E	ENE	SE
-Speed (MPH)	25	60	59	54	53	46	49	48	63	63	48	62	58	63
-Year		1966	1972	1964	1970	1974	1972	1965	1971	1961	1965	1968	1974	AUG 1971
Peak Gust														
-Direction	1	NW	WNW	ENE	S	NE		SE	SW	WSW	NNE	E	WNW	ENE
-Speed (mph)	1	49	60	87	54	52		52	46	78	58	53	54	87
-Date		1985	1985	1984	1985	1985		1984	1985	1985	1984	1985	1985	MAR 1984

See Reference Notes to this table on the following page.

TABLE 2 PRECIPITATION (inches)   ATLANTIC CITY, NEW JERSEY STATE MARINA

YEAR	JAN	FEB	MAR	APR	MAY	JUNE	JULY	AUG	SEP	OCT	NOV	DEC	ANNUAL
#1958	5.24	6.86	7.57	4.33	4.42	4.38	9.57	9.53	3.85	6.71	2.09	2.62	67.17
1959	1.18	1.88	3.94	3.83	3.50	1.14	15.69	5.08	0.39	5.47	4.82	3.49	50.41
1960	2.21	4.01	2.07	2.13	2.32	1.47	5.87	3.93	3.50	2.51	1.53	3.64	35.19
1961	3.04	4.10	5.48	2.86	3.05	1.29	3.36	2.12	3.36	3.49	3.76	3.25	39.16
1962	4.83	3.25	4.56	3.92	1.49	4.28	2.81	4.37	2.96	2.38	4.46	2.79	42.10
1963	2.90	2.35	5.30	1.38	1.94	2.13	3.68	2.08	3.12	0.02	5.46	2.20	32.56
1964	5.84	4.04	2.65	6.62	1.30	1.21	1.37	0.93	5.55	2.35	1.21	2.62	35.69
1965	3.48	1.99	3.28	2.30	2.39	1.87	4.92	5.50	1.57	0.61	0.82	1.32	30.05
1966	3.12	4.84	0.68	2.58	4.62	2.22	2.70	4.11	4.35	2.81	1.44	4.58	38.05
1967	0.87	2.98	2.82	2.78	3.33	1.33	2.62	14.77	2.01	2.64	1.09	4.93	42.17
1968	2.09	1.41	4.05	1.13	5.04	6.24	1.26	1.07	2.17	2.36	3.08	3.85	33.75
1969	2.24	2.92	2.63	3.76	1.43	1.64	11.11	2.46	2.46	0.51	4.56	6.55	43.96
1970	1.54	3.39	3.26	4.77	2.46	4.60	3.72	1.24	0.47	3.35	4.37	2.30	35.47
1971	2.89	4.31	1.84	1.35	2.14	0.61	4.20	10.37	5.47	3.60	4.29	2.85	43.92
1972	3.25	3.76	3.87	3.97	5.10	5.21	3.73	1.46	4.35	4.43	7.57	4.98	51.68
1973	2.87	3.51	2.64	4.03	2.83	7.28	3.21	1.06	2.96	1.80	1.33	5.23	38.75
1974	2.78	2.35	4.33	1.75	2.44	2.01	2.57	6.31	2.56	1.59	1.00	3.31	33.00
1975	4.29	3.08	3.39	3.94	2.73	3.72	5.63	2.05	3.82	1.74	2.93	2.57	39.89
1976	3.66	2.98	0.91	0.82	2.19	1.59	1.74	5.01	3.08	4.94	0.95	2.42	30.29
1977	2.48	2.10	2.30	1.89	0.90	1.78	2.17	4.10	1.81	3.05	6.77	5.12	34.47
1978	5.59	1.11	3.82	1.33	6.00	2.07	4.20	4.01	0.87	1.35	2.48	3.95	36.78
1979	6.96	6.29	3.28	4.00	3.55	3.12	2.21	5.93	3.68	2.07	3.80	1.56	46.45
1980	2.97	0.79	6.39	4.38	1.47	2.73	2.25	1.41	2.08	3.63	3.11	1.13	32.34
1981	0.35	4.08	1.98	4.96	2.16	2.49	0.57	3.75	2.09	2.98	1.28	3.83	30.52
1982	2.81	2.03	2.48	3.98	2.24	2.88	4.14	2.08	1.16	1.11	4.05	3.04	32.00
1983	2.22	2.82	5.63	5.17	3.75	2.50	0.71	2.43	1.88	3.41	6.64	3.47	40.63
1984	2.38	4.32	6.50	4.46	6.01	1.44	4.24	1.03	2.08	1.53	2.28	1.53	37.80
1985	2.04	1.73	2.58	0.97	5.01	2.39	3.37	3.43	2.07	1.08	2.16	0.71	27.54
Record Mean	3.08	3.19	3.58	3.19	3.06	2.70	4.06	4.05	2.70	2.63	3.19	3.21	38.64

TABLE 3 AVERAGE TEMPERATURE (deg. F)   ATLANTIC CITY, NEW JERSEY STATE MARINA

YEAR	JAN	FEB	MAR	APR	MAY	JUNE	JULY	AUG	SEP	OCT	NOV	DEC	ANNUAL
#1958	33.8	29.8	40.3	51.2	57.8	65.8	74.6	73.3	66.5	56.4	49.0	29.6	52.3
1959	31.9	34.4	40.3	49.7	61.1	71.5	73.7	75.9	70.8	60.5	46.5	40.7	54.8
1960	36.9	37.2	33.0	52.9	60.3	69.4	72.4	73.8	68.0	57.7	48.5	32.3	53.5
1961	28.8	36.4	42.0	46.8	55.4	66.9	74.7	74.3	72.3	60.4	50.3	34.1	53.9
1962	35.2	34.5	40.1	50.5	58.8	65.0	72.2	74.4	65.5	57.6	44.6	31.9	52.7
1963	32.5	30.7	44.7	52.4	59.2	69.7	75.7	71.6	62.8	58.8	49.9	31.9	53.3
1964	36.6	34.6	42.6	48.9	61.9	69.7	73.6	72.4	69.2	56.3	50.6	39.4	54.7
1965	31.8	34.4	39.6	48.3	62.8	68.0	73.0	73.3	69.0	55.8	47.2	39.5	53.6
1966	31.8	34.3	42.6	48.2	58.3	68.4	76.6	73.8	67.1	56.4	49.3	37.7	53.7
1967	38.9	32.5	39.1	51.2	55.4	68.6	75.3	73.5	66.0	58.3	45.2	40.6	53.7
1968	32.0	32.5	44.8	52.6	60.2	70.6	76.5	76.2	71.0	60.2	48.4	35.6	55.1
1969	32.4	33.6	38.6	52.8	61.4	70.8	75.2	75.4	69.3	58.6	47.8	37.9	54.5
1970	29.5	36.7	41.0	51.1	61.4	69.3	72.5	74.9	68.3	58.5	48.3	38.2	54.1
1971	31.5	38.1	40.9	48.7	56.9	67.9	71.9	71.0	67.9	60.0	46.2	44.9	53.8
1972	38.6	36.6	40.4	48.3	58.1	64.5	73.1	72.0	67.5	54.3	46.4	43.6	53.6
1973	39.0	38.4	48.4	54.0	57.2	68.8	73.7	75.1	69.8	60.4	48.9	40.8	56.2
1974	39.1	34.1	43.6	52.5	59.8	67.5	75.6	73.4	65.8	53.3	47.4	39.5	54.3
1975	37.4	36.0	39.4	45.9	61.3	68.8	73.5	74.6	64.7	59.3	50.8	38.5	54.2
1976	32.7	42.0	45.0	53.4	58.0	69.1	71.8	71.4	65.9	53.1	41.6	34.5	53.2
1977	24.4	35.6	45.7	52.2	60.9	67.7	75.3	76.1	71.1	56.6	48.9	36.9	54.2
1978	31.2	27.9	39.8	49.6	55.3	65.8	70.6	74.4	65.0	55.2	48.2	39.3	51.9
1979	33.1	27.4	43.3	50.5	59.3	66.6	72.8	73.1	67.5	56.2	51.4	41.9	53.6
1980	35.5	32.7	40.7	52.2	62.3	67.0	72.9	75.1	69.4	55.6		36.7	
1981	29.9	39.1	41.4	53.2	60.7	71.4	77.5	74.7	67.3	54.6	46.9	37.9	54.5
1982	29.1	35.8	42.2	48.7	60.2	66.1	72.8	69.6	66.0	56.3	48.8	40.7	53.0
1983	34.7	39.0	46.0	51.3	59.0	67.5	76.0	74.6	68.7	57.6	48.3	35.4	54.9
1984	30.4	41.3	39.6	51.1	60.7	72.8	73.4	77.4	67.4	62.8	47.6	47.2	56.0
1985	30.6	36.9	46.4	54.1	62.0	68.4	75.0	74.4	70.4	61.7	54.4	36.4	55.9
Record Mean	33.2	35.0	41.8	50.8	59.5	68.3	74.0	73.9	67.9	57.5	48.2	38.3	54.0
Max	39.5	41.5	48.0	57.4	65.5	74.3	79.8	79.7	73.8	64.1	54.7	44.8	60.3
Min	26.9	28.6	35.6	44.1	53.5	62.3	68.1	68.1	61.9	51.0	41.7	31.7	47.8

REFERENCE NOTES FOR TABLES 1, 2, 3 and 6   **(ATLANTIC CITY, NJ)**

GENERAL

T - TRACE AMOUNT
BLANK ENTRIES DENOTE MISSING/UNREPORTED DATA.
# INDICATES A STATION OR INSTRUMENT RELOCATION.

SPECIFIC

TABLE 1

(a) - LENGTH OF RECORD IN YEARS. ALTHOUGH INDIVIDUAL MONTHS MAY BE MISSING.
* LESS THAN .05

NORMALS — BASED ON THE 1951-1980 RECORD PERIOD.
EXTREMES — DATES ARE THE MOST RECENT OCCURRENCE.
WIND DIR. — NUMERALS SHOW TENS OF DEGREES CLOCKWISE FROM TRUE NORTH.
  "00" INDICATES CALM.
RESULTANT WIND DIRECTIONS ARE GIVEN TO WHOLE DEGREES.

EXCEPTIONS

TABLE 1

1. WINDS ARE FASTEST OBSERVED 1 - MINUTE WINDS WITH DIRECTIONS IN COMPASS POINTS.

TABLES 2, 3, and 6

RECORD MEANS ARE THROUGH THE CURRENT YEAR, BEGINNING IN   1959 FOR TEMPERATURE
        1959 FOR PRECIPITATION
        1959 FOR SNOWFALL

* FOR REPRESENTATIVE SNOWFALL DATA FOR THE ATLANTIC CITY AREA, CONSULT THE ANNUAL ISSUE OF THE ATLANTIC CITY AIRPORT OFFICE, ATLANTIC CITY NEW JERSEY, N.A.F.E.C.

TABLE 4   HEATING DEGREE DAYS Base 65 deg. F   ATLANTIC CITY, NEW JERSEY STATE MARINA

SEASON	JULY	AUG	SEP	OCT	NOV	DEC	JAN	FEB	MAR	APR	MAY	JUNE	TOTAL
1959-60	0	0	39	215	551	709	865	800	988	368	157	12	4704
1960-61	0	0	16	233	488	1007	1115	795	706	540	290	21	5211
1961-62	0	0	10	155	435	798	919	847	768	432	202	39	4605
1962-63	0	0	61	234	608	951	999	954	622	377	193	8	5007
1963-64	1	7	112	189	447	1020	873	874	690	482	160	22	4877
1964-65	0	0	17	272	426	785	1022	851	778	492	126	38	4807
1965-66	0	8	23	284	528	783	1023	854	688	498	217	26	4932
1966-67	0	0	47	263	463	835	800	903	796	405	293	8	4813
1967-68	0	0	49	226	591	751	1016	937	618	371	156	1	4716
1968-69	0	0	1	180	490	903	1005	872	810	358	141	2	4762
1969-70	0	0	25	222	509	833	1093	789	736	411	142	3	4763
1970-71	0	0	33	204	493	823	1029	749	738	480	246	30	4825
1971-72	0	2	27	158	560	615	812	818	755	494	209	47	4497
1972-73	2	0	23	328	553	658	796	738	508	330	245	9	4190
1973-74	0	0	16	163	476	742	798	860	655	379	173	24	4286
1974-75	0	0	57	358	524	784	848	805	786	567	151	11	4891
1975-76	0	0	49	188	422	814	994	664	612	356	212	29	4340
1976-77	0	3	38	366	696	936	1253	816	590	384	154	36	5272
1977-78	0	0	11	267	477	862	1043	1031	776	455	302	41	5265
1978-79	6	0	62	297	497	791	979	1045	666	427	173	22	4965
1979-80	5	3	27	275	404	708	906	931	746	376	115	43	4539
1980-81	0	0	19	288			871	720	723	350	160	3	1080
1981-82	0	0	51	315	538	834	1106	813	698	482	155	39	5031
1982-83	1	10	36	280	486	749	931	722	584	405	179	17	4400
1983-84	1	9	63	236	496	911	1065	681	781	406	139	5	4793
1984-85	0	0	56	96	514	543	1059	783	571	336	119	16	4093
1985-86	0	0	19	132	313	876							

TABLE 5   COOLING DEGREE DAYS Base 65 deg. F   ATLANTIC CITY, NEW JERSEY STATE MARINA

YEAR	JAN	FEB	MAR	APR	MAY	JUNE	JULY	AUG	SEP	OCT	NOV	DEC	TOTAL
1969	0	0	0	0	38	181	322	327	160	32	0	0	1060
1970	0	0	0	0	37	138	240	315	141	11	0	0	882
1971	0	0	0	0	2	122	219	194	336	7	2	0	882
1972	0	0	0	0	0	39	258	224	103	2	0	0	626
1973	0	0	0	3	9	233	278	319	168	29	0	0	1039
1974	0	0	0	10	18	125	334	266	91	1	0	0	845
1975	0	0	0	0	42	131	269	313	50	16	0	0	821
1976	0	0	0	14	2	160	219	210	69	2	0	0	676
1977	0	0	0	6	37	105	324	355	200	10	0	0	1037
1978	0	0	0	0	9	72	186	297	69	0	0	0	633
1979	0	0	0	0	5	74	252	263	111	8	0	0	713
1980	0	0	0	0	39	109	250	319	159	2	0	0	
1981	0	0	0	3	33	200	392	305	125	1	0	0	1059
1982	0	0	0	0	13	77	247	160	70	20	3	0	590
1983	0	0	0	2	0	96	349	313	181	14	0	0	955
1984	0	0	0	0	12	246	269	393	135	36	0	0	1091
1985	0	0	2	12	34	124	315	298	188	35	0	0	1008

TABLE 6   SNOWFALL (inches)   ATLANTIC CITY, NEW JERSEY STATE MARINA

SEASON	JULY	AUG	SEP	OCT	NOV	DEC	JAN	FEB	MAR	APR	MAY	JUNE	TOTAL
Record Mean													

See Reference Notes, relative to all above tables, on preceding page.

Terrain in vicinity of the station is flat and rather marshy. To the northwest are ridges oriented roughly in a south-southwest to north-northeast direction. They rise to an elevation of about 200 feet at 4.5 to 5 miles and to 500 to 600 feet at 7 to 8 miles. All winds between west-northwest and north-northwest are downslope and therefore are subject to some adiabatic temperature increase. This effect is evident in the rapid improvement which normally occurs with shift of wind to westerly, following a coastal storm or frontal passage. The drying effect of the downslope winds accounts for the relatively few local thunderstorms occurring at the station, compared to areas to the west. Easterly winds, particularly southeasterly, moderate the temperature because of the influence of the Atlantic Ocean.

Temperature falls of 5 to 15 degrees, depending on the season, are not uncommon when the wind backs from southwesterly to southeasterly. Periods of very hot weather, lasting as long as a week, are associated with a west-southwest air flow which has a long trajectory over land. Extremes of cold are related to rapidly moving outbreaks of cold air traveling southeastward from the Hudson Bay region. Temperatures of zero or below occur in one winter out of four, but are much more common several miles to the west of the station. Average dates of the last occurrence in spring and the first occurrence in autumn of temperatures as low as 32 degrees are in mid-April and the end of October or early November. Areas to the west of the station experience a growing season at least a month shorter than that at the airport.

A considerable amount of precipitation is realized from the Northeasters of the Atlantic coast. These storms, more typical of the fall and winter, generally last for a period of two days and commonly produce between 1 and 2 inches of precipitation. Storms producing 4 inches or more of snow occur from two to five times a winter. Snowstorms producing 8 inches or more have occurred in about one-half the winters. As many as three such storms have been experienced in one winter. The frequency and intensity of snow storms and the duration of snow cover increase dramatically within a few miles to the west of the station.

TABLE 1   NORMALS, MEANS AND EXTREMES

NEWARK, NEW JERSEY

LATITUDE: 40°42'N   LONGITUDE: 74°10'W   ELEVATION: FT. GRND   7 BARO 00029   TIME ZONE: EASTERN   WBAN: 14734

	(a)	JAN	FEB	MAR	APR	MAY	JUNE	JULY	AUG	SEP	OCT	NOV	DEC	YEAR
TEMPERATURE °F:														
Normals														
-Daily Maximum		38.2	40.3	49.1	61.3	71.6	80.6	85.6	84.0	76.9	66.0	54.0	42.3	62.5
-Daily Minimum		24.2	25.3	33.3	42.9	53.0	62.4	67.9	67.0	59.4	48.3	39.0	28.6	45.9
-Monthly		31.2	32.8	41.2	52.1	62.3	71.5	76.8	75.5	68.2	57.2	46.5	35.5	54.2
Extremes														
-Record Highest	44	74	76	89	93	98	102	105	103	105	92	85	72	105
-Year		1950	1949	1945	1976	1962	1952	1966	1948	1953	1949	1950	1982	JUL 1966
-Record Lowest	44	-8	-7	6	16	33	43	52	45	35	28	15	-1	-8
-Year		1985	1943	1943	1982	1947	1945	1945	1982	1947	1969	1955	1980	JAN 1985
NORMAL DEGREE DAYS:														
Heating (base 65°F)		1045	902	738	387	140	0	0	0	36	254	555	915	4972
Cooling (base 65°F)		0	0	0	0	56	199	366	326	132	12	0	0	1091
% OF POSSIBLE SUNSHINE														
MEAN SKY COVER (tenths)														
Sunrise - Sunset	39	6.4	6.3	6.3	6.3	6.5	6.2	6.2	6.0	5.7	5.5	6.3	6.4	6.2
MEAN NUMBER OF DAYS:														
Sunrise to Sunset														
-Clear	43	8.0	7.4	7.8	7.3	6.3	7.0	6.7	7.9	9.8	10.7	7.7	8.1	94.8
-Partly Cloudy	43	7.7	7.6	8.6	9.0	10.7	10.6	12.3	11.5	8.9	8.5	8.3	7.9	111.5
-Cloudy	43	15.3	13.3	14.7	13.7	14.0	12.4	12.0	11.6	11.3	11.7	14.0	15.0	159.0
Precipitation														
.01 inches or more	44	11.1	9.5	11.5	10.8	12.0	10.2	9.8	9.1	8.3	7.8	10.3	11.0	121.4
Snow,Ice pellets														
1.0 inches or more	44	2.2	1.9	1.3	0.2	0.0	0.0	0.0	0.0	0.0	0.0	0.2	1.5	7.3
Thunderstorms	44	0.2	0.2	1.0	1.6	3.7	4.8	5.8	4.6	2.2	1.0	0.4	0.2	25.7
Heavy Fog Visibility														
1/4 mile or less	44	2.3	1.9	1.5	1.1	1.8	1.2	0.5	0.6	0.9	2.0	2.0	1.8	17.7
Temperature °F														
-Maximum														
90° and above	20	0.0	0.0	0.0	0.2	0.9	3.8	8.1	6.3	1.5	0.0	0.0	0.0	20.7
32° and below	20	11.1	5.9	1.0	0.1	0.0	0.0	0.0	0.0	0.0	0.0	0.1	3.8	22.0
-Minimum														
32° and below	20	24.6	20.7	12.4	2.0	0.0	0.0	0.0	0.0	0.0	0.0	0.7	6.1	19.0	85.5
0° and below	20	0.6	0.3	0.0	0.0	0.0	0.0	0.0	0.0	0.0	0.0	0.0	0.1	1.0
AVG. STATION PRESS.(mb)	12	1016.4	1016.6	1015.3	1013.6	1013.7	1014.3	1014.4	1016.1	1017.5	1018.3	1017.6	1017.8	1016.0
RELATIVE HUMIDITY (%)														
Hour 01	20	71	69	66	64	72	73	72	76	77	76	73	73	72
Hour 07	20	74	73	69	65	70	71	71	75	78	79	78	75	73
Hour 13 (Local Time)	20	59	55	51	47	51	53	51	54	55	54	57	60	54
Hour 19	20	64	61	57	53	58	59	58	62	64	64	65	66	61
PRECIPITATION (inches):														
Water Equivalent														
-Normal		3.13	3.05	4.15	3.57	3.59	2.94	3.85	4.30	3.66	3.09	3.59	3.42	42.34
-Maximum Monthly	44	10.10	4.94	11.14	11.14	10.22	6.40	8.65	11.84	10.28	8.20	11.53	9.47	11.84
-Year		1979	1979	1983	1983	1984	1975	1984	1955	1944	1943	1977	1983	AUG 1955
-Minimum Monthly	44	0.45	1.22	1.10	0.90	0.52	0.07	0.89	0.50	0.95	0.21	0.51	0.27	0.07
-Year		1981	1968	1981	1963	1964	1949	1966	1976	1951	1963	1955	1955	JUN 1949
-Maximum in 24 hrs	32	3.59	2.45	2.66	3.73	4.22	2.31	3.40	7.84	5.27	3.04	7.22	2.77	7.84
-Year		1979	1961	1978	1984	1979	1973	1971	1971	1971	1973	1977	1983	AUG 1971
Snow,Ice pellets														
-Maximum Monthly	44	27.4	26.1	26.0	13.8	T					0.3	3.1	29.1	29.1
-Year		1978	1979	1956	1982	1977					1952	1967	1947	DEC 1947
-Maximum in 24 hrs	44	17.8	20.0	17.6	12.8	T					0.3	3.1	26.0	26.0
-Year		1978	1961	1956	1982	1977					1952	1967	1947	DEC 1947
WIND:														
Mean Speed (mph)	41	11.3	11.6	12.0	11.4	10.0	9.4	8.9	8.7	9.0	9.4	10.2	10.8	10.2
Prevailing Direction														
through 1963		NE	NW	NW	WNW	SW	SW	SW	SW	SW	SW	SW	SW	SW
Fastest Obs. 1 Min.														
-Direction	37	30	23	27	27	32	26	18	09	05	11	09	32	09
-Speed (MPH)	37	52	46	43	50	50	58	45	46	51	48	82	55	82
-Year		1964	1965	1950	1951	1963	1984	1950	1955	1960	1954	1950	1962	NOV 1950
Peak Gust														
-Direction	2	NW	NW	W	N	W	W	W	N	W	NW	W	W	W
-Speed (mph)	2	48	58	56	53	52	83	59	68	67	39	47	56	83
-Date		1985	1984	1985	1985	1984	1984	1984	1985	1985	1984	1984	1985	JUN 1984

See Reference Notes to this table on the following page.

NEWARK, NEW JERSEY

TABLE 2 PRECIPITATION (inches) NEWARK, NEW JERSEY

YEAR	JAN	FEB	MAR	APR	MAY	JUNE	JULY	AUG	SEP	OCT	NOV	DEC	ANNUAL
1956	1.50	4.47	6.29	3.04	2.45	3.31	4.99	3.08	2.10	3.83	3.98	3.22	42.26
1957	1.77	2.77	2.73	5.45	1.87	1.54	1.51	2.64	3.80	2.13	3.13	5.74	35.08
1958	3.55	4.36	4.31	6.41	3.76	3.66	4.53	2.39	3.60	5.48	1.97	1.45	45.47
1959	2.40	1.89	4.09	2.05	1.47	3.60	2.49	6.94	2.45	4.22	3.94	4.17	39.71
1960	2.79	4.11	2.98	3.09	4.11	1.07	6.39	4.61	6.23	2.79	2.41	3.89	44.47
1961	3.34	3.97	4.96	5.28	3.35	2.46	7.95	4.22	1.49	2.06	2.64	3.65	45.37
1962	2.56	4.25	3.35	3.44	1.46	3.89	2.34	5.73	3.33	3.72	4.39	2.39	40.85
1963	2.19	2.16	3.92	0.90	2.37	2.01	2.24	1.93	3.94	0.21	5.68	1.97	29.52
1964	5.12	2.59	2.27	5.56	0.52	3.09	4.74	0.50	1.30	1.55	2.08	4.10	33.42
1965	2.86	2.91	2.81	2.60	1.23	1.23	1.73	2.87	2.20	2.31	1.48	1.86	26.09
1966	2.29	4.41	1.12	3.01	4.86	0.49	0.89	3.08	7.86	3.78	3.06	3.01	37.86
1967	1.15	3.00	5.86	2.84	3.57	3.31	7.53	5.53	1.35	2.87	2.35	4.65	44.01
1968	1.71	1.22	3.59	2.24	6.28	4.37	1.87	2.41	2.48	2.02	4.38	4.32	36.89
1969	1.47	2.68	3.53	3.51	2.73	2.53	7.11	2.24	6.63	1.75	2.80	4.97	41.95
1970	0.87	3.29	3.42	3.52	2.64	2.41	3.68	3.91	1.83	2.36	4.41	2.05	34.39
1971	2.74	4.44	3.29	1.35	3.65	1.48	6.98	10.63	7.88	2.96	3.86	1.51	50.77
1972	2.26	4.01	3.09	3.08	6.02	4.70	2.30	1.03	4.83	8.42	4.10		49.86
1973	3.65	3.39	3.63	5.77	3.56	4.03	3.63	3.36	3.39	3.35	1.29	7.24	46.29
1974	2.84	1.44	4.11	2.37	3.49	3.60	1.31	1.17	5.76	1.85	0.80	4.02	38.76
1975	3.99	2.56	2.94	2.29	3.27	6.40	8.02	4.36	9.00	3.24	3.67	2.91	52.65
1976	5.04	2.52	2.33	2.50	4.12	1.54	3.91	2.98	2.50	5.07	0.51	2.17	35.19
1977	1.55	2.77	5.67	3.16	1.31	3.89	1.51	4.29	3.99	3.53	11.53	4.77	47.97
1978	7.76	2.26	4.58	2.60	7.97	2.05	4.99	7.30	4.23	1.64	2.66	5.37	53.41
1979	10.10	4.94	3.65	3.66	7.78	2.73	3.39	4.38	5.72	4.58	3.09	2.08	56.10
1980	1.66	1.28	9.13	7.28	2.61	3.27	2.78	0.92	1.87	3.37	3.71	0.63	38.51
1981	0.45	4.81	1.10	3.15	3.88	2.61	4.51	0.57	3.42	3.47	1.75	5.32	35.04
1982	6.77	2.36	2.82	6.20	2.96	5.28	2.86	2.78	2.39	1.68	3.16	1.32	40.58
1983	4.37	3.03	11.14	4.22	2.81	1.59	3.46	2.93		5.80	5.54	9.47	65.50
1984	2.78	4.57	6.96	6.36	10.22	4.77	8.65	1.74	2.46	3.93	2.88	3.69	59.01
1985	1.22	2.58	1.59	1.17	4.23	4.29	4.52	2.58	4.19	1.29	8.32	1.31	37.29
Record Mean	3.30	2.93	4.06	3.64	3.78	3.34	3.88	4.07	3.75	3.03	3.55	3.37	42.72

TABLE 3 AVERAGE TEMPERATURE (deg. F) NEWARK, NEW JERSEY

YEAR	JAN	FEB	MAR	APR	MAY	JUNE	JULY	AUG	SEP	OCT	NOV	DEC	ANNUAL
1956	32.0	36.0	36.9	47.8	58.9	71.9	73.1	73.7	64.0	57.2	46.1	39.9	53.1
1957	28.0	36.4	42.0	53.3	62.9	73.0	77.3	73.0	69.5	55.6	48.3	39.1	54.9
1958	31.7	27.8	40.4	53.0	59.2	67.5	76.7	74.9	67.4	55.0	47.4	29.2	52.5
1959	31.4	31.6	40.3	53.5	65.7	71.6	76.7	77.6	71.5	59.9	45.5	38.0	55.3
1960	34.4	36.3	33.9	54.8	62.2	72.3	74.9	75.1	67.9	57.3	48.5	30.0	54.0
1961	26.6	35.8	41.2	48.6	59.7	71.9	77.3	74.5	67.4	59.5	47.4	33.8	54.4
1962	30.8	30.3	42.0	52.5	64.3	72.5	73.9	72.9	64.7	57.3	43.5	31.1	53.0
1963	29.6	27.6	42.5	52.6	61.1	72.0	77.0	74.0	64.0	61.2	49.7	29.3	53.4
1964	34.3	31.9	42.6	49.1	65.4	71.2	76.0	73.9	68.9	55.9	49.4	35.9	54.6
#1965	28.3	32.4	39.0	50.0	67.3	71.6	75.7	74.5	68.4	54.0	44.4	38.8	53.7
1966	30.4	33.2	41.7	48.2	59.3	73.8	79.6	76.5	66.6	55.5	48.9	36.5	54.2
1967	36.9	29.4	37.6	50.9	54.3	72.0	74.2	73.5	66.6	56.4	42.2	38.3	52.7
1968	27.8	29.9	43.1	54.0	59.6	69.7	78.2	76.9	70.7	59.7	45.7	32.5	54.0
1969	31.3	31.3	38.8	54.6	64.1	72.8	74.2	77.3	67.5	56.2	45.5	33.1	53.9
1970	24.2	33.0	39.0	51.9	64.6	76.0	77.2	77.3	70.6	59.5	49.1	35.3	54.4
1971	27.3	35.2	41.2	51.4	60.6	74.8	77.8	76.0	71.8	63.2	46.2	41.4	55.6
1972	35.4	31.3	40.5	50.0	63.0	68.8	77.9	75.9	69.8	53.3	44.8	39.7	54.2
1973	35.5	33.3	48.6	54.2	60.4	74.6	78.7	79.6	71.0	60.3	48.8	39.4	57.0
1974	35.4	31.9	43.4	56.5	62.7	70.1	77.1	76.5	66.6	53.9	47.5	38.9	55.0
1975	36.9	35.1	39.7	47.3	65.8	71.6	76.9	75.1	64.3	59.1	51.7	35.4	54.9
1976	26.8	39.3	44.0	55.2	61.1	73.6	74.9	74.4	66.5	52.6	39.9	29.1	53.1
1977	20.9	32.8	46.8	53.7	65.4	70.3	78.2	75.1	68.0	54.5	47.1	33.3	53.9
1978	27.2	25.5	38.5	51.0	60.5	71.6	75.1	76.7	66.1	57.5	48.8	38.1	53.1
1979	32.5	23.5	46.2	52.0	64.5	69.4	77.0	76.6	69.1	56.5	51.8	40.2	54.9
1980	34.0	30.8	38.9	52.6	65.9	70.2	78.9	78.6	70.8	55.0	42.9	30.4	54.1
1981	24.1	37.6	40.2	55.3	64.0	74.6	79.3	75.1	67.2	53.1	46.0	34.6	54.3
1982	24.2	36.2	41.8	50.6	63.2	67.9	78.4	72.5	66.7	56.9	48.8	42.8	54.2
1983	35.0	35.9	44.7	52.2	60.8	73.5	79.6	77.6	70.6	57.8	47.8	34.2	55.8
1984	27.8	40.8	36.5	52.2	62.2	75.0	76.6	77.3	65.4	62.3	45.3	44.0	55.2
1985	24.9	33.5	44.5	57.0	67.1	69.4	76.3	75.6	70.2	58.5	49.5	33.3	55.0
Record Mean	31.2	32.7	40.9	51.5	62.1	71.2	76.5	74.8	67.7	56.9	46.2	35.2	53.9
Max	38.3	40.3	49.0	60.8	71.7	80.6	85.6	83.7	76.7	66.1	54.0	42.2	62.4
Min	24.1	25.0	32.7	42.2	52.5	61.8	67.4	66.0	58.6	47.7	38.3	28.1	45.4

REFERENCE NOTES FOR TABLES 1, 2, 3 and 6   (NEWARK, NJ)

GENERAL

T - TRACE AMOUNT
BLANK ENTRIES DENOTE MISSING/UNREPORTED DATA.
# INDICATES A STATION OR INSTRUMENT RELOCATION.

SPECIFIC

TABLE 1

(a) - LENGTH OF RECORD IN YEARS. ALTHOUGH
INDIVIDUAL MONTHS MAY BE MISSING.

＊ LESS THAN .05

NORMALS — BASED ON THE 1951-1980 RECORD PERIOD.
EXTREMES — DATES ARE THE MOST RECENT OCCURRENCE.
WIND DIR. — NUMERALS SHOW TENS OF DEGREES
CLOCKWISE FROM TRUE NORTH.
''00'' INDICATES CALM.
RESULTANT WIND DIRECTIONS ARE GIVEN TO WHOLE DEGREES.

EXCEPTIONS

TABLES 2, 3, and 6

RECORD MEANS ARE THROUGH THE CURRENT YEAR,
BEGINNING IN   1931 FOR TEMPERATURE
1931 FOR PRECIPITATION
1942 FOR SNOWFALL

TABLE 4 HEATING DEGREE DAYS Base 65 deg. F   NEWARK, NEW JERSEY

SEASON	JULY	AUG	SEP	OCT	NOV	DEC	JAN	FEB	MAR	APR	MAY	JUNE	TOTAL
1956-57	9	2	117	244	565	767	1140	795	705	367	132	17	4860
1957-58	0	0	54	289	495	796	1025	1036	752	355	188	25	5015
1958-59	0	0	37	323	520	1103	1036	928	761	343	101	27	5179
1959-60	0	0	39	233	578	832	940	827	959	322	114	0	4844
1960-61	0	0	14	247	485	1079	1185	814	729	487	169	3	5212
1961-62	0	0	21	200	526	960	1052	963	705	393	120	7	4947
1962-63	0	7	81	250	640	1046	1091	1041	691	368	164	4	5383
1963-64	0	0	108	139	454	1100	946	955	687	473	88	21	4971
#1964-65	1	0	40	278	461	895	1133	905	799	442	55	17	5026
1965-66	0	11	50	339	610	807	1066	882	717	500	212	14	5208
1966-67	0	0	63	286	480	876	864	991	842	425	331	5	5163
1967-68	0	1	58	285	677	823	1148	1012	676	325	167	12	5184
1968-69	0	0	6	193	573	1003	1039	938	804	317	101	2	4976
1969-70	0	0	49	284	575	984	1255	892	796	390	97	5	5327
1970-71	0	0	24	199	472	914	1160	827	732	402	155	7	4892
1971-72	0	1	12	95	569	724	909	969	757	444	93	19	4592
1972-73	0	0	22	356	599	776	906	882	504	339	163	1	4548
1973-74	0	0	18	166	479	787	909	921	661	273	127	12	4353
1974-75	0	0	62	341	521	802	864	832	775	524	84	6	4811
1975-76	0	1	59	195	400	913	1177	738	645	338	141	17	4624
1976-77	0	4	56	381	745	1107	1361	895	563	352	89	24	5577
1977-78	0	0	50	319	527	975	1168	1099	814	411	190	13	5566
1978-79	6	0	66	239	481	830	1001	1155	577	386	68	11	4820
1979-80	2	4	28	289	393	763	953	987	802	366	62	24	4673
1980-81	0	0	28	314	654	1066	1261	762	764	290	96	0	5235
1981-82	0	0	52	360	563	934	1258	802	712	433	85	42	5241
1982-83	0	13	36	267	493	679	923	810	622	395	162	5	4405
1983-84	0	0	52	249	510	949	1144	696	874	366	128	9	4977
1984-85	0	0	83	114	584	745	1235	877	641	268	62	15	4624
1985-86	0	0	21	212	462	971							

TABLE 5 COOLING DEGREE DAYS Base 65 deg. F   NEWARK, NEW JERSEY

YEAR	JAN	FEB	MAR	APR	MAY	JUNE	JULY	AUG	SEP	OCT	NOV	DEC	TOTAL
1969	0	0	0	15	80	243	293	390	131	17	0	0	1169
1970	0	0	0	4	94	187	384	387	201	33	0	0	1290
1971	0	0	0	0	25	307	403	350	222	46	12	0	1365
1972	0	0	3	4	41	142	406	347	175	3	0	0	1121
1973	0	0	0	20	26	296	432	459	205	28	0	0	1466
1974	0	0	0	28	64	172	381	361	115	1	3	0	1125
1975	0	0	0	0	117	211	375	321	46	20	10	0	1100
1976	0	0	0	50	30	281	317	305	110	6	0	0	1099
1977	0	0	6	18	111	191	414	321	146	1	0	0	1208
1978	0	0	0	0	59	217	325	367	105	15	0	0	1088
1979	0	0	0	2	59	147	381	372	158	34	3	0	1156
1980	0	0	0	0	97	187	435	427	209	10	0	0	1365
1981	0	0	0	6	75	293	446	319	124	0	0	0	1263
1982	0	0	0	6	39	136	421	249	95	24	12	0	982
1983	0	0	0	19	39	268	458	396	226	36	0	0	1442
1984	0	0	0	2	47	316	365	388	102	36	0	0	1256
1985	0	0	11	36	134	152	357	335	183	19	3	0	1230

TABLE 6 SNOWFALL (inches)   NEWARK, NEW JERSEY

SEASON	JULY	AUG	SEP	OCT	NOV	DEC	JAN	FEB	MAR	APR	MAY	JUNE	TOTAL
1956-57	0.0	0.0	0.0	0.0	T	1.3	7.8	8.8	2.1	4.1	0.0	0.0	24.1
1957-58	0.0	0.0	0.0	T	T	16.2	6.3	16.3	19.5	T	0.0	0.0	58.3
1958-59	0.0	0.0	0.0	0.0	T	5.3	2.2	1.0	8.2	1.1	0.0	0.0	17.8
1959-60	0.0	0.0	0.0	0.0	0.4	9.1	3.6	5.2	19.0	T	0.0	0.0	37.3
1960-61	0.0	0.0	0.0	T	0.0	24.0	22.2	23.3	4.0	T	0.0	0.0	73.5
1961-62	0.0	0.0	0.0	0.0	0.8	13.2	1.0	13.1	1.5	T	0.0	0.0	29.6
1962-63	0.0	0.0	0.0	T	0.3	7.8	7.5	3.6	2.5	T	0.0	0.0	21.7
1963-64	0.0	0.0	0.0	0.0	T	10.7	13.5	15.0	4.0	T	0.0	0.0	43.2
1964-65	0.0	0.0	0.0	0.0	0.0	3.9	16.1	1.8	4.6	0.7	0.0	0.0	27.1
1965-66	0.0	0.0	0.0	T	0.0	T	10.2	8.6	T	0.0	0.0	0.0	18.8
1966-67	0.0	0.0	0.0	0.0	0.0	12.6	1.3	25.4	18.0	T	0.0	0.0	57.3
1967-68	0.0	0.0	0.0	0.0	3.1	3.9	4.6	0.6	1.7	0.0	0.0	0.0	13.9
1968-69	0.0	0.0	0.0	0.0	0.4	1.1	1.1	16.5	5.9	0.0	0.0	0.0	28.6
1969-70	0.0	0.0	0.0	0.0	T	8.5	9.1	5.5	4.3	T	0.0	0.0	27.4
1970-71	0.0	0.0	0.0	0.0	0.0	2.9	13.2	1.1	4.2	2.2	0.0	0.0	23.6
1971-72	0.0	0.0	0.0	0.0	T	0.4	3.1	12.3	1.0	T	0.0	0.0	16.8
1972-73	0.0	0.0	0.0	T	T	T	0.7	0.6	0.6	T	0.0	0.0	1.9
1973-74	0.0	0.0	0.0	0.0	0.0	2.1	6.8	8.1	3.1	0.3	0.0	0.0	20.4
1974-75	0.0	0.0	0.0	0.0	T	1.2	1.4	12.7	1.1	T	0.0	0.0	16.4
1975-76	0.0	0.0	0.0	0.0	T	2.4	7.2	6.1	4.2	T	0.0	0.0	19.9
1976-77	0.0	0.0	0.0	0.0	T	6.7	10.8	5.8	1.7	T	T	0.0	25.0
1977-78	0.0	0.0	0.0	0.0	1.5	0.2	27.4	25.3	10.5	T	0.0	0.0	64.9
1978-79	0.0	0.0	0.0	0.0	2.6	T	9.4	26.1	T	T	0.0	0.0	38.1
1979-80	0.0	0.0	0.0	T	0.0	3.7	2.5	1.8	6.3	T	0.0	0.0	14.3
1980-81	0.0	0.0	0.0	0.0	0.4	3.1	6.9	T	9.1	0.0	0.0	0.0	19.5
1981-82	0.0	0.0	0.0	0.0	T	3.4	12.3	0.5	0.8	13.8	0.0	0.0	30.8
1982-83	0.0	0.0	0.0	0.0	T	2.9	2.3	21.5	0.2	4.1	0.0	0.0	31.0
1983-84	0.0	0.0	0.0	0.0	1.2	13.7	0.3	11.3	T	0.0	0.0	0.0	28.9
1984-85	0.0	0.0	0.0	0.0	T	6.8	8.9	7.4	0.1	0.0	0.0	0.0	23.2
1985-86	0.0	0.0	0.0	0.0	0.6	4.6							
Record Mean	0.0	0.0	0.0	T	0.5	6.1	7.5	8.3	4.9	0.8	T	0.0	28.1

See Reference Notes, relative to all above tables, on preceding page.

ALBUQUERQUE, NEW MEXICO

The Albuquerque metropolitan area is largely situated in the Rio Grande Valley and on the mesas and piedmont slopes which rise either side of the valley floor. The Rio Grande flows from north to south through the area. The Sandia and Manzano Mountains rise abruptly at the eastern edge of the city with Tijeras Canyon separating the two ranges. West of the city the land gradually rises to the Continental Divide, some 90 miles away.

The climate of Albuquerque is best described as arid continental with abundant sunshine, low humidity, scant precipitation, and a wide yet tolerable seasonal range of temperatures. Sunny days and low humidity are renowned features of the climate. More than three-fourths of the daylight hours have sunshine, even in the winter months. The air is normally dry and muggy days are rare. The combination of dry air and plentiful solar radiation allows widespread use of energy-efficient devices such as evaporative coolers and solar collectors.

Precipitation within the valley area is adequate only for native desert vegetation and deep-rooted imports. However, irrigation supports successful farming and fruit growing in the Rio Grande Valley. On the east slopes of the Sandias and Manzanos, precipitation is sufficient for thick stands of timber and good grass cover.

Meager amounts of precipitation fall in the winter, much of it as snow. Snowfalls of an inch or more occur about four times a year in the Rio Grande Valley, while the mountains receive substantial snowfall on occasion. Snow seldom remains on the ground more than 24 hours in the city proper. However, snow cover on the east slopes of the Sandias is sufficient for skiing during most winters.

Nearly half of the annual precipitation in Albuquerque results from afternoon and evening thunderstorms during the summer. Thunderstorm frequency increases rapidly around July 1st, peaks during August, then tapers off by the end of September. Thunderstorms are usually brief, sometimes produce heavy rainfall, and often lower afternoon temperatures noticeably. Hailstorms are infrequent and tornadoes rare.

Temperatures in Albuquerque are those characteristic of a dry, high altitude, continental climate. The average daily range of temperature is relatively high, but extreme temperatures are rare. High temperatures during the winter are near 50 degrees with only a few days on which the temperature fails to rise above the freezing mark. In the summer, daytime maxima are about 90 degrees, but with the large daily range, the nights usually are comfortably cool.

The average number of days between the last freezing temperature in spring and the first freeze in fall varies widely across the Albuquerque metropolitan area. The growing season in Albuquerque and adjacent suburbs ranges from around 170 days in the Rio Grande Valley to about 200 days in parts of the northeast section of the city.

Sustained winds of 12 mph or less occur approximately 80 percent of the time at the Albuquerque International Airport, while sustained winds greater than 25 mph have a frequency less than 3 percent. Late winter and spring storms along with occasional east winds out of Tijeras Canyon are the main sources of strong wind conditions. Blowing dust, the least attractive feature of the climate, often accompanies the occasional strong winds of winter and spring.

TABLE 1   NORMALS, MEANS AND EXTREMES

ALBUQUERQUE, NEW MEXICO

LATITUDE: 35°03'N   LONGITUDE: 106°37'W   ELEVATION: FT. GRND 5311 BARO 05313   TIME ZONE: MOUNTAIN   WBAN: 23050

	(a)	JAN	FEB	MAR	APR	MAY	JUNE	JULY	AUG	SEP	OCT	NOV	DEC	YEAR
TEMPERATURE °F:														
Normals														
-Daily Maximum		47.2	52.9	60.7	70.6	79.9	90.6	92.8	89.4	83.0	71.7	57.2	48.0	70.3
-Daily Minimum		22.3	25.9	31.7	39.5	48.6	58.4	64.7	62.8	54.9	43.1	30.7	23.2	42.1
-Monthly		34.8	39.4	46.2	55.0	64.3	74.5	78.8	76.1	69.0	57.4	44.0	35.6	56.2
Extremes														
-Record Highest	46	69	75	85	89	98	105	105	101	100	91	77	72	105
-Year		1971	1972	1971	1965	1951	1980	1980	1979	1979	1979	1975	1958	JUN 1980
-Record Lowest	46	-17	-5	8	19	28	40	52	52	37	25	-7	3	-17
-Year		1971	1951	1948	1980	1975	1980	1985	1968	1971	1980	1976	1974	JAN 1971
NORMAL DEGREE DAYS:														
Heating (base 65°F)		936	717	583	302	81	0	0	0	12	242	630	911	4414
Cooling (base 65°F)		0	0	0	0	59	285	428	344	132	6	0	0	1254
% OF POSSIBLE SUNSHINE	46	72	73	73	77	80	83	76	76	79	79	77	72	76
MEAN SKY COVER (tenths)														
Sunrise - Sunset	46	4.8	4.9	5.1	4.5	4.1	3.3	4.5	4.3	3.6	3.5	4.0	4.6	4.3
MEAN NUMBER OF DAYS:														
Sunrise to Sunset														
-Clear	46	12.9	11.3	11.2	12.8	14.7	17.9	12.0	13.8	16.8	17.5	15.1	13.9	170.0
-Partly Cloudy	46	7.8	7.8	10.0	9.4	10.2	8.6	14.3	12.4	7.7	7.6	7.7	7.5	111.2
-Cloudy	46	10.3	9.2	9.7	7.7	6.1	3.6	4.7	4.8	5.5	5.9	7.2	9.5	84.1
Precipitation														
.01 inches or more	46	3.9	4.0	4.5	3.3	4.3	3.7	8.8	9.3	5.7	4.8	3.3	4.0	59.8
Snow, Ice pellets														
1.0 inches or more	46	1.0	0.8	0.7	0.2	0.*	0.0	0.0	0.0	0.0	0.0	0.4	0.9	4.1
Thunderstorms	46	0.1	0.3	0.9	1.5	3.8	4.9	11.2	11.0	4.7	2.4	0.6	0.2	41.7
Heavy Fog Visibility														
1/4 mile or less	46	1.2	1.0	0.6	0.2	0.*	0.*	0.1	0.*	0.1	0.3	0.6	1.4	5.6
Temperature °F														
-Maximum														
90° and above	25	0.0	0.0	0.0	0.0	2.6	17.4	24.0	16.9	4.3	0.2	0.0	0.0	65.3
32° and below	25	2.5	0.6	0.2	0.0	0.0	0.0	0.0	0.0	0.0	0.0	0.2	1.5	5.0
-Minimum														
32° and below	25	28.9	23.5	16.8	5.0	0.3	0.0	0.0	0.0	0.0	2.2	16.3	28.5	121.6
0° and below	25	0.5	0.0	0.0	0.0	0.0	0.0	0.0	0.0	0.0	0.0	0.1	0.0	0.6
AVG. STATION PRESS.(mb)	13	838.5	837.9	834.6	835.5	836.0	837.9	840.4	840.8	840.2	839.8	838.8	839.2	838.3
RELATIVE HUMIDITY (%)														
Hour 05	25	71	65	56	48	48	45	60	65	62	62	65	70	60
Hour 11 (Local Time)	25	51	44	34	26	25	23	34	39	40	38	42	50	37
Hour 17	25	41	32	25	18	18	17	27	30	31	30	36	43	29
Hour 23	25	62	52	44	35	34	32	47	52	52	50	55	61	48
PRECIPITATION (inches):														
Water Equivalent														
-Normal		0.41	0.40	0.52	0.40	0.46	0.51	1.30	1.51	0.85	0.86	0.38	0.52	8.12
-Maximum Monthly	46	1.32	1.42	2.18	1.97	3.07	1.71	3.33	3.30	1.99	3.08	1.45	1.85	3.33
-Year		1978	1948	1973	1942	1941	1967	1968	1967	1940	1972	1940	1959	JUL 1968
-Minimum Monthly	46	T	T	T	T	T	T	0.08	T	0.00	0.00	0.00	0.00	0.00
-Year		1970	1984	1966	1972	1945	1975	1980	1962	1957	1952	1949	1981	DEC 1981
-Maximum in 24 hrs	46	0.87	0.51	1.11	1.66	1.14	1.64	1.77	1.75	1.92	1.80	0.76	1.35	1.92
-Year		1962	1981	1973	1969	1969	1952	1961	1980	1955	1969	1940	1958	SEP 1955
Snow, Ice pellets														
-Maximum Monthly	46	9.5	8.2	13.9	8.1	1.0				T	0.9	9.3	14.7	14.7
-Year		1973	1964	1973	1973	1979				1971	1979	1940	1959	DEC 1959
-Maximum in 24 hrs	46	5.1	4.2	10.7	6.6	1.0				T	0.9	5.5	14.2	14.2
-Year		1973	1946	1973	1973	1979				1971	1979	1946	1958	DEC 1958
WIND:														
Mean Speed (mph)	46	8.1	8.8	10.2	11.1	10.5	10.0	9.1	8.2	8.6	8.3	7.9	7.7	9.0
Prevailing Direction														
through 1963		N	N	SE	S	S	S	SE	SE	SE	SE	N	N	SE
Fastest Mile														
-Direction	44	E	NW	NW	S	W	SE	E	SE	SE	N	NW	SE	SE
-Speed (MPH)	44	61	68	80	72	72	82	68	61	62	66	57	90	90
-Year		1949	1944	1943	1946	1950	1946	1945	1951	1945	1959	1948	1943	DEC 1943
Peak Gust														
-Direction	2	E	W	SW	SW	S	E	S	E	W	W	SW	E	SW
-Speed (mph)	2	41	63	58	64	54	48	58	47	61	51	52	46	64
-Date		1985	1984	1984	1984	1985	1985	1984	1984	1985	1984	1985	1984	APR 1984

See Reference Notes to this table on the following page.

TABLE 2 PRECIPITATION (inches)   ALBUQUERQUE, NEW MEXICO

YEAR	JAN	FEB	MAR	APR	MAY	JUNE	JULY	AUG	SEP	OCT	NOV	DEC	ANNUAL
1956	0.46	0.49	T	T	0.18	0.43	1.49	0.62	0.02	0.34	0.03	0.00	4.06
1957	0.78	0.59	0.52	0.38	0.35	0.04	2.48	1.32	T	2.59	1.24	0.32	10.61
#1958	0.21	0.27	1.71	0.62	0.43	0.22	0.14	1.74	1.34	1.72	0.37	1.35	10.12
1959	0.17	0.04	0.42	0.43	0.80	0.78	0.73	2.79	0.36	1.70	0.07	1.85	10.14
1960	0.34	0.38	0.44	0.19	0.71	0.91	0.47	0.78	0.56	2.88	0.07	0.39	8.12
1961	0.23	0.10	0.61	0.73	0.01	0.11	2.70	1.69	1.09	0.47	0.48	0.65	8.87
1962	1.01	0.11	0.18	0.07	0.01	0.19	1.24	T	0.71	0.75	0.61	0.51	5.39
1963	0.29	0.24	0.55	0.14	0.03	0.11	1.43	3.00	0.63	0.76	0.29	T	7.47
1964	0.07	1.12	0.13	0.61	0.35	T	1.87	0.98	1.57	0.04	0.21	0.49	7.44
1965	0.47	0.60	0.49	0.49	0.19	0.99	1.65	0.61	1.18	0.89	0.33	1.42	9.31
1966	0.42	0.30	T	0.04	0.02	1.66	1.63	1.06	1.04	0.54	0.09	0.01	6.81
1967	0.01	0.44	0.25	T	0.04	1.71	0.61	3.30	0.79	0.18	0.15	0.56	8.04
1968	0.01	0.98	1.48	0.51	0.99	0.05	3.33	1.49	0.30	0.12	0.59	0.82	10.67
1969	0.08	0.34	0.41	1.76	1.31	0.59	0.94	0.95	1.08	2.37	0.01	0.72	10.56
1970	T	0.27	0.42	0.05	0.33	0.40	1.22	2.24	0.79	0.25	0.08	0.23	6.28
1971	0.27	0.21	0.03	0.78	0.16	0.02	1.05	1.44	1.15	0.67	1.40		8.05
1972	0.12	0.12	0.08	T	0.18	0.55	1.00	2.93	1.00	3.08	0.69	0.36	10.11
1973	0.85	0.33	2.18	0.91	0.66	1.37	1.80	1.19	1.13	0.35	0.08	0.03	10.88
1974	0.88	0.11	0.85	0.14	0.01	0.22	2.40	0.79	1.58	1.96	0.38	0.51	9.83
1975	0.26	0.99	0.95	0.10	0.66	T	1.43	1.40	1.66	T	0.28	0.28	8.01
1976	0.00	0.40	0.09	0.31	0.82	0.60	1.32	0.73	0.45	0.03	0.24	0.20	5.19
1977	0.88	0.13	0.63	1.07	0.10	0.04	0.69	2.28	0.78	0.76	0.42	0.13	7.91
1978	1.32	1.02	0.54	0.05	0.69	1.05	0.24	2.49	0.59	1.22	1.00	0.76	10.97
1979	1.07	0.62	0.14	0.24	2.48	1.02	0.80	1.13	0.40	0.27	0.91	0.87	10.35
1980	0.87	0.58	0.60	0.60	0.56	0.01	0.08	2.61	1.83	0.09	0.30	0.74	8.87
1981	0.05	0.67	0.80	0.30	0.53	0.35	1.07	1.68	0.41	1.43	0.37	0.00	7.66
1982	0.32	0.20	0.84	0.05	0.52	0.09	1.32	1.09	1.34	0.26	0.60	0.78	7.41
1983	1.10	0.71	0.61	0.02	0.32	1.21	0.55	0.27	0.91	1.20	0.44	0.42	7.76
1984	0.33	T	0.62	0.50	0.16	0.48	1.13	2.70	1.13	3.04	0.63	1.36	12.08
1985	0.49	0.54	0.70	1.69	1.12	0.53	1.16	0.49	1.53	2.15	0.19	0.16	10.75
Record Mean	0.40	0.37	0.45	0.54	0.63	0.58	1.38	1.36	0.93	0.85	0.43	0.46	8.37

TABLE 3 AVERAGE TEMPERATURE (deg. F)   ALBUQUERQUE, NEW MEXICO

YEAR	JAN	FEB	MAR	APR	MAY	JUNE	JULY	AUG	SEP	OCT	NOV	DEC	ANNUAL
1956	41.1	36.0	49.2	55.3	69.6	78.9	78.4	76.3	74.1	60.3	41.0	36.3	58.1
1957	39.9	48.1	47.3	54.2	61.9	74.8	79.1	76.0	70.3	56.6	40.3	38.5	57.3
#1958	35.3	43.5	42.8	53.0	68.6	78.5	79.6	78.9	69.2	56.9	46.0	41.5	57.8
1959	34.3	39.5	45.9	57.0	65.1	77.0	78.6	76.2	70.6	56.8	43.9	38.2	56.9
#1960	33.0	36.5	49.9	57.9	64.7	76.6	78.7	78.4	71.8	56.5	46.5	33.9	57.1
1961	33.9	40.6	47.0	54.5	65.9	75.8	76.7	75.2	65.6	56.8	40.3	34.1	55.5
1962	31.6	42.3	41.2	58.1	64.1	72.7	76.3	77.6	69.4	58.1	46.9	36.9	56.3
1963	29.4	40.5	45.2	57.7	68.0	74.6	76.4	75.9	72.5	61.5	45.7	34.8	57.3
1964	30.0	29.1	41.5	51.7	65.8	73.6	78.2	76.8	69.3	59.4	43.7	35.5	54.5
1965	38.8	39.4	44.6	54.8	61.7	69.4	77.9	75.4	66.6	58.0	48.4	35.8	55.9
1966	30.1	33.2	45.6	54.6	67.2	72.8	79.8	75.7	68.4	56.8	46.7	34.3	55.4
1967	33.2	40.5	52.0	57.8	63.8	71.5	79.2	74.5	68.4	58.2	46.1	32.4	56.5
1968	36.8	43.3	46.7	53.4	62.7	75.2	76.1	72.4	68.0	58.3	42.8	30.0	55.5
1969	38.0	38.5	41.1	57.4	66.2	73.6	80.2	79.0	70.0	53.8	41.4	39.1	56.6
1970	34.5	42.8	44.1	52.5	66.2	72.7	79.6	77.8	67.5	52.6	44.5	36.4	56.0
1971	33.6	38.9	47.7	53.3	61.7	73.8	78.1	78.1	66.4	53.8	45.2	31.9	54.8
1972	36.1	42.5	53.6	56.9	64.0	73.7	78.6	74.1	68.1	57.6	40.1	35.0	56.7
1973	31.8	35.9	45.1	50.2	62.7	73.5	78.4	78.0	67.5	56.4	44.6	34.0	54.8
1974	33.6	37.9	52.8	56.4	68.5	80.1	77.0	72.7	66.1	58.1	45.0	32.0	56.7
1975	30.8	38.0	45.0	49.9	61.0	73.0	76.8	76.1	66.3	56.5	42.6	35.6	54.3
1976	33.2	43.3	44.3	54.6	62.8	73.4	77.0	75.0	68.0	53.1	40.6	33.0	54.9
1977	29.8	40.7	43.2	56.5	64.2	75.5	78.6	77.4	69.4	58.9	46.4	40.4	56.8
1978	36.8	39.3	50.2	57.7	60.5	75.5	81.6	75.5	69.1	60.3	47.5	34.3	57.4
1979	32.9	41.1	48.4	56.9	63.7	73.3	80.6	77.1	72.3	61.5	41.0	37.7	57.2
1980	40.2	44.2	46.1	52.1	61.1	77.2	82.7	77.4	69.9	54.5	43.5	40.5	57.4
1981	38.0	42.9	46.2	59.0	64.5	77.0	79.8	76.4	69.7	55.7	47.0	40.5	58.0
1982	35.9	39.4	47.4	56.1	63.0	74.8	79.1	77.4	69.5	54.8	42.9	34.4	56.2
1983	35.0	39.7	49.2	50.2	63.0	73.4	80.4	79.4	73.4	58.3	45.1	36.7	56.8
1984	34.1	40.1	46.8	52.8	69.9	73.6	78.9	75.7	68.8	51.6	43.7	35.6	56.0
1985	33.8	38.3	47.5	57.4	64.0	74.1	77.1	76.6	65.9	57.5	45.4	37.6	56.3
Record Mean	34.5	39.6	46.4	54.8	63.8	73.5	77.4	75.3	68.5	56.8	43.9	35.3	55.9
Max	47.0	52.9	60.8	69.9	79.0	89.1	91.3	88.8	82.3	71.0	57.5	47.6	69.8
Min	22.0	26.3	32.0	39.7	48.6	57.9	63.5	61.8	54.6	42.5	30.3	23.0	41.9

REFERENCE NOTES FOR TABLES 1, 2, 3 and 6   (ALBUQUERQUE, NM)

GENERAL

T - TRACE AMOUNT
BLANK ENTRIES DENOTE MISSING/UNREPORTED DATA.
# INDICATES A STATION OR INSTRUMENT RELOCATION.

SPECIFIC

TABLE 1

(a) - LENGTH OF RECORD IN YEARS. ALTHOUGH
        INDIVIDUAL MONTHS MAY BE MISSING.
   * LESS THAN .05

NORMALS — BASED ON THE 1951-1980 RECORD PERIOD.
EXTREMES — DATES ARE THE MOST RECENT OCCURRENCE.
WIND DIR. — NUMERALS SHOW TENS OF DEGREES
        CLOCKWISE FROM TRUE NORTH.
        "00" INDICATES CALM.
RESULTANT WIND DIRECTIONS ARE GIVEN TO WHOLE DEGREES.

EXCEPTIONS

TABLES 2, 3, and 6

RECORD MEANS ARE THROUGH THE CURRENT YEAR,
BEGINNING IN   1893 FOR TEMPERATURE
                  1893 FOR PRECIPITATION
                  1940 FOR SNOWFALL

TABLE 4   HEATING DEGREE DAYS Base 65 deg. F   ALBUQUERQUE, NEW MEXICO

SEASON	JULY	AUG	SEP	OCT	NOV	DEC	JAN	FEB	MAR	APR	MAY	JUNE	TOTAL
1956-57	0	0	0	195	715	880	771	468	543	317	123	3	4015
#1957-58	0	0	3	269	739	817	914	596	679	354	24	0	4395
1958-59	0	0	35	249	562	722	948	587	249	60	0		4118
#1959-60	0	0	20	250	628	823	987	819	462	211	82	0	4282
1960-61	0	0	3	256	548	956	956	678	551	308	65	0	4321
1961-62	0	0	43	248	731	951	1030	629	730	214	78	2	4656
1962-63	0	0	22	208	534	863	1098	680	605	219	7	2	4238
1963-64	0	0	0	124	573	931	1076	1036	722	391	85	3	4941
1964-65	0	2	20	173	632	909	805	709	624	300	128	24	4326
1965-66	0	0	56	217	492	895	1074	882	595	305	53	0	4569
1966-67	0	0	15	247	541	942	980	682	396	211	109	0	4123
1967-68	0	0	13	220	557	1003	870	623	559	343	107	8	4303
1968-69	2	0	12	208	660	1080	831	735	735	228	84	0	4575
1969-70	0	0	1	348	701	795	938	612	644	367	63	11	4480
1970-71	0	0	58	380	605	878	968	725	533	343	122	5	4617
1971-72	0	0	101	341	587	1022	889	648	346	244	76	0	4254
1972-73	0	3	14	244	740	925	1020	811	607	440	113	3	4920
1973-74	0	0	43	257	606	955	963	754	373	255	29	4	4239
1974-75	0	2	68	212	593	1020	1051	748	614	449	143	6	4906
1975-76	0	0	47	256	664	905	979	622	634	304	99	1	4511
1976-77	0	0	35	367	726	985	1084	675	669	250	61	0	4852
1977-78	0	0	1	192	551	757	870	713	454	215	175	2	3930
1978-79	0	0	20	167	521	945	988	665	509	241	100	12	4168
1979-80	0	0	23	148	715	840	763	595	577	379	139	2	4181
1980-81	0	0	6	335	640.	752	827	611	575	197	62	2	4007
1981-82	0	0	3	280	534	754	895	709	538	268	94	0	4075
1982-83	0	0	23	314	658	941	922	703	556	439	127	0	4683
1983-84	0	0	11	198	592	875	948	714	559	362	22	3	4284
1984-85	0	0	51	411	631	903	960	744	536	220	74	7	4537
1985-86	0	0	61	228	581	842							

TABLE 5   COOLING DEGREE DAYS Base 65 deg. F   ALBUQUERQUE, NEW MEXICO

YEAR	JAN	FEB	MAR	APR	MAY	JUNE	JULY	AUG	SEP	OCT	NOV	DEC	TOTAL
1969	0	0	0	6	127	263	478	442	158	7	0	0	1481
1970	0	0	0	0	105	246	461	405	141	4	0	0	1362
1971	0	0	5	0	26	277	414	282	149	0	0	0	1153
1972	0	0	0	5	52	267	428	294	113	23	0	0	1182
1973	0	0	0	0	48	267	422	409	124	0	0	0	1270
1974	0	0	0	5	144	464	380	247	107	6	0	0	1353
1975	0	0	0	0	25	256	372	351	96	0	0	0	1100
1976	0	0	0	0	38	260	382	319	137	5	0	0	1141
1977	0	0	0	0	44	324	427	392	141	7	0	0	1335
1978	0	0	0	4	41	324	521	330	151	27	0	0	1398
1979	0	0	0	5	67	269	491	382	249	45	0	0	1508
1980	0	0	0	0	27	375	557	392	160	15	0	0	1526
1981	0	0	0	28	51	368	470	360	152	1	0	0	1430
1982	0	0	0	6	38	301	441	394	163	4	0	0	1347
1983	0	0	0	1	72	260	484	450	267	1	0	0	1535
1984	0	0	0	4	179	266	441	340	169	1	0	0	1400
1985	0	0	0	0	51	289	383	368	97	0	0	0	1188

TABLE 6   SNOWFALL (inches)   ALBUQUERQUE, NEW MEXICO

SEASON	JULY	AUG	SEP	OCT	NOV	DEC	JAN	FEB	MAR	APR	MAY	JUNE	TOTAL
1956-57	0.0	0.0	0.0	0.0	T	0.0	1.5	T	T	T	0.0	0.0	1.5
1957-58	0.0	0.0	0.0	T	T	2.9	2.3	0.5	7.3	T	0.0	0.0	13.0
1958-59	0.0	0.0	0.0	T	2.2	14.2	1.7	0.6	4.8	T	0.0	0.0	23.5
1959-60	0.0	0.0	0.0	T	0.0	14.7	1.8	0.8	2.8	0.0	0.0	0.0	20.1
1960-61	0.0	0.0	0.0	0.0	0.0	2.5	0.6	0.4	3.0	T	0.0	0.0	6.5
1961-62	0.0	0.0	0.0	T	3.4	2.4	4.0	T	0.2	0.0	0.0	0.0	10.0
1962-63	0.0	0.0	0.0	0.0	T	1.0	2.5	0.8	2.5	0.0	0.0	0.0	6.8
1963-64	0.0	0.0	0.0	0.0	T	T	0.5	8.2	1.3	T	0.0	0.0	10.0
1964-65	0.0	0.0	0.0	0.0	T	0.3	1.4	3.6	T	T	0.0	0.0	5.3
1965-66	0.0	0.0	0.0	0.0	T	3.0	5.4	1.0	0.0	T	0.0	0.0	9.4
1966-67	0.0	0.0	0.0	0.0	T	T	T	1.1	1.1	T	0.0	0.0	2.1
1967-68	0.0	0.0	0.0	0.2	1.0	2.8	T	2.0	1.4	T	0.0	0.0	7.4
1968-69	0.0	0.0	0.0	0.0	T	7.4	T	1.8	5.5	T	0.0	0.0	14.7
1969-70	0.0	0.0	0.0	0.0	T	1.1	T	2.7	3.3	0.0	0.0	0.0	7.1
1970-71	0.0	0.0	0.0	0.5	T	0.5	3.0	2.3	0.5	T	0.0	0.0	6.8
1971-72	0.0	0.0	T	T	T	1.2	1.1	1.1	0.0	T	0.0	0.0	9.1
1972-73	0.0	0.0	0.0	T	2.9	1.2	9.5	1.8	13.9	8.1	0.0	0.0	37.4
1973-74	0.0	0.0	0.0	0.3	0.6	0.1	9.3	0.6	2.0	0.0	0.0	0.0	12.9
1974-75	0.0	0.0	0.0	0.0	T	4.9	0.9	6.7	3.8	0.2	0.0	0.0	16.5
1975-76	0.0	0.0	0.0	0.0	0.2	2.9	0.0	T	0.5	0.2	0.0	0.0	3.8
1976-77	0.0	0.0	0.0	T	2.4	1.2	8.4	1.4	2.3	2.6	0.0	0.0	18.3
1977-78	0.0	0.0	0.0	0.0	0.0	0.0	6.0	3.4	2.0	0.5	0.1	0.0	11.5
1978-79	0.0	0.0	0.0	0.0	T	1.0	2.6	6.0	T	T	1.0	0.0	11.1
1979-80	0.0	0.0	0.0	0.9	0.8	2.7	T	0.9	3.1	T	T	0.0	8.4
1980-81	0.0	0.0	0.0	T	2.8	7.4	0.5	2.6	0.9	T	0.0	0.0	14.2
1981-82	0.0	0.0	0.0	0.0	0.0	0.0	3.6	1.2	0.7	T	0.0	0.0	5.5
1982-83	0.0	0.0	0.0	0.0	0.9	3.3	7.3	4.2	1.0	T	T	0.0	16.7
1983-84	0.0	0.0	0.0	0.0	0.8	0.8	4.1	T	0.1	3.0	0.0	0.0	8.8
1984-85	0.0	0.0	0.0	T	T	3.4	2.0	2.9	0.6	0.0	0.0	0.0	8.9
1985-86	0.0	0.0	0.0	0.0	0.7	0.9							
Record Mean	0.0	0.0	T	T	1.1	2.6	2.5	2.0	1.9	0.5	T	0.0	10.7

See Reference Notes, relative to all above tables, on preceding page.

Clayton is located on the high plains of northeastern New Mexico some 90 miles southeast of the eastern slope of the Rocky Mountains. The climate is semi-arid. Nearly 80 percent of the rainfall occurs from May through October in sudden thunderstorms which form over the mountains northwest of Clayton and drift southeastward. This makes the growing of small grains and the raising of range cattle profitable. Native grasses in the area remain nutritious even in winter.

The climate of Clayton is characteristic of that found in the higher-altitude sections of the continental southwest. Temperatures are mostly moderate. While daytime temperatures in summer are moderately warm, 90 degrees or slightly higher about half the time in July, hot days recording temperatures of 100 degrees or more, only occur about once a year. Minimum temperatures range from the teens in January to the 60s in July. With clear nocturnal skies and an altitude of about 5,000 feet, summer nights in Clayton are usually comfortable for sleeping. While winter minima are generally below freezing, zero temperatures only occur about three times a year.

From June through August nearly all precipitation is from scattered thunderstorms. As late summer and fall give way to winter, showery precipitation becomes less frequent and copious. Occasional winter snows, caused in part by upslope movement of air from the Gulf of Mexico, supply some winter moisture.

Blizzards are rare but high winds and cold temperatures frequent winter storms and can produce blizzard conditions. These storms may often close highways and, unless proper precautions are taken, they may also cause loss of life and livestock.

Based on the 1951-1980 period, the average first occurrence of 32 degrees Fahrenheit in the fall is October 16 and the average last occurrence in the spring is May 1.

TABLE 1   NORMALS, MEANS AND EXTREMES

CLAYTON, NEW MEXICO

LATITUDE: 36°27'N   LONGITUDE: 103°09'W   ELEVATION: FT. GRND 4969 BARO 04972   TIME ZONE: MOUNTAIN   WBAN: 23051

	(a)	JAN	FEB	MAR	APR	MAY	JUNE	JULY	AUG	SEP	OCT	NOV	DEC	YEAR
TEMPERATURE °F:														
Normals														
-Daily Maximum		47.4	50.6	56.3	65.9	74.5	84.4	87.8	85.6	78.4	69.2	55.8	49.6	67.1
-Daily Minimum		18.6	22.1	26.4	36.0	45.8	55.3	60.5	58.9	51.2	40.1	27.7	21.5	38.7
-Monthly		33.0	36.4	41.4	51.0	60.2	69.8	74.2	72.3	64.8	54.7	41.8	35.5	52.9
Extremes														
-Record Highest	41	78	81	85	91	95	104	102	102	99	90	85	83	104
-Year		1956	1963	1971	1965	1984	1968	1964	1944	1948	1979	1980	1955	JUN 1968
-Record Lowest	41	-21	-17	-11	9	23	37	45	45	26	17	-10	-14	-21
-Year		1959	1951	1948	1945	1967	1964	1958	1964	1985	1969	1976	1983	JAN 1959
NORMAL DEGREE DAYS:														
Heating (base 65°F)		992	801	732	420	177	25	0	0	80	327	699	915	5168
Cooling (base 65°F)		0	0	0	0	25	172	288	230	74	8	0	0	797
% OF POSSIBLE SUNSHINE														
MEAN SKY COVER (tenths)														
Sunrise - Sunset	26	5.1	5.0	5.3	5.0	5.1	4.1	4.8	4.3	3.8	3.5	4.4	4.6	4.6
MEAN NUMBER OF DAYS:														
Sunrise to Sunset														
-Clear	28	12.5	11.2	11.3	11.7	11.4	13.5	12.3	15.0	17.2	17.8	14.8	13.9	162.5
-Partly Cloudy	28	7.0	7.3	8.8	8.8	9.3	10.4	11.9	9.7	6.0	6.6	6.0	7.0	98.7
-Cloudy	28	11.5	9.6	10.9	9.5	10.4	6.2	6.8	6.3	6.8	6.6	9.2	10.1	103.9
Precipitation														
.01 inches or more	39	3.5	3.4	4.8	5.2	8.1	7.8	10.3	8.7	5.4	3.9	3.3	2.9	67.3
Snow, Ice pellets														
1.0 inches or more	37	1.5	1.2	1.6	0.6	0.1	0.0	0.0	0.0	0.1	0.3	0.9	1.2	7.5
Thunderstorms	19	0.1	0.3	0.7	2.2	8.3	10.5	13.4	11.4	5.3	1.2	0.3	0.1	53.8
Heavy Fog Visibility														
1/4 mile or less	19	0.7	1.4	2.1	1.0	0.9	0.6	0.6	0.3	0.8	0.7	0.9	0.8	10.9
Temperature °F														
-Maximum														
90° and above	40	0.0	0.0	0.0	0.1	1.0	8.6	13.4	9.5	2.9	0.1	0.0	0.0	35.6
32° and below	40	4.8	3.2	1.9	0.2	0.0	0.0	0.0	0.0	0.0	0.2	1.5	3.4	15.2
-Minimum														
32° and below	40	29.4	24.9	23.3	9.5	0.9	0.0	0.0	0.0	0.3	4.9	20.8	28.1	142.1
0° and below	40	1.5	0.5	0.2	0.0	0.0	0.0	0.0	0.0	0.0	0.0	0.1	0.6	2.8
AVG. STATION PRESS. (mb)														
RELATIVE HUMIDITY (%)														
Hour 05	38	64	65	65	65	71	71	76	77	73	64	64	63	68
Hour 11 (Local Time)	39	43	42	39	34	39	38	41	42	41	36	40	42	40
Hour 17	32	47	42	37	31	36	35	41	42	40	40	47	52	41
Hour 23	4	65	63	62	54	67	68	61	65	60	66	61	67	63
PRECIPITATION (inches):														
Water Equivalent														
-Normal		0.27	0.28	0.59	1.05	2.23	1.74	2.53	2.43	1.48	0.75	0.48	0.29	14.12
-Maximum Monthly	41	1.06	1.07	2.30	4.67	6.77	4.51	7.77	5.73	5.22	4.55	2.08	1.10	7.77
-Year		1960	1948	1957	1944	1949	1950	1950	1981	1960	1984	1978	1960	JUL 1950
-Minimum Monthly	41	T	T	0.01	0.06	0.27	0.19	0.66	0.28	T	T	T	0.00	0.00
-Year		1970	1950	1966	1974	1974	1955	1946	1983	1956	1980	1966	1957	DEC 1957
-Maximum in 24 hrs	41	0.71	0.75	0.90	2.42	4.67	2.86	2.69	4.41	3.48	3.90	1.58	0.71	4.67
-Year		1960	1953	1959	1980	1954	1984	1982	1963	1960	1965	1961	1947	MAY 1954
Snow, Ice pellets														
-Maximum Monthly	41	12.0	9.5	16.0	10.9	8.0				5.0	8.0	14.8	11.0	16.0
-Year		1983	1978	1984	1955	1978				1984	1984	1961	1978	MAR 1984
-Maximum in 24 hrs	39	7.2	7.5	9.0	10.9	7.6				5.0	6.0	12.7	8.1	12.7
-Year		1983	1953	1973	1955	1978				1984	1984	1961	1958	NOV 1961
WIND:														
Mean Speed (mph)	3	11.7	11.7	12.6	14.5	12.6	12.4	10.6	9.6	11.4	10.9	11.9	12.3	11.8
Prevailing Direction														
Fastest Obs. 1 Min.														
-Direction (!!!)														
-Speed (MPH)														
-Year														
Peak Gust														
-Direction (!!!)														
-Speed (mph)														
-Date														

See Reference Notes to this table on the following page.

TABLE 2 — PRECIPITATION (inches) CLAYTON, NEW MEXICO

YEAR	JAN	FEB	MAR	APR	MAY	JUNE	JULY	AUG	SEP	OCT	NOV	DEC	ANNUAL
1956	0.10	0.12	0.16	0.24	2.87	1.00	4.23	3.01	T	0.15	0.04	T	11.92
1957	0.20	0.06	2.30	1.50	4.07	0.78	1.22	3.34	0.33	1.51	0.79	0.00	16.10
1958	0.26	0.46	1.81	1.69	3.83	0.47	5.15	0.83	1.60	0.10	0.36	0.64	17.20
1959	0.72	0.10	1.57	0.53	2.57	0.78	1.30	3.88	0.49	1.19	0.15	0.68	13.96
1960	1.06	0.48	0.32	0.82	0.49	2.66	2.84	1.38	5.22	2.97	T	1.10	19.34
1961	0.04	0.19	0.55	0.89	0.57	1.71	2.35	2.59	3.01	1.35	1.84	0.10	15.19
1962	0.45	0.03	0.18	0.50	1.34	1.91	4.42	0.54	0.61	0.01	0.32	0.12	10.43
1963	T	0.50	0.09	0.25	2.00	2.23	1.17	5.60	0.74	0.20	T	0.08	12.86
1964	T	0.56	0.08	0.34	0.88	1.00	1.33	2.32	1.38	0.01	0.80	0.23	8.93
1965	0.10	0.50	0.28	0.62	1.75	3.96	3.64	2.93	2.97	3.91	T	0.70	21.36
1966	0.28	0.07	0.01	0.43	0.79	3.20	3.82	3.21	2.76	0.69	T	0.17	15.43
1967	0.07	0.05	0.09	0.97	2.68	3.02	1.61	2.25	2.43	0.12	0.22	0.33	13.84
1968	0.17	0.08	0.66	1.00	0.75	2.09	2.56	2.12	0.73	0.02	0.31	0.20	10.69
1969	T	0.65	1.48	0.93	4.38	3.23	2.21	5.17	3.59	3.15	0.27	0.64	25.70
1970	T	0.18	0.83	1.33	0.69	0.79	2.11	1.97	1.88	0.47	0.11	T	10.36
1971	0.13	0.33	0.38	1.16	2.27	2.95	3.31	2.49	0.87	0.90	0.87	0.52	16.18
1972	0.06	0.01	0.44	0.20	2.60	3.08	4.58	3.33	3.05	0.58	1.05	0.17	19.15
1973	0.42	0.22	2.29	2.65	0.86	1.04	2.00	1.31	0.99	0.40	0.10	0.74	13.02
1974	0.33	0.10	0.24	0.06	0.27	2.84	2.85	2.45	0.71	1.32	0.30	0.13	11.60
1975	0.39	0.17	0.14	0.63	1.77	1.55	3.80	2.06	1.71	T	1.02	0.06	13.30
1976	0.10	0.18	0.13	0.82	2.39	0.67	3.83	1.04	3.32	0.22	0.25	0.03	12.98
1977	0.20	0.04	0.15	3.00	1.47	0.80	1.39	2.44	0.45	0.01	0.19	0.08	10.22
1978	0.27	0.61	0.09	0.26	3.63	2.66	1.14	1.41	0.58	0.18	2.08	0.39	13.30
1979	0.74	0.03	0.86	1.36	2.41	2.73	3.38	1.01	1.31	0.41	0.31	0.30	14.85
1980	0.29	0.04	0.79	3.11	2.70	2.03	0.88	1.25	0.74	T	0.44	0.24	12.51
1981	0.07	0.03	1.27	0.46	1.57	3.33	3.01	5.73	0.63	0.81	0.45	0.24	17.60
1982	0.04	0.18	0.20	0.37	2.32	2.99	6.67	2.09	2.29	0.43	0.33	0.56	18.47
1983	0.78	0.58	0.72	0.69	3.35	2.12	2.12	1.39	0.28	0.36	0.33	0.45	11.40
1984	0.34	0.19	1.36	0.91	0.75	4.38	2.58	3.95	0.74	4.55	0.12	0.41	20.28
1985	0.43	0.10	1.14	1.89	1.72	0.77	1.40	1.78	2.49	2.85	0.15	0.04	14.76
Record Mean	0.30	0.30	0.62	1.03	2.38	1.88	2.83	2.47	1.46	0.94	0.42	0.29	14.94

TABLE 3 — AVERAGE TEMPERATURE (deg. F) CLAYTON, NEW MEXICO

YEAR	JAN	FEB	MAR	APR	MAY	JUNE	JULY	AUG	SEP	OCT	NOV	DEC	ANNUAL
1956	35.8	33.8	43.5	50.0	64.8	73.6	73.0	72.3	68.5	57.1	39.6	38.5	54.2
1957	32.2	43.1	40.2	46.4	56.5	67.9	76.1	72.4	63.4	52.7	37.1	40.3	52.4
1958	34.3	36.5	32.8	46.7	62.6	70.2	72.0	74.5	66.1	55.2	43.2	36.4	52.6
1959	29.7	35.9	40.3	49.6	60.5	71.5	72.9	73.1	63.9	51.7	39.6	37.0	52.1
1960	30.8	27.8	41.0	53.6	61.0	70.8	71.7	73.4	65.7	53.8	43.6	32.4	52.1
1961	34.5	36.2	41.8	49.2	60.3	69.5	72.2	71.7	60.3	53.9	37.0	32.1	51.6
1962	29.1	39.8	40.1	52.5	64.9	67.2	72.9	75.0	57.9	44.7	37.6	35.3	53.9
1963	25.5	37.8	44.3	55.5	63.5	70.2	77.2	73.0	68.6	61.4	45.6	31.9	54.5
1964	34.5	29.1	38.6	50.2	62.7	70.8	73.4	73.4	64.6	56.3	42.5	34.9	52.9
1965	38.6	34.0	33.7	55.0	60.4	66.8	73.7	70.9	61.2	55.6	48.0	38.6	53.0
1966	28.7	32.5	45.4	49.8	60.9	68.8	76.9	68.8	64.5	53.1	46.2	32.1	52.3
1967	37.1	36.8	47.4	54.7	58.0	67.0	72.8	69.0	63.4	56.2	43.1	30.3	53.0
1968	34.5	35.9	43.0	48.7	57.5	70.8	71.2	71.2	57.6	56.2	43.3	33.5	52.4
1969	38.9	37.1	33.3	53.2	61.2	65.6	75.8	73.9	65.9	47.7	42.5	36.1	52.6
1970	32.3	39.5	36.5	48.6	62.0	68.3	71.7	73.4	61.6	48.5	42.6	37.8	52.2
1971	34.2	34.5	42.6	50.1	58.4	71.5	71.7	69.3	61.6	53.5	43.1	34.8	52.1
1972	34.4	39.9	48.4	55.4	59.9	70.1	70.6	69.8	62.5	53.1	43.0	34.9	52.1
1973	31.6	36.1	41.0	44.4	57.4	68.7	72.8	73.5	62.5	56.4	46.0	34.9	52.1
1974	31.7	38.0	47.7	52.3	65.5	70.6	74.5	68.6	62.6	55.8	42.1	32.1	53.2
1975	32.9	33.9	40.8	48.8	58.7	68.6	71.6	73.2	62.2	56.1	40.9	38.7	52.2
1976	34.5	42.8	41.6	53.4	57.6	69.3	72.7	72.4	64.8	48.8	38.4	36.2	52.7
1977	30.0	39.9	42.4	52.9	62.3	72.8	75.7	72.9	68.5	56.3	43.5	38.8	54.7
1978	27.7	30.2	44.6	54.3	58.3	69.4	76.8	71.6	65.8	55.6	42.2	29.4	52.2
1979	23.3	36.0	43.5	50.9	56.9	67.5	73.9	70.2	66.4	56.5	37.5	37.8	51.7
1980	33.1#	37.6	40.7	48.6	57.1	72.8	75.0	74.1	66.4	54.3	46.7	42.8	54.1
1981	38.3	39.2	43.9	57.7	59.8	73.1	75.0	70.4	65.7	54.1	47.3	37.2	55.1
1982	34.9	33.7	43.3	50.5	59.3	66.1	72.9	72.9	66.0	53.3	40.8	33.6	52.3
1983	35.4	36.4	42.1	45.7	55.8	65.8	75.3	76.5	68.4	56.2	43.6	23.7	52.1
1984	30.9	37.6	40.2	47.7	61.8	70.1	75.2	72.3	62.7	49.3	43.2	37.2	52.6
1985	29.5	34.1	43.6	54.9	62.3	70.2	75.2	73.1	62.1	53.4	40.2	32.6	52.6
Record Mean	33.0	36.4	41.5	51.1	60.0	69.6	74.0	72.5	65.0	54.9	42.0	35.4	53.0
Max	47.2	50.5	56.3	65.9	74.3	84.0	87.7	85.8	78.7	69.3	56.0	49.3	67.1
Min	18.8	22.2	26.6	36.2	45.6	55.2	60.3	59.1	51.3	40.5	27.9	21.5	38.8

REFERENCE NOTES FOR TABLES 1, 2, 3 and 6   (CLAYTON, NM)

GENERAL

T - TRACE AMOUNT
BLANK ENTRIES DENOTE MISSING/UNREPORTED DATA.
# INDICATES A STATION OR INSTRUMENT RELOCATION.

SPECIFIC

TABLE 1

(a) - LENGTH OF RECORD IN YEARS. ALTHOUGH INDIVIDUAL MONTHS MAY BE MISSING.

* LESS THAN .05

NORMALS — BASED ON THE 1951-1980 RECORD PERIOD.
EXTREMES — DATES ARE THE MOST RECENT OCCURRENCE.
WIND DIR. — NUMERALS SHOW TENS OF DEGREES CLOCKWISE FROM TRUE NORTH. "00" INDICATES CALM.
RESULTANT WIND DIRECTIONS ARE GIVEN TO WHOLE DEGREES.

EXCEPTIONS

TABLE 1

1. THUNDERSTORMS AND HEAVY FOG ARE THROUGH 1964 AND MAY BE INCOMPLETE, DUE TO PART-TIME OPERATIONS.
2. MEAN SKY COVER, AND DAYS CLEAR-PARTLY CLOUDY-CLOUDY ARE THROUGH 1974.

TABLES 2, 3, and 6

RECORD MEANS ARE THROUGH THE CURRENT YEAR, BEGINNING IN   1944 FOR TEMPERATURE
1944 FOR PRECIPITATION
1944 FOR SNOWFALL

TABLE 4   HEATING DEGREE DAYS Base 65 deg. F   CLAYTON, NEW MEXICO

SEASON	JULY	AUG	SEP	OCT	NOV	DEC	JAN	FEB	MAR	APR	MAY	JUNE	TOTAL
1956-57	0	22	26	246	755	815	1010	607	763	551	260	46	5101
1957-58	0	0	85	387	828	761	943	793	991	544	120	24	5476
1958-59	7	0	75	309	648	880	1089	807	760	454	157	17	5203
1959-60	0	0	98	410	753	864	1052	1072	737	336	141	13	5476
1960-61	15	0	63	338	634	1002	939	800	710	467	181	18	5167
1961-62	2	0	172	338	831	1013	1104	700	766	369	74	32	5401
1962-63	0	0	67	226	601	843	1218	757	636	285	108	22	4763
1963-64	0	1	15	127	578	1020	942	1035	811	439	134	24	5126
1964-65	0	2	106	270	670	926	811	865	965	312	165	39	5131
1965-66	2	0	167	292	502	812	1121	905	603	450	165	17	5036
1966-67	0	22	57	361	558	1012	859	784	538	306	249	22	4768
1967-68	0	31	85	286	650	1066	940	837	673	482	238	10	5298
1968-69	11	13	68	247	734	969	802	773	973	348	142	72	5152
1969-70	0	0	28	530	668	886	1005	708	875	483	125	58	5366
1970-71	6	0	151	506	667	835	950	849	690	439	215	0	5308
1971-72	25	8	180	351	649	930	946	719	511	293	171	0	4783
1972-73	25	14	91	376	921	1018	1025	805	735	612	250	24	5896
1973-74	4	0	119	263	561	928	1025	749	528	377	70	31	4655
1974-75	2	11	180	285	680	1010	987	861	742	479	199	42	5478
1975-76	0	0	158	270	718	811	938	637	716	341	228	13	4830
1976-77	0	2	80	497	790	884	1079	696	693	355	94	0	5170
1977-78	0	3	17	268	637	806	1148	968	627	314	236	36	5060
1978-79	0	8	68	291	678	1100	1286	806	660	416	252	62	5627
1979-80	0	7	69	275	818	836	981	791	750	486	245	16	5274
1980-81	0	0	64	333	675	682	821	716	646	228	170	10	4345
1981-82	0	6	50	332	525	856	924	873	663	434	179	52	4894
1982-83	4	0	60	354	718	966	910	795	703	571	287	66	5434
1983-84	0	0	63	271	636	1276	1050	786	764	512	143	9	5510
1984-85	0	1	150	481	650	858	1092	858	655	296	126	26	5193
1985-86	0	4	183	355	739	995							

TABLE 5   COOLING DEGREE DAYS Base 65 deg. F   CLAYTON, NEW MEXICO

YEAR	JAN	FEB	MAR	APR	MAY	JUNE	JULY	AUG	SEP	OCT	NOV	DEC	TOTAL
1969	0	0	0	0	33	96	344	282	61	4	0	0	820
1970	0	0	0	0	36	162	297	270	95	3	0	0	863
1971	0	0	4	0	17	200	239	149	82	0	0	0	691
1972	0	0	0	13	19	160	208	169	71	15	0	0	655
1973	0	0	0	0	21	143	254	273	51	5	0	0	747
1974	0	0	0	4	92	205	303	130	31	3	0	0	768
1975	0	0	0	0	10	156	210	260	82	1	0	0	719
1976	0	0	0	0	7	152	244	240	81	3	0	0	727
1977	0	0	0	0	16	241	342	256	130	3	0	0	988
1978	0	0	0	0	34	177	372	221	100	6	0	0	910
1979	0	0	0	1	6	145	283	173	117	17	0	0	742
1980	0	0	0	0	8	257	433	290	113	7	2	0	1110
1981	0	0	0	18	17	258	318	180	77	1	0	0	869
1982	0	0	0	4	8	90	255	251	97	6	0	0	705
1983	0	0	0	0	9	98	325	363	172	6	0	0	973
1984	0	0	0	0	50	170	282	233	88	2	0	0	825
1985	0	0	0	1	49	188	321	263	100	0	0	0	922

TABLE 6   SNOWFALL (inches)   CLAYTON, NEW MEXICO

SEASON	JULY	AUG	SEP	OCT	NOV	DEC	JAN	FEB	MAR	APR	MAY	JUNE	TOTAL
1956-57	0.0	0.0	0.0	0.0	0.4	T	2.4	0.1	14.9	0.6	0.0	0.0	18.4
1957-58	0.0	0.0	0.0	T	8.0	0.0	2.7	5.4	16.0	5.7	0.0	0.0	37.8
1958-59	0.0	0.0	0.0	T	4.0	8.1	7.3	1.0	14.0	3.8	0.0	0.0	38.2
1959-60	0.0	0.0	0.0	0.4	1.0	5.1	4.4	5.1	1.4	T	0.0	0.0	17.4
1960-61	0.0	0.0	0.0	2.6	T	9.0	0.7	2.6	3.1	2.3	0.0	0.0	20.3
1961-62	0.0	0.0	0.0	0.0	14.8	1.2	3.0	0.6	0.8	T	0.0	0.0	20.4
1962-63	0.0	0.0	0.0	0.0	1.2	0.8	T	5.5	0.6	0.0	0.0	0.0	8.1
1963-64	0.0	0.0	0.0	T	0.0	1.6	T	7.9	1.2	1.7	0.0	0.0	12.4
1964-65	0.0	0.0	0.0	T	1.4	2.7	2.0	5.3	5.0	T	T	0.0	16.4
1965-66	0.0	0.0	0.0	0.0	0.0	5.2	6.1	1.5	0.6	T	0.0	0.0	13.4
1966-67	0.0	0.0	0.0	1.7	0.0	2.2	1.4	0.6	1.1	0.0	T	0.0	7.0
1967-68	0.0	0.0	0.0	0.7	2.2	7.0	1.0	1.6	7.0	0.6	0.0	0.0	19.5
1968-69	0.0	0.0	0.0	2.3	2.1	T	3.8	15.1	3.6	0.6	0.0	0.0	27.5
1969-70	0.0	0.0	0.0	3.9	1.8	3.6	T	3.8	11.9	3.2	0.0	0.0	28.2
1970-71	0.0	0.0	T	5.2	1.5	T	2.3	6.3	4.4	T	0.0	0.0	20.5
1971-72	0.0	0.0	1.1		0.5	6.6	1.5	0.3	6.3	1.3	0.0	0.0	17.6
1972-73	0.0	0.0	0.0	1.0	13.5	1.7	5.7	2.2	14.4	5.6	T	0.0	44.1
1973-74	0.0	0.0	T	T	1.6	5.8	5.8	1.0	1.0	T	0.0	0.0	17.4
1974-75	0.0	0.0	0.0	0.0	0.6	2.5	6.5	3.3	1.0	1.3	0.0	0.0	15.2
1975-76	0.0	0.0	0.0	0.0	2.3	2.3	1.5	2.0	1.0	0.0	0.0	0.0	8.7
1976-77	0.0	0.0	T	T	3.9	0.5	3.8	1.0	0.6	2.0	0.0	0.0	11.8
1977-78	0.0	0.0	0.0	0.0	1.3	0.2	4.7	9.5	1.8	1.0	8.0	0.0	26.5
1978-79	0.0	0.0	0.0	0.0	4.5	11.0	6.9	0.2	0.6	6.1	1.7	0.0	31.0
1979-80	0.0	0.0	0.0	2.7	2.3	3.8	1.9	1.0	5.6	4.0	0.0	0.0	21.3
1980-81	0.0	0.0	0.0	0.0	6.5	1.5	1.3	0.7	1.0	0.5	T	0.0	11.5
1981-82	0.0	0.0	0.0	T	T	4.0	0.8	5.3	3.0	0.8	T	0.0	13.9
1982-83	0.0	0.0	0.0	0.0	4.2	8.6	12.0	8.3	9.7	8.2	0.2	0.0	51.2
1983-84	0.0	0.0	0.0	0.0	1.0	7.6	6.4	3.0	16.0	T	0.0	0.0	34.0
1984-85	0.0	0.0	5.0	8.0	1.3	5.0	9.8	0.8	5.0	0.0	0.0	0.0	34.9
1985-86	0.0	0.0	T	0.0	1.3	1.5							
Record Mean	0.0	0.0	0.2	0.6	2.7	3.3	3.7	3.1	5.0	1.8	0.3	0.0	20.7

See Reference Notes, relative to all above tables, on preceding page.

The climate at Roswell conforms to the basic trend of the four seasons, but shows certain deviations related to geography. Higher landmasses almost surround the valley location, with a long, gradual descent from points southwest through west and north. The topography acts to modify air masses, especially the cold outbreaks in wintertime. Downslope warming of air, as well as air interchange within a tempering environment, often prevents sharp cooling. Moreover, the elevation of 3,600 feet is high enough to moderate the heat and humidity compared to locations to the south and east.

Summer moves into a wet phase that delivers the most important rain of the year. Rather frequent showers and thunderstorms from June through September account for over half of the annual precipitation. Storm clouds that build up from the heat of the day, overspread the sky on many afternoons, retarding a further rise in temperature. At the same time, relative humidity shows moderation, ranging from about 70 percent in early morning to 30 percent in the mid-afternoon. Temperatures are quite warm on most summer days with readings of 100 degrees or higher occurring on 10 days in an average year.

Rainfall tapers off markedly in the fall with decline in storm activity. This leaves usually agreeable conditions because of low wind movement and mostly clear skies. Frosty nights alternate with warm days. Relative humidity reaches rather low levels in autumn, but dryness is not as rigorous as in the spring.

In winter, sub-freezing at night is tempered by considerable warming during the day. Zero or lower temperatures occur on only one day in an average winter. Sub-zero cold spells are of short duration. Winter is the season of least precipitation.

Spring ushers in the driest season of the year with respect to relative humidity. Wind movement shows a large increase, especially from the plateau areas of the west. Most of the 60 days a year with winds of 25 mph or more occur from February to May. Destructive storms seldom strike the city, but minor damage results from thundersqualls or hailstorms about once a year. Rain is most erratic in spring, ranging from none of consequence in some years, to excessive amounts in others.

TABLE 1 NORMALS, MEANS AND EXTREMES

ROSWELL, NEW MEXICO

LATITUDE: 33°18'N   LONGITUDE: 104°32'W   ELEVATION: FT. GRND 3649 BARO 03653   TIME ZONE: MOUNTAIN   WBAN: 23009

	(a)	JAN	FEB	MAR	APR	MAY	JUNE	JULY	AUG	SEP	OCT	NOV	DEC	YEAR
TEMPERATURE °F:														
Normals														
-Daily Maximum		55.4	60.4	67.7	76.9	85.0	93.1	93.7	91.3	84.9	75.8	63.1	56.7	75.3
-Daily Minimum		27.4	31.4	37.9	46.8	55.6	64.8	69.0	67.0	59.6	47.5	35.0	28.2	47.5
-Monthly		41.4	45.9	52.8	61.9	70.3	79.0	81.3	79.2	72.3	61.7	49.0	42.5	61.4
Extremes														
-Record Highest	13	82	84	90	92	101	109	105	104	100	95	86	81	109
-Year		1975	1979	1978	1978	1978	1981	1980	1980	1983	1979	1980	1981	JUN 1981
-Record Lowest	13	-9	3	18	23	34	51	59	57	42	25	4	-8	-9
-Year		1979	1985	1980	1973	1975	1983	1973	1976	1973	1980	1976	1978	JAN 1979
NORMAL DEGREE DAYS:														
Heating (base 65°F)		732	535	386	134	11	0	0	0	10	143	477	698	3126
Cooling (base 65°F)		0	0	8	41	176	420	508	440	229	41	0	0	1863
% OF POSSIBLE SUNSHINE	7	60	68	75	77	80	83	77	73	72	77	73	71	74
MEAN SKY COVER (tenths)														
Sunrise - Sunset	9	5.3	4.7	4.4	4.2	4.3	3.5	4.6	4.5	4.8	3.6	3.9	4.2	4.3
MEAN NUMBER OF DAYS:														
Sunrise to Sunset														
-Clear	9	11.3	12.1	14.2	14.3	13.4	17.2	11.1	12.8	12.8	18.1	15.8	14.9	168.1
-Partly Cloudy	9	8.8	7.8	9.1	8.9	11.8	9.9	14.7	12.2	7.9	5.9	7.1	9.0	113.0
-Cloudy	9	10.9	8.3	7.7	6.8	5.8	2.9	5.2	6.0	9.3	7.0	7.1	7.1	84.1
Precipitation														
.01 inches or more	13	4.8	3.3	2.6	3.1	3.8	4.8	5.5	7.2	7.2	4.5	3.0	3.3	53.0
Snow,Ice pellets														
1.0 inches or more	13	1.1	1.0	0.3	0.3	0.0	0.0	0.0	0.0	0.0	0.1	0.7	0.9	4.4
Thunderstorms	13	0.2	0.0	0.5	2.0	4.4	6.6	7.2	7.5	3.9	2.0	0.5	0.0	34.8
Heavy Fog Visibility 1/4 mile or less	13	3.3	2.8	0.5	0.2	0.5	0.2	0.2	0.2	0.8	2.2	2.8	2.2	15.8
Temperature °F														
-Maximum														
90° and above	13	0.0	0.0	0.1	0.2	7.5	20.9	24.5	21.2	9.2	1.0	0.0	0.0	84.5
32° and below	13	3.0	0.7	0.1	0.0	0.0	0.0	0.0	0.0	0.0	0.0	0.5	2.2	6.5
-Minimum														
32° and below	13	24.8	18.2	6.9	2.2	0.0	0.0	0.0	0.0	0.0	1.1	11.2	25.0	89.5
0° and below	13	0.3	0.0	0.0	0.0	0.0	0.0	0.0	0.0	0.0	0.0	0.0	0.3	0.6
AVG. STATION PRESS. (mb)	13	891.0	890.1	886.6	887.2	886.8	888.2	890.6	891.0	891.2	891.2	890.7	891.2	889.6
RELATIVE HUMIDITY (%)														
Hour 05	13	71	65	56	53	59	63	67	72	75	70	67	66	65
Hour 11	13	53	45	34	30	33	35	42	45	50	46	46	46	42
Hour 17 (Local Time)	13	43	33	24	23	25	26	32	37	42	37	40	40	34
Hour 23	13	64	55	43	41	44	46	53	58	63	60	58	59	54
PRECIPITATION (inches):														
Water Equivalent														
-Normal		0.24	0.28	0.27	0.37	0.77	0.91	1.38	2.17	1.72	0.99	0.33	0.27	9.70
-Maximum Monthly	13	0.85	1.06	1.48	2.48	3.35	4.55	6.27	6.48	6.58	3.81	1.57	1.62	6.58
-Year		1980	1975	1973	1985	1981	1981	1981	1974	1980	1974	1984	1982	SEP 1980
-Minimum Monthly	13	0.04	T	0.00	0.02	T	0.03	0.01	0.34	0.15	T	T	0.00	0.00
-Year		1984	1984	1980	1978	1974	1974	1980	1985	1979	1980	1979	1976	MAR 1980
-Maximum in 24 hrs	13	0.64	0.37	1.43	2.24	1.77	3.05	4.91	3.94	2.71	1.69	1.33	0.64	4.91
-Year		1982	1975	1973	1985	1981	1981	1981	1977	1980	1983	1984	1982	JUL 1981
Snow,Ice pellets														
-Maximum Monthly	13	7.3	7.5	4.8	5.3	T					4.2	12.3	8.0	12.3
-Year		1980	1973	1984	1983	1978	1973	1973	1973	1973	1976	1980	1982	NOV 1980
-Maximum in 24 hrs	13	7.3	5.4	4.8	4.0	T	T				3.1	6.3	5.0	7.3
-Year		1980	1979	1984	1983	1985	1985	1973	1973	1973	1976	1980	1978	JAN 1980
WIND:														
Mean Speed (mph)	13	7.9	8.6	10.5	10.4	9.8	9.6	8.5	7.8	8.1	8.0	7.8	7.6	8.7
Prevailing Direction														
Fastest Mile														
-Direction (!!!)	9	NW	NW	NW	SW	NW	NW	NE	NW	NE	22	NE	SW	NW
-Speed (MPH)	9	47	56	52	48	60	73	42	44	40	44	65	58	73
-Year		1973	1977	1973	1977	1977	1975	1977	1974	1976	1975	1975	1977	JUN 1975
Peak Gust														
-Direction (!!!)														
-Speed (mph)														
-Date														

See Reference Notes to this table on the following page.

TABLE 2 PRECIPITATION (inches)   ROSWELL, NEW MEXICO

YEAR	JAN	FEB	MAR	APR	MAY	JUNE	JULY	AUG	SEP	OCT	NOV	DEC	ANNUAL
1956	0.02	1.42	0.03	0.03	0.40	0.04	0.54	1.13	0.16	0.54	T	0.04	4.35
1957	0.09	0.64	0.80	0.31	0.43	0.06	0.87	1.23	1.18	2.91	0.80	0.00	9.32
1958	1.57	0.84	1.93	0.84	0.77	0.20	0.66	1.27	3.56	0.98	0.19	0.25	13.06
1959	0.02	0.10	0.03	0.59	1.44	0.82	2.98	1.87	0.16	0.52	0.24	0.75	9.52
1960	1.26	0.43	0.04	T	1.03	1.24	3.31	0.16	0.45	3.53	T	2.12	13.57
1961	0.68	0.04	0.81	0.02	0.44	0.62	1.08	1.37	0.44	0.44	1.62	0.29	7.85
1962	0.38	0.51	0.12	0.09	0.21	0.97	3.44	1.31	3.51	0.50	0.62	0.15	11.81
1963	0.44	0.77	0.00	0.16	0.88	0.60	0.21	2.26	0.62	0.15	0.05	0.16	6.30
1964	0.80	1.25	0.15	0.02	0.30	1.10	0.17	0.57	2.05	T	0.33	0.24	6.98
1965	0.12	0.84	0.21	0.38	0.35	1.09	1.50	0.83	0.76	0.05	0.08	0.47	6.68
1966	0.53	0.03	0.25	1.97	0.54	2.35	0.15	2.89	0.97	T	T	T	9.68
1967	0.00	0.20	0.07	T	0.11	3.55	0.97	4.00	0.85	0.02	0.22	1.07	11.06
1968	1.50	1.17	1.93	0.06	0.57	0.60	0.50	2.67	0.10	0.41	1.11	0.22	15.84
1969	0.01	0.47	1.14	0.44	0.10	0.35	1.32	0.71	2.67	4.34	T	1.78	13.33
1970	0.01	0.28	0.51	0.02	0.48	2.72	2.07	0.52	0.97	0.78	0.09	0.18	8.63
1971	0.18	0.23	0.11	0.26	T	0.18	1.88	3.62	1.57	0.76	0.45	0.80	10.04
#1972	0.20	0.00	0.03	0.00	0.16	2.06	5.43	3.35	3.25	1.27	0.49	0.26	16.50
1973	0.73	0.92	1.48	0.15	0.73	0.97	2.26	1.27	2.55	0.51	0.01	0.02	11.60
1974	0.24	0.01	0.11	0.50	T	0.03	0.31	6.48	6.47	3.81	0.09	0.60	18.65
1975	0.20	1.06	0.27	0.29	0.13	0.57	2.75	1.28	2.83	0.16	T	0.05	9.59
1976	0.12	0.22	0.24	0.79	0.82	1.55	2.44	1.98	2.29	0.69	0.41	0.00	11.55
1977	0.07	0.36	0.27	1.25	2.43	0.25	0.46	4.45	0.29	0.62	0.48	0.02	10.95
1978	0.50	0.48	0.39	0.02	1.81	4.31	0.52	3.49	3.58	1.47	1.25	0.43	18.25
1979	0.41	0.44	0.13	0.32	1.25	1.56	1.44	2.28	0.15	0.18	T	0.37	8.53
1980	0.85	0.19	0.00	1.06	0.85	0.29	0.01	2.45	6.58	T	0.77	0.15	13.20
1981	0.27	0.17	0.10	0.79	3.35	4.55	6.27	4.73	2.70	1.02	0.25	0.13	24.33
1982	0.66	0.20	0.12	0.41	0.20	0.76	1.03	0.93	2.00	0.20	0.92	1.62	9.05
1983	0.50	0.22	0.11	0.64	0.93	0.67	0.37	0.80	0.42	3.43	1.52	0.42	10.03
1984	0.04	T	0.46	0.03	1.62	4.52	0.85	5.03	1.04	2.74	1.57	0.85	18.75
1985	0.37	0.04	0.70	2.48	2.22	2.58	2.71	0.34	1.93	0.98	0.12	0.07	14.54
Record Mean	0.42	0.45	0.51	0.71	1.07	1.42	1.96	1.87	1.89	1.14	0.57	0.50	12.51

TABLE 3 AVERAGE TEMPERATURE (deg. F)   ROSWELL, NEW MEXICO

YEAR	JAN	FEB	MAR	APR	MAY	JUNE	JULY	AUG	SEP	OCT	NOV	DEC	ANNUAL
1956	42.1	37.4	52.5	57.7	71.5	81.1	80.5	78.3	73.5	63.1	44.6	41.7	60.3
1957	41.6	52.3	51.5	57.0	66.6	77.9	82.6	79.4	70.4	58.1	44.8	43.7	60.5
1958	39.5	45.0	46.3	58.3	70.3	81.3	83.1	80.5	71.3	58.6	49.1	40.5	60.3
1959	39.2	43.6	49.7	59.9	70.0	78.8	79.3	80.8	73.1	59.0	45.0	42.4	60.1
#1960	37.5	40.0	51.1	62.4	69.0	80.0	78.9	78.9	71.2	59.8	47.8	29.3	58.8
1961	34.3	42.4	49.9	58.0	69.3	76.4	78.1	76.8	69.6	59.1	40.4	40.0	57.9
1962	33.7	48.8	47.4	60.2	70.5	75.4	78.2	77.4	69.9	58.4	46.0	40.2	58.8
1963	31.9	43.0	50.3	61.3	69.6	76.7	81.9	78.5	72.4	63.2	47.9	35.1	59.3
1964	34.7	35.3	48.6	58.2	69.8	76.7	81.7	80.4	72.1	60.9	48.1	39.7	58.8
1965	44.2	40.1	45.3	62.0	69.0	77.0	80.9	76.7	70.7	59.0	52.5	42.0	60.0
1966	30.8	38.5	51.6	59.8	70.0	76.6	84.3	76.5	70.9	56.8	50.6	39.5	58.8
1967	41.2	44.3	57.2	65.5	68.5	75.3	80.0	75.0	68.8	59.3	47.5	35.3	59.8
1968	39.1	43.7	47.7	55.5	66.7	77.0	75.6	75.6	67.2	60.3	46.3	38.2	57.6
#1969	45.0	44.6	43.1	61.9	69.1	77.1	82.9	81.4	71.9	58.1	47.1	41.0	60.3
1970	37.0	46.1	46.6	58.0	67.7	74.3	80.3	78.3	71.0	55.7	48.5	43.4	58.9
1971	41.3	42.9	52.9	58.5	68.6	78.1	79.7	74.9	69.3	59.4	48.4	40.8	59.6
#1972	40.6	45.7	56.2	63.4	68.3	77.5	79.0	76.0	70.8	59.1	42.9	39.5	59.9
1973	39.0	43.6	52.3	54.5	66.2	75.3	78.8	77.6	69.7	61.4	51.1	41.9	59.3
1974	40.4	43.3	57.6	61.2	73.5	79.2	80.0	74.5	64.4	59.2	47.5	39.5	60.1
1975	41.3	42.3	49.9	56.6	65.9	76.7	77.0	78.3	68.5	62.5	49.5	43.4	59.3
1976	40.4	53.3	54.1	63.7	68.7	79.3	78.6	80.3	71.2	56.2	42.7	39.3	60.7
1977	38.6	48.2	52.1	62.3	73.3	81.6	84.2	83.0	78.1	64.1	53.1	47.0	63.9
1978	36.0	43.6	55.6	66.2	71.5	79.3	83.4	78.0	69.2	60.3	49.0	37.2	60.8
1979	34.9	43.6	50.5	60.6	67.5	75.1	81.0	76.5	72.2	63.8	44.6	40.6	59.3
1980	39.9	44.6	51.1	57.7	68.0	83.5	85.4	78.7	71.6	57.9	43.8	44.0	60.5
1981	41.6	46.2	51.7	63.5	68.5	79.0	79.9	75.7	69.9	59.8	53.7	45.3	61.3
1982	38.9	42.7	53.4	60.5	68.0	76.7	80.9	80.5	73.6	59.4	47.0	37.8	60.0
1983	38.9	45.5	52.8	54.7	66.3	76.5	81.6	81.4	76.1	62.9	51.8	35.2	60.3
1984	38.5	46.5	51.7	60.2	72.5	75.8	79.1	76.2	69.3	57.7	48.5	42.1	59.8
1985	37.0	43.2	54.2	63.4	70.1	75.8	79.4	80.4	70.7	61.1	53.2	40.3	60.7
Record Mean	39.5	44.0	51.2	59.6	68.2	76.7	79.2	77.8	70.8	59.8	47.8	40.1	59.5
Max	54.5	59.7	67.5	75.9	83.9	92.0	92.7	91.3	84.8	75.2	63.4	54.9	74.7
Min	24.4	28.2	34.8	43.2	52.4	61.3	65.6	64.2	56.8	44.5	32.2	25.2	44.4

REFERENCE NOTES FOR TABLES 1, 2, 3 and 6   (ROSWELL, NM)

GENERAL

T - TRACE AMOUNT
BLANK ENTRIES DENOTE MISSING/UNREPORTED DATA.
# INDICATES A STATION OR INSTRUMENT RELOCATION.

SPECIFIC

TABLE 1

(a) - LENGTH OF RECORD IN YEARS. ALTHOUGH
INDIVIDUAL MONTHS MAY BE MISSING.
 * LESS THAN .05

NORMALS — BASED ON THE 1951-1980 RECORD PERIOD.
EXTREMES — DATES ARE THE MOST RECENT OCCURRENCE.
WIND DIR. — NUMERALS SHOW TENS OF DEGREES
CLOCKWISE FROM TRUE NORTH.
"00" INDICATES CALM.
RESULTANT WIND DIRECTIONS ARE GIVEN TO WHOLE DEGREES.

EXCEPTIONS

TABLE 1

1. PERCENT OF POSSIBLE SUNSHINE, SKY COVER, AND
DAYS CLEAR-PARTLY CLOUDY-CLOUDY ARE THROUGH
1981.
2. FASTEST MILE WINDS ARE THROUGH MAY 1982.

TABLES 2, 3, and 6

RECORD MEANS ARE THROUGH THE CURRENT YEAR,
BEGINNING IN   1895 FOR TEMPERATURE
1895 FOR PRECIPITATION
1949 FOR SNOWFALL

TABLE 4   HEATING DEGREE DAYS Base 65 deg. F   ROSWELL, NEW MEXICO

SEASON	JULY	AUG	SEP	OCT	NOV	DEC	JAN	FEB	MAR	APR	MAY	JUNE	TOTAL
1956-57	0	0	0	107	603	715	720	353	409	240	33	6	3186
1957-58	0	0	10	222	600	655	783	552	572	226	16	0	3636
1958-59	0	0	38	217	471	753	789	594	466	212	11	0	3551
1959-60	0	0	14	191	593	693	848	716	426	120	28	0	3629
#1960-61	0	0	10	178	507	1102	942	627	463	234	17	1	4081
1961-62	0	0	21	194	730	768	964	447	537	160	19	0	3840
1962-63	0	0	8	210	564	764	1020	610	450	131	22	0	3779
1963-64	0	0	4	59	508	921	936	853	501	209	18	2	4011
1964-65	0	0	21	145	501	782	637	693	606	126	22	0	3533
1965-66	0	0	34	200	372	708	1052	734	407	168	33	0	3708
1966-67	0	5	5	257	425	788	731	577	244	59	52	0	3143
1967-68	0	2	9	205	517	915	796	609	529	283	52	0	3917
#1968-69	0	0	26	171	603	823	615	564	672	109	38	2	3623
1969-70	0	0	0	257	529	738	861	523	565	214	55	17	3759
1970-71	0	0	56	298	490	662	729	612	384	206	30	0	3467
1971-72	0	0	82	183	495	743	750	551	273	109	25	0	3211
#1972-73	0	0	13	216	660	785	800	594	386	310	73	0	3837
1973-74	0	0	27	130	410	710	755	599	233	151	11	0	3026
1974-75	0	0	104	190	518	784	729	630	462	269	36	1	3723
1975-76	0	0	48	104	462	662	755	334	334	90	36	0	2825
1976-77	0	0	18	275	663	789	811	463	393	109	0	0	3521
1977-78	0	0	0	68	349	551	895	593	294	45	39	3	2837
1978-79	0	0	55	174	468	856	929	591	444	161	51	2	3731
1979-80	0	0	27	115	602	748	773	587	425	222	53	0	3552
1980-81	0	0	19	234	630	645	719	521	404	100	21	0	3293
1981-82	0	0	11	183	333	603	803	616	357	160	27	0	3093
1982-83	0	0	2	195	530	834	802	541	372	322	44	2	3644
1983-84	0	0	7	115	389	920	817	529	404	164	19	0	3364
1984-85	0	0	56	227	488	704	863	604	332	85	10	1	3370
1985-86	0	0	46	131	344	759							

TABLE 5   COOLING DEGREE DAYS Base 65 deg. F   ROSWELL, NEW MEXICO

YEAR	JAN	FEB	MAR	APR	MAY	JUNE	JULY	AUG	SEP	OCT	NOV	DEC	TOTAL
1969	0	0	0	21	174	372	560	516	215	52	0	0	1910
1970	0	0	0	9	146	303	482	420	244	16	0	0	1620
1971	0	0	14	17	146	402	463	315	220	17	0	0	1594
#1972	0	0	7	68	133	382	438	350	197	41	0	0	1616
1973	0	0	0	2	118	316	434	400	174	25	0	0	1469
1974	0	0	11	45	283	433	473	299	95	16	0	0	1655
1975	0	0	0	19	67	361	379	416	159	33	3	0	1437
1976	0	0	7	57	159	436	429	478	211	9	0	0	1786
1977	0	0	0	34	269	505	602	564	408	46	0	0	2428
1978	0	0	8	91	244	438	576	410	187	37	0	0	1991
1979	0	0	0	35	133	311	506	364	251	84	0	0	1684
1980	0	0	0	8	152	564	636	430	223	21	1	0	2035
1981	0	0	0	53	133	430	467	336	166	31	0	0	1622
1982	0	0	6	35	124	357	499	488	266	27	0	0	1802
1983	0	0	0	21	89	353	523	512	346	58	0	0	1902
1984	0	0	2	26	256	331	443	356	189	6	0	0	1609
1985	0	0	8	44	175	332	457	485	224	18	0	0	1743

TABLE 6   SNOWFALL (inches)   ROSWELL, NEW MEXICO

SEASON	JULY	AUG	SEP	OCT	NOV	DEC	JAN	FEB	MAR	APR	MAY	JUNE	TOTAL
1956-57	0.0	0.0	0.0	0.0	T	T	0.0	0.0	T	T	0.0	0.0	T
1957-58	0.0	0.0	0.0	0.0	3.3	0.0	5.5	0.8	0.9	1.4	0.0	0.0	11.9
1958-59	0.0	0.0	0.0	0.0	T	2.5	0.2	T	0.3	T	0.0	0.0	3.0
1959-60	0.0	0.0	0.0	0.0	0.0	3.1	4.5	4.4	T	0.0	0.0	0.0	12.0
1960-61	0.0	0.0	0.0	0.0	0.0	20.0	5.4	0.4	1.4	0.0	0.0	0.0	27.2
1961-62	0.0	0.0	0.0	0.0	12.2	0.3	1.2	T	1.2	0.0	0.0	0.0	14.9
1962-63	0.0	0.0	0.0	0.0	0.6	0.1	4.5	4.8	0.0	0.0	0.0	0.0	10.0
1963-64	0.0	0.0	0.0	0.0	0.0	1.0	1.1	11.3	0.9	0.0	0.0	0.0	14.3
1964-65	0.0	0.0	0.0	0.0	0.0	1.9	1.2	7.8	1.7	0.0	0.0	0.0	12.6
1965-66	0.0	0.0	0.0	0.0	0.0	2.7	5.2	T	T	0.0	0.0	0.0	7.9
1966-67	0.0	0.0	0.0	0.0	0.0	T	0.0	2.0	0.6	0.0	0.0	0.0	2.6
1967-68	0.0	0.0	0.0	0.0	0.8	6.4	2.9	4.0	9.9	0.0	0.0	0.0	24.0
1968-69	0.0	0.0	0.0	0.0	8.5	0.0	T	0.0	10.9	0.0	0.0	0.0	19.4
1969-70	0.0	0.0	0.0	0.0	T	13.9	T	2.0	4.9	0.0	0.0	0.0	20.8
1970-71	0.0	0.0	0.0	0.2	0.0	0.0	2.0	2.4	1.0	0.0	0.0	0.0	5.6
1971-72	0.0	0.0	0.0	0.0	0.0	5.5	1.9	0.0	0.0	0.0	0.0	0.0	7.4
#1972-73	0.0	0.0	0.0	0.6	2.6	4.8	6.8	7.5	0.5	1.1	0.0	0.0	23.9
1973-74	0.0	0.0	0.0	0.0	0.0	T	1.0	T	0.2	0.0	0.0	0.0	1.2
1974-75	0.0	0.0	0.0	0.0	0.0	1.4	1.5	3.7	1.4	0.0	0.0	0.0	8.0
1975-76	0.0	0.0	0.0	0.0	T	0.3	1.2	0.0	2.4	0.0	0.0	0.0	3.9
1976-77	0.0	0.0	0.0	4.2	4.1	0.0	0.4	0.4	0.9	0.0	0.0	0.0	10.0
1977-78	0.0	0.0	0.0	0.0	0.0	0.0	3.2	2.3	0.7	0.0	T	0.0	6.2
1978-79	0.0	0.0	0.0	0.0	0.5	7.7	0.7	6.4	0.0	0.0	0.0	0.0	15.3
1979-80	0.0	0.0	0.0	0.4	0.0	3.8	7.3	4.0	0.0	3.0	0.0	0.0	18.5
1980-81	0.0	0.0	0.0	T	12.3	0.6	5.3	0.0	T	0.0	0.0	0.0	18.2
1981-82	0.0	0.0	0.0	0.0	0.0	0.0	6.9	2.6	1.2	T	0.0	0.0	10.7
1982-83	0.0	0.0	0.0	0.0	7.1	8.0	2.4	1.0	T	5.3	0.0	0.0	23.8
1983-84	0.0	0.0	0.0	0.0	0.0	3.5	0.5	T	4.8		0.0	0.0	
1984-85	0.0	0.0	0.0	0.0	0.0	5.4	1.9	0.3	0.0	0.0	0.0	0.0	7.6
1985-86	0.0	0.0	0.0	0.0	0.0	1.1							
Record Mean	0.0	0.0	0.0	0.1	1.5	2.8	2.8	2.4	1.3	0.3	T	0.0	11.3

See Reference Notes, relative to all above tables, on preceding page.

Albany is located on the west bank of the Hudson River some 150 miles north of New York City, and 8 miles south of the confluence of the Mohawk and Hudson Rivers. The river-front portion of the city is only a few feet above sea level, and there is a tidal effect upstream to Troy. Eleven miles west of Albany the Helderberg escarpment rises to 1,800 feet. Between it and the Hudson River the valley floor is gently rolling, ranging some 200 to 500 feet above sea level. East of the city there is more rugged terrain 5 or 6 miles wide with elevations of 300 to 600 feet. Farther to the east the terrain rises more sharply. It reaches a north-south range of hills 12 miles east of Albany with elevations ranging to 2,000 feet.

The climate at Albany is primarily continental in character, but is subjected to some modification by the Atlantic Ocean. The moderating effect on temperatures is more pronounced during the warmer months than in winter when outbursts of cold air sweep down from Canada. In the warmer seasons, temperatures rise rapidly in the daytime. However, temperatures also fall rapidly after sunset so that the nights are relatively cool. Occasionally there are extended periods of oppressive heat up to a week or more in duration.

Winters are usually cold and sometimes fairly severe. Maximum temperatures during the colder winters are often below freezing and nighttime lows are frequently below 10 degrees. Sub-zero readings occur about twelve times a year. Snowfall throughout the area is quite variable and snow flurries are quite frequent during the winter. Precipitation is sufficient to serve the economy of the region in most years, and only occasionally do periods of drought exist. Most of the rainfall in the summer is from thunderstorms. Tornadoes are quite rare and hail is not usually of any consequence.

Wind velocities are moderate. The north-south Hudson River Valley has a marked effect on the lighter winds and in the warm months, average wind direction is usually southerly. Destructive winds rarely occur.

The area enjoys one of the highest percentages of sunshine in the entire state. Seldom does the area experience long periods of cloudy days and long periods of smog are rare.

Based on the 1951-1980 period, the average first occurrence of 32 degrees Fahrenheit in the fall is September 29 and the average last occurrence in the spring is May 7.

TABLE 1 — NORMALS, MEANS AND EXTREMES

ALBANY, NEW YORK

LATITUDE: 42°45'N   LONGITUDE: 73°48'W   ELEVATION: FT. GRND   275 BARO   296   TIME ZONE: EASTERN   WBAN: 14735

	(a)	JAN	FEB	MAR	APR	MAY	JUNE	JULY	AUG	SEP	OCT	NOV	DEC	YEAR
TEMPERATURE °F:														
Normals														
-Daily Maximum		30.2	32.7	42.5	57.6	69.5	78.3	83.2	80.7	72.8	61.5	47.8	34.6	57.6
-Daily Minimum		11.9	14.0	24.6	35.5	45.4	55.0	59.6	57.6	49.6	39.4	30.8	18.2	36.8
-Monthly		21.1	23.4	33.5	46.5	57.5	66.7	71.4	69.2	61.2	50.5	39.3	26.4	47.2
Extremes														
-Record Highest	39	62	67	85	92	94	99	100	99	100	89	82	71	100
-Year		1974	1976	1977	1976	1981	1952	1953	1955	1953	1963	1950	1984	JUL 1953
-Record Lowest	39	-28	-21	-21	10	26	36	40	34	24	16	5	-22	-28
-Year		1971	1973	1948	1965	1968	1980	1978	1982	1947	1969	1972	1969	JAN 1971
NORMAL DEGREE DAYS:														
Heating (base 65°F)		1361	1165	973	552	252	38	7	15	149	450	771	1194	6927
Cooling (base 65°F)		0	0	0	0	19	89	206	145	35	0	0	0	494
% OF POSSIBLE SUNSHINE	47	45	51	53	53	55	59	64	60	57	51	36	38	52
MEAN SKY COVER (tenths)														
Sunrise - Sunset	47	7.0	6.9	7.0	6.9	6.9	6.5	6.2	6.2	6.0	6.3	7.5	7.4	6.7
MEAN NUMBER OF DAYS:														
Sunrise to Sunset														
-Clear	47	5.5	5.4	6.0	5.8	5.2	5.3	6.1	6.9	7.8	7.7	3.7	4.9	70.2
-Partly Cloudy	47	8.1	7.4	7.9	8.1	9.2	11.1	12.9	11.6	10.0	9.3	7.8	6.8	110.2
-Cloudy	47	17.4	15.4	17.1	16.1	16.6	13.6	12.0	12.5	12.2	14.0	18.5	19.3	184.8
Precipitation														
.01 inches or more	39	12.4	10.6	12.1	12.2	13.1	11.2	10.2	10.1	9.5	8.7	11.9	12.5	134.4
Snow, Ice pellets														
1.0 inches or more	39	3.9	3.3	2.7	0.7	0.1	0.0	0.0	0.0	0.0	0.*	1.1	4.0	15.8
Thunderstorms	47	0.1	0.2	0.5	1.3	3.6	5.6	6.7	4.7	2.4	0.9	0.3	0.1	26.4
Heavy Fog Visibility 1/4 mile or less	47	1.1	0.9	1.2	0.8	1.4	1.3	1.4	2.6	3.8	4.5	1.7	1.8	22.5
Temperature °F														
-Maximum														
90° and above	20	0.0	0.0	0.0	0.1	0.3	1.5	3.5	1.6	0.5	0.0	0.0	0.0	7.6
32° and below	20	18.6	13.1	4.0	0.3	0.0	0.0	0.0	0.0	0.0	0.0	1.3	11.8	49.0
-Minimum														
32° and below	20	29.6	26.0	24.5	13.4	2.3	0.0	0.0	0.0	0.6	8.9	18.0	27.6	150.9
0° and below	20	7.6	4.6	0.4	0.0	0.0	0.0	0.0	0.0	0.0	0.0	0.0	2.5	15.1
AVG. STATION PRESS. (mb)	13	1006.1	1007.2	1005.1	1004.0	1003.9	1004.4	1004.7	1006.6	1007.7	1008.4	1007.2	1007.3	1006.1
RELATIVE HUMIDITY (%)														
Hour 01	20	76	73	71	70	78	83	84	87	87	82	79	78	79
Hour 07	20	77	76	74	70	75	79	81	86	88	85	81	80	79
Hour 13 (Local Time)	20	63	59	54	48	53	56	55	58	59	57	63	66	58
Hour 19	20	71	66	61	55	60	64	63	70	75	72	73	74	67
PRECIPITATION (inches):														
Water Equivalent														
-Normal		2.39	2.26	3.01	2.94	3.31	3.29	3.00	3.34	3.23	2.93	3.04	3.00	35.74
-Maximum Monthly	39	6.44	5.02	5.90	7.95	8.96	7.36	6.96	7.33	7.89	8.83	8.07	6.73	8.96
-Year		1978	1981	1977	1983	1953	1973	1975	1950	1960	1955	1972	1973	MAY 1953
-Minimum Monthly	39	0.42	0.36	0.26	1.14	1.05	0.65	0.49	0.73	0.40	0.20	0.91	0.64	0.20
-Year		1980	1968	1981	1963	1980	1964	1968	1947	1964	1963	1978	1958	OCT 1963
-Maximum in 24 hrs	39	1.91	1.50	2.11	2.20	2.17	3.48	2.70	4.52	3.66	2.70	2.01	4.02	4.52
-Year		1978	1975	1977	1968	1968	1952	1960	1971	1960	1976	1959	1948	AUG 1971
Snow, Ice pellets														
-Maximum Monthly	39	40.8	34.5	34.7	17.7	1.6					2.0	24.6	57.5	57.5
-Year		1978	1962	1956	1982	1977					1952	1972	1969	DEC 1969
-Maximum in 24 hrs	39	21.2	17.9	17.0	17.5	1.6					2.0	21.9	18.3	21.9
-Year		1983	1958	1984	1982	1977					1952	1971	1966	NOV 1971
WIND:														
Mean Speed (mph)	47	9.8	10.3	10.7	10.6	9.0	8.2	7.5	7.0	7.4	8.0	9.0	9.3	8.9
Prevailing Direction through 1963		WNW	WNW	WNW	WNW	S	S	S	S	S	S	S	S	S
Fastest Mile														
-Direction	45	W	NW	W	W	W	NW	NW	W	S	NW	E	W	NW
-Speed (MPH)	45	57	71	55	49	50	57	43	40	48	45	70	54	71
-Year		1952	1953	1941	1954	1950	1971	1938	1981	1961	1939	1950	1944	FEB 1953
Peak Gust														
-Direction	2	W	W	W	SE	W	W	S	NW	W	S	W	NW	W
-Speed (mph)	2	55	56	51	44	40	43	35	33	47	44	44	56	56
-Date		1985	1985	1984	1985	1984	1984	1985	1984	1984	1985	1984	1985	FEB 1985

See Reference Notes to this table on the following page.

TABLE 2 PRECIPITATION (inches) ALBANY, NEW YORK

YEAR	JAN	FEB	MAR	APR	MAY	JUNE	JULY	AUG	SEP	OCT	NOV	DEC	ANNUAL
1956	2.25	3.30	4.76	2.62	3.08	1.86	2.76	2.76	4.71	1.19	2.30	3.59	35.18
1957	1.16	1.02	1.54	2.25	4.94	2.36	2.17	1.66	1.61	2.57	2.24	4.12	27.64
1958	4.12	2.71	2.45	3.10	2.11	2.06	4.47	1.21	3.68	3.07	0.64	2.77	32.39
1959	2.75	1.59	3.20	2.46	2.09	1.75	1.67	2.10	1.88	5.60	4.44	2.86	32.39
1960	2.51	2.85	1.69	4.19	3.33	2.70	5.89	3.20	1.58	7.89	1.17	1.35	38.35
1961	1.47	2.47	3.11	3.09	4.44	2.97	4.78	4.76	2.47	1.22	2.98	1.96	35.72
1962	2.05	3.65	1.70	3.25	1.40	1.15	2.12	2.60	3.45	3.58	2.11	2.24	29.30
1963	2.38	1.84	3.45	1.14	1.90	2.94	1.20	2.49	2.69	0.20	4.15	1.86	26.24
1964	3.35	1.63	2.93	2.17	1.31	0.65	1.29	1.29	2.55	0.54	1.45	3.28	21.55
1965	1.95	1.92	1.73	2.38	1.22	1.91	3.52	4.32	3.76	2.37	1.89	0.97	27.94
1966	2.29	2.71	3.63	1.46	2.35	2.95	3.88	1.44	5.61	2.22	1.79	3.04	33.37
1967	1.22	1.76	2.56	3.69	3.36	2.85	3.38	2.17	2.23	3.48	2.68	3.90	33.28
1968	1.48	0.36	2.62	2.64	4.79	4.38	0.49	1.77	1.49	2.18	5.48	4.60	32.28
1969	2.13	1.66	1.32	3.51	2.64	5.30	5.08	2.18	2.06	1.55	6.51	3.89	39.50
1970	0.81	1.98	2.87	3.01	1.78	3.14	1.93	3.35	3.79	2.49	1.48	3.89	30.52
1971	1.78	4.10	3.11	2.00	3.48	2.81	3.89	7.04	2.40	2.09	3.78	3.09	39.57
1972	1.21	3.04	4.05	3.63	5.98	6.84	3.10	1.48	1.99	3.60	8.07	4.19	47.18
1973	2.16	1.34	1.99	4.47	5.45	7.36	1.68	2.89	1.33	2.07	1.27	6.73	38.74
1974	2.04	2.12	3.10	2.80	3.47	3.31	4.84	3.53	5.37	1.49	3.83	2.57	38.47
1975	2.75	3.58	2.72	2.18	2.96	3.80	6.96	5.98	4.57	5.88	2.89	2.78	47.05
1976	3.78	2.60	3.57	3.63	4.89	5.37	2.60	5.04	2.61	5.65	1.41	1.39	42.54
1977	1.51	2.63	5.90	3.41	2.29	2.87	2.31	3.66	6.66	4.00	4.85	4.21	44.30
1978	6.44	0.88	1.99	1.68	1.96	4.60	4.04	3.06	1.87	2.95	0.91	3.08	33.46
1979	6.37	1.71	1.83	3.89	4.13	1.94	2.78	2.67	4.05	3.42	3.41	0.94	37.14
1980	0.42	0.89	4.44	3.02	1.05	4.90	2.69	6.45	2.24	2.27	2.99	1.23	32.59
1981	0.59	5.02	0.26	1.99	2.44	2.78	3.50	1.76	3.45	3.55	1.56	3.54	30.44
1982	3.18	2.14	3.23	2.46	2.60	6.48	2.43	2.01	1.42	0.99	3.80	1.33	32.07
1983	3.73	2.03	5.33	7.95	6.26	1.95	1.34	3.41	2.28	2.18	4.73	5.10	46.29
1984	1.28	2.98	3.04	4.29	7.92	1.74	3.97	3.25	1.53	2.50	2.15	2.48	37.13
1985	0.81	1.18	3.67	1.44	2.71	4.12	1.86	2.23	3.07	1.81	5.00	2.05	29.95
Record Mean	2.47	2.36	2.76	2.78	3.33	3.69	3.67	3.51	3.28	3.03	2.93	2.65	36.48

TABLE 3 AVERAGE TEMPERATURE (deg. F) ALBANY, NEW YORK

YEAR	JAN	FEB	MAR	APR	MAY	JUNE	JULY	AUG	SEP	OCT	NOV	DEC	ANNUAL
1956	23.5	26.1	27.9	42.6	53.5	67.3	69.0	69.0	57.7	49.8	39.9	30.4	46.4
1957	16.1	28.8	35.5	48.9	57.0	71.0	71.1	67.3	62.3	50.2	41.7	33.2	48.6
1958	22.1	18.9	36.0	49.1	54.4	62.1	70.2	70.2	61.5	48.9	39.9	19.2	46.2
1959	21.1	20.2	32.2	48.0	60.7	67.6	74.2	73.5	65.6	51.4	38.4	29.9	48.5
1960	22.7	28.0	26.5	49.6	60.4	67.1	69.5	69.9	62.8	49.4	41.5	21.6	47.4
1961	15.3	25.5	33.0	43.9	55.4	67.0	71.6	69.0	68.7	53.9	39.4	27.7	47.6
1962	21.9	20.5	34.7	47.0	59.8	68.1	69.0	69.0	58.7	50.2	35.0	23.0	46.4
1963	20.3	17.2	33.6	45.8	56.6	67.3	72.0	66.4	55.5	44.4		18.1	46.2
#1964	23.9	22.3	34.9	45.7	61.7	66.5	74.4	66.2	61.0	49.2	41.4	27.9	47.9
1965	18.1	22.3	31.2	42.2	59.6	66.0	68.9	69.4	63.6	51.2	37.6	30.8	46.8
1966	21.5	23.3	34.3	44.0	53.9	67.4	72.2	69.2	58.0	48.5	42.3	27.3	46.8
1967	27.0	18.0	29.0	43.5	50.4	69.9	71.6	69.3	61.3	51.0	34.8	28.9	46.2
1968	14.7	21.1	37.1	51.1	54.9	66.7	72.7	68.6	63.7	53.3	38.5	23.5	47.2
1969	20.9	24.7	31.1	47.6	56.3	66.0	69.7	70.6	62.4	49.0	39.7	21.6	46.6
1970	9.7	23.1	32.0	48.7	60.5	65.9	72.0	69.6	63.3	52.9	41.9	21.6	46.8
1971	13.9	25.4	30.6	42.3	54.9	66.3	68.4	66.8	64.8	54.7	36.9	30.0	46.3
1972	22.9	21.1	30.5	41.2	59.5	63.6	70.9	67.2	60.7	45.7	35.1	28.9	45.6
1973	27.0	22.0	41.9	48.8	55.3	68.7	72.9	72.9	61.5	51.0	39.9	28.2	49.1
1974	23.3	21.3	32.4	48.1	54.1	65.0	69.3	67.9	58.3	44.4	38.6	28.9	46.0
1975	25.7	24.9	30.8	40.7	61.9	66.1	72.8	70.0	59.4	53.3	45.5	26.1	48.0
1976	16.0	31.5	36.7	49.7	55.0	69.4	68.5	67.4	59.0	46.5	34.9	21.4	46.3
1977	15.5	24.5	40.0	46.8	60.2	64.6	71.7	67.8	61.4	49.7	42.6	26.7	47.6
1978	21.5	18.2	30.8	43.4	58.4	64.4	68.9	69.2	56.8	48.6	38.6	28.7	45.6
1979	22.1	14.4	38.9	45.4	60.0	66.0	72.5	69.0	61.2	50.2	41.4	31.4	47.9
1980	24.1	19.8	33.3	48.0	59.5	63.3	72.2	70.7	62.6	47.4	34.8	19.9	46.3
1981	14.0	33.1	34.7	48.1	58.9	66.7	69.3	68.5	58.8	44.8	37.7	25.7	46.7
1982	14.3	23.4	32.8	44.3	59.5	69.1	70.1	65.5	60.5	50.6	43.0	33.7	46.7
1983	24.3	26.8	37.6	46.7	54.9	67.2	72.2	69.8		49.6	39.2	24.0	47.9
1984	18.1	32.4	29.0	47.6	53.2	66.4	68.9	71.8	60.2	53.8	40.3	33.8	48.0
1985	19.9	26.8	37.3	49.7	60.0	62.2	70.7	68.7	63.3	50.2	40.1	24.5	47.8
Record Mean	22.6	23.9	33.6	46.5	58.4	67.4	70.1	70.1	62.6	51.3	39.6	27.6	48.0
Max	31.0	32.5	42.1	56.3	69.1	77.9	82.8	80.4	72.8	61.0	47.2	35.0	57.3
Min	14.3	15.2	25.1	36.6	47.7	56.9	61.7	59.7	52.3	41.5	32.0	20.1	38.6

REFERENCE NOTES FOR TABLES 1, 2, 3 and 6   (ALBANY, NY)

GENERAL

T – TRACE AMOUNT
BLANK ENTRIES DENOTE MISSING/UNREPORTED DATA.
# INDICATES A STATION OR INSTRUMENT RELOCATION.

SPECIFIC

TABLE 1

(a) – LENGTH OF RECORD IN YEARS, ALTHOUGH INDIVIDUAL MONTHS MAY BE MISSING.
* LESS THAN .05

NORMALS — BASED ON THE 1951-1980 RECORD PERIOD.
EXTREMES — DATES ARE THE MOST RECENT OCCURRENCE.
WIND DIR. — NUMERALS SHOW TENS OF DEGREES CLOCKWISE FROM TRUE NORTH.
   "00" INDICATES CALM.
RESULTANT WIND DIRECTIONS ARE GIVEN TO WHOLE DEGREES.

EXCEPTIONS

TABLES 2, 3, and 6

RECORD MEANS ARE THROUGH THE CURRENT YEAR, BEGINNING IN   1874 FOR TEMPERATURE
   1826 FOR PRECIPITATION
   1947 FOR SNOWFALL

TABLE 4 HEATING DEGREE DAYS Base 65 deg. F ALBANY, NEW YORK

SEASON	JULY	AUG	SEP	OCT	NOV	DEC	JAN	FEB	MAR	APR	MAY	JUNE	TOTAL
1956-57	22	16	262	464	749	1066	1511	1006	908	491	264	36	6795
1957-58	3	30	152	450	691	981	1323	1286	892	476	322	118	6724
1958-59	8	8	142	499	745	1415	1354	1249	1010	502	220	68	7220
1959-60	0	8	142	423	789	1080	1307	1067	1184	457	152	32	6641
1960-61	12	11	100	480	696	1338	1530	1101	987	627	314	47	7243
1961-62	11	18	79	335	761	1152	1327	1240	933	552	211	27	6646
1962-63	6	23	207	451	894	1297	1381	1331	968	571	261	62	7452
1963-64	18	29	251	296	612	1446	1266	1233	927	571	145	78	6872
#1964-65	1	48	169	484	702	1141	1449	1193	1041	679	197	68	7172
1965-66	11	49	120	421	817	1051	1342	1162	948	623	347	57	6948
1966-67	3	5	216	502	673	1163	1169	1312	1111	639	447	11	7251
1967-68	0	19	153	429	899	1112	1557	1269	857	412	304	46	7057
1968-69	7	45	76	359	787	1281	1360	1122	1043	518	284	55	6937
1969-70	13	22	137	491	749	1339	1708	1168	1016	495	165	75	7378
1970-71	3	7	127	377	686	1336	1580	1104	1059	672	315	50	7316
1971-72	20	45	109	311	838	1080	1298	1269	1060	707	175	97	7009
1972-73	16	38	154	590	890	1113	1168	1198	709	486	299	47	6708
1973-74	2	3	200	431	750	1136	1285	1216	1005	511	343	54	6936
1974-75	17	14	227	631	786	1113	1212	1115	1053	722	145	88	7123
1975-76	9	19	173	357	580	1199	1511	964	871	472	315	43	6504
1976-77	7	40	196	564	895	1345	1526	1127	764	545	205	85	7299
1977-78	7	51	156	471	666	1179	1340	1306	1051	642	245	84	7198
1978-79	43	19	256	503	784	1119	1324	1414	803	579	188	63	7095
1979-80	19	37	163	468	619	1036	1259	1303	974	503	190	106	6677
1980-81	0	7	140	539	900	1393	1575	885	930	502	235	30	7136
1981-82	8	22	204	622	816	1209	1564	1160	992	617	182	87	7483
1982-83	20	65	156	436	657	1255	1062	843		539	312	58	6372
1983-84	5	24	150	479	766	1265	1448	939	1109	517	363	60	7125
1984-85	12	8	170	344	737	959	1389	1062	852	458	184	106	6281
1985-86	7	16	123	452	740	1246							

TABLE 5 COOLING DEGREE DAYS Base 65 deg. F ALBANY, NEW YORK

YEAR	JAN	FEB	MAR	APR	MAY	JUNE	JULY	AUG	SEP	OCT	NOV	DEC	TOTAL
1969	0	0	0	2	23	95	165	203	68	0	0	0	556
1970	0	0	0	12	36	107	225	160	83	7	0	0	630
1971	0	0	0	9	12	98	132	107	109	1	0	0	456
1972	0	0	0	0	12	58	208	112	31	0	0	0	421
1973	0	0	0	7	6	164	248	255	71	2	0	0	753
1974	0	0	0	11	12	59	157	111	35	0	1	0	386
1975	0	0	0	0	58	97	248	180	12	0	2	0	597
1976	0	0	0	19	11	184	120	120	22	0	0	0	476
1977	0	0	0	0	8	66	79	222	146	53	0	0	574
1978	0	0	0	0	47	70	169	154	16	0	0	0	456
1979	0	0	0	0	39	99	258	168	55	17	0	0	636
1980	0	0	0	0	28	63	230	189	73	0	0	0	583
1981	0	0	0	2	53	87	149	137	25	0	0	0	453
1982	0	0	0	0	19	31	184	88	29	0	4	0	355
1983	0	0	0	0	8	134	236	179	86	6	0	0	649
1984	0	0	0	0	3	107	140	226	35	3	0	0	514
1985	0	0	0	5	37	27	191	140	80	2	0	0	482

TABLE 6 SNOWFALL (inches) ALBANY, NEW YORK

SEASON	JULY	AUG	SEP	OCT	NOV	DEC	JAN	FEB	MAR	APR	MAY	JUNE	TOTAL
1956-57	0.0	0.0	0.0	0.0	7.6	28.1	16.7	2.7	6.8	8.4	0.0	0.0	70.3
1957-58	0.0	0.0	0.0	0.2	T	7.0	26.1	26.0	15.1	T	0.0	0.0	74.4
1958-59	0.0	0.0	0.0	T	4.6	6.4	12.2	10.9	29.1	0.0	0.0	0.0	63.2
1959-60	0.0	0.0	0.0	0.0	0.9	6.9	20.4	13.5	18.3	0.1	0.0	0.0	60.1
1960-61	0.0	0.0	0.0	T	0.1	16.6	17.2	10.9	21.7	6.2	T	0.0	72.7
1961-62	0.0	0.0	0.0	0.0	3.6	14.4	2.3	34.5	3.2	4.6	0.0	0.0	62.6
1962-63	0.0	0.0	0.0	T	1.6	11.3	24.5	15.5	18.4	T	0.0	0.0	71.3
1963-64	0.0	0.0	0.0	0.0	T	21.0	27.3	21.4	7.3	T	0.0	0.0	77.0
1964-65	0.0	0.0	0.0	T	T	11.2	20.4	3.7	8.4	2.1	0.0	0.0	45.8
1965-66	0.0	0.0	0.0	T	0.5	2.7	28.8	24.5	9.2	T	1.4	0.0	67.1
1966-67	0.0	0.0	0.0	0.0	T	29.4	5.7	16.3	26.2	3.1	0.2	0.0	80.9
1967-68	0.0	0.0	0.0	T	9.0	17.8	8.0	1.8	5.6	0.0	0.0	0.0	42.2
1968-69	0.0	0.0	0.0	0.0	13.5	18.1	6.3	20.7	4.6	0.2	0.0	0.0	63.3
1969-70	0.0	0.0	0.0	T	3.2	57.5	7.2	7.4	11.2	1.2	T	0.0	87.7
1970-71	0.0	0.0	0.0	T	T	43.8	15.2	17.6	32.0	3.9	0.0	0.0	112.5
1971-72	0.0	0.0	0.0	0.0	24.0	10.1	8.5	24.8	15.9	6.0	0.0	0.0	89.3
1972-73	0.0	0.0	0.0	T	24.6	22.5	11.2	12.5	T	0.1	0.0	0.0	70.9
1973-74	0.0	0.0	0.0	0.0	0.1	18.9	10.0	12.4	5.6	11.3	0.0	0.0	58.3
1974-75	0.0	0.0	0.0	T	2.2	12.5	14.0	21.2	2.9	1.8	0.0	0.0	54.6
1975-76	0.0	0.0	0.0	0.0	3.6	16.4	15.0	4.4	14.8	T	T	0.0	54.2
1976-77	0.0	0.0	0.0	T	5.7	7.8	22.1	17.9	15.2	0.3	1.6	0.0	70.6
1977-78	0.0	0.0	0.0	0.0	8.4	19.8	40.8	15.8	7.4	0.2	T	0.0	92.4
1978-79	0.0	0.0	0.0	T	3.4	19.9	26.5	4.6	0.9	8.2	0.0	0.0	63.5
1979-80	0.0	0.0	0.0	T	0.0	5.8	0.6	10.2	10.8	0.0	0.0	0.0	27.4
1980-81	0.0	0.0	0.0	0.0	11.8	12.8	11.9	6.9	1.5	T	0.0	0.0	44.9
1981-82	0.0	0.0	0.0	0.0	1.1	31.4	18.2	9.6	19.1	17.7	0.0	0.0	97.1
1982-83	0.0	0.0	0.0	0.0	0.6	5.5	27.5	17.4	9.2	14.7	0.1	0.0	75.0
1983-84	0.0	0.0	0.0	0.0	1.7	11.6	16.5	7.2	28.2	T	0.0	0.0	65.2
1984-85	0.0	0.0	0.0	0.0	2.2	11.7	8.4	10.1	8.7	0.2	0.0	0.0	41.3
1985-86	0.0	0.0	0.0	0.0	11.8	11.5							
Record Mean	0.0	0.0	0.0	0.1	4.4	15.7	15.8	14.2	11.8	3.1	0.1	0.0	65.1

See Reference Notes, relative to all above tables, on preceding page.

Binghamton, in south central New York lies in a comparatively narrow valley at the confluence of the Susquehanna and Chenango Rivers. Within a radius of 5 miles, hills rise to elevations of 1,400–1,600 feet above mean sea level. In the spring, melting snow, sometimes supplemented by rainfall, occasionally causes flooding in the city and along the streams. Less frequently, heavy rains in the warmer months produce some flooding.

The climate of Binghamton is representative of the humid area of the north-eastern United States and is primarily continental in type. The area, being adjacent to the so-called St. Lawrence Valley storm track, and also subject to cold air masses approaching from the west and north, has a variable climate, characterized by frequent and rapid changes. Furthermore, diurnal and seasonal changes assist in the production of an invigorating climate. In the warmer months, it is seldom that either high temperatures or humidity become depressing to humans. As a rule, the temperature rises rapidly to moderate daytime levels with readings of 90 degrees or above only a few days in any month. Summer nights are sufficiently cool to provide favorable sleeping conditions and relief from the heat of the day.

Winters are usually cold, but not commonly severe. Highest daytime temperatures average in the high 20s to low 30s, while the lowest nighttime readings average from the mid-teens to low 20s. Ordinarily a few sub-zero readings may be expected in January and February, with a lesser number in November, December, and March. The transitional seasons, spring and autumn, are the most variable of the year.

Most of the precipitation in the Binghamton area derives from moisture laden air transported from the Gulf of Mexico and cyclonic systems moving northward along the Atlantic coast. The annual rainfall is rather evenly distributed over the year. However, the greatest average monthly amounts occur during the growing season, April through September. As a rule, rainfall is ample for good crop growth and comes mostly in the form of thunderstorms. Ordinarily, the requirements for water supplies are adequately met by the precipitation that is received. Annual snowfall is around 50 inches in Binghamton and above 85 inches at Edwin A. Link Field, some 10 miles to the NNW, and about 700 feet higher in elevation. Most of the snow falls during the normal winter months. However, heavy snows can occur as early as November and as late as April. Being adjacent to the track of storms that move through the St. Lawrence Valley, and being under the influence of winds that sweep across Lakes Erie and Ontario to the interior of the state, the area is subject to much cloudiness and winter snow flurries.

Furthermore, the combination of a valley location and surrounding hills produces numerous advection fogs which also reduce the amount of sunshine received.

For the most part, the winds at Binghamton have northerly and westerly components. Tornadoes, although rare, have struck in the Binghamton area.

The growing season averages 150 to 160 days. Usually the last spring frost occurs during early May, and the first frost in autumn during early October.

TABLE 1   NORMALS, MEANS AND EXTREMES

BINGHAMTON, NEW YORK

LATITUDE: 42°13'N   LONGITUDE: 75°59'W   ELEVATION: FT. GRND 1590 BARO 01618   TIME ZONE: EASTERN   WBAN: 04725

	(a)	JAN	FEB	MAR	APR	MAY	JUNE	JULY	AUG	SEP	OCT	NOV	DEC	YEAR
TEMPERATURE °F:														
Normals														
-Daily Maximum		28.0	29.6	38.7	53.5	64.9	73.9	78.4	76.4	68.9	57.6	44.4	32.4	53.9
-Daily Minimum		14.3	15.1	24.0	35.1	45.5	54.6	59.4	57.9	50.6	40.5	31.3	19.9	37.4
-Monthly		21.1	22.4	31.4	44.3	55.2	64.3	68.9	67.2	59.8	49.0	37.9	26.1	45.6
Extremes														
-Record Highest	34	63	66	82	85	89	94	96	94	96	82	77	65	96
-Year		1967	1954	1977	1985	1982	1952	1983	1985	1953	1963	1982	1984	JUL 1983
-Record Lowest	34	-20	-15	-6	9	25	33	39	37	25	17	3	-18	-20
-Year		1957	1979	1980	1982	1978	1980	1963	1965	1974	1976	1976	1980	JAN 1957
NORMAL DEGREE DAYS:														
Heating (base 65°F)		1358	1193	1042	621	313	79	16	35	178	493	813	1203	7344
Cooling (base 65°F)		0	0	0	0	9	58	137	104	22	0	0	0	330
% OF POSSIBLE SUNSHINE	34	37	42	44	51	56	61	65	60	56	48	30	28	48
MEAN SKY COVER (tenths)														
Sunrise - Sunset	34	8.0	7.9	7.7	7.2	7.1	6.7	6.4	6.6	6.5	6.8	8.1	8.4	7.3
MEAN NUMBER OF DAYS:														
Sunrise to Sunset														
-Clear	34	2.4	2.6	3.7	5.2	4.7	4.7	5.4	5.1	5.7	6.0	2.6	2.0	50.0
-Partly Cloudy	34	6.9	6.2	7.1	6.9	9.0	11.3	13.1	11.5	10.3	8.2	5.7	6.0	102.3
-Cloudy	34	21.7	19.5	20.2	18.0	17.4	13.9	12.5	14.4	14.0	16.8	21.7	23.1	213.0
Precipitation														
.01 inches or more	34	16.9	14.6	15.8	13.7	13.3	12.1	10.4	10.7	9.9	11.6	14.8	17.8	161.6
Snow, Ice pellets														
1.0 inches or more	34	5.9	4.7	3.7	1.2	0.1	0.0	0.0	0.0	0.0	0.1	2.3	5.8	23.9
Thunderstorms	34	0.1	0.1	1.1	2.2	3.5	6.6	6.5	5.2	2.8	1.1	0.3	0.2	29.8
Heavy Fog Visibility														
1/4 mile or less	34	3.3	3.2	5.1	4.0	4.1	4.3	4.5	5.0	5.2	4.2	4.5	4.8	52.3
Temperature °F														
-Maximum														
90° and above	34	0.0	0.0	0.0	0.0	0.0	0.2	1.0	0.6	0.2	0.0	0.0	0.0	2.1
32° and below	34	20.8	15.9	8.8	0.8	0.0	0.0	0.0	0.0	0.0	0.1	3.5	16.1	66.0
-Minimum														
32° and below	34	29.9	26.4	25.5	12.2	1.4	0.0	0.0	0.0	0.2	5.6	17.3	27.4	146.0
0° and below	34	3.9	2.6	0.3	0.0	0.0	0.0	0.0	0.0	0.0	0.0	0.0	1.3	8.1
AVG. STATION PRESS.(mb)	13	955.3	956.6	955.5	955.6	956.4	957.7	958.7	960.2	960.3	960.2	958.1	957.0	957.6
RELATIVE HUMIDITY (%)														
Hour 01	34	79	77	76	72	75	80	81	84	85	80	80	81	79
Hour 07	34	80	79	79	76	78	82	84	88	90	85	82	82	82
Hour 13 (Local Time)	34	71	67	63	56	56	58	57	60	63	62	69	74	63
Hour 19	34	74	71	68	61	61	66	65	69	73	69	74	78	69
PRECIPITATION (inches):														
Water Equivalent														
-Normal		2.54	2.33	2.94	3.07	3.19	3.60	3.48	3.35	3.32	3.00	3.04	2.92	36.78
-Maximum Monthly	34	6.39	4.36	6.00	8.57	6.46	9.46	7.40	7.48	9.66	9.43	7.52	6.11	9.66
-Year		1979	1971	1980	1983	1968	1960	1956	1959	1977	1955	1972	1983	SEP 1977
-Minimum Monthly	34	0.76	0.51	0.69	0.98	0.78	0.98	0.83	0.61	0.61	0.26	1.01	0.94	0.26
-Year		1970	1968	1981	1985	1962	1979	1955	1953	1961	1963	1960	1960	OCT 1963
-Maximum in 24 hrs	34	1.80	2.16	1.95	2.86	1.88	3.19	3.24	3.19	3.57	3.88	2.66	2.81	3.88
-Year		1958	1966	1964	1980	1984	1972	1976	1959	1985	1955	1972	1983	OCT 1955
Snow, Ice pellets														
-Maximum Monthly	34	41.0	44.3	33.5	22.9	3.4				T	2.6	24.4	59.6	59.6
-Year		1978	1972	1971	1983	1966				1970	1972	1954	1969	DEC 1969
-Maximum in 24 hrs	34	18.4	23.0	15.8	11.5	3.4				T	2.4	10.1	15.6	23.0
-Year		1964	1961	1971	1960	1966				1970	1972	1953	1969	FEB 1961
WIND:														
Mean Speed (mph)	34	11.7	11.7	11.9	11.5	10.1	9.2	8.4	8.3	8.8	9.8	10.9	11.4	10.3
Prevailing Direction through 1963		WSW	SSE	NW	WNW	NNW	NNW	WSW	SSW	SSW	WSW	NNW	WSW	WSW
Fastest Mile														
-Direction (!!!)	34	NW	NW	NW	W	W	NW	NW	N	NW	S	SE	NW	S
-Speed (MPH)	34	59	66	61	52	54	60	60	58	42	72	57	59	72
-Year		1959	1956	1955	1956	1954	1957	1983	1958	1960	1954	1951	1956	OCT 1954
Peak Gust														
-Direction (!!!)	2	NW	E	SW	W	SW	NW	NW	NW	NW	W	NW	SW	E
-Speed (mph)	2	38	52	44	52	52	38	46	45	36	39	40	48	52
-Date	2	1985	1985	1985	1985	1984	1984	1985	1984	1984	1984	1985	1985	FEB 1985

See Reference Notes to this table on the following page.

TABLE 2   PRECIPITATION (inches)   BINGHAMTON, NEW YORK

YEAR	JAN	FEB	MAR	APR	MAY	JUNE	JULY	AUG	SEP	OCT	NOV	DEC	ANNUAL
1956	1.97	3.87	5.11	3.84	3.13	3.17	7.40	4.41	3.01	2.58	1.30	4.05	43.84
1957	1.74	1.31	1.47	4.12	3.14	3.30	4.24	1.48	2.69	2.63	1.71	2.60	30.43
1958	4.31	2.93	4.29	3.01	2.87	4.91	4.43	2.69	2.47	2.47	1.03	3.23	37.64
1959	2.59	1.50	2.34	2.67	2.17	4.61	2.49	7.48	0.68	7.62	4.84	3.24	42.23
1960	2.36	3.98	1.56	3.02	4.45	9.46	2.63	4.27	5.57	1.78	1.01	0.94	41.03
1961	1.40	3.37	2.82	4.68	3.13	3.91	5.08	4.87	0.61	1.22	3.12	1.52	35.73
1962	2.43	2.27	2.31	3.34	0.78	2.57	1.88	2.93	3.45	7.15	1.61	1.89	32.61
1963	1.79	1.69	2.86	2.43	4.73	4.24	2.51	3.39	1.87	0.26	4.15	2.53	32.45
1964	3.00	2.00	4.56	5.09	2.01	1.22	4.80	1.85	0.66	1.06	3.18	1.33	31.33
1965	4.15	1.55	2.23	2.88	1.73	1.90	2.23	4.25	3.72	3.02	2.30	1.69	31.65
1966	3.19	3.62	2.86	2.51	2.56	2.90	1.45	1.41	3.26	1.40	2.87	2.75	30.78
1967	1.60	1.60	2.78	2.09	5.04	2.90	3.45	4.92	3.11	3.29	4.45	2.49	37.72
1968	2.26	0.51	3.25	1.61	6.46	6.96	1.66	2.26	5.49	3.14	5.62	3.17	42.39
1969	2.00	0.97	0.69	2.78	1.60	4.00	4.32	1.96	1.84	2.25	3.71	4.85	30.97
1970	0.76	2.22	2.41	3.58	3.03	1.15	4.50	3.97	3.85	2.07	2.45	3.58	33.57
1971	1.68	4.36	2.77	2.02	3.31	1.73	4.60	2.10	1.66	1.89	3.17	4.16	33.45
1972	1.29	3.74	3.79	2.83	5.17	9.18	1.68	3.79	2.03	2.17	7.52	4.85	48.04
1973	1.60	1.95	2.01	3.74	3.29	2.93	1.93	2.40	3.12	2.72	1.85	5.81	33.35
1974	2.20	1.70	4.05	2.02	3.19	3.48	4.23	1.69	2.98	0.98	3.51	3.05	33.08
1975	2.55	3.97	2.16	1.77	4.42	2.90	6.13	5.33	8.41	3.48	2.49	3.22	46.83
1976	3.69	2.88	2.78	2.69	2.53	4.42	6.40	6.79	3.85	6.30	1.12	1.71	45.16
1977	1.68	1.54	5.11	2.73	1.72	3.17	3.27	2.95	9.66	4.76	5.10	4.84	46.53
1978	6.06	1.26	2.36	1.92	2.55	3.85	2.54	4.61	1.16	3.57	1.29	3.16	34.33
1979	6.39	1.67	2.73	3.13	4.26	0.98	1.45	2.44	5.70	2.46	3.70	1.83	36.74
1980	1.08	1.08	6.00	5.48	1.54	5.68	2.09	1.58	2.81	2.86	2.96	1.60	34.76
1981	0.89	3.88	0.69	3.18	1.94	3.42	1.99	1.99	3.40	4.72	1.67	2.49	30.26
1982	3.40	2.26	2.61	2.29	3.89	7.09	1.87	2.94	1.86	0.93	4.04	1.90	35.08
1983	2.56	1.50	2.57	8.57	4.05	4.08	2.20	3.21	1.53	2.61	3.58	6.11	42.57
1984	1.59	3.34	2.19	5.07	6.09	2.65	5.44	3.07	1.92	1.58	3.55	3.15	39.64
1985	1.30	1.30	3.63	0.98	2.69	2.61	4.14	2.72	4.76	2.47	4.63	2.19	33.42
Record Mean	2.45	2.31	2.85	3.23	3.31	3.63	3.44	3.33	3.24	2.97	3.11	2.94	36.81

TABLE 3   AVERAGE TEMPERATURE (deg. F)   BINGHAMTON, NEW YORK

YEAR	JAN	FEB	MAR	APR	MAY	JUNE	JULY	AUG	SEP	OCT	NOV	DEC	ANNUAL
1956	21.0	25.0	26.7	40.6	50.9	65.1	65.5	66.1	55.9	50.3	38.0	31.1	44.7
1957	17.4	27.0	32.7	45.4	54.1	67.8	67.5	64.6	60.6	47.0	30.6	30.6	46.2
1958	21.2	17.0	30.2	44.6	51.8	58.6	68.3	67.6	59.6	48.3	39.5	19.5	43.8
1959	21.5	23.1	30.8	46.6	59.0	65.3	71.7	71.7	64.4	51.1	36.1	30.5	47.6
1960	25.7	26.1	22.8	48.2	56.8	64.5	66.8	68.0	62.2	48.6	40.6	20.8	45.9
1961	16.8	25.6	32.0	39.8	52.3	64.6	66.8	66.9	65.5	53.2	35.1	25.1	45.6
1962	19.8	18.9	31.0	44.3	59.1	64.5	66.1	66.9	56.9	49.2	34.8	21.6	44.4
1963	19.2	16.6	35.5	45.6	53.6	65.4	67.0	62.8	54.7	54.8	41.0	19.3	44.6
1964	24.0	18.9	31.9	43.0	58.4	62.0	70.3	65.2	61.2	48.3	42.9	28.0	46.2
1965	19.8	22.7	28.7	40.3	59.2	63.1	66.7	66.6	60.8	46.8	36.9	30.4	45.2
1966	19.1	22.8	33.2	41.5	50.6	65.9	71.2	69.4	56.8	46.9	41.1	26.5	45.4
1967	29.0	18.5	29.4	43.6	47.6	68.2	67.8	66.2	59.1	48.6	32.9	28.9	45.0
1968	17.1	17.3	34.5	48.4	51.3	62.9	69.4	67.1	61.8	49.9	36.7	23.6	45.0
1969	22.1	23.0	29.9	47.0	56.3	63.8	67.0	67.6	60.1	48.0	37.9	23.5	45.5
1970	16.5	25.0	28.9	45.0	57.0	62.8	67.9	67.2	60.6	50.9	39.2	24.2	45.5
1971	15.6	23.6	28.4	40.7	54.3	66.3	65.9	64.1	62.8	55.4	35.8	30.9	45.3
1972	24.7	20.6	28.2	40.4	58.8	62.2	71.5	65.8	59.9	42.9	33.6	29.4	44.8
1973	25.8	20.1	40.0	45.4	51.5	66.3	70.2	69.7	59.7	52.6	39.8	29.0	47.5
1974	27.3	22.2	32.1	48.7	52.5	63.2	69.7	68.7	56.9	44.9	37.3	27.9	46.0
1975	24.9	24.5	27.2	37.0	62.0	66.5	72.8	67.4	56.5	51.4	44.6	26.3	46.0
1976	16.9	30.3	35.5	47.0	52.8	67.0	65.7	65.5	57.3	43.7	31.1	20.3	44.4
1977	12.0	24.1	38.0	47.4	59.0	61.8	70.0	66.5	60.3	46.7	40.2	24.6	45.9
1978	19.1	15.0	27.6	41.5	58.6	63.7	68.4	69.5	58.3	47.6	39.2	27.9	44.7
1979	21.2	13.6	37.9	42.5	55.0	62.1	69.0	65.7	58.7	48.4	42.6	31.2	45.6
1980	23.5	20.2	31.8	46.5	58.8	62.0	70.4	70.6	62.2	44.9	33.7	21.4	45.5
1981	15.3	30.7	32.3	47.5	57.3	65.2	69.4	67.8	56.4	46.3	38.2	26.5	46.2
1982	14.9	24.4	32.3	42.2	59.7	62.1	70.0	64.4	61.0	50.6	42.1	34.4	46.5
1983	24.7	28.0	35.9	43.2	53.4	66.8	72.3	70.9	63.7	50.6	39.7	23.8	47.8
1984	19.7	33.1	25.9	45.2	53.0	66.4	68.3	70.0	58.1	54.0	38.3	34.2	47.2
1985	18.9	26.1	35.3	49.1	58.9	61.2	68.8	67.4	61.3	50.0	39.9	22.4	46.6
Record Mean	20.5	23.3	31.5	44.5	55.4	64.3	69.1	67.3	59.9	49.2	38.3	26.4	45.8
Max	27.6	30.6	39.0	53.6	65.1	74.0	78.6	76.6	69.0	57.8	44.9	32.6	54.1
Min	13.3	15.9	24.0	35.3	45.7	54.7	59.5	58.0	50.7	40.5	31.7	20.2	37.5

REFERENCE NOTES FOR TABLES 1, 2, 3 and 6   (BINGHAMPTON, NY)

GENERAL

T - TRACE AMOUNT
BLANK ENTRIES DENOTE MISSING/UNREPORTED DATA.
# INDICATES A STATION OR INSTRUMENT RELOCATION.

SPECIFIC

TABLE 1

(a) - LENGTH OF RECORD IN YEARS. ALTHOUGH
      INDIVIDUAL MONTHS MAY BE MISSING.
  # LESS THAN .05

NORMALS — BASED ON THE 1951-1980 RECORD PERIOD.
EXTREMES — DATES ARE THE MOST RECENT OCCURRENCE.
WIND DIR. — NUMERALS SHOW TENS OF DEGREES
            CLOCKWISE FROM TRUE NORTH.
            "00" INDICATES CALM.
RESULTANT WIND DIRECTIONS ARE GIVEN TO WHOLE DEGREES.

EXCEPTIONS

TABLES 2, 3, and 6

RECORD MEANS ARE THROUGH THE CURRENT YEAR,
BEGINNING IN   1952 FOR TEMPERATURE
                 1952 FOR PRECIPITATION
                 1952 FOR SNOWFALL

TABLE 4 HEATING DEGREE DAYS Base 65 deg. F   BINGHAMTON, NEW YORK

SEASON	JULY	AUG	SEP	OCT	NOV	DEC	JAN	FEB	MAR	APR	MAY	JUNE	TOTAL
1956-57	51	44	290	448	804	1043	1471	1058	995	586	343	51	7184
1957-58	42	68	166	552	770	1057	1351	1341	1072	604	407	204	7634
1958-59	23	27	191	513	759	1405	1339	1167	1053	544	240	94	7355
1959-60	6	11	143	433	859	1065	1210	1120	1305	503	253	75	6983
1960-61	36	12	114	502	724	1365	1490	1099	1016	748	394	81	7581
1961-62	38	37	117	362	825	1228	1392	1288	1049	639	246	73	7294
1962-63	21	39	261	478	900	1339	1414	1348	911	583	356	68	7718
1963-64	53	94	306	315	714	1409	1265	1333	1018	654	231	151	7543
1964-65	9	61	165	510	657	1142	1392	1176	1119	734	211	140	7316
1965-66	41	76	184	558	836	1063	1417	1173	981	696	447	88	7560
1966-67	8	16	252	552	709	1186	1297	1107	1098	634	531	27	7417
1967-68	26	40	178	505	953	1113	1477	1375	940	493	416	119	7635
1968-69	29	62	106	465	843	1279	1323	1171	1081	536	292	100	7287
1969-70	30	40	195	525	808	1280	1499	1116	1113	599	261	124	7590
1970-71	20	31	179	431	769	1258	1525	1154	1130	723	339	56	7615
1971-72	45	85	144	296	872	1050	1243	1280	1133	732	205	116	7201
1972-73	19	56	173	680	937	1097	1210	1250	770	584	411	47	7234
1973-74	5	27	200	374	749	1109	1160	1192	1016	495	384	90	6801
1974-75	14	5	251	615	825	1144	1237	1127	1164	833	147	65	7427
1975-76	0	39	250	415	604	1192	1487	999	910	554	374	50	6874
1976-77	31	62	241	653	1013	1382	1636	1137	831	537	237	130	7890
1977-78	15	61	166	560	736	1246	1415	1392	1150	699	248	100	7788
1978-79	53	7	208	533	765	1143	1351	1437	832	667	325	127	7448
1979-80	37	73	204	522	669	1040	1278	1294	1022	548	214	144	7045
1980-81	7	11	143	617	929	1349	1537	951	1006	522	266	56	7394
1981-82	19	21	215	573	799	1185	1549	1135	1006	679	178	102	7461
1982-83	19	72	147	439	685	944	1241	1031	897	649	356	68	6548
1983-84	10	13	130	447	750	1271	1399	917	1205	588	373	65	7168
1984-85	19	19	222	341	795	947	1424	1083	914	488	213	129	6594
1985-86	13	31	169	458	747	1315							

TABLE 5 COOLING DEGREE DAYS Base 65 deg. F   BINGHAMTON, NEW YORK

YEAR	JAN	FEB	MAR	APR	MAY	JUNE	JULY	AUG	SEP	OCT	NOV	DEC	TOTAL
1969	0	0	0	0	30	71	101	127	55	0	0	0	384
1970	0	0	0	7	22	66	117	109	52	0	0	0	373
1971	0	0	0	0	13	101	82	61	84	5	0	0	346
1972	0	0	0	0	19	36	228	90	27	0	0	0	400
1973	0	0	0	4	0	92	175	180	50	0	0	0	501
1974	0	0	0	12	5	42	168	128	17	0	0	0	372
1975	0	0	0	0	59	119	251	121	1	1	0	0	552
1976	0	0	0	21	3	119	61	82	15	0	0	0	301
1977	0	0	3	14	56	39	174	115	33	0	0	0	434
1978	0	0	0	0	58	68	165	154	14	0	0	0	459
1979	0	0	0	0	23	47	167	98	22	14	0	0	371
1980	0	0	0	0	30	61	183	191	66	1	0	0	532
1981	0	0	0	4	30	68	161	115	22	0	0	0	400
1982	0	0	0	0	20	23	179	60	35	0	8	0	325
1983	0	0	0	1	0	129	240	205	98	7	0	0	680
1984	0	0	0	0	9	111	128	182	21	4	0	0	455
1985	0	0	0	17	31	21	138	116	67	0	0	0	390

TABLE 6 SNOWFALL (inches)   BINGHAMTON, NEW YORK

SEASON	JULY	AUG	SEP	OCT	NOV	DEC	JAN	FEB	MAR	APR	MAY	JUNE	TOTAL
1956-57	0.0	0.0	T	T	7.3	16.4	24.6	12.2	13.3	16.4	T	0.0	90.2
1957-58	0.0	0.0	0.0	0.7	1.4	4.7	34.1	34.6	29.7	3.0	0.0	0.0	108.2
1958-59	0.0	0.0	0.0	0.4	8.0	15.2	20.0	7.6	25.7	T	T	0.0	76.9
1959-60	0.0	0.0	0.0	T	3.9	15.1	14.2	39.4	16.1	12.7	0.0	0.0	101.4
1960-61	0.0	0.0	0.0	0.8	0.4	22.6	21.1	27.8	13.8	13.2	T	0.0	99.7
1961-62	0.0	0.0	0.0	T	11.8	7.6	21.1	7.4	2.8	7.4	0.4	0.0	65.6
1962-63	0.0	0.0	T	0.7	6.9	21.7	19.1	22.6	20.1	3.5	1.0	0.0	95.6
1963-64	0.0	0.0	0.0	T	3.5	27.1	32.6	27.5	12.0	0.6	0.0	0.0	103.3
1964-65	0.0	0.0	0.0	T	T	15.0	24.9	6.8	14.3	12.7	0.0	0.0	73.7
1965-66	0.0	0.0	0.0	1.3	4.5	5.4	36.6	16.8	11.7	2.0	3.4	0.0	81.7
1966-67	0.0	0.0	0.0	T	0.1	25.4	13.2	19.4	26.3	3.9	0.2	0.0	88.5
1967-68	0.0	0.0	0.0	T	18.6	15.3	11.0	11.7	7.0	T	0.0	0.0	63.6
1968-69	0.0	0.0	0.0	1.1	15.2	11.9	10.0	12.4	1.3	0.1	T	0.0	52.0
1969-70	0.0	0.0	0.0	0.3	9.7	59.6	14.0	12.9	14.1	3.4	0.0	0.0	114.0
1970-71	0.0	0.0	T	0.4	1.7	17.2	17.0	18.7	33.5	0.8	T	0.0	108.6
1971-72	0.0	0.0	0.0	0.0	17.9	8.1	11.0	44.3	17.5	7.4	0.0	0.0	106.6
1972-73	0.0	0.0	0.0	2.6	16.1	21.5	6.8	11.8	2.5	4.6	1.8	0.0	67.7
1973-74	0.0	0.0	0.0	T	1.7	26.9	10.7	14.2	21.2	10.1	T	0.0	84.8
1974-75	0.0	0.0	0.0	0.5	7.5	13.7	15.7	16.2	9.1	4.4	0.0	0.0	67.1
1975-76	0.0	0.0	0.0	T	4.2	18.3	29.7	10.7	11.6	1.8	T	0.0	76.3
1976-77	0.0	0.0	0.0	0.1	10.4	12.3	22.8	10.7	15.5	0.5	2.1	0.0	74.4
1977-78	0.0	0.0	0.0	T	13.5	31.8	41.0	17.5	10.1	1.4	0.0	0.0	115.3
1978-79	0.0	0.0	T	T	6.2	24.4	26.3	11.0	2.1	7.3	0.0	0.0	77.3
1979-80	0.0	0.0	0.0	0.1	1.5	16.7	7.0	13.2	17.9	0.4	0.0	0.0	56.8
1980-81	0.0	0.0	0.0	0.5	13.9	12.9	16.1	8.8	7.1	T	0.0	0.0	59.3
1981-82	0.0	0.0	0.0	T	4.0	22.6	15.8	12.0	15.5	11.7	0.0	0.0	81.6
1982-83	0.0	0.0	0.0	0.5	3.2	13.3	23.2	8.6	9.1	22.9	0.2	0.0	81.0
1983-84	0.0	0.0	T	10.6	7.1	18.6	9.7	24.9	T	T	0.0	0.0	70.9
1984-85	0.0	0.0	0.0	0.0	6.8	10.5	23.5	10.5	9.7	1.5	0.0	0.0	62.5
1985-86	0.0	0.0	0.0	0.0	1.3	19.9							
Record Mean	0.0	0.0	T	0.4	7.7	18.5	19.6	17.4	14.6	5.1	0.3	0.0	83.6

See Reference Notes, relative to all above tables, on preceding page.

The country surrounding Buffalo is comparatively low and level to the west. To the east and south the land is gently rolling, rising to pronounced hills within 12 to 18 miles, and to 1,000 feet above the level of Lake Erie about 35 miles south-southeast of the city. An escarpment of 50 to 100 feet lies east-west 1-1/2 miles to the north. The eastern end of Lake Erie is 9 miles to the west-southwest, while Lake Ontario lies 25 miles to the north, the two being connected by the Niagara River, which flows north-northwestward from the end of Lake Erie.

Buffalo is located near the mean position of the polar front. Its weather is varied and changeable, characteristic of the latitude. Wide seasonal swings of temperature from hot to cold are tempered appreciably by the proximity of Lakes Erie and Ontario. Lake Erie lies to the southwest, the direction of the prevailing wind. Wind flow throughout the year is somewhat higher due to this exposure. The vigorous interplay of warm and cold air masses during the winter and early spring months causes one or more windstorms. Precipitation is moderate and fairly evenly divided throughout the twelve months.

The spring season is more cloudy and cooler than points not affected by the cold lake. Spring growth of vegetation is retarded, protecting it from late spring frosts. With heavy winter ice accumulations in the lake, typical spring conditions are delayed until late May or early June.

Summer comes suddenly in mid-June. Lake breezes temper the extreme heat of the summer season. Temperatures of 90 degrees and above are infrequent. There is more summer sunshine here than in any other section of the state. Due to the stabilizing effects of Lake Erie, thunderstorms are relatively infrequent. Most of them are caused by frontal action. To the north and south of the city thunderstorms occur more often.

Autumn has long, dry periods and is frost free usually until mid-October. Cloudiness increases in November, continuing mostly cloudy throughout the winter and early spring. Snow flurries off the lake begin in mid-November or early December. Outbreaks of Arctic air in December and throughout the winter months produce locally heavy snowfalls from the lake. At the same time, temperatures of well below zero over Canada and the midwest are raised 10 to 30 degrees in crossing the lakes. Only on rare occasions do polar air masses drop southward from eastern Hudson Bay across Lake Ontario without appreciable warming.

TABLE 1 **NORMALS, MEANS AND EXTREMES**

BUFFALO, NEW YORK

LATITUDE: 42°56'N   LONGITUDE: 78°44'W   ELEVATION: FT. GRND   705 BARO   00715   TIME ZONE: EASTERN   WBAN: 14733

	(a)	JAN	FEB	MAR	APR	MAY	JUNE	JULY	AUG	SEP	OCT	NOV	DEC	YEAR
TEMPERATURE °F:														
Normals														
-Daily Maximum		30.0	31.4	40.4	54.4	65.9	75.6	80.2	78.2	71.4	60.2	47.0	35.0	55.8
-Daily Minimum		17.0	17.5	25.6	36.3	46.3	56.4	61.2	59.6	52.7	42.7	33.6	22.5	39.3
-Monthly		23.5	24.5	33.0	45.4	56.1	66.0	70.7	68.9	62.0	51.5	40.3	28.8	47.5
Extremes														
-Record Highest	42	72	65	81	88	90	95	94	99	98	87	80	74	99
-Year		1950	1981	1945	1985	1977	1957	1968	1948	1953	1951	1961	1982	AUG 1948
-Record Lowest	42	-16	-20	-7	12	26	35	43	38	32	20	9	-10	-20
-Year		1982	1961	1984	1982	1947	1945	1945	1982	1963	1965	1971	1980	FEB 1961
NORMAL DEGREE DAYS:														
Heating (base 65°F)		1287	1134	992	588	294	53	9	25	130	423	741	1122	6798
Cooling (base 65°F)		0	0	0	0	18	83	186	146	43	0	0	0	476
% OF POSSIBLE SUNSHINE	42	32	38	45	52	58	65	68	64	59	50	29	26	49
MEAN SKY COVER (tenths)														
Sunrise - Sunset	42	8.4	8.2	7.6	7.1	6.8	6.2	6.0	6.2	6.3	6.7	8.3	8.5	7.2
MEAN NUMBER OF DAYS:														
Sunrise to Sunset														
-Clear	42	1.4	2.0	3.7	4.9	5.6	6.3	7.0	6.8	6.5	6.5	2.1	1.2	53.9
-Partly Cloudy	42	6.2	5.6	7.4	8.2	9.6	11.6	13.0	11.8	10.0	8.3	5.4	6.1	103.2
-Cloudy	42	23.5	20.7	19.9	16.9	15.7	12.1	11.0	12.4	13.5	16.2	22.5	23.7	208.1
Precipitation														
.01 inches or more	42	20.2	16.9	16.2	14.3	12.5	10.3	9.8	10.6	10.8	11.4	15.9	19.9	169.0
Snow, Ice pellets														
1.0 inches or more	42	7.5	5.5	3.8	1.0	0.*	0.0	0.0	0.0	0.0	0.1	3.1	6.2	27.3
Thunderstorms	42	0.1	0.2	1.2	2.4	2.9	5.0	5.6	6.1	3.7	1.5	1.1	0.5	30.5
Heavy Fog Visibility 1/4 mile or less	42	1.6	1.7	2.5	2.4	2.5	1.3	0.9	0.9	1.0	1.5	1.4	1.3	19.0
Temperature °F														
-Maximum														
90° and above	25	0.0	0.0	0.0	0.0	0.*	0.6	1.2	0.3	0.*	0.0	0.0	0.0	2.2
32° and below	25	18.7	15.8	7.4	0.7	0.0	0.0	0.0	0.0	0.0	0.0	2.0	12.8	57.5
-Minimum														
32° and below	25	28.9	25.7	24.4	11.0	0.7	0.0	0.0	0.0	0.*	3.3	14.0	25.9	134.0
0° and below	25	2.5	1.7	0.2	0.0	0.0	0.0	0.0	0.0	0.0	0.0	0.0	0.6	5.0
AVG. STATION PRESS.(mb)	13	990.0	991.4	989.4	989.1	988.7	989.4	990.2	991.8	992.2	992.6	991.0	990.7	990.5
RELATIVE HUMIDITY (%)														
Hour 01	25	78	79	78	75	76	79	79	83	82	80	80	80	79
Hour 07	25	79	79	80	76	75	77	78	83	83	82	81	81	80
Hour 13 (Local Time)	25	73	70	67	58	56	57	54	58	60	60	70	74	63
Hour 19	25	77	75	73	64	62	61	60	67	72	73	77	78	70
PRECIPITATION (inches):														
Water Equivalent														
-Normal		3.02	2.40	2.97	3.06	2.89	2.72	2.96	4.16	3.37	2.93	3.62	3.42	37.52
-Maximum Monthly	42	6.88	5.80	5.59	5.90	6.39	6.86	6.43	10.67	8.99	9.13	9.75	8.02	10.67
-Year		1982	1960	1976	1961	1953	1984	1963	1977	1977	1954	1985	1977	AUG 1977
-Minimum Monthly	42	1.03	0.81	1.20	1.27	1.21	0.11	0.99	1.10	0.77	0.30	1.44	0.69	0.11
-Year		1946	1968	1967	1946	1965	1955	1972	1948	1964	1963	1944	1943	JUN 1955
-Maximum in 24 hrs	42	2.57	2.31	2.14	1.71	2.03	3.04	3.38	3.88	4.94	3.49	2.51	2.16	4.94
-Year		1982	1954	1954	1977	1957	1968	1963	1963	1979	1945	1949	1945	SEP 1979
Snow, Ice pellets														
-Maximum Monthly	42	68.3	54.2	29.2	15.0	2.0	T			T	3.1	31.3	68.4	68.4
-Year		1977	1958	1959	1975	1945	1980			1956	1972	1976	1985	DEC 1985
-Maximum in 24 hrs	42	25.3	19.4	15.8	6.8	2.0	T			T	2.5	19.9	24.3	25.3
-Year		1982	1984	1954	1975	1945	1980			1956	1972	1955	1945	JAN 1982
WIND:														
Mean Speed (mph)	46	14.3	13.8	13.5	12.8	11.5	11.0	10.4	9.8	10.4	11.2	12.7	13.4	12.1
Prevailing Direction through 1963		WSW	SW	SW	SW	SW	SW	SW	SW	S	S	S	WSW	SW
Fastest Mile														
-Direction	39	SW	SW	W	W	SW	NW	NW	SW	SW	SW	SW	S	SW
-Speed (MPH)	39	91	70	68	67	63	56	59	56	59	63	66	60	91
-Year		1950	1946	1959	1957	1950	1954	1953	1944	1954	1954	1948	1945	JAN 1950
Peak Gust														
-Direction	2	SW	SW	SW	W	SW	W	SW	NW	SW	SW	W	SW	W
-Speed (mph)	2	71	51	59	74	55	47	41	40	47	46	61	66	74
-Date		1985	1985	1985	1985	1984	1985	1985	1984	1984	1984	1985	1985	APR 1985

See Reference Notes to this table on the following page.

TABLE 2 PRECIPITATION (inches)   BUFFALO, NEW YORK

YEAR	JAN	FEB	MAR	APR	MAY	JUNE	JULY	AUG	SEP	OCT	NOV	DEC	ANNUAL
1956	2.12	3.33	5.25	3.69	4.61	0.99	3.54	5.89	3.95	0.86	2.88	4.82	41.93
1957	4.92	1.65	2.23	4.86	4.03	2.92	2.97	1.11	3.84	1.12	5.30	3.74	38.69
1958	3.81	4.78	1.36	3.95	1.79	2.91	1.49	3.16	4.59	4.93	1.40	1.66	34.39
1959	6.47	3.45	3.87	3.21	2.10	1.94	1.98	4.78	2.52	4.93	3.58	3.98	42.81
1960	3.90	5.80	2.35	2.34	4.05	2.48	1.89	3.75	1.20	1.89	2.94	2.57	35.16
1961	1.41	2.63	2.59	5.90	3.01	3.66	3.02	4.03	2.53	2.41	3.30	2.62	37.11
1962	2.78	2.65	1.23	2.25	2.36	2.80	1.89	3.00	3.14	1.90	1.78	2.77	28.55
1963	1.51	1.03	2.19	2.77	2.22	0.61	6.43	8.04	1.20	0.30	5.07	1.83	33.20
1964	2.12	1.09	3.72	3.36	2.91	1.55	2.57	5.02	0.77	1.89	2.58	2.58	29.67
1965	3.27	2.99	1.97	1.99	1.21	1.50	3.69	4.12	2.37	5.07	4.69	2.60	35.47
1966	3.74	2.11	2.78	2.06	1.36	1.97	4.92	3.60	2.65	0.93	4.50	2.25	32.87
1967	1.18	1.39	1.20	2.60	3.69	2.50	1.57	4.04	6.36	4.78	3.13	2.16	34.60
1968	2.18	0.81	2.67	1.78	3.30	4.45	1.19	5.33	5.63	3.03	4.47	3.42	38.26
1969	3.85	0.97	1.62	4.16	3.75	3.51	3.83	2.48	2.04	2.77	4.09	3.09	36.16
1970	2.06	1.74	1.72	2.54	2.87	2.55	4.02	2.01	4.55	4.20	3.20	3.25	34.71
1971	1.46	3.03	2.07	1.48	1.56	4.25	4.50	4.43	1.88	1.57	3.07	3.61	32.91
1972	2.17	3.44	3.99	2.99	3.64	6.06	0.99	4.19	3.06	2.96	4.28	3.86	41.63
1973	2.03	1.98	3.27	3.56	2.99	1.68	3.68	2.98	1.44	4.27	4.07	4.89	36.84
1974	2.44	2.19	3.19	3.15	3.36	3.86	1.80	3.64	2.42	1.75	5.38	3.13	36.31
1975	2.11	2.93	2.92	1.86	3.31	3.65	2.34	8.49	2.44	1.13	2.77	4.58	38.53
1976	3.19	3.43	5.59	4.01	4.70	3.36	5.65	1.65	5.39	3.61	2.11	3.83	46.52
1977	3.38	1.59	2.42	3.60	1.39	2.79	3.64	10.67	8.99	2.61	4.45	8.02	53.55
1978	6.29	1.36	1.72	1.84	3.95	2.42	1.48	3.51	4.40	3.72	1.55	3.50	35.74
1979	5.43	2.03	2.48	3.16	1.63	2.18	3.51	6.26	5.61	3.88	4.14	3.43	43.74
1980	1.97	1.08	4.05	2.43	1.60	5.82	3.55	2.55	4.53	4.69	2.36	2.65	38.31
1981	1.11	3.50	1.70	3.09	2.56	3.68	5.05	3.13	4.24	3.31	2.22	2.87	36.46
1982	6.88	1.28	2.64	2.33	3.66	3.14	1.50	4.62	3.37	2.06	6.31	3.32	41.11
1983	1.44	1.30	3.20	2.55	3.28	2.99	2.01	3.51	2.11	4.62	5.19	7.30	39.50
1984	1.54	3.59	1.77	2.53	4.67	6.86	1.37	4.16	3.73	0.87	2.66	3.67	37.42
1985	4.27	3.34	4.42	1.33	3.46	3.21	1.81	4.63	1.20	3.73	9.75	4.85	46.00
Record Mean	3.11	2.68	2.76	2.68	2.89	2.83	2.90	3.24	3.06	3.04	3.31	3.28	35.78

TABLE 3 AVERAGE TEMPERATURE (deg. F)   BUFFALO, NEW YORK

YEAR	JAN	FEB	MAR	APR	MAY	JUNE	JULY	AUG	SEP	OCT	NOV	DEC	ANNUAL
1956	24.6	28.3	30.6	43.0	53.1	66.4	68.9	69.0	58.9	55.3	42.0	34.2	47.9
1957	20.6	30.8	35.9	48.8	54.7	69.1	69.8	67.3	62.4	50.1	40.8	34.8	48.7
1958	25.1	20.4	34.5	48.2	54.0	61.9	71.3	68.9	62.1	52.2	41.8	22.3	46.9
1959	22.3	24.2	31.0	46.0	60.2	66.8	72.8	74.8	66.8	52.6	38.6	32.3	49.1
#1960	26.5	27.2	24.1	48.4	58.1	65.8	69.1	69.0	64.3	50.6	42.8	21.6	47.3
1961	18.5	26.5	34.2	39.8	53.1	63.4	69.7	69.6	68.6	54.5	40.8	29.7	47.4
1962	22.6	21.3	32.5	44.9	60.9	64.9	68.2	68.1	58.6	51.5	37.1	25.1	46.3
1963	18.9	18.8	35.4	44.2	52.9	66.7	70.2	64.3	57.1	57.1	43.6	23.4	46.1
1964	29.3	23.5	34.0	46.9	59.2	65.7	73.1	64.9	60.9	48.1	42.1	29.5	48.1
1965	23.6	25.8	30.0	41.2	59.6	64.3	67.6	67.8	63.5	47.8	40.0	34.3	47.1
1966	20.4	24.9	34.7	43.3	52.2	67.4	71.4	68.5	58.7	48.8	41.5	28.6	46.7
1967	29.8	20.6	30.9	46.1	50.1	72.5	71.2	68.1	60.7	51.9	36.3	33.0	47.6
1968	19.9	20.7	35.7	49.2	53.4	64.8	71.2	69.4	66.1	53.5	40.7	26.8	47.6
1969	25.0	24.6	30.9	46.8	54.4	64.4	70.5	71.2	62.2	51.0	39.1	24.8	47.1
1970	17.6	24.8	30.1	46.9	57.3	66.0	71.0	70.2	64.0	54.5	41.6	27.4	47.6
1971	20.9	27.0	29.8	41.8	54.5	67.6	68.7	67.8	62.8	58.7	39.1	33.5	47.9
1972	25.5	22.0	30.1	41.1	59.1	62.6	71.0	67.7	62.8	46.2	36.0	30.8	46.3
1973	27.6	22.9	42.4	46.9	54.5	68.2	72.3	71.8	61.7	54.3	40.8	29.0	49.4
1974	27.1	22.3	33.0	46.2	53.1	65.6	69.9	69.9	59.6	49.2	40.2	31.7	47.3
1975	30.1	29.1	30.8	39.3	62.1	68.0	72.3	69.7	58.3	53.1	46.9	28.3	49.0
1976	19.7	31.8	37.2	46.5	53.4	68.4	67.8	67.5	60.1	46.3	34.1	22.0	46.3
1977	13.8	24.6	39.8	47.0	60.3	64.4	72.0	68.1	62.6	49.6	43.3	27.9	47.8
1978	20.4	15.5	28.2	42.5	57.4	65.1	70.4	70.3	60.8	49.5	40.4	30.4	45.9
1979	20.5	14.9	38.2	44.3	56.9	66.5	71.3	67.5	61.9	50.7	43.5	33.4	47.5
1980	25.8	21.2	31.8	46.1	58.1	61.9	71.7	72.6	62.4	48.7	39.4	25.3	47.1
1981	19.3	32.9	33.9	47.2	56.4	66.2	71.8	70.0	60.0	48.2	40.4	29.0	48.0
1982	17.2	23.2	32.5	41.6	61.0	62.2	71.8	65.0	61.6	52.6	43.0	37.5	47.5
1983	27.0	29.6	36.7	43.6	53.9	67.6	74.2	71.2	63.7	51.7	40.8	22.7	48.6
1984	20.4	33.8	27.1	47.7	52.9	67.8	70.3	70.3	58.5	53.2	39.0	35.6	48.1
1985	21.1	24.8	35.6	49.5	59.5	62.7	69.7	69.2	64.2	52.5	42.0	25.6	48.0
Record Mean	24.7	24.6	32.5	43.5	54.8	64.8	70.3	69.0	62.5	51.7	40.0	29.5	47.3
Max	31.1	31.5	39.7	51.7	63.3	72.5	78.2	76.9	70.5	59.1	46.3	35.3	54.7
Min	18.2	17.6	25.2	35.3	46.3	57.0	62.5	61.0	54.4	44.2	33.8	23.6	39.9

REFERENCE NOTES FOR TABLES 1, 2, 3 and 6   (BUFFALO, NY)

GENERAL

T - TRACE AMOUNT
BLANK ENTRIES DENOTE MISSING/UNREPORTED DATA.
# INDICATES A STATION OR INSTRUMENT RELOCATION.

SPECIFIC

TABLE 1

(a) - LENGTH OF RECORD IN YEARS. ALTHOUGH INDIVIDUAL MONTHS MAY BE MISSING.
* LESS THAN .05

NORMALS — BASED ON THE 1951-1980 RECORD PERIOD.
EXTREMES — DATES ARE THE MOST RECENT OCCURRENCE.
WIND DIR. — NUMERALS SHOW TENS OF DEGREES CLOCKWISE FROM TRUE NORTH. "00" INDICATES CALM.
RESULTANT WIND DIRECTIONS ARE GIVEN TO WHOLE DEGREES.

EXCEPTIONS

TABLES 2, 3, and 6

RECORD MEANS ARE THROUGH THE CURRENT YEAR, BEGINNING IN   1874 FOR TEMPERATURE
1871 FOR PRECIPITATION
1944 FOR SNOWFALL

TABLE 4 HEATING DEGREE DAYS Base 65 deg. F BUFFALO, NEW YORK

SEASON	JULY	AUG	SEP	OCT	NOV	DEC	JAN	FEB	MAR	APR	MAY	JUNE	TOTAL
1956-57	12	13	218	300	687	947	1370	951	894	508	321	38	6259
1957-58	13	26	140	456	717	930	1231	1242	940	498	343	111	6647
1958-59	2	19	124	395	687	1318	1317	1134	1050	562	216	63	6887
1959-60	1	1	111	388	784	1004	1188	1088	1258	494	235	48	6600
#1960-61	16	10	77	441	660	1339	1435	1072	952	749	385	90	7226
1961-62	30	17	76	323	722	1089	1310	1216.	1002	609	195	66	6655
1962-63	9	26	213	415	832	1231	1420	1288	907	618	370	57	7386
1963-64	20	72	240	241	635	1282	1099	1198	955	535	204	98	6579
1964-65	5	68	176	518	680	1097	1277	1092	1080	706	186	100	6985
1965-66	23	46	122	525	742	942	1374	1114	931	648	401	68	6936
1966-67	7	19	199	495	700	1124	1086	1239	1047	560	457	4	6937
1967-68	12	26	162	403	853	985	1393	1281	901	469	352	84	6921
1968-69	11	29	58	374	722	1180	1233	1125	1052	540	325	102	6751
1969-70	13	16	147	433	769	1240	1459	1121	1076	552	255	66	7147
1970-71	6	6	93	328	695	1161	1361	1057	1085	691	327	36	6846
1971-72	11	29	87	202	771	971	1218	1237	1070	707	187	112	6602
1972-73	16	33	113	574	860	1054	1152	1173	696	542	318	24	6555
1973-74	2	14	171	326	720	1107	1167	1187	989	553	365	51	6652
1974-75	2	0	187	483	738	1024	1077	1001	1053	764	175	32	6536
1975-76	3	15	197	368	535	1134	1400	958	853	557	358	40	6418
1976-77	15	35	180	573	921	1328	1580	1123	775	544	207	90	7371
1977-78	5	40	110	473	646	1146	1376	1378	1130	670	282	81	7337
1978-79	14	3	154	472	732	1067	1371	1400	823	619	285	65	7005
1979-80	16	35	134	455	636	1224	1265	1208	1022	559	240	142	6685
1980-81	2	0	128	498	759	1224	1411	895	956	527	269	33	6702
1981-82	6	11	170	514	732	1108	1476	1163	1002	698	147	95	7122
1982-83	4	65	140	382	656	848	1172	987	868	636	342	71	6171
1983-84	5	10	125	418	722	1304	1378	899	1167	519	385	35	6967
1984-85	11	22	210	360	774	905	1354	1120	902	476	196	95	6425
1985-86	8	12	114	378	685	1215							

TABLE 5 COOLING DEGREE DAYS Base 65 deg. F BUFFALO, NEW YORK

YEAR	JAN	FEB	MAR	APR	MAY	JUNE	JULY	AUG	SEP	OCT	NOV	DEC	TOTAL
1969	0	0	0	0	1	88	192	212	69	6	0	0	568
1970	0	0	0	16	21	108	197	173	72	12	0	0	599
1971	0	0	0	0	9	119	136	122	107	15	0	0	508
1972	0	0	0	0	12	48	210	123	57	0	0	0	450
1973	0	0	0	6	2	126	233	230	78	3	0	0	678
1974	0	0	0	0	7	71	163	158	29	0	0	0	428
1975	0	0	0	0	0	90	129	238	171	3	0	0	634
1976	0	0	0	8	7	149	109	119	40	0	0	0	432
1977	0	0	0	0	12	68	78	228	142	45	0	1	574
1978	0	0	0	0	52	91	189	173	35	0	0	0	540
1979	0	0	0	6	40	118	217	120	49	20	0	0	570
1980	0	0	0	0	32	56	217	242	58	2	0	0	607
1981	0	0	0	2	13	78	225	173	55	0	0	0	546
1982	0	0	0	3	31	18	221	74	45	2	0	2	396
1983	0	0	0	0	5	157	300	214	90	15	0	0	781
1984	0	0	0	5	16	123	183	193	23	1	0	0	544
1985	0	0	0	18	32	32	161	151	96	0	1	0	491

TABLE 6 SNOWFALL (inches) BUFFALO, NEW YORK

SEASON	JULY	AUG	SEP	OCT	NOV	DEC	JAN	FEB	MAR	APR	MAY	JUNE	TOTAL
1956-57	0.0	0.0	T	0.0	16.2	35.6	36.6	3.1	10.2	12.0	0.0	0.0	113.7
1957-58	0.0	0.0	0.0	T	19.2	4.8	31.1	54.2	11.2	4.2	0.0	0.0	124.7
1958-59	0.0	0.0	0.0	0.0	13.5	19.0	38.4	14.0	29.2	0.4	T	0.0	114.5
1959-60	0.0	0.0	0.0	T	12.2	14.2	18.3	49.5	19.5	1.9	T	0.0	115.6
1960-61	0.0	0.0	0.0	1.0	15.9	31.9	23.5	11.9	5.1	13.1	T	0.0	102.4
1961-62	0.0	0.0	0.0	T	5.6	30.2	26.2	28.2	6.7	4.5	0.0	0.0	101.4
1962-63	0.0	0.0	0.0	2.0	2.5	30.2	31.5	15.5	7.7	0.3	0.1	0.0	89.8
1963-64	0.0	0.0	0.0	0.0	3.1	24.0	13.7	14.6	12.8	3.3	0.0	0.0	71.5
1964-65	0.0	0.0	0.0	T	5.4	15.2	19.2	9.4	17.5	4.2	0.0	0.0	70.9
1965-66	0.0	0.0	0.0	1.2	12.2	7.0	48.0	15.2	11.4	3.2	0.1	0.0	98.3
1966-67	0.0	0.0	0.0	0.0	10.0	12.1	11.6	19.8	10.8	0.6	1.2	0.0	66.1
1967-68	0.0	0.0	0.0	T	19.7	10.4	19.1	11.7	10.6	0.1	0.0	0.0	71.6
1968-69	0.0	0.0	0.0	T	11.6	11.7	31.2	12.8	8.0	3.1	0.0	0.0	78.4
1969-70	0.0	0.0	0.0	1.0	22.1	23.4	38.0	21.9	12.6	1.5	T	0.0	120.5
1970-71	0.0	0.0	0.0	0.0	2.6	32.3	17.2	19.4	22.6	2.9	0.0	0.0	97.0
1971-72	0.0	0.0	0.0	0.0	18.7	12.9	27.6	31.4	14.1	5.2	0.0	0.0	109.9
1972-73	0.0	0.0	0.0	3.1	18.9	19.8	9.9	16.1	8.5	2.4	0.1	0.0	78.8
1973-74	0.0	0.0	0.0	0.0	3.0	23.1	19.7	22.8	12.9	7.1	0.0	0.0	88.7
1974-75	0.0	0.0	0.0	T	22.1	23.6	11.0	16.3	7.6	15.0	0.0	0.0	95.6
1975-76	0.0	0.0	0.0	T	5.5	27.3	21.6	8.3	17.3	2.5	T	0.0	82.5
1976-77	0.0	0.0	0.0	0.2	31.3	60.7	68.3	22.7	13.5	2.2	.5	0.0	199.4
1977-78	0.0	0.0	0.0	T	15.0	53.4	56.5	21.7	5.8	1.8	0.1	0.0	154.3
1978-79	0.0	0.0	0.0	T	3.0	10.1	42.6	28.3	4.6	8.7	0.0	0.0	97.3
1979-80	0.0	0.0	0.0	T	12.6	19.7	14.4	11.7	13.9	0.3	T	T	68.4
1980-81	0.0	0.0	0.0	T	6.7	21.6	14.4	5.0	13.2	T	0.0	0.0	60.9
1981-82	0.0	0.0	0.0	T	1.8	24.8	53.2	12.7	9.0	10.9	0.0	0.0	112.4
1982-83	0.0	0.0	0.0	0.0	15.8	12.9	9.0	5.5	6.9	2.3	T	0.0	52.4
1983-84	0.0	0.0	0.0	T	17.7	52.0	13.4	32.5	16.0	0.9	T	0.0	132.5
1984-85	0.0	0.0	0.0	0.0	1.4	11.2	65.9	20.9	6.3	1.5	0.0	0.0	107.2
1985-86	0.0	0.0	0.0	0.0	5.2	68.4							
Record Mean	0.0	0.0	T	0.2	12.2	23.7	24.9	18.0	11.6	3.2	0.1	T	94.0

See Reference Notes, relative to all above tables, on preceding page.

New York City, in area exceeding 300 square miles, is located on the Atlantic coastal plain at the mouth of the Hudson River. The terrain is laced with numerous waterways, all but one of the five boroughs in the city are situated on islands. Elevations range from less than 50 feet over most of Manhattan, Brooklyn, and Queens to almost 300 feet in northern Manhattan and the Bronx, and over 400 feet in Staten Island. Extensive suburban areas on Long Island, and in Connecticut, New York State and New Jersey border the city on the east, north, and west. About 30 miles to the west and northwest, hills rise to about 1,500 feet and to the north in upper Westchester County to 800 feet. To the southwest and to the east are the low-lying land areas of the New Jersey coastal plain and of Long Island, bordering on the Atlantic.

The New York Metropolitan area is close to the path of most storm and frontal systems which move across the North American continent. Therefore, weather conditions affecting the city most often approach from a westerly direction. New York City can thus experience higher temperatures in summer and lower ones in winter than would otherwise be expected in a coastal area. However, the frequent passage of weather systems often helps reduce the length of both warm and cold spells, and is also a major factor in keeping periods of prolonged air stagnation to a minimum.

Although continental influence predominates, oceanic influence is by no means absent. During the summer local sea breezes, winds blowing onshore from the cool water surface, often moderate the afternoon heat. The effect of the sea breeze diminishes inland. On winter mornings, ocean temperatures which are warm relative to the land reinforce the effect of the city heat island and low temperatures are often 10-20 degrees lower in the inland suburbs than in the central city. The relatively warm water temperatures also delay the advent of winter snows. Conversely, the lag in warming of water temperatures keeps spring temperatures relatively cool. One year-round measure of the ocean influence is the small average daily variation in temperature.

Precipitation is moderate and distributed fairly evenly throughout the year. Most of the rainfall from May through October comes from thunderstorms, usually of brief duration and sometimes intense. Heavy rains of long duration associated with tropical storms occur infrequently in late summer or fall. For the other months of the year precipitation is more likely to be associated with widespread storm areas, so that day-long rain, snow or a mixture of both is more common. Coastal storms, occurring most often in the fall and winter months, produce on occasion considerable amounts of precipitation and have been responsible for record rains, snows, and high winds.

The average annual precipitation is reasonably uniform within the city but is higher in the northern and western suburbs and less on eastern Long Island. Annual snowfall totals also show a consistent increase to the north and west of the city with lesser amounts along the south shores and the eastern end of Long Island, reflecting the influence of the ocean waters.

Local Climatological Data is published for three locations in New York City, Central Park, La Guardia Airport, and John F. Kennedy International Airport. Other nearby locations for which it is published are Newark, New Jersey, and Bridgeport, Connecticut.

Based on the 1951-1980 period, the average first occurrence of 32 degrees Fahrenheit in the fall is November 11 and the average last occurrence in the spring is April 1.

TABLE 1   NORMALS, MEANS AND EXTREMES

NEW YORK, CENTRAL PARK, NEW YORK

LATITUDE: 40°47'N   LONGITUDE: 73°58'W   ELEVATION: FT. GRND   132 BARO   00087   TIME ZONE: EASTERN   WBAN: 94728

	(a)	JAN	FEB	MAR	APR	MAY	JUNE	JULY	AUG	SEP	OCT	NOV	DEC	YEAR
TEMPERATURE °F:														
Normals														
-Daily Maximum		38.0	40.1	48.6	61.1	71.5	80.1	85.3	83.7	76.4	65.6	53.6	42.1	62.2
-Daily Minimum		25.6	26.6	34.1	43.8	53.3	62.7	68.2	67.1	60.1	49.9	40.8	30.3	46.9
-Monthly		31.8	33.4	41.4	52.5	62.4	71.4	76.8	75.4	68.3	57.8	47.2	36.2	54.5
Extremes														
-Record Highest	117	72	75	86	96	99	101	106	104	102	94	84	72	106
-Year		1950	1985	1945	1976	1962	1966	1936	1918	1953	1941	1950	1982	JUL 1936
-Record Lowest	117	-6	-15	3	12	32	44	52	50	39	28	5	-13	-15
-Year		1882	1934	1872	1923	1891	1945	1943	1982	1912	1936	1875	1917	FEB 1934
NORMAL DEGREE DAYS:														
Heating (base 65°F)		1029	885	732	378	134	7	0	0	36	240	534	893	4868
Cooling (base 65°F)		0	0	0	0	56	199	363	322	135	14	0	0	1089
% OF POSSIBLE SUNSHINE	100	50	55	56	59	61	64	65	64	63	61	52	49	58
MEAN SKY COVER (tenths)														
Sunrise - Sunset	42	6.0	5.8	5.7	6.0	5.7	5.6	5.5	5.5	5.2	4.9	5.8	5.9	5.6
MEAN NUMBER OF DAYS:														
Sunrise to Sunset														
-Clear	42	8.1	8.3	8.8	7.6	8.0	8.0	8.5	9.2	10.6	11.8	9.0	8.9	106.7
-Partly Cloudy	42	9.2	8.7	10.1	10.5	12.4	12.4	13.0	12.1	10.0	9.7	9.5	9.1	126.7
-Cloudy	42	13.7	11.2	12.1	11.9	10.7	9.6	9.5	9.7	9.4	9.5	11.5	13.0	131.8
Precipitation														
.01 inches or more	116	11.2	9.8	11.5	10.6	11.1	10.1	10.4	9.7	8.3	8.3	9.2	10.4	120.7
Snow,Ice pellets														
1.0 inches or more	115	2.2	2.3	1.5	0.2	0.0	0.0	0.0	0.0	0.0	0.0	0.3	1.6	8.1
Thunderstorms	24	0.1	0.3	0.8	1.0	2.5	3.8	3.8	3.7	1.3	0.8	0.3	0.1	18.5
Heavy Fog Visibility														
1/4 mile or less														
Temperature °F														
-Maximum														
90° and above	72	0.0	0.0	0.0	0.1	0.9	3.2	6.5	4.5	1.5	0.1	0.0	0.0	16.7
32° and below	72	8.8	5.7	1.3	0.*	0.0	0.0	0.0	0.0	0.0	0.0	0.2	4.9	21.0
-Minimum														
32° and below	72	22.8	20.2	12.5	1.5	0.*	0.0	0.0	0.0	0.0	0.3	4.6	18.2	80.2
0° and below	72	0.2	0.2	0.0	0.0	0.0	0.0	0.0	0.0	0.0	0.0	0.0	0.1	0.5
AVG. STATION PRESS.(mb)	10	1013.9	1013.7	1012.3	1011.5	1011.3	1012.3	1012.5	1014.2	1014.7	1015.1	1014.5	1014.5	1013.4
RELATIVE HUMIDITY (%)														
Hour 01	49	65	64	64	64	70	73	74	76	76	72	69	67	70
Hour 07	61	68	68	67	67	71	74	75	78	79	76	73	69	72
Hour 13 (Local Time)	61	60	58	55	51	53	55	55	57	57	55	59	60	56
Hour 19	61	60	59	57	56	60	61	63	66	66	63	63	62	61
PRECIPITATION (inches):														
Water Equivalent														
-Normal		3.21	3.13	4.22	3.75	3.76	3.23	3.77	4.03	3.66	3.41	4.14	3.81	44.12
-Maximum Monthly	116	10.52	6.87	10.41	8.77	9.74	9.78	11.89	10.86	16.85	13.31	12.41	9.98	16.85
-Year		1979	1869	1980	1874	1984	1903	1889	1955	1882	1903	1972	1973	SEP 1882
-Minimum Monthly	116	0.58	0.46	0.90	0.95	0.30	0.02	0.49	0.24	0.21	0.14	0.34	0.25	0.02
-Year		1981	1895	1885	1881	1903	1949	1910	1964	1884	1963	1976	1955	JUN 1949
-Maximum in 24 hrs	73	3.91	3.04	4.25	4.22	4.88	4.74	3.60	5.78	8.30	11.17	8.09	3.21	11.17
-Year		1979	1973	1876	1984	1968	1884	1971	1971	1882	1903	1977	1909	OCT 1903
Snow,Ice pellets														
-Maximum Monthly	117	27.4	27.9	30.5	13.5	T					0.8	19.0	29.6	30.5
-Year		1925	1934	1896	1875	1977					1925	1898	1947	MAR 1896
-Maximum in 24 hrs	117	13.6	17.6	18.1	10.2	T					0.8	10.0	26.4	26.4
-Year		1978	1983	1941	1915	1977					1925	1898	1947	DEC 1947
WIND:														
Mean Speed (mph)	58	10.7	10.8	11.0	10.5	8.8	8.1	7.6	7.6	8.1	8.9	9.9	10.4	9.4
Prevailing Direction														
through 1963		NW	NW	NW	NW	SW	SW	SW	SW	SW	SW	NW	NW	SW
Fastest Mile														
-Direction	31	NW	NE	NW	NW	NE	SW	WSW	NE	NE	SE	NE	NW	NE
-Speed (MPH)	31	47	47	60	47	38	49	43	36	44	46	70	47	70
-Year		1959	1961	1950	1981	1967	1952	1980	1954	1960	1980	1950	1980	NOV 1950
Peak Gust														
-Direction	2	SW	NE	NE	SE	WNW	S	SW	NNW	W	NE	WNW	W	NE
-Speed (mph)	2	37	51	63	46	43	37	35	32	52	33	44	40	63
-Date		1985	1984	1984	1984	1984	1984	1984	1984	1985	1984	1984	1985	MAR 1984

See reference Notes to this table on the following page.

TABLE 2 PRECIPITATION (inches) NEW YORK, CENTRAL PARK, NEW YORK

YEAR	JAN	FEB	MAR	APR	MAY	JUNE	JULY	AUG	SEP	OCT	NOV	DEC	ANNUAL
1956	1.54	4.18	5.03	2.78	2.20	2.99	3.13	2.56	2.32	3.61	2.62	3.29	36.25
1957	1.70	2.43	1.99	4.51	3.67	1.85	1.47	2.87	3.01	3.27	4.46	5.26	36.49
1958	3.79	2.98	3.19	6.14	3.25	2.55	3.68	4.44	4.44	5.46	1.85	1.25	40.94
1959	2.34	1.69	3.77	1.91	1.33	4.20	4.28	4.45	1.11	4.83	4.22	4.64	38.77
1960	2.40	4.43	2.96	3.05	2.97	1.74	8.29	6.26	5.38	2.82	3.05	3.04	46.39
1961	1.88	3.96	4.23	5.08	3.60	2.86	4.92	3.13	1.70	2.21	2.71	3.04	39.32
1962	2.62	3.74	2.97	3.00	1.26	3.73	1.67	5.71	3.10	3.15	3.94	2.26	37.15
1963	1.93	2.55	3.61	1.27	2.16	2.72	2.19	3.21	3.95	0.14	8.24	2.31	34.28
1964	4.62	2.93	2.57	5.09	0.57	2.67	4.17	0.24	1.69	1.73	2.55	4.16	32.99
1965	3.09	3.66	2.49	2.90	1.58	1.33	1.27	2.73	1.70	2.16	1.46	1.72	26.09
1966	2.63	4.96	0.94	2.69	4.26	1.17	1.25	1.89	8.82	4.64	3.47	3.18	39.90
1967	1.39	2.68	5.97	3.45	4.08	4.64	6.99	5.94	1.84	3.47	2.59	6.08	49.12
1968	2.04	1.13	4.79	2.82	7.06	6.15	2.63	2.88	1.97	2.20	5.75	4.15	43.57
1969	1.10	3.05	3.73	3.99	2.67	3.16	7.37	2.53	8.32	1.97	3.58	7.07	48.54
1970	0.66	4.52	4.18	3.48	3.34	2.27	2.19	2.47	1.74	2.48	5.14	2.82	35.29
1971	2.67	5.33	3.80	2.95	4.24	2.31	7.20	9.37	7.36	4.14	5.64	1.76	56.77
1972	2.41	5.90	4.55	3.92	8.39	9.30	4.54	1.92	1.33	6.27	12.41	6.09	67.03
1973	4.53	4.55	3.60	8.05	4.51	4.55	5.89	3.08	2.75	3.92	1.82	9.98	57.23
1974	3.80	1.49	5.76	3.83	4.29	3.29	1.33	5.99	8.05	2.59	0.94	6.33	47.69
1975	4.76	3.33	3.32	3.04	3.38	7.58	11.77	3.05	9.32	3.70	4.33	3.63	61.21
1976	5.78	3.13	2.99	2.80	4.77	2.31	1.42	6.52	3.15	5.31	0.34	2.29	41.28
1977	2.25	2.51	7.41	3.75	1.71	3.83	1.60	4.57	4.75	5.03	12.26	5.06	54.73
1978	8.27	1.59	2.73	2.38	9.15	1.69	4.48	5.26	4.06	1.50	2.85	5.61	49.81
1979	10.52	4.58	4.40	4.04	6.23	1.56	1.76	4.27	4.83	3.87	3.38	2.69	52.13
1980	1.72	1.04	10.41	8.26	2.33	3.84	5.26	1.16	1.98	3.86	4.11	0.58	44.55
1981	0.58	6.04	1.19	3.42	3.56	2.71	6.21	0.59	3.45	3.49	1.69	5.18	38.11
1982	6.46	2.37	2.56	5.67	2.43	5.12	3.14	4.66	1.77	2.31	3.44	1.47	41.40
1983													
1984	1.87	4.86	6.30	6.62	9.74	5.76	7.03	1.38	2.51	3.63	4.07	3.26	57.03
1985	1.00	2.41	1.91	1.41	5.72	4.41	4.41	2.58	4.75	1.30	8.09	0.83	38.82
Record Mean	3.41	3.40	3.86	3.51	3.58	3.45	4.24	4.24	3.70	3.48	3.47	3.45	43.80

TABLE 3 AVERAGE TEMPERATURE (deg. F) NEW YORK, CENTRAL PARK, NEW YORK

YEAR	JAN	FEB	MAR	APR	MAY	JUNE	JULY	AUG	SEP	OCT	NOV	DEC	ANNUAL
1956	32.0	36.6	37.4	48.2	58.7	71.4	72.9	74.2	64.8	58.1	46.7	40.9	53.5
1957	28.5	37.3	41.9	53.2	63.1	74.3	77.7	73.6	69.7	56.2	49.4	40.2	55.5
1958	31.9	27.4	40.3	52.9	59.1	67.2	76.1	75.2	67.6	55.5	47.9	29.4	52.5
1959	31.1	32.1	40.1	53.8	66.4	71.2	76.3	77.5	72.3	59.8	45.8	38.4	55.4
1960	33.9	36.3	33.3	54.1	62.6	71.8	74.6	74.9	68.0	58.1	49.7	30.9	54.0
1961	27.7	36.7	41.5	49.0	59.9	72.3	78.1	76.4	73.6	61.1	48.8	35.5	55.1
1962	32.6	31.8	43.1	53.3	64.5	72.5	74.0	72.4	64.9	57.4	43.2	31.5	53.4
1963	30.1	28.3	43.7	53.7	61.1	70.9	76.4	72.1	63.1	61.8	50.4	31.2	53.6
1964	35.7	32.9	43.1	49.7	65.4	71.6	75.4	72.9	67.2	55.0	49.4	36.4	54.6
1965	29.7	33.9	40.0	50.6	66.4	70.3	74.3	73.2	67.5	57.3	46.8	40.5	54.2
1966	32.2	35.1	42.7	49.7	61.6	75.4	79.7	76.9	66.5	56.2	48.9	35.7	55.1
1967	37.4	29.2	37.6	49.6	55.2	72.8	75.3	73.9	66.7	57.2	42.5	38.2	53.0
1968	26.7	28.9	43.3	55.0	59.6	69.7	77.3	76.0	70.6	60.5	46.9	34.3	54.1
1969	31.8	32.6	40.1	55.9	65.3	73.1	74.8	77.4	69.0	57.7	46.4	33.4	54.8
1970	25.1	33.0	38.7	52.1	64.0	70.9	77.1	77.6	70.8	58.9	48.5	34.4	54.3
1971	27.0	35.1	40.1	50.8	61.4	74.2	77.8	75.9	71.6	62.7	45.1	40.8	55.2
1972	35.1	31.4	39.8	50.1	63.3	67.9	77.2	75.6	69.5	53.5	44.4	38.5	53.8
1973	35.5	32.5	46.4	53.4	59.5	73.4	77.4	77.6	69.5	60.2	48.3	39.0	56.1
1974	35.3	31.7	42.1	55.2	61.0	69.0	77.2	76.4	66.7	54.1	48.2	39.4	54.7
1975	37.3	35.8	40.2	47.9	65.8	70.5	75.8	74.4	64.2	59.2	52.3	35.9	54.9
1976	27.4	39.9	44.4	55.0	60.2	73.2	74.8	74.3	66.6	52.9	41.7	29.9	53.4
1977	22.1	33.5	46.8	53.7	65.0	70.2	79.0	75.7	68.2	54.9	47.3	35.7	54.3
1978	28.0	27.2	39.0	51.6	61.5	71.3	74.4	76.0	65.0	54.9	47.8	38.9	53.0
1979	33.6	25.5	46.9	52.6	65.3	69.2	76.9	76.8	70.5	57.3	52.5	41.1	55.7
1980	33.7	31.4	41.2	54.5	65.6	70.3	79.3	80.3	70.8	55.2	44.6	32.5	55.0
1981	26.3	39.3	42.3	56.2	64.8	73.0	78.5	76.0	67.6	54.4	47.7	36.5	55.2
1982	26.1	35.3	42.0	52.3	64.1	68.6	77.9	77.9	68.3	58.5	50.4	42.8	54.9
1983	34.5	36.4	44.0	52.3	60.2	73.4	79.5	77.7	71.8	57.9	48.9	35.2	56.0
1984	29.9	40.6	36.7	51.9	61.6	76.7	74.7	76.7	65.9	61.8	47.3	43.8	55.5
1985	28.8	36.6	45.8	55.5	65.3	68.6	76.2	75.4	70.5	59.5	50.0	34.2	55.5
Record Mean	31.9	32.9	40.9	51.4	62.0	70.9	76.3	74.8	68.1	57.8	46.9	35.8	54.2
Max	38.2	39.7	48.4	59.8	71.0	79.6	84.7	83.0	76.3	65.7	53.3	41.8	61.8
Min	25.5	26.1	33.4	43.0	53.0	62.3	67.8	66.6	59.8	49.9	40.4	29.8	46.5

REFERENCE NOTES FOR TABLES 1, 2, 3 and 6   (NEW YORK, NY)

GENERAL

T - TRACE AMOUNT
BLANK ENTRIES DENOTE MISSING/UNREPORTED DATA.
# INDICATES A STATION OR INSTRUMENT RELOCATION.

SPECIFIC

TABLE 1

(a) - LENGTH OF RECORD IN YEARS. ALTHOUGH INDIVIDUAL MONTHS MAY BE MISSING.

* LESS THAN .05

NORMALS — BASED ON THE 1951-1980 RECORD PERIOD.
EXTREMES — DATES ARE THE MOST RECENT OCCURRENCE.
WIND DIR. — NUMERALS SHOW TENS OF DEGREES
　　　　　　CLOCKWISE FROM TRUE NORTH.
　　　　　　''00'' INDICATES CALM.
RESULTANT WIND DIRECTIONS ARE GIVEN TO WHOLE DEGREES.

EXCEPTIONS

TABLE 1

1. MEAN SKY COVER, AND DAYS CLEAR - PARTLY CLOUDY -CLOUDY ARE THROUGH 1966.
2. PERCENT OF POSSIBLE SUNSHINE, AND MEAN WIND SPEED ARE THROUGH 1976.
3. FASTEST MILE WINDS ARE THROUGH MARCH 1977 AND FEBRUARY 1980 THROUGH OCTOBER 1981.
4. RELATIVE HUMIDITY IS THROUGH 1980.
5. LIQUID PRECIPITATION FOR 1983 IS NOT CONSIDERED IN DETERMINING EXTREMES.

TABLES 2, 3, and 6

RECORD MEANS ARE THROUGH THE CURRENT YEAR,
BEGINNING IN　　1912 FOR TEMPERATURE
　　　　　　　　1869 FOR PRECIPITATION
　　　　　　　　1869 FOR SNOWFALL

TABLE 4 HEATING DEGREE DAYS Base 65 deg. F   NEW YORK, CENTRAL PARK, NEW YORK

SEASON	JULY	AUG	SEP	OCT	NOV	DEC	JAN	FEB	MAR	APR	MAY	JUNE	TOTAL
1956-57	11	3	98	225	545	739	1124	771	708	367	126	7	4724
1957-58	0	0	48	268	461	761	1019	1045	759	358	193	29	4941
1958-59	0	0	38	313	507	1099	1040	915	765	341	87	32	5137
1959-60	0	0	30	224	572	820	954	827	975	336	101	0	4839
1960-61	0	0	10	229	452	1049	1149	788	719	473	172	6	5047
1961-62	0	0	20	168	490	907	997	921	675	370	123	12	4683
1962-63	1	10	78	243	646	1032	1074	1021	653	337	161	9	5265
1963-64	0	0	125	134	431	1040	902	927	669	454	90	23	4795
1964-65	3	0	63	308	461	879	1088	867	765	426	64	30	4954
1965-66	0	13	54	239	538	755	1007	830	685	451	166	9	4747
1966-67	0	0	63	270	475	901	849	999	843	462	305	5	5172
1967-68	0	4	55	264	671	825	1179	1042	668	292	170	15	5185
1968-69	0	0	3	183	538	944	1023	902	768	285	74	0	4720
1969-70	0	0	28	240	551	974	1227	890	809	387	109	6	5221
1970-71	0	0	27	210	490	940	1173	830	764	419	135	9	4997
1971-72	0	0	14	106	596	743	920	965	775	445	94	26	4684
1972-73	2	0	25	355	611	812	907	903	572	362	188	2	4739
1973-74	0	0	29	162	493	800	913	925	704	309	165	27	4527
1974-75	1	0	59	333	502	789	852	812	764	507	86	11	4716
1975-76	0	3	62	193	387	898	1163	723	630	360	167	18	4604
1976-77	0	4	44	373	692	1082	1322	877	560	354	100	27	5435
1977-78	0	0	56	307	524	903	1140	1051	797	394	179	13	5364
1978-79	5	0	75	311	510	802	969	1100	769	369	55	14	4764
1979-80	4	4	20	271	373	734	963	969	731	310	67	22	4468
1980-81	0	0	31	305	602	1000	1194	715	698	264	78	3	4890
1981-82	0	0	48	320	513	876	1198	825	707	413	74	36	5010
1982-83	0	5	24	229	446	679	936	793	644	393	161	3	4313
1983-84	0	0	34	249	480	914	1082	698	870	389	137	9	4862
1984-85	0	0	69	114	525	654	1113	789	596	305	79	24	4268
1985-86	0	0	17	188	448	947							

TABLE 5   COOLING DEGREE DAYS Base 65 deg. F   NEW YORK, CENTRAL PARK, NEW YORK

YEAR	JAN	FEB	MAR	APR	MAY	JUNE	JULY	AUG	SEP	OCT	NOV	DEC	TOTAL
1969	0	0	0	20	88	250	310	392	154	20	0	0	1234
1970	0	0	0	8	86	190	385	398	207	30	0	0	1304
1971	0	0	0	0	29	290	404	347	218	40	7	0	1335
1972	0	0	0	5	47	118	384	338	169	3	0	0	1064
1973	0	0	0	20	23	260	390	401	171	22	2	0	1289
1974	0	0	0	19	47	155	385	360	115	1	6	0	1088
1975	0	0	0	0	120	185	341	299	43	22	15	0	1025
1976	0	0	0	65	24	270	310	299	103	5	0	0	1076
1977	0	0	3	22	110	189	442	338	159	0	0	0	1263
1978	0	0	0	0	77	209	301	348	81	4	0	0	1020
1979	0	0	0	4	71	149	378	376	192	43	5	0	1218
1980	0	0	0	1	94	188	448	480	213	11	0	0	1435
1981	0	0	0	4	78	252	425	347	129	0	0	0	1235
1982	0	0	0	7	55	152	405	266	129	36	16	0	1066
1983	0	0	0	19	16	259	460	404	244	35	0	0	1437
1984	0	0	0	3	39	301	306	367	106	26	0	0	1148
1985	0	0	8	28	95	139	353	329	189	21	5	0	1167

TABLE 6   SNOWFALL (inches)   NEW YORK, CENTRAL PARK, NEW YORK

SEASON	JULY	AUG	SEP	OCT	NOV	DEC	JAN	FEB	MAR	APR	MAY	JUNE	TOTAL
1956-57	0.0	0.0	0.0	0.0	T	0.9	8.9	7.0	2.6	2.5	0.0	0.0	21.9
1957-58	0.0	0.0	0.0	0.0	0.0	8.7	9.2	10.7	15.9	0.2	0.0	0.0	44.7
1958-59	0.0	0.0	0.0	0.0	T	3.8	1.5	0.4	6.7	0.6	0.0	0.0	13.0
1959-60	0.0	0.0	0.0	0.0	0.5	15.8	2.5	1.9	18.5	0.0	0.0	0.0	39.2
1960-61	0.0	0.0	0.0	T	0.0	18.6	16.7	18.2	1.2	T	0.0	0.0	54.7
1961-62	0.0	0.0	0.0	0.0	T	7.7	0.6	9.6	0.2	T	0.0	0.0	18.1
1962-63	0.0	0.0	0.0	T	T	4.5	5.3	3.7	2.8	T	0.0	0.0	16.3
1963-64	0.0	0.0	0.0	0.0	T	11.3	13.3	14.1	6.0	T	0.0	0.0	44.7
1964-65	0.0	0.0	0.0	0.0	0.0	3.1	14.8	2.5	2.8	1.2	0.0	0.0	24.4
1965-66	0.0	0.0	0.0	T	0.0	T	11.6	9.8	T	0.0	0.0	0.0	21.4
1966-67	0.0	0.0	0.0	0.0	0.0	9.1	1.4	23.6	17.4	T	0.0	0.0	51.5
1967-68	0.0	0.0	0.0	0.0	3.2	5.5	3.6	1.1	6.1	0.0	0.0	0.0	19.5
1968-69	0.0	0.0	0.0	0.0	T	7.0	1.0	16.6	5.6	0.0	0.0	0.0	30.2
1969-70	0.0	0.0	0.0	0.0	T	6.8	8.4	6.4	4.0	T	0.0	0.0	25.6
1970-71	0.0	0.0	0.0	0.0	0.0	2.4	11.4	T	1.3	0.4	0.0	0.0	15.5
1971-72	0.0	0.0	0.0	0.0	T	T	2.8	17.8	2.3	T	0.0	0.0	22.9
1972-73	0.0	0.0	0.0	T	T	1.8	0.8	0.2	T	0.0	0.0	0.0	2.8
1973-74	0.0	0.0	0.0	0.0	0.0	7.8	7.8	9.4	3.2	0.3	0.0	0.0	23.5
1974-75	0.0	0.0	0.0	0.0	0.1	0.1	2.0	10.6	0.3	T	0.0	0.0	13.1
1975-76	0.0	0.0	0.0	0.0	T	2.3	5.6	5.0	4.4	T	0.0	0.0	17.3
1976-77	0.0	0.0	0.0	0.0	T	5.1	13.0	5.8	0.6	T	T	0.0	24.5
1977-78	0.0	0.0	0.0	0.0	0.2	0.4	20.3	23.0	6.8	T	0.0	0.0	50.7
1978-79	0.0	0.0	0.0	0.0	2.2	0.5	6.6	20.1	T	T	0.0	0.0	29.4
1979-80	0.0	0.0	0.0	T	0.0	3.5	2.0	2.7	4.6	T	0.0	0.0	12.8
1980-81	0.0	0.0	0.0	0.0	T	2.8	8.0	T	8.6	0.0	0.0	0.0	19.4
1981-82	0.0	0.0	0.0	0.0	0.0	2.1	11.8	0.4	0.7	9.6	0.0	0.0	24.6
1982-83	0.0	0.0	0.0	0.0	0.0	3.0	1.9	23.5	T	0.8	0.0	0.0	29.2
1983-84	0.0	0.0	0.0	0.0	T	1.6	11.7	0.2	11.9	0.0	0.0	0.0	25.4
1984-85	0.0	0.0	0.0	0.0	T	5.5	8.4	10.0	0.2	T	0.0	0.0	24.1
1985-86	0.0	0.0	0.0	0.0	T	0.9							
Record Mean	0.0	0.0	0.0	T	0.9	5.6	7.6	8.7	5.1	0.9	T	0.0	28.8

See Reference Notes, relative to all above tables, on preceding page.

New York City, in area exceeding 300 square miles, is located on the Atlantic coastal plain at the mouth of the Hudson River. The terrain is laced with numerous waterways, all but one of the five boroughs in the city are situated on islands. Elevations range from less than 50 feet over most of Manhattan, Brooklyn, and Queens to almost 300 feet in northern Manhattan and the Bronx, and over 400 feet in Staten Island. Extensive suburban areas on Long Island, and in Connecticut, New York State and New Jersey border the city on the east, north, and west. About 30 miles to the west and northwest, hills rise to about 1,500 feet and to the north in upper Westchester County to 800 feet. To the southwest and to the east are the low-lying land areas of the New Jersey coastal plain and of Long Island, bordering on the Atlantic.

The New York Metropolitan area is close to the path of most storm and frontal systems which move across the North American continent. Therefore, weather conditions affecting the city most often approach from a westerly direction. New York City can thus experience higher temperatures in summer and lower ones in winter than would otherwise be expected in a coastal area. However, the frequent passage of weather systems often helps reduce the length of both warm and cold spells, and is also a major factor in keeping periods of prolonged air stagnation to a minimum.

Although continental influence predominates, oceanic influence is by no means absent. During the summer local sea breezes, winds blowing onshore from the cool water surface, often moderate the afternoon heat. The effect of the sea breeze diminishes inland. On winter mornings, ocean temperatures which are warm relative to the land reinforce the effect of the city heat island and low temperatures are often 10–20 degrees lower in the inland suburbs than in the central city. The relatively warm water temperatures also delay the advent of winter snows. Conversely, the lag in warming of water temperatures keeps spring temperatures relatively cool. One year-round measure of the ocean influence is the small average daily variation in temperature.

Precipitation is moderate and distributed fairly evenly throughout the year. Most of the rainfall from May through October comes from thunderstorms. It is therefore usually of brief duration and sometimes intense. Heavy rains of long duration associated with tropical storms occur infrequently in late summer or fall. For the other months of the year precipitation is more likely to be associated with widespread storm areas, so that day-long rain, snow or a mixture of both is more common. Precipitation accompanying winter storms sometimes starts as snow, later changes to rain, and perhaps briefly back to snow before ending. Coastal storms, occurring most often in the fall and winter months, produce on occasion considerable amounts of precipitation and have been responsible for record rains, snows, and high winds.

The average annual precipitation and snowfall totals are reasonably uniform within the city but show a consistent increase to the north and west with lesser amounts along the south shores and the eastern end of Long Island, reflecting the influence of the ocean waters. Relative humidity averages about the same over the metropolitan area except again that the immediate coastal areas are more humid than inland locations.

Local Climatological Data is published for three locations in New York City, Central Park, La Guardia Airport, and John F. Kennedy International Airport. Other nearby locations for which it is published are Newark, New Jersey, and Bridgeport, Connecticut.

TABLE 1   NORMALS, MEANS AND EXTREMES

J.F.K INTERNATIONAL AIRPORT N.Y.C, NEW YORK

LATITUDE: 40°39'N   LONGITUDE: 73°47'W   ELEVATION: FT. GRND   13 BARO   32   TIME ZONE: EASTERN   WBAN: 94789

	(a)	JAN	FEB	MAR	APR	MAY	JUNE	JULY	AUG	SEP	OCT	NOV	DEC	YEAR
TEMPERATURE °F:														
Normals														
–Daily Maximum		37.5	39.1	46.6	58.3	67.7	76.9	82.7	81.7	75.2	64.7	53.2	41.8	60.5
–Daily Minimum		25.1	25.9	33.2	42.3	51.7	61.0	67.2	66.3	59.2	48.7	39.6	29.6	45.8
–Monthly		31.3	32.5	39.9	50.3	59.7	69.0	75.0	74.0	67.2	56.7	46.4	35.7	53.1
Extremes														
–Record Highest	25	65	67	77	90	99	99	104	100	98	84	77	70	104
–Year		1974	1976	1985	1977	1969	1964	1966	1983	1983	1967	1982	1982	JUL 1966
–Record Lowest	25	-2	-2	7	20	34	45	55	46	40	25	19	2	-2
–Year		1985	1963	1967	1982	1966	1967	1979	1965	1963	1961	1976	1983	JAN 1985
NORMAL DEGREE DAYS:														
Heating (base 65°F)		1045	910	775	441	192	23	0	0	47	270	558	908	5169
Cooling (base 65°F)		0	0	0	0	28	143	310	279	113	13	0	0	886
% OF POSSIBLE SUNSHINE														
MEAN SKY COVER (tenths)														
Sunrise – Sunset	27	6.1	6.1	6.2	6.0	6.2	6.0	5.9	5.8	5.4	5.4	6.2	6.3	6.0
MEAN NUMBER OF DAYS:														
Sunrise to Sunset														
–Clear	27	8.4	7.9	7.9	8.2	6.9	7.3	7.5	7.5	10.1	10.8	7.9	7.7	98.1
–Partly Cloudy	27	8.6	7.6	9.4	9.3	11.4	11.0	12.6	13.1	9.6	9.1	8.3	8.2	118.3
–Cloudy	27	14.0	12.8	13.7	12.5	12.7	11.7	10.9	10.4	10.4	11.0	13.7	15.1	148.9
Precipitation														
.01 inches or more	27	10.3	9.6	11.3	10.3	11.3	9.9	8.9	9.0	8.1	7.5	10.0	11.4	117.7
Snow, Ice pellets														
1.0 inches or more	27	2.6	2.0	1.2	0.1	0.0	0.0	0.0	0.0	0.0	0.0	0.1	1.1	7.1
Thunderstorms	27	0.1	0.3	1.0	1.6	2.8	3.8	4.4	4.8	1.7	1.1	0.4	0.1	22.3
Heavy Fog Visibility														
1/4 mile or less	27	2.7	2.7	3.7	2.7	4.3	3.9	2.5	1.3	1.1	2.3	2.2	2.6	32.1
Temperature °F														
–Maximum														
90° and above	24	0.0	0.0	0.0	0.*	0.3	1.8	4.1	3.0	0.9	0.0	0.0	0.0	10.2
32° and below	24	10.1	6.0	1.0	0.*	0.0	0.0	0.0	0.0	0.0	0.0	0.*	4.1	21.3
–Minimum														
32° and below	24	24.1	20.7	11.7	1.6	0.0	0.0	0.0	0.0	0.0	0.4	4.3	17.7	80.4
0° and below	24	0.2	0.1	0.0	0.0	0.0	0.0	0.0	0.0	0.0	0.0	0.0	0.0	0.3
AVG. STATION PRESS.(mb)	13	1016.4	1017.0	1015.1	1014.1	1014.1	1014.7	1014.9	1016.7	1017.7	1018.6	1017.5	1017.7	1016.2
RELATIVE HUMIDITY (%)														
Hour 01	24	68	68	68	70	77	80	78	79	80	76	73	71	74
Hour 07 (Local Time)	24	70	71	70	69	73	75	75	78	80	78	76	73	74
Hour 13	24	59	58	56	55	59	61	59	59	59	56	59	61	58
Hour 19	24	63	63	63	64	69	72	72	72	72	69	67	66	68
PRECIPITATION (inches):														
Water Equivalent														
–Normal		2.93	3.20	3.99	3.76	3.40	2.98	3.56	4.10	3.51	2.98	3.73	3.62	41.76
–Maximum Monthly	38	8.33	5.48	8.17	9.51	8.59	9.20	8.48	17.41	9.65	6.41	9.51	6.16	17.41
–Year		1979	1960	1980	1983	1984	1984	1969	1955	1975	1958	1972	1969	AUG 1955
–Minimum Monthly	38	0.21	1.06	0.95	1.12	0.38	T	0.46	0.42	0.70	0.09	0.32	0.87	T
–Year		1956	1980	1981	1963	1955	1949	1954	1972	1951	1963	1976	1985	JUN 1949
–Maximum in 24 hrs	38	3.25	2.87	2.40	3.31	2.88	6.27	5.92	6.59	5.83	3.42	4.09	2.46	6.59
–Year		1979	1958	1977	1980	1968	1984	1984	1955	1960	1972	1972	1974	AUG 1955
Snow, Ice pellets														
–Maximum Monthly	27	20.1	25.3	21.1	8.2	T					0.5	2.1	16.4	25.3
–Year		1978	1961	1960	1982	1967					1962	1967	1960	FEB 1961
–Maximum in 24 hrs	27	14.2	21.7	8.1	8.0	T					0.5	2.1	8.2	21.7
–Year		1978	1983	1967	1982	1967					1962	1967	1960	FEB 1983
WIND:														
Mean Speed (mph)	27	13.6	13.9	14.0	13.4	11.9	10.9	10.6	10.3	10.6	11.2	12.5	13.0	12.2
Prevailing Direction														
Fastest Obs. 1 Min.														
–Direction (!!!)	22	26	25	28	29	16	29	22	30	28	26	05	06	26
–Speed (MPH)	22	52	46	44	44	44	41	40	46	46	44	44	46	52
–Year		1966	1967	1971	1982	1973	1981	1984	1965	1985	1980	1968	1974	JAN 1966
Peak Gust														
–Direction (!!!)	2	N	NW	NE	S	W	NW	SW	S	W	W	SW	W	NE
–Speed (mph)	2	45	60	61	46	41	49	54	32	58	33	47	45	61
–Date		1985	1985	1984	1985	1984	1984	1984	1984	1985	1985	1984	1985	MAR 1984

See Reference Notes to this table on the following page.

TABLE 2 PRECIPITATION (inches)   J.F.K INTERNATIONAL AIRPORT N.Y.C. NEW YORK

YEAR	JAN	FEB	MAR	APR	MAY	JUNE	JULY	AUG	SEP	OCT	NOV	DEC	ANNUAL
1956	0.21	5.21	4.40	3.38	2.95	3.54	5.45	2.59	2.50	2.54	5.37	3.90	42.04
1957	1.76	2.19	2.97	6.36	2.60	1.96	1.41	2.81	2.54	3.46	2.74	5.02	35.82
1958	3.95	4.42	4.01	5.90	4.41	2.98	3.91	3.69	4.02	6.41	2.27	1.72	47.69
1959	2.10	1.87	4.41	2.07	2.93	3.61	4.87	3.71	0.87	2.89	3.49	4.02	36.84
1960	2.96	5.48	3.65	2.49	3.03	1.33	8.08	4.80	9.60	2.28	2.92	3.24	49.86
1961	2.43	4.17	5.23	6.29	4.07	1.86	7.34	3.56	3.08	2.87	2.29	2.88	46.07
1962	2.39	4.93	2.96	2.43	1.45	4.27	2.66	4.78	2.85	2.81	5.29	2.41	39.23
1963	2.05	3.35	4.56	1.12	2.64	1.91	2.81	2.65	5.01	0.09	7.89	1.80	35.88
1964	4.59	2.84	2.34	6.60	0.62	2.22	4.65	0.87	2.58	2.03	2.02	5.34	36.70
1965	2.99	2.79	2.74	3.18	1.61	1.12	2.45	2.11	0.96	1.85	1.36	2.22	25.38
1966	3.18	4.08	1.35	3.24	3.51	0.21	0.69	2.99	7.47	4.92	2.29	2.78	36.71
1967	1.47	2.48	5.48	3.67	5.12	3.39	4.98	7.36	1.44	2.13	5.70	1.89	45.11
1968	1.99	1.73	4.90	1.97	5.34	4.16	2.58	2.78	2.54	1.85	6.84	4.55	41.23
1969	1.10	3.72	2.03	4.26	1.72	2.85	8.48	2.39	3.80	2.48	6.16	1.70	40.69
1970	0.56	3.33	4.03	3.15	2.01	3.04	0.54	2.57	1.61	1.93	3.84	2.55	29.16
1971	2.31	4.76	3.17	2.90	2.61	1.71	3.19	4.35	3.72	3.28	4.07	1.46	37.53
1972	1.74	4.89	4.09	3.32	5.02	6.70	2.60	0.42	1.49	5.00	9.51	4.26	49.04
1973	3.37	2.61	3.72	6.98	3.97	5.25	2.64	0.84	3.03	2.26	1.41	5.96	42.04
1974	2.83	1.41	4.60	2.33	2.69	2.38	1.29	4.24	5.97	2.19	1.10	6.07	37.10
1975	5.05	3.57	3.50	3.13	3.52	7.06	7.86	3.54	9.65	3.25	4.02	3.03	57.18
1976	4.55	3.16	2.30	2.26	3.56	3.25	3.50	8.30	2.24	4.06	0.32	2.02	40.24
1977	2.67	2.41	4.70	3.62	2.29	3.25	2.30	4.85	6.69	4.41	6.59	4.92	48.70
1978	7.74	1.74	2.38	2.10	6.44	1.24	3.50	3.58	3.41	1.68	1.86	5.29	40.96
1979	8.33	4.58	3.77	3.03	5.02	3.42	1.66	5.65	4.13	3.53	2.00	2.09	47.21
1980	1.55	1.06	8.17	7.53	2.77	3.57	3.93	0.85	1.32	2.79	3.98	0.90	38.42
1981	0.49	4.33	0.95	2.77	2.80	4.28	5.32	1.05	2.67	3.74	1.77	4.40	34.57
1982	4.83	2.08	2.61	4.40	3.43	5.08	2.67	2.15	1.15	1.29	2.85	1.52	34.06
1983	4.14	2.79	6.66	9.51	3.56	2.10	3.69	5.84	3.95	5.12	5.62	6.14	59.12
1984	1.60	4.55	5.99	5.82	8.59	9.20	6.66	1.47	2.02	1.91	2.47	2.94	53.22
1985	1.06	2.13	2.17	2.12	4.77	4.56	2.63	1.80	3.29	0.93	6.68	0.87	33.01
Record Mean	2.86	3.20	3.94	3.92	3.58	3.27	3.65	3.87	3.39	2.92	3.76	3.56	41.92

TABLE 3 AVERAGE TEMPERATURE (deg. F)   J.F.K INTERNATIONAL AIRPORT N.Y.C. NEW YORK

YEAR	JAN	FEB	MAR	APR	MAY	JUNE	JULY	AUG	SEP	OCT	NOV	DEC	ANNUAL
1956	30.4	35.9	36.6	46.0	55.8	70.1	72.4	73.7	64.9	57.1	46.2	40.7	52.5
#1957	28.3	36.5	40.3	49.6	60.7	71.6	76.3	72.4	67.3	55.9	49.3	40.0	54.0
1958	31.6	28.1	40.4	51.1	57.9	66.1	75.9	74.6	67.2	55.4	48.7	30.3	52.4
1959	31.6	31.9	39.8	51.7	62.9	70.3	75.6	76.8	71.3	59.5	45.3	38.7	54.6
1960	34.3	36.3	32.9	52.6	61.4	70.9	74.2	74.2	67.4	57.4	49.2	31.4	53.4
#1961	27.6	35.0	40.2	47.5	58.4	68.7	76.3	75.3	73.0	59.5	47.0	34.2	53.6
1962	30.5	30.2	39.7	48.9	62.3	71.4	73.3	71.5	64.0	55.9	42.5	32.0	52.0
1963	29.9	28.7	41.6	52.0	59.7	70.9	75.5	72.4	63.2	59.1	49.6	29.9	52.7
1964	33.5	31.5	39.9	47.1	63.1	70.4	74.5	72.2	68.3	55.8	48.8	36.5	53.5
1965	28.5	32.3	38.6	48.3	61.8	67.9	72.2	72.3	66.1	55.6	45.4	38.6	52.3
1966	31.2	33.8	41.9	47.5	56.8	70.0	77.0	74.5	66.2	55.2	47.7	35.6	53.1
1967	36.4	27.8	34.9	46.4	51.7	66.7	73.3	71.4	64.4	55.5	41.1	36.7	50.5
1968	25.6	27.9	40.2	50.5	56.8	68.1	77.0	75.8	70.1	58.2	46.6	33.0	52.5
1969	31.3	30.8	38.2	52.3	61.7	71.2	75.0	77.4	68.5	58.1	47.2	35.2	53.9
1970	26.7	33.3	37.5	50.0	61.7	70.3	77.0	77.4	70.0	58.3	48.6	35.0	53.8
1971	27.9	34.2	38.9	47.8	59.5	72.1	77.2	76.5	71.8	63.5	47.2	42.8	55.0
1972	36.4	33.5	40.4	49.8	61.2	68.0	77.0	74.8	69.3	54.5	46.5	41.2	54.4
1973	36.6	35.1	47.7	54.2	57.2	69.2	74.0	75.3	67.2	57.7	45.9	37.2	54.8
1974	35.1	31.6	42.7	52.5	59.6	67.8	76.6	75.7	66.9	54.4	47.9	39.8	54.2
1975	37.9	35.9	39.9	47.1	62.3	69.0	75.0	74.4	64.1	60.0	52.8	36.6	54.6
1976	28.1	38.7	43.1	53.5	58.6	70.2	72.9	72.7	65.6	52.8	40.7	30.1	52.3
1977	22.0	32.1	43.7	51.0	61.7	67.4	75.1	73.3	66.8	53.7	47.2	34.3	52.3
1978	29.2	26.8	38.7	50.6	59.5	70.7	74.5	77.5	66.6	57.3	48.8	39.3	53.3
1979	33.2	25.0	45.4	51.5	62.8	66.9	74.7	73.6	67.1	54.9	48.3	38.1	53.5
1980	32.2	29.2	38.4	51.9	64.2	68.1	76.5	78.0	69.9	56.3	45.4	33.7	53.6
1981	26.5	38.7	42.2	53.7	60.6	70.5	77.3	75.4	68.1	54.6	47.7	37.5	54.4
1982	26.7	35.8	40.9	49.2	61.8	66.9	75.7	72.7	67.1	57.6	50.4	43.6	54.1
1983	35.8	36.2	44.6	51.5	58.0	71.6	78.7	76.7	71.0	58.8	49.6	35.9	55.7
1984	29.3	39.8	36.2	51.1	61.0	73.9	74.8	77.6	66.1	61.7	47.4	43.9	55.2
1985	28.8	34.8	45.1	54.2	63.1	68.4	74.8	73.9	69.2	58.3	50.1	34.0	54.6
Record Mean	31.2	33.4	40.5	50.5	60.2	69.6	75.5	74.7	67.7	57.3	47.1	36.6	53.7
Max	37.4	40.0	47.3	58.2	68.0	77.4	83.0	82.0	75.2	64.8	53.7	42.6	60.8
Min	25.0	26.8	33.8	42.8	52.4	61.8	67.9	67.3	60.1	49.7	40.5	30.6	46.6

REFERENCE NOTES FOR TABLES 1, 2, 3 and 6   (NEW YORK [JFK INT. AIRPORT], NY)

GENERAL

T - TRACE AMOUNT
BLANK ENTRIES DENOTE MISSING/UNREPORTED DATA.
# INDICATES A STATION OR INSTRUMENT RELOCATION.

SPECIFIC

TABLE 1

(a) - LENGTH OF RECORD IN YEARS. ALTHOUGH INDIVIDUAL MONTHS MAY BE MISSING.

* LESS THAN .05

NORMALS — BASED ON THE 1951-1980 RECORD PERIOD.
EXTREMES — DATES ARE THE MOST RECENT OCCURRENCE.
WIND DIR. — NUMERALS SHOW TENS OF DEGREES CLOCKWISE FROM TRUE NORTH. "00" INDICATES CALM.
RESULTANT WIND DIRECTIONS ARE GIVEN TO WHOLE DEGREES.

EXCEPTIONS

TABLE 1

1. FROM AUGUST 1966-MARCH 1969, MAXIMUM 24-HOUR PRECIPITATION IS FOR CALENDAR DAY (MIDNIGHT TO MIDNIGHT) INSTEAD OF FOR ANY CONSECUTIVE 24 HOURS.

2. MAXIMUM 24-HOUR SNOW IS FOR CALENDAR DAY THROUGH MARCH 1969.

TABLES 2, 3, and 6

RECORD MEANS ARE THROUGH THE CURRENT YEAR, BEGINNING IN   1951 FOR TEMPERATURE
1951 FOR PRECIPITATION
1951 FOR SNOWFALL

TABLE 4 HEATING DEGREE DAYS Base 65 deg. F   J.F.K INTERNATIONAL AIRPORT N.Y.C, NEW YORK

SEASON	JULY	AUG	SEP	OCT	NOV	DEC	JAN	FEB	MAR	APR	MAY	JUNE	TOTAL
1959-60	0	0	36	225	583	809	945	827	972	371	125	1	4894
#1960-61	0	0	16	242	467	1026	1151	831	760	517	199	10	5219
1961-62	0	0	28	191	535	950	1064	967	777	477	144	10	5143
1962-63	0	8	84	282	667	1020	1079	1010	721	383	181	6	5441
1963-64	0	1	113	190	457	1080	969	968	773	530	124	24	5229
1964-65	0	0	42	282	478	878	1126	909	809	496	146	26	5192
1965-66	0	14	58	289	584	810	1041	868	707	521	256	14	5162
1966-67	0	0	60	295	513	904	883	1033	930	554	408	22	5602
1967-68	0	4	72	305	710	871	1214	1066	761	427	245	19	5694
1968-69	0	0	3	221	545	983	1038	950	822	374	144	0	5080
1969-70	0	0	37	232	526	917	1180	881	844	444	134	3	5198
1970-71	0	0	25	219	487	921	1144	858	803	508	180	9	5154
1971-72	0	0	11	83	532	685	881	904	755	449	123	9	4432
1972-73	0	0	21	321	549	732	876	831	530	325	242	13	4440
1973-74	0	0	64	234	568	856	920	929	686	373	183	26	4839
1974-75	0	0	47	322	508	775	833	809	770	531	135	13	4743
1975-76	0	3	60	171	367	874	1140	756	672	348	195	28	4614
1976-77	0	6	53	373	726	1076	1326	916	651	418	147	34	5726
1977-78	0	7	68	344	530	947	1104	1062	805	422	197	7	5493
1978-79	2	0	62	237	481	791	978	1115	603	397	88	19	4773
1979-80	2	6	45	313	493	827	1010	1030	823	387	106	39	5081
1980-81	0	0	25	266	585	962	1184	731	700	334	159	12	4958
1981-82	0	0	46	313	511	844	1182	812	741	468	110	44	5071
1982-83	0	6	24	245	434	655	901	801	625	400	212	8	4311
1983-84	0	2	38	224	455	891	1101	726	886	409	134	7	4873
1984-85	0	0	67	119	522	647	1118	841	606	328	95	18	4361
1985-86	0	0	26	213	441	954							

TABLE 5   COOLING DEGREE DAYS Base 65 deg. F   J.F.K INTERNATIONAL AIRPORT N.Y.C, NEW YORK

YEAR	JAN	FEB	MAR	APR	MAY	JUNE	JULY	AUG	SEP	OCT	NOV	DEC	TOTAL
1969	0	0	0	1	48	192	313	390	147	26	0	0	1117
1970	0	0	0	0	42	168	378	389	182	17	0	0	1176
1971	0	0	0	0	16	231	384	365	221	45	8	0	1270
1972	0	0	0	0	11	107	378	309	159	4	0	0	968
1973	0	0	0	9	7	145	287	324	137	14	0	0	923
1974	0	0	0	7	24	116	366	342	111	1	1	0	968
1975	0	0	0	0	58	137	319	301	41	24	7	0	887
1976	0	0	0	9	5	187	250	253	80	2	0	0	786
1977	0	0	0	5	51	114	321	268	127	0	0	0	886
1978	0	0	0	0	37	185	305	391	119	5	0	0	1042
1979	0	0	0	0	26	83	310	283	115	7	0	0	824
1980	0	0	0	0	48	139	365	412	178	6	0	0	1148
1981	0	0	0	3	30	184	390	327	143	0	0	0	1077
1982	0	0	0	0	16	107	336	252	92	21	4	0	828
1983	0	0	0	0	1	213	431	371	223	36	0	0	1275
1984	0	0	0	10	16	285	310	396	106	21	1	0	1135
1985	0	0	0	10	44	126	312	284	157	14	0	0	947

TABLE 6   SNOWFALL (inches)   J.F.K INTERNATIONAL AIRPORT N.Y.C, NEW YORK

SEASON	JULY	AUG	SEP	OCT	NOV	DEC	JAN	FEB	MAR	APR	MAY	JUNE	TOTAL
1959-60	0.0	0.0	0.0	0.0	0.6	6.7	3.2	2.6	21.1	T	0.0	0.0	34.2
1960-61	0.0	0.0	0.0	0.0	0.0	16.4	16.7	25.3	T	T	0.0	0.0	58.4
1961-62	0.0	0.0	0.0	0.0	T	6.1	2.3	9.9	0.9	T	0.0	0.0	19.2
1962-63	0.0	0.0	0.0	0.5	T	6.2	4.4	4.4	1.5	T	0.0	0.0	17.0
1963-64	0.0	0.0	0.0	0.0	T	7.1	14.9	10.0	2.4	T	0.0	0.0	34.4
1964-65	0.0	0.0	0.0	0.0	T	3.0	17.4	4.1	3.6	1.4	0.0	0.0	29.5
1965-66	0.0	0.0	0.0	T	T	T	10.1	5.5	T	T	0.0	0.0	15.6
1966-67	0.0	0.0	0.0	0.0	T	8.8	2.8	19.9	15.5	T	T	0.0	47.0
1967-68	0.0	0.0	0.0	0.0	2.1	4.3	5.1	1.7	4.2	0.0	0.0	0.0	17.4
1968-69	0.0	0.0	0.0	0.0	T	4.4	0.6	22.4	3.4	0.0	0.0	0.0	30.8
1969-70	0.0	0.0	0.0	0.0	T	6.3	5.5	5.3	2.4	0.0	0.0	0.0	19.5
1970-71	0.0	0.0	0.0	0.0	T	2.3	11.6	0.2	2.8	3.2	0.0	0.0	20.1
1971-72	0.0	0.0	0.0	0.0	T	0.1	1.7	10.3	2.8	T	0.0	0.0	14.9
1972-73	0.0	0.0	0.0	0.0	T	T	0.8	0.8	T	T	0.0	0.0	1.6
1973-74	0.0	0.0	0.0	0.0	0.0	0.5	6.7	10.5	4.1	T	0.0	0.0	21.8
1974-75	0.0	0.0	0.0	0.0	0.2	T	0.6	8.8	0.9	T	0.0	0.0	10.5
1975-76	0.0	0.0	0.0	0.0	T	1.2	6.9	7.3	5.3	0.0	0.0	0.0	20.7
1976-77	0.0	0.0	0.0	0.0	T	5.8	13.4	3.5	T	0.0	0.0	0.0	22.7
1977-78	0.0	0.0	0.0	0.0	0.6	1.0	20.1	18.1	8.7	0.0	0.0	0.0	48.5
1978-79	0.0	0.0	0.0	0.0	1.7	0.3	7.4	17.6	0.2	T	0.0	0.0	27.2
1979-80	0.0	0.0	0.0	T	0.0	2.5	3.0	2.3	3.2	T	0.0	0.0	11.0
1980-81	0.0	0.0	0.0	0.0	0.2	1.7	7.7	T	6.9	0.0	0.0	0.0	16.5
1981-82	0.0	0.0	0.0	0.0	T	3.1	12.5	0.8	0.3	8.2	0.0	0.0	24.9
1982-83	0.0	0.0	0.0	0.0	0.0	4.8	1.0	24.7	0.1	1.5	0.0	0.0	32.1
1983-84	0.0	0.0	0.0	0.0	T	1.2	9.9	T	10.9	T	0.0	0.0	22.0
1984-85	0.0	0.0	0.0	0.0	T	5.5	12.4	9.0	0.4	T	0.0	0.0	27.3
1985-86	0.0	0.0	0.0	0.0	T	2.8							
Record Mean	0.0	0.0	0.0	T	0.2	3.8	7.4	8.4	4.0	0.5	T	0.0	24.4

See Reference Notes, relative to all above tables, on preceding page.

New York City, in area exceeding 300 square miles, is located on the Atlantic coastal plain at the mouth of the Hudson River. The terrain is laced with numerous waterways, all but one of the five boroughs in the city are situated on islands. Elevations range from less than 50 feet over most of Manhattan, Brooklyn, and Queens to almost 300 feet in northern Manhattan and the Bronx, and over 400 feet in Staten Island. Extensive suburban areas on Long Island, and in Connecticut, New York State and New Jersey border the city on the east, north, and west. About 30 miles to the west and northwest, hills rise to about 1,500 feet and to the north in upper Westchester County to 800 feet. To the southwest and to the east are the low-lying land areas of the New Jersey coastal plain and of Long Island, bordering on the Atlantic.

The New York Metropolitan area is close to the path of most storm and frontal systems which move across the North American continent. Therefore, weather conditions affecting the city most often approach from a westerly direction. New York City can thus experience higher temperatures in summer and lower ones in winter than would otherwise be expected in a coastal area. However, the frequent passage of weather systems often helps reduce the length of both warm and cold spells, and is also a major factor in keeping periods of prolonged air stagnation to a minimum.

Although continental influence predominates, oceanic influence is by no means absent. During the summer local sea breezes, winds blowing onshore from the cool water surface, often moderate the afternoon heat. The effect of the sea breeze diminishes inland. On winter mornings, ocean temperatures which are warm relative to the land reinforce the effect of the city heat island and low temperatures are often 10–20 degrees lower in the inland suburbs than in the central city. The relatively warm water temperatures also delay the advent of winter snows. Conversely, the lag in warming of water temperatures keeps spring temperatures relatively cool. One year-round measure of the ocean influence is the small average daily variation in temperature.

Precipitation is moderate and distributed fairly evenly throughout the year. Most of the rainfall from May through October comes from thunderstorms. It is therefore usually of brief duration and sometimes intense. Heavy rains of long duration associated with tropical storms occur infrequently in late summer or fall. For the other months of the year precipitation is more likely to be associated with widespread storm areas, so that day-long rain, snow or a mixture of both is more common. Precipitation accompanying winter storms sometimes starts as snow, later changes to rain, and perhaps briefly back to snow before ending. Coastal storms, occurring most often in the fall and winter months, produce on occasion considerable amounts of precipitation and have been responsible for record rains, snows, and high winds.

The average annual precipitation and snowfall totals are reasonably uniform within the city but show a consistent increase to the north and west with lesser amounts along the south shores and the eastern end of Long Island, reflecting the influence of the ocean waters. Relative humidity averages about the same over the metropolitan area except again that the immediate coastal areas are more humid than inland locations.

Local Climatological Data is published for three locations in New York City, Central Park, La Guardia Airport, and John F. Kennedy International Airport. Other nearby locations for which it is published are Newark, New Jersey, and Bridgeport, Connecticut.

TABLE 1   NORMALS, MEANS AND EXTREMES

NEW YORK, LA GUARDIA FIELD, NEW YORK

LATITUDE: 40°46'N   LONGITUDE: 73°54'W   ELEVATION: FT. GRND   11 BARO   00039   TIME ZONE: EASTERN   WBAN: 14732

	(a)	JAN	FEB	MAR	APR	MAY	JUNE	JULY	AUG	SEP	OCT	NOV	DEC	YEAR
TEMPERATURE °F:														
Normals														
-Daily Maximum		37.4	39.2	47.3	59.6	69.7	78.7	83.9	82.3	75.2	64.5	52.9	41.5	61.0
-Daily Minimum		26.1	27.3	34.6	44.2	53.7	63.2	68.9	68.2	61.2	50.5	41.2	30.8	47.5
-Monthly		31.8	33.3	41.0	51.9	61.7	71.0	76.4	75.3	68.2	57.5	47.0	36.2	54.3
Extremes														
-Record Highest	24	68	73	80	91	95	97	107	97	96	85	79	69	107
-Year		1967	1985	1985	1976	1962	1964	1966	1985	1983	1967	1974	1982	JUL 1966
-Record Lowest	24	-3	-2	8	22	38	46	56	51	44	30	18	-1	-3
-Year		1985	1963	1980	1982	1983	1972	1979	1982	1974	1969	1976	1980	JAN 1985
NORMAL DEGREE DAYS:														
Heating (base 65°F)		1029	888	744	393	149	8	0	0	35	246	537	893	4922
Cooling (base 65°F)		0	0	0	0	47	188	353	319	131	13	0	0	1051
% OF POSSIBLE SUNSHINE														
MEAN SKY COVER (tenths)														
Sunrise - Sunset	37	6.4	6.3	6.3	6.2	6.4	6.0	5.9	5.8	5.6	5.5	6.4	6.5	6.1
MEAN NUMBER OF DAYS:														
Sunrise to Sunset														
-Clear	37	7.7	7.6	7.3	7.5	6.3	7.4	7.4	7.9	9.5	10.4	7.4	7.8	94.2
-Partly Cloudy	37	8.4	7.4	9.3	9.4	11.2	11.1	12.8	12.2	9.5	9.1	8.6	8.7	117.5
-Cloudy	37	14.9	13.3	14.4	13.1	13.5	11.5	10.9	10.9	11.0	11.6	14.0	14.5	153.5
Precipitation														
.01 inches or more	45	11.2	9.5	11.2	10.7	11.4	9.9	9.2	9.2	8.1	7.8	10.0	10.7	118.8
Snow,Ice pellets														
1.0 inches or more	41	2.2	2.1	1.2	0.1	0.0	0.0	0.0	0.0	0.0	0.*	0.1	1.3	7.1
Thunderstorms	37	0.2	0.2	0.9	1.9	3.2	3.9	5.0	4.7	2.2	0.9	0.4	0.1	23.7
Heavy Fog Visibility														
1/4 mile or less	37	1.7	1.4	1.4	1.3	1.6	1.2	0.8	0.4	0.3	0.9	0.7	1.2	12.9
Temperature °F														
-Maximum														
90° and above	23	0.0	0.0	0.0	0.*	0.5	2.6	5.4	3.4	1.3	0.0	0.0	0.0	13.3
32° and below	23	11.3	6.9	1.1	0.*	0.0	0.0	0.0	0.0	0.0	0.0	0.1	4.4	24.0
-Minimum														
32° and below	23	23.6	20.0	10.7	1.3	0.0	0.0	0.0	0.0	0.0	0.1	3.1	16.3	75.1
0° and below	23	0.2	0.1	0.0	0.0	0.0	0.0	0.0	0.0	0.0	0.0	0.0	0.*	0.3
AVG. STATION PRESS.(mb)	13	1016.3	1017.0	1015.0	1013.9	1013.9	1014.3	1014.6	1016.3	1017.5	1018.5	1017.4	1017.7	1016.0
RELATIVE HUMIDITY (%)														
Hour 01	23	63	62	63	63	69	72	71	73	73	70	68	66	68
Hour 07	23	65	64	66	65	70	72	72	75	75	73	71	68	70
Hour 13 (Local Time)	23	57	55	54	50	53	55	52	55	56	55	58	60	55
Hour 19	23	59	57	57	55	59	60	60	63	64	62	62	62	60
PRECIPITATION (inches):														
Water Equivalent														
-Normal		3.11	3.08	4.10	3.76	3.46	3.15	3.67	4.32	3.48	3.24	3.77	3.68	42.82
-Maximum Monthly	45	8.68	5.76	8.73	11.51	9.27	8.15	12.33	16.05	9.63	9.09	9.92	7.70	16.05
-Year		1979	1960	1953	1983	1984	1972	1975	1955	1975	1943	1972	1973	AUG 1955
-Minimum Monthly	45	0.51	0.92	0.87	0.99	0.43	0.03	0.69	0.24	0.62	0.06	0.31	0.31	0.03
-Year		1981	1978	1966	1985	1964	1949	1954	1964	1941	1963	1976	1955	JUN 1949
-Maximum in 24 hrs	45	3.55	2.90	3.25	3.06	3.02	3.73	3.82	7.11	4.52	3.58	4.46	3.44	7.11
-Year		1979	1941	1953	1984	1968	1984	1971	1955	1969	1972	1977	1941	AUG 1955
Snow,Ice pellets														
-Maximum Monthly	41	18.3	26.4	18.9	8.2	T					1.2	4.0	26.8	26.8
-Year		1948	1983	1958	1982	1977					1962	1953	1947	DEC 1947
-Maximum in 24 hrs	41	11.0	22.0	15.3	8.2	T					1.2	4.0	22.8	22.8
-Year		1978	1983	1960	1982	1977					1962	1953	1947	DEC 1947
WIND:														
Mean Speed (mph)	37	13.9	14.0	14.2	13.1	11.6	10.9	10.4	10.3	11.0	11.7	12.7	13.5	12.3
Prevailing Direction														
through 1963		WNW	WNW	NW	NW	NE	S	S	S	S	SW	WNW	WNW	S
Fastest Obs. 1 Min.														
-Direction (!!)	6	18	07	13	32	30	34	36	30	30	10	31	32	10
-Speed (MPH)	6	37	39	44	39	32	35	35	31	46	52	35	37	52
-Year		1980	1984	1980	1982	1985	1984	1983	1980	1985	1980	1980	1980	OCT 1980
Peak Gust														
-Direction (!!)	2	NW	SE	NE	SE	S	N	SW	SW	NW	N	NW	NW	NW
-Speed (mph)	2	48	55	62	52	44	51	46	48	64	41	44	54	64
-Date		1985	1985	1984	1984	1984	1984	1984	1985	1985	1985	1984	1985	SEP 1985

See reference Notes to this table on the following page.

TABLE 2 PRECIPITATION (inches)   NEW YORK, LA GUARDIA FIELD, NEW YORK

YEAR	JAN	FEB	MAR	APR	MAY	JUNE	JULY	AUG	SEP	OCT	NOV	DEC	ANNUAL
1956	1.73	4.84	4.76	3.31	2.59	2.76	4.15	3.06	2.11	4.94	2.89	4.07	41.21
1957	1.86	2.20	2.43	5.70	2.92	1.64	1.32	2.43	4.40	3.73	4.17	5.25	38.05
1958	5.23	5.47	4.06	7.24	4.72	2.61	3.89	4.08	6.88	1.97	1.56		51.34
1959	2.58	2.01.	4.79	2.38	1.70	4.76	3.37	4.17	1.42	4.88	4.13	5.00	41.19
1960	2.77	5.76	3.36	2.81	3.39	1.99	9.27	5.88	7.61	2.43	2.71	3.19	51.17
1961	3.20	4.10	5.10	7.36	3.97	2.76	5.47	3.37	1.93	2.05	2.49	3.52	45.32
1962	2.59	4.08	2.61	2.69	1.03	1.38	5.09	3.02	2.38	3.96	2.05		35.15
1963	2.18	2.46	3.25	1.60	1.78	1.43	1.79	4.24	3.53	0.06	7.32	2.13	31.77
1964	4.22	2.39	2.14	5.12	0.43	2.85	4.56	0.24	1.78	1.68	1.65	4.25	31.31
1965	2.93	3.24	1.94	2.75	1.35	1.14	1.34	2.02	1.26	1.44	1.29	1.47	22.17
1966	2.36	4.29	0.87	2.31	3.78	1.44	1.12	1.74	5.07	3.73	2.44	2.54	31.69
1967	1.06	2.36	5.77	2.91	3.52	3.13	5.02	7.03	1.39	2.66	2.39	5.82	43.06
1968	1.67	1.37	4.77	2.59	4.67	6.16	2.39	3.19	1.84	1.95	5.13	3.80	39.53
1969	0.93	2.59	3.06	4.57	2.52	4.20	8.35	2.64	7.35	1.69	3.06	1.86	46.26
1970	0.76	3.24	3.16	2.36	2.02	2.36	1.18	2.68	1.70	1.45	4.59	1.86	27.36
1971	2.27	4.10	3.08	2.08	3.37	2.21	6.32	8.38	4.93	3.32	4.58	1.55	46.19
1972	2.02	4.80	4.12	3.19	6.85	8.15	3.47	0.94	1.65	4.91	9.92	4.66	54.68
1973	3.47	3.02	3.47	7.17	4.21	6.79	4.90	2.32	2.63	2.73	1.54	7.70	49.95
1974	2.78	1.28	4.45	2.94	3.82	2.51	1.43	6.07	6.90	2.03	1.01	6.38	41.60
1975	4.50	3.20	2.81	2.77	3.79	7.19	12.33	3.79	9.63	3.36	3.92	3.69	60.79
1976	5.42	2.94	2.30	2.45	3.87	2.74	1.00	5.95	2.82	4.62	0.31	2.23	36.65
1977	2.02	2.13	6.57	2.93	1.82	3.56	1.53	4.48	4.80	5.84	8.30	4.86	48.84
1978	6.11	0.92	2.14	1.95	8.15	1.30	3.79	4.01	3.93	1.41	2.24	4.90	40.85
1979	8.68	4.28	3.76	3.55	4.32	1.51	1.37	4.80	4.07	3.83	3.04	2.57	45.78
1980	1.94	0.95	8.65	6.55	2.14	3.43	4.74	1.32	1.16	3.15	4.17	0.61	38.81
1981	0.51	5.42	1.11	3.01	3.32	2.32	5.73	0.31	2.99	3.21	1.63	4.65	34.21
1982	4.81	2.25	2.39	4.14	2.03	4.70	2.97	3.11	1.41	1.65	3.19	1.42	34.07
1983	4.14	2.90	8.22	11.51	3.77	1.95	3.41	2.67	3.47	7.32	4.85	6.63	60.84
1984	1.51	4.31	5.19	5.26	9.27	6.85	5.75	1.19	2.65	3.01	3.13	2.58	50.70
1985	0.76	1.81	1.81	0.99	5.18	4.48	5.77	2.80	4.23	1.18	7.00	0.63	36.64
Record Mean	3.05	3.05	3.99	3.72	3.70	3.25	4.00	4.14	3.37	3.12	3.82	3.65	42.86

TABLE 3 AVERAGE TEMPERATURE (deg. F)   NEW YORK, LA GUARDIA FIELD, NEW YORK

YEAR	JAN	FEB	MAR	APR	MAY	JUNE	JULY	AUG	SEP	OCT	NOV	DEC	ANNUAL
1956	32.8	37.1	38.0	48.4	58.9	72.4	74.2	75.0	65.8	58.6	47.6	41.6	54.2
1957	29.1	37.8	42.4	53.2	63.4	74.5	77.6	74.0	70.3	56.9	50.0	41.1	55.8
1958	32.3	28.0	40.7	52.7	59.1	67.5	76.9	75.6	68.4	56.0	48.8	30.0	53.0
1959	31.6	32.1	40.1	53.4	65.1	71.6	76.9	78.1	72.5	58.8	46.1	39.2	55.5
1960	34.5	36.6	33.9	54.1	62.6	72.2	75.3	75.1	68.0	58.5	50.4	31.8	54.4
1961	27.9	36.7	40.9	48.6	58.6	71.5	76.0	75.3	73.6	60.0	48.1	35.4	54.4
#1962	32.2	30.8	41.1	51.5	63.6	72.7	73.8	72.6	64.7	57.5	43.6	31.2	52.9
1963	29.3	28.0	41.5	52.2	60.6	70.9	76.2	73.0	65.3	61.4	50.8	30.5	53.3
1964	34.4	32.6	42.8	49.7	65.6	71.3	76.1	73.8	68.1	55.4	49.9	36.7	54.7
1965	30.2	34.2	40.2	50.7	67.1	72.5	76.6	75.7	70.3	55.9	45.9	39.6	54.9
1966	31.8	34.5	42.0	48.6	59.0	74.0	80.8	77.7	68.3	56.6	49.2	36.8	55.0
1967	38.2	30.6	37.8	49.7	55.0	71.8	75.3	73.7	66.5	57.2	42.7	38.6	53.1
1968	26.8	29.1	42.3	54.0	59.4	69.2	77.2	75.9	70.6	59.8	46.4	33.5	53.7
1969	30.9	31.4	39.1	53.9	62.9	70.3	73.2	76.0	67.7	56.5	46.1	33.8	53.5
1970	26.0	32.7	38.4	51.1	62.6	70.5	77.1	77.5	70.4	59.2	49.6	36.2	54.3
1971	28.8	36.0	39.8	49.6	59.4	72.1	76.1	74.8	70.9	62.4	44.5	40.0	54.5
1972	34.2	30.6	37.8	47.6	59.7	65.8	74.6	73.6	67.8	52.2	43.1	38.5	52.1
1973	35.1	32.5	45.0	53.0	58.9	72.7	76.8	77.5	69.3	60.1	48.7	39.4	55.7
1974	35.6	31.9	41.7	54.2	59.9	68.9	76.8	75.7	66.2	53.8	47.8	39.5	54.3
1975	37.1	35.5	39.6	47.1	64.3	69.7	75.7	74.1	64.5	59.4	52.6	37.1	54.7
1976	28.8	40.7	44.8	55.6	61.5	74.1	75.8	75.0	67.5	54.5	43.0	30.7	54.3
1977	22.3	33.2	45.7	52.6	64.1	69.4	77.2	74.8	67.2	54.8	47.3	35.7	53.7
1978	28.8	26.9	38.6	50.3	59.6	70.0	73.8	75.4	65.3	55.7	48.2	38.0	52.5
1979	31.6	23.0	44.2	49.9	63.0	68.9	77.1	75.5	68.2	56.0	50.2	39.0	53.9
1980	32.7	30.7	40.2	53.2	64.7	70.1	78.7	78.2	70.4	55.7	43.1	30.8	54.0
1981	24.7	38.4	41.3	54.3	63.8	72.7	78.2	75.7	65.9	53.9	47.0	36.6	54.4
1982	25.3	35.1	41.1	50.2	63.3	66.8	76.3	72.4	66.9	57.4	49.0	41.8	53.8
1983	34.4	35.0	43.2	51.7	58.9	72.5	78.3	76.7	70.7	57.4	48.0	35.1	55.1
1984	29.3	39.2	35.3	50.3	61.3	73.6	73.5	76.3	65.7	62.5	46.4	43.3	54.7
1985	28.4	35.6	44.9	53.9	64.6	68.8	76.6	75.8	70.7	59.5	50.6	34.7	55.3
Record Mean	31.8	33.6	41.3	51.7	61.8	71.2	76.6	75.3	68.4	58.1	47.5	36.4	54.5
Max	37.7	39.9	48.5	59.7	70.0	79.2	84.3	82.6	75.6	64.9	53.5	42.1	61.5
Min	25.8	27.2	34.1	43.6	53.5	63.1	68.9	67.9	61.3	51.2	41.5	30.7	47.4

REFERENCE NOTES FOR TABLES 1, 2, 3 and 6   (NEW YORK [LAGUARDIA FIELD], NY)

GENERAL

T - TRACE AMOUNT
BLANK ENTRIES DENOTE MISSING/UNREPORTED DATA.
# INDICATES A STATION OR INSTRUMENT RELOCATION.

SPECIFIC

TABLE 1

(a) - LENGTH OF RECORD IN YEARS. ALTHOUGH
INDIVIDUAL MONTHS MAY BE MISSING.
x LESS THAN .05

NORMALS — BASED ON THE 1951-1980 RECORD PERIOD.
EXTREMES — DATES ARE THE MOST RECENT OCCURRENCE.
WIND DIR. — NUMERALS SHOW TENS OF DEGREES
CLOCKWISE FROM TRUE NORTH.
"00" INDICATES CALM.
RESULTANT WIND DIRECTIONS ARE GIVEN TO WHOLE DEGREES.

EXCEPTIONS

TABLES 2, 3, and 6

RECORD MEANS ARE THROUGH THE CURRENT YEAR,
BEGINNING IN   1941 FOR TEMPERATURE
1941 FOR PRECIPITATION
1945 FOR SNOWFALL

TABLE 4 HEATING DEGREE DAYS Base 65 deg. F NEW YORK, LA GUARDIA FIELD, NEW YORK

SEASON	JULY	AUG	SEP	OCT	NOV	DEC	JAN	FEB	MAR	APR	MAY	JUNE	TOTAL
1956-57	8	1	80	205	517	716	1104	755	696	369	120	3	4574
1957-58	0	0	40	247	446	734	997	1032	746	364	193	26	4825
1958-59	0	0	26	296	478	1076	1029	916	765	351	107	27	5071
1959-60	0	0	32	244	561	791	937	817	959	334	98	0	4773
1960-61	0	0	11	215	432	1022	1142	784	739	490	202	3	5040
#1961-62	0	0	22	181	501	910	1006	952	735	412	129	7	4855
1962-63	0	7	80	237	633	1039	1100	1028	720	378	169	8	5399
1963-64	0	0	83	136	421	1063	944	938	680	452	88	16	4821
1964-65	2	0	52	290	444	872	1072	858	763	426	47	14	4840
1965-66	0	5	26	280	565	780	1021	846	705	484	214	17	4943
1966-67	0	0	39	256	469	866	823	960	839	454	308	5	5019
1967-68	0	1	56	262	662	812	1181	1034	698	323	176	14	5219
1968-69	0	0	3	196	554	969	1051	933	795	336	111	2	4950
1969-70	0	0	49	272	562	959	1204	896	821	414	118	6	5301
1970-71	0	0	26	200	456	886	1117	807	774	455	179	11	4911
1971-72	0	1	17	108	612	768	945	987	837	520	164	43	5002
1972-73	4	0	37	389	653	813	920	904	613	371	195	4	4903
1973-74	0	0	30	167	485	785	903	919	713	331	186	27	4546
1974-75	0	0	63	341	512	783	858	819	778	529	105	14	4802
1975-76	0	4	56	188	372	857	1118	695	618	320	134	14	4376
1976-77	0	2	36	324	654	1061	1316	883	591	377	115	28	5387
1977-78	0	0	59	309	524	901	1119	1061	812	436	203	16	5440
1978-79	5	0	72	287	498	831	1031	1173	637	446	94	15	5089
1979-80	2	5	32	295	440	802	997	988	764	350	74	21	4770
1980-81	0	0	24	292	651	1052	1241	739	728	316	97	4	5144
1981-82	0	0	63	338	532	875	1222	832	737	443	79	49	5170
1982-83	0	6	36	253	482	712	942	832	669	402	187	5	4526
1983-84	0	2	48	259	505	919	1102	740	913	436	147	9	5080
1984-85	0	0	74	101	552	666	1127	820	620	338	87	20	4405
1985-86	0	0	17	183	428	934							

TABLE 5 COOLING DEGREE DAYS Base 65 deg. F NEW YORK, LA GUARDIA FIELD, NEW YORK

YEAR	JAN	FEB	MAR	APR	MAY	JUNE	JULY	AUG	SEP	OCT	NOV	DEC	TOTAL
1969	0	0	0	8	55	186	258	346	135	14	0	0	1002
1970	0	0	0	2	54	180	382	395	196	27	0	0	1236
1971	0	0	0	0	12	229	352	308	199	32	6	0	1138
1972	0	0	0	2	8	77	311	274	129	0	0	0	801
1973	0	0	0	14	16	242	372	397	164	24	0	0	1229
1974	0	0	0	16	35	149	371	338	104	0	4	0	1017
1975	0	0	0	0	87	160	338	293	46	20	9	0	953
1976	0	0	0	44	31	293	344	322	119	5	0	0	1158
1977	0	0	0	10	93	166	387	311	130	0	0	0	1097
1978	0	0	0	44	172	286	326	86	5	0	0	0	919
1979	0	0	0	0	41	138	382	335	131	22	0	0	1049
1980	0	0	0	0	71	180	430	413	192	8	0	0	1294
1981	0	0	0	0	66	243	413	336	111	0	0	0	1169
1982	0	0	0	5	32	112	358	239	100	25	6	0	877
1983	0	0	0	9	4	233	417	368	222	31	0	0	1284
1984	0	0	0	0	39	274	271	355	100	31	0	0	1070
1985	0	0	1	11	81	139	368	340	193	20	3	0	1156

TABLE 6 SNOWFALL (inches) NEW YORK, LA GUARDIA FIELD, NEW YORK

SEASON	JULY	AUG	SEP	OCT	NOV	DEC	JAN	FEB	MAR	APR	MAY	JUNE	TOTAL
1956-57	0.0	0.0	0.0	0.0	T	1.2	8.1	5.7	1.5	5.4	0.0	0.0	21.9
1957-58	0.0	0.0	0.0	0.0	T	11.0	8.8	12.3	18.9	0.5	0.0	0.0	51.5
1958-59	0.0	0.0	0.0	0.0	T	3.0	2.2	0.8	10.8	1.3	0.0	0.0	18.1
1959-60	0.0	0.0	0.0	0.0	0.2	8.6	2.0	3.9	18.8	T	0.0	0.0	33.5
1960-61	0.0	0.0	0.0	T	0.0	15.3	15.5	21.3	3.8	0.6	T	0.0	56.5
1961-62	0.0	0.0	0.0	0.0	0.4	5.2	0.4	11.3	0.7	T	0.0	0.0	18.0
1962-63	0.0	0.0	0.0	1.2	1.1	3.5	6.9	3.0	2.4	T	0.0	0.0	18.1
1963-64	0.0	0.0	0.0	0.0	T	11.6	10.2	8.8	3.2	T	0.0	0.0	33.8
1964-65	0.0	0.0	0.0	0.0	0.0	3.1	11.9	2.3	3.1	0.7	0.0	0.0	21.1
1965-66	0.0	0.0	0.0	0.0	T	T	T	9.3	9.8	T	T	0.0	19.1
1966-67	0.0	0.0	0.0	0.0	0.0	7.6	20.0	14.9	T	0.0	0.0	0.0	43.4
1967-68	0.0	0.0	0.0	0.0	2.2	3.4	5.2	1.5	2.2	0.0	0.0	0.0	14.5
1968-69	0.0	0.0	0.0	0.0	T	4.9	1.4	18.5	4.2	0.0	0.0	0.0	28.8
1969-70	0.0	0.0	0.0	0.0	0.0	8.1	7.4	0.3	4.6	T	0.0	0.0	24.3
1970-71	0.0	0.0	0.0	0.0	0.0	1.9	10.4	0.3	2.3	1.0	0.0	0.0	15.9
1971-72	0.0	0.0	0.0	0.0	T	T	2.2	17.2	2.7	0.1	0.0	0.0	22.2
1972-73	0.0	0.0	0.0	T	T	T	0.9	1.0	T	T	0.0	0.0	1.9
1973-74	0.0	0.0	0.0	0.0	0.0	2.2	6.4	7.8	2.2	0.3	0.0	0.0	18.9
1974-75	0.0	0.0	0.0	0.0	0.0	0.4	1.8	9.1	0.4	T	0.0	0.0	11.7
1975-76	0.0	0.0	0.0	0.0	T	3.0	6.1	4.8	2.8	0.0	0.0	0.0	16.7
1976-77	0.0	0.0	0.0	0.0	T	4.9	10.9	5.8	0.4	T	T	0.0	22.0
1977-78	0.0	0.0	0.0	0.0	T	0.3	16.6	18.7	7.9	0.0	0.0	0.0	43.5
1978-79	0.0	0.0	0.0	0.0	2.3	0.2	6.0	17.4	T	T	0.0	0.0	25.9
1979-80	0.0	0.0	0.0	T	0.0	3.2	2.3	1.6	3.2	0.0	0.0	0.0	10.3
1980-81	0.0	0.0	0.0	0.0	0.2	1.8	7.7	T	6.4	0.0	0.0	0.0	16.1
1981-82	0.0	0.0	0.0	0.0	T	3.6	13.1	0.4	0.3	8.2	0.0	0.0	25.6
1982-83	0.0	0.0	0.0	0.0	0.0	2.1	1.7	26.4	T	T	0.0	0.0	30.2
1983-84	0.0	0.0	0.0	0.0	T	1.6	9.8	T	12.7	0.0	0.0	0.0	24.1
1984-85	0.0	0.0	0.0	0.0	T	5.5	8.3	8.8	0.3	T	0.0	0.0	22.9
1985-86	0.0	0.0	0.0	0.0	0.4	0.9							
Record Mean	0.0	0.0	0.0	T	0.4	5.1	6.7	8.4	4.5	0.7	T	0.0	26.0

See Reference Notes, relative to all above tables, on preceding page.

Rochester is located at the mouth of the Genesee River at about the mid point of the south shore of Lake Ontario. The river flows northward from northwest Pennsylvania and empties into Lake Ontario. The land slopes from a lakeshore elevation of 246 feet to over 1,000 feet some 20 miles south. The airport is located just south of the city.

Lake Ontario plays a major role in the Rochester weather. In the summer its cooling effect inhibits the temperature from rising much above the low to mid 90s. In the winter the modifying temperature effect prevents temperatures from falling below −15 degrees most of the time, although temperatures at locations more than 15 miles inland do drop below −30 degrees.

The lake plays a major role in winter snowfall distribution. Well inland from the lake and toward the airport, the seasonal snowfall is usually less than in the area north of the airport and toward the lakeshore where wide variations occur. This is due to what is called the lake effect. Snowfalls of one to two feet or more in 24 hours are common near the lake in winter due the lake effect alone. The lake rarely freezes over because of its depth. The area is also prone to other heavy snowstorms and blizzards because of its proximity to the paths of low pressure systems coming up the east coast, out of the Ohio Valley, or, to a lesser extent, from the Alberta area. The climate is favorable for winter sports activities with a continuous snow cover likely from December through March.

Moisture in the air from the lake enhances the climatic conditions for fruit growing. Apples, peaches, pears, cantaloupes, plums, cherries, and grapes are grown abundantly in Greater Rochester and the Western Finger Lakes Region.

Precipitation is rather evenly distributed throughout the year. Excessive rains occur infrequently but may be caused by slowly moving thunderstorms, slowly moving or stalled major low pressure systems, or by hurricanes and tropical storms that move inland. Hail occurs occasionally and heavy fog is rare.

The growing season averages 150 to 180 days. The years first frost usually occurs in late September and the last frost typically occurs in mid−May.

TABLE 1   NORMALS, MEANS AND EXTREMES

ROCHESTER, NEW YORK

LATITUDE: 43°07'N   LONGITUDE: 77°40'W   ELEVATION: FT. GRND   547 BARO   00547   TIME ZONE: EASTERN   WBAN: 14768

	(a)	JAN	FEB	MAR	APR	MAY	JUNE	JULY	AUG	SEP	OCT	NOV	DEC	YEAR
TEMPERATURE °F:														
Normals														
-Daily Maximum		30.8	32.2	41.2	56.0	67.7	77.7	82.3	80.1	72.8	61.5	48.0	35.5	57.2
-Daily Minimum		16.3	16.7	25.3	36.1	46.0	55.7	60.3	58.7	51.6	41.8	33.2	22.3	38.7
-Monthly		23.6	24.5	33.3	46.0	56.9	66.7	71.3	69.4	62.2	51.7	40.6	28.9	47.9
Extremes														
-Record Highest	45	74	67	84	93	94	100	98	99	99	91	81	72	100
-Year		1950	1947	1945	1970	1974	1953	1978	1948	1953	1951	1950	1982	JUN 1953
-Record Lowest	45	-16	-19	-6	13	26	35	42	36	28	20	5	-16	-19
-Year		1957	1979	1980	1982	1979	1949	1963	1965	1947	1972	1971	1942	FEB 1979
NORMAL DEGREE DAYS:														
Heating (base 65°F)		1283	1137	983	570	274	41	10	23	132	412	732	1116	6713
Cooling (base 65°F)		0	0	0	0	23	92	205	163	48	0	0	0	531
% OF POSSIBLE SUNSHINE	45	36	41	49	54	59	66	69	66	60	49	30	30	51
MEAN SKY COVER (tenths)														
Sunrise - Sunset	45	8.2	8.0	7.4	6.7	6.6	6.0	5.7	5.9	6.1	6.6	8.2	8.4	7.0
MEAN NUMBER OF DAYS:														
Sunrise to Sunset														
-Clear	45	2.0	2.2	4.4	6.2	5.9	7.2	8.2	7.8	7.2	6.7	2.2	2.0	62.0
-Partly Cloudy	45	6.8	6.8	8.3	8.2	9.7	10.9	12.6	11.6	10.7	8.3	6.2	5.5	105.5
-Cloudy	45	22.1	19.3	18.4	15.6	15.4	11.8	10.3	11.6	12.1	16.0	21.6	23.6	197.7
Precipitation														
.01 inches or more	45	17.6	15.7	14.7	13.2	12.0	10.4	9.5	10.0	10.6	11.3	14.9	17.7	157.5
Snow, Ice pellets														
1.0 inches or more	45	7.4	6.8	4.1	1.0	0.*	0.0	0.0	0.0	0.0	0.*	1.9	6.3	27.7
Thunderstorms	45	0.1	0.1	0.9	2.1	3.6	5.2	6.4	5.7	3.0	1.0	0.4	0.2	28.9
Heavy Fog Visibility														
1/4 mile or less	45	1.0	0.6	1.4	1.1	1.1	1.2	0.6	0.9	1.3	1.8	0.7	1.1	12.7
Temperature °F														
-Maximum														
90° and above	22	0.0	0.0	0.0	0.*	0.3	1.7	4.5	2.0	0.7	0.0	0.0	0.0	9.3
32° and below	22	17.9	14.5	6.6	0.5	0.0	0.0	0.0	0.0	0.0	0.0	1.5	10.9	52.0
-Minimum														
32° and below	22	28.8	25.2	23.2	11.1	1.1	0.0	0.0	0.0	0.*	4.2	14.6	25.5	133.7
0° and below	22	3.5	2.4	0.3	0.0	0.0	0.0	0.0	0.0	0.0	0.0	0.0	0.7	6.8
AVG. STATION PRESS.(mb)	13	995.9	997.3	995.3	994.7	994.3	994.8	995.6	997.2	997.8	998.5	996.9	996.7	996.2
RELATIVE HUMIDITY (%)														
Hour 01	22	76	77	76	75	78	82	82	86	87	82	80	80	80
Hour 07 (Local Time)	22	77	78	78	77	77	80	82	87	88	84	82	81	81
Hour 13	22	69	67	63	55	54	56	53	57	61	61	69	74	62
Hour 19	22	74	72	69	61	60	60	59	67	74	74	77	78	69
PRECIPITATION (inches):														
Water Equivalent														
-Normal		2.30	2.32	2.53	2.64	2.58	2.78	2.48	3.20	2.66	2.54	2.65	2.59	31.27
-Maximum Monthly	45	5.79	5.07	5.42	4.90	6.62	6.77	9.70	6.00	6.30	7.85	6.99	5.05	9.70
-Year		1978	1950	1942	1944	1974	1980	1947	1984	1977	1955	1985	1944	JUL 1947
-Minimum Monthly	45	0.81	0.74	0.47	1.28	0.36	0.22	0.98	0.76	0.28	0.23	0.44	0.62	0.22
-Year		1946	1968	1958	1971	1977	1963	1955	1951	1960	1976	1958	1958	JUN 1963
-Maximum in 24 hrs	45	1.64	2.43	2.21	1.99	3.85	2.86	2.94	2.39	3.54	2.98	3.13	1.60	3.85
-Year		1966	1950	1942	1943	1974	1950	1947	1968	1979	1980	1945	1978	MAY 1974
Snow, Ice pellets														
-Maximum Monthly	45	60.4	64.8	40.3	20.2	2.0			T	T	1.4	17.6	46.1	64.8
-Year		1978	1958	1959	1979	1945			1965	1956	1960	1983	1981	FEB 1958
-Maximum in 24 hrs	45	18.2	22.8	17.6	8.3	2.0			T	T	1.4	11.2	19.1	22.8
-Year		1966	1978	1959	1979	1945			1965	1956	1960	1953	1978	FEB 1978
WIND:														
Mean Speed (mph)	45	11.8	11.5	11.2	10.8	9.3	8.6	8.0	7.7	8.1	8.8	10.2	10.9	9.7
Prevailing Direction														
through 1963		WSW	WSW	WSW	WSW	WSW	SW	SW	SW	SW	SW	WSW	WSW	WSW
Fastest Mile														
-Direction	45	W	W	W	SW	SW	SW	W	NE	SW	NW	E	W	W
-Speed (MPH)	45	73	66	65	60	63	61	56	59	59	50	59	56	73
-Year		1950	1956	1956	1979	1950	1949	1956	1955	1949	1941	1950	1949	JAN 1950
Peak Gust														
-Direction	2	W	NE	W	SW	SW	SW	SW	N	W	SW	SW	SW	SW
-Speed (mph)	2	53	44	52	67	41	35	36	47	44	38	60	51	67
-Date		1985	1984	1984	1984	1984	1984	1984	1984	1984	1985	1985	1985	APR 1984

See Reference Notes to this table on the following page.

TABLE 2 PRECIPITATION (inches)   ROCHESTER, NEW YORK

YEAR	JAN	FEB	MAR	APR	MAY	JUNE	JULY	AUG	SEP	OCT	NOV	DEC	ANNUAL
1956	2.09	3.23	3.41	4.14	3.83	0.97	2.91	5.67	3.58	0.94	1.47	2.16	34.40
1957	2.62	1.56	1.75	2.77	2.98	3.63	2.24	2.46	0.96	0.97	1.81	1.92	25.67
1958	2.92	4.20	0.47	3.10	1.89	4.58	4.17	2.56	3.55	4.95	2.79	0.62	35.80
1959	3.87	2.70	2.28	1.91	2.21	2.06	1.63	2.72	2.72	4.07	1.84	4.64	32.23
1960	2.82	4.46	1.64	1.87	3.72	3.05	1.39	4.05	0.28	1.72	0.73	1.03	26.76
1961	1.10	3.21	2.70	4.07	2.96	3.78	2.41	3.09	0.39	1.58	3.99	1.23	30.51
1962	1.84	2.87	1.25	2.62	2.70	3.02	1.89	3.53	4.01	1.95	1.99	1.58	29.25
1963	1.24	1.43	2.53	2.74	2.33	0.22	2.72	3.26	1.17	0.23	4.32	1.90	24.09
1964	1.99	0.89	2.89	3.54	2.77	1.13	1.52	2.74	0.58	0.76	1.60	2.04	22.45
1965	3.05	2.24	2.07	1.85	0.50	0.64	1.46	2.94	1.98	2.82	3.40	2.21	25.16
1966	4.10	2.51	1.42	2.04	1.26	1.85	2.77	2.14	2.47	0.68	3.12	1.75	26.11
1967	0.94	1.67	1.31	1.69	2.74	1.57	2.68	4.64	3.84	4.35	2.89	1.52	29.84
1968	1.91	0.74	2.38	1.33	2.84	2.84	1.42	5.95	1.86	2.89	4.28	3.31	31.75
1969	2.46	0.91	1.16	3.48	2.25	4.69	1.83	1.82	1.77	1.69	3.42	3.66	29.14
1970	1.80	2.28	1.49	2.58	3.03	3.74	4.91	3.88	2.49	3.96	3.50	4.12	37.78
1971	2.66	4.21	3.43	1.28	1.71	3.52	5.59	3.18	1.79	1.34	1.96	3.49	34.16
1972	1.50	3.96	2.19	2.68	3.32	6.56	1.43	3.14	3.84	2.25	4.83	2.58	38.28
1973	1.28	1.70	2.92	3.21	2.68	2.84	1.14	1.94	1.41	2.67	3.82	3.62	29.23
1974	1.75	2.06	3.61	2.60	6.62	2.59	2.82	3.64	3.48	1.34	3.23	2.86	36.60
1975	1.83	2.82	2.74	1.43	2.85	5.35	1.18	2.31	3.15	1.83	1.35	3.76	30.60
1976	2.33	1.67	3.54	3.81	2.63	3.37	5.15	3.04	2.13	4.73	0.44	1.48	34.32
1977	1.49	0.97	2.18	2.49	0.36	1.33	3.26	5.65	6.30	2.64	3.78	4.65	35.10
1978	5.79	2.40	1.48	2.25	2.03	1.30	2.17	2.66	3.63	2.56	1.14	4.35	31.76
1979	4.18	2.40	1.76	3.78	3.14	1.85	3.16	2.05	5.32	2.60	1.80	2.86	34.90
1980	1.11	1.16	3.83	2.35	1.49	6.77	1.90	3.44	3.57	3.73	2.52	2.45	34.32
1981	1.24	3.13	1.04	1.95	2.27	2.70	4.60	4.44	5.37	3.29	2.18	2.78	34.99
1982	4.16	1.01	1.73	1.63	1.77	3.92	3.13	3.00	3.57	1.79	3.95	2.17	31.83
1983	1.43	1.23	2.45	3.50	3.44	2.40	1.13	5.43	1.56	3.26	4.91	4.47	35.21
1984	1.62	2.97	2.08	3.05	5.47	1.67	1.90	6.00	3.34	0.76	1.47	3.31	33.64
1985	2.49	1.78	3.47	1.30	2.08	2.63	1.86	1.11	2.49	2.34	6.99	1.46	30.00
Record Mean	2.49	2.37	2.62	2.57	2.90	2.98	3.06	2.88	2.75	2.75	2.70	2.59	32.66

TABLE 3 AVERAGE TEMPERATURE (deg. F)   ROCHESTER, NEW YORK

YEAR	JAN	FEB	MAR	APR	MAY	JUNE	JULY	AUG	SEP	OCT	NOV	DEC	ANNUAL
1956	23.1	26.9	28.7	42.2	52.4	66.6	68.2	68.2	58.0	53.8	41.0	32.7	46.8
1957	19.4	29.9	34.4	48.4	55.4	69.4	69.9	66.6	61.2	49.4	40.7	34.2	48.3
1958	23.9	19.4	33.6	47.9	54.3	61.4	70.4	68.2	61.5	51.2	41.5	21.4	46.2
1959	21.9	22.6	30.2	46.6	59.4	67.5	72.5	74.5	66.6	51.9	37.9	30.9	48.5
1960	24.6	26.1	22.4	48.6	58.7	65.6	68.5	67.7	64.5	50.2	43.2	23.5	47.0
1961	19.4	27.5	34.3	41.7	53.9	65.4	70.8	69.9	68.5	55.5	40.6	29.6	48.1
1962	22.9	21.4	33.5	45.7	62.4	66.0	67.6	68.9	58.9	51.8	36.9	26.0	46.8
#1963	19.3	17.9	35.7	46.2	54.3	67.1	70.9	65.5	56.7	57.3	43.7	21.7	46.4
1964	28.3	24.0	34.4	47.2	60.3	65.8	73.8	66.2	61.2	48.7	42.3	29.6	48.5
1965	21.6	25.7	29.5	41.3	60.1	64.4	67.5	68.4	63.8	48.8	39.8	34.6	47.1
1966	22.6	25.5	36.6	44.4	53.0	73.0	68.0	69.4	59.2	50.1	43.5	30.8	48.0
1967	31.1	21.4	32.5	46.4	49.1	70.4	69.4	67.6	60.4	52.1	36.3	32.3	47.4
1968	19.8	20.8	36.0	49.5	53.2	64.7	70.7	70.7	64.8	53.8	41.2	27.7	47.6
1969	25.2	26.0	32.0	47.6	55.6	65.0	70.7	72.0	63.6	50.9	40.1	25.1	47.8
1970	18.0	23.7	30.8	48.4	59.5	68.1	72.2	70.1	63.0	53.9	41.4	25.5	47.9
1971	19.5	26.9	28.5	41.1	54.0	67.5	68.8	68.8	67.4	59.0	38.4	30.9	47.6
1972	26.0	23.4	30.3	42.2	60.4	65.2	73.0	69.8	64.3	47.6	37.0	33.1	47.7
1973	28.7	22.2	42.5	48.0	56.1	70.7	73.4	73.0	62.5	55.0	43.0	31.2	50.5
1974	27.1	22.5	33.0	49.4	53.9	65.7	71.3	70.6	59.2	47.6	39.5	31.4	47.6
1975	29.5	28.4	31.6	39.3	63.2	67.1	73.0	70.0	58.4	53.1	47.2	27.8	49.0
1976	19.8	33.3	37.2	48.6	55.4	69.2	69.2	68.4	60.9	47.5	35.4	23.6	47.4
1977	15.5	25.4	39.8	47.9	60.7	64.8	72.9	72.9	62.7	49.4	43.6	28.4	48.3
1978	22.9	16.2	29.7	43.6	60.2	67.4	72.6	71.6	62.4	51.0	41.0	30.0	47.4
1979	21.5	13.7	38.6	44.0	56.4	66.2	72.3	67.1	61.3	50.3	43.0	32.3	47.2
1980	21.5	19.7	32.4	47.8	60.0	63.1	72.9	74.3	63.7	48.8	38.8	24.7	47.5
1981	15.7	32.3	34.5	48.0	57.2	67.3	71.9	69.4	59.8	47.3	39.9	28.7	47.7
1982	16.1	23.0	33.5	43.2	60.9	63.6	72.0	66.1	62.8	52.7	43.4	37.4	47.9
1983	27.4	29.1	37.2	43.9	53.8	66.7	73.8	70.7	63.9	52.9	40.7	25.1	48.8
1984	20.4	33.2	26.5	47.5	52.6	66.8	69.2	72.0	60.6	54.9	40.7	35.9	48.0
1985	21.9	25.6	36.7	49.6	58.6	61.7	68.8	68.7	63.8	51.0	41.4	25.0	47.7
Record Mean	24.5	24.6	32.8	45.0	56.8	66.4	71.3	69.3	62.6	51.4	39.8	29.0	47.8
Max	31.4	31.9	40.4	54.0	66.7	76.5	81.2	78.9	72.2	60.2	46.6	35.2	56.3
Min	17.6	17.2	25.2	36.1	46.8	56.3	61.4	59.7	53.0	42.6	33.1	22.7	39.3

REFERENCE NOTES FOR TABLES 1, 2, 3 and 6   (ROCHESTER, NY)

GENERAL

T - TRACE AMOUNT
BLANK ENTRIES DENOTE MISSING/UNREPORTED DATA.
# INDICATES A STATION OR INSTRUMENT RELOCATION.

SPECIFIC

TABLE 1

(a) - LENGTH OF RECORD IN YEARS. ALTHOUGH INDIVIDUAL MONTHS MAY BE MISSING.

* LESS THAN .05

NORMALS — BASED ON THE 1951-1980 RECORD PERIOD.
EXTREMES — DATES ARE THE MOST RECENT OCCURRENCE.
WIND DIR. — NUMERALS SHOW TENS OF DEGREES
CLOCKWISE FROM TRUE NORTH.
"00" INDICATES CALM.
RESULTANT WIND DIRECTIONS ARE GIVEN TO WHOLE DEGREES.

EXCEPTIONS

TABLES 2, 3, and 6

RECORD MEANS ARE THROUGH THE CURRENT YEAR,
BEGINNING IN   1872 FOR TEMPERATURE
                1829 FOR PRECIPITATION
                1941 FOR SNOWFALL

TABLE 4 HEATING DEGREE DAYS Base 65 deg. F   ROCHESTER, NEW YORK

SEASON	JULY	AUG	SEP	OCT	NOV	DEC	JAN	FEB	MAR	APR	MAY	JUNE	TOTAL
1956-57	19	27	239	342	715	993	1408	979	944	520	311	48	6545
1957-58	17	49	158	477	721	948	1268	1269	965	509	337	131	6849
1958-59	5	31	139	428	700	1344	1327	1182	1072	545	235	67	7075
1959-60	0	3	126	407	806	1052	1244	1122	1313	491	212	56	6832
1960-61	22	24	83	452	647	1280	1410	1045	945	692	366	64	7030
1961-62	28	21	82	302	728	1092	1296	1215	969	589	181	52	6555
#1962-63	10	21	209	409	836	1202	1412	1309	902	556	332	51	7249
1963-64	24	52	253	247	615	1335	1132	1181	939	526	187	98	6589
1964-65	3	48	169	500	674	1090	1337	1096	1095	706	192	114	7024
1965-66	24	53	122	496	748	937	1307	1098	871	612	382	57	6707
1966-67	0	16	190	458	639	1053	1046	1216	998	551	486	8	6661
1967-68	25	28	164	401	856	1005	1399	1275	894	463	357	78	6945
1968-69	15	36	67	369	710	1150	1229	1087	1013	517	302	86	6581
1969-70	18	10	126	437	737	1231	1448	1149	1053	506	209	49	6973
1970-71	3	11	126	349	699	1218	1405	1059	1122	707	342	49	7090
1971-72	17	33	74	194	792	1048	1200	1199	1071	677	161	78	6544
1972-73	7	24	92	534	833	982	1118	1189	690	519	279	17	6284
1973-74	2	14	162	305	653	1040	1167	1187	983	475	352	59	6399
1974-75	1	1	209	535	755	1034	1096	1017	1031	764	139	52	6634
1975-76	4	14	194	365	525	1146	1395	914	858	507	300	35	6257
1976-77	11	27	173	538	879	1279	1524	1103	777	523	204	89	7127
1977-78	9	44	113	477	634	1127	1298	1360	1087	634	220	63	7066
1978-79	5	1	136	428	711	1077	1342	1432	813	626	310	79	6960
1979-80	13	37	155	468	655	1006	1264	1306	1003	510	195	125	6737
1980-81	1	0	108	498	782	1243	1522	908	938	507	260	26	6793
1981-82	6	12	201	546	748	1119	1510	1171	972	648	162	67	7162
1982-83	10	54	113	377	643	847	1161	998	854	627	347	78	6109
1983-84	9	8	121	387	723	1228	1376	917	1187	520	395	50	6921
1984-85	14	7	162	307	724	897	1330	1097	869	471	217	119	6214
1985-86	15	23	121	429	700	1231							

TABLE 5 COOLING DEGREE DAYS Base 65 deg. F   ROCHESTER, NEW YORK

YEAR	JAN	FEB	MAR	APR	MAY	JUNE	JULY	AUG	SEP	OCT	NOV	DEC	TOTAL
1969	0	0	0	0	18	94	202	233	92	7	0	0	646
1970	0	0	0	16	47	148	235	176	69	13	0	0	704
1971	0	0	0	0	10	133	143	159	155	15	0	0	615
1972	0	0	0	0	24	94	261	179	79	0	0	0	637
1973	0	0	0	15	6	194	267	269	96	4	0	0	851
1974	0	0	0	0	13	14	88	204	181	40	0	0	540
1975	0	0	0	0	89	121	257	178	5	6	0	0	656
1976	0	0	0	24	9	189	150	138	55	1	0	0	566
1977	0	0	3	16	80	88	260	164	50	0	1	0	662
1978	0	0	0	0	77	141	245	212	66	3	0	0	744
1979	0	0	0	1	52	121	244	112	49	18	0	0	597
1980	0	0	0	0	46	73	253	294	76	2	0	0	744
1981	0	0	0	5	23	102	228	156	50	0	0	0	564
1982	0	0	0	3	40	30	232	95	52	3	1	0	456
1983	0	0	0	0	7	136	289	192	96	20	0	0	740
1984	0	0	0	1	14	113	152	233	35	1	0	0	549
1985	0	0	0	15	23	27	139	145	90	0	0	0	439

TABLE 6 SNOWFALL (inches)   ROCHESTER, NEW YORK

SEASON	JULY	AUG	SEP	OCT	NOV	DEC	JAN	FEB	MAR	APR	MAY	JUNE	TOTAL
1956-57	0.0	0.0	T	0.0	10.0	17.7	25.6	8.0	11.7	6.2	T	0.0	79.2
1957-58	0.0	0.0	0.0	1.4	9.9	6.9	37.9	64.8	8.8	1.1	T	0.0	130.8
1958-59	0.0	0.0	0.0	0.0	13.5	12.4	46.8	27.4	40.3	0.2	T	0.0	140.6
1959-60	0.0	0.0	0.0	T	11.6	25.4	32.7	58.3	33.0	0.7	0.0	0.0	161.7
1960-61	0.0	0.0	0.0	1.4	2.8	17.9	23.6	18.8	19.5	5.4	T	0.0	89.4
1961-62	0.0	0.0	0.0	0.0	7.5	6.0	11.7	28.2	4.6	7.6	T	0.0	65.6
1962-63	0.0	0.0	0.0	0.8	6.0	14.2	23.7	22.9	6.7	1.1	1.0	0.0	76.4
1963-64	0.0	0.0	0.0	0.0	4.4	34.6	20.2	13.1	16.1	3.6	0.0	0.0	92.0
1964-65	0.0	0.0	0.0	0.0	5.1	11.6	26.6	10.3	15.6	1.9	0.0	0.0	71.1
1965-66	0.0	T	0.0	0.9	8.0	6.0	60.2	21.0	6.2	0.9	T	0.0	103.2
1966-67	0.0	0.0	0.0	0.0	3.0	14.4	12.7	27.6	16.0	T	0.3	0.0	74.0
1967-68	0.0	0.0	0.0	T	10.0	6.9	24.2	20.4	15.2	T	0.0	0.0	76.7
1968-69	0.0	0.0	0.0	T	8.6	22.2	25.6	17.8	4.6	1.0	0.0	0.0	79.8
1969-70	0.0	0.0	0.0	T	5.8	42.0	37.9	27.7	4.9	1.3	T	0.0	119.6
1970-71	0.0	0.0	0.0	0.2	3.6	44.2	34.1	29.7	29.7	1.2	0.0	0.0	142.7
1971-72	0.0	0.0	0.0	0.0	11.2	13.8	18.1	35.7	19.0	7.3	0.0	0.0	105.1
1972-73	0.0	0.0	0.0	0.2	16.9	22.7	8.9	18.4	4.4	1.5	T	0.0	73.0
1973-74	0.0	0.0	0.0	0.3	4.2	23.4	14.4	26.6	22.3	8.2	T	0.0	99.1
1974-75	0.0	0.0	0.0	0.3	4.6	26.5	10.8	23.2	10.9	14.9	0.0	0.0	91.2
1975-76	0.0	0.0	0.0	T	1.8	28.3	29.9	8.8	15.2	1.8	0.4	0.0	86.2
1976-77	0.0	0.0	0.0	0.5	6.5	24.5	30.2	15.0	13.0	1.8	0.6	0.0	92.1
1977-78	0.0	0.0	0.0	T	12.7	35.2	60.4	40.7	7.5	4.2	0.2	0.0	160.9
1978-79	0.0	0.0	0.0	T	3.3	30.9	36.8	39.1	8.2	20.2	0.0	0.0	138.5
1979-80	0.0	0.0	0.0	0.2	1.2	12.2	13.1	24.0	21.2	0.3	0.0	0.0	72.2
1980-81	0.0	0.0	0.0	T	8.4	31.8	31.5	9.3	12.0	1.4	0.0	0.0	94.4
1981-82	0.0	0.0	0.0	0.1	2.4	46.1	43.6	14.9	8.9	12.4	0.0	0.0	128.4
1982-83	0.0	0.0	0.0	T	3.0	11.6	10.2	13.6	9.3	12.2	T	0.0	59.9
1983-84	0.0	0.0	0.0	0.0	17.6	19.6	23.4	27.8	29.1	0.5	T	0.0	118.0
1984-85	0.0	0.0	0.0	0.0	1.6	11.6	36.8	26.1	8.4	2.6	0.0	0.0	87.1
1985-86	0.0	0.0	0.0	0.0	7.6	18.3							
Record Mean	0.0	T	T	0.1	6.7	19.6	23.3	22.3	14.3	3.5	0.1	0.0	90.0

See Reference Notes, relative to all above tables, on preceding page.

Syracuse is located approximately at the geographical center of the state. Gently rolling terrain stretches northward for about 30 miles to the eastern end of Lake Ontario. Oneida Lake is about 8 miles northeast of Syracuse. Approximately 5 miles south of the city, hills rise to 1,500 feet. Immediately to the west, the terrain is gently rolling with elevations 500 to 800 feet above sea level.

The climate of Syracuse is primarily continental in character and comparatively humid. Nearly all cyclonic systems moving from the interior of the country through the St. Lawrence Valley will affect the Syracuse area. Seasonal and diurnal changes are marked and produce an invigorating climate.

In the summer and in portions of the transitional seasons, temperatures usually rise rapidly during the daytime to moderate levels and as a rule fall rapidly after sunset. The nights are relatively cool and comfortable. There are only a few days in a year when atmospheric humidity causes great personal discomfort.

Winters are usually cold and are sometimes severe in part. Daytime temperatures average in the low 30s with nighttime lows in the teens. Low winter temperatures below −25 degrees have been recorded. The autumn, winter, and spring seasons display marked variability.

Based on the 1951–1980 period, the average first occurrence of 32 degrees Fahrenheit in the fall is October 16 and the average last occurrence in the spring is April 28.

Precipitation in the Syracuse area is derived principally from cyclonic storms which pass from the interior of the country through the St. Lawrence Valley. Lake Ontario provides the source of significant winter precipitation. The lake is quite deep and never freezes so cold air flowing over the lake is quickly saturated and produces the cloudiness and snow squalls which are a well-known feature of winter weather in the Syracuse area.

The area enjoys sufficient precipitation in most years to meet the needs of agriculture and water supplies. The precipitation is uncommonly well distributed, averaging about 3 inches per month throughout the year. Snowfall is moderately heavy with an average just over 100 inches. There are about 30 days per year with thunderstorms, mostly during the warmer months.

Wind velocities are moderate, but during the winter months there are numerous days with sufficient winds to cause blowing and drifting snow.

During December, January, and February there is much cloudiness. Syracuse receives only about one-third of possible sunshine during winter months. Approximately two-thirds of possible sunshine is received during the warm months.

TABLE 1   NORMALS, MEANS AND EXTREMES

SYRACUSE, NEW YORK

LATITUDE: 43°07'N   LONGITUDE: 76°07'W   ELEVATION: FT. GRND   410 BARO   00420   TIME ZONE: EASTERN   WBAN: 14771

	(a)	JAN	FEB	MAR	APR	MAY	JUNE	JULY	AUG	SEP	OCT	NOV	DEC	YEAR
TEMPERATURE °F:														
Normals														
-Daily Maximum		30.6	32.2	41.4	56.2	67.9	77.2	81.6	79.6	72.3	60.9	47.9	35.3	56.9
-Daily Minimum		15.0	15.8	25.2	36.0	46.0	55.4	60.3	58.9	51.8	41.7	33.3	21.3	38.4
-Monthly		22.8	24.0	33.3	46.1	57.0	66.3	71.0	69.3	62.0	51.3	40.6	28.3	47.7
Extremes														
-Record Highest	36	70	69	85	89	96	98	97	97	97	87	81	70	98
-Year		1967	1981	1977	1962	1977	1953	1962	1965	1953	1963	1950	1966	JUN 1953
-Record Lowest	36	-26	-26	-16	9	25	35	45	40	28	19	5	-22	-26
-Year		1966	1979	1950	1972	1966	1966	1976	1965	1965	1976	1976	1980	FEB 1979
NORMAL DEGREE DAYS:														
Heating (base 65°F)		1308	1148	983	567	269	47	12	25	133	425	732	1138	6787
Cooling (base 65°F)		0	0	0	0	21	86	195	158	46	0	0	0	506
% OF POSSIBLE SUNSHINE	36	34	39	46	51	54	59	64	59	53	44	25	25	46
MEAN SKY COVER (tenths)														
Sunrise - Sunset	36	8.1	7.9	7.5	6.8	6.6	6.2	5.8	6.2	6.2	6.7	8.3	8.4	7.1
MEAN NUMBER OF DAYS:														
Sunrise to Sunset														
-Clear	36	2.7	2.8	4.5	6.4	6.0	7.3	7.9	6.9	6.9	6.6	2.3	2.3	62.6
-Partly Cloudy	36	6.4	6.2	6.8	6.8	10.0	10.4	12.2	11.2	10.4	7.8	5.4	4.7	98.4
-Cloudy	36	21.9	19.3	19.8	16.8	15.0	12.3	10.9	12.9	12.7	16.7	22.3	24.0	204.3
Precipitation														
.01 inches or more	36	19.0	15.8	17.0	14.1	13.0	11.0	10.7	10.9	10.9	11.8	16.3	19.2	169.6
Snow, Ice pellets														
1.0 inches or more	36	8.6	7.4	5.2	1.1	0.1	0.0	0.0	0.0	0.0	0.2	2.8	7.7	33.1
Thunderstorms	36	0.2	0.2	0.8	1.9	3.1	5.6	6.3	5.6	2.6	1.0	0.6	0.1	27.9
Heavy Fog Visibility														
1/4 mile or less	36	0.7	0.7	0.8	0.6	0.7	0.6	0.4	0.8	0.9	1.2	0.6	0.8	8.7
Temperature °F														
-Maximum														
90° and above	22	0.0	0.0	0.0	0.0	0.4	1.2	3.5	1.5	0.4	0.0	0.0	0.0	6.9
32° and below	22	18.1	14.1	5.9	0.4	0.0	0.0	0.0	0.0	0.0	0.0	1.5	11.7	51.6
-Minimum														
32° and below	22	28.9	25.3	23.8	12.3	1.0	0.0	0.0	0.0	0.1	5.0	14.3	26.2	136.8
0° and below	22	5.0	3.0	0.7	0.0	0.0	0.0	0.0	0.0	0.0	0.0	0.0	1.5	10.2
AVG. STATION PRESS.(mb)	13	1001.3	1002.6	1000.6	999.7	999.4	999.8	1000.4	1002.2	1003.0	1003.7	1002.3	1002.3	1001.4
RELATIVE HUMIDITY (%)														
Hour 01	22	76	76	76	74	78	83	84	87	86	82	80	80	80
Hour 07 (Local Time)	22	77	77	78	75	76	80	81	87	88	84	82	80	80
Hour 13	22	68	65	61	53	55	57	56	60	62	61	68	72	62
Hour 19	22	75	72	68	59	60	63	63	70	76	75	77	78	70
PRECIPITATION (inches):														
Water Equivalent														
-Normal		2.61	2.65	3.11	3.34	3.16	3.63	3.76	3.77	3.29	3.14	3.45	3.20	39.11
-Maximum Monthly	36	5.77	5.38	6.84	8.12	7.41	12.30	9.52	8.41	8.81	8.29	6.79	5.50	12.30
-Year		1978	1951	1955	1976	1976	1972	1974	1956	1975	1955	1972	1983	JUN 1972
-Minimum Monthly	36	1.02	0.80	1.01	1.22	0.75	1.10	0.90	1.33	0.75	0.21	1.25	1.73	0.21
-Year		1970	1978	1981	1985	1977	1969	1962	1980	1964	1963	1978	1958	OCT 1963
-Maximum in 24 hrs	36	1.47	1.99	1.34	2.85	3.13	3.88	4.07	4.27	4.14	3.60	2.09	2.18	4.27
-Year		1958	1961	1974	1976	1969	1972	1974	1954	1975	1955	1967	1952	AUG 1954
Snow, Ice pellets														
-Maximum Monthly	36	72.2	72.6	40.3	16.4	1.2					4.4	25.9	52.5	72.6
-Year		1978	1958	1984	1983	1973					1952	1976	1969	FEB 1958
-Maximum in 24 hrs	36	24.5	21.4	14.7	7.1	1.2					2.4	12.1	15.6	24.5
-Year		1966	1961	1971	1975	1973					1974	1973	1978	JAN 1966
WIND:														
Mean Speed (mph)	36	10.9	11.0	11.1	10.9	9.3	8.5	8.3	7.9	8.5	9.1	10.3	10.6	9.7
Prevailing Direction														
through 1963		WSW	WNW	WNW	WNW	WNW	WNW	WNW	WSW	S	WSW	WSW	WSW	WNW
Fastest Mile														
-Direction	36	W	W	SE	NW	NW	NW	NW	NW	W	SE	E	W	SE
-Speed (MPH)	36	60	62	56	52	50	49	47	43	52	63	59	52	63
-Year		1974	1967	1956	1957	1964	1961	1982	1958	1962	1954	1950	1962	OCT 1954
Peak Gust														
-Direction	2	W	SW	SW	W	W	S	NW	W	NW	W	SE	W	W
-Speed (mph)	2	51	56	53	61	49	43	53	33	39	40	49	52	61
-Date		1985	1985	1985	1985	1985	1984	1985	1984	1984	1985	1984	1985	APR 1985

See Reference Notes to this table on the following page.

TABLE 2 PRECIPITATION (inches) SYRACUSE, NEW YORK

YEAR	JAN	FEB	MAR	APR	MAY	JUNE	JULY	AUG	SEP	OCT	NOV	DEC	ANNUAL
1956	2.28	3.90	4.63	3.35	3.03	1.71	3.75	8.41	4.27	1.28	2.38	3.02	42.01
1957	2.19	1.76	2.01	2.60	2.88	4.18	6.13	3.45	2.28	0.93	1.86	2.90	33.17
1958	4.46	5.28	1.31	3.36	3.70	5.24	3.63	2.06	4.89	3.37	3.76	1.73	42.79
1959	4.59	2.22	2.93	2.51	1.97	2.53	2.54	4.48	0.93	7.15	4.34	5.01	41.20
1960	3.11	4.90	2.48	2.94	3.96	1.86	1.03	2.69	2.93	2.77	1.68	1.86	32.21
1961	2.30	4.14	4.22	3.74	2.40	3.68	5.08	1.78	1.21	3.59	2.99	2.45	37.58
1962	2.87	2.96	1.96	3.57	1.05	1.10	2.74	4.63	1.99	3.30	2.22	2.25	30.64
1963	1.85	2.05	2.79	2.22	2.84	2.49	1.21	3.59	0.85	0.21	5.65	2.06	27.81
1964	2.18	1.13	3.83	3.66	2.31	1.41	2.15	3.09	0.75	1.52	2.20	2.87	27.10
1965	2.28	2.92	1.63	3.53	1.61	2.04	1.34	1.95	3.60	2.70	2.97	1.92	28.39
1966	3.98	2.96	2.27	3.05	1.79	2.73	2.09	2.64	4.75	0.90	2.05	3.93	33.14
1967	1.47	1.49	1.34	2.11	3.33	1.56	6.33	5.00	2.73	3.52	4.48	2.66	36.02
1968	2.08	1.10	3.13	2.40	3.46	6.14	3.77	4.17	3.43	5.81	4.07	4.67	44.23
1969	3.37	1.49	1.08	3.95	4.34	3.74	0.90	1.77	1.13	2.30	4.56	3.42	32.05
1970	1.02	1.84	2.45	3.68	2.79	2.93	4.42	4.07	4.33	3.84	3.53	3.33	38.23
1971	1.90	4.07	2.90	2.19	3.40	3.26	6.49	4.01	2.56	1.62	3.52	3.26	39.18
1972	1.10	2.87	2.49	4.03	6.19	12.30	3.45	3.76	4.12	4.36	6.79	3.95	55.41
1973	1.85	1.71	3.45	6.91	5.58	7.07	3.62	2.97	4.57	3.81	6.73	4.38	52.65
1974	2.08	1.70	4.34	3.09	5.78	4.67	9.52	4.60	4.45	1.58	4.95	3.47	50.23
1975	2.54	3.05	2.67	2.01	2.74	4.08	9.32	5.35	8.81	3.69	3.54	4.10	51.90
1976	2.79	2.71	4.62	8.12	7.41	7.42	5.24	6.73	3.27	6.53	1.53	1.80	58.17
1977	1.84	1.62	3.47	3.04	0.75	3.30	4.76	4.93	6.54	4.75	5.31	4.33	44.64
1978	5.77	0.80	3.08	1.87	1.90	3.58	2.78	3.31	3.93	2.68	1.25	4.12	35.07
1979	4.70	2.54	2.73	3.89	3.07	2.33	2.33	3.69	5.25	2.91	3.25	1.84	38.53
1980	1.47	1.38	4.34	3.33	1.34	4.45	2.57	1.33	3.40	2.56	2.64	3.27	32.08
1981	1.34	2.72	1.01	2.04	2.61	1.89	2.68	2.63	5.58	6.66	3.09	2.96	35.21
1982	3.59	1.26	2.63	1.71	2.87	4.64	3.83	2.60	4.22	0.72	4.52	2.55	35.14
1983	1.92	1.07	2.30	6.34	3.33	1.50	2.31	2.80	2.98	1.98	4.30	5.50	36.33
1984	1.30	2.88	2.39	3.16	4.97	2.02	3.66	5.17	2.61	1.95	3.48	4.38	37.97
1985	2.49	1.55	2.61	1.22	3.39	2.80	2.75	1.44	3.88	3.39	5.18	1.80	32.50
Record Mean	2.69	2.51	3.08	3.09	3.02	3.56	3.45	3.32	3.07	3.03	3.00	2.96	36.79

TABLE 3 AVERAGE TEMPERATURE (deg. F) SYRACUSE, NEW YORK

YEAR	JAN	FEB	MAR	APR	MAY	JUNE	JULY	AUG	SEP	OCT	NOV	DEC	ANNUAL
1956	22.6	27.0	28.5	42.9	53.4	67.0	68.2	69.2	58.8	52.7	41.2	33.3	47.1
1957	18.8	30.4	35.1	48.4	56.0	70.4	70.4	67.1	62.6	50.1	42.0	34.2	48.8
1958	21.4	19.5	34.3	48.3	54.5	62.5	70.5	69.1	61.9	50.4	41.9	19.4	46.1
1959	21.0	21.8	31.1	47.6	60.0	66.3	73.5	74.1	67.3	52.3	38.6	31.1	48.8
1960	24.7	27.0	24.4	49.5	59.8	66.6	69.4	68.7	63.8	50.0	43.8	23.7	47.6
1961	18.8	25.9	33.2	42.9	55.5	66.5	71.8	70.4	69.5	55.8	40.7	29.1	48.3
1962	24.1	21.5	34.6	47.3	62.3	68.3	69.2	69.1	59.7	51.2	35.5	24.5	47.3
#1963	20.8	18.4	34.2	45.6	54.7	66.1	71.7	66.1	57.2	56.3	44.9	20.8	46.4
1964	25.9	23.7	35.1	46.5	61.6	66.0	73.2	67.5	61.6	49.5	43.6	29.6	48.6
1965	20.5	24.6	30.3	42.3	59.4	65.7	67.5	69.1	62.8	47.9	38.7	32.1	46.6
1966	19.0	23.1	34.4	42.6	51.2	65.7	71.1	70.4	59.5	49.5	42.9	29.2	46.6
1967	30.6	19.2	31.6	44.9	50.2	66.5	67.7	66.6	60.7	51.8	37.6	32.7	46.9
1968	18.5	21.1	33.2	48.1	53.6	64.9	69.9	68.8	64.8	52.5	40.0	27.2	46.9
1969	24.3	23.6	30.4	46.9	55.7	64.9	69.7	71.3	63.8	51.1	40.4	23.8	47.1
1970	16.1	24.1	31.7	47.2	57.5	63.5	69.7	68.0	61.4	52.3	41.7	25.4	46.5
1971	18.5	26.5	31.2	42.8	55.8	69.0	69.0	67.1	65.6	56.6	36.9	33.2	47.6
1972	26.4	22.9	29.4	40.5	58.5	64.5	72.9	69.2	63.5	46.5	37.0	30.8	46.9
1973	28.4	21.4	42.6	46.8	54.3	69.6	72.7	73.5	62.0	53.7	40.7	29.5	49.6
1974	26.0	21.6	32.3	48.8	54.1	65.6	69.1	68.9	59.1	46.5	40.6	30.4	46.9
1975	29.4	28.1	31.7	39.9	62.9	67.1	71.7	68.2	57.1	53.2	46.6	27.6	48.6
1976	18.1	32.5	36.6	48.4	54.2	66.7	66.0	68.0	59.8	46.9	35.8	22.6	46.3
1977	15.7	26.0	40.1	48.2	60.3	62.7	70.8	67.3	62.5	50.6	44.0	27.3	48.0
1978	21.3	17.6	29.4	42.2	58.3	64.8	71.9	71.7	59.9	49.6	40.3	30.6	46.4
1979	22.4	12.9	39.1	45.1	58.6	66.0	71.7	67.9	61.4	50.9	44.5	33.4	47.8
1980	25.6	19.8	32.4	47.8	59.8	66.0	72.5	73.8	63.4	48.8	37.6	22.6	47.3
1981	15.0	33.7	36.4	50.0	59.2	68.0	73.3	70.4	61.6	47.9	39.0	29.0	48.6
1982	14.8	25.1	33.2	43.9	59.4	65.1	70.4	65.3	60.6	50.4	43.9	34.1	47.0
1983	23.4	26.4	35.7	44.3	53.7	66.7	72.0	69.0	62.5	50.3	39.0	22.5	47.1
1984	18.7	32.0	24.5	46.0	52.4	65.4	68.0	68.8	57.7	52.2	38.3	33.5	46.5
1985	22.0	27.3	36.3	47.8	59.5	62.0	69.8	68.9	63.5	51.4	41.2	26.0	48.0
Record Mean	23.6	24.1	33.4	45.4	56.9	66.1	71.1	69.2	62.2	51.4	40.2	28.2	47.6
Max	31.4	32.1	41.4	54.7	67.1	76.2	81.1	78.9	71.8	60.4	47.2	35.0	56.4
Min	15.9	16.1	25.4	36.1	46.7	56.0	61.1	59.4	52.5	42.3	33.2	21.4	38.8

REFERENCE NOTES FOR TABLES 1, 2, 3 and 6   (SYRACUSE, NY)

GENERAL

T - TRACE AMOUNT
BLANK ENTRIES DENOTE MISSING/UNREPORTED DATA.
# INDICATES A STATION OR INSTRUMENT RELOCATION.

SPECIFIC

TABLE 1

(a) - LENGTH OF RECORD IN YEARS. ALTHOUGH
INDIVIDUAL MONTHS MAY BE MISSING.

* LESS THAN .05

NORMALS — BASED ON THE 1951-1980 RECORD PERIOD.
EXTREMES — DATES ARE THE MOST RECENT OCCURRENCE.
WIND DIR. — NUMERALS SHOW TENS OF DEGREES
CLOCKWISE FROM TRUE NORTH.
"00" INDICATES CALM.
RESULTANT WIND DIRECTIONS ARE GIVEN TO WHOLE DEGREES.

EXCEPTIONS

TABLES 2, 3, and 6

RECORD MEANS ARE THROUGH THE CURRENT YEAR,
BEGINNING IN   1902 FOR TEMPERATURE
1902 FOR PRECIPITATION
1950 FOR SNOWFALL

TABLE 4 HEATING DEGREE DAYS Base 65 deg. F   SYRACUSE, NEW YORK

SEASON	JULY	AUG	SEP	OCT	NOV	DEC	JAN	FEB	MAR	APR	MAY	JUNE	TOTAL
1956-57	21	9	225	373	710	979	1428	965	920	507	300	37	6474
1957-58	10	38	131	457	683	949	1345	1265	943	496	330	114	6761
1958-59	7	17	133	451	683	1408	1357	1201	1045	516	222	66	7106
1959-60	0	4	115	393	786	1045	1241	1096	1255	468	176	47	6626
1960-61	11	12	83	460	630	1273	1427	1091	977	656	317	44	6981
1961-62	14	13	70	292	724	1107	1262	1211	937	552	172	25	6379
#1962-63	6	16	185	422	875	1250	1364	1300	948	575	319	70	7330
1963-64	14	43	240	276	597	1365	1205	1190	922	546	160	89	6647
1964-65	2	30	154	475	636	1087	1370	1069	1069	676	215	132	6971
1965-66	27	50	144	521	782	1011	1422	1168	945	665	429	80	7244
1966-67	6	7	186	473	652	1104	1059	1275	1030	595	453	13	6853
1967-68	24	35	154	407	812	995	1438	1266	979	501	345	83	7039
1968-69	27	41	54	391	745	1163	1256	1152	1063	536	306	103	6837
1969-70	22	20	134	425	730	1269	1508	1139	1027	536	244	115	7169
1970-71	7	27	150	388	692	1222	1437	1069	1040	658	295	50	7035
1971-72	13	51	96	256	840	980	1189	1216	1098	731	204	84	6758
1972-73	9	23	98	567	833	1053	1128	1217	687	547	325	31	6518
1973-74	2	12	164	344	723	1094	1200	1206	1004	493	339	52	6633
1974-75	16	3	202	565	726	1069	1100	1026	1026	749	138	46	6666
1975-76	3	32	230	357	545	1154	1449	936	872	509	329	47	6463
1976-77	24	45	179	556	869	1303	1520	1086	767	511	209	111	7180
1977-78	14	60	121	444	624	1162	1348	1322	1097	677	252	92	7213
1978-79	10	1	184	470	735	1062	1315	1457	796	591	242	74	6937
1979-80	19	39	146	454	607	971	1215	1302	1007	511	194	115	6580
1980-81	3	0	120	496	814	1307	1544	869	882	446	221	27	6729
1981-82	2	4	145	523	775	1110	1552	1114	978	626	183	79	7091
1982-83	13	57	152	449	628	951	1280	1073	902	615	351	67	6538
1983-84	11	25	140	457	769	1312	1432	949	1246	563	386	68	7358
1984-85	16	33	227	390	797	971	1329	1048	882	514	193	109	6509
1985-86	10	18	121	415	702	1200							

TABLE 5 COOLING DEGREE DAYS Base 65 deg. F   SYRACUSE, NEW YORK

YEAR	JAN	FEB	MAR	APR	MAY	JUNE	JULY	AUG	SEP	OCT	NOV	DEC	TOTAL
1969	0	0	0	0	22	94	183	222	102	1	0	0	624
1970	0	0	0	8	22	74	160	127	51	3	0	0	445
1971	0	0	0	0	17	145	145	124	117	4	0	0	552
1972	0	0	0	0	9	78	262	160	61	0	0	0	570
1973	0	0	0	7	0	177	249	281	79	2	0	0	795
1974	0	0	0	14	6	77	148	128	31	1	0	0	405
1975	0	0	0	0	80	114	221	138	1	1	0	0	555
1976	0	0	0	16	2	141	84	83	31	0	0	0	357
1977	0	0	1	12	71	47	202	138	49	0	0	0	520
1978	0	0	0	0	49	92	231	215	36	0	0	0	623
1979	0	0	0	2	50	109	232	134	46	22	0	0	595
1980	0	0	0	0	41	62	243	279	80	1	0	0	706
1981	0	0	3	4	47	125	264	180	49	0	0	0	672
1982	0	0	0	0	18	25	186	72	25	0	3	0	329
1983	0	0	0	0	2	125	236	155	70	7	0	0	595
1984	0	0	0	0	4	88	119	154	14	1	0	0	380
1985	0	0	0	7	30	26	165	144	87	0	0	0	459

TABLE 6 SNOWFALL (inches)   SYRACUSE, NEW YORK

SEASON	JULY	AUG	SEP	OCT	NOV	DEC	JAN	FEB	MAR	APR	MAY	JUNE	TOTAL
1956-57	0.0	0.0	0.0	0.0	7.7	16.8	25.2	6.7	12.7	7.0	T	0.0	76.1
1957-58	0.0	0.0	0.0	T	5.0	13.5	35.6	72.6	12.5	1.9	0.0	0.0	141.1
1958-59	0.0	0.0	0.0	T	22.1	22.5	48.8	17.8	26.0	T	0.0	0.0	137.2
1959-60	0.0	0.0	0.0	T	16.6	11.6	31.6	50.5	22.9	1.6	0.0	0.0	134.8
1960-61	0.0	0.0	0.0	2.0	2.7	27.8	37.3	25.8	26.5	8.4	T	0.0	130.5
1961-62	0.0	0.0	0.0	T	9.5	22.3	13.6	25.0	1.2	5.7	0.0	0.0	77.3
1962-63	0.0	0.0	0.0	2.8	11.0	33.8	22.2	28.3	15.8	1.8	0.8	0.0	116.5
1963-64	0.0	0.0	0.0	T	4.0	28.4	18.8	16.1	15.2	1.3	0.0	0.0	83.8
1964-65	0.0	0.0	0.0	0.3	4.0	18.3	31.8	24.9	13.3	4.7	0.0	0.0	97.3
1965-66	0.0	0.0	0.0	1.8	2.7	7.1	71.0	27.0	7.8	0.5	0.9	0.0	118.8
1966-67	0.0	0.0	0.0	0.0	T	33.0	18.3	21.0	10.4	0.3	T	0.0	83.0
1967-68	0.0	0.0	0.0	T	14.4	14.4	18.5	23.2	10.7	T	0.0	0.0	81.2
1968-69	0.0	0.0	0.0	0.8	16.5	25.4	24.5	21.3	9.4	T	0.0	0.0	97.9
1969-70	0.0	0.0	0.0	1.7	9.7	52.5	21.7	25.8	12.7	1.2	0.2	0.0	125.5
1970-71	0.0	0.0	0.0	0.8	7.0	51.9	30.3	25.2	37.2	4.8	0.0	0.0	157.2
1971-72	0.0	0.0	0.0	0.0	16.7	18.3	18.2	50.0	22.7	7.8	0.0	0.0	133.7
1972-73	0.0	0.0	0.0	0.3	15.8	29.8	11.9	13.3	3.6	5.3	1.2	0.0	81.2
1973-74	0.0	0.0	0.0	T	20.6	24.4	15.5	23.7	31.2	7.8	T	0.0	123.2
1974-75	0.0	0.0	0.0	2.8	4.8	26.2	11.8	27.3	20.6	12.0	0.0	0.0	105.5
1975-76	0.0	0.0	0.0	T	2.8	27.0	35.8	12.7	16.6	0.9	T	0.0	95.8
1976-77	0.0	0.0	0.0	0.3	25.9	25.7	52.3	24.4	13.5	1.9	1.0	0.0	145.0
1977-78	0.0	0.0	0.0	0.0	11.3	40.1	72.2	26.1	11.1	0.4	T	0.0	161.2
1978-79	0.0	0.0	0.0	T	3.9	40.9	27.9	20.7	14.9	10.2	0.0	0.0	118.5
1979-80	0.0	0.0	0.0	0.1	1.5	13.8	24.5	32.8	20.5	0.2	0.0	0.0	93.4
1980-81	0.0	0.0	0.0	T	7.3	28.8	23.4	8.5	10.6	0.4	0.0	0.0	79.0
1981-82	0.0	0.0	0.0	0.5	12.1	37.3	48.2	11.6	14.4	13.0	0.0	0.0	137.1
1982-83	0.0	0.0	0.0	T	1.9	10.9	20.3	8.2	8.3	16.4	T	0.0	66.0
1983-84	0.0	0.0	0.0	0.0	7.6	24.2	21.8	19.7	40.3	T	0.0	0.0	113.6
1984-85	0.0	0.0	0.0	0.0	5.0	23.4	57.3	21.6	7.1	2.0	0.0	0.0	116.4
1985-86	0.0	0.0	0.0	0.0	8.0	28.2							
Record Mean	0.0	0.0	0.0	0.5	9.1	26.0	28.7	25.1	17.2	3.7	0.1	0.0	110.3

See Reference Notes, relative to all above tables, on preceding page.

The city of Asheville is located on both banks of the French Broad River, near the center of the French Broad Basin. Upstream from Asheville, the valley runs south for 18 miles and then curves toward the south-southwest. Downstream from the city, the valley is oriented toward the north-northwest. Two miles upstream from the principal section of Asheville, the Swannanoa River joins the French Broad from the east. The entire valley is known as the Asheville Plateau, having an average elevation near 2,200 feet above sea level, and is flanked by mountain ridges to the east and west, whose peaks range from 2,000 to 4,400 feet above the valley floor. At the Carolina-Tennessee border, about 25 miles north-northwest of Asheville, a relatively high ridge of mountains blocks the northern end of the valley. Thirty miles south, the Blue Ridge Mountains form an escarpment, having a general elevation of about 2,700 feet above sea level. The tallest peaks near Asheville are Mt. Mitchell, 6,684 feet above sea level, 20 miles northeast of the city, and Big Pisgah Mountain, 5,721 feet above sea level, 16 miles to the southwest.

Asheville has a temperate, but invigorating, climate. Considerable variation in temperature often occurs from day to day in summer, as well as during the other seasons.

While the office was located in the city, the combination of roof exposure conditions and a smoke blanket, caused by inversions in temperature in the valley on quiet nights, resulted in higher early morning temperatures at City Office sites than were experienced nearer ground level in nearby rural areas. The growing season in this area is of sufficient length for commercial crops, the average length of freeze-free period being about 195 days. The average last occurrence in spring of a temperature 32 degrees or lower is mid-April and the average first occurrence in fall of 32 degrees is late October.

The orientation of the French Broad Valley appears to have a pronounced influence on the wind direction. Prevailing winds are from the northwest during all months of the year. Also, the shielding effect of the nearby mountain barriers apparently has a direct bearing on the annual amount of precipitation received in this vicinity. In an area northwest of Asheville, the average annual precipitation is the lowest in North Carolina. Precipitation increases sharply in all other directions, especially to the south and southwest.

Destructive events caused directly by meteorological conditions are infrequent. The most frequent, occurring at approximately 12-year intervals, are floods on the French Broad River. These floods are usually associated with heavy rains caused by storms moving out of the Gulf of Mexico. Snowstorms which have seriously disrupted normal life in this community are infrequent. Hailstorms that cause property damage are extremely rare.

TABLE 1 — NORMALS, MEANS AND EXTREMES

ASHEVILLE, NORTH CAROLINA

LATITUDE: 35°26'N   LONGITUDE: 82°33'W   ELEVATION: FT. GRND 2140 BARO 02161   TIME ZONE: EASTERN   WBAN: 03812

	(a)	JAN	FEB	MAR	APR	MAY	JUNE	JULY	AUG	SEP	OCT	NOV	DEC	YEAR
TEMPERATURE °F:														
Normals														
–Daily Maximum		47.5	50.6	58.4	68.6	75.6	81.4	84.0	83.5	77.9	68.7	58.6	50.3	67.1
–Daily Minimum		26.0	27.6	34.4	42.7	51.0	58.2	62.4	61.6	55.8	43.3	34.2	28.2	43.8
–Monthly		36.8	39.1	46.4	55.7	63.3	69.8	73.2	72.6	66.8	56.0	46.4	39.3	55.4
Extremes														
–Record Highest	21	78	77	83	89	91	96	96	100	92	85	81	78	100
–Year		1975	1977	1985	1972	1969	1969	1983	1983	1975	1981	1974	1971	AUG 1983
–Record Lowest	21	–16	–2	9	23	29	35	46	43	30	21	8	–7	–16
–Year		1985	1967	1980	1985	1971	1966	1967	1968	1967	1976	1970	1983	JAN 1985
NORMAL DEGREE DAYS:														
Heating (base 65°F)		874	725	577	283	114	23	0	0	57	286	558	797	4294
Cooling (base 65°F)		0	0	0	0	61	167	254	239	114	7	0	0	842
% OF POSSIBLE SUNSHINE	21	56	61	62	66	61	64	58	54	56	60	58	58	60
MEAN SKY COVER (tenths)														
Sunrise - Sunset	21	6.0	5.8	6.0	5.6	6.1	6.1	6.5	6.3	6.2	5.3	5.5	5.8	5.9
MEAN NUMBER OF DAYS:														
Sunrise to Sunset														
–Clear	21	9.7	9.4	8.8	9.9	7.8	6.3	4.6	5.2	7.0	11.8	10.7	10.7	101.8
–Partly Cloudy	21	7.0	6.3	8.7	8.7	9.5	12.2	13.5	13.7	10.3	7.8	7.2	6.8	111.9
–Cloudy	21	14.2	12.6	13.5	11.4	13.7	11.5	13.0	12.1	12.7	11.4	12.1	13.5	151.6
Precipitation														
.01 inches or more	21	10.3	9.3	11.3	9.2	11.9	11.3	12.5	12.0	9.3	8.2	9.2	9.5	123.9
Snow, Ice pellets														
1.0 inches or more	21	1.6	1.5	0.9	0.1	0.0	0.0	0.0	0.0	0.0	0.0	0.2	0.5	4.8
Thunderstorms	21	0.4	0.8	2.3	3.2	7.4	8.0	9.2	8.9	3.4	1.0	0.7	0.3	45.7
Heavy Fog Visibility 1/4 mile or less	21	4.0	3.0	2.2	2.6	5.4	8.1	10.3	14.9	12.1	8.5	4.4	4.8	80.5
Temperature °F														
–Maximum														
90° and above	21	0.0	0.0	0.0	0.0	0.*	1.8	3.8	2.3	0.3	0.0	0.0	0.0	8.2
32° and below	21	3.8	1.6	0.3	0.0	0.0	0.0	0.0	0.0	0.0	0.0	0.1	0.9	6.6
–Minimum														
32° and below	21	24.0	21.3	13.7	4.1	0.3	0.0	0.0	0.0	0.*	3.8	13.4	20.8	101.4
0° and below	21	0.6	0.*	0.0	0.0	0.0	0.0	0.0	0.0	0.0	0.0	0.0	0.1	0.8
AVG. STATION PRESS. (mb)	13	941.0	941.0	939.8	940.6	940.4	942.2	943.4	944.2	943.8	944.1	943.1	942.4	942.2
RELATIVE HUMIDITY (%)														
Hour 01	21	81	78	79	78	90	94	95	97	96	92	86	83	87
Hour 07	21	85	84	85	86	92	95	96	98	97	94	89	86	91
Hour 13 (Local Time)	21	60	56	54	51	58	60	65	64	64	58	57	59	59
Hour 19	21	69	63	60	56	67	70	76	79	81	76	70	70	70
PRECIPITATION (inches):														
Water Equivalent														
–Normal		3.48	3.60	5.13	3.84	4.19	4.20	4.43	4.79	3.96	3.29	3.29	3.51	47.71
–Maximum Monthly	21	7.47	7.02	9.86	7.26	8.83	6.54	9.92	11.28	9.12	7.05	7.76	8.48	11.28
–Year		1978	1982	1975	1979	1973	1972	1982	1967	1977	1971	1979	1973	AUG 1967
–Minimum Monthly	21	0.45	0.44	0.77	0.25	1.59	1.47	0.63	0.52	0.16	0.30	1.19	0.16	0.16
–Year		1981	1978	1985	1976	1985	1985	1978	1981	1984	1978	1981	1965	SEP 1984
–Maximum in 24 hrs	21	2.95	3.47	5.13	3.06	4.95	3.54	4.02	4.12	3.41	2.95	4.03	2.66	5.13
–Year		1978	1982	1968	1973	1973	1972	1969	1967	1975	1977	1977	1973	MAR 1968
Snow, Ice pellets														
–Maximum Monthly	21	17.6	25.5	13.0	3.0	T					T	9.6	16.3	25.5
–Year		1966	1969	1969	1982	1979					1977	1968	1971	FEB 1969
–Maximum in 24 hrs	21	8.0	11.7	10.9	3.0	T					T	5.7	16.3	16.3
–Year		1983	1969	1969	1982	1979					1977	1968	1971	DEC 1971
WIND:														
Mean Speed (mph)	21	9.8	9.6	9.5	8.8	7.1	6.1	5.8	5.4	5.6	6.7	8.2	8.9	7.6
Prevailing Direction														
Fastest Obs. 1 Min.														
–Direction	21	34	34	35	22	34	36	35	34	32	33	32	34	34
–Speed (MPH)	21	40	60	46	44	40	40	43	40	40	35	40	44	60
–Year		1975	1972	1969	1970	1971	1977	1966	1973	1980	1972	1974	1965	FEB 1972
Peak Gust														
–Direction	2	NW	N	SW	N	N	N	SW	NW	SW	N	SE	N	NW
–Speed (mph)	2	49	39	45	40	37	37	36	36	37	33	36	44	49
–Date		1984	1984	1984	1985	1984	1985	1985	1985	1985	1984	1984	1984	JAN 1984

See Reference Notes to this table on the following page.

TABLE 2 PRECIPITATION (inches) ASHEVILLE, NORTH CAROLINA

YEAR	JAN	FEB	MAR	APR	MAY	JUNE	JULY	AUG	SEP	OCT	NOV	DEC	ANNUAL
1956	1.02	4.87	3.53	4.33	1.65	2.21	5.11	1.38	7.31	2.67	1.34	2.96	38.38
1957	4.02	3.13	1.57	6.76	1.73	6.65	1.43	0.98	4.26	2.18	6.22	3.36	42.29
1958	2.89	3.03	3.05	4.04	4.02	1.15	3.91	1.56	1.94	1.83	1.88	3.93	33.23
1959	2.96	1.51	3.37	4.70	7.33	1.53	3.38	3.41	8.33	5.09	1.01	1.97	44.59
1960	2.79	4.19	3.97	2.76	2.11	3.05	2.85	7.24	2.21	4.26	0.68	1.49	37.60
1961	1.45	5.18	3.19	2.98	3.04	4.44	2.54	8.13	1.07	2.36	4.85	6.09	45.32
1962	4.46	3.58	4.13	3.25	2.83	6.20	3.24	3.47	2.40	2.40	1.66		40.02
1963	1.73	1.76	7.66	3.02	2.53	2.71	2.93	3.83	3.64	T	4.42	2.44	36.67
#1964	2.83	3.58	5.13	5.21	0.94	0.80	3.29	8.88	5.37	8.46	2.88		49.88
1965	2.16	4.60	5.10	2.62	3.33	4.12	4.47	4.03	4.69	2.92	1.30	0.16	39.50
1966	3.37	6.56	2.59	5.47	4.73	2.46	3.24	7.73	4.55	5.37	3.32	2.36	51.75
1967	2.02	2.20	2.86	1.11	6.79	4.45	6.90	11.28	2.53	3.30	2.54	6.13	52.11
1968	2.93	0.62	6.65	2.37	2.92	5.06	7.18	3.31	2.64	5.02	2.98	3.10	44.78
1969	2.64	5.08	4.01	3.53	3.32	3.82	7.53	6.47	3.04	2.63	1.91	4.63	48.61
1970	1.75	2.42	2.62	2.96	1.72	2.72	5.02	2.46	1.17	5.55	1.83	2.72	32.94
1971	2.53	4.93	3.48	2.06	3.54	5.00	4.57	3.03	3.80	7.05	2.84	4.32	48.05
1972	3.57	2.02	3.19	1.49	6.63	6.54	4.66	1.88	5.29	4.44	4.42	3.89	48.02
1973	4.26	4.23	8.91	5.71	8.83	3.87	6.95	4.57	3.12	2.41	3.57	8.48	64.91
1974	3.44	4.24	3.18	4.99	5.58	3.73	3.93	7.34	4.13	1.28	4.22	2.38	48.44
1975	3.86	4.56	9.86	0.61	8.17	2.12	3.31	3.63	7.53	3.94	4.89	4.44	56.92
1976	3.51	2.20	4.96	0.25	8.67	5.51	3.18	4.23	3.50	5.59	1.58	4.05	47.23
1977	2.09	1.02	7.29	4.05	3.96	5.11	4.44	3.68	9.12	3.79	6.88	2.43	50.45
1978	7.47	0.44	5.22	2.97	4.65	2.29	0.63	6.91	2.57	0.30	2.49	4.32	40.26
1979	6.81	5.14	5.72	7.26	5.35	2.20	5.52	3.63	5.60	1.40	7.76	1.05	57.44
1980	2.85	0.53	8.26	4.77	4.54	4.68	2.21	2.38	4.36	2.62	3.04	0.59	40.83
1981	0.45	4.80	3.24	2.07	7.50	4.41	2.06	0.52	1.36	2.19	1.19	4.79	34.58
1982	5.41	7.02	1.92	3.62	3.78	3.98	9.92	1.73	1.33	3.48	4.59	4.04	50.82
1983	3.39	5.63	6.27	5.27	3.48	3.71	1.06	0.95	5.66	4.43	4.77	8.30	52.92
1984	2.36	6.43	4.82	4.05	6.62	3.69	5.88	5.02	0.16	2.73	2.61	1.34	45.71
1985	2.95	4.74	0.77	2.74	1.59	1.47	4.37	7.04	1.25	3.41	4.91	0.70	35.94
Record Mean	3.32	3.78	4.81	3.33	5.03	3.85	4.50	4.37	3.69	3.52	3.51	3.53	47.25

TABLE 3 AVERAGE TEMPERATURE (deg. F) ASHEVILLE, NORTH CAROLINA

YEAR	JAN	FEB	MAR	APR	MAY	JUNE	JULY	AUG	SEP	OCT	NOV	DEC	ANNUAL
1956	34.6	44.7	47.2	54.0	65.7	70.0	73.6	73.5	64.9	59.0	45.2	49.3	56.8
1957	39.8	46.8	45.5	59.0	65.8	72.5	74.0	72.1	69.0	52.0	47.7	41.2	57.1
1958	32.2	29.7	42.9	55.5	64.8	71.2	74.3	73.4	66.4	55.5	50.1	37.0	54.4
1959	36.5	42.1	44.6	57.3	66.5	69.9	74.4	75.4	67.6	59.9	45.1	40.7	56.7
1960	39.0	36.6	33.5	58.2	61.1	71.1	73.4	74.1	68.1	58.9	47.4	34.7	54.7
1961	33.5	44.1	50.1	50.4	60.0	68.6	72.5	72.0	68.4	55.3	50.4	39.7	55.5
1962	37.4	45.8	43.4	52.9	69.4	70.2	73.7	72.3	65.4	58.3	46.0	34.4	55.7
1963	34.4	34.2	51.0	57.6	63.6	69.2	71.4	72.0	66.2	59.6	47.8	31.4	54.8
#1964	38.1	34.7	46.3	57.3	65.3	72.8	72.9	71.7	66.0	53.0	51.4	42.9	56.1
1965	37.0	37.9	42.4	57.4	65.8	66.8	72.2	71.1	66.8	53.8	46.4	40.3	54.8
1966	30.1	36.2	43.5	52.0	60.1	66.1	71.1	69.9	62.9	51.7	45.0	37.6	52.2
1967	38.7	35.0	49.8	57.6	59.7	66.8	68.5	68.5	63.5	53.5	42.7	41.8	53.6
1968	34.3	32.4	46.6	54.8	61.0	69.5	73.1	74.1	64.1	56.7	46.0	36.2	54.0
1969	36.7	37.8	41.3	56.7	65.2	73.1	75.9	70.7	65.8	56.2	44.0	36.5	55.0
1970	30.9	39.1	46.8	57.6	63.7	70.1	74.4	73.1	70.7	59.0	46.0	42.8	56.2
1971	36.5	39.5	43.1	55.2	61.1	72.4	72.5	72.2	69.5	61.8	45.8	47.7	56.5
1972	42.1	37.6	46.5	55.8	61.2	66.5	72.3	72.5	69.0	55.0	45.5	45.2	55.7
1973	37.5	38.5	52.7	52.8	60.3	71.0	74.1	74.2	70.0	58.8	49.7	39.8	56.5
1974	48.2	40.5	51.1	54.9	64.2	66.7	72.9	72.3	65.7	54.6	47.4	40.3	56.6
1975	41.7	42.5	44.8	54.1	66.0	68.8	72.4	72.9	65.2	57.3	48.2	38.6	56.1
1976	33.6	45.3	50.6	54.9	59.5	68.1	71.2	70.2	63.1	51.5	41.2	36.2	53.8
1977	24.8	37.4	50.7	58.2	64.6	69.7	75.7	73.8	66.1	54.3	49.3	36.8	55.3
1978	29.3	33.4	45.9	56.8	62.0	71.1	73.4	74.1	70.0	55.7	51.8	41.0	55.4
1979	34.2	35.8	50.0	55.9	64.3	68.8	72.2	73.2	66.5	55.3	49.2	42.0	55.6
1980	40.5	35.1	46.2	56.5	64.8	71.7	77.5	74.8	70.2	54.7	47.1	39.6	56.5
1981	33.3	39.9	44.9	60.1	60.7	74.3	75.0	71.7	66.1	54.5	48.2	35.8	55.4
1982	32.3	41.2	50.0	53.6	67.3	71.5	74.6	71.7	64.5	56.3	47.1	44.9	56.3
1983	36.7	38.8	46.7	51.1	61.6	69.0	75.7	76.5	66.5	57.5	47.3	36.4	55.3
1984	34.0	40.5	44.8	51.7	59.9	70.0	70.6	71.6	62.8	62.7	43.1	46.3	54.8
1985	30.5	38.3	48.1	56.6	62.6	69.8	72.2	70.9	64.2	60.5	56.0	34.7	55.4
Record Mean	35.4	38.2	47.0	55.5	62.6	69.6	73.2	72.4	66.3	56.2	47.0	40.0	55.3
Max	46.0	49.7	59.2	68.4	74.8	81.0	83.5	82.8	77.2	68.4	58.8	51.0	66.7
Min	24.7	26.7	34.7	42.5	50.4	58.1	62.9	61.9	55.4	44.0	35.1	29.0	43.8

REFERENCE NOTES FOR TABLES 1, 2, 3 and 6   **(ASHEVILLE, NC)**

GENERAL

T - TRACE AMOUNT
BLANK ENTRIES DENOTE MISSING/UNREPORTED DATA.
# INDICATES A STATION OR INSTRUMENT RELOCATION.

SPECIFIC

TABLE 1

(a) - LENGTH OF RECORD IN YEARS. ALTHOUGH
INDIVIDUAL MONTHS MAY BE MISSING.
* LESS THAN .05

NORMALS — BASED ON THE 1951-1980 RECORD PERIOD.
EXTREMES — DATES ARE THE MOST RECENT OCCURRENCE.
WIND DIR. — NUMERALS SHOW TENS OF DEGREES
CLOCKWISE FROM TRUE NORTH.
"00" INDICATES CALM.
RESULTANT WIND DIRECTIONS ARE GIVEN TO WHOLE DEGREES.

EXCEPTIONS

TABLES 2, 3, and 6

RECORD MEANS ARE THROUGH THE CURRENT YEAR,
BEGINNING IN 1965 FOR TEMPERATURE
1965 FOR PRECIPITATION
1965 FOR SNOWFALL

TABLE 4 HEATING DEGREE DAYS Base 65 deg. F   ASHEVILLE, NORTH CAROLINA

SEASON	JULY	AUG	SEP	OCT	NOV	DEC	JAN	FEB	MAR	APR	MAY	JUNE	TOTAL
1956-57	0	3	82	197	586	481	774	505	601	235	60	3	3527
1957-58	0	0	60	398	509	731	1010	981	679	288	70	4	4730
1958-59	0	0	41	292	449	860	877	635	624	249	58	3	4088
1959-60	0	0	21	204	591	747	800	818	969	226	181	2	4559
1960-61	0	0	32	194	522	931	971	582	453	444	169	18	4316
1961-62	2	0	49	295	433	778	851	531	664	371	27	1	4002
1962-63	0	0	91	224	565	679	943	857	426	241	98	6	4392
1963-64	4	0	51	164	513	1038	826	873	572	251	59	8	4359
#1964-65	0	20	46	372	399	679	863	751	691	232	23	27	4103
1965-66	0	7	39	344	550	759	1075	800	660	383	149	42	4808
1966-67	1	1	87	405	593	838	810	834	465	226	185	51	4496
1967-68	7	2	158	351	660	713	947	939	566	306	150	7	4806
1968-69	0	20	42	258	563	884	873	755	729	246	70	9	4449
1969-70	0	8	59	280	623	875	1050	720	557	236	86	3	4497
1970-71	0	0	29	194	565	682	875	707	672	290	137	0	4151
1971-72	0	0	6	129	576	530	704	790	569	294	116	35	3749
1972-73	3	0	8	304	578	605	846	737	374	362	158	0	3975
1973-74	0	0	7	205	473	772	516	680	423	299	83	24	3482
1974-75	0	0	65	316	519	760	715	624	619	331	46	7	4002
1975-76	0	0	77	232	498	812	966	566	439	296	168	33	4087
1976-77	2	3	83	411	706	884	1239	768	437	198	66	25	4822
1977-78	0	0	14	331	466	868	1101	878	586	241	139	0	4624
1978-79	0	0	12	283	390	741	951	810	457	268	71	18	4001
1979-80	5	0	44	299	468	707	753	861	573	258	65	2	4035
1980-81	0	0	37	315	533	778	978	696	615	152	152	0	4256
1981-82	0	1	57	326	499	897	1006	659	458	333	38	13	4274
1982-83	0	0	74	274	531	616	872	725	562	410	127	13	4204
1983-84	0	0	84	229	527	882	955	706	618	391	176	9	4577
1984-85	1	0	107	91	648	576	1064	737	520	249	109	19	4121
1985-86	0	6	111	156	266	932							

TABLE 5 COOLING DEGREE DAYS Base 65 deg. F   ASHEVILLE, NORTH CAROLINA

YEAR	JAN	FEB	MAR	APR	MAY	JUNE	JULY	AUG	SEP	OCT	NOV	DEC	TOTAL
1969	0	0	0	4	85	262	343	196	92	15	0	0	997
1970	0	0	0	22	52	159	296	259	206	17	0	0	1011
1971	0	0	0	3	25	232	238	231	149	37	8	0	923
1972	0	0	0	24	6	84	236	237	134	0	1	0	722
1973	0	0	0	1	16	190	288	292	163	19	0	0	969
1974	0	0	0	3	65	82	254	234	92	1	0	0	731
1975	0	0	0	11	82	124	237	252	89	0	0	0	795
1976	0	0	0	0	5	135	198	170	35	2	0	0	545
1977	0	0	0	2	59	173	340	279	146	7	1	0	1007
1978	0	0	0	2	53	188	266	292	168	4	0	0	973
1979	0	0	0	1	55	141	234	261	96	4	0	0	792
1980	0	0	0	8	64	210	396	311	198	4	0	0	1191
1981	0	0	0	10	25	286	316	213	98	7	0	0	955
1982	0	0	0	0	117	206	305	215	64	16	0	0	923
1983	0	0	0	0	25	139	335	362	141	5	0	0	1007
1984	0	0	0	0	25	165	180	211	49	27	0	0	657
1985	0	0	5	2	43	170	229	194	90	25	4	0	762

TABLE 6 SNOWFALL (inches)   ASHEVILLE, NORTH CAROLINA

SEASON	JULY	AUG	SEP	OCT	NOV	DEC	JAN	FEB	MAR	APR	MAY	JUNE	TOTAL
1956-57	0.0	0.0	0.0	0.0	T	0.2	0.8	0.1	2.5	T	0.0	0.0	3.6
1957-58	0.0	0.0	0.0	T	0.5	1.5	1.1	12.8	1.5	0.0	T	0.0	17.4
1958-59	0.0	0.0	0.0	0.0	T	2.3	1.5	T	T	0.3	0.0	0.0	4.1
1959-60	0.0	0.0	0.0	0.0	T	1.0	2.0	8.5	28.9	0.2	T	0.0	40.6
1960-61	0.0	0.0	0.0	0.0	T	0.2	0.8	3.0	T	T	0.0	0.0	4.0
1961-62	0.0	0.0	0.0	T	T	1.2	13.3	T	8.3	T	0.0	0.0	22.8
1962-63	0.0	0.0	0.0	0.0	0.3	4.2	T	4.5	T	T	0.0	0.0	9.0
1963-64	0.0	0.0	0.0	0.0	1.2	8.9	1.7	13.9	0.1	0.0	0.0	0.0	25.8
#1964-65	0.0	0.0	0.0	0.0	T	T	5.5	4.3	5.0	0.0	0.0	0.0	14.8
1965-66	0.0	0.0	0.0	0.0	T	T	17.6	6.2	0.2	T	0.0	0.0	24.0
1966-67	0.0	0.0	0.0	0.0	1.3	0.8	1.5	4.2	T	0.0	0.0	0.0	7.8
1967-68	0.0	0.0	0.0	0.0	0.0	1.9	7.2	6.0	0.1	0.0	0.0	0.0	15.2
1968-69	0.0	0.0	0.0	0.0	9.6	T	0.1	25.5	13.0	0.0	0.0	0.0	48.2
1969-70	0.0	0.0	0.0	0.0	T	10.9	4.8	1.1	0.5	0.0	0.0	0.0	17.3
1970-71	0.0	0.0	0.0	0.0	0.0	6.1	0.1	0.1	8.9	0.2	0.0	0.0	15.4
1971-72	0.0	0.0	0.0	0.0	0.6	16.3	T	7.4	7.4	0.0	0.0	0.0	31.7
1972-73	0.0	0.0	0.0	0.0	0.6	T	7.1	0.5	1.0	T	0.0	0.0	9.2
1973-74	0.0	0.0	0.0	0.0	0.0	3.0	T	0.3	1.1	T	0.0	0.0	4.4
1974-75	0.0	0.0	0.0	0.0	3.1	3.0	0.4	4.3	3.7	T	0.0	0.0	14.5
1975-76	0.0	0.0	0.0	0.0	5.0	0.4	1.6	3.5	T	0.0	0.0	0.0	10.5
1976-77	0.0	0.0	0.0	0.0	0.1	0.3	11.9	0.7	0.0	0.0	0.0	0.0	13.0
1977-78	0.0	0.0	0.0	T	T	1.5	9.7	5.3	5.3	T	0.0	0.0	21.8
1978-79	0.0	0.0	0.0	0.0	0.0	T	5.2	17.8	T	0.0	T	0.0	23.0
1979-80	0.0	0.0	0.0	0.0	T	T	2.1	6.3	5.4	T	0.0	0.0	13.8
1980-81	0.0	0.0	0.0	0.0	T	T	4.7	T	9.9	0.0	0.0	0.0	14.6
1981-82	0.0	0.0	0.0	0.0	T	2.0	8.6	8.1	0.1	3.0	0.0	0.0	21.8
1982-83	0.0	0.0	0.0	0.0	0.0	0.4	10.5	9.3	4.5	2.0	0.0	0.0	26.7
1983-84	0.0	0.0	0.0	0.0	T	T	0.2	2.9	T	0.0	0.0	0.0	3.1
1984-85	0.0	0.0	0.0	0.0	0.0	T	4.5	3.1	0.1	0.4	0.0	0.0	8.1
1985-86	0.0	0.0	0.0	0.0	0.0	0.4							
Record Mean	0.0	0.0	0.0	T	1.0	2.2	4.9	5.6	3.2	0.3	T	0.0	17.1

See Reference Notes, relative to all above tables, on preceding page.

CAPE HATTERAS, NORTH CAROLINA

Hatteras Island is the largest and easternmost Island in North Carolina. The average elevation of the Island is less than 10 feet above mean sea level. It is separated from the mainland by the Pamlico Sound and is part of a chain of islands known as the Outer Banks. The Island is narrow, ranging from a few hundred yards wide to a few miles wide and is about 54 miles long. Much of the island is a National Park and waterfowl reserve.

The Weather Office is located in the village of Buxton about 1 mile west–northwest of the famous Cape Hatteras Lighthouse. Weather observations have been taken continuously since 1874 from locations all within 10 miles of the present stations location.

With its maritime climate, Cape Hatteras is very humid, with cooler summers and warmer winters than mainland North Carolina. Ninety degree temperatures are rare in summer, as are the teens in winter. The average first occurrence of freezing temperatures is early December, and the average last occurrence is late February.

Average rainfall is greater than any other coastal station in the state. Rainfall is rather evenly distributed throughout the year, with the maximum during July, August, and September. Snowfall is rare and generally light, usually melting as it falls.

Winter storms frequently breed offshore where the warm waters of the Gulf Stream and the southermost penetration of the Labrador Current meet some 20 to 50 miles off the coast. Late summer and fall tracks of tropical cyclones occasionally threaten the island. These storms produce strong winds, heavy rains and tidal flooding from both the ocean and Pamlico Sound. Many ships have been lost near, or wrecked on, the beaches of the island and attest to the fury of wind and wave.

More than a million tourists visit the island each year. The proximity of the gulfstream, natural beaches, excellent surf, and offshore fishing make Cape Hatteras a preferred place for vacationers, sportsmen, and campers. The surfing conditions are said to be the best on the east coast.

TABLE 1 — NORMALS, MEANS AND EXTREMES

CAPE HATTERAS, NORTH CAROLINA

LATITUDE: 35°16'N   LONGITUDE: 75°33'W   ELEVATION: FT. GRND   7 BARO 00025   TIME ZONE: EASTERN   WBAN: 93729

	(a)	JAN	FEB	MAR	APR	MAY	JUNE	JULY	AUG	SEP	OCT	NOV	DEC	YEAR
TEMPERATURE °F:														
Normals														
–Daily Maximum		52.6	53.5	58.8	67.2	74.1	80.5	84.4	84.4	80.5	71.7	63.6	56.4	69.0
–Daily Minimum		37.6	37.7	43.3	51.1	59.7	67.5	71.9	72.0	67.9	58.1	48.3	40.9	54.7
–Monthly		45.1	45.6	51.0	59.2	66.9	74.0	78.2	78.2	74.2	64.9	56.0	48.7	61.8
Extremes														
–Record Highest	28	75	76	79	89	88	95	95	94	92	86	81	77	95
–Year		1985	1971	1977	1985	1962	1978	1969	1968	1959	1971	1982	JUN 1978	
–Record Lowest	28	6	14	19	26	39	44	54	56	45	32	22	12	6
–Year		1985	1958	1967	1972	1971	1966	1972	1979	1970	1979	1967	1983	JAN 1985
NORMAL DEGREE DAYS:														
Heating (base 65°F)		617	543	437	186	37	0	0	0	0	76	276	510	2682
Cooling (base 65°F)		0	0	6	12	96	270	409	409	276	72	6	0	1556
% OF POSSIBLE SUNSHINE	23	49	53	60	66	62	62	62	63	62	58	56	48	58
MEAN SKY COVER (tenths)														
Sunrise – Sunset	28	6.1	6.0	6.0	5.4	6.0	6.2	6.6	6.1	5.6	5.5	5.4	5.9	5.9
MEAN NUMBER OF DAYS:														
Sunrise to Sunset														
–Clear	28	9.3	9.0	9.8	10.3	8.3	7.2	6.3	7.9	9.5	10.9	10.5	10.0	108.7
–Partly Cloudy	28	6.8	5.4	7.1	8.6	9.5	10.4	9.8	10.2	9.4	7.7	7.9	7.1	100.0
–Cloudy	28	14.9	13.9	14.2	11.1	13.2	12.4	15.0	12.9	11.0	12.5	11.6	13.9	156.5
Precipitation														
.01 inches or more	28	11.0	10.4	10.5	8.4	10.5	9.2	11.9	10.4	8.9	9.4	9.0	9.6	119.2
Snow, Ice pellets														
1.0 inches or more	28	0.2	0.3	0.1	0.0	0.0	0.0	0.0	0.0	0.0	0.0	0.0	0.*	0.7
Thunderstorms	28	0.8	1.5	2.0	3.1	5.4	5.1	8.6	7.4	3.4	2.0	1.5	1.0	41.9
Heavy Fog Visibility														
1/4 mile or less	28	2.7	2.8	2.5	1.3	1.2	0.3	0.2	0.4	0.3	0.8	1.3	2.4	15.9
Temperature °F														
–Maximum														
90° and above	28	0.0	0.0	0.0	0.0	0.0	0.6	1.3	1.6	0.4	0.0	0.0	0.0	3.9
32° and below	28	0.9	0.4	0.0	0.0	0.0	0.0	0.0	0.0	0.0	0.0	0.0	0.2	1.5
–Minimum														
32° and below	28	11.3	9.7	3.7	0.4	0.0	0.0	0.0	0.0	0.0	0.*	1.3	6.7	33.1
0° and below	28	0.0	0.0	0.0	0.0	0.0	0.0	0.0	0.0	0.0	0.0	0.0	0.0	0.0
AVG. STATION PRESS.(mb)	13	1018.5	1018.5	1016.7	1016.6	1015.6	1016.2	1016.9	1017.7	1017.6	1018.7	1019.1	1019.6	1017.7
RELATIVE HUMIDITY (%)														
Hour 01	28	78	79	79	81	86	87	89	88	85	82	81	79	83
Hour 07	28	80	80	80	78	82	83	85	86	84	83	82	81	82
Hour 13 (Local Time)	28	68	64	62	59	65	68	70	69	67	66	64	66	66
Hour 19	28	77	76	76	75	79	80	82	82	80	79	78	78	79
PRECIPITATION (inches):														
Water Equivalent														
–Normal		4.72	4.11	3.97	3.21	4.09	4.22	5.36	6.11	5.78	4.83	4.84	4.48	55.72
–Maximum Monthly	28	9.72	8.45	9.29	7.10	11.44	10.80	9.99	11.73	12.78	15.05	16.20	8.63	16.20
–Year		1979	1983	1983	1962	1972	1962	1965	1976	1979	1985	1985	1962	NOV 1985
–Minimum Monthly	28	1.75	1.42	0.98	0.59	0.61	0.38	0.45	0.99	0.73	0.53	1.23	0.64	0.38
–Year		1981	1976	1967	1976	1962	1978	1958	1983	1978	1984	1973	1985	JUN 1978
–Maximum in 24 hrs	28	5.00	2.92	2.91	5.60	3.55	6.63	5.53	8.11	5.46	7.67	7.69	3.59	8.11
–Year		1979	1970	1982	1963	1984	1962	1967	1962	1979	1983	1985	1979	AUG 1962
Snow, Ice pellets														
–Maximum Monthly	28	3.5	4.4	8.5								T	2.5	8.5
–Year		1962	1978	1960								1970	1970	MAR 1960
–Maximum in 24 hrs	28	3.5	4.4	7.3								T	2.5	7.3
–Year		1962	1978	1980								1970	1970	MAR 1980
WIND:														
Mean Speed (mph)	28	12.4	12.6	12.2	12.1	11.1	10.8	10.3	9.7	10.8	11.4	11.2	11.7	11.4
Prevailing Direction through 1963		NNE	NNE	SW	SW	SW	SSW	SW	SW	NE	NNE	NNE	NNE	NNE
Fastest Obs. 1 Min.														
–Direction (!!!)	15	35	16	02	18	18	29	22	16	11	02	12	22	11
–Speed (MPH)	15	32	44	48	40	35	37	26	41	60	44	39	39	60
–Year		1983	1985	1980	1983	1972	1972	1982	1971	1985	1982	1985	1972	SEP 1985
Peak Gust														
–Direction (!!!)	2	N	SW	SW	SE	SW	SW	S	SW	SW	NE	SW	NW	SW
–Speed (mph)	2	46	58	58	49	46	43	39	40	87	55	59	39	87
–Date		1984	1984	1984	1984	1984	1984	1984	1985	1985	1985	1985	1985	SEP 1985

See Reference Notes to this table on the following page

TABLE PRECIPITATION (inches)   CAPE HATTERAS, NORTH CAROLINA

YEAR	JAN	FEB	MAR	APR	MAY	JUNE	JULY	AUG	SEP	OCT	NOV	DEC	ANNUAL
1956	4.41	6.48	2.80	3.54	2.04	5.51	5.35	5.16	4.12	9.89	0.67	1.21	51.18
#1957	2.60	3.33	3.34	0.36	5.81	6.03	5.60	6.30	6.16	7.19	6.64	6.40	58.48
1958	4.13	3.28	5.21	3.63	4.19	5.57	0.45	6.30	3.98	5.39	3.46	3.16	53.47
1959	2.53	4.09	7.82	3.32	1.28	1.76	8.74	3.21	3.24	4.66	3.46	5.58	50.43
1960	6.75	6.11	4.68	4.88	5.61	7.13	7.25	3.93	3.24	3.17	1.43	4.62	58.80
1961	3.93	3.02	3.05	3.20	4.36	4.01	2.47	8.32	3.43	2.32	2.16	2.40	42.67
1962	6.96	1.65	5.72	7.10	0.61	10.80	3.24	11.68	3.45	14.63	8.63	8.63	83.22
1963	4.05	7.48	1.93	6.44	4.42	7.80	2.87	4.59	4.82	6.11	3.70	3.96	58.17
1964	6.27	5.86	1.11	4.32	2.63	1.87	6.29	10.62	5.51	7.41	2.08	3.19	57.16
1965	1.95	5.21	2.94	1.20	0.92	6.72	9.99	2.80	2.90	3.19	1.46	2.24	41.52
1966	9.07	4.24	2.32	1.21	7.51	6.80	2.16	3.61	6.41	5.04	1.23	6.58	56.18
1967	4.08	4.51	0.98	1.20	5.48	1.04	8.36	8.65	8.73	2.83	2.38	4.71	52.95
1968	5.62	4.06	1.93	4.07	3.63	7.65	1.80	6.47	9.58	4.21	2.64	5.13	55.13
1969	3.22	3.50	4.87	3.37	0.87	2.31	6.69	6.77	4.41	1.41	3.60	5.27	46.29
1970	5.61	5.92	5.31	4.25	2.86	4.66	6.16	4.21	4.07	2.94	8.28	2.44	56.71
1971	4.53	3.76	4.70	3.73	5.52	1.58	6.48	9.74	8.59	11.24	2.40	2.07	64.34
1972	6.42	5.66	2.96	0.72	11.44	1.83	8.62	1.78	2.74	7.26	7.29	4.84	61.56
1973	3.99	4.75	6.15	3.35	4.79	4.29	2.56	3.02	3.04	1.34	1.23	8.52	47.03
1974	2.02	4.07	3.48	2.38	4.31	8.31	2.49	11.04	2.94	8.99	1.72	5.14	56.89
1975	4.38	2.95	1.89	4.37	4.05	4.40	6.69	1.91	5.21	4.04	3.71	4.87	48.47
1976	3.22	1.42	2.73	0.59	5.20	4.71	4.31	11.73	8.58	4.62	3.08	6.90	57.09
1977	4.53	1.64	4.01	1.89	1.81	3.07	2.48	2.74	2.70	9.13	12.00	5.99	51.99
1978	7.61	2.85	5.70	5.73	6.04	0.38	4.47	1.04	0.73	3.15	6.77	3.32	47.79
1979	9.72	4.67	2.96	2.68	6.66	4.04	6.30	3.33	12.78	2.04	8.46	5.19	68.83
1980	7.76	2.38	8.94	4.03	0.87	5.49	3.47	1.48	5.09	2.58	5.29	4.48	51.86
1981	1.75	1.99	2.36	1.84	4.90	3.91	9.32	11.34	1.31	2.69	2.93	7.20	51.54
1982	5.85	7.03	6.04	5.85	0.90	5.50	6.76	2.14	4.53	3.90	2.50	5.84	56.84
1983	9.27	8.45	9.29	3.40	0.81	4.87	1.75	0.99	6.15	9.46	4.58	6.10	65.12
1984	2.34	4.28	2.99	4.26	7.27	3.12	5.23	4.73	3.81	0.53	4.95	1.69	45.20
1985	3.67	6.28	2.28	0.67	3.77	3.92	3.46	7.17	4.06	15.05	16.20	0.64	67.17
Record Mean	4.41	4.14	4.21	3.46	3.82	4.42	5.69	5.84	5.49	4.86	4.13	4.67	55.14

TABLE 3 AVERAGE TEMPERATURE (deg. F)   CAPE HATTERAS, NORTH CAROLINA

YEAR	JAN	FEB	MAR	APR	MAY	JUNE	JULY	AUG	SEP	OCT	NOV	DEC	ANNUAL
1956	39.7	50.6	51.2	58.2	66.4	74.7	79.0	77.7	73.0	67.4	56.7	56.7	62.6
#1957	45.3	50.6	50.8	63.3	69.1	76.4	77.6	76.3	76.3	62.3	57.9	51.5	63.1
1958	41.5	38.7	45.4	56.1	65.7	71.6	80.4	79.4	73.0	63.8	59.1	44.7	60.0
1959	43.6	48.7	51.5	60.5	70.0	75.2	78.7	79.0	75.8	68.3	57.6	50.1	63.3
1960	46.4	46.8	42.5	62.4	68.2	75.2	77.7	78.8	74.3	64.9	57.9	42.7	61.5
1961	42.1	49.1	54.4	56.4	63.0	72.7	79.1	78.2	75.8	62.4	56.4	49.4	61.6
1962	46.1	47.7	46.7	58.7	68.4	74.3	77.2	77.5	72.6	65.6	53.6	44.7	61.1
1963	42.9	41.3	54.7	58.4	63.4	72.6	76.4	76.8	70.5	63.9	56.5	41.9	59.9
1964	46.8	44.2	52.1	59.1	66.6	74.5	77.3	76.1	73.7	62.3	58.7	52.0	61.9
1965	45.8	45.5	47.5	55.0	68.9	73.2	77.5	78.5	74.4	63.1	56.0	48.2	61.1
1966	43.4	44.6	49.4	56.8	63.9	70.9	78.1	77.6	73.7	64.8	55.2	46.9	60.4
1967	48.5	45.0	50.4	56.8	62.8	70.5	77.6	76.9	68.8	64.2	51.4	51.6	60.4
1968	41.8	38.7	52.2	58.6	64.8	74.9	78.6	80.9	73.0	66.6	57.4	45.6	61.1
1969	42.5	43.5	45.4	60.0	66.3	75.1	78.9	76.8	73.0	67.1	52.1	46.1	60.6
1970	37.7	44.2	49.5	59.5	67.5	75.2	77.5	78.5	74.9	67.5	54.8	48.2	61.3
1971	42.5	47.3	49.4	54.7	64.9	74.6	78.3	77.7	75.6	70.9	56.0	56.2	62.3
1972	52.9	47.0	52.2	56.3	66.0	71.5	77.9	78.3	73.8	64.8	56.7	54.2	62.6
1973	46.1	43.4	55.2	60.5	67.3	76.5	77.4	78.8	75.9	66.7	57.4	52.6	63.3
1974	55.5	47.8	56.3	61.8	68.5	74.2	77.3	77.6	75.2	61.5	55.2	49.9	63.4
1975	51.1	51.5	52.5	55.5	69.1	76.5	77.6	80.2	76.2	68.3	59.0	48.7	63.9
1976	43.7	51.7	57.3	60.0	67.2	73.6	77.7	75.9	71.8	62.2	49.9	46.5	61.5
1977	35.8	44.3	55.2	62.9	69.6	74.6	80.8	79.5	77.3	63.4	59.4	48.8	62.6
1978	42.0	35.7	50.1	60.2	65.1	73.8	78.0	80.6	76.3	63.5	61.3	51.2	61.5
1979	45.4	41.0	50.7	58.9	67.6	71.6	77.4	77.9	75.2	63.8	58.6	49.1	61.4
1980	45.8	39.0	49.3	61.1	68.6	73.1	79.3	79.1	76.0	62.2	52.0	47.1	61.1
1981	36.2	46.1	45.4	62.6	65.4	78.3	78.6	76.5	71.5	62.9	52.9	44.4	60.1
1982	41.3	47.6	52.1	57.4	68.8	73.3	78.3	77.4	72.8	63.7	59.7	47.5	62.2
1983	44.9	45.4	52.7	57.4	66.8	72.5	79.5	78.7	74.3	66.7	55.9	47.5	61.9
1984	41.7	48.8	48.5	57.7	68.2	75.5	77.8	79.4	72.9	68.8	55.4	55.7	62.5
1985	41.0	45.3	55.1	63.7	69.8	75.2	79.6	78.2	75.8	71.4	66.6	47.7	64.1
Record Mean	46.1	46.4	51.7	59.0	67.6	74.5	78.2	78.2	74.6	65.8	56.4	48.8	62.3
Max	52.8	53.2	58.4	65.7	73.7	80.0	83.6	83.5	79.8	71.4	62.7	55.4	68.3
Min	39.4	39.5	44.9	52.3	61.5	69.0	72.9	72.9	69.3	60.1	50.1	42.2	56.2

REFERENCE NOTES FOR TABLES 1, 2, 3 and 6   (CAPE HATTERAS, NC)

GENERAL

T - TRACE AMOUNT
BLANK ENTRIES DENOTE MISSING/UNREPORTED DATA.
# INDICATES A STATION OR INSTRUMENT RELOCATION.

SPECIFIC

TABLE 1

(a) - LENGTH OF RECORD IN YEARS. ALTHOUGH INDIVIDUAL MONTHS MAY BE MISSING.

\# - LESS THAN .05

NORMALS — BASED ON THE 1951-1980 RECORD PERIOD.
EXTREMES — DATES ARE THE MOST RECENT OCCURRENCE.
WIND DIR. — NUMERALS SHOW TENS OF DEGREES CLOCKWISE FROM TRUE NORTH. "00" INDICATES CALM.
RESULTANT WIND DIRECTIONS ARE GIVEN TO WHOLE DEGREES.

EXCEPTIONS

TABLES 2, 3, and 6

RECORD MEANS ARE THROUGH THE CURRENT YEAR, BEGINNING IN   1882 FOR TEMPERATURE
     1875 FOR PRECIPITATION
     1958 FOR SNOWFALL

TABLE 4 HEATING DEGREE DAYS Base 65 deg. F   CAPE HATTERAS, NORTH CAROLINA

SEASON	JULY	AUG	SEP	OCT	NOV	DEC	JAN	FEB	MAR	APR	MAY	JUNE	TOTAL
#1956-57	0	0	2	7	264	259	609	395	433	126	44	0	2139
1957-58	0	0	1	108	226	413	723	729	603	270	35	6	3114
1958-59	0	0	0	77	180	623	658	450	412	164	7	3	2574
1959-60	0	0	0	41	251	456	570	521	694	116	27	0	2676
1960-61	0	0	0	81	219	682	700	440	326	262	109	3	2822
1961-62	0	0	0	124	281	477	579	478	563	208	34	0	2744
1962-63	0	0	4	99	335	623	678	658	314	216	112	0	3039
1963-64	0	0	8	77	251	732	558	596	395	194	58	2	2871
1964-65	0	0	3	136	194	409	588	538	537	296	15	0	2716
1965-66	0	0	0	135	267	514	665	565	478	253	104	23	3004
1966-67	0	0	0	77	296	562	505	551	449	245	107	15	2807
1967-68	0	0	11	77	399	418	708	758	391	196	70	0	3028
1968-69	0	0	0	73	228	594	686	594	601	166	41	0	2983
1969-70	0	0	1	51	390	581	842	573	472	167	44	0	3121
1970-71	0	0	10	40	299	515	690	486	476	300	65	0	2881
1971-72	0	0	0	6	289	278	379	517	393	267	39	16	2184
1972-73	0	0	0	61	257	335	578	597	302	151	50	0	2331
1973-74	0	0	0	31	246	389	293	476	276	143	28	0	1882
1974-75	0	0	3	141	306	459	425	378	383	289	26	0	2410
1975-76	0	0	0	32	206	498	653	379	244	190	52	4	2258
1976-77	0	0	1	135	448	566	900	575	309	104	27	0	3065
1977-78	0	0	0	103	196	504	704	813	455	152	60	0	2987
1978-79	0	0	0	90	123	426	601	663	439	198	32	0	2572
1979-80	0	0	0	112	208	487	587	750	477	145	33	0	2799
1980-81	0	0	0	132	385	549	884	524	600	109	82	0	3265
1981-82	0	0	15	104	360	630	727	482	392	237	44	0	2991
1982-83	0	0	0	86	176	341	613	544	373	238	50	2	2423
1983-84	0	0	10	55	272	535	715	466	507	215	47	0	2822
1984-85	0	0	2	22	296	282	736	547	305	113	19	0	2322
1985-86	0	0	0	1	48	532							

TABLE 5 COOLING DEGREE DAYS Base 65 deg. F   CAPE HATTERAS, NORTH CAROLINA

YEAR	JAN	FEB	MAR	APR	MAY	JUNE	JULY	AUG	SEP	OCT	NOV	DEC	TOTAL
1969	0	0	0	22	89	312	439	371	248	123	8	0	1612
1970	0	0	0	7	127	314	393	422	313	122	0	0	1698
1971	0	0	0	1	67	294	419	402	325	194	26	10	1738
1972	7	0	2	12	74	218	406	422	271	63	17	8	1500
1973	0	1	5	25	128	352	394	434	334	126	25	9	1833
1974	7	0	13	54	146	283	390	399	318	42	20	0	1672
1975	1	7	3	20	160	353	398	479	345	142	32	0	1940
1976	0	0	11	48	127	268	401	344	214	53	0	0	1466
1977	0	1	10	49	177	295	499	453	374	57	37	6	1958
1978	0	0	0	13	68	271	411	493	345	48	21	7	1677
1979	0	0	2	23	121	203	394	407	313	69	24	0	1556
1980	0	0	0	35	153	252	449	445	335	54	3	0	1726
1981	0	0	1	45	104	406	429	361	218	46	1	0	1611
1982	0	0	1	15	168	259	419	393	239	53	24	8	1579
1983	0	1	0	18	110	236	458	433	295	112	8	0	1671
1984	0	0	0	6	153	323	403	452	244	146	13	2	1742
1985	0	2	4	82	176	317	461	418	330	207	104	3	2104

TABLE 6 SNOWFALL (inches)   CAPE HATTERAS, NORTH CAROLINA

SEASON	JULY	AUG	SEP	OCT	NOV	DEC	JAN	FEB	MAR	APR	MAY	JUNE	TOTAL
#1956-57	0.0	0.0	0.0	0.0	0.0	T	T	0.0	0.0	0.0	0.0	0.0	T
1957-58	0.0	0.0	0.0	0.0	0.0	T	T	4.0	0.0	0.0	0.0	0.0	4.0
1958-59	0.0	0.0	0.0	0.0	0.0	0.0	T	T	T	0.0	0.0	0.0	T
1959-60	0.0	0.0	0.0	0.0	T	0.0	T	0.0	8.5	0.0	0.0	0.0	8.5
1960-61	0.0	0.0	0.0	0.0	0.0	T	T	0.0	0.0	0.0	0.0	0.0	T
1961-62	0.0	0.0	0.0	0.0	0.0	0.0	3.5	T	T	0.0	0.0	0.0	3.5
1962-63	0.0	0.0	0.0	0.0	0.0	T	T	1.3	0.0	0.0	0.0	0.0	1.3
1963-64	0.0	0.0	0.0	0.0	0.0	0.0	0.0	T	0.0	0.0	0.0	0.0	T
1964-65	0.0	0.0	0.0	0.0	0.0	0.0	0.3	0.0	0.0	0.0	0.0	0.0	0.3
1965-66	0.0	0.0	0.0	0.0	0.0	0.0	T	0.0	0.0	0.0	0.0	0.0	T
1966-67	0.0	0.0	0.0	0.0	0.0	T	0.0	0.0	0.0	0.0	0.0	0.0	T
1967-68	0.0	0.0	0.0	0.0	0.0	T	T	3.5	T	0.0	0.0	0.0	3.5
1968-69	0.0	0.0	0.0	0.0	0.0	0.2	T	0.0	0.0	0.0	0.0	0.0	0.2
1969-70	0.0	0.0	0.0	0.0	0.0	0.0	1.4	1.2	T	0.0	0.0	0.0	2.6
1970-71	0.0	0.0	0.0	0.0	T	2.5	T	0.0	0.0	0.0	0.0	0.0	2.5
1971-72	0.0	0.0	0.0	0.0	0.0	0.0	T	T	T	0.0	0.0	0.0	T
1972-73	0.0	0.0	0.0	0.0	0.0	0.0	1.9	3.3	0.0	0.0	0.0	0.0	5.2
1973-74	0.0	0.0	0.0	0.0	0.0	T	0.0	0.0	0.0	0.0	0.0	0.0	T
1974-75	0.0	0.0	0.0	0.0	0.0	0.0	0.0	0.0	T	0.0	0.0	0.0	T
1975-76	0.0	0.0	0.0	0.0	0.0	0.0	0.0	T	0.0	0.0	0.0	0.0	T
1976-77	0.0	0.0	0.0	0.0	0.0	T	0.0	1.0	0.0	0.0	0.0	0.0	1.0
1977-78	0.0	0.0	0.0	0.0	0.0	0.0	T	4.4	T	0.0	0.0	0.0	4.4
1978-79	0.0	0.0	0.0	0.0	0.0	0.0	T	1.1	0.0	0.0	0.0	0.0	1.1
1979-80	0.0	0.0	0.0	0.0	0.0	0.0	3.0	T	7.3	0.0	0.0	0.0	10.3
1980-81	0.0	0.0	0.0	0.0	0.0	T	2.9	0.0	0.0	0.0	0.0	0.0	2.9
1981-82	0.0	0.0	0.0	0.0	0.0	0.0	T	0.0	0.0	0.0	0.0	0.0	T
1982-83	0.0	0.0	0.0	0.0	0.0	0.0	0.0	T	0.0	0.0	0.0	0.0	T
1983-84	0.0	0.0	0.0	0.0	0.0	0.2	0.0	T	0.0	0.0	0.0	0.0	0.2
1984-85	0.0	0.0	0.0	0.0	0.0	0.0	0.1	0.0	0.0	0.0	0.0	0.0	0.1
1985-86	0.0	0.0	0.0	0.0	0.0	T							
Record Mean	0.0	0.0	0.0	0.0	T	0.1	0.5	0.7	0.6	0.0	0.0	0.0	1.8

See Reference Notes, relative to all above tables, on preceding page.

CHARLOTTE, NORTH CAROLINA

Charlotte is located in the Piedmont of the Carolinas, a transitional area of rolling country between the mountains to the west and the Coastal Plain to the east. The mountains are to the northwest about 80 miles from Charlotte. The general elevation of the area around Charlotte is about 730 feet. The Atlantic ocean is about 160 miles southeast.

The mountains have a moderating effect on winter temperatures, causing appreciable warming of cold air from the northwest winds. The ocean is too far away to have any immediate effect on summer temperatures but in winter an occasional general and sustained flow of air from the warm ocean waters results in considerable warming.

Charlotte enjoys a moderate climate, characterized by cool winters and quite warm summers. Temperatures fall as low as the freezing point on a little over one-half of the days in the winter months. Winter weather is changeable, with occasional cold periods, but extreme cold is rare. Snow is infrequent, and the first snowfall of the season usually comes in late November or December. Heavy snowfalls have occurred, but any appreciable accumulation of snow on the ground for more than a day or two is rare.

Summers are long and quite warm, with afternoon temperatures frequently in the low 90s. The growing season is also long, the average length of the freeze-free period being 216 days. On the average, the last occurrence in spring with a temperature of 32 degrees is early April. In the fall the average first occurrence of 32 degrees is early November.

Rainfall is generally rather evenly distributed throughout the year, the driest weather usually coming in the fall. Summer rainfall comes principally from thunderstorms with occasional dry spells of one to three weeks duration.

Hurricanes which strike the Carolina coast may produce heavy rain but seldom cause dangerous winds.

TABLE 1   NORMALS, MEANS AND EXTREMES

CHARLOTTE, NORTH CAROLINA

LATITUDE: 35°13'N   LONGITUDE: 80°56'W   ELEVATION: FT. GRND 737 BARO 738   TIME ZONE: EASTERN   WBAN: 13881

	(a)	JAN	FEB	MAR	APR	MAY	JUNE	JULY	AUG	SEP	OCT	NOV	DEC	YEAR
TEMPERATURE °F:														
Normals														
-Daily Maximum		50.3	53.6	61.6	72.1	79.1	85.2	88.3	87.6	81.7	71.7	61.7	52.6	70.5
-Daily Minimum		30.7	32.1	39.1	48.4	57.2	64.7	68.7	68.2	62.3	49.6	39.7	32.6	49.4
-Monthly		40.5	42.9	50.4	60.3	68.2	75.0	78.5	77.9	72.0	60.7	50.7	42.6	60.0
Extremes														
-Record Highest	46	78	81	90	93	100	103	103	103	104	98	85	77	104
-Year		1952	1977	1945	1960	1941	1954	1952	1983	1954	1954	1961	1971	SEP 1954
-Record Lowest	46	-5	5	4	24	32	45	53	53	39	24	11	2	-5
-Year		1985	1958	1980	1960	1963	1972	1961	1965	1967	1962	1950	1962	JAN 1985
NORMAL DEGREE DAYS:														
Heating (base 65°F)		760	619	459	155	50	0	0	0	10	166	429	694	3342
Cooling (base 65°F)		0	0	7	14	149	304	419	400	220	33	0	0	1546
% OF POSSIBLE SUNSHINE	35	56	60	63	69	68	70	67	69	67	67	60	58	65
MEAN SKY COVER (tenths)														
Sunrise - Sunset	36	6.2	6.0	6.0	5.6	6.1	6.0	6.2	5.8	5.7	4.8	5.4	5.9	5.8
MEAN NUMBER OF DAYS:														
Sunrise to Sunset														
-Clear	37	9.2	8.8	9.2	9.8	8.0	7.4	6.5	7.7	9.4	13.2	11.7	10.4	111.3
-Partly Cloudy	37	6.4	6.3	8.2	8.7	9.9	11.2	11.4	12.7	9.3	7.5	6.4	6.0	104.1
-Cloudy	37	15.3	13.1	13.6	11.5	13.1	11.5	13.1	10.6	11.3	10.3	11.9	14.6	149.9
Precipitation														
.01 inches or more	46	10.1	9.6	11.2	8.8	9.7	9.8	11.4	9.4	7.1	6.8	7.6	9.5	111.2
Snow, Ice pellets														
1.0 inches or more	46	0.7	0.5	0.3	0.0	0.0	0.0	0.0	0.0	0.0	0.0	0.*	0.2	1.7
Thunderstorms	46	0.5	0.9	1.8	3.0	5.7	7.6	9.8	7.1	3.0	1.2	0.6	0.4	41.7
Heavy Fog Visibility														
1/4 mile or less	46	3.7	2.8	2.4	1.4	1.0	1.1	1.1	1.5	2.1	2.0	3.0	4.2	26.5
Temperature °F														
-Maximum														
90° and above	25	0.0	0.0	0.0	0.3	1.8	6.0	11.5	10.0	4.0	0.0	0.0	0.0	33.6
32° and below	25	1.9	0.4	0.*	0.0	0.0	0.0	0.0	0.0	0.0	0.0	0.0	0.4	2.8
-Minimum														
32° and below	25	19.9	16.8	8.0	1.2	0.*	0.0	0.0	0.0	0.0	0.9	7.2	16.1	70.2
0° and below	25	0.1	0.0	0.0	0.0	0.0	0.0	0.0	0.0	0.0	0.0	0.0	0.0	0.1
AVG. STATION PRESS. (mb)	13	991.4	991.2	989.0	989.1	988.2	989.3	990.2	991.2	991.3	992.4	992.4	992.6	990.7
RELATIVE HUMIDITY (%)														
Hour 01	25	72	67	69	69	78	81	84	84	83	80	76	74	76
Hour 07	25	78	76	79	78	83	85	88	89	90	87	83	79	83
Hour 13 (Local Time)	25	56	52	50	47	53	56	58	58	57	54	53	57	54
Hour 19	25	61	54	52	49	58	62	66	67	67	67	63	63	61
PRECIPITATION (inches):														
Water Equivalent														
-Normal		3.80	3.81	4.83	3.27	3.64	3.57	3.92	3.75	3.59	2.72	2.86	3.40	43.16
-Maximum Monthly	46	7.44	7.59	8.76	7.64	12.48	8.26	9.12	9.98	10.89	8.33	8.68	7.49	12.48
-Year		1962	1979	1980	1958	1975	1961	1941	1948	1945	1976	1985	1983	MAY 1975
-Minimum Monthly	46	0.45	0.74	0.58	0.30	0.11	0.67	0.53	0.61	0.02	T	0.46	0.43	T
-Year		1981	1978	1985	1976	1941	1954	1983	1972	1954	1953	1973	1965	OCT 1953
-Maximum in 24 hrs	46	3.57	2.92	3.83	3.20	3.67	3.77	3.00	4.52	4.74	5.34	3.27	2.87	5.34
-Year		1962	1973	1977	1962	1975	1949	1949	1978	1959	1976	1985	1972	OCT 1976
Snow, Ice pellets														
-Maximum Monthly	46	11.7	14.9	19.3	0.1							2.5	7.5	19.3
-Year		1962	1979	1960	1982							1968	1971	MAR 1960
-Maximum in 24 hrs	46	10.2	12.0	10.3	0.1							2.5	7.5	12.0
-Year		1965	1969	1983	1982							1968	1971	FEB 1969
WIND:														
Mean Speed (mph)	36	7.9	8.4	8.9	8.8	7.5	6.9	6.6	6.4	6.7	7.0	7.2	7.4	7.5
Prevailing Direction														
through 1963		SW	NE	SW	S	SW	SW	SW	S	NE	NNE	SSW	SW	SW
Fastest Obs. 1 Min.														
-Direction (!!!)	6	31	07	25	25	02	31	36	25	08	21	32	30	21
-Speed (MPH)	6	25	32	29	29	25	30	23	24	29	37	30	35	37
-Year		1982	1984	1985	1982	1985	1980	1980	1983	1985	1979	1979	1979	OCT 1979
Peak Gust														
-Direction (!!!)	2	NW	NE	SW	NW	N	N	SW	E	NE	NW	W	W	NE
-Speed (mph)	2	43	53	49	41	46	45	35	32	52	31	37	39	53
-Date		1985	1984	1984	1985	1985	1985	1985	1984	1985	1984	1984	1985	FEB 1984

See Reference Notes to this table on the following pages.

CHARLOTTE, NORTH CAROLINA

TABLE 2 PRECIPITATION (inches) CHARLOTTE, NORTH CAROLINA

YEAR	JAN	FEB	MAR	APR	MAY	JUNE	JULY	AUG	SEP	OCT	NOV	DEC	ANNUAL
1956	1.24	5.85	3.74	3.98	3.98	1.22	3.16	2.47	5.06	1.92	1.31	3.20	37.13
1957	2.73	4.83	3.06	3.33	5.29	5.29	1.40	3.23	4.74	2.44	7.70	2.19	46.23
1958	4.27	3.35	2.77	7.64	2.34	2.22	6.21	3.96	0.96	1.55	1.31	3.17	39.75
1959	3.02	2.87	4.32	5.36	3.25	1.62	8.33	7.92	6.11	6.13	1.92	2.81	53.66
1960	5.74	6.31	4.93	3.20	3.51	6.56	1.74	3.25	1.48	3.74	0.60	2.99	44.05
1961	2.21	6.68	4.40	4.05	3.19	8.26	1.43	4.07	0.06	0.73	2.73	6.03	43.84
1962	7.44	4.27	5.47	4.82	1.84	4.25	1.96	2.76	5.06	1.57	5.33	3.01	47.78
1963	2.71	3.15	5.28	2.69	2.91	2.01	2.19	3.55	3.70	0.02	3.46	3.04	34.71
1964	5.68	4.29	4.18	4.52	0.89	2.11	5.39	4.83	2.32	7.20	2.05	3.18	46.64
1965	2.17	2.69	6.59	3.32	1.96	5.30	6.19	4.87	0.88	1.76	3.19	0.43	39.35
1966	3.79	5.59	2.13	1.73	3.61	2.58	1.86	2.36	5.42	2.29	1.19	2.58	35.13
1967	2.76	3.79	2.48	1.57	4.70	3.08	9.23	5.13	1.60	0.83	2.51	5.53	43.21
1968	4.95	0.87	4.96	2.73	2.45	6.63	5.13	0.88	1.46	3.02	4.68	2.29	40.05
1969	1.93	5.19	4.04	3.49	2.03	2.32	3.48	5.14	4.83	1.33	1.14	4.87	39.79
1970	1.70	3.66	2.93	2.04	2.69	2.62	5.73	4.03	0.55	5.12	1.58	3.14	35.79
1971	3.40	5.19	4.87	2.79	4.47	5.09	4.99	2.89	2.68	6.88	2.96	2.24	48.45
1972	4.47	3.21	2.59	1.75	5.61	3.01	6.59	0.61	3.77	1.37	5.42	5.83	44.23
1973	4.14	4.44	6.96	2.13	4.31	4.91	3.62	2.31	3.15	2.38	0.46	5.32	44.13
1974	5.22	4.90	3.30	3.26	4.49	2.32	4.16	6.35	6.50	0.46	4.50	3.82	49.28
1975	6.14	3.50	7.62	1.69	12.48	1.86	7.58	4.48	6.51	3.58	2.83	3.82	62.09
1976	1.89	1.13	4.36	0.30	4.26	3.84	2.26	0.90	5.55	8.33	3.37	5.60	41.79
1977	2.73	1.48	8.45	2.05	3.16	3.12	0.82	2.44	6.35	4.74	4.20	1.97	41.51
1978	6.80	0.74	4.97	2.69	4.91	4.19	4.03	8.11	1.16	1.18	2.81	3.13	44.72
#1979	5.31	7.59	3.79	6.47	4.54	4.72	4.74	1.27	9.69	2.95	4.61	1.36	57.04
1980	4.67	1.31	8.76	2.31	3.59	2.27	2.63	1.94	5.37	1.67	3.77	0.83	39.12
1981	0.45	3.63	2.12	0.67	4.27	1.81	6.61	2.67	3.42	3.94	0.87	6.23	36.69
1982	4.30	4.87	1.58	3.84	4.97	4.16	4.19	2.03	0.64	3.83	3.05	4.23	41.69
1983	2.53	5.50	6.07	2.66	2.14	3.77	0.53	3.61	0.74	2.43	4.05	7.49	41.52
1984	4.09	5.90	5.89	4.50	4.78	2.95	5.96	3.95	1.74	0.75	2.08	2.40	44.99
1985	5.20	4.05	0.58	1.90	5.14	5.46	4.14	7.35	0.74	5.16	8.68	0.92	49.32
Record Mean	3.81	3.97	4.35	3.26	3.57	3.94	4.74	4.59	3.30	2.96	2.73	3.69	44.90

TABLE 3 AVERAGE TEMPERATURE (deg. F) CHARLOTTE, NORTH CAROLINA

YEAR	JAN	FEB	MAR	APR	MAY	JUNE	JULY	AUG	SEP	OCT	NOV	DEC	ANNUAL
1956	41.0	47.9	51.2	58.7	70.0	76.8	79.7	79.8	70.0	63.4	50.2	53.1	61.8
1957	44.2	49.8	50.5	64.1	70.3	77.0	80.0	78.0	73.4	57.6	53.0	45.1	61.9
1958	37.3	36.8	46.8	60.6	70.2	76.3	79.5	78.6	72.5	59.5	54.4	39.6	59.4
1959	40.9	45.8	49.2	62.1	71.9	77.0	78.9	79.8	73.1	63.1	50.9	44.4	61.4
#1960	42.5	42.6	39.7	63.2	67.9	76.4	78.9	79.7	73.6	63.7	51.8	38.6	59.9
1961	38.2	46.4	52.9	53.0	63.8	72.8	77.3	76.9	75.4	62.1	56.3	42.7	59.8
1962	39.4	46.7	48.2	58.4	74.5	74.7	78.6	77.0	70.1	62.6	48.0	38.0	59.8
1963	37.3	37.5	54.1	62.9	68.8	75.3	77.9	78.9	70.1	63.6	52.7	35.6	59.6
1964	41.5	40.1	51.0	60.6	69.8	77.7	76.9	75.0	70.9	55.7	54.8	45.3	59.9
1965	41.5	44.6	46.8	62.4	73.4	73.3	77.8	77.8	73.1	59.6	51.4	44.0	60.5
1966	36.0	42.2	50.2	58.7	67.6	74.0	79.6	77.2	70.7	58.0	50.3	41.5	58.9
1967	43.3	40.0	53.9	62.7	63.7	73.1	76.3	75.2	67.0	60.0	47.8	46.3	59.1
1968	37.9	38.7	51.6	59.0	64.8	73.8	77.0	79.3	71.4	60.6	49.4	38.6	58.5
1969	38.1	39.7	45.1	60.8	67.9	77.3	80.8	75.6	69.0	60.4	47.7	39.5	58.5
1970	34.7	41.1	48.6	61.2	68.2	74.4	79.5	78.9	75.9	63.6	49.7	44.7	60.0
1971	39.9	41.4	46.3	58.3	65.0	76.5	76.2	75.8	73.1	64.7	49.0	50.6	59.7
1972	44.6	40.4	49.6	58.5	64.7	71.3	77.0	77.1	72.7	59.0	48.8	46.5	59.2
1973	39.6	40.0	54.1	57.6	64.7	74.7	78.5	76.8	74.6	62.1	52.7	41.7	59.8
1974	49.8	44.0	55.2	60.0	68.5	72.2	76.9	75.9	68.4	58.3	49.9	42.1	60.1
1975	45.2	45.0	50.4	59.0	70.5	75.0	76.2	78.9	70.8	63.5	53.5	42.4	60.9
1976	39.3	50.2	54.7	59.9	65.0	72.4	76.9	76.1	69.5	55.5	44.1	39.1	58.6
1977	30.1	42.1	54.9	62.5	69.8	75.2	82.4	79.7	74.4	58.7	53.6	42.5	60.5
1978	36.9	36.7	49.4	61.5	66.7	76.2	78.9	79.0	74.2	60.1	56.2	44.8	60.0
#1979	36.7	37.9	53.7	60.2	67.7	71.6	76.6	78.4	70.4	59.3	52.9	43.7	59.1
1980	41.4	38.9	47.4	60.0	68.2	74.0	80.0	80.9	74.8	58.3	48.8	42.6	59.6
1981	36.1	44.2	48.5	64.2	65.1	78.2	78.7	75.1	70.2	58.0	51.3	39.3	59.1
1982	36.0	45.4	52.7	57.6	72.0	75.0	78.4	76.3	70.0	60.5	52.3	47.9	60.3
1983	38.8	41.8	50.6	54.8	66.7	73.9	80.7	80.2	71.1	61.0	50.0	39.6	59.1
1984	38.3	45.4	50.0	56.5	65.9	76.4	76.1	77.1	69.0	67.8	47.8	50.0	60.0
1985	35.6	43.1	54.0	61.4	68.0	75.8	77.4	75.7	70.6	64.8	58.6	39.7	60.4
Record Mean	41.5	43.7	51.0	60.0	68.8	75.5	78.7	77.5	72.1	61.5	51.0	43.7	60.4
Max	50.5	53.4	61.5	71.0	79.5	86.1	88.3	86.9	81.7	71.8	61.1	52.1	70.3
Min	32.5	34.0	40.6	48.9	58.0	65.7	69.0	68.1	62.5	51.1	40.8	34.1	50.4

REFERENCE NOTES FOR TABLES 1, 2, 3 and 6 (CHARLOTTE, NC)

GENERAL

T - TRACE AMOUNT
BLANK ENTRIES DENOTE MISSING/UNREPORTED DATA.
# INDICATES A STATION OR INSTRUMENT RELOCATION.

SPECIFIC

TABLE 1

(a) - LENGTH OF RECORD IN YEARS. ALTHOUGH INDIVIDUAL MONTHS MAY BE MISSING.

* LESS THAN .05

NORMALS — BASED ON THE 1951-1980 RECORD PERIOD.
EXTREMES — DATES ARE THE MOST RECENT OCCURRENCE.
WIND DIR. — NUMERALS SHOW TENS OF DEGREES CLOCKWISE FROM TRUE NORTH. "00" INDICATES CALM.
RESULTANT WIND DIRECTIONS ARE GIVEN TO WHOLE DEGREES.

EXCEPTIONS

TABLES 2, 3, and 6

RECORD MEANS ARE THROUGH THE CURRENT YEAR, BEGINNING IN 1879 FOR TEMPERATURE
1878 FOR PRECIPITATION
1940 FOR SNOWFALL

TABLE 4 HEATING DEGREE DAYS Base 65 deg. F   CHARLOTTE, NORTH CAROLINA

SEASON	JULY	AUG	SEP	OCT	NOV	DEC	JAN	FEB	MAR	APR	MAY	JUNE	TOTAL
1956-57	0	0	46	87	449	365	643	420	442	123	35	3	2613
1957-58	0	0	37	229	361	611	849	783	559	165	28	0	3622
1958-59	0	0	2	178	331	783	741	528	482	145	13	0	3203
1959-60	0	0	2	153	422	633	691	644	776	124	72	0	3517
#1960-61	0	0	4	112	388	809	825	513	383	376	74	8	3492
1961-62	0	0	8	133	323	685	787	509	512	230	5	0	3192
1962-63	0	0	36	156	504	829	851	766	333	136	50	0	3661
1963-64	0	0	24	81	361	904	721	720	434	178	39	0	3462
1964-65	0	5	14	287	307	604	721	565	554	128	0	7	3192
1965-66	0	0	7	194	403	643	892	631	455	213	46	8	3492
1966-67	0	0	2	231	437	724	664	695	343	128	122	14	3360
1967-68	0	0	40	177	509	572	835	756	413	197	64	0	3563
1968-69	0	0	0	175	462	811	829	702	610	141	45	0	3775
1969-70	0	0	22	180	511	784	933	665	502	154	42	0	3793
1970-71	0	0	12	111	456	622	770	655	574	208	73	0	3481
1971-72	0	0	4	72	490	444	625	710	472	226	47	13	3103
1972-73	0	0	3	195	490	564	780	693	341	241	74	0	3381
1973-74	0	0	2	124	361	715	463	583	320	182	36	0	2786
1974-75	0	0	50	216	454	699	609	554	444	212	12	0	3250
1975-76	0	0	13	103	347	694	789	425	321	192	68	12	2964
1976-77	0	0	5	299	621	799	1075	636	306	120	36	7	3904
1977-78	0	0	3	215	356	690	862	785	473	140	72	0	3596
1978-79	0	0	7	155	255	620	873	750	350	163	30	4	3207
#1979-80	1	0	14	197	357	655	726	750	538	171	27	0	3436
1980-81	0	0	26	230	482	689	890	574	508	87	74	0	3560
1981-82	0	2	18	228	405	790	891	544	376	235	0		3495
1982-83	0	0	20	191	380	529	806	645	441	308	46	1	3367
1983-84	0	0	44	149	445	780	820	562	459	266	74	2	3601
1984-85	0	0	47	31	517	458	905	607	358	150	32	0	3105
1985-86	0	0	28	85	198	777							

TABLE 5 COOLING DEGREE DAYS Base 65 deg. F   CHARLOTTE, NORTH CAROLINA

YEAR	JAN	FEB	MAR	APR	MAY	JUNE	JULY	AUG	SEP	OCT	NOV	DEC	TOTAL
1969	0	0	0	22	143	373	498	336	151	42	0	0	1565
1970	0	0	0	47	148	289	453	438	346	71	0	0	1792
1971	0	0	0	12	83	354	354	341	251	69	16	5	1485
1972	0	0	0	40	43	208	377	385	241	17	10	0	1321
1973	0	0	11	25	72	292	426	373	296	39	0	0	1534
1974	0	0	20	38	152	223	375	346	158	15	10	0	1337
1975	0	0	0	40	188	307	356	436	196	67	11	0	1601
1976	0	2	11	43	76	238	374	351	146	14	0	0	1255
1977	0	2	0	50	191	319	545	465	294	27	20	0	1913
1978	0	0	0	41	132	343	438	440	287	13	1	1	1696
#1979	0	0	4	28	122	208	369	419	183	30	3	0	1366
1980	0	0	0	28	133	275	473	497	328	26	0	0	1760
1981	0	0	2	69	87	403	431	323	182	21	0	0	1518
1982	0	0	3	22	230	308	424	357	179	59	6	5	1593
1983	0	0	1	6	107	275	494	476	234	32	0	0	1625
1984	0	0	0	18	110	350	348	381	176	124	6	0	1513
1985	0	0	24	48	132	330	395	338	204	87	11	0	1569

TABLE 6 SNOWFALL (inches)   CHARLOTTE, NORTH CAROLINA

SEASON	JULY	AUG	SEP	OCT	NOV	DEC	JAN	FEB	MAR	APR	MAY	JUNE	TOTAL
1956-57	0.0	0.0	0.0	0.0	T	0.0	T	T	0.0	0.0	0.0	0.0	T
1957-58	0.0	0.0	0.0	0.0	0.0	0.0	1.0	2.2	0.0	0.0	0.0	0.0	3.2
1958-59	0.0	0.0	0.0	0.0	0.0	3.0	T	T	1.3	0.0	0.0	0.0	4.3
1959-60	0.0	0.0	0.0	0.0	0.0	T	0.0	3.3	19.3	0.0	0.0	0.0	22.6
1960-61	0.0	0.0	0.0	0.0	0.0	0.3	1.2	0.8	T	0.0	0.0	0.0	2.3
1961-62	0.0	0.0	0.0	0.0	0.0	T	11.7	T	1.4	0.0	0.0	0.0	13.1
1962-63	0.0	0.0	0.0	0.0	T	T	T	2.5	0.0	0.0	0.0	0.0	2.5
1963-64	0.0	0.0	0.0	0.0	0.0	1.1	0.6	0.2	T	0.0	0.0	0.0	1.9
1964-65	0.0	0.0	0.0	0.0	T	0.0	11.2	1.0	T	0.0	0.0	0.0	12.2
1965-66	0.0	0.0	0.0	0.0	0.0	0.0	9.6	1.1	0.0	0.0	0.0	0.0	10.7
1966-67	0.0	0.0	0.0	0.0	0.0	T	5.4	T	0.0	0.0	0.0	0.0	5.4
1967-68	0.0	0.0	0.0	0.0	0.0	0.0	4.3	2.4	T	0.0	0.0	0.0	6.7
1968-69	0.0	0.0	0.0	0.0	2.5	T	T	13.2	3.0	0.0	0.0	0.0	18.7
1969-70	0.0	0.0	0.0	0.0	0.0	T	4.5	T	0.0	0.0	0.0	0.0	4.5
1970-71	0.0	0.0	0.0	0.0	0.0	2.7	0.1	0.8	5.7	0.0	0.0	0.0	9.3
1971-72	0.0	0.0	0.0	0.0	T	7.5	0.0	0.5	3.7	0.0	0.0	0.0	11.7
1972-73	0.0	0.0	0.0	0.0	T	T	5.5	1.0	0.0	0.0	0.0	0.0	6.5
1973-74	0.0	0.0	0.0	0.0	0.0	5.4	0.0	T	1.4	0.0	0.0	0.0	6.8
1974-75	0.0	0.0	0.0	0.0	0.0	T	0.0	1.0	0.2	0.0	0.0	0.0	1.2
1975-76	0.0	0.0	0.0	0.0	0.5	T	T	T	0.0	0.0	0.0	0.0	0.5
1976-77	0.0	0.0	0.0	0.0	T	T	3.4	0.0	0.0	0.0	0.0	0.0	3.4
1977-78	0.0	0.0	0.0	0.0	0.0	T	T	0.7	4.8	0.0	0.0	0.0	5.5
1978-79	0.0	0.0	0.0	0.0	0.0	0.0	0.4	14.9	0.0	0.0	0.0	0.0	15.3
#1979-80	0.0	0.0	0.0	0.0	0.0	0.0	0.4	7.3	6.8	0.0	0.0	0.0	14.5
1980-81	0.0	0.0	0.0	0.0	0.0	0.0	0.3	1.8	T	T	0.0	0.0	2.1
1981-82	0.0	0.0	0.0	0.0	0.0	T	4.8	5.9	0.0	0.1	0.0	0.0	10.8
1982-83	0.0	0.0	0.0	0.0	0.0	0.0	0.8	1.5	10.3	0.0	0.0	0.0	12.6
1983-84	0.0	0.0	0.0	0.0	T	T	T	5.9	0.0	0.0	0.0	0.0	5.9
1984-85	0.0	0.0	0.0	0.0	0.0	T	1.7	T	0.0	0.0	0.0	0.0	1.7
1985-86	0.0	0.0	0.0	0.0	0.0	T							
Record Mean	0.0	0.0	0.0	0.0	0.1	0.6	2.0	1.9	1.4	T	0.0	0.0	5.9

See Reference Notes, relative to all above tables, on preceding page.

The Greensboro–High Point–Winston–Salem Regional Airport is located in the west–central part of Guilford County, in the northern Piedmont section of North Carolina. The location is near the headwaters of the Haw and Deep Rivers, both branches of the Cape Fear River system. A few miles west is a ridge beyond which lies the Yadkin River Basin. To the north, across a similar ridge, the waters of the Dan River flow northeastward into the Roanoke. West, beyond the Yadkin River Basin, the land gradually rises into the Brushy Mountains. To the northwest, other outcroppings southeast of the Blue Ridge rise into peaks occasionally exceeding 2,500 feet. The Blue Ridge proper forms a northeast–southwest barrier with heights occasionally exceeding 3,000 feet.

Winter temperatures and rainfall are both modified by the mountain barrier, but to a lesser extent than in areas closer to the Appalachian Range. Shallow cold air masses from the west tend to be stopped or turned aside by the mountains, while deeper masses are lifted over the range, losing moisture and warming during the passage. For this reason the lowest temperatures recorded in Forsyth and Guilford Counties usually occur when clear, cold air drifts southward, east of the Appalachian Range. The summer temperatures vary with the cloudiness and shower activity, but are generally mild.

Northwesterly winds seldom bring heavy or prolonged winter rain or snow. Flurries of light snow may fall when cold air blows across the mountains, but the heavier winter precipitation comes with winds blowing from northeast through east and south to southwest. When moist winds blowing from an easterly or southerly direction meet cold air moving out of the north or northwest in the vicinity of North Carolina, snow, sleet, or glaze may occur. Glazing is more common here than in most of North Carolina, but only occasionally becomes severe enough to do much damage in the northern Piedmont area.

Seasonal snowfall has a wide range and there have been a few winters with only a trace of snow. Snow seldom stays on the ground more than a few days.

Summer precipitation is largely from thunderstorms, mostly local in character. The frequency of these showers and the amount of rain received varies greatly from year to year and from place to place. Sizeable areas are sometimes without significant rain in late spring or early summer for two or more weeks, while other areas in the vicinity may be well watered.

Damaging storms are infrequent in the Northern Piedmont area. The highest winds to occur have been associated with thunderstorms, and were of brief duration. Hail is reported within Guilford and Forsyth Counties each year. The occurrence of tornadoes is rare. Hurricanes have produced heavy rainfall here, but no winds of destructive force.

Based on the 1951–1980 period, the average first occurrence of 32 degrees Fahrenheit in the fall is October 27 and the average last occurrence in the spring is April 11.

TABLE 1 NORMALS, MEANS AND EXTREMES

GREENSBORO, NORTH CAROLINA

LATITUDE: 36°05'N   LONGITUDE: 79°57'W   ELEVATION: FT. GRND 897 BARO 00908   TIME ZONE: EASTERN   WBAN: 13723

	(a)	JAN	FEB	MAR	APR	MAY	JUNE	JULY	AUG	SEP	OCT	NOV	DEC	YEAR
TEMPERATURE °F:														
Normals														
-Daily Maximum		47.6	50.8	59.3	70.7	77.9	84.2	87.4	86.2	80.4	70.1	59.9	50.4	68.7
-Daily Minimum		27.3	29.0	36.5	45.9	55.0	62.6	66.9	66.3	59.3	46.7	37.1	29.9	46.9
-Monthly		37.5	39.9	47.9	58.3	66.5	73.4	77.2	76.3	69.8	58.4	48.5	40.2	57.8
Extremes														
-Record Highest	57	78	81	90	94	98	102	102	101	100	95	85	78	102
-Year		1975	1977	1945	1930	1941	1954	1977	1932	1954	1954	1974	1971	JUL 1977
-Record Lowest	57	-8	-4	5	21	32	42	48	47	35	20	10	0	-8
-Year		1985	1936	1960	1943	1963	1977	1933	1946	1942	1962	1970	1962	JAN 1985
NORMAL DEGREE DAYS:														
Heating (base 65°F)		853	703	533	215	73	0	0	0	12	221	495	769	3874
Cooling (base 65°F)		0	0	6	14	120	259	378	350	159	17	0	0	1303
% OF POSSIBLE SUNSHINE	57	52	57	60	64	65	65	62	63	63	65	59	55	61
MEAN SKY COVER (tenths)														
Sunrise - Sunset	57	6.2	6.0	6.0	5.7	5.9	5.9	6.2	5.9	5.5	4.7	5.4	6.0	5.8
MEAN NUMBER OF DAYS:														
Sunrise to Sunset														
-Clear	57	8.9	8.7	9.1	9.5	8.2	6.7	6.4	7.5	10.1	13.6	11.3	10.2	110.1
-Partly Cloudy	57	6.8	6.5	8.1	8.5	10.8	12.6	12.5	12.5	8.7	7.3	6.7	6.3	107.5
-Cloudy	57	15.3	13.0	13.8	11.9	11.9	10.7	12.1	11.0	11.2	10.1	12.0	14.5	147.6
Precipitation														
.01 inches or more	57	10.3	9.7	11.2	9.5	10.4	10.5	12.4	10.7	7.7	7.1	8.3	9.4	117.0
Snow, Ice pellets														
1.0 inches or more	57	0.9	0.7	0.5	0.0	0.0	0.0	0.0	0.0	0.0	0.0	0.*	0.4	2.6
Thunderstorms	57	0.5	0.8	1.9	3.2	6.7	8.3	10.8	8.3	3.2	1.1	0.4	0.2	45.5
Heavy Fog Visibility														
1/4 mile or less	57	4.8	3.3	2.8	1.7	1.9	1.4	1.9	2.2	2.9	2.6	3.2	4.1	32.8
Temperature °F														
-Maximum														
90° and above	22	0.0	0.0	0.0	0.4	1.2	5.4	10.1	8.2	2.3	0.0	0.0	0.0	27.6
32° and below	22	3.0	1.0	0.1	0.0	0.0	0.0	0.0	0.0	0.0	0.0	0.*	0.9	5.1
-Minimum														
32° and below	22	22.8	19.8	10.4	2.1	0.0	0.0	0.0	0.0	0.0	1.5	9.9	18.3	84.9
0° and below	22	0.2	0.0	0.0	0.0	0.0	0.0	0.0	0.0	0.0	0.0	0.0	0.0	0.2
AVG. STATION PRESS.(mb)	13	986.7	986.6	984.7	984.7	984.1	985.3	986.2	987.2	987.4	988.3	987.9	987.8	986.4
RELATIVE HUMIDITY (%)														
Hour 01	22	75	70	70	71	82	85	88	88	87	84	77	76	79
Hour 07	22	80	76	78	78	83	86	89	90	90	88	83	80	83
Hour 13 (Local Time)	22	56	51	51	48	55	57	60	60	58	56	53	56	55
Hour 19	22	64	57	54	52	62	66	69	71	71	72	66	66	64
PRECIPITATION (inches):														
Water Equivalent														
-Normal		3.51	3.37	3.88	3.16	3.37	3.93	4.27	4.19	3.64	3.18	2.59	3.38	42.47
-Maximum Monthly	57	8.24	7.04	8.76	6.19	8.35	7.99	12.72	12.53	13.26	9.60	8.26	6.44	13.26
-Year		1937	1929	1975	1936	1982	1965	1984	1939	1947	1959	1985	1973	SEP 1947
-Minimum Monthly	57	0.66	0.73	0.67	0.55	0.37	0.32	0.98	0.71	T	0.26	0.35	0.33	T
-Year		1981	1978	1985	1942	1936	1933	1953	1972	1985	1963	1981	1955	SEP 1985
-Maximum in 24 hrs	57	3.06	3.00	3.07	2.70	3.11	4.91	4.43	4.47	7.49	6.24	3.32	3.60	7.49
-Year		1936	1934	1932	1944	1978	1972	1944	1949	1947	1954	1962	1958	SEP 1947
Snow, Ice pellets														
-Maximum Monthly	57	22.9	16.3	21.3	T							5.9	14.3	22.9
-Year		1966	1979	1960	1983							1968	1930	JAN 1966
-Maximum in 24 hrs	57	14.0	9.3	11.1	T							5.0	14.3	14.3
-Year		1940	1979	1960	1983							1968	1930	DEC 1930
WIND:														
Mean Speed (mph)	57	8.1	8.6	9.1	8.8	7.6	6.9	6.5	6.2	6.6	7.0	7.5	7.6	7.5
Prevailing Direction														
through 1963		SW	SW	SW	SW	SW	SW	SW	SW	NE	NE	SW	SW	SW
Fastest Obs. 1 Min.														
-Direction (!!!)	5	33	09	31	31	31	28	35	30	32	30	29	29	35
-Speed (MPH)	5	24	29	27	30	27	23	50	25	29	46	29	30	50
-Year		1985	1984	1981	1982	1982	1981	1983	1983	1983	1985	1983	1983	JUL 1984
Peak Gust														
-Direction (!!!)	2	W	E	W	W	SW	NW	N	N	NW	NW	SW	SW	NW
-Speed (mph)	2	38	48	38	35	33	32	58	26	24	60	28	32	60
-Date		1984	1984	1985	1984	1984	1985	1984	1985	1984	1985	1985	1985	OCT 1985

See Reference Notes to this table on the following page.

TABLE 2 PRECIPITATION (inches) GREENSBORO, NORTH CAROLINA

YEAR	JAN	FEB	MAR	APR	MAY	JUNE	JULY	AUG	SEP	OCT	NOV	DEC	ANNUAL
1956	1.04	5.32	2.54	3.69	4.18	2.66	4.71	4.76	7.19	2.89	1.93	3.14	44.05
1957	5.09	3.95	2.15	3.79	1.96	5.33	2.14	3.28	5.27	1.54	2.49	4.37	43.17
1958	3.36	2.87	3.03	6.11	2.98	3.70	5.28	4.63	0.64	2.02	1.84	4.37	40.83
1959	3.01	3.32	3.10	4.37	3.07	2.47	9.81	3.69	5.55	9.60	1.57	2.86	52.42
1960	4.44	5.81	4.78	4.30	4.68	0.95	4.24	3.78	2.32	4.04	0.66	2.54	42.54
1961	2.35	3.43	4.86	3.95	4.31	6.21	3.30	7.17	0.20	1.02	1.70	5.30	43.80
1962	5.77	4.74	4.76	2.89	2.18	7.08	2.93	1.19	5.66	0.65	6.33	4.94	49.12
1963	2.41	2.96	6.46	2.49	2.92	3.54	3.49	2.34	4.01	0.26	4.77	1.54	37.19
1964	4.48	4.09	2.75	2.93	0.56	3.07	6.61	6.75	1.03	8.15	1.57	3.20	45.19
1965	2.27	2.80	5.04	3.07	1.00	7.99	9.08	3.91	4.31	3.72	1.29	0.64	45.12
1966	4.48	5.44	1.53	0.97	4.15	4.04	2.01	3.01	3.59	3.25	1.55	2.26	36.28
1967	1.84	2.76	1.21	2.03	4.03	2.11	3.11	5.43	2.39	1.23	1.40	6.34	33.88
1968	3.74	0.88	3.97	2.64	3.92	1.62	2.60	2.51	0.58	4.60	3.55	2.25	32.86
1969	2.01	3.01	3.72	2.30	3.17	7.86	4.11	3.95	3.44	1.89	1.41	4.93	41.80
1970	1.66	3.25	3.81	4.52	2.76	3.59	2.56	9.13	0.70	5.57	2.87	2.37	42.79
1971	1.80	4.46	3.32	2.74	5.37	2.99	3.88	4.22	2.51	6.71	2.58	1.61	42.19
1972	2.60	3.71	2.65	2.72	6.24	6.37	2.54	0.71	3.72	1.82	4.87	4.49	42.44
1973	3.50	3.53	5.69	4.31	5.71	4.88	5.30	2.72	1.15	2.31	1.24	6.44	46.78
1974	4.58	3.13	3.32	2.15	5.69	2.72	1.71	7.14	7.57	0.81	2.42	4.23	45.47
1975	5.31	3.15	8.76	2.00	6.23	1.71	12.35	1.48	7.29	2.62	1.81	3.80	56.51
1976	2.25	1.47	2.42	0.62	2.94	3.91	1.87	2.63	2.35	8.61	1.69	3.92	34.68
1977	2.27	1.69	2.79	2.25	0.70	1.54	1.52	4.03	6.35	4.67	3.11	3.30	34.22
1978	7.70	0.73	4.38	4.13	5.09	3.72	9.23	9.60	2.70	1.25	3.17	4.45	56.15
1979	5.87	5.03	2.95	3.21	4.46	2.72	4.30	1.13	13.08	2.64	4.78	1.12	51.29
1980	4.00	1.77	5.04	3.24	3.23	3.44	2.68	2.14	2.95	2.13	2.58	0.78	33.98
1981	0.66	3.61	2.59	1.13	2.78	2.92	8.99	2.60	6.35	3.63	0.35	5.60	41.21
1982	2.91	4.64	2.39	2.72	8.35	7.28	3.46	1.46	1.22	4.65	2.73	3.56	45.37
1983	1.31	3.82	5.75	5.16	4.12	5.15	2.78	1.44	2.68	4.09	5.37	4.38	46.05
1984	3.51	5.25	5.31	2.77	5.83	3.10	12.72	5.55	1.12	1.16	2.02	1.90	50.24
1985	4.34	3.77	0.67	0.94	4.32	2.25	4.31	4.86	T	1.92	8.26	1.46	37.10
Record Mean	3.34	3.34	3.75	3.12	3.66	3.77	4.71	4.25	3.54	2.95	2.83	3.26	42.50

TABLE 3 AVERAGE TEMPERATURE (deg. F) GREENSBORO, NORTH CAROLINA

YEAR	JAN	FEB	MAR	APR	MAY	JUNE	JULY	AUG	SEP	OCT	NOV	DEC	ANNUAL
1956	36.4	43.5	47.0	55.3	66.7	73.8	76.9	77.0	67.5	60.4	46.5	48.8	58.3
1957	39.1	45.9	47.0	61.3	68.0	74.7	76.9	74.6	71.0	54.4	49.8	42.5	58.8
1958	34.2	33.3	43.1	57.4	66.4	73.4	77.4	76.1	68.6	56.9	50.9	35.4	56.2
1959	37.0	42.2	46.6	58.9	69.6	74.0	76.9	78.0	69.7	60.2	47.1	41.4	58.5
1960	39.7	38.7	35.1	60.7	64.7	73.9	75.7	77.2	69.9	59.6	48.6	34.8	56.6
1961	34.7	43.2	50.9	51.8	62.9	71.7	76.1	75.3	72.0	58.3	52.1	39.5	57.4
1962	35.8	42.4	44.5	55.4	71.6	72.6	75.3	75.8	62.8	60.9	46.8	36.0	57.1
#1963	35.2	33.9	51.3	59.6	65.3	71.8	74.0	74.6	65.8	59.8	49.6	32.3	56.1
1964	38.1	37.7	48.7	58.4	67.6	75.7	76.4	74.0	67.3	55.4	52.5	42.6	57.8
1965	38.9	40.6	44.3	58.6	72.6	72.8	75.7	77.0	71.6	57.9	50.1	43.2	58.6
1966	33.1	39.2	48.7	55.9	66.7	73.5	79.0	76.4	69.8	57.7	50.0	40.5	57.6
1967	43.0	38.2	53.0	63.3	64.0	73.7	77.1	75.9	66.6	58.5	46.8	45.1	58.8
1968	35.6	36.3	53.4	60.2	65.0	72.9	77.0	78.1	66.9	60.2	48.9	37.5	57.9
1969	36.4	40.1	44.3	59.5	67.3	76.2	80.4	75.7	68.6	59.3	47.0	38.5	57.8
1970	33.4	40.0	47.8	60.9	67.7	76.0	79.5	75.4	73.0	60.0	47.8	41.1	58.6
1971	35.2	38.7	43.7	54.4	62.5	74.7	75.1	73.3	71.0	63.0	47.7	48.7	57.3
1972	42.3	37.8	47.1	56.2	63.0	68.7	76.1	75.5	70.8	54.7	46.2	44.1	56.9
1973	36.1	37.1	52.2	55.4	62.7	75.4	76.5	76.0	72.3	59.8	49.7	38.2	57.6
1974	45.8	41.1	52.0	58.0	66.1	70.7	76.6	76.5	67.9	57.5	49.5	41.8	58.6
1975	42.9	44.1	46.2	56.7	69.1	75.2	77.1	80.0	70.1	62.0	52.0	40.1	59.6
1976	35.9	49.3	53.8	60.5	64.2	71.2	75.9	73.1	66.9	53.2	41.8	37.0	56.9
1977	26.7	39.1	53.3	61.2	69.1	73.6	80.4	78.2	71.7	55.1	49.8	38.2	58.1
1978	31.3	31.6	45.2	57.0	63.1	73.4	76.4	77.1	71.4	57.2	52.1	42.2	56.5
1979	35.1	32.8	50.9	59.1	66.2	69.8	75.7	76.5	69.1	58.0	52.3	41.5	57.3
1980	38.4	36.0	46.5	58.2	66.0	70.8	84.7	78.7	73.1	56.1	46.6	40.2	57.4
1981	33.5	42.3	45.7	61.9	63.2	77.9	78.5	73.6	68.1	55.6	48.9	36.4	57.1
1982	32.7	42.8	49.2	54.9	70.1	72.1	75.0	68.0	65.9	57.6	49.1	45.5	57.9
1983	36.0	38.6	48.4	53.0	63.8	71.6	78.0	78.2	69.1	58.3	48.7	36.5	56.7
1984	35.2	44.1	46.0	54.3	75.0	74.2	75.4	65.7	65.5	45.5	48.3	57.7	57.7
1985	32.8	40.6	51.7	61.0	66.8	73.1	75.7	73.4	67.9	61.1	56.2	37.6	58.2
Record Mean	38.2	40.4	48.0	57.8	66.5	74.0	77.1	76.0	69.8	58.8	48.4	40.0	57.9
Max	48.1	51.2	59.5	70.0	78.2	84.9	87.3	86.1	80.5	70.6	59.8	50.2	68.9
Min	28.2	29.5	36.5	45.5	54.8	63.1	66.9	65.9	59.2	46.9	37.0	29.9	46.9

REFERENCE NOTES FOR TABLES 1, 2, 3 and (GREENSBORO, NC)

GENERAL

T - TRACE AMOUNT
BLANK ENTRIES DENOTE MISSING/UNREPORTED DATA.
# INDICATES A STATION OR INSTRUMENT RELOCATION.

SPECIFIC

TABLE 1

(a) - LENGTH OF RECORD IN YEARS. ALTHOUGH INDIVIDUAL MONTHS MAY BE MISSING.
* LESS THAN .05

NORMALS — BASED ON THE 1951-1980 RECORD PERIOD.
EXTREMES — DATES ARE THE MOST RECENT OCCURRENCE.
WIND DIR. — NUMERALS SHOW TENS OF DEGREES CLOCKWISE FROM TRUE NORTH.
"00" INDICATES CALM.
RESULTANT WIND DIRECTIONS ARE GIVEN TO WHOLE DEGREES.

EXCEPTIONS

TABLE 1

1. FASTEST MILE WINDS ARE THROUGH FEBRUARY 1980.

TABLES 2, 3, and 6

RECORD MEANS ARE THROUGH THE CURRENT YEAR, BEGINNING IN 1929 FOR TEMPERATURE
1929 FOR PRECIPITATION
1929 FOR SNOWFALL

TABLE 4   HEATING DEGREE DAYS Base 65 deg. F   GREENSBORO, NORTH CAROLINA

SEASON	JULY	AUG	SEP	OCT	NOV	DEC	JAN	FEB	MAR	APR	MAY	JUNE	TOTAL
1956-57	0	0	75	172	551	493	796	528	547	193	48	8	3411
1957-58	0	1	52	324	448	693	946	882	675	244	57	0	4322
1958-59	0	0	21	253	424	912	860	632	565	208	33	2	3910
1959-60	0	0	26	222	534	726	776	756	919	182	112	0	4253
1960-61	0	0	21	182	485	929	935	607	444	410	100	10	4123
1961-62	0	2	23	215	413	783	898	625	630	310	15	0	3914
#1962-63	0	0	75	180	537	893	918	865	420	198	80	1	4167
1963-64	0	2	64	163	456	1008	826	785	502	234	61	3	4104
1964-65	0	2	28	329	369	688	803	676	687	215	0	11	3758
1965-66	0	3	15	230	441	670	979	714	501	285	71	12	3921
1966-67	0	0	14	237	438	751	676	742	376	130	111	14	3489
1967-68	0	0	49	218	539	610	904	828	374	170	69	2	3763
1968-69	0	1	0	180	476	848	878	689	637	175	48	0	3932
1969-70	0	0	30	211	536	816	973	695	525	160	59	0	4005
1970-71	0	1	19	179	511	736	916	728	653	312	120	0	4175
1971-72	0	0	9	101	523	501	696	783	396	282	76	30	3551
1972-73	1	0	9	314	562	642	889	774	396	292	111	0	3990
1973-74	0	0	4	179	453	823	586	664	403	232	59	0	3403
1974-75	0	0	53	244	477	711	676	579	578	256	26	0	3600
1975-76	0	0	31	121	391	765	895	454	357	199	74	20	3307
1976-77	0	1	19	370	687	860	1184	723	358	159	42	11	4414
1977-78	0	0	7	307	459	821	1039	927	605	245	131	4	4545
1978-79	0	0	16	238	381	700	921	896	440	198	52	8	3850
1979-80	1	2	25	239	377	719	817	834	565	212	65	7	3863
1980-81	0	0	31	278	544	762	970	630	590	135	116	0	4056
1981-82	0	1	39	299	477	879	997	616	484	304	14	0	4110
1982-83	0	0	38	253	467	602	894	734	505	358	97	9	3957
1983-84	0	0	74	222	482	876	913	600	580	331	107	4	4189
1984-85	0	0	96	53	580	512	992	679	425	160	50	6	3553
1985-86	0	1	0	65	151	263	843						

TABLE 5   COOLING DEGREE DAYS Base 65 deg. F   GREENSBORO, NORTH CAROLINA

YEAR	JAN	FEB	MAR	APR	MAY	JUNE	JULY	AUG	SEP	OCT	NOV	DEC	TOTAL
1969	0	0	0	18	125	345	486	339	145	41	0	0	1499
1970	0	0	0	45	151	338	456	335	264	32	0	0	1621
1971	0	0	0	0	50	298	321	261	195	47	13	5	1190
1972	0	0	3	23	21	147	352	328	188	2	7	0	1071
1973	0	0	6	11	46	319	363	343	229	25	0	0	1342
1974	0	0	8	28	101	176	366	333	144	15	18	0	1189
1975	0	0	1	15	159	313	379	473	192	37	7	0	1576
1976	0	4	14	71	55	212	344	257	82	9	0	0	1048
1977	0	2	1	50	175	275	483	417	214	9	10	0	1636
1978	0	0	0	11	80	263	364	384	216	4	0	0	1322
1979	0	0	6	26	95	159	340	365	157	28	4	0	1180
1980	0	0	0	16	102	189	424	434	281	12	0	0	1458
1981	0	0	0	47	69	394	427	276	141	15	0	0	1369
1982	0	0	0	7	179	216	382	317	136	33	0	2	1272
1983	0	0	0	5	68	210	409	415	202	20	0	0	1329
1984	0	0	0	16	66	309	292	328	122	79	0	0	1212
1985	0	1	18	47	109	259	339	268	159	36	7	0	1243

TABLE 6   SNOWFALL (inches)   GREENSBORO, NORTH CAROLINA

SEASON	JULY	AUG	SEP	OCT	NOV	DEC	JAN	FEB	MAR	APR	MAY	JUNE	TOTAL
1956-57	0.0	0.0	0.0	0.0	0.2	T	0.4	T	T	0.0	0.0	0.0	0.6
1957-58	0.0	0.0	0.0	0.0	0.0	T	2.4	6.3	T	T	0.0	0.0	8.7
1958-59	0.0	0.0	0.0	0.0	0.0	5.9	4.7	T	T	T	0.0	0.0	10.6
1959-60	0.0	0.0	0.0	0.0	0.0	T	1.8	8.6	21.3	T	0.0	0.0	31.7
1960-61	0.0	0.0	0.0	0.0	0.0	T	0.7	1.8	T	T	0.0	0.0	2.5
1961-62	0.0	0.0	0.0	0.0	T	T	14.4	0.4	10.2	0.0	0.0	0.0	25.0
1962-63	0.0	0.0	0.0	0.0	0.0	0.8	T	4.9	T	0.0	0.0	0.0	5.7
1963-64	0.0	0.0	0.0	0.0	T	2.6	1.5	3.3	T	0.0	0.0	0.0	7.4
1964-65	0.0	0.0	0.0	0.0	0.3	T	9.1	3.3	0.2	0.0	0.0	0.0	12.9
1965-66	0.0	0.0	0.0	0.0	0.0	T	22.9	3.1	0.0	0.0	0.0	0.0	26.0
1966-67	0.0	0.0	0.0	0.0	T	2.6	2.7	0.0	T	0.0	0.0	0.0	6.1
1967-68	0.0	0.0	0.0	0.0	T	T	6.0	3.2	T	0.0	0.0	0.0	9.2
1968-69	0.0	0.0	0.0	0.0	5.9	T	0.3	0.7	10.7	0.0	0.0	0.0	17.6
1969-70	0.0	0.0	0.0	0.0	0.0	1.6	6.4	T	T	0.0	0.0	0.0	8.0
1970-71	0.0	0.0	0.0	0.0	0.0	3.1	0.4	0.7	3.5	T	0.0	0.0	7.7
1971-72	0.0	0.0	0.0	0.0	0.9	4.2	T	4.8	3.1	0.0	0.0	0.0	13.0
1972-73	0.0	0.0	0.0	0.0	T	0.0	6.0	T	1.5	0.0	0.0	0.0	7.5
1973-74	0.0	0.0	0.0	0.0	0.0	8.3	0.0	T	4.5	0.0	0.0	0.0	12.8
1974-75	0.0	0.0	0.0	0.0	0.0	T	T	2.0	2.0	0.0	0.0	0.0	4.0
1975-76	0.0	0.0	0.0	0.0	0.0	T	T	0.3	0.0	0.0	0.0	0.0	0.3
1976-77	0.0	0.0	0.0	0.0	T	1.2	8.1	1.2	0.0	0.0	0.0	0.0	10.5
1977-78	0.0	0.0	0.0	0.0	T	0.7	0.4	5.9	3.6	0.0	0.0	0.0	10.6
1978-79	0.0	0.0	0.0	0.0	T	T	T	16.3	0.0	0.0	0.0	0.0	16.3
1979-80	0.0	0.0	0.0	0.0	0.0	0.0	3.5	7.0	7.9	0.0	0.0	0.0	18.4
1980-81	0.0	0.0	0.0	0.0	T	0.5	2.4	0.0	7.6	0.0	0.0	0.0	10.5
1981-82	0.0	0.0	0.0	0.0	0.0	0.1	8.8	4.8	T	T	0.0	0.0	13.7
1982-83	0.0	0.0	0.0	0.0	0.0	0.2	0.4	5.7	T	0.0	0.0	0.0	6.3
1983-84	0.0	0.0	0.0	0.0	T	T	T	3.8	T	0.0	0.0	0.0	3.8
1984-85	0.0	0.0	0.0	0.0	0.0	0.0	5.1	T	0.0	0.0	0.0	0.0	5.1
1985-86	0.0	0.0	0.0	0.0	0.0	0.3							
Record Mean	0.0	0.0	0.0	0.0	0.1	1.3	3.2	2.4	1.9	T	0.0	0.0	8.9

See Reference Notes, relative to all above tables, on preceding page.

The Raleigh–Durham Airport is located in the zone of transition between the Coastal Plain and the Piedmont Plateau. The surrounding terrain is rolling, with an average elevation of around 400 feet, the range over a 10-mile radius is roughly between 200 and 550 feet. Being centrally located between the mountains on the west and the coast on the south and east, the Raleigh–Durham area enjoys a favorable climate. The mountains form a partial barrier to cold air masses moving eastward from the interior of the nation. As a result, there are few days in the heart of the winter season when the temperature falls below 20 degrees. Tropical air is present over the eastern and central sections of North Carolina during much of the summer season, bringing warm temperatures and rather high humidities to the Raleigh–Durham area. Afternoon temperatures reach 90 degrees or higher on about one-fourth of the days in the middle of summer, but reach 100 degrees less than once per year. Even in the hottest weather, early morning temperatures almost always drop into the lower 70s.

Rainfall is well distributed throughout the year as a whole. July and August have the greatest amount of rainfall, and October and November the least. There are times in spring and summer when soil moisture is scanty. This usually results from too many days between rains rather than from a shortage of total rainfall, but occasionally the accumulated total during the growing season falls short of plant needs. Most summer rain is produced by thunderstorms, which may occasionally be accompanied by strong winds, intense rains, and hail. The Raleigh–Durham area is far enough from the coast so that the bad weather effects of coastal storms are reduced. While snow and sleet usually occur each year, excessive accumulations of snow are rare.

From September 1887 to December 1950, the office was located in the downtown areas of Raleigh. The various buildings occupied were within an area of three blocks. All thermometers were exposed on the roof, and this, plus the smoke over the city, had an effect on the temperature record of that period. Lowest temperatures at the city office were frequently from 2 to 5 degrees higher than those recorded in surrounding rural areas. Maximum temperatures in the city were generally a degree or two lower. These observations are supported by a period of simultaneous record from the Municipal Airport and the city office location between 1937 and 1940.

From September 1946 to May 1954, simultaneous records were kept at a surface location on the North Carolina State College campus in Raleigh, and at the Raleigh–Durham Airport 10 1/2 air miles to the northwest.

Based on the 1951–1980 period, the average first occurrence of 32 degrees Fahrenheit in the fall is October 27 and the average last occurrence in the spring is April 11.

TABLE 1 — NORMALS, MEANS AND EXTREMES

RALEIGH, NORTH CAROLINA

LATITUDE: 35°52'N   LONGITUDE: 78°47'W   ELEVATION: FT. GRND   416 BARO   415   TIME ZONE: EASTERN   WBAN: 13722

	(a)	JAN	FEB	MAR	APR	MAY	JUNE	JULY	AUG	SEP	OCT	NOV	DEC	YEAR
TEMPERATURE °F:														
Normals														
-Daily Maximum		50.1	52.8	61.0	72.3	79.0	85.2	88.2	87.1	81.6	71.6	61.8	52.7	70.3
-Daily Minimum		29.1	30.3	37.7	46.5	55.3	62.6	67.1	66.8	60.4	47.7	38.1	31.2	47.7
-Monthly		39.6	41.5	49.4	59.4	67.2	73.9	77.7	77.0	71.0	59.7	50.0	42.0	59.0
Extremes														
-Record Highest	41	79	84	92	95	97	104	105	101	104	98	88	79	105
-Year		1952	1977	1945	1980	1953	1954	1952	1983	1954	1954	1950	1978	JUL 1952
-Record Lowest	41	-9	5	11	23	31	38	48	46	37	19	11	4	-9
-Year		1985	1971	1980	1985	1977	1977	1975	1965	1983	1962	1970	1983	JAN 1985
NORMAL DEGREE DAYS:														
Heating (base 65°F)		787	655	496	181	53	0	0	0	9	187	450	713	3531
Cooling (base 65°F)		0	0	9	16	121	270	394	372	189	23	0	0	1394
% OF POSSIBLE SUNSHINE	31	54	59	62	64	60	61	61	61	60	61	59	55	60
MEAN SKY COVER (tenths)														
Sunrise - Sunset	36	6.1	5.9	5.9	5.6	6.0	5.8	6.1	5.9	5.6	5.1	5.3	5.8	5.7
MEAN NUMBER OF DAYS:														
Sunrise to Sunset														
-Clear	37	9.2	8.9	9.5	9.8	8.2	7.7	7.3	7.8	10.1	12.6	11.7	10.3	113.0
-Partly Cloudy	37	6.9	6.3	7.6	8.9	10.1	11.8	11.7	12.4	8.8	7.1	7.3	6.8	105.6
-Cloudy	37	14.8	13.1	13.9	11.3	12.8	10.5	12.0	10.8	11.2	11.3	11.0	13.9	146.6
Precipitation														
.01 inches or more	41	10.0	9.6	10.3	8.9	10.3	9.2	11.2	9.8	7.5	7.1	8.2	9.0	111.3
Snow, Ice pellets														
1.0 inches or more	41	1.0	0.7	0.4	0.*	0.0	0.0	0.0	0.0	0.0	0.0	0.1	0.3	2.4
Thunderstorms	41	0.5	0.8	1.9	3.3	6.3	7.1	10.4	8.0	3.6	1.5	0.7	0.3	44.5
Heavy Fog Visibility 1/4 mile or less	36	3.6	2.8	2.2	1.6	2.5	2.1	2.9	3.4	3.5	3.8	3.3	3.6	35.3
Temperature °F														
-Maximum														
90° and above	21	0.0	0.0	0.0	0.6	1.1	5.9	10.9	9.7	3.1	0.0	0.0	0.0	31.2
32° and below	21	2.5	0.5	0.1	0.0	0.0	0.0	0.0	0.0	0.0	0.0	0.*	0.7	3.9
-Minimum														
32° and below	21	21.5	18.1	10.0	2.3	0.*	0.0	0.0	0.0	0.0	1.5	10.0	17.1	80.7
0° and below	21	0.2	0.0	0.0	0.0	0.0	0.0	0.0	0.0	0.0	0.0	0.0	0.0	0.2
AVG. STATION PRESS.(mb)	13	1003.0	1002.9	1000.8	1000.5	999.7	1000.6	1001.3	1002.3	1002.6	1003.9	1003.9	1004.0	1002.1
RELATIVE HUMIDITY (%)														
Hour 01	21	73	69	71	73	84	87	88	89	88	85	78	75	80
Hour 07	21	79	76	79	80	86	87	90	92	92	89	84	80	85
Hour 13 (Local Time)	21	55	51	49	45	54	57	59	60	59	54	52	55	54
Hour 19	21	63	57	56	53	66	68	71	75	76	76	67	67	66
PRECIPITATION (inches):														
Water Equivalent														
-Normal		3.55	3.43	3.69	2.91	3.67	3.66	4.38	4.44	3.29	2.73	2.87	3.14	41.76
-Maximum Monthly	41	7.52	6.00	7.78	6.10	7.67	9.38	10.05	10.49	12.94	7.53	8.22	6.65	12.94
-Year		1954	1983	1983	1978	1974	1973	1945	1955	1945	1971	1948	1983	SEP 1945
-Minimum Monthly	41	0.87	1.00	1.03	0.23	0.92	0.55	0.80	0.81	0.23	0.44	0.61	0.25	0.23
-Year		1981	1968	1985	1976	1964	1981	1953	1950	1985	1963	1973	1965	SEP 1985
-Maximum in 24 hrs	41	3.11	3.22	3.70	4.04	4.40	3.44	3.89	5.20	5.16	4.10	4.70	3.18	5.20
-Year		1984	1973	1983	1978	1957	1967	1952	1955	1944	1954	1963	1958	AUG 1955
Snow, Ice pellets														
-Maximum Monthly	41	14.4	17.2	14.0	1.8							2.6	10.6	17.2
-Year		1955	1979	1960	1983							1975	1958	FEB 1979
-Maximum in 24 hrs	41	9.0	10.4	9.3	1.8							2.6	9.1	10.4
-Year		1966	1979	1969	1983							1975	1958	FEB 1979
WIND:														
Mean Speed (mph)	36	8.6	8.9	9.4	9.0	7.6	6.9	6.6	6.3	6.8	7.1	7.7	8.0	7.7
Prevailing Direction through 1963		SW	SW	SW	SW	SW	SW	SW	NE	NE	NNE	SW	SW	SW
Fastest Obs. 1 Min.														
-Direction	32	27	12	32	14	20	33	23	33	23	29	32	21	29
-Speed (MPH)	32	41	44	44	40	54	39	69	46	35	35	35	35	73
-Year		1971	1984	1967	1961	1972	1977	1962	1969	1972	1954	1969	1968	OCT 1954
Peak Gust														
-Direction	2	NW	SE	N	W	W	NW	NW	SW	N	SW	W	NW	SE
-Speed (mph)	2	45	62	46	43	55	39	39	43	33	28	38	37	62
-Date		1985	1984	1984	1984	1984	1984	1984	1985	1984	1985	1984	1985	FEB 1984

See Reference Notes to this table on the following page.

TABLE 2 PRECIPITATION (inches) RALEIGH, NORTH CAROLINA

YEAR	JAN	FEB	MAR	APR	MAY	JUNE	JULY	AUG	SEP	OCT	NOV	DEC	ANNUAL
1956	1.05	4.70	3.37	4.74	1.88	4.43	6.83	1.31	3.74	3.18	1.38	3.28	39.89
1957	2.82	4.40	2.25	1.66	5.86	6.03	2.15	4.85	4.58	1.71	6.53	3.42	46.26
1958	3.43	3.24	2.55	5.02	6.45	4.40	4.82	6.23	1.18	2.98	1.03	4.24	45.57
1959	1.77	3.38	3.22	5.83	2.53	2.59	8.83	5.92	3.52	6.53	2.58	2.25	48.95
1960	4.59	5.26	4.94	4.57	5.07	2.53	7.47	5.43	3.05	2.29	0.88	1.92	48.00
1961	2.88	5.75	4.37	2.23	2.94	4.05	3.10	6.52	1.25	1.16	2.10	4.76	41.11
1962	6.56	2.74	4.85	3.22	1.37	6.37	7.07	1.98	3.72	0.99	7.19	2.21	48.27
1963	2.96	3.45	3.80	1.77	4.05	1.72	3.51	2.10	2.77	0.44	7.06	3.28	36.91
1964	3.66	4.11	2.93	3.39	0.92	3.41	4.06	5.68	5.29	3.95	1.38	4.13	42.91
1965	1.47	2.40	4.08	1.51	2.20	8.32	5.54	3.00	2.65	1.77	1.23	0.25	34.42
1966	5.42	4.76	1.81	2.02	4.95	3.68	0.91	5.79	3.58	2.01	2.06	2.61	39.60
1967	1.64	3.80	1.62	3.02	4.15	4.57	3.49	6.22	1.74	2.26	2.14	4.93	39.58
1968	2.88	1.00	2.22	3.03	3.82	1.74	5.15	2.50	1.77	5.15	3.59	2.75	35.60
1969	1.55	3.60	3.95	1.43	2.85	4.81	4.40	6.31	6.21	1.01	3.31		41.52
1970	2.26	3.47	4.04	2.07	3.36	0.87	5.64	4.47	1.20	4.47	1.59	2.57	36.01
1971	3.28	3.85	3.69	2.59	4.68	2.79	4.56	6.26	2.91	7.53	1.81	1.69	45.64
1972	1.97	4.13	2.50	1.92	5.34	4.16	6.80	4.17	5.80	3.96	5.98	5.01	51.74
1973	2.67	5.50	4.06	4.40	3.99	9.38	3.12	4.60	1.13	0.60	0.61	6.38	46.44
1974	4.39	2.87	3.34	1.32	7.67	4.02	1.56	4.82	3.71	1.23	1.79	4.02	40.74
1975	6.09	2.85	6.26	1.64	3.84	1.66	6.74	2.11	5.77	1.23	4.60	4.04	46.83
1976	3.07	1.54	3.17	0.23	4.74	2.55	1.00	1.52	5.99	3.97	1.89	4.04	33.71
1977	2.82	2.13	5.63	1.89	3.94	0.84	0.89	4.12	3.86	5.06	2.22	3.70	37.10
1978	7.03	1.43	4.40	6.10	4.20	4.06	3.63	1.86	1.37	1.46	4.17	3.26	42.97
#1979	5.71	5.55	2.69	2.63	4.71	3.27	4.84	1.66	6.76	1.88	4.73	0.94	45.37
1980	4.39	1.91	5.87	1.97	2.33	4.89	2.11	1.87	3.76	2.25	2.87	1.42	35.64
1981	0.87	3.02	2.35	1.03	4.28	0.55	5.69	5.34	2.70	4.64	0.95	4.96	36.38
1982	3.43	4.97	3.02	3.33	4.20	8.39	3.34	1.83	1.55	3.93	2.34	4.02	44.35
1983	1.79	6.00	7.78	3.54	5.89	3.09	1.10	1.81	2.13	3.59	3.86	6.65	47.23
1984	4.93	5.65	5.40	4.45	5.43	4.08	9.20	1.13	2.31	0.73	1.64	2.32	46.27
1985	4.83	4.44	1.03	0.64	3.95	2.87	6.28	3.73	0.23	1.75	7.61	0.81	38.17
Record Mean	3.42	3.71	3.73	3.22	3.93	4.16	5.24	4.98	3.65	2.88	2.63	3.28	44.83

TABLE 3 AVERAGE TEMPERATURE (deg. F) RALEIGH, NORTH CAROLINA

YEAR	JAN	FEB	MAR	APR	MAY	JUNE	JULY	AUG	SEP	OCT	NOV	DEC	ANNUAL
1956	38.9	45.9	48.8	57.2	67.5	74.9	77.8	77.2	69.1	61.9	48.8	51.0	59.9
1957	40.7	47.1	48.7	63.0	68.1	75.6	77.9	75.4	72.4	55.5	51.7	44.8	60.1
1958	35.9	35.5	44.6	59.1	67.3	72.9	79.4	76.8	69.0	58.3	53.3	36.9	57.4
1959	39.6	43.7	48.6	60.7	70.1	74.3	77.5	78.6	71.0	61.7	49.1	43.3	59.9
1960	41.6	40.5	37.6	62.1	66.1	75.0	76.1	77.6	70.8	60.5	50.8	36.6	57.9
1961	36.5	44.9	52.7	53.6	64.2	72.8	77.6	76.6	73.2	59.1	52.4	41.3	58.7
1962	38.8	43.9	45.5	57.4	72.0	73.5	75.7	75.8	68.2	61.8	47.7	38.0	58.2
1963	36.7	36.1	53.4	60.5	65.4	72.7	76.0	76.3	67.1	60.5	51.0	34.7	57.5
#1964	41.2	39.4	50.6	59.6	68.1	75.6	76.4	74.9	69.3	55.2	52.7	43.3	58.9
1965	40.5	42.9	44.6	58.3	71.9	71.9	76.2	77.2	72.7	58.8	50.9	43.8	59.1
1966	35.8	42.3	49.7	57.0	66.5	73.4	79.1	77.7	71.0	58.6	50.8	41.9	58.7
1967	44.8	38.9	53.7	62.5	63.9	72.0	76.9	76.3	67.5	59.4	46.8	46.3	59.1
1968	37.1	36.4	52.6	58.1	63.9	73.6	76.9	79.2	70.9	62.0	51.7	38.8	58.4
1969	37.4	40.9	43.4	58.7	66.2	74.7	78.0	73.8	67.5	58.6	46.1	37.7	56.9
1970	32.9	38.9	46.2	59.4	66.0	72.7	76.1	75.3	73.3	61.0	49.4	42.7	57.8
1971	37.0	42.1	45.1	56.4	64.4	75.4	76.6	75.3	72.0	64.8	48.7	50.3	59.0
1972	44.7	40.0	49.5	58.0	64.3	69.9	77.1	75.6	70.4	57.4	48.1	46.2	58.5
1973	39.3	40.0	54.8	57.9	64.5	75.1	76.5	76.6	73.2	62.2	54.6	42.5	59.8
1974	49.3	42.9	54.4	60.2	67.4	71.8	76.5	76.0	69.0	56.5	48.7	43.2	59.7
1975	43.8	43.7	47.1	55.8	68.2	73.7	76.5	77.8	71.1	62.2	53.5	42.0	59.6
1976	37.6	50.1	56.2	60.7	66.8	74.7	78.6	75.4	69.7	55.7	42.5	37.1	58.7
1977	26.6	39.3	53.3	62.6	68.4	73.3	80.6	78.1	72.8	56.4	51.8	39.9	58.6
1978	35.3	33.1	48.2	59.2	66.0	75.8	78.0	79.9	73.7	59.5	55.1	44.9	59.1
1979	39.1	36.5	52.2	59.7	69.8	75.4	77.2	77.1	71.2	59.9	52.0	43.4	58.6
1980	40.6	36.5	46.5	62.0	69.8	75.0	78.9	79.6	74.9	58.7	49.1	40.7	59.3
1981	33.4	43.9	46.2	61.7	64.0	78.9	80.8	74.6	68.3	57.2	50.7	39.7	58.3
1982	35.5	45.5	51.7	57.4	71.0	74.7	79.1	76.5	70.5	60.6	51.9	44.5	60.2
1983	38.1	40.7	50.7	55.1	65.4	72.5	79.1	79.1	70.7	60.4	50.9	39.5	58.5
1984	36.3	45.7	47.2	55.9	65.5	75.5	74.9	76.6	67.5	66.3	47.1	49.7	59.0
1985	34.0	41.9	52.7	62.0	67.3	73.9	76.8	75.2	69.7	63.7	58.4	39.4	59.6
Record Mean	41.2	42.8	50.4	59.3	67.9	75.2	78.2	77.1	71.6	60.8	50.7	42.8	59.8
Max	50.6	52.8	61.3	70.9	79.0	85.6	88.0	86.7	81.4	71.5	61.2	52.3	70.1
Min	31.7	32.8	39.5	47.7	56.8	64.7	68.4	67.5	61.7	50.1	40.2	33.2	49.5

REFERENCE NOTES FOR TABLES 1, 2, 3 and 6   (RALEIGH, NC)

GENERAL

T - TRACE AMOUNT
BLANK ENTRIES DENOTE MISSING/UNREPORTED DATA.
# INDICATES A STATION OR INSTRUMENT RELOCATION.

SPECIFIC

TABLE 1

(a) - LENGTH OF RECORD IN YEARS. ALTHOUGH
      INDIVIDUAL MONTHS MAY BE MISSING.
 * LESS THAN .05

NORMALS — BASED ON THE 1951-1980 RECORD PERIOD.
EXTREMES — DATES ARE THE MOST RECENT OCCURRENCE.
WIND DIR. — NUMERALS SHOW TENS OF DEGREES
            CLOCKWISE FROM TRUE NORTH.
            "00" INDICATES CALM.
RESULTANT WIND DIRECTIONS ARE GIVEN TO WHOLE DEGREES.

EXCEPTIONS

TABLES 2, 3, and 6

RECORD MEANS ARE THROUGH THE CURRENT YEAR,
BEGINNING IN   1887 FOR TEMPERATURE
               1887 FOR PRECIPITATION
               1945 FOR SNOWFALL

TABLE 4 HEATING DEGREE DAYS Base 65 deg. F   RALEIGH, NORTH CAROLINA

SEASON	JULY	AUG	SEP	OCT	NOV	DEC	JAN	FEB	MAR	APR	MAY	JUNE	TOTAL
1956-57	0	0	58	129	489	428	748	494	499	162	59	5	3071
1957-58	0	1	38	295	399	619	893	819	623	210	41	1	3939
1958-59	0	0	23	217	355	864	780	591	504	179	26	4	3543
1959-60	0	0	23	189	473	667	721	704	844	159	87	2	3869
1960-61	0	0	11	170	422	874	878	559	393	367	83	8	3765
1961-62	0	0	19	197	403	730	810	584	598	261	11	0	3613
1962-63	0	0	61	166	510	832	870	803	354	193	86	4	3879
1963-64	0	0	57	149	413	933	731	735	445	205	54	0	3722
#1964-65	0	1	25	305	365	665	754	616	628	232	0	15	3606
1965-66	0	5	7	208	416	650	898	629	473	263	72	12	3633
1966-67	0	0	11	216	425	709	623	723	361	151	105	13	3337
1967-68	0	0	35	199	539	574	855	824	391	213	87	0	3717
1968-69	0	0	0	151	396	805	848	667	667	195	61	0	3790
1969-70	0	0	45	222	561	841	989	725	576	200	74	0	4233
1970-71	0	0	22	154	460	684	863	636	611	258	87	0	3775
1971-72	0	0	3	61	496	456	623	718	478	237	51	16	3139
1972-73	0	0	9	238	504	576	790	692	334	231	88	0	3462
1973-74	0	0	2	126	312	690	481	614	346	187	48	0	2806
1974-75	0	0	44	268	501	668	651	589	553	293	34	0	3601
1975-76	0	0	17	117	351	705	843	426	300	194	52	6	3011
1976-77	0	0	7	302	668	857	1183	715	358	132	49	14	4285
1977-78	0	0	4	283	411	768	914	883	514	196	83	0	4056
1978-79	0	0	7	184	292	627	793	792	398	183	43	8	3327
1979-80	0	0	13	196	394	661	753	820	564	130	33	0	3564
1980-81	0	0	16	225	477	747	973	583	579	149	99	0	3848
1981-82	0	4	31	253	425	776	907	538	411	244	15	0	3604
1982-83	0	0	14	182	392	542	828	675	438	305	79	7	3462
1983-84	0	0	59	180	417	784	882	553	545	283	83	5	3791
1984-85	0	0	63	42	530	468	954	644	395	146	42	4	3288
1985-86	0	0	36	96	207	789							

TABLE 5 COOLING DEGREE DAYS Base 65 deg. F   RALEIGH, NORTH CAROLINA

YEAR	JAN	FEB	MAR	APR	MAY	JUNE	JULY	AUG	SEP	OCT	NOV	DEC	TOTAL
1969	0	0	0	12	105	295	413	283	128	28	0	0	1264
1970	0	0	0	41	113	236	350	327	278	38	0	0	1383
1971	0	0	0	6	78	320	364	329	222	61	17	5	1402
1972	0	0	5	37	36	170	382	336	177	6	6	2	1157
1973	0	0	24	26	81	310	363	365	254	48	7	0	1478
1974	0	0	25	51	130	210	363	347	169	9	21	0	1325
1975	0	0	3	22	141	269	337	421	209	38	12	0	1452
1976	0	3	31	71	116	304	428	330	157	19	0	0	1459
1977	0	2	4	68	162	272	490	414	245	25	19	0	1701
1978	0	0	2	30	120	330	412	468	275	21	3	10	1671
1979	0	0	6	28	105	159	332	384	205	46	10	0	1275
1980	0	0	0	45	190	306	441	460	321	38	6	0	1807
1981	0	0	2	56	75	425	497	309	139	19	0	0	1522
1982	0	0	3	24	208	299	443	363	183	53	5	7	1588
1983	0	0	0	16	97	238	441	447	239	42	0	0	1520
1984	0	0	0	16	108	324	311	366	143	90	0	0	1358
1985	0	3	20	65	121	277	373	323	181	64	14	0	1441

TABLE 6 SNOWFALL (inches)   RALEIGH, NORTH CAROLINA

SEASON	JULY	AUG	SEP	OCT	NOV	DEC	JAN	FEB	MAR	APR	MAY	JUNE	TOTAL
1956-57	0.0	0.0	0.0	0.0	T	T	T	T	T	0.0	0.0	0.0	T
1957-58	0.0	0.0	0.0	0.0	0.0	1.0	3.0	3.9	T	0.0	0.0	0.0	7.9
1958-59	0.0	0.0	0.0	0.0	0.0	10.6	2.9	T	T	0.0	0.0	0.0	13.5
1959-60	0.0	0.0	0.0	0.0	T	T	T	4.7	14.0	0.0	0.0	0.0	18.7
1960-61	0.0	0.0	0.0	0.0	0.0	0.0	2.3	1.5	0.0	0.0	0.0	0.0	3.8
1961-62	0.0	0.0	0.0	0.0	0.0	T	10.1	0.4	4.3	0.0	0.0	.0.0	14.8
1962-63	0.0	0.0	0.0	0.0	1.3	T	0.1	6.9	0.0	0.0	0.0	.0.0	8.3
1963-64	0.0	0.0	0.0	0.0	0.0	T	T	0.4	3.1	T	0.0	0.0	3.5
1964-65	0.0	0.0	0.0	0.0	0.4	0.0	9.7	3.4	T	0.0	0.0	0.0	13.5
1965-66	0.0	0.0	0.0	0.0	0.0	0.0	12.3	T	0.0	0.0	0.0	0.0	12.3
1966-67	0.0	0.0	0.0	0.0	T	1.0	0.5	9.1	0.0	0.0	0.0	0.0	10.6
1967-68	0.0	0.0	0.0	0.0	T	1.4	3.0	1.3	T	0.0	0.0	0.0	5.7
1968-69	0.0	0.0	0.0	0.0	1.2	0.7	T	0.8	9.3	0.0	0.0	0.0	12.0
1969-70	0.0	0.0	0.0	0.0	0.0	0.0	2.0	T	T	0.0	0.0	0.0	2.0
1970-71	0.0	0.0	0.0	0.0	0.0	0.6	T	T	5.3	0.0	0.0	0.0	5.9
1971-72	0.0	0.0	0.0	0.0	T	3.7	0.0	1.4	2.6	0.0	0.0	0.0	7.7
1972-73	0.0	0.0	0.0	0.0	T	0.0	6.4	4.5	0.4	0.0	0.0	0.0	11.3
1973-74	0.0	0.0	0.0	0.0	0.0	2.8	0.0	T	2.9	0.0	0.0	0.0	5.7
1974-75	0.0	0.0	0.0	0.0	0.0	T	T	T	0.6	0.0	0.0	0.0	0.6
1975-76	0.0	0.0	0.0	0.0	2.6	T	0.4	T	0.0	0.0	0.0	0.0	3.0
1976-77	0.0	0.0	0.0	0.0	T	T	2.1	1.5	0.0	0.0	0.0	0.0	3.6
1977-78	0.0	0.0	0.0	0.0	T	T	T	9.0	1.6	0.0	0.0	0.0	10.6
1978-79	0.0	0.0	0.0	0.0	0.0	0.0	0.4	17.2	T	0.0	0.0	0.0	17.6
#1979-80	0.0	0.0	0.0	0.0	0.0	0.0	2.2	5.0	11.1	0.0	0.0	0.0	18.3
1980-81	0.0	0.0	0.0	0.0	0.0	3.1	2.6	0.0	T	0.0	0.0	0.0	5.7
1981-82	0.0	0.0	0.0	0.0	0.0	T	6.0	0.6	0.0	0.0	0.0	0.0	6.6
1982-83	0.0	0.0	0.0	0.0	0.0	T	T	2.7	7.3	1.8	0.0	0.0	11.8
1983-84	0.0	0.0	0.0	0.0	0.0	0.0	T	6.9	T	0.0	0.0	0.0	6.9
1984-85	0.0	0.0	0.0	0.0	0.0	0.0	4.1	T	0.0	0.0	0.0	0.0	4.1
1985-86	0.0	0.0	0.0	0.0	0.0	T							
Record Mean	0.0	0.0	0.0	0.0	0.1	0.8	2.5	2.5	1.6	T	0.0	0.0	7.5

See Reference Notes, relative to all above tables, on preceding page.

Wilmington is located in the tidewater section of southeastern North Carolina, near the Atlantic Ocean. The city proper is built adjacent to the east bank of the Cape Fear River. Because of the curvature of the coastline in this area, the ocean lies about 5 miles east and about 20 miles south. The surrounding terrain is typical of coastal Carolina. It is low-lying with an average elevation of less than 40 feet, and is characterized by level to gently rolling land with rivers, creeks, and lakes that frequently have considerable swamp or marshland adjoining them. Large wooded areas alternate with cultivated fields.

The maritime location makes the climate of Wilmington unusually mild for its latitude. All wind directions from the east-northeast through southwest have some moderating effects on temperatures throughout the year, because the ocean is relatively warm in winter and cool in summer. The daily range in temperatures is moderate compared to a continental type of climate. As a rule, summers are quite warm and humid, but excessive heat is rare. Sea breezes, arriving early in the afternoon, tend to alleviate the heat further inland. Long-term averages show afternoon temperatures reach 90 degrees or higher on one-third of the days in midsummer, but several years may pass without 100 degree weather. During the colder part of the year, numerous outbreaks of polar air masses reach the Atlantic Coast, causing sharp drops in temperatures. However, these cold outbreaks are significantly moderated by the long trajectories from the source regions, the effects of passing over the Appalachian Range, and the warming effects of the ocean air. As a result, most winters are short and quite mild. Even in the most severe cold spells, the temperature usually remains above zero. Normally, the temperature fails to rise above the freezing point during a 24-hour period only once each winter.

Rainfall in this area is usually ample and well-distributed throughout the year, the greatest amount occurring in the summer. Summer rainfall comes principally from thunderstorms, and is therefore usually of short duration, but often heavy and unevenly distributed. Thunderstorms occur about one out of three days from June through August. Winter rain is more likely to be of the slow, steady type, lasting one or two days. Generally, the winter rain is evenly distributed and associated with slow-moving, low-pressure systems. Seldom is there a winter without a few flakes of snow, but several years may pass without a measurable amount, and appreciable accumulation on the ground is rare. Hail occurs less than once a year. Sunshine is abundant, with the area receiving about two-thirds of the sunshine hours possible at its latitude.

Because of these many factors, the growing season is long, averaging 244 days, but records show the range is from 180 days to as long as 302 days. This area is exceptionally good for floriculture. Agricultural pursuits, principally field-grown flowers, nursery plantings, and vegetables, are an important part of the economy. Some types of plants continue to grow throughout the year.

In common with most Atlantic Coastal localities, the area is subject to the effects of coastal storms and occasional hurricanes which produce high winds, above normal tides, and heavy rains.

TABLE 1 NORMALS, MEANS AND EXTREMES

WILMINGTON, NORTH CAROLINA

LATITUDE: 34°16'N   LONGITUDE: 77°54'W   ELEVATION: FT. GRND   30 BARO   34   TIME ZONE: EASTERN   WBAN: 13748

	(a)	JAN	FEB	MAR	APR	MAY	JUNE	JULY	AUG	SEP	OCT	NOV	DEC	YEAR
TEMPERATURE °F:														
Normals														
–Daily Maximum		55.9	58.1	64.8	74.3	80.9	86.1	89.3	88.6	83.9	75.2	66.8	59.1	73.6
–Daily Minimum		35.3	36.6	43.3	51.8	60.4	67.1	71.3	70.8	65.7	53.7	43.9	37.2	53.1
–Monthly		45.6	47.4	54.0	63.0	70.7	76.6	80.3	79.7	74.8	64.5	55.4	48.2	63.3
Extremes														
–Record Highest	34	82	85	89	95	98	104	102	102	98	95	87	81	104
–Year		1975	1962	1974	1967	1953	1952	1977	1954	1975	1954	1974	1984	JUN 1952
–Record Lowest	34	5	11	9	30	40	48	59	55	44	27	20	9	5
–Year		1985	1958	1980	1983	1981	1983	1972	1982	1981	1962	1970	1983	JAN 1985
NORMAL DEGREE DAYS:														
Heating (base 65°F)		607	498	350	94	10	0	0	0	0	94	295	521	2469
Cooling (base 65°F)		6	5	12	37	187	348	474	456	294	78	7	0	1904
% OF POSSIBLE SUNSHINE	34	57	59	63	70	67	66	63	63	62	64	65	60	63
MEAN SKY COVER (tenths)														
Sunrise – Sunset	34	6.0	5.9	5.8	5.3	5.9	6.1	6.5	6.2	6.0	5.1	5.0	5.7	5.8
MEAN NUMBER OF DAYS:														
Sunrise to Sunset														
–Clear	34	9.8	9.4	9.9	11.0	8.4	7.1	5.6	7.1	8.4	12.6	12.3	10.6	112.2
–Partly Cloudy	34	6.3	5.9	7.9	7.9	10.4	10.9	12.0	12.0	9.6	7.2	7.3	7.0	104.5
–Cloudy	34	14.9	13.0	13.2	11.1	12.2	12.0	13.4	11.9	11.9	11.2	10.4	13.4	148.5
Precipitation														
.01 inches or more	34	10.5	9.6	10.3	8.0	9.7	10.0	13.2	11.9	9.4	7.2	7.6	9.1	116.4
Snow,Ice pellets														
1.0 inches or more	34	0.2	0.2	0.1	0.0	0.0	0.0	0.0	0.0	0.0	0.0	0.0	0.1	0.6
Thunderstorms	34	0.4	1.1	2.2	3.1	5.4	7.4	11.1	9.0	3.9	1.2	0.7	0.6	46.1
Heavy Fog Visibility														
1/4 mile or less	34	2.6	1.7	2.3	1.6	1.9	1.7	0.8	1.3	2.6	2.8	2.8	2.8	24.8
Temperature °F														
–Maximum														
90° and above	22	0.0	0.0	0.0	0.9	2.1	7.6	15.2	13.0	4.5	0.1	0.0	0.0	43.4
32° and below	22	0.6	0.1	0.*	0.0	0.0	0.0	0.0	0.0	0.0	0.0	0.0	0.*	0.8
–Minimum														
32° and below	22	14.5	11.7	3.8	0.3	0.0	0.0	0.0	0.0	0.0	0.1	3.3	10.7	44.3
0° and below	22	0.0	0.0	0.0	0.0	0.0	0.0	0.0	0.0	0.0	0.0	0.0	0.0	0.0
AVG. STATION PRESS.(mb)	13	1018.2	1018.0	1015.9	1015.8	1014.5	1015.2	1016.0	1016.9	1016.6	1018.0	1018.5	1019.1	1016.9
RELATIVE HUMIDITY (%)														
Hour 01	22	78	76	79	79	87	88	89	91	90	87	83	79	84
Hour 07	22	79	78	81	79	84	85	86	90	90	88	85	81	84
Hour 13 (Local Time)	22	56	52	52	48	56	60	64	64	62	57	53	56	57
Hour 19	22	70	65	66	63	71	73	77	80	81	80	76	74	73
PRECIPITATION (inches):														
Water Equivalent														
–Normal		3.64	3.44	4.04	2.98	4.22	5.65	7.44	6.64	5.71	2.97	3.19	3.43	53.35
–Maximum Monthly	34	7.08	8.74	8.09	8.21	9.12	12.87	15.12	14.06	18.94	9.81	7.87	6.57	18.94
–Year		1964	1983	1983	1961	1956	1962	1966	1981	1984	1964	1972	1982	SEP 1984
–Minimum Monthly	34	1.09	1.01	0.93	0.33	1.13	0.89	1.65	1.66	1.07	0.17	0.49	0.48	0.17
–Year		1981	1976	1967	1957	1983	1984	1961	1968	1981	1953	1973	1955	OCT 1953
–Maximum in 24 hrs	34	3.08	3.20	3.31	3.52	4.95	7.73	5.63	5.10	8.24	4.34	4.82	3.88	8.24
–Year		1982	1983	1960	1961	1963	1966	1966	1981	1958	1964	1969	1980	SEP 1958
Snow,Ice pellets														
–Maximum Monthly	34	2.9	12.5	6.6								T	4.0	12.5
–Year		1965	1973	1980								1976	1970	FEB 1973
–Maximum in 24 hrs	34	2.8	11.7	5.7								T	4.0	11.7
–Year		1965	1973	1980								1976	1970	FEB 1973
WIND:														
Mean Speed (mph)	34	9.2	10.0	10.3	10.5	9.4	8.5	8.0	7.5	8.0	8.2	8.2	8.6	8.9
Prevailing Direction through 1963		N	NW	SSW	SSW	SSW	SSW	SSW	SW	N	N	N	N	N
Fastest Obs. 1 Min.														
–Direction (!!!)	5	06	21	29	21	21	30	22	24	08	01	10	06	29
–Speed (MPH)	5	29	31	58	38	28	24	33	35	46	35	29	31	58
–Year		1983	1984	1981	1982	1985	1983	1983	1982	1984	1982	1985	1980	MAR 1981
Peak Gust														
–Direction (!!!)	2	NW	SW	SW	S	SW	SW	SW	SW	E	N	SW	NW	E
–Speed (mph)	2	40	49	64	38	45	33	46	35	74	32	43	44	74
–Date		1985	1984	1984	1985	1985	1985	1984	1985	1984	1984	1984	1984	SEP 1984

See Reference Notes to this table on the following page.

TABLE 2 PRECIPITATION (inches)   WILMINGTON, NORTH CAROLINA

YEAR	JAN	FEB	MAR	APR	MAY	JUNE	JULY	AUG	SEP	OCT	NOV	DEC	ANNUAL
1956	1.55	4.36	2.13	2.67	9.12	4.36	3.31	5.01	2.31	7.25	0.52	1.30	43.89
1957	2.32	2.31	5.54	0.33	3.29	4.54	3.59	5.08	13.25	1.26	3.75	4.84	50.10
1958	3.73	3.30	4.98	6.20	4.21	6.44	4.34	8.26	10.10	5.68	0.89	5.31	63.44
1959	2.11	5.04	6.48	2.74	1.36	7.58	10.90	1.41	3.00	5.11	4.72	3.28	53.73
1960	3.63	4.75	7.44	3.15	5.52	3.24	10.28	4.53	9.06	2.03	1.85	2.64	58.12
1961	2.12	2.99	4.52	8.21	3.70	11.79	1.65	9.21	4.25	1.19	1.85	0.78	52.26
1962	5.98	2.27	5.01	5.12	2.35	12.87	5.86	4.98	5.48	0.76	5.01	1.99	57.68
1963	2.54	4.20	0.94	1.22	8.68	3.86	11.74	2.73	3.87	4.83	2.72		51.35
1964	7.08	6.17	2.89	1.32	5.11	5.02	7.45	3.69	3.58	9.81	1.17	4.54	57.83
1965	1.29	4.83	5.72	2.40	2.76	9.38	12.09	10.58	2.90	3.44	1.12	0.81	57.32
1966	6.32	5.54	2.89	1.48	7.50	9.78	15.12	5.67	6.09	1.10	0.70	3.44	65.63
1967	3.89	4.14	0.93	1.15	2.58	6.03	5.78	8.13	1.55	1.07	2.24	4.67	42.16
1968	3.71	1.44	0.97	3.50	2.30	2.52	9.31	1.66	1.24	4.46	4.40	2.26	37.77
1969	2.80	2.53	4.60	3.41	7.32	9.31	13.46	4.69	2.59	0.95	5.23	3.86	60.75
1970	1.98	2.45	7.19	1.37	3.92	4.48	5.71	13.98	2.23	4.12	2.22	3.22	52.87
1971	4.97	3.51	4.57	3.46	2.29	5.60	8.08	10.42	5.04	6.05	2.22	1.49	57.70
1972	4.27	4.57	3.36	0.78	3.50	3.50	6.36	4.91	6.73	0.69	7.87	5.29	51.83
1973	4.37	4.85	3.48	5.32	3.48	10.70	9.33	6.42	5.60	1.52	0.49	5.60	61.16
1974	2.81	4.23	2.30	2.18	5.15	5.37	8.36	13.33	5.09	1.12	3.14	3.72	56.80
1975	5.06	5.09	3.15	3.90	3.11	7.11	9.76	4.52	6.06	3.14	3.00	5.20	59.10
1976	2.93	1.01	2.57	0.91	4.33	12.74	8.28	9.53	4.79	3.31	2.14	5.37	57.91
1977	2.94	1.83	5.66	1.43	6.75	4.26	3.25	5.85	4.88	5.73	6.33	2.83	51.74
1978	6.68	1.33	2.94	3.82	3.46	2.34	7.38	4.77	1.58	1.01	3.71	4.68	43.70
#1979	6.23	4.21	4.82	5.60	5.27	4.74	2.82	2.48	15.17	0.38	2.01	2.35	56.08
1980	4.16	1.52	6.03	1.33	4.65	2.46	5.95	3.21	5.97	1.62	1.87	5.81	44.58
1981	1.09	3.08	3.02	1.43	5.02	2.45	5.23	14.06	1.07	1.39	0.78	5.76	44.38
1982	5.50	6.67	1.84	4.03	2.04	7.59	8.59	3.67	7.08	2.56	1.32	6.57	57.46
1983	4.90	8.74	8.09	2.09	1.13	6.71	5.53	5.63	5.59	1.02	4.49	5.20	59.12
1984	2.58	4.82	4.43	3.23	6.45	0.89	9.01	4.79	18.94	0.49	1.16	1.32	58.11
1985	2.01	5.08	1.66	0.71	2.76	4.56	10.34	3.63	2.75	2.43	6.74	1.35	44.02
Record Mean	3.35	3.52	3.61	2.82	3.75	5.21	7.40	6.67	5.54	3.11	2.57	3.30	50.85

TABLE 3 AVERAGE TEMPERATURE (deg. F)   WILMINGTON, NORTH CAROLINA

YEAR	JAN	FEB	MAR	APR	MAY	JUNE	JULY	AUG	SEP	OCT	NOV	DEC	ANNUAL
1956	42.1	52.8	53.8	61.0	69.7	76.8	81.3	79.6	72.4	66.8	53.7	57.1	63.9
1957	46.4	52.6	53.0	64.3	71.0	77.6	79.0	77.5	76.5	60.6	49.3	43.1	63.7
1958	41.0	40.9	49.1	61.4	69.9	75.6	82.1	79.9	73.8	62.1	58.2	48.3	61.4
1959	44.5	49.6	53.0	62.8	72.4	77.1	79.1	80.2	74.5	67.1	54.7	48.3	63.6
1960	45.9	46.4	43.7	64.9	70.8	77.1	79.0	79.7	74.5	65.4	55.5	41.6	62.1
1961	41.9	51.0	58.6	57.9	67.3	75.7	81.4	79.5	76.3	62.4	56.3	48.8	63.1
1962	45.2	51.6	50.5	60.8	73.5	75.8	79.6	78.8	72.0	65.8	52.7	43.8	62.5
#1963	43.0	42.2	57.7	63.7	68.2	75.9	78.0	78.8	70.2	64.3	55.4	42.2	61.7
1964	46.7	45.8	54.2	63.0	71.3	77.9	78.3	78.1	74.8	62.0	59.2	52.2	63.6
1965	46.2	48.3	51.7	62.4	74.4	76.2	79.1	80.2	75.4	64.8	57.5	48.8	63.8
1966	43.2	48.0	53.3	62.2	69.6	74.8	80.9	80.1	75.8	66.8	56.1	47.8	63.2
1967	50.4	47.5	58.0	66.0	71.3	76.3	81.1	80.1	72.0	64.9	53.0	52.1	64.4
1968	42.2	40.3	54.5	63.2	69.6	78.3	80.5	81.9	76.5	65.2	54.7	45.0	62.4
1969	43.2	45.1	50.0	63.1	68.8	77.4	80.8	76.9	72.2	65.5	52.5	44.9	61.7
1970	39.2	45.5	53.6	64.0	70.8	76.2	80.2	79.7	76.7	67.4	55.1	49.9	63.2
1971	44.8	48.0	50.2	60.4	68.8	78.0	79.8	78.8	75.4	70.8	55.1	56.6	63.9
1972	52.0	47.0	53.8	61.3	67.2	72.7	80.1	79.3	74.7	64.3	54.8	51.9	63.3
1973	46.5	45.3	58.8	60.8	69.5	76.6	79.8	80.1	77.2	67.3	58.2	50.2	64.2
1974	58.8	49.5	49.5	64.4	70.7	75.5	79.1	79.3	74.4	61.7	54.8	49.4	64.8
1975	51.0	52.1	54.4	62.3	73.8	78.7	79.9	82.4	77.0	68.2	59.7	49.4	65.7
1976	45.1	55.3	60.8	63.9	69.6	76.0	81.7	77.7	74.0	60.5	49.7	46.4	63.4
1977	35.8	45.6	57.9	65.9	71.4	77.7	82.6	80.5	77.8	62.0	58.8	47.6	63.6
1978	41.9	38.5	51.3	64.2	69.8	77.1	80.3	82.1	76.8	63.8	60.2	49.4	63.1
1979	43.7	43.2	53.9	65.6	71.4	74.7	80.4	80.6	74.5	63.3	57.8	47.8	63.1
1980	47.2	41.7	50.4	63.2	70.5	76.9	82.9	82.2	78.7	62.9	51.9	45.0	62.8
1981	37.5	46.6	49.9	63.9	66.9	79.8	80.7	76.9	71.4	60.9	52.3	44.4	60.9
1982	41.5	50.7	55.8	58.6	71.1	75.8	79.1	77.7	72.0	62.0	57.3	52.6	62.8
1983	42.6	45.1	53.7	57.4	67.3	75.0	82.8	81.5	74.8	66.1	55.1	46.9	62.4
1984	42.1	50.4	52.9	61.0	70.0	77.3	77.9	78.3	71.0	68.7	55.0	57.5	63.5
1985	43.2	50.4	59.1	66.3	71.8	78.7	81.4	78.9	75.0	71.1	66.6	47.0	65.8
Record Mean	47.0	48.4	54.6	62.1	70.1	76.7	79.7	79.0	74.5	64.8	55.5	48.5	63.4
Max	56.4	58.0	64.3	71.8	79.2	85.1	87.6	86.9	82.8	74.4	65.6	58.2	72.5
Min	37.6	38.8	44.8	52.3	61.0	68.2	71.8	71.0	66.1	55.3	45.4	38.8	54.3

REFERENCE NOTES FOR TABLES 1, 2, 3 and 6   (WILMINGTON, NC)

GENERAL

T - TRACE AMOUNT
BLANK ENTRIES DENOTE MISSING/UNREPORTED DATA.
# INDICATES A STATION OR INSTRUMENT RELOCATION.

SPECIFIC

TABLE 1

(a) - LENGTH OF RECORD IN YEARS. ALTHOUGH INDIVIDUAL MONTHS MAY BE MISSING.

* LESS THAN .05

NORMALS — BASED ON THE 1951-1980 RECORD PERIOD.
EXTREMES — DATES ARE THE MOST RECENT OCCURRENCE.
WIND DIR. — NUMERALS SHOW TENS OF DEGREES CLOCKWISE FROM TRUE NORTH. "00" INDICATES CALM.
RESULTANT WIND DIRECTIONS ARE GIVEN TO WHOLE DEGREES.

EXCEPTIONS

TABLE 1

1. FASTEST MILE WIND IS THROUGH JANUARY 1980.

TABLES 2, 3, and 6

RECORD MEANS ARE THROUGH THE CURRENT YEAR, BEGINNING IN   1874 FOR TEMPERATURE
1871 FOR PRECIPITATION
1952 FOR SNOWFALL

TABLE 4 HEATING DEGREE DAYS Base 65 deg. F   WILMINGTON, NORTH CAROLINA

SEASON	JULY	AUG	SEP	OCT	NOV	DEC	JAN	FEB	MAR	APR	MAY	JUNE	TOTAL
1956-57	0	0	12	29	353	265	571	351	366	121	35	0	2103
1957-58	0	0	4	156	277	482	736	668	488	153	15	0	2979
1958-59	0	0	2	126	221	674	627	534	370	131	9	0	2587
1959-60	0	0	1	76	325	512	584	534	657	93	21	0	2803
1960-61	0	0	0	88	286	721	708	394	230	243	29	7	2706
1961-62	0	0	0	119	303	499	609	387	449	176	4	3	2549
#1962-63	0	0	22	113	362	652	675	634	243	135	56	0	2892
1963-64	0	0	17	67	283	701	559	548	335	142	13	0	2665
1964-65	0	0	0	145	184	415	575	465	408	131	1	0	2324
1965-66	0	0	0	102	217	494	671	470	360	147	35	2	2498
1966-67	0	0	0	57	276	530	442	485	252	79	15	7	2143
1967-68	0	0	4	88	362	402	701	709	342	106	20	0	2734
1968-69	0	0	0	103	312	609	670	554	458	102	27	0	2835
1969-70	0	0	1	83	376	614	793	539	348	113	16	0	2883
1970-71	0	0	10	43	295	461	620	469	450	160	27	0	2535
1971-72	0	0	0	11	318	276	404	516	340	167	29	4	2065
1972-73	0	0	0	81	299	399	567	546	217	154	30	0	2293
1973-74	0	0	0	39	224	459	211	428	209	113	24	0	1707
1974-75	0	0	5	149	333	473	429	362	338	160	0	0	2249
1975-76	0	0	0	40	199	478	613	294	183	122	32	0	1961
1976-77	0	0	0	172	455	569	899	541	241	78	27	1	2983
1977-78	0	0	0	144	231	537	709	736	419	89	30	0	2895
1978-79	0	0	0	91	154	489	653	606	346	60	5	0	2404
1979-80	0	0	2	110	239	525	543	670	445	120	21	0	2675
1980-81	0	0	0	128	388	613	846	510	465	101	64	0	3115
1981-82	0	0	13	149	374	630	726	396	290	204	17	2	2799
1982-83	0	0	3	145	241	386	688	552	345	231	46	2	2639
1983-84	0	0	12	66	299	551	703	417	371	147	37	1	2604
1984-85	0	0	16	18	306	243	682	420	228	80	11	0	2004
1985-86	0	0	4	17	60	559							

TABLE 5 COOLING DEGREE DAYS Base 65 deg. F   WILMINGTON, NORTH CAROLINA

YEAR	JAN	FEB	MAR	APR	MAY	JUNE	JULY	AUG	SEP	OCT	NOV	DEC	TOTAL
1969	0	2	0	51	152	381	496	377	225	107	10	0	1801
1970	0	0	0	90	205	343	482	463	370	125	3	3	2084
1971	0	0	0	31	152	397	465	433	319	199	28	23	2047
1972	5	0	2	64	101	241	477	453	298	66	19	2	1728
1973	1	0	31	34	174	354	465	472	372	116	24	8	2051
1974	25	0	52	101	208	321	443	449	294	53	32	0	1978
1975	1	5	17	84	276	417	469	545	369	145	45	0	2373
1976	2	17	58	94	183	336	525	403	277	39	2	0	1936
1977	0	4	28	114	234	387	553	487	391	56	53	3	2310
1978	0	0	0	68	187	372	482	536	359	63	17	13	2097
1979	0	2	9	84	210	298	487	490	293	65	28	0	1966
1980	0	1	0	71	198	364	560	540	419	70	2	0	2225
1981	0	0	3	74	131	449	491	375	213	32	2	0	1770
1982	0	3	11	18	211	331	445	400	218	58	24	9	1728
1983	0	0	2	10	122	307	560	517	313	107	10	0	1948
1984	0	0	4	35	201	378	405	418	204	138	12	18	1813
1985	11	17	57	126	229	420	513	437	309	211	112	10	2452

TABLE 6 SNOWFALL (inches)   WILMINGTON, NORTH CAROLINA

SEASON	JULY	AUG	SEP	OCT	NOV	DEC	JAN	FEB	MAR	APR	MAY	JUNE	TOTAL
1970-71	0.0	0.0	0.0	0.0	T	4.0	T	0.0	T	0.0	0.0	0.0	4.0
1971-72	0.0	0.0	0.0	0.0	0.0	T	0.0	T	0.0	0.0	0.0	0.0	T
1972-73	0.0	0.0	0.0	0.0	0.0	0.0	1.9	12.5	0.0	0.0	0.0	0.0	14.4
1973-74	0.0	0.0	0.0	0.0	0.0	0.4	0.0	0.0	0.0	0.0	0.0	0.0	0.4
1974-75	0.0	0.0	0.0	0.0	0.0	0.0	0.0	T	0.0	0.0	0.0	0.0	T
1975-76	0.0	0.0	0.0	0.0	0.0	0.0	T	0.0	0.0	0.0	0.0	0.0	T
1976-77	0.0	0.0	0.0	0.0	T	T	0.6	0.3	0.0	0.0	0.0	0.0	0.9
1977-78	0.0	0.0	0.0	0.0	0.0	T	T	T	0.0	0.0	0.0	0.0	T
1978-79	0.0	0.0	0.0	0.0	0.0	0.0	T	0.2	0.0	0.0	0.0	0.0	0.2
#1979-80	0.0	0.0	0.0	0.0	0.0	0.0	0.0	0.3	6.6	0.0	0.0	0.0	6.9
1980-81	0.0	0.0	0.0	0.0	0.0	0.0	1.0	0.0	0.0	0.0	0.0	0.0	1.0
1981-82	0.0	0.0	0.0	0.0	0.0	0.0	T	0.1	0.0	0.0	0.0	0.0	0.1
1982-83	0.0	0.0	0.0	0.0	0.0	0.0	0.0	4.2	0.0	0.0	0.0	0.0	4.2
1983-84	0.0	0.0	0.0	0.0	0.0	0.0	T	T	0.0	0.0	0.0	0.0	T
1984-85	0.0	0.0	0.0	0.0	0.0	T	T	0.0	0.0	0.0	0.0	0.0	T
1985-86	0.0	0.0	0.0	0.0	0.0	0.0							
Record Mean	0.0	0.0	0.0	0.0	T	0.3	0.3	0.6	0.5	0.0	0.0	0.0	1.8

See Reference Notes, relative to all above tables, on preceding page.

Bismarck, the State Capital and County Seat of Burleigh County, is located in south-central North Dakota, near the center of North America. It is on the east bank of the Missouri River in a shallow basin 7 miles wide and 11 miles long.

The Weather Service Forecast Office is located at the Municipal Airport approximately 2 miles southeast of city center. It is almost entirely surrounded by low-lying hills. The closest hills, 3 miles to the north, and other hills 5 miles to the southeast, are about 200 to 300 feet high. West across the Missouri River the land is more hilly and 300 to 600 feet higher.

The climate is semi-arid, typically continental in character, and invigorating. Summers are warm, but there are not many hot days, and very few hot and humid days. Winters tend to be long and quite cold, but there are plenty of mild days to make winter weather pleasant much of the time. Sunshine is abundant, averaging 2,700 hours out of a possible 4,470 hours.

More than 75 percent of annual precipitation falls during the six month period from April through September, and nearly 50 percent during May, June, and July. Snow has been reported in all months except July and August. Three inches or more can be expected on about three days each year.

Most summer precipitation occurs during thunderstorms in the late afternoon and evening. Thunderstorms occur on about 34 days each year, accompanied by hail on two or three of the days. A damaging hailstorm is experienced about once every ten years. Tornadoes are rare, but damaging winds occasionally occur with the heavier thunderstorms.

The winter season usually begins in late November and continues until late March. Winter precipitation is nearly all in the form of snow and is often associated with strong winds and low temperatures. This combination produces winter storms and occasional blizzards that must never be taken lightly. A severe blizzard lasting two or three days may be expected every few years. But several times each winter storms lasting a few hours occur in which drifting snow can make travel difficult and even block roads. A stalled motorist can be in serious trouble if he is not prepared with adequate winter clothing and some kind of emergency provisions. A motorist must never leave his vehicle in a blinding snowstorm as he can easily become lost.

The temperature range from summer to winter is very large and typical of the Northern Great Plains. The average freeze-free period is 134 days, from mid-May to late September.

TABLE 1 NORMALS, MEANS AND EXTREMES

BISMARCK, NORTH DAKOTA

LATITUDE: 46°46'N   LONGITUDE: 100°45'W   ELEVATION: FT. GRND 1647 BARO 01655   TIME ZONE: CENTRAL   WBAN: 24011

	(a)	JAN	FEB	MAR	APR	MAY	JUNE	JULY	AUG	SEP	OCT	NOV	DEC	YEAR
TEMPERATURE °F:														
Normals														
-Daily Maximum		17.5	25.2	36.4	54.2	67.7	76.8	84.4	83.3	71.4	59.3	39.4	25.9	53.5
-Daily Minimum		-4.2	3.7	15.6	30.8	42.0	51.8	56.4	54.2	43.2	32.8	17.7	4.8	29.1
-Monthly		6.7	14.4	26.0	42.5	54.9	64.3	70.4	68.8	57.3	46.0	28.6	15.4	41.3
Extremes														
-Record Highest	46	62	68	81	93	98	100	109	109	105	95	75	65	109
-Year		1981	1958	1946	1980	1941	1961	1973	1941	1959	1963	1978	1979	JUL 1973
-Record Lowest	46	-44	-39	-31	-12	15	30	35	33	11	5	-30	-43	-44
-Year		1950	1982	1948	1975	1967	1969	1971	1982	1974	1960	1985	1967	JAN 1950
NORMAL DEGREE DAYS:														
Heating (base 65°F)		1807	1414	1209	675	324	100	18	57	255	586	1092	1538	9075
Cooling (base 65°F)		0	0	0	0	10	79	186	174	24	0	0	0	473
% OF POSSIBLE SUNSHINE	46	54	54	59	59	62	64	75	72	65	58	44	47	59
MEAN SKY COVER (tenths)														
Sunrise - Sunset	46	6.7	6.9	7.0	6.8	6.5	6.1	4.8	4.9	5.5	5.9	6.9	6.8	6.2
MEAN NUMBER OF DAYS:														
Sunrise to Sunset														
-Clear	46	6.9	5.3	5.4	5.8	6.3	7.3	11.6	11.7	10.2	9.5	6.3	6.6	93.0
-Partly Cloudy	46	7.6	7.7	8.4	8.7	10.5	10.5	12.7	11.1	8.8	7.7	6.9	6.9	107.4
-Cloudy	46	16.6	15.3	17.1	15.5	14.3	12.2	6.7	8.2	11.0	13.8	16.8	17.4	164.9
Precipitation														
.01 inches or more	46	7.8	6.8	8.1	8.1	9.8	11.8	8.8	8.5	7.0	5.8	6.1	7.9	96.4
Snow, Ice pellets														
1.0 inches or more	46	2.6	2.0	2.4	1.2	0.3	0.0	0.0	0.0	0.1	0.4	1.8	2.1	12.9
Thunderstorms	46	0.0	0.*	0.1	1.0	3.7	8.7	9.4	7.9	2.6	0.6	0.*	0.0	34.1
Heavy Fog Visibility														
1/4 mile or less	46	1.0	1.5	1.5	0.8	0.4	0.5	0.7	0.7	0.7	1.1	1.3	1.4	11.6
Temperature °F														
-Maximum														
90° and above	26	0.0	0.0	0.0	0.1	0.5	2.0	7.8	8.5	1.9	0.1	0.0	0.0	20.8
32° and below	26	23.9	17.3	10.8	0.8	0.1	0.0	0.0	0.0	0.0	0.5	8.9	21.8	84.1
-Minimum														
32° and below	26	30.8	28.1	28.7	17.5	4.1	0.1	0.0	0.0	2.8	14.3	28.2	31.0	185.6
0° and below	26	18.6	11.2	4.2	0.2	0.0	0.0	0.0	0.0	0.0	0.0	2.5	13.3	50.0
AVG. STATION PRESS.(mb)	13	958.1	957.4	954.6	955.0	953.4	953.1	954.9	955.1	956.1	955.8	956.7	956.8	955.6
RELATIVE HUMIDITY (%)														
Hour 00	26	73	76	77	72	70	78	74	71	73	72	77	77	74
Hour 06 (Local Time)	26	73	76	79	79	78	84	83	83	82	79	80	77	79
Hour 12 (Local Time)	26	67	67	63	52	48	53	47	45	49	52	63	70	56
Hour 18	26	69	69	60	47	44	48	42	39	45	51	65	73	54
PRECIPITATION (inches):														
Water Equivalent														
-Normal		0.51	0.45	0.70	1.51	2.23	3.01	2.05	1.69	1.38	0.81	0.51	0.51	15.36
-Maximum Monthly	46	1.29	1.21	3.19	5.46	5.18	8.29	5.24	5.05	6.93	4.30	2.56	0.95	8.29
-Year		1969	1979	1975	1975	1965	1947	1969	1944	1977	1982	1944	1967	JUN 1947
-Minimum Monthly	46	0.02	0.03	0.09	T	0.28	0.50	0.18	0.03	0.02	0.05	T	T	T
-Year		1940	1985	1981	1952	1984	1974	1968	1971	1948	1968	1963	1944	NOV 1963
-Maximum in 24 hrs	46	0.67	0.73	1.30	1.97	2.54	3.25	2.33	2.68	3.02	1.81	0.99	0.59	3.25
-Year		1952	1958	1950	1964	1985	1947	1969	1965	1977	1980	1944	1960	JUN 1947
Snow, Ice pellets														
-Maximum Monthly	46	25.0	25.6	31.1	18.7	10.3	T			5.0	7.6	24.2	17.2	31.1
-Year		1982	1979	1975	1984	1950	1969			1984	1946	1985	1977	MAR 1975
-Maximum in 24 hrs	46	7.9	8.9	15.5	11.9	11.0	T			4.8	4.9	9.4	7.6	15.5
-Year		1952	1979	1966	1984	1967	1969			1984	1946	1985	1950	MAR 1966
WIND:														
Mean Speed (mph)	46	10.0	9.9	11.0	12.1	11.8	10.5	9.2	9.5	10.0	10.1	9.9	9.5	10.3
Prevailing Direction														
through 1963		WNW	WNW	WNW	WNW	SSE	WNW	SSE	E	WNW	WNW	WNW	WNW	WNW
Fastest Obs. 1 Min.														
-Direction (!!!)	6	33	34	31	33	27	28	32	30	31	31	35	31	30
-Speed (MPH)	6	43	35	43	41	53	44	46	54	33	39	35	45	54
-Year		1980	1984	1984	1984	1980	1981	1982	1980	1985	1980	1982	1985	AUG 1980
Peak Gust														
-Direction (!!!)	2	NW	NW	NW	SW	NW	NW	NW	SW	W	W	NW	NW	NW
-Speed (mph)	2	51	49	55	56	56	58	53	61	48	52	48	63	63
-Date		1985	1984	1984	1985	1985	1985	1984	1984	1985	1984	1984	1985	DEC 1985

See Reference Notes to this table on the following page.

BISMARCK, NORTH DAKOTA

TABLE 2   PRECIPITATION (inches)   BISMARCK, NORTH DAKOTA

YEAR	JAN	FEB	MAR	APR	MAY	JUNE	JULY	AUG	SEP	OCT	NOV	DEC	ANNUAL
1956	0.97	0.18	1.21	0.11	3.83	2.36	2.78	2.93	0.55	0.25	1.43	0.26	16.86
1957	0.43	0.23	0.25	1.61	2.77	2.56	1.58	1.58	0.74	1.59	0.42	0.38	14.14
1958	0.43	1.09	0.31	0.76	1.02	3.57	1.74	0.71	0.41	1.79	1.72	0.34	12.59
1959	0.29	0.41	0.24	0.37	2.21	2.54	0.41	1.15	1.57	1.44	1.34	0.39	12.36
1960	0.35	0.15	0.67	0.41	2.89	3.40	0.68	3.81	0.29	0.10	0.40	0.89	14.04
1961	0.05	0.56	0.11	1.70	0.80	1.78	1.50	0.43	2.81	0.50	0.01	0.88	11.13
1962	0.54	0.31	0.71	0.55	4.80	2.92	2.40	1.04	0.79	0.31	0.30	0.24	14.91
1963	0.32	0.48	0.29	1.98	2.61	5.45	2.76	1.39	0.72	1.18	T	0.81	17.99
1964	0.40	0.22	0.52	2.90	0.90	5.71	2.18	0.76	0.66	0.10	0.43	0.65	15.43
1965	0.54	0.21	0.46	1.88	5.18	2.57	3.35	3.39	3.02	0.46	0.18	0.29	21.53
1966	0.29	0.41	1.85	1.23	0.91	3.60	4.27	2.25	0.86	0.59	0.29	0.18	16.73
1967	0.85	0.68	0.28	2.79	1.67	0.85	0.29	1.21	2.09	1.72	0.16	0.95	13.54
1968	0.30	0.12	0.95	1.30	2.51	6.52	0.18	4.56	1.20	0.05	0.56	0.78	19.03
1969	1.29	1.17	0.15	0.77	1.57	2.01	5.24	0.92	0.49	0.30	0.06	0.79	14.76
1970	0.46	0.34	0.55	4.05	2.32	3.78	1.63	0.13	1.47	0.69	1.33	0.16	16.91
1971	0.78	0.41	0.30	1.06	1.34	3.86	1.04	0.03	1.63	3.74	0.59	0.47	15.25
1972	0.68	0.41	1.32	1.81	3.16	1.81	1.60	1.70	0.29	1.60	0.11	0.67	15.16
1973	0.07	0.08	0.55	0.87	1.77	1.15	1.24	0.30	2.32	1.40	0.46	0.83	11.04
1974	0.11	0.29	0.40	2.23	2.68	0.50	1.11	1.52	0.23	0.74	0.36	0.50	10.66
1975	0.53	0.47	3.19	5.46	1.81	4.60	2.63	0.64	0.57	0.71	0.19	0.70	21.50
1976	0.52	0.27	0.52	2.84	1.08	2.66	0.46	0.32	1.59	0.21	0.15	0.55	11.17
1977	0.55	0.33	0.71	0.13	1.09	2.35	1.39	1.92	6.93	1.00	1.36	0.78	18.54
1978	0.14	0.41	0.30	2.00	4.65	1.79	2.58	1.21	1.80	0.45	1.13	0.49	16.95
1979	0.54	1.21	1.15	0.94	1.07	0.76	3.22	1.73	0.82	0.10	0.05	0.22	11.81
1980	0.70	0.28	0.32	0.43	1.08	1.67	3.16	5.03	1.11	2.31	0.09	0.21	16.39
1981	0.12	0.42	0.09	0.58	0.90	1.67	3.69	3.32	1.88	0.52	0.79	0.48	14.46
1982	0.75	0.40	1.08	0.76	3.71	2.04	2.17	1.52	0.45	4.30	0.41	0.48	18.07
1983	0.23	0.44	0.51	1.46	2.95	2.04	2.04	0.87	1.03	0.75	0.73	0.48	13.14
1984	0.38	0.31	1.65	3.65	0.28	3.58	0.81	0.87	0.94	0.99	0.73	0.58	14.77
1985	0.27	0.03	0.80	1.77	4.13	1.80	0.55	4.61	1.29	1.33	0.91	0.35	17.84
Record Mean	0.46	0.45	0.87	1.52	2.23	3.23	2.19	1.79	1.33	0.95	0.56	0.52	16.08

TABLE 3   AVERAGE TEMPERATURE (deg. F)   BISMARCK, NORTH DAKOTA

YEAR	JAN	FEB	MAR	APR	MAY	JUNE	JULY	AUG	SEP	OCT	NOV	DEC	ANNUAL
1956	4.7	9.1	23.6	37.4	54.9	70.1	67.9	67.7	55.5	49.0	30.4	20.6	40.9
1957	1.9	13.8	27.8	41.7	55.1	62.8	75.3	69.6	56.1	45.5	31.0	24.2	42.1
1958	18.1	10.9	28.5	44.6	59.6	60.3	66.7	71.6	60.2	47.5	29.2	12.6	42.5
#1959	3.2	6.2	33.1	43.3		68.5	73.1	72.8	57.1	40.6	22.9	26.1	41.6
1960	11.0	14.4	20.1	45.2	56.2	63.2	74.2	70.7	59.5	47.0	28.9	16.9	42.3
1961	18.4	19.0	36.1	40.3	54.3	69.1	69.5	73.7	52.4	47.9	30.8	10.0	43.5
1962	10.7	10.3	23.2	44.6	55.0	65.4	67.4	70.2	57.6	49.5	35.5	20.2	42.5
1963	3.0	16.5	36.1	43.9	55.3	68.2	71.9	69.1	60.9	55.0	33.2	10.6	43.7
1964	15.9	21.9	22.8	45.1	56.8	63.0	71.7	64.9	53.9	47.2	25.3	3.5	41.0
1965	2.5	10.2	14.2	42.0	54.5	63.0	69.4	66.8	46.3	49.3	30.1	23.6	39.5
1966	-3.0	10.4	33.0	38.0	53.4	64.2	72.6	64.3	58.7	46.1	21.3	15.9	39.6
1967	11.2	9.2	28.1	38.3	49.3	61.1	69.0	67.4	60.4	45.0	28.3	12.8	40.0
1968	7.1	11.9	32.8	42.8	49.8	61.2	67.6	64.6	57.5	45.1	32.0	11.4	40.3
1969	-0.8	14.7	17.5	46.1	55.9	59.5	70.4	73.8	61.9	40.6	32.5	17.6	40.8
1970	4.0	15.6	18.8	39.6	54.3	67.3	71.0	69.9	58.1	44.4	27.3	11.1	40.1
1971	1.5	12.9	28.2	43.8	52.4	65.7	66.2	71.6	56.8	45.1	30.1	10.9	40.4
1972	4.8	9.1	27.5	42.4	58.1	65.2	66.7	70.7	57.0	41.4	27.6	8.1	39.9
1973	15.5	21.9	39.4	40.9	54.1	63.3	68.4	72.0	54.6	46.8	22.6	9.8	42.5
1974	6.5	15.9	27.1	42.3	50.5	62.8	73.8	63.0	52.9	45.9	28.9	17.5	40.6
1975	12.9	12.5	22.5	35.5	53.9	63.2	72.8	66.8	55.0	47.0	31.5	19.1	41.1
1976	11.9	27.2	27.3	46.5	55.6	67.1	71.2	72.9	60.1	41.4	26.4	14.8	43.5
1977	-1.6	23.9	33.8	49.1	61.7	65.6	70.2	62.0	57.0	46.2	26.4	10.5	42.1
1978	-1.7	9.0	27.9	42.7	56.9	63.0	68.8	68.2	61.8	46.0	22.7	10.8	39.6
1979	-2.2	-0.7	22.2	35.1	50.0	64.3	70.2	65.7	59.4	44.6	25.9	23.7	38.2
1980	8.7	15.4	24.8	48.7	60.1	66.2	71.9	65.7	58.1	45.6	35.9	18.5	43.3
1981	19.5	23.1	34.9	46.5	54.1	60.7	71.2	70.1	59.3	46.9	36.9	14.7	44.8
1982	-4.3	8.7	23.5	40.0	54.4	59.2	70.3	67.5	56.7	44.5	24.4	22.1	39.0
1983	23.0	27.4	30.4	39.9	50.9	62.3	72.3	74.8	56.3	44.8	30.9	-1.2	42.7
1984	15.5	28.0	25.4	41.7	51.9	63.1	70.1	71.6	52.6	45.5	30.3	8.6	42.0
1985	6.2	13.2	31.6	46.0	59.1	59.3	69.9	64.3	53.6	44.3	14.0	9.1	39.2
Record Mean	8.1	12.7	25.6	43.0	54.7	64.0	70.6	68.5	57.8	45.6	28.4	15.2	41.2
Max	18.6	23.2	35.8	54.6	66.9	75.8	83.6	82.1	71.1	58.0	38.7	25.2	52.8
Min	-2.5	2.2	15.4	31.4	42.4	52.2	57.5	54.9	44.5	33.2	18.1	5.2	29.5

REFERENCE NOTES FOR TABLES 1, 2, 3 and 6   (BISMARCK, ND)

GENERAL

T - TRACE AMOUNT
BLANK ENTRIES DENOTE MISSING/UNREPORTED DATA.
# INDICATES A STATION OR INSTRUMENT RELOCATION.

SPECIFIC

TABLE 1

(a) - LENGTH OF RECORD IN YEARS. ALTHOUGH INDIVIDUAL MONTHS MAY BE MISSING.

* LESS THAN .05

NORMALS — BASED ON THE 1951-1980 RECORD PERIOD.
EXTREMES — DATES ARE THE MOST RECENT OCCURRENCE.
WIND DIR. — NUMERALS SHOW TENS OF DEGREES CLOCKWISE FROM TRUE NORTH. "00" INDICATES CALM.
RESULTANT WIND DIRECTIONS ARE GIVEN TO WHOLE DEGREES.

EXCEPTIONS

TABLES 2, 3, and 6

RECORD MEANS ARE THROUGH THE CURRENT YEAR, BEGINNING IN
1875 FOR TEMPERATURE
1875 FOR PRECIPITATION
1940 FOR SNOWFALL

TABLE 4 HEATING DEGREE DAYS Base 65 deg. F BISMARCK, NORTH DAKOTA

SEASON	JULY	AUG	SEP	OCT	NOV	DEC	JAN	FEB	MAR	APR	MAY	JUNE	TOTAL
1956-57	17	48	287	485	1032	1369	1958	1432	1145	692	307	93	8865
1957-58	0	44	264	605	1012	1257	1444	1515	1123	606	204	166	8240
1958-59	49	39	185	535	1066	1621	1914	1645	980	643	397	56	9130
#1959-60	4	13	273	749	1256	1197	1668	1460	1390	587	290	83	8970
1960-61	7	13	228	551	1076	1488	1437	1283	892	733	344	36	8088
1961-62	11	3	385	524	1017	1701	1681	1529	1289	606	304	59	9109
1962-63	22	28	228	474	877	1382	1925	1351	890	628	308	26	8139
1963-64	3	26	139	311	945	1683	1520	1243	1302	588	268	123	8151
1964-65	5	108	342	542	1180	1909	1937	1533	1501	684	324	83	10148
1965-66	14	57	556	479	1040	1279	2106	1525	987	804	368	118	9333
1966-67	4	93	228	577	1307	1517	1664	1562	1138	795	491	129	9505
1967-68	55	43	164	614	1095	1613	1795	1535	990	660	462	139	9165
1968-69	34	82	238	612	982	1658	2040	1405	1461	563	314	169	9558
1969-70	6	3	149	748	967	1463	1892	1376	1423	758	331	34	9150
1970-71	8	24	263	632	1123	1667	1967	1458	1134	631	383	41	9331
1971-72	52	20	271	608	1043	1672	1868	1618	1157	671	243	65	9288
1972-73	41	39	256	724	1116	1762	1528	1201	787	714	332	93	8593
1973-74	12	4	316	557	1263	1709	1816	1369	1167	674	440	120	9447
1974-75	6	119	357	582	1078	1466	1612	1467	1309	880	345	112	9333
1975-76	9	39	295	550	1002	1419	1642	1087	1163	550	288	65	8109
1976-77	11	6	193	726	1152	1551	2063	1147	956	473	164	45	8487
1977-78	17	111	246	574	1153	1687	2066	1564	1143	663	257	105	9586
1978-79	18	35	217	585	1263	1679	2084	1837	1320	891	463	81	10473
1979-80	11	55	191	626	1166	1273	1744	1435	1238	492	211	58	8500
1980-81	5	53	217	594	867	1438	1402	1173	925	547	335	144	7700
1981-82	11	11	191	554	836	1556	2151	1577	1279	742	310	176	9394
1982-83	5	73	291	626	1210	1321	1294	1048	1066	746	434	128	8242
1983-84	6	0	300	620	1015	2049	1532	1067	1225	690	421	94	9019
1984-85	17	27	373	602	1033	1749	1821	1450	1027	562	194	192	9047
1985-86	21	85	345	636	1526	1731							

TABLE 5 COOLING DEGREE DAYS Base 65 deg. F BISMARCK, NORTH DAKOTA

YEAR	JAN	FEB	MAR	APR	MAY	JUNE	JULY	AUG	SEP	OCT	NOV	DEC	TOTAL
1969	0	0	0	0	39	11	179	282	60	0	0	0	571
1970	0	0	0	0	5	109	199	182	60	0	0	0	555
1971	0	0	0	0	0	67	94	230	32	0	0	0	423
1972	0	0	0	0	37	76	98	226	24	0	0	0	461
1973	0	0	0	0	1	47	125	227	11	0	0	0	411
1974	0	0	0	0	0	61	287	61	0	0	0	0	409
1975	0	0	0	0	6	64	257	104	2	0	0	0	433
1976	0	0	0	0	3	137	212	257	54	0	0	0	663
1977	0	0	0	3	70	72	186	25	11	0	0	0	367
1978	0	0	0	0	13	50	141	143	128	0	0	0	475
1979	0	0	0	0	6	67	178	86	28	0	0	0	365
1980	0	0	0	8	64	103	227	84	16	0	0	0	502
1981	0	0	0	0	3	24	210	177	27	0	0	0	441
1982	0	0	0	0	3	8	175	159	49	0	0	0	394
1983	0	0	0	0	3	53	243	312	47	0	0	0	658
1984	0	0	0	0	21	44	180	241	11	4	0	0	501
1985	0	0	0	0	16	26	180	67	8	0	0	0	297

TABLE 6 SNOWFALL (inches) BISMARCK, NORTH DAKOTA

SEASON	JULY	AUG	SEP	OCT	NOV	DEC	JAN	FEB	MAR	APR	MAY	JUNE	TOTAL
1956-57	0.0	0.0	0.0	0.0	5.4	2.3	6.5	4.4	3.3	2.2	0.0	0.0	24.1
1957-58	0.0	0.0	T	T	2.9	5.7	3.7	8.6	5.1	T	0.2	0.0	26.2
1958-59	0.0	0.0	T	T	16.1	4.1	8.2	5.9	4.5	2.9	T	0.0	41.7
1959-60	0.0	0.0	T	3.4	15.5	4.9	6.0	2.2	11.7	2.2	2.8	0.0	48.7
1960-61	0.0	0.0	0.0	T	5.6	11.9	0.6	7.7	0.8	6.5	T	0.0	33.1
1961-62	0.0	0.0	T	T	0.1	5.8	6.0	4.8	5.7	1.2	0.0	0.0	23.6
1962-63	0.0	0.0	0.0	T	0.4	2.6	3.7	4.5	2.6	7.6	T	0.0	21.4
1963-64	0.0	0.0	0.0	0.0	T	9.4	4.7	2.8	5.8	2.7	0.0	0.0	25.4
1964-65	0.0	0.0	0.0	1.0	6.7	10.7	10.2	3.2	7.5	7.2	8.0	0.0	54.5
1965-66	0.0	0.0	3.6	0.0	0.4	2.3	4.4	5.4	22.7	4.9	0.6	0.0	44.3
1966-67	0.0	0.0	0.0	0.4	5.1	2.5	12.7	13.9	4.1	14.8	8.2	0.0	61.7
1967-68	0.0	0.0	0.0	T	2.6	4.3	1.8	7.4	3.2	2.3	0.0	T	35.1
1968-69	0.0	0.0	0.0	T	2.4	13.0	16.0	17.4	3.2	0.3	T	T	52.3
1969-70	0.0	0.0	0.0	0.0	0.9	13.7	7.0	4.5	7.6	18.2	0.0	0.0	52.3
1970-71	0.0	0.0	0.0	2.1	13.4	2.8	12.7	6.1	3.3	1.8	T	0.0	42.2
1971-72	0.0	0.0	0.0	3.1	6.4	8.7	13.3	5.8	8.5	0.9	T	0.0	46.7
1972-73	0.0	0.0	T	5.8	1.1	10.2	0.9	1.2	1.4	0.5	0.0	0.0	21.1
1973-74	0.0	0.0	0.0	0.6	5.5	11.5	1.4	4.5	3.4	0.0	0.0	0.0	27.5
1974-75	0.0	0.0	T	T	3.6	6.9	6.3	6.0	31.1	4.8	0.0	0.0	58.7
1975-76	0.0	0.0	0.0	T	2.2	8.6	5.3	2.7	6.5	0.8	T	0.0	26.1
1976-77	0.0	0.0	0.0	1.9	3.3	7.7	8.3	5.1	1.6	0.7	0.0	0.0	28.6
1977-78	0.0	0.0	0.0	3.5	16.2	17.2	4.1	11.1	7.6	4.0	0.0	0.0	63.7
1978-79	0.0	0.0	0.0	T	16.6	10.0	10.3	25.6	11.6	7.3	1.6	0.0	83.0
1979-80	0.0	0.0	0.0	T	1.0	1.6	12.6	5.6	5.7	0.1	0.0	0.0	26.6
1980-81	0.0	0.0	0.0	1.1	0.2	2.1	3.4	3.6	T	T	0.0	0.0	11.7
1981-82	0.0	0.0	0.0	0.7	6.3	10.4	25.0	10.3	22.5	5.1	0.0	0.0	80.3
1982-83	0.0	0.0	0.0	4.8	7.2	2.9	2.2	5.3	9.3	0.5	T	0.0	32.2
1983-84	0.0	0.0	T	0.0	6.6	10.6	4.1	5.9	20.6	18.7	T	0.0	66.5
1984-85	0.0	0.0	5.0	1.3	2.0	14.9	3.9	0.5	12.4	0.2	0.0	0.0	40.2
1985-86	0.0	0.0	T	3.2	24.2	7.1							
Record Mean	0.0	0.0	0.3	1.3	5.5	6.9	7.2	6.3	8.5	3.8	0.9	T	40.6

See Reference Notes, relative to all above tables, on preceding page.

Moorhead, Minnesota, and Fargo are twin cities in the Red River Valley of the north. The Red River of the north flows northward between the two cities and is a part of the Hudson Bay drainage area. The Red River is approximately 2 miles east of the airport at its nearest point and has no significant effect on the weather. In recent years, spring floods due to melting snow have been common. Summer floods caused by heavy rains are infrequent.

The surrounding terrain is flat and open. Northerly winds blowing up the valley occasionally causing low cloudiness and fog. However, this upslope cloudiness is very infrequent. Aside from this, there are no pronounced climatic differences due to geographical features in the immediate area.

The summers are generally comfortable with very few days of hot and humid weather. Nights, with few exceptions, are comfortably cool. The winter months are cold and dry with temperatures rising above freezing only on an average of six days each month, and nighttime lows dropping below zero approximately half of the time.

Precipitation is the most important climatic factor in the area. The Red River Valley lies in an area where lighter amounts fall to the west and heavier amounts to the east. Seventy-five percent of the precipitation occurs during the growing season (April to September) and is often accompanied by electrical storms and heavy falls in a short time. Winter precipitation is light, indicating that heavy snowfall is the exception rather than the rule. The first light snow in the fall occasionally falls in September, but usually very little, if any, occurs until October or November. The latest fall is generally in April.

With the flat terrain, surface friction has little effect on the wind in the area and this fact has led to the legendary Dakota blizzards. Strong winds with even light snowfall cause much drifting and blowing snow, reducing visibility to near zero. Fortunately, these conditions occur only several times during the winter months.

TABLE 1   NORMALS, MEANS AND EXTREMES

FARGO, NORTH DAKOTA

LATITUDE: 46°54'N   LONGITUDE: 96°48'W   ELEVATION: FT. GRND   896 BARO   00911   TIME ZONE: CENTRAL   WBAN: 14914

	(a)	JAN	FEB	MAR	APR	MAY	JUNE	JULY	AUG	SEP	OCT	NOV	DEC	YEAR
TEMPERATURE °F:														
Normals														
-Daily Maximum		13.7	20.5	33.2	52.5	68.1	76.9	82.7	81.1	69.8	57.7	37.0	21.3	51.2
-Daily Minimum		-5.1	1.5	14.8	31.6	43.0	53.5	58.4	56.4	45.7	34.9	19.4	4.0	29.8
-Monthly		4.3	11.0	24.0	42.0	55.5	65.2	70.6	68.8	57.8	46.3	28.2	12.7	40.5
Extremes														
-Record Highest	33	52	66	78	100	98	99	102	106	102	93	73	57	106
-Year		1981	1958	1967	1980	1964	1959	1980	1976	1959	1963	1978	1962	AUG 1976
-Record Lowest	33	-35	-34	-23	-7	20	30	36	33	19	7	-24	-32	-35
-Year		1977	1962	1980	1975	1966	1969	1967	1982	1965	1976	1985	1967	JAN 1977
NORMAL DEGREE DAYS:														
Heating (base 65°F)		1882	1512	1271	687	311	86	17	36	236	580	1104	1621	9343
Cooling (base 65°F)		0	0	0	0	19	92	191	154	20	0	0	0	476
% OF POSSIBLE SUNSHINE	43	50	56	57	59	60	60	71	68	59	55	40	43	57
MEAN SKY COVER (tenths)														
Sunrise - Sunset	40	6.7	6.7	7.0	6.7	6.4	6.2	5.0	5.2	5.8	6.2	7.2	7.0	6.3
MEAN NUMBER OF DAYS:														
Sunrise to Sunset														
-Clear	43	6.6	6.2	5.4	6.4	6.8	6.5	10.5	10.6	8.7	8.7	5.4	6.1	87.9
-Partly Cloudy	43	7.6	7.3	8.9	8.7	9.8	10.9	13.3	11.7	9.1	8.2	6.4	7.4	109.1
-Cloudy	43	16.8	14.8	16.7	15.0	14.4	12.6	7.2	8.7	12.2	14.1	18.2	17.5	168.2
Precipitation														
.01 inches or more	43	8.7	7.0	7.9	8.2	10.0	10.7	9.5	9.0	8.0	6.6	6.0	8.1	99.6
Snow, Ice pellets														
1.0 inches or more	43	2.4	1.6	2.2	1.1	0.*	0.0	0.0	0.0	0.0	0.3	1.7	2.2	11.6
Thunderstorms	43	0.0	0.*	0.3	1.3	3.7	7.4	8.6	6.8	3.0	1.0	0.1	0.*	32.2
Heavy Fog Visibility														
1/4 mile or less	43	0.7	1.7	1.8	0.6	0.4	0.6	0.7	1.1	0.8	1.0	1.4	1.8	12.6
Temperature °F														
-Maximum														
90° and above	26	0.0	0.0	0.0	0.1	0.4	1.6	5.2	5.4	1.3	0.*	0.0	0.0	14.0
32° and below	26	28.0	21.9	13.5	1.3	0.*	0.0	0.0	0.0	0.0	0.5	10.3	25.8	101.5
-Minimum														
32° and below	26	31.0	28.2	27.9	16.5	4.5	0.*	0.0	0.0	1.9	12.3	26.7	30.8	179.8
0° and below	26	19.9	13.2	5.0	0.2	0.0	0.0	0.0	0.0	0.0	0.0	2.1	14.0	54.3
AVG. STATION PRESS.(mb)	13	986.0	985.5	982.4	982.0	980.0	979.1	981.1	981.6	982.5	982.4	983.8	984.6	982.6
RELATIVE HUMIDITY (%)														
Hour 00	26	73	76	80	74	68	76	78	76	78	76	79	77	76
Hour 06 (Local Time)	26	73	76	82	80	77	82	85	86	85	82	82	77	81
Hour 12	26	70	72	71	59	50	57	55	55	58	60	70	74	63
Hour 18	26	72	73	71	55	46	52	51	50	55	61	73	75	61
PRECIPITATION (inches):														
Water Equivalent														
-Normal		0.55	0.42	0.83	1.90	2.24	3.06	3.34	2.67	1.87	1.29	0.79	0.63	19.59
-Maximum Monthly	44	1.36	1.74	2.27	4.24	7.30	9.40	8.42	8.52	6.13	7.03	4.58	2.19	9.40
-Year		1950	1979	1983	1942	1977	1975	1952	1944	1957	1982	1977	1951	JUN 1975
-Minimum Monthly	44	0.09	0.03	0.03	0.02	0.46	0.58	0.42	0.18	0.13	0.08	0.04	0.04	0.02
-Year		1961	1954	1958	1980	1976	1972	1950	1984	1974	1972	1967	1958	APR 1980
-Maximum in 24 hrs	44	0.81	1.22	1.16	1.91	4.10	4.02	3.93	4.72	3.97	3.22	1.99	0.87	4.72
-Year		1980	1946	1950	1963	1977	1975	1952	1943	1957	1982	1977	1960	AUG 1943
Snow, Ice pellets														
-Maximum Monthly	44	30.0	19.5	18.7	12.8	1.0				0.6	8.1	24.3	20.3	30.0
-Year		1982	1979	1975	1970	1950				1942	1951	1985	1951	JAN 1982
-Maximum in 24 hrs	44	17.4	11.2	10.4	8.6	1.0				0.6	7.8	12.6	8.0	17.4
-Year		1982	1951	1975	1970	1950				1942	1951	1977	1967	JAN 1982
WIND:														
Mean Speed (mph)	43	12.8	12.6	13.2	14.1	13.1	11.8	10.6	11.1	12.0	12.8	13.0	12.3	12.4
Prevailing Direction														
through 1963		SSE	N	N	N	N	SSE	S	SSE	SSE	SSE	S	S	SSE
Fastest Mile														
-Direction	43	SE	W	N	NW	NW	NW	S	NW	N	NW	N	N	NW
-Speed (MPH)	43	62	56	56	68	72	115	60	71	88	57	66	58	115
-Year		1968	1957	1960	1964	1960	1959	1965	1955	1960	1960	1956	1957	JUN 1959
Peak Gust														
-Direction	2	NW	N	SE	S	NW	SE	W	NW	SE	S	SE	N	NW
-Speed (mph)	2	59	59	53	44	58	52	53	46	45	49	48	55	59
-Date		1985	1984	1985	1984	1985	1984	1985	1985	1984	1985	1985	1985	JAN 1985

See Reference Notes to this table on the following page.

TABLE 2 PRECIPITATION (inches) FARGO, NORTH DAKOTA

YEAR	JAN	FEB	MAR	APR	MAY	JUNE	JULY	AUG	SEP	OCT	NOV	DEC	ANNUAL
1956	0.72	0.18	0.91	1.45	1.63	2.85	3.58	1.75	0.83	1.14	1.60	0.31	16.95
1957	0.31	0.23	0.08	1.90	2.00	5.47	4.72	1.68	6.13	1.64	0.82	0.05	25.03
1958	0.16	0.15	0.03	1.39	1.41	5.08	5.74	0.71	0.72	1.58	0.04		20.94
1959	0.13	0.20	0.08	0.66	1.88	4.78	4.26	3.02	0.77	1.61	0.31	0.53	18.23
1960	0.17	0.06	0.11	3.62	1.71	1.82	2.40	3.91	1.26	1.34	1.63	1.01	19.04
1961	0.09	0.18	0.38	2.27	2.71	1.36	3.00	1.02	4.44	1.70	0.06	0.57	17.78
1962	1.07	0.97	1.08	1.51	5.95	2.78	5.92	3.25	2.42	0.86	0.54	0.30	26.65
1963	0.13	0.33	0.51	2.67	2.61	1.69	0.66	4.41	1.19	0.23	0.12	0.39	14.94
1964	0.54	0.27	0.92	3.76	0.87	4.85	0.77	2.85	1.70	0.10	0.72	0.91	18.26
1965	0.10	0.14	1.36	3.04	3.06	3.10	4.81	2.55	3.50	0.55	0.79	1.01	24.01
1966	0.40	0.26	1.92	1.78	1.27	2.91	4.01	3.80	0.54	1.40	0.18	0.50	18.97
1967	1.03	0.21	0.34	4.14	1.00	2.54	0.60	0.41	0.31	1.06	0.04	1.36	13.04
1968	0.37	0.27	1.29	4.09	2.08	3.94	1.49	1.61	2.23	1.75	0.37	1.11	20.60
1969	1.27	0.46	0.54	1.55	2.36	2.03	5.92	0.38	1.55	1.51	0.14	0.81	18.52
1970	0.10	0.20	1.52	2.30	2.83	2.63	0.43	1.24	3.61	1.61	0.96	0.47	17.90
1971	0.81	0.34	0.56	1.10	2.68	3.51	2.80	0.92	4.30	4.42	0.83	0.59	22.86
1972	0.94	0.61	0.74	0.96	3.52	0.58	2.78	3.45	1.22	1.25	0.22	1.51	17.78
1973	0.12	0.13	1.25	0.70	1.65	1.78	3.60	3.85	4.98	1.54	0.90	1.02	21.52
1974	0.35	0.36	0.71	3.40	4.03	0.90	4.75	6.46	0.13	3.10	0.48	0.32	24.99
1975	1.32	0.27	1.48	3.24	1.45	9.40	2.42	2.90	1.24	1.76	0.64	0.18	26.30
1976	1.25	0.35	1.00	1.19	0.46	2.34	0.63	0.41	0.55	0.16	0.26	0.24	8.84
1977	0.65	1.24	1.72	0.84	7.30	1.64	5.36	2.53	3.21	2.46	4.58	0.75	32.28
1978	0.16	0.18	0.43	1.15	1.78	4.40	2.92	3.79	0.92	0.13	1.11	0.47	17.44
1979	0.44	1.74	2.00	3.04	2.02	2.92	3.38	0.90	0.31	2.60	0.48	0.14	19.97
1980	1.23	0.57	0.62	0.02	0.64	2.68	0.76	4.24	2.52	1.06	0.47	0.30	15.11
1981	0.11	0.49	0.67	0.61	3.46	2.56	3.21	1.76	1.11	2.36	0.40	0.85	17.59
1982	1.32	0.54	1.25	0.45	1.82	1.61	2.64	1.12	1.12	7.03	1.13	0.17	20.20
1983	0.46	0.21	2.27	0.42	2.00	2.34	4.16	2.56	1.63	1.62	1.04	0.96	19.67
1984	0.79	0.90	1.12	1.68	0.61	5.38	0.64	0.18	1.23	6.76	0.18	0.90	20.37
1985	0.20	0.18	1.35	0.60	5.03	1.44	3.91	2.30	1.39	1.12	1.06	0.59	19.17
Record Mean	0.63	0.60	0.92	1.94	2.47	3.39	3.14	2.76	1.93	1.58	0.86	0.66	20.87

TABLE 3 AVERAGE TEMPERATURE (deg. F) FARGO, NORTH DAKOTA

YEAR	JAN	FEB	MAR	APR	MAY	JUNE	JULY	AUG	SEP	OCT	NOV	DEC	ANNUAL
1956	6.1	5.8	19.5	35.0	53.3	70.7	68.0	68.5	55.7	51.7	30.6	16.7	40.1
1957	2.5	12.5	27.5	42.6	56.1	62.7	75.4	69.0	56.1	46.5	29.0	21.5	41.8
1958	17.4	12.4	30.9	45.2	58.2	59.7	67.4	68.9	59.9	48.8	29.7	10.5	42.4
#1959	3.7	9.5	30.9	43.2	55.2	68.2	72.2	72.5	58.9	40.7	20.5	25.9	41.8
1960	8.1	11.3	15.5	41.2	56.6	63.3	71.4	70.6	59.9	47.5	29.9	10.0	40.5
1961	6.8	18.7	34.5	37.8	53.2	68.7	70.5	73.5	56.0	47.9	30.1	10.7	42.4
1962	4.4	5.6	22.7	39.9	56.6	66.2	68.8	71.6	57.6	51.1	35.0	16.6	41.3
1963	2.0	10.1	29.0	43.5	54.3	68.8	73.6	70.3	62.2	57.3	34.7	8.6	42.9
1964	15.6	18.9	20.8	46.4	61.1	67.2	74.0	64.9	55.4	45.2	29.2	3.8	41.9
1965	-1.2	7.1	13.7	41.6	54.4	63.9	68.5	66.5	48.9	47.3	26.0	19.2	38.0
1966	-6.4	6.2	29.7	37.2	51.4	66.1	73.8	65.5	58.2	44.0	23.1	11.9	38.4
1967	9.3	3.8	26.8	38.3	49.7	62.6	67.9	66.7	60.8	44.0	29.0	15.1	39.5
1968	7.7	9.9	34.1	43.9	52.9	64.1	69.6	68.1	59.3	46.7	31.0	8.9	41.4
1969	-1.6	12.8	15.3	45.4	54.8	57.3	68.4	72.4	59.0	40.3	30.5	15.6	39.2
1970	0.6	10.9	18.6	39.1	51.7	67.9	71.8	69.7	59.7	46.8	27.8	9.3	39.5
1971	-0.7	12.7	27.6	44.5	54.2	67.5	65.1	68.3	58.5	47.4	29.6	12.2	40.6
1972	2.7	4.1	23.9	41.0	59.8	66.9	68.4	70.5	56.9	42.4	28.6	3.8	39.1
1973	10.3	16.3	36.0	41.4	54.2	64.7	68.3	71.7	54.8	50.2	25.1	10.1	41.9
1974	1.7	9.6	22.9	42.2	51.1	64.4	73.7	64.3	53.4	47.5	29.2	20.9	40.1
1975	12.3	10.1	18.4	35.9	56.6	65.2	74.3	68.1	55.4	49.4	31.1	14.8	41.0
1976	7.7	21.4	23.0	47.0	55.8	68.5	71.8	73.6	60.0	39.5	23.2	6.9	41.5
1977	-3.3	17.5	32.0	49.5	66.5	66.6	72.2	62.5	57.9	47.1	25.6	6.5	41.7
1978	-1.4	3.4	23.5	42.5	59.1	64.7	69.5	69.1	63.6	46.4	22.8	7.3	39.2
1979	-4.2	-1.5	20.4	36.0	50.4	65.4	71.9	67.3	62.0	42.6	24.5	20.7	38.0
1980	6.6	8.3	20.7	49.0	61.4	65.7	71.9	67.5	56.9	42.4	33.1	12.7	41.6
1981	11.8	19.6	33.5	45.6	55.5	62.8	71.1	69.6	57.4	44.5	35.4	8.7	43.0
1982	-7.0	8.9	22.9	40.7	58.1	59.1	70.9	68.5	57.5	45.7	24.1	20.9	39.2
1983	16.1	21.8	29.9	40.2	52.1	66.1	73.5	72.9	56.7	44.4	31.3	-0.3	42.1
1984	9.7	24.9	23.4	45.6	54.2	65.8	70.6	73.3	54.4	47.4	29.7	9.6	42.4
1985	5.1	10.9	32.9	46.6	60.2	60.0	69.0	64.5	53.9	44.6	15.4	3.9	38.9
Record Mean	5.0	9.8	24.3	42.3	54.9	64.6	70.1	68.0	57.9	45.6	27.4	12.5	40.2
Max	14.5	19.4	33.4	52.9	67.0	75.9	81.9	80.1	69.7	56.4	36.1	21.3	50.7
Min	-4.6	0.2	15.1	31.7	42.8	53.3	58.2	55.9	46.1	34.8	18.7	3.8	29.7

REFERENCE NOTES FOR TABLES 1, 2, 3 and 6   (FARGO, ND)

GENERAL

T - TRACE AMOUNT
BLANK ENTRIES DENOTE MISSING/UNREPORTED DATA.
# INDICATES A STATION OR INSTRUMENT RELOCATION.

SPECIFIC

TABLE 1

(a) - LENGTH OF RECORD IN YEARS. ALTHOUGH
      INDIVIDUAL MONTHS MAY BE MISSING.
* LESS THAN .05

NORMALS — BASED ON THE 1951-1980 RECORD PERIOD.
EXTREMES — DATES ARE THE MOST RECENT OCCURRENCE.
WIND DIR. — NUMERALS SHOW TENS OF DEGREES
            CLOCKWISE FROM TRUE NORTH.
            "00" INDICATES CALM.
RESULTANT WIND DIRECTIONS ARE GIVEN TO WHOLE DEGREES.

EXCEPTIONS

TABLES 2, 3, and 6

RECORD MEANS ARE THROUGH THE CURRENT YEAR,
BEGINNING IN   1881 FOR TEMPERATURE
                1881 FOR PRECIPITATION
                1943 FOR SNOWFALL

TABLE 4 HEATING DEGREE DAYS Base 65 deg. F   FARGO, NORTH DAKOTA

SEASON	JULY	AUG	SEP	OCT	NOV	DEC	JAN	FEB	MAR	APR	MAY	JUNE	TOTAL
1956-57	14	39	291	404	1025	1494	1935	1465	1159	668	286	88	8868
1957-58	0	35	274	568	1074	1343	1470	1471	1049	587	237	182	8290
1958-59	38	56	174	498	1054	1684	1903	1555	1051	649	342	63	9067
#1959-60	8	8	226	746	1329	1205	1762	1550	1526	708	276	87	9431
1960-61	23	9	223	538	1046	1701	1803	1292	936	811	375	41	8798
1961-62	2	9	301	527	1039	1681	1880	1660	1304	749	258	61	9471
1962-63	11	5	232	439	892	1492	1957	1112	1535	640	339	45	8699
1963-64	5	14	127	262	901	1748	1528	1331	1364	553	177	68	8078
1964-65	5	94	304	604	1065	1902	2051	1618	1582	695	324	66	10310
1965-66	19	59	477	544	1161	1415	2216	1649	1089	829	430	86	9974
1966-67	0	84	230	644	1253	1642	1723	1716	1178	795	480	96	9841
1967-68	65	49	158	645	1073	1544	1773	1592	952	628	375	79	8933
1968-69	31	66	201	565	1015	1736	2066	1460	1536	582	348	230	9836
1969-70	20	10	229	757	1028	1522	1996	1511	1431	773	407	55	9739
1970-71	16	26	251	560	1109	1723	2035	1461	1153	609	333	30	9306
1971-72	57	36	241	535	1052	1630	1931	1763	1269	718	231	60	9523
1972-73	25	41	261	695	1089	1897	1688	1361	893	701	328	79	9058
1973-74	32	3	309	451	1187	1698	1963	1550	1298	676	431	86	9684
1974-75	3	91	345	537	1066	1362	1630	1535	1438	867	265	79	9218
1975-76	14	22	284	492	1012	1550	1774	1257	1296	533	285	55	8574
1976-77	13	9	227	788	1247	1797	2119	1327	1015	466	74	30	9112
1977-78	7	95	211	549	1178	1817	2061	1721	1284	668	209	90	9890
1978-79	15	39	179	571	1262	1788	2147	1863	1377	861	457	64	10623
1979-80	3	45	139	689	1209	1367	1808	1644	1363	493	206	61	9027
1980-81	3	35	267	696	951	1616	1645	1266	971	574	298	84	8406
1981-82	14	10	250	627	881	1742	2236	1570	1298	725	222	187	9762
1982-83	0	66	257	589	1219	1359	1513	1206	1082	738	390	74	8493
1983-84	16	2	301	631	1004	2023	1714	1154	1280	576	344	52	9097
1984-85	15	13	339	541	1053	1715	1853	1514	988	550	172	179	8932
1985-86	13	72	329	625	1487	1895							

TABLE 5 COOLING DEGREE DAYS Base 65 deg. F   FARGO, NORTH DAKOTA

YEAR	JAN	FEB	MAR	APR	MAY	JUNE	JULY	AUG	SEP	OCT	NOV	DEC	TOTAL
1969	0	0	0	0	39	5	131	249	59	0	0	0	483
1970	0	0	0	0	2	146	233	180	95	3	0	0	659
1971	0	0	0	0	3	109	65	142	56	0	0	0	375
1972	0	0	0	0	74	125	135	217	23	0	0	0	574
1973	0	0	0	0	0	76	140	219	13	0	0	0	448
1974	0	0	0	0	9	75	281	75	3	1	0	0	444
1975	0	0	0	0	11	92	308	126	1	15	0	0	553
1976	0	0	0	0	4	164	228	283	83	4	0	0	766
1977	0	0	0	6	129	86	235	23	8	0	0	0	487
1978	0	0	0	0	31	86	165	176	146	0	0	0	604
1979	0	0	0	0	12	85	225	124	58	0	0	0	504
1980	0	0	0	18	102	89	222	119	31	1	0	0	582
1981	0	0	0	0	9	25	212	159	26	0	0	0	431
1982	0	0	0	2	11	20	189	179	39	0	0	0	440
1983	0	0	0	0	2	113	288	252	55	0	0	0	710
1984	0	0	0	0	18	81	196	279	24	4	0	0	602
1985	0	0	0	6	31	35	143	63	4	0	0	0	282

TABLE 6 SNOWFALL (inches)   FARGO, NORTH DAKOTA

SEASON	JULY	AUG	SEP	OCT	NOV	DEC	JAN	FEB	MAR	APR	MAY	JUNE	TOTAL
1956-57	0.0	0.0	T	0.0	0.5	3.2	3.3	4.4	0.7	2.8	0.0	0.0	14.9
1957-58	0.0	0.0	T		1.6	1.2	2.8	1.9	1.8		0.0	0.0	9.3
1958-59	0.0	0.0	0.0	6.5	0.9	2.0	3.5	1.2	T		0.0	0.0	14.2
1959-60	0.0	0.0	0.0	0.7	3.6	1.5	6.2	1.2	2.4	7.9	0.2	0.0	23.7
1960-61	0.0	0.0	0.0	T	11.0	3.6	1.6	2.3	T	2.8	T	0.0	21.3
1961-62	0.0	0.0	0.0	T	0.5	6.5	10.9	10.1	10.7	2.0	0.0	0.0	40.7
1962-63	0.0	0.0	0.0	0.2	3.1	2.5	1.3	3.9	5.3	6.0	0.1	0.0	22.4
1963-64	0.0	0.0	0.0	0.0	0.4	4.3	8.4	6.2	8.4	9.2	0.0	0.0	36.9
1964-65	0.0	0.0	0.0		2.7	1.7	11.9	1.2	13.1	2.5	T	0.0	33.1
1965-66	0.0	0.0	T	0.0	6.1	6.1	5.0	1.1	15.4	5.0	0.0	0.0	38.7
1966-67	0.0	0.0	0.0	1.2	1.3	3.4	15.1	2.6	4.9	5.0	T	0.0	33.5
1967-68	0.0	0.0	0.0	0.7	10.3	4.0	2.4	2.7	0.4	11.6	0.0	0.0	32.1
1968-69	0.0	0.0	0.0	0.8	3.9	11.4	14.5	7.8	3.0		0.0	0.0	41.4
1969-70	0.0	0.0	0.0	2.0	1.9	9.5	12.8	9.1	4.8	1.1	0.0	0.0	41.2
1970-71	0.0	0.0	0.0	0.9	6.4	8.3	15.1	4.8	1.8	1.0	0.0	0.0	38.3
1971-72	0.0	0.0	0.0	3.8	2.3	16.5	10.9	10.7	7.1	2.4	0.0	0.0	53.7
1972-73	0.0	0.0	T		3.8	18.5	1.7	1.4	1.4	2.4	0.0	0.0	30.9
1973-74	0.0	0.0	0.0		3.9	12.3	6.1	7.1	10.5	2.7	0.0	0.0	42.6
1974-75	0.0	0.0	0.0	0.4	1.0	5.1	18.3	5.9	18.7	3.7	T	0.0	53.1
1975-76	0.0	0.0	0.0	0.4	3.8	1.9	14.0	6.2	14.0	T	0.1	0.0	40.4
1976-77	0.0	0.0	0.0	0.1	2.7	5.5	12.6	10.7	4.6	2.1	0.0	0.0	38.3
1977-78	0.0	0.0	0.0	T	24.2	7.2	4.6	3.9	7.1	2.8	0.0	0.0	49.8
1978-79	0.0	0.0	T		8.5	11.7	7.8	19.5	4.3	2.7	0.8	0.0	55.3
1979-80	0.0	0.0	0.0	1.4	6.0	1.5	17.3	7.2	6.5	T	0.0	0.0	39.9
1980-81	0.0	0.0	0.0	0.5	1.1	4.6	2.1	4.5	0.3	T	0.0	0.0	13.1
1981-82	0.0	0.0	T	2.3	2.2	9.9	30.0	10.9	14.0	0.2	0.0	0.0	69.5
1982-83	0.0	0.0	0.0	0.0	6.8	0.3	3.8	2.0	7.4	2.9	T	0.0	23.2
1983-84	0.0	0.0	T		5.3	11.8	11.5	3.1	7.7	0.5	0.0	0.0	39.9
1984-85	0.0	0.0	T	T	1.4	7.4	3.7	3.1	12.6	T	0.0	0.0	28.2
1985-86	0.0	0.0	0.0	T	24.3	10.4							
Record Mean	0.0	0.0	T	0.7	5.1	6.9	7.8	5.7	6.8	3.2	0.1	0.0	36.3

See Reference Notes, relative to all above tables, on preceding page.

Williston lies in a flat valley at the junction of the Missouri River and Little Muddy Creek. The surrounding country is rolling. Hills to the east are highest, ranging from 250 to 300 feet in height at a distance of 5 to 7 miles. Across the Missouri River to the south, the bluffs are about 225 feet high at 4 miles distance.

Great extremes of temperatures are encountered, winters being cold, while summer days are usually warm. In winter, temperatures below zero are common and lows of −50 degrees have been recorded. When temperatures are lowest, however, the air is generally dry, with little or no wind and the weather is fine and invigorating. At the other extreme, temperatures above 100 degrees have been reached in all months from May to September. The low humidity that generally prevails on the hottest summer days keeps them from becoming oppressive.

The climate of Williston and vicinity is continental, semi-arid, characterized by marked season changes. Winter is the relatively dry season, with only about 1/2 inch of monthly precipitation occurring from November to February. There is considerably less than the average amount of snowfall for similar locations in the United States. Ice crystals, which rarely yield more than a trace of precipitation, are common in the cold months. Although snow has been observed every month except July and August, there is usually very little from April to November. Accumulated winter snow remains unmelted on the ground until about March. Summer precipitation is variable from year to year. The amount of rain occurring during the growing period is the most important element of climate for agricultural interests in the vicinity of Williston. Generally, considerably more precipitation occurs in the spring and summer months than in winter, but even so, the rainfall is just adequate for successful farming operations in normal years. A series of dry years, in addition to causing failure of crops, may result in erosion of the fertile topsoil by winds.

The growing season averages 131 days. It has ranged from 94 to 172 days during the period of record.

Clear and partly cloudy skies, nearly equally distributed, occur about 70 percent of the time. Heavy fog occurs on the average about ten times a year. Because of the northern latitude of Williston, it enjoys long hours of daylight in the spring and summer. Relatively little cloudiness occurs then, so that the duration of sunshine averages about two-thirds of the possible amount. These conditions are conducive to rapid growth of vegetation, making successful agricultural pursuits possible in spite of the relatively short growing season.

Summer storms are generally in the form of thunderstorms or rain showers, occasionally accompanied by hail and squally winds. Tornadoes are rare in this area. In the winter, cold waves and occasionally blizzard conditions occur. Cold waves result when extremely cold air advances southward from northwestern Canada. In blizzard conditions the advancing cold wave is accompanied by winds of gale force and the air is filled with fine, wind-driven snow. In extreme instances in the country, it becomes impossible for persons to ascertain their bearings or to remain alive many hours without shelter in such storms.

TABLE 1 — NORMALS, MEANS AND EXTREMES

WILLISTON, NORTH DAKOTA

LATITUDE: 64°49'N   LONGITUDE: 103°38'W   ELEVATION: FT. GRND 1900 BARO 1903   TIME ZONE: CENTRAL   WBAN: 94014

	(a)	JAN	FEB	MAR	APR	MAY	JUNE	JULY	AUG	SEP	OCT	NOV	DEC	YEAR
TEMPERATURE °F:														
Normals														
—Daily Maximum		17.5	25.6	36.3	54.0	67.5	76.5	84.0	82.5	70.2	58.2	38.1	25.5	53.0
—Daily Minimum		-4.3	3.5	14.3	29.6	41.5	51.1	56.0	53.7	43.0	32.0	17.0	4.3	28.5
—Monthly		6.6	14.6	25.3	41.8	54.5	63.8	70.0	68.1	56.6	45.1	27.6	14.9	40.7
Extremes														
—Record Highest	24	53	58	78	92	106	103	109	107	104	93	73	58	109
—Year		1981	1984	1978	1980	1980	1979	1980	1983	1983	1963	1981	1979	JUL 1980
—Record Lowest	24	-40	-41	-28	-15	17	30	34	35	17	0	-24	-50	-50
—Year		1966	1962	1962	1975	1980	1969	1967	1982	1974	1984	1985	1983	DEC 1983
NORMAL DEGREE DAYS:														
Heating (base 65°F)		1810	1411	1231	696	335	108	23	58	277	617	1122	1553	9241
Cooling (base 65°F)		0	0	0	0	10	72	178	155	25	0	0	0	440
% OF POSSIBLE SUNSHINE	22	52	57	61	61	63	67	76	75	65	59	44	48	61
MEAN SKY COVER (tenths)														
Sunrise - Sunset	24	6.9	7.0	6.6	6.7	6.5	6.0	4.7	4.9	5.7	6.2	6.8	6.8	6.2
MEAN NUMBER OF DAYS:														
Sunrise to Sunset														
—Clear	24	5.5	4.9	7.0	5.8	5.8	6.9	11.5	11.5	9.1	8.6	6.6	6.3	89.5
—Partly Cloudy	24	8.7	7.5	8.4	9.1	11.2	11.6	12.7	11.5	8.6	7.8	7.3	7.2	111.8
—Cloudy	24	16.8	15.9	15.6	15.1	14.0	11.5	6.8	8.0	12.3	14.6	16.0	17.5	163.9
Precipitation														
.01 inches or more	24	8.5	6.5	7.6	7.7	9.6	10.3	8.5	7.0	7.0	5.4	6.2	8.5	92.7
Snow, Ice pellets														
1.0 inches or more	24	2.3	1.8	2.0	1.6	0.1	0.0	0.0	0.0	0.2	0.5	1.7	2.8	13.0
Thunderstorms	21	0.0	0.0	0.1	0.7	2.8	7.4	8.2	5.9	2.0	0.0	0.0	0.0	27.2
Heavy Fog Visibility 1/4 mile or less	21	0.7	1.1	1.0	1.0	0.3	0.3	0.3	0.1	0.6	1.1	1.5	1.2	9.2
Temperature °F														
—Maximum														
70° and above	24	0.0	0.0	0.0	0.1	0.9	2.1	8.6	9.2	1.8	0.*	0.0	0.0	22.7
32° and below	24	23.6	16.7	10.4	1.0	0.0	0.0	0.0	0.0	0.0	0.8	10.0	21.5	84.0
—Minimum														
32° and below	24	30.8	27.9	29.3	18.3	4.1	0.2	0.0	0.0	3.0	15.3	28.6	31.0	188.5
0° and below	24	18.2	11.5	4.3	0.1	0.0	0.0	0.0	0.0	0.0	0.*	3.6	14.2	52.1
AVG. STATION PRESS. (mb)	13	948.8	947.7	945.2	945.9	944.3	944.2	946.1	946.2	947.3	946.6	947.5	947.3	946.4
RELATIVE HUMIDITY (%)														
Hour 03	18	77	79	79	71	67	71	66	64	69	73	79	80	73
Hour 09	24	77	80	82	80	77	81	79	78	80	80	81	79	80
Hour 15 (Local Time)	24	72	71	65	52	47	50	45	44	51	55	68	73	58
Hour 21	24	74	71	62	48	42	45	38	36	44	53	70	76	55
PRECIPITATION (inches):														
Water Equivalent														
—Normal		0.55	0.50	0.57	1.29	1.85	2.68	1.83	1.42	1.37	0.74	0.50	0.55	13.85
—Maximum Monthly		1.42	1.48	2.26	3.31	7.38	5.92	6.20	3.38	3.06	3.56	1.14	1.43	7.38
—Year		1967	1967	1975	1967	1965	1964	1963	1968	1973	1971	1975	1982	MAY 1965
—Minimum Monthly	24	0.03	0.07	0.01	0.03	0.15	0.87	0.49	0.07	0.11	T	0.04	0.09	T
—Year		1973	1983	1966	1983	1980	1983	1976	1971	1963	1965	1969	1979	OCT 1965
—Maximum in 24 hrs	24	0.51	0.43	0.92	2.04	2.05	2.20	5.03	2.45	2.24	2.21	0.80	0.88	5.03
—Year		1980	1962	1985	1967	1965	1964	1963	1972	1971	1971	1974	1982	JUL 1963
Snow, Ice pellets														
—Maximum Monthly	24	24.3	16.5	30.9	22.2	15.5				4.0	14.2	14.1	15.2	30.9
—Year		1982	1972	1975	1970	1983				1984	1985	1975	1978	MAR 1975
—Maximum in 24 hrs	24	8.4	4.3	9.7	12.1	14.6				4.0	10.5	7.9	10.1	14.6
—Year		1976	1962	1985	1984	1983				1984	1985	1975	1978	MAY 1983
WIND:														
Mean Speed (mph)	21	9.9	9.8	10.4	11.3	11.3	10.2	9.4	9.7	10.1	10.2	9.2	9.8	10.1
Prevailing Direction through 1963		W	NE	NW	SE	SE	SE	SE	SW	SW	SW	SW	SW	SW
Fastest Obs. 1 Min.														
—Direction (!!!)	6	32	32	27	25	27	28	35	28	29	34	30	31	27
—Speed (MPH)	6	41	33	44	40	46	39	40	44	38	41	35	45	46
—Year		1983	1981	1982	1981	1984	1980	1984	1985	1985	1982	1985	1985	MAY 1984
Peak Gust														
—Direction (!!!)	2	NW	SE	NW	N	W	NW	N	W	W	NW	NW	NW	NW
—Speed (mph)	2	48	44	53	62	51	58	55	55	51	43	46	63	63
—Date		1985	1985	1984	1984	1985	1985	1984	1985	1985	1985	1985	1985	DEC 1985

See Reference Notes to this table on the following page

TABLE 2 PRECIPITATION (inches) WILLISTON, NORTH DAKOTA

YEAR	JAN	FEB	MAR	APR	MAY	JUNE	JULY	AUG	SEP	OCT	NOV	DEC	ANNUAL
1956	0.14	0.10	0.76	0.11	1.86	3.24	2.20	1.03	0.58	0.40	0.27	0.51	11.20
1957	0.60	0.19	0.44	1.69	0.90	2.38	1.83	0.76	0.71	0.74	0.84	0.30	11.38
1958	0.22	1.06	0.28	0.33	0.06	1.94	0.81	0.70	1.11	1.11	1.50	0.51	8.79
1959	0.23	0.51	0.17	0.30	0.78	4.93	0.17	1.08	3.74	1.38	1.71	0.55	15.55
1960	0.61	0.39	0.28	1.11	1.73	5.22	0.89	2.98	0.05	0.02	0.64	0.45	14.37
1961	0.09	1.25	0.12	2.39	0.58	0.43	1.80	0.14	2.49	0.07	0.09	0.50	9.95
#1962	0.47	0.95	0.60	0.38	3.50	3.14	4.10	1.97	0.43	2.08	0.38	0.27	18.27
1963	0.35	0.53	0.70	2.33	2.05	2.05	6.20	1.62	0.11	T	0.27	0.30	17.46
1964	0.45	0.24	0.94	1.08	1.10	5.92	2.24	1.14	0.40	0.27	0.65	1.28	15.71
1965	0.71	0.10	0.48	0.99	7.38	2.03	1.52	1.23	2.20	T	0.21	0.61	17.46
1966	0.42	0.31	0.01	0.94	1.75	0.92	1.92	1.80	0.40	0.36	0.30	0.33	9.46
1967	1.42	1.48	0.70	3.31	0.18	0.91	1.05	0.67	1.15	1.15	0.56	1.28	12.68
1968	0.53	0.16	0.10	0.78	1.05	3.68	0.63	3.38	0.85	0.31	0.37	1.09	12.93
1969	1.03	0.60	0.45	0.93	0.42	2.73	3.47	0.54	1.14	0.79	0.04	0.48	12.62
1970	0.56	0.10	0.59	3.20	3.00	2.07	3.11	0.53	2.16	0.56	0.81	0.49	17.18
1971	0.79	0.15	0.45	0.91	2.28	3.88	0.07	1.92	2.48	3.56	0.40	0.42	17.31
1972	0.43	1.37	0.91	1.91	4.10	1.18	2.14	3.21	1.32	0.43	0.15	0.80	17.95
1973	0.03	0.12	0.52	1.79	1.06	2.30	1.01	0.64	3.06	0.28	0.75	0.56	12.12
1974	0.47	0.32	0.91	0.77	3.73	1.88	1.29	2.46	0.17	1.08	0.13	0.30	13.51
1975	0.14	0.13	2.26	2.81	0.86	3.41	2.35	1.50	1.77	1.29	1.14	1.20	18.86
1976	0.58	0.21	0.74	1.08	0.51	2.45	0.49	1.06	0.96	0.28	0.31	0.49	9.16
1977	0.97	0.30	0.02	0.53	2.94	1.30	1.28	1.55	2.30	0.41	0.97	0.86	13.43
1978	0.30	0.52	0.44	0.65	4.17	2.53	2.62	0.51	2.09	0.18	0.89	0.98	15.88
1979	0.12	0.81	1.01	1.69	1.82	2.93	1.78	0.33	0.38	0.23	0.06	0.09	11.25
1980	0.58	0.28	0.16	0.40	0.15	1.80	0.54	1.83	2.24	1.62	0.48	0.72	10.80
1981	0.08	0.21	0.18	0.39	1.18	3.58	2.47	0.87	0.57	0.51	0.51	0.36	10.91
1982	1.28	0.45	1.34	1.02	1.79	2.77	1.17	1.96	1.86	1.91	0.07	1.43	17.05
1983	0.35	0.07	1.07	0.03	1.98	0.87	2.12	0.87	0.51	0.30	0.78	0.57	9.52
1984	0.81	0.12	0.90	1.48	0.47	1.87	1.87	0.93	0.65	0.49	0.33	0.27	10.56
1985	0.13	0.24	1.07	1.53	1.08	1.12	1.40	1.96	1.08	1.86	0.58	0.64	12.69
Record Mean	0.54	0.44	0.68	1.16	1.85	3.13	1.92	1.44	1.15	0.79	0.55	0.56	14.21

TABLE 3 AVERAGE TEMPERATURE (deg. F) WILLISTON, NORTH DAKOTA

YEAR	JAN	FEB	MAR	APR	MAY	JUNE	JULY	AUG	SEP	OCT	NOV	DEC	ANNUAL
1956	6.8	12.6	26.2	37.6	54.1	68.5	67.2	67.9	56.3	47.4	32.1	19.6	41.4
1957	1.7	15.2	29.3	41.0	56.7	62.6	75.5	68.1	57.3	43.8	31.0	25.6	42.3
1958	22.3	14.7	25.5	45.4	60.4	60.5	66.1	72.3	58.5	48.1	27.9	16.2	43.2
1959	5.7	5.9	34.5	42.9	51.9	66.7	72.3	69.7	55.0	39.3	23.2	27.5	41.2
1960	10.2	12.7	21.6	42.6	55.9	62.0	74.1	69.6	60.2	47.6	29.1	19.1	42.1
1961	19.2	22.4	35.3	38.7	55.4	71.1	71.3	74.9	51.9	46.7	29.9	9.0	43.8
#1962	10.7	8.8	21.0	45.2	51.8	65.3	65.5	68.3	56.6	48.8	36.1	21.4	41.6
1963	2.7	18.8	35.6	42.5	53.6	65.0	72.5	69.2	63.4	55.0	31.4	12.7	43.5
1964	16.2	24.4	21.3	45.4	57.4	62.9	72.7	66.3	53.5	46.4	23.6	0.0	40.8
1965	3.0	9.3	12.4	40.8	52.9	63.4	70.4	67.7	45.4	48.9	26.7	20.9	38.5
1966	-2.7	9.6	32.8	36.8	54.8	63.5	73.0	65.3	60.1	45.7	21.6	17.9	39.9
1967	11.7	10.5	23.3	38.9	51.1	62.5	70.3	69.7	62.0	45.2	29.7	13.1	40.7
1968	8.8	15.6	35.7	42.0	52.0	61.4	69.4	64.8	58.0	45.3	30.6	7.9	40.9
1969	-4.2	12.0	20.8	49.0	55.0	57.8	67.9	72.8	60.8	39.4	32.5	18.6	40.2
1970	3.6	16.9	21.6	37.9	52.5	67.9	72.5	70.0	56.4	41.6	25.3	10.4	39.7
1971	1.6	14.9	28.2	43.1	54.9	64.8	66.3	73.1	56.0	43.0	29.9	9.6	40.4
1972	3.3	8.2	25.9	41.3	56.3	65.8	64.0	69.3	52.6	39.4	28.5	6.7	38.4
1973	16.9	21.5	37.2	41.2	54.8	64.4	68.4	72.8	55.5	47.2	20.2	13.0	42.8
1974	6.8	19.9	26.0	43.0	50.3	65.1	74.3	63.2	53.0	47.0	29.7	23.0	41.8
1975	12.0	12.3	21.9	34.7	52.9	62.1	73.0	65.3	55.0	44.4	28.4	15.7	39.8
1976	10.8	22.6	24.6	46.2	56.3	65.0	70.5	70.5	59.2	40.9	24.3	14.5	42.1
1977	-0.3	25.1	33.9	49.1	62.3	66.4	71.1	61.6	56.1	46.2	24.4	8.3	42.0
1978	-2.4	7.7	27.8	43.0	57.2	63.8	68.6	67.5	60.2	46.4	20.8	8.3	39.1
1979	-2.2	1.7	23.0	34.6	49.2	65.0	70.3	66.9	61.1	46.7	25.1	23.8	38.8
1980	7.7	14.9	25.6	49.7	60.6	67.1	73.2	65.7	58.1	46.6	33.2	16.2	43.6
1981	21.8	23.3	38.1	49.1	59.2	63.8	73.8	74.3	61.8	45.4	34.6	16.1	46.8
1982	-5.8	11.8	26.3	41.8	54.9	64.3	71.7	70.3	56.4	44.4	24.0	16.0	39.7
1983	20.0	25.0	30.0	41.2	51.1	63.7	74.9	78.0	56.3	47.2	30.3	-4.5	42.8
1984	15.2	30.9	28.3	47.4	54.5	63.4	73.0	73.7	50.2	41.0	27.2	5.5	42.5
1985	7.0	13.2	32.4	47.0	58.9	58.8	71.2	64.6	51.2	42.8	13.2	11.2	39.3
Record Mean	7.8	12.1	25.0	42.7	54.2	63.3	69.9	67.8	56.5	44.7	27.5	14.6	40.5
Max	18.0	22.5	35.3	54.4	66.4	75.1	83.0	81.5	69.7	56.9	37.4	24.3	52.1
Min	-2.5	1.8	14.7	31.0	41.9	51.5	56.8	54.2	43.4	32.5	17.6	4.9	29.0

REFERENCE NOTES FOR TABLES 1, 2, 3 and 6   (WILLISTON, ND)

GENERAL

T - TRACE AMOUNT
BLANK ENTRIES DENOTE MISSING/UNREPORTED DATA.
# INDICATES A STATION OR INSTRUMENT RELOCATION.

SPECIFIC

TABLE 1

(a) - LENGTH OF RECORD IN YEARS. ALTHOUGH
INDIVIDUAL MONTHS MAY BE MISSING.
* LESS THAN .05

NORMALS — BASED ON THE 1951-1980 RECORD PERIOD.
EXTREMES — DATES ARE THE MOST RECENT OCCURRENCE.
WIND DIR. — NUMERALS SHOW TENS OF DEGREES
CLOCKWISE FROM TRUE NORTH.
"00" INDICATES CALM.
RESULTANT WIND DIRECTIONS ARE GIVEN TO WHOLE DEGREES.

EXCEPTIONS

TABLE 1

1. PRIOR TO MARCH 1967, THUNDERSTORMS AND HEAVY
FOG MAY BE INCOMPLETE, DUE TO PART-TIME OPERA-
TIONS.

TABLES 2, 3, and 6

RECORD MEANS ARE THROUGH THE CURRENT YEAR,
BEGINNING IN   1879 FOR TEMPERATURE
1879 FOR PRECIPITATION
1962 FOR SNOWFALL

TABLE 4 HEATING DEGREE DAYS Base 65 deg. F WILLISTON, NORTH DAKOTA

SEASON	JULY	AUG	SEP	OCT	NOV	DEC	JAN	FEB	MAR	APR	MAY	JUNE	TOTAL
1956-57	40	58	264	537	981	1401	1963	1391	1098	711	262	103	8809
1957-58	0	59	235	668	1014	1216	1320	1407	1217	582	189	173	8080
1958-59	55	36	227	517	1104	1508	1839	1653	941	658	403	65	9006
1959-60	10	18	312	788	1153	1695	1511	1342	917	667	292	117	9154
1960-61	10	26	209	532	1070	1417	1414	1186	917	781	321	22	7905
#1961-62	8	7	396	563	1046	1732	1681	1572	1358	587	399	65	9414
1962-63	45	44	249	494	858	1346	1936	1292	905	671	368	48	8256
1963-64	0	24	93	314	1001	1620	1507	1171	1347	581	265	123	8046
1964-65	0	86	349	567	1236	2017	1920	1557	1624	717	370	87	10530
1965-66	3	62	580	494	1142	1358	2098	1547	992	839	323	128	9566
1966-67	0	94	190	593	1297	1455	1649	1525	1285	775	435	99	9397
1967-68	44	27	151	605	1056	1607	1741	1425	900	683	397	146	8782
1968-69	26	90	225	603	1028	1769	2149	1361	1480	475	326	223	9755
1969-70	25	10	166	786	966	1431	1903	1340	1337	807	382	30	9183
1970-71	2	17	286	719	1182	1689	1964	1400	1135	649	310	62	9415
1971-72	42	5	283	674	1063	1713	1913	1644	1209	702	291	50	9589
1972-73	75	25	380	789	1089	1805	1801	1213	1204	708	316	78	8821
1973-74	14	1	291	546	1338	1608	1801	1258	1204	653	448	80	9242
1974-75	1	100	354	553	1052	1294	1635	1470	1329	904	371	117	9180
1975-76	1	60	299	632	1092	1522	1675	1226	1247	556	267	87	8664
1976-77	10	11	203	740	1215	1563	2024	1112	959	469	149	46	8501
1977-78	8	131	267	577	1212	1759	2091	1600	1148	655	252	91	9791
1978-79	21	39	213	569	1322	1756	2085	1774	1294	906	495	76	10550
1979-80	7	42	145	560	1190	1272	1775	1447	1216	460	237	43	8394
1980-81	5	52	221	567	947	1511	1330	1162	825	473	205	70	7368
1981-82	4	9	149	600	903	1509	2198	1485	1191	692	322	79	9141
1982-83	2	46	278	633	1223	1516	1389	1115	1076	708	436	101	8523
1983-84	1	1	300	544	1031	2158	1538	983	1131	520	343	92	8642
1984-85	0	25	444	741	1128	1845	1793	1446	1001	535	210	201	9369
1985-86	21	73	415	683	1553	1664							

TABLE 5 COOLING DEGREE DAYS Base 65 deg. F WILLISTON, NORTH DAKOTA

YEAR	JAN	FEB	MAR	APR	MAY	JUNE	JULY	AUG	SEP	OCT	NOV	DEC	TOTAL
1969	0	0	0	0	23	14	119	258	50	0	0	0	464
1970	0	0	0	0	3	123	244	176	34	0	0	0	580
1971	0	0	0	0	3	62	90	262	23	0	0	0	440
1972	0	0	0	0	28	80	52	167	15	0	0	0	342
1973	0	0	0	0	7	72	125	250	13	0	0	0	467
1974	0	0	0	2	0	92	295	51	0	0	0	0	440
1975	0	0	0	0	2	39	253	77	5	0	0	0	376
1976	0	0	0	0	4	95	185	185	37	1	0	0	507
1977	0	0	0	0	74	96	204	30	6	0	0	0	410
1978	0	0	0	0	19	60	142	125	75	0	0	0	421
1979	0	0	0	0	12	83	180	107	33	0	0	0	415
1980	0	0	0	7	107	114	265	84	22	4	0	0	603
1981	0	0	0	2	31	41	285	301	60	0	0	0	720
1982	0	0	0	0	13	64	217	215	29	0	0	0	538
1983	0	0	0	0	8	68	314	413	45	0	0	0	848
1984	0	0	0	0	21	54	251	304	8	0	0	0	638
1985	0	0	0	0	26	21	218	68	6	0	0	0	339

TABLE 6 SNOWFALL (inches) WILLISTON, NORTH DAKOTA

SEASON	JULY	AUG	SEP	OCT	NOV	DEC	JAN	FEB	MAR	APR	MAY	JUNE	TOTAL
1956-57	0.0	0.0	T	0.0	0.2	5.2	7.1	2.1	3.2	2.0	0.0	0.0	19.8
1957-58	0.0	0.0	0.0	1.5	5.7	4.2	1.6	11.2	2.9	T	0.0	0.0	27.1
1958-59	0.0	0.0	T	0.8	14.1	5.1	2.5	5.2	1.3	0.8	0.2	0.0	30.0
1959-60	0.0	0.0	0.0	10.0	14.7	3.3	6.0	3.4	1.8	4.2	0.0	0.0	43.4
1960-61	0.0	0.0	0.0	0.0	4.5	4.5	0.9	11.5	0.3	11.4	0.1	0.0	33.2
#1961-62	0.0	0.0	1.5	0.5	0.5	5.8	4.7	9.5	5.4	2.7	0.0	0.0	30.6
1962-63	0.0	0.0	0.0	5.0	2.4	2.6	3.5	4.1	3.6	10.5	T	0.0	31.7
1963-64	0.0	0.0	0.0	0.0	2.6	3.0	5.0	2.4	9.5	5.4	0.0	0.0	27.9
1964-65	0.0	0.0	0.0	1.3	6.2	12.8	7.1	0.9	5.0	1.6	1.4	0.0	36.3
1965-66	0.0	0.0	3.0	0.0	2.2	6.7	4.3	3.3	0.1	6.2	0.5	0.0	26.3
1966-67	0.0	0.0	T	2.0	3.4	3.7	16.6	15.0	6.2	12.4	1.1	0.0	60.4
1967-68	0.0	0.0	0.0	T	1.6	5.6	8.9	1.2	0.3	2.9	T	0.0	20.5
1968-69	0.0	0.0	0.0	T	2.6	11.6	10.7	6.0	4.5	T	T	0.0	35.4
1969-70	0.0	0.0	0.0	1.5	T	5.3	5.9	1.0	7.6	22.2	0.0	0.0	43.5
1970-71	0.0	0.0	T	T	6.4	5.6	10.5	2.5	1.6	1.9	0.7	0.0	29.2
1971-72	0.0	0.0	T	1.8	4.3	5.4	6.5	16.5	6.4	2.7	0.0	0.0	43.6
1972-73	0.0	0.0	3.0	3.9	1.5	11.5	0.6	1.8	2.3	1.6	0.0	0.0	26.2
1973-74	0.0	0.0	0.0	T	6.4	5.5	3.4	5.7	9.3	1.4	T	0.0	31.7
1974-75	0.0	0.0	0.0	0.1	2.2	4.4	6.3	1.3	30.9	10.3	0.0	0.0	52.8
1975-76	0.0	0.0	0.0	4.5	14.1	14.1	10.3	2.0	14.6	0.4	0.0	0.0	60.0
1976-77	0.0	0.0	0.0	1.9	2.9	6.9	11.1	3.1	T	2.0	0.0	0.0	27.9
1977-78	0.0	0.0	0.0	T	8.4	10.0	3.9	5.7	4.6	1.4	T	0.0	34.0
1978-79	0.0	0.0	0.0	T	9.4	15.2	2.9	14.1	10.3	10.1	T	0.0	62.0
1979-80	0.0	0.0	0.0	T	1.2	0.3	10.4	7.9	2.1	3.5	T	0.0	25.4
1980-81	0.0	0.0	0.0	1.1	3.3	10.4	1.2	3.1	T	0.0	T	0.0	19.1
1981-82	0.0	0.0	0.0	0.3	4.6	5.7	24.3	8.2	17.0	9.8	0.5	0.0	70.4
1982-83	0.0	0.0	T	T	0.8	13.4	1.7	1.6	9.2	0.1	15.5	0.3	42.3
1983-84	0.0	0.0	1.0	T	6.0	9.2	7.2	0.5	7.9	13.5	0.3	0.0	45.6
1984-85	0.0	0.0	4.0	4.5	3.5	6.5	1.8	3.1	11.4	1.5	0.0	0.0	36.3
1985-86	0.0	0.0	0.1	14.2	11.3	9.2							
Record Mean	0.0	0.0	0.5	1.8	4.5	7.7	6.9	5.0	7.1	5.2	0.8	0.0	39.4

See Reference Notes, relative to all above tables, on preceding page.

The station at the Akron-Canton Airport is located about midway between Akron and Canton, a few miles south of the crest separating the Lake Erie and Muskingum River drainage areas. Precipitation at the station and southward drains through the Muskingum River into the Ohio, while northward of the crest the Cuyahoga and other streams flow into Lake Erie. The terrain is rolling with highest elevations near 1,300 feet above sea level and many small lakes provide water for local industry as well as recreational facilities for the densely populated region. The area is mainly industrial, agricultural operations having diminished rapidly in recent years.

Lake Erie has considerable influence on the area weather, tempering cold air masses during the late fall and winter, as well as contributing to the formation of brief, but heavy snow squalls until the lake freezes over.

The arrival of spring is late in this area, but has the good effect of retarding plant growth and allowing growing of normally frost-susceptible fruits. Summers are moderately warm, but quite humid, while the months of September, October, and sometimes November are usually pleasant although with considerable morning fog. The average last occurrence of freezing temperatures in spring is the end of April, and the first occurrence in fall is late October. In past years, growing seasons for most vegetation has varied from 120 to 211 days. Temperatures and occurences of frost vary widely over the area because of the hilly terrain. Due to the influence of Lake Erie, snowfall is usually much heavier north of the station.

TABLE 1   NORMALS, MEANS AND EXTREMES

AKRON, OHIO

LATITUDE: 40°55'N   LONGITUDE: 81°26'W   ELEVATION: FT. GRND 1208 BARO 01242   TIME ZONE: EASTERN   WBAN: 14895

	(a)	JAN	FEB	MAR	APR	MAY	JUNE	JULY	AUG	SEP	OCT	NOV	DEC	YEAR
TEMPERATURE °F:														
Normals														
−Daily Maximum		32.9	35.6	45.8	59.2	69.8	78.7	82.3	80.9	74.3	62.6	48.9	37.5	59.0
−Daily Minimum		17.2	18.8	27.5	38.0	47.7	56.8	61.0	59.9	53.2	42.4	33.0	23.0	39.9
−Monthly		25.1	27.2	36.7	48.6	58.8	67.8	71.7	70.4	63.8	52.5	41.0	30.3	49.5
Extremes														
−Record Highest	37	70	68	79	85	92	100	100	98	99	86	80	76	100
−Year		1950	1961	1977	1985	1978	1952	1980	1953	1953	1952	1961	1982	JUL 1980
−Record Lowest	37	−24	−13	−3	10	24	32	43	41	32	20	−1	−15	−24
−Year		1985	1979	1980	1964	1966	1972	1979	1982	1956	1952	1958	1983	JAN 1985
NORMAL DEGREE DAYS:														
Heating (base 65°F)		1237	1058	877	492	228	37	0	14	108	394	720	1076	6241
Cooling (base 65°F)		0	0	0	0	36	121	208	181	72	7	0	0	625
% OF POSSIBLE SUNSHINE														
MEAN SKY COVER (tenths)														
Sunrise – Sunset	37	8.0	7.7	7.5	7.1	6.7	6.2	5.9	5.9	5.8	6.0	7.7	8.1	6.9
MEAN NUMBER OF DAYS:														
Sunrise to Sunset														
−Clear	37	3.1	3.5	4.3	4.9	6.3	6.6	7.1	7.8	8.9	9.2	4.1	2.8	68.5
−Partly Cloudy	37	5.9	6.0	6.6	7.5	9.2	11.2	13.0	11.9	9.1	7.6	6.2	5.7	99.9
−Cloudy	37	22.1	18.7	20.1	17.6	15.5	12.2	10.9	11.3	12.0	14.2	19.7	22.5	196.8
Precipitation														
.01 inches or more	37	16.1	14.5	16.2	14.2	12.9	10.9	10.9	9.8	9.2	10.1	13.7	15.6	154.0
Snow,Ice pellets														
1.0 inches or more	37	4.1	3.1	2.8	0.7	0.*	0.0	0.0	0.0	0.0	0.2	1.4	2.9	15.1
Thunderstorms	37	0.3	0.4	2.2	3.9	5.6	7.4	8.0	6.2	3.5	1.4	0.7	0.3	39.8
Heavy Fog Visibility														
1/4 mile or less	37	2.9	2.8	2.2	1.5	1.6	1.5	1.7	2.6	2.3	2.3	2.0	3.1	26.6
Temperature °F														
−Maximum														
90° and above	22	0.0	0.0	0.0	0.0	0.1	1.5	3.2	1.3	0.5	0.0	0.0	0.0	6.5
32° and below	22	15.7	11.4	3.6	0.2	0.0	0.0	0.0	0.0	0.0	0.0	1.4	9.0	41.3
−Minimum														
32° and below	22	28.4	23.8	21.2	9.2	1.0	0.*	0.0	0.0	0.0	3.4	14.2	24.5	125.7
0° and below	22	3.7	2.3	0.1	0.0	0.0	0.0	0.0	0.0	0.0	0.0	0.0	0.7	6.8
AVG. STATION PRESS.(mb)	13	972.2	972.8	970.7	970.9	970.6	971.7	972.9	974.2	974.5	974.6	973.4	972.9	972.6
RELATIVE HUMIDITY (%)														
Hour 01	22	75	74	71	68	73	78	81	83	82	76	75	75	76
Hour 07 (Local Time)	22	77	76	75	74	76	80	83	87	86	81	78	77	79
Hour 13	22	68	64	60	53	54	56	56	59	59	57	65	69	60
Hour 19	22	71	67	63	57	58	60	62	67	70	67	70	73	65
PRECIPITATION (inches):														
Water Equivalent														
−Normal		2.56	2.18	3.37	3.26	3.55	3.27	4.01	3.31	2.96	2.24	2.54	2.65	35.90
−Maximum Monthly	37	8.70	5.24	8.83	6.46	9.60	7.06	11.43	8.19	7.85	8.42	9.39	4.89	11.43
−Year		1950	1956	1964	1981	1956	1970	1958	1974	1979	1954	1985	1951	JUL 1958
−Minimum Monthly	37	0.71	0.48	1.04	0.91	1.05	1.01	1.56	0.49	0.20	0.45	0.62	0.31	0.20
−Year		1961	1978	1958	1971	1977	1967	1965	1970	1960	1953	1976	1955	SEP 1960
−Maximum in 24 hrs	37	2.99	2.57	3.29	1.98	3.18	2.87	4.18	3.00	6.30	2.77	2.66	1.66	6.30
−Year		1959	1959	1964	1957	1985	1970	1958	1975	1979	1954	1985	1974	SEP 1979
Snow,Ice pellets														
−Maximum Monthly	37	37.5	21.1	20.9	15.8	3.2				T	6.8	22.3	29.4	37.5
−Year		1978	1984	1960	1961	1966				1965	1952	1950	1974	JAN 1978
−Maximum in 24 hrs	37	10.9	12.3	10.7	5.9	3.2				T	3.9	7.4	17.9	17.9
−Year		1966	1984	1973	1961	1966				1965	1952	1950	1974	DEC 1974
WIND:														
Mean Speed (mph)	37	11.6	11.2	11.6	11.0	9.2	8.4	7.6	7.3	8.0	9.1	10.9	11.4	9.8
Prevailing Direction through 1963		SW	NW	NW	SW	SW	S	SW	SW	S	S	S	S	S
Fastest Obs. 1 Min.														
−Direction (!!!)	24	22	25	22	30	32	36	25	29	36	22	20	24	25
−Speed (MPH)	24	44	51	40	40	46	35	35	35	40	30	35	40	51
−Year		1978	1962	1964	1968	1967	1971	1983	1980	1962	1983	1963	1968	FEB 1962
Peak Gust														
−Direction (!!!)	2	SW	N	W	SW	NW	W	SW	W	W	SW	W	SW	SW
−Speed (mph)	2	47	39	51	60	41	40	44	38	38	35	40	58	60
−Date		1985	1984	1985	1984	1985	1985	1985	1985	1984	1985	1985	1985	APR 1984

See Reference Notes to this table on the following page.

TABLE 2 PRECIPITATION (inches) AKRON, OHIO

YEAR	JAN	FEB	MAR	APR	MAY	JUNE	JULY	AUG	SEP	OCT	NOV	DEC	ANNUAL
1956	1.44	5.24	4.23	3.51	9.60	3.68	5.12	6.12	2.61	1.31	1.07	3.13	47.06
1957	1.87	1.29	1.58	5.23	3.10	4.86	2.60	0.77	3.33	2.03	2.03	3.94	32.14
1958	2.19	1.05	1.04	4.20	3.24	3.57	11.43	3.07	3.84	0.64	2.32	0.79	37.38
1959	5.46	4.15	2.38	3.43	4.51	1.95	4.91	3.79	3.98	4.79	3.18	2.71	45.24
1960	2.63	2.32	1.63	1.79	3.15	4.86	4.15	2.60	0.20	1.02	1.46	1.27	27.08
1961	0.71	2.94	3.29	5.07	2.64	2.41	6.39	2.14	2.17	3.10	1.59	1.59	34.35
1962	2.57	2.15	1.83	1.84	1.84	1.38	3.81	1.95	3.82	2.72	2.11	2.44	28.98
1963	1.08	1.08	4.72	2.82	2.07	2.45	1.95	2.25	1.55	0.48	1.86	1.48	23.79
1964	2.19	1.81	8.83	6.06	3.29	1.17	3.54	5.65	0.58	1.26	1.90	4.54	40.82
1965	5.25	2.66	2.68	1.57	2.55	2.04	1.56	4.16	3.12	3.80	2.41	1.75	33.55
1966	2.03	2.20	1.95	3.32	2.30	1.92	2.82	2.95	2.01	5.05	2.11	2.11	29.64
1967	1.14	2.47	3.07	3.76	4.43	1.01	2.11	3.01	3.69	1.98	2.80	2.41	31.88
1968	2.88	0.48	2.71	2.35	6.54	2.40	4.13	1.61	3.25	1.38	3.86	3.88	35.47
1969	2.57	0.88	1.91	3.88	3.78	2.94	6.08	1.68	1.51	2.15	2.85	2.73	32.96
1970	1.48	1.84	2.46	3.94	3.30	7.06	4.22	0.49	3.46	4.24	3.11	2.63	38.23
1971	1.75	3.89	2.41	0.91	2.65	4.61	2.65	2.86	3.06	1.07	2.02	4.32	32.20
1972	1.33	2.35	4.09	3.96	2.88	3.20	7.26	6.47	1.54	4.27	3.33	4.88	43.88
1973	1.67	2.33	4.36	3.45	3.97	3.62	3.88	2.01	2.09	4.60	2.43	2.38	36.79
1974	2.93	1.85	5.68	2.50	4.85	2.74	2.42	8.19	1.81	1.51	3.41	4.32	42.21
1975	3.79	3.58	3.45	1.76	3.57	2.76	2.22	7.47	5.27	2.57	1.67	2.88	40.99
1976	3.16	2.99	3.49	1.64	1.15	3.94	9.00	2.01	3.90	2.31	0.62	1.42	35.63
1977	1.24	1.13	3.62	4.12	1.05	5.43	4.75	4.58	3.72	2.04	4.19	4.09	39.96
1978	3.69	0.48	2.54	2.21	4.41	3.31	2.76	4.14	1.98	2.58	1.06	4.20	33.36
1979	3.17	2.11	1.74	3.58	5.06	2.48	1.96	4.28	7.85	1.51	3.13	2.44	39.31
1980	1.58	1.64	4.71	2.97	5.42	3.76	5.43	4.94	1.76	1.86	1.94	2.15	38.16
1981	0.80	4.63	1.95	6.46	4.84	5.76	4.13	1.47	3.20	1.66	1.92	3.35	40.17
1982	4.71	1.80	3.81	1.18	3.13	4.23	2.13	1.47	2.98	0.85	4.94	3.45	34.68
1983	1.58	1.38	3.76	5.13	4.53	2.03	3.38	3.06	2.73	3.22	4.16	3.43	38.39
1984	1.15	3.02	3.00	2.82	5.44	1.64	3.02	3.77	2.62	2.66	3.41	2.58	35.13
1985	1.33	1.80	4.64	1.14	6.49	2.72	2.56	3.47	0.72	1.55	9.39	2.80	38.61
Record Mean	2.56	2.21	3.14	3.13	3.82	3.61	3.91	3.37	3.18	2.44	2.56	2.54	36.49

TABLE 3 AVERAGE TEMPERATURE (deg. F) AKRON, OHIO

YEAR	JAN	FEB	MAR	APR	MAY	JUNE	JULY	AUG	SEP	OCT	NOV	DEC	ANNUAL
1956	25.5	30.1	34.2	44.8	56.2	67.0	69.7	69.5	58.7	55.8	40.3	36.4	49.0
1957	21.5	31.4	37.1	49.9	58.7	68.8	70.6	69.3	62.7	48.9	40.5	33.4	49.4
1958	25.2	21.0	34.6	49.5	56.9	63.2	71.8	68.5	62.8	51.7	42.1	20.6	47.3
1959	22.6	28.9	34.8	48.6	62.8	67.9	72.3	74.2	66.9	52.4	37.3	33.4	50.2
1960	30.2	27.4	24.1	52.5	57.1	66.7	68.7	71.5	66.4	52.8	42.4	21.9	48.5
1961	21.5	32.7	39.3	42.7	53.8	65.6	71.0	71.3	68.8	55.0	41.4	29.0	49.4
1962	23.9	26.6	34.8	48.3	65.7	69.8	70.4	71.3	60.9	53.6	40.7	24.4	49.2
#1963	19.2	19.0	39.4	48.6	56.4	68.3	71.2	66.9	61.3	59.0	42.7	20.5	47.7
1964	28.8	25.0	37.4	49.9	61.3	68.0	72.9	67.1	63.4	50.0	44.6	31.9	50.0
1965	25.2	26.2	32.3	47.3	64.6	66.9	69.8	69.2	66.1	50.5	42.3	36.4	49.7
1966	22.0	28.4	38.7	46.0	53.9	69.4	73.5	69.7	61.5	50.5	41.7	31.0	48.9
1967	32.3	24.9	38.2	50.6	53.2	72.4	70.7	68.2	60.9	53.2	36.7	34.2	49.6
1968	23.0	23.4	40.1	52.2	57.1	69.8	72.9	72.9	66.4	54.6	42.9	30.4	50.5
1969	27.5	30.4	33.9	50.1	59.0	65.9	72.1	70.2	62.5	52.2	39.1	26.9	49.2
1970	19.6	26.5	33.4	50.1	62.1	68.8	71.9	71.5	67.6	56.3	42.7	32.0	50.2
1971	22.6	30.4	34.3	46.3	57.3	71.6	71.0	68.3	66.7	58.7	38.5	35.7	50.1
1972	25.7	24.0	32.6	45.4	59.2	63.2	71.2	69.8	64.5	49.6	39.6	35.6	48.4
1973	29.8	27.5	47.5	49.9	57.2	71.0	73.1	72.5	65.7	57.1	45.3	32.9	52.5
1974	30.7	27.1	39.3	51.8	58.1	66.3	71.9	71.1	60.1	50.7	43.5	31.5	50.2
1975	32.0	31.2	36.0	43.9	63.2	69.5	72.7	73.2	58.9	53.9	47.5	33.4	51.3
1976	23.0	36.8	43.3	49.2	55.2	68.4	68.4	66.3	59.8	46.6	33.2	23.6	47.9
1977	11.4	27.0	44.7	54.0	65.7	66.0	74.8	71.3	64.8	49.9	43.7	28.3	50.1
1978	19.1	15.9	32.2	47.0	58.6	67.8	70.5	72.1	68.2	50.1	42.5	32.6	48.0
1979	19.8	16.8	41.9	46.6	56.7	66.2	69.3	69.2	63.1	50.8	42.6	33.1	48.0
1980	26.8	23.3	34.9	47.4	60.5	63.9	72.6	74.5	65.8	48.6	39.4	29.5	48.9
1981	21.3	31.9	36.0	51.9	58.5	69.8	71.7	70.6	62.8	50.4	43.5	30.7	49.9
1982	20.1	27.7	37.6	45.4	66.1	64.7	73.4	68.2	62.4	55.0	45.0	40.0	50.5
1983	28.8	32.8	41.7	47.3	56.7	64.7	68.4	75.0	73.8	65.4	44.5	24.7	51.1
1984	22.2	36.2	30.0	48.5	54.7	70.5	69.8	70.1	60.2	56.7	40.1	37.1	49.7
1985	19.7	24.8	40.7	53.6	60.6	63.1	70.7	69.3	65.3	54.2	45.5	24.3	49.3
Record Mean	26.4	27.3	36.5	48.0	59.1	68.2	72.3	70.6	64.2	52.8	40.8	30.2	49.7
Max	34.0	35.4	45.5	58.2	69.8	78.8	83.0	81.2	74.7	62.6	48.4	37.0	59.1
Min	18.8	19.1	27.6	37.8	48.3	57.6	61.5	60.0	53.7	42.9	33.1	23.4	40.3

REFERENCE NOTES FOR TABLES 1, 2, 3 and 6   (AKRON, OH)

GENERAL

T - TRACE AMOUNT
BLANK ENTRIES DENOTE MISSING/UNREPORTED DATA.
# INDICATES A STATION OR INSTRUMENT RELOCATION.

SPECIFIC

TABLE 1

(a) - LENGTH OF RECORD IN YEARS. ALTHOUGH
      INDIVIDUAL MONTHS MAY BE MISSING.

 * LESS THAN .05

NORMALS — BASED ON THE 1951-1980 RECORD PERIOD.
EXTREMES — DATES ARE THE MOST RECENT OCCURRENCE.
WIND DIR. — NUMERALS SHOW TENS OF DEGREES
            CLOCKWISE FROM TRUE NORTH.
            "00" INDICATES CALM.
RESULTANT WIND DIRECTIONS ARE GIVEN TO WHOLE DEGREES.

EXCEPTIONS

TABLES 2, 3, and 6

RECORD MEANS ARE THROUGH THE CURRENT YEAR,
BEGINNING IN   1887 FOR TEMPERATURE
                1887 FOR PRECIPITATION
                1949 FOR SNOWFALL

TABLE 4 HEATING DEGREE DAYS Base 65 deg. F AKRON, OHIO

SEASON	JULY	AUG	SEP	OCT	NOV	DEC	JAN	FEB	MAR	APR	MAY	JUNE	TOTAL
1956-57	5	27	209	279	735	877	1344	935	858	479	219	36	6003
1957-58	11	12	133	491	730	973	1225	1227	936	461	256	98	6553
1958-59	5	29	116	407	680	1373	1306	1006	931	484	157	62	6556
1959-60	0	0	91	403	824	970	1075	1083	1261	392	260	38	6397
1960-61	15	0	49	373	672	1328	1341	897	787	664	351	80	6557
1961-62	14	3	74	311	701	1109	1265	1067	928	518	104	24	6118
#1962-63	11	10	165	356	722	1249	1412	1282	786	490	275	45	6803
1963-64	22	30	134	189	661	1372	1117	1154	849	447	157	68	6200
1964-65	2	46	121	458	605	1020	1226	1081	1005	524	94	56	6238
1965-66	8	40	82	442	675	878	1329	1019	807	568	344	46	6238
1966-67	3	10	148	445	690	1044	1006	1116	825	433	368	8	6096
1967-68	17	22	146	368	841	949	1296	1198	763	379	246	19	6244
1968-69	9	19	48	338	655	1067	1156	963	441	215	81		5951
1969-70	1	6	133	401	768	1172	1402	1073	972	458	161	33	6580
1970-71	10	2	61	281	664	1015	1309	962	946	554	248	9	6061
1971-72	0	18	68	202	786	899	1214	1182	999	581	184	110	6243
1972-73	25	16	84	470	755	905	1082	1043	538	455	243	1	5617
1973-74	2	12	78	240	587	988	1059	1059	790	402	241	45	5503
1974-75	0	0	169	438	637	1029	1015	940	895	626	122	38	5909
1975-76	2	1	183	344	519	971	1293	811	646	484	308	24	5606
1976-77	15	56	171	564	948	1272	1656	1057	627	354	105	82	6907
1977-78	4	21	68	458	632	1130	1413	1366	1013	533	247	51	6936
1978-79	9	0	58	456	670	997	1397	1348	712	550	288	62	6547
1979-80	35	28	118	444	664	981	1179	1202	925	520	168	95	6359
1980-81	1	3	79	500	763	1093	1351	920	889	384	222	15	6220
1981-82	6	6	138	444	638	1058	1386	1037	844	583	54	53	6247
1982-83	5	32	129	318	596	770	1113	896	715	527	261	45	5407
1983-84	10	0	103	338	608	1244	1321	831	1078	493	323	7	6356
1984-85	7	16	180	254	740	860	1397	1119	746	371	174	82	5946
1985-86	3	5	116	329	578	1254							

TABLE 5 COOLING DEGREE DAYS Base 65 deg. F AKRON, OHIO

YEAR	JAN	FEB	MAR	APR	MAY	JUNE	JULY	AUG	SEP	OCT	NOV	DEC	TOTAL
1969	0	0	0	3	36	115	225	179	63	11	0	0	632
1970	0	0	0	17	79	152	231	210	146	18	0	0	853
1971	0	0	0	0	15	217	196	130	123	13	0	0	694
1972	0	0	0	9	13	63	223	169	77	0	0	0	545
1973	0	0	0	0	5	188	259	252	104	5	0	0	822
1974	0	0	0	13	33	90	222	196	29	1	1	0	585
1975	0	0	0	0	74	179	250	260	5	7	0	0	775
1976	0	0	0	16	14	139	132	101	22	0	0	0	424
1977	0	0	5	30	131	118	314	223	71	0	0	0	892
1978	0	0	0	54	143	185	226	159	0	0	0	767
1979	0	0	0	5	36	106	173	166	68	10	0	0	564
1980	0	0	0	0	37	69	244	306	109	0	0	0	765
1981	0	0	0	1	26	166	223	187	78	1	0	0	682
1982	0	0	0	3	97	52	272	140	59	14	3	3	643
1983	0	0	0	2	11	153	326	279	122	5	0	0	898
1984	0	0	0	3	12	179	163	180	43	4	0	0	584
1985	0	0	0	36	44	32	186	149	134	1	1	0	583

TABLE 6 SNOWFALL (inches) AKRON, OHIO

SEASON	JULY	AUG	SEP	OCT	NOV	DEC	JAN	FEB	MAR	APR	MAY	JUNE	TOTAL
1956-57	0.0	0.0	0.0	0.0	2.5	16.6	9.3	3.4	6.8	7.3	0.0	0.0	45.9
1957-58	0.0	0.0	0.0	2.1	0.6	2.3	8.5	10.8	6.4	1.0	0.0	0.0	31.7
1958-59	0.0	0.0	0.0	0.0	6.7	10.1	15.0	3.6	13.7	0.3	0.0	0.0	49.4
1959-60	0.0	0.0	0.0	0.0	8.4	7.5	1.9	16.5	20.9	1.7	T	0.0	56.9
1960-61	0.0	0.0	0.0	T	5.6	25.4	17.3	13.4	4.0	15.8	T	0.0	81.5
1961-62	0.0	0.0	0.0	T	3.4	7.7	8.9	12.7	14.7	3.8	0.0	0.0	51.2
1962-63	0.0	0.0	0.0	1.8	T	24.3	14.3	18.6	15.5	0.3	0.4	0.0	75.2
1963-64	0.0	0.0	0.0	T	4.9	11.8	17.6	16.5	6.6	0.8	0.0	0.0	58.2
1964-65	0.0	0.0	0.0	T	5.1	9.2	15.9	8.9	10.9	T	0.0	0.0	50.0
1965-66	0.0	0.0	T	0.1	0.9	1.8	16.3	6.8	4.0	4.8	3.2	0.0	37.9
1966-67	0.0	0.0	0.0	T	9.7	6.8	4.9	19.7	13.8	0.8	T	0.0	55.7
1967-68	0.0	0.0	0.0	T	8.8	4.4	16.6	6.6	4.1	0.1	0.0	0.0	40.6
1968-69	0.0	0.0	0.0	0.2	2.4	11.1	4.8	5.8	5.9	T	0.0	0.0	30.2
1969-70	0.0	0.0	0.0	0.5	4.7	17.1	7.0	10.5	10.5	0.2	T	0.0	53.4
1970-71	0.0	0.0	0.0	0.0	0.7	4.4	9.0	15.7	15.1	T	T	0.0	44.9
1971-72	0.0	0.0	0.0	0.0	10.8	2.1	7.5	14.7	10.5	1.1	0.0	0.0	46.7
1972-73	0.0	0.0	0.0	0.5	2.9	6.5	3.9	9.9	10.8	2.6	T	0.0	37.1
1973-74	0.0	0.0	0.0	0.0	0.7	8.8	8.3	8.7	8.0	3.9	T	0.0	38.4
1974-75	0.0	0.0	0.0	3.5	6.0	29.4	8.8	10.5	11.9	2.6	0.0	0.0	72.7
1975-76	0.0	0.0	0.0	0.0	3.9	4.8	25.1	9.8	4.8	2.7	T	0.0	51.1
1976-77	0.0	0.0	0.0	0.4	7.0	7.4	19.5	5.6	1.0	1.5	0.0	0.0	42.4
1977-78	0.0	0.0	0.0	T	10.0	19.3	37.5	9.1	6.0	0.1	0.0	0.0	82.0
1978-79	0.0	0.0	0.0	0.0	1.1	4.5	21.9	14.5	3.1	1.6	0.0	0.0	46.7
1979-80	0.0	0.0	0.0	0.4	1.6	5.1	8.5	12.2	6.0	0.4	T	0.0	34.2
1980-81	0.0	0.0	0.0	T	7.0	11.0	13.0	9.3	12.0	T	0.0	0.0	52.3
1981-82	0.0	0.0	0.0	T	1.4	19.0	17.8	5.3	10.4	7.8	0.0	0.0	61.7
1982-83	0.0	0.0	0.0	T	2.4	14.2	4.7	8.9	7.6	1.0	0.0	0.0	38.8
1983-84	0.0	0.0	0.0	0.0	4.4	8.8	10.5	21.1	12.5	0.2	0.0	0.0	57.5
1984-85	0.0	0.0	0.0	0.0	2.6	7.8	20.9	12.4	1.2	5.3	0.0	0.0	50.2
1985-86	0.0	0.0	0.0	0.0	T	10.7							
Record Mean	0.0	0.0	T	0.5	4.8	10.3	11.8	9.9	8.9	2.2	0.1	0.0	48.4

See Reference Notes, relative to all above tables, on preceding page.

Greater Cincinnati Airport is located on a gently rolling plateau about 12 miles southwest of downtown Cincinnati and 2 miles south of the Ohio River at its nearest point. The river valley is rather narrow and steep-sided varying from 1 to 3 miles in width and the river bed is 500 feet below the level of the airport.

The climate is continental with a rather wide range of temperatures from winter to summer. A precipitation maximum occurs during winter and spring with a late summer and fall minimum. On the average, the maximum snowfall occurs during January, although the heaviest 24-hour amounts have been recorded during late November and February.

The heaviest precipitation, as well as the precipitation of the longest duration, is normally associated with low pressure disturbances moving in a general southwest to northeast direction through the Ohio valley and south of the Cincinnati area.

Summers are warm and rather humid. The temperature will reach 100 degrees or more in 1 year out of 3. However, the temperature will reach 90 degrees or higher on about 19 days each year. Winters are moderately cold with frequent periods of extensive cloudiness.

The freeze free period lasts on the average 187 days from mid-April to the latter part of October.

TABLE 1   NORMALS, MEANS AND EXTREMES

CINCINNATI, (GREATER CINCINNATI AIRPORT) OHIO

LATITUDE: 39°03'N   LONGITUDE: 84°40'W   ELEVATION: FT. GRND 869 BARO 00888   TIME ZONE: EASTERN   WBAN: 93814

	(a)	JAN	FEB	MAR	APR	MAY	JUNE	JULY	AUG	SEP	OCT	NOV	DEC	YEAR
TEMPERATURE °F:														
Normals														
-Daily Maximum		37.3	41.2	51.5	64.5	74.2	82.3	85.8	84.8	78.7	66.7	52.6	41.9	63.5
-Daily Minimum		20.4	23.0	32.0	42.4	51.7	60.5	64.9	63.3	56.3	43.9	34.1	25.7	43.2
-Monthly		28.9	32.1	41.8	53.5	63.0	71.4	75.3	74.1	67.5	55.3	43.4	33.8	53.3
Extremes														
-Record Highest	24	69	73	83	89	93	97	101	102	98	88	80	75	102
-Year		1967	1972	1981	1976	1962	1971	1983	1962	1964	1963	1982	1982	AUG 1962
-Record Lowest	24	-25	-11	-11	17	27	39	47	43	33	16	1	-12	-25
-Year		1977	1982	1980	1964	1963	1972	1963	1965	1983	1962	1976	1983	JAN 1977
NORMAL DEGREE DAYS:														
Heating (base 65°F)		1119	921	719	350	143	12	0	0	52	316	648	967	5247
Cooling (base 65°F)		0	0	0	5	81	204	322	282	127	16	0	0	1037
% OF POSSIBLE SUNSHINE	2	35	47	39	55	56	62	66	65	75	44	35	35	51
MEAN SKY COVER (tenths)														
Sunrise - Sunset	34	7.5	7.3	7.4	7.0	6.7	6.2	6.1	5.8	5.6	5.7	7.1	7.5	6.6
MEAN NUMBER OF DAYS:														
Sunrise to Sunset														
-Clear	34	5.0	5.3	5.0	5.6	5.9	6.9	7.4	8.1	9.8	10.5	6.1	5.1	80.4
-Partly Cloudy	34	6.3	5.7	6.8	7.5	9.5	10.3	11.7	12.0	8.9	7.2	5.9	5.9	97.6
-Cloudy	34	19.8	17.3	19.2	16.9	15.6	12.8	11.9	10.9	11.4	13.3	18.1	20.0	187.2
Precipitation														
.01 inches or more	38	12.2	11.0	13.3	12.6	11.5	10.6	10.0	9.0	7.8	8.3	11.0	12.2	129.5
Snow, Ice pellets														
1.0 inches or more	38	2.4	1.8	1.3	0.2	0.0	0.0	0.0	0.0	0.0	0.*	0.6	1.2	7.4
Thunderstorms	38	0.7	0.9	2.4	4.2	5.7	7.1	8.3	7.5	3.2	1.4	1.1	0.4	42.9
Heavy Fog Visibility														
1/4 mile or less	22	2.5	2.1	1.6	0.9	1.2	1.4	2.0	2.9	3.8	2.9	1.8	2.6	25.6
Temperature °F														
-Maximum														
90° and above	23	0.0	0.0	0.0	0.0	0.4	4.2	7.5	4.9	2.2	0.0	0.0	0.0	19.1
32° and below	23	13.4	8.4	1.4	0.0	0.0	0.0	0.0	0.0	0.0	0.0	0.7	6.6	30.5
-Minimum														
32° and below	23	26.8	23.2	16.4	4.5	0.3	0.0	0.0	0.0	0.0	3.2	13.0	22.1	109.5
0° and below	23	3.7	2.0	0.1	0.0	0.0	0.0	0.0	0.0	0.0	0.0	0.0	1.1	7.0
AVG. STATION PRESS.(mb)	13	987.2	986.9	984.0	984.2	983.3	984.4	985.6	986.6	987.1	987.8	987.1	987.3	986.0
RELATIVE HUMIDITY (%)														
Hour 01	23	75	74	73	71	77	80	83	84	82	77	75	76	77
Hour 07	23	78	77	78	76	80	82	85	88	88	83	80	79	81
Hour 13 (Local Time)	23	68	64	60	54	55	56	57	58	57	56	64	69	60
Hour 19	23	69	65	61	55	57	59	61	64	67	65	69	72	64
PRECIPITATION (inches):														
Water Equivalent														
-Normal		3.13	2.73	3.95	3.58	3.84	4.09	4.28	2.97	2.91	2.54	3.12	3.00	40.14
-Maximum Monthly	38	9.43	6.72	12.18	7.19	9.48	7.36	8.36	7.71	8.61	8.60	7.51	6.46	12.18
-Year		1950	1955	1964	1970	1968	1977	1962	1982	1979	1983	1985	1978	MAR 1964
-Minimum Monthly	38	0.57	0.25	1.14	1.04	1.13	0.95	1.18	0.31	0.18	0.25	0.43	0.51	0.18
-Year		1981	1978	1960	1971	1964	1965	1951	1953	1963	1963	1949	1976	SEP 1963
-Maximum in 24 hrs	38	4.33	2.32	5.21	2.72	3.71	3.45	3.47	3.12	4.54	4.47	3.36	2.96	5.21
-Year		1959	1950	1964	1950	1956	1974	1953	1957	1979	1985	1948	1948	MAR 1964
Snow, Ice pellets														
-Maximum Monthly	38	31.5	13.3	13.0	3.7	T					1.7	12.1	12.5	31.5
-Year		1978	1971	1968	1977	1966					1962	1966	1960	JAN 1978
-Maximum in 24 hrs	38	8.1	9.3	9.8	3.6	T					1.7	9.0	7.3	9.8
-Year		1978	1966	1968	1977	1966					1962	1966	1984	MAR 1968
WIND:														
Mean Speed (mph)	38	10.7	10.4	11.1	10.7	8.7	7.9	7.2	6.8	7.4	8.2	9.6	10.3	9.1
Prevailing Direction														
through 1963		SSW	SSW	SSW	SSW	SSW	SSW	SSW	SSW	SSW	SSW	SSW	SSW	SSW
Fastest Obs. 1 Min.														
-Direction	22	28	29	25	25	22	24	34	31	24	27	20	21	25
-Speed (MPH)	22	46	40	44	46	37	40	35	37	32	35	35	40	46
-Year		1976	1967	1977	1982	1983	1971	1980	1983	1975	1967	1968	1973	APR 1982
Peak Gust														
-Direction	2	W	S	SW	S	NW	NW	NW	W	W	W	NW	W	S
-Speed (mph)	2	48	53	58	61	37	40	43	38	38	39	43	49	61
-Date		1985	1984	1985	1985	1985	1985	1984	1984	1984	1985	1984	1985	APR 1985

See reference Notes to this table on the following page.

TABLE 2 PRECIPITATION (inches)   CINCINNATI, (GREATER CINCINNATI AIRPORT) OHIO

YEAR	JAN	FEB	MAR	APR	MAY	JUNE	JULY	AUG	SEP	OCT	NOV	DEC	ANNUAL
1956	2.50	5.92	4.19	4.09	7.04	4.98	3.90	2.91	2.29	1.54	1.91	2.56	43.83
1957	3.08	2.79	1.44	5.62	5.74	2.68	2.44	3.74	3.20	2.34	6.25	5.98	45.30
1958	2.34	0.61	1.73	3.49	5.33	5.19	7.54	2.52	3.67	2.67	0.68		37.23
1959	7.61	3.18	2.52	2.33	2.92	5.85	2.77	1.80	1.88	3.05	4.04	3.03	40.98
1960	2.48	3.49	1.14	1.32	3.22	6.16	3.15	5.21	1.07	4.00	2.33	1.73	35.30
1961	1.87	3.56	4.76	2.81	7.31	2.28	5.11	1.07	0.97	1.88	3.36	3.39	38.37
1962	3.98	5.58	4.25	1.26	3.64	4.40	8.36	2.14	2.67	2.87	2.37	1.36	42.88
1963	2.04	1.09	9.91	2.01	2.73	1.59	4.34	1.83	0.18	0.25	0.94	1.08	27.99
1964	2.88	1.98	12.18	6.73	1.13	4.32	2.56	2.25	1.65	0.59	2.69	4.92	43.88
1965	3.11	5.07	2.86	4.90	1.46	0.95	4.42	3.24	6.06	3.81	1.26	1.19	38.33
1966	3.84	3.73	1.22	5.38	2.42	2.52	4.06	4.31	3.13	0.57	4.18	3.31	38.67
1967	0.75	1.86	3.63	3.66	5.64	1.72	4.99	0.77	1.79	2.65	3.84	4.01	35.24
1968	1.79	0.64	4.24	3.47	9.48	2.43	7.50	2.26	2.16	1.35	3.21	4.01	42.54
1969	4.64	1.27	1.42	3.59	2.05	4.91	3.28	2.15	2.87	1.53	3.67	2.59	33.97
1970	1.27	1.68	4.71	7.19	1.88	5.73	3.47	2.96	3.87	2.44	2.29	3.32	40.81
1971	2.47	5.89	2.55	1.04	3.31	5.18	3.70	3.45	6.56	1.44	1.68	3.39	40.66
1972	1.96	2.20	3.68	5.89	6.02	2.41	1.50	2.64	5.96	2.55	6.26	4.23	45.30
1973	1.79	1.58	6.11	5.81	3.46	6.27	7.16	2.62	2.63	4.39	4.95	2.66	49.43
1974	3.65	1.63	4.39	5.08	5.53	4.38	3.82	5.75	4.44	1.07	4.19	2.83	46.76
1975	4.05	3.38	6.76	4.16	3.11	5.09	1.62	1.97	3.64	4.59	2.50	3.36	44.23
1976	3.00	2.37	2.14	1.21	1.80	5.94	2.33	4.36	1.95	3.85	0.83	0.51	30.29
1977	1.90	1.29	4.52	4.16	1.53	7.36	1.90	5.45	1.80	3.74	3.90	4.00	41.55
1978	4.52	0.25	1.99	2.28	5.30	6.63	6.86	4.41	0.43	5.03	2.67	6.46	46.83
1979	3.68	3.77	2.05	4.90	4.00	5.92	5.49	4.80	8.61	1.77	4.86	2.91	52.76
1980	2.26	1.04	4.50	1.96	4.59	4.13	5.51	4.19	1.83	3.28	2.58	1.26	37.13
1981	0.57	3.86	1.72	5.05	5.07	3.34	3.66	2.15	1.47	2.33	2.94	2.39	34.55
1982	7.17	1.17	4.67	2.18	4.60	3.61	2.44	7.71	1.27	0.99	5.08	4.25	45.14
1983	1.56	1.14	2.02	4.84	8.89	2.22	1.96	3.23	1.22	8.60	4.20	2.84	42.72
1984	0.75	2.40	3.61	4.88	4.82	2.11	2.57	3.30	3.50	3.85	6.00	4.21	42.00
1985	1.68	2.25	6.90	1.34	6.18	4.55	3.59	2.02	0.76	5.83	7.51	1.52	44.13
Record Mean	3.33	2.77	3.91	3.56	3.99	4.02	4.09	3.04	2.80	2.74	3.45	3.02	40.73

TABLE 3 AVERAGE TEMPERATURE (deg. F)   CINCINNATI, (GREATER CINCINNATI AIRPORT) OHIO

YEAR	JAN	FEB	MAR	APR	MAY	JUNE	JULY	AUG	SEP	OCT	NOV	DEC	ANNUAL
1956	29.3	37.6	41.7	50.0	63.3	72.1	74.3	73.6	64.5	60.8	42.6	42.0	54.3
1957	27.8	38.3	42.3	56.1	64.0	73.3	76.3	74.5	67.3	53.1	43.6	38.7	54.6
1958	29.3	25.4	37.8	53.8	63.1	69.0	74.3	73.1	67.3	56.1	46.4	27.0	51.9
1959	29.3	35.6	41.1	55.6	67.9	72.6	76.7	78.5	71.0	56.8	40.6	37.7	55.3
1960	34.5	30.6	29.7	57.7	60.9	70.9	74.4	76.3	70.5	57.1	46.1	28.8	53.1
1961	27.8	38.7	46.5	48.4	59.3	70.1	75.4	74.4	71.8	58.1		33.5	54.0
#1962	28.2	34.9	40.7	52.3	68.4	72.4	74.3	75.1	63.5	56.2	42.8	28.5	53.1
1963	22.2	24.4	44.8	53.8	58.9	69.2	72.3	70.5	65.7	61.8	45.2	22.9	51.0
1964	32.1	30.2	43.1	56.1	65.8	71.9	74.2	73.7	67.1	52.7	47.1	35.5	54.1
1965	30.6	32.9	36.6	53.3	67.1	71.9	73.2	72.5	67.1	53.4	44.2	38.8	53.5
1966	23.8	31.5	42.9	50.1	60.4	73.1	78.8	74.5	66.5	51.0	44.7	33.6	52.6
1967	35.4	27.5	44.8	55.4	60.5	73.0	72.9	71.1	65.3	55.5	40.3	36.8	53.2
1968	26.8	28.1	45.2	55.5	60.7	72.9	75.7	75.3	67.4	55.9	45.6	33.7	53.6
1969	30.2	33.7	37.3	55.2	65.0	71.9	77.3	74.0	66.5	55.7	40.3	30.6	53.1
1970	24.0	30.6	39.0	56.8	66.8	72.2	75.5	75.1	73.7	58.6	45.0	38.0	54.6
1971	28.3	33.3	40.2	53.1	60.8	75.8	74.8	72.4	70.3	61.4	43.2	41.2	54.6
1972	30.2	29.5	40.6	52.5	63.1	65.3	75.5	73.4	68.7	51.7	40.3	36.1	52.3
1973	31.7	33.0	50.6	51.1	58.4	72.3	74.8	73.7	69.6	59.0	43.3	33.6	54.4
1974	35.7	33.9	45.9	54.3	61.9	67.7	75.2	74.3	61.6	53.2	44.4	34.4	53.5
1975	34.0	37.1	39.8	50.6	67.6	72.3	74.6	76.9	62.4	57.5	49.1	37.4	54.9
1976	27.1	41.7	47.0	56.1	60.0	71.5	73.6	71.2	63.6	48.8	34.9	27.6	52.0
1977	12.0	29.8	46.5	56.3	68.2	68.7	77.9	73.9	69.8	52.3	45.4	28.7	52.5
1978	18.4	18.2	36.4	53.4	60.1	72.2	75.0	73.0	70.8	52.6	46.4	36.0	51.0
1979	21.3	21.4	46.6	50.9	60.2	69.4	73.4	72.3	66.1	54.3	44.2	36.2	51.4
1980	29.9	24.0	38.5	50.3	64.9	70.1	76.6	76.5	68.6	50.6	41.5	32.9	52.0
1981	24.1	34.2	40.1	58.1	59.8	72.3	75.9	73.6	65.2	53.9	43.7	29.0	52.5
1982	23.9	30.6	44.3	49.6	68.1	67.4	77.0	71.3	66.9	59.3	48.4	42.9	54.1
1983	31.6	35.3	44.7	49.4	59.0	71.6	79.2	78.3	67.2	55.7	44.7	24.6	53.4
1984	23.7	38.2	34.4	51.1	58.9	74.1	72.2	74.3	65.7	61.5	42.0	42.4	53.2
1985	22.7	29.4	47.5	57.9	65.4	69.9	75.1	72.5	67.5	59.3	49.9	26.2	53.6
Record Mean	28.9	32.7	42.0	53.5	63.0	71.6	75.5	74.1	67.3	55.9	43.7	33.9	53.5
Max	37.1	41.6	51.5	64.2	74.0	82.3	85.9	84.8	78.3	66.7	52.6	41.8	63.4
Min	20.7	23.7	32.4	42.7	52.0	60.8	65.1	63.4	56.2	45.1	34.8	26.0	43.6

REFERENCE NOTES FOR TABLES 1, 2, 3 and 6   (CINCINNATI, OH)

GENERAL

T - TRACE AMOUNT
BLANK ENTRIES DENOTE MISSING/UNREPORTED DATA.
# INDICATES A STATION OR INSTRUMENT RELOCATION.

SPECIFIC

TABLE 1

(a) - LENGTH OF RECORD IN YEARS. ALTHOUGH
INDIVIDUAL MONTHS MAY BE MISSING.

* LESS THAN .05

NORMALS — BASED ON THE 1951-1980 RECORD PERIOD.
EXTREMES — DATES ARE THE MOST RECENT OCCURRENCE.
WIND DIR. — NUMERALS SHOW TENS OF DEGREES
CLOCKWISE FROM TRUE NORTH.
"00" INDICATES CALM.
RESULTANT WIND DIRECTIONS ARE GIVEN TO WHOLE DEGREES.

EXCEPTIONS

TABLES 2, 3, and 6

RECORD MEANS ARE THROUGH THE CURRENT YEAR,
BEGINNING IN   1948 FOR TEMPERATURE
                    1948 FOR PRECIPITATION
                    1948 FOR SNOWFALL

TABLE 4   HEATING DEGREE DAYS Base 65 deg. F   CINCINNATI, (GREATER CINCINNATI AIRPORT) OHIO

SEASON	JULY	AUG	SEP	OCT	NOV	DEC	JAN	FEB	MAR	APR	MAY	JUNE	TOTAL
1956-57	0	8	90	138	663	704	1147	740	698	334	113	0	4635
1957-58	0	1	47	361	636	809	1097	1104	832	336	100	15	5338
1958-59	0	2	49	282	561	1171	1102	815	736	292	76	9	5095
1959-60	1	0	37	294	723	841	937	989	1090	262	187	4	5365
1960-61	0	0	18	252	559	1116	1146	729	568	497	196	23	5104
#1961-62	0	0	47	226	622	970	1132	838	747	402	51	4	5039
1962-63	2	2	129	311	660	1126	1321	1132	619	359	202	25	5888
1963-64	1	6	75	119	586	1296	1011	1005	671	273	64	26	5133
1964-65	0	14	68	376	529	907	1058	894	872	350	43	0	5111
1965-66	2	15	64	362	614	803	1270	931	681	442	181	15	5380
1966-67	0	0	70	428	602	969	910	1043	622	305	175	9	5133
1967-68	0	2	82	313	736	867	1178	1063	609	284	151	6	5291
1968-69	0	4	32	316	576	965	1072	872	852	298	84	29	5100
1969-70	0	0	66	309	737	1058	1265	954	797	264	91	2	5543
1970-71	4	0	28	212	591	829	1133	881	758	351	157	0	4944
1971-72	0	0	31	128	649	731	1073	1024	751	374	116	79	4956
1972-73	2	3	24	404	733	891	1025	888	445	425	206	3	5049
1973-74	0	1	27	217	583	964	901	863	591	332	161	31	4671
1974-75	0	0	147	371	614	942	953	774	778	435	50	10	5074
1975-76	5	0	142	244	474	848	1168	670	558	321	176	5	4611
1976-77	0	4	72	498	893	1157	1640	980	571	276	67	36	6194
1977-78	0	2	32	391	586	1118	1440	1303	880	346	207	10	6315
1978-79	0	0	21	381	552	891	1348	1216	563	425	179	15	5591
1979-80	1	14	60	346	616	887	1080	1182	814	434	92	24	5550
1980-81	0	0	48	446	697	988	1261	858	768	230	191	6	5493
1981-82	0	0	87	344	634	1107	1268	956	635	460	28	19	5538
1982-83	0	1	56	244	505	682	1029	825	627	466	199	21	4655
1983-84	1	0	89	288	600	1247	1274	773	939	425	219	4	5859
1984-85	0	0	101	128	684	692	1306	992	543	256	72	22	4796
1985-86	0	0	78	212	450	1195							

TABLE 5   COOLING DEGREE DAYS Base 65 deg. F   CINCINNATI, (GREATER CINCINNATI AIRPORT) OHIO

YEAR	JAN	FEB	MAR	APR	MAY	JUNE	JULY	AUG	SEP	OCT	NOV	DEC	TOTAL
1969	0	0	0	9	92	240	389	285	116	31	0	0	1162
1970	0	0	0	28	151	225	336	320	292	21	0	0	1373
1971	0	0	0	0	33	335	313	236	195	25	0	0	1137
1972	0	0	0	7	65	96	334	268	142	0	0	0	912
1973	0	0	4	13	7	228	310	278	170	36	0	0	1046
1974	0	0	8	17	70	121	323	297	50	12	2	0	900
1975	0	0	0	0	8	138	236	309	376	72	19	0	1161
1976	0	0	6	59	28	207	276	202	39	5	0	0	822
1977	0	0	7	22	171	152	407	285	181	4	5	0	1234
1978	0	0	0	4	63	231	315	255	200	2	0	0	1070
1979	0	0	2	8	38	154	271	248	102	22	0	0	845
1980	0	0	0	0	98	187	364	363	166	5	0	0	1183
1981	0	0	1	31	34	234	343	275	99	9	0	0	1026
1982	0	0	0	5	129	99	381	203	120	73	13	8	1031
1983	0	0	4	18	225	448	417	161	8	0	0		1285
1984	0	0	0	13	38	289	233	295	130	29	0	0	1027
1985	0	0	6	47	93	174	318	241	162	41	5	0	1087

TABLE 6   SNOWFALL (inches)   CINCINNATI, (GREATER CINCINNATI AIRPORT) OHIO

SEASON	JULY	AUG	SEP	OCT	NOV	DEC	JAN	FEB	MAR	APR	MAY	JUNE	TOTAL
1956-57	0.0	0.0	0.0	0.0	3.8	1.2	5.8	1.5	1.2	0.1	0.0	0.0	13.6
1957-58	0.0	0.0	0.0	T	0.4	1.5	6.9	2.6	7.7	T	0.0	0.0	19.1
1958-59	0.0	0.0	0.0	0.0	8.3	2.8	8.0	1.6	7.9	0.6	0.0	0.0	29.2
1959-60	0.0	0.0	0.0	T	2.3	3.9	2.5	12.1	12.8	T	1.4	0.0	33.6
1960-61	0.0	0.0	0.0	0.0	T	12.5	6.9	13.3	1.4	T	0.0	0.0	35.5
1961-62	0.0	0.0	0.0	0.0	0.9	8.1	3.9	9.0	5.2	1.7	0.0	0.0	28.8
1962-63	0.0	0.0	0.0	1.7	0.8	4.7	9.6	6.4	0.1	0.0	0.0	0.0	23.3
1963-64	0.0	0.0	0.0	0.0	2.3	7.6	15.3	6.7	1.3	T	0.0	0.0	33.2
1964-65	0.0	0.0	0.0	0.0	1.7	0.9	9.3	5.9	6.2	T	0.0	0.0	24.0
1965-66	0.0	0.0	0.0	0.0	T	2.5	6.8	9.5	1.2	0.1	T	0.0	20.1
1966-67	0.0	0.0	0.0	0.0	12.1	1.6	2.5	7.3	8.1	0.0	0.0	0.0	31.6
1967-68	0.0	0.0	0.0	T	3.1	3.1	8.7	3.6	13.0	0.0	0.0	0.0	31.5
1968-69	0.0	0.0	0.0	0.0	0.4	2.0	1.0	0.3	2.8	0.0	0.0	0.0	6.5
1969-70	0.0	0.0	0.0	0.0	1.1	6.0	6.7	4.3	12.0	T	0.0	0.0	30.1
1970-71	0.0	0.0	0.0	0.0	0.4	0.4	3.2	13.3	9.7	T	0.0	0.0	27.0
1971-72	0.0	0.0	0.0	0.0	3.0	T	1.6	10.8	0.1	0.5	0.0	0.0	16.0
1972-73	0.0	0.0	0.0	T	6.5	1.9	0.5	3.1	3.9	1.8	0.0	0.0	17.7
1973-74	0.0	0.0	0.0	0.0	0.4	2.0	1.7	3.2	3.4	0.5	0.0	0.0	11.2
1974-75	0.0	0.0	0.0	T	5.0	5.6	2.0	2.2	6.8	0.2	0.0	0.0	21.8
1975-76	0.0	0.0	0.0	0.0	3.7	1.7	8.0	0.1	0.6	0.0	0.0	0.0	14.1
1976-77	0.0	0.0	0.0	0.0	2.7	1.0	30.3	4.2	5.4	3.7	0.0	0.0	47.3
1977-78	0.0	0.0	0.0	0.0	4.0	6.3	31.5	4.6	7.5	0.0	0.0	0.0	53.9
1978-79	0.0	0.0	0.0	0.0	T	0.7	17.5	11.7	0.6	0.1	0.0	0.0	30.6
1979-80	0.0	0.0	0.0	0.0	1.0	1.0	8.3	11.9	7.9	T	0.0	0.0	30.1
1980-81	0.0	0.0	0.0	T	1.2	3.7	4.0	2.6	2.5	0.0	0.0	0.0	14.0
1981-82	0.0	0.0	0.0	T	0.3	10.9	7.1	3.9	0.5	1.5	0.0	0.0	24.2
1982-83	0.0	0.0	0.0	0.0	T	T	0.8	5.5	0.3	T	0.0	0.0	6.6
1983-84	0.0	0.0	0.0	0.0	T	1.7	4.1	6.7	4.1	0.0	0.0	0.0	16.6
1984-85	0.0	0.0	0.0	0.0	1.4	7.3	12.2	9.5	0.4	1.7	0.0	0.0	32.5
1985-86	0.0	0.0	0.0	0.0	0.0	5.0							
Record Mean	0.0	0.0	0.0	T	2.4	3.8	7.6	5.4	4.4	0.5	T	0.0	24.1

See Reference Notes, relative to all above tables, on preceding page.

Cleveland is on the south shore of Lake Erie in northeast Ohio. The metropolitan area has a lake frontage of 31 miles. The surrounding terrain is generally level except for an abrupt ridge on the eastern edge of the city which rises some 500 feet above the shore terrain. The Cuyahoga River, which flows through a rather deep but narrow north-south valley, bisects the city.

Local climate is continental in character but with strong modifying influences by Lake Erie. West to northerly winds blowing off Lake Erie tend to lower daily high temperatures in summer and raise temperatures in winter. Temperatures at Hopkins Airport which is 5 miles south of the lakeshore average from 2-4 degrees higher than the lakeshore in summer, while overnight low temperatures average from 2-4 degrees lower than the lakefront during all seasons.

In this area, summers are moderately warm and humid with occasional days when temperatures exceed 90 degrees. Winters are relatively cold and cloudy with an average of 5 days with sub-zero temperatures. Weather changes occur every few days from the passing of cold fronts.

The daily range in temperature is usually greatest in late summer and least in winter. Annual extremes in temperature normally occur soon after late June and December. Maximum temperatures below freezing occur most often in December, January, and February. Temperatures of 100 degrees or higher are rare. On the average, freezing temperatures in fall are first recorded in October while the last freezing temperature in spring normally occurs in April.

As is characteristic of continental climates, precipitation varies widely from year to year. However, it is normally abundant and well distributed throughout the year with spring being the wettest season. Showers and thunderstorms account for most of the rainfall during the growing season. Thunderstorms are most frequent from April through August. Snowfall may fluctuate widely. Mean annual snowfall increases from west to east in Cuyahoga County ranging from about 45 inches in the west to more than 90 inches in the extreme east.

Damaging winds of 50 mph or greater are usually associated with thunderstorms. Tornadoes, one of the most destructive of all atmospheric storms, occasionally occur in Cuyahoga County.

TABLE 1

NORMALS, MEANS AND EXTREMES

CLEVELAND, OHIO

LATITUDE: 41°25'N   LONGITUDE: 81°52'W   ELEVATION: FT. GRND   777 BARO   00779   TIME ZONE: EASTERN   WBAN: 14820

	(a)	JAN	FEB	MAR	APR	MAY	JUNE	JULY	AUG	SEP	OCT	NOV	DEC	YEAR
TEMPERATURE °F:														
Normals														
-Daily Maximum		32.5	34.8	44.8	57.9	68.5	78.0	81.7	80.3	74.2	62.7	49.3	37.5	58.5
-Daily Minimum		18.5	19.9	28.4	38.3	47.9	57.2	61.4	60.5	54.0	43.6	34.3	24.6	40.7
-Monthly		25.5	27.4	36.6	48.1	58.2	67.6	71.6	70.4	64.1	53.2	41.8	31.1	49.6
Extremes														
-Record Highest	44	73	69	83	88	92	101	103	102	101	90	82	77	103
-Year		1950	1961	1945	1942	1959	1944	1941	1948	1953	1946	1950	1982	JUL 1941
-Record Lowest	44	-19	-15	-5	10	25	31	41	38	32	22	3	-11	-19
-Year		1963	1963	1984	1964	1966	1972	1968	1982	1942	1969	1976	1976	JAN 1963
NORMAL DEGREE DAYS:														
Heating (base 65°F)		1225	1053	880	507	244	33	8	11	99	371	696	1051	6178
Cooling (base 65°F)		0	0	0	0	33	111	213	178	72	5	0	0	612
% OF POSSIBLE SUNSHINE	42	31	37	44	52	58	65	67	63	60	53	31	26	49
MEAN SKY COVER (tenths)														
Sunrise - Sunset	44	8.3	7.9	7.6	7.0	6.6	6.1	5.6	5.7	5.9	6.2	8.0	8.4	6.9
MEAN NUMBER OF DAYS:														
Sunrise to Sunset														
-Clear	44	2.8	3.1	4.4	5.2	6.0	6.7	8.6	8.7	8.5	8.3	3.1	2.4	67.7
-Partly Cloudy	44	4.8	5.5	6.5	7.9	10.1	11.3	11.9	11.1	9.4	7.9	5.9	4.9	97.3
-Cloudy	44	23.5	19.7	20.2	16.9	14.9	12.0	10.5	11.2	12.1	14.8	21.0	23.7	200.3
Precipitation														
.01 inches or more	44	16.5	14.3	15.6	14.2	13.0	10.9	10.1	9.7	9.8	10.9	14.5	16.4	155.9
Snow, Ice pellets														
1.0 inches or more	44	4.3	3.9	3.1	0.7	0.*	0.0	0.0	0.0	0.0	0.2	1.8	4.2	18.3
Thunderstorms	44	0.1	0.5	1.8	3.7	5.0	6.9	6.4	5.3	3.4	1.5	1.0	0.3	35.9
Heavy Fog Visibility														
1/4 mile or less	44	1.5	1.6	1.8	1.2	1.3	0.6	0.5	1.0	0.5	1.0	0.6	1.1	12.7
Temperature °F														
-Maximum														
90° and above	25	0.0	0.0	0.0	0.0	0.2	1.6	3.4	1.7	0.7	0.0	0.0	0.0	7.5
32° and below	25	16.9	13.3	5.2	0.3	0.0	0.0	0.0	0.0	0.0	0.0	1.1	10.5	47.3
-Minimum														
32° and below	25	28.4	24.4	21.1	9.4	1.0	0.*	0.0	0.0	0.0	2.6	12.8	24.6	124.3
0° and below	25	3.6	2.4	0.1	0.0	0.0	0.0	0.0	0.0	0.0	0.0	0.0	0.8	6.8
AVG. STATION PRESS.(mb)	13	988.4	989.0	986.6	986.6	986.0	986.8	987.8	989.2	989.7	990.2	989.0	988.9	988.2
RELATIVE HUMIDITY (%)														
Hour 01	25	75	75	74	72	76	79	81	83	81	76	75	75	77
Hour 07	25	77	77	78	75	77	78	81	85	84	80	77	76	79
Hour 13 (Local Time)	25	70	68	63	56	57	57	57	60	60	59	66	70	62
Hour 19	25	72	71	69	61	59	61	61	66	69	68	71	73	67
PRECIPITATION (inches):														
Water Equivalent														
-Normal		2.47	2.20	2.99	3.32	3.30	3.49	3.37	3.38	2.92	2.45	2.76	2.75	35.40
-Maximum Monthly	44	7.01	4.64	6.07	6.61	6.04	9.06	6.47	8.96	6.75	9.50	8.80	5.60	9.50
-Year		1950	1950	1954	1961	1947	1972	1969	1975	1981	1954	1985	1951	OCT 1954
-Minimum Monthly	44	0.36	0.48	0.78	1.18	1.00	1.17	1.21	0.53	0.74	0.61	0.80	0.71	0.36
-Year		1961	1978	1958	1946	1963	1967	1982	1969	1964	1976	1958	1958	JAN 1961
-Maximum in 24 hrs	44	2.33	2.33	2.76	2.24	3.73	4.00	2.87	3.07	2.38	3.44	2.73	2.06	4.00
-Year		1959	1959	1948	1961	1955	1972	1969	1947	1979	1954	1985	1974	JUN 1972
Snow, Ice pellets														
-Maximum Monthly	44	42.8	27.1	26.3	14.5	2.1				T	8.0	22.3	30.3	42.8
-Year		1978	1984	1954	1943	1974				1976	1962	1950	1962	JAN 1978
-Maximum in 24 hrs	44	10.5	11.5	14.9	11.6	2.1				T	6.7	15.0	12.2	15.0
-Year		1978	1984	1954	1982	1974				1976	1962	1950	1974	NOV 1950
WIND:														
Mean Speed (mph)	44	12.4	12.1	12.4	11.8	10.2	9.4	8.7	8.4	9.1	10.0	11.9	12.3	10.7
Prevailing Direction														
through 1963		SW	S	W	S	S	S	S	S	S	S	S	S	S
Fastest Obs. 1 Min.														
-Direction	8	22	05	25	23	20	23	23	30	22	23	23	21	22
-Speed (MPH)	8	53	32	37	44	42	37	36	35	29	37	35	43	53
-Year		1978	1984	1985	1982	1983	1982	1983	1980	1981	1983	1978	1982	JAN 1978
Peak Gust														
-Direction	2	SW	NE	W	SW	NW	NW	SW	NW	NW	SW	SE	W	SW
-Speed (mph)	2	55	49	55	69	48	51	37	38	35	37	46	51	69
-Date		1985	1984	1985	1984	1984	1984	1985	1984	1984	1984	1985	1985	APR 1984

See Reference Notes to this table on the following page.

CLEVELAND, OHIO

TABLE 2 PRECIPITATION (inches)   CLEVELAND, OHIO

YEAR	JAN	FEB	MAR	APR	MAY	JUNE	JULY	AUG	SEP	OCT	NOV	DEC	ANNUAL
#1956	2.75	3.23	4.60	3.61	5.49	4.14	3.74	5.14	1.56	0.85	1.14	2.70	38.95
1957	2.03	1.40	1.50	5.78	4.60	5.85	3.25	2.30	2.62	2.64	2.27	3.38	37.62
1958	1.49	0.96	0.78	3.31	2.14	4.67	4.82	3.98	3.62	4.76	0.71		32.87
1959	4.61	4.24	2.14	4.70	4.71	1.82	4.32	1.85	3.07	3.57	2.96	2.39	40.38
1960	2.33	1.87	2.06	2.78	3.56	3.01	3.33	6.95	1.84	1.02	2.07	1.09	31.91
1961	0.36	3.23	3.20	6.61	1.31	2.95	4.30	4.28	2.35	2.15	2.78	1.84	35.36
1962	2.83	1.85	1.73	1.78	1.91	2.95	3.42	1.30	4.39	3.60	2.77	3.05	31.58
1963	1.06	0.73	2.83	2.41	1.00	1.93	1.88	1.70	2.00	0.71	1.33	1.05	18.63
1964	1.45	1.49	5.21	4.87	3.02	2.06	3.37	3.82	0.74	1.78	0.92	2.67	31.40
1965	4.45	3.00	1.66	1.83	2.29	3.05	3.01	3.58	2.53	2.55	1.89	2.07	31.91
1966	1.53	2.31	2.26	3.61	2.21	1.83	3.89	3.48	1.66	1.18	5.16	2.84	31.96
1967	0.97	2.35	2.08	3.12	3.82	1.17	1.90	1.85	2.08	2.11	2.88	2.46	26.79
1968	3.27	0.79	2.07	2.25	4.08	2.32	3.58	1.82	3.36	2.90	4.35	3.94	34.73
1969	2.84	0.75	1.82	4.49	5.73	4.61	6.47	0.53	4.92	1.90	2.86	2.46	39.38
1970	1.28	1.35	2.32	2.64	2.95	4.98	4.14	0.92	3.16	3.98	3.69	2.25	33.66
1971	1.35	3.69	2.01	1.24	3.29	3.79	3.72	0.91	4.27	1.61	2.02	3.90	31.80
1972	1.95	2.01	2.97	3.40	3.74	9.06	4.44	6.38	4.91	1.64	4.58	3.26	48.34
1973	1.62	2.40	3.48	3.40	4.79	6.72	2.94	3.11	2.69	3.95	2.62	3.53	41.25
1974	2.56	2.43	3.88	3.64	4.78	3.57	1.90	3.29	3.06	1.19	4.72	4.86	39.88
1975	3.06	3.20	3.47	1.31	3.23	4.10	2.54	8.96	3.35	1.73	2.09	3.77	40.81
1976	3.38	3.97	3.11	2.17	2.94	3.64	3.48	3.50	3.71	2.54	0.80	1.57	34.81
1977	1.29	1.38	4.49	3.56	1.02	4.91	3.94	3.92	2.52	1.93	3.62	3.51	36.09
1978	3.67	0.48	2.17	3.02	3.01	3.30	2.40	3.58	3.68	3.23	1.19	2.96	32.69
1979	2.61	2.74	2.33	3.09	4.77	3.47	3.76	4.46	3.66	1.79	3.16	4.00	39.84
1980	1.18	1.27	3.61	2.65	3.13	2.69	4.77	4.38	3.51	2.38	1.29	2.10	32.61
1981	0.76	2.72	1.61	4.62	2.19	4.68	5.31	2.61	6.75	2.33	1.99	3.44	39.01
1982	4.00	1.41	3.77	1.62	2.65	5.01	1.21	2.66	4.82	0.93	5.17	3.68	36.93
1983	1.08	0.77	3.54	4.48	4.17	3.45	4.16	3.15	2.87	4.14	5.89	2.92	40.62
1984	1.25	3.82	3.80	2.29	5.95	3.40	3.35	5.51	2.43	2.20	3.95	3.38	41.33
1985	1.78	2.60	4.97	1.38	3.45	2.93	3.23	4.01	2.05	3.45	8.80	2.63	41.28
Record Mean	2.49	2.33	2.90	2.80	3.17	3.36	3.40	3.04	3.14	2.58	2.73	2.50	34.45

TABLE 3 AVERAGE TEMPERATURE (deg. F)   CLEVELAND, OHIO

YEAR	JAN	FEB	MAR	APR	MAY	JUNE	JULY	AUG	SEP	OCT	NOV	DEC	ANNUAL
#1956	27.6	31.3	36.0	45.9	57.8	68.7	71.7	71.7	61.2	58.7	42.7	38.0	50.9
1957	23.0	33.0	38.1	50.7	59.1	70.3	72.1	70.2	64.9	51.8	42.8	36.1	51.0
1958	28.2	23.2	35.5	51.4	58.7	64.4	73.5	70.4	64.9	55.1	44.5	23.3	49.4
1959	24.0	29.1	36.5	49.8	64.3	70.0	73.3	76.1	68.9	54.3	39.1	36.0	51.8
#1960	31.3	29.5	24.0	51.0	56.8	65.8	67.6	69.8	65.0	51.0	42.4	22.9	48.1
1961	21.6	31.6	40.0	43.4	54.3	65.0	71.3	70.9	68.5	56.8	42.5	29.7	49.7
1962	23.9	26.0	33.8	47.4	65.0	67.9	69.4	70.2	62.0	54.6	42.4	25.9	49.0
1963	18.0	17.5	39.1	48.4	54.7	67.8	71.3	66.7	60.3	54.9	43.9	21.9	47.4
1964	29.8	25.5	37.1	48.8	60.7	67.5	72.2	67.5	63.5	49.2	44.2	32.2	49.8
1965	27.2	28.0	31.7	45.6	63.2	66.9	69.0	68.7	67.4	51.3	42.5	37.4	49.9
1966	21.9	26.7	37.8	46.2	54.2	68.9	72.7	68.8	60.9	50.3	42.8	30.5	48.5
1967	32.4	25.9	37.2	49.7	52.3	71.7	69.7	68.8	61.7	54.0	38.6	34.7	49.7
1968	23.0	22.6	37.6	49.4	54.4	66.5	70.0	71.7	63.9	52.1	42.4	30.0	48.6
1969	25.4	27.9	34.3	49.4	58.6	65.5	72.4	71.8	63.6	51.9	40.2	27.2	49.0
1970	18.9	27.2	33.8	50.2	62.7	69.8	71.9	69.9	66.0	54.4	41.6	32.2	49.9
1971	21.4	27.9	31.6	43.2	56.5	71.0	69.5	68.9	67.7	59.9	41.4	38.1	49.8
1972	27.3	25.7	34.8	46.0	58.6	62.7	71.4	68.9	63.8	49.1	39.6	34.5	48.5
1973	30.4	27.9	46.5	50.2	56.7	70.4	72.6	73.2	66.3	57.7	44.6	34.3	52.6
1974	32.0	27.8	39.6	51.3	56.4	66.2	72.2	70.4	59.9	51.2	42.9	31.7	50.1
1975	31.9	30.4	34.6	41.8	62.3	69.8	71.3	72.3	58.6	53.8	47.0	32.0	50.5
1976	21.6	36.0	45.0	49.1	55.3	69.5	71.6	68.4	61.1	48.1	33.7	23.7	48.6
1977	11.0	25.0	42.7	51.4	61.8	63.3	73.1	69.6	65.6	52.6	45.4	29.2	49.3
1978	20.1	16.8	32.4	47.0	59.4	69.0	72.2	73.0	69.2	53.2	44.2	33.7	49.2
1979	22.0	19.1	42.9	46.6	56.9	66.9	71.1	71.5	65.0	52.4	42.3	33.7	49.2
1980	25.5	21.9	33.6	46.1	58.5	64.0	72.3	73.2	64.7	47.9	39.4	28.5	48.0
1981	20.1	31.5	36.0	50.6	55.7	68.2	71.3	70.0	62.4	50.0	42.6	30.6	49.1
1982	19.8	25.2	37.1	44.6	64.9	64.1	73.6		62.7	55.3	45.4	40.5	50.1
1983	30.7	33.9	40.8	47.1	55.7	69.0	75.2	73.7	65.1	53.4	43.9	23.2	51.0
1984	20.7	34.5	28.4	46.8	54.0	66.7	70.6	70.6	61.1	56.3	40.9	36.6	49.0
1985	20.8	25.2	40.3	53.6	60.4	62.7	71.1	68.9	64.9	54.0	46.0	24.3	49.4
Record Mean	26.9	27.7	36.0	47.2	58.2	67.8	72.1	70.6	64.6	53.7	41.6	31.2	49.8
Max	33.8	35.0	43.8	55.8	67.1	76.6	80.6	79.0	73.1	61.9	48.4	37.4	57.7
Min	20.0	20.4	28.3	38.5	49.3	59.0	63.6	62.2	56.0	45.4	34.8	25.0	41.9

REFERENCE NOTES FOR TABLES 1, 2, 3 and 6   (CLEVELAND, OH)

GENERAL

T - TRACE AMOUNT
BLANK ENTRIES DENOTE MISSING/UNREPORTED DATA.
# INDICATES A STATION OR INSTRUMENT RELOCATION.

SPECIFIC

TABLE 1

(a) - LENGTH OF RECORD IN YEARS. ALTHOUGH
      INDIVIDUAL MONTHS MAY BE MISSING.
  * LESS THAN .05

NORMALS — BASED ON THE 1951-1980 RECORD PERIOD.
EXTREMES — DATES ARE THE MOST RECENT OCCURRENCE.
WIND DIR. — NUMERALS SHOW TENS OF DEGREES
      CLOCKWISE FROM TRUE NORTH.
      "00" INDICATES CALM.
RESULTANT WIND DIRECTIONS ARE GIVEN TO WHOLE DEGREES.

EXCEPTIONS

TABLES 2, 3, and 6

RECORD MEANS ARE THROUGH THE CURRENT YEAR,
BEGINNING IN   1871 FOR TEMPERATURE
                    1871 FOR PRECIPITATION
                    1942 FOR SNOWFALL

TABLE 4 HEATING DEGREE DAYS Base 65 deg. F   CLEVELAND, OHIO

SEASON	JULY	AUG	SEP	OCT	NOV	DEC	JAN	FEB	MAR	APR	MAY	JUNE	TOTAL
1956-57	4	13	159	198	665	828	1295	891	825	471	221	29	5599
1957-58	0	6	97	403	656	891	1137	1164	907	413	218	73	5965
1958-59	0	21	86	316	608	1288	1261	998	877	451	133	43	6082
#1959-60	0	0	78	348	769	894	1039	1023	1262	430	265	58	6166
1960-61	38	7	67	427	671	1299	1336	930	771	640	341	95	6622
1961-62	19	3	74	258	668	1085	1264	1085	958	547	124	43	6128
1962-63	10	17	151	331	674	1206	1452	1327	793	500	328	52	6841
1963-64	30	32	152	191	627	1326	1084	1139	859	481	179	80	6180
1964-65	3	46	117	483	617	1009	1165	1032	1025	576	130	64	6267
1965-66	24	49	67	418	671	852	1328	1067	837	562	346	53	6274
1966-67	6	15	162	452	655	1063	1000	1087	858	461	393	17	6169
1967-68	21	19	137	351	784	934	1295	1224	845	459	328	59	6456
1968-69	26	34	93	414	672	1080	1220	1032	946	471	234	100	6322
1969-70	1	7	121	406	736	1166	1425	1052	960	462	154	39	6529
1970-71	9	12	86	332	696	1009	1344	1032	1031	650	277	16	6494
1971-72	9	13	63	168	704	828	1160	1133	930	564	196	124	5892
1972-73	32	27	95	485	752	937	1067	1033	569	450	254	3	5704
1973-74	3	9	73	234	605	946	1015	1035	777	419	280	49	5445
1974-75	2	5	176	423	660	1026	1021	962	934	691	154	38	6092
1975-76	5	4	187	345	532	1015	1336	836	614	493	309	25	5701
1976-77	0	25	150	519	932	1286	1672	1113	689	423	166	115	7090
1977-78	4	26	60	378	592	1103	1387	1343	1005	534	218	43	6693
1978-79	7	2	43	362	620	965	1328	1281	680	552	290	60	6190
1979-80	20	11	87	403	670	967	1218	1244	967	561	223	103	6474
1980-81	3	2	97	521	763	1125	1385	935	894	430	298	30	6483
1981-82	11	11	145	458	664	1059	1393	1109	860	608	78	75	6471
1982-83	5	42	136	310	586	760	1056	864	742	533	294	56	5384
1983-84	7	0	116	362	628	1291	1366	878	1126	544	347	19	6684
1984-85	16	17	174	270	716	877	1364	1110	757	370	187	99	5957
1985-86	2	7	118	338	565	1255							

TABLE 5 COOLING DEGREE DAYS Base 65 deg. F   CLEVELAND, OHIO

YEAR	JAN	FEB	MAR	APR	MAY	JUNE	JULY	AUG	SEP	OCT	NOV	DEC	TOTAL
1969	0	0	0	10	41	120	237	223	84	10	0	0	725
1970	0	0	0	22	89	189	230	171	121	10	0	0	832
1971	0	0	0	0	22	198	158	143	152	19	0	0	692
1972	0	0	0	1	5	63	239	157	64	0	0	0	529
1973	0	0	0	13	7	168	244	273	119	17	0	0	841
1974	0	0	0	14	18	91	231	180	30	3	2	0	569
1975	0	0	0	0	75	187	206	241	6	5	0	0	720
1976	0	0	3	23	14	167	214	138	39	2	0	0	600
1977	0	0	4	22	74	73	262	175	84	0	9	0	703
1978	0	0	0	0	53	170	237	256	177	3	0	0	896
1979	0	0	0	6	42	122	213	218	93	21	0	0	715
1980	0	0	0	0	27	83	235	263	97	0	0	0	705
1981	0	0	0	4	16	132	214	175	73	0	0	0	614
1982	0	0	0	3	84	54	278	140	73	17	6	6	661
1983	0	0	0	5	12	185	327	277	127	12	0	0	945
1984	0	0	0	3	13	159	139	197	60	5	0	0	576
1985	0	0	0	38	52	34	201	131	122	4	2	0	584

TABLE 6 SNOWFALL (inches)   CLEVELAND, OHIO

SEASON	JULY	AUG	SEP	OCT	NOV	DEC	JAN	FEB	MAR	APR	MAY	JUNE	TOTAL
1956-57	0.0	0.0	0.0	0.0	7.4	11.4	12.8	3.1	8.4	13.2	T	0.0	56.3
1957-58	0.0	0.0	0.0	2.5	2.5	7.5	4.4	9.4	4.6	0.2	0.0	0.0	31.1
1958-59	0.0	0.0	0.0	T	7.4	9.7	14.0	5.8	14.6	T	T	0.0	51.5
1959-60	0.0	0.0	0.0	T	7.2	5.3	2.8	14.6	19.5	0.5	T	0.0	49.9
1960-61	0.0	0.0	0.0	T	4.7	14.1	6.0	6.8	1.4	5.1	T	0.0	38.1
1961-62	0.0	0.0	0.0	T	0.9	4.3	6.0	16.2	8.9	1.0	0.0	0.0	37.3
1962-63	0.0	0.0	0.0	8.0	T	30.3	12.4	13.4	10.4	0.3	0.1	0.0	74.9
1963-64	0.0	0.0	0.0	0.0	0.1	14.1	16.9	15.7	8.5	0.5	0.0	0.0	55.8
1964-65	0.0	0.0	0.0	T	1.0	8.7	13.6	15.6	12.9	0.4	0.0	0.0	52.2
1965-66	0.0	0.0	0.0	T	1.2	1.2	15.3	10.1	7.0	2.5	T	0.0	37.3
1966-67	0.0	0.0	0.0	0.0	8.8	10.9	2.0	18.5	7.3	0.1	0.0	0.0	47.6
1967-68	0.0	0.0	0.0	0.1	9.1	8.3	14.5	8.9	7.7	0.2	T	0.0	43.3
1968-69	0.0	0.0	0.0	T	6.8	8.3	5.8	5.6	9.0	1.5	T	0.0	37.0
1969-70	0.0	0.0	0.0	0.6	6.6	17.4	10.5	6.6	11.5	0.2	T	0.0	53.4
1970-71	0.0	0.0	T	T	5.2	6.0	8.6	14.3	16.6	0.7	0.0	0.0	51.4
1971-72	0.0	0.0	0.0	0.0	5.3	1.9	15.0	14.8	6.3	2.3	0.0	0.0	45.6
1972-73	0.0	0.0	0.0	5.5	7.8	15.2	9.8	20.4	8.3	0.9	0.6	0.0	68.5
1973-74	0.0	0.0	0.0	T	3.3	13.8	8.9	16.9	7.1	6.4	2.1	0.0	58.5
1974-75	0.0	0.0	0.0	1.6	5.3	24.1	9.7	9.9	15.2	1.2	0.0	0.0	67.0
1975-76	0.0	0.0	0.0	0.0	5.6	13.1	21.5	6.8	5.8	1.6	T	0.0	54.4
1976-77	0.0	0.0	T	1.6	8.9	16.3	21.1	9.6	4.2	1.7	0.0	0.0	63.4
1977-78	0.0	0.0	0.0	T	9.7	23.1	42.8	10.8	3.5	0.2	0.0	0.0	90.1
1978-79	0.0	0.0	0.0	T	1.9	2.5	15.1	16.0	2.4	0.4	0.0	0.0	38.3
1979-80	0.0	0.0	0.0	0.2	0.5	4.0	11.3	19.2	3.5	T	T	0.0	38.7
1980-81	0.0	0.0	0.0	T	5.4	13.5	15.0	9.7	16.9	T	0.0	0.0	60.5
1981-82	0.0	0.0	0.0	4.0	2.9	27.1	28.1	7.6	17.6	13.2	0.0	0.0	100.5
1982-83	0.0	0.0	0.0	T	2.2	6.3	6.5	8.3	11.3	3.4	0.0	0.0	38.0
1983-84	0.0	0.0	0.0	0.0	7.1	13.0	12.9	27.1	19.3	T	0.0	0.0	79.4
1984-85	0.0	0.0	0.0	0.0	4.0	8.9	25.5	18.2	1.2	5.9	0.0	0.0	63.7
1985-86	0.0	0.0	0.0	0.0	T	23.4							
Record Mean	0.0	0.0	T	0.7	5.3	11.7	12.4	11.6	9.9	2.3	0.1	0.0	54.0

See Reference Notes, relative to all above tables, on preceding page.

Columbus is located in the center of the state and in the drainage area of the Ohio River. The airport is located at the eastern boundary of the city approximately 7 miles from the center of the business district.

Four nearly parallel streams run through or adjacent to the city. The Scioto River is the principal stream and flows from the northwest into the center of the city and then flows straight south toward the Ohio River. The Olentangy River runs almost due south and empties into the Scioto just west of the business district. Two minor streams run through portions of Columbus or skirt the eastern and southern fringes of the area. They are Alum Creek and Big Walnut Creek. Alum Creek empties into the Big Walnut southeast of the city and the Big Walnut empties into the Scioto a few miles downstream. The Scioto and Olentangy are gorge-like in character with very little flood plain and the two creeks have only a little more flood plain or bottomland.

The narrow valleys associated with the streams flowing through the city supply the only variation in the micro-climate of the area. The city proper shows the typical metropolitan effect with shrubs and flowers blossoming earlier than in the immediate surroundings and in retarding light frost on clear quiet nights. Many small areas to the southeast and to the north and northeast show marked effects of air drainage as evidenced by the frequent formation of shallow ground fog at daybreak during the summer and fall months and the higher frequency of frost in the spring and fall.

The average occurrence of the last freezing temperature in the spring within the city proper is mid-April, and the first freeze in the fall is very late October, but in the immediate surroundings there is much variation. For example, at Valley Crossing located at the southeastern outskirts of the city, the average occurrence of the last 32 degree temperature in the spring is very early May, while the first 32 degree temperature in the fall is mid-October.

The records show a high frequency of calm or very low wind speeds during the late evening and early morning hours, from June through September. The rolling landscape is conducive to air drainage and from the Weather Service location at the airport the air drainage is toward the northwest with the wind direction indicated as southeast. Air drainage takes place at speeds generally 4 mph or less and frequently provides the only perceptible breeze during the night.

Columbus is located in the area of changeable weather. Air masses from central and northwest Canada frequently invade this region. Air from the Gulf of Mexico often reachs central Ohio during the summer and to a much lesser extent in the fall and winter. There are also occasional weather changes brought about by cool outbreaks from the Hudson Bay region of Canada, especially during the spring months. At infrequent intervals the general circulation will bring showers or snow to Columbus from the Atlantic. Although Columbus does not have a wet or dry season as such, the month of October usually has the least amount of precipitation.

TABLE 1   NORMALS, MEANS AND EXTREMES

COLUMBUS, OHIO

LATITUDE: 40°00'N   LONGITUDE: 82°53'W   ELEVATION: FT. GRND 813 BARO 816   TIME ZONE: EASTERN   WBAN: 14821

	(a)	JAN	FEB	MAR	APR	MAY	JUNE	JULY	AUG	SEP	OCT	NOV	DEC	YEAR
TEMPERATURE °F:														
Normals														
-Daily Maximum		34.7	38.1	49.3	62.3	72.6	81.3	84.4	83.0	76.9	65.0	50.7	39.4	61.5
-Daily Minimum		19.4	21.5	30.6	40.5	50.2	59.0	63.2	61.7	54.6	42.8	33.5	24.7	41.8
-Monthly		27.1	29.8	40.0	51.4	61.4	70.2	73.8	72.3	65.8	53.9	42.1	32.0	51.6
Extremes														
-Record Highest	46	74	73	85	89	94	102	100	101	100	90	80	76	102
-Year		1950	1957	1945	1948	1941	1944	1955	1983	1951	1951	1950	1982	JUN 1944
-Record Lowest	46	-19	-13	-6	14	25	35	43	39	31	20	5	-12	-19
-Year		1985	1977	1984	1982	1966	1972	1972	1965	1963	1962	1976	1983	JAN 1985
NORMAL DEGREE DAYS:														
Heating (base 65°F)		1175	986	775	408	178	19	0	5	78	355	687	1020	5686
Cooling (base 65°F)		0	0	0	0	66	175	273	235	102	11	0	0	862
% OF POSSIBLE SUNSHINE	34	36	42	44	51	56	60	60	60	61	55	37	30	49
MEAN SKY COVER (tenths)														
Sunrise - Sunset	36	7.8	7.5	7.4	7.0	6.6	6.2	6.0	5.9	5.6	5.7	7.4	7.8	6.7
MEAN NUMBER OF DAYS:														
Sunrise to Sunset														
-Clear	36	4.2	4.0	4.9	5.4	6.0	6.3	6.6	7.1	9.5	10.3	4.9	3.9	73.0
-Partly Cloudy	36	6.3	6.2	6.7	7.9	10.4	11.3	13.4	12.9	9.1	7.3	6.6	6.1	104.2
-Cloudy	36	20.5	18.0	19.5	16.7	14.6	12.5	11.0	11.1	11.5	13.3	18.4	20.9	188.1
Precipitation														
.01 inches or more	46	13.3	11.5	14.0	12.9	12.7	11.0	10.8	9.4	8.3	8.8	11.5	12.7	136.8
Snow, Ice pellets														
1.0 inches or more	37	3.0	2.1	1.5	0.2	0.0	0.0	0.0	0.0	0.0	0.*	0.8	1.9	9.6
Thunderstorms	46	0.4	0.5	2.2	4.1	6.4	8.0	8.1	6.4	3.0	1.2	0.9	0.3	41.4
Heavy Fog Visibility														
1/4 mile or less	36	2.0	1.6	1.1	0.6	0.9	1.1	1.2	1.8	1.8	1.5	1.3	1.5	16.5
Temperature °F														
-Maximum														
90° and above	26	0.0	0.0	0.0	0.0	0.5	3.2	5.3	3.0	1.4	0.0	0.0	0.0	13.4
32° and below	26	14.8	10.1	3.0	0.1	0.0	0.0	0.0	0.0	0.0	0.0	1.2	9.2	38.4
-Minimum														
32° and below	26	27.5	23.9	18.9	6.8	0.7	0.0	0.0	0.0	0.1	3.9	14.0	23.9	119.8
0° and below	26	3.3	1.7	0.2	0.0	0.0	0.0	0.0	0.0	0.0	0.0	0.0	0.8	6.0
AVG. STATION PRESS. (mb)	13	988.5	988.5	985.8	985.9	985.0	986.0	987.1	988.3	988.9	989.4	988.8	988.7	987.6
RELATIVE HUMIDITY (%)														
Hour 01	26	74	73	69	69	76	80	82	84	82	78	77	76	77
Hour 07 (Local Time)	26	76	75	73	74	79	81	84	87	87	81	80	78	80
Hour 13	26	67	64	57	52	54	55	56	58	57	55	64	69	59
Hour 19	26	69	66	60	54	57	58	60	63	65	64	69	72	63
PRECIPITATION (inches):														
Water Equivalent														
-Normal		2.75	2.18	3.23	3.41	3.76	4.01	4.01	3.70	2.76	1.91	2.64	2.61	36.97
-Maximum Monthly	46	8.29	4.60	9.59	6.36	9.11	9.75	9.46	8.63	6.76	5.24	10.67	5.07	10.67
-Year		1950	1981	1964	1964	1968	1958	1958	1979	1979	1954	1985	1951	NOV 1985
-Minimum Monthly	46	0.53	0.29	0.61	0.67	0.95	0.71	0.48	0.58	0.51	0.11	0.60	0.46	0.11
-Year		1944	1978	1941	1971	1977	1984	1940	1951	1963	1963	1976	1955	OCT 1963
-Maximum in 24 hrs	38	4.81	2.15	3.40	2.37	2.72	2.93	3.82	3.79	4.86	1.87	2.47	1.74	4.86
-Year		1959	1975	1964	1957	1968	1958	1969	1972	1979	1965	1985	1978	SEP 1979
Snow, Ice pellets														
-Maximum Monthly	38	34.4	16.4	13.5	7.1	T				T	1.3	15.2	17.3	34.4
-Year		1978	1979	1962	1973	1966				1967	1962	1950	1960	JAN 1978
-Maximum in 24 hrs	38	7.5	8.9	8.6	6.3	T				T	1.3	8.2	8.7	8.9
-Year		1978	1971	1962	1973	1966				1967	1962	1950	1960	FEB 1971
WIND:														
Mean Speed (mph)	36	10.3	10.1	10.7	10.1	8.5	7.6	6.9	6.5	6.7	7.7	9.5	9.9	8.7
Prevailing Direction														
through 1963		SSW	NW	SSW	WNW	S	SSW	SSW	NNW	S	S	S	W	S
Fastest Mile														
-Direction	29	W	W	NW	W	NW	NW	W	SW	N	SW	N	SW	NW
-Speed (MPH)	29	56	57	63	56	54	47	51	43	38	43	61	47	63
-Year		1959	1956	1955	1970	1964	1966	1976	1979	1963	1979	1952	1971	MAR 1955
Peak Gust														
-Direction	2	W	E	W	S	W	SW	N	W	S	NW	W	W	W
-Speed (mph)	2	51	44	53	52	52	40	47	56	38	37	39	48	56
-Date		1985	1984	1985	1985	1985	1984	1985	1984	1985	1985	1985	1985	AUG 1984

See Reference Notes to this table on the following page.

TABLE 2 PRECIPITATION (inches) COLUMBUS, OHIO

YEAR	JAN	FEB	MAR	APR	MAY	JUNE	JULY	AUG	SEP	OCT	NOV	DEC	ANNUAL
1956	2.13	4.33	3.49	4.06	4.85	3.22	4.92	2.23	1.37	1.15	0.97	3.21	35.93
1957	1.94	1.87	1.61	5.70	5.41	5.20	2.98	0.89	2.99	1.27	3.23	4.68	37.77
1958	1.87	0.82	1.50	4.09	3.54	9.75	9.46	2.81	3.01	1.27	1.56	0.69	40.37
1959	7.44	3.08	1.45	3.60	4.15	1.55	3.23	1.55	1.55	3.33	3.71	2.27	36.91
1960	2.31	2.57	1.11	1.60	5.82	2.47	4.54	2.48	0.83	2.55	1.60	1.62	29.50
1961	0.65	2.90	4.83	4.58	2.90	3.49	4.61	2.73	1.05	1.18	3.49	2.42	34.83
1962	3.17	3.46	2.43	1.33	2.31	2.26	3.59	3.62	2.06	2.94	1.76	0.55	31.24
1963	1.39	1.01	7.14	3.27	1.61	1.25	2.90	3.67	0.51	0.11	0.80	0.85	24.51
1964	1.82	1.68	9.59	6.36	1.95	5.71	2.97	3.19	1.66	0.38	1.81	4.09	41.21
1965	2.70	3.76	2.90	5.90	4.00	2.42	3.76	4.62	6.18	3.98	1.19	1.24	42.65
1966	2.87	2.59	1.04	4.89	3.13	1.28	5.91	4.90	3.56	0.79	4.05	3.33	38.34
1967	0.78	2.46	4.40	3.29	4.59	2.92	4.22	1.51	2.63	1.39	3.22	2.55	33.96
1968	2.22	0.38	3.01	2.20	9.11	2.96	2.80	3.08	1.77	2.59	4.26	3.40	37.78
1969	3.40	1.17	1.32	3.10	3.04	8.19	7.65	3.25	1.40	1.52	3.87	2.30	40.21
1970	1.60	1.68	3.04	5.52	5.37	5.65	3.73	3.94	3.95	2.07	2.88	2.50	41.93
1971	1.57	3.16	2.70	0.67	3.66	4.16	4.22	2.81	3.08	1.32	1.73	4.61	33.69
1972	1.40	1.74	2.86	3.74	6.56	3.98	2.60	7.96	5.13	1.74	4.40	3.49	45.60
1973	2.46	1.29	3.43	3.72	3.36	8.77	4.07	4.97	2.82	3.29	5.37	2.70	46.25
1974	2.40	2.30	4.38	2.66	3.29	5.04	1.14	4.88	3.32	1.51	3.39	2.68	36.99
1975	3.21	3.47	4.10	2.71	3.17	3.53	2.04	4.51	5.46	2.29	1.54	3.01	39.04
1976	3.15	2.03	2.17	1.44	1.41	4.52	5.12	5.08	2.54	2.86	0.60	0.93	31.85
1977	1.57	1.02	3.88	4.04	0.95	4.02	2.52	4.76	3.48	2.57	3.77	3.54	36.12
1978	5.89	0.29	2.98	3.02	4.15	3.65	1.81	5.23	1.16	2.39	1.56	5.01	37.14
1979	3.32	2.88	1.01	4.01	3.27	4.23	8.06	8.63	6.76	1.26	3.91	1.83	49.17
1980	1.69	1.38	3.77	1.59	4.56	5.17	4.58	6.26	1.86	2.53	2.07	1.96	37.42
1981	0.70	4.60	1.11	5.38	6.50	5.73	4.14	1.41	2.28	1.40	1.65	2.88	37.78
1982	4.77	1.49	3.99	1.90	4.68	3.37	3.90	1.02	4.25	0.92	5.19	3.84	39.32
1983	1.20	0.74	1.69	5.58	5.06	4.59	2.80	2.23	1.91	4.45	5.00	3.16	38.41
1984	1.07	1.97	3.89	3.10	4.93	0.71	3.15	2.96	1.48	2.91	4.41	2.84	33.42
1985	1.31	1.67	3.78	0.73	4.96	1.41	6.88	2.34	1.18	1.93	10.67	1.81	38.67
Record Mean	2.89	2.44	3.97	3.16	3.72	3.70	3.73	3.26	2.60	2.22	2.80	2.59	37.09

TABLE 3 AVERAGE TEMPERATURE (deg. F) COLUMBUS, OHIO

YEAR	JAN	FEB	MAR	APR	MAY	JUNE	JULY	AUG	SEP	OCT	NOV	DEC	ANNUAL
1956	27.7	35.6	39.5	48.5	61.3	71.1	73.8	73.3	63.4	59.4	43.1	40.8	53.1
1957	25.8	36.5	40.7	54.1	62.7	72.3	75.1	73.3	66.3	51.5	43.4	36.8	53.2
1958	28.4	24.9	37.7	52.4	61.4	67.2	74.4	72.6	66.1	54.4	43.9	23.4	50.6
#1959	26.0	32.7	38.8	52.6	66.8	71.5	75.4	77.3	66.9	55.1	39.5	36.4	53.5
1960	34.1	28.8	28.4	55.3	59.4	69.7	71.8	74.9	68.3	55.1	44.4	25.1	51.3
1961	23.5	36.9	45.3	47.1	56.5	67.2	73.7	73.1	70.0	55.3	41.8	29.7	51.8
1962	24.6	28.8	36.9	48.7	67.2	72.3	71.3	71.6	61.5	54.4	40.8	24.9	50.2
1963	21.7	22.4	42.8	51.1	58.8	70.8	73.9	69.9	63.7	59.9	44.6	21.6	50.1
1964	30.1	27.7	41.1	53.6	63.4	70.3	74.4	71.2	65.0	51.1	45.1	33.9	52.3
1965	28.3	29.1	34.8	51.4	68.1	69.4	72.0	71.0	68.3	54.4	43.3	36.6	52.2
1966	22.2	29.4	41.8	49.4	56.1	71.4	75.6	71.3	63.2	50.7	42.4	32.6	50.5
1967	34.8	25.8	40.4	52.6	55.4	71.8	71.7	68.3	64.3	51.9	36.7	34.3	50.3
1968	23.6	25.7	43.3	53.1	58.5	71.5	74.1	73.0	65.5	54.1	44.0	31.1	51.5
1969	27.0	31.4	35.3	51.5	61.7	68.5	74.1	71.2	63.8	53.9	39.3	26.8	50.4
1970	20.6	28.5	36.9	53.4	64.9	69.8	73.5	72.4	68.9	55.6	42.4	34.3	51.8
1971	24.3	30.8	36.6	49.0	58.0	73.5	70.5	68.9	67.8	59.8	40.4	38.4	51.5
1972	28.2	27.7	37.0	48.8	60.8	63.6	71.9	70.1	64.6	49.6	40.5	36.1	49.9
1973	31.2	31.4	50.4	51.1	59.5	72.6	74.3	74.2	58.6	58.1	33.7	34.0	54.3
1974	33.2	31.2	44.6	54.3	60.8	67.7	74.4	74.0	62.2	52.8	44.5	34.0	52.8
1975	32.5	33.4	37.3	46.7	66.6	72.4	75.1	77.3	62.7	54.7	47.5	33.5	53.3
1976	24.0	37.4	46.5	50.9	58.1	70.5	72.0	68.3	61.7	47.5	33.9	24.8	49.6
1977	11.4	26.5	45.6	54.8	66.8	67.5	76.2	72.0	68.2	52.0	45.1	29.5	51.3
1978	19.0	16.6	34.5	50.6	59.6	70.4	73.5	73.2	69.8	51.5	44.4	34.4	49.8
1979	21.4	19.3	44.3	50.1	60.5	69.6	71.8	71.9	65.1	53.3	43.6	35.1	50.5
1980	29.3	25.2	37.2	49.5	62.4	67.4	75.9	75.9	60.3	50.8	40.8	32.5	51.3
1981	23.3	34.0	40.2	55.8	59.5	70.9	71.9	70.4	62.3	51.1	40.9	30.6	50.9
1982	21.2	29.2	40.4	46.4	66.8	65.8	74.4	69.2	63.5	56.2	45.4	40.4	51.6
1983	29.9	34.0	43.3	48.4	57.6	69.4	76.7	76.2	67.1	54.5	44.0	24.8	52.2
1984	23.3	37.4	32.3	50.0	57.6	73.1	71.2	72.9	63.1	59.4	40.6	39.5	51.7
1985	21.7	26.0	43.7	56.3	62.6	66.9	72.7	71.2	66.6	57.3	48.2	26.0	51.6
Record Mean	28.7	30.6	40.0	51.1	61.8	70.8	74.7	72.8	66.4	54.8	42.3	32.4	52.2
Max	36.2	38.6	49.0	61.3	72.4	81.2	85.1	83.1	77.1	65.1	50.4	39.5	61.6
Min	21.2	22.6	31.0	40.9	51.2	60.3	64.2	62.5	55.8	44.6	34.1	25.2	42.8

REFERENCE NOTES FOR TABLES 1, 2, 3 and 6   (COLUMBUS, OH)

GENERAL

T - TRACE AMOUNT
BLANK ENTRIES DENOTE MISSING/UNREPORTED DATA.
# INDICATES A STATION OR INSTRUMENT RELOCATION.

SPECIFIC

TABLE 1

(a) - LENGTH OF RECORD IN YEARS. ALTHOUGH
INDIVIDUAL MONTHS MAY BE MISSING.
* LESS THAN .05

NORMALS — BASED ON THE 1951-1980 RECORD PERIOD.
EXTREMES — DATES ARE THE MOST RECENT OCCURRENCE.
WIND DIR. — NUMERALS SHOW TENS OF DEGREES
CLOCKWISE FROM TRUE NORTH.
"00" INDICATES CALM.
RESULTANT WIND DIRECTIONS ARE GIVEN TO WHOLE DEGREES.

EXCEPTIONS

TABLES 2, 3, and 6

RECORD MEANS ARE THROUGH THE CURRENT YEAR,
BEGINNING IN   1879 FOR TEMPERATURE
1879 FOR PRECIPITATION
1948 FOR SNOWFALL

TABLE 4 HEATING DEGREE DAYS Base 65 deg. F COLUMBUS, OHIO

SEASON	JULY	AUG	SEP	OCT	NOV	DEC	JAN	FEB	MAR	APR	MAY	JUNE	TOTAL
1956-57	0	8	114	173	651	745	1211	792	745	379	138	9	4965
1957-58	0	0	73	413	640	867	1131	1115	841	383	141	28	5632
1958-59	1	7	68	326	632	1284	1201	898	809	366	98	34	5724
#1959-60	0	0	62	336	758	878	950	1043	1128	315	205	7	5682
1960-61	1	0	33	306	612	1229	1282	782	601	536	267	53	5702
1961-62	2	2	60	297	689	1087	1249	1005	866	496	76	2	5831
1962-63	8	3	165	342	721	1238	1334	1186	683	424	208	18	6330
1963-64	6	14	93	174	603	1336	1076	1075	735	340	113	40	5605
1964-65	0	28	90	424	589	957	1132	999	932	403	39	26	5619
1965-66	4	28	53	330	645	873	1325	989	710	465	286	29	5737
1966-67	1	5	106	440	671	999	929	1093	756	374	300	9	5683
1967-68	7	23	164	406	843	942	1276	1133	667	351	205	6	6023
1968-69	6	20	57	362	624	1043	1173	933	916	402	143	54	5733
1969-70	0	2	107	359	763	1175	1017	861	861	365	119	13	6150
1970-71	10	0	58	297	674	944	1256	950	871	475	230	4	5769
1971-72	3	5	52	181	733	815	1133	1077	860	482	146	101	5588
1972-73	22	18	77	473	727	889	1041	934	444	427	184	0	5236
1973-74	0	3	35	219	589	963	977	940	628	332	178	31	4895
1974-75	0	0	130	374	609	954	999	878	850	542	73	18	5427
1975-76	0	0	110	321	520	973	1263	791	570	440	229	4	5221
1976-77	1	25	118	537	925	1241	1659	1071	601	324	91	64	6657
1977-78	1	17	36	394	594	1091	1420	1346	938	424	223	23	6507
1978-79	0	0	38	411	610	943	1346	1270	637	449	185	18	5907
1979-80	11	16	83	376	632	920	1099	1148	855	458	133	53	5784
1980-81	0	0	46	435	717	1000	1286	864	761	287	195	14	5605
1981-82	8	5	141	429	713	1061	1351	997	758	556	45	33	6097
1982-83	3	19	107	304	585	759	1081	863	669	493	239	30	5152
1983-84	6	0	83	325	626	1236	1284	796	1007	447	254	3	6067
1984-85	6	3	143	182	727	782	1339	1086	654	286	134	35	5377
1985-86	0	2	96	249	500	1202							

TABLE 5 COOLING DEGREE DAYS Base 65 deg. F COLUMBUS, OHIO

YEAR	JAN	FEB	MAR	APR	MAY	JUNE	JULY	AUG	SEP	OCT	NOV	DEC	TOTAL
1969	0	0	0	3	45	165	290	203	76	18	0	0	800
1970	0	0	0	22	125	166	281	237	179	13	0	0	1023
1971	0	0	0	0	21	266	181	135	144	24	0	0	771
1972	0	0	0	1	24	67	245	183	71	0	0	0	591
1973	0	0	3	14	17	236	295	292	160	25	0	0	1042
1974	0	0	4	20	58	117	296	286	52	3	0	0	836
1975	0	0	0	0	1	130	248	320	389	48	10	1	1147
1976	0	0	0	3	23	23	174	223	135	25	2	0	608
1977	0	0	0	8	24	151	148	354	242	139	0	7	1073
1978	0	0	0	0	59	190	270	261	188	0	0	0	968
1979	0	0	0	7	54	163	230	239	93	22	0	0	808
1980	0	0	0	0	61	132	343	344	151	3	0	0	1034
1981	0	0	0	16	32	198	231	181	64	4	0	0	726
1982	0	0	0	4	111	66	301	154	67	39	7	4	753
1983	0	0	1	2	17	167	377	355	152	9	0	0	1080
1984	0	0	0	8	30	253	205	254	94	14	0	0	858
1985	0	0	2	32	64	97	245	201	152	19	2	0	814

TABLE 6 SNOWFALL (inches) COLUMBUS, OHIO

SEASON	JULY	AUG	SEP	OCT	NOV	DEC	JAN	FEB	MAR	APR	MAY	JUNE	TOTAL
1956-57	0.0	0.0	0.0	0.0	2.6	5.0	10.2	3.0	3.4	0.8	T	0.0	25.0
1957-58	0.0	0.0	0.0	T	T	8.5	2.3	4.8	5.9	T	0.0	0.0	21.5
1958-59	0.0	0.0	0.0	0.0	6.2	7.5	6.5	1.8	5.1	0.1	0.0	0.0	27.2
1959-60	0.0	0.0	0.0	0.0	3.5	5.5	0.5	9.7	10.8	T	0.0	0.0	30.0
1960-61	0.0	0.0	0.0	0.0	0.8	17.3	7.6	13.1	0.7	2.7	0.0	0.0	42.2
1961-62	0.0	0.0	0.0	0.0	1.3	8.8	3.9	8.8	13.5	1.1	0.0	0.0	37.4
1962-63	0.0	0.0	0.0	1.3	T	9.5	10.1	8.7	2.4	T	0.0	0.0	32.0
1963-64	0.0	0.0	0.0	0.0	1.2	7.3	12.3	11.3	2.9	T	0.0	0.0	35.0
1964-65	0.0	0.0	0.0	0.0	1.0	3.2	10.2	7.6	8.6	T	0.0	0.0	30.6
1965-66	0.0	0.0	0.0	0.0	0.2	1.1	7.6	6.7	1.2	0.7	T	0.0	17.5
1966-67	0.0	0.0	0.0	T	10.4	6.4	2.8	15.6	11.4	0.0	0.0	0.0	46.6
1967-68	0.0	0.0	T	T	6.5	5.2	11.6	2.8	6.1	T	0.0	0.0	32.2
1968-69	0.0	0.0	0.0	0.0	1.0	6.8	2.5	2.5	1.9	0.5	0.0	0.0	15.2
1969-70	0.0	0.0	0.0	0.0	1.8	9.7	18.4	3.2	10.3	0.9	0.0	0.0	44.3
1970-71	0.0	0.0	0.0	0.0	0.9	1.4	6.5	12.3	12.3	T	0.0	0.0	33.4
1971-72	0.0	0.0	0.0	0.0	5.0	5.8	6.6	5.0	5.0	0.6	0.0	0.0	23.6
1972-73	0.0	0.0	0.0	T	6.3	2.8	4.4	1.8	2.1	7.1	0.0	0.0	24.5
1973-74	0.0	0.0	0.0	0.0	T	6.4	2.3	5.0	4.5	0.3	0.0	0.0	18.5
1974-75	0.0	0.0	0.0	T	0.3	7.4	8.1	3.7	2.6	T	0.0	0.0	22.1
1975-76	0.0	0.0	0.0	0.0	1.1	2.9	12.4	1.8	1.0	T	0.0	0.0	19.2
1976-77	0.0	0.0	0.0	T	3.1	4.6	18.1	6.7	0.3	0.1	0.0	0.0	32.9
1977-78	0.0	0.0	0.0	0.0	2.2	7.5	34.4	4.5	5.5	T	0.0	0.0	54.1
1978-79	0.0	0.0	0.0	T	1.3	1.8	17.3	16.4	0.8	0.3	0.0	0.0	37.9
1979-80	0.0	0.0	0.0	0.0	0.1	0.2	7.0	8.1	1.2	T	0.0	0.0	16.6
1980-81	0.0	0.0	0.0	T	8.0	7.3	7.8	3.7	3.3	0.0	0.0	0.0	30.1
1981-82	0.0	0.0	0.0	0.0	1.9	9.8	11.8	3.7	3.2	4.7	0.0	0.0	35.1
1982-83	0.0	0.0	0.0	0.0	T	1.5	2.6	4.5	2.8	0.1	0.0	0.0	11.5
1983-84	0.0	0.0	0.0	0.0	0.5	5.7	9.0	10.8	9.8	0.3	0.0	0.0	36.1
1984-85	0.0	0.0	0.0	0.0	0.9	7.3	21.9	12.5	T	0.8	0.0	0.0	43.4
1985-86	0.0	0.0	0.0	0.0	0.0	8.6							
Record Mean	0.0	0.0	T	T	2.6	5.8	8.9	6.3	4.5	0.7	T	0.0	28.8

See Reference Notes, relative to all above tables, on preceding page.

Dayton is located near the center of the Miami River Valley, which is a nearly flat plain, 50 to 200 feet below the general elevation of the adjacent rolling country. Three Miami River tributaries, the Mad River, the Stillwater River, and Wolf Creek converge, fanwise, from the north to join the master stream within the city limits of Dayton. Heavy rains in March 1913 caused the worst flood disaster in the history of the Miami Valley. During the flood more than 400 people lost their lives and property damage amounted to $100 million. After the 1913 flood, dams were built on the streams north of Dayton, forming retarding basins. No floods have occurred at Dayton since the construction of these dams.

The elevation of the city of Dayton is about 750 feet. Terrain north of the city slopes gradually upward to about 1,100 feet at Indian Lake. Ten miles southeast of Indian Lake, near Bellefontaine, is the highest point in the state, with an elevation of about 1,550 feet. South of the city, the terrain slopes gradually downward to about 450 feet where the Miami River empties into the Ohio River.

Precipitation, which is rather evenly distributed throughout the year, and moderate temperatures help to make the Miami Valley a rich agricultural region. High relative humidities during much of the year cause some discomfort to people with allergies. Temperatures of zero or below will be experienced in about four years out of five, while 100 degrees or higher will be recorded in about one year out of five. Extreme temperatures are usually of short duration. The downward slope of about 700 feet in the 163 miles of the Miami River may have some moderating influence on the winter temperatures in the Miami Valley.

The average last occurrence in the spring of freezing temperatures is mid–April, and the average first occurrence in the autumn is late October.

Cold, polar air, flowing across the Great Lakes, causes much cloudiness during the winter, and is accompanied by frequent snow flurries. These add little to the total snowfall.

TABLE 1   NORMALS, MEANS AND EXTREMES

DAYTON, OHIO

LATITUDE: 39°54'N   LONGITUDE: 84°12'W   ELEVATION: FT. GRND   995 BARO   01005   TIME ZONE: EASTERN   WBAN: 93815

	(a)	JAN	FEB	MAR	APR	MAY	JUNE	JULY	AUG	SEP	OCT	NOV	DEC	YEAR
TEMPERATURE °F:														
Normals														
-Daily Maximum		34.5	38.0	48.6	62.0	72.4	81.6	84.9	83.4	77.1	65.1	50.5	39.3	61.5
-Daily Minimum		18.8	21.2	30.3	41.0	51.2	60.4	64.3	62.6	55.5	43.9	33.7	24.3	42.3
-Monthly		26.6	29.6	39.5	51.5	61.8	71.0	74.6	73.0	66.3	54.5	42.1	31.8	51.9
Extremes														
-Record Highest	42	71	71	82	89	93	99	102	100	101	89	79	72	102
-Year		1950	1976	1981	1962	1962	1944	1954	1964	1954	1951	1975	1982	JUL 1954
-Record Lowest	42	-24	-16	-7	15	27	40	44	40	32	21	-2	-15	-24
-Year		1985	1951	1980	1972	1947	1972	1972	1965	1974	1962	1958	1983	JAN 1985
NORMAL DEGREE DAYS:														
Heating (base 65°F)		1190	991	791	405	171	15	0	0	68	342	687	1029	5689
Cooling (base 65°F)		0	0	0	0	72	195	301	252	110	17	0	0	947
% OF POSSIBLE SUNSHINE	42	41	45	49	53	60	67	68	68	66	59	41	37	55
MEAN SKY COVER (tenths)														
Sunrise - Sunset	42	7.6	7.3	7.4	7.0	6.7	6.2	5.8	5.7	5.5	5.6	7.3	7.6	6.6
MEAN NUMBER OF DAYS:														
Sunrise to Sunset														
-Clear	42	4.8	5.2	4.6	5.5	6.0	6.6	7.7	8.6	9.7	10.6	5.2	4.3	79.0
-Partly Cloudy	42	6.1	5.9	7.3	7.3	9.5	10.3	12.5	11.8	9.0	7.6	6.8	6.5	100.6
-Cloudy	42	20.1	17.2	19.0	17.2	15.5	13.1	10.8	10.5	11.2	12.8	18.0	20.2	185.7
Precipitation														
.01 inches or more	42	13.0	10.8	13.0	12.8	12.1	10.4	10.1	9.5	8.1	8.7	11.3	12.1	131.9
Snow, Ice pellets														
1.0 inches or more	42	2.6	1.9	1.7	0.2	0.0	0.0	0.0	0.0	0.0	0.*	0.6	2.0	9.0
Thunderstorms	42	0.5	0.5	2.5	4.4	6.2	7.4	7.2	6.1	3.1	1.4	0.8	0.3	40.4
Heavy Fog Visibility 1/4 mile or less	42	3.6	2.7	1.9	0.8	1.2	1.0	1.3	1.8	1.6	1.6	1.9	3.3	23.0
Temperature °F														
-Maximum														
90° and above	22	0.0	0.0	0.0	0.0	0.4	3.6	7.0	3.5	1.4	0.0	0.0	0.0	15.9
32° and below	22	14.9	9.8	2.6	0.1	0.0	0.0	0.0	0.0	0.0	0.0	0.9	7.6	35.9
-Minimum														
32° and below	22	27.7	23.6	18.8	6.0	0.4	0.0	0.0	0.0	0.*	3.5	14.0	23.5	117.5
0° and below	22	3.6	2.1	0.2	0.0	0.0	0.0	0.0	0.0	0.0	0.0	0.0	0.9	6.9
AVG. STATION PRESS.(mb)	13	981.9	981.9	979.2	979.3	978.6	979.7	980.9	982.0	982.5	982.9	982.0	982.0	981.1
RELATIVE HUMIDITY (%)														
Hour 01	22	74	74	74	71	73	76	78	81	80	76	76	77	76
Hour 07	22	77	77	78	75	76	79	81	85	86	81	80	79	80
Hour 13 (Local Time)	22	69	66	62	55	54	54	55	57	57	57	66	71	60
Hour 19	22	70	69	65	58	57	57	57	58	63	65	71	75	64
PRECIPITATION (inches):														
Water Equivalent														
-Normal		2.57	2.11	3.08	3.43	3.69	3.81	3.37	3.10	2.39	2.01	2.64	2.51	34.71
-Maximum Monthly	42	9.86	4.57	7.65	6.69	7.76	10.89	7.21	8.03	5.69	5.54	8.07	5.07	10.89
-Year		1950	1950	1964	1947	1957	1958	1955	1974	1965	1983	1985	1951	JUN 1958
-Minimum Monthly	42	0.30	0.14	1.07	0.56	1.55	0.32	0.47	0.34	0.27	0.10	0.48	0.36	0.10
-Year		1981	1947	1966	1962	1964	1962	1974	1967	1963	1944	1949	1955	OCT 1944
-Maximum in 24 hrs	42	4.30	2.79	2.87	3.10	2.93	3.76	3.29	3.62	2.60	2.31	2.93	1.71	4.30
-Year		1959	1959	1964	1977	1968	1981	1955	1974	1981	1983	1955	1945	JAN 1959
Snow, Ice pellets														
-Maximum Monthly	42	40.2	17.5	13.8	4.9	T					2.0	12.7	15.6	40.2
-Year		1978	1979	1984	1974	1966					1962	1950	1960	JAN 1978
-Maximum in 24 hrs	42	12.2	7.7	11.3	4.7	T					2.0	10.0	7.6	12.2
-Year		1978	1984	1968	1974	1966					1962	1950	1974	JAN 1978
WIND:														
Mean Speed (mph)	42	11.7	11.6	12.1	11.7	9.8	9.0	8.0	7.5	8.2	9.0	11.1	11.4	10.1
Prevailing Direction through 1963		S	WNW	WNW	SSW	SSW	SSW	SSW	SSW	SSW	SSW	S	SSW	SSW
Fastest Mile														
-Direction (!!!)	41	W	SW	W	S	SW	NW	W	SW	W	SW	SW	SW	NW
-Speed (MPH)	41	59	66	70	60	61	78	52	49	65	56	57	47	78
-Year		1959	1946	1948	1947	1970	1950	1972	1947	1950	1954	1952	1982	JUN 1950
Peak Gust														
-Direction (!!!)	2	W	NW	W	SW	W	SW	NW	E	S	W	S	W	SW
-Speed (mph)	2	46	43	58	62	35	60	32	28	37	36	40	59	62
-Date		1985	1985	1985	1985	1984	1984	1984	1984	1985	1985	1985	1985	APR 1985

See reference Notes to this table on the following page.

TABLE 2 PRECIPITATION (inches) DAYTON, OHIO

YEAR	JAN	FEB	MAR	APR	MAY	JUNE	JULY	AUG	SEP	OCT	NOV	DEC	ANNUAL
1956	1.56	2.96	3.36	3.41	4.02	2.29	4.47	2.75	1.15	1.07	2.77	3.05	32.86
1957	1.80	2.31	1.42	5.81	7.76	5.29	3.28	0.36	2.75	2.07	3.22	4.04	40.11
1958	1.81	0.57	1.50	4.80	4.26	10.89	6.34	2.29	4.10	1.68	2.47	0.54	41.25
1959	6.85	4.14	2.60	3.58	4.40	1.16	4.10	3.59	1.61	2.57	3.17	2.40	40.17
1960	1.97	2.30	1.21	1.18	3.66	4.51	3.04	1.89	0.90	1.86	1.91	1.76	26.19
1961	1.13	3.99	5.39	5.15	2.81	3.45	3.86	3.16	4.37	1.59	2.61	2.59	40.10
1962	2.90	2.42	2.34	0.56	3.94	0.32	6.78	2.14	0.96	2.23	2.56	0.59	27.74
1963	0.96	0.48	6.80	3.09	2.45	3.30	3.13	2.29	0.27	0.17	0.69	0.58	24.21
1964	1.60	1.52	7.65	5.98	1.55	3.15	2.15	1.12	1.04	0.53	1.95	3.33	31.57
1965	2.32	2.80	2.33	4.54	2.33	1.44	2.24	1.35	5.69	2.76	1.20	1.46	30.46
1966	2.33	2.64	1.07	2.63	2.37	2.63	3.95	3.49	3.63	1.28	3.44	2.81	32.27
1967	0.69	1.38	3.77	4.24	6.72	3.98	2.42	0.34	1.09	2.23	2.99	3.58	33.43
1968	1.67	0.27	2.22	1.44	7.33	3.66	3.61	2.89	2.74	1.38	3.66	3.48	34.35
1969	3.75	0.73	1.41	2.74	4.20	5.90	5.71	3.39	1.13	0.98	3.14	1.83	34.91
1970	1.23	1.16	2.08	5.60	2.68	3.01	2.64	1.04	1.29	2.91	1.63	1.95	27.22
1971	1.64	3.66	1.84	1.00	4.20	2.39	4.09	3.11	3.78	2.17	1.55	3.80	33.23
1972	1.47	0.85	2.46	3.77	4.34	3.04	2.08	3.13	4.64	2.22	5.00	2.81	35.81
1973	1.63	1.28	4.64	3.45	3.10	5.72	3.76	3.93	0.69	3.28	3.86	3.14	38.48
1974	2.67	2.04	3.59	2.97	5.20	4.50	0.47	8.03	3.68	0.98	3.44	2.86	40.43
1975	3.59	3.95	3.94	3.99	2.28	2.23	5.50	4.75	3.30	2.45	2.18	3.53	41.09
1976	3.01	1.73	2.97	1.80	1.90	5.32	0.95	1.88	1.71	2.82	0.87	0.67	25.63
1977	1.64	1.78	3.50	5.13	2.02	2.46	1.73	3.59	2.73	3.85	2.67	4.47	35.57
1978	4.72	0.24	2.45	3.89	2.85	4.66	3.83	6.98	0.43	2.47	2.03	4.45	39.00
1979	3.29	2.85	1.35	3.62	2.90	4.34	4.43	7.95	3.51	2.03	2.12	4.30	43.30
1980	2.16	1.69	4.43	3.55	5.06	9.54	2.98	4.60	1.45	2.25	1.81	1.44	40.96
1981	0.30	3.37	1.18	5.06	4.76	6.32	5.08	3.51	5.06	2.79	2.80	3.46	43.69
1982	6.03	1.82	5.54	1.95	4.80	4.05	1.46	6.42	1.40	1.42	4.10	3.72	42.71
1983	1.39	0.65	2.67	4.73	4.43	5.73	3.57	1.16	0.88	5.54	4.21	2.89	37.85
1984	1.15	2.67	3.63	3.92	4.29	1.87	2.37	2.06	3.30	3.52	3.38	3.83	35.99
1985	1.56	2.26	4.85	1.56	4.43	2.27	2.69	2.50	0.98	2.39	8.07	2.25	35.81
Record Mean	2.91	2.13	3.37	3.41	3.73	3.85	3.31	3.12	2.66	2.44	2.80	2.58	36.31

TABLE 3 AVERAGE TEMPERATURE (deg. F) DAYTON, OHIO

YEAR	JAN	FEB	MAR	APR	MAY	JUNE	JULY	AUG	SEP	OCT	NOV	DEC	ANNUAL
1956	26.3	33.9	38.5	47.8	61.2	71.3	73.3	73.0	63.4	59.6	41.6	39.3	52.4
1957	23.9	35.1	39.8	53.2	61.7	71.6	74.9	73.6	65.9	51.5	41.8	36.3	52.4
1958	27.0	23.5	36.0	51.6	61.0	66.5	73.4	72.0	65.6	54.6	43.7	22.7	49.8
1959	25.2	31.9	38.4	52.0	65.9	72.1	75.1	77.1	69.0	54.5	38.1	35.7	52.9
1960	31.8	28.4	26.7	54.7	59.0	69.4	72.6	75.3	69.5	55.6	43.3	24.4	50.9
1961	23.5	35.8	43.7	45.6	56.8	68.5	73.9	72.8	70.2	55.9	41.8	29.6	51.5
1962	25.2	29.0	36.9	50.8	68.1	71.3	72.5	72.5	62.4	56.1	41.2	26.1	51.0
#1963	20.3	22.4	42.6	51.9	58.9	70.9	72.4	68.5	64.2	62.1	46.1	21.7	50.2
1964	31.5	30.0	41.0	54.2	65.4	72.7	75.1	72.8	67.1	52.3	46.1	33.5	53.5
1965	29.2	30.7	34.7	51.3	69.4	73.1	73.8	73.1	68.3	53.1	42.3	36.2	52.8
1966	21.8	30.3	42.2	49.7	57.8	73.8	77.5	72.3	64.3	51.3	42.2	33.0	51.4
1967	33.9	25.1	40.2	52.3	56.5	74.3	73.5	70.9	63.7	54.7	37.6	34.7	51.5
1968	24.7	25.5	41.9	52.9	58.3	71.3	73.9	73.3	65.5	53.9	43.8	30.5	51.3
1969	26.7	31.6	35.5	52.7	62.4	68.1	75.2	73.0	66.4	55.4	39.6	28.9	51.3
1970	20.7	29.3	37.6	54.2	65.1	72.2	75.0	73.6	69.3	56.0	42.0	35.5	52.6
1971	23.9	30.0	37.8	51.4	59.5	74.6	72.7	71.1	68.2	62.9	42.8	38.8	52.8
1972	27.5	28.2	36.7	50.3	62.0	66.8	74.7	72.2	65.8	50.4	39.1	33.6	50.6
1973	31.3	31.2	48.6	50.5	58.7	73.5	75.8	74.5	70.2	58.7	45.0	31.0	54.1
1974	32.4	29.5	42.5	52.7	60.1	67.3	75.5	73.7	60.5	52.5	44.4	33.0	52.0
1975	33.3	33.9	37.9	47.1	65.9	71.6	73.1	74.5	62.0	55.8	48.6	33.8	53.1
1976	24.1	38.6	46.4	52.3	59.1	70.5	73.1	68.8	61.7	47.6	35.2	26.4	50.3
1977	11.6	27.5	45.6	55.7	68.8	69.2	77.5	72.8	68.3	51.8	44.8	29.0	51.9
1978	18.7	16.9	34.8	51.7	60.4	71.6	73.3	71.1	69.4	52.0	44.9	34.4	50.0
1979	20.6	19.1	44.3	50.3	61.5	70.3	73.0	71.6	64.7	52.6	43.2	35.1	50.5
1980	28.0	22.9	35.7	48.3	62.3	68.1	76.8	77.0	68.8	49.3	39.3	31.2	50.7
1981	23.3	32.9	39.3	54.9	57.3	71.9	75.1	72.8	63.4	52.1	43.6	29.7	51.4
1982	20.9	27.6	40.0	46.6	67.9	66.7	75.0	70.7	63.9	55.4	43.9	39.6	51.6
1983	29.1	34.4	42.0	46.9	56.7	69.9	76.4	75.8	66.2	54.1	43.1	21.9	51.4
1984	21.2	36.3	30.9	48.6	57.2	72.4	69.9	71.5	62.5	58.8	39.6	38.1	50.6
1985	19.4	25.3	43.6	56.2	62.8	67.0	73.0	70.8	66.3	57.4	47.6	23.8	51.1
Record Mean	28.1	30.8	40.0	51.3	61.7	70.9	74.9	73.2	66.5	55.3	42.5	32.0	52.3
Max	35.7	38.8	49.0	61.4	71.9	81.2	85.2	83.3	77.0	65.4	50.6	39.2	61.6
Min	20.4	22.7	31.0	41.2	51.4	60.7	64.7	63.0	56.0	45.2	34.4	24.8	43.0

REFERENCE NOTES FOR TABLES 1, 2, 3 and 6   (DAYTON, OH)

GENERAL

T - TRACE AMOUNT
BLANK ENTRIES DENOTE MISSING/UNREPORTED DATA.
# INDICATES A STATION OR INSTRUMENT RELOCATION.

SPECIFIC

TABLE 1

(a) - LENGTH OF RECORD IN YEARS. ALTHOUGH
INDIVIDUAL MONTHS MAY BE MISSING.
* LESS THAN .05

NORMALS — BASED ON THE 1951-1980 RECORD PERIOD.
EXTREMES — DATES ARE THE MOST RECENT OCCURRENCE.
WIND DIR. — NUMERALS SHOW TENS OF DEGREES
CLOCKWISE FROM TRUE NORTH.
"00" INDICATES CALM.
RESULTANT WIND DIRECTIONS ARE GIVEN TO WHOLE DEGREES.

EXCEPTIONS

TABLE 1

1. FASTEST MILE WINDS ARE THROUGH APRIL 1983.

TABLES 2, 3, and 6

RECORD MEANS ARE THROUGH THE CURRENT YEAR,
BEGINNING IN 1912 FOR TEMPERATURE
1912 FOR PRECIPITATION
1944 FOR SNOWFALL

TABLE 4 HEATING DEGREE DAYS Base 65 deg. F   DAYTON, OHIO

SEASON	JULY	AUG	SEP	OCT	NOV	DEC	JAN	FEB	MAR	APR	MAY	JUNE	TOTAL
1956-57	0	8	106	171	697	790	1267	829	775	402	152	12	5209
1957-58	0	0	77	410	689	882	1170	1155	893	401	148	38	5863
1958-59	2	7	69	323	639	1302	1227	920	820	383	110	17	5819
1959-60	0	0	62	351	800	901	1021	1054	1180	336	213	9	5927
1960-61	1	0	28	297	641	1252	1279	811	653	580	259	32	5833
1961-62	0	2	58	282	686	1088	1227	1000	865	460	70	21	5759
#1962-63	5	4	152	300	707	1196	1380	1186	688	395	203	20	6236
1963-64	1	23	79	120	561	1335	1032	1010	740	326	74	17	5318
1964-65	0	24	75	386	559	968	1104	957	930	405	31	4	5443
1965-66	0	19	62	371	677	886	1331	965	701	456	240	16	5724
1966-67	0	3	88	420	674	984	959	1112	763	378	277	1	5659
1967-68	3	13	115	336	816	934	1241	1139	708	356	218	10	5889
1968-69	6	14	56	366	629	1062	1181	933	908	371	55	55	5720
1969-70	0	1	71	314	755	1115	1369	995	843	337	115	11	5926
1970-71	5	0	48	278	685	907	1265	975	836	403	193	1	5596
1971-72	0	0	50	97	661	806	1155	1058	870	438	130	67	5332
1972-73	15	6	63	443	771	967	1038	941	502	444	204	0	5394
1973-74	0	2	31	227	593	1045	1004	986	691	379	198	42	5198
1974-75	0	0	164	384	613	988	977	868	835	530	84	22	5465
1975-76	2	0	130	292	492	957	1259	760	570	403	203	2	5070
1976-77	0	25	121	535	886	1191	1651	1042	600	309	63	36	6459
1977-78	0	7	45	403	605	1108	1428	1339	927	395	205	15	6477
1978-79	1	11	47	396	597	942	1372	1281	634	445	166	15	5907
1979-80	7	23	91	397	649	923	1145	1217	901	495	139	49	6036
1980-81	0	0	44	487	764	1043	1289	894	792	305	252	9	5879
1981-82	0	2	119	398	638	1089	1362	1039	766	546	27	22	6008
1982-83	1	9	102	320	630	782	1105	851	706	538	258	28	5330
1983-84	8	0	104	345	649	1331	1351	827	1051	493	263	2	6424
1984-85	7	8	142	191	756	824	1406	1104	660	294	128	34	5554
1985-86	0	1	107	255	516	1271							

TABLE 5 COOLING DEGREE DAYS Base 65 deg. F   DAYTON, OHIO

YEAR	JAN	FEB	MAR	APR	MAY	JUNE	JULY	AUG	SEP	OCT	NOV	DEC	TOTAL
1969	0	0	0	6	64	154	327	254	118	22	0	0	945
1970	0	0	0	22	125	234	322	275	185	5	0	0	1168
1971	0	0	0	3	28	298	247	196	153	39	0	0	964
1972	0	0	0	4	42	130	324	237	94	0	0	0	831
1973	0	0	0	15	18	260	343	306	191	39	0	0	1172
1974	0	0	0	15	52	118	331	277	36	3	3	0	835
1975	0	0	0	2	119	228	257	302	46	13	7	0	974
1976	0	0	2	30	27	175	259	147	28	1	0	0	669
1977	0	0	4	36	188	167	393	255	153	0	3	0	1199
1978	0	0	0	1	66	219	265	210	184	0	0	0	945
1979	0	0	1	12	63	179	260	234	87	18	0	0	854
1980	0	0	0	0	61	148	373	378	164	6	0	0	1130
1981	0	0	1	11	17	221	321	251	76	5	0	0	903
1982	0	0	0	0	123	78	316	191	75	29	3	1	816
1983	0	0	1		8	180	369	339	148	10	0	0	1055
1984	0	0	0	5	25	233	169	218	72	7	0	0	729
1985	0	0	3	36	63	101	253	190	152	25	1	0	824

TABLE 6 SNOWFALL (inches)   DAYTON, OHIO

SEASON	JULY	AUG	SEP	OCT	NOV	DEC	JAN	FEB	MAR	APR	MAY	JUNE	TOTAL
1956-57	0.0	0.0	0.0	0.0	2.0	4.0	9.0	2.1	0.2	0.5	T	0.0	17.8
1957-58	0.0	0.0	0.0	0.5	T	2.8	5.6	3.0	11.5	T	0.0	0.0	23.4
1958-59	0.0	0.0	0.0	0.0	6.7	5.9	9.9	2.7	8.7	T	0.0	0.0	33.9
1959-60	0.0	0.0	0.0	0.0	2.9	4.7	1.1	8.8	13.3	0.5	T	0.0	31.3
1960-61	0.0	0.0	0.0	0.0	0.9	15.6	8.5	9.0	0.5	3.2	0.0	0.0	37.7
1961-62	0.0	0.0	0.0	0.0	2.0	8.4	4.7	12.1	10.7	0.2	0.0	0.0	38.1
1962-63	0.0	0.0	0.0	2.0	T	6.3	7.1	7.5	9.9	T	0.0	0.0	32.8
1963-64	0.0	0.0	0.0	0.0	1.1	7.8	19.4	12.5	4.0	T	0.0	0.0	44.8
1964-65	0.0	0.0	0.0	0.0	2.0	2.1	2.1	5.3	9.8	T	0.0	0.0	27.9
1965-66	0.0	0.0	0.0	0.0	T	2.0	5.2	8.4	2.8	1.7	T	0.0	20.1
1966-67	0.0	0.0	0.0	0.0	9.8	7.5	3.6	8.7	7.8	T	0.0	0.0	37.4
1967-68	0.0	0.0	0.0	T	6.2	4.0	12.1	2.4	13.8	T	0.0	0.0	38.5
1968-69	0.0	0.0	0.0	T	1.4	3.7	2.5	0.9	3.8	0.3	0.0	0.0	12.6
1969-70	0.0	0.0	0.0	0.0	2.3	10.1	12.0	6.4	10.5	T	0.0	0.0	41.3
1970-71	0.0	0.0	0.0	0.0	3.6	0.7	4.6	9.8	11.0	T	0.0	0.0	29.7
1971-72	0.0	0.0	0.0	0.0	3.4	1.0	6.6	9.6	1.2	1.2	0.0	0.0	23.0
1972-73	0.0	0.0	0.0	T	2.5	5.2	2.4	3.6	4.4	3.6	0.0	0.0	21.7
1973-74	0.0	0.0	0.0	0.0	T	11.6	2.2	6.3	3.5	4.9	0.0	0.0	28.3
1974-75	0.0	0.0	0.0	T	6.6	8.8	8.2	6.4	11.1	T	0.0	0.0	41.1
1975-76	0.0	0.0	0.0	0.0	T	4.7	3.0	11.6	1.3	T	0.0	0.0	21.8
1976-77	0.0	0.0	0.0	T	2.6	3.9	20.2	10.6	1.2	0.3	0.0	0.0	38.8
1977-78	0.0	0.0	0.0	T	2.4	8.5	40.2	3.6	8.0	0.0	0.0	0.0	62.7
1978-79	0.0	0.0	0.0	0.0	T	1.1	20.5	17.5	0.2	T	0.0	0.0	39.3
1979-80	0.0	0.0	0.0	T	0.4	0.2	7.3	14.6	2.4	T	0.0	0.0	24.9
1980-81	0.0	0.0	0.0	T	5.3	3.3	3.3	5.7	2.0	0.0	0.0	0.0	19.6
1981-82	0.0	0.0	0.0	T	0.8	14.7	13.3	5.1	4.9	4.1	0.0	0.0	42.9
1982-83	0.0	0.0	0.0	0.0	0.1	0.3	1.5	2.5	0.9	0.2	0.0	0.0	5.5
1983-84	0.0	0.0	0.0	T	0.0	5.9	9.2	12.2	13.8	T	0.0	0.0	41.1
1984-85	0.0	0.0	0.0	3.2	5.3	5.3	14.0	12.6	0.3	2.3	0.0	0.0	37.7
1985-86	0.0	0.0	0.0	0.0	0.0	7.5							
Record Mean	0.0	0.0	0.0	0.1	2.4	5.9	8.3	6.2	5.4	0.7	T	0.0	29.0

See Reference Notes, relative to all above tables, on preceding page.

Mansfield is in the north central highlands at the geographical and climatological junction of central Ohio, northwest Ohio, and northeast Ohio. The station is on a plateau 3 miles north of the city of Mansfield and surrounded by rolling open farmland. The general elevation ranges from around 1,300 to 1,400 feet above sea level with the 1,000-foot contour east to west some 15 miles to the north. The climate is continental, with the modifying effects of Lake Erie most pronounced in winter. Lake Erie is just 38 miles due north.

The lake influence, plus the elevation, produce cloudy skies and considerable snow shower activity from late November into April with any wind flow from northwest through northeast. Because of this, any windshift with a cold frontal passage in winter does not bring the clearing skies, indeed, more snow is often measured from the flurry activity behind the front than from the pre-frontal conditions. A frozen Lake Erie will allow clearing skies, but an open lake dictates overcast and snow flurries. Usually the lake is open enough to set off the flurries and cloudy conditions. The major snow producer will be an intense storm moving out of the southwest with the Gulf of Mexico moisture available. Snow cover is almost constant from December through March due to almost daily snow flurries, but the depth of cover is rarely more than 8 inches. Daytime winter temperatures are not above the freezing mark too often.

Spring is a short period of rapid transition from hard winter to summer conditions. April usually brings abundant shower activity and the crops and vegetation get a quick start.

Summer is a pleasant season with low humidities and no extremely high temperatures. Rarely does the temperature climb above the 90 degree point. Thunderstorms average about once every three days during the season from June through September. Highest winds are associated with the heavier thunderstorms, and while hail does not occur often, it is of major concern to the applegrowers in the area. Flooding problems are confined to the flash-flood type on the small streams in the area.

The growing season is normally about 153 days. Autumn usually produces many clear warm days and cool invigorating nights. Ground fog is at a maximum incidence during the autumn. Little rainfall occurs to interfere with harvest time and county fair time.

TABLE 1 NORMALS, MEANS AND EXTREMES

MANSFIELD, OHIO

LATITUDE: 40°49'N   LONGITUDE: 82°31'W   ELEVATION: FT. GRND 1295 BARO 01300   TIME ZONE: EASTERN   WBAN: 14891

	(a)	JAN	FEB	MAR	APR	MAY	JUNE	JULY	AUG	SEP	OCT	NOV	DEC	YEAR
TEMPERATURE °F:														
Normals														
-Daily Maximum		32.2	35.0	45.4	58.8	69.3	78.4	82.1	80.7	74.4	62.8	48.5	36.8	58.7
-Daily Minimum		17.4	19.2	27.9	38.4	48.2	57.5	61.8	60.5	53.8	43.0	33.0	22.9	40.3
-Monthly		24.8	27.1	36.7	48.6	58.8	68.0	72.0	70.6	64.1	52.9	40.8	29.9	49.5
Extremes														
-Record Highest	26	63	67	76	85	92	97	96	95	93	85	78	73	97
-Year		1972	1961	1983	1960	1962	1971	1965	1964	1964	1963	1968	1982	JUN 1971
-Record Lowest	26	-22	-11	-6	8	25	38	44	40	34	20	2	-14	-22
-Year		1985	1982	1980	1982	1966	1980	1965	1965	1974	1976	1976	1983	JAN 1985
NORMAL DEGREE DAYS:														
Heating (base 65°F)		1246	1058	877	492	231	35	0	10	105	381	726	1088	6249
Cooling (base 65°F)		0	0	0	0	39	125	220	184	78	6	0	0	652
% OF POSSIBLE SUNSHINE														
MEAN SKY COVER (tenths)														
Sunrise - Sunset	26	7.9	7.6	7.6	7.0	6.6	6.0	5.8	5.9	5.9	6.1	7.7	8.1	6.9
MEAN NUMBER OF DAYS:														
Sunrise to Sunset														
-Clear	26	3.7	3.7	4.5	5.3	6.6	6.9	7.3	8.0	8.7	8.9	4.0	3.4	71.0
-Partly Cloudy	26	6.3	6.5	6.8	7.6	9.5	11.6	13.4	11.1	8.8	7.8	5.8	5.2	100.4
-Cloudy	26	21.1	18.2	19.7	17.0	14.9	11.5	10.2	11.9	12.5	14.3	20.2	22.4	193.9
Precipitation														
.01 inches or more	26	12.9	12.0	14.8	13.2	12.9	10.7	9.6	9.8	8.9	9.5	12.6	13.9	140.8
Snow,Ice pellets														
1.0 inches or more	26	3.8	3.7	2.3	0.5	0.*	0.0	0.0	0.0	0.0	0.0	1.0	3.0	14.4
Thunderstorms	26	0.2	0.5	2.3	4.0	5.2	6.7	7.1	6.2	3.4	0.9	0.9	0.2	37.6
Heavy Fog Visibility														
1/4 mile or less	26	2.6	2.8	3.6	1.8	2.3	1.5	1.5	2.7	2.2	2.5	2.5	4.2	30.1
Temperature °F														
-Maximum														
90° and above	20	0.0	0.0	0.0	0.0	0.9	3.2	1.1	0.6	0.0	0.0	0.0	5.9	
32° and below	20	17.5	12.4	4.6	0.3	0.0	0.0	0.0	0.0	0.0	0.0	1.6	11.2	47.6
-Minimum														
32° and below	20	28.4	24.0	20.6	8.5	0.9	0.0	0.0	0.0	0.0	0.0	3.3	14.5	25.0	125.1
0° and below	20	4.2	2.5	0.3	0.0	0.0	0.0	0.0	0.0	0.0	0.0	0.0	1.1	8.1
AVG. STATION PRESS.(mb)														
RELATIVE HUMIDITY (%)														
Hour 01	10	78	77	76	72	76	81	82	83	82	77	79	79	79
Hour 07	19	80	79	79	76	77	80	82	86	86	81	81	82	81
Hour 13 (Local Time)	19	71	68	62	56	57	58	57	60	60	59	67	74	62
Hour 19	13	75	72	68	61	61	64	64	68	72	69	74	77	69
PRECIPITATION (inches):														
Water Equivalent														
-Normal		2.25	1.86	3.00	3.55	3.75	3.44	3.73	3.25	3.04	1.94	2.66	2.40	34.87
-Maximum Monthly	26	4.53	4.28	7.04	6.58	6.57	10.00	8.06	7.60	6.85	4.89	12.82	4.78	12.82
-Year		1982	1961	1964	1964	1985	1981	1969	1974	1972	1978	1985	1978	NOV 1985
-Minimum Monthly	26	0.41	0.29	1.16	0.75	1.50	1.25	0.94	0.63	0.73	0.43	0.70	0.74	0.29
-Year		1981	1978	1960	1971	1961	1984	1975	1970	1963	1963	1976	1976	FEB 1978
-Maximum in 24 hrs	26	1.70	2.79	2.45	2.34	2.33	3.72	5.06	3.85	4.10	1.84	2.32	1.85	5.06
-Year		1979	1961	1964	1979	1965	1978	1969	1972	1979	1983	1985	1978	JUL 1969
Snow,Ice pellets														
-Maximum Monthly	26	42.1	19.1	14.8	13.4	1.4				T	0.6	7.4	18.0	42.1
-Year		1978	1984	1963	1982	1966				1970	1962	1980	1962	JAN 1978
-Maximum in 24 hrs	26	12.0	10.0	7.5	6.4	1.4				T	0.6	3.2	12.3	12.3
-Year		1968	1985	1960	1982	1966				1970	1962	1966	1974	DEC 1974
WIND:														
Mean Speed (mph)	11	13.4	12.6	12.5	12.3	10.3	9.9	8.4	8.4	9.0	10.6	11.9	12.7	11.0
Prevailing Direction														
Fastest Obs. 1 Min.														
-Direction (!!!)	19	24	24	25	33	22	23	25	31	28	25	21	18	33
-Speed (MPH)	19	46	44	37	46	35	40	37	37	28	30	37	46	46
-Year		1971	1967	1982	1981	1983	1973	1983	1979	1973	1981	1981	1971	APR 1981
Peak Gust														
-Direction (!!!)	2	SW	SW	S	S	SW	S	W	SW	SW	SW	W	SW	S
-Speed (mph)	2	56	48	61	59	45	40	55	37	45	40	51	53	61
-Date		1985	1985	1985	1984	1984	1984	1985	1985	1984	1985	1985	1985	MAR 1985

See reference Notes to this table on the following page.

TABLE 2 PRECIPITATION (inches) MANSFIELD, OHIO

YEAR	JAN	FEB	MAR	APR	MAY	JUNE	JULY	AUG	SEP	OCT	NOV	DEC	ANNUAL
1960	2.54	2.67	1.16	2.07	3.31	4.40	3.04	2.75	1.12	1.72	1.92	1.10	27.80
1961	0.65	4.28	4.47	5.21	1.50	3.99	4.05	3.20	4.49	1.23	2.92	2.38	38.37
1962	2.93	1.52	1.78	1.31	1.87	3.17	3.50	3.39	3.79	1.98	2.42	1.54	29.20
1963	0.91	0.68	4.78	2.30	1.69	1.77	3.08	3.36	0.73	0.43	1.34	0.74	21.81
1964	1.45	1.33	7.04	6.58	4.16	3.03	2.16	2.44	1.13	1.09	1.49	2.88	34.78
1965	3.01	2.64	2.28	3.02	3.69	1.84	1.42	3.30	2.11	2.98	2.47	1.88	30.64
1966	1.55	2.17	1.44	2.73	2.84	1.88	3.82	4.66	1.95	1.23	4.33	2.57	31.17
1967	0.74	1.33	3.21	2.61	4.50	1.63	3.29	1.08	4.61	1.75	2.38	2.76	29.89
1968	2.27	0.48	2.40	1.84	6.27	3.94	3.13	1.56	2.26	0.78	4.01	3.17	32.11
1969	2.41	0.74	1.25	5.50	6.15	3.68	8.06	2.36	2.70	1.82	3.25	1.55	39.47
1970	1.02	0.93	1.27	5.27	5.17	0.63	5.67	1.86	4.65	2.69	3.24	2.39	34.79
1971	1.40	2.58	1.78	0.75	4.24	1.29	3.79	2.18	2.41	1.15	1.11	3.42	26.10
1972	0.95	1.15	2.49	5.98	3.07	3.75	3.11	5.74	6.85	1.52	4.30	2.79	41.70
1973	1.83	1.39	4.21	3.35	4.15	7.07	4.35	4.00	1.38	3.59	3.19	2.85	41.36
1974	3.59	1.59	4.14	2.47	5.65	1.90	2.93	7.60	3.41	0.79	5.15	2.64	41.86
1975	3.33	3.74	2.69	1.95	4.01	2.36	0.94	5.63	4.97	2.19	1.59	2.92	36.32
1976	2.69	2.99	3.96	1.37	2.35	4.34	2.55	3.16	3.61	2.43	0.70	0.74	30.89
1977	0.79	1.17	4.26	3.25	1.99	3.69	4.93	3.40	4.56	1.75	3.04	4.15	36.98
1978	3.76	0.29	2.29	4.70	4.20	4.98	1.70	3.20	1.10	4.89	1.96	4.78	37.85
1979	3.29	1.74	1.81	5.63	5.25	4.82	2.55	6.78	5.64	1.39	4.93	2.28	46.11
1980	1.30	1.44	5.72	3.34	4.11	6.31	5.01	4.41	2.45	1.98	2.01	2.15	40.23
1981	0.41	3.41	1.71	5.62	4.49	10.00	2.04	5.08	4.41	2.62	1.53	3.50	44.82
1982	4.53	1.76	5.18	1.93	3.73	6.09	4.08	5.09	2.52	0.99	6.82	4.68	47.40
1983	1.67	1.16	2.43	5.80	5.68	2.92	4.61	4.51	5.16	4.60	6.49	3.37	48.40
1984	1.22	2.91	4.68	5.48	6.23	1.25	3.55	4.43	3.78	3.71	4.15	4.41	45.80
1985	1.49	2.09	6.17	1.42	6.57	3.13	3.00	6.90	1.63	2.46	12.82	3.17	50.85
Record Mean	1.99	1.85	3.25	3.52	4.11	3.66	3.48	3.88	3.21	2.07	3.44	2.72	37.18

TABLE 3 AVERAGE TEMPERATURE (deg. F) MANSFIELD, OHIO

YEAR	JAN	FEB	MAR	APR	MAY	JUNE	JULY	AUG	SEP	OCT	NOV	DEC	ANNUAL
1960	29.2	26.5	23.9	52.4	57.2	66.4	69.1	71.8	66.1	52.8	42.5	22.9	48.4
1961	21.0	33.1	40.1	42.6	54.2	66.0	71.0	70.9	68.1	55.1	40.7	27.7	49.2
1962	23.4	25.8	34.1	48.0	65.1	68.9	69.7	69.7	60.1	53.0	39.9	23.2	48.4
1963	17.8	18.7	39.0	47.9	55.7	68.3	71.2	66.6	61.3	61.0	43.4	19.9	47.6
1964	28.9	25.3	36.6	49.1	61.9	68.0	72.4	68.3	63.6	49.9	43.9	30.6	49.9
#1965	25.8	25.5	30.6	45.9	62.9	65.2	68.0	67.2	63.9	48.8	40.6	35.2	48.3
1966	20.3	27.1	37.5	45.5	52.6	68.2	72.0	68.1	59.7	49.3	40.8	29.0	47.5
1967	30.7	23.8	37.4	50.2	53.3	70.9	70.2	67.6	60.1	52.6	36.0	33.1	48.8
1968	22.8	23.4	39.7	51.3	55.5	68.9	72.3	73.0	66.4	54.4	42.9	29.7	50.0
1969	26.2	29.8	35.0	51.9	61.4	67.4	74.6	73.0	64.8	54.2	38.7	26.0	50.3
1970	20.1	27.6	34.4	50.6	62.0	69.9	73.7	73.6	69.1	56.4	42.3	32.3	51.0
1971	22.4	28.9	34.1	46.0	56.4	71.3	70.3	68.6	67.8	60.0	40.7	37.6	50.4
1972	27.9	27.3	36.2	47.7	61.4	66.0	73.5	70.5	65.7	49.6	38.6	33.6	49.8
1973	30.0	27.8	47.5	50.7	57.9	71.8	73.5	73.6	67.4	58.2	45.1	32.7	53.0
1974	32.0	28.2	37.4	51.9	58.2	66.3	72.7	71.7	59.9	51.5	41.7	30.2	50.2
1975	29.8	29.7	33.9	42.2	62.4	68.5	72.9	72.6	58.7	54.1	45.8	29.9	50.0
1976	20.8	34.6	42.3	49.1	55.3	67.6	69.2	66.0	59.2	45.3	31.3	20.7	46.8
1977	8.7	23.6	42.0	51.3	63.3	63.5	74.1	70.1	65.8	50.5	43.4	27.4	48.6
1978	18.3	15.4	31.6	47.2	58.1	68.4	71.2	71.8	68.5	50.3	41.9	31.1	47.8
1979	18.6	16.4	40.8	46.2	56.6	66.6	68.6	68.2	62.8	50.8	40.9	31.9	47.4
1980	25.7	21.4	33.9	46.9	58.3	62.8	72.7	73.9	65.2	49.0	39.5	30.4	48.3
1981	22.3	32.4	37.0	51.8	57.1	69.7	72.8	70.5	62.8	51.3	43.4	28.5	50.0
1982	18.6	24.8	36.1	43.2	64.1	68.4	70.9	67.2	60.9	54.6	44.1	39.5	48.9
1983	29.2	33.1	40.6	47.5	56.5	69.7	75.6	74.9	65.5	53.9	44.4	22.3	51.1
1984	20.2	34.5	28.7	48.5	55.9	72.2	70.1	71.1	61.0	58.3	41.0	38.2	50.0
1985	21.0	25.2	42.0	54.4	60.7	63.0	70.7	68.6	64.8	54.4	45.2	23.1	49.4
Record Mean	23.5	26.6	36.7	48.4	58.6	67.6	71.7	70.3	63.8	53.0	41.5	29.5	49.3
Max	31.0	34.5	45.4	58.5	69.0	77.9	81.7	80.1	73.6	62.6	49.0	36.5	58.3
Min	16.0	18.6	27.9	38.3	48.2	57.3	61.6	60.5	54.0	43.5	33.9	22.5	40.2

REFERENCE NOTES FOR TABLES 1, 2, 3 and 6 (MANSFIELD, OH)

GENERAL

T - TRACE AMOUNT
BLANK ENTRIES DENOTE MISSING/UNREPORTED DATA.
# INDICATES A STATION OR INSTRUMENT RELOCATION.

SPECIFIC

TABLE 1

(a) - LENGTH OF RECORD IN YEARS. ALTHOUGH
 INDIVIDUAL MONTHS MAY BE MISSING.

 * LESS THAN .05

NORMALS — BASED ON THE 1951-1980 RECORD PERIOD.
EXTREMES — DATES ARE THE MOST RECENT OCCURRENCE.
WIND DIR. — NUMERALS SHOW TENS OF DEGREES
 CLOCKWISE FROM TRUE NORTH.
 "00" INDICATES CALM.
RESULTANT WIND DIRECTIONS ARE GIVEN TO WHOLE DEGREES.

EXCEPTIONS

TABLE 1

1. MEAN WIND SPEED IS THROUGH 1979.

TABLES 2, 3, and 6

RECORD MEANS ARE THROUGH THE CURRENT YEAR,
BEGINNING IN 1960 FOR TEMPERATURE
 1960 FOR PRECIPITATION
 1960 FOR SNOWFALL

TABLE 4   HEATING DEGREE DAYS Base 65 deg. F   MANSFIELD, OHIO

SEASON	JULY	AUG	SEP	OCT	NOV	DEC	JAN	FEB	MAR	APR	MAY	JUNE	TOTAL
1959-60							1101	1106	1268	397	250	38	
1960-61	15	0	60	375	670	1299	1356	889	764	668	336	82	6514
1961-62	15	4	83	309	725	1149	1284	1092	952	526	111	29	6279
1962-63	9	11	181	374	746	1290	1458	1292	796	510	302	45	7014
1963-64	17	39	131	151	640	1391	1111	1146	872	474	144	64	6180
#1964-65	2	37	127	462	627	1061	1208	1098	1059	567	119	76	6443
1965-66	23	59	115	500	728	921	1380	1054	845	579	381	54	6639
1966-67	7	20	189	480	723	1108	1060	1148	852	442	364	13	6406
1967-68	21	28	171	391	865	982	1300	1201	778	406	292	25	6460
1968-69	9	17	48	354	655	1086	1196	982	923	396	168	70	5904
1969-70	0	0	89	347	784	1204	1386	1043	940	444	159	27	6423
1970-71	6	0	53	275	674	1006	1316	1004	955	563	275	16	6143
1971-72	3	12	54	170	722	840	1143	1085	886	515	143	71	5644
1972-73	9	16	70	471	783	966	1078	1037	534	453	219	1	5637
1973-74	1	8	51	223	588	995	1017	1022	847	402	234	56	5444
1974-75	0	0	175	413	695	1073	1085	983	956	676	142	45	6243
1975-76	2	4	188	343	570	1078	1367	870	699	494	307	30	5952
1976-77	8	56	185	606	1006	1369	1743	1155	708	421	128	112	7497
1977-78	9	24	63	442	643	1159	1442	1380	1028	530	259	39	7018
1978-79	12	6	63	449	686	1044	1433	1353	744	557	290	54	6691
1979-80	38	39	118	453	719	1018	1214	1256	956	535	217	115	6678
1980-81	4	3	86	490	757	1067	1318	908	859	389	259	12	6152
1981-82	5	7	128	418	637	1127	1434	1120	887	645	79	106	6593
1982-83	8	41	161	325	624	786	1102	886	751	521	266	39	5510
1983-84	8	0	106	347	613	1319	1363	880	1123	493	298	5	6575
1984-85	9	9	172	214	713	825	1361	1108	706	344	177	88	5726
1985-86	2	11	128	323	587	1293							

TABLE 5   COOLING DEGREE DAYS Base 65 deg. F   MANSFIELD, OHIO

YEAR	JAN	FEB	MAR	APR	MAY	JUNE	JULY	AUG	SEP	OCT	NOV	DEC	TOTAL
1969	0	0	0	10	63	148	303	255	91	19	0	0	889
1970	0	0	0	17	71	181	284	271	184	16	0	0	1024
1971	0	0	0	0	16	211	175	131	147	25	0	0	705
1972	0	0	0	0	35	109	279	198	102	0	0	0	723
1973	0	0	0	11	8	210	272	282	129	19	0	0	931
1974	0	0	0	16	30	103	246	215	27	2	1	0	640
1975	0	0	0	0	68	160	254	247	8	9	0	0	746
1976	0	0	0	21	12	117	145	91	16	0	0	0	402
1977	0	0	0	18	84	72	299	190	96	0	1	0	760
1978	0	0	0	0	50	146	209	226	174	0	0	0	805
1979	0	0	0	4	38	109	157	146	60	17	0	0	531
1980	0	0	0	0	17	57	249	285	102	1	0	0	711
1981	0	0	0	2	20	159	255	186	68	3	0	0	693
1982	0	0	0	1	59	33	197	118	47	11	3	3	472
1983	0	0	0	4	12	185	344	313	127	9	0	0	994
1984	0	0	0	6	22	228	176	205	57	14	0	0	708
1985	0	0	0	36	53	36	187	129	129	0	0	0	570

TABLE 6   SNOWFALL (inches)   MANSFIELD, OHIO

SEASON	JULY	AUG	SEP	OCT	NOV	DEC	JAN	FEB	MAR	APR	MAY	JUNE	TOTAL
1959-60					3.7	12.6	1.5	14.6	14.8	1.0	T	0.0	41.3
1960-61	0.0	0.0	0.0	0.0	0.4	5.8	5.4	14.8	7.0	1.4	T	0.0	34.8
1961-62	0.0	0.0	0.0	0.6	T	18.0	12.4	11.6	14.8	0.9	T	0.0	58.3
1962-63	0.0	0.0	0.0	T	T	7.9	17.0	16.1	8.3	1.2	0.0	0.0	50.5
1963-64	0.0	0.0	0.0	T	1.9	8.2	11.3	10.8	10.2	T	0.0	0.0	42.4
1964-65	0.0	0.0	0.0	T	1.4	0.8	8.0	5.8	3.6	2.7	0.0	0.0	22.3
1965-66	0.0	0.0	0.0	T	T	6.3	8.0	5.8	3.6	2.7	0.0	0.0	
1966-67	0.0	0.0	0.0	0.0	6.2	6.3	1.4	12.8	11.2	T	0.0	0.0	37.9
1967-68	0.0	0.0	0.0	T	3.5	8.2	21.3	8.1	8.1	0.3	0.0	0.0	49.5
1968-69	0.0	0.0	0.0	T	1.9	7.5	4.0	7.4	5.5	2.1	T	0.0	28.4
1969-70	0.0	0.0	0.0	0.0	3.7	11.7	10.7	5.0	4.3	T	0.0	0.0	35.4
1970-71	0.0	0.0	T	0.0	1.7	5.4	7.2	15.9	13.5	T	0.0	0.0	43.7
1971-72	0.0	0.0	0.0	0.0	2.8	1.7	7.1	9.9	5.9	1.4	0.0	0.0	28.8
1972-73	0.0	0.0	0.0	T	4.4	9.1	3.2	9.0	3.3	1.3	T	0.0	30.3
1973-74	0.0	0.0	0.0	0.2	11.0	8.1	11.0	6.3	2.7	T	0.0	0.0	39.3
1974-75	0.0	0.0	0.0	0.0	5.2	16.8	10.1	12.4	8.9	0.7	T	0.0	54.1
1975-76	0.0	0.0	0.0	0.0	2.7	6.2	14.0	10.8	2.2	1.7	T	0.0	37.6
1976-77	0.0	0.0	0.0	T	4.7	12.8	24.8	10.0	2.0	1.4	0.0	0.0	55.7
1977-78	0.0	0.0	0.0	T	4.2	16.8	42.1	9.6	4.9	T	0.0	0.0	77.6
1978-79	0.0	0.0	0.0	T	0.3	1.3	20.5	10.9	1.3	0.3	T	0.0	34.6
1979-80	0.0	0.0	0.0	T	0.4	5.5	6.8	12.6	1.7	0.6	T	0.0	27.6
1980-81	0.0	0.0	0.0	T	7.4	7.5	9.9	7.1	11.0	T	0.0	0.0	43.4
1981-82	0.0	0.0	0.0	T	3.3	16.1	16.5	6.6	11.0	13.4	0.0	0.0	66.9
1982-83	0.0	0.0	0.0	0.0	1.0	3.3	5.9	5.9	3.9	0.9	0.0	0.0	16.6
1983-84	0.0	0.0	0.0	0.0	3.7	9.7	11.6	19.1	13.4	0.2	0.0	0.0	57.7
1984-85	0.0	0.0	0.0	0.0	0.7	7.1	20.3	18.2	0.5	2.2	0.0	0.0	49.0
1985-86	0.0	0.0	0.0	0.0	T	13.3							
Record Mean	0.0	0.0	T	T	2.5	9.0	11.7	11.0	6.8	1.6	0.1	0.0	42.7

See Reference Notes, relative to all above tables, on preceding page.

Toledo is located on the western end of Lake Erie at the mouth of the Maumee River. Except for a bank up from the river about 30 feet, the terrain is generally level with only a slight slope toward the river and Lake Erie. The city has quite a diversified industrial section and excellent harbor facilities, making it a large transportation center for rail, water, and motor freight. Generally rich agricultural land is found in the surrounding area, especially up the Maumee Valley toward the Indiana state line.

Rainfall is usually sufficient for general agriculture. The terrain is level and drainage rather poor, therefore, a little less than the normal precipitation during the growing season is better than excessive amounts. Snowfall is generally light in this area, distributed throughout the winter from November to March with frequent thaws.

The nearness of Lake Erie and the other Great Lakes has a moderating effect on the temperature, and extremes are seldom recorded. On average, only fifteen days a year experience temperatures of 90 degrees or higher, and only eight days when it drops to zero or lower. The growing season averages 160 days, but has ranged from over 220 to less than 125 days.

Humidity is rather high throughout the year in this area, and there is an excessive amount of cloudiness. In the winter months the sun shines during only about 30 percent of the daylight hours. December and January, the cloudiest months, sometimes have as little as 16 percent of the possible hours of sunshine.

Severe windstorms, causing more than minor damage, occur infrequently. There are on the average twenty-three days per year having a sustained wind velocity of 32 mph or more.

Flooding in the Toledo area is produced by several factors. Heavy rains of 1 inch or more will cause a sudden rise in creeks and drainage ditches to the point of overflow. The western shores of Lake Erie are subject to flooding when the lake level is high and prolonged periods of east to northeast winds prevail.

TABLE 1 — NORMALS, MEANS AND EXTREMES

TOLEDO, OHIO

LATITUDE: 41°36'N   LONGITUDE: 83°48'W   ELEVATION: FT. GRND 669 BARO 00694   TIME ZONE: EASTERN   WBAN: 94830

	(a)	JAN	FEB	MAR	APR	MAY	JUNE	JULY	AUG	SEP	OCT	NOV	DEC	YEAR
TEMPERATURE °F:														
Normals														
-Daily Maximum		30.7	34.0	44.6	59.1	70.5	79.9	83.4	81.8	75.1	63.3	47.9	35.5	58.8
-Daily Minimum		15.5	17.5	26.1	36.5	46.6	56.0	60.2	58.4	51.2	40.1	30.6	20.6	38.3
-Monthly		23.1	25.8	35.4	47.8	58.5	68.0	71.8	70.1	63.2	51.7	39.3	28.1	48.5
Extremes														
-Record Highest	30	62	68	80	88	95	99	101	98	98	91	78	68	101
-Year		1967	1957	1963	1977	1962	1971	1977	1964	1978	1963	1968	1982	JUL 1977
-Record Lowest	30	-20	-14	-6	8	25	32	43	34	26	15	2	-15	-20
-Year		1984	1982	1984	1982	1974	1972	1972	1982	1974	1976	1958	1983	JAN 1984
NORMAL DEGREE DAYS:														
Heating (base 65°F)		1299	1098	918	516	237	39	0	16	113	419	771	1144	6570
Cooling (base 65°F)		0	0	0	0	38	129	215	174	59	7	0	0	622
% OF POSSIBLE SUNSHINE	30	42	47	50	53	60	64	67	63	61	55	38	34	53
MEAN SKY COVER (tenths)														
Sunrise - Sunset	30	7.5	7.3	7.3	6.9	6.4	6.0	5.7	5.7	5.9	6.1	7.6	7.9	6.7
MEAN NUMBER OF DAYS:														
Sunrise to Sunset														
-Clear	30	4.5	4.6	4.8	5.9	6.3	6.7	7.7	8.1	8.7	8.4	3.7	3.1	72.6
-Partly Cloudy	30	6.8	6.9	7.2	7.7	10.9	12.0	13.5	12.2	9.4	8.6	7.1	6.5	108.7
-Cloudy	30	19.7	16.7	19.0	16.4	13.8	11.3	9.8	10.7	12.0	14.0	19.2	21.4	183.9
Precipitation														
.01 inches or more	30	13.6	11.1	13.5	12.7	12.1	10.3	9.4	9.1	10.0	9.0	11.7	14.4	136.9
Snow,Ice pellets														
1.0 inches or more	30	2.9	2.5	2.1	0.6	0.0	0.0	0.0	0.0	0.0	0.0	1.2	2.8	12.1
Thunderstorms	30	0.1	0.5	2.0	4.0	5.0	7.4	7.1	6.3	3.8	1.1	0.7	0.2	38.2
Heavy Fog Visibility 1/4 mile or less	30	1.5	2.0	1.6	0.9	0.8	0.9	0.8	1.9	1.7	2.2	1.7	2.2	18.1
Temperature °F														
-Maximum														
90° and above	30	0.0	0.0	0.0	0.0	0.6	3.3	4.6	2.9	1.3	0.*	0.0	0.0	12.7
32° and below	30	18.6	13.2	4.7	0.2	0.0	0.0	0.0	0.0	0.0	0.0	2.2	12.5	51.3
-Minimum														
32° and below	30	29.6	25.9	23.3	11.1	1.7	0.*	0.0	0.0	0.4	6.5	17.7	26.6	142.9
0° and below	30	4.7	3.2	0.2	0.0	0.0	0.0	0.0	0.0	0.0	0.0	0.0	1.8	9.9
AVG. STATION PRESS.(mb)	13	992.4	993.0	990.3	990.1	989.3	990.0	991.2	992.5	993.1	993.6	992.5	992.5	991.7
RELATIVE HUMIDITY (%)														
Hour 01	30	75	74	75	76	78	82	84	88	87	81	81	81	80
Hour 07	30	79	79	81	79	80	82	86	90	91	85	83	83	83
Hour 13 (Local Time)	30	69	66	62	55	53	54	55	59	58	57	67	74	61
Hour 19	30	74	71	66	59	56	58	61	67	71	69	74	78	67
PRECIPITATION (inches):														
Water Equivalent														
-Normal		1.99	1.80	2.64	3.04	2.90	3.49	3.26	3.19	2.53	1.94	2.41	2.59	31.78
-Maximum Monthly	30	4.61	4.43	5.70	6.10	5.13	8.48	6.75	8.47	8.10	3.79	6.86	6.81	8.48
-Year		1965	1976	1985	1977	1968	1981	1969	1965	1972	1981	1982	1967	JUN 1981
-Minimum Monthly	30	0.27	0.27	0.58	0.88	0.96	1.49	0.68	0.40	0.58	0.28	0.55	0.54	0.27
-Year		1961	1969	1964	1962	1964	1984	1974	1976	1963	1964	1976	1958	FEB 1969
-Maximum in 24 hrs	30	1.78	1.61	2.60	3.43	1.96	3.21	4.39	2.42	3.97	1.92	3.17	3.53	4.39
-Year		1959	1981	1985	1977	1970	1978	1969	1972	1972	1981	1982	1967	JUL 1969
Snow,Ice pellets														
-Maximum Monthly	30	30.8	14.4	15.0	12.0	T				T	0.9	17.9	24.2	30.8
-Year		1978	1967	1977	1957	1984				1967	1980	1966	1977	JAN 1978
-Maximum in 24 hrs	30	10.4	7.7	7.8	9.8	T				T	0.9	8.3	13.9	13.9
-Year		1978	1981	1977	1957	1984				1967	1980	1966	1974	DEC 1974
WIND:														
Mean Speed (mph)	30	10.9	10.6	11.0	10.9	9.6	8.4	7.3	7.1	7.6	8.7	10.1	10.5	9.4
Prevailing Direction through 1963		WSW	WSW	WSW	E	WSW	SW	WSW	SW	SSW	WSW	WSW	SW	WSW
Fastest Mile														
-Direction	30	W	SW	W	SW	W	W	NW	W	NW	SW	SW	SW	SW
-Speed (MPH)	30	47	56	56	72	45	50	54	47	47	40	65	45	72
-Year		1972	1967	1957	1956	1957	1969	1970	1965	1969	1956	1957	1971	APR 1956
Peak Gust														
-Direction	2	SW	NE	W	W	NW	NW	NW	NW	SW	SW	SW	W	W
-Speed (mph)	2	39	49	54	58	48	43	37	51	41	43	48	53	58
-Date		1985	1984	1985	1984	1985	1984	1984	1985	1984	1985	1985	1985	APR 1984

See Reference Notes to this table on the following page.

TABLE 2   PRECIPITATION (inches)   TOLEDO, OHIO

YEAR	JAN	FEB	MAR	APR	MAY	JUNE	JULY	AUG	SEP	OCT	NOV	DEC	ANNUAL
1956	1.49	2.18	3.39	3.77	4.73	3.65	2.98	5.81	0.76	0.37	2.11	2.52	33.76
1957	1.66	1.47	1.11	4.24	2.19	3.61	2.03	2.17	2.06	3.34	1.94	3.61	31.03
1958	1.23	0.82	0.58	2.00	2.29	3.04	6.71	5.08	2.06	0.73	3.20	0.54	28.28
1959	3.89	2.74	3.15	3.67	3.95	2.37	3.13	2.04	2.04	3.72	3.09	2.37	37.24
1960	2.94	3.13	1.01	1.62	3.11	4.86	3.62	2.12	2.27	0.99	1.02	0.55	27.24
1961	0.27	2.64	3.09	4.94	2.15	2.70	2.78	2.02	2.86	0.86	2.04	1.29	27.64
1962	2.46	2.17	1.74	0.88	2.83	2.22	4.28	1.38	3.39	2.08	1.70	1.23	26.36
1963	0.93	0.81	3.14	2.17	2.63	4.31	1.98	2.22	0.58	0.67	0.83	1.78	22.05
1964	1.87	0.95	4.88	3.49	0.96	1.89	1.58	3.80	1.61	0.28	0.77	2.20	24.28
1965	4.61	1.96	1.77	2.07	3.80	2.57	2.03	8.47	4.93	3.28	1.75	3.61	40.85
1966	0.46	1.46	1.82	2.81	1.88	3.42	3.73	4.60	1.17	0.97	4.63	5.12	32.07
1967	1.29	2.12	1.72	2.77	2.28	1.92	3.95	0.81	2.14	3.07	2.85	6.81	31.73
1968	1.91	1.29	2.26	3.01	5.13	3.40	4.50	1.45	1.52	1.11	3.52	3.97	33.07
1969	3.70	0.27	1.54	3.64	3.74	4.82	6.75	1.15	2.70	1.58	3.81	2.10	35.80
1970	1.09	0.89	2.61	4.26	4.05	4.59	5.99	3.00	5.78	2.00	2.09	1.49	37.84
1971	0.82	2.59	1.34	1.08	2.33	2.64	2.77	1.10	1.84	1.77	1.17	3.73	23.18
1972	1.42	0.77	2.33	3.74	2.63	4.09	2.77	4.47	8.10	1.46	3.55	3.08	38.41
1973	1.63	1.05	4.20	1.79	2.85	6.51	3.17	1.18	1.09	2.76	3.27	3.17	32.67
1974	2.27	2.00	2.93	2.55	4.18	3.31	0.68	1.61	1.41	0.70	3.57	3.41	28.62
1975	2.57	2.57	1.90	2.34	3.83	4.21	4.99	5.52	2.70	2.42	2.17	3.35	38.57
1976	2.80	4.43	3.56	2.79	1.72	3.70	2.08	0.40	3.68	2.14	0.55	0.93	28.78
1977	1.29	1.99	4.43	6.10	1.53	3.48	1.83	5.79	4.27	1.77	2.72	3.56	38.76
1978	3.14	0.54	2.34	3.74	2.48	5.34	1.86	1.67	3.19	1.65	2.48	3.31	31.74
1979	1.24	0.70	2.55	4.03	3.15	4.23	3.96	4.71	2.90	2.02	4.25	2.46	36.20
1980	0.74	0.96	3.65	3.13	2.93	3.26	4.49	5.89	1.63	1.79	0.97	2.48	31.92
1981	0.48	3.27	0.63	3.54	2.38	8.48	3.72	2.28	6.05	3.79	0.84	2.93	38.39
1982	3.61	1.15	3.74	1.53	2.61	2.01	1.97	1.38	2.03	1.14	6.86	3.48	31.51
1983	0.88	0.59	1.86	4.28	3.98	4.06	3.39	2.15	1.42	3.59	5.56	3.91	35.67
1984	0.99	1.18	2.95	5.15	3.48	1.49	2.30	3.87	2.02	1.75	2.74	3.22	31.14
1985	2.02	3.23	5.70	1.40	1.85	2.90	3.86	4.30	2.53	3.05	5.89	1.62	38.35
Record Mean	2.14	1.91	2.61	2.86	3.15	3.47	2.99	2.89	2.66	2.28	2.43	2.40	31.78

TABLE 3   AVERAGE TEMPERATURE (deg. F)   TOLEDO, OHIO

YEAR	JAN	FEB	MAR	APR	MAY	JUNE	JULY	AUG	SEP	OCT	NOV	DEC	ANNUAL
1956	23.9	28.8	33.7	45.6	57.8	69.8	72.1	71.0	59.8	56.8	40.0	35.1	49.5
1957	20.1	31.4	36.8	48.7	57.9	68.6	72.6	70.4	62.9	49.6	40.0	34.3	49.5
1958	25.8	22.2	35.9	49.5	58.2	63.9	72.2	69.6	63.2	53.4	42.0	21.1	48.1
#1959	21.0	26.8	35.3	48.7	63.7	70.0	73.2	76.1	66.7	49.6	34.8	33.1	49.9
1960	28.7	27.5	25.7	50.3	57.3	65.3	69.8	71.6	66.4	52.2	40.9	21.9	48.2
1961	22.4	31.2	39.1	41.6	54.6	65.2	70.1	69.8	67.6	53.9	38.9	26.5	48.4
1962	21.8	24.8	35.0	48.3	65.6	69.6	69.6	70.6	61.2	54.2	38.3	22.7	48.5
1963	14.9	18.1	37.0	47.5	55.3	67.9	72.7	67.4	61.7	59.1	43.1	19.6	47.0
1964	28.5	25.9	36.0	48.6	62.6	68.8	73.8	67.8	62.5	47.5	42.3	27.8	49.2
1965	24.7	25.8	29.8	46.8	63.5	66.9	68.5	67.2	63.7	49.1	39.5	34.1	48.3
1966	20.2	27.0	38.1	45.6	53.8	70.8	74.1	69.1	61.1	50.4	40.4	27.7	48.2
1967	29.1	23.0	34.8	49.0	52.4	71.1	69.1	65.4	58.4	51.2	34.8	31.6	47.5
1968	22.1	24.5	38.6	49.5	55.9	69.6	71.3	74.0	65.2	52.0	40.6	26.6	49.2
1969	21.6	27.4	33.2	49.3	58.4	64.6	71.8	71.4	62.9	50.5	37.6	25.7	47.9
1970	16.2	24.3	31.8	48.9	61.1	67.2	71.1	69.6	64.4	54.0	35.8	28.9	48.1
1971	20.3	27.9	32.9	46.1	56.4	71.3	69.0	69.0	66.8	59.0	37.6	33.6	49.1
1972	23.4	24.4	34.1	46.1	60.4	63.9	71.4	68.4	62.2	47.2	37.7	30.3	47.5
1973	28.2	25.2	44.1	48.3	55.7	70.1	72.3	71.3	64.4	55.7	41.9	27.5	50.4
1974	26.1	23.2	36.2	48.8	56.0	65.4	72.5	71.5	59.6	49.5	40.4	28.9	48.2
1975	29.2	28.3	33.3	42.7	62.5	69.0	70.8	72.0	57.4	51.9	45.3	28.9	49.3
1976	19.8	32.8	41.6	49.3	56.0	69.3	72.2	68.2	60.5	45.6	32.3	19.9	47.3
1977	9.6	24.3	41.6	53.3	63.6	65.0	74.6	69.3	65.0	49.3	41.0	24.7	48.4
1978	16.7	11.8	28.7	45.8	58.9	67.6	70.9	70.4	68.0	49.8	40.3	30.1	46.6
1979	17.6	15.1	38.7	45.5	57.9	67.7	67.7	68.8	63.0	51.3	40.0	32.1	47.4
1980	24.3	21.4	32.4	46.8	59.5	65.5	73.6	73.3	63.8	46.8	37.4	26.0	47.6
1981	17.6	28.5	36.5	49.9	55.4	68.4	71.7	69.8	61.3	47.7	39.6	27.4	47.8
1982	15.8	20.2	33.4	42.7	64.4	64.3	72.6	67.5	61.9	52.7	41.8	36.6	47.8
1983	27.6	30.5	37.9	44.2	54.8	67.9	74.7	73.8	64.2	51.9	41.3	20.0	49.1
1984	16.6	33.0	27.6	46.8	54.4	71.2	69.8	71.2	60.8	55.2	38.7	34.0	48.3
1985	19.5	22.6	39.3	53.5	61.6	64.8	73.2	69.1	64.0	53.3	43.9	22.3	48.9
Record Mean	25.4	26.8	35.8	47.5	59.0	68.7	73.1	71.1	64.3	52.9	40.4	29.4	49.5
Max	32.5	34.3	44.1	57.0	68.9	78.6	83.1	80.9	74.2	62.3	47.9	36.1	58.3
Min	18.2	19.2	27.5	38.1	49.0	58.8	63.1	61.2	54.4	43.5	32.9	22.8	40.7

REFERENCE NOTES FOR TABLES 1, 2, 3 and 6   (TOLEDO, OH)

GENERAL

T - TRACE AMOUNT
BLANK ENTRIES DENOTE MISSING/UNREPORTED DATA.
# INDICATES A STATION OR INSTRUMENT RELOCATION.

SPECIFIC

TABLE 1

(a) - LENGTH OF RECORD IN YEARS. ALTHOUGH
       INDIVIDUAL MONTHS MAY BE MISSING.

  * LESS THAN .05

NORMALS — BASED ON THE 1951-1980 RECORD PERIOD.
EXTREMES — DATES ARE THE MOST RECENT OCCURRENCE.
WIND DIR. — NUMERALS SHOW TENS OF DEGREES
              CLOCKWISE FROM TRUE NORTH.
              "00" INDICATES CALM.
RESULTANT WIND DIRECTIONS ARE GIVEN TO WHOLE DEGREES.

EXCEPTIONS

TABLES 2, 3, and 6

RECORD MEANS ARE THROUGH THE CURRENT YEAR,
BEGINNING IN   1874 FOR TEMPERATURE
                    1871 FOR PRECIPITATION
                    1956 FOR SNOWFALL

TABLE 4   HEATING DEGREE DAYS Base 65 deg. F   TOLEDO, OHIO

SEASON	JULY	AUG	SEP	OCT	NOV	DEC	JAN	FEB	MAR	APR	MAY	JUNE	TOTAL
1956-57	1	13	190	256	745	923	1385	934	868	509	236	45	6105
1957-58	0	9	145	471	741	944	1210	1193	897	465	233	70	6378
1958-59	4	19	117	367	681	1354	1359	1064	914	485	136	31	6531
#1959-60	0	0	98	470	899	983	1113	1078	1211	451	244	62	6609
1960-61	7	0	64	393	716	1327	1313	938	795	698	322	92	6665
1961-62	27	12	87	344	775	1188	1333	1120	924	510	125	17	6462
1962-63	1	2	169	348	795	1305	1549	1310	859	520	317	55	7230
1963-64	9	33	120	199	646	1399	1121	1128	920	493	146	58	6272
1964-65	1	55	150	536	676	1149	1242	1088	1084	538	123	51	6693
1965-66	27	57	129	486	759	953	1381	1058	828	579	348	30	6635
1966-67	1	15	147	451	729	1150	1105	1170	929	479	390	16	6582
1967-68	27	57	206	437	900	1031	1324	1168	814	458	284	27	6733
1968-69	8	17	71	424	726	1184	1340	1047	976	470	239	107	6609
1969-70	3	7	126	446	818	1213	1507	1130	1022	495	176	70	7013
1970-71	14	11	118	345	749	1111	1379	1035	987	561	272	22	6604
1971-72	18	12	78	197	813	966	1283	1169	952	560	158	95	6301
1972-73	28	36	134	543	810	1073	1135	1106	639	499	285	3	6291
1973-74	3	16	114	289	686	1157	1197	1166	885	483	295	71	6362
1974-75	2	0	190	478	730	1108	1104	1021	974	664	148	45	6464
1975-76	7	6	227	406	585	1110	1393	927	717	497	277	16	6168
1976-77	1	33	162	596	976	1393	1708	1135	718	381	135	91	7329
1977-78	3	29	71	481	713	1241	1490	1484	1121	573	243	43	7492
1978-79	11	11	74	466	732	1076	1461	1390	808	577	259	42	6907
1979-80	16	33	121	440	724	1009	1258	1256	1005	542	199	83	6686
1980-81	0	3	113	560	822	1206	1464	1015	879	450	309	24	6845
1981-82	7	15	169	529	754	1160	1522	1250	972	665	81	76	7200
1982-83	3	47	148	386	690	871	1154	958	833	624	311	55	6080
1983-84	8	0	127	407	705	1389	1494	920	1151	545	341	9	7096
1984-85	11	15	173	297	782	951	1404	1182	791	368	158	58	6190
1985-86	0	16	138	356	626	1316							

TABLE 5   COOLING DEGREE DAYS Base 65 deg. F   TOLEDO, OHIO

YEAR	JAN	FEB	MAR	APR	MAY	JUNE	JULY	AUG	SEP	OCT	NOV	DEC	TOTAL
1969	0	0	0	7	43	101	220	215	69	1	0	0	656
1970	0	0	0	19	62	142	210	159	107	10	0	0	709
1971	0	0	0	1	13	219	148	143	138	18	0	0	680
1972	0	0	0	0	22	67	236	148	55	0	0	0	528
1973	0	0	0	5	3	163	237	222	103	9	0	0	742
1974	0	0	0	4	25	91	243	206	34	5	0	0	608
1975	0	0	0	0	79	172	197	230	7	7	0	0	692
1976	0	0	0	31	10	155	230	137	34	2	0	0	599
1977	0	0	0	37	95	99	309	167	77	0	0	0	784
1978	0	0	0	0	58	128	200	184	170	1	0	0	741
1979	0	0	0	0	46	127	182	158	67	22	0	0	602
1980	0	0	0	3	35	106	275	265	84	4	0	0	772
1981	0	0	1	2	17	132	220	170	64	0	0	0	606
1982	0	0	0	0	68	61	245	132	62	11	0	0	579
1983	0	0	0	4	2	148	311	279	109	11	0	0	864
1984	0	0	0	5	17	203	168	214	51	1	0	0	659
1985	0	0	0	29	60	58	263	147	116	0	0	0	673

TABLE 6   SNOWFALL (inches)   TOLEDO, OHIO

SEASON	JULY	AUG	SEP	OCT	NOV	DEC	JAN	FEB	MAR	APR	MAY	JUNE	TOTAL
1956-57	0.0	0.0	0.0	0.0	3.5	7.7	13.2	2.2	2.6	12.0	0.0	0.0	41.2
1957-58	0.0	0.0	0.0	T	T	4.2	8.1	7.1	3.5	T	0.0	0.0	22.9
1958-59	0.0	0.0	0.0	0.0	3.1	6.2	10.2	1.1	3.5	T	0.0	0.0	24.1
1959-60	0.0	0.0	0.0	0.0	8.3	4.6	3.0	14.2	11.2	T	T	0.0	41.3
1960-61	0.0	0.0	0.0	0.0	0.5	5.9	3.4	3.3	0.6	7.6	T	0.0	21.3
1961-62	0.0	0.0	0.0	0.0	0.9	8.9	6.9	14.2	9.6	T	0.0	0.0	40.5
1962-63	0.0	0.0	0.0	0.2	T	13.2	12.4	9.4	5.6	0.3	0.0	0.0	41.1
1963-64	0.0	0.0	0.0	0.0	T	8.1	8.6	12.4	11.6	0.1	0.0	0.0	40.8
1964-65	0.0	0.0	0.0	0.0	3.6	7.4	9.0	12.6	10.2	1.4	0.0	0.0	44.2
1965-66	0.0	0.0	0.0	0.0	0.1	0.9	5.3	4.5	7.9	1.1	T	0.0	19.8
1966-67	0.0	0.0	0.0	0.0	17.9	13.6	4.1	14.4	9.8	0.8	0.0	0.0	60.6
1967-68	0.0	0.0	T	T	1.9	5.1	10.4	5.6	11.2	0.2	0.0	0.0	34.4
1968-69	0.0	0.0	0.0	0.0	1.8	8.2	9.2	2.5	4.9	1.5	0.0	0.0	28.1
1969-70	0.0	0.0	0.0	T	5.7	19.0	14.2	7.7	8.3	4.5	0.0	0.0	59.4
1970-71	0.0	0.0	0.0	0.0	3.6	8.5	8.5	8.0	5.2	T	0.0	0.0	33.4
1971-72	0.0	0.0	0.0	0.0	5.7	1.4	10.1	7.6	3.3	1.8	0.0	0.0	29.9
1972-73	0.0	0.0	0.0	0.2	5.0	7.7	3.0	11.6	4.0	T	0.0	0.0	31.5
1973-74	0.0	0.0	0.0	T	0.2	13.8	7.5	11.6	2.9	1.1	0.0	0.0	37.1
1974-75	0.0	0.0	0.0	T	2.8	23.9	5.4	5.5	5.3	1.8	0.0	0.0	44.7
1975-76	0.0	0.0	0.0	0.0	5.7	12.2	14.5	8.4	4.0	1.3	0.0	0.0	46.1
1976-77	0.0	0.0	0.0	T	1.3	11.1	17.2	8.7	15.0	0.6	0.0	0.0	53.9
1977-78	0.0	0.0	0.0	0.0	6.6	24.2	30.8	9.0	2.5	T	0.0	0.0	73.1
1978-79	0.0	0.0	0.0	0.0	2.8	2.3	7.6	5.1	1.2	4.0	0.0	0.0	23.0
1979-80	0.0	0.0	0.0	0.0	1.6	1.5	4.1	6.4	3.4	0.5	T	0.0	17.5
1980-81	0.0	0.0	0.0	0.9	3.5	11.6	6.9	11.2	3.6	0.0	0.0	0.0	37.7
1981-82	0.0	0.0	0.0	T	0.8	14.9	18.4	14.3	10.7	9.1	0.0	0.0	68.2
1982-83	0.0	0.0	0.0	T	2.2	1.2	0.7	4.1	3.6	0.7	T	0.0	12.5
1983-84	0.0	0.0	0.0	0.0	3.4	13.4	12.2	6.3	9.8	T	T	0.0	45.1
1984-85	0.0	0.0	0.0	0.0	2.4	5.1	14.0	12.4	2.6	2.0	0.0	0.0	38.5
1985-86	0.0	0.0	0.0	0.0	2.5	8.7							
Record Mean	0.0	0.0	T	T	3.2	9.1	9.7	8.3	6.2	1.8	T	0.0	38.4

See Reference Notes, relative to all above tables, on preceding page.

YOUNGSTOWN, OHIO

The Youngstown Municipal Airport is located in northeastern Ohio approximately 8 miles north of the city of Youngstown in Trumbull County. Airport elevation is 1,178 feet, about 200 feet higher than most communities in the Mahoning and Shenango River Valleys. There are numerous natural and man-made lakes in the region, including Lake Erie, 45 miles to the north. Drainage from the area flows southward through the Mahoning and Shenango Rivers which join to form the Beaver River at New Castle, Pennsylvania. The Beaver empties into the Ohio River at Rochester, Pennsylvania.

This entire area experiences frequent outbreaks of cold Canadian air masses which may be modified by passage over Lake Erie. This effect produces widespread cloudiness especially during the cool months of the year. The winter months are characterized by persistent cloudiness and intermittent snow flurries. The daily temperature range during most winter days is quite small. During most winters, the bulk of the snow falls as flurries of 2 inches or less per occurrence, although several snowstorms per year will produce amounts in the 4- to 10-inch range.

Destructive storms seldom occur, and tornadoes are not common. During recent years flood control projects have all but eliminated the threat of serious river flooding. Flash flooding of small streams and creeks rarely affects residential areas. Certain communities have well known areas of urban flooding during periods of prolonged heavy thunderstorms.

The climate of the Youngstown district has had an important role in the growth and development of this industrial area. Temperatures seldom reach extreme values especially during the summer months. However, high humidity during most days of the year tends to accentuate the temperature. Rainfall, reasonably well distributed throughout the year, provides a more than adequate supply of water for agriculture, industrial, and residential use.

Based on the 1951-1980 period, the average first occurrence of 32 degrees Fahrenheit in the fall is October 14 and the average last occurrence in the spring is May 6.

TABLE 1 — NORMALS, MEANS AND EXTREMES

YOUNGSTOWN, OHIO

LATITUDE: 41°16'N   LONGITUDE: 80°40'W   ELEVATION: FT. GRND 1178 BARO 01199   TIME ZONE: EASTERN   WBAN: 14852

	(a)	JAN	FEB	MAR	APR	MAY	JUNE	JULY	AUG	SEP	OCT	NOV	DEC	YEAR
TEMPERATURE °F:														
Normals														
–Daily Maximum		31.4	33.8	44.0	57.9	68.6	77.5	81.2	79.8	73.2	61.4	47.7	36.0	57.7
–Daily Minimum		16.9	18.0	26.5	36.9	46.0	55.1	59.0	58.2	51.5	41.5	32.7	22.6	38.7
–Monthly		24.1	25.9	35.3	47.4	57.3	66.3	70.1	69.0	62.4	51.5	40.2	29.3	48.2
Extremes														
–Record Highest	42	71	67	80	87	92	99	100	97	99	87	80	76	100
–Year		1950	1961	1977	1976	1962	1952	1954	1953	1954	1953	1961	1982	JUL 1954
–Record Lowest	42	-20	-14	-10	11	24	30	42	32	29	20	1	-12	-20
–Year		1985	1979	1980	1950	1970	1972	1968	1982	1957	1969	1976	1983	JAN 1985
NORMAL DEGREE DAYS:														
Heating (base 65°F)		1265	1095	921	528	267	57	7	19	130	423	744	1104	6560
Cooling (base 65°F)		0	0	0	0	28	96	166	143	52	0	0	0	485
% OF POSSIBLE SUNSHINE														
MEAN SKY COVER (tenths)														
Sunrise – Sunset	39	8.3	7.9	7.7	7.2	6.7	6.2	6.0	6.0	6.1	6.3	7.9	8.4	7.0
MEAN NUMBER OF DAYS:														
Sunrise to Sunset														
–Clear	42	2.8	3.2	4.3	5.0	6.1	6.5	7.1	7.5	8.0	8.4	3.3	2.5	64.5
–Partly Cloudy	42	5.1	5.4	6.2	7.4	8.5	11.2	13.0	11.8	9.8	7.5	5.5	5.4	97.2
–Cloudy	42	23.0	19.7	20.5	17.6	16.4	12.3	10.9	11.7	12.2	15.1	21.2	23.2	203.6
Precipitation														
.01 inches or more	42	17.0	14.9	16.0	14.4	13.0	11.5	10.4	10.1	10.1	10.7	14.8	17.6	160.6
Snow,Ice pellets														
1.0 inches or more	42	4.3	3.8	3.3	0.8	0.*	0.0	0.0	0.0	0.0	0.2	1.6	4.5	18.6
Thunderstorms	42	0.3	0.4	1.7	3.4	4.5	6.8	6.7	5.6	3.2	1.2	0.7	0.3	34.7
Heavy Fog Visibility														
1/4 mile or less	42	2.3	2.2	1.9	1.6	2.0	2.3	2.7	3.5	3.2	2.0	2.1	2.8	28.5
Temperature °F														
–Maximum														
90° and above	42	0.0	0.0	0.0	0.0	0.1	1.3	2.7	2.0	0.5	0.0	0.0	0.0	6.6
32° and below	42	16.5	12.3	5.6	0.3	0.0	0.0	0.0	0.0	0.0	0.*	2.4	12.7	49.8
–Minimum														
32° and below	42	28.3	25.2	22.6	10.9	1.6	0.*	0.0	0.*	0.1	4.2	15.9	25.7	134.5
0° and below	42	3.1	2.3	0.2	0.0	0.0	0.0	0.0	0.0	0.0	0.0	0.0	0.9	6.5
AVG. STATION PRESS.(mb)	13	973.8	974.5	972.6	972.7	972.4	973.5	974.6	975.9	976.2	976.4	975.0	974.6	974.3
RELATIVE HUMIDITY (%)														
Hour 01	37	79	78	77	75	78	83	84	86	85	80	79	80	80
Hour 07	38	80	80	79	77	78	82	84	88	88	84	82	81	82
Hour 13 (Local Time)	38	72	68	64	56	54	56	55	57	58	58	67	73	62
Hour 19	38	75	71	68	61	60	63	63	68	73	71	73	77	69
PRECIPITATION (inches):														
Water Equivalent														
–Normal		2.69	2.23	3.29	3.46	3.29	3.53	4.04	3.47	3.10	2.65	2.82	2.76	37.33
–Maximum Monthly	42	7.64	5.26	6.20	6.43	9.87	6.97	7.41	7.86	6.17	8.59	9.11	5.52	9.87
–Year		1950	1950	1964	1957	1946	1957	1958	1956	1945	1954	1985	1971	MAY 1946
–Minimum Monthly	42	0.73	0.60	1.34	1.01	0.78	1.35	1.57	0.51	0.27	0.43	0.94	0.88	0.27
–Year		1985	1978	1960	1982	1977	1952	1957	1957	1960	1953	1976	1958	SEP 1960
–Maximum in 24 hrs	42	2.79	2.76	2.47	1.75	2.85	3.14	3.82	2.86	4.02	4.31	3.00	2.28	4.31
–Year		1959	1959	1954	1957	1946	1983	1967	1980	1979	1954	1985	1979	OCT 1954
Snow,Ice pellets														
–Maximum Monthly	42	36.0	22.7	23.2	12.2	5.4				T	7.4	30.6	23.0	36.0
–Year		1978	1967	1965	1961	1966				1983	1962	1950	1963	JAN 1978
–Maximum in 24 hrs	42	17.5	13.4	10.7	6.4	5.4				T	4.9	20.7	14.8	20.7
–Year		1948	1984	1983	1961	1966				1983	1962	1950	1944	NOV 1950
WIND:														
Mean Speed (mph)	36	11.8	11.4	11.6	11.1	9.7	8.7	7.9	7.5	8.3	9.4	11.1	11.6	10.0
Prevailing Direction														
through 1963		SW	W	W	SW	SW	SW	SW	SW	SSW	SSW	SW	SW	SW
Fastest Obs. 1 Min.														
–Direction (!!!)	36	25	27	25	25	25	23	27	27	36	23	25	28	27
–Speed (MPH)	36	48	58	55	49	40	45	58	44	40	44	52	40	58
–Year		1959	1956	1959	1957	1956	1949	1959	1956	1960	1959	1957	1970	JUL 1959
Peak Gust														
–Direction (!!!)	2	SW	W	W	W	W	NW	W	NE	NW	SW	W	SW	W
–Speed (mph)	2	48	39	53	64	44	58	46	30	52	39	45	58	64
–Date		1985	1985	1985	1985	1984	1984	1985	1985	1984	1984	1985	1985	APR 1985

See Reference Notes to this table on the following page

TABLE 2 PRECIPITATION (inches) YOUNGSTOWN, OHIO

YEAR	JAN	FEB	MAR	APR	MAY	JUNE	JULY	AUG	SEP	OCT	NOV	DEC	ANNUAL
1956	1.59	4.83	4.45	5.25	5.79	4.91	7.07	7.86	1.90	0.90	1.54	2.49	48.58
1957	2.15	1.71	2.13	6.43	3.17	6.97	1.57	0.51	2.64	2.43	2.22	3.75	35.68
1958	2.25	1.18	1.34	3.22	2.58	4.78	7.41	3.39	4.45	0.82	3.28	0.88	35.58
1959	5.34	4.25	2.50	4.15	2.45	3.76	4.40	2.34	3.04	3.66	3.21	3.29	42.39
1960	2.60	3.15	1.34	1.35	4.57	2.97	4.80	4.63	0.27	1.43	1.38	1.01	29.50
1961	0.82	2.84	3.14	4.99	1.94	5.64	5.87	2.18	2.46	2.88	2.98	0.99	36.73
1962	2.41	2.07	1.96	2.31	2.13	2.47	2.40	1.31	3.00	3.62	2.38	1.99	28.05
1963	1.12	0.80	3.68	2.91	2.11	2.10	2.07	3.11	0.94	0.45	3.24	1.26	23.79
1964	2.03	1.36	6.20	5.97	2.96	3.55	4.18	3.20	0.94	1.83	1.76	4.58	38.56
1965	4.80	3.21	3.31	2.56	2.98	2.95	3.42	4.10	4.02	4.14	2.65	2.33	40.47
1966	2.73	1.93	1.75	4.34	1.81	1.88	2.52	5.05	1.59	1.40	5.52	2.49	33.01
1967	1.26	2.53	3.36	2.52	3.77	2.14	5.51	2.02	3.64	3.26	2.94	1.92	34.87
1968	3.75	0.71	3.24	2.34	5.25	3.16	2.06	5.08	2.88	4.94	4.14	4.00	41.55
1969	2.56	0.83	3.72	3.91	4.07	3.57	3.51	0.51	1.95	2.80	2.67	2.84	33.33
1970	1.39	1.93	2.22	2.85	3.22	3.52	3.48	1.08	1.80	5.20	3.24	2.90	32.83
1971	2.08	3.56	2.36	1.57	2.58	2.46	2.19	2.78	2.74	1.40	3.26	5.52	32.50
1972	1.35	2.28	3.22	3.97	2.41	4.43	3.66	1.11	4.38	0.81	4.29	3.11	35.02
1973	1.80	1.79	3.72	4.39	5.10	3.51	2.97	2.83	2.96	2.89	1.71	3.04	36.71
1974	2.78	1.61	4.48	3.05	4.67	3.95	3.20	7.21	3.36	1.67	3.98	3.03	42.99
1975	3.14	3.12	3.22	1.61	5.94	2.75	2.35	6.06	5.04	2.59	1.91	2.89	40.62
1976	3.20	3.24	4.01	1.64	1.61	4.07	7.16	2.41	5.16	2.43	0.94	1.66	37.53
1977	1.53	1.19	4.46	4.06	0.78	5.75	5.72	4.64	5.05	2.25	4.55	3.75	43.73
1978	4.52	0.60	1.64	3.01	4.49	3.88	3.98	2.72	4.23	4.34	1.36	4.22	38.99
1979	2.95	2.03	1.94	4.02	4.40	2.16	3.60	4.72	5.57	1.63	2.67	3.95	39.64
1980	1.73	1.26	5.17	2.16	2.74	3.77	6.20	7.74	4.37	2:08	1.88	1.52	40.62
1981	0.75	3.85	2.04	4.84	3.99	3.21	3.39	2.01	4.12	2.34	1.70	2.77	35.01
1982	4.27	1.50	3.24	1.01	4.37	5.20	1.73	1.75	2.16	0.59	4.93	2.96	33.71
1983	1.20	1.19	3.57	5.28	3.00	4.42	2.59	2.31	3.98	3.55	3.90	3.70	38.69
1984	1.20	2.40	2.90	2.43	5.84	3.72	4.18	3.84	2.53	2.28	3.55	2.92	37.79
1985	0.73	1.22	5.77	1.76	3.06	2.88	6.01	3.07	2.40	1.34	9.11	2.07	39.42
Record Mean	2.76	2.28	3.35	3.49	3.74	3.56	3.85	3.34	3.10	2.60	3.12	2.77	37.95

TABLE 3 AVERAGE TEMPERATURE (deg. F) YOUNGSTOWN, OHIO

YEAR	JAN	FEB	MAR	APR	MAY	JUNE	JULY	AUG	SEP	OCT	NOV	DEC	ANNUAL
1956	25.4	29.9	33.9	44.9	55.5	66.7	69.4	69.6	58.5	55.2	40.9	36.6	48.9
1957	21.6	31.2	37.3	49.5	58.2	68.4	69.9	67.9	62.3	49.1	40.8	33.7	49.2
1958	26.1	20.6	34.2	49.5	56.4	61.9	71.0	68.1	62.0	51.1	42.3	20.8	47.0
#1959	22.6	28.3	33.9	48.5	61.9	66.7	70.8	73.3	66.0	51.8	36.9	33.5	49.5
1960	29.9	27.3	24.3	52.2	56.1	65.8	67.9	69.9	64.9	51.1	42.5	22.3	47.8
1961	21.3	31.8	38.6	42.1	52.8	65.0	71.1	70.5	67.2	54.3	40.4	27.6	48.5
1962	22.3	25.2	34.0	46.9	63.6	67.4	68.5	69.2	58.6	52.3	39.0	23.5	47.5
1963	18.3	17.4	38.4	47.9	54.6	66.2	69.5	64.8	57.8		42.7	20.6	46.4
1964	27.9	23.4	36.5	48.8	60.9	66.8	71.1	66.3	62.1	49.0	44.6	31.1	49.0
1965	23.9	25.5	30.8	45.0	63.1	65.4	67.3	65.3		49.4	41.7	35.4	48.3
1966	20.7	27.3	37.6	44.6	52.2	67.7	72.4	68.2	59.1	49.0	40.5	28.3	47.3
1967	29.5	22.0	35.1	48.0	50.2	70.2	69.6	65.7	58.1	50.6	34.6	31.8	47.0
1968	20.3	19.5	36.9	48.9	52.4	65.0	70.1	70.5	64.5	52.4	41.1	27.6	47.5
1969	24.7	26.8	31.5	48.5	56.8	64.3	69.8	69.1	61.4	50.9	38.0	24.6	47.2
1970	17.8	24.5	31.5	48.4	61.2	65.8	70.0	68.9	65.1	53.7	40.9	30.1	48.1
1971	20.7	27.1	30.3	43.4	55.2	69.6	68.1	67.5	66.0	57.9	38.5	35.8	48.3
1972	26.5	23.5	31.6	44.4	58.3	60.8	69.1	67.6	60.9	46.1	36.7	33.3	46.6
1973	27.4	24.9	44.5	47.8	54.4	69.2	71.1	71.0	63.5	54.8	42.8	31.4	50.2
1974	30.0	25.4	36.5	49.8	56.1	65.3	69.7	69.1	58.3	49.3	41.7	30.5	48.5
1975	30.2	29.2	33.4	40.9	62.0	68.6	71.1	71.2	58.7	52.5	46.2	30.5	49.6
1976	19.9	34.6	43.2	50.2	55.4	68.9	68.7	66.1	59.5	45.9	32.4	22.1	47.3
1977	10.3	24.8	42.3	50.7	62.6	63.4	72.3	69.4	65.6	51.0	43.2	28.0	48.6
1978	19.5	15.6	31.5	45.5	57.3	65.4	68.4	70.4	65.9	50.0	42.0	32.4	47.0
1979	20.7	17.2	39.6	46.1	56.0	65.3	68.4	68.1	62.0	50.9	43.0	33.0	47.5
1980	24.4	20.9	32.2	44.9	58.0	61.5	69.4	71.0	63.1	46.1	37.2	25.5	46.2
1981	18.4	29.4	35.9	51.3	58.2	70.2	72.0	68.1	60.6	48.9	40.7	29.2	48.5
1982	19.2	25.6	35.2	43.9	64.2	62.4	71.1	65.9	60.8	55.5	45.5	40.3	49.1
1983	29.9	33.2	40.4	46.9	53.3	66.3	74.1	74.2	63.8	51.8	42.0	24.0	50.1
1984	19.9	35.5	29.4	48.7	54.9	69.2	68.8	71.0	60.4	56.2	39.3	36.2	49.1
1985	19.2	24.1	39.1	53.4	59.6	62.2	68.3	67.9	64.0	52.9	45.2	24.4	48.4
Record Mean	24.5	26.6	35.8	47.5	57.5	66.4	70.3	69.3	62.5	52.0	40.5	29.4	48.5
Max	31.8	34.6	44.7	58.0	68.8	77.7	81.5	80.2	73.3	62.1	48.0	36.2	58.1
Min	17.1	18.6	26.8	36.9	46.2	55.2	59.2	58.3	51.6	42.0	32.9	22.6	39.0

REFERENCE NOTES FOR TABLES 1, 2, 3 and 6   (YOUNGSTOWN, OH)

GENERAL

T - TRACE AMOUNT
BLANK ENTRIES DENOTE MISSING/UNREPORTED DATA.
# INDICATES A STATION OR INSTRUMENT RELOCATION.

SPECIFIC

TABLE 1

(a) - LENGTH OF RECORD IN YEARS. ALTHOUGH
INDIVIDUAL MONTHS MAY BE MISSING.
* LESS THAN .05

NORMALS — BASED ON THE 1951-1980 RECORD PERIOD.
EXTREMES — DATES ARE THE MOST RECENT OCCURRENCE.
WIND DIR. — NUMERALS SHOW TENS OF DEGREES
CLOCKWISE FROM TRUE NORTH.
"00" INDICATES CALM.
RESULTANT WIND DIRECTIONS ARE GIVEN TO WHOLE DEGREES.

EXCEPTIONS

TABLES 2, 3, and 6

RECORD MEANS ARE THROUGH THE CURRENT YEAR,
BEGINNING IN   1943 FOR TEMPERATURE
1943 FOR PRECIPITATION
1943 FOR SNOWFALL

TABLE 4 HEATING DEGREE DAYS Base 65 deg. F   YOUNGSTOWN, OHIO

SEASON	JULY	AUG	SEP	OCT	NOV	DEC	JAN	FEB	MAR	APR	MAY	JUNE	TOTAL
1956-57	8	24	214	298	719	874	1339	939	854	492	234	40	6035
1957-58	12	20	148	486	721	961	1200	1238	949	462	276	130	6603
1958-59	1	40	131	428	675	1362	1309	1023	958	489	172	75	6663
#1959-60	3	4	107	423	833	970	1081	1085	1253	400	284	48	6491
1960-61	24	2	60	426	670	1316	1345	922	812	681	385	85	6728
1961-62	18	7	89	329	731	1150	1319	1109	952	559	140	36	6439
1962-63	18	23	222	397	773	1277	1442	1326	817	515	330	63	7203
1963-64	29	55	197	231	661	1368	1145	1198	875	482	174	93	6508
1964-65	8	57	140	490	603	1044	1266	1098	1053	589	123	80	6551
1965-66	24	66	100	481	694	908	1367	1049	844	605	398	65	6601
1966-67	5	19	211	494	728	1130	1093	1200	919	508	453	18	6778
1967-68	34	48	214	445	908	1023	1378	1315	865	476	384	91	7181
1968-69	32	39	73	400	710	1150	1243	1066	1032	491	275	116	6627
1969-70	13	17	167	440	803	1242	1457	1128	1036	509	180	77	7069
1970-71	20	19	101	347	716	1077	1367	1055	1070	642	314	18	6746
1971-72	16	26	76	221	788	898	1186	1198	1028	611	208	157	6413
1972-73	51	39	154	580	842	972	1157	1116	630	512	324	6	6383
1973-74	8	17	120	308	659	1037	1077	1103	878	465	288	60	6020
1974-75	8	10	216	480	692	1064	1069	995	972	717	149	48	6420
1975-76	8	8	189	389	559	1063	1391	873	672	468	313	22	5955
1976-77	18	59	183	584	969	1321	1692	1119	703	449	148	127	7372
1977-78	9	28	65	429	652	1142	1142	1378	1030	582	273	86	7077
1978-79	31	2	82	456	680	1004	1363	1332	781	568	310	74	6683
1979-80	35	37	141	445	652	987	1270	1270	1008	596	233	139	6793
1980-81	11	12	117	576	826	1220	1438	989	894	408	232	16	6739
1981-82	12	21	189	491	721	1101	1414	1098	920	628	81	99	6775
1982-83	11	65	161	302	587	763	1083	885	757	543	359	81	5597
1983-84	13	0	143	406	662	1265	1393	850	1101	486	323	18	6660
1984-85	16	13	182	271	763	885	1417	1137	796	383	200	111	6174
1985-86	10	17	141	365	589	1255							

TABLE 5 COOLING DEGREE DAYS Base 65 deg. F   YOUNGSTOWN, OHIO

YEAR	JAN	FEB	MAR	APR	MAY	JUNE	JULY	AUG	SEP	OCT	NOV	DEC	TOTAL
1969	0	0	0	1	26	101	170	151	67	8	0	0	524
1970	0	0	0	16	71	110	182	148	112	4	0	0	643
1971	0	0	0	0	16	164	119	109	114	10	0	0	532
1972	0	0	0	0	8	38	184	125	39	0	0	0	394
1973	0	0	0	4	2	137	204	210	81	0	0	0	638
1974	0	0	0	15	21	74	162	143	21	0	1	0	437
1975	0	0	0	0	65	163	204	207	8	9	0	0	656
1976	0	0	0	30	21	145	137	100	28	0	0	0	461
1977	0	0	4	28	80	86	246	171	89	0	4	0	708
1978	0	0	0	0	37	102	142	176	117	0	0	0	574
1979	0	0	0	7	38	95	147	139	61	14	0	0	501
1980	0	0	0	0	23	41	153	203	66	0	0	0	486
1981	0	0	0	6	30	179	212	124	55	0	0	0	606
1982	0	0	0	2	66	28	209	98	42	16	10	4	475
1983	0	0	0	5	5	127	299	292	117	4	0	0	849
1984	0	0	0	6	18	151	141	207	53	6	0	0	582
1985	0	0	0	41	39	33	119	111	118	0	1	0	462

TABLE 6 SNOWFALL (inches)   YOUNGSTOWN, OHIO

SEASON	JULY	AUG	SEP	OCT	NOV	DEC	JAN	FEB	MAR	APR	MAY	JUNE	TOTAL
1956-57	0.0	0.0	T	0.0	3.2	15.5	17.1	9.5	13.0	11.5	0.0	0.0	69.8
1957-58	0.0	0.0	0.0	1.7	2.3	7.4	13.3	13.2	11.0	0.6	0.0	0.0	49.5
1958-59	0.0	0.0	0.0	T	9.1	11.4	15.8	4.8	15.0	T	0.0	0.0	56.1
1959-60	0.0	0.0	0.0	T	7.3	9.9	8.3	20.8	20.6	2.6	T	0.0	69.5
1960-61	0.0	0.0	0.0	0.3	3.3	17.7	14.3	9.2	2.6	12.2	T	0.0	59.6
1961-62	0.0	0.0	0.0	0.0	4.0	10.8	9.9	16.0	19.6	5.2	0.0	0.0	65.5
1962-63	0.0	0.0	0.0	7.4	0.3	22.4	17.8	16.5	16.1	1.3	0.2	0.0	82.0
1963-64	0.0	0.0	0.0	T	8.4	23.0	20.5	17.2	8.2	1.6	0.0	0.0	78.9
1964-65	0.0	0.0	0.0	T	3.8	14.2	16.7	12.2	23.2	0.5	0.0	0.0	70.6
1965-66	0.0	0.0	0.0	0.4	3.1	3.9	23.4	6.7	6.1	7.2	5.4	0.0	56.2
1966-67	0.0	0.0	0.0	T	7.4	12.5	8.1	22.7	13.9	2.4	T	0.0	67.0
1967-68	0.0	0.0	0.0	0.6	12.7	8.0	14.0	10.1	4.9	T	0.0	0.0	50.3
1968-69	0.0	0.0	0.0	T	5.8	21.6	8.6	11.8	9.3	T	T	0.0	57.1
1969-70	0.0	0.0	0.0	0.4	6.0	21.2	15.2	13.4	10.0	0.6	0.0	0.0	66.8
1970-71	0.0	0.0	T	T	1.7	15.6	16.0	18.0	22.0	0.4	0.0	0.0	73.7
1971-72	0.0	0.0	0.0	0.0	20.1	7.2	9.0	19.0	9.3	2.1	0.0	0.0	66.7
1972-73	0.0	0.0	0.0	0.4	8.2	8.3	5.3	11.8	8.5	1.8	T	0.0	44.3
1973-74	0.0	0.0	0.0	0.0	3.3	7.0	9.8	9.2	9.2	4.7	0.0	0.0	43.2
1974-75	0.0	0.0	0.0	0.9	5.3	20.4	6.6	10.2	13.5	1.8	0.0	0.0	58.7
1975-76	0.0	0.0	0.0	0.0	3.7	10.2	19.7	7.9	5.0	0.6	T	0.0	47.1
1976-77	0.0	0.0	0.0	T	8.9	10.9	22.0	10.5	3.3	0.6	0.0	0.0	56.2
1977-78	0.0	0.0	0.0	T	5.6	18.6	36.0	9.6	3.6	T	0.0	0.0	73.4
1978-79	0.0	0.0	0.0	0.0	2.0	5.7	16.2	9.6	2.8	T	0.0	0.0	36.3
1979-80	0.0	0.0	0.0	T	1.6	7.4	9.2	6.2	8.4	T	T	0.0	32.8
1980-81	0.0	0.0	0.0	0.1	5.9	11.0	13.8	8.2	10.1	T	0.0	0.0	49.1
1981-82	0.0	0.0	0.0	0.4	1.5	17.6	13.5	5.4	13.1	10.6	0.0	0.0	62.1
1982-83	0.0	0.0	0.0	T	4.1	10.7	3.4	7.6	11.9	1.7	0.0	0.0	39.4
1983-84	0.0	0.0	T	0.0	2.8	7.9	12.4	22.3	18.0	T	0.0	0.0	63.4
1984-85	0.0	0.0	0.0	0.0	4.6	9.4	21.2	9.1	1.7	5.8	0.0	0.0	51.8
1985-86	0.0	0.0	0.0	0.0	T	16.4							
Record Mean	0.0	0.0	T	0.4	6.0	12.5	13.2	11.1	10.5	2.4	0.1	0.0	56.1

See Reference Notes, relative to all above tables, on preceding page.

Oklahoma City is located along the North Canadian River, a frequently nearly-dry stream, at the geographic center of the state. It is not quite 1,000 miles south of the Canadian Border and a little less than 500 miles north of the Gulf of Mexico. The surrounding country is gently rolling with the nearest hills or low mountains, the Arbuckles, 80 miles south. The elevation ranges around 1,250 feet above sea level.

Although some influence is exerted at times by warm, moist air currents from the Gulf of Mexico, the climate of Oklahoma City falls mainly under continental controls characteristic of the Great Plains Region. The continental effect produces pronounced daily and seasonal temperature changes and considerable variation in seasonal and annual precipitation. Summers are long and usually hot. Winters are comparatively mild and short.

During the year, temperatures of 100 degrees or more occur on an average of 10 days, but have occurred on as many as 50 days or more. While summers are usually hot, the discomforting effect of extreme heat is considerably mitigated by low humidity and the prevalence of a moderate southerly breeze. Approximately one winter in three has temperatures of zero or lower.

The length of the growing season varies from 180 to 251 days. Average date of last freeze is early April and average date of first freeze is early November. Freezes have occurred in early October.

During an average year, skies are clear approximately 40 percent of the time, partly cloudy 25 percent, and cloudy 35 percent of the time. The city is almost smoke-free as a result of favorable atmospheric conditions and the almost exclusive use of natural gas for heating. Flying conditions are generally very good with flight by visual flight rules possible about 96 percent of the time.

Summer rainfall comes mainly from showers and thunderstorms. Winter precipitation is generally associated with frontal passages. Measurable precipitation has occurred on as many as 122 days and as few as 55 days during the year. The seasonal distribution of precipitation is normally 12 percent in winter, 34 percent in spring, 30 percent in summer, and 24 percent in fall. The The period with the least number of days with precipitation is November through January, and the month with the most rainy days is May. Thunderstorms occur most often in late spring and early summer. Large hail and/or destructive winds on occasion accompany these thunderstorms.

Snowfall averages less than 10 inches per year and seldom remains on the ground very long. Occasional brief periods of freezing rain and sleet storms occur.

Heavy fogs are infrequent. Prevailing winds are southerly except in January and February when northerly breezes predominate.

TABLE 1 NORMALS, MEANS AND EXTREMES

OKLAHOMA CITY, OKLAHOMA

LATITUDE: 35°24'N   LONGITUDE: 97°36'W   ELEVATION: FT. GRND 1285 BARO 01283   TIME ZONE: CENTRAL   WBAN: 13967

	(a)	JAN	FEB	MAR	APR	MAY	JUNE	JULY	AUG	SEP	OCT	NOV	DEC	YEAR
TEMPERATURE °F:														
Normals														
-Daily Maximum		46.6	52.2	61.0	71.7	79.0	87.6	93.5	92.8	84.7	74.3	59.9	50.7	71.2
-Daily Minimum		25.2	29.4	37.1	48.6	57.7	66.3	70.6	69.4	61.9	50.2	37.6	29.1	48.6
-Monthly		35.9	40.8	49.0	60.2	68.3	77.0	82.1	81.1	73.3	62.3	48.8	39.9	59.9
Extremes														
-Record Highest	32	79	84	93	100	104	105	108	110	102	96	87	86	110
-Year		1967	1981	1967	1972	1985	1980	1980	1980	1985	1972	1980	1955	AUG 1980
-Record Lowest	32	-4	-3	3	20	37	47	53	51	37	22	11	-3	-4
-Year		1959	1979	1960	1957	1981	1954	1971	1956	1985	1957	1959	1983	JAN 1959
NORMAL DEGREE DAYS:														
Heating (base 65°F)		902	678	506	184	41	0	0	0	15	145	486	778	3735
Cooling (base 65°F)		0	0	13	40	147	360	530	499	264	61	0	0	1914
% OF POSSIBLE SUNSHINE	31	59	60	63	66	67	74	79	79	72	69	60	59	67
MEAN SKY COVER (tenths)														
Sunrise - Sunset	37	5.9	5.8	5.9	5.7	5.8	5.0	4.4	4.3	4.6	4.6	5.1	5.5	5.2
MEAN NUMBER OF DAYS:														
Sunrise to Sunset														
-Clear	37	10.3	9.1	9.6	9.5	8.9	10.8	14.5	14.9	13.6	14.3	11.9	11.7	139.3
-Partly Cloudy	37	6.1	6.9	8.0	7.7	9.8	10.5	9.4	9.6	8.0	6.9	6.7	6.1	95.8
-Cloudy	37	14.6	12.2	13.4	12.7	12.2	8.6	7.1	6.5	8.4	9.8	11.3	13.2	130.1
Precipitation														
.01 inches or more	46	5.5	6.3	7.2	7.8	10.0	8.5	6.4	6.4	6.7	6.4	5.2	5.3	81.7
Snow, Ice pellets														
1.0 inches or more	46	0.9	1.0	0.4	0.0	0.0	0.0	0.0	0.0	0.0	0.0	0.2	0.6	3.1
Thunderstorms	46	0.5	1.4	3.0	5.5	9.1	8.7	6.2	6.3	4.5	3.1	1.3	0.6	50.1
Heavy Fog Visibility 1/4 mile or less	37	4.0	3.3	1.9	1.0	0.6	0.4	0.3	0.3	0.8	1.6	2.1	3.3	19.6
Temperature °F														
-Maximum														
90° and above	20	0.0	0.0	0.2	0.3	1.7	11.4	22.9	22.8	9.4	0.9	0.0	0.0	69.5
32° and below	20	6.2	2.6	0.1	0.0	0.0	0.0	0.0	0.0	0.0	0.0	0.1	2.5	11.4
-Minimum														
32° and below	20	23.8	17.6	8.0	1.1	0.0	0.0	0.0	0.0	0.0	0.6	8.6	20.8	80.6
0° and below	20	0.4	0.2	0.0	0.0	0.0	0.0	0.0	0.0	0.0	0.0	0.0	0.2	0.8
AVG. STATION PRESS.(mb)	13	973.6	971.8	967.4	967.6	966.5	967.7	969.4	969.6	970.3	971.1	971.3	972.4	969.9
RELATIVE HUMIDITY (%)														
Hour 00	20	73	72	68	69	76	76	70	70	74	72	74	72	72
Hour 06 (Local Time)	20	78	78	76	77	82	83	79	79	82	79	79	77	79
Hour 12	20	60	57	53	52	57	56	49	49	54	52	56	58	54
Hour 18	20	60	54	49	49	54	54	45	45	52	54	60	61	53
PRECIPITATION (inches):														
Water Equivalent														
-Normal		0.96	1.29	2.07	2.91	5.50	3.87	3.04	2.40	3.41	2.71	1.53	1.20	30.89
-Maximum Monthly	46	5.68	3.71	7.16	10.78	12.07	9.94	8.44	6.77	9.64	13.18	5.46	8.14	13.18
-Year		1949	1985	1948	1947	1982	1979	1959	1966	1970	1983	1964	1984	OCT 1983
-Minimum Monthly	46	T	T	T	0.62	0.33	0.63	T	0.25	T	T	T	0.03	T
-Year		1976	1947	1940	1971	1942	1952	1983	1978	1948	1958	1949	1955	JUL 1983
-Maximum in 24 hrs	46	3.10	2.21	3.44	3.80	5.63	4.55	5.75	3.34	7.68	8.95	2.01	2.55	8.95
-Year		1982	1978	1944	1970	1970	1985	1981	1966	1970	1983	1961	1984	OCT 1983
Snow, Ice pellets														
-Maximum Monthly	46	17.3	12.0	13.9	0.7						T	7.5	8.2	17.3
-Year		1949	1978	1968	1957						1967	1972	1960	JAN 1949
-Maximum in 24 hrs	46	8.0	6.1	8.4	0.7						T	5.5	5.7	8.4
-Year		1944	1979	1948	1957						1967	1972	1984	MAR 1948
WIND:														
Mean Speed (mph)	37	12.9	13.3	14.6	14.5	12.7	12.3	10.9	10.6	11.2	12.0	12.4	12.6	12.5
Prevailing Direction through 1963		N	N	SSE	SSE	SSE	SSE	SSE	SSE	SSE	SSE	S	S	SSE
Fastest Mile														
-Direction	30	W	W	S	NW	SE	NNW	NW	S	NW	S	S	NW	NNW
-Speed (MPH)	30	63	61	61	75	72	87	73	56	54	65	66	56	87
-Year		1960	1960	1954	1960	1960	1951	1960	1953	1961	1951	1958	1961	JUN 1951
Peak Gust														
-Direction	2	N	N	SW	NW	SE	N	S	W	N	W	NW	N	NW
-Speed (mph)	2	51	46	62	66	51	53	46	51	44	52	47	40	66
-Date		1985	1984	1985	1985	1985	1985	1985	1985	1985	1984	1984	1984	APR 1985

See Reference Notes to this table on the following page.

TABLE 2 PRECIPITATION (inches) OKLAHOMA CITY, OKLAHOMA

YEAR	JAN	FEB	MAR	APR	MAY	JUNE	JULY	AUG	SEP	OCT	NOV	DEC	ANNUAL
1956	0.26	1.31	0.56	2.43	5.07	2.46	2.19	0.86	0.20	3.76	1.49	1.97	22.56
1957	1.49	0.86	2.19	7.32	8.67	8.60	0.63	1.08	5.58	2.65	2.25	0.72	42.04
1958	1.29	0.83	3.39	2.22	3.10	7.17	2.02	3.56	3.87	T	0.60	1.11	29.16
1959	0.31	0.69	1.40	2.81	7.60	2.93	8.44	4.10	7.90	6.08	0.94	3.26	46.46
1960	1.47	2.12	0.85	3.01	6.79	5.68	7.69	2.89	1.62	4.66	0.24	2.48	39.50
1961	0.15	1.98	3.35	0.73	1.92	3.86	4.82	2.91	7.37	2.86	3.81	1.04	34.80
1962	1.45	1.02	0.80	2.16	2.64	7.84	1.71	2.26	3.08	2.43	1.34	0.76	27.49
1963	0.21	0.22	3.21	2.77	1.91	2.35	6.19	1.61	1.91	2.05	2.45	0.89	25.77
1964	0.83	2.17	1.30	2.06	5.21	0.77	2.01	4.91	2.96	0.84	5.46	0.62	29.14
1965	0.98	0.85	0.86	3.24	2.14	3.65	1.57	3.37	3.94	1.00	0.06	2.51	24.17
1966	1.05	2.39	1.30	3.68	0.88	2.63	2.38	6.77	2.82	0.37	0.84	0.45	25.56
1967	0.77	0.20	2.49	5.71	4.25	2.27	1.21	1.40	3.15	2.92	0.40	1.04	25.81
1968	2.19	1.02	2.84	3.03	8.40	2.39	1.41	3.75	2.64	2.40	4.11	1.33	35.51
1969	0.20	1.93	3.01	1.66	3.99	4.92	1.42	2.38	6.51	1.58	0.06	1.44	29.10
1970	0.32	0.29	2.09	5.33	6.53	2.45	1.30	0.80	9.64	3.29	1.03	0.26	33.33
1971	0.75	1.95	0.07	0.62	2.68	5.15	4.13	2.13	4.25	2.62	0.29	2.79	27.43
1972	0.21	0.43	1.13	3.10	4.03	1.36	3.22	1.82	2.04	7.17	2.28	0.84	27.63
1973	3.39	0.31	6.76	2.32	3.61	6.31	3.38	1.36	8.00	2.81	0.47	0.47	41.77
1974	0.10	2.68	3.12	4.66	5.01	3.36	0.48	4.42	6.24	5.57	2.34	1.47	39.45
1975	1.99	1.90	1.72	1.92	8.76	4.82	7.71	0.60	1.92	0.84	1.77	1.30	35.25
1976	T	0.33	3.09	2.94	4.36	0.88	1.38	1.46	1.53	1.78	0.12	0.19	18.06
1977	0.32	1.40	1.30	2.88	7.97	2.00	4.10	3.08	1.20	2.41	1.59	0.34	28.59
1978	1.26	3.23	1.32	1.65	10.12	4.04	3.75	0.25	0.96	1.02	2.88	0.70	31.18
1979	1.55	0.63	2.73	2.78	7.29	9.94	5.62	3.78	0.72	1.58	1.93	2.57	41.12
1980	1.69	1.29	1.38	2.16	9.00	2.52	0.42	0.60	2.21	0.99	0.51	1.58	24.35
1981	0.19	1.15	2.87	2.97	2.73	7.49	6.45	3.61	1.48	7.70	2.11	0.20	38.95
1982	3.68	0.98	1.63	1.92	12.07	4.06	2.11	1.13	2.86	1.03	2.78	1.94	36.19
1983	2.62	1.71	2.51	2.34	6.88	3.18	T	3.18	0.90	13.18	1.90	0.70	39.10
1984	0.35	1.16	4.70	1.79	1.62	3.48	0.30	2.35	1.01	6.64	2.05	8.14	33.59
1985	0.92	3.71	6.60	5.35	1.49	8.34	1.33	2.63	4.59	5.23	3.73	0.26	44.18
Record Mean	1.26	1.26	2.20	3.20	5.15	3.99	2.74	2.66	3.13	2.95	1.90	1.47	31.89

TABLE 3 AVERAGE TEMPERATURE (deg. F) OKLAHOMA CITY, OKLAHOMA

YEAR	JAN	FEB	MAR	APR	MAY	JUNE	JULY	AUG	SEP	OCT	NOV	DEC	ANNUAL
1956	36.1	40.6	50.5	57.9	72.6	78.1	83.0	84.4	76.6	66.3	47.1	41.7	61.3
1957	33.2	45.4	46.7	55.9	65.7	74.2	83.5	80.7	69.3	57.3	45.6	44.3	58.5
1958	39.4	35.5	41.0	56.1	69.4	76.8	80.1	79.9	73.2	61.9	51.4	36.5	58.4
1959	33.0	40.0	49.6	58.5	70.4	75.7	77.4	80.1	72.6	58.3	43.5	44.2	58.6
1960	36.3	34.5	39.7	61.7	66.5	77.3	78.2	79.3	74.5	64.4	51.4	37.4	58.4
1961	36.3	42.7	52.2	59.2	67.7	74.1	79.1	78.4	70.6	63.0	46.5	36.7	58.9
1962	32.3	45.1	48.1	58.3	74.3	75.3	82.2	82.5	71.4	64.7	49.4	40.6	60.4
1963	28.3	40.0	53.8	64.8	70.3	78.7	83.1	82.0	75.1	71.1	52.3	33.0	61.1
1964	40.1	38.5	47.1	64.1	70.0	77.4	85.3	80.8	72.3	60.3	49.9	37.9	60.3
#1965	38.8	39.5	40.7	65.5	71.3	78.0	84.6	80.6	73.9	62.8	56.3	48.8	61.8
1966	33.8	38.6	52.8	57.8	68.7	77.9	86.4	78.8	70.5	60.9	54.3	37.8	59.9
1967	41.8	41.7	56.4	65.4	66.9	77.4	79.7	79.3	70.9	62.6	49.4	39.8	60.9
1968	36.6	36.4	50.7	58.1	64.7	74.8	79.8	80.0	70.5	62.1	46.2	38.1	58.1
1969	38.8	42.3	42.0	60.4	67.7	75.0	83.9	80.2	73.3	57.6	48.7	40.4	59.2
1970	31.8	42.7	45.0	60.2	69.2	76.3	82.1	83.6	75.1	57.9	46.0	43.9	59.5
1971	36.9	39.1	49.1	60.4	67.3	78.6	80.7	77.2	73.3	63.4	49.0	42.2	59.8
1972	34.9	42.1	53.4	63.2	67.6	79.0	79.8	80.4	75.8	61.1	43.4	34.4	59.6
1973	33.3	39.8	52.5	56.0	66.8	75.2	79.7	79.8	70.6	64.3	53.1	39.3	59.2
1974	35.0	44.4	54.8	60.0	71.5	74.1	82.7	78.5	65.5	63.5	49.3	39.6	59.9
1975	40.3	36.5	46.1	58.7	67.4	75.1	78.0	80.1	68.3	63.4	50.7	41.8	58.8
1976	39.0	52.2	52.4	61.6	63.6	74.8	79.8	81.3	72.6	56.5	43.9	38.8	59.7
1977	29.2	45.9	54.1	62.5	70.0	79.6	83.0	80.7	78.0	62.7	50.9	40.0	61.4
1978	26.3	29.4	49.1	64.5	68.1	77.3	87.0	82.6	79.7	64.7	50.4	36.9	59.7
1979	25.4	31.5	51.2	58.1	65.8	75.2	81.0	80.0	73.1	65.7	46.5	43.3	58.1
1980	38.2	38.2	46.3	56.7	69.0	81.4	88.3	88.0	76.3	61.1	50.3	41.9	61.3
1981	37.7	43.9	51.9	65.6	65.7	78.4	84.2	78.8	74.1	60.1	50.3	39.1	60.8
1982	35.3	37.7	52.7	57.5	68.2	72.2	81.0	84.1	74.5	62.7	48.6	43.2	59.8
1983	38.6	42.6	48.8	54.0	64.6	73.4	81.6	84.0	74.9	62.7	50.4	25.8	58.5
1984	34.0	45.4	46.4	56.5	68.4	78.6	81.6	82.6	71.5	61.6	49.7	43.0	59.9
1985	30.6	37.2	53.0	62.7	70.0	76.0	80.9	81.3	73.1	61.2	46.1	35.1	58.9
Record Mean	36.8	40.7	49.9	60.8	68.1	76.9	81.6	81.3	73.8	62.5	49.4	39.8	60.1
Max	46.8	51.5	61.3	71.2	78.2	87.2	92.4	92.4	84.8	73.7	60.0	49.6	70.8
Min	26.8	29.9	38.4	49.1	57.9	66.7	70.8	70.1	62.8	51.3	38.7	29.9	49.4

REFERENCE NOTES FOR TABLES 1, 2, 3 and 6   (OKLAHOMA CITY, OK)

GENERAL

T - TRACE AMOUNT
BLANK ENTRIES DENOTE MISSING/UNREPORTED DATA.
# INDICATES A STATION OR INSTRUMENT RELOCATION.

SPECIFIC

TABLE 1

(a) - LENGTH OF RECORD IN YEARS. ALTHOUGH INDIVIDUAL MONTHS MAY BE MISSING.
  * LESS THAN .05

NORMALS — BASED ON THE 1951-1980 RECORD PERIOD.
EXTREMES — DATES ARE THE MOST RECENT OCCURRENCE.
WIND DIR. — NUMERALS SHOW TENS OF DEGREES CLOCKWISE FROM TRUE NORTH. "00" INDICATES CALM.
RESULTANT WIND DIRECTIONS ARE GIVEN TO WHOLE DEGREES.

EXCEPTIONS

TABLE 1

1. FASTEST MILE WIND IS THOROUGH OCTOBER 1981.

TABLES 2, 3, and 6

RECORD MEANS ARE THROUGH THE CURRENT YEAR, BEGINNING IN   1891 FOR TEMPERATURE
 1891 FOR PRECIPITATION
 1940 FOR SNOWFALL

TABLE 4 HEATING DEGREE DAYS Base 65 deg. F   OKLAHOMA CITY, OKLAHOMA

SEASON	JULY	AUG	SEP	OCT	NOV	DEC	JAN	FEB	MAR	APR	MAY	JUNE	TOTAL
1956-57	0	0	0	72	531	713	978	542	560	284	72	9	3761
1957-58	0	0	17	263	572	633	789	818	737	268	24	3	4124
1958-59	0	0	17	158	410	876	983	692	481	248	27	0	3892
1959-60	0	0	24	206	640	644	880	877	777	134	73	0	4255
1960-61	0	0	2	96	407	852	885	618	395	247	49	0	3551
1961-62	0	0	40	106	547	868	1006	560	524	230	5	0	3886
1962-63	0	0	19	129	458	750	1131	692	372	101	50	0	3702
1963-64	0	0	7	23	381	988	761	762	547	116	27	6	3618
1964-65	0	0	24	150	454	831	808	706	745	71	2	0	3791
#1965-66	0	0	28	129	262	496	961	734	388	223	62	0	3283
1966-67	0	0	6	166	338	837	713	647	307	71	77	0	3162
1967-68	0	0	27	155	464	773	872	826	444	215	71	0	3847
1968-69	0	0	0	152	561	829	808	629	708	158	38	2	3885
1969-70	0	0	0	274	481	752	1022	620	615	187	31	12	3994
1970-71	0	0	18	254	559	651	866	718	492	163	36	0	3757
1971-72	0	0	59	88	475	702	923	660	365	144	46	0	3462
1972-73	0	0	23	225	640	940	975	701	380	283	55	0	4222
1973-74	0	0	37	99	362	787	922	573	330	168	8	0	3286
1974-75	0	0	56	88	463	784	763	792	583	235	29	0	3793
1975-76	0	0	64	126	430	713	801	367	406	128	100	0	3135
1976-77	0	0	19	306	629	805	1103	529	338	107	7	0	3843
1977-78	0	0	0	115	420	766	1192	990	493	90	64	0	4130
1978-79	0	0	2	89	437	866	1221	932	434	217	81	0	4279
1979-80	0	0	2	92	551	669	823	771	572	249	24	0	3753
1980-81	0	0	23	180	444	710	839	587	400	69	69	0	3321
1981-82	0	0	22	189	434	797	913	759	382	248	25	13	3782
1982-83	0	0	14	156	490	671	809	622	496	345	96	9	3708
1983-84	0	0	25	117	439	1207	955	561	572	263	45	0	4184
1984-85	0	0	75	162	462	676	1059	773	377	108	10	0	3702
1985-86	0	0	63	146	562	921							

TABLE 5 COOLING DEGREE DAYS Base 65 deg. F   OKLAHOMA CITY, OKLAHOMA

YEAR	JAN	FEB	MAR	APR	MAY	JUNE	JULY	AUG	SEP	OCT	NOV	DEC	TOTAL
1969	0	0	0	29	128	310	593	477	255	52	0	0	1844
1970	0	0	1	47	169	357	536	582	328	38	0	0	2058
1971	0	0	4	31	117	416	493	388	313	45	3	0	1810
1972	0	2	11	97	133	429	470	483	351	109	0	0	2085
1973	0	0	0	19	119	312	465	462	216	83	11	0	1687
1974	0	0	22	26	217	280	553	426	80	47	0	0	1651
1975	0	0	0	1	53	108	310	410	476	170	83	4	1615
1976	0	1	23	33	62	300	468	512	253	50	0	0	1702
1977	0	1	8	37	170	445	565	491	395	49	2	0	2163
1978	0	0	8	80	165	378	690	553	450	87	7	0	2418
1979	0	0	10	18	112	314	505	471	252	121	2	0	1805
1980	0	0	0	7	155	498	729	721	366	65	11	2	2554
1981	0	4	0	94	98	409	603	435	304	47	0	0	1994
1982	0	0	9	28	130	234	503	598	305	90	3	1	1901
1983	0	0	0	20	91	266	523	599	329	54	8	0	1890
1984	0	0	0	16	159	414	521	551	279	64	5	0	2009
1985	0	0	12	43	172	336	501	512	313	38	0	0	1927

TABLE 6 SNOWFALL (inches)   OKLAHOMA CITY, OKLAHOMA

SEASON	JULY	AUG	SEP	OCT	NOV	DEC	JAN	FEB	MAR	APR	MAY	JUNE	TOTAL
1956-57	0.0	0.0	0.0	0.0	0.0	1.6	T	T	0.1	0.7	0.0	0.0	2.4
1957-58	0.0	0.0	0.0	T	T	0.0	1.4	0.5	6.8	0.0	0.0	0.0	8.7
1958-59	0.0	0.0	0.0	0.0	2.6	7.0	3.3	0.5	1.2	0.0	0.0	0.0	14.6
1959-60	0.0	0.0	0.0	0.0	0.3	T	8.2	5.4	3.2	0.0	0.0	0.0	17.1
1960-61	0.0	0.0	0.0	0.0	0.0	8.2	1.9	7.2	0.7	T	0.0	0.0	18.0
1961-62	0.0	0.0	0.0	0.0	0.0	0.8	8.5	0.8	0.0	0.0	0.0	0.0	10.1
1962-63	0.0	0.0	0.0	0.0	T	1.2	0.5	1.6	0.0	0.0	0.0	0.0	3.3
1963-64	0.0	0.0	0.0	0.0	0.0	2.5	T	0.6	1.5	0.0	0.0	0.0	4.6
1964-65	0.0	0.0	0.0	0.0	0.0	T	2.0	0.9	0.1	0.0	0.0	0.0	3.0
1965-66	0.0	0.0	0.0	0.0	0.0	5.8	T	2.3	0.0	0.0	0.0	0.0	8.1
1966-67	0.0	0.0	0.0	0.0	T	0.6	T	0.4	0.0	0.0	0.0	0.0	1.1
1967-68	0.0	0.0	0.0	T	1.2	1.2	0.4	7.7	13.9	0.0	0.0	0.0	24.4
1968-69	0.0	0.0	0.0	0.0	1.4	1.3	T	1.3	8.2	0.0	0.0	0.0	12.2
1969-70	0.0	0.0	0.0	0.0	0.0	4.2	0.4	T	2.3	T	0.0	0.0	6.9
1970-71	0.0	0.0	0.0	0.0	T	T	T	5.1	0.7	0.0	0.0	0.0	5.8
1971-72	0.0	0.0	0.0	0.0	0.5	5.2	0.8	4.9	0.0	0.0	0.0	0.0	11.4
1972-73	0.0	0.0	0.0	0.0	7.5	1.4	8.3	0.2	T	T	0.0	0.0	17.4
1973-74	0.0	0.0	0.0	0.0	0.0	0.6	0.7	1.0	0.5	0.0	0.0	0.0	2.8
1974-75	0.0	0.0	0.0	0.0	1.0	2.0	0.6	0.9	0.1	0.0	0.0	0.0	4.6
1975-76	0.0	0.0	0.0	0.0	0.7	3.9	T	0.3	0.0	0.0	0.0	0.0	4.9
1976-77	0.0	0.0	0.0	0.0	0.3	T	2.8	0.4	0.0	0.0	0.0	0.0	3.5
1977-78	0.0	0.0	0.0	0.0	T	T	8.4	12.0	T	0.0	0.0	0.0	20.4
1978-79	0.0	0.0	0.0	0.0	0.0	3.3	4.0	6.1	0.0	0.0	0.0	0.0	13.4
1979-80	0.0	0.0	0.0	0.0	T	T	T	1.8	T	0.0	0.0	0.0	1.8
1980-81	0.0	0.0	0.0	0.0	4.0	0.0	T	T	0.0	0.0	0.0	0.0	4.0
1981-82	0.0	0.0	0.0	0.0	0.0	T	1.0	3.9	2.5	0.0	0.0	0.0	7.4
1982-83	0.0	0.0	0.0	0.0	T	T	5.1	4.3	T	0.0	0.0	0.0	9.4
1983-84	0.0	0.0	0.0	0.0	T	1.9	5.6	2.0	T	0.0	0.0	0.0	9.5
1984-85	0.0	0.0	0.0	0.0	T	6.1	1.5	2.3	0.0	0.0	0.0	0.0	9.9
1985-86	0.0	0.0	0.0	0.0	T	2.9							
Record Mean	0.0	0.0	0.0	T	0.5	1.6	2.9	2.5	1.5	T	0.0	0.0	9.0

See Reference Notes, relative to all above tables, on preceding page.

The city of Tulsa lies along the Arkansas River at an elevation of 700 feet above sea level. The surrounding terrain is gently rolling.

At latitude 36 degrees, Tulsa is far enough north to escape the long periods of heat in summer, yet far enough south to miss the extreme cold of winter. The influence of warm moist air from the Gulf of Mexico is often noted, due to the high humidity, but the climate is essentially continental characterized by rapid changes in temperature. Generally the winter months are mild. Temperatures occasionally fall below zero but only last a very short time. Temperatures of 100 degrees or higher are often experienced from late July to early September, but are usually accompanied by low relative humidity and a good southerly breeze. The fall season is long with a great number of pleasant, sunny days and cool, bracing nights.

Rainfall is ample for most agricultural pursuits and is distributed favorably throughout the year. Spring is the wettest season, having an abundance of rain in the form of showers and thunderstorms.

The steady rains of fall are a contrast to the spring and summer showers and provide a good supply of moisture and more ideal conditions for the growth of winter grains and pastures. The greatest amounts of snow are received in January and early March. The snow is usually light and only remains on the ground for brief periods.

The average date of the last 32 degree temperature occurrence is late March and the average date of the first 32 degree occurrence is early November. The average growing season is 216 days.

The Tulsa area is occasionally subjected to large hail and violent windstorms which occur mostly during spring and early summer, although occurrences have been noted throughout the year.

Prevailing surface winds are southerly during most of the year. Heavy fogs are infrequent. Sunshine is abundant. The prevalence of good flying weather throughout the year has contributed to the development of Tulsa as an aviation center.

TABLE 1 NORMALS, MEANS AND EXTREMES

TULSA, OKLAHOMA

LATITUDE: 36°12'N LONGITUDE: 95°54'W ELEVATION: FT. GRND 650 BARO 00669 TIME ZONE: CENTRAL WBAN: 13968

	(a)	JAN	FEB	MAR	APR	MAY	JUNE	JULY	AUG	SEP	OCT	NOV	DEC	YEAR
TEMPERATURE °F:														
Normals														
-Daily Maximum		45.6	51.9	60.8	72.4	79.7	87.9	93.9	93.0	85.0	74.9	60.2	50.3	71.3
-Daily Minimum		24.8	29.5	37.7	49.5	58.5	67.5	72.4	70.3	62.5	50.3	38.1	29.3	49.2
-Monthly		35.2	40.7	49.3	61.0	69.1	77.7	83.2	81.7	73.8	62.6	49.2	39.8	60.3
Extremes														
-Record Highest	47	79	86	96	102	96	103	112	110	109	98	87	80	112
-Year		1950	1962	1974	1972	1985	1953	1954	1970	1939	1979	1945	1966	JUL 1954
-Record Lowest	47	-8	-7	-3	22	35	49	51	52	35	26	10	-3	-8
-Year		1947	1979	1948	1957	1961	1954	1971	1967	1984	1952	1976	1963	JAN 1947
NORMAL DEGREE DAYS:														
Heating (base 65°F)		924	680	500	168	40	0	0	0	18	146	474	781	3731
Cooling (base 65°F)		0	0	14	45	167	381	564	518	282	72	0	0	2043
% OF POSSIBLE SUNSHINE	43	52	55	56	57	58	65	73	72	65	64	56	53	61
MEAN SKY COVER (tenths)														
Sunrise - Sunset	43	6.0	5.8	6.0	6.0	6.0	5.4	4.6	4.4	4.8	4.7	5.3	5.8	5.4
MEAN NUMBER OF DAYS:														
Sunrise to Sunset														
-Clear	47	9.1	8.9	8.9	8.1	8.1	9.0	12.8	13.6	12.6	13.4	11.3	10.2	126.1
-Partly Cloudy	47	7.2	6.5	8.2	8.8	10.3	11.4	10.9	10.6	8.1	7.6	6.8	7.4	103.8
-Cloudy	47	14.7	12.9	13.9	13.0	12.6	9.6	7.3	6.8	9.2	10.0	11.9	13.4	135.4
Precipitation														
.01 inches or more	47	6.3	7.0	8.2	8.9	10.6	8.9	6.4	6.7	7.1	6.7	6.2	6.6	89.6
Snow, Ice pellets														
1.0 inches or more	47	1.4	1.0	0.4	0.*	0.0	0.0	0.0	0.0	0.0	0.0	0.2	0.6	3.5
Thunderstorms	47	0.7	1.3	3.3	6.1	9.1	8.4	5.8	6.2	5.0	3.1	1.4	0.8	51.3
Heavy Fog Visibility 1/4 mile or less	47	2.0	1.7	0.9	0.2	0.4	0.3	0.2	0.1	0.6	1.0	1.2	1.6	10.1
Temperature °F														
-Maximum														
90° and above	25	0.0	0.0	0.3	0.7	2.2	12.8	24.4	22.2	9.9	1.8	0.0	0.0	74.3
32° and below	25	6.4	2.5	0.3	0.0	0.0	0.0	0.0	0.0	0.0	0.0	0.1	3.2	12.5
-Minimum														
32° and below	25	24.8	17.9	8.7	0.6	0.0	0.0	0.0	0.0	0.0	0.3	7.8	20.4	80.5
0° and below	25	0.7	0.1	0.0	0.0	0.0	0.0	0.0	0.0	0.0	0.0	0.0	0.3	1.1
AVG. STATION PRESS. (mb)	13	996.8	994.7	989.9	989.7	988.5	989.5	990.9	991.2	992.3	993.4	993.8	995.3	992.2
RELATIVE HUMIDITY (%)														
Hour 00	25	72	70	68	69	78	79	72	73	79	76	75	74	74
Hour 06 (Local Time)	25	78	77	76	78	86	86	82	84	87	83	80	79	81
Hour 12	25	60	56	53	52	58	59	53	53	58	53	58	60	56
Hour 18	25	59	55	50	49	56	57	49	49	57	55	60	62	55
PRECIPITATION (inches):														
Water Equivalent														
-Normal		1.35	1.74	3.14	4.15	5.14	4.57	3.51	3.01	4.37	3.41	2.56	1.82	38.77
-Maximum Monthly	47	6.65	5.73	11.94	9.23	18.00	11.17	10.88	7.47	18.81	16.51	7.57	8.70	18.81
-Year		1949	1985	1973	1947	1943	1948	1961	1942	1971	1941	1946	1984	SEP 1971
-Minimum Monthly	47	T	0.40	0.08	0.51	1.33	0.53	0.03	0.21	T	T	0.01	0.16	T
-Year		1943	1947	1971	1950	1945	1963	1954	1945	1948	1952	1949	1950	OCT 1952
-Maximum in 24 hrs	47	2.25	4.34	2.67	4.58	9.27	5.01	7.54	4.16	6.39	5.80	5.14	3.27	9.27
-Year		1946	1985	1969	1964	1984	1941	1963	1942	1940	1983	1974	1984	MAY 1984
Snow, Ice pellets														
-Maximum Monthly	47	12.7	10.1	11.8	1.7						T	5.6	9.9	12.7
-Year		1979	1960	1968	1957						1967	1972	1958	JAN 1979
-Maximum in 24 hrs	47	9.0	6.3	9.8	1.7						T	4.0	8.8	9.8
-Year		1944	1944	1968	1957						1967	1972	1954	MAR 1968
WIND:														
Mean Speed (mph)	37	10.6	11.0	12.2	12.2	10.7	10.1	9.1	8.9	9.2	9.8	10.3	10.4	10.4
Prevailing Direction through 1963		N	N	SSE	S	S	S	S	SSE	SSE	SSE	S	S	S
Fastest Obs. 1 Min.														
-Direction	8	02	23	21	29	21	23	36	36	17	18	26	36	29
-Speed (MPH)	8	35	30	32	52	37	35	29	32	29	35	33	29	52
-Year		1985	1984	1980	1982	1985	1983	1982	1979	1982	1985	1981	1979	APR 1982
Peak Gust														
-Direction	2	N	SW	S	SW	SW	W	N	SW	NW	S	NW	N	W
-Speed (mph)	2	45	46	41	52	52	53	45	33	38	47	52	38	53
-Date		1985	1984	1984	1984	1985	1985	1985	1984	1985	1985	1985	1984	JUN 1985

See Reference Notes to this table on the following page.

TABLE 2 PRECIPITATION (inches)   TULSA, OKLAHOMA

YEAR	JAN	FEB	MAR	APR	MAY	JUNE	JULY	AUG	SEP	OCT	NOV	DEC	ANNUAL
1956	0.98	1.15	1.23	2.31	5.13	2.03	1.36	0.30	0.87	2.72	2.78	2.38	23.24
1957	1.60	1.46	3.01	8.74	9.80	7.25	0.50	3.79	5.33	1.58	2.09	1.44	46.59
1958	1.78	0.86	6.14	4.39	3.69	3.08	3.33	3.61	3.61	0.21	1.39	0.82	32.88
1959	0.77	1.69	3.02	1.46	6.66	3.30	9.85	2.13	7.42	9.08	1.48	2.53	49.39
1960	1.22	2.65	1.06	3.63	8.91	1.65	9.01	1.87	0.89	3.53	0.63	3.01	38.06
1961	0.66	2.86	3.30	1.49	9.09	6.36	10.88	3.16	7.37	0.86	3.18	2.18	51.39
1962	1.33	1.44	3.24	3.40	1.69	5.52	4.83	3.10	10.50	3.92	0.36	0.41	41.46
1963	0.98	0.42	2.84	2.21	2.49	0.53	10.60	3.28	2.01	0.18	2.28	0.98	28.80
1964	0.63	2.17	3.96	5.87	4.77	5.79	1.80	6.14	3.33	1.24	6.90	1.67	44.27
1965	1.56	1.45	0.73	3.00	3.91	3.76	3.39	3.72	4.59	0.26	0.03	4.29	30.69
1966	0.69	2.35	0.86	4.84	1.86	2.56	2.00	4.59	2.68	1.39	0.51	2.53	26.86
1967	1.51	0.65	1.42	5.09	5.34	4.60	6.88	0.57	4.89	3.75	1.09	1.12	36.91
1968	3.26	1.08	3.49	4.40	3.56	4.08	1.37	1.90	2.80	2.64	5.19	2.01	35.78
1969	1.63	1.34	3.25	1.56	1.98	6.40	1.08	3.24	1.67	5.86	0.32	1.62	29.95
1970	0.41	0.57	2.05	5.66	4.20	4.60	0.13	1.85	6.73	5.83	0.84	1.15	34.02
1971	1.37	4.18	0.08	1.37	6.59	3.27	3.34	1.86	18.81	7.99	1.21	6.34	56.41
1972	0.17	0.49	0.91	4.45	2.43	2.69	2.68	5.16	2.95	7.58	5.00	1.03	35.54
1973	3.39	0.74	11.94	7.22	5.30	7.69	6.47	4.70	6.56	6.16	3.39	6.93	69.88
1974	0.79	3.17	2.62	3.65	6.94	7.88	0.55	5.30	11.78	6.40	7.30	2.88	59.26
1975	2.61	3.44	5.45	2.20	7.22	6.75	2.14	3.52	3.34	1.47	3.53	3.04	44.71
1976	0.21	0.84	3.95	8.27	6.75	1.87	4.37	1.17	2.60	2.65	0.68	0.55	33.91
1977	1.43	1.57	5.58	2.05	5.72	6.69	2.00	4.86	5.57	2.75	2.31	0.93	41.46
1978	0.81	2.84	2.99	7.14	9.28	6.06	0.36	1.37	0.13	0.95	5.48	0.78	38.19
1979	2.07	0.81	3.97	4.47	6.15	8.90	2.68	4.77	0.28	2.20	5.60	0.45	42.35
1980	2.07	1.32	3.59	3.44	7.23	5.57	0.09	2.34	3.47	2.05	0.79	1.37	33.33
1981	0.69	1.63	1.67	1.90	6.70	3.31	6.22	2.47	3.11	6.73	2.25	0.20	36.88
1982	3.58	0.67	1.04	1.28	9.30	4.13	1.65	1.42	2.95	1.22	4.61	3.39	35.24
1983	2.95	1.98	2.19	3.88	6.85	1.47	0.58	0.65	2.11	9.33	2.14	0.61	34.74
1984	1.00	1.95	6.72	2.44	11.25	1.72	0.48	1.96	2.77	6.98	2.80	8.70	48.77
1985	1.24	5.74	5.39	5.62	4.19	7.63	2.38	1.91	3.29	6.26	6.27	1.39	51.30
Record Mean	1.66	1.66	2.90	4.07	5.37	4.67	3.08	3.13	3.75	3.50	2.51	1.91	38.21

TABLE 3 AVERAGE TEMPERATURE (deg. F)   TULSA, OKLAHOMA

YEAR	JAN	FEB	MAR	APR	MAY	JUNE	JULY	AUG	SEP	OCT	NOV	DEC	ANNUAL
1956	35.6	42.2	51.7	58.0	72.9	78.8	85.2	86.8	78.0	68.0	47.9	42.5	62.3
1957	34.0	46.7	48.2	58.0	68.0	76.3	85.9	82.0	70.9	58.6	47.7	46.0	60.2
1958	38.6	35.1	42.3	59.0	71.0	78.7	81.4	80.7	74.6	62.3	53.5	37.3	59.5
1959	33.6	41.4	51.5	60.5	71.5	77.1	78.3	81.7	74.0	59.7	44.5	44.9	59.9
#1960	37.1	35.0	38.4	60.7	65.8	76.5	79.1	79.7	75.4	65.6	51.9	38.5	58.6
1961	34.3	42.7	52.0	57.6	65.9	73.8	78.6	77.3	70.7	62.6	46.2	35.1	58.1
1962	30.2	42.5	46.1	57.6	75.3	75.2	81.5	81.4	70.7	64.9	49.8	39.9	59.6
1963	28.6	38.5	54.8	65.6	70.9	81.7	85.0	82.7	76.0	72.2	52.3	31.3	61.6
1964	40.9	39.4	46.9	65.7	71.7	78.2	84.9	80.7	73.0	60.1	52.7	39.1	61.1
1965	39.8	39.4	40.5	66.2	72.6	78.1	83.2	81.7	75.3	63.3	55.7	46.8	61.9
1966	32.8	39.0	52.7	57.9	67.6	77.3	87.3	78.5	70.3	59.5	53.4	37.2	59.5
1967	39.6	38.9	55.1	64.3	65.2	76.5	77.5	76.0	69.2	61.2	48.2	39.7	59.3
1968	36.0	37.0	49.8	59.3	65.5	76.8	80.6	80.9	71.9	62.2	46.7	36.8	58.6
1969	36.9	41.4	42.7	61.6	69.9	75.0	85.9	80.8	74.9	59.5	48.1	39.3	59.7
1970	29.7	41.9	44.6	60.7	70.7	76.9	82.8	84.8	74.5	58.9	45.6	42.5	59.4
1971	36.5	39.0	49.9	60.2	66.7	79.4	80.0	79.0	73.0	65.0	50.1	43.7	60.2
1972	34.8	41.8	53.0	62.8	68.0	79.5	80.4	81.7	75.5	60.9	43.6	33.7	59.6
1973	34.0	39.8	54.3	58.2	67.5	76.7	81.2	79.3	72.3	65.0	53.4	38.1	60.0
1974	34.1	43.7	55.2	61.8	72.1	73.8	85.4	78.3	64.7	63.0	49.1	39.7	60.1
1975	39.9	36.9	45.3	60.4	69.1	76.0	81.2	82.2	67.2	63.2	50.8	40.1	59.5
1976	37.2	51.1	51.9	61.5	63.0	75.0	81.4	79.7	72.8	56.1	43.1	37.1	59.2
1977	26.9	46.6	55.0	64.4	72.6	81.0	84.8	81.7	76.6	62.2	51.1	39.0	61.7
1978	24.9	29.4	47.5	63.5	68.3	77.6	87.8	84.3	80.6	63.5	51.6	38.0	59.7
1979	23.1	30.2	52.4	61.0	68.7	77.8	83.4	81.8	74.7	66.2	47.5	44.4	59.3
1980	38.6	37.1	48.3	61.1	70.6	82.5	91.7	89.7	78.3	61.5	50.5	42.3	62.7
1981	37.6	43.6	53.3	68.0	65.9	80.0	85.9	79.4	73.9	60.9	51.4	38.5	61.5
1982	33.6	38.2	55.3	59.3	72.9	74.7	84.2	85.3	74.6	63.4	50.6	44.4	61.4
1983	39.1	42.9	49.0	55.4	67.0	76.6	84.7	88.1	77.4	64.5	52.9	26.7	60.8
1984	34.4	46.4	48.3	58.0	67.5	80.1	82.0	82.7	71.5	63.8	50.4	44.7	60.8
1985	30.2	35.9	54.7	63.3	70.6	75.8	82.9	81.7	74.6	63.1	47.8	34.5	59.6
Record Mean	36.6	41.3	50.3	60.8	68.7	77.7	82.8	81.9	74.2	62.9	49.8	40.0	60.5
Max	46.9	52.4	61.9	72.1	79.4	88.2	94.0	93.7	85.9	75.0	61.0	50.2	71.7
Min	26.3	30.2	38.6	49.4	58.0	67.2	71.5	70.2	62.5	50.7	38.6	29.9	49.4

REFERENCE NOTES FOR TABLES 1, 2, 3 and 6   (TULSA, OK)

GENERAL

T - TRACE AMOUNT
BLANK ENTRIES DENOTE MISSING/UNREPORTED DATA.
# INDICATES A STATION OR INSTRUMENT RELOCATION.

SPECIFIC

TABLE 1

(a) - LENGTH OF RECORD IN YEARS. ALTHOUGH
    INDIVIDUAL MONTHS MAY BE MISSING.
  * LESS THAN .05

NORMALS — BASED ON THE 1951-1980 RECORD PERIOD.
EXTREMES — DATES ARE THE MOST RECENT OCCURRENCE.
WIND DIR. — NUMERALS SHOW TENS OF DEGREES
          CLOCKWISE FROM TRUE NORTH.
          "00" INDICATES CALM.
RESULTANT WIND DIRECTIONS ARE GIVEN TO WHOLE DEGREES.

EXCEPTIONS

TABLES 2, 3, and 6

RECORD MEANS ARE THROUGH THE CURRENT YEAR,
BEGINNING IN   1906 FOR TEMPERATURE
              1888 FOR PRECIPITATION
              1939 FOR SNOWFALL

TABLE 4 HEATING DEGREE DAYS Base 65 deg. F   TULSA, OKLAHOMA

SEASON	JULY	AUG	SEP	OCT	NOV	DEC	JAN	FEB	MAR	APR	MAY	JUNE	TOTAL
1956-57	0	0	0	41	509	689	957	505	513	235	36	3	3488
1957-58	0	0	5	225	515	582	814	834	696	194	23	0	3888
1958-59	0	0	18	146	345	853	969	652	416	196	21	0	3616
#1959-60	0	0	14	170	614	618	858	864	818	161	88	0	4205
1960-61	0	0	0	88	393	814	943	619	394	281	70	0	3602
1961-62	0	0	37	129	563	918	1072	628	586	250	3	0	4186
1962-63	0	0	21	117	450	771	1119	736	343	90	46	0	3693
1963-64	0	0	5	20	376	1038	739	734	553	87	19	1	3572
1964-65	0	0	18	162	385	797	775	711	752	73	1	0	3674
1965-66	0	0	15	122	283	556	992	722	395	227	67	0	3379
1966-67	0	0	13	202	376	859	776	724	362	100	101	0	3513
1967-68	0	0	37	184	498	781	892	803	479	184	67	0	3925
1968-69	0	0	1	160	543	864	863	652	680	127	24	2	3916
1969-70	0	0	0	241	498	789	1088	642	625	174	26	18	4101
1970-71	0	0	18	217	577	692	878	721	463	176	37	0	3779
1971-72	0	0	53	60	446	653	932	670	373	161	47	0	3395
1972-73	0	0	19	183	634	964	954	700	321	233	42	0	4050
1973-74	0	0	24	95	343	824	951	591	341	137	5	0	3311
1974-75	0	0	74	94	473	777	773	780	610	205	19	0	3805
1975-76	0	0	57	146	429	762	855	402	407	126	109	0	3293
1976-77	0	0	16	317	648	858	1173	511	309	99	1	0	3932
1977-78	0	0	1	118	412	801	1236	989	541	110	67	0	4275
1978-79	0	0	0	121	406	834	1293	972	391	164	47	0	4228
1979-80	0	0	0	90	525	632	812	801	513	154	22	0	3549
1980-81	0	0	13	172	438	703	843	598	360	48	58	0	3233
1981-82	0	0	23	178	402	817	967	747	322	208	11	5	3680
1982-83	0	0	23	146	437	635	794	611	492	321	50	0	3509
1983-84	0	0	19	89	378	1179	941	533	509	229	47	0	3924
1984-85	0	0	73	130	438	628	1073	809	330	103	7	0	3591
1985-86	0	0	46	111	510	936							

TABLE 5 COOLING DEGREE DAYS Base 65 deg. F   TULSA, OKLAHOMA

YEAR	JAN	FEB	MAR	APR	MAY	JUNE	JULY	AUG	SEP	OCT	NOV	DEC	TOTAL
1969	0	0	0	29	185	312	656	496	307	77	0	0	2062
1970	0	0	0	50	209	382	555	619	307	39	0	0	2161
1971	0	0	3	40	97	442	471	444	298	65	7	0	1867
1972	0	6	11	99	144	446	487	524	339	64	0	0	2120
1973	0	0	0	35	124	357	508	452	249	101	5	0	1831
1974	0	0	47	48	232	270	641	419	71	40	2	0	1770
1975	0	0	9	77	156	335	509	542	192	97	12	0	1929
1976	0	6	7	28	52	307	520	461	256	48	0	0	1685
1977	0	1	6	84	248	486	619	525	327	38	0	0	2334
1978	0	0	7	73	180	388	713	605	476	79	14	0	2535
1979	0	0	9	48	167	388	577	527	298	137	6	0	2157
1980	0	0	0	43	200	533	833	774	419	69	6	4	2881
1981	0	5	4	145	96	456	658	452	296	57	1	0	2170
1982	0	0	28	44	266	300	601	637	319	106	10	5	2316
1983	0	0	3	40	120	353	615	725	396	80	20	0	2352
1984	0	0	0	25	132	464	534	556	272	100	9	2	2094
1985	0	0	19	59	185	333	564	523	340	57	0	0	2080

TABLE 6 SNOWFALL (inches)   TULSA, OKLAHOMA

SEASON	JULY	AUG	SEP	OCT	NOV	DEC	JAN	FEB	MAR	APR	MAY	JUNE	TOTAL
1956-57	0.0	0.0	0.0	0.0	0.0	0.4	1.3	T	0.1	1.7	0.0	0.0	3.5
1957-58	0.0	0.0	0.0	T	T	T	4.6	1.0	11.6	0.0	0.0	0.0	17.2
1958-59	0.0	0.0	0.0	0.0	2.0	9.9	4.6	2.5	T	T	0.0	0.0	19.0
1959-60	0.0	0.0	0.0	0.0	0.2	0.0	6.1	10.1	3.0	0.0	0.0	0.0	19.4
1960-61	0.0	0.0	0.0	0.0	0.0	1.0	T	5.2	T	0.0	0.0	0.0	6.2
1961-62	0.0	0.0	0.0	0.0	T	2.0	2.7	1.8	0.0	0.0	0.0	0.0	6.5
1962-63	0.0	0.0	0.0	0.0	T	1.0	0.6	1.5	0.0	0.0	0.0	0.0	3.1
1963-64	0.0	0.0	0.0	0.0	0.0	4.0	1.1	T	8.8	0.0	0.0	0.0	13.9
1964-65	0.0	0.0	0.0	0.0	0.0	T	0.3	1.5	0.6	0.0	0.0	0.0	2.4
1965-66	0.0	0.0	0.0	0.0	0.0	T	4.3	5.1	T	0.0	0.0	0.0	9.4
1966-67	0.0	0.0	0.0	0.0	T	3.1	0.7	1.1	0.0	0.0	0.0	0.0	5.8
1967-68	0.0	0.0	0.0	T	0.9	1.6	0.6	2.1	11.8	0.0	0.0	0.0	17.0
1968-69	0.0	0.0	0.0	0.0	T	1.4	T	5.3	1.3	0.0	0.0	0.0	8.0
1969-70	0.0	0.0	0.0	0.0	0.0	5.8	4.7	T	9.9	T	0.0	0.0	20.4
1970-71	0.0	0.0	0.0	0.0	T	0.0	T	6.5	T	0.0	0.0	0.0	6.5
1971-72	0.0	0.0	0.0	0.0	2.0	1.7	0.8	4.9	T	0.0	0.0	0.0	8.7
1972-73	0.0	0.0	0.0	0.0	5.6	1.7	4.3	2.2	0.0	0.3	0.0	0.0	14.1
1973-74	0.0	0.0	0.0	0.0	T	1.8	T	T	T	0.0	0.0	0.0	1.8
1974-75	0.0	0.0	0.0	0.0	1.7	T	T	3.0	1.8	T	0.0	0.0	6.5
1975-76	0.0	0.0	0.0	0.0	0.8	1.3	T	T	T	0.0	0.0	0.0	2.1
1976-77	0.0	0.0	0.0	0.0	0.5	T	10.5	0.3	0.0	0.0	0.0	0.0	11.3
1977-78	0.0	0.0	0.0	0.0	T	5.4	6.3	T	0.0	0.0	0.0	0.0	11.7
1978-79	0.0	0.0	0.0	0.0	0.0	2.8	12.7	3.4	0.0	T	0.0	0.0	18.9
1979-80	0.0	0.0	0.0	0.0	T	0.0	0.4	3.8	T	0.0	0.0	0.0	4.2
1980-81	0.0	0.0	0.0	0.0	T	0.0	T	0.9	T	0.0	0.0	0.0	0.9
1981-82	0.0	0.0	0.0	0.0	0.0	T	0.3	5.6	T	0.0	0.0	0.0	5.9
1982-83	0.0	0.0	0.0	0.0	T	3.8	1.4	T	0.0	0.0	0.0	0.0	5.2
1983-84	0.0	0.0	0.0	0.0	T	3.0	4.6	0.2	T	0.0	0.0	0.0	7.8
1984-85	0.0	0.0	0.0	0.0	0.0	6.6	3.3	4.3	0.0	0.0	0.0	0.0	14.2
1985-86	0.0	0.0	0.0	0.0	T	2.5							
Record Mean	0.0	0.0	0.0	T	0.4	1.5	3.2	2.5	1.5	T	0.0	0.0	9.1

See Reference Notes, relative to all above tables, on preceding page.

ASTORIA, OREGON

Astoria is ringed by low mountains on the north, east, and south. On the west, the area is open to the Pacific Ocean at the mouth of the Columbia River. North of the station, 8 to 12 miles distant, the Washington hills rise to 1,000 to 1,200 feet. Maximum visibility is 19 miles north-northeastward to the Willapa Hills. East-northeastward 2 to 4 miles, the Astoria hills rise to 600 feet. East-southeastward 4 to 14 miles, consecutively, rise other ridges of the Coast Ranges, and southeastward is the most prominent landmark, Saddle Mountain, 3,283 feet high. Forests cover most of the uplands. From Seaside northward to the south bank of the Columbia are 18 miles of sandy beaches, and a 2 to 3 mile wide stretch of dune lands.

The airport sits by the south bank of the Columbia estuary, west of Youngs Bay, on the flood plain or tidal flats. Low dikes prevent flooding and increase the bog-like characteristics of the area. When air temperature falls below water temperature, fog forms easily, or rolls in from the ocean, river, or bay. This usually begins from late afternoon to early morning, and may persist well into the following day. During the summer months, sea breezes commonly blow up the river by noon and stop the diurnal rise in temperature. In winter, cold air may funnel down the Columbia from the interior.

Weather hazards occasionally occur. For flying, the greatest are fog and gales. Even with moderate surface velocities, wind and turbulence at 800 feet may be severe enough to upset a heavy plane. Even in fair weather, wind and wave may combine to produce a type of breaker known as the widow-maker and swamp a boat. Heavy rains inundate lowlands, and high tides aggravated by gales may push seawater across highways or up beaches. Rains may cause earthslides, mostly in highway cuts. Lightning strikes are rare. Showers of ice pellets may briefly whiten the ground during many of the months. Occasionally in winter there may be rather brief periods of freezing temperatures, with snow or ice.

The climate is generally healthful, except for dampness and a lack of sunshine in winter. Heat waves are uncommon and usually brief. Soil leaching necessitates supplementary mineral diets for both animals and plants.

TABLE 1 — NORMALS, MEANS AND EXTREMES

ASTORIA, OREGON

LATITUDE: 46°09'N   LONGITUDE: 123°53'W   ELEVATION: FT. GRND   8 BARO 00013   TIME ZONE: PACIFIC   WBAN: 94224

	(a)	JAN	FEB	MAR	APR	MAY	JUNE	JULY	AUG	SEP	OCT	NOV	DEC	YEAR
TEMPERATURE °F:														
Normals														
– Daily Maximum		46.8	50.6	51.9	55.5	60.2	63.9	67.9	68.6	67.8	61.4	53.5	48.8	58.1
– Daily Minimum		35.4	37.1	36.9	39.7	44.1	49.2	52.2	52.6	49.2	44.3	39.7	37.3	43.1
– Monthly		41.1	43.9	44.4	47.6	52.2	56.5	60.0	60.6	58.5	52.9	46.6	43.0	50.6
Extremes														
– Record Highest	32	65	72	73	83	87	93	100	96	95	85	71	64	100
– Year		1961	1968	1979	1956	1985	1955	1961	1981	1972	1980	1970	1980	JUL 1961
– Record Lowest	32	11	19	22	29	30	37	39	39	33	26	15	6	6
– Year		1980	1979	1971	1968	1954	1980	1971	1973	1983	1971	1955	1972	DEC 1972
NORMAL DEGREE DAYS:														
Heating (base 65°F)		741	591	639	522	397	252	158	143	199	375	552	679	5248
Cooling (base 65°F)		0	0	0	0	0	0	7	7	0	0	0	0	14
% OF POSSIBLE SUNSHINE														
MEAN SKY COVER (tenths)														
Sunrise – Sunset	32	8.4	8.3	8.1	7.9	7.6	7.7	6.7	6.6	6.3	7.2	8.0	8.4	7.6
MEAN NUMBER OF DAYS:														
Sunrise to Sunset														
– Clear	32	2.9	2.9	3.0	3.2	3.0	3.2	5.7	6.4	8.2	5.4	3.3	2.5	49.5
– Partly Cloudy	32	3.8	3.5	5.0	6.0	8.7	7.2	10.2	9.5	7.2	6.6	5.1	4.3	77.1
– Cloudy	32	24.3	21.8	23.0	20.8	19.3	19.6	15.1	15.2	14.6	19.0	21.7	24.3	238.6
Precipitation														
.01 inches or more	32	21.8	19.7	20.5	18.4	15.0	13.3	7.2	8.4	10.8	16.4	20.3	22.8	194.6
Snow, Ice pellets														
1.0 inches or more	32	0.8	0.1	0.3	0.*	0.0	0.0	0.0	0.0	0.0	0.0	0.1	0.4	1.7
Thunderstorms	32	0.6	0.3	0.5	0.6	0.3	0.3	0.4	0.4	0.9	1.0	1.1	1.0	7.3
Heavy Fog Visibility 1/4 mile or less	32	3.9	2.9	2.5	2.2	1.7	1.5	2.0	4.2	5.6	6.8	3.8	4.0	41.1
Temperature °F														
– Maximum														
90° and above	32	0.0	0.0	0.0	0.0	0.0	0.0	0.1	0.1	0.2	0.0	0.0	0.0	0.5
32° and below	32	0.7	0.0	0.0	0.0	0.0	0.0	0.0	0.0	0.0	0.0	0.2	0.6	1.6
– Minimum														
32° and below	32	9.9	6.3	6.4	2.1	0.1	0.0	0.0	0.0	0.0	0.5	4.8	8.1	38.1
0° and below	32	0.0	0.0	0.0	0.0	0.0	0.0	0.0	0.0	0.0	0.0	0.0	0.0	0.0
AVG. STATION PRESS. (mb)	13	1017.7	1015.4	1015.1	1017.8	1018.2	1018.4	1018.3	1017.1	1016.4	1017.1	1015.3	1017.3	1017.0
RELATIVE HUMIDITY (%)														
Hour 04	32	86	87	87	88	89	90	89	91	90	89	87	86	88
Hour 10 (Local Time)	32	84	83	78	74	73	76	75	77	76	81	82	84	79
Hour 16	32	78	74	71	69	70	71	69	70	70	73	77	80	73
Hour 22	32	85	85	85	85	85	85	86	87	87	88	86	86	
PRECIPITATION (inches):														
Water Equivalent														
– Normal		11.29	7.81	7.26	4.60	2.84	2.43	1.04	1.56	3.11	6.21	9.88	11.57	69.60
– Maximum Monthly	32	18.94	21.89	13.47	8.04	6.60	5.48	4.39	5.22	6.93	12.56	16.75	16.57	21.89
– Year		1954	1961	1956	1955	1960	1954	1983	1968	1978	1975	1983	1955	FEB 1961
– Minimum Monthly	32	0.69	2.60	0.93	1.33	0.37	0.75	0.01	0.08	0.04	1.01	1.45	2.67	0.01
– Year		1985	1973	1965	1956	1982	1965	1960	1970	1975	1978	1976	1985	JUL 1960
– Maximum in 24 hrs	32	5.12	2.86	2.66	2.26	1.82	2.42	1.98	1.65	2.63	3.71	3.48	3.61	5.12
– Year		1982	1961	1956	1965	1979	1968	1974	1968	1953	1982	1959	1974	JAN 1982
Snow, Ice pellets														
– Maximum Monthly	32	26.3	4.0	6.7	1.1	T				T	T	4.6	19.0	26.3
– Year		1969	1962	1966	1975	1985				1972	1984	1985	1964	JAN 1969
– Maximum in 24 hrs	32	10.8	4.0	5.9	1.0	T				T	T	4.3	7.2	10.8
– Year		1971	1962	1960	1975	1985				1972	1984	1985	1964	JAN 1971
WIND:														
Mean Speed (mph)	32	9.0	8.9	8.9	8.6	8.4	8.4	8.6	7.9	7.5	7.6	8.6	9.1	8.5
Prevailing Direction through 1963		E	ESE	SE	WNW	NW	NW	NW	NW	SE	SE	SE	ESE	SE
Fastest Obs. 1 Min.														
– Direction (!!)	32	17	19	19	20	22	20	29	23	20	20	20	25	17
– Speed (MPH)	32	55	47	44	52	37	29	28	28	35	44	46	52	55
– Year		1971	1979	1964	1962	1975	1962	1979	1961	1959	1962	1962	1961	JAN 1971
Peak Gust														
– Direction (!!)	2	S	S	S	S	S	S	SW	S	S	SW	S	S	S
– Speed (mph)	2	46	59	62	43	43	31	33	35	39	44	59	47	62
– Date		1984	1985	1985	1985	1985	1985	1985	1984	1984	1984	1984	1984	MAR 1985

See reference Notes to this table on the following page.

TABLE 2 PRECIPITATION (inches)   ASTORIA, OREGON

YEAR	JAN	FEB	MAR	APR	MAY	JUNE	JULY	AUG	SEP	OCT	NOV	DEC	ANNUAL
1956	17.09	9.32	13.47	1.33	1.43	4.64	0.18	2.15	3.76	11.37	2.57	9.02	76.33
1957	4.76	6.90	9.73	3.94	2.82	3.31	1.63	1.34	0.82	5.43	7.68	11.97	60.33
1958	9.61	10.96	4.62	7.03	1.03	2.80	0.09	0.52	1.94	7.33	14.14	12.17	72.24
1959	13.24	8.04	7.88	4.40	3.45	3.77	0.91	0.92	5.56	6.48	11.40	8.36	74.41
1960	10.09	8.47	7.40	5.92	6.60	1.87	0.01	1.84	1.69	7.33	13.91	6.12	71.25
1961	9.03	21.89	10.69	5.47	2.90	1.10	0.50	1.30	1.45	7.32	8.34	10.40	80.39
1962	6.53	5.61	5.18	7.44	2.88	1.87	0.34	2.49	3.50	7.40	14.21	6.78	64.23
1963	4.76	6.44	6.13	5.76	1.91	1.80	1.52	1.20	2.20	9.58	13.16	9.12	63.58
1964	18.50	4.06	7.41	3.59	2.27	2.70	2.59	2.21	2.73	2.61	11.15	13.67	73.49
1965	16.59	6.77	0.93	5.47	2.74	0.75	0.46	1.95	0.51	3.97	11.82	11.78	63.74
1966	8.61	5.53	8.79	2.90	2.18	2.13	0.54	1.01	2.18	5.83	10.00	14.07	63.77
1967	14.95	6.07	8.38	5.52	1.37	1.37	0.22	0.19	3.07	11.06	5.94	9.04	66.95
1968	9.57	9.57	10.42	4.22	3.91	4.81	1.23	5.22	4.60	8.03	11.96	13.85	87.39
1969	12.02	5.67	3.16	3.84	3.92	3.63	0.56	0.62	6.55	5.28	5.77	11.69	62.71
1970	14.46	5.29	4.28	7.74	1.92	1.19	0.31	0.08	3.65	5.80	9.86	15.93	70.51
1971	16.69	6.67	9.96	4.09	2.30	2.97	1.55	1.14	4.65	6.34	9.08	13.83	79.27
1972	10.62	8.58	10.04	6.82	1.22	0.92	2.01	0.37	4.72	1.96	6.90	13.28	67.44
1973	5.72	2.60	5.71	2.38	3.16	4.26	0.07	0.46	4.19	5.92	14.93	15.75	65.15
1974	12.47	8.38	10.73	4.88	4.37	2.33	4.20	0.29	0.67	1.85	8.95	13.84	72.96
1975	15.21	8.03	5.66	3.90	2.41	1.99	0.22	2.82	0.04	12.56	12.28.	15.66	80.78
1976	11.67	7.86	7.17	3.55	2.20	1.27	2.46	2.55	1.58	2.96	1.45	4.20	48.92
1977	3.20	5.22	9.74	1.65	6.00	1.36	0.44	3.85	5.44	4.38	12.37	14.34	67.99
1978	8.66	5.43	4.40	6.35	4.75	3.07	0.90	2.61	6.93	1.01	8.43	4.99	57.53
1979	3.83	11.76	4.52	4.38	4.19	1.82	0.92	0.81	4.35	8.46	7.87	13.18	66.09
1980	7.21	9.60	6.31	4.85	1.45	1.57	0.74	1.24	2.51	2.79	12.02	12.44	62.63
1981	2.63	8.69	5.80	7.30	2.97	5.47	1.06	0.62	2.77	8.67	10.66	11.80	68.44
1982	13.98	10.87	7.19	6.52	0.37	1.22	0.75	0.63	3.72	8.31	12.14	12.14	75.32
1983	13.52	8.66	8.84	4.26	3.59	4.53	4.39	1.14	1.83	1.87	16.75	9.44	78.82
1984	6.60	8.34	5.90	5.02	5.34	3.90	0.05	0.52	3.16	8.10	15.19	6.51	68.63
1985	0.69	4.09	7.00	2.95	1.90	3.09	0.78	1.11	3.23	8.11	5.96	2.67	40.89
Record Mean	10.23	7.85	7.20	4.88	2.84	2.67	1.15	1.42	3.11	6.40	10.28	11.09	69.13

TABLE 3 AVERAGE TEMPERATURE (deg. F)   ASTORIA, OREGON

YEAR	JAN	FEB	MAR	APR	MAY	JUNE	JULY	AUG	SEP	OCT	NOV	DEC	ANNUAL
1956	41.9	38.4	42.8	48.5	53.9	55.2	60.9	59.9	57.2	50.3	44.8	42.4	49.7
1957	35.9	42.4	45.6	48.8	54.2	57.8	59.0	60.2	62.0	53.1	45.0	50.3	50.8
1958	45.9	49.3	44.8	48.5	55.6	60.9	62.6	62.0	58.9	55.1	46.4	47.6	53.1
1959	43.7	43.0	45.0	49.4	52.2	56.8	60.3	58.8	56.4	53.6	46.6	42.2	50.7
1960	40.3	43.7	44.8	49.1	51.0	56.3	59.0	59.8	57.3	53.5	46.9	43.5	50.5
1961	47.3	46.6	45.8	47.5	52.6	58.7	61.9	61.3	56.4	51.0	43.7	42.3	51.3
1962	40.1	43.7	43.3	48.3	50.3	54.9	58.2	60.5	58.6	53.4	48.8	44.6	50.4
1963	36.9	49.7	45.0	48.0	50.8	55.8	60.3	61.1	61.2	54.9	44.8	44.5	51.5
1964	43.7	42.3	44.3	45.8	49.5	55.2	59.2	59.3	57.4	53.8	44.8	40.0	49.6
1965	42.3	44.2	47.2	49.3	50.9	56.2	60.2	61.7	57.0	55.3	48.8	40.3	51.1
1966	42.3	42.6	43.7	49.0	50.5	56.2	59.6	60.3	59.2	51.7	47.8	45.5	50.7
1967	43.8	43.2	42.2	44.0	51.7	58.3	61.1	62.8	60.5	52.9	48.1	41.2	50.8
1968	42.0	47.5	46.7	46.1	52.2	56.1	61.0	60.1	60.6	47.1	39.5	50.6
1969	34.3	40.3	44.3	46.8	53.5	59.6	58.6	58.7	57.2	51.0	46.5	44.1	49.6
1970	41.8	46.3	44.9	45.3	51.4	56.6	59.3	60.2	56.6	51.0	47.1	40.8	50.1
1971	40.7	42.2	42.0	46.5	51.4	54.6	59.4	62.4	57.9	51.3	46.8	41.2	49.7
1972	40.0	43.2	47.8	46.6	54.7	58.3	62.2	63.4	57.6	52.3	47.3	38.6	51.0
1973	40.5	44.3	44.4	48.9	53.4	56.6	59.7	57.6	57.9	51.2	44.3	45.2	50.3
1974	38.9	42.0	44.7	48.0	50.5	56.1	59.3	62.1	60.9	51.9	48.0	45.6	50.7
1975	42.9	43.1	43.8	44.9	52.8	55.6	60.7	59.7	59.2	50.9	46.2	44.7	50.4
1976	44.1	42.6	43.7	48.7	52.8	55.9	61.1	61.7	60.8	53.2	48.0	43.5	51.3
1977	39.9	46.8	44.6	48.9	50.2	56.8	58.7	62.7	57.0	51.7	44.6	43.9	50.5
1978	44.0	46.0	47.0	48.8	52.3	59.4	60.9	61.5	58.1	54.3	41.7	37.6	50.9
1979	35.3	41.8	47.3	49.2	53.8	56.8	62.4	61.8	61.4	55.0	46.9	47.4	51.6
1980	38.7	46.4	44.8	49.5	52.3	55.4	60.7	58.8	58.9	55.0	50.1	47.5	51.5
1981	48.2	46.6	48.1	49.5	53.4	57.7	60.4	62.4	59.1	52.7	49.0	45.3	52.7
1982	41.7	44.0	45.2	46.8	52.2	57.8	59.3	61.3	59.7	55.5	46.6	43.5	51.2
1983	47.3	50.2	50.9	50.2	55.7	59.8	61.2	61.7	56.5	50.4	48.6	37.9	52.5
1984	43.5	45.4	48.4	46.7	50.8	55.6	59.6	59.9	58.4	50.3	46.7	39.6	50.4
1985	39.3	40.3	42.8	48.4	52.5	56.8	60.6	59.4	56.2	50.7	39.2	39.3	49.7
Record Mean	41.5	44.2	45.0	47.7	52.3	56.8	60.1	60.6	58.3	52.5	46.5	42.8	50.7
Max	47.4	50.9	52.5	55.5	60.1	64.0	67.8	68.7	67.6	61.0	53.3	48.6	58.1
Min	35.6	37.4	37.4	39.8	44.4	49.5	52.3	52.5	49.1	44.1	39.7	36.9	43.2

REFERENCE NOTES FOR TABLES 1, 2, 3 and 6   (ASTORIA, OR)

GENERAL

T - TRACE AMOUNT
BLANK ENTRIES DENOTE MISSING/UNREPORTED DATA.
# INDICATES A STATION OR INSTRUMENT RELOCATION.

SPECIFIC

TABLE 1

(a) - LENGTH OF RECORD IN YEARS. ALTHOUGH
INDIVIDUAL MONTHS MAY BE MISSING.
* LESS THAN 05

NORMALS — BASED ON THE 1951-1980 RECORD PERIOD.
EXTREMES — DATES ARE THE MOST RECENT OCCURRENCE.
WIND DIR. — NUMERALS SHOW TENS OF DEGREES
CLOCKWISE FROM TRUE NORTH.
"00" INDICATES CALM.
RESULTANT WIND DIRECTIONS ARE GIVEN TO WHOLE DEGREES.

EXCEPTIONS

TABLES 2, 3, and 6

RECORD MEANS ARE THROUGH THE CURRENT YEAR,
BEGINNING IN   1954 FOR TEMPERATURE
1954 FOR PRECIPITATION
1954 FOR SNOWFALL

TABLE 4   HEATING DEGREE DAYS Base 65 deg. F   ASTORIA, OREGON

SEASON	JULY	AUG	SEP	OCT	NOV	DEC	JAN	FEB	MAR	APR	MAY	JUNE	TOTAL
1956-57	138	150	228	448	602	696	896	625	594	476	327	209	5389
1957-58	181	145	100	359	561	613	585	435	620	492	285	128	4504
1958-59	86	91	190	303	552	531	654	611	615	469	428	256	4720
1959-60	149	188	253	343	546	703	756	611	615	520	378	186	5317
1960-61	178	155	235	349	537	661	541	511	590	493	451	299	5491
1961-62	105	110	254	428	632	697	765	590	612	502	367	267	5019
1962-63	203	133	190	353	478	628	864	422	635	502	477	285	5092
1963-64	140	116	121	307	510	629	653	652	567	466	432	259	5257
1964-65	174	179	222	345	598	768	696	576	542	466	443	267	5166
1965-66	144	98	235	295	476	760	697	621	658	472	443	267	5157
1966-67	160	142	169	405	514	595	651	603	697	623	404	194	5157
1967-68	118	68	135	369	500	731	705	500	557	560	387	260	4890
1968-69	124	151	216	438	531	782	944	687	633	540	351	161	5558
1969-70	193	188	226	423	551	637	710	518	584	617	416	250	5313
1970-71	170	142	264	428	531	743	746	632	703	549	415	310	5633
1971-72	165	85	212	420	538	728	768	623	526	545	313	196	5119
1972-73	99	65	230	388	525	810	755	574	635	473	357	246	5157
1973-74	164	225	213	421	614	611	803	639	623	503	445	257	5518
1974-75	174	97	130	398	502	596	678	606	653	598	373	273	5078
1975-76	133	155	183	429	558	624	642	640	651	483	374	269	5141
1976-77	117	105	128	361	503	660	774	504	628	475	453	238	4946
1977-78	190	82	231	403	606	646	643	524	551	482	389	161	4908
1978-79	125	110	202	324	693	843	910	641	543	467	338	243	5439
1979-80	91	92	111	305	537	540	807	529	617	459	388	279	4755
1980-81	140	186	183	309	439	533	513	507	518	458	354	211	4351
1981-82	139	107	173	372	474	604	714	581	610	538	391	212	4915
1982-83	171	112	152	288	546	661	543	407	430	438	282	149	4179
1983-84	109	98	249	445	485	834	658	559	509	541	433	276	5196
1984-85	160	155	199	452	542	780	788	686	683	493	383	252	5573
1985-86	131	172	259	438	769	788							

TABLE 5   COOLING DEGREE DAYS Base 65 deg. F   ASTORIA, OREGON

YEAR	JAN	FEB	MAR	APR	MAY	JUNE	JULY	AUG	SEP	OCT	NOV	DEC	TOTAL
1969	0	0	0	0	1	4	0	0	0	0	0	0	5
1970	0	0	0	0	0	7	0	0	1	0	0	0	8
1971	0	0	0	0	0	2	0	11	4	2	0	0	19
1972	0	0	0	0	0	1	19	23	16	0	0	0	59
1973	0	0	0	0	0	0	6	0	7	0	0	0	13
1974	0	0	0	0	0	0	5	11	13	0	0	0	29
1975	0	0	0	0	1	0	5	0	17	0	0	0	23
1976	0	0	0	0	0	3	4	8	8	1	0	0	24
1977	0	0	0	0	0	0	0	18	0	0	0	0	18
1978	0	0	0	0	0	0	6	8	0	0	0	0	14
1979	0	0	0	0	0	0	15	0	11	1	0	0	27
1980	0	0	0	0	0	0	13	0	7	8	0	0	28
1981	0	0	0	0	0	0	4	36	2	0	0	0	42
1982	0	0	0	0	0	6	0	4	1	1	0	0	12
1983	0	0	0	0	0	0	0	4	1	0	0	0	5
1984	0	0	0	0	0	0	3	1	5	0	0	0	9
1985	0	0	0	0	3	11	1	6	0	0	0	0	21

TABLE 6   SNOWFALL (inches)   ASTORIA, OREGON

SEASON	JULY	AUG	SEP	OCT	NOV	DEC	JAN	FEB	MAR	APR	MAY	JUNE	TOTAL
1956-57	0.0	0.0	0.0	0.0	0.0	2.2	4.2	0.2	0.0	0.0	0.0	0.0	6.6
1957-58	0.0	0.0	0.0	0.0	0.0	0.0	0.0	0.0	0.0	0.0	0.0	0.0	0.0
1958-59	0.0	0.0	0.0	0.0	0.0	0.0	1.1	T	T	0.0	0.0	0.0	1.1
1959-60	0.0	0.0	0.0	0.0	0.0	0.0	1.1	0.0	5.9	0.0	0.0	0.0	7.0
1960-61	0.0	0.0	0.0	0.0	T	0.0	0.0	T	0.4	0.0	0.0	0.0	0.4
1961-62	0.0	0.0	0.0	0.0	0.0	T	T	4.0	1.6	0.0	0.0	0.0	5.6
1962-63	0.0	0.0	0.0	0.0	0.0	0.0	0.3	0.0	0.0	T	0.0	0.0	0.3
1963-64	0.0	0.0	0.0	0.0	0.0	0.0	T	0.0	0.0	0.0	0.0	0.0	T
1964-65	0.0	0.0	0.0	0.0	T	19.0	10.7	0.0	0.0	0.0	0.0	0.0	29.7
1965-66	0.0	0.0	0.0	0.0	0.0	3.4	0.7	T	6.7	T	0.0	0.0	10.8
1966-67	0.0	0.0	0.0	0.0	0.0	0.8	T	3.7	0.0	0.0	0.0	0.0	4.5
1967-68	0.0	0.0	0.0	0.0	0.0	4.5	26.3	1.0	0.0	0.0	0.0	0.0	31.8
1968-69	0.0	0.0	0.0	0.0	0.0	0.0	0.0	0.0	0.0	T	0.0	0.0	T
1969-70	0.0	0.0	0.0	0.0	0.0	1.8	16.1	0.2	1.5	0.2	T	0.0	19.8
1970-71	0.0	0.0	0.0	T	0.0	3.4	1.7	T	T	0.2	0.0	0.0	5.3
1971-72	0.0	0.0	0.0	T	0.0	3.4	1.7	T	T	T	0.0	0.0	5.4
1972-73	0.0	0.0	T	0.0	0.0	0.0	0.4	T	1.7	0.0	T	0.0	2.1
1973-74	0.0	0.0	0.0	0.0	T	T	0.4	0.2	T	1.1	T	0.0	1.6
1974-75	0.0	0.0	0.0	0.0	T	T	0.3	0.1	T	T	0.0	0.0	1.2
1975-76	0.0	0.0	0.0	T	1.1	T	T	T	T	T	0.0	0.0	T
1976-77	0.0	0.0	0.0	0.0	0.0	0.1	T	T	T	0.0	0.0	0.0	0.1
1977-78	0.0	0.0	0.0	0.0	0.0	0.1	2.5	0.3	T	0.0	0.0	0.0	2.9
1978-79	0.0	0.0	0.0	T	0.3	0.2	0.5	1.7	0.5	T	0.0	0.0	3.2
1979-80	0.0	0.0	0.0	0.0	T	T	0.0	0.0	T	T	0.0	0.0	T
1980-81	0.0	0.0	0.0	0.0	0.0	T	6.1	T	T	T	0.0	0.0	6.1
1981-82	0.0	0.0	0.0	0.0	T	0.0	T	T	T	T	T	0.0	T
1982-83	0.0	0.0	0.0	0.0	T	T	0.0	0.0	T	T	T	0.0	0.0
1983-84	0.0	0.0	0.0	0.0	T	T	0.0	0.0	1.1	T	T	0.0	1.1
1984-85	0.0	0.0	0.0	T	T	0.0	0.0	0.0	T	T	T	0.0	
1985-86	0.0	0.0	0.0	0.0	4.6	0.4							
Record Mean	0.0	0.0	T	T	0.3	1.4	2.5	0.3	0.7	T	T	0.0	5.2

See Reference Notes, relative to all above tables, on preceding page.

Burns is located near the center of the high plateau area that comprises much of central Oregon. The crest of the Cascade Range is some 135 miles to the west, the Steens Mountains about 45 miles to the southeast. A prong of the Blue Mountains to the north approaches within 40 miles of Burns with lower hills reaching the city limits. Approximately 30 miles east, a number of low hills separate this area from the Malheur Valley. From the Blue Mountains in the north to the southern border of Oregon and between the foothills of the Cascades in the west to the chain of lesser mountains to the east is a series of shallow valleys, each with its own small creek or creeks. These are separated by slightly higher, gently rolling bench lands which in turn are cut here and there by rough, rocky canyons, and with numerous buttes or small mesas interspersed through the area, rising from 500 to 1,500 feet above the general terrain. Well distributed over much of central Oregon south of Burns are a number of large, relatively shallow landlocked lakes and marshes.

Burns, in common with most of the Great Basin area of the Western Plateau, has a semi-arid climate. Maritime air moving in from the Pacific Ocean is greatly modified by its passage over the Coastal and Cascade Mountain Ranges that lie between Burns and the coast. In its ascent over them much of its precipitable moisture has been given up and annual precipitation totals here are small and humidities generally low. This makes for an abundance of sunshine and a rather wide range between daily maximum and minimum temperatures. Nighttime frosts may occur any month in the year. An occasional continental air mass moving from northern Canada southward through this area, instead of in its usual trajectory along the east slope of the Rocky Mountains, will produce even lower humidities along with fairly strong winds and more extreme temperatures. The average total precipitation for the entire high plateau region of Oregon is between 9 and 12 inches. About one-third of this falls in the form of snow. There are several thunderstorms each year. Occasionally these are accompanied by hail, but seldom do either cause appreciable damage. No tornadoes have ever been reported in the immediate area of Burns, but funnel clouds have been sighted.

Approximately 90 percent of the agricultural income in Harney County, of which Burns is the county seat, is derived from beef cattle and sheep, with far the greater portion coming from cattle. The very low nightly temperatures during much of the year necessitate the raising of only very hardy, frost-resistant crops with a short growing season. The low annual precipitation total limits cultivation to valley bottom lands. Through the use of spring runoff water and some reservoir storage for summer irrigation, about 120,000 acres of native grasslands have been developed into native grass hay meadows and about 40,000 additional acres are in cultivated crops, principally the small hardy grains, with some 6,000 acres of alfalfa. The rest of the more than 6,000,000 acres in Harney County is in range with about 400,000 acres of the rangeland covered by pine forests along the northern border.

TABLE 1   NORMALS, MEANS AND EXTREMES

BURNS, OREGON

LATITUDE: 43°35'N   LONGITUDE: 119°03'W   ELEVATION: FT. GRND 4141 BARO 4149   TIME ZONE: PACIFIC   WBAN: 94185

	(a)	JAN	FEB	MAR	APR	MAY	JUNE	JULY	AUG	SEP	OCT	NOV	DEC	YEAR
TEMPERATURE °F:														
Normals														
-Daily Maximum		37.3	44.0	47.1	55.9	65.2	74.0	84.4	81.9	74.2	62.1	46.6	36.8	59.1
-Daily Minimum		17.7	23.0	25.8	30.2	38.9	46.9	54.4	52.5	43.6	33.7	24.6	18.2	34.1
-Monthly		27.5	33.5	36.5	43.0	52.0	60.5	69.4	67.2	58.9	47.9	35.6	27.5	46.6
Extremes														
-Record Highest	5	53	54	60	81	90	95	100	98	93	86	67	54	100
-Year		1981	1983	1981	1981	1983	1985	1980	1984	1981	1980	1980	1981	JUL 1980
-Record Lowest	5	-27	-28	4	16	22	23	28	28	17	7	-13	-20	-28
-Year		1982	1985	1985	1981	1985	1980	1981	1980	1984	1980	1985	1984	FEB 1985
NORMAL DEGREE DAYS:														
Heating (base 65°F)		1163	882	884	657	405	177	18	48	215	530	882	1163	7024
Cooling (base 65°F)		0	0	0	0	5	42	155	116	32	0	0	0	350
% OF POSSIBLE SUNSHINE														
MEAN SKY COVER (tenths)														
Sunrise - Sunset	5	7.7	7.1	7.2	6.5	6.0	5.0	3.0	3.3	4.2	6.1	7.7	7.0	5.9
MEAN NUMBER OF DAYS:														
Sunrise to Sunset														
-Clear	5	3.2	5.2	5.8	5.4	7.4	11.0	20.0	18.2	14.2	8.6	3.8	5.4	108.2
-Partly Cloudy	5	8.0	6.0	5.2	11.2	10.8	9.6	6.8	8.8	8.0	7.6	6.0	7.4	95.4
-Cloudy	5	19.8	17.0	20.0	13.4	12.8	9.4	4.2	4.0	7.8	14.8	19.4	18.2	160.8
Precipitation														
.01 inches or more	5	8.4	10.2	13.6	8.0	8.4	7.4	3.0	4.8	4.8	8.4	15.2	13.0	105.2
Snow,Ice pellets														
1.0 inches or more	5	2.4	2.0	1.6	0.4	0.0	0.0	0.0	0.0	0.0	0.2	3.2	5.0	14.8
Thunderstorms														
Heavy Fog Visibility														
1/4 mile or less														
Temperature °F														
-Maximum														
90° and above	5	0.0	0.0	0.0	0.0	0.2	0.8	7.2	6.2	0.8	0.0	0.0	0.0	15.2
32° and below	5	15.0	8.2	0.4	0.0	0.0	0.0	0.0	0.0	0.0	0.0	6.6	16.0	46.2
-Minimum														
32° and below	5	30.4	26.6	27.4	20.4	10.2	2.6	0.2	0.2	9.8	24.4	26.8	30.8	209.8
0° and below	5	5.8	1.8	0.0	0.0	0.0	0.0	0.0	0.0	0.0	0.0	2.2	7.4	17.2
AVG. STATION PRESS.(mb)														
RELATIVE HUMIDITY (%)														
Hour 04	1	91	90	86	83	86	79	65	56	63	84	89	89	80
Hour 10 (Local Time)	4	83	79	69	51	48	47	37	37	42	56	75	81	59
Hour 16	1	77	67	55	40	35	40	24	22	30	43		72	
Hour 22														
PRECIPITATION (inches):														
Water Equivalent														
-Normal		1.55	0.99	0.89	0.67	0.78	0.58	0.23	0.45	0.46	0.79	1.18	1.56	10.13
-Maximum Monthly	5	1.15	2.13	3.66	1.10	2.14	0.90	1.09	1.16	1.12	1.43	2.73	3.88	3.88
-Year		1982	1983	1983	1983	1981	1982	1983	1984	1985	1982	1984	1981	DEC 1981
-Minimum Monthly	5	0.09	0.49	0.61	0.34	0.64	0.11	0.23	0.03	0.05	0.81	0.59	0.73	0.03
-Year		1985	1985	1980	1985	1982	1985	1981	1980	1984	1985	1980	1985	AUG 1980
-Maximum in 24 hrs	5	0.36	0.60	0.68	0.54	0.82	0.42	0.89	1.01	0.76	0.87	0.69	0.67	1.01
-Year		1982	1982	1983	1982	1983	1984	1985	1984	1985	1982	1981	1983	AUG 1984
Snow,Ice pellets														
-Maximum Monthly	5	13.8	9.4	13.5	3.2	1.4	0.6				3.6	17.4	26.8	26.8
-Year		1982	1984	1985	1982	1983	1981				1984	1983	1983	DEC 1983
-Maximum in 24 hrs	5	3.9	3.2	6.4	1.7	0.7	0.6				1.7	7.0	6.5	7.0
-Year		1982	1985	1985	1980	1983	1981				1985	1983	1981	NOV 1983
WIND:														
Mean Speed (mph)														
Prevailing Direction														
Fastest Obs. 1 Min.														
-Direction (!!!)														
-Speed (MPH)														
-Year														
Peak Gust														
-Direction (!!!)														
-Speed (mph)														
-Date														

See Reference Notes to this table on the following page

TABLE 2 PRECIPITATION (inches) BURNS, OREGON

YEAR	JAN	FEB	MAR	APR	MAY	JUNE	JULY	AUG	SEP	OCT	NOV	DEC	ANNUAL
1956	3.97	0.74	0.37	0.15	1.29	0.47	0.59	0.43	T	3.37	0.23	1.04	12.65
1957	1.37	2.31	2.16	0.73	1.32	0.61	0.10	0.11	0.60	1.84	0.72	1.58	13.45
1958	2.37	2.27	0.90	1.09	0.24	1.58	1.13	1.00	0.34	0.74	0.76		12.84
1959	0.99	0.91	1.29	0.24	1.37	0.06	0.19	0.84	1.93	1.02	0.13	0.67	9.64
1960	2.08	2.09	2.00	0.43	1.61	T	0.68	0.26	0.54	0.20	2.07	0.81	12.77
1961	0.69	1.22	1.76	0.21	0.82	0.83	0.36	0.12	0.61	0.92	0.99	2.24	10.77
1962	0.77	2.13	1.22	0.23	1.32	0.04	0.01	0.09	0.33	3.68	1.23	1.02	12.07
1963	1.42	1.15	1.15	2.10	2.31	0.83	0.28	0.25	1.20	0.32	1.86	1.02	13.89
1964	2.09	0.25	0.85	0.58	0.78	1.93	0.53	1.21	0.06	0.26	2.02	5.47	16.03
1965	2.40	0.22	0.19	1.28	0.75	1.93	0.52	1.10	T	0.14	1.75	0.82	11.10
1966	0.79	0.73	0.77	0.15	0.18	1.15	0.22	0.63	0.80	0.33	2.03	1.98	9.76
1967	2.15	0.42	2.44	1.31	0.71	0.84	0.02	0.01	0.32	1.22	0.81	1.03	11.28
1968	0.99	1.63	0.13	0.09	1.14	0.21	T	1.73	0.03	0.73	2.22	1.83	10.73
1969	3.17	1.47	0.31	0.45	0.30	1.33	T	T	0.22	1.03	3.10	1.84	11.84
1970	5.73	0.72	0.45	0.23	0.66	0.43	T	0.02	0.00	0.22	1.21	3.20	2.53	15.40
1971	1.77	0.46	2.02	0.33	2.02	0.98	1.14	T	0.57	0.51	1.62	2.06	13.48
1972	2.07	0.97	1.83	0.17	0.31	0.40	0.01	0.33	0.63	0.62	2.69	2.11	10.29
1973	1.37	0.45	0.43	0.15	1.06	0.04	T	0.31	0.00	0.35	0.11	2.45	10.20
1974	0.85	0.82	2.42	0.72	0.09	0.01	0.59	0.10	0.07	1.35	0.82	1.36	7.42
1975	1.54	1.78	1.71	1.22	0.05	0.38	0.58	0.19	0.07	1.35	0.82	1.64	11.33
1976	1.49	1.46	0.45	0.59	0.37	0.66	0.55	1.97	0.51	0.11	0.30	T	8.46
1977	0.69	0.45	0.27	0.16	1.67	0.41	0.51	0.29	1.11	0.44	2.09	2.31	10.40
1978	2.54	1.42	1.56	1.88	0.18	0.44	0.45	0.56	1.11	0.00	1.21	0.54	11.89
1979	2.96	1.74	0.99	0.82	0.81	0.21	0.03	1.99	0.56	1.03	1.93	1.62	14.69
#1980	1.58	2.01	0.61	0.57	1.08	0.85	0.26	0.03	0.56	0.88	0.59	1.36	10.38
1981	0.79	0.91	1.62	0.84	2.14	0.47	0.23	0.18	0.54	1.34	2.62	3.88	15.56
1982	1.15	1.70	0.78	0.90	0.64	0.90	0.41	0.35	1.06	1.43	1.09	2.56	12.97
1983	1.04	2.13	3.66	1.10	1.54	0.31	1.09	0.70	0.06	1.08	2.44	3.09	18.24
1984	0.21	0.76	1.71	0.80	1.04	0.88	0.49	1.16	0.05	1.02	2.73	0.87	11.72
1985	0.09	0.49	1.03	0.34	1.37	0.11	0.92	0.09	1.12	0.81	0.97	0.73	8.07
Record Mean	0.66	1.20	1.76	0.80	1.35	0.53	0.63	0.50	0.57	1.14	1.97	2.23	13.31

TABLE 3 AVERAGE TEMPERATURE (deg. F) BURNS, OREGON

YEAR	JAN	FEB	MAR	APR	MAY	JUNE	JULY	AUG	SEP	OCT	NOV	DEC	ANNUAL
1956	26.0	23.5	36.7	46.9	54.9	57.7	70.0	65.6	58.7	45.7	36.2	29.6	46.0
1957	18.2	30.6	38.5	45.2	53.8	61.2	66.6	65.0	61.3	45.6	34.2	31.3	46.0
1958	28.1	36.8	35.5	43.2	60.2	61.3	68.8	71.0	57.1	51.7	37.8	35.0	48.9
1959	32.8	32.7	37.7	47.6	48.0	62.4	71.3	66.0	54.9	48.4	37.2	26.7	47.2
1960	22.7	27.7	39.2	44.0	49.0	62.8	73.2	63.5	60.3	47.3	34.5	28.4	46.1
1961	30.1	36.1	37.7	44.2	50.4	66.3	70.6	70.9	53.1	45.9	33.6	26.2	47.1
1962	21.0	29.3	34.7	47.7	48.9	60.0	66.3	63.8	60.5	48.2	37.9	32.8	45.9
1963	25.3	40.3	37.2	39.9	54.5	57.0	63.4	66.4	62.8	50.5	36.5	27.6	46.8
1964	25.6	27.0	34.1	41.3	49.4	57.0	67.1	63.0	55.6	49.9	33.6	29.1	44.4
1965	28.6	34.6	37.3	45.7	49.6	60.5	67.1	63.8	53.3	51.1	37.6	26.2	46.3
1966	27.1	29.4	38.4	45.8	55.6	57.5	66.5	67.0	60.4	47.1	38.7	28.7	46.9
1967	31.1	34.9	35.0	37.9	52.7	61.1	72.0	72.3	62.8	46.2	38.1	25.5	47.5
1968	27.0	38.9	40.2	40.6	51.7	61.1	72.4	62.2	57.4	45.4	35.9	28.3	46.8
1969	25.9	26.2	34.1	44.3	56.8	61.0	68.5	67.0	59.9	42.3	37.0	31.1	46.2
1970	29.7	36.9	37.3	37.8	52.3	63.0	69.9	66.5	52.9	43.9	36.8	22.9	46.1
1971	29.9	33.1	33.4	42.7	52.1	56.8	68.6	70.8	52.0	43.9	34.8	23.9	45.2
1972	23.4	30.7	41.6	40.5	54.7	62.2	67.3	68.1	52.1	48.3	35.2	20.7	45.4
1973	25.6	34.2	36.9	44.0	55.1	60.6	70.8	66.2	57.7	46.5	34.6	32.4	47.0
1974	26.6	32.9	36.8	42.8	50.2	67.5	67.5	67.0	61.5	48.2	36.5	28.4	47.0
1975	25.1	30.2	33.9	38.1	50.5	58.6	70.9	63.5	61.8	45.4	33.6	30.8	45.2
1976	27.6	30.3	34.1	41.9	53.9	64.5	68.7	61.2	60.4	48.2	40.9	29.8	46.2
1977	19.2	35.3	34.7	48.4	47.1	66.0	67.2	68.9	56.3	48.6	34.7	31.8	46.5
1978	30.1	31.8	43.6	42.9	49.1	58.8	68.3	64.6	54.6	50.6	32.8	21.8	45.7
1979	15.8	31.2	39.2	44.2	53.6	61.4	69.3	66.8	62.3	51.0	32.0	29.8	46.4
#1980	25.2	35.1	35.1	45.2	50.9	54.2	65.8	61.0	55.6	44.5	35.8	29.0	44.8
1981	30.1	29.2	37.6	44.6	49.2	56.6	64.1	66.5	55.5	42.3	35.6	29.1	45.0
1982	17.6	24.7	35.3	39.1	49.1	57.8	64.9	64.9	52.9	43.5	29.0	22.8	41.8
1983	28.7	33.4	39.1	40.9	51.6	55.0	61.1	64.6	53.9	47.2	35.7	21.7	44.4
1984	17.0	23.1	35.8	40.7	50.0	54.4	67.0	64.2	52.7	39.7	33.6	19.0	41.5
1985	16.9	23.0	31.3	46.3	51.1	59.1	70.7	61.4	48.9	41.8	20.9	11.9	40.3
Record Mean	22.0	26.6	35.8	39.0	50.2	56.6	65.6	64.3	52.8	42.9	31.0	20.9	42.3
Max	31.7	36.8	46.2	48.7	63.9	72.4	83.3	83.4	70.4	58.2	41.4	30.3	55.6
Min	12.3	16.5	25.4	29.3	36.4	40.7	47.8	45.3	35.1	27.6	20.5	11.5	29.0

REFERENCE NOTES FOR TABLES 1, 2, 3 and 6   (BURNS, OR)

GENERAL

T - TRACE AMOUNT
BLANK ENTRIES DENOTE MISSING/UNREPORTED DATA.
# INDICATES A STATION OR INSTRUMENT RELOCATION.

SPECIFIC

TABLE 1

(a) - LENGTH OF RECORD IN YEARS. ALTHOUGH INDIVIDUAL MONTHS MAY BE MISSING.

* LESS THAN .05

NORMALS — BASED ON THE 1951-1980 RECORD PERIOD.
EXTREMES — DATES ARE THE MOST RECENT OCCURRENCE.
WIND DIR. — NUMERALS SHOW TENS OF DEGREES
CLOCKWISE FROM TRUE NORTH.
"00" INDICATES CALM.
RESULTANT WIND DIRECTIONS ARE GIVEN TO WHOLE DEGREES.

EXCEPTIONS

TABLES 2, 3, and 6

RECORD MEANS ARE THROUGH THE CURRENT YEAR, BEGINNING IN   1939 FOR TEMPERATURE
1939 FOR PRECIPITATION
1937 FOR SNOWFALL

TABLE 4 HEATING DEGREE DAYS Base 65 deg. F   BURNS, OREGON

SEASON	JULY	AUG	SEP	OCT	NOV	DEC	JAN	FEB	MAR	APR	MAY	JUNE	TOTAL
1963-64	66	49	109	449	847	1152	1214	1094	950	709	477	240	7356
1964-65	40	121	278	465	934	1108	1120	846	852	573	469	150	6956
1965-66	50	92	345	423	815	1194	1165	995	817	569	298	224	6987
1966-67	45	62	162	548	781	1116	1045	836	921	806	381	161	6864
1967-68	1	5	94	574	800	1221	1173	748	762	724	405	158	6665
1968-69	11	162	234	598	866	1129	1205	1079	952	615	250	148	7249
1969-70	29	38	182	697	836	1044	1085	782	852	809	386	158	6898
1970-71	12	6	357	645	841	1298	1077	889	973	665	390	248	7401
1971-72	68	31	381	647	900	1267	1287	989	716	730	330	114	7460
1972-73	47	41	394	513	888	1369	1211	857	864	624	313	189	7310
1973-74	19	62	229	570	904	1002	1184	895	866	662	453	84	6930
1974-75	60	60	115	515	849	1127	1231	967	960	801	440	199	7324
1975-76	39	107	123	599	934	1054	1150	998	949	687	339	262	7241
1976-77	25	150	154	512	717	1083	1413	825	928	493	551	51	6902
1977-78	47	86	278	499	903	1020	1075	923	656	655	484	189	6815
1978-79	34	122	310	443	961	1335	1517	941	792	618	350	150	7573
#1979-80	33	45	103	434	984	1082	1227	862	918	587	430	321	7026
1980-81	52	130	274	627	868	1110	1074	996	844	607	487	255	7324
1981-82	80	46	287	696	875	1106	1461	1123	914	770	484	224	8066
1982-83	86	76	360	659	1075	1303	1116	882	793	720	424	294	7788
1983-84	154	76	328	544	872	1339	1481	1210	898	719	464	324	8409
1984-85	21	45	363	778	937	1421	1487	1170	1037	555	421	190	8425
1985-86	14	131	475	715	1314	1639							

TABLE 5 COOLING DEGREE DAYS Base 65 deg. F   BURNS, OREGON

YEAR	JAN	FEB	MAR	APR	MAY	JUNE	JULY	AUG	SEP	OCT	NOV	DEC	TOTAL
1970	0	0	0	0	0	105	171	151	0	0	0	0	427
1971	0	0	0	0	0	8	187	219	0	0	0	0	414
1972	0	0	0	0	14	38	124	141	16	0	0	0	333
1973	0	0	0	0	17	63	206	106	15	0	0	0	407
1974	0	0	0	0	2	114	145	127	18	0	0	0	406
1975	0	0	0	0	0	11	230	71	35	0	0	0	347
1976	0	0	0	0	1	15	150	37	24	0	0	0	227
1977	0	0	0	0	3	87	120	214	27	0	0	0	451
1978	0	0	0	0	0	12	142	118	5	0	0	0	277
1979	0	0	0	0	3	48	172	111	27	6	0	0	367
#1980	0	0	0	0	0	2	81	11	0	0	0	0	94
1981	0	0	0	0	0	11	61	102	13	0	0	0	187
1982	0	0	0	0	0	13	88	62	3	0	0	0	166
1983	0	0	0	0	16	0	41	71	0	0	0	0	128
1984	0	0	0	0	3	14	89	48	4	0	0	0	158
1985	0	0	0	0	0	18	199	27	0	0	0	0	244

TABLE 6 SNOWFALL (inches)   BURNS, OREGON

SEASON	JULY	AUG	SEP	OCT	NOV	DEC	JAN	FEB	MAR	APR	MAY	JUNE	TOTAL
1956-57	0.0	0.0	0.0	1.5	2.0	1.5	17.5	6.4	4.3	T	0.0	0.0	33.2
1957-58	0.0	0.0	0.0	T	1.7	10.7	16.1	6.5	7.0	5.0	0.0	0.0	47.0
1958-59	0.0	0.0	0.0	T	4.9	8.3	9.2	6.6	0.5	1.5	T		31.0
1959-60	0.0	0.0	0.0	T	0.8	5.9	21.2	8.7	10.2	0.1	0.3	0.0	47.2
1960-61	0.0	T	0.0	T	5.7	3.0	1.5	4.9	9.9	0.6	T	0.0	25.6
1961-62	0.0	0.0	0.0	1.3	4.8	18.8	11.2	9.4	11.3	T	T	T	56.8
1962-63	0.0	0.0	0.0	T	1.3	0.2	11.0	0.9	4.5	8.8	T	0.0	26.7
1963-64	0.0	0.0	0.0	T	5.1	8.2	20.6	2.8	8.6	2.8	0.9	0.0	49.0
1964-65	0.0	0.0	0.0	0.0	11.3	15.6	24.9	1.8	2.7	1.4	T	0.0	57.7
1965-66	0.0	0.0	0.0	0.0	0.0	4.2	17.8	8.0	5.1	T	T	T	35.1
1966-67	0.0	0.0	0.0	1.4	2.4	9.8	15.7	6.5	23.1	3.6	0.4	0.0	62.9
1967-68	0.0	0.0	0.0	0.0	6.3	12.0	4.6	2.2	1.7	0.2	0.0	0.0	25.9
1968-69	0.0	0.0	0.0	T	4.8	18.3	27.3	12.1	1.7	0.2	0.0	0.0	64.4
1969-70	0.0	0.0	0.0	1.1	0.9	10.3	14.6	0.1	3.3	0.2	2.3	0.0	32.8
1970-71	0.0	0.0	0.0	4.9	6.1	22.1	5.3	4.4	7.2	1.6	T	0.0	51.6
1971-72	0.0	0.0	0.2	1.3	7.2	22.0	11.7	8.9	6.1	0.2	0.0	0.0	57.6
1972-73	0.0	0.0	T	0.3	1.2	10.2	10.1	4.0	0.8	0.7	T	0.4	27.7
1973-74	0.0	0.0	0.0	T	11.9	11.7	4.5	6.8	8.6	2.0	0.7	0.0	46.2
1974-75	0.0	0.0	0.0	0.0	0.1	9.9	15.9	13.0	13.0	7.1	T	0.0	59.0
1975-76	0.0	0.0	0.0	4.4	5.2	9.9	11.8	10.5	1.0	1.0	0.0	0.0	43.8
1976-77	0.0	0.0	0.0	0.0	T	T	7.0	4.1	3.3	0.4	1.4	0.0	16.2
1977-78	0.0	0.0	0.0	T	14.6	12.7	13.3	9.4	3.6	5.8	T	0.0	59.4
1978-79	0.0	0.0	T	0.0	1.2	6.0	22.0	16.8	2.5	1.2	0.9	0.0	50.6
#1979-80	0.0	0.0	0.0	T	11.0	13.7	5.5	7.0	4.9	1.9	T		44.0
1980-81	0.0	0.0	0.0	0.3	1.0	5.2	4.8	1.9	1.6	T	T	0.6	15.4
1981-82	0.0	0.0	0.0	T	4.5	24.4	13.8	2.3	3.4	3.2	0.2	T	51.8
1982-83	0.0	0.0	0.0	T	2.1	16.0	7.1	7.2	8.3	0.3	1.4	0.0	42.4
1983-84	0.0	0.0	0.0	0.0	17.4	26.8	4.2	9.4	2.7	1.5	T	0.0	62.0
1984-85	0.0	0.0	0.0	3.6	9.2	7.2	1.1	5.8	13.5	1.6	0.0	0.0	42.0
1985-86	0.0	0.0	0.0	1.7	13.8	7.4							
Record Mean	0.0	0.0	0.0	1.1	9.4	16.4	6.2	5.3	5.9	1.3	0.3	0.1	46.0

See Reference Notes, relative to all above tables, on preceding page.

Eugene is located at the upper or southern end of the fertile Willamette Valley. Mahlon Sweet Field, location of the National Weather Service Office, is 9 miles northwest of the city center. The Cascade Mountains to the east and the Coast Range to the west bound the valley, and low hills to the south nearly close it, but northward the level valley floor broadens rapidly. Hills of the rolling, wooded Coast Range begin about 5 miles west of the airport and rise to elevations of 1,500 to 2,500 feet midway between Eugene and the Pacific Ocean lying 50 miles to the west. About 10 miles east, the Coburg Hills, rising to an elevation of 2,500 feet, obscure the snow-covered peaks of the Cascade Range, which reach elevations of 10,000 feet about 75 miles away. Abundant moisture and moderate temperatures result in rapid growth of evergreen timber and lumbering is a major industry.

The Willamette River passes about 5 miles east of the airport. The Fern Ridge flood control reservoir, with a normal area of 9,360 acres, begins about 2 miles southwest. These two water areas are the main sources of local fog, but numerous small creeks and low places, which fill with water in the wet season, also produce considerable fog. The Coast Range acts as a barrier to coastal fog, but active storms cross these ridges with little hindrance. The Cascade Range blocks westward passage of all but the strongest continental air masses, but when air does flow into the valley from the east, dry, hot weather develops in summer causing an extreme fire hazard. In winter this situation causes clear, sunny days and cool, frosty nights.

The centers of low barometric pressure, with which rain is associated, generally pass inland north of Eugene and as a result southwest winds with speeds of 10 to 20 mph usually accompany rainfall. Heavier storms bring winds of 30 to 40 mph and occasional southwest winds exceeding 50 mph are experienced. Fair weather in both summer and winter is most often accompanied by calm nights and daytime northerly winds increasing to speeds of 5 to 15 mph in the afternoon.

The first fall rains usually arrive during the second or third week of September, after which rain gradually increases until about the first of January and then slowly decreases to the latter part of June. July and August are normally very dry, occasionally passing without rainfall. When snow occurs, it frequently melts on contact with the ground or within a few hours, but occasionally an accumulation of a few inches will persist as a ground covering for several days. Snowfall for a winter season exceeds 5 inches in about one-third of the years.

Temperatures are so largely controlled by maritime air from the Pacific that long periods of extremely hot or severely cold weather never occur. Temperatures of 95 degrees or higher have occurred only in the months of June, July, August, and September, and average three days a year.

Based on the 1951–1980 period, the average first occurrence of 32 degrees Fahrenheit in the fall is October 25 and the average last occurrence in the spring is April 24.

TABLE 1   **NORMALS, MEANS AND EXTREMES**

EUGENE, OREGON

LATITUDE: 44°07'N   LONGITUDE: 123°13'W   ELEVATION: FT. GRND   359 BARO   00364 • TIME ZONE: PACIFIC   WBAN: 24221

	(a)	JAN	FEB	MAR	APR	MAY	JUNE	JULY	AUG	SEP	OCT	NOV	DEC	YEAR
TEMPERATURE °F:														
Normals														
–Daily Maximum		46.3	51.4	55.0	60.5	67.2	74.2	82.6	81.3	76.4	64.6	52.8	47.3	63.3
–Daily Minimum		33.8	35.5	36.5	38.7	42.9	48.0	51.0	51.1	47.7	42.0	37.8	35.3	41.7
–Monthly		40.0	43.5	45.8	49.6	55.0	61.1	66.8	66.2	62.0	53.3	45.3	41.3	52.5
Extremes														
–Record Highest	43	67	71	77	86	92	100	105	108	101	94	76	68	108
–Year		1975	1968	1978	1957	1983	1961	1961	1981	1944	1980	1975	1979	AUG 1981
–Record Lowest	43	-4	-3	20	27	28	32	39	38	32	19	12	-12	-12
–Year		1957	1950	1956	1983	1954	1976	1973	1969	1983	1971	1978	1972	DEC 1972
NORMAL DEGREE DAYS:														
Heating (base 65°F)		772	602	595	462	307	145	44	57	126	363	591	735	4799
Cooling (base 65°F)		0	0	0	0	0	28	100	94	39	0	0	0	261
% OF POSSIBLE SUNSHINE														
MEAN SKY COVER (tenths)														
Sunrise – Sunset	41	8.5	8.4	8.0	7.5	6.8	6.3	3.8	4.6	5.1	7.1	8.4	8.9	6.9
MEAN NUMBER OF DAYS:														
Sunrise to Sunset														
–Clear	43	2.2	2.3	3.0	4.0	5.8	7.4	16.3	13.7	11.8	4.7	1.6	1.5	74.1
–Partly Cloudy	43	4.0	4.6	6.0	7.1	8.6	8.0	7.9	8.8	8.3	8.8	5.6	3.8	81.4
–Cloudy	43	24.8	21.4	22.0	19.0	16.5	14.7	6.8	8.5	9.9	17.6	22.7	25.7	209.7
Precipitation														
.01 inches or more	43	17.5	15.3	16.6	12.6	10.0	6.9	2.2	3.9	6.0	11.5	16.4	18.6	137.7
Snow, Ice pellets														
1.0 inches or more	43	1.2	0.1	0.3	0.0	0.0	0.0	0.0	0.0	0.0	0.0	0.1	0.5	2.2
Thunderstorms	42	0.2	0.1	0.2	0.5	0.8	0.7	0.5	0.6	0.5	0.2	0.2	0.2	4.7
Heavy Fog Visibility														
1/4 mile or less	42	8.6	6.5	3.7	2.2	1.3	0.9	0.4	0.9	4.6	11.4	9.5	9.2	59.1
Temperature °F														
–Maximum														
90° and above	43	0.0	0.0	0.0	0.0	0.1	1.2	6.3	4.7	2.3	0.*	0.0	0.0	14.7
32° and below	43	1.8	0.2	0.0	0.0	0.0	0.0	0.0	0.0	0.0	0.0	0.2	0.8	3.0
–Minimum														
32° and below	43	15.3	9.6	7.3	2.9	0.5	0.*	0.0	0.0	0.1	2.2	7.6	11.4	56.9
0° and below	43	0.1	0.*	0.0	0.0	0.0	0.0	0.0	0.0	0.0	0.0	0.0	0.1	0.2
AVG. STATION PRESS. (mb)	11	1006.5	1004.2	1003.0	1005.3	1004.8	1004.4	1004.0	1003.2	1003.3	1004.8	1003.6	1005.8	1004.4
RELATIVE HUMIDITY (%)														
Hour 04	28	91	92	91	90	91	90	87	88	89	94	93	92	91
Hour 10	41	87	85	78	70	66	62	56	60	65	80	87	89	74
Hour 16 (Local Time)	41	80	72	64	57	53	49	38	40	43	63	78	84	60
Hour 22	30	90	90	86	83	81	79	72	74	77	88	92	91	84
PRECIPITATION (inches):														
Water Equivalent														
–Normal		8.39	5.12	5.11	2.76	1.97	1.24	0.27	0.95	1.45	3.47	6.82	8.49	46.04
–Maximum Monthly	43	14.83	12.28	12.46	6.89	4.44	4.76	2.63	5.79	3.45	12.66	20.48	20.99	20.99
–Year		1964	1983	1974	1982	1960	1952	1947	1968	1978	1950	1973	1964	DEC 1964
–Minimum Monthly	43	0.31	0.86	0.79	0.49	0.26	T	0.00	0.00	T	0.29	1.20	1.24	0.00
–Year		1985	1964	1965	1985	1982	1951	1967	1967	1975	1978	1956	1976	JUL 1967
–Maximum in 24 hrs	43	4.88	4.81	2.44	2.25	2.37	2.36	1.39	1.92	1.68	3.85	4.53	5.15	5.15
–Year		1974	1984	1963	1971	1972	1952	1947	1983	1981	1955	1960	1981	DEC 1981
Snow, Ice pellets														
–Maximum Monthly	43	47.1	4.8	10.8	T	T	T			T	T	6.0	10.2	47.1
–Year		1969	1949	1951	1985	1984	1981			1971	1984	1955	1964	JAN 1969
–Maximum in 24 hrs	43	22.9	2.5	4.9	T	T	T			T	T	5.0	6.3	22.9
–Year		1969	1971	1951	1985	1984	1981			1971	1984	1955	1972	JAN 1969
WIND:														
Mean Speed (mph)	33	7.9	7.9	8.4	7.7	7.4	7.4	8.0	7.5	7.4	6.6	7.3	7.7	7.6
Prevailing Direction														
Fastest Obs. 1 Min.														
–Direction (!!!)	29	20	20	18	18	25	23	36	11	20	18	23	25	18
–Speed (MPH)	29	58	54	48	44	46	29	26	32	32	63	46	40	63
–Year		1961	1961	1963	1972	1961	1961	1960	1979	1959	1962	1957	1961	OCT 1962
Peak Gust														
–Direction (!!!)	2	S	S	W	W	SW	N	SW	N	S	SW	S	N	W
–Speed (mph)	2	25	44	45	41	45	32	36	39	30	35	45	32	45
–Date		1984	1984	1985	1984	1985	1985	1984	1985	1985	1985	1984	1985	MAR 1985

See Reference Notes to this table on the following pages.

TABLE 2 PRECIPITATION (inches) EUGENE, OREGON

YEAR	JAN	FEB	MAR	APR	MAY	JUNE	JULY	AUG	SEP	OCT	NOV	DEC	ANNUAL
1956	10.89	5.29	4.83	0.97	2.17	1.05	T	0.47	0.74	8.33	1.20	4.56	40.50
1957	2.98	5.87	8.66	2.16	3.11	0.85	0.19	0.54	1.03	2.81	12.82		44.15
1958	9.35	10.14	2.32	3.18	1.53	2.14	T	0.03	1.48	2.34	7.87	4.01	44.39
1959	12.39	5.41	3.61	0.90	2.16	0.80	0.32	0.06	0.88	2.23	1.42	2.76	32.94
1960	4.83	6.54	9.81	3.29	4.44	0.07	T	0.47	0.75	1.65	12.02	3.21	47.08
1961	5.25	11.58	8.42	1.48	2.87	0.76	0.32	0.34	1.12	4.31	9.36	7.37	53.18
1962	1.39	4.53	7.07	3.06	1.92	0.66	T	1.18	1.93	6.33	6.34	3.03	37.44
1963	2.55	5.27	7.17	5.23	1.40	1.40	0.32	0.17	2.19	2.39	7.87	3.09	41.60
1964	14.83	0.86	4.53	1.28	0.90	1.29	0.54	0.27	0.73	1.03	10.70	20.99	57.95
1965	9.92	1.60	0.79	2.92	0.90	0.44	0.24	0.90	T	2.43	7.43	7.69	35.26
1966	10.97	1.83	5.95	0.52	0.54	0.23	0.45	0.02	1.44	1.93	9.10	8.31	41.29
1967	10.33	2.13	2.90	3.02	1.59	1.70	0.00	0.00	1.83	5.10	3.22	4.99	36.81
1968	7.53	6.32	3.91	1.06	2.86	0.96	0.02	5.79	2.57	5.99	7.02	12.52	56.55
1969	12.67	3.21	2.74	2.92	2.38	3.13	T	T	2.05	4.22	2.74	11.68	47.74
1970	14.38	3.37	2.45	3.04	0.80	0.76	T	T	1.19	4.30	8.23	11.57	50.09
1971	10.83	5.11	7.99	4.49	2.52	3.10	T	1.33	3.04	3.32	9.48	9.46	60.67
1972	12.48	6.52	7.63	5.80	2.94	1.91	0.02	1.70	2.71	1.31	3.76	10.78	57.56
1973	6.20	2.09	5.18	1.67	0.86	1.35	T	0.80	2.48	2.37	20.48	11.82	55.30
1974	12.80	8.42	12.46	2.47	1.12	0.37	1.37	0.42	0.08	1.59	6.42	9.26	56.78
1975	6.93	6.79	7.56	2.90	2.16	0.88	1.18	2.09	T	5.69	8.47	7.12	51.77
1976	9.82	7.66	6.23	1.89	0.94	0.21	0.42	2.04	1.13	1.87	1.33	1.24	34.78
1977	1.11	5.05	4.66	1.47	2.84	0.97	0.11	1.70	2.39	2.87	9.14	14.60	46.91
1978	9.05	3.25	1.68	6.56	2.12	0.74	0.72	2.17	3.45	0.29	6.61	2.86	39.50
1979	2.98	9.52	3.12	4.71	2.61	0.56	0.41	3.46	2.32	8.12	6.09	7.38	51.28
1980	7.45	4.68	5.11	4.20	1.39	2.06	0.39	0.02	0.75	1.90	8.66	14.73	51.34
1981	2.13	4.35	4.16	2.69	3.27	3.51	0.08	T	3.15	5.42	9.51	17.63	55.90
1982	9.31	8.14	4.88	6.89	0.26	1.92	0.54	0.72	2.81	3.95	7.07	13.53	60.02
1983	6.75	12.28	10.58	3.35	1.81	1.78	1.77	3.19	0.54	1.36	13.13	7.47	64.01
1984	2.11	9.58	6.36	5.41	3.91	3.88	0.27	0.03	0.94	6.05	18.67	4.56	61.77
1985	0.31	5.15	5.65	0.49	1.53	2.51	1.37	0.04	2.13	4.83	6.31	3.51	33.83
Record Mean	7.40	5.43	5.19	2.78	1.96	1.42	0.38	0.85	1.47	3.86	7.27	7.94	45.95

TABLE 3 AVERAGE TEMPERATURE (deg. F) EUGENE, OREGON

YEAR	JAN	FEB	MAR	APR	MAY	JUNE	JULY	AUG	SEP	OCT	NOV	DEC	ANNUAL
1956	41.8	37.2	44.2	50.4	56.6		66.6	65.2	61.2	50.9	41.9	40.8	51.2
1957	32.1	42.8	46.9	51.9	57.8	61.6	64.1	63.3	64.8	52.4	43.0	42.7	52.0
1958	42.9	48.8	44.3	50.6	60.0	63.5	71.0	69.8	61.9	53.6	47.1	46.1	55.0
1959	43.3	42.8	46.3	51.0	53.6	61.1	68.2	65.5	59.3	54.5	44.4	40.9	52.6
1960	37.8	42.4	46.8	50.4	53.6	62.0	68.5	63.8	61.6	53.3	45.4	38.9	52.1
1961	43.6	46.7	46.9	50.2	54.9	63.6	66.8	68.6	58.5	52.3	41.7	42.0	53.0
1962	37.5	42.3	44.6	52.1	52.0	59.9	64.9	64.7	62.7	52.6	47.8	42.5	52.0
#1963	34.4	49.3	45.7	47.6	56.0	60.1	62.9	66.7	66.0	55.4	47.1	42.2	52.8
1964	42.4	41.6	45.1	47.9	53.2	59.8	66.4	65.6	61.0	53.2	43.2	40.5	51.9
1965	40.8	44.3	47.8	51.8	53.6	60.7	68.0	67.4	60.8	55.5	49.4	39.2	53.3
1966	41.3	42.1	46.6	52.5	55.2	62.6	66.7	67.2	64.6	53.5	47.5	45.1	53.7
1967	44.4	44.6	45.0	45.9	55.7	64.8	68.8	72.0	66.0	54.4	46.9	42.4	54.2
1968	41.5	48.9	49.1	48.5	55.5	61.6	68.7	66.4	62.2	52.9	48.2	41.0	53.7
1969	36.9	40.7	47.6	50.2	60.1	65.3	66.5	64.6	61.3	50.6	45.6	43.1	52.7
1970	42.5	44.5	44.8	46.9	56.1	65.4	69.9	67.6	60.2	52.6	45.6	40.0	53.2
1971	40.1	41.1	43.1	48.3	54.5	57.8	66.7	68.2	59.2	49.5	43.5	38.7	50.9
1972	37.9	43.5	48.6	46.6	56.9	60.9	69.0	69.1	59.0	51.1	47.0	34.3	52.0
1973	38.3	44.5	44.9	49.9	55.9	61.7	67.7	64.0	62.7	52.7	44.3	43.4	52.5
1974	39.3	43.2	49.5	52.6	56.1	63.9	67.9	68.9	67.6	55.0	48.7	45.7	54.9
1975	45.9	43.5	44.7	46.4	55.5	61.1	68.5	64.5	66.1	55.0	46.2	43.0	53.4
1976	42.0	41.9	44.3	48.8	54.6	58.1	67.7	66.1	63.8	54.0	47.6	37.7	52.2
1977	38.9	44.5	44.9	50.7	51.3	61.0	64.2	69.7	59.4	52.5	43.5	43.0	52.0
1978	42.5	44.2	48.4	48.1	52.7	62.5	66.4	66.1	59.6	53.0	39.3	35.2	51.5
1979	31.5	42.9	48.8	50.4	55.5	60.0	67.3	65.5	63.6	56.5	42.4	43.0	52.3
1980	35.7	42.9	45.5	50.8	53.2	57.5	66.8	63.5	61.9	54.1	46.4	42.3	51.7
1981	40.4	43.4	45.9	49.9	54.4	59.5	64.8	68.2	62.0	50.2	45.4	41.8	52.2
1982	36.2	41.3	44.1	46.4	53.9	62.2	64.4	66.5	60.9	52.2	40.8	39.3	50.7
1983	41.7	45.1	48.3	49.1	56.1	58.6	63.0	66.1	59.1	52.1	46.6	36.3	51.9
1984	40.5	44.2	48.5	47.9	53.2	58.3	65.9	66.0	60.8	50.8	43.8	37.3	51.4
1985	34.2	40.6	42.8	52.0	54.9	61.4	69.9	65.8	58.3	51.8	37.6	33.0	50.2
Record Mean	39.2	43.3	45.8	49.7	55.3	61.0	66.6	66.2	61.8	52.9	45.0	40.7	52.3
Max	45.6	51.3	55.0	60.6	67.5	73.9	82.3	81.3	76.3	64.0	52.5	46.8	63.1
Min	32.7	35.2	36.5	38.8	43.0	48.0	50.9	51.0	47.3	41.8	37.6	34.6	41.5

REFERENCE NOTES FOR TABLES 1, 2, 3 and 6   (EUGENE, OR)

GENERAL

T - TRACE AMOUNT
BLANK ENTRIES DENOTE MISSING/UNREPORTED DATA.
# INDICATES A STATION OR INSTRUMENT RELOCATION.

SPECIFIC

TABLE 1

(a) - LENGTH OF RECORD IN YEARS. ALTHOUGH INDIVIDUAL MONTHS MAY BE MISSING.

* LESS THAN .05

NORMALS — BASED ON THE 1951-1980 RECORD PERIOD.
EXTREMES — DATES ARE THE MOST RECENT OCCURRENCE.
WIND DIR. — NUMERALS SHOW TENS OF DEGREES CLOCKWISE FROM TRUE NORTH. "00" INDICATES CALM.
RESULTANT WIND DIRECTIONS ARE GIVEN TO WHOLE DEGREES.

EXCEPTIONS

TABLES 2, 3, and 6

RECORD MEANS ARE THROUGH THE CURRENT YEAR, BEGINNING IN 1943 FOR TEMPERATURE / 1943 FOR PRECIPITATION / 1943 FOR SNOWFALL

TABLE 4 HEATING DEGREE DAYS Base 65 deg. F   EUGENE, OREGON

SEASON	JULY	AUG	SEP	OCT	NOV	DEC	JAN	FEB	MAR	APR	MAY	JUNE	TOTAL
1956-57	46	50	111	428	689	745	1011	613	551	392	216	104	4956
1957-58	42	58	51	382	652	682	675	451	613	426	154	92	4278
1958-59	0	2	116	349	531	575	667	615	574	413	348	127	4317
1959-60	38	25	167	317	612	740	834	648	556	432	347	106	4822
1960-61	21	84	117	350	581	802	658	507	556	436	309	98	4519
1961-62	35	12	195	391	695	708	845	629	625	379	396	156	5066
1962-63	71	45	80	379	507	691	943	437	589	515	288	169	4714
#1963-64	71	25	44	289	530	699	691	671	610	505	360	151	4646
1964-65	40	45	120	282	646	753	744	572	522	389	350	138	4601
1965-66	31	17	132	288	461	791	728	634	565	368	295	108	4418
1966-67	25	31	53	352	517	611	633	566	614	564	285	64	4315
1967-68	5	1	28	322	538	693	722	461	486	487	287	123	4153
1968-69	9	39	100	366	497	738	867	673	536	438	160	44	4467
1969-70	17	44	134	440	576	673	692	568	557	534	272	75	4582
1970-71	11	11	135	374	576	769	767	664	671	496	318	212	5004
1971-72	60	23	169	474	638	811	834	617	503	547	256	123	5055
1972-73	23	27	202	424	536	945	825	567	615	444	276	132	5016
1973-74	29	77	90	374	614	666	790	604	472	364	270	84	4434
1974-75	29	16	25	305	480	589	585	595	623	548	294	138	4227
1975-76	18	56	40	302	557	676	707	664	636	478	316	210	4660
1976-77	10	38	58	340	518	840	801	567	615	422	417	126	4752
1977-78	68	20	178	385	637	675	690	576	508	503	375	102	4717
1978-79	52	48	165	366	763	919	1029	614	496	432	288	159	5331
1979-80	33	19	50	257	673	671	901	599	599	421	361	219	4838
1980-81	29	65	101	344	551	697	752	599	585	446	321	164	4654
1981-82	64	33	121	450	580	710	887	657	642	551	335	124	5154
1982-83	67	20	137	392	719	791	716	553	511	470	287	189	4852
1983-84	78	15	171	393	544	879	753	596	502	504	359	207	5001
1984-85	33	19	145	440	628	851	948	674	678	383	311	128	5238
1985-86	5	43	197	405	817	983							

TABLE 5 COOLING DEGREE DAYS Base 65 deg. F   EUGENE, OREGON

YEAR	JAN	FEB	MAR	APR	MAY	JUNE	JULY	AUG	SEP	OCT	NOV	DEC	TOTAL
1969	0	0	0	0	12	60	72	37	33	0	0	0	214
1970	0	0	0	0	4	93	173	99	1	0	0	0	370
1971	0	0	0	0	0	3	119	129	2	0	0	0	253
1972	0	0	0	0	11	6	156	159	26	0	0	0	358
1973	0	0	0	0	2	38	119	50	26	0	0	0	235
1974	0	0	0	0	3	57	127	143	110	0	0	0	440
1975	0	0	0	0	3	25	135	47	82	1	0	0	293
1976	0	0	0	0	3	12	101	82	30	3	0	0	231
1977	0	0	0	0	0	10	52	177	12	0	0	0	251
1978	0	0	0	0	2	34	101	88	10	0	0	0	235
1979	0	0	0	0	0	16	111	40	15	0	0	0	182
1980	0	0	0	0	0	0	91	23	16	12	0	0	142
1981	0	0	0	1	0	4	63	138	38	0	0	0	244
1982	0	0	0	0	0	51	59	73	20	0	0	0	203
1983	0	0	0	0	17	3	23	56	0	0	0	0	99
1984	0	0	0	0	2	14	69	55	23	5	0	0	168
1985	0	0	0	0	4	28	166	73	2	0	0	0	273

TABLE 6 SNOWFALL (inches)   EUGENE, OREGON

SEASON	JULY	AUG	SEP	OCT	NOV	DEC	JAN	FEB	MAR	APR	MAY	JUNE	TOTAL
1956-57	0.0	0.0	0.0	T	0.0	5.2	7.9	T	T	0.0	0.0	0.0	13.1
1957-58	0.0	0.0	0.0	0.0	0.0	T	0.0	0.0	T	0.0	0.0	0.0	T
1958-59	0.0	0.0	0.0	0.0	T	0.0	1.2	T	T	0.0	0.0	0.0	1.2
1959-60	0.0	0.0	0.0	0.0	0.0	T	T	0.0	3.9	0.0	0.0	0.0	3.9
1960-61	0.0	0.0	0.0	0.0	0.0	0.0	0.0	0.0	0.0	0.0	0.0	0.0	0.0
1961-62	0.0	0.0	0.0	0.0	2.0	0.0	2.0	T	1.9	T	0.0	0.0	5.9
1962-63	0.0	0.0	0.0	0.0	0.0	0.0	3.9	0.0	T	T	0.0	0.0	3.9
1963-64	0.0	0.0	0.0	0.0	0.0	0.0	T	0.0	T	0.0	0.0	0.0	T
1964-65	0.0	0.0	0.0	0.0	T	10.2	4.9	T	T	0.0	0.0	0.0	15.1
1965-66	0.0	0.0	0.0	0.0	0.0	4.8	0.1	0.3	2.5	0.0	0.0	0.0	7.7
1966-67	0.0	0.0	0.0	0.0	0.0	0.0	T	0.7	T	T	0.0	0.0	0.7
1967-68	0.0	0.0	0.0	0.0	0.0	1.1	3.6	0.0	0.0	0.0	0.0	0.0	4.7
1968-69	0.0	0.0	0.0	T	2.3	0.0	47.1	T	0.0	0.0	0.0	0.0	49.4
1969-70	0.0	0.0	0.0	0.0	0.0	0.1	18.9	4.5	0.8	T	0.0	0.0	24.3
1970-71	0.0	0.0	0.0	0.0	T	0.1	T	T	T	0.0	0.0	0.0	2.9
1971-72	0.0	0.0	T	T	0.0	1.2	1.2	T	0.5	T	0.0	0.0	2.9
1972-73	0.0	0.0	0.0	0.0	0.0	9.9	1.1	0.0	T	T	0.0	0.0	11.0
1973-74	0.0	0.0	0.0	1.5	T	T	0.4	T	T	0.0	0.0	0.0	1.9
1974-75	0.0	0.0	0.0	0.0	0.0	T	4.0	T	T	T	0.0	0.0	4.1
1975-76	0.0	0.0	0.0	0.0	T	2.8	T	T	1.3	0.0	0.0	0.0	4.1
1976-77	0.0	0.0	0.0	0.0	0.0	0.0	0.0	T	T	0.0	T	0.0	T
1977-78	0.0	0.0	0.0	0.0	0.0	0.2	0.0	T	0.0	T	0.0	0.0	0.2
1978-79	0.0	0.0	0.0	0.0	1.0	T	T	T	0.0	T	0.0	0.0	1.0
1979-80	0.0	0.0	0.0	0.0	0.0	0.0	T	0.0	T	T	0.0	T	T
1980-81	0.0	0.0	0.0	0.0	0.0	0.0	T	0.0	T	0.0	0.0	0.0	T
1981-82	0.0	0.0	0.0	0.0	T	T	5.0	2.1	T	T	0.0	0.0	7.1
1982-83	0.0	0.0	0.0	0.0	0.0	T	0.0	0.0	T	T	0.0	0.0	0.0
1983-84	0.0	0.0	0.0	0.0	T	3.7	T	T	T	T	0.0	0.0	3.7
1984-85	0.0	0.0	0.0	T	0.0	T	T	0.2	T	T	0.0	0.0	0.2
1985-86	0.0	0.0	0.0	0.0	2.3	4.7							
Record Mean	0.0	0.0	T	T	0.3	1.2	4.2	0.4	0.6	T	T	T	6.6

See Reference Notes, relative to all above tables, on preceding page.

Medford is located in a mountain valley formed by the famous Rogue River and one of its tributaries, Bear Creek. The major portion of the valley ranges in elevation from 1,300 to 1,400 feet above sea level. Mountains surround the valley on all sides, to the east the Cascades, ranging up to 9,500 feet, to the south the Siskiyous, ranging up to 7,600 feet, and to the west and north, the Coast Range and Umpqua Divide, ranging up to 5,500 feet above sea level. The valley exits to the ocean 80 miles westward through the narrow canyon of the Rogue River.

Medford has a moderate climate of marked seasonal characteristics. Late fall, winter, and early spring months are damp, cloudy, and cool under the influence of marine air. Late spring, summer, and early fall are warm, dry, and sunny, due to the dry continental nature of the prevailing winds aloft that cross this area.

The rain shadow afforded by the Siskiyous and Coast Range results in a relatively light annual rainfall, most of which falls during the winter season. Summertime rainfall is brought by thunderstorm activity. Snowfall is quite heavy in the surrounding mountains during the winter, providing excellent skiing. The mountains provide irrigation water storage which is necessary for production of most commercial crops during the dry summer. Valley snowfall is light. Individual accumulations of snow seldom last more than 24 hours and present little hindrance to transportation on the valley floor.

Few extremes of temperature occur. High temperatures in the summer months average slightly below 90 degrees. High temperatures are always accompanied by low humidity, and hot days give way to cool nights as cool air drains down the mountain slopes into the valley. The length of the growing season is 170 days, from late April to mid-October. The last date of 32 degrees in the spring normally occurs in mid-June and the first date of 32 degrees in the fall occurs in mid-September.

Valley winds are usually very light, prevailing from the north or northwest much of the year. Winds exceeding 10 mph during the winter months nearly always come from the southerly quadrant. Highest velocities are reached when a well developed storm off the northern California coast causes a foehn or chinook wind off the Siskiyou Mountains to the south, speeds to 50 mph are common, and gusts to 70 mph have been recorded occasionally. Summer thunderstorms produce gusty winds to 40 or 50 mph which may come from any direction.

Fog often fills the lower portion of the valley during the winter and early spring months, when rapid clearing of the sky after a storm allows nocturnal cooling of the entrapped, moist air to the saturation point. Duration of the fog is seldom more than three days. Geographical and meteorological conditions contribute to a smoke problem during the fall, winter and early spring months. Smoke, from local sources, occasionally reduces visibility to 1 to 3 miles under stable conditions.

TABLE 1   NORMALS, MEANS AND EXTREMES

MEDFORD, OREGON

LATITUDE: 42°22'N   LONGITUDE: 122°52'W   ELEVATION: FT. GRND 1298 BARO 1304   TIME ZONE: PACIFIC   WBAN: 24225

	(a)	JAN	FEB	MAR	APR	MAY	JUNE	JULY	AUG	SEP	OCT	NOV	DEC	YEAR
TEMPERATURE °F:														
Normals														
-Daily Maximum		45.0	52.9	57.1	63.8	72.2	81.0	90.7	88.8	82.8	68.7	52.6	44.2	66.7
-Daily Minimum		30.2	31.9	33.9	36.8	42.7	49.3	54.2	53.4	47.4	39.6	34.5	31.2	40.4
-Monthly		37.6	42.4	45.5	50.3	57.5	65.2	72.5	71.1	65.1	54.2	43.5	37.7	53.5
Extremes														
-Record Highest	56	71	77	86	92	100	109	115	114	107	99	75	72	115
-Year		1981	1968	1930	1947	1941	1961	1946	1981	1955	1980	1970	1962	JUL 1946
-Record Lowest	56	-3	6	16	21	28	31	38	39	29	18	10	-6	-6
-Year		1930	1950	1956	1936	1968	1952	1976	1962	1950	1971	1978	1972	DEC 1972
NORMAL DEGREE DAYS:														
Heating (base 65°F)		849	633	605	441	245	85	6	19	92	335	642	846	4798
Cooling (base 65°F)		0	0	0	0	12	91	239	208	95	0	0	0	645
% OF POSSIBLE SUNSHINE														
MEAN SKY COVER (tenths)														
Sunrise - Sunset	56	8.2	7.7	7.3	6.7	5.7	4.7	2.1	2.5	3.3	5.6	7.8	8.6	5.9
MEAN NUMBER OF DAYS:														
Sunrise to Sunset														
-Clear	56	2.6	3.4	5.2	5.9	9.3	12.6	23.3	21.5	17.7	10.0	3.6	1.9	117.1
-Partly Cloudy	56	4.9	5.7	6.5	8.4	8.8	8.3	5.2	5.8	6.6	8.4	5.9	4.0	78.3
-Cloudy	56	23.5	19.1	19.3	15.6	12.8	9.2	2.6	3.7	5.7	12.6	20.6	25.1	169.9
Precipitation														
.01 inches or more	56	13.5	11.4	11.7	9.2	8.1	5.3	1.5	2.1	4.2	7.9	12.3	14.5	101.6
Snow, Ice pellets														
1.0 inches or more	56	1.2	0.5	0.3	0.1	0.0	0.0	0.0	0.0	0.0	0.*	0.1	0.5	2.7
Thunderstorms	56	0.*	0.2	0.2	0.8	1.7	1.8	1.5	1.2	0.9	0.2	0.*	0.*	8.6
Heavy Fog Visibility														
1/4 mile or less	56	11.8	5.8	1.6	0.4	0.3	0.2	0.*	0.1	0.5	4.6	10.5	13.5	49.4
Temperature °F														
-Maximum														
90° and above	24	0.0	0.0	0.0	0.0	1.7	7.5	18.3	16.5	8.1	0.8	0.0	0.0	53.0
32° and below	24	1.2	0.0	0.0	0.0	0.0	0.0	0.0	0.0	0.0	0.0	0.*	2.2	3.4
-Minimum														
32° and below	24	20.2	15.2	12.3	7.4	1.0	0.0	0.0	0.0	0.2	4.5	10.5	16.1	87.3
0° and below	24	0.*	0.0	0.0	0.0	0.0	0.0	0.0	0.0	0.0	0.0	0.0	0.2	0.2
AVG. STATION PRESS.(mb)	13	971.9	970.1	968.4	970.1	969.5	969.1	968.4	968.0	968.5	970.6	970.6	972.5	969.8
RELATIVE HUMIDITY (%)														
Hour 04	24	90	88	86	84	82	78	73	74	79	87	91	90	84
Hour 10	24	88	84	73	63	56	48	44	48	54	72	87	89	67
Hour 16 (Local Time)	24	72	59	50	45	39	33	26	28	31	46	69	76	48
Hour 22	24	87	82	75	69	65	59	50	53	61	78	87	88	71
PRECIPITATION (inches):														
Water Equivalent														
-Normal		3.42	2.12	1.85	1.07	1.19	0.67	0.25	0.46	0.75	1.68	2.89	3.49	19.84
-Maximum Monthly	56	6.67	5.67	5.54	3.07	4.58	3.49	1.63	2.83	4.22	9.16	8.62	12.72	12.72
-Year		1936	1983	1957	1965	1945	1931	1966	1976	1977	1950	1942	1964	DEC 1964
-Minimum Monthly	56	0.19	0.21	0.29	0.16	T	0.00	0.00	0.00	0.00	T	0.01	0.36	0.00
-Year		1984	1964	1969	1949	1982	1951	1970	1981	1974	1936	1936	1976	AUG 1981
-Maximum in 24 hrs	56	3.17	2.96	1.61	1.05	1.67	1.96	1.07	1.13	3.09	2.92	2.99	3.75	3.75
-Year		1943	1956	1972	1965	1956	1931	1966	1945	1977	1950	1953	1964	DEC 1964
Snow, Ice pellets														
-Maximum Monthly	56	22.6	11.6	8.1	4.2	T					1.3	11.4	12.2	22.6
-Year		1930	1956	1956	1953	1984					1956	1955	1972	JAN 1930
-Maximum in 24 hrs	56	9.3	5.2	7.9	4.2	T					1.3	8.5	4.2	9.3
-Year		1971	1956	1956	1953	1984					1956	1977	1964	JAN 1971
WIND:														
Mean Speed (mph)	36	4.1	4.6	5.3	5.7	5.7	5.9	5.8	5.3	4.5	3.7	3.6	3.6	4.8
Prevailing Direction														
through 1963		SSE	S	NNW	WNW	WNW	WNW	WNW	WNW	WNW	S	N	N	WNW
Fastest Obs. 1 Min.														
-Direction (!!)	36	23	25	16	14	12	18	07	16	13	20	19	14	16
-Speed (MPH)	36	50	46	55	35	38	33	44	48	40	40	40	44	55
-Year		1950	1958	1952	1965	1966	1958	1958	1956	1982	1962	1981	1965	MAR 1952
Peak Gust														
-Direction (!!)	2	N	W	SE	SW	SW	SW	NW	W	W	NW	SE	SE	SE
-Speed (mph)	2	17	32	38	37	39	37	25	23	36	24	52	56	56
-Date		1984	1985	1984	1984	1985	1985	1984	1985	1984	1984	1984	1985	DEC 1985

See Reference Notes to this table on the following page.

TABLE 2 PRECIPITATION (inches) MEDFORD, OREGON

YEAR	JAN	FEB	MAR	APR	MAY	JUNE	JULY	AUG	SEP	OCT	NOV	DEC	ANNUAL
1956	5.88	4.95	1.31	0.64	4.18	0.80	0.94	0.32	0.64	5.89	0.91	2.32	28.78
1957	1.70	2.99	5.54	0.36	1.10	0.03	T	0.80		1.64	2.28	3.92	20.52
1958	5.63	5.37	1.83	0.40	1.01	2.72	1.35	0.14	0.28	0.42	1.63	2.51	23.29
1959	1.99	2.78	0.88	0.59	1.40	0.27	0.00	0.28	0.29	0.61	0.16	1.17	10.42
1960	2.35	4.12	4.40	0.67	1.97	T	0.09	0.03	0.18	0.38	4.70	1.71	20.60
1961	1.12	2.74	3.05	0.96	1.86	0.34	0.10	0.15	0.93	2.38	3.42	2.60	19.65
1962	1.69	1.05	1.55	0.81	0.80	0.15	T	1.00	0.76	6.27	4.37	4.68	23.13
1963	1.75	2.47	0.88	2.25	2.23	0.92	0.15	0.26	0.26	1.40	5.25	1.05	18.87
1964	5.60	0.21	2.71	0.37	0.82	0.79	0.97	0.10	0.15	0.90	3.75	12.72	29.09
1965	4.30	0.70	0.41	3.07	0.31	1.05	0.03	1.52	T	0.46	2.56	3.71	18.12
1966	4.80	0.37	1.70	0.45	0.20	0.37	1.63	0.19	1.88	0.76	5.89	2.80	21.04
1967	5.44	1.14	2.08	1.72	0.96	0.27	T	T	0.28	2.34	1.04	3.40	18.67
1968	1.86	2.95	0.90	0.38	1.05	0.06	T	1.33	0.32	0.62	3.04	2.78	15.29
1969	6.16	1.46	0.29	0.60	1.62	1.31	0.02	0.00	0.62	2.46	0.49	5.44	20.47
1970	6.19	1.70	1.13	1.44	0.34	0.59	0.00	0.34	0.22	1.39	6.57	3.36	23.27
1971	3.68	1.43	2.72	1.34	1.13	0.97	0.07	0.28	1.24	0.61	3.43	2.45	19.35
1972	3.55	2.49	3.62	0.94	1.61	1.59	T	0.36	0.52	1.21	1.50	3.23	20.62
1973	1.98	0.54	1.58	0.76	0.45	0.06	0.04	0.03	0.64	2.79	7.01	3.02	18.90
1974	4.32	2.78	3.26	1.70	0.22	T	0.10	T	0.00	1.17	1.13	3.91	18.59
1975	2.64	2.64	3.97	1.27	0.24	0.38	0.22	0.54	0.65	2.21	1.85	2.74	19.35
1976	1.62	2.21	1.13	1.67	0.11	0.04	0.84	2.83	0.90	0.18	0.43	0.36	12.32
1977	1.17	0.67	1.12	0.81	2.37	0.53	0.23	0.36	4.22	0.96	4.91	4.81	22.16
1978	1.53	2.45	2.03	1.26	1.59	1.02	0.54	1.46	1.68	0.01	1.50	0.66	15.73
1979	2.81	1.54	0.83	2.24	1.42	0.55	0.02	0.63	0.32	3.98	3.17	2.73	20.24
1980	2.59	1.78	1.27	1.75	0.69	1.22	0.02	0.00	0.18	1.52	2.28	2.59	15.89
1981	0.54	1.72	1.23	0.55	1.17	0.47	0.41	0.00	0.52	1.23	6.05	8.02	21.91
1982	1.43	3.64	2.30	0.87	T	0.85	0.07	0.03	0.97	1.60	2.17	5.31	19.24
1983	0.92	5.67	3.21	1.12	0.81	0.66	0.59	2.21	2.05	1.21	4.97	6.73	30.15
1984	0.19	2.50	2.05	1.11	0.39	0.79	0.16	0.40	0.51	1.93	6.56	1.96	18.55
1985	0.23	1.58	1.22	0.39	1.00	0.37	T	0.02	1.53	1.50	2.02	0.83	10.69
Record Mean	2.75	2.12	1.67	1.18	1.16	0.83	0.25	0.32	0.73	1.66	2.88	3.24	18.79

TABLE 3 AVERAGE TEMPERATURE (deg. F) MEDFORD, OREGON

YEAR	JAN	FEB	MAR	APR	MAY	JUNE	JULY	AUG	SEP	OCT	NOV	DEC	ANNUAL
1956	40.6	37.3	44.0	53.5	59.1	61.8	73.2	69.3	64.0	51.3	40.1	34.4	52.4
1957	33.9	44.2	46.8	52.1	59.7	67.3	69.3	67.5	67.4	51.8	40.7	38.2	53.3
1958	39.3	47.7	43.9	50.8	63.6	66.0	75.3	75.8	63.8	57.7	45.6	43.1	56.1
1959	41.3	41.7	46.4	53.8	55.3	65.6	74.9	70.3	62.4	56.0	43.4	34.6	53.8
1960	39.5	42.3	49.2	51.1	55.1	68.2	75.5	70.0	66.9	54.7	44.4	37.7	54.5
#1961	40.3	44.8	46.2	51.1	55.6	69.2	73.0	75.1	60.6	52.9	41.6	35.6	53.8
1962	31.5	40.7	44.6	53.6	54.3	63.5	70.9	68.8	65.4	51.3	40.6	38.1	52.2
1963	30.8	46.1	43.9	46.1	57.0	63.1	66.7	70.4	71.1	55.2	42.6	35.9	52.4
1964	35.2	37.9	43.6	47.7	54.9	63.3	70.9	69.6	61.1	55.9	41.1	41.5	51.9
1965	39.0	42.7	48.7	53.4	56.5	64.5	72.9	70.5	62.2	57.4	46.6	35.0	54.1
1966	38.2	39.9	46.8	53.3	59.4	63.8	68.7	71.3	65.2	53.5	45.7	40.6	53.9
1967	39.6	40.5	44.4	45.8	58.9	68.4	75.5	78.0	69.2	53.6	45.5	35.3	54.6
1968	36.6	47.6	48.1	49.3	57.9	68.8	74.1	69.8	68.2	56.7	44.8	38.2	55.0
1969	35.7	40.9	45.5	49.5	62.4	67.9	72.3	69.9	64.9	49.6	42.5	39.7	53.4
1970	42.5	44.0	47.0	46.2	57.9	68.7	75.8	73.7	62.7	53.1	47.4	38.1	54.8
1971	36.1	40.8	44.9	48.4	55.8	61.4	73.1	74.1	62.6	51.1	41.5	36.1	52.2
1972	36.0	43.2	49.8	48.0	57.9	65.4	74.9	73.0	60.8	52.4	44.8	31.5	53.2
1973	37.9	46.2	45.0	52.1	62.4	67.5	75.2	72.0	66.9	53.4	45.8	43.6	55.7
1974	38.7	39.7	44.8	47.9	56.4	67.7	71.3	73.0	59.7	54.0	44.4	38.9	53.9
1975	36.6	41.8	43.3	45.1	57.4	63.9	72.3	68.4	68.9	51.5	41.3	39.3	52.5
1976	38.8	40.2	42.8	48.0	57.3	61.7	71.5	67.3	65.5	54.1	43.6	31.7	51.9
1977	34.0	44.0	43.0	53.0	53.3	69.7	72.0	76.0	63.8	53.1	42.3	42.4	53.6
1978	44.0	44.5	50.5	49.7	56.4	66.4	72.6	70.9	59.1	56.9	39.2	33.4	53.6
1979	35.5	41.9	49.2	51.1	57.6	65.9	72.6	70.3	68.3	59.1	42.4	39.8	54.5
1980	38.7	45.6	44.8	52.2	56.3	61.3	73.1	68.9	65.4	55.8	43.8	38.1	53.7
1981	41.7	42.1	46.9	52.8	57.2	65.5	71.2	74.9	66.6	52.2	45.9	42.8	55.0
1982	34.6	41.2	44.5	48.8	58.0	67.1	71.6	71.9	62.8	54.9	42.1	38.9	53.0
1983	39.7	41.6	48.9	49.1	59.6	64.2	68.6	73.0	63.6	54.3	43.7	40.4	54.2
1984	38.8	43.0	48.9	49.3	58.3	63.3	73.8	71.8	65.2	51.5	44.5	36.3	53.7
1985	37.6	42.1	44.1	56.2	57.9	68.2	76.4	71.0	61.0	53.1	38.1	36.1	53.5
Record Mean	37.6	42.4	46.4	51.4	58.0	65.1	72.2	71.2	64.5	54.0	43.8	38.2	53.7
Max	45.1	52.6	58.2	64.8	72.7	80.8	90.1	89.1	82.0	68.4	53.3	44.8	66.8
Min	30.1	32.2	34.5	37.9	43.2	49.4	54.2	53.3	47.0	39.6	34.2	31.5	40.6

REFERENCE NOTES FOR TABLES 1, 2, 3 and 6   (MEDFORD, OR)

GENERAL

T - TRACE AMOUNT
BLANK ENTRIES DENOTE MISSING/UNREPORTED DATA.
# INDICATES A STATION OR INSTRUMENT RELOCATION.

SPECIFIC

TABLE 1

(a) - LENGTH OF RECORD IN YEARS. ALTHOUGH
      INDIVIDUAL MONTHS MAY BE MISSING.
* LESS THAN .05

NORMALS — BASED ON THE 1951-1980 RECORD PERIOD.
EXTREMES — DATES ARE THE MOST RECENT OCCURRENCE.
WIND DIR. — NUMERALS SHOW TENS OF DEGREES
            CLOCKWISE FROM TRUE NORTH.
            "00" INDICATES CALM.
RESULTANT WIND DIRECTIONS ARE GIVEN TO WHOLE DEGREES.

EXCEPTIONS

TABLES 2, 3, and 6

RECORD MEANS ARE THROUGH THE CURRENT YEAR,
BEGINNING IN   1912 FOR TEMPERATURE
                  1912 FOR PRECIPITATION
                  1930 FOR SNOWFALL

TABLE 4 HEATING DEGREE DAYS Base 65 deg. F MEDFORD, OREGON

SEASON	JULY	AUG	SEP	OCT	NOV	DEC	JAN	FEB	MAR	APR	MAY	JUNE	TOTAL
1956-57	9	21	69	421	739	945	957	578	557	386	188	27	4897
1957-58	6	12	44	405	723	823	790	477	646	420	98	66	4510
1958-59	1	0	90	234	576	671	727	647	570	333	302	63	4214
1959-60	3	18	134	274	642	934	786	653	482	410	299	23	4658
#1960-61	0	35	39	318	612	839	759	558	575	411	289	35	4470
1961-62	9	1	144	369	696	905	1032	671	625	337	325	88	5202
1962-63	13	11	57	417	622	827	1051	524	645	560	255	122	5104
1963-64	20	7	6	306	665	893	915	777	655	513	314	86	5157
1964-65	8	23	120	284	710	723	802	621	498	346	267	78	4480
1965-66	4	7	106	231	544	923	823	696	557	345	182	93	4511
1966-67	26	11	53	352	571	749	781	678	634	569	206	50	4680
1967-68	0	0	10	344	578	913	873	500	515	466	218	48	4465
1968-69	0	31	56	252	599	826	900	670	599	459	133	37	4562
1969-70	0	9	56	471	670	780	691	583	549	558	227	68	4674
1970-71	0	0	89	363	520	827	890	674	617	489	280	121	4870
1971-72	16	12	108	424	698	889	891	624	465	505	237	54	4923
1972-73	2	11	178	385	599	1031	835	519	616	379	139	48	4742
1973-74	0	12	28	352	569	654	809	703	620	504	267	30	4548
1974-75	20	6	9	333	611	803	873	642	663	590	249	86	4885
1975-76	18	18	14	416	706	789	807	712	679	503	238	127	5027
1976-77	9	31	48	338	636	1023	952	581	675	358	262	17	5027
1977-78	14	4	105	360	673	694	646	571	444	451	262	44	4268
1978-79	4	15	180	244	768	973	905	639	483	410	223	58	4902
1979-80	8	2	5	203	668	776	807	556	620	376	266	121	4408
1980-81	1	9	56	313	630	827	715	635	553	369	244	57	4409
1981-82	20	0	89	392	566	682	935	660	628	479	223	63	4737
1982-83	13	8	114	310	682	803	775	566	491	474	235	68	4539
1983-84	26	3	77	322	632	757	806	633	491	467	219	108	4541
1984-85	0	0	79	416	608	881	842	635	642	257	225	37	4622
1985-86	0	4	135	368	801	890							

TABLE 5 COOLING DEGREE DAYS Base 65 deg. F MEDFORD, OREGON

YEAR	JAN	FEB	MAR	APR	MAY	JUNE	JULY	AUG	SEP	OCT	NOV	DEC	TOTAL
1969	0	0	0	0	58	131	235	168	70	0	0	0	662
1970	0	0	0	0	15	185	343	278	28	3	0	0	852
1971	0	0	0	0	2	16	275	302	44	2	0	0	641
1972	0	0	0	0	27	70	315	265	58	0	0	0	735
1973	0	0	0	0	64	132	323	233	92	0	0	0	844
1974	0	0	0	0	7	120	224	262	155	0	0	0	768
1975	0	0	0	0	21	60	250	129	140	2	0	0	602
1976	0	0	0	0	6	34	215	110	66	5	0	0	436
1977	0	0	0	2	2	165	239	352	77	0	0	0	837
1978	0	0	0	0	4	93	244	205	11	0	0	0	557
1979	0	0	0	0	2	95	251	172	108	30	0	0	658
1980	0	0	0	0	2	16	257	134	78	35	0	0	522
1981	0	0	0	11	8	78	216	313	144	0	0	0	770
1982	0	0	0	0	14	136	223	228	56	1	0	0	658
1983	0	0	0	0	74	49	144	259	43	0	0	0	569
1984	0	0	0	0	20	64	280	219	90	5	0	0	678
1985	0	0	0	0	10	141	360	196	21	5	0	0	733

TABLE 6 SNOWFALL (inches) MEDFORD, OREGON

SEASON	JULY	AUG	SEP	OCT	NOV	DEC	JAN	FEB	MAR	APR	MAY	JUNE	TOTAL
1956-57	0.0	0.0	0.0	1.3	T	2.1	2.2	0.0	T	0.0	0.0	0.0	5.6
1957-58	0.0	0.0	0.0	0.0	0.0	1.5	T	T	T	0.0	0.0	0.0	1.5
1958-59	0.0	0.0	0.0	0.0	0.8	0.0	T	0.3	0.0	0.0	0.0	0.0	1.1
1959-60	0.0	0.0	0.0	0.0	0.0	2.2	1.4	T	T	T	0.0	0.0	3.6
1960-61	0.0	0.0	0.0	0.0	3.5	T	T	T	T	3.2	0.0	0.0	6.7
1961-62	0.0	0.0	0.0	0.0	1.7	3.6	8.8	T	4.7	0.6	0.0	0.0	18.8
1962-63	0.0	0.0	0.0	0.0	0.0	0.0	1.2	0.0	0.8	T	0.0	0.0	1.8
1963-64	0.0	0.0	0.0	0.0	0.8	T	9.3	13.2	T	T	T	0.0	23.3
1964-65	0.0	0.0	0.0	0.0	T	9.0	3.4	T	2.0	0.0	0.0	0.0	14.4
1965-66	0.0	0.0	0.0	0.0	T							0.0	6.5
1966-67	0.0	0.0	0.0	0.0	T	T	1.1	3.4	0.4	1.6	T	0.0	6.5
1967-68	0.0	0.0	0.0	0.0	0.0	4.3	6.4	0.0	0.0	0.0	0.0	0.0	10.7
1968-69	0.0	0.0	0.0	0.0	T	2.0	13.7	3.4	T	0.0	0.0	0.0	19.1
1969-70	0.0	0.0	0.0	0.0	T	0.4	0.4	1.5	0.8	T	0.0	0.0	2.9
1970-71	0.0	0.0	0.0	0.0	T	0.1	10.5	2.3	5.0	T	0.0	0.0	17.9
1971-72	0.0	0.0	0.0	T	T	2.8	0.3	0.9	2.6	T	T	0.0	6.6
1972-73	0.0	0.0	0.0	0.0	T	12.2	2.7	T	0.0	T	T	0.0	14.9
1973-74	0.0	0.0	0.0	0.0	0.0	T	T	T	5.2	0.3	T	0.0	5.5
1974-75	0.0	0.0	0.0	0.0	0.0	5.4	6.6	1.9	T	0.4	T	0.0	14.3
1975-76	0.0	0.0	0.0	0.0	0.4	2.3	2.1	1.0	0.2	T	T	0.0	6.0
1976-77	0.0	0.0	0.0	0.0	0.0	0.1	1.0	T	0.0	T	0.0	0.0	1.1
1977-78	0.0	0.0	0.0	0.0	8.5	0.2	0.0	0.1	0.0	T	0.0	0.0	8.8
1978-79	0.0	0.0	0.0	0.0	T	T	0.6	T	T	T	0.0	0.0	0.6
1979-80	0.0	0.0	0.0	0.0	T	0.1	0.3	T	0.1	T	0.0	0.0	0.5
1980-81	0.0	0.0	0.0	0.0	0.0	0.4	T	0.0	T	T	0.0	0.0	0.4
1981-82	0.0	0.0	0.0	0.0	0.0	3.0	4.1	T	1.8	T	0.0	0.0	8.9
1982-83	0.0	0.0	0.0	0.0	T	T	T	0.0	0.0	T	0.0	0.0	T
1983-84	0.0	0.0	0.0	T	T	3.4	T	T	0.0	T	0.0	0.0	3.4
1984-85	0.0	0.0	0.0	T	0.1	0.9	1.8	0.4	2.0	T	0.0	0.0	5.2
1985-86	0.0	0.0	0.0	0.0	0.5	T							
Record Mean	0.0	0.0	0.0	T	0.5	1.4	3.5	1.3	0.8	0.2	T	0.0	7.8

See Reference Notes, relative to all above tables, on preceding page.

Pendleton is located in the southeastern part of the Columbia Basin, that low country of northern Oregon and central and eastern Washington which is almost entirely surrounded by mountains. This Basin is bounded on the south by the high country of central Oregon, on the north by the mountains of western Canada, on the west by the Cascade Range and on the east by the Blue Mountains and the north Idaho plateau. The gorge in the Cascades through which the Columbia River reaches the Pacific is the most important break in the barriers surrounding this basin. These physical features have important influences on the general climate of Pendleton and the surrounding territory.

The Weather Service Office at Pendleton Airport is located in rolling country which slopes generally upward toward the Blue Mountains about 15 miles to the east and southeast. The Columbia River approaches the area from the northwest to its junction with the Walla Walla River at an elevation of 351 feet and some 25 miles north of Pendleton, then turns southwestward to be joined a few miles below by the Umatilla River. Both the Walla Walla and Umatilla Rivers have their sources in the Blue Mountains and flow westward to the Columbia. The observation station is at an elevation of nearly 1,500 feet, about 3 miles northwest of downtown Pendleton. The city of Pendleton lies in the shallow east-west valley of the Umatilla River, approximately 400 feet lower than the airport.

Precipitation in the Pendleton area is definitely seasonal in occurrence with an average of only 10 percent of the annual total occurring in the three-month period, July-September. Most precipitation reaching this area accompanies cyclonic storms moving in from the Pacific Ocean. These storms reach their greatest intensity and frequency from October through April. The Cascade Range west of the Columbia Basin reduces the amount of precipitation received from the Pacific cyclonic storms. This influence is felt, particularly, in the desert area of the central part of the Basin. A gradual rise in elevation from the Columbia River to the foothills of the Blue Mountains again results in increased precipitation. This increase supplies sufficient moisture for productive wheat, pea, and stock raising activity in the area surrounding Pendleton.

The lighter summertime precipitation usually accompanies thunderstorms which often move into the area from the south or southwest. On occasion, these storms are quite intense, causing flash flooding with resultant heavy property damage and even loss of life.

Seasonal temperature extremes are usually quite moderate for the latitude. The last occurrence in spring of temperatures as low as 32 degrees is mid-April, and the average last occurrence in the fall of 32 degrees is late October. At the city station, where cool air settles in the valley on still nights, temperatures of 32 degrees have been recorded later in the spring and earlier in the fall. Under usual atmospheric conditions, air from the Pacific, with moderate temperature characteristics, moves across the Cascades or through the Columbia Gorge resulting in mild temperatures in the Pendleton area. When this flow of air from the west is impeded by slow-moving high pressure systems over the interior of the continent, temperature conditions sometimes become rather severe, hot in summer and cold in winter. During the summer or early fall, if a stagnant high predominates to the north or east of Pendleton, the hot, dry conditions may prove detrimental to crops during late May and June, and cause fire danger in the forest and grassland areas during late summer and early fall. During winter, coldest temperatures occur when air from a cold high pressure system in central Canada moves southwestward across the Rockies and flows down into the Columbia Basin. Under this condition the heavy cold air sometimes remains at low levels in the Basin for several days while warmer air from the Pacific flows above it, causing comparatively mild temperatures at higher elevations. Extreme winter temperatures are not particularly common in the Pendleton area. Below zero readings are recorded in approximately 60 percent of winters. Maximum temperatures usually reach 100 degrees or slightly higher on a few days during the summer.

TABLE 1   NORMALS, MEANS AND EXTREMES

PENDLETON, OREGON

LATITUDE: 45°41'N   LONGITUDE: 118°51'W   ELEVATION: FT. GRND 1482 BARO 01507   TIME ZONE: PACIFIC   WBAN: 24155

	(a)	JAN	FEB	MAR	APR	MAY	JUNE	JULY	AUG	SEP	OCT	NOV	DEC	YEAR
TEMPERATURE °F:														
Normals														
-Daily Maximum		39.4	46.9	53.4	61.4	70.6	79.6	88.9	85.9	77.1	63.7	48.7	42.5	63.2
-Daily Minimum		26.3	31.8	34.4	39.2	46.1	52.9	58.6	57.5	50.5	41.3	33.4	29.5	41.8
-Monthly		32.9	39.4	43.9	50.3	58.4	66.3	73.8	71.7	63.8	52.5	41.0	36.0	52.5
Extremes														
-Record Highest	50	68	69	79	91	99	108	110	113	102	92	77	67	113
-Year		1974	1982	1964	1977	1936	1961	1939	1961	1955	1980	1975	1980	AUG 1961
-Record Lowest	50	-22	-18	10	18	25	36	42	40	30	11	-12	-19	-22
-Year		1957	1950	1955	1936	1954	1966	1971	1980	1970	1935	1985	1983	JAN 1957
NORMAL DEGREE DAYS:														
Heating (base 65°F)		998	717	654	441	220	75	7	27	120	388	717	899	5263
Cooling (base 65°F)		0	0	0	0	16	111	280	235	84	0	0	0	726
% OF POSSIBLE SUNSHINE														
MEAN SKY COVER (tenths)														
Sunrise - Sunset	40	8.4	8.1	7.3	6.8	6.1	5.4	2.9	3.4	4.2	5.8	7.8	8.3	6.2
MEAN NUMBER OF DAYS:														
Sunrise to Sunset														
-Clear	50	2.4	2.6	4.9	5.5	7.6	9.7	19.9	18.2	14.7	9.9	3.6	2.6	101.8
-Partly Cloudy	50	5.3	5.5	7.4	9.3	10.7	10.1	7.5	7.7	8.0	7.8	6.4	4.6	90.2
-Cloudy	50	23.3	20.1	18.7	15.2	12.7	10.2	3.6	5.1	7.3	13.3	19.9	23.8	173.3
Precipitation														
.01 inches or more	50	12.5	11.0	10.7	9.0	7.8	6.7	2.6	3.2	4.5	7.6	11.3	13.0	99.9
Snow,Ice pellets														
1.0 inches or more	50	2.7	1.0	0.3	0.1	0.0	0.0	0.0	0.0	0.0	0.*	0.6	1.4	6.2
Thunderstorms	48	0.0	0.*	0.2	0.7	1.8	1.9	1.8	2.1	1.1	0.3	0.1	0.*	10.0
Heavy Fog Visibility														
1/4 mile or less	48	7.2	4.6	1.7	0.3	0.3	0.1	0.0	0.*	0.3	1.1	6.1	8.1	29.8
Temperature °F														
-Maximum														
90° and above	50	0.0	0.0	0.0	0.*	0.8	4.5	14.7	10.5	2.7	0.*	0.0	0.0	33.3
32° and below	50	9.7	2.7	0.2	0.0	0.0	0.0	0.0	0.0	0.0	0.0	2.2	7.0	21.8
-Minimum														
32° and below	50	21.3	15.6	9.7	2.6	0.1	0.0	0.0	0.0	0.1	2.5	12.8	18.8	83.5
0° and below	50	1.9	0.6	0.0	0.0	0.0	0.0	0.0	0.0	0.0	0.0	0.1	0.5	3.1
AVG. STATION PRESS.(mb)	12	966.5	964.2	961.0	962.7	962.1	961.7	961.8	961.3	963.0	964.7	963.8	966.2	963.3
RELATIVE HUMIDITY (%)														
Hour 04	44	81	79	73	71	69	66	54	54	62	73	80	81	70
Hour 10 (Local Time)	46	77	71	59	51	47	42	34	37	43	56	73	78	56
Hour 16	46	75	65	49	41	37	32	23	26	33	47	70	78	48
Hour 22	43	80	77	69	63	58	52	38	41	51	67	78	81	63
PRECIPITATION (inches):														
Water Equivalent														
-Normal		1.73	1.11	1.06	0.99	1.09	0.70	0.30	0.55	0.58	0.95	1.48	1.66	12.20
-Maximum Monthly	50	3.92	3.03	2.82	2.78	3.02	2.70	1.26	2.58	2.34	2.79	3.76	4.68	4.68
-Year		1970	1940	1983	1978	1962	1947	1948	1977	1941	1947	1973	1973	DEC 1973
-Minimum Monthly	50	0.21	0.07	0.24	0.01	0.03	0.12	T	0.00	T	T	0.04	0.27	0.00
-Year		1949	1964	1941	1956	1964	1940	1967	1969	1974	1978	1939	1965	AUG 1969
-Maximum in 24 hrs	50	1.29	1.09	1.33	1.04	1.52	1.49	1.19	1.48	1.23	1.88	1.35	1.25	1.88
-Year		1956	1959	1983	1978	1972	1947	1948	1977	1981	1982	1971	1978	OCT 1982
Snow,Ice pellets														
-Maximum Monthly	50	41.6	15.8	4.9	2.2	T					3.2	14.9	26.6	41.6
-Year		1950	1936	1971	1975	1978					1973	1985	1983	JAN 1950
-Maximum in 24 hrs	50	13.3	9.7	4.0	2.2	T					3.2	8.0	9.9	13.3
-Year		1950	1949	1970	1975	1978					1973	1977	1948	JAN 1950
WIND:														
Mean Speed (mph)	32	8.1	8.7	9.7	10.3	9.9	10.1	9.2	8.9	8.7	7.9	7.8	8.1	9.0
Prevailing Direction														
through 1963		SE	SE	W	W	W	W	WNW	SE	SE	SE	SE	SE	SE
Fastest Obs. 1 Min.														
-Direction	30	27	25	29	27	27	29	28	27	27	25	27	29	27
-Speed (MPH)	30	49	54	63	77	48	62	46	40	47	49	62	63	77
-Year		1962	1955	1956	1960	1959	1956	1968	1961	1954	1959	1959	1959	APR 1960
Peak Gust														
-Direction	2	W	W	W	W	W	W	SW	W	W	W	SW	W	W
-Speed (mph)	2	40	47	63	47	49	47	44	46	56	47	52	40	63
-Date		1984	1985	1984	1985	1985	1984	1985	1984	1984	1985	1984	1984	MAR 1984

See reference Notes to this table on the following page.

TABLE 2 PRECIPITATION (inches) PENDLETON, OREGON

YEAR	JAN	FEB	MAR	APR	MAY	JUNE	JULY	AUG	SEP	OCT	NOV	DEC	ANNUAL
1956	2.97	0.81	0.54	0.01	2.82	0.56	0.27	1.00	T	1.39	0.34	1.27	11.98
1957	1.81	0.88	2.31	0.90	2.21	0.67	0.11	0.01	0.71	1.79	1.10	1.85	14.35
1958	2.26	1.58	1.41	2.45	1.09	0.59	0.02	0.02	0.35	0.05	1.88	2.53	14.23
1959	2.45	1.90	0.92	0.51	0.79	1.19	0.41	0.42	2.03	0.65	0.31	0.62	12.20
1960	1.10	0.99	1.95	0.80	1.57	0.52	0.02	0.72	0.49	1.22	1.70	0.63	11.71
1961	0.47	2.46	2.25	1.30	0.94	0.28	0.08	0.09	0.17	0.70	1.46	1.27	11.47
1962	0.70	0.72	1.14	0.76	3.02	0.15	T	0.43	0.79	1.62	1.46	1.46	12.17
1963	1.40	1.67	0.37	1.86	0.65	0.21	0.32	0.30	0.70	0.44	2.03	1.29	11.24
1964	1.07	0.07	0.66	0.34	0.03	0.64	0.64	0.21	0.15	0.80	1.93	3.23	10.14
1965	3.08	0.37	0.29	0.65	0.57	1.10	0.51	1.21	0.23	0.19	1.95	0.27	10.42
1966	2.19	0.83	0.96	0.08	0.07	0.55	0.79	0.17	0.43	0.75	2.09	2.65	11.56
1967	1.59	0.15	0.89	1.05	0.56	0.41	T	T	0.40	0.64	0.63	0.45	6.77
1968	0.59	1.82	0.47	0.17	0.66	0.89	0.17	0.61	0.57	1.03	2.06	2.19	11.23
1969	2.88	0.88	0.57	2.05	1.40	0.86	0.02	0.00	0.42	1.13	0.36	1.88	12.45
1970	3.92	1.48	0.99	0.63	0.32	0.57	0.08	0.03	0.78	0.81	1.78	0.80	12.19
1971	0.84	0.69	1.11	1.15	1.41	1.73	0.32	0.14	1.03	0.70	2.73	2.59	14.44
1972	0.96	1.08	1.47	0.68	1.97	0.80	0.58	0.36	0.16	0.58	0.70	2.31	11.65
1973	0.50	0.19	0.43	0.27	0.67	0.15	0.01	0.08	1.34	1.71	3.76	4.68	14.69
1974	0.79	1.57	0.81	2.13	0.26	0.19	0.90	T	T	0.29	1.00	1.59	9.53
1975	3.53	1.30	0.65	0.97	0.30	0.28	0.73	0.67	0.00	1.80	0.84	1.98	13.05
1976	1.77	1.00	1.65	1.09	0.92	0.33	0.16	1.77	0.18	0.54	0.19	0.44	10.04
1977	0.48	0.64	1.51	0.18	1.87	0.37	0.06	2.58	1.17	0.51	2.00	2.42	13.79
1978	2.82	1.60	1.03	2.78	0.63	0.76	0.77	2.21	0.92	T	2.37	1.86	17.75
1979	1.43	1.72	1.18	1.17	0.39	0.21	0.09	1.40	0.30	1.68	1.83	0.62	12.02
1980	2.48	1.39	1.60	0.59	2.14	1.12	0.77	0.03	0.59	1.22	0.84	1.20	13.97
1981	0.89	1.35	1.43	1.20	1.59	1.53	0.94	0.03	1.31	0.86	1.91	2.31	15.35
1982	1.54	0.77	1.22	0.84	0.31	0.63	0.51	0.24	1.47	2.67	0.34	2.20	12.74
1983	0.86	1.57	2.82	0.70	0.73	1.44	0.52	0.56	0.46	0.84	1.67	3.42	15.59
1984	0.53	1.74	1.83	1.70	1.02	1.13	0.06	0.44	0.39	1.02	2.14	0.92	12.92
1985	0.44	1.33	1.13	0.37	0.44	0.69	0.34	0.26	2.10	0.89	2.11	1.27	11.37
Record Mean	1.57	1.32	1.22	1.04	1.10	0.97	0.33	0.45	0.73	1.10	1.53	1.64	13.01

TABLE 3 AVERAGE TEMPERATURE (deg. F) PENDLETON, OREGON

YEAR	JAN	FEB	MAR	APR	MAY	JUNE	JULY	AUG	SEP	OCT	NOV	DEC	ANNUAL
1956	34.0	28.4	43.5	54.0	61.2	63.2	75.9	71.5	65.3	50.9	37.4	36.7	51.8
1957	19.0	36.1	43.7	52.7	61.5	67.9	71.1	69.0	66.3	49.1	41.0	41.3	51.5
1958	37.9	46.8	42.8	49.9	64.9	69.8	77.6	76.9	63.5	54.7	41.9	38.4	55.4
1959	36.0	37.7	45.1	51.7	55.4	66.2	74.6	69.7	61.0	53.5	37.6	33.6	51.9
#1960	25.4	39.0	44.0	51.7	55.9	67.4	78.3	69.1	64.3	53.8	42.5	30.6	51.8
1961	37.1	45.5	46.3	50.2	57.4	71.6	75.8	76.9	59.5	50.0	36.3	36.9	53.7
1962	32.3	39.8	43.5	53.8	54.0	65.1	72.9	69.8	65.0	52.7	45.2	39.3	52.8
1963	27.0	42.8	45.4	48.0	58.9	65.8	68.6	71.7	67.6	55.5	44.4	32.7	52.4
1964	40.3	41.1	43.5	48.5	56.8	64.9	72.2	68.4	62.7	51.9	40.1	32.9	51.9
1965	35.4	41.1	40.6	52.4	58.0	66.0	73.8	71.5	60.1	57.5	44.7	36.9	53.2
1966	38.3	39.7	45.4	52.0	59.9	64.6	71.0	72.4	66.6	54.0	45.1	41.1	54.2
1967	42.4	42.5	43.5	45.8	58.5	58.7	76.4	79.7	69.9	54.9	41.1	36.8	55.1
1968	37.3	42.8	48.5	49.6	59.7	67.5	77.2	70.9	65.9	50.8	42.6	33.3	53.8
1969	22.0	35.5	44.6	49.8	61.7	69.8	72.5	70.1	64.7	49.3	43.2	34.9	51.5
1970	32.3	39.5	43.5	45.8	57.8	68.7	74.8	72.5	56.3	47.4	40.4	35.7	51.2
1971	40.0	39.8	40.5	49.4	60.5	63.3	76.1	76.8	59.1	51.4	43.7	36.9	53.1
1972	34.0	37.4	47.8	47.6	60.9	68.3	74.7	76.2	61.1	51.1	42.6	27.1	52.4
1973	31.3	38.4	45.8	50.3	61.3	66.6	75.3	71.7	64.0	52.8	42.6	41.5	53.5
1974	30.4	43.8	46.4	51.7	57.3	71.1	73.3	75.5	67.5	54.8	44.8	40.6	54.8
1975	37.1	39.0	45.2	47.5	59.3	65.8	78.4	70.1	67.0	54.3	42.3	40.5	53.9
1976	39.2	37.9	42.8	50.2	58.8	63.7	73.4	67.7	67.3	53.1	42.7	35.9	52.7
1977	26.3	41.5	44.2	55.3	55.1	69.1	70.3	74.9	58.5	50.0	38.3	34.8	51.5
1978	32.2	39.3	45.7	48.0	54.4	66.3	72.2	69.4	60.5	51.7	33.5	29.5	50.2
1979	15.3	37.7	46.0	50.4	59.5	66.6	72.8	70.6	65.5	54.3	34.7	38.2	51.0
1980	25.6	36.1	41.3	51.9	56.4	60.4	72.1	66.9	63.3	51.3	42.0	39.2	50.6
1981	36.2	38.9	45.7	50.4	56.0	61.6	69.2	74.3	63.8	50.6	44.2	37.2	52.3
1982	35.0	38.1	43.5	47.6	56.8	67.6	71.1	71.5	60.7	50.7	37.3	35.7	51.3
1983	40.8	43.8	47.8	49.0	58.9	62.7	68.4	72.7	58.9	52.5	45.9	23.2	52.1
1984	34.6	39.7	46.8	48.2	54.7	62.1	72.9	72.2	60.4	49.1	41.8	30.4	51.1
1985	26.3	33.5	43.2	53.1	58.5	65.6	77.4	68.1	57.0	50.3	26.5	19.5	48.3
Record Mean	32.3	38.0	44.9	51.3	58.5	65.6	73.2	71.4	62.9	52.4	41.2	35.3	52.3
Max	39.3	46.1	55.1	63.5	71.7	79.6	89.7	87.4	77.3	64.6	49.5	41.9	63.8
Min	25.3	29.8	34.7	39.0	45.2	51.5	56.7	55.4	48.5	40.2	32.8	28.7	40.7

REFERENCE NOTES FOR TABLES 1, 2, 3 and 6   (PENDLETON, OR)

GENERAL

T - TRACE AMOUNT
BLANK ENTRIES DENOTE MISSING/UNREPORTED DATA.
# INDICATES A STATION OR INSTRUMENT RELOCATION.

SPECIFIC

TABLE 1

(a) - LENGTH OF RECORD IN YEARS. ALTHOUGH INDIVIDUAL MONTHS MAY BE MISSING.
* LESS THAN .05

NORMALS — BASED ON THE 1951-1980 RECORD PERIOD.
EXTREMES — DATES ARE THE MOST RECENT OCCURRENCE.
WIND DIR. — NUMERALS SHOW TENS OF DEGREES CLOCKWISE FROM TRUE NORTH. "00" INDICATES CALM.
RESULTANT WIND DIRECTIONS ARE GIVEN TO WHOLE DEGREES.

EXCEPTIONS

TABLES 2, 3, and 6

RECORD MEANS ARE THROUGH THE CURRENT YEAR, BEGINNING IN   1900 FOR TEMPERATURE
1900 FOR PRECIPITATION
1936 FOR SNOWFALL

TABLE 4 HEATING DEGREE DAYS Base 65 deg. F   PENDLETON, OREGON

SEASON	JULY	AUG	SEP	OCT	NOV	DEC	JAN	FEB	MAR	APR	MAY	JUNE	TOTAL
1956-57	1	22	63	429	824	872	1424	804	651	372	132	28	5622
1957-58	0	4	61	483	712	729	832	502	682	444	91	20	4560
1958-59	1	0	100	326	686	814	891	757	609	393	295	53	4925
1959-60	11	13	149	351	816	966	1221	748	643	392	284	40	5634
#1960-61	0	43	72	347	670	1061	860	540	573	434	237	25	4862
1961-62	0	0	167	458	855	863	1006	699	660	333	334	86	5461
1962-63	15	16	67	375	585	789	1170	613	598	504	205	61	4998
1963-64	8	9	57	308	611	993	760	688	662	487	257	70	4910
1964-65	6	30	92	397	742	990	912	664	750	357	223	44	5207
1965-66	15	19	153	226	602	865	820	702	597	382	194	75	4650
1966-67	19	10	48	333	590	736	691	621	659	571	218	21	4517
1967-68	0	0	24	306	711	866	850	638	505	462	174	46	4582
1968-69	0	15	73	434	664	977	1327	820	623	450	140	32	5555
1969-70	0	11	79	480	646	926	1007	707	660	568	228	83	5395
1970-71	0	1	260	540	731	903	767	698	755	460	169	95	5379
1971-72	11	9	182	428	633	868	955	793	528	515	171	29	5122
1972-73	5	4	165	422	663	1170	1036	738	588	434	169	73	5467
1973-74	1	16	97	372	666	721	1064	589	573	391	241	29	4760
1974-75	8	0	39	313	600	750	857	721	609	517	194	57	4665
1975-76	0	12	43	332	673	751	791	782	679	436	206	89	4794
1976-77	4	42	31	363	660	896	1192	653	639	299	301	26	5106
1977-78	20	35	200	461	792	927	1011	714	593	504	322	46	5625
1978-79	7	41	146	403	936	1094	1533	757	582	432	184	62	6177
1979-80	12	0	43	326	902	823	1210	829	728	388	267	141	5669
1980-81	4	33	88	438	681	794	886	724	593	435	275	126	5077
1981-82	20	1	128	440	617	855	919	747	662	515	256	72	5232
1982-83	22	7	171	435	825	901	741	588	528	470	242	95	5025
1983-84	42	1	180	381	569	1292	935	729	558	496	316	134	5633
1984-85	4	0	182	490	692	1065	1196	876	665	351	224	65	5810
1985-86	4	22	242	452	1149	1402							

TABLE 5 COOLING DEGREE DAYS Base 65 deg. F   PENDLETON, OREGON

YEAR	JAN	FEB	MAR	APR	MAY	JUNE	JULY	AUG	SEP	OCT	NOV	DEC	TOTAL
1969	0	0	0	0	45	183	238	177	76	0	0	0	719
1970	0	0	0	0	11	201	313	243	4	0	0	0	772
1971	0	0	0	0	36	50	363	379	12	13	0	0	853
1972	0	0	0	0	50	134	314	358	55	0	0	0	911
1973	0	0	0	0	63	137	327	232	72	0	0	0	831
1974	0	0	0	0	9	219	272	332	122	4	0	0	958
1975	0	0	0	0	27	88	423	179	109	8	0	0	834
1976	0	0	0	0	20	53	270	129	103	3	0	0	578
1977	0	0	0	16	3	152	190	348	16	0	0	0	725
1978	0	0	0	0	1	93	236	182	16	0	0	0	528
1979	0	0	0	0	21	114	261	186	65	3	0	0	650
1980	0	0	0	2	5	13	232	101	44	20	0	0	417
1981	0	0	0	4	2	28	155	297	101	0	0	0	587
1982	0	0	0	0	7	158	219	215	47	0	0	0	646
1983	0	0	0	0	60	32	155	246	6	0	0	0	499
1984	0	0	0	0	7	55	256	231	51	3	0	0	603
1985	0	0	0	0	28	91	394	127	7	0	0	0	647

TABLE 6 SNOWFALL (inches)   PENDLETON, OREGON

SEASON	JULY	AUG	SEP	OCT	NOV	DEC	JAN	FEB	MAR	APR	MAY	JUNE	TOTAL
1956-57	0.0	0.0	0.0	T	T	1.2	22.8	6.9	0.7	0.0	0.0	0.0	31.6
1957-58	0.0	0.0	0.0	0.0	0.0	T	1.2	T	3.1	0.0	0.0	0.0	4.3
1958-59	0.0	0.0	0.0	0.0	1.1	1.1	5.3	13.4	T	T	0.0	0.0	20.9
1959-60	0.0	0.0	0.0	0.0	0.9	1.6	10.9	0.4	4.3	T	0.0	0.0	18.1
1960-61	0.0	0.0	0.0	0.0	0.0	0.4	T	T	0.1	T	0.0	0.0	0.5
1961-62	0.0	0.0	0.0	T	9.2	5.6	8.2	1.3	0.7	T	0.0	0.0	25.0
1962-63	0.0	0.0	0.0	0.0	T	T	12.1	0.3	T	1.0	0.0	0.0	13.4
1963-64	0.0	0.0	0.0	0.0	T	2.0	T	0.3	0.5	T	0.0	0.0	2.8
1964-65	0.0	0.0	0.0	0.0	0.2	4.9	7.0	0.5	2.5	0.0	0.0	0.0	15.1
1965-66	0.0	0.0	0.0	0.0	T	0.1	11.5	1.6	0.2	T	0.0	0.0	13.4
1966-67	0.0	0.0	0.0	T	0.4	2.3	2.3	0.4	3.3	T	0.0	0.0	9.3
1967-68	0.0	0.0	0.0	0.0	0.8	2.5	1.3	T	T	T	0.0	0.0	4.6
1968-69	0.0	0.0	0.0	0.0	T	11.9	27.4	2.7	T	0.0	0.0	0.0	42.0
1969-70	0.0	0.0	0.0	0.0	T	3.5	9.9	3.8	1.3	T	0.0	0.0	18.5
1970-71	0.0	0.0	0.0	0.0	1.6	2.3	4.0	0.6	4.9	T	0.0	0.0	13.4
1971-72	0.0	0.0	0.0	1.9	T	11.8	3.6	6.2	0.1	1.1	0.0	0.0	24.7
1972-73	0.0	0.0	0.0	T	T	12.6	2.2	5.9	T	0.1	0.0	0.0	20.8
1973-74	0.0	0.0	0.0	3.2	9.1	5.3	2.6	0.5	T	T	0.0	0.0	20.7
1974-75	0.0	0.0	0.0	0.0	T	T	16.6	3.3	T	2.2	T	0.0	22.1
1975-76	0.0	0.0	0.0	0.0	5.2	3.0	0.3	0.3	0.1	0.0	0.0	0.0	8.9
1976-77	0.0	0.0	0.0	0.0	0.0	1.0	3.1	0.5	0.4	0.0	0.0	0.0	5.0
1977-78	0.0	0.0	0.0	0.0	8.5	11.5	6.1	T	3.9	0.0	T	0.0	30.0
1978-79	0.0	0.0	0.0	0.0	9.0	7.4	14.7	2.2	T	0.0	0.0	0.0	33.3
1979-80	0.0	0.0	0.0	0.0	4.3	T	16.6	0.9	3.9	0.0	0.0	0.0	25.7
1980-81	0.0	0.0	0.0	0.0	2.0	2.7	3.6	1.2	0.0	0.0	0.0	0.0	9.5
1981-82	0.0	0.0	0.0	0.0	0.6	5.1	5.7	1.5	1.9	T	0.0	0.0	14.8
1982-83	0.0	0.0	0.0	0.0	T	T	1.6	0.2	0.9	0.0	0.0	0.0	2.7
1983-84	0.0	0.0	0.0	0.0	T	26.6	1.0	1.2	T	T	0.0	0.0	28.8
1984-85	0.0	0.0	0.0	0.0	0.0	6.2	0.8	12.7	0.6	0.0	0.0	0.0	20.3
1985-86	0.0	0.0	0.0	0.0	14.9	9.1							
Record Mean	0.0	0.0	0.0	0.1	1.9	4.2	7.5	3.4	0.9	0.1	T	0.0	18.2

See Reference Notes, relative to all above tables, on preceding page.

The Portland Weather Service Office is located 6 miles north-northeast of downtown Portland. Portland is situated about 65 miles inland from the Pacific Coast and midway between the northerly oriented low coast range on the west and the higher Cascade range on the east, each about 30 miles distant. The airport lies on the south bank of the Columbia River. The coast range provides limited shielding from the Pacific Ocean. The Cascade range provides a steep slope for orographic lift of moisture-laden westerly winds and consequent moderate rainfall, and also forms a barrier from continental air masses originating over the interior Columbia Basin. Airflow is usually northwesterly in Portland in spring and summer and southeasterly in fall and winter. The Portland Airport location is drier than most surrounding localities.

Portland has a very definite winter rainfall climate. Approximately 88 percent of the annual total occurs in the months of October through May, 9 percent in June and September, while only 3 percent comes in July and August. Precipitation is mostly rain, as on the average there are only five days each year with measurable snow. Snowfalls are seldom more than a couple of inches, and generally last only a few days.

The winter season is marked by relatively mild temperatures, cloudy skies and rain with southeasterly surface winds predominating. Summer produces pleasantly mild temperatures, northwesterly winds and very little precipitation. Fall and spring are transitional in nature. Fall and early winter are times with most frequent fog.

At all times, incursions of marine air are a frequent moderating influence. Outbreaks of continental high pressure from east of the Cascade Mountains produce strong easterly flow through the Columbia Gorge into the Portland area. In winter this brings the coldest weather with the extremes of low temperature registered in the cold air mass. Freezing rain and ice glaze are sometimes transitional effects. Temperatures below zero are very infrequent. In summer, hot, dry continental air brings the highest temperatures. Temperatures above 100 degrees are infrequent, but 90 degrees or higher are reached every year, but seldom persist for more than two or three days.

Destructive storms are infrequent in the Portland area. Surface winds seldom exceed gale force and rarely in the period of record have winds reached higher than 75 mph. Thunderstorms occur about once a month through the spring and summer months. Heavy downpours are infrequent but gentle rains occur almost daily during winter months.

Most rural areas around Portland are farmed for berries, green beans, and vegetables for fresh market and processing. The long growing season with mild temperatures and ample moisture favors local nursery and seed industries.

Based on the 1951-1980 period, the average first occurrence of 32 degrees Fahrenheit in the fall is November 7 and the average last occurrence in the spring is April 3.

TABLE 1   NORMALS, MEANS AND EXTREMES

PORTLAND, OREGON

LATITUDE: 45°36'N   LONGITUDE: 122°36'W   ELEVATION: FT. GRND   21 BARO   00027   TIME ZONE: PACIFIC   WBAN: 24229

	(a)	JAN	FEB	MAR	APR	MAY	JUNE	JULY	AUG	SEP	OCT	NOV	DEC	YEAR
TEMPERATURE °F:														
Normals														
-Daily Maximum		44.3	50.4	54.5	60.2	66.9	72.7	79.5	78.6	74.2	63.9	52.3	46.4	62.0
-Daily Minimum		33.5	36.0	37.4	40.6	46.4	52.2	55.8	55.8	51.1	44.6	38.6	35.4	44.0
-Monthly		38.9	43.2	46.0	50.4	56.7	62.5	67.7	67.2	62.7	54.3	45.5	40.9	53.0
Extremes														
-Record Highest	45	62	70	80	87	100	100	107	107	101	90	73	64	107
-Year		1984	1968	1947	1957	1983	1982	1965	1981	1944	1980	1975	1980	AUG 1981
-Record Lowest	45	-2	-3	19	29	29	39	43	44	34	26	13	6	-3
-Year		1950	1950	1955	1955	1954	1966	1955	1980	1965	1971	1985	1964	FEB 1950
NORMAL DEGREE DAYS:														
Heating (base 65°F)		809	610	592	438	263	118	35	51	111	332	585	747	4691
Cooling (base 65°F)		0	0	0	0	6	43	119	122	42	0	0	0	332
% OF POSSIBLE SUNSHINE	36	27	37	47	53	58	55	70	66	61	42	29	22	47
MEAN SKY COVER (tenths)														
Sunrise - Sunset	37	8.4	8.3	8.1	7.7	7.2	6.8	4.7	5.2	5.6	7.2	8.2	8.7	7.2
MEAN NUMBER OF DAYS:														
Sunrise to Sunset														
-Clear	37	2.9	2.5	3.2	3.8	4.8	6.2	13.1	11.1	10.0	5.3	3.1	2.0	67.9
-Partly Cloudy	37	3.5	3.6	4.4	5.8	7.3	7.4	8.8	9.5	8.0	7.0	4.1	3.1	72.4
-Cloudy	37	24.6	22.1	23.4	20.4	19.0	16.5	9.1	10.4	12.0	18.7	22.9	25.9	225.0
Precipitation														
.01 inches or more	45	18.2	16.4	17.1	14.0	11.6	9.5	3.7	5.1	7.9	12.9	17.9	19.0	153.3
Snow,Ice pellets														
1.0 inches or more	45	1.2	0.2	0.1	0.0	0.*	0.0	0.0	0.0	0.0	0.0	0.2	0.5	2.3
Thunderstorms	45	0.*	0.1	0.5	0.8	1.4	0.9	0.8	1.0	0.7	0.4	0.3	0.*	7.1
Heavy Fog Visibility														
1/4 mile or less	43	4.3	3.7	2.3	1.1	0.2	0.1	0.1	0.2	3.0	7.6	6.2	4.9	33.6
Temperature °F														
-Maximum														
90° and above	45	0.0	0.0	0.0	0.0	0.2	1.2	3.6	3.4	1.5	0.*	0.0	0.0	10.0
32° and below	45	2.6	0.3	0.*	0.0	0.0	0.0	0.0	0.0	0.0	0.0	0.3	0.8	4.0
-Minimum														
32° and below	45	13.6	8.2	5.1	1.1	0.1	0.0	0.0	0.0	0.0	0.6	5.5	9.4	43.6
0° and below	45	0.*	0.*	0.0	0.0	0.0	0.0	0.0	0.0	0.0	0.0	0.0	0.0	*
AVG. STATION PRESS.(mb)	13	1018.5	1016.1	1014.9	1017.0	1016.8	1016.5	1016.1	1015.2	1015.4	1016.9	1016.1	1018.2	1016.5
RELATIVE HUMIDITY (%)														
Hour 04	45	86	86	86	86	85	84	82	83	87	90	88	87	86
Hour 10 (Local Time)	45	82	80	73	68	66	65	61	64	67	79	82	84	73
Hour 16	45	76	68	60	55	53	49	45	46	49	63	74	79	60
Hour 22	45	83	81	78	75	73	71	68	70	75	84	84	85	77
PRECIPITATION (inches):														
Water Equivalent														
-Normal		6.16	3.93	3.61	2.31	2.08	1.47	0.46	1.13	1.61	3.05	5.17	6.41	37.39
-Maximum Monthly	45	12.83	9.46	7.52	4.72	4.57	4.06	2.68	4.53	3.98	8.04	11.57	11.12	12.83
-Year		1953	1949	1957	1955	1945	1984	1983	1968	1982	1947	1942	1968	JAN 1953
-Minimum Monthly	45	0.06	0.78	1.10	0.53	0.46	0.03	0.00	T	T	0.36	0.77	1.38	0.00
-Year		1985	1964	1965	1956	1982	1951	1967	1970	1975	1978	1976	1976	JUL 1967
-Maximum in 24 hrs	45	2.61	2.00	1.83	1.47	1.47	1.82	1.09	1.54	2.38	2.18	2.62	2.59	2.62
-Year		1974	1982	1943	1962	1968	1958	1978	1977	1982	1941	1973	1977	NOV 1973
Snow,Ice pellets														
-Maximum Monthly	45	41.4	13.2	12.9	T	0.6	T			T	0.2	8.2	15.7	41.4
-Year		1950	1949	1951	1985	1953	1981			1949	1950	1955	1968	JAN 1950
-Maximum in 24 hrs	45	10.6	3.2	7.7	T	0.5	T			T	0.2	7.4	8.0	10.6
-Year		1950	1962	1951	1985	1953	1981			1949	1950	1977	1964	JAN 1950
WIND														
Mean Speed (mph)	37	10.0	9.1	8.3	7.4	7.0	7.1	7.6	7.1	6.5	6.5	8.7	9.6	7.9
Prevailing Direction														
through 1963		ESE	ESE	ESE	NW	NW	NW	NW	NW	NW	ESE	ESE	ESE	ESE
Fastest Mile														
-Direction	35	S	SW	S	S	SW	SW	SW	SW	S	S	SW	S	S
-Speed (MPH)	35	54	61	57	60	42	40	33	29	61	88	56	57	88
-Year		1951	1958	1963	1957	1960	1958	1983	1961	1963	1962	1961	1951	OCT 1962
Peak Gust														
-Direction	2	E	SW	SW	N	SW	E	S	E	E	S	SW	E	E
-Speed (mph)	2	39	43	43	35	41	31	29	28	39	35	52	53	53
-Date		1984	1984	1985	1984	1985	1985	1984	1985	1985	1985	1984	1985	DEC 1985

See Reference Notes to this table on the following page.

TABLE 2   PRECIPITATION (inches)   PORTLAND, OREGON

YEAR	JAN	FEB	MAR	APR	MAY	JUNE	JULY	AUG	SEP	OCT	NOV	DEC	ANNUAL
1956	11.66	3.03	4.30	0.53	2.50	2.03	0.01	2.56	1.12	5.10	1.47	3.64	37.95
1957	2.23	4.14	7.52	1.84	1.97	0.73	0.19	0.69	0.49	3.53	3.07	6.15	32.55
1958	6.56	5.13	2.20	3.33	1.35	3.04	T	0.02	1.05	1.49	6.39	5.06	35.62
1959	7.57	4.18	3.22	0.92	2.89	2.38	0.56	0.09	2.81	3.51	3.30	3.08	34.51
1960	3.93	4.00	4.77	3.33	3.37	0.52	T	1.00	1.37	2.39	8.63	2.61	35.92
1961	4.50	8.92	6.04	3.59	2.80	0.47	0.42	1.07	0.64	2.89	4.67	5.94	41.95
1962	1.58	3.43	4.25	3.15	2.56	0.78	0.06	1.49	1.66	3.31	2.59	2.59	34.18
1963	2.27	3.48	4.69	3.78	2.74	1.71	1.17	0.87	0.75	3.04	5.64	3.60	33.74
1964	9.51	0.78	2.30	1.56	1.04	1.96	0.68	0.90	1.61	0.84	6.78	9.92	37.88
1965	7.44	2.22	1.10	2.20	1.31	0.83	0.44	0.73	0.01	2.03	5.64	7.34	31.29
1966	5.74	1.70	4.71	0.85	0.91	1.02	1.19	0.59	1.70	3.06	5.50	6.89	33.86
1967	6.21	2.02	4.31	2.17	1.02	1.01	0.00	T	0.76	4.72	2.27	4.75	29.24
1968	4.58	6.64	2.68	1.91	3.63	2.20	0.14	4.53	2.20	5.03	6.23	11.12	50.89
1969	7.60	3.14	1.13	2.28	1.61	2.99	0.14	0.04	3.86	3.02	3.18	8.12	37.11
1970	11.81	4.77	2.58	2.94	1.55	0.49	0.05	T	1.10	2.85	5.72	7.49	41.35
1971	7.09	3.36	4.87	2.72	1.00	1.76	0.26	0.95	3.53	2.37	5.76	8.05	41.72
1972	5.71	4.08	5.41	2.98	2.23	0.68	0.56	0.67	3.06	0.87	3.78	8.79	38.82
1973	3.69	1.94	2.45	1.33	1.43	1.45	0.06	1.41	3.29	3.14	9.93	41.67	41.67
1974	8.51	4.61	5.65	1.76	1.74	0.80	2.01	0.07	0.21	2.14	6.73	6.05	40.28
1975	8.43	4.75	3.45	1.88	1.35	1.13	0.43	2.10	T	4.76	4.10	6.68	39.06
1976	5.14	4.92	2.93	2.34	2.29	0.78	0.66	3.29	0.73	1.48	0.77	1.38	26.71
1977	1.07	2.49	3.50	1.04	4.30	0.83	0.39	3.26	3.33	2.28	5.56	8.98	37.03
1978	4.85	3.28	1.49	3.96	3.17	1.69	1.36	2.05	2.07	0.36	3.83	2.51	30.62
1979	2.55	6.53	2.51	2.47	2.41	0.64	0.25	1.18	1.75	4.85	3.38	7.23	35.75
1980	8.51	4.01	3.11	2.58	2.19	2.50	0.19	0.39	1.56	1.18	6.47	9.72	42.41
1981	1.47	3.86	2.33	1.79	2.25	3.23	0.24	0.15	1.86	4.12	4.62	8.37	34.29
1982	6.31	5.98	2.38	3.56	0.46	1.66	0.94	1.66	3.98	4.44	3.51	8.16	43.04
1983	6.23	7.78	6.80	1.87	1.30	1.95	2.68	2.29	0.39	1.95	8.65	5.30	47.19
1984	2.01	3.93	3.19	3.20	3.41	4.06	T	0.09	1.46	3.85	9.74	2.56	37.50
1985	0.06	1.79	3.08	1.07	1.52	2.34	0.55	0.48	2.76	2.75	3.89	2.19	22.48
Record Mean	5.51	4.11	3.61	2.25	2.09	1.63	0.54	0.99	1.73	3.29	5.57	6.16	37.47

TABLE 3   AVERAGE TEMPERATURE (deg. F)   PORTLAND, OREGON

YEAR	JAN	FEB	MAR	APR	MAY	JUNE	JULY	AUG	SEP	OCT	NOV	DEC	ANNUAL
1956	39.9	35.9	44.2	52.7	59.1	59.0	68.0	65.4	60.8	51.7	42.6	40.7	51.7
1957	31.2	41.7	46.2	51.8	58.9	62.1	65.6	64.5	65.3	52.8	44.0	43.9	52.3
1958	43.6	48.6	45.5	50.9	61.8	65.2	70.6	70.3	62.3	55.2	45.5	43.6	55.2
1959	40.4	42.7	46.8	52.3	55.7	62.8	68.7	65.5	60.3	54.2	42.8	39.6	52.7
1960	35.4	42.3	45.0	50.6	54.0	63.0	69.1	65.4	61.5	54.5	46.5	38.7	52.2
1961	43.6	47.2	47.7	50.1	56.6	65.3	69.4	70.3	59.5	53.3	42.9	40.9	53.9
1962	38.6	43.0	45.3	52.6	53.9	61.6	66.0	66.0	63.3	55.1	47.6	42.6	53.0
1963	35.0	47.8	45.2	48.6	56.6	60.0	63.1	66.1	65.1	54.2	46.0	38.0	52.1
1964	40.9	40.1	43.8	46.7	52.6	58.7	64.5	63.7	58.4	53.2	41.0	37.0	50.1
1965	40.2	43.4	47.6	51.8	54.3	61.8	69.7	68.6	60.4	57.5	49.7	39.5	53.7
1966	40.3	42.8	47.2	50.6	57.0	62.4	66.4	67.5	64.4	53.7	47.1	44.2	53.7
1967	43.6	43.7	44.0	46.9	57.1	65.9	69.3	72.9	66.4	54.8	46.2	40.4	54.3
1968	39.3	48.2	48.2	48.0	56.3	62.2	68.6	65.8	61.5	52.1	46.5	37.3	52.8
1969	31.9	39.7	46.5	50.1	59.5	66.5	66.5	66.1	63.4	53.3	47.2	42.9	52.8
1970	40.6	46.0	46.9	48.4	57.0	65.9	69.2	68.2	60.7	53.0	47.1	40.0	53.6
1971	40.4	42.8	43.7	49.6	56.8	60.2	69.2	71.6	60.8	52.4	45.5	40.4	52.8
1972	39.2	43.8	49.8	48.0	60.2	64.0	70.9	71.7	61.2	53.0	48.2	37.4	53.9
1973	39.0	44.9	47.9	52.3	59.4	63.9	70.3	65.9	64.4	54.3	44.2	44.7	54.2
1974	38.0	43.0	47.2	51.3	55.7	64.4	67.1	68.9	67.3	55.2	48.1	44.1	54.2
1975	41.5	41.2	45.0	47.3	55.7	61.9	69.0	65.3	65.7	53.5	46.0	42.7	53.1
1976	42.2	42.1	44.4	50.3	56.6	60.4	67.2	65.5	64.2	54.7	47.0	39.5	52.8
1977	35.7	44.6	45.5	52.9	54.8	63.9	66.3	71.7	66.0	53.8	43.3	42.0	52.9
1978	40.1	44.7	49.1	50.5	54.7	65.1	68.4	67.6	60.9	54.7	39.1	35.3	52.5
1979	30.7	42.9	50.8	53.1	60.1	65.1	70.5	68.6	66.3	58.1	45.0	44.4	54.6
1980	35.1	42.5	46.3	53.8	57.3	60.7	68.9	66.4	63.8	56.0	48.5	44.0	53.6
1981	43.9	44.0	48.8	52.5	57.5	61.8	67.5	72.2	64.9	53.3	48.8	42.7	54.8
1982	39.7	43.6	48.5	49.0	57.6	66.0	67.5	68.6	63.2	54.9	44.4	41.7	53.7
1983	44.4	47.3	50.7	52.7	60.4	62.8	66.5	69.1	61.5	54.2	49.3	36.4	54.6
1984	42.2	45.9	51.1	50.4	56.4	62.2	69.1	69.4	63.7	52.9	46.7	38.3	54.0
1985	36.1	41.1	45.8	53.9	58.3	64.4	74.1	69.3	60.8	52.7	37.3	33.0	52.2
Record Mean	38.6	43.2	46.4	50.9	57.0	60.9	66.4	66.1	62.8	54.0	45.5	40.8	52.7
Max	44.3	50.4	54.9	60.5	67.2	70.8	77.9	77.2	74.1	63.3	52.2	46.1	61.6
Min	32.9	36.0	37.9	41.2	46.9	50.9	54.9	54.9	51.4	44.8	38.8	35.4	43.8

REFERENCE NOTES FOR TABLES 1, 2, 3 and 6   (PORTLAND, OR)

GENERAL

T - TRACE AMOUNT
BLANK ENTRIES DENOTE MISSING/UNREPORTED DATA.
# INDICATES A STATION OR INSTRUMENT RELOCATION.

SPECIFIC

TABLE 1

(a) - LENGTH OF RECORD IN YEARS. ALTHOUGH
INDIVIDUAL MONTHS MAY BE MISSING.
* LESS THAN .05

NORMALS — BASED ON THE 1951-1980 RECORD PERIOD.
EXTREMES — DATES ARE THE MOST RECENT OCCURRENCE.
WIND DIR. — NUMERALS SHOW TENS OF DEGREES
CLOCKWISE FROM TRUE NORTH.
"00" INDICATES CALM.
RESULTANT WIND DIRECTIONS ARE GIVEN TO WHOLE DEGREES.

EXCEPTIONS

TABLES 2, 3, and 6

RECORD MEANS ARE THROUGH THE CURRENT YEAR,
BEGINNING IN   1941 FOR TEMPERATURE
                  1941 FOR PRECIPITATION
                  1941 FOR SNOWFALL

TABLE 4   HEATING DEGREE DAYS Base 65 deg. F   PORTLAND, OREGON

SEASON	JULY	AUG	SEP	OCT	NOV	DEC	JAN	FEB	MAR	APR	MAY	JUNE	TOTAL
1956-57	22	47	120	407	662	746	1041	648	578	390	190	95	4946
1957-58	25	35	45	370	624	644	656	451	599	420	128	62	4059
1958-59	3	5	105	295	579	657	755	618	559	375	280	89	4320
1959-60	26	32	142	326	659	781	910	653	612	424	337	78	4980
1960-61	18	52	114	316	548	807	654	494	530	442	261	66	4302
1961-62	11	4	169	359	656	740	811	613	605	364	339	118	4789
1962-63	49	19	71	299	515	687	925	477	606	486	272	168	4574
1963-64	72	24	41	329	561	830	739	718	650	539	380	182	5065
1964-65	67	81	191	358	711	860	761	599	533	388	325	113	4987
1965-66	22	16	139	227	451	786	759	620	545	425	249	99	4338
1966-67	27	16	56	345	531	635	655	591	647	535	246	50	4334
1967-68	3	0	29	306	558	758	789	482	515	500	261	110	4311
1968-69	17	43	123	395	544	852	1022	703	570	442	178	51	4940
1969-70	17	22	85	357	526	678	751	526	553	493	246	71	4325
1970-71	14	14	130	369	530	771	757	615	615	454	253	149	4709
1971-72	33	5	123	388	578	756	793	607	466	501	174	61	4485
1972-73	10	6	153	363	497	848	799	560	525	378	202	89	4430
1973-74	6	47	59	326	618	624	832	610	545	403	282	72	4424
1974-75	32	16	29	301	500	640	722	660	615	523	240	127	4405
1975-76	24	41	48	354	565	686	698	658	632	437	258	155	4556
1976-77	15	41	47	319	536	783	901	564	596	358	340	68	4568
1977-78	40	19	131	339	644	707	764	561	485	430	317	58	4495
1978-79	29	26	134	312	772	915	1058	615	434	351	162	57	4865
1979-80	8	2	19	214	592	631	920	647	575	329	232	125	4294
1980-81	15	25	64	284	485	644	650	583	494	372	229	108	3953
1981-82	23	5	76	355	478	687	780	596	502	472	229	71	4274
1982-83	22	10	99	307	614	715	635	492	435	363	184	81	3957
1983-84	27	2	109	325	463	880	701	546	425	430	269	115	4292
1984-85	9	2	80	377	539	820	893	664	588	327	213	62	4574
1985-86	0	7	124	373	826	982							

TABLE 5   COOLING DEGREE DAYS Base 65 deg. F   PORTLAND, OREGON

YEAR	JAN	FEB	MAR	APR	MAY	JUNE	JULY	AUG	SEP	OCT	NOV	DEC	TOTAL
1969	0	0	0	0	13	102	74	65	43	0	0	0	297
1970	0	0	0	0	8	106	150	120	8	2	0	0	394
1971	0	0	0	0	5	14	170	217	7	3	0	0	416
1972	0	0	0	0	27	39	200	221	44	0	0	0	531
1973	0	0	0	0	34	65	178	81	45	0	0	0	403
1974	0	0	0	0	1	60	102	144	102	0	0	0	409
1975	0	0	0	0	12	39	157	57	75	2	0	0	342
1976	0	0	0	0	4	23	89	66	30	4	0	0	216
1977	0	0	0	0	4	42	90	233	10	0	0	0	375
1978	0	0	0	0	3	69	141	112	18	0	0	0	343
1979	0	0	0	0	18	65	183	124	65	7	0	0	462
1980	0	0	0	1	0	2	141	75	35	12	0	0	266
1981	0	0	0	3	4	16	109	232	82	0	0	0	446
1982	0	0	0	0	4	107	103	127	50	0	0	0	391
1983	0	0	0	0	48	23	80	137	12	0	0	0	300
1984	0	0	0	0	10	34	140	144	47	6	0	0	381
1985	0	0	0	0	11	53	291	145	5	0	0	0	505

TABLE 6   SNOWFALL (inches)   PORTLAND, OREGON

SEASON	JULY	AUG	SEP	OCT	NOV	DEC	JAN	FEB	MAR	APR	MAY	JUNE	TOTAL
1956-57	0.0	0.0	0.0	0.0	0.0	2.6	4.2	1.4	T	0.0	0.0	0.0	8.2
1957-58	0.0	0.0	0.0	0.0	0.0	T	0.0	0.0	0.0	0.0	0.0	0.0	T
1958-59	0.0	0.0	0.0	0.0	T	0.0	0.9	2.0	T	0.0	0.0	0.0	2.9
1959-60	0.0	0.0	0.0	0.0	0.0	T	10.3	T	2.6	0.0	T	0.0	12.9
1960-61	0.0	0.0	0.0	0.0	0.0	0.0	0.0	T	T	T	0.0	0.0	T
1961-62	0.0	0.0	0.0	0.0	0.0	1.0	0.7	3.8	0.1	0.0	0.0	0.0	5.6
1962-63	0.0	0.0	0.0	0.0	0.0	0.0	5.0	0.0	0.0	T	0.0	0.0	5.0
1963-64	0.0	0.0	0.0	0.0	T	0.0	T	0.0	T	0.0	0.0	0.0	T
1964-65	0.0	0.0	0.0	0.0	T	11.0	T	0.0	0.3	0.0	T	0.0	11.3
1965-66	0.0	0.0	0.0	0.0	0.0	T	T	T	0.6	0.0	0.0	0.0	0.6
1966-67	0.0	0.0	0.0	0.0	0.0	0.0	T	T	T	T	0.0	0.0	T
1967-68	0.0	0.0	0.0	0.0	0.0	5.7	5.2	0.0	0.0	T	0.0	0.0	10.9
1968-69	0.0	0.0	0.0	0.0	0.0	15.7	18.3	T	0.0	0.0	0.0	0.0	34.0
1969-70	0.0	0.0	0.0	0.0	0.0	0.0	T	T	T	0.0	0.0	0.0	T
1970-71	0.0	0.0	0.0	0.0	T	1.4	6.9	1.7	T	T	0.0	0.0	10.0
1971-72	0.0	0.0	0.0	T	0.0	4.6	0.4	T	T	T	0.0	0.0	5.0
1972-73	0.0	0.0	0.0	0.0	0.0	6.1	0.4	T	T	0.0	0.0	0.0	6.5
1973-74	0.0	0.0	0.0	0.0	T	0.0	T	T	T	T	0.0	0.0	T
1974-75	0.0	0.0	0.0	0.0	0.0	T	T	0.1	T	T	T	0.0	0.1
1975-76	0.0	0.0	0.0	0.0	T	T	T	T	T	T	0.0	0.0	T
1976-77	0.0	0.0	0.0	0.0	0.0	7.0	0.0	T	0.0	0.0	T	0.0	7.7
1977-78	0.0	0.0	0.0	0.0	0.0	7.6	T	0.0	0.0	0.1	T	0.0	8.4
1978-79	0.0	0.0	0.0	0.0	3.0	2.4	1.9	1.1	T	T	T	0.0	12.4
1979-80	0.0	0.0	0.0	0.0	0.0	0.0	12.4	T	T	T	T	T	T
1980-81	0.0	0.0	0.0	0.0	T	0.0	0.0	T	T	0.0	0.0	0.0	T
1981-82	0.0	0.0	0.0	0.0	0.0	2.0	2.1	T	T	T	0.0	0.0	4.1
1982-83	0.0	0.0	0.0	0.0	T	0.0	0.0	0.0	T	T	0.0	0.0	T
1983-84	0.0	0.0	0.0	0.0	0.0	2.3	0.1	0.1	T	0.0	0.0	0.0	2.4
1984-85	0.0	0.0	0.0	0.0	0.0	2.8	T	4.8	T	T	0.0	0.0	7.6
1985-86	0.0	0.0	0.0	0.0	3.4	1.6							
Record Mean	0.0	0.0	T	T	0.5	1.5	3.8	0.7	0.5	T	T	T	7.0

See Reference Notes, relative to all above tables, on preceding page.

Salem is located in the middle Willamette Valley some 60 airline miles east of the Pacific Ocean. The valley here is approximately 50 miles wide with the city about equidistant from the valley walls formed by the Coast Range on the west and the Cascade Range on the east.

The usual movement of very moist maritime air masses from the Pacific Ocean inland over the Coast Range produces, near its crest, some of the heaviest yearly rainfall in the United States. Annual totals of nearly 170 inches have been recorded in the mountains. From the ridge crest of the Coast Range, approximately 3,000 feet above sea level, there is a gradual decrease of rainfall downslope to the valley floor where annual totals are between 35 and 45 inches. As these marine conditioned air masses continue to move farther inland they are forced to ascend the west slopes of the Cascades to approximately 5,000 feet above sea level and again rainfall amounts substantially increase with elevation.

Most of this precipitation in both the valley and its bordering mountain ranges occurs during the winter. At Salem, 70 percent of the annual total occurs during the five months of November through March while only 6 percent occurs during the three summer months, with practically all of it falling in the form of rain. In the immediate area, there are only three or four days a year with measurable amounts of snow. Its depth on the ground rarely exceeds 2 or 3 inches, and it usually melts in a day or two. The few thunderstorms that occur each year are not generally severe and seldom do they, or the hail that occasionally accompanies them, cause any serious damage. A tornado in the immediate metropolitan area has never been recorded.

The seasonal difference in temperatures is much less marked than that of precipitation. There is a range of about 28 degrees between the temperature for January, the coldest month, and July, the warmest. Highs of 100 degrees or more seldom occur, and only in a few years since records began in 1892, have 0 degree or lower temperatures been observed. There is an average growing season of six and a half months.

The mild temperatures, long growing season, and plentiful supply of moisture are ideal for a wide variety of crops. In dollar value of agricultural returns, this is the most productive area in Oregon. Large orchards of sweet cherries are grown and processed here for maraschino cherries. Hops, filberts, walnuts, cane, and strawberries each contribute many millions of dollars to the annual farm income. A wide variety of vegetables is raised for both the fresh market and to support a large number of processing plants located in Salem. This climate is also suitable for the production of a number of specialty crops including mint, several seed crops, and nursery stock, particularly roses and ornamental shrubs.

Based on the 1951–1980 period, the average first occurrence of 32 degrees Fahrenheit in the fall is October 22 and the average last occurrence in the spring is May 5.

TABLE 1   NORMALS, MEANS AND EXTREMES

SALEM, OREGON

LATITUDE: 44°55'N   LONGITUDE: 123°00'W   ELEVATION: FT. GRND   196 BARO   00199   TIME ZONE: PACIFIC   WBAN: 24232

	(a)	JAN	FEB	MAR	APR	MAY	JUNE	JULY	AUG	SEP	OCT	NOV	DEC	YEAR
TEMPERATURE °F:														
Normals														
-Daily Maximum		45.7	51.1	54.6	60.3	67.3	73.9	82.2	81.2	76.2	64.5	52.6	47.0	63.1
-Daily Minimum		32.8	34.3	35.0	37.4	42.3	47.8	50.3	50.7	47.0	41.4	36.8	34.4	40.9
-Monthly		39.3	42.7	44.8	48.9	54.8	60.9	66.3	66.0	61.6	53.0	44.7	40.7	52.0
Extremes														
-Record Highest	48	65	72	80	88	100	102	108	108	103	93	72	66	108
-Year		1984	1968	1947	1957	1983	1982	1941	1981	1944	1970	1970	1980	AUG 1981
-Record Lowest	48	-10	-4	12	23	25	32	37	36	26	23	9	-12	-12
-Year		1950	1950	1971	1968	1954	1976	1962	1980	1972	1971	1955	1972	DEC 1972
NORMAL DEGREE DAYS:														
Heating (base 65°F)		797	624	626	483	316	152	46	65	131	372	609	753	4974
Cooling (base 65°F)		0	0	0	0	0	29	87	93	29	0	0	0	238
% OF POSSIBLE SUNSHINE														
MEAN SKY COVER (tenths)														
Sunrise - Sunset	41	8.3	8.2	7.9	7.4	6.9	6.4	4.1	4.7	5.1	6.9	8.1	8.7	6.9
MEAN NUMBER OF DAYS:														
Sunrise to Sunset														
-Clear	48	2.8	3.0	3.5	4.2	5.8	7.0	15.1	13.5	11.2	5.8	2.9	2.0	76.7
-Partly Cloudy	48	4.6	4.8	6.1	7.2	7.7	8.1	8.6	8.6	8.4	8.3	5.2	3.7	81.3
-Cloudy	48	23.6	20.5	21.4	18.6	17.5	14.9	7.3	8.9	10.4	17.0	21.9	25.3	207.3
Precipitation														
.01 inches or more	48	18.0	16.4	17.1	13.7	10.8	8.0	3.0	4.2	7.1	12.8	17.9	19.2	148.2
Snow,Ice pellets														
1.0 inches or more	48	1.0	0.2	0.2	0.0	0.0	0.0	0.0	0.0	0.0	0.0	0.1	0.5	2.1
Thunderstorms	48	0.1	0.1	0.2	0.5	0.9	0.5	0.8	0.8	0.9	0.4	0.2	0.1	5.4
Heavy Fog Visibility														
1/4 mile or less	48	6.6	4.2	2.1	0.7	0.4	0.1	0.*	0.3	2.0	6.9	6.7	7.2	37.4
Temperature °F														
-Maximum														
90° and above	23	0.0	0.0	0.0	0.0	0.1	1.8	6.1	5.7	2.0	0.1	0.0	0.0	15.8
32° and below	23	1.4	0.*	0.0	0.0	0.0	0.0	0.0	0.0	0.0	0.0	0.3	1.3	3.0
-Minimum														
32° and below	23	14.6	11.7	10.6	6.9	1.1	0.*	0.0	0.0	0.3	3.3	8.6	12.5	69.6
0° and below	23	0.0	0.0	0.0	0.0	0.0	0.0	0.0	0.0	0.0	0.0	0.0	0.2	0.2
AVG. STATION PRESS.(mb)	13	1012.1	1009.7	1008.9	1011.0	1010.9	1010.6	1010.1	1009.2	1009.2	1010.7	1009.9	1012.0	1010.4
RELATIVE HUMIDITY (%)														
Hour 04	23	86	87	87	87	86	86	85	86	87	90	90	87	87
Hour 10	23	83	82	76	70	66	62	57	59	64	77	85	85	72
Hour 16 (Local Time)	23	75	69	61	57	53	50	40	41	46	61	76	80	59
Hour 22	23	85	84	81	79	76	74	68	71	78	86	88	86	80
PRECIPITATION (inches):														
Water Equivalent														
-Normal		7.05	4.56	4.31	2.41	1.95	1.23	0.35	0.76	1.59	3.33	5.71	7.10	40.35
-Maximum Monthly	48	15.40	12.31	8.56	5.18	4.58	4.19	2.63	4.17	3.98	11.17	15.23	12.40	15.40
-Year		1953	1949	1983	1955	1942	1984	1983	1968	1971	1947	1973	1964	JAN 1953
-Minimum Monthly	48	0.24	0.78	0.87	0.39	0.18	0.01	0.00	T	0.00	0.37	0.84	1.26	0.00
-Year		1985	1964	1965	1939	1947	1951	1967	1984	1975	1978	1939	1976	SEP 1975
-Maximum in 24 hrs	48	3.07	3.16	3.03	2.22	1.84	1.73	1.27	1.25	1.86	2.84	2.82	2.72	3.16
-Year		1972	1949	1943	1971	1963	1985	1983	1943	1951	1955	1950	1964	FEB 1949
Snow,Ice pellets														
-Maximum Monthly	48	32.8	8.4	10.9	0.1	T				T	T	6.1	14.6	32.8
-Year		1950	1962	1951	1972	1984				1981	1974	1977	1972	JAN 1950
-Maximum in 24 hrs	48	10.8	5.4	8.5	0.1	T				T	T	6.1	9.4	10.8
-Year		1943	1962	1960	1972	1984				1981	1974	1977	1972	JAN 1943
WIND:														
Mean Speed (mph)	37	8.1	7.7	7.9	7.1	6.5	6.5	6.5	6.2	6.1	6.1	7.4	8.0	7.0
Prevailing Direction														
through 1963		S	S	S	S	S	S	N	NW	S	S	S	S	S
Fastest Obs. 1 Min.														
-Direction (!!)	36	18	18	19	18	25	18	24	29	18	18	18	23	18
-Speed (MPH)	36	40	46	40	44	28	25	26	24	31	58	44	45	58
-Year		1966	1958	1971	1962	1975	1974	1979	1969	1969	1962	1981	1953	OCT 1962
Peak Gust														
-Direction (!!)	2	S	S	S	S	SW	W	N	SW	NW	S	S	S	S
-Speed (mph)	2	31	47	41	39	44	29	26	28	32	40	38	44	47
-Date		1984	1984	1985	1984	1985	1985	1985	1985	1984	1985	1984	1985	FEB 1984

See Reference Notes to this table on the following page.

TABLE 2 PRECIPITATION (inches) SALEM, OREGON

YEAR	JAN	FEB	MAR	APR	MAY	JUNE	JULY	AUG	SEP	OCT	NOV	DEC	ANNUAL
1956	12.68	5.42	5.91	0.64	1.61	1.20	T	0.37	0.87	6.50	1.03	2.94	39.17
1957	2.57	4.93	8.16	2.02	2.77	2.10	0.18	0.27	0.96	2.23	7.18	4.71	39.25
1958	8.80	7.04	2.50	3.71	1.38	2.53	T	0.03	1.00	2.10	2.06	3.97	35.38
1959	11.15	4.98	4.45	1.12	2.09	1.41	0.50	0.02	0.65	2.75	9.45	3.24	41.11
1960	4.41	5.41	6.99	3.50	3.59	0.47	T		0.65	2.75	2.06	9.74	41.11
1961	4.79	10.82	8.19	3.19	2.44	0.30	0.96	0.28	0.91	3.18	4.42	6.64	46.12
1962	1.11	3.97	3.03	2.11	0.69	T		0.70	1.53	4.55	8.54	3.01	34.89
1963	2.80	3.34	6.51	4.07	3.70	0.85	0.91	0.09	1.41	3.59	6.52	3.85	37.64
1964	11.19	0.78	3.55	1.28	0.59	1.73	0.45	0.41	0.74	0.93	8.44	12.40	42.49
1965	8.15	1.57	0.87	2.41	1.16	1.11	0.19	0.99	0.13	2.20	7.00	7.95	33.73
1966	6.60	2.24	6.08	1.07	0.78	0.58	0.53	0.40	1.66	2.06	5.88	7.32	35.20
1967	7.29	2.06	3.84	2.02	1.87	0.69	0.00	T	0.84	5.08	3.30	5.45	32.44
1968	6.37	7.73	3.32	1.47	3.46	1.29	0.39	4.17	2.48	6.14	6.49	11.05	54.36
1969	8.61	3.24	3.55	1.63	0.89	2.94	0.05	0.05	3.58	4.44	3.21	9.23	40.38
1970	13.47	4.46	1.92	2.63	1.36	0.85	0.01	T	1.81	3.25	7.18	9.74	46.68
1971	6.49	4.34	6.93	4.05	1.89	2.47	0.01	1.49	3.98	3.09	6.27	8.18	49.19
1972	7.98	4.68	4.96	3.79	2.40	0.69	0.12	0.14	2.07	0.70	3.77	8.70	40.00
1973	5.64	1.62	3.50	1.69	1.11	1.48		0.80	2.80	2.79	15.23	11.08	47.74
1974	10.89	5.56	7.95	1.48	0.90	0.41	1.80	0.11	0.28	2.15	7.42	6.94	45.89
1975	4.96	4.68	4.22	2.20	1.66	0.81	0.51	1.96		5.51	6.06	6.07	38.64
1976	5.47	6.92	3.66	2.00	1.33	1.04	0.67	1.89	1.13	1.51	1.13	1.26	28.01
1977	0.88	2.83	3.54	0.62	3.76	0.73	0.26	1.70	2.36	2.37	6.19	8.73	33.76
1978	5.67	3.54	1.23	3.50	2.97	0.48	1.07	2.56	2.64	0.37	4.50	2.64	31.17
1979	2.84	7.19	2.17	2.82	2.20	0.65	0.30	0.70	2.19	6.06	3.83	6.95	37.90
1980	6.58	4.04	3.48	3.58	1.53	1.99	0.22	0.04	1.05	1.45	5.24	10.44	39.64
1981	2.09	3.29	3.45	2.06	2.19	3.33	0.15	T	2.78	4.63	7.54	9.73	41.24
1982	6.11	6.02	2.81	3.80	0.73	1.36	0.33	0.41	1.83	3.04	4.92	9.26	40.62
1983	6.00	10.57	8.56	2.72	2.12	2.48	2.63	2.09	0.32	1.31	10.06	6.49	55.35
1984	2.34	5.50	4.40	3.74	3.32	4.19	0.03	T	1.19	4.31	12.70	3.71	45.43
1985	0.24	3.44	3.80	0.93	0.65	2.42	0.32	0.20	1.25	3.17	4.81	2.51	23.74
Record Mean	6.17	5.05	4.41	2.34	1.98	1.40	0.42	0.67	1.54	3.58	6.08	6.85	40.49

TABLE 3 AVERAGE TEMPERATURE (deg. F) SALEM, OREGON

YEAR	JAN	FEB	MAR	APR	MAY	JUNE	JULY	AUG	SEP	OCT	NOV	DEC	ANNUAL
1956	41.1	37.0	43.9	51.3	58.6	58.5	67.8	65.2	61.7	51.7	42.4	40.1	51.6
1957	33.0	42.7	46.9	51.8	58.4	61.7	64.7	63.8	65.3	53.3	42.9	43.1	52.3
1958	42.9	48.3	45.0	50.2	61.0	65.0	71.2	70.3	62.4	55.4	46.9	45.5	55.4
1959	42.8	42.9	46.0	50.8	54.0	61.6	68.0	65.5	60.2	54.4	44.3	39.6	52.5
1960	36.6	41.6	44.4	49.6	52.5	60.7	67.6	64.8	62.4	54.8	46.9	40.1	51.8
1961	43.0	46.1	46.4	49.1	54.0	63.3	67.5	69.3	58.3	52.6	41.2	41.0	52.6
#1962	37.1	40.8	43.9	50.9	51.4	59.3	64.3	64.0	62.5	53.8	46.7	39.9	51.3
1963	33.7	47.7	44.4	47.0	55.2	59.1	61.9	65.3	65.7	53.8	46.1	39.9	51.7
1964	41.7	40.3	43.7	48.1	53.7	60.3	66.4	63.8	58.8	53.2	42.0	39.6	51.0
1965	40.5	43.1	46.6	50.3	52.2	58.9	67.2	66.7	59.6	55.3	47.9	37.8	52.2
1966	39.6	40.6	44.6	50.5	54.9	61.8	65.2	66.9	63.2	52.6	47.1	42.0	52.6
1967	44.1	42.6	43.1	44.6	54.4	64.2	67.6	71.6	64.9	54.3	46.4	42.0	53.3
1968	39.7	48.1	47.3	46.9	54.5	59.6	66.1	65.2	60.5	51.5	45.3	39.0	51.8
1969	33.0	40.3	46.3	49.1	58.9	65.4	64.9	63.5	61.1	49.9	44.1	41.9	51.5
1970	41.6	44.9	45.5	45.8	54.2	64.4	67.2	65.4	58.3	50.7	43.6	39.0	51.7
1971	39.1	39.4	41.9	47.1	53.3	58.2	67.3	69.1	59.1	50.6	44.5	39.5	50.8
1972	38.3	43.2	47.9	46.0	56.5	60.0	67.6	68.2	57.6	50.1	46.4	35.2	51.4
1973	38.8	43.9	44.6	49.1	55.8	61.1	67.5	63.6	62.1	51.9	44.0	43.4	52.2
1974	37.6	41.2	46.1	49.3	52.9	61.8	65.8	67.4	65.8	53.0	46.4	43.0	52.5
1975	42.6	41.9	43.6	45.3	54.3	59.9	67.1	63.1	63.5	51.6	43.9	41.9	51.6
1976	40.7	40.2	42.5	47.2	53.2	56.9	65.3	64.2	62.0	52.1	45.9	38.8	50.8
1977	37.7	44.5	44.5	50.6	51.8	61.8	64.5	70.1	59.0	52.1	44.6	43.2	52.0
1978	41.4	45.3	48.2	48.9	54.8	65.2	68.8	68.3	61.0	54.1	39.4	34.8	52.5
1979	31.2	42.7	49.0	50.5	56.8	61.8	67.9	65.7	62.9	54.7	41.8	44.8	52.5
1980	35.6	43.2	44.6	50.7	53.7	57.5	65.5	62.8	62.5	53.6	47.3	42.9	51.6
1981	40.6	43.1	46.6	50.6	54.5	59.0	65.1	68.7	61.9	50.4	44.6	43.2	52.3
1982	38.6	41.6	44.5	45.9	54.7	65.1	65.7	66.4	61.4	53.4	42.4	41.1	51.7
1983	43.3	46.1	48.9	50.1	57.3	60.3	64.3	67.8	60.3	52.7	49.3	38.4	53.3
1984	43.2	44.2	48.7	47.8	52.9	57.4	65.4	66.8	61.3	51.4	44.4	37.5	51.8
1985	34.3	39.6	42.8	51.4	55.2	61.2	70.2	65.4	58.4	51.2	37.3	33.3	50.0
Record Mean	38.9	42.8	45.5	49.6	55.5	61.1	66.7	66.3	61.8	53.0	44.8	40.8	52.2
Max	45.6	51.1	55.2	60.9	67.9	74.0	82.3	81.5	76.2	64.0	52.5	47.1	63.2
Min	32.2	34.5	35.7	38.3	43.0	48.2	51.0	51.1	47.4	42.0	37.1	34.5	41.2

REFERENCE NOTES FOR TABLES 1, 2, 3 and 6 (SALEM, OR)

GENERAL

T - TRACE AMOUNT
BLANK ENTRIES DENOTE MISSING/UNREPORTED DATA.
# INDICATES A STATION OR INSTRUMENT RELOCATION.

SPECIFIC

TABLE 1

(a) - LENGTH OF RECORD IN YEARS. ALTHOUGH INDIVIDUAL MONTHS MAY BE MISSING.
  * LESS THAN .05

NORMALS — BASED ON THE 1951-1980 RECORD PERIOD.
EXTREMES — DATES ARE THE MOST RECENT OCCURRENCE.
WIND DIR. — NUMERALS SHOW TENS OF DEGREES CLOCKWISE FROM TRUE NORTH.
  "00" INDICATES CALM.
RESULTANT WIND DIRECTIONS ARE GIVEN TO WHOLE DEGREES.

EXCEPTIONS

TABLES 2, 3, and 6

RECORD MEANS ARE THROUGH THE CURRENT YEAR, BEGINNING IN 1938 FOR TEMPERATURE
1938 FOR PRECIPITATION
1938 FOR SNOWFALL

TABLE 4 HEATING DEGREE DAYS Base 65 deg. F SALEM, OREGON

SEASON	JULY	AUG	SEP	OCT	NOV	DEC	JAN	FEB	MAR	APR	MAY	JUNE	TOTAL
1956-57	33	47	102	407	670	765	986	618	554	393	201	103	4879
1957-58	39	45	47	359	657	674	680	463	614	436	142	68	4224
1958-59	0	2	110	296	539	593	682	612	579	417	333	119	4282
1959-60	33	31	147	320	615	782	873	673	634	457	384	135	5084
1960-61	32	59	106	309	538	765	675	524	570	473	334	112	4497
#1961-62	26	9	199	382	706	737	857	673	647	419	414	171	5240
1962-63	91	49	86	377	542	713	964	475	630	531	302	185	4945
1963-64	102	35	43	340	559	770	715	710	658	501	346	135	4914
1964-65	41	79	181	357	684	783	751	607	565	433	390	188	5059
1965-66	46	26	166	293	505	836	781	679	629	425	308	114	4808
1966-67	44	24	74	375	530	637	642	620	669	604	323	73	4615
1967-68	11	2	47	323	549	706	778	481	541	536	318	167	4459
1968-69	47	49	142	413	581	860	987	685	575	469	195	49	5052
1969-70	45	70	137	459	621	710	719	555	596	571	329	94	4906
1970-71	33	45	195	438	635	797	796	710	706	528	355	201	5439
1971-72	61	18	169	440	607	784	821	625	523	560	267	148	5023
1972-73	34	42	239	454	552	919	803	586	626	468	288	141	5152
1973-74	30	85	106	398	624	665	845	658	580	463	371	123	4948
1974-75	48	28	47	364	550	674	688	641	656	584	327	164	4771
1975-76	37	84	79	409	624	708	743	709	693	529	356	243	5214
1976-77	43	63	98	395	565	807	842	569	629	424	400	108	4943
1977-78	57	23	181	392	606	670	727	547	513	476	316	51	4559
1978-79	17	20	129	328	763	928	1038	619	486	430	249	121	5128
1979-80	24	21	76	315	691	620	905	627	627	423	344	217	4890
1980-81	51	82	87	357	525	682	746	610	566	426	318	176	4626
1981-82	59	33	124	447	607	676	814	649	629	568	317	85	5008
1982-83	42	27	128	356	671	732	668	524	491	439	253	135	4466
1983-84	59	3	135	376	463	821	667	597	501	511	367	231	4731
1984-85	44	12	124	418	610	846	947	706	684	403	299	132	5225
1985-86	5	51	196	423	824	975							

TABLE 5 COOLING DEGREE DAYS Base 65 deg. F SALEM, OREGON

YEAR	JAN	FEB	MAR	APR	MAY	JUNE	JULY	AUG	SEP	OCT	NOV	DEC	TOTAL
1969	0	0	0	0	12	66	50	28	27	0	0	0	183
1970	0	0	0	0	2	82	106	66	0	1	0	0	257
1971	0	0	0	0	0	5	140	152	2	0	0	0	299
1972	0	0	0	0	10	6	122	148	21	0	0	0	307
1973	0	0	0	0	6	34	115	47	29	0	0	0	231
1974	0	0	0	0	0	34	78	108	78	0	0	0	298
1975	0	0	0	0	1	16	109	30	41	0	0	0	197
1976	0	0	0	0	0	6	59	48	16	3	0	0	132
1977	0	0	0	0	0	19	50	187	5	0	0	0	261
1978	0	0	0	0	6	63	143	130	16	0	0	0	358
1979	0	0	0	0	0	29	121	48	20	1	0	0	219
1980	0	0	0	0	0	0	73	21	17	9	0	0	120
1981	0	0	0	1	0	2	69	156	38	0	0	0	266
1982	0	0	0	0	5	91	68	82	27	0	0	0	273
1983	0	0	0	0	21	4	46	94	3	0	0	0	168
1984	0	0	0	0	1	10	64	73	19	5	0	0	172
1985	0	0	0	0	1	26	171	71	1	0	0	0	270

TABLE 6 SNOWFALL (inches) SALEM, OREGON

SEASON	JULY	AUG	SEP	OCT	NOV	DEC	JAN	FEB	MAR	APR	MAY	JUNE	TOTAL
1956-57	0.0	0.0	0.0	0.0	0.0	8.2	2.4	T	0.0	0.0	0.0	0.0	10.6
1957-58	0.0	0.0	0.0	0.0	0.0	0.0	0.0	T	0.0	0.0	0.0	0.0	T
1958-59	0.0	0.0	0.0	0.0	T	0.0	3.4	2.1	0.0	0.0	0.0	0.0	5.5
1959-60	0.0	0.0	0.0	0.0	0.0	0.0	6.9	T	8.5	0.0	0.0	0.0	15.4
1960-61	0.0	0.0	0.0	0.0	0.0	T	0.0	0.0	T	T	0.0	0.0	T
1961-62	0.0	0.0	0.0	0.0	0.0	T	2.5	8.4	2.6	0.0	0.0	0.0	13.5
1962-63	0.0	0.0	0.0	0.0	0.0	0.0	6.6	0.0	0.0	T	0.0	0.0	6.6
1963-64	0.0	0.0	0.0	0.0	0.0	0.0	T	0.0	T	0.0	0.0	0.0	T
1964-65	0.0	0.0	0.0	0.0	T	5.7	T	0.0	0.0	0.0	0.0	0.0	5.7
1965-66	0.0	0.0	0.0	0.0	0.0	7.4	1.7	0.2	0.6	0.0	0.0	0.0	9.9
1966-67	0.0	0.0	0.0	0.0	0.0	T	T	T	T	T	0.0	0.0	T
1967-68	0.0	0.0	0.0	0.0	0.0	1.0	9.3	0.0	0.0	T	0.0	0.0	10.3
1968-69	0.0	0.0	0.0	0.0	T	12.6	21.9	0.4	0.0	0.0	0.0	0.0	34.9
1969-70	0.0	0.0	0.0	0.0	0.0	0.0	0.2	0.0	T	T	0.0	0.0	0.2
1970-71	0.0	0.0	0.0	T	T	0.9	11.1	8.1	1.4	T	0.0	0.0	21.5
1971-72	0.0	0.0	0.0	0.0	0.0	2.2	0.4	T	0.3	0.1	0.0	0.0	3.0
1972-73	0.0	0.0	0.0	0.0	0.0	14.6	0.7	0.0	T	T	0.0	0.0	15.3
1973-74	0.0	0.0	0.0	0.0	0.5	0.0	T	T	T	0.0	0.0	0.0	0.5
1974-75	0.0	0.0	0.0	T	0.0	0.0	1.6	T	T	T	T	0.0	1.6
1975-76	0.0	0.0	0.0	0.0	T	1.5	0.0	0.0	0.3	T	0.0	0.0	1.8
1976-77	0.0	0.0	0.0	0.0	0.0	0.0	T	0.0	T	0.0	T	0.0	T
1977-78	0.0	0.0	0.0	0.0	6.1	0.0	0.0	0.0	0.0	T	T	0.0	6.1
1978-79	0.0	0.0	0.0	0.0	1.5	2.9	2.2	0.6	0.0	T	0.0	0.0	7.2
1979-80	0.0	0.0	0.0	0.0	0.0	T	1.8	T	T	T	0.0	0.0	1.8
1980-81	0.0	0.0	0.0	0.0	0.0	T	0.0	0.0	0.0	0.0	0.0	0.0	T
1981-82	0.0	0.0	T	0.0	T	T	4.4	0.6	T	T	0.0	0.0	5.0
1982-83	0.0	0.0	0.0	0.0	0.0	0.0	0.0	0.0	T	T	0.0	0.0	T
1983-84	0.0	0.0	0.0	0.0	T	4.3	T	0.0	0.0	T	0.0	0.0	4.3
1984-85	0.0	0.0	0.0	0.0	0.0	0.3	T	4.6	T	T	0.0	0.0	4.9
1985-86	0.0	0.0	0.0	0.0	2.0	6.5							
Record Mean	0.0	0.0	T	T	0.3	1.5	3.2	0.8	0.6	T	T	0.0	6.5

See Reference Notes, relative to all above tables, on preceding page.

The National Weather Service Office, Sexton Summit, is located at the peak of one of the minor mountains in the coast range of southwestern Oregon. The terrain is mostly covered with coniferous timber and slopes away steeply in all directions to valleys some 2,500 feet below the station. This small mountain is a typical example of the general character of the terrain for a distance of 50 miles or more in all directions. Many of the peaks are higher than Sexton Mountain, but few are over 5,000 feet above sea level. The Pacific Ocean lies 60 miles to the west.

The station is exposed to free circulating air that has not been held in close proximity with the ground and this has a stabilizing effect on the temperature. The extremes of both high and low temperature lack about 10 degrees of equaling those of the surrounding valleys.

Clouds often form over the valleys at an elevation below the station, and in drifting by, cause much of the fog recorded at all times of the year. Stratiform types based below the station sometimes cause the station to remain obscured for several consecutive days during the winter. At other times the station is entirely above such a layer and records clear skies while the valleys are overcast.

Precipitation occurs very frequently during the winter months and usually takes the form of rain in the valleys and snow on the elevations 3,000 feet above sea level. Large accumulations of snow are very rare as the temperature usually rises well above freezing between storms. Only the highest mountains in the area remain snow covered for more than a few days.

Persistent accumulations of rime form in midwinter due to fog at temperatures slightly below freezing. This occurs only near the tops of the mountains and the temperature in the valleys is usually several degrees above freezing at the time.

The winds accompanying a storm in winter occasionally reach damaging velocity at this exposed position, but in the more sheltered areas wind damage is a rarity.

The general character of the climate is mild with some six months of very fair weather in the summer and much cloudiness and rain in the winter. The structure of the soil is such that the precipitation drains away very rapidly and the ground becomes firm and dry in a few minutes after a rain.

Based on the 1951–1980 period, the average first occurrence of 32 degrees Fahrenheit in the fall is October 22 and the average last occurrence in the spring is May 26.

TABLE 1 NORMALS, MEANS AND EXTREMES

SEXTON SUMMIT, OREGON

LATITUDE: 42°37'N   LONGITUDE: 123°22'W   ELEVATION: FT. GRND 3836 BARO 03841   TIME ZONE: PACIFIC   WBAN: 24235

	(a)	JAN	FEB	MAR	APR	MAY	JUNE	JULY	AUG	SEP	OCT	NOV	DEC	YEAR
TEMPERATURE °F:														
Normals														
-Daily Maximum		39.9	43.0	44.4	51.0	59.3	66.9	75.8	74.3	69.6	58.6	46.8	41.6	55.9
-Daily Minimum		30.1	31.8	30.8	33.1	38.9	45.1	51.9	51.8	49.6	43.2	36.0	31.8	39.5
-Monthly		35.0	37.4	37.6	42.0	49.1	56.0	63.9	63.0	59.6	50.9	41.4	36.7	47.7
Extremes														
-Record Highest	41	66	67	74	82	89	95	100	97	97	87	77	66	100
-Year		1985	1977	1966	1947	1983	1961	1946	1981	1955	1980	1966	1980	JUL 1946
-Record Lowest	41	-2	9	15	19	22	27	36	36	26	20	11	2	-2
-Year		1962	1962	1971	1972	1965	1966	1981	1947	1945	1971	1955	1972	JAN 1962
NORMAL DEGREE DAYS:														
Heating (base 65°F)		930	773	849	687	493	283	112	140	217	445	708	874	6511
Cooling (base 65°F)		0	0	0	0	0	16	77	81	55	8	0	0	237
% OF POSSIBLE SUNSHINE														
MEAN SKY COVER (tenths)														
Sunrise - Sunset	28	7.7	7.6	7.5	6.8	6.0	4.7	2.1	2.7	3.5	5.7	7.4	7.6	5.8
MEAN NUMBER OF DAYS:														
Sunrise to Sunset														
-Clear	28	5.0	4.5	5.2	5.9	9.6	13.4	23.5	21.0	17.4	10.9	5.8	5.0	127.0
-Partly Cloudy	28	4.7	4.3	4.9	7.5	7.0	6.9	4.8	6.1	5.9	6.3	4.6	4.4	67.3
-Cloudy	28	21.3	19.4	21.0	16.6	14.4	9.7	2.8	3.9	6.8	13.8	19.6	21.7	170.9
Precipitation														
.01 inches or more	41	16.4	14.9	16.7	11.7	9.4	5.9	2.0	2.8	5.3	10.1	15.5	17.0	127.6
Snow, Ice pellets														
1.0 inches or more	41	5.5	4.6	5.7	3.1	0.7	0.*	0.0	0.0	0.*	0.5	2.9	4.8	27.8
Thunderstorms	21	0.*	0.1	0.1	0.3	1.3	1.1	0.7	0.9	0.7	0.2	0.0	0.0	5.5
Heavy Fog Visibility														
1/4 mile or less	21	18.8	17.2	19.0	14.2	13.6	10.4	5.5	5.6	7.3	13.1	16.9	17.8	159.3
Temperature °F														
-Maximum														
90° and above	41	0.0	0.0	0.0	0.0	0.0	0.3	0.9	1.1	0.2	0.0	0.0	0.0	2.5
32° and below	41	6.6	3.5	2.9	0.7	0.1	0.0	0.0	0.0	0.0	0.1	1.8	5.5	21.2
-Minimum														
32° and below	41	18.7	16.0	20.0	15.9	6.7	0.5	0.0	0.0	0.2	3.3	11.5	17.8	110.6
0° and below	41	0.*	0.0	0.0	0.0	0.0	0.0	0.0	0.0	0.0	0.0	0.0	0.0	*
AVG. STATION PRESS. (mb)	8	883.5	882.8	881.5	883.8	884.5	885.2	885.8	885.4	884.8	885.2	884.0	884.8	884.3
RELATIVE HUMIDITY (%)														
Hour 04	7	80	79	84	84	74	73	65	67	66	70	81	79	75
Hour 10 (Local Time)	36	77	77	78	73	69	67	60	61	59	66	76	78	70
Hour 16	19	72	71	68	63	53	49	40	42	45	58	75	76	59
Hour 22	7	79	78	80	81	70	58	60	60	63	69	80	79	72
PRECIPITATION (inches):														
Water Equivalent														
-Normal		7.00	4.35	4.02	2.14	1.86	0.96	0.27	0.70	1.31	3.15	5.96	6.42	38.14
-Maximum Monthly	44	19.03	10.34	8.97	7.54	7.26	4.69	3.78	4.23	4.26	13.87	24.09	20.03	24.09
-Year		1970	1961	1960	1963	1963	1947	1947	1976	1977	1950	1973	1964	NOV 1973
-Minimum Monthly	44	0.18	0.81	0.26	0.37	0.11	0.00	0.00	0.00	0.00	0.03	0.57	0.49	0.00
-Year		1984	1964	1965	1946	1982	1960	1980	1981	1974	1978	1959	1976	AUG 1981
-Maximum in 24 hrs	41	5.98	3.96	2.64	2.03	3.07	2.24	2.23	1.66	2.27	5.20	4.74	4.43	5.98
-Year		1974	1961	1972	1965	1963	1978	1947	1983	1977	1950	1973	1964	JAN 1974
Snow, Ice pellets														
-Maximum Monthly	44	106.7	57.8	45.5	43.5	12.8	1.5		T	3.1	17.3	42.6	80.3	106.7
-Year		1969	1956	1948	1948	1953	1954		1976	1971	1956	1973	1971	JAN 1969
-Maximum in 24 hrs	44	34.1	22.6	18.3	13.6	8.2	1.5		T	3.1	8.8	10.0	33.5	34.1
-Year		1969	1959	1945	1948	1953	1954		1976	1971	1956	1960	1965	JAN 1969
WIND:														
Mean Speed (mph)	6	12.8	12.5	11.7	10.6	10.2	11.1	11.8	10.9	11.1	11.9	14.1	13.3	11.8
Prevailing Direction														
through 1963		S	S	S	S	S	N	N	N	N	S	S	S	S
Fastest Obs. 1 Min.														
-Direction (!!!)	11	16	19	21	20	20	18	35	20	12	14	19	20	20
-Speed (MPH)	11	60	53	60	50	44	42	39	45	45	51	59	63	63
-Year		1969	1972	1977	1971	1975	1968	1973	1972	1971	1967	1968	1970	DEC 1970
Peak Gust														
-Direction (!!!)														
-Speed (mph)														
-Date														

See Reference Notes to this table on the following page.

TABLE 2 PRECIPITATION (inches)   SEXTON SUMMIT, OREGON

YEAR	JAN	FEB	MAR	APR	MAY	JUNE	JULY	AUG	SEP	OCT	NOV	DEC	ANNUAL
1956	10.00	5.26	2.49	0.60	2.82	1.01	1.20	0.07	0.69	7.36	0.90	3.30	35.70
1957	3.49	5.14	7.09	1.12	2.52	0.24	0.38	0.04	2.45	3.30	3.33	7.12	36.22
1958	8.42	9.36	3.58	1.66	1.16	2.85	0.24	0.02	0.76	1.70	2.90	3.46	36.11
1959	8.47	3.66	5.17	1.97	1.45	0.68	T	0.07	2.81	1.52	0.57	2.69	24.58
1960	4.51	8.27	8.97	1.83	5.56	0.00	0.01	0.33	0.36	2.13	10.83	3.90	46.70
1961	4.84	10.34	8.89	1.76	3.78	1.13	0.35	0.27	1.33	5.34	11.78	4.44	54.25
1962	1.81	5.97	3.80	1.82	1.81	0.38	0.00	2.21	1.68	9.77	6.87	4.92	41.04
1963	3.95	6.73	5.17	7.54	7.26	1.47	0.35	T	0.54	2.99	11.75	2.83	50.58
1964	10.31	0.81	3.35	0.75	0.65	1.42	0.62	0.01	0.27	0.89	8.18	20.03	47.29
1965	6.41	1.14	0.26	4.40	0.51	0.45	0.05	1.33	0.00	0.89	4.70	6.95	27.09
1966	5.73	1.39	4.17	1.13	0.19	0.84	0.95	0.46	1.82	0.93	9.78	5.53	32.92
1967	5.66	1.21	3.40	1.74	1.07	0.28	0.00	0.00	0.62	2.51	1.33	2.96	20.78
1968	5.41	4.23	2.02	0.71	1.37	0.18	0.06	2.42	0.90	3.81	5.15	7.60	33.86
1969	13.10	2.52	0.95	1.76	2.67	1.65	0.19	0.00	1.25	4.62	1.39	14.05	44.15
1970	19.03	2.88	2.43	2.52	0.63	0.73	0.00	0.00	0.33	3.28	11.80	7.53	51.16
1971	9.44	3.80	7.36	3.75	1.57	1.69	0.15	0.59	2.19	1.31	9.48	8.46	50.81
1972	10.02	7.07	7.17	2.87	1.38	0.93	T	0.90	0.99	2.46	3.32	9.26	45.57
1973	6.65	2.52	4.30	1.02	0.44	0.43	0.14	0.10	3.10	6.40	24.09	5.89	58.45
1974	17.33	5.11	7.26	1.96	0.49	0.19	0.46	0.02	0.00	2.36	3.38	5.89	44.45
1975	4.17	7.22	7.34	2.45	0.67	0.32	0.38	0.80	0.70	5.00	5.10	4.56	38.71
1976	1.76	2.63	3.11	1.29	0.58	0.34	1.32	4.23	1.26	0.59	0.74	0.49	18.34
1977	1.02	2.35	2.98	1.01	4.06	0.34	0.22	0.66	4.26	1.82	5.17	5.46	29.35
1978	3.42	4.51	2.89	1.89	1.11	2.80	0.23	2.09	3.74	0.03	2.46	1.12	26.29
1979	3.18	4.76	1.86	3.76	2.96	0.64	0.49	1.21	0.89	6.59	4.57	5.61	36.52
1980	4.19	3.43	2.99	2.99	0.96	1.02	0.00	0.00	0.17	3.06	4.21	6.61	29.63
1981	1.46	3.09	3.06	0.96	1.43	1.27	0.16	0.00	1.64	4.99	8.27	10.04	36.37
1982	2.97	3.58	3.64	3.98	0.11	1.48	0.06	0.15	1.43	5.09	3.67	6.54	32.70
1983	3.84	8.49	6.98	1.90	0.95	0.76	1.81	2.89	0.23	1.67	8.74	6.72	44.98
1984	0.18	5.79	3.21	2.16	1.06	0.76	0.04	0.40	0.72	2.52	12.13	2.67	31.64
1985	0.38	2.38	2.93	0.73	0.60	1.41	0.21	0.11	1.22	0.90	2.41	1.56	14.84
Record Mean	5.90	4.18	3.84	2.07	1.86	1.13	0.35	0.61	1.19	3.45	5.92	5.84	36.35

TABLE 3 AVERAGE TEMPERATURE (deg. F)   SEXTON SUMMIT, OREGON

YEAR	JAN	FEB	MAR	APR	MAY	JUNE	JULY	AUG	SEP	OCT	NOV	DEC	ANNUAL
1956	33.6	32.0	36.5	45.9	52.1	53.5	65.8	61.8	60.2	46.1	48.1	40.5	48.0
1957	29.8	40.0	38.0	45.1	51.2	58.1	60.1	63.7	59.3	41.5	38.6	43.1	47.6
1958	38.3	39.3	35.6	42.5	56.8	57.7	68.9	68.8	57.0	55.4	42.7	43.1	50.5
1959	35.1	33.7	37.8	45.8	46.2	55.6	65.8	61.9	52.2	50.8	37.7	37.4	47.4
1960	32.1	33.7	40.7	42.0	44.7	60.0	67.5	61.3	61.6	51.4	38.1	40.6	47.8
1961	43.0	37.0	36.0	42.1	45.5	61.0	65.0	67.3	53.8	39.7	35.2	42.9	47.9
1962	36.9	35.9	35.9	46.4	43.7	55.4	64.0	60.8	61.2	51.1	41.4	42.9	48.0
1963	37.9	44.6	36.6	36.2	50.3	52.7	56.8	62.5	62.4	48.0	40.3	33.4	47.6
1964	33.4	38.2	35.6	39.6	46.8	53.4	61.3	61.7	57.4	56.1	38.0	33.4	46.2
1965	36.8	38.6	43.1	42.5	46.9	55.1	65.5	65.7	57.5	41.2	41.2	33.1	48.3
1966	33.6	33.7	39.3	47.6	51.1	54.8	59.8	63.7	59.9	52.2	41.8	37.0	47.9
1967	35.2	40.4	34.6	34.7	51.2	59.5	65.0	71.5	63.1	49.3	44.1	32.8	48.5
1968	33.7	44.0	40.2	41.0	47.3	57.1	65.5	60.8	60.8	48.8	39.4	31.1	47.5
1969	29.4	31.1	39.5	40.9	55.6	58.0	61.8	61.8	59.2	46.8	47.1	38.6	47.5
1970	37.0	42.5	39.1	37.3	50.1	60.7	66.4	66.0	56.8	50.1	41.0	31.6	48.2
1971	35.6	36.2	34.0	39.8	47.9	53.5	65.5	66.3	56.2	56.1	39.5	30.1	46.0
1972	32.3	36.9	43.0	38.3	51.9	57.5	68.1	66.2	56.8	49.3	39.8	30.3	47.5
1973	32.6	40.1	35.0	44.8	53.8	55.8	65.0	61.6	58.2	46.9	33.7	35.4	46.9
1974	30.7	33.3	36.3	39.5	45.9	57.5	61.3	64.4	67.7	52.0	39.8	33.5	46.8
1975	36.3	34.6	34.0	36.0	51.5	55.3	64.1	60.3	66.3	45.4	37.2	38.6	46.7
1976	40.4	35.8	35.9	40.3	49.6	53.1	64.0	59.7	60.4	55.5	50.0	44.0	49.0
1977	40.1	43.6	34.9	47.8	44.1	60.4	62.4	69.3	54.2	49.4	38.0	37.2	48.4
1978	38.0	38.2	44.8	39.1	46.9	57.3	63.8	61.2	52.4	56.1	38.6	32.2	47.4
1979	35.5	34.4	44.0	42.9	51.5	57.9	63.8	60.3	62.1	52.6	42.2	44.7	49.0
1980	36.9	41.1	36.7	43.7	47.9	50.9	63.7	60.3	60.3	54.1	43.3	44.7	48.6
1981	43.0	40.8	39.8	44.5	48.1	55.5	61.6	67.3	60.1	48.2	41.5	37.1	49.0
1982	33.6	37.3	36.7	40.5	50.0	57.3	61.2	62.8	54.0	48.2	38.5	36.2	46.4
1983	39.2	37.5	39.4	40.4	51.0	53.1	56.6	61.3	55.2	52.7	37.3	33.6	46.4
1984	41.4	38.3	41.2	39.6	48.9	52.9	64.1	61.3	59.2	44.5	36.3	33.1	46.9
1985	40.9	35.7	36.0	48.5	48.6	58.7	67.6	62.7	53.7	47.9	32.7	44.2	48.1
Record Mean	35.4	37.3	37.7	42.4	49.4	55.9	63.5	63.0	59.0	50.2	40.9	36.6	47.6
Max	40.5	42.9	44.4	51.3	59.7	66.8	75.4	74.5	69.1	57.8	46.2	41.4	55.8
Min	30.3	31.7	31.0	33.5	39.1	45.0	51.5	51.6	49.0	42.6	35.5	31.8	39.4

REFERENCE NOTES FOR TABLES 1, 2, 3 and 6   (SEXTON SUMMIT, OR)

GENERAL

T - TRACE AMOUNT
BLANK ENTRIES DENOTE MISSING/UNREPORTED DATA.
# INDICATES A STATION OR INSTRUMENT RELOCATION.

SPECIFIC

TABLE 1

(a) - LENGTH OF RECORD IN YEARS. ALTHOUGH INDIVIDUAL MONTHS MAY BE MISSING.
* LESS THAN .05

NORMALS — BASED ON THE 1951-1980 RECORD PERIOD.
EXTREMES — DATES ARE THE MOST RECENT OCCURRENCE.
WIND DIR. — NUMERALS SHOW TENS OF DEGREES CLOCKWISE FROM TRUE NORTH. "00" INDICATES CALM.
RESULTANT WIND DIRECTIONS ARE GIVEN TO WHOLE DEGREES.

EXCEPTIONS

TABLE 1

1. THUNDERSTORMS AND HEAVY FOG ARE THROUGH 1964 AND MAY BE INCOMPLETE, DUE TO PART-TIME OPERATIONS.
2. MEAN WIND SPEEDS ARE THROUGH 1964.
3. FASTEST OBSERVED WINDS ARE INCOMPLETE FROM 1965 THROUGH AUGUST 1977. PEAK GUST WINDS WERE REPORTED SEPTEMBER 1977-FEBRUARY 1980 AND NOT INCLUDED IN EXTREMES.
4. RELATIVE HUMIDITY HOURS 04 AND 22 ARE FOR YEARS 1967, 1968, 1972, 1973, 1975, 1976, AND 1978.

TABLES 2, 3, and 6

RECORD MEANS ARE THROUGH THE CURRENT YEAR, BEGINNING IN   1945 FOR TEMPERATURE
1942 FOR PRECIPITATION
1942 FOR SNOWFALL

TABLE 4 HEATING DEGREE DAYS Base 65 deg. F SEXTON SUMMIT, OREGON

SEASON	JULY	AUG	SEP	OCT	NOV	DEC	JAN	FEB	MAR	APR	MAY	JUNE	TOTAL
1956-57	97	135	164	582	499	749	1083	693	828	593	428	214	6065
1957-58	157	184	120	584	702	811	823	714	901	668	269	265	6198
1958-59	22	15	265	319	662	671	918	870	834	569	580	291	6016
1959-60	114	134	389	437	549	839	1010	901	743	684	623	179	6602
1960-61	50	194	155	420	800	748	676	781	892	680	593	200	6189
1961-62	84	50	334	513	752	914	863	808	894	552	652	290	6706
1962-63	125	163	163	424	701	680	834	566	875	859	461	381	6232
1963-64	254	112	144	521	733	694	973	770	905	757	560	355	6778
1964-65	162	165	241	303	804	972	868	732	670	668	552	297	6434
1965-66	99	134	227	251	711	982	967	872	792	515	436	323	6309
1966-67	180	116	194	397	690	861	916	684	935	902	434	209	6518
1967-68	73	25	113	479	619	992	963	601	762	712	540	270	6149
1968-69	78	201	184	498	763	1045	1095	944	782	718	465	235	6862
1969-70	130	132	212	557	531	812	862	624	796	822	528	341	6178
1970-71	48	39	270	479	713	1030	906	799	954	750	528	341	6857
1971-72	123	74	276	569	756	1072	1008	808	679	795	412	236	6808
1972-73	40	111	288	479	749	1068	999	689	925	600	364	300	6612
1973-74	107	151	244	553	931	913	1059	881	881	758	586	251	7315
1974-75	185	96	54	412	748	971	883	846	952	861	431	304	6743
1975-76	109	166	71	610	827	813	757	840	895	736	472	363	6659
1976-77	86	203	161	301	441	644	766	592	928	510	640	151	5423
1977-78	126	74	329	477	805	854	832	745	618	771	557	254	6442
1978-79	135	189	377	283	784	1012	908	849	643	657	416	240	6493
1979-80	122	135	141	405	676	762	863	684	869	632	524	416	6229
1980-81	99	152	172	386	643	621	674	672	774	611	521	280	5605
1981-82	155	90	237	513	700	859	965	770	871	730	459	264	6613
1982-83	169	138	334	515	788	888	796	763	787	732	466	357	6733
1983-84	269	151	292	394	823	965	723	767	729	754	504	369	6740
1984-85	84	113	205	628	850	981	740	813	892	488	502	221	6517
1985-86	34	132	346	529	964	637							

TABLE 5 COOLING DEGREE DAYS Base 65 deg. F SEXTON SUMMIT, OREGON

YEAR	JAN	FEB	MAR	APR	MAY	JUNE	JULY	AUG	SEP	OCT	NOV	DEC	TOTAL
1969	0	0	0	0	33	34	37	37	45	0	0	0	186
1970	0	0	0	0	10	110	98	80	33	26	0	0	357
1971	0	0	0	0	4	3	147	124	18	18	0	0	314
1972	0	0	0	0	13	16	143	155	49	0	0	0	376
1973	0	0	0	0	24	28	117	54	48	0	0	0	271
1974	0	0	0	0	0	32	76	83	141	16	0	0	348
1975	0	0	0	0	20	20	89	27	118	9	0	0	283
1976	0	0	0	0	3	13	61	47	29	13	0	0	166
1977	0	0	0	0	0	18	50	212	11	0	0	0	291
1978	0	0	0	0	1	28	106	79	8	11	0	0	233
1979	0	0	0	0	5	35	88	11	63	26	0	0	228
1980	0	0	0	0	0	0	64	13	37	56	0	0	170
1981	0	0	0	3	3	1	53	168	93	1	0	0	322
1982	0	0	0	0	3	40	58	79	11	0	0	0	191
1983	0	0	0	0	39	5	15	42	4	0	0	0	105
1984	0	0	0	0	13	11	63	53	35	0	0	0	175
1985	0	0	0	0	0	38	121	65	13	3	0	0	240

TABLE 6 SNOWFALL (inches) SEXTON SUMMIT, OREGON

SEASON	JULY	AUG	SEP	OCT	NOV	DEC	JAN	FEB	MAR	APR	MAY	JUNE	TOTAL
1956-57	0.0	0.0	0.0	17.3	2.0	7.7	30.8	5.6	22.6	4.5	0.9	0.0	91.4
1957-58	0.0	0.0	0.0	6.7	5.3	31.1	12.7	12.0	34.8	12.6	0.0	0.0	115.2
1958-59	0.0	0.0	0.0	T	6.7	10.1	8.8	41.8	6.1	0.6	2.7	0.0	76.8
1959-60	0.0	0.0	0.0	0.0	0.0	13.4	23.5	20.8	20.7	4.9	4.7	0.0	88.0
1960-61	0.0	0.0	0.0	0.0	14.4	1.8	1.0	13.4	32.0	9.2	5.2	0.0	77.0
1961-62	0.0	0.0	0.0	7.8	10.9	10.2	14.8	8.7	24.1	2.4	1.0	T	79.9
1962-63	0.0	0.0	0.0	0.0	6.4	0.2	4.3	0.0	14.3	35.2	1.0	T	61.4
1963-64	0.0	0.0	0.0	0.0	4.7	3.2	45.3	9.6	36.4	4.9	5.8	0.0	109.9
1964-65	0.0	0.0	0.0	0.0	32.2	38.5	38.5	3.6	1.4	11.1	T	1.2	126.4
1965-66	0.0	0.0	0.0	T	7.5	64.0	17.4	10.0	17.2	T	T	1.2	117.3
1966-67	0.0	0.0	0.0	2.4	10.0	19.7	25.9	12.7	25.0	12.5	1.2	0.0	109.4
1967-68	0.0	0.0	0.0	0.0	5.7	25.5	25.5	T	5.1	1.3	0.4	0.0	63.5
1968-69	0.0	0.0	0.1	1.6	17.6	39.6	106.7	16.0	8.3	4.9	0.5	0.9	196.2
1969-70	0.0	0.0	0.0	T	0.5	13.3	5.8	9.8	6.7	18.3	2.8	0.0	57.2
1970-71	0.0	0.0	0.0	2.8	8.2	41.2	29.0	33.1	31.9	31.0	T	0.5	180.6
1971-72	0.0	0.0	3.1	3.5	7.9	80.3	22.0	32.8	32.3	4.3	0.3	T	172.8
1972-73	0.0	0.0	T	7.2	4.6	21.3	25.1	1.8	32.3	8.6	1.5	0.0	96.9
1973-74	0.0	0.0	T	T	42.6	29.5	7.0	41.9	35.8	20.6	1.7	T	166.9
1974-75	0.0	0.0	0.0	0.0	0.5	20.6	24.3	43.9	34.0	20.6	1.7	0.0	145.6
1975-76	0.0	0.0	0.0	9.3	17.7	17.7	6.8	11.5	28.7	5.5	0.4	0.0	97.6
1976-77	0.0	T	0.0	0.0	0.0	1.5	9.7	18.5	27.2	1.0	5.5	0.4	63.4
1977-78	0.0	0.0	0.0	T	15.4	16.0	13.0	2.1	14.7	6.4	7.4	0.4	64.3
1978-79	0.0	0.0	T	0.2	2.3	4.0	8.2	18.5	1.7	10.8	1.3	T	48.7
1979-80	0.0	0.0	0.0	T	0.0	8.6	14.3	8.3	T	10.8	0.9	0.6	58.7
1980-81	0.0	0.0	0.0	T	T	2.3	9.2	4.1	8.3	10.8	0.9	0.6	37.1
1981-82	0.0	0.0	0.0	2.5	8.8	11.7	29.8	3.6	29.1	27.8	0.1	0.0	113.4
1982-83	0.0	0.0	0.0	T	14.7	14.2	5.2	28.1	15.8	7.9	4.2	0.0	90.1
1983-84	0.0	0.0	0.0	0.0	16.0	24.2	T	31.6	7.7	8.9	0.5	0.0	88.9
1984-85	0.0	0.0	0.0	2.3	15.1	21.7	3.5	12.0	25.8	3.3	0.8	0.0	84.5
1985-86	0.0	0.0	0.0	0.5	14.7	3.2							
Record Mean	0.0	T	0.1	1.6	9.9	20.0	21.7	15.8	19.5	9.2	2.1	0.1	100.0

See Reference Notes, relative to all above tables, on preceding page.

Guam is the largest and southernmost of the Mariana Islands. The Philippine Sea lies to the west and the Pacific Ocean to the east. The island is 28 miles long, 4 to 8 miles wide, and is oriented north northeast and south southwest. Located 1,500 miles east of Manila and 3,000 miles west of Honolulu, Guam serves as an important stopping place for aircraft and ships. It also has long been an important American military base. Outside of the activities of the Federal and Territorial governments, the most important single industry is agriculture.

Guam is shaped like a bow tie, and in correspondence with this shape, there are three topographic regions. The northern portion of the island is a limestone plateau that is bounded by steep cliffs that either fall directly to the sea or to narrow beaches. The surface of the plateau is 300 to 600 feet above sea level. The southern portion of the island is mountainous with several peaks that rise above 1,000 feet. The highest of these is Mount Lamlam which reaches 1,334 feet. The third major region, the narrow waist between the northern and southern regions, is quite low, generally less than 200 feet above sea level.

The National Weather Service station on Guam is located on the western side of the northern plateau. The ocean is 1 1/2 miles to the west, 9 miles to the north and east, and 5 miles to the southeast. The weather instruments at the station are well exposed in the center of an open field that is 40 acres in area. The trade winds reach the station after rising sharply up the 500-foot cliffs on the eastern side of the island and flowing 9 miles on an easy downslope grade across the surface of the northern plateau.

The climate of Guam is almost uniformly warm and humid throughout the year. Afternoon temperatures are typically in the middle or high 80s and nighttime temperatures typically fall to the low 70s or high 60s. Relative humidity commonly ranges from around 65 to 75 percent in the afternoon to 85 to 100 percent at night. Though temperature and humidity vary only slightly throughout the year, rainfall and wind conditions vary markedly, and it is these latter variations that really define the seasons.

There are two primary seasons and two secondary seasons on Guam. The primary seasons are the four-month dry season, which extends from January through April, and the four-month rainy season which extends from mid-July to mid-November. The secondary seasons are May to mid-July and mid-November through December. These are transitional seasons that may be either rainy or dry depending upon the nature of the particular year. On the average, about 15 percent of the annual rainfall occurs during the dry season and 55 percent during the rainy season.

At all times of the year the dominant winds on Guam are the trade winds which blow from the east or northeast. The trades are strongest and most constant during the dry season, when wind speeds of 15 to 25 mph are very common. During the rainy season there is often a breakdown of the trades, and on some days the weather may be dominated by westerly-moving storm systems that bring heavy showers or steady, and sometimes torrential, rain. Occasionally there are typhoons, and these bring not only tremendous rains, but also violent winds that may cause a surge of water onto low-lying coastal areas. Typhoons have passed sufficiently close to Guam to produce high winds and heavy rains in every month, but their most frequent occurrence is during the latter half of the year.

TABLE 1 — NORMALS, MEANS AND EXTREMES

GUAM, PACIFIC

LATITUDE: 13°33'N   LONGITUDE: 144°50' E   ELEVATION: FT. GRND   361 BARO   00365   TIME ZONE: 150E MER   WBAN: 41415

	(a)	JAN	FEB	MAR	APR	MAY	JUNE	JULY	AUG	SEP	OCT	NOV	DEC	YEAR
TEMPERATURE °F:														
Normals														
-Daily Maximum		83.4	83.4	84.4	85.7	86.5	86.9	86.4	86.0	85.8	85.7	85.3	84.2	85.3
-Daily Minimum		71.0	71.1	71.2	72.4	72.9	73.1	72.5	72.2	72.2	72.4	73.1	72.6	72.2
-Monthly		77.2	77.3	77.8	79.1	79.7	80.0	79.5	79.1	79.0	79.1	79.2	78.4	78.8
Extremes														
-Record Highest	28	87	88	89	90	91	92	94	91	95	91	89	89	95
-Year		1978	1960	1965	1960	1984	1983	1983	1968	1957	1957	1980	1966	SEP 1957
-Record Lowest	28	56	59	54	59	62	63	66	63	61	64	62	61	54
-Year		1978	1959	1965	1965	1960	1983	1983	1983	1958	1982	1957	1980	MAR 1965
NORMAL DEGREE DAYS:														
Heating (base 65°F)		0	0	0	0	0	0	0	0	0	0	0	0	0
Cooling (base 65°F)		378	344	397	423	456	450	450	437	420	437	426	415	5033
% OF POSSIBLE SUNSHINE	23	54	56	62	63	59	55	44	39	38	39	43	46	50
MEAN SKY COVER (tenths)														
Sunrise - Sunset	13	7.4	7.8	7.4	7.5	7.5	8.1	8.5	9.0	8.7	8.2	7.7	7.6	8.0
MEAN NUMBER OF DAYS:														
Sunrise to Sunset														
-Clear	13	1.3	0.9	1.5	0.8	0.3	0.3	0.1	0.0	0.2	0.4	0.6	0.9	7.3
-Partly Cloudy	13	13.4	10.6	14.4	13.5	14.8	10.2	8.5	4.4	5.8	10.2	13.1	13.3	132.3
-Cloudy	13	16.3	16.8	15.2	15.6	15.8	19.5	22.4	26.6	24.0	20.5	16.3	16.8	225.7
Precipitation														
.01 inches or more	29	20.2	18.2	18.7	19.8	19.8	23.7	25.7	25.3	25.8	25.4	25.1	23.3	271.0
Snow, Ice pellets														
1.0 inches or more	29	0.0	0.0	0.0	0.0	0.0	0.0	0.0	0.0	0.0	0.0	0.0	0.0	0.0
Thunderstorms	12	0.6	0.0	0.0	0.5	1.1	1.8	4.4	5.8	6.6	3.5	2.2	0.3	26.8
Heavy Fog Visibility														
1/4 mile or less	12	0.0	0.0	0.0	0.0	0.0	0.0	0.0	0.0	0.0	0.0	0.0	0.2	0.2
Temperature °F														
-Maximum														
90° and above	28	0.0	0.0	0.0	0.*	1.2	2.3	1.2	0.3	0.3	0.3	0.0	0.0	5.6
32° and below	28	0.0	0.0	0.0	0.0	0.0	0.0	0.0	0.0	0.0	0.0	0.0	0.0	0.0
-Minimum														
32° and below	28	0.0	0.0	0.0	0.0	0.0	0.0	0.0	0.0	0.0	0.0	0.0	0.0	0.0
0° and below	28	0.0	0.0	0.0	0.0	0.0	0.0	0.0	0.0	0.0	0.0	0.0	0.0	0.0
AVG. STATION PRESS.(mb)														
RELATIVE HUMIDITY (%)														
Hour 04	3	88	88	88	90	92	92	94	96	96	95	91	88	92
Hour 10 (Local Time)	18	77	76	75	74	73	75	79	81	81	80	80	78	77
Hour 16	18	75	74	73	72	73	75	78	80	80	80	81	78	77
Hour 22	17	87	86	86	86	88	90	92	94	94	93	91	88	90
PRECIPITATION (inches):														
Water Equivalent														
-Normal		5.43	4.76	4.19	4.14	6.41	5.53	10.31	13.92	14.25	13.87	8.98	6.07	97.86
-Maximum Monthly	29	20.39	14.79	16.94	19.55	40.13	14.61	20.00	25.66	27.13	26.05	18.14	16.19	40.13
-Year		1976	1980	1971	1963	1976	1985	1972	1974	1982	1979	1957	1963	MAY 1976
-Minimum Monthly	29	1.31	0.67	0.59	0.50	0.90	0.80	4.74	3.87	6.79	6.63	2.08	2.22	0.50
-Year		1983	1960	1965	1965	1959	1983	1957	1957	1969	1976	1973	1977	APR 1965
-Maximum in 24 hrs	29	6.26	9.24	3.55	6.37	27.00	4.55	5.17	8.29	7.48	10.14	7.26	6.09	27.00
-Year		1976	1980	1972	1974	1976	1958	1969	1984	1965	1979	1957	1963	MAY 1976
Snow, Ice pellets														
-Maximum Monthly	29													
-Year														
-Maximum in 24 hrs	29													
-Year														
WIND:														
Mean Speed (mph)	6	8.2	10.2	9.0	8.9	8.3	6.3	5.1	4.8	4.7	6.2	7.8	9.1	7.4
Prevailing Direction														
through 1963		E	NE	E	E	E	E	E	E	E	E	E	E	E
Fastest Mile														
-Direction (!!!)	27	NE	NE	NE	SW	NE	E	E	SE	E	W	NE	E	NE
-Speed (MPH)	27	37	36	32	64	76	32	34	43	35	44	80	35	80
-Year		1979	1962	1977	1963	1976	1976	1963	1979	1976	1968	1962	1985	NOV 1962
Peak Gust														
-Direction (!!!)	2	NE	NE	NE	SE	SE	SE	SW	SE	NE	E	E	E	E
-Speed (mph)	2	39	35	35	41	29	36	29	35	36	48	72	46	72
-Date		1985	1985	1985	1985	1985	1985	1985	1985	1985	1984	1984	1985	NOV 1984

See Reference Notes to this table on the following page.

TABLE 2 PRECIPITATION (inches) GUAM, PACIFIC

YEAR	JAN	FEB	MAR	APR	MAY	JUNE	JULY	AUG	SEP	OCT	NOV	DEC	ANNUAL
1956										8.16	13.45	10.44	
1957	5.51	3.77	3.00	3.13	1.83	3.46	4.74	13.77	14.79	20.59	18.14	2.96	95.69
1958	7.91	2.23	1.70	2.44	3.87	11.53	12.89	11.22	14.43	12.12	5.80	4.62	90.76
1959	3.45	2.27	2.08	3.22	0.90	2.27	6.15	11.56	18.39	10.76	11.71	6.31	79.07
1960	2.71	0.67	1.73	1.44	5.69	7.54	7.93	14.82	9.35	11.78	11.30	6.23	81.19
1961	7.66	2.74	4.85	4.28	5.37	5.90	8.10	16.61	15.40	19.17	5.31	4.35	99.74
1962	2.24	9.17	1.66	8.66	6.68	7.97	14.98	14.12	21.86	14.95	13.09	13.08	128.46
1963	9.33	9.47	2.46	19.55	12.56	7.11	11.25	11.85	13.47	18.06	6.88	16.19	138.18
1964	1.99	4.11	4.44	7.48	20.69	7.49	7.67	8.92	19.30	14.62	6.05	6.92	109.68
1965	10.29	1.02	0.59	0.50	1.69	5.35	15.67	3.87	22.27	6.89	6.97	3.13	78.24
1966	2.13	1.69	1.71	1.31	2.51	5.72	6.74	11.39	22.28	6.96	7.26	4.76	74.46
1967	5.26	4.45	11.28	8.97	3.76	8.37	12.62	23.07	20.28	14.20	11.75	2.51	126.52
1968	5.75	8.48	3.71	1.69	3.16	6.17	12.36	14.61	11.82	13.31	11.92	3.36	96.34
1969	2.79	1.12	0.97	1.36	2.72	1.52	13.83	9.30	6.79	25.32	10.10	10.11	85.93
1970	11.93	3.60	2.68	1.22	2.96	5.60	8.99	10.42	13.70	11.65	9.77	5.71	88.23
1971	5.25	5.78	16.94	5.74	22.68	5.06	16.06	20.92	12.21	8.63	6.75	3.14	129.16
1972	4.86	6.40	11.29	3.35	3.02	7.95	20.00	18.09	16.09	7.09	4.83	6.35	109.32
1973	2.10	2.94	1.57	1.90	3.01	4.85	9.55	6.65	9.05	17.82	2.08	5.35	66.87
1974	4.27	3.13	12.25	14.83	10.46	9.32	25.66	9.96	10.00	8.03	6.20	12.00	126.11
1975	9.68	1.15	2.25	3.67	1.33	2.46	11.13	22.54	8.86	10.68	12.35	6.42	92.52
1976	20.39	13.56	9.04	4.37	40.13	6.40	13.38	24.40	10.79	6.63	7.91	8.91	165.91
1977	3.12	3.01	4.98	2.29	6.88	5.10	7.25	6.09	17.32	16.67	12.38	2.22	87.31
1978	2.44	4.99	0.85	2.65	3.48	7.89	8.91	19.63	10.01	10.49	14.10	4.80	90.24
1979	5.60	1.85	4.20	1.89	1.48	4.28	11.83	12.75	8.34	26.05	6.78	5.67	90.72
1980	2.07	14.79	3.42	2.92	8.60	7.96	10.93	9.42	24.34	12.02	7.76	6.14	110.37
1981	9.05	2.44	3.88	6.61	5.68	6.11	13.18	17.29	9.73	14.92	15.57	7.20	111.66
1982	2.92	8.77	2.46	0.99	6.35	8.77	13.61	6.32	27.13	11.86	9.34	6.33	104.85
1983	1.31	1.21	3.34	1.83	1.10	0.80	6.15	10.88	14.62	10.17	10.52	5.13	67.06
1984	3.19	4.19	4.02	1.56	3.10	6.69	6.58	24.05	17.41	9.40	12.93	5.89	99.01
1985	8.16	3.69	5.53	5.62	11.95	14.61	13.23	15.97	18.06	8.33	4.96	8.40	118.51
Record Mean	5.63	4.58	4.44	4.33	7.08	6.39	10.86	14.35	15.11	13.14	9.39	6.15	101.45

TABLE 3 AVERAGE TEMPERATURE (deg. F) GUAM, PACIFIC

YEAR	JAN	FEB	MAR	APR	MAY	JUNE	JULY	AUG	SEP	OCT	NOV	DEC	ANNUAL
1956										78.6	79.3	78.1	
1957	76.8	77.1	77.4	79.0	79.6	80.5	79.6	79.4	79.7	78.9	78.1	78.3	78.7
1958	77.4	76.7	77.1	79.7	80.0	79.6	79.2	80.1	79.3	80.3	79.8	78.6	79.1
1959	77.7	78.1	77.9	79.0	80.1	80.5	80.1	73.1	79.3	79.5	80.0	79.4	79.2
1960	77.7	78.6	77.9	81.2	79.9	80.0	79.7	79.2	79.7	79.0	80.1	79.1	79.3
1961	78.5	79.1	78.8	79.5	80.3	79.9	78.8	78.2	78.5	78.8	77.9	78.7	78.9
1962	77.3	77.5	79.2	78.5	79.9	79.0	79.0	78.9	79.2	79.4	79.9	78.6	79.0
1963	76.4	77.4	78.1	78.5	79.2	79.9	79.1	80.7	80.0	79.0	79.6	78.8	78.9
1964	78.0	77.4	77.8	79.2	79.3	80.0	79.0	78.7	78.7	79.0	79.2	78.0	78.7
1965	76.9	77.4	76.4	78.1	80.5	79.4	78.7	78.8	78.7	78.5	78.7	78.2	78.3
1966	77.0	77.0	77.5	80.0	79.9	80.4	80.6	79.2	78.9	79.1	79.1	78.7	79.0
1967	77.5	76.6	75.9	78.0	79.3	79.5	79.1	78.8	78.9	77.9	79.4	76.8	78.1
1968	77.3	75.6	77.2	78.6	79.1	79.4	79.5	78.5	78.4	78.0	77.7		78.1
1969	76.9	76.2	77.6	79.9	80.2	80.7	80.1	79.0	79.8	78.7	80.0	79.3	79.1
1970	76.9	77.7	78.7	79.0	79.8	80.2	78.9	79.0	79.0	79.5	79.6	79.0	79.0
1971	78.1	78.1	78.5	79.7	79.6	79.6	78.6	78.1	79.0	78.8	79.1	79.2	78.9
1972	76.7	76.5	77.3	77.7	79.0	79.4	79.4	78.8	79.4	79.2	79.4	78.3	78.4
1973	75.5	76.9	78.3	79.6	80.1	80.2	79.9	79.5	79.4	78.7	79.5	79.0	78.9
1974	77.4	78.7	77.6	79.1	79.0	79.0	78.9	78.9	79.1	79.7	79.5	78.7	78.8
1975	77.9	77.5	77.2	79.2	79.4	80.9	78.4	78.0	78.8	78.8	78.3	78.2	78.5
1976	76.1	76.6	77.4	78.4	79.1	79.5	79.4	79.0	78.6	79.5	78.8	77.6	78.3
1977	76.4	77.0	77.7	78.3	78.9	80.1	80.0	80.0	79.0	78.9	79.2	78.6	78.7
1978	76.7	76.6	78.1	79.8	81.0	79.9	79.9	79.1	79.7	79.7	78.9	78.1	79.0
1979	77.4	77.0	77.7	78.9	80.2	81.1	79.7	79.7	78.8	79.2	78.9	78.0	78.8
1980	76.4	77.4	77.6	79.3	79.5	79.7	79.6	80.2	79.4	79.8	80.4	78.0	78.9
1981	78.3	77.0	78.0		80.1	80.1	79.4	79.0	79.7	79.8	79.3		79.1
1982	77.4	76.8	77.4	78.7	79.6	79.7	78.8	78.6	78.6	77.5	78.5	77.3	78.2
1983	75.1	75.1	75.6	77.7	79.4	80.7	80.0	78.5	79.1	80.1	79.0	78.0	78.3
1984	76.8	76.2	77.6	79.1	79.9	79.4	78.8	78.5	78.4	79.6	78.9	78.6	78.5
1985	77.7	78.3	78.8	78.9	79.9	78.8	78.9	78.5	78.7	78.9	79.3	78.5	78.7
Record Mean	77.1	77.2	77.7	79.0	79.7	79.9	79.4	79.0	79.1	79.1	79.2	78.4	78.7
Max	83.4	83.4	84.4	85.6	86.6	86.9	86.3	85.9	86.0	85.8	85.2	84.2	85.3
Min	70.8	70.9	71.0	72.3	72.7	72.9	72.4	72.1	72.2	72.3	73.1	72.6	72.1

REFERENCE NOTES FOR TABLES 1, 2, 3 and 6 (GUAM, PACIFIC)

GENERAL

T - TRACE AMOUNT
BLANK ENTRIES DENOTE MISSING/UNREPORTED DATA.
# INDICATES A STATION OR INSTRUMENT RELOCATION.

SPECIFIC

TABLE 1

(a) - LENGTH OF RECORD IN YEARS. ALTHOUGH INDIVIDUAL MONTHS MAY BE MISSING.

* LESS THAN .05

NORMALS — BASED ON THE 1951-1980 RECORD PERIOD.
EXTREMES — DATES ARE THE MOST RECENT OCCURRENCE.
WIND DIR. — NUMERALS SHOW TENS OF DEGREES
 CLOCKWISE FROM TRUE NORTH.
 "00" INDICATES CALM.
RESULTANT WIND DIRECTIONS ARE GIVEN TO WHOLE DEGREES.

EXCEPTIONS

TABLE 1

1. THUNDERSTORMS AND HEAVY FOG ARE FOR 1957-1964 AND 1970-1972 AND MAY BE INCOMPLETE, DUE TO PART-TIME OPERATIONS
2. MEAN WIND SPEED IS THROUGH 1964 AND 1970-1971.
3. MEAN SKY COVER, AND DAYS CLEAR-PARTLY CLOUDY-CLOUDY ARE 1968-1970 AND 1976 TO DATE
4. RELATIVE HUMIDITY HOUR 04 IS 1970-1972.

TABLES 2, 3, and 6

RECORD MEANS ARE THROUGH THE CURRENT YEAR,
BEGINNING IN 1957 FOR TEMPERATURE
 1957 FOR PRECIPITATION

TABLE 4   HEATING DEGREE DAYS Base 65 deg. F   GUAM, PACIFIC

SEASON	JULY	AUG	SEP	OCT	NOV	DEC	JAN	FEB	MAR	APR	MAY	JUNE	TOTAL
1983-84	0	0	0	0	0	0	0	0	0	0	0	0	0
1984-85	0	0	0	0	0	0	0	0	0	0	0	0	0
1985-86	0	0	0	0	0	0							

TABLE 5   COOLING DEGREE DAYS Base 65 deg. F   GUAM, PACIFIC

YEAR	JAN	FEB	MAR	APR	MAY	JUNE	JULY	AUG	SEP	OCT	NOV	DEC	TOTAL
1969	375	321	400	454	479	475	476	442	451	431	454	451	5209
1970	373	362	431	425	465	463	437	441	428	460	445	437	5167
1971	411	372	427	448	461	443	429	412	429	437	429	449	5147
1972	370	343	387	390	440	437	449	436	438	448	440	419	4997
1973	331	341	417	444	478	463	470	456	439	434	442	442	5157
1974	393	389	396	430	461	427	427	434	427	465	444	430	5123
1975	409	357	383	434	454	487	422	409	422	435	405	418	5035
1976	350	346	392	411	445	442	454	444	414	456	421	398	4973
1977	363	341	401	405	438	459	474	474	428	442	435	424	5084
1978	368	329	413	451	504	454	466	445	449	463	421	413	5176
1979	391	344	403	423	480	491	461	435	425	447	422	408	5130
1980	358	367	397	434	456	445	457	476	437	465	466	412	5170
1981	416	342	410		475	461	454	440	448	467	437	442	
1982	393	335	390	420	459	446	432	430	414	393	412	389	4913
1983	320	289	337	386	451	480	494	425	455	474	427	411	4949
1984	374	333	396	428	467	437	433	413	408	461	424	429	5003
1985	401	378	436	424	430	423	437	425	427	436	434	427	5078

TABLE 6   SNOWFALL (inches)   GUAM, PACIFIC

SEASON	JULY	AUG	SEP	OCT	NOV	DEC	JAN	FEB	MAR	APR	MAY	JUNE	TOTAL
1971-72	0.0	0.0	0.0	0.0	0.0	0.0	0.0	0.0	0.0	0.0	0.0	0.0	0.0
1972-73	0.0	0.0	0.0	0.0	0.0	0.0	0.0	0.0	0.0	0.0	0.0	0.0	0.0
1973-74	0.0	0.0	0.0	0.0	0.0	0.0	0.0	0.0	0.0	0.0	0.0	0.0	0.0
1974-75	0.0	0.0	0.0	0.0	0.0	0.0	0.0	0.0	0.0	0.0	0.0	0.0	0.0
1975-76	0.0	0.0	0.0	0.0	0.0	0.0	0.0	0.0	0.0	0.0	0.0	0.0	0.0
1976-77	0.0	0.0	0.0	0.0	0.0	0.0	0.0	0.0	0.0	0.0	0.0	0.0	0.0
1977-78	0.0	0.0	0.0	0.0	0.0	0.0	0.0	0.0	0.0	0.0	0.0	0.0	0.0
1978-79	0.0	0.0	0.0	0.0	0.0	0.0	0.0	0.0	0.0	0.0	0.0	0.0	0.0
1979-80	0.0	0.0	0.0	0.0	0.0	0.0	0.0	0.0	0.0	0.0	0.0	0.0	0.0
1980-81	0.0	0.0	0.0	0.0	0.0	0.0	0.0	0.0	0.0	0.0	0.0	0.0	0.0
1981-82	0.0	0.0	0.0	0.0	0.0	0.0	0.0	0.0	0.0	0.0	0.0	0.0	0.0
1982-83	0.0	0.0	0.0	0.0	0.0	0.0	0.0	0.0	0.0	0.0	0.0	0.0	0.0
1983-84	0.0	0.0	0.0	0.0	0.0	0.0	0.0	0.0	0.0	0.0	0.0	0.0	0.0
1984-85	0.0	0.0	0.0	0.0	0.0	0.0	0.0	0.0	0.0	0.0	0.0	0.0	0.0
1985-86	0.0	0.0	0.0	0.0	0.0								
Record Mean	0.0	0.0	0.0	0.0	0.0	0.0	0.0	0.0	0.0	0.0	0.0	0.0	0.0

See Reference Notes, relative to all above tables, on preceding page.

Johnston Island, an atoll oriented NE-SW, lies near the heart of the trade wind zone in the east-central tropical Pacific about 900 statute miles SW of Honolulu, Hawaii. The maximum elevation is less than 20 feet and prior to 1960 the land area was less than 250 acres. Extensive dredging and filling extended nearly all of the original shore line and increased the land area of Johnston Island to over 600 acres by 1970. Shallow waters extend well outward in all directions. Three small islands, each with an area of about one half square mile in 1970, built up by deposit of dredging spoil, lie 1 to 2 miles offshore to the north and east. There is no appreciable variation in climatic conditions from one to another part of the island. Due to the original vegetation having been removed during the course of various construction projects, there is very little vegetation on the island. A few palm trees have been planted and there are some seeded grass areas.

Surface trade winds are dominant at all times of the year. Winds from between NE and E (inclusive) are experienced 62 percent or more of the time in every month, with the annual average being 85 percent (10-year average). Very broadly, two seasons can be distinguished. The first extends from December through March. During each of the four months of this "winter" season the wind is from trade directions (NE through E) less than 80 percent of the time, with a low frequency value of 62 percent for January. Correspondingly, this is the season when light variable winds are most frequent. December-March is also the rainier of the two seasons, with a mean of 11 inches as contrasted with a mean of only 15 inches for the "summer" season, which is eight months long. And winter rains unlike summer ones, are frequently associated with organized disturbances, such as easterly waves, the passage of troughs aloft, or weak cold-front passages.

During the summer season--from April through November--the rainfall is dominantly in the form of trade wind showers. However, in April and again in September-October-November, identifiable organized disturbances are not uncommon, and these may bring very heavy rains, as when a tropical depression or tropical storm passes close by. Observations from weather satellites indicate that tropical storms in the Johnston Island area, although infrequent, may not be as unusual as was once supposed. But until August 19, 1972, there was no reliable evidence that the island had ever experienced a true hurricane (winds of 64 knots or more). On that date, however, the center of Hurricane CELESTE, which had first appeared off the southern coast of Mexico nearly two weeks earlier and 3,000 nautical miles to the east, passed only about 25 miles to the northeast of Johnston. Although the island had been entirely evacuated as a precaution against the possible excape of toxic gases stored there, automatic weather instruments recorded hurricane-force winds from the northern quadrant between 3:54 a.m. and 9:18 a.m. on the 19th, with a fastest mile of 105 miles an hour (by far the highest ever observed on Johnston), minimum sea level pressure of 29.04 inches and 6.21 inches of rain. Damage to the island and to its facilities appears to have been minor, although details were classified by the military.

Temperatures are slightly higher at Johnston Island during the summer than during the winter. Thus the monthly means during July-October are 80°-81°; whereas during December-March they are 76°-78°. The mean diurnal range is 7°-8°, with little seasonal variation. Humidities are generally high throughout the year. Conditions from terminal aircraft operations are excellent as compared with most locations. However, visibility is occasionally a limiting factor because of torrential showers, especially during the winter season.

NORMALS, MEANS AND EXTREMES

Elev. 17 feet m.s.l.

Temperatures °F

Month	Normal Daily maximum	Normal Daily minimum	Normal Monthly	Extremes Record highest	Year	Extremes Record lowest	Year
(a)	28			28		28	
JAN	80.5	72.6	76.7	88	1970	65	1975
FEB	80.4	72.7	76.6	87	1970	64	1962
MAR	80.8	72.9	76.9	95	1982	67	1973
APR	81.6	73.9	77.7	86	1981	65	1977
MAY	82.6	75.3	79.0	89	1982	68	1965
JUN	83.7	76.3	80.0	88	1982	69	1965
JUL	84.3	76.5	80.6	89	1982	70	1965
AUG	84.7	77.4	81.1	90	1982	71	1959
SEP	84.8	77.2	81.0	89	1982	71	1975
OCT	84.4	76.6	80.7	89	1983	66	1961
NOV	82.8	75.5	79.2	89	1969	63	1961
DEC	81.3	73.9	77.6	86	1983	62	1964
YEAR	82.6	75.1	78.9	95 SEP 1982		62 DEC 1964	

Precipitation in inches

Month	Normal	Water equiv. Max monthly	Year	Water equiv. Min monthly	Year	Max in 24 hrs.	Year
(a)		28		28		25	
JAN	2.35	12.93	1954	0.29	1963	2.65	1959
FEB	1.88	5.18	1966	0.12	1963	2.12	1966
MAR	2.72	12.38	1968	0.36	1966	3.82	1958
APR	2.41	12.41	1964	0.31	1966	6.20	1974
MAY	1.81	12.41	1963	0.11	1963	4.57	1966
JUN	0.92	2.11	1974	0.07	1974	1.14	1974
JUL	1.05	2.96	1976	0.37	1976	0.83	1978
AUG	2.11	15.80	1959	0.30	1959	3.51	1977
SEP	2.09	7.12	1957	0.33	1957	3.51	1961
OCT	3.02	12.72	1961	0.57	1954	9.51	1961
NOV	3.06	13.80	1979	0.55	1957	5.35	1979
DEC	3.10	13.04	1964	0.22	1956	9.23	1956
YEAR	26.52	15.80 AUG 1959		0.07 JUN 1974		9.51 OCT 1961	

Snow, Ice pellets

All months: Maximum monthly 0.0, Maximum in 24 hrs. 0.0, Year 0.0. (Length of record 28)

Relative humidity pct. / Wind / Sky

Month	RH Hr 01	RH Hr 07	RH Hr 13	RH Hr 19	Wind Mean speed m.p.h.	Fastest mile Speed m.p.h.	Direction	Year	Pct. of possible sunshine	Mean sky cover sunrise to sunset, tenths
(Yrs)	22	22	21	19	22	20	20		21	23
JAN	76	76	68	74	14.4	46	NE	1972	70	4.5
FEB	76	77	68	75	15.8	43	E	1973	74	4.6
MAR	79	79	70	77	16.9	49	SE	1968	75	5.0
APR	79	79	71	77	16.7	40	NE	1974	70	6.2
MAY	80	78	70	77	15.9	44	E	1979	74	6.3
JUN	79	77	69	76	16.1	40	SE	1964	78	5.7
JUL	79	78	70	77	16.0	36	E	1975	79	5.8
AUG	80	79	71	78	15.4	105	NW	1972	76	6.2
SEP	81	80	72	79	14.4	48	NW	1968	73	6.3
OCT	81	80	72	79	16.6	42	SE	1979	66	6.1
NOV	80	80	73	79	16.0	49	NE	1959	59	6.1
DEC	78	78	72	77	16.9	48	E	1966	62	5.0
YEAR	79	78	71	77	15.8	105	NW	AUG 1972	71	5.7

Mean number of days

Month	Clear	Partly cloudy	Cloudy	Precip .01 inch or more	Snow 1.0 inch or more	Thunderstorms	Heavy fog, visibility ¼ mile or less	Max 90° and above	Max 32° and below	Min 32° and below	Min 0° and below	Avg. station pressure mb.
(Yrs)	23	23	23	27	27	24	27	27	27	27	27	10
JAN	13	13	5	11	0	*	0	0	0	0	0	1012.3
FEB	13	13	5	11	0	*	0	0	0	0	0	1013.3
MAR	11	14	6	15	0	*	0	0	0	0	0	1013.9
APR	6	13	12	13	0	*	0	0	0	0	0	1014.0
MAY	6	13	12	13	0	*	0	0	0	0	0	1014.0
JUN	7	15	8	12	0	*	0	0	0	0	0	1013.6
JUL	7	15	9	9	0	*	0	0	0	0	0	1013.2
AUG	6	14	11	12	0	*	0	0	0	0	0	1012.6
SEP	5	14	11	14	0	1	0	0	0	0	0	1012.1
OCT	7	12	12	14	0	*	0	0	0	0	0	1012.2
NOV	7	11	12	15	0	*	0	0	0	0	0	1012.0
DEC	11	13	7	16	0	*	0	0	0	0	0	1012.8
YEAR	97	159	109	161	0	*	0	0	0	0	0	1013.0

FOOTNOTES

(a) Length of record, years, through the current year unless otherwise noted, based on January data.

(b) 70° and above at Alaskan stations.

* Less than one half.

T Trace.

BLANK entries denote missing or unreported data.

NORMALS - Based on record for the 1951-1980 period.
MEANS - Length of record in (a) is for complete data years.
EXTREMES - Length of record in (a) may be for other than complete or consecutive data years. Date is the most recent in cases of multiple occurrence.

WIND DIRECTION - Numerals indicate tens of degrees clockwise from true north. 00 indicates calm.

FASTEST MILE WIND - Speed is fastest observed 1-minute value when direction is in tens of degrees.

NORMALS, MEANS, AND EXTREMES TABLE NOTE(S):
1. Fastest mile wind data is through 1979.

Precipitation

Year	Jan	Feb	Mar	Apr	May	June	July	Aug	Sept	Oct	Nov	Dec	Annual
1952				1.04	1.20		0.52	0.53	2.95	2.77	0.98	0.67	
1953	0.38	0.80	2.76	1.19	1.26	0.89	1.11	0.34	0.33	1.70	0.55	1.55	12.86
1954	12.93	0.33	0.67	1.28	1.30	0.65	1.52	2.82	1.04	0.57			
1955	2.78	1.56	1.27	1.57	1.62	1.25	0.88	1.40	2.79	4.41	6.71	6.68	32.92
#1956	9.17	3.35	5.73	6.63	0.72	1.12	1.98	1.56	1.25	2.15	1.50	0.22	35.38
1957	2.99	5.08	0.87	1.06	0.28	0.31	0.50	0.30		0.89	0.66	4.44	
1958	1.39	0.67	1.31	0.79						3.89	3.65	0.62	
1959	6.42	1.54	1.05	3.46	0.49	0.88	1.02	15.80	2.02	1.61	1.03	5.64	40.96
1960	1.19	0.41	5.27	1.35	1.04	0.79	0.89	1.67	3.83	6.98	1.18	2.12	27.72
#1961	0.64	1.10	0.77	3.17	0.40	0.84	0.85	3.37	1.15	12.72	3.03	1.74	29.78
1962	1.95	3.91									1.76	0.62	
1963	0.29	0.37	2.32	7.63	12.41	2.11	1.20	3.30	2.08	1.10	1.68	0.98	35.47
1964	1.03	1.18	3.76	8.29	3.57	1.42	0.87	0.96	1.06	2.33	1.63	13.04	39.14
1965	1.56	0.97	1.05	0.57	4.47	1.08	0.62	1.46	1.79	3.26	1.65	2.24	20.72
1966	0.71	5.34	0.36	0.31	0.11	0.42	1.25	2.13	0.49	8.28	4.31	1.21	24.92
1967	0.87	2.54	6.69	0.51	2.09	1.85	2.96	1.53	7.12	4.96	5.34	2.99	39.45
1968	0.83	0.36	12.38	2.63	5.56	1.48	0.71	6.14	2.68	2.91	1.28	5.31	42.27
1969	1.25	4.53	1.49	1.41	0.81	0.75	0.56	1.29	1.60	1.26	0.57	1.59	17.11
1970	0.32	1.04	0.66	1.27	1.09	0.52	1.59	0.37	0.67	3.38	3.87	8.27	23.05
1971	1.47	0.34	6.82	1.33	0.49	1.84	0.73	1.80	2.46	1.81	1.53	3.57	24.19
1972	3.29	2.71	1.55	3.89	6.07	1.03	0.82	6.91	1.38	3.27	1.56	1.75	34.23
1973	0.82	0.88	2.31	0.87	0.82	0.66	0.50	0.65	3.44	2.29	7.41	4.47	25.12
1974	0.33	1.45	3.79	7.25	0.87	0.07	0.89	2.25	2.38	0.98	1.09	1.49	22.84
1975	1.49	1.10	1.74	0.51	0.73	0.32	0.47	0.33	0.64	1.42	1.13	0.52	10.40
1976	6.41	3.62	4.11	1.18	0.69	0.94	0.37	0.77	1.96	2.40	2.14	1.85	26.44
1977	0.47	0.39	1.13	2.58	1.14	0.90	0.72	1.74	5.64	2.14	13.12	1.77	31.74
1978	0.59	0.26	1.53	0.56	1.01	1.08	2.01	4.08	2.89	2.10	2.58	5.39	24.08
1979	1.51	1.62	0.87	1.07	1.17	0.85	1.54	0.67	0.88	2.27	13.80	4.52	30.77
1980	0.34	1.11	0.84	2.39	1.02	0.66	1.58	1.10	0.86	2.86	4.33	0.26	17.35
1981	0.88	1.26	1.16	1.89	0.63	0.39	1.00	0.96	1.49	5.27	5.39	3.21	23.53
1982	3.29	0.90	1.65	1.21	0.67	1.17	1.50	4.32	3.49	1.18	8.59	2.96	30.93
1983	0.57	0.12	0.77	2.13	1.37	0.67	0.61	0.67	0.86	2.73	2.40	1.09	13.99
RECORD MEAN	1.82	1.60	2.73	2.48	1.94	0.96	1.08	2.50	2.12	3.25	3.61	3.14	27.23

Average Temperature

Year	Jan	Feb	Mar	Apr	May	June	July	Aug	Sept	Oct	Nov	Dec	Annual
1952				78.5	78.1		81.1	82.0	81.2	80.7	79.6	78.6	
1953	77.9	77.4	77.0	78.1	79.1	80.4	80.9	81.6	82.0	81.0	79.6	78.6	
1954	76.7	77.2	75.6	78.6	79.0	81.2	81.8	82.4	82.4	82.0	79.9	77.8	79.4
1955	76.9	75.3	75.3	77.1	77.9	79.3	79.8	80.2	80.2	79.4	79.1	78.0	78.2
#1956	77.2	76.5	77.0	76.0	78.2	79.4	80.8	81.4	81.4	81.2			
1957													
1958											78.9	77.8	
1959	76.0	75.5	77.3	77.2	78.7	80.5	81.0	80.8	81.2	81.1	79.6	76.8	78.8
1960	76.6	76.4	76.6	77.8	79.0	80.9	81.2	81.5	81.6	81.0	80.0	77.9	79.2
#1961	77.2	78.2	78.6	78.4	78.6	78.7	79.8	79.4	79.4	78.4	78.2	76.7	78.5
1962	76.4	76.1									79.1	77.4	
1963	77.1	77.6	77.0	78.8	77.8	79.2	80.3	81.0	80.9	79.6	79.6	77.8	79.0
1964	77.0	75.8	76.2	77.3	78.6	80.0	80.1	80.5	79.9	79.5	78.3	77.3	78.4
1965	75.7	74.3	75.2	76.9	78.4	78.8	79.7	80.4	81.2	80.7	79.8	77.3	74.2
1966	76.6	75.0	77.0	77.1	79.8	81.0	81.6	81.8	82.0	80.5	79.5	78.0	79.2
1967	77.0	76.9	76.3	78.0	80.1	80.8	81.5	82.5	81.0	81.3	78.9	78.1	79.3
1968	76.7	77.1	77.0	78.3	80.0	81.4	82.0	81.7	82.1	82.1	81.2	79.4	79.9
1969	75.5	77.2	77.1	77.5	79.3	80.8	81.6	81.5	81.2	80.9	80.5	78.2	79.3
1970	78.5	77.9	77.5	77.8	79.8	80.4	80.5	81.3	80.9	80.4	78.4	76.9	79.2
1971	75.8	77.3	76.0	77.2	78.5	78.9	79.9	80.2	81.1	80.1	79.0	77.0	78.4
1972	76.6	76.2	77.8	78.4	78.3	79.6	80.0	81.0	81.2	80.4	79.2	76.4	78.8
1973	75.5	75.4	75.9	76.7	77.7	79.3	79.8	80.5	80.1	79.8	78.4	77.2	78.0
1974	77.8	77.1	77.1	78.0	78.9	80.7	80.5	80.8	81.1	81.2	79.3	77.3	79.2
1975	76.5	76.1	76.0	77.0	78.1	78.7	79.4	80.1	80.2	79.7	78.5	77.3	78.2
1976	75.9	76.1	74.9	76.3	77.8	79.0	79.2	80.1	80.5	80.1	78.4	77.9	78.0
1977	77.1	77.8	76.8	76.6	79.0	79.3	80.4	81.6	81.3	81.7	79.5	78.1	79.1
1978	77.4	77.8	77.6	78.3	79.4	80.8	81.6	81.9	81.4	81.1	79.0	76.8	79.3
1979	75.7	75.7	77.1	77.3	79.1	80.0	80.9	81.9	82.5	82.0	78.9	78.7	79.2
1980	77.8	77.7	79.1	79.5	80.5	81.6	82.3	82.6	83.0	81.8	80.3	79.7	80.5
1981	79.2	78.7	78.9	79.7	80.4	81.5	82.0	83.0	83.3	82.3	80.1	77.6	80.6
1982	76.2	77.2	77.5	79.7	80.6	82.2	83.1	83.3	82.5	82.2	79.8	76.6	80.1
1983	77.3	77.6	77.6	78.7	79.0	81.1	81.5	82.1	82.0	80.9	79.8	78.2	79.6
RECORD MEAN	76.8	76.7	77.0	77.8	79.0	80.1	80.8	81.3	81.3	80.8	79.3	77.6	79.0
MAX	80.7	80.5	80.9	81.6	82.7	83.9	84.5	85.0	85.1	84.6	82.9	81.3	82.8
MIN	72.9	72.8	73.0	73.9	75.2	76.3	77.0	77.5	77.4	77.0	75.6	73.9	75.2

Heating Degree Days

Season	July	Aug	Sept	Oct	Nov	Dec	Jan	Feb	Mar	Apr	May	June	Total

Cooling Degree Days

Year	Jan	Feb	Mar	Apr	May	June	July	Aug	Sept	Oct	Nov	Dec	Total
1969	333	349	382	382	454	483	522	517	494	498	475	413	5302
1970	426	365	393	393	468	471	488	511	483	488	407	378	5271
1971	342	350	349	374	423	424	473	478	487	476	427	382	4982
1972	366	334	403	408	420	447	469		492	484	436	361	
1973	330	298	345	358	400	436	467	490	458	463	411	386	4842
1974	406	346	383	396	439	478	487	494	492	507	434	403	5265
1975	361	319	347	368	413	420	457	476	461	461	410	388	4881
1976	345	328	315	345	404	426	446	478	473	476	407	405	4848
1977	384	366	373	355	440	436	485	520	496	528	443	415	5241
1978	391	366	397	403	454	484	523	529	499	504	428	377	5355
1979	338	304	394	377	442	459	500	531	532	534	424	431	5256
1980	404	373	442	442	490	503	541	552	545	528	466	459	5745
1981	446	390	438	448	485	504	538	565	555	540	461	403	5773
1982	356	348	395	449	490	523	566	579	531	542	452	366	5596
1983	386	355	399	418	439	487	518	538	517	503	445	417	5422

Snowfall

Season	July	Aug	Sept	Oct	Nov	Dec	Jan	Feb	Mar	Apr	May	June	Total

Koror is one of the islands of the Palau Group, which lies in the extreme western Carolines and along the eastern side of the Philippine Sea. It is the administrative center of the Republic of Palau of the U.S. Trust Territory of the Pacific Islands. Like most of the other islands of the group, Koror is hilly and is surrounded by a lagoon whose outer border is a barrier reef. Koror is not an isolated island, but is immediately adjacent to the islands of Malakal and Arakabesan, to which it is joined by causeways. Babelthuap, the largest of the Palau Islands, lies 1 mile north-northeast and is connected to Korror by bridge. The highest hill within 3 miles of the Weather Station on Koror has an elevation of 610 feet.

Precipitation is heavy at Koror. Rainfall of 150 inches or more in a year is not uncommon. Rainfall is variable for each month and from year to year. Normal monthly precipitation exceeds 10 inches and during some years, each month has received at at least 15 inches.

During the period December through March, the northeast trades prevail. Winds are generally light to moderate. Precipitation, heavy during December and January, decreases sharply when the Intertropical Convergence Zone moves well south of the island. February, March, and April are the driest months of the year. During April, the frequency of northeast winds decreases, and there is an increase in the frequency of east winds. In May the winds are predominantly from southeast, to northeast.

Usually during June the Intertropical Convergence Zone moves northward across Koror, bringing with it heavy rainfall and thunderstorms that may yield 1 inch of rain in 15 to 30 minutes. The convergence zone remains in the vicinity of Koror, though most commonly toward the north, from July through January, with heavy rainfall persisting. During November, Koror is usually near the heart of the zone. Calm to light variable winds and continued heavy showers are the rule.

TABLE 1 NORMALS, MEANS AND EXTREMES

KOROR ISLAND, PACIFIC

LATITUDE: 7°20'N   LONGITUDE: 134°29' E   ELEVATION: FT. GRND   94 BARO   98   TIME ZONE: 135E MER   WBAN: 40309

	(a)	JAN	FEB	MAR	APR	MAY	JUNE	JULY	AUG	SEP	OCT	NOV	DEC	YEAR
TEMPERATURE °F:														
Normals														
-Daily Maximum		86.7	86.7	87.4	88.1	88.3	87.8	87.2	87.1	87.5	87.9	88.3	87.5	87.5
-Daily Minimum		74.9	74.9	75.2	75.8	76.1	75.6	75.3	75.5	75.8	75.9	75.9	75.5	75.5
-Monthly		80.8	80.8	81.3	82.0	82.2	81.7	81.3	81.3	81.7	81.9	82.1	81.5	81.5
Extremes														
-Record Highest	36	91	92	91	93	93	95	93	93	92	92	92	92	95
-Year		1981	1954	1978	1954	1972	1976	1968	1947	1979	1974	1980	1973	JUN 1976
-Record Lowest	36	70	71	69	69	72	70	70	70	70	71	70	71	69
-Year		1974	1972	1953	1979	1985	1948	1980	1984	1978	1981	1949	1984	APR 1979
NORMAL DEGREE DAYS:														
Heating (base 65°F)		0	0	0	0	0	0	0	0	0	0	0	0	0
Cooling (base 65°F)		490	442	505	510	533	501	505	505	501	524	513	512	6041
% OF POSSIBLE SUNSHINE	25	56	58	66	64	57	47	49	47	55	49	54	51	54
MEAN SKY COVER (tenths)														
Sunrise - Sunset	33	9.0	8.9	8.6	8.5	8.7	9.2	9.1	9.3	9.1	8.9	8.8	9.0	8.9
MEAN NUMBER OF DAYS:														
Sunrise to Sunset														
-Clear	33	0.2	0.2	0.6	0.6	0.3	0.1	0.2	0.1	0.2	0.4	0.2	0.2	3.2
-Partly Cloudy	33	4.8	4.8	6.4	7.2	6.7	3.5	3.8	3.4	4.3	5.4	5.6	4.8	60.8
-Cloudy	33	25.9	23.3	24.0	22.2	24.0	26.3	27.0	27.5	25.6	25.2	24.2	26.1	301.2
Precipitation														
.01 inches or more	34	22.2	19.2	20.1	19.3	23.9	24.8	23.9	23.2	21.2	22.8	22.1	24.4	267.3
Snow, Ice pellets														
1.0 inches or more	34	0.0	0.0	0.0	0.0	0.0	0.0	0.0	0.0	0.0	0.0	0.0	0.0	0.0
Thunderstorms	34	1.1	1.0	1.1	2.0	3.4	3.6	2.9	3.7	4.0	4.4	3.8	2.7	33.5
Heavy Fog Visibility														
1/4 mile or less	34	0.0	0.0	0.0	0.0	0.0	0.0	0.0	0.0	0.0	0.0	0.0	0.0	0.0
Temperature °F														
-Maximum														
90° and above	34	1.3	0.9	2.1	5.6	10.6	6.1	3.1	2.5	4.2	6.9	9.7	4.6	57.5
32° and below	34	0.0	0.0	0.0	0.0	0.0	0.0	0.0	0.0	0.0	0.0	0.0	0.0	0.0
-Minimum														
32° and below	34	0.0	0.0	0.0	0.0	0.0	0.0	0.0	0.0	0.0	0.0	0.0	0.0	0.0
0° and below	34	0.0	0.0	0.0	0.0	0.0	0.0	0.0	0.0	0.0	0.0	0.0	0.0	0.0
AVG. STATION PRESS.(mb)	10	1005.8	1006.5	1006.6	1006.0	1005.7	1005.7	1005.6	1005.7	1005.8	1005.4	1005.0	1005.4	1005.8
RELATIVE HUMIDITY (%)														
Hour 03	18	90	90	90	91	92	93	91	91	89	91	92	91	91
Hour 09	34	81	81	79	78	80	81	81	81	80	80	80	81	80
Hour 15 (Local Time)	34	76	75	74	74	77	78	78	78	77	77	77	77	77
Hour 21	34	87	86	87	87	89	89	89	87	87	88	89	89	88
PRECIPITATION (inches):														
Water Equivalent														
-Normal		11.14	8.02	8.19	10.21	13.09	14.69	16.32	15.20	13.06	13.68	10.95	12.95	147.50
-Maximum Monthly	36	28.13	22.46	21.98	27.69	27.46	29.17	34.82	24.66	23.16	22.47	22.06	21.10	34.82
-Year		1974	1978	1972	1979	1954	1981	1962	1952	1985	1974	1958	1975	JUL 1962
-Minimum Monthly	36	2.11	0.64	1.71	1.65	5.73	5.91	4.14	6.89	1.04	6.77	4.70	6.28	0.64
-Year		1973	1983	1983	1948	1983	1976	1964	1981	1982	1951	1951	1985	FEB 1983
-Maximum in 24 hrs	36	13.86	8.42	6.17	16.95	9.86	6.38	8.27	8.18	8.47	6.18	6.09	6.47	16.95
-Year		1974	1980	1953	1979	1982	1981	1981	1962	1949	1957	1985	1974	APR 1979
Snow, Ice pellets														
-Maximum Monthly	36													
-Year														
-Maximum in 24 hrs	36													
-Year														
WIND:														
Mean Speed (mph)	23	8.5	8.8	8.6	7.7	6.5	5.9	6.7	6.8	6.9	7.2	6.4	7.4	7.3
Prevailing Direction														
through 1963		NE	ENE	NE	ENE	E	E	NW	SW	W	W	NE	ENE	NE
Fastest Mile														
-Direction (!!!)	25	NW	SE	S	SW	SW	SW	W	SW	N	SW	SE	N	S
-Speed (MPH)	25	43	30	73	60	35	40	34	34	34	36	59	50	73
-Year		1975	1968	1967	1976	1976	1967	1969	1978	1969	1968	1964	1972	MAR 1967
Peak Gust														
-Direction (!!!)	2	NW	NE	NE	NE	S	NW	SW	NW	SW	SW	SW	SE	NW
-Speed (mph)	2	45	33	33	33	33	43	38	44	39	43	38	39	45
-Date		1985	1985	1984	1985	1985	1985	1985	1984	1985	1985	1984	1984	JAN 1985

See Reference Notes to this table on the following page.

TABLE 2 PRECIPITATION (inches) KOROR ISLAND, PACIFIC

YEAR	JAN	FEB	MAR	APR	MAY	JUNE	JULY	AUG	SEP	OCT	NOV	DEC	ANNUAL
1956	24.45	5.50	8.80	15.63	22.78	15.71	12.84	18.95	14.00	11.98	12.88	15.63	179.15
1957	4.74	2.42	6.72	16.42	9.68	12.60	16.69	21.57	16.12	19.22	4.70	7.62	138.50
1958	8.42	2.69	2.93	8.09	17.18	9.73	15.74	20.27	18.11	9.95	22.06	12.62	147.79
1959	6.28	3.49	10.82	12.96	13.09	12.26	22.53	14.88	13.12	16.64	15.41	10.41	151.89
1960	15.16	10.20	3.75	11.66	11.45	14.16	13.41	11.36	20.79	15.04	10.12	14.11	151.21
1961	15.79	10.13	6.25	6.72	20.14	22.67	13.68	17.37	11.61	18.55	7.81	13.23	163.95
1962	16.78	9.38	7.65	6.84	18.30	8.59	34.82	19.45	20.79	9.09	6.84	14.04	172.57
1963	18.63	8.79	8.41	3.39	12.01	13.99	11.19	14.24	14.01	10.23	13.40	11.45	135.02
1964	7.27	16.10	6.98	7.46	18.32	12.12	4.14	15.73	7.06	10.19	13.75	11.45	130.57
1965	6.40	13.03	14.60	7.11	9.78	19.85	30.57	16.76	12.98	9.68	6.16	12.67	159.59
1966	8.23	3.31	10.02	15.27	9.08	10.44	23.68	9.69	7.20	16.66	11.76	14.93	140.27
1967	18.86	3.74	7.20	4.63	9.91	17.19	12.20	17.11	6.75	17.06	11.79	12.23	138.67
1968	8.02	15.54	8.59	7.56	9.51	7.01	16.43	11.00	12.25	11.46	8.15	15.95	131.47
1969	6.14	2.93	5.11	6.83	11.72	16.79	28.21	12.39	14.00	9.81	9.06	6.32	129.31
1970	6.23	5.78	4.83	3.21	8.25	12.38	12.61	15.77	8.51	12.99	9.17	14.82	114.55
1971	13.54	10.68	11.09	8.32	16.31	19.61	14.49	10.40	13.98	19.59	10.26	11.06	159.33
1972	10.78	10.83	21.98	7.08	9.49	20.68	11.15	15.88	14.90	9.85	9.96	7.54	150.12
1973	2.11	1.24	2.95	11.29	10.16	13.78	12.79	11.35	12.18	19.14	16.56	9.87	123.42
1974	28.13	7.98	13.75	10.86	8.10	9.72	21.16	13.75	14.80	22.47	15.73	18.54	184.99
1975	17.29	2.82	6.69	10.00	9.01	16.24	22.86	8.28	17.24	11.52	11.18	21.10	154.23
1976	7.80	7.27	8.05	20.09	8.66	5.91	8.08	16.64	7.72	12.49	6.30	16.54	125.55
1977	5.18	5.30	3.60	4.48	11.36	11.15	20.72	19.20	12.65	10.63	7.79		119.44
1978	10.34	22.46	6.02	8.98	12.52	16.04	9.13	20.36	10.85	20.06	17.66	10.33	164.75
1979	6.98	6.47	7.96	27.69	11.26	22.84	17.79	11.69	12.25	17.00	11.57		160.08
1980	8.72	16.01	5.53	18.80	10.02	19.50	12.40	15.26	13.60	17.11	12.17	19.95	169.07
1981	11.32	15.00	4.49	3.00	9.66	29.17	21.14	6.89	14.30	14.30	9.81		152.85
1982	5.79	6.81	9.90	9.45	19.12	22.41	19.40	10.94	1.04	8.82	9.92	13.71	137.31
1983	3.44	0.64	1.71	3.12	5.73	18.48	21.20	17.96	11.73	14.23	11.40	10.48	120.12
1984	18.57	10.81	13.58	7.23	10.85	16.49	12.82	17.47	10.39	15.94	9.19	9.42	152.76
1985	13.22	13.88	5.38	11.82	10.41	25.25	17.55	14.32	23.16	8.64	13.61	6.28	163.52
Record Mean	10.85	8.16	8.20	9.70	12.94	15.81	16.64	15.12	13.05	13.64	11.05	12.63	147.78

TABLE 3 AVERAGE TEMPERATURE (deg. F) KOROR ISLAND, PACIFIC

YEAR	JAN	FEB	MAR	APR	MAY	JUNE	JULY	AUG	SEP	OCT	NOV	DEC	ANNUAL
1956	79.7	81.1	81.3	81.2	81.3	81.5	80.8	80.9	81.0	81.2	81.1	80.8	81.0
1957	81.2	81.0	81.6	81.4	82.2	81.6	81.0	81.2	81.8	82.2	82.3	81.7	81.6
1958	81.6	81.4	81.5	82.0	82.9	82.9	82.0	81.8	81.3	82.7	81.4	81.8	82.0
1959	80.6	80.2	81.0	81.6	81.8	81.8	80.6	80.9	81.7	81.5	81.9	81.3	81.2
1960	80.5	80.6	81.4	81.8	82.0	81.5	82.1	82.1	81.8	82.2	82.4	81.6	81.7
1961	80.4	81.1	81.5	82.1	81.7	80.5	80.9	80.9	81.3	80.6	81.7	80.7	81.1
1962	80.2	80.5	80.6	82.0	81.5	81.8	80.7	81.2	81.0	82.6	82.7	81.3	81.3
1963	80.7	80.8	81.1	82.4	82.0	81.9	81.2	81.0	81.9	81.9	81.6	81.6	81.5
1964	80.8	79.7	81.3	81.7	82.1	81.5	82.0	80.5	82.0	81.5	81.0	81.0	81.3
1965	80.9	80.0	80.1	81.4	81.4	80.6	78.8	80.9	81.4	81.8	82.3	80.9	80.9
1966	79.6	80.1	80.9	81.8	82.5	81.6	81.1	81.7	82.6	82.1	82.1	81.6	81.4
1967	80.1	81.4	81.2	82.4	82.5	82.0	81.2	81.2	80.6	82.5	81.6	81.0	81.5
1968	81.0	80.0	81.3	82.2	82.7	83.2	82.0	81.9	82.1	82.5	82.5	81.2	81.9
1969	80.7	81.1	81.7	82.4	82.7	82.5	80.8	81.9	81.3	82.2	82.1	81.8	81.8
1970	81.5	81.9	82.0	82.6	82.4	82.4	82.1	82.0	82.0	82.3	82.2	81.6	82.1
1971	81.3	81.3	82.1	82.9	82.3	81.1	80.7	81.5	82.1	80.9	82.0	82.0	81.7
1972	80.8	80.5	80.9	81.8	81.7	82.2	81.6	81.8	81.8	82.5	81.9	81.9	81.7
1973	81.2	81.6	81.9	82.3	82.7	82.7	81.9	82.5	82.2	81.9	82.2	82.3	82.1
1974	81.0	81.7	81.1	81.6	82.7	81.8	81.4	82.1	82.3	82.0	82.1	81.6	81.8
1975	80.8	81.4	81.7	82.0	82.2	81.2	80.8	81.6	81.6	81.8	82.3	81.0	81.5
1976	81.9	80.9	81.4	81.7	82.5	82.0	82.0	80.7	81.8	82.4	81.8	81.6	81.8
1977	81.5	81.2	81.8	82.5	83.0	82.5	81.1	81.2	82.2	82.7	83.0	82.5	82.1
1978	81.3	80.1	82.0	81.9	82.6	81.8	81.8	80.8	80.9	81.4	81.7	81.8	81.6
1979	81.4	81.3	81.3	81.3	82.4	81.0	81.1	81.6	82.1	82.0	81.5	81.6	81.6
1980	80.8	80.2	81.4	81.8	82.5	81.8	81.1	80.9	81.6	82.3	82.3	81.7	81.7
1981	81.1	80.4	81.2	82.0	82.6	80.9	81.2	81.7	81.6	81.2	82.5	82.1	81.5
1982	81.7	81.7	81.1	81.7	81.8	81.5	81.1	81.1	81.1	81.8	82.3	81.7	81.7
1983	80.5	81.2	81.7	82.2	83.0	81.9	81.1	82.0	82.0	82.0	82.4	82.0	81.7
1984	81.1	80.9	81.7	82.3	82.7	81.2	81.5	81.0	81.0	80.9	82.2	82.4	81.7
1985	80.9	81.3	82.1	81.6	82.4	80.5	80.0	80.7	80.5	81.8	82.5	82.4	81.4
Record Mean	80.9	80.9	81.3	82.0	82.3	81.6	81.2	81.3	81.7	81.8	82.2	81.6	81.6
Max	86.8	86.8	87.5	88.1	88.4	87.7	87.1	87.1	87.6	87.8	88.4	87.6	87.6
Min	74.9	74.9	75.2	75.8	76.1	75.5	75.3	75.5	75.8	75.8	75.9	75.5	75.5

REFERENCE NOTES FOR TABLES 1, 2, 3 and 6 (KOROR ISLAND, PACIFIC)

GENERAL

T - TRACE AMOUNT
BLANK ENTRIES DENOTE MISSING/UNREPORTED DATA.
# INDICATES A STATION OR INSTRUMENT RELOCATION.

SPECIFIC

TABLE 1

(a) - LENGTH OF RECORD IN YEARS. ALTHOUGH INDIVIDUAL MONTHS MAY BE MISSING.
* LESS THAN .05

NORMALS — BASED ON THE 1951-1980 RECORD PERIOD.
EXTREMES — DATES ARE THE MOST RECENT OCCURRENCE.
WIND DIR. — NUMERALS SHOW TENS OF DEGREES CLOCKWISE FROM TRUE NORTH.
"00" INDICATES CALM.
RESULTANT WIND DIRECTIONS ARE GIVEN TO WHOLE DEGREES.

EXCEPTIONS

TABLE 1

1. THUNDERSTORMS AND HEAVY FOG, APRIL 1973 TO DATE MAY BE INCOMPLETE, DUE TO PART-TIME OPERATIONS

TABLES 2, 3, and 6

RECORD MEANS ARE THROUGH THE CURRENT YEAR, BEGINNING IN   1952 FOR TEMPERATURE
1948 FOR PRECIPITATION

TABLE 4   HEATING DEGREE DAYS Base 65 deg. F   KOROR ISLAND, PACIFIC

SEASON	JULY	AUG	SEP	OCT	NOV	DEC	JAN	FEB	MAR	APR	MAY	JUNE	TOTAL
1983-84	0	0	0	0	0	0	0	0	0	0	0	0	0
1984-85	0	0	0	0	0	0	0	0	0	0	0	0	0
1985-86	0	0	0	0	0	0							

TABLE 5   COOLING DEGREE DAYS Base 65 deg. F   KOROR ISLAND, PACIFIC

YEAR	JAN	FEB	MAR	APR	MAY	JUNE	JULY	AUG	SEP	OCT	NOV	DEC	TOTAL
1969	493	457	522	530	557	532	496	529	495	541	521	527	6200
1970	519	481	536	536	546	526	537	534	527	538	527	523	6330
1971	514	463	537	539	539	488	495	517	521	499	519	536	6167
1972	499	456	502	512	553	509	539	523	510	551	528	530	6212
1973	506	470	530	525	555	536	531	548	523	529	522	547	6322
1974	504	476	504	505	555	513	517	540	526	535	524	519	6218
1975	498	467	524	515	541	493	493	524	504	527	527	503	6116
1976	530	467	515	504	549	517	533	495	510	547	543	523	6233
1977	516	460	527	530	563	532	507	509	524	551	550	551	6320
1978	513	429	535	513	556	510	542	495	484	511	508	529	6125
1979	516	463	511	497	547	488	506	524	520	533	522	520	6147
1980	496	449	514	509	546	509	508	500	503	545	548	543	6170
1981	504	437	507	514	553	485	506	523	504	508	532	536	6109
1982	527	474	505	508	525	501	507	507	530	525	530	535	6162
1983	489	461	524	525	564	511	505	537	516	535	530	535	6232
1984	503	471	523	525	554	493	517	502	510	500	539	547	6184
1985	500	466	536	507	547	471	473	493	473	526	530	545	6067

TABLE 6   SNOWFALL (inches)   KOROR ISLAND, PACIFIC

SEASON	JULY	AUG	SEP	OCT	NOV	DEC	JAN	FEB	MAR	APR	MAY	JUNE	TOTAL
1971-72	0.0	0.0	0.0	0.0	0.0	0.0	0.0	0.0	0.0	0.0	0.0	0.0	0.0
1972-73	0.0	0.0	0.0	0.0	0.0	0.0	0.0	0.0	0.0	0.0	0.0	0.0	0.0
1973-74	0.0	0.0	0.0	0.0	0.0	0.0	0.0	0.0	0.0	0.0	0.0	0.0	0.0
1974-75	0.0	0.0	0.0	0.0	0.0	0.0	0.0	0.0	0.0	0.0	0.0	0.0	0.0
1975-76	0.0	0.0	0.0	0.0	0.0	0.0	0.0	0.0	0.0	0.0	0.0	0.0	0.0
1976-77	0.0	0.0	0.0	0.0	0.0	0.0	0.0	0.0	0.0	0.0	0.0	0.0	0.0
1977-78	0.0	0.0	0.0	0.0	0.0	0.0	0.0	0.0	0.0	0.0	0.0	0.0	0.0
1978-79	0.0	0.0	0.0	0.0	0.0	0.0	0.0	0.0	0.0	0.0	0.0	0.0	0.0
1979-80	0.0	0.0	0.0	0.0	0.0	0.0	0.0	0.0	0.0	0.0	0.0	0.0	0.0
1980-81	0.0	0.0	0.0	0.0	0.0	0.0	0.0	0.0	0.0	0.0	0.0	0.0	0.0
1981-82	0.0	0.0	0.0	0.0	0.0	0.0	0.0	0.0	0.0	0.0	0.0	0.0	0.0
1982-83	0.0	0.0	0.0	0.0	0.0	0.0	0.0	0.0	0.0	0.0	0.0	0.0	0.0
1983-84	0.0	0.0	0.0	0.0	0.0	0.0	0.0	0.0	0.0	0.0	0.0	0.0	0.0
1984-85	0.0	0.0	0.0	0.0	0.0	0.0	0.0	0.0	0.0	0.0	0.0	0.0	0.0
1985-86	0.0	0.0	0.0	0.0	0.0	0.0							
Record Mean	0.0	0.0	0.0	0.0	0.0	0.0	0.0	0.0	0.0	0.0	0.0	0.0	0.0

See Reference Notes, relative to all above tables, on preceding page.

Kwajalein Island, although only 3 miles in length and 1/2 mile wide, is the largest of the fringing reef islands composing Kwajalein Atoll. Kwajalein Atoll, spanning some 70 miles, is one of the largest coral atolls in the world. The land surface of the island, which has very little effect on the climate of the locality, has an average elevation of less than 10 feet above sea level. The highest points on the island are 12 to 15 feet above sea level.

Kwajalein, located less than 700 miles north of the equator, has a tropical marine climate characterized by relatively high annual rainfall and warm to hot, humid weather throughout the year.

Temperatures are very uniform from day to day and month to month. Because of the low latitude, there are only slight seasonal variations in the length of daylight period and the altitude of the sun at Kwajalein. As a result, the variation of the amount of solar energy received is small. The small variation in solar energy and the marine influence are the principal reasons for the uniform temperatures in the area. The range of normal temperature between the coldest month and the warmest month averages about 2 degrees.

The principal rainfall season extends from May through November. Light, easterly winds, almost constant cloudiness, and frequent moderate to heavy showers prevail during the wet season.

The dry season includes the period December through April, and is characterized not so much by lack of showers as by light showers of short duration. In this season the trade winds are persistent, blowing from the northeast 15 to 20 knots almost continuously. Cloudiness is at a minimum, and the sky is less than one-half covered most of the time, but clear skies are rare.

Severe storms with attendant damaging winds are rare in the vicinity of Kwajalein. During the wet season, however, small, weak depressions may form near the island. Some of these intensify and a few eventually develop into typhoons after moving westward away from the island. These depressions cause heavy rainfall in the Kwajalein Atoll.

The relative humidity is uniformly high throughout the year, and is slightly higher in the wet season than in the dry season. The combination of high humidity and proximity of the salt water ocean presents a corrosion problem.

TABLE 1 — NORMALS, MEANS AND EXTREMES

KWAJALEIN, MARSHALL ISLANDS, PACIFIC

LATITUDE: °44'N   LONGITUDE: 167°44'E   ELEVATION: FT. GRND   8 BARO 00011   TIME ZONE: 180E MER   WBAN: 40604

	(a)	JAN	FEB	MAR	APR	MAY	JUNE	JULY	AUG	SEP	OCT	NOV	DEC	YEAR
TEMPERATURE °F:														
Normals														
-Daily Maximum		85.7	86.4	87.0	86.5	86.3	86.4	86.4	86.8	86.9	86.7	86.5	86.0	86.5
-Daily Minimum		76.8	76.9	77.4	77.1	76.9	76.9	76.9	77.0	77.1	77.0	77.1	77.2	77.0
-Monthly		81.3	81.7	82.2	81.8	81.6	81.7	81.7	81.9	82.0	81.8	81.8	81.6	81.7
Extremes														
-Record Highest	33	91	92	91	92	93	92	94	95	93	97	92	90	97
-Year		1953	1954	1954	1953	1953	1958	1953	1958	1958	1958	1953	1955	OCT 1958
-Record Lowest	33	68	71	70	71	71	71	70	71	68	71	70	69	68
-Year		1979	1963	1985	1964	1968	1976	1954	1971	1984	1967	1975	1963	SEP 1984
NORMAL DEGREE DAYS:														
Heating (base 65°F)		0	0	0	0	0	0	0	0	0	0	0	0	0
Cooling (base 65°F)		505	468	533	504	515	501	518	524	510	524	504	515	6121
% OF POSSIBLE SUNSHINE														
MEAN SKY COVER (tenths)														
Sunrise - Sunset	24	7.8	7.6	7.7	8.3	8.4	8.3	8.5	8.2	8.4	8.6	8.3	8.0	8.2
MEAN NUMBER OF DAYS:														
Sunrise to Sunset														
-Clear	24	2.9	3.5	3.0	1.9	1.0	1.4	1.2	1.6	1.6	1.2	1.3	2.6	23.2
-Partly Cloudy	24	8.4	7.6	8.8	6.3	7.1	7.3	6.6	7.5	6.2	6.1	7.8	7.4	87.1
-Cloudy	24	19.7	17.2	19.2	21.8	22.9	21.3	23.2	21.9	22.2	23.8	21.0	21.0	255.0
Precipitation														
.01 inches or more	33	14.8	13.1	14.4	17.5	21.1	22.4	23.4	23.3	22.5	23.3	22.8	18.8	237.3
Snow, Ice pellets														
1.0 inches or more	33	0.0	0.0	0.0	0.0	0.0	0.0	0.0	0.0	0.0	0.0	0.0	0.0	0.0
Thunderstorms	25	0.2	0.1	0.1	0.4	0.6	0.8	1.4	1.2	1.9	1.4	1.2	0.3	9.6
Heavy Fog Visibility														
1/4 mile or less	25	0.0	0.0	0.0	0.0	0.0	0.0	0.0	0.0	0.0	0.0	0.0	0.0	0.0
Temperature °F														
-Maximum														
90° and above	33	0.1	0.3	0.8	0.6	1.3	0.9	1.5	2.2	2.8	2.3	0.5	0.1	13.5
32° and below	33	0.0	0.0	0.0	0.0	0.0	0.0	0.0	0.0	0.0	0.0	0.0	0.0	0.0
-Minimum														
32° and below	33	0.0	0.0	0.0	0.0	0.0	0.0	0.0	0.0	0.0	0.0	0.0	0.0	0.0
0° and below	33	0.0	0.0	0.0	0.0	0.0	0.0	0.0	0.0	0.0	0.0	0.0	0.0	0.0
AVG. STATION PRESS.(mb)	9	1009.0	1009.1	1009.3	1009.4	1009.6	1009.3	1008.9	1009.0	1009.2	1008.8	1008.3	1008.4	1009.0
RELATIVE HUMIDITY (%)														
Hour 00	25	79	79	80	82	84	84	84	84	83	83	83	81	82
Hour 06 (Local Time)	25	80	79	80	83	85	84	85	84	84	83	83	81	83
Hour 12	25	71	69	70	74	77	77	77	77	76	76	76	74	75
Hour 18	25	75	73	74	77	80	80	79	78	79	79	79	77	78
PRECIPITATION (inches):														
Water Equivalent														
-Normal		4.91	2.97	5.17	7.60	11.24	10.11	10.30	10.30	10.94	12.24	10.97	7.96	104.71
-Maximum Monthly	41	15.66	9.05	24.33	20.29	26.86	19.61	17.33	17.46	21.16	20.05	19.51	30.38	30.38
-Year		1951	1976	1951	1971	1980	1955	1958	1979	1972	1964	1957	1950	DEC 1950
-Minimum Monthly	41	0.48	0.04	0.16	0.20	0.53	3.56	3.53	5.38	5.32	5.04	3.51	1.90	0.04
-Year		1977	1977	1975	1983	1984	1984	1984	1981	1965	1969	1973	1971	FEB 1977
-Maximum in 24 hrs	25	6.46	4.60	3.77	5.41	8.35	4.69	5.68	5.35	4.69	6.53	7.24	17.15	17.15
-Year		1979	1985	1968	1971	1980	1971	1983	1979	1972	1968	1975	1972	DEC 1972
Snow, Ice pellets														
-Maximum Monthly	41													
-Year														
-Maximum in 24 hrs	41													
-Year														
WIND:														
Mean Speed (mph)	19	16.4	16.6	16.0	14.8	13.6	12.7	10.8	9.7	9.2	10.1	12.1	16.3	13.2
Prevailing Direction through 1963		ENE	ENE	ENE	ENE	ENE	ENE	E	ENE	E	E	ENE	ENE	ENE
Fastest Obs. 1 Min.														
-Direction (!!!)	25	22	18	07	08	11	13	09	08	19	20	13	11	22
-Speed (MPH)	25	55	35	36	37	44	35	41	43	35	40	49	40	55
-Year		1969	1963	1981	1964	1967	1985	1962	1970	1967	1968	1982	1961	JAN 1969
Peak Gust														
-Direction (!!!)	2	E	E	E	SE	NE	E	SE	E	E	NE	NE	NE	NE
-Speed (mph)	2	44	38	41	45	40	40	40	40	43	53	48	44	53
-Date		1985	1985	1985	1985	1985	1984	1984	1985	1985	1985	1984	1984	OCT 1985

See reference Notes to this table on the following page.

TABLE 2 PRECIPITATION (inches)   KWAJALEIN, MARSHALL ISLANDS, PACIFIC

YEAR	JAN	FEB	MAR	APR	MAY	JUNE	JULY	AUG	SEP	OCT	NOV	DEC	ANNUAL
1956	4.36	1.52	9.42	6.00	11.63	11.45	9.13	10.37	8.57	9.60	9.42	3.91	95.38
1957	4.49	5.95	0.81	1.18	10.99	6.29	4.84	7.08	5.51	7.27	9.01	2.90	82.45
1958	6.74	1.07	1.93	5.72	6.28	9.45	17.33	7.77	5.51	12.20	9.01	7.92	90.93
1959	1.15	4.87	1.93	6.21	4.27	7.52	8.79	7.50	12.41	8.33	19.26	6.48	88.72
1960	1.32	1.13	6.13	10.69	6.98	11.19	8.55	12.25	11.89	9.00	19.20	3.67	102.00
1961	3.29	1.95	4.32	4.85	13.71	7.84	9.98	10.60	15.53	10.29	13.61	8.50	104.47
1962	3.50	4.84	3.88	5.76	12.49	5.75	10.67	7.26	11.75	7.79	14.17	4.68	92.54
1963	6.22	6.32	4.56	5.02	7.47	13.76	9.80	15.09	5.61	19.86	8.81	14.11	116.63
1964	0.90	3.07	3.44	14.13	24.18	9.59	10.67	11.71	19.76	20.05	9.27	12.60	139.37
1965	3.85	1.01	0.18	0.24	3.20	12.55	12.03	8.05	5.32	9.95	9.17	4.49	70.04
1966	2.97	0.40	4.00	11.65	13.72	10.42	7.38	15.03	12.55	7.80	9.19	21.08	116.19
1967	5.14	4.42	17.19	4.61	10.35	14.14	9.64	6.11	10.91	11.18	7.71	7.17	108.57
1968	5.79	3.12	10.28	11.63	11.80	11.56	12.01	12.15	14.75	10.34	9.57		120.25
1969	6.11	0.43	1.40	6.07	8.13	12.56	12.81	5.74	9.91	5.04	8.46	2.68	79.34
1970	0.82	0.43	0.73	2.99	6.74	8.06	6.84	11.94	13.83	19.12	6.89	5.16	83.55
1971	5.35	3.63	4.28	20.29	18.66	11.54	16.43	11.19	12.52	15.65	4.86	1.90	126.30
1972	10.02	6.20	3.88	8.13	14.26	11.84	12.33	9.44	21.16	11.75	3.69	22.70	135.40
1973	0.52	0.52	1.20	4.53	7.92	8.60	11.29	7.51	8.33	3.51	13.84	5.21	70.59
1974	8.74	7.90	5.82	18.98	11.84	9.05	5.80	17.21	10.28	12.62	13.84	8.50	130.58
1975	1.05	0.24	0.16	6.45	8.17	13.99	11.74	11.22	11.60	15.62	15.61	5.69	101.54
1976	4.18	9.05	6.73	12.31	11.03	11.24	8.79	8.94	13.50	10.43	5.81	2.46	104.47
1977	0.48	0.04	4.71	5.79	12.36	7.93	13.01	7.58	8.17	14.09	8.76	7.16	90.08
1978	13.86	6.31	4.59	5.30	15.81	7.11	8.52	5.48	8.75	11.47	16.16	3.14	106.50
1979	10.73	2.81	1.06	12.15	2.74	6.19	9.29	17.46	8.44	12.52	17.61	8.27	109.29
1980	8.45	2.82	1.20	8.55	26.86	9.29	13.75	13.02	14.42	11.91	9.58	4.27	124.12
1981	1.91	1.29	7.94	5.10	1.74	8.36	11.97	5.38	12.22	10.76	15.74	5.56	87.97
1982	2.16	2.20	1.93	6.50	8.88	13.60	16.14	13.73	13.55	7.10	11.10	4.28	101.17
1983	0.89	0.68	0.36	0.20	1.76	3.83	13.27	10.74	14.44	10.49	11.84	4.24	79.78
1984	3.20	4.07	1.06	2.12	0.53	3.56	3.53	7.54	7.10	8.48	11.84	6.32	59.35
1985	2.28	6.97	5.75	12.46	8.05	7.66	7.61	11.18	7.64	9.99	11.45	5.06	96.10
Record Mean	4.17	2.82	5.18	6.89	9.83	9.56	10.14	10.31	11.09	11.84	11.37	8.17	101.38

TABLE 3 AVERAGE TEMPERATURE (deg. F)   KWAJALEIN, MARSHALL ISLANDS, PACIFIC

YEAR	JAN	FEB	MAR	APR	MAY	JUNE	JULY	AUG	SEP	OCT	NOV	DEC	ANNUAL
1956	80.3	80.3	80.4	81.3	81.5	81.4	82.5	82.5	82.4	82.4	82.6	82.1	81.6
1957	81.0	80.6	81.9	81.5	81.9	81.8	83.0	83.2	82.7	83.3	81.7	81.6	82.0
1958	80.8	81.2	81.4	82.0	82.2	81.7	82.3	83.6	84.7	84.3	82.7	82.5	82.4
1959	80.9	80.0	80.9	80.8	82.2	82.9	83.4	84.2	82.9	82.9	81.3	81.2	82.0
1960	80.3	80.7	81.5	80.4	81.3	81.6	81.7	81.8	82.1	82.0	81.0	82.1	81.4
1961	81.5	82.1	82.1	82.1	81.1	81.2	81.3	81.4	81.2	81.7	81.1	80.8	81.4
1962	81.3	81.6	82.1	81.9	81.2	81.7	81.4	81.7	81.3	82.1	81.0	82.1	81.6
1963	82.2	81.2	81.6	82.3	82.6	82.1	81.6	82.0	82.8	81.7	82.1	81.0	81.9
1964	80.8	80.8	81.5	80.7	81.0	81.5	81.1	81.1	81.0	80.5	80.6	80.4	80.9
1965	80.5	81.0	82.6	82.8	82.2	81.1	80.9	81.9	82.2	81.9	80.9	81.3	81.6
1966	80.7	81.8	82.1	81.3	81.6	81.6	82.4	82.4	83.1	82.6	82.0	80.3	81.8
1967	81.0	81.5	80.4	81.9	81.7	81.4	81.8	82.1	82.3	81.6	82.0	81.2	81.6
1968	81.0	81.2	80.4	80.8	81.3	81.1	82.1	82.1	82.9	82.9	82.4	82.3	81.6
1969	80.8	82.2	82.7	82.2	82.0	82.3	81.8	83.0	82.4	83.1	83.0	82.6	82.4
1970	82.4	83.3	83.5	83.5	82.6	82.5	82.0	81.9	81.6	80.8	81.8	82.1	82.3
1971	81.6	81.7	82.1	80.4	80.8	80.9	80.8	81.1	81.0	81.0	81.5	81.6	81.2
1972	80.8	81.3	82.3	81.4	82.4	81.9	82.1	82.5	82.0	81.9	82.4	81.3	81.9
1973	81.1	81.8	83.4	82.7	82.0	82.1	81.9	81.9	81.9	81.5	82.5	81.9	82.0
1974	81.6	81.7	82.1	80.8	81.7	81.6	82.4	81.5	81.7	81.7	81.9	81.7	81.7
1975	81.8	82.4	83.3	82.8	81.3	81.1	80.7	80.8	80.9	80.3	80.8	81.0	81.4
1976	80.4	80.5	81.2	81.3	81.1	81.2	81.6	81.7	81.4	81.5	81.7	81.5	81.3
1977	81.2	82.3	82.1	81.9	81.7	82.1	81.8	82.9	83.3	82.4	82.7	82.4	82.2
1978	81.6	81.5	82.5	82.0	81.2	81.7	81.6	82.3	82.6	82.5	81.6	82.0	81.9
1979	81.1	81.4	82.8	81.5	82.4	82.4	82.5	82.1	83.0	82.9	82.4	82.0	82.2
1980	81.9	82.0	83.3	82.0	81.4	82.4	82.3	82.9	82.3	82.8	82.0	82.5	82.3
1981	82.3	82.3	82.3	82.3	83.3	82.2	81.9	82.6	81.9	83.0	82.2	81.9	82.2
1982	81.6	82.1	82.6	82.7	82.5	82.4	81.8	82.3	81.8	82.3	82.7	80.9	82.2
1983	80.1	81.1	82.0	82.7	83.6	83.5	82.6	82.4	82.2	82.5	81.5	82.1	82.2
1984	82.0	82.2	83.0	83.1	83.6	82.9	82.7	81.8	82.3	82.6	82.0	82.5	82.6
1985	82.2	82.2	82.9	81.4	82.3	82.0	82.1	81.9	82.8	82.9	82.0	82.2	82.2
Record Mean	81.4	91.7	82.1	82.0	82.0	82.1	82.1	82.3	82.4	82.3	82.0	81.8	82.0
Max	85.5	86.0	86.5	86.3	86.4	86.6	86.8	87.1	87.3	87.1	86.4	85.8	86.5
Min	77.3	77.3	77.7	77.7	77.6	77.6	77.4	77.5	77.5	77.5	77.5	77.7	77.5

REFERENCE NOTES FOR TABLES 1, 2, 3 and 6   (KWAJALEIN, MARSHALL IS., PAC.)

GENERAL

T - TRACE AMOUNT
BLANK ENTRIES DENOTE MISSING/UNREPORTED DATA.
# INDICATES A STATION OR INSTRUMENT RELOCATION.

SPECIFIC

TABLE 1

(a) - LENGTH OF RECORD IN YEARS. ALTHOUGH
      INDIVIDUAL MONTHS MAY BE MISSING.
 * LESS THAN .05

NORMALS — BASED ON THE 1951-1980 RECORD PERIOD.
EXTREMES — DATES ARE THE MOST RECENT OCCURRENCE.
WIND DIR. — NUMERALS SHOW TENS OF DEGREES
           CLOCKWISE FROM TRUE NORTH.
           "00" INDICATES CALM.
RESULTANT WIND DIRECTIONS ARE GIVEN TO WHOLE DEGREES.

EXCEPTIONS

TABLES 2, 3, and 6

RECORD MEANS ARE THROUGH THE CURRENT YEAR,
BEGINNING IN   1948 FOR TEMPERATURE
                 1945 FOR PRECIPITATION

TABLE 4   HEATING DEGREE DAYS Base 65 deg. F   KWAJALEIN, MARSHALL ISLANDS, PACIFIC

SEASON	JULY	AUG	SEP	OCT	NOV	DEC	JAN	FEB	MAR	APR	MAY	JUNE	TOTAL
1983-84	0	0	0	0	0	0	0	0	0	0	0	0	0
1984-85	0	0	0	0	0	0	0	0	0	0	0	0	0
1985-86	0	0	0	0	0	0							

TABLE 5   COOLING DEGREE DAYS Base 65 deg. F   KWAJALEIN, MARSHALL ISLANDS, PACIFIC

YEAR	JAN	FEB	MAR	APR	MAY	JUNE	JULY	AUG	SEP	OCT	NOV	DEC	TOTAL
1969	498	487	556	524	532	523	528	564	528	566	542	552	6400
1970	547	518	580	562	552	532	534	530	506	497	510	536	6404
1971	519	470	538	469	498	484	494	509	488	505	528	521	5995
1972	497	477	545	496	548	515	536	549	518	530	534	513	6252
1973	509	477	574	538	533	520	526	533	513	520	529	525	6306
1974	521	476	538	480	526	503	545	518	509	523	517	525	6181
1975	526	495	572	539	516	493	496	498	482	480	479	502	6078
1976	483	458	511	496	507	491	522	525	499	521	507	518	6038
1977	507	491	537	514	526	518	526	564	555	547	537	549	6371
1978	525	465	548	519	507	509	521	542	536	552	507	535	6266
1979	502	467	561	504	544	526	546	536	547	559	530	532	6354
1980	528	501	576	515	517	529	542	557	526	560	519	551	6421
1981	545	491	541	527	573	525	532	551	515	565	523	530	6418
1982	521	486	554	535	548	528	530	544	513	541	502	535	6333
1983	478	454	531	538	582	561	552	546	521	548	502	535	6348
1984	535	504	565	551	582	546	557	526	527	554	518	549	6514
1985	541	500	533	501	542	513	542	529	540	564	516	539	6360

TABLE 6   SNOWFALL (inches)   KWAJALEIN, MARSHALL ISLANDS, PACIFIC

SEASON	JULY	AUG	SEP	OCT	NOV	DEC	JAN	FEB	MAR	APR	MAY	JUNE	TOTAL
1971-72	0.0	0.0	0.0	0.0	0.0	0.0	0.0	0.0	0.0	0.0	0.0	0.0	0.0
1972-73	0.0	0.0	0.0	0.0	0.0	0.0	0.0	0.0	0.0	0.0	0.0	0.0	0.0
1973-74	0.0	0.0	0.0	0.0	0.0	0.0	0.0	0.0	0.0	0.0	0.0	0.0	0.0
1974-75	0.0	0.0	0.0	0.0	0.0	0.0	0.0	0.0	0.0	0.0	0.0	0.0	0.0
1975-76	0.0	0.0	0.0	0.0	0.0	0.0	0.0	0.0	0.0	0.0	0.0	0.0	0.0
1976-77	0.0	0.0	0.0	0.0	0.0	0.0	0.0	0.0	0.0	0.0	0.0	0.0	0.0
1977-78	0.0	0.0	0.0	0.0	0.0	0.0	0.0	0.0	0.0	0.0	0.0	0.0	0.0
1978-79	0.0	0.0	0.0	0.0	0.0	0.0	0.0	0.0	0.0	0.0	0.0	0.0	0.0
1979-80	0.0	0.0	0.0	0.0	0.0	0.0	0.0	0.0	0.0	0.0	0.0	0.0	0.0
1980-81	0.0	0.0	0.0	0.0	0.0	0.0	0.0	0.0	0.0	0.0	0.0	0.0	0.0
1981-82	0.0	0.0	0.0	0.0	0.0	0.0	0.0	0.0	0.0	0.0	0.0	0.0	0.0
1982-83	0.0	0.0	0.0	0.0	0.0	0.0	0.0	0.0	0.0	0.0	0.0	0.0	0.0
1983-84	0.0	0.0	0.0	0.0	0.0	0.0	0.0	0.0	0.0	0.0	0.0	0.0	0.0
1984-85	0.0	0.0	0.0	0.0	0.0	0.0	0.0	0.0	0.0	0.0	0.0	0.0	0.0
1985-86	0.0	0.0	0.0	0.0	0.0								
Record Mean	0.0	0.0	0.0	0.0	0.0	0.0	0.0	0.0	0.0	0.0	0.0	0.0	0.0

See Reference Notes, relative to all above tables, on preceding page.

The station at Majuro is located on the southeastern end of the Majuro Atoll. This atoll is approximately 160 square miles in area with a lagoon of about 150 square miles. The lagoon is oblong, 22 miles long and about 4 miles wide. Dalap Island, on which the station is located, is oriented roughly east-west.

The climate of Majuro is predominately a trade-wind climate with the trade winds prevailing throughout the year. Tropical storms are very rare.

Minor storms of the easterly wave type are quite common from March to April and October to November. The trades are frequently locally interrupted during the summer months by the movement of the zone of intertropical convergence across the area.

Rainfall is heavy, with the wettest months being October and November. Precipitation is generally of the shower type, however, continuous rain is not uncommon.

One of the outstanding features of the climate is the extremely consistent temperature regime. The range between the coolest and the warmest months averages less than 1 degree. The average daily range is less than 9 degrees. Nighttime minima are generally 2-4 degrees warmer than the average daily minimum because lowest temperatures usually occur during heavy showers in the daytime.

Skies at Majuro are quite cloudy. Cumuliform clouds are predominant but altostratus-altocumulus and cirriform clouds are also present most of the time.

TABLE 1   NORMALS, MEANS AND EXTREMES

MAJURO, MARSHALL ISLANDS, PACIFIC

LATITUDE: 7°05'N   LONGITUDE: 171°23'E   ELEVATION: FT. GRND   10 BARO   8   TIME ZONE: 180E MER   WBAN: 40710

	(a)	JAN	FEB	MAR	APR	MAY	JUNE	JULY	AUG	SEP	OCT	NOV	DEC	YEAR
TEMPERATURE °F:														
Normals														
-Daily Maximum		84.7	85.1	85.3	85.2	85.4	85.5	85.5	85.9	86.0	86.0	85.6	85.0	85.4
-Daily Minimum		76.7	77.0	76.9	76.5	76.6	76.4	76.4	76.6	76.5	76.5	76.6	76.8	76.6
-Monthly		80.7	81.1	81.1	80.8	81.0	81.0	81.0	81.3	81.3	81.3	81.1	80.9	81.0
Extremes														
-Record Highest	30	89	88	89	89	89	89	90	91	90	91	90	90	91
-Year		1979	1984	1984	1983	1983	1983	1980	1969	1977	1958	1979	1979	AUG 1969
-Record Lowest	30	69	70	70	70	70	70	70	71	72	70	70	70	69
-Year		1958	1985	1982	1985	1985	1958	1985	1981	1985	1984	1984	1984	JAN 1958
NORMAL DEGREE DAYS:														
Heating (base 65°F)		0	0	0	0	0	0	0	0	0	0	0	0	0
Cooling (base 65°F)		487	451	499	477	496	480	496	505	489	505	483	493	5861
% OF POSSIBLE SUNSHINE	25	63	64	66	56	57	54	55	60	59	54	53	54	58
MEAN SKY COVER (tenths)														
Sunrise - Sunset	29	8.6	8.3	8.4	8.6	8.6	8.7	8.6	8.4	8.5	8.6	8.7	8.7	8.6
MEAN NUMBER OF DAYS:														
Sunrise to Sunset														
-Clear	29	1.0	1.1	1.3	0.8	0.7	0.4	0.6	0.7	1.0	0.9	0.6	0.6	9.5
-Partly Cloudy	29	6.5	7.5	6.9	6.4	6.9	6.4	6.2	8.1	6.2	6.7	6.6	6.3	80.8
-Cloudy	29	23.6	19.7	22.7	22.8	23.4	23.2	24.2	22.2	22.8	23.4	22.9	24.1	274.9
Precipitation														
.01 inches or more	31	16.8	15.7	18.0	21.3	23.6	24.2	24.4	23.5	22.5	23.9	23.1	21.9	258.9
Snow,Ice pellets														
1.0 inches or more	31	0.0	0.0	0.0	0.0	0.0	0.0	0.0	0.0	0.0	0.0	0.0	0.0	0.0
Thunderstorms	18	0.3	0.6	0.7	0.6	0.9	1.9	1.6	1.8	2.8	2.2	2.1	1.1	16.6
Heavy Fog Visibility														
1/4 mile or less	19	0.0	0.0	0.0	0.0	0.0	0.0	0.0	0.0	0.0	0.0	0.0	0.0	0.0
Temperature °F														
-Maximum														
90° and above	31	0.0	0.0	0.0	0.0	0.0	0.0	0.*	0.2	0.3	0.4	0.*	0.*	0.9
32° and below	31	0.0	0.0	0.0	0.0	0.0	0.0	0.0	0.0	0.0	0.0	0.0	0.0	0.0
-Minimum														
32° and below	31	0.0	0.0	0.0	0.0	0.0	0.0	0.0	0.0	0.0	0.0	0.0	0.0	0.0
0° and below	31	0.0	0.0	0.0	0.0	0.0	0.0	0.0	0.0	0.0	0.0	0.0	0.0	0.0
AVG. STATION PRESS.(mb)	11	1008.9	1009.3	1009.6	1009.3	1009.6	1009.3	1009.1	1009.3	1009.3	1008.8	1008.2	1008.3	1009.1
RELATIVE HUMIDITY (%)														
Hour 00	29	80	80	81	83	84	84	83	82	82	82	82	82	82
Hour 06 (Local Time)	30	81	80	81	84	85	84	84	84	83	83	83	82	83
Hour 12	30	75	74	74	77	78	78	77	77	76	76	77	76	76
Hour 18	29	77	77	77	78	80	81	80	79	78	79	80	79	79
PRECIPITATION (inches):														
Water Equivalent														
-Normal		7.99	6.37	8.96	11.91	12.32	12.04	12.65	11.61	13.09	15.24	13.47	11.52	137.17
-Maximum Monthly	31	21.97	18.34	18.51	31.10	22.23	17.63	18.69	16.77	21.11	24.26	23.56	24.80	31.10
-Year		1961	1957	1955	1971	1956	1975	1964	1985	1964	1955	1978	1968	APR 1971
-Minimum Monthly	31	0.78	0.40	0.66	1.97	1.49	5.40	5.34	5.33	6.42	7.11	4.53	2.28	0.40
-Year		1973	1970	1983	1983	1983	1984	1961	1959	1984	1969	1972	1957	FEB 1970
-Maximum in 24 hrs	31	9.57	6.28	8.14	6.63	5.86	7.39	5.39	4.38	5.76	8.74	10.01	17.88	17.88
-Year		1961	1957	1972	1973	1962	1983	1971	1981	1982	1974	1957	1972	DEC 1972
Snow,Ice pellets														..
-Maximum Monthly	31													
-Year														
-Maximum in 24 hrs	31													
-Year														
WIND:														
Mean Speed (mph)	24	12.8	13.7	13.2	12.1	11.1	10.0	8.5	7.3	7.1	7.5	8.9	12.5	10.4
Prevailing Direction														
through 1963		ENE	ENE	ENE	ENE	ENE	ENE	ENE	ENE	E	E	E	ENE	ENE
Fastest Mile														
-Direction (!!!)	27	NE	E	NE	E	E	NE	E	NE	E	E	SW	E	SW
-Speed (MPH)	27	38	35	36	35	38	38	34	32	36	38	45	38	45
-Year		1967	1962	1959	1963	1962	1964	1973	1970	1973	1985	1982	1973	NOV 1982
Peak Gust														
-Direction (!!!)	2	E	E	E	E	E	E	E	E	E	E	SW	E	E
-Speed (mph)	2	39	39	37	35	36	40	35	38	39	47	39	39	47
-Date		1985	1984	1984	1984	1984	1984	1984	1984	1984	1985	1984	1984	OCT 1985

See Reference Notes to this table on the following page

TABLE 2 PRECIPITATION (inches) MAJURO, MARSHALL ISLANDS, PACIFIC

YEAR	JAN	FEB	MAR	APR	MAY	JUNE	JULY	AUG	SEP	OCT	NOV	DEC	ANNUAL
1956	9.36	14.75	14.33	11.02	22.23	11.50	15.27	16.72	15.45	20.59	18.48	3.44	173.14
1957	5.90	18.34	2.83	7.21	8.03	11.03	9.10	12.87	12.52	14.24	20.83	2.28	125.18
1958	9.78	0.96	7.77	8.03	8.53	14.30	14.55	9.05	6.88	18.71	13.53	8.45	120.54
1959	1.07	9.47	8.72	12.69	6.35	14.17	11.00	5.33	16.36	11.49	19.92	14.00	130.57
1960	9.17	3.60	11.17	23.41	14.27	13.22	14.10	14.59	16.93	9.71	16.32	6.54	153.03
1961	21.97	6.50	4.24	8.50	8.34	13.90	11.50	11.31	11.14	11.50	12.04	16.91	131.69
1962	17.55	5.15	11.48	5.95	12.01	7.54	11.02	8.91	21.03	16.36	22.69	11.71	151.40
1963	17.46	9.57	12.43	6.21	11.31	11.96	11.69	10.76	6.83	13.13	11.60	8.57	131.52
1964	1.40	6.99	7.23	11.46	22.02	11.16	18.69	15.58	21.11	22.79	16.85	7.42	162.70
1965	9.85	5.32	1.98	4.69	7.93	11.45	14.85	6.92	15.46	14.71	12.12	9.55	114.83
1966	3.79	4.42	5.80	16.03	6.64	9.40	14.94	6.52	13.95	13.53	12.24	19.44	128.70
1967	11.88	9.72	12.46	7.64	4.93	10.98	13.87	7.99	13.78	15.16	11.16	6.48	126.05
1968	5.38	3.49	11.12	8.86	9.33	16.07	11.39	11.50	9.77	12.06	11.97	24.80	135.74
1969	8.22	2.35	16.17	17.21	8.78	13.01	16.65	10.24	15.65	7.11	11.68	7.21	134.28
1970	5.62	0.40	1.73	2.87	9.23	10.66	7.73	11.24	11.75	12.64	6.68	8.40	88.95
1971	8.21	5.74	9.80	31.10	19.86	13.42	15.49	14.92	7.93	18.06	9.46	8.40	162.39
1972	9.58	7.11	15.45	9.17	14.96	14.88	14.76	10.84	18.96	14.06	4.53	23.36	157.66
1973	0.78	1.84	11.05	14.59	14.33	12.23	7.29	13.86	12.78	13.79	14.21	7.24	123.99
1974	11.09	8.07	7.18	15.67	12.84	13.66	12.48	13.69	10.44	19.90	9.29	14.49	148.80
1975	5.20	3.21	7.77	12.76	10.58	17.63	14.23	16.35	16.51	18.29	15.28	13.95	151.76
1976	8.57	9.42	15.68	19.41	15.28	9.43	16.78	8.36	17.66	8.95	12.70	2.77	145.01
1977	2.39	0.77	2.60	10.62	17.21	8.37	10.88	11.15	9.72	17.59	11.85	18.88	122.03
1978	3.60	5.25	3.39	12.65	13.90	10.70	10.26	8.86	9.73	20.56	23.56	14.35	142.80
1979	6.78	2.77	7.14	11.75	7.91	13.23	6.67	13.03	6.54	15.04	11.33	7.10	109.29
1980	8.11	9.70	5.05	7.03	11.34	6.73	8.48	13.69	12.85	9.25	5.35	10.56	108.34
1981	0.90	4.34	17.40	10.20	9.04	5.43	16.53	12.24	6.71	7.28	14.61	14.47	119.15
1982	12.63	9.72	13.29	4.68	11.46	16.98	14.66	11.72	18.94	8.17	19.08	3.17	144.50
1983	0.83	0.98	0.66	1.97	1.49	14.45	12.58	6.05	11.25	13.47	9.84	12.74	86.31
1984	16.12	16.83	1.29	3.87	4.18	5.40	9.35	9.20	6.42	14.77	13.31	14.95	115.69
1985	8.70	16.56	4.59	15.38	9.67	14.67	13.18	16.77	8.03	18.06	12.81	11.30	149.72
Record Mean	7.96	6.90	8.72	11.16	11.31	12.00	12.63	11.45	12.64	14.68	13.43	11.23	134.10

TABLE 3 AVERAGE TEMPERATURE (deg. F) MAJURO, MARSHALL ISLANDS, PACIFIC

YEAR	JAN	FEB	MAR	APR	MAY	JUNE	JULY	AUG	SEP	OCT	NOV	DEC	ANNUAL
1956	80.5	80.9	81.1	81.1	80.6	80.7	81.0	80.5	80.9	80.8	81.3	81.9	81.0
1957	80.8	80.6	82.0	81.6	81.8	81.3	81.4	81.7	81.7	81.8	81.5	81.9	81.5
1958	80.3	81.8	81.1	81.0	80.5	80.6	81.0	81.7	81.7	81.6	81.7	81.3	81.2
1959	81.2	80.9	81.5	80.5	82.0	81.1	81.3	82.2	81.4	81.4	81.1	80.6	81.3
1960	81.0	81.6	81.5	80.2	80.6	80.6	81.1	81.2	81.7	82.3	81.8	81.7	81.3
1961	81.3	81.8	82.6	81.7	81.5	81.2	81.5	81.3	81.1	81.9	81.4	80.9	81.5
1962	80.9	81.6	80.7	81.7	82.2	81.3	81.2	81.8	81.0	81.5	80.7	81.5	81.3
1963	80.5	80.5	80.7	82.1	82.3	81.8	81.7	82.2	82.8	81.7	81.8	81.3	81.6
1964	81.8	81.6	81.3	81.4	81.1	80.7	80.8	80.8	80.5	80.7	80.7	80.8	81.0
1965	80.2	80.5	81.5	81.3	81.0	81.2	80.7	82.1	81.3	81.4	81.0	80.8	81.1
1966	81.0	81.3	81.3	80.7	81.7	81.7	81.8	82.5	81.8	81.9	81.2	80.9	81.5
1967	81.0	80.8	80.2	81.2	82.0	81.3	81.4	82.2	82.0	81.4	81.2	81.5	81.4
1968	81.1	81.4	80.3	80.6	80.8	81.1	80.9	81.8	81.1	81.1	80.6	81.0	81.0
1969	80.1	81.1	81.0	80.7	81.5	81.2	80.5	81.6	81.5	82.4	82.0	81.3	81.2
1970	81.2	82.0	82.0	82.1	81.5	80.7	81.1	80.8	81.1	80.6	81.2	80.6	81.3
1971	80.6	80.9	80.9	79.5	80.0	80.2	80.5	80.1	80.8	80.5	81.1	80.5	80.5
1972	80.2	80.9	80.8	80.8	81.2	81.5	80.9	81.2	81.3	80.9	81.7	80.8	81.0
1973	80.9	81.8	81.6	81.3	80.6	80.8	80.9	80.8	80.2	80.5	80.7	81.1	80.9
1974	79.9	80.8	.80.8	80.5	80.8	80.6	80.7	81.0	80.9	81.0	80.9	80.3	80.7
1975	80.4	81.0	80.7	80.2	80.5	79.7	79.7	79.9	80.0	78.8	79.4	79.5	80.1
1976	79.4	79.4	79.6	79.5	80.0	80.0	80.0	80.7	80.4	81.4	80.2	80.1	80.1
1977	80.3	81.3	81.5	80.5	80.1	81.2	80.9	81.2	82.4	81.1	81.1	81.1	81.1
1978	81.2	81.1	80.9	80.5	80.9	80.4	81.5	81.6	81.2	80.6	80.3	81.0	81.0
1979	81.2	81.0	81.6	79.7	80.9	81.5	81.5	80.9	82.0	82.1	81.8	81.7	81.3
1980	81.4	81.3	81.3	81.6	81.5	81.9	81.5	81.5	81.7	82.2	81.9	81.0	81.6
1981	81.4	81.5	81.0	80.9	81.4	82.0	80.8	81.4	82.0	82.0	81.0	80.6	81.3
1982	80.5	80.7	80.5	81.8	81.3	81.3	81.1	81.3	81.4	82.0	81.4	80.2	81.1
1983	80.1	80.5	81.4	82.2	83.0	81.4	81.3	82.2	81.8	80.9	81.0	80.3	81.3
1984	80.6	80.6	82.0	81.9	80.3	80.5	81.1	81.2	80.8	80.8	80.9	80.9	80.0
1985	80.8	80.3	80.7	79.8	81.1	80.4	80.7	80.4	81.5	81.4	81.5	80.9	80.8
Record Mean	80.7	81.0	81.1	80.9	81.1	81.0	81.0	81.3	81.3	81.3	81.1	80.8	81.1
Max	84.8	85.2	85.4	85.3	85.6	85.6	85.6	86.0	86.2	86.1	85.7	85.1	85.6
Min	76.6	76.8	76.8	76.5	76.6	76.3	76.3	76.5	76.5	76.4	76.5	76.6	76.5

REFERENCE NOTES FOR TABLES 1, 2, 3 and 6 (MAJURO, MARSHALL IS., PACIFIC)

GENERAL

T - TRACE AMOUNT
BLANK ENTRIES DENOTE MISSING/UNREPORTED DATA.
# INDICATES A STATION OR INSTRUMENT RELOCATION.

SPECIFIC

TABLE 1

(a) - LENGTH OF RECORD IN YEARS. ALTHOUGH INDIVIDUAL MONTHS MAY BE MISSING.
* LESS THAN .05

NORMALS — BASED ON THE 1951-1980 RECORD PERIOD.
EXTREMES — DATES ARE THE MOST RECENT OCCURRENCE.
WIND DIR. — NUMERALS SHOW TENS OF DEGREES
CLOCKWISE FROM TRUE NORTH.
"00" INDICATES CALM.
RESULTANT WIND DIRECTIONS ARE GIVEN TO WHOLE DEGREES.

EXCEPTIONS

TABLE 1

1. THUNDERSTORMS AND HEAVY FOG ARE THROUGH 1972.

TABLES 2, 3, and 6

RECORD MEANS ARE THROUGH THE CURRENT YEAR,
BEGINNING IN 1955 FOR TEMPERATURE
1955 FOR PRECIPITATION

TABLE 4   HEATING DEGREE DAYS Base 65 deg. F   MAJURO, MARSHALL ISLANDS, PACIFIC

SEASON	JULY	AUG	SEP	OCT	NOV	DEC	JAN	FEB	MAR	APR	MAY	JUNE	TOTAL
1983-84	0	0	0	0	0	0	0	0	0	0	0	0	0
1984-85	0	0	0	0	0	0							
1985-86	0	0	0	0	0	0							

TABLE 5   COOLING DEGREE DAYS Base 65 deg. F   MAJURO, MARSHALL ISLANDS, PACIFIC

YEAR	JAN	FEB	MAR	APR	MAY	JUNE	JULY	AUG	SEP	OCT	NOV	DEC	TOTAL
1969	475	458	502	477	519	496	486	520	503	544	515	513	6008
1970	512	483	536	521	518	474	507	497	491	485	491	488	6007
1971	494	454	500	443	473	462	486	476	480	487	492	489	5736
1972	480	466	498	483	509	504	502	508	499	500	506	498	5953
1973	498	476	523	494	490	480	500	493	464	484	479	505	5886
1974	469	449	497	468	498	498	475	492	503	486	505	487	5810
1975	483	456	494	464	490	447	460	468	458	436	439	458	5553
1976	451	424	459	442	470	460	473	491	469	514	465	477	5595
1977	478	464	518	475	476	495	501	510	528	506	492	507	5950
1978	508	456	517	482	486	482	483	518	506	509	475	480	5902
1979	507	454	519	447	501	504	520	500	512	537	510	527	6038
1980	519	478	513	507	519	516	516	518	509	542	515	502	6154
1981	515	470	501	483	516	517	496	512	517	534	489	488	6038
1982	485	447	487	511	511	497	507	511	498	534	498	478	5964
1983	476	439	519	522	566	500	511	537	511	499	487	482	6049
1984	492	459	533	512	518	464	490	507	494	495	475	499	5938
1985	499	434	492	453	507	469	495	486	503	515	500	501	5854

TABLE 6   SNOWFALL (inches)   MAJURO, MARSHALL ISLANDS, PACIFIC

SEASON	JULY	AUG	SEP	OCT	NOV	DEC	JAN	FEB	MAR	APR	MAY	JUNE	TOTAL
1971-72	0.0	0.0	0.0	0.0	0.0	0.0	0.0	0.0	0.0	0.0	0.0	0.0	0.0
1972-73	0.0	0.0	0.0	0.0	0.0	0.0	0.0	0.0	0.0	0.0	0.0	0.0	0.0
1973-74	0.0	0.0	0.0	0.0	0.0	0.0	0.0	0.0	0.0	0.0	0.0	0.0	0.0
1974-75	0.0	0.0	0.0	0.0	0.0	0.0	0.0	0.0	0.0	0.0	0.0	0.0	0.0
1975-76	0.0	0.0	0.0	0.0	0.0	0.0	0.0	0.0	0.0	0.0	0.0	0.0	0.0
1976-77	0.0	0.0	0.0	0.0	0.0	0.0	0.0	0.0	0.0	0.0	0.0	0.0	0.0
1977-78	0.0	0.0	0.0	0.0	0.0	0.0	0.0	0.0	0.0	0.0	0.0	0.0	0.0
1978-79	0.0	0.0	0.0	0.0	0.0	0.0	0.0	0.0	0.0	0.0	0.0	0.0	0.0
1979-80	0.0	0.0	0.0	0.0	0.0	0.0	0.0	0.0	0.0	0.0	0.0	0.0	0.0
1980-81	0.0	0.0	0.0	0.0	0.0	0.0	0.0	0.0	0.0	0.0	0.0	0.0	0.0
1981-82	0.0	0.0	0.0	0.0	0.0	0.0	0.0	0.0	0.0	0.0	0.0	0.0	0.0
1982-83	0.0	0.0	0.0	0.0	0.0	0.0	0.0	0.0	0.0	0.0	0.0	0.0	0.0
1983-84	0.0	0.0	0.0	0.0	0.0	0.0	0.0	0.0	0.0	0.0	0.0	0.0	0.0
1984-85	0.0	0.0	0.0	0.0	0.0	0.0	0.0	0.0	0.0	0.0	0.0	0.0	0.0
1985-86	0.0	0.0	0.0	0.0	0.0								
Record Mean	0.0	0.0	0.0	0.0	0.0	0.0	0.0	0.0	0.0	0.0	0.0	0.0	0.0

See Reference Notes, relative to all above tables, on preceding page.

Pago Pago Airport is located on the southeastern coast of the island of Tutuila in the American Samoa group, approximately 2,600 miles south-southwest of Hawaii, 1,600 miles north-northeast of New Zealand, and 4,500 miles southwest of California. Tutuila is a long, narrow island lying southwest-northeast, with a land area of 76 square miles, a greatest length of just over 20 miles, and a width ranging from 1 to 2 miles in the eastern half and from 2 to 5 miles in the western. It is volcanic in origin, extremely mountainous, and nearly surrounded by a coral reef. The principal ridge extends the length of the island, reaching a maximum height of 2,141 feet, at Matafao peak, near the central portion of the long axis. Vegetation is moderately dense, with many coconut, banana, and other tropical fruit trees, grass, and low-growing brush. The orientation of Tutuila is such that winds from the east-northeast clockwise to south approach Pago Pago Airport directly from the ocean without being deflected by the terrain, while winds from other directions may be considerably disturbed by topography.

Samoa has a maritime climate with abundant rain and warm, humid days and nights. Rainfall, usually falling as showers, is about 125 inches a year at the airport, but varies greatly over small distances because of topography. Thus, Pago Pago, less than 4 miles north of the airport and at the head of a hill-encircled harbor open to the prevailing wind, receives nearly 200 inches a year. The crest of the range receives well above 250 inches. In most years, the airport records about 300 days with a trace or more of rain and about 175 with .1 inch or more.

The driest months are June through September (southern winter) and the wettest, December through March (southern summer). However, the seasonal rainfall may vary widely in individual years, and heavy showers and long rainy periods can occur in any month. Flooding rains are not unknown. Some of these have been associated with hurricanes and tropical storms, but they have occurred at other times as well.

June, July and August are the coolest months and January, February, and March, the warmest. Afternoon temperatures ordinarily reach the upper 80s in summer and the mid 80s in winter, while nighttime temperatures fall to the mid 70s in the summer and low 70s in winter. The highest temperatures recorded at the airport are in the low 90s and lowest near 60.

The prevailing winds throughout the year are the easterly trades. These tend to be more directly from the east in December through March, but predominantly from east-southeast and southeast during the rest of the year. The trade winds are also less prevalent in summer than in winter. As the foregoing suggests, the trades are interrupted more often in summer than in winter. These interruptions are sometimes associated with the proximity of small tropical storms, of bands of converging winds, or of low pressure systems higher in the atmosphere, all of which help make summer the rainy season. At other times, the absence of the trades is marked by periods of light and variable winds and by land and sea breezes. Westerly to northerly winds, in particular, are more frequent then. These are strong at times, but are often quite light, and may then reflect the nighttime drainage of cooled air from the mountains west and north of the airport.

Thunderstorms are less frequent than might be expected, considering the moistness and instability of the tropical air mass which usually overlies Samoa.

TABLE 1 NORMALS, MEANS AND EXTREMES

PAGO PAGO, AMERICAN SAMOA

LATITUDE: 14°20'S   LONGITUDE: 170°43' W   ELEVATION: FT. GRND   12 BARO   00015   TIME ZONE: 165W MER   WBAN: 61705

	(a)	JAN	FEB	MAR	APR	MAY	JUNE	JULY	AUG	SEP	OCT	NOV	DEC	YEAR
TEMPERATURE °F:														
Normals														
-Daily Maximum		86.5	86.7	86.9	86.4	85.2	84.4	83.4	83.4	84.5	85.0	85.5	86.1	85.3
-Daily Minimum		74.9	74.8	75.1	74.6	74.5	75.0	74.3	73.9	74.2	75.0	75.1	75.1	74.7
-Monthly		80.7	80.8	81.0	80.5	79.8	79.7	78.8	78.7	79.3	80.0	80.3	80.6	80.0
Extremes														
-Record Highest	26	92	92	92	92	90	90	89	90	89	90	92	92	92
-Year		1969	1983	1966	1980	1972	1966	1961	1981	1983	1966	1966	1985	DEC 1985
-Record Lowest	26	67	67	67	68	66	64	62	64	63	67	67	67	62
-Year		1965	1965	1965	1979	1974	1965	1964	1979	1970	1973	1964	1964	JUL 1964
NORMAL DEGREE DAYS:														
Heating (base 65°F)		0	0	0	0	0	0	0	0	0	0	0	0	0
Cooling (base 65°F)		487	442	496	465	462	441	431	425	432	465	459	484	5489
% OF POSSIBLE SUNSHINE	18	41	42	47	38	35	34	45	49	55	42	41	39	42
MEAN SKY COVER (tenths)														
Sunrise - Sunset	19	8.2	8.2	7.8	7.7	7.5	7.1	7.1	6.9	6.9	7.6	7.8	7.9	7.6
MEAN NUMBER OF DAYS:														
Sunrise to Sunset														
-Clear	19	0.7	0.7	0.9	1.3	1.2	1.8	2.0	2.2	2.4	1.1	1.0	1.3	16.6
-Partly Cloudy	19	8.5	8.3	11.5	11.2	13.0	14.2	15.0	15.8	14.2	12.4	10.7	9.9	144.6
-Cloudy	19	21.7	19.3	18.6	17.6	16.8	14.0	14.0	13.0	13.4	17.5	18.3	19.8	204.1
Precipitation														
.01 inches or more	19	24.3	21.9	23.1	21.7	20.6	18.6	18.1	17.9	17.6	22.1	20.5	23.1	249.7
Snow,Ice pellets														
1.0 inches or more	19	0.0	0.0	0.0	0.0	0.0	0.0	0.0	0.0	0.0	0.0	0.0	0.0	0.0
Thunderstorms	19	2.5	2.8	2.9	3.4	2.9	1.1	0.4	0.7	0.8	2.4	3.0	3.4	26.4
Heavy Fog Visibility														
1/4 mile or less	19	0.0	0.0	0.0	0.0	0.0	0.0	0.0	0.0	0.0	0.0	0.0	0.0	0.0
Temperature °F														
-Maximum														
90° and above	26	2.2	2.3	2.8	1.7	0.2	0.*	0.0	0.*	0.0	0.1	0.3	1.9	11.7
32° and below	26	0.0	0.0	0.0	0.0	0.0	0.0	0.0	0.0	0.0	0.0	0.0	0.0	0.0
-Minimum														
32° and below	26	0.0	0.0	0.0	0.0	0.0	0.0	0.0	0.0	0.0	0.0	0.0	0.0	0.0
0° and below	26	0.0	0.0	0.0	0.0	0.0	0.0	0.0	0.0	0.0	0.0	0.0	0.0	0.0
AVG. STATION PRESS.(mb)	13	1007.8	1009.0	1009.2	1010.1	1011.1	1011.9	1012.4	1012.7	1012.7	1011.7	1009.9	1008.2	1010.6
RELATIVE HUMIDITY (%)														
Hour 01	17	89	89	89	90	87	85	84	83	85	86	87	88	87
Hour 07	17	89	89	90	90	88	85	84	84	84	85	85	86	87
Hour 13 (Local Time)	17	76	76	76	76	76	76	75	74	73	76	76	75	75
Hour 19	17	82	82	83	85	85	82	81	81	81	83	82	82	82
PRECIPITATION (inches):														
Water Equivalent														
-Normal		12.78	12.53	11.38	11.25	10.72	8.56	6.51	7.08	6.69	11.05	11.20	14.21	123.96
-Maximum Monthly	26	22.14	32.66	31.84	24.46	20.48	12.82	19.59	16.46	25.29	21.48	25.67	26.95	32.66
-Year		1973	1968	1961	1967	1966	1961	1962	1978	1972	1980	1978	1984	FEB 1968
-Minimum Monthly	26	4.65	5.91	4.01	4.13	1.61	2.62	0.72	1.19	0.99	2.10	1.36	2.84	0.72
-Year		1968	1966	1976	1982	1983	1978	1974	1974	1974	1975	1965	1982	JUL 1974
-Maximum in 24 hrs	26	8.74	9.01	6.48	10.79	6.68	5.94	8.65	6.18	10.28	7.50	5.75	8.97	10.79
-Year		1973	1968	1981	1975	1976	1961	1962	1978	1972	1967	1977	1970	APR 1975
Snow,Ice pellets														
-Maximum Monthly	26													
-Year														
-Maximum in 24 hrs	26													
-Year														
WIND:														
Mean Speed (mph)	18	8.7	8.4	8.2	8.1	9.9	12.0	12.6	12.6	12.0	12.1	10.2	8.8	10.3
Prevailing Direction														
Fastest Obs. 1 Min.														
-Direction (!!)	6	34	35	02	07	22	08	08	12	10	10	09	32	34
-Speed (MPH)	6	46	39	33	35	31	33	31	31	32	32	30	37	46
-Year		1985	1984	1982	1985	1982	1984	1985	1985	1983	1981	1985	1983	JAN 1985
Peak Gust														
-Direction (!!)	2	N	N	N	E	N	NE	E	SE	E	E	E	NW	N
-Speed (mph)	2	67	56	52	41	44	39	48	43	38	39	38	56	67
-Date		1985	1984	1984	1985	1985	1984	1985	1985	1985	1985	1985	1984	JAN 1985

See Reference Notes to this table on the following pages.

TABLE 2 PRECIPITATION (inches) PAGO PAGO, AMERICAN SAMOA

YEAR	JAN	FEB	MAR	APR	MAY	JUNE	JULY	AUG	SEP	OCT	NOV	DEC	ANNUAL
1960		11.47	10.36		9.42	9.33	14.37	10.15	2.62	10.08	12.01	13.29	
1961	8.33	7.50	31.84	4.73	6.27	12.82	3.17	4.54	2.99	14.39	12.65	8.61	117.84
1962	11.53	19.57	8.06	14.25	11.20	10.58	19.59	12.03	5.12	13.97	24.02	156.38
1963	16.97	14.38	15.51	16.32	11.25	5.51	2.01	8.06	4.74	16.24	10.09	6.00	127.08
1964	14.92		10.95	11.72	10.53	9.47	10.34	7.69	14.92	7.54	16.68	16.42	
1965	9.84	12.73	6.01		17.69	12.12	2.06	3.75	1.95	19.67	1.36	12.95	
1966	16.38	5.91	10.83	10.42	20.48	9.17	6.40	7.62	8.46	18.63	5.76	16.21	136.27
1967	6.97	8.60	4.95	24.46	9.97	11.74	9.00	13.02	4.34	15.92	15.30	16.36	140.63
1968	4.65	32.66	11.69	12.60	7.00	2.71	3.56	4.57	8.01	18.88	4.27	6.24	116.84
1969	17.95	11.71	11.93	15.95	4.91	3.96	11.42	7.35	2.26	10.05	9.35	16.68	123.52
1970	9.43	8.87	21.62	13.58	8.78	12.16	4.68	8.10	7.69	5.69	7.98	26.38	134.96
1971	11.48	14.33	7.69	11.70	4.77	12.03	3.16	5.24	2.61	7.25	9.24	6.67	96.17
1972	17.77	11.17	9.45	6.50	6.55	3.67	7.35	5.61	25.29	10.44	6.48	20.30	130.58
1973	22.14	8.44	4.94	9.74	7.93	6.09	8.12	8.58	5.15	16.19	18.93	14.90	131.15
1974	9.26	9.46	9.67	8.80	8.07	5.40	0.72	1.19	0.99	2.44	12.73	9.16	77.89
1975	22.13	9.88	6.57	22.77	17.88	5.21	8.17	2.95	5.57	2.10	7.69	14.15	125.07
1976	5.69	13.36	4.01	10.31	18.98	7.11	8.73	4.09	5.17	3.50	12.19	21.38	114.52
1977	7.29	6.30	11.64	6.98	20.21	5.19	2.92	2.40	3.12	10.06	14.81	6.86	97.78
1978	21.36	9.52	25.64	9.24	8.71	2.62	3.48	16.46	4.71	10.04	25.67	15.62	153.07
1979	12.70	14.43	9.03	9.53	4.23	7.86	8.83	7.19	9.12	11.73	7.89	8.41	110.95
1980	9.73	7.47	14.84	6.52	19.50	11.90	4.14	13.97	15.47	21.48	7.43	13.44	145.89
1981	14.25	14.23	25.37	22.00	9.25	9.47	10.19	10.64	4.04	16.38	10.08	19.58	165.48
1982	9.09	30.25	7.68	4.13	7.90	4.22	6.96	16.20	6.30	4.38	6.35	2.84	106.30
1983	9.45	12.09	6.07	10.97	1.61	2.71	1.12	2.11	7.78	7.17	13.23	13.05	87.36
1984	9.70	8.03	19.34	6.70	4.79	7.44	1.86	4.25	5.23	15.21	8.08	26.95	117.58
1985	16.71	8.27	5.00	18.66	10.41	12.60	3.84	5.76	9.56	10.39	9.24	6.27	116.71
Record Mean	12.63	12.43	11.95	12.02	10.32	7.81	6.39	7.44	6.66	11.24	10.75	13.95	123.60

TABLE 3 AVERAGE TEMPERATURE (deg. F) PAGO PAGO, AMERICAN SAMOA

YEAR	JAN	FEB	MAR	APR	MAY	JUNE	JULY	AUG	SEP	OCT	NOV	DEC	ANNUAL
1960	80.7	79.8	80.2	81.3	80.4	80.0	77.7	78.0	79.1	79.1	81.3	81.1	79.9
1961	82.3	81.4	79.5	81.6	80.5	79.2	78.6	78.9	80.0	80.0	80.0	80.6	80.1
1962	80.2	80.7	80.7	80.3	79.4	78.9	79.3	79.7	80.2	80.0	80.3	80.2	79.9
1963	81.1	81.4	80.9	81.8	80.8	80.2	80.3	79.7	79.1	80.0	81.2	81.7	80.7
1964	81.9	82.0	82.4	82.0	79.7	79.5	78.0	79.1	78.7	80.1	80.4	79.1	80.3
1965	79.9	79.1	79.7	79.6	79.2	78.0	78.2	78.2	78.7	79.7	80.6	80.8	79.3
1966	81.3	82.1	82.3	81.6	80.1	79.8	79.6	78.4	79.5	80.0	81.2	80.5	80.5
1967	80.4	80.0	81.7	80.0	79.1	77.9	77.4	77.6	79.6	78.8	79.0	80.0	79.3
1968	81.0	80.1	79.7	78.4	79.3	79.8	78.5	77.4	79.1	79.5	80.8	81.7	79.6
1969	81.7	81.1	81.2	80.6	79.8	80.5	78.4	78.3	78.7	80.4	80.1	80.5	80.1
1970	81.0	81.7	80.1	81.0	81.2	80.0	79.6	78.3	78.4	79.8	79.5	78.7	79.9
1971	80.0	79.4	79.5	79.3	79.2	78.5	78.1	78.4	78.4	79.3	80.0	80.1	79.2
1972	79.2	79.7	80.9	81.2	79.9	80.1	78.4	79.5	79.3	80.2	80.8	82.6	80.2
1973	81.6	82.6	82.9	82.2	81.3	80.4	79.9	79.3	80.3	78.8	79.6	80.0	80.7
1974	80.6	78.9	79.9	79.6	78.4	79.2	79.2	78.1	80.5	80.8	80.2	80.0	79.6
1975	80.5	80.6	81.1	79.5	80.0	79.2	79.3	79.3	79.6	81.1	80.3	79.4	80.0
1976	79.6	79.8	80.6	80.5	79.1	78.8	78.5	78.8	78.9	80.8	80.9	80.1	79.7
1977	81.9	82.3	81.4	80.9	79.4	80.5	79.1	79.1	78.5	80.0	80.8	82.1	80.5
1978	80.9	81.6	80.3	80.3	80.5	79.8	78.9	79.5	79.5	80.1	79.3	80.9	80.2
1979	81.0	81.4	82.3	80.1	81.2	81.1	78.4	78.8	80.5	81.0	80.4	80.5	80.5
1980	81.1	81.8	82.2	82.4	79.6	80.6	79.2	79.4	80.2	79.7	81.8	81.6	80.8
1981	81.7	80.8	80.6	80.2	80.6	79.1	79.0	79.2	80.6	79.8	80.8	80.9	80.3
1982	82.0	80.4	82.3	81.8	80.8	80.4	78.7	79.0	79.2	80.8	82.6	81.1	80.5
1983	82.1	83.5	81.9	81.1	81.1	79.7	78.7	78.3	80.2	79.6	80.4	81.1	80.5
1984	80.7	81.5	81.3	81.5	81.8	80.6	79.0	79.8	79.8	79.6	80.4	81.4	80.7
1985	80.3	81.6	82.2	80.8	80.4	80.0	79.5	79.9	79.7	80.9	80.5	81.7	80.6
Record Mean	80.9	81.0	81.1	80.8	80.1	79.7	78.9	78.8	79.4	79.9	80.4	80.7	80.2
Max	86.7	86.8	87.0	86.6	85.3	84.3	83.4	83.4	84.4	84.8	85.6	86.2	85.4
Min	75.1	75.1	75.1	74.9	74.9	75.0	74.3	74.1	74.4	75.0	75.2	75.2	74.9

REFERENCE NOTES FOR TABLES 1, 2, 3 and 6 (PAGO PAGO, AMERICAN SAMOA)

GENERAL

T - TRACE AMOUNT
BLANK ENTRIES DENOTE MISSING/UNREPORTED DATA.
# INDICATES A STATION OR INSTRUMENT RELOCATION.

SPECIFIC

TABLE 1

(a) - LENGTH OF RECORD IN YEARS. ALTHOUGH INDIVIDUAL MONTHS MAY BE MISSING.
* LESS THAN .05

NORMALS — BASED ON THE 1951-1980 RECORD PERIOD.
EXTREMES — DATES ARE THE MOST RECENT OCCURRENCE.
WIND DIR. — NUMERALS SHOW TENS OF DEGREES CLOCKWISE FROM TRUE NORTH.
     "00" INDICATES CALM.
RESULTANT WIND DIRECTIONS ARE GIVEN TO WHOLE DEGREES.

EXCEPTIONS

TABLE 1

1. TEMPERATURES ARE BASED ON DAILY EXTREMES OF HOURLY VALUES APRIL 1964 THROUGH MARCH 1966.

TABLES 2, 3, and 6

RECORD MEANS ARE THROUGH THE CURRENT YEAR,
BEGINNING IN 1960 FOR TEMPERATURE
             1960 FOR PRECIPITATION

TABLE 4 HEATING DEGREE DAYS Base 65 deg. F   PAGO PAGO, AMERICAN SAMOA

SEASON	JULY	AUG	SEP	OCT	NOV	DEC	JAN	FEB	MAR	APR	MAY	JUNE	TOTAL
1983-84	0	0	0	0	0	0	0	0	0	0	0	0	0
1984-85	0	0	0	0	0	0	0	0	0	0	0	0	0
1985-86	0	0	0	0	0	0							

TABLE 5 COOLING DEGREE DAYS Base 65 deg. F   PAGO PAGO, AMERICAN SAMOA

YEAR	JAN	FEB	MAR	APR	MAY	JUNE	JULY	AUG	SEP	OCT	NOV	DEC	TOTAL
1969	526	459	511	476	464	472	421	416	419	487	458	487	5596
1970	500	474	474	486	508	456	458	423	408	467	441	430	5525
1971	474	410	454	438	448	410	412	429	410	448	458	475	5266
1972	446	433	503	490	468	461	421	456	436	480	480	551	5625
1973	525	500	561	522	513	464	469	450	465	436	446	473	5824
1974	488	396	469	447	421	435	448	414	472	497	462	471	5420
1975	485	445	506	452	473	431	452	450	444	507	464	454	5563
1976	462	437	492	470	444	422	427	435	426	496	483	477	5471
1977	531	491	516	487	454	474	445	445	411	472	480	537	5743
1978	501	472	483	467	486	448	439	455	445	474	437	500	5607
1979	502	465	540	458	509	487	424	436	469	502	470	487	5749
1980	507	495	543	530	457	475	446	453	463	465	513	523	5870
1981	526	450	493	465	490	432	439	447	473	467	481	502	5665
1982	533	438	543	512	497	469	433	442	434	496	463	506	5766
1983	539	524	529	493	504	449	429	419	463	456	467	516	5788
1984	495	487	514	504	528	472	442	462	445	459	487	482	5777
1985	481	470	543	480	487	458	457	468	449	500	474	526	5793

TABLE 6 SNOWFALL (inches)   PAGO PAGO, AMERICAN SAMOA

SEASON	JULY	AUG	SEP	OCT	NOV	DEC	JAN	FEB	MAR	APR	MAY	JUNE	TOTAL
1971-72	0.0	0.0	0.0	0.0	0.0	0.0	0.0	0.0	0.0	0.0	0.0	0.0	0.0
1972-73	0.0	0.0	0.0	0.0	0.0	0.0	0.0	0.0	0.0	0.0	0.0	0.0	0.0
1973-74	0.0	0.0	0.0	0.0	0.0	0.0	0.0	0.0	0.0	0.0	0.0	0.0	0.0
1974-75	0.0	0.0	0.0	0.0	0.0	0.0	0.0	0.0	0.0	0.0	0.0	0.0	0.0
1975-76	0.0	0.0	0.0	0.0	0.0	0.0	0.0	0.0	0.0	0.0	0.0	0.0	0.0
1976-77	0.0	0.0	0.0	0.0	0.0	0.0	0.0	0.0	0.0	0.0	0.0	0.0	0.0
1977-78	0.0	0.0	0.0	0.0	0.0	0.0	0.0	0.0	0.0	0.0	0.0	0.0	0.0
1978-79	0.0	0.0	0.0	0.0	0.0	0.0	0.0	0.0	0.0	0.0	0.0	0.0	0.0
1979-80	0.0	0.0	0.0	0.0	0.0	0.0	0.0	0.0	0.0	0.0	0.0	0.0	0.0
1980-81	0.0	0.0	0.0	0.0	0.0	0.0	0.0	0.0	0.0	0.0	0.0	0.0	0.0
1981-82	0.0	0.0	0.0	0.0	0.0	0.0	0.0	0.0	0.0	0.0	0.0	0.0	0.0
1982-83	0.0	0.0	0.0	0.0	0.0	0.0	0.0	0.0	0.0	0.0	0.0	0.0	0.0
1983-84	0.0	0.0	0.0	0.0	0.0	0.0	0.0	0.0	0.0	0.0	0.0	0.0	0.0
1984-85	0.0	0.0	0.0	0.0	0.0	0.0	0.0	0.0	0.0	0.0	0.0	0.0	0.0
1985-86	0.0	0.0	0.0	0.0	0.0	0.0							
Record Mean	0.0	0.0	0.0	0.0	0.0	0.0	0.0	0.0	0.0	0.0	0.0	0.0	0.0

See Reference Notes, relative to all above tables, on preceding page.

Pohnpei, about 129 square miles in area, is a nearly circular island of volcanic origin, encircled by coral barrier reefs, and covered with lush, tropical vegetation. The island, located less than 500 miles north of the equator, rises from the Pacific Ocean to an elevation of 2,595 feet, the highest point in the Caroline Islands. The topography is a complicated system of ridges and valleys, interlaced with small rivers and intermittent streams, and covered with tall grasses, tropical trees and flowers, and the coconut palms which are the backbone of the island economy. The interior of the island is covered by a rain forest which acts as a watershed area supplying fresh water the year-round.

The Weather Station lies a little north of the center of a bowl-shaped valley, about 3 miles south of the Pacific Ocean. Encircling it on a radius of 1 to 3 miles are volcanic outcroppings which rise rapidly from the ocean to an average elevation of about 2,100 feet in the east, south, and southwest, but to only about 700 feet to the west and northwest. With the exception of the cliff area in the northwest, the vegetation is lush and extremely dense. The valley is relatively level compared to the rest of the island. A small stream, oriented northeast-southwest lies about one-eighth of a mile to the east of the station.

From about November to June the climate of Pohnpei is chiefly influenced by the northeasterly trade winds. Wind speeds in the vicinity of the weather station, Pohnpei, are reduced somewhat due to the surrounding terrain. By about April the trades begin to diminish in strength, and by July have given way to the lighter and more variable winds of the doldrums. Between July and November the island is frequently under the influence of the Intertropical Convergence Zone (ITCZ – also called the Intertropical Front) which has moved northward into the area. This is also the season when moist southerly winds and tropical disturbances, many of them associated with the ITCZ, are most frequent and when humidities are often oppressively high.

Rainfall at Pohnpei is heavy and frequent throughout the year, averaging 192 inches annually. The wettest period is April and May. Measurable rain (.01 inch or more) falls on about 300 days a year.

The temperature is remarkably uniform throughout the year, with only slightly more than 1 degree separating the averages of the warmest and coolest months. High temperatures normally range in the mid to upper 80s and lows in the low to mid 70s. Temperature extremes above 90 degrees and below 70 degrees have occurred in every month of the year. Humidities are usually high throughout the year.

On most days, cumulus clouds predominate and usually cover more than eight-tenths of the sky. Days are normally cloudier than nights. High clouds, such as cirrus or cirrostratus often form and are obscurred by clouds at lower altitudes. Clouds at middle heights, usually altocumulus, sometimes combined with altostratus, occur quite frequently, especially if there are tropical disturbances in the vicinity.

Although Pohnpei is located within the spawning grounds of typhoons, the major typhoon tracks of the Western Pacific lie well to the north and west. Typhoons have caused extensive damage to crops and buildings on the island on several occasions.

The steep slopes surrounding the Weather Station reduce the wind speed. They can cause gentle up and downslope air currents which augment the land and sea breezes. The temperture and humidity are also greatly influenced by the reduced circulation and dense vegetation.

TABLE 1 NORMALS, MEANS AND EXTREMES

POHNPEI (PONAPE), EASTERN CAROLINE ISLANDS, PACIFIC

LATITUDE: 6°58'N   LONGITUDE: 158°13' E   ELEVATION: FT. GRND   123 BARO   00126   TIME ZONE: 165E MER   WBAN: 40504

	(a)	JAN	FEB	MAR	APR	MAY	JUNE	JULY	AUG	SEP	OCT	NOV	DEC	YEAR
TEMPERATURE °F:														
Normals														
-Daily Maximum		86.2	86.2	86.7	86.7	87.0	87.2	87.4	87.8	87.8	87.9	87.7	86.8	87.1
-Daily Minimum		75.4	75.6	75.4	74.9	74.5	73.9	73.0	72.7	72.7	72.7	73.4	74.8	74.1
-Monthly		80.8	80.9	81.1	80.8	80.8	80.6	80.2	80.3	80.3	80.3	80.6	80.8	80.6
Extremes														
-Record Highest	35	92	92	95	92	92	92	92	93	96	94	93	93	96
-Year		1980	1950	1950	1983	1983	1983	1981	1981	1950	1957	1979	1979	SEP 1950
-Record Lowest	35	66	67	69	68	69	69	68	68	68	68	66	66	66
-Year		1974	1976	1976	1972	1964	1964	1950	1974	1978	1955	1969	1977	DEC 1977
NORMAL DEGREE DAYS:														
Heating (base 65°F)		0	0	0	0	0	0	0	0	0	0	0	0	0
Cooling (base 65°F)		490	445	499	474	490	468	471	474	459	474	468	490	5702
% OF POSSIBLE SUNSHINE	27	42	43	48	42	41	42	44	46	47	41	43	38	43
MEAN SKY COVER (tenths)														
Sunrise - Sunset	33	8.9	8.9	8.9	9.0	8.8	8.8	8.6	8.5	8.6	8.6	8.6	8.9	8.8
MEAN NUMBER OF DAYS:														
Sunrise to Sunset														
-Clear	33	0.2	0.2	0.2	0.5	0.2	0.4	0.5	0.6	0.9	0.6	0.5	0.3	5.1
-Partly Cloudy	33	4.9	4.8	4.7	4.3	5.2	5.0	6.2	7.3	5.8	6.8	6.2	5.4	66.8
-Cloudy	33	25.9	23.2	26.0	25.2	25.6	24.6	24.3	23.1	23.3	23.6	23.3	25.3	293.4
Precipitation														
.01 inches or more	34	21.8	19.9	22.1	24.6	27.3	27.3	27.0	26.5	24.4	25.1	25.1	24.9	295.9
Snow,Ice pellets														
1.0 inches or more	34	0.0	0.0	0.0	0.0	0.0	0.0	0.0	0.0	0.0	0.0	0.0	0.0	0.0
Thunderstorms	34	0.9	0.8	1.1	1.4	2.2	2.2	3.0	2.5	3.0	3.1	3.1	1.7	25.0
Heavy Fog Visibility														
1/4 mile or less	34	0.0	0.0	0.0	0.0	0.0	0.0	0.0	0.0	0.0	0.0	0.0	0.0	0.0
Temperature °F														
-Maximum														
90° and above	34	0.9	0.9	2.6	3.1	5.7	4.3	6.1	7.8	9.1	10.1	7.6	3.2	61.4
32° and below	34	0.0	0.0	0.0	0.0	0.0	0.0	0.0	0.0	0.0	0.0	0.0	0.0	0.0
-Minimum														
32° and below	34	0.0	0.0	0.0	0.0	0.0	0.0	0.0	0.0	0.0	0.0	0.0	0.0	0.0
0° and below	34	0.0	0.0	0.0	0.0	0.0	0.0	0.0	0.0	0.0	0.0	0.0	0.0	0.0
AVG. STATION PRESS.(mb)														
RELATIVE HUMIDITY (%)														
Hour 05	18	84	83	86	89	91	93	95	96	96	95	93	88	91
Hour 11 (Local Time)	34	76	76	76	78	80	80	79	79	78	78	79	78	78
Hour 17 (Local Time)	34	78	76	78	80	81	80	80	79	79	79	80	79	79
Hour 23	34	83	82	84	87	89	90	93	93	93	93	91	86	89
PRECIPITATION (inches):														
Water Equivalent														
-Normal		11.31	11.06	14.51	18.98	20.36	17.39	17.69	17.13	15.95	15.99	16.91	15.65	192.93
-Maximum Monthly	35	26.67	19.76	25.30	38.65	38.43	24.88	37.20	32.74	29.53	22.25	31.79	33.35	38.65
-Year		1962	1964	1976	1959	1980	1969	1965	1976	1972	1975	1957	1975	APR 1959
-Minimum Monthly	35	1.89	1.05	1.52	2.03	2.21	9.60	9.40	10.06	6.57	7.98	4.55	2.40	1.05
-Year		1983	1977	1983	1983	1983	1963	1973	1965	1979	1972	1963	1957	FEB 1977
-Maximum in 24 hrs	35	5.04	8.91	7.13	8.86	7.72	6.26	7.07	13.25	9.35	6.55	22.48	8.04	22.48
-Year		1954	1968	1967	1977	1980	1980	1965	1976	1982	1985	1957	1975	NOV 1957
Snow,Ice pellets														
-Maximum Monthly	35													
-Year														
-Maximum in 24 hrs	35													
-Year														
WIND:														
Mean Speed (mph)	14	8.6	9.5	8.4	7.3	6.6	5.6	4.9	4.5	4.7	4.8	5.6	7.7	6.5
Prevailing Direction														
through 1963		NE	NE	NE	NE	NE	NE	E	ESE	S	ESE	NE	NE	NE
Fastest Mile														
-Direction (!!!)	27	SE	NE	E	NE	SW	NE	NE	NE	E	SW	SW	NE	E
-Speed (MPH)	27	28	26	36	26	29	26	26	32	26	28	29	35	36
-Year		1965	1967	1959	1973	1972	1975	1975	1973	1970	1972	1967	1973	MAR 1959
Peak Gust														
-Direction (!!!)	2	NE	NE	NE	E	SE	SE	NE	SE	NE	NE	E	NE	E
-Speed (mph)	2	36	41	39	48	37	43	35	35	24	29	45	38	48
-Date		1984	1985	1985	1985	1984	1985	1984	1984	1985	1985	1984	1984	APR 1985

See Reference Notes to this table on the following page.

TABLE 2 PRECIPITATION (inches) POHNPEI (PONAPE), EASTERN CAROLINE ISLANDS, PACIFIC

YEAR	JAN	FEB	MAR	APR	MAY	JUNE	JULY	AUG	SEP	OCT	NOV	DEC	ANNUAL
#1956	17.61	6.51	18.33	28.08	23.40	13.41	16.90	17.43	13.29	15.40	15.68	19.98	206.02
1957	11.23	14.58	9.93	6.53	15.23	20.58	15.72	17.57	16.79	8.62	31.79	2.40	170.97
1958	9.72	15.71	17.00	23.59	20.40	16.62	21.28	11.23	14.89	20.56	23.16	9.35	203.51
1959	3.91	15.37	22.03	38.65	21.26	15.61	19.12	12.73	18.42	13.28	11.31	26.11	217.80
1960	13.82	12.45	15.81	25.72	22.08	18.36	12.01	16.02	13.11	16.18	18.88	20.85	205.29
1961	16.60	17.86	17.47	11.59	22.21	18.11	15.51	17.52	20.69	14.07	18.29	16.47	206.39
1962	26.67	16.04	11.04	11.94	22.41	11.89	13.60	18.54	22.91	18.57	28.39	13.60	215.60
1963	20.99	16.37	17.06	12.44	19.12	9.60	13.73	18.23	13.12	20.68	4.55	9.08	174.97
1964	3.59	19.76	14.03	16.02	12.69	13.16	14.33	16.47	15.44	11.02	12.66	18.22	167.39
1965	11.74	11.12	6.36	14.28	14.18	18.73	37.20	10.06	24.67	15.27	15.50	14.20	193.31
1966	15.84	1.71	14.77	6.07	21.27	11.87	24.22	10.18	10.65	16.61	18.74	18.12	170.05
1967	10.21	18.83	21.82	20.90	13.46	11.48	14.02	20.32	19.51	16.37	22.44	12.60	201.96
1968	9.88	13.60	24.47	21.09	16.99	10.54	22.32	17.36	17.58	14.65	9.32	13.33	191.13
1969	6.39	9.47	10.71	22.35	17.75	24.88	28.90	13.28	18.01	12.86	16.20	15.00	195.80
1970	7.98	7.14	7.35	14.58	15.73	16.80	12.40	16.10	15.74	20.40	15.62	19.93	169.77
1971	16.64	13.12	19.98	17.37	24.59	23.62	22.69	16.67	9.89	19.11	12.02	8.67	204.37
1972	10.52	11.79	14.66	20.39	33.46	9.73	36.31	12.38	29.53	7.98	7.30	7.71	201.76
1973	3.31	3.70	7.71	24.38	16.28	23.40	9.40	17.85	18.15	21.32	10.38	18.16	174.04
1974	10.66	16.56	18.81	22.24	16.74	21.27	21.57	18.77	8.84	17.30	20.79	12.32	205.87
1975	6.61	4.26	15.99	16.79	17.50	18.83	15.60	11.26	12.61	22.25	17.22	33.35	192.27
1976	6.02	12.76	25.30	20.18	24.39	20.99	13.04	32.74	24.11	16.94	26.34	13.48	236.29
1977	4.45	1.05	12.65	15.93	26.17	14.70	16.97	18.72	10.88	20.00	16.14	4.95	162.61
1978	16.38	6.18	6.17	18.91	12.82	19.27	10.28	13.62	11.44	16.97	13.19	14.00	159.23
1979	8.16	6.65	11.98	23.38	18.76	23.85	17.07	22.35	6.57	17.57	20.09	19.58	196.01
1980	14.01	18.63	8.11	15.10	38.43	23.21	15.87	15.79	15.32	11.70	7.96	10.07	194.20
1981	14.59	13.75	8.55	10.47	23.00	17.92	16.20	13.41	15.04	15.43	16.46	17.47	182.29
1982	14.06	16.35	12.94	16.59	22.67	17.28	23.00	17.99	25.58	8.42	9.94	16.05	200.87
1983	1.89	1.72	1.52	2.03	2.21	15.91	24.55	14.29	12.36	20.05	20.99	16.10	133.62
1984	23.24	13.24	9.43	5.85	6.73	17.49	10.49	11.49	12.51	16.75	17.79	13.89	158.90
1985	14.83	12.91	5.15	24.22	13.04	13.80	14.62	14.02	12.48	18.81	19.95	20.11	183.94
Record Mean	11.57	11.05	13.53	18.05	19.29	17.11	17.63	16.42	16.05	16.16	16.88	15.79	189.54

TABLE 3 AVERAGE TEMPERATURE (deg. F) POHNPEI (PONAPE), EASTERN CAROLINE ISLANDS, PACIFIC

YEAR	JAN	FEB	MAR	APR	MAY	JUNE	JULY	AUG	SEP	OCT	NOV	DEC	ANNUAL
#1956	80.2	80.6	80.4	79.7	80.1	79.9	79.2	79.6	79.5	80.0	80.8	80.7	80.1
1957	80.8	80.4	81.8	81.5	81.8	81.1	81.0	80.5	81.0	81.4	80.0	82.4	81.1
1958	81.2	80.0	80.3	81.0	80.7	80.5	80.1	80.6	81.2	81.1	80.3	80.8	80.7
1959	80.7	79.6	80.6	80.2	80.5	80.5	80.1	80.3	80.3	80.6	81.0	80.6	80.4
1960	80.7	80.6	80.6	80.4	80.3	80.8	80.5	79.8	80.5	80.2	81.0	80.8	80.5
1961	80.5	80.9	80.9	81.6	80.2	80.0	79.7	80.1	79.7	80.1	79.8	80.6	80.4
1962	80.1	80.9	81.0	81.0	80.8	80.8	80.3	79.7	80.1	80.1	80.0	81.0	80.5
1963	79.5	80.3	80.3	80.4	81.0	80.4	80.3	80.5	80.2	80.1	81.1	80.5	80.5
1964	81.8	81.6	80.9	80.8	80.3	79.7	79.6	79.7	79.9	79.6	80.0	80.3	80.3
1965	80.4	80.6	80.9	80.3	80.7	80.4	78.9	80.0	79.9	80.4	80.6	80.6	80.3
1966	80.1	81.4	80.7	81.7	80.6	81.0	80.4	81.1	80.5	80.5	80.3	80.4	80.7
1967	81.5	80.6	80.4	80.0	81.1	80.4	80.1	80.0	79.9	79.8	81.2	80.6	80.5
1968	81.1	80.6	80.2	79.9	80.0	80.1	79.5	80.4	79.9	79.9	80.1	80.6	80.2
1969	80.2	79.1	80.2	80.8	81.1	80.7	79.7	79.6	79.6	80.0	79.5	80.7	80.7
1970	80.8	82.0	81.9	81.6	81.6	81.6	80.7	80.1	80.1	79.4	79.3	81.3	80.7
1971	80.5	79.4	79.5	79.7	79.8	79.5	79.5	79.4	80.3	79.9	80.6	81.2	80.0
1972	80.6	80.8	80.8	80.0	80.0	81.0	79.6	79.9	79.9	80.1	81.1	81.5	80.5
1973	81.0	80.7	82.2	81.0	81.5	81.3	81.5	80.9	80.6	80.4	81.6	81.1	81.2
1974	79.8	81.5	80.9	80.8	81.1	79.6	79.8	80.2	80.2	81.0	80.7	80.8	80.5
1975	81.8	82.3	81.0	81.5	80.8	80.2	79.8	80.2	80.5	79.6	80.2	79.6	80.6
1976	81.6	80.2	81.1	80.1	80.2	80.3	79.8	79.6	80.9	80.8	80.8	80.5	80.4
1977	81.0	82.6	81.9	82.0	81.3	81.5	81.2	81.6	81.5	81.0	81.0	81.7	81.5
1978	81.1	81.3	82.6	81.4	82.0	80.6	80.8	80.9	80.9	80.8	81.6	81.3	81.3
1979	81.7	81.5	81.7	80.6	81.6	81.7	81.4	80.7	81.8	81.6	81.7	81.1	81.4
1980	81.9	81.7	82.8	82.4	80.9	81.7	81.3	81.3	81.0	81.3	81.9	82.3	81.7
1981	81.7	81.7	82.2	82.0	82.3	81.5	81.2	81.6	81.9	81.8	81.8	81.8	81.8
1982	81.4	81.7	81.7	81.8	81.2	81.6	80.3	80.6	80.4	80.7	81.0	82.1	81.2
1983	81.4	81.7	82.4	83.4	84.4	82.8	81.5	81.5	81.7	81.3	81.9	81.5	82.1
1984	80.8	81.5	82.7	83.0	83.3	81.4	81.6	81.1	81.1	81.2	80.8	81.8	81.7
1985	81.3	81.9	82.3	81.0	82.1	81.1	80.5	80.8	81.3	80.9	81.4	81.5	81.3
Record Mean	80.9	81.0	81.3	81.0	81.1	80.7	80.3	80.4	80.4	80.5	80.7	81.0	80.8
Max	86.3	86.4	86.9	87.0	87.3	87.3	87.5	87.9	88.0	88.0	87.8	87.0	87.3
Min	75.4	75.6	75.6	75.0	74.8	74.1	73.0	72.8	72.8	72.9	73.5	75.0	74.2

REFERENCE NOTES FOR TABLES 1, 2, 3 and 6   (POHNPEI [PONAPE], E. CAR. IS. PAC.)

GENERAL

T - TRACE AMOUNT
BLANK ENTRIES DENOTE MISSING/UNREPORTED DATA.
# INDICATES A STATION OR INSTRUMENT RELOCATION.

SPECIFIC

TABLE 1

(a) - LENGTH OF RECORD IN YEARS. ALTHOUGH INDIVIDUAL MONTHS MAY BE MISSING.
* LESS THAN .05

NORMALS — BASED ON THE 1951-1980 RECORD PERIOD.
EXTREMES — DATES ARE THE MOST RECENT OCCURRENCE.
WIND DIR. — NUMERALS SHOW TENS OF DEGREES
            CLOCKWISE FROM TRUE NORTH.
            "00" INDICATES CALM.
RESULTANT WIND DIRECTIONS ARE GIVEN TO WHOLE DEGREES.

EXCEPTIONS

TABLE 1

1. MEAN WIND SPEED IS THROUGH 1972.
2. THUNDERSTORMS AND HEAVY FOG ARE APRIL 1973 TO DATE AND MAY BE INCOMPLETE, DUE TO PART-TIME OPERATIONS

TABLES 2, 3, and 6

RECORD MEANS ARE THROUGH THE CURRENT YEAR, BEGINNING IN   1952 FOR TEMPERATURE
                                                            1950 FOR PRECIPITATION

TABLE 4   HEATING DEGREE DAYS Base 65 deg. F   POHNPEI (PONAPE), EASTERN CAROLINE ISLANDS, PACIFIC

SEASON	JULY	AUG	SEP	OCT	NOV	DEC	JAN	FEB	MAR	APR	MAY	JUNE	TOTAL
1983-84	0	0	0	0	0	0	0	0	0	0	0	0	0
1984-85	0	0	0	0	0	0							
1985-86	0	0	0	0	0	0							

TABLE 5   COOLING DEGREE DAYS Base 65 deg. F   POHNPEI (PONAPE), EASTERN CAROLINE ISLANDS, PACIFIC

YEAR	JAN	FEB	MAR	APR	MAY	JUNE	JULY	AUG	SEP	OCT	NOV	DEC	TOTAL
1969	478	401	477	480	506	479	462	460	450	474	444	492	5603
1970	494	480	534	508	521	477	472	476	442	452	455	513	5824
1971	491	408	454	450	465	442	457	452	467	468	476	506	5536
1972	492	465	499	458	472	488	461	469	455	477	493	521	5750
1973	504	447	541	486	520	495	519	500	473	485	505	507	5982
1974	464	467	501	480	506	446	464	489	463	476	476	497	5729
1975	525	491	501	503	498	464	467	481	471	461	459	458	5779
1976	522	445	507	459	477	470	479	464	444	499	481	498	5745
1977	504	498	530	518	511	504	507	522	499	505	490	525	6113
1978	508	459	554	501	536	476	497	513	484	498	479	523	6028
1979	524	465	525	477	520	506	516	494	512	524	508	503	6074
1980	531	488	557	530	502	507	514	512	487	515	511	544	6198
1981	526	474	540	519	546	503	509	521	513	530	511	531	6223
1982	512	472	527	508	509	507	481	491	469	495	486	537	5994
1983	513	473	548	562	609	544	519	519	509	511	512	516	6335
1984	497	488	559	547	576	499	520	504	490	504	479	528	6191
1985	513	481	544	487	535	489	490	496	495	501	500	519	6050

TABLE 6   SNOWFALL (inches)   POHNPEI (PONAPE), EASTERN CAROLINE ISLANDS, PACIFIC

SEASON	JULY	AUG	SEP	OCT	NOV	DEC	JAN	FEB	MAR	APR	MAY	JUNE	TOTAL
1971-72	0.0	0.0	0.0	0.0	0.0	0.0	0.0	0.0	0.0	0.0	0.0	0.0	0.0
1972-73	0.0	0.0	0.0	0.0	0.0	0.0	0.0	0.0	0.0	0.0	0.0	0.0	0.0
1973-74	0.0	0.0	0.0	0.0	0.0	0.0	0.0	0.0	0.0	0.0	0.0	0.0	0.0
1974-75	0.0	0.0	0.0	0.0	0.0	0.0	0.0	0.0	0.0	0.0	0.0	0.0	0.0
1975-76	0.0	0.0	0.0	0.0	0.0	0.0	0.0	0.0	0.0	0.0	0.0	0.0	0.0
1976-77	0.0	0.0	0.0	0.0	0.0	0.0	0.0	0.0	0.0	0.0	0.0	0.0	0.0
1977-78	0.0	0.0	0.0	0.0	0.0	0.0	0.0	0.0	0.0	0.0	0.0	0.0	0.0
1978-79	0.0	0.0	0.0	0.0	0.0	0.0	0.0	0.0	0.0	0.0	0.0	0.0	0.0
1979-80	0.0	0.0	0.0	0.0	0.0	0.0	0.0	0.0	0.0	0.0	0.0	0.0	0.0
1980-81	0.0	0.0	0.0	0.0	0.0	0.0	0.0	0.0	0.0	0.0	0.0	0.0	0.0
1981-82	0.0	0.0	0.0	0.0	0.0	0.0	0.0	0.0	0.0	0.0	0.0	0.0	0.0
1982-83	0.0	0.0	0.0	0.0	0.0	0.0	0.0	0.0	0.0	0.0	0.0	0.0	0.0
1983-84	0.0	0.0	0.0	0.0	0.0	0.0	0.0	0.0	0.0	0.0	0.0	0.0	0.0
1984-85	0.0	0.0	0.0	0.0	0.0	0.0	0.0	0.0	0.0	0.0	0.0	0.0	0.0
1985-86	0.0	0.0	0.0	0.0	0.0								
Record Mean	0.0	0.0	0.0	0.0	0.0	0.0	0.0	0.0	0.0	0.0	0.0	0.0	0.0

See Reference Notes, relative to all above tables, on preceding page.

Truk Atoll, located in the Western Pacific about 500 miles north of the equator, is comprised of numerous small coral and high volcanic islands scattered over a large lagoon and surrounded by a 125 mile-long barrier reef. Moen Island, roughly triangular in shape and 7.3 square miles in area, is the second largest of the 19 volcanic islands. Its highest point is Mount Teroken, which rises to an elevation of 1,214 feet, and has dense, tropical forests covering its upper reaches and small streams descending its slopes.

The Weather Station is situated on northwestern Moen about 300 feet east of Truk Lagoon and immediately south of the approach end of the airport runway.

A steep-sided 754 foot hill lies a short distance northeast of the Weather Station, and a northeast to southwest trending mountain ridge, of which Mount Teroken is part, to the east. In funneling between these high points, the northeasterly trade winds are deflected to north-northeasterly, and at times generate local eddies with a southerly component near the Weather Station.

From about November to June the climate of Truk is chiefly influenced by the northeasterly trade winds, with average monthly speeds of 8 to 12 miles per hour. By about April, however, the trades begin to weaken, and by July have given way to the lighter and more variable winds of the doldrums. Between July and November the island is frequently under the influence of the Intertropical Convergence Zone which has moved northward into the area. This is also the season when moist southerly winds and tropical disturbances, many of them associated with the ITCZ, are most frequent and when humidities are often oppressively high.

Rainfall at Truk averages about 140 inches a year. A relatively dry period occurs from January to March, when monthly averages run below 8.50 inches. February is the driest month. Rainfall varies widely from year to year and seasonally. Annual totals at Truk have been as low as 120 inches and as high as 180 inches, while even during the ordinarily drier season of January to March, the monthly rainfall has been as much as 24 inches in individual years. However, it has also been less than 1 inch, so that extended dry spells are not uncommon.

The temperature is remarkably uniform throughout the year, with less than one-half degree separating the averages of the warmest and coolest months. With highs in the mid 80s and lows in the mid 70s, the daily range is about 10 degrees. Temperatures below 70 degrees are rare. Every month has had a high temperature of at least 90 degrees. Humidities are high throughout the year.

Although the major typhoon tracks of the Western Pacific lie well to the north and west of Truk, several of the storms have passed close to or over the island within recent years. The storms have caused widespread damage to buildings and to coconut palms and other crops, as well as wave damage to shorelines and coastal structures.

TABLE 1   # NORMALS, MEANS AND EXTREMES

TRUK, EASTERN CAROLINE IS., PACIFIC

LATITUDE: 7°27'N   LONGITUDE: 151°50' E   ELEVATION: FT. GRND   5 BARO 00008   TIME ZONE: 150E MER   WBAN: 40505

	(a)	JAN	FEB	MAR	APR	MAY	JUNE	JULY	AUG	SEP	OCT	NOV	DEC	YEAR
TEMPERATURE °F:														
Normals														
—Daily Maximum		85.4	85.5	85.8	86.2	86.6	86.9	86.7	87.0	87.1	87.0	86.8	86.0	86.4
—Daily Minimum		76.9	76.9	77.0	76.8	76.4	76.0	75.2	75.1	75.3	75.5	76.1	76.8	76.2
—Monthly		81.2	81.2	81.4	81.5	81.5	81.5	81.0	81.1	81.2	81.3	81.5	81.4	81.3
Extremes														
—Record Highest	35	91	91	94	92	94	93	92	92	93	92	91	91	94
—Year		1969	1946	1946	1982	1946	1957	1984	1981	1981	1981	1983	1981	MAR 1946
—Record Lowest	35	70	70	71	71	70	70	70	70	68	66	71	70	66
—Year		1985	1963	1968	1967	1980	1965	1974	1968	1973	1980	1980	1980	OCT 1980
NORMAL DEGREE DAYS:														
Heating (base 65°F)		0	0	0	0	0	0	0	0	0	0	0	0	0
Cooling (base 65°F)		502	454	508	495	512	495	496	499	486	505	495	508	5955
% OF POSSIBLE SUNSHINE	25	56	59	61	56	52	50	53	55	51	47	50	51	53
MEAN SKY COVER (tenths)														
Sunrise - Sunset	34	9.1	9.3	9.2	9.1	9.1	9.2	9.1	9.1	9.1	9.1	9.1	9.3	9.1
MEAN NUMBER OF DAYS:														
Sunrise to Sunset														
—Clear	34	0.3	0.2	0.3	0.3	0.4	0.2	0.2	0.1	0.3	0.4	0.1	0.1	2.9
—Partly Cloudy	34	3.6	3.0	3.3	4.2	4.2	3.9	4.5	4.9	4.0	4.1	4.0	3.5	47.4
—Cloudy	34	27.1	25.0	27.4	25.6	26.4	25.9	26.2	26.0	25.7	26.5	25.9	27.3	315.0
Precipitation														
.01 inches or more	34	18.9	16.3	18.4	20.5	25.1	24.1	24.1	24.4	22.5	23.7	23.9	23.0	264.8
Snow, Ice pellets														
1.0 inches or more	34	0.0	0.0	0.0	0.0	0.0	0.0	0.0	0.0	0.0	0.0	0.0	0.0	0.0
Thunderstorms	34	0.9	0.3	0.9	1.3	1.9	1.7	1.6	1.3	1.8	2.1	2.2	1.7	17.6
Heavy Fog Visibility 1/4 mile or less	34	0.0	0.0	0.0	0.0	0.0	0.0	0.0	0.0	0.0	0.0	0.0	0.0	0.0
Temperature °F														
—Maximum														
90° and above	34	0.1	0.0	0.1	0.6	2.0	1.7	2.5	3.0	2.9	2.1	1.2	0.4	16.6
32° and below	34	0.0	0.0	0.0	0.0	0.0	0.0	0.0	0.0	0.0	0.0	0.0	0.0	0.0
—Minimum														
32° and below	34	0.0	0.0	0.0	0.0	0.0	0.0	0.0	0.0	0.0	0.0	0.0	0.0	0.0
0° and below	34	0.0	0.0	0.0	0.0	0.0	0.0	0.0	0.0	0.0	0.0	0.0	0.0	0.0
AVG. STATION PRESS. (mb)	10	1008.9	1009.5	1009.6	1009.3	1009.2	1009.4	1009.1	1009.4	1009.3	1008.8	1008.3	1008.7	1009.1
RELATIVE HUMIDITY (%)														
Hour 04	18	82	81	82	85	87	88	90	90	89	89	87	84	86
Hour 10 (Local Time)	34	76	76	76	78	79	79	79	79	79	79	79	78	78
Hour 16	34	76	75	75	77	79	79	78	77	77	78	78	79	77
Hour 22	34	81	80	81	83	85	86	87	87	87	87	86	83	84
PRECIPITATION (inches):														
Water Equivalent														
—Normal		8.36	6.67	9.11	12.76	15.64	12.37	14.32	14.04	13.23	14.68	12.07	12.59	145.84
—Maximum Monthly	36	19.19	13.44	24.02	23.38	28.39	21.72	32.99	25.96	21.17	24.71	26.12	34.89	34.89
—Year		1981	1974	1967	1956	1976	1950	1962	1979	1955	1979	1962	1959	DEC 1959
—Minimum Monthly	36	0.96	0.56	1.95	3.28	3.80	6.10	2.65	5.37	5.35	4.17	1.88	3.24	0.56
—Year		1959	1983	1983	1983	1983	1966	1984	1949	1984	1972	1982	1977	FEB 1983
—Maximum in 24 hrs	36	6.78	6.59	8.21	6.70	11.13	7.61	10.07	4.91	6.24	6.55	10.41	14.92	14.92
—Year		1985	1970	1972	1981	1976	1972	1962	1963	1978	1968	1962	1959	DEC 1959
Snow, Ice pellets														
—Maximum Monthly	36													
—Year														
—Maximum in 24 hrs	36													
—Year														
WIND:														
Mean Speed (mph)	23	10.7	11.4	10.8	9.7	8.6	7.4	7.5	7.2	7.7	7.8	7.9	9.6	8.9
Prevailing Direction through 1963		NNE	NNE	NNE	NNE	NNE	NNE	SE	S	SW	S	NNE	NNE	NNE
Fastest Mile														
—Direction (!!)	24	NW	S	NE	NE	S	SW	NW	W	SW	NW	N	W	S
—Speed (MPH)	24	37	31	34	40	78	40	41	38	50	41	45	39	78
—Year		1985	1962	1978	1971	1971	1972	1962	1979	1972	1979	1962	1979	MAY 1971
Peak Gust														
—Direction (!!)	2	NE	SE	NE	NE	E	SE	SW	SE	S	W	NE	NE	E
—Speed (mph)	2	46	41	43	38	52	49	38	41	39	40	36	47	52
—Date		1984	1984	1984	1984	1985	1985	1985	1984	1985	1984	1984	1984	MAY 1985

See Reference Notes to this table on the following page.

TABLE 2   PRECIPITATION (inches)   TRUK, EASTERN CAROLINE IS., PACIFIC

YEAR	JAN	FEB	MAR	APR	MAY	JUNE	JULY	AUG	SEP	OCT	NOV	DEC	ANNUAL
1956	10.59	8.79	12.21	23.38	17.11	12.41	15.76	10.92	15.75	9.83	12.33	14.53	163.61
1957	14.22	5.04	4.68	11.15	11.97	10.96	10.71	9.64	12.39	11.97	13.41	3.43	119.57
1958	6.01	4.19	13.46	18.81	16.22	9.73	13.71	15.03	11.21	6.91	13.66	7.51	136.45
1959	0.96	8.23	9.11	16.12	20.08	7.97	13.78	11.47	14.53	11.36	11.30	34.89	159.80
1960	12.50	6.30	7.43	10.86	14.15	11.09	9.45	13.19	8.49	11.14	16.00	18.10	138.70
1961	9.12	7.62	6.01	11.28	22.36	12.71	19.24	17.86	17.12	17.89	7.82	15.24	164.27
1962	7.91	9.85	12.77	6.98	18.33	12.27	32.99	16.51	14.64	9.77	26.12	11.04	179.18
1963	11.27	7.35	5.44	7.41	8.54	7.64	14.01	18.35	16.88	16.61	7.08	9.48	130.06
1964	2.00	10.80	2.44	12.29	18.45	9.99	13.55	12.47	16.88	15.80	7.85	17.73	140.25
1965	13.86	6.70	15.30	8.17	10.79	12.17	25.19	16.97	9.13	16.97	9.53	4.26	137.61
1966	4.61	1.70	7.57	7.53	13.62	6.10	20.11	12.39	12.49	8.88	8.44	18.21	121.65
1967	8.04	6.38	24.02	17.21	15.17	14.13	19.90	17.36	7.82	15.80	15.55	14.17	175.55
1968	8.67	9.23	13.91	20.50	10.00	14.20	15.75	7.77	12.37	11.78	6.91	24.51	155.60
1969	1.22	1.44	3.82	11.28	19.26	14.91	16.38	14.29	12.26	10.38	16.00	10.39	131.63
1970	14.80	11.80	2.40	10.43	18.60	13.99	7.49	12.98	10.75	19.04	8.43	13.57	144.28
1971	8.25	8.63	9.85	10.20	15.33	13.92	13.18	10.63	14.54	16.40	5.20	8.04	134.17
1972	9.83	10.65	18.49	15.79	16.68	14.73	16.68	11.58	13.60	4.17	7.49	9.26	148.95
1973	1.36	2.30	4.59	8.83	8.96	10.31	13.78	13.18	11.34	21.16	8.59	17.60	122.00
1974	10.23	13.44	19.75	11.59	13.47	14.83	12.49	10.72	14.33	20.14	14.91	8.84	164.74
1975	3.71	3.86	11.17	4.25	17.91	16.12	7.35	13.72	12.02	12.24	17.44	9.99	129.78
1976	10.57	9.37	5.70	17.80	28.39	12.26	11.55	14.74	15.14	15.22	16.09	6.41	163.24
1977	6.44	1.98	8.31	11.47	11.67	7.07	9.11	14.20	13.94	16.21	12.45	3.24	116.09
1978	5.73	2.29	4.85	8.17	13.25	14.01	8.40	14.37	14.98	21.21	12.99	12.47	128.81
1979	7.69	4.39	7.83	20.32	13.91	19.02	9.02	25.96	10.44	24.71	20.97	7.36	171.62
1980	13.91	5.96	8.14	6.55	18.26	12.88	17.26	12.78	8.50	18.91	2.98	18.44	144.57
1981	19.19	4.41	7.24	12.54	6.04	14.46	8.44	10.76	11.35	14.48	18.20	18.20	136.73
1982	7.04	5.92	11.21	8.67	14.68	11.99	11.55	10.70	9.38	6.76	1.88	4.61	104.39
1983	5.16	0.56	1.95	3.28	3.80	9.28	23.09	12.84	9.75	15.32	12.08	15.17	112.28
1984	12.92	10.10	8.27	7.03	11.06	7.47	2.65	14.88	5.35	18.06	14.58	6.84	119.21
1985	16.99	7.85	3.64	16.39	11.67	10.21	12.24	9.73	15.81	7.47	8.70	12.91	133.61
Record Mean	8.63	6.23	8.45	12.01	14.89	12.47	14.00	13.54	13.05	14.15	11.60	13.02	142.06

TABLE 3   AVERAGE TEMPERATURE (deg. F)   TRUK, EASTERN CAROLINE IS., PACIFIC

YEAR	JAN	FEB	MAR	APR	MAY	JUNE	JULY	AUG	SEP	OCT	NOV	DEC	ANNUAL
1956	80.7	81.2	80.4	80.5	80.1	80.7	80.2	80.6	80.7	81.3	81.7	81.6	80.8
1957	81.6	81.4	82.3	81.8	82.2	81.6	81.2	81.3	81.1	81.1	81.9	81.9	81.6
1958	81.3	81.6	81.7	81.6	82.1	82.0	81.5	81.7	81.8	82.5	81.9	82.1	81.8
1959	82.0	81.4	81.1	80.8	81.5	82.5	81.4	80.9	81.4	81.4	81.4	81.4	81.5
1960	80.6	81.4	81.7	81.7	81.3	81.2	81.9	81.3	81.4	81.5	81.4	81.1	81.4
1961	81.7	81.7	81.6	82.0	81.6	81.2	80.5	80.7	80.7	80.7	80.7	80.8	81.2
1962	81.3	81.1	81.5	82.0	81.4	80.9	80.2	80.5	80.8	81.3	80.7	81.1	81.1
1963	80.0	80.5	80.7	81.2	81.1	81.4	80.8	80.9	80.9	81.0	81.5	80.9	80.9
1964	81.8	80.7	81.7	81.4	81.1	81.2	80.5	80.2	80.2	80.8	81.4	80.2	80.9
1965	80.2	80.3	80.0	80.4	80.8	80.1	79.0	80.7	80.5	80.6	80.9	81.3	80.4
1966	80.1	81.3	81.4	82.0	81.3	81.1	81.1	81.3	81.2	81.4	81.5	80.8	81.2
1967	81.3	81.3	80.7	80.5	81.5	80.8	80.4	80.1	80.8	81.0	81.6	80.8	80.9
1968	81.0	81.0	80.8	80.7	81.4	81.5	80.2	81.1	80.9	81.2	80.8	80.8	80.9
1969	80.7	80.2	81.4	81.0	81.7	81.6	80.7	81.0	81.0	81.4	81.9	81.8	81.2
1970	81.5	81.9	82.6	82.4	81.8	81.8	81.8	81.6	81.7	81.1	82.1	81.9	81.8
1971	81.5	81.3	81.5	81.6	81.2	81.0	80.4	80.7	81.1	81.3	81.7	81.7	81.3
1972	80.9	80.5	81.1	81.0	81.5	81.4	80.7	80.7	81.3	81.6	82.1	81.4	81.2
1973	81.6	81.2	82.2	81.8	82.2	82.4	81.6	81.8	81.8	81.0	82.2	81.8	81.8
1974	81.2	81.2	81.6	81.8	82.0	81.4	81.2	81.3	81.5	81.4	81.8	81.7	81.5
1975	81.6	81.7	81.7	82.2	81.4	81.2	81.0	81.1	81.1	81.4	80.5	81.2	81.3
1976	81.2	80.8	81.3	80.9	80.9	80.9	81.3	80.7	80.5	81.7	81.6	81.7	81.1
1977	81.3	81.6	81.4	82.1	82.0	82.5	81.7	81.9	81.9	82.1	82.2	81.8	81.8
1978	81.3	81.7	82.4	82.1	82.1	82.0	82.4	82.1	81.9	81.8	82.0	81.8	82.0
1979	81.9	82.2	81.8	81.8	81.9	82.3	81.8	81.3	82.4	81.3	81.5	81.8	81.8
1980	81.2	81.5	81.8	82.5	82.0	81.8	80.8	81.1	81.5	80.4	81.3	81.2	81.4
1981	81.0	82.0	82.2	82.6	83.0	82.0	82.7	82.5	82.4	82.3	82.4	82.4	82.3
1982	82.7	82.2	81.8	82.7	82.4	82.0	82.1	81.9	82.2	82.0	82.1	82.1	82.2
1983	80.8	81.7	81.8	82.8	84.0	83.4	81.8	82.5	82.6	82.5	82.7	82.4	82.4
1984	81.4	81.7	82.5	83.3	83.6	83.8	81.5	81.6	82.6	81.9	82.7	83.1	82.5
1985	81.5	82.6	82.9	82.1	82.8	82.7	81.7	81.9	81.6	82.4	82.9	82.7	82.3
Record Mean	81.2	81.3	81.6	81.7	81.7	81.5	81.2	81.1	81.3	81.3	81.6	81.5	81.4
Max	85.5	85.6	86.0	86.4	86.8	86.9	86.9	87.0	87.1	87.1	86.9	86.1	86.5
Min	76.8	76.9	77.1	76.9	76.5	76.1	75.4	75.2	75.4	75.5	76.2	76.9	76.3

REFERENCE NOTES FOR TABLES 1, 2, 3 and 6 (TRUK, E. CAROLINA IS., PACIFIC)

GENERAL

T - TRACE AMOUNT
BLANK ENTRIES DENOTE MISSING/UNREPORTED DATA.
# INDICATES A STATION OR INSTRUMENT RELOCATION.

SPECIFIC

TABLE 1

(a) - LENGTH OF RECORD IN YEARS. ALTHOUGH INDIVIDUAL MONTHS MAY BE MISSING.

 *   LESS THAN .05

NORMALS — BASED ON THE 1951-1980 RECORD PERIOD.
EXTREMES — DATES ARE THE MOST RECENT OCCURRENCE.
WIND DIR. — NUMERALS SHOW TENS OF DEGREES
           CLOCKWISE FROM TRUE NORTH.
           ''00'' INDICATES CALM.
RESULTANT WIND DIRECTIONS ARE GIVEN TO WHOLE DEGREES.

EXCEPTIONS

TABLE 1

1. THUNDERSTORMS AND HEAVY FOG ARE FOR APRIL 1973 TO DECEMBER 1978 AND SEPTEMBER 1979 TO DATE. THESE DATA MAY BE INCOMPLETE, DUE TO PART-TIME OPERATIONS

TABLES 2, 3, and 6

RECORD MEANS ARE THROUGH THE CURRENT YEAR, BEGINNING IN   1952 FOR TEMPERATURE
                    1948 FOR PRECIPITATION

TABLE 4 __   __

TABLE 5 __   __

TABLE 6   SNOWFALL (inches)   TRUK, EASTERN CAROLINE IS., PACIFIC

SEASON	JULY	AUG	SEP	OCT	NOV	DEC	JAN	FEB	MAR	APR	MAY	JUNE	TOTAL
1971-72	0.0	0.0	0.0	0.0	0.0	0.0	0.0	0.0	0.0	0.0	0.0	0.0	0.0
1972-73	0.0	0.0	0.0	0.0	0.0	0.0	0.0	0.0	0.0	0.0	0.0	0.0	0.0
1973-74	0.0	0.0	0.0	0.0	0.0	0.0	0.0	0.0	0.0	0.0	0.0	0.0	0.0
1974-75	0.0	0.0	0.0	0.0	0.0	0.0	0.0	0.0	0.0	0.0	0.0	0.0	0.0
1975-76	0.0	0.0	0.0	0.0	0.0	0.0	0.0	0.0	0.0	0.0	0.0	0.0	0.0
1976-77	0.0	0.0	0.0	0.0	0.0	0.0	0.0	0.0	0.0	0.0	0.0	0.0	0.0
1977-78	0.0	0.0	0.0	0.0	0.0	0.0	0.0	0.0	0.0	0.0	0.0	0.0	0.0
1978-79	0.0	0.0	0.0	0.0	0.0	0.0	0.0	0.0	0.0	0.0	0.0	0.0	0.0
1979-80	0.0	0.0	0.0	0.0	0.0	0.0	0.0	0.0	0.0	0.0	0.0	0.0	0.0
1980-81	0.0	0.0	0.0	0.0	0.0	0.0	0.0	0.0	0.0	0.0	0.0	0.0	0.0
1981-82	0.0	0.0	0.0	0.0	0.0	0.0	0.0	0.0	0.0	0.0	0.0	0.0	0.0
1982-83	0.0	0.0	0.0	0.0	0.0	0.0	0.0	0.0	0.0	0.0	0.0	0.0	0.0
1983-84	0.0	0.0	0.0	0.0	0.0	0.0	0.0	0.0	0.0	0.0	0.0	0.0	0.0
1984-85	0.0	0.0	0.0	0.0	0.0	0.0	0.0	0.0	0.0	0.0	0.0	0.0	0.0
1985-86	0.0	0.0	0.0	0.0	0.0								
Record Mean	0.0	0.0	0.0	0.0	0.0	0.0	0.0	0.0	0.0	0.0	0.0	0.0	0.0

See Reference Notes, relative to all above tables, on preceding page.

Wake is a coral atoll, a coral and shell accumulation atop the rim of an extinct underwater volcano. The central lagoon is the former crater and the rim is three small islands, Wake, Peale, and Wilkes, totalling less than three square miles in area and approximately nine miles in outside perimeter. The average height above sea level is only about 12 feet. The highest terrain, 15 to 20 feet, consists of coral rock and shell ridges parallel to the shore and inland some 10 to 50 feet from the sand beaches. The terrain slopes gently from the ridges to the lagoon, and includes several tidal flats. There are no other islands within several hundred miles. The terrain does not appreciably affect the climate.

The climate of Wake is maritime and is chiefly controlled by the easterly trade winds which dominate the island throughout the year. Occasionally during late fall, winter, and early spring, polar outbreaks reach the island and are marked by temperature drops of several degrees, increased cloudiness, and light to moderate showers of short duration. The winds during these outbreaks swing to the northerly directions, and may reach gust velocities of 40 mph. The unusual weather lasts only a few days but may produce temperatures in the low 60s.

Frequent tropical disturbances (low pressure systems, aloft or at the surface) approach the island from the southeast quadrant during the late summer and early fall months. These systems bring periods of light wind, high temperatures, high humidities, and moderate to heavy rain showers. When the system is vigorous and close to the island, winds strengthen and showers may be prolonged to several hours duration. Although typhoons in the area are not unusual, only rarely have typhoon-force winds occurred at Wake since observations began in 1935.

Clouds are predominately cumulus types with little difference in amounts from day to night. High cloudiness, rare during winter months, occurs in association with fall and summer tropical disturbances.

Showers account for most of the precipitation and occur most frequently between midnight and sunrise. The sky is seldom completely overcast or completely clear. Thunderstorms are infrequent, occurring mainly with tropical disturbances. The occurrence of hail and fog are virtually unknown to Wake. Although rare, the visibility can be reduced by volcanic haze, carried from the Hawaiian Islands by the Easterly Trades.

TABLE 1   NORMALS, MEANS AND EXTREMES

WAKE ISLAND, PACIFIC

LATITUDE: 19°17'N   LONGITUDE: 166°39'E   ELEVATION: FT. GRND   11 BARO 00016   TIME ZONE: 180E MER   WBAN: 41606

	(a)	JAN	FEB	MAR	APR	MAY	JUNE	JULY	AUG	SEP	OCT	NOV	DEC	YEAR
TEMPERATURE °F:														
Normals														
–Daily Maximum		81.7	81.7	82.8	83.5	85.3	87.5	88.0	88.1	88.0	87.1	85.1	83.1	85.2
–Daily Minimum		72.1	71.5	72.5	73.0	74.7	76.4	76.9	77.0	77.3	76.7	75.6	73.7	74.8
–Monthly		76.9	76.6	77.7	78.3	80.0	82.0	82.5	82.6	82.7	81.9	80.3	78.4	80.0
Extremes														
–Record Highest	38	89	88	90	91	92	93	94	95	95	92	90	91	95
–Year		1982	1985	1985	1970	1984	1980	1985	1984	1985	1985	1984	1980	AUG 1984
–Record Lowest	38	65	65	65	65	69	71	69	68	69	68	65	64	64
–Year		1958	1972	1977	1985	1976	1978	1980	1958	1975	1961	1950	1954	DEC 1954
NORMAL DEGREE DAYS:														
Heating (base 65°F)		0	0	0	0	0	0	0	0	0	0	0	0	0
Cooling (base 65°F)		369	325	394	399	465	510	543	546	531	524	462	415	5483
% OF POSSIBLE SUNSHINE	16	69	74	77	72	74	76	72	70	70	69	63	66	71
MEAN SKY COVER (tenths)														
Sunrise – Sunset	36	4.7	4.5	4.7	5.0	5.2	5.3	6.5	6.7	6.4	5.9	4.8	4.3	5.3
MEAN NUMBER OF DAYS:														
Sunrise to Sunset														
–Clear	36	11.9	12.0	12.1	10.6	10.1	9.3	6.3	5.7	6.4	8.1	11.7	14.1	118.3
–Partly Cloudy	36	13.3	11.3	13.7	13.9	14.2	13.3	11.1	11.4	11.2	12.3	12.9	12.6	151.1
–Cloudy	36	5.6	4.9	5.1	5.6	6.7	7.3	13.6	13.9	12.3	10.7	5.4	4.3	95.6
Precipitation														
.01 inches or more	38	10.7	9.6	12.1	14.2	14.7	15.7	19.2	19.2	18.8	19.4	14.9	12.5	181.0
Snow,Ice pellets														
1.0 inches or more	38	0.0	0.0	0.0	0.0	0.0	0.0	0.0	0.0	0.0	0.0	0.0	0.0	0.0
Thunderstorms	37	0.0	0.0	0.0	0.2	0.2	0.5	1.1	1.4	1.2	1.4	0.4	0.3	6.5
Heavy Fog Visibility														
1/4 mile or less	38	0.0	0.0	0.0	0.0	0.0	0.0	0.0	0.0	0.0	0.0	0.0	0.0	0.0
Temperature °F														
–Maximum														
90° and above	38	0.0	0.0	0.1	0.3	1.2	2.8	5.5	7.6	6.7	3.7	0.2	0.1	28.2
32° and below	38	0.0	0.0	0.0	0.0	0.0	0.0	0.0	0.0	0.0	0.0	0.0	0.0	0.0
–Minimum														
32° and below	38	0.0	0.0	0.0	0.0	0.0	0.0	0.0	0.0	0.0	0.0	0.0	0.0	0.0
0° and below	38	0.0	0.0	0.0	0.0	0.0	0.0	0.0	0.0	0.0	0.0	0.0	0.0	0.0
AVG. STATION PRESS.(mb)	13	1013.8	1014.5	1014.6	1015.4	1014.6	1014.1	1013.1	1012.3	1012.3	1012.7	1013.5	1014.2	1013.8
RELATIVE HUMIDITY (%)														
Hour 00	38	76	77	79	79	81	81	81	82	81	81	79	77	80
Hour 06 (Local Time)	36	77	78	80	81	83	83	83	83	83	82	80	78	81
Hour 12	38	65	65	67	68	69	68	70	71	71	71	69	67	68
Hour 18	38	70	69	71	73	73	72	73	74	75	76	76	73	73
PRECIPITATION (inches):														
Water Equivalent														
–Normal		1.17	1.20	1.94	2.09	1.87	2.34	3.84	5.46	5.66	4.76	2.77	1.76	34.86
–Maximum Monthly	38	4.25	3.49	6.91	8.36	4.55	5.42	10.74	27.71	17.66	15.45	8.54	6.87	27.71
–Year		1974	1976	1976	1985	1970	1978	1970	1981	1952	1948	1950	1975	AUG 1981
–Minimum Monthly	38	0.06	0.15	0.36	0.31	0.39	0.86	0.78	1.16	0.80	1.04	0.32	0.28	0.06
–Year		1965	1968	1952	1970	1979	1958	1965	1974	1983	1982	1966	1980	JAN 1965
–Maximum in 24 hrs	38	2.31	2.05	2.88	5.67	2.58	3.69	4.20	7.83	15.00	3.39	6.24	2.63	15.00
–Year		1974	1976	1981	1985	1970	1953	1967	1981	1952	1965	1954	1975	SEP 1952
Snow,Ice pellets														
–Maximum Monthly	38													
–Year														
–Maximum in 24 hrs	38													
–Year														
WIND:														
Mean Speed (mph)	35	13.3	13.3	14.5	15.6	14.1	12.4	12.6	11.9	12.4	14.0	15.8	14.8	13.7
Prevailing Direction														
through 1963		ENE	ENE	ENE	ENE	ENE	E	E	E	E	ENE	ENE	ENE	ENE
Fastest Obs. 1 Min.														
–Direction (!!)	5	06	08	06	06	09	09	10	19	16	08	07	03	19
–Speed (MPH)	5	35	30	35	38	30	28	41	51	30	33	40	39	51
–Year		1983	1983	1985	1982	1985	1985	1983	1981	1982	1981	1985	1985	AUG 1981
Peak Gust														
–Direction (!!)	2	NE	NE	E	NE	E	S	S	E	NE	E	E	NE	E
–Speed (mph)	2	35	41	41	45	31	37	39	45	35	40	52	44	52
–Date		1984	1984	1985	1984	1984	1984	1984	1985	1985	1984	1985	1985	NOV 1985

See Reference Notes to this table on the following page.

TABLE 2 PRECIPITATION (inches) WAKE ISLAND, PACIFIC

YEAR	JAN	FEB	MAR	APR	MAY	JUNE	JULY	AUG	SEP	OCT	NOV	DEC	ANNUAL
1956	1.92	0.27	1.86	0.74	1.53	1.52	4.94	2.58	2.05	6.26	5.53	0.96	30.16
1957	0.88	0.39	2.99	1.80	2.53	2.25	1.78	8.71	5.17	2.11	1.35	0.91	30.87
1958	0.80	0.42	0.67	2.01	1.38	0.86	4.13	2.22	4.33	5.34	3.09	0.76	26.01
1959	1.02	0.57	0.55	2.22	2.07	2.10	2.37	3.00	3.37	2.36	2.12	3.09	24.84
1960	0.41	0.73	0.52	1.94	0.95	3.74	3.10	3.20	4.93	6.70	4.48	0.65	31.35
1961	0.83	0.56	3.47	1.09	0.50	1.09	4.94	8.06	10.81	8.19	3.32	0.76	43.62
1962	0.49	1.63	1.00	0.58	1.02	5.19	8.49	8.54	6.30	11.70	1.23	2.44	48.61
1963	0.55	0.32	1.09	2.86	3.66	3.37	1.38	4.38	5.54	3.12	1.74		30.10
1964	0.92	0.61	1.08	3.68	0.42	5.25	4.46	5.85	6.92	2.80	2.53	1.67	36.19
1965	0.06	1.08	1.38	3.31	2.23	1.41	0.78	4.02	1.44	8.31	1.92	0.51	26.45
1966	0.67	1.34	0.64	1.30	3.20	1.59	1.70	4.85	5.15	2.53	0.32	0.62	23.91
1967	1.17	1.90	6.14	3.80	2.25	4.18	10.56	8.71	16.60	3.12	2.08	1.85	62.36
1968	1.20	0.15	0.74	6.22	3.65	2.26	3.48	6.10	7.12	4.29	2.03		40.93
1969	1.15	0.74	0.91	1.62	1.38	1.64	1.15	1.63	4.18	2.41	2.92	4.74	24.47
1970	0.95	1.11	1.66	0.31	4.55	0.98	10.74	5.44	2.88	3.26	1.61	1.91	35.79
1971	1.79	3.34	1.62	3.01	2.38	3.52	2.40	6.51	17.08	5.00	5.01	0.98	52.64
1972	1.75	0.75	1.77	1.21	0.67	1.30	3.38	1.44	8.05	3.36	1.77	4.15	29.60
1973	2.17	1.71	1.41	0.89	1.99	1.95	1.23	7.42	3.12	2.73	2.25	0.93	27.80
1974	4.25	2.13	0.59	1.93	0.96	1.74	9.25	1.16	4.94	3.18	1.88	1.73	33.74
1975	0.88	2.12	1.44	0.91	2.19	0.96	3.08	4.48	5.73	3.48	2.82	6.87	34.96
1976	1.19	3.49	6.91	1.53	1.29	1.81	3.18	2.69	2.24	1.40	2.44	0.64	28.81
1977	0.43	0.87	5.55	1.63	1.60	1.83	2.40	12.39	4.37	10.33	2.83	1.36	45.59
1978	1.00	0.33	1.90	4.14	2.35	5.42	2.57	5.09	7.58	3.09	4.21	1.16	38.84
1979	2.28	1.45	1.53	1.48	0.39	1.50	1.77	14.20	4.58	3.95	3.09	1.17	37.39
1980	0.38	1.54	1.11	2.02	1.50	1.00	2.71	3.92	3.85	4.81	3.08	0.28	26.20
1981	0.84	1.39	5.19	1.26	1.19	2.23	7.23	27.71	2.63	3.11	2.67	1.38	56.83
1982	0.14	3.15	3.14	4.04	2.20	2.20	1.28	2.72	4.50	1.04	2.01	0.45	26.87
1983	0.65	0.27	0.50	4.28	0.62	0.99	2.21	1.72	0.80	3.03	2.26	1.11	18.44
1984	2.11	2.51	1.20	3.18	1.61	4.24	5.51	1.62	2.60	8.33	4.00	1.00	37.91
1985	0.82	1.12	1.54	8.36	2.01	1.90	3.57	7.06	2.05	1.09	2.18	0.52	32.22
Record Mean	1.16	1.42	1.88	2.21	1.90	2.16	4.08	6.51	5.48	4.87	2.86	1.82	36.34

TABLE 3 AVERAGE TEMPERATURE (deg. F) WAKE ISLAND, PACIFIC

YEAR	JAN	FEB	MAR	APR	MAY	JUNE	JULY	AUG	SEP	OCT	NOV	DEC	ANNUAL
1956	75.9	76.1	77.4	78.0	79.4	81.3	81.6	82.5	83.4	81.8	80.2	78.1	79.6
1957	75.8	75.7	77.3	79.0	80.1	81.8	82.8	83.3	83.4	82.2	80.3	78.3	80.0
1958	76.4	76.4	77.5	78.0	79.3	81.2	81.2	82.3	82.9	81.8	80.4	79.5	79.7
1959	75.9	76.4	78.1	76.7	78.6	81.8	83.0	83.5	84.0	83.4	80.3	78.5	80.0
1960	77.8	77.9	78.8	79.2	80.9	82.7	82.6	83.0	82.8	81.2	80.3	80.0	80.6
1961	78.7	79.1	78.1	77.4	79.2	81.7	81.3	82.2	81.5	80.4	79.5	78.1	79.8
1962	77.5	76.6	77.4	79.2	80.2	81.6	81.9	81.9	82.1	80.6	80.3	78.2	79.8
1963	78.1	78.2	79.4	76.5	79.8	81.8	82.8	82.8	82.8	81.7	81.8	79.6	80.4
1964	78.1	77.5	78.3	77.8	80.2	82.6	81.9	82.1	82.2	80.4	79.9	76.3	79.8
1965	75.6	76.3	76.8	76.5	77.6	79.4	81.4	81.5	82.3	80.8	78.3	76.6	78.6
1966	75.6	75.8	76.6	78.0	78.8	81.7	82.9	82.1	82.5	81.7	78.9	78.1	79.4
1967	77.4	76.3	76.1	77.4	80.1	82.2	82.4	81.0	82.3	81.3	79.7	79.8	79.8
1968	77.3	75.9	77.2	77.6	80.1	82.5	82.8	83.4	82.9	82.7	81.9	78.0	80.2
1969	77.0	77.0	79.2	80.4	80.8	81.8	83.4	84.7	83.7	83.4	80.6	79.1	80.9
1970	77.7	78.3	80.9	82.6	82.9	84.2	82.2	82.7	83.5	83.0	81.4	79.4	81.6
1971	77.5	76.3	76.3	79.4	81.5	82.5	83.1	82.4	80.7	81.5	80.2	76.6	80.0
1972	74.5	74.5	75.6	75.6	79.7	81.9	82.4	83.3	82.9	82.0	80.6	77.4	79.2
1973	76.5	75.8	78.4	78.4	80.1	81.6	83.2	81.0	82.2	81.1	79.9	78.7	79.7
1974	76.6	76.5	75.9	77.0	78.6	80.3	80.0	81.4	80.4	81.3	79.7	78.3	78.9
1975	76.3	75.5	77.0	77.3	79.7	82.3	81.9	82.2	81.5	81.8	80.3	77.8	79.5
1976	77.4	76.1	77.5	78.3	79.3	82.3	82.4	82.7	82.9	82.5	81.0	79.1	80.1
1977	76.3	76.9	76.6	77.6	80.4	82.3	82.8	82.0	82.8	81.0	80.0	77.7	79.7
1978	76.8	76.8	77.0	79.3	80.1	81.6	82.3	82.3	84.1	84.0	81.2	79.7	80.5
1979	77.4	77.4	79.1	80.0	81.8	83.3	85.2	81.5	83.1	83.4	79.4	78.5	80.9
1980	77.6	76.1	78.3	79.7	80.7	82.8	84.3	83.5	84.6	82.1	81.6	80.2	81.0
1981	79.3	78.6	80.2	81.4	82.1	84.0	83.9	82.7	84.3	82.8	82.8	81.7	82.0
1982	79.4	78.1	79.2	77.3	79.7	81.2	84.0	84.4	84.5	83.8	80.9	78.4	80.9
1983	75.7	76.7	77.3	77.4	79.7	81.3	82.8	83.7	84.3	84.4	82.5	81.2	80.6
1984	79.5	80.1	81.0	82.1	83.7	83.8	84.1	85.8	86.1	83.2	83.1	80.5	82.8
1985	79.2	79.7	79.6	80.0	82.6	83.1	84.2	83.3	84.9	84.1	80.8	79.0	81.7
Record Mean	77.2	77.0	77.9	78.5	80.1	81.9	82.4	82.7	82.8	82.0	80.6	78.7	80.1
Max	81.8	81.9	82.8	83.5	85.2	87.2	87.7	88.0	88.0	87.0	85.2	83.2	85.1
Min	72.5	72.1	72.9	73.4	75.0	76.7	77.1	77.3	77.6	76.9	76.0	74.2	75.1

REFERENCE NOTES FOR TABLES 1, 2, 3 and 6   (WAKE ISLAND, PACIFIC)

GENERAL

T - TRACE AMOUNT
BLANK ENTRIES DENOTE MISSING/UNREPORTED DATA.
# INDICATES A STATION OR INSTRUMENT RELOCATION.

SPECIFIC

TABLE 1

(a) - LENGTH OF RECORD IN YEARS. ALTHOUGH
INDIVIDUAL MONTHS MAY BE MISSING.
* LESS THAN .05

NORMALS — BASED ON THE 1951-1980 RECORD PERIOD.
EXTREMES — DATES ARE THE MOST RECENT OCCURRENCE.
WIND DIR. — NUMERALS SHOW TENS OF DEGREES
CLOCKWISE FROM TRUE NORTH.
"00" INDICATES CALM.
RESULTANT WIND DIRECTIONS ARE GIVEN TO WHOLE DEGREES.

EXCEPTIONS

TABLES 2, 3, and 6

RECORD MEANS ARE THROUGH THE CURRENT YEAR,
BEGINNING IN   1948 FOR TEMPERATURE
1936 FOR PRECIPITATION

TABLE 4   HEATING DEGREE DAYS Base 65 deg. F   WAKE ISLAND, PACIFIC

SEASON	JULY	AUG	SEP	OCT	NOV	DEC	JAN	FEB	MAR	APR	MAY	JUNE	TOTAL
1983-84	0	0	0	0	0	0	0	0	0	0	0	0	0
1984-85	0	0	0	0	0	0	0	0	0	0	0	0	0
1985-86	0	0	0	0	0	0							

TABLE 5   COOLING DEGREE DAYS Base 65 deg. F   WAKE ISLAND, PACIFIC

YEAR	JAN	FEB	MAR	APR	MAY	JUNE	JULY	AUG	SEP	OCT	NOV	DEC	TOTAL
1969	379	338	448	467	495	510	572	616	568	578	472	445	5888
1970	401	378	497	535	562	585	540	554	561	565	500	452	6130
1971	394	324	426	440	517	531	568	547	479	518	463	368	5575
1972	302	279	335	323	463	513	545	571	545	535	477	391	5279
1973	367	309	421	407	475	502	572	504	520	505	455	432	5469
1974	368	329	343	370	431	464	474	516	470	512	445	416	5138
1975	357	300	378	375	460	525	530	539	505	528	463	402	5362
1976	388	331	391	407	450	525	545	554	542	549	486	445	5613
1977	359	341	366	384	484	524	560	534	541	502	457	399	5451
1978	374	338	378	436	476	505	544	599	577	588	495	464	5774
1979	392	354	443	458	527	556	632	518	550	576	436	426	5868
1980	400	326	418	446	493	542	603	580	593	540	506	477	5924
1981	451	386	479	500	535	578	594	557	587	562	540	523	6292
1982	453	378	444	373	465	492	596	606	590	591	484	423	5895
1983	341	334	390	381	462	496	557	586	587	608	534	511	5787
1984	457	444	502	516	586	572	599	655	639	570	552	489	6581
1985	450	415	462	455	554	548	601	574	606	600	482	442	6189

TABLE 6   SNOWFALL (inches)   WAKE ISLAND, PACIFIC

SEASON	JULY	AUG	SEP	OCT	NOV	DEC	JAN	FEB	MAR	APR	MAY	JUNE	TOTAL
1971-72	0.0	0.0	0.0	0.0	0.0	0.0	0.0	0.0	0.0	0.0	0.0	0.0	0.0
1972-73	0.0	0.0	0.0	0.0	0.0	0.0	0.0	0.0	0.0	0.0	0.0	0.0	0.0
1973-74	0.0	0.0	0.0	0.0	0.0	0.0	0.0	0.0	0.0	0.0	0.0	0.0	0.0
1974-75	0.0	0.0	0.0	0.0	0.0	0.0	0.0	0.0	0.0	0.0	0.0	0.0	0.0
1975-76	0.0	0.0	0.0	0.0	0.0	0.0	0.0	0.0	0.0	0.0	0.0	0.0	0.0
1976-77	0.0	0.0	0.0	0.0	0.0	0.0	0.0	0.0	0.0	0.0	0.0	0.0	0.0
1977-78	0.0	0.0	0.0	0.0	0.0	0.0	0.0	0.0	0.0	0.0	0.0	0.0	0.0
1978-79	0.0	0.0	0.0	0.0	0.0	0.0	0.0	0.0	0.0	0.0	0.0	0.0	0.0
1979-80	0.0	0.0	0.0	0.0	0.0	0.0	0.0	0.0	0.0	0.0	0.0	0.0	0.0
1980-81	0.0	0.0	0.0	0.0	0.0	0.0	0.0	0.0	0.0	0.0	0.0	0.0	0.0
1981-82	0.0	0.0	0.0	0.0	0.0	0.0	0.0	0.0	0.0	0.0	0.0	0.0	0.0
1982-83	0.0	0.0	0.0	0.0	0.0	0.0	0.0	0.0	0.0	0.0	0.0	0.0	0.0
1983-84	0.0	0.0	0.0	0.0	0.0	0.0	0.0	0.0	0.0	0.0	0.0	0.0	0.0
1984-85	0.0	0.0	0.0	0.0	0.0	0.0	0.0	0.0	0.0	0.0	0.0	0.0	0.0
1985-86	0.0	0.0	0.0	0.0	0.0	0.0							
Record Mean	0.0	0.0	0.0	0.0	0.0	0.0	0.0	0.0	0.0	0.0	0.0	0.0	0.0

See Reference Notes, relative to all above tables, on preceding page.

The Yap group consists of four large islands and ten small islands surrounded by a coral reef. These islands were formed by land upheaval and are not, therefore, of volcanic or of coral origin. The soil is clay-like and contains considerable rock. The islands are mostly low, rolling grass-covered hills.

The lowlands, which occupy the southwesterly end of Yap, are covered with dense jungle growth and are marsh-like, except during the dry spells that occur from time to time, particularly during the early months of the year. The terrain around the station is level for about 1/2 mile. A ridge about 3 1/4 miles to the northwest rises 224 feet above sea level and slopes northeastward toward the highest hill on the island, elevation 585 feet. Other small hills lie approximately 3 miles northeast and 3 miles southwest of the station. Lush vegetation, interspersed with sparse pandanas and a few coconut trees, is visible in all directions. The ocean itself cannot be seen from the station, although it lies only about 3/4 of a mile away to the east and south and about 1 mile to the west.

During northern summer, the Intertropical Convergence Zone lies near Yap, particularly as it moves northward in July and southward again in October. At such times showers and light variable winds predominate, interspersed with heavier showers or thunderstorms, occasionally accompanied by strong, shifting winds. Thunderstorms are relatively infrequent, averaging two per month from August through December and fifteen for the year as a whole.

Tropical cyclones affect the area much less often than they do the Pacific further to the northwest. June to December are the months of greatest frequency. Fully-developed typhoons are uncommon near Yap. Most of them pass to the north and then move westward to northwestward away from the island.

Yap is under the influence of the northeast trade winds for eight months of the year, November through June. From July through October the prevailing wind is southwesterly, with frequent periods of calm and light variable winds. This is also the wettest season with monthly rainfall exceeding 13 inches. The nearest approach to a dry season is February through April, when the monthly rainfall is less than 7 inches.

Temperature varies much less seasonally than between day and night. Thus, the warmest and coolest months differ by less than 2 degrees in average temperature, as compared with a difference of nearly 12 degrees between the warmest and coolest times of day.

Humidity is higher and clear skies more frequent during the night and early morning than during the day. Cloudless days are rare. A common daily sequence from May through December is to have the late morning fair weather clouds build up in late afternoon into towering cumulus that give rise to evening and early morning showers. Visibility in such showers is seldom less than 5 miles.

Despite their relatively small size and low relief, the islands nevertheless appear to be large and high enough to cause local differences in temperature, wind, humidity, and, perhaps, rainfall.

TABLE 1   NORMALS, MEANS AND EXTREMES

YAP ISLAND, PACIFIC

LATITUDE: 9°29'N   LONGITUDE: 138°05' E   ELEVATION: FT. GRND   44 BARO   00047   TIME ZONE: 135E MER   WBAN: 40308

	(a)	JAN	FEB	MAR	APR	MAY	JUNE	JULY	AUG	SEP	OCT	NOV	DEC	YEAR
TEMPERATURE °F:														
Normals														
-Daily Maximum		85.5	85.8	86.5	87.4	87.5	87.4	87.1	86.9	87.2	87.4	87.2	86.2	86.8
-Daily Minimum		75.0	75.2	75.4	75.9	75.8	75.4	74.9	74.7	74.8	74.9	75.4	75.4	75.2
-Monthly		80.3	80.5	81.0	81.7	81.7	81.4	81.0	80.8	81.0	81.2	81.3	80.8	81.0
Extremes														
-Record Highest	37	90	92	90	97	93	94	96	93	93	93	93	94	97
-Year		1963	1950	1983	1951	1984	1958	1950	1957	1956	1957	1956	1955	APR 1951
-Record Lowest	37	69	68	69	70	69	65	70	67	69	69	69	65	65
-Year		1983	1978	1976	1983	1974	1975	1982	1975	1974	1971	1980	1973	JUN 1975
NORMAL DEGREE DAYS:														
Heating (base 65°F)		0	0	0	0	0	0	0	0	0	0	0	0	0
Cooling (base 65°F)		474	434	496	501	518	492	496	490	480	502	489	490	5862
% OF POSSIBLE SUNSHINE	26	58	63	68	70	62	53	50	47	50	47	56	54	57
MEAN SKY COVER (tenths)														
Sunrise - Sunset	37	8.7	8.7	8.6	8.3	8.3	8.9	9.1	9.4	9.1	8.9	8.7	8.6	8.8
MEAN NUMBER OF DAYS:														
Sunrise to Sunset														
-Clear	36	0.8	0.6	0.7	1.0	0.9	0.3	0.6	0.1	0.4	0.6	0.6	0.7	7.3
-Partly Cloudy	36	6.8	6.6	7.3	8.7	9.2	5.3	3.9	3.1	5.1	6.1	7.2	7.8	77.0
-Cloudy	36	23.4	21.1	23.0	20.3	20.9	24.4	26.5	27.8	24.6	24.3	22.2	22.5	281.0
Precipitation														
.01 inches or more	37	20.5	17.3	17.8	17.2	21.4	23.6	24.6	24.2	23.2	23.9	22.8	22.4	258.8
Snow,Ice pellets														
1.0 inches or more	37	0.0	0.0	0.0	0.0	0.0	0.0	0.0	0.0	0.0	0.0	0.0	0.0	0.0
Thunderstorms	37	0.5	0.3	0.3	0.8	1.2	1.4	1.7	1.7	2.4	2.5	1.9	1.6	16.4
Heavy Fog Visibility														
1/4 mile or less	37	0.0	0.0	0.0	0.0	0.0	0.0	0.0	0.0	0.0	0.0	0.0	0.0	0.0
Temperature °F														
-Maximum														
90° and above	37	0.2	0.2	0.3	2.8	4.8	4.1	3.2	3.3	3.7	4.2	2.5	0.7	29.8
32° and below	37	0.0	0.0	0.0	0.0	0.0	0.0	0.0	0.0	0.0	0.0	0.0	0.0	0.0
-Minimum														
32° and below	37	0.0	0.0	0.0	0.0	0.0	0.0	0.0	0.0	0.0	0.0	0.0	0.0	0.0
0° and below	37	0.0	0.0	0.0	0.0	0.0	0.0	0.0	0.0	0.0	0.0	0.0	0.0	0.0
AVG. STATION PRESS.(mb)	10	1007.8	1008.5	1008.5	1008.1	1007.6	1007.5	1007.1	1007.1	1007.2	1006.8	1006.5	1007.3	1007.5
RELATIVE HUMIDITY (%)														
Hour 03	20	87	86	86	87	89	91	92	91	90	92	90	88	89
Hour 09 (Local Time)	37	78	77	76	75	77	79	80	81	80	80	79	79	78
Hour 15 (Local Time)	37	76	74	74	74	76	77	77	78	79	79	78	78	77
Hour 21	37	86	84	84	85	87	89	90	89	89	90	89	87	87
PRECIPITATION (inches):														
Water Equivalent														
-Normal		7.92	5.54	6.28	6.56	9.96	11.39	14.24	14.64	13.18	12.56	9.84	10.07	122.18
-Maximum Monthly	37	23.08	13.36	16.46	18.15	18.23	32.01	34.71	29.44	19.57	21.16	20.66	17.05	34.71
-Year		1955	1962	1950	1956	1964	1982	1969	1953	1949	1961	1960	1954	JUL 1969
-Minimum Monthly	37	1.25	0.27	1.38	0.91	1.77	3.40	4.99	5.13	6.41	2.59	1.96	3.65	0.27
-Year		1983	1983	1958	1977	1984	1951	1949	1973	1984	1976	1957	1957	FEB 1983
-Maximum in 24 hrs	37	10.45	5.94	5.09	6.57	10.06	18.75	7.44	7.57	8.35	5.32	8.91	6.66	18.75
-Year		1958	1962	1963	1962	1967	1982	1969	1970	1978	1961	1960	1985	JUN 1982
Snow,Ice pellets														
-Maximum Monthly	37													
-Year														
-Maximum in 24 hrs	37													
-Year														
WIND:														
Mean Speed (mph)	20	10.0	10.4	10.0	9.1	8.0	6.6	6.4	6.4	6.8	6.6	7.4	8.9	8.1
Prevailing Direction														
through 1963		NE	NE	NE	NE	NE	NE	SW	SW	SW	W	NE	NE	NE
Fastest Mile														
-Direction (!!)	26	S	NE	NE	NW	E	NE	W	W	SW	SW	SW	SW	SW
-Speed (MPH)	26	35	68	50	34	33	37	45	55	46	56	72	72	72
-Year		1985	1970	1961	1962	1962	1964	1979	1962	1977	1979	1960	1960	NOV 1960
Peak Gust														
-Direction (!!)	2	NE	E	E	NE	NE	E	S	NW	W	SW	SW	E	SW
-Speed (mph)	2	38	37	35	33	31	41	30	37	29	43	38	36	43
-Date		1984	1985	1985	1985	1985	1985	1985	1984	1985	1985	1984	1985	OCT 1985

See reference Notes to this table on the following page.

TABLE 2 PRECIPITATION (inches)   YAP ISLAND, PACIFIC

YEAR	JAN	FEB	MAR	APR	MAY	JUNE	JULY	AUG	SEP	OCT	NOV	DEC	ANNUAL
1956	7.78	3.63	8.74	18.15	12.17	11.45	19.56	11.41	12.13	14.26	11.97	14.27	145.52
1957	15.10	4.70	8.53	5.27	6.16	11.48	13.82	11.80	17.60	9.33	1.96	3.65	109.40
1958	13.70	1.46	1.38	2.92	4.68	10.21	15.84	6.07	12.80	9.74	18.53	5.20	102.53
1959	7.75	7.99	9.07	4.38	11.54	4.69	18.95	11.61	11.18	11.61	10.34	11.16	120.27
1960	7.78	6.23	4.22	6.30	12.70	9.46	11.45	11.96	10.63	18.07	20.66	8.13	127.59
1961	11.65	5.66	11.15	4.75	18.08	12.33	12.70	17.25	15.10	21.16	4.42	11.27	145.52
1962	8.53	13.36	7.82	15.95	14.43	7.96	19.44	17.32	12.23	9.65	7.41	15.01	149.11
1963	11.26	12.20	11.13	4.20	7.14	8.77	13.49	28.20	10.25	16.67	7.47	10.17	140.95
1964	2.37	6.47	4.01	7.61	18.23	6.74	9.44	16.72	12.55	11.69	6.19	10.98	113.00
1965	3.32	6.00	7.63	4.25	8.12	10.88	26.47	12.39	17.73	8.42	12.02	3.69	120.92
1966	4.98	1.29	2.31	1.86	6.71	12.52	17.98	9.02	9.59	7.11	8.84	9.97	92.18
1967	12.02	6.25	5.37	11.76	16.00	16.71	14.14	16.45	11.72	12.80	10.44	7.48	141.14
#1968	10.77	8.04	3.72	1.82	3.94	5.76	14.24	10.90	10.66	11.21	3.59	8.34	92.99
1969	4.10	1.24	2.08	3.03	7.69	8.78	34.71	11.58	17.03	11.48	9.76	8.32	119.80
1970	4.64	6.17	4.67	3.04	9.76	8.76	8.80	25.45	11.04	12.31	9.56	8.15	112.35
1971	10.42	10.11	13.48	12.25	12.84	13.94	14.12	12.15	13.87	15.15	10.26	9.71	148.30
1972	6.03	10.42	14.21	8.97	5.33	10.18	9.20	11.09	17.60	5.64	9.35	5.14	113.16
1973	2.14	1.00	1.54	5.62	5.98	12.35	10.11	5.13	17.64	14.92	10.57	7.03	94.03
1974	11.84	4.27	9.99	10.07	9.77	14.30	14.40	12.33	9.48	19.11	18.85	13.30	147.71
1975	19.48	1.20	3.12	10.73	9.09	10.67	8.38	11.90	11.25	12.67	6.79	10.93	116.21
1976	7.36	3.19	8.76	6.77	12.52	13.30	11.43	16.29	13.44	2.59	8.88	9.97	114.50
1977	3.94	2.18	2.42	0.91	10.36	7.49	17.21	13.99	18.73	5.76	9.47	11.64	104.10
1978	4.22	5.25	2.04	5.38	4.87	12.89	8.67	18.52	19.17	18.10	11.09	8.98	119.18
1979	3.88	3.16	7.06	3.98	8.82	21.07	14.44	19.57	9.59	12.18	13.40	12.40	124.49
1980	2.32	4.60	6.42	7.72	10.57	13.52	17.84	9.52	12.71	13.41	7.20	14.52	120.35
1981	12.90	8.00	2.89	1.10	5.05	10.77	14.54	13.61	19.03	14.22	10.12	11.01	127.24
1982	7.30	12.58	7.50	2.62	10.49	32.01	13.04	14.26	13.93	9.34	4.95	7.01	135.03
1983	1.25	0.27	2.76	1.36	3.59	6.98	16.14	16.59	12.59	8.37	13.56	5.38	88.84
1984	5.33	9.59	3.90	2.21	1.77	12.38	9.59	15.33	6.41	17.29	12.03	5.44	101.27
1985	14.46	3.27	6.70	8.83	6.81	18.65	11.52	15.49	17.34	10.31	5.79	14.32	133.49
Record Mean	7.94	5.54	6.02	6.01	9.16	12.06	13.78	14.39	13.48	12.53	9.82	9.84	120.58

TABLE 3 AVERAGE TEMPERATURE (deg. F)   YAP ISLAND, PACIFIC

YEAR	JAN	FEB	MAR	APR	MAY	JUNE	JULY	AUG	SEP	OCT	NOV	DEC	ANNUAL
1956	80.2	80.1	80.9	81.1	81.8	81.8	81.7	81.6	81.3	81.4	80.8	81.3	81.3
1957	80.3	80.1	81.4	81.9	82.7	83.1	81.2	81.8	82.0	82.2	82.5	81.7	81.7
1958	80.2	80.4	81.4	81.9	82.6	82.5	81.2	82.4	81.4	82.3	81.4	82.3	81.7
1959	81.3	80.9	81.3	82.1	82.0	82.8	81.0	80.4	80.9	81.6	82.3	81.7	81.5
1960	81.3	81.3	81.4	81.4	82.0	82.0	81.0	80.9	81.6	80.3	81.9	80.6	81.4
1961	80.6	81.4	81.1	82.1	81.2	80.7	80.1	80.3	80.6	79.8	80.1	80.8	80.7
1962	80.7	80.7	80.9	80.6	81.4	81.4	80.8	80.3	81.1	82.0	81.6	79.9	81.0
1963	80.2	80.0	80.9	81.7	81.7	81.8	81.0	80.7	81.3	80.8	81.5	81.1	81.1
1964	80.5	80.0	80.7	81.4	81.1	81.0	81.4	80.7	81.7	81.5	81.6	80.7	81.0
1965	81.1	80.5	79.9	81.2	82.0	81.0	79.0	80.7	80.8	81.3	81.0	81.2	80.8
1966	79.6	80.3	80.5	82.1	82.2	81.6	81.1	81.2	81.5	82.2	81.6	80.4	81.2
1967	79.8	80.3	80.1	80.7	81.5	80.9	80.7	80.6	81.7	81.3	80.9	80.5	80.8
#1968	79.9	80.0	81.1	81.0	81.8	81.2	80.6	80.5	80.8	80.4	81.2	80.4	80.7
1969	80.1	79.8	80.5	81.6	81.4	81.9	80.5	80.4	80.1	80.9	80.7	80.8	80.7
1970	80.8	81.0	81.5	82.1	81.7	81.8	81.7	80.6	80.9	81.1	81.4	81.2	81.4
1971	80.3	80.3	81.5	81.1	81.0	80.5	79.9	80.2	80.8	80.0	80.6	80.9	80.6
1972	80.2	80.3	80.4	81.1	81.0	81.0	82.9	81.0	81.0	81.3	81.0	81.0	81.1
1973	79.7	81.1	81.7	82.6	82.2	81.7	80.9	81.0	80.6	80.0	81.2	80.5	81.1
1974	79.3	81.0	80.7	81.2	80.8	80.3	80.7	81.2	80.6	80.5	80.7	80.6	80.6
1975	80.1	80.8	80.8	80.8	80.7	79.9	80.2	79.6	79.7	80.6	80.7	79.9	80.3
1976	79.2	79.9	79.9	79.8	81.0	80.0	79.5	79.5	79.5	81.2	80.6	80.1	80.0
1977	80.0	80.5	81.4	82.3	81.2	81.1	80.1	80.7	80.3	81.3	81.2	80.8	80.9
1978	80.2	79.7	81.2	81.6	82.2	81.3	81.4	80.4	80.2	79.8	81.0	81.1	80.8
1979	80.5	80.2	80.8	81.7	81.5	81.7	80.1	80.2	81.4	81.0	80.4	80.5	80.9
1980	80.7	80.2	81.0	81.7	81.7	80.9	80.6	80.6	81.4	80.3	80.9	80.5	80.9
1981	79.7	80.0	80.6	81.9	82.5	80.6	80.0	80.6	81.5	81.2	81.5	81.7	80.8
1982	80.9	80.6	80.9	82.1	81.3	80.5	80.5	80.3	80.1	80.4	81.1	80.3	80.8
1983	79.3	80.0	80.4	81.0	82.2	81.8	80.8	81.0	81.1	81.8	81.5	81.5	81.5
1984	80.7	80.5	81.3	82.6	83.8	81.8	81.3	80.6	81.4	80.6	81.1	81.8	81.5
1985	80.6	81.4	81.5	81.7	82.1	81.2	80.6	80.5	81.0	81.1	81.7	81.0	81.2
Record Mean	80.3	80.6	81.0	81.9	81.9	81.4	80.8	80.6	81.0	81.2	81.3	80.9	81.1
Max	85.7	86.0	86.7	87.6	87.8	87.4	87.1	87.0	87.2	87.4	87.3	86.4	87.0
Min	75.0	75.1	75.3	75.8	75.9	75.3	74.8	74.7	74.8	74.9	75.3	75.4	75.2

REFERENCE NOTES FOR TABLES 1, 2, 3 and 6   (YAP ISLAND, PACIFIC)

GENERAL

T - TRACE AMOUNT
BLANK ENTRIES DENOTE MISSING/UNREPORTED DATA.
# INDICATES A STATION OR INSTRUMENT RELOCATION.

SPECIFIC

TABLE 1

(a) · LENGTH OF RECORD IN YEARS. ALTHOUGH
INDIVIDUAL MONTHS MAY BE MISSING.
* LESS THAN .05

NORMALS — BASED ON THE 1951-1980 RECORD PERIOD.
EXTREMES — DATES ARE THE MOST RECENT OCCURRENCE.
WIND DIR. — NUMERALS SHOW TENS OF DEGREES
CLOCKWISE FROM TRUE NORTH.
"00" INDICATES CALM.
RESULTANT WIND DIRECTIONS ARE GIVEN TO WHOLE DEGREES.

EXCEPTIONS

TABLE 1

1. JUNE 1970 TO DATE, THUNDERSTORMS AND HEAVY FOG
MAY BE INCOMPLETE, DUE TO PART-TIME OPERATIONS

TABLES 2, 3, and 6

RECORD MEANS ARE THROUGH THE CURRENT YEAR,
BEGINNING IN   1949 FOR TEMPERATURE
                 1949 FOR PRECIPITATION

TABLE 4 HEATING DEGREE DAYS Base 65 deg. F   YAP ISLAND, PACIFIC

SEASON	JULY	AUG	SEP	OCT	NOV	DEC	JAN	FEB	MAR	APR	MAY	JUNE	TOTAL
1983-84	0	0	0	0	0	0	0	0	0	0	0	0	0
1984-85	0	0	0	0	0	0	0	0	0	0	0	0	0
1985-86	0	0	0	0	0	0	0	0	0	0	0	0	0

TABLE 5 COOLING DEGREE DAYS Base 65 deg. F   YAP ISLAND, PACIFIC

YEAR	JAN	FEB	MAR	APR	MAY	JUNE	JULY	AUG	SEP	OCT	NOV	DEC	TOTAL
1969	476	422	487	502	517	510	491	480	460	500	480	498	5823
1970	496	454	521	519	526	512	528	492	485	507	499	510	6049
1971	480	443	520	490	504	469	468	478	481	470	474	498	5775
1972	478	449	483	491	503	503	561	505	490	509	488	496	5956
1973	461	457	525	535	538	509	500	502	476	472	495	487	5957
1974	452	455	497	493	501	466	494	511	474	489	477	490	5799
1975	474	451	498	482	491	451	479	459	448	489	477	469	5668
1976	449	437	469	450	502	454	453	455	439	508	474	472	5562
1977	474	440	516	528	508	490	478	493	466	512	493	496	5894
1978	476	417	508	508	498	513	486	462	467	488	507	5868
1979	490	434	499	510	518	500	475	477	501	505	493	485	5887
1980	496	449	503	510	522	483	489	513	467	498	480	486	5896
1981	458	428	490	514	548	475	472	491	500	509	501	525	5911
1982	500	442	502	520	511	471	489	479	463	484	490	480	5831
1983	451	426	485	484	540	510	496	506	490	530	495	519	5932
1984	495	454	514	534	591	508	511	490	500	489	491	529	6106
1985	486	465	518	506	536	494	488	485	491	506	510	504	5989

TABLE 6 SNOWFALL (inches)   YAP ISLAND, PACIFIC

SEASON	JULY	AUG	SEP	OCT	NOV	DEC	JAN	FEB	MAR	APR	MAY	JUNE	TOTAL
1971-72	0.0	0.0	0.0	0.0	0.0	0.0	0.0	0.0	0.0	0.0	0.0	0.0	0.0
1972-73	0.0	0.0	0.0	0.0	0.0	0.0	0.0	0.0	0.0	0.0	0.0	0.0	0.0
1973-74	0.0	0.0	0.0	0.0	0.0	0.0	0.0	0.0	0.0	0.0	0.0	0.0	0.0
1974-75	0.0	0.0	0.0	0.0	0.0	0.0	0.0	0.0	0.0	0.0	0.0	0.0	0.0
1975-76	0.0	0.0	0.0	0.0	0.0	0.0	0.0	0.0	0.0	0.0	0.0	0.0	0.0
1976-77	0.0	0.0	0.0	0.0	0.0	0.0	0.0	0.0	0.0	0.0	0.0	0.0	0.0
1977-78	0.0	0.0	0.0	0.0	0.0	0.0	0.0	0.0	0.0	0.0	0.0	0.0	0.0
1978-79	0.0	0.0	0.0	0.0	0.0	0.0	0.0	0.0	0.0	0.0	0.0	0.0	0.0
1979-80	0.0	0.0	0.0	0.0	0.0	0.0	0.0	0.0	0.0	0.0	0.0	0.0	0.0
1980-81	0.0	0.0	0.0	0.0	0.0	0.0	0.0	0.0	0.0	0.0	0.0	0.0	0.0
1981-82	0.0	0.0	0.0	0.0	0.0	0.0	0.0	0.0	0.0	0.0	0.0	0.0	0.0
1982-83	0.0	0.0	0.0	0.0	0.0	0.0	0.0	0.0	0.0	0.0	0.0	0.0	0.0
1983-84	0.0	0.0	0.0	0.0	0.0	0.0	0.0	0.0	0.0	0.0	0.0	0.0	0.0
1984-85	0.0	0.0	0.0	0.0	0.0	0.0	0.0	0.0	0.0	0.0	0.0	0.0	0.0
1985-86	0.0	0.0	0.0	0.0	0.0								
Record Mean	0.0	0.0	0.0	0.0	0.0	0.0	0.0	0.0	0.0	0.0	0.0	0.0	0.0

See Reference Notes, relative to all above tables, on preceding page.

ALLENTOWN, PENNSYLVANIA

Allentown is located in the east central section of the state and in the Lehigh River valley. Twelve miles to the north is Blue Mountain, a ridge from 1,000 to 1,800 feet in height. The South Mountain, 500 to 1,000 feet high, fringes the southern edge of the city. Otherwise the country is generally rolling with numerous small streams. A modified climate prevails. Temperatures are usually moderate and precipitation generally ample and dependable with the largest amounts occurring during the summer months when precipitation is generally showery. General climatological features of the area are slightly modified by the mountain ranges so that at times during the winter there is a temperature difference of 10 to 15 degrees between Allentown and Philadelphia, only 50 miles to the south.

The growing season averages 177 days, and generally ranges from 170 to 185 days. It begins late in April and ends late in October. The average occurrence of the last temperature of 32 degrees in the spring is late April, and the average first fall minimum of 32 degrees is mid–October.

Maximum temperatures during most years are not excessively high and temperatures above 100 degrees are seldom recorded. However, the average humidity in the valley is quite high, and combined with the normal summer temperatures, causes periods of discomfort.

Winters in the valley are comparatively mild. Minimum temperatures during December, January, and February are usually below freezing, but below zero temperatures are seldom recorded.

Seasonal snowfall is quite variable. Freezing rain is a common problem throughout the Lehigh Valley. Snowstorms producing 10 inches or more occur an average of once in two years. The accumulation of snowfall over the drainage area of the Lehigh River to the north of Allentown, combined with spring rains, frequently presents a flood threat to the city and surrounding area. The valley is also subject to torrential rains that cause quick rises in the river and feeder creeks.

The area is seldom subject to destructive storms of large extent. Heavy thunderstorms and tornadoes occasionally cause damage over limited areas. An exception to the usual weather was the storm that battered the east coast on November 25, 1950, when gusts of 88 mph were observed at the station.

TABLE 1   NORMALS, MEANS AND EXTREMES

ALLENTOWN, PENNSYLVANIA

LATITUDE: 40°39'N   LONGITUDE: 75°26'W   ELEVATION: FT. GRND   387 BARO   00379   TIME ZONE: EASTERN   WBAN: 14737

	(a)	JAN	FEB	MAR	APR	MAY	JUNE	JULY	AUG	SEP	OCT	NOV	DEC	YEAR
TEMPERATURE °F:														
Normals														
-Daily Maximum		34.9	37.8	47.6	61.0	71.1	80.1	84.6	82.4	75.3	64.2	51.3	39.2	60.8
-Daily Minimum		19.5	20.9	29.2	39.0	48.8	58.3	63.0	61.6	54.1	42.6	33.6	23.8	41.2
-Monthly		27.2	29.4	38.4	50.0	60.0	69.2	73.8	72.0	64.7	53.4	42.5	31.5	51.0
Extremes														
-Record Highest	42	72	76	86	93	97	100	105	100	99	90	81	72	105
-Year		1950	1985	1945	1976	1962	1966	1966	1955	1980	1951	1950	1984	JUL 1966
-Record Lowest	42	-12	-7	-1	16	28	39	48	43	30	22	11	-8	-12
-Year		1961	1967	1967	1982	1947	1972	1976	1982	1947	1969	1976	1950	JAN 1961
NORMAL DEGREE DAYS:														
Heating (base 65°F)		1172	1000	822	450	190	12	0	6	85	364	675	1039	5815
Cooling (base 65°F)		0	0	0	0	35	138	273	226	79	0	0	0	751
% OF POSSIBLE SUNSHINE	1	35	50	45	53			40	47	59	45	42	35	
MEAN SKY COVER (tenths)														
Sunrise - Sunset	41	6.7	6.4	6.4	6.4	6.5	6.0	5.9	5.8	5.7	5.6	6.6	6.7	6.2
MEAN NUMBER OF DAYS:														
Sunrise to Sunset														
-Clear	42	6.9	7.1	7.5	6.9	6.6	7.2	8.3	8.7	9.4	10.5	6.6	7.0	92.7
-Partly Cloudy	42	8.2	7.9	8.6	9.2	10.7	11.3	12.1	11.1	9.0	8.7	8.1	7.5	112.5
-Cloudy	42	16.0	13.3	14.8	14.0	13.7	11.5	10.7	11.1	11.5	11.9	15.3	16.5	160.1
Precipitation														
.01 inches or more	42	11.0	9.6	11.4	11.3	12.3	10.6	10.1	9.8	8.9	8.0	10.1	11.0	124.1
Snow,Ice pellets														
1.0 inches or more	42	2.4	2.2	1.5	0.3	0.0	0.0	0.0	0.0	0.0	0.*	0.5	2.0	8.9
Thunderstorms	42	0.2	0.3	0.8	2.2	4.4	6.0	7.0	6.1	3.1	1.0	0.7	0.1	32.0
Heavy Fog Visibility 1/4 mile or less	42	2.7	2.5	2.5	1.5	2.0	1.2	1.0	1.8	2.7	3.0	2.7	3.0	26.6
Temperature °F														
-Maximum														
90° and above	42	0.0	0.0	0.0	0.1	0.5	3.5	6.7	3.9	1.2	0.*	0.0	0.0	16.0
32° and below	42	11.8	7.7	1.5	0.*	0.0	0.0	0.0	0.0	0.0	0.0	0.3	7.4	28.9
-Minimum														
32° and below	42	28.2	25.0	20.4	6.5	0.3	0.0	0.0	0.0	0.1	3.6	14.2	25.8	124.0
0° and below	42	1.4	0.6	0.*	0.0	0.0	0.0	0.0	0.0	0.0	0.0	0.0	0.4	2.5
AVG. STATION PRESS.(mb)	13	1003.5	1004.1	1002.2	1001.3	1001.2	1001.9	1002.5	1004.1	1004.9	1005.7	1004.7	1004.8	1003.4
RELATIVE HUMIDITY (%)														
Hour 01	35	74	73	71	71	77	81	82	85	86	83	79	77	78
Hour 07 (Local Time)	35	76	76	75	74	77	80	81	86	88	86	82	79	80
Hour 13	35	62	59	54	50	52	54	52	55	57	56	60	63	56
Hour 19	35	69	65	60	57	59	60	61	66	70	69	70	71	65
PRECIPITATION (inches):														
Water Equivalent														
-Normal		3.35	3.02	3.88	3.93	3.57	3.45	4.13	4.44	4.03	3.05	3.73	3.73	44.31
-Maximum Monthly	42	8.42	5.44	7.21	10.09	10.62	8.58	10.42	12.10	8.72	6.84	9.69	7.89	12.10
-Year		1979	1971	1953	1952	1984	1972	1969	1955	1985	1955	1972	1973	AUG 1955
-Minimum Monthly	42	0.67	1.01	0.97	0.61	0.09	0.34	0.42	0.93	0.94	0.15	0.68	0.39	0.09
-Year		1981	1980	1981	1985	1964	1949	1955	1980	1967	1963	1976	1955	MAY 1964
-Maximum in 24 hrs	42	2.47	2.05	3.08	2.52	3.36	3.55	4.54	5.84	7.85	2.96	3.40	2.86	7.85
-Year		1949	1966	1952	1968	1984	1967	1969	1982	1985	1955	1972	1983	SEP 1985
Snow,Ice pellets														
-Maximum Monthly	42	24.1	29.5	30.5	13.4	T					1.4	7.8	28.4	30.5
-Year		1966	1983	1958	1982	1977					1972	1967	1966	MAR 1958
-Maximum in 24 hrs	42	16.0	25.2	17.5	11.4	T					1.4	6.4	13.3	25.2
-Year		1961	1983	1958	1982	1977					1972	1968	1966	FEB 1983
WIND:														
Mean Speed (mph)	36	10.7	11.1	11.7	11.0	9.1	8.1	7.2	6.8	7.3	8.2	9.7	10.1	9.2
Prevailing Direction through 1963		W	WNW	WNW	W	WSW	SW	WSW	SW	SW	WSW	W	W	W
Fastest Obs. 1 Min.														
-Direction (!!!)	37	29	25	29	29	30	27	27	23	25	14	11	29	27
-Speed (MPH)	37	55	58	58	60	58	81	55	58	46	49	55	52	81
-Year		1959	1956	1955	1963	1964	1964	1951	1949	1956	1954	1950	1956	JUN 1964
Peak Gust														
-Direction (!!!)	2	W	E	W	W	SW	NW	W	W	NW	NE	E	W	W
-Speed (mph)	2	47	51	51	51	46	41	66	39	56	37	39	54	66
-Date		1985	1985	1985	1985	1984	1985	1984	1984	1985	1984	1985	1985	JUL 1984

See Reference Notes to this table on the following page

TABLE 2 PRECIPITATION (inches) ALLENTOWN, PENNSYLVANIA

YEAR	JAN	FEB	MAR	APR	MAY	JUNE	JULY	AUG	SEP	OCT	NOV	DEC	ANNUAL
1956	1.68	4.90	4.88	3.42	3.83	4.85	6.16	3.56	3.27	2.72	2.41	4.74	46.42
1957	1.87	2.16	2.31	6.49	2.35	3.88	1.05	1.39	2.06	2.85	3.28	5.52	35.21
1958	4.30	4.86	4.64	5.44	3.05	2.11	4.66	2.15	5.72	3.78	2.89	1.04	44.64
1959	2.58	2.41	3.24	2.84	1.39	3.32	5.87	5.68	4.15	3.95	2.43	3.76	41.62
1960	3.66	3.68	2.58	3.45	3.60	3.38	6.64	6.31	7.54	1.08	2.30	2.59	46.81
1961	3.32	3.17	4.47	4.36	2.36	2.09	7.00	2.70	1.53	0.80	3.63	3.84	39.27
1962	2.90	4.59	3.14	3.97	1.26	3.53	1.61	6.50	4.28	3.05	4.13	2.62	41.58
1963	3.00	2.55	3.97	1.18	1.67	1.79	2.48	3.17	6.59	0.15	6.30	2.19	35.04
1964	4.67	3.29	1.95	7.01	0.09	3.82	2.29	1.26	2.17	1.37	2.34	4.48	34.74
1965	2.70	3.21	2.69	2.10	1.34	1.81	1.58	2.29	5.92	3.46	1.66	1.79	30.55
1966	4.20	4.84	1.44	3.75	2.53	0.89	1.91	2.68	6.23	2.77	3.05	3.99	38.28
1967	1.76	1.98	6.68	2.61	4.44	4.10	5.85	5.00	0.94	2.30	3.00	5.13	43.79
1968	2.54	1.31	3.15	3.14	6.74	3.59	1.60	2.74	2.64	2.99	4.32	3.67	38.43
1969	1.47	1.61	1.93	3.52	1.90	3.20	10.42	3.33	3.87	1.62	2.59	6.43	41.89
1970	0.69	2.40	3.98	4.65	2.97	2.91	5.40	1.69	2.96	5.06	6.08	2.80	41.59
1971	2.73	5.44	3.52	1.65	5.30	2.66	3.04		3.71	3.57	6.31	1.32	48.50
1972	2.79	4.05	3.37	2.78	6.20	8.58	5.85	2.26	1.27	3.59	9.69	5.42	55.85
1973	3.58	2.65	2.80	5.94	5.51	5.36	2.16	2.36	6.18	2.07	1.67	7.89	48.17
1974	3.66	1.60	5.38	3.98	3.85	4.31	1.40	9.42	6.67	1.58	1.82	4.52	48.19
1975	5.17	3.57	3.49	3.50	4.25	5.09	7.71	4.23	7.60	4.84	3.37	2.72	55.54
1976	5.03	2.80	2.64	2.16	3.12	3.57	2.11	5.09	4.60	5.70	0.68	2.40	39.90
1977	1.42	2.54	5.90	4.23	1.75	3.66	4.33	5.79	4.10	4.21	6.13	5.54	49.60
1978	7.42	1.60	3.94	1.55	5.52	3.15	3.64	8.95	2.28	1.96	2.02	3.96	45.99
1979	8.42	4.49	2.35	3.74	4.85	1.20	5.57	2.07	7.19	4.36	2.99	2.48	49.71
1980	0.81	1.01	5.84	4.39	3.38	2.90	1.50	0.93	1.99	3.09	3.23	0.75	29.82
1981	0.67	4.79	0.97	3.43	4.82	5.72	2.71	1.60	2.74	2.82	1.63	3.18	35.08
1982	3.48	2.63	1.99	4.73	4.02	7.16	2.86	8.43	1.76	1.56	3.40	1.38	43.40
1983	2.84	3.21	5.35	7.87	4.21	4.68	2.02	2.38	1.41	4.83	6.29	7.61	52.70
1984	1.53	3.64	3.82	5.08	10.62	4.20	9.26	4.17	1.10	4.26	1.66	3.07	52.41
1985	0.84	2.28	1.56	0.61	5.47	5.06	7.61	3.46	8.72	1.97	7.74	1.71	47.03
Record Mean	3.16	2.93	3.63	3.87	4.09	3.80	4.32	4.26	3.91	2.84	3.71	3.62	44.15

TABLE 3 AVERAGE TEMPERATURE (deg. F) ALLENTOWN, PENNSYLVANIA

YEAR	JAN	FEB	MAR	APR	MAY	JUNE	JULY	AUG	SEP	OCT	NOV	DEC	ANNUAL
1956	29.2	32.9	34.9	46.7	56.8	69.7	69.9	70.8	60.9	53.9	41.7	36.8	50.3
1957	25.1	33.0	39.2	51.5	59.9	71.9	73.6	69.9	65.9	51.5	43.9	35.5	51.7
1958	28.6	24.6	37.4	51.5	57.7	64.6	74.7	71.9	63.4	52.2	42.6	24.4	49.4
1959	28.0	29.2	37.7	52.3	63.8	70.6	74.3	73.6	68.2	56.5	41.5	34.1	52.5
1960	31.6	33.1	29.5	54.1	58.8	68.9	71.1	72.7	64.8	53.0	43.3	24.2	50.4
1961	21.1	30.6	39.0	45.9	57.6	69.3	73.8	72.0	70.8	55.5	44.4	29.8	50.8
1962	26.4	27.0	39.0	50.3	63.0	70.4	71.5	70.5	60.4	53.5	39.7	26.3	49.8
1963	24.0	22.0	39.5	50.5	58.6	69.5	74.1	69.7	60.9	57.4	46.6	24.9	49.8
1964	29.2	27.5	39.1	47.4	62.2	68.8	73.9	69.6	65.8	50.8	44.8	32.8	51.0
1965	24.3	30.2	36.2	47.2	64.3	68.3	72.4	71.1	65.8	50.9	41.3	33.9	50.5
1966	26.2	28.4	39.0	45.7	56.5	71.1	77.0	74.0	62.4	49.8	43.5	30.3	50.3
1967	31.3	24.3	35.1	48.8	52.8	70.6	72.7	69.1	61.3	51.1	37.3	32.6	48.9
1968	21.7	25.5	40.6	52.0	56.2	67.2	74.5	73.8	66.5	55.6	42.3	29.6	50.4
1969	28.0	30.2	37.3	52.6	62.4	70.0	72.2	72.0	63.6	51.5	40.9	28.4	50.8
1970	19.2	27.9	34.4	48.4	60.8	67.5	73.8	72.9	67.3	55.5	44.0	30.7	50.2
1971	22.0	30.0	35.7	47.7	56.9	70.2	72.6	70.8	67.7	58.9	40.9	36.5	50.8
1972	30.4	26.8	37.2	47.1	61.4	66.6	74.9	71.1	64.8	48.2	39.9	35.0	50.2
1973	30.7	29.3	44.3	50.0	56.2	71.6	75.3	75.3	65.6	55.6	45.0	35.1	52.8
1974	31.3	30.2	40.3	52.5	59.1	66.6	73.3	72.4	62.5	49.6	43.5	35.2	51.4
1975	31.6	32.0	37.6	45.5	64.1	69.2	73.7	72.5	61.3	56.9	48.8	32.7	52.1
1976	24.1	36.6	41.8	52.3	58.3	72.6	71.6	71.0	63.4	50.2	37.6	27.0	50.5
1977	18.6	30.8	45.4	52.4	63.9	68.7	75.5	73.0	66.7	53.3	44.2	30.0	51.9
1978	24.9	22.3	36.7	50.5	60.8	70.8	72.4	73.3	64.2	52.9	44.3	33.5	50.6
1979	28.0	20.5	44.3	49.5	62.1	68.0	73.2	72.8	65.3	52.5	47.8	36.2	51.7
1980	30.7	27.3	38.6	52.9	63.8	68.2	77.3	78.2	70.3	54.4	41.4	29.6	52.7
1981	23.2	35.3	38.4	51.7	61.2	71.6	74.7	70.6	63.6	49.8	43.3	30.6	51.2
1982	19.9	29.9	37.3	47.2	62.5	65.7	73.5	68.3	64.1	54.0	45.5	38.5	50.5
1983	30.2	32.0	41.9	49.2	58.3	70.0	76.4	75.5	66.5	54.3	43.5	28.6	52.2
1984	23.0	36.7	33.7	49.3	58.7	72.0	72.0	74.1	62.4	59.5	43.4	38.4	52.0
1985	24.9	32.2	42.3	54.2	63.6	67.3	73.3	71.9	66.2	54.2	46.1	29.4	52.1
Record Mean	27.2	29.6	38.7	49.9	60.1	69.2	74.0	72.1	64.7	53.8	42.7	31.5	51.2
Max	34.9	38.1	48.1	60.9	71.3	80.1	84.8	82.6	75.4	64.7	51.6	39.2	61.0
Min	19.4	21.2	29.3	38.8	48.9	58.2	63.1	61.5	54.0	42.9	33.8	23.7	41.3

REFERENCE NOTES FOR TABLES 1, 2, 3 and 6   (ALLENTOWN, PA)

GENERAL

T - TRACE AMOUNT
BLANK ENTRIES DENOTE MISSING/UNREPORTED DATA.
# INDICATES A STATION OR INSTRUMENT RELOCATION.

SPECIFIC

TABLE 1

(a) - LENGTH OF RECORD IN YEARS. ALTHOUGH
INDIVIDUAL MONTHS MAY BE MISSING.
* LESS THAN .05

NORMALS — BASED ON THE 1951-1980 RECORD PERIOD.
EXTREMES — DATES ARE THE MOST RECENT OCCURRENCE.
WIND DIR. — NUMERALS SHOW TENS OF DEGREES
CLOCKWISE FROM TRUE NORTH.
"00" INDICATES CALM.
RESULTANT WIND DIRECTIONS ARE GIVEN TO WHOLE DEGREES.

EXCEPTIONS

TABLES 2, 3, and 6

RECORD MEANS ARE THROUGH THE CURRENT YEAR,
BEGINNING IN   1944 FOR TEMPERATURE
1944 FOR PRECIPITATION
1944 FOR SNOWFALL

TABLE 4 HEATING DEGREE DAYS Base 65 deg. F   ALLENTOWN, PENNSYLVANIA

SEASON	JULY	AUG	SEP	OCT	NOV	DEC	JAN	FEB	MAR	APR	MAY	JUNE	TOTAL
1956-57	12	9	178	335	594	864	1230	892	792	420	184	30	5640
1957-58	0	8	93	410	626	907	1122	1125	849	402	232	62	5836
1958-59	0	2	100	397	667	1256	1141	996	838	372	133	30	5932
1959-60	0	0	67	311	698	950	1031	921	1089	348	193	11	5619
1960-61	0	0	55	366	646	1257	1353	959	801	568	239	16	6260
1961-62	0	1	56	293	620	1084	1193	1056	800	454	148	6	5711
1962-63	1	7	174	352	753	1189	1266	1198	783	430	206	16	6375
1963-64	3	9	157	234	546	1235	1103	1079	795	523	132	47	5863
1964-65	1	15	78	436	599	990	1256	969	889	526	92	44	5895
1965-66	0	23	81	432	705	957	1197	1017	801	573	272	29	6087
1966-67	0	0	136	463	640	1070	1038	1134	921	484	375	12	6273
1967-68	0	20	146	431	825	1000	1336	1139	750	386	266	34	6333
1968-69	0	7	33	293	670	1091	1141	967	852	369	133	14	5570
1969-70	3	5	125	420	717	1129	1409	1032	940	490	169	35	6474
1970-71	0	0	64	292	621	1056	1326	969	900	512	250	24	6014
1971-72	0	8	52	191	719	877	1064	1102	853	531	127	54	5578
1972-73	4	11	69	513	745	924	1055	995	633	446	275	6	5676
1973-74	0	2	72	290	593	922	1036	968	755	385	216	32	5271
1974-75	2	0	123	471	638	917	1027	919	841	577	98	19	5632
1975-76	0	7	122	257	482	995	1262	820	713	414	219	28	5319
1976-77	1	9	101	455	817	1174	1428	950	604	384	127	35	6085
1977-78	0	10	56	356	623	1077	1238	1190	871	430	189	15	6055
1978-79	9	0	94	368	614	971	1142	1241	633	460	125	29	5686
1979-80	13	22	76	386	509	884	1058	1087	808	356	87	41	5327
1980-81	0	0	26	326	702	1092	1290	823	818	392	163	6	5638
1981-82	0	3	100	465	643	1058	1389	975	851	528	110	50	6172
1982-83	5	28	82	347	575	815	1071	920	708	477	222	23	5273
1983-84	0	1	96	336	639	1121	1295	813	963	464	211	12	5951
1984-85	0	1	135	180	642	820	1240	914	702	346	106	28	5114
1985-86	0	3	85	324	559	1093							

TABLE 5 COOLING DEGREE DAYS Base 65 deg. F   ALLENTOWN, PENNSYLVANIA

YEAR	JAN	FEB	MAR	APR	MAY	JUNE	JULY	AUG	SEP	OCT	NOV	DEC	TOTAL
1969	0	0	0	5	58	172	232	230	88	6	0	0	791
1970	0	0	0	2	44	116	278	249	141	6	0	0	836
1971	0	0	0	0	4	184	239	194	140	10	4	0	775
1972	0	0	0	0	18	106	318	208	69	0	0	0	719
1973	0	0	0	3	7	212	322	331	103	4	0	0	982
1974	0	0	0	16	40	86	266	236	52	0	0	0	696
1975	0	0	0	0	79	154	279	247	18	13	1	0	791
1976	0	0	0	37	20	262	214	203	59	2	0	0	797
1977	0	0	4	13	99	155	331	263	115	1	5	0	986
1978	0	0	0	0	65	196	247	264	76	0	0	0	811
1979	0	0	0	1	42	126	274	268	92	7	1	0	811
1980	0	0	0	0	59	147	388	413	194	6	0	0	1207
1981	0	0	0	0	50	210	307	183	61	0	0	0	811
1982	0	0	0	1	37	76	274	137	63	13	0	0	601
1983	0	0	0	11	21	179	361	337	148	13	0	0	1070
1984	0	0	0	0	25	226	260	291	65	18	0	0	885
1985	0	0	2	27	66	100	262	222	125	1	0	0	805

TABLE 6 SNOWFALL (inches)   ALLENTOWN, PENNSYLVANIA

SEASON	JULY	AUG	SEP	OCT	NOV	DEC	JAN	FEB	MAR	APR	MAY	JUNE	TOTAL
1956-57	0.0	0.0	0.0	0.0	T	1.1	7.3	7.8	1.9	3.1	0.0	0.0	21.2
1957-58	0.0	0.0	0.0	T	T	6.9	5.1	21.1	30.5	T	0.0	0.0	63.6
1958-59	0.0	0.0	0.0	0.0	0.8	4.5	3.8	1.4	11.7	1.3	0.0	0.0	23.5
1959-60	0.0	0.0	0.0	0.0	T	6.2	3.9	8.3	21.3	T	0.0	0.0	39.7
1960-61	0.0	0.0	0.0	T	0.0	16.3	22.1	18.8	5.0	2.9	0.0	0.0	65.1
1961-62	0.0	0.0	0.0	0.0	2.9	13.3	0.9	14.7	4.6	0.2	0.0	0.0	36.4
1962-63	0.0	0.0	0.0	0.2	3.5	13.9	12.2	12.6	4.5	T	0.0	0.0	47.1
1963-64	0.0	0.0	0.0	0.0	T	11.0	14.5	22.4	6.8	1.7	0.0	0.0	54.7
1964-65	0.0	0.0	0.0	0.0	T	2.2	11.7	1.9	5.1	T	0.0	0.0	22.6
1965-66	0.0	0.0	0.0	T	T	T	24.1	15.5	T	0.0	0.0	0.0	39.7
1966-67	0.0	0.0	0.0	0.0	T	28.4	0.7	19.7	17.4	1.0	0.0	0.0	67.2
1967-68	0.0	0.0	0.0	0.0	7.8	8.4	3.0	0.6	0.1	0.0	0.0	0.0	19.9
1968-69	0.0	0.0	0.0	0.0	7.2	3.7	1.4	11.8	8.4	0.0	0.0	0.0	32.5
1969-70	0.0	0.0	0.0	0.0	T	20.0	11.7	4.3	14.8	T	0.0	0.0	46.9
1970-71	0.0	0.0	0.0	T	T	6.8	11.2	4.4	6.8	0.5	0.0	0.0	29.7
1971-72	0.0	0.0	0.0	6.1	1.4	4.5	1.9	1.0	1.4	0.4	0.0	0.0	28.6
1972-73	0.0	0.0	0.0	1.4	2.3	0.6	1.9	4.3	0.6	0.1	0.0	0.0	7.4
1973-74	0.0	0.0	0.0	0.0	T	7.2	7.2	4.3	0.6	T	0.0	0.0	19.4
1974-75	0.0	0.0	0.0	0.9	T	3.2	8.3	9.6	2.1	T	0.0	0.0	23.2
1975-76	0.0	0.0	0.0	T	T	1.6	6.3	4.3	6.7	T	0.0	0.0	18.9
1976-77	0.0	0.0	0.0	0.0	0.4	6.6	11.5	8.4	1.5	0.3	0.0	0.0	27.9
1977-78	0.0	0.0	0.0	T	2.5	3.9	20.9	19.9	8.1	T	0.0	0.0	55.6
1978-79	0.0	0.0	0.0	T	2.9	9.1	21.0	T	T	0.0	0.0	0.0	33.9
1979-80	0.0	0.0	0.0	1.0	T	4.5	1.0	5.9	9.5	T	0.0	0.0	21.9
1980-81	0.0	0.0	0.0	0.0	4.2	3.6	9.5	1.3	6.9	0.0	0.0	0.0	25.5
1981-82	0.0	0.0	0.0	0.0	T	8.9	12.4	5.2	4.0	13.4	0.0	0.0	43.9
1982-83	0.0	0.0	0.0	0.0	T	2.3	8.8	29.5	0.6	4.6	0.0	0.0	45.8
1983-84	0.0	0.0	0.0	0.0	T	5.0	12.8	0.6	15.9	T	0.0	0.0	34.3
1984-85	0.0	0.0	0.0	0.0	T	5.7	6.7	11.5	0.3	T	0.0	0.0	24.2
1985-86	0.0	0.0	0.0	0.0	T	2.7							
Record Mean	0.0	0.0	0.0	0.1	1.3	6.7	8.3	9.2	5.9	0.8	T	0.0	32.4

See Reference Notes, relative to all above tables, on preceding page.

Erie is located on the southeast shore of Lake Erie and observations are made at Erie International Airport, which is 6 miles southwest of the center of the city and about 1 mile from the lake shore. The terrain rises gradually in a series of ridges paralleling the shoreline to 500 feet above the lake level 3 to 4 miles inland and to 1,000 feet about 15 miles inland. Snowfall from instability showers moving southward off the lake usually increases due to the upslope terrain. Snowfall is somewhat higher south of the city than along the lake shore.

During the winter months, the many cold air masses moving south from Canada are modified by the relatively warm waters of Lake Erie. However, the temperature difference between air and water produces an excess of cloudiness and frequent snow from November through March.

Spring weather is quite variable in Erie, but generally cloudy and cool. Proximity to the lake frequently prevents killing frosts that occur inland. This has led to the establishment of numerous vineyards and orchards in a narrow belt along the shore. Summer heat waves are tempered by cool lake breezes that may reach several miles inland, and days with temperatures above 90 degrees are infrequent. Summer thunderstorms are usually less destructive in Erie than inland areas because of the stabilizing effects of Lake Erie.

Autumn, with long dry periods and an abundance of sunshine, is usually the most pleasant period of the year in Erie. The growing season is extended by the influence of the warmer waters of the lake. Precipitation is well distributed throughout the year, although the number of days with measurable amounts varies considerably from a low average of about one day in three for the period June through September to about one-half of the days from November through March, when snow flurries and squalls move in from the lake.

TABLE 1   NORMALS, MEANS AND EXTREMES

ERIE, PENNSYLVANIA

LATITUDE: 42°05'N   LONGITUDE: 80°11'W   ELEVATION: FT. GRND 731 BARO 744   TIME ZONE: EASTERN   WBAN: 14860

	(a)	JAN	FEB	MAR	APR	MAY	JUNE	JULY	AUG	SEP	OCT	NOV	DEC	YEAR
TEMPERATURE °F:														
Normals														
-Daily Maximum		30.9	32.2	41.1	53.7	64.6	74.0	78.2	77.0	71.0	60.1	47.1	35.7	55.5
-Daily Minimum		18.0	17.7	25.8	36.1	45.4	55.2	59.9	59.4	53.1	43.2	34.3	24.2	39.4
-Monthly		24.5	25.0	33.5	44.9	55.0	64.6	69.1	68.2	62.0	51.7	40.7	30.0	47.4
Extremes														
-Record Highest	32	65	67	79	85	89	92	94	92	94	88	80	75	94
-Year		1985	1957	1977	1958	1975	1964	1968	1978	1959	1963	1961	1982	JUL 1968
-Record Lowest	32	-16	-17	-9	12	26	32	44	37	33	24	7	-6	-17
-Year		1985	1979	1980	1982	1970	1972	1963	1982	1974	1975	1976	1983	FEB 1979
NORMAL DEGREE DAYS:														
Heating (base 65°F)		1256	1120	977	603	323	80	17	28	130	420	729	1085	6768
Cooling (base 65°F)		0	0	0	0	13	68	144	127	43	7	0	0	402
% OF POSSIBLE SUNSHINE														
MEAN SKY COVER (tenths)														
Sunrise - Sunset	30	8.6	8.0	7.5	6.8	6.4	5.7	5.5	5.8	6.2	6.7	8.4	9.0	7.1
MEAN NUMBER OF DAYS:														
Sunrise to Sunset														
-Clear	30	1.7	2.5	4.5	5.9	6.8	8.2	8.8	8.7	6.9	6.4	2.5	1.3	64.3
-Partly Cloudy	30	4.2	6.1	7.2	7.8	10.3	11.3	12.6	11.1	9.6	7.9	4.3	3.3	95.6
-Cloudy	30	25.1	19.7	19.3	16.3	13.9	10.5	9.7	11.2	13.5	16.7	23.2	26.4	205.4
Precipitation														
.01 inches or more	32	18.6	14.9	15.3	13.9	12.0	10.2	9.5	10.9	10.7	12.4	16.2	19.1	163.6
Snow,Ice pellets														
1.0 inches or more	30	7.8	4.9	3.5	0.8	0.0	0.0	0.0	0.0	0.0	0.2	2.5	6.8	26.5
Thunderstorms	30	0.2	0.5	1.5	3.0	4.0	6.0	6.7	7.0	4.2	2.4	1.7	0.4	37.5
Heavy Fog Visibility														
1/4 mile or less	30	1.0	1.6	2.7	1.8	2.0	0.9	0.4	0.5	0.3	0.4	0.9	0.8	13.4
Temperature °F														
-Maximum														
90° and above	20	0.0	0.0	0.0	0.0	0.0	0.3	0.5	0.2	0.1	0.0	0.0	0.0	1.1
32° and below	20	18.1	14.9	7.7	0.6	0.0	0.0	0.0	0.0	0.0	0.0	1.5	10.6	53.4
-Minimum														
32° and below	20	28.5	25.1	23.7	11.9	1.3	0.1	0.0	0.0	0.0	2.0	12.1	25.0	129.7
0° and below	20	2.6	2.3	0.3	0.0	0.0	0.0	0.0	0.0	0.0	0.0	0.0	0.4	5.6
AVG. STATION PRESS.(mb)	13	990.0	991.0	988.8	988.6	988.1	988.8	989.8	991.3	991.6	992.1	990.7	990.5	990.1
RELATIVE HUMIDITY (%)														
Hour 01	20	75	77	75	72	77	80	81	82	81	75	75	76	77
Hour 07	20	77	78	77	74	77	80	81	84	83	77	76	76	78
Hour 13 (Local Time)	20	73	71	66	61	63	65	64	65	66	64	69	73	67
Hour 19	20	74	75	71	63	64	65	65	69	74	73	74	75	70
PRECIPITATION (inches):														
Water Equivalent														
-Normal		2.49	2.12	2.91	3.49	3.28	3.72	3.28	3.85	3.89	3.37	3.74	3.25	39.39
-Maximum Monthly	32	4.59	5.21	6.78	7.11	5.83	7.74	7.70	11.06	10.65	9.87	10.40	5.63	11.06
-Year		1959	1981	1976	1961	1984	1957	1970	1977	1977	1954	1985	1977	AUG 1977
-Minimum Monthly	32	0.87	0.57	0.63	1.63	1.45	0.85	0.65	0.58	1.45	1.13	1.52	1.38	0.57
-Year		1981	1978	1960	1975	1962	1963	1978	1978	1960	1963	1978	1960	FEB 1978
-Maximum in 24 hrs	30	1.51	2.16	1.87	2.53	2.23	2.80	3.22	3.53	6.11	4.35	3.67	2.39	6.11
-Year		1959	1961	1965	1977	1969	1957	1970	1980	1979	1954	1985	1979	SEP 1979
Snow,Ice pellets														
-Maximum Monthly	31	62.4	32.1	26.8	17.2	0.1					4.0	36.3	59.9	62.4
-Year		1978	1972	1971	1957	1963					1954	1967	1985	JAN 1978
-Maximum in 24 hrs	31	12.7	17.8	12.0	10.0	0.2					2.3	23.0	13.1	23.0
-Year		1985	1979	1965	1957	1963					1974	1956	1983	NOV 1956
WIND:														
Mean Speed (mph)	31	13.3	12.2	12.2	11.7	10.0	9.6	9.0	9.0	9.9	11.2	13.0	13.6	11.2
Prevailing Direction through 1963		WSW	WSW	NNE	WSW	WSW	S	S	S	S	SSE	SSW	SSW	S
Fastest Obs. 1 Min.														
-Direction (!!!)	28	20	29	14	21	25	36	32	24	17	24	30	17	14
-Speed (MPH)	28	53	52	55	46	37	37	46	35	45	43	39	38	55
-Year		1978	1972	1960	1965	1964	1975	1959	1970	1965	1983	1968	1964	MAR 1960
Peak Gust														
-Direction (!!!)	2	W	NE	W	SW	W	W	S	SW	S	W	W	W	W
-Speed (mph)	2	53	44	52	55	46	41	38	53	38	44	49	60	60
-Date		1985	1984	1985	1985	1984	1985	1985	1984	1985	1985	1985	1985	DEC 1985

See Reference Notes to this table on the following pages.

TABLE 2 PRECIPITATION (inches)   ERIE, PENNSYLVANIA

YEAR	JAN	FEB	MAR	APR	MAY	JUNE	JULY	AUG	SEP	OCT	NOV	DEC	ANNUAL
1956	1.31	2.45	3.74	3.65	5.59	1.58	7.11	6.61	1.79	1.19	5.60	2.81	43.43
1957	2.93	1.73	1.97	6.33	3.29	7.74	1.75	1.43	4.63	1.98	2.26	2.20	38.24
1958	2.86	1.52	0.68	3.03	1.60	4.84	4.65	5.64	6.01	2.21	4.03	2.73	39.80
1959	4.59	3.08	3.14	3.38	2.55	3.40	2.31	0.58	5.85	6.29	4.76	3.58	43.51
1960	3.09	2.55	0.63	2.18	3.66	2.60	2.95	2.03	1.45	2.21	2.68	1.38	27.41
1961	1.13	4.03	2.13	7.11	2.05	3.90	3.25	3.12	2.16	2.56	2.22	2.84	36.50
1962	3.30	1.84	1.17	1.84	1.45	4.98	2.39	4.56	4.94	4.02	2.13	4.28	36.90
1963	1.51	0.95	2.84	2.94	2.02	0.85	2.71	2.62	1.46	1.13	5.91	3.17	28.11
1964	2.35	1.39	3.81	3.67	3.75	2.26	3.42	4.60	1.97	2.96	3.08	3.21	36.47
1965	3.86	3.00	3.55	1.66	3.40	3.75	3.48	2.68	2.24	3.84	4.27	2.68	38.41
1966	2.01	1.92	3.47	4.26	2.03	3.82	2.08	3.52	3.24	1.73	5.34	4.32	37.74
1967	0.90	1.67	1.67	4.21	3.15	2.49	4.79	4.48	4.25	2.77	5.19	2.21	37.78
1968	2.98	1.13	1.69	2.64	3.06	2.63	2.34	2.76	2.93	3.40	4.71	4.06	34.33
1969	2.95	0.73	1.43	5.27	5.49	4.88	3.65	1.75	1.99	3.02	2.99	2.43	36.58
1970	1.44	2.09	1.61	2.48	3.23	2.65	7.70	2.03	7.08	3.52	4.87	2.80	41.50
1971	1.63	2.05	1.85	1.81	2.20	2.83	2.67	3.46	3.51	3.48	4.51	4.06	34.06
1972	1.94	2.77	3.95	2.64	4.69	7.50	2.91	3.01	5.37	1.72	3.36	3.69	43.55
1973	1.72	2.00	3.18	2.71	4.57	6.28	1.55	3.96	2.22	3.56	2.83	3.46	38.04
1974	2.45	1.93	5.02	4.89	4.44	5.33	1.11	2.97	3.34	1.57	5.20	3.58	41.83
1975	2.99	3.30	3.58	1.63	2.36	4.36	3.22	8.93	2.52	2.42	3.17	4.55	43.03
1976	3.07	5.01	6.78	2.71	4.18	3.98	4.35	1.40	3.87	4.14	2.12	2.24	43.85
1977	1.05	1.44	3.45	5.21	2.53	4.39	6.67	11.06	10.65	3.37	6.25	5.63	61.70
1978	3.61	0.57	1.79	2.45	2.76	2.68	0.65	4.56	4.83	5.81	1.52	3.81	35.04
1979	3.50	2.15	2.50	4.71	3.70	3.58	4.27	5.09	8.44	6.63	5.84	4.90	55.31
1980	1.57	1.29	4.11	3.79	2.23	4.83	5.42	6.76	5.48	6.51	2.56	2.49	47.04
1981	0.87	5.21	1.58	6.09	2.13	4.84	3.04	3.85	4.26	5.04	2.22	2.84	41.97
1982	3.85	1.24	3.50	1.81	3.06	6.02	4.40	2.20	4.07	2.74	5.33	3.34	41.56
1983	1.49	1.07	3.63	2.93	3.91	3.84	5.52	4.74	5.27	3.77	6.11	3.97	46.25
1984	1.65	2.42	1.91	2.63	5.83	4.49	1.94	2.09	5.29	1.82	3.62	4.10	37.79
1985	2.56	2.75	5.08	1.76	2.94	3.50	4.97	1.66	2.22	5.20	10.40	2.83	45.87
Record Mean	2.67	2.39	2.80	3.10	3.36	3.47	3.30	3.28	3.65	3.52	3.56	2.89	37.98

TABLE 3 AVERAGE TEMPERATURE (deg. F)   ERIE, PENNSYLVANIA

YEAR	JAN	FEB	MAR	APR	MAY	JUNE	JULY	AUG	SEP	OCT	NOV	DEC	ANNUAL
1956	25.7	28.9	31.7	43.2	54.1	65.6	68.6	68.9	59.2	55.3	42.3	36.0	48.3
1957	22.9	30.8	35.1	48.1	55.3	67.9	68.8	66.5	69.7	49.7	42.3	35.8	48.8
1958	26.2	21.4	33.4	47.2	54.4	61.2	70.2	67.9	63.0	53.2	43.8	23.2	46.8
1959	23.3	26.3	32.5	47.2	60.1	67.2	71.8	75.5	67.6	53.7	39.9	34.2	50.0
1960	29.8	28.5	24.6	49.9	56.6	66.1	68.4	69.9	66.3	52.1	44.8	25.5	48.6
1961	22.8	29.4	37.4	42.0	53.6	64.7	70.8	70.4	68.5	56.2	43.2	30.9	49.2
1962	23.9	24.5	33.3	46.5	62.6	66.8	69.5	69.0	60.6	54.1	40.4	26.9	48.2
1963	19.8	18.2	37.7	46.8	53.9	66.6	70.8	66.7	59.6	58.7	45.0	25.3	47.5
1964	30.0	24.8	36.6	47.2	60.1	66.6	72.5	66.6	62.7	50.0	45.0	33.2	49.6
#1965	25.8	26.6	29.5	41.6	61.0	64.5	67.5	68.2	66.3	50.5	42.4	36.0	48.4
1966	22.8	26.6	37.6	44.5	51.6	67.5	71.8	69.3	61.1	51.3	43.3	32.3	48.3
1967	32.8	24.2	34.6	47.6	49.6	71.1	68.8	66.6	60.0	53.1	36.9	34.0	48.3
1968	23.5	20.7	36.4	48.3	52.6	65.0	70.1	70.6	65.9	55.1	42.4	30.5	48.4
1969	27.0	25.7	31.8	47.2	54.7	62.6	69.0	69.6	61.7	50.0	38.3	25.8	47.0
1970	16.8	23.1	28.4	45.1	58.2	63.9	68.9	68.0	62.8	53.0	40.4	30.8	46.6
1971	22.0	27.3	30.0	40.6	51.9	66.0	66.9	64.7	64.1	57.0	39.3	35.9	47.2
1972	26.9	22.9	29.8	40.0	55.6	59.8	68.4	66.1	61.1	47.2	37.3	33.3	45.7
1973	29.6	23.8	42.2	46.0	52.5	66.2	69.7	69.6	62.6	54.8	42.8	31.7	49.3
1974	30.0	24.6	35.4	47.6	53.0	63.2	67.7	68.1	58.7	48.3	41.0	31.0	47.4
1975	29.9	28.4	32.1	38.8	60.3	67.4	71.8	69.4	58.5	52.7	46.6	30.7	48.9
1976	20.9	32.8	38.9	45.8	52.5	67.0	66.4	67.0	60.5	48.1	34.2	23.9	46.4
1977	12.5	25.0	40.6	47.8	58.4	61.9	70.3	67.1	62.5	49.4	43.7	28.3	47.3
1978	19.8	14.5	27.9	41.8	56.0	65.2	69.5	71.1	62.3	49.3	40.6	31.0	45.7
1979	19.5	14.1	36.5	42.2	53.9	63.5	67.6	69.5	64.3	53.6	43.7	34.5	46.9
1980	26.6	20.9	31.0	43.3	55.7	61.3	69.6	72.2	63.7	48.1	38.9	27.5	46.6
1981	19.6	30.3	33.1	47.4	53.9	66.0	72.1	70.6	61.9	49.7	41.6	31.4	48.1
1982	19.2	23.4	34.4	42.5	60.9	61.3	70.9	66.2	62.5	54.2	44.8	40.3	48.4
1983	30.6	31.6	38.5	45.3	55.0	67.0	73.0	72.4	65.7	54.4	45.2	25.9	50.4
1984	21.6	35.5	28.5	47.0	53.8	64.6	69.6	72.0	61.8	56.6	42.4	37.3	49.5
1985	21.7	25.9	37.4	51.4	59.8	63.4	70.2	70.7	66.9	54.6	46.3	27.2	49.6
Record Mean	26.9	26.5	34.4	45.4	56.5	66.3	71.2	69.8	63.9	53.2	41.9	31.6	49.0
Max	33.5	33.7	41.9	53.4	64.9	74.4	78.9	77.4	71.6	60.6	48.1	37.4	56.3
Min	20.3	19.3	26.9	37.3	48.1	58.3	63.4	62.2	56.2	45.8	35.7	25.8	41.6

REFERENCE NOTES FOR TABLES 1, 2, 3 and 6   (ERIE, PA)

GENERAL

T - TRACE AMOUNT
BLANK ENTRIES DENOTE MISSING/UNREPORTED DATA.
# INDICATES A STATION OR INSTRUMENT RELOCATION.

SPECIFIC

TABLE 1

(a) - LENGTH OF RECORD IN YEARS. ALTHOUGH INDIVIDUAL MONTHS MAY BE MISSING.
 *   LESS THAN .05

NORMALS — BASED ON THE 1951-1980 RECORD PERIOD.
EXTREMES — DATES ARE THE MOST RECENT OCCURRENCE.
WIND DIR. — NUMERALS SHOW TENS OF DEGREES CLOCKWISE FROM TRUE NORTH.
  ''00'' INDICATES CALM.
RESULTANT WIND DIRECTIONS ARE GIVEN TO WHOLE DEGREES.

EXCEPTIONS

TABLES 2, 3, and 6

RECORD MEANS ARE THROUGH THE CURRENT YEAR, BEGINNING IN   1847 FOR TEMPERATURE
                     1847 FOR PRECIPITATION
                     1955 FOR SNOWFALL

TABLE 4 HEATING DEGREE DAYS Base 65 deg. F ERIE, PENNSYLVANIA

SEASON	JULY	AUG	SEP	OCT	NOV	DEC	JAN	FEB	MAR	APR	MAY	JUNE	TOTAL
1956-57	12	19	202	298	675	891	1298	951	920	526	314	51	6157
1957-58	21	38	155	467	675	898	1196	1215	973	529	330	140	6637
1958-59	3	34	112	365	629	1287	1286	1077	1002	530	222	69	6616
1959-60	1	0	91	361	748	947	1086	1050	1245	464	277	52	6322
1960-61	25	6	49	397	597	1221	1302	989	848	681	370	96	6581
1961-62	23	10	66	276	648	1051	1266	1125	977	578	168	59	6247
1962-63	7	27	173	338	731	1174	1394	1306	839	543	350	64	6946
1963-64	20	28	173	221	594	1222	1077	1158	874	530	195	91	6183
1964-65	3	53	140	461	596	978	1209	1072	1093	696	174	105	6580
#1965-66	26	44	76	441	673	892	1302	1068	842	609	417	64	6454
1966-67	5	10	151	420	644	1006	993	1137	941	518	471	18	6314
1967-68	25	32	164	374	835	955	1280	1277	881	494	377	81	6775
1968-69	20	26	55	333	671	1063	1169	1093	1020	528	324	145	6447
1969-70	20	20	169	462	796	1211	1488	1169	1129	605	235	103	7407
1970-71	18	18	125	366	731	1056	1324	1048	1078	723	404	68	6959
1971-72	32	55	97	242	763	895	1172	1217	1082	742	287	184	6768
1972-73	55	47	144	546	820	976	1090	1146	697	568	379	40	6508
1973-74	9	18	141	309	658	1022	1128	1077	915	521	373	101	6272
1974-75	32	9	204	511	713	1046	1081	1016	1013	781	203	51	6660
1975-76	3	14	197	381	546	1057	1361	931	804	580	386	51	6311
1976-77	28	42	163	519	920	1269	1620	1116	751	519	249	126	7322
1977-78	9	51	110	477	636	1131	1395	1408	1143	688	303	84	7435
1978-79	15	2	125	481	723	1046	1406	1424	874	678	369	106	7249
1979-80	36	17	84	378	634	939	1188	1275	1045	644	299	161	6700
1980-81	12	2	108	520	775	1156	1400	967	982	523	345	42	6832
1981-82	3	6	153	467	694	1034	1413	1159	942	672	158	123	6824
1982-83	11	47	126	336	605	762	1062	933	814	585	312	67	5660
1983-84	10	0	89	336	591	1207	1338	847	1123	539	360	25	6465
1984-85	11	5	141	262	673	853	1337	1088	846	423	202	87	5928
1985-86	5	3	75	316	558	1164							

TABLE 5 COOLING DEGREE DAYS Base 65 deg. F ERIE, PENNSYLVANIA

YEAR	JAN	FEB	MAR	APR	MAY	JUNE	JULY	AUG	SEP	OCT	NOV	DEC	TOTAL
1969	0	0	0	4	10	81	150	170	76	5	0	0	496
1970	0	0	0	13	28	76	145	119	62	3	0	0	446
1971	0	0	0	0	7	105	96	55	75	3	0	0	341
1972	0	0	0	0	2	32	165	88	35	1	0	0	322
1973	0	0	0	6	0	84	163	167	76	1	0	0	497
1974	0	0	0	5	9	55	120	114	23	0	0	0	326
1975	0	0	0	0	64	132	221	156	9	6	0	0	588
1976	0	0	2	10	5	98	81	110	33	2	0	0	339
1977	0	0	0	9	52	42	179	121	44	0	1	0	450
1978	0	0	0	0	32	94	162	198	48	0	0	0	534
1979	0	0	0	3	29	69	125	163	71	32	0	0	492
1980	0	0	0	0	20	56	162	234	75	4	0	0	551
1981	0	0	0	2	8	81	230	187	65	0	0	0	573
1982	0	0	0	4	35	19	201	90	59	5	0	4	425
1983	0	0	0	0	7	135	261	235	117	15	0	0	770
1984	0	0	0	7	18	145	161	227	52	3	0	0	613
1985	0	0	0	21	45	45	172	187	140	4	3	0	617

TABLE 6 SNOWFALL (inches) ERIE, PENNSYLVANIA

SEASON	JULY	AUG	SEP	OCT	NOV	DEC	JAN	FEB	MAR	APR	MAY	JUNE	TOTAL
1956-57	0.0	0.0	0.0	0.0	34.8	10.2	25.6	6.7	13.5	17.2	T	0.0	108.0
1957-58	0.0	0.0	0.0	1.2	7.9	4.7	28.9	12.7	6.6	2.4	0.0	0.0	64.4
1958-59	0.0	0.0	0.0	0.0	15.7	45.0	24.7	7.6	15.5	T	T	0.0	108.5
1959-60	0.0	0.0	0.0	0.0	16.0	12.8	7.3	22.6	11.3	2.6	T	0.0	72.6
1960-61	0.0	0.0	0.0	T	5.3	17.4	23.7	13.6	0.7	8.9	T	0.0	69.6
1961-62	0.0	0.0	0.0	T	0.8	11.9	9.5	14.4	4.5	1.4	0.0	0.0	42.5
1962-63	0.0	0.0	0.0	3.0	0.3	30.5	18.7	12.3	10.9	0.1	0.1	0.0	75.9
1963-64	0.0	0.0	0.0	0.0	6.5	56.0	25.8	19.7	7.4	0.5	0.0	0.0	115.9
1964-65	0.0	0.0	0.0	T	19.9	13.5	16.8	13.7	26.8	2.2	0.0	0.0	92.9
1965-66	0.0	0.0	0.0	T	18.4	5.3	29.9	8.7	17.7	4.8	T	0.0	84.8
1966-67	0.0	0.0	0.0	0.0	7.2	31.4	6.0	13.5	8.8	1.7	T	0.0	68.6
1967-68	0.0	0.0	0.0	T	36.3	13.3	24.2	24.6	9.1	0.4	0.0	0.0	107.9
1968-69	0.0	0.0	0.0	T	9.8	18.8	26.1	13.6	9.4	2.0	0.0	0.0	79.7
1969-70	0.0	0.0	0.0	1.2	11.4	21.9	21.9	21.4	8.6	0.9	0.0	0.0	85.6
1970-71	0.0	0.0	0.0	T	18.6	23.6	26.8	21.4	26.8	2.8	0.0	0.0	120.0
1971-72	0.0	0.0	0.0	T	14.6	9.3	27.3	32.1	7.9	1.1	0.0	0.0	92.3
1972-73	0.0	0.0	0.0	0.6	6.6	18.5	7.9	12.6	5.7	1.8	T	0.0	53.7
1973-74	0.0	0.0	0.0	T	0.4	10.8	21.2	17.9	13.2	5.1	T	0.0	68.6
1974-75	0.0	0.0	0.0	2.3	2.9	20.1	17.7	12.6	10.0	1.1	0.0	0.0	66.7
1975-76	0.0	0.0	0.0	0.0	2.1	24.7	24.8	6.8	7.5	0.1	T	0.0	66.0
1976-77	0.0	0.0	0.0	3.1	33.9	27.6	29.0	13.0	4.5	0.4	0.0	0.0	111.5
1977-78	0.0	0.0	0.0	0.0	13.1	51.0	62.4	12.6	3.6	0.1	0.0	0.0	142.8
1978-79	0.0	0.0	0.0	0.0	1.8	10.6	33.4	27.8	2.2	0.7	0.0	0.0	76.5
1979-80	0.0	0.0	0.0	0.1	22.2	4.8	9.2	12.2	6.7	T	0.0	0.0	55.2
1980-81	0.0	0.0	0.0	4.9	21.0	27.9	22.5	13.1			0.0	0.0	89.4
1981-82	0.0	0.0	0.0	T	0.4	17.3	27.1	9.0	7.3	10.2	0.0	0.0	71.3
1982-83	0.0	0.0	0.0	T	9.2	8.9	6.7	9.2	5.6	1.6	0.0	0.0	41.2
1983-84	0.0	0.0	0.0	0.0	1.4	41.1	18.7	27.2	21.6	T	0.0	0.0	110.0
1984-85	0.0	0.0	0.0	0.0	4.7	16.7	57.2	19.0	3.5	5.2	0.0	0.0	106.3
1985-86	0.0	0.0	0.0	0.0	1.9	59.9							
Record Mean	0.0	0.0	0.0	0.4	10.8	21.8	22.8	15.4	10.3	2.6	T	0.0	84.0

See Reference Notes, relative to all above tables, on preceding page.

Harrisburg, the capital of Pennsylvania, is situated on the east bank of the Susquehanna River. It is in the Great Valley formed by the eastern foothills of the Appalachian Chain, and about 60 miles southeast of the Commonwealths geographic center. It is nestled in a saucer-like bowl, 10 miles south of Blue Mountain, which serves as a barrier to the severe winter climate experienced 50 to 100 miles to the north and west. Although the severity of the winter climate is lessened, the city lies a little too far inland to derive the full benefits of the coastal climate.

Air masses change with some regularity, and any one condition does not persist for many days in succession. The mountain barrier occasionally prevents cold waves from reaching the Great Valley. The city is favorably located to receive precipitation produced when warm, maritime air from the Atlantic Ocean is forced upslope to cross the Blue Ridge Mountains.

The growing season in the Harrisburg area is about 192 days. Prolonged dry spells occur on occasion. Flood stage on the Susquehanna River occurs on the average of about every three years in Harrisburg, but serious flooding is much less frequent. About one-third of all floods have occurred during the month of March. Tropical hurricanes rarely reach Harrisburg with destructive winds, but have produced rainfalls in excess of 15 inches.

TABLE 1   NORMALS, MEANS AND EXTREMES

HARRISBURG, PENNSYLVANIA

LATITUDE: 40°13'N   LONGITUDE: 76°51'W   ELEVATION: FT. GRND   338 BARO   00339   TIME ZONE: EASTERN   WBAN: 14751

	(a)	JAN	FEB	MAR	APR	MAY	JUNE	JULY	AUG	SEP	OCT	NOV	DEC	YEAR
TEMPERATURE °F:														
Normals														
–Daily Maximum		36.7	39.5	49.6	62.9	73.0	81.8	86.2	84.4	77.2	65.4	52.4	40.6	62.5
–Daily Minimum		22.1	23.5	31.5	41.5	51.0	60.5	65.3	64.2	56.6	44.6	35.4	26.2	43.5
–Monthly		29.4	31.5	40.5	52.2	62.0	71.2	75.8	74.3	66.9	55.0	43.9	33.4	53.0
Extremes														
–Record Highest	47	73	75	86	93	97	100	107	101	102	97	84	75	107
–Year		1950	1985	1945	1985	1942	1966	1966	1944	1953	1941	1950	1984	JUL 1966
–Record Lowest	47	–9	–5	5	19	31	40	49	45	30	23	13	–8	–9
–Year		1985	1979	1984	1982	1966	1980	1945	1976	1963	1969	1955	1960	JAN 1985
NORMAL DEGREE DAYS:														
Heating (base 65°F)		1104	938	756	384	150	12	0	0	58	320	633	980	5335
Cooling (base 65°F)		0	0	0	0	57	198	335	291	115	10	0	0	1006
% OF POSSIBLE SUNSHINE	47	49	55	57	59	60	64	69	68	63	57	46	45	58
MEAN SKY COVER (tenths)														
Sunrise – Sunset	35	6.7	6.5	6.6	6.5	6.5	6.1	6.0	5.9	5.7	5.8	6.8	6.9	6.3
MEAN NUMBER OF DAYS:														
Sunrise to Sunset														
–Clear	47	6.9	6.8	7.0	6.4	6.1	6.7	7.3	8.3	9.0	9.9	6.2	6.2	86.8
–Partly Cloudy	47	7.4	7.2	8.2	8.7	10.2	11.1	11.4	10.7	9.4	7.9	8.1	7.6	107.8
–Cloudy	47	16.7	14.3	15.9	14.9	14.7	12.2	12.3	12.0	11.5	13.3	15.7	17.2	170.6
Precipitation														
.01 inches or more	7	10.9	10.3	12.1	12.4	13.6	11.0	9.1	10.0	7.1	9.9	10.6	10.4	127.4
Snow,Ice pellets														
1.0 inches or more	7	3.0	2.6	1.3	0.6	0.0	0.0	0.0	0.0	0.0	0.0	0.3	1.9	9.6
Thunderstorms	47	0.2	0.2	1.1	2.3	5.2	6.3	7.0	5.3	2.9	0.8	0.5	0.2	31.9
Heavy Fog Visibility														
1/4 mile or less	47	2.3	2.3	1.7	1.0	0.9	0.6	0.6	0.8	1.6	2.7	1.8	2.4	18.8
Temperature °F														
–Maximum														
90° and above	47	0.0	0.0	0.0	0.3	1.0	4.6	8.7	6.1	2.0	0.1	0.0	0.0	22.7
32° and below	47	9.8	6.1	1.4	0.0	0.0	0.0	0.0	0.0	0.0	0.0	0.2	5.4	23.0
–Minimum														
32° and below	47	26.5	23.1	17.3	3.6	0.1	0.0	0.0	0.0	0.1	1.9	10.9	22.9	106.3
0° and below	47	0.7	0.2	0.0	0.0	0.0	0.0	0.0	0.0	0.0	0.0	0.0	0.2	1.1
AVG. STATION PRESS.(mb)	13	1005.8	1006.3	1004.1	1003.2	1002.8	1003.5	1004.1	1005.6	1006.7	1007.6	1006.7	1006.9	1005.3
RELATIVE HUMIDITY (%)														
Hour 01	43	69	68	67	67	73	79	79	82	80	79	73	70	74
Hour 07	43	71	71	71	70	74	77	78	82	84	82	76	73	76
Hour 13 (Local Time)	42	58	55	52	49	52	53	52	55	55	55	57	58	54
Hour 19	43	63	61	57	54	57	60	60	65	68	67	65	64	62
PRECIPITATION (inches):														
Water Equivalent														
–Normal		2.96	2.73	3.50	3.19	3.67	3.63	3.32	3.29	3.60	2.73	3.24	3.23	39.09
–Maximum Monthly	7	8.01	5.93	5.47	7.96	6.29	8.12	4.67	4.11	6.62	4.21	6.23	7.57	8.12
–Year		1979	1981	1980	1983	1985	1982	1981	1981	1979	1983	1985	1983	JUN 1982
–Minimum Monthly	7	0.43	0.82	1.02	0.45	1.86	2.50	0.97	1.51	1.06	1.34	0.96	0.77	0.43
–Year		1981	1980	1981	1985	1981	1980	1983	1980	1980	1985	1981	1980	JAN 1981
–Maximum in 24 hrs	5	2.09	1.84	1.80	1.25	2.91	2.28	1.90	1.65	3.03	2.19	2.16	0.86	3.03
–Year		1979	1985	1980	1980	1984	1984	1981	1981	1979	1980	1984	1979	SEP 1979
Snow,Ice pellets														
–Maximum Monthly	7	18.8	28.8	14.9	10.2						T	4.0	12.5	28.8
–Year		1982	1983	1984	1982						1982	1980	1981	FEB 1983
–Maximum in 24 hrs	5	3.5	14.2	7.0	2.6						T	4.0	5.6	14.2
–Year		1984	1979	1984	1985						1979	1980	1981	FEB 1979
WIND:														
Mean Speed (mph)	43	8.4	9.1	9.7	9.2	7.7	6.8	6.3	5.9	6.1	6.5	7.8	8.1	7.6
Prevailing Direction														
through 1963		WNW	WNW	WNW	WNW	W	W	W	W	WNW	W	WNW	WNW	WNW
Fastest Obs. 1 Min.														
–Direction	5	27	27	28	29	29	33	07	35	30	25	27	31	33
–Speed (MPH)	5	44	31	37	35	47	58	25	46	27	30	29	46	58
–Year		1978	1982	1985	1979	1980	1980	1980	1979	1984	1981	1984	1978	JUN 1980
Peak Gust														
–Direction														
–Speed (mph)														
–Date														

See Reference Notes to this table on the following pages.

TABLE 2 PRECIPITATION (inches)   HARRISBURG, PENNSYLVANIA

YEAR	JAN	FEB	MAR	APR	MAY	JUNE	JULY	AUG	SEP	OCT	NOV	DEC	ANNUAL
1956	1.45	3.47	4.41	2.66	2.43	2.66	4.57	3.48	3.24	3.52	1.97	4.26	38.12
1957	1.64	2.64	1.97	3.89	0.84	3.38	1.68	0.93	2.19	1.78	2.81	5.23	28.98
1958	4.61	3.61	3.67	3.21	3.08	2.76	2.46	4.20	1.30	3.85	0.66		36.70
1959	2.66	1.92	2.57	3.18	1.51	4.81	3.20	5.54	0.91	4.17	2.14	3.01	35.62
1960	2.10	3.30	2.52	2.59	7.41	4.17	4.73	1.85	5.34	1.34	1.39	2.41	39.15
1961	3.46	3.07	4.19	4.56	2.03	1.93	6.60	5.49	1.24	0.92	3.56	3.42	40.47
1962	2.15	4.33	2.70	2.81	2.96	4.03	1.20	4.18	3.59	4.21	4.10	3.32	39.58
1963	2.19	1.83	3.86	1.52	2.66	2.36	1.97	2.55	2.82	0.04	5.93	2.36	30.09
1964	4.78	3.12	2.94	4.91	0.51	4.20	2.25	3.07	1.77	1.92	1.87	3.11	34.45
1965	2.70	3.29	3.61	1.25	2.38	2.60	3.99	2.12	3.65	1.63	0.87		31.19
1966	3.57	4.44	1.88	3.44	0.98	0.07	0.81	1.53	6.12	2.12	3.56	3.08	31.60
1967	1.81	1.54	5.26	2.58	4.32	1.90	5.96	5.61	1.80	3.15	2.89	4.27	41.09
1968	1.32	0.53	3.40	2.43	6.55	2.25	1.94	1.77	5.18	2.34	3.38	2.12	33.21
1969	1.06	1.70	2.20	2.13	1.56	2.54	9.72	2.07	2.32	1.63	3.29	6.46	36.68
1970	0.88	3.25	3.64	5.03	2.39	5.80	6.34	2.97	2.12	3.20	4.59	3.50	43.71
1971	2.70	5.62	2.67	1.04	5.30	1.80	2.84	7.77	1.94	2.85	4.96	1.93	41.42
1972	2.65	5.00	2.68	4.10	5.56	18.55	2.26	2.52	1.41	2.03	7.20	5.31	59.27
1973	3.24	2.50	2.00	6.23	3.34	2.27	2.18	2.19	5.73	2.47	1.04	6.52	43.81
1974	3.82	1.36	4.64	3.21	4.38	3.69	2.79	4.13	6.79	1.25	2.30	4.59	42.95
1975	4.12	3.10	3.78	2.80	5.25	6.51	3.13	1.83	14.97	2.62	2.92	3.19	54.22
1976	4.34	1.88	1.28	1.63	5.42	2.42	5.50	3.28	4.79	9.87	0.79	1.96	45.31
1977	1.44	1.75	6.10	4.48	1.00	3.17	3.01	0.93	3.73	3.66	5.61	4.82	39.70
1978	7.44	1.35	3.94	1.97	5.67	5.16	4.35	3.60	1.64	2.51	2.13	3.95	43.71
#1979	8.01	4.74	1.93	3.60	4.66	2.62	3.14	3.24	6.62	3.91	2.67	1.46	46.60
1980	0.90	0.82	5.47	4.27	4.58	2.50	1.59	1.51	1.06	2.94	3.65	0.77	30.06
1981	0.43	5.93	1.02	2.77	1.86	4.66	4.67	4.11	2.20	3.76	0.96	2.41	34.78
1982	3.63	1.92	2.20	4.17	4.89	8.12	2.90	2.47	2.87	1.82	3.37	1.56	39.92
1983	2.26	3.38	4.86	7.96	5.36	2.81	0.97	2.50	1.40	4.21	5.29	7.57	48.57
1984	1.12	4.51	5.36	4.46	6.20	6.36	3.76	2.75	1.49	1.98	3.78	2.28	44.05
1985	1.06	2.91	2.78	0.45	6.29	3.07	2.50	2.14	3.76	1.34	6.23	1.28	33.81
Record Mean	2.80	2.70	3.25	3.07	3.70	3.67	3.58	3.65	3.15	2.89	2.74	2.94	38.15

TABLE 3 AVERAGE TEMPERATURE (deg. F)   HARRISBURG, PENNSYLVANIA

YEAR	JAN	FEB	MAR	APR	MAY	JUNE	JULY	AUG	SEP	OCT	NOV	DEC	ANNUAL
1956	32.0	35.4	38.2	49.1	59.7	71.0	72.0	72.4	62.5	55.2	43.7	40.2	52.6
1957	28.0	35.9	41.3	54.3	62.7	73.6	75.0	72.3	67.8	52.7	44.9	37.3	53.8
1958	30.3	26.8	38.5	53.9	61.1	66.6	76.0	72.9	65.8	54.2	45.0	26.8	51.5
1959	29.5	32.5	39.8	54.8	66.0	72.2	76.0	76.0	67.4	58.2	42.0	36.0	54.4
1960	33.8	34.1	32.0	56.7	59.9	70.5	72.9	75.0	67.4	55.1	45.1	25.6	52.3
1961	25.1	33.6	41.5	48.1	59.1	69.9	75.1	73.5	72.3	57.2	46.0	31.1	52.7
1962	28.0	28.2	39.3	52.0	64.9	72.1	74.9	73.0	62.8	55.7	40.1	28.2	51.7
1963	26.2	25.2	43.2	54.4	61.6	72.3	76.4	70.1	61.5	57.1	45.8	26.9	51.7
1964	30.2	28.9	40.5	49.6	64.3	71.4	76.1	71.5	66.2	52.0	45.7	33.9	52.5
1965	26.4	32.5	37.5	50.1	66.5	71.7	75.6	74.6	69.8	53.4	42.2	35.7	53.0
1966	26.3	31.5	43.8	49.0	61.4	74.6	79.8	78.3	66.1	54.7	45.1	31.7	53.6
1967	33.6	26.9	38.8	53.0	55.2	73.9	74.2	71.8	65.3	53.3	39.8	34.8	51.7
1968	24.0	29.2	43.7	53.7	58.0	70.3	77.3	76.7	67.6	56.1	44.9	31.3	52.7
1969	29.9	33.3	38.6	53.4	63.2	72.4	75.3	73.4	65.6	53.8	42.7	31.4	52.8
1970	22.9	30.4	36.7	51.4	65.1	70.7	76.2	75.6	71.6	58.7	46.0	34.8	53.4
1971	26.4	33.2	38.6	50.4	60.0	72.5	75.5	72.2	69.9	60.6	43.5	40.3	53.6
1972	34.5	30.2	40.0	49.6	62.3	67.7	76.2	74.0	68.4	52.1	41.8	36.8	52.8
1973	33.7	31.7	45.2	50.9	57.8	73.4	75.5	76.1	68.6	57.4	46.9	36.2	54.5
1974	34.8	32.3	42.8	56.2	63.0	70.5	77.1	76.1	64.5	51.6	44.9	35.4	54.2
1975	33.3	32.4	38.3	47.4	64.8	70.7	75.3	76.2	63.3	57.7	50.0	34.2	53.7
1976	27.6	39.6	44.2	54.5	59.4	73.4	72.3	72.4	64.8	51.5	39.6	30.0	52.4
1977	20.1	30.4	46.0	54.1	64.5	68.6	75.9	74.3	68.3	52.7	46.1	31.6	52.7
1978	26.2	22.8	38.6	51.0	61.5	69.5	73.1	76.9	67.6	54.4	46.7	36.5	52.1
1979	29.2	22.6	45.1	50.4	62.2	68.5	73.4	72.9	65.5	52.3	46.8	37.6	52.2
1980	30.3	29.1	38.9	52.8	63.3	67.8	76.3	76.1	67.7	51.5	39.9	29.6	51.9
1981	23.7	34.6	38.7	53.7	61.9	71.7	75.7	72.2	63.9	50.7	44.7	31.9	51.9
1982	22.8	30.9	38.6	47.6	62.2	65.1	74.0	70.5	65.3	55.1	41.4	41.4	51.8
1983	33.0	33.4	42.7	49.3	58.4	69.1	75.9	74.9	66.2	53.6	43.9	28.7	52.4
1984	24.8	36.6	33.7	48.0	58.2	72.9	74.3	75.8	64.4	61.5	43.3	41.5	53.0
1985	27.9	34.4	44.5	56.9	65.2	69.4	75.9	74.1	69.2	57.2	47.9	31.0	54.5
Record Mean	30.1	31.1	40.2	50.6	62.1	70.7	75.2	73.3	66.6	55.1	43.6	33.2	52.7
Max	36.9	38.4	48.5	59.5	72.1	80.5	84.8	82.7	76.0	64.4	51.4	40.0	61.3
Min	23.2	23.7	31.9	41.7	52.1	60.9	65.6	63.9	57.1	45.8	35.8	26.4	44.0

REFERENCE NOTES FOR TABLES 1, 2, 3 and 6   (HARRISBURG, PA)

GENERAL

T - TRACE AMOUNT
BLANK ENTRIES DENOTE MISSING/UNREPORTED DATA.
# INDICATES A STATION OR INSTRUMENT RELOCATION.

SPECIFIC

TABLE 1

(a) - LENGTH OF RECORD IN YEARS. ALTHOUGH INDIVIDUAL MONTHS MAY BE MISSING.
* LESS THAN .05

NORMALS — BASED ON THE 1951-1980 RECORD PERIOD.
EXTREMES — DATES ARE THE MOST RECENT OCCURRENCE.
WIND DIR. — NUMERALS SHOW TENS OF DEGREES CLOCKWISE FROM TRUE NORTH.
"00" INDICATES CALM.
RESULTANT WIND DIRECTIONS ARE GIVEN TO WHOLE DEGREES.

EXCEPTIONS

TABLE 1

1. MAXIMUM 24-HOUR PRECIPITATION AND SNOW ARE THROUGH 1981.
2. PERCENT OF POSSIBLE SUNSHINE IS FROM THE FEDERAL OFFICE BUILDING COMMENCING DECEMBER 1983.

TABLES 1, 2, and 3

COMMENCING JANUARY 12, 1979 FOR PRECIPITATION AND JUNE 13, 1984 FOR TEMPERATURE, DATA ARE FROM THE FEDERAL OFFICE BUILDING, 3.5 MILES NORTHWEST OF THE AIRPORT.

TABLES 2, 3, and 6

RECORD MEANS ARE THROUGH THE CURRENT YEAR, BEGINNING IN   1889 FOR TEMPERATURE
1889 FOR PRECIPITATION
1939 FOR SNOWFALL

TABLE 4 HEATING DEGREE DAYS Base 65 deg. F   HARRISBURG, PENNSYLVANIA

SEASON	JULY	AUG	SEP	OCT	NOV	DEC	JAN	FEB	MAR	APR	MAY	JUNE	TOTAL
1956-57	7	0	146	295	613	762	1139	809	730	358	137	13	5009
1957-58	0	0	76	376	594	855	1070	1065	815	339	148	42	5380
1958-59	0	0	62	336	596	1177	1095	904	773	307	103	23	5376
1959-60	0	0	59	280	683	895	961	885	1017	292	171	3	5246
1960-61	0	0	23	307	593	1216	1232	720	871	509	204	22	5697
1961-62	0	2	41	244	569	1042	1139	1025	791	415	120	2	5390
1962-63	0	3	126	297	712	1132	1198	1107	666	325	146	4	5716
1963-64	1	7	149	242	570	1173	1072	1041	753	460	93	34	5595
1964-65	0	4	71	399	570	958	1192	902	847	441	66	27	5477
1965-66	0	7	44	357	677	900	1191	932	648	472	166	12	5406
1966-67	0	0	69	312	592	1024	966	1061	806	364	306	3	5503
1967-68	0	5	85	369	750	931	1266	1032	654	331	216	12	5651
1968-69	0	3	18	277	593	1038	1083	882	809	345	123	8	5179
1969-70	0	0	89	352	662	1036	1299	962	872	407	93	13	5785
1970-71	0	0	39	213	564	931	1192	884	812	432	165	12	5244
1971-72	0	3	40	146	643	759	940	1001	766	457	108	44	4907
1972-73	0	1	25	395	686	865	964	926	607	422	227	7	5125
1973-74	0	1	34	238	534	887	931	910	683	289	133	2	4642
1974-75	0	0	94	414	600	911	977	903	818	520	97	13	5347
1975-76	0	0	87	232	445	951	1150	730	639	354	193	23	4804
1976-77	0	5	75	418	756	1075	1387	966	588	340	106	32	5748
1977-78	0	5	35	377	562	1029	1196	1175	810	414	173	17	5793
1978-79	14	0	48	321	544	876	1104	1182	611	435	123	26	5284
1979-80	12	14	71	393	536	844	1070	1033	799	361	103	48	5284
1980-81	0	0	57	411	761	1091	1277	844	809	339	147	6	5742
1981-82	0	1	94	437	599	1021	1304	948	812	518	128	61	5923
1982-83	7	12	67	318	520	725	985	876	686	468	221	25	4910
1983-84	0	2	103	362	628	1117	1238	817	962	502	240	11	5982
1984-85	0	0	105	131	627	724	1143	849	627	292	87	16	4601
1985-86	0	0	41	237	508	1049							

TABLE 5 COOLING DEGREE DAYS Base 65 deg. F   HARRISBURG, PENNSYLVANIA

YEAR	JAN	FEB	MAR	APR	MAY	JUNE	JULY	AUG	SEP	OCT	NOV	DEC	TOTAL
1969	0	0	0	3	71	235	327	267	113	11	0	0	1027
1970	0	0	0	7	102	191	353	333	246	24	0	0	1256
1971	0	0	0	0	17	241	331	236	195	17	6	0	1043
1972	0	0	0	0	33	133	357	287	137	3	0	0	950
1973	0	0	0	8	13	266	334	352	148	9	0	0	1130
1974	0	0	0	34	79	176	401	381	88	1	3	0	1163
1975	0	0	0	0	0	97	192	325	354	44	12	2	1026
1976	0	0	0	47	22	283	233	240	73	3	0	0	901
1977	0	0	0	4	19	95	147	347	297	140	1	3	1053
1978	0	0	0	0	69	162	273	377	133	1	0	0	1015
1979	0	0	0	5	43	138	279	264	92	7	0	0	828
1980	0	0	0	0	57	138	355	350	145	0	0	0	1045
1981	0	0	0	6	60	213	339	232	66	0	0	0	916
1982	0	0	0	2	48	70	307	191	83	19	4	0	724
1983	0	0	0	6	22	154	343	315	146	13	0	0	999
1984	0	0	0	0	34	256	292	342	95	35	0	1	1055
1985	0	0	0	55	99	157	345	290	173	4	0	0	1123

TABLE 6 SNOWFALL (inches)   HARRISBURG, PENNSYLVANIA

SEASON	JULY	AUG	SEP	OCT	NOV	DEC	JAN	FEB	MAR	APR	MAY	JUNE	TOTAL
1956-57	0.0	0.0	0.0	0.0	1.0	0.7	10.8	16.4	2.4	0.7	0.0	0.0	32.0
1957-58	0.0	0.0	0.0	T	T	7.0	4.8	13.0	16.8	T	0.0	0.0	41.6
1958-59	0.0	0.0	0.0	0.0	1.7	2.4	6.6	2.8	8.8	3.1	0.0	0.0	25.4
1959-60	0.0	0.0	0.0	0.0	T	8.0	0.7	9.5	22.6	0.1	0.0	0.0	40.9
1960-61	0.0	0.0	0.0	T	0.0	22.1	34.0	18.7	4.0	2.5	0.0	0.0	81.3
1961-62	0.0	0.0	0.0	0.0	3.7	18.8	2.3	15.9	10.9	T	0.0	0.0	51.6
1962-63	0.0	0.0	0.0	T	4.5	16.6	4.5	14.2	6.1	T	T	0.0	50.5
1963-64	0.0	0.0	0.0	0.0	T	15.8	19.4	30.2	9.0	0.3	0.0	0.0	74.7
1964-65	0.0	0.0	0.0	0.0	T	1.4	13.4	1.7	15.1	0.2	0.0	0.0	31.8
1965-66	0.0	0.0	0.0	T	T	T	24.8	16.8	1.0	T	T	0.0	42.6
1966-67	0.0	0.0	0.0	0.0	0.2	19.9	1.6	16.5	10.2	T	0.0	0.0	48.4
1967-68	0.0	0.0	0.0	0.0	9.7	13.0	3.0	1.8	3.5	0.0	0.0	0.0	31.0
1968-69	0.0	0.0	0.0	0.0	3.0	0.2	1.2	16.8	3.8	0.0	0.0	0.0	25.0
1969-70	0.0	0.0	0.0	T	2.1	28.3	9.8	7.5	12.9	T	0.0	0.0	60.6
1970-71	0.0	0.0	0.0	0.0	0.0	10.9	11.6	5.1	5.3	T	0.0	0.0	32.9
1971-72	0.0	0.0	0.0	0.0	8.8		2.6	21.6	0.5	1.1	0.0	0.0	34.6
1972-73	0.0	0.0	0.0	1.2	5.9	0.5	T	5.7	T	T	0.0	0.0	13.3
1973-74	0.0	0.0	0.0	0.0	T	15.3	0.5	5.5	5.5	T	0.0	0.0	27.8
1974-75	0.0	0.0	0.0	0.0	T	0.5	11.3	13.1	6.1	T	0.0	0.0	31.0
1975-76	0.0	0.0	0.0	0.0	T	2.2	3.6	2.5	10.0	0.0	0.0	0.0	18.3
1976-77	0.0	0.0	0.0	0.0	1.4	5.1	12.2	4.5	0.2	T	T	0.0	23.4
1977-78	0.0	0.0	0.0	T	1.5	5.5	33.5	21.1	9.0	0.0	0.0	0.0	70.6
#1978-79	0.0	0.0	0.0	0.0	4.0	0.3	9.2	26.0	T	T	0.0	0.0	39.5
1979-80	0.0	0.0	0.0	T	T	0.2	3.8	2.7	7.9	0.0	0.0	0.0	14.6
1980-81	0.0	0.0	0.0	0.0	4.0	4.3	5.5	4.4	6.7	0.0	0.0	0.0	24.9
1981-82	0.0	0.0	0.0	0.0	0.8	12.5	18.8	8.4	7.8	10.2	0.0	0.0	58.5
1982-83	0.0	0.0	0.0	0.0	T	T	1.1	4.4	28.8	1.3	0.0	0.0	36.0
1983-84	0.0	0.0	0.0	0.0	T	4.6	9.7	2.3	14.9	T	0.0	0.0	31.5
1984-85	0.0	0.0	0.0	0.0	1.9	2.6	10.6	11.4	T	3.6	0.0	0.0	30.1
1985-86	0.0	0.0	0.0	0.0	T	5.6							
Record Mean	0.0	0.0	0.0	0.1	2.0	7.1	9.5	9.6	6.4	0.5	T	0.0	35.2

See Reference Notes, relative to all above tables, on preceding page.

The Appalachian Mountains to the west and the Atlantic Ocean to the east have a moderating effect on climate. Periods of very high or very low temperatures seldom last for more than three or four days. Temperatures below zero or above 100 degrees are a rarity. On occasion, the area becomes engulfed with maritime air during the summer months, and high humidity adds to the discomfort of seasonably warm temperatures.

Precipitation is fairly evenly distributed throughout the year with maximum amounts during the late summer months. Much of the summer rainfall is from local thunderstorms and amounts vary in different areas of the city. This is due, in part, to the higher elevations to the west and north. Snowfall amounts are often considerably larger in the northern suburbs than in the central and southern parts of the city. In many cases, the precipitation will change from snow to rain within the city. Single storms of 10 inches or more occur about every five years.

The prevailing wind direction for the summer months is from the southwest, while northwesterly winds prevail during the winter. The annual prevailing direction is from the west—southwest. Destructive velocities are comparatively rare and occur mostly in gustiness during summer thunderstorms. High winds occurring in the winter months, as a rule, come with the advance of cold air after the passage of a deep low pressure system. Only rarely have hurricanes in the vicinity caused widespread damage, primarily because of flooding.

Flood stages in the Schuylkill River normally occur about twice a year. Flood stages seldom last over 12 hours and usually occur after excessive thunderstorms. Flooding rarely occurs on the Delaware River.

TABLE 1   NORMALS, MEANS AND EXTREMES

PHILADELPHIA, PENNSYLVANIA

LATITUDE: 39°53'N   LONGITUDE: 75°15'W   ELEVATION: FT. GRND   5 BARO 00063   TIME ZONE: EASTERN   WBAN: 13739

	(a)	JAN	FEB	MAR	APR	MAY	JUNE	JULY	AUG	SEP	OCT	NOV	DEC	YEAR
TEMPERATURE °F:														
Normals														
-Daily Maximum		38.6	41.1	50.5	63.2	73.0	81.7	86.1	84.6	77.8	66.5	54.5	43.0	63.4
-Daily Minimum		23.8	25.0	33.1	42.6	52.5	61.5	66.8	66.0	58.6	46.5	37.1	28.0	45.1
-Monthly		31.2	33.0	41.8	52.9	62.8	71.6	76.5	75.3	68.2	56.5	45.8	35.5	54.3
Extremes														
-Record Highest	44	74	74	87	94	96	100	104	101	100	96	81	72	104
-Year		1950	1985	1945	1976	1962	1964	1966	1955	1953	1941	1974	1984	JUL 1966
-Record Lowest	44	-7	-4	7	19	28	44	51	45	35	25	15	1	-7
-Year		1984	1961	1984	1982	1966	1984	1966	1965	1963	1969	1976	1983	JAN 1984
NORMAL DEGREE DAYS:														
Heating (base 65°F)		1048	893	719	363	127	0	0	0	33	273	576	915	4947
Cooling (base 65°F)		0	0	0	0	59	202	357	319	129	9	0	0	1075
% OF POSSIBLE SUNSHINE	43	50	53	55	56	56	62	62	62	60	59	52	49	56
MEAN SKY COVER (tenths)														
Sunrise - Sunset	45	6.6	6.3	6.4	6.4	6.5	6.2	6.0	5.9	5.6	5.6	6.3	6.5	6.2
MEAN NUMBER OF DAYS:														
Sunrise to Sunset														
-Clear	45	7.4	7.2	7.6	7.1	6.2	6.9	7.2	8.3	9.8	10.5	7.2	7.5	93.0
-Partly Cloudy	45	7.5	7.3	8.1	8.9	10.4	11.3	11.8	11.0	9.0	8.4	8.9	8.4	111.0
-Cloudy	45	16.1	13.7	15.3	14.0	14.4	11.8	12.1	11.6	11.2	12.2	13.8	15.0	161.2
Precipitation														
.01 inches or more	45	11.0	9.2	11.1	10.7	11.4	10.2	9.2	9.1	7.9	7.6	9.4	10.1	116.9
Snow,Ice pellets														
1.0 inches or more	45	2.0	1.6	1.1	0.1	0.0	0.0	0.0	0.0	0.0	0.*	0.2	0.9	6.0
Thunderstorms	45	0.2	0.3	1.0	2.0	4.3	5.4	5.6	5.2	2.3	0.7	0.5	0.2	27.5
Heavy Fog Visibility 1/4 mile or less	45	2.8	2.4	1.8	1.3	1.4	1.2	0.9	1.1	1.6	3.4	2.5	2.7	23.0
Temperature °F														
-Maximum														
90° and above	26	0.0	0.0	0.0	0.4	0.5	3.5	7.5	5.8	2.0	0.0	0.0	0.0	19.7
32° and below	26	9.7	6.0	1.0	0.0	0.0	0.0	0.0	0.0	0.0	0.0	0.*	4.4	21.2
-Minimum														
32° and below	26	26.5	22.7	14.5	2.8	0.*	0.0	0.0	0.0	0.0	1.6	9.0	21.4	98.4
0° and below	26	0.5	0.2	0.0	0.0	0.0	0.0	0.0	0.0	0.0	0.0	0.0	0.0	0.7
AVG. STATION PRESS.(mb)	13	1017.1	1017.5	1015.3	1014.3	1013.9	1014.6	1014.9	1016.4	1017.5	1018.7	1017.9	1018.2	1016.3
RELATIVE HUMIDITY (%)														
Hour 01	26	70	68	68	68	76	80	81	82	82	80	75	72	75
Hour 07	26	72	71	71	70	75	77	79	81	83	83	78	74	76
Hour 13 (Local Time)	26	59	56	53	48	53	54	54	55	55	54	56	59	55
Hour 19	26	65	60	58	54	59	61	63	65	68	69	66	67	63
PRECIPITATION (inches):														
Water Equivalent														
-Normal		3.18	2.81	3.86	3.47	3.18	3.92	3.88	4.10	3.42	2.83	3.32	3.45	41.42
-Maximum Monthly	43	8.86	6.44	7.01	8.12	7.41	7.88	8.33	9.70	8.78	5.21	9.06	7.37	9.70
-Year		1978	1979	1980	1983	1948	1973	1969	1955	1960	1943	1972	1983	AUG 1955
-Minimum Monthly	43	0.45	0.96	0.68	0.52	0.47	0.11	0.64	0.49	0.44	0.09	0.32	0.25	0.09
-Year		1955	1980	1966	1985	1964	1949	1957	1964	1968	1963	1976	1955	OCT 1963
-Maximum in 24 hrs	39	2.70	1.96	2.39	2.76	3.18	4.62	4.26	5.68	5.45	3.85	3.99	2.04	5.68
-Year		1979	1966	1968	1970	1984	1973	1969	1971	1960	1980	1977	1978	AUG 1971
Snow,Ice pellets														
-Maximum Monthly	43	23.4	27.6	13.4	4.3	T					2.1	8.8	18.8	27.6
-Year		1978	1979	1958	1971	1963					1979	1953	1966	FEB 1979
-Maximum in 24 hrs	43	13.2	21.3	10.0	4.3	T					2.1	8.7	14.6	21.3
-Year		1961	1983	1958	1971	1963					1979	1953	1960	FEB 1983
WIND:														
Mean Speed (mph)	45	10.3	11.0	11.4	10.9	9.6	8.7	8.0	7.8	8.2	8.8	9.6	10.0	9.5
Prevailing Direction through 1963		WNW	NW	N	SW	WSW	WSW	WSW	SW	SW	WSW	WSW	WNW	WSW
Fastest Mile														
-Direction	45	NE	NW	NW	SW	SW	NW	SW	E	NE	SW	SW	NW	NW
-Speed (MPH)	45	61	59	56	59	56	73	49	67	49	66	60	47	73
-Year		1958	1956	1955	1958	1957	1958	1980	1955	1960	1954	1958	1958	JUN 1958
Peak Gust														
-Direction	2	W	W	W	NW	NW	NW	W	NW	W	NW	E	W	NW
-Speed (mph)	2	40	41	48	39	67	46	51	28	53	35	35	48	67
-Date		1985	1985	1985	1985	1984	1985	1985	1984	1985	1985	1985	1985	MAY 1984

See Reference Notes to this table on the following page.

TABLE 2 PRECIPITATION (inches) PHILADELPHIA, PENNSYLVANIA

YEAR	JAN	FEB	MAR	APR	MAY	JUNE	JULY	AUG	SEP	OCT	NOV	DEC	ANNUAL
1956	2.30	4.64	4.65	2.68	3.84	3.86	4.61	2.79	3.75	3.47	5.71	3.70	46.00
1957	1.67	2.81	3.24	4.22	1.21	2.41	0.64	3.38	3.10	2.05	2.98	4.49	32.20
1958	3.53	4.64	4.97	4.19	3.65	5.13	5.98	6.20	2.55	3.85	2.05	1.13	47.87
1959	2.03	1.60	3.55	2.25	0.80	5.28	7.48	3.73	1.33	3.41	3.29	3.62	38.37
1960	3.11	3.44	1.96	2.92	3.65	0.71	5.52	3.19	8.78	2.79	1.92	3.16	41.15
1961	3.16	3.13	5.17	4.82	3.38	2.95	5.96	3.42	2.41	1.83	2.04	2.78	41.05
1962	2.95	3.51	3.91	3.69	1.85	7.40	2.30	6.58	2.77	0.95	4.60	2.11	42.62
1963	2.31	2.19	3.94	1.13	1.06	2.88	3.13	3.35	6.44	0.09	6.67	1.76	34.95
1964	3.92	2.83	1.94	5.27	0.47	0.21	3.83	0.49	2.42	1.73	5.13	1.64	29.88
1965	2.35	2.18	3.19	2.33	1.23	2.85	3.22	4.05	3.02	2.02	1.05	1.85	29.34
1966	2.82	4.30	0.68	4.35	2.95	0.41	2.35	1.63	8.70	5.12	2.36	4.33	40.00
1967	1.67	1.82	4.53	2.17	3.49	4.12	7.11	7.08	2.96	2.00	1.99	5.88	44.82
1968	2.90	1.40	4.98	1.57	5.17	5.89	2.00	1.24	0.44	3.15	4.17	2.54	35.45
1969	1.57	1.88	1.92	1.68	3.30	7.31	8.33	2.66	4.38	1.13	1.97	7.23	43.36
1970	0.74	2.08	3.83	6.12	2.57	4.60	2.75	3.99	0.82	3.66	4.71	3.27	39.14
1971	2.13	5.43	2.58	1.84	4.10	1.01	4.84	9.61	5.83	3.84	5.37	1.21	47.79
1972	2.34	5.09	2.69	4.08	4.11	5.79	2.62	3.76	1.12	3.77	9.06	5.20	49.63
1973	3.93	2.96	3.52	6.68	4.14	7.88	2.39	2.03	3.39	2.16	0.64	6.34	46.06
1974	2.95	2.14	4.91	2.77	3.21	4.43	2.08	3.83	4.68	1.93	0.81	4.04	37.78
1975	4.00	2.91	4.68	2.97	4.99	7.57	6.32	2.21	7.21	3.24	3.14	2.89	52.13
1976	4.50	1.66	2.38	2.06	4.35	3.42	4.04	2.17	2.44	4.30	0.32	1.63	33.27
1977	2.61	1.33	4.19	5.59	0.70	5.33	1.47	8.70	3.44	3.11	7.76	5.19	49.42
1978	8.86	1.35	4.31	1.76	6.01	1.75	5.27	6.04	1.59	1.20	2.20	5.61	45.95
1979	8.74	6.44	2.43	4.08	3.98	4.34	3.95	5.95	4.89	3.84	2.48	1.67	52.79
1980	2.27	0.96	7.01	4.79	3.22	1.73	6.58	0.80	2.79	5.03	2.85	0.77	38.80
1981	0.50	2.94	1.61	3.60	4.53	4.40	4.54	5.11	2.83	2.68	0.95	4.14	37.83
1982	4.45	3.16	2.66	6.06	4.47	5.76	1.94	2.20	2.32	1.94	3.67	1.80	40.43
1983	2.81	3.53	6.70	8.12	7.03	2.75	0.68	2.57	3.45	3.69	5.71	7.37	54.41
1984	2.22	2.81	6.14	4.25	6.87	2.85	6.99	3.28	1.96	2.56	1.56	2.17	43.66
1985	1.55	2.44	1.95	0.52	4.99	1.88	4.66	2.82	5.78	1.54	6.09	0.98	35.20
Record Mean	3.22	3.07	3.56	3.36	3.45	3.63	4.09	4.46	3.39	2.80	3.12	3.20	41.35

TABLE 3 AVERAGE TEMPERATURE (deg. F) PHILADELPHIA, PENNSYLVANIA

YEAR	JAN	FEB	MAR	APR	MAY	JUNE	JULY	AUG	SEP	OCT	NOV	DEC	ANNUAL
1956	32.1	37.6	38.8	49.9	60.9	72.3	73.6	74.1	65.2	57.7	46.3	41.7	54.2
1957	29.2	37.4	42.3	54.2	63.5	74.9	76.6	73.2	69.2	54.1	47.8	39.4	55.1
1958	31.8	28.2	39.9	54.2	61.2	67.8	77.4	73.4	65.9	55.7	46.6	29.4	52.6
#1959	31.6	33.3	41.5	54.8	66.4	72.2	75.9	76.5	70.8	60.1	45.2	38.2	55.5
1960	34.2	35.4	32.7	56.7	61.2	70.6	73.3	74.5	67.3	54.8	45.5	27.6	52.8
1961	25.0	34.0	43.1	49.8	58.6	69.9	75.6	73.5	71.5	55.7	45.2	31.0	52.7
1962	30.0	30.4	40.5	52.0	64.1	71.7	72.0	72.0	63.1	56.3	42.1	31.0	52.1
1963	27.5	26.5	42.9	52.5	60.2	70.4	76.0	71.2	62.8	57.1	48.0	27.9	51.9
1964	33.0	31.8	42.7	50.8	65.1	72.4	76.6	72.2	67.2	52.6	47.1	37.5	54.1
1965	29.2	33.3	37.6	49.0	65.5	70.0	74.1	73.1	69.2	53.7	44.2	37.0	53.0
1966	29.1	31.5	42.5	47.8	59.5	72.1	77.9	74.8	65.2	53.1	46.8	35.5	53.0
1967	36.0	29.0	38.5	51.7	55.9	72.1	76.6	75.1	67.0	56.8	42.8	38.5	53.3
1968	28.9	30.4	44.4	54.6	59.7	71.2	77.1	77.8	69.4	58.1	45.6	32.3	54.1
1969	29.8	32.0	39.7	55.3	64.6	73.4	75.1	75.2	67.2	55.0	44.4	33.5	53.8
1970	24.5	33.1	38.3	51.5	64.9	71.6	76.9	76.7	72.0	60.1	48.2	35.8	54.5
1971	27.8	36.1	40.7	51.6	60.9	74.3	75.3	75.3	71.6	63.5	46.1	41.6	55.6
1972	35.1	32.4	40.7	49.7	63.6	68.7	77.1	76.0	69.2	52.7	43.6	39.9	54.1
1973	34.4	33.6	47.2	53.4	60.3	74.6	76.9	78.8	70.7	59.2	48.0	38.6	56.4
1974	35.9	31.7	43.3	55.8	62.4	70.3	76.9	76.8	68.1	54.8	48.5	39.4	55.3
1975	37.3	35.8	41.2	48.7	66.6	72.2	76.6	77.1	66.6	61.2	52.7	36.9	56.1
1976	28.7	40.9	46.3	56.6	62.7	75.2	75.3	74.8	67.3	52.5	39.9	30.3	54.2
1977	20.0	33.6	48.8	57.2	65.8	68.6	77.8	76.2	69.9	54.3	46.4	32.6	54.3
1978	28.0	24.7	39.0	50.6	61.4	72.6	75.6	79.2	68.5	55.5	47.9	38.6	53.5
1979	32.5	23.0	47.0	52.3	66.4	69.1	75.2	75.5	68.5	54.9	50.1	38.2	54.5
1980	31.8	29.7	40.2	54.7	65.4	70.6	78.5	80.0	72.2	54.9	43.2	32.5	54.5
1981	25.3	37.9	40.0	54.7	62.6	72.0	76.9	74.9	66.8	53.1	45.6	34.6	53.7
1982	24.7	34.4	41.7	50.2	65.9	68.7	76.9	73.5	67.6	56.9	48.4	41.3	54.2
1983	34.1	34.0	43.7	51.0	62.1	72.0	77.9	77.1	69.0	56.6	46.7	33.2	54.8
1984	26.2	38.7	35.5	50.2	60.2	72.0	73.9	75.2	64.7	61.2	44.4	41.9	53.8
1985	27.3	35.3	44.6	55.5	64.5	68.8	75.4	74.1	69.1	59.3	51.3	33.3	54.9
Record Mean	32.5	33.7	41.7	52.3	63.0	71.7	76.6	74.8	68.4	57.3	46.3	36.1	54.6
Max	39.6	41.2	50.0	61.7	72.6	81.0	85.4	83.2	77.0	66.0	53.9	43.0	62.9
Min	25.5	26.2	33.4	42.9	53.4	62.4	67.8	66.4	59.8	48.6	38.6	29.2	46.2

REFERENCE NOTES FOR TABLES 1, 2, 3 and 6 **(PHILADELPHIA, PA)**

GENERAL

T - TRACE AMOUNT
BLANK ENTRIES DENOTE MISSING/UNREPORTED DATA.
# INDICATES A STATION OR INSTRUMENT RELOCATION.

SPECIFIC

TABLE 1

(a) - LENGTH OF RECORD IN YEARS. ALTHOUGH
INDIVIDUAL MONTHS MAY BE MISSING.
* LESS THAN .05

NORMALS — BASED ON THE 1951-1980 RECORD PERIOD.
EXTREMES — DATES ARE THE MOST RECENT OCCURRENCE.
WIND DIR. — NUMERALS SHOW TENS OF DEGREES
CLOCKWISE FROM TRUE NORTH.
"00" INDICATES CALM.
RESULTANT WIND DIRECTIONS ARE GIVEN TO WHOLE DEGREES.

EXCEPTIONS

TABLES 2, 3, and 6

RECORD MEANS ARE THROUGH THE CURRENT YEAR,
BEGINNING IN 1874 FOR TEMPERATURE
1872 FOR PRECIPITATION
1943 FOR SNOWFALL

TABLE 4   HEATING DEGREE DAYS Base 65 deg. F   PHILADELPHIA, PENNSYLVANIA

SEASON	JULY	AUG	SEP	OCT	NOV	DEC	JAN	FEB	MAR	APR	MAY	JUNE	TOTAL
1956-57	2	2	107	222	556	718	1103	765	695	346	126	15	4657
1957-58	0	0	64	335	510	786	1023	1024	770	328	136	27	5003
1958-59	0	0	64	296	547	1096	1030	884	721	311	91	24	5064
#1959-60	0	0	46	246	589	823	947	850	997	274	129	5	4906
1960-61	0	0	25	312	577	1150	1232	862	672	452	214	15	5511
1961-62	0	0	45	283	593	1049	1078	963	748	408	133	7	5307
1962-63	0	4	109	272	681	1048	1159	1072	680	375	175	12	5587
1963-64	0	7	118	242	502	1144	985	955	685	424	76	13	5151
1964-65	0	2	51	377	532	847	1107	883	839	475	66	26	5205
1965-66	0	18	41	342	614	862	1110	931	693	509	207	21	5348
1966-67	0	0	83	362	538	908	893	1001	817	396	280	6	5284
1967-68	0	0	55	271	660	814	1112	995	633	305	170	7	5022
1968-69	0	0	14	234	576	1008	1084	918	782	290	84	2	4992
1969-70	0	0	54	316	611	970	1247	890	821	399	92	0	5400
1970-71	0	0	29	191	499	899	1145	802	746	394	140	3	4848
1971-72	0	0	17	79	576	719	920	941	748	450	86	26	4562
1972-73	0	0	22	378	635	775	940	874	547	359	176	1	4707
1973-74	0	0	18	196	507	810	897	926	667	292	128	11	4452
1974-75	0	0	46	313	500	786	852	812	732	483	66	4	4594
1975-76	0	0	45	152	372	866	1120	692	572	307	119	13	4258
1976-77	0	2	42	387	743	1069	1390	873	505	258	73	36	5378
1977-78	0	0	24	328	558	998	1139	1121	797	423	161	10	5559
1978-79	5	0	41	296	507	811	999	1170	556	378	38	17	4818
1979-80	4	7	28	324	439	823	1021	1016	763	301	72	17	4815
1980-81	0	0	22	320	646	999	1222	752	768	309	129	4	5171
1981-82	0	0	58	364	576	936	1243	850	714	440	50	25	5256
1982-83	0	8	31	277	497	730	951	861	653	423	128	2	4561
1983-84	0	0	70	283	540	981	1196	756	911	438	181	13	5369
1984-85	0	0	92	138	613	709	1161	824	627	306	89	9	4568
1985-86	0	0	38	187	407	975							

TABLE 5   COOLING DEGREE DAYS Base 65 deg. F   PHILADELPHIA, PENNSYLVANIA

YEAR	JAN	FEB	MAR	APR	MAY	JUNE	JULY	AUG	SEP	OCT	NOV	DEC	TOTAL
1969	0	0	0	9	77	259	319	323	126	15	0	0	1128
1970	0	0	0	3	100	204	376	367	247	46	0	0	1343
1971	0	0	0	0	19	292	394	326	223	37	14	0	1305
1972	0	0	3	0	47	143	381	344	153	3	0	0	1074
1973	0	0	0	16	35	294	404	435	193	23	0	0	1400
1974	0	0	0	24	55	179	373	372	145	5	12	0	1165
1975	0	0	0	0	0	121	224	366	380	98	42	12	1243
1976	0	0	0	64	58	326	326	315	115	7	0	0	1211
1977	0	0	10	32	104	150	402	355	175	3	6	0	1237
1978	0	0	0	0	57	244	338	447	153	8	0	0	1247
1979	0	0	6	5	90	146	357	339	137	16	1	0	1097
1980	0	0	0	0	89	194	428	470	244	10	0	0	1435
1981	0	0	0	9	62	224	373	315	119	1	0	0	1103
1982	0	0	0	3	85	142	376	280	115	31	5	0	1037
1983	0	0	11	43	217	409	380	199	27	0	0	0	1286
1984	0	0	0	0	39	260	283	324	90	30	0	0	1026
1985	0	0	0	27	81	133	330	291	166	19	0	0	1047

TABLE 6   SNOWFALL (inches)   PHILADELPHIA, PENNSYLVANIA

SEASON	JULY	AUG	SEP	OCT	NOV	DEC	JAN	FEB	MAR	APR	MAY	JUNE	TOTAL
1956-57	0.0	0.0	0.0	0.0	T	0.2	4.7	1.8	1.2	T	0.0	0.0	7.9
1957-58	0.0	0.0	0.0	T	T	7.8	3.7	16.9	13.4	T	0.0	0.0	41.8
1958-59	0.0	0.0	0.0	0.0	T	0.3	3.3	T	1.5	T	0.0	0.0	5.1
1959-60	0.0	0.0	0.0	0.0	T	5.7	0.8	3.1	12.2	T	0.0	0.0	21.8
1960-61	0.0	0.0	0.0	0.0	0.0	17.5	19.7	11.8	0.1	T	0.0	0.0	49.1
1961-62	0.0	0.0	0.0	0.0	3.2	5.2	1.1	12.5	7.2	T	0.0	0.0	29.2
1962-63	0.0	0.0	0.0	T	T	9.5	6.1	4.7	0.2	0.0	T	0.0	20.5
1963-64	0.0	0.0	0.0	0.0	T	8.0	7.4	12.4	5.1	T	0.0	0.0	32.9
1964-65	0.0	0.0	0.0	0.0	T	2.6	11.9	2.2	6.5	3.0	0.0	0.0	26.2
1965-66	0.0	0.0	0.0	T	0.0	T	16.0	11.4	T	T	0.0	0.0	27.4
1966-67	0.0	0.0	0.0	0.0	T	18.8	0.6	18.4	6.4	0.1	0.0	0.0	44.3
1967-68	0.0	0.0	0.0	0.0	4.9	5.6	1.5	1.7	2.2	0.0	0.0	0.0	15.9
1968-69	0.0	0.0	0.0	0.0	0.4	3.1	1.9	9.5	8.8	0.0	0.0	0.0	23.7
1969-70	0.0	0.0	0.0	0.0	0.2	7.5	7.5	2.7	2.4	T	0.0	0.0	20.3
1970-71	0.0	0.0	0.0	T	0.0	1.1	7.7	0.8	4.4	4.3	0.0	0.0	18.3
1971-72	0.0	0.0	0.0	0.0	T	0.1	3.2	8.2	0.3	0.4	0.0	0.0	12.2
1972-73	0.0	0.0	0.0	T	T	T	T	T	T	T	0.0	0.0	T
1973-74	0.0	0.0	0.0	0.0	T	4.6	4.1	12.1	T	T	0.0	0.0	20.8
1974-75	0.0	0.0	0.0	0.0	T	0.8	3.9	6.6	2.3	T	0.0	0.0	13.6
1975-76	0.0	0.0	0.0	0.0	0.0	1.1	6.4	3.1	6.9	0.0	0.0	0.0	17.5
1976-77	0.0	0.0	0.0	0.0	T	2.8	15.7	0.2	T	0.0	0.0	0.0	18.7
1977-78	0.0	0.0	0.0	0.0	0.2	0.2	23.4	19.0	12.1	T	0.0	0.0	54.9
1978-79	0.0	0.0	0.0	0.0	2.5	T	10.1	27.6	T	T	0.0	0.0	40.2
1979-80	0.0	0.0	0.0	2.1	T	4.9	6.1	0.4	7.4	T	0.0	0.0	20.9
1980-81	0.0	0.0	0.0	0.0	0.2	1.4	5.0	T	8.8	0.0	0.0	0.0	15.4
1981-82	0.0	0.0	0.0	0.0	T	2.8	14.0	3.5	1.1	4.0	0.0	0.0	25.4
1982-83	0.0	0.0	0.0	0.0	0.0	6.8	0.2	26.1	0.9	1.9	0.0	0.0	35.9
1983-84	0.0	0.0	0.0	0.0	0.8	T	10.5	T	10.3	T	0.0	0.0	21.6
1984-85	0.0	0.0	0.0	0.0	T	0.2	11.9	4.4	T	T	0.0	0.0	16.5
1985-86	0.0	0.0	0.0	0.0	0.0	1.5							
Record Mean	0.0	0.0	0.0	T	0.6	3.7	6.5	6.7	3.9	0.3	T	0.0	21.8

See Reference Notes, relative to all above tables, on preceding page.

Pittsburgh lies at the foothills of the Allegheny Mountains at the confluence of the Allegheny and Monongahela Rivers which form the Ohio. The city is a little over 100 miles southeast of Lake Erie. It has a humid continental type of climate modified only slightly by its nearness to the Atlantic Seaboard and the Great Lakes.

The predominant winter air masses influencing the climate of Pittsburgh have a polar continental source in Canada and move in from the Hudson Bay region or the Canadian Rockies. During the summer, frequent invasions of air from the Gulf of Mexico bring warm humid weather. Occasionally, Gulf air reaches as far north as Pittsburgh during the winter and produces intermittent periods of thawing. The last spring temperature of 32 degrees usually occurs in late April and the first in late October. The average growing season is about 180 days. There is a wide variation in the time of the first and last frosts over a radius of 25 miles from the center of Pittsburgh due to terrain differences.

Precipitation is distributed well throughout the year. During the winter months about a fourth of the precipitation occurs as snow and there is about a 50 percent chance of measurable precipitation on any day. Thunderstorms occur normally during all months, except midwinter, and have a maximum frequency in midsummer. The first appreciable snowfall generally occurs in late November and usually the last occurs early in April. Snow lies on the ground in the suburbs on an average of about 33 days during the year.

Seven months of the year, April through October, have sunshine more than 50 percent of the possible time. During the remaining five months cloudiness is heavier because the track of migratory storms from west to east is closer to the area and because of the frequent periods of cloudy, showery weather associated with northwest winds from across the Great Lakes. Cold air drainage induced by the many hills leads to the frequent formation of early morning fog which may be quite persistent in the river valleys during the colder months.

Rising of the tributary streams cause occasional flooding at Pittsburgh. Serious inconvenience is occasioned by the Ohio River reaching the flood stage of 25 feet about once each year. Significant flooding, or a 30-foot stage, occurs about once each three years.

TABLE 1

NORMALS, MEANS AND EXTREMES

PITTSBURGH, GRTR. PITT. AIRPORT PENNSYLVANIA

LATITUDE: 40°30'N   LONGITUDE: 80°13'W   ELEVATION: FT. GRND 1137 BARO 01213   TIME ZONE: EASTERN   WBAN: 94823

	(a)	JAN	FEB	MAR	APR	MAY	JUNE	JULY	AUG	SEP	OCT	NOV	DEC	YEAR
TEMPERATURE °F:														
Normals														
-Daily Maximum		34.1	36.8	47.6	60.7	70.8	79.1	82.7	81.1	74.8	62.9	49.8	38.4	59.9
-Daily Minimum		19.2	20.7	29.4	39.4	48.5	57.1	61.3	60.1	53.3	42.1	33.3	24.3	40.7
-Monthly		26.6	28.8	38.5	50.0	59.7	68.1	72.0	70.6	64.1	52.5	41.5	31.4	50.3
Extremes														
-Record Highest	33	69	69	80	87	91	96	99	97	97	87	82	74	99
-Year		1985	1954	1977	1970	1962	1971	1954	1953	1954	1959	1961	1982	JUL 1954
-Record Lowest	33	-18	-12	-1	14	26	34	42	39	31	16	-1	-12	-18
-Year		1985	1979	1980	1982	1970	1972	1963	1982	1959	1965	1958	1983	JAN 1985
NORMAL DEGREE DAYS:														
Heating (base 65°F)		1187	1014	822	447	201	28	0	13	101	393	702	1042	5950
Cooling (base 65°F)		0	0	0	0	37	121	222	186	74	5	0	0	645
% OF POSSIBLE SUNSHINE	33	33	38	44	48	52	57	59	56	58	52	38	29	47
MEAN SKY COVER (tenths)														
Sunrise - Sunset	33	8.1	7.8	7.6	7.2	6.9	6.4	6.3	6.3	6.1	6.3	7.7	8.2	7.1
MEAN NUMBER OF DAYS:														
Sunrise to Sunset														
-Clear	33	2.9	3.4	4.1	4.5	5.2	5.2	5.3	6.3	7.7	7.9	3.9	2.6	59.0
-Partly Cloudy	33	6.0	5.8	6.8	8.2	9.2	11.6	13.2	11.8	10.3	8.5	6.2	5.8	103.4
-Cloudy	33	22.1	19.0	20.1	17.4	16.6	13.2	12.5	12.9	12.0	14.6	19.9	22.6	202.9
Precipitation														
.01 inches or more	33	16.5	14.1	16.1	13.5	12.3	11.5	10.8	9.8	9.2	10.6	13.2	16.5	154.1
Snow, Ice pellets														
1.0 inches or more	33	3.9	3.0	2.3	0.4	0.1	0.0	0.0	0.0	0.0	0.1	1.1	2.7	13.6
Thunderstorms	33	0.1	0.3	1.8	3.3	5.0	6.7	7.0	5.6	3.1	1.3	0.6	0.3	35.3
Heavy Fog Visibility														
1/4 mile or less	33	1.4	1.2	1.0	0.9	1.2	1.1	1.7	2.1	2.4	1.8	1.5	1.7	17.8
Temperature °F														
-Maximum														
90° and above	26	0.0	0.0	0.0	0.0	0.2	1.4	2.5	1.2	0.7	0.0	0.0	0.0	6.1
32° and below	26	15.5	11.1	4.0	0.1	0.0	0.0	0.0	0.0	0.0	0.0	1.4	10.4	42.5
-Minimum														
32° and below	26	27.5	24.2	19.8	8.4	1.0	0.0	0.0	0.0	0.0	3.9	14.4	24.7	124.0
0° and below	26	2.8	1.7	0.1	0.0	0.0	0.0	0.0	0.0	0.0	0.0	0.*	0.8	5.4
AVG. STATION PRESS. (mb)	13	973.0	973.4	971.3	971.4	971.1	972.4	973.5	974.8	975.1	975.4	974.2	973.7	973.3
RELATIVE HUMIDITY (%)														
Hour 01	25	72	70	69	66	72	77	80	82	81	76	75	74	75
Hour 07 (Local Time)	25	75	74	74	72	75	78	82	85	85	81	78	76	78
Hour 13	25	65	62	58	50	52	52	53	56	56	54	62	67	57
Hour 19	25	67	63	59	52	54	57	57	59	65	62	68	70	62
PRECIPITATION (inches):														
Water Equivalent														
-Normal		2.86	2.40	3.58	3.28	3.54	3.30	3.83	3.31	2.80	2.49	2.34	2.57	36.30
-Maximum Monthly	33	6.25	5.98	6.10	7.61	6.36	8.20	7.43	7.56	5.42	8.20	11.05	5.24	11.05
-Year		1978	1956	1967	1964	1968	1981	1958	1975	1972	1954	1985	1978	NOV 1985
-Minimum Monthly	33	0.77	0.51	1.14	0.48	1.21	0.90	1.82	0.78	0.28	0.16	0.90	0.40	0.16
-Year		1981	1969	1969	1971	1965	1967	1965	1957	1985	1963	1976	1955	OCT 1963
-Maximum in 24 hrs	33	1.47	2.30	2.00	2.15	2.44	1.93	2.97	3.06	2.25	3.56	1.97	1.76	3.56
-Year		1982	1975	1964	1964	1971	1955	1971	1956	1975	1954	1985	1978	OCT 1954
Snow, Ice pellets														
-Maximum Monthly	33	40.2	24.2	21.3	7.2	3.1					1.8	11.0	21.2	40.2
-Year		1978	1972	1960	1985	1966					1972	1958	1974	JAN 1978
-Maximum in 24 hrs	33	14.0	12.3	14.7	4.8	3.1					1.8	10.5	12.5	14.7
-Year		1966	1960	1962	1985	1966					1972	1958	1974	MAR 1962
WIND:														
Mean Speed (mph)	33	10.7	10.6	10.9	10.5	8.9	8.0	7.3	6.9	7.4	8.4	9.8	10.4	9.1
Prevailing Direction through 1963		WSW	WSW	WSW	WSW	WSW	WSW	WSW	WSW	WSW	WSW	WSW	WSW	WSW
Fastest Obs. 1 Min.														
-Direction	33	23	26	25	27	25	27	25	29	02	27	29	25	26
-Speed (MPH)	33	52	58	48	46	42	40	51	46	32	35	45	48	58
-Year		1978	1967	1954	1974	1957	1957	1956	1963	1960	1959	1969	1968	FEB 1967
Peak Gust														
-Direction														
-Speed (mph)														
-Date														

See Reference Notes to this table on the following page.

TABLE 2 PRECIPITATION (inches) PITTSBURGH, GRTR. PITT. AIRPORT PENNSYLVANIA

YEAR	JAN	FEB	MAR	APR	MAY	JUNE	JULY	AUG	SEP	OCT	NOV	DEC	ANNUAL
1956	1.90	5.98	5.28	4.31	5.90	4.19	4.25	5.07	1.93	1.50	1.03	3.35	44.69
1957	1.65	1.35	2.02	4.58	2.73	4.07	3.97	0.78	4.06	1.74	2.50	4.22	33.67
1958	3.17	1.11	1.87	3.42	4.82	2.74	7.43	3.71	4.52	0.97	2.47	1.10	37.33
1959	3.99	2.15	2.11	3.33	2.56	3.70	4.25	4.04	1.34	5.94	2.43	2.78	38.62
1960	3.01	3.16	2.06	1.37	5.62	2.72	3.46	3.55	1.84	1.64	1.22	1.64	31.29
1961	1.95	3.13	3.48	5.21	2.80	4.21	5.53	2.11	1.98	2.58	3.41	1.71	38.10
1962	2.33	3.55	3.85	3.03	1.87	1.82	2.44	2.57	4.69	2.11	1.53	1.83	31.62
1963	1.96	2.09	5.28	2.39	1.57	2.40	3.45	2.31	1.40	0.16	2.54	1.24	26.79
1964	2.55	1.73	4.96	7.61	1.77	3.84	4.48	1.79	0.74	1.42	2.74	4.26	37.89
1965	3.84	2.98	3.16	1.79	1.21	2.31	1.82	3.26	4.07	2.82	2.35	0.63	30.24
1966	4.52	3.23	1.88	3.73	2.76	1.72	2.70	5.13	1.92	1.38	3.39	1.70	34.06
1967	1.06	2.54	6.10	4.41	5.21	0.90	4.54	2.67	2.05	2.05	3.07	2.22	36.38
1968	2.83	0.79	4.53	2.33	6.36	2.38	2.36	3.97	3.08	2.13	2.07	3.24	36.07
1969	2.02	0.51	1.14	2.91	1.89	3.74	4.52	2.96	0.91	2.59	2.44	3.95	29.58
1970	1.61	1.92	3.35	3.09	4.36	4.61	3.89	1.55	2.77	4.80	2.64	3.29	37.88
1971	2.29	4.04	3.20	0.48	3.87	1.41	6.82	1.23	3.86	0.84	1.94	3.24	33.22
1972	1.84	3.64	3.68	4.37	1.38	5.08	2.98	1.79	5.42	2.15	4.70	3.04	40.07
1973	2.03	1.80	3.86	4.69	5.87	3.12	2.16	3.40	3.56	4.45	2.65	2.15	39.74
1974	3.47	2.10	3.72	3.26	5.35	5.08	3.30	2.93	4.42	1.12	3.06	4.02	41.83
1975	3.34	4.64	4.62	2.27	1.84	4.58	4.48	7.56	5.06	3.46	1.77	2.90	46.42
1976	3.25	1.74	4.45	1.24	1.99	3.37	4.72	1.25	3.30	3.76	0.90	1.81	31.78
1977	2.06	0.87	4.12	3.26	2.57	2.85	3.38	2.66	3.13	2.44	2.59	3.27	33.20
1978	6.25	0.54	1.65	2.25	4.26	4.11	2.15	3.65	2.64	3.42	1.62	5.24	37.78
1979	4.80	3.12	1.32	3.17	4.49	1.73	4.31	6.84	3.60	2.46	2.43	2.29	40.56
1980	1.56	1.32	5.65	2.94	4.32	4.34	6.76	5.10	1.29	2.42	2.38	1.38	39.46
1981	0.77	4.20	2.12	4.92	2.04	8.20	3.82	0.98	4.13	1.82	1.50	3.00	37.50
1982	4.44	1.93	3.52	1.44	3.98	3.05	2.36	1.97	2.80	0.40	3.33	2.79	32.01
1983	1.19	1.58	3.50	4.33	5.24	4.82	3.32	3.13	2.42	3.67	3.94	4.27	41.41
1984	1.40	2.05	2.32	3.72	5.22	1.98	3.01	5.15	0.84	3.45	3.14	3.04	35.32
1985	1.43	1.45	3.37	1.64	5.80	2.26	4.06	2.64	0.28	2.27	11.05	2.26	38.51
Record Mean	2.88	2.45	3.29	3.09	3.38	3.70	3.97	3.20	2.62	2.49	2.45	2.73	36.26

TABLE 3 AVERAGE TEMPERATURE (deg. F) PITTSBURGH, GRTR. PITT. AIRPORT PENNSYLVANIA

YEAR	JAN	FEB	MAR	APR	MAY	JUNE	JULY	AUG	SEP	OCT	NOV	DEC	ANNUAL
1956	27.7	34.0	37.2	47.0	58.5	67.7	70.3	70.8	60.4	56.7	42.0	39.4	51.0
1957	25.0	34.6	39.7	52.7	60.9	70.4	71.8	70.0	64.8	49.6	42.7	35.3	51.5
1958	27.4	22.7	36.1	51.2	58.6	64.2	73.0	69.6	63.1	52.3	44.0	23.0	48.8
#1959	25.3	31.8	37.0	51.3	63.9	68.6	72.7	74.9	68.3	53.8	34.8	51.8	51.8
1960	30.7	28.7	26.0	54.0	57.5	66.5	68.6	71.2	66.1	53.5	43.1	23.4	49.1
1961	22.2	32.3	41.3	44.0	55.2	65.1	70.5	71.2	68.5	55.3	42.8	31.3	50.0
1962	26.2	28.3	36.5	48.4	65.3	69.3	70.1	70.8	58.6	53.3	41.1	24.1	49.4
1963	21.1	19.3	40.7	49.0	56.5	67.2	70.8	67.7	61.3	58.8	43.7	22.4	48.2
1964	31.4	27.0	40.0	51.7	62.7	67.9	72.3	67.1	63.7	50.4	45.5	34.0	51.1
1965	28.2	28.4	35.2	49.0	65.9	66.9	69.9	69.1	64.7	48.1	41.3	37.5	50.4
1966	23.1	30.3	40.9	47.9	56.1	70.4	75.6	71.1	61.3	50.8	42.8	31.4	50.1
1967	32.3	25.6	40.2	52.2	54.3	73.0	71.5	68.8	61.1	52.5	36.8	34.8	50.3
1968	23.4	22.2	40.4	51.2	54.7	66.9	72.4	71.8	64.8	52.2	41.3	27.6	49.1
1969	26.7	29.5	34.3	51.7	60.2	69.3	72.7	69.7	63.0	52.9	39.2	26.7	49.7
1970	20.7	27.7	35.5	52.5	63.9	68.2	71.6	71.6	67.8	54.9	42.2	32.1	50.7
1971	23.7	30.4	34.3	46.0	56.6	71.4	70.2	69.6	59.5	40.4	38.8	30.8	50.8
1972	29.6	26.5	36.4	48.5	61.8	63.8	71.2	70.6	65.3	48.4	39.3	37.2	49.9
1973	29.7	28.8	48.3	49.3	56.4	70.9	73.2	73.2	66.5	56.1	44.1	33.3	52.5
1974	34.0	29.9	41.2	51.8	58.3	65.2	73.1	72.8	62.2	52.4	43.9	32.5	51.4
1975	32.6	32.1	36.3	44.3	63.0	67.8	72.8	73.0	58.8	53.3	46.3	32.9	51.1
1976	23.5	37.2	45.2	50.6	55.6	68.4	67.4	65.3	59.9	45.9	33.1	23.9	48.0
1977	11.4	26.9	43.7	50.8	63.0	63.8	71.8	68.1	64.7	50.5	45.6	31.1	49.3
1978	22.6	20.9	36.9	51.0	60.2	69.4	73.0	71.4	66.2	49.1	43.0	32.7	49.7
1979	21.4	18.0	43.1	49.7	59.1	67.7	70.3	69.6	63.4	50.9	44.7	34.6	49.4
1980	26.9	24.2	35.6	48.1	60.3	66.2	75.0	74.5	67.1	49.5	38.6	28.6	49.5
1981	20.5	31.4	35.6	51.9	58.4	68.8	72.1	69.7	61.9	49.4	40.3	29.4	49.1
1982	20.9	28.4	38.4	45.3	64.7	63.7	72.4	68.2	63.4	54.4	41.7	39.9	50.4
1983	30.0	32.6	40.7	47.1	55.8	67.8	73.0	72.8	64.4	53.0	43.5	25.4	50.5
1984	23.2	36.4	32.2	49.2	55.3	69.7	68.5	70.8	61.4	55.8	40.2	39.3	50.4
1985	22.1	27.7	42.1	55.0	60.6	64.2	70.5	69.6	65.3	55.2	47.1	27.4	50.6
Record Mean	29.9	31.0	39.9	51.0	61.8	70.3	74.3	72.5	66.3	54.9	43.1	33.4	52.4
Max	37.5	39.2	48.9	61.1	72.3	80.5	84.3	82.4	76.4	64.6	50.9	40.4	61.5
Min	22.3	22.8	30.8	40.8	51.2	60.0	64.2	62.6	56.3	45.2	35.3	26.4	43.2

REFERENCE NOTES FOR TABLES 1, 2, 3 and 6   (PITTSBURGH, PA)

GENERAL

T - TRACE AMOUNT
BLANK ENTRIES DENOTE MISSING/UNREPORTED DATA.
# INDICATES A STATION OR INSTRUMENT RELOCATION.

SPECIFIC

TABLE 1

(a) - LENGTH OF RECORD IN YEARS. ALTHOUGH
INDIVIDUAL MONTHS MAY BE MISSING.
  * LESS THAN .05

NORMALS — BASED ON THE 1951-1980 RECORD PERIOD.
EXTREMES — DATES ARE THE MOST RECENT OCCURRENCE.
WIND DIR. — NUMERALS SHOW TENS OF DEGREES
        CLOCKWISE FROM TRUE NORTH.
        "00" INDICATES CALM.
RESULTANT WIND DIRECTIONS ARE GIVEN TO WHOLE DEGREES.

EXCEPTIONS

TABLE 1

1. TEMPERATURE DATA MAY BE SUSPECT NOVEMBER 1977
    THROUGH JULY 1978 DUE TO INTERMITTENT INSTRUMENT
    MALFUNCTION.

TABLES 2, 3, and 6

RECORD MEANS ARE THROUGH THE CURRENT YEAR,
BEGINNING IN   1875 FOR TEMPERATURE
        1872 FOR PRECIPITATION
        1953 FOR SNOWFALL

TABLE 4 HEATING DEGREE DAYS Base 65 deg. F   PITTSBURGH, GRTR. PITT. AIRPORT PENNSYLVANIA

SEASON	JULY	AUG	SEP	OCT	NOV	DEC	JAN	FEB	MAR	APR	MAY	JUNE	TOTAL
1956-57	4	19	180	250	683	785	1234	843	778	409	173	22	5380
1957-58	4	10	101	469	662	912	1159	1180	888	415	207	83	6090
#1958-59	4	20	117	393	625	1295	1224	924	861	407	130	54	6054
1959-60	1	0	73	374	776	927	1055	1045	1203	357	250	32	6093
1960-61	15	0	47	352	648	1284	1321	907	729	627	311	73	6314
1961-62	17	2	71	302	666	1039	1197	1020	873	513	100	18	5818
1962-63	11	12	216	365	707	1263	1354	1273	747	479	271	45	6743
1963-64	21	22	139	196	634	1315	1035	1095	769	395	116	63	5800
1964-65	1	51	99	447	577	954	1134	1018	920	476	63	53	5793
1965-66	9	40	99	518	702	848	1293	963	741	510	289	34	6046
1966-67	2	6	156	435	659	1035	1007	1097	765	387	332	4	5885
1967-68	10	13	146	391	840	931	1284	1232	758	406	313	60	6384
1968-69	8	31	54	400	703	1152	1181	988	944	394	182	35	6072
1969-70	0	8	127	383	770	1183	1370	1039	908	390	127	31	6336
1970-71	5	1	69	318	678	1013	1277	961	949	562	264	6	6103
1971-72	1	6	41	184	729	807	1093	1112	881	489	128	96	5567
1972-73	20	11	63	508	767	853	1087	1006	508	474	264	2	5563
1973-74	2	8	55	274	621	729	957	978	729	403	223	54	5282
1974-75	0	0	124	384	630	1001	997	916	881	617	116	48	5714
1975-76	0	0	192	362	554	989	1278	801	605	453	301	24	5559
1976-77	15	59	159	587	953	1268	1655	1060	658	436	138	102	7090
1977-78	11	41	78	442	583	1043	1307	1229	860	412	209	38	6253
1978-79	4	3	80	485	656	993	1346	1311	671	458	219	38	6264
1979-80	23	26	111	438	601	935	1175	1177	906	500	172	71	6135
1980-81	0	5	48	476	787	1117	1372	936	904	391	223	18	6277
1981-82	3	10	159	475	736	1098	1361	1017	819	586	82	67	6413
1982-83	9	23	119	336	605	770	1080	904	746	535	280	44	5451
1983-84	10	2	126	365	639	1223	1293	823	1008	471	305	16	6281
1984-85	12	7	165	214	734	790	1322	1038	701	334	163	65	5545
1985-86	3	9	116	300	531	1160							

TABLE 5 COOLING DEGREE DAYS Base 65 deg. F   PITTSBURGH, GRTR. PITT. AIRPORT PENNSYLVANIA

YEAR	JAN	FEB	MAR	APR	MAY	JUNE	JULY	AUG	SEP	OCT	NOV	DEC	TOTAL
1969	0	0	0	2	42	170	245	162	72	14	0	0	707
1970	0	0	0	21	100	133	215	213	162	11	0	0	855
1971	0	0	0	0	13	204	171	158	153	17	0	0	716
1972	0	0	0	0	34	68	219	192	76	0	0	0	589
1973	0	0	0	10	5	185	264	269	108	5	0	0	846
1974	0	0	0	13	19	66	258	247	45	5	4	0	657
1975	0	0	0	0	0	60	137	248	257	12	7	0	721
1976	0	0	1	25	14	134	99	73	12	0	0	0	358
1977	0	0	3	14	83	75	231	141	72	0	4	0	623
1978	0	0	0	0	69	178	260	207	122	0	0	0	836
1979	0	0	0	9	41	125	193	175	70	7	0	0	620
1980	0	0	0	0	34	115	317	306	118	0	0	0	890
1981	0	0	0	5	25	139	230	160	72	0	0	0	631
1982	0	0	0	0	79	33	246	127	77	15	3	0	580
1983	0	0	0	3	3	135	263	251	115	0	0	0	770
1984	0	0	0	3	12	165	127	194	63	13	0	0	577
1985	0	0	0	41	33	49	181	160	130	4	0	0	598

TABLE 6 SNOWFALL (inches)   PITTSBURGH, GRTR. PITT. AIRPORT PENNSYLVANIA

SEASON	JULY	AUG	SEP	OCT	NOV	DEC	JAN	FEB	MAR	APR	MAY	JUNE	TOTAL
1956-57	0.0	0.0	0.0	0.0	3.4	11.2	9.3	4.3	7.3	2.2	0.0	0.0	37.7
1957-58	0.0	0.0	0.0	1.5	1.0	4.9	12.9	7.4	8.7	1.5	0.0	0.0	37.9
1958-59	0.0	0.0	0.0	0.0	11.0	8.0	13.7	2.2	7.6	3.1	0.0	0.0	45.6
1959-60	0.0	0.0	0.0	0.0	4.3	11.5	2.2	21.8	21.3	1.1	0.0	0.0	62.2
1960-61	0.0	0.0	0.0	0.1	2.6	20.8	22.7	22.5	1.4	5.9	T	0.0	76.0
1961-62	0.0	0.0	0.0	T	2.2	5.6	4.0	8.6	19.1	3.6	0.0	0.0	43.1
1962-63	0.0	0.0	0.0	1.8	T	11.9	12.7	20.4	4.5	0.3	1.8	0.0	53.4
1963-64	0.0	0.0	0.0	T	5.8	16.4	20.3	13.7	6.1	0.3	0.0	0.0	62.6
1964-65	0.0	0.0	0.0	T	1.6	6.1	6.1	10.6	10.4	13.3	0.2	0.0	42.2
1965-66	0.0	0.0	0.0	0.2	0.2	1.8	24.6	6.9	8.5	2.7	3.1	0.0	48.0
1966-67	0.0	0.0	0.0	0.0	5.1	7.8	4.5	21.7	20.0	0.5	0.0	0.0	59.6
1967-68	0.0	0.0	0.0	T	10.1	7.9	15.4	6.1	11.0	T	0.0	0.0	50.5
1968-69	0.0	0.0	0.0	T	2.7	13.3	6.5	4.0	3.9	0.0	T	0.0	30.4
1969-70	0.0	0.0	0.0	0.4	7.9	20.6	12.6	13.0	16.1	0.1	0.0	0.0	70.7
1970-71	0.0	0.0	0.0	T	0.1	10.1	12.1	20.6	16.8	0.2	0.0	0.0	59.9
1971-72	0.0	0.0	0.0	0.0	10.5	0.7	4.9	24.2	9.8	1.8	0.0	0.0	51.9
1972-73	0.0	0.0	0.0	1.8	6.1	2.9	3.4	6.1	4.6	1.4	T	0.0	26.3
1973-74	0.0	0.0	0.0	0.0	0.8	4.8	4.9	2.2	2.3	1.6	T	0.0	16.6
1974-75	0.0	0.0	0.0	T	2.6	21.2	10.1	13.9	9.8	1.1	0.0	0.0	58.7
1975-76.	0.0	0.0	0.0	0.0	1.9	3.8	21.8	3.3	4.3	0.5	0.0	0.0	35.6
1976-77	0.0	0.0	0.0	T	6.6	7.9	26.5	6.4	0.9	1.3	T	0.0	49.6
1977-78	0.0	0.0	0.0	T	3.3	9.1	40.2	5.4	4.0	0.2	0.0	0.0	62.2
1978-79	0.0	0.0	0.0	0.0	2.3	3.2	18.2	13.7	2.0	1.4	0.0	0.0	40.8
1979-80	0.0	0.0	0.0	T	1.1	1.1	7.8	6.2	7.9	T	0.0	0.0	24.1
1980-81	0.0	0.0	0.0	T	9.7	6.3	12.5	11.9	7.6	T	0.0	0.0	48.0
1981-82	0.0	0.0	0.0	T	0.6	11.5	13.4	3.6	12.2	3.8	0.0	0.0	45.1
1982-83	0.0	0.0	0.0	T	0.1	8.8	3.9	12.0	4.3	1.0	0.0	0.0	30.1
1983-84	0.0	0.0	0.0	0.0	6.1	10.5	10.8	11.4	10.4	T	0.0	0.0	49.2
1984-85	0.0	0.0	0.0	0.0	1.5	4.8	14.6	8.1	0.2	7.2	0.0	0.0	36.4
1985-86	0.0	0.0	0.0	0.0	T	15.3							
Record Mean	0.0	0.0	0.0	0.2	3.7	8.5	12.2	10.0	8.4	1.6	0.1	0.0	44.7

See Reference Notes, relative to all above tables, on preceding page.

The Wilkes-Barre Scranton National Weather Service Office is located about midway between the two cities, at the southwest end of the crescent-shaped Lackawanna River Valley. The river flows through this valley and empties into the Susquehanna River and the Wyoming Valley a few miles west of the airport. The surrounding mountains protect both cities and the airport from high winds. They influence the temperature and precipitation during both summer and winter, causing wide departures in both within a few miles of the station. Because of the proximity of the mountains, the climate is relatively cool in summer with frequent shower and thunderstorm activity, usually of brief duration. The winter temperatures in the valley are not severe. The occurrence of sub-zero temperatures and severe snowstorms is infrequent. A high percentage of the winter precipitation occurs as rain.

Although severe snowstorms are infrequent, when they do occur they approach blizzard conditions. High winds cause huge drifts and normal routines are disrupted for several days.

While the incidence of tornadoes is very low, Wilkes-Barre has occasionally been hit with these storms which caused loss of life and great property damage.

The area has felt the effects of tropical storms. Considerable wind damage has occasionally occurred, but the most devastating damage has come from flooding caused by the large amounts of precipitation deposited by the storms. The worst natural disaster to hit the region was the result of the flooding caused by a hurricane.

TABLE 1

NORMALS, MEANS AND EXTREMES

AVOCA, WILKES-BARRE – SCRANTON PENNSYLVANIA

LATITUDE: 41°20'N   LONGITUDE: 75°44'W   ELEVATION: FT. GRND 930 BARO 00959   TIME ZONE: EASTERN   WBAN: 14777

	(a)	JAN	FEB	MAR	APR	MAY	JUNE	JULY	AUG	SEP	OCT	NOV	DEC	YEAR
TEMPERATURE °F:														
Normals														
-Daily Maximum		32.1	34.4	44.1	58.2	69.1	77.8	82.1	80.0	72.7	61.4	48.2	36.3	58.0
-Daily Minimum		18.2	19.2	28.1	38.4	48.1	56.9	61.4	60.0	52.8	42.0	33.6	23.1	40.1
-Monthly		25.1	26.8	36.1	48.3	58.6	67.3	71.8	70.0	62.8	51.7	40.9	29.7	49.1
Extremes														
-Record Highest	30	67	71	83	92	93	97	101	94	95	84	80	67	101
-Year		1967	1985	1977	1976	1962	1964	1966	1983	1983	1959	1982	1984	JUL 1966
-Record Lowest	30	-14	-16	-4	14	27	34	43	38	30	19	9	-7	-16
-Year		1985	1979	1967	1982	1974	1972	1979	1982	1974	1972	1976	1980	FEB 1979
NORMAL DEGREE DAYS:														
Heating (base 65°F)		1234	1070	896	501	227	34	7	10	117	417	723	1094	6330
Cooling (base 65°F)		0	0	0	0	29	106	218	165	51	0	0	0	569
% OF POSSIBLE SUNSHINE	30	43	47	49	53	56	60	62	60	55	51	36	35	51
MEAN SKY COVER (tenths)														
Sunrise – Sunset	30	7.5	7.3	7.2	6.8	6.7	6.2	6.1	6.1	6.1	6.3	7.6	7.8	6.8
MEAN NUMBER OF DAYS:														
Sunrise to Sunset														
-Clear	30	4.2	4.4	5.2	6.4	5.8	7.0	6.3	6.9	7.2	8.0	3.8	3.8	69.0
-Partly Cloudy	30	7.3	6.9	7.5	7.4	9.6	10.9	12.8	11.9	10.0	8.4	6.6	6.5	105.9
-Cloudy	30	19.5	16.9	18.3	16.2	15.6	12.2	11.9	12.2	12.8	14.6	19.5	20.7	190.4
Precipitation														
.01 inches or more	30	12.2	11.1	13.1	12.3	12.8	11.9	11.3	10.8	9.5	9.5	12.1	13.2	139.8
Snow, Ice pellets														
1.0 inches or more	30	3.3	2.9	2.8	0.7	0.*	0.0	0.0	0.0	0.0	0.*	0.9	2.6	13.3
Thunderstorms	30	0.1	0.3	1.0	2.2	3.6	6.0	7.1	5.0	2.7	1.0	0.4	0.2	29.5
Heavy Fog Visibility 1/4 mile or less	30	2.0	2.0	2.1	1.5	1.3	1.2	1.5	1.9	2.6	2.2	1.9	2.4	22.7
Temperature °F														
-Maximum														
90° and above	30	0.0	0.0	0.0	0.1	0.3	1.7	2.8	1.7	0.5	0.0	0.0	0.0	7.0
32° and below	30	16.6	12.1	3.9	0.2	0.0	0.0	0.0	0.0	0.0	0.0	1.3	11.2	45.3
-Minimum														
32° and below	30	28.4	24.8	22.1	8.9	0.8	0.0	0.0	0.0	0.2	4.2	13.8	25.5	128.6
0° and below	30	2.3	1.5	0.1	0.0	0.0	0.0	0.0	0.0	0.0	0.0	0.0	0.6	4.5
AVG. STATION PRESS.(mb)	13	981.9	982.9	981.2	980.7	980.8	981.8	982.5	984.1	984.8	985.1	983.7	983.4	982.7
RELATIVE HUMIDITY (%)														
Hour 01	30	72	71	69	66	72	79	80	83	83	78	75	75	75
Hour 07 (Local Time)	30	75	75	73	71	75	82	84	87	87	83	79	77	79
Hour 13	30	66	64	59	52	52	56	56	58	60	59	65	68	60
Hour 19	30	67	64	60	54	55	61	62	66	69	66	69	70	64
PRECIPITATION (inches):														
Water Equivalent														
-Normal		2.27	2.05	2.63	3.01	3.16	3.42	3.39	3.47	3.36	2.78	2.98	2.54	35.06
-Maximum Monthly	30	6.48	8.06	4.83	9.56	7.33	7.22	6.81	5.23	7.83	8.12	7.69	6.58	9.56
-Year		1979	1981	1977	1983	1972	1982	1969	1965	1985	1976	1972	1983	APR 1983
-Minimum Monthly	30	0.39	0.30	0.49	1.17	0.77	0.27	1.23	1.23	0.82	0.03	0.80	0.35	0.03
-Year		1980	1968	1981	1975	1959	1966	1972	1980	1964	1963	1976	1958	OCT 1963
-Maximum in 24 hrs	30	1.89	3.11	2.20	3.80	2.58	3.61	2.33	3.18	6.52	3.27	2.91	2.86	6.52
-Year		1978	1981	1964	1983	1972	1973	1969	1966	1985	1976	1972	1983	SEP 1985
Snow, Ice pellets														
-Maximum Monthly	30	28.8	22.0	29.7	26.7	2.4				T	4.4	22.5	33.9	33.9
-Year		1978	1964	1967	1983	1977				1956	1962	1971	1969	DEC 1969
-Maximum in 24 hrs	30	20.1	13.3	15.5	12.2	2.4				T	4.4	20.5	12.4	20.5
-Year		1964	1961	1960	1983	1977				1956	1962	1971	1969	NOV 1971
WIND:														
Mean Speed (mph)	30	9.0	9.1	9.4	9.6	8.6	7.8	7.3	7.0	7.4	7.9	8.6	8.9	8.4
Prevailing Direction through 1963		SW	SW	NW	SW	WSW	SW	WSW	SW	SW	WSW	WSW	SW	SW
Fastest Mile														
-Direction	30	SE	W	S	NW	SW	W	NW	NE	SW	E	S	SW	W
-Speed (MPH)	30	47	60	49	47	46	43	42	50	38	40	45	47	60
-Year		1977	1956	1970	1957	1980	1956	1960	1956	1957	1980	1957	1957	FEB 1956
Peak Gust														
-Direction	2	NW	NW	SE	SW	NW	NW	SW	W	NW	SW	SE	W	SW
-Speed (mph)	2	40	48	53	64	44	37	36	48	41	39	40	51	64
-Date		1985	1985	1984	1985	1984	1984	1985	1984	1984	1984	1984	1985	APR 1985

See Reference Notes to this table on the following page.

TABLE 2 PRECIPITATION (inches) AVOCA, WILKES-BARRE – SCRANTON PENNSYLVANIA

YEAR	JAN	FEB	MAR	APR	MAY	JUNE	JULY	AUG	SEP	OCT	NOV	DEC	ANNUAL
1956	1.01	3.64	2.48	3.16	2.25	4.68	3.18	3.03	2.54	1.35	1.31	3.24	31.87
1957	1.49	0.89	1.66	5.81	1.44	4.01	2.05	1.38	2.18	2.54	2.42	4.04	29.91
1958	2.42	3.12	2.18	3.59	3.40	3.41	3.57	4.30	7.11	3.37	2.78	0.35	39.60
1959	3.11	2.23	1.77	3.76	0.77	3.23	4.46	3.00	1.09	4.25	3.77	2.71	34.15
1960	2.94	2.64	1.49	3.40	4.32	3.33	5.55	3.54	7.78	1.22	1.77	1.01	38.99
1961	1.82	1.92	2.87	2.48	3.30	3.80	5.96	3.98	0.99	1.58	3.57	2.52	34.79
1962	2.67	3.04	1.98	2.59	0.84	1.71	1.48	4.05	3.14	5.46	2.80	2.20	31.96
1963	2.06	1.95	2.19	1.30	2.44	1.93	4.03	2.04	1.69	0.03	4.62	1.94	26.22
1964	3.40	2.03	3.54	3.82	0.98	5.00	1.23	2.85	0.82	1.13	1.86	3.67	30.33
1965	2.07	1.90	1.83	2.63	2.51	1.22	1.30	5.23	3.13	1.80	1.43	1.30	26.35
1966	1.66	2.31	1.60	2.91	3.32	0.27	1.89	4.76	2.70	2.04	2.97	2.00	28.43
1967	1.11	0.89	3.91	2.09	4.41	4.48	3.61	5.20	2.13	2.58	2.31	2.45	35.17
1968	2.03	0.30	2.73	2.37	4.64	4.82	1.23	1.43	4.20	1.65	3.40	1.95	30.75
1969	0.64	0.96	1.45	3.04	2.42	4.58	6.81	4.47	1.92	2.20	3.77	3.42	35.68
1970	0.52	2.41	2.33	3.07	2.75	2.56	5.19	2.72	3.03	2.24	2.87	1.85	31.54
1971	1.54	3.92	1.93	1.29	3.38	2.44	5.73	4.88	1.93	3.00	3.55	2.08	35.67
1972	2.05	2.42	4.00	3.31	7.33	7.04	1.23	1.64	1.57	3.30	7.69	3.61	45.19
1973	2.13	1.28	1.79	4.38	3.80	5.99	3.87	2.61	3.62	1.97	1.50	6.07	39.01
1974	2.66	1.48	4.75	2.71	1.89	3.85	2.80	3.50	6.85	1.07	2.26	3.40	37.22
1975	2.78	3.26	2.52	1.17	4.01	5.64	3.85	2.78	6.10	3.29	3.00	1.84	40.24
1976	3.25	2.14	2.18	2.27	3.24	5.43	3.20	2.57	3.81	8.12	0.80	1.50	38.51
1977	0.88	1.82	4.83	3.98	1.72	3.16	3.44	4.23	5.97	5.27	3.98	3.44	42.72
1978	5.33	0.93	2.30	1.67	4.30	2.48	2.16	3.28	3.06	3.35	1.02	3.09	32.97
1979	6.48	2.44	1.52	3.69	5.16	2.54	2.97	2.05	5.84	3.68	3.17	1.70	41.24
1980	0.39	0.69	3.72	2.35	2.37	4.36	3.76	1.43	1.43	2.17	2.83	1.24	26.54
1981	0.63	8.06	0.49	3.54	3.00	3.45	4.27	1.75	2.74	3.50	1.84	2.13	35.40
1982	2.71	2.28	2.55	3.48	3.52	7.22	3.32	3.42	1.10	0.84	3.44	1.52	35.40
1983	1.17	1.46	3.28	9.56	3.28	4.81	2.76	1.77	2.12	2.73	3.71	6.58	43.23
1984	1.11	2.92	2.42	4.09	6.70	4.75	5.12	2.81	1.36	2.30	2.63	2.36	38.57
1985	0.61	1.58	2.24	2.00	6.10	3.00	6.09	2.62	7.83	1.92	4.47	1.96	40.42
Record Mean	2.33	2.30	2.83	3.11	3.28	3.78	4.02	3.56	3.16	2.89	2.76	2.64	36.66

TABLE 3 AVERAGE TEMPERATURE (deg. F) AVOCA, WILKES-BARRE – SCRANTON PENNSYLVANIA

YEAR	JAN	FEB	MAR	APR	MAY	JUNE	JULY	AUG	SEP	OCT	NOV	DEC	ANNUAL
1956	25.4	30.0	31.5	44.9	54.7	67.1	67.5	67.9	58.0	52.1	40.4	34.2	47.8
1957	21.6	31.1	36.2	49.1	57.3	69.3	69.6	66.2	62.6	49.2	41.9	33.6	49.0
1958	25.3	21.0	34.2	49.0	56.5	62.2	72.3	69.6	61.1	49.8	41.1	22.4	47.1
#1959	24.7	25.7	35.4	50.4	62.8	68.6	72.7	73.4	67.4	54.0	37.8	32.1	50.4
1960	28.9	29.5	25.3	52.4	57.7	67.2	68.8	70.5	62.8	50.5	41.4	21.9	48.1
1961	18.9	28.9	35.1	42.0	56.1	67.1	72.0	71.0	68.4	52.9	40.1	29.6	48.6
1962	24.9	24.6	36.6	49.0	62.6	68.5	70.3	70.4	59.5	51.6	36.6	24.0	48.2
1963	23.2	19.3	38.2	49.1	57.8	67.4	71.5	68.3	60.0	56.7	45.1	24.3	48.4
1964	27.9	23.5	36.9	46.6	62.0	67.4	74.1	68.2	63.8	49.4	44.1	31.6	49.6
1965	23.0	28.1	34.2	45.3	63.7	68.0	71.0	69.5	64.5	49.6	40.4	33.5	49.2
1966	23.8	27.2	39.0	46.0	55.7	70.1	75.0	72.4	60.2	50.2	43.4	30.5	49.4
1967	32.6	23.5	34.9	49.1	52.7	71.1	71.7	68.9	61.6	51.2	36.6	31.7	48.8
1968	20.0	24.1	40.4	52.9	56.3	67.5	73.7	71.0	64.4	53.4	40.7	27.2	49.3
1969	26.0	26.6	34.4	50.5	59.9	68.2	70.9	69.9	62.7	50.6	39.2	25.7	48.7
1970	17.7	25.7	31.8	48.4	60.4	64.9	70.8	69.6	64.2	52.8	41.5	27.3	47.9
1971	19.5	27.2	33.0	45.0	55.0	68.6	69.9	68.3	65.5	56.4	38.8	34.1	48.5
1972	27.6	24.3	32.8	43.5	60.0	62.9	71.5	69.6	61.9	45.3	36.0	32.9	47.4
1973	28.5	24.5	41.9	47.5	53.4	68.3	71.6	71.6	62.4	53.2	41.5	31.3	49.7
1974	27.8	24.2	34.7	49.1	56.2	64.4	70.8	68.5	61.2	48.5	43.0	34.4	48.6
1975	31.8	32.1	35.8	43.9	64.9	68.8	73.7	71.2	59.9	55.8	47.8	31.1	51.4
1976	22.1	35.0	41.0	51.3	56.7	70.4	69.2	69.1	60.9	48.0	37.0	22.9	48.6
1977	15.0	26.9	40.5	48.9	60.0	63.8	71.1	68.7	62.7	48.5	43.1	29.6	48.2
1978	24.4	19.2	33.1	46.0	58.7	65.2	69.4	71.0	60.5	50.7	40.5	29.1	47.3
1979	24.2	16.0	40.7	46.4	58.6	65.1	70.7	70.7	63.2	51.8	45.9	35.6	49.1
1980	27.8	24.2	35.9	51.0	61.8	65.3	73.2	75.2	66.1	49.8	37.6	25.7	49.5
1981	19.5	34.9	36.2	51.0	59.8	68.5	72.2	70.1	62.1	49.1	41.1	29.1	49.5
1982	18.7	27.8	36.0	46.3	61.2	64.3	71.0	66.2	62.4	52.3	44.3	36.7	48.9
1983	27.3	29.4	38.8	45.9	55.7	67.6	72.4	71.6	64.8	52.5	42.8	27.1	49.7
1984	23.0	35.9	30.9	48.3	57.5	69.3	71.6	72.6	61.7	58.0	40.8	37.3	50.6
1985	21.5	29.8	39.1	51.4	60.6	63.8	70.1	69.0	64.0	52.7	44.5	26.7	49.5
Record Mean	26.6	27.4	36.7	48.0	59.0	67.4	72.1	70.0	63.1	52.3	41.2	30.3	49.5
Max	33.9	35.2	45.2	57.8	69.7	77.9	82.5	80.2	73.3	62.0	48.6	36.9	58.6
Min	19.3	19.5	28.2	38.1	48.4	56.9	61.6	59.8	52.9	42.5	33.7	23.6	40.4

REFERENCE NOTES FOR TABLES 1, 2, 3 and 6 (WILKES-BARRE/SCRANTON, PA)

GENERAL

T - TRACE AMOUNT
BLANK ENTRIES DENOTE MISSING/UNREPORTED DATA.
# INDICATES A STATION OR INSTRUMENT RELOCATION.

SPECIFIC

TABLE 1

(a) - LENGTH OF RECORD IN YEARS. ALTHOUGH
INDIVIDUAL MONTHS MAY BE MISSING.
* LESS THAN .05

NORMALS — BASED ON THE 1951-1980 RECORD PERIOD.
EXTREMES — DATES ARE THE MOST RECENT OCCURRENCE.
WIND DIR. — NUMERALS SHOW TENS OF DEGREES
CLOCKWISE FROM TRUE NORTH.
"00" INDICATES CALM.
RESULTANT WIND DIRECTIONS ARE GIVEN TO WHOLE DEGREES.

EXCEPTIONS

TABLES 2, 3, and 6

RECORD MEANS ARE THROUGH THE CURRENT YEAR,
BEGINNING IN 1901 FOR TEMPERATURE
1901 FOR PRECIPITATION
1956 FOR SNOWFALL

TABLE 4 HEATING DEGREE DAYS Base 65 deg. F AVOCA, WILKES-BARRE – SCRANTON PENNSYLVANIA

SEASON	JULY	AUG	SEP	OCT	NOV	DEC	JAN	FEB	MAR	APR	MAY	JUNE	TOTAL
1956-57	28	29	241	394	736	943	1337	946	886	485	264	41	6330
1957-58	11	37	139	479	685	966	1222	1225	946	478	268	118	6574
1958-59	0	15	157	463	711	1316	1243	1095	911	429	167	55	6562
#1959-60	0	4	94	372	810	1012	1109	1024	1225	393	221	29	6293
1960-61	8	4	92	446	700	1331	1422	1005	922	684	293	44	6951
1961-62	3	7	80	370	743	1092	1236	1125	874	508	155	16	6209
1962-63	3	10	199	414	848	1289	1260	1272	820	475	236	45	6871
1963-64	11	20	175	253	588	1254	1140	1198	866	544	139	69	6257
1964-65	0	31	110	475	621	1028	1293	1027	951	582	100	55	6273
1965-66	7	35	99	471	731	971	1270	1053	799	565	300	41	6342
1966-67	1	0	170	451	642	1063	998	1154	923	475	379	12	6268
1967-68	5	17	139	427	844	1023	1388	1181	754	357	261	42	6438
1968-69	0	28	55	359	725	1165	1204	1068	943	434	188	38	6207
1969-70	6	15	134	440	765	1212	1459	1094	1022	500	176	76	6899
1970-71	2	5	109	371	695	1162	1404	1051	985	593	292	32	6701
1971-72	5	22	90	263	783	951	1152	1175	992	634	168	109	6344
1972-73	21	20	125	603	860	988	1124	1131	704	521	354	25	6476
1973-74	2	11	140	368	699	1036	1145	1135	934	480	291	65	6306
1974-75	5	2	155	503	655	941	1024	913	902	627	88	25	5840
1975-76	0	8	158	291	509	922	1322	985	737	451	265	31	5679
1976-77	8	25	155	519	834	1297	1546	1058	756	487	206	90	6981
1977-78	14	37	119	505	653	1090	1252	1279	984	562	240	73	6808
1978-79	38	2	153	436	728	1103	1257	1370	747	552	221	66	6673
1979-80	34	31	120	420	568	1000	1144	1175	895	414	137	94	5932
1980-81	1	0	82	466	813	1211	1407	835	886	416	195	19	6331
1981-82	2	5	132	485	706	1105	1426	1034	896	554	147	68	6560
1982-83	17	55	112	390	619	1158	990	992	805	569	292	41	5920
1983-84	7	11	119	392	659	1169	1297	837	1052	493	247	34	6317
1984-85	7	6	148	219	719	852	1342	981	799	421	162	78	5734
1985-86	4	11	127	376	610	1181							

TABLE 5 COOLING DEGREE DAYS Base 65 deg. F AVOCA, WILKES-BARRE – SCRANTON PENNSYLVANIA

YEAR	JAN	FEB	MAR	APR	MAY	JUNE	JULY	AUG	SEP	OCT	NOV	DEC	TOTAL
1969	0	0	0	7	34	141	193	178	73	1	0	0	627
1970	0	0	0	10	40	81	189	159	92	2	0	0	573
1971	0	0	0	0	16	145	160	132	113	4	3	0	573
1972	0	0	0	0	20	53	232	167	42	0	0	0	514
1973	0	0	0	4	0	132	212	223	72	7	0	0	650
1974	0	0	0	10	28	52	194	117	46	0	0	0	447
1975	0	0	0	0	91	146	278	207	13	14	0	0	749
1976	0	0	0	46	16	198	145	159	37	0	0	0	601
1977	0	0	2	13	57	62	208	162	59	0	2	0	565
1978	0	0	0	0	52	84	181	194	26	0	0	0	537
1979	0	0	0	2	32	78	218	214	75	15	0	0	634
1980	0	0	0	0	42	107	263	322	122	3	0	0	859
1981	0	0	0	4	42	131	231	172	55	0	0	0	635
1982	0	0	0	1	34	55	208	98	41	3	5	0	445
1983	0	0	0	4	12	125	243	224	118	9	0	0	735
1984	0	0	0	0	20	165	218	248	58	12	0	0	721
1985	0	0	0	20	32	47	169	142	104	0	0	0	514

TABLE 6 SNOWFALL (inches) AVOCA, WILKES-BARRE – SCRANTON PENNSYLVANIA

SEASON	JULY	AUG	SEP	OCT	NOV	DEC	JAN	FEB	MAR	APR	MAY	JUNE	TOTAL
1956-57	0.0	0.0	T	0.0	2.7	5.6	11.1	3.3	6.7	8.0	0.0	0.0	37.4
1957-58	0.0	0.0	0.0	T	0.3	6.0	8.1	19.5	18.6	1.1	0.0	0.0	53.6
1958-59	0.0	0.0	0.0	T	2.4	3.8	7.2	3.9	11.5	0.4	0.0	0.0	29.2
1959-60	0.0	0.0	0.0	0.5	0.7	6.2	15.4	22.3	3.6	8.4	0.0	0.0	57.2
1960-61	0.0	0.0	0.0	0.4	0.7	19.0	19.0	16.9	9.3	8.4	0.0	0.0	73.7
1961-62	0.0	0.0	0.0	0.0	2.4	12.8	2.7	11.2	2.8	3.0	T	0.0	34.9
1962-63	0.0	0.0	0.0	4.4	4.6	12.9	15.2	18.0	8.3	0.5	0.2	0.0	64.1
1963-64	0.0	0.0	0.0	T	1.4	13.1	27.9	22.0	9.7	0.6	0.0	0.0	74.7
1964-65	0.0	0.0	0.0	0.2	0.1	4.1	11.5	2.3	9.9	3.8	0.0	0.0	31.9
1965-66	0.0	0.0	0.0	0.6	0.5	1.3	20.7	17.5	3.8	0.7	0.6	0.0	45.7
1966-67	0.0	0.0	0.0	T	T	17.7	8.3	14.4	29.7	4.8	0.4	0.0	75.3
1967-68	0.0	0.0	0.0	T	7.2	13.8	6.0	2.9	2.7	0.0	0.0	0.0	32.6
1968-69	0.0	0.0	0.0	T	10.9	5.8	2.7	14.8	2.5	0.0	0.0	0.0	36.7
1969-70	0.0	0.0	0.0	T	2.7	33.9	9.5	9.5	20.6	0.6	T	0.0	76.8
1970-71	0.0	0.0	0.0	0.3	T	12.3	15.2	12.1	15.6	1.6	0.0	0.0	57.1
1971-72	0.0	0.0	0.0	0.0	22.5	4.1	5.9	19.7	6.6	3.8	0.0	0.0	62.6
1972-73	0.0	0.0	0.0	0.8	7.9	3.9	5.0	3.1	1.9	0.2	0.4	0.0	23.2
1973-74	0.0	0.0	0.0	T	0.4	16.0	12.8	4.5	15.7	2.8	0.0	0.0	52.2
1974-75	0.0	0.0	0.0	0.2	2.2	5.2	13.7	15.2	5.5	1.2	0.0	0.0	43.2
1975-76	0.0	0.0	0.0	0.0	1.3	3.7	13.0	10.2	7.5	0.5	0.0	0.0	36.2
1976-77	0.0	0.0	0.0	0.0	6.0	6.7	15.7	13.0	11.2	1.4	2.4	0.0	56.4
1977-78	0.0	0.0	0.0	0.6	8.7	9.8	28.8	18.2	6.5	0.9	0.0	0.0	73.5
1978-79	0.0	0.0	0.0	T	4.1	7.9	12.7	14.3	1.1	4.4	0.0	0.0	44.5
1979-80	0.0	0.0	0.0	T	T	5.5	1.4	8.1	10.5	T	0.0	0.0	25.5
1980-81	0.0	0.0	0.0	T	8.6	8.0	11.1	7.0	5.8	T	0.0	0.0	40.5
1981-82	0.0	0.0	0.0	T	1.0	14.2	14.1	13.5	8.7	8.1	0.0	0.0	59.6
1982-83	0.0	0.0	0.0	0.0	0.5	7.4	8.4	12.3	3.8	26.7	0.0	0.0	59.1
1983-84	0.0	0.0	0.0	0.0	3.1	2.7	11.2	4.0	18.4	0.0	0.0	0.0	39.4
1984-85	0.0	0.0	0.0	0.0	3.0	9.2	10.8	9.1	1.4	1.8	0.0	0.0	35.3
1985-86	0.0	0.0	0.0	0.0	1.7	13.4							
Record Mean	0.0	0.0	T	0.3	3.6	9.6	11.4	11.4	9.9	3.4	0.1	0.0	49.6

See Reference Notes, relative to all above tables, on preceding page.

The climate of the Lycoming valley is favorably influenced by the lower elevation of the area compared to the surrounding terrain. Since the prevailing winds reach the area from the southwest to the north, there is a slight moderating effect on winter extremes of cold. Radiation cooling on clear nights is somewhat more frequent than in adjacent areas. Deep valley fogs occasionally persist until nearly midday. Cold air drainage from the surrounding hills is experienced during several nights but the cool temperatures are often modified by the proximity of the river and adjacent damp areas. The winters are milder than those experienced to the west and cold spells are frequently interrupted by incursions of warmer coastal weather. In summer the air frequently becomes trapped in the valley and higher temperatures and humidities result, generally benefitting the local agriculture.

The long irregular range south of the river forms an effective barrier to free air movement. Moderate or strong south to southwest winds are deflected to southeast or south winds in crossing the range and the air becomes quite turbulent with distinct wave effects. Banner clouds with one to three rolls are frequently observed with low overcasts.

An average growing season of 168 days extends from April 29 to October 14. Snowfall in the valley is generally uniform but varies considerably with the rise in terrain, particularly to the north and south. Snow depth on the ridge 2 miles south of the observation point is frequently double the amount at the station.

TABLE 1 — NORMALS, MEANS AND EXTREMES

WILLIAMSPORT, PENNSYLVANIA

LATITUDE: 41°15'N   LONGITUDE: 76°55'W   ELEVATION: FT. GRND   524 BARO   00543   TIME ZONE: EASTERN   WBAN: 14778

	(a)	JAN	FEB	MAR	APR	MAY	JUNE	JULY	AUG	SEP	OCT	NOV	DEC	YEAR
TEMPERATURE °F:														
Normals														
-Daily Maximum		34.1	36.8	46.8	60.6	71.2	79.7	83.7	82.0	74.5	63.0	49.6	38.1	60.0
-Daily Minimum		18.3	19.5	28.4	38.5	48.0	56.8	61.3	60.3	53.2	41.6	33.2	23.3	40.2
-Monthly		26.2	28.1	37.6	49.5	59.6	68.3	72.5	71.2	63.9	52.3	41.4	30.7	50.1
Extremes														
-Record Highest	41	69	71	86	92	95	102	100	100	102	91	83	67	102
-Year		1967	1985	1977	1976	1969	1952	1966	1955	1953	1951	1950	1984	SEP 1953
-Record Lowest	41	-17	-13	-1	15	28	36	43	38	28	20	8	-15	-17
-Year		1977	1971	1984	1982	1966	1945	1965	1965	1947	1972	1976	1950	JAN 1977
NORMAL DEGREE DAYS:														
Heating (base 65°F)		1203	1030	849	462	196	30	0	7	101	398	708	1063	6047
Cooling (base 65°F)		0	0	0	0	29	129	237	196	68	0	0	0	659
% OF POSSIBLE SUNSHINE														
MEAN SKY COVER (tenths)														
Sunrise - Sunset	41	7.3	7.0	6.9	6.7	6.7	6.3	6.2	6.3	6.5	6.4	7.5	7.5	6.8
MEAN NUMBER OF DAYS:														
Sunrise to Sunset														
-Clear	41	5.0	5.5	6.0	6.4	5.6	6.3	6.2	5.7	5.9	7.2	4.2	4.2	68.2
-Partly Cloudy	41	7.6	6.8	7.8	7.8	10.1	11.8	13.2	13.3	11.1	9.1	6.6	6.9	112.1
-Cloudy	41	18.4	16.0	17.2	15.9	15.3	12.0	11.6	12.0	13.0	14.7	19.2	19.9	185.0
Precipitation														
.01 inches or more	41	12.2	11.3	12.9	13.2	13.5	11.8	11.5	11.0	9.7	10.0	12.4	12.8	142.2
Snow, Ice pellets														
1.0 inches or more	41	3.2	3.1	2.1	0.3	0.0	0.0	0.0	0.0	0.0	0.*	0.9	2.6	12.2
Thunderstorms	24	0.2	0.3	1.0	1.8	4.0	6.9	7.7	6.2	3.2	0.9	0.5	0.3	33.0
Heavy Fog Visibility														
1/4 mile or less	24	1.3	1.3	1.3	1.2	2.3	2.9	3.5	4.6	7.9	6.1	3.0	1.9	37.2
Temperature °F														
-Maximum														
90° and above	41	0.0	0.0	0.0	0.2	0.6	2.8	5.2	2.8	0.9	0.*	0.0	0.0	12.5
32° and below	41	13.3	8.2	2.0	0.*	0.0	0.0	0.0	0.0	0.0	0.0	0.7	8.0	32.3
-Minimum														
32° and below	41	28.0	24.8	21.5	7.8	0.7	0.0	0.0	0.0	0.2	4.3	14.8	25.6	127.7
0° and below	41	2.5	1.4	0.1	0.0	0.0	0.0	0.0	0.0	0.0	0.0	0.0	0.8	4.7
AVG. STATION PRESS. (mb)	13	998.4	999.0	997.0	996.2	995.9	996.7	997.4	999.0	999.9	1000.6	999.5	999.5	998.3
RELATIVE HUMIDITY (%)														
Hour 01	34	74	73	73	73	80	87	88	90	90	86	80	77	81
Hour 07	40	76	76	76	75	80	84	87	90	92	88	82	78	82
Hour 13 (Local Time)	40	62	58	54	49	52	54	55	58	59	57	62	64	57
Hour 19	39	68	64	60	55	58	62	64	69	74	72	71	71	66
PRECIPITATION (inches):														
Water Equivalent														
-Normal		2.88	2.83	3.66	3.53	3.66	3.88	3.92	3.26	3.57	3.22	3.63	3.24	41.28
-Maximum Monthly	41	8.25	8.42	5.96	7.03	9.45	16.80	8.30	7.67	10.02	8.14	8.09	7.36	16.80
-Year		1978	1981	1980	1983	1946	1972	1958	1955	1975	1976	1972	1973	JUN 1972
-Minimum Monthly	41	0.52	0.57	0.86	0.99	0.80	0.66	0.99	0.96	0.50	0.19	0.83	0.84	0.19
-Year		1985	1968	1981	1946	1964	1966	1955	1951	1964	1963	1976	1955	OCT 1963
-Maximum in 24 hrs	41	2.46	2.72	2.52	2.71	4.15	8.66	2.53	3.46	4.60	4.38	3.46	3.29	8.66
-Year		1978	1971	1964	1977	1946	1972	1958	1950	1975	1955	1956	1983	JUN 1972
Snow, Ice pellets														
-Maximum Monthly	41	38.1	34.3	29.5	13.8	0.2					1.0	13.7	35.5	38.1
-Year		1978	1972	1967	1982	1977					1977	1953	1969	JAN 1978
-Maximum in 24 hrs	41	23.1	20.4	13.9	8.8	0.2					1.0	12.1	16.5	23.1
-Year		1964	1972	1967	1982	1977					1977	1953	1969	JAN 1964
WIND:														
Mean Speed (mph)	24	9.2	9.2	9.5	9.4	8.0	7.0	6.4	6.0	6.2	6.8	8.2	8.7	7.9
Prevailing Direction through 1963		W	WNW	W	W	W	W	W	W	WNW	W	W	W	W
Fastest Obs. 1 Min.														
-Direction (!!!)	32	27	14	11	18	18	29	20	29	16	11	09	16	20
-Speed (MPH)	32	66	60	58	62	55	62	78	60	59	75	77	58	78
-Year		1951	1951	1954	1950	1953	1951	1951	1949	1951	1954	1950	1953	JUL 1951
Peak Gust														
-Direction (!!!)	2	W	W	W	W	NW	SW	NW	NW	W	W	W	W	W
-Speed (mph)	2	48	49	58	60	48	55	48	49	35	44	46	55	60
-Date		1985	1985	1985	1985	1984	1984	1985	1984	1984	1985	1984	1985	APR 1985

See Reference Notes to this table on the following pages.

TABLE 2 PRECIPITATION (inches) WILLIAMSPORT, PENNSYLVANIA

YEAR	JAN	FEB	MAR	APR	MAY	JUNE	JULY	AUG	SEP	OCT	NOV	DEC	ANNUAL
1956	1.48	4.15	3.28	3.08	3.33	3.02	7.17	4.09	2.60	3.53	4.75	4.28	44.76
1957	1.59	2.20	2.87	6.30	1.93	5.74	1.65	1.25	3.80	1.50	3.12	5.05	37.00
1958	4.95	3.52	3.74	4.26	2.66	3.98	8.30	3.03	4.04	1.73	3.25	0.90	44.36
1959	3.49	1.85	3.55	4.38	1.45	4.24	4.33	2.73	4.09	6.96	4.56	4.18	45.81
1960	2.21	4.81	2.95	3.46	7.24	3.57	4.64	2.99	6.35	1.78	1.45	1.80	43.25
1961	2.35	3.17	3.75	5.46	2.94	5.11	5.49	3.69	0.89	1.86	3.65	2.53	40.89
1962	3.10	4.05	4.80	3.00	2.61	2.11	2.54	4.55	3.25	5.31	2.47	2.87	40.66
1963	2.09	1.73	3.14	1.34	4.91	2.48	3.12	2.28	2.29	0.19	5.08	2.41	31.06
1964	4.95	3.11	3.87	5.56	0.80	2.60	3.37	4.18	0.50	1.10	2.24	2.66	34.94
1965	2.88	2.27	3.21	3.07	1.88	1.75	1.70	5.03	4.04	3.98	1.69	0.93	32.43
1966	3.32	4.58	1.90	2.83	4.16	0.66	3.90	2.45	5.22	1.43	5.77	1.92	38.14
1967	1.35	1.49	5.36	2.88	5.50	3.24	6.03	4.97	2.20	4.50	3.63	2.71	43.86
1968	2.49	0.57	1.10	1.05	4.93	5.20	1.39	1.47	6.41	3.72	4.50	2.96	37.79
1969	1.24	0.72	1.49	2.78	1.83	4.38	6.46	3.09	1.81	3.04	5.08	5.63	37.55
1970	0.95	3.76	2.95	3.28	3.23	3.32	5.45	3.00	1.91	5.43	5.08	4.12	42.48
1971	2.54	6.50	2.69	1.17	2.88	1.89	5.68	2.48	3.33	3.88	3.18	2.87	39.09
1972	2.49	4.84	3.96	4.22	6.79	16.80	3.79	1.71	1.61	2.18	8.09	4.79	61.27
1973	2.73	2.74	3.47	5.56	5.18	5.86	3.99	1.95	5.50	3.17	2.24	7.36	49.75
1974	3.03	1.90	5.78	2.99	3.04	5.51	2.67	3.05	4.84	0.86	2.57	4.50	40.74
1975	3.53	4.34	3.07	1.53	6.54	5.25	3.66	4.43	10.02	2.98	2.83	3.24	51.42
1976	2.73	1.90	2.75	1.90	3.81	5.31	4.80	3.18	3.21	8.14	0.83	1.74	40.30
1977	1.35	1.74	5.74	6.79	1.02	3.51	3.59	2.29	7.30	4.22	5.45	4.22	47.22
1978	8.25	0.86	2.70	2.01	6.36	4.60	3.99	3.56	2.84	2.76	1.78	3.68	43.39
1979	6.36	3.20	3.18	2.73	4.00	2.23	4.24	5.54	5.83	4.08	4.07	1.90	47.36
1980	0.88	0.77	5.96	6.72	1.71	3.50	1.82	1.82	3.32	3.19	2.68	1.18	33.55
1981	0.68	8.42	0.86	2.93	2.07	5.51	4.09	1.02	2.10	4.03	1.46	1.95	35.12
1982	3.37	1.93	2.13	2.96	3.53	9.23	1.81	1.00	2.59	0.59	3.00	1.35	33.49
1983	1.79	1.87	3.05	7.03	4.51	5.55	3.38	3.68	1.88	3.01	5.89	6.54	48.18
1984	1.09	4.83	3.63	4.58	4.89	5.15	6.42	3.60	0.62	2.78	4.76	2.34	44.69
1985	0.52	1.12	2.81	1.18	3.98	2.49	3.47	4.28	2.58	1.83	5.99	1.42	31.67
Record Mean	2.67	2.85	3.49	3.53	3.80	3.99	4.02	3.31	3.37	3.09	3.76	3.13	41.01

TABLE 3 AVERAGE TEMPERATURE (deg. F) WILLIAMSPORT, PENNSYLVANIA

YEAR	JAN	FEB	MAR	APR	MAY	JUNE	JULY	AUG	SEP	OCT	NOV	DEC	ANNUAL
1956	28.7	31.8	35.0	47.8	56.9	70.0	69.9	70.0	60.5	53.6	41.9	35.8	50.1
1957	24.1	33.1	39.0	51.8	60.3	71.6	72.7	69.8	64.5	51.1	42.6	35.2	51.3
1958	27.8	23.9	37.4	50.9	58.1	63.2	73.9	70.3	62.8	51.6	41.6	23.6	48.8
1959	26.6	28.3	35.6	50.3	63.5	69.6	73.3	74.3	67.3	55.0	39.3	32.8	51.3
1960	31.1	32.3	28.5	54.5	57.5	68.1	69.8	72.0	64.9	51.2	22.1	22.1	49.5
1961	20.6	30.0	38.2	45.4	56.7	68.3	72.6	71.2	70.0	54.9	42.6	30.3	50.0
1962	26.1	26.8	37.3	49.7	63.3	68.7	70.6	70.3	59.3	52.5	38.8	24.7	49.0
1963	23.2	20.7	38.8	50.2	58.0	68.4	72.2	68.3	59.7	56.7	45.1	25.4	48.9
1964	27.9	26.8	38.5	47.0	62.6	68.4	74.7	68.8	64.5	51.4	43.5	31.5	50.5
1965	22.3	27.3	33.5	44.5	61.5	65.2	68.9	68.1	63.4	49.1	41.4	34.7	48.3
1966	25.7	29.0	39.5	46.6	55.9	66.3	72.2	70.9	59.4	47.6	41.3	30.1	48.7
1967	31.4	23.5	34.3	50.4	52.8	71.2	71.4	69.4	63.3	51.8	37.9	33.5	49.3
1968	23.0	26.4	41.0	52.0	56.3	66.3	72.4	71.9	64.5	53.9	42.9	30.5	50.1
1969	27.3	28.2	35.5	50.0	60.9	69.5	71.7	71.1	64.2	51.7	41.8	28.5	50.0
1970	18.5	26.5	34.2	49.9	62.5	67.4	73.2	72.4	67.4	56.4	45.1	32.8	50.6
1971	23.0	30.4	36.6	48.4	58.5	70.6	71.8	70.4	68.2	58.8	39.6	35.1	51.0
1972	28.6	24.2	34.9	46.0	61.0	64.0	72.1	70.1	63.1	46.6	37.7	34.3	48.6
1973	28.6	25.9	43.4	48.9	54.9	70.4	71.2	71.1	63.6	53.4	42.4	31.7	50.5
1974	29.7	27.5	38.1	50.9	57.2	65.7	71.8	71.3	61.5	49.0	43.1	34.1	50.0
1975	30.6	31.2	37.2	43.2	61.5	66.9	72.1	70.9	59.5	54.4	47.2	31.8	50.6
1976	22.7	35.4	42.1	52.9	58.3	71.6	70.5	70.2	62.3	49.4	36.3	26.1	49.8
1977	14.9	28.9	44.7	53.2	63.8	67.0	73.8	72.1	67.2	50.4	45.3	28.4	50.8
1978	22.8	20.5	34.5	48.3	60.7	68.4	72.1	74.8	65.0	52.5	43.7	32.5	49.7
1979	26.7	19.7	42.3	48.3	60.9	67.0	72.6	71.0	62.8	50.8	44.8	35.0	50.2
1980	28.0	25.7	36.9	51.3	61.4	65.7	73.8	75.4	66.8	49.6	37.1	25.0	49.7
1981	20.3	33.2	36.5	51.9	62.5	70.8	74.4	72.4	64.5	51.3	42.2	30.2	50.9
1982	19.3	28.4	35.7	46.1	62.1	64.5	71.1	66.6	63.1	52.0	43.8	36.9	49.1
1983	29.3	31.5	41.0	47.9	56.3	66.3	72.2	71.1	62.0	49.9	40.1	26.7	49.5
1984	21.0	34.7	30.4	47.5	55.9	68.7	70.6	72.2	60.6	56.7	40.3	35.7	49.5
1985	23.4	29.9	40.1	53.3	61.0	64.4	71.0	70.1	65.3	52.9	45.7	27.9	50.4
Record Mean	26.0	28.5	37.8	49.4	59.4	68.0	72.4	71.0	63.7	52.7	41.5	30.5	50.1
Max	33.9	37.1	47.2	60.4	71.0	79.4	83.6	81.8	74.4	63.4	49.8	37.8	60.0
Min	18.1	19.9	28.4	38.4	47.8	56.6	61.2	60.1	53.0	41.9	33.3	23.3	40.2

REFERENCE NOTES FOR TABLES 1, 2, 3 and 6 **(WILLIAMSPORT, PA)**

GENERAL

T - TRACE AMOUNT
BLANK ENTRIES DENOTE MISSING/UNREPORTED DATA.
# INDICATES A STATION OR INSTRUMENT RELOCATION.

SPECIFIC

TABLE 1

(a) - LENGTH OF RECORD IN YEARS. ALTHOUGH
      INDIVIDUAL MONTHS MAY BE MISSING.
  * LESS THAN .05

NORMALS — BASED ON THE 1951-1980 RECORD PERIOD.
EXTREMES — DATES ARE THE MOST RECENT OCCURRENCE.
WIND DIR. — NUMERALS SHOW TENS OF DEGREES
            CLOCKWISE FROM TRUE NORTH.
            "00" INDICATES CALM.
RESULTANT WIND DIRECTIONS ARE GIVEN TO WHOLE DEGREES.

EXCEPTIONS

TABLES 2, 3, and 6

RECORD MEANS ARE THROUGH THE CURRENT YEAR,
BEGINNING IN   1945 FOR TEMPERATURE
                   1945 FOR PRECIPITATION
                   1945 FOR SNOWFALL

TABLE 4 HEATING DEGREE DAYS Base 65 deg. F   WILLIAMSPORT, PENNSYLVANIA

SEASON	JULY	AUG	SEP	OCT	NOV	DEC	JAN	FEB	MAR	APR	MAY	JUNE	TOTAL
1956-57	10	11	176	346	688	899	1263	889	799	418	185	20	5704
1957-58	0	11	109	424	664	917	1146	1146	847	416	216	100	5996
1958-59	1	7	116	408	693	1274	1184	1021	905	432	138	33	6212
1959-60	0	1	88	343	766	992	1042	941	1122	346	229	17	5887
1960-61	4	2	60	422	698	1325	1369	976	822	581	264	28	6551
1961-62	0	5	73	308	666	1069	1198	1063	851	479	137	8	5857
1962-63	1	7	200	389	777	1242	1288	1235	803	442	237	30	6651
1963-64	8	19	179	250	593	1220	1143	1102	813	532	119	60	6038
1964-65	0	20	89	417	639	1032	1317	1047	970	611	137	80	6359
1965-66	17	41	121	484	702	937	1213	1002	783	548	283	60	6191
1966-67	5	4	196	531	703	1075	1035	1157	946	437	371	5	6465
1967-68	5	12	109	406	807	973	1296	1113	739	385	265	50	6160
1968-69	4	22	46	337	656	1063	1162	1022	909	444	171	24	5860
1969-70	2	5	106	412	690	1125	1433	1072	945	453	127	30	6400
1970-71	0	0	56	271	591	990	1297	961	871	495	212	16	5760
1971-72	0	7	58	201	757	921	1122	1176	923	562	141	90	5958
1972-73	12	19	91	565	810	942	1125	1089	664	479	306	16	6118
1973-74	5	16	119	354	672	1027	1087	1047	828	427	258	36	5876
1974-75	1	0	142	488	648	950	1059	940	852	648	155	44	5927
1975-76	1	9	168	324	529	1023	1305	850	704	393	217	19	5542
1976-77	0	12	121	478	855	1201	1547	1008	624	366	119	38	6369
1977-78	4	15	46	444	588	1126	1300	1239	938	495	190	28	6413
1978-79	9	0	76	381	633	998	1179	1264	696	497	167	40	5940
1979-80	15	19	131	438	601	919	1141	1134	863	405	139	77	5882
1980-81	0	0	65	472	830	1232	1378	886	877	391	130	7	6268
1981-82	0	0	97	416	677	1070	1408	1020	903	558	123	57	6329
1982-83	15	45	108	393	632	865	1101	934	736	511	271	50	5661
1983-84	14	8	159	461	741	1182	1360	873	1069	517	288	30	6702
1984-85	6	6	171	257	733	901	1282	978	765	377	151	59	5686
1985-86	4	6	108	369	590	1142							

TABLE 5 COOLING DEGREE DAYS Base 65 deg. F   WILLIAMSPORT, PENNSYLVANIA

YEAR	JAN	FEB	MAR	APR	MAY	JUNE	JULY	AUG	SEP	OCT	NOV	DEC	TOTAL
1969	0	0	0	0	50	167	216	201	90	5	0	0	729
1970	0	0	0	8	58	109	262	237	136	12	0	0	822
1971	0	0	0	1	19	189	221	178	160	15	2	0	785
1972	0	0	0	0	25	67	236	183	40	0	0	0	551
1973	0	0	0	4	0	186	205	213	83	1	0	0	692
1974	0	0	0	11	26	63	217	204	43	0	0	0	564
1975	0	0	0	0	54	108	231	197	10	4	0	0	604
1976	0	0	0	35	20	223	173	177	47	0	0	0	675
1977	0	0	0	18	91	106	282	241	119	0	5	0	862
1978	0	0	0	0	63	134	237	310	83	0	0	0	827
1979	0	0	0	3	48	105	262	212	73	4	0	0	707
1980	0	0	0	0	34	104	279	333	128	0	0	0	878
1981	0	0	0	5	60	186	301	236	90	0	0	0	878
1982	0	0	0	0	42	48	213	103	56	2	5	0	469
1983	0	0	0	3	8	99	242	205	76	0	0	0	633
1984	0	0	0	0	9	147	186	236	44	7	0	0	629
1985	0	0	2	30	36	48	194	170	121	1	0	0	602

TABLE 6 SNOWFALL (inches)   WILLIAMSPORT, PENNSYLVANIA

SEASON	JULY	AUG	SEP	OCT	NOV	DEC	JAN	FEB	MAR	APR	MAY	JUNE	TOTAL
1956-57	0.0	0.0	0.0	0.0	5.0	6.5	9.3	9.5	14.5	5.9	T	0.0	50.7
1957-58	0.0	0.0	0.0	T	0.2	3.0	9.9	18.8	17.0	0.2	0.0	0.0	49.1
1958-59	0.0	0.0	0.0	0.0	3.3	5.4	9.6	2.5	16.4	1.9	0.0	0.0	39.1
1959-60	0.0	0.0	0.0	0.0	T	8.8	2.0	15.6	23.2	2.3	0.0	0.0	51.9
1960-61	0.0	0.0	0.0	T	T	19.7	20.5	19.6	12.5	7.9	0.0	0.0	80.2
1961-62	0.0	0.0	0.0	0.0	2.5	11.3	6.1	19.7	9.5	0.5	T	0.0	49.6
1962-63	0.0	0.0	0.0	0.3	3.1	17.7	11.9	13.6	9.2	T	T	0.0	55.8
1963-64	0.0	0.0	0.0	T	1.2	15.3	33.5	25.2	0.9	0.1	0.0	0.0	76.2
1964-65	0.0	0.0	0.0	T	T	0.7	9.4	11.7	3.0	0.1	T	0.0	24.9
1965-66	0.0	0.0	0.0	T	T	0.7	25.6	11.7	1.0	0.1	T	0.0	39.1
1966-67	0.0	0.0	0.0	0.0	0.3	11.6	3.0	17.6	29.5	2.7	T	0.0	64.7
1967-68	0.0	0.0	0.0	0.0	9.8	10.4	2.6	3.0	3.2	0.0	0.0	0.0	29.0
1968-69	0.0	0.0	0.0	0.0	7.5	2.1	0.8	6.7	1.2	0.2	T	0.0	18.3
1969-70	0.0	0.0	0.0	0.0	1.2	35.5	12.2	13.5	20.0	0.2	0.0	0.0	82.6
1970-71	0.0	0.0	0.0	T	0.2	14.8	14.2	15.5	16.7	T	0.0	0.0	61.4
1971-72	0.0	0.0	0.0	0.0	9.3	1.6	6.0	34.3	6.7	1.6	0.0	0.0	59.5
1972-73	0.0	0.0	0.0	0.3	12.8	5.1	1.5	8.3	2.6	T	0.0	0.0	30.6
1973-74	0.0	0.0	0.0	0.0	0.4	15.6	7.8	4.8	5.7	6.3	0.0	0.0	40.6
1974-75	0.0	0.0	0.0	T	T	4.6	15.4	14.9	3.8	0.1	0.0	0.0	38.8
1975-76	0.0	0.0	0.0	0.0	0.2	4.6	11.4	5.2	7.5	T	0.0	0.0	28.9
1976-77	0.0	0.0	0.0	T	3.8	6.0	16.0	7.3	8.5	0.3	0.2	0.0	42.1
1977-78	0.0	0.0	0.0	1.0	4.6	17.6	38.1	14.3	6.8	1.2	0.0	0.0	83.6
1978-79	0.0	0.0	0.0	T	3.9	4.8	10.5	17.7	T	0.4	0.0	0.0	37.3
1979-80	0.0	0.0	0.0	T	T	4.8	1.9	5.2	8.6	T	0.0	0.0	20.5
1980-81	0.0	0.0	0.0	T	12.0	6.8	10.8	5.5	6.5	T	0.0	0.0	41.6
1981-82	0.0	0.0	0.0	T	T	8.1	14.3	10.4	7.9	13.8	0.0	0.0	54.5
1982-83	0.0	0.0	0.0	T	T	1.4	6.3	7.6	0.5	1.8	0.0	0.0	17.6
1983-84	0.0	0.0	0.0	0.0	T	5.3	11.7	1.6	22.1	T	0.0	0.0	40.7
1984-85	0.0	0.0	0.0	0.0	T	8.9	8.7	6.4	0.6	0.4	0.0	0.0	25.0
1985-86	0.0	0.0	0.0	0.0	T	6.5							
Record Mean	0.0	0.0	0.0	T	3.0	8.9	10.4	10.5	8.4	1.3	T	0.0	42.7

See Reference Notes, relative to all above tables, on preceding page.

Block Island has an area of nearly 7,000 acres and is formed from glacial terminal moraine material. It is located in the Atlantic Ocean 12 miles east-northeast of Long Island and the same distance south of Charleston, RI. The climate is typically maritime, but conditions of extreme cold or heat on the mainland are also felt on the island. Temperatures have ranged from below zero in winter to above 90 degrees in summer but these are rare occurrences.

Summers are usually dry, but high monthly rainfall totals do occur. The island is too small to contribute to the development of thunderstorms. The greatest amounts of rainfall occur from storms moving in from the ocean. Fog occurs on one out of four days in the early summer, when the ocean is relatively cold.

Based on the 1951–1980 period, the average first occurrence of 32 degrees Fahrenheit in the fall is November 11 and the average last occurrence in the spring is April 10.

Winters are distinguished for their comparative mildness because of the ocean influence. Sea water temperatures are always somewhat above freezing. Since the surface winds are usually from the east when snow begins, it soon changes to rain or melts rapidly.

The ocean moderates the temperature of the air as it moves from the mainland over the island in summer as well as winter. Winds, unimpeded by mainland topography, can reach as high as 40 mph when anticyclonic conditions prevail on the mainland during the winter. The most pronounced winds come during frequent winter gales and summer or fall tropical storms moving up the coast.

TABLE 1   NORMALS, MEANS AND EXTREMES

BLOCK ISLAND, RHODE ISLAND

LATITUDE: 41°10'N   LONGITUDE: 71°35'W   ELEVATION: FT. GRND 110 BARO 109   TIME ZONE: EASTERN   WBAN: 94793

	(a)	JAN	FEB	MAR	APR	MAY	JUNE	JULY	AUG	SEP	OCT	NOV	DEC	YEAR
TEMPERATURE °F:														
Normals														
-Daily Maximum		37.2	36.9	42.8	51.8	60.7	69.8	76.0	75.8	69.7	60.8	51.5	41.9	56.2
-Daily Minimum		25.0	25.1	31.4	38.9	47.6	56.9	63.6	63.8	57.9	48.9	40.2	29.6	44.1
-Monthly		31.1	31.0	37.1	45.4	54.2	63.4	69.8	69.8	63.8	54.9	45.9	35.8	50.2
Extremes														
-Record Highest	33	57	62	74	92	82	90	91	91	87	77	70	64	92
-Year		1962	1976	1977	1976	1969	1952	1972	1973	1983	1967	1956	1953	APR 1976
-Record Lowest	33	-2	-2	8	18	34	41	51	45	42	30	20	-4	-4
-Year		1968	1961	1967	1982	1972	1967	1979	1982	1973	1976	1957	1962	DEC 1962
NORMAL DEGREE DAYS:														
Heating (base 65°F)		1051	952	865	588	335	83	7	5	75	316	573	905	5755
Cooling (base 65°F)		0	0	0	0	0	35	155	154	39	0	0	0	383
% OF POSSIBLE SUNSHINE														
MEAN SKY COVER (tenths)														
Sunrise - Sunset	19	6.5	6.0	5.8	6.3	6.2	6.3	6.8	6.6	5.9	5.4	6.6	6.2	6.2
MEAN NUMBER OF DAYS:														
Sunrise to Sunset														
-Clear	19	8.0	7.9	9.0	8.1	7.2	8.5	7.1	7.4	9.9	11.4	6.6	6.7	97.8
-Partly Cloudy	19	8.6	7.9	8.5	8.4	10.8	9.8	10.9	10.9	9.2	7.7	9.8	10.2	112.7
-Cloudy	19	14.4	12.5	13.5	13.5	13.1	11.7	13.0	12.6	10.9	11.9	13.6	14.1	154.8
Precipitation														
.01 inches or more	30	10.2	9.3	10.8	10.1	10.0	8.6	7.2	8.0	7.3	7.6	10.3	11.4	110.8
Snow,Ice pellets														
1.0 inches or more	26	1.7	1.6	1.7	0.2	0.0	0.0	0.0	0.0	0.0	0.0	0.1	1.2	6.4
Thunderstorms	15	0.2	0.3	0.3	1.4	1.8	1.9	3.9	3.6	1.4	1.1	0.4	0.1	16.4
Heavy Fog Visibility														
1/4 mile or less	15	3.6	3.7	4.7	8.2	10.1	10.0	12.4	10.7	5.4	3.7	3.2	3.1	78.8
Temperature °F														
-Maximum														
90° and above	33	0.0	0.0	0.0	0.*	0.0	0.*	0.1	0.1	0.0	0.0	0.0	0.0	0.2
32° and below	33	9.4	7.1	1.4	0.*	0.0	0.0	0.0	0.0	0.0	0.0	0.1	4.1	22.1
-Minimum														
32° and below	33	24.3	21.4	16.1	2.7	0.0	0.0	0.0	0.0	0.0	0.2	4.6	17.6	86.9
0° and below	33	0.1	0.1	0.0	0.0	0.0	0.0	0.0	0.0	0.0	0.0	0.0	0.1	0.4
AVG. STATION PRESS.(mb)														
RELATIVE HUMIDITY (%)														
Hour 01														
Hour 07	14	73	73	75	79	80	84	87	86	84	80	76	72	79
Hour 13 (Local Time)	14	65	65	65	65	67	69	72	71	70	66	65	65	67
Hour 19														
PRECIPITATION (inches):														
Water Equivalent														
-Normal		3.53	3.38	3.98	3.55	3.37	2.28	2.71	4.06	3.51	3.21	3.99	4.34	41.91
-Maximum Monthly	31	6.74	6.88	8.52	9.21	6.09	8.66	6.15	9.73	11.51	8.74	8.06	8.12	11.51
-Year		1958	1971	1959	1983	1984	1982	1959	1954	1961	1955	1969	1967	SEP 1961
-Minimum Monthly	31	0.27	0.80	1.16	0.83	0.72	T	0.39	0.16	0.33	0.81	0.89	0.83	T
-Year		1970	1980	1966	1985	1955	1957	1952	1984	1971	1984	1984	1955	JUN 1957
-Maximum in 24 hrs	32	4.06	2.86	3.63	2.73	3.67	4.30	3.61	4.86	8.52	6.63	3.96	4.39	8.52
-Year		1962	1972	1968	1983	1984	1981	1978	1953	1960	1955	1969	1967	SEP 1960
Snow,Ice pellets														
-Maximum Monthly	28	30.0	16.9	24.1	2.0						T	2.5	10.4	30.0
-Year		1978	1961	1956	1973						1970	1955	1963	JAN 1978
-Maximum in 24 hrs	28	21.7	16.9	11.5	2.0						T	2.5	6.3	21.7
-Year		1978	1961	1960	1973						1970	1955	1960	JAN 1978
WIND:														
Mean Speed (mph)	1	10.3	12.9	13.4	10.3	10.5	9.9	13.1	6.8	5.9	7.9	12.3	6.6	10.0
Prevailing Direction														
Fastest Obs. 1 Min.														
-Direction	6	27	36	05	13	01	21	30	04	16	25	27	25	16
-Speed (MPH)	6	36	46	45	28	25	25	28	29	53	36	36	40	53
-Year		1980	1983	1984	1983	1985	1983	1983	1983	1985	1980	1983	1983	SEP 1985
Peak Gust														
-Direction														
-Speed (mph)														
-Date														

See reference Notes to this table on the following page.

TABLE 2 PRECIPITATION (inches) BLOCK ISLAND, RHODE ISLAND

YEAR	JAN	FEB	MAR	APR	MAY	JUNE	JULY	AUG	SEP	OCT	NOV	DEC	ANNUAL
1956	3.59	4.62	5.47	1.99	2.72	1.50	4.35	1.63	1.86	2.30	3.06	3.49	36.58
1957	1.79	1.27	3.18	3.70	2.40	T	1.84	5.62	1.58	2.42	2.84	5.62	32.26
1958	6.74	4.44	5.09	5.54	5.25	2.12	4.85	5.53	5.50	3.75	3.01	2.20	54.02
1959	1.61	3.97	8.52	3.64	2.62	4.24	6.15	3.55	0.38	5.21	4.49	5.37	49.75
1960	4.28	5.67	1.91	4.07	4.37	2.47	4.14	1.05	11.36	2.65	3.84	6.37	52.18
1961	2.35	3.68	2.92	6.24	5.38	0.94	2.33	4.06	11.51	3.17	5.51	4.15	52.24
1962	5.07	6.44	2.52	6.11	1.59	6.81	1.97	6.74	4.24	7.31	7.49	3.30	59.59
1963	3.39	3.52	3.45	1.15	3.56	0.94	3.13	4.96	1.62	1.49	5.77	3.24	38.22
1964	3.37	3.54	2.99	5.82	1.41	1.32	1.89	0.26	3.73	2.73	1.29	6.61	34.96
1965	2.92	2.00	1.57	2.70	1.76	1.76	1.75	4.09	1.70	1.14	1.27	1.36	24.08
1966	3.23	2.29	1.16	1.32	5.62	1.55	1.18	1.41	4.74	2.40	2.89	2.49	30.28
1967	1.90	2.18	5.28	3.46	5.98	2.48	4.09	2.75	2.67	1.32	4.15	8.12	44.38
1968	2.37	1.20	6.69	1.26	2.78	4.50	0.78	1.28	0.65	2.74	5.41	5.38	35.04
1969	0.88	3.89	3.44	2.46	2.34	1.72	4.02	2.94	3.87	2.60	8.06	7.31	43.53
1970	0.27	3.63	4.74	2.99	1.69	3.09	2.63	5.35	2.79	3.83	6.20	2.97	40.18
1971	2.48	6.88	3.07	3.33	4.57	0.60	3.11	1.24	0.33	2.44	5.15	1.97	35.17
1972	2.18	5.85	4.95	3.98	4.73	6.20	1.87	0.99	6.53	2.16	7.88	5.86	53.18
1973	2.30	2.21	3.37	7.78	4.15	3.21	5.29	3.81	3.08	3.33	1.23	6.66	46.42
1974	3.85	2.12	3.58	3.17	3.17	2.88	2.10	2.90	2.73	1.91	1.36	4.71	34.48
1975	6.02	4.26	4.01	3.24	4.42	5.05	1.34	4.39	5.06		4.69	3.40	
1976	5.59	2.49	3.63	1.26	1.93	0.78	1.78	8.98	1.92	4.75	1.04	3.03	37.18
1977	2.59	2.19	4.00	3.61	1.66	4.44	1.82	4.89	4.10	4.85	2.22	6.08	42.45
1978	8.05	1.25	2.46	1.31	5.75	0.75		3.04	3.70	2.28	1.78	5.51	
1979	8.83	3.76	1.07	4.02	4.69	1.35		4.23	2.66		2.83	1.98	
1980	0.80	0.80	8.05	3.95	1.70	1.69	1.62	2.37	0.82	3.18	3.20	2.29	30.47
1981	0.74	4.81	1.34	3.68	1.64	6.15	1.87	1.69	1.96	2.77	2.34	5.19	34.18
1982	4.13	1.82	3.20	4.05	2.57	8.66	1.36	2.74	3.19	1.47	3.44	2.27	38.90
1983	3.27	4.02	5.59	9.21	1.97	2.03	0.67	1.98	1.12	2.53	7.20	3.17	42.76
1984	2.68	4.87	5.11	3.36	6.09	4.25	5.64	0.16	1.86	2.69	0.89	2.23	39.83
1985		1.86	2.61	0.83	5.55	4.45	1.78	3.00	1.98	0.84	5.58	0.93	29.66
Record Mean	3.54	3.47	3.88	3.58	3.28	2.73	2.75	3.50	3.03	3.22	3.74	3.76	40.49

TABLE 3 AVERAGE TEMPERATURE (deg. F) BLOCK ISLAND, RHODE ISLAND

YEAR	JAN	FEB	MAR	APR	MAY	JUNE	JULY	AUG	SEP	OCT	NOV	DEC	ANNUAL
1956	32.1	33.5	34.2	42.5	49.8	62.9	67.1	69.0	61.9	54.8	45.8	39.9	49.5
1957	28.2	34.3	38.6	46.5	54.8	66.1	69.3	67.7	65.1	54.6	48.4	41.8	51.3
1958	33.3	27.0	38.0	46.9	52.1	59.8	68.7	69.1	62.7	52.7	47.0	30.1	48.9
1959	30.9	29.5	36.8	46.1	56.0	62.1	68.9	69.8	66.2	56.1	45.3	38.1	50.5
1960	33.0	35.6	32.4	46.3	55.1	64.8	68.5	68.4	63.2	54.7	48.0	31.7	50.1
1961	26.6	31.9	36.4	43.6	51.4	61.6	69.1	69.3	67.7	56.6	45.6	34.6	49.6
1962	30.9	28.5	36.4	45.1	54.7	64.3	67.2	68.5	62.3	53.8	43.3	31.0	48.8
1963	30.1	26.6	37.0	45.3	52.3	65.3	69.4	67.9	59.7	56.2	48.8	30.0	48.9
1964	32.9	29.5	37.0	43.3	54.9	62.8	68.4	66.7	63.0	53.2	46.3	36.2	49.5
1965	28.1	29.7	35.0	42.5	54.6	61.0	68.5	69.6	62.9	51.7	42.9	36.4	48.6
1966	29.3	31.5	37.3	42.2	51.0	62.3	70.1	69.8	62.4	52.8	47.3	36.3	49.3
1967	35.8	28.7	33.2	42.2	48.5	61.6	61.6	68.8	61.5	55.5	41.6	36.8	48.7
1968	27.2	26.5	37.6	46.6	53.0	62.2	69.7	69.7	65.3	57.4	45.3	32.8	49.5
1969	30.7	30.6	34.7	45.7	54.8	64.9	69.0	71.4	64.3	54.6	45.8	34.4	50.1
1970	24.3	30.6	35.3	44.6	54.9	62.1	70.8	71.4	64.3	55.7	47.4	33.7	49.6
1971	26.8	31.8	36.3	42.9	53.3	63.6	70.8	70.1	67.4	59.7	44.1	38.8	50.4
1972	33.7	30.9	36.0	42.5	53.5	61.6	71.4	70.3	64.3	55.6	43.7	38.7	49.9
1973	33.0	31.6	41.6	46.7	54.4	65.2	70.6	72.0	63.9	56.1	45.6	39.6	51.7
1974	34.4	30.4	38.5	47.1	52.7	62.9	69.6	72.0	64.5	51.3	46.6	39.1	50.8
1975	36.5	33.1	36.7	42.9	55.9	64.0	71.7	71.3	62.9		50.5	34.3	
1976	27.5	37.8	40.9	52.0	58.1	69.4	68.8		62.6	51.1	40.0	29.7	50.5
1977	22.6	28.6	40.1	46.4	56.0	63.5	71.1	70.9	64.1	55.1	47.9	35.8	50.2
1978	29.2	26.4	34.2	43.9	53.3	63.8	67.9	71.0	61.3	54.1	46.4	38.0	49.1
1979	33.1	22.1	40.0	45.3	57.2	63.2	71.0	70.6	65.2	54.7	51.4	40.8	51.2
1980	33.3	29.9	38.2	48.0	57.1	62.3	72.7	71.9	64.8	52.8	42.9	32.1	50.6
1981	23.9	35.1	35.8	46.2	55.0	65.0	70.8	68.3	62.4	51.9	44.9	35.8	49.6
1982	26.3	32.3	37.1	46.2	54.7	61.6	71.6	67.4	62.2	54.6	48.8	41.6	50.1
1983	34.2	34.7	40.7	47.5	54.2	65.5	71.5	69.9	67.1	56.3	49.2	35.2	52.2
1984	30.1	37.9	34.6	45.9	55.9	66.0	69.7	72.5	63.6	58.2	47.2	43.9	52.1
1985		32.8	39.3	47.8	56.3	62.6	70.5	70.8	65.9	57.2	50.3	36.9	51.7
Record Mean	31.4	30.9	36.7	44.7	53.6	62.6	69.1	69.1	64.3	55.4	46.0	36.3	50.0
Max	37.4	36.7	42.2	50.6	59.7	68.7	75.0	74.6	69.9	60.9	51.4	42.1	55.8
Min	25.4	25.0	31.1	38.8	47.5	56.6	63.2	63.5	58.8	49.9	40.6	30.4	44.2

REFERENCE NOTES FOR TABLES 1, 2, 3 and 6   (BLOCK ISLAND, RI)

GENERAL

T - TRACE AMOUNT
BLANK ENTRIES DENOTE MISSING/UNREPORTED DATA.
# INDICATES A STATION OR INSTRUMENT RELOCATION.

SPECIFIC

TABLE 1

(a) - LENGTH OF RECORD IN YEARS. ALTHOUGH INDIVIDUAL MONTHS MAY BE MISSING.

 * LESS THAN .05

NORMALS — BASED ON THE 1951-1980 RECORD PERIOD.
EXTREMES — DATES ARE THE MOST RECENT OCCURRENCE.
WIND DIR. — NUMERALS SHOW TENS OF DEGREES CLOCKWISE FROM TRUE NORTH.
    "00" INDICATES CALM.
RESULTANT WIND DIRECTIONS ARE GIVEN TO WHOLE DEGREES.

EXCEPTIONS

TABLE 1

1. THUNDERSTORMS AND HEAVY FOG MAY BE INCOMPLETE, DUE TO PART-TIME OPERATIONS.
2. THUNDERSTORMS, HEAVY FOG, AND RELATIVE HUMIDITY ARE THROUGH 1964.
3. MEAN SKY COVER, AND DAYS CLEAR - PARTLY CLOUDY - CLOUDY ARE THROUGH 1968 AND JANUARY-JULY 1974.
3. DAYS OF SNOW 1 INCH OR MORE ARE THROUGH 1977.
4. SNOW DATA IS THROUGH DECEMBER 1978.

TABLES 2, 3, and 6

RECORD MEANS ARE THROUGH THE CURRENT YEAR, BEGINNING IN   1881 FOR TEMPERATURE
    1881 FOR PRECIPITATION
    1951 - 1978 FOR SNOWFALL

TABLE 4 HEATING DEGREE DAYS Base 65 deg. F   BLOCK ISLAND, RHODE ISLAND

SEASON	JULY	AUG	SEP	OCT	NOV	DEC	JAN	FEB	MAR	APR	MAY	JUNE	TOTAL
1956-57	15	11	130	306	571	772	1131	853	812	548	307	60	5516
1957-58	3	14	67	316	491	714	974	1058	830	537	394	158	5556
1958-59	6	3	86	374	535	1070	1050	988	868	561	276	112	5929
1959-60	2	10	73	282	585	825	984	844	1001	556	298	37	5497
1960-61	2	5	79	309	504	1021	1182	920	880	634	415	100	6051
1961-62	7	4	39	257	577	932	1052	1015	881	591	313	65	5733
1962-63	11	11	106	344	645	1047	1073	1071	860	585	384	92	6229
1963-64	7	9	162	266	478	1076	988	1024	860	644	309	96	5919
1964-65	9	14	99	359	555	884	1136	982	923	669	320	132	6082
1965-66	1	20	96	405	657	877	1099	931	854	675	429	105	6149
1966-67	0	0	101	371	523	883	898	1012	977	676	507	105	6053
1967-68	2	9	115	294	695	865	1166	1109	846	543	365	103	6112
1968-69	7	10	40	239	583	992	1060	959	931	572	318	45	5756
1969-70	5	3	74	319	568	943	1255	957	915	601	307	95	6042
1970-71	0	0	70	288	524	963	1180	927	883	659	357	73	5924
1971-72	0	12	31	162	619	805	962	982	895	669	350	105	5592
1972-73	4	4	53	409	636	808	984	926	716	542	325	37	5444
1973-74	0	2	97	272	577	782	943	961	816	532	377	86	5445
1974-75	6	0	73	417	546	797	874	887	870	657	280	66	5473
1975-76	0	6	76		427	944	1154	783	740	402	217	29	
1976-77	3	14	87	425	742	1088	1014	764	549	278	70	6340
1977-78	0	3	84	301	508	899	1104	1073	946	624	358	72	5972
1978-79	16	5	125	331	550	827	985	1196	769	583	235	68	5690
1979-80	18	11	70	315	402	745	976	1011	823	505	242	74	5192
1980-81	0	1	68	372	656	1014	1268	831	897	559	305	30	6001
1981-82	0	13	93	400	597	899	1191	908	860	650	310	114	6035
1982-83	5	28	96	317	479	720	949	842	748	519	329	46	5078
1983-84	1	5	57	274	468	915	1073	779	934	565	276	40	5387
1984-85	0	0	81	201	525	645		897	790	509	263	82	
1985-86	7	2	45	237	435	864							

TABLE 5 COOLING DEGREE DAYS Base 65 deg. F   BLOCK ISLAND, RHODE ISLAND

YEAR	JAN	FEB	MAR	APR	MAY	JUNE	JULY	AUG	SEP	OCT	NOV	DEC	TOTAL
1969	0	0	0	0	8	52	136	209	60	3	0	0	468
1970	0	0	0	0	0	12	186	208	55	5	0	0	466
1971	0	0	0	0	0	38	185	177	111	2	0	0	513
1972	0	0	0	0	0	8	208	174	38	0	0	0	428
1973	0	0	0	0	2	48	184	225	73	4	0	0	536
1974	0	0	0	0	0	32	157	227	64	0	0	0	480
1975	0	0	0	0	4	40	217	208	18		0	0	
1976	0	0	0	18	10	166	128	138	21	0	0	0	481
1977	0	0	0	0	3	33	197	193	65	0	0	0	491
1978	0	0	0	0	0	42	113	198	25	0	0	0	378
1979	0	0	0	0	0	22	211	190	84	2	0	0	509
1980	0	0	0	0	0	27	245	222	70	0	0	0	564
1981	0	0	0	0	0	36	188	123	25	0	0	0	372
1982	0	0	0	0	0	16	215	109	18	1	0	0	359
1983	0	0	0	0	0	70	211	163	125	10	0	0	579
1984	0	0	0	0	0	78	153	239	45	0	0	0	515
1985		0	0	0	0	17	184	189	82	3	0	0	475

TABLE 6 SNOWFALL (inches)   BLOCK ISLAND, RHODE ISLAND

SEASON	JULY	AUG	SEP	OCT	NOV	DEC	JAN	FEB	MAR	APR	MAY	JUNE	TOTAL
1956-57	0.0	0.0	0.0	0.0	T	2.2	2.2	2.8	4.2	1.1	0.0	0.0	12.5
1957-58	0.0	0.0	0.0	0.0	T	1.5	T	15.4	18.1	T	0.0	0.0	35.0
1958-59	0.0	0.0	0.0	0.0	T	4.4	0.6	1.8	7.8	0.3	0.0	0.0	14.9
1959-60	0.0	0.0	0.0	0.0	T	4.2	7.9	1.9	13.7	T	0.0	0.0	27.7
1960-61	0.0	0.0	0.0	0.0	0.0	7.0	9.0	16.9	5.4	T	0.0	0.0	38.3
1961-62	0.0	0.0	0.0	0.0	0.0	0.0	4.2	3.8	12.0	0.8	0.0	0.0	20.8
1962-63	0.0	0.0	0.0	T	0.3	7.2	5.2	2.5	3.0	0.5	0.0	0.0	18.7
1963-64	0.0	0.0	0.0	0.0	0.1	10.4	5.9	12.1	1.9	0.0	0.0	0.0	30.4
1964-65	0.0	0.0	0.0	0.0	0.0	1.1	21.5	2.2	2.6	1.3	0.0	0.0	28.7
1965-66	0.0	0.0	0.0	0.0	T	T	8.5	4.6	2.0	0.0	0.0	0.0	15.1
1966-67	0.0	0.0	0.0	0.0	0.0	3.2	0.5	15.6	19.3	T	0.0	0.0	38.6
1967-68	0.0	0.0	0.0	0.0	0.4	3.7	8.8	2.8	4.4	0.0	0.0	0.0	20.1
1968-69	0.0	0.0	0.0	0.0	0.0	0.1	T	8.3	8.5	0.0	0.0	0.0	16.9
1969-70	0.0	0.0	0.0	0.0	T	1.5	1.4	2.2	11.7	T	0.0	0.0	16.8
1970-71	0.0	0.0	0.0	T	0.0	4.2	3.5	0.5	T	0.0	0.0	0.0	8.2
1971-72	0.0	0.0	0.0	0.0	T	1.0	2.2	5.0	0.2	0.5	0.0	0.0	8.9
1972-73	0.0	0.0	0.0	0.0	0.0	0.9	0.3	2.1	1.4	2.0	0.0	0.0	6.7
1973-74	0.0	0.0	0.0	0.0	0.0	T	8.8	8.5	T	0.6	0.0	0.0	17.9
1974-75	0.0	0.0	0.0	0.0	T	2.5	10.3	7.8	T	0.0	0.0	20.6
1975-76	0.0	0.0	0.0	0.0	0.0	1.8	9.3	2.8	7.6	0.0	0.0	0.0	21.5
1976-77	0.0	0.0	0.0	0.0	0.0	5.5	4.6	8.3	3.0	0.0	0.0	0.0	21.4
1977-78	0.0	0.0	0.0	0.0	T	0.8	30.0	11.1	9.8	0.0	0.0	0.0	51.7
1978-79	0.0	0.0	0.0	0.0	T								
1979-80													
1980-81													
1981-82													
1982-83													
1983-84													
1984-85													
1985-86		0.0	0.0										
Record Mean	0.0	0.0	0.0	T	0.2	2.9	5.1	6.3	5.9	0.3	0.0	0.0	20.6

See Reference Notes, relative to all above tables, on preceding page.

The proximity to Narragansett Bay and the Atlantic Ocean plays an important part in determining the climate for Providence and vicinity. In winter, the temperatures are modified considerably, and many major snowstorms change to rain before reaching the area. In summer, many days that could be uncomfortably warm are cooled by refreshing sea breezes. At other times of the year, sea fog may be advected in over land by onshore winds. In fact, most cases of dense fog are produced this way, but the number of such days is few, averaging two or three days per month. In early fall, severe coastal storms of tropical origin sometimes bring destructive winds to this area. Even at other times of the year, it is usually coastal storms which produce the severest weather.

The temperature for the entire year averages around 50 degrees with 70 degree temperatures common from near the end of May to the latter part of September. During this period, there may be several days reaching 90 degrees or more. Temperatures of 100 degrees and more are rare.

Freezing temperatures occur on the average about 125 days per year. They become a common daily occurrence in the latter part of November, and become less frequent near the end of March. The average date for the last freeze in spring is mid-April, while the average date for the first freeze in fall is late October, making the growing season about 195 days in length. Sub-zero weather in winter seldom occurs, averaging less than one day for December and one or two days each for January and February.

Measurable precipitation occurs on about one day out of every three, and is fairly evenly distributed throughout the year. There is usually no definite dry season, but occasionally droughts do occur.

Thunderstorms are responsible for much of the rainfall from May through August. They usually produce heavy, and sometimes even excessive amounts of rainfall. However, since their duration is relatively short, damage is ordinarily light. The thunderstorms of summer are frequently accompanied by extremely gusty winds, which may result in some damage to property.

The first measurable snowfall of winter usually comes toward the end of November, and the last in spring is about the middle of March. Winters with over 50 inches of snow are not common. The area normally receives less than 25 inches. The month of greatest snowfall is usually February, but January and March are close seconds. It is unusual for the ground to remain well covered with snow for any long period of time.

TABLE 1   NORMALS, MEANS AND EXTREMES

PROVIDENCE, RHODE ISLAND

LATITUDE: 41°44'N   LONGITUDE: 71°26'W   ELEVATION: FT. GRND 51 BARO 00058   TIME ZONE: EASTERN   WBAN: 14765

	(a)	JAN	FEB	MAR	APR	MAY	JUNE	JULY	AUG	SEP	OCT	NOV	DEC	YEAR
TEMPERATURE °F:														
Normals														
–Daily Maximum		36.4	37.7	45.5	57.5	67.6	76.6	81.7	80.3	73.1	63.2	51.9	40.5	59.3
–Daily Minimum		20.0	20.9	29.2	38.3	47.6	57.0	63.3	61.9	53.8	43.1	34.8	24.1	41.2
–Monthly		28.2	29.3	37.4	47.9	57.6	66.8	72.5	71.1	63.5	53.2	43.4	32.3	50.3
Extremes														
–Record Highest	32	66	72	78	98	94	96	100	104	100	86	78	70	104
–Year		1974	1985	1977	1976	1964	1980	1980	1975	1983	1979	1974	1984	AUG 1975
–Record Lowest	32	-13	-7	1	14	29	41	49	40	33	20	14	-10	-13
–Year		1976	1979	1967	1954	1956	1980	1979	1965	1980	1976	1972	1980	JAN 1976
NORMAL DEGREE DAYS:														
Heating (base 65°F)		1141	1000	856	513	239	31	0	6	94	366	648	1014	5908
Cooling (base 65°F)		0	0	0	0	10	85	235	195	49	0	0	0	574
% OF POSSIBLE SUNSHINE	32	57	57	57	57	57	60	64	60	61	60	50	52	58
MEAN SKY COVER (tenths)														
Sunrise - Sunset	32	6.2	6.3	6.6	6.5	6.7	6.3	6.3	6.2	5.8	5.5	6.3	6.2	6.2
MEAN NUMBER OF DAYS:														
Sunrise to Sunset														
–Clear	32	9.9	7.9	8.6	7.9	6.7	6.9	7.2	8.3	9.7	11.0	8.4	8.3	100.8
–Partly Cloudy	32	6.7	7.5	7.6	8.2	10.0	10.2	12.0	10.5	8.3	7.9	6.7	7.8	103.4
–Cloudy	32	14.4	12.8	14.9	13.8	14.3	12.8	11.8	12.3	12.1	12.1	14.9	14.8	161.0
Precipitation														
.01 inches or more	32	11.1	10.1	11.8	10.9	11.4	10.8	8.7	9.5	8.2	8.3	11.0	12.3	124.2
Snow, Ice pellets														
1.0 inches or more	32	2.8	2.3	2.0	0.3	0.*	0.0	0.0	0.0	0.0	0.1	0.3	2.2	10.0
Thunderstorms	32	0.2	0.2	0.6	1.3	2.7	3.7	4.3	3.8	1.7	1.0	0.8	0.2	20.3
Heavy Fog Visibility														
1/4 mile or less	32	2.0	2.1	2.1	2.0	2.3	2.2	1.9	1.4	1.8	3.2	2.2	1.9	25.1
Temperature °F														
–Maximum														
90° and above	22	0.0	0.0	0.0	0.1	0.6	1.9	3.7	2.2	1.0	0.0	0.0	0.0	9.6
32° and below	22	12.5	8.1	1.4	0.*	0.0	0.0	0.0	0.0	0.0	0.0	0.3	6.2	28.5
–Minimum														
32° and below	22	28.2	24.2	19.8	6.0	0.2	0.0	0.0	0.0	0.0	3.8	12.3	24.7	119.2
0° and below	22	1.6	0.7	0.0	0.0	0.0	0.0	0.0	0.0	0.0	0.0	0.0	0.4	2.7
AVG. STATION PRESS. (mb)	13	1013.5	1014.4	1012.7	1011.9	1012.4	1012.7	1012.8	1014.8	1015.9	1016.5	1015.1	1015.0	1014.0
RELATIVE HUMIDITY (%)														
Hour 01	22	68	67	68	69	77	82	81	83	83	79	75	72	75
Hour 07	22	70	69	70	67	72	75	76	79	81	79	77	74	74
Hour 13 (Local Time)	22	56	54	53	47	52	56	55	56	55	53	57	58	54
Hour 19	22	63	61	60	58	63	67	66	70	72	70	69	67	66
PRECIPITATION (inches):														
Water Equivalent														
–Normal		4.06	3.72	4.29	3.95	3.48	2.79	3.01	4.04	3.54	3.75	4.22	4.47	45.32
–Maximum Monthly	32	11.66	7.20	8.84	12.74	8.38	11.08	8.08	11.12	7.92	11.89	11.01	10.75	12.74
–Year		1979	1984	1983	1983	1984	1982	1976	1955	1961	1962	1983	1969	APR 1983
–Minimum Monthly	32	0.50	1.16	0.56	1.48	0.71	0.39	1.00	0.71	0.77	1.62	0.81	0.58	0.39
–Year		1970	1980	1981	1966	1964	1957	1970	1984	1959	1964	1976	1955	JUN 1957
–Maximum in 24 hrs	32	3.34	3.14	4.53	4.45	5.17	5.03	4.83	6.71	4.89	6.63	4.18	3.85	6.71
–Year		1962	1978	1968	1983	1984	1984	1976	1979	1961	1962	1983	1969	AUG 1979
Snow, Ice pellets														
–Maximum Monthly	32	28.7	30.9	31.6	7.6	7.0					2.5	4.1	19.8	31.6
–Year		1965	1962	1956	1982	1977					1979	1980	1963	MAR 1956
–Maximum in 24 hrs	32	10.8	27.6	16.9	7.6	7.0					2.5	4.1	11.9	27.6
–Year		1978	1978	1960	1982	1977					1979	1980	1981	FEB 1978
WIND:														
Mean Speed (mph)	32	11.3	11.6	12.2	12.2	10.9	10.0	9.5	9.3	9.4	9.6	10.5	10.9	10.6
Prevailing Direction														
through 1963		NW	NNW	WNW	SW	S	SW	SW	SSW	SW	NW	SW	WNW	SW
Fastest Obs. 1 Min.														
–Direction	32	20	16	18	20	20	20	34	11	18	14	18	14	11
–Speed (MPH)	32	46	46	60	51	42	40	35	90	58	41	52	48	90
–Year		1978	1972	1959	1956	1956	1957	1964	1954	1960	1954	1957	1957	AUG 1954
Peak Gust														
–Direction	2	NW	SE	NE	SE	S	S	SW	N	S	SW	SW	W	S
–Speed (mph)	2	39	53	52	54	40	46	43	32	81	49	45	48	81
–Date		1984	1985	1984	1984	1984	1985	1984	1985	1985	1985	1985	1985	SEP 1985

See Reference Notes to this table on the following page.

TABLE 2 PRECIPITATION (inches) PROVIDENCE, RHODE ISLAND

YEAR	JAN	FEB	MAR	APR	MAY	JUNE	JULY	AUG	SEP	OCT	NOV	DEC	ANNUAL
1956	4.92	4.60	5.51	3.08	1.43	1.57	4.92	0.91	3.10	3.74	3.62	5.27	42.67
1957	2.17	1.68	3.29	4.46	0.93	0.39	1.41	2.51	0.87	2.52	3.99	5.86	30.08
1958	7.12	2.95	3.45	7.21	4.05	3.15	6.29	5.15	5.02	3.08	2.58	1.49	51.54
1959	2.27	3.67	6.04	3.83	1.46	4.83	4.01	3.53	0.77	4.71	3.85	4.17	43.14
1960	3.02	5.63	2.48	2.94	3.79	1.26	4.61	1.06	5.98	2.24	2.77	4.30	40.08
1961	3.52	4.68	4.16	7.32	5.21	1.48	2.76	3.86	7.92	2.39	3.10	3.16	49.56
1962	4.70	5.16	1.93	3.85	2.14	5.52	1.62	2.73	3.67	11.89	4.49	2.63	50.33
1963	3.40	3.15	3.78	1.62	4.69	3.54	3.35	1.56	4.10	1.63	6.53	2.15	39.50
1964	5.65	3.15	2.26	5.34	0.71	2.34	2.63	2.38	3.95	2.11	2.43	5.46	38.41
1965	3.46	3.77	1.72	2.43	1.08	1.91	1.28	1.90	1.64	2.75	2.08	1.42	25.44
1966	3.40	4.30	2.40	1.48	3.85	2.31	2.77	3.37	5.23	2.60	3.93	3.04	38.68
1967	1.60	2.51	5.49	4.19	7.27	2.72	3.95	3.24	3.17	2.25	2.75	7.36	46.50
1968	3.50	1.31	7.83	1.49	3.54	4.74	1.49	1.61	1.14	1.79	6.22	6.70	41.36
1969	2.23	4.30	3.10	3.95	2.41	1.23	2.98	2.58	3.09	1.62	6.35	10.75	44.59
1970	0.50	5.34	4.75	3.91	3.03	4.25	1.00	6.59	1.79	4.41	5.31	4.54	45.42
1971	2.01	5.36	3.81	2.31	3.83	1.64	3.48	3.03	2.54	2.88	5.16	2.37	38.42
1972	1.85	5.19	6.70	3.71	5.73	6.83	4.25	2.98	7.31	4.36	8.45	7.70	65.06
1973	3.06	3.55	2.78	7.16	3.99	3.48	2.92	5.17	3.04	3.17	2.29	7.63	48.24
1974	4.45	3.04	4.51	2.86	2.74	3.28	1.64	3.10	6.15	2.79	1.56	4.54	40.66
1975	6.78	3.29	3.07	2.99	2.06	4.73	3.51	2.19	6.15	4.66	6.29	5.11	50.83
1976	6.38	2.91	3.44	2.00	2.53	1.60	8.08	7.01	1.57	6.52	0.81	3.47	46.32
1977	3.90	2.87	5.62	3.35	3.43	3.92	2.04	2.12	5.60	6.90	3.24	5.85	48.84
1978	9.01	3.20	3.10	2.53	5.27	1.97	2.63	6.46	1.82	3.22	2.61	5.19	47.01
1979	11.66	4.08	2.21	5.12	7.62	1.44	1.65	10.09	4.08	3.94	4.49	1.81	58.19
1980	1.40	1.16	8.11	6.18	1.78	3.85	2.03	1.99	0.90	3.41	3.73	1.57	36.11
1981	0.77	4.79	0.56	4.10	1.92	2.31	3.75	2.65	2.58	3.38	3.20	6.36	36.37
1982	6.09	3.08	3.76	3.64	1.61	11.08	3.51	3.67	3.61	3.08	4.32	1.81	49.26
1983	4.32	4.81	8.84	12.74	4.67	1.41	2.14	2.71	2.16	4.50	11.01	7.71	67.52
1984	2.00	7.20	5.77	4.30	8.38	4.09	5.16	0.71	1.77	4.25	1.95	3.16	48.74
1985	1.18	1.57	3.08	1.65	4.76	4.70	2.88	8.57	1.69	1.78	7.14	1.42	40.42
Record Mean	3.72	3.32	3.79	3.69	3.27	3.07	3.10	3.69	3.28	3.17	3.80	3.86	41.77

TABLE 3 AVERAGE TEMPERATURE (deg. F) PROVIDENCE, RHODE ISLAND

YEAR	JAN	FEB	MAR	APR	MAY	JUNE	JULY	AUG	SEP	OCT	NOV	DEC	ANNUAL
1956	30.4	32.3	33.0	43.9	53.4	67.1	70.5	70.5	60.1	52.1	43.9	36.2	49.4
1957	22.7	33.9	38.9	49.5	58.8	70.5	72.4	67.8	64.8	52.9	45.5	38.4	51.3
1958	30.7	24.7	38.4	47.8	54.9	62.5	71.5	70.7	63.0	51.5	44.9	25.3	48.8
1959	28.2	26.8	36.4	49.0	64.6	72.5	73.7	66.8	54.1	43.3	34.5	50.9	50.4
1960	29.1	35.1	31.5	48.7	58.5	67.5	71.0	70.8	62.5	52.6	45.7	27.4	50.0
1961	23.7	30.6	37.0	45.4	55.2	67.5	72.2	71.0	60.0	55.8	43.6	32.1	50.3
1962	28.5	26.6	37.7	48.6	56.6	66.6	68.5	69.0	60.8	52.2	40.9	28.6	48.7
#1963	28.9	26.3	38.9	48.3	57.1	67.5	72.8	69.4	59.6	56.8	46.9	24.7	49.8
1964	30.6	27.9	37.8	46.1	60.3	66.3	71.7	66.6	62.5	52.6	44.6	33.2	50.0
1965	25.0	28.4	36.1	45.4	60.0	67.0	71.6	71.6	64.1	52.1	41.1	35.4	49.8
1966	28.8	29.9	38.8	43.8	54.4	67.4	72.7	70.7	61.3	51.4	45.5	32.8	49.8
1967	33.7	25.7	33.3	44.8	51.2	66.8	72.8	70.6	62.7	53.7	39.5	34.5	49.1
1968	24.5	24.6	38.1	49.9	55.7	65.0	73.1	70.6	64.8	55.9	42.4	30.2	49.6
1969	28.8	28.7	35.0	49.7	57.7	68.2	71.5	74.3	64.0	53.3	42.4	30.5	50.4
1970	19.6	29.3	35.0	47.8	58.2	65.6	74.1	72.3	64.4	54.2	44.7	28.5	49.5
1971	22.9	30.9	36.7	45.9	58.1	69.0	74.3	73.0	68.7	59.2	40.4	35.0	51.2
1972	30.8	28.0	36.4	44.3	57.7	64.9	72.6	70.6	65.1	49.6	40.9	34.3	49.6
1973	31.1	29.6	43.7	50.0	56.7	70.3	73.6	75.0	63.4	54.2	43.8	38.3	52.5
1974	31.6	29.0	38.7	50.6	55.6	65.3	72.6	72.6	63.2	48.2	43.7	35.7	50.5
1975	34.1	30.4	35.5	44.5	61.4	66.0	74.3	71.4	61.0	55.3	48.0	32.1	51.2
1976	23.5	35.5	39.0	52.6	58.0	70.0	70.0	70.0	61.6	48.7	37.9	25.4	49.4
1977	20.9	29.8	43.6	50.6	61.2	66.7	74.3	73.0	64.1	52.9	45.9	31.6	51.2
1978	25.1	22.1	33.8	46.7	57.8	68.0	71.7	71.3	59.5	51.5	42.3	33.4	48.6
1979	30.0	19.7	40.4	46.8	60.3	65.1	73.5	70.2	64.0	53.0	48.3	37.3	50.7
1980	29.7	26.8	37.1	49.5	60.3	64.5	74.8	73.4	64.9	49.8	41.0	28.5	50.0
1981	20.3	37.4	38.7	51.4	58.5	69.4	75.6	70.0	62.4	49.1	43.0	31.1	50.6
1982	21.5	31.5	38.8	47.8	58.9	63.9	73.6	69.2	64.1	53.2	47.5	38.6	50.7
1983	31.4	32.9	40.4	49.9	56.9	70.2	76.6	74.3	69.6	55.3	46.0	32.5	53.0
1984	26.4	37.1	33.8	47.6	57.4	69.1	71.5	73.5	62.1	56.3	43.6	37.9	51.4
1985	22.5	32.1	40.8	51.0	60.2	64.8	73.0	71.1	65.2	54.6	45.9	30.4	51.0
Record Mean	28.9	29.4	37.7	47.7	58.0	66.6	72.8	71.0	63.8	53.8	43.4	32.6	50.5
Max	36.7	37.5	45.9	56.9	67.7	76.6	82.0	80.1	73.2	63.2	51.4	40.2	59.3
Min	21.0	21.3	29.5	38.5	48.2	57.3	63.6	61.9	54.5	44.4	35.4	25.0	41.7

REFERENCE NOTES FOR TABLES 1, 2, 3 and 6   (PROVIDENCE, RI)

GENERAL

T - TRACE AMOUNT
BLANK ENTRIES DENOTE MISSING/UNREPORTED DATA.
# INDICATES A STATION OR INSTRUMENT RELOCATION.

SPECIFIC

TABLE 1

(a) - LENGTH OF RECORD IN YEARS. ALTHOUGH
INDIVIDUAL MONTHS MAY BE MISSING.

* LESS THAN .05

NORMALS — BASED ON THE 1951-1980 RECORD PERIOD.
EXTREMES — DATES ARE THE MOST RECENT OCCURRENCE.
WIND DIR. — NUMERALS SHOW TENS OF DEGREES
CLOCKWISE FROM TRUE NORTH.
"00" INDICATES CALM.
RESULTANT WIND DIRECTIONS ARE GIVEN TO WHOLE DEGREES.

EXCEPTIONS

TABLES 2, 3, and 6

RECORD MEANS ARE THROUGH THE CURRENT YEAR,
BEGINNING IN   1905 FOR TEMPERATURE
1905 FOR PRECIPITATION
1954 FOR SNOWFALL

TABLE 4 HEATING DEGREE DAYS Base 65 deg. F   PROVIDENCE, RHODE ISLAND

SEASON	JULY	AUG	SEP	OCT	NOV	DEC	JAN	FEB	MAR	APR	MAY	JUNE	TOTAL
1956-57	13	13	194	394	630	885	1304	866	804	463	222	35	5823
1957-58	0	22	93	368	576	817	1056	1124	816	509	309	114	5804
1958-59	2	6	108	415	595	1225	1134	1063	879	473	180	88	6168
1959-60	0	6	101	354	644	942	1106	861	1034	480	196	30	5754
1960-61	0	8	117	382	573	1159	1274	955	864	581	302	26	6241
1961-62	0	6	49	284	633	1013	1123	1068	840	489	274	40	5819
1962-63	12	14	152	389	717	1125	1115	1076	805	493	252	46	6196
#1963-64	6	8	173	259	536	1242	1059	1066	835	560	188	52	5984
1964-65	9	24	125	377	605	981	1231	1018	891	581	182	67	6091
1965-66	3	29	99	395	711	907	1115	975	806	630	326	57	6053
1966-67	1	1	135	417	577	994	963	1093	976	598	424	48	6227
1967-68	0	7	103	356	761	937	1246	1166	827	447	281	74	6205
1968-69	2	16	59	295	672	1072	1117	1010	923	509	214	60	6330
1969-70	2	4	119	365	673	1065	1399	996	924	566	212	34	6097
1970-71	0	0	102	342	602	1124	1298	868	949	447	281	74	6097
1971-72	0	7	42	181	736	922	1054	1064	879	616	226	55	5782
1972-73	8	10	64	473	717	945	1044	984	653	451	268	16	5633
1973-74	2	3	125	331	632	819	1028	1003	808	433	313	62	5559
1974-75	0	0	114	512	634	899	951	962	907	606	160	64	5809
1975-76	0	13	132	298	506	1013	1283	850	798	403	223	39	5558
1976-77	2	23	124	501	806	1219	1361	983	653	434	176	51	6333
1977-78	0	6	103	368	568	1030	1231	1192	964	542	238	26	6268
1978-79	8	8	180	412	673	970	1075	1261	755	540	162	52	6096
1979-80	11	25	94	380	496	849	1088	1104	857	459	158	93	5614
1980-81	0	1	120	465	715	1125	1379	769	808	405	228	13	6028
1981-82	0	20	119	486	651	1044	1343	932	802	510	190	91	6188
1982-83	1	26	78	363	518	809	1038	892	755	449	254	13	5196
1983-84	0	4	62	323	563	1001	1190	802	961	513	236	36	5691
1984-85	1	0	125	270	637	832	1309	914	743	417	177	63	5488
1985-86	0	6	78	321	567	1065							

TABLE 5 COOLING DEGREE DAYS Base 65 deg. F   PROVIDENCE, RHODE ISLAND

YEAR	JAN	FEB	MAR	APR	MAY	JUNE	JULY	AUG	SEP	OCT	NOV	DEC	TOTAL
1969	0	0	0	0	23	125	211	299	94	9	0	0	761
1970	0	0	0	0	13	86	289	237	91	14	0	0	730
1971	0	0	0	0	3	157	296	263	158	8	5	0	890
1972	0	0	0	1	7	60	248	190	76	1	0	0	583
1973	0	0	0	8	17	181	272	318	84	3	0	0	883
1974	0	0	0	7	27	79	242	244	66	0	1	0	666
1975	0	0	0	0	55	100	300	218	16	4	1	0	694
1976	0	0	0	40	13	196	163	183	33	3	0	0	631
1977	0	0	0	5	68	108	295	260	85	0	0	0	821
1978	0	0	0	0	25	126	224	211	24	0	0	0	610
1979	0	0	0	0	26	59	279	190	74	12	0	0	640
1980	0	0	0	0	21	84	312	272	122	0	0	0	811
1981	0	0	0	2	33	152	335	183	47	0	0	0	752
1982	0	0	0	0	11	64	276	165	59	3	2	0	580
1983	0	0	0	1	8	177	367	298	206	30	0	0	1087
1984	0	0	0	0	6	164	206	272	47	7	0	0	702
1985	0	0	0	5	34	65	256	203	90	5	0	0	658

TABLE 6 SNOWFALL (inches)   PROVIDENCE, RHODE ISLAND

SEASON	JULY	AUG	SEP	OCT	NOV	DEC	JAN	FEB	MAR	APR	MAY	JUNE	TOTAL
1956-57	0.0	0.0	0.0	0.0	T	6.1	14.8	3.5	10.5	2.1	0.0	0.0	37.0
1957-58	0.0	0.0	0.0	T	0.2	0.3	7.5	11.0	14.4	3.7	0.0	0.0	37.1
1958-59	0.0	0.0	0.0	0.0	T	7.5	2.2	5.2	11.0	T	0.0	0.0	25.9
1959-60	0.0	0.0	0.0	0.0	T	5.1	9.9	1.6	21.6	T	0.0	0.0	38.2
1960-61	0.0	0.0	0.0	T	T	14.1	17.4	18.7	12.2	0.3	0.0	0.0	62.7
1961-62	0.0	0.0	0.0	T	1.7	9.2	4.9	30.9	T	0.3	0.0	0.0	47.0
1962-63	0.0	0.0	0.0	1.6	1.7	6.6	5.3	5.2	9.4	T	0.0	0.0	29.8
1963-64	0.0	0.0	0.0	T	T	19.8	12.5	14.8	2.5	T	0.0	0.0	49.6
1964-65	0.0	0.0	0.0	0.0	T	6.9	28.7	2.7	6.6	T	0.0	0.0	44.9
1965-66	0.0	0.0	0.0	0.0	T	1.9	16.1	9.1	8.1	T	0.0	0.0	35.2
1966-67	0.0	0.0	0.0	0.0	T	7.2	1.3	23.1	24.9	1.6	0.0	0.0	58.1
1967-68	0.0	0.0	0.0	0.0	0.8	19.0	13.5	4.6	5.1	0.0	0.0	0.0	43.0
1968-69	0.0	0.0	0.0	0.0	0.1	2.7	0.5	26.7	6.0	0.0	0.0	0.0	36.0
1969-70	0.0	0.0	0.0	0.0	T	15.4	6.5	6.8	15.7	1.1	0.0	0.0	45.5
1970-71	0.0	0.0	0.0	T	0.0	17.8	11.0	5.0	6.1	1.9	0.0	0.0	41.8
1971-72	0.0	0.0	0.0	0.0	T	4.7	2.2	13.7	8.7	0.7	0.0	0.0	30.0
1972-73	0.0	0.0	0.0	0.7	0.3	4.1	2.0	3.2	0.6	0.4	0.0	0.0	11.3
1973-74	0.0	0.0	0.0	0.0	T	T	15.1	11.1	0.4	1.3	0.0	0.0	27.9
1974-75	0.0	0.0	0.0	0.0	0.3	2.1	2.0	18.2	2.2	0.2	0.0	0.0	25.0
1975-76	0.0	0.0	0.0	T	1.2	7.5	15.6	3.7	9.5	0.0	0.0	0.0	37.5
1976-77	0.0	0.0	0.0	0.0	T	9.8	14.0	11.4	4.4	T	7.0	0.0	46.6
1977-78	0.0	0.0	0.0	0.0	1.3	3.9	20.5	28.6	15.9	T	0.0	0.0	70.2
1978-79	0.0	0.0	0.0	0.0	2.3	2.4	6.0	5.5	T	1.1	0.0	0.0	17.3
1979-80	0.0	0.0	0.0	2.5	0.0	T	0.6	3.8	5.3	0.0	0.0	0.0	12.2
1980-81	0.0	0.0	0.0	0.0	4.1	3.6	12.9	0.6	0.3	0.0	0.0	0.0	21.5
1981-82	0.0	0.0	0.0	T	T	16.4	13.4	4.3	5.7	7.6	0.0	0.0	47.4
1982-83	0.0	0.0	0.0	0.0	0.0	7.3	3.8	21.3	T	T	0.0	0.0	32.4
1983-84	0.0	0.0	0.0	0.0	T	4.5	17.9	T	13.7	T	0.0	0.0	36.1
1984-85	0.0	0.0	0.0	0.0	T	2.0	9.8	10.0	0.6	T	0.0	0.0	22.4
1985-86	0.0	0.0	0.0	0.0	1.8	2.6							
Record Mean	0.0	0.0	0.0	0.2	0.6	6.9	9.9	10.0	8.0	0.8	0.2	0.0	36.6

See Reference Notes, relative to all above tables, on preceding page.

Charleston is a peninsula city bounded on the west and south by the Ashley River, on the east by the Cooper River, and on the southeast by a spacious harbor. Weather records for the airport are from a site some 10 miles inland. The terrain is generally level, ranging in elevation from sea level to 20 feet on the peninsula, with gradual increases in elevation toward inland areas. The soil is sandy to sandy loam with lesser amounts of loam. The drainage varies from good to poor. Because of the very low elevation, a considerable portion of this community and the nearby coastal islands are vulnerable to tidal flooding.

The climate is temperate, modified considerably by the nearness to the ocean. The marine influence is noticeable during winter when the low temperatures are sometimes 10-15 degrees higher on the peninsula than at the airport. By the same token, high temperatures are generally a few degrees lower on the peninsula. The prevailing winds are northerly in the fall and winter, southerly in the spring and summer.

Summer is warm and humid. Temperatures of 100 degrees or more are infrequent. High temperatures are generally several degrees lower along the coast than inland due to the cooling effect of the sea breeze. Summer is the rainiest season with 41 percent of the annual total. The rain, except during occasional tropical storms, generally occurs as showers or thunderstorms.

The fall season passes through the warm Indian Summer period to the pre-winter cold spells which begin late in November. From late September to early November the weather is mostly sunny and temperature extremes are rare. Late summer and early fall is the period of maximum threat to the South Carolina coast from hurricanes.

The winter months, December through February, are mild with periods of rain. However, the winter rainfall is generally of a more uniform type. There is some chance of a snow flurry, with the best probability of its occurrence in January, but a significant amount is rarely measured. An average winter would experience less than one cold wave and severe freeze. Temperatures of 20 degrees or less on the peninsula and along the coast are very unusual.

The most spectacular time of the year, weatherwise, is spring with its rapid changes from windy and cold in March to warm and pleasant in May. Severe local storms are more likely to occur in spring than in summer.

The average occurrence of the first freeze in the fall is early December, and the average last freeze is late February, giving an average growing season of about 294 days.

TABLE 1   # NORMALS, MEANS AND EXTREMES

CHARLESTON, SOUTH CAROLINA

LATITUDE: 32°54'N   LONGITUDE: 80°02'W   ELEVATION: FT. GRND 40 BARO 47   TIME ZONE: EASTERN   WBAN: 13880

	(a)	JAN	FEB	MAR	APR	MAY	JUNE	JULY	AUG	SEP	OCT	NOV	DEC	YEAR
TEMPERATURE °F:														
Normals														
-Daily Maximum		58.8	61.2	68.0	76.0	82.9	87.0	89.4	88.8	84.6	76.8	68.7	61.4	75.3
-Daily Minimum		36.9	38.4	45.3	52.5	61.4	68.0	71.6	71.2	66.7	54.7	44.6	38.5	54.2
-Monthly		47.9	49.8	56.7	64.3	72.2	77.5	80.5	80.0	75.7	65.8	56.7	50.0	64.7
Extremes														
-Record Highest	43	83	86	90	93	98	103	101	102	99	94	88	83	103
-Year		1950	1962	1974	1985	1953	1944	1977	1954	1944	1954	1961	1972	JUN 1944
-Record Lowest	43	6	12	15	29	36	50	58	56	42	27	15	8	6
-Year		1985	1973	1980	1944	1963	1972	1952	1979	1967	1976	1950	1962	JAN 1985
NORMAL DEGREE DAYS:														
Heating (base 65°F)		543	434	286	69	6	0	0	0	0	76	262	471	2147
Cooling (base 65°F)		13	9	29	48	229	378	481	465	321	101	13	6	2093
% OF POSSIBLE SUNSHINE	26	58	62	67	71	71	68	67	65	63	66	63	59	65
MEAN SKY COVER (tenths)														
Sunrise - Sunset	36	6.2	5.9	6.0	5.4	6.0	6.3	6.6	6.2	6.2	5.2	5.1	5.9	5.9
MEAN NUMBER OF DAYS:														
Sunrise to Sunset														
-Clear	37	9.0	9.1	9.1	11.2	7.9	6.2	4.9	5.9	6.6	11.5	12.4	9.7	103.5
-Partly Cloudy	37	6.5	6.5	8.3	7.7	10.9	11.3	12.2	13.1	10.7	8.4	6.4	7.1	109.2
-Cloudy	37	15.5	12.6	13.6	11.1	12.2	12.5	13.9	12.0	12.6	11.1	11.2	14.2	152.5
Precipitation														
.01 inches or more	43	9.7	8.9	10.3	7.3	9.2	10.8	13.7	12.3	9.3	6.0	6.9	8.5	113.0
Snow, Ice pellets														
1.0 inches or more	43	0.0	0.1	0.*	0.0	0.0	0.0	0.0	0.0	0.0	0.0	0.0	0.*	0.2
Thunderstorms	43	0.7	1.1	2.3	2.7	6.9	9.8	13.2	11.0	5.1	1.3	0.7	0.6	55.5
Heavy Fog Visibility														
1/4 mile or less	36	4.2	2.0	2.4	2.1	2.1	1.7	0.8	1.4	1.7	2.8	3.7	3.6	28.6
Temperature °F														
-Maximum														
90° and above	43	0.0	0.0	0.*	0.7	3.8	10.8	15.3	13.9	5.0	0.3	0.0	0.0	49.8
32° and below	43	0.2	0.1	0.*	0.0	0.0	0.0	0.0	0.0	0.0	0.0	0.0	0.1	0.3
-Minimum														
32° and below	43	11.4	8.3	3.0	0.2	0.0	0.0	0.0	0.0	0.0	0.0	0.1	3.5	9.5	36.0
0° and below	43	0.0	0.0	0.0	0.0	0.0	0.0	0.0	0.0	0.0	0.0	0.0	0.0	0.0
AVG. STATION PRESS.(mb)	13	1018.2	1017.7	1015.6	1015.7	1014.1	1014.7	1015.7	1016.4	1015.8	1017.5	1018.3	1019.2	1016.6
RELATIVE HUMIDITY (%)														
Hour 01	43	80	78	81	83	88	89	90	91	90	88	84	82	85
Hour 07 (Local Time)	43	83	81	83	83	85	86	88	90	90	89	86	83	86
Hour 13	43	55	51	50	49	54	59	63	63	62	56	52	55	56
Hour 19	43	71	67	66	66	71	74	77	79	80	79	76	73	73
PRECIPITATION (inches):														
Water Equivalent														
-Normal		3.33	3.37	4.38	2.58	4.41	6.54	7.33	6.50	4.94	2.92	2.18	3.11	51.59
-Maximum Monthly	43	6.68	6.35	11.11	9.50	9.28	27.24	18.46	16.99	17.31	9.12	7.35	7.09	27.24
-Year		1966	1983	1983	1958	1957	1973	1964	1974	1945	1959	1972	1953	JUN 1973
-Minimum Monthly	43	0.63	0.33	0.99	0.01	0.68	0.96	1.76	0.73	0.53	0.08	0.48	0.66	0.01
-Year		1950	1947	1963	1972	1944	1970	1980	1971	1980	1966	1984	APR 1972
-Maximum in 24 hrs	43	2.49	3.28	6.63	4.10	6.23	10.10	5.81	5.77	8.84	5.77	5.24	3.40	10.10
-Year		1983	1944	1959	1958	1967	1973	1960	1964	1945	1944	1969	1978	JUN 1973
Snow, Ice pellets														
-Maximum Monthly	43	1.0	7.1	2.0	T							T	3.8	7.1
-Year		1977	1973	1969	1985							1950	1980	FEB 1973
-Maximum in 24 hrs	43	0.8	5.9	2.0	T							T	3.8	5.9
-Year		1966	1973	1969	1985							1950	1980	FEB 1973
WIND:														
Mean Speed (mph)	36	9.2	10.0	10.1	9.8	8.7	8.4	7.9	7.4	7.9	8.1	8.1	8.6	8.7
Prevailing Direction														
through 1963		SW	NNE	SSW	SSW	S	S	SW	SW	NNE	NNE	N	NNE	NNE
Fastest Obs. 1 Min.														
-Direction	10	20	29	31	19	25	03	14	03	11	18	15	24	03
-Speed (MPH)	10	40	37	37	36	32	40	37	32	38	30	37	39	40
-Year		1978	1976	1984	1983	1978	1981	1985	1981	1979	1976	1985	1975	JUN 1981
Peak Gust														
-Direction	2	SW	W	NW	E	W	W	SE	SE	SW	NE	SE	W	NW
-Speed (mph)	2	35	44	49	40	49	41	46	44	37	33	48	39	49
-Date		1985	1984	1984	1985	1984	1985	1985	1985	1985	1985	1985	1984	MAR 1984

See Reference Notes to this table on the following pages.

CHARLESTON, SOUTH CAROLINA

TABLE 2 PRECIPITATION (inches) CHARLESTON, SOUTH CAROLINA

YEAR	JAN	FEB	MAR	APR	MAY	JUNE	JULY	AUG	SEP	OCT	NOV	DEC	ANNUAL
1956	2.56	2.91	2.45	2.39	4.62	7.69	10.60	5.97	4.81	3.89	0.88	0.95	49.72
1957	1.43	3.35	5.53	1.70	9.28	6.83	8.32	5.86	8.44	1.83	4.71	3.57	60.85
1958	7.20	4.04	5.78	9.50	4.43	12.09	7.74	7.87	5.54	3.31	0.52	4.15	72.17
1959	4.20	5.99	11.11	2.39	4.91	1.36	8.47	5.45	9.60	9.12	2.07	2.49	67.16
1960	5.81	4.06	2.72	2.05	2.26	6.76	14.10	3.15	7.60	0.98	0.79	2.13	52.41
1961	1.77	4.15	5.83	5.78	5.23	6.84	6.67	6.07	2.97	1.71	1.59	1.43	50.04
1962	3.59	1.28	7.88	2.52	1.81	16.07	5.93	11.54	3.81	3.25	2.01	1.46	61.15
1963	3.14	4.90	0.99	2.68	1.70	14.34	7.94	6.34	2.75	2.27	4.74	3.18	54.97
1964	6.53	6.32	4.40	2.72	4.77	6.95	18.46	7.67	4.11	7.53	0.52	3.01	72.99
1965	1.69	5.49	7.99	3.10	1.32	6.68	8.90	5.90	4.66	6.23	0.87	1.20	54.03
1966	6.68	4.61	2.65	2.83	7.71	6.03	11.48	3.45	2.20	1.70	0.48	3.76	53.58
1967	4.93	3.12	2.81	0.84	8.91	4.06	9.19	3.74	1.39	0.52	1.35	2.79	43.65
1968	2.25	1.27	1.35	1.99	2.54	9.41	7.77	4.42	1.66	6.86	2.67	3.65	45.84
1969	1.19	2.05	5.14	3.46	2.34	4.88	5.34	11.84	5.37	1.50	5.49	3.52	52.12
1970	2.51	2.86	7.72	1.34	3.78	0.96	5.93	10.64	2.53	4.08	0.67	2.90	45.92
1971	5.45	4.71	4.05	4.11	4.15	4.07	6.04	16.32	0.53	7.22	1.61	2.28	60.54
1972	4.13	5.18	2.52	0.01	5.67	5.29	1.76	4.52	1.82	0.25	7.35	4.36	42.86
1973	4.59	5.57	6.15	2.55	1.83	27.24	3.60	6.66	7.93	0.63	0.84	4.58	72.17
1974	1.42	2.96	3.04	0.86	4.82	9.45	3.09	16.99	4.80	0.40	3.78	3.00	54.61
1975	4.92	3.54	4.54	3.74	5.06	5.96	9.34	7.18	5.16	1.97	1.43	3.35	56.19
1976	1.62	0.95	2.33	0.62	8.87	5.59	4.48	5.22	6.03	4.10	3.57	5.12	48.50
1977	2.72	1.38	5.31	0.45	4.66	2.12	3.86	8.13	2.48	2.49	1.76	5.88	41.24
1978	4.31	1.82	3.25	1.97	4.68	3.42	6.19	4.01	5.06	0.18	1.87	4.13	40.89
1979	3.43	3.04	3.01	3.81	8.09	2.23	8.35	0.88	15.36	3.87	3.29	2.62	57.98
1980	3.99	1.25	7.99	3.43	5.85	3.15	6.97	0.73	2.60	1.52	2.19	1.25	40.92
1981	0.93	2.23	2.38	1.87	4.02	6.04	12.66	9.30	1.27	1.95	1.06	5.73	49.44
1982	2.18	3.64	1.26	6.51	3.04	9.16	5.40	4.10	3.92	2.42	1.19	4.20	47.02
1983	4.86	6.35	11.11	3.57	0.75	2.37	8.89	2.90	3.50	2.36	3.08	4.35	54.09
1984	5.12	3.51	5.63	6.30	6.89	2.96	4.87	1.96	5.27	1.67	1.39	0.66	46.23
1985	0.87	2.70	1.50	1.12	2.79	7.02	12.06	8.48	2.53	4.58	5.49	1.21	50.35
Record Mean	3.01	3.28	3.71	2.78	3.60	5.23	7.25	6.61	4.98	3.10	2.27	2.90	48.72

TABLE 3 AVERAGE TEMPERATURE (deg. F) CHARLESTON, SOUTH CAROLINA

YEAR	JAN	FEB	MAR	APR	MAY	JUNE	JULY	AUG	SEP	OCT	NOV	DEC	ANNUAL
1956	45.1	56.8	56.9	63.3	73.0	76.9	79.9	79.3	71.9	65.8	53.8	57.0	65.0
1957	50.1	56.6	55.3	64.2	71.6	77.4	78.9	78.9	75.9	61.8	58.2	50.2	64.9
1958	42.9	43.0	52.3	64.3	70.9	76.7	81.2	79.9	76.1	62.5	59.8	46.4	64.5
1959	46.3	53.4	54.8	64.3	73.1	75.9	77.9	78.3	73.8	69.0	56.5	49.8	64.4
#1960	49.2	48.4	46.3	65.8	72.2	77.5	80.1	80.1	75.9	67.8	57.9	43.7	63.7
1961	44.3	52.6	61.3	59.7	69.5	76.7	80.6	79.0	76.8	63.7	51.0	51.0	64.5
1962	47.1	56.2	53.2	62.1	75.9	76.3	80.8	79.0	73.9	66.5	53.5	45.6	64.2
1963	45.3	45.3	60.9	65.8	71.6	79.2	79.9	80.9	73.8	67.2	57.3	43.8	64.3
1964	48.5	47.5	57.6	66.0	73.1	80.0	78.0	79.6	75.4	62.5	52.9	52.9	65.1
1965	48.4	51.2	54.5	64.5	75.3	76.1	79.0	79.6	75.9	65.2	57.6	49.3	64.7
1966	43.9	49.6	53.9	63.2	70.6	73.7	79.6	80.0	75.8	67.4	56.3	48.9	63.6
1967	50.6	48.5	60.0	67.1	71.3	75.6	79.9	79.5	69.9	62.6	53.8	53.4	64.4
1968	43.5	41.8	55.7	65.7	71.7	77.8	81.5	81.5	75.1	67.1	53.7	46.5	63.5
1969	44.6	45.2	50.6	63.3	69.9	80.3	82.7	77.5	73.7	68.2	53.4	46.5	63.0
1970	41.0	48.5	57.1	66.3	72.7	78.4	82.5	80.6	77.3	67.5	54.1	52.1	64.8
1971	49.9	49.8	51.9	62.7	71.1	80.4	80.3	79.6	77.1	70.8	57.1	58.8	65.8
1972	54.8	48.9	56.9	64.0	69.8	74.0	80.1	80.4	76.4	67.3	56.8	55.7	65.5
1973	48.0	46.1	61.0	61.5	71.7	78.4	82.6	81.0	79.6	69.6	61.4	51.3	66.0
1974	61.8	51.5	62.0	63.6	74.0	75.4	78.2	79.3	75.0	61.8	55.5	51.0	65.8
1975	53.8	54.7	56.9	62.3	75.1	78.5	79.2	81.5	76.8	69.0	59.3	49.7	66.4
1976	44.8	55.9	62.3	64.0	70.3	75.8	81.0	77.2	73.9	61.4	55.0	48.8	63.9
1977	38.7	46.3	60.6	66.4	72.8	81.2	83.8	81.4	78.7	63.5	61.3	5.0	65.4
1978	43.5	42.7	55.2	66.5	72.0	78.6	81.1	81.3	77.4	65.7	63.3	52.7	65.0
1979	45.4	46.8	57.4	64.9	72.4	75.9	82.0	81.4	76.5	66.0	59.4	48.7	64.7
1980	48.7	45.9	54.6	64.3	71.4	79.8	82.4	82.1	79.8	65.0	55.4	47.5	64.6
1981	41.6	50.8	54.3	67.5	70.8	82.7	83.5	80.3	74.8	64.1	55.4	46.2	64.3
1982	45.1	51.5	59.2	61.8	72.2	78.8	81.2	80.0	74.5	65.1	60.9	57.0	65.6
1983	45.6	49.0	56.4	61.0	71.7	76.9	82.8	82.9	75.5	68.9	57.4	48.8	64.8
1984	46.1	52.8	57.6	64.0	71.7	78.8	79.9	81.1	73.0	71.3	54.2	57.2	65.6
1985	42.6	50.5	60.7	67.8	73.6	79.6	80.9	79.9	75.8	72.2	67.3	47.9	66.6
Record Mean	49.7	51.4	57.5	64.6	72.5	78.6	81.1	80.5	76.4	67.2	57.8	51.0	65.7
Max	58.5	60.4	66.5	73.7	80.9	86.3	88.5	87.7	83.7	75.7	67.1	60.0	74.1
Min	40.8	42.3	48.4	55.5	64.1	70.9	73.7	73.3	69.2	58.7	48.5	42.1	57.3

REFERENCE NOTES FOR TABLES 1, 2, 3 and 6   (CHARLESTON, SC)

GENERAL

T - TRACE AMOUNT
BLANK ENTRIES DENOTE MISSING/UNREPORTED DATA.
# INDICATES A STATION OR INSTRUMENT RELOCATION.

SPECIFIC

TABLE 1

(a) - LENGTH OF RECORD IN YEARS. ALTHOUGH
INDIVIDUAL MONTHS MAY BE MISSING.

* LESS THAN .05

NORMALS — BASED ON THE 1951-1980 RECORD PERIOD.
EXTREMES — DATES ARE THE MOST RECENT OCCURRENCE.
WIND DIR. — NUMERALS SHOW TENS OF DEGREES
CLOCKWISE FROM TRUE NORTH.
"00" INDICATES CALM.
RESULTANT WIND DIRECTIONS ARE GIVEN TO WHOLE DEGREES.

EXCEPTIONS

TABLES 2, 3, and 6

RECORD MEANS ARE THROUGH THE CURRENT YEAR,
BEGINNING IN   1874 FOR TEMPERATURE
                1871 FOR PRECIPITATION
                1943 FOR SNOWFALL

TABLE 4 HEATING DEGREE DAYS Base 65 deg. F   CHARLESTON, SOUTH CAROLINA

SEASON	JULY	AUG	SEP	OCT	NOV	DEC	JAN	FEB	MAR	APR	MAY	JUNE	TOTAL
1956-57	0	0	12	38	347	260	465	245	307	90	29	0	1793
1957-58	0	0	5	130	224	453	674	609	387	91	15	0	2588
1958-59	0	0	0	113	183	568	574	330	314	100	3	0	2185
#1959-60	0	0	0	38	279	465	483	474	582	65	12	0	2398
1960-61	0	0	0	62	216	651	635	347	155	181	11	2	2260
1961-62	0	0	3	92	251	443	553	263	363	160	0	0	2128
1962-63	0	0	8	95	339	597	604	545	172	69	29	0	2458
1963-64	0	0	2	33	233	650	504	500	240	61	4	0	2227
1964-65	0	0	0	135	153	385	507	392	340	95	0	0	2007
1965-66	0	0	0	95	225	481	644	427	338	119	15	2	2346
1966-67	0	0	0	51	280	498	442	460	194	50	29	4	2008
1967-68	0	0	15	112	334	367	657	667	307	67	4	0	2530
1968-69	0	0	0	80	334	564	624	551	444	90	6	0	2693
1969-70	0	0	1	50	349	567	735	454	249	82	9	0	2496
1970-71	0	0	11	42	324	465	424	404	392	127	16	0	2205
1971-72	0	0	0	13	261	220	317	463	249	113	8	0	1644
1972-73	0	0	0	33	268	302	520	524	167	141	18	0	1973
1973-74	0	0	0	34	158	428	131	378	150	114	2	0	1395
1974-75	0	0	5	136	299	432	350	294	273	152	0	0	1941
1975-76	0	0	0	40	221	466	624	265	146	94	15	3	1874
1976-77	0	0	0	159	418	501	808	516	186	58	17	0	2663
1977-78	0	0	0	112	175	459	663	616	309	52	18	0	2404
1978-79	0	0	0	57	83	399	602	505	241	70	2	0	1959
1979-80	0	0	0	68	203	500	495	555	321	82	17	0	2241
1980-81	0	0	0	80	287	537	719	393	333	55	16	0	2420
1981-82	0	0	3	88	291	577	611	372	214	132	3	0	2291
1982-83	0	0	0	102	154	276	596	440	264	146	2	0	1980
1983-84	0	0	4	24	230	500	578	347	240	92	16	0	2031
1984-85	0	0	9	13	337	249	692	418	183	47	4	0	1952
1985-86	0	0	2	16	54	526							

TABLE 5 COOLING DEGREE DAYS Base 65 deg. F   CHARLESTON, SOUTH CAROLINA

YEAR	JAN	FEB	MAR	APR	MAY	JUNE	JULY	AUG	SEP	OCT	NOV	DEC	TOTAL
1969	0	0	1	46	165	465	555	394	268	158	6	0	2058
1970	0	0	9	126	253	410	552	487	384	130	3	1	2355
1971	2	5	5	65	215	469	480	457	369	199	31	32	2329
1972	7	3	5	93	163	275	475	488	351	110	30	22	2022
1973	1	0	50	42	233	412	554	501	445	184	56	10	2488
1974	41	7	63	80	288	319	417	450	312	46	18	3	2044
1975	8	13	26	74	318	414	449	516	361	171	58	0	2408
1976	2	9	73	70	187	329	502	384	274	52	2	1	1885
1977	0	1	54	107	263	493	588	518	417	71	71	1	2584
1978	0	0	13	106	242	414	505	514	378	86	40	21	2319
1979	0	2	9	71	241	335	533	514	354	105	40	0	2204
1980	0	9	7	69	221	407	549	539	451	87	5	1	2345
1981	0	0	9	138	199	539	582	481	307	66	9	0	2330
1982	0	2	42	42	232	420	510	475	293	111	36	34	2197
1983	0	0	6	32	217	362	559	567	322	149	10	4	2228
1984	0	1	19	67	228	420	471	509	254	212	21	10	2212
1985	7	17	57	136	276	445	501	470	332	245	129	5	2620

TABLE 6 SNOWFALL (inches)   CHARLESTON, SOUTH CAROLINA

SEASON	JULY	AUG	SEP	OCT	NOV	DEC	JAN	FEB	MAR	APR	MAY	JUNE	TOTAL
1970-71	0.0	0.0	0.0	0.0	0.0	0.0	T	T	T	0.0	0.0	0.0	T
1971-72	0.0	0.0	0.0	0.0	0.0	0.0	0.0	0.0	0.0	0.0	0.0	0.0	0.0
1972-73	0.0	0.0	0.0	0.0	0.0	0.0	0.0	7.1	0.0	0.0	0.0	0.0	7.1
1973-74	0.0	0.0	0.0	0.0	0.0	T	0.0	0.0	0.0	0.0	0.0	0.0	T
1974-75	0.0	0.0	0.0	0.0	0.0	0.0	0.0	0.0	0.0	0.0	0.0	0.0	0.0
1975-76	0.0	0.0	0.0	0.0	0.0	0.0	0.4	0.0	0.0	0.0	0.0	0.0	0.4
1976-77	0.0	0.0	0.0	0.0	0.0	0.0	1.0	0.3	0.0	0.0	0.0	0.0	1.3
1977-78	0.0	0.0	0.0	0.0	0.0	0.0	T	0.4	T	0.0	0.0	0.0	0.4
1978-79	0.0	0.0	0.0	0.0	0.0	0.0	0.0	1.8	0.0	0.0	0.0	0.0	1.8
1979-80	0.0	0.0	0.0	0.0	0.0	0.0	0.0	T	1.3	0.0	0.0	0.0	1.3
1980-81	0.0	0.0	0.0	0.0	0.0	3.8	0.0	0.0	0.0	0.0	0.0	0.0	3.8
1981-82	0.0	0.0	0.0	0.0	0.0	0.0	T	0.0	0.0	0.0	0.0	0.0	T
1982-83	0.0	0.0	0.0	0.0	0.0	0.0	0.0	0.0	T	0.0	0.0	0.0	T
1983-84	0.0	0.0	0.0	0.0	0.0	0.0	T	0.0	0.0	0.0	0.0	0.0	T
1984-85	0.0	0.0	0.0	0.0	0.0	0.0	T	0.0	0.0	T	0.0	0.0	T
1985-86	0.0	0.0	0.0	0.0	0.0	0.0							
Record Mean	0.0	0.0	0.0	0.0	T	0.1	0.1	0.3	0.1	T	0.0	0.0	0.6

See Reference Notes, relative to all above tables, on preceding page.

COLUMBIA, SOUTH CAROLINA

Columbia is centrally located within the state of South Carolina and lies on the Congaree River near the confluence of the Broad and Saluda Rivers. The surrounding terrain is rolling, sloping from about 350 feet above sea level in northern Columbia to about 200 feet in the southeastern part of the city.

The climate in the Columbia area is relatively temperate. The Appalachian Mountain chain, some 150 miles to the northwest, frequently retards the approach of unseasonable cold weather in the winter. The terrain offers little moderating effect on the summer heat.

Long summers are prevalent with warm weather usually lasting from sometime in May into September. In summer the Bermuda high is the greatest single weather factor influencing the area. This permanent high more or less blocks the entry of cold fronts so that many stall before reaching central South Carolina. Also, the southwestern flow around the offshore Bermuda high pressure supplies moisture for the many summer thunderstorms. There are relatively few breaks in the heat during midsummer. The typical summer has about six days with 100 degrees or more. Thunderstorm activity usually shows a decided increase during June, decreasing about the first of September. About once or twice a year, passing tropical storms produce strong winds and heavy rains. The incidence of these storms is greatest in September, although they represent a possible threat from midsummer to late fall. Damage from tropical storms is usually minor in the Columbia area.

Fall is the most pleasant time of the year. Rainfall during the late fall is at an annual minimum, while the sunshine is at a relative maximum. Winters are mild with the cold weather usually lasting from late November to mid-March. The winter weather at Columbia is largely made up of polar air outbreaks that reach this area in a much modified form. On rare occasions in winter, Arctic air masses push southward as far as central South Carolina and cause some of the coldest temperatures. Disruption of activities from snowfall is unusual, in fact, more than three days of sustained snow cover is rare.

Spring is the most changeable season of the year. The temperature varies from an occasional cold snap in March to generally warm and pleasant in May. While tornadoes are infrequent, they occur most often in the spring. Hailstorms are not frequent, with the annual incidence at a maximum in spring and early summer. The average occurrence of the last spring freeze is very late March, and the first fall freeze is early November, for a growing period of about 218 days.

TABLE 1   NORMALS, MEANS AND EXTREMES

COLUMBIA, SOUTH CAROLINA

LATITUDE: 33°57'N   LONGITUDE: 81°07'W   ELEVATION: FT. GRND   213 BARO   00245   TIME ZONE: EASTERN   WBAN: 13883

	(a)	JAN	FEB	MAR	APR	MAY	JUNE	JULY	AUG	SEP	OCT	NOV	DEC	YEAR
TEMPERATURE °F:														
Normals														
-Daily Maximum		56.2	59.5	67.1	77.0	83.8	89.2	91.9	91.0	85.5	76.5	67.1	58.8	75.3
-Daily Minimum		33.2	34.6	41.9	50.5	59.1	66.1	70.1	69.4	63.9	50.3	40.6	34.7	51.2
-Monthly		44.7	47.0	54.5	63.8	71.5	77.7	81.0	80.2	74.7	63.4	53.9	46.8	63.3
Extremes														
-Record Highest	38	84	84	91	94	101	107	107	107	101	101	90	83	107
-Year		1975	1977	1974	1970	1953	1954	1952	1983	1954	1954	1961	1978	AUG 1983
-Record Lowest	38	-1	5	4	26	34	44	54	53	40	23	12	4	-1
-Year		1985	1973	1980	1983	1963	1984	1951	1969	1967	1952	1970	1958	JAN 1985
NORMAL DEGREE DAYS:														
Heating (base 65°F)		637	508	346	87	22	0	0	0	0	123	339	567	2629
Cooling (base 65°F)		8	6	20	51	223	381	496	471	297	74	6	0	2033
% OF POSSIBLE SUNSHINE	32	57	61	65	69	68	68	67	67	66	66	64	60	65
MEAN SKY COVER (tenths)														
Sunrise - Sunset	37	6.1	5.8	5.9	5.2	5.7	5.8	6.2	5.7	5.6	4.8	5.0	5.7	5.6
MEAN NUMBER OF DAYS:														
Sunrise to Sunset														
-Clear	38	9.4	9.1	9.2	11.4	9.7	8.1	6.5	8.5	9.6	13.6	12.6	10.6	118.4
-Partly Cloudy	38	6.1	6.6	8.1	7.6	9.9	11.4	12.6	12.8	9.0	7.1	6.3	6.2	103.6
-Cloudy	38	15.5	12.6	13.7	10.9	11.4	10.5	11.9	9.7	11.4	10.3	11.2	14.2	143.3
Precipitation														
.01 inches or more	38	10.1	9.6	10.6	8.2	9.0	9.5	11.8	10.5	7.4	6.1	7.1	9.2	109.1
Snow, Ice pellets														
1.0 inches or more	38	0.2	0.2	0.1	0.0	0.0	0.0	0.0	0.0	0.0	0.0	0.0	0.*	0.5
Thunderstorms	38	0.7	1.5	2.7	3.7	6.4	9.1	12.8	9.9	3.9	1.4	0.8	0.3	53.1
Heavy Fog Visibility														
1/4 mile or less	37	2.9	2.2	1.7	1.3	1.5	1.5	1.7	2.5	2.8	2.8	3.0	3.6	27.6
Temperature °F														
-Maximum														
90° and above	19	0.0	0.0	0.1	1.9	4.9	13.7	20.3	16.2	8.4	0.5	0.0	0.0	65.9
32° and below	19	0.6	0.1	0.1	0.0	0.0	0.0	0.0	0.0	0.0	0.0	0.0	0.1	0.9
-Minimum														
32° and below	19	17.4	14.6	6.5	1.1	0.0	0.0	0.0	0.0	0.0	0.0	1.0	8.5	13.9	63.0
0° and below	19	0.1	0.0	0.0	0.0	0.0	0.0	0.0	0.0	0.0	0.0	0.0	0.0	0.1
AVG. STATION PRESS. (mb)	13	1011.7	1011.1	1008.8	1008.8	1007.4	1008.3	1009.1	1009.9	1009.8	1011.4	1011.9	1012.6	1010.1
RELATIVE HUMIDITY (%)														
Hour 01	19	78	76	76	77	85	87	87	91	90	88	84	80	83
Hour 07	19	82	81	84	84	87	87	89	92	92	91	88	84	87
Hour 13 (Local Time)	19	55	49	48	44	50	53	56	57	56	52	50	54	52
Hour 19	19	65	57	54	50	59	62	68	72	73	74	71	69	65
PRECIPITATION (inches):														
Water Equivalent														
-Normal		4.38	3.99	5.16	3.59	3.85	4.45	5.35	5.56	4.23	2.55	2.51	3.50	49.12
-Maximum Monthly	38	9.26	8.68	10.89	6.85	8.85	14.81	13.87	16.72	8.78	12.09	7.20	8.54	16.72
-Year		1978	1961	1973	1979	1967	1973	1959	1949	1953	1959	1957	1981	AUG 1949
-Minimum Monthly	38	0.84	0.87	0.56	0.81	0.29	1.26	0.57	1.02	0.07	T	0.41	0.32	T
-Year		1981	1976	1985	1976	1951	1955	1977	1976	1963	1963	1973	1955	OCT 1963
-Maximum in 24 hrs	38	2.82	3.69	3.59	3.66	5.57	5.44	5.81	7.66	6.23	5.46	2.57	3.18	7.66
-Year		1968	1962	1960	1956	1967	1973	1959	1949	1953	1964	1976	1970	AUG 1949
Snow, Ice pellets														
-Maximum Monthly	38	3.5	16.0	4.1								T	9.1	16.0
-Year		1982	1973	1980								1976	1958	FEB 1973
-Maximum in 24 hrs	38	3.3	15.7	4.1								T	8.8	15.7
-Year		1982	1973	1980								1976	1958	FEB 1973
WIND:														
Mean Speed (mph)	37	7.2	7.6	8.2	8.3	7.0	6.6	6.3	5.8	6.2	6.2	6.4	6.7	6.9
Prevailing Direction														
through 1963		SW	SW	SW	SW	SW	SW	SW	SW	NE	NE	SW	WSW	SW
Fastest Obs. 1 Min.														
-Direction (!!!)	32	28	20	27	27	23	23	35	16	11	21	35	28	27
-Speed (MPH)	32	46	40	60	40	46	40	40	44	38	27	35	35	60
-Year		1964	1966	1954	1961	1958	1957	1965	1961	1959	1968	1967	1975	MAR 1954
Peak Gust														
-Direction (!!!)	2	W	W	S	W	SW	NW	N	NE	NE	N	SW	W	N
-Speed (mph)	2	48	44	49	46	37	49	40	30	32	54	38	49	54
-Date		1985	1985	1984	1985	1984	1984	1984	1984	1984	1985	1984	1984	OCT 1985

See Reference Notes to this table on the following page.

TABLE 2 PRECIPITATION (inches) COLUMBIA, SOUTH CAROLINA

YEAR	JAN	FEB	MAR	APR	MAY	JUNE	JULY	AUG	SEP	OCT	NOV	DEC	ANNUAL
1956	1.73	5.45	3.99	5.31	1.92	1.70	3.68	1.77	7.94	1.83	0.66	2.44	38.42
1957	2.48	1.30	4.59	2.25	6.71	1.86	1.15	4.12	6.74	1.80	7.20	2.44	42.64
1958	4.09	3.87	4.47	5.89	3.79	3.61	8.70	0.76	2.65	0.58	0.58	3.85	44.19
1959	2.94	4.99	6.28	2.64	5.79	2.67	13.87	4.52	7.12	12.09	0.67	2.42	66.00
1960	7.15	5.53	6.17	3.91	1.47	2.37	4.79	5.52	3.94	1.71	0.68	2.37	45.61
1961	2.93	8.68	5.75	5.52	2.98	1.95	5.70	14.94	1.46	0.82	1.01	3.21	54.95
1962	6.49	5.18	4.40	3.21	2.32	4.78	2.67	3.10	2.85	0.89	4.53	2.27	42.69
1963	5.38	3.94	3.28	4.18	2.87	4.84	2.48	3.98	T	4.20	5.05	4.21	42.11
1964	6.34	5.33	6.16	3.60	2.63	2.97	10.32	9.97	6.93	10.34	1.36	4.58	70.53
1965	1.43	5.33	7.68	3.99	1.46	8.20	4.33	2.34	5.99	2.34	1.77	0.64	52.55
1966	7.22	4.54	2.23	3.58	6.14	3.66	2.87	3.22	2.02	2.47	1.05	3.31	42.31
1967	2.79	4.36	3.08	3.72	8.85	7.27	11.16	2.38	0.62	3.71	2.59	4.18	54.71
1968	5.94	1.14	1.92	4.52	4.17	5.41	9.28	1.11	2.40	4.31	5.21	3.26	48.67
1969	2.64	3.03	5.16	4.57	3.28	4.70	4.31	2.93	3.17	1.20	4.51	40.67
1970	3.28	2.58	8.42	0.91	4.50	2.05	4.74	7.13	3.72	8.18	1.43	4.55	51.49
1971	4.55	5.23	9.53	4.31	2.71	7.46	11.13	10.68	3.44	2.35	2.90	69.32
1972	7.62	3.58	3.79	1.16	6.41	6.10	9.31	2.87	2.51	1.15	5.62	5.39	55.51
1973	5.25	5.75	10.89	4.47	4.04	14.81	3.19	6.92	4.47	0.71	0.41	6.66	67.57
1974	6.16	4.49	2.36	2.97	3.40	4.50	4.40	6.20	4.44	0.02	4.47	4.61	48.02
1975	4.26	6.43	5.41	4.59	7.88	2.85	9.91	3.16	3.32	0.88	2.23	5.03	55.95
1976	3.58	0.87	5.24	0.81	4.63	11.67	6.55	5.74	5.21	5.13	7.54	57.99
1977	4.20	1.22	6.34	0.91	0.89	2.20	0.57	10.73	1.51	4.81	2.10	3.69	39.17
1978	9.26	1.28	3.49	4.28	3.09	4.73	2.10	4.45	4.09	0.79	2.98	1.82	42.36
1979	5.19	8.10	3.53	6.85	6.47	5.48	7.28	4.05	7.86	1.76	3.89	1.51	61.97
1980	4.72	1.88	10.72	2.02	4.51	2.27	1.24	3.29	7.25	1.58	1.72	1.33	42.53
1981	0.84	4.08	2.25	1.87	3.38	5.28	5.42	4.65	0.39	1.90	1.47	8.54	40.07
1982	3.74	4.39	1.65	6.44	2.92	4.23	9.98	5.88	3.32	1.47	2.62	3.72	50.36
1983	3.66	5.38	7.35	5.68	0.70	2.85	0.73	3.36	3.25	2.22	3.63	6.58	45.39
1984	3.99	4.88	5.54	3.75	4.29	6.47	8.69	3.23	0.67	1.03	0.78	1.75	45.07
1985	3.27	7.15	0.56	1.29	3.13	3.96	7.47	5.65	0.07	8.44	5.98	0.88	47.85
Record Mean	3.41	3.90	3.95	3.25	3.29	4.20	5.59	5.51	3.77	2.47	2.86	3.24	45.44

TABLE 3 AVERAGE TEMPERATURE (deg. F) COLUMBIA, SOUTH CAROLINA

YEAR	JAN	FEB	MAR	APR	MAY	JUNE	JULY	AUG	SEP	OCT	NOV	DEC	ANNUAL
1956	42.7	52.0	54.7	62.4	73.7	79.4	82.5	81.9	72.5	66.0	52.8	56.2	64.8
1957	47.9	55.1	53.5	65.6	72.4	79.2	81.1	79.1	76.3	60.0	55.7	47.7	64.5
1958	40.0	40.8	49.9	63.6	71.9	78.2	81.9	81.8	75.1	61.2	57.5	42.2	62.0
1959	44.6	50.1	53.0	65.6	73.6	78.3	80.3	81.7	74.2	66.4	53.5	47.1	64.0
1960	45.3	45.6	43.8	65.3	70.8	79.0	81.2	81.1	75.9	66.5	54.2	41.3	62.5
1961	41.3	50.1	58.1	58.6	68.8	77.1	81.2	80.6	76.2	61.9	58.7	47.0	63.1
1962	44.8	52.5	51.3	60.8	76.7	77.3	81.6	80.6	73.9	65.4	52.3	43.7	63.4
1963	41.3	42.0	59.0	65.3	70.9	77.5	79.9	81.7	72.3	64.3	54.3	39.6	62.4
1964	44.1	43.7	54.8	64.2	72.4	80.5	78.8	78.4	73.8	59.1	58.2	49.8	63.2
1965	45.7	48.3	52.2	65.1	76.1	75.5	79.9	80.5	75.4	62.8	54.7	47.3	63.6
#1966	40.3	46.9	51.8	63.1	70.2	75.0	81.6	79.2	73.8	62.8	53.4	45.7	62.0
1967	46.9	43.0	57.1	65.6	67.0	73.3	77.6	75.7	67.3	60.2	49.1	49.6	61.0
1968	41.1	40.1	54.8	63.8	69.6	77.9	80.5	82.9	72.4	64.0	53.6	42.7	62.0
1969	42.7	43.5	51.1	62.8	70.6	78.3	81.1	77.5	72.8	65.3	50.8	44.1	62.1
1970	38.3	47.5	57.1	67.6	73.0	78.8	83.4	81.3	78.6	64.7	50.7	50.1	64.3
1971	45.4	46.3	50.0	61.3	69.8	80.3	80.0	80.1	77.1	69.8	53.8	56.4	64.2
1972	50.9	45.7	53.9	63.1	68.4	73.9	79.2	79.1	75.2	62.7	53.5	51.3	63.1
1973	44.9	44.0	58.6	60.0	67.8	77.2	80.7	80.6	78.6	66.7	59.0	49.5	64.0
1974	59.2	50.9	61.7	65.3	74.7	77.7	81.3	78.7	72.5	61.1	54.3	48.0	65.5
1975	51.6	52.0	54.0	63.2	74.8	77.8	79.0	80.5	75.3	67.2	56.1	46.3	64.8
1976	43.5	55.5	61.0	64.4	69.1	74.6	79.6	79.0	72.4	58.6	47.7	44.0	62.5
1977	35.8	46.1	59.6	66.2	72.5	78.3	82.5	78.6	75.4	59.5	56.4	44.4	62.9
1978	37.3	38.4	51.4	63.0	69.2	77.0	79.9	80.3	77.3	63.9	59.9	49.5	62.3
1979	43.4	42.9	55.8	62.4	69.0	73.0	78.5	79.7	73.7	62.0	55.0	44.7	61.7
1980	44.2	41.4	49.8	61.5	69.0	75.7	81.9	80.5	76.8	60.2	51.7	44.7	61.5
1981	38.7	46.7	51.2	66.9	68.0	81.4	80.9	76.1	71.4	61.8	53.2	43.3	61.7
1982	40.7	49.8	56.5	60.2	72.0	77.4	80.3	77.9	71.8	65.0	55.0	51.3	62.9
1983	40.9	45.3	53.6	57.8	69.9	75.3	83.3	82.7	73.2	63.7	52.3	43.3	61.8
1984	41.7	48.0	53.6	59.6	70.1	79.1	78.7	79.2	71.7	70.4	50.6	53.4	63.0
1985	39.5	47.0	57.5	65.3	71.5	78.7	78.7	79.9	77.7	72.4	68.4	62.5	42.9	63.6
Record Mean	45.9	47.8	55.1	63.3	71.7	78.3	80.7	79.8	74.9	64.3	54.3	47.1	63.7
Max	56.0	58.5	66.3	75.1	83.1	89.1	90.8	89.6	84.9	75.7	65.7	57.4	74.4
Min	35.8	37.1	43.9	51.6	60.2	67.6	70.6	69.9	64.9	52.9	42.9	36.8	52.9

REFERENCE NOTES FOR TABLES 1, 2, 3 and 6   (COLUMBIA, SC)

GENERAL

T - TRACE AMOUNT
BLANK ENTRIES DENOTE MISSING/UNREPORTED DATA.
# INDICATES A STATION OR INSTRUMENT RELOCATION.

SPECIFIC

TABLE 1

(a) - LENGTH OF RECORD IN YEARS. ALTHOUGH
INDIVIDUAL MONTHS MAY BE MISSING.

* LESS THAN .05

NORMALS — BASED ON THE 1951-1980 RECORD PERIOD.
EXTREMES — DATES ARE THE MOST RECENT OCCURRENCE.
WIND DIR. — NUMERALS SHOW TENS OF DEGREES
CLOCKWISE FROM TRUE NORTH.
"00" INDICATES CALM.
RESULTANT WIND DIRECTIONS ARE GIVEN TO WHOLE DEGREES.

EXCEPTIONS

TABLES 2, 3, and 6

RECORD MEANS ARE THROUGH THE CURRENT YEAR,
BEGINNING IN   1880 FOR TEMPERATURE
1880 FOR PRECIPITATION
1948 FOR SNOWFALL

TABLE 4 HEATING DEGREE DAYS Base 65 deg. F   COLUMBIA, SOUTH CAROLINA

SEASON	JULY	AUG	SEP	OCT	NOV	DEC	JAN	FEB	MAR	APR	MAY	JUNE	TOTAL
1956-57	0	0	23	45	372	283	531	283	350	102	31	0	2020
1957-58	0	0	20	160	287	526	766	671	463	113	19	0	3025
1958-59	0	0	3	148	249	702	630	412	370	95	5	0	2614
1959-60	0	0	1	98	350	550	603	557	655	84	38	0	2936
1960-61	0	0	0	85	317	728	728	412	238	227	15	3	2753
1961-62	0	0	7	123	272	550	620	360	430	183	0	0	2545
1962-63	0	0	14	112	378	653	726	641	208	86	41	0	2859
1963-64	0	0	14	70	311	779	640	612	325	112	17	0	2880
1964-65	0	0	3	204	216	473	592	466	398	93	0	0	2445
1965-66	0	0	2	138	303	543	759	502	408	128	19	1	2803
#1966-67	0	0	0	127	351	596	554	605	270	67	71	10	2651
1967-68	0	0	32	164	469	732	717	480	327	106	20	0	3047
1968-69	0	0	0	121	339	684	683	594	430	71	13	0	2935
1969-70	0	0	2	86	419	640	823	466	252	73	12	0	2791
1970-71	0	0	6	85	420	462	602	519	462	155	27	0	2738
1971-72	0	0	0	13	354	292	429	555	341	136	12	1	2133
1972-73	0	0	0	98	350	423	618	583	219	179	47	0	2517
1973-74	0	0	0	57	205	477	199	394	161	99	7	0	1599
1974-75	0	0	18	163	342	521	417	365	344	136	0	0	2306
1975-76	0	0	0	56	296	577	662	280	176	97	26	4	2174
1976-77	0	0	0	219	512	645	901	525	211	69	20	0	3102
1977-78	0	0	1	192	277	629	850	741	418	109	37	0	3254
1978-79	0	0	0	89	162	492	664	613	294	107	27	2	2450
1979-80	0	0	6	138	314	616	641	680	466	147	27	0	3035
1980-81	0	0	14	178	394	622	809	506	424	59	43	0	3049
1981-82	0	0	13	131	349	665	748	425	283	179	5	0	2798
1982-83	0	0	18	162	306	440	739	546	352	230	24	1	2818
1983-84	0	0	21	89	373	666	717	485	355	190	31	2	2929
1984-85	0	0	12	22	444	355	792	503	262	79	9	0	2478
1985-86	0	0	16	45	110	676							

TABLE 5 COOLING DEGREE DAYS Base 65 deg. F   COLUMBIA, SOUTH CAROLINA

YEAR	JAN	FEB	MAR	APR	MAY	JUNE	JULY	AUG	SEP	OCT	NOV	DEC	TOTAL
1969	0	0	5	81	191	406	570	392	242	104	1	0	1992
1970	2	0	12	158	266	420	576	513	421	82	0	6	2456
1971	0	2	6	51	185	465	470	473	369	168	25	32	2246
1972	1	0	5	86	126	274	450	445	313	35	14	4	1753
1973	0	0	28	33	143	375	494	490	413	116	32	6	2130
1974	29	8	65	116	316	388	511	432	249	51	29	1	2195
1975	8	6	11	94	308	390	442	488	317	129	37	0	2230
1976	2	10	58	87	160	295	460	442	231	30	0	0	1775
1977	0	3	53	110	257	407	549	429	318	29	25	0	2180
1978	0	0	1	56	172	369	466	478	376	62	15	17	2012
1979	0	0	13	36	157	253	425	461	276	50	22	0	1693
1980	0	0	0	47	159	328	530	487	376	35	2	0	1964
1981	0	0	6	119	143	497	499	349	213	41	0	2	1869
1982	0	4	26	39	229	377	484	406	226	74	11	20	1896
1983	0	0	6	21	184	319	573	558	270	58	0	0	1989
1984	0	0	10	33	194	430	436	448	219	199	18	3	1990
1985	9	2	38	94	215	417	468	402	244	155	45	0	2089

TABLE 6 SNOWFALL (inches)   COLUMBIA, SOUTH CAROLINA

SEASON	JULY	AUG	SEP	OCT	NOV	DEC	JAN	FEB	MAR	APR	MAY	JUNE	TOTAL
1956-57	0.0	0.0	0.0	0.0	0.0	0.0	T	0.0	T	0.0	0.0	0.0	T
1957-58	0.0	0.0	0.0	0.0	0.0	T	T	1.4	0.0	0.0	0.0	0.0	1.4
1958-59	0.0	0.0	0.0	0.0	0.0	9.1	0.0	0.0	1.0	0.0	0.0	0.0	10.1
1959-60	0.0	0.0	0.0	0.0	0.0	T	T	T	3.2	0.0	0.0	0.0	3.2
1960-61	0.0	0.0	0.0	0.0	0.0	T	1.4	0.9	0.0	0.0	0.0	0.0	2.3
1961-62	0.0	0.0	0.0	0.0	0.0	0.0	T	T	0.0	0.0	0.0	0.0	T
1962-63	0.0	0.0	0.0	0.0	0.0	0.0	0.0	T	0.0	0.0	0.0	0.0	T
1963-64	0.0	0.0	0.0	0.0	0.0	T	1.1	0.0	0.0	0.0	0.0	0.0	1.1
1964-65	0.0	0.0	0.0	0.0	0.0	0.0	1.0	0.0	0.0	0.0	0.0	0.0	1.0
1965-66	0.0	0.0	0.0	0.0	0.0	0.0	1.0	T	0.0	0.0	0.0	0.0	T
1966-67	0.0	0.0	0.0	0.0	0.0	T	0.0	3.4	0.0	0.0	0.0	0.0	3.4
1967-68	0.0	0.0	0.0	0.0	0.0	0.0	1.7	1.0	0.0	0.0	0.0	0.0	2.7
1968-69	0.0	0.0	0.0	0.0	T	0.0	T	0.8	0.0	0.0	0.0	0.0	0.8
1969-70	0.0	0.0	0.0	0.0	0.0	T	1.8	T	0.0	0.0	0.0	0.0	1.8
1970-71	0.0	0.0	0.0	0.0	0.0	T	T	T	1.7	0.0	0.0	0.0	1.7
1971-72	0.0	0.0	0.0	0.0	0.0	0.8	0.0	0.0	0.0	0.0	0.0	0.0	0.8
1972-73	0.0	0.0	0.0	0.0	0.0	0.0	2.2	16.0	0.0	0.0	0.0	0.0	18.2
1973-74	0.0	0.0	0.0	0.0	0.0	T	0.0	0.0	T	0.0	0.0	0.0	T
1974-75	0.0	0.0	0.0	0.0	0.0	0.0	0.0	T	0.0	0.0	0.0	0.0	T
1975-76	0.0	0.0	0.0	0.0	0.0	0.0	0.0	0.0	0.0	0.0	0.0	0.0	T
1976-77	0.0	0.0	0.0	0.0	T	0.0	T	0.0	0.0	0.0	0.0	0.0	T
1977-78	0.0	0.0	0.0	0.0	0.0	0.0	T	0.5	0.0	0.0	0.0	0.0	0.5
1978-79	0.0	0.0	0.0	0.0	0.0	T	0.0	5.5	0.0	0.0	0.0	0.0	5.5
1979-80	0.0	0.0	0.0	0.0	0.0	0.0	0.0	3.0	4.1	0.0	0.0	0.0	7.1
1980-81	0.0	0.0	0.0	0.0	0.0	0.3	T	0.0	0.0	0.0	0.0	0.0	0.3
1981-82	0.0	0.0	0.0	0.0	0.0	T	3.5	0.8	0.0	0.0	0.0	0.0	4.3
1982-83	0.0	0.0	0.0	0.0	0.0	0.0	0.1	T	0.4	0.0	0.0	0.0	0.5
1983-84	0.0	0.0	0.0	0.0	0.0	0.0	T	T	0.0	0.0	0.0	0.0	T
1984-85	0.0	0.0	0.0	0.0	0.0	0.0	T	0.0	0.0	0.0	0.0	0.0	T
1985-86	0.0	0.0	0.0	0.0	0.0	0.0							
Record Mean	0.0	0.0	0.0	0.0	T	0.3	0.4	0.9	0.3	0.0	0.0	0.0	1.8

See Reference Notes, relative to all above tables, on preceding page.

This station, three miles south of Greer, South Carolina, is located in the Piedmont section, on the eastern slope of the Southern Appalachian Mountains. It is rolling country with the first ridge of the mountains about 20 miles to the northwest and the main ridge about 55 miles to the northwest. These mountains usually protect this area from the full force of the cold air masses which move southeastward from central Canada during the winter months.

At present, the National Weather Service Office is located at the Greenville-Spartanburg Jet Age Airport, on a level with, or slightly higher than, most of the surrounding countryside. No bodies of water are nearby. Temperatures are quite consistent with those in Greer, Greenville, and Spartanburg.

The elevation of the area, ranging from 800 to 1,100 feet is conducive to cool nights, especially during the summer months. Winters are quite pleasant, with the temperature remaining below freezing throughout the daylight hours only a few times during a normal year. There are usually two freezing rainstorms each winter and two or three small snowstorms.

Rainfall in this section is usually abundant and spread quite evenly through the months. Droughts have been experienced, but are usually of short duration.

The mountain ridges, which lie in a northeast-southwest direction, appear to have a definite overall influence on the direction of the wind. The prevailing directions are northeast and southwest, divided almost evenly, with fall and winter favoring northeast and spring and summer favoring southwest. Destructive winds occur occasionally, while tornadoes are infrequent in this vicinity.

In the southern two-thirds of Greenville and Spartanburg Counties, including the cities of the same names, the average occurrence of the last temperature of 32 degrees in spring is late March and the average occurrence of the first in fall is early November, giving an average growing season of 225 days. In a normal year some flowering shrubs bloom through the winter. In the higher elevations in the northern thirds of these counties, the growing season begins about one month later and ends about one month earlier.

TABLE 1 **NORMALS, MEANS AND EXTREMES**

GREENVILLE-SPARTANBURG (GREER), SOUTH CAROLINA

LATITUDE: 34°54'N   LONGITUDE: 82°13' W   ELEVATION: FT. GRND   857 BARO   00951   TIME ZONE: EASTERN   WBAN: 03870

	(a)	JAN	FEB	MAR	APR	MAY	JUNE	JULY	AUG	SEP	OCT	NOV	DEC	YEAR
TEMPERATURE °F:														
Normals														
-Daily Maximum		51.0	54.5	62.5	72.6	79.7	85.4	88.2	87.5	81.7	72.2	62.1	53.5	70.9
-Daily Minimum		31.2	32.6	39.4	48.3	56.9	64.2	68.2	67.4	61.7	49.1	39.6	33.2	49.3
-Monthly		41.1	43.5	51.0	60.5	68.3	74.8	78.2	77.5	71.7	60.7	50.9	43.4	60.1
Extremes														
-Record Highest	23	79	79	88	91	97	100	101	103	96	88	85	76	103
-Year		1975	1982	1967	1980	1967	1985	1983	1983	1975	1981	1974	1984	AUG 1983
-Record Lowest	23	-6	8	11	25	33	40	54	52	36	25	12	5	-6
-Year		1966	1967	1980	1983	1971	1972	1979	1968	1967	1976	1970	1985	JAN 1966
NORMAL DEGREE DAYS:														
Heating (base 65°F)		741	599	442	154	41	0	0	0	7	162	423	670	3239
Cooling (base 65°F)		0	0	8	19	143	297	409	388	208	29	0	0	1501
% OF POSSIBLE SUNSHINE	23	56	61	65	66	61	61	59	61	63	65	60	56	61
MEAN SKY COVER (tenths)														
Sunrise - Sunset	23	5.8	5.6	5.6	5.4	5.8	5.8	6.2	5.6	5.5	4.6	5.1	5.7	5.6
MEAN NUMBER OF DAYS:														
Sunrise to Sunset														
-Clear	23	11.0	10.5	10.3	10.9	8.8	8.3	6.1	8.7	10.0	14.0	12.7	11.5	122.8
-Partly Cloudy	23	5.9	5.6	8.0	8.1	9.7	11.1	12.9	12.2	8.9	6.8	6.0	5.7	100.9
-Cloudy	23	14.2	12.2	12.7	11.0	12.5	10.6	12.0	10.0	11.1	10.2	11.3	13.8	141.5
Precipitation														
.01 inches or more	23	11.0	9.3	11.3	9.0	11.4	10.1	12.3	9.7	8.1	7.4	8.7	10.0	118.2
Snow,Ice pellets														
1.0 inches or more	23	0.8	0.7	0.3	0.0	0.0	0.0	0.0	0.0	0.0	0.0	0.*	0.1	2.0
Thunderstorms	23	0.5	0.7	2.7	2.9	6.3	6.4	10.0	6.7	3.1	0.9	0.7	0.6	41.7
Heavy Fog Visibility														
1/4 mile or less	23	4.7	3.3	3.2	2.3	1.7	1.3	2.3	3.0	2.2	1.9	3.6	5.2	34.7
Temperature °F														
-Maximum														
90° and above	23	0.0	0.0	0.0	0.2	1.4	6.7	11.3	8.6	2.6	0.0	0.0	0.0	30.7
32° and below	23	1.4	0.3	0.*	0.0	0.0	0.0	0.0	0.0	0.0	0.0	0.0	0.3	2.0
-Minimum														
32° and below	23	19.4	16.3	7.3	1.0	0.0	0.0	0.0	0.0	0.0	0.7	7.2	15.3	67.1
0° and below	23	0.1	0.0	0.0	0.0	0.0	0.0	0.0	0.0	0.0	0.0	0.0	0.0	0.1
AVG. STATION PRESS.(mb)	13	984.0	983.8	981.7	981.8	980.9	982.1	983.1	983.9	984.0	985.1	985.0	985.1	983.4
RELATIVE HUMIDITY (%)														
Hour 01	23	72	69	69	70	79	82	85	86	85	82	76	74	77
Hour 07	23	76	74	76	77	83	84	87	89	88	85	81	78	82
Hour 13 (Local Time)	23	55	51	49	48	53	55	59	59	58	53	53	55	54
Hour 19	23	62	57	54	52	61	63	68	69	72	70	66	66	63
PRECIPITATION (inches):														
Water Equivalent														
-Normal		4.21	4.39	5.87	4.35	4.22	4.77	4.08	3.66	4.35	3.49	3.21	3.93	50.53
-Maximum Monthly	23	7.19	7.43	11.37	11.30	8.89	9.59	13.57	7.51	11.65	10.24	7.52	8.45	13.57
-Year		1979	1971	1980	1964	1972	1969	1984	1967	1975	1964	1985	1983	JUL 1984
-Minimum Monthly	23	0.29	0.53	1.13	0.69	1.09	1.29	0.80	1.16	0.27	0.24	1.34	0.37	0.24
-Year		1981	1978	1985	1976	1965	1981	1977	1963	1978	1974	1973	1965	OCT 1974
-Maximum in 24 hrs	23	3.30	3.57	4.45	3.76	3.58	4.80	3.89	4.49	6.21	4.53	2.83	3.00	6.21
-Year		1982	1984	1963	1963	1972	1980	1964	1967	1973	1977	1964	1972	SEP 1973
Snow,Ice pellets														
-Maximum Monthly	23	9.1	12.3	9.3	0.1							1.9	11.4	12.3
-Year		1966	1979	1983	1983							1968	1971	FEB 1979
-Maximum in 24 hrs	23	5.7	8.2	9.3	0.1							1.9	11.4	11.4
-Year		1965	1979	1983	1983							1968	1971	DEC 1971
WIND:														
Mean Speed (mph)	23	7.2	7.8	8.0	7.7	6.8	6.3	5.8	5.5	5.9	6.4	6.5	7.2	6.8
Prevailing Direction														
through 1963		NE	NE	SW	SW	NE	NE	WSW	NE	NE	NE	SW	NE	NE
Fastest Mile														
-Direction (!!!)	23	SW	SW	W	SW	SW	NW	NE	N	NE	NE	NE	NE	NE
-Speed (MPH)	23	44	44	38	44	36	35	52	34	31	31	34	47	52
-Year		1967	1966	1963	1970	1967	1969	1966	1973	1964	1964	1974	1963	JUL 1966
Peak Gust														
-Direction (!!!)	2	SW	SW	SW	SW	N	N	SW	NE	NE	NE	W	W	SW
-Speed (mph)	2	44	45	51	69	38	40	45	39	35	35	45	38	69
-Date		1985	1985	1984	1985	1984	1985	1984	1984	1985	1985	1984	1985	APR 1985

See Reference Notes to this table on the following page.

TABLE 2 PRECIPITATION (inches)   GREENVILLE-SPARTANBURG (GREER), SOUTH CAROLINA

YEAR	JAN	FEB	MAR	APR	MAY	JUNE	JULY	AUG	SEP	OCT	NOV	DEC	ANNUAL
1963	3.93	3.29	9.66	5.95	3.06	4.73	2.46	1.16	4.68	0.24	4.19	3.78	47.09
1964	5.44	4.67	7.11	11.30	1.59	8.07	7.44	6.64	0.93	10.24	3.36	3.62	70.41
1965	2.39	5.22	7.60	4.93	1.09	8.62	3.13	3.57	2.32	3.60	2.82	0.37	45.66
1966	4.64	6.78	3.26	2.53	3.06	3.84	2.98	5.01	7.98	3.78	1.93	3.15	48.94
1967	3.97	3.32	1.98	2.36	4.97	4.87	3.86	7.51	2.05	2.35	3.50	7.40	48.14
1968	4.12	1.00	3.68	2.40	3.93	5.71	6.92	1.31	3.04	2.82	5.07	3.18	43.18
1969	3.94	5.24	4.56	7.18	1.93	9.59	3.17	6.53	3.68	2.38	2.24	4.60	55.04
1970	1.74	3.74	3.45	2.94	3.13	3.60	2.31	3.59	1.34	7.02	1.77	2.88	37.51
1971	3.33	7.43	5.52	3.09	5.72	2.19	5.64	2.44	3.28	9.51	4.22	3.79	56.16
1972	6.14	3.04	4.59	2.28	8.89	8.16	4.18	3.21	2.20	3.44	5.31	6.68	58.12
1973	4.33	4.88	8.73	4.04	5.59	3.87	3.70	2.03	7.56	0.98	1.34	7.55	54.60
1974	4.24	4.90	3.26	4.06	5.45	3.78	3.23	4.03	3.76	0.24	4.81	2.50	44.26
1975	5.42	5.78	8.64	1.14	7.81	5.39	4.79	3.21	11.65	7.45	3.98	3.07	68.33
1976	4.49	2.15	7.30	0.69	8.10	2.81	5.75	2.09	8.28	8.49	2.75	6.21	59.11
1977	3.53	2.00	8.47	3.23	2.71	2.88	0.80	4.99	9.44	6.39	4.43	3.55	52.42
1978	6.93	0.53	6.09	2.97	4.84	3.51	6.77	2.98	0.27	0.81	1.93	3.39	41.02
1979	7.19	6.11	4.19	10.15	5.69	3.74	8.66	4.34	7.50	3.33	3.91	1.25	66.06
1980	4.28	1.19	11.37	3.47	5.92	6.72	1.05	3.33	5.82	2.83	4.11	0.64	50.73
1981	0.29	3.86	3.22	0.88	4.15	1.29	5.30	1.17	2.08	4.40	1.66	7.19	35.49
1982	6.27	5.21	2.77	4.57	6.18	3.32	12.52	1.66	1.44	3.07	4.17	5.02	56.20
1983	2.70	5.26	6.26	4.66	5.80	4.67	1.13	3.27	3.59	3.05	5.29	8.45	54.13
1984	3.04	7.04	5.67	4.76	8.30	3.07	13.57	4.00	1.34	2.28	2.60	2.22	57.89
1985	4.94	4.29	1.13	1.31	2.42	2.85	6.96	5.93	1.62	4.55	7.52	1.44	44.96
Record Mean	4.23	4.21	5.59	3.95	4.80	4.66	5.06	3.65	4.17	4.05	3.60	4.00	51.98

TABLE 3 AVERAGE TEMPERATURE (deg. F)   GREENVILLE-SPARTANBURG (GREER), SOUTH CAROLINA

YEAR	JAN	FEB	MAR	APR	MAY	JUNE	JULY	AUG	SEP	OCT	NOV	DEC	ANNUAL
1963	39.7	39.1	55.4	64.0	69.1	74.9	77.5	78.5	70.5	64.1	52.8	36.4	60.2
1964	41.7	40.7	51.9	61.2	71.3	77.6	77.6	76.7	72.4	57.8	55.9	45.3	60.8
1965	42.8	44.2	47.8	61.7	73.1	72.7	77.5	78.1	72.4	59.6	52.3	45.0	60.6
1966	36.5	42.8	50.1	59.1	66.9	74.2	77.9	75.7	69.9	58.7	51.6	43.0	58.8
1967	44.0	41.0	56.5	64.6	66.1	72.6	75.2	74.2	66.8	59.7	48.2	47.2	59.7
1968	39.0	38.3	53.3	60.2	65.4	74.2	77.0	78.7	69.8	60.9	49.5	39.0	58.8
1969	39.5	40.9	45.9	60.7	67.6	76.3	80.6	74.6	69.8	60.9	48.8	41.4	58.9
1970	35.4	43.3	51.7	63.3	68.8	75.2	79.3	76.9	74.8	63.0	48.7	44.7	60.5
1971	39.5	41.9	46.3	59.2	65.3	76.2	76.7	75.8	72.9	64.9	48.9	50.4	59.8
1972	44.6	40.9	50.7	59.5	64.9	70.9	77.0	76.3	73.1	59.2	49.3	48.1	59.5
1973	41.1	42.0	54.5	57.8	65.8	74.6	78.2	76.9	74.5	61.8	51.9	41.8	60.1
1974	51.4	43.9	55.9	58.8	68.0	72.2	78.0	77.2	69.9	58.8	49.6	43.4	60.6
1975	45.8	46.8	48.5	58.7	70.5	74.4	76.5	78.6	70.0	62.6	53.0	42.7	60.7
1976	38.0	50.6	54.0	59.2	63.3	71.4	77.2	75.3	68.6	55.6	44.6	39.9	58.1
1977	30.7	42.4	54.5	62.4	71.5	77.5	81.9	78.7	74.4	59.9	54.8	42.8	61.0
1978	35.3	36.7	49.2	60.8	65.5	75.4	77.8	77.9	73.9	60.4	55.6	43.5	59.3
1979	37.3	39.2	54.2	61.4	69.6	72.0	74.7	76.6	70.5	59.2	52.7	45.4	59.4
1980	44.0	39.3	47.6	59.1	68.2	74.5	81.5	79.4	73.6	57.4	49.3	43.4	59.8
1981	38.1	43.3	49.1	64.2	66.3	80.3	80.1	75.4	69.3	58.0	50.8	39.3	59.5
1982	36.1	45.8	53.3	56.7	70.6	74.2	77.6	75.2	69.8	61.0	52.7	48.3	60.1
1983	39.2	42.1	51.4	54.7	66.3	73.1	79.7	79.5	69.0	59.4	49.2	39.2	58.6
1984	38.8	45.6	50.3	56.2	66.9	77.7	75.6	76.3	69.0	67.5	47.8	50.1	60.2
1985	36.2	44.6	54.2	61.8	67.9	76.7	76.4	75.1	69.6	64.3	58.6	38.9	60.4
Record Mean	39.8	42.4	51.5	60.2	67.8	74.7	77.8	76.9	71.1	60.6	51.2	43.5	59.8
Max	49.7	53.3	63.3	72.3	79.0	85.0	87.5	86.4	81.0	71.8	62.2	53.4	70.4
Min	29.8	31.5	39.8	48.1	56.5	64.4	68.2	67.3	61.1	49.4	40.1	33.5	49.1

REFERENCE NOTES FOR TABLES 1, 2, 3 and 6 (GREENVILLE, SPARTANSBURG, SC)

GENERAL

T - TRACE AMOUNT
BLANK ENTRIES DENOTE MISSING/UNREPORTED DATA.
# INDICATES A STATION OR INSTRUMENT RELOCATION.

SPECIFIC

TABLE 1

(a) - LENGTH OF RECORD IN YEARS. ALTHOUGH
INDIVIDUAL MONTHS MAY BE MISSING.

* LESS THAN .05

NORMALS — BASED ON THE 1951-1980 RECORD PERIOD.
EXTREMES — DATES ARE THE MOST RECENT OCCURRENCE.
WIND DIR. — NUMERALS SHOW TENS OF DEGREES
CLOCKWISE FROM TRUE NORTH.
"00" INDICATES CALM.
RESULTANT WIND DIRECTIONS ARE GIVEN TO WHOLE DEGREES.

EXCEPTIONS

TABLES 2, 3, and 6

RECORD MEANS ARE THROUGH THE CURRENT YEAR,
BEGINNING IN   1963 FOR TEMPERATURE
                  1963 FOR PRECIPITATION
                  1963 FOR SNOWFALL

TABLE 4 HEATING DEGREE DAYS Base 65 deg. F GREENVILLE-SPARTANBURG (GREER), SOUTH CAROLINA

SEASON	JULY	AUG	SEP	OCT	NOV	DEC	JAN	FEB	MAR	APR	MAY	JUNE	TOTAL
1962-63							774	720	306	108	53	0	
1963-64	0	0	20	75	362	880	714	698	401	165	20	0	3335
1964-65	0	2	0	237	281	604	681	572	527	147	0	6	3057
1965-66	0	0	6	186	375	615	877	616	454	194	42	5	3370
1966-67	0	0	6	202	394	675	643	663	276	88	85	14	3046
1967-68	0	0	41	181	498	547	799	769	368	169	54	0	3426
1968-69	0	1	1	168	461	801	785	668	585	142	44	0	3656
1969-70	0	0	19	165	481	725	910	600	405	104	28	0	3437
1970-71	0	0	10	120	480	620	782	643	572	188	73	0	3498
1971-72	0	0	2	74	492	449	625	694	434	205	48	14	3037
1972-73	0	0	4	189	476	519	734	641	323	229	61	0	3176
1973-74	0	0	1	132	387	712	419	584	290	205	54	0	2784
1974-75	0	0	39	205	464	663	590	505	507	227	9	0	3209
1975-76	0	0	24	118	361	683	828	415	345	190	85	15	3064
1976-77	0	0	11	297	605	773	1057	626	319	117	19	0	3824
1977-78	0	0	2	184	322	681	916	787	486	143	88	0	3609
1978-79	0	0	6	152	276	658	851	715	336	129	15	3	3141
1979-80	1	0	9	193	362	600	647	737	532	199	27	0	3307
1980-81	0	0	27	246	462	663	829	602	491	91	50	0	3461
1981-82	0	0	26	229	422	791	887	533	365	249	12	0	3514
1982-83	0	0	24	177	370	515	796	631	414	307	43	1	3278
1983-84	0	0	59	179	467	792	807	553	452	271	67	1	3648
1984-85	0	0	37	32	508	454	884	565	341	141	32	0	2994
1985-86	0	0	38	101	194	800							

TABLE 5 COOLING DEGREE DAYS Base 65 deg. F GREENVILLE-SPARTANBURG (GREER), SOUTH CAROLINA

YEAR	JAN	FEB	MAR	APR	MAY	JUNE	JULY	AUG	SEP	OCT	NOV	DEC	TOTAL
1969	0	0	0	19	133	344	493	305	169	46	0	0	1509
1970	0	0	0	59	154	313	454	378	310	67	0	0	1735
1971	0	0	0	20	88	342	369	345	246	77	15	3	1505
1972	0	0	0	48	49	197	380	356	253	15	9	1	1308
1973	0	0	6	21	92	297	414	375	291	36	0	0	1532
1974	1	0	17	26	153	222	408	386	193	19	10	0	1435
1975	0	0	0	45	185	290	363	427	180	50	7	0	1547
1976	0	2	11	22	40	213	387	322	126	12	0	0	1135
1977	0	0	1	46	226	383	528	431	289	34	20	0	1958
1978	0	0	0	23	109	319	403	404	282	19	0	0	1559
1979	0	0	8	28	164	215	312	368	180	20	1	0	1296
1980	0	0	0	27	135	292	519	454	294	16	0	0	1737
1981	0	0	4	75	96	467	475	327	158	19	0	0	1621
1982	0	1	8	6	193	282	396	324	174	58	5	3	1450
1983	0	0	1	3	91	252	465	457	185	15	0	0	1469
1984	0	0	0	13	132	387	335	356	164	116	2	0	1505
1985	0	0	13	50	132	360	359	320	183	88	9	0	1514

TABLE 6 SNOWFALL (inches) GREENVILLE-SPARTANBURG (GREER), SOUTH CAROLINA

SEASON	JULY	AUG	SEP	OCT	NOV	DEC	JAN	FEB	MAR	APR	MAY	JUNE	TOTAL
1962-63							0.0	2.7	0.0	0.0	0.0	0.0	
1963-64	0.0	0.0	0.0	0.0	0.0	2.1	0.5	1.2	0.0	0.0	0.0	0.0	3.8
1964-65	0.0	0.0	0.0	0.0	0.0	0.0	7.7	1.0	1.3	0.0	0.0	0.0	10.0
1965-66	0.0	0.0	0.0	0.0	0.0	0.0	9.1	3.4	0.0	0.0	0.0	0.0	12.5
1966-67	0.0	0.0	0.0	0.0	0.0	0.0	0.5	1.5	0.0	0.0	0.0	0.0	2.0
1967-68	0.0	0.0	0.0	0.0	0.0	T	T	0.9	1.8	0.0	0.0	0.0	2.7
1968-69	0.0	0.0	0.0	0.0	1.9	T	T	6.9	1.9	0.0	0.0	0.0	10.7
1969-70	0.0	0.0	0.0	0.0	0.0	T	2.0	0.0	0.0	0.0	0.0	0.0	2.0
1970-71	0.0	0.0	0.0	0.0	0.0	2.9	T	1.0	6.6	0.0	0.0	0.0	10.5
1971-72	0.0	0.0	0.0	0.0	T	11.4	0.0	3.9	1.8	0.0	0.0	0.0	17.1
1972-73	0.0	0.0	0.0	0.0	T	0.0	4.6	0.0	0.0	0.0	0.0	0.0	4.6
1973-74	0.0	0.0	0.0	0.0	T	T	0.0	0.0	0.4	T	0.0	0.0	0.4
1974-75	0.0	0.0	0.0	0.0	0.0	T	T	1.3	T	0.0	0.0	0.0	1.3
1975-76	0.0	0.0	0.0	0.0	T	T	T	0.0	0.0	0.0	0.0	0.0	T
1976-77	0.0	0.0	0.0	0.0	0.1	0.0	6.9	0.0	0.0	0.0	0.0	0.0	7.0
1977-78	0.0	0.0	0.0	0.0	0.0	T	T	1.1	2.8	0.0	0.0	0.0	3.9
1978-79	0.0	0.0	0.0	0.0	0.0	0.0	0.9	12.3	0.0	0.0	0.0	0.0	13.2
1979-80	0.0	0.0	0.0	0.0	0.0	3.2	7.6	3.6	T	0.0	0.0	0.0	14.4
1980-81	0.0	0.0	0.0	0.0	0.0	T	1.0	0.0	T	0.0	0.0	0.0	1.0
1981-82	0.0	0.0	0.0	0.0	0.0	T	7.0	5.1	0.0	0.0	0.0	0.0	12.1
1982-83	0.0	0.0	0.0	0.0	0.0	0.0	4.4	3.0	9.3	0.1	0.0	0.0	16.8
1983-84	0.0	0.0	0.0	0.0	0.0	T	T	0.6	0.0	0.0	0.0	0.0	0.6
1984-85	0.0	0.0	0.0	0.0	0.0		1.2	0.3	0.0	0.0	0.0	0.0	1.5
1985-86	0.0	0.0	0.0	0.0	0.0	0.0							
Record Mean	0.0	0.0	0.0	0.0	0.1	0.7	2.2	2.4	1.2	T	0.0	0.0	6.6

See Reference Notes, relative to all above tables, on preceding page.

Aberdeen is located in the northeast quarter of South Dakota, approximately 200 miles south of the geographical center of the North American continent. The surrounding area, extensively cultivated, is the bed of glacial Lake Dakota, which is by far the largest flat area in South Dakota. The lake bed slopes gently to the south. The elevation of Aberdeen at the northern end of the lake bed is 1,300 feet. The elevation at the southern end, some 30 miles distant is 1,280 feet. Low hills rim the area on the east and west. These hills effect ceilings, visibility, and precipitation, which are a hazard to private aircraft operating in the area during periods of marginal weather. Principal drainage for the area is through the southward flowing, meandering James River with its associated meandering rivers and creeks.

Located near the center of the North American land mass, the climate is continental with distinct seasons. Frequent and rapid weather changes occur during all seasons of the year as migratory storms sweep through the area. The winters are cold and dry. Sub-zero minimum temperatures may set in as early as late November, although temperatures of zero and below are generally not recorded until mid-December. Lowest temperatures of the winter generally occur in the period from mid-January to mid-February. During the coldest periods the days are generally sunny with light winds, and these conditions partially moderate the discomfort experienced at such low temperatures. Some days of the winter will be extremely unpleasant with temperatures near or below zero and brisk winds. Heavy snowfalls rarely occur during the first two-thirds of the winter season, with heaviest snowfalls developing during late February and early March as temperatures moderate. Blizzards are infrequent, many winters will pass without a single occurrence of this type of weather phenomenon. However, difficult driving conditions occur several times during most winters during periods of weather termed ground blizzards by residents of the state. These ground blizzards may develop at wind speeds as low as 12 mph, depending on the condition of the snow pack, wind direction with respect to the highway, etc. Depending on the wind speed, the ground blizzard may restrict visibility only a few inches to as much as 5 feet above the highway surface, completely obscuring the highway surface, but with excellent visibility prevailing above the low blanket of wind-driven snow.

Spring is a very short and transitional period, the shortest season of the four distinct seasons, and one marked by very rapid weather changes. Cool to quite cold nights prevail into mid-May, although afternoon temperatures may be quite warm, as high as the mid-80s. By mid-May temperatures below the freezing point rarely occur and frost is rarely experienced after the end of May. Precipitation increases markedly during the spring, 42 percent of the total annual precipitation normally being recorded in the three month period from April through June.

Summers are pleasant with a maximum of sunshine, warm days, and generally cool and comfortable nights. Temperatures of 100 degrees or above may occur several times during the summer season, but low humidities, brisk winds during the heat of the day, and rapid cooling during the evening hours, which generally occur during the periods of elevated temperatures, markedly moderate the physical discomfort normally experienced at these high temperatures. Thunderstorms occur frequently. In June and August, thunderstorms are more likely to occur during the early evening and nighttime hours. During July, thunderstorms are approximately equally distributed throughout the 24 hours of the day. Hail, occurring in connection with thunderstorm activity, is most likely in late May and early June.

Autumn is most pleasant with mild days, cool nights, ample sunshine, and declining occurrences and amounts of precipitation. The first frost may be expected by late September, although it may occur as early as late August. By mid-October, the temperatures during the night will be near or below freezing, and late October temperatures are normally in the 20 degree range. The growing season is about 132 days.

TABLE 1   NORMALS, MEANS AND EXTREMES

ABERDEEN, SOUTH DAKOTA

LATITUDE: 45°27'N   LONGITUDE: 98°26'W   ELEVATION: FT. GRND 1296 BARO 01315   TIME ZONE: CENTRAL   WBAN: 14929

	(a)	JAN	FEB	MAR	APR	MAY	JUNE	JULY	AUG	SEP	OCT	NOV	DEC	YEAR
TEMPERATURE °F:														
Normals														
-Daily Maximum		18.9	26.1	37.9	56.5	69.6	78.6	85.3	84.4	73.4	61.3	41.0	26.5	55.0
-Daily Minimum		-2.2	5.3	17.5	32.6	43.9	54.0	58.9	56.5	45.5	34.1	19.4	6.4	31.0
-Monthly		8.4	15.7	27.7	44.5	56.8	66.3	72.1	70.5	59.5	47.7	30.2	16.5	43.0
Extremes														
-Record Highest	24	60	62	82	97	96	103	110	112	103	96	78	62	112
-Year		1981	1976	1963	1980	1969	1963	1966	1965	1970	1963	1975	1969	AUG 1965
-Record Lowest	24	-35	-37	-29	-2	19	33	39	35	20	11	-27	-39	-39
-Year		1972	1971	1962	1975	1961	1964	1971	1965	1965	1976	1964	1967	DEC 1967
NORMAL DEGREE DAYS:														
Heating (base 65°F)		1758	1380	1156	612	274	73	16	20	197	536	1044	1504	8570
Cooling (base 65°F)		0	0	0	0	19	112	236	190	32	0	0	0	589
% OF POSSIBLE SUNSHINE														
MEAN SKY COVER (tenths)														
Sunrise - Sunset	19	6.4	6.8	6.9	6.8	6.4	5.8	4.7	4.8	5.3	5.9	6.9	6.8	6.1
MEAN NUMBER OF DAYS:														
Sunrise to Sunset														
-Clear	19	8.1	6.4	6.3	6.6	7.1	8.4	11.8	11.8	10.6	9.9	6.2	7.2	100.5
-Partly Cloudy	19	7.2	6.2	7.4	7.1	10.1	10.2	12.7	10.9	8.3	7.7	6.8	6.6	101.2
-Cloudy	19	15.8	15.6	17.3	16.4	13.8	11.4	6.5	8.2	11.1	13.3	16.9	17.2	163.6
Precipitation														
.01 inches or more	54	6.6	5.7	7.1	8.2	9.6	10.4	8.3	7.9	6.2	5.1	5.5	6.3	86.9
Snow, Ice pellets														
1.0 inches or more	54	2.6	2.1	2.1	0.9	0.1	0.0	0.0	0.0	0.0	0.3	1.6	2.0	11.7
Thunderstorms	14	0.0	0.0	0.4	1.9	4.9	9.6	9.0	6.1	2.9	1.4	0.0	0.0	36.1
Heavy Fog Visibility														
1/4 mile or less	14	1.6	2.1	2.4	1.4	0.5	0.9	0.9	0.9	1.1	1.1	2.5	2.9	18.3
Temperature °F														
-Maximum														
90° and above	24	0.0	0.0	0.0	0.2	0.3	2.7	9.5	9.0	2.2	0.2	0.0	0.0	24.0
32° and below	24	24.4	17.3	10.7	0.3	0.0	0.0	0.0	0.0	0.0	0.1	8.1	21.9	82.8
-Minimum														
32° and below	24	30.9	27.9	27.2	15.1	3.4	0.0	0.0	0.0	2.3	13.3	27.3	30.9	178.1
0° and below	24	17.3	10.0	3.3	0.*	0.0	0.0	0.0	0.0	0.0	0.0	2.1	12.0	44.8
AVG. STATION PRESS.(mb)	7	971.0	970.8	966.7	967.4	965.7	965.4	967.2	966.8	968.5	968.8	969.6	969.8	968.1
RELATIVE HUMIDITY (%)														
Hour 00	13	73	76	79	75	72	77	75	72	74	74	80	80	76
Hour 06	17	74	77	81	81	80	84	83	85	84	82	82	80	81
Hour 12 (Local Time)	19	69	70	68	56	52	57	52	50	52	56	68	72	60
Hour 18	19	71	72	72	66	52	47	53	47	48	56	70	75	58
PRECIPITATION (inches):														
Water Equivalent														
-Normal		0.49	0.64	0.99	1.94	2.56	3.22	2.42	1.95	1.50	1.01	0.60	0.47	17.79
-Maximum Monthly	54	2.23	2.06	3.45	5.13	6.36	8.88	7.71	6.62	4.51	5.14	2.36	1.86	8.88
-Year		1937	1952	1977	1953	1949	1939	1972	1942	1941	1982	1977	1935	JUN 1939
-Minimum Monthly	54	0.01	0.00	0.04	0.33	0.28	0.37	0.30	0.06	0.05	0.00	T	T	0.00
-Year		1961	1932	1971	1952	1948	1974	1975	1947	1979	1952	1980	1943	OCT 1952
-Maximum in 24 hrs	54	1.12	1.02	3.00	2.28	3.81	5.20	3.46	2.72	2.60	2.38	1.30	0.90	5.20
-Year		1939	1958	1937	1938	1949	1978	1983	1930	1967	1982	1977	1935	JUN 1978
Snow, Ice pellets														
-Maximum Monthly	54	26.2	25.1	27.9	24.4	2.0				0.1	5.5	24.7	18.5	27.9
-Year		1937	1969	1975	1970	1943				1965	1970	1936	1935	MAR 1975
-Maximum in 24 hrs	54	10.0	14.3	13.0	15.0	2.0				0.1	5.0	9.0	8.0	15.0
-Year		1937	1951	1937	1970	1943				1965	1932	1953	1935	APR 1970
WIND:														
Mean Speed (mph)	15	11.3	11.4	12.4	13.2	12.3	10.5	9.4	10.3	10.7	11.2	10.9	10.8	11.2
Prevailing Direction														
Fastest Obs. 1 Min.														
-Direction (!!!)	16	34	06	35	36	16	19	32	34	32	33	34	36	32
-Speed (MPH)	16	58	52	52	49	46	43	58	40	41	37	41	40	58
-Year		1967	1967	1966	1968	1965	1967	1973	1985	1977	1968	1982	1972	JUL 1973
Peak Gust														
-Direction (!!!)	2	NW	N	S	S	NW	NW	NW	N	N	NW	NW	NW	S
-Speed (mph)	2	46	56	51	60	56	49	54	58	49	47	48	59	60
-Date		1985	1984	1985	1985	1985	1985	1984	1985	1985	1984	1984	1985	APR 1985

See Reference Notes to this table on the following pages.

TABLE 2 PRECIPITATION (inches)   ABERDEEN, SOUTH DAKOTA

YEAR	JAN	FEB	MAR	APR	MAY	JUNE	JULY	AUG	SEP	OCT	NOV	DEC	ANNUAL
1956	1.22	0.58	0.94	0.74	2.01	3.06	2.95	2.87	0.82	1.52	1.30	0.40	18.41
1957	0.23	0.52	0.32	3.91	4.95	4.05	2.50	1.29	1.60	2.26	0.52	0.53	22.68
1958	0.23	1.29	0.52	2.04	1.46	2.54	1.61	0.93	0.08	0.27	1.17	0.29	12.43
1959	0.22	0.85	0.15	0.42	3.40	0.50	1.74	1.48	3.15	1.54	0.88	0.73	15.06
1960	1.01	0.52	0.79	1.27	2.18	1.59	0.97	4.21	0.69	0.56	0.49	0.68	14.96
1961	0.01	0.39	0.15	1.12	3.61	1.62	2.08	1.15	2.09	2.24	0.02	0.59	15.07
1962	0.50	1.41	0.87	1.08	5.74	5.74	3.14	0.75	2.44	1.06	0.46	0.18	23.37
1963	0.46	0.30	0.33	1.15	3.92	2.10	3.93	2.55	1.64	0.50	0.30	0.23	17.41
1964	0.36	0.31	0.89	3.19	2.68	2.69	3.47	4.04	0.56	T	0.24	0.39	18.82
1965	0.15	0.09	0.71	2.93	3.01	2.11	2.22	1.26	3.25	0.37	0.65	0.38	17.13
1966	0.16	0.52	2.34	0.92	0.91	4.53	4.66	2.77	1.25	1.86	0.59	0.03	20.54
1967	0.50	0.42	0.09	3.38	0.58	5.13	1.06	1.40	3.41	0.65	0.15	0.47	17.24
1968	0.15	0.10	2.74	4.81	1.66	3.24	2.20	1.07	2.82	0.60	0.83	0.94	21.16
1969	1.11	1.54	0.36	0.92	3.38	0.38	4.07	0.74	0.38	0.57	0.14	0.71	17.29
1970	0.21	0.17	1.33	3.42	1.49	2.11	2.04	0.33	1.67	1.24	1.39	0.19	15.59
1971	0.35	0.41	0.04	1.86	1.70	4.29	3.75	1.39	2.30	2.80	1.39	0.47	20.75
1972	0.39	0.39	1.02	1.05	4.86	1.80	7.71	0.52	0.25	1.26	0.26	0.88	20.39
1973	0.21	0.15	1.29	1.02	1.00	1.37	1.77	1.17	3.51	1.28	0.66	0.44	13.87
1974	0.05	0.26	0.74	2.10	3.77	0.94	1.72	0.94	0.15	0.54	0.12	0.05	10.81
1975	0.82	0.29	2.71	2.60	2.26	5.28	0.30	1.50	1.73	1.23	0.29	0.30	19.31
1976	0.81	0.52	0.70	1.27	0.52	1.41	0.50	0.66	0.86	0.32	0.01	0.30	7.88
1977	0.34	0.91	3.45	0.90	2.82	1.99	2.16	2.48	4.36	1.24	2.36	0.83	23.84
1978	0.15	0.23	0.46	2.25	4.23	7.30	2.17	4.04	0.60	0.08	0.77	0.11	22.39
1979	1.01	0.73	1.60	2.93	1.93	4.99	2.56	1.24	0.05	1.18	0.02	0.14	18.38
1980	0.51	0.44	0.88	1.15	1.64	2.53	0.80	5.93	0.92	1.44	T	0.14	16.38
1981	0.12	0.20	2.00	0.12	1.60	2.10	3.97	2.91					13.02
1982										5.14	0.59	0.09	
1983	0.16	0.26	2.65	0.69	1.66	3.47	6.46	2.21	1.55	0.81	0.60	0.55	21.07
1984	0.47	0.70	1.94	2.39	1.13	5.65	2.64	2.23	0.84	2.93	0.06	0.61	21.59
1985	0.23	0.08	1.82	0.63	3.41	1.76	2.38	2.71	2.71	0.87	1.60	0.57	18.77
Record Mean	0.55	0.58	1.16	1.98	2.43	3.52	2.59	2.11	1.53	1.17	0.67	0.52	18.79

TABLE 3 AVERAGE TEMPERATURE (deg. F)   ABERDEEN, SOUTH DAKOTA

YEAR	JAN	FEB	MAR	APR	MAY	JUNE	JULY	AUG	SEP	OCT	NOV	DEC	ANNUAL
1956	8.3	6.9	24.3	37.4	55.8	71.4	69.1	68.4	57.3	50.7	30.4	21.6	41.9
1957	4.9	16.9	29.3	43.3	54.0	63.3	75.7	69.9	57.0	47.0	31.2	25.1	43.2
1958	18.6	13.3	30.4	44.1	58.7	60.2	67.8	71.4	61.4	48.3	32.1	12.9	43.3
1959	6.5	8.2	33.3	44.0	55.3	70.2	72.4	74.1	59.5	41.6	22.1	27.3	42.9
#1960	10.5	10.9	15.2	43.7	56.0	63.2	73.2	68.2	56.7	46.8	29.6	14.7	40.8
1961	11.5	16.1	34.6	38.7	52.7	67.9	70.8	73.6	55.6	48.2	30.6	10.3	42.6
1962	8.8	11.2	21.9	44.8	58.6	65.3	67.2	70.2	57.2	49.8	34.9	19.1	42.5
1963	2.9	16.9	36.3	45.0	55.6	66.7	73.4	71.0	62.3	55.9	33.6	9.4	44.4
1964	16.3	24.3	23.0	46.4	58.5	66.7	75.1	66.4	49.7	50.2	29.2	25.4	42.9
1965	2.1	11.2	19.3	43.7	56.2	65.1	71.7	71.7	59.0	46.7	25.5	19.3	41.1
1966	-0.1	9.2	34.0	38.9	53.8	67.1	76.4	66.6	59.0	46.7	25.5	19.3	41.4
1967	15.1	11.7	33.1	43.2	52.7	64.2	69.2	68.0	61.7	46.3	31.2	16.5	42.7
1968	10.4	15.8	36.3	44.0	51.3	65.0	69.9	68.2	58.9	46.7	32.4	13.8	42.7
1969	4.4	17.9	16.2	46.4	57.5	59.9	70.0	73.1	63.2	42.6	33.6	13.8	42.0
1970	2.0	15.6	23.1	41.2	56.7	68.2	72.5	72.1	62.8	48.3	30.5	13.8	42.3
1971	5.4	15.2	31.4	47.9	55.6	70.3	68.2	73.0	59.7	48.5	32.2	12.6	43.4
1972	6.3	9.6	29.2	44.8	60.7	67.2	69.5	71.6	60.3	44.2	30.3	10.5	42.0
1973	17.0	24.3	39.5	45.6	57.5	68.2	72.9	75.6	57.9	50.6	27.0	12.2	45.7
1974	6.3	18.2	30.6	46.0	53.9	65.7	77.4	68.1	57.2	50.2	32.5	22.9	44.1
1975	17.1	14.7	24.9	40.6	59.8	68.1	78.3	72.3	58.9	51.9	34.0	17.6	44.9
1976	12.0	26.8	30.5	49.7	57.0	69.6	75.0	75.9	62.1	43.0	26.5	14.6	45.2
1977	1.4	24.2	33.6	51.9	64.9	67.7	74.1	65.7	59.6	46.5	27.2	11.1	44.0
1978	-1.7	6.0	25.3	42.9	58.2	65.6	70.1	70.7	65.1	47.4	24.2	10.8	40.4
1979	-1.2	2.5	24.6	40.7	52.6	65.3	70.5	68.3	63.2	45.9	29.2	25.6	40.6
1980	13.0	14.7	25.6	49.9	59.5	67.0	72.9	68.9	59.5	45.3	36.2	20.0	44.4
1981	19.4	23.9	35.6	49.9	56.3	65.3	74.4	72.1					33.1
1982										46.6	27.7	25.0	
1983	23.1	28.0	33.3	41.5	53.6	65.2	75.2	76.5	60.8	47.3	32.2	-0.6	44.7
1984	15.7	30.3	27.9	47.0	54.4	65.6	71.3	72.7	55.4	48.6	32.5	13.0	44.5
1985	8.5	16.3	36.0	50.3	61.2	61.3	72.6	67.3	56.5	45.3	16.4	7.8	41.6
Record Mean	9.7	15.9	28.7	44.9	57.0	66.5	73.1	71.1	60.0	47.9	30.1	16.6	43.5
Max	20.3	26.4	38.7	57.0	69.9	78.9	86.8	85.2	74.0	61.3	40.6	26.5	55.5
Min	-0.9	5.4	18.6	32.7	44.0	54.0	59.4	57.0	46.1	34.5	19.7	6.7	31.4

REFERENCE NOTES FOR TABLES 1, 2, 3 and 6   (ABERDEEN, SD)

GENERAL

T - TRACE AMOUNT
BLANK ENTRIES DENOTE MISSING/UNREPORTED DATA.
# INDICATES A STATION OR INSTRUMENT RELOCATION.

SPECIFIC

TABLE 1

(a) - LENGTH OF RECORD IN YEARS. ALTHOUGH INDIVIDUAL MONTHS MAY BE MISSING.

  * LESS THAN .05

NORMALS — BASED ON THE 1951-1980 RECORD PERIOD.
EXTREMES — DATES ARE THE MOST RECENT OCCURRENCE.
WIND DIR. — NUMERALS SHOW TENS OF DEGREES CLOCKWISE FROM TRUE NORTH. "00" INDICATES CALM.
RESULTANT WIND DIRECTIONS ARE GIVEN TO WHOLE DEGREES.

EXCEPTIONS

TABLE 1

1. PRIOR TO 1948, MAXIMUM 24-HOUR PRECIPITATION IS FOR OBSERVATIONAL DAY (A 24-HOUR PERIOD OTHER THAN MIDNIGHT-TO-MIDNIGHT) INSTEAD OF ANY CONSECUTIVE 24 HOURS.
2. MEAN WIND SPEED, THUNDERSTORMS, AND HEAVY FOG ARE THROUGH 1977.
3. TOTAL BREAK IN RECORD SEPTEMBER 1981 THROUGH SEPTEMBER 1982.
4. FASTEST OBSERVED WINDS ARE THROUGH JUNE 1978 AND OCTOBER 1982 TO DATE.

TABLES 2, 3, and 6

RECORD MEANS ARE THROUGH THE CURRENT YEAR, BEGINNING IN
  1930 FOR TEMPERATURE
  1930 FOR PRECIPITATION
  1930 FOR SNOWFALL

TABLE 4 HEATING DEGREE DAYS Base 65 deg. F   ABERDEEN, SOUTH DAKOTA

SEASON	JULY	AUG	SEP	OCT	NOV	DEC	JAN	FEB	MAR	APR	MAY	JUNE	TOTAL
1956-57	11	31	247	445	1037	1333	1768	1326	1103	649	350	81	8381
1957-58	0	33	249	562	1014	1227	1434	1365	1070	627	220	188	7989
1958-59	33	40	172	518	985	1557	1724	1540	979	629	341	51	8569
1959-60	8	5	227	723	1282	1167	1640	1544	1498	637	297	84	9112
#1960-61	14	41	313	561	1050	1538	1605	1346	938	793	388	59	8646
1961-62	3	2	332	515	1021	1691	1739	1505	1327	604	199	49	8987
1962-63	33	16	235	469	893	1415	1927	1342	880	590	296	29	8125
1963-64	0	21	111	295	932	1722	1507	1173	1296	553	232	67	7909
1964-65	2	77	274	560	1139	1789	1951	1505	1413	635	279	52	9676
1965-66	5	38	456	447	1069	1217	2017	1560	950	774	355	67	8955
1966-67	0	60	212	560	1181	1409	1544	1491	982	647	395	56	8537
1967-68	35	39	134	575	1008	1500	1690	1421	883	623	423	77	8408
1968-69	21	60	207	562	970	1580	1882	1312	1505	550	273	169	9091
1969-70	7	4	126	688	935	1438	1954	1379	1294	706	266	43	8840
1970-71	11	9	187	513	1029	1583	1846	1388	1035	515	287	20	8423
1971-72	32	15	220	504	977	1621	1821	1603	1100	597	201	39	8730
1972-73	19	35	176	638	1033	1688	1484	1135	783	580	238	27	7836
1973-74	2	0	239	442	1132	1630	1819	1305	1058	563	350	73	8613
1974-75	0	43	250	455	967	1302	1477	1401	1234	725	184	44	8082
1975-76	2	6	208	411	921	1465	1639	1101	1063	454	245	39	7554
1976-77	3	2	165	676	1149	1557	1973	1135	968	395	83	23	8129
1977-78	2	39	172	569	1128	1669	2070	1650	1228	659	226	74	9486
1978-79	13	23	142	540	1221	1678	2051	1750	1245	727	389	68	9847
1979-80	6	41	124	585	1068	1215	1608	1452	1213	463	221	31	8027
1980-81	1	24	192	601	860	1388	1408	1146	907	447	273	43	7290
1981-82	13	1	0	0	0	0							
1982-83				562	1111	1232	1291	1032	976	697	351	89	
1983-84	4	0	198	540	975	2030	1522	1001	1147	533	332	44	8326
1984-85	12	15	312	514	967	1608	1747	1361	891	444	158	145	8174
1985-86	2	42	292	608	1453	1770							

TABLE 5 COOLING DEGREE DAYS Base 65 deg. F   ABERDEEN, SOUTH DAKOTA

YEAR	JAN	FEB	MAR	APR	MAY	JUNE	JULY	AUG	SEP	OCT	NOV	DEC	TOTAL
1969	0	0	0	0	48	17	170	265	75	0	0	0	575
1970	0	0	0	0	14	141	253	235	127	3	0	0	773
1971	0	0	0	7	1	186	138	269	67	0	0	0	668
1972	0	0	0	0	73	112	166	245	43	0	0	0	639
1973	0	0	0	2	10	129	252	335	33	0	0	0	761
1974	0	0	0	0	13	99	393	145	20	3	0	0	673
1975	0	0	0	0	27	142	420	241	34	13	0	0	877
1976	0	0	0	0	5	193	322	345	85	1	0	0	951
1977	0	0	0	9	85	109	291	67	19	0	0	0	580
1978	0	0	0	0	21	100	179	206	152	0	0	0	658
1979	0	0	0	0	11	83	183	150	75	0	0	0	502
1980	0	0	0	16	55	98	251	150	33	0	0	0	603
1981	0	0	0	2	11	57	314	230			0	0	614
1982										0	0	0	
1983	0	0	0	0	3	100	329	363	79	0	0	0	874
1984	0	0	0	0	12	70	210	259	31	15	0	0	597
1985	0	0	0	9	48	41	244	121	46	0	0	0	509

TABLE 6 SNOWFALL (inches)   ABERDEEN, SOUTH DAKOTA

SEASON	JULY	AUG	SEP	OCT	NOV	DEC	JAN	FEB	MAR	APR	MAY	JUNE	TOTAL
1956-57	0.0	0.0	0.0	0.0	4.7	1.7	2.1	6.6	3.6	6.8	0.0	0.0	25.5
1957-58	0.0	0.0	0.0	0.0	T	8.8	3.5	1.0	4.0	3.5	0.0	0.0	20.8
1958-59	0.0	0.0	0.0	0.0	2.3	2.2	2.5	10.3	1.0	T	0.0	0.0	18.3
1959-60	0.0	0.0	0.0	1.0	8.6	5.7	9.7	3.3	8.8	7.9	0.0	0.0	45.0
1960-61	0.0	0.0	0.0	0.0	6.9	6.6	0.1	3.9	0.6	0.3	0.0	0.0	18.4
1961-62	0.0	0.0	0.0	0.0	0.1	6.3	5.6	11.5	6.6	T	0.0	0.0	30.1
1962-63	0.0	0.0	0.0	0.0	1.6	T	4.5	2.4	3.5	6.2	0.0	0.0	18.2
1963-64	0.0	0.0	0.0	0.0	2.2	4.1	3.0	3.0	5.5	T	0.0	0.0	17.8
1964-65	0.0	0.0	0.0	T	4.5	8.6	4.1	1.9	10.4	10.9	T	0.0	40.4
1965-66	0.0	0.0	0.1	0.0	5.8	0.7	3.3	5.7	18.9	7.2	0.3	0.0	42.0
1966-67	0.0	0.0	0.0	T	4.5	0.7	7.9	8.8	1.2	1.3	0.5	0.0	24.9
1967-68	0.0	0.0	0.0	1.4	1.3	9.1	3.1	1.4	3.2	15.5	T	0.0	35.0
1968-69	0.0	0.0	0.0	0.3	3.7	12.7	13.7	2.8	25.1	0.0	0.0	0.0	58.3
1969-70	0.0	0.0	0.0	0.3	1.1	12.0	4.8	3.7	6.8	24.4	0.1	0.0	53.2
1970-71	0.0	0.0	0.0	5.5	8.9	3.9	10.2	6.2	1.3	3.5	0.0	0.0	39.5
1971-72	0.0	0.0	T	2.9	9.1	9.9	8.2	7.0	6.0	0.4	0.0	0.0	43.5
1972-73	0.0	0.0	T	1.3	4.2	15.0	4.3	2.0	T	2.7	0.0	0.0	29.5
1973-74	0.0	0.0	0.0	T	4.2	10.7	1.6	6.6	4.2	T	0.0	0.0	27.3
1974-75	0.0	0.0	0.0	0.0	1.1	1.1	14.6	4.7	27.9	0.5	0.0	0.0	49.9
1975-76	0.0	0.0	0.0	1.9	1.5	5.1	13.6	8.5	6.8	0.4	0.3	0.0	38.1
1976-77	0.0	0.0	0.0	2.4	0.2	6.3	5.3	11.0	9.0	1.0	0.0	0.0	35.2
1977-78	0.0	0.0	0.0	0.2	20.2	12.5	2.9	5.0	5.8	1.5	0.0	0.0	48.1
1978-79	0.0	0.0	0.0	0.0	8.8	2.1	15.1	10.5	7.2	7.0	0.9	0.0	51.6
1979-80	0.0	0.0	0.0	3.2	0.7	1.0	5.3	6.4	10.7	1.5	0.0	0.0	28.8
1980-81	0.0	0.0	0.0	0.4	0.0	2.3	1.8	2.2	1.6	T	0.0	0.0	8.3
1981-82	0.0	0.0	0.0	0.0	0.0	0.0							
1982-83				0.0	0.0	0.0	0.9	2.2	7.9	2.3	0.0	0.0	
1983-84	0.0	0.0	0.0	T	5.5	5.2	2.0	3.8	17.3	0.1	0.0	0.0	33.9
1984-85	0.0	0.0	T	1.0	0.2	5.6	3.7	1.0	18.4	0.1	0.0	0.0	30.0
1985-86	0.0	0.0	0.0	T	22.4	9.6							
Record Mean	0.0	0.0	T	0.7	4.7	6.3	7.0	6.6	7.1	3.5	0.2	0.0	36.1

See Reference Notes, relative to all above tables, on preceding page.

HURON, SOUTH DAKOTA

Located on the west bank of the James River at about the middle of the river valley, Huron has a climate that can be classified as continental with frequent daily temperature fluctuations and distinct seasons. The seasons have varied quite markedly from year to year. Agriculture is the main industry in the area.

Winter is characteristically cold and dry with storms of short duration. Normal temperatures for the season are in the middle teens and precipitation is mainly in the form of snow. Seasonal snowfall has varied from under 9 inches to over 75 inches. Wintertime storms of blizzard proportions are infrequent, but they do occur. These blizzards are characterized by strong winds, low temperatures, snow, and very poor visibility. Many mild days can be expected during the winter since about 39 percent of the daily maximum temperatures are above the freezing mark.

Spring is characterized by marked upward trends in both precipitation and temperature with the moisture amounts increasing some three to four fold over winter. Nearly one-third, over 5 inches, of the annual total precipitation usually falls during the spring months. In early spring some of the precipitation falls as snow. As a consequence, the month of March has a slightly higher snowfall average than any of the winter months.

Based on the 1951–1980 period, the average first occurrence of 32 degrees Fahrenheit in the fall is September 29 and the average last occurrence in the spring is May 11.

Summers are hot but not extreme. Temperatures of 100 degrees or higher usually occur three or four times a year, but nights are normally cool and comfortable. Summertime precipitation is mainly in the form of showers and thunderstorms. Hail occurs about three times a year in the summertime thunderstorms. Sunshine also reaches its maximum during the summer months, when the sun shines during more than 70 percent of the daylight hours.

Autumn, with a relatively slow drop in temperature and steadily lessening amounts of rainfall, is a delightful season with mildly warm days, cool nights, and plentiful sunshine.

The terrain around Huron is exceptionally level and flat. Even the James River has a slope of only 4 to 6 inches per mile. However, floods in the Huron area have generally been of minor importance because the flood plain is agricultural land, and is not developed for homes or businesses. Heavy rains from summertime thunderstorms may, at times, cause some local flooding problems in the city. Moderate to fresh winds occur quite frequently during the daytime in all seasons of the year. Winds are normally from the north in the winter and from the south in the summer. The unusually level terrain helps to accentuate this tendency toward windy days.

TABLE 1   NORMALS, MEANS AND EXTREMES

HURON, SOUTH DAKOTA

LATITUDE: 44°23'N   LONGITUDE: 98°13'W   ELEVATION: FT. GRND 1281 BARO 01285   TIME ZONE: CENTRAL   WBAN: 14936

	(a)	JAN	FEB	MAR	APR	MAY	JUNE	JULY	AUG	SEP	OCT	NOV	DEC	YEAR
TEMPERATURE °F:														
Normals														
-Daily Maximum		21.9	28.6	39.4	57.7	70.2	80.3	87.3	85.5	74.7	62.2	43.2	29.0	56.7
-Daily Minimum		0.4	7.6	19.2	33.5	44.2	55.0	60.5	58.4	46.9	35.5	21.1	9.0	32.6
-Monthly		11.2	18.1	29.3	45.6	57.2	67.7	73.9	72.0	60.8	48.9	32.2	19.0	44.6
Extremes														
-Record Highest	44	63	71	89	97	99	106	112	110	106	102	77	62	112
-Year		1944	1958	1943	1980	1959	1979	1966	1965	1970	1963	1962	1969	JUL 1966
-Record Lowest	44	-35	-39	-24	-2	17	32	37	36	19	9	-21	-30	-39
-Year		1970	1962	1960	1975	1976	1964	1971	1964	1974	1976	1964	1985	FEB 1962
NORMAL DEGREE DAYS:														
Heating (base 65°F)		1668	1313	1104	582	259	62	15	14	171	505	984	1426	8103
Cooling (base 65°F)		0	0	0	0	20	143	294	231	45	5	0	0	738
% OF POSSIBLE SUNSHINE	44	58	60	59	60	65	69	77	74	68	63	52	50	63
MEAN SKY COVER (tenths)														
Sunrise - Sunset	42	6.6	6.7	7.0	6.6	6.3	5.7	4.6	4.7	5.1	5.5	6.6	6.8	6.0
MEAN NUMBER OF DAYS:														
Sunrise to Sunset														
-Clear	46	8.0	6.4	5.7	6.3	7.3	8.7	12.3	12.4	11.8	10.8	7.0	7.2	103.9
-Partly Cloudy	46	7.5	7.6	7.7	8.7	9.8	11.2	12.3	11.2	8.2	8.4	7.5	7.1	107.2
-Cloudy	46	15.5	14.3	17.6	15.0	13.9	10.2	6.4	7.3	10.0	11.8	15.4	16.8	154.2
Precipitation														
.01 inches or more	46	6.3	6.6	8.3	9.0	10.1	10.9	8.7	8.3	6.9	5.9	5.9	5.8	92.7
Snow, Ice pellets														
1.0 inches or more	42	2.2	2.6	2.5	0.8	0.1	0.0	0.0	0.0	0.0	0.2	1.6	1.9	12.0
Thunderstorms	46	0.*	0.*	0.4	2.5	5.4	8.8	9.5	8.0	3.9	1.5	0.1	0.*	40.2
Heavy Fog Visibility														
1/4 mile or less	46	1.7	1.8	2.0	0.7	0.7	0.5	0.7	1.0	1.0	1.1	1.5	2.1	14.7
Temperature °F														
-Maximum														
90° and above	26	0.0	0.0	0.0	0.2	0.5	3.7	12.3	10.2	3.0	0.2	0.0	0.0	30.1
32° and below	26	21.7	15.4	7.9	0.3	0.0	0.0	0.0	0.0	0.0	0.1	6.3	19.2	70.8
-Minimum														
32° and below	26	30.8	27.4	26.2	13.3	2.8	0.*	0.0	0.0	1.6	11.1	26.1	30.8	170.2
0° and below	26	15.3	9.0	2.5	0.*	0.0	0.0	0.0	0.0	0.0	0.0	1.3	9.2	37.4
AVG. STATION PRESS.(mb)	13	971.9	970.9	967.4	967.1	965.9	965.7	967.4	967.6	968.8	969.0	969.8	970.5	968.5
RELATIVE HUMIDITY (%)														
Hour 00	26	74	78	80	77	76	81	78	78	79	76	78	77	78
Hour 06 (Local Time)	26	73	78	83	83	84	87	86	88	87	82	82	78	83
Hour 12	26	67	69	66	55	54	57	52	52	54	55	64	69	60
Hour 18	26	70	71	65	52	51	53	47	47	52	57	68	74	59
PRECIPITATION (inches):														
Water Equivalent														
-Normal		0.42	0.77	1.22	1.99	2.72	3.31	2.26	2.01	1.36	1.39	0.69	0.52	18.66
-Maximum Monthly	46	1.93	3.87	5.89	5.39	7.69	11.49	5.10	5.47	3.57	6.44	3.01	1.53	11.49
-Year		1975	1962	1977	1968	1962	1984	1982	1956	1965	1946	1947	1968	JUN 1984
-Minimum Monthly	46	0.04	0.03	0.12	0.24	0.50	0.67	0.42	0.14	0.10	T	T	T	T
-Year		1942	1968	1939	1981	1940	1950	1941	1976	1952	1952	1939	1943	OCT 1952
-Maximum in 24 hrs	46	1.58	2.16	2.82	2.40	3.49	5.48	2.31	4.15	2.60	4.20	2.00	1.24	5.48
-Year		1944	1962	1985	1970	1945	1967	1985	1956	1950	1961	1972	1949	JUN 1967
Snow, Ice pellets														
-Maximum Monthly	46	27.7	39.9	33.9	12.8	1.6				T	7.3	32.7	26.0	39.9
-Year		1975	1962	1975	1970	1954				1965	1976	1985	1968	FEB 1962
-Maximum in 24 hrs	46	12.3	17.5	18.3	8.9	1.6				T	5.0	10.0	10.7	18.3
-Year		1982	1962	1985	1957	1954				1965	1970	1953	1955	MAR 1985
WIND:														
Mean Speed (mph)	46	11.5	11.5	12.5	13.6	12.4	11.3	10.5	10.7	11.4	11.4	11.8	11.2	11.6
Prevailing Direction														
through 1963		SSE	NW	SSE	SSE	SSE	SSE	SSE	SSE	SSE	SSE	SSE	SSE	SSE
Fastest Mile														
-Direction (!!!)	45	NW	NW	NW	SE	NW	SE	NW	NW	NW	W	NW	NW	NW
-Speed (MPH)	45	57	56	68	73	70	65	77	72	64	72	73	59	77
-Year		1967	1955	1953	1955	1942	1960	1957	1968	1951	1949	1954	1963	JUL 1957
Peak Gust														
-Direction (!!!)	2	NW	N	NE	S	NW	SW	SW	W	NE	NW	S	NW	SW
-Speed (mph)	2	49	52	51	64	60	63	71	63	54	47	46	52	71
-Date		1984	1984	1985	1985	1985	1985	1985	1984	1984	1984	1984	1985	JUL 1985

See Reference Notes to this table on the following page.

TABLE 2 PRECIPITATION (inches)   HURON, SOUTH DAKOTA

YEAR	JAN	FEB	MAR	APR	MAY	JUNE	JULY	AUG	SEP	OCT	NOV	DEC	ANNUAL
1956	0.46	0.29	1.35	1.23	1.25	0.96	3.47	5.47	1.06	1.08	1.26	0.20	18.08
1957	0.17	0.53	0.25	2.41	5.92	4.05	1.13	1.54	1.69	2.89	1.94	0.05	22.57
1958	0.05	1.58	0.65	2.20	1.44	2.04	3.14	0.45	0.98	0.19	0.49	0.37	13.58
1959	0.32	0.55	0.16	0.49	3.12	3.59	1.64	4.01	2.27	1.90	0.92	1.08	20.05
1960	0.55	0.76	0.67	1.55	2.19	3.35	0.61	3.76	2.37	0.76	0.87	0.95	18.39
1961	0.19	0.41	0.37	0.46	4.36	4.54	2.29	0.59	1.69	4.66	0.07	0.29	19.92
1962	0.22	3.87	2.37	2.02	7.69	5.67	4.89	2.07	2.05	0.55	0.08	0.23	31.71
1963	0.63	0.35	1.19	1.70	3.11	2.60	4.40	0.88	1.31	1.59	0.84	0.48	19.08
1964	0.14	0.20	1.06	2.78	1.29	1.61	0.69	1.28	0.86	0.06	0.19	0.69	10.85
1965	0.22	0.51	1.21	2.42	4.33	3.49	0.68	1.72	3.57	0.35	0.45	0.57	19.52
1966	0.32	0.82	1.35	1.95	1.13	2.42	4.24	4.05	2.69	3.12	0.40	0.06	22.55
1967	0.39	0.77	0.19	0.87	0.56	8.30	0.69	1.03	1.55	0.29	0.06	0.47	15.17
1968	0.29	0.03	0.16	5.39	1.02	7.23	2.05	3.45	2.66	1.93	0.43	1.53	26.17
1969	0.65	1.94	0.28	1.45	3.84	2.38	1.57	0.67	2.11	0.09	0.59		20.09
1970	0.29	0.08	2.10	4.69	2.62	3.33	3.45	0.61	1.49	1.44	1.66	0.44	22.20
1971	0.18	0.99	0.15	2.22	1.99	2.87	0.91	1.85	1.78	3.16	2.04	0.41	18.55
1972	0.43	0.65	1.21	3.80	6.86	1.25	4.93	0.69	0.30	2.66	2.44	1.24	26.46
1973	0.74	0.42	2.16	1.17	2.04	0.82	2.11	1.61	2.69	2.12	0.94	0.56	17.38
1974	0.10	0.68	1.62	1.07	4.26	2.00	1.29	1.26	0.15	0.49	0.08	0.03	13.03
1975	1.93	0.21	2.71	3.31	1.98	3.44	0.46	2.50	0.82	0.81	0.15	0.06	18.38
1976	0.64	0.63	0.73	1.10	0.62	1.97	2.05	0.14	1.43	1.26	0.01	0.39	10.97
1977	0.18	1.73	5.89	2.13	1.83	1.31	1.50	2.57	2.29	1.59	1.60	0.52	23.14
1978	0.13	0.49	0.35	3.11	3.66	2.20	2.49	1.81	0.99	0.13	0.27	0.26	15.89
1979	0.89	0.29	1.96	2.23	1.64	2.37	2.61	1.50	0.53	1.97	0.22	0.04	15.95
1980	0.25	0.49	0.81	0.82	1.68	5.31	3.21	3.94	0.53	0.91	0.08	0.12	18.15
1981	0.12	0.12	1.90	0.24	0.68	1.48	2.24	4.63	1.39	1.28	1.20	0.51	15.79
1982	1.36	0.17	1.60	1.29	4.58	1.25	5.10	2.85	2.32	3.49	0.60	0.43	25.04
1983	0.08	0.21	2.34	1.14	1.37	4.62	2.22	0.45	2.55	1.27	1.81	0.48	18.54
1984	0.36	0.69	1.50	2.83	2.52	11.49	1.81	3.55	0.94	3.36	0.13	0.55	29.73
1985	0.40	0.10	3.86	1.57	2.03	2.36	4.67	2.57	3.28	1.06	2.46	0.59	24.95
Record Mean	0.52	0.59	1.12	2.09	2.69	3.49	2.55	2.18	1.57	1.36	0.67	0.53	19.34

TABLE 3 AVERAGE TEMPERATURE (deg. F)   HURON, SOUTH DAKOTA

YEAR	JAN	FEB	MAR	APR	MAY	JUNE	JULY	AUG	SEP	OCT	NOV	DEC	ANNUAL
1956	9.1	9.1	26.6	39.6	58.6	75.7	71.9	71.0	59.9	53.9	32.5	25.2	44.4
1957	8.0	20.5	31.1	44.4	55.5	65.6	77.3	72.3	59.3	48.8	33.0	28.7	45.4
1958	23.0	15.6	28.9	46.1	60.9	63.5	69.6	74.6	63.6	51.7	34.9	16.5	45.8
#1959	8.9	10.3	36.0	46.7	58.1	73.1	74.4	76.4	61.1	43.5	29.5	15.6	45.2
1960	12.2	11.5	15.0	45.5	57.2	65.0	75.2	73.2	61.2	48.5	32.1	15.6	42.7
1961	9.3	19.4	36.3	39.6	54.2	68.5	72.2	75.3	58.1	50.3	32.5	13.5	44.1
1962	11.0	12.0	20.7	45.0	59.6	66.4	69.9	70.6	57.7	50.7	36.9	21.3	43.5
1963	5.1	20.1	39.7	47.9	57.3	70.9	76.0	72.3	65.8	59.4	36.4	11.4	46.9
1964	20.9	26.2	25.9	49.5	60.4	69.2	78.6	69.4	61.2	49.9	29.3	10.1	45.9
1965	5.4	14.3	21.1	45.6	57.9	65.5	73.0	71.0	50.5	50.7	32.4	28.4	43.0
1966	4.5	15.1	35.9	41.3	55.6	68.7	78.3	66.9	58.9	47.5	28.6	20.8	43.5
1967	19.5	14.8	36.3	46.3	53.0	66.2	70.3	70.3	62.7	48.2	33.8	20.9	45.2
1968	15.4	19.3	38.6	45.5	52.2	67.7	71.4	70.2	59.7	48.4	33.9	16.0	44.8
1969	7.8	20.6	20.7	49.0	57.0	60.4	71.7	71.7	62.7	42.7	34.8	21.7	43.4
1970	6.5	22.5	27.4	43.6	58.1	68.2	74.8	73.7	63.2	48.1	33.0	15.1	44.5
1971	7.6	15.7	28.5	45.4	54.3	71.4	69.4	74.0	60.0	49.4	33.3	16.8	43.8
1972	9.6	12.9	31.5	44.8	60.0	66.1	69.7	70.6	60.5	44.8	31.5	15.4	43.1
1973	18.9	23.9	39.8	45.5	56.2	67.0	72.1	75.5	58.0	52.6	31.6	16.5	46.4
1974	13.1	24.4	34.2	47.9	54.2	66.0	79.2	69.0	58.4	51.5	33.0	23.1	46.2
1975	16.7	15.7	24.9	39.8	59.0	67.0	78.6	73.6	57.9	50.8	32.6	21.4	44.8
1976	14.5	28.3	31.1	48.1	55.0	69.2	74.7	75.2	61.5	42.1	25.4	16.1	45.1
1977	5.3	25.4	33.9	52.8	65.4	70.2	76.7	67.5	61.1	48.4	31.3	15.5	46.2
1978	0.3	7.1	29.3	44.0	57.5	65.9	70.7	71.2	66.1	47.4	29.1	14.5	42.0
1979	0.3	6.1	28.3	42.4	53.1	68.9	75.2	69.8	63.9	46.2	31.6	28.8	42.9
1980	16.9	18.5	29.7	49.1	60.3	68.5	73.4	70.6	61.7	46.5	37.0	22.0	46.2
1981	22.1	25.6	36.3	52.6	57.8	69.4	77.5	72.2	63.6	49.3	39.7	17.9	48.7
1982	3.5	18.9	33.2	44.5	59.6	63.6	75.4	72.7	61.2	49.0	31.4	26.3	44.9
1983	25.6	30.1	35.0	42.4	54.3	66.3	76.5	78.9	63.1	49.5	34.0	2.8	46.5
1984	18.8	31.1	29.3	47.1	55.8	67.2	74.0	74.5	56.9	49.2	35.4	18.1	46.5
1985	13.2	20.2	37.2	51.6	62.0	63.2	72.5	66.9	58.3	47.0	18.7	8.6	43.3
Record Mean	12.4	17.0	30.2	46.0	57.1	67.2	73.4	71.3	61.3	48.8	31.9	19.0	44.6
Max	23.0	27.4	40.4	58.0	69.5	79.1	86.2	84.3	74.6	61.6	42.7	29.0	56.3
Min	1.8	6.5	19.9	34.0	44.7	55.2	60.6	58.3	48.0	35.9	21.1	9.0	32.9

REFERENCE NOTES FOR TABLES 1, 2, 3 and 6   **(HURON, SD)**

GENERAL

T - TRACE AMOUNT
BLANK ENTRIES DENOTE MISSING/UNREPORTED DATA.
# INDICATES A STATION OR INSTRUMENT RELOCATION.

SPECIFIC

TABLE 1

(a) - LENGTH OF RECORD IN YEARS: ALTHOUGH
      INDIVIDUAL MONTHS MAY BE MISSING.
  * LESS THAN .05

NORMALS — BASED ON THE 1951-1980 RECORD PERIOD.
EXTREMES — DATES ARE THE MOST RECENT OCCURRENCE.
WIND DIR. — NUMERALS SHOW TENS OF DEGREES
            CLOCKWISE FROM TRUE NORTH.
            ''00'' INDICATES CALM.
RESULTANT WIND DIRECTIONS ARE GIVEN TO WHOLE DEGREES.

EXCEPTIONS

TABLES 2, 3, and 6

RECORD MEANS ARE THROUGH THE CURRENT YEAR,
BEGINNING IN   1882 FOR TEMPERATURE
                  1882 FOR PRECIPITATION
                  1940 FOR SNOWFALL

TABLE 4 HEATING DEGREE DAYS Base 65 deg. F   HURON, SOUTH DAKOTA

SEASON	JULY	AUG	SEP	OCT	NOV	DEC	JAN	FEB	MAR	APR	MAY	JUNE	TOTAL
1956-57	1	19	186	345	970	1230	1766	1240	1045	611	304	41	7758
1957-58	0	18	190	501	950	1118	1296	1379	1112	564	164	114	7406
1958-59	14	18	117	426	895	1500	1738	1525	890	543	259	20	7945
#1959-60	3	0	186	656	1203	1096	1634	1545	1546	583	259	53	8764
1960-61	4	1	220	505	982	1523	1725	1269	879	758	341	54	8261
1961-62	0	2	276	452	971	1592	1675	1480	1367	601	175	60	8651
1962-63	20	13	225	443	835	1351	1860	1251	781	507	258	12	7556
1963-64	0	17	57	205	849	1658	1361	1117	1207	465	192	52	7180
1964-65	1	55	179	463	1068	1699	1846	1415	1352	579	229	40	8926
1965-66	0	29	431	436	971	1127	1875	1394	894	707	307	51	8222
1966-67	0	62	208	535	1083	1362	1403	1404	884	556	393	38	7928
1967-68	38	24	118	521	933	1361	1318	1318	812	577	392	56	7683
1968-69	13	35	190	511	926	1515	1769	1239	1366	473	280	162	8479
1969-70	0	6	118	684	899	1340	1814	1186	1158	635	225	40	8105
1970-71	7	2	181	522	952	1541	1777	1375	1127	579	325	16	8404
1971-72	32	8	211	478	946	1486	1714	1510	1032	600	210	57	8284
1972-73	26	38	171	621	997	1530	1428	1144	776	578	271	31	7611
1973-74	5	0	227	380	995	1499	1607	1131	948	506	346	73	7717
1974-75	1	36	218	414	954	1293	1489	1373	1239	756	195	67	8035
1975-76	4	3	236	447	966	1346	1561	1057	1043	502	306	39	7510
1976-77	5	4	173	706	1181	1514	1849	1101	958	374	61	12	7938
1977-78	0	27	129	506	1002	1530	2007	1620	1099	624	247	89	8880
1978-79	12	27	132	539	1070	1560	2006	1647	1130	668	374	35	9200
1979-80	0	32	122	574	997	1117	1485	1343	1088	489	193	29	7469
1980-81	1	19	160	578	835	1326	1322	1098	882	376	243	12	6852
1981-82	5	1	105	480	752	1456	1908	1287	976	609	173	85	7837
1982-83	0	19	168	489	1002	1193	1215	970	923	673	329	83	7064
1983-84	2	0	166	473	924	1926	1430	978	1101	528	294	36	7858
1984-85	4	7	277	493	883	1448	1598	1250	856	403	141	123	7483
1985-86	7	46	267	553	1383	1746							

TABLE 5 COOLING DEGREE DAYS Base 65 deg. F   HURON, SOUTH DAKOTA

YEAR	JAN	FEB	MAR	APR	MAY	JUNE	JULY	AUG	SEP	OCT	NOV	DEC	TOTAL
1969	0	0	0	0	36	31	214	219	56	0	0	0	556
1970	0	0	0	0	18	143	320	279	136	4	0	0	900
1971	0	0	0	0	1	217	175	293	69	0	0	0	755
1972	0	0	0	0	62	94	181	222	46	0	0	0	605
1973	0	0	0	0	4	98	233	331	22	3	0	0	691
1974	0	0	0	0	18	109	448	166	29	1	0	0	771
1975	0	0	0	4	17	136	432	276	33	12	0	0	910
1976	0	0	0	0	2	170	313	328	75	3	0	0	891
1977	0	0	0	14	79	176	369	113	44	0	0	0	795
1978	0	0	0	0	21	119	196	227	173	0	0	0	736
1979	0	0	0	0	13	162	324	188	96	0	0	0	783
1980	0	0	0	18	56	139	269	199	56	10	0	0	757
1981	0	0	0	9	25	153	399	232	70	0	0	0	888
1982	0	0	0	0	14	50	329	263	57	0	0	0	713
1983	0	0	0	0	4	130	366	437	113	0	0	0	1050
1984	0	0	0	0	13	107	291	306	41	9	0	0	767
1985	0	0	0	7	55	73	246	112	71	0	0	0	564

TABLE 6 SNOWFALL (inches)   HURON, SOUTH DAKOTA

SEASON	JULY	AUG	SEP	OCT	NOV	DEC	JAN	FEB	MAR	APR	MAY	JUNE	TOTAL
1956-57	0.0	0.0	0.0	0.0	4.2	3.4	4.1	10.5	3.7	10.0	0.0	0.0	35.9
1957-58	0.0	0.0	0.0	T	3.5	0.8	1.0	5.4	9.3	1.4	0.0	0.0	21.4
1958-59	0.0	0.0	0.0	0.0	4.0	5.3	5.6	11.6	0.4	0.9	0.0	0.0	27.8
1959-60	0.0	0.0	T	1.5	12.2	5.9	6.7	6.1	10.8	3.9	0.0	0.0	47.1
1960-61	0.0	0.0	0.0	0.0	7.8	7.0	3.1	5.7	3.6	1.2	0.0	0.0	28.4
1961-62	0.0	0.0	0.0	T	0.9	5.1	3.7	39.9	24.9	3.2	0.0	0.0	77.7
1962-63	0.0	0.0	0.0	0.9	T	3.0	10.4	3.4	6.4	2.2	0.0	0.0	26.3
1963-64	0.0	0.0	0.0	0.0	2.0	6.7	3.8	4.3	12.8	0.7	0.0	0.0	30.3
1964-65	0.0	0.0	0.0	0.0	3.6	10.1	3.6	10.7	17.2	1.2	T	0.0	46.4
1965-66	0.0	0.0	T	0.0	2.3	0.2	6.7	8.4	14.4	6.3	0.5	0.0	38.8
1966-67	0.0	0.0	0.0	T	4.3	1.7	6.7	17.5	2.1	0.6	0.7	0.0	33.6
1967-68	0.0	0.0	0.0	T	0.8	4.3	4.5	0.7	2.5	5.3	0.3	0.0	18.4
1968-69	0.0	0.0	T	3.4	26.0	13.2	22.3	1.6	15.0	12.8	T	0.0	67.4
1969-70	0.0	0.0	0.0	1.4	1.4	12.2	7.3	1.6	16.0	1.9	T	0.0	51.7
1970-71	0.0	0.0	0.0	5.3	1.9	7.8	4.6	16.0	1.9	T	0.0	0.0	37.5
1971-72	0.0	0.0	0.0	3.8	11.8	10.4	13.5	0.9	0.1	1.0	0.0	0.0	46.7
1972-73	0.0	0.0	0.0	1.3	10.9	13.8	10.7	6.2	0.1	T	T	0.0	43.0
1973-74	0.0	0.0	0.0	T	6.1	10.6	8.0	6.0	6.0	0.0	0.0	0.0	32.1
1974-75	0.0	0.0	0.0	0.0	2.0	1.3	27.7	5.9	33.9	0.7	0.0	0.0	71.5
1975-76	0.0	0.0	0.0	T	2.9	1.6	10.9	6.6	13.0	T	T	0.0	35.0
1976-77	0.0	0.0	0.0	7.3	0.4	5.8	3.5	9.8	9.1	0.3	0.0	0.0	36.2
1977-78	0.0	0.0	0.0	T	7.8	6.5	4.2	9.0	4.8	0.8	T	0.0	33.1
1978-79	0.0	0.0	0.0	0.0	3.5	5.9	23.8	5.9	2.6	4.0	T	0.0	45.7
1979-80	0.0	0.0	0.0	2.5	3.2	0.4	1.5	6.0	8.6	T	0.0	0.0	22.2
1980-81	0.0	0.0	0.0	1.5	0.0	3.0	1.6	2.3	1.2	T	0.0	0.0	10.4
1981-82	0.0	0.0	0.0	0.1	8.0	7.4	23.8	1.8	9.5	9.1	0.0	0.0	59.7
1982-83	0.0	0.0	0.0	2.7	2.1	4.4	0.7	0.5	11.8	5.1	0.0	0.0	27.3
1983-84	0.0	0.0	0.0	0.0	16.9	6.9	4.3	6.8	19.1	T	0.0	0.0	54.0
1984-85	0.0	0.0	0.0	T	0.2	8.1	5.5	1.8	26.7	2.0	0.0	0.0	44.3
1985-86	0.0	0.0	0.0	T	32.7	11.8							
Record Mean	0.0	0.0	T	0.8	5.2	6.2	7.1	8.5	9.4	2.3	0.1	0.0	39.6

See Reference Notes, relative to all above tables, on preceding page.

Rapid City, which is not far from the geographical center of North America, experiences the large temperature ranges, both daily and seasonal, that are typical of semi−arid continental climates.

The city is surrounded by contrasting landforms, with the forested Black Hills rising immediately west of the city, and rolling prairie extending out in the other directions. From 40 to 70 miles southeast lie the eroded Badlands. The Black Hills, many of which are more than 5,000 feet above sea level, with a number of peaks above 7,000 feet, exert a pronounced influence on the climate of this area. The rolling land to the east of the city is cut by the valleys of the Box Elder and Rapid Creeks, which flow generally east−southeastward. The station is located on the north slope of the irrigated Rapid Valley. An east−west ridge 200 to 300 feet higher than the airport separates the station from the Box Elder Creek Valley.

The principal agricultural products in the area are cattle and wheat, and ranchers and farmers are dependent on the current weather forecasts, which are at times of vital interest in the protection of livestock.

Although the annual precipitation is light at lower elevations, the distribution is beneficial to agriculture with the greatest amounts occurring during the growing season. The heaviest snows are expected in the spring, which helps to furnish moisture for the early maturing crops such as wheat, while heavy winter snows at the higher elevations provide irrigation water for the fertile valleys.

Summer days are normally warm with cool, comfortable nights. Nearly all of the summer precipitation occurs as thunderstorms. Hail is often associated with the more severe thunderstorms, with resultant damage to vegetation as well as other fragile material in the path of the storms. Autumn, which begins soon after the first of September, is characterized by mild, balmy days, and cool, invigorating mornings and evenings. Autumn weather usually extends into November and often into December.

Temperatures for the winter months of December, January, and February are among the warmest in South Dakota due to the protection of the Black Hills, the frequent occurrence of Chinook winds, and the fact that the winter tracks of arctic air masses usually pass east of Rapid City. Rapid City has become the retirement home for many farmers and ranchers from the western half of the state because of the cool summer nights and the relatively mild winters.

Snowfall is normally light with the greatest monthly average of about 8 inches occurring in March. Cold waves can be expected occasionally, and one or more blizzards may occur each winter.

Spring is characterized by unsettled conditions. Wide variations usually occur in temperatures, and snows may fall as late as May.

Based on the 1951−1980 period, the average first occurrence of 32 degrees Fahrenheit in the fall is September 29 and the average last occurrence in the spring is May 7.

TABLE 1   # NORMALS, MEANS AND EXTREMES

RAPID CITY, SOUTH DAKOTA

LATITUDE: 44°03'N   LONGITUDE: 103°04'W   ELEVATION: FT. GRND 3162 BARO 03169   TIME ZONE: MOUNTAIN   WBAN: 24090

	(a)	JAN	FEB	MAR	APR	MAY	JUNE	JULY	AUG	SEP	OCT	NOV	DEC	YEAR
TEMPERATURE °F:														
Normals														
-Daily Maximum		32.4	37.4	44.2	57.0	68.1	77.9	86.5	85.7	75.4	63.2	46.7	37.4	59.3
-Daily Minimum		9.2	14.6	21.0	32.1	43.0	52.5	58.7	57.0	46.4	36.1	23.0	14.8	34.0
-Monthly		20.8	26.0	32.6	44.5	55.5	65.2	72.6	71.3	60.9	49.7	34.9	26.1	46.7
Extremes														
-Record Highest	43	74	74	82	92	98	106	110	106	104	94	77	75	110
-Year		1953	1954	1946	1980	1969	1961	1973	1947	1978	1963	1965	1965	JUL 1973
-Record Lowest	43	-27	-22	-17	1	18	31	39	38	18	10	-19	-27	-27
-Year		1950	1971	1962	1975	1950	1951	1959	1966	1985	1972	1959	1983	DEC 1983
NORMAL DEGREE DAYS:														
Heating (base 65°F)		1370	1092	1004	612	298	101	21	24	188	482	903	1206	7301
Cooling (base 65°F)		0	0	0	0	7	107	257	223	65	8	0	0	667
% OF POSSIBLE SUNSHINE	43	55	59	61	60	58	62	72	73	68	65	54	53	62
MEAN SKY COVER (tenths)														
Sunrise - Sunset	43	6.5	6.5	6.7	6.6	6.4	5.6	4.4	4.3	4.5	5.0	6.2	6.2	5.7
MEAN NUMBER OF DAYS:														
Sunrise to Sunset														
-Clear	43	7.6	6.1	6.3	5.8	6.5	8.9	13.1	13.9	13.3	12.5	7.7	8.0	109.7
-Partly Cloudy	43	7.6	8.5	9.0	9.4	11.0	11.3	12.6	11.7	8.6	7.8	8.6	8.1	114.3
-Cloudy	43	15.9	13.7	15.6	14.7	13.5	9.8	5.3	5.4	8.1	10.7	13.7	14.9	141.3
Precipitation														
.01 inches or more	43	6.8	7.1	8.7	9.3	11.9	12.5	9.2	7.9	6.4	4.9	5.7	6.1	96.4
Snow,Ice pellets														
1.0 inches or more	35	1.3	2.3	2.6	1.9	0.2	0.*	0.0	0.0	0.1	0.4	1.7	1.7	12.3
Thunderstorms	43	0.0	0.0	0.1	1.1	5.8	10.8	11.8	8.9	3.4	0.5	0.*	0.*	42.3
Heavy Fog Visibility														
1/4 mile or less	43	1.5	1.9	2.5	1.7	0.9	1.1	0.6	0.6	0.4	0.6	1.9	2.1	15.7
Temperature °F														
-Maximum														
90° and above	43	0.0	0.0	0.0	0.*	0.5	3.0	11.3	11.5	3.5	0.2	0.0	0.0	30.0
32° and below	43	13.5	10.0	7.0	0.8	0.*	0.0	0.0	0.0	0.0	0.3	5.2	11.0	48.0
-Minimum														
32° and below	43	30.1	26.9	27.5	14.7	2.7	0.1	0.0	0.0	1.7	10.1	24.5	29.9	168.3
0° and below	43	8.6	4.4	2.1	0.0	0.0	0.0	0.0	0.0	0.0	0.0	1.1	4.3	20.5
AVG. STATION PRESS.(mb)	13	904.8	904.1	901.5	902.7	902.5	903.4	905.3	905.2	905.9	905.3	904.6	904.0	904.1
RELATIVE HUMIDITY (%)														
Hour 05	35	68	71	74	72	74	77	73	70	67	65	68	68	71
Hour 11 (Local Time)	35	60	59	56	48	49	52	45	42	41	42	52	60	51
Hour 17	35	64	61	54	45	46	48	41	37	38	45	58	65	50
Hour 23	35	68	71	71	67	70	73	65	61	59	61	67	68	67
PRECIPITATION (inches):														
Water Equivalent														
-Normal		0.42	0.62	1.02	1.96	2.63	3.26	2.12	1.44	1.03	0.81	0.51	0.45	16.27
-Maximum Monthly	43	1.77	2.46	3.02	5.16	7.35	7.00	6.13	4.83	3.94	3.82	2.22	1.65	7.35
-Year		1944	1953	1945	1967	1946	1968	1969	1982	1946	1982	1985	1975	MAY 1946
-Minimum Monthly	43	0.01	0.06	0.12	0.27	0.33	0.64	0.60	0.10	0.03	T	0.03	0.04	T
-Year		1952	1985	1964	1954	1966	1973	1965	1943	1975	1960	1945	1957	OCT 1960
-Maximum in 24 hrs	43	1.26	1.00	2.19	3.01	3.40	4.01	2.51	2.60	2.13	2.49	1.09	1.04	4.01
-Year		1944	1953	1945	1946	1965	1963	1944	1982	1966	1982	1944	1975	JUN 1963
Snow,Ice pellets														
-Maximum Monthly	43	24.0	23.7	30.7	30.6	11.6	3.6			2.0	10.2	33.6	17.9	33.6
-Year		1949	1953	1950	1970	1950	1951			1970	1971	1985	1975	NOV 1985
-Maximum in 24 hrs	43	16.3	10.0	14.9	16.0	13.4	3.6			2.0	7.6	9.4	9.8	16.3
-Year		1944	1953	1973	1970	1967	1951			1970	1971	1977	1975	JAN 1944
WIND:														
Mean Speed (mph)	35	10.8	11.1	12.7	13.3	12.3	10.8	10.1	10.3	11.1	11.2	10.8	10.5	11.3
Prevailing Direction														
through 1963		NNW	NNW	NNW	NNW	NNW	NNW	NNW	NNW	NNW	NNW	NNW	NNW	NNW
Fastest Mile														
-Direction	32	SW	NW	W	NW	NW	NW	SW	NW	NW	NW	NW	W	SW
-Speed (MPH)	32	65	62	65	61	57	57	66	65	61	59	61	60	66
-Year		1967	1977	1953	1978	1976	1975	1959	1951	1964	1971	1977	1955	JUL 1959
Peak Gust														
-Direction	2	NW	NW	NW	NW	NW	N	NW	NW	S	NW	NW	NW	NW
-Speed (mph)	2	62	61	62	62	55	55	72	69	70	60	60	63	72
-Date	2	1985	1984	1984	1984	1985	1985	1984	1985	1985	1985	1984	1985	JUL 1984

See Reference Notes to this table on the following page.

TABLE 2 PRECIPITATION (inches) RAPID CITY, SOUTH DAKOTA

YEAR	JAN	FEB	MAR	APR	MAY	JUNE	JULY	AUG	SEP	OCT	NOV	DEC	ANNUAL
1956	0.47	0.52	0.75	1.69	2.09	1.32	2.27	3.15	0.30	0.10	1.48	0.36	14.50
1957	0.40	0.29	0.66	2.06	7.05	3.43	2.01	1.03	0.69	1.29	0.39	0.04	19.34
1958	0.43	0.84	1.08	3.03	1.04	3.86	4.49	0.26	0.22	0.22	0.42	0.37	16.07
1959	0.39	0.63	0.60	1.89	3.25	2.79	0.67	0.36	2.37	0.22	1.18	0.26	14.61
1960	0.16	0.85	0.78	1.21	2.23	2.19	0.70	1.24	0.68	T	0.39	0.80	11.23
1961	0.10	0.22	0.75	1.53	1.29	0.76	2.11	0.43	0.98	0.94	0.36	0.51	9.98
1962	0.51	0.98	1.28	0.69	6.90	4.01	4.53	1.03	0.67	1.63	0.08	0.19	22.50
1963	1.03	0.92	1.60	3.80	1.18	5.47	2.03	1.32	1.21	0.77	0.12	0.32	19.77
1964	0.35	0.83	0.63	1.24	2.52	4.69	0.77	1.87	0.69	0.50	0.30	0.78	15.17
1965	0.61	0.22	0.46	1.50	6.97	3.56	0.60	1.46	1.46	0.57	0.15	0.12	17.68
1966	0.24	1.00	1.78	2.50	0.33	1.31	3.93	3.24	2.84	1.50	0.95	0.79	20.41
1967	0.47	0.59	0.82	5.16	3.20	6.78	1.07	0.95	2.10	0.28	0.28	0.89	22.59
1968	0.43	0.35	0.21	1.82	1.68	7.00	2.44	2.46	0.89	0.14	0.46	0.58	18.46
1969	0.11	0.73	0.66	1.60	2.20	2.04	6.13	0.31	0.35	0.85	0.35	0.57	15.90
1970	0.73	0.54	1.48	4.63	2.41	2.16	1.04	0.67	1.57	1.23	0.80	0.61	17.87
1971	1.18	1.00	1.25	2.86	3.70	1.92	1.46	0.52	2.32	2.02	0.79	0.15	19.17
1972	0.22	0.44	0.47	2.78	3.28	4.11	1.67	2.49	0.24	0.77	0.38	0.34	17.19
1973	0.11	0.31	2.71	2.69	2.37	0.64	1.46	0.74	1.44	1.38	0.73	0.54	15.12
1974	0.16	0.30	0.34	1.55	1.32	1.10	0.68	1.37	0.88	1.18	0.12	0.12	9.12
1975	1.05	0.35	2.45	1.37	1.23	5.63	1.57	0.87	0.03	0.69	0.57	1.65	17.46
1976	0.28	0.47	0.33	2.70	2.74	4.81	1.05	1.31	0.28	0.21	0.61	0.41	15.20
1977	0.83	0.25	2.63	1.57	2.49	1.76	2.98	1.79	2.86	1.06	0.82	0.36	19.40
1978	0.19	0.84	0.40	2.19	3.12	2.01	4.08	1.42	0.26	0.26	0.63	0.25	15.57
1979	0.49	0.33	0.47	0.31	1.17	3.60	4.11	2.32	0.07	0.90	0.15	0.07	13.99
1980	0.20	0.51	0.86	1.13	1.58	3.56	4.75	1.78	0.48	2.28	0.57	0.66	17.18
1981	0.14	0.09	0.12	0.32	2.81	1.89	4.47	1.74	0.16	1.81	0.23	0.35	14.13
1982	0.39	0.37	1.35	0.69	6.50	2.89	1.81	4.83	2.69	3.82	0.27	0.36	25.97
1983	0.34	0.18	0.84	1.00	2.18	3.01	1.94	2.39	0.33	1.74	1.07	0.47	15.49
1984	0.10	0.18	0.69	3.10	1.57	4.72	1.57	1.00	0.74	0.67	0.51	0.38	15.23
1985	0.46	0.06	1.55	0.32	1.24	1.58	1.03	1.86	1.57	0.98	2.22	0.77	13.64
Record Mean	0.43	0.45	1.00	1.92	3.07	3.30	2.31	1.63	1.22	1.03	0.61	0.52	17.48

TABLE 3 AVERAGE TEMPERATURE (deg. F) RAPID CITY, SOUTH DAKOTA

YEAR	JAN	FEB	MAR	APR	MAY	JUNE	JULY	AUG	SEP	OCT	NOV	DEC	ANNUAL
1956	23.1	24.1	34.1	39.1	56.8	72.8	72.0	69.5	63.0	53.4	33.2	29.9	47.6
1957	10.8	27.6	33.7	41.0	54.0	62.7	75.3	71.6	59.2	47.5	35.4	34.9	46.1
1958	31.9	22.0	30.7	44.0	61.8	62.4	67.1	74.7	63.8	52.3	37.2	27.0	47.9
1959	20.6	21.8	37.9	43.3	53.7	70.1	73.8	75.7	58.8	44.5	27.9	34.2	46.8
1960	24.0	19.1	28.0	47.5	56.9	65.7	76.8	72.4	63.6	51.8	36.9	25.0	47.3
1961	27.3	33.2	39.4	41.8	55.8	73.1	77.2	71.2	55.6	49.9	34.9	23.0	48.6
1962	19.3	24.2	30.2	49.1	58.3	65.3	69.2	71.2	60.3	52.9	40.6	30.1	47.4
1963	12.3	30.2	39.7	44.5	57.0	68.1	74.4	73.8	66.5	59.4	39.6	22.7	49.0
1964	28.4	28.1	29.9	46.3	58.5	64.9	76.7	69.5	59.0	51.7	33.7	18.9	47.1
#1965	25.4	25.6	21.2	47.0	54.7	72.1	70.7		49.7	54.8	38.2	33.3	46.4
1966	14.2	19.3	35.0	38.7	56.0	65.3	76.6	65.9	61.3	48.6	34.6	26.7	45.2
1967	27.9	27.5	35.2	44.1	49.8	60.0	70.6	70.6	63.4	50.1	34.6	22.5	46.3
1968	23.4	27.4	40.4	42.6	50.8	63.0	69.2	67.0	59.8	51.0	35.5	18.5	45.7
1969	16.0	25.4	27.8	49.6	57.5	59.8	71.0	73.8	64.8	40.6	38.7	28.2	46.1
1970	17.8	30.8	27.5	39.2	55.6	65.6	72.7	73.8	59.1	44.1	34.2	21.9	45.2
1971	17.5	22.6	32.6	46.0	53.6	66.4	68.0	73.9	57.4	46.1	34.5	23.3	45.2
1972	17.3	24.0	36.5	43.6	55.2	64.7	65.6	69.1	59.4	43.8	31.0	18.6	44.1
1973	26.8	28.9	37.5	42.9	53.7	64.7	70.7	74.3	57.1	51.3	32.3	25.4	47.1
1974	21.9	32.5	37.6	47.2	53.8	66.6	77.3	67.7	57.6	51.7	36.4	28.9	48.3
1975	23.7	17.6	27.6	40.5	54.0	62.2	74.7	70.3	58.5	49.2	33.8	29.4	45.1
1976	23.9	34.6	34.4	46.9	55.4	64.3	72.8	72.6	63.3	45.4	31.7	26.8	47.7
1977	12.7	34.6	36.4	49.5	60.3	69.0	73.4	66.0	61.0	48.8	33.3	21.9	47.3
1978	11.0	15.3	35.4	45.2	55.1	65.1	71.1	69.7	66.2	50.5	29.2	17.1	44.2
1979	7.4	16.8	35.4	44.4	53.6	65.4	70.4	68.5	66.3	50.9	33.0	33.2	45.4
1980	21.0	27.1	31.5	48.9	57.4	67.1	74.9	68.6	61.6	48.8	39.3	30.3	48.0
1981	32.6	29.4	40.0	51.5	54.9	64.9	72.0	70.0	63.5	47.7	40.4	25.8	49.4
1982	11.9	23.9	33.0	42.1	53.3	59.7	70.7	70.2	58.7	47.2	32.7	28.6	44.3
1983	32.1	37.3	36.4	40.7	52.0	63.1	73.6	78.0	60.8	49.4	34.9	8.1	47.2
1984	28.0	36.1	34.3	43.8	53.6	62.8	72.2	74.7	57.2	47.2	37.5	21.4	47.4
1985	21.6	23.8	35.9	52.0	61.8	62.1	74.6	69.0	55.6	47.3	16.0	21.0	45.1
Record Mean	22.5	25.4	33.1	45.0	55.0	64.5	72.3	70.9	60.7	49.3	35.8	26.3	46.7
Max	34.1	36.8	44.5	56.8	66.7	76.3	85.4	84.3	74.1	62.1	47.3	37.4	58.8
Min	11.0	14.0	21.7	33.1	43.3	52.6	59.2	57.4	47.2	36.5	24.2	15.2	34.6

REFERENCE NOTES FOR TABLES 1, 2, 3 and 6 (RAPID CITY, SD)

GENERAL

T - TRACE AMOUNT
BLANK ENTRIES DENOTE MISSING/UNREPORTED DATA.
# INDICATES A STATION OR INSTRUMENT RELOCATION.

SPECIFIC

TABLE 1

(a) - LENGTH OF RECORD IN YEARS. ALTHOUGH INDIVIDUAL MONTHS MAY BE MISSING.

* LESS THAN .05

NORMALS — BASED ON THE 1951-1980 RECORD PERIOD.
EXTREMES — DATES ARE THE MOST RECENT OCCURRENCE.
WIND DIR. — NUMERALS SHOW TENS OF DEGREES CLOCKWISE FROM TRUE NORTH. "00" INDICATES CALM.
RESULTANT WIND DIRECTIONS ARE GIVEN TO WHOLE DEGREES.

EXCEPTIONS

TABLES 2, 3, and 6

RECORD MEANS ARE THROUGH THE CURRENT YEAR, BEGINNING IN 1900 FOR TEMPERATURE
1900 FOR PRECIPITATION
1943 FOR SNOWFALL

TABLE 4 HEATING DEGREE DAYS Base 65 deg. F   RAPID CITY, SOUTH DAKOTA

SEASON	JULY	AUG	SEP	OCT	NOV	DEC	JAN	FEB	MAR	APR	MAY	JUNE	TOTAL
1956-57	1	37	119	364	950	1083	1078	1042	966	714	336	106	7390
1957-58	0	22	189	553	876	924	1019	1201	1057	622	129	121	6713
1958-59	30	7	127	412	826	1168	1374	1206	833	643	352	47	7025
1959-60	8	1	260	630	1107	949	1266	1325	1140	520	262	58	7526
1960-61	1	12	154	407	837	1234	1158	884	787	690	314	25	6503
1961-62	2	0	327	469	898	1295	1414	1139	1121	483	213	60	7421
1962-63	4	36	165	372	721	1073	1634	968	776	607	253	28	6637
1963-64	4	4	59	205	755	1305	1129	1064	1081	555	237	85	6483
1964-65	0	64	198	410	931	1419	1218	1098	1351	534	327	55	7605
#1965-66	3	27	461	312	797	978	1573	1273	923	784	285	103	7519
1966-67	0	83	158	504	905	1180	1143	1042	917	622	473	162	7189
1967-68	31	19	129	464	902	1311	1285	1083	755	669	432	107	7187
1968-69	26	46	176	428	881	1438	1516	1102	1145	458	264	176	7656
1969-70	0	0	62	750	781	1135	1461	950	1157	767	293	74	7430
1970-71	2	0	245	644	918	1330	1463	1182	996	563	344	35	7722
1971-72	43	2	267	578	909	1284	1473	1182	877	634	322	74	7645
1972-73	74	43	193	649	1010	1436	1181	1008	847	659	350	83	7533
1973-74	16	0	246	416	972	1223	1329	905	840	530	343	87	6907
1974-75	1	42	242	407	849	1112	1274	1318	1151	728	343	119	7586
1975-76	3	16	206	493	929	1096	1269	878	940	535	295	98	6758
1976-77	3	6	132	606	991	1177	1616	846	877	459	165	17	6895
1977-78	1	48	163	494	944	1330	1669	1383	912	588	312	91	7935
1978-79	17	40	111	443	1068	1480	1781	1348	910	614	362	82	8256
1979-80	3	25	64	433	952	982	1359	1094	1032	483	251	54	6732
1980-81	1	18	144	510	763	1070	998	993	765	402	311	65	6040
1981-82	21	7	108	531	730	1209	1646	1146	985	682	358	170	7593
1982-83	7	21	226	545	962	1119	1012	772	880	723	407	113	6787
1983-84	8	0	208	474	896	1762	1139	832	948	626	366	101	7360
1984-85	0	0	268	546	820	1344	1341	1148	895	393	146	144	7045
1985-86	8	27	327	544	1466	1358							

TABLE 5 COOLING DEGREE DAYS Base 65 deg. F   RAPID CITY, SOUTH DAKOTA

YEAR	JAN	FEB	MAR	APR	MAY	JUNE	JULY	AUG	SEP	OCT	NOV	DEC	TOTAL
1969	0	0	0	0	35	29	193	280	63	0	0	0	600
1970	0	0	0	0	7	96	248	279	78	5	0	0	713
1971	0	0	0	0	0	81	142	284	43	0	0	0	550
1972	0	0	0	0	21	70	98	178	34	0	0	0	401
1973	0	0	0	0	7	80	202	295	15	0	0	0	599
1974	0	0	0	1	3	143	390	132	28	0	0	0	697
1975	0	0	0	0	9	41	314	188	19	12	0	0	583
1976	0	0	0	0	4	83	251	248	87	3	0	0	676
1977	0	0	0	0	25	143	270	85	51	0	0	0	574
1978	0	0	0	0	9	99	214	193	154	0	0	0	669
1979	0	0	0	4	15	99	179	141	110	2	0	0	550
1980	0	0	0	6	25	123	315	136	48	14	0	0	667
1981	0	0	0	3	5	67	243	170	74	0	0	0	562
1982	0	0	0	0	3	18	189	190	41	0	0	0	441
1983	0	0	0	0	9	62	282	407	88	0	0	0	848
1984	0	0	0	0	18	41	234	309	42	2	0	0	646
1985	0	0	0	10	53	64	312	158	51	0	0	0	648

TABLE 6 SNOWFALL (inches)   RAPID CITY, SOUTH DAKOTA

SEASON	JULY	AUG	SEP	OCT	NOV	DEC	JAN	FEB	MAR	APR	MAY	JUNE	TOTAL
1956-57	0.0	0.0	0.0	0.2	12.1	5.3	4.1	3.9	7.0	16.2	0.0	0.0	48.8
1957-58	0.0	0.0	T	T	2.1	0.2	4.2	8.4	7.1	8.8	0.0	0.0	30.8
1958-59	0.0	0.0	0.0	.T	2.8	4.1	3.7	6.8	1.2	13.9	T	0.0	32.5
1959-60	0.0	0.0	T	T	12.6	3.0	2.0	8.7	7.4	4.3	0.0	0.0	38.0
1960-61	0.0	0.0	T	.T	1.8	7.1	1.0	2.2	3.8	5.2	T	0.0	21.1
1961-62	0.0	0.0	0.2	3.7	2.7	5.5	4.9	6.9	7.9	0.1	0.0	0.0	31.9
1962-63	0.0	0.0	T	T	0.4	1.6	10.8	12.0	17.3	7.4	0.0	0.0	49.5
1963-64	0.0	0.0	0.0	T	0.3	3.0	3.5	8.9	5.7	4.2	0.0	0.0	25.6
1964-65	0.0	0.0	0.0	0.0	2.6	7.8	4.3	1.9	4.6	5.5	8.8	0.0	35.5
1965-66	0.0	0.0	1.5	0.0	0.8	0.0	2.4	10.0	16.5	10.0	0.4	0.0	42.3
1966-67	0.0	0.0	T	2.1	6.4	8.5	4.0	7.4	7.6	14.0	3.1	0.0	53.1
1967-68	0.0	0.0	0.0	0.0	2.2	4.0	4.0	3.4	1.3	13.7	T	0.0	33.7
1968-69	0.0	0.0	0.0	0.0	2.7	7.5	1.2	7.9	6.3	5.2	T	0.5	31.3
1969-70	0.0	0.0	0.0	4.1	2.9	5.5	5.5	6.0	14.8	30.6	T	0.0	69.6
1970-71	0.0	0.0	2.0	5.7	4.5	7.8	13.2	15.7	13.8	5.3	0.0	0.0	68.0
1971-72	0.0	0.0	0.0	10.2	6.9	2.0	2.6	8.0	1.2	1.0	T	0.0	31.9
1972-73	0.0	0.0	0.0	0.7	3.6	6.0	2.0	3.1	16.9	1.9	0.0	0.0	34.2
1973-74	0.0	0.0	0.0	1.4	9.7	5.6	2.0	4.0	1.7	4.7	0.0	0.0	29.1
1974-75	0.0	0.0	0.0	T	1.5	1.4	13.7	6.1	27.4	1.3	0.0	0.0	51.4
1975-76	0.0	0.0	0.0	4.6	8.4	17.9	2.0	6.5	4.3	3.5	0.0	0.0	47.2
1976-77	0.0	0.0	0.0	0.5	7.0	6.1	11.1	2.6	26.0	3.4	0.0	0.0	56.7
1977-78	0.0	0.0	0.0	0.8	10.6	7.4	3.1	15.0	3.0	2.6	T	0.0	42.5
1978-79	0.0	0.0	0.0	T	9.8	4.0	6.1	4.4	2.6	1.8	2.8	0.0	31.5
1979-80	0.0	0.0	0.0	T	1.8	0.3	3.0	10.1	8.6	5.4	0.0	0.0	29.2
1980-81	0.0	0.0	0.0	T	6.9	1.4	1.2	1.3	T	T	0.0	0.0	16.9
1981-82	0.0	0.0	0.0	1.6	1.2	3.8	6.2	5.0	11.5	5.5	0.0	0.0	34.8
1982-83	0.0	0.0	0.0	1.4	1.2	4.0	2.9	0.3	6.5	4.3	4.3	0.0	24.9
1983-84	0.0	0.0	0.3	0.9	6.9	7.1	1.9	2.5	6.1	22.1	0.2	0.0	48.0
1984-85	0.0	0.0	1.3	0.7	2.0	4.9	3.8	0.7	16.2	0.4	0.0	T	30.0
1985-86	0.0	0.0	1.4	0.6	33.6	10.2							
Record Mean	0.0	0.0	0.2	1.5	5.0	4.8	5.1	6.1	9.1	6.0	0.8	0.1	38.7

See Reference Notes, relative to all above tables, on preceding page.

Sioux Falls is located in the Big Sioux River Valley in southeast South Dakota. The surrounding terrain is gently rolling. The land slopes upward for about 100 miles north and northwest to an elevation about 400 feet higher than the city. To the southeast, the land slopes downward 200 to 300 feet over the same distance. Little change in elevation occurs in the remaining directions.

The climate is of the continental type. There are frequent weather changes from day to day or week to week as the locality is visited by differing air masses. Cold air masses arrive from the interior of Canada, cool, dry air from the northern Pacific, warm, moist air from the Gulf of Mexico, or hot, dry air from the southwest.

Temperatures fluctuate frequently as cold air masses move in very rapidly. During the late fall and winter, cold fronts accompanied by strong, gusty winds drop temperatures by 20 to 30 degrees in a 24–hour period. Severe cold spells usually last only a few days. The winter months of December through February have experienced cold spells with average temperatures under 8 degrees and more than 60 consecutive days below 32 degrees.

Temperatures of 100 degrees and above occur about one in every three years, and will most likely happen in July. Summer nights are usually comfortable with temperatures below 70 degrees.

Rainfall is heavier during the spring and summer with lighter amounts in winter. Nearly 64 percent of the normal yearly precipitation falls during the growing season of April through August.

One or two very heavy snows usually fall each winter. Eight to 12 inches of snow may fall in 24 hours. There have been a few snows in excess of 15 inches and almost 30 inches have fallen during a severe winter storm. Strong winds often cause drifting snow, and blizzard conditions may block highways for a day or so.

Southerly winds prevail from late spring to early fall with northwest winds the remainder of the year. Strong winds of 70 mph with gusts to 90 mph have occurred.

Thunderstorms are frequent during the late spring and summer with June and July the most active months. The thunderstorms usually occur during the late afternoon and evening with a secondary peak of activity between 2 and 5 in the morning. Some of the most severe thunderstorms with damaging winds, hail and an occasional tornado, occur most frequently June.

There is occasional flooding in the lower areas of Sioux Falls along the Big Sioux River and Skunk Creek. Runoff from the melting snow in the spring often causes substantial rises in the rivers. A diversion canal around Sioux Falls has reduced the threat of damaging floods.

Based on the 1951–1980 period, the average first occurrence of 32 degrees Fahrenheit in the fall is October 1 and the average last occurrence in the spring is May 10.

TABLE 1   NORMALS, MEANS AND EXTREMES

SIOUX FALLS, SOUTH DAKOTA

LATITUDE: 43°34'N   LONGITUDE: 96°44'W   ELEVATION: FT. GRND 1418 BARO 01429   TIME ZONE: CENTRAL   WBAN: 14944

	(a)	JAN	FEB	MAR	APR	MAY	JUNE	JULY	AUG	SEP	OCT	NOV	DEC	YEAR
TEMPERATURE °F:														
Normals														
-Daily Maximum		22.9	29.3	40.1	58.1	70.5	80.3	86.2	83.9	73.5	62.1	43.7	29.3	56.7
-Daily Minimum		1.9	8.9	20.6	34.6	45.7	56.3	61.8	59.7	48.5	36.7	22.3	10.1	33.9
-Monthly		12.4	19.1	30.4	46.4	58.1	68.3	74.0	71.8	61.0	49.4	33.0	19.7	45.3
Extremes														
-Record Highest	40	66	70	87	94	100	101	108	108	104	94	76	61	108
-Year		1981	1982	1968	1962	1967	1974	1947	1973	1976	1963	1978	1984	AUG 1973
-Record Lowest	40	-36	-31	-23	5	17	33	38	34	22	9	-17	-26	-36
-Year		1970	1962	1948	1982	1967	1969	1971	1950	1974	1972	1964	1983	JAN 1970
NORMAL DEGREE DAYS:														
Heating (base 65°F)		1631	1285	1073	561	240	52	14	15	161	489	960	1404	7885
Cooling (base 65°F)		0	0	0	0	29	154	293	226	41	6	0	0	749
% OF POSSIBLE SUNSHINE														
MEAN SKY COVER (tenths)														
Sunrise - Sunset	40	6.4	6.6	6.9	6.5	6.3	5.7	4.8	4.9	5.1	5.5	6.6	6.7	6.0
MEAN NUMBER OF DAYS:														
Sunrise to Sunset														
-Clear	40	8.1	6.9	6.1	6.9	7.3	8.6	11.7	12.0	11.7	11.2	7.1	7.3	104.8
-Partly Cloudy	40	7.8	6.7	7.6	8.2	9.9	11.2	12.1	10.4	7.9	7.5	7.3	7.0	103.5
-Cloudy	40	15.2	14.6	17.3	14.9	13.8	10.3	7.3	8.6	10.4	12.3	15.6	16.7	156.9
Precipitation														
.01 inches or more	40	6.2	6.5	8.7	9.2	10.4	10.9	9.3	8.9	8.1	6.2	6.0	6.3	96.8
Snow, Ice pellets														
1.0 inches or more	40	2.1	2.2	2.6	0.6	0.0	0.0	0.0	0.0	0.0	0.1	1.5	2.1	11.2
Thunderstorms	40	0.0	0.1	0.8	2.9	5.8	9.0	9.3	8.1	5.4	2.0	0.4	0.1	44.0
Heavy Fog Visibility 1/4 mile or less	40	2.5	3.0	2.8	1.0	0.7	0.6	0.6	1.1	1.3	1.7	2.8	3.7	21.6
Temperature °F														
-Maximum														
90° and above	22	0.0	0.0	0.0	0.3	0.5	3.5	10.6	7.9	1.7	0.0	0.0	0.0	24.6
32° and below	22	22.5	15.7	8.2	0.4	0.0	0.0	0.0	0.0	0.0	0.*	6.5	19.8	73.1
-Minimum														
32° and below	22	30.9	27.5	26.0	12.1	2.4	0.0	0.0	0.0	1.5	11.3	25.2	30.7	167.6
0° and below	22	14.6	8.5	1.8	0.0	0.0	0.0	0.0	0.0	0.0	0.0	1.4	8.1	34.4
AVG. STATION PRESS.(mb)	13	967.1	966.2	962.9	962.7	962.0	962.2	964.1	964.5	965.2	965.1	965.3	965.8	964.4
RELATIVE HUMIDITY (%)														
Hour 00	22	74	78	79	74	71	75	74	77	78	75	79	79	76
Hour 06	22	75	78	81	81	80	82	82	84	85	81	82	79	81
Hour 12 (Local Time)	22	67	68	64	56	53	56	53	55	57	56	65	71	60
Hour 18	22	71	70	64	53	49	51	50	52	56	59	70	75	60
PRECIPITATION (inches):														
Water Equivalent														
-Normal		0.50	0.93	1.58	2.36	3.21	3.70	2.71	3.13	2.79	1.57	0.92	0.72	24.12
-Maximum Monthly	40	1.71	4.05	3.60	5.79	7.29	8.43	7.79	9.09	6.34	5.73	2.95	2.62	9.09
-Year		1969	1962	1977	1984	1965	1984	1948	1975	1966	1973	1983	1968	AUG 1975
-Minimum Monthly	40	0.05	0.05	0.14	0.17	0.61	1.02	0.25	0.53	0.29	T	0.02	0.04	T
-Year		1958	1985	1967	1969	1981	1976	1947	1970	1956	1952	1980	1979	OCT 1952
-Maximum in 24 hrs	40	1.61	2.00	1.96	2.64	3.92	4.32	3.04	4.59	4.02	4.54	1.62	1.44	4.59
-Year		1960	1962	1956	1953	1972	1957	1982	1975	1966	1973	1972	1955	AUG 1975
Snow, Ice pellets														
-Maximum Monthly	40	19.6	48.4	31.5	18.4	0.2				0.9	5.1	21.9	41.1	48.4
-Year		1969	1962	1951	1983	1954				1985	1970	1985	1968	FEB 1962
-Maximum in 24 hrs	40	11.8	26.0	18.9	9.0	0.2				0.9	5.0	11.8	16.6	26.0
-Year		1960	1962	1956	1957	1954				1985	1970	1979	1968	FEB 1962
WIND:														
Mean Speed (mph)	37	11.0	11.1	12.5	13.3	11.9	10.7	9.7	9.8	10.3	10.8	11.5	10.7	11.1
Prevailing Direction through 1963		NW	NW	NW	NW	S	S	S	S	S	S	NW	NW	S
Fastest Obs. 1 Min.														
-Direction	37	31	30	02	25	11	23	36	29	16	27	36	36	23
-Speed (MPH)	37	45	44	60	48	46	70	69	52	50	60	52	46	70
-Year		1975	1972	1950	1955	1950	1952	1982	1956	1953	1949	1975	1982	JUN 1952
Peak Gust														
-Direction	2	NW	NW	E	SW	NW	W	NW	SW	S	NW	NW	NW	SW
-Speed (mph)	2	46	55	52	64	53	52	64	40	43	53	47	51	64
-Date		1985	1984	1985	1985	1985	1985	1984	1985	1985	1984	1984	1985	APR 1985

See Reference Notes to this table on the following page.

TABLE 2 PRECIPITATION (inches)   SIOUX FALLS, SOUTH DAKOTA

YEAR	JAN	FEB	MAR	APR	MAY	JUNE	JULY	AUG	SEP	OCT	NOV	DEC	ANNUAL
1956	0.58	0.38	2.19	1.29	1.31	6.86	4.02	4.09	0.29	0.62	0.87	0.24	22.74
1957	0.12	0.26	1.84	3.15	4.89	5.41	2.80	3.76	2.66	1.24	1.66	0.40	28.19
1958	0.05	1.77	0.95	2.55	0.81	1.48	3.52	0.61	2.11	0.08	1.15	0.25	15.33
1959	0.16	0.92	0.53	0.36	7.28	2.02	0.39	7.47	4.62	3.29	1.89	0.88	29.81
1960	1.65	0.22	1.90	3.64	5.08	3.97	0.72	5.17	2.21	0.67	1.51	0.77	27.51
1961	0.25	0.92	1.14	1.04	4.67	3.86	2.16	1.79	2.36	2.66	1.40	0.80	23.05
1962	0.29	4.05	1.72	1.70	6.07	3.98	5.50	2.77	3.58	0.46	0.16	0.19	30.47
1963	0.90	0.53	1.16	1.25	2.00	2.51	6.45	0.94	2.40	1.65	0.36	0.85	21.00
1964	0.34	0.08	2.12	4.03	1.29	1.68	4.03	3.87	4.06	0.09	0.38	0.59	22.56
1965	0.24	1.46	1.09	3.35	7.29	4.91	1.49	1.29	4.91	1.05	0.26	0.60	27.94
1966	0.54	0.99	0.70	1.71	1.94	2.68	1.54	2.12	6.34	1.43	0.20	0.65	20.84
1967	0.75	0.44	0.14	3.90	0.72	4.26	0.53	3.46	0.87	0.39	0.03	0.91	16.40
1968	0.33	0.09	0.61	4.34	2.69	4.10	2.37	1.70	4.01	4.57	0.39	2.62	27.82
1969	1.71	2.55	1.09	0.17	2.43	4.85	2.73	5.07	2.41	2.05	0.34	1.18	26.58
1970	0.37	0.10	2.03	3.75	4.83	3.81	2.98	0.53	3.14	3.13	2.17	0.54	27.38
1971	0.13	0.90	0.85	1.59	1.06	6.10	2.92	0.71	3.23	3.06	2.45	0.64	23.64
1972	0.18	0.40	0.97	2.73	7.25	2.09	3.49	2.65	1.75	1.78	1.89	1.25	26.43
1973	0.43	0.43	3.52	2.12	1.93	2.38	3.50	1.05	5.61	5.73	1.01	0.48	28.19
1974	0.13	0.30	1.65	1.33	3.11	2.79	1.27	5.16	0.58	0.34	0.27	0.10	17.03
1975	1.35	0.22	1.95	2.45	1.66	4.48	0.62	9.09	1.35	0.49	2.25	0.19	26.10
1976	0.41	0.48	1.60	2.15	1.02	1.02	1.53	1.31	0.76	0.71	0.07	0.36	11.42
1977	0.19	0.83	3.60	2.17	3.17	1.73	3.64	5.63	5.63	2.36	1.80	0.87	31.62
1978	0.47	0.33	0.56	3.98	3.47	2.91	4.79	3.08	2.45	0.14	0.48	0.51	23.17
1979	1.14	0.41	3.47	2.75	4.90	3.01	3.13	4.35	4.03	3.30	1.72	0.04	32.25
1980	0.18	0.47	0.70	0.77	2.52	2.17	1.63	2.92	0.79	1.36	0.02	0.29	13.82
1981	0.12	0.33	1.86	0.58	0.61	3.90	3.89	2.28	0.50	2.45	1.21	0.38	18.11
1982	0.76	0.13	1.17	1.87	4.72	1.18	4.60	5.23	3.49	5.18	2.94	1.99	33.26
1983	0.52	0.22	3.35	2.88	2.92	6.75	1.82	2.00	1.92	0.71	2.95	0.73	26.77
1984	0.37	1.10	1.83	5.79	2.95	8.43	1.63	0.76	1.62	4.11	0.03	1.02	29.64
1985	0.45	0.05	2.37	5.18	3.29	2.52	2.70	4.07	3.34	0.75	1.97	0.47	27.16
Record Mean	0.60	0.78	1.44	2.50	3.52	4.17	3.00	3.11	2.67	1.62	1.05	0.73	25.19

TABLE 3 AVERAGE TEMPERATURE (deg. F)   SIOUX FALLS, SOUTH DAKOTA

YEAR	JAN	FEB	MAR	APR	MAY	JUNE	JULY	AUG	SEP	OCT	NOV	DEC	ANNUAL
1956	11.2	13.6	27.6	40.9	60.1	75.1	70.8	71.0	61.2	56.1	32.8	24.7	45.4
1957	9.2	23.4	31.4	45.2	56.0	66.7	78.6	71.1	59.1	48.3	33.3	29.4	46.0
1958	23.9	16.9	31.3	46.3	61.8	64.4	70.3	74.6	63.9	52.8	34.8	17.6	46.6
1959	11.8	13.8	35.2	46.8	58.7	72.4	74.3	76.0	61.2	44.7	25.0	29.3	45.8
1960	14.1	14.2	17.7	45.9	58.0	65.6	73.2	72.8	62.8	50.7	35.9	21.0	44.4
1961	14.2	23.6	35.6	41.9	55.0	68.8	72.2	74.0	59.3	51.3	33.4	16.0	45.5
1962	12.8	17.2	24.6	45.7	63.0	67.6	70.7	71.3	59.4	52.7	38.6	22.7	45.6
#1963	6.7	19.8	38.9	49.4	59.4	73.5	75.8	70.8	64.3	58.1	39.2	13.2	47.4
1964	23.0	26.9	26.6	48.1	61.6	70.1	77.3	68.3	59.3	48.7	32.7	15.5	46.5
1965	12.0	13.0	21.7	45.9	59.0	65.5	70.8	69.4	51.5	51.7	32.8	28.4	43.5
1966	5.1	17.3	37.7	41.1	54.3	68.5	78.7	68.8	59.9	48.3	19.4	19.4	44.2
1967	17.6	15.3	37.1	46.0	53.1	66.3	71.7	69.9	61.3	47.2	32.7	21.9	45.0
1968	16.8	18.9	39.8	47.8	52.6	69.6	71.8	72.7	61.0	49.4	33.5	16.2	45.9
1969	8.8	19.0	20.5	48.4	59.6	61.4	73.5	73.3	62.3	43.4	33.4	19.5	43.6
1970	4.7	19.4	26.4	44.8	59.9	69.0	73.3	72.5	61.9	46.2	32.0	15.6	43.8
1971	8.3	18.8	31.2	47.8	55.3	71.3	69.3	73.0	61.5	51.7	33.3	17.4	44.9
1972	8.9	11.5	31.5	43.8	59.1	66.7	70.2	70.5	59.5	44.4	31.9	14.8	42.7
1973	18.5	23.1	39.7	46.4	56.9	68.9	74.4	76.7	60.1	52.9	34.7	17.1	47.5
1974	12.8	24.3	34.5	48.9	55.7	65.9	79.2	67.6	57.7	51.4	33.9	23.6	46.3
1975	17.1	16.5	25.5	41.4	61.1	67.6	78.5	72.7	57.0	52.1	33.5	20.4	45.3
1976	15.6	29.5	34.8	51.0	57.1	71.5	76.7	75.2	62.9	44.3	27.6	16.6	46.9
1977	4.7	25.6	36.0	53.8	66.8	71.1	77.0	68.1	61.9	47.7	31.3	16.3	46.7
1978	1.9	8.6	30.5	44.3	57.7	66.6	70.9	71.1	66.7	47.8	31.3	14.6	42.7
1979	1.8	7.5	28.2	44.1	55.2	67.8	74.0	69.9	64.3	48.2	30.2	27.1	43.2
1980	18.1	18.1	31.7	50.1	58.4	68.8	74.2	71.1	62.0	45.5	36.7	21.2	46.3
1981	22.3	26.0	38.4	53.5	58.3	70.4	75.5	71.1	63.0	48.7	39.5	18.0	48.8
1982	3.8	20.2	32.8	43.9	59.6	62.6	74.4	71.3	60.0	48.8	30.3	24.5	44.3
1983	20.1	26.2	33.1	41.5	55.1	66.7	77.1	78.3	63.5	49.6	34.9	2.1	45.7
1984	17.4	27.8	24.6	45.8	55.9	68.1	73.6	74.2	57.2	50.4	36.2	20.4	46.0
1985	13.4	19.9	37.8	52.8	62.7	64.2	71.5	66.3	58.2	46.6	20.7	9.5	43.6
Record Mean	14.7	20.5	32.0	47.1	58.8	68.3	74.3	71.8	61.8	50.1	33.3	20.2	46.0
Max	24.6	30.4	41.8	58.8	71.0	80.1	86.3	83.7	73.7	61.9	43.1	29.5	57.1
Min	4.7	10.5	22.2	35.4	46.5	56.5	62.2	60.0	49.9	38.2	23.5	10.9	35.0

REFERENCE NOTES FOR TABLES 1, 2, 3 and 6   (SIOUX FALLS, SD)

GENERAL

T - TRACE AMOUNT
BLANK ENTRIES DENOTE MISSING/UNREPORTED DATA.
# INDICATES A STATION OR INSTRUMENT RELOCATION.

SPECIFIC

TABLE 1

(a) - LENGTH OF RECORD IN YEARS. ALTHOUGH
INDIVIDUAL MONTHS MAY BE MISSING.

* LESS THAN .05

NORMALS — BASED ON THE 1951-1980 RECORD PERIOD.
EXTREMES — DATES ARE THE MOST RECENT OCCURRENCE.
WIND DIR. — NUMERALS SHOW TENS OF DEGREES
CLOCKWISE FROM TRUE NORTH.
"00" INDICATES CALM.
RESULTANT WIND DIRECTIONS ARE GIVEN TO WHOLE DEGREES.

EXCEPTIONS

TABLES 2, 3, and 6

RECORD MEANS ARE THROUGH THE CURRENT YEAR,
BEGINNING IN   1921 FOR TEMPERATURE
1891 FOR PRECIPITATION
1946 FOR SNOWFALL

TABLE 4 HEATING DEGREE DAYS Base 65 deg. F   SIOUX FALLS, SOUTH DAKOTA

SEASON	JULY	AUG	SEP	OCT	NOV	DEC	JAN	FEB	MAR	APR	MAY	JUNE	TOTAL
1956-57	0	27	165	284	957	1243	1729	1157	1039	591	292	37	7521
1957-58	0	13	189	509	943	1099	1271	1341	1036	554	148	97	7200
1958-59	7	17	104	397	901	1465	1646	1430	918	540	236	27	7688
1959-60	3	0	192	623	1194	1100	1572	1469	1462	576	233	38	8462
1960-61	3	2	173	439	867	1357	1566	1153	906	689	321	43	7519
1961-62	0	5	236	422	943	1512	1616	1333	1245	593	117	42	8064
1962-63	18	5	191	391	786	1305	1809	1261	803	467	203	7	7246
#1963-64	0	24	92	222	767	1603	1299	1101	1181	509	174	47	7019
1964-65	0	55	210	499	965	1528	1639	1452	1340	568	195	41	8492
1965-66	2	37	405	408	964	1129	1856	1331	838	709	343	49	8071
1966-67	0	41	188	508	1016	1406	1462	1385	858	563	399	37	7863
1967-68	36	32	141	549	963	1328	1490	1331	776	511	380	53	7590
1968-69	14	14	152	489	938	1511	1742	1283	1373	493	217	143	8369
1969-70	2	2	117	668	941	1404	1867	1274	1192	603	206	46	8322
1970-71	16	3	195	577	986	1526	1754	1288	1041	511	297	17	8211
1971-72	29	12	177	409	945	1470	1737	1551	1031	628	236	50	8275
1972-73	29	43	194	631	984	1555	1440	1166	775	549	256	19	7641
1973-74	1	0	178	373	902	1481	1616	1133	939	477	303	85	7488
1974-75	2	39	240	419	927	1279	1476	1351	1217	701	157	51	7859
1975-76	3	5	271	407	937	1377	1528	1023	929	418	261	23	7182
1976-77	1	0	142	643	1114	1495	1867	1097	891	341	38	4	7633
1977-78	0	22	125	530	1000	1503	1957	1576	1063	613	252	78	8719
1978-79	11	18	104	525	1004	1554	1960	1607	1134	622	321	40	8900
1979-80	0	28	103	512	1038	1168	1448	1349	1027	464	243	32	7412
1980-81	1	14	157	602	841	1354	1318	1087	816	353	230	5	6778
1981-82	5	5	114	497	758	1452	1899	1247	991	632	174	105	7879
1982-83	0	28	188	494	1035	1249	1385	1082	982	699	317	67	7526
1983-84	1	0	160	476	894	1947	1468	1072	1247	569	285	23	8142
1984-85	2	12	265	451	857	1376	1594	1260	836	378	124	101	7256
1985-86	5	44	269	562	1327	1721							

TABLE 5 COOLING DEGREE DAYS Base 65 deg. F   SIOUX FALLS, SOUTH DAKOTA

YEAR	JAN	FEB	MAR	APR	MAY	JUNE	JULY	AUG	SEP	OCT	NOV	DEC	TOTAL
1969	0	0	0	0	58	43	270	265	41	4	0	0	681
1970	0	0	0	6	54	172	280	241	108	2	0	0	863
1971	0	0	0	4	1	211	167	268	79	2	0	0	732
1972	0	0	0	0	60	108	197	219	35	0	0	0	619
1973	0	0	0	0	13	143	300	370	38	8	0	0	872
1974	0	0	0	2	23	118	450	126	26	6	0	0	751
1975	0	0	0	0	44	134	428	252	36	16	0	0	910
1976	0	0	0	2	24	226	370	324	85	9	0	0	1040
1977	0	0	0	11	101	197	382	123	40	0	0	0	854
1978	0	0	0	0	32	133	202	213	163	0	0	0	743
1979	0	0	0	3	25	131	289	189	87	0	0	0	724
1980	0	0	0	23	46	153	290	207	74	6	0	0	799
1981	0	0	0	14	27	175	342	203	61	0	0	0	822
1982	0	0	0	4	12	42	298	230	45	0	0	0	631
1983	0	0	0	15	122	381	420	120		2	0	0	1060
1984	0	0	0	0	10	120	276	305	36	4	0	0	751
1985	0	0	0	19	59	85	212	93	72	0	0	0	540

TABLE 6 SNOWFALL (inches)   SIOUX FALLS, SOUTH DAKOTA

SEASON	JULY	AUG	SEP	OCT	NOV	DEC	JAN	FEB	MAR	APR	MAY	JUNE	TOTAL
1956-57	0.0	0.0	0.0	T	4.1	3.8	2.2	4.5	19.0	10.7	0.0	0.0	44.3
1957-58	0.0	0.0	0.0	T	1.6	0.1	0.5	2.9	8.2	T	0.0	0.0	13.3
1958-59	0.0	0.0	0.0	0.0	2.8	4.5	2.4	15.4	3.6	1.0	0.0	0.0	29.7
1959-60	0.0	0.0	0.0	0.6	17.0	3.7	12.3	3.0	20.4	4.6	0.0	0.0	61.6
1960-61	0.0	0.0	0.0	T	0.1	6.4	3.4	11.6	12.0	T	0.0	0.0	33.5
1961-62	0.0	0.0	T	0.0	1.0	7.7	3.8	48.4	14.7	4.2	0.0	0.0	79.8
1962-63	0.0	0.0	0.0	0.7	0.0	1.7	11.0	6.8	2.8	2.3	0.0	0.0	25.3
1963-64	0.0	0.0	0.0	T	0.0	11.6	3.5	0.7	9.1	T	0.0	0.0	24.9
1964-65	0.0	0.0	0.0	0.0	3.9	6.5	3.3	21.1	13.3	0.4	0.0	0.0	48.5
1965-66	0.0	0.0	0.0	0.0	1.6	0.1	6.2	1.8	4.6	0.9	T	0.0	15.2
1966-67	0.0	0.0	0.0	0.4	2.3	9.3	6.0	11.8	1.0	T	0.1	0.0	30.9
1967-68	0.0	0.0	0.0	0.5	0.4	2.5	4.6	0.9	T	0.1	0.1	0.0	9.1
1968-69	0.0	0.0	0.0	0.7	1.9	41.1	19.6	28.5	2.9	0.0	0.0	0.0	94.7
1969-70	0.0	0.0	0.0	T	2.9	15.2	5.9	1.5	19.3	3.0	T	0.0	47.8
1970-71	0.0	0.0	0.0	5.1	0.5	8.5	1.9	1.7	4.7	0.3	0.0	0.0	22.7
1971-72	0.0	0.0	0.0	T	4.4	10.6	1.8	10.6	5.0	0.8	0.0	0.0	33.2
1972-73	0.0	0.0	0.0	0.2	2.5	6.2	5.5	7.4	0.3	T	0.0	0.0	22.1
1973-74	0.0	0.0	0.0	T	0.1	11.6	1.9	2.9	5.1	3.3	0.0	0.0	24.9
1974-75	0.0	0.0	0.0	0.0	2.5	1.1	18.3	3.1	17.9	T	0.0	0.0	42.9
1975-76	0.0	0.0	0.0	T	13.2	2.5	6.5	6.6	7.9	T	0.1	0.0	36.8
1976-77	0.0	0.0	0.0	3.4	0.7	6.7	2.7	1.9	13.5	T	0.0	0.0	28.9
1977-78	0.0	0.0	0.0	0.9	8.6	8.0	7.6	6.5	5.5	0.6	0.0	0.0	37.7
1978-79	0.0	0.0	0.0	0.0	2.9	9.5	19.0	7.3	13.2	1.5	0.0	0.0	53.4
1979-80	0.0	0.0	0.0	T	15.2	0.3	2.9	5.5	5.8	T	0.0	0.0	29.7
1980-81	0.0	0.0	0.0	0.8	0.0	4.6	1.4	3.1	T	0.5	0.0	0.0	10.8
1981-82	0.0	0.0	0.0	0.8	8.9	5.5	16.9	1.0	1.8	7.5	0.0	0.0	42.4
1982-83	0.0	0.0	0.0	3.3	4.1	17.6	4.7	3.6	18.8	18.4	0.0	0.0	70.5
1983-84	0.0	0.0	0.0	T	19.0	13.7	5.0	11.9	19.4	6.0	0.0	0.0	75.0
1984-85	0.0	0.0	T	T	4.7	7.4	0.7	16.1	2.5	0.0	0.0		31.4
1985-86	0.0	0.0	0.9	0.2	21.9	9.1							
Record Mean	0.0	0.0	T	0.5	5.0	7.6	6.5	8.3	10.2	2.2	T	0.0	40.5

See Reference Notes, relative to all above tables, on preceding page.

The Weather Service Office is located an almost equal distance of 15 miles in the middle of a geographical triangle between the cities of Bristol, Tennessee–Virginia, Kingsport and Johnson City, Tennessee, and is more commonly known as the Tri–City Area. This location is situated in the extreme upper East Tennessee Valley. The terrain immediately surrounding the station ranges from gently rolling on the east and south to very hilly on the west and north. Mountain ranges begin about 10 miles to the southeast and about 15 miles to the west and north, with many peaks and ridges rising to 4,000 feet, and some to 6,000 feet toward the southeast.

This section does not lie directly within any of the principal storm tracks that cross the country, but comes under the influence of storm centers that pass along the Gulf Coast and then up the Atlantic Coast toward the northeast. Being quite varied, the topography has considerable influence on the weather. Moist air from the east is forced up the slopes of the mountains causing much of the moisture to be precipitated before the air mass reaches the Bristol area. The same process occurs to a lesser extent when air masses move over the smaller mountain ranges to the west and north. The maximum monthly precipitation occurs in July, usually from afternoon and early evening thunderstorms. A second maximum of precipitation occurs in the late winter months, due mainly to moist air associated with storm centers to the south or northeast. Annual precipitation amounts recorded in mountainous sections to the east and southeast are almost double what they are in the immediate vicinity.

Lowest temperatures normally occur during the early morning hours, but rise rapidly during the morning hours. Periods of cold weather are generally associated with air flow from winter storm centers near the northeast coast. Periods of unusually high temperatures occur most frequently when Gulf air associated with the Bermuda high pressure system dominates the area.

Snowfall seldom occurs before November and rarely remains on the ground for more than a few days. However, mountains to the east and south of the station are frequently well blanketed with snow for much longer periods of time.

Agricultural activies within this area include such staple crops as tobacco, beans, and hay which are raised in such amounts as to be important commercially. The last freezing temperature in spring normally occurs in late April, and the first in autumn around mid–October. The growing season of 180 days, usually coupled with ample sunshine and rainfall, permits a second planting and harvesting of some staple crops.

TABLE 1 — NORMALS, MEANS AND EXTREMES

BRISTOL, JOHNSON CITY, KINGSPORT, TENNESSEE

LATITUDE: 36°29'N   LONGITUDE: 82°24'W   ELEVATION: FT. GRND 1507 BARO 01558   TIME ZONE: EASTERN   WBAN: 13877

	(a)	JAN	FEB	MAR	APR	MAY	JUNE	JULY	AUG	SEP	OCT	NOV	DEC	YEAR
TEMPERATURE °F:														
Normals														
-Daily Maximum		44.5	48.4	57.6	68.3	76.3	82.9	85.5	85.2	80.4	69.5	57.3	48.1	67.0
-Daily Minimum		25.5	27.4	34.9	43.8	52.5	60.1	64.3	63.4	57.2	44.7	35.1	28.3	44.8
-Monthly		35.0	37.9	46.3	56.0	64.4	71.5	74.9	74.3	68.8	57.1	46.2	38.2	55.9
Extremes														
-Record Highest	40	79	80	85	88	92	97	102	98	100	90	81	78	102
-Year		1950	1977	1954	1957	1969	1952	1952	1983	1954	1954	1974	1951	JUL 1952
-Record Lowest	40	-21	-5	-2	21	30	38	45	47	34	20	5	-9	-21
-Year		1985	1958	1980	1982	1963	1966	1947	1979	1983	1962	1950	1962	JAN 1985
NORMAL DEGREE DAYS:														
Heating (base 65°F)		930	759	580	273	111	10	0	0	35	263	564	831	4356
Cooling (base 65°F)		0	0	0	6	93	205	307	288	149	18	0	0	1066
% OF POSSIBLE SUNSHINE														
MEAN SKY COVER (tenths)														
Sunrise - Sunset	37	7.2	6.8	6.8	6.3	6.2	6.0	6.3	5.9	5.5	5.2	6.3	6.8	6.3
MEAN NUMBER OF DAYS:														
Sunrise to Sunset														
-Clear	48	5.9	5.9	6.6	7.5	6.9	6.1	5.6	6.9	9.9	12.3	8.6	6.9	89.1
-Partly Cloudy	48	6.8	7.0	7.8	8.7	10.6	12.4	12.5	13.8	9.9	8.1	7.1	7.2	111.9
-Cloudy	48	18.3	15.4	16.6	13.8	13.5	11.5	12.8	10.4	10.2	10.6	14.3	16.9	164.3
Precipitation														
.01 inches or more	40	14.1	11.9	13.2	11.2	11.6	11.1	12.1	10.5	7.8	8.1	10.6	11.4	133.4
Snow, Ice pellets														
1.0 inches or more	42	1.8	1.3	0.7	0.*	0.0	0.0	0.0	0.0	0.0	0.0	0.2	1.0	5.0
Thunderstorms	42	0.3	0.9	2.0	3.4	6.4	8.4	9.6	7.5	3.6	1.0	0.5	0.2	43.6
Heavy Fog Visibility														
1/4 mile or less	42	3.4	2.6	1.3	1.6	3.4	3.7	4.9	7.3	5.4	4.9	2.7	3.1	44.4
Temperature °F														
-Maximum														
90° and above	24	0.0	0.0	0.0	0.0	0.3	2.3	4.1	2.8	2.0	0.0	0.0	0.0	11.5
32° and below	24	6.2	3.3	0.4	0.0	0.0	0.0	0.0	0.0	0.0	0.0	0.3	2.9	13.0
-Minimum														
32° and below	24	23.8	20.4	13.8	3.6	0.1	0.0	0.0	0.0	0.0	2.6	12.6	20.6	97.6
0° and below	24	1.1	0.3	0.*	0.0	0.0	0.0	0.0	0.0	0.0	0.0	0.0	0.2	1.7
AVG. STATION PRESS.(mb)	13	964.1	963.8	961.9	962.4	961.9	963.5	964.5	965.5	965.3	965.9	965.3	965.0	964.1
RELATIVE HUMIDITY (%)														
Hour 01	24	76	73	71	71	82	86	88	89	87	83	79	77	80
Hour 07	24	79	78	78	80	88	90	91	92	92	88	83	81	85
Hour 13 (Local Time)	24	63	58	52	49	55	57	62	61	57	54	57	62	57
Hour 19	24	66	60	55	51	60	63	68	70	69	64	65	67	63
PRECIPITATION (inches):														
Water Equivalent														
-Normal		3.56	3.43	4.29	3.46	3.61	3.46	4.19	3.23	3.00	2.50	2.98	3.53	41.24
-Maximum Monthly	40	9.18	7.29	9.56	5.85	9.71	6.68	9.73	7.07	7.09	5.65	5.90	6.75	9.73
-Year		1957	1956	1955	1970	1950	1957	1949	1966	1972	1959	1948	1961	JUL 1949
-Minimum Monthly	40	1.37	0.75	1.31	0.21	1.31	1.10	0.79	0.82	0.50	0.07	1.07	0.21	0.07
-Year		1981	1968	1985	1976	1966	1980	1957	1954	1985	1963	1953	1965	OCT 1963
-Maximum in 24 hrs	40	2.34	1.87	3.35	2.66	3.26	3.10	2.90	3.07	3.61	3.65	2.55	2.95	3.65
-Year		1950	1954	1973	1977	1984	1954	1946	1982	1972	1964	1957	1969	OCT 1964
Snow, Ice pellets														
-Maximum Monthly	48	22.1	20.4	27.9	5.6	T					T	18.1	12.9	27.9
-Year		1966	1979	1960	1983	1963					1977	1952	1963	MAR 1960
-Maximum in 24 hrs	42	9.7	10.7	13.0	5.6	T					T	16.2	9.6	16.2
-Year		1955	1969	1960	1983	1963					1977	1952	1969	NOV 1952
WIND:														
Mean Speed (mph)	31	6.6	6.8	7.4	7.1	5.3	4.8	4.2	3.9	4.3	4.7	5.7	6.0	5.6
Prevailing Direction														
through 1963		WSW	NE	WNW	WSW	WSW	NE	WSW	NE	NE	NE	W	WSW	NE
Fastest Obs. 1 Min.														
-Direction (!!!)	30	25	25	25	25	32	28	23	34	31	28	29	24	32
-Speed (MPH)	30	40	46	40	41	50	39	40	46	29	35	35	40	50
-Year		1965	1961	1952	1977	1951	1975	1961	1962	1967	1965	1965	1968	MAY 1951
Peak Gust														
-Direction (!!!)	2	W	W	SW	W	NW	NW	N	W	SW	W	W	W	NW
-Speed (mph)	2	46	45	52	52	46	55	47	40	54	32	36	48	55
-Date		1985	1984	1984	1985	1984	1985	1985	1985	1984	1985	1984	1984	JUN 1985

See Reference Notes to this table on the following page.

TABLE 2 PRECIPITATION (inches)   BRISTOL, JOHNSON CITY, KINGSPORT, TENNESSEE

YEAR	JAN	FEB	MAR	APR	MAY	JUNE	JULY	AUG	SEP	OCT	NOV	DEC	ANNUAL
1956	2.36	7.29	4.60	4.87	3.18	1.39	4.34	2.60	2.44	1.40	1.78	4.88	41.13
1957	9.18	5.13	1.33	4.38	2.39	6.68	0.79	5.30	5.05	1.12	5.38	4.45	51.18
1958	2.17	3.84	2.76	3.98	2.35	1.35	0.58	5.35	1.32	0.58	2.47	3.14	40.54
1959	2.93	2.79	4.06	4.56	2.33	1.35	1.57	4.10	2.56	5.65	4.73	3.87	40.50
1960	2.24	2.81	3.34	1.73	2.06	3.95	6.93	5.12	1.13	4.42	1.57	2.46	37.76
1961	2.27	6.04	4.12	2.72	2.31	4.46	3.16	4.66	1.00	4.05	2.80	6.75	44.34
1962	5.15	5.38	3.36	3.60	1.69	3.35	3.27	2.22	6.19	2.27	4.35	2.00	43.60
1963	3.20	2.27	7.52	1.38	4.64	4.38	3.26	3.73	1.40	0.07	4.73	2.00	38.58
1964	3.25	4.11	4.66	4.65	2.27	1.14	3.73	3.79	3.14	4.95	1.88	2.48	40.05
1965	3.23	2.77	5.82	3.42	2.75	3.45	3.45	2.65	1.78	2.16	1.85	0.21	37.26
1966	3.30	3.50	2.05	5.04	1.31	1.77	6.21	7.07	4.73	3.50	3.71	3.07	45.26
1967	2.00	4.14	3.02	2.80	6.29	2.14	3.68	3.68	3.78	1.99	3.73	5.62	42.87
1968	3.58	0.75	4.95	4.12	3.83	2.89	2.89	2.32	0.93	2.51	1.89	2.13	32.79
1969	2.72	5.72	2.14	2.11	2.06	5.24	8.18	2.11	4.31	1.54	1.86	5.78	43.77
1970	2.89	2.75	1.60	5.85	1.60	2.48	2.95	4.28	1.67	3.29	1.33	3.40	34.09
1971	3.83	3.59	3.42	3.38	4.38	5.41	5.94	2.32	2.91	3.18	2.45	2.43	43.24
1972	4.87	4.81	2.86	2.92	5.92	2.97	3.18	2.70	7.09	4.57	2.85	6.43	51.17
1973	1.80	2.04	7.18	3.58	6.46	2.99	5.66	2.60	1.68	2.29	4.52	5.07	45.87
1974	5.33	4.07	4.34	3.96	8.66	4.01	1.78	1.95	3.12	2.34	3.29	3.53	46.38
1975	4.19	3.13	9.22	3.67	4.29	3.19	2.46	3.87	5.02	1.54	3.14	3.33	47.05
1976	2.60	3.07	3.24	0.21	3.31	5.62	2.14	2.27	4.44	5.30	1.41	3.80	37.41
1977	1.90	1.01	4.86	5.43	1.91	3.77	2.13	3.09	2.36	4.89	4.91	2.72	38.98
1978	4.22	0.89	3.18	2.44	4.50	5.67	4.78	3.71	3.04	0.58	2.64	4.89	40.54
1979	5.29	3.58	3.16	3.68	3.35	3.55	6.12	2.80	3.89	2.19	4.44	1.66	43.71
1980	3.91	1.39	5.68	3.56	2.69	1.10	3.82	2.54	3.01	2.09	2.10	1.38	33.27
1981	1.37	2.59	1.94	5.10	4.51	4.28	6.24	3.05	4.17	2.51	1.95	3.20	40.91
1982	4.07	5.07	3.35	2.30	2.55	5.52	9.14	4.70	5.53	2.54	4.12	2.89	51.78
1983	1.67	2.14	1.73	4.44	4.83	4.60	3.29	5.05	1.88	2.18	2.74	4.15	38.70
1984	1.79	4.50	2.73	2.85	7.42	3.86	4.63	1.23	1.43	1.14	2.61	1.76	35.95
1985	3.21	3.40	1.31	2.08	2.85	4.35	4.38	3.09	0.50	3.02	5.87	1.17	35.23
Record Mean	3.47	3.45	3.87	3.21	3.64	3.60	4.90	3.53	2.94	2.31	2.89	3.32	41.14

TABLE 3 AVERAGE TEMPERATURE (deg. F)   CHATTANOOGA, TENNESSEE

YEAR	JAN	FEB	MAR	APR	MAY	JUNE	JULY	AUG	SEP	OCT	NOV	DEC	ANNUAL
1956	38.1	48.7	51.4	58.3	70.8	75.3	79.3	79.1	69.7	63.0	47.7	51.4	61.1
1957	42.6	50.5	49.4	63.0	69.9	76.5	79.2	77.7	72.7	56.4	50.5	43.6	61.0
1958	36.5	33.9	46.4	59.3	68.5	75.9	78.8	77.5	71.3	59.4	51.6	38.5	58.1
1959	38.0	44.1	47.4	60.3	70.7	74.5	79.3	79.1	73.1	63.2	47.0	43.8	60.0
1960	41.9	38.9	39.0	61.4	64.7	76.2	79.3	78.8	72.8	61.2	48.9	36.4	58.3
#1961	34.8	47.4	52.8	56.7	65.8	73.7	78.9	78.1	74.0	59.7	53.0	42.1	59.7
1962	38.5	49.5	47.9	56.5	75.2	76.1	79.1	79.8	72.2	63.2	48.6	37.1	60.3
1963	32.6	34.8	55.0	63.2	69.2	74.7	76.7	78.3	72.1	64.8	51.0	34.5	58.9
1964	39.9	39.9	50.2	62.9	69.9	77.0	76.8	76.7	72.6	57.2	52.4	43.9	60.0
1965	40.9	40.7	45.9	62.5	70.4	73.3	76.8	76.9	71.9	58.3	51.4	41.5	59.2
1966	33.1	41.3	49.0	58.3	64.8	73.8	78.9	75.5	69.5	56.8	49.5	40.4	57.6
1967	41.0	38.2	53.9	63.5	64.9	73.0	72.8	73.6	67.5	59.4	46.9	45.8	58.4
1968	37.1	34.4	49.8	57.9	65.8	75.3	79.3	80.9	72.5	61.9	49.7	38.9	58.6
1969	38.3	42.2	44.6	61.2	67.1	75.1	78.6	74.3	68.3	56.5	43.6	35.8	57.1
1970	32.0	38.9	46.8	60.2	67.6	75.1	79.7	78.1	75.6	62.4	47.0	44.2	59.0
1971	38.8	39.9	45.2	58.9	65.1	78.3	77.1	78.0	75.5	66.4	48.7	49.4	60.1
1972	43.4	40.0	48.8	59.2	63.9	70.2	74.5	75.0	71.9	56.8	46.0	43.1	57.7
1973	36.0	37.5	55.4	57.3	64.8	76.4	79.0	77.6	75.8	62.7	52.3	41.5	59.7
1974	49.3	43.3	54.5	57.1	65.8	68.4	75.7	75.8	68.3	57.0	47.6	40.7	58.6
1975	42.7	44.0	45.6	56.1	68.0	72.8	78.0	77.1	68.4	60.7	49.0	39.4	58.5
1976	34.6	48.6	53.8	59.4	63.5	72.8	76.6	76.6	69.1	56.3	42.7	37.4	57.6
1977	28.5	40.5	53.6	63.1	70.5	77.6	83.4	81.8	75.3	57.8	53.0	40.6	60.5
1978	30.3	35.5	49.7	62.6	68.1	76.5	81.2	78.7	75.6	59.0	54.9	44.3	59.7
1979	33.7	38.8	52.1	60.1	67.2	74.1	76.5	78.0	72.4	60.2	50.2	41.3	58.7
1980	40.7	36.8	47.8	57.8	67.2	73.8	82.2	82.2	74.0	57.2	49.8	41.2	59.3
1981	35.5	43.4	48.1	64.7	64.6	78.0	81.4	77.3	69.7	58.6	50.8	38.2	59.2
1982	36.0	43.4	53.4	55.9	70.2	74.2	79.6	76.7	70.0	60.5	50.9	48.0	59.9
1983	38.8	42.8	50.5	54.3	65.9	73.5	80.4	81.9	71.2	60.8	48.5	37.0	58.8
1984	35.5	43.2	48.2	56.8	64.4	77.4	76.2	76.6	68.0	66.7	46.5	49.0	59.0
1985	32.8	40.2	52.9	60.8	67.6	74.9	78.4	76.6	69.4	64.7	58.8	36.7	59.5
Record Mean	40.8	43.5	51.1	60.3	68.3	75.8	78.7	77.8	72.5	61.3	50.0	42.5	60.3
Max	49.5	53.0	61.4	71.3	79.4	86.3	88.6	87.7	82.8	72.5	60.3	51.3	70.4
Min	32.1	33.9	40.8	49.2	57.2	65.3	68.8	68.0	62.1	50.1	39.8	33.6	50.1

REFERENCE NOTES FOR TABLES 1, 2, 3 and 6   (BRISTOL, JOHNSON CITY, KINGSPORT, TN)

GENERAL

T - TRACE AMOUNT
BLANK ENTRIES DENOTE MISSING/UNREPORTED DATA.
# INDICATES A STATION OR INSTRUMENT RELOCATION.

SPECIFIC

TABLE 1

(a) - LENGTH OF RECORD IN YEARS. ALTHOUGH INDIVIDUAL MONTHS MAY BE MISSING.

* LESS THAN .05

NORMALS — BASED ON THE 1951-1980 RECORD PERIOD.
EXTREMES — DATES ARE THE MOST RECENT OCCURRENCE.
WIND DIR. — NUMERALS SHOW TENS OF DEGREES CLOCKWISE FROM TRUE NORTH.
     "00" INDICATES CALM.
RESULTANT WIND DIRECTIONS ARE GIVEN TO WHOLE DEGREES.

EXCEPTIONS

TABLES 2, 3, and 6

RECORD MEANS ARE THROUGH THE CURRENT YEAR, BEGINNING IN
     1938 FOR TEMPERATURE
     1938 FOR PRECIPITATION
     1938 FOR SNOWFALL

TABLE 4   HEATING DEGREE DAYS Base 65 deg. F   BRISTOL, JOHNSON CITY, KINGSPORT, TENNESSEE

SEASON	JULY	AUG	SEP	OCT	NOV	DEC	JAN	FEB	MAR	APR	MAY	JUNE	TOTAL
1956-57	0	5	79	132	595	539	831	524	578	221	50	0	3554
1957-58	0	0	45	351	496	738	1001	1032	678	282	65	0	4688
1958-59	0	0	22	264	488	948	959	646	640	252	45	3	4267
1959-60	0	0	7	217	597	777	822	815	968	201	144	0	4548
#1960-61	0	0	8	198	550	999	1031	586	441	423	138	12	4386
1961-62	2	0	27	263	446	795	906	572	622	383	18	0	4034
1962-63	1	0	71	228	602	983	1062	920	454	239	78	1	4639
1963-64	5	0	28	88	531	1148	952	878	622	269	55	5	4581
1964-65	0	12	26	349	436	719	874	772	656	223	17	5	4089
1965-66	0	3	21	292	518	803	1096	755	547	307	109	20	4471
1966-67	0	0	23	303	533	843	821	868	455	218	216	29	4309
1967-68	12	2	107	271	662	745	979	1004	553	276	137	1	4749
1968-69	0	4	19	245	590	920	947	813	814	300	113	9	4774
1969-70	0	0	49	293	702	940	1155	784	553	201	66	1	4744
1970-71	0	0	26	147	538	737	903	780	676	296	135	0	4238
1971-72	0	0	0	87	560	537	743	844	610	307	86	36	3810
1972-73	8	0	10	281	559	652	898	779	334	333	152	0	4006
1973-74	0	0	12	169	483	792	552	700	424	257	86	14	3489
1974-75	0	0	72	326	527	785	755	616	621	328	29	1	4060
1975-76	0	0	46	188	451	768	970	509	402	297	134	9	3774
1976-77	0	0	46	409	776	973	1321	845	459	184	61	21	5095
1977-78	0	0	19	359	455	913	1207	1005	620	286	130	4	4938
1978-79	0	0	0	268	411	786	1065	920	511	283	88	11	4343
1979-80	3	3	23	323	483	804	832	990	670	285	76	4	4496
1980-81	0	0	31	337	586	882	1099	752	686	161	137	0	4671
1981-82	0	0	83	346	570	942	1042	687	465	351	32	0	4518
1982-83	0	0	66	256	494	667	922	745	570	436	140	19	4315
1983-84	0	0	98	242	583	928	995	709	627	357	199	8	4746
1984-85	0	0	91	73	652	599	1154	819	518	243	88	19	4256
1985-86	0	2	76	124	283	960							

TABLE 5   COOLING DEGREE DAYS Base 65 deg. F   BRISTOL, JOHNSON CITY, KINGSPORT, TENNESSEE

YEAR	JAN	FEB	MAR	APR	MAY	JUNE	JULY	AUG	SEP	OCT	NOV	DEC	TOTAL
1969	0	0	0	0	60	261	363	255	93	14	0	0	1046
1970	0	0	0	28	106	220	373	314	270	45	0	0	1356
1971	0	0	0	5	32	273	285	276	228	44	10	2	1155
1972	0	0	0	11	22	113	279	274	160	3	0	0	862
1973	0	0	0	5	26	249	309	267	212	32	1	0	1101
1974	0	0	0	9	73	102	286	254	73	5	1	0	803
1975	0	0	0	12	101	190	310	353	136	7	0	0	1109
1976	0	0	2	14	24	185	262	212	54	0	0	0	753
1977	0	0	1	23	124	213	374	309	163	2	8	0	1217
1978	0	0	0	6	53	204	299	302	224	4	0	0	1092
1979	0	0	0	1	59	113	194	259	85	9	0	0	720
1980	0	0	0	6	61	179	359	377	215	5	0	0	1202
1981	0	0	0	27	30	309	325	226	87	1	0	0	1005
1982	0	0	0	0	136	199	322	232	93	36	0	0	1018
1983	0	0	2	5	21	163	294	307	131	3	0	0	926
1984	0	0	0	2	35	236	202	253	83	38	0	0	849
1985	0	0	5	11	53	159	262	201	114	22	0	0	827

TABLE 6   SNOWFALL (inches)   BRISTOL, JOHNSON CITY, KINGSPORT, TENNESSEE

SEASON	JULY	AUG	SEP	OCT	NOV	DEC	JAN	FEB	MAR	APR	MAY	JUNE	TOTAL
1956-57	0.0	0.0	0.0	0.0	0.3	0.5	1.0	T	2.7	T	0.0	0.0	4.5
1957-58	0.0	0.0	0.0	T	0.4	0.7	1.7	12.6	0.1	T	0.0	0.0	15.5
1958-59	0.0	0.0	0.0	0.0	T	4.8	4.2	T	2.0	T	0.0	0.0	11.0
1959-60	0.0	0.0	0.0	0.0	0.4	4.1	6.6	12.0	27.9	T	0.0	0.0	51.0
1960-61	0.0	0.0	0.0	0.0	0.4	0.8	4.3	7.4	1.0	0.2	0.0	0.0	14.1
1961-62	0.0	0.0	0.0	T	T	3.2	10.8	T	8.8	0.6	0.0	0.0	23.4
1962-63	0.0	0.0	0.0	T	0.1	11.2	4.1	8.0	0.3	0.0	T	0.0	23.7
1963-64	0.0	0.0	0.0	0.0	6.7	12.9	8.9	11.0	0.5	0.0	0.0	0.0	40.0
1964-65	0.0	0.0	0.0	0.0	0.7	T	7.5	9.0	5.5	0.0	0.0	0.0	22.7
1965-66	0.0	0.0	0.0	T	T	0.1	22.1	0.3	T	T	0.0	0.0	22.5
1966-67	0.0	0.0	0.0	0.0	1.9	5.8	4.6	4.8	T	0.0	0.0	0.0	17.1
1967-68	0.0	0.0	0.0	0.0	T	4.1	12.1	6.3	1.0	0.0	0.0	0.0	23.5
1968-69	0.0	0.0	0.0	T	2.9	1.0	5.9	15.1	4.4	0.0	0.0	0.0	29.3
1969-70	0.0	0.0	0.0	0.0	1.0	9.7	10.1	7.8	0.6	0.0	0.0	0.0	29.2
1970-71	0.0	0.0	0.0	0.0	0.4	4.2	4.5	9.9	6.3	T	0.0	0.0	25.3
1971-72	0.0	0.0	0.0	0.0	0.7	4.3	T	5.3	2.4	T	0.0	0.0	12.7
1972-73	0.0	0.0	0.0	0.0	2.1	0.1	4.2	1.2	T	T	0.0	0.0	7.6
1973-74	0.0	0.0	0.0	0.0	0.0	3.3	0.0	0.9	T	T	0.0	0.0	4.2
1974-75	0.0	0.0	0.0	0.0	0.5	3.1	1.8	0.3	3.4	T	0.0	0.0	9.1
1975-76	0.0	0.0	0.0	0.0	T	T	0.5	3.2	T	0.0	0.0	0.0	3.7
1976-77	0.0	0.0	0.0	0.0	1.2	6.2	12.7	2.4	0.4	T	0.0	0.0	22.9
1977-78	0.0	0.0	0.0	T	1.6	0.4	13.6	6.7	3.9	T	0.0	0.0	26.2
1978-79	0.0	0.0	0.0	0.0	0.0	T	9.2	20.4	0.8	0.0	0.0	0.0	30.4
1979-80	0.0	0.0	0.0	0.0	T	0.4	7.8	5.4	3.6	0.0	0.0	0.0	17.2
1980-81	0.0	0.0	0.0	0.0	T	T	7.3	1.1	1.1	0.0	0.0	0.0	9.5
1981-82	0.0	0.0	0.0	0.0	T	7.3	3.6	2.3	2.6	0.6	0.0	0.0	16.4
1982-83	0.0	0.0	0.0	0.0	0.0	6.3	3.4	5.5	1.1	5.6	0.0	0.0	21.9
1983-84	0.0	0.0	0.0	0.0	T	0.3	4.6	3.8	1.2	T	0.0	0.0	9.9
1984-85	0.0	0.0	0.0	0.0	0.0	0.3	9.7	6.0	T	T	0.0	0.0	16.0
1985-86	0.0	0.0	0.0	0.0	0.0	3.2							
Record Mean	0.0	0.0	0.0	T	1.1	2.8	5.2	4.5	2.4	0.2	T	0.0	16.1

See Reference Notes, relative to all above tables, on preceding page.

CHATTANOOGA, TENNESSEE

Chattanooga is located in the southern portion of the Great Valley of Tennessee, an area of the Tennessee River between the Cumberland Mountains to the west and the Appalachian Mountains to the east. Local topography is complex with a number of minor valleys and ridges giving a local relief of as much as 500 feet. The Tennessee River approaches Chattanooga from the northeast and forms a loop southwest to west to northwest of the city at an elevation of about 630 feet above mean sea level. Most of the city lies on the south side of the river. On the north and southwest sides, the terrain rises abruptly to about 1,200 feet above the river. This complex topography results in marked variations in air drainage, wind, and minimum temperatures within short distances. In winter the Cumberland Mountains have a moderating influence on the local climate by retarding the flow of cold air from the north and west.

Chattanooga enjoys a moderate climate, characterized by cool winters and quite warm summers. Because of the sheltering effect of the mountains, winter temperatures average about 30 degrees warmer than at stations on the southern Cumberland Plateau section of the state. Winter weather is changeable and alternates between cool spells with an occasional cold period, but extreme cold is rare. Temperatures fall as low as the freezing point on a little over one-half of the winter days, but temperatures below zero rarely occur. Snowfall from year to year is highly variable with some winters having little or none. Heavy snowfalls have occurred, but any accumulation of snow seldom remains on the ground more than a few days. Ice storms of freezing rain or glaze are not uncommon, occasionally midwinter icing becomes severe enough to do some damage in the area.

Summer temperatures are either in the high 80s or low 90s and temperatures over 100 degrees are unusual. Most afternoon temperatures are modified by thunderstorms. Temperatures frequently plunge 10 to 15 degrees in a matter of minutes during one of these showers.

Precipitation in the Chattanooga area is well distributed throughout the year with the greater amounts in wintertime when cyclonic storms from the Gulf of Mexico reach the area with greater intensity and frequency. A second peak rainfall period generally occurs in July, principally from thunderstorms that move into the area from the south and southwest. During any year there are usually a few of these storms that can be classified as severe, with hail and damaging winds.

The growing season averages 228 days. The average occurrence of last freezing temperature in spring is early April and the average first freezing temperature in the fall is early November.

Spring and autumn are very enjoyable seasons in Chattanooga, with many days being nearly ideal in temperature. To many, the fall months of September, October, and November are the most pleasant. Rainfall is at a minimum, sunshine at a relative maximum and the temperature range is comfortable.

TABLE 1   NORMALS, MEANS AND EXTREMES

CHATTANOOGA, TENNESSEE

LATITUDE: 35°02'N   LONGITUDE: 85°12'W   ELEVATION: FT. GRND 665 BARO 00681   TIME ZONE: EASTERN   WBAN: 13882

	(a)	JAN	FEB	MAR	APR	MAY	JUNE	JULY	AUG	SEP	OCT	NOV	DEC	YEAR
TEMPERATURE °F:														
Normals														
-Daily Maximum		48.2	52.6	60.9	72.6	79.8	86.4	89.3	88.8	83.0	72.3	60.3	51.4	70.5
-Daily Minimum		29.2	31.2	38.6	47.5	55.7	63.8	68.1	67.4	61.5	47.7	37.5	31.5	48.3
-Monthly		38.7	41.9	49.8	60.0	67.8	75.1	78.7	78.1	72.3	60.0	48.9	41.5	59.4
Extremes														
-Record Highest	46	78	79	87	93	99	104	106	105	102	94	84	78	106
-Year		1949	1977	1963	1942	1941	1952	1952	1947	1954	1954	1961	1951	JUL 1952
-Record Lowest	46	-10	1	8	26	34	41	51	50	36	22	4	-2	-10
-Year		1985	1958	1960	1973	1971	1972	1972	1946	1967	1952	1950	1983	JAN 1985
NORMAL DEGREE DAYS:														
Heating (base 65°F)		815	644	478	171	56	6	0	0	11	190	483	729	3583
Cooling (base 65°F)		0	0	7	24	142	309	425	406	230	35	0	0	1578
% OF POSSIBLE SUNSHINE	55	43	49	52	60	65	65	61	62	64	62	53	44	57
MEAN SKY COVER (tenths)														
Sunrise - Sunset	55	6.7	6.5	6.4	5.9	5.8	5.7	6.0	5.7	5.4	4.9	5.7	6.6	5.9
MEAN NUMBER OF DAYS:														
Sunrise to Sunset														
-Clear	55	7.3	7.6	7.9	9.0	8.9	8.1	6.7	8.3	10.0	13.1	10.3	8.0	105.4
-Partly Cloudy	55	6.5	6.0	7.6	8.1	10.3	11.9	12.9	12.9	9.7	7.7	6.7	6.4	106.6
-Cloudy	55	17.2	14.7	15.5	12.9	11.8	10.0	11.3	9.8	10.3	10.1	13.0	16.7	153.2
Precipitation														
.01 inches or more	55	11.7	10.5	12.1	9.7	10.1	10.5	11.9	9.9	7.7	6.8	8.8	10.8	120.5
Snow,Ice pellets														
1.0 inches or more	55	0.8	0.5	0.2	0.*	0.0	0.0	0.0	0.0	0.0	0.0	0.*	0.3	1.9
Thunderstorms	55	1.2	1.8	3.7	4.7	7.2	9.4	11.4	9.3	3.8	1.4	1.2	0.6	55.5
Heavy Fog Visibility														
1/4 mile or less	55	3.1	1.7	1.8	1.7	2.4	2.1	2.1	2.8	3.8	6.1	4.0	3.3	34.9
Temperature °F														
-Maximum														
90° and above	45	0.0	0.0	0.0	0.1	2.6	10.4	15.9	13.8	4.8	0.3	0.0	0.0	47.9
32° and below	45	2.1	1.0	0.1	0.0	0.0	0.0	0.0	0.0	0.0	0.0	0.*	0.8	4.0
-Minimum														
32° and below	45	19.6	15.6	9.1	1.5	0.0	0.0	0.0	0.0	0.0	0.0	1.2	10.2	17.8	75.0
0° and below	45	0.2	0.0	0.0	0.0	0.0	0.0	0.0	0.0	0.0	0.0	0.0	0.*	0.2
AVG. STATION PRESS.(mb)	13	995.5	994.6	991.9	992.0	990.6	991.8	992.6	993.3	993.5	994.9	995.2	995.9	993.5
RELATIVE HUMIDITY (%)														
Hour 01	45	79	77	76	77	86	88	88	89	89	88	82	81	83
Hour 07	55	81	80	81	81	85	86	89	91	90	89	84	83	85
Hour 13 (Local Time)	55	62	57	53	49	53	54	57	57	56	53	56	62	56
Hour 19	55	67	59	55	51	57	59	64	66	66	67	66	69	62
PRECIPITATION (inches):														
Water Equivalent														
-Normal		5.20	4.69	6.31	4.57	4.00	3.32	4.55	3.41	4.30	2.92	4.19	5.14	52.60
-Maximum Monthly	46	12.28	11.03	16.32	11.92	9.21	9.40	11.78	7.54	14.18	9.91	13.59	13.68	16.32
-Year		1947	1944	1980	1964	1979	1949	1979	1975	1977	1949	1948	1961	MAR 1980
-Minimum Monthly	46	1.13	0.62	1.17	0.44	0.54	0.87	0.20	0.56	0.34	0.24	0.93	0.86	0.20
-Year		1961	1941	1967	1942	1941	1968	1957	1963	1941	1963	1953	1965	JUL 1957
-Maximum in 24 hrs	46	4.44	3.93	6.53	3.36	3.47	4.85	5.77	3.70	6.62	3.52	4.56	5.25	6.62
-Year		1949	1958	1973	1983	1979	1949	1979	1941	1977	1977	1948	1942	SEP 1977
Snow,Ice pellets														
-Maximum Monthly	55	8.4	10.4	10.1	1.0	T					T	2.8	9.1	10.4
-Year		1966	1960	1960	1971	1944					1954	1950	1963	FEB 1960
-Maximum in 24 hrs	55	8.4	8.7	6.0	1.0	T					T	2.8	8.9	8.9
-Year		1966	1960	1960	1971	1944					1954	1950	1963	DEC 1963
WIND:														
Mean Speed (mph)	45	7.1	7.5	8.0	7.5	6.0	5.4	5.1	4.6	4.9	5.0	6.1	6.5	6.2
Prevailing Direction														
through 1963		S	S	S	S	S	S	S	S	S	S	S	S	S
Fastest Obs. 1 Min.														
-Direction (!!!)	10	31	25	24	18	29	34	35	05	15	31	18	31	05
-Speed (MPH)	10	30	37	30	32	33	30	25	37	26	24	24	28	37
-Year		1978	1976	1978	1977	1977	1981	1980	1979	1979	1981	1979	1976	AUG 1979
Peak Gust														
-Direction (!!!)	2	W	W	W	SW	W	S	NW	NW	SW	SE	W	W	SW
-Speed (mph)	2	36	33	36	41	35	38	31	31	26	39	32	32	41
-Date		1984	1985	1984	1985	1984	1984	1984	1984	1984	1984	1984	1984	APR 1985

See Reference Notes to this table on the following pages.

TABLE 2 PRECIPITATION (inches) CHATTANOOGA, TENNESSEE

YEAR	JAN	FEB	MAR	APR	MAY	JUNE	JULY	AUG	SEP	OCT	NOV	DEC	ANNUAL
1956	2.60	9.48	3.80	5.52	3.01	1.42	3.49	4.50	1.43	3.85	2.11	5.12	46.33
1957	9.12	6.38	2.38	4.55	2.16	3.68	0.20	2.00	12.19	2.70	10.36	4.55	60.27
1958	4.07	3.18	4.88	5.64	2.46	4.07	5.90	5.09	4.41	0.86	2.03	1.81	44.40
1959	6.11	2.90	3.83	4.30	4.64	2.00	1.44	3.58	3.33	5.63	4.49	4.50	46.75
1960	4.96	4.02	6.36	2.69	1.09	3.59	2.72	2.52	5.92	3.77	2.68	4.19	44.51
1961	1.13	9.27	6.55	3.01	5.46	6.99	3.85	3.55	0.61	1.63	3.61	13.68	59.34
1962	7.79	9.47	4.42	4.46	1.68	5.31	4.89	1.17	3.48	3.38	6.06	3.28	55.39
1963	4.81	2.22	12.83	6.35	3.06	4.87	10.82	0.56	1.27	0.24	5.13	5.05	57.21
1964	5.98	4.51	10.44	11.92	4.28	4.06	4.99	4.87	2.06	3.53	4.03	3.59	64.26
1965	3.51	4.46	9.43	2.87	1.65	4.92	3.13	3.55	4.47	1.03	4.48	0.86	44.36
1966	3.67	6.50	2.99	2.93	4.00	1.03	4.51	4.16	6.40	4.77	2.65	3.58	47.19
1967	3.02	4.10	1.17	3.39	6.02	1.63	10.56	2.67	2.08	4.91	4.49	7.89	51.93
1968	4.98	0.94	3.70	3.72	3.93	0.87	1.92	2.90	2.82	2.52	2.81	3.98	35.09
1969	6.84	5.82	2.96	4.20	2.36	3.61	2.32	6.20	4.70	1.72	2.11	7.98	50.82
1970	2.45	1.80	4.08	7.36	2.76	4.69	3.88	2.92	2.92	5.56	2.03	4.30	44.75
1971	5.19	8.50	6.71	2.26	2.17	2.71	7.28	1.91	3.68	2.11	1.93	6.56	51.01
1972	8.64	4.53	5.72	3.46	6.47	3.95	7.53	5.31	3.19	4.80	3.81	7.09	64.50
1973	5.09	3.76	13.80	5.17	6.37	4.58	6.93	2.58	5.38	3.69	5.87	8.34	71.56
1974	8.00	5.68	3.44	3.09	6.53	1.19	4.74	6.10	1.30	1.71	5.04	4.51	51.33
1975	5.63	6.34	10.44	1.03	4.45	3.92	2.18	7.54	10.31	6.28	5.00	4.98	68.10
1976	4.32	2.10	5.41	1.02	5.86	6.76	5.57	1.38	3.35	3.55	2.92	5.54	47.78
1977	3.14	2.98	8.39	9.10	2.55	2.44	1.01	3.53	14.18	6.21	8.66	2.37	64.56
1978	5.79	0.72	4.71	4.78	4.17	2.59	0.80	4.17	1.62	0.32	2.84	7.57	40.08
1979	5.44	3.85	4.77	7.49	9.21	2.32	11.78	4.50	7.75	2.45	7.35	1.66	68.57
1980	4.69	2.16	16.32	4.78	4.25	2.67	1.28	0.96	4.62	1.56	4.68	0.90	48.87
1981	1.91	5.07	4.44	3.81	3.38	3.70	3.99	4.21	2.74	2.80	4.85	5.02	45.92
1982	8.29	5.95	4.80	4.48	1.44	2.76	4.79	6.42	1.34	1.98	7.65	8.05	57.95
1983	2.65	4.52	3.66	6.40	6.65	3.29	2.52	1.34	1.15	2.64	9.57	8.34	52.73
1984	3.25	4.40	3.75	6.28	6.57	2.52	6.32	2.48	0.51	6.38	3.44	1.84	47.74
1985	3.45	4.95	2.27	2.81	4.36	2.22	3.10	4.93	0.93	4.83	3.34	2.36	39.55
Record Mean	5.18	4.92	5.93	4.65	3.98	3.88	4.63	3.77	3.29	3.05	3.82	5.08	52.18

TABLE 3 AVERAGE TEMPERATURE (deg. F) BRISTOL, JOHNSON CITY, KINGSPORT, TENNESSEE

YEAR	JAN	FEB	MAR	APR	MAY	JUNE	JULY	AUG	SEP	OCT	NOV	DEC	ANNUAL
1956	32.2	44.5	46.9	53.4	66.8	71.3	74.9	74.1	65.6	61.3	44.9	47.5	57.0
1957	38.0	46.1	46.1	60.2	67.8	74.7	75.8	73.4	53.5	48.3	40.9	40.9	57.9
1958	32.5	28.1	42.9	55.6	65.5	72.2	76.6	74.1	67.8	56.2	48.7	34.2	54.5
1959	33.8	41.6	44.1	56.8	68.7	71.8	76.7	77.3	69.9	59.3	45.0	39.7	57.1
1960	38.3	36.7	33.5	59.1	62.9	72.3	74.6	75.8	71.0	59.4	46.4	32.5	55.2
#1961	31.6	43.9	50.6	51.1	61.0	69.8	74.1	74.1	70.9	56.5	50.7	39.2	56.1
1962	35.5	44.4	44.7	52.5	71.2	72.3	74.0	74.4	67.3	58.6	44.7	33.0	54.1
1963	30.6	31.9	50.2	57.6	64.2	70.4	71.1	72.5	68.4	62.4	47.1	27.8	54.5
1964	34.1	34.6	44.7	56.4	65.9	74.2	75.2	73.2	68.6	54.0	50.2	41.6	56.1
1965	36.5	37.3	43.6	57.9	68.1	71.2	74.5	73.9	71.3	55.7	47.6	38.9	56.4
1966	29.5	37.9	47.1	54.9	62.9	71.9	75.4	73.3	66.4	55.0	47.0	37.7	54.9
1967	38.2	33.7	50.2	57.7	58.7	68.9	69.8	70.7	62.6	56.1	42.7	40.8	54.2
1968	33.2	30.1	46.9	55.8	61.3	70.8	75.5	67.1	57.6	45.2	35.2	34.6	54.6
1969	34.3	35.8	38.6	54.7	63.0	73.1	76.4	73.0	66.2	55.7	41.4	34.5	53.9
1970	27.6	36.8	47.0	59.0	66.1	72.1	76.8	75.0	72.9	61.5	46.8	41.2	56.9
1971	35.6	36.9	43.0	55.1	61.5	73.9	74.0	73.7	72.3	63.4	46.5	47.5	57.0
1972	40.8	35.7	45.1	54.9	62.7	67.4	73.6	73.7	69.8	55.9	46.1	43.7	55.8
1973	35.8	36.9	53.9	53.9	60.7	73.1	74.7	73.4	71.4	60.3	48.8	39.3	56.8
1974	47.0	39.8	51.2	56.4	64.3	67.7	74.0	73.0	64.8	54.5	47.2	39.5	56.6
1975	40.4	42.8	44.7	54.2	67.1	71.1	74.8	76.1	67.8	59.0	49.8	40.0	57.3
1976	33.5	47.3	51.9	55.3	61.2	70.7	71.6	71.6	65.1	51.6	38.8	33.3	54.5
1977	22.1	34.5	50.0	59.4	66.8	71.2	76.8	74.8	69.6	53.3	49.9	35.3	55.3
1978	25.8	28.8	44.8	55.5	62.3	71.5	74.4	74.5	72.3	56.3	51.0	39.5	54.7
1979	30.4	31.9	48.3	55.4	63.8	68.2	70.9	73.1	66.8	54.6	48.7	38.8	54.2
1980	37.9	30.6	43.2	55.4	64.2	70.6	76.4	76.9	70.9	54.1	45.3	36.3	55.1
1981	29.3	37.9	42.6	60.4	61.3	75.1	75.3	72.1	65.0	53.7	45.8	34.5	54.4
1982	31.2	40.2	49.8	53.1	68.2	71.4	75.1	72.3	65.7	57.6	48.3	43.3	56.3
1983	35.1	38.2	46.4	50.5	60.9	69.5	74.2	74.7	65.9	57.1	45.3	34.8	54.4
1984	32.7	40.3	44.5	53.0	59.4	72.4	71.3	72.9	64.5	63.6	43.0	45.5	55.3
1985	27.6	35.5	48.2	57.0	63.8	69.5	73.2	71.2	66.1	61.4	55.3	33.9	55.2
Record Mean	35.5	38.7	46.5	55.6	64.3	71.9	74.8	73.9	68.2	57.6	46.9	38.3	55.9
Max	44.9	49.3	57.9	68.3	76.3	83.3	85.3	84.8	79.9	70.1	57.5	48.2	67.1
Min	26.0	28.0	35.0	42.9	52.3	60.4	64.2	63.0	56.4	45.1	35.0	28.4	44.7

REFERENCE NOTES FOR TABLES 1, 2, 3 and 6   (CHATTANOOGA, TN)

GENERAL

T - TRACE AMOUNT
BLANK ENTRIES DENOTE MISSING/UNREPORTED DATA.
# INDICATES A STATION OR INSTRUMENT RELOCATION.

SPECIFIC

TABLE 1

(a) - LENGTH OF RECORD IN YEARS. ALTHOUGH
INDIVIDUAL MONTHS MAY BE MISSING.
* LESS THAN .05

NORMALS — BASED ON THE 1951-1980 RECORD PERIOD.
EXTREMES — DATES ARE THE MOST RECENT OCCURRENCE.
WIND DIR. — NUMERALS SHOW TENS OF DEGREES
CLOCKWISE FROM TRUE NORTH.
"00" INDICATES CALM.
RESULTANT WIND DIRECTIONS ARE GIVEN TO WHOLE DEGREES.

EXCEPTIONS

TABLES 2, 3, and 6

RECORD MEANS ARE THROUGH THE CURRENT YEAR,
BEGINNING IN   1930 FOR TEMPERATURE
1930 FOR PRECIPITATION
1930 FOR SNOWFALL

TABLE 4 HEATING DEGREE DAYS Base 65 deg. F   CHATTANOOGA, TENNESSEE

SEASON	JULY	AUG	SEP	OCT	NOV	DEC	JAN	FEB	MAR	APR	MAY	JUNE	TOTAL
1956-57	0	0	15	96	514	416	688	398	475	143	37	0	2782
1957-58	0	0	19	269	429	657	876	865	570	178	40	0	3903
1958-59	0	0	8	174	403	819	829	580	540	171	25	0	3549
1959-60	0	0	1	148	536	651	709	748	801	140	129	0	3863
#1960-61	0	0	6	160	477	877	928	486	373	279	57	2	3645
1961-62	0	0	9	178	372	703	813	430	523	268	5	0	3301
1962-63	0	0	26	140	486	860	998	839	314	117	40	0	3820
1963-64	0	0	10	51	414	940	773	722	458	120	16	0	3504
1964-65	0	1	5	255	378	646	741	674	587	130	2	0	3419
1965-66	0	0	15	215	399	721	982	656	486	231	63	10	3778
1966-67	0	0	7	252	459	756	737	743	351	94	93	8	3500
1967-68	1	0	45	200	538	587	857	881	465	227	58	0	3859
1968-69	0	0	1	169	452	805	821	634	628	141	48	1	3700
1969-70	0	0	20	263	636	897	1017	729	556	194	47	1	4360
1970-71	0	0	13	119	535	636	807	698	606	206	73	0	3693
1971-72	0	0	0	45	496	478	661	721	496	215	62	10	3184
1972-73	0	0	10	253	567	674	890	767	294	246	80	0	3781
1973-74	0	0	3	143	374	722	480	604	322	242	65	16	2971
1974-75	0	0	28	250	517	746	686	581	593	290	14	0	3705
1975-76	0	0	53	152	474	785	934	470	340	180	87	2	3477
1976-77	0	0	11	269	661	851	1123	679	353	111	17	0	4075
1977-78	0	0	5	233	359	753	1072	821	467	114	60	0	3884
1978-79	0	0	0	192	299	637	962	725	390	152	42	0	3399
1979-80	0	0	0	167	438	728	745	813	526	215	38	0	3670
1980-81	0	0	24	246	450	736	907	600	515	70	81	0	3629
1981-82	0	0	31	211	422	823	890	597	371	275	13	0	3633
1982-83	0	0	29	202	418	528	807	616	447	321	42	0	3410
1983-84	0	0	38	146	487	863	910	626	515	246	100	2	3933
1984-85	0	0	27	40	550	488	990	687	381	158	28	4	3353
1985-86	0	0	32	67	198	869							

TABLE 5 COOLING DEGREE DAYS Base 65 deg. F   CHATTANOOGA, TENNESSEE

YEAR	JAN	FEB	MAR	APR	MAY	JUNE	JULY	AUG	SEP	OCT	NOV	DEC	TOTAL
1969	0	0	0	33	119	311	429	298	125	7	0	0	1322
1970	0	0	0	57	133	313	462	412	337	44	0	0	1758
1971	0	0	0	27	81	406	384	411	324	96	16	2	1747
1972	0	0	0	51	35	177	303	316	224	8	4	0	1118
1973	0	0	1	22	79	350	443	400	334	79	1	0	1709
1974	0	0	3	11	97	124	338	343	132	8	2	0	1058
1975	0	0	0	29	115	238	411	382	162	27	1	0	1365
1976	0	0	1	19	49	241	342	363	142	8	0	0	1165
1977	0	0	9	60	194	384	578	532	320	15	3	0	2095
1978	0	0	0	48	164	350	507	432	325	14	2	5	1847
1979	0	0	0	12	119	279	361	410	227	24	0	0	1432
1980	0	0	0	8	111	272	562	540	300	10	1	0	1804
1981	0	0	0	68	77	400	513	385	178	21	2	0	1644
1982	0	0	19	9	184	283	460	368	187	68	1	7	1586
1983	0	0	2	6	78	264	485	529	230	22	0	0	1616
1984	0	0	0	8	86	379	356	367	124	102	0	0	1422
1985	0	0	12	38	118	310	422	366	171	64	16	0	1517

TABLE 6 SNOWFALL (inches)   CHATTANOOGA, TENNESSEE

SEASON	JULY	AUG	SEP	OCT	NOV	DEC	JAN	FEB	MAR	APR	MAY	JUNE	TOTAL
1956-57	0.0	0.0	0.0	0.0	T	T	T	0.0	T	0.0	0.0	0.0	T
1957-58	0.0	0.0	0.0	0.0	T	T	T	6.7	T	0.0	0.0	0.0	6.7
1958-59	0.0	0.0	0.0	0.0	0.0	1.4	0.1	T	T	0.0	0.0	0.0	1.5
1959-60	0.0	0.0	0.0	0.0	T	T	1.3	10.4	10.1	0.0	0.0	0.0	21.8
1960-61	0.0	0.0	0.0	0.0	0.0	T	0.2	T	T	0.0	0.0	0.0	0.2
1961-62	0.0	0.0	0.0	0.0	0.0	1.8	5.2	T	T	0.0	0.0	0.0	7.0
1962-63	0.0	0.0	0.0	0.0	0.0	T	0.4	3.0	0.0	0.0	0.0	0.0	3.4
1963-64	0.0	0.0	0.0	0.0	T	9.1	3.4	0.1	0.0	0.0	0.0	0.0	12.6
1964-65	0.0	0.0	0.0	0.0	T	0.0	1.8	0.9	0.7	0.0	0.0	0.0	3.4
1965-66	0.0	0.0	0.0	0.0	0.0	0.0	8.4	T	T	0.0	0.0	0.0	8.4
1966-67	0.0	0.0	0.0	0.0	T	T	T	1.8	T	0.0	0.0	0.0	1.8
1967-68	0.0	0.0	0.0	0.0	0.0	T	5.6	4.9	T	0.0	0.0	0.0	10.5
1968-69	0.0	0.0	0.0	0.0	T	T	7.3	0.8	0.2	0.0	0.0	0.0	8.3
1969-70	0.0	0.0	0.0	0.0	T	5.1	7.9	T	T	0.0	0.0	0.0	13.0
1970-71	0.0	0.0	0.0	0.0	0.0	3.4	0.6	0.6	4.5	1.0	0.0	0.0	10.1
1971-72	0.0	0.0	0.0	0.0	T	T	T	1.0	0.0	0.0	0.0	0.0	1.0
1972-73	0.0	0.0	0.0	0.0	T	0.0	2.3	T	T	0.0	0.0	0.0	2.3
1973-74	0.0	0.0	0.0	0.0	0.0	0.2	0.0	T	T	0.0	0.0	0.0	0.2
1974-75	0.0	0.0	0.0	0.0	0.0	T	T	T	T	0.0	0.0	0.0	T
1975-76	0.0	0.0	0.0	0.0	T	T	0.1	T	0.0	0.0	0.0	0.0	0.1
1976-77	0.0	0.0	0.0	0.0	T	T	3.7	T	0.4	0.0	0.0	0.0	3.7
1977-78	0.0	0.0	0.0	0.0	T	T	2.5	0.7	0.4	0.0	0.0	0.0	3.6
1978-79	0.0	0.0	0.0	0.0	0.0	0.0	0.5	8.7	0.0	0.0	0.0	0.0	9.2
1979-80	0.0	0.0	0.0	0.0	T	0.0	T	7.4	0.4	0.0	0.0	0.0	7.8
1980-81	0.0	0.0	0.0	0.0	T	0.0	T	T	T	0.0	0.0	0.0	T
1981-82	0.0	0.0	0.0	0.0	0.0	0.1	2.9	2.3	T	0.0	0.0	0.0	5.3
1982-83	0.0	0.0	0.0	0.0	0.0	T	1.0	0.9	0.6	0.0	0.0	0.0	2.5
1983-84	0.0	0.0	0.0	0.0	0.0	T	0.6	1.8	T	0.0	0.0	0.0	2.4
1984-85	0.0	0.0	0.0	0.0	0.0	T	3.8	1.1	0.0	0.0	0.0	0.0	4.9
1985-86	0.0	0.0	0.0	0.0	0.0	0.3							
Record Mean	0.0	0.0	0.0	T	0.1	0.6	1.7	1.3	0.4	T	T	0.0	4.1

See Reference Notes, relative to all above tables, on preceding page.

Knoxville is located in a broad valley between the Cumberland Mountains, which lie northwest of the city, and the Great Smoky Mountains, which lie southeast of the city. These two mountain ranges exercise a marked influence upon the climate of the valley. The Cumberland Mountains, to the northwest, serve to retard and weaken the force of the cold winter air which frequently penetrates far south of the latitude of Knoxville over the plains areas to the west of the mountains.

The mountains also serve to modify the hot summer winds which are common to the plains to the west. In addition, they serve as a fixed incline plane which lifts the warm, moist air flowing northward from the Gulf of Mexico and thereby increases the frequency of afternoon thunderstorms. Relief from extremely high temperatures which such thunderstorms produce serves to reduce the number of extremely warm days in the valley.

July is usually the warmest month of the year. The coldest weather usually occurs during the month of January. Sudden great temperature changes occur infrequently. This again is due mainly to the retarding effect of the mountains. Summer nights are nearly always comfortable.

Rainfall is ample for agricultural purposes and is favorably distributed during the year for most crops. Precipitation is greatest in the wintertime. Another peak period occurs during the late spring and summer months. The period of lowest rainfall occurs during the fall. A cumulative total of approximately 12 inches of snow falls annually. However, this usually comes in amounts of less than 4 inches at one time. It is unusual for snow to remain on the ground in measurable amounts longer than one week.

The topography also has a pronounced effect upon the prevailing wind direction. Daytime winds usually have a southwesterly component, while nighttime winds usually move from the northeast. The winds are relatively light and tornadoes are extremely rare.

TABLE 1 NORMALS, MEANS AND EXTREMES

KNOXVILLE, TENNESSEE

LATITUDE: 35°49'N   LONGITUDE: 83°59'W   ELEVATION: FT. GRND   905 BARO   00918   TIME ZONE: EASTERN   WBAN: 13891

	(a)	JAN	FEB	MAR	APR	MAY	JUNE	JULY	AUG	SEP	OCT	NOV	DEC	YEAR
TEMPERATURE °F:														
Normals														
-Daily Maximum		46.9	51.2	60.1	71.0	78.3	84.6	87.2	86.9	81.7	70.9	59.1	50.3	69.0
-Daily Minimum		29.5	31.7	39.3	48.2	56.5	64.0	68.0	67.1	61.2	48.1	38.4	31.9	48.7
-Monthly		38.2	41.5	49.7	59.6	67.4	74.3	77.6	77.0	71.5	59.5	48.8	41.1	58.8
Extremes														
-Record Highest	44	77	83	86	92	94	102	103	102	103	91	84	80	103
-Year		1950	1977	1963	1942	1962	1944	1952	1944	1954	1953	1948	1982	SEP 1954
-Record Lowest	44	-24	-2	1	23	34	43	51	49	36	25	5	-6	-24
-Year		1985	1958	1980	1985	1963	1956	1961	1946	1967	1962	1950	1983	JAN 1985
NORMAL DEGREE DAYS:														
Heating (base 65°F)		831	658	483	181	63	0	0	0	14	201	486	741	3658
Cooling (base 65°F)		0	0	8	19	137	283	391	372	209	30	0	0	1449
% OF POSSIBLE SUNSHINE	43	40	47	52	62	63	64	62	62	60	60	49	40	55
MEAN SKY COVER (tenths)														
Sunrise - Sunset	43	7.1	6.7	6.7	6.1	6.1	5.8	6.0	5.6	5.6	5.0	6.2	6.7	6.1
MEAN NUMBER OF DAYS:														
Sunrise to Sunset														
-Clear	43	6.1	6.6	7.1	8.0	8.0	7.5	6.6	8.7	9.5	12.3	8.8	7.1	96.1
-Partly Cloudy	43	6.6	6.1	7.5	8.7	9.7	12.6	12.9	11.9	9.6	7.8	6.6	6.8	106.8
-Cloudy	43	18.3	15.5	16.4	13.3	13.3	10.0	11.5	10.5	10.9	10.9	14.7	17.1	162.3
Precipitation														
.01 inches or more	43	12.4	11.3	12.8	11.1	11.0	10.3	11.5	9.7	8.3	7.7	10.0	10.8	127.0
Snow, Ice pellets														
1.0 inches or more	43	1.4	1.1	0.5	0.1	0.0	0.0	0.0	0.0	0.0	0.0	0.2	0.6	3.9
Thunderstorms	43	0.8	1.4	3.2	4.2	6.5	8.2	9.7	6.8	3.0	1.4	1.0	0.5	46.8
Heavy Fog Visibility														
1/4 mile or less	43	3.0	1.7	1.6	1.1	2.3	2.1	2.1	3.4	4.0	4.9	3.0	2.6	31.8
Temperature °F														
-Maximum														
90° and above	25	0.0	0.0	0.0	0.0	0.8	4.1	9.0	6.9	2.6	0.0	0.0	0.0	23.4
32° and below	25	3.8	1.4	0.1	0.0	0.0	0.0	0.0	0.0	0.0	0.0	0.1	1.4	6.8
-Minimum														
32° and below	25	20.6	17.2	8.3	1.4	0.0	0.0	0.0	0.0	0.0	0.6	7.7	17.3	73.1
0° and below	25	0.5	0.*	0.0	0.0	0.0	0.0	0.0	0.0	0.0	0.0	0.0	0.1	0.7
AVG. STATION PRESS.(mb)	13	984.4	983.7	981.2	981.5	980.4	981.7	982.5	983.3	983.4	984.6	984.6	984.9	983.0
RELATIVE HUMIDITY (%)														
Hour 01	25	77	73	71	71	82	86	87	88	87	84	79	77	80
Hour 07	25	80	79	79	80	86	88	90	92	91	89	83	82	85
Hour 13 (Local Time)	25	64	59	55	52	57	59	62	61	60	56	59	63	59
Hour 19	25	66	59	56	52	60	63	66	67	67	64	66	68	63
PRECIPITATION (inches):														
Water Equivalent														
-Normal		4.65	4.18	5.49	3.87	3.71	3.95	4.33	3.02	2.99	2.73	3.78	4.59	47.29
-Maximum Monthly	44	11.74	9.38	10.42	7.20	10.98	7.58	10.09	8.88	8.61	6.67	10.36	11.63	11.74
-Year		1954	1944	1975	1970	1974	1969	1967	1942	1944	1949	1948	1961	JAN 1954
-Minimum Monthly	44	1.05	0.74	1.98	0.39	0.74	0.20	0.70	0.77	0.42	T	0.97	0.45	T
-Year		1981	1968	1985	1976	1970	1944	1957	1954	1985	1963	1942	1965	OCT 1963
-Maximum in 24 hrs	44	3.89	2.89	4.85	3.65	3.40	3.57	4.69	3.25	5.08	2.44	4.06	4.89	5.08
-Year		1946	1956	1973	1977	1984	1972	1942	1959	1944	1961	1948	1969	SEP 1944
Snow, Ice pellets														
-Maximum Monthly	44	15.1	23.3	20.2	7.0	T					T	18.2	12.2	23.3
-Year		1962	1960	1960	1971	1945					1962	1952	1963	FEB 1960
-Maximum in 24 hrs	44	12.0	17.5	12.1	7.0	T					T	18.2	8.9	18.2
-Year		1962	1960	1942	1971	1945					1962	1952	1969	NOV 1952
WIND:														
Mean Speed (mph)	43	8.0	8.4	8.9	8.8	7.1	6.6	6.1	5.6	5.7	5.8	6.9	7.4	7.1
Prevailing Direction														
through 1963		NE	NE	NE	WSW	SW	SW	WSW	NE	NE	NE	NE	NE	NE
Fastest Obs. 1 Min.														
-Direction	11	27	26	22	23	19	07	36	32	30	30	26	25	22
-Speed (MPH)	11	35	32	36	32	31	35	35	29	23	28	29	35	36
-Year		1978	1977	1977	1983	1985	1975	1985	1983	1984	1983	1982	1977	MAR 1977
Peak Gust														
-Direction	1	W	SW	NW	W	SW	NE	NE	N	NW	W	NW	W	W
-Speed (mph)	1	39	44	46	59	47	51	46	32	32	29	35	38	59
-Date		1984	1984	1984	1985	1985	1985	1985	1985	1984	1985	1984	1984	APR 1985

See Reference Notes to this table on the following page.

TABLE 2 PRECIPITATION (inches)   KNOXVILLE, TENNESSEE

YEAR	JAN	FEB	MAR	APR	MAY	JUNE	JULY	AUG	SEP	OCT	NOV	DEC	ANNUAL
1956	3.18	8.29	4.58	5.07	3.27	2.69	4.54	1.45	1.66	2.59	1.98	8.06	47.36
1957	9.63	7.86	2.26	4.56	3.53	5.86	0.70	4.07	6.98	3.43	7.45	5.16	61.49
1958	2.13	2.81	4.29	5.41	4.78	2.01	5.57	2.83	1.49	0.68	3.06	2.19	37.25
1959	4.09	4.06	4.12	4.17	3.53	2.78	2.97	4.91	0.93	5.00	4.68	4.03	45.27
1960	3.39	3.00	4.78	1.83	1.12	4.89	2.36	3.72	4.58	4.34	2.55	3.61	40.17
1961	2.55	7.82	7.80	2.50	4.18	4.54	3.49	3.04	0.50	3.41	3.04	11.64	54.51
1962	6.22	7.96	4.13	3.84	1.88	5.05	7.82	2.08	5.13	2.20	4.60	3.67	54.58
1963	4.20	3.54	9.92	4.51	3.32	3.64	4.92	4.92	2.37	T	5.23	2.86	48.14
1964	4.71	4.09	5.75	6.98	3.96	0.97	3.70	5.75	1.10	2.10	2.80	4.02	45.93
1965	3.94	3.25	9.31	4.16	3.13	4.84	3.49	2.18	2.48	1.08	2.50	0.45	40.81
1966	3.88	4.68	2.72	2.88	2.92	2.27	5.44	4.09	4.41	4.80	5.12	2.66	45.87
1967	2.67	4.58	4.08	2.00	4.10	6.53	10.09	4.06	2.70	2.33	5.57	6.95	55.66
1968	4.13	0.74	4.78	4.12	3.01	3.97	2.57	1.29	2.53	3.40	2.00	3.22	35.76
1969	4.11	5.54	2.89	2.41	1.33	7.58	3.51	6.72	3.03	1.56	2.56	7.74	48.98
1970	3.04	2.86	3.18	7.20	0.74	4.26	3.11	4.89	2.75	5.33	1.40	4.67	43.43
1971	5.03	4.93	4.21	3.87	3.78	3.73	8.76	3.05	3.41	1.98	2.21	5.48	50.44
1972	7.35	4.19	4.98	2.54	4.49	5.02	6.76	1.61	4.70	5.99	3.36	7.02	58.01
1973	3.24	2.59	10.24	5.15	5.71	5.26	4.38	2.31	3.28	3.48	5.01	7.38	58.03
1974	7.05	5.24	6.15	5.77	10.98	2.70	2.92	3.14	3.33	2.35	5.18	4.52	59.33
1975	4.66	4.68	10.42	2.43	2.98	2.43	2.25	1.61	3.28	4.02	2.92	3.59	45.27
1976	3.86	2.18	5.22	0.39	5.53	3.46	3.75	1.98	2.87	5.33	3.45	4.42	42.44
1977	2.55	1.52	6.08	6.96	1.16	6.49	1.08	5.78	6.91	4.04	5.06	3.30	50.93
1978	5.22	1.01	4.42	4.10	3.44	5.27	5.06	2.44	1.26	0.82	3.62	5.91	42.57
1979	6.18	4.17	4.21	4.30	7.21	3.80	9.47	2.29	2.64	1.97	5.73	1.92	53.89
1980	5.54	1.78	8.72	3.30	3.80	1.94	3.57	2.34	2.38	1.53	3.78	1.78	40.46
1981	1.05	3.62	2.83	4.84	3.02	5.53	2.03	3.48	6.09	4.15	3.01	4.14	43.79
1982	6.03	4.88	6.36	3.26	5.52	3.93	6.60	2.68	2.68	2.66	5.21	4.89	54.70
1983	1.58	2.90	1.99	5.88	5.42	3.26	3.18	3.89	0.95	3.34	4.40	5.69	42.48
1984	2.26	4.42	3.79	3.37	10.14	4.34	9.03	1.72	0.85	3.26	2.87	2.49	48.54
1985	3.17	4.11	1.98	2.86	1.60	4.77	2.63	4.07	0.42	3.04	5.39	2.36	36.40
Record Mean	4.58	4.52	5.15	4.14	3.83	4.07	4.53	3.65	2.81	2.67	3.42	4.37	47.74

TABLE 3 AVERAGE TEMPERATURE (deg. F)   KNOXVILLE, TENNESSEE

YEAR	JAN	FEB	MAR	APR	MAY	JUNE	JULY	AUG	SEP	OCT	NOV	DEC	ANNUAL
1956	36.1	47.3	50.0	56.6	69.9	74.6	77.5	78.4	69.1	63.6	47.2	51.4	60.1
1957	41.4	49.6	48.8	62.8	69.9	76.7	79.6	77.7	72.7	56.6	43.9	50.9	60.9
1958	35.6	31.9	45.9	59.0	68.2	75.7	78.5	77.2	72.1	59.8	51.4	37.4	57.8
1959	37.3	43.7	47.1	60.0	71.5	75.0	78.5	79.3	73.2	62.4	46.7	42.6	59.8
1960	40.8	38.6	36.8	61.3	65.9	74.8	78.4	78.2	73.1	61.8	48.8	35.2	57.8
#1961	34.2	46.7	51.6	52.7	62.1	70.7	74.4	74.2	72.4	57.9	51.7	39.9	57.4
1962	37.0	47.2	46.8	54.7	74.0	74.4	78.3	78.2	70.9	62.5	46.4	37.0	59.0
1963	33.4	35.0	55.1	60.8	65.9	72.4	73.6	75.5	69.4	62.0	47.3	30.2	56.7
1964	36.3	36.1	47.7	60.5	66.9	75.1	75.7	75.7	70.0	55.5	52.3	43.5	57.8
1965	40.1	39.7	46.8	62.4	72.3	73.9	78.0	77.5	73.5	59.0	51.7	43.0	59.9
1966	34.0	41.6	50.4	58.9	66.1	74.3	79.3	75.0	68.9	57.5	49.1	40.0	57.9
1967	41.9	38.1	54.7	62.7	63.5	73.2	71.8	73.2	65.6	58.2	44.2	44.6	57.6
1968	37.1	33.3	49.9	59.0	65.3	73.6	77.2	78.4	69.1	58.8	48.1	37.0	57.0
1969	36.5	39.4	42.1	59.5	67.3	74.5	78.9	74.5	69.3	59.1	45.5	37.7	57.0
1970	30.6	39.3	48.7	61.2	68.6	73.4	78.5	77.1	75.3	62.8	48.1	44.1	59.0
1971	38.5	39.4	45.6	58.5	63.9	75.9	75.4	76.0	73.3	65.3	48.7	50.4	59.2
1972	42.5	39.4	48.2	59.5	64.9	70.8	75.5	75.9	71.9	56.8	47.5	44.8	58.1
1973	38.0	39.9	55.8	56.4	63.5	74.8	76.7	76.2	73.8	62.4	52.4	41.1	59.2
1974	49.3	43.1	55.2	59.8	68.1	70.5	77.7	76.5	69.8	58.6	50.8	41.9	60.1
1975	43.3	45.5	46.6	57.5	70.1	74.6	77.6	78.8	68.9	60.1	50.2	41.1	59.5
1976	34.3	48.2	52.6	58.9	63.4	73.9	76.2	74.9	67.7	54.5	43.2	37.5	57.1
1977	27.2	41.4	55.0	62.9	69.8	75.3	80.3	78.9	74.0	57.4	52.6	40.2	59.6
1978	29.4	34.6	49.0	61.1	66.6	75.4	78.7	77.1	74.8	58.2	54.4	42.8	58.5
1979	33.0	36.7	53.1	59.9	66.9	72.8	75.0	77.6	72.2	58.3	51.4	42.2	58.3
1980	41.6	36.6	47.6	59.1	68.1	75.2	82.1	81.7	73.7	56.0	47.3	40.1	59.1
1981	33.4	41.9	46.8	63.5	63.9	77.1	81.4	78.3	70.2	58.9	49.4	38.7	58.6
1982	33.8	43.1	53.0	55.2	69.3	72.6	77.6	76.7	70.0	59.5	50.9	46.4	59.0
1983	38.0	41.1	49.7	53.2	65.4	72.5	78.5	79.6	70.3	60.2	47.4	36.4	57.7
1984	35.3	43.4	47.7	57.3	62.8	74.8	73.3	74.1	67.1	67.6	46.3	47.5	58.1
1985	29.4	37.1	50.2	58.6	65.7	72.5	76.5	75.1	68.5	63.7	56.8	34.8	57.4
Record Mean	38.9	41.7	49.4	58.8	67.3	74.8	77.7	76.8	71.4	59.9	48.3	40.7	58.8
Max	47.4	51.0	59.6	69.6	78.1	85.0	87.5	86.6	81.8	71.0	58.3	49.2	68.8
Min	30.4	32.4	39.2	47.9	56.4	64.5	67.9	66.9	60.9	48.7	38.3	32.1	48.8

REFERENCE NOTES FOR TABLES 1, 2, 3 and 6   (KNOXVILLE, TN)

GENERAL

T - TRACE AMOUNT
BLANK ENTRIES DENOTE MISSING/UNREPORTED DATA.
# INDICATES A STATION OR INSTRUMENT RELOCATION.

SPECIFIC

TABLE 1

(a) - LENGTH OF RECORD IN YEARS. ALTHOUGH
INDIVIDUAL MONTHS MAY BE MISSING.
* LESS THAN .05

NORMALS — BASED ON THE 1951-1980 RECORD PERIOD.
EXTREMES — DATES ARE THE MOST RECENT OCCURRENCE.
WIND DIR. — NUMERALS SHOW TENS OF DEGREES
CLOCKWISE FROM TRUE NORTH.
"00" INDICATES CALM.
RESULTANT WIND DIRECTIONS ARE GIVEN TO WHOLE DEGREES.

EXCEPTIONS

TABLES 2, 3, and 6

RECORD MEANS ARE THROUGH THE CURRENT YEAR,
BEGINNING IN   1871 FOR TEMPERATURE
1871 FOR PRECIPITATION
1943 FOR SNOWFALL

TABLE 4   HEATING DEGREE DAYS Base 65 deg. F   KNOXVILLE, TENNESSEE

SEASON	JULY	AUG	SEP	OCT	NOV	DEC	JAN	FEB	MAR	APR	MAY	JUNE	TOTAL
1956-57	0	0	31	75	524	415	728	425	496	168	33	0	2895
1957-58	0	0	23	264	430	648	902	920	586	188	53	0	4014
1958-59	0	0	5	169	412	847	854	589	550	182	30	0	3638
1959-60	0	0	0	157	547	686	744	758	869	158	108	0	4027
#1960-61	0	0	3	.142	481	917	947	507	400	377	112	6	3892
1961-62	0	0	15	222	405	771	862	492	556	324	5	0	3652
1962-63	0	0	32	156	548	861	973	835	312	176	64	0	3957
1963-64	0	0	18	101	525	1072	884	834	532	160	44	1	4171
1964-65	0	8	9	308	374	657	765	701	559	125	0	0	3506
1965-66	0	0	13	200	393	674	956	650	452	217	59	9	3623
1966-67	0	0	9	232	472	767	709	748	330	112	121	8	3508
1967-68	1	0	63	229	617	626	860	912	467	191	66	0	4032
1968-69	0	0	5	215	501	859	879	710	704	175	50	4	4102
1969-70	0	0	11	200	579	840	1063	712	496	139	45	0	4085
1970-71	0	0	13	109	497	640	813	710	596	209	76	0	3663
1971-72	0	0	0	53	496	450	692	735	516	211	47	11	3211
1972-73	0	0	9	251	522	622	828	697	283	272	97	0	3581
1973-74	0	0	10	135	373	734	481	606	304	190	41	10	2884
1974-75	0	0	23	215	439	713	666	542	563	257	9	0	3427
1975-76	0	0	42	172	437	733	944	485	380	203	91	0	3487
1976-77	0	0	19	320	651	845	1166	658	319	110	33	2	4123
1977-78	0	0	3	242	374	764	1097	846	487	148	74	0	4035
1978-79	0	0	0	210	310	681	985	786	376	167	46	0	3561
1979-80	0	0	0	220	407	700	715	815	529	185	32	1	3604
1980-81	0	0	23	284	523	761	974	641	556	104	94	0	3960
1981-82	0	0	32	196	461	809	959	606	384	295	20	0	3762
1982-83	0	0	30	228	416	577	829	662	472	356	50	3	3623
1983-84	0	0	51	163	523	878	912	619	530	240	139	5	4060
1984-85	0	0	44	30	556	536	1095	776	459	208	51	13	3768
1985-86	0	0	44	86	250	927							

TABLE 5   COOLING DEGREE DAYS Base 65 deg. F   KNOXVILLE, TENNESSEE

YEAR	JAN	FEB	MAR	APR	MAY	JUNE	JULY	AUG	SEP	OCT	NOV	DEC	TOTAL
1969	0	0	0	15	125	294	438	304	147	23	0	0	1346
1970	0	0	0	33	162	258	427	384	326	46	0	0	1636
1971	0	0	0	20	50	333	328	350	252	69	13	3	1418
1972	0	0	0	53	51	193	332	347	223	4	2	0	1205
1973	0	0	4	19	56	298	369	355	282	60	2	0	1445
1974	0	0	9	39	142	179	399	365	171	19	17	0	1340
1975	0	0	0	39	173	293	397	434	168	26	0	0	1530
1976	0	3	1	26	48	273	356	314	107	5	0	0	1133
1977	0	5	14	57	188	317	483	440	283	14	10	0	1811
1978	0	0	0	37	130	319	432	384	302	8	0	0	1612
1979	0	0	15	20	111	242	317	399	224	23	4	0	1355
1980	0	0	0	16	136	315	538	525	290	12	0	0	1832
1981	0	0	0	62	65	373	512	421	193	14	1	0	1641
1982	0	0	17	6	157	233	396	372	187	65	2	8	1443
1983	0	0	4	6	71	235	425	462	217	23	0	0	1443
1984	0	0	0	12	77	306	263	290	116	120	2	0	1186
1985	0	0	11	20	81	247	363	322	155	53	11	0	1263

TABLE 6   SNOWFALL (inches)   KNOXVILLE, TENNESSEE

SEASON	JULY	AUG	SEP	OCT	NOV	DEC	JAN	FEB	MAR	APR	MAY	JUNE	TOTAL
1956-57	0.0	0.0	0.0	0.0	0.6	0.4	T	0.0	3.1	T	0.0	0.0	4.1
1957-58	0.0	0.0	0.0	0.0	T	0.6	0.3	10.3	T	0.0	0.0	0.0	11.2
1958-59	0.0	0.0	0.0	0.0	T	4.3	2.5	T	0.3	0.0	0.0	0.0	7.1
1959-60	0.0	0.0	0.0	0.0	0.6	3.7	8.9	23.3	20.2	T	0.0	0.0	56.7
1960-61	0.0	0.0	0.0	0.0	0.0	2.0	5.4	3.4	0.1	T	0.0	0.0	10.9
1961-62	0.0	0.0	0.0	0.0	T	4.5	15.1	T	1.9	T	0.0	0.0	21.5
1962-63	0.0	0.0	0.0	T		6.9	4.2	7.2	T	0.0	0.0	0.0	18.3
1963-64	0.0	0.0	0.0	0.0	5.9	12.2	8.5	5.2	T	0.0	0.0	0.0	31.8
1964-65	0.0	0.0	0.0	0.0	T	T	6.9	6.9	4.4	0.0	0.0	0.0	17.6
1965-66	0.0	0.0	0.0	0.0	T	T	14.2	T	0.5	0.0	0.0	0.0	14.7
1966-67	0.0	0.0	0.0	0.0	T	3.1	2.2	5.7	0.0	0.0	0.0	0.0	11.0
1967-68	0.0	0.0	0.0	0.0	0.0	T	5.9	4.7	1.9	0.0	0.0	0.0	12.5
1968-69	0.0	0.0	0.0	0.0	0.3	T	0.5	7.7	5.2	0.0	0.0	0.0	13.7
1969-70	0.0	0.0	0.0	0.0	T	8.9	12.9	4.6	T	0.0	0.0	0.0	26.4
1970-71	0.0	0.0	0.0	0.0	T	6.3	1.3	6.4	3.6	7.0	0.0	0.0	24.6
1971-72	0.0	0.0	0.0	0.0	T	T	0.4	4.2	6.7	0.0	0.0	0.0	11.3
1972-73	0.0	0.0	0.0	0.0	T	T	9.0	T	1.9	T	0.0	0.0	10.9
1973-74	0.0	0.0	0.0	0.0	0.0	2.0	T	T	1.8	0.0	0.0	0.0	3.8
1974-75	0.0	0.0	0.0	0.0	T	1.6	1.3	T	2.5	0.0	0.0	0.0	5.4
1975-76	0.0	0.0	0.0	0.0	T	T	0.3	2.8	0.0	0.0	0.0	0.0	3.1
1976-77	0.0	0.0	0.0	0.0	1.0	2.1	7.9	0.2	0.0	0.0	0.0	0.0	11.2
1977-78	0.0	0.0	0.0	0.0	0.1	0.6	11.3	5.5	1.8	0.0	0.0	0.0	19.3
1978-79	0.0	0.0	0.0	0.0	0.0	T	4.7	18.4	T	T	0.0	0.0	23.1
1979-80	0.0	0.0	0.0	0.0	T	T	0.5	11.0	3.5	T	0.0	0.0	15.0
1980-81	0.0	0.0	0.0	0.0	T	T	5.0	2.5	T	0.0	0.0	0.0	7.5
1981-82	0.0	0.0	0.0	0.0	T	0.1	4.4	0.1	1.1	T	0.0	0.0	5.7
1982-83	0.0	0.0	0.0	0.0	0.0	3.0	1.1	3.4	0.7	2.0	0.0	0.0	10.2
1983-84	0.0	0.0	0.0	0.0	0.0	0.7	2.8	3.5	T	0.0	0.0	0.0	7.0
1984-85	0.0	0.0	0.0	0.0	0.0	T	14.2	8.3	0.0	0.0	0.0	0.0	22.5
1985-86	0.0	0.0	0.0	0.0	0.0	0.4							
Record Mean	0.0	0.0	0.0	T	0.7	1.7	4.2	4.0	1.7	0.2	T	0.0	12.5

See Reference Notes, relative to all above tables, on preceding page.

Topography varies from the level alluvial area in east–central Arkansas to the slightly rolling area in northwestern Mississippi and southwestern Tennessee.

Agricultural interests are varied, with major crops being cotton, corn, hay, soybeans, peaches, apples, and a considerable number of vegetables. The climate is quite favorable for dairy interests, and for the raising of cattle and hogs.

The growing season is about 230 days in length. The average date for the last occurrence of temperatures as low as 32 degrees is late March. The average date of the first temperature of 32 degrees or below is early November.

Precipitation of nearly 50 inches per year is fairly well distributed. Crops and pastures receive, on the average, an adequate supply of moisture during the growing season, with lesser amounts during the fall harvesting period.

Sunshine averages slightly over 70 percent of the possible amount during the growing season. Relative humidity averages about 70 percent for the year.

Memphis, although not in the normal paths of storms coming from the Gulf or from western Canada, is affected by both, and thereby has comparatively frequent changes in weather. Extremely high or low temperatures, however, are relatively rare.

TABLE 1   NORMALS, MEANS AND EXTREMES

MEMPHIS, TENNESSEE

LATITUDE: 35°03'N   LONGITUDE: 90°00'W   ELEVATION: FT. GRND 258 BARO 271   TIME ZONE: CENTRAL   WBAN: 13893

	(a)	JAN	FEB	MAR	APR	MAY	JUNE	JULY	AUG	SEP	OCT	NOV	DEC	YEAR
TEMPERATURE °F:														
Normals														
-Daily Maximum		48.3	53.0	61.4	72.9	81.0	88.4	91.5	90.3	84.3	74.5	61.4	52.3	71.6
-Daily Minimum		30.9	34.1	41.9	52.2	60.9	68.9	72.6	70.8	64.1	51.3	41.1	34.3	51.9
-Monthly		39.6	43.5	51.7	62.5	71.0	78.7	82.1	80.6	74.2	62.9	51.3	43.3	61.8
Extremes														
-Record Highest	44	78	81	85	91	99	104	108	105	103	95	85	81	108
-Year		1972	1962	1963	1952	1977	1954	1980	1943	1954	1954	1955	1982	JUL 1980
-Record Lowest	44	-4	-11	12	29	38	48	52	48	36	25	9	-13	-13
-Year		1985	1951	1943	1944	1944	1966	1947	1946	1949	1952	1950	1963	DEC 1963
NORMAL DEGREE DAYS:														
Heating (base 65°F)		787	602	433	126	25	0	0	0	9	137	415	673	3207
Cooling (base 65°F)		0	0	20	54	211	411	530	484	285	72	0	0	2067
% OF POSSIBLE SUNSHINE	35	50	54	56	64	69	74	74	75	69	70	58	50	64
MEAN SKY COVER (tenths)														
Sunrise - Sunset	37	6.8	6.4	6.5	6.0	5.9	5.3	5.3	5.0	5.0	4.6	5.6	6.4	5.7
MEAN NUMBER OF DAYS:														
Sunrise to Sunset														
-Clear	35	7.9	7.8	8.0	8.8	8.7	10.1	10.4	11.8	12.5	14.3	10.1	8.8	119.2
-Partly Cloudy	35	5.9	5.8	6.6	7.1	9.7	11.1	11.8	11.4	7.7	7.1	6.4	5.9	96.5
-Cloudy	35	17.2	14.7	16.4	14.1	12.6	8.8	8.8	7.8	9.9	9.6	13.6	16.3	149.5
Precipitation														
.01 inches or more	35	10.1	9.5	11.0	10.4	9.2	8.3	8.7	8.0	7.1	6.2	8.6	9.7	106.8
Snow,Ice pellets														
1.0 inches or more	35	0.9	0.6	0.3	0.0	0.0	0.0	0.0	0.0	0.0	0.0	0.*	0.2	1.9
Thunderstorms	35	1.9	2.3	4.7	6.6	6.7	7.2	8.2	6.3	3.4	1.9	2.3	1.6	53.1
Heavy Fog Visibility														
1/4 mile or less	35	2.2	1.3	0.8	0.3	0.2	0.2	0.3	0.4	0.7	1.1	1.3	1.8	10.6
Temperature °F														
-Maximum														
90° and above	44	0.0	0.0	0.0	0.*	3.0	14.5	21.5	18.1	7.5	0.6	0.0	0.0	65.2
32° and below	44	3.5	1.2	0.2	0.0	0.0	0.0	0.0	0.0	0.0	0.0	0.1	1.3	6.3
-Minimum														
32° and below	44	18.5	12.8	5.4	0.3	0.0	0.0	0.0	0.0	0.0	0.4	6.0	14.6	57.9
0° and below	44	0.1	0.*	0.0	0.0	0.0	0.0	0.0	0.0	0.0	0.0	0.0	0.1	0.2
AVG. STATION PRESS.(mb)	13	1011.5	1010.0	1006.0	1005.9	1004.3	1005.4	1006.3	1006.9	1007.5	1009.0	1009.5	1010.7	1007.7
RELATIVE HUMIDITY (%)														
Hour 00	46	75	73	71	71	76	78	79	80	80	77	74	74	76
Hour 06 (Local Time)	46	78	78	76	78	82	82	84	86	86	83	80	78	81
Hour 12	46	63	59	56	54	55	56	57	57	57	56	51	61	57
Hour 18	46	67	62	57	55	57	57	59	60	62	61	63	67	61
PRECIPITATION (inches):														
Water Equivalent														
-Normal		4.61	4.33	5.44	5.77	5.06	3.58	4.03	3.74	3.62	2.37	4.17	4.85	51.57
-Maximum Monthly	35	12.21	9.39	12.08	12.29	11.58	6.88	8.84	9.65	7.61	7.75	9.56	13.81	13.81
-Year		1951	1956	1975	1955	1953	1951	1959	1978	1958	1984	1983	1982	DEC 1982
-Minimum Monthly	35	0.84	1.12	1.50	2.05	0.83	0.04	0.43	0.43	0.19	0.75	1.05	T	T
-Year		1961	1980	1966	1965	1977	1953	1954	1953	1953	1963	1965	1955	OCT 1963
-Maximum in 24 hrs	35	3.89	3.63	5.95	4.35	4.94	4.76	4.71	4.04	4.63	3.40	4.72	5.42	5.95
-Year		1974	1966	1975	1985	1958	1980	1980	1978	1957	1981	1983	1978	MAR 1975
Snow,Ice pellets														
-Maximum Monthly	35	12.4	8.3	17.3	T							1.5	14.3	17.3
-Year		1985	1985	1968	1971							1976	1963	MAR 1968
-Maximum in 24 hrs	35	8.1	5.8	16.1	T							1.2	14.3	16.1
-Year		1985	1960	1968	1971							1976	1963	MAR 1968
WIND:														
Mean Speed (mph)	37	10.3	10.3	11.1	10.6	8.9	8.0	7.5	7.0	7.5	7.8	9.2	9.9	9.0
Prevailing Direction														
through 1963		S	S	S	S	S	S	S	S	E	S	S	S	S
Fastest Obs. 1 Min.														
-Direction	13	34	32	16	24	34	02	34	20	36	28	23	30	24
-Speed (MPH)	13	35	35	40	46	40	40	35	35	35	39	40	36	46
-Year		1976	1984	1973	1979	1979	1977	1982	1980	1972	1984	1979	1972	APR 1979
Peak Gust														
-Direction														
-Speed (mph)														
-Date														

See reference Notes to this table on the following page.

TABLE 2   PRECIPITATION (inches)   MEMPHIS, TENNESSEE

YEAR	JAN	FEB	MAR	APR	MAY	JUNE	JULY	AUG	SEP	OCT	NOV	DEC	ANNUAL
1956	4.14	9.39	2.54	6.74	3.41	3.71	3.05	2.48	0.57	2.36	2.58	3.79	44.76
1957	11.39	4.65	4.37	9.31	7.10	6.64	4.95	7.48	3.64	8.56	4.45	4.31	76.85
1958	2.75	3.11	4.04	6.77	8.88	4.00	5.55	4.05	7.61	1.03	3.81	2.33	53.93
1959	4.34	5.21	3.56	3.73	1.85	4.82	8.84	4.56	2.73	1.57	2.66	5.33	49.20
1960	3.74	3.08	4.20	3.65	4.20	5.34	1.30	7.84	4.05	4.11	3.62	4.30	49.43
1961	0.84	6.89	7.13	4.65	4.40	1.49	3.97	1.71	1.28	8.06	8.56	0.66	49.64
1962	4.19	4.22	4.80	3.62	0.84	5.71	3.94	4.18	5.28	2.57	2.31	1.35	43.01
1963	1.28	2.91	6.17	5.60	3.77	4.33	4.38	2.15	2.05	T	2.72	3.31	38.68
1964	3.73	3.50	7.34	11.03	3.28	1.39	6.14	5.76	2.74	2.21	2.59	7.97	57.68
1965	4.79	6.78	5.35	2.05	7.42	0.98	1.60	3.98	7.38	0.54	0.75	1.17	42.79
1966	2.84	6.88	1.50	5.42	5.69	0.52	2.18	4.28	3.23	1.92	1.57	5.21	41.24
1967	2.23	2.33	4.65	4.46	6.38	1.70	6.01	5.17	1.86	2.38	1.90	7.37	46.44
1968	5.57	1.98	6.52	5.15	5.21	3.76	2.69	1.61	5.58	2.87	4.89	6.04	51.87
1969	3.14	3.20	2.63	8.29	1.34	1.60	1.92	6.62	0.90	1.24	4.19	7.05	42.12
1970	1.16	3.87	5.32	7.08	3.70	1.78	4.99	5.76	3.80	6.20	2.62	3.71	49.99
1971	2.15	7.21	3.64	2.89	3.90	3.82	2.90	6.00	3.42	0.06	1.49	6.71	44.19
1972	4.73	2.23	4.80	3.51	4.55	5.50	4.89	1.94	5.46	3.92	8.05	9.37	58.95
1973	4.62	3.62	7.63	9.44	6.23	1.00	4.49	4.88	5.06	3.37	8.49	5.35	64.18
1974	8.90	4.65	3.40	6.34	7.76	6.30	6.33	4.78	3.45	2.67	4.96	5.03	64.57
1975	4.65	5.53	12.08	4.98	8.72	2.42	2.26	2.03	2.62	2.69	7.77	2.93	58.68
1976	2.85	4.41	7.68	2.41	4.73	4.06	3.82	0.86	5.40	5.66	1.83	1.79	45.50
1977	2.57	1.99	4.13	5.42	0.83	3.38	3.41	1.62	6.43	2.02	6.01	3.39	41.20
1978	8.13	1.31	4.05	2.14	8.14	4.45	3.89	9.65	1.52	1.82	5.56	13.12	63.78
1979	5.98	5.66	6.60	11.47	7.78	4.93	3.12	5.92	4.49	2.60	7.42	4.92	70.89
1980	3.23	1.12	10.86	7.53	4.43	5.75	4.73	1.23	5.32	3.14	5.23	1.86	54.43
1981	1.38	3.66	4.98	3.67	7.06	2.93	1.71	4.21	0.61	5.83	2.12	1.84	40.00
1982	6.61	4.16	4.47	6.76	5.50	6.68	4.13	3.11	1.92	5.23	6.43	13.81	68.81
1983	2.32	2.61	3.66	8.84	9.58	3.50	3.83	0.61	1.52	2.94	9.56	8.68	57.65
1984	1.88	4.37	6.07	5.24	9.06	1.12	4.59	5.00	1.96	7.75	5.85	4.35	57.24
1985	3.78	4.10	4.96	6.51	2.23	4.55	3.50	3.50	4.03	3.36	3.87	3.27	47.66
Record Mean	4.93	4.30	5.26	5.14	4.41	3.67	3.42	3.27	2.97	2.86	4.31	4.71	49.26

TABLE 3   AVERAGE TEMPERATURE (deg. F)   MEMPHIS, TENNESSEE

YEAR	JAN	FEB	MAR	APR	MAY	JUNE	JULY	AUG	SEP	OCT	NOV	DEC	ANNUAL
1956	38.7	47.8	51.5	60.4	73.2	77.3	82.6	82.5	72.9	65.7	50.6	49.4	62.7
1957	39.3	50.1	50.4	64.4	71.7	78.3	80.4	79.0	72.0	59.6	51.3	47.8	62.0
1958	37.6	34.0	45.6	60.4	71.1	78.0	81.4	79.9	73.5	62.1	54.0	39.1	59.8
1959	39.5	44.4	51.7	61.5	74.4	77.4	79.3	81.1	75.7	64.5	46.6	45.4	61.8
#1960	41.9	40.8	40.2	64.8	68.5	77.8	81.8	81.6	76.1	63.8	50.8	37.2	60.4
1961	36.1	47.2	55.1	58.2	66.7	75.1	80.2	77.8	74.6	62.5	49.6	42.4	60.4
1962	36.2	50.3	47.3	58.8	76.8	77.1	81.3	80.8	73.4	65.9	50.2	39.9	61.5
1963	34.4	37.6	56.7	64.2	71.1	78.6	80.1	80.2	73.4	68.6	53.4	31.5	60.8
1964	41.1	40.2	51.7	64.0	71.9	79.3	80.6	78.6	72.9	58.9	54.4	44.3	61.5
1965	43.4	42.9	44.0	66.4	74.7	78.1	81.8	80.2	73.7	60.8	55.8	45.8	62.3
1966	34.2	42.2	52.8	60.6	67.9	76.6	84.7	76.9	70.8	58.1	53.7	41.4	60.0
1967	42.2	39.2	56.7	66.5	68.8	78.3	77.7	76.0	70.3	62.4	49.1	45.0	61.0
1968	38.9	37.3	50.9	62.7	69.7	79.5	80.8	82.0	71.3	62.2	50.9	41.7	60.7
1969	41.2	43.3	45.0	62.9	72.1	78.8	84.9	79.3	72.9	62.9	49.0	40.1	61.1
1970	35.3	41.7	48.7	65.0	72.3	77.3	79.7	81.2	77.9	61.7	49.8	46.4	61.4
1971	39.6	43.5	48.4	60.5	66.6	80.6	80.6	78.6	76.1	69.3	50.9	50.7	62.1
1972	42.3	44.7	52.2	63.1	69.7	77.5	79.4	79.7	75.9	61.4	45.5	40.0	60.9
1973	38.6	40.5	57.3	59.8	68.3	81.0	83.2	79.6	76.1	67.6	57.3	43.3	62.7
1974	45.7	45.6	58.7	61.8	72.1	74.7	82.5	79.2	68.5	62.4	53.3	45.2	62.5
1975	45.9	46.2	49.9	61.9	73.5	78.8	81.1	81.2	70.9	65.8	53.8	44.1	62.8
1976	39.5	53.8	58.5	63.6	65.6	76.4	81.5	78.9	73.0	58.9	45.5	41.9	61.5
1977	30.7	45.1	58.6	66.9	76.4	81.9	84.7	82.6	79.0	62.2	55.1	44.1	64.0
1978	32.7	35.0	50.3	66.3	70.9	79.8	83.8	80.9	77.7	62.5	57.7	44.0	61.8
1979	30.9	38.5	54.3	63.0	70.0	77.2	82.6	80.9	73.4	65.8	50.7	45.5	61.1
1980	43.2	39.5	49.4	60.9	72.5	80.9	88.8	87.2	80.5	62.7	53.3	45.9	63.8
1981	40.9	47.3	54.3	70.2	70.0	82.5	84.6	81.8	74.0	62.5	53.8	40.9	63.6
1982	36.6	40.5	55.5	58.5	74.5	78.0	85.0	82.9	73.3	63.7	53.4	49.5	62.6
1983	40.4	45.3	51.9	56.8	68.2	77.2	83.6	84.9	75.2	66.2	53.1	34.7	61.5
1984	35.9	47.6	51.1	61.0	69.5	80.9	80.9	79.8	71.9	68.4	50.9	53.8	62.6
1985	32.4	40.6	57.7	65.0	71.8	78.9	82.2	80.2	73.6	67.2	57.6	36.9	62.0
Record Mean	40.8	44.0	52.4	62.3	70.4	78.3	81.3	80.2	74.2	63.5	51.8	43.5	61.9
Max	48.7	52.4	61.3	71.5	79.6	87.3	90.1	89.0	83.4	73.6	60.8	51.5	70.8
Min	32.9	35.5	43.5	53.0	61.3	69.3	72.6	71.3	64.9	53.5	42.7	35.6	53.0

REFERENCE NOTES FOR TABLES 1, 2, 3 and 6   (MEMPHIS, TN)

GENERAL

T - TRACE AMOUNT
BLANK ENTRIES DENOTE MISSING/UNREPORTED DATA.
# INDICATES A STATION OR INSTRUMENT RELOCATION.

SPECIFIC

TABLE 1

(a) - LENGTH OF RECORD IN YEARS. ALTHOUGH
INDIVIDUAL MONTHS MAY BE MISSING.

* LESS THAN .05

NORMALS — BASED ON THE 1951-1980 RECORD PERIOD.
EXTREMES — DATES ARE THE MOST RECENT OCCURRENCE.
WIND DIR. — NUMERALS SHOW TENS OF DEGREES
CLOCKWISE FROM TRUE NORTH.
"00" INDICATES CALM.
RESULTANT WIND DIRECTIONS ARE GIVEN TO WHOLE DEGREES.

EXCEPTIONS

TABLES 2, 3, and 6

RECORD MEANS ARE THROUGH THE CURRENT YEAR,
BEGINNING IN   1875 FOR TEMPERATURE
1872 FOR PRECIPITATION
1951 FOR SNOWFALL

TABLE 4 HEATING DEGREE DAYS Base 65 deg. F   MEMPHIS, TENNESSEE

SEASON	JULY	AUG	SEP	OCT	NOV	DEC	JAN	FEB	MAR	APR	MAY	JUNE	TOTAL
1956-57	0	0	0	48	439	482	788	414	447	145	25	0	2788
1957-58	0	0	6	191	407	526	844	862	592	154	36	0	3618
1958-59	0	0	11	138	344	795	781	570	410	155	15	0	3219
1959-60	0	0	0	111	547	602	713	691	763	91	69	0	3587
#1960-61	0	0	0	104	425	853	892	490	309	249	70	2	3394
1961-62	0	0	8	148	467	690	887	410	543	229	5	0	3387
1962-63	0	0	14	107	436	772	944	760	291	119	34	0	3477
1963-64	0	0	10	33	352	1034	739	713	406	91	13	0	3391
1964-65	0	0	13	202	335	634	666	614	645	72	0	0	3181
1965-66	0	0	22	162	275	590	947	634	382	175	47	5	3239
1966-67	0	0	9	227	344	732	707	715	301	60	35	0	3130
1967-68	0	0	34	144	469	613	803	795	444	114	21	0	3437
1968-69	0	0	1	149	423	716	733	601	608	103	12	0	3346
1969-70	0	0	0	151	473	768	917	648	500	97	20	0	3574
1970-71	0	0	7	150	455	571	781	593	509	171	50	0	3287
1971-72	0	0	0	13	432	435	698	582	391	146	27	0	2733
1972-73	0	0	12	172	583	766	809	679	237	200	32	0	3490
1973-74	0	0	8	67	244	665	599	535	235	150	1	0	2504
1974-75	0	0	28	121	367	607	591	521	463	180	2	0	2880
1975-76	0	0	40	90	352	643	783	326	238	100	58	0	2630
1976-77	0	0	0	231	581	708	1056	547	212	61	4	0	3400
1977-78	0	0	0	123	313	640	995	835	454	74	47	0	3481
1978-79	0	0	0	116	230	643	1049	734	345	121	23	0	3261
1979-80	0	0	0	76	426	598	669	733	478	156	7	0	3143
1980-81	0	0	5	146	362	586	739	492	342	18	23	0	2713
1981-82	0	0	9	153	331	739	873	680	324	215	2	0	3326
1982-83	0	0	20	134	352	500	759	543	406	273	25	0	3012
1983-84	0	0	27	73	368	935	894	499	426	162	24	0	3408
1984-85	0	0	37	48	423	367	1004	683	254	100	6	0	2922
1985-86	0	0	17	54	257	864							

TABLE 5 COOLING DEGREE DAYS Base 65 deg. F   MEMPHIS, TENNESSEE

YEAR	JAN	FEB	MAR	APR	MAY	JUNE	JULY	AUG	SEP	OCT	NOV	DEC	TOTAL
1969	0	0	0	48	240	422	627	449	243	91	0	0	2120
1970	3	0	0	104	251	375	463	509	400	51	6	2	2164
1971	0	0	3	46	107	474	489	426	349	154	14	0	2062
1972	0	0	3	95	179	383	449	464	346	66	6	0	1991
1973	0	0	4	48	143	486	571	458	350	156	19	0	2235
1974	6	0	46	59	228	299	550	445	138	46	23	0	1840
1975	8	0	3	93	272	421	507	510	224	121	23	2	2184
1976	0	7	44	64	84	349	519	438	247	48	0	0	1800
1977	0	0	23	123	362	516	619	551	426	41	20	0	2681
1978	0	0	6	122	235	452	590	501	387	46	18	0	2357
1979	0	0	19	68	184	394	553	499	259	108	4	0	2088
1980	0	0	0	40	249	480	744	695	476	80	18	2	2784
1981	0	5	20	181	184	532	614	527	285	80	2	0	2430
1982	0	1	36	26	305	399	563	623	275	100	14	25	2367
1983	0	0	7	32	131	373	584	622	338	116	17	0	2220
1984	0	1	4	51	169	482	502	462	249	162	5	23	2110
1985	0	6	30	107	224	425	540	478	285	129	42	0	2266

TABLE 6 SNOWFALL (inches)   MEMPHIS, TENNESSEE

SEASON	JULY	AUG	SEP	OCT	NOV	DEC	JAN	FEB	MAR	APR	MAY	JUNE	TOTAL
1956-57	0.0	0.0	0.0	0.0	T	0.0	T	T	T	0.0	0.0	0.0	T
1957-58	0.0	0.0	0.0	0.0	T	1.0	T	4.8	0.0	0.0	0.0	0.0	5.8
1958-59	0.0	0.0	0.0	0.0	T	4.0	T	0.6	0.0	0.0	0.0	0.0	4.6
1959-60	0.0	0.0	0.0	0.0	T	6.5	T	7.6	6.4	0.0	0.0	0.0	20.5
1960-61	0.0	0.0	0.0	0.0	0.0	1.2	T	0.2	0.0	0.0	0.0	0.0	1.4
1961-62	0.0	0.0	0.0	0.0	T	5.0	T	T	0.0	0.0	0.0	0.0	5.0
1962-63	0.0	0.0	0.0	0.0	0.0	3.0	1.2	0.8	0.0	0.0	0.0	0.0	5.0
1963-64	0.0	0.0	0.0	0.0	0.0	14.3	0.5	T	T	0.0	0.0	0.0	14.8
1964-65	0.0	0.0	0.0	0.0	T	T	3.3	4.6	0.0	0.0	0.0	0.0	7.9
1965-66	0.0	0.0	0.0	0.0	0.0	0.0	12.2	T	T	0.0	0.0	0.0	12.2
1966-67	0.0	0.0	0.0	0.0	T	0.7	0.6	T	0.3	0.0	0.0	0.0	1.6
1967-68	0.0	0.0	0.0	0.0	0.0	T	2.0	4.5	17.3	0.0	0.0	0.0	23.8
1968-69	0.0	0.0	0.0	0.0	0.0	T	T	T	T	0.0	0.0	0.0	T
1969-70	0.0	0.0	0.0	0.0	T	0.1	3.3	0.2	T	T	0.0	0.0	3.6
1970-71	0.0	0.0	0.0	0.0	T	1.0	T	6.7	1.6	T	0.0	0.0	9.3
1971-72	0.0	0.0	0.0	0.0	0.8	0.0	0.3	0.1	T	0.0	0.0	0.0	1.2
1972-73	0.0	0.0	0.0	0.0	T	T	1.4	T	0.0	0.0	0.0	0.0	1.4
1973-74	0.0	0.0	0.0	0.0	0.0	0.2	0.9	0.5	T	0.0	0.0	0.0	1.6
1974-75	0.0	0.0	0.0	T	0.0	0.2	3.9	0.5	1.4	0.0	0.0	0.0	6.0
1975-76	0.0	0.0	0.0	0.0	0.0	0.1	0.3	T	0.0	0.0	0.0	0.0	0.4
1976-77	0.0	0.0	0.0	0.0	1.5	0.3	3.5	T	0.0	0.0	0.0	0.0	5.3
1977-78	0.0	0.0	0.0	0.0	0.0	0.0	4.3	3.2	T	0.0	0.0	0.0	7.5
1978-79	0.0	0.0	0.0	0.0	0.0	T	3.0	7.4	0.0	0.0	0.0	0.0	10.4
1979-80	0.0	0.0	0.0	0.0	0.0	0.0	1.3	1.5	0.8	0.0	0.0	0.0	3.6
1980-81	0.0	0.0	0.0	0.0	T	T	T	T	0.0	0.0	0.0	0.0	T
1981-82	0.0	0.0	0.0	0.0	0.0	T	4.5	0.7	1.2	0.0	0.0	0.0	6.4
1982-83	0.0	0.0	0.0	0.0	0.0	T	7.3	0.2	0.5	0.0	0.0	0.0	7.5
1983-84	0.0	0.0	0.0	0.0	0.0	0.8	2.0	T	0.5	0.0	0.0	0.0	3.3
1984-85	0.0	0.0	0.0	0.0	0.0	T	12.4	8.3	0.0	0.0	0.0	0.0	20.7
1985-86	0.0	0.0	0.0	0.0	0.0	T							
Record Mean	0.0	0.0	0.0	0.0	0.1	0.7	2.6	1.5	1.0	T	0.0	0.0	5.9

See Reference Notes, relative to all above tables, on preceding page.

The city of Nashville is located on the Cumberland River, in the northwestern corner of the Central Basin of middle Tennessee near the escarpment of the Highland Rim. The Rim, as it is called, rises to the height of 300 to 400 feet above the mean elevation of the basin, forming an amphitheater about the city from the southwest to the southeast, with the south being more or less open but undulating.

Temperatures are moderate, with great extremes of either heat or cold rarely occurring, yet there are changes of sufficient amplitude and frequency to give variety.

Based on the 1951–1980 period, the average first occurrence of 32 degrees Fahrenheit in the fall is October 29 and the average last occurrence in the spring is April 5.

Humidity is an important phase of climate in relation to bodily health and comfort. The Nashville records show that the average relative humidity is moderate as compared with the general conditions east of the Mississippi River and south of the Ohio.

Nashville is not in the most frequented path of general storms that cross the country, however, it is in the zone of moderate frequency of thunderstorms. The thunderstorm season usually begins in the latter part of March and continues through September.

TABLE 1   NORMALS, MEANS AND EXTREMES

NASHVILLE, TENNESSEE

LATITUDE: 36°07'N   LONGITUDE: 86°41'W   ELEVATION: FT. GRND 590 BARO 630   TIME ZONE: CENTRAL   WBAN: 13897

	(a)	JAN	FEB	MAR	APR	MAY	JUNE	JULY	AUG	SEP	OCT	NOV	DEC	YEAR
TEMPERATURE °F:														
Normals														
-Daily Maximum		46.3	50.7	59.6	71.2	79.2	86.7	89.8	89.0	83.2	72.3	59.2	50.4	69.8
-Daily Minimum		27.8	30.1	38.3	48.1	56.9	64.8	69.0	67.8	61.3	48.0	38.0	31.3	48.5
-Monthly		37.0	40.4	49.0	59.7	68.1	75.8	79.4	78.4	72.3	60.2	48.6	40.9	59.1
Extremes														
-Record Highest	46	78	84	86	90	97	106	107	104	105	94	84	79	107
-Year		1972	1962	1982	1980	1941	1952	1952	1954	1954	1953	1971	1982	JUL 1952
-Record Lowest	46	-17	-13	2	23	34	42	51	47	36	26	-1	-7	-17
-Year		1985	1951	1980	1982	1976	1966	1947	1946	1983	1952	1950	1962	JAN 1985
NORMAL DEGREE DAYS:														
Heating (base 65°F)		865	689	510	186	55	0	0	0	19	193	492	747	3756
Cooling (base 65°F)		0	0	14	24	151	328	446	415	238	45	0	0	1661
% OF POSSIBLE SUNSHINE	43	41	47	51	58	61	66	63	64	63	62	50	42	56
MEAN SKY COVER (tenths)														
Sunrise - Sunset	45	7.1	6.7	6.6	6.1	6.0	5.6	5.7	5.3	5.2	4.9	6.1	6.7	6.0
MEAN NUMBER OF DAYS:														
Sunrise to Sunset														
-Clear	44	6.3	6.9	7.5	8.2	8.3	8.3	8.1	10.1	10.8	12.6	8.8	7.3	103.1
-Partly Cloudy	44	6.2	6.0	6.8	8.4	9.9	12.4	13.0	11.8	8.8	8.3	6.8	6.8	105.1
-Cloudy	44	18.5	15.4	16.7	13.5	12.8	9.3	9.9	9.0	10.4	10.0	14.4	17.0	157.0
Precipitation														
.01 inches or more	44	11.3	10.7	12.2	11.0	10.7	9.4	10.2	8.9	7.8	7.0	9.4	10.9	119.5
Snow,Ice pellets														
1.0 inches or more	44	1.4	1.2	0.5	0.*	0.0	0.0	0.0	0.0	0.0	0.0	0.1	0.6	3.8
Thunderstorms	44	1.3	1.7	4.2	5.5	7.4	8.3	9.9	7.8	3.7	1.6	1.6	1.2	54.1
Heavy Fog Visibility														
1/4 mile or less	44	2.6	1.3	1.1	0.5	1.0	0.9	1.2	1.7	1.9	2.1	1.7	1.8	17.6
Temperature °F														
-Maximum														
90° and above	20	0.0	0.0	0.0	0.1	0.9	9.1	15.9	10.9	5.2	0.2	0.0	0.0	42.2
32° and below	20	5.9	2.5	0.2	0.0	0.0	0.0	0.0	0.0	0.0	0.0	0.2	1.9	10.6
-Minimum														
32° and below	20	22.8	17.7	9.4	1.5	0.0	0.0	0.0	0.0	0.0	1.1	8.3	17.0	77.8
0° and below	20	1.0	0.3	0.0	0.0	0.0	0.0	0.0	0.0	0.0	0.0	0.0	0.1	1.4
AVG. STATION PRESS.(mb)	13	998.9	997.8	994.6	994.7	993.4	994.5	995.2	996.1	996.6	997.9	998.0	998.8	996.4
RELATIVE HUMIDITY (%)														
Hour 00	20	75	73	71	72	82	84	86	86	86	81	77	76	79
Hour 06	20	79	79	78	80	86	87	90	91	90	86	82	80	84
Hour 12 (Local Time)	20	64	59	54	52	56	55	58	59	59	55	60	63	58
Hour 18	20	66	60	55	54	59	59	63	64	65	61	64	66	61
PRECIPITATION (inches):														
Water Equivalent														
-Normal		4.49	4.03	5.58	4.47	4.56	3.70	3.82	3.40	3.71	2.58	3.52	4.63	48.49
-Maximum Monthly	46	13.92	10.31	12.35	8.41	11.04	9.37	7.75	8.31	11.44	6.13	9.04	13.63	13.92
-Year		1950	1956	1975	1984	1983	1960	1950	1942	1979	1959	1945	1978	JAN 1950
-Minimum Monthly	46	1.16	0.64	1.39	1.15	0.69	0.78	0.71	0.69	0.28	T	0.54	0.98	T
-Year		1970	1968	1966	1948	1941	1952	1954	1968	1956	1963	1949	1985	OCT 1963
-Maximum in 24 hrs	46	4.40	4.04	4.66	3.29	4.27	4.91	3.56	5.34	6.68	3.75	3.74	5.12	6.68
-Year		1946	1948	1975	1979	1984	1960	1950	1963	1979	1975	1973	1978	SEP 1979
Snow,Ice pellets														
-Maximum Monthly	46	18.8	18.9	16.1	1.1						T	9.2	13.2	18.9
-Year		1948	1979	1960	1971						1954	1950	1963	FEB 1979
-Maximum in 24 hrs	46	7.5	8.3	8.8	1.1						T	9.2	10.2	10.2
-Year		1966	1979	1951	1971						1954	1950	1963	DEC 1963
WIND:														
Mean Speed (mph)	44	9.2	9.3	10.0	9.5	7.7	7.1	6.4	6.1	6.4	6.7	8.4	9.0	8.0
Prevailing Direction														
through 1963		S	S	S	S	S	S	S	S	S	S	S	S	S
Fastest Obs. 1 Min.														
-Direction	10	34	32	18	31	36	30	34	02	34	16	15	17	36
-Speed (MPH)	10	32	35	37	35	41	35	33	40	33	28	39	29	41
-Year		1985	1980	1985	1982	1984	1981	1978	1983	1977	1979	1984	1984	MAY 1984
Peak Gust														
-Direction	2	W	W	W	S	NW	W	W	NW	W	E	W	SW	W
-Speed (mph)	2	46	44	51	45	55	47	37	46	44	33	60	38	60
-Date		1984	1984	1984	1985	1984	1984	1985	1985	1985	1984	1985	1985	NOV 1985

See Reference Notes to this table on the following page.

TABLE 2 PRECIPITATION (inches) NASHVILLE, TENNESSEE

YEAR	JAN	FEB	MAR	APR	MAY	JUNE	JULY	AUG	SEP	OCT	NOV	DEC	ANNUAL
1956	5.67	10.31	4.08	4.23	2.87	2.42	1.94	1.89	0.28	1.99	1.73	6.53	43.94
1957	9.39	5.24	2.85	2.79	8.23	7.06	1.83	2.55	4.09	3.89	6.31	5.84	60.07
1958	2.60	1.61	3.92	6.35	2.72	3.30	6.15	5.10	3.04	1.24	3.38	1.49	40.90
1959	3.26	4.60	3.71	2.49	5.20	2.41	3.90	2.54	3.73	6.13	4.96	5.04	47.97
1960	3.22	5.40	4.37	2.04	3.16	9.37	1.46	1.72	3.96	1.38	2.72	3.62	42.42
1961	1.44	5.33	6.52	4.50	4.36	2.96	5.34	2.62	0.35	1.12	3.87	6.44	44.85
1962	6.51	9.07	5.89	6.91	1.87	7.29	1.97	2.45	8.03	2.29	3.37	1.92	57.57
1963	1.60	2.83	10.03	3.37	2.47	3.09	5.33	7.63	3.43	T	2.43	2.15	44.36
1964	3.70	3.26	5.92	5.86	5.04	1.21	2.16	2.16	2.65	1.83	3.67	5.15	45.01
1965	2.98	4.71	6.13	5.72	3.12	2.74	3.32	2.53	5.02	0.57	1.82	1.01	39.67
1966	3.93	3.63	1.39	5.08	3.99	1.09	2.70	5.29	3.87	2.50	2.76	5.69	41.92
1967	1.62	1.78	4.44	3.40	6.98	4.23	7.46	2.06	1.93	1.57	3.87	5.88	45.22
1968	3.50	0.64	4.47	3.57	6.28	2.26	6.87	0.69	2.76	3.92	5.39	3.58	43.93
1969	4.96	4.48	2.12	6.03	4.81	3.34	5.33	2.27	2.06	2.01	1.83	8.03	47.27
1970	1.16	4.36	3.87	6.81	5.90	6.73	3.61	2.99	2.76	2.94	2.20	3.60	46.93
1971	2.66	4.70	2.95	3.34	2.93	3.47	5.00	5.87	2.11	1.27	1.18	5.17	40.65
1972	5.15	3.45	4.34	3.58	3.52	2.54	6.40	4.30	3.71	4.06	5.22	8.14	54.41
1973	3.40	3.63	9.88	7.00	5.72	4.80	7.67	1.79	1.56	3.32	7.78	3.23	59.78
1974	9.45	3.01	5.25	3.97	5.04	6.80	2.10	4.13	10.44	1.47	6.23	2.81	60.70
1975	4.67	5.22	12.35	3.55	6.52	2.22	2.96	4.69	5.42	5.86	3.00	4.12	60.58
1976	4.11	2.28	5.32	1.53	6.19	4.72	4.01	8.05	5.08	5.17	1.30	1.81	49.57
1977	2.53	3.27	5.83	7.87	1.65	4.29	1.15	4.65	5.04	4.22	5.96	4.25	50.71
1978	5.95	1.57	4.88	2.42	8.03	1.46	4.03	3.81	1.37	2.28	4.01	13.63	53.44
1979	7.13	4.01	4.92	7.80	8.18	2.79	4.27	4.59	11.44	3.97	5.98	5.04	70.12
1980	2.59	1.38	7.27	3.67	6.14	2.89	3.53	1.24	1.09	1.17	2.55	1.40	34.92
1981	1.60	3.83	3.38	4.78	3.05	8.05	3.49	3.10	1.37	2.82	3.83	2.38	41.68
1982	6.50	4.80	3.00	4.36	4.19	2.28	5.47	3.46	3.23	1.91	3.87	6.36	49.43
1983	2.56	2.93	3.44	6.80	11.04	3.93	1.71	1.36	0.45	2.77	6.98	7.75	51.72
1984	1.79	2.38	5.14	8.41	9.68	4.49	6.63	2.42	0.97	6.00	6.20	2.38	56.49
1985	3.02	3.30	2.70	2.91	2.65	1.53	2.00	3.91	2.52	1.59	3.81	0.98	30.92
Record Mean	4.68	4.10	5.16	4.28	4.09	3.80	3.97	3.41	3.27	2.51	3.56	3.98	46.81

TABLE 3 AVERAGE TEMPERATURE (deg. F) NASHVILLE, TENNESSEE

YEAR	JAN	FEB	MAR	APR	MAY	JUNE	JULY	AUG	SEP	OCT	NOV	DEC	ANNUAL
1956	35.9	46.9	49.5	57.7	70.2	76.1	81.3	79.8	71.4	64.0	47.7	50.3	60.9
1957	37.5	47.4	48.3	63.1	69.6	77.3	79.1	77.5	72.5	56.6	44.8	44.8	60.2
1958	35.4	31.0	44.0	58.4	68.2	75.0	80.1	77.7	71.8	59.6	51.7	36.2	57.4
1959	37.0	42.1	47.8	59.7	71.3	74.9	79.8	79.8	74.0	61.8	45.1	43.1	59.7
1960	40.0	35.6	36.3	61.0	65.7	75.0	78.3	79.5	73.6	61.6	47.9	34.7	57.4
1961	33.2	46.9	52.2	54.5	63.6	72.4	77.4	76.4	73.5	59.8	49.6	41.1	58.4
1962	35.3	46.2	45.3	56.0	75.2	75.0	79.1	79.7	69.3	62.9	47.9	35.0	58.9
1963	30.9	34.2	53.8	61.6	67.3	75.9	77.6	76.8	70.0	65.5	49.3	30.5	57.8
1964	38.9	37.2	49.0	62.6	69.7	77.7	78.7	77.2	71.3	56.7	51.7	42.7	59.4
#1965	40.0	39.8	41.9	60.9	71.7	75.1	78.6	79.0	74.1	59.3	51.9	44.8	59.7
1966	32.3	41.3	50.5	58.8	65.3	75.1	82.3	76.5	69.5	57.0	50.7	40.2	58.3
1967	42.3	37.5	57.0	63.9	66.3	76.3	75.7	72.6	66.8	59.3	44.3	42.6	58.7
1968	34.0	32.4	47.8	58.7	66.3	74.9	77.8	79.5	69.9	59.5	48.7	38.1	57.3
1969	37.2	39.8	42.6	60.9	68.9	77.2	82.7	78.1	70.8	60.6	46.5	37.3	58.6
1970	32.4	38.4	46.8	61.3	68.4	73.6	77.2	79.2	76.9	61.0	47.5	43.2	58.8
1971	35.6	38.7	44.6	57.9	63.3	77.5	76.8	76.4	74.4	66.7	49.9	49.2	59.3
1972	41.8	41.8	50.2	60.5	67.4	73.0	77.3	77.2	75.7	60.2	47.3	42.8	59.6
1973	38.0	39.7	56.8	56.4	64.2	76.1	78.7	78.1	76.2	66.2	54.7	40.5	60.5
1974	45.4	41.8	54.9	58.6	70.0	71.4	78.0	77.6	67.5	59.4	50.0	42.6	59.8
1975	43.4	44.6	47.3	58.5	70.4	76.0	78.5	79.1	67.9	62.4	52.1	42.8	60.2
1976	36.7	50.5	55.9	59.9	64.0	73.3	76.4	74.4	66.8	53.9	40.9	36.6	57.5
1977	24.5	40.6	53.9	63.0	71.9	77.2	82.2	79.5	74.0	57.0	50.8	38.6	59.4
1978	27.6	29.2	46.9	61.0	66.7	76.2	80.5	78.7	75.5	57.4	53.6	42.4	58.0
1979	29.7	33.4	50.7	57.8	66.3	73.7	77.6	77.0	70.5	60.3	48.6	41.5	57.3
1980	39.7	35.7	46.3	57.3	67.8	75.5	82.8	81.7	76.0	57.8	48.5	41.0	59.2
1981	35.5	42.6	47.5	64.0	64.2	77.5	79.8	76.6	68.1	60.4	49.9	38.3	58.7
1982	34.0	39.5	52.5	54.6	71.0	73.3	79.8	76.1	69.6	61.1	51.4	48.2	59.3
1983	38.8	42.7	50.3	54.5	64.8	75.5	80.5	83.2	73.7	62.4	49.9	34.0	59.2
1984	32.2	43.4	46.1	58.2	64.2	77.4	76.1	76.5	68.6	66.7	46.0	49.6	58.8
1985	27.8	36.5	53.2	61.9	68.4	75.7	80.2	77.2	70.8	64.4	56.9	34.2	58.9
Record Mean	38.5	40.9	49.6	59.4	68.1	76.2	79.4	78.3	72.1	61.0	49.0	41.0	59.5
Max	47.1	50.0	59.4	69.8	78.4	86.2	89.2	88.1	82.6	72.1	58.6	49.6	69.3
Min	29.9	31.7	39.7	49.0	57.7	66.1	69.6	68.4	61.6	49.8	39.3	32.5	49.6

REFERENCE NOTES FOR TABLES 1, 2, 3 and 6   (NASHVILLE, TN)

GENERAL

T - TRACE AMOUNT
BLANK ENTRIES DENOTE MISSING/UNREPORTED DATA.
# INDICATES A STATION OR INSTRUMENT RELOCATION.

SPECIFIC

TABLE 1

(a) - LENGTH OF RECORD IN YEARS. ALTHOUGH
INDIVIDUAL MONTHS MAY BE MISSING.
* LESS THAN .05

NORMALS — BASED ON THE 1951-1980 RECORD PERIOD.
EXTREMES — DATES ARE THE MOST RECENT OCCURRENCE.
WIND DIR. — NUMERALS SHOW TENS OF DEGREES
CLOCKWISE FROM TRUE NORTH.
"00" INDICATES CALM.
RESULTANT WIND DIRECTIONS ARE GIVEN TO WHOLE DEGREES.

EXCEPTIONS

TABLES 2, 3, and 6

RECORD MEANS ARE THROUGH THE CURRENT YEAR,
BEGINNING IN   1871 FOR TEMPERATURE
1871 FOR PRECIPITATION
1942 FOR SNOWFALL

TABLE 4 HEATING DEGREE DAYS Base 65 deg. F   NASHVILLE, TENNESSEE

SEASON	JULY	AUG	SEP	OCT	NOV	DEC	JAN	FEB	MAR	APR	MAY	JUNE	TOTAL
1956-57	0	0	10	73	512	450	845	487	511	175	45	0	3108
1957-58	0	0	10	270	488	619	911	950	643	204	56	1	4152
1958-59	0	0	16	187	420	886	861	634	526	191	35	0	3766
1959-60	0	0	0	170	587	674	769	845	886	171	121	0	4223
1960-61	0	0	0	148	509	931	977	501	393	336	98	4	3897
1961-62	0	0	23	186	471	734	916	523	605	303	5	0	3766
1962-63	0	0	43	152	508	924	1050	855	358	170	63	0	4123
1963-64	0	0	28	48	465	1062	803	797	490	139	23	0	3855
#1964-65	0	3	18	265	398	685	768	698	711	168	-3	0	3717
1965-66	0	0	26	198	386	618	1007	657	452	234	78	7	3663
1966-67	0	0	13	255	423	763	697	763	300	106	69	1	3390
1967-68	0	3	58	216	615	688	952	941	531	205	64	1	4274
1968-69	0	0	4	220	484	825	855	700	692	149	38	3	3970
1969-70	0	0	10	201	551	854	1005	737	556	156	51	0	4121
1970-71	0	0	13	159	522	671	902	733	624	227	101	0	3952
1971-72	0	0	0	39	462	483	713	667	454	193	36	6	3053
1972-73	0	0	10	168	533	682	830	702	261	275	83	0	3544
1973-74	0	0	8	84	316	753	601	641	320	227	28	3	2981
1974-75	0	0	48	196	464	685	665	567	547	241	6	0	3419
1975-76	0	0	68	138	398	683	870	417	303	183	94	0	3154
1976-77	0	0	31	349	718	872	1250	679	350	129	28	1	4407
1977-78	0	0	3	255	425	813	1152	996	556	164	92	0	4456
1978-79	0	0	1	240	338	695	1088	877	449	213	57	0	3958
1979-80	0	0	5	180	487	723	777	848	571	240	38	0	3869
1980-81	0	0	9	259	487	739	909	621	537	97	96	0	3754
1981-82	0	0	42	175	445	820	956	707	416	309	8	0	3878
1982-83	0	0	30	194	413	537	806	620	458	322	71	0	3451
1983-84	0	0	45	121	447	956	1009	621	578	220	106	0	4103
1984-85	0	0	59	63	564	473	1146	794	383	145	25	6	3658
1985-86	0	0	30	91	264	948							

TABLE 5 COOLING DEGREE DAYS Base 65 deg. F   NASHVILLE, TENNESSEE

YEAR	JAN	FEB	MAR	APR	MAY	JUNE	JULY	AUG	SEP	OCT	NOV	DEC	TOTAL
1969	0	0	0	33	165	378	554	416	191	74	0	0	1811
1970	0	0	0	56	163	265	386	446	374	40	1	1	1732
1971	0	0	0	23	55	380	374	360	293	101	16	1	1603
1972	0	0	1	62	117	250	385	387	341	24	6	1	1573
1973	0	0	14	25	61	339	432	412	351	128	8	0	1770
1974	0	0	16	39	191	203	410	399	130	30	22	0	1440
1975	3	0	3	55	183	341	424	444	164	62	19	0	1698
1976	0	1	28	36	68	257	363	299	92	10	0	0	1154
1977	0	0	11	74	253	371	543	458	281	13	4	0	2008
1978	0	0	1	50	152	344	489	432	324	13	2	0	1807
1979	0	0	11	5	103	264	393	381	175	44	0	0	1376
1980	0	0	0	17	131	322	562	527	344	44	1	0	1948
1981	0	0	1	71	81	383	464	366	145	42	0	0	1553
1982	0	0	37	4	199	256	470	352	177	84	12	21	1612
1983	0	0	9	12	69	320	488	568	315	49	2	0	1832
1984	0	0	0	21	87	382	352	364	173	121	0	1	1501
1985	0	2	24	59	137	335	479	386	206	79	29	0	1736

TABLE 6 SNOWFALL (inches)   NASHVILLE, TENNESSEE

SEASON	JULY	AUG	SEP	OCT	NOV	DEC	JAN	FEB	MAR	APR	MAY	JUNE	TOTAL
1956-57	0.0	0.0	0.0	0.0	T	T	T	T	3.1	T	0.0	0.0	3.1
1957-58	0.0	0.0	0.0	0.0	1.2	0.3	2.0	6.6	0.1	0.0	0.0	0.0	10.2
1958-59	0.0	0.0	0.0	0.0	T	T	1.0	2.0	T	0.0	0.0	0.0	3.0
1959-60	0.0	0.0	0.0	0.0	T	T	7.4	15.0	16.1	T	0.0	0.0	38.5
1960-61	0.0	0.0	0.0	0.0	0.0	0.2	3.0	1.6	0.2	0.0	0.0	0.0	5.0
1961-62	0.0	0.0	0.0	0.0	T	2.0	2.2	1.1	1.4	T	0.0	0.0	6.7
1962-63	0.0	0.0	0.0	0.0	T	8.2	6.8	8.7	0.0	0.0	0.0	0.0	23.7
1963-64	0.0	0.0	0.0	0.0	T	13.2	5.0	4.2	T	0.0	0.0	0.0	22.4
1964-65	0.0	0.0	0.0	0.0	T	0.0	1.2	2.9	3.4	0.0	0.0	0.0	7.5
1965-66	0.0	0.0	0.0	0.0	T	0.0	11.4	T	T	0.0	0.0	0.0	11.4
1966-67	0.0	0.0	0.0	0.0	7.2	4.3	1.2	T	T	0.0	0.0	0.0	12.7
1967-68	0.0	0.0	0.0	0.0	0.0	8.4	7.2	2.9	8.5	0.0	0.0	0.0	27.0
1968-69	0.0	0.0	0.0	0.0	T	T	5.2	6.9	4.8	0.0	0.0	0.0	16.9
1969-70	0.0	0.0	0.0	0.0	0.3	3.8	5.6	3.0	T	0.0	0.0	0.0	12.7
1970-71	0.0	0.0	0.0	0.0	T	1.2	1.2	6.5	3.0	1.1	0.0	0.0	13.0
1971-72	0.0	0.0	0.0	0.0	0.1	T	0.4	0.5	0.9	T	0.0	0.0	1.9
1972-73	0.0	0.0	0.0	0.0	0.1	0.6	4.8	T	0.2	0.1	0.0	0.0	5.8
1973-74	0.0	0.0	0.0	0.0	0.0	2.4	T	0.3	T	0.0	0.0	0.0	2.7
1974-75	0.0	0.0	0.0	0.0	T	2.1	4.2	T	T	0.0	0.0	0.0	6.3
1975-76	0.0	0.0	0.0	0.0	T	T	1.1	2.3	T	0.0	0.0	0.0	3.4
1976-77	0.0	0.0	0.0	0.0	1.2	1.8	18.5	T	0.0	T	0.0	0.0	21.5
1977-78	0.0	0.0	0.0	0.0	T	0.1	12.9	9.8	2.4	0.0	0.0	0.0	25.2
1978-79	0.0	0.0	0.0	0.0	0.0	T	8.0	18.9	0.6	0.0	0.0	0.0	27.5
1979-80	0.0	0.0	0.0	0.0	T	T	0.3	6.6	3.1	0.0	0.0	0.0	10.0
1980-81	0.0	0.0	0.0	0.0	T	T	1.2	1.7	T	0.0	0.0	0.0	2.9
1981-82	0.0	0.0	0.0	0.0	0.0	0.2	4.8	3.7	1.0	0.0	0.0	0.0	9.7
1982-83	0.0	0.0	0.0	0.0	0.0	0.4	0.3	0.8	T	0.0	0.0	0.0	1.5
1983-84	0.0	0.0	0.0	0.0	0.0	0.7	5.3	3.7	T	0.0	0.0	0.0	9.7
1984-85	0.0	0.0	0.0	0.0	0.0	0.8	9.8	8.0	0.0	0.0	0.0	0.0	18.6
1985-86	0.0	0.0	0.0	0.0	0.0	0.5							
Record Mean	0.0	0.0	0.0	0.0	0.5	1.7	4.2	3.3	1.5	0.1	0.0	0.0	11.3

See Reference Notes, relative to all above tables, on preceding page.

Oak Ridge is located in a broad valley between the Cumberland Mountains, which lie to the northwest of the area, and the Great Smoky Mountains, to the southeast. These mountain ranges are oriented northeast–southwest and the valley between is corrugated by broken ridges 300 to 500 feet high and oriented parallel to the main valley. During periods of light winds, daytime winds are usually southwesterly, nighttime winds northeasterly. Wind velocities are somewhat decreased by the ridges. Tornadoes rarely occur in the valley between the Cumberlands and the Great Smokies. In winter the Cumberland Mountains have a moderating influence on the local climate by retarding the flow of cold air from the north and west.

Temperatures of 100 degrees or more have occurred during less than one–half of the years of the period of record, and temperatures of zero or below are rare. Summer nights are seldom oppressively hot and humid.

Precipitation is more than adequate for agriculture and is normally well distributed through the year for agricultural purposes.

Occasionally there is sufficient dry weather in late summer or early fall to cause small damage to crops and pastures and to create conditions favorable for destructive forest fires. Winter and early spring are the seasons of heaviest precipitation. A few of the larger monthly precipitation amounts recorded have occurred in the normally drier fall months.

Light snow usually occurs in all of the months from November through March, but the total monthly snowfall is often only a trace. Snowfalls sufficiently heavy to interfere with traffic and outdoor activities occur infrequently.

Based on the 1951–1980 period, the average first occurrence of 32 degrees Fahrenheit in the fall is October 27 and the average last occurrence in the spring is April 11.

TABLE 1   NORMALS, MEANS AND EXTREMES

OAK RIDGE, TENNESSEE

LATITUDE: 36°06'N   LONGITUDE: 84°11'W   ELEVATION: FT. GRND   880 BARO   TIME ZONE: EASTERN   WBAN: 03841

	(a)	JAN	FEB	MAR	APR	MAY	JUNE	JULY	AUG	SEP	OCT	NOV	DEC	YEAR
TEMPERATURE °F:														
Normals														
-Daily Maximum		45.7	50.2	59.0	70.5	78.1	84.6	87.2	86.7	81.3	70.4	58.1	48.9	68.4
-Daily Minimum		27.7	29.3	36.7	45.6	54.1	61.8	65.9	65.2	59.1	46.0	36.3	30.2	46.5
-Monthly		36.7	39.8	47.9	58.0	66.1	73.2	76.6	76.0	70.2	58.2	47.2	39.5	57.4
Extremes														
-Record Highest	38	75	79	85	91	93	101	105	103	102	90	83	78	105
-Year		1952	1977	1982	1970	1962	1954	1952	1983	1954	1954	1961	1982	JUL 1952
-Record Lowest	38	-17	1	1	21	30	39	50	51	33	21	0	-7	-17
-Year		1985	1965	1980	1985	1976	1977	1967	1979	1967	1952	1950	1983	JAN 1985
NORMAL DEGREE DAYS:														
Heating (base 65°F)		877	706	536	217	82	8	0	0	25	234	534	787	4006
Cooling (base 65°F)		0	0	5	10	117	254	360	341	181	26	0	0	1294
% OF POSSIBLE SUNSHINE														
MEAN SKY COVER (tenths)														
Sunrise - Sunset	31	6.7	6.5	6.6	5.9	5.9	5.5	5.9	5.4	5.6	5.0	6.1	6.5	6.0
MEAN NUMBER OF DAYS:														
Sunrise to Sunset														
-Clear	33	7.2	7.7	7.9	9.6	9.8	9.5	8.7	10.1	10.4	13.6	9.1	8.4	112.1
-Partly Cloudy	33	6.1	5.3	6.5	6.5	8.1	10.5	10.5	10.5	8.1	6.7	6.7	5.7	91.2
-Cloudy	33	17.6	15.2	16.6	13.9	13.1	9.9	11.8	10.4	11.5	10.8	14.3	16.9	162.0
Precipitation														
.01 inches or more	37	12.6	11.4	12.8	10.9	10.8	10.4	12.0	10.3	8.2	7.9	10.0	10.9	128.1
Snow, Ice pellets														
1.0 inches or more	33	1.4	1.2	0.3	0.0	0.0	0.0	0.0	0.0	0.0	0.0	0.0	0.5	3.5
Thunderstorms	17	0.7	1.7	2.6	4.2	7.4	8.0	10.8	9.1	3.4	1.4	1.2	0.8	51.3
Heavy Fog Visibility														
1/4 mile or less	15	1.3	1.1	0.8	0.9	1.5	1.4	3.1	4.2	3.8	7.6	5.7	2.2	33.7
Temperature °F														
-Maximum														
90° and above	37	0.0	0.0	0.0	0.1	1.3	5.9	10.9	8.2	3.1	0.1	0.0	0.0	29.6
32° and below	37	3.1	1.4	0.2	0.0	0.0	0.0	0.0	0.0	0.0	0.0	0.1	1.6	6.4
-Minimum														
32° and below	37	20.9	17.5	12.1	2.6	0.1	0.0	0.0	0.0	0.0	1.9	12.1	19.7	86.9
0° and below	37	0.4	0.0	0.0	0.0	0.0	0.0	0.0	0.0	0.0	0.0	0.*	0.1	0.5
AVG. STATION PRESS. (mb)														
RELATIVE HUMIDITY (%)														
Hour 01														
Hour 07 (Local Time)														
Hour 13														
Hour 19														
PRECIPITATION (inches):														
Water Equivalent														
-Normal		5.25	4.60	6.21	4.41	4.23	4.26	5.21	3.75	3.80	2.89	4.50	5.65	54.76
-Maximum Monthly	38	13.27	10.47	12.24	9.71	10.70	8.09	19.27	10.46	9.10	6.95	12.22	10.31	19.27
-Year		1954	1956	1975	1956	1984	1962	1967	1960	1957	1972	1948	1956	JUL 1967
-Minimum Monthly	38	0.93	0.84	2.13	0.88	0.80	0.85	1.55	0.54	0.41	T	1.37	0.67	T
-Year		1981	1968	1957	1976	1970	1980	1970	1953	1961	1963	1949	1965	OCT 1963
-Maximum in 24 hrs	38	4.25	2.94	4.74	6.24	4.41	3.70	4.91	7.48	3.76	2.66	5.29	5.12	7.48
-Year		1954	1954	1973	1977	1973	1969	1967	1960	1982	1976	1973	1969	AUG 1960
Snow, Ice pellets														
-Maximum Monthly	38	9.6	17.2	21.0	0.3						T	6.5	14.8	21.0
-Year		1966	1979	1960	1971						1962	1950	1963	MAR 1960
-Maximum in 24 hrs	38	8.3	9.1	12.0	0.3						T	6.5	10.8	12.0
-Year		1962	1960	1960	1971						1962	1950	1963	MAR 1960
WIND:														
Mean Speed (mph)	16	4.8	5.0	5.3	5.7	4.5	4.2	3.9	3.7	3.8	3.6	4.1	4.5	4.4
Prevailing Direction														
through 1961		SW	ENE	SW	SW	SW	SW	SW	E	E	E	E	SW	SW
Fastest Mile														
-Direction (!!!)														
-Speed (MPH)	22	59	47	48	50	46	50	50	53	38	39	45	50	59
-Year		1959	1967	1962	1959	1973	1975	1961	1964	1959	1967	1968	1964	JAN 1959
Peak Gust														
-Direction (!!!)														
-Speed (mph)														
-Date														

See Reference Notes to this table on the following page.

TABLE 2 PRECIPITATION (inches) OAK RIDGE, TENNESSEE

YEAR	JAN	FEB	MAR	APR	MAY	JUNE	JULY	AUG	SEP	OCT	NOV	DEC	ANNUAL
1956	4.57	10.47	6.44	9.71	4.44	2.28	7.90	2.08	2.91	3.80	2.23	10.31	67.14
1957	10.08	8.60	2.13	4.55	2.45	4.80	2.72	1.82	9.10	4.16	10.07	7.40	67.88
1958	2.52	2.57	3.66	6.62	3.17	1.91	4.46	3.46	0.41	0.41	3.22	2.43	37.43
1959	5.81	4.39	4.28	4.35	3.24	3.85	3.23	5.63	0.56	4.05	5.78	5.37	50.54
1960	3.75	3.59	5.23	2.09	1.97	7.00	3.76	10.46	4.64	4.55	2.72	4.56	54.32
1961	1.86	7.88	7.33	3.61	4.39	7.16	7.06	4.02	0.41	2.91	4.41	9.86	60.90
1962	5.93	9.01	5.88	3.94	2.64	8.09	4.28	2.87	6.42	3.28	5.44	3.31	61.09
1963	2.83	2.86	9.05	3.70	3.24	3.62	7.51	3.17	1.43	T	4.30	2.99	44.70
1964	4.81	4.33	6.32	7.13	2.84	0.86	5.05	3.41	3.14	2.61	5.62	4.94	49.94
1965	4.17	2.89	9.92	3.86	3.27	5.75	7.54	2.76	3.01	1.08	2.92	0.67	47.84
1966	3.72	5.73	2.29	5.50	3.52	2.65	2.57	4.73	4.20	3.85	4.74	3.72	47.22
1967	3.78	3.77	6.11	2.62	4.77	6.40	19.27	2.22	3.27	3.61	5.01	7.94	68.77
1968	4.47	0.84	5.26	5.21	4.01	3.56	2.80	1.25	3.47	2.23	2.24	4.28	39.62
1969	4.30	4.84	2.24	2.86	2.76	5.07	5.27	4.17	4.08	1.83	8.54	49.58
1970	2.96	3.74	4.06	9.24	0.80	5.90	1.55	9.10	2.54	6.07	1.78	4.47	52.21
1971	4.88	4.68	4.61	4.73	7.17	2.25	8.22	2.57	3.32	1.75	2.25	6.90	53.33
1972	7.32	4.83	5.99	2.94	5.81	5.56	5.22	1.77	4.28	6.95	5.02	9.20	64.89
1973	4.21	3.42	11.43	5.66	10.43	6.94	5.54	2.25	3.41	3.36	10.78	8.90	76.33
1974	9.62	4.72	7.22	3.78	7.98	2.00	1.99	4.31	4.18	1.78	5.04	4.96	57.58
1975	5.88	5.91	12.24	2.36	4.02	5.60	2.88	2.67	5.79	5.55	3.42	4.36	60.68
1976	5.47	2.60	6.00	0.88	7.24	4.80	4.03	4.89	4.56	5.91	2.27	4.68	53.33
1977	2.52	1.89	5.42	8.50	1.67	6.68	4.11	4.40	8.67	4.97	9.12	4.82	62.77
1978	5.99	1.13	4.69	3.21	4.50	3.97	4.89	5.75	1.70	0.39	5.54	6.65	48.41
1979	7.60	4.30	5.01	5.25	9.32	3.73	12.92	5.49	3.74	1.93	5.77	2.24	67.30
1980	6.24	1.51	9.20	3.89	3.01	0.85	2.39	2.07	3.31	1.08	4.55	2.02	40.12
1981	0.93	4.69	3.59	4.58	2.65	4.50	2.42	3.11	3.90	4.89	3.20	4.12	42.58
1982	6.70	5.42	6.20	2.79	2.88	2.15	7.01	4.61	5.31	1.95	7.81	7.22	60.05
1983	1.75	4.38	2.57	6.40	6.90	2.53	2.41	1.28	2.07	4.58	5.85	6.95	47.67
1984	2.62	3.92	4.81	4.16	10.70	4.70	8.72	1.87	1.87	6.19	4.37	2.59	56.52
1985	2.88	3.73	2.60	2.23	3.83	5.08	6.43	8.50	1.56	3.08	4.49	2.11	46.52
Record Mean	5.20	4.71	5.85	4.24	4.39	4.08	5.41	3.82	3.53	3.06	4.68	5.47	54.45

TABLE 3 AVERAGE TEMPERATURE (deg. F) OAK RIDGE, TENNESSEE

YEAR	JAN	FEB	MAR	APR	MAY	JUNE	JULY	AUG	SEP	OCT	NOV	DEC	ANNUAL
1956	35.0	47.6	50.4	57.3	69.1	73.5	77.0	77.0	69.5	64.1	48.4	51.0	60.0
1957	41.4	49.1	48.9	61.6	69.7	77.0	77.6	77.6	73.6	57.0	50.4	43.1	60.7
1958	35.7	32.7	44.9	57.6	67.9	75.2	78.7	75.3	69.5	58.7	50.0	36.6	56.9
1959	35.9	42.9	44.6	59.0	70.6	72.9	77.7	78.2	72.7	62.1	47.4	41.3	58.7
1960	40.1	37.3	35.9	58.8	63.5	73.0	77.1	77.1	71.9	60.6	47.2	34.6	56.4
1961	33.3	46.5	51.8	53.9	62.7	71.6	73.9	75.1	72.2	57.6	50.8	40.4	57.5
1962	36.0	45.6	46.0	53.9	72.5	73.2	76.1	76.0	68.3	60.8	46.1	34.4	57.4
1963	31.2	32.5	52.3	59.6	65.8	73.4	73.7	75.4	69.4	62.0	47.8	30.5	56.1
1964	36.9	35.4	47.5	60.7	67.6	75.4	74.9	74.3	68.8	55.1	51.3	41.7	57.5
1965	38.5	38.9	45.1	60.7	68.6	71.4	75.5	75.8	71.1	56.1	49.5	41.4	57.7
1966	33.1	39.7	48.8	57.0	64.6	72.5	78.4	75.0	68.5	55.5	47.8	38.3	56.6
1967	40.5	36.3	52.9	61.3	62.7	72.0	72.1	72.8	64.9	57.8	44.3	44.2	56.8
1968	36.8	32.7	48.9	58.2	65.2	73.9	77.3	78.3	68.1	58.7	47.3	36.4	56.8
1969	35.7	38.7	41.2	59.4	67.3	74.6	76.1	74.8	68.5	58.1	44.8	36.5	56.6
1970	29.8	38.0	47.0	60.3	67.7	72.1	76.7	76.5	73.9	61.2	46.9	42.5	57.7
1971	36.8	39.1	44.4	57.6	63.5	75.6	75.0	74.8	72.6	64.0	46.7	47.8	58.2
1972	40.6	37.4	46.5	58.0	63.9	69.8	75.2	75.5	72.1	57.3	47.0	44.2	57.3
1973	37.4	38.2	55.0	55.4	62.0	74.0	76.8	75.3	72.8	62.1	50.0	39.3	58.2
1974	47.1	41.6	53.6	57.7	65.9	67.7	75.4	74.2	65.6	54.1	45.8	39.6	57.2
1975	41.1	43.0	44.6	54.9	68.2	72.0	74.7	76.2	66.0	58.5	48.4	39.4	57.3
1976	32.6	45.7	49.9	56.2	60.5	70.9	73.5	72.3	65.0	52.3	39.7	34.6	54.4
1977	24.8	36.7	50.8	59.2	66.4	72.0	75.3	75.3	70.3	53.6	49.9	36.1	56.0
1978	27.3	31.1	46.1	57.2	64.1	72.5	75.7	75.6	70.8	55.7	51.7	39.7	55.8
1979	31.1	35.3	50.0	57.2	64.4	71.5	71.5	74.9	66.9	56.5	47.9	39.5	56.0
1980	38.8	34.3	44.5	55.4	65.7	72.8	80.6	78.9	73.7	55.4	46.6	38.3	57.1
1981	32.3	39.9	44.1	61.2	62.3	75.4	77.8	74.1	66.8	55.0	46.0	34.3	55.8
1982	31.6	41.2	50.7	52.6	69.3	71.6	77.5	73.8	66.8	57.6	47.4	43.0	57.0
1983	35.4	38.6	47.7	51.2	63.8	71.5	79.2	79.2	69.4	59.4	45.7	34.4	56.2
1984	33.2	41.3	44.9	54.7	60.7	74.0	73.5	75.1	66.5	65.5	43.1	45.8	56.5
1985	28.0	35.3	47.8	57.3	64.2	71.0	74.8	73.1	67.0	61.9	54.5	30.9	55.5
Record Mean	36.5	40.0	47.8	58.0	65.9	73.2	76.6	75.7	69.7	58.6	47.1	39.3	57.4
Max	45.7	50.7	59.3	70.5	78.1	84.7	87.2	86.4	80.9	70.7	58.1	48.7	68.4
Min	27.4	29.4	36.3	45.4	53.7	61.6	65.9	64.9	58.4	46.5	36.1	29.8	46.3

REFERENCE NOTES FOR TABLES 1, 2, 3 and 6 (OAK RIDGE, TN)

GENERAL

T - TRACE AMOUNT
BLANK ENTRIES DENOTE MISSING/UNREPORTED DATA.
# INDICATES A STATION OR INSTRUMENT RELOCATION.

SPECIFIC

TABLE 1

(a) - LENGTH OF RECORD IN YEARS. ALTHOUGH
      INDIVIDUAL MONTHS MAY BE MISSING.
* LESS THAN .05

NORMALS — BASED ON THE 1951-1980 RECORD PERIOD.
EXTREMES — DATES ARE THE MOST RECENT OCCURRENCE.
WIND DIR. — NUMERALS SHOW TENS OF DEGREES
            CLOCKWISE FROM TRUE NORTH.
            "00" INDICATES CALM.
RESULTANT WIND DIRECTIONS ARE GIVEN TO WHOLE DEGREES.

EXCEPTIONS

TABLE 1

1. PEAK GUST WINDS ARE THROUGH 1979.
2. THUNDERSTORMS AND HEAVY FOG ARE THROUGH 1984
   AND MAY BE INCOMPLETE, DUE TO PART-TIME OPERA-
   TIONS
3. MEAN WIND SPEED IS THROUGH 1984.

TABLES 2, 3, and 6

RECORD MEANS ARE THROUGH THE CURRENT YEAR,
BEGINNING IN   1948 FOR TEMPERATURE
                1948 FOR PRECIPITATION
                1949 FOR SNOWFALL

TABLE 4   HEATING DEGREE DAYS Base 65 deg. F   OAK RIDGE, TENNESSEE

SEASON	JULY	AUG	SEP	OCT	NOV	DEC	JAN	FEB	MAR	APR	MAY	JUNE	TOTAL
1956-57	0	0	20	69	493	427	723	438	491	178	38	0	2877
1957-58	0	0	12	254	432	673	901	897	617	231	48	0	4065
1958-59	0	0	22	194	448	875	894	611	625	211	36	0	3916
1959-60	0	0	0	153	527	727	764	797	896	206	141	0	4211
1960-61	0	0	4	165	528	935	976	513	400	350	105	5	3981
1961-62	0	0	22	237	435	755	891	536	583	339	19	0	3817
1962-63	0	0	58	185	562	945	1042	905	387	209	73	0	4366
1963-64	1	0	24	102	508	1064	865	855	539	161	38	0	4157
1964-65	0	6	29	310	405	715	815	724	610	166	9	2	3791
1965-66	0	0	22	279	460	726	984	702	496	267	79	12	4027
1966-67	0	0	15	288	511	820	754	797	378	138	135	10	3846
1967-68	2	0	70	243	616	638	870	930	494	216	70	0	4149
1968-69	0	0	11	219	523	878	902	731	728	181	48	7	4228
1969-70	0	0	22	227	602	876	1088	751	550	172	52	2	4342
1970-71	0	0	19	134	536	689	870	717	633	233	87	0	3918
1971-72	0	0	0	73	551	526	748	792	568	244	67	17	3586
1972-73	0	0	10	241	536	642	848	745	304	301	122	0	3749
1973-74	0	0	11	139	415	789	548	651	347	228	65	26	3219
1974-75	0	0	61	330	568	781	736	610	629	318	24	0	4057
1975-76	0	0	79	203	490	786	816	556	643	262	148	5	3988
1976-77	1	0	45	389	752	936	1242	788	437	169	60	23	4842
1977-78	0	0	10	352	448	890	1161	946	577	232	110	2	4728
1978-79	0	0	0	281	393	780	1043	826	462	230	86	2	4103
1979-80	0	0	7	267	507	783	804	883	629	288	63	3	4234
1980-81	0	0	23	304	545	821	1009	695	643	145	120	0	4305
1981-82	0	0	68	307	562	944	1030	662	450	367	29	0	4419
1982-83	0	0	62	263	522	678	910	734	529	413	78	11	4200
1983-84	0	0	57	181	572	942	977	682	616	304	173	5	4509
1984-85	0	0	58	47	648	588	1140	824	525	239	84	20	4173
1985-86	0	0	61	123	311	1049							

TABLE 5   COOLING DEGREE DAYS Base 65 deg. F   OAK RIDGE, TENNESSEE

YEAR	JAN	FEB	MAR	APR	MAY	JUNE	JULY	AUG	SEP	OCT	NOV	DEC	TOTAL
1969	0	0	0	23	129	300	443	311	136	22	0	0	1364
1970	0	0	0	37	140	219	371	361	291	25	0	0	1444
1971	0	0	0	16	46	323	316	312	232	47	9	0	1301
1972	0	0	0	41	40	168	322	330	229	5	2	0	1137
1973	0	0	2	16	34	277	374	325	254	56	0	0	1338
1974	0	0	1	20	103	116	330	295	86	2	0	0	953
1975	0	0	0	23	133	216	306	353	114	11	0	0	1156
1976	0	1	0	6	17	187	273	234	53	0	0	0	771
1977	0	0	3	4	109	241	387	324	177	7	4	0	1256
1978	0	0	0	3	89	234	342	335	240	1	0	0	1244
1979	0	0	2	5	74	202	285	314	151	11	0	0	1044
1980	0	0	0	6	92	244	491	435	291	13	0	0	1572
1981	0	0	0	39	40	319	404	286	126	3	0	0	1217
1982	0	0	15	2	167	207	392	279	120	44	0	3	1229
1983	0	0	0	5	47	211	405	445	196	12	0	0	1321
1984	0	0	0	3	47	283	275	318	106	69	0	0	1101
1985	0	0	2	16	63	207	313	258	125	31	3	0	1018

TABLE 6   SNOWFALL (inches)   OAK RIDGE, TENNESSEE

SEASON	JULY	AUG	SEP	OCT	NOV	DEC	JAN	FEB	MAR	APR	MAY	JUNE	TOTAL
1956-57	0.0	0.0	0.0	0.0	T	T	T	0.0	0.6	0.0	0.0	0.0	0.6
1957-58	0.0	0.0	0.0	T	T	T	T	3.8	T	0.0	0.0	0.0	3.8
1958-59	0.0	0.0	0.0	0.0	T	1.1	1.3	T	0.1	0.0	0.0	0.0	2.5
1959-60	0.0	0.0	0.0	0.0	0.2	1.1	7.8	11.3	21.0	0.0	0.0	0.0	41.4
1960-61	0.0	0.0	0.0	0.0	0.0	2.4	3.5	1.7	0.4	0.0	0.0	0.0	8.0
1961-62	0.0	0.0	0.0	0.0	0.0	5.6	8.7	0.2	2.4	T	0.0	0.0	16.9
1962-63	0.0	0.0	0.0	T	T	6.9	2.6	7.7	T	0.0	0.0	0.0	17.2
1963-64	0.0	0.0	0.0	0.0	0.1	14.8	0.3	3.1	T	0.0	0.0	0.0	18.3
1964-65	0.0	0.0	0.0	0.0	T	T	4.3	5.8	1.2	0.0	0.0	0.0	11.3
1965-66	0.0	0.0	0.0	0.0	0.0	T	9.6	T	0.3	0.0	0.0	0.0	9.9
1966-67	0.0	0.0	0.0	0.0	1.0	4.4	4.2	8.2	T	0.0	0.0	0.0	17.8
1967-68	0.0	0.0	0.0	0.0	0.0	0.2	3.9	2.7	0.5	0.0	0.0	0.0	7.3
1968-69	0.0	0.0	0.0	0.0	0.4	0.5	2.8	10.2	0.9	0.0	0.0	0.0	14.8
1969-70	0.0	0.0	0.0	0.0	T	8.1	8.6	4.6	0.3	0.0	0.0	0.0	21.6
1970-71	0.0	0.0	0.0	0.0	T	5.4	1.7	1.9	2.6	0.3	0.0	0.0	11.9
1971-72	0.0	0.0	0.0	0.0	0.2	0.7	0.4	5.0	4.0	0.0	0.0	0.0	10.3
1972-73	0.0	0.0	0.0	0.0	0.0	0.4	T	8.0	0.2	0.6	0.2	0.0	9.4
1973-74	0.0	0.0	0.0	0.0	0.0	0.6	0.0	0.3	0.4	0.0	0.0	0.0	1.3
1974-75	0.0	0.0	0.0	0.0	0.4	1.5	1.1	T	3.1	T	0.0	0.0	6.1
1975-76	0.0	0.0	0.0	0.0	0.1	T	2.2	3.9	0.0	0.0	0.0	0.0	6.2
1976-77	0.0	0.0	0.0	0.0	T	1.3	8.7	T	T	0.0	0.0	0.0	10.0
1977-78	0.0	0.0	0.0	0.0	0.0	1.0	8.6	6.2	3.4	T	0.0	0.0	19.2
1978-79	0.0	0.0	0.0	0.0	0.0	T	4.2	17.2	0.5	0.0	0.0	0.0	21.9
1979-80	0.0	0.0	0.0	0.0	T	T	1.0	7.9	3.2	T	0.0	0.0	12.1
1980-81	0.0	0.0	0.0	0.0	T	T	5.3	0.2	0.1	0.0	0.0	0.0	5.6
1981-82	0.0	0.0	0.0	0.0	T	0.9	4.4	0.5	1.8	T	0.0	0.0	7.6
1982-83	0.0	0.0	0.0	0.0	0.0	2.8	1.4	4.8	T	0.2	0.0	0.0	9.2
1983-84	0.0	0.0	0.0	0.0	T	0.3	4.6	5.1	0.2	0.0	0.0	0.0	10.2
1984-85	0.0	0.0	0.0	0.0	0.0	0.1	9.4	3.6	0.0	0.0	0.0	0.0	13.1
1985-86	0.0	0.0	0.0	0.0	0.0	T							
Record Mean	0.0	0.0	0.0	T	0.4	1.7	3.5	3.3	1.3	T	0.0	0.0	10.3

See Reference Notes, relative to all above tables, on preceding page.

Abilene is located in north central Texas. The station elevation is 1,750 feet above sea level. Topography of the area includes rolling plains, treeless except for mesquite, broken by low hills to the south and west. The land rises gently to the east and southeast. Regional agricultural products are mainly cattle, dry-land cotton, and feed crops.

Abilene is on the boundary between the humid east Texas climate and the semi-arid west and north Texas climate. The rainfall pattern is typical of the Great Plains. Most precipitation occurs from April to October and is usually associated with thunderstorms. Severe storms are infrequent, occurring mostly in the spring.

The large range of high and low temperatures, characteristic of the Great Plains, extends south to the Abilene area. High daytime temperatures prevail in the summer, but are normally broken by thunderstorms about five times a month. Rapid cooling after sunset results in pleasant nights with low summertime temperatures in the upper 60s and low 70s. High summer temperatures are usually associated with fair skies, southwesterly winds, and low humidities.

Rapid wintertime temperature changes occur when cold, dry, arctic air replaces warm moist tropical air. Drops in temperature of 20 to 30 degrees in one hour are not unusual. However, cold weather periods are short lived. Fair, mild weather is typical.

South is the prevailing wind direction, and southerly winds are frequently high and persist for several days. Strong northerly winds often occur during the passage of cold fronts. Dusty conditions are infrequent, occurring mostly with westerly winds. Dust storm frequency and intensity depend on soil conditions in eastern New Mexico, west Texas, and the Texas Panhandle.

Based on the 1951-1980 period, the average first occurrence of 32 degrees Fahrenheit in the fall is November 13 and the average last occurrence in the spring is March 25.

TABLE 1

NORMALS, MEANS AND EXTREMES

ABILENE, TEXAS

LATITUDE: 32°25'N   LONGITUDE: 99°41'W   ELEVATION: FT. GRND 1784 BARO 01789   TIME ZONE: CENTRAL   WBAN: 13962

	(a)	JAN	FEB	MAR	APR	MAY	JUNE	JULY	AUG	SEP	OCT	NOV	DEC	YEAR
TEMPERATURE °F:														
Normals														
-Daily Maximum		55.5	60.3	68.6	77.6	84.1	91.8	95.4	94.5	87.1	77.6	64.8	58.4	76.3
-Daily Minimum		31.2	35.5	42.6	52.8	60.8	69.0	72.7	71.7	64.9	54.1	42.0	34.3	52.6
-Monthly		43.4	47.9	55.6	65.2	72.5	80.4	84.1	83.1	76.0	65.8	53.4	46.4	64.5
Extremes														
-Record Highest	46	89	90	97	99	107	109	110	109	106	103	92	89	110
-Year		1943	1940	1974	1948	1967	1980	1978	1943	1952	1979	1980	1955	JUL 1978
-Record Lowest	46	-9	-7	7	25	36	47	55	55	35	28	14	2	-9
-Year		1947	1985	1943	1973	1979	1964	1940	1961	1942	1957	1976	1983	JAN 1947
NORMAL DEGREE DAYS:														
Heating (base 65°F)		673	479	321	98	11	0	0	0	10	91	361	577	2621
Cooling (base 65°F)		0	0	29	104	244	465	592	561	340	119	13	0	2467
% OF POSSIBLE SUNSHINE	37	62	65	69	71	70	78	79	77	69	71	68	65	70
MEAN SKY COVER (tenths)														
Sunrise - Sunset	46	5.7	5.6	5.4	5.2	5.4	4.5	4.4	4.4	4.5	4.4	4.8	5.2	5.0
MEAN NUMBER OF DAYS:														
Sunrise to Sunset														
-Clear	46	10.7	10.2	11.2	11.5	10.2	12.9	14.4	14.5	14.0	14.7	13.5	12.1	149.8
-Partly Cloudy	46	6.4	6.2	7.4	7.7	10.2	10.3	9.8	9.8	7.8	7.0	6.0	6.5	95.1
-Cloudy	46	13.9	11.9	12.4	10.8	10.6	6.8	6.8	6.8	8.2	9.3	10.5	12.4	120.3
Precipitation														
.01 inches or more	46	5.1	5.1	4.6	6.3	7.9	6.0	4.7	5.4	5.9	5.6	4.4	4.3	65.4
Snow, Ice pellets														
1.0 inches or more	46	0.7	0.5	0.2	0.0	0.0	0.0	0.0	0.0	0.0	0.0	0.2	0.3	1.9
Thunderstorms	46	0.5	1.2	2.8	5.2	7.9	5.9	4.7	5.2	3.2	2.8	1.3	0.7	41.5
Heavy Fog Visibility														
1/4 mile or less	46	1.3	1.3	0.6	0.4	0.5	0.1	0.*	0.*	0.3	0.7	1.0	1.0	7.2
Temperature °F														
-Maximum														
90° and above	22	0.0	0.0	0.5	2.3	7.3	20.0	26.5	25.0	12.5	2.2	0.1	0.0	96.5
32° and below	22	2.3	0.8	0.*	0.0	0.0	0.0	0.0	0.0	0.0	0.0	0.1	1.1	4.4
-Minimum														
32° and below	22	17.9	12.4	4.4	0.5	0.0	0.0	0.0	0.0	0.0	0.1	4.9	14.1	54.2
0° and below	22	0.*	0.*	0.0	0.0	0.0	0.0	0.0	0.0	0.0	0.0	0.0	0.0	0.1
AVG. STATION PRESS. (mb)	13	956.1	954.5	950.5	950.5	949.3	950.8	952.8	952.9	953.2	954.2	954.4	955.4	952.9
RELATIVE HUMIDITY (%)														
Hour 00	22	67	66	61	63	68	64	57	60	67	67	69	66	65
Hour 06 (Local Time)	22	73	72	70	72	78	77	71	73	77	76	74	71	74
Hour 12	22	55	53	48	46	51	49	44	47	53	52	53	52	50
Hour 18	22	50	45	40	40	44	41	38	40	48	49	52	51	45
PRECIPITATION (inches):														
Water Equivalent														
-Normal		0.97	0.96	1.08	2.35	3.25	2.52	2.11	2.47	3.06	2.32	1.32	0.85	23.26
-Maximum Monthly	46	4.35	2.84	5.16	6.80	13.19	9.60	7.15	8.18	11.03	10.68	4.60	3.08	13.19
-Year		1968	1940	1979	1966	1957	1961	1968	1969	1974	1981	1968	1984	MAY 1957
-Minimum Monthly	46	T	0.04	0.03	T	0.15	0.03	T	T	T	0.00	0.00	T	0.00
-Year		1967	1962	1963	1961	1956	1954	1970	1943	1956	1952	1949	1972	OCT 1952
-Maximum in 24 hrs	46	2.18	1.74	2.23	3.75	2.85	3.66	3.74	6.30	6.70	6.08	2.43	2.30	6.70
-Year		1961	1940	1977	1957	1969	1959	1960	1978	1961	1981	1975	1946	SEP 1961
Snow, Ice pellets														
-Maximum Monthly	46	13.5	8.4	7.3	T						T	8.1	7.8	13.5
-Year		1973	1956	1970	1980						1980	1968	1983	JAN 1973
-Maximum in 24 hrs	46	7.5	4.5	6.1	T						T	5.3	4.2	7.5
-Year		1973	1979	1970	1980						1980	1976	1946	JAN 1973
WIND:														
Mean Speed (mph)	41	11.9	12.7	14.1	14.1	13.1	13.0	10.9	10.5	10.6	11.1	11.7	11.9	12.1
Prevailing Direction through 1963		S	S	S	SSE	SSE	SSE	SSE	SSE	SSE	SSE	S	SSW	SSE
Fastest Obs. 1 Min.														
-Direction (!!!)	5	26	35	27	33	25	36	36	03	32	31	19	18	31
-Speed (MPH)	5	35	36	41	40	39	35	39	29	28	46	35	31	46
-Year		29	1984	1984	1983	1980	1982	1981	1984	1983	1984	1982	1984	OCT 1984
Peak Gust														
-Direction (!!!)	2	N	N	W	W	W	N	SE	NE	S	NW	NW	S	W
-Speed (mph)	2	48	52	61	56	63	41	40	38	36	58	39	44	63
-Date		1985	1984	1984	1984	1985	1985	1985	1984	1985	1984	1985	1985	MAY 1985

See Reference Notes to this table on the following page

TABLE 2 — PRECIPITATION (inches) — ABILENE, TEXAS

YEAR	JAN	FEB	MAR	APR	MAY	JUNE	JULY	AUG	SEP	OCT	NOV	DEC	ANNUAL
1956	0.69	1.39	0.35	2.32	0.15	0.12	0.87	0.55	T	1.91	0.45	0.98	9.78
1957	0.54	2.15	0.54	6.52	13.19	2.30	0.62	0.47	2.92	4.35	2.45	0.52	36.57
1958	1.39	1.63	1.37	3.50	3.45	2.08	3.12	2.06	4.23	3.72	0.71	0.23	27.49
1959	0.04	0.73	0.42	1.88	2.34	7.22	4.93	0.72	0.77	5.79	0.35	2.12	27.31
1960	2.14	0.92	0.27	2.42	1.93	1.95	4.51	2.68	1.43	3.99	0.02	2.83	25.09
1961	3.99	1.52	0.65	T	1.29	9.60	4.35	1.41	7.86	1.67	2.91	0.30	35.55
1962	0.07	0.04	1.20	1.37	1.55	7.65	4.54	1.48	5.12	2.27	1.06	0.77	27.12
1963	0.01	0.34	0.03	1.53	6.21	1.63	0.28	3.56	0.67	0.39	2.67	0.44	17.76
1964	2.36	2.47	1.26	2.76	1.53	1.55	2.58	5.87	2.11	0.50	2.19	0.17	25.35
1965	1.13	1.48	0.59	2.78	5.04	3.70	0.69	1.59	1.86	2.67	1.85	0.97	24.35
1966	0.85	0.40	0.40	6.80	0.46	1.16	0.12	5.08	4.08	2.09	0.28	0.05	21.77
1967	T	0.24	1.01	0.49	4.39	2.85	1.32	0.54	5.44	1.74	2.89	2.43	23.34
1968	4.35	1.93	2.12	3.32	2.40	1.09	7.15	0.89	0.53	0.14	4.60	0.12	28.64
1969	0.81	1.28	2.08	2.49	7.76	2.42	2.15	8.18	4.10	1.98	1.11	2.48	36.84
1970	0.04	1.84	2.41	2.56	3.68	1.94	T	2.41	1.18	0.08			17.92
1971	0.01	0.57	0.04	2.44	2.17	1.78	1.85	6.92	5.33	2.43	0.76	1.81	26.11
1972	0.29	0.11	0.23	1.04	2.47	3.67	1.22	4.92	3.15	4.49	0.37	T	21.96
1973	3.52	1.61	3.28	1.10	1.38	2.21	2.68	0.30	4.65	3.15	0.22	0.01	24.11
1974	0.18	0.57	0.72	3.44	0.72	1.03	2.20	5.61	11.03	5.06	1.75	1.14	33.45
1975	0.83	1.86	1.49	0.52	4.43	1.68	1.93	2.47	2.27	0.84	2.67	1.22	22.21
1976	0.02	0.08	0.25	3.70	1.04	0.68	4.27	1.47	3.97	6.24	0.63	0.18	22.53
1977	1.22	0.08	2.35	3.11	0.44	1.71	1.82	2.54	0.09	2.15	0.57	0.19	16.27
1978	0.62	1.26	0.17	1.00	1.50	1.24	0.72	6.70	2.36	1.49	0.94	0.28	18.28
1979	0.72	1.10	5.16	1.72	1.86	2.89	1.55	1.55	0.01	0.53	0.65	2.62	20.36
1980	0.77	0.72	0.69	0.17	5.00	1.14	0.24	1.62	6.30	0.71	1.62	1.69	20.67
1981	1.20	1.06	2.36	3.52	1.48	2.73	1.69	0.63	1.74	10.68	0.43	0.25	27.77
1982	1.00	1.20	0.43	0.38	6.87	3.98	1.50	1.12	1.09	0.74	1.72	1.28	21.31
1983	2.28	0.12	1.95	0.54	1.42	3.86	2.57	0.10	0.87	3.25	1.77	0.77	19.50
1984	0.98	0.46	0.47	0.20	0.42	1.70	0.97	3.24	3.55	5.15	2.11	3.08	22.33
1985	0.53	1.50	2.86	0.90	4.03	1.78	1.71	3.66	1.28	2.43	1.56	0.03	22.27
Record Mean	0.94	0.99	1.21	2.38	3.78	2.65	2.02	2.20	2.64	2.66	1.29	1.18	23.94

TABLE 3 — AVERAGE TEMPERATURE (deg. F) — ABILENE, TEXAS

YEAR	JAN	FEB	MAR	APR	MAY	JUNE	JULY	AUG	SEP	OCT	NOV	DEC	ANNUAL
1956	43.4	47.7	58.7	64.1	76.9	84.7	86.2	85.1	79.3	69.8	52.0	49.0	66.4
1957	44.0	53.3	54.3	61.9	68.2	77.8	85.7	84.6	73.5	62.6	48.9	49.8	63.7
1958	44.1	43.6	47.3	61.2	72.0	81.1	84.5	83.5	75.5	63.1	54.7	42.6	62.8
1959	42.5	46.9	56.0	63.1	74.7	79.0	79.8	82.1	78.3	63.1	47.8	48.1	63.5
1960	44.2	42.7	49.8	65.9	72.3	82.8	83.4	82.8	77.3	67.0	57.0	40.9	63.9
1961	40.6	48.0	58.0	65.4	74.2	76.2	79.2	79.7	72.9	65.2	50.0	44.9	62.9
1962	39.7	54.5	53.0	64.0	77.3	77.3	82.8	84.3	75.2	69.1	55.0	46.6	64.9
#1963	38.7	47.1	60.7	70.3	75.0	79.5	86.1	85.6	78.8	73.0	57.4	41.4	66.1
1964	46.2	44.9	57.2	68.8	75.3	81.3	87.0	85.3	76.1	66.0	56.7	48.6	66.1
1965	48.3	45.8	48.1	68.1	72.3	79.3	85.7	84.7	79.2	66.5	62.3	51.4	66.0
1966	39.3	44.4	56.9	63.2	71.0	79.9	86.9	80.4	75.2	64.1	59.5	44.6	63.8
1967	48.0	46.6	63.5	71.9	72.5	81.9	82.8	80.2	71.1	63.7	54.2	43.7	65.0
1968	43.9	43.6	53.1	61.7	71.4	79.3	80.7	82.5	74.4	69.5	52.6	45.5	63.2
1969	48.0	48.0	47.5	65.7	70.4	78.4	87.4	85.6	75.9	61.9	53.4	47.7	64.2
1970	40.8	48.8	51.4	65.1	70.3	79.2	84.4	83.8	78.1	63.4	51.2	51.5	64.0
1971	46.4	48.7	55.1	64.5	72.4	80.4	84.4	76.9	73.3	66.0	55.0	48.9	64.3
1972	43.7	49.2	59.5	69.4	70.6	80.6	80.8	78.9	75.3	63.7	46.4	42.1	63.3
1973	37.5	44.4	57.2	58.6	70.6	77.4	81.2	82.5	73.8	66.5	58.2	46.8	62.9
1974	42.8	51.2	62.6	65.6	76.0	80.8	84.4	78.7	66.3	64.6	51.7	44.4	64.1
1975	44.8	45.1	54.1	62.9	70.2	78.5	79.7	81.3	70.8	65.6	54.3	46.1	62.8
1976	42.4	56.0	56.5	64.8	68.2	78.9	77.7	81.6	73.5	56.9	45.6	42.5	62.1
1977	35.7	50.9	56.8	62.4	73.4	81.3	83.8	83.4	82.6	66.9	54.6	48.7	65.0
1978	34.5	38.7	55.0	70.3	76.3	83.3	89.0	82.3	77.3	66.9	55.3	43.3	64.3
1979	35.6	44.9	56.7	64.2	69.8	79.0	83.7	82.2	77.7	71.6	50.9	47.7	63.7
1980	45.2	47.9	54.8	63.9	72.1	84.4	89.4	86.1	77.7	64.9	52.6	49.3	65.7
1981	44.8	49.4	55.1	68.3	71.6	80.3	85.8	83.3	78.0	66.2	57.4	46.8	65.8
1982	44.8	44.7	58.3	62.9	72.1	78.8	84.1	85.6	77.6	65.9	54.4	45.8	64.6
1983	42.6	47.0	54.9	60.6	70.8	76.7	82.8	85.2	77.9	69.1	57.0	34.2	63.2
1984	39.5	49.9	55.3	63.6	75.3	83.1	83.7	83.1	73.1	63.8	52.7	48.8	64.3
1985	37.2	44.4	57.5	66.7	73.8	77.9	81.6	84.9	75.7	65.7	54.6	41.4	63.5
Record Mean	44.0	47.9	56.0	64.9	72.3	80.0	83.5	83.0	76.1	65.9	53.9	46.1	64.5
Max	55.4	59.8	68.4	77.1	83.6	91.2	94.6	94.2	87.0	77.3	65.2	57.2	75.9
Min	32.5	35.9	43.6	52.8	60.9	68.8	72.3	71.8	65.1	54.5	42.6	34.9	53.0

REFERENCE NOTES FOR TABLES 1, 2, 3 and 6 (ABILENE, TX)

GENERAL

T - TRACE AMOUNT
BLANK ENTRIES DENOTE MISSING/UNREPORTED DATA.
# INDICATES A STATION OR INSTRUMENT RELOCATION.

SPECIFIC

TABLE 1

(a) - LENGTH OF RECORD IN YEARS ALTHOUGH INDIVIDUAL MONTHS MAY BE MISSING.
* LESS THAN .05

NORMALS — BASED ON THE 1951-1980 RECORD PERIOD.
EXTREMES — DATES ARE THE MOST RECENT OCCURRENCE.
WIND DIR. — NUMERALS SHOW TENS OF DEGREES CLOCKWISE FROM TRUE NORTH.
  "00" INDICATES CALM.
RESULTANT WIND DIRECTIONS ARE GIVEN TO WHOLE DEGREES.

EXCEPTIONS

TABLE 1

1. FASTEST MILE WINDS ARE THROUGH JANUARY 1980.

TABLES 2, 3, and 6

RECORD MEANS ARE THROUGH THE CURRENT YEAR, BEGINNING IN
  1886 FOR TEMPERATURE
  1986 FOR PRECIPITATION
  1940 FOR SNOWFALL

TABLE 4 HEATING DEGREE DAYS Base 65 deg. F   ABILENE, TEXAS

SEASON	JULY	AUG	SEP	OCT	NOV	DEC	JAN	FEB	MAR	APR	MAY	JUNE	TOTAL
1956-57	0	0	0	36	396	489	647	328	333	155	36	6	2426
1957-58	0	0	6	141	487	466	642	590	544	160	16	0	3052
1958-59	0	0	6	133	319	690	689	507	299	183	8	0	2834
1959-60	0	0	8	118	518	520	639	642	482	81	40	0	3048
1960-61	0	0	3	55	261	739	749	473	241	114	10	3	2648
1961-62	0	0	20	71	449	618	777	310	379	115	8	0	2747
1962-63	0	0	2	68	299	562	805	502	207	51	15	0	2511
#1963-64	0	0	2	6	251	725	573	575	258	54	6	3	2453
1964-65	0	0	2	44	293	505	519	538	522	59	0	0	2482
1965-66	0	0	4	67	117	424	791	570	281	122	42	0	2418
1966-67	0	0	0	119	198	633	532	508	154	14	23	0	2181
1967-68	0	0	17	118	321	654	648	611	366	134	15	0	2884
1968-69	0	0	0	42	387	599	520	468	541	53	15	1	2626
1969-70	0	0	0	192	350	528	748	449	422	92	31	1	2813
1970-71	0	0	15	147	414	411	569	455	326	96	22	0	2455
1971-72	0	0	44	44	312	492	651	464	207	49	12	0	2275
1972-73	0	0	13	158	554	702	844	574	235	228	43	0	3351
1973-74	0	0	6	60	227	557	680	389	181	78	7	0	2185
1974-75	0	0	60	67	400	631	620	553	351	144	7	0	2833
1975-76	0	0	55	97	327	578	692	276	296	68	39	0	2428
1976-77	0	0	11	276	575	690	901	393	262	89	0	0	3197
1977-78	0	0	0	56	314	500	938	730	337	37	34	0	2946
1978-79	0	0	5	56	308	667	906	567	272	84	55	0	2920
1979-80	0	0	0	43	429	530	605	496	316	118	10	0	2547
1980-81	0	0	16	116	389	485	619	439	312	51	20	0	2447
1981-82	0	0	7	116	237	496	617	570	253	149	22	0	2467
1982-83	0	0	0	111	340	589	686	497	325	209	23	4	2784
1983-84	0	0	9	37	261	947	781	431	310	107	12	0	2895
1984-85	0	0	65	108	375	501	853	572	259	52	3	0	2788
1985-86	0	0	28	70	328	725							

TABLE 5 COOLING DEGREE DAYS Base 65 deg. F   ABILENE, TEXAS

YEAR	JAN	FEB	MAR	APR	MAY	JUNE	JULY	AUG	SEP	OCT	NOV	DEC	TOTAL
1969	0	0	5	83	190	412	699	644	332	105	12	0	2482
1970	2	0	5	104	202	433	607	590	411	106	6	1	2467
1971	0	4	26	88	256	469	607	376	297	83	20	0	2226
1972	0	12	41	188	189	474	497	436	330	128	1	0	2296
1973	0	0	0	44	224	379	508	550	277	111	32	0	2125
1974	0	7	113	102	354	480	607	430	104	62	8	0	2267
1975	0	0	20	89	175	411	462	513	236	120	13	0	2039
1976	0	21	40	67	145	424	401	521	275	31	0	0	1925
1977	0	4	13	17	267	496	590	579	534	122	7	1	2630
1978	0	0	35	201	392	555	751	542	379	121	23	1	3000
1979	0	9	20	67	209	428	586	539	253	11	0		2512
1980	0	7	6	91	237	588	762	660	405	119	21	5	2901
1981	0	6	8	154	230	464	653	575	403	162	15	1	2671
1982	1	8	52	92	249	422	600	645	386	147	28	1	2631
1983	0	0	21	84	208	365	558	634	403	171	30	0	2474
1984	0	0	20	71	339	550	589	570	315	80	13	4	2551
1985	0	0	32	107	281	393	519	621	356	97	25	0	2431

TABLE 6 SNOWFALL (inches)   ABILENE, TEXAS

SEASON	JULY	AUG	SEP	OCT	NOV	DEC	JAN	FEB	MAR	APR	MAY	JUNE	TOTAL
1956-57	0.0	0.0	0.0	0.0	0.0	0.0	T	0.0	4.0	0.0	0.0	0.0	T
1957-58	0.0	0.0	0.0	0.0	3.5	0.0	4.7	1.8	4.0	0.0	0.0	0.0	14.0
1958-59	0.0	0.0	0.0	0.0	0.0	4.0	T	1.5	T	0.0	0.0	0.0	5.5
1959-60	0.0	0.0	0.0	0.0	T	0.0	T	1.2	T	0.0	0.0	0.0	1.2
1960-61	0.0	0.0	0.0	0.0	0.0		3.9	4.7	0.0	0.0	0.0	0.0	8.6
1961-62	0.0	0.0	0.0	0.0	T	0.0	0.4	T	5.7	0.0	0.0	0.0	6.1
1962-63	0.0	0.0	0.0	0.0	T	T	0.7	2.5	T	0.0	0.0	0.0	3.2
1963-64	0.0	0.0	0.0	0.0	0.0	0.7	2.0	4.3	T	0.0	0.0	0.0	7.0
1964-65	0.0	0.0	0.0	0.0	0.0	T	0.0	T	T	0.0	0.0	0.0	T
1965-66	0.0	0.0	0.0	0.0	0.0	0.0	6.6	1.5	0.0	0.0	0.0	0.0	8.1
1966-67	0.0	0.0	0.0	0.0	0.0	T	0.0	T	T	0.0	0.0	0.0	T
1967-68	0.0	0.0	0.0	T	0.0	4.0	0.0	1.6	2.5	0.0	0.0	0.0	8.1
1968-69	0.0	0.0	0.0	0.0	8.1	0.0	0.0	T	0.5	0.0	0.0	0.0	8.6
1969-70	0.0	0.0	0.0	0.0	T	3.1	0.2	0.0	7.3	0.0	0.0	0.0	10.6
1970-71	0.0	0.0	0.0	0.0	0.0	0.0	1.2	0.4	0.0	0.0	0.0	0.0	1.6
1971-72	0.0	0.0	0.0	0.0	0.0	T	1.0	0.0	T	0.0	0.0	0.0	1.0
1972-73	0.0	0.0	0.0	0.0	0.1	0.0	13.5	3.0	0.0	0.0	0.0	0.0	16.6
1973-74	0.0	0.0	0.0	0.0	0.0	T	T	0.2	T	0.0	0.0	0.0	0.2
1974-75	0.0	0.0	0.0	0.0	0.0	T	6.0	1.6	T	0.0	0.0	0.0	7.6
1975-76	0.0	0.0	0.0	0.0	0.0	2.6	0.0	0.0	T	0.0	0.0	0.0	2.6
1976-77	0.0	0.0	0.0	0.0	5.6	0.0	3.9	0.0	0.0	0.0	0.0	0.0	9.5
1977-78	0.0	0.0	0.0	0.0	0.0	0.0	1.7	1.4	1.0	0.0	0.0	0.0	4.1
1978-79	0.0	0.0	0.0	0.0	0.0	0.9	0.5	5.9	0.0	0.0	0.0	0.0	7.3
1979-80	0.0	0.0	0.0	0.0	0.0	0.3	T	2.2	T	T	0.0	0.0	2.5
1980-81	0.0	0.0	0.0	T	3.7	0.0	1.9	0.0	0.0	0.0	0.0	0.0	5.6
1981-82	0.0	0.0	0.0	0.0	0.0	0.0	5.0	1.2	0.0	0.0	0.0	0.0	6.2
1982-83	0.0	0.0	0.0	0.0	T	1.4	10.8	T	T	0.0	0.0	0.0	12.2
1983-84	0.0	0.0	0.0	0.0	0.0	7.8	T	0.4	0.8	0.0	0.0	0.0	9.0
1984-85	0.0	0.0	0.0	0.0	0.0	0.0	6.1	1.9	0.0	0.0	0.0	0.0	8.0
1985-86	0.0	0.0	0.0	0.0	0.0	0.2							
Record Mean	0.0	0.0	0.0	T	0.5	0.6	2.1	1.1	0.6	T	0.0	0.0	4.9

See Reference Notes, relative to all above tables, on preceding page.

The station is located 7 statute miles east northeast of the downtown post office in a region of rather flat topography. The Canadian River flows eastward 18 miles north of the station, with its bed about 800 feet below the plains. The Prairie Dog Town Fork of the Red River flows southeastward about 15 miles south of the station where it enters the Palo Duro Canyon, which is about 1,000 feet deep. There are numerous shallow Playa lakes, often dry, over the area, and the nearly treeless grasslands slope downward to the east. The terrain gradually rises to the west and northwest.

Three-fourths of the total annual precipitation falls from April through September, occurring from thunderstorm activity. Snow usually melts within a few days after it falls. Heavier snowfalls of 10 inches or more, usually with near blizzard conditions, average once every 5 years and last 2 to 3 days.

The Amarillo area is subject to rapid and large temperature changes, especially during the winter months when cold fronts from the northern Rocky Mountain and Plains states sweep across the area. Temperature drops of 50 to 60 degrees within a 12-hour period are not uncommon. Temperature drops of 40 degrees have occurred within a few minutes.

Humidity averages are low, occasionally dropping below 20 percent in the spring. Low humidity moderates the effect of high summer afternoon temperatures, permits evaporative cooling systems to be very effective, and provides many pleasant evenings and nights.

Severe local storms are infrequent, although a few thunderstorms with damaging hail, lightning, and wind in a very localized area occur most years, usually in spring and summer. These storms are often accompanied by very heavy rain, which produces local flooding, particularly of roads and streets. Tornadoes are rare.

Based on the 1951-1980 period, the average first occurrence of 32 degrees Fahrenheit in the fall is October 29 and the average last occurrence in the spring is April 14.

TABLE 1   NORMALS, MEANS AND EXTREMES

AMARILLO, TEXAS

LATITUDE: 35°14'N   LONGITUDE: 101°42'W   ELEVATION: FT. GRND 3604 BARO 03591   TIME ZONE: CENTRAL   WBAN: 23047

	(a)	JAN	FEB	MAR	APR	MAY	JUNE	JULY	AUG	SEP	OCT	NOV	DEC	YEAR
TEMPERATURE °F:														
Normals														
-Daily Maximum		49.1	53.1	60.8	71.0	79.1	88.2	91.4	89.6	82.4	72.7	58.7	51.8	70.7
-Daily Minimum		21.7	26.1	32.0	42.0	51.9	61.5	66.2	64.5	56.9	45.5	32.1	24.8	43.8
-Monthly		35.4	39.6	46.4	56.5	65.5	74.8	78.8	77.1	69.7	59.1	45.4	38.3	57.2
Extremes														
-Record Highest	45	81	88	94	98	102	108	105	106	102	95	87	81	108
-Year		1950	1963	1971	1965	1953	1953	1981	1944	1983	1954	1980	1955	JUN 1953
-Record Lowest	45	-11	-14	-3	14	28	42	53	49	-30	21	0	-7	-14
-Year		1984	1951	1948	1945	1954	1955	1950	1956	1984	1980	1976	1983	FEB 1951
NORMAL DEGREE DAYS:														
Heating (base 65°F)		918	711	577	271	92	6	0	0	25	215	588	828	4231
Cooling (base 65°F)		0	0	0	16	108	303	428	372	166	35	0	0	1428
% OF POSSIBLE SUNSHINE	44	68	69	71	73	72	77	78	77	73	74	71	67	73
MEAN SKY COVER (tenths)														
Sunrise - Sunset	44	5.2	5.2	5.3	5.0	5.1	4.3	4.5	4.3	4.1	4.0	4.4	4.9	4.7
MEAN NUMBER OF DAYS:														
Sunrise to Sunset														
-Clear	44	12.3	10.5	11.4	11.7	11.0	13.1	13.0	14.7	15.3	16.2	14.5	12.9	156.7
-Partly Cloudy	44	7.3	7.7	8.7	8.8	10.4	11.2	12.4	10.1	7.2	7.0	6.8	7.6	105.1
-Cloudy	44	11.4	10.0	10.9	9.6	9.5	5.7	5.6	6.2	7.5	7.9	8.8	10.5	103.4
Precipitation														
.01 inches or more	44	4.1	4.4	4.7	5.0	8.2	8.2	8.3	8.3	5.8	5.0	3.3	3.8	69.0
Snow,Ice pellets														
1.0 inches or more	44	1.4	1.1	0.7	0.2	0.0	0.0	0.0	0.0	0.0	0.1	0.5	0.8	4.8
Thunderstorms	44	0.1	0.5	1.5	3.4	8.5	9.4	9.5	8.8	4.0	2.5	0.6	0.2	49.1
Heavy Fog Visibility														
1/4 mile or less	44	3.4	4.2	3.3	2.0	2.0	0.7	0.5	0.7	2.1	2.4	2.8	2.8	26.9
Temperature °F														
-Maximum														
90° and above	24	0.0	0.0	0.1	0.8	4.8	13.2	22.0	17.0	7.0	1.1	0.0	0.0	65.9
32° and below	24	5.0	2.7	0.8	0.*	0.0	0.0	0.0	0.0	0.0	0.*	0.8	3.5	12.9
-Minimum														
32° and below	24	27.5	22.0	14.9	3.3	0.1	0.0	0.0	0.0	0.2	1.8	14.2	26.5	110.6
0° and below	24	1.3	0.3	0.0	0.0	0.0	0.0	0.0	0.0	0.0	0.0	0.*	0.4	2.0
AVG. STATION PRESS.(mb)	13	891.9	891.0	887.8	888.7	888.7	890.3	892.8	893.0	892.9	892.6	891.6	891.8	891.1
RELATIVE HUMIDITY (%)														
Hour 00	24	65	64	58	55	62	64	60	66	69	64	66	65	63
Hour 06	24	71	71	68	67	73	76	72	77	79	72	73	70	72
Hour 12 (Local Time)	24	51	49	42	38	42	44	41	45	49	44	48	48	45
Hour 18	24	48	42	35	31	36	39	37	42	44	42	49	49	41
PRECIPITATION (inches):														
Water Equivalent														
-Normal		0.46	0.57	0.87	1.08	2.79	3.50	2.70	2.95	1.72	1.39	0.58	0.49	19.10
-Maximum Monthly	45	2.33	1.83	3.99	3.74	9.81	10.73	7.59	7.55	5.02	7.64	2.26	4.52	10.73
-Year		1968	1948	1973	1942	1951	1965	1960	1974	1950	1941	1961	1959	JUN 1965
-Minimum Monthly	45	T	T	T	T	0.04	0.01	0.12	0.28	0.03	0.00	T	T	0.00
-Year		1976	1943	1950	1964	1984	1953	1946	1983	1977	1952	1960	1976	OCT 1952
-Maximum in 24 hrs	45	1.74	1.28	2.27	1.99	6.75	6.15	4.74	4.26	3.42	3.45	1.29	3.11	6.75
-Year		1968	1971	1973	1985	1951	1960	1982	1945	1941	1948	1971	1943	MAY 1951
Snow,Ice pellets														
-Maximum Monthly	45	14.5	17.3	14.7	6.4	0.5				0.3	3.9	13.6	8.5	17.3
-Year		1983	1971	1961	1947	1978				1984	1976	1952	1943	FEB 1971
-Maximum in 24 hrs	45	9.8	13.5	9.8	5.1	0.5				0.3	3.2	12.2	7.4	13.5
-Year		1983	1971	1957	1947	1978				1984	1976	1952	1971	FEB 1971
WIND:														
Mean Speed (mph)	44	13.0	14.1	15.5	15.5	14.7	14.3	12.6	12.1	13.0	13.0	13.1	13.0	13.7
Prevailing Direction through 1963		SW	SW	SW	SW	S	S	S	S	S	SW	SW	SW	SW
Fastest Obs. 1 Min.														
-Direction	11	36	36	34	27	36	31	36	30	36	31	26	28	31
-Speed (MPH)	11	41	47	58	45	41	40	41	37	39	58	44	37	58
-Year		1975	1984	1977	1981	1983	1981	1979	1981	1983	1979	1975	1977	OCT 1979
Peak Gust														
-Direction	2	W	NW	W	N	W	NE	SE	S	NE	N	W	W	NW
-Speed (mph)	2	46	62	54	52	45	56	38	41	51	58	48	44	62
-Date		1985	1984	1985	1984	1985	1984	1985	1985	1985	1985	1985	1984	FEB 1984

See Reference Notes to this table on the following page.

TABLE 2   PRECIPITATION (inches)   AMARILLO, TEXAS

YEAR	JAN	FEB	MAR	APR	MAY	JUNE	JULY	AUG	SEP	OCT	NOV	DEC	ANNUAL
1956	0.09	1.10	0.03	0.23	1.99	2.03	2.82	0.79	0.48	0.38	T	T	9.94
1957	0.33	1.11	2.82	2.69	4.36	0.53	0.13	4.85	0.88	2.57	0.94	0.03	21.24
1958	1.05	0.58	2.36	1.74	2.45	4.22	6.16	2.08	1.60	0.15	0.60	0.30	23.29
1959	0.16	0.06	0.26	1.18	4.82	2.19	2.85	2.24	2.29	2.10	0.14	4.52	22.81
1960	1.30	0.95	1.66	1.66	0.82	9.85	7.59	3.15	4.22	4.82	T	0.65	36.67
1961	0.12	0.27	2.55	0.24	3.40	3.42	4.10	3.14	1.87	0.91	2.26	0.16	22.44
1962	0.47	0.39	0.02	1.48	1.76	10.16	7.51	3.29	2.66	0.85	0.53	0.64	29.76
1963	0.06	0.67	0.28	0.47	3.66	3.60	2.04	3.93	0.43	1.54	0.29		17.30
1964	T	1.37	0.03	T	1.69	1.90	0.94	5.69	3.95	0.08	1.53	0.79	17.97
1965	0.55	0.47	0.72	0.23	1.88	10.73	1.54	1.71	0.79	1.02	0.07	0.38	20.09
1966	0.43	0.69	0.01	0.87	0.19	4.62	1.37	3.77	2.40	0.29	0.08	0.19	14.91
1967	T	0.15	0.42	1.95	1.40	2.55	3.70	1.81	2.47	1.61	0.28	0.51	16.85
1968	2.33	0.73	0.45	0.93	2.84	1.68	2.96	3.35	0.62	0.90	0.92	0.26	17.97
1969	0.02	0.50	1.15	0.30	2.93	4.09	2.55	4.51	2.77	2.56	0.34	0.83	22.55
1970	0.02	0.02	2.10	1.33	0.23	1.54	1.39	1.27	0.34	1.06	0.26	T	9.56
1971	0.10	1.65	0.10	0.77	0.91	4.17	1.75	3.33	4.70	2.59	2.08	0.89	23.04
1972	0.21	0.11	0.11	0.03	2.81	3.87	2.59	1.73	0.71	1.66	1.19	0.32	15.34
1973	0.56	0.42	3.99	1.88	1.43	0.84	4.08	2.31	1.22	1.05	0.10	0.17	18.05
1974	0.33	0.24	0.60	0.04	4.06	3.33	1.31	7.55	1.65	3.44	0.12	0.42	23.09
1975	0.28	1.33	0.51	1.02	2.47	4.15	5.19	3.97	0.76	0.33	0.92	0.15	21.08
1976	T	0.10	0.79	1.65	1.36	2.94	1.77	1.78	4.28	1.14	0.43	T	16.24
1977	0.64	0.53	0.24	2.74	4.01	2.06	3.14	4.94	0.03	0.26	0.32	0.27	19.18
1978	0.63	0.80	0.21	0.55	5.76	6.50	1.82	1.61	2.42	0.97	0.47	0.27	22.01
1979	0.92	0.28	1.46	1.29	3.94	3.19	2.03	5.08	0.52	1.28	0.40	0.07	20.46
1980	0.85	0.55	1.38	0.82	2.88	1.30	0.65	1.80	1.55	0.42	0.84	0.35	13.39
1981	0.11	0.23	1.87	0.90	2.11	1.04	2.73	5.22	3.47	1.79	1.50	0.03	21.00
1982	0.15	0.39	0.52	0.43	1.96	4.75	6.23	0.55	1.37	0.71	0.75	0.79	18.60
1983	1.78	1.19	0.98	0.83	2.85	1.76	0.74	0.28	0.37	3.23	0.33	0.64	14.98
1984	0.56	0.37	0.98	1.18	0.04	6.76	0.83	2.28	0.95	3.19	1.09	1.00	19.23
1985	0.99	0.77	1.49	2.79	0.86	3.08	2.07	1.67	4.96	3.07	0.39	0.26	22.40
Record Mean	0.54	0.67	0.86	1.40	2.86	3.09	2.66	2.97	2.10	1.66	0.83	0.67	20.31

TABLE 3   AVERAGE TEMPERATURE (deg. F)   AMARILLO, TEXAS

YEAR	JAN	FEB	MAR	APR	MAY	JUNE	JULY	AUG	SEP	OCT	NOV	DEC	ANNUAL
1956	37.5	35.3	47.5	53.8	70.1	78.0	77.6	77.2	73.0	62.7	43.7	42.1	58.2
1957	36.6	46.4	45.0	52.4	60.7	72.3	81.2	77.1	66.4	56.0	40.7	42.7	56.4
1958	36.8	37.7	37.0	51.1	66.7	76.0	78.0	78.2	70.5	58.1	46.7	37.3	56.2
1959	32.3	39.2	45.9	55.0	66.0	74.6	75.5	78.4	70.2	54.8	42.4	40.3	56.2
1960	31.4	31.4	42.3	58.3	65.1	74.3	75.5	76.2	69.5	58.6	47.1	34.9	55.4
#1961	34.6	39.3	48.5	56.2	67.3	73.9	76.4	77.2	67.6	59.5	41.3	36.5	56.5
1962	31.7	45.5	46.7	57.2	71.7	72.4	78.0	78.3	70.4	61.5	47.8	40.9	58.5
1963	28.9	41.7	50.4	61.8	69.0	74.1	81.3	77.8	72.3	66.2	49.5	32.5	58.8
1964	37.2	32.1	44.8	57.9	68.9	76.5	81.9	78.1	69.6	60.6	47.5	39.1	57.9
1965	41.0	37.6	38.4	59.4	66.9	72.1	78.6	76.9	68.5	59.5	52.5	42.9	57.8
1966	27.8	34.1	50.0	54.5	66.0	74.2	82.4	73.5	68.5	57.7	51.6	35.5	56.3
1967	40.7	40.6	53.4	60.6	63.4	73.7	77.2	75.0	68.3	60.9	46.6	35.4	58.0
1968	38.3	38.0	48.7	55.3	63.4	74.9	77.0	77.0	69.1	62.1	45.4	36.9	57.2
1969	41.4	40.4	38.8	59.9	67.2	72.8	82.6	80.2	71.1	54.5	46.7	38.4	57.8
1970	33.2	43.2	41.7	56.0	68.8	74.5	80.7	79.1	70.9	54.0	46.4	43.4	57.7
1971	37.8	38.8	48.3	55.7	64.9	75.9	76.9	72.2	67.1	57.2	45.6	37.6	56.5
1972	35.7	40.9	51.7	59.5	63.2	73.6	74.5	73.9	66.9	57.9	36.9	32.9	55.8
1973	33.4	39.5	47.1	50.3	62.9	75.1	78.1	78.8	68.7	62.0	50.8	38.6	57.1
1974	35.0	42.2	53.1	60.2	71.5	75.0	79.3	73.6	62.9	59.0	47.5	36.1	57.3
1975	37.1	35.3	45.0	54.9	64.5	73.3	75.5	76.7	65.8	61.1	45.8	40.4	56.3
1976	36.9	47.8	46.5	56.8	60.2	72.2	74.8	75.1	66.3	49.9	38.5	37.5	55.2
1977	30.1	43.8	48.7	57.8	67.3	78.1	80.7	77.5	74.3	60.7	47.4	39.9	58.9
1978	29.0	30.1	47.2	61.4	63.7	75.4	80.7	76.1	70.7	59.4	46.5	32.7	56.0
1979	24.9	40.0	46.6	54.5	62.2	70.9	77.1	73.6	69.8	60.1	40.5	38.8	54.9
1980	34.9	37.7	43.9	52.4	61.9	78.3	82.9	78.5	70.7	56.6	42.7	41.4	56.8
1981	37.9	42.2	48.8	63.1	65.6	78.5	81.3	74.4	69.0	56.6	49.1	40.1	58.9
1982	37.1	35.9	47.3	53.8	63.3	72.2	78.7	78.7	71.1	57.3	45.6	36.1	56.4
1983	33.4	36.2	45.3	50.9	60.3	70.4	80.0	81.0	73.6	60.5	47.9	24.7	55.3
1984	31.6	40.3	44.0	51.8	64.6	74.6	75.6	75.3	65.7	56.9	47.0	40.5	55.9
1985	31.5	37.2	49.5	60.0	68.1	75.1	75.1	79.6	68.9	56.9	43.5	34.0	57.0
Record Mean	36.2	39.2	46.9	56.3	64.8	74.1	78.0	76.8	69.8	58.9	46.3	37.9	57.1
Max	48.9	52.3	60.8	70.2	77.8	87.0	90.3	89.1	82.3	71.8	59.2	50.2	70.0
Min	23.4	26.0	33.0	42.3	51.8	61.2	65.7	64.6	57.2	45.9	33.3	25.5	44.2

REFERENCE NOTES FOR TABLES 1, 2, 3 and 6   (AMARILLO, TX)

GENERAL

T - TRACE AMOUNT
BLANK ENTRIES DENOTE MISSING/UNREPORTED DATA.
# INDICATES A STATION OR INSTRUMENT RELOCATION.

SPECIFIC

TABLE 1

(a) -   LENGTH OF RECORD IN YEARS. ALTHOUGH
        INDIVIDUAL MONTHS MAY BE MISSING.
  *   LESS THAN .05

NORMALS — BASED ON THE 1951-1980 RECORD PERIOD.
EXTREMES — DATES ARE THE MOST RECENT OCCURRENCE.
WIND DIR. — NUMERALS SHOW TENS OF DEGREES
          CLOCKWISE FROM TRUE NORTH.
          "00" INDICATES CALM.
RESULTANT WIND DIRECTIONS ARE GIVEN TO WHOLE DEGREES.

EXCEPTIONS

TABLES 2, 3, and 6

RECORD MEANS ARE THROUGH THE CURRENT YEAR,
BEGINNING IN   1892 FOR TEMPERATURE
                  1892 FOR PRECIPITATION
                  1942 FOR SNOWFALL

TABLE 4 HEATING DEGREE DAYS Base 65 deg. F   AMARILLO, TEXAS

SEASON	JULY	AUG	SEP	OCT	NOV	DEC	JAN	FEB	MAR	APR	MAY	JUNE	TOTAL
1956-57	0	9	7	127	634	702	871	515	616	381	140	26	4028
1957-58	0	0	45	303	721	683	867	758	861	412	66	5	4721
1958-59	0	0	37	233	546	852	1008	716	584	321	74	3	4374
1959-60	0	0	46	311	671	758	1033	965	696	214	80	0	4774
#1960-61	3	0	27	210	530	929	935	714	507	303	54	7	4219
1961-62	0	0	63	189	702	879	1026	540	563	254	23	15	4254
1962-63	0	0	23	159	508	741	1114	648	450	139	61	7	3850
1963-64	0	0	11	42	458	1000	855	949	618	235	53	7	4228
1964-65	0	0	48	149	517	795	740	762	820	194	56	7	4088
1965-66	0	0	73	191	376	676	1145	857	456	312	99	1	4186
1966-67	0	23	22	250	395	908	743	675	356	159	135	9	3675
1967-68	0	1	23	185	544	911	820	777	504	290	109	0	4164
1968-69	0	0	6	159	583	865	726	681	808	176	69	24	4097
1969-70	0	0	1	374	543	817	981	602	713	280	50	33	4394
1970-71	0	0	47	359	544	661	837	726	524	290	89	0	4077
1971-72	0	0	120	245	575	843	900	692	409	203	106	0	4094
1972-73	13	2	48	274	833	987	972	706	548	438	113	0	4934
1973-74	0	0	56	154	420	813	922	633	368	190	17	0	3573
1974-75	0	0	119	199	571	890	853	826	612	323	67	16	4476
1975-76	0	0	104	158	569	757	861	492	567	253	171	0	3932
1976-77	0	0	59	464	790	846	1075	592	499	215	22	0	4562
1977-78	0	0	1	150	522	768	1107	972	551	136	139	6	4352
1978-79	0	0	33	197	574	992	1236	697	561	319	144	30	4783
1979-80	0	2	28	186	727	806	926	788	649	373	146	0	4631
1980-81	0	0	35	280	662	723	832	630	496	111	65	0	3834
1981-82	0	0	26	271	469	765	856	809	539	340	96	2	4173
1982-83	0	0	23	252	575	888	972	800	603	421	171	32	4737
1983-84	0	0	40	175	506	1241	1028	709	642	390	56	0	4787
1984-85	0	0	125	262	531	752	1034	769	474	169	37	5	4158
1985-86	0	0	111	249	640	957							

TABLE 5 COOLING DEGREE DAYS Base 65 deg. F   AMARILLO, TEXAS

YEAR	JAN	FEB	MAR	APR	MAY	JUNE	JULY	AUG	SEP	OCT	NOV	DEC	TOTAL
1969	0	0	0	30	147	264	550	478	190	54	0	0	1713
1970	0	0	0	14	176	327	495	446	231	24	0	0	1713
1971	0	0	13	18	94	334	381	232	189	10	0	0	1271
1972	0	0	4	44	59	265	313	285	182	60	0	0	1212
1973	0	0	0	3	53	310	414	435	173	67	1	0	1456
1974	0	0	6	52	228	306	449	274	61	20	0	0	1396
1975	0	0	0	24	61	273	330	367	134	46	0	0	1235
1976	0	0	0	13	32	223	312	318	110	5	0	0	1013
1977	0	0	0	10	100	399	493	389	287	22	0	0	1700
1978	0	0	6	33	108	324	497	351	209	28	0	0	1556
1979	0	0	0	12	64	218	380	276	177	41	0	0	1168
1980	0	0	0	4	58	408	562	429	211	26	0	0	1698
1981	0	0	1	59	87	410	512	299	154	16	0	0	1538
1982	0	0	0	9	52	225	437	432	213	22	0	0	1390
1983	0	0	0	2	31	201	473	502	306	41	3	0	1559
1984	0	0	0	1	121	298	331	325	151	17	0	0	1244
1985	0	0	2	27	143	315	473	458	235	6	0	0	1659

TABLE 6 SNOWFALL (inches)   AMARILLO, TEXAS

SEASON	JULY	AUG	SEP	OCT	NOV	DEC	JAN	FEB	MAR	APR	MAY	JUNE	TOTAL
1956-57	0.0	0.0	0.0	0.0	T	T	1.6	1.6	11.1	T	0.0	0.0	14.3
1957-58	0.0	0.0	0.0	T	2.9	0.0	5.2	5.2	8.6	T	0.0	0.0	21.9
1958-59	0.0	0.0	0.0	0.0	2.1	2.7	1.7	0.9	1.4	T	0.0	0.0	8.8
1959-60	0.0	0.0	0.0	0.0	0.9	T	12.9	5.9	0.5	0.0	0.0	0.0	20.2
1960-61	0.0	0.0	0.0	0.0	T	3.6	1.8	3.1	14.7	0.3	0.0	0.0	23.5
1961-62	0.0	0.0	0.0	0.0	4.9	0.5	2.3	0.1	T	0.0	0.0	0.0	7.8
1962-63	0.0	0.0	0.0	0.0	7.6	0.8	0.5	4.8	T	T	0.0	0.0	13.7
1963-64	0.0	0.0	0.0	0.0	0.0	3.6	T	17.3	T	T	0.0	0.0	20.9
1964-65	0.0	0.0	0.0	0.0	T	0.5	5.6	3.8	2.3	0.0	0.0	0.0	12.2
1965-66	0.0	0.0	0.0	0.0	0.0	1.4	7.2	0.4	0.1	0.0	0.0	0.0	9.1
1966-67	0.0	0.0	0.0	T	T	1.8	T	1.1	0.4	0.0	0.0	0.0	3.3
1967-68	0.0	0.0	0.0	T	2.1	6.1	0.6	8.0	2.9	0.0	0.0	0.0	19.7
1968-69	0.0	0.0	0.0	0.0	0.3	1.8	0.2	2.9	10.3	0.0	0.0	0.0	15.5
1969-70	0.0	0.0	0.0	0.0	T	7.1	0.1	T	14.1	1.8	T	0.0	23.1
1970-71	0.0	0.0	0.0	3.9	2.6	T	0.8	17.3	1.0	T	0.0	0.0	25.6
1971-72	0.0	0.0	0.0	0.0	T	7.9	1.6	2.0	0.6	0.0	0.0	0.0	12.1
1972-73	0.0	0.0	0.0	0.4	9.9	2.8	7.0	4.2	0.5	5.7	0.0	0.0	30.5
1973-74	0.0	0.0	0.0	0.0	0.4	1.5	3.6	1.2	T	0.0	0.0	0.0	6.7
1974-75	0.0	0.0	0.0	0.0	T	1.5	3.8	11.7	1.2	T	0.0	0.0	18.2
1975-76	0.0	0.0	0.0	0.0	0.4	1.8	T	T	4.2	0.0	0.0	0.0	6.4
1976-77	0.0	0.0	0.0	3.9	4.3	T	6.4	3.7	T	T	0.0	0.0	18.3
1977-78	0.0	0.0	0.0	0.0	1.3	0.2	8.2	9.4	1.5	T	0.5	0.0	21.1
1978-79	0.0	0.0	0.0	0.0	2.0	3.3	1.9	1.9	T	T	0.0	0.0	13.1
1979-80	0.0	0.0	0.0	1.6	2.2	0.3	1.4	5.2	1.0	T	0.0	0.0	11.7
1980-81	0.0	0.0	0.0	0.0	8.6	T	1.1	0.1	0.1	0.0	0.0	0.0	9.9
1981-82	0.0	0.0	0.0	0.0	T	0.3	1.1	4.7	1.7	T	0.0	0.0	7.8
1982-83	0.0	0.0	0.0	0.0	4.5	1.9	14.5	13.0	8.1	5.9	0.0	0.0	47.9
1983-84	0.0	0.0	0.0	0.0	T	5.2	7.0	3.5	2.5	0.0	0.0	0.0	18.2
1984-85	0.0	0.0	0.3	0.0	0.1	0.3	7.4	0.3	0.2	0.0	0.0	0.0	8.6
1985-86	0.0	0.0	T	0.0	T	2.8							
Record Mean	0.0	0.0	T	0.2	1.8	2.2	3.9	3.8	2.5	0.6	T	0.0	14.9

See Reference Notes, relative to all above tables, on preceding page.

Austin, capital of Texas, is located on the Colorado River where the stream crosses the Balcones escarpment separating the Texas Hill Country from the Blackland Prairies to the east. Elevations within the city vary from 400 feet to nearly 1,000 feet above sea level. Native trees include cedar, oak, walnut, mesquite, and pecan.

The climate of Austin is humid subtropical with hot summers. Winters are mild, with below freezing temperatures occurring on an average of about 25 days each year. Rather strong northerly winds, accompanied by sharp drops in temperature, frequently occur during the winter months in connection with cold fronts, but cold spells are usually of short duration, seldom lasting more than two days. Daytime temperatures in summer are hot, but summer nights are usually pleasant.

Precipitation is fairly evenly distributed throughout the year, with heaviest amounts occurring in late spring. A secondary rainfall peak occurs in September, primarily because of tropical cyclones that migrate out of the Gulf of Mexico. Precipitation from April through September usually results from thunderstorms, with fairly large amounts of rain falling within short periods of time. While thunderstorms and heavy rains may occur in all months of the year, most of the winter precipitation consists of light rain. Snow is insignificant as a source of moisture, and usually melts as rapidly as it falls. The city may experience several seasons in succession with no measurable snowfall.

Prevailing winds are southerly, however in winter, northerly winds are about as frequent as those from the south. Destructive winds and damaging hailstorms are infrequent. On rare occasions dissipating tropical storms produce strong winds and heavy rains in the area. Blowing dust occurs occasionally in spring, but visibility rarely drops substantially, and then only for a few hours.

The average length of the warm season (freeze-free period) is 273 days. The average occurrence of the last temperature of 32 degrees in spring is early March and the average occurrence of the first temperature of 32 degrees is late November.

TABLE 1   NORMALS, MEANS AND EXTREMES

AUSTIN, TEXAS

LATITUDE: 30°18'N   LONGITUDE: 97°42'W   ELEVATION: FT. GRND   587 BARO   590   TIME ZONE: CENTRAL   WBAN: 13958

	(a)	JAN	FEB	MAR	APR	MAY	JUNE	JULY	AUG	SEP	OCT	NOV	DEC	YEAR
TEMPERATURE °F:														
Normals														
-Daily Maximum		59.4	64.1	71.7	79.0	84.7	91.6	95.4	95.3	89.3	80.8	69.2	62.8	78.6
-Daily Minimum		38.8	42.2	49.3	58.3	65.1	71.5	73.9	73.7	69.1	58.7	48.1	41.4	57.5
-Monthly		49.1	53.2	60.5	68.7	74.9	81.6	84.7	84.5	79.2	69.8	58.7	52.1	68.1
Extremes														
-Record Highest	44	90	93	98	98	100	105	109	106	104	97	91	90	109
-Year		1971	1954	1971	1982	1984	1980	1954	1984	1985	1979	1951	1955	JUL 1954
-Record Lowest	44	-2	7	18	35	43	53	64	61	41	32	20	10	-2
-Year		1949	1951	1948	1973	1954	1970	1970	1967	1942	1957	1976	1983	JAN 1949
NORMAL DEGREE DAYS:														
Heating (base 65°F)		505	347	203	41	0	0	0	0	0	37	221	406	1760
Cooling (base 65°F)		12	16	63	152	307	498	611	605	426	186	32	6	2914
% OF POSSIBLE SUNSHINE	44	49	52	55	54	58	69	76	75	67	64	56	51	61
MEAN SKY COVER (tenths)														
Sunrise - Sunset	44	6.3	6.0	6.1	6.3	6.1	5.2	4.7	4.7	5.0	4.8	5.3	5.9	5.5
MEAN NUMBER OF DAYS:														
Sunrise to Sunset														
-Clear	44	8.9	8.5	8.5	7.7	6.8	8.3	11.8	11.7	10.8	12.5	11.2	10.2	116.9
-Partly Cloudy	44	5.9	6.4	8.0	7.3	11.4	15.0	13.5	13.8	10.8	9.2	6.8	6.0	114.0
-Cloudy	44	16.2	13.4	14.6	15.0	12.8	6.8	5.7	5.5	8.4	9.2	12.1	14.8	134.3
Precipitation														
.01 inches or more	44	7.9	7.7	7.2	7.5	8.7	6.2	4.8	5.2	7.1	6.6	7.0	7.0	82.8
Snow,Ice pellets														
1.0 inches or more	44	0.2	0.2	0.*	0.0	0.0	0.0	0.0	0.0	0.0	0.0	0.1	0.0	0.5
Thunderstorms	44	0.8	2.0	3.2	4.7	6.9	4.7	3.9	4.8	4.0	2.8	1.6	1.2	40.7
Heavy Fog Visibility														
1/4 mile or less	44	4.5	3.1	2.7	1.4	1.0	0.5	0.3	0.4	0.8	2.1	2.8	4.3	23.6
Temperature °F														
-Maximum														
90° and above	24	0.*	0.*	0.7	1.5	6.3	20.7	27.7	28.2	16.4	3.5	0.0	0.0	105.0
32° and below	24	0.8	0.2	0.0	0.0	0.0	0.0	0.0	0.0	0.0	0.0	0.0	0.3	1.2
-Minimum														
32° and below	24	9.7	5.0	1.0	0.0	0.0	0.0	0.0	0.0	0.0	0.0	1.0	5.8	22.5
0° and below	24	0.0	0.0	0.0	0.0	0.0	0.0	0.0	0.0	0.0	0.0	0.0	0.0	0.0
AVG. STATION PRESS.(mb)	13	998.2	996.9	992.2	991.9	990.4	991.8	993.3	993.2	993.2	995.1	996.1	997.5	994.2
RELATIVE HUMIDITY (%)														
Hour 00	24	72	71	71	75	80	79	74	73	78	75	76	73	75
Hour 06	24	78	79	79	82	88	88	87	86	86	83	82	79	83
Hour 12 (Local Time)	24	60	58	56	58	60	56	50	50	56	55	58	59	56
Hour 18	24	57	52	49	53	57	53	46	46	55	55	59	58	53
PRECIPITATION (inches):														
Water Equivalent														
-Normal		1.60	2.49	1.68	3.11	4.19	3.06	1.89	2.24	3.60	3.38	2.20	2.06	31.50
-Maximum Monthly	44	7.94	6.39	6.03	9.93	9.98	14.96	10.54	8.90	8.11	12.31	7.91	5.91	14.96
-Year		1968	1958	1983	1957	1965	1981	1979	1974	1942	1960	1946	1944	JUN 1981
-Minimum Monthly	44	0.04	0.28	T	0.06	0.81	T	0.00	0.00	0.00	0.07	T	T	0.00
-Year		1971	1954	1972	1984	1960	1960	1962	1952	1947	1952	1970	1950	JUL 1962
-Maximum in 24 hrs	44	3.44	3.73	2.69	3.86	5.66	6.50	5.46	4.68	6.74	7.22	5.09	4.02	7.22
-Year		1965	1958	1980	1942	1979	1964	1961	1945	1973	1960	1974	1953	OCT 1960
Snow,Ice pellets														
-Maximum Monthly	44	7.5	6.0	2.0								2.0	T	7.5
-Year		1985	1966	1965								1980	1983	JAN 1985
-Maximum in 24 hrs	44	7.0	6.0	2.0								2.0	T	7.0
-Year		1944	1966	1965								1980	1983	JAN 1944
WIND:														
Mean Speed (mph)	44	9.8	10.2	10.9	10.6	9.7	9.3	8.4	7.9	8.0	8.1	9.0	9.2	9.2
Prevailing Direction through 1963		S	S	S	SSE	SSE	S	S	S	S	S	S	S	S
Fastest Obs. 1 Min.														
-Direction (!!!)	6	35	34	33	33	32	35	28	07	03	35	30	35	33
-Speed (MPH)	6	33	39	36	40	35	35	40	35	33	33	36	30	40
-Year		1985	1984	1984	1983	1980	1982	1980	1984	1981	1985	1983	1979	APR 1983
Peak Gust														
-Direction (!!!)	2	N	NW	NW	N	N	SE	S	N	N	N	N	W	NW
-Speed (mph)	2	52	55	56	41	47	38	40	44	41	46	40	41	56
-Date		1985	1984	1984	1984	1985	1985	1985	1985	1985	1985	1984	1984	MAR 1984

See reference Notes to this table on the following page.

TABLE 2 PRECIPITATION (inches) AUSTIN, TEXAS

YEAR	JAN	FEB	MAR	APR	MAY	JUNE	JULY	AUG	SEP	OCT	NOV	DEC	ANNUAL
1956	1.65	1.74	0.26	0.56	3.12	0.94	0.11	1.21	0.09	0.84	2.13	2.76	15.41
1957	0.55	3.14	4.58	9.93	7.38	5.25	1.10	T	6.43	8.79	2.95	1.20	51.30
1958	3.09	6.39	2.55	4.24	3.67	2.89	3.42	0.68	5.18	5.18	0.87	1.15	41.02
1959	0.42	2.30	0.23	4.35	1.66	3.30	3.49	4.80	4.37	5.98	1.95	2.11	34.96
1960	1.03	2.36	1.37	1.01	0.81	4.26	2.41	2.60	1.68	12.31	1.90	4.08	35.82
1961	1.27	4.85	0.67	0.10	1.03	11.43	8.40	0.40	3.68	0.91	2.82	0.91	36.47
1962	0.56	0.63	1.19	4.04	1.06	8.21	0.00	4.58	4.75	4.07	0.92	3.47	33.48
1963	0.59	2.83	0.22	3.51	1.32	2.10	0.58	0.88	1.50	0.78	1.57	1.42	17.30
1964	2.57	1.47	1.95	1.47	1.87	7.54	0.65	2.09	6.29	3.74	2.45	0.88	32.97
1965	4.09	5.06	1.30	1.91	1.91	9.98	0.89	0.37	1.32	5.46	3.26	4.28	40.57
1966	1.58	3.23	0.50	3.74	3.13	1.53	0.47	6.21	3.22	0.60	0.11	0.87	25.19
1967	0.25	1.52	1.09	4.44	3.35	T	1.15	3.71	5.71	4.55	4.36	3.41	33.54
1968	7.94	1.64	2.09	1.87	8.75	3.10	3.11	0.74	3.42	0.60	4.91	0.55	38.72
1969	0.40	4.18	3.26	5.04	4.78	2.66	0.12	5.78	1.17	2.65	0.79	4.29	33.59
1970	1.83	5.70	2.47	1.36	8.18	0.29	0.66	1.00	3.82	5.22	T	0.11	30.64
1971	0.04	0.69	0.79	1.07	1.37	1.68	1.23	5.69	2.13	3.02	3.02	4.22	24.95
1972	1.48	0.31	T	1.46	7.88	2.20	2.55	2.53	1.55	2.96	2.62	0.53	26.07
1973	3.42	2.05	2.92	3.09	1.38	4.70	2.95	0.06	7.44	11.11	0.58	0.76	40.46
1974	2.74	0.36	1.34	1.79	5.88	0.21	0.61	8.90	1.58	3.45	7.35	2.00	36.21
1975	1.11	2.30	0.80	3.86	8.16	7.07	2.25	2.54	3.62	2.54	0.52	2.04	36.81
1976	1.16	1.11	2.11	8.13	6.05	3.19	4.71	0.80	3.80	5.93	1.78	2.48	41.25
1977	2.25	2.58	2.18	6.08	1.24	1.22	0.21	0.06	3.10	1.19	1.69	0.34	22.14
1978	0.88	1.95	0.84	1.72	5.78	2.98	1.19	1.49	4.44	1.38	5.48	2.84	30.97
1979	2.11	3.54	3.76	2.98	7.29	0.83	10.54	0.61	1.40	0.45	0.59	3.40	37.50
1980	0.85	2.33	2.20	2.20	5.43	0.31	0.28	1.18	5.66	1.29	3.41	1.24	27.38
1981	1.61	1.18	3.05	0.81	9.02	14.96	3.39	0.91	2.65	7.04	0.72	0.39	45.73
1982	0.85	0.80	1.39	4.17	5.68	2.99	0.13	0.77	1.88	2.66	3.19	2.12	26.63
1983	1.88	2.84	6.03	0.16	5.33	3.84	2.85	2.21	2.83	2.82	2.66	0.53	33.98
1984	1.66	1.00	2.49	0.06	1.27	1.69	1.44	0.45	0.79	10.34	1.88	3.23	26.30
1985	1.34	2.10	1.84	2.39	1.65	5.64	1.53	0.37	3.98	5.84	4.75	1.06	32.49
Record Mean	2.02	2.39	2.24	3.45	4.30	2.91	2.22	2.14	3.59	3.18	2.40	2.44	33.28

TABLE 3 AVERAGE TEMPERATURE (deg. F) AUSTIN, TEXAS

YEAR	JAN	FEB	MAR	APR	MAY	JUNE	JULY	AUG	SEP	OCT	NOV	DEC	ANNUAL
1956	50.8	55.8	61.8	69.1	78.7	84.4	87.0	85.9	81.7	73.7	57.3	56.6	70.2
1957	50.5	60.9	59.5	66.0	73.5	80.9	86.3	86.0	77.0	65.3	55.9	56.3	68.2
1958	48.7	48.8	54.6	66.3	76.0	83.2	85.1	85.8	79.5	67.5	60.7	48.5	67.0
1959	47.7	52.9	59.9	65.1	77.1	82.6	84.2	83.5	80.1	69.5	52.2	53.7	67.4
1960	49.3	47.9	54.3	69.7	73.5	83.4	84.7	83.3	79.3	71.5	60.7	48.5	67.2
#1961	46.0	54.8	63.7	67.5	76.3	79.6	81.6	81.8	78.2	69.8	56.5	52.2	67.3
1962	45.5	60.4	57.9	67.7	76.6	80.3	85.4	86.8	79.7	73.3	59.6	51.0	68.7
1963	43.6	51.6	62.9	72.9	76.9	82.9	86.4	87.5	82.0	75.1	62.6	45.1	69.3
1964	50.3	48.4	60.7	70.8	76.6	81.3	85.2	86.0	79.9	67.6	62.5	52.2	68.5
1965	53.4	49.4	53.4	71.4	74.7	80.7	84.2	84.3	80.5	66.6	64.1	56.0	68.2
1966	45.3	48.9	60.1	69.1	73.9	79.9	84.7	82.2	78.2	68.6	64.2	51.0	67.2
1967	50.9	51.4	67.3	75.9	75.3	83.7	85.2	82.6	75.2	67.7	58.9	49.4	68.6
1968	47.6	47.2	57.3	68.1	74.2	79.5	82.2	84.1	76.0	72.0	55.8	50.4	66.2
1969	53.6	52.6	53.9	69.0	73.7	80.7	86.3	84.9	79.7	68.6	57.3	53.8	67.9
1970	44.0	53.1	55.2	68.7	71.9	79.9	83.8	85.8	79.2	66.7	57.4	58.7	67.1
1971	54.6	54.9	62.1	68.2	76.2	83.7	85.9	81.2	79.3	72.2	60.5	56.4	69.6
1972	51.0	55.7	66.1	73.5	73.5	81.7	82.3	83.1	82.4	70.9	52.8	48.3	68.5
1973	45.9	50.3	63.7	63.1	73.7	78.4	82.8	82.5	79.7	71.1	65.1	52.4	67.4
1974	49.0	56.3	66.5	68.9	77.7	80.1	84.4	81.3	72.3	69.5	56.9	50.7	67.8
1975	53.4	52.2	59.7	67.8	73.7	79.9	82.0	82.6	75.8	70.0	60.4	52.4	67.5
1976	49.5	61.3	62.0	68.0	70.7	79.8	80.2	83.4	78.1	61.3	53.6	48.5	66.2
1977	41.6	54.1	61.3	67.0	75.3	82.2	85.1	86.8	83.8	71.6	61.7	53.6	68.7
1978	40.7	45.0	58.5	69.8	77.3	82.3	86.4	84.8	79.0	69.8	60.6	49.7	67.0
1979	40.4	48.5	61.1	67.0	71.5	79.8	83.1	82.5	77.6	73.5	56.5	53.0	66.2
1980	51.2	52.8	61.4	66.7	75.1	84.6	87.9	86.0	81.7	68.4	57.5	53.7	68.9
1981	50.6	53.4	59.3	72.4	74.6	81.4	84.9	85.5	79.7	72.2	63.7	53.7	69.3
1982	51.9	51.0	64.7	67.3	75.6	81.6	86.5	87.4	81.8	70.7	58.5	52.3	69.1
1983	47.4	51.3	57.8	64.6	72.7	78.4	82.7	84.4	78.3	72.0	62.4	41.9	66.2
1984	46.3	55.5	63.4	71.1	78.0	82.6	86.1	86.0	76.3	70.5	58.7	58.4	69.3
1985	44.1	49.7	64.3	70.0	76.3	80.7	82.9	87.0	79.3	71.7	63.0	49.1	68.2
Record Mean	49.8	53.3	60.6	68.0	74.7	81.5	84.2	84.3	79.1	69.9	59.2	51.7	68.0
Record Max	60.1	64.1	71.6	78.5	84.7	91.6	94.7	95.1	89.4	80.9	69.8	61.9	78.5
Record Min	39.6	42.6	49.6	57.4	64.7	71.4	73.6	73.5	68.7	58.9	48.6	41.4	57.5

REFERENCE NOTES FOR TABLES 1, 2, 3 and 6   (AUSTIN, TX)

GENERAL

T - TRACE AMOUNT
BLANK ENTRIES DENOTE MISSING/UNREPORTED DATA.
# INDICATES A STATION OR INSTRUMENT RELOCATION.

SPECIFIC

TABLE 1

(a) - LENGTH OF RECORD IN YEARS. ALTHOUGH
INDIVIDUAL MONTHS MAY BE MISSING.
  * LESS THAN .05

NORMALS — BASED ON THE 1951-1980 RECORD PERIOD.
EXTREMES — DATES ARE THE MOST RECENT OCCURRENCE.
WIND DIR. — NUMERALS SHOW TENS OF DEGREES
CLOCKWISE FROM TRUE NORTH.
    "00" INDICATES CALM.
RESULTANT WIND DIRECTIONS ARE GIVEN TO WHOLE DEGREES.

EXCEPTIONS

TABLES 2, 3, and 6

RECORD MEANS ARE THROUGH THE CURRENT YEAR,
BEGINNING IN   1898 FOR TEMPERATURE
                    1956 FOR PRECIPITATION
                    1942 FOR SNOWFALL

TABLE 4   HEATING DEGREE DAYS Base 65 deg. F   AUSTIN, TEXAS

SEASON	JULY	AUG	SEP	OCT	NOV	DEC	JAN	FEB	MAR	APR	MAY	JUNE	TOTAL
1956-57	0	0	0	0	250	285	454	168	185	79	16	0	1437
1957-58	0	0	0	98	298	275	499	457	319	53	0	0	1999
1958-59	0	0	0	74	184	505	530	346	183	112	0	0	1934
1959-60	0	0	0	38	393	348	492	495	350	17	12	0	2145
#1960-61	0	0	0	26	182	512	582	299	116	80	0	0	1797
1961-62	0	0	0	21	270	404	598	161	243	48	0	0	1745
1962-63	0	0	0	20	172	429	657	374	109	27	4	0	1792
1963-64	0	0	0	0	137	610	451	475	154	20	0	0	1947
1964-65	0	0	0	17	158	415	370	431	367	12	0	0	1770
1965-66	0	0	2	58	76	288	611	449	168	24	11	0	1687
1966-67	0	0	0	31	109	453	446	380	77	0	2	0	1498
1967-68	0	0	7	41	192	477	538	510	265	37	1	0	2068
1968-69	0	0	0	7	294	446	368	345	346	6	1	0	1813
1969-70	0	0	0	69	253	339	653	330	309	47	10	0	2010
1970-71	0	0	3	75	240	231	328	300	171	52	0	0	1400
1971-72	0	0	3	0	184	271	434	298	66	6	0	0	1262
1972-73	0	0	0	46	375	513	586	405	72	128	3	0	2128
1973-74	0	0	0	8	83	387	496	258	108	36	0	0	1376
1974-75	0	0	5	11	271	441	377	356	204	53	0	0	1718
1975-76	0	0	0	1	27	204	404	476	158	173	19	3	1465
1976-77	0	0	0	170	402	506	719	308	148	20	0	0	2273
1977-78	0	0	0	0	16	136	354	750	561	225	15	7	2064
1978-79	0	0	0	0	5	186	475	754	460	144	37	10	2071
1979-80	0	0	0	16	278	372	425	357	168	47	0	0	1663
1980-81	0	0	0	69	262	361	443	340	196	6	0	0	1677
1981-82	0	0	0	44	86	349	424	400	147	81	2	0	1533
1982-83	0	0	0	37	245	408	538	376	229	93	4	0	1930
1983-84	0	0	4	7	154	721	573	280	123	16	0	0	1878
1984-85	0	0	18	34	221	228	643	428	99	19	0	0	1690
1985-86	0	0	7	12	134	490							

TABLE 5   COOLING DEGREE DAYS Base 65 deg. F   AUSTIN, TEXAS

YEAR	JAN	FEB	MAR	APR	MAY	JUNE	JULY	AUG	SEP	OCT	NOV	DEC	TOTAL
1969	25	5	8	132	276	476	668	620	450	186	46	1	2893
1970	8	1	10	165	235	451	589	651	435	134	22	43	2744
1971	14	20	86	157	355	570	656	511	436	229	53	13	3100
1972	6	34	111	269	272	510	547	571	529	239	16	3	3107
1973	1	0	34	78	277	407	557	550	447	205	93	4	2653
1974	6	21	160	158	398	463	606	513	229	157	34	6	2751
1975	21	0	47	141	277	454	536	552	329	187	72	22	2638
1976	0	57	85	115	186	451	479	577	401	61	6	0	2418
1977	0	7	38	90	324	526	629	681	572	227	40	7	3141
1978	5	6	31	167	396	526	670	621	427	161	60	7	3077
1979	2	7	29	102	216	449	570	549	385	285	29	8	2631
1980	5	11	62	106	322	598	718	659	510	182	45	19	3237
1981	2	23	27	233	304	501	627	641	446	272	54	5	3135
1982	23	14	144	157	337	506	672	701	509	219	61	18	3361
1983	0	0	13	90	250	406	556	608	409	227	80	11	2650
1984	0	13	80	204	413	535	629	659	364	206	40	30	3173
1985	0	9	85	174	358	480	562	687	443	228	82	4	3112

TABLE 6   SNOWFALL (inches)   AUSTIN, TEXAS

SEASON	JULY	AUG	SEP	OCT	NOV	DEC	JAN	FEB	MAR	APR	MAY	JUNE	TOTAL
1970-71	0.0	0.0	0.0	0.0	0.0	0.0	0.0	T	0.0	0.0	0.0	0.0	T
1971-72	0.0	0.0	0.0	0.0	0.0	T	T	0.0	0.0	0.0	0.0	0.0	T
1972-73	0.0	0.0	0.0	0.0	0.0	0.0	0.8	0.9	0.0	0.0	0.0	0.0	1.7
1973-74	0.0	0.0	0.0	0.0	0.0	0.0	0.0	0.0	0.0	0.0	0.0	0.0	0.0
1974-75	0.0	0.0	0.0	0.0	0.0	0.0	T	T	0.0	0.0	0.0	0.0	T
1975-76	0.0	0.0	0.0	0.0	0.0	0.0	0.0	0.0	0.0	0.0	0.0	0.0	0.0
1976-77	0.0	0.0	0.0	0.0	T	0.0	0.0	T	0.0	0.0	0.0	0.0	T
1977-78	0.0	0.0	0.0	0.0	0.0	0.0	T	T	T	0.0	0.0	0.0	T
1978-79	0.0	0.0	0.0	0.0	0.0	T	T	T	0.0	0.0	0.0	0.0	T
1979-80	0.0	0.0	0.0	0.0	T	0.0	T	T	0.0	0.0	0.0	0.0	T
1980-81	0.0	0.0	0.0	0.0	2.0	0.0	T	0.0	0.0	0.0	0.0	0.0	2.0
1981-82	0.0	0.0	0.0	0.0	0.0	0.0	2.0	T	T	0.0	0.0	0.0	2.0
1982-83	0.0	0.0	0.0	0.0	0.0	0.0	0.0	0.0	0.0	0.0	0.0	0.0	0.0
1983-84	0.0	0.0	0.0	0.0	0.0	T	0.0	0.0	0.0	0.0	0.0	0.0	T
1984-85	0.0	0.0	0.0	0.0	0.0	0.0	7.5	1.2	0.0	0.0	0.0	0.0	8.7
1985-86	0.0	0.0	0.0	0.0	0.0	0.0							
Record Mean	0.0	0.0	0.0	0.0	0.1	T	0.6	0.4	T	0.0	0.0	0.0	1.1

See Reference Notes, relative to all above tables, on preceding page.

Brownsville is located at the southern tip of Texas. It is the largest city in the four county area referred to as the Lower Rio Grande Valley or just the Valley.

The Gulf of Mexico, located about 18 miles east, is the dominant influence on local weather. Prevailing southeast breezes off the Gulf provide a humid but generally mild climate. Winds are frequently strong and gusty in the spring.

Brownsville weather is generally favorable for outdoor activities and the Valley is a popular tourist area, especially for Winter Texans who come to enjoy the mild winters. High temperatures range mostly in the 70s and 80s from October through April, with lows in the 50s and 60s during the same period. For the remainder of the year highs are frequently in the 90s with lows in the 70s.

Temperature extremes are rare but do occur. Temperatures in the 90s have occurred in every month of the year, with 100 degree readings noted as early as March and as late as September. Temperatures of 100 degrees or more are associated with west winds bringing hot dry air out of Mexico. Very hot temperatures are often moderated by a cooling sea breeze from the Gulf during the afternoon hours.

Located about 150 miles north of the tropics, cold weather in Brownsville is infrequent and of short duration. Some winters pass without a single day with freezing temperatures. This climate permits year around gardening and cultivation of citrus and other cold sensitive tropical and sub-tropical plants. Damaging cold comes from frigid air masses, called northers or arctic outbreaks, plunging south from Canada or the Arctic. The worst of these can drop temperatures well below freezing for several hours, and a few have produced readings in the teens. Fortunately such events are very rare since they are disasterous to the local economy.

Rainfall is not well distributed. Heaviest rains occur in May through June and mid August through mid October. Extended periods of cool rainy weather, called overrunning, can occur in winter. Torrential rains of 10 to 20 inches or more may accompany tropical storms or hurricanes that occasionally move over the area in summer or fall. Rainy spells may be followed by long dry periods. Irrigation is required to ensure production of corps such as cotton, grains, and vegetables. Snow and freezing rain or drizzle are so rare that years may pass between occurrences.

Brownsville is blessed by having little severe weather. Damaging hail or winds from heavy thunderstorms are generally limited to the Spring season and many years may elapse between occurrences. Tornadoes are even more rare. Tropical storms and hurricanes from the Gulf are a threat each summer and fall, but again, damaging storms are quite rare.

TABLE 1   NORMALS, MEANS AND EXTREMES

BROWNSVILLE, TEXAS

LATITUDE: 25°54'N   LONGITUDE: 97°26'W   ELEVATION: FT. GRND   19 BARO   00022   TIME ZONE: CENTRAL   WBAN: 12919

	(a)	JAN	FEB	MAR	APR	MAY	JUNE	JULY	AUG	SEP	OCT	NOV	DEC	YEAR
TEMPERATURE °F:														
Normals														
-Daily Maximum		69.7	72.5	77.5	83.2	87.0	90.5	92.6	92.8	89.8	84.4	77.0	71.9	82.4
-Daily Minimum		50.8	53.0	59.5	66.6	71.3	74.7	75.6	75.4	73.1	66.1	58.3	52.6	64.8
-Monthly		60.3	62.8	68.5	74.9	79.2	82.6	84.1	84.1	81.5	75.3	67.7	62.3	73.6
Extremes														
-Record Highest	47	93	94	106	102	102	101	102	102	104	96	94	94	106
-Year		1971	1982	1984	1984	1974	1942	1939	1962	1947	1977	1969	1977	MAR 1984
-Record Lowest	47	19	22	32	38	52	60	68	63	55	44	33	20	19
-Year		1962	1951	1980	1980	1970	1975	1937	1967	1942	1939	1976	1983	JAN 1962
NORMAL DEGREE DAYS:														
Heating (base 65°F)		216	135	53	0	0	0	0	0	0	0	55	150	609
Cooling (base 65°F)		70	73	164	297	440	528	592	592	492	322	136	66	3772
% OF POSSIBLE SUNSHINE	43	42	48	52	57	64	73	80	76	67	64	51	43	60
MEAN SKY COVER (tenths)														
Sunrise - Sunset	43	6.9	6.5	6.8	6.6	6.1	5.2	4.8	5.0	5.3	4.9	5.7	6.6	5.9
MEAN NUMBER OF DAYS:														
Sunrise to Sunset														
-Clear	43	6.3	7.1	6.3	5.1	6.0	8.4	11.3	10.5	9.2	11.4	9.3	7.0	97.9
-Partly Cloudy	43	6.9	6.1	8.0	10.6	14.3	15.8	14.3	14.0	13.1	12.4	9.3	7.8	132.4
-Cloudy	43	17.7	15.1	16.7	14.3	10.7	5.8	5.4	6.6	7.7	7.2	11.4	16.2	135.0
Precipitation														
.01 inches or more	43	7.6	6.2	4.3	3.7	4.8	5.6	4.9	6.9	10.2	6.5	5.7	6.5	73.0
Snow,Ice pellets														
1.0 inches or more	43	0.0	0.0	0.0	0.0	0.0	0.0	0.0	0.0	0.0	0.0	0.0	0.0	0.0
Thunderstorms	43	0.6	0.7	0.6	2.3	3.5	2.7	2.7	4.5	4.6	2.0	0.9	0.5	25.4
Heavy Fog Visibility														
1/4 mile or less	43	5.9	4.6	3.6	2.4	1.0	0.2	0.1	0.2	0.3	0.7	2.9	5.4	27.4
Temperature °F														
-Maximum														
90° and above	19	0.2	0.2	1.6	4.0	10.6	22.9	26.4	27.2	18.1	5.9	0.4	0.1	117.5
32° and below	19	0.0	0.0	0.0	0.0	0.0	0.0	0.0	0.0	0.0	0.0	0.0	0.1	0.1
-Minimum														
32° and below	19	1.4	0.4	0.1	0.0	0.0	0.0	0.0	0.0	0.0	0.0	0.0	0.6	2.5
0° and below	19	0.0	0.0	0.0	0.0	0.0	0.0	0.0	0.0	0.0	0.0	0.0	0.0	0.0
AVG. STATION PRESS.(mb)	13	1018.7	1017.7	1013.1	1012.4	1010.7	1012.4	1014.1	1013.8	1013.1	1015.3	1016.6	1018.3	1014.7
RELATIVE HUMIDITY (%)														
Hour 00	19	86	85	85	84	86	86	85	85	86	85	84	85	85
Hour 06	19	88	88	87	87	89	89	90	90	89	88	86	87	88
Hour 12 (Local Time)	19	68	62	60	59	60	59	55	56	61	60	60	64	60
Hour 18	19	74	68	67	67	68	65	62	63	68	70	71	74	68
PRECIPITATION (inches):														
Water Equivalent														
-Normal		1.25	1.55	0.50	1.57	2.15	2.70	1.51	2.83	5.24	3.54	1.44	1.16	25.44
-Maximum Monthly	46	5.11	10.25	4.27	6.62	9.12	13.06	9.43	9.56	20.18	17.12	6.26	9.45	20.18
-Year		1945	1958	1941	1977	1982	1942	1976	1975	1984	1958	1957	1940	SEP 1984
-Minimum Monthly	46	T	T	T	T	T	0.01	T	0.02	0.07	0.34	0.01	T	T
-Year		1956	1954	1971	1984	1978	1955	1982	1974	1959	1961	1949	1969	APR 1984
-Maximum in 24 hrs	46	2.95	4.98	2.59	5.20	4.56	8.18	4.25	5.48	12.19	6.67	3.64	5.69	12.19
-Year		1958	1958	1981	1977	1969	1942	1976	1980	1967	1954	1957	1940	SEP 1967
Snow,Ice pellets														
-Maximum Monthly	46	T	T	T								T	T	T
-Year		1985	1973	1943								1976	1966	JAN 1985
-Maximum in 24 hrs	46	T	T	T								T	T	T
-Year		1985	1973	1943								1976	1966	JAN 1985
WIND:														
Mean Speed (mph)	43	11.5	12.1	13.5	14.0	13.2	12.3	11.4	10.4	9.5	9.6	10.7	10.8	11.6
Prevailing Direction through 1963		SSE	SSE	SE	SE	SE	SE	SE	SE	SE	SE	SSE	NNW	SE
Fastest Obs. 1 Min.														
-Direction (!!!)	6	36	23	33	14	03	35	18	19	15	17	17	17	19
-Speed (MPH)	6	31	36	41	35	32	35	32	48	30	30	39	35	48
-Year		1982	1983	1980	1985	1985	1982	1984	1980	1985	1984	1983	1981	AUG 1980
Peak Gust														
-Direction (!!!)	2	N	NW	S	SE	NW	S	S	SE	N	S	S	S	NW
-Speed (mph)	2	46	46	45	54	60	39	33	37	41	38	40	40	60
-Date	2	1985	1984	1985	1985	1985	1985	1985	1984	1984	1984	1985	1984	MAY 1985

See Reference Notes to this table on the following page.

TABLE 2 PRECIPITATION (inches) BROWNSVILLE, TEXAS

YEAR	JAN	FEB	MAR	APR	MAY	JUNE	JULY	AUG	SEP	OCT	NOV	DEC	ANNUAL
1956	T	1.94	0.29	4.75	0.48	4.02	0.05	0.57	3.25	0.67	0.67	0.05	16.74
1957	0.35	5.20	1.97	1.85	2.24	6.34	0.10	3.45	1.34	2.03	6.26	1.27	32.40
1958	3.98	10.25	0.84	0.06	1.71	0.77	1.68	1.28	6.18	17.12	1.61	2.03	47.51
1959	2.98	2.92	0.60	2.62	0.07	4.90	1.09	1.07	0.07	3.21	2.71	0.61	22.85
1960	0.65	1.51	1.10	2.52	1.78	3.19	0.34	2.83	4.88	3.67	1.18	2.65	26.30
1961	1.95	0.30	0.04	2.68	0.47	0.48	0.91	3.10	13.30	0.34	1.63	0.77	25.97
1962	0.60	0.05	1.12	1.21	3.20	3.98	T	0.13	1.98	1.04	0.84	1.84	15.99
1963	0.23	0.65	0.08	0.05	4.51	2.99	0.76	0.14	4.31	1.03	1.23	2.74	18.72
1964	0.35	1.32	0.26	1.05	3.12	0.59	0.22	0.11	4.94	0.40	0.60	2.76	15.72
1965	0.51	1.57	0.13	0.10	1.16	0.07	1.83	3.12	5.40	0.74	2.60	3.98	21.21
1966	3.45	0.83	0.53	0.80	6.05	2.18	1.58	0.58	0.52	7.45	0.02	0.68	24.67
1967	1.69	1.05	0.44	T	0.64	1.51	0.58	5.95	19.26	2.82	0.63	1.10	35.67
1968	3.08	0.80	1.08	1.27	5.90	4.71	1.20	2.78	4.48	3.39	0.85	0.10	29.64
1969	0.51	1.81	0.82	0.24	6.69	0.08	0.23	7.74	2.80	3.25	3.18	T	27.35
1970	4.12	0.22	0.34	2.22	3.28	1.64	2.35	1.79	6.95	3.28	0.14	0.12	26.45
1971	0.22	0.30	T	1.82	2.47	3.44	1.08	2.64	10.78	3.11	0.74	1.08	27.68
1972	1.30	2.72	1.14	1.54	2.02	8.52	5.16	0.90	4.22	3.33	1.25	0.74	32.84
1973	2.07	4.74	0.13	0.69	1.24	7.57	0.59	2.80	4.61	3.77	0.88	0.53	29.62
1974	0.65	0.01	0.20	0.52	1.82	4.59	1.02	0.02	4.93	2.60	0.65	0.78	17.79
1975	0.60	0.09	0.01	0.01	2.22	2.19	4.78	9.56	4.77	0.51	1.66	2.17	28.57
1976	0.48	0.03	1.28	5.71	4.95	0.80	9.43	3.35	2.85	8.45	2.49	1.32	41.14
1977	1.24	1.37	0.12	6.62	0.76	4.73	0.27	1.27	2.84	2.87	4.07	0.14	26.30
1978	1.94	1.29	0.01	2.39	T	2.25	0.39	3.20	8.28	4.45	0.82	1.86	26.88
1979	1.43	1.10	0.14	3.91	0.59	1.52	2.10	5.25	8.84	1.18	0.12	2.04	28.22
1980	1.05	1.74	0.28	0.01	1.78	0.02	1.46	7.29	1.48	2.26	2.50	1.90	21.77
1981	1.79	0.76	3.47	0.34	5.88	2.29	2.65	4.47	5.05	2.47	0.33	0.75	30.25
1982	0.04	0.75	0.19	4.08	9.12	0.18	T	1.04	2.42	1.63	3.11	2.70	25.26
1983	1.10	2.62	0.61	T	1.41	1.78	6.11	2.34	8.61	2.53	0.52	0.48	28.11
1984	4.79	0.42	0.13	T	6.18	2.44	1.59	1.80	20.18	0.93	0.02	1.85	40.33
1985	1.49	0.54	0.40	1.91	4.21	6.47	4.18	2.10	6.04	4.04	1.02	0.42	32.82
Record Mean	1.36	1.30	1.00	1.45	2.61	2.71	1.86	2.62	5.63	3.18	1.72	1.54	26.98

TABLE 3 AVERAGE TEMPERATURE (deg. F) BROWNSVILLE, TEXAS

YEAR	JAN	FEB	MAR	APR	MAY	JUNE	JULY	AUG	SEP	OCT	NOV	DEC	ANNUAL
1956	63.0	65.7	68.6	74.2	78.8	80.3	81.9	82.9	79.4	76.6	64.4	64.4	73.4
1957	65.5	69.1	69.7	74.4	79.7	82.5	84.9	84.7	80.9	75.0	66.6	64.4	74.8
1958	58.0	61.7	63.8	74.2	77.5	83.4	83.7	84.7	81.7	72.8	68.3	58.8	72.4
1959	56.8	61.5	64.3	70.7	80.3	81.9	83.5	84.9	83.9	76.7	62.8	62.7	72.5
1960	61.7	60.6	64.3	73.6	76.9	82.9	85.0	84.4	78.7	78.5	69.6	58.6	72.9
1961	55.5	62.3	71.5	72.4	79.7	82.8	83.4	82.9	80.5	74.9	67.4	63.9	73.1
1962	55.9	70.5	66.5	73.4	78.2	82.2	83.9	85.8	83.3	79.6	68.1	60.8	74.0
1963	56.3	60.1	70.8	78.5	79.3	83.7	84.2	85.0	82.4	76.4	69.7	54.4	73.4
1964	59.6	58.5	67.5	77.3	80.3	81.6	84.9	86.0	83.2	72.2	70.7	60.3	73.5
1965	63.0	61.3	64.2	76.3	79.0	82.7	83.5	81.8	81.8	71.5	72.1	64.9	73.5
#1966	54.3	56.7	65.5	74.1	76.2	80.0	83.3	84.4	81.2	72.9	68.9	61.1	71.6
1967	58.5	60.9	70.1	80.1	80.9	83.2	85.3	82.0	77.7	72.8	69.9	61.3	73.6
1968	57.8	57.5	64.0	74.2	80.2	83.3	83.6	84.3	81.6	77.7	68.0	64.7	73.1
1969	64.2	65.7	63.4	75.9	78.8	83.8	87.2	84.9	81.8	77.3	66.9	65.3	74.6
1970	57.6	64.0	66.0	76.6	76.5	82.2	84.0	85.7	82.7	75.2	66.2	69.8	73.9
1971	66.5	67.2	72.1	74.3	79.5	82.8	83.4	83.0	81.4	76.9	69.3	67.3	75.3
1972	66.0	64.7	72.4	77.4	77.6	80.6	81.3	82.2	82.0	77.5	63.1	59.5	73.7
1973	54.5	58.7	70.4	71.9	77.8	81.1	83.1	81.5	81.6	76.4	63.0	62.0	72.9
1974	61.4	63.9	73.1	75.6	81.6	80.9	81.9	84.9	80.0	73.6	66.5	61.2	73.7
1975	61.7	64.0	71.0	76.7	81.8	82.6	82.4	82.4	78.3	74.8	66.2	61.3	73.7
1976	59.1	65.6	71.0	74.2	75.8	81.6	81.3	82.3	81.7	70.2	60.8	57.4	71.8
1977	55.0	61.9	69.0	73.7	80.2	82.6	84.6	85.9	84.3	77.1	69.8	64.4	74.0
1978	54.6	55.9	66.8	74.8	83.3	85.0	86.8	85.4	82.3	75.9	71.2	62.9	73.8
1979	56.3	59.5	69.0	76.1	77.0	82.1	85.5	84.5	77.7	76.6	65.3	59.7	72.5
1980	63.9	60.6	69.6	72.0	82.0	86.9	87.5	84.9	84.9	73.6	63.1	61.6	74.2
1981	59.7	62.9	68.1	77.0	80.3	84.3	85.3	85.6	81.7	77.6	70.9	64.8	74.9
1982	62.7	61.1	71.7	76.0	80.1	85.7	86.9	86.3	83.4	77.0	68.4	62.4	75.1
1983	59.8	62.8	68.2	73.0	79.8	83.4	84.5	85.6	81.7	76.2	71.5	55.4	73.5
1984	55.9	62.4	69.7	76.3	79.3	82.8	84.2	84.5	79.4	79.3	68.8	70.3	74.4
1985	54.4	59.2	71.4	75.9	80.6	82.5	82.9	84.9	82.1	76.7	73.0	60.4	73.7
Record Mean	60.2	63.9	68.4	74.4	79.0	82.7	83.8	84.2	81.3	75.5	67.9	62.1	73.6
Max	69.3	72.4	77.3	82.8	86.9	90.7	92.3	92.9	89.7	84.7	77.0	71.3	82.3
Min	51.0	53.6	59.5	66.0	71.1	74.6	75.3	75.4	72.8	66.3	58.8	52.8	64.8

REFERENCE NOTES FOR TABLES 1, 2, 3 and 6   (BROWNSVILLE, TX)

GENERAL

T - TRACE AMOUNT
BLANK ENTRIES DENOTE MISSING/UNREPORTED DATA.
# INDICATES A STATION OR INSTRUMENT RELOCATION.

SPECIFIC

TABLE 1

(a) - LENGTH OF RECORD IN YEARS. ALTHOUGH INDIVIDUAL MONTHS MAY BE MISSING.

* LESS THAN .05

NORMALS — BASED ON THE 1951-1980 RECORD PERIOD.
EXTREMES — DATES ARE THE MOST RECENT OCCURRENCE.
WIND DIR. — NUMERALS SHOW TENS OF DEGREES CLOCKWISE FROM TRUE NORTH. "00" INDICATES CALM.
RESULTANT WIND DIRECTIONS ARE GIVEN TO WHOLE DEGREES.

EXCEPTIONS

TABLES 2, 3, and 6

RECORD MEANS ARE THROUGH THE CURRENT YEAR, BEGINNING IN   1882 FOR TEMPERATURE
1870 FOR PRECIPITATION
1940 FOR SNOWFALL

TABLE 4 HEATING DEGREE DAYS Base 65 deg. F BROWNSVILLE, TEXAS

SEASON	JULY	AUG	SEP	OCT	NOV	DEC	JAN	FEB	MAR	APR	MAY	JUNE	TOTAL
1956-57	0	0	0	0	108	89	102	30	25	9	0	0	363
1957-58	0	0	0	18	102	94	219	117	102	0	0	0	652
1958-59	0	0	0	32	53	212	268	153	84	27	0	0	829
1959-60	0	0	0	3	169	107	173	159	104	5	0	0	720
1960-61	0	0	0	2	41	229	302	123	16	14	0	0	727
1961-62	0	0	0	1	37	138	292	23	79	7	0	0	577
1962-63	0	0	0	0	28	168	292	185	17	0	0	0	690
1963-64	0	0	0	0	48	342	204	190	39	0	0	0	823
1964-65	0	0	0	5	52	215	127	136	138	0	0	0	673
1965-66	0	0	0	8	8	76	333	239	65	1	0	0	730
#1966-67	0	0	0	4	48	185	238	151	32	0	0	0	658
1967-68	0	0	1	4	8	174	262	218	126	4	0	0	797
1968-69	0	0	0	0	73	96	141	45	115	0	0	0	470
1969-70	0	0	0	0	92	69	256	72	76	4	1	0	570
1970-71	0	0	1	5	86	46	130	76	38	6	0	0	388
1971-72	0	0	0	0	16	67	112	93	11	2	0	0	301
1972-73	0	0	0	0	144	217	339	190	4	27	0	0	921
1973-74	0	0	0	0	16	130	190	102	40	5	0	0	483
1974-75	0	0	0	3	77	169	177	90	31	5	0	0	552
1975-76	0	0	0	5	74	180	215	74	22	0	0	0	570
1976-77	0	0	0	21	177	256	319	123	43	3	0	0	942
1977-78	0	0	0	0	25	101	342	286	66	6	0	0	826
1978-79	0	0	0	0	35	152	305	203	41	0	0	0	736
1979-80	0	0	0	0	99	204	108	180	51	13	0	0	655
1980-81	0	0	0	34	141	166	181	117	38	0	0	0	677
1981-82	0	0	0	14	23	101	183	187	54	17	0	0	579
1982-83	0	0	0	0	82	145	187	85	18	11	0	0	528
1983-84	0	0	0	0	22	365	299	138	39	2	0	0	865
1984-85	0	0	0	2	51	55	349	213	21	0	0	0	691
1985-86	0	0	0	1	21	184							

TABLE 5 COOLING DEGREE DAYS Base 65 deg. F BROWNSVILLE, TEXAS

YEAR	JAN	FEB	MAR	APR	MAY	JUNE	JULY	AUG	SEP	OCT	NOV	DEC	TOTAL
1969	123	71	72	335	437	570	695	624	511	390	156	87	4071
1970	35	48	116	356	364	522	597	648	540	330	128	202	3886
1971	185	142	266	294	459	541	578	566	498	375	153	143	4200
1972	148	91	246	382	398	475	512	542	517	395	95	53	3854
1973	19	21	178	242	404	490	585	519	504	365	309	74	3710
1974	83	75	299	329	520	485	534	624	458	275	130	59	3871
1975	79	68	223	363	526	536	548	546	408	315	176	69	3857
1976	41	97	216	283	343	507	511	547	509	190	59	24	3327
1977	15	44	174	272	476	535	535	658	585	384	177	88	4023
1978	32	38	129	308	572	605	680	638	524	342	228	92	4188
1979	45	56	173	339	380	517	645	614	388	368	118	46	3689
1980	83	62	198	230	536	661	707	623	602	309	91	67	4169
1981	22	64	139	365	482	586	635	646	507	414	204	102	4166
1982	119	85	267	357	474	626	684	667	555	381	190	73	4478
1983	29	29	124	254	466	560	607	645	506	354	222	73	3869
1984	24	70	190	349	451	541	604	611	441	451	175	224	4131
1985	28	57	226	334	492	533	560	623	518	370	269	47	4057

TABLE 6 SNOWFALL (inches) BROWNSVILLE, TEXAS

SEASON	JULY	AUG	SEP	OCT	NOV	DEC	JAN	FEB	MAR	APR	MAY	JUNE	TOTAL
1970-71	0.0	0.0	0.0	0.0	0.0	0.0	0.0	0.0	0.0	0.0	0.0	0.0	0.0
1971-72	0.0	0.0	0.0	0.0	0.0	0.0	0.0	0.0	0.0	0.0	0.0	0.0	0.0
1972-73	0.0	0.0	0.0	0.0	0.0	0.0	0.0	T	0.0	0.0	0.0	0.0	T
1973-74	0.0	0.0	0.0	0.0	0.0	0.0	0.0	0.0	0.0	0.0	0.0	0.0	0.0
1974-75	0.0	0.0	0.0	0.0	0.0	0.0	0.0	0.0	0.0	0.0	0.0	0.0	0.0
1975-76	0.0	0.0	0.0	0.0	0.0	0.0	0.0	0.0	0.0	0.0	0.0	0.0	T
1976-77	0.0	0.0	0.0	0.0	T	0.0	0.0	0.0	0.0	0.0	0.0	0.0	0.0
1977-78	0.0	0.0	0.0	0.0	0.0	0.0	0.0	0.0	0.0	0.0	0.0	0.0	0.0
1978-79	0.0	0.0	0.0	0.0	0.0	0.0	0.0	0.0	0.0	0.0	0.0	0.0	0.0
1979-80	0.0	0.0	0.0	0.0	0.0	0.0	0.0	0.0	0.0	0.0	0.0	0.0	0.0
1980-81	0.0	0.0	0.0	0.0	0.0	0.0	0.0	0.0	0.0	0.0	0.0	0.0	0.0
1981-82	0.0	0.0	0.0	0.0	0.0	0.0	0.0	0.0	0.0	0.0	0.0	0.0	0.0
1982-83	0.0	0.0	0.0	0.0	0.0	0.0	0.0	0.0	0.0	0.0	0.0	0.0	0.0
1983-84	0.0	0.0	0.0	0.0	0.0	0.0	T	0.0	0.0	0.0	0.0	0.0	T
1984-85	0.0	0.0	0.0	0.0	0.0	0.0	T	0.0	0.0	0.0	0.0	0.0	
1985-86	0.0	0.0	0.0	0.0	0.0	0.0							
Record Mean	0.0	0.0	0.0	0.0	T	T	T	T	T	0.0	0.0	0.0	T

See Reference Notes, relative to all above tables, on preceding page.

Corpus Christi is located on Corpus Christi Bay, an inlet of the Gulf of Mexico, in south Texas. The climatic conditions vary between the humid subtropical region to the northeast along the Texas coast and the semi-arid region to the west and southwest. Temperatures at the International Airport, which is about 7 miles west of downtown Corpus Christi, may be substantially different than those in the city during calm winter mornings and during summer afternoon sea breezes.

Peak rainfall months are May and September. Winter months have the least amounts of rainfall. The hurricane season from June to November can greatly effect the rainfall totals. Dry periods frequently occur. Several months during the years of record have had no rainfall, or only a trace. Snow falls on an average of about one day every two years.

There is little change in the day-to-day weather of the summer months, except for an occasional rainshower or a tropical storm in the area. High temperatures range in the high 80s to mid 90s, except for brief periods in the high 90s. The sea breeze during the afternoon and evening hours moderates the summer heat. Low temperatures are usually in the mid 70s. Mornings are generally warm. Summertime temperatures rarely reach 100 degrees near the bay, but occasionally do in most other parts of the city. Temperatures above 100 degrees are frequent about 30 to 60 miles to the west and southwest. Summertime afternoons are more pleasant than mornings because they are usually clear and windy. In the summer season the region receives nearly 80 percent of the possible sunshine.

The fall months of September and October are essentially an extension of the summer months. November is a transition to the conditions of the coming winter months, with greater temperature extremes, stronger winds, and the first occurrences of northers. The winter months are relatively mild, but with temperatures sufficiently low to be stimulating. Temperatures below 32 degrees seldom occur near the bay, but are more frequent inland. January is the coldest month with a prevailing northerly wind. The most extreme cold weather, in which daytime highs do not exceed 32 degrees, does not occur more than once every three or four years. The earliest occurrence of a temperature below 32 degrees is in early November and the latest occurrence in the spring is mid to late March.

Relative humidity, because of the nearness of the Gulf of Mexico, is high throughout the year. However, during the afternoons the humidity usually drops to between 50 and 60 percent.

Severe tropical storms average about one every ten years. Lesser strength storms average about one every five years. The city of Corpus Christi has a feature not found in most other coastal cities. A bluff rises 30 to 40 feet above the level of the lowlands area near the bay. This serves as a natural protection from high water. Protection for the main city is now furnished by sea walls.

Chief hurricane months are August and September, although tropical storms have occurred as early as June and as late as October. The majority of the storms pass either to the south or east of the city. Tornadoes are an infrequent occurrence in the area, and hail occurs only about once a year.

TABLE 1   NORMALS, MEANS AND EXTREMES

CORPUS CHRISTI, TEXAS

LATITUDE: 27°46'N   LONGITUDE: 97°30'W   ELEVATION: FT. GRND   41 BARO 00056   TIME ZONE: CENTRAL   WBAN: 12924

	(a)	JAN	FEB	MAR	APR	MAY	JUNE	JULY	AUG	SEP	OCT	NOV	DEC	YEAR
TEMPERATURE °F:														
Normals														
-Daily Maximum		66.5	69.9	76.1	82.1	86.7	91.2	94.2	94.1	90.1	83.9	75.1	69.3	81.6
-Daily Minimum		46.1	48.7	55.7	63.9	69.5	74.1	75.6	75.8	72.8	64.1	54.9	48.8	62.5
-Monthly		56.3	59.3	65.9	73.0	78.1	82.7	84.9	85.0	81.5	74.0	65.0	59.0	72.0
Extremes														
-Record Highest	47	91	98	101	102	103	101	104	103	103	98	95	91	104
-Year		1971	1940	1984	1984	1984	1980	1939	1962	1977	1950	1949	1977	JUL 1939
-Record Lowest	47	14	18	24	33	47	58	64	64	50	40	29	14	14
-Year		1962	1951	1980	1980	1970	1975	1967	1967	1942	1964	1969	1983	DEC 1983
NORMAL DEGREE DAYS:														
Heating (base 65°F)		310	209	97	7	0	0	0	0	0	11	116	220	970
Cooling (base 65°F)		40	50	125	247	406	531	617	620	495	290	116	37	3574
% OF POSSIBLE SUNSHINE	43	45	51	55	57	61	74	82	78	69	68	57	47	62
MEAN SKY COVER (tenths)														
Sunrise - Sunset	43	6.8	6.4	6.7	6.7	6.4	5.2	4.8	4.9	5.2	4.7	5.7	6.5	5.8
MEAN NUMBER OF DAYS:														
Sunrise to Sunset														
-Clear	43	6.9	7.6	6.7	5.6	5.7	8.9	10.8	11.3	10.0	12.5	9.3	8.1	103.3
-Partly Cloudy	43	6.9	6.0	7.5	9.2	12.3	14.6	14.5	12.8	11.9	10.3	8.7	6.6	121.2
-Cloudy	43	17.2	14.7	16.8	15.2	13.0	6.6	5.7	6.9	8.2	8.2	12.0	16.3	140.7
Precipitation														
.01 inches or more	46	8.2	6.9	5.6	4.8	6.4	5.8	4.8	5.8	9.1	6.8	5.8	6.6	76.5
Snow,Ice pellets														
1.0 inches or more	46	0.*	0.*	0.0	0.0	0.0	0.0	0.0	0.0	0.0	0.0	0.0	0.0	0.1
Thunderstorms	46	0.8	1.1	1.4	2.3	4.6	3.3	2.8	3.8	4.8	2.4	1.1	0.8	29.2
Heavy Fog Visibility														
1/4 mile or less	43	5.9	5.0	4.1	2.6	1.0	0.2	0.2	0.2	0.3	1.1	3.5	5.3	29.4
Temperature °F														
-Maximum														
90° and above	21	0.1	0.*	1.0	1.7	4.7	18.6	27.1	26.5	16.2	5.0	0.1	0.1	101.1
32° and below	21	0.0	0.*	0.0	0.0	0.0	0.0	0.0	0.0	0.0	0.0	0.0	0.1	0.2
-Minimum														
32° and below	21	3.5	1.5	0.3	0.0	0.0	0.0	0.0	0.0	0.0	0.0	0.3	1.4	7.0
0° and below	21	0.0	0.0	0.0	0.0	0.0	0.0	0.0	0.0	0.0	0.0	0.0	0.0	0.0
AVG. STATION PRESS.(mb)	13	1018.5	1017.3	1012.4	1011.9	1010.2	1011.7	1013.4	1013.1	1012.6	1014.9	1016.1	1018.0	1014.2
RELATIVE HUMIDITY (%)														
Hour 00	21	85	84	84	86	89	89	87	86	86	86	85	83	86
Hour 06	21	87	87	87	89	92	92	92	92	90	90	88	86	89
Hour 12 (Local Time)	21	69	65	62	63	66	63	57	59	63	61	62	64	63
Hour 18	21	71	66	65	68	71	68	63	64	68	69	71	71	68
PRECIPITATION (inches):														
Water Equivalent														
-Normal		1.63	1.55	0.84	1.99	3.05	3.36	1.96	3.51	6.15	3.19	1.55	1.40	30.18
-Maximum Monthly	47	10.78	8.11	4.80	8.04	9.38	13.35	11.92	14.79	20.33	11.02	8.53	7.80	20.33
-Year		1958	1982	1974	1956	1968	1973	1976	1980	1967	1981	1947	1960	SEP 1967
-Minimum Monthly	47	0.03	T	T	T	T	0.03	0.00	0.10	0.49	0.00	T	0.01	0.00
-Year		1971	1976	1971	1984	1961	1980	1957	1952	1981	1952	1949	1950	JUL 1957
-Maximum in 24 hrs	43	6.38	4.85	2.67	7.19	4.65	5.65	4.61	8.92	8.76	7.25	3.44	3.86	8.92
-Year		1958	1982	1945	1956	1968	1978	1981	1980	1967	1960	1947	1960	AUG 1980
Snow,Ice pellets														
-Maximum Monthly	47	1.2	1.1	T								T	T	1.2
-Year		1940	1973	1962								1979	1983	JAN 1940
-Maximum in 24 hrs	47	1.1	1.1	T								T	T	1.1
-Year		1940	1973	1962								1979	1983	FEB 1973
WIND:														
Mean Speed (mph)	43	12.1	13.0	14.1	14.4	12.9	11.9	11.5	10.9	10.3	10.2	11.5	11.4	12.0
Prevailing Direction														
through 1963		SSE	SSE	SSE	SE	SE	SE	SSE	SSE	SE	SE	SSE	SSE	SSE
Fastest Obs. 1 Min.														
-Direction (!!!)	9	32	15	14	32	32	12	02	11	04	35	17	18	11
-Speed (MPH)	9	37	41	38	40	44	31	46	55	38	35	39	38	55
-Year		1979	1981	1983	1983	1985	1983	1980	1980	1979	1985	1983	1982	AUG 1980
Peak Gust														
-Direction (!!!)	2	N	NW	S	SE	NW	SE	NE	S	NE	E	NW	S	NW
-Speed (mph)	2	52	58	48	44	60	41	37	41	52	53	44	44	60
-Date		1985	1984	1984	1984	1985	1984	1984	1984	1985	1984	1984	1984	MAY 1985

See Reference Notes to this table on the following page.

TABLE 2 PRECIPITATION (inches) CORPUS CHRISTI, TEXAS

YEAR	JAN	FEB	MAR	APR	MAY	JUNE	JULY	AUG	SEP	OCT	NOV	DEC	ANNUAL
1956	0.43	0.85	0.09	8.04	3.60	0.62	0.98	1.33	1.00	2.76	1.13	0.90	21.73
1957	0.14	1.48	2.74	2.53	4.82	5.34	0.00	2.12	2.42	0.40	5.24	0.77	28.00
1958	10.78	5.24	0.64	0.37	0.81	0.75	1.13	1.33	8.42	8.43	0.84	3.88	42.62
1959	1.74	4.53	0.31	1.39	4.49	5.69	2.29	2.41	7.73	0.76	1.52	38.44
#1960	1.56	1.07	1.97	3.26	1.93	3.77	1.42	7.06	1.61	10.66	2.24	7.80	44.35
1961	2.38	2.08	0.08	3.78	T	5.64	4.37	3.30	3.14	0.05	1.09	0.53	26.44
1962	0.22	0.06	0.41	1.18	0.24	2.93	T	0.90	5.37	0.39	1.13	2.66	15.49
1963	0.19	1.36	0.09	0.31	0.85	2.35	0.49	2.99	0.92	2.61	1.64	0.86	14.66
1964	1.61	1.53	1.14	0.08	4.39	0.38	2.25	0.50	6.98	0.19	0.21	2.45	21.71
1965	0.86	4.41	0.78	0.80	4.01	1.99	1.25	2.09	1.36	1.96	3.14	25.29
1966	2.12	1.15	0.69	5.03	7.23	4.35	1.23	4.15	2.84	0.85	0.07	0.18	29.89
1967	2.63	2.38	0.08	0.23	1.83	0.35	1.05	5.36	20.33	2.86	0.28	0.84	38.22
1968	2.11	2.42	0.90	0.82	9.38	8.36	5.43	0.62	6.34	3.68	1.34	0.13	41.53
1969	0.35	2.92	0.49	2.89	2.07	0.13	0.03	2.83	2.05	2.85	5.09	1.87	23.57
1970	1.79	1.01	1.55	0.15	3.92	9.16	1.72	7.32	8.51	3.13	0.81	0.40	39.47
1971	0.03	0.22	T	2.29	4.55	1.24	0.31	8.32	12.17	3.96	0.44	3.42	36.95
1972	1.23	3.41	1.44	1.53	5.99	3.65	2.82	3.74	9.49	0.46	2.48	0.17	36.41
1973	2.18	1.42	0.16	1.73	0.58	13.35	0.52	5.63	7.58	9.95	0.31	0.12	43.53
1974	1.98	T	4.80	0.08	4.24	2.43	0.19	1.05	3.88	3.57	1.76	0.83	24.81
1975	1.94	0.42	0.05	0.08	1.67	1.31	4.05	4.84	6.70	2.02	0.90	1.21	25.19
1976	0.15	T	0.15	3.68	5.95	0.76	11.92	0.86	2.54	6.81	4.27	2.30	39.39
1977	3.11	1.72	0.96	6.00	1.96	3.56	1.15	0.39	0.87	4.73	1.74	0.06	26.25
1978	2.01	0.84	0.03	2.20	1.68	12.04	3.92	0.81	10.83	2.46	0.50	1.82	39.14
1979	3.93	0.83	1.55	3.69	4.28	3.23	3.52	2.53	13.77	0.41	0.28	1.02	39.04
1980	1.24	1.01	0.31	0.34	2.82	0.03	1.47	14.79	6.01	1.18	3.16	0.33	32.69
1981	2.55	1.91	2.37	0.98	8.64	3.02	5.98	5.79	0.49	11.02	0.12	1.15	44.02
1982	0.07	8.11	0.46	1.01	4.17	0.72	0.01	0.64	0.55	1.70	4.33	0.70	22.47
1983	0.75	3.27	3.03	T	2.77	2.50	8.78	2.67	7.04	3.99	1.53	0.58	36.91
1984	5.91	0.39	0.19	T	2.22	0.23	0.25	0.90	3.03	6.49	1.71	0.92	22.24
1985	2.68	2.86	1.82	3.54	2.87	3.99	1.04	2.88	8.39	3.40	1.62	1.61	36.70
Record Mean	1.56	1.60	1.36	1.85	3.17	2.81	1.99	2.48	4.79	2.71	1.82	1.63	27.78

TABLE 3 AVERAGE TEMPERATURE (deg. F) CORPUS CHRISTI, TEXAS

YEAR	JAN	FEB	MAR	APR	MAY	JUNE	JULY	AUG	SEP	OCT	NOV	DEC	ANNUAL
1956	58.6	62.8	65.6	71.4	77.2	82.2	83.8	84.5	81.1	76.3	62.1	60.7	72.2
1957	60.0	66.7	66.4	72.1	77.8	81.4	84.5	85.1	79.5	72.3	63.2	61.5	72.5
1958	55.0	57.0	60.2	71.9	77.4	84.1	85.5	85.6	81.3	71.8	65.6	55.4	70.9
1959	52.8	57.6	63.5	68.1	78.4	81.6	83.6	83.2	81.9	74.2	58.8	59.1	70.2
#1960	56.0	55.4	59.9	71.1	74.9	82.1	84.2	84.2	79.5	75.8	66.5	55.3	70.4
1961	51.6	59.5	68.5	70.1	78.7	81.6	83.4	82.9	80.2	73.4	62.6	59.6	71.0
1962	50.5	67.6	63.0	72.0	77.8	82.2	85.4	86.4	82.8	78.7	64.7	56.4	72.3
1963	50.1	56.3	68.7	76.6	78.8	83.1	84.6	85.1	81.9	75.5	67.6	51.0	71.6
#1964	56.9	53.7	64.1	74.6	78.8	81.3	84.2	85.6	80.9	70.3	67.8	57.6	71.3
1965	59.1	56.4	59.8	73.6	78.4	82.2	84.3	83.1	82.6	70.1	71.2	62.6	72.0
1966	50.5	54.4	64.5	71.9	75.6	79.5	83.6	84.7	80.7	72.6	67.0	58.2	70.3
1967	56.1	57.4	69.2	77.3	78.5	82.8	84.6	81.4	77.2	71.0	65.7	56.1	71.5
1968	53.2	52.4	60.1	71.7	76.9	79.9	82.0	83.6	79.1	74.5	62.6	58.9	69.6
1969	59.1	60.2	57.8	70.8	75.5	82.1	86.6	85.3	81.3	74.2	62.9	61.2	71.4
1970	51.2	59.0	60.0	73.3	75.4	80.4	82.6	84.2	80.9	72.6	63.5	66.9	70.9
1971	62.3	62.0	68.7	70.9	78.4	82.9	84.9	82.7	81.1	76.6	67.1	64.9	73.5
1972	61.6	61.1	70.2	76.7	77.0	82.0	83.9	83.8	83.6	74.0	60.7	56.2	72.8
1973	51.7	56.0	68.8	69.3	77.1	81.1	84.9	82.8	81.8	76.2	72.9	59.7	71.9
1974	57.9	61.5	70.7	73.6	80.9	81.0	83.8	85.3	78.5	74.0	64.2	57.5	72.4
1975	59.7	60.1	67.2	75.0	80.5	83.4	84.5	83.3	79.0	74.4	66.5	58.5	72.7
1976	56.7	64.4	68.8	73.5	74.4	81.6	81.2	83.0	81.4	67.3	57.3	54.8	70.4
1977	50.3	58.6	66.0	71.5	78.8	82.9	84.6	86.7	85.5	75.7	67.5	61.9	72.5
1978	49.2	51.7	63.9	72.3	81.0	83.5	85.6	84.9	81.9	73.2	68.8	59.1	71.3
1979	51.3	57.0	67.9	74.3	76.6	81.7	85.5	84.9	78.8	75.4	63.3	58.8	71.3
1980	60.0	58.1	67.6	69.4	77.8	83.8	85.8	83.2	82.4	71.5	59.9	57.7	71.4
1981	55.3	58.4	63.7	74.4	77.0	82.5	83.4	83.7	80.7	75.2	68.3	59.4	71.9
1982	57.4	55.9	66.8	71.1	76.7	82.7	84.6	84.9	82.3	73.9	64.5	58.4	71.6
1983	55.1	57.9	63.7	69.0	76.0	81.2	83.1	84.2	79.4	74.1	67.7	49.3	70.1
1984	51.4	59.2	66.4	73.2	77.1	81.6	83.3	83.7	78.3	76.5	64.9	66.4	71.8
1985	49.2	54.1	67.6	72.3	77.6	80.2	82.0	84.1	80.7	75.2	69.0	55.5	70.6
Record Mean	56.2	59.0	65.0	71.5	76.8	81.1	83.3	83.4	80.6	73.9	64.9	58.3	71.2
Max	64.4	67.3	72.8	78.4	83.2	88.0	90.4	90.7	87.6	81.8	73.2	66.7	78.7
Min	48.0	50.6	57.2	64.6	70.4	74.2	76.1	76.2	73.5	66.0	56.7	50.0	63.6

REFERENCE NOTES FOR TABLES 1, 2, 3 and 6   (CORPUS CHRISTI,TX)

GENERAL

T - TRACE AMOUNT
BLANK ENTRIES DENOTE MISSING/UNREPORTED DATA.
# INDICATES A STATION OR INSTRUMENT RELOCATION.

SPECIFIC

TABLE 1

(a) - LENGTH OF RECORD IN YEARS. ALTHOUGH
INDIVIDUAL MONTHS MAY BE MISSING.

* LESS THAN .05

NORMALS — BASED ON THE 1951-1980 RECORD PERIOD.
EXTREMES — DATES ARE THE MOST RECENT OCCURRENCE.
WIND DIR. — NUMERALS SHOW TENS OF DEGREES
CLOCKWISE FROM TRUE NORTH.
"00" INDICATES CALM.
RESULTANT WIND DIRECTIONS ARE GIVEN TO WHOLE DEGREES.

EXCEPTIONS

TABLES 2, 3, and 6

RECORD MEANS ARE THROUGH THE CURRENT YEAR,
BEGINNING IN   1887 FOR TEMPERATURE
1887 FOR PRECIPITATION
1940 FOR SNOWFALL

TABLE 4   HEATING DEGREE DAYS Base 65 deg. F   CORPUS CHRISTI, TEXAS

SEASON	JULY	AUG	SEP	OCT	NOV	DEC	JAN	FEB	MAR	APR	MAY	JUNE	TOTAL
1956-57	0	0	0	0	156	180	216	64	53	22	0	0	691
1957-58	0	0	0	31	161	157	307	232	176	7	0	0	1071
1958-59	0	0	0	38	93	292	382	227	105	39	0	0	1176
1959-60	0	0	0	5	253	190	299	286	189	10	1	0	1233
#1960-61	0	0	0	12	78	330	408	166	39	35	0	0	1088
1961-62	0	0	0	4	123	220	448	51	142	10	0	0	998
1962-63	0	0	0	4	85	274	466	271	36	0	0	0	1136
#1963-64	0	0	0	0	69	440	263	324	87	7	0	0	1190
1964-65	0	0	0	18	70	289	219	252	222	0	0	0	1070
1965-66	0	0	0	22	14	125	450	300	79	12	1	0	1003
1966-67	0	0	0	5	68	267	304	222	42	0	0	0	908
1967-68	0	0	0	17	63	299	373	362	202	9	0	0	1325
1968-69	0	0	0	3	144	219	237	146	226	0	0	0	975
1969-70	0	0	0	10	160	152	442	168	166	29	2	0	1129
1970-71	0	0	2	12	125	93	174	151	64	26	0	0	647
1971-72	0	0	0	0	51	112	188	166	22	2	0	0	541
1972-73	0	0	0	8	188	295	415	262	23	44	0	0	1235
1973-74	0	0	0	0	20	204	264	151	56	13	0	0	708
1974-75	0	0	0	4	117	260	227	167	77	10	0	0	862
1975-76	0	0	0	4	107	247	275	94	45	1	0	0	773
1976-77	0	0	0	59	253	311	455	192	65	6	0	0	1341
1977-78	0	0	0	4	56	160	502	382	101	17	0	0	1222
1978-79	0	0	0	1	57	236	445	256	38	1	0	0	1034
1979-80	0	0	0	6	131	233	195	247	74	30	0	0	916
1980-81	0	0	0	43	204	254	294	221	90	0	0	0	1106
1981-82	0	0	0	24	34	207	292	275	92	40	0	0	964
1982-83	0	0	0	10	134	245	308	195	88	40	0	0	1020
1983-84	0	0	0	6	66	511	415	199	69	7	0	0	1273
1984-85	0	0	3	5	106	93	485	320	48	6	0	0	1066
1985-86	0	0	2	3	50	304							

TABLE 5   COOLING DEGREE DAYS Base 65 deg. F   CORPUS CHRISTI, TEXAS

YEAR	JAN	FEB	MAR	APR	MAY	JUNE	JULY	AUG	SEP	OCT	NOV	DEC	TOTAL
1969	62	16	11	180	332	521	677	636	496	300	103	40	3374
1970	20	8	47	287	329	466	552	599	485	255	89	160	3297
1971	98	74	187	207	423	542	625	556	489	367	121	116	3805
1972	93	61	187	360	380	520	594	588	561	368	68	28	3808
1973	13	14	149	179	381	489	629	560	511	353	265	48	3591
1974	52	60	243	276	498	486	587	637	410	289	97	32	3667
1975	70	35	154	319	486	557	608	573	422	300	161	52	3737
1976	25	84	170	265	299	505	510	563	498	136	31	3	3089
1977	4	18	104	206	435	545	616	677	621	342	136	71	3775
1978	18	16	74	243	502	561	647	622	511	264	177	56	3691
1979	26	39	140	289	366	507	639	624	421	333	89	47	3520
1980	45	52	159	405	405	573	649	570	530	247	56	34	3486
1981	4	41	57	288	378	535	580	589	476	349	141	41	3479
1982	63	24	158	229	370	540	613	627	525	289	128	48	3614
1983	10	6	55	166	349	494	568	605	438	295	152	31	3169
1984	0	39	119	257	382	502	573	588	411	367	108	141	3487
1985	5	19	136	230	399	461	532	595	480	327	179	19	3382

TABLE 6   SNOWFALL (inches)   CORPUS CHRISTI, TEXAS

SEASON	JULY	AUG	SEP	OCT	NOV	DEC	JAN	FEB	MAR	APR	MAY	JUNE	TOTAL
1970-71	0.0	0.0	0.0	0.0	0.0	0.0	0.0	0.0	0.0	0.0	0.0	0.0	0.0
1971-72	0.0	0.0	0.0	0.0	0.0	0.0	0.0	0.0	0.0	0.0	0.0	0.0	0.0
1972-73	0.0	0.0	0.0	0.0	0.0	0.0	0.2	1.1	0.0	0.0	0.0	0.0	1.3
1973-74	0.0	0.0	0.0	0.0	0.0	0.0	0.0	0.0	0.0	0.0	0.0	0.0	0.0
1974-75	0.0	0.0	0.0	0.0	0.0	0.0	0.0	0.0	0.0	0.0	0.0	0.0	0.0
1975-76	0.0	0.0	0.0	0.0	0.0	0.0	0.0	0.0	0.0	0.0	0.0	0.0	0.0
1976-77	0.0	0.0	0.0	0.0	T	0.0	0.0	0.0	0.0	0.0	0.0	0.0	T
1977-78	0.0	0.0	0.0	0.0	0.0	0.0	T	0.0	0.0	0.0	0.0	0.0	T
1978-79	0.0	0.0	0.0	0.0	0.0	0.0	0.0	0.0	0.0	0.0	0.0	0.0	0.0
1979-80	0.0	0.0	0.0	0.0	T	0.0	0.0	0.0	0.0	0.0	0.0	0.0	T
1980-81	0.0	0.0	0.0	0.0	0.0	0.0	T	0.0	0.0	0.0	0.0	0.0	T
1981-82	0.0	0.0	0.0	0.0	0.0	0.0	0.0	0.0	0.0	0.0	0.0	0.0	0.0
1982-83	0.0	0.0	0.0	0.0	0.0	0.0	0.0	0.0	0.0	0.0	0.0	0.0	0.0
1983-84	0.0	0.0	0.0	0.0	0.0	T	0.0	0.0	0.0	0.0	0.0	0.0	T
1984-85	0.0	0.0	0.0	0.0	0.0	0.0	T	T	0.0	0.0	0.0	0.0	T
1985-86	0.0	0.0	0.0	0.0	0.0	0.0							
Record Mean	0.0	0.0	0.0	0.0	T	T	0.1	T	T	0.0	0.0	0.0	0.1

See Reference Notes, relative to all above tables, on preceding page.

The Dallas–Fort Worth Metroplex is located in North Central Texas, approximately 250 miles north of the Gulf of Mexico. It is near the headwaters of the Trinity River, which lie in the upper margins of the Coastal Plain. The rolling hills in the area range from 500 to 800 feet in elevation.

The Dallas–Fort Worth climate is humid subtropical with hot summers. It is also continental, characterized by a wide annual temperature range. Precipitation also varies considerably, ranging from less than 20 to more than 50 inches.

Winters are mild, but northers occur about three times each month, and often are accompanied by sudden drops in temperature. Periods of extreme cold that occasionally occur are short-lived, so that even in January mild weather occurs frequently.

The highest temperatures of summer are associated with fair skies, westerly winds and low humidities. Characteristically, hot spells in summer are broken into three-to-five day periods by thunderstorm activity. There are only a few nights each summer when the low temperature exceeds 80 degrees. Summer daytime temperatures frequently exceed 100 degrees. Air conditioners are recommended for maximum comfort indoors and while traveling via automobile.

Throughout the year, rainfall occurs more frequently during the night. Usually, periods of rainy weather last for only a day or two, and are followed by several days with fair skies. A large part of the annual precipitation results from thunderstorm activity, with occasional heavy rainfall over brief periods of time. Thunderstorms occur throughout the year, but are most frequent in the spring. Hail falls on about two or three days a year, ordinarily with only slight and scattered damage. Windstorms occurring during thunderstorm activity are sometimes destructive. Snowfall is rare.

The average length of the warm season (freeze-free period) in the Dallas–Fort Worth Metroplex is about 249 days. The average last occurrence of 32 degrees or below is mid March and the average first occurrence of 32 degrees or below is in late November.

TABLE 1   NORMALS, MEANS AND EXTREMES

DALLAS – FORT WORTH, TEXAS

LATITUDE: 32°54'N   LONGITUDE: 97°02'W   ELEVATION: FT. GRND   551 BARO   00575   TIME ZONE: CENTRAL   WBAN: 03927

	(a)	JAN	FEB	MAR	APR	MAY	JUNE	JULY	AUG	SEP	OCT	NOV	DEC	YEAR
TEMPERATURE °F:														
Normals														
-Daily Maximum		54.0	59.1	67.2	76.8	84.4	93.2	97.8	97.3	89.7	79.5	66.2	58.1	76.9
-Daily Minimum		33.9	37.8	44.9	55.0	62.9	70.8	74.7	73.7	67.5	56.3	44.9	37.4	55.0
-Monthly		44.0	48.5	56.0	65.9	73.7	82.0	86.3	85.5	78.6	67.9	55.5	47.8	66.0
Extremes														
-Record Highest	32	88	88	96	95	103	113	110	108	106	102	89	88	113
-Year		1969	1959	1974	1972	1985	1980	1980	1964	1985	1979	1955	1955	JUN 1980
-Record Lowest	32	4	7	15	30	41	51	59	56	43	29	20	5	4
-Year		1964	1985	1980	1973	1978	1964	1972	1967	1984	1980	1959	1983	JAN 1964
NORMAL DEGREE DAYS:														
Heating (base 65°F)		651	469	313	85	0	0	0	0	0	56	300	533	2407
Cooling (base 65°F)		0	7	37	112	275	510	660	636	408	146	18	0	2809
% OF POSSIBLE SUNSHINE	7	53	56	58	64	65	72	80	78	73	59	58	58	65
MEAN SKY COVER (tenths)														
Sunrise – Sunset	32	6.1	5.7	5.9	6.0	5.8	4.8	4.2	4.2	4.7	4.7	5.2	5.5	5.2
MEAN NUMBER OF DAYS:														
Sunrise to Sunset														
-Clear	32	9.8	9.9	9.6	8.7	8.4	11.2	15.3	15.2	13.0	13.4	12.1	11.5	138.2
-Partly Cloudy	32	5.7	5.7	7.5	8.0	10.9	11.5	9.3	10.0	8.5	7.5	6.0	6.3	97.0
-Cloudy	32	15.5	12.7	13.9	13.3	11.7	7.3	6.4	5.8	8.4	10.0	11.9	13.2	130.1
Precipitation														
.01 inches or more	32	7.0	6.4	7.3	8.1	8.6	5.9	4.8	4.6	6.8	6.1	5.7	6.0	77.5
Snow, Ice pellets														
1.0 inches or more	32	0.7	0.4	0.1	0.0	0.0	0.0	0.0	0.0	0.0	0.0	0.*	0.1	1.3
Thunderstorms	32	0.9	1.7	4.2	6.0	7.3	5.8	4.8	4.5	3.5	2.8	1.6	1.0	44.1
Heavy Fog Visibility														
1/4 mile or less	32	2.7	1.7	1.1	0.7	0.4	0.1	0.0	0.*	0.1	0.9	1.5	2.5	11.7
Temperature °F														
-Maximum														
90° and above	22	0.0	0.0	0.2	0.6	3.8	19.6	27.8	26.9	14.2	2.6	0.0	0.0	95.8
32° and below	22	1.9	0.7	0.0	0.0	0.0	0.0	0.0	0.0	0.0	0.0	0.0	0.8	3.4
-Minimum														
32° and below	22	16.0	10.0	2.7	0.1	0.0	0.0	0.0	0.0	0.0	0.*	2.5	10.0	41.3
0° and below	22	0.0	0.0	0.0	0.0	0.0	0.0	0.0	0.0	0.0	0.0	0.0	0.0	0.0
AVG. STATION PRESS.(mb)	13	999.7	998.0	993.4	993.2	991.6	992.8	994.2	994.1	994.8	996.4	997.2	998.6	995.3
RELATIVE HUMIDITY (%)														
Hour 00	22	73	71	71	73	79	73	66	67	74	74	74	73	72
Hour 06 (Local Time)	22	79	79	80	82	87	85	80	80	84	82	81	79	82
Hour 12	22	61	59	57	57	60	55	48	50	56	55	57	59	56
Hour 18	22	59	54	51	53	57	50	44	44	53	56	58	58	53
PRECIPITATION (inches):														
Water Equivalent														
-Normal		1.65	1.93	2.42	3.63	4.27	2.59	2.00	1.76	3.31	2.47	1.76	1.67	29.46
-Maximum Monthly	32	3.60	6.20	6.39	12.19	13.66	7.85	11.13	6.85	9.52	14.18	6.23	6.99	14.18
-Year		1968	1965	1968	1957	1982	1981	1973	1970	1964	1981	1964	1971	OCT 1981
-Minimum Monthly	32	0.13	0.15	0.10	0.59	0.99	0.40	0.09	T	0.09	T	0.20	0.17	T
-Year		1976	1963	1972	1983	1977	1964	1965	1980	1984	1975	1970	1981	AUG 1980
-Maximum in 24 hrs	32	2.39	4.06	4.39	4.55	4.86	3.11	3.76	4.05	4.76	5.91	2.83	3.10	5.91
-Year		1975	1965	1977	1957	1965	1966	1975	1976	1965	1959	1964	1971	OCT 1959
Snow, Ice pellets														
-Maximum Monthly	32	12.1	13.5	2.5								5.0	2.6	13.5
-Year		1964	1978	1962								1976	1963	FEB 1978
-Maximum in 24 hrs	32	12.1	7.5	2.5								4.8	2.5	12.1
-Year		1964	1978	1962								1976	1963	JAN 1964
WIND:														
Mean Speed (mph)	32	11.1	11.9	13.0	12.7	11.1	10.8	9.5	9.0	9.4	9.7	10.7	11.1	10.8
Prevailing Direction														
through 1963		S	S	S	S	S	S	S	S	S	S	S	S	S
Fastest Obs. 1 Min.														
-Direction	32	36	36	29	32	14	32	36	36	11	27	34	32	36
-Speed (MPH)	32	55	51	55	55	55	52	65	73	53	44	50	53	73
-Year		1985	1962	1954	1970	1955	1955	1961	1959	1961	1957	1957	1968	AUG 1959
Peak Gust														
-Direction	2	N	S	S	W	S	E	S	NW	N	S	NW	NW	NW
-Speed (mph)	2	66	54	53	48	53	53	36	81	43	41	55	43	81
-Date		1985	1985	1985	1984	1985	1984	1985	1985	1984	1985	1984	1985	AUG 1985

See Reference Notes to this table on the following page.

TABLE 2 PRECIPITATION (inches) DALLAS – FORT WORTH, TEXAS

YEAR	JAN	FEB	MAR	APR	MAY	JUNE	JULY	AUG	SEP	OCT	NOV	DEC	ANNUAL
1956	1.34	2.54	0.11	3.12	3.83	0.88	0.38	0.23	0.23	1.20	2.61	2.08	18.55
1957	1.72	1.77	4.18	12.19	12.64	3.96	0.65	0.12	3.23	3.53	4.72	1.78	50.49
1958	1.70	0.84	5.49	8.63	1.50	0.67	3.69	3.64	5.10	1.07	2.26	1.09	35.68
1959	0.36	1.61	2.31	0.92	3.27	5.27	3.27	0.93	2.40	9.22	1.74	2.84	34.14
1960	2.29	2.16	0.74	1.67	1.89	1.72	3.96	2.76	1.82	0.49	1.25	4.22	24.97
1961	3.29	2.20	2.95	2.23	1.06	5.93	2.32	0.02	2.92	2.82	2.72	2.12	30.58
1962	1.00	2.01	1.80	5.66	1.58	6.94	6.36	3.22	3.79	4.15	0.99		41.43
1963	0.86	0.15	0.48	6.20	2.52	0.57	2.28	2.73	1.70	0.23	1.29	1.45	20.46
1964	3.53	1.17	3.35	2.71	2.85	0.40	0.25	2.43	9.52	0.62	6.23	1.25	34.31
1965	2.77	6.20	1.45	2.15	8.97	1.50	0.09	2.26	5.04	1.97	2.43	1.73	36.56
1966	1.68	2.84	1.38	10.74	3.13	5.47	0.72	3.38	4.23	1.48	0.53	1.17	39.29
1967	0.28	0.32	2.09	3.84	4.02	0.72	2.20	0.48	5.94	4.19	0.92	2.30	27.30
1968	3.60	1.48	6.39	2.41	6.02	3.50	1.88	2.71	2.53	2.18	4.58	1.20	38.48
1969	1.26	1.99	3.62	3.40	7.12	0.63	0.77	2.56	4.55	5.82	1.22	2.75	35.69
1970	0.72	4.78	3.49	4.68	3.62	0.61	0.94	6.85	6.25	2.95	0.20	1.01	36.10
1971	0.19	1.32	0.34	2.76	1.88	0.83	3.60	5.70	3.24	7.64	1.77	6.99	36.26
1972	1.09	0.26	0.10	3.25	2.35	1.50	0.59	0.81	2.42	6.89	2.36	0.61	22.23
1973	3.26	1.92	2.28	6.06	3.18	5.88	11.13	0.01	7.16	6.85	0.83		50.62
1974	1.79	1.01	0.80	2.51	6.00	5.44	0.67	4.19	6.04	5.93	3.32	1.93	39.63
1975	3.34	3.72	1.67	3.40	6.88	1.95	5.06	0.30	0.87	T	0.42	1.49	29.10
1976	0.13	0.52	2.29	5.71	6.03	1.40	3.83	4.75	5.02	3.46	0.50	1.99	35.63
1977	2.39	1.68	5.88	4.31	0.99	0.69	2.20	2.33	1.72	2.96	1.79	0.25	27.19
1978	1.41	3.33	2.66	1.34	8.01	0.77	0.33	1.53	0.93	0.55	2.73	0.78	24.37
1979	3.35	1.52	6.33	2.03	5.90	1.36	1.94	2.47	3.38	0.43	2.72		32.42
1980	2.52	0.84	1.24	2.23	3.01	1.25	0.71	T	6.54	1.08	1.23	1.43	22.08
1981	0.58	1.44	3.39	2.69	6.24	7.85	1.81	2.32	2.40	14.18	1.53	0.17	44.60
1982	2.33	1.89	1.71	2.71	13.66	4.28	2.73	0.52	0.58	3.36	4.22	2.76	40.75
1983	2.55	1.25	4.36	0.59	5.83	2.07	1.56	5.55	0.22	4.04	2.22	0.83	31.07
1984	1.07	3.11	4.92	1.41	3.04	2.79	0.43	1.47	0.09	6.50	2.97	6.09	33.89
1985	0.81	2.62	3.70	3.75	2.13	3.78	2.40	0.53	3.35	3.91	3.11	0.61	30.70
Record Mean	1.80	1.98	2.40	3.80	4.70	2.92	2.21	2.22	2.85	3.06	2.24	1.95	32.13

TABLE 3 AVERAGE TEMPERATURE (deg. F) DALLAS – FORT WORTH, TEXAS

YEAR	JAN	FEB	MAR	APR	MAY	JUNE	JULY	AUG	SEP	OCT	NOV	DEC	ANNUAL
1956	44.9	50.1	58.4	64.8	77.7	83.3	89.0	88.0	80.9	71.0	52.7	50.0	67.6
1957	43.7	54.1	52.6	61.6	71.5	80.2	87.4	85.1	75.1	62.3	51.6	51.7	64.8
1958	44.9	43.6	48.7	61.5	73.1	82.3	85.7	85.0	78.4	65.8	57.4	43.6	64.2
1959	42.5	48.9	56.6	63.4	75.1	80.7	83.1	84.5	79.7	66.1	49.5	50.2	65.0
1960	44.1	43.0	49.1	67.3	72.0	82.9	84.6	84.0	79.6	70.0	57.5	43.7	64.8
1961	40.9	50.3	59.3	64.0	73.1	77.9	82.3	82.7	76.9	67.3	52.7	45.8	64.5
1962	39.6	53.3	54.0	64.3	77.6	79.9	85.5	85.6	77.1	70.4	55.5	47.2	65.8
#1963	37.8	46.4	61.2	70.4	75.0	83.1	87.4	87.3	79.2	73.5	58.4	40.3	66.7
1964	43.8	43.8	55.6	66.8	73.2	81.0	87.1	85.3	76.9	63.6	57.6	47.1	65.2
1965	47.1	45.8	47.0	68.4	72.9	79.9	86.3	84.1	79.4	66.6	62.9	52.8	66.1
1966	40.3	45.4	56.3	63.8	70.8	79.6	86.3	82.7	75.8	65.0	60.7	45.3	64.4
1967	48.3	46.8	63.3	71.1	71.4	81.4	82.9	83.1	74.1	65.4	55.6	47.0	65.9
1968	44.4	44.2	54.6	63.4	72.4	79.5	81.0	83.5	74.6	67.8	53.9	47.4	63.9
1969	49.0	50.0	49.8	65.4	71.9	79.8	87.9	84.2	77.1	65.3	55.1	49.9	65.5
1970	40.6	48.6	52.1	66.2	71.7	79.1	84.0	85.8	78.2	65.1	54.7	53.6	65.0
1971	46.7	49.2	55.6	64.0	70.5	82.9	84.4	79.5	77.1	70.1	57.0	52.2	65.8
1972	45.0	51.5	62.1	70.1	72.7	81.4	83.1	84.7	80.8	67.5	50.1	44.0	66.1
1973	42.5	47.9	60.0	60.7	71.7	79.3	83.9	82.9	76.1	68.3	59.8	48.4	65.1
1974	43.6	52.3	62.9	65.8	75.7	78.7	86.1	82.9	70.9	69.2	55.5	47.2	65.9
1975	49.0	46.6	53.8	64.7	72.4	80.9	83.6	84.8	75.6	66.9	57.3	49.0	65.6
1976	45.0	58.4	59.4	64.9	68.6	78.8	82.1	84.2	76.1	60.2	49.5	45.0	64.3
1977	34.7	49.4	57.2	66.8	77.4	84.1	87.1	84.4	81.6	66.7	56.4	47.6	66.2
1978	33.8	36.7	54.1	67.1	73.1	82.3	88.4	84.6	80.2	68.9	57.7	46.1	64.4
1979	35.4	42.2	56.7	64.4	69.7	81.0	84.5	82.5	77.0	70.8	52.9	49.4	63.9
1980	45.5	46.6	54.2	63.1	75.0	87.0	92.0	88.5	80.3	65.4	54.9	49.4	66.8
1981	44.6	48.9	55.7	69.2	70.5	80.3	85.9	83.4	76.2	66.1	57.5	47.3	65.4
1982	44.6	44.5	59.8	62.5	72.5	79.2	84.6	86.7	78.1	67.0	55.6	49.2	65.4
1983	43.4	48.5	54.5	60.6	69.5	77.3	83.6	84.3	77.1	67.8	57.3	34.8	63.3
1984	39.3	50.9	56.3	63.7	73.7	82.5	85.5	85.8	76.1	67.0	54.6	52.6	65.7
1985	37.8	45.0	60.8	67.2	74.0	80.2	84.4	87.6	77.7	67.6	56.3	42.3	65.1
Record Mean	45.0	48.6	56.9	65.2	72.7	80.9	84.6	84.7	77.8	67.7	56.0	47.5	65.6
Max	55.3	59.3	67.9	75.8	82.7	91.1	95.0	95.3	88.2	78.5	66.5	57.4	76.1
Min	34.7	37.9	45.8	54.5	62.6	70.7	74.2	74.0	67.4	56.8	45.5	37.4	55.1

REFERENCE NOTES FOR TABLES 1, 2, 3 and 6   (DALLAS/FT WORTH, TX)

GENERAL

T - TRACE AMOUNT
BLANK ENTRIES DENOTE MISSING/UNREPORTED DATA.
# INDICATES A STATION OR INSTRUMENT RELOCATION.

SPECIFIC

TABLE 1

(a) - LENGTH OF RECORD IN YEARS. ALTHOUGH
INDIVIDUAL MONTHS MAY BE MISSING.
* LESS THAN .05

NORMALS — BASED ON THE 1951-1980 RECORD PERIOD.
EXTREMES — DATES ARE THE MOST RECENT OCCURRENCE.
WIND DIR. — NUMERALS SHOW TENS OF DEGREES
CLOCKWISE FROM TRUE NORTH.
"00" INDICATES CALM.
RESULTANT WIND DIRECTIONS ARE GIVEN TO WHOLE DEGREES.

EXCEPTIONS

TABLES 2, 3, and 6

RECORD MEANS ARE THROUGH THE CURRENT YEAR,
BEGINNING IN   1899 FOR TEMPERATURE
1899 FOR PRECIPITATION
1954 FOR SNOWFALL

TABLE 4 HEATING DEGREE DAYS Base 65 deg. F   DALLAS – FORT WORTH, TEXAS

SEASON	JULY	AUG	SEP	OCT	NOV	DEC	JAN	FEB	MAR	APR	MAY	JUNE	TOTAL
1956-57	0	0	0	21	374	468	656	311	381	142	21	0	2374
1957-58	0	0	0	144	402	410	615	592	496	138	9	0	2806
1958-59	0	0	0	82	247	659	690	452	276	143	4	0	2553
1959-60	0	0	0	68	470	450	643	639	499	54	32	0	2855
1960-61	0	0	0	45	253	658	740	409	209	132	9	2	2457
1961-62	0	0	0	50	381	590	781	328	345	107	2	0	2584
#1962-63	0	0	0	46	280	545	839	517	185	34	13	0	2459
1963-64	0	0	0	4	227	760	651	608	285	65	1	0	2601
1964-65	0	0	6	81	260	550	550	530	551	36	0	0	2564
1965-66	0	0	2	60	103	376	760	542	274	84	26	0	2227
1966-67	0	0	0	79	182	627	514	503	146	15	21	0	2087
1967-68	0	0	13	80	282	548	631	598	330	100	2	0	2584
1968-69	0	0	0	47	348	540	492	416	468	49	6	4	2370
1969-70	0	0	0	116	306	463	756	455	404	63	21	1	2585
1970-71	0	0	7	105	316	369	564	440	307	97	19	0	2224
1971-72	0	0	12	7	270	389	615	398	143	26	1	0	1861
1972-73	0	0	3	96	446	644	690	475	155	182	12	0	2703
1973-74	0	0	1	36	182	509	656	352	173	70	1	0	1980
1974-75	0	0	20	16	296	546	489	508	355	112	0	0	2342
1975-76	0	0	4	33	266	500	616	217	222	48	20	0	1926
1976-77	0	0	0	214	459	614	931	431	241	37	41	0	2927
1977-78	0	0	0	55	257	536	962	786	346	54	29	0	3037
1978-79	0	0	0	27	247	578	911	635	261	78	6	0	2766
1979-80	0	0	0	34	370	478	597	530	339	102	6	0	2456
1980-81	0	0	18	99	330	486	625	448	284	26	23	0	2339
1981-82	0	0	10	116	228	541	625	569	232	140	9	0	2470
1982-83	0	0	1	94	316	495	663	454	324	186	21	2	2556
1983-84	0	0	12	52	269	933	789	401	281	89	11	0	2837
1984-85	0	0	38	66	322	389	837	558	171	37	0	0	2418
1985-86	0	0	19	53	285	696							

TABLE 5 COOLING DEGREE DAYS Base 65 deg. F   DALLAS – FORT WORTH, TEXAS

YEAR	JAN	FEB	MAR	APR	MAY	JUNE	JULY	AUG	SEP	OCT	NOV	DEC	TOTAL
1969	3	3	3	67	228	453	715	602	372	133	17	0	2596
1970	5	0	7	108	236	433	595	653	409	115	12	22	2595
1971	0	6	21	71	195	546	606	456	382	171	36	1	2491
1972	2	14	57	185	249	498	569	618	480	183	4	0	2859
1973	1	0	6	60	230	435	593	559	342	146	33	0	2405
1974	0	2	115	101	341	419	660	563	202	153	20	2	2578
1975	0	0	15	107	236	483	580	620	331	189	39	9	2609
1976	0	32	59	52	138	421	537	602	338	72	0	0	2251
1977	0	0	7	94	391	581	693	626	505	112	6	2	3017
1978	0	0	16	125	301	524	733	614	462	153	37	0	2965
1979	0	1	9	67	179	489	613	551	366	220	14	0	2509
1980	0	0	11	52	320	668	844	737	485	117	35	10	3279
1981	0	5	5	158	200	467	654	577	352	155	8	0	2581
1982	1	2	77	71	252	433	614	679	403	160	40	10	2742
1983	0	0	7	61	171	382	582	626	381	145	46	0	2401
1984	0	0	20	60	288	531	644	652	376	135	16	12	2734
1985	0	5	51	108	287	460	608	706	408	139	29	0	2801

TABLE 6 SNOWFALL (inches)   DALLAS – FORT WORTH, TEXAS

SEASON	JULY	AUG	SEP	OCT	NOV	DEC	JAN	FEB	MAR	APR	MAY	JUNE	TOTAL
1956-57	0.0	0.0	0.0	0.0	0.0	0.0	T	0.0	0.0	0.0	0.0	0.0	T
1957-58	0.0	0.0	0.0	0.0	T	0.0	T	0.9	T	0.0	0.0	0.0	0.9
1958-59	0.0	0.0	0.0	0.0	0.0	0.3	0.2	2.0	0.0	0.0	0.0	0.0	2.5
1959-60	0.0	0.0	0.0	0.0	0.0	0.0	0.9	1.0	T	0.0	0.0	0.0	1.9
1960-61	0.0	0.0	0.0	0.0	0.0	0.0	3.5	2.9	0.0	0.0	0.0	0.0	6.4
1961-62	0.0	0.0	0.0	0.0	0.0	0.0	0.0	2.6	T	2.5	0.0	0.0	5.1
1962-63	0.0	0.0	0.0	0.0	0.0	0.0	T	T	T	0.1	0.0	0.0	0.1
1963-64	0.0	0.0	0.0	0.0	0.0	0.0	2.6	12.1	0.2	0.4	0.0	0.0	15.3
1964-65	0.0	0.0	0.0	0.0	0.0	0.0	T	0.0	T	0.0	0.0	0.0	T
1965-66	0.0	0.0	0.0	0.0	0.0	0.0	4.4	2.9	0.0	0.0	0.0	0.0	7.3
1966-67	0.0	0.0	0.0	0.0	0.0	0.0	0.0	T	T	0.0	0.0	0.0	T
1967-68	0.0	0.0	0.0	0.0	0.0	0.0	0.4	2.6	T	0.0	0.0	0.0	3.0
1968-69	0.0	0.0	0.0	0.0	T	0.0	0.0	0.0	0.8	0.0	0.0	0.0	0.8
1969-70	0.0	0.0	0.0	0.0	0.0	T	T	T	1.6	0.0	0.0	0.0	1.6
1970-71	0.0	0.0	0.0	0.0	0.0	0.0	0.0	T	T	0.0	0.0	0.0	T
1971-72	0.0	0.0	0.0	0.0	T	0.0	T	T	0.0	0.0	0.0	0.0	T
1972-73	0.0	0.0	0.0	0.0	T	1.4	2.3	T	0.0	0.0	0.0	0.0	3.7
1973-74	0.0	0.0	0.0	0.0	0.0	0.0	T	3.7	0.0	0.0	0.0	0.0	3.7
1974-75	0.0	0.0	0.0	0.0	0.0	T	0.4	0.0	T	0.0	0.0	0.0	0.4
1975-76	0.0	0.0	0.0	0.0	0.0	5.0	5.4	0.0	0.0	0.0	0.0	0.0	10.4
1976-77	0.0	0.0	0.0	0.0	0.0	0.0	4.1	13.5	T	0.0	0.0	0.0	17.6
1977-78	0.0	0.0	0.0	0.0	0.0	0.8	1.8	0.7	0.0	0.0	0.0	0.0	3.3
1978-79	0.0	0.0	0.0	0.0	0.0	0.0	0.0	1.6	0.0	0.0	0.0	0.0	1.6
1979-80	0.0	0.0	0.0	0.0	0.0	0.0	T	T	0.0	0.0	0.0	0.0	T
1980-81	0.0	0.0	0.0	0.0	0.0	0.0	0.8	T	0.0	0.0	0.0	0.0	0.8
1981-82	0.0	0.0	0.0	0.0	0.0	0.0	T	T	0.0	0.0	0.0	0.0	T
1982-83	0.0	0.0	0.0	0.0	0.0	2.0	0.0	0.0	0.0	0.0	0.0	0.0	2.0
1983-84	0.0	0.0	0.0	0.0	0.0	0.0	3.4	1.7	0.0	0.0	0.0	0.0	5.1
1984-85	0.0	0.0	0.0	0.0	0.0	T							
1985-86	0.0	0.0	0.0	0.0	0.0								
Record Mean	0.0	0.0	0.0	0.0	0.2	0.2	1.5	1.1	0.2	0.0	0.0	0.0	3.2

See Reference Notes, relative to all above tables, on preceding page.

Del Rio is located on the Rio Grande River, on the western tip of the Balcones escarpment, in southwest Texas. Elevation is near 1,000 feet and varies little within the city but rises to 2,300 feet in the northern part of the county. Regional agriculture is chiefly wool and mohair production to the north and west of Del Rio and garden crops to the southeast. Lake Amistad, a reservoir of 65,000 surface acres, lies 10 miles west of Del Rio.

The climate of Del Rio is semi—arid continental. Annual precipitation is insufficient for dry farming. However, San Felipe Springs and the Rio Grande provide adequate water for irrigation farming. Over 80 percent of the average annual precipitation occurs from April through October. During this period, rainfall is chiefly in the form of showers and thunderstorms, often as heavy downpours, resulting in flash flooding. The small amount of precipitation for November through March usually falls as steady light rain.

Hail occurs in the vicinity of Del Rio about once per year and reaches severe proportions about once every five years. Sleet or snow falls on an average of once a year, but frequently melts as it falls. A snowfall heavy enough to blanket the ground only occurs about once every four or five years, and seldom remains more than 24 hours.

Temperature averages indicate mild winters and quite warm summers. Cold periods in winter are ushered in by strong, dry, dusty north, and northwest winds known as northers, and temperature drops of as much as 25 degrees in a few hours are not uncommon. Cold weather periods usually do not last more than two or three days. Temperatures as low as 32 degrees have occurred as early as October and as late as March. Normal occurrences of the earliest freezing temperature in autumn and the latest in spring are early December and mid February, which results in an average growing season of 300 days. Hot weather is rather persistent from late May to mid September and temperatures above 100 degrees have been recorded as early as March and as late as October. Low humidity and fresh breezes tend to alleviate uncomfortable conditions usually associated with high temperatures. The mean early morning humidity is about 79 percent, and the mean afternoon humidity is near 44 percent.

Clear to partly cloudy skies predominate, and even in the more cloudy winter months, the mean number of cloudy days are less than the number of clear days.

TABLE 1 NORMALS, MEANS AND EXTREMES

DEL RIO, TEXAS

LATITUDE: 29°22'N LONGITUDE: 100°50'W ELEVATION: FT. GRND 1026 BARO 01030 TIME ZONE: CENTRAL WBAN: 22010

	(a)	JAN	FEB	MAR	APR	MAY	JUNE	JULY	AUG	SEP	OCT	NOV	DEC	YEAR
TEMPERATURE °F:														
Normals														
-Daily Maximum		63.2	68.6	76.5	84.2	89.1	95.1	97.7	97.0	91.7	82.4	71.2	64.8	81.8
-Daily Minimum		38.3	42.5	50.2	59.4	66.1	72.1	74.2	73.6	68.9	59.2	47.5	40.1	57.7
-Monthly		50.8	55.5	63.4	71.8	77.6	83.6	86.0	85.3	80.3	70.8	59.4	52.5	69.7
Extremes														
-Record Highest	23	89	91	101	106	106	108	108	109	105	106	93	90	109
-Year		1972	1980	1971	1984	1979	1980	1980	1969	1983	1979	1980	1977	AUG 1969
-Record Lowest	23	15	14	21	33	45	55	64	64	48	34	22	13	13
-Year		1982	1985	1980	1973	1970	1970	1976	1966	1970	1980	1976	1983	DEC 1983
NORMAL DEGREE DAYS:														
Heating (base 65°F)		450	282	145	15	0	0	0	0	0	27	203	388	1510
Cooling (base 65°F)		10	18	95	219	391	558	651	629	459	207	35	0	3272
% OF POSSIBLE SUNSHINE														
MEAN SKY COVER (tenths)														
Sunrise - Sunset	16	5.8	5.5	5.5	6.1	6.4	5.5	4.9	5.0	5.4	4.9	5.2	5.6	5.5
MEAN NUMBER OF DAYS:														
Sunrise to Sunset														
-Clear	16	10.1	10.3	10.6	8.4	6.5	7.9	11.8	11.0	9.2	12.4	11.8	11.3	121.2
-Partly Cloudy	16	6.6	6.7	7.6	7.4	8.9	13.3	11.3	12.1	11.0	8.4	6.9	5.6	105.8
-Cloudy	16	14.3	11.3	12.8	14.2	15.6	8.8	7.9	7.9	9.8	10.1	11.4	14.2	138.3
Precipitation														
.01 inches or more	22	4.8	4.6	4.7	5.5	7.0	4.8	4.2	4.3	6.7	5.1	4.4	4.7	60.8
Snow,Ice pellets														
1.0 inches or more	22	0.1	0.1	0.*	0.0	0.0	0.0	0.0	0.0	0.0	0.0	0.0	0.0	0.3
Thunderstorms	17	0.4	0.4	1.6	4.5	7.6	4.2	3.8	4.2	4.1	1.9	0.9	0.4	34.1
Heavy Fog Visibility														
1/4 mile or less	17	3.2	1.5	0.9	0.2	0.2	0.0	0.0	0.2	0.9	2.9	4.4	14.5
Temperature °F														
-Maximum														
90° and above	22	0.0	0.2	1.8	6.2	13.3	24.5	28.1	28.1	19.3	3.8	0.2	0.*	125.5
32° and below	22	0.1	0.*	0.0	0.0	0.0	0.0	0.0	0.0	0.0	0.0	0.0	0.1	0.3
-Minimum														
32° and below	22	7.5	3.8	0.7	0.0	0.0	0.0	0.0	0.0	0.0	0.0	1.0	4.7	17.7
0° and below	22	0.0	0.0	0.0	0.0	0.0	0.0	0.0	0.0	0.0	0.0	0.0	0.0	0.0
AVG. STATION PRESS.(mb)	7	982.9	981.7	976.7	976.3	974.6	976.5	977.7	977.9	977.8	980.1	981.5	982.2	978.8
RELATIVE HUMIDITY (%)														
Hour 00	16	66	62	57	62	69	65	59	62	70	71	71	68	65
Hour 06 (Local Time)	16	76	74	72	77	83	82	79	80	84	82	80	76	79
Hour 12	16	55	52	48	52	57	55	52	54	57	57	57	54	54
Hour 18	16	46	40	36	40	46	42	39	41	49	50	52	49	44
PRECIPITATION (inches):														
Water Equivalent														
-Normal		0.51	0.89	0.63	1.85	1.99	1.72	1.69	1.60	2.73	2.24	0.80	0.55	17.20
-Maximum Monthly	23	1.44	2.45	2.66	7.51	5.15	5.54	13.18	6.10	15.79	11.33	3.36	2.44	15.79
-Year		1984	1982	1979	1981	1980	1981	1976	1971	1964	1969	1969	1984	SEP 1964
-Minimum Monthly	23	T	0:00	T	0.13	0.20	T	0.04	T	0.36	0.00	T	T	0.00
-Year		1971	1974	1971	1983	1973	1974	1970	1985	1981	1979	1973	1973	OCT 1979
-Maximum in 24 hrs	19	0.93	1.49	1.73	4.57	1.86	2.57	6.34	3.81	5.52	7.60	2.74	1.37	7.60
-Year		1985	1975	1979	1969	1964	1966	1975	1969	1970	1969	1978	1974	OCT 1969
Snow,Ice pellets														
-Maximum Monthly	22	9.8	2.7	3.0							T	T	T	9.8
-Year		1985	1973	1965							1967	1976	1969	JAN 1985
-Maximum in 24 hrs	19	8.6	2.7	3.0							T	T	T	8.6
-Year		1985	1973	1965							1967	1976	1969	JAN 1985
WIND:														
Mean Speed (mph)	16	8.8	9.5	10.9	11.0	10.7	11.4	10.9	10.2	9.2	9.1	8.5	8.4	9.9
Prevailing Direction														
Fastest Obs. 1 Min.														
-Direction (!!!)	16	30	30	33	31	32	36	32	14	31	33	35	29	14
-Speed (MPH)	16	37	37	52	37	48	38	38	60	35	46	35	43	60
-Year		1973	1974	1964	1969	1970	1968	1969	1970	1976	1967	1972	1969	AUG 1970
Peak Gust														
-Direction (!!!)														
-Speed (mph)														
-Date														

See Reference Notes to this table on the following page.

TABLE 2 PRECIPITATION (inches)   DEL RIO, TEXAS

YEAR	JAN	FEB	MAR	APR	MAY	JUNE	JULY	AUG	SEP	OCT	NOV	DEC	ANNUAL
1956	0.28	0.34	T	0.05	0.03	0.03	T	0.25	1.07	2.20	0.02	0.07	4.34
#1957	0.18	0.94	1.63	6.49	10.23	1.22	T	T	2.76	1.97	1.51	0.82	27.75
1958	3.25	1.67	1.19	0.28	6.17	3.94	0.33	0.88	5.47	4.48	0.24	0.15	28.05
1959	0.08	0.65	T	2.25	3.02	6.65	3.11	0.45	0.69	5.37	0.46	0.72	23.45
1960	0.50	2.17	0.50	0.49	0.51	0.10	4.81	2.52	2.68	3.20	0.33	1.81	19.62
1961	1.56	0.54	0.08	1.30	0.49	7.17	3.51	2.97	0.87	3.21	0.75	0.29	22.74
1962	0.15	0.27	0.17	1.81	0.48	1.11	0.29	0.05	2.10	2.91	0.26	0.31	9.91
#1963	T	1.95	T	1.26	2.72	1.49	0.09	T	1.91	0.81	0.63	0.67	11.53
1964	0.42	0.76	0.45	1.20	2.28	0.03	0.34	1.16	15.79	2.47	T	0.25	25.15
1965	0.16	1.72	0.33	3.13	2.34	2.03	0.38	1.64	2.61	0.43	1.26	1.09	17.12
1966	0.94	1.11	0.48	4.36	2.23	3.37	0.05	2.10	1.45	1.26	T	0.04	17.39
1967	0.02	0.29	0.58	1.11	0.65	0.36	0.12	1.00	7.02	1.39	0.66	0.76	13.96
1968	0.66	1.53	1.15	2.53	1.03	1.72	3.07	0.25	1.20	1.84	2.19	T	17.17
1969	1.04	1.17	0.60	5.46	2.36	0.20	1.47	4.27	0.78	11.33	3.36	1.18	33.22
1970	0.42	1.61	0.57	0.14	2.85	1.38	0.04	1.43	9.87	0.06	T	0.12	18.49
1971	T	0.29	T	2.16	0.59	4.87	0.45	6.10	0.50	2.36	0.83	0.25	18.40
1972	0.63	0.08	0.52	1.58	2.61	1.82	2.60	5.74	0.92	0.38	0.71	0.01	17.60
1973	0.76	1.36	1.19	2.08	0.20	3.22	2.43	1.08	2.86	4.78	T	T	19.96
1974	0.05	0.00	1.39	1.39	1.42	T	0.09	3.37	4.98	2.81	0.62	1.47	17.59
1975	0.51	1.92	0.06	1.06	2.17	1.12	8.26	0.27	1.24	1.36	0.10	0.53	18.60
1976	0.15	0.01	0.14	2.45	2.82	0.46	13.18	0.43	1.10	3.20	0.89	1.51	26.34
1977	0.89	1.23	0.87	2.66	2.27	1.23	0.06	0.10	1.41	4.96	0.67	0.01	16.36
1978	0.07	0.36	0.15	2.09	3.46	2.02	2.15	1.14	2.76	0.95	3.35	0.76	19.26
1979	0.29	1.57	2.66	1.53	0.62	4.33	0.25	0.72	0.68	0.00	0.60	0.78	14.03
1980	0.22	0.44	0.30	0.44	5.15	0.32	0.28	2.06	1.86	0.06	2.00	1.02	14.15
1981	0.67	0.21	1.64	7.51	1.94	5.54	0.56	2.18	0.36	6.64	0.10	0.06	27.41
1982	0.12	2.45	0.13	0.47	1.59	2.37	0.86	0.12	1.10	0.17	1.77	0.79	11.94
1983	0.70	1.01	1.07	0.13	1.19	1.07	0.32	0.79	1.35	5.09	1.42	0.06	14.20
1984	1.44	0.24	0.06	0.55	2.09	0.45	0.90	0.51	2.78	3.02	1.03	2.44	15.51
1985	1.25	0.51	0.81	2.51	2.20	3.91	1.71	T	2.81	0.63	1.05	0.03	17.42
Record Mean	0.63	0.85	0.78	1.66	2.34	2.14	1.77	1.49	2.57	2.32	0.95	0.70	18.20

TABLE 3 AVERAGE TEMPERATURE (deg. F)   DEL RIO, TEXAS

YEAR	JAN	FEB	MAR	APR	MAY	JUNE	JULY	AUG	SEP	OCT	NOV	DEC	ANNUAL
1956	51.2	56.6	65.2	71.0	81.3	87.0	87.2	87.2	80.8	74.8	58.1	54.5	71.3
#1957	52.5	63.3	64.1	67.9	74.5	82.0	87.7	87.1	79.6	69.6	56.0	55.1	70.0
1958	50.6	52.7	57.9	69.6	76.9	83.5	85.6	86.3	80.0	67.9	59.5	50.7	68.4
1959	49.2	54.9	62.4	67.4	78.2	83.0	83.2	84.7	82.0	71.2	53.2	53.9	68.6
1960	51.6	51.9	58.8	71.8	77.7	86.3	86.3	84.4	80.3	74.2	61.4	48.6	69.4
1961	47.6	56.1	66.2	71.1	79.5	82.7	83.2	83.1	81.1	71.0	53.8	53.8	69.4
1962	46.8	63.4	61.4	72.0	79.8	83.3	87.9	88.8	82.6	76.6	61.9	52.3	71.4
#1963	46.8	54.4	66.9	76.0	77.3	83.7	86.5	88.1	82.0	74.2	61.9	45.8	70.3
1964	49.2	50.7	63.8	74.4	79.2	84.9	88.2	87.0	79.8	67.5	62.4	51.3	69.9
1965	54.5	57.0	57.4	73.6	75.9	81.9	85.8	84.5	81.3	69.1	65.4	56.4	69.7
1966	46.8	51.0	62.3	71.0	75.3	80.2	86.9	83.2	79.7	68.5	63.7	50.9	68.3
1967	51.5	54.2	68.8	76.1	79.1	85.2	86.8	83.3	76.2	68.2	60.8	49.5	70.0
1968	49.8	50.4	58.0	69.4	77.4	82.3	83.5	85.5	77.4	73.1	58.5	51.3	68.0
1969	54.5	56.6	57.3	71.1	75.5	84.4	88.4	86.0	79.9	68.5	57.2	54.0	69.5
1970	47.1	54.9	57.5	70.1	73.8	81.2	85.4	85.3	78.7	66.9	57.3	57.7	68.0
1971	56.1	56.8	64.9	70.2	78.3	80.1	82.0	78.6	78.5	71.1	61.7	56.5	69.6
1972	53.6	58.1	68.8	75.3	74.9	82.2	83.0	80.8	80.5	71.9	55.9	50.9	69.7
1973	47.7	51.2	65.9	68.1	77.9	80.1	82.5	83.7	80.2	71.4	64.2	53.0	68.8
1974	53.1	56.9	68.8	72.2	80.6	83.7	85.9	83.6	73.9	70.1	58.0	50.9	69.8
1975	53.1	55.2	63.8	70.9	76.4	82.6	81.2	83.4	76.1	70.8	61.4	53.5	69.1
1976	50.1	63.0	66.5	70.3	72.9	82.5	78.0	81.5	78.8	63.0	53.2	49.2	67.4
1977	46.4	55.4	62.8	68.4	75.6	82.9	85.8	88.3	85.8	72.4	61.7	55.5	70.1
1978	46.5	49.9	63.4	73.9	80.0	83.8	87.7	84.6	78.5	69.5	61.5	51.1	69.2
1979	44.2	52.2	62.9	70.2	75.8	79.6	87.4	85.0	80.8	75.4	57.3	53.9	68.8
1980	54.1	56.1	63.6	70.8	78.2	87.2	90.1	85.0	83.6	70.7	57.6	53.5	70.9
1981	52.2	55.3	61.6	71.4	76.3	80.5	84.3	84.9	80.4	72.1	63.0	54.5	69.7
1982	50.9	52.2	64.9	71.0	76.0	84.3	86.2	87.3	81.2	72.2	57.6	51.8	69.8
1983	51.1	55.1	62.5	69.0	78.0	81.9	86.3	86.2	81.1	72.9	62.4	44.6	69.3
1984	47.4	57.0	65.7	72.8	79.7	84.3	85.8	86.7	78.8	71.8	59.9	57.9	70.6
1985	45.4	51.8	64.2	71.1	79.1	81.1	83.6	87.3	80.7	71.6	64.6	49.7	69.2
Record Mean	51.2	55.8	63.3	70.9	77.2	83.0	85.2	85.1	80.0	70.9	59.8	52.4	69.6
Max	62.3	67.4	75.3	82.5	87.9	93.4	95.7	95.8	90.2	81.4	70.5	63.3	80.5
Min	40.1	44.2	51.3	59.3	66.5	72.5	74.6	74.4	69.8	60.4	49.0	41.5	58.6

REFERENCE NOTES FOR TABLES 1, 2, 3 and 6   (DEL RIO, TX)

GENERAL

T - TRACE AMOUNT
BLANK ENTRIES DENOTE MISSING/UNREPORTED DATA.
# INDICATES A STATION OR INSTRUMENT RELOCATION.

SPECIFIC

TABLE 1

(a) - LENGTH OF RECORD IN YEARS. ALTHOUGH INDIVIDUAL MONTHS MAY BE MISSING.
  * LESS THAN .05

NORMALS — BASED ON THE 1951-1980 RECORD PERIOD.
EXTREMES — DATES ARE THE MOST RECENT OCCURRENCE.
WIND DIR. — NUMERALS SHOW TENS OF DEGREES CLOCKWISE FROM TRUE NORTH.
  "00" INDICATES CALM.
RESULTANT WIND DIRECTIONS ARE GIVEN TO WHOLE DEGREES.

EXCEPTIONS

TABLE 1

1. MAXIMUM 24-HOUR PRECIPITATION AND SNOW, WINDS, MEAN SKY COVER, DAYS CLEAR-PARTLY CLOUDY-CLOUDY, THUNDERSTORMS, AND HEAVY FOG ARE THROUGH 1979.

TABLES 2, 3, and 6

RECORD MEANS ARE THROUGH THE CURRENT YEAR, BEGINNING IN   1906 FOR TEMPERATURE
  1906 FOR PRECIPITATION
  1964 FOR SNOWFALL

TABLE 4 HEATING DEGREE DAYS Base 65 deg. F   DEL RIO, TEXAS

SEASON	JULY	AUG	SEP	OCT	NOV	DEC	JAN	FEB	MAR	APR	MAY	JUNE	TOTAL
1956-57	0	0	0	3	216	327	382	117	87	64	2	0	1198
#1957-58	0	0	0	3	216	326	438	340	224	17	4	0	1568
1958-59	0	0	0	57	285	302	486	288	133	69	0	0	1620
1959-60	0	0	0	68	184	439	419	375	233	19	1	0	1738
1960-61	0	0	0	20	366	337	530	260	67	33	0	0	1613
1961-62	0	0	0	17	137	499	559	96	163	21	0	0	1492
#1962-63	0	0	0	9	246	347	559	302	71	13	0	0	1547
1963-64	0	0	0	2	112	388	484	405	91	15	0	0	1497
1964-65	0	0	0	21	152	417	326	397	268	4	0	0	1585
1965-66	0	0	3	16	41	263	554	388	128	8	9	0	1410
1966-67	0	0	0	26	93	434	417	299	54	0	0	0	1323
1967-68	0	0	1	36	159	475	466	419	246	28	0	0	1830
1968-69	0	0	0	0	225	419	324	240	266	1	0	0	1475
1969-70	0	0	0	62	273	330	551	277	242	40	7	0	1782
1970-71	0	0	24	69	233	224	278	239	107	35	0	0	1209
1971-72	0	0	9	6	135	263	350	228	34	2	0	0	1027
1972-73	0	0	0	21	283	429	531	381	31	72	0	0	1748
1973-74	0	0	0	1	83	362	363	231	53	11	0	0	1104
1974-75	0	0	6	9	223	431	375	274	114	22	0	0	1454
1975-76	0	0	0	16	180	361	456	125	76	3	5	0	1222
1976-77	0	0	0	127	348	481	568	269	122	13	0	0	1928
1977-78	0	0	0	6	126	294	564	422	111	3	0	0	1526
1978-79	0	0	1	8	147	426	639	363	125	8	0	0	1717
1979-80	0	0	0	10	233	346	336	279	119	26	0	0	1349
1980-81	0	0	0	51	268	353	388	280	119	9	0	0	1468
1981-82	0	0	0	33	92	319	436	359	125	42	0	0	1406
1982-83	0	0	0	20	221	409	423	269	103	66	0	0	1511
1983-84	0	0	0	2	141	625	539	240	82	9	0	0	1638
1984-85	0	0	10	22	214	222	603	367	96	28	0	0	1562
1985-86	0	0	8	9	85	466							

TABLE 5 COOLING DEGREE DAYS Base 65 deg. F   DEL RIO, TEXAS

YEAR	JAN	FEB	MAR	APR	MAY	JUNE	JULY	AUG	SEP	OCT	NOV	DEC	TOTAL
1969	6	13	32	193	330	590	735	660	455	175	46	0	3235
1970	3	0	15	200	286	496	642	637	443	135	11	0	2868
1971	10	19	114	196	421	461	534	429	423	201	43	8	2859
1972	3	37	161	317	315	521	566	496	473	241	18	1	3149
1973	0	0	68	170	410	460	550	586	463	207	69	0	2983
1974	2	10	178	237	489	565	654	584	278	177	23	2	3199
1975	13	1	83	205	363	534	508	582	345	202	78	13	2927
1976	0	74	130	168	255	534	410	521	419	73	3	0	2587
1977	0	7	63	121	334	543	649	733	630	239	34	6	3359
1978	0	7	71	274	473	571	710	614	415	159	49	1	3344
1979	0	9	66	172	340	444	701	625	483	337	24	9	3210
1980	6	25	81	206	417	671	785	629	567	233	53	5	3678
1981	0	15	24	210	356	472	603	624	470	258	40	0	3072
1982	7	9	129	228	348	587	662	698	493	253	68	5	3487
1983	0	0	33	193	409	511	669	664	489	253	71	0	3292
1984	0	12	111	249	463	585	651	682	431	241	40	12	3477
1985	0	6	78	218	443	491	580	698	481	222	81	0	3298

TABLE 6 SNOWFALL (inches)   DEL RIO, TEXAS

SEASON	JULY	AUG	SEP	OCT	NOV	DEC	JAN	FEB	MAR	APR	MAY	JUNE	TOTAL
1956-57	0.0	0.0	0.0	0.0	0.0	0.0	0.0	0.0	0.0	0.0	0.0	0.0	0.0
#1957-58	0.0	0.0	0.0	0.0	0.1	0.0	0.0	0.0	0.0	0.0	0.0	0.0	0.1
1958-59	0.0	0.0	0.0	0.0	0.0	0.0	T	0.0	0.0	0.0	0.0	0.0	T
1959-60	0.0	0.0	0.0	0.0	0.0	0.0	0.0	0.5	0.0	0.0	0.0	0.0	0.5
1960-61	0.0	0.0	0.0	0.0	0.0	0.0	T	T	0.0	0.0	0.0	0.0	T
1961-62	0.0	0.0	0.0	0.0	0.0	0.0	T	0.0	T	0.0	0.0	0.0	T
#1962-63	0.0	0.0	0.0	0.0	0.0	0.0	0.0	T	0.0	0.0	0.0	0.0	T
1963-64	0.0	0.0	0.0	0.0	0.0	0.0	0.0	T	3.0	0.0	0.0	0.0	3.0
1964-65	0.0	0.0	0.0	0.0	0.0	0.0	T	2.2	0.0	0.0	0.0	0.0	2.2
1965-66	0.0	0.0	0.0	0.0	0.0	0.0	0.6	0.0	0.0	0.0	0.0	0.0	0.6
1966-67	0.0	0.0	0.0	0.0	0.0	0.0	T	0.0	0.0	0.0	0.0	0.0	T
1967-68	0.0	0.0	0.0	T	0.0	T	0.0	0.0	0.0	0.0	0.0	0.0	0.0
1968-69	0.0	0.0	0.0	0.0	0.0	0.0	0.0	0.0	0.0	0.0	0.0	0.0	T
1969-70	0.0	0.0	0.0	0.0	0.0	T	0.0	T	0.0	0.0	0.0	0.0	0.0
1970-71	0.0	0.0	0.0	0.0	0.0	0.0	0.0	0.0	0.0	0.0	0.0	0.0	T
1971-72	0.0	0.0	0.0	0.0	0.0	0.0	T	2.7	0.0	0.0	0.0	0.0	2.7
1972-73	0.0	0.0	0.0	0.0	0.0	0.0	T	0.0	0.0	0.0	0.0	0.0	0.0
1973-74	0.0	0.0	0.0	0.0	0.0	0.0	0.0	0.0	0.0	0.0	0.0	0.0	0.0
1974-75	0.0	0.0	0.0	0.0	0.0	0.0	T	0.0	0.0	0.0	0.0	0.0	T
1975-76	0.0	0.0	0.0	0.0	0.0	0.0	0.0	0.0	0.0	0.0	0.0	0.0	T
1976-77	0.0	0.0	0.0	0.0	T	0.0	0.0	0.0	0.0	0.0	0.0	0.0	0.0
1977-78	0.0	0.0	0.0	0.0	0.0	0.0	0.0	0.0	0.0	0.0	0.0	0.0	0.0
1978-79	0.0	0.0	0.0	0.0	0.0	0.0	0.0	0.0	0.0	0.0	0.0	0.0	T
1979-80	0.0	0.0	0.0	0.0	0.0	0.0	0.0	T	0.0	0.0	0.0	0.0	T
1980-81	0.0	0.0	0.0	0.0	0.0	0.0	0.2	0.0	0.0	0.0	0.0	0.0	0.2
1981-82	0.0	0.0	0.0	0.0	0.0	0.0	1.0	T	0.0	0.0	0.0	0.0	1.0
1982-83	0.0	0.0	0.0	0.0	0.0	0.0	0.0	T	0.0	0.0	0.0	0.0	T
1983-84	0.0	0.0	0.0	0.0	0.0	0.0	T	0.0	0.0	0.0	0.0	0.0	T
1984-85	0.0	0.0	0.0	0.0	0.0	0.0	9.8	T	0.0	0.0	0.0	0.0	9.8
1985-86	0.0	0.0	0.0	0.0	0.0	0.0							
Record Mean	0.0	0.0	0.0	T	T	T	0.5	0.3	0.1	0.0	0.0	0.0	0.9

See Reference Notes, relative to all above tables, on preceding page.

The city of El Paso is located in the extreme west point of Texas at an elevation of about 3,700 feet . The National Weather Service station is located on a mesa about 200 feet higher than the city. The climate of the region is characterized by an abundance of sunshine throughout the year, high daytime summer temperatures, very low humidity, scanty rainfall, and a relatively mild winter season. The Franklin Mountains begin within the city limits and extend northward for about 16 miles. Peaks of these mountains range from 4,687 to 7,152 feet above sea level.

Rainfall throughout the year is light, insufficient for any growth except desert vegetation. Irrigation is necessary for crops, gardens, and lawns. Dry periods lasting several months are not unusual. Almost half of the precipitation occurs in the three-month period, July through September, from brief but often heavy thunderstorms. Small amounts of snow fall nearly every winter, but snow cover rarely amounts to more than an inch and seldom remains on the ground for more than a few hours.

Daytime summer temperatures are high, frequently above 90 degrees and occasionally above 100 degrees. Summer nights are usually comfortable, with temperatures in the 60s. It should be noted that when temperatures are high the relative humidity is generally quite low. A 20-year tabulation of observations with temperatures above 90 degrees shows that in April, May, and June the humidity averaged from 10 to 14 percent, while in July, August, and September it averaged 22 to 24 percent. This low humidity aids the efficiency of evaporative air coolers, which are widely used in homes and public buildings and are quite effective in cooling the air to comfortable temperatures.

Winter daytime temperatures are mild. At night they drop below freezing about half the time in December and January. The flat, irrigated land of the Rio Grande Valley in the vicinity of El Paso is noticeably cooler, particularly at night, than the airport or the city proper, both in summer and winter. This results in more comfortable temperatures in summer but increases the severity of freezes in winter. The cooler air in the Valley also causes marked short-period fluctuations of temperature and dewpoint at the airport with changes in wind direction, especially during the early morning hours.

Dust and sandstorms are the most unpleasant features of the weather in El Paso. While wind velocities are not excessively high, the soil surface is dry and loose and natural vegetation is sparse, so moderately strong winds raise considerable dust and sand. A tabulation of duststorms for a period of 20 years shows that they are most frequent in March and April, and comparatively rare in the period July through December. prevailing winds are from the north in winter and the south in summer.

TABLE 1   NORMALS, MEANS AND EXTREMES

EL PASO, TEXAS

LATITUDE: 31°48'N   LONGITUDE: 106°24'W   ELEVATION: FT. GRND 3918 BARO 03932   TIME ZONE: MOUNTAIN   WBAN: 23044

	(a)	JAN	FEB	MAR	APR	MAY	JUNE	JULY	AUG	SEP	OCT	NOV	DEC	YEAR
TEMPERATURE °F:														
Normals														
–Daily Maximum		57.9	62.7	69.6	78.7	87.1	95.9	95.3	93.0	87.5	78.5	65.7	58.2	77.5
–Daily Minimum		30.4	34.1	40.5	48.5	56.6	65.7	69.6	67.5	60.6	48.7	37.0	30.6	49.2
–Monthly		44.2	48.4	55.0	63.6	71.8	80.8	82.5	80.3	74.1	63.6	51.4	44.4	63.3
Extremes														
–Record Highest	46	80	83	88	98	104	111	112	108	104	96	87	80	112
–Year		1970	1972	1971	1965	1951	1978	1979	1980	1982	1979	1983	1973	JUL 1979
–Record Lowest	46	-8	8	14	23	31	48	57	56	41	25	1	5	-8
–Year		1962	1985	1971	1983	1967	1979	1985	1973	1945	1970	1976	1953	JAN 1962
NORMAL DEGREE DAYS:														
Heating (base 65°F)		645	465	318	93	0	0	0	0	0	96	408	639	2664
Cooling (base 65°F)		0	0	11	51	218	474	543	474	273	52	0	0	2096
% OF POSSIBLE SUNSHINE	43	77	82	85	87	89	89	80	81	82	83	82	78	83
MEAN SKY COVER (tenths)														
Sunrise – Sunset	43	4.7	4.1	4.3	3.6	3.2	2.8	4.5	4.2	3.5	3.2	3.5	4.2	3.8
MEAN NUMBER OF DAYS:														
Sunrise to Sunset														
–Clear	43	13.9	14.0	14.8	16.5	18.8	20.0	12.3	14.0	17.7	19.0	17.3	15.3	193.4
–Partly Cloudy	43	7.4	7.4	8.0	8.0	7.9	7.2	13.0	12.3	7.2	6.6	6.4	7.5	99.0
–Cloudy	43	9.8	6.8	8.1	5.5	4.3	2.8	5.7	4.7	5.2	5.4	6.3	8.3	72.9
Precipitation														
.01 inches or more	46	4.1	2.8	2.4	1.7	2.1	3.3	7.5	7.4	5.2	4.2	2.7	3.6	47.0
Snow, Ice pellets														
1.0 inches or more	46	0.5	0.3	0.*	0.1	0.0	0.0	0.0	0.0	0.0	0.*	0.3	0.4	1.8
Thunderstorms	46	0.2	0.4	0.4	1.0	2.6	4.5	10.2	10.0	4.0	1.9	0.3	0.2	35.8
Heavy Fog Visibility														
1/4 mile or less	46	0.7	0.2	0.1	0.*	0.*	0.0	0.0	0.0	0.1	0.2	0.3	0.5	2.2
Temperature °F														
–Maximum														
90° and above	25	0.0	0.0	0.0	1.8	12.9	26.1	27.4	24.1	11.8	1.6	0.0	0.0	105.8
32° and below	25	0.4	0.1	0.0	0.0	0.0	0.0	0.0	0.0	0.0	0.0	0.1	0.1	0.7
–Minimum														
32° and below	25	18.8	12.3	4.9	0.9	0.*	0.0	0.0	0.0	0.0	0.3	7.4	18.5	63.2
0° and below	25	0.1	0.0	0.0	0.0	0.0	0.0	0.0	0.0	0.0	0.0	0.0	0.0	0.1
AVG. STATION PRESS. (mb)	13	883.5	882.7	879.6	879.8	879.4	880.3	882.3	882.9	882.8	883.3	883.3	884.0	882.0
RELATIVE HUMIDITY (%)														
Hour 05	25	65	55	46	39	41	44	61	64	67	63	61	64	56
Hour 11 (Local Time)	25	45	36	29	23	22	25	37	40	44	38	39	44	35
Hour 17	25	36	26	21	16	16	18	28	32	34	30	33	36	27
Hour 23	25	56	44	34	28	28	31	45	50	53	52	52	55	44
PRECIPITATION (inches):														
Water Equivalent														
–Normal		0.38	0.45	0.32	0.19	0.24	0.56	1.60	1.21	1.42	0.73	0.33	0.39	7.82
–Maximum Monthly	46	1.84	1.69	2.26	1.42	1.92	3.18	5.53	5.57	6.68	4.31	1.63	2.61	6.68
–Year		1949	1973	1958	1983	1941	1984	1968	1984	1974	1945	1961	1982	SEP 1974
–Minimum Monthly	46	0.00	0.00	T	0.00	0.00	T	0.04	T	T	0.00	0.00	0.00	0.00
–Year		1967	1943	1982	1978	1962	1980	1978	1962	1959	1952	1964	1955	APR 1978
–Maximum in 24 hrs	46	0.61	0.87	1.72	1.08	1.23	1.45	2.63	2.30	2.52	1.77	1.19	1.05	2.63
–Year		1960	1956	1941	1966	1941	1966	1968	1984	1958	1945	1943	1946	JUL 1968
Snow, Ice pellets														
–Maximum Monthly	46	8.3	8.9	7.3	16.5						1.0	12.7	18.2	18.2
–Year		1949	1956	1958	1983						1980	1976	1982	DEC 1982
–Maximum in 24 hrs	46	4.8	7.2	7.3	8.8						1.0	7.8	7.1	8.8
–Year		1981	1956	1958	1983						1980	1961	1951	APR 1983
WIND:														
Mean Speed (mph)	43	8.6	9.4	11.4	11.4	10.6	9.5	8.4	7.9	7.8	7.7	8.1	8.1	9.1
Prevailing Direction through 1963		N	N	WSW	WSW	WSW	S	SSE	S	S	N	N	N	N
Fastest Obs. 1 Min.														
–Direction	10	28	26	30	30	25	22	11	13	08	27	24	24	26
–Speed (MPH)	10	40	48	48	42	35	40	35	35	29	35	40	42	48
–Year		1976	1977	1977	1975	1979	1976	1981	1979	1979	1978	1975	1975	FEB 1977
Peak Gust														
–Direction	2	W	NW	W	W	SW	N	E	S	NW	W	W	SW	W
–Speed (mph)	2	46	53	55	55	48	48	43	40	46	45	44	32	55
–Date		1985	1984	1985	1984	1985	1985	1984	1984	1984	1984	1984	1984	MAR 1985

See Reference Notes to this table on the following page.

TABLE 2   PRECIPITATION (inches)   EL PASO, TEXAS

YEAR	JAN	FEB	MAR	APR	MAY	JUNE	JULY	AUG	SEP	OCT	NOV	DEC	ANNUAL
1956	0.35	1.06	T	0.05	T	1.19	1.10	0.61	0.43	0.01	T	0.64	5.44
1957	0.24	0.46	0.33	0.09	0.10	0.02	2.64	4.11	0.11	2.34	0.74	0.02	11.20
1958	0.74	1.11	2.26	0.05	0.40	1.66	1.36	1.14	6.29	1.98	0.20	T	17.19
1959	0.21	T	0.07	0.15	0.30	0.46	0.40	2.39	T	0.58	0.14	0.29	4.99
1960	0.72	0.37	0.21	0.02	0.04	0.76	3.61	0.77	0.01	0.77	0.11	1.73	9.12
1961	0.41	T	0.29	0.01	T	0.27	2.18	1.40	0.69	0.18	1.63	0.63	7.69
1962	0.94	0.58	0.24	0.10	0.00	T	1.82	T	3.54	0.55	0.21	0.30	8.28
1963	0.13	0.53	T	T	0.71	0.05	0.52	1.03	0.64	0.55	0.76	T	4.92
1964	T	T	0.99	0.08	0.02	T	0.18	0.76	2.40	0.40	0.00	0.52	5.35
1965	0.19	0.59	0.03	0.01	0.11	0.66	0.17	0.49	2.12	0.18	0.12	0.74	5.41
1966	0.38	0.20	T	1.08	0.04	2.67	1.17	1.85	1.79	0.01	0.01	0.04	9.24
1967	0.00	0.04	0.17	0.03	0.05	1.41	0.84	0.54	1.54	0.09	0.23	0.78	5.72
1968	0.47	1.11	0.85	0.10	T	0.03	5.53	1.71	0.53	0.11	1.35	0.23	12.02
1969	0.05	0.08	0.17	T	0.28	T	1.14	0.28	0.43	0.59	0.63	0.69	4.34
1970	0.03	0.55	0.47	T	0.71	0.73	1.41	0.41	1.01	0.68	T	0.06	6.06
1971	0.17	0.04	0.00	0.42	T	0.01	2.34	1.59	0.96	1.07	0.14	0.50	7.24
1972	0.44	T	T	0.00	0.04	1.62	0.71	2.59	1.60	1.25	0.33	0.42	9.00
1973	1.23	1.69	0.60	0.00	0.29	0.71	2.12	0.73	0.01	0.07	0.08	T	7.53
1974	0.27	T	0.36	0.12	0.05	0.36	2.21	0.63	6.68	1.90	0.50	0.87	13.95
1975	0.70	0.59	0.19	T	0.03	T	1.11	0.45	2.18	0.25	T	0.71	6.21
1976	0.26	0.52	T	0.30	0.74	0.50	3.17	0.23	1.70	1.20	1.20	0.32	10.14
1977	0.57	T	0.17	0.09	0.06	0.04	1.09	1.36	0.16	1.65	0.05	0.26	5.50
1978	0.44	0.47	0.07	0.00	0.57	1.46	0.04	2.18	4.14	2.28	0.45	0.47	12.57
1979	0.77	0.68	T	0.28	0.24	0.03	0.98	2.16	0.41	T	0.04	0.25	5.84
1980	0.54	0.73	0.25	0.31	0.08	T	T	0.21	1.76	1.90	0.95	0.54	7.31
1981	1.10	0.36	0.39	0.65	0.72	0.64	2.08	5.26	0.52	0.53	0.30	0.08	12.63
1982	0.34	0.55	T	0.05	0.19	0.18	1.00	0.48	5.28	T	0.29	2.61	10.97
1983	0.35	0.60	0.45	1.42	0.05	0.23	0.43	0.97	1.51	1.48	0.34	0.16	7.99
1984	0.31	0.00	0.44	0.01	0.59	3.18	0.69	5.57	0.58	3.12	0.51	1.17	16.17
1985	0.95	0.19	0.59	0.07	0.01	0.10	1.32	1.46	1.47	1.82	0.13	0.05	8.16
Record Mean	0.45	0.41	0.33	0.25	0.34	0.61	1.64	1.49	1.31	0.84	0.43	0.50	8.58

TABLE 3   AVERAGE TEMPERATURE (deg. F)   EL PASO, TEXAS

YEAR	JAN	FEB	MAR	APR	MAY	JUNE	JULY	AUG	SEP	OCT	NOV	DEC	ANNUAL
1956	49.7	44.9	57.5	61.8	75.7	83.9	82.0	79.9	77.4	67.5	48.4	44.6	64.4
1957	49.3	57.6	57.0	62.4	69.5	81.9	83.9	80.5	74.5	62.4	49.0	46.4	64.5
1958	42.9	51.6	51.2	63.3	74.2	83.5	84.5	82.6	74.0	62.5	53.3	47.0	64.2
1959	47.9	49.3	54.0	65.0	73.9	83.0	82.9	81.0	77.9	66.1	50.3	46.3	64.8
#1960	42.4	46.3	59.7	66.3	73.7	85.0	81.9	82.5	75.6	63.5	52.9	38.8	64.0
1961	41.0	47.9	56.3	64.1	74.1	80.8	82.8	80.5	74.5	63.5	47.8	46.3	63.3
1962	40.5	53.5	50.9	67.3	74.1	80.0	81.3	84.0	73.9	64.9	53.7	45.2	64.1
1963	40.5	49.2	56.4	66.5	74.9	81.1	84.9	80.3	76.2	66.4	53.5	42.7	64.4
1964	39.3	40.8	53.6	63.2	73.9	81.5	84.5	82.6	75.7	63.4	51.8	44.0	62.8
1965	48.0	46.7	52.1	65.3	71.8	78.2	84.2	81.1	74.0	63.0	56.8	45.7	63.9
1966	40.1	42.9	56.4	65.1	74.5	79.5	83.6	78.7	73.4	62.1	54.5	42.4	62.8
1967	41.6	48.0	59.6	65.1	70.9	79.1	83.0	78.6	73.3	63.7	53.1	41.5	63.1
1968	42.4	50.0	53.0	61.4	73.0	81.0	79.1	76.5	72.4	65.0	51.0	41.3	62.2
1969	48.6	48.4	49.4	65.8	72.1	81.5	84.9	85.7	77.1	67.7	52.5	48.6	65.2
1970	46.9	52.3	55.6	63.5	72.2	79.7	82.8	81.4	74.2	59.5	51.9	48.0	64.0
1971	44.6	48.4	58.1	62.6	72.1	81.1	82.3	77.0	73.5	62.5	52.1	44.7	63.2
1972	45.2	52.3	61.2	65.2	69.8	78.3	82.2	77.7	72.9	65.7	48.8	46.6	63.8
1973	42.9	47.0	52.4	57.7	68.7	76.6	79.7	79.2	75.9	63.3	53.8	45.7	61.9
1974	44.3	44.8	59.6	65.0	75.6	82.8	79.6	77.0	69.4	63.0	49.9	41.3	62.7
1975	43.1	48.9	55.1	59.7	69.8	80.9	79.8	80.9	72.2	64.0	51.5	44.2	62.5
1976	42.3	52.6	55.9	64.1	69.5	79.4	78.2	78.7	70.5	58.5	44.8	41.9	61.4
1977	44.7	47.3	49.7	61.5	70.7	81.5	82.2	82.5	77.4	64.3	53.7	49.6	63.8
#1978	45.4	48.8	58.6	66.1	73.6	83.4	84.5	80.0	72.3	63.6	55.8	44.6	64.7
1979	41.0	47.1	53.1	63.6	70.1	78.4	85.1	78.8	74.2	66.2	47.9	43.1	62.4
1980	46.8	50.6	54.1	60.6	70.5	86.3	87.2	82.4	75.6	60.3	49.2	48.5	64.3
1981	45.2	50.3	57.2	64.8	73.6	82.6	83.6	79.5	75.9	64.6	54.3	48.7	65.1
1982	42.3	48.7	57.7	64.4	69.6	80.9	84.2	83.5	77.1	64.6	53.2	43.3	64.1
1983	41.6	49.5	54.6	56.3	68.9	77.3	82.9	81.8	78.7	66.6	54.3	45.5	63.2
1984	44.4	47.0	55.7	62.0	75.0	79.5	81.1	80.4	72.9	61.4	51.6	45.7	63.1
1985	40.0	45.6	55.2	64.2	72.1	79.0	79.4	80.6	72.8	61.4	52.9	43.1	62.2
Record Mean	44.6	49.2	55.2	63.7	72.1	80.8	82.0	80.3	74.8	64.5	52.4	45.2	63.8
Max	57.2	62.5	69.2	77.6	86.0	94.6	94.1	92.1	86.9	77.7	65.7	57.6	76.8
Min	32.0	35.9	41.9	49.7	58.2	66.9	69.9	68.5	62.7	51.3	39.1	32.8	50.7

REFERENCE NOTES FOR TABLES 1, 2, 3 and 6   (EL PASO, TX)

GENERAL

T - TRACE AMOUNT
BLANK ENTRIES DENOTE MISSING/UNREPORTED DATA.
# INDICATES A STATION OR INSTRUMENT RELOCATION.

SPECIFIC

TABLE 1

(a) - LENGTH OF RECORD IN YEARS. ALTHOUGH INDIVIDUAL MONTHS MAY BE MISSING.
* LESS THAN .05

NORMALS — BASED ON THE 1951-1980 RECORD PERIOD.
EXTREMES — DATES ARE THE MOST RECENT OCCURRENCE.
WIND DIR. — NUMERALS SHOW TENS OF DEGREES CLOCKWISE FROM TRUE NORTH.
"00" INDICATES CALM.
RESULTANT WIND DIRECTIONS ARE GIVEN TO WHOLE DEGREES.

EXCEPTIONS

TABLES 2, 3, and 6

RECORD MEANS ARE THROUGH THE CURRENT YEAR, BEGINNING IN   1887 FOR TEMPERATURE
1879 FOR PRECIPITATION
1940 FOR SNOWFALL

TABLE 4   HEATING DEGREE DAYS Base 65 deg. F   EL PASO, TEXAS

SEASON	JULY	AUG	SEP	OCT	NOV	DEC	JAN	FEB	MAR	APR	MAY	JUNE	TOTAL
1956-57	0	0	0	65	494	624	478	205	242	102	19	0	2229
1957-58	0	0	4	127	476	570	677	370	422	116	3	0	2765
1958-59	0	0	16	120	343	549	523	434	335	86	3	0	2409
1959-60	0	0	0	33	439	573	691	534	176	58	10	0	2514
#1960-61	0	0	0	86	357	806	735	473	266	97	1	0	2821
1961-62	0	0	0	82	513	575	754	318	433	36	3	0	2714
1962-63	0	0	0	65	333	608	753	438	279	41	0	0	2517
1963-64	0	0	0	17	341	683	789	695	354	99	3	0	2981
1964-65	0	0	0	76	391	643	521	504	397	54	7	2	2595
1965-66	0	0	4	107	240	592	769	612	264	59	4	0	2651
1966-67	0	0	2	126	307	695	718	469	173	56	25	0	2571
1967-68	0	0	2	106	352	720	691	415	377	128	0	0	2791
1968-69	0	0	0	61	414	728	503	464	477	43	24	0	2714
1969-70	0	0	0	62	371	504	556	348	286	94	33	0	2254
1970-71	0	0	39	180	388	519	625	457	254	110	6	0	2578
1971-72	0	0	31	112	381	624	607	364	126	56	3	0	2304
1972-73	0	0	3	87	480	563	679	499	384	218	31	0	2944
1973-74	0	0	7	90	336	592	636	558	178	79	5	0	2481
1974-75	0	0	41	107	445	728	672	445	309	188	13	0	2948
1975-76	0	0	20	66	399	640	696	351	278	91	26	0	2567
1976-77	0	0	7	214	601	709	623	492	469	138	3	0	3256
1977-78	0	0	0	56	328	472	603	449	200	57	22	0	2187
#1978-79	0	0	16	106	272	625	735	494	362	118	26	1	2755
1979-80	0	0	24	56	505	670	555	410	331	157	19	0	2727
1980-81	0	0	2	203	467	503	607	405	233	82	2	0	2504
1981-82	0	0	0	93	313	499	697	449	237	82	17	0	2387
1982-83	0	0	0	88	344	668	720	430	316	284	23	0	2873
1983-84	0	0	0	52	317	599	633	514	285	126	8	0	2534
1984-85	0	0	18	144	404	592	768	537	302	71	5	0	2841
1985-86	0	0	10	125	358	670							

TABLE 5   COOLING DEGREE DAYS Base 65 deg. F   EL PASO, TEXAS

YEAR	JAN	FEB	MAR	APR	MAY	JUNE	JULY	AUG	SEP	OCT	NOV	DEC	TOTAL
1969	0	0	1	71	250	501	627	647	372	155	2	0	2626
1970	0	0	2	57	263	448	559	514	321	19	0	0	2183
1971	0	0	45	45	235	492	543	375	293	43	0	0	2071
1972	0	4	15	70	159	404	543	401	247	120	0	0	1963
1973	0	0	0	7	152	355	459	448	340	44	7	0	1812
1974	0	0	19	84	338	540	459	378	181	54	0	0	2053
1975	0	0	9	35	170	482	469	502	241	41	1	0	1950
1976	0	0	4	71	170	441	418	434	179	18	0	0	1735
1977	0	0	0	39	186	500	540	552	380	43	0	0	2240
#1978	0	0	8	98	295	559	612	474	238	70	2	0	2356
1979	0	0	0	84	190	414	630	432	308	99	0	0	2157
1980	0	0	0	34	198	646	693	546	329	63	0	0	2509
1981	0	2	2	84	275	534	586	455	333	89	0	2	2362
1982	0	0	14	70	167	484	602	583	371	82	0	0	2373
1983	0	0	0	32	151	374	564	527	417	108	6	0	2179
1984	0	0	2	44	324	441	507	482	260	38	8	0	2106
1985	0	0	9	55	233	428	457	488	252	18	0	0	1940

TABLE 6   SNOWFALL (inches)   EL PASO, TEXAS

SEASON	JULY	AUG	SEP	OCT	NOV	DEC	JAN	FEB	MAR	APR	MAY	JUNE	TOTAL
1956-57	0.0	0.0	0.0	0.0	0.0	2.3	T	0.0	T	0.0	0.0	0.0	2.3
1957-58	0.0	0.0	0.0	0.0	4.1	0.0	2.0	0.0	7.3	T	0.0	0.0	13.4
1958-59	0.0	0.0	0.0	0.0	0.8	0.0	T	T	T	T	0.0	0.0	0.8
1959-60	0.0	0.0	0.0	0.0	0.0	T	T	2.1	T	T	0.0	0.0	2.1
1960-61	0.0	0.0	0.0	0.0	0.0	10.1	0.0	T	T	T	0.0	0.0	10.1·
1961-62	0.0	0.0	0.0	0.0	7.8	T	1.6	0.0	T	0.0	0.0	0.0	9.4
1962-63	0.0	0.0	0.0	0.0	0.0	0.0	0.8	5.3	0.0	0.0	0.0	0.0	6.1
1963-64	0.0	0.0	0.0	0.0	0.0	T	0.0	T	T	T	0.0	0.0	T
1964-65	0.0	0.0	0.0	0.0	0.0	T	0.0	0.3	T	T	0.0	0.0	0.3
1965-66	0.0	0.0	0.0	0.0	0.0	T	0.4	T	T	T	0.0	0.0	0.4
1966-67	0.0	0.0	0.0	0.0	0.0	T	T	0.0	T	0.0	0.0	0.0	T
1967-68	0.0	0.0	0.0	0.0	0.0	5.6	2.1	2.3	0.8	0.0	0.0	0.0	10.8
1968-69	0.0	0.0	0.0	0.0	7.0	T	0.0	0.0	T	0.0	0.0	0.0	7.0
1969-70	0.0	0.0	0.0	0.0	6.0	1.9	0.3	1.7	T	0.0	0.0	0.0	9.9
1970-71	0.0	0.0	0.0	0.0	0.0	0.0	3.8	T	0.0	T	0.0	0.0	3.8
1971-72	0.0	0.0	0.0	0.0	0.0	1.0	2.5	0.0	0.0	0.0	0.0	0.0	3.5
1972-73	0.0	0.0	0.0	0.0	0.0	T	5.3	4.6	T	0.0	0.0	0.0	9.9
1973-74	0.0	0.0	0.0	0.0	0.0	0.0	T	0.0	0.0	0.0	0.0	0.0	T
1974-75	0.0	0.0	0.0	0.0	0.0	5.3	2.2	0.2	2.0	0.0	0.0	0.0	9.7
1975-76	0.0	0.0	0.0	0.0	T	1.0	1.0	0.0	T	T	0.0	0.0	1.0
1976-77	0.0	0.0	0.0	T	12.7	2.0	T	T	0.0	T	0.0	0.0	14.7
1977-78	0.0	0.0	0.0	0.0	0.0	0.0	T	0.0	0.0	0.0	0.0	0.0	T
1978-79	0.0	0.0	0.0	0.0	T	T	T	1.2	0.0	0.0	0.0	0.0	1.2
1979-80	0.0	0.0	0.0	0.0	0.0	0.0	T	3.6	0.0	2.0	0.0	0.0	5.6
1980-81	0.0	0.0	0.0	1.0	4.0	0.0	4.8	0.0	0.0	0.0	0.0	0.0	9.8
1981-82	0.0	0.0	0.0	0.0	0.0	0.0	T	T	0.0	0.0	0.0	0.0	T
1982-83	0.0	0.0	0.0	0.0	0.3	18.2	T	0.0	T	16.5	0.0	0.0	35.0
1983-84	0.0	0.0	0.0	0.0	T	T	0.0	6.1	0.0	0.0	0.0	0.0	6.1
1984-85	0.0	0.0	0.0	0.0	T	2.9	5.4	1.1	0.0	0.0	0.0	0.0	9.4
1985-86	0.0	0.0	0.0	0.0	0.0	0.9							
Record Mean	0.0	0.0	0.0	T	1.0	1.3	1.4	0.8	0.5	0.4	0.0	0.0	5.4

See Reference Notes, relative to all above tables, on preceding page.

The city of Galveston is located on Galveston Island off the southeast coast of Texas. The island is about 2 3/4 miles across at the widest point and 29 miles long. It is bounded on the southeast by the Gulf of Mexico and on the northwest by Galveston Bay, which is about 3 miles wide at this point. The climate of the Galveston area is predominantly marine, with periods of modified continental influence during the colder months, when cold fronts from the northwest sometimes reach the coast.

Because of its coastal location and relatively low latitude, cold fronts which do reach the area are very seldom severe and temperatures below 32 degrees are recorded on an average only four times a year. Normal monthly high temperatures range from about 60 degrees in January to nearly 88 degrees in August. Lows range from about 48 degrees in January to the upper 70s during the summer season.

High humidities prevail throughout the year. Annual precipitation averages about 42 inches. Rainfall during the summer months may vary greatly on different parts of the island, as most of the rain in this season is from local thunderstorm activity. Hail is rare because the necessary strong vertical lifting is usually absent. There have been several instances when a monthly rainfall total amounted to only a trace, but these have been offset in the means by many monthly totals in excess of 15 inches. Winter precipitation comes mainly from frontal activity and from low stratus clouds, which produce slow, steady rains.

The island has been subject at infrequent intervals to major tropical storms of hurricane force.

TABLE 1 NORMALS, MEANS AND EXTREMES

GALVESTON, TEXAS

LATITUDE: 29°18'N LONGITUDE: 94°48'W ELEVATION: FT. GRND 7 BARO 00069 TIME ZONE: CENTRAL WBAN: 12944

	(a)	JAN	FEB	MAR	APR	MAY	JUNE	JULY	AUG	SEP	OCT	NOV	DEC	YEAR
TEMPERATURE °F:														
Normals														
-Daily Maximum		59.2	60.9	66.4	73.3	79.8	85.1	87.3	87.5	84.6	77.6	68.3	62.3	74.4
-Daily Minimum		47.9	50.2	56.5	64.9	71.6	77.2	79.1	78.8	75.4	67.7	57.6	51.2	64.8
-Monthly		53.5	55.5	61.5	69.1	75.7	81.2	83.2	83.2	80.0	72.7	63.0	56.8	69.6
Extremes														
-Record Highest	115	77	83	85	92	94	99	101	100	96	94	85	80	101
-Year		1969	1932	1879	1953	1984	1918	1932	1924	1927	1952	1886	1918	JUL 1932
-Record Lowest	115	11	8	26	38	52	57	66	67	52	41	26	14	8
-Year		1886	1899	1980	1938	1954	1903	1910	1966	1942	1925	1911	1983	FEB 1899
NORMAL DEGREE DAYS:														
Heating (base 65°F)		376	282	160	19	0	0	0	0	0	10	139	267	1253
Cooling (base 65°F)		23	18	48	142	332	486	564	564	450	248	79	13	2967
% OF POSSIBLE SUNSHINE	94	48	51	55	60	67	75	72	71	68	71	59	49	62
MEAN SKY COVER (tenths)														
Sunrise - Sunset														
MEAN NUMBER OF DAYS:														
Sunrise to Sunset														
-Clear														
-Partly Cloudy														
-Cloudy														
Precipitation														
.01 inches or more	114	10.0	8.6	7.8	6.3	6.1	6.5	8.5	9.1	9.3	6.5	7.7	9.8	96.2
Snow, Ice pellets														
1.0 inches or more	114	0.*	0.*	0.0	.0.0	0.0	0.0	0.0	0.0	0.0	0.0	0.0	0.0	*
Thunderstorms														
Heavy Fog Visibility														
1/4 mile or less														
Temperature °F														
-Maximum														
90° and above	100	0.0	0.0	0.0	0.*	0.1	1.3	3.8	5.2	1.6	0.1	0.0	0.0	12.2
32° and below	115	0.1	0.1	0.0	0.0	0.0	0.0	0.0	0.0	0.0	0.0	0.0	0.*	0.2
-Minimum														
32° and below	115	2.0	0.8	0.1	0.0	0.0	0.0	0.0	0.0	0.0	0.0	0.1	0.7	3.7
0° and below	115	0.0	0.0	0.0	0.0	0.0	0.0	0.0	0.0	0.0	0.0	0.0	0.0	0.0
AVG. STATION PRESS.(mb)														
RELATIVE HUMIDITY (%)														
Hour 00	44	83	82	84	85	83	80	80	78	78	75	81	82	81
Hour 06	96	85	84	85	86	84	81	81	81	81	80	83	85	83
Hour 12 (Local Time)	66	77	74	74	75	73	70	70	69	68	65	72	76	72
Hour 18	96	80	77	79	80	77	73	73	73	74	71	77	79	76
PRECIPITATION (inches):														
Water Equivalent														
-Normal		2.96	2.34	2.10	2.62	3.30	3.48	3.77	4.40	5.82	2.60	3.23	3.62	40.24
-Maximum Monthly	115	10.39	8.29	9.49	11.04	10.79	15.49	18.74	19.08	26.01	17.78	16.18	10.28	26.01
-Year		1899	1881	1973	1904	1975	1919	1900	1915	1885	1871	1940	1887	SEP 1885
-Minimum Monthly	115	0.02	0.09	0.06	T	T	T	T	0.00	0.04	T	0.03	0.23	0.00
-Year		1909	1954	1953	1984	1978	1907	1962	1902	1924	1952	1903	1889	AUG 1902
-Maximum in 24 hrs	115	5.38	6.55	8.10	9.23	7.71	12.56	14.35	10.86	11.65	14.10	9.01	5.43	14.35
-Year		1923	1952	1973	1904	1975	1961	1900	1981	1961	1901	1940	1964	JUL 1900
Snow, Ice pellets														
-Maximum Monthly	115	2.5	15.4	T									0.2	15.4
-Year		1973	1895	1932									1924	FEB 1895
-Maximum in 24 hrs	115	2.5	15.4	T									0.2	15.4
-Year		1973	1895	1932									1924	FEB 1895
WIND:														
Mean Speed (mph)	93	11.6	11.8	11.9	12.1	11.5	10.7	9.8	9.4	10.1	10.3	11.2	11.3	11.0
Prevailing Direction														
Fastest Mile														
-Direction (!!!)	114	S	N	SE	NW	W	SE	NW	E	NE	N	N	NW	NE
-Speed (MPH)	114	53	60	50	68	66	62	68	91	100	70	54	50	100
-Year		1915	1927	1952	1983	1959	1921	1943	1915	1900	1985	1950	1954	SEP 1900
Peak Gust														
-Direction (!!!)														
-Speed (mph)														
-Date														

See Reference Notes to this table on the following pages.

GALVESTON, TEXAS

TABLE 2   PRECIPITATION (inches)   GALVESTON, TEXAS

YEAR	JAN	FEB	MAR	APR	MAY	JUNE	JULY	AUG	SEP	OCT	NOV	DEC	ANNUAL
1956	2.44	1.87	1.14	2.35	1.00	0.90	0.05	1.86	1.84	1.46	0.78	6.15	21.84
1957	1.54	2.03	7.73	3.76	0.94	4.89	0.46	2.98	5.62	3.00	3.17	1.19	37.31
1958	3.50	1.66	0.95	0.84	0.46	0.23	1.69	6.59	13.77	0.96	1.39	2.67	34.71
1959	1.22	7.54	0.23	1.94	2.31	3.16	10.68	5.76	3.62	2.09	0.76	3.23	42.54
1960	0.58	2.29	0.93	0.83	0.26	2.35	0.75	6.97	1.21	6.47	3.88	5.37	31.89
1961	4.02	1.84	0.26	2.25	0.09	14.76	6.71	4.48	15.41	0.23	10.80	3.03	63.88
1962	1.24	0.79	1.59	2.43	0.82	8.72	T	2.20	1.59	3.29	3.59	6.03	32.29
1963	1.49	2.62	0.07	0.34	0.40	4.31	1.94	1.41	9.30	0.05	5.52	2.80	30.25
1964	4.13	3.39	2.53	0.37	1.72	1.29	5.79	5.67	4.38	1.00	1.49	7.12	38.88
1965	1.65	1.77	0.36	0.55	2.57	2.45	1.24	4.88	4.02	1.91	4.26	9.00	34.66
1966	5.52	4.50	0.70	5.41	10.34	4.86	1.25	5.56	7.99	4.03	0.93	1.99	53.08
1967	1.60	2.26	1.55	1.69	3.33	0.81	8.73	5.74	2.21	3.18	1.64	2.50	35.24
1968	6.14	1.70	1.93	3.93	4.82	13.03	2.26	3.17	3.95	2.28	4.73	3.22	51.16
1969	1.59	3.54	2.30	6.03	6.97	3.45	4.76	5.75	0.28	1.90	0.91	4.31	41.79
1970	1.47	1.33	6.40	3.47	4.33	5.14	3.07	14.31	5.14	1.48	0.96		48.47
1971	0.18	4.12	0.64	1.77	1.31	0.82	1.98	3.61	10.21	2.37	2.29	6.67	35.97
1972	3.68	1.87	1.11	0.76	6.37	0.50	3.44	1.40	7.85	3.03	6.30	3.64	39.95
1973	3.19	2.62	9.49	10.41	1.61	4.01	2.44	10.25	6.80	6.44	0.57	2.64	60.47
1974	3.28	0.83	2.79	0.91	7.84	2.03	2.18	8.08	3.31	3.57	4.10	4.34	43.26
1975	3.36	1.79	0.96	2.19	10.79	3.63	1.28	6.58	4.61	4.83	4.56	3.96	48.54
1976	1.00	0.31	0.85	2.53	2.16	4.28	5.63	1.43	8.27	3.85	4.02	7.73	42.06
1977	3.41	1.07	1.39	6.74	0.67	2.83	0.68	6.45	4.91	2.95	9.51	1.46	42.07
1978	8.88	1.81	0.38	1.62	T	3.55	4.23	0.63	3.55	0.30	2.99	1.34	29.28
1979	3.48	2.80	3.34	4.91	3.45	0.79	17.48	4.47	10.86	2.75	2.05	2.97	59.35
1980	6.85	0.83	3.93	0.57	7.44	0.44	3.85	2.35	5.05	0.28	1.25	1.74	34.58
1981	1.88	1.30	0.52	0.33	4.31	10.66	4.87	13.38	2.49	3.03	1.76	2.25	46.78
1982	1.80	4.00	1.24	1.81	3.38	4.51	0.24	1.80	1.29	1.20	6.61	6.38	34.26
1983	4.24	3.37	2.91	0.24	1.52	4.94	7.28	11.18	11.58	1.24	2.76	2.64	53.90
1984	3.19	1.19	1.36	T	5.07	0.86	1.82	3.08	4.97	8.97	2.41	2.72	35.64
1985	2.88	5.42	5.63	1.20	2.01	3.41	7.08	2.46	2.93	3.41	2.24	2.57	41.24
Record Mean	3.39	2.75	2.62	2.93	3.43	3.83	4.07	4.48	5.66	3.79	3.60	3.88	44.44

TABLE 3   AVERAGE TEMPERATURE (deg. F)   GALVESTON, TEXAS

YEAR	JAN	FEB	MAR	APR	MAY	JUNE	JULY	AUG	SEP	OCT	NOV	DEC	ANNUAL
1956	55.5	59.1	62.2	67.7	77.8	81.1	83.1	83.4	80.8	75.3	63.1	60.8	70.8
1957	58.9	62.7	62.3	68.4	76.3	81.1	84.6	83.7	78.1	69.8	63.1	59.2	70.7
1958	51.9	51.2	57.9	68.8	77.3	83.2	84.6	83.6	79.5	71.0	64.3	54.0	68.9
1959	51.3	55.4	60.8	62.3	77.1	81.3	82.1	82.3	81.1	73.2	57.6	56.8	68.8
1960	52.5	51.1	56.8	68.6	73.5	81.1	84.4	82.0	79.7	74.0	65.2	53.4	68.5
1961	49.2	56.5	64.8	66.4	74.9	79.0	82.4	80.7	78.8	72.6	61.2	58.2	68.7
1962	49.9	62.1	60.3	68.3	76.5	80.4	84.0	84.1	81.7	75.3	62.7	54.8	70.0
1963	48.9	51.7	64.0	72.4	77.2	81.4	83.6	83.6	80.0	76.4	65.5	49.3	69.5
1964	52.1	51.5	59.9	69.4	76.7	80.6	82.3	83.8	80.5	69.8	65.8	55.5	69.0
1965	56.9	55.0	57.0	71.1	76.3	81.7	83.7	82.8	80.7	70.8	70.3	60.7	70.6
1966	50.5	53.3	60.6	68.8	75.2	79.6	83.4	81.6	79.2	70.8	66.3	56.0	68.7
1967	53.3	54.6	65.4	75.0	75.6	81.7	81.5	80.7	77.8	72.3	64.5	57.5	70.0
1968	52.6	50.8	57.4	69.7	75.7	80.3	82.3	84.4	79.0	74.2	62.0	55.8	68.7
1969	55.4	56.5	55.8	69.6	75.5	81.4	84.3	84.4	81.2	73.9	63.2	59.1	70.0
1970	48.9	56.1	58.9	69.7	74.1	81.0	82.7	83.6	80.0	69.7	60.6	61.8	69.0
1971	57.0	57.1	60.6	68.1	75.0	81.5	83.4	82.5	79.8	75.7	65.2	62.2	70.7
1972	59.0	58.1	66.0	72.5	76.3	81.8	82.2	84.1	82.8	73.4	58.8	55.7	70.9
1973	51.1	53.0	64.8	65.5	74.3	79.9	83.4	82.0	80.6	76.2	70.4	57.3	69.9
1974	57.1	58.8	65.8	70.4	77.2	80.7	83.0	82.6	76.0	72.8	63.0	56.9	70.3
1975	58.7	58.1	62.6	68.2	77.0	80.9	83.8	84.1	77.4	72.9	64.1	55.2	70.2
1976	53.3	60.5	63.5	70.0	73.1	80.0	81.4	82.5	79.7	65.3	54.6	52.8	68.0
1977	46.6	55.1	62.2	70.0	76.1	81.0	82.7	82.6	81.4	72.8	65.1	57.8	69.4
1978	45.3	47.0	58.3	68.4	76.7	82.6	83.9	84.5	81.4	73.4	66.7	57.4	68.8
1979	47.0	51.9	61.6	68.6	73.5	81.0	82.3	82.4	76.0	73.2	58.9	55.0	67.7
1980	56.3	52.9	60.0	67.1	75.7	81.8	84.3	84.2	82.8	70.8	59.0	55.0	69.2
1981	52.5	53.7	61.0	72.1	74.8	82.0	84.7	84.8	80.4	74.3	66.0	57.2	70.3
1982	53.2	53.1	63.3	67.4	76.3	82.4	84.4	84.5	81.0	71.5	63.4	59.1	69.9
1983	53.2	55.9	61.3	65.9	75.1	81.3	84.7	84.4	79.2	74.8	67.4	49.8	69.4
1984	49.3	57.2	61.5	68.8	75.6	81.1	84.5	82.3	76.8	75.1	63.2	62.4	69.8
1985	48.1	50.0	63.2	70.3	76.3	79.9	81.7	83.7	79.9	73.0	68.1	54.3	69.0
Record Mean	53.8	56.1	62.0	69.0	75.6	81.3	83.3	83.3	80.0	73.0	63.3	56.8	69.8
Record Max	59.3	61.5	67.0	73.5	80.0	85.6	87.7	87.8	84.7	77.9	68.6	62.1	74.7
Record Min	48.2	50.7	56.9	64.5	71.2	77.1	78.8	78.7	75.3	68.0	58.0	51.4	64.9

REFERENCE NOTES FOR TABLES 1, 2, 3 and 6   (GALVESTON, TX)

GENERAL

T - TRACE AMOUNT
BLANK ENTRIES DENOTE MISSING/UNREPORTED DATA.
# INDICATES A STATION OR INSTRUMENT RELOCATION.

SPECIFIC

TABLE 1

(a) - LENGTH OF RECORD IN YEARS. ALTHOUGH INDIVIDUAL MONTHS MAY BE MISSING.

* LESS THAN .05

NORMALS — BASED ON THE 1951-1980 RECORD PERIOD.
EXTREMES — DATES ARE THE MOST RECENT OCCURRENCE.
WIND DIR. — NUMERALS SHOW TENS OF DEGREES CLOCKWISE FROM TRUE NORTH. "00" INDICATES CALM.
RESULTANT WIND DIRECTIONS ARE GIVEN TO WHOLE DEGREES.

EXCEPTIONS

TABLE 1

1. FASTEST MILE WIND OF 100 MPH WAS RECORDED AT 1815H SEPTEMBER 8, 1900 JUST BEFORE THE ANEMOMETER BLEW AWAY, MAXIMUM VELOCITY ESTIMATED 120 MPH NE BETWEEN 1930H AND 2030H
2. MEAN WIND SPEED IS THROUGH 1964.
3. RELATIVE HUMIDITY IS THROUGH 1983

TABLES 2, 3, and 6

RECORD MEANS ARE THROUGH THE CURRENT YEAR, BEGINNING IN 1874 FOR TEMPERATURE
1871 FOR PRECIPITATION
1871 FOR SNOWFALL

TABLE 4   HEATING DEGREE DAYS Base 65 deg. F   GALVESTON, TEXAS

SEASON	JULY	AUG	SEP	OCT	NOV	DEC	JAN	FEB	MAR	APR	MAY	JUNE	TOTAL
1956-57	0	0	0	0	147	154	204	104	106	32	2	0	749
1957-58	0	0	0	49	145	175	401	379	220	13	0	0	1382
1958-59	0	0	0	35	102	334	418	263	134	43	0	0	1329
1959-60	0	0	0	10	253	248	384	400	264	11	3	0	1573
1960-61	0	0	0	15	74	366	484	235	51	55	1	0	1281
1961-62	0	0	0	7	148	232	461	97	166	24	0	0	1135
1962-63	0	0	0	4	99	311	494	366	5	5	0	0	1354
1963-64	0	0	0	0	80	480	393	384	154	11	0	0	1502
1964-65	0	0	0	5	102	293	254	276	249	2	0	0	1181
1965-66	0	0	0	5	12	152	447	320	138	9	0	0	1083
1966-67	0	0	0	16	63	298	356	287	61	0	0	0	1081
1967-68	0	0	3	12	78	243	381	404	233	8	0	0	1362
1968-69	0	0	0	4	142	283	296	236	281	0	0	0	1242
1969-70	0	0	0	9	127	189	490	244	185	25	3	0	1272
1970-71	0	0	0	31	161	130	250	223	143	30	0	0	968
1971-72	0	0	0	0	91	131	210	211	53	2	0	0	698
1972-73	0	0	0	23	225	289	424	329	39	73	0	0	1402
1973-74	0	0	0	1	23	249	251	181	61	10	0	0	776
1974-75	0	0	0	2	121	254	208	201	120	28	0	0	934
1975-76	0	0	0	4	126	309	355	128	82	0	0	0	1004
1976-77	0	0	0	96	313	373	563	274	102	3	0	0	1724
1977-78	0	0	0	6	63	238	606	501	204	16	2	0	1636
1978-79	0	0	0	0	72	258	550	361	120	12	2	0	1375
1979-80	0	0	0	7	200	302	268	348	158	28	0	0	1311
1980-81	0	0	0	43	222	306	382	314	127	0	0	0	1394
1981-82	0	0	0	30	45	242	358	333	109	51	0	0	1168
1982-83	0	0	0	23	109	201	359	249	129	49	0	0	1119
1983-84	0	0	2	2	56	491	478	228	108	17	0	0	1382
1984-85	0	0	2	1	120	109	515	412	81	2	0	0	1242
1985-86	0	0	1	11	41	333							

TABLE 5   COOLING DEGREE DAYS Base 65 deg. F   GALVESTON, TEXAS

YEAR	JAN	FEB	MAR	APR	MAY	JUNE	JULY	AUG	SEP	OCT	NOV	DEC	TOTAL
1969	3	4	4	145	333	500	602	609	493	289	81	11	3074
1970	0	0	3	174	292	488	557	583	456	181	36	39	2809
1971	8	5	15	130	316	500	574	549	448	340	103	49	3037
1972	30	20	91	234	360	510	540	597	540	294	48	8	3272
1973	1	0	41	93	296	454	576	533	476	354	191	15	3030
1974	10	14	94	178	385	481	561	553	335	253	68	7	2939
1975	19	11	53	132	379	483	588	599	378	254	106	11	3013
1976	0	6	40	156	259	457	516	550	445	113	7	0	2549
1977	0	4	21	160	350	488	553	555	499	254	75	21	2980
1978	1	2	5	124	373	535	595	613	499	273	128	25	3173
1979	0	0	22	128	271	488	544	547	339	268	22	0	2629
1980	2	1	12	99	339	512	604	602	542	228	52	0	2993
1981	0	0	12	218	311	520	617	620	465	327	80	7	3177
1982	1	6	63	132	357	524	608	611	486	230	66	29	3113
1983	3	0	19	83	320	496	617	609	435	314	135	25	3056
1984	0	7	7	138	338	489	610	545	364	324	71	38	2931
1985	0	0	31	168	358	456	523	587	457	269	137	9	2995

TABLE 6   SNOWFALL (inches)   GALVESTON, TEXAS

SEASON	JULY	AUG	SEP	OCT	NOV	DEC	JAN	FEB	MAR	APR	MAY	JUNE	TOTAL
1970-71	0.0	0.0	0.0	0.0	0.0	0.0	T	0.0	0.0	0.0	0.0	0.0	T
1971-72	0.0	0.0	0.0	0.0	0.0	0.0	0.0	0.0	0.0	0.0	0.0	0.0	0.0
1972-73	0.0	0.0	0.0	0.0	0.0	0.0	2.5	1.6	0.0	0.0	0.0	0.0	4.1
1973-74	0.0	0.0	0.0	0.0	0.0	0.0	0.0	0.0	0.0	0.0	0.0	0.0	0.0
1974-75	0.0	0.0	0.0	0.0	0.0	0.0	0.0	0.0	0.0	0.0	0.0	0.0	0.0
1975-76	0.0	0.0	0.0	0.0	0.0	0.0	0.0	0.0	0.0	0.0	0.0	0.0	0.0
1976-77	0.0	0.0	0.0	0.0	0.0	0.0	0.0	0.0	0.0	0.0	0.0	0.0	0.0
1977-78	0.0	0.0	0.0	0.0	0.0	0.0	T	0.0	0.0	0.0	0.0	0.0	T
1978-79	0.0	0.0	0.0	0.0	0.0	0.0	0.0	0.0	0.0	0.0	0.0	0.0	0.0
1979-80	0.0	0.0	0.0	0.0	0.0	0.0	0.0	T	0.0	0.0	0.0	0.0	T
1980-81	0.0	0.0	0.0	0.0	0.0	0.0	0.0	0.0	0.0	0.0	0.0	0.0	0.0
1981-82	0.0	0.0	0.0	0.0	0.0	0.0	T	0.0	0.0	0.0	0.0	0.0	T
1982-83	0.0	0.0	0.0	0.0	0.0	0.0	0.0	0.0	0.0	0.0	0.0	0.0	0.0
1983-84	0.0	0.0	0.0	0.0	0.0	0.0	0.0	0.0	0.0	0.0	0.0	0.0	0.0
1984-85	0.0	0.0	0.0	0.0	0.0	0.0	T	0.0	0.0	0.0	0.0	0.0	T
1985-86	0.0	0.0	0.0	0.0	0.0	0.0							
Record Mean	0.0	0.0	0.0	0.0	0.0	T	T	0.2	T	0.0	0.0	0.0	0.2

See Reference Notes, relative to all above tables, on preceding page.

Houston, the largest city in Texas, is located in the flat Coastal Plains, about 50 miles from the Gulf of Mexico and about 25 miles from Galveston Bay. The climate is predominantly marine. The terrain includes numerous small streams and bayous which, together with the nearness to Galveston Bay, favor the development of both ground and advective fogs. Prevailing winds are from the southeast and south, except in January, when frequent passages of high pressure areas bring invasions of polar air and prevailing northerly winds.

Temperatures are moderated by the influence of winds from the Gulf, which result in mild winters. Another effect of the nearness of the Gulf is abundant rainfall, except for rare extended dry periods. Polar air penetrates the area frequently enough to provide variability in the weather.

Records of sky cover for daylight hours indicate about one-fourth of the days per year as clear, with a high number of clear days in October and November. Cloudy days are relatively frequent from December to May and partly cloudy days are the more frequent for June through September. Sunshine averages nearly 60 percent of the possible amount for the year ranging from 42 percent in January to 67 percent in June.

Heavy fog occurs on an average of 16 days a year and light fog occurs about 62 days a year in the city. The frequency of heavy fog is considerably higher at William P. Hobby Airport and at Intercontinental Airport.

Destructive windstorms are fairly infrequent, but both thundersqualls and tropical storms occasionally pass through the area.

TABLE 1 NORMALS, MEANS AND EXTREMES

HOUSTON, TEXAS

LATITUDE: 29°58'N   LONGITUDE: 95°21'W   ELEVATION: FT. GRND   96 BARO   00122   TIME ZONE: CENTRAL   WBAN: 12960

	(a)	JAN	FEB	MAR	APR	MAY	JUNE	JULY	AUG	SEP	OCT	NOV	DEC	YEAR
TEMPERATURE °F:														
Normals														
-Daily Maximum		61.9	65.7	72.1	79.0	85.1	90.9	93.6	93.1	88.7	81.9	71.6	65.2	79.1
-Daily Minimum		40.8	43.2	49.8	58.3	64.7	70.2	72.5	72.1	68.1	57.5	48.6	42.7	57.4
-Monthly		51.4	54.5	61.0	68.7	74.9	80.6	83.1	82.6	78.4	69.7	60.1	54.0	68.2
Extremes														
-Record Highest	16	84	85	90	92	95	103	104	107	102	94	89	83	107
-Year		1975	1982	1974	1981	1978	1980	1980	1980	1985	1981	1978	1978	AUG 1980
-Record Lowest	16	12	20	22	31	44	52	62	62	48	33	19	11	11
-Year		1982	1985	1980	1973	1978	1970	1972	1970	1975	1976	1976	1983	DEC 1983
NORMAL DEGREE DAYS:														
Heating (base 65°F)		442	314	175	32	0	0	0	0	0	36	201	349	1549
Cooling (base 65°F)		20	20	51	143	307	468	561	546	402	181	54	8	2761
% OF POSSIBLE SUNSHINE	16	43	50	47	51	58	64	66	64	62	59	52	55	56
MEAN SKY COVER (tenths)														
Sunrise - Sunset	16	7.0	6.3	6.9	6.6	6.1	5.5	5.7	5.6	5.6	5.3	5.6	6.5	6.1
MEAN NUMBER OF DAYS:														
Sunrise to Sunset														
-Clear	16	7.6	7.8	6.3	7.3	7.1	8.8	7.4	6.8	8.9	10.8	9.9	8.4	96.9
-Partly Cloudy	16	5.3	6.1	6.6	6.6	10.8	12.4	15.4	16.1	10.6	8.8	7.4	5.8	111.6
-Cloudy	16	18.2	14.4	18.2	16.2	13.2	8.8	8.1	8.1	10.6	11.5	12.6	16.9	156.8
Precipitation														
.01 inches or more	16	10.3	7.6	9.9	7.1	8.2	7.8	9.6	9.7	9.6	8.1	8.6	8.4	104.9
Snow,Ice pellets														
1.0 inches or more	16	0.1	0.2	0.0	0.0	0.0	0.0	0.0	0.0	0.0	0.0	0.0	0.0	0.3
Thunderstorms	16	1.8	1.4	3.8	3.6	6.8	6.9	10.8	10.8	7.0	3.9	2.6	1.9	61.1
Heavy Fog Visibility														
1/4 mile or less	16	5.6	4.1	3.5	3.2	1.7	0.8	0.3	0.6	1.5	3.3	4.1	4.8	33.4
Temperature °F														
-Maximum														
90° and above	16	0.0	0.0	0.1	0.4	4.4	19.1	26.5	25.9	13.4	2.4	0.0	0.0	92.1
32° and below	16	0.3	0.1	0.0	0.0	0.0	0.0	0.0	0.0	0.0	0.0	0.0	0.3	0.6
-Minimum														
32° and below	16	8.8	5.4	1.3	0.1	0.0	0.0	0.0	0.0	0.0	0.0	1.8	5.6	22.8
0° and below	16	0.0	0.0	0.0	0.0	0.0	0.0	0.0	0.0	0.0	0.0	0.0	0.0	0.0
AVG. STATION PRESS.(mb)	13	1017.2	1016.1	1011.7	1011.3	1009.5	1010.8	1012.3	1012.0	1011.7	1014.0	1015.1	1016.6	1013.2
RELATIVE HUMIDITY (%)														
Hour 00	16	83	83	83	86	87	87	87	89	90	89	86	83	86
Hour 06	16	86	87	87	89	92	92	93	93	93	91	89	86	90
Hour 12 (Local Time)	16	64	60	60	59	59	59	58	59	61	57	59	61	60
Hour 18	16	69	60	61	62	62	61	63	64	68	71	73	71	65
PRECIPITATION (inches):														
Water Equivalent														
-Normal		3.21	3.25	2.68	4.24	4.69	4.06	3.33	3.66	4.93	3.67	3.38	3.66	44.76
-Maximum Monthly	16	7.68	5.38	8.52	10.92	14.39	13.46	8.10	9.42	11.35	16.05	8.91	7.33	16.05
-Year		1974	1985	1972	1976	1970	1973	1979	1983	1976	1984	1982	1971	OCT 1984
-Minimum Monthly	16	0.36	0.38	1.21	0.43	0.79	0.26	1.42	1.14	0.80	0.05	1.54	0.64	0.05
-Year		1971	1976	1971	1983	1977	1970	1971	1985	1978	1978	1970	1973	OCT 1978
-Maximum in 24 hrs	16	2.56	2.22	7.47	8.16	5.11	6.61	3.99	6.83	7.98	9.31	3.62	3.43	9.31
-Year		1984	1985	1972	1976	1981	1973	1973	1981	1976	1984	1981	1971	OCT 1984
Snow,Ice pellets														
-Maximum Monthly	16	2.0	2.8									T		2.8
-Year		1973	1973									1979		FEB 1973
-Maximum in 24 hrs	16	2.0	1.4									T		2.0
-Year		1973	1980									1979		JAN 1973
WIND:														
Mean Speed (mph)	16	8.2	8.7	9.5	9.3	8.2	7.7	6.8	6.1	6.9	6.9	7.8	8.1	7.8
Prevailing Direction														
through 1963		NNW	SSE	SSE	SSE	SSE	SSE	S	SSE	SSE	ESE	SSE	SSE	SSE
Fastest Obs. 1 Min.														
-Direction	16	31	26	10	14	23	30	10	08	05	32	33	31	08
-Speed (MPH)	16	32	46	35	45	46	45	46	51	37	35	37	35	51
-Year		1978	1984	1979	1978	1983	1973	1969	1983	1982	1973	1972	1973	AUG 1983
Peak Gust														
-Direction	2	N	W	NW	SE	SW	S	E	NE	W	34	34	SW	W
-Speed (mph)	2	38	61	44	46	44	52	47	46	31	41	41	39	61
-Date		1985	1984	1984	1984	1985	1985	1984	1985	1985	1985	1985	1984	FEB 1984

See Reference Notes to this table on the following page.

TABLE 2 PRECIPITATION (inches)   HOUSTON, TEXAS

YEAR	JAN	FEB	MAR	APR	MAY	JUNE	JULY	AUG	SEP	OCT	NOV	DEC	ANNUAL
1956	3.96	2.42	0.78	1.84	3.32	4.07	0.38	0.38	1.88	2.66	1.52	5.11	28.32
1957	0.91	2.46	11.42	8.07	1.87	3.96	0.84	3.23	8.29	11.53	6.70	1.83	61.11
1958	4.69	3.59	1.47	2.58	2.40	0.47	6.04	3.16	2.03	1.03	0.78	5.34	43.93
1959	2.70	11.33	1.58	7.76	6.20	2.78	9.67	8.45	4.76	5.76	1.90	5.34	68.23
1960	2.05	3.93	0.38	1.42	0.90	14.66	2.34	7.42	0.61	7.32	3.69	8.97	53.69
1961	4.44	3.88	1.84	2.42	3.59	11.11	10.07	4.17	7.89	0.05	10.20	3.31	62.97
1962	1.73	0.71	0.94	4.81	1.15	7.40	0.07	2.77	3.97	3.12	5.68	4.78	37.13
1963	3.09	2.60	0.55	0.92	0.62	7.79	2.08	1.85	1.94	0.30	5.72	4.83	32.29
1964	2.89	4.97	2.24	1.63	2.25	1.89	1.68	2.61	6.76	2.35	4.28	5.57	39.12
1965	1.87	3.27	0.81	0.95	6.53	3.06	1.57	2.29	3.56	3.09	4.82	6.15	37.97
1966	4.46	7.75	2.20	7.98	11.21	4.42	1.45	7.11	4.01	5.45	1.56	1.53	59.13
1967	2.41	2.17	1.83	4.42	2.54	0.17	7.77	1.60	4.84	3.18	0.50	5.02	36.45
1968	8.02	1.99	2.92	3.02	13.24	11.18	6.49	2.90	3.87	3.91	2.71	1.19	61.44
#1969	2.74	5.31	3.18	3.34	4.73	1.51	3.89	2.67	6.08	3.30	2.13	4.38	43.26
1970	1.93	2.52	5.08	2.21	14.39	0.26	2.28	2.03	6.22	9.09	1.54	0.64	48.19
1971	0.36	2.11	1.21	2.14	3.41	2.42	1.42	6.95	5.17	3.49	1.82	7.33	37.83
1972	3.30	1.20	8.52	2.85	6.99	3.02	2.76	3.90	6.23	3.34	6.49	2.20	50.80
1973	5.00	3.40	3.68	7.15	4.22	13.46	6.77	3.73	9.38	9.31	1.59	2.47	70.16
1974	7.68	0.55	4.20	1.68	5.61	0.59	1.75	6.94	4.51	4.53	7.90	3.35	49.29
1975	1.97	2.63	3.19	4.80	7.57	7.50	5.48	5.72	0.80	5.62	2.08	3.61	50.97
1976	1.39	0.38	1.53	10.92	5.80	2.63	3.93	1.59	11.35	5.83	3.05	6.22	54.62
1977	2.67	1.70	1.95	4.34	0.79	3.55	2.69	4.45	3.92	0.82	5.17	2.89	34.94
1978	7.15	3.07	1.70	0.57	4.15	9.37	2.35	3.66	4.27	0.05	5.99	2.60	44.93
1979	6.30	5.23	2.88	7.79	3.78	1.88	8.10	4.57	9.83	2.80	1.78	4.03	58.97
1980	6.09	2.54	5.39	2.05	5.63	0.92	1.57	1.40	6.00	4.03	2.12	1.25	38.99
1981	2.32	2.21	1.74	2.69	8.75	9.65	4.43	7.01	2.91	6.96	5.26	2.05	55.98
1982	1.82	1.59	1.55	2.28	6.87	1.10	4.32	1.90	0.98	6.64	8.91	4.91	42.87
1983	2.00	3.97	3.85	0.43	7.29	5.37	5.23	9.42	7.23	1.56	3.17	3.69	53.21
1984	3.99	4.37	2.41	0.56	3.13	1.99	3.43	3.52	3.87	16.05	2.28	2.59	48.19
1985	2.10	5.38	4.52	4.31	1.57	5.29	4.93	1.14	4.67	6.54	4.84	3.85	49.14
Record Mean	3.67	3.10	2.71	3.40	4.78	4.38	4.17	4.15	4.75	4.07	4.06	4.05	47.29

TABLE 3 AVERAGE TEMPERATURE (deg. F)   HOUSTON, TEXAS

YEAR	JAN	FEB	MAR	APR	MAY	JUNE	JULY	AUG	SEP	OCT	NOV	DEC	ANNUAL
1956	54.3	59.2	62.6	67.7	77.3	80.8	84.3	83.8	80.2	73.3	59.9	59.0	70.2
1957	57.9	62.9	61.2	68.6	76.5	80.5	83.8	83.8	76.9	67.9	57.3	70.0	70.0
1958	50.3	49.9	57.8	69.2	77.2	83.4	84.8	84.2	79.7	69.6	62.5	52.6	68.4
1959	50.3	56.3	60.3	66.7	76.9	82.3	82.3	82.3	80.1	72.2	55.8	55.0	68.4
#1960	51.8	49.7	57.4	69.8	73.9	82.4	84.7	82.4	79.5	73.2	64.3	52.8	68.5
1961	49.5	57.9	66.8	68.0	77.0	80.4	82.6	82.2	79.7	70.5	59.8	56.3	69.2
1962	49.3	63.9	58.9	68.1	75.7	79.8	84.1	85.9	81.3	74.6	60.1	54.8	69.7
1963	48.3	52.6	64.9	74.5	77.5	82.0	84.3	84.3	80.4	75.0	64.1	47.1	69.6
1964	51.6	49.4	60.4	70.2	75.8	80.4	83.7	84.4	79.3	67.8	65.0	55.9	68.7
1965	56.0	55.1	58.7	73.5	77.1	83.0	85.1	83.5	81.2	69.4	69.1	58.9	70.9
1966	48.4	52.8	61.5	70.6	75.9	80.0	84.3	82.4	79.6	69.7	64.1	54.2	68.7
1967	54.9	54.5	67.6	75.7	75.9	82.6	82.1	81.1	77.6	70.5	64.1	55.4	70.2
1968	52.7	50.4	59.0	71.1	76.3	80.5	82.5	84.0	78.6	73.5	59.7	55.7	68.7
#1969	56.7	56.9	56.1	70.3	75.4	80.0	84.4	83.2	78.2	71.0	58.5	55.2	68.8
1970	46.7	53.9	56.9	69.3	72.1	78.4	81.4	83.1	78.0	66.7	56.9	60.5	67.1
1971	56.7	55.7	59.4	66.6	74.1	80.3	83.9	80.4	78.3	72.0	60.0	59.9	68.9
1972	56.5	55.2	64.3	71.2	73.7	80.8	80.3	79.6	69.8	69.8	54.2	52.0	68.2
1973	47.4	51.4	63.7	64.6	72.8	79.2	83.0	79.5	78.1	71.8	67.3	53.5	67.7
1974	55.0	56.2	66.8	67.2	76.9	79.9	82.8	81.6	74.6	70.6	60.2	54.6	68.9
1975	56.9	55.4	61.1	68.3	75.9	80.0	81.5	81.1	74.9	69.5	59.7	52.7	68.1
1976	50.6	60.1	62.2	67.7	70.5	78.4	80.5	81.4	76.2	60.6	51.8	49.2	65.8
1977	42.7	53.8	60.9	66.9	74.5	81.0	82.4	83.1	80.0	69.2	61.8	53.7	67.5
1978	40.8	45.1	57.3	67.6	76.0	80.4	83.7	83.1	79.3	68.9	64.7	52.9	66.5
1979	44.1	51.7	62.4	68.7	73.1	79.8	82.6	81.5	75.6	70.7	55.7	52.4	66.5
1980	55.0	53.7	60.9	66.2	77.3	85.1	87.5	86.6	83.2	67.8	58.0	55.2	69.7
1981	51.4	55.4	60.9	74.3	75.3	82.7	84.4	84.4	78.6	72.3	64.4	54.5	69.9
1982	52.9	52.1	64.9	67.8	75.3	83.0	85.4	84.1	79.3	69.5	60.9	55.4	69.2
1983	50.1	52.5	58.3	64.0	73.4	79.0	82.2	82.6	76.6	70.1	63.1	45.7	66.5
1984	47.0	54.0	61.9	67.8	74.9	78.6	81.8	82.9	77.4	74.2	60.0	63.4	68.7
1985	45.7	49.6	64.7	70.0	75.6	81.0	81.6	84.2	79.8	72.5	67.0	51.0	68.6
Record Mean	50.0	53.5	61.7	68.0	74.4	80.4	82.9	82.6	78.2	69.8	60.3	54.2	67.9
Max	60.2	64.9	72.5	78.5	84.7	90.5	93.3	92.9	88.1	81.2	71.6	65.3	78.6
Min	39.7	42.0	50.8	57.5	64.2	70.3	72.4	72.2	68.2	58.5	48.9	43.1	57.3

REFERENCE NOTES FOR TABLES 1, 2, 3 and 6   (HOUSTON, TX)

GENERAL

T - TRACE AMOUNT
BLANK ENTRIES DENOTE MISSING/UNREPORTED DATA.
# INDICATES A STATION OR INSTRUMENT RELOCATION.

SPECIFIC

TABLE 1

(a) - LENGTH OF RECORD IN YEARS. ALTHOUGH INDIVIDUAL MONTHS MAY BE MISSING.

* LESS THAN .05

NORMALS — BASED ON THE 1951-1980 RECORD PERIOD.
EXTREMES — DATES ARE THE MOST RECENT OCCURRENCE.
WIND DIR. — NUMERALS SHOW TENS OF DEGREES CLOCKWISE FROM TRUE NORTH. "00" INDICATES CALM.
RESULTANT WIND DIRECTIONS ARE GIVEN TO WHOLE DEGREES.

EXCEPTIONS

TABLES 2, 3, and 6

RECORD MEANS ARE THROUGH THE CURRENT YEAR, BEGINNING IN   1969 FOR TEMPERATURE
1933 FOR PRECIPITATION
1935 FOR SNOWFALL

TABLE 4   HEATING DEGREE DAYS Base 65 deg. F   HOUSTON, TEXAS

SEASON	JULY	AUG	SEP	OCT	NOV	DEC	JAN	FEB	MAR	APR	MAY	JUNE	TOTAL
1956-57	0	0	0	0	204	220	256	129	138	45	3	0	995
1957-58	0	0	0	71	185	245	449	417	234	30	0	0	1631
1958-59	0	0	0	48	145	375	452	254	163	58	0	0	1495
1959-60	0	0	0	7	309	306	416	441	260	8	1	0	1748
#1960-61	0	0	0	17	104	390	473	227	56	61	0	0	1328
1961-62	0	0	0	16	184	287	480	99	209	32	0	0	1307
1962-63	0	0	0	8	157	313	515	351	88	7	0	0	1439
1963-64	0	0	0	0	108	551	413	446	158	19	0	0	1695
1964-65	0	0	0	31	114	315	300	284	250	0	0	0	1294
1965-66	0	0	2	20	23	212	516	334	144	12	0	0	1263
1966-67	0	0	0	27	99	359	342	298	68	0	0	0	1193
1967-68	0	0	3	18	99	312	390	415	229	17	0	0	1483
#1968-69	0	0	0	5	199	297	284	234	281	1	0	2	1303
1969-70	0	0	0	29	238	304	579	309	252	51	12	0	1774
1970-71	0	0	0	72	274	209	298	273	219	72	3	0	1420
1971-72	0	0	2	6	195	194	315	295	85	17	0	0	1109
1972-73	0	0	2	50	320	410	540	379	75	117	5	0	1898
1973-74	0	0	0	8	74	364	330	273	95	60	0	0	1204
1974-75	0	0	0	15	196	336	290	270	179	48	0	0	1334
1975-76	0	0	0	26	217	399	441	178	155	26	7	0	1449
1976-77	0	0	0	0	173	398	484	687	312	25	17	0	2245
1977-78	0	0	0	34	150	365	752	553	250	33	17	0	2154
1978-79	0	0	0	22	111	393	646	376	135	23	2	0	1708
1979-80	0	0	0	27	297	389	308	350	169	45	0	0	1585
1980-81	0	0	0	67	255	323	416	291	144	6	1	0	1503
1981-82	0	0	0	50	82	326	409	363	143	79	1	0	1453
1982-83	0	0	0	53	175	328	457	346	219	96	0	0	1674
1983-84	0	0	6	27	138	606	549	325	150	45	2	0	1848
1984-85	0	0	6	12	204	144	591	432	91	22	0	0	1502
1985-86	0	0	5	17	76	434							

TABLE 5   COOLING DEGREE DAYS Base 65 deg. F   HOUSTON, TEXAS

YEAR	JAN	FEB	MAR	APR	MAY	JUNE	JULY	AUG	SEP	OCT	NOV	DEC	TOTAL
#1969	35	13	11	167	328	456	608	569	402	222	48	9	2868
1970	18	3	7	188	238	409	513	567	423	131	38	76	2611
1971	48	18	55	126	292	466	594	485	409	229	52	44	2818
1972	58	20	71	208	275	480	480	482	447	206	17	12	2756
1973	1	4	41	111	253	434	564	458	401	225	151	12	2655
1974	24	33	158	132	374	454	558	519	295	196	60	18	2821
1975	47	8	61	155	342	455	514	505	303	174	68	24	2656
1976	5	43	75	110	182	408	490	520	341	42	9	0	2225
1977	0	5	44	91	302	487	547	565	456	173	58	23	2751
1978	10	5	19	120	369	471	584	568	437	150	108	25	2866
1979	7	13	62	142	261	454	552	519	324	211	26	6	2577
1980	4	31	49	86	388	610	705	677	553	162	52	26	3343
1981	1	28	23	295	330	538	606	609	413	285	71	7	3206
1982	39	11	147	170	329	547	641	599	437	199	60	40	3219
1983	0	0	18	76	268	427	541	554	362	196	87	18	2547
1984	0	13	64	135	315	415	527	562	384	302	62	100	2879
1985	0	6	87	180	335	487	521	602	456	257	143	9	3083

TABLE 6   SNOWFALL (inches)   HOUSTON, TEXAS

SEASON	JULY	AUG	SEP	OCT	NOV	DEC	JAN	FEB	MAR	APR	MAY	JUNE	TOTAL
1970-71	0.0	0.0	0.0	0.0	0.0	0.0	T	0.0	0.0	0.0	0.0	0.0	T
1971-72	0.0	0.0	0.0	0.0	0.0	0.0	T	0.0	0.0	0.0	0.0	0.0	T
1972-73	0.0	0.0	0.0	0.0	0.0	0.0	2.0	2.8	0.0	0.0	0.0	0.0	4.8
1973-74	0.0	0.0	0.0	0.0	0.0	0.0	0.0	0.0	0.0	0.0	0.0	0.0	0.0
1974-75	0.0	0.0	0.0	0.0	0.0	0.0	T	0.0	0.0	0.0	0.0	0.0	T
1975-76	0.0	0.0	0.0	0.0	0.0	0.0	0.0	0.0	0.0	0.0	0.0	0.0	0.0
1976-77	0.0	0.0	0.0	0.0	T	0.0	0.0	0.0	0.0	0.0	0.0	0.0	T
1977-78	0.0	0.0	0.0	0.0	0.0	0.0	0.4	0.0	0.0	0.0	0.0	0.0	0.4
1978-79	0.0	0.0	0.0	0.0	0.0	0.0	T	0.0	0.0	0.0	0.0	0.0	T
1979-80	0.0	0.0	0.0	0.0	0.0	T	0.0	1.4	0.0	0.0	0.0	0.0	1.4
1980-81	0.0	0.0	0.0	0.0	0.0	0.0	T	T	0.0	0.0	0.0	0.0	T
1981-82	0.0	0.0	0.0	0.0	0.0	0.0	T	0.0	0.0	0.0	0.0	0.0	T
1982-83	0.0	0.0	0.0	0.0	0.0	0.0	0.0	0.0	0.0	0.0	0.0	0.0	0.0
1983-84	0.0	0.0	0.0	0.0	0.0	0.0	0.0	0.0	0.0	0.0	0.0	0.0	0.0
1984-85	0.0	0.0	0.0	0.0	0.0	0.0	1.4	0.3	0.0	0.0	0.0	0.0	1.7
1985-86	0.0	0.0	0.0	0.0	0.0	0.0							
Record Mean	0.0	0.0	0.0	0.0	T	T	0.2	0.2	T	0.0	0.0	0.0	0.4

See Reference Notes, relative to all above tables, on preceding page.

Lubbock is located on a plateau area of Northwestern Texas that is referred to locally as the South Plains Region. The general elevation of the area is about 3,250 feet. The Region is a major part of the Llano Estacado (staked plains). The latter, which includes a large portion of Northwest Texas, is bounded on the east and southeast by an erosional escarpment that is usually referred to as the Cap Rock. The Llano Estacado extends southwestward into the upper Pecos Valley and westward into eastern New Mexico.

The South Plains are predominately flat, but contain numerous small playas (or clay lined depressions) and small stream valleys. During the rainy months the playas collect run-off water and form small lakes or ponds. The stream valleys drain into the major rivers of West Texas, but throughout most of the year these streams carry only very light flows.

The escarpment, or Cap Rock, is the primary terrain feature that causes a noticeable distortion of the smooth wind flow patterns across the South Plains. The most noticeable influence is on southeasterly winds as they are deflected upward along the face of the escarpment.

The Lubbock area is the heart of the largest cotton-producing section of Texas. Grain sorghum production and cattle feeding make significant contributions to the agroeconomy of the area. Irrigation from underground sources is often used as a supplement to natural rainfall to improve crop yields. The soils of the region are sandy clay loams which consist of limy clays, silts, and sands of a reddish hue.

The area is semi-arid, transitional between the desert conditions on the west and the humid climates to the east and southeast. The greatest monthly rainfall totals occur from May through September when warm moist tropical air may be carried into the area from the Gulf of Mexico. This air mass often brings moderate to heavy afternoon and evening thunderstorms, accompanied by hail. Precipitation across the area is characterized by its variability. The monthly precipitation extremes range from trace amounts in several isolated months to 14 inches.

Snow may occur from late October until April. Each snowfall is generally light and seldom remains on the ground for more than two or three days at any one period.

High winds are associated primarily with intense thunderstorms and at times may cause significant damage to structures. Winds in excess of 25 mph occasionally occur for periods of 12 hours or longer. These prolonged winds are generally associated with late winter and springtime low-pressure centers. Spring winds often bring widespread dust causing discomfort to residents for periods of several hours.

Overall, the climate of the region is rated as pleasant. Most periods of disagreeable weather are of short duration. They generally occur from the winter months into the early summer months.

The summer heat is not considered oppressive. One moderating factor is a variable, but usually gentle, wind. Intrusions of dry air from the west often reduce any discomfort from the summer heat and lower temperatures into the 60s.

The average first occurrence of temperatures below 32 degrees Fahrenheit in the fall is the first of November and the average last occurrence in the spring is in mid April.

TABLE 1
NORMALS, MEANS AND EXTREMES

LUBBOCK, TEXAS

LATITUDE: 33°39'N   LONGITUDE: 101°49'W   ELEVATION: FT. GRND 3254 BARO 03258   TIME ZONE: CENTRAL   WBAN: 23042

	(a)	JAN	FEB	MAR	APR	MAY	JUNE	JULY	AUG	SEP	OCT	NOV	DEC	YEAR
TEMPERATURE °F:														
Normals														
-Daily Maximum		53.3	57.3	65.1	74.8	82.8	90.8	91.9	90.1	83.6	74.7	62.1	55.5	73.5
-Daily Minimum		24.3	27.9	35.2	45.8	55.2	64.3	67.6	65.7	58.7	47.3	34.8	27.4	46.2
-Monthly		38.8	42.6	50.2	60.3	69.0	77.6	79.8	77.9	71.2	61.0	48.5	41.5	59.8
Extremes														
-Record Highest	39	83	87	94	96	104	108	108	106	103	98	86	81	108
-Year		1972	1979	1971	1972	1947	1980	1983	1966	1948	1979	1980	1958	JUL 1983
-Record Lowest	39	-16	-8	2	22	30	44	51	52	33	23	-1	0	-16
-Year		1963	1960	1948	1948	1967	1947	1952	1956	1983	1980	1957	1983	JAN 1963
NORMAL DEGREE DAYS:														
Heating (base 65°F)		812	627	470	178	33	0	0	0	15	157	495	729	3516
Cooling (base 65°F)		0	0	11	37	157	378	459	400	201	33	0	0	1676
% OF POSSIBLE SUNSHINE	13	64	69	72	74	73	78	78	80	72	73	68	66	72
MEAN SKY COVER (tenths)														
Sunrise - Sunset	37	5.3	5.0	5.1	4.8	4.9	4.1	4.4	4.2	4.3	3.9	4.4	4.8	4.6
MEAN NUMBER OF DAYS:														
Sunrise to Sunset														
-Clear	39	12.5	11.2	11.6	12.8	11.4	13.6	14.1	15.5	14.7	16.5	14.6	13.5	162.1
-Partly Cloudy	39	6.4	7.4	8.7	8.3	11.3	10.9	11.0	9.7	7.8	6.7	7.1	7.1	102.4
-Cloudy	39	12.1	9.7	10.6	8.9	8.4	5.4	5.9	5.8	7.5	7.8	8.4	10.3	100.7
Precipitation														
.01 inches or more	39	3.8	3.9	3.9	4.4	7.4	6.9	6.8	6.5	5.6	5.2	3.4	3.7	61.6
Snow, Ice pellets														
1.0 inches or more	39	0.8	1.0	0.5	0.1	0.0	0.0	0.0	0.0	0.0	0.1	0.3	0.5	3.3
Thunderstorms	39	0.1	0.5	1.8	3.4	8.6	9.0	7.3	6.9	4.1	2.8	0.8	0.2	45.6
Heavy Fog Visibility														
1/4 mile or less	39	2.9	2.8	1.6	1.0	1.2	0.3	0.2	0.3	1.1	1.7	2.5	1.7	17.3
Temperature °F														
-Maximum														
90° and above	38	0.0	0.0	0.1	1.6	8.1	18.3	22.0	19.4	8.9	0.9	0.0	0.0	79.1
32° and below	38	3.3	1.2	0.4	0.0	0.0	0.0	0.0	0.0	0.0	0.0	0.2	1.5	6.7
-Minimum														
32° and below	38	25.8	19.6	11.1	2.0	0.1	0.0	0.0	0.0	0.0	0.0	12.0	23.8	95.3
0° and below	38	0.4	0.1	0.0	0.0	0.0	0.0	0.0	0.0	0.0	0.0	0.*	0.1	0.6
AVG. STATION PRESS. (mb)	13	905.2	904.2	900.8	901.4	900.9	902.6	904.9	905.1	905.1	905.3	904.7	905.0	903.8
RELATIVE HUMIDITY (%)														
Hour 00	38	66	63	56	54	61	62	61	64	68	67	65	65	63
Hour 06	38	73	72	68	68	75	77	74	77	79	78	74	72	74
Hour 12 (Local Time)	38	50	49	41	40	43	44	47	48	51	48	47	47	46
Hour 18	38	47	41	33	32	36	36	39	41	45	45	47	47	41
PRECIPITATION (inches):														
Water Equivalent														
-Normal		0.38	0.57	0.90	1.08	2.59	2.81	2.34	2.20	2.06	1.81	0.59	0.43	17.76
-Maximum Monthly	39	4.05	2.51	3.23	3.48	7.80	7.95	7.20	8.85	6.62	10.80	2.67	1.95	10.80
-Year		1949	1961	1958	1957	1949	1967	1976	1966	1974	1983	1968	1982	OCT 1983
-Minimum Monthly	39	0.00	T	T	0.10	0.10	0.32	T	0.05	T	0.00	0.00	T	0.00
-Year		1967	1955	1972	1961	1962	1973	1970	1960	1954	1952	1960	1973	JAN 1967
-Maximum in 24 hrs	39	1.56	2.15	1.80	2.18	5.14	5.70	3.25	3.78	2.80	5.82	1.57	1.12	5.82
-Year		1983	1961	1973	1982	1949	1967	1985	1966	1965	1983	1968	1959	OCT 1983
Snow, Ice pellets														
-Maximum Monthly	37	25.3	16.8	14.3	5.3						7.5	21.4	9.9	25.3
-Year		1983	1956	1958	1983						1976	1980	1960	JAN 1983
-Maximum in 24 hrs	37	16.3	12.1	10.0	4.5						4.7	10.8	6.3	16.3
-Year		1983	1961	1969	1983						1976	1980	1960	JAN 1983
WIND:														
Mean Speed (mph)	36	12.1	13.4	14.8	14.9	14.2	13.7	11.2	9.9	10.5	11.1	11.6	11.9	12.4
Prevailing Direction through 1963		SW	SW	SW	SW	S	S	S	S	S	S	WSW	WSW	S
Fastest Obs. 1 Min.														
-Direction (!!)	37	28	25	34	25	36	05	25	16	36	25	25	25	36
-Speed (MPH)	37	59	58	69	58	70	63	64	46	45	65	59	58	70
-Year		1965	1960	1957	1956	1952	1955	1950	1956	1953	1957	1955	1957	MAY 1952
Peak Gust														
-Direction (!!)	2	NE	N	W	S	SW	S	SW	N	NW	N	NW	W	S
-Speed (mph)	2	46	64	64	71	55	67	39	52	39	49	53	40	71
-Date		1985	1984	1984	1985	1985	1985	1985	1985	1985	1985	1985	1985	APR 1985

See Reference Notes to this table on the following page

TABLE 2 PRECIPITATION (inches) LUBBOCK, TEXAS

YEAR	JAN	FEB	MAR	APR	MAY	JUNE	JULY	AUG	SEP	OCT	NOV	DEC	ANNUAL
1956	0.01	1.59	T	0.36	1.80	3.26	0.69	1.06	0.03	1.73	T	0.30	10.83
1957	0.08	0.73	0.98	3.48	6.43	4.96	1.54	0.32	0.51	4.20	1.27	0.06	24.56
1958	1.35	0.33	3.23	1.97	2.94	0.71	2.65	0.21	2.90	0.94	0.34	0.02	17.59
1959	0.08	0.07	T	1.28	2.15	7.25	1.30	0.72	0.89	0.98	0.02	1.47	16.21
1960	0.66	0.94	0.61	0.26	1.16	5.72	5.37	0.05	0.34	5.83	0.00	1.25	22.19
1961	0.56	2.51	1.34	0.10	2.05	4.03	4.06	1.78	0.18	0.55	1.31	0.35	18.82
1962	0.26	0.02	0.10	1.20	0.10	2.56	4.85	1.31	4.17	2.66	0.45	0.67	18.35
1963	0.06	0.54	0.73	0.25	6.79	2.10	0.37	2.67	0.78	0.59	1.13	0.20	16.21
1964	0.45	0.16	0.64	0.11	1.67	5.00	0.82	1.14	2.46	0.30	0.57	0.90	14.22
1965	0.08	0.35	0.22	0.41	1.63	1.44	2.14	0.62	5.68	1.06	0.02	0.50	14.15
1966	0.52	0.06	0.13	3.03	0.67	2.27	0.57	8.85	2.18	T	0.11	0.03	18.42
1967	0.00	0.14	2.09	0.95	3.45	7.95	3.29	0.71	0.98	0.45	0.11	0.52	20.64
1968	0.94	0.82	2.77	0.58	2.01	1.81	3.14	2.72	0.67	0.81	2.67	0.48	19.42
1969	T	1.13	1.77	1.14	3.88	1.41	2.99	2.59	4.93	7.76	0.77	0.82	29.19
1970	T	0.11	2.15	0.26	4.30	1.36	T	1.18	1.80	1.34	0.05	0.08	12.63
1971	T	0.81	0.21	1.36	2.44	2.25	0.76	4.15	5.22	1.79	0.43	0.81	20.23
1972	0.16	0.13	T	0.35	3.20	5.37	4.47	5.40	2.95	1.75	0.97	0.32	25.07
1973	1.44	1.26	1.90	1.40	0.43	0.32	4.16	0.36	0.73	0.89	T	T	12.89
1974	0.08	0.01	1.56	0.82	1.23	1.11	2.22	5.14	6.62	3.89	0.89	0.44	24.01
1975	0.41	1.53	0.04	0.45	2.74	1.80	4.32	2.21	2.61	0.06	1.18	0.34	17.69
1976	T	0.03	0.24	1.76	1.19	2.46	7.20	1.99	3.28	1.39	0.56	0.01	20.11
1977	0.24	0.38	0.82	2.90	2.46	2.28	1.13	4.31	0.49	1.11	0.02	0.01	16.15
1978	0.59	1.39	0.23	0.21	3.20	1.93	0.15	0.34	3.29	1.06	1.11	0.17	13.67
1979	0.33	0.85	2.95	1.17	4.00	3.69	1.84	3.81	0.21	0.59	0.09	1.29	20.82
1980	0.54	0.38	0.19	1.13	3.46	1.78	0.20	1.64	3.55	0.19	2.29	0.51	15.86
1981	0.32	0.67	1.19	2.05	1.25	0.79	3.35	5.41	1.78	5.34	0.64	0.20	22.99
1982	0.05	0.39	0.44	2.53	4.54	4.99	2.08	1.08	1.29	0.48	1.18	1.95	21.00
1983	2.75	0.32	0.55	0.77	1.23	1.79	0.41	0.32	0.39	10.80	0.54	0.36	20.23
1984	0.03	0.17	0.23	0.23	0.45	4.32	0.53	3.72	0.15	1.74	1.87	1.18	14.62
1985	0.38	0.27	1.19	0.48	2.97	4.51	3.94	0.63	4.73	3.60	0.27	0.18	23.15
Record Mean	0.52	0.53	0.82	1.09	2.80	2.82	2.25	2.09	2.06	2.03	0.61	0.46	18.07

TABLE 3 AVERAGE TEMPERATURE (deg. F) LUBBOCK, TEXAS

YEAR	JAN	FEB	MAR	APR	MAY	JUNE	JULY	AUG	SEP	OCT	NOV	DEC	ANNUAL
1956	40.5	38.5	51.8	58.4	72.7	80.0	79.9	78.2	73.7	63.7	45.3	44.0	60.6
1957	40.1	49.7	49.6	55.8	64.8	75.1	81.4	79.0	68.6	58.3	42.7	44.1	59.1
1958	36.8	40.6	41.4	55.4	68.9	79.3	80.9	79.0	71.9	59.5	49.9	39.6	58.6
1959	37.9	42.5	49.0	58.6	70.0	76.6	77.2	79.2	72.9	58.4	44.4	42.3	59.1
1960	37.5	36.0	45.9	62.1	68.9	78.7	77.7	78.9	72.5	62.1	50.8	36.6	59.0
1961	35.6	41.5	50.5	60.3	69.5	75.0	76.3	76.0	69.9	61.2	44.7	39.3	58.3
1962	34.4	48.4	47.6	60.4	74.6	74.8	78.7	78.2	70.8	62.7	50.5	42.4	60.3
1963	32.8	42.5	53.0	64.3	70.5	76.2	81.9	79.0	72.9	65.8	50.7	36.7	60.5
1964	38.8	36.5	49.2	61.5	71.3	76.7	82.0	79.9	70.9	62.1	50.5	42.8	60.2
#1965	44.3	40.4	42.8	63.6	71.1	77.5	81.7	78.0	72.9	63.5	57.3	47.0	61.7
1966	34.2	40.5	55.7	60.4	69.8	78.5	85.4	76.8	71.6	60.9	55.6	38.0	60.6
1967	40.7	41.5	55.8	64.1	66.0	75.2	77.0	74.9	68.6	60.5	49.3	36.8	59.2
1968	40.7	38.5	48.5	55.3	65.5	74.5	75.3	74.2	66.7	61.9	46.6	39.9	57.3
1969	44.9	43.3	41.5	61.8	68.4	76.9	83.3	79.6	70.2	56.4	48.6	43.1	59.9
1970	37.5	45.5	46.2	58.5	68.7	75.7	80.0	79.0	71.5	56.7	48.7	45.1	59.5
1971	41.4	42.7	51.0	59.7	68.8	78.4	80.0	73.4	70.4	61.9	50.8	43.7	60.2
1972	41.3	45.9	56.3	65.7	67.4	76.9	76.5	74.4	71.0	59.6	42.4	38.0	59.6
1973	34.9	40.2	51.4	55.1	67.8	76.9	77.3	77.7	70.3	63.9	53.5	42.2	59.3
1974	41.3	45.6	58.5	63.1	75.1	78.6	80.4	74.4	64.0	59.9	44.8	41.0	60.8
1975	40.8	41.0	49.5	58.7	67.6	77.4	75.4	77.5	67.5	62.5	50.3	42.8	59.2
1976	39.6	51.2	52.1	62.4	66.0	77.1	75.0	77.9	70.0	54.1	42.6	40.3	59.0
1977	34.5	46.2	51.9	60.4	71.8	79.3	80.3	79.5	77.4	63.3	52.3	45.0	61.8
1978	32.1	33.9	51.9	65.1	70.1	79.0	82.9	78.2	72.0	62.1	49.9	38.1	59.6
1979	31.7	41.9	52.9	61.8	69.2	76.7	81.8	77.4	73.2	64.6	45.7	42.5	59.9
1980	40.3	44.3	51.3	59.0	68.7	83.1	84.3	80.6	73.0	60.2	46.0	45.7	61.3
1981	41.7	44.7	51.3	64.0	68.1	79.6	81.7	76.2	70.5	59.7	53.4	44.0	61.2
1982	39.7	42.4	49.3	59.3	67.8	73.5	80.1	81.0	74.4	60.6	48.6	38.3	60.0
1983	32.5	42.8	51.0	54.8	65.9	74.0	80.6	80.4	74.7	63.6	52.4	31.7	58.7
1984	37.9	45.3	49.4	58.0	71.2	76.9	78.3	78.2	69.3	59.5	49.5	43.9	59.8
1985	35.6	41.9	52.3	63.0	70.0	75.5	79.7	81.5	71.4	61.2	49.8	38.0	60.0
Record Mean	38.4	42.7	50.1	60.2	68.8	77.3	79.8	78.1	71.3	61.1	48.7	41.2	59.8
Max	52.5	57.3	64.9	74.7	82.7	90.6	92.0	90.3	83.7	74.6	62.5	55.0	73.4
Min	24.3	28.1	35.2	45.7	55.0	64.1	67.5	65.8	58.8	47.6	34.9	27.4	46.2

REFERENCE NOTES FOR TABLES 1, 2, 3 and 6   (LUBBOCK, TX)

GENERAL

T - TRACE AMOUNT
BLANK ENTRIES DENOTE MISSING/UNREPORTED DATA.
# INDICATES A STATION OR INSTRUMENT RELOCATION.

SPECIFIC

TABLE 1

(a) - LENGTH OF RECORD IN YEARS. ALTHOUGH INDIVIDUAL MONTHS MAY BE MISSING.

* LESS THAN .05

NORMALS — BASED ON THE 1951-1980 RECORD PERIOD.
EXTREMES — DATES ARE THE MOST RECENT OCCURRENCE.
WIND DIR. — NUMERALS SHOW TENS OF DEGREES CLOCKWISE FROM TRUE NORTH. "00" INDICATES CALM.
RESULTANT WIND DIRECTIONS ARE GIVEN TO WHOLE DEGREES.

EXCEPTIONS

TABLES 2, 3, and 6

RECORD MEANS ARE THROUGH THE CURRENT YEAR, BEGINNING IN
1947 FOR TEMPERATURE
1947 FOR PRECIPITATION
1949 FOR SNOWFALL

TABLE 4 HEATING DEGREE DAYS Base 65 deg. F LUBBOCK, TEXAS

SEASON	JULY	AUG	SEP	OCT	NOV	DEC	JAN	FEB	MAR	APR	MAY	JUNE	TOTAL
1956-57	0	0	1	101	584	646	766	421	473	289	67	22	3370
1957-58	0	0	19	228	661	640	868	677	726	295	33	0	4147
1958-59	0	0	24	197	446	779	835	622	490	246	35	1	3675
1959-60	0	0	31	223	609	695	846	833	589	142	57	0	4025
1960-61	0	0	8	135	421	872	906	653	444	210	30	5	3684
1961-62	0	0	24	150	602	790	942	457	535	176	13	3	3692
1962-63	0	0	18	143	428	692	988	627	378	92	36	0	3402
1963-64	0	0	11	27	425	872	808	821	483	139	13	4	3603
1964-65	0	0	35	111	433	677	635	683	682	107	12	0	3375
#1965-66	0	0	22	94	235	551	946	677	307	166	48	0	3046
1966-67	0	8	3	170	283	827	748	652	289	85	84	3	3152
1967-68	0	4	17	190	463	866	748	762	507	293	73	4	3927
1968-69	0	1	31	151	543	770	614	601	723	126	45	7	3612
1969-70	0	0	3	312	486	673	846	539	577	198	47	21	3702
1970-71	0	0	50	286	484	608	723	618	438	186	30	0	3423
1971-72	0	0	83	124	421	656	728	549	275	94	40	0	2970
1972-73	2	0	23	220	672	831	928	688	416	312	58	0	4150
1973-74	0	0	31	99	340	703	726	536	223	127	11	0	2796
1974-75	0	0	105	166	500	753	744	664	474	233	18	4	3661
1975-76	0	0	76	122	436	680	780	397	392	116	57	0	3056
1976-77	0	0	30	333	665	760	939	520	397	153	1	0	3798
1977-78	0	0	0	80	373	611	1014	864	419	75	64	0	3500
1978-79	0	0	31	129	447	827	1023	640	369	134	45	9	3654
1979-80	0	0	9	104	570	690	756	592	436	205	48	0	3410
1980-81	0	0	21	190	565	590	712	563	420	108	33	0	3202
1981-82	0	0	16	202	341	643	777	635	351	198	45	2	3210
1982-83	0	0	0	189	485	821	1001	613	430	326	68	13	3946
1983-84	0	0	18	107	371	1025	836	566	477	216	20	0	3636
1984-85	0	0	80	194	460	646	908	639	392	104	26	3	3452
1985-86	0	0	65	133	448	832							

TABLE 5 COOLING DEGREE DAYS Base 65 deg. F LUBBOCK, TEXAS

YEAR	JAN	FEB	MAR	APR	MAY	JUNE	JULY	AUG	SEP	OCT	NOV	DEC	TOTAL
1969	0	0	0	36	160	370	575	461	165	56	0	0	1823
1970	0	0	0	11	167	348	498	440	249	32	2	0	1747
1971	0	0	14	35	157	408	476	269	252	34	4	0	1649
1972	0	0	14	120	119	362	366	297	211	63	0	0	1552
1973	0	0	0	19	153	364	387	401	197	71	2	0	1594
1974	0	0	30	78	330	413	491	300	80	18	0	0	1740
1975	0	0	0	49	107	383	332	392	155	49	0	0	1467
1976	0	1	0	45	93	369	317	406	187	4	0	0	1422
1977	0	0	0	19	221	435	482	455	379	36	0	0	2027
1978	0	0	12	85	230	426	562	417	248	46	0	0	2026
1979	0	0	0	44	183	371	524	394	264	98	0	0	1878
1980	0	0	0	32	172	549	605	488	272	48	3	0	2169
1981	0	0	3	84	137	443	524	356	187	42	0	0	1776
1982	0	5	11	34	141	265	476	504	289	58	0	0	1783
1983	0	0	2	27	101	291	490	486	313	70	2	0	1782
1984	0	0	0	13	219	363	420	416	217	30	1	0	1679
1985	0	0	6	52	187	324	462	518	264	22	0	0	1835

TABLE 6 SNOWFALL (inches) LUBBOCK, TEXAS

SEASON	JULY	AUG	SEP	OCT	NOV	DEC	JAN	FEB	MAR	APR	MAY	JUNE	TOTAL
1956-57	0.0	0.0	0.0	0.0	T	0.0	T	T	2.4	0.1	0.0	0.0	2.5
1957-58	0.0	0.0	0.0	T	3.5	0.0	8.1	1.2	14.3	0.0	0.0	0.0	27.1
1958-59	0.0	0.0	0.0	0.0	0.1	0.2	0.4	1.0	T	0.0	0.0	0.0	1.7
1959-60	0.0	0.0	0.0	0.0	T	T	3.6	9.8	0.2	0.0	0.0	0.0	13.6
1960-61	0.0	0.0	0.0	0.0	0.0	9.9	3.9	13.7	6.9	T	0.0	0.0	34.4
1961-62	0.0	0.0	0.0	0.0	0.5	0.7	0.3	T	0.1	0.0	0.0	0.0	1.6
1962-63	0.0	0.0	0.0	0.0	T	1.4	2.3	8.0	T	0.0	0.0	0.0	11.7
1963-64	0.0	0.0	0.0	0.0	1.5	0.0	T	3.3	0.2	0.0	0.0	0.0	5.0
1964-65	0.0	0.0	0.0	0.0	0.0	0.1	0.2	2.0	1.2	0.0	0.0	0.0	3.5
1965-66	0.0	0.0	0.0	0.0	0.0	0.7	4.9	0.4	0.0	0.0	0.0	0.0	6.0
1966-67	0.0	0.0	0.0	0.0	T	T	0.0	0.7	T	0.0	0.0	0.0	0.7
1967-68	0.0	0.0	0.0	0.2	T	3.0	0.2	3.8	9.7	0.0	0.0	0.0	16.9
1968-69	0.0	0.0	0.0	0.0	1.2	0.6	0.0	0.1	11.7	0.0	0.0	0.0	13.6
1969-70	0.0	0.0	0.0	0.0	3.0	4.8	T	0.4	6.2	0.0	0.0	0.0	14.4
1970-71	0.0	0.0	0.0	0.0	0.1	0.0	T	5.2	2.1	0.0	0.0	0.0	7.4
1971-72	0.0	0.0	0.0	0.0	T	6.5	2.2	1.0	T	0.0	0.0	0.0	9.7
1972-73	0.0	0.0	0.0	T	5.9	0.4	9.4	9.6	0.3	0.0	0.0	0.0	25.6
1973-74	0.0	0.0	0.0	0.0	T	0.0	T	T	0.2	0.0	0.0	0.0	0.2
1974-75	0.0	0.0	0.0	0.0	0.0	1.1	1.7	3.8	0.7	T	0.0	0.0	7.3
1975-76	0.0	0.0	0.0	0.0	0.0	3.4	T	T	T	T	0.0	0.0	3.4
1976-77	0.0	0.0	0.0	7.5	9.1	0.1	2.0	1.7	T	0.0	0.0	0.0	20.4
1977-78	0.0	0.0	0.0	0.0	T	T	5.7	10.2	0.7	0.0	0.0	0.0	16.6
1978-79	0.0	0.0	0.0	0.0	0.5	1.7	0.8	8.6	0.6	0.0	0.0	0.0	12.2
1979-80	0.0	0.0	0.0	0.0	0.1	5.0	3.6	2.9	T	1.4	0.0	0.0	13.0
1980-81	0.0	0.0	0.0	0.0	21.4	T	3.5	0.1	T	0.0	0.0	0.0	25.0
1981-82	0.0	0.0	0.0	0.0	0.0	0.0	0.5	2.7	T	0.0	0.0	0.0	3.2
1982-83	0.0	0.0	0.0	0.0	0.4	6.8	25.3	3.4	T	5.3	0.0	0.0	41.2
1983-84	0.0	0.0	0.0	0.0	0.0	2.3	0.5	1.7	T	T	0.0	0.0	4.5
1984-85	0.0	0.0	0.0	0.0	T	0.1	2.0	T	T	0.0	0.0	0.0	2.1
1985-86	0.0	0.0	0.0	0.0	0.0	1.7							
Record Mean	0.0	0.0	0.0	0.2	1.3	1.6	2.5	3.1	1.6	0.2	0.0	0.0	10.5

See Reference Notes, relative to all above tables, on preceding page.

The Midland-Odessa region is on the southern extension of the South Plains of Texas. The terrain is level with only slight occasional undulations.

The climate is typical of a semi-arid region. The vegetation of the area consists mostly of native grasses and a few trees, mostly of the mesquite variety.

Most of the annual precipitation in the area comes as a result of very violent spring and early summer thunderstorms. These are usually accompanied by excessive rainfall, over limited areas, and sometimes hail. Due to the flat nature of the countryside, local flooding occurs, but is of short duration. Tornadoes are occasionally sighted.

During the late winter and early spring months, blowing dust occurs frequently. The flat plains of the area with only grass as vegetation offer little resistance to the strong winds. The sky is occasionally obscured by dust but in most storms visibilities range from 1 to 3 miles.

Daytime temperatures are quite hot in the summer, but there is a large diurnal range of temperature and most nights are comfortable. The temperature drops below 32 degrees in the fall about mid-November and the last temperature below 32 degrees in spring comes early in April.

Winters are characterized by frequent cold periods followed by rapid warming. Cold frontal passages are followed by chilly weather for two or three days. Cloudiness is at a minimum. Summers are hot and dry with numerous small convective showers.

The prevailing wind direction in this area is from the southeast. This, together with the upslope of the terrain from the same direction, causes occasional low cloudiness and drizzle during winter and spring months. Snow is infrequent. Maximum temperatures during the summer months frequently are from 2 to 6 degrees cooler than those at places 100 miles southeast, due to the cooling effect of the upslope winds.

Very low humidities are conducive to personal comfort, because even though summer afternoon temperatures are frequently above 90 degrees, the low humidity with resultant rapid evaporation, has a cooling effect. The climate of the area is generally quite pleasant with the most disagreeable weather concentrated in the late winter and spring months.

TABLE 1   NORMALS, MEANS AND EXTREMES

MIDLAND – ODESSA, TEXAS

LATITUDE: 31°57'N   LONGITUDE: 102°11'W   ELEVATION: FT. GRND 2851 BARO 02865   TIME ZONE: CENTRAL   WBAN: 23023

	(a)	JAN	FEB	MAR	APR	MAY	JUNE	JULY	AUG	SEP	OCT	NOV	DEC	YEAR
TEMPERATURE °F:														
Normals														
-Daily Maximum		57.6	62.1	69.8	78.8	86.0	93.0	94.2	93.1	86.4	77.7	65.5	59.7	77.0
-Daily Minimum		29.7	33.3	40.2	49.4	58.2	66.6	69.2	68.0	61.9	51.1	39.0	32.2	49.9
-Monthly		43.7	47.7	55.0	64.1	72.1	79.8	81.7	80.6	74.2	64.4	52.3	46.0	63.4
Extremes														
-Record Highest	38	84	87	95	99	107	109	106	107	107	100	88	85	109
-Year		1974	1950	1971	1972	1953	1951	1983	1964	1953	1979	1963	1954	JUN 1951
-Record Lowest	38	-8	-11	9	20	34	47	53	56	40	27	13	6	-11
-Year		1962	1985	1980	1973	1970	1983	1978	1967	1983	1980	1976	1983	FEB 1985
NORMAL DEGREE DAYS:														
Heating (base 65°F)		660	484	329	102	8	0	0	0	7	94	385	589	2658
Cooling (base 65°F)		0	0	19	75	228	444	518	484	283	75	0	0	2126
% OF POSSIBLE SUNSHINE	5	64	69	71	75	78	78	84	78	80	66	72	67	74
MEAN SKY COVER (tenths)														
Sunrise - Sunset	37	5.2	4.9	4.9	4.6	4.6	3.9	4.4	4.3	4.3	3.9	4.3	4.7	4.5
MEAN NUMBER OF DAYS:														
Sunrise to Sunset														
-Clear	37	12.5	11.6	13.1	13.6	13.5	15.1	13.6	14.3	14.5	16.8	14.7	14.2	167.4
-Partly Cloudy	37	6.3	7.0	7.6	7.7	9.3	9.4	10.8	10.5	7.7	6.2	6.5	6.3	95.4
-Cloudy	37	12.2	9.6	10.3	8.7	8.2	5.5	6.6	6.2	7.8	8.0	8.8	10.5	102.5
Precipitation														
.01 inches or more	38	3.8	3.8	2.5	3.4	5.7	4.9	5.2	5.4	5.5	4.8	3.1	3.1	51.1
Snow,Ice pellets														
1.0 inches or more	38	0.7	0.5	0.1	0.0	0.0	0.0	0.0	0.0	0.0	0.0	0.2	0.3	1.8
Thunderstorms	37	0.1	0.4	1.1	2.9	6.6	5.8	6.1	5.9	3.6	2.5	0.6	0.4	36.0
Heavy Fog Visibility														
1/4 mile or less	37	3.4	2.8	0.9	0.5	0.5	0.1	0.1	0.1	0.6	1.3	2.6	2.9	15.6
Temperature °F														
-Maximum														
90° and above	22	0.0	0.0	0.3	2.9	11.2	21.0	26.4	22.6	11.3	1.9	0.0	0.0	97.5
32° and below	22	1.6	0.3	0.*	0.0	0.0	0.0	0.0	0.0	0.0	0.0	0.1	0.9	3.0
-Minimum														
32° and below	22	19.9	13.9	5.6	0.7	0.0	0.0	0.0	0.0	0.0	0.1	6.1	17.1	63.5
0° and below	22	0.*	0.1	0.0	0.0	0.0	0.0	0.0	0.0	0.0	0.0	0.0	0.0	0.1
AVG. STATION PRESS.(mb)	13	918.3	917.3	913.6	913.8	913.0	914.5	916.6	916.9	917.0	917.7	917.5	918.2	916.2
RELATIVE HUMIDITY (%)														
Hour 00	22	64	62	53	52	60	59	55	59	68	71	69	64	61
Hour 06 (Local Time)	22	71	72	65	66	75	76	71	73	79	80	76	71	73
Hour 12	22	48	44	35	34	38	41	40	43	50	47	46	44	43
Hour 18	22	42	35	28	27	31	32	33	36	44	44	45	43	37
PRECIPITATION (inches):														
Water Equivalent														
-Normal		0.42	0.58	0.51	0.84	2.05	1.44	1.72	1.60	2.08	1.41	0.60	0.45	13.70
-Maximum Monthly	38	3.66	1.79	2.86	2.85	4.99	3.93	7.73	4.43	9.70	5.93	2.32	2.80	9.70
-Year		1949	1969	1970	1949	1959	1949	1975	1974	1980	1985	1968	1979	SEP 1980
-Minimum Monthly	38	0.00	0.01	T	0.00	0.08	0.05	T	0.17	0.08	0.00	0.00	T	0.00
-Year		1967	1971	1984	1964	1953	1951	1983	1967	1979	1952	1950	1958	JAN 1967
-Maximum in 24 hrs	32	1.15	1.22	2.20	1.62	4.75	2.54	5.99	2.41	2.75	3.59	2.16	1.58	5.99
-Year		1958	1965	1970	1979	1968	1969	1961	1965	1980	1985	1975	1979	JUL 1961
Snow,Ice pellets														
-Maximum Monthly	38	9.0	3.9	5.9	0.5						T	7.2	6.4	9.0
-Year		1985	1973	1970	1983						1980	1980	1982	JAN 1985
-Maximum in 24 hrs	32	6.8	3.9	5.0	0.5						T	5.7	3.5	6.8
-Year		1974	1985	1970	1983						1980	1980	1982	JAN 1974
WIND:														
Mean Speed (mph)	32	10.3	11.3	12.8	12.9	12.4	12.3	10.6	10.0	10.2	10.1	10.2	10.1	11.1
Prevailing Direction														
through 1963		S	SW	S	SSE	SSE	SSE	SSE	SE	SSE	S	S	SW	SSE
Fastest Obs. 1 Min.														
-Direction (!!!)	32	27	25	27	30	20	24	05	20	23	14	30	26	25
-Speed (MPH)	32	41	67	53	53	52	58	58	40	53	46	45	47	67
-Year		1965	1960	1984	1983	1958	1966	1979	1979	1985	1983	1973	1977	FEB 1960
Peak Gust														
-Direction (!!!)	2	NE	NW	W	W	26	SW	W	N	SW	N	SW	S	SW
-Speed (mph)	2	47	52	74	67	52	61	67	45	82	48	46	38	82
-Date		1985	1984	1984	1984	1985	1985	1984	1984	1985	1985	1985	1984	SEP 1985

See reference Notes to this table on the following page.

TABLE 2 PRECIPITATION (inches)   MIDLAND – ODESSA, TEXAS

YEAR	JAN	FEB	MAR	APR	MAY	JUNE	JULY	AUG	SEP	OCT	NOV	DEC	ANNUAL
1956	0.16	0.16	0.04	1.15	1.41	1.40	1.71	1.73	0.17	1.42	T	0.27	9.62
1957	0.39	1.15	0.09	0.44	4.30	2.55	0.40	1.19	1.56	1.78	1.56	0.17	15.58
1958	1.97	1.57	0.83	0.78	3.24	0.50	0.89	3.84	3.42	2.40	0.88	T	20.32
1959	0.06	0.62	0.16	0.39	4.99	1.89	4.46	0.90	2.48	2.75	0.68	1.01	20.39
1960	1.17	0.60	0.17	0.28	1.55	0.42	3.33	2.59	0.14	2.29	0.16	1.88	14.58
1961	1.33	0.43	1.12	0.02	2.63	2.96	6.73	0.21	3.12	0.03	1.63	0.33	20.54
1962	0.17	0.03	0.48	0.79	0.14	1.83	2.12	2.66	4.06	0.87	0.38	0.58	14.11
1963	0.01	0.75	T	0.95	2.62	2.61	1.17	1.24	0.39	0.17	0.77	0.28	10.96
1964	0.40	0.27	0.64	0.00	2.14	0.40	0.24	0.58	2.19	0.19	0.20	0.40	7.65
1965	0.06	1.71	0.06	0.33	3.04	0.90	0.06	3.23	0.78	1.41	0.29	0.29	12.16
1966	0.48	0.20	0.41	2.44	2.86	2.91	0.17	3.37	3.29	0.80	0.04	0.02	16.99
1967	0.00	0.03	1.45	0.44	0.56	1.17	1.64	0.17	1.59	0.20	0.44	0.78	8.47
1968	1.07	1.06	1.62	1.59	4.96	1.42	1.01	2.10	1.39	0.19	2.32	0.18	18.51
1969	0.02	1.79	0.51	2.10	3.28	2.68	0.74	0.32	1.53	2.52	0.85	0.60	16.94
1970	0.01	1.40	2.86	0.29	0.18	0.99	0.23	0.87	2.00	0.78	T	0.11	9.72
1971	T	0.01	T	0.81	2.77	1.07	0.79	3.27	2.55	0.67	0.42	0.24	12.60
1972	0.37	0.04	T	0.29	1.20	1.70	0.58	3.80	1.05	2.26	0.22	0.15	11.66
1973	0.93	1.63	1.73	1.38	0.87	0.34	2.47	0.25	1.09	0.59	T	0.00	11.28
1974	0.54	0.28	0.18	1.17	0.34	0.75	0.26	4.43	6.16	5.42	0.72	0.25	20.50
1975	0.81	0.69	0.08	0.09	3.44	2.05	7.73	1.68	3.44	0.70	2.16	0.37	23.24
1976	0.03	0.18	0.30	2.02	1.08	2.14	3.56	0.84	1.25	1.84	0.26	0.08	13.58
1977	0.63	0.55	0.61	1.22	0.60	0.81	0.40	0.37	0.09	1.27	0.02	0.27	6.84
1978	0.22	0.34	T	0.09	1.97	1.15	2.51	1.01	5.02	2.51	2.27	0.00	17.29
1979	0.16	0.26	0.81	1.63	1.14	2.98	3.05	2.17	0.08	0.97	T	2.80	16.05
1980	0.49	0.29	T	0.86	1.85	1.59	T	0.93	9.70	0.12	0.78	1.15	17.76
1981	0.56	0.67	0.56	2.15	2.21	0.46	0.40	3.33	1.68	5.53	T	0.05	17.60
1982	0.42	0.16	0.01	1.08	3.16	1.45	3.24	0.67	1.43	1.01	0.62	1.40	14.65
1983	1.14	0.26	0.21	0.06	0.50	0.19	T	0.43	0.62	4.64	1.62	0.33	10.00
1984	0.41	0.17	T	0.01	3.40	2.15	0.84	2.45	1.89	1.99	2.31	0.83	16.45
1985	0.84	0.59	0.63	0.21	0.99	2.23	0.90	0.81	3.15	5.93	0.08	0.05	16.41
Record Mean	0.64	0.58	0.46	0.85	2.01	1.61	1.74	1.57	2.02	1.68	0.61	0.58	14.35

TABLE 3 AVERAGE TEMPERATURE (deg. F)   MIDLAND – ODESSA, TEXAS

YEAR	JAN	FEB	MAR	APR	MAY	JUNE	JULY	AUG	SEP	OCT	NOV	DEC	ANNUAL
1956	45.2	46.4	57.0	63.0	76.4	83.0	82.5	81.2	76.6	67.6	50.2	47.5	64.7
1957	44.9	54.2	55.4	61.0	69.5	77.9	83.3	82.4	73.4	61.9	47.2	48.5	63.3
1958	42.0	44.5	48.0	60.2	71.8	81.5	84.2	82.0	74.7	61.5	53.6	43.4	62.3
#1959	42.7	46.8	53.9	62.1	73.3	78.8	78.1	81.2	76.3	63.0	47.1	46.8	62.5
1960	43.3	42.6	51.3	65.3	72.3	82.4	80.6	80.8	75.5	66.9	54.7	40.4	63.0
1961	40.2	47.9	55.4	64.4	73.7	77.5	78.0	78.3	73.3	65.6	48.7	44.6	62.3
1962	39.1	54.1	52.4	65.0	77.4	78.9	82.2	83.0	74.5	65.6	54.9	45.6	64.5
#1963	39.8	47.4	58.7	69.5	74.4	79.2	83.5	82.2	76.2	71.1	56.0	40.2	64.8
1964	43.6	42.7	55.6	67.1	75.7	83.5	86.9	87.2	76.5	66.6	56.6	49.0	65.9
1965	47.6	43.2	47.7	67.4	71.7	79.0	82.8	79.6	74.9	63.9	57.4	49.7	63.7
1966	38.0	42.7	57.2	61.9	70.4	78.9	84.1	78.9	72.8	61.2	56.8	43.4	62.1
1967	43.6	46.4	60.5	68.7	72.3	79.9	80.7	77.7	71.1	64.0	54.0	41.9	63.4
1968	44.3	44.7	53.5	60.8	70.8	77.5	78.5	79.2	72.6	67.0	49.6	43.3	61.8
1969	47.6	48.1	47.5	65.2	69.3	78.2	84.1	82.7	74.2	61.0	51.4	47.6	63.1
1970	42.0	47.6	49.8	61.5	69.9	77.5	82.2	79.9	73.4	61.9	53.1	50.7	62.4
1971	47.4	47.2	55.8	62.8	72.2	78.4	80.7	74.4	71.6	64.0	53.3	47.8	63.0
1972	44.2	49.3	60.0	69.7	70.4	78.9	79.2	75.7	72.5	62.5	46.4	43.5	62.7
1973	38.9	43.8	54.4	57.1	68.8	77.0	78.1	78.9	72.0	65.0	56.6	46.7	61.4
1974	45.2	49.5	62.7	66.1	77.5	81.1	83.4	79.5	67.8	64.3	53.6	45.6	64.7
1975	45.9	48.0	54.9	63.0	71.8	80.3	77.8	79.0	70.6	64.9	53.5	46.7	63.1
1976	43.9	56.2	57.0	64.6	68.5	78.1	75.8	78.8	71.7	58.4	46.6	44.3	62.0
1977	40.5	50.2	54.8	61.6	76.4	82.1	83.6	85.4	82.1	67.0	55.0	49.4	65.7
1978	38.2	43.8	58.1	67.8	72.1	78.1	81.9	78.9	71.4	61.3	52.3	42.6	62.2
1979	35.9	45.3	55.4	62.6	70.7	77.4	82.1	79.0	75.1	68.1	49.5	46.4	62.3
1980	44.9	46.3	52.4	59.1	69.1	81.9	84.1	80.8	72.7	60.2	47.2	46.2	62.1
1981	43.2	48.6	52.4	63.9	70.2	79.1	83.6	80.2	75.7	65.6	57.0	47.8	64.0
1982	43.6	45.9	57.5	63.8	70.7	79.1	82.3	82.8	77.4	64.2	52.2	43.4	63.6
1983	41.8	48.0	55.9	59.5	71.5	79.3	83.7	84.1	77.8	67.8	55.0	37.5	63.5
1984	40.1	48.2	55.0	63.6	74.1	79.8	80.9	80.2	71.5	61.7	50.7	48.0	62.8
1985	37.5	44.3	56.7	66.9	74.8	77.3	81.2	84.1	74.2	63.7	54.6	41.1	63.0
Record Mean	43.3	47.8	55.0	63.9	72.1	79.7	81.5	80.8	74.4	64.9	52.5	45.8	63.4
Max	56.6	61.8	69.7	78.5	85.8	92.8	94.2	93.3	86.6	77.8	65.8	58.8	76.8
Min	29.9	33.7	40.2	49.3	58.4	66.5	68.8	68.3	62.2	51.9	39.2	32.1	50.0

REFERENCE NOTES FOR TABLES 1, 2, 3 and 6   (MIDLAND-ODESSA, TX)

GENERAL

T - TRACE AMOUNT
BLANK ENTRIES DENOTE MISSING/UNREPORTED DATA.
# INDICATES A STATION OR INSTRUMENT RELOCATION.

SPECIFIC

TABLE 1

(a) - LENGTH OF RECORD IN YEARS. ALTHOUGH
INDIVIDUAL MONTHS MAY BE MISSING.

* LESS THAN .05

NORMALS — BASED ON THE 1951-1980 RECORD PERIOD.
EXTREMES — DATES ARE THE MOST RECENT OCCURRENCE.
WIND DIR. — NUMERALS SHOW TENS OF DEGREES
CLOCKWISE FROM TRUE NORTH.
"00" INDICATES CALM.
RESULTANT WIND DIRECTIONS ARE GIVEN TO WHOLE DEGREES.

EXCEPTIONS

TABLES 2, 3, and 6

RECORD MEANS ARE THROUGH THE CURRENT YEAR,
BEGINNING IN   1930 FOR TEMPERATURE
1930 FOR PRECIPITATION
1949 FOR SNOWFALL

TABLE 4 HEATING DEGREE DAYS Base 65 deg. F MIDLAND – ODESSA, TEXAS

SEASON	JULY	AUG	SEP	OCT	NOV	DEC	JAN	FEB	MAR	APR	MAY	JUNE	TOTAL
1956-57	0	0	0	54	439	536	617	301	293	175	25	10	2450
1957-58	0	0	2	155	530	504	709	566	522	176	16	0	3180
1958-59	0	0	10	161	338	664	686	501	341	179	6	0	2886
#1959-60	0	0	9	110	534	556	666	643	426	84	28	0	3056
1960-61	0	0	2	64	307	758	761	474	293	124	5	4	2792
1961-62	0	0	12	66	481	625	796	309	394	96	7	0	2786
#1962-63	0	0	2	74	299	595	773	487	233	45	19	0	2527
1963-64	0	0	1	3	274	761	656	643	287	63	5	0	2693
1964-65	0	0	6	34	283	493	535	604	529	52	5	0	2541
1965-66	0	0	11	96	232	468	831	621	258	124	46	0	2687
1966-67	0	3	4	159	242	663	654	516	176	23	22	0	2462
1967-68	0	0	9	95	324	708	634	582	360	143	14	0	2869
1968-69	0	0	0	60	456	665	535	467	534	52	30	0	2799
1969-70	0	0	0	216	405	532	706	482	463	137	48	5	2994
1970-71	0	0	41	150	348	441	540	493	297	119	15	0	2444
1971-72	0	0	51	71	350	528	639	446	176	37	12	0	2310
1972-73	0	0	19	158	551	660	804	587	320	249	55	0	3403
1973-74	0	0	15	81	252	561	607	429	121	68	1	0	2135
1974-75	0	0	44	65	344	594	582	468	314	135	2	3	2551
1975-76	0	0	40	82	340	560	652	258	261	73	38	0	2304
1976-77	0	0	10	220	545	635	755	409	320	117	0	0	3011
1977-78	0	0	0	47	295	478	823	586	245	49	30	0	2553
1978-79	0	0	36	138	374	689	893	546	297	106	34	3	3116
1979-80	0	0	3	63	459	566	619	535	384	184	34	0	2847
1980-81	0	0	20	173	531	575	669	452	383	99	20	0	2922
1981-82	0	0	0	109	234	527	655	535	245	112	21	0	2438
1982-83	0	0	0	128	378	661	711	469	281	224	19	3	2874
1983-84	0	0	9	56	304	844	768	481	313	100	7	0	2882
1984-85	0	0	66	135	426	521	843	573	277	53	2	1	2897
1985-86	0	0	30	87	308	734							

TABLE 5 COOLING DEGREE DAYS Base 65 deg. F MIDLAND – ODESSA, TEXAS

YEAR	JAN	FEB	MAR	APR	MAY	JUNE	JULY	AUG	SEP	OCT	NOV	DEC	TOTAL
1969	0	0	0	66	170	402	597	559	286	97	5	0	2182
1970	0	0	0	37	205	389	542	467	301	59	0	1	2001
1971	0	0	19	59	245	410	491	301	256	47	3	0	1831
1972	0	0	31	187	183	423	446	338	249	88	0	0	1945
1973	0	0	0	18	178	368	414	437	229	88	5	0	1737
1974	0	0	58	109	394	488	575	455	136	50	5	0	2270
1975	0	0	10	80	220	466	404	439	217	86	5	0	1927
1976	0	10	21	68	155	398	344	434	219	22	0	0	1671
1977	0	0	9	21	359	522	584	638	518	112	1	0	2764
1978	0	0	38	139	260	401	531	438	233	30	0	0	2070
1979	0	0	5	43	216	383	539	442	314	167	0	0	2109
1980	0	0	0	17	166	514	600	498	260	33	2	0	2090
1981	0	0	0	71	188	428	583	480	327	135	0	0	2212
1982	0	3	20	84	204	429	543	556	374	111	1	0	2325
1983	0	0	5	64	227	439	588	599	401	151	11	0	2485
1984	0	0	10	69	296	447	500	482	267	39	2	0	2112
1985	0	1	26	115	311	376	505	599	310	56	3	0	2302

TABLE 6 SNOWFALL (inches) MIDLAND – ODESSA, TEXAS

SEASON	JULY	AUG	SEP	OCT	NOV	DEC	JAN	FEB	MAR	APR	MAY	JUNE	TOTAL
1956-57	0.0	0.0	0.0	0.0	T	T	0.0	T	1.0	0.0	0.0	0.0	T
1957-58	0.0	0.0	0.0	0.0	1.1	0.0	3.7	T	1.0	0.0	0.0	0.0	5.8
1958-59	0.0	0.0	0.0	T	0.0	T	0.3	0.3	T	0.0	0.0	0.0	0.6
1959-60	0.0	0.0	0.0	0.0	T	T	T	2.8	T	0.0	0.0	0.0	2.8
1960-61	0.0	0.0	0.0	0.0	0.0	4.8	2.0	2.2	T	0.0	0.0	0.0	9.0
1961-62	0.0	0.0	0.0	0.0	T	T	1.6	T	3.5	0.0	0.0	0.0	5.1
1962-63	0.0	0.0	0.0	0.0	0.0	0.7	T	2.7	0.0	0.0	0.0	0.0	3.4
1963-64	0.0	0.0	0.0	0.0	0.0	T	T	0.4	T	0.0	0.0	0.0	0.4
1964-65	0.0	0.0	0.0	0.0	0.0	T	0.0	3.5	1.0	0.0	0.0	0.0	4.5
1965-66	0.0	0.0	0.0	0.0	0.0	T	3.6	T	0.0	0.0	0.0	0.0	3.6
1966-67	0.0	0.0	0.0	0.0	0.0	0.0	0.0	T	0.0	0.0	0.0	0.0	T
1967-68	0.0	0.0	0.0	0.0	0.0	1.2	0.8	1.7	0.2	0.0	0.0	0.0	3.9
1968-69	0.0	0.0	0.0	0.0	4.5	T	0.0	0.0	1.5	0.0	0.0	0.0	6.0
1969-70	0.0	0.0	0.0	0.0	1.3	0.5	0.1	1.7	5.9	0.0	0.0	0.0	9.5
1970-71	0.0	0.0	0.0	0.0	0.0	0.0	T	0.1	T	0.0	0.0	0.0	0.1
1971-72	0.0	0.0	0.0	0.0	0.0	T	2.4	T	T	0.0	0.0	0.0	2.4
1972-73	0.0	0.0	0.0	0.0	0.6	T	2.9	3.9	0.0	T	0.0	0.0	7.4
1973-74	0.0	0.0	0.0	0.0	0.0	0.0	7.0	0.7	0.0	0.0	0.0	0.0	7.7
1974-75	0.0	0.0	0.0	0.0	0.0	0.1	0.4	1.6	T	0.0	0.0	0.0	2.1
1975-76	0.0	0.0	0.0	0.0	0.0	1.3	0.3	T	0.0	0.0	0.0	0.0	1.6
1976-77	0.0	0.0	0.0	0.0	5.1	T	0.7	0.0	0.0	0.0	0.0	0.0	5.8
1977-78	0.0	0.0	0.0	0.0	0.0	T	1.1	T	T	0.0	0.0	0.0	1.1
1978-79	0.0	0.0	0.0	0.0	0.0	1.1	T	2.0	0.0	0.0	0.0	0.0	3.1
1979-80	0.0	0.0	0.0	0.0	0.0	2.1	0.1	3.2	0.0	T	0.0	0.0	5.4
1980-81	0.0	0.0	0.0	T	7.2	T	4.4	0.0	0.0	0.0	0.0	0.0	11.6
1981-82	0.0	0.0	0.0	0.0	0.0	0.0	3.8	T	0.0	0.0	0.0	0.0	3.8
1982-83	0.0	0.0	0.0	0.0	0.1	6.4	5.3	T	T	0.5	0.0	0.0	12.3
1983-84	0.0	0.0	0.0	0.0	0.0	2.1	T	T	T	0.0	0.0	0.0	2.1
1984-85	0.0	0.0	0.0	0.0	0.0	T	9.0	0.7	0.0	0.0	0.0	0.0	9.7
1985-86	0.0	0.0	0.0	0.0	0.0	0.5							
Record Mean	0.0	0.0	0.0	T	0.5	0.6	1.7	0.9	0.3	T	0.0	0.0	4.1

See Reference Notes, relative to all above tables, on preceding page.

PORT ARTHUR, TEXAS

Port Arthur is located on the flat Coastal Plain in the extreme southeast corner of Texas. The climate is a mixture of tropical and temperate zone conditions.

Sea breezes prevent extremely high temperatures in the summer, except on rare occasions. The area lies far enough south so that cold air masses modify in severity but still provide freezing temperatures up to six times a year.

High humidity is the result of fairly evenly distributed high normal rainfall and prevailing southerly winds from the Gulf of Mexico.

Cloudy, rainy weather is most common in the winter. Only slightly more than half the winters record even a trace of sleet or snow. Heavy rainfall in summer occurs in short duration thunderstorms and in infrequent tropical storms.

Slow moving systems in the spring and fall often result in three to five days of stormy weather and heavy rain. The lightest precipitation usually occurs in March and October. Funnel clouds and waterspouts are common near the coast. The area enjoys approximately 60 percent of possible sunshine.

Fog, most frequent in midwinter and early spring, is rare in summer. It usually dissipates before noon, but occasionally under stagnant conditions lasts a day or two. Along the immediate coast, fog usually does not form until daybreak, but inland it may form before midnight.

The average wind movement is near 11 mph. Except for severe storms and tropical disturbances, wind seldom exceeds 45 mph. It exceeds 30 mph on only about 40 days in any one year.

The climate is favorable for outdoor activities throughout the year. The abundant rainfall, moderate temperatures, and the short period of temperatures below freezing are particularly favorable for farming and livestock production. Heaviest rain usually falls in the summer when needed for rice. The comparatively dry harvest season simplifies the gathering of rice and feed crops. Cattle on the open range of the coastal marshes need little supplemental feeding or protection. Improved pastures are easily provided because of the moderate temperatures and abundant rainfall.

TABLE 1 — NORMALS, MEANS AND EXTREMES

PORT ARTHUR, TEXAS

LATITUDE: 29°57'N   LONGITUDE: 94°01'W   ELEVATION: FT. GRND 16 BARO 21   TIME ZONE: CENTRAL   WBAN: 12917

	(a)	JAN	FEB	MAR	APR	MAY	JUNE	JULY	AUG	SEP	OCT	NOV	DEC	YEAR
TEMPERATURE °F:														
Normals														
-Daily Maximum		61.7	65.4	71.8	78.5	85.0	90.5	92.5	92.2	88.6	81.5	71.4	65.0	78.7
-Daily Minimum		42.1	44.3	51.0	59.4	66.1	71.8	73.7	73.3	69.8	58.9	49.8	44.4	58.7
-Monthly		51.9	54.9	61.4	69.0	75.6	81.2	83.1	82.8	79.2	70.2	60.6	54.7	68.7
Extremes														
-Record Highest	32	81	84	87	93	97	100	103	107	100	95	88	84	107
-Year		1982	1962	1974	1955	1977	1954	1980	1962	1980	1977	1978	1978	AUG 1962
-Record Lowest	32	14	20	25	36	46	56	61	62	45	35	22	15	14
-Year		1962	1981	1980	1980	1954	1984	1967	1967	1967	1957	1976	1983	JAN 1962
NORMAL DEGREE DAYS:														
Heating (base 65°F)		431	306	167	23	0	0	0	0	0	33	190	327	1477
Cooling (base 65°F)		25	23	56	143	329	486	561	552	426	194	58	8	2861
% OF POSSIBLE SUNSHINE	26	42	52	52	52	64	69	65	63	62	67	57	47	58
MEAN SKY COVER (tenths)														
Sunrise - Sunset	32	7.0	6.3	6.7	6.6	6.0	5.3	5.9	5.8	5.5	4.9	5.6	6.3	6.0
MEAN NUMBER OF DAYS:														
Sunrise to Sunset														
-Clear	32	6.6	8.0	6.8	6.3	6.8	8.6	5.7	6.7	8.8	11.8	9.9	8.8	94.8
-Partly Cloudy	32	6.3	5.8	7.5	7.7	12.3	14.1	15.8	15.3	11.7	10.7	7.6	6.3	120.9
-Cloudy	32	18.2	14.5	16.7	16.0	12.0	7.3	9.5	9.1	9.5	8.4	12.5	15.8	149.5
Precipitation														
.01 inches or more	32	9.6	8.5	8.1	6.7	7.3	7.5	11.2	12.3	9.9	6.3	7.7	9.2	104.3
Snow, Ice pellets														
1.0 inches or more	32	0.*	0.1	0.*	0.0	0.0	0.0	0.0	0.0	0.0	0.0	0.0	0.0	0.1
Thunderstorms	32	2.2	2.5	3.3	3.6	5.7	7.2	13.6	12.3	6.9	3.0	2.6	2.2	65.1
Heavy Fog Visibility														
1/4 mile or less	32	7.6	5.8	5.9	3.1	1.3	0.3	0.2	0.2	0.8	3.1	4.9	6.3	39.5
Temperature °F														
-Maximum														
90° and above	25	0.0	0.0	0.0	0.*	2.7	17.0	23.8	22.9	12.0	1.6	0.0	0.0	80.0
32° and below	25	0.2	0.0	0.0	0.0	0.0	0.0	0.0	0.0	0.0	0.0	0.0	0.1	0.3
-Minimum														
32° and below	25	7.0	3.7	1.0	0.0	0.0	0.0	0.0	0.0	0.0	0.0	0.9	4.2	16.8
0° and below	25	0.0	0.0	0.0	0.0	0.0	0.0	0.0	0.0	0.0	0.0	0.0	0.0	0.0
AVG. STATION PRESS.(mb)	12	1020.3	1019.1	1015.0	1014.8	1012.9	1014.5	1015.6	1015.3	1014.9	1017.2	1018.3	1019.9	1016.5
RELATIVE HUMIDITY (%)														
Hour 00	23	86	85	86	87	90	91	93	92	90	88	87	87	89
Hour 06 (Local Time)	25	88	87	88	90	92	93	94	94	92	91	89	89	91
Hour 12	25	69	63	63	64	64	63	66	66	65	59	62	67	64
Hour 18	25	76	70	69	70	70	70	72	72	74	73	76	77	72
PRECIPITATION (inches):														
Water Equivalent														
-Normal		4.18	3.71	2.93	4.05	4.50	3.96	5.37	5.45	6.13	3.63	4.33	4.55	52.79
-Maximum Monthly	32	9.57	11.76	9.35	15.30	11.39	14.05	18.71	17.26	21.96	15.09	10.84	17.98	21.96
-Year		1961	1959	1979	1973	1983	1961	1959	1966	1980	1970	1977	1982	SEP 1980
-Minimum Monthly	32	0.60	0.36	0.06	0.35	0.10	0.76	0.63	0.98	0.50	0.00	0.15	1.32	0.00
-Year		1971	1954	1955	1965	1978	1956	1968	1968	1963	1967	1954		OCT 1963
-Maximum in 24 hrs	32	4.92	5.05	6.04	10.09	7.66	10.20	10.56	8.45	17.16	8.06	7.26	9.98	17.16
-Year		1961	1965	1979	1973	1983	1961	1979	1966	1980	1970	1961	1982	SEP 1980
Snow, Ice pellets														
-Maximum Monthly	32	3.0	4.4	1.4								T	T	4.4
-Year		1973	1960	1968								1976	1983	FEB 1960
-Maximum in 24 hrs	32	3.0	4.4	1.4								T	T	4.4
-Year		1973	1960	1968								1976	1983	FEB 1960
WIND:														
Mean Speed (mph)	32	11.0	11.5	11.9	12.0	10.3	8.9	7.7	7.4	8.6	8.9	10.2	10.6	9.9
Prevailing Direction through 1963		N	S	S	S	S	S	S	S	NE	N	N	N	S
Fastest Obs. 1 Min.														
-Direction (!!)	6	15	24	08	30	29	14	17	13	08	31	20	15	29
-Speed (MPH)	6	37	35	38	46	52	37	39	40	29	32	36	32	52
-Year		1983	1984	1979	1981	1983	1981	1979	1983	1980	1985	1983	1983	MAY 1983
Peak Gust														
-Direction (!!)	2	N	SW	NW	SE	NW	S	S	E	SE	NW	NW	S	NW
-Speed (mph)	2	41	54	48	46	62	38	45	46	39	47	39	47	62
-Date		1985	1984	1984	1984	1985	1985	1984	1985	1985	1985	1984	1984	MAY 1985

See reference Notes to this table on the following page.

TABLE 2 PRECIPITATION (inches) PORT ARTHUR, TEXAS

YEAR	JAN	FEB	MAR	APR	MAY	JUNE	JULY	AUG	SEP	OCT	NOV	DEC	ANNUAL
1956	3.50	4.83	2.81	3.01	6.39	2.04	0.63	2.32	0.64	1.87	2.64	12.47	43.15
1957	0.64	0.74	8.83	7.05	0.64	11.01	6.00	2.54	14.04	6.59	5.18	3.01	66.27
1958	3.91	3.82	1.62	1.97	2.69	1.93	5.65	2.59	14.59	1.72	2.74	3.32	46.55
1959	2.97	11.76	0.85	6.68	4.01	3.84	18.71	5.91	2.49	5.85	2.73	4.12	69.92
1960	1.65	5.29	0.64	3.51	0.55	1.63	6.03	14.48	2.69	6.16	2.82	7.37	52.82
1961	9.57	4.36	1.77	2.79	1.80	14.05	6.09	3.91	4.79	3.04	10.42	4.77	67.36
1962	2.10	0.72	1.32	2.21	0.86	5.82	1.48	5.89	5.23	3.07	6.13	4.11	38.94
1963	6.14	3.97	0.21	1.10	0.59	2.88	4.49	2.00	18.15	0.00	9.90	2.89	52.32
1964	4.28	3.10	3.64	2.11	6.02	1.62	6.16	4.96	5.57	0.07	4.19	5.06	46.78
1965	1.59	5.41	2.96	0.35	3.15	2.41	3.21	1.91	4.62	0.78	2.29	5.78	34.46
1966	6.22	5.68	0.65	5.86	9.63	3.50	3.97	17.26	3.17	6.49	4.85	3.39	70.67
1967	1.92	2.06	0.91	5.86	5.85	1.06	4.20	8.63	3.35	2.13	0.15	4.55	40.67
1968	5.48	2.56	2.51	7.38	3.20	12.17	6.15	0.98	6.01	3.11	5.69	2.74	57.98
1969	1.19	4.14	3.67	5.38	5.69	2.08	10.14	3.06	1.50	1.74	1.52	8.33	48.44
1970	1.78	2.88	4.32	2.47	8.54	0.96	1.55	3.07	7.37	15.09	1.57	2.03	51.63
1971	0.60	4.31	1.15	1.76	3.47	1.86	4.31	10.56	5.07	2.16	2.64	7.53	45.42
1972	8.33	1.83	3.93	5.42	5.92	3.43	3.46	5.80	6.50	2.88	6.11	3.95	58.96
1973	4.76	3.00	7.24	15.30	5.47	5.80	8.72	7.16	11.44	5.11	2.01	2.90	78.91
1974	8.81	2.37	4.88	5.33	7.61	1.43	3.75	4.39	2.83	3.69	5.84	4.53	55.46
1975	5.46	1.60	1.75	3.77	5.30	4.97	7.56	6.86	2.80	4.68	3.26	3.60	51.61
1976	3.64	0.44	2.69	1.89	6.39	6.16	3.99	2.57	4.47	5.16	5.54	6.11	49.05
1977	4.93	1.06	2.23	4.79	3.41	3.37	2.11	7.20	2.87	4.06	10.84	2.01	48.88
1978	6.66	2.32	0.44	0.36	0.10	7.47	3.41	5.11	3.29	T	4.80	3.72	37.68
1979	5.41	4.21	9.35	7.35	4.39	3.96	4.07	9.45	15.68	6.62	3.20	2.51	76.20
1980	4.66	1.92	6.83	0.93	7.76	0.76	2.24	1.18	21.96	7.27	4.82	2.22	62.55
1981	3.33	3.26	2.09	3.16	2.78	9.14	10.75	1.63	2.72	5.68	2.16	3.85	50.55
1982	2.04	3.38	3.12	5.63	9.36	9.30	5.88	2.47	1.00	4.15	7.46	17.98	71.77
1983	5.78	3.97	4.57	0.51	11.39	8.10	6.53	14.35	11.89	0.33	4.24	6.49	78.15
1984	6.69	4.27	1.28	1.03	10.43	1.05	3.83	3.91	4.14	14.94	6.44	2.03	60.04
1985	3.44	6.85	4.74	1.13	5.15	3.17	6.91	11.25	7.39	11.60	3.85	2.74	68.22
Record Mean	4.43	3.70	3.21	3.70	4.78	4.52	6.20	5.42	5.28	3.66	4.02	4.94	53.87

TABLE 3 AVERAGE TEMPERATURE (deg. F) PORT ARTHUR, TEXAS

YEAR	JAN	FEB	MAR	APR	MAY	JUNE	JULY	AUG	SEP	OCT	NOV	DEC	ANNUAL
1956	52.9	58.7	61.2	67.1	77.6	79.7	82.9	82.4	78.7	71.9	59.3	59.4	69.3
1957	57.7	62.5	60.7	68.9	76.5	81.1	83.8	82.4	76.2	66.6	61.1	56.3	69.5
1958	49.2	49.1	57.9	68.9	76.2	82.8	84.0	82.8	79.2	69.3	61.2	51.2	67.7
1959	49.7	54.9	59.4	66.4	77.0	81.3	81.3	81.6	79.8	71.3	55.7	53.9	67.7
#1960	50.5	48.9	55.6	69.4	72.9	82.3	83.6	81.4	78.7	71.7	62.3	51.0	67.4
1961	47.3	56.7	65.4	65.4	74.2	78.9	81.6	80.6	79.1	69.3	59.2	55.8	67.8
1962	47.3	62.5	58.1	67.5	76.1	80.3	84.1	85.6	80.4	72.9	58.8	52.3	68.3
1963	46.4	51.1	64.4	72.6	76.1	82.4	84.6	84.3	80.0	74.2	46.2	48.8	68.8
1964	50.4	49.2	59.9	69.2	76.6	81.1	82.1	83.4	79.1	66.6	54.7		68.0
1965	55.0	54.4	57.1	72.7	77.3	81.2	83.8	82.4	80.7	68.4	58.1	57.4	69.9
1966	48.0	52.9	59.8	69.9	75.8	80.2	84.3	81.7	78.9	68.3	64.2	53.5	68.1
1967	51.8	53.4	63.9	73.2	74.7	81.9	81.7	80.9	76.4	69.0	55.5	55.8	68.8
1968	51.0	48.0	58.2	69.9	74.6	79.5	81.3	83.2	76.6	70.4	58.9	53.8	67.1
1969	54.6	54.6	54.7	69.2	74.9	81.0	84.9	83.1	78.4	72.0	59.7	55.9	68.6
1970	47.5	55.4	59.7	71.5	74.7	80.9	84.7	85.5	82.0	70.1	59.3	62.1	69.4
1971	57.5	56.6	60.7	69.0	75.2	81.9	81.7	79.3	77.8	71.8	59.3	59.8	69.2
1972	55.3	55.3	55.3	69.6	73.3	80.2	79.3	80.2	79.4	69.6	54.0	52.5	67.6
1973	47.5	51.0	63.5	63.7	72.6	79.4	81.7	79.6	78.9	72.1	66.5	52.4	67.4
1974	56.4	55.1	66.2	68.6	76.5	78.9	81.7	81.1	74.9	68.7	59.0	53.0	68.3
1975	56.2	55.2	60.0	66.9	75.9	79.9	81.4	81.2	75.5	70.3	62.0	52.0	68.0
1976	49.8	59.9	62.8	68.7	72.1	78.7	81.2	81.2	78.3	61.9	52.5	50.6	66.5
1977	44.5	55.3	63.2	68.4	77.4	83.6	85.0	84.9	83.1	71.8	63.8	55.3	69.7
1978	43.8	47.2	59.8	70.2	79.0	83.3	85.1	85.3	81.8	72.0	67.4	56.8	69.3
1979	45.8	52.5	62.5	69.5	72.9	80.3	81.8	81.7	76.5	70.6	56.4	52.8	66.9
1980	55.2	52.7	61.4	65.6	76.4	82.9	85.1	84.2	82.6	67.5	57.8	54.4	68.8
1981	50.3	53.8	61.2	72.9	73.7	82.0	83.2	82.7	78.6	71.0	63.4	55.1	68.8
1982	52.8	52.4	65.0	68.0	75.6	81.6	82.7	82.7	78.6	70.0	61.8	57.6	69.0
1983	50.4	54.0	58.9	63.8	73.3	79.1	82.4	82.4	76.7	70.3	63.7	48.7	69.1
1984	48.7	56.0	62.2	68.6	74.7	79.7	81.2	81.7	77.0	74.8	60.8	63.7	69.1
1985	48.1	53.1	67.0	72.1	76.2	81.0	82.7	81.8	78.7	72.4	66.6	50.8	69.3
Record Mean	52.3	55.5	61.3	68.7	75.2	81.1	82.7	82.7	78.9	70.8	60.7	54.7	68.7
Max	60.8	64.3	70.1	76.9	83.3	89.0	90.9	91.0	87.3	80.5	70.1	63.5	77.3
Min	43.7	46.7	52.6	60.5	67.1	73.1	74.5	74.4	70.5	61.1	51.3	45.9	60.1

TABLE 4 HEATING DEGREE DAYS Base 65 deg. F   PORT ARTHUR, TEXAS

SEASON	JULY	AUG	SEP	OCT	NOV	DEC	JAN	FEB	MAR	APR	MAY	JUNE	TOTAL
1956-57	0	0	0	4	231	203	260	127	150	50	5	0	1030
1957-58	0	0	0	83	181	269	478	440	232	34	0	0	1717
1958-59	0	0	0	46	170	422	469	289	183	59	0	0	1638
1959-60	0	0	0	20	313	340	454	461	304	15	6	0	1913
#1960-61	0	0	0	25	122	443	542	240	81	99	0	0	1552
1961-62	0	0	0	33	198	306	543	110	230	42	0	0	1462
1962-63	0	0	0	20	193	387	573	384	102	12	0	0	1671
1963-64	0	0	0	4	141	579	447	452	172	22	0	0	1817
1964-65	0	0	0	52	135	337	322	298	270	3	0	0	1417
1965-66	0	0	0	43	34	250	529	336	180	19	0	0	1391
1966-67	0	0	0	46	115	383	417	325	111	0	2	0	1399
1967-68	0	0	10	33	133	313	435	485	244	18	0	0	1671
1968-69	0	0	0	20	207	343	331	289	317	4	0	0	1511
1969-70	0	0	0	22	212	281	545	293	180	22	3	0	1558
1970-71	0	0	0	28	213	168	269	254	167	43	0	0	1142
1971-72	0	0	0	6	208	199	317	291	109	20	0	0	1150
1972-73	0	0	0	57	340	390	532	389	77	118	2	0	1905
1973-74	0	0	0	13	83	384	288	288	85	36	0	0	1177
1974-75	0	0	0	20	230	384	297	276	180	67	0	0	1454
1975-76	0	0	0	16	209	413	464	170	128	16	0	0	1416
1976-77	0	0	0	149	375	439	626	268	110	16	0	0	1983
1977-78	0	0	0	16	112	321	662	497	184	14	0	0	1806
1978-79	0	0	0	4	75	296	591	360	125	12	0	0	1463
1979-80	0	0	0	23	278	382	306	377	151	44	0	0	1561
1980-81	0	0	0	70	248	339	448	315	140	3	3	0	1566
1981-82	0	0	3	59	94	315	402	357	124	59	0	0	1413
1982-83	0	0	0	43	154	266	446	301	203	91	1	0	1505
1983-84	0	0	1	27	122	524	500	265	129	31	1	0	1600
1984-85	0	0	5	8	176	125	514	342	34	6	0	0	1210
1985-86	0	0	3	14	78	438							

TABLE 5 COOLING DEGREE DAYS Base 65 deg. F   PORT ARTHUR, TEXAS

YEAR	JAN	FEB	MAR	APR	MAY	JUNE	JULY	AUG	SEP	OCT	NOV	DEC	TOTAL
1969	17	4	5	139	316	484	621	569	408	247	60	6	2876
1970	13	2	23	225	309	484	617	644	518	194	49	85	3163
1971	42	24	42	167	325	517	527	451	388	223	42	43	2791
1972	21	18	52	164	262	461	449	479	440	209	17	9	2581
1973	0	2	38	84	244	440	524	459	426	242	137	6	2602
1974	27	15	130	153	362	423	524	505	305	143	58	20	2665
1975	28	9	61	130	344	454	516	508	335	186	71	18	2660
1976	2	29	66	133	226	418	512	512	404	61	8	0	2371
1977	0	4	62	125	391	562	627	621	551	235	82	28	3288
1978	11	8	29	177	439	555	627	635	510	227	154	49	3421
1979	4	17	54	156	250	464	529	523	352	203	27	11	2590
1980	8	27	47	69	359	546	633	603	534	152	37	15	3030
1981	0	8	28	245	278	515	570	557	367	250	53	14	2885
1982	27	9	129	155	335	504	555	554	416	205	67	46	3002
1983	0	0	20	64	264	429	547	546	361	200	88	21	2540
1984	0	9	50	147	309	449	508	524	369	317	56	94	2832
1985	0	13	104	228	354	489	527	573	422	253	134	5	3102

TABLE 6 SNOWFALL (inches)   PORT ARTHUR, TEXAS

SEASON	JULY	AUG	SEP	OCT	NOV	DEC	JAN	FEB	MAR	APR	MAY	JUNE	TOTAL
1970-71	0.0	0.0	0.0	0.0	0.0	0.0	0.0	0.0	0.0	0.0	0.0	0.0	0.0
1971-72	0.0	0.0	0.0	0.0	0.0	0.0	0.0	0.0	0.0	0.0	0.0	0.0	0.0
1972-73	0.0	0.0	0.0	0.0	0.0	0.0	3.0	0.4	0.0	0.0	0.0	0.0	3.4
1973-74	0.0	0.0	0.0	0.0	0.0	0.0	0.0	0.0	0.0	0.0	0.0	0.0	0.0
1974-75	0.0	0.0	0.0	0.0	0.0	0.0	0.0	0.0	0.0	0.0	0.0	0.0	T
1975-76	0.0	0.0	0.0	0.0	0.0	0.0	T	0.0	0.0	0.0	0.0	0.0	T
1976-77	0.0	0.0	0.0	0.0	T	0.0	0.0	0.0	0.0	0.0	0.0	0.0	T
1977-78	0.0	0.0	0.0	0.0	0.0	0.0	T	0.0	0.0	0.0	0.0	0.0	T
1978-79	0.0	0.0	0.0	0.0	0.0	0.0	T	0.0	0.0	0.0	0.0	0.0	T
1979-80	0.0	0.0	0.0	0.0	0.0	0.0	0.0	0.7	0.0	0.0	0.0	0.0	0.7
1980-81	0.0	0.0	0.0	0.0	0.0	0.0	0.0	T	0.0	0.0	0.0	0.0	T
1981-82	0.0	0.0	0.0	0.0	0.0	0.0	0.2	0.0	0.0	0.0	0.0	0.0	0.2
1982-83	0.0	0.0	0.0	0.0	0.0	0.0	0.0	0.0	0.0	0.0	0.0	0.0	0.0
1983-84	0.0	0.0	0.0	0.0	0.0	T	0.0	0.0	0.0	0.0	0.0	0.0	T
1984-85	0.0	0.0	0.0	0.0	0.0	0.0	0.4	T	0.0	0.0	0.0	0.0	0.4
1985-86	0.0	0.0	0.0	0.0	0.0	0.0							
Record Mean	0.0	0.0	0.0	0.0	T	T	0.1	0.3	T	0.0	0.0	0.0	0.4

See Reference Notes, relative to all above tables, on preceding page.

San Angelo is located near the center of Texas at the northern edge of the Edwards Plateau. Ground elevation ranges from about 1,700 to 2,700 feet above sea level. Topography varies from level and slightly rolling to broken. The climate is generally classified as semi-arid or steppe, but has some humid temperate characteristics. Warm, dry weather predominates, although changes may be rapid and frequent with the passage of cold fronts or northers.

High temperatures of summer are associated with fair skies, south to southwest winds and dry air. Low humidities, however, are conducive to personal comfort because of rapid evaporation. Rapid temperature drops occur after sunset, and most nights are pleasant with lows in the upper 60s and lower 70s. Rapid temperature drops occur in the winter as cold polar air invades the region. Temperature drops of 20 to 30 degrees in a short time are not uncommon. Cold polar outbreaks have produced record low temperatures of zero or below throughout the area.

The rainfall is typical of the Great Plains. Much of the rainfall occurs from thunderstorm activity, and wide variations in annual precipitation occur from year to year. Heavy rainfall occurs in April, May, June, September and October. Also, in the late summer months, heavy precipitation may occur when tropical disturbances move inland over south Texas and pass near the San Angelo area.

The prevailing wind direction is from the south, and winds are frequently high and persistent for several days. Dusty conditions are infrequent and occur in early spring when west or northwest winds predominate. The frequency and intensity of the dust storms are dependent on soil conditions in the Texas Panhandle and in New Mexico.

Agriculture in the region consists of cattle, sheep, and goat raising. Cotton, from dry-land and irrigated fields, maize, corn, melons, truck farming, and pecan production are also important crops.

TABLE 1 — NORMALS, MEANS AND EXTREMES

SAN ANGELO, TEXAS

LATITUDE: 31°22'N   LONGITUDE: 100°30'W   ELEVATION: FT. GRND 1903 BARO 01909   TIME ZONE: CENTRAL   WBAN: 23034

	(a)	JAN	FEB	MAR	APR	MAY	JUNE	JULY	AUG	SEP	OCT	NOV	DEC	YEAR
TEMPERATURE °F:														
Normals														
-Daily Maximum		58.7	63.3	71.5	80.2	86.3	93.4	96.5	95.4	88.0	79.2	67.2	61.2	78.4
-Daily Minimum		32.2	36.1	43.4	53.4	61.5	69.3	72.0	71.2	65.0	54.0	42.0	34.7	52.9
-Monthly		45.5	49.7	57.5	66.8	73.9	81.3	84.3	83.3	76.5	66.6	54.6	48.0	65.7
Extremes														
-Record Highest	38	90	90	97	103	107	110	111	107	107	100	93	91	111
-Year		1969	1957	1974	1972	1984	1969	1960	1969	1952	1951	1980	1954	JUL 1960
-Record Lowest	38	5	-1	8	25	35	48	58	55	38	29	13	-1	-1
-Year		1982	1985	1980	1973	1967	1964	1968	1961	1984	1980	1979	1983	FEB 1985
NORMAL DEGREE DAYS:														
Heating (base 65°F)		605	428	274	73	5	0	0	0	0	75	326	527	2313
Cooling (base 65°F)		0	0	42	127	281	492	598	567	350	125	14	0	2596
% OF POSSIBLE SUNSHINE														
MEAN SKY COVER (tenths)														
Sunrise - Sunset	37	5.5	5.2	5.2	5.1	5.2	4.4	4.3	4.1	4.6	4.4	4.6	5.0	4.8
MEAN NUMBER OF DAYS:														
Sunrise to Sunset														
-Clear	37	11.7	11.1	11.7	11.3	10.9	13.6	14.6	15.2	13.5	14.9	13.6	12.7	155.0
-Partly Cloudy	37	6.2	6.3	7.8	7.8	10.0	10.3	9.9	10.0	8.3	7.3	6.3	6.9	97.1
-Cloudy	37	13.1	10.8	11.5	10.9	10.1	6.1	6.5	5.8	8.2	8.8	10.1	11.4	113.2
Precipitation														
.01 inches or more	38	4.8	4.4	3.8	4.9	6.9	5.1	4.5	4.8	6.2	5.0	3.7	3.6	57.6
Snow, Ice pellets														
1.0 inches or more	38	0.6	0.3	0.1	0.0	0.0	0.0	0.0	0.0	0.0	0.0	0.2	0.1	1.2
Thunderstorms	38	0.4	0.9	2.1	4.3	6.9	4.9	4.2	4.8	3.6	2.6	0.8	0.4	35.9
Heavy Fog Visibility 1/4 mile or less	38	1.3	1.1	0.6	0.3	0.3	0.1	0.*	0.*	0.2	0.7	1.5	1.4	7.6
Temperature °F														
-Maximum														
90° and above	25	0.*	0.0	1.0	5.1	11.8	21.3	27.1	26.6	12.9	2.7	0.1	0.0	108.7
32° and below	25	1.6	0.4	0.0	0.0	0.0	0.0	0.0	0.0	0.0	0.0	0.1	0.6	2.7
-Minimum														
32° and below	25	18.3	11.4	4.0	0.6	0.0	0.0	0.0	0.0	0.0	0.1	4.7	14.4	53.4
0° and below	25	0.0	0.*	0.0	0.0	0.0	0.0	0.0	0.0	0.0	0.0	0.0	0.*	0.1
AVG. STATION PRESS. (mb)	13	952.0	950.6	946.6	946.4	945.2	946.8	948.6	948.7	949.0	950.2	950.6	951.5	948.9
RELATIVE HUMIDITY (%)														
Hour 00	25	69	66	60	61	68	66	59	62	72	74	74	71	67
Hour 06 (Local Time)	25	76	75	71	73	80	80	76	77	83	83	80	78	78
Hour 12 (Local Time)	25	53	49	44	43	48	49	43	45	54	53	52	51	49
Hour 18	25	48	42	36	36	41	41	37	39	49	51	53	51	44
PRECIPITATION (inches):														
Water Equivalent														
-Normal		0.64	0.84	0.79	1.75	2.52	1.88	1.22	1.85	3.04	2.05	0.97	0.64	18.19
-Maximum Monthly	38	3.65	2.86	5.00	5.10	7.10	6.01	7.21	8.13	11.00	8.68	3.55	3.48	11.00
-Year		1961	1958	1953	1977	1957	1982	1959	1971	1980	1981	1968	1984	SEP 1980
-Minimum Monthly	38	0.00	0.01	T	0.10	0.26	0.07	T	T	T	0.00	0.00	T	0.00
-Year		1967	1974	1972	1967	1962	1956	1970	1959	1983	1952	1950	1973	JAN 1967
-Maximum in 24 hrs	38	2.49	1.92	4.65	3.32	2.56	2.86	2.95	3.00	6.25	5.11	2.16	2.71	6.25
-Year		1961	1958	1953	1971	1975	1961	1959	1971	1980	1959	1975	1984	SEP 1980
Snow, Ice pellets														
-Maximum Monthly	38	9.0	5.8	3.1	T							8.8	1.9	9.0
-Year		1978	1973	1962	1980							1968	1967	JAN 1978
-Maximum in 24 hrs	38	7.4	4.1	3.1	T							5.8	1.4	7.4
-Year		1978	1966	1962	1980							1968	1967	JAN 1978
WIND:														
Mean Speed (mph)	36	10.2	10.9	12.3	12.1	11.3	11.2	9.7	9.1	9.1	9.3	9.9	9.9	10.4
Prevailing Direction through 1963		SW	SSW	SSW	S	S	S	S	S	S	S	SW	W	S
Fastest Obs. 1 Min.														
-Direction (!!)	37	27	29	27	28	02	02	03	02	34	29	29	30	28
-Speed (MPH)	37	44	48	58	75	60	57	40	43	52	60	66	43	75
-Year		1960	1960	1961	1969	1963	1955	1981	1949	1981	1960	1958	1965	APR 1969
Peak Gust														
-Direction (!!)	2	NE	S	W	W	NW	W	NE	E	E	S	S	S	NW
-Speed (mph)	2	39	36	54	52	58	48	39	49	39	37	38	36	58
-Date		1985	1985	1984	1984	1985	1984	1984	1985	1985	1985	1985	1984	MAY 1985

See Reference Notes to this table on the following page

TABLE 2 PRECIPITATION (inches) SAN ANGELO, TEXAS

YEAR	JAN	FEB	MAR	APR	MAY	JUNE	JULY	AUG	SEP	OCT	NOV	DEC	ANNUAL
1956	0.71	0.17	0.05	1.44	0.48	0.07	0.21	0.75	0.48	2.14	0.42	0.49	7.41
1957	0.55	0.98	0.30	4.00	7.10	1.74	0.05	0.53	2.38	2.55	1.65	0.33	22.16
1958	1.45	2.86	1.04	0.95	2.87	1.14	0.21	2.07	3.20	1.86	0.66	0.04	18.35
1959	T	0.53	0.11	2.40	1.59	4.17	7.21	T	9.20	5.29	1.30	2.06	33.86
1960	1.95	0.38	0.41	1.24	2.05	0.46	1.53	0.51	0.69	3.70	0.21	1.61	14.74
1961	3.65	0.76	0.26	0.29	2.63	5.82	0.81	1.02	2.95	1.94	1.15	0.11	21.39
1962	0.02	0.11	0.40	2.67	0.26	0.74	1.96	0.59	1.85	0.72	0.75	0.46	10.53
1963	0.06	0.82	T	0.96	3.89	2.85	T	1.55	1.08	0.32	2.13	0.29	13.95
1964	0.95	1.27	0.83	1.90	0.41	0.16	0.47	1.35	2.80	0.63	1.14	0.27	12.18
1965	0.53	2.54	0.12	0.12	3.65	2.01	0.26	2.71	1.18	2.11	0.39	0.63	16.25
1966	0.49	0.58	0.27	2.16	2.10	0.77	0.09	5.05	2.41	1.90	T	T	15.82
1967	0.00	0.17	0.78	0.10	1.91	2.42	0.59	1.43	7.21	1.82	2.41	1.14	19.98
1968	2.32	1.58	1.88	2.26	1.98	4.71	1.23	1.87	1.81	0.11	3.55	T	23.30
1969	0.02	1.07	1.76	4.38	3.20	0.26	0.62	2.96	5.74	6.59	1.20	2.24	30.04
1970	0.14	1.31	2.36	1.10	1.83	0.39	T	0.47	4.28	0.94	T	0.06	12.88
1971	T	0.42	T	3.89	1.42	1.72	1.84	8.13	3.59	2.09	0.27	0.88	24.25
1972	0.48	0.09	T	0.57	4.89	4.21	0.17	3.66	5.30	3.22	0.31	0.03	22.93
1973	1.86	1.82	1.18	1.67	1.43	2.75	1.06	0.36	4.22	2.03	0.03	T	18.41
1974	0.12	0.01	0.48	1.86	2.43	0.16	1.54	4.77	6.11	4.82	1.40	1.40	25.10
1975	0.37	1.28	0.09	0.86	5.26	2.32	2.46	0.38	2.25	3.37	2.19	0.55	21.58
1976	0.04	0.42	0.30	3.41	1.77	0.41	3.51	1.41	4.66	5.11	0.58	0.18	21.80
1977	0.61	0.26	0.99	5.10	2.14	0.47	0.52	0.38	0.27	1.38	0.68	0.15	12.95
1978	0.61	1.17	0.41	0.73	1.83	2.81	0.41	2.93	1.35	0.42	1.75	0.25	14.67
1979	0.19	1.83	2.25	1.90	0.67	2.15	1.30	2.18	0.06	0.93	T	2.70	16.16
1980	0.84	0.79	0.71	0.52	4.72	3.14	0.33	3.30	11.00	0.01	2.53	2.20	30.09
1981	1.17	0.68	2.81	3.51	4.70	1.97	2.68	1.23	2.72	8.68	T	0.02	30.17
1982	1.06	1.53	0.40	0.83	4.17	6.01	0.35	0.46	0.09	1.26	1.11	0.91	18.18
1983	2.06	0.42	1.20	0.81	0.52	3.72	1.18	0.03	T	3.47	1.78	0.07	15.26
1984	2.38	0.54	0.49	0.23	0.54	2.82	0.60	0.26	2.99	3.74	1.09	3.48	19.16
1985	0.67	0.38	1.69	0.42	4.78	3.55	1.13	0.24	2.84	5.64	0.48	0.01	21.83
Record Mean	0.87	0.88	0.92	1.88	2.92	2.01	1.68	1.99	2.86	2.28	1.14	1.01	20.45

TABLE 3 AVERAGE TEMPERATURE (deg. F) SAN ANGELO, TEXAS

YEAR	JAN	FEB	MAR	APR	MAY	JUNE	JULY	AUG	SEP	OCT	NOV	DEC	ANNUAL
1956	46.3	49.9	59.5	65.3	76.9	84.0	84.5	84.3	78.0	69.5	52.8	49.5	66.7
1957	45.8	55.7	57.5	62.8	69.4	78.2	85.1	84.7	73.6	62.8	50.0	51.0	64.7
1958	45.1	45.3	49.0	62.3	72.1	80.9	83.9	83.0	75.3	62.9	55.2	45.0	63.3
1959	44.9	48.6	57.1	63.9	75.0	79.6	78.9	81.8	78.5	64.6	48.3	48.3	64.1
#1960	46.1	44.9	52.1	66.8	73.5	85.5	85.9	84.8	79.3	71.2	58.0	43.4	66.0
1961	42.2	50.2	58.9	67.1	76.8	79.0	79.8	80.4	74.8	66.0	50.3	46.2	64.3
1962	41.5	55.8	54.5	66.3	78.4	81.1	86.4	86.9	78.9	72.1	56.9	47.8	67.2
1963	41.8	48.8	62.6	73.2	77.0	81.1	87.1	85.2	78.2	71.0	56.4	40.8	66.9
1964	45.1	45.4	60.6	70.3	77.3	83.1	86.9	86.6	77.1	64.7	56.9	48.9	66.9
1965	49.7	46.6	49.7	70.5	74.0	81.4	85.7	82.7	78.5	65.8	61.2	51.5	66.5
1966	40.7	45.5	59.0	66.3	73.6	81.0	87.0	81.0	74.1	62.8	59.3	45.2	64.6
1967	46.5	48.5	64.8	72.5	74.4	83.3	84.5	83.0	72.6	66.1	57.0	45.2	66.5
1968	46.3	45.8	55.0	62.2	74.5	79.3	82.5	84.1	74.4	69.4	53.3	47.2	64.5
1969	52.1	51.6	51.6	67.0	71.4	82.0	87.5	84.9	75.2	64.5	50.4	50.5	66.1
1970	43.9	50.8	52.9	66.2	71.3	78.9	83.3	85.4	78.3	64.0	54.4	54.6	65.4
1971	50.2	52.0	59.6	66.7	75.2	80.9	82.8	75.4	74.3	66.8	50.0	50.2	65.8
1972	47.4	52.4	62.4	71.7	70.8	81.1	83.3	79.9	78.2	66.2	50.4	46.2	65.8
1973	41.0	46.4	59.0	60.6	73.0	78.9	83.3	82.6	75.1	67.0	60.7	49.4	64.7
1974	47.0	52.4	64.9	67.9	76.0	80.9	84.0	79.6	68.9	66.4	55.0	46.8	65.8
1975	46.5	48.0	56.1	64.8	71.8	79.7	78.6	80.6	71.6	65.3	54.7	48.0	63.8
1976	43.9	56.7	58.1	65.5	68.8	79.7	76.2	80.7	73.8	58.6	47.5	45.6	62.9
1977	40.4	51.5	57.9	65.0	75.5	82.3	84.8	86.2	83.5	69.2	57.1	52.3	67.2
1978	38.5	43.8	56.3	70.6	75.7	80.3	85.1	79.5	74.6	64.7	55.5	44.2	64.1
1979	36.2	46.7	57.2	65.4	72.1	78.0	80.3	80.4	75.2	69.9	50.8	47.4	63.5
1980	46.1	49.2	55.4	63.7	71.8	82.6	86.9	84.4	76.9	64.5	53.0	49.9	65.4
1981	45.8	50.2	54.2	66.4	70.7	77.8	82.1	80.0	74.3	64.6	55.6	48.7	64.2
1982	46.0	45.9	59.1	63.5	69.6	77.9	83.2	84.6	77.4	66.0	54.3	45.8	64.5
1983	45.1	48.8	56.6	62.2	72.5	76.6	82.2	82.1	77.4	70.2	58.4	36.3	64.0
1984	40.4	50.3	57.4	65.1	76.0	82.3	81.5	82.8	72.5	64.5	53.0	50.2	64.7
1985	38.8	46.2	59.3	68.1	75.1	77.5	80.5	84.5	75.7	66.9	57.9	43.6	64.5
Record Mean	45.4	49.5	57.2	65.8	73.3	80.8	83.6	83.1	76.1	66.3	54.3	47.4	65.2
Max	58.5	63.2	71.6	79.9	86.4	93.2	96.1	95.7	88.3	79.1	67.4	60.4	78.3
Min	32.2	35.8	42.6	51.8	60.3	68.4	71.0	70.5	63.9	53.4	41.2	34.3	52.1

REFERENCE NOTES FOR TABLES 1, 2, 3 and 6 (SAN ANGELO, TX)

GENERAL

T - TRACE AMOUNT
BLANK ENTRIES DENOTE MISSING/UNREPORTED DATA.
# INDICATES A STATION OR INSTRUMENT RELOCATION.

SPECIFIC

TABLE 1

(a) - LENGTH OF RECORD IN YEARS. ALTHOUGH
INDIVIDUAL MONTHS MAY BE MISSING.
* LESS THAN .05

NORMALS — BASED ON THE 1951-1980 RECORD PERIOD.
EXTREMES — DATES ARE THE MOST RECENT OCCURRENCE.
WIND DIR. — NUMERALS SHOW TENS OF DEGREES
CLOCKWISE FROM TRUE NORTH.
"00" INDICATES CALM.
RESULTANT WIND DIRECTIONS ARE GIVEN TO WHOLE DEGREES.

EXCEPTIONS

TABLES 2, 3, and 6

RECORD MEANS ARE THROUGH THE CURRENT YEAR,
BEGINNING IN   1907 FOR TEMPERATURE
1867 FOR PRECIPITATION
1948 FOR SNOWFALL

TABLE 4   HEATING DEGREE DAYS Base 65 deg. F   SAN ANGELO, TEXAS

SEASON	JULY	AUG	SEP	OCT	NOV	DEC	JAN	FEB	MAR	APR	MAY	JUNE	TOTAL
1956-57	0	0	0	33	365	473	594	273	239	147	30	5	2159
1957-58	0	0	0	138	451	424	611	546	487	142	14	0	2813
1958-59	0	0	2	143	297	616	614	458	264	156	6	0	2556
#1959-60	0	0	4	87	499	512	576	580	406	69	31	0	2764
1960-61	0	0	0	39	232	662	701	411	215	85	1	1	2347
1961-62	0	0	8	66	437	575	719	265	331	83	6	0	2490
1962-63	0	0	0	36	239	526	711	451	161	34	7	0	2165
1963-64	0	0	3	4	269	747	609	562	169	30	5	2	2400
1964-65	0	0	2	65	277	492	469	514	471	31	0	0	2321
1965-66	0	0	0	76	144	416	747	538	221	55	30	0	2227
1966-67	0	0	0	127	192	614	568	459	113	2	19	0	2094
1967-68	0	0	5	76	249	608	577	550	304	127	1	0	2497
1968-69	0	0	0	38	362	542	399	372	416	33	9	0	2171
1969-70	0	0	0	122	335	441	654	391	376	92	44	0	2455
1970-71	0	0	26	126	320	326	453	359	211	81	9	0	1911
1971-72	0	0	23	33	289	450	539	382	138	26	3	0	1883
1972-73	0	0	3	97	438	577	738	514	183	191	20	0	2761
1973-74	0	0	1	33	170	476	554	352	115	53	1	0	1755
1974-75	0	0	33	37	308	560	568	470	288	101	1	0	2366
1975-76	0	0	32	75	320	519	645	251	237	48	30	0	2157
1976-77	0	0	3	220	518	595	757	377	227	45	0	0	2742
1977-78	0	0	0	44	249	390	818	588	286	31	21	0	2427
1978-79	0	0	3	74	294	642	886	510	251	60	34	0	2754
1979-80	0	0	0	57	430	542	578	453	300	105	11	0	2476
1980-81	0	0	5	101	374	464	588	411	330	62	17	0	2352
1981-82	0	0	8	133	281	496	582	532	220	124	20	0	2396
1982-83	0	0	0	105	338	586	611	448	268	163	13	3	2535
1983-84	0	0	7	20	245	883	754	422	258	74	5	0	2668
1984-85	0	0	59	90	365	455	804	522	223	52	1	0	2571
1985-86	0	0	22	50	232	655							

TABLE 5   COOLING DEGREE DAYS Base 65 deg. F   SAN ANGELO, TEXAS

YEAR	JAN	FEB	MAR	APR	MAY	JUNE	JULY	AUG	SEP	OCT	NOV	DEC	TOTAL
1969	6	1	6	99	213	513	707	620	311	113	23	0	2612
1970	6	1	7	135	248	422	575	639	432	101	10	9	2585
1971	0	4	49	139	332	482	557	328	308	96	23	0	2318
1972	0	20	65	232	191	488	574	468	406	140	7	2	2593
1973	0	0	6	67	274	425	573	554	314	104	49	0	2366
1974	3	4	120	147	349	481	597	460	159	87	15	0	2422
1975	3	0	18	103	220	448	427	493	238	90	18	0	2058
1976	0	12	29	66	155	449	354	491	272	27	0	0	1855
1977	0	6	14	53	333	523	621	662	562	183	17	1	2975
1978	0	0	22	204	360	467	634	454	302	72	13	0	2528
1979	0	2	17	77	263	394	579	487	309	217	12	0	2357
1980	0	0	9	73	229	535	685	610	368	90	19	2	2620
1981	0	1	2	112	202	390	536	474	290	129	6	0	2142
1982	0	4	43	88	170	392	572	616	378	142	24	0	2429
1983	0	0	13	88	253	359	539	535	390	191	55	0	2423
1984	0	2	30	86	352	528	517	558	291	82	10	4	2460
1985	0	1	53	153	318	381	486	610	352	113	26	0	2493

TABLE 6   SNOWFALL (inches)   SAN ANGELO, TEXAS

SEASON	JULY	AUG	SEP	OCT	NOV	DEC	JAN	FEB	MAR	APR	MAY	JUNE	TOTAL
1970-71	0.0	0.0	0.0	0.0	0.0	0.0	0.0	T	T	0.0	0.0	0.0	T
1971-72	0.0	0.0	0.0	0.0	0.0	0.0	3.3	0.0	T	0.0	0.0	0.0	3.3
1972-73	0.0	0.0	0.0	0.0	T	T	7.7	5.8	0.2	0.0	0.0	0.0	13.5
1973-74	0.0	0.0	0.0	0.0	0.0	0.0	T	T	0.2	0.0	0.0	0.0	0.2
1974-75	0.0	0.0	0.0	0.0	0.0	0.1	1.5	T	0.0	0.0	0.0	0.0	1.6
1975-76	0.0	0.0	0.0	0.0	0.0	0.0	T	0.0	0.0	0.0	0.0	0.0	T
1976-77	0.0	0.0	0.0	0.0	3.0	0.0	0.1	0.0	0.0	0.0	0.0	0.0	3.1
1977-78	0.0	0.0	0.0	0.0	0.0	0.0	9.0	0.6	0.0	0.0	0.0	0.0	9.6
1978-79	0.0	0.0	0.0	0.0	0.0	1.0	T	T	0.0	0.0	0.0	0.0	1.0
1979-80	0.0	0.0	0.0	0.0	0.0	0.0	T	1.0	0.0	T	0.0	0.0	1.0
1980-81	0.0	0.0	0.0	0.0	2.3	0.0	1.8	T	0.0	0.0	0.0	0.0	4.1
1981-82	0.0	0.0	0.0	0.0	0.0	0.0	6.8	T	0.0	0.0	0.0	0.0	6.8
1982-83	0.0	0.0	0.0	0.0	0.0	T	2.9	0.0	0.0	0.0	0.0	0.0	2.9
1983-84	0.0	0.0	0.0	0.0	0.0	0.3	T	0.9	T	0.0	0.0	0.0	1.2
1984-85	0.0	0.0	0.0	0.0	T	T	8.4	0.4	0.0	0.0	0.0	0.0	8.8
1985-86	0.0	0.0	0.0	0.0	0.0	T							
Record Mean	0.0	0.0	0.0	0.0	0.6	0.1	1.6	0.7	0.2	T	0.0	0.0	3.3

See Reference Notes, relative to all above tables, on preceding page.

The city of San Antonio is located in the south-central portion of Texas on the Balcones escarpment. Northwest of the city, the terrain slopes upward to the Edwards Plateau and to the southeast it slopes downward to the Gulf Coastal Plains. Soils are blackland clay and silty loam on the Plains and thin limestone soils on the Edwards Plateau.

The location of San Antonio on the edge of the Gulf Coastal Plains is influenced by a modified subtropical climate, predominantly continental during the winter months and marine during the summer months. Temperatures range from 50 degrees in January to the middle 80s in July and August. While the summer is hot, with daily temperatures above 90 degrees over 80 percent of the time, extremely high temperatures are rare. Mild weather prevails during much of the winter months, with below-freezing temperatures occurring on an average of about 20 days each year.

San Antonio is situated between a semi-arid area to the west and the coastal area of heavy precipitation to the east. The normal annual rainfall of nearly 28 inches is sufficient for the production of most crops. Precipitation is fairly well distributed throughout the year with the heaviest amounts occurring during May and September. The precipitation from April through September usually occurs from thunderstorms. Large amounts of precipitation may fall during short periods of time. Most of the winter precipitation occurs as light rain or drizzle. Thunderstorms and heavy rains have occurred in all months of the year. Hail of damaging intensity seldom occurs but light hail is frequent with the springtime thunderstorms. Measurable snow occurs only once in three or four years. Snowfall of 2 to 4 inches occurs about every ten years.

Northerly winds prevail during most of the winter, and strong northerly winds occasionally occur during storms called northers. Southeasterly winds from the Gulf of Mexico also occur frequently during winter and are predominant in summer.

Since San Antonio is located only 140 miles from the Gulf of Mexico, tropical storms occasionally affect the city with strong winds and heavy rains. One of the fastest winds recorded, 74 mph, occurred as a tropical storm moved inland east of the city in August 1942.

Relative humidity is above 80 percent during the early morning hours most of the year, dropping to near 50 percent in the late afternoon.

San Antonio has about 50 percent of the possible amount of sunshine during the winter months and more than 70 percent during the summer months. Skies are clear to partly cloudy more than 60 percent of the time and cloudy less than 40 percent. Air carried over San Antonio by southeasterly winds is lifted orographically, causing low stratus clouds to develop frequently during the later part of the night. These clouds usually dissipate around noon, and clear skies prevail a high percentage of the time during the afternoon.

The first occurrence of 32 degrees Fahrenheit is in late November and the average last occurrence is in early March.

TABLE 1   # NORMALS, MEANS AND EXTREMES

SAN ANTONIO, TEXAS

LATITUDE: 29°32'N   LONGITUDE: 98°28'W   ELEVATION: FT. GRND 788 BARO 00796   TIME ZONE: CENTRAL   WBAN: 12921

	(a)	JAN	FEB	MAR	APR	MAY	JUNE	JULY	AUG	SEP	OCT	NOV	DEC	YEAR
TEMPERATURE °F:														
Normals														
-Daily Maximum		61.7	66.3	73.7	80.3	85.5	91.8	94.9	94.6	89.3	81.5	70.7	64.6	79.6
-Daily Minimum		39.0	42.4	49.8	58.8	65.5	72.0	74.3	73.7	69.4	58.9	48.2	41.4	57.8
-Monthly		50.4	54.4	61.8	69.6	75.5	81.9	84.6	84.2	79.3	70.2	59.5	53.0	68.7
Extremes														
-Record Highest	44	89	92	100	100	101	105	106	106	103	98	91	90	106
-Year		1971	1959	1971	1984	1967	1980	1954	1962	1985	1979	1962	1955	AUG 1962
-Record Lowest	44	0	6	19	33	43	53	62	61	41	33	21	9	0
-Year		1949	1951	1980	1983	1984	1964	1967	1966	1942	1980	1976	1983	JAN 1949
NORMAL DEGREE DAYS:														
Heating (base 65°F)		463	319	178	28	0	0	0	0	0	41	199	378	1606
Cooling (base 65°F)		10	19	78	166	326	507	608	595	432	202	34	6	2983
% OF POSSIBLE SUNSHINE	43	48	52	57	55	56	67	74	73	67	63	55	50	60
MEAN SKY COVER (tenths)														
Sunrise - Sunset	43	6.3	6.1	6.2	6.4	6.4	5.5	5.0	4.9	5.2	5.0	5.5	5.9	5.7
MEAN NUMBER OF DAYS:														
Sunrise to Sunset														
-Clear	43	8.9	8.4	8.4	7.3	6.2	7.0	9.2	10.4	9.5	11.6	10.7	9.9	107.5
-Partly Cloudy	43	6.2	6.0	7.3	7.4	11.3	15.4	15.3	14.6	12.2	9.8	6.9	6.1	118.4
-Cloudy	43	15.9	13.9	15.3	15.2	13.5	7.7	6.5	6.0	8.3	9.6	12.5	15.0	139.3
Precipitation														
.01 inches or more	43	8.1	7.7	7.1	7.5	8.3	6.0	4.3	5.3	7.1	6.5	6.5	7.3	81.5
Snow,Ice pellets														
1.0 inches or more	43	0.1	0.1	0.0	0.0	0.0	0.0	0.0	0.0	0.0	0.0	0.0	0.0	0.2
Thunderstorms	43	0.9	1.4	2.4	4.0	6.5	4.3	3.4	4.2	4.0	2.6	1.8	0.8	36.3
Heavy Fog Visibility														
1/4 mile or less	43	5.3	3.1	2.5	1.4	0.7	0.1	0.1	0.*	0.2	1.5	3.0	4.7	22.7
Temperature °F														
-Maximum														
90° and above	43	0.0	0.1	0.9	2.2	8.4	21.6	28.3	28.0	16.8	4.1	0.1	0.*	110.4
32° and below	43	0.3	0.1	0.0	0.0	0.0	0.0	0.0	0.0	0.0	0.0	0.0	0.1	0.5
-Minimum														
32° and below	43	8.7	5.0	1.5	0.0	0.0	0.0	0.0	0.0	0.0	0.0	1.9	6.0	23.1
0° and below	43	0.*	0.0	0.0	0.0	0.0	0.0	0.0	0.0	0.0	0.0	0.0	0.0	*
AVG. STATION PRESS.(mb)	13	991.7	990.5	986.0	985.7	984.2	985.8	987.3	987.3	987.2	988.9	989.9	991.2	988.0
RELATIVE HUMIDITY (%)														
Hour 00	43	76	75	72	76	81	80	75	74	77	77	76	76	76
Hour 06 (Local Time)	43	80	80	79	82	87	88	87	86	86	84	81	80	83
Hour 12	43	59	57	54	56	59	56	51	51	55	54	55	57	55
Hour 18	43	57	52	47	51	55	51	45	46	52	53	56	57	52
PRECIPITATION (inches):														
Water Equivalent														
-Normal		1.55	1.86	1.33	2.73	3.67	3.03	1.92	2.69	3.75	2.88	2.34	1.38	29.13
-Maximum Monthly	43	8.52	6.43	4.19	9.32	11.24	10.44	8.19	11.14	15.78	9.56	6.01	4.51	15.78
-Year		1968	1965	1957	1957	1972	1973	1942	1974	1946	1942	1977	1965	SEP 1946
-Minimum Monthly	43	0.04	0.03	0.03	0.11	0.17	0.01	T	0.00	0.06	T	T	0.03	0.00
-Year		1971	1954	1961	1984	1961	1967	1984	1952	1947	1952	1966	1950	AUG 1952
-Maximum in 24 hrs	43	3.18	2.34	2.36	4.88	6.53	6.18	6.97	5.57	7.28	5.29	4.87	2.89	7.28
-Year		1968	1965	1945	1977	1972	1951	1958	1950	1973	1942	1977	1944	SEP 1973
Snow,Ice pellets														
-Maximum Monthly	43	15.9	3.5	T								0.3	0.2	15.9
-Year		1985	1966	1978								1957	1964	JAN 1985
-Maximum in 24 hrs	43	13.2	3.5	T								0.3	0.2	13.2
-Year		1985	1966	1978								1957	1964	JAN 1985
WIND:														
Mean Speed (mph)	43	9.2	9.8	10.6	10.6	10.2	10.2	9.3	8.6	8.6	8.6	8.9	8.7	9.5
Prevailing Direction														
through 1963		N	NE	SE	SE	SE	SE	SSE	SE	SE	N	N	N	SE
Fastest Obs. 1 Min.														
-Direction	9	31	31	36	35	02	34	09	20	18	02	33	32	09
-Speed (MPH)	9	35	42	35	39	43	35	48	37	42	31	37	30	48
-Year		1979	1984	1980	1979	1983	1984	1979	1984	1977	1977	1983	1984	JUL 1979
Peak Gust														
-Direction	2	N	NW	NW	N	N	NW	SE	SW	SE	N	NW	NW	NW
-Speed (mph)	2	51	56	46	37	38	51	40	49	41	43	38	48	56
-Date		1985	1984	1984	1985	1985	1984	1985	1984	1984	1985	1985	1984	FEB 1984

See reference Notes to this table on the following page.

TABLE 2 PRECIPITATION (inches) SAN ANTONIO, TEXAS

YEAR	JAN	FEB	MAR	APR	MAY	JUNE	JULY	AUG	SEP	OCT	NOV	DEC	ANNUAL
1956	0.81	0.85	0.27	0.49	3.07	0.27	0.53	3.94	0.62	1.23	1.13	1.10	14.31
1957	0.51	2.53	4.19	9.32	8.22	3.49	0.73	0.21	11.10	4.71	2.90	0.92	48.83
1958	4.57	3.88	1.08	1.32	1.98	3.39	7.39	0.45	8.36	5.43	0.77	1.07	39.69
1959	0.52	2.50	0.13	2.55	2.43	1.32	1.48	3.05	1.72	5.11	2.17	1.52	24.50
1960	0.76	1.22	1.65	2.08	1.21	2.70	1.31	5.96	0.76	7.84	1.30	2.97	29.76
1961	0.68	1.79	0.03	0.32	0.17	7.87	7.04	0.15	2.24	3.39	2.09	0.70	26.47
1962	0.48	0.90	0.91	4.02	1.31	2.44	0.13	1.57	2.69	2.19	4.97	2.29	23.90
1963	0.27	3.59	0.21	1.88	3.03	2.28	0.03	0.63	1.11	2.75	1.93	0.94	18.65
1964	3.40	1.89	1.73	1.16	1.79	4.88	0.02	5.19	4.15	1.64	4.81	1.22	31.88
1965	2.40	6.43	2.30	1.97	8.18	2.42	0.08	1.65	3.13	2.69	0.89	4.51	36.65
1966	1.47	2.30	1.14	3.20	3.53	1.78	0.06	4.28	2.13	1.11	T	0.44	21.44
1967	0.18	0.48	2.18	0.94	2.22	0.01	2.12	3.17	11.16	2.00	3.42	1.38	29.26
1968	8.52	1.85	1.27	1.92	2.82	2.63	1.53	0.94	2.99	0.69	4.58	0.66	30.40
1969	1.76	2.90	2.35	2.46	4.61	2.32	0.36	4.19	1.32	5.85	1.02	2.28	31.42
1970	1.10	2.66	1.98	1.13	7.30	0.89	0.91	0.95	4.35	1.31	0.01	0.15	22.74
1971	0.04	0.81	0.04	1.39	1.52	2.74	1.05	9.42	4.57	4.62	2.74	2.86	31.80
1972	1.35	0.40	0.13	1.94	11.24	2.86	3.13	4.24	1.40	1.99	2.37	0.44	31.49
1973	2.77	2.76	1.58	5.41	2.73	10.44	6.91	1.29	13.09	4.85	0.29	0.16	52.28
1974	1.36	0.04	0.94	2.18	4.28	1.02	1.28	11.14	3.85	4.09	5.39	1.43	37.00
1975	1.04	3.30	0.52	2.69	6.91	4.60	1.06	1.28	0.51	2.25	0.03	1.48	25.67
1976	0.56	0.13	1.20	5.67	5.80	1.61	5.39	2.09	3.79	8.48	2.46	1.95	39.13
1977	3.10	0.91	0.88	8.80	1.62	2.26	0.10	0.06	3.47	6.01	0.32	0.64	29.64
1978	0.68	1.76	1.71	3.62	2.45	3.96	1.43	4.97	8.86	0.55	4.91	1.09	35.99
1979	4.07	1.38	3.55	5.34	1.98	5.59	7.38	2.09	0.86	0.11	1.43	2.86	36.64
1980	0.72	0.74	0.98	1.67	6.42	0.52	0.26	2.64	5.05	1.09	3.53	0.61	24.23
1981	2.06	0.96	1.96	2.21	6.43	8.71	0.25	2.41	1.36	8.61	0.72	0.69	36.37
1982	0.72	1.28	0.69	1.23	6.42	1.37	0.14	0.55	0.87	2.84	4.54	2.31	22.96
1983	1.48	1.54	3.89	0.18	4.37	1.27	2.43	2.00	3.86	1.64	3.06	0.39	26.11
1984	1.87	0.54	1.91	0.11	3.76	1.40	T	3.04	1.06	5.94	2.91	3.41	25.95
1985	2.68	1.91	2.85	3.27	2.47	8.20	5.80	0.45	4.80	3.91	5.00	0.09	41.43
Record Mean	1.61	1.64	1.66	2.92	3.46	2.90	2.03	2.39	3.21	2.54	1.97	1.63	27.99

TABLE 3 AVERAGE TEMPERATURE (deg. F) SAN ANTONIO, TEXAS

YEAR	JAN	FEB	MAR	APR	MAY	JUNE	JULY	AUG	SEP	OCT	NOV	DEC	ANNUAL
1956	52.1	57.1	63.3	69.7	78.9	84.7	85.5	84.8	80.0	73.0	57.8	56.9	70.3
1957	53.3	62.9	61.9	66.3	73.4	81.0	85.7	86.0	77.8	66.9	57.1	55.5	69.0
1958	50.1	50.0	55.8	67.7	75.1	82.8	84.6	84.8	79.2	67.7	60.8	50.1	67.4
1959	48.9	54.0	60.0	65.9	77.1	82.9	84.0	84.4	81.4	70.4	53.3	54.0	68.0
1960	50.1	49.9	56.1	69.7	74.1	83.3	84.2	83.6	78.6	73.3	62.2	50.1	67.9
1961	47.9	55.9	65.7	68.5	78.5	81.3	82.6	82.5	80.5	71.1	58.0	54.2	68.9
1962	45.9	62.8	59.1	69.7	77.9	82.3	86.9	87.5	80.9	75.5	60.4	52.2	70.1
1963	46.2	52.6	65.6	74.6	77.7	83.4	85.4	85.7	81.1	74.1	62.4	45.7	69.6
1964	51.0	49.8	61.5	70.5	77.6	82.4	86.3	86.2	80.0	66.4	62.6	52.3	68.9
1965	54.4	49.8	54.9	71.6	75.0	81.6	84.9	84.0	80.7	66.8	64.5	55.5	68.6
1966	45.4	49.8	60.0	68.6	73.5	78.8	84.2	81.9	77.5	67.0	63.0	50.7	66.7
1967	50.2	51.8	66.9	76.6	76.6	84.5	85.3	82.7	75.5	66.9	60.5	51.0	69.0
1968	49.8	48.3	58.0	68.1	75.3	80.5	82.7	84.2	76.0	72.2	56.4	50.7	66.8
1969	52.5	53.6	54.9	69.0	73.5	81.2	86.8	85.7	79.6	69.8	58.1	55.1	68.3
1970	45.6	54.8	56.8	70.2	72.9	80.7	84.0	85.7	81.1	67.7	58.0	60.1	68.1
1971	56.0	57.4	64.6	69.4	78.1	83.6	85.9	81.2	80.1	73.9	63.2	57.2	70.9
1972	52.8	56.7	66.3	73.7	72.8	80.3	82.2	82.1	82.0	71.9	54.0	50.3	68.8
1973	47.2	51.9	66.1	66.0	74.7	79.2	83.2	82.1	79.3	72.5	65.8	52.2	68.4
1974	51.0	56.5	67.9	69.7	77.3	79.4	83.0	81.2	72.3	68.2	57.3	50.9	67.9
1975	53.2	53.5	61.4	68.4	73.5	80.0	80.9	81.7	76.0	71.1	60.3	53.1	67.8
1976	49.6	61.2	63.8	68.9	71.3	79.8	79.8	81.6	77.5	61.1	52.1	49.9	66.4
1977	44.1	52.8	61.8	66.9	74.8	81.5	84.9	84.7	82.3	71.2	61.4	53.4	68.3
1978	43.4	46.4	59.6	68.9	77.1	82.7	86.1	83.1	78.5	69.3	62.4	51.7	67.5
1979	43.7	52.4	63.3	69.7	73.9	80.9	84.7	83.1	78.7	74.7	58.2	55.4	68.2
1980	52.6	53.7	61.5	67.6	76.1	85.1	88.1	85.3	83.7	70.7	58.3	55.0	69.8
1981	50.8	53.7	60.7	72.9	75.3	81.5	84.2	84.7	78.9	71.9	62.4	53.0	69.2
1982	50.8	49.7	63.1	66.9	74.5	81.6	85.5	86.0	80.1	69.3	59.4	52.4	68.3
1983	48.9	52.1	58.7	65.2	73.6	79.2	82.9	84.5	78.5	70.9	62.5	43.0	66.7
1984	46.7	54.1	64.2	69.7	77.1	82.8	85.0	84.7	77.6	71.2	58.8	59.6	69.3
1985	44.2	50.5	64.1	69.4	76.7	80.2	82.2	85.5	79.4	71.7	64.4	49.9	68.2
Record Mean	51.6	55.1	62.2	69.3	75.4	81.6	84.1	84.1	79.3	70.8	60.5	53.7	69.0
Max	62.1	66.0	73.4	80.0	85.4	91.5	94.4	94.8	89.4	81.7	71.0	64.1	79.5
Min	41.1	44.2	51.0	58.6	65.4	71.6	73.5	73.4	69.2	59.9	49.9	43.2	58.4

REFERENCE NOTES FOR TABLES 1, 2, 3 and 6   (SAN ANTONIO, TX)

GENERAL

T - TRACE AMOUNT
BLANK ENTRIES DENOTE MISSING/UNREPORTED DATA.
# INDICATES A STATION OR INSTRUMENT RELOCATION.

SPECIFIC

TABLE 1

(a) - LENGTH OF RECORD IN YEARS. ALTHOUGH
      INDIVIDUAL MONTHS MAY BE MISSING.
  * LESS THAN .05

NORMALS — BASED ON THE 1951-1980 RECORD PERIOD.
EXTREMES — DATES ARE THE MOST RECENT OCCURRENCE.
WIND DIR. — NUMERALS SHOW TENS OF DEGREES
            CLOCKWISE FROM TRUE NORTH.
            "00" INDICATES CALM.
RESULTANT WIND DIRECTIONS ARE GIVEN TO WHOLE DEGREES.

EXCEPTIONS

TABLES 2, 3, and 6

RECORD MEANS ARE THROUGH THE CURRENT YEAR,
BEGINNING IN   1885 FOR TEMPERATURE
                 1885 FOR PRECIPITATION
                 1943 FOR SNOWFALL

TABLE 4   HEATING DEGREE DAYS Base 65 deg. F   SAN ANTONIO, TEXAS

SEASON	JULY	AUG	SEP	OCT	NOV	DEC	JAN	FEB	MAR	APR	MAY	JUNE	TOTAL
1956-57	0	0	0	3	229	271	370	126	132	71	7	0	1209
1957-58	0	0	0	71	270	292	453	419	289	22	0	0	1816
1958-59	0	0	0	75	171	451	497	311	179	90	0	0	1774
1959-60	0	0	0	27	374	336	467	433	291	14	7	0	1949
1960-61	0	0	0	17	138	457	523	272	82	63	0	0	1552
1961-62	0	0	0	19	223	351	586	108	206	27	0	0	1520
1962-63	0	0	0	9	164	393	575	349	87	17	3	0	1597
1963-64	0	0	0	0	141	592	428	434	143	23	0	0	1761
1964-65	0	0	0	41	155	414	346	419	327	13	0	0	1715
1965-66	0	0	2	62	64	301	607	426	182	39	5	0	1688
1966-67	0	0	0	57	131	456	470	366	80	0	0	0	1560
1967-68	0	0	8	48	164	429	477	478	254	39	0	0	1897
1968-69	0	0	0	9	278	437	394	319	315	5	3	0	1760
1969-70	0	0	0	52	253	299	599	282	266	45	7	0	1803
1970-71	0	0	1	72	247	201	282	239	134	52	1	0	1229
1971-72	0	0	1	0	129	266	382	263	61	7	0	0	1109
1972-73	0	0	0	29	334	457	551	362	29	94	1	0	1857
1973-74	0	0	0	4	85	391	437	257	74	39	0	0	1287
1974-75	0	0	2	19	260	433	389	316	152	41	0	0	1612
1975-76	0	0	1	21	214	394	472	166	143	11	2	0	1424
1976-77	0	0	0	160	382	461	643	336	144	32	0	0	2158
1977-78	0	0	0	19	138	360	667	521	192	27	4	0	1928
1978-79	0	0	0	12	152	413	657	356	109	20	4	0	1723
1979-80	0	0	0	15	243	306	386	333	163	42	0	0	1488
1980-81	0	0	0	62	245	331	437	332	157	10	0	0	1574
1981-82	0	0	2	52	112	368	445	430	171	77	2	0	1659
1982-83	0	0	0	49	237	404	490	356	208	99	1	0	1844
1983-84	0	0	5	20	154	681	563	315	120	21	2	0	1881
1984-85	0	0	9	28	228	203	635	406	109	26	0	0	1644
1985-86	0	0	10	9	112	467							

TABLE 5   COOLING DEGREE DAYS Base 65 deg. F   SAN ANTONIO, TEXAS

YEAR	JAN	FEB	MAR	APR	MAY	JUNE	JULY	AUG	SEP	OCT	NOV	DEC	TOTAL
1969	11	5	8	133	273	494	683	652	448	207	53	2	2969
1970	3	3	20	208	259	478	592	651	493	163	40	57	2967
1971	14	36	130	189	414	564	658	509	459	281	81	31	3366
1972	11	31	105	276	252	465	542	539	515	249	12	6	3003
1973	8	0	69	129	310	431	570	536	437	242	114	0	2846
1974	11	22	171	188	387	439	568	506	229	124	34	5	2684
1975	29	1	51	151	273	457	502	524	337	217	80	30	2652
1976	3	62	113	136	202	451	467	521	383	45	0	0	2383
1977	0	3	52	98	311	502	620	618	525	218	38	5	2990
1978	3	7	30	152	384	537	660	567	410	154	79	11	2994
1979	3	13	65	166	285	482	619	570	418	322	42	13	2998
1980	11	14	61	127	355	614	725	635	567	245	51	26	3431
1981	3	24	30	255	324	502	603	619	424	273	41	3	3101
1982	11	5	117	142	304	504	645	659	459	191	72	19	3128
1983	0	0	21	111	276	435	560	611	417	207	84	8	2730
1984	0	8	101	169	383	541	625	618	394	230	46	44	3159
1985	0	8	85	165	368	462	539	641	450	223	101	5	3047

TABLE 6   SNOWFALL (inches)   SAN ANTONIO, TEXAS

SEASON	JULY	AUG	SEP	OCT	NOV	DEC	JAN	FEB	MAR	APR	MAY	JUNE	TOTAL
1956-57	0.0	0.0	0.0	0.0	0.0	0.0	T	0.0	0.0	0.0	0.0	0.0	T
1957-58	0.0	0.0	0.0	0.0	0.3	0.0	0.0	1.2	T	0.0	0.0	0.0	1.5
1958-59	0.0	0.0	0.0	0.0	0.0	0.0	0.0	0.0	T	0.0	0.0	0.0	T
1959-60	0.0	0.0	0.0	0.0	T	0.0	0.0	T	0.0	0.0	0.0	0.0	T
1960-61	0.0	0.0	0.0	0.0	0.0	T	T	T	0.0	0.0	0.0	0.0	T
1961-62	0.0	0.0	0.0	0.0	0.0	0.0	0.0	0.0	0.0	0.0	0.0	0.0	0.0
1962-63	0.0	0.0	0.0	0.0	0.0	0.0	0.0	T	0.0	0.0	0.0	0.0	T
1963-64	0.0	0.0	0.0	0.0	0.0	T	T	2.0	0.0	0.0	0.0	0.0	2.0
1964-65	0.0	0.0	0.0	0.0	0.0	0.2	0.0	T	0.0	0.0	0.0	0.0	0.2
1965-66	0.0	0.0	0.0	0.0	0.0	0.0	T	3.5	0.0	0.0	0.0	0.0	3.5
1966-67	0.0	0.0	0.0	0.0	0.0	0.0	T	T	0.0	0.0	0.0	0.0	T
1967-68	0.0	0.0	0.0	0.0	0.0	0.0	0.0	T	0.0	0.0	0.0	0.0	T
1968-69	0.0	0.0	0.0	0.0	0.0	0.0	0.0	0.0	0.0	0.0	0.0	0.0	0.0
1969-70	0.0	0.0	0.0	0.0	0.0	T	T	T	0.0	0.0	0.0	0.0	T
1970-71	0.0	0.0	0.0	0.0	0.0	0.0	0.0	0.0	0.0	0.0	0.0	0.0	0.0
1971-72	0.0	0.0	0.0	0.0	0.0	0.0	T	0.0	0.0	0.0	0.0	0.0	T
1972-73	0.0	0.0	0.0	0.0	0.0	T	0.8	2.1	0.0	0.0	0.0	0.0	2.9
1973-74	0.0	0.0	0.0	0.0	0.0	0.0	0.0	0.0	0.0	0.0	0.0	0.0	0.0
1974-75	0.0	0.0	0.0	0.0	0.0	T	T	0.0	0.0	0.0	0.0	0.0	T
1975-76	0.0	0.0	0.0	0.0	0.0	0.0	0.0	0.0	0.0	0.0	0.0	0.0	T
1976-77	0.0	0.0	0.0	0.0	T	T	0.0	0.0	0.0	0.0	0.0	0.0	T
1977-78	0.0	0.0	0.0	0.0	0.0	0.0	0.0	T	0.0	0.0	0.0	0.0	T
1978-79	0.0	0.0	0.0	0.0	0.0	0.0	T	0.0	0.0	0.0	0.0	0.0	T
1979-80	0.0	0.0	0.0	0.0	T	0.0	0.0	T	0.0	0.0	0.0	0.0	T
1980-81	0.0	0.0	0.0	0.0	T	0.0	T	0.0	0.0	0.0	0.0	0.0	T
1981-82	0.0	0.0	0.0	0.0	0.0	0.0	0.5	0.0	0.0	0.0	0.0	0.0	0.5
1982-83	0.0	0.0	0.0	0.0	0.0	0.0	0.0	0.0	0.0	0.0	0.0	0.0	0.0
1983-84	0.0	0.0	0.0	0.0	0.0	0.0	0.0	0.0	0.0	0.0	0.0	0.0	0.0
1984-85	0.0	0.0	0.0	0.0	0.0	0.0	15.9	T	0.0	0.0	0.0	0.0	15.9
1985-86	0.0	0.0	0.0	0.0	0.0	0.0							
Record Mean	0.0	0.0	0.0	0.0	T	T	0.6	0.2	T	0.0	0.0	0.0	0.8

See Reference Notes, relative to all above tables, on preceding page.

The city of Victoria is located in the south-central Texas Coastal Plain. The climate is classified as humid subtropical. Summers are hot with about 100 days with temperatures of 90 degrees or above. However, pleasant sea breezes from the nearby Gulf of Mexico make the high temperatures bearable.

Spring is characterized by mild days, brisk winds, and occasional showers and thunderstorms. Strong southeast winds begin in March, diminish in April and May, and become pleasant sea breezes in the first half of June. Thunderstorm activity increases through March and April, reaching a peak in May. Considerable cloudiness is the rule, with almost 50 percent of the days in the spring having overcast or nearly overcast skies.

The sea breeze diminishes during the summer, and at times fails altogether, and some hot nights are experienced in late June, July, and early August. High summer humidity gives way to clear, drier air in late August. Nighttime temperatures drop to pleasant levels. Thunderstorms continue, and lawns and fields remain green.

The first norther usually arrives near the beginning of fall, in late September. October and November are ideal fall months with long periods of clear days with mild temperatures and cool nights. The amount of rainfall decreases.

The winter season weather conditions alternate between clear, cold, dry periods and cloudy, mild, drizzly days as fronts move down from the north. The temperature drops below 32 degrees on an average of about a dozen mornings per year.

The normal rainfall of about 36 inches is well distributed throughout the year, with the heaviest falls coming during the growing season. Some of the smaller streams dry up in the late summer, and during occasional periods of general drought some of the larger streams may reach pool stage.

The area is subject to occasional tropical disturbances during summer and fall. Destructive winds and torrential rains may occur in these storms. Approximately 50 days per year have thunderstorms, but hail is infrequent. Destructive storms with tornados are rare.

TABLE 1   NORMALS, MEANS AND EXTREMES

VICTORIA, TEXAS

LATITUDE: 28°51'N   LONGITUDE: 96°55'W   ELEVATION: FT. GRND   104 BARO   00107   TIME ZONE: CENTRAL   WBAN: 12912

	(a)	JAN	FEB	MAR	APR	MAY	JUNE	JULY	AUG	SEP	OCT	NOV	DEC	YEAR
TEMPERATURE °F:														
Normals														
-Daily Maximum		63.6	67.1	73.8	80.2	85.6	90.8	93.7	93.7	89.3	82.8	73.0	66.7	80.0
-Daily Minimum		43.1	45.9	52.8	61.5	67.7	73.1	75.2	74.7	70.9	61.0	51.5	45.4	60.2
-Monthly		53.4	56.5	63.3	70.8	76.7	82.0	84.5	84.2	80.1	71.9	62.3	56.0	70.1
Extremes														
-Record Highest	25	88	90	95	98	101	100	104	107	102	95	92	88	107
-Year		1971	1962	1984	1963	1964	1980	1964	1962	1985	1962	1963	1964	AUG 1962
-Record Lowest	25	14	19	21	35	49	59	62	63	49	36	24	14	14
-Year		1982	1985	1980	1980	1978	1984	1967	1967	1984	1980	1976	1983	DEC 1983
NORMAL DEGREE DAYS:														
Heating (base 65°F)		386	268	140	18	0	0	0	0	0	20	150	291	1273
Cooling (base 65°F)		27	30	87	195	363	510	605	595	453	234	69	16	3184
% OF POSSIBLE SUNSHINE														
MEAN SKY COVER (tenths)														
Sunrise - Sunset	24	7.0	6.4	7.0	7.1	6.7	5.8	5.7	5.8	5.7	5.2	5.7	6.8	6.2
MEAN NUMBER OF DAYS:														
Sunrise to Sunset														
-Clear	24	6.5	7.6	6.2	5.3	4.9	6.0	6.6	6.6	7.9	11.1	9.7	7.3	85.6
-Partly Cloudy	24	6.5	5.9	6.8	7.3	11.1	15.1	15.6	15.3	12.2	10.0	7.3	5.9	118.9
-Cloudy	24	18.0	14.7	18.0	17.4	15.0	8.9	8.8	9.2	10.0	10.0	13.0	17.9	160.8
Precipitation														
.01 inches or more	24	8.1	6.5	6.9	6.0	7.2	7.2	7.6	8.8	10.2	6.8	6.5	7.9	89.5
Snow, Ice pellets														
1.0 inches or more	24	0.1	0.*	0.0	0.0	0.0	0.0	0.0	0.0	0.0	0.0	0.0	0.0	0.1
Thunderstorms	24	0.9	1.5	2.4	3.2	5.9	5.4	6.6	9.0	7.9	4.0	1.8	1.0	49.3
Heavy Fog Visibility														
1/4 mile or less	24	7.0	5.2	5.1	4.3	1.9	0.6	0.1	0.3	1.2	3.1	5.9	6.4	41.1
Temperature °F														
-Maximum														
90° and above	24	0.0	0.*	0.3	0.8	5.3	20.0	27.3	27.6	16.3	4.0	0.1	0.0	101.9
32° and below	24	0.2	0.*	0.0	0.0	0.0	0.0	0.0	0.0	0.0	0.0	0.0	0.2	0.4
-Minimum														
32° and below	24	6.0	2.6	0.5	0.0	0.0	0.0	0.0	0.0	0.0	0.0	0.7	3.1	12.9
0° and below	24	0.0	0.0	0.0	0.0	0.0	0.0	0.0	0.0	0.0	0.0	0.0	0.0	0.0
AVG. STATION PRESS. (mb)	13	1016.5	1015.4	1010.8	1010.3	1008.6	1010.0	1011.6	1011.3	1010.9	1013.1	1014.4	1016.0	1012.4
RELATIVE HUMIDITY (%)														
Hour 00	21	84	83	83	84	87	87	87	86	87	86	86	84	85
Hour 06 (Local Time)	24	86	86	86	88	90	91	91	91	91	90	88	87	89
Hour 12	24	65	60	59	60	61	59	56	57	61	57	59	63	60
Hour 18	24	68	62	61	63	66	64	60	62	66	66	68	70	65
PRECIPITATION (inches):														
Water Equivalent														
-Normal		1.87	2.24	1.34	2.61	4.47	4.53	2.58	3.33	6.24	3.31	2.24	2.14	36.90
-Maximum Monthly	25	5.21	5.42	5.51	9.43	14.08	12.68	10.47	7.30	19.05	10.16	8.68	6.97	19.05
-Year		1979	1969	1985	1969	1968	1973	1983	1974	1978	1981	1982	1975	SEP 1978
-Minimum Monthly	25	0.02	0.27	0.18	0.09	1.00	T	0.07	0.34	1.11	0.34	0.02	0.36	T
-Year		1971	1974	1971	1984	1971	1980	1982	1965	1982	1964	1981	1972	JUN 1980
-Maximum in 24 hrs	25	3.65	2.63	2.65	8.57	8.45	9.30	4.22	6.14	8.51	5.05	6.63	6.12	9.30
-Year		1980	1969	1985	1969	1972	1977	1983	1964	1967	1981	1982	1975	JUN 1977
Snow, Ice pellets														
-Maximum Monthly	25	2.1	1.0									0.2	T	2.1
-Year		1985	1973									1976	1969	JAN 1985
-Maximum in 24 hrs	25	2.1	1.0									0.2	T	2.1
-Year		1985	1973									1976	1969	JAN 1985
WIND:														
Mean Speed (mph)	24	10.6	11.0	11.7	11.9	10.7	9.6	8.8	8.3	8.6	8.9	9.8	10.2	10.0
Prevailing Direction														
Fastest Mile														
-Direction (!!!)	21		N	N	E	NW	S			N	W	N	N	
-Speed (MPH)	21	55	54	59	60	63	62	99	67	54	60	48	52	99
-Year		1963	1981	1970	1969	1975	1971	1963	1963	1976	1977	1983	1973	JUL 1963
Peak Gust														
-Direction (!!!)	2	N	N	W	SE	NW	NE	S	SE	E	N	N	N	NE
-Speed (mph)	2	46	54	43	51	51	56	43	40	39	54	39	41	56
-Date		1985	1984	1984	1984	1984	1985	1985	1984	1985	1985	1985	1984	JUN 1985

See Reference Notes to this table on the following pages.

TABLE 2 PRECIPITATION (inches) VICTORIA, TEXAS

YEAR	JAN	FEB	MAR	APR	MAY	JUNE	JULY	AUG	SEP	OCT	NOV	DEC	ANNUAL
1956	1.14	0.80	0.64	1.83	2.50	1.18	0.70	0.41	1.92	4.23	0.30	2.30	17.95
1957	0.66	3.19	7.91	7.05	6.06	5.03	0.11	1.13	5.11	3.10	7.38	0.83	47.56
1958	4.02	5.26	1.25	0.93	0.93	0.76	0.88	0.45	13.43	8.08	1.12	3.90	41.01
1959	0.31	5.32	0.38	1.73	3.24	3.39	1.03	5.31	3.31	6.58	1.05	3.57	35.22
1960	0.52	2.47	1.20	0.78	2.11	5.26	4.14	7.81	0.76	17.25	3.80	4.18	50.28
#1961	1.99	2.73	0.55	2.37	0.70	5.99	6.82	2.20	6.55	0.82	4.88	0.54	36.14
1962	0.46	0.31	0.83	4.38	1.28	4.09	1.17	1.22	5.82	1.47	2.44	2.42	25.89
1963	0.22	1.50	0.25	0.84	1.58	4.89	1.48	3.02	1.23	1.25	3.91	1.88	22.05
1964	2.03	2.49	2.10	0.50	3.09	3.77	1.81	6.84	7.64	0.34	0.90	1.81	33.32
1965	2.07	2.86	0.80	0.91	3.34	4.88	0.29	0.34	4.34	4.04	2.07	4.91	30.85
1966	3.71	2.30	0.39	4.02	4.77	3.60	5.14	6.05	2.39	0.99	0.87	1.24	35.47
1967	2.31	1.41	0.49	1.72	2.38	T	1.26	3.03	14.52	4.92	1.31	0.55	33.90
1968	3.27	2.03	1.41	0.81	14.08	10.67	3.75	1.92	4.28	1.91	4.24	0.95	49.32
1969	0.43	5.42	2.90	9.43	2.09	2.33	1.74	3.49	6.71	3.57	3.05	3.48	44.64
1970	2.74	1.68	4.44	2.38	9.93	2.87	3.51	1.44	7.41	2.63	0.10	0.65	39.78
1971	0.02	1.10	0.18	1.75	1.00	5.23	0.27	4.01	12.03	6.87	1.37	2.23	36.06
1972	1.74	0.72	1.55	0.35	11.24	3.17	7.30	4.38	5.97	3.44	2.19	0.36	42.41
1973	2.40	2.75	1.04	4.73	1.22	12.68	2.89	2.55	7.20	6.26	0.80	1.13	45.65
1974	2.89	0.27	1.75	0.90	11.16	3.33	0.99	7.30	5.84	2.88	3.43	2.60	43.34
1975	0.96	0.46	0.36	0.89	6.73	7.68	3.71	2.38	1.93	3.88	1.01	6.97	36.96
1976	0.77	0.39	1.45	5.90	2.61	1.32	5.75	2.76	7.61	6.18	3.05	5.46	43.25
1977	2.39	2.56	1.10	3.90	2.26	12.21	0.76	2.53	3.20	4.21	3.64	0.45	39.21
1978	3.43	2.87	0.54	1.95	1.05	4.88	2.51	2.08	19.05	0.57	2.88	1.27	43.08
1979	5.21	2.32	1.69	5.16	6.66	4.03	6.94	2.07	10.50	1.65	0.89	2.18	49.30
1980	4.52	1.78	1.97	0.48	8.16	T	0.41	5.65	6.07	0.90	1.81	0.79	32.54
1981	2.22	1.01	1.40	1.42	8.39	9.29	4.37	4.23	1.22	10.16	0.02	1.37	45.10
1982	0.39	5.38	0.23	1.40	8.61	0.06	0.07	1.78	8.68	4.07	8.68	0.75	32.53
1983	1.64	3.79	4.21	0.24	1.76	2.96	10.47	1.88	4.80	7.00	3.14	0.52	42.41
1984	3.02	1.34	1.74	0.09	4.02	2.05	1.02	4.16	1.87	8.52	2.16	3.93	33.92
1985	3.37	1.97	5.51	8.56	1.03	6.97	1.26	1.88	3.29	2.03	1.74	2.38	39.99
Record Mean	2.23	2.15	2.03	2.68	4.18	3.55	3.27	2.92	4.39	3.52	2.47	2.44	35.83

TABLE 3 AVERAGE TEMPERATURE (deg. F) VICTORIA, TEXAS

YEAR	JAN	FEB	MAR	APR	MAY	JUNE	JULY	AUG	SEP	OCT	NOV	DEC	ANNUAL
1956	55.6	60.5	63.6	70.7	78.2	83.0	85.0	85.3	81.6	75.9	60.5	59.5	71.6
1957	57.1	64.5	63.0	69.4	76.0	81.0	84.8	85.1	78.1	68.7	60.4	58.4	70.5
1958	52.1	53.1	58.5	70.3	77.0	83.4	85.1	86.4	80.3	70.0	63.6	52.4	69.4
1959	51.5	56.0	61.6	66.8	77.8	82.1	82.9	83.0	79.9	72.5	56.7	56.9	69.0
1960	53.2	51.9	58.0	70.5	74.1	82.4	84.0	82.5	78.3	73.3	64.4	52.4	68.8
#1961	49.7	57.4	66.2	68.0	76.6	80.7	82.3	81.7	79.6	71.3	60.3	56.7	69.2
1962	48.6	63.8	60.5	69.8	76.7	80.8	84.5	86.5	81.3	76.0	61.7	54.9	70.4
1963	47.5	53.8	64.4	75.6	78.3	82.8	84.4	85.2	81.3	75.5	65.4	48.8	70.4
1964	54.3	51.8	62.3	72.5	79.8	83.1	85.5	84.5	80.0	68.6	66.0	56.1	70.4
1965	56.9	54.5	58.3	73.5	77.3	82.0	85.0	84.7	81.8	68.8	68.7	59.3	70.9
1966	48.9	52.6	62.8	71.5	75.3	80.0	83.8	83.6	80.0	70.7	66.1	55.3	69.2
1967	54.5	55.5	67.8	76.1	76.8	83.5	84.6	82.7	77.4	70.8	65.4	54.8	70.8
1968	52.1	50.7	60.0	71.2	76.8	80.8	82.7	84.1	78.4	74.1	60.2	56.0	68.9
1969	57.0	57.7	56.9	70.9	74.9	81.9	87.0	86.0	80.5	72.7	62.0	58.7	70.5
1970	49.3	56.7	59.0	71.4	73.3	80.0	82.8	84.7	79.9	69.3	60.3	62.7	69.1
1971	59.7	58.0	64.2	69.1	77.0	82.8	85.2	82.5	79.7	74.5	64.2	61.8	71.6
1972	57.9	57.9	67.5	74.7	76.1	82.1	82.9	83.5	82.9	73.5	57.3	53.0	70.8
1973	49.6	54.4	67.7	67.2	76.3	80.7	85.3	82.8	80.7	74.9	60.7	57.1	70.6
1974	55.5	59.6	69.3	71.7	79.5	81.2	84.6	83.9	76.3	73.3	62.2	56.1	71.1
1975	58.8	58.2	65.6	71.7	78.4	82.5	84.1	83.9	78.1	72.3	63.2	55.5	71.0
1976	53.2	62.5	65.2	70.9	73.3	80.7	81.0	82.5	78.5	64.0	54.4	51.9	68.2
1977	46.0	56.2	64.4	69.3	76.9	82.0	84.1	85.0	83.2	72.8	64.9	57.5	70.2
1978	45.2	48.4	60.7	70.0	78.6	82.4	84.8	84.1	80.6	71.4	64.6	54.3	68.8
1979	45.9	52.7	64.7	70.7	73.5	80.6	83.0	82.8	76.9	72.7	57.6	54.3	68.0
1980	55.7	54.9	62.8	67.2	77.1	84.1	86.9	84.1	82.4	69.4	58.5	56.0	69.9
1981	53.1	56.1	62.0	73.7	75.6	82.2	84.0	83.3	78.8	72.6	65.4	56.2	70.2
1982	53.8	53.2	65.2	69.5	75.9	83.2	85.9	86.1	81.8	71.9	62.7	57.4	70.5
1983	52.4	55.1	61.1	67.0	75.2	80.7	82.8	84.0	77.7	72.1	65.3	46.6	68.4
1984	50.3	57.4	64.0	71.6	76.5	82.2	85.2	84.2	78.3	75.3	62.7	65.0	71.1
1985	47.3	52.5	66.7	71.4	78.5	82.0	83.5	86.2	81.1	73.8	67.4	52.7	70.3
Record Mean	54.5	57.4	63.9	71.8	76.6	82.1	84.4	84.7	80.3	72.6	63.0	56.3	70.6
Max	65.0	68.4	74.7	80.7	86.2	91.7	94.3	95.1	90.6	84.1	74.2	66.9	81.0
Min	43.9	46.4	53.1	63.0	66.9	72.5	74.4	74.2	70.1	61.0	51.8	45.6	60.2

REFERENCE NOTES FOR TABLES 1, 2, 3 and 6 (VICTORIA, TX)

GENERAL

T - TRACE AMOUNT
BLANK ENTRIES DENOTE MISSING/UNREPORTED DATA.
# INDICATES A STATION OR INSTRUMENT RELOCATION.

SPECIFIC

TABLE 1

(a) - LENGTH OF RECORD IN YEARS. ALTHOUGH INDIVIDUAL MONTHS MAY BE MISSING.
* LESS THAN .05

NORMALS — BASED ON THE 1951-1980 RECORD PERIOD.
EXTREMES — DATES ARE THE MOST RECENT OCCURRENCE.
WIND DIR. — NUMERALS SHOW TENS OF DEGREES CLOCKWISE FROM TRUE NORTH.
"00" INDICATES CALM.
RESULTANT WIND DIRECTIONS ARE GIVEN TO WHOLE DEGREES.

EXCEPTIONS

TABLE 1

1. THUNDERSTORMS AND HEAVY FOG MAY BE IN-COMPLETE, DUE TO PART-TIME OPERATIONS

TABLES 2, 3, and 6

RECORD MEANS ARE THROUGH THE CURRENT YEAR, BEGINNING IN 1904 FOR TEMPERATURE
1893 FOR PRECIPITATION
1962 FOR SNOWFALL

TABLE 4 HEATING DEGREE DAYS Base 65 deg. F VICTORIA, TEXAS

SEASON	JULY	AUG	SEP	OCT	NOV	DEC	JAN	FEB	MAR	APR	MAY	JUNE	TOTAL
1956-57	0	0	0	0	180	212	281	108	102	44	3	0	930
1957-58	0	0	0	63	199	224	392	338	223	9	0	0	1448
1958-59	0	0	0	51	122	385	426	263	139	66	0	0	1452
1959-60	0	0	0	14	295	253	377	378	244	7	3	0	1571
#1960-61	0	0	0	17	106	395	466	233	67	57	0	0	1341
1961-62	0	0	0	15	168	287	501	90	178	20	0	0	1259
1962-63	0	0	0	7	131	313	538	318	68	3	0	0	1378
1963-64	0	0	0	0	95	499	332	376	113	16	0	0	1431
1964-65	0	0	0	22	102	321	275	294	255	0	0	0	1269
1965-66	0	0	2	28	25	202	499	341	122	16	3	0	1238
1966-67	0	0	0	17	83	327	350	267	50	0	0	0	1094
1967-68	0	0	4	15	83	332	414	409	208	16	0	0	1481
1968-69	0	0	0	6	188	287	281	208	261	1	0	0	1232
1969-70	0	0	0	14	174	202	498	230	202	34	5	0	1359
1970-71	0	0	0	44	189	149	228	222	120	45	0	0	997
1971-72	0	0	2	0	96	152	270	229	50	2	0	0	801
1972-73	0	0	0	14	264	376	476	301	20	64	0	0	1515
1973-74	0	0	0	0	26	254	310	183	70	17	0	0	860
1974-75	0	0	0	7	155	286	244	206	92	14	0	0	1004
1975-76	0	0	0	6	165	316	364	120	95	2	0	0	1068
1976-77	0	0	0	111	325	398	585	246	82	8	0	0	1755
1977-78	0	0	0	0	8	86	261	620	467	166	23	1	1632
1978-79	0	0	0	0	3	121	349	591	352	79	13	6	1514
1979-80	0	0	0	20	243	329	299	309	135	39	0	0	1374
1980-81	0	0	0	59	231	303	362	272	120	2	0	0	1349
1981-82	0	0	0	46	64	284	375	337	126	47	0	0	1279
1982-83	0	0	0	21	155	270	390	274	151	60	0	0	1321
1983-84	0	0	3	12	99	580	451	230	88	14	0	0	1477
1984-85	0	0	8	6	139	123	544	358	60	5	0	0	1243
1985-86	0	0	4	6	69	385							

TABLE 5 COOLING DEGREE DAYS Base 65 deg. F VICTORIA, TEXAS

YEAR	JAN	FEB	MAR	APR	MAY	JUNE	JULY	AUG	SEP	OCT	NOV	DEC	TOTAL
1969	38	10	15	184	313	512	687	659	472	263	89	14	3256
1970	19	6	23	234	269	458	561	617	452	184	55	84	2962
1971	70	35	104	172	380	540	630	548	452	303	78	58	3370
1972	58	28	134	299	350	519	562	582	541	285	37	13	3408
1973	3	10	109	138	358	477	638	556	476	313	201	15	3294
1974	20	38	207	225	458	494	613	594	346	271	75	17	3358
1975	61	19	118	223	421	535	597	591	401	240	116	29	3351
1976	7	54	107	186	263	478	505	550	414	88	16	1	2669
1977	4	8	74	144	376	517	599	625	551	259	89	37	3283
1978	10	8	39	177	428	528	622	598	476	205	118	23	3232
1979	7	18	78	190	277	476	563	559	362	266	28	5	2829
1980	15	24	73	114	383	582	684	599	526	202	41	29	3272
1981	0	32	35	268	333	523	595	576	421	285	83	18	3169
1982	40	13	140	191	344	552	655	661	513	244	93	41	3487
1983	7	0	36	126	323	479	557	598	393	238	112	20	2889
1984	1	21	95	219	363	524	632	602	413	332	74	128	3404
1985	4	15	117	204	422	517	581	667	493	287	146	12	3465

TABLE 6 SNOWFALL (inches) VICTORIA, TEXAS

SEASON	JULY	AUG	SEP	OCT	NOV	DEC	JAN	FEB	MAR	APR	MAY	JUNE	TOTAL
1970-71	0.0	0.0	0.0	0.0	0.0	0.0	0.0	0.0	0.0	0.0	0.0	0.0	0.0
1971-72	0.0	0.0	0.0	0.0	0.0	0.0	0.0	0.0	0.0	0.0	0.0	0.0	0.0
1972-73	0.0	0.0	0.0	0.0	0.0	0.0	1.2	1.0	0.0	0.0	0.0	0.0	2.2
1973-74	0.0	0.0	0.0	0.0	0.0	0.0	0.0	0.0	0.0	0.0	0.0	0.0	0.0
1974-75	0.0	0.0	0.0	0.0	0.0	0.0	T	0.0	0.0	0.0	0.0	0.0	T
1975-76	0.0	0.0	0.0	0.0	0.0	0.0	0.0	0.0	0.0	0.0	0.0	0.0	0.0
1976-77	0.0	0.0	0.0	0.0	0.2	0.0	0.0	0.0	0.0	0.0	0.0	0.0	0.2
1977-78	0.0	0.0	0.0	0.0	0.0	0.0	0.0	0.0	0.0	0.0	0.0	0.0	0.0
1978-79	0.0	0.0	0.0	0.0	0.0	0.0	0.0	0.0	0.0	0.0	0.0	0.0	0.0
1979-80	0.0	0.0	0.0	0.0	T	0.0	0.0	0.0	0.0	0.0	0.0	0.0	T
1980-81	0.0	0.0	0.0	0.0	0.0	0.0	T	0.0	0.0	0.0	0.0	0.0	T
1981-82	0.0	0.0	0.0	0.0	0.0	0.0	T	0.0	0.0	0.0	0.0	0.0	T
1982-83	0.0	0.0	0.0	0.0	0.0	0.0	0.0	0.0	0.0	0.0	0.0	0.0	0.0
1983-84	0.0	0.0	0.0	0.0	0.0	0.0	T	0.0	0.0	0.0	0.0	0.0	T
1984-85	0.0	0.0	0.0	0.0	0.0	0.0	2.1	T	0.0	0.0	0.0	0.0	2.1
1985-86	0.0	0.0	0.0	0.0	0.0	0.0							
Record Mean	0.0	0.0	0.0	0.0	T	T	0.1	T	0.0	0.0	0.0	0.0	0.2

See Reference Notes, relative to all above tables, on preceding page.

One of the major cities of Texas, Waco is located in the rich agricultural region of the Brazos River Valley in North Central Texas. The city lies on the edge of the gently rolling Blackland Prairies. To the west lies the rolling to hilly Grand Prairie. Waco is a commercial hub with an economy based on industry, education and agriculture. Baylor University, founded in 1845, is located here. Regional agriculture includes chiefly cattle, poultry, sorghum, cotton and corn. Soils are black waxy, loam and sandy types. Lake Waco, a reservoir of 7,260 surface acres, lies within the Waco city limits, with the north shoreline approximately 0.8 mile south of the Municipal Airport.

The climate of Waco is humid subtropical with hot summers. It is a continental type climate characterized by extreme variations in temperature. Tropical maritime air masses predominate throughout the late spring, summer and early fall months, while Polar air masses frequent the area in winter. In an average year, April and May are the wettest months, while the July–August period is the driest. Most warm season rainfall occurs from thunderstorm activity. Consequently, considerable spatial variation in amounts occur.

Winters are mild. Cold fronts moving down from the High Plains often are accompanied by strong, gusty, northerly winds and sharp drops in temperature. Cold spells are of short duration, rarely lasting longer than 2 or 3 days before a rapid warming occurs. Winter precipitation is closely associated with frontal activity, and may fall as rain, freezing rain, sleet or snow. During most years, snowfall is of little or no consequence.

Daytime temperatures are hot in summer, particularly in July and August. The highest temperatures are associated with fair skies, light winds, and comparatively low humidities. There is little variety in the day-to-day weather during July and August. Air conditioning is recommended for maximum comfort indoors or while traveling.

The spring and fall seasons are very pleasant at Waco. Temperatures are comfortable. Cloudiness and showers are more frequent in the spring than in the fall. The average first occurrence of 32 degrees Fahrenheit is late November and the average last occurrence is in mid March.

TABLE 1 NORMALS, MEANS AND EXTREMES

WACO, TEXAS

LATITUDE: 31°37'N   LONGITUDE: 97°13'W   ELEVATION: FT. GRND   501 BARO   00499   TIME ZONE: CENTRAL   WBAN: 13959

	(a)	JAN	FEB	MAR	APR	MAY	JUNE	JULY	AUG	SEP	OCT	NOV	DEC	YEAR
TEMPERATURE °F:														
Normals														
-Daily Maximum		56.6	61.6	69.5	77.6	84.2	92.1	96.5	96.7	89.7	80.3	67.9	60.3	77.8
-Daily Minimum		35.7	39.4	46.6	56.5	64.2	71.5	75.2	74.5	68.6	57.2	46.1	38.5	56.2
-Monthly		46.2	50.5	58.0	67.1	74.2	81.8	85.8	85.6	79.2	68.8	57.0	49.4	67.0
Extremes														
-Record Highest	43	88	90	100	101	102	109	108	112	106	101	92	91	112
-Year		1971	1954	1971	1963	1985	1980	1957	1969	1985	1979	1948	1955	AUG 1969
-Record Lowest	43	-5	4	15	27	37	52	61	60	40	29	17	7	-5
-Year		1949	1985	1948	1975	1981	1964	1970	1967	1983	1980	1976	1983	JAN 1949
NORMAL DEGREE DAYS:														
Heating (base 65°F)		591	415	257	71	0	0	0	0	0	46	265	481	2126
Cooling (base 65°F)		8	9	43	134	288	507	648	639	426	164	25	0	2891
% OF POSSIBLE SUNSHINE														
MEAN SKY COVER (tenths)														
Sunrise - Sunset	40	6.3	6.0	5.9	6.1	6.0	4.9	4.4	4.3	4.8	4.7	5.2	5.7	5.4
MEAN NUMBER OF DAYS:														
Sunrise to Sunset														
-Clear	42	9.0	9.0	9.6	8.6	8.2	10.6	14.0	13.8	12.9	13.1	11.7	10.8	131.2
-Partly Cloudy	42	6.2	5.8	6.8	7.0	10.2	12.2	10.5	11.6	8.9	8.2	5.9	5.9	99.2
-Cloudy	42	15.8	13.5	14.6	14.4	12.6	7.2	6.5	5.6	8.3	9.7	12.4	14.4	134.9
Precipitation														
.01 inches or more	42	7.3	7.2	7.4	7.5	8.6	6.0	4.3	5.0	6.3	5.9	6.4	5.9	77.7
Snow, Ice pellets														
1.0 inches or more	42	0.3	0.2	0.*	0.0	0.0	0.0	0.0	0.0	0.0	0.0	0.0	0.*	0.6
Thunderstorms	42	1.2	2.5	3.8	5.5	7.7	5.4	4.1	4.4	3.9	3.0	2.0	1.4	44.8
Heavy Fog Visibility														
1/4 mile or less	42	2.9	2.2	1.1	0.8	0.5	0.2	0.*	0.0	0.2	0.9	1.9	2.6	13.4
Temperature °F														
-Maximum														
90° and above	22	0.0	0.0	0.5	1.0	6.7	22.1	28.1	28.5	16.8	3.4	0.0	0.0	107.2
32° and below	22	1.1	0.2	0.0	0.0	0.0	0.0	0.0	0.0	0.0	0.0	0.0	0.5	1.8
-Minimum														
32° and below	22	13.8	8.3	2.0	0.1	0.0	0.0	0.0	0.0	0.0	0.*	2.1	9.3	35.7
0° and below	22	0.0	0.0	0.0	0.0	0.0	0.0	0.0	0.0	0.0	0.0	0.0	0.0	0.0
AVG. STATION PRESS.(mb)	13	1002.6	1001.0	996.3	996.1	994.5	995.8	997.2	997.2	997.5	999.3	1000.2	1001.7	998.3
RELATIVE HUMIDITY (%)														
Hour 00	22	77	75	74	75	78	74	67	67	75	76	77	76	74
Hour 06	22	82	82	81	84	86	84	80	81	84	84	83	82	83
Hour 12 (Local Time)	22	64	60	59	60	61	54	48	48	55	56	59	61	57
Hour 18	22	62	56	53	55	57	50	43	44	53	57	62	63	55
PRECIPITATION (inches):														
Water Equivalent														
-Normal		1.69	2.04	1.99	3.79	4.73	2.58	1.78	1.95	3.18	3.06	2.24	1.92	30.95
-Maximum Monthly	43	5.83	4.55	6.84	13.37	15.00	12.06	8.58	8.91	7.29	10.51	6.24	7.03	15.00
-Year		1961	1944	1945	1957	1965	1961	1971	1974	1970	1984	1952	1960	MAY 1965
-Minimum Monthly	43	0.03	0.17	0.04	0.12	0.72	0.27	T	T	0.00	0.00	0.13	0.04	0.00
-Year		1971	1972	1956	1983	1945	1953	1963	1952	1956	1952	1970	1950	SEP 1956
-Maximum in 24 hrs	43	2.24	3.03	3.07	5.09	7.18	4.21	4.49	4.80	4.57	5.72	4.26	3.11	7.18
-Year		1961	1977	1946	1957	1953	1947	1973	1958	1957	1974	1952	1945	MAY 1953
Snow, Ice pellets														
-Maximum Monthly	43	7.0	4.8	1.0								0.8	2.0	7.0
-Year		1949	1966	1962								1980	1946	JAN 1949
-Maximum in 24 hrs	43	7.0	4.8	1.0								0.8	2.0	7.0
-Year		1949	1966	1962								1980	1946	JAN 1949
WIND:														
Mean Speed (mph)	36	11.6	12.0	13.1	12.9	11.9	11.7	10.7	9.9	9.6	10.0	10.8	11.2	11.3
Prevailing Direction														
through 1963		S	S	S	S	S	S	S	S	S	S	S	S	S
Fastest Obs. 1 Min.														
-Direction (!!!)	37	32	36	27	36	36	09	36	05	32	34	29	32	09
-Speed (MPH)	37	49	58	65	62	60	69	60	60	60	52	62	52	69
-Year		1952	1954	1952	1953	1952	1961	1953	1951	1952	1960	1953	1954	JUN 1961
Peak Gust														
-Direction (!!!)	2	N	SW	S	S	NW	S	SE	NW	NE	W	SW	N	SW
-Speed (mph)	2	54	58	53	41	41	51	49	54	43	54	41	43	58
-Date		1985	1985	1985	1985	1985	1985	1985	1985	1985	1984	1984	1985	FEB 1985

See Reference Notes to this table on the following page.

TABLE 2 PRECIPITATION (inches) WACO, TEXAS

YEAR	JAN	FEB	MAR	APR	MAY	JUNE	JULY	AUG	SEP	OCT	NOV	DEC	ANNUAL
1956	2.75	1.25	0.04	0.65	4.25	0.37	0.34	0.14	0.00	1.08	2.34	1.94	15.15
1957	1.53	2.41	5.58	13.37	7.60	1.03	0.61	0.50	5.69	5.40	4.77	0.42	48.91
1958	2.49	3.27	0.96	3.60	5.58	2.72	0.68	7.34	4.68	1.24	1.81	0.94	35.31
1959	0.29	2.72	0.22	2.74	1.96	5.74	1.84	2.14	3.49	7.36	4.06	33.81
1960	1.79	1.91	0.69	1.67	2.38	2.73	1.20	3.00	1.12	5.25	0.86	7.03	29.63
1961	5.83	3.86	1.76	1.05	1.34	12.06	7.68	0.90	2.28	1.28	3.53	1.14	42.71
1962	0.65	1.41	0.81	2.95	2.19	4.69	0.07	0.72	3.92	2.37	1.48	1.34	22.60
1963	0.49	0.51	2.75	2.76	5.70	1.83	T	1.28	0.83	1.09	3.22	1.10	19.57
1964	3.15	1.29	2.85	2.85	1.48	4.32	0.23	2.30	5.49	0.79	3.67	0.90	29.32
1965	2.97	4.41	3.13	1.35	15.00	0.82	0.12	0.64	3.80	2.91	4.64	2.31	42.10
1966	1.15	2.00	0.77	9.37	2.55	2.34	0.46	4.09	4.73	0.01	0.13	0.45	28.05
1967	0.37	0.77	1.93	8.80	4.17	0.36	1.59	0.43	4.31	3.80	3.53	2.91	32.97
1968	3.47	1.78	2.15	2.94	7.12	5.11	0.98	0.87	5.61	1.46	4.77	0.45	36.71
1969	0.45	2.10	3.62	3.98	3.98	0.44	0.05	2.23	4.70	5.53	1.93	2.49	31.50
1970	0.76	3.82	4.33	1.80	2.24	0.92	0.10	2.47	7.29	3.37	0.13	0.83	28.06
1971	0.03	1.33	0.64	2.79	0.75	0.99	8.58	2.16	0.74	3.26	1.64	4.30	27.21
1972	2.54	0.17	0.09	2.04	1.95	3.05	4.14	1.79	2.57	6.62	2.87	0.78	28.61
1973	3.48	1.52	3.04	5.41	3.16	7.10	4.89	0.36	5.24	9.36	0.66	0.76	44.98
1974	1.33	1.21	0.61	3.30	2.21	1.07	0.73	8.91	6.80	6.76	2.54	1.92	37.39
1975	1.40	2.94	1.09	4.58	13.21	2.84	2.77	0.72	2.27	2.40	0.40	1.83	36.45
1976	1.74	0.33	1.60	6.54	4.99	3.23	7.40	0.24	5.66	5.19	0.66	2.50	40.08
1977	1.83	3.67	2.90	7.38	1.71	0.78	0.21	1.96	0.26	1.50	2.43	0.15	24.78
1978	1.03	2.83	2.46	2.37	2.88	1.54	0.26	1.42	0.53	1.87	4.57	2.01	23.77
1979	1.99	1.97	4.31	1.38	9.68	4.70	5.01	4.58	2.51	1.94	0.33	3.97	42.37
1980	2.26	1.83	2.13	4.03	4.47	0.34	0.01	0.16	2.03	0.76	2.29	2.70	23.01
1981	0.85	1.80	2.97	1.21	3.69	7.23	0.16	2.14	3.61	8.41	0.99	0.50	33.56
1982	1.86	1.41	3.79	2.46	5.54	3.93	2.61	0.18	0.15	1.43	3.57	2.11	29.04
1983	1.28	2.97	3.46	0.12	3.83	0.95	3.41	2.54	0.71	1.55	2.90	0.46	24.18
1984	0.82	1.00	3.45	0.54	3.71	1.75	1.82	1.33	1.32	10.51	3.05	5.07	34.37
1985	0.76	1.91	2.57	3.37	1.59	5.62	0.68	0.68	6.38	3.84	3.53	2.02	32.95
Record Mean	2.01	2.24	2.68	3.96	4.54	3.05	2.02	2.12	2.95	2.82	2.50	2.49	33.37

TABLE 3 AVERAGE TEMPERATURE (deg. F) WACO, TEXAS

YEAR	JAN	FEB	MAR	APR	MAY	JUNE	JULY	AUG	SEP	OCT	NOV	DEC	ANNUAL
1956	47.4	52.5	59.8	66.4	78.2	83.8	87.5	87.0	81.3	72.6	54.9	53.1	68.7
1957	46.6	57.3	56.3	64.2	73.6	81.2	87.4	86.4	76.7	64.5	54.1	54.3	66.9
1958	46.8	46.5	51.9	64.2	75.0	82.4	85.8	85.1	78.4	66.7	59.3	46.0	65.7
1959	45.3	50.2	58.1	63.8	76.4	80.7	83.4	83.9	79.4	67.4	50.4	52.1	65.9
1960	46.9	45.7	52.5	69.1	73.3	82.7	84.8	84.0	79.8	71.1	59.3	46.0	66.3
1961	43.3	53.2	61.6	66.1	75.0	78.3	81.4	82.3	78.0	68.6	54.5	49.3	66.0
1962	43.0	57.1	56.1	66.5	77.1	79.9	86.0	88.0	80.4	72.3	57.8	49.1	67.8
#1963	40.9	48.7	63.1	72.0	76.2	83.7	87.5	87.4	80.8	74.9	59.5	41.2	68.0
1964	46.7	45.5	57.4	68.7	75.5	82.6	87.9	86.8	78.7	65.8	60.9	50.6	67.3
1965	51.4	48.2	51.2	71.9	75.6	82.6	87.2	86.0	79.9	66.0	63.2	52.4	68.0
1966	42.1	46.2	55.6	65.6	73.6	81.4	88.6	84.2	78.4	68.1	63.4	47.6	66.2
1967	48.5	48.7	65.3	74.0	73.2	84.9	86.9	85.2	76.8	68.4	58.8	49.4	68.4
1968	46.7	47.0	56.1	64.8	73.1	79.9	82.9	85.6	76.3	70.7	55.9	49.1	65.7
1969	51.1	51.3	51.8	68.0	75.2	82.7	90.5	88.1	80.5	68.8	56.8	50.5	67.9
1970	41.7	50.0	52.7	68.0	72.8	80.8	85.1	88.1	79.6	66.3	55.9	66.5
1971	51.7	52.9	60.8	67.8	75.1	84.0	85.9	80.5	79.1	71.2	58.5	53.4	68.4
1972	47.3	52.9	63.6	71.0	73.3	82.7	82.2	84.4	81.3	67.7	50.8	44.1	66.8
1973	42.3	47.2	60.4	61.1	71.6	77.3	83.8	84.4	79.1	69.9	63.2	50.0	65.8
1974	45.3	53.4	64.6	67.4	77.2	80.1	86.8	84.4	71.8	69.8	56.7	46.2	67.0
1975	48.5	47.0	54.5	64.1	72.0	79.7	82.0	84.6	77.2	70.5	59.8	50.5	65.9
1976	46.8	59.0	60.5	66.2	70.7	80.4	82.2	85.6	77.9	66.8	50.1	45.5	65.4
1977	38.0	52.7	59.8	67.2	76.5	84.6	89.1	87.9	85.7	71.4	60.2	51.7	68.7
1978	37.8	42.0	56.4	70.2	76.5	83.7	89.2	86.3	81.4	69.5	57.3	46.5	66.4
1979	36.4	44.5	58.4	64.9	69.5	78.9	82.3	81.8	75.5	70.4	52.4	49.3	63.7
1980	46.7	48.7	56.0	62.5	73.5	85.2	89.2	87.5	81.2	65.7	55.9	51.4	67.0
1981	46.6	50.1	55.7	69.9	72.6	81.7	85.8	83.9	77.5	68.9	58.8	48.7	66.7
1982	46.8	47.5	61.8	64.4	75.0	80.5	86.3	88.2	80.4	68.8	57.9	50.4	67.3
1983	45.5	48.9	55.4	61.8	71.5	79.2	84.3	85.6	77.4	69.8	59.3	38.3	64.7
1984	41.8	51.5	58.8	65.5	76.2	82.4	86.0	86.6	76.6	69.1	56.0	54.9	67.1
1985	40.4	46.2	61.5	68.4	76.8	82.1	85.3	87.4	77.5	69.4	59.7	44.9	66.6
Record Mean	47.2	50.0	58.7	66.9	74.7	82.3	85.8	85.7	79.2	68.9	57.5	49.4	67.2
Max	57.6	60.3	70.0	77.7	84.8	92.6	96.3	96.5	89.8	80.2	68.3	59.8	77.8
Min	36.7	39.7	47.3	56.1	64.5	72.1	75.2	74.9	68.5	57.6	46.7	38.9	56.5

REFERENCE NOTES FOR TABLES 1, 2, 3 and 6   (WACO, TX)

GENERAL

T - TRACE AMOUNT
BLANK ENTRIES DENOTE MISSING/UNREPORTED DATA.
# INDICATES A STATION OR INSTRUMENT RELOCATION.

SPECIFIC

TABLE 1

(a) - LENGTH OF RECORD IN YEARS. ALTHOUGH
INDIVIDUAL MONTHS MAY BE MISSING.

* LESS THAN .05

NORMALS — BASED ON THE 1951-1980 RECORD PERIOD.
EXTREMES — DATES ARE THE MOST RECENT OCCURRENCE.
WIND DIR. — NUMERALS SHOW TENS OF DEGREES
CLOCKWISE FROM TRUE NORTH.
"00" INDICATES CALM.
RESULTANT WIND DIRECTIONS ARE GIVEN TO WHOLE DEGREES.

EXCEPTIONS

TABLES 2, 3, and 6

RECORD MEANS ARE THROUGH THE CURRENT YEAR,
BEGINNING IN   1897 FOR TEMPERATURE
1897 FOR PRECIPITATION
1944 FOR SNOWFALL

TABLE 4 HEATING DEGREE DAYS Base 65 deg. F   WACO, TEXAS

SEASON	JULY	AUG	SEP	OCT	NOV	DEC	JAN	FEB	MAR	APR	MAY	JUNE	TOTAL
1956-57	0	0	0	15	314	378	568	238	270	107	16	0	1906
1957-58	0	0	0	113	340	327	559	514	401	80	7	0	2341
1958-59	0	0	0	71	208	577	604	411	231	132	0	0	2234
1959-60	0	0	0	53	448	396	561	562	402	29	13	0	2464
1960-61	0	0	0	36	206	586	665	333	159	101	0	0	2086
1961-62	0	0	0	37	326	489	674	234	282	67	1	0	2110
1962-63	0	0	0	26	218	484	741	452	136	31	5	0	2093
#1963-64	0	0	0	2	208	726	562	559	235	42	2	0	2336
1964-65	0	0	0	45	196	458	427	465	424	15	0	0	2030
1965-66	0	0	2	72	100	393	706	522	291	75	13	0	2174
1966-67	0	0	0	52	126	557	511	450	114	4	14	0	1828
1967-68	0	0	6	46	196	480	560	519	288	79	0	0	2174
1968-69	0	0	0	22	294	484	438	382	405	15	1	0	2041
1969-70	0	0	0	75	278	444	719	414	.372	60	11	0	2373
1970-71	0	0	4	87	286	288	411	349	188	50	5	0	1668
1971-72	0	0	6	5	230	360	543	368	109	19	2	0	1642
1972-73	0	0	4	89	426	638	696	492	149	170	13	0	2677
1973-74	0	0	0	22	118	460	605	322	145	52	0	0	1724
1974-75	0	0	13	14	274	578	507	495	343	120	0	0	2344
1975-76	0	0	2	27	230	462	558	200	204	38	6	0	1727
1976-77	0	0	0	205	446	596	827	345	175	21	2	0	2617
1977-78	0	0	0	22	173	417	839	639	284	19	19	0	2412
1978-79	0	0	0	23	256	573	880	570	217	59	30	0	2608
1979-80	0	0	0	43	384	483	562	475	297	119	6	0	2369
1980-81	0	0	2	98	305	427	561	417	285	27	13	0	2135
1981-82	0	0	6	75	194	498	566	486	202	103	3	0	2133
1982-83	0	0	1	73	265	453	599	443	298	159	9	0	2300
1983-84	0	0	11	38	220	823	710	390	221	73	6	0	2492
1984-85	0	0	23	41	293	319	752	527	157	41	0	0	2153
1985-86	0	0	14	38	207	619							

TABLE 5 COOLING DEGREE DAYS Base 65 deg. F   WACO, TEXAS

YEAR	JAN	FEB	MAR	APR	MAY	JUNE	JULY	AUG	SEP	OCT	NOV	DEC	TOTAL
1969	12	2	1	111	325	537	796	723	470	200	40	0	3217
1970	4	0	0	155	260	483	629	723	447	136	20	35	2892
1971	6	14	66	142	325	576	655	489	436	206	40	7	2962
1972	3	24	72	203	266	539	541	608	499	178	6	0	2939
1973	0	0	12	60	225	375	587	611	429	182	72	0	2553
1974	0	5	138	128	382	461	683	610	225	169	31	2	2834
1975	2	0	26	100	227	448	533	618	376	206	83	19	2638
1976	0	33	72	81	193	469	542	644	395	73	5	0	2507
1977	0	7	20	94	366	593	754	719	628	224	34	13	3452
1978	2	0	21	181	383	568	757	666	496	169	30	6	3279
1979	0	1	21	61	177	426	544	529	320	220	17	1	2317
1980	0	6	25	49	278	611	758	705	498	126	39	12	3107
1981	0	5	4	184	256	509	648	594	384	207	15	0	2806
1982	8	0	112	93	321	470	666	725	471	197	62	9	3134
1983	0	0	7	68	217	431	604	648	390	193	56	3	2617
1984	0	4	39	95	360	529	658	677	379	174	28	14	2957
1985	0	7	56	150	370	518	637	699	397	177	55	3	3069

TABLE 6 SNOWFALL (inches)   WACO, TEXAS

SEASON	JULY	AUG	SEP	OCT	NOV	DEC	JAN	FEB	MAR	APR	MAY	JUNE	TOTAL
1956-57	0.0	0.0	0.0	0.0	0.0	0.0	T	0.0	0.0	0.0	0.0	0.0	T
1957-58	0.0	0.0	0.0	0.0	T	0.0	0.0	T	T	0.0	0.0	0.0	T
1958-59	0.0	0.0	0.0	0.0	0.0	T	0.0	T	0.0	0.0	0.0	0.0	T
1959-60	0.0	0.0	0.0	0.0	T	0.0	T	0.4	0.0	0.0	0.0	0.0	0.4
1960-61	0.0	0.0	0.0	0.0	0.0	T	2.0	0.3	0.0	0.0	0.0	0.0	2.3
1961-62	0.0	0.0	0.0	0.0	0.0	T	T	T	1.0	0.0	0.0	0.0	1.0
1962-63	0.0	0.0	0.0	0.0	0.0	0.0	T	1.0	T	0.0	0.0	0.0	1.0
1963-64	0.0	0.0	0.0	0.0	0.0	1.0	5.6	1.0	T	0.0	0.0	0.0	7.6
1964-65	0.0	0.0	0.0	0.0	0.0	T	0.0	2.0	0.2	0.0	0.0	0.0	2.2
1965-66	0.0	0.0	0.0	0.0	0.0	0.0	2.6	4.8	0.0	0.0	0.0	0.0	7.4
1966-67	0.0	0.0	0.0	0.0	0.0	0.0	0.0	T	0.0	0.0	0.0	0.0	T
1967-68	0.0	0.0	0.0	0.0	0.0	0.0	T	1.0	T	0.0	0.0	0.0	1.0
1968-69	0.0	0.0	0.0	0.0	T	0.0	T	T	0.0	0.0	0.0	0.0	T
1969-70	0.0	0.0	0.0	0.0	T	T	T	T	0.0	0.0	0.0	0.0	T
1970-71	0.0	0.0	0.0	0.0	0.0	0.0	T	T	0.0	0.0	0.0	0.0	T
1971-72	0.0	0.0	0.0	0.0	0.0	0.0	T	T	0.0	0.0	0.0	0.0	T
1972-73	0.0	0.0	0.0	0.0	T	T	2.6	0.0	0.0	0.0	0.0	0.0	2.6
1973-74	0.0	0.0	0.0	0.0	0.0	T	T	0.0	0.0	0.0	0.0	0.0	T
1974-75	0.0	0.0	0.0	0.0	0.0	T	1.0	0.8	0.0	0.0	0.0	0.0	1.8
1975-76	0.0	0.0	0.0	0.0	0.0	0.0	0.0	0.0	0.0	0.0	0.0	0.0	T
1976-77	0.0	0.0	0.0	0.0	0.8	0.0	1.8	0.0	0.0	0.0	0.0	0.0	2.6
1977-78	0.0	0.0	0.0	0.0	T	0.0	2.2	0.2	T	0.0	0.0	0.0	2.4
1978-79	0.0	0.0	0.0	0.0	0.0	0.2	T	T	0.0	0.0	0.0	0.0	0.2
1979-80	0.0	0.0	0.0	0.0	T	0.0	0.0	T	0.0	0.0	0.0	0.0	T
1980-81	0.0	0.0	0.0	0.0	0.8	0.0	T	1.0	0.0	0.0	0.0	0.0	1.8
1981-82	0.0	0.0	0.0	0.0	0.0	0.0	6.0	T	T	0.0	0.0	0.0	6.0
1982-83	0.0	0.0	0.0	0.0	0.0	0.0	T	T	0.0	0.0	0.0	0.0	T
1983-84	0.0	0.0	0.0	0.0	0.0	0.0	T	0.0	T	0.0	0.0	0.0	T
1984-85	0.0	0.0	0.0	0.0	0.0	0.0	1.0	1.9	0.0	0.0	0.0	0.0	2.9
1985-86	0.0	0.0	0.0	0.0	0.0	0.0							
Record Mean	0.0	0.0	0.0	0.0	T	0.1	0.9	0.4	0.1	0.0	0.0	0.0	1.5

See Reference Notes, relative to all above tables, on preceding page.

Wichita Falls is located in the West Cross Timbers subdivision of the North Central Plains of Texas, about 10 miles south of the Red River and 400 miles northwest of the nearest portion of the Gulf of Mexico. The topography is gently rolling mesquite plain, and the elevation of the area is about 1,000 feet.

This region lies between the humid subtropical climate of east Texas and a continental climate to the north and west. The climate of Wichita Falls is classified as continental. It is characterized by rapid changes in temperature, large daily and annual temperature extremes, and by rather erratic rainfall.

The area lies in the path of polar air masses which move down from the north during the winter season. With the passage of cold fronts or northers in the fall and winter, abrupt drops in temperature of as much as 20 to 30 degrees within an hour sometimes occur. While the area is subject to a wide range of temperature, winters are on the whole relatively mild. January, the coldest month, has an average temperature around 40 degrees. Sub-zero temperatures occur about once every five years.

The summers in Wichita Falls are generally of the continental climate type, characterized by low humidity and windy conditions. Temperatures over 100 degrees are frequent during the common periods of hot weather. July and August, the hottest months, have average temperatures in the middle 80s.

The normal rainfall is nearly 27 inches per year, but the distribution is erratic to such an extent that prolonged dry periods are common. Several lakes in the area provide water for domestic, industrial, and irrigation purposes. The greater part of the rainfall comes in the form of showers rather than general rains. Over 75 percent of the annual moisture occurs during the period from late March to mid November, but dry periods of three to four weeks are to be expected during this time almost every year. While the dry conditions materially affect agriculture in this region, complete crop failure seldom results. Moderate flooding along Holliday Creek and the Wichita River, which run through the city, occur about once in each ten-year period. Snowfall, measuring an inch or more, occurs on average only two days a year.

Wind speeds average over 11 mph, and southerly winds prevail. Rather strong winds are observed in all months. Even though strong, gusty winds occur frequently, severe duststorms are rare. Most severe dust observed in the area is blown in from the north and west.

The area around Wichita Falls enjoys excellent aviation weather. Flying activities are possible on all but a very few days of the year. Approximately 95 percent of the time the ceiling is 1,000 feet or more with visibility of 3 miles or more.

TABLE 1 NORMALS, MEANS AND EXTREMES

WICHITA FALLS, TEXAS

LATITUDE: 33°58'N   LONGITUDE: 98°29'W   ELEVATION: FT. GRND   994 BARO   01007   TIME ZONE: CENTRAL   WBAN: 13966

	(a)	JAN	FEB	MAR	APR	MAY	JUNE	JULY	AUG	SEP	OCT	NOV	DEC	YEAR
TEMPERATURE °F:														
Normals														
-Daily Maximum		52.3	58.0	66.7	76.8	84.1	93.2	98.5	97.3	88.7	78.2	64.4	56.2	76.2
-Daily Minimum		28.2	32.6	39.9	50.6	59.4	68.2	72.5	71.2	63.7	52.0	39.6	31.6	50.8
-Monthly		40.3	45.3	53.3	63.7	71.8	80.7	85.5	84.3	76.2	65.1	52.0	43.9	63.5
Extremes														
-Record Highest	39	87	92	100	102	105	117	114	113	108	102	89	88	117
-Year		1969	1979	1971	1972	1985	1980	1980	1964	1977	1979	1965	1954	JUN 1980
-Record Lowest	39	-5	-8	9	24	36	51	54	54	38	25	14	1	-8
-Year		1966	1985	1965	1975	1979	1983	1970	1962	1984	1957	1950	1983	FEB 1985
NORMAL DEGREE DAYS:														
Heating (base 65°F)		766	552	388	118	18	0	0	0	14	105	396	654	3011
Cooling (base 65°F)		0	0	26	79	229	471	639	598	350	108	6	0	2506
% OF POSSIBLE SUNSHINE														
MEAN SKY COVER (tenths)														
Sunrise - Sunset	40	5.7	5.5	5.6	5.4	5.3	4.5	4.2	4.0	4.2	4.3	4.8	5.3	4.9
MEAN NUMBER OF DAYS:														
Sunrise to Sunset														
-Clear	42	10.7	9.8	10.7	11.0	10.8	13.0	14.6	15.8	15.1	15.0	12.9	12.3	151.6
-Partly Cloudy	42	6.5	6.6	7.6	7.4	9.3	10.5	9.6	9.3	7.2	7.3	6.5	6.2	94.0
-Cloudy	42	13.8	11.9	12.7	11.6	10.9	6.5	6.7	5.9	7.8	8.8	10.6	12.4	119.6
Precipitation														
.01 inches or more	42	5.0	5.4	5.9	6.6	8.9	6.4	5.2	5.3	6.0	5.9	4.6	4.8	70.0
Snow, Ice pellets														
1.0 inches or more	42	0.8	0.7	0.3	0.0	0.0	0.0	0.0	0.0	0.0	0.0	0.1	0.3	2.2
Thunderstorms	42	0.9	1.5	3.2	5.5	9.4	7.1	5.6	5.5	3.8	3.3	1.6	0.8	48.3
Heavy Fog Visibility 1/4 mile or less	42	2.3	2.3	1.1	0.8	0.5	0.2	0.*	0.1	0.5	1.0	1.5	2.1	12.4
Temperature °F														
-Maximum														
90° and above	25	0.0	0.*	0.9	2.4	8.4	21.0	28.5	27.0	14.6	3.9	0.0	0.0	106.9
32° and below	25	4.0	1.5	0.1	0.0	0.0	0.0	0.0	0.0	0.0	0.0	0.*	1.7	7.4
-Minimum														
32° and below	25	22.4	15.1	6.4	0.7	0.0	0.0	0.0	0.0	0.0	0.2	6.0	17.9	68.7
0° and below	25	0.1	0.*	0.0	0.0	0.0	0.0	0.0	0.0	0.0	0.0	0.0	0.0	0.2
AVG. STATION PRESS. (mb)	13	983.6	981.7	977.0	977.0	975.5	976.7	978.4	978.5	979.3	980.6	981.2	982.5	979.3
RELATIVE HUMIDITY (%)														
Hour 00	25	73	73	69	72	76	74	65	67	76	74	76	74	72
Hour 06	25	80	80	78	80	85	84	77	79	85	84	83	81	81
Hour 12 (Local Time)	25	57	55	50	49	52	50	42	45	53	52	54	56	51
Hour 18	25	57	51	45	47	50	46	38	41	51	54	60	60	50
PRECIPITATION (inches):														
Water Equivalent														
-Normal		0.93	1.00	1.82	2.99	4.34	2.85	2.00	2.14	3.41	2.61	1.42	1.22	26.73
-Maximum Monthly	42	4.48	3.43	3.89	8.50	13.22	8.38	11.86	7.61	10.23	7.86	5.69	5.03	13.22
-Year		1968	1945	1973	1957	1982	1945	1950	1971	1980	1972	1957	1984	MAY 1982
-Minimum Monthly	42	T	0.10	T	0.35	0.01	0.26	T	T	T	T	0.00	0.02	0.00
-Year		1976	1976	1956	1980	1966	1980	1943	1943	1983	1949	1949	1950	NOV 1949
-Maximum in 24 hrs	42	2.02	3.00	2.61	4.09	5.70	5.36	3.93	4.62	6.22	5.61	2.58	2.40	6.22
-Year		1968	1981	1978	1967	1975	1985	1950	1971	1980	1959	1968	1984	SEP 1980
Snow, Ice pellets														
-Maximum Monthly	42	11.9	11.8	6.5	0.8						T	3.9	7.1	11.9
-Year		1966	1978	1947	1973						1967	1957	1983	JAN 1966
-Maximum in 24 hrs	42	8.1	4.5	5.2	0.8						T	3.9	5.6	8.1
-Year		1985	1958	1958	1973						1967	1957	1983	JAN 1985
WIND:														
Mean Speed (mph)	37	11.4	12.0	13.5	13.4	12.3	12.2	11.0	10.4	10.7	10.7	11.4	11.3	11.7
Prevailing Direction through 1963		N	N	S	S	SSE	S	S	S	SE	S	S	S	S
Fastest Obs. 1 Min.														
-Direction (!!)	37	32	29	27	14	34	27	34	34	01	29	32	29	27
-Speed (MPH)	37	49	57	59	52	58	60	46	55	53	60	56	55	60
-Year		1949	1952	1950	1950	1949	1954	1958	1949	1973	1949	1949	1949	JUN 1954
Peak Gust														
-Direction (!!)	2	N	N	NW	W	SE	S	S	SE	SW	N	NW	N	W
-Speed (mph)	2	59	61	61	66	52	51	39	48	60	52	46	37	66
-Date		1985	1984	1984	1984	1985	1984	1984	1985	1985	1985	1985	1984	APR 1984

See Reference Notes to this table on the following page.

TABLE 2 AVERAGE TEMPERATURE (deg. F) WICHITA FALLS, TEXAS

YEAR	JAN	FEB	MAR	APR	MAY	JUNE	JULY	AUG	SEP	OCT	NOV	DEC	ANNUAL
1956	40.4	45.4	55.5	62.7	76.6	83.6	87.9	87.5	80.2	69.4	50.8	47.3	65.6
1957	39.2	49.5	50.3	59.9	69.1	78.4	87.5	85.4	73.7	60.9	48.6	47.4	62.5
1958	42.7	40.7	44.5	59.7	71.9	81.6	84.0	85.4	77.3	64.8	55.3	41.3	62.4
1959	38.4	45.4	54.8	63.7	75.2	80.0	81.0	84.2	76.8	61.9	46.6	47.5	63.0
#1960	40.8	39.5	45.6	64.9	69.8	81.7	83.5	84.4	78.9	67.4	55.6	39.5	62.6
1961	39.3	46.0	55.7	62.1	72.6	77.9	83.7	82.0	74.7	65.0	48.1	41.5	62.4
1962	36.5	50.5	51.3	62.0	76.9	77.2	84.0	85.8	73.8	68.7	54.3	46.1	63.9
1963	35.3	44.6	57.5	68.1	74.0	82.6	86.6	85.2	78.5	73.7	56.1	38.2	65.1
1964	43.6	42.9	53.2	66.8	73.1	80.4	88.7	86.0	75.1	63.0	54.3	42.1	64.1
1965	42.8	44.7	44.0	67.2	72.6	77.8	86.2	83.4	77.2	63.4	57.0	48.9	63.8
1966	34.8	42.1	57.1	61.7	71.8	81.3	88.2	81.3	72.8	63.2	58.0	41.4	62.8
1967	45.6	45.5	62.9	69.8	69.6	80.8	83.0	82.7	73.1	65.1	52.6	43.0	64.5
1968	42.2	41.0	53.7	61.3	69.7	79.6	82.7	83.5	73.7	66.4	50.7	42.9	62.3
1969	43.9	45.4	45.0	63.6	71.0	79.2	88.8	83.8	75.4	59.8	51.2	44.3	62.6
1970	36.1	46.0	47.6	63.2	71.6	79.7	84.8	85.0	76.3	60.6	49.0	47.7	62.3
1971	40.6	44.8	52.5	63.0	71.6	83.8	86.4	79.9	75.5	67.9	53.0	46.1	63.7
1972	39.4	45.2	56.8	67.0	70.0	82.0	83.3	82.3	77.1	63.4	45.8	39.9	62.7
1973	37.7	44.1	58.0	57.7	68.3	76.5	82.8	83.2	74.4	67.7	57.7	44.8	62.7
1974	40.5	49.9	60.6	64.5	76.2	79.9	86.6	81.3	68.0	65.4	51.7	43.1	64.0
1975	43.9	42.4	51.2	62.3	71.6	79.3	81.8	82.7	71.0	65.7	54.0	45.3	62.6
1976	42.5	55.3	55.9	64.2	67.5	78.0	81.0	84.2	73.9	57.6	46.4	41.1	62.3
1977	31.7	48.4	55.8	63.3	72.9	82.7	85.8	83.2	81.5	65.5	53.3	44.0	64.0
1978	30.1	33.3	52.8	66.5	72.1	80.8	89.9	82.5	79.8	65.5	53.3	40.3	62.3
1979	30.7	38.1	54.4	62.1	70.0	79.5	85.1	82.4	75.6	68.6	49.8	46.7	61.9
1980	42.8	42.4	49.9	61.1	70.9	84.8	91.9	88.8	77.9	62.8	51.2	46.5	64.3
1981	44.0	48.6	54.0	68.3	69.6	80.8	86.9	82.7	77.4	63.1	53.2	44.5	64.4
1982	40.8	40.3	54.9	59.3	71.5	76.9	82.3	85.7	76.5	65.2	52.4	45.1	62.6
1983	41.2	44.7	52.3	57.8	67.9	75.7	84.6	86.4	78.1	66.3	54.1	30.5	61.6
1984	37.2	47.1	51.4	60.1	72.7	82.9	84.8	84.4	73.9	63.3	51.9	46.2	63.0
1985	34.8	40.8	55.9	64.2	72.2	78.1	83.0	85.4	75.9	64.7	50.7	38.6	62.0
Record Mean	41.0	46.1	54.0	64.1	71.9	80.9	85.4	84.9	76.8	66.2	52.8	44.1	64.0
Max	52.6	58.3	67.0	76.8	84.0	93.0	97.8	97.5	89.0	78.6	64.8	55.6	76.2
Min	29.4	33.9	41.1	51.3	59.8	68.8	72.9	72.3	64.7	53.7	40.7	32.6	51.8

TABLE 3 PRECIPITATION (inches) WICHITA FALLS, TEXAS

YEAR	JAN	FEB	MAR	APR	MAY	JUNE	JULY	AUG	SEP	OCT	NOV	DEC	ANNUAL
1956	0.39	1.11	T	1.05	7.06	1.51	0.68	2.65	T	4.61	0.75	1.50	21.31
1957	1.17	1.62	2.38	8.50	12.03	3.37	0.69	0.05	1.71	3.83	5.69	0.56	41.60
1958	1.80	0.55	2.54	2.51	4.43	1.82	3.35	0.37	0.42	0.84	0.57	0.40	19.60
1959	0.21	0.15	0.57	1.53	3.61	5.93	4.65	2.03	1.49	7.27	1.27	4.26	32.97
1960	1.25	1.28	1.12	1.85	3.02	2.36	2.91	1.05	4.11	6.03	T	3.80	28.78
1961	0.34	1.03	3.35	1.71	1.82	6.53	1.22	1.58	4.21	1.48	3.24	1.03	27.54
1962	0.27	0.26	1.30	4.61	1.21	8.29	3.81	0.96	5.26	2.15	1.79	1.23	31.14
1963	0.21	0.52	2.36	1.13	3.10	0.43	4.46	1.77	2.32	0.37	2.53	0.89	20.55
1964	1.35	1.30	1.32	3.22	8.43	2.16	0.38	3.74	6.26	0.35	2.74	0.60	31.85
1965	1.73	0.83	0.66	2.45	5.92	3.88	0.35	2.29	2.58	3.11	0.01	0.62	24.43
1966	1.10	1.41	1.76	7.97	0.01	0.28	3.13	5.05	7.45	0.59	0.60	0.27	29.62
1967	0.03	0.20	1.30	5.84	6.50	1.84	3.18	0.22	3.93	1.98	0.36	1.44	26.82
1968	4.48	1.08	2.58	1.80	4.41	1.19	3.29	2.35	2.21	2.40	4.05	0.72	30.56
1969	0.57	2.82	3.18	1.87	5.04	3.11	0.34	3.95	4.61	3.16	0.54	2.42	31.61
1970	0.02	1.10	2.95	3.00	0.94	1.48	0.10	0.36	4.10	1.39	0.22	0.41	16.07
1971	0.54	1.37	0.12	0.74	1.06	0.63	1.83	7.61	5.13	4.21	0.91	3.37	27.52
1972	0.17	0.44	0.40	2.39	2.90	1.75	1.09	1.78	2.20	7.86	1.86	0.52	23.36
1973	2.74	1.08	3.89	3.51	0.74	5.17	4.51	0.05	5.31	1.93	1.68	0.08	30.69
1974	0.29	1.50	1.71	6.15	1.81	1.40	0.86	3.28	5.92	3.47	0.53	0.88	27.80
1975	0.91	1.71	1.57	2.94	9.66	1.69	3.41	3.46	3.65	0.82	1.31	1.86	32.99
1976	T	0.10	1.90	5.39	2.28	2.18	3.98	2.42	8.06	6.02	0.30	1.00	33.63
1977	1.09	1.28	2.08	2.51	3.78	2.53	1.49	4.10	0.39	0.69	0.93	0.06	20.93
1978	0.41	1.96	3.53	0.61	3.51	3.10	0.27	4.16	2.16	1.07	2.14	0.65	23.57
1979	2.04	0.71	3.43	2.71	4.69	6.07	1.74	3.57	0.06	1.16	2.20	2.00	30.38
1980	1.57	0.63	0.76	0.35	6.67	0.26	0.03	0.26	10.23	1.65	1.57	1.94	25.92
1981	0.14	3.30	1.86	3.95	3.54	4.59	1.33	2.29	1.49	7.83	0.86	0.30	31.48
1982	1.66	1.03	2.04	3.38	13.22	7.41	0.92	0.71	2.06	1.99	2.73	2.22	39.37
1983	1.87	0.90	2.06	2.19	3.21	5.05	0.19	0.19	T	7.79	0.91	0.89	25.25
1984	0.17	0.79	1.48	0.62	1.44	1.78	0.92	3.07	0.80	6.24	3.32	5.03	25.66
1985	1.08	2.61	3.77	6.15	1.69	7.07	0.21	1.75	1.46	3.69	1.11	0.11	30.70
Record Mean	1.10	1.27	1.76	2.73	4.15	3.27	1.98	2.07	2.93	2.92	1.43	1.37	26.99

REFERENCE NOTES FOR TABLES 1, 2, 3 and 6   (WICHITA FALLS, TX)

GENERAL

T - TRACE AMOUNT
BLANK ENTRIES DENOTE MISSING/UNREPORTED DATA.
# INDICATES A STATION OR INSTRUMENT RELOCATION.

SPECIFIC

TABLE 1

(a) - LENGTH OF RECORD IN YEARS. ALTHOUGH
INDIVIDUAL MONTHS MAY BE MISSING.

* LESS THAN .05

NORMALS — BASED ON THE 1951-1980 RECORD PERIOD.
EXTREMES — DATES ARE THE MOST RECENT OCCURRENCE.
WIND DIR. — NUMERALS SHOW TENS OF DEGREES
CLOCKWISE FROM TRUE NORTH.
"00" INDICATES CALM.
RESULTANT WIND DIRECTIONS ARE GIVEN TO WHOLE DEGREES.

EXCEPTIONS

TABLES 2, 3, and 6

RECORD MEANS ARE THROUGH THE CURRENT YEAR,
BEGINNING IN   1924 FOR TEMPERATURE
1924 FOR PRECIPITATION
1944 FOR SNOWFALL

WICHITA FALLS, TEXAS

TABLE 4 HEATING DEGREE DAYS Base 65 deg. F — WICHITA FALLS, TEXAS

SEASON	JULY	AUG	SEP	OCT	NOV	DEC	JAN	FEB	MAR	APR	MAY	JUNE	TOTAL
1956-57	0	0	0	37	425	549	793	430	449	197	34	5	2919
1957-58	0	0	4	190	491	539	688	677	630	183	18	0	3420
1958-59	0	0	4	120	308	724	818	540	329	166	4	0	3013
#1959-60	0	0	9	139	551	537	744	732	606	84	52	0	3454
#1960-61	0	0	1	61	294	786	790	525	300	167	16	0	2940
1961-62	0	0	4	62	499	723	875	408	427	139	3	0	3140
1962-63	0	0	2	68	315	577	914	565	281	56	22	0	2800
1963-64	0	0	1	8	278	825	655	634	366	82	6	0	2855
1964-65	0	0	12	86	352	706	679	563	647	47	2	0	3094
1965-66	0	0	13	120	241	494	929	633	275	139	33	0	2877
1966-67	0	2	0	119	247	723	593	538	167	30	41	0	2460
1967-68	0	0	14	93	366	674	699	687	357	133	23	0	3046
1968-69	0	0	0	68	431	676	646	543	609	91	14	4	3082
1969-70	0	0	1	235	411	634	891	526	534	131	19	4	3386
1970-71	0	0	20	205	474	529	749	560	395	131	15	0	3078
1971-72	0	0	37	25	364	578	787	574	273	77	12	0	2727
1972-73	0	0	8	193	571	771	839	577	213	234	47	0	3453
1973-74	0	0	10	42	248	621	752	418	212	88	1	0	2392
1974-75	0	0	36	54	398	673	644	628	430	157	0	0	3020
1975-76	0	0	52	87	338	606	690	295	318	73	40	0	2499
1976-77	0	0	7	265	552	731	1023	459	281	80	1	0	3399
1977-78	0	0	0	64	345	645	1078	881	394	73	50	0	3530
1978-79	0	0	1	76	364	758	1056	750	343	130	47	0	3525
1979-80	0	0	0	57	464	563	682	649	460	162	18	0	3055
1980-81	0	0	32	139	416	567	644	455	339	50	33	0	2675
1981-82	0	0	11	144	355	629	745	689	331	220	25	0	3149
1982-83	0	0	4	116	387	612	732	564	387	259	43	4	3108
1983-84	0	0	13	60	342	1059	855	509	420	181	17	0	3456
1984-85	0	0	65	129	400	582	930	673	301	67	1	0	3148
1985-86	0	0	31	83	431	812							

TABLE 5 COOLING DEGREE DAYS Base 65 deg. F — WICHITA FALLS, TEXAS

YEAR	JAN	FEB	MAR	APR	MAY	JUNE	JULY	AUG	SEP	OCT	NOV	DEC	TOTAL
1969	0	0	0	57	206	438	742	591	320	79	3	0	2436
1970	0	0	4	85	233	452	622	627	365	76	3	0	2467
1971	0	0	14	78	226	573	672	470	357	124	11	0	2525
1972	0	4	26	145	174	517	572	543	377	147	2	0	2507
1973	0	0	3	21	155	351	559	570	298	132	35	0	2124
1974	0	0	85	80	356	454	676	512	135	73	6	0	2377
1975	0	0	10	81	210	436	529	558	237	116	13	0	2190
1976	0	21	39	54	124	395	500	601	282	44	0	0	2060
1977	0	2	2	34	254	537	652	570	502	88	3	0	2644
1978	0	0	27	128	281	481	777	550	449	98	17	0	2808
1979	0	2	18	49	208	444	628	551	324	176	14	1	2415
1980	0	0	0	52	204	603	843	745	427	80	7	2	2963
1981	0	5	6	159	179	480	687	558	389	90	5	0	2558
1982	0	3	24	59	234	364	542	648	354	130	18	3	2379
1983	0	0	1	49	140	331	614	672	411	107	18	0	2343
1984	0	0	2	42	262	545	620	609	338	87	13	3	2521
1985	0	0	27	53	230	401	562	638	365	81	11	0	2368

TABLE 6 SNOWFALL (inches) — WICHITA FALLS, TEXAS

SEASON	JULY	AUG	SEP	OCT	NOV	DEC	JAN	FEB	MAR	APR	MAY	JUNE	TOTAL
1956-57	0.0	0.0	0.0	0.0	0.0	0.0	1.2	0.0	T	T	0.0	0.0	1.2
1957-58	0.0	0.0	0.0	0.0	3.9	0.0	0.4	4.8	5.2	0.0	0.0	0.0	14.3
1958-59	0.0	0.0	0.0	0.0	T	4.7	1.0	2.1	T	0.0	0.0	0.0	7.8
1959-60	0.0	0.0	0.0	0.0	T	T	3.6	2.2	0.7	0.0	0.0	0.0	6.5
1960-61	0.0	0.0	0.0	0.0	0.0	3.4	1.1	5.7	0.0	0.0	0.0	0.0	10.2
1961-62	0.0	0.0	0.0	0.0	0.0	T	0.0	1.2	0.8	0.0	0.0	0.0	2.0
1962-63	0.0	0.0	0.0	0.0	0.0	0.6	0.2	4.1	T	0.0	0.0	0.0	4.9
1963-64	0.0	0.0	0.0	0.0	0.0	1.3	1.2	1.7	1.3	0.0	0.0	0.0	5.5
1964-65	0.0	0.0	0.0	0.0	0.0	0.0	T	T	1.4	0.0	0.0	0.0	1.4
1965-66	0.0	0.0	0.0	0.0	0.0	0.0	11.9	0.8	0.0	0.0	0.0	0.0	12.7
1966-67	0.0	0.0	0.0	0.0	T	T	0.0	T	0.6	0.0	0.0	0.0	0.6
1967-68	0.0	0.0	0.0	T	T	T	T	8.6	3.8	0.0	0.0	0.0	12.4
1968-69	0.0	0.0	0.0	0.0	T	T	T	1.7	3.0	0.0	0.0	0.0	4.7
1969-70	0.0	0.0	0.0	0.0	T	5.4	T	T	2.3	T	0.0	0.0	7.7
1970-71	0.0	0.0	0.0	0.0	0.0	0.0	T	3.0	0.0	0.0	0.0	0.0	3.0
1971-72	0.0	0.0	0.0	0.0	T	4.2	0.4	1.9	0.0	0.0	0.0	0.0	6.5
1972-73	0.0	0.0	0.0	0.0	2.1	T	3.1	4.6	T	0.8	0.0	0.0	10.6
1973-74	0.0	0.0	0.0	0.0	0.0	0.5	T	T	0.2	0.0	0.0	0.0	0.5
1974-75	0.0	0.0	0.0	0.0	0.6	0.0	T	1.9	T	0.0	0.0	0.0	2.7
1975-76	0.0	0.0	0.0	0.0	0.5	2.3	T	T	0.0	0.0	0.0	0.0	2.8
1976-77	0.0	0.0	0.0	0.0	3.7	T	5.9	0.0	0.0	0.0	0.0	0.0	9.6
1977-78	0.0	0.0	0.0	0.0	T	0.0	1.9	11.8	T	0.0	0.0	0.0	13.7
1978-79	0.0	0.0	0.0	0.0	0.0	3.0	6.0	1.6	0.0	0.0	0.0	0.0	10.6
1979-80	0.0	0.0	0.0	0.0	0.0	T	T	3.6	T	0.0	0.0	0.0	3.6
1980-81	0.0	0.0	0.0	0.0	2.7	0.0	T	T	0.0	0.0	0.0	0.0	2.7
1981-82	0.0	0.0	0.0	0.0	0.0	0.0	1.3	3.2	T	0.0	0.0	0.0	4.5
1982-83	0.0	0.0	0.0	0.0	T	1.1	7.3	T	0.0	0.0	0.0	0.0	8.4
1983-84	0.0	0.0	0.0	0.0	0.0	7.1	0.3	0.0	0.0	0.0	0.0	0.0	7.4
1984-85	0.0	0.0	0.0	0.0	0.0	0.4	8.7	0.2	T	0.0	0.0	0.0	9.3
1985-86	0.0	0.0	0.0	0.0	0.0	0.9							
Record Mean	0.0	0.0	0.0	T	0.4	1.0	2.0	1.8	0.7	T	0.0	0.0	6.0

See Reference Notes, relative to all above tables, on preceding page.

Milford is located in Beaver County in the west-central portion of the state. The city is situated in a flat to gently sloping valley 15 to 20 miles in width. The Mineral Mountains, 10 miles to the east of the station, and the San Francisco Range, 15 miles to the northwest, rise about 5,000 feet above the valley floor.

The station is in the Sevier River Basin, and drainage is toward the north. The Beaver River just to the east extends north-south through the valley, but no significant body of water is reached by it. The river is dry most of the time due to the low annual rainfall in the area, and to the Minersville Reservoir 6 miles east of Minersville, which regulates the flow of water in the stream. Water for the irrigation of agricultural land in the valley is obtained from the diversion of surface water from this reservoir and from numerous deep wells.

The climate is temperate and dry. Irrigation water is necessary for the economic production of most crops.

Snowfall is rather evenly distributed during the season. The snow is usually light and powdery with below average moisture content. January, the coldest month of the year, has the greatest average monthly total.

Relative humidity is rather low during the summer months. It increases considerably in the change from summer to winter, and winters are cold and uncomfortable. In four out of five years the temperature can be expected to drop to -10 degrees or lower.

Summers are characterized by warm days and cool nights. Temperatures of 100 degrees or more occur about once in every two years.

Diurnal heating is a factor in producing strong southerly winds during the spring and summer months. Winter winds may cause considerable drifting snow, with resultant hazards to stock and transportation in the area.

Low pressure storm systems are rare during the summer months. Precipitation during this period occurs as showers or thundershowers and rainfall amounts from these storms are quite variable. As winter approaches, the number of atmospheric disturbances increases.

Based on the 1951-1980 period, the average first occurrence of 32 degrees Fahrenheit in the fall is September 22 and the average last occurrence in the spring is May 25.

TABLE 1   NORMALS, MEANS AND EXTREMES

MILFORD, UTAH

LATITUDE: 38°26'N   LONGITUDE: 113°01'W   ELEVATION: FT. GRND 5028 BARO 05033   TIME ZONE: MOUNTAIN   WBAN: 23176

	(a)	JAN	FEB	MAR	APR	MAY	JUNE	JULY	AUG	SEP	OCT	NOV	DEC	YEAR
TEMPERATURE °F:														
Normals														
-Daily Maximum		39.4	45.3	52.7	62.2	73.1	84.8	92.9	89.9	81.2	67.8	51.5	41.5	65.2
-Daily Minimum		13.4	18.8	23.6	30.4	38.6	46.8	55.6	54.2	43.9	32.6	22.0	14.8	32.9
-Monthly		26.4	32.0	38.2	46.3	55.9	65.8	74.3	72.1	62.5	50.2	36.8	28.1	49.0
Extremes														
-Record Highest	37	66	73	78	87	94	105	104	102	98	90	76	65	105
-Year		1975	1963	1972	1959	1967	1970	1985	1958	1950	1963	1965	1979	JUN 1970
-Record Lowest	37	-28	-27	-14	9	17	24	38	34	23	-2	-13	-32	-32
-Year		1949	1949	1966	1975	1975	1976	1982	1968	1984	1971	1958	1972	DEC 1972
NORMAL DEGREE DAYS:														
Heating (base 65°F)		1197	921	831	561	295	68	0	9	126	456	846	1141	6451
Cooling (base 65°F)		0	0	0	0	13	92	288	229	54	0	0	0	676
% OF POSSIBLE SUNSHINE	12	56	64	60	68	73	83	77	81	81	75	62	60	70
MEAN SKY COVER (tenths)														
Sunrise - Sunset	37	6.1	6.1	5.9	5.4	5.0	3.4	3.8	3.7	3.1	3.7	5.3	5.8	4.8
MEAN NUMBER OF DAYS:														
Sunrise to Sunset														
-Clear	37	8.8	7.3	9.2	10.0	11.3	17.4	15.6	16.2	18.6	16.7	10.8	9.8	151.5
-Partly Cloudy	37	8.1	8.4	8.3	9.4	10.3	7.7	10.4	10.2	7.3	7.1	7.8	8.0	102.9
-Cloudy	37	14.1	12.5	13.5	10.6	9.4	4.9	5.0	4.7	4.1	7.2	11.4	13.2	110.8
Precipitation														
.01 inches or more	37	7.0	6.6	8.4	6.2	5.5	3.2	5.2	5.8	3.9	4.4	4.7	5.9	66.8
Snow, Ice pellets														
1.0 inches or more	37	2.7	2.2	3.3	1.8	0.5	0.0	0.0	0.0	0.1	0.5	1.4	2.3	14.9
Thunderstorms	17	0.1	0.3	0.4	1.5	3.4	2.9	8.8	9.8	3.9	1.3	0.3	0.1	32.8
Heavy Fog Visibility														
1/4 mile or less	17	1.4	1.2	0.5	0.2	0.2	0.0	0.0	0.1	0.1	0.1	1.3	1.8	6.7
Temperature °F														
-Maximum														
90° and above	37	0.0	0.0	0.0	0.0	0.6	9.7	23.9	18.0	4.7	0.*	0.0	0.0	56.9
32° and below	37	8.3	3.8	0.9	0.*	0.0	0.0	0.0	0.0	0.0	0.1	1.6	6.4	21.1
-Minimum														
32° and below	37	29.9	26.3	26.9	18.5	5.9	0.6	0.0	0.0	2.5	15.5	25.8	29.8	181.8
0° and below	37	5.4	2.1	0.4	0.0	0.0	0.0	0.0	0.0	0.0	0.*	0.5	3.1	11.6
AVG. STATION PRESS. (mb)	10	847.1	846.8	843.0	843.8	843.9	845.2	847.2	847.3	847.4	848.5	847.6	848.4	846.3
RELATIVE HUMIDITY (%)														
Hour 05														
Hour 11 (Local Time)	24	66	62	45	34	28	23	24	27	28	35	51	65	41
Hour 17	18	60	54	37	28	23	17	20	21	20	28	46	60	35
Hour 23														
PRECIPITATION (inches):														
Water Equivalent														
-Normal		0.69	0.74	0.99	0.96	0.73	0.42	0.61	0.71	0.69	0.73	0.69	0.63	8.59
-Maximum Monthly	37	1.86	1.67	2.00	2.28	2.26	2.43	1.78	3.75	3.64	2.61	2.21	2.45	3.75
-Year		1980	1976	1981	1973	1981	1967	1985	1984	1982	1972	1978	1966	AUG 1984
-Minimum Monthly	37	0.05	0.01	0.00	0.02	T	T	0.01	0.03	T	0.00	0.06	0.05	0.00
-Year		1972	1972	1972	1955	1963	1979	1958	1957	1979	1952	1956	1976	MAR 1972
-Maximum in 24 hrs	37	0.87	0.74	1.16	1.06	1.37	1.18	1.25	1.73	1.53	1.20	1.01	1.03	1.73
-Year		1954	1953	1953	1973	1981	1982	1985	1984	1982	1951	1963	1972	AUG 1984
Snow, Ice pellets														
-Maximum Monthly	37	29.8	18.8	29.4	24.4	11.4				8.4	17.4	14.0	30.6	30.6
-Year		1949	1971	1985	1973	1975				1965	1971	1963	1972	DEC 1972
-Maximum in 24 hrs	37	11.8	9.2	20.4	11.2	8.6				6.7	6.4	8.5	16.3	20.4
-Year		1957	1971	1985	1973	1975				1965	1971	1978	1949	MAR 1985
WIND:														
Mean Speed (mph)														
Prevailing Direction														
Fastest Mile														
-Direction (!!!)	13	SW	W	SW	SW	SW	SW	NW	SW	SW	NW	SW	SW	SW
-Speed (MPH)	13	44	56	52	52	57	57	45	52	47	45	56	53	57
-Year		1981	1983	1976	1974	1984	1976	1980	1976	1985	1975	1973	1973	MAY 1984
Peak Gust														
-Direction (!!!)	2	N	SW	SW	SW	SW	SW	S	NE	SW	S	SW	SW	SW
-Speed (mph)	2	39	45	56	63	72	54	58	55	60	60	60	48	72
-Date		1984	1984	1985	1985	1984	1984	1984	1984	1985	1984	1984	1984	MAY 1984

See reference Notes to this table on the following page.

TABLE 2 PRECIPITATION (inches)   MILFORD, UTAH

YEAR	JAN	FEB	MAR	APR	MAY	JUNE	JULY	AUG	SEP	OCT	NOV	DEC	ANNUAL
1956	0.51	1.04	0.30	1.35	0.51	0.73	0.29	0.03	T	0.11	0.06	0.13	5.06
1957	1.27	0.21	0.92	1.43	1.19	0.66	0.18	0.03	T	1.24	1.13	0.15	8.41
1958	0.64	0.89	1.15	0.74	1.19	T	0.01	0.47	0.48	0.07	0.77	0.07	6.48
1959	0.24	1.34	0.90	0.19	0.34	0.07	0.10	0.51	0.33	0.35	0.27	1.00	5.64
1960	0.38	0.71	1.26	0.63	0.48	0.01	0.12	0.18	0.88	0.94	0.93	0.26	6.78
1961	0.42	0.28	1.45	1.21	0.30	T	0.26	0.84	1.66	0.54	0.47	0.49	7.92
1962	0.39	1.50	0.95	0.21	0.98	0.24	0.36	0.13	1.21	0.55	0.19	0.44	7.15
1963	0.19	0.77	0.83	0.98	T	1.23	0.05	0.59	1.24	0.40	2.10	0.18	8.56
1964	0.45	0.38	1.03	1.80	0.62	1.00	0.13	0.42	0.80	0.09	1.38	0.56	8.66
1965	0.59	1.01	0.68	1.31	0.85	0.62	0.34	0.79	1.69	0.03	1.23	1.66	10.80
1966	0.57	0.88	0.42	0.59	0.28	0.11	0.15	0.40	0.25	0.99	0.34	2.45	7.43
1967	1.07	0.20	1.10	0.41	0.55	2.43	1.42	0.32	2.60	0.22	0.52	0.83	11.67
1968	0.17	0.87	1.08	1.33	0.46	0.15	0.97	0.98	0.12	0.93	0.27	0.83	8.16
1969	1.63	1.27	1.17	1.76	0.27	1.22	1.36	0.64	0.44	0.90	0.63	0.34	11.63
1970	0.23	0.35	0.86	1.09	0.08	0.28	1.38	0.76	1.33	0.27	0.74	0.45	7.82
1971	0.25	1.30	0.31	0.91	1.06	0.15	0.48	1.59	0.49	2.47	0.20	1.33	10.54
1972	0.05	0.01	0.00	0.96	0.02	0.66	0.04	2.52	0.62	2.61	0.95	2.21	10.65
1973	0.71	0.61	1.65	2.28	0.96	0.89	0.49	0.67	0.15	0.77	0.87	0.44	10.49
1974	1.51	0.36	0.45	0.82	0.01	T	1.03	0.23	0.19	1.29	0.38	0.16	6.43
1975	0.35	0.50	1.40	0.60	1.87	0.70	1.37	0.74	0.02	0.35	0.76	0.46	9.12
1976	0.37	1.67	1.06	1.18	0.34	0.07	1.08	0.15	0.43	1.13	0.48	0.05	8.01
1977	0.21	0.10	0.77	0.39	1.89	0.20	1.09	1.57	0.20	0.27	0.17	0.46	7.32
1978	1.56	0.80	2.00	1.49	1.44	T	0.46	0.48	1.70	0.33	2.21	0.59	13.06
1979	1.31	0.66	1.73	0.63	0.96	T	0.05	0.70	T	0.48	0.57	0.46	7.55
1980	1.86	1.50	1.21	0.54	1.01	0.02	0.94	0.61	2.06	1.55	0.75	0.22	12.27
1981	0.39	0.29	2.00	1.09	2.26	0.32	1.57	0.67	2.12	1.09	0.85	0.35	13.00
1982	0.60	0.20	1.29	0.40	0.57	1.38	0.68	1.88	3.64	1.23	0.94	0.81	13.62
1983	0.36	0.67	1.33	1.19	1.05	0.09	0.24	1.72	0.95	0.89	1.24	0.76	10.49
1984	0.61	0.73	1.59	1.11	0.57	0.77	1.69	3.75	0.03	1.33	0.53	1.20	13.91
1985	0.42	0.47	1.76	1.42	0.46	0.49	1.78	0.09	1.42	1.12	1.17	0.61	11.21
Record Mean	0.65	0.76	1.05	0.87	0.72	0.42	0.77	0.79	0.64	0.84	0.66	0.72	8.90

TABLE 3 AVERAGE TEMPERATURE (deg. F)   MILFORD, UTAH

YEAR	JAN	FEB	MAR	APR	MAY	JUNE	JULY	AUG	SEP	OCT	NOV	DEC	ANNUAL
1956	36.2	24.5	40.0	47.5	59.3	69.5	74.0	71.1	66.3	50.2	30.8	27.5	49.7
1957	26.1	36.2	41.7	47.2	53.7	66.3	73.7	73.3	61.4	49.7	31.3	32.8	49.4
1958	26.7	39.4	35.8	44.3	59.7	67.5	72.5	75.4	63.3	52.0	33.7	34.6	50.4
1959	31.3	31.4	39.0	49.5	54.6	70.0	75.8	72.7	60.4	50.3	37.6	28.1	50.0
1960	21.0	29.4	43.2	48.8	56.8	70.0	76.4	72.3	66.9	50.5	39.6	28.6	50.3
1961	26.6	35.9	38.7	46.8	56.7	70.8	75.2	72.9	57.8	48.0	33.3	28.4	49.3
1962	19.3	31.8	34.4	50.6	54.7	64.0	70.9	71.5	64.1	52.1	41.2	27.4	48.5
1963	22.7	37.2	37.5	44.1	60.2	63.6	74.3	73.5	66.0	56.3	38.9	26.1	50.0
1964	24.2	25.7	32.6	44.9	54.3	63.3	75.1	71.8	60.1	53.5	31.6	32.7	47.5
1965	31.6	30.9	37.2	47.4	52.8	61.8	72.2	69.9	56.3	51.9	42.3	29.1	48.6
1966	24.9	28.9	38.6	47.6	59.8	66.1	74.9	73.4	64.1	49.3	41.5	26.0	49.6
1967	21.1	33.6	42.5	42.8	54.4	62.1	75.3	75.7	64.2	52.3	41.8	19.6	48.8
1968	24.1	39.5	41.5	42.7	54.6	67.2	74.3	67.9	61.0	50.0	37.9	25.2	48.9
1969	33.0	32.4	35.1	48.2	61.2	63.1	74.9	77.1	66.7	44.1	36.0	31.8	50.3
1970	31.0	37.8	37.0	41.7	57.1	66.1	75.1	75.9	59.2	46.0	40.0	30.1	49.7
1971	29.8	30.4	38.4	46.7	52.9	66.1	75.7	74.2	58.7	45.0	34.8	25.1	48.2
1972	29.1	35.7	44.3	47.6	56.7	68.2	75.7	70.5	61.0	49.4	34.9	15.8	49.1
1973	14.3	29.0	35.9	42.7	56.8	64.5	72.5	72.2	58.9	49.2	37.5	30.4	47.0
1974	20.9	26.9	43.3	44.3	59.4	69.3	73.3	70.5	62.8	49.9	37.9	25.5	48.7
1975	24.2	30.2	36.4	39.5	49.9	62.4	73.1	69.2	62.7	47.8	33.4	28.9	46.5
1976	26.2	34.4	32.3	44.7	58.1	65.3	74.7	68.6	62.9	47.4	37.6	26.6	48.1
1977	25.5	32.7	33.6	49.4	52.3	68.7	73.6	72.2	63.4	51.9	38.8	35.8	49.8
1978	33.4	36.4	45.0	46.9	52.7	65.5	73.2	69.8	59.7	52.7	34.2	23.2	49.4
1979	19.3	29.3	39.1	47.2	55.7	64.4	73.4	70.1	65.7	53.1	32.8	30.8	48.4
1980	31.4	36.8	36.3	47.0	52.5	65.1	74.7	70.7	62.3	48.4	38.4	34.6	49.8
1981	29.9	34.7	38.8	50.4	54.7	69.1	76.0	72.5	64.8	46.4	40.3	33.3	50.9
1982	25.0	30.3	39.2	44.8	54.7	62.6	71.5	73.5	60.8	45.0	35.6	29.6	47.6
1983	32.0	37.0	41.0	41.7	51.5	64.2	71.7	72.1	65.2	51.2	37.7	30.9	49.7
1984	19.1	25.1	37.2	43.2	59.4	64.6	73.3	71.8	62.4	44.1	36.8	22.7	46.6
1985	24.5	25.2	37.2	49.5	57.7	68.3	74.8	71.5	57.8	48.8	35.2	21.3	47.7
Record Mean	25.5	32.0	39.2	47.2	56.3	65.6	74.2	72.1	62.3	49.8	37.0	27.9	49.1
Max	38.1	44.8	53.5	63.0	73.4	84.4	92.4	89.8	80.9	67.3	52.0	41.0	65.1
Min	12.9	19.1	24.9	31.4	39.2	46.8	55.9	54.3	43.6	32.3	22.0	14.7	33.1

REFERENCE NOTES FOR TABLES 1, 2, 3 and 6   (MILFORD, UT)

GENERAL

T - TRACE AMOUNT
BLANK ENTRIES DENOTE MISSING/UNREPORTED DATA.
# INDICATES A STATION OR INSTRUMENT RELOCATION.

SPECIFIC

TABLE 1

(a) - LENGTH OF RECORD IN YEARS. ALTHOUGH
INDIVIDUAL MONTHS MAY BE MISSING.
* LESS THAN .05

NORMALS — BASED ON THE 1951-1980 RECORD PERIOD.
EXTREMES — DATES ARE THE MOST RECENT OCCURRENCE.
WIND DIR. — NUMERALS SHOW TENS OF DEGREES
CLOCKWISE FROM TRUE NORTH.
"00" INDICATES CALM.
RESULTANT WIND DIRECTIONS ARE GIVEN TO WHOLE DEGREES.

EXCEPTIONS

TABLE 1

1. THUNDERSTORMS AND HEAVY FOG ARE THROUGH 1964
AND MAY BE INCOMPLETE, DUE TO PART-TIME OPERA-
TIONS.

2. RELATIVE HUMIDITY HOUR 17 IS THROUGH 1966, HOUR
11 IS THROUGH 1971 PLUS 1974.

TABLES 2, 3, and 6

RECORD MEANS ARE THROUGH THE CURRENT YEAR,
BEGINNING IN   1909 FOR TEMPERATURE
1909 FOR PRECIPITATION
1949 FOR SNOWFALL

TABLE 4 HEATING DEGREE DAYS Base 65 deg. F   MILFORD, UTAH

SEASON	JULY	AUG	SEP	OCT	NOV	DEC	JAN	FEB	MAR	APR	MAY	JUNE	TOTAL
1956-57	0	0	53	453	1017	1156	1203	802	716	531	347	68	6346
1957-58	0	9	127	471	1006	993	1181	709	897	613	184	34	6224
1958-59	0	0	131	399	930	936	1040	796	459	316	21		5963
1959-60	0	0	182	450	814	1136	1358	1026	669	479	264	6	6384
1960-61	0	14	31	442	756	1121	1184	808	809	540	252	34	5991
1961-62	0	0	223	519	945	1126	1407	924	941	425	310	97	6917
1962-63	4	12	67	392	707	1159	1305	774	843	619	149	81	6112
1963-64	0	4	29	265	776	1201	1258	1137	997	596	325	104	6692
1964-65	0	29	147	355	996	995	1029	949	855	523	374	109	6361
1965-66	0	3	264	399	674	1103	1239	1005	811	516	161	58	6233
1966-67	0	7	75	478	700	1205	1354	872	688	662	336	131	6508
1967-68	0	0	88	390	688	1260	1403	731	720	660	322	65	6327
1968-69	1	35	179	429	806	1227	983	907	920	499	125	87	6198
1969-70	0	0	32	637	863	1022	1047	755	861	689	244	86	6236
1970-71	0	0	215	583	746	1072	1088	962	819	540	367	78	6470
1971-72	0	0	224	612	899	1229	1107	847	631	516	265	5	6335
1972-73	0	6	132	470	898	1523	1570	1003	895	660	243	111	7511
1973-74	0	8	194	486	815	1063	1362	1062	667	614	188	41	6500
1974-75	0	4	124	460	807	1219	1258	969	880	755	464	107	7047
1975-76	0	7	100	527	944	1113	1197	880	1006	602	207	93	6676
1976-77	0	3	84	539	814	1183	1217	899	967	462	391	13	6572
1977-78	0	16	110	399	782	898	973	794	611	533	377	42	5535
1978-79	1	21	206	375	920	1288	1409	993	800	531	284	101	6929
1979-80	0	8	28	361	958	1050	1036	810	884	532	379	58	6104
1980-81	0	8	107	507	788	935	1084	843	805	433	316	58	5884
1981-82	0	6	49	569	734	976	1233	965	795	600	313	98	6338
1982-83	19	0	174	612	875	1015	1184	775	737	660	417	87	6555
1983-84	7	0	75	420	813	1051	1417	1150	854	650	180	84	6701
1984-85	0	0	132	641	842	1309	1252	1109	856	458	223	58	6880
1985-86	0	3	231	495	889	1349							

TABLE 5 COOLING DEGREE DAYS Base 65 deg. F   MILFORD, UTAH

YEAR	JAN	FEB	MAR	APR	MAY	JUNE	JULY	AUG	SEP	OCT	NOV	DEC	TOTAL
1969	0	0	0	0	14	35	314	383	89	0	0	0	835
1970	0	0	0	0	7	124	322	347	46	0	0	0	846
1971	0	0	0	0	0	118	336	291	41	0	0	0	786
1972	0	0	0	0	15	103	338	185	18	0	0	0	659
1973	0	0	0	0	0	102	239	241	16	0	0	0	598
1974	0	0	0	0	20	172	266	184	66	0	0	0	708
1975	0	0	0	0	0	37	260	141	37	0	0	0	475
1976	0	0	0	0	0	55	306	121	30	0	0	0	512
1977	0	0	0	0	2	131	273	249	67	0	0	0	722
1978	0	0	0	0	0	64	264	176	54	0	0	0	558
1979	0	0	0	0	4	89	267	174	54	1	0	0	589
1980	0	0	0	0	0	70	307	191	28	0	0	0	596
1981	0	0	0	0	2	187	349	245	49	0	0	0	832
1982	0	0	0	0	0	65	230	270	53	0	0	0	618
1983	0	0	0	0	4	69	222	229	92	0	0	0	616
1984	0	0	0	0	14	78	264	216	59	0	0	0	631
1985	0	0	0	0	1	163	310	209	21	0	0	0	704

TABLE 6 SNOWFALL (inches)   MILFORD, UTAH

SEASON	JULY	AUG	SEP	OCT	NOV	DEC	JAN	FEB	MAR	APR	MAY	JUNE	TOTAL
1956-57	0.0	0.0	0.0	1.3	0.4	2.4	18.3	0.1	2.7	7.4	0.2	0.0	32.8
1957-58	0.0	0.0	0.0	T	11.8	T	11.9	4.8	10.2	3.9	T	0.0	42.6
1958-59	0.0	0.0	0.0	0.0	5.5	0.7	1.9	12.3	6.0	T	0.8	0.0	27.2
1959-60	0.0	0.0	0.0	0.7	0.0	8.8	4.5	8.0	3.6	0.3	T	0.0	25.9
1960-61	0.0	0.0	0.0	T	2.9	3.0	3.2	1.7	10.5	3.8	0.0	0.0	25.1
1961-62	0.0	0.0	0.0	4.2	4.4	2.5	7.9	9.6	12.5	0.0	2.1	0.0	43.2
1962-63	0.0	0.0	0.0	T	2.7	1.5	2.2	8.6	7.2	8.7	0.0	0.0	30.9
1963-64	0.0	0.0	0.0	0.0	14.0	2.2	6.4	4.4	14.3	16.0	3.2	0.0	60.5
1964-65	0.0	0.0	0.0	0.0	12.6	5.1	6.1	12.0	4.3	8.4	T	0.0	48.5
1965-66	0.0	0.0	8.4	0.0	2.4	13.2	9.1	9.2	6.6	3.6	0.0	0.0	52.5
1966-67	0.0	0.0	0.0	3.2	4.1	16.0	16.0	1.7	12.9	5.9	1.4	0.0	61.2
1867-68	0.0	0.0	0.0	0.0	5.3	14.7	2.8	3.8	9.0	8.3	2.4	0.0	46.3
1968-69	0.0	0.0	0.0	T	3.1	11.6	5.0	14.7	19.3	5.8	0.0	0.0	59.5
1969-70	0.0	0.0	0.0	4.2	1.4	0.8	0.8	1.6	10.8	6.5	1.1	0.0	33.9
1970-71	0.0	0.0	0.0	1.5	0.5	3.9	4.7	18.8	4.6	7.2	T	0.0	41.2
1971-72	0.0	0.0	1.1	17.4	2.8	8.7	1.0	0.1	0.0	10.8	0.0	0.0	41.9
1972-73	0.0	0.0	0.0	5.2	7.0	30.6	9.5	1.6	20.2	24.4	3.3	0.0	101.8
1973-74	0.0	0.0	0.0	3.2	6.4	6.5	20.3	6.0	0.8	5.1	0.0	0.0	48.3
1974-75	0.0	0.0	0.0	0.8	0.1	2.6	6.2	5.6	16.3	5.6	11.4	0.0	48.6
1975-76	0.0	0.0	0.0	0.6	9.3	5.9	5.8	15.9	14.7	9.5	0.0	0.0	61.7
1976-77	0.0	0.0	0.0	T	T	0.9	1.1	1.0	9.8	4.6	1.5	0.0	18.9
1977-78	0.0	0.0	0.0	0.0	1.6	2.0	13.6	5.7	2.2	3.4	7.4	0.0	35.9
1978-79	0.0	0.0	0.4	T	11.7	4.8	19.0	7.2	7.9	5.3	2.1	0.0	58.4
1979-80	0.0	0.0	0.0	T	0.8	3.8	12.4	3.0	10.9	0.5	0.5	0.0	31.9
1980-81	0.0	0.0	0.0	2.6	2.7	2.3	2.6	2.6	17.3	1.6	0.4	0.0	32.1
1981-82	0.0	0.0	0.0	0.8	8.6	3.6	9.0	3.1	7.5	3.8	0.2	0.0	36.6
1982-83	0.0	0.0	1.8	2.0	3.8	6.3	3.1	2.9	12.4	11.9	4.7	0.0	48.9
1983-84	0.0	0.0	0.0	0.0	13.6	13.6	10.0	3.9	12.9	8.0	0.0	0.0	64.2
1984-85	0.0	0.0	T	3.2	5.2	13.6	3.9	8.6	29.4	7.9	T	0.0	71.8
1985-86	0.0	0.0	0.0	T	9.2	4.6							
Record Mean	0.0	0.0	0.3	1.7	4.7	6.8	8.6	6.9	9.4	5.4	1.3	0.0	45.0

See Reference Notes, relative to all above tables, on preceding page.

Salt Lake City is located in a northern Utah valley surrounded by mountains on three sides and the Great Salt Lake to the northwest. The city varies in altitude from near 4,200 to 5,000 feet above sea level.

The Wasatch Mountains to the east have peaks to nearly 12,000 feet above sea level. Their orographic effects cause more precipitation in the eastern part of the city than over the western part.

The Oquirrh Mountains to the southwest of the city have several peaks to above 10,000 feet above sea level. The Traverse Mountain Range at the south end of the Salt Lake Valley rises to above 6,000 feet above sea level. These mountain ranges help to shelter the valleys from storms from the southwest in the winter, but are instrumental in developing thunderstorms which can drift over the valley in the summer.

Besides the mountain ranges, the most influential natural condition affecting the climate of Salt Lake City is the Great Salt Lake. This large inland body of water, which never freezes over due to its high salt content, can moderate the temperatures of cold winter winds blowing from the northwest and helps drive a lake/valley wind system. The warmer lake water during the winter and spring also contributes to increased precipitation in the valley downwind from the lake. The combination of the Great Salt Lake and the Wasatch Mountains often enhances storm precipitation in the valley.

Salt Lake City normally has a semi-arid continental climate with four well-defined seasons. Summers are characterized by hot, dry weather, but the high temperatures are usually not oppressive since the relative humidity is generally low and the nights usually cool. July is the hottest month with temperature readings in the 90s.

The mean diurnal temperature range is about 30 degrees in the summer and 18 degrees during the winter. Temperatures above 102 degrees in the summer or colder than −10 degrees in the winter are likely to occur one season out of four.

Winters are cold, but usually not severe. Mountains to the north and east act as a barrier to frequent invasions of cold continental air. The average annual snowfall is under 60 inches at the airport but much higher amounts fall in higher bench locations. Heavy fog can develop under temperature inversions in the winter and persist for several days.

Precipitation, generally light during the summer and early fall, is heavy in the spring when storms from the Pacific Ocean are moving through the area more frequently than at any other season of the year.

Winds are usually light, although occasional high winds have occurred in every month of the year, particularly in March.

The growing season is over five months in length. Yard and garden foilage generally are making good growth by mid-April. The last freezing temperature in the spring averages late April and the first freeze of the fall is mid-October.

TABLE 1 — NORMALS, MEANS AND EXTREMES

SALT LAKE CITY, UTAH

LATITUDE: 40°46'N   LONGITUDE: 111°58'W   ELEVATION: FT. GRND 4221 BARO 04224   TIME ZONE: MOUNTAIN   WBAN: 24127

	(a)	JAN	FEB	MAR	APR	MAY	JUNE	JULY	AUG	SEP	OCT	NOV	DEC	YEAR
TEMPERATURE °F:														
Normals														
-Daily Maximum		37.4	43.7	51.5	61.1	72.4	83.3	93.2	90.0	80.0	66.7	50.2	38.9	64.0
-Daily Minimum		19.7	24.4	29.9	37.2	45.2	53.3	61.8	59.7	50.0	39.3	29.2	21.6	39.3
-Monthly		28.6	34.0	40.7	49.2	58.8	68.3	77.5	74.8	65.0	53.0	39.7	30.3	51.7
Extremes														
-Record Highest	57	62	69	78	85	93	104	107	104	100	89	75	67	107
-Year		1982	1972	1960	1985	1984	1979	1960	1979	1979	1963	1967	1969	JUL 1960
-Record Lowest	57	-22	-30	2	14	25	35	40	37	27	16	-14	-21	-30
-Year		1949	1933	1966	1936	1965	1962	1968	1965	1965	1971	1955	1932	FEB 1933
NORMAL DEGREE DAYS:														
Heating (base 65°F)		1128	865	753	474	220	53	0	0	97	377	759	1076	5802
Cooling (base 65°F)		0	0	0	0	28	152	388	311	97	5	0	0	981
% OF POSSIBLE SUNSHINE	47	45	54	63	67	72	79	83	83	82	72	53	42	66
MEAN SKY COVER (tenths)														
Sunrise - Sunset	50	7.3	7.1	6.7	6.3	5.7	4.3	3.6	3.6	3.6	4.6	6.2	7.2	5.5
MEAN NUMBER OF DAYS:														
Sunrise to Sunset														
-Clear	57	5.5	5.2	7.0	7.0	9.2	13.7	16.8	15.7	16.9	14.1	8.6	6.2	125.9
-Partly Cloudy	57	6.6	7.0	8.3	9.3	10.4	10.0	9.8	10.8	8.0	7.8	7.2	6.5	101.8
-Cloudy	57	18.8	16.1	15.7	13.7	11.4	6.3	4.4	4.5	5.1	9.1	14.2	18.3	137.5
Precipitation														
.01 inches or more	57	9.9	8.8	9.9	9.5	8.0	5.5	4.6	5.6	5.3	6.4	7.6	9.3	90.5
Snow, Ice pellets														
1.0 inches or more	57	4.1	3.2	3.1	1.4	0.2	0.0	0.0	0.0	0.1	0.3	2.0	3.9	18.3
Thunderstorms	57	0.2	0.7	1.3	2.1	5.0	5.5	7.0	7.9	4.2	2.0	0.5	0.3	36.8
Heavy Fog Visibility														
1/4 mile or less	57	4.2	2.3	0.4	0.1	0.*	0.0	0.0	0.0	0.0	0.*	0.9	3.6	11.5
Temperature °F														
-Maximum														
90° and above	26	0.0	0.0	0.0	0.0	0.7	8.7	24.3	19.3	3.5	0.0	0.0	0.0	56.6
32° and below	26	10.8	3.9	0.7	0.0	0.0	0.0	0.0	0.0	0.0	0.*	0.6	8.7	24.8
-Minimum														
32° and below	26	27.1	22.8	17.3	7.3	0.8	0.0	0.0	0.0	0.4	5.4	17.7	27.1	126.1
0° and below	26	2.0	0.3	0.0	0.0	0.0	0.0	0.0	0.0	0.0	0.0	0.0	0.6	3.0
AVG. STATION PRESS.(mb)	13	874.5	873.4	869.1	869.5	869.1	870.1	871.5	871.5	872.2	873.5	873.1	874.7	871.8
RELATIVE HUMIDITY (%)														
Hour 05	26	78	77	71	67	65	60	52	55	62	69	74	78	67
Hour 11 (Local Time)	26	71	64	53	44	39	31	27	30	35	44	58	70	47
Hour 17 (Local Time)	26	69	59	47	39	33	26	21	23	29	41	59	71	43
Hour 23	26	78	76	68	62	58	50	42	45	54	67	73	78	63
PRECIPITATION (inches):														
Water Equivalent														
-Normal		1.35	1.33	1.72	2.21	1.47	0.97	0.72	0.92	0.89	1.14	1.22	1.37	15.31
-Maximum Monthly	57	3.14	3.22	3.97	4.90	4.76	2.93	2.57	3.66	7.04	3.91	2.63	4.37	7.04
-Year		1940	1936	1983	1944	1977	1947	1982	1968	1982	1981	1985	1983	SEP 1982
-Minimum Monthly	57	0.09	0.12	0.10	0.45	T	0.01	T	T	T	0.00	0.01	0.08	0.00
-Year		1961	1946	1956	1981	1934	1946	1963	1944	1951	1952	1939	1976	OCT 1952
-Maximum in 24 hrs	57	1.36	1.05	1.83	2.41	2.03	1.88	2.35	1.96	2.30	1.76	1.13	1.82	2.41
-Year		1953	1958	1944	1957	1942	1948	1962	1932	1982	1984	1954	1972	APR 1957
Snow, Ice pellets														
-Maximum Monthly	57	32.3	27.9	41.9	26.4	7.5	T			4.0	20.4	27.2	35.2	41.9
-Year		1937	1969	1977	1974	1975	1984			1971	1984	1985	1972	MAR 1977
-Maximum in 24 hrs	57	10.7	8.8	15.4	16.2	6.4	T			4.0	18.4	11.0	18.1	18.4
-Year		1980	1984	1944	1974	1975	1984			1971	1984	1930	1972	OCT 1984
WIND:														
Mean Speed (mph)	56	7.6	8.2	9.3	9.6	9.4	9.4	9.5	9.7	9.1	8.5	7.9	7.5	8.8
Prevailing Direction														
through 1963		SSE	SE	SSE	SE	SE	SSE	SSE	SSE	SE	SE	SSE	SSE	SSE
Fastest Mile														
-Direction	50	NW	SE	NW	NW	NW	W	W	SW	W	NW	NW	S	NW
-Speed (MPH)	50	59	56	71	57	57	63	49	58	61	67	63	54	71
-Year		1980	1954	1954	1964	1953	1963	1936	1946	1952	1950	1937	1955	MAR 1954
Peak Gust														
-Direction	2	NW	S	NE	NW	S	NW	S	SW	SW	NW	S	SW	NW
-Speed (mph)	2	21	48	45	54	60	45	55	43	48	63	52	36	63
-Date		1985	1985	1985	1984	1984	1984	1984	1985	1985	1985	1985	1985	OCT 1985

See Reference Notes to this table on the following page.

SALT LAKE CITY, UTAH

TABLE 2   PRECIPITATION (inches)   SALT LAKE CITY, UTAH

YEAR	JAN	FEB	MAR	APR	MAY	JUNE	JULY	AUG	SEP	OCT	NOV	DEC	ANNUAL
1956	2.39	1.10	0.10	1.50	2.33	0.63	0.08	0.03	0.14	1.86	0.56	1.67	12.39
1957	1.37	0.72	2.18	3.24	3.37	1.47	0.31	1.69	0.33	0.78	1.37	1.61	18.44
1958	0.87	2.20	2.19	2.92	0.30	0.04	0.05	0.23	0.25	T	1.13	0.54	10.72
1959	1.60	1.41	0.79	1.61	2.05	1.38	0.19	1.76	1.66	0.22	0.10	1.05	13.82
1960	0.96	1.58	2.45	1.11	0.73	0.39	0.02	1.33	0.49	1.19	1.73	0.39	12.37
1961	0.09	2.06	1.85	0.95	0.24	0.09	0.54	1.20	1.10	1.60	1.15	0.88	11.75
1962	0.84	1.43	2.34	2.98	2.12	0.49	2.52	0.26	0.27	0.93	0.44	0.28	14.90
1963	0.53	0.67	2.11	3.86	0.23	1.67	T	0.54	1.08	1.05	1.56	0.79	14.09
1964	0.94	0.35	2.26	2.69	2.77	2.61	0.26	0.17	0.13	0.45	1.42	3.82	17.87
1965	2.13	1.13	0.14	2.30	2.02	1.87	1.50	2.08	1.93	0.39	1.13	1.81	18.43
1966	0.41	1.19	1.21	1.43	0.51	0.07	0.33	0.22	0.83	1.18	0.75	0.86	8.99
1967	2.05	0.67	1.94	2.08	2.15	2.73	1.14	0.07	0.73	0.66	1.64	1.64	16.52
1968	0.46	2.32	2.21	2.82	2.18	1.58	0.09	3.66	0.56	1.64	1.32	2.27	21.11
1969	1.69	2.84	0.57	1.38	0.18	2.83	1.51	0.34	0.18	1.96	0.92	1.69	16.09
1970	1.24	0.94	1.01	3.25	0.89	1.63	0.86	0.57	2.80	1.61	2.27	2.80	19.87
1971	1.06	2.13	1.01	2.16	1.34	0.64	0.94	2.15	1.75	3.23	1.03	1.35	18.79
1972	1.22	0.48	1.18	3.62	0.14	0.15	0.06	0.21	1.36	2.74	3.22	1.53	15.74
1973	1.49	0.91	2.67	1.64	1.74	0.19	1.07	1.16	4.07	0.67	2.52	2.26	20.39
1974	1.80	1.65	0.97	4.57	0.39	0.28	0.18	0.32	0.03	2.03	0.90	1.34	14.46
1975	1.28	1.24	3.44	2.46	2.58	1.81	0.28	0.10	0.08	1.91	1.71	1.03	17.92
1976	0.63	1.90	1.18	2.47	0.99	1.24	1.55	0.82	0.16	0.57	0.03	0.08	12.34
1977	0.76	0.64	3.10	0.59	4.76	0.06	0.61	1.85	1.85	0.83	1.20	1.42	17.67
1978	2.33	1.96	3.47	2.90	1.57	0.06	0.06	0.92	2.51	T	1.73	0.58	18.09
1979	0.72	1.05	0.80	1.04	0.84	0.35	0.40	0.63	0.05	1.29	0.98	0.55	8.70
1980	2.87	2.25	2.46	0.89	2.70	0.42	1.34	0.26	0.72	1.74	1.17	0.37	17.19
1981	0.64	0.81	2.11	0.45	3.68	1.03	0.33	0.23	0.48	3.91	1.03	1.89	16.59
1982	1.08	0.53	2.39	1.63	1.86	0.66	2.57	0.56	7.04	1.87	0.75	1.92	22.86
1983	1.19	1.36	3.97	1.63	2.58	0.62	1.02	2.64	1.03	1.62	2.23	4.37	24.26
1984	0.50	0.95	1.76	4.43	1.17	1.86	1.72	1.49	1.72	3.70	1.45	0.80	21.55
1985	0.91	0.85	1.80	0.64	2.95	1.30	0.85	0.03	1.98	1.61	2.63	1.42	16.97
Record Mean	1.28	1.34	1.87	1.99	1.77	0.90	0.64	0.87	0.96	1.44	1.37	1.39	15.81

TABLE 3   AVERAGE TEMPERATURE (deg. F)   SALT LAKE CITY, UTAH

YEAR	JAN	FEB	MAR	APR	MAY	JUNE	JULY	AUG	SEP	OCT	NOV	DEC	ANNUAL
1956	35.5	26.4	42.4	49.9	60.4	69.6	76.4	72.7	67.0	52.1	34.3	30.0	51.4
1957	26.0	40.3	43.1	48.4	57.2	66.8	77.0	76.5	64.6	52.2	34.5	33.4	51.7
1958	30.7	41.7	39.2	47.2	65.1	70.6	74.9	77.8	65.4	55.3	38.8	36.2	53.6
#1959	33.9	36.1	41.4	51.6	55.7	71.5	77.2	74.1	62.0	51.0	37.7	27.8	51.7
1960	25.6	28.7	42.5	50.8	58.7	71.0	81.2	74.2	68.2	51.6	40.4	29.6	51.9
1961	28.7	38.1	42.9	50.1	60.8	74.7	79.8	77.8	60.0	50.0	35.3	28.2	52.2
1962	20.5	31.4	35.1	52.8	58.7	68.2	75.9	73.1	65.8	55.3	41.6	28.4	50.6
1963	19.5	38.6	39.4	44.3	60.7	63.3	77.8	77.9	67.8	57.8	38.9	24.4	50.9
1964	21.9	25.8	32.0	45.6	55.8	63.2	77.5	71.9	61.5	53.0	37.8	33.3	48.3
1965	31.0	33.0	36.8	51.0	54.7	64.8	75.0	70.9	57.5	54.5	46.1	30.2	50.5
1966	30.6	29.4	41.6	49.6	62.7	69.2	80.1	74.1	67.5	49.9	43.1	29.2	52.3
1967	29.4	37.5	44.2	46.1	56.3	64.6	78.4	78.6	66.7	52.4	43.0	25.1	51.8
1968	24.5	38.2	44.7	45.4	56.4	67.5	78.3	69.4	61.4	51.7	38.5	26.8	50.2
1969	32.2	28.7	38.4	50.4	64.0	64.8	76.6	77.6	69.7	47.7	39.5	32.4	51.9
1970	34.6	40.4	40.6	44.2	58.8	67.6	76.6	77.7	59.0	47.1	42.6	29.2	51.5
1971	32.4	34.9	40.4	48.2	56.6	67.5	76.4	76.9	59.8	47.5	37.6	26.9	50.4
1972	29.8	37.8	46.9	48.1	60.5	71.9	77.2	75.8	63.9	53.6	39.4	22.7	52.3
1973	19.6	32.3	41.8	47.6	61.6	70.2	76.6	76.6	61.5	54.1	40.5	33.4	51.3
1974	26.7	31.4	41.6	48.1	58.8	73.4	79.2	74.2	66.5	54.7	43.4	31.7	52.8
1975	27.4	35.5	41.1	44.3	54.3	64.8	78.8	73.4	65.4	53.4	37.3	32.9	50.7
1976	27.9	34.1	38.1	49.3	62.2	67.6	72.3	66.4	66.4	51.0	29.4	29.4	51.6
1977	26.8	35.8	37.7	54.1	55.0	73.2	77.3	75.0	66.4	55.6	42.5	37.9	53.2
1978	36.3	39.8	48.0	50.2	56.0	69.2	78.0	74.0	64.0	55.5	41.0	26.8	53.3
1979	22.1	32.5	43.2	51.1	60.2	70.1	78.9	75.6	71.4	56.7	36.5	32.9	52.6
1980	33.7	36.0	41.5	52.7	57.0	67.5	77.0	74.1	66.3	52.6	41.3	33.6	52.8
1981	32.1	38.3	44.1	53.4	57.6	69.6	78.2	78.0	68.5	50.5	44.3	36.4	54.3
1982	29.8	32.3	43.3	46.5	56.7	68.0	75.3	78.4	64.0	48.8	38.1	29.9	51.0
1983	35.2	39.4	44.6	45.9	55.8	67.7	76.6	77.8	67.8	56.0	43.0	31.9	53.5
1984	23.8	25.8	40.1	48.5	61.6	67.3	78.5	77.2	66.5	49.5	42.7	29.9	51.0
1985	24.2	25.6	40.8	55.7	63.9	72.5	80.7	76.5	62.7	53.1	37.4	27.7	51.7
Record Mean	28.1	33.2	40.8	49.2	58.3	68.2	77.3	75.5	65.2	53.1	40.5	31.5	51.8
Max	36.2	41.6	50.4	60.0	70.3	81.6	91.2	89.0	78.5	64.9	50.1	39.3	62.8
Min	20.0	24.8	31.2	38.3	46.2	54.8	63.3	61.9	51.8	41.2	30.9	23.6	40.7

REFERENCE NOTES FOR TABLES 1, 2, 3 and 6   (SALT LAKE CITY, UT)

GENERAL

T - TRACE AMOUNT
BLANK ENTRIES DENOTE MISSING/UNREPORTED DATA.
# INDICATES A STATION OR INSTRUMENT RELOCATION.

SPECIFIC

TABLE 1

(a) - LENGTH OF RECORD IN YEARS. ALTHOUGH INDIVIDUAL MONTHS MAY BE MISSING.

*   LESS THAN .05

NORMALS — BASED ON THE 1951-1980 RECORD PERIOD.
EXTREMES — DATES ARE THE MOST RECENT OCCURRENCE.
WIND DIR. — NUMERALS SHOW TENS OF DEGREES
       CLOCKWISE FROM TRUE NORTH.
       "00" INDICATES CALM.
RESULTANT WIND DIRECTIONS ARE GIVEN TO WHOLE DEGREES.

EXCEPTIONS

TABLES 2, 3, and 6

RECORD MEANS ARE THROUGH THE CURRENT YEAR,
BEGINNING IN   1874 FOR TEMPERATURE
                1874 FOR PRECIPITATION
                1929 FOR SNOWFALL

TABLE 4 HEATING DEGREE DAYS Base 65 deg. F   SALT LAKE CITY, UTAH

SEASON	JULY	AUG	SEP	OCT	NOV	DEC	JAN	FEB	MAR	APR	MAY	JUNE	TOTAL
1956-57	0	12	58	398	915	1077	1203	685	667	489	238	70	5812
1957-58	0	1	101	401	909	971	1053	649	794	527	106	13	5525
1958-59	0	0	116	300	778	887	956	804	722	404	296	34	5297
#1959-60	1	0	170	423	814	1142	1212	1049	692	419	227	14	6163
1960-61	0	19	16	419	730	1088	1116	743	678	440	183	14	5446
1961-62	0	0	207	461	881	1132	1373	936	921	369	220	75	6575
1962-63	3	17	59	322	695	1128	1403	731	787	614	135	98	5992
1963-64	0	1	18	243	777	1252	1331	1130	1016	576	303	125	6772
1964-65	0	44	134	365	808	975	1046	889	869	414	316	61	5921
1965-66	0	20	239	317	564	1069	1058	989	717	456	140	40	5609
1966-67	0	4	57	460	649	1101	1097	763	638	564	287	76	5696
1967-68	0	0	57	387	653	1228	1246	772	622	583	276	57	5881
1968-69	3	49	166	407	786	1174	1009	1010	818	433	75	67	5997
1969-70	1	0	17	530	759	1003	935	681	754	619	218	69	5586
1970-71	0	0	218	550	667	1103	1002	836	754	499	258	55	5942
1971-72	0	0	201	535	817	1176	1085	783	556	499	168	2	5822
1972-73	0	0	110	347	761	1307	1400	909	711	515	135	67	6262
1973-74	1	0	140	333	732	975	1181	935	603	502	214	41	5657
1974-75	0	5	54	316	638	1025	1157	819	734	613	334	92	5787
1975-76	0	1	62	365	825	989	1144	890	826	464	112	67	5745
1976-77	0	7	37	432	689	1096	1175	813	838	333	304	0	5724
1977-78	0	11	73	282	670	835	880	697	522	433	293	36	4732
1978-79	0	12	144	284	714	1178	1327	902	666	414	196	57	5894
1979-80	0	0	7	270	846	987	964	835	723	371	250	77	5330
1980-81	0	10	57	379	704	965	1013	742	641	346	233	46	5136
1981-82	0	0	34	444	614	879	1087	909	668	548	259	62	5504
1982-83	7	0	134	495	800	1080	916	710	624	569	314	36	5685
1983-84	6	0	49	276	650	1018	1269	1130	763	493	157	76	5887
1984-85	0	0	98	480	662	1084	1260	1097	740	285	109	17	5832
1985-86	0	0	140	360	821	1151							

TABLE 5 COOLING DEGREE DAYS Base 65 deg. F   SALT LAKE CITY, UTAH

YEAR	JAN	FEB	MAR	APR	MAY	JUNE	JULY	AUG	SEP	OCT	NOV	DEC	TOTAL
1969	0	0	0	1	53	68	366	398	164	0	0	0	1050
1970	0	0	0	0	32	152	365	398	46	0	0	0	993
1971	0	0	0	0	5	136	361	374	50	0	0	0	926
1972	0	0	0	0	34	213	386	340	85	0	0	0	1058
1973	0	0	0	0	38	226	370	367	44	3	0	0	1048
1974	0	0	0	2	31	303	446	298	108	3	0	0	1191
1975	0	0	0	0	9	89	439	269	80	14	0	0	900
1976	0	0	0	0	34	151	431	237	87	3	0	0	943
1977	0	0	0	12	2	254	389	328	123	0	0	0	1108
1978	0	0	0	0	21	167	411	299	120	0	0	0	1018
1979	0	0	0	2	54	214	439	336	208	21	0	0	1274
1980	0	0	0	9	10	159	399	301	99	1	0	0	978
1981	0	0	0	3	12	190	412	409	145	2	0	0	1173
1982	0	0	0	0	11	158	338	423	109	0	0	0	1039
1983	0	0	0	0	37	123	370	405	138	4	0	0	1077
1984	0	0	0	3	58	153	426	383	147	4	0	0	1174
1985	0	0	0	11	78	249	493	364	79	0	0	0	1274

TABLE 6 SNOWFALL (inches)   SALT LAKE CITY, UTAH

SEASON	JULY	AUG	SEP	OCT	NOV	DEC	JAN	FEB	MAR	APR	MAY	JUNE	TOTAL
1956-57	0.0	0.0	0.0	10.4	7.0	11.8	18.4	0.3	4.2	5.1	T	0.0	57.2
1957-58	0.0	0.0	0.0	0.0	6.3	11.1	7.6	9.5	15.7	15.5	0.0	0.0	65.7
1958-59	0.0	0.0	0.0	T	10.4	2.9	11.8	0.4	15.2	1.3	0.0	0.0	42.0
1959-60	0.0	0.0	0.0	T	0.6	10.6	13.5	18.2	11.4	0.7	1.0	0.0	56.0
1960-61	0.0	0.0	0.0	T	5.0	4.5	0.1	16.0	4.9	0.8	T	0.0	31.3
1961-62	0.0	0.0	0.0	8.3	11.1	8.5	15.6	10.1	25.3	1.6	0.0	0.0	80.5
1962-63	0.0	0.0	0.0	T	2.4	0.9	7.4	0.5	16.0	14.3	0.0	0.0	41.5
1963-64	0.0	0.0	0.0	T	7.6	12.9	18.2	8.0	33.5	1.9	5.3	0.0	87.4
1964-65	0.0	0.0	0.0	0.0	6.7	6.2	15.7	9.5	1.1	2.4	5.3	0.0	46.9
1965-66	0.0	0.0	2.2	2.6	2.6	12.8	5.9	18.1	17.4	2.8	0.0	0.0	61.8
1966-67	0.0	0.0	0.0	3.6	3.2	8.7	30.4	4.5	11.8	11.4	1.0	0.0	74.6
1967-68	0.0	0.0	0.0	0.0	4.2	27.1	6.8	13.6	8.4	14.2	T	0.0	74.3
1968-69	0.0	0.0	0.0	T	8.7	33.3	13.7	27.9	5.4	0.2	0.0	0.0	89.2
1969-70	0.0	0.0	0.0	0.1	5.6	2.8	3.9	5.2	23.6	0.0	0.0	0.0	57.2
1970-71	0.0	0.0	0.0	0.3	0.7	25.8	13.6	8.7	8.9	1.7	1.4	0.0	61.1
1971-72	0.0	0.0	4.0	16.6	5.4	10.5	7.6	1.4	15.0	0.0	0.0	0.0	78.2
1972-73	0.0	0.0	0.0	6.0	1.1	35.2	20.9	3.6	17.8	2.6	0.0	0.0	87.2
1973-74	0.0	0.0	0.0	1.3	19.5	19.6	20.1	17.2	6.7	26.4	T	0.0	110.8
1974-75	0.0	0.0	0.0	T	T	8.8	12.5	7.9	22.8	13.1	7.5	0.0	72.6
1975-76	0.0	0.0	0.0	0.1	18.0	11.8	8.6	15.8	18.7	3.5	0.0	T	76.5
1976-77	0.0	0.0	0.0	0.0	T	1.2	8.6	3.2	41.9	4.8	0.6	0.0	60.3
1977-78	0.0	0.0	0.0	0.2	8.5	8.2	15.6	15.5	6.2	2.5	4.6	0.0	61.3
1978-79	0.0	0.0	1.0	0.0	17.4	8.7	13.8	12.4	3.6	7.7	T	0.0	64.6
1979-80	0.0	0.0	0.0	0.0	4.6	8.5	24.5	2.9	19.9	1.2	T	T	61.6
1980-81	0.0	0.0	0.0	T	3.9	3.3	8.9	2.7	11.1	0.3	T	T	30.2
1981-82	0.0	0.0	0.0	4.4	2.4	11.5	15.3	4.5	10.2	9.5	T	T	57.8
1982-83	0.0	0.0	0.0	0.2	1.0	20.1	6.2	13.3	1.0	0.0	5.0	0.0	55.8
1983-84	0.0	0.0	0.0	0.0	5.9	34.2	7.6	18.5	6.7	25.1	T	T	98.0
1984-85	0.0	0.0	T	20.4	6.6	12.9	12.7	11.4	8.0	0.7	0.0	0.0	72.7
1985-86	0.0	0.0	0.0	T	27.2	14.7							
Record Mean	0.0	0.0	0.1	1.4	6.6	12.4	13.2	9.5	10.3	5.2	0.7	T	59.3

See Reference Notes, relative to all above tables, on preceding page.

BURLINGTON, VERMONT

Burlington is located on the eastern shore of Lake Champlain at the widest part of the lake. About 35 miles to the west lie the highest peaks of the Adirondacks, while the foothills of the Green Mountains begin 10 miles to the east and southeast.

Its northerly latitude assures the variety and vigor of a true New England climate, while thanks to the modifying influence of the lake, the many rapid and marked weather changes are tempered in severity. Due to its location in the path of the St. Lawrence Valley storm track and the lake effects, the city is one of the cloudiest in the United States.

Lake Champlain exercises a tempering influence on the local temperature. During the winter months and prior to the lake freezing, temperatures along the lake shore are often 5-10 degrees warmer than at the airport 3 1/2 miles inland. At the airport the average occurrence of the last freeze in spring is around May 10th and that of the first in fall is early October, giving a growing season of 145 days. This location is justly proud of its delightful summer weather. On average, there are few days a year with maxima of 90 degrees or higher. This moderate summer heat gives way to a cooler, but none the less pleasant fall period, usually extending well into October. High pressure systems moving down rapidly from central Canada or Hudson Bay produce the coldest temperatures during the winter months, but extended periods of very cold weather are rare.

Precipitation, although generally plentiful and well distributed throughout the year, is less in the Champlain Valley than in other areas of Vermont due to the shielding effect of the mountain barriers to the east and west. The heaviest rainfall usually occurs during summer thunderstorms, but excessively heavy rainfall is quite uncommon. Droughts are infrequent.

Because of the trend of the Champlain Valley between the Adirondack and Green Mountain ranges, most winds have a northerly or southerly component. The prevailing direction most of the year is from the south. Winds of damaging force are very uncommon.

Smoke pollution is nearly non-existent since there is no concentration of heavy industry here, however, haze has been on the increase over the years due to the large increase in industry to the north and south. During the spring and fall months, fog occasionally forms along the Winooski River to the north and east and may drift over the airport with favorable winds. In spite of the high percentage of cloudiness, periods of low aircraft ceilings and visibilities are usually of short duration, allowing this area to have one of the highest percentages of flying weather in New England.

TABLE 1 **NORMALS, MEANS AND EXTREMES**

BURLINGTON, VERMONT

LATITUDE: 44°28'N LONGITUDE: 73°09'W ELEVATION: FT. GRND 332 BARO 00350 TIME ZONE: EASTERN WBAN: 14742

	(a)	JAN	FEB	MAR	APR	MAY	JUNE	JULY	AUG	SEP	OCT	NOV	DEC	YEAR
TEMPERATURE °F:														
Normals														
-Daily Maximum		25.4	27.3	37.7	52.6	66.4	75.9	80.5	77.6	68.8	57.0	43.6	30.3	53.6
-Daily Minimum		7.7	8.8	20.8	32.7	44.0	54.0	58.6	56.6	48.7	38.7	29.6	14.9	34.6
-Monthly		16.6	18.1	29.3	42.7	55.2	65.0	69.6	67.1	58.8	47.9	36.6	22.6	44.1
Extremes														
-Record Highest	42	63	62	84	91	93	96	99	101	94	85	75	65	101
-Year		1950	1981	1946	1976	1977	1946	1977	1944	1945	1949	1948	1982	AUG 1944
-Record Lowest	42	-30	-30	-20	2	24	33	39	35	25	15	-2	-26	-30
-Year		1957	1979	1948	1972	1966	1965	1962	1976	1963	1972	1958	1980	FEB 1979
NORMAL DEGREE DAYS:														
Heating (base 65°F)		1500	1313	1110	669	326	64	23	50	202	530	852	1314	7953
Cooling (base 65°F)		0	0	0	0	22	61	165	115	16	0	0	0	379
% OF POSSIBLE SUNSHINE	42	41	47	50	49	55	58	64	60	54	48	30	32	49
MEAN SKY COVER (tenths)														
Sunrise - Sunset	42	7.6	7.3	7.2	7.1	7.0	6.8	6.4	6.4	6.5	6.9	8.3	8.1	7.1
MEAN NUMBER OF DAYS:														
Sunrise to Sunset														
-Clear	42	4.3	4.4	5.6	5.1	5.0	4.8	5.3	6.0	6.0	6.1	2.5	2.8	57.8
-Partly Cloudy	42	6.6	6.4	6.8	7.5	9.1	10.7	13.0	11.7	9.9	7.9	5.1	5.8	100.7
-Cloudy	42	20.1	17.5	18.6	17.3	17.0	14.5	12.7	13.3	14.0	17.0	22.3	22.4	206.7
Precipitation														
.01 inches or more	42	14.3	11.6	13.1	12.4	13.4	12.5	11.9	12.4	11.6	11.4	14.1	15.0	153.7
Snow,Ice pellets														
1.0 inches or more	42	5.2	4.7	3.7	1.2	0.1	0.0	0.0	0.0	0.0	0.*	2.0	5.4	22.2
Thunderstorms	42	0.*	0.0	0.3	0.9	2.4	5.2	6.3	5.4	2.2	0.6	0.3	0.*	23.8
Heavy Fog Visibility														
1/4 mile or less	42	0.9	1.1	1.2	1.3	0.9	1.1	0.8	1.4	2.5	2.0	1.3	1.2	15.7
Temperature °F														
-Maximum														
90° and above	21	0.0	0.0	0.0	0.1	0.6	1.1	2.5	0.9	0.*	0.0	0.0	0.0	5.2
32° and below	21	22.8	18.2	9.3	0.6	0.0	0.0	0.0	0.0	0.0	0.*	4.7	17.3	73.0
-Minimum														
32° and below	21	29.9	26.4	25.9	16.0	3.0	0.0	0.0	0.0	0.6	8.6	18.9	27.6	156.9
0° and below	21	11.0	8.4	2.2	0.0	0.0	0.0	0.0	0.0	0.0	0.0	0.0	4.7	26.3
AVG. STATION PRESS.(mb)	13	1003.2	1004.5	1002.6	1001.5	1001.5	1001.7	1002.0	1004.3	1005.2	1005.9	1004.4	1004.4	1003.5
RELATIVE HUMIDITY (%)														
Hour 01	20	69	71	72	73	76	81	82	84	85	79	77	75	77
Hour 07 (Local Time)	20	70	73	74	73	74	78	78	83	86	81	78	75	77
Hour 13	20	63	63	59	54	51	56	53	57	62	62	67	69	60
Hour 19	20	66	65	63	58	57	62	60	67	74	71	73	72	66
PRECIPITATION (inches):														
Water Equivalent														
-Normal		1.85	1.73	2.20	2.77	2.96	3.64	3.43	3.87	3.20	2.81	2.80	2.43	33.69
-Maximum Monthly	42	4.69	5.38	3.58	6.55	6.31	7.69	6.12	11.54	8.18	6.22	6.85	5.95	11.54
-Year		1978	1981	1972	1983	1983	1973	1972	1955	1945	1959	1983	1973	AUG 1955
-Minimum Monthly	42	0.49	0.21	0.38	0.93	0.29	1.09	1.23	0.72	0.87	0.50	0.63	0.62	0.21
-Year		1981	1978	1965	1966	1977	1949	1979	1957	1948	1963	1952	1960	FEB 1978
-Maximum in 24 hrs	42	1.53	1.93	1.62	2.16	2.26	2.83	2.69	3.59	3.26	2.17	1.80	2.60	3.59
-Year		1978	1981	1971	1968	1955	1972	1985	1955	1983	1983	1959	1950	AUG 1955
Snow,Ice pellets														
-Maximum Monthly	42	42.4	34.3	33.1	21.3	3.9				T	5.1	19.2	56.7	56.7
-Year		1978	1958	1971	1983	1966				1963	1969	1971	1970	DEC 1970
-Maximum in 24 hrs	42	14.5	16.5	15.6	15.6	3.5				T	5.1	10.1	17.0	17.0
-Year		1961	1966	1971	1983	1966				1963	1969	1958	1978	DEC 1978
WIND:														
Mean Speed (mph)	42	9.5	9.2	9.4	9.3	8.8	8.3	7.9	7.4	8.1	8.6	9.5	9.8	8.8
Prevailing Direction														
through 1963		S	S	N	S	S	S	S	S	S	S	S	S	S
Fastest Mile														
-Direction	40	S	SE	S	S	SW	SW	NW	N	S	SE	SE	S	SE
-Speed (MPH)	40	57	50	56	52	46	42	46	54	45	70	72	47	72
-Year		1950	1965	1951	1948	1983	1957	1974	1955	1947	1954	1950	1950	NOV 1950
Peak Gust														
-Direction	2	S	S	NW	NW	S	NW	S	S	NW	S	S	SE	SE
-Speed (mph)	2	41	43	41	40	39	39	31	39	36	39	45	46	46
-Date		1984	1985	1984	1985	1984	1984	1985	1984	1985	1985	1984	1985	DEC 1985

See Reference Notes to this table on the following page.

TABLE 2 PRECIPITATION (inches)   BURLINGTON, VERMONT

YEAR	JAN	FEB	MAR	APR	MAY	JUNE	JULY	AUG	SEP	OCT	NOV	DEC	ANNUAL
1956	1.92	1.78	2.37	2.47	4.74	2.92	4.06	2.00	3.91	1.58	1.67	1.80	31.22
1957	1.20	0.90	0.95	2.11	2.95	7.35	5.34	0.72	3.27	1.55	3.07	3.93	33.34
1958	3.74	2.21	1.06	2.84	2.93	3.77	3.89	3.06	4.66	1.89	0.82	3.88	34.75
1959	2.72	1.98	1.31	1.57	1.49	3.41	1.81	4.38	2.03	6.22	5.07	2.55	34.54
1960	1.24	1.98	1.42	2.64	3.64	3.51	3.37	1.53	4.90	4.04	1.96	0.62	30.85
1961	0.93	1.65	1.56	3.96	2.63	3.71	4.98	3.24	2.69	2.50	2.31	1.75	31.91
1962	1.07	1.36	1.86	2.59	2.24	2.66	5.93	3.46	3.56	3.28	2.75	1.73	32.49
1963	1.14	1.22	2.35	2.52	2.37	1.90	2.79	5.11	1.42	0.50	3.95	0.96	26.23
1964	2.27	0.63	2.64	2.11	4.67	3.00	2.87	4.10	1.49	2.20	2.10	1.63	29.71
1965	0.60	0.93	0.38	2.16	1.05	4.08	2.91	6.27	3.19	3.32	2.65	1.47	29.01
1966	2.02	2.49	2.63	0.93	2.49	2.63	1.92	4.46	3.33	1.41	1.41	2.82	28.54
1967	1.65	0.77	0.51	3.77	3.19	3.12	4.60	3.79	3.06	3.03	2.12	2.61	32.22
1968	1.26	1.28	3.23	3.54	2.43	3.66	2.70	2.36	2.06	2.73	4.37	3.12	32.74
1969	2.43	0.94	1.93	2.93	3.10	4.01	2.40	3.71	1.88	1.62	4.98	4.59	34.52
1970	0.65	1.95	2.01	2.78	3.14	4.38	1.92	3.44	3.93	2.66	2.35	3.77	32.98
1971	1.24	2.98	2.71	2.65	2.97	2.29	4.29	4.85	1.63	2.16	2.29	1.93	31.99
1972	0.93	1.69	3.58	2.26	2.83	6.52	6.12	2.35	1.69	2.60	4.10	3.43	38.10
1973	2.13	1.55	2.09	3.80	5.38	7.69	3.02	5.41	5.02	1.93	2.31	5.95	46.28
1974	1.90	1.54	2.73	3.47	4.61	4.45	3.70	2.60	3.23	0.78	3.60	2.08	34.69
1975	2.20	2.01	2.86	1.71	1.17	2.47	3.77	2.85	4.12	3.85	3.14	2.36	32.51
1976	2.99	2.85	2.35	2.54	5.86	4.04	3.05	4.69	3.77	4.34	1.63	1.97	40.08
1977	1.61	1.78	2.97	3.13	0.29	2.06	3.34	6.27	6.33	5.02	4.22	3.42	40.44
1978	4.69	0.21	2.98	2.51	2.16	4.36	3.50	1.82	2.07	3.72	0.95	2.11	31.08
1979	4.50	0.60	2.15	3.61	3.12	1.39	1.23	3.42	3.84	2.31	3.89	1.50	31.56
1980	0.61	0.67	2.44	2.39	1.61	1.92	6.11	3.83	4.41	2.48	2.92	1.50	30.89
1981	0.49	5.38	1.32	3.05	3.76	3.07	3.22	5.58	6.24	5.26	2.73	2.03	42.13
1982	2.74	1.43	2.31	2.63	1.95	4.95	3.07	3.55	2.12	2.31	3.59	1.69	32.34
1983	3.09	1.66	2.60	6.55	6.31	1.49	3.92	4.31	3.77	4.38	6.85	5.23	50.16
1984	0.81	2.73	1.72	4.25	5.27	1.70	5.11	3.30	2.81	1.89	3.08	3.14	35.81
1985	1.46	1.26	2.46	1.90	3.53	3.76	4.42	2.67	3.30	3.31	3.68	1.59	33.34
Record Mean	1.82	1.68	2.16	2.54	3.03	3.48	3.62	3.48	3.35	2.92	2.78	2.12	32.97

TABLE 3 AVERAGE TEMPERATURE (deg. F)   BURLINGTON, VERMONT

YEAR	JAN	FEB	MAR	APR	MAY	JUNE	JULY	AUG	SEP	OCT	NOV	DEC	ANNUAL
1956	19.6	23.0	23.7	40.8	50.0	64.9	66.5	66.7	55.9	49.4	37.9	26.3	43.7
1957	11.7	25.1	31.4	45.4	54.0	68.2	68.4	64.1	60.9	48.6	39.4	29.6	45.6
1958	19.5	12.9	32.7	45.4	51.9	59.8	68.0	68.0	59.1	47.3	37.8	13.1	43.0
#1959	17.2	13.8	27.8	45.0	58.3	65.5	72.5	71.3	63.3	48.0	35.1	26.0	45.3
1960	16.5	24.5	23.9	43.9	60.9	64.7	66.9	66.0	59.5	45.4	39.2	19.7	44.3
1961	9.2	18.6	26.5	38.5	51.1	63.5	68.2	66.8	65.4	49.6	36.5	23.9	43.2
1962	15.5	13.7	29.0	41.9	55.7	64.6	64.0	65.5	55.7	46.3	31.6	20.1	42.0
1963	16.6	10.7	26.0	40.3	52.4	65.5	70.3	62.5	53.5	51.7	40.3	12.9	41.9
#1964	21.9	17.2	31.1	43.3	58.6	63.0	69.6	62.9	55.8	45.3	35.7	24.8	44.1
1965	13.7	20.0	27.4	39.3	57.2	62.9	65.1	66.0	58.7	46.8	32.9	28.6	43.2
1966	15.5	17.9	30.3	41.3	51.2	65.3	69.6	67.4	55.7	47.0	40.6	23.4	43.7
1967	23.9	11.7	25.3	40.7	47.6	67.6	70.1	67.0	58.1	48.8	32.9	25.5	43.3
1968	8.4	11.0	29.6	46.2	51.7	60.8	68.7	63.7	61.0	49.8	32.2	17.9	41.8
1969	16.5	18.5	24.3	41.4	51.3	64.1	67.8	68.6	57.9	45.8	36.3	18.6	42.6
1970	3.6	16.9	25.8	42.6	54.1	63.7	70.6	68.4	60.1	50.5	39.0	14.3	42.5
1971	9.7	20.0	24.1	37.3	54.5	64.9	68.9	67.1	63.5	53.8	33.5	24.3	43.5
1972	21.1	17.0	24.8	35.6	56.2	63.1	65.3	65.3	58.5	42.4	22.5	22.4	42.4
1973	21.5	14.6	37.1	44.6	53.6	66.9	70.6	72.1	58.5	49.3	37.2	27.4	46.1
1974	18.7	15.6	29.2	44.4	51.3	66.5	70.2	69.1	58.2	43.4	28.5	20.1	44.3
1975	23.6	20.7	28.0	37.1	62.3	66.4	74.6	69.1	58.0	50.4	42.1	20.1	46.1
1976	11.1	24.6	33.4	47.4	54.7	69.2	68.5	65.7	57.0	43.7	33.0	16.3	43.7
1977	11.1	20.5	37.6	45.3	60.0	64.7	69.6	67.5	58.7	46.6	40.0	22.4	45.3
1978	15.1	9.5	26.0	38.7	60.1	63.9	69.4	68.3	55.2	46.4	34.8	25.2	42.7
1979	18.0	7.5	36.9	43.5	58.5	65.3	72.2	65.9	58.6	48.1	41.3	29.0	45.4
1980	21.2	17.6	31.1	46.5	58.9	64.4	70.6	70.7	57.9	45.0	32.3	15.0	44.2
1981	8.9	32.9	33.5	46.7	58.2	66.1	71.1	67.1	59.3	44.9	36.9	25.3	45.9
1982	9.6	19.1	30.3	43.4	57.3	60.7	69.5	65.9	62.3	50.1	42.3	31.9	45.2
1983	21.0	22.3	33.0	42.3	52.9	66.3	71.3	68.6	62.9	48.2	38.1	22.4	45.8
1984	16.5	28.7	21.9	44.7	52.3	66.0	70.3	71.1	57.2	50.0	38.4	30.3	45.6
1985	13.4	22.5	31.6	44.3	55.8	61.7	69.6	67.5	60.3	49.1	36.9	21.3	44.5
Record Mean	17.7	18.6	29.5	42.8	55.4	64.8	69.8	67.3	59.5	48.6	36.7	23.3	44.5
Max	26.2	27.4	37.8	51.9	65.6	74.8	79.8	77.1	69.0	57.3	43.4	30.6	53.4
Min	9.1	9.8	21.2	33.6	45.2	54.8	59.7	57.6	50.1	39.9	29.9	16.0	35.6

REFERENCE NOTES FOR TABLES 1, 2, 3 and 6   (BURLINGTON, VT)

GENERAL

T - TRACE AMOUNT
BLANK ENTRIES DENOTE MISSING/UNREPORTED DATA.
# INDICATES A STATION OR INSTRUMENT RELOCATION.

SPECIFIC

TABLE 1

(a) - LENGTH OF RECORD IN YEARS. ALTHOUGH
      INDIVIDUAL MONTHS MAY BE MISSING.
 *   LESS THAN .05

NORMALS — BASED ON THE 1951-1980 RECORD PERIOD.
EXTREMES — DATES ARE THE MOST RECENT OCCURRENCE.
WIND DIR. — NUMERALS SHOW TENS OF DEGREES
            CLOCKWISE FROM TRUE NORTH.
            "00" INDICATES CALM.
RESULTANT WIND DIRECTIONS ARE GIVEN TO WHOLE DEGREES.

EXCEPTIONS

TABLE 1

1. FASTEST MILE WIND IS THROUGH 1983.

TABLES 2, 3, and 6

RECORD MEANS ARE THROUGH THE CURRENT YEAR,
BEGINNING IN   1893 FOR TEMPERATURE
                1884 FOR PRECIPITATION
                1944 FOR SNOWFALL

TABLE 4 HEATING DEGREE DAYS Base 65 deg. F   BURLINGTON, VERMONT

SEASON	JULY	AUG	SEP	OCT	NOV	DEC	JAN	FEB	MAR	APR	MAY	JUNE	TOTAL
1956-57	41	46	298	480	807	1196	1649	1113	1035	582	350	72	7669
1957-58	17	82	167	503	764	1092	1404	1456	994	580	405	169	7633
1958-59	20	21	191	543	810	1603	1478	1430	1145	594	260	97	8192
#1959-60	5	16	177	525	889	1200	1497	1169	1267	624	147	81	7597
1960-61	28	55	195	600	767	1399	1728	1297	1187	788	454	91	8589
1961-62	34	47	104	468	849	1266	1529	1433	1109	690	324	73	7926
1962-63	71	54	296	571	997	1386	1491	1204	731	385	84	8784	
1963-64	30	118	343	411	735	1609	1327	1380	1046	677	229	124	8029
#1964-65	12	102	280	601	872	1238	1585	1256	1159	766	257	136	8264
1965-66	43	80	236	558	956	1125	1531	1313	1069	707	432	91	8141
1966-67	17	26	280	551	725	1285	1269	1490	1225	722	533	29	8152
1967-68	11	35	223	495	957	1216	1751	1561	1089	562	407	140	8447
1968-69	32	104	127	472	979	1451	1496	1298	1256	700	422	107	8444
1969-70	41	41	244	589	856	1434	1906	1342	1208	663	341	105	8770
1970-71	10	36	174	444	773	1567	1710	1257	1263	821	336	83	8474
1971-72	12	49	131	344	938	1254	1357	1387	1239	872	281	113	7977
1972-73	26	69	212	694	982	1310	1344	1410	855	608	345	86	7941
1973-74	10	17	256	480	825	1160	1431	1378	1101	618	430	37	7743
1974-75	2	6	224	665	858	1128	1276	1236	1141	831	152	82	7601
1975-76	0	45	208	448	681	1385	1669	1168	973	545	331	50	7503
1976-77	20	68	254	654	954	1505	1667	1240	842	590	223	89	8106
1977-78	24	53	207	564	740	1314	1539	1547	1202	781	225	90	8286
1978-79	49	38	295	571	897	1227	1452	1610	866	641	224	90	7960
1979-80	23	65	213	528	703	1107	1350	1371	1043	550	204	91	7248
1980-81	10	3	240	611	976	1545	1738	894	969	544	239	43	7812
1981-82	13	36	204	617	837	1224	1716	1277	1069	643	255	133	8024
1982-83	30	54	124	455	676	1021	1356	1188	983	675	367	77	7006
1983-84	19	36	148	518	803	1317	1500	1044	1331	602	395	68	7781
1984-85	6	24	241	460	792	1068	1592	1185	1029	615	296	118	7426
1985-86	11	42	169	489	835	1344							

TABLE 5 COOLING DEGREE DAYS Base 65 deg. F   BURLINGTON, VERMONT

YEAR	JAN	FEB	MAR	APR	MAY	JUNE	JULY	AUG	SEP	OCT	NOV	DEC	TOTAL
1969	0	0	0	0	2	86	134	160	38	0	0	0	420
1970	0	0	0	0	11	75	189	150	36	1	0	0	462
1971	0	0	0	0	17	87	138	118	90	4	0	0	454
1972	0	0	0	0	14	64	169	81	30	0	0	0	358
1973	0	0	0	3	0	149	187	243	68	0	0	0	650
1974	0	0	0	5	9	89	171	140	27	1	0	0	442
1975	0	0	0	0	75	131	306	181	5	1	0	0	699
1976	0	0	0	24	19	185	135	97	23	0	0	0	483
1977	0	0	0	7	75	86	174	138	27	0	0	0	507
1978	0	0	0	0	79	64	194	146	6	0	0	0	489
1979	0	0	0	2	29	106	253	101	27	13	0	0	531
1980	0	0	0	0	24	78	189	184	34	0	0	0	509
1981	0	0	0	2	35	85	211	110	39	0	0	0	482
1982	0	0	0	1	24	11	179	90	51	0	0	0	356
1983	0	0	0	0	0	121	223	155	92	6	0	0	597
1984	0	0	0	0	7	106	175	217	15	3	0	0	523
1985	0	0	0	0	15	25	160	123	34	0	0	0	357

TABLE 6 SNOWFALL (inches)   BURLINGTON, VERMONT

SEASON	JULY	AUG	SEP	OCT	NOV	DEC	JAN	FEB	MAR	APR	MAY	JUNE	TOTAL
1956-57	0.0	0.0	T	T	10.3	22.0	12.2	1.3	8.6	1.0	T	0.0	55.4
1957-58	0.0	0.0	T	T	1.8	12.0	33.7	34.3	12.3	0.8	T	0.0	94.9
1958-59	0.0	0.0	0.0	T	11.2	9.3	29.2	24.4	13.1	0.1	T	0.0	87.3
1959-60	0.0	0.0	0.0	T	6.8	22.4	16.0	19.3	8.0	0.9	0.0	0.0	73.4
1960-61	0.0	0.0	0.0	T	0.7	9.1	15.9	2.4	16.5	7.0	T	0.0	51.6
1961-62	0.0	0.0	0.0	T	5.9	21.1	6.9	24.3	15.0	3.6	0.0	0.0	76.8
1962-63	0.0	0.0	0.0	0.1	4.3	16.8	12.8	15.8	21.5	1.3	T	0.0	72.6
1963-64	0.0	0.0	0.0	T	4.4	14.8	7.5	8.8	14.4	6.5	0.0	0.0	56.4
1964-65	0.0	0.0	0.0	0.1	1.2	23.0	11.8	4.3	7.9	1.1	0.0	0.0	49.4
1965-66	0.0	0.0	0.0	0.4	12.4	11.9	41.3	28.5	8.3	4.9	3.9	0.0	111.6
1966-67	0.0	0.0	0.0	T	2.4	36.2	20.5	12.6	6.1	4.7	2.6	0.0	85.1
1967-68	0.0	0.0	0.0	T	10.3	17.1	18.4	24.8	14.5	T	0.0	0.0	85.1
1968-69	0.0	0.0	0.0	T	18.8	28.6	15.8	17.0	12.4	3.7	0.0	0.0	96.3
1969-70	0.0	0.0	0.0	5.1	10.5	50.8	11.1	13.8	10.5	2.4	0.4	0.0	104.6
1970-71	0.0	0.0	0.0	0.1	2.7	56.7	17.1	23.1	33.1	12.6	0.0	0.0	145.4
1971-72	0.0	0.0	0.0	0.0	19.2	19.3	14.3	25.1	21.8	9.2	0.0	0.0	108.9
1972-73	0.0	0.0	0.0	T	12.2	39.0	11.4	18.5	2.3	6.3	0.0	0.0	89.7
1973-74	0.0	0.0	0.0	0.1	2.6	24.1	21.5	9.9	20.9	16.8	0.0	0.0	95.9
1974-75	0.0	0.0	0.0	0.1	11.5	16.8	14.8	22.0	12.4	13.3	0.0	0.0	90.9
1975-76	0.0	0.0	0.0	T	5.3	16.0	28.3	20.4	18.8	0.9	T	0.0	89.7
1976-77	0.0	0.0	0.0	0.9	13.3	11.5	24.2	16.4	9.6	1.8	T	0.0	77.7
1977-78	0.0	0.0	0.0	0.0	16.0	22.6	42.4	4.0	12.5	1.9	T	0.0	99.4
1978-79	0.0	0.0	0.0	T	5.7	24.1	37.9	6.6	1.6	8.4	0.0	0.0	84.3
1979-80	0.0	0.0	0.0	1.5	0.4	17.5	3.0	11.6	16.8	0.3	0.0	0.0	39.6
1980-81	0.0	0.0	0.0	T	12.2	17.5	8.7	11.9	13.3	1.1	0.0	0.0	64.7
1981-82	0.0	0.0	0.0	T	3.9	32.8	19.4	8.3	13.0	4.1	0.0	0.0	81.5
1982-83	0.0	0.0	0.0	T	0.8	5.0	22.5	18.3	11.9	21.3	0.7	0.0	80.5
1983-84	0.0	0.0	0.0	T	4.7	14.4	15.2	13.7	16.1	0.4	T	0.0	64.5
1984-85	0.0	0.0	0.0	0.0	6.0	29.3	25.9	10.9	16.6	2.7	0.0	0.0	91.4
1985-86	0.0	0.0	0.0	T	4.6	21.3							
Record Mean	0.0	0.0	T	0.2	6.9	19.5	18.6	16.6	12.5	3.8	0.2	0.0	78.2

See Reference Notes, relative to all above tables, on preceding page.

Lynchburg is situated in the valley of the James River, and on the eastern edge of the Blue Ridge Mountains. The terrain is definitely hilly, with sheltered valleys which are visited by early autumn and late spring frosts. The climate is usually a pleasant one, being neither too hot in the summer, nor too cold in the winter. Rainfall is fairly evenly distributed throughout the year, but there is a distinct summertime rainfall, occasioned by afternoon thunderstorms.

Spring makes itself felt in March, when the mean monthly temperature increases about 7 degrees over the February temperature. Autumn rapidly comes in October, which shows about a 10 degree drop below the September mean. The approaching autumn season brings periods of two to three days of cloudy, cool weather, with high humidity and light rain or drizzle. In midwinter, however, after the passage of a cold front, dry invigorating air, with clear skies, is the rule in Lynchburg. There are occasional snow showers, but the mountains to the immediate west act as a barrier and shelter the area from many storms and high winds.

The mountains also act as a barrier to extremely cold weather. Temperatures have fallen below zero only on a few days, and 100 degree heat is almost as rare, although this mark has been exceeded in the months of May through September.

Great variation in temperature is quite frequently noted during clear, still nights in the winter months. On some such nights, differences of as much as 10-15 degrees occur between the low valleys and the higher terrain.

Based on the 1951-1980 period, the average first occurrence of 32 degrees Fahrenheit in the fall is October 23 and the average last occurrence in the spring is April 13.

TABLE 1   NORMALS, MEANS AND EXTREMES

LYNCHBURG, VIRGINIA

LATITUDE: 37°20'N   LONGITUDE: 79°12'W   ELEVATION: FT. GRND   921 BARO   00971   TIME ZONE: EASTERN   WBAN: 13733

	(a)	JAN	FEB	MAR	APR	MAY	JUNE	JULY	AUG	SEP	OCT	NOV	DEC	YEAR
TEMPERATURE °F:														
Normals														
-Daily Maximum		44.4	47.1	56.2	68.1	75.8	82.5	86.1	85.0	78.8	68.1	57.2	47.5	66.4
-Daily Minimum		25.9	27.6	35.1	44.6	53.2	60.6	65.1	64.5	57.9	46.1	36.7	29.1	45.5
-Monthly		35.2	37.4	45.7	56.4	64.5	71.6	75.6	74.8	68.3	57.1	47.0	38.3	56.0
Extremes														
-Record Highest	41	76	79	87	92	93	100	103	102	101	93	83	78	103
-Year		1952	1985	1945	1985	1969	1945	1954	1983	1954	1951	1974	1984	JUL 1954
-Record Lowest	41	-10	0	7	20	31	40	50	45	36	21	8	-4	-10
-Year		1985	1965	1965	1985	1966	1977	1963	1965	1983	1969	1970	1983	JAN 1985
NORMAL DEGREE DAYS:														
Heating (base 65°F)		927	773	598	263	97	7	0	0	32	258	540	828	4323
Cooling (base 65°F)		0	0	0	5	81	205	332	304	134	13	0	0	1074
% OF POSSIBLE SUNSHINE	41	51	56	59	61	63	66	62	62	61	60	55	52	59
MEAN SKY COVER (tenths)														
Sunrise - Sunset	39	6.1	6.0	6.0	5.8	6.0	5.7	5.9	5.6	5.4	4.9	5.5	5.8	5.7
MEAN NUMBER OF DAYS:														
Sunrise to Sunset														
-Clear	39	9.1	8.7	8.7	8.7	8.2	8.0	7.9	9.1	10.6	13.3	10.6	10.4	113.3
-Partly Cloudy	39	7.0	7.1	9.1	9.0	10.3	11.8	11.3	11.4	8.7	6.8	7.2	6.7	106.4
-Cloudy	39	14.9	12.5	13.2	12.3	12.5	10.2	11.8	10.5	10.7	10.9	12.2	13.9	145.6
Precipitation														
.01 inches or more	41	11.2	9.4	11.1	9.8	11.4	10.0	11.4	10.1	8.0	7.9	9.1	9.5	118.9
Snow, Ice pellets														
1.0 inches or more	41	1.6	1.6	0.8	0.1	0.0	0.0	0.0	0.0	0.0	0.*	0.2	1.0	5.4
Thunderstorms	23	0.3	0.4	1.2	2.9	6.4	7.1	9.3	7.8	3.5	0.9	0.5	0.1	40.5
Heavy Fog Visibility 1/4 mile or less	23	3.9	3.9	2.7	2.6	3.5	2.3	2.6	3.4	3.7	2.6	3.8	3.9	39.0
Temperature °F														
-Maximum														
90° and above	22	0.0	0.0	0.0	0.4	0.5	4.0	7.6	6.7	2.6	0.0	0.0	0.0	21.9
32° and below	22	5.5	2.5	0.2	0.0	0.0	0.0	0.0	0.0	0.0	0.0	0.1	2.1	10.5
-Minimum														
32° and below	22	23.8	21.0	12.5	3.0	0.2	0.0	0.0	0.0	0.0	2.5	10.4	19.5	92.8
0° and below	22	0.6	0.*	0.0	0.0	0.0	0.0	0.0	0.0	0.0	0.0	0.0	0.*	0.7
AVG. STATION PRESS. (mb)	10	984.0	984.3	982.8	982.8	981.8	983.2	983.9	985.2	984.9	985.6	985.7	984.9	984.1
RELATIVE HUMIDITY (%)														
Hour 01	22	71	71	72	72	80	83	86	89	88	85	78	74	79
Hour 07 (Local Time)	22	71	71	72	72	80	83	86	89	88	85	78	74	79
Hour 13	22	53	49	48	45	52	55	58	57	56	53	51	54	53
Hour 19	15	60	55	55	51	62	67	70	73	73	71	61	63	63
PRECIPITATION (inches):														
Water Equivalent														
-Normal		3.06	2.93	3.69	2.90	3.65	3.47	3.85	3.69	3.23	3.36	2.92	3.16	39.91
-Maximum Monthly	41	7.97	5.70	9.24	6.67	9.07	8.58	10.30	11.36	9.22	11.40	8.77	7.15	11.40
-Year		1978	1972	1975	1983	1971	1972	1984	1952	1979	1976	1985	1973	OCT 1976
-Minimum Monthly	41	0.49	0.54	0.74	0.28	1.36	0.65	1.15	0.93	0.02	0.38	0.90	0.33	0.02
-Year		1981	1978	1966	1976	1957	1980	1977	1963	1978	1974	1960	1965	SEP 1978
-Maximum in 24 hrs	41	2.28	2.87	2.47	3.67	3.47	6.27	4.82	4.47	3.70	4.98	2.65	3.03	6.27
-Year		1978	1984	1975	1978	1960	1972	1984	1967	1979	1954	1951	1948	JUN 1972
Snow, Ice pellets														
-Maximum Monthly	41	31.8	19.2	24.9	4.8						2.4	11.6	17.9	31.8
-Year		1966	1979	1960	1971						1979	1968	1966	JAN 1966
-Maximum in 24 hrs	41	10.9	14.6	13.4	4.8						2.4	6.7	12.7	14.6
-Year		1966	1983	1969	1971						1979	1968	1969	FEB 1983
WIND:														
Mean Speed (mph)	24	8.8	8.8	9.3	9.2	7.7	6.9	6.6	6.3	6.9	7.4	7.9	7.9	7.8
Prevailing Direction through 1963		SW	SW	SW	SW	SW	SW	SW	N	N	N	SW	SW	SW
Fastest Mile														
-Direction (!!)	41	W	S	S	NE	N	SW	NW	W	NE	N	NW	SE	N
-Speed (MPH)	41	45	50	43	43	56	56	43	48	40	41	43	45	56
-Year		1971	1961	1950	1978	1958	1951	1954	1977	1956	1954	1949	1950	MAY 1958
Peak Gust														
-Direction (!!)	2	NW	NE	S	W	SW	W	W	NW	NE	NE	SE	NW	W
-Speed (mph)	2	43	44	47	41	44	52	41	46	31	35	38	47	52
-Date		1985	1984	1984	1985	1984	1985	1984	1985	1985	1985	1985	1985	JUN 1985

See Reference Notes to this table on the following page.

TABLE 2 PRECIPITATION (inches)   LYNCHBURG, VIRGINIA

YEAR	JAN	FEB	MAR	APR	MAY	JUNE	JULY	AUG	SEP	OCT	NOV	DEC	ANNUAL
1956	0.76	3.64	2.52	3.22	2.37	1.65	6.40	1.97	3.67	3.20	2.26	2.88	34.54
1957	3.99	4.31	1.11	3.77	1.36	5.30	2.05	2.40	7.48	1.93	4.36	3.09	41.15
1958	3.32	2.80	3.57	3.06	5.45	2.40	4.72	4.45	1.49	1.72	1.38	3.58	37.94
1959	2.01	1.57	3.40	4.26	1.57	1.85	4.56	4.76	4.91	3.61	2.63	2.49	37.62
1960	2.39	4.64	3.25	2.72	6.02	2.15	2.33	3.22	4.26	2.26	0.90	1.77	35.91
1961	1.06	5.23	4.15	2.72	3.06	6.12	2.28	4.12	4.05	4.55	2.46	4.89	44.69
1962	3.50	2.88	5.17	2.44	1.64	4.92	5.17	2.41	2.45	1.50	4.97	2.57	39.62
1963	1.92	1.86	4.56	1.15	1.87	1.87	1.75	0.93	3.83	0.79	5.10	1.51	26.56
1964	4.82	4.46	1.99	2.14	1.64	0.67	6.07	2.02	1.26	2.06	3.08	3.52	33.73
1965	2.18	3.35	4.35	1.42	2.21	2.04	1.83	2.87	2.31	4.39	1.00	0.33	28.28
1966	3.62	4.51	0.74	1.70	3.16	1.36	2.97	2.58	6.22	4.69	2.65	3.18	37.38
1967	2.18	2.13	2.61	1.45	3.69	2.54	5.41	11.27	1.17	1.83	1.28	5.95	41.51
1968	2.35	0.64	3.28	2.48	3.58	1.49	2.58	1.53	1.74	3.65	3.32	1.78	28.42
1969	2.30	2.56	4.32	2.21	2.37	3.27	4.06	3.03	2.29	0.87	5.37	5.37	33.89
1970	1.16	2.39	2.36	2.82	2.12	0.74	4.72	3.62	0.78	5.25	4.01	2.90	32.87
1971	1.73	4.45	2.64	2.29	9.07	2.85	4.48	3.02	1.85	7.40	3.11	1.38	44.27
1972	3.25	5.70	1.76	2.81	8.04	8.08	8.45	1.21	3.10	6.71	4.32	5.11	59.71
1973	3.05	2.94	6.44	4.57	3.81	4.17	6.64	4.44	2.53	3.95	1.04	7.15	50.73
1974	3.94	2.05	2.96	3.11	4.86	2.93	2.84	6.00	5.55	0.38	1.74	3.68	40.04
1975	3.82	2.66	9.24	2.31	6.54	4.78	4.42	4.68	7.42	2.43	2.96	4.34	55.60
1976	2.70	1.98	4.07	0.28	5.32	7.34	1.29	1.07	5.06	11.40	1.38	2.83	44.72
1977	1.76	0.61	2.71	4.84	1.50	3.39	1.15	5.73	2.25	5.77	5.70	3.29	38.70
1978	7.97	0.54	3.84	5.33	6.69	3.36	4.83	3.89	0.02	0.84	3.22	4.06	44.59
1979	6.65	4.99	3.78	3.26	3.10	5.28	4.50	3.42	9.22	3.77	3.18	1.13	52.28
1980	4.63	1.07	5.03	3.99	3.03	0.65	3.61	1.34	1.79	2.35	2.85	0.56	30.90
1981	0.49	3.81	1.81	2.44	1.66	5.24	5.02	4.21	3.61	2.80	0.93	3.90	35.92
1982	3.59	4.41	2.41	3.28	3.67	6.45	5.07	2.60	2.25	2.75	2.85	2.26	41.59
1983	1.14	3.70	4.13	6.67	2.83	1.64	1.46	2.62	1.66	5.99	5.14	5.48	42.46
1984	1.54	5.39	6.45	4.27	4.21	2.74	10.30	5.98	2.15	2.18	1.99	2.01	49.21
1985	2.77	3.41	0.96	1.83	5.97	1.45	3.61	7.10	0.06	3.30	8.77	1.11	40.34
Record Mean	3.28	3.01	3.61	3.08	3.53	3.82	4.10	4.07	3.27	3.16	2.69	3.15	40.76

TABLE 3 AVERAGE TEMPERATURE (deg. F)   LYNCHBURG, VIRGINIA

YEAR	JAN	FEB	MAR	APR	MAY	JUNE	JULY	AUG	SEP	OCT	NOV	DEC	ANNUAL
1956	36.4	41.7	45.3	54.1	64.5	72.1	76.1	74.8	66.3	58.4	45.6	47.7	56.9
1957	35.5	43.0	46.4	59.8	66.4	73.2	75.8	73.6	68.8	53.3	48.6	41.3	57.1
1958	33.5	31.6	41.2	55.9	63.3	70.9	77.3	73.7	67.8	50.2	33.7	54.6
1959	36.0	40.4	45.5	57.9	68.0	73.5	75.8	77.7	69.4	59.1	45.5	41.5	57.5
1960	37.8	36.8	34.7	60.1	63.0	72.4	75.6	76.6	68.9	58.3	48.5	33.0	55.5
1961	32.4	40.4	49.3	50.8	61.2	70.7	75.4	74.9	71.2	58.4	49.6	37.5	56.0
1962	34.3	38.9	42.7	54.2	68.7	71.0	72.6	73.4	64.9	59.6	45.2	33.8	55.0
#1963	33.2	31.6	50.0	59.2	63.8	71.8	74.8	74.1	66.8	61.1	49.5	30.3	55.5
1964	36.9	35.9	45.9	56.2	66.0	73.6	74.8	72.4	66.1	52.7	50.2	40.8	55.9
1965	34.9	36.0	41.5	54.3	68.8	69.4	73.9	73.9	69.4	54.7	47.3	41.6	55.5
1966	31.3	36.4	47.0	52.2	64.0	71.8	77.9	73.6	66.8	55.4	48.2	37.5	55.2
1967	40.5	34.3	46.0	57.7	58.7	71.0	72.8	72.1	63.1	54.3	41.9	40.7	54.4
1968	32.0	32.4	49.9	56.0	62.1	72.1	76.0	77.0	68.0	57.8	47.4	34.0	55.4
1969	32.2	36.4	40.7	51.4	65.4	72.3	76.2	72.8	65.9	56.0	44.4	34.0	54.5
1970	29.4	36.3	41.1	56.0	65.9	73.0	75.6	73.6	73.1	60.0	47.2	39.3	55.9
1971	33.5	38.7	43.1	54.8	62.1	73.4	74.6	72.6	70.0	62.6	44.9	47.3	56.5
1972	39.1	36.3	46.4	56.2	63.1	67.7	75.2	74.1	69.4	54.8	45.1	43.1	55.9
1973	36.0	36.0	50.7	54.0	60.9	73.4	74.7	75.4	70.8	59.4	49.8	37.3	56.6
1974	44.2	38.6	48.9	56.9	63.8	68.0	74.4	74.3	65.7	55.1	47.2	39.0	56.3
1975	38.7	39.7	42.3	52.9	66.9	71.4	73.9	76.5	66.2	59.2	50.5	37.3	56.3
1976	32.7	45.8	50.5	57.6	61.8	70.4	74.1	73.2	66.4	53.2	40.8	34.3	55.1
1977	23.0	36.8	51.5	58.8	67.1	70.1	78.4	76.0	70.0	54.6	49.6	36.2	56.0
1978	29.2	30.1	44.1	57.2	63.6	72.4	74.5	77.3	71.0	55.6	50.1	39.7	55.4
1979	31.9	27.9	49.3	56.3	63.9	69.2	74.3	75.0	67.9	55.2	50.1	40.7	55.2
1980	36.0	33.7	43.8	56.9	65.8	69.4	77.8	78.6	72.6	55.2	45.2	38.7	56.1
1981	31.3	40.8	43.3	59.4	62.5	75.7	76.4	73.3	66.3	53.8	47.0	34.0	55.3
1982	27.8	38.6	45.5	53.8	67.5	72.0	78.0	75.7	73.5	68.8	50.0	44.4	56.3
1983	35.9	37.7	46.9	52.4	62.1	72.1	77.4	78.4	68.7	57.5	48.0	34.9	56.0
1984	32.8	43.3	43.6	53.7	63.3	74.7	74.3	74.9	65.6	65.5	45.4	47.4	56.9
1985	32.3	39.2	49.6	61.2	65.9	71.5	74.9	73.4	68.5	60.3	54.6	35.6	57.3
Record Mean	37.0	38.9	46.7	56.2	65.5	73.1	76.7	75.2	69.1	58.1	47.5	39.0	56.9
Max	45.9	48.4	57.0	67.5	76.7	83.6	86.9	85.1	79.4	69.1	57.7	48.0	67.1
Min	28.0	29.4	36.3	44.8	54.3	62.5	66.5	65.3	58.7	47.1	37.3	30.1	46.7

REFERENCE NOTES FOR TABLES 1, 2, 3 and 6   (LYNCHBURG, VA)

GENERAL

T - TRACE AMOUNT
BLANK ENTRIES DENOTE MISSING/UNREPORTED DATA.
# INDICATES A STATION OR INSTRUMENT RELOCATION.

SPECIFIC

TABLE 1

(a) - LENGTH OF RECORD IN YEARS. ALTHOUGH
      INDIVIDUAL MONTHS MAY BE MISSING.
  * LESS THAN .05

NORMALS — BASED ON THE 1951-1980 RECORD PERIOD.
EXTREMES — DATES ARE THE MOST RECENT OCCURRENCE.
WIND DIR. — NUMERALS SHOW TENS OF DEGREES
            CLOCKWISE FROM TRUE NORTH.
            "00" INDICATES CALM.
RESULTANT WIND DIRECTIONS ARE GIVEN TO WHOLE DEGREES.

EXCEPTIONS

TABLE 1

1. THUNDERSTORMS AND HEAVY FOG ARE THROUGH 1966
   AND MAY BE INCOMPLETE, DUE TO PART-TIME OPERA-
   TIONS PRIOR TO AUGUST 1962.

TABLES 2, 3, and 6

RECORD MEANS ARE THROUGH THE CURRENT YEAR,
BEGINNING IN   1875 FOR TEMPERATURE
                1872 FOR PRECIPITATION
                1945 FOR SNOWFALL

TABLE 4 HEATING DEGREE DAYS Base 65 deg. F   LYNCHBURG, VIRGINIA

SEASON	JULY	AUG	SEP	OCT	NOV	DEC	JAN	FEB	MAR	APR	MAY	JUNE	TOTAL
1956-57	0	5	92	216	575	527	912	610	569	238	69	17	3830
1957-58	0	6	65	355	488	728	969	929	730	289	100	7	4666
1958-59	0	0	34	264	444	964	891	683	595	230	53	11	4169
1959-60	0	0	37	250	577	722	837	809	933	206	146	1	4518
1960-61	0	0	29	226	492	989	1005	681	489	439	137	17	4504
1961-62	0	2	37	208	476	846	945	723	684	342	50	0	4313
#1962-63	1	0	105	212	588	957	979	931	461	217	107	8	4566
1963-64	0	2	58	140	456	1070	865	834	585	282	75	23	4390
1964-65	0	5	48	374	438	743	926	803	724	321	15	27	4424
1965-66	0	9	33	317	522	718	1038	795	554	392	102	22	4502
1966-67	0	0	46	294	496	844	752	855	583	237	212	17	4336
1967-68	0	0	97	330	688	744	1018	941	468	268	117	1	4672
1968-69	0	2	7	233	523	953	1009	793	745	239	74	15	4593
1969-70	0	0	58	294	612	954	1095	798	732	278	79	0	4900
1970-71	0	0	21	183	528	789	968	729	673	303	119	3	4316
1971-72	0	0	13	110	613	546	796	826	571	282	73	36	3866
1972-73	2	0	20	310	593	671	893	435	495	339	158	0	4230
1973-74	0	0	9	192	448	851	641	736	495	265	105	16	3758
1974-75	0	0	74	304	537	799	808	701	696	362	45	11	4337
1975-76	0	0	60	185	432	853	994	550	443	273	132	23	3945
1976-77	0	2	34	370	719	947	1294	783	417	217	58	26	4867
1977-78	0	0	12	325	464	887	1103	972	637	237	125	6	4768
1978-79	0	0	25	289	439	777	1018	1033	489	265	83	11	4429
1979-80	0	6	27	320	444	746	892	901	652	255	65	19	4327
1980-81	0	0	28	312	589	807	1039	670	666	197	131	2	4441
1981-82	0	0	55	353	535	954	1144	731	597	331	29	1	4730
1982-83	0	1	33	229	446	637	893	758	554	380	127	9	4067
1983-84	0	0	73	243	502	927	991	622	656	348	111	5	4478
1984-85	0	0	91	59	580	539	1006	716	482	170	49	11	3703
1985-86	1	0	53	167	307	901							

TABLE 5 COOLING DEGREE DAYS Base 65 deg. F   LYNCHBURG, VIRGINIA

YEAR	JAN	FEB	MAR	APR	MAY	JUNE	JULY	AUG	SEP	OCT	NOV	DEC	TOTAL
1969	0	0	0	9	94	239	355	247	91	18	0	0	1053
1970	0	0	0	17	116	249	338	272	271	37	0	0	1300
1971	0	0	0	3	36	263	304	242	170	44	16	5	1083
1972	0	0	0	25	20	126	324	290	159	2	3	0	949
1973	0	0	3	16	34	260	308	330	188	24	0	0	1163
1974	0	0	3	30	75	113	298	296	103	7	10	0	935
1975	0	0	0	6	108	209	282	365	103	14	4	0	1091
1976	0	0	3	57	37	190	289	259	80	13	0	0	928
1977	0	0	5	38	129	184	424	349	167	9	8	0	1313
1978	0	0	0	9	88	237	305	389	212	5	0	0	1245
1979	0	0	11	12	54	143	298	322	121	21	0	0	982
1980	0	0	0	15	99	159	403	432	262	18	0	0	1388
1981	0	0	0	35	59	331	361	265	101	11	0	0	1163
1982	0	0	0	3	113	183	339	272	153	49	3	4	1119
1983	0	0	0	10	45	227	394	422	193	17	0	1	1308
1984	0	0	0	17	65	296	265	314	115	81	0	1	1154
1985	0	1	13	65	83	215	316	268	163	29	2	0	1155

TABLE 6 SNOWFALL (inches)   LYNCHBURG, VIRGINIA

SEASON	JULY	AUG	SEP	OCT	NOV	DEC	JAN	FEB	MAR	APR	MAY	JUNE	TOTAL
1956-57	0.0	0.0	0.0	0.0	T	T	4.6	0.8	0.0	2.8	0.0	0.0	8.2
1957-58	0.0	0.0	0.0	T	T	1.4	3.3	13.0	6.3	0.0	0.0	0.0	24.0
1958-59	0.0	0.0	0.0	0.0	T	4.1	7.8	T	T	0.6	0.0	0.0	12.5
1959-60	0.0	0.0	0.0	0.0	T	0.7	5.6	14.0	24.9	0.0	0.0	0.0	45.2
1960-61	0.0	0.0	0.0	0.0	0.0	1.2	8.8	8.1	T	T	0.0	0.0	18.1
1961-62	0.0	0.0	0.0	0.0	0.6	2.1	17.3	2.0	23.2	0.0	0.0	0.0	45.2
1962-63	0.0	0.0	0.0	T	0.8	7.0	9.3	T	0.0	0.0	0.0	0.0	17.1
1963-64	0.0	0.0	0.0	T	T	5.5	6.3	11.8	T	0.0	0.0	0.0	23.6
1964-65	0.0	0.0	0.0	0.0	0.4	T	8.3	3.7	2.8	0.0	0.0	0.0	15.2
1965-66	0.0	0.0	0.0	0.0	0.0	0.1	31.8	6.3	T	0.0	0.0	0.0	38.2
1966-67	0.0	0.0	0.0	0.0	T	17.9	4.6	14.7	T	0.0	0.0	0.0	37.2
1967-68	0.0	0.0	0.0	0.0	T	4.0	3.9	3.6	T	0.0	0.0	0.0	11.5
1968-69	0.0	0.0	0.0	0.0	11.6	T	1.0	4.1	18.2	0.0	0.0	0.0	34.9
1969-70	0.0	0.0	0.0	0.0	T	12.7	5.8	1.6	T	T	0.0	0.0	20.1
1970-71	0.0	0.0	0.0	0.0	1.8	9.3	1.7	2.3	7.8	4.8	0.0	0.0	27.7
1971-72	0.0	0.0	0.0	0.0	5.9	T	T	15.2	T	T	0.0	0.0	21.1
1972-73	0.0	0.0	0.0	0.0	0.5	0.0	2.3	1.7	5.3	0.2	0.0	0.0	10.0
1973-74	0.0	0.0	0.0	0.0	0.0	6.6	0.0	5.9	T	0.0	0.0	0.0	12.5
1974-75	0.0	0.0	0.0	0.0	0.6	0.6	6.5	4.4	3.8	0.0	0.0	0.0	15.9
1975-76	0.0	0.0	0.0	0.0	0.0	0.1	T	1.2	0.6	0.0	0.0	0.0	1.9
1976-77	0.0	0.0	0.0	0.0	T	2.8	10.0	0.3	0.0	0.0	0.0	0.0	13.1
1977-78	0.0	0.0	0.0	0.0	0.7	0.3	10.8	5.9	8.1	0.0	0.0	0.0	25.8
1978-79	0.0	0.0	0.0	0.0	T	T	4.0	19.2	0.0	0.0	0.0	0.0	23.2
1979-80	0.0	0.0	0.0	2.4	0.0	T	14.2	7.5	8.9	0.0	0.0	0.0	33.0
1980-81	0.0	0.0	0.0	0.0	0.0	T	3.8	0.0	3.6	0.0	0.0	0.0	7.4
1981-82	0.0	0.0	0.0	0.0	3.7	5.9	7.6	15.1	0.2	2.5	0.0	0.0	35.0
1982-83	0.0	0.0	0.0	0.0	0.0	6.8	0.5	18.6	T	T	0.0	0.0	25.9
1983-84	0.0	0.0	0.0	0.0	T	0.0	3.8	0.1	0.4	T	0.0	0.0	4.3
1984-85	0.0	0.0	0.0	0.0	0.0	0.0	6.4	T	0.5	0.0	0.0	0.0	6.9
1985-86	0.0	0.0	0.0	0.0	0.0	1.2							
Record Mean	0.0	0.0	0.0	0.1	0.8	3.0	5.4	5.4	3.6	0.3	0.0	0.0	18.5

See Reference Notes, relative to all above tables, on preceding page.

The city of Norfolk, Virginia, is located near the coast and the southern border of the state. It is almost surrounded by water, with the Chesapeake Bay immediately to the north, Hampton Roads to the west, and the Atlantic Ocean only 18 miles to the east. It is traversed by numerous rivers and waterways and its average elevation above sea level is 13 feet. There are no nearby hilly areas and the land is low and level throughout the city. The climate is generally marine. The geographic location of the city with respect to the principal storm tracks, is especially favorable, being south of the average path of storms originating in the higher latitudes and north of the usual tracks of hurricanes and other tropical storms.

The winters are usually mild, while the autumn and spring seasons usually are delightful. Summers, though warm and long, frequently are tempered by cool periods, often associated with northeasterly winds off the Atlantic. Temperatures of 100 degrees or higher occur infrequently. Extreme cold waves seldom penetrate the area and temperatures of zero or below are almost nonexistent. Winters pass, on occasion, without a measurable amount of snowfall. Most of the snowfall in Norfolk is light and generally melts within 24 hours.

Based on the 1951-1980 period, the average first occurrence of 32 degrees Fahrenheit in the fall is November 17 and the average last occurrence in the spring is March 23.

TABLE 1   NORMALS, MEANS AND EXTREMES

NORFOLK, VIRGINIA

LATITUDE: 36°54'N   LONGITUDE: 76°12'W   ELEVATION: FT. GRND   24 BARO   00044   TIME ZONE: EASTERN   WBAN: 13737

	(a)	JAN	FEB	MAR	APR	MAY	JUNE	JULY	AUG	SEP	OCT	NOV	DEC	YEAR
TEMPERATURE °F:														
Normals														
–Daily Maximum		48.1	49.9	57.5	68.2	75.7	83.2	86.9	85.7	80.2	69.8	60.8	51.9	68.2
–Daily Minimum		31.7	32.3	39.4	48.1	57.2	65.3	69.9	69.6	64.2	52.8	43.0	35.0	50.7
–Monthly		39.9	41.1	48.5	58.2	66.5	74.3	78.4	77.7	72.2	61.3	51.9	43.5	59.4
Extremes														
–Record Highest	37	78	81	88	97	97	101	103	104	99	95	86	80	104
–Year		1970	1976	1985	1960	1956	1964	1952	1980	1983	1954	1974	1978	AUG 1980
–Record Lowest	37	–3	8	18	28	36	45	54	49	45	27	20	7	–3
–Year		1985	1965	1980	1982	1966	1967	1979	1982	1967	1976	1950	1983	JAN 1985
NORMAL DEGREE DAYS:														
Heating (base 65°F)		778	669	512	219	53	0	0	0	9	146	393	667	3446
Cooling (base 65°F)		0	0	0	15	96	282	415	394	225	31	0	0	1458
% OF POSSIBLE SUNSHINE	21	56	59	63	65	65	68	64	65	64	59	58	57	62
MEAN SKY COVER (tenths)														
Sunrise – Sunset	37	6.2	6.1	6.1	5.8	6.1	5.8	6.0	5.7	5.7	5.5	5.5	6.0	5.9
MEAN NUMBER OF DAYS:														
Sunrise to Sunset														
–Clear	37	9.2	8.6	9.0	8.9	7.9	7.6	7.6	8.3	9.3	11.4	10.7	9.4	107.9
–Partly Cloudy	37	6.4	6.0	7.4	9.2	9.7	11.5	11.6	11.9	9.5	7.0	7.9	7.3	105.3
–Cloudy	37	15.4	13.7	14.6	11.8	13.4	10.8	11.8	10.9	11.2	12.6	11.4	14.4	152.0
Precipitation														
.01 inches or more	37	10.4	10.0	11.0	9.9	9.9	9.0	11.3	10.3	7.8	7.8	8.0	9.1	114.5
Snow, Ice pellets														
1.0 inches or more	37	0.8	0.7	0.3	0.*	0.0	0.0	0.0	0.0	0.0	0.0	0.0	0.4	2.2
Thunderstorms	37	0.4	0.6	1.8	2.7	4.9	5.7	8.5	7.1	2.7	1.4	0.5	0.4	36.7
Heavy Fog Visibility 1/4 mile or less	37	2.2	2.7	1.8	1.4	1.9	1.2	0.6	1.2	1.3	2.4	2.0	2.3	21.1
Temperature °F														
–Maximum														
90° and above	37	0.0	0.0	0.0	0.5	1.3	6.3	11.0	8.7	3.0	0.1	0.0	0.0	30.8
32° and below	37	2.8	1.3	0.2	0.0	0.0	0.0	0.0	0.0	0.0	0.0	0.0	1.0	5.3
–Minimum														
32° and below	37	17.3	14.5	6.0	0.4	0.0	0.0	0.0	0.0	0.0	0.2	3.2	13.6	55.1
0° and below	37	0.*	0.0	0.0	0.0	0.0	0.0	0.0	0.0	0.0	0.0	0.0	0.0	*
AVG. STATION PRESS.(mb)	13	1017.9	1017.9	1015.8	1015.3	1014.5	1015.1	1015.7	1016.9	1017.4	1018.8	1018.6	1018.9	1016.9
RELATIVE HUMIDITY (%)														
Hour 01	37	72	72	71	72	80	83	84	86	84	82	76	73	78
Hour 07	37	74	74	73	72	77	79	82	84	83	83	79	75	78
Hour 13 (Local Time)	37	59	56	53	49	56	57	59	61	61	60	56	58	57
Hour 19	37	67	65	62	60	66	67	70	74	75	75	69	68	68
PRECIPITATION (inches):														
Water Equivalent														
–Normal		3.72	3.28	3.86	2.87	3.75	3.45	5.15	5.33	4.35	3.41	2.88	3.17	45.22
–Maximum Monthly	37	6.47	6.23	7.80	7.25	10.12	9.72	13.73	11.19	13.80	10.12	7.01	6.10	13.80
–Year		1979	1983	1978	1984	1979	1963	1975	1967	1979	1971	1951	1983	SEP 1979
–Minimum Monthly	37	1.05	0.86	1.34	0.43	1.48	0.37	0.77	0.74	0.36	0.57	0.49	0.79	0.36
–Year		1981	1950	1967	1985	1965	1954	1983	1975	1958	1984	1965	1985	SEP 1958
–Maximum in 24 hrs	37	3.80	2.71	3.18	2.93	3.41	6.85	5.64	11.40	6.79	4.38	3.35	2.76	11.40
–Year		1967	1983	1958	1984	1980	1963	1969	1964	1959	1971	1952	1983	AUG 1964
Snow, Ice pellets														
–Maximum Monthly	37	14.2	18.9	13.7	1.2							0.6	14.7	18.9
–Year		1966	1980	1980	1964							1950	1958	FEB 1980
–Maximum in 24 hrs	37	9.1	12.4	9.9	1.2							0.6	11.4	12.4
–Year		1973	1980	1980	1964							1950	1958	FEB 1980
WIND:														
Mean Speed (mph)	37	11.5	11.9	12.4	11.8	10.4	9.6	8.9	8.8	9.6	10.4	10.6	11.0	10.6
Prevailing Direction through 1963		SW	NNE	SW	SW	SW	SW	SW	SW	NE	NE	SW	SW	SW
Fastest Obs. 1 Min.														
–Direction	13	23	36	22	06	02	30	34	35	30	04	10	26	04
–Speed (MPH)	13	39	44	46	39	38	46	46	46	46	48	39	37	48
–Year		1978	1973	1973	1978	1979	1977	1973	1979	1985	1982	1985	1985	OCT 1982
Peak Gust														
–Direction	2	NW	E	NW	SW	NW	W	SW	W	NW	N	E	W	NW
–Speed (mph)	2	52	56	47	56	66	47	52	46	67	43	55	53	67
–Date		1985	1984	1984	1985	1984	1985	1985	1985	1985	1984	1985	1985	SEP 1985

See Reference Notes to this table on the following page.

TABLE 2 PRECIPITATION (inches)   NORFOLK, VIRGINIA

YEAR	JAN	FEB	MAR	APR	MAY	JUNE	JULY	AUG	SEP	OCT	NOV	DEC	ANNUAL
1956	2.49	5.72	3.09	3.81	2.57	2.07	6.63	5.04	3.70	8.06	1.51	2.76	47.45
1957	4.11	4.74	4.06	1.86	1.94	3.56	2.18	5.96	7.18	3.44	4.81	4.53	48.37
1958	4.16	3.55	6.41	3.28	4.99	6.27	10.48	0.36	4.88	4.10	3.07	4.10	57.78
1959	1.52	1.84	3.96	5.03	1.98	2.49	7.49	2.81	10.73	8.78	1.85	2.65	51.13
1960	3.85	3.57	2.54	1.40	4.43	2.57	4.78	6.49	5.88	2.76	1.10	2.37	41.74
1961	3.52	4.56	3.59	2.74	7.77	6.70	1.69	7.42	1.62	4.12	1.20	3.74	48.67
1962	5.56	2.50	4.32	4.49	3.68	3.03	9.05	4.09	3.07	4.48	4.06	3.80	52.13
1963	3.36	3.75	2.98	1.29	1.55	9.72	2.01	2.40	6.84	1.21	5.31	2.85	43.27
1964	4.56	4.56	2.26	2.38	1.56	2.58	7.33	10.58	12.26	5.55	1.14	2.95	57.71
1965	2.73	2.53	2.83	2.24	1.48	4.69	3.46	3.08	0.77	1.29	0.49	1.08	26.67
1966	4.86	3.83	1.50	1.68	5.95	1.82	4.26	5.24	3.39	1.25	1.05	3.13	37.96
1967	5.44	3.56	1.34	1.31	3.25	1.37	7.21	11.19	3.02	0.93	1.75	4.84	45.21
1968	3.62	2.01	4.76	3.17	2.16	3.07	4.23	2.04	1.51	4.44	3.56	3.14	37.71
1969	2.26	2.16	4.88	2.07	2.05	4.13	12.70	5.28	2.72	3.18	2.97	3.93	48.33
1970	2.27	3.97	3.37	3.19	2.58	4.10	5.33	2.04	1.72	1.30	2.34	3.01	35.22
1971	4.03	3.59	3.88	2.18	4.46	2.16	4.81	4.63	5.46	10.12	1.44	1.47	47.73
1972	2.94	3.50	2.55	2.15	3.35	4.93	4.65	1.60	6.91	4.09	5.44	4.12	46.23
1973	2.54	3.21	4.44	3.44	3.62	5.93	4.19	7.92	0.86	1.37	1.90	5.83	45.50
1974	3.52	2.98	5.16	3.34	3.74	4.76	5.47	8.33	4.40	1.23	1.22	3.81	47.96
1975	4.18	4.18	5.72	4.19	3.37	1.16	13.73	0.74	4.82	3.19	1.63	3.62	50.53
1976	2.51	1.50	2.21	0.99	3.74	1.59	5.19	2.62	3.51	2.90	2.38	3.22	32.36
1977	3.33	2.23	4.05	2.20	3.86	2.41	2.70	4.57	3.00	6.09	5.41	3.92	43.77
1978	6.32	1.91	7.80	2.90	5.64	7.84	4.19	1.66	1.17	1.50	4.40	2.31	47.64
1979	6.47	5.01	5.13	7.00	10.12	2.97	4.19	1.79	13.80	1.74	5.26	0.98	64.96
1980	4.54	2.91	4.40	3.25	5.17	1.39	1.85	4.54	1.47	4.21	2.01	2.64	38.38
1981	1.05	2.26	1.88	2.26	2.75	5.00	5.10	6.87	3.18	3.28	1.78	5.77	41.18
1982	3.35	5.81	3.04	1.71	3.07	4.22	5.83	6.51	3.63	4.25	3.43	4.30	49.15
1983	2.21	6.23	4.55	6.13	3.52	3.84	0.77	3.07	4.52	5.29	3.24	6.10	49.47
1984	2.77	4.66	5.09	7.25	6.23	1.50	7.66	2.25	1.94	0.57	2.68	2.22	44.82
1985	3.98	3.53	2.02	0.43	3.23	6.81	6.14	1.89	6.36	3.92	5.71	0.79	44.44
Record Mean	3.29	3.38	3.76	3.22	3.73	4.02	5.67	5.27	3.89	3.14	2.66	3.22	45.25

TABLE 3 AVERAGE TEMPERATURE (deg. F)   NORFOLK, VIRGINIA

YEAR	JAN	FEB	MAR	APR	MAY	JUNE	JULY	AUG	SEP	OCT	NOV	DEC	ANNUAL
1956	37.2	45.9	48.1	56.6	64.3	75.3	78.7	77.3	70.8	63.5	51.1	52.2	60.1
1957	39.7	44.6	48.4	62.3	68.1	77.7	78.6	75.7	74.8	58.2	53.9	46.7	60.7
1958	37.3	36.4	43.6	57.8	66.7	71.8	81.7	77.8	70.7	61.4	55.6	37.2	58.2
#1959	39.7	43.7	49.4	60.2	69.7	76.4	78.7	79.4	73.6	64.7	51.1	44.8	60.9
1960	41.7	41.4	39.2	63.7	68.3	75.5	77.8	80.0	72.8	61.8	52.7	36.5	59.3
1961	35.0	43.6	53.1	55.5	63.6	72.4	80.6	77.7	75.3	60.6	53.2	41.5	59.4
1962	38.9	41.0	43.7	56.4	68.0	73.4	75.2	76.0	69.1	61.8	49.8	37.2	57.6
1963	37.1	35.7	53.4	60.3	64.6	73.2	77.0	76.7	66.7	60.1	52.0	35.2	57.7
1964	42.3	39.9	49.5	55.2	66.4	74.5	77.3	74.4	70.8	57.7	54.6	46.5	59.1
1965	39.7	41.0	44.4	54.5	69.7	72.8	76.7	77.1	73.6	59.0	51.6	43.5	58.6
1966	35.8	38.6	47.9	54.5	63.6	71.6	78.0	74.9	69.9	59.3	50.5	41.4	57.1
1967	45.7	39.9	47.7	58.0	61.2	71.2	76.3	75.2	66.5	58.7	45.9	44.0	57.5
1968	34.8	34.0	50.0	55.1	64.6	74.9	78.0	80.5	71.5	63.1	52.9	41.4	58.4
1969	38.5	39.8	44.7	59.8	67.4	77.1	79.2	76.2	71.1	62.2	48.1	40.9	58.9
1970	33.9	39.3	44.7	56.7	67.5	74.9	76.9	78.0	74.7	63.7	51.7	47.0	59.1
1971	38.6	44.7	46.9	55.9	65.0	76.0	77.2	75.7	73.2	66.7	52.4	52.3	60.4
1972	46.4	43.2	49.0	56.4	63.5	70.5	77.6	75.8	71.9	59.2	51.6	49.2	59.5
1973	40.5	39.7	52.3	58.5	66.9	76.9	78.5	75.0	71.4	64.2	53.5	46.2	60.9
1974	48.6	43.4	53.1	60.8	66.8	72.8	78.3	77.4	71.4	58.7	53.5	46.0	60.9
1975	46.0	45.4	47.4	52.7	68.3	77.0	78.6	79.6	72.3	63.4	55.7	43.2	60.8
1976	38.9	49.9	53.4	61.9	66.3	75.9	78.2	75.9	71.1	57.7	45.9	41.4	59.7
1977	29.2	41.5	54.7	61.9	68.2	74.3	81.4	81.0	76.3	60.5	54.8	43.5	60.6
1978	37.0	32.6	46.1	57.2	65.6	74.1	76.1	80.5	73.2	60.8	56.0	45.3	58.7
1979	39.4	33.3	49.1	58.1	66.7	70.4	77.1	78.5	72.8	60.4	56.4	44.9	58.9
1980	40.3	34.7	46.5	58.6	67.8	73.9	80.9	80.9	76.1	60.4	49.9	42.3	59.4
1981	32.7	43.1	45.4	61.2	65.1	78.3	79.8	75.1	70.7	59.6	50.7	41.0	58.6
1982	35.4	42.0	48.8	55.0	69.4	73.4	78.6	75.3	70.0	60.2	54.4	48.8	59.3
1983	40.2	40.8	51.0	55.7	65.8	73.0	80.3	79.0	72.8	62.7	52.6	41.6	59.6
1984	35.5	46.7	45.4	55.6	67.8	76.2	76.7	78.4	70.5	66.9	49.9	50.9	60.0
1985	34.9	40.4	51.8	62.0	68.8	74.2	78.2	77.2	73.4	65.9	60.3	41.2	60.7
Record Mean	41.1	42.1	48.9	57.5	66.7	74.6	77.6	77.4	72.4	62.2	52.2	43.7	59.8
Max	48.9	50.4	57.7	66.8	75.7	83.2	86.8	85.1	79.7	69.8	60.1	51.3	68.0
Min	33.3	33.8	40.0	48.1	57.7	66.0	70.4	70.0	65.1	54.5	44.2	36.0	51.6

REFERENCE NOTES FOR TABLES 1, 2, 3 and 6   (NORFOLK, VA)

GENERAL

T - TRACE AMOUNT
BLANK ENTRIES DENOTE MISSING/UNREPORTED DATA.
# INDICATES A STATION OR INSTRUMENT RELOCATION.

SPECIFIC

TABLE 1

(a) - LENGTH OF RECORD IN YEARS. ALTHOUGH
   INDIVIDUAL MONTHS MAY BE MISSING.
  * LESS THAN .05

NORMALS — BASED ON THE 1951-1980 RECORD PERIOD.
EXTREMES — DATES ARE THE MOST RECENT OCCURRENCE.
WIND DIR. — NUMERALS SHOW TENS OF DEGREES
   CLOCKWISE FROM TRUE NORTH.
   "00" INDICATES CALM.
RESULTANT WIND DIRECTIONS ARE GIVEN TO WHOLE DEGREES.

EXCEPTIONS

TABLE 1

1. PERCENT OF POSSIBLE SUNSHINE IS THROUGH 1980.

TABLES 2, 3, and 6

RECORD MEANS ARE THROUGH THE CURRENT YEAR,
BEGINNING IN   1875 FOR TEMPERATURE
   1871 FOR PRECIPITATION
   1949 FOR SNOWFALL

TABLE 4   HEATING DEGREE DAYS Base 65 deg. F   NORFOLK, VIRGINIA

SEASON	JULY	AUG	SEP	OCT	NOV	DEC	JAN	FEB	MAR	APR	MAY	JUNE	TOTAL
1956-57	0	0	21	69	422	390	778	564	508	173	59	1	2985
1957-58	0	0	9	214	329	558	852	795	655	242	44	7	3705
#1958-59	0	0	7	132	282	855	778	591	478	180	37	4	3344
1959-60	0	0	10	132	424	621	714	679	796	152	37	1	3566
1960-61	0	0	0	139	368	877	921	595	385	318	110	16	3729
1961-62	0	0	7	155	368	722	800	668	655	288	50	0	3713
1962-63	0	0	37	148	449	854	859	815	357	202	108	0	3829
1963-64	0	0	44	156	384	920	697	719	482	303	108	4	3817
1964-65	0	0	4	232	312	575	780	667	635	320	29	15	3569
1965-66	0	6	1	195	398	657	897	734	527	330	121	21	3887
1966-67	0	0	22	191	437	725	588	699	533	244	157	21	3617
1967-68	0	0	36	211	566	644	928	895	471	294	88	0	4133
1968-69	0	0	0	124	361	726	814	697	624	192	44	0	3582
1969-70	0	0	8	131	469	741	960	714	622	263	57	0	3965
1970-71	0	0	16	93	393	552	812	567	555	269	69	0	3326
1971-72	0	0	3	27	391	390	572	628	494	272	81	11	2869
1972-73	0	0	4	197	406	486	752	703	403	217	47	0	3215
1973-74	0	0	0	83	353	575	504	599	377	183	63	0	2737
1974-75	0	0	16	213	371	584	584	547	541	382	47	0	3285
1975-76	0	0	6	98	290	671	804	443	362	186	62	6	2928
1976-77	0	0	0	245	566	726	1104	657	330	150	40	1	3819
1977-78	0	0	0	158	321	661	860	902	580	235	72	3	3792
1978-79	0	0	3	162	268	614	787	879	499	213	52	5	3482
1979-80	0	0	0	190	272	616	759	872	564	196	58	2	3529
1980-81	0	0	11	181	449	699	994	610	605	159	96	0	3804
1981-82	0	0	12	189	423	739	907	636	495	303	21	0	3725
1982-83	0	4	6	177	334	498	762	674	426	295	85	3	3264
1983-84	0	0	27	126	370	718	908	522	601	281	54	3	3610
1984-85	0	0	16	37	450	432	928	686	421	172	21	0	3163
1985-86	0	0	6	61	162	731							

TABLE 5   COOLING DEGREE DAYS Base 65 deg. F   NORFOLK, VIRGINIA

YEAR	JAN	FEB	MAR	APR	MAY	JUNE	JULY	AUG	SEP	OCT	NOV	DEC	TOTAL
1969	0	0	0	42	125	369	446	357	199	49	0	0	1587
1970	1	0	0	19	140	303	374	412	311	60	0	0	1620
1971	0	3	0	3	76	336	383	343	259	87	23	5	1518
1972	0	0	8	20	40	183	398	343	217	22	10	2	1243
1973	0	0	16	27	112	363	420	424	307	64	17	1	1751
1974	3	0	16	64	124	244	419	390	213	26	32	0	1531
1975	2	3	0	22	157	366	429	460	233	55	17	0	1744
1976	1	13	11	102	110	337	417	347	193	27	0	0	1558
1977	0	4	16	66	145	289	515	502	347	24	22	0	1930
1978	0	0	0	9	96	286	352	487	257	36	3	9	1535
1979	0	0	11	13	112	171	385	426	239	54	22	0	1433
1980	0	0	0	11	153	274	499	497	351	45	1	1	1832
1981	0	0	0	51	103	407	468	320	189	29	0	0	1567
1982	0	0	1	8	166	257	428	331	164	39	21	4	1419
1983	0	0	0	21	115	250	481	440	265	62	4	0	1638
1984	0	0	0	5	146	345	368	426	188	102	5	2	1587
1985	0	5	20	91	146	284	419	382	267	97	28	0	1739

TABLE 6   SNOWFALL (inches)   NORFOLK, VIRGINIA

SEASON	JULY	AUG	SEP	OCT	NOV	DEC	JAN	FEB	MAR	APR	MAY	JUNE	TOTAL
1956-57	0.0	0.0	0.0	0.0	T	T	4.1	T	T	T	0.0	0.0	4.1
1957-58	0.0	0.0	0.0	0.0	0.0	1.1	3.0	0.4	T	0.0	0.0	0.0	4.5
1958-59	0.0	0.0	0.0	0.0	0.0	14.7	1.1	0.0	T	0.0	0.0	0.0	15.8
1959-60	0.0	0.0	0.0	0.0	T	T	3.9	7.9	0.0	0.0	0.0	0.0	11.8
1960-61	0.0	0.0	0.0	0.0	0.0	0.7	3.1	1.2	0.0	0.0	0.0	0.0	5.0
1961-62	0.0	0.0	0.0	0.0	T	T	11.9	0.8	1.2	T	0.0	0.0	13.9
1962-63	0.0	0.0	0.0	0.0	0.0	4.5	1.9	7.5	T	0.0	0.0	0.0	13.9
1963-64	0.0	0.0	0.0	0.0	0.0	T	5.8	1.0	1.2	0.0	0.0	0.0	8.0
1964-65	0.0	0.0	0.0	0.0	T	0.0	10.6	3.9	T	0.0	0.0	0.0	14.5
1965-66	0.0	0.0	0.0	0.0	0.0	T	14.2	0.5	0.0	0.0	0.0	0.0	14.7
1966-67	0.0	0.0	0.0	0.0	T	1.0	4.2	5.1	T	0.0	0.0	0.0	10.3
1967-68	0.0	0.0	0.0	0.0	T	2.0	1.5	2.9	0.9	0.0	0.0	0.0	7.3
1968-69	0.0	0.0	0.0	0.0	T	3.8	T	0.8	1.9	0.0	0.0	0.0	6.5
1969-70	0.0	0.0	0.0	0.0	0.0	T	3.0	2.8	T	0.0	0.0	0.0	5.8
1970-71	0.0	0.0	0.0	0.0	T	T	T	2.4	4.2	T	0.0	0.0	6.6
1971-72	0.0	0.0	0.0	0.0	T	T	0.0	1.8	T	T	0.0	0.0	1.8
1972-73	0.0	0.0	0.0	0.0	T	T	9.1	4.7	T	0.0	0.0	0.0	13.8
1973-74	0.0	0.0	0.0	0.0	0.0	1.4	T	0.9	7.5	0.0	0.0	0.0	9.8
1974-75	0.0	0.0	0.0	0.0	0.0	T	0.3	T	0.8	T	0.0	0.0	1.1
1975-76	0.0	0.0	0.0	0.0	0.0	T	T	T	T	0.0	0.0	0.0	T
1976-77	0.0	0.0	0.0	0.0	T	1.0	4.7	1.4	0.0	0.0	0.0	0.0	7.1
1977-78	0.0	0.0	0.0	0.0	0.0	T	1.3	9.2	2.3	0.0	0.0	0.0	12.8
1978-79	0.0	0.0	0.0	0.0	0.0	0.0	1.0	12.7	T	0.0	0.0	0.0	13.7
1979-80	0.0	0.0	0.0	0.0	0.0	0.0	9.3	18.9	13.7	0.0	0.0	0.0	41.9
1980-81	0.0	0.0	0.0	0.0	0.0	T	T	0.0	0.3	0.0	0.0	0.0	0.3
1981-82	0.0	0.0	0.0	0.0	0.0	1.8	4.2	0.1	T	T	0.0	0.0	6.1
1982-83	0.0	0.0	0.0	0.0	0.0	0.4	T	3.0	T	0.0	0.0	0.0	3.4
1983-84	0.0	0.0	0.0	0.0	0.0	T	T	5.2	T	0.0	0.0	0.0	5.2
1984-85	0.0	0.0	0.0	0.0	0.0	T	4.3	0.0	T	0.0	0.0	0.0	4.3
1985-86	0.0	0.0	0.0	0.0	0.0	T							
Record Mean	0.0	0.0	0.0	0.0	T	1.0	3.0	2.7	1.2	T	0.0	0.0	7.9

See Reference Notes, relative to all above tables, on preceding page.

Richmond is located in east-central Virginia at the head of navigation on the James River and along a line separating the Coastal Plains (Tidewater Virginia) from the Piedmont. The Blue Ridge Mountains lie about 90 miles to the west and the Chesapeake Bay 60 miles to the east. Elevations range from a few feet above sea level along the river to a little over 300 feet in parts of the western section of the city.

The climate might be classified as modified continental. Summers are warm and humid and winters generally mild. The mountains to the west act as a partial barrier to outbreaks of cold, continental air in winter. The cold winter air is delayed long enough to be modified, then further warmed as it subsides in its approach to Richmond. The open waters of the Chesapeake Bay and Atlantic Ocean contribute to the humid summers and mild winters. The coldest weather normally occurs in late December and January, when low temperatures usually average in the upper 20s, and the high temperatures in the upper 40s. Temperatures seldom lower to zero, but there have been several occurrences of below zero temperatures. Summertime high temperatures above 100 degrees are not uncommon, but do not occur every year.

Precipitation is rather uniformly distributed throughout the year. However, dry periods lasting several weeks do occur, especially in autumn when long periods of pleasant, mild weather are most common. There is considerable variability in total monthly amounts from year to year. Snow usually remains on the ground only one or two days at a time. Ice storms (freezing rain or glaze) are not uncommon, but they are seldom severe enough to do any considerable damage. A notable exception was the spectacular glaze storm of January 27-28, 1943, when nearly 1 inch of ice accumulation caused heavy damage to trees and overhead transmission lines.

The James River reaches tidewater at Richmond where flooding may occur in every month of the year, most frequently in March and least in July. Hurricanes and tropical storms have been responsible for most of the flooding during the summer and early fall months. Hurricanes passing near Richmond have produced record rainfalls. In 1955, three hurricanes brought record rainfall to Richmond within a six-week period. The most noteworthy of these were Hurricanes Connie and Diane that brought heavy rains five days apart.

Damaging storms occur mainly from snow and freezing rain in winter and from hurricanes, tornadoes, and severe thunderstorms in other seasons. Damage may be from wind, flooding, or rain, or from any combination of these. Tornadoes are infrequent but some notable occurrences have been observed within the Richmond area.

Based on the 1951-1980 period, the average first occurrence of 32 degrees Fahrenheit in the fall is October 26 and the average last occurrence in the spring is April 10.

TABLE 1 — NORMALS, MEANS AND EXTREMES

RICHMOND, VIRGINIA

LATITUDE: 37°30'N   LONGITUDE: 77°20'W   ELEVATION: FT. GRND 164 BARO 00178   TIME ZONE: EASTERN   WBAN: 13740

	(a)	JAN	FEB	MAR	APR	MAY	JUNE	JULY	AUG	SEP	OCT	NOV	DEC	YEAR
TEMPERATURE °F:														
Normals														
-Daily Maximum		46.7	49.6	58.5	70.6	77.9	84.8	88.4	87.1	81.0	70.5	60.5	50.2	68.8
-Daily Minimum		26.5	28.1	35.8	45.1	54.2	62.2	67.2	66.4	59.3	46.7	37.3	29.6	46.5
-Monthly		36.6	38.9	47.2	57.9	66.1	73.5	77.8	76.8	70.2	58.6	48.9	39.9	57.7
Extremes														
-Record Highest	56	80	83	93	96	100	104	105	102	103	99	86	80	105
-Year		1950	1932	1938	1985	1941	1952	1977	1983	1954	1941	1974	1971	JUL 1977
-Record Lowest	56	-12	-10	11	23	31	40	51	46	35	21	10	-1	-12
-Year		1940	1936	1960	1985	1956	1967	1965	1934	1974	1962	1933	1942	JAN 1940
NORMAL DEGREE DAYS:														
Heating (base 65°F)		880	731	552	226	65	0	0	0	24	221	483	778	3960
Cooling (base 65°F)		0	0	0	13	99	258	397	366	180	23	0	0	1336
% OF POSSIBLE SUNSHINE	35	53	58	60	65	65	68	66	66	64	60	58	53	61
MEAN SKY COVER (tenths)														
Sunrise - Sunset	40	6.4	6.1	6.3	6.1	6.3	6.1	6.2	6.0	5.7	5.5	5.7	6.2	6.0
MEAN NUMBER OF DAYS:														
Sunrise to Sunset														
-Clear	40	8.3	8.7	8.4	7.9	7.0	6.6	6.9	7.4	9.5	11.3	9.6	9.4	100.9
-Partly Cloudy	40	6.8	6.2	8.2	9.3	10.2	12.0	11.4	11.6	8.5	7.2	7.7	6.4	105.5
-Cloudy	40	15.9	13.4	14.7	12.7	13.8	11.4	12.6	11.9	11.9	12.6	12.7	15.1	158.8
Precipitation														
.01 inches or more	48	10.3	9.1	10.8	9.2	10.8	9.5	11.1	9.9	7.9	7.4	8.3	8.8	113.1
Snow,Ice pellets														
1.0 inches or more	48	1.4	1.1	0.7	0.1	0.0	0.0	0.0	0.0	0.0	0.0	0.1	0.6	4.0
Thunderstorms	48	0.2	0.4	1.5	2.5	5.5	6.6	8.7	6.8	3.1	1.0	0.5	0.3	37.1
Heavy Fog Visibility 1/4 mile or less	56	2.8	2.1	1.7	1.7	1.9	1.6	2.3	2.6	3.1	3.4	2.3	2.9	28.2
Temperature °F														
-Maximum														
90° and above	56	0.0	0.0	0.1	0.9	2.7	9.3	13.4	11.0	4.6	0.4	0.0	0.0	42.3
32° and below	56	3.3	1.5	0.2	0.0	0.0	0.0	0.0	0.0	0.0	0.0	0.*	1.7	6.7
-Minimum														
32° and below	56	21.6	18.9	10.2	2.2	0.1	0.0	0.0	0.0	0.0	1.8	10.2	20.3	85.2
0° and below	56	0.4	0.1	0.0	0.0	0.0	0.0	0.0	0.0	0.0	0.0	0.0	0.*	0.5
AVG. STATION PRESS.(mb)	13	1012.3	1012.4	1010.2	1009.5	1008.8	1009.6	1010.1	1011.4	1012.1	1013.3	1013.1	1013.3	1011.3
RELATIVE HUMIDITY (%)														
Hour 01	51	77	74	73	74	83	87	88	90	90	87	80	77	82
Hour 07	51	80	79	78	75	80	82	85	89	90	89	84	81	83
Hour 13 (Local Time)	51	57	52	49	45	51	54	57	57	56	53	51	55	53
Hour 19	51	69	63	58	55	65	67	71	75	78	77	70	70	68
PRECIPITATION (inches):														
Water Equivalent														
-Normal		3.23	3.13	3.57	2.90	3.55	3.60	5.14	5.01	3.52	3.74	3.29	3.39	44.07
-Maximum Monthly	48	7.97	5.97	8.65	5.92	8.87	9.24	18.87	14.10	10.98	9.39	7.64	7.07	18.87
-Year		1978	1979	1984	1984	1972	1938	1945	1955	1975	1971	1959	1973	JUL 1945
-Minimum Monthly	48	0.64	0.48	0.94	0.64	0.87	0.38	0.51	0.52	0.26	0.30	0.36	0.40	0.26
-Year		1981	1978	1966	1963	1965	1980	1983	1943	1978	1963	1965	1980	SEP 1978
-Maximum in 24 hrs	48	3.31	2.67	2.54	2.60	3.08	4.61	5.73	8.79	4.02	6.50	4.07	3.16	8.79
-Year		1962	1979	1984	1978	1981	1963	1969	1955	1985	1961	1956	1958	AUG 1955
Snow,Ice pellets														
-Maximum Monthly	48	28.5	21.4	19.7	2.0						T	7.3	12.5	28.5
-Year		1940	1983	1960	1940						1979	1953	1958	JAN 1940
-Maximum in 24 hrs	48	21.6	16.8	12.1	2.0						T	7.3	7.5	21.6
-Year		1940	1983	1962	1940						1979	1953	1966	JAN 1940
WIND:														
Mean Speed (mph)	37	8.0	8.6	9.0	8.8	7.7	7.2	6.7	6.3	6.5	6.9	7.4	7.6	7.6
Prevailing Direction through 1963		S	NNE	W	S	SSW	S	SSW	S	S	NNE	S	SW	S
Fastest Mile														
-Direction	32	NW	SW	SE	NW	N	NW	NW	W	SE	SE	NW	SW	SE
-Speed (MPH)	32	43	45	42	40	45	52	56	54	45	68	38	40	68
-Year		1971	1951	1952	1972	1962	1952	1955	1964	1952	1954	1977	1968	OCT 1954
Peak Gust														
-Direction	1	NW	S	W	W	SW	N	S	NW	N	SW	SE	W	SW
-Speed (mph)	1	48	37	43	48	55	46	53	35	36	31	43	41	55
-Date		1985	1985	1985	1984	1984	1985	1985	1985	1985	1984	1985	1985	MAY 1984

See Reference Notes to this table on the following page.

TABLE 2 PRECIPITATION (inches) RICHMOND, VIRGINIA

YEAR	JAN	FEB	MAR	APR	MAY	JUNE	JULY	AUG	SEP	OCT	NOV	DEC	ANNUAL
1956	1.65	3.57	3.06	2.75	4.35	3.28	10.32	2.28	2.96	4.92	6.11	3.98	49.23
1957	3.36	5.29	2.82	2.25	2.75	3.92	1.80	7.46	3.43	5.35	5.30	6.88	50.61
1958	2.96	4.38	3.81	4.35	5.79	6.09	3.27	9.77	1.90	5.35	1.43	4.43	53.53
1959	1.31	1.87	2.92	4.32	2.44	3.45	12.85	5.75	3.30	3.25	7.64	2.24	51.34
1960	2.13	4.56	3.29	3.57	3.59	0.91	7.34	7.20	6.21	3.31	0.85	3.04	46.00
1961	2.57	5.39	4.02	1.73	4.83	6.49	2.85	3.90	1.64	8.78	1.81	5.05	49.06
1962	5.95	3.00	4.87	3.80	4.08	5.57	5.65	2.37	3.46	0.50	6.73	2.64	48.62
1963	1.55	2.98	5.62	0.64	2.39	7.01	0.52	3.75	3.20	0.30	6.70	2.80	37.46
1964	4.16	4.46	2.61	2.71	1.14	2.40	6.46	9.88	2.56	3.62	1.98	3.05	45.03
1965	2.51	2.77	3.68	2.13	0.87	3.39	6.33	0.81	4.81	1.38	0.36	0.72	29.76
1966	4.58	3.80	0.94	2.18	2.58	2.54	4.07	1.31	5.06	4.81	1.31	3.07	36.25
1967	1.50	3.35	2.34	1.32	3.71	3.58	5.00	6.65	0.95	1.00	1.76	6.48	37.64
1968	2.53	0.98	4.00	2.93	3.13	2.89	3.41	3.71	1.78	1.59	3.87	2.28	33.10
1969	2.04	3.95	3.95	2.60	3.32	4.36	13.90	9.31	3.89	1.88	1.87	5.26	56.33
1970	1.32	2.37	3.70	2.84	1.84	1.12	4.74	1.69	1.02	1.55	3.10	3.00	28.29
1971	1.84	4.37	2.68	1.76	6.82	4.10	4.40	3.73	2.35	9.39	2.76	0.75	44.95
1972	1.43	5.15	2.11	3.35	8.87	8.82	3.84	3.84	3.35	7.89	5.82	2.91	59.34
1973	2.66	3.11	3.44	4.58	3.56	2.45	3.64	4.34	1.82	2.56	1.27	7.07	40.50
1974	3.21	2.54	3.79	1.58	3.02	1.80	2.25	6.84	4.83	0.39	1.23	4.22	35.70
1975	5.71	2.96	8.04	2.78	2.59	4.00	12.29	2.31	10.98	3.10	2.04	4.51	61.31
1976	3.39	1.35	2.14	1.08	3.76	2.85	2.63	1.35	4.78	6.99	1.88	2.56	34.76
1977	2.22	1.34	2.67	2.33	3.99	1.25	4.20	6.15	2.16	7.88	4.32	5.57	44.08
1978	7.97	0.48	5.67	4.31	3.92	5.26	4.24	5.93	0.26	1.21	4.57	3.80	47.62
1979	6.16	5.97	2.59	3.97	3.80	2.42	4.36	7.08	9.76	3.87	5.50	1.64	57.12
1980	6.05	1.01	5.49	4.28	4.68	0.38	5.18	2.15	2.37	6.96	2.18	0.40	41.13
1981	0.64	2.76	1.52	2.96	6.62	3.69	4.01	2.89	2.70	2.36	0.68	5.04	35.87
1982	2.76	4.44	3.74	2.97	3.48	3.97	9.21	4.39	2.55	2.90	2.70	3.37	46.48
1983	1.59	3.95	6.04	5.21	2.50	5.46	0.51	0.97	3.05	4.02	5.63	4.50	43.43
1984	3.98	3.97	8.65	5.92	4.52	2.01	3.55	4.58	1.86	2.14	3.34	1.52	46.04
1985	3.54	3.20	1.80	0.65	2.36	4.01	5.31	10.58	4.97	5.09	6.99	0.58	49.08
Record Mean	3.12	3.03	3.61	2.95	3.70	3.68	5.46	4.95	3.61	3.41	3.26	3.15	43.93

TABLE 3 AVERAGE TEMPERATURE (deg. F) RICHMOND, VIRGINIA

YEAR	JAN	FEB	MAR	APR	MAY	JUNE	JULY	AUG	SEP	OCT	NOV	DEC	ANNUAL
1956	36.0	43.0	46.3	55.5	65.0	74.7	77.8	76.5	67.9	60.9	47.6	48.9	58.3
1957	35.2	43.3	47.2	61.5	67.8	76.2	78.4	74.6	71.9	54.6	50.3	43.0	58.7
1958	34.6	33.8	42.3	57.5	65.7	71.3	80.2	76.4	69.1	58.7	51.2	33.4	56.2
1959	37.5	41.6	47.3	59.3	69.4	74.8	77.9	79.0	70.8	61.4	47.3	41.6	59.0
1960	38.8	39.3	35.9	61.8	64.9	73.7	76.3	77.5	69.3	57.1	50.1	34.6	56.6
1961	33.5	42.2	50.8	53.0	63.6	72.8	78.5	77.1	73.5	58.1	50.1	37.1	57.5
1962	36.6	39.7	45.0	57.5	70.6	74.0	74.8	74.6	66.2	60.5	47.2	36.1	56.9
1963	35.9	33.3	50.8	59.2	64.0	72.0	76.1	75.7	65.5	58.6	50.1	32.4	56.1
1964	38.1	37.2	47.6	55.4	66.4	73.1	75.8	73.1	67.1	53.4	51.5	42.9	56.8
1965	35.6	38.8	43.0	53.9	69.6	70.7	74.9	75.9	70.7	56.1	48.2	41.3	56.6
1966	31.1	37.7	47.5	52.8	63.1	71.4	74.6	74.6	67.2	55.5	49.5	38.0	55.4
1967	40.9	34.8	46.6	58.8	60.7	72.1	76.6	75.5	65.7	57.2	44.0	41.9	56.2
1968	33.9	34.2	52.0	58.8	64.7	74.7	78.9	78.9	70.9	61.9	51.3	37.0	58.1
1969	33.9	36.8	42.3	57.6	65.5	75.7	78.3	75.1	68.1	58.5	46.8	35.5	56.2
1970	30.1	37.1	42.9	58.2	69.1	75.7	78.3	78.0	74.8	62.9	49.9	40.4	58.1
1971	33.8	39.5	44.5	55.0	63.3	74.7	76.6	75.3	71.4	64.6	48.5	48.0	57.9
1972	40.7	37.6	47.2	56.2	64.6	70.1	77.1	75.2	70.1	55.8	45.9	44.7	57.4
1973	37.6	38.5	52.6	57.9	65.1	76.0	77.4	77.5	72.3	60.6	51.3	40.8	59.0
1974	45.8	40.1	50.4	59.9	65.8	70.6	76.9	75.7	67.4	55.4	48.5	41.7	58.2
1975	40.7	41.4	45.3	52.9	67.7	73.6	76.0	78.8	69.3	62.5	53.6	40.0	58.5
1976	35.1	48.5	52.6	60.5	65.2	74.6	77.3	75.7	68.7	54.4	42.7	36.7	57.7
1977	25.3	40.5	53.7	61.1	68.2	73.0	81.4	79.8	74.2	57.3	52.3	39.5	58.9
1978	33.4	30.3	44.5	57.3	65.5	74.7	77.5	80.1	72.9	58.3	52.5	42.5	57.5
1979	36.4	28.6	51.1	58.4	67.1	70.8	76.9	77.8	71.0	58.3	53.3	42.3	57.7
1980	38.8	36.0	47.4	61.1	68.3	72.8	80.0	80.7	74.7	56.9	46.2	38.6	58.5
1981	31.2	42.2	44.6	60.6	64.1	77.9	79.6	75.1	69.4	56.4	49.1	38.0	57.3
1982	31.6	41.7	49.1	55.9	70.4	73.4	78.6	75.0	69.8	59.2	51.9	46.1	58.6
1983	37.8	39.1	50.4	56.1	66.1	75.6	79.4	77.7	68.8	58.1	49.0	36.2	57.9
1984	32.6	44.5	43.6	55.8	65.4	77.7	76.0	77.0	67.5	66.1	46.6	47.7	58.4
1985	32.6	40.2	49.7	62.0	68.0	74.3	79.0	77.5	70.8	62.6	56.6	37.8	59.3
Record Mean	37.2	39.3	47.2	57.2	66.3	74.1	77.7	76.4	70.1	58.8	48.9	39.8	57.8
Max	47.0	50.0	58.6	69.7	78.1	85.2	88.1	86.6	80.9	70.7	60.4	49.9	68.8
Min	27.3	28.6	35.7	44.7	54.4	62.9	67.3	66.2	59.2	47.0	37.4	29.6	46.7

REFERENCE NOTES FOR TABLES 1, 2, 3 and 6   (RICHMOND, VA)

GENERAL

T - TRACE AMOUNT
BLANK ENTRIES DENOTE MISSING/UNREPORTED DATA.
# INDICATES A STATION OR INSTRUMENT RELOCATION.

SPECIFIC

TABLE 1

(a) - LENGTH OF RECORD IN YEARS. ALTHOUGH
      INDIVIDUAL MONTHS MAY BE MISSING.
* LESS THAN .05

NORMALS — BASED ON THE 1951-1980 RECORD PERIOD.
EXTREMES — DATES ARE THE MOST RECENT OCCURRENCE.
WIND DIR. — NUMERALS SHOW TENS OF DEGREES
            CLOCKWISE FROM TRUE NORTH.
            "00" INDICATES CALM.
RESULTANT WIND DIRECTIONS ARE GIVEN TO WHOLE DEGREES.

EXCEPTIONS

TABLES 2, 3, and 6

RECORD MEANS ARE THROUGH THE CURRENT YEAR,
BEGINNING IN 1930 FOR TEMPERATURE
              1930 FOR PRECIPITATION
              1938 FOR SNOWFALL

TABLE 4 HEATING DEGREE DAYS Base 65 deg. F RICHMOND, VIRGINIA

SEASON	JULY	AUG	SEP	OCT	NOV	DEC	JAN	FEB	MAR	APR	MAY	JUNE	TOTAL
1956-57	0	3	75	146	519	493	915	600	546	201	63	5	3566
1957-58	0	0	50	317	434	674	936	871	696	254	70	6	4308
1958-59	0	0	24	212	409	973	847	650	542	212	44	10	3923
1959-60	0	0	36	217	530	717	807	737	894	184	88	2	4212
1960-61	0	0	24	257	439	936	971	632	461	390	106	7	4223
1961-62	0	0	27	218	459	860	875	702	623	276	32	0	4072
1962-63	0	0	73	175	526	891	897	882	434	218	102	1	4199
1963-64	0	0	71	197	439	1004	826	801	537	306	74	12	4267
1964-65	0	0	32	352	402	676	909	726	674	339	17	34	4161
1965-66	0	6	25	275	498	726	1043	759	538	371	133	27	4401
1966-67	0	0	47	293	466	833	738	841	560	230	171	17	4196
1967-68	0	0	64	256	623	708	956	887	416	191	86	0	4187
1968-69	0	0	0	161	403	864	957	783	695	237	66	0	4166
1969-70	0	0	45	221	541	907	1076	778	677	231	51	0	4527
1970-71	0	0	12	124	445	756	960	709	627	295	104	3	4035
1971-72	0	0	11	69	512	526	748	788	554	286	58	21	3573
1972-73	0	0	17	285	513	588	843	735	394	247	79	0	3701
1973-74	0	0	5	163	414	744	589	691	455	204	75	5	3345
1974-75	0	0	62	310	513	715	746	654	604	368	44	1	4017
1975-76	0	0	27	121	356	770	917	480	386	227	78	11	3373
1976-77	0	1	15	332	660	869	1227	680	366	176	42	7	4375
1977-78	0	0	4	259	401	784	974	964	627	235	88	5	4341
1978-79	0	0	16	214	366	694	876	1011	439	218	44	4	3882
1979-80	0	0	8	242	353	698	806	835	541	135	47	2	3667
1980-81	0	0	14	267	557	813	1042	633	626	171	107	0	4230
1981-82	0	1	29	273	473	834	1029	645	486	280	6	1	4057
1982-83	0	6	10	213	399	585	836	718	445	282	69	2	3565
1983-84	0	1	86	236	475	887	994	589	657	282	93	3	4303
1984-85	0	0	73	57	546	531	997	692	484	177	35	5	3597
1985-86	0	0	31	114	257	838							

TABLE 5 COOLING DEGREE DAYS Base 65 deg. F RICHMOND, VIRGINIA

YEAR	JAN	FEB	MAR	APR	MAY	JUNE	JULY	AUG	SEP	OCT	NOV	DEC	TOTAL
1969	0	0	0	21	90	328	417	321	147	26	0	0	1350
1970	0	0	0	35	185	328	418	410	313	67	0	0	1756
1971	0	0	0	0	56	297	367	327	209	62	22	5	1345
1972	0	0	7	30	52	180	381	326	178	9	8	0	1171
1973	0	0	13	42	91	338	391	395	231	32	9	2	1544
1974	0	0	10	58	106	180	377	340	141	21	26	0	1259
1975	0	0	0	16	135	267	348	433	165	51	18	0	1433
1976	0	8	9	99	91	307	389	337	133	12	0	0	1385
1977	0	0	22	66	148	258	513	467	289	24	27	0	1814
1978	0	0	0	12	112	302	393	475	263	15	0	1	1573
1979	0	0	16	30	117	188	374	404	195	42	9	0	1375
1980	0	0	0	1	25	157	243	472	494	313	23	1	1729
1981	0	0	1	45	89	395	458	319	169	16	0	0	1492
1982	0	0	0	13	181	259	428	323	157	43	13	7	1424
1983	0	0	0	23	108	325	452	405	207	27	0	0	1547
1984	0	0	0	10	114	392	346	381	154	100	0	2	1499
1985	0	4	20	94	139	290	441	392	213	51	10	0	1654

TABLE 6 SNOWFALL (inches) RICHMOND, VIRGINIA

SEASON	JULY	AUG	SEP	OCT	NOV	DEC	JAN	FEB	MAR	APR	MAY	JUNE	TOTAL
1956-57	0.0	0.0	0.0	0.0	T	T	5.8	0.7	T	1.2	0.0	0.0	7.7
1957-58	0.0	0.0	0.0	0.0	0.0	2.9	3.7	7.7	6.3	0.0	0.0	0.0	20.6
1958-59	0.0	0.0	0.0	0.0	0.0	12.5	2.4	0.0	0.0	0.0	0.0	0.0	14.9
1959-60	0.0	0.0	0.0	0.0	T	1.2	2.5	8.9	19.7	0.0	0.0	0.0	32.3
1960-61	0.0	0.0	0.0	0.0	0.0	1.7	7.2	5.0	T	0.0	0.0	0.0	13.9
1961-62	0.0	0.0	0.0	0.0	T	0.9	20.6	1.2	16.2	0.0	0.0	0.0	38.9
1962-63	0.0	0.0	0.0	0.0	0.9	8.1	1.6	6.3	T	0.0	0.0	0.0	16.9
1963-64	0.0	0.0	0.0	0.0	T	0.4	5.7	10.2	7.0	1.2	0.0	0.0	24.5
1964-65	0.0	0.0	0.0	0.0	0.4	0.0	12.4	6.6	1.0	0.0	0.0	0.0	20.4
1965-66	0.0	0.0	0.0	0.0	0.0	0.0	26.2	3.0	0.0	0.0	0.0	0.0	29.2
1966-67	0.0	0.0	0.0	0.0	0.2	12.2	6.3	17.1	0.0	0.0	0.0	0.0	35.8
1967-68	0.0	0.0	0.0	0.0	T	5.6	2.3	2.4	2.8	0.0	0.0	0.0	13.1
1968-69	0.0	0.0	0.0	1.2	2.8	T	T	11.9	0.0	0.0	0.0	0.0	15.9
1969-70	0.0	0.0	0.0	0.0	0.0	1.8	5.4	0.4	0.0	0.0	0.0	0.0	7.6
1970-71	0.0	0.0	0.0	0.0	0.0	0.9	3.3	2.0	8.4	0.6	0.0	0.0	15.2
1971-72	0.0	0.0	0.0	0.0	T	0.0	T	13.7	0.0	T	0.0	0.0	13.7
1972-73	0.0	0.0	0.0	T	0.6	0.0	4.3	0.4	1.4	0.0	0.0	0.0	6.7
1973-74	0.0	0.0	0.0	0.0	0.0	9.9	T	5.0	0.0	0.0	0.0	0.0	14.9
1974-75	0.0	0.0	0.0	0.0	T	0.0	2.7	2.9	0.4	0.0	0.0	0.0	6.0
1975-76	0.0	0.0	0.0	0.0	0.0	T	0.2	T	1.0	0.0	0.0	0.0	1.2
1976-77	0.0	0.0	0.0	0.0	1.0	1.7	11.1	T	0.0	0.0	0.0	0.0	13.8
1977-78	0.0	0.0	0.0	0.0	T	T	1.3	5.1	5.0	0.0	0.0	0.0	11.4
1978-79	0.0	0.0	0.0	0.0	0.0	0.0	0.7	19.5	T	0.0	0.0	0.0	20.2
1979-80	0.0	0.0	0.0	T	0.0	T	16.6	7.0	15.0	0.0	0.0	0.0	38.6
1980-81	0.0	0.0	0.0	0.0	0.0	0.2	0.6	0.0	0.2	0.0	0.0	0.0	1.0
1981-82	0.0	0.0	0.0	0.0	T	1.9	8.3	10.8	T	0.2	0.0	0.0	21.2
1982-83	0.0	0.0	0.0	0.0	0.0	7.9	0.1	21.4	0.0	T	0.0	0.0	29.4
1983-84	0.0	0.0	0.0	0.0	0.1	T	1.1	2.8	0.3	0.0	0.0	0.0	4.3
1984-85	0.0	0.0	0.0	0.0	0.0	0.0	8.3	T	T	0.0	0.0	0.0	8.3
1985-86	0.0	0.0	0.0	0.0	0.0	1.3							
Record Mean	0.0	0.0	0.0	T	0.4	2.0	5.1	4.2	2.7	0.1	0.0	0.0	14.5

See Reference Notes, relative to all above tables, on preceding page.

The climate of Roanoke is relatively mild. Roanoke is nestled among mountains which interrupt the Great Valley, extending from northernmost Virginia southwestward into east Tennessee. This location, at a point where the valley is pinched between the Blue Ridges and the Alleghenies, offers a natural barrier to the winter cold as it moves southward. It is also far enough inland that hurricanes lose much of their destructive force before reaching Roanoke. Finally, the rough terrain is an inhospitable breeding ground for tornadic activity. The elevation in the vicinity usually produces cool summer nights that make a light cover comfortable for sleeping. Although past records show extremes over 100 degrees and below zero, many years pass without either extreme being threatened.

Roanoke is located near the headwaters of the Roanoke River, which flows in a general southeasterly direction. Numerous creeks and small streams from nearby mountainous areas empty into the Roanoke River. The usual low water stage is 1 to 1.5 feet, and flood stage is 10 feet. Some low-lying streets in Roanoke and nearby Salem have to be blocked off during 7 to 8 foot stages, but damage is minor until the river overflows its banks. The highest stage on record exceeds 19 feet. Damage has been widespread on occasion and has amounted to several million dollars in the city of Roanoke alone.

The growing season averages 190 days. The average date of the last freezing temperature in spring is mid-April and the average date of the first freezing date in the fall is late October.

Rainfall is well apportioned throughout the year. Droughts are so infrequent that quoting actual records would be difficult. Snow usually falls each winter, ranging from only a trace to more than 60 inches.

TABLE 1   NORMALS, MEANS AND EXTREMES

ROANOKE, VIRGINIA

LATITUDE: 37°19'N   LONGITUDE: 79°58'W   ELEVATION: FT. GRND 1149 BARO 01193   TIME ZONE: EASTERN   WBAN: 13741

	(a)	JAN	FEB	MAR	APR	MAY	JUNE	JULY	AUG	SEP	OCT	NOV	DEC	YEAR
TEMPERATURE °F:														
Normals														
-Daily Maximum		44.8	48.0	56.9	68.2	76.4	83.0	86.7	85.5	79.4	68.6	57.4	47.8	66.9
-Daily Minimum		26.2	27.8	35.3	44.3	53.0	60.1	64.6	63.8	57.0	44.9	36.3	28.7	45.2
-Monthly		35.5	37.9	46.1	56.3	64.7	71.6	75.7	74.7	68.2	56.8	46.9	38.3	56.0
Extremes														
-Record Highest	38	78	80	86	95	96	100	104	105	101	93	83	76	105
-Year		1952	1985	1968	1957	1962	1959	1954	1983	1954	1951	1950	1984	AUG 1983
-Record Lowest	38	-11	1	10	20	31	39	48	43	34	22	9	-4	-11
-Year		1985	1970	1980	1985	1966	1977	1983	1965	1983	1976	1950	1983	JAN 1985
NORMAL DEGREE DAYS:														
Heating (base 65°F)		915	759	586	268	99	12	0	0	38	267	543	828	4315
Cooling (base 65°F)		0	0	0	7	89	210	332	301	134	12	0	0	1085
% OF POSSIBLE SUNSHINE														
MEAN SKY COVER (tenths)														
Sunrise - Sunset	37	6.3	6.2	6.3	6.0	6.1	6.0	6.1	5.8	5.6	5.2	5.9	6.1	6.0
MEAN NUMBER OF DAYS:														
Sunrise to Sunset														
-Clear	38	8.3	7.9	7.8	8.6	7.5	7.2	6.6	7.9	9.8	12.6	8.9	8.8	101.9
-Partly Cloudy	38	7.7	7.2	9.0	8.7	10.4	11.5	12.8	12.6	8.8	7.1	8.6	8.1	112.5
-Cloudy	38	15.0	13.1	14.2	12.7	13.1	11.3	11.6	10.6	11.4	11.3	12.5	14.1	150.9
Precipitation														
.01 inches or more	38	10.3	9.5	11.2	10.1	11.9	10.1	11.7	10.7	8.2	7.8	9.0	8.8	119.3
Snow, Ice pellets														
1.0 inches or more	38	1.8	1.9	1.1	0.1	0.0	0.0	0.0	0.0	0.0	0.*	0.6	1.2	6.6
Thunderstorms	38	0.1	0.3	1.1	2.9	6.1	6.6	8.5	7.1	2.6	1.1	0.3	0.1	36.8
Heavy Fog Visibility														
1/4 mile or less	38	2.9	2.8	1.9	1.2	1.8	1.1	1.5	1.5	2.5	2.2	2.2	2.5	24.0
Temperature °F														
-Maximum														
90° and above	21	0.0	0.0	0.0	0.4	0.6	4.7	9.0	7.4	2.6	0.0	0.0	0.0	24.8
32° and below	21	5.6	3.0	0.2	0.0	0.0	0.0	0.0	0.0	0.0	0.0	0.2	2.4	11.4
-Minimum														
32° and below	21	23.9	20.6	12.6	2.5	0.1	0.0	0.0	0.0	0.0	2.5	10.5	19.7	92.4
0° and below	21	0.5	0.0	0.0	0.0	0.0	0.0	0.0	0.0	0.0	0.0	0.0	0.1	0.6
AVG. STATION PRESS.(mb)	13	975.7	975.8	974.0	973.9	973.6	975.0	975.9	976.9	977.3	977.9	977.1	976.9	975.8
RELATIVE HUMIDITY (%)														
Hour 01	21	67	64	63	63	76	83	84	86	86	80	72	69	74
Hour 07 (Local Time)	21	70	68	69	70	78	81	83	86	88	83	76	72	77
Hour 13	21	53	49	48	46	52	54	55	56	56	54	52	54	52
Hour 19	21	58	54	50	48	57	62	64	67	68	65	60	61	60
PRECIPITATION (inches):														
Water Equivalent														
-Normal		2.83	3.19	3.69	3.09	3.51	3.34	3.45	3.91	3.14	3.48	2.59	2.93	39.15
-Maximum Monthly	38	6.12	7.17	7.80	7.95	8.42	7.55	7.85	9.54	9.18	9.72	12.36	7.10	12.36
-Year		1978	1960	1975	1983	1950	1972	1949	1984	1979	1976	1985	1948	NOV 1985
-Minimum Monthly	38	0.29	0.56	0.43	0.48	1.27	1.46	0.45	1.16	0.44	0.27	0.44	0.18	0.18
-Year		1981	1968	1966	1976	1951	1960	1977	1965	1968	1963	1960	1965	DEC 1965
-Maximum in 24 hrs	38	2.71	2.62	3.02	5.57	3.99	3.98	2.72	5.22	3.80	6.41	6.63	3.40	6.63
-Year		1968	1984	1983	1978	1973	1972	1966	1985	1979	1968	1985	1948	NOV 1985
Snow, Ice pellets														
-Maximum Monthly	38	41.2	27.6	30.3	7.3	T				T	1.0	13.8	22.6	41.2
-Year		1966	1960	1960	1971	1963				1953	1957	1968	1966	JAN 1966
-Maximum in 24 hrs	38	13.7	18.4	17.4	7.3	T				T	1.0	10.0	16.4	18.4
-Year		1966	1983	1960	1971	1963				1953	1957	1968	1969	FEB 1983
WIND:														
Mean Speed (mph)	37	9.8	10.0	10.5	10.0	8.0	7.0	6.7	6.2	6.2	7.0	8.5	8.9	8.2
Prevailing Direction through 1963		WNW	SE	WNW	SE	SE	SE	W	SE	SE	SE	NW	NW	SE
Fastest Obs. 1 Min.														
-Direction	24	30	31	32	32	36	28	34	30	31	34	34	30	32
-Speed (MPH)	24	53	40	52	58	46	46	46	44	29	35	52	40	58
-Year		1964	1972	1967	1963	1962	1966	1980	1975	1971	1963	1963	1970	APR 1963
Peak Gust														
-Direction	2	NW	W	NW	NW	W	W	N	W	W	W	W	NW	NW
-Speed (mph)	2	54	47	52	47	41	40	37	37	35	33	44	51	54
-Date		1985	1984	1985	1985	1984	1985	1984	1985	1984	1985	1985	1984	JAN 1985

See reference Notes to this table on the following page.

TABLE 2 PRECIPITATION (inches) ROANOKE, VIRGINIA

YEAR	JAN	FEB	MAR	APR	MAY	JUNE	JULY	AUG	SEP	OCT	NOV	DEC	ANNUAL
1956	0.60	4.87	2.22	3.12	1.58	1.72	2.85	2.58	5.74	6.80	1.84	1.91	35.83
1957	3.74	5.31	1.48	4.55	2.23	5.35	1.13	1.20	7.06	2.37	3.39	2.46	40.27
1958	3.66	3.40	4.76	2.83	5.00	1.88	6.21	3.28	1.56	1.12	1.66	4.15	39.51
1959	2.37	1.04	3.38	4.27	3.17	2.82	2.81	4.49	4.95	3.97	1.94	2.18	37.39
1960	1.92	7.17	3.96	2.19	5.91	1.46	2.06	4.69	3.30	3.62	0.44	2.23	38.95
1961	1.61	4.50	4.17	3.86	2.05	3.79	1.59	5.14	1.79	3.35	3.62	5.07	40.54
1962	2.54	3.22	4.76	1.69	2.70	3.08	5.34	3.55	3.66	2.22	5.11	3.35	41.22
1963	1.10	1.97	3.89	0.87	1.94	2.61	2.68	1.81	2.87	0.27	3.80	1.86	25.67
1964	5.20	5.33	1.81	3.54	1.79	1.97	3.37	3.54	2.70	2.80	3.09	2.73	37.87
1965	3.72	3.53	3.93	1.70	3.83	1.83	5.50	1.16	1.97	3.42	0.95	0.18	31.72
1966	4.26	4.78	0.43	2.59	3.49	1.54	3.00	4.73	7.25	4.04	1.73	3.25	41.09
1967	1.25	2.51	4.51	1.67	3.95	3.19	4.05	6.36	1.78	2.42	1.30	4.84	37.83
1968	3.33	0.56	3.03	2.73	1.89	2.11	2.95	5.36	0.44	8.06	2.82	1.80	35.08
1969	1.86	2.74	3.35	1.40	1.58	4.95	5.35	5.41	2.79	2.22	1.40	5.54	38.59
1970	1.31	2.36	1.95	2.80	1.51	4.87	3.28	6.40	1.99	7.51	3.67	2.81	40.46
1971	1.21	5.13	2.28	2.55	7.50	4.84	5.23	4.46	3.87	6.75	2.18	0.83	46.83
1972	2.49	4.80	1.76	3.31	6.00	7.55	4.89	2.62	4.79	5.63	3.18	4.62	51.64
1973	2.60	2.95	5.92	5.39	5.58	3.65	5.10	3.34	1.84	4.28	1.79	5.60	48.04
1974	3.33	2.13	3.12	1.86	3.76	3.71	4.93	3.04	0.77	1.28	3.16	2.93	34.02
1975	3.59	3.05	7.80	2.04	6.65	1.54	5.15	5.68	6.46	3.01	1.77	3.67	50.41
1976	2.16	1.27	4.54	0.48	6.13	1.20	1.24	2.20	3.17	9.72	1.31	2.59	40.01
1977	1.46	0.73	2.61	3.50	1.52	2.41	0.45	2.29	2.71	4.70	6.46	2.49	31.33
1978	6.12	0.65	5.92	7.54	4.85	2.05	4.83	6.33	0.52	0.78	2.55	3.15	45.29
1979	5.27	5.37	3.38	3.99	2.65	5.78	3.97	3.37	9.18	3.56	3.77	1.13	51.42
1980	4.10	0.67	5.41	5.51	2.66	1.81	5.18	2.87	1.66	2.30	1.78	0.60	34.55
1981	0.29	2.43	2.30	1.75	4.56	2.49	2.86	1.32	4.52	3.90	0.68	3.79	30.89
1982	3.76	4.75	2.33	2.01	4.83	4.99	3.98	5.20	2.67	4.13	3.65	2.53	44.83
1983	1.28	4.12	6.41	7.95	3.17	2.38	1.67	2.23	1.52	7.73	4.26	5.61	48.33
1984	1.35	4.85	4.30	3.97	4.49	2.34	4.17	9.54	2.69	1.42	2.67	1.84	43.63
1985	2.45	3.64	1.80	1.75	6.89	2.08	4.18	8.67	1.26	3.77	12.36	0.85	49.70
Record Mean	2.75	3.22	3.63	3.17	3.86	3.41	3.61	4.33	3.13	3.51	2.91	3.03	40.58

TABLE 3 AVERAGE TEMPERATURE (deg. F) ROANOKE, VIRGINIA

YEAR	JAN	FEB	MAR	APR	MAY	JUNE	JULY	AUG	SEP	OCT	NOV	DEC	ANNUAL
1956	35.4	42.0	45.9	53.6	65.7	72.3	76.3	75.7	66.8	58.9	45.6	47.9	57.2
1957	35.7	43.4	45.4	59.8	67.6	73.7	76.5	73.7	68.5	53.5	48.1	41.1	57.2
1958	33.4	31.8	41.3	55.9	64.6	71.8	75.0	75.0	68.2	57.0	48.2	34.2	55.2
1959	36.0	40.9	45.5	58.5	68.8	73.8	77.1	79.1	70.3	59.6	44.8	41.1	58.0
1960	38.0	36.6	34.0	60.2	63.2	73.1	75.9	77.7	69.7	58.8	47.7	32.4	55.6
1961	33.2	41.1	50.1	51.0	61.5	71.3	76.6	75.6	71.9	58.3	49.3	37.2	56.4
1962	35.7	40.2	43.5	53.8	69.9	72.4	74.2	74.4	64.8	59.4	45.0	33.3	55.6
1963	32.9	32.0	49.7	58.6	64.4	72.9	75.0	73.9	66.0	60.5	48.6	30.5	55.4
1964	37.3	35.6	46.6	56.9	67.4	74.6	75.9	74.0	66.8	54.1	51.1	40.0	56.7
1965	36.0	36.4	41.9	54.9	69.2	69.7	73.9	74.2	69.7	56.0	47.5	41.5	55.9
1966	30.6	35.1	46.7	51.5	63.3	71.1	77.9	73.0	66.1	54.6	47.2	37.8	54.6
1967	40.8	34.8	48.5	59.0	59.0	70.3	73.4	72.6	63.2	54.7	43.4	41.3	55.1
1968	33.0	33.8	51.3	56.5	62.4	71.3	75.6	75.2	66.7	57.4	48.1	33.7	55.4
1969	32.3	37.2	40.7	57.3	65.0	73.1	76.0	73.5	66.7	56.6	44.5	33.6	54.7
1970	29.8	36.9	42.5	56.5	66.8	72.9	76.1	74.1	72.6	55.7	46.3	40.0	56.2
1971	33.9	39.1	43.6	55.5	61.5	72.6	73.5	72.6	69.8	62.1	45.4	45.3	56.3
1972	39.6	36.0	45.6	55.5	62.0	67.7	74.3	73.2	67.6	52.2	45.4	44.5	55.3
1973	37.3	35.8	51.1	53.8	61.1	73.8	75.5	76.0	70.8	59.2	49.3	38.2	56.8
1974	45.3	39.0	50.8	57.2	64.3	68.1	74.5	73.5	64.8	55.0	47.3	38.9	56.6
1975	39.6	40.4	42.2	52.9	66.5	71.5	74.5	77.0	66.6	60.0	51.1	38.8	56.8
1976	33.4	46.7	50.8	57.4	62.3	70.0	74.0	72.1	65.2	50.3	40.2	34.7	54.7
1977	23.6	36.6	52.5	58.9	68.1	71.1	79.7	77.8	70.4	53.6	48.5	35.9	56.4
1978	27.8	29.5	44.2	56.3	63.8	72.7	76.0	77.5	71.4	54.0	49.5	39.5	55.2
1979	31.4	29.4	48.6	55.7	63.8	69.3	73.5	74.0	66.3	54.6	49.6	40.9	54.8
1980	37.5	34.0	43.5	56.7	64.8	70.0	78.1	76.8	71.5	55.5	45.4	38.7	56.1
1981	32.0	38.2	43.1	58.6	60.9	74.0	76.1	73.4	66.1	53.7	45.2	32.8	54.5
1982	28.7	38.1	44.8	50.8	67.5	70.1	75.5	72.3	65.8	57.4	47.8	42.4	55.1
1983	35.6	36.5	47.0	52.4	61.0	70.1	77.0	77.9	68.1	57.1	47.1	34.9	55.4
1984	33.6	43.6	43.3	54.2	63.1	70.1	74.2	77.0	64.1	64.3	46.0	47.1	56.8
1985	31.3	38.6	50.3	61.9	67.1	72.7	76.5	73.8	68.4	60.3	55.4	35.1	57.6
Record Mean	35.5	38.1	45.9	56.8	64.7	72.0	75.9	74.8	67.8	57.3	46.9	38.3	56.2
Max	44.5	48.0	56.5	67.8	76.3	83.4	86.8	85.6	78.8	68.6	57.2	47.5	66.8
Min	26.4	28.2	35.3	45.7	53.1	60.5	65.0	63.9	56.7	46.0	36.6	29.1	45.6

REFERENCE NOTES FOR TABLES 1, 2, 3 and 6 (ROANOKE, VA)

GENERAL

T - TRACE AMOUNT
BLANK ENTRIES DENOTE MISSING/UNREPORTED DATA.
# INDICATES A STATION OR INSTRUMENT RELOCATION.

SPECIFIC

TABLE 1

(a) - LENGTH OF RECORD IN YEARS. ALTHOUGH INDIVIDUAL MONTHS MAY BE MISSING.
* LESS THAN .05

NORMALS — BASED ON THE 1951-1980 RECORD PERIOD.
EXTREMES — DATES ARE THE MOST RECENT OCCURRENCE.
WIND DIR. — NUMERALS SHOW TENS OF DEGREES CLOCKWISE FROM TRUE NORTH.
"00" INDICATES CALM.
RESULTANT WIND DIRECTIONS ARE GIVEN TO WHOLE DEGREES.

EXCEPTIONS

TABLES 2, 3, and 6

RECORD MEANS ARE THROUGH THE CURRENT YEAR, BEGINNING IN
1948 FOR TEMPERATURE
1948 FOR PRECIPITATION
1948 FOR SNOWFALL

TABLE 4 HEATING DEGREE DAYS Base 65 deg. F   ROANOKE, VIRGINIA

SEASON	JULY	AUG	SEP	OCT	NOV	DEC	JAN	FEB	MAR	APR	MAY	JUNE	TOTAL
1956-57	0	6	86	202	573	526	901	599	600	246	52	10	3801
1957-58	0	3	67	349	498	735	974	920	727	282	82	3	4640
1958-59	0	0	30	256	436	948	893	668	597	218	49	8	4103
1959-60	0	0	28	238	602	734	829	819	955	206	142	0	4553
1960-61	0	0	19	209	511	1001	981	663	465	433	127	15	4424
1961-62	0	0	36	211	485	856	904	688	658	354	34	1	4227
1962-63	0	0	100	221	594	975	989	917	470	228	89	5	4588
1963-64	0	1	64	150	485	1063	848	852	566	265	62	15	4371
#1964-65	0	9	41	338	409	770	892	794	710	305	21	19	4308
1965-66	0	9	38	287	520	722	1058	831	565	409	118	25	4582
1966-67	0	0	51	317	527	837	740	838	505	202	206	28	4251
1967-68	0	0	105	321	643	728	984	902	423	255	111	3	4475
1968-69	0	12	11	244	502	963	1005	772	746	227	77	9	4568
1969-70	0	0	53	274	607	967	1087	783	691	269	66	2	4799
1970-71	1	1	24	194	553	769	959	719	658	285	143	3	4309
1971-72	0	0	12	117	595	603	780	837	594	305	99	31	3973
1972-73	10	1	33	391	582	628	852	812	428	344	144	1	4226
1973-74	0	0	12	196	461	826	607	722	440	255	96	14	3629
1974-75	0	0	84	308	539	801	783	683	699	361	60	12	4330
1975-76	0	0	59	173	415	809	973	523	438	271	126	22	3809
1976-77	0	3	47	452	735	945	1275	786	385	203	56	23	4910
1977-78	0	0	18	350	496	896	1147	989	637	261	112	4	4910
1978-79	0	0	29	335	461	784	1037	992	512	279	88	15	4532
1979-80	3	9	49	329	458	738	848	893	656	266	78	14	4341
1980-81	0	0	30	301	582	807	1016	744	672	212	158	3	4525
1981-82	0	0	58	357	589	991	1121	746	618	421	51	3	4955
1982-83	0	6	61	264	509	695	904	792	551	381	157	14	4334
1983-84	1	0	87	246	531	924	966	614	664	336	123	6	4498
1984-85	0	0	120	74	565	549	1041	734	471	168	39	11	3772
1985-86	0	0	55	171	282	918							

TABLE 5 COOLING DEGREE DAYS Base 65 deg. F   ROANOKE, VIRGINIA

YEAR	JAN	FEB	MAR	APR	MAY	JUNE	JULY	AUG	SEP	OCT	NOV	DEC	TOTAL
1969	0	0	0	2	86	256	347	272	107	21	0	0	1091
1970	0	0	0	22	128	246	351	291	261	35	0	0	1334
1971	0	0	0	4	40	239	270	245	163	35	13	1	1010
1972	0	0	0	26	13	118	305	262	118	0	.2	0	844
1973	0	0	3	15	29	269	331	346	192	23	0	0	1208
1974	0	0	7	27	82	117	303	267	83	3	12	0	901
1975	0	0	0	5	108	213	301	379	114	27	5	0	1152
1976	0	0	8	53	49	179	285	230	58	1	0	0	863
1977	0	0	7	28	158	214	461	400	185	4	7	0	1464
1978	0	0	0	7	79	243	347	391	228	2	0	0	1297
1979	0	0	10	7	56	150	273	296	97	12	1	0	902
1980	0	0	0	21	78	171	412	374	231	13	0	0	1300
1981	0	0	2	26	39	278	350	267	97	13	0	0	1072
1982	0	0	0	0	137	165	332	242	90	35	0	0	1001
1983	0	0	0	11	41	172	382	407	188	10	0	0	1211
1984	0	0	0	20	71	290	260	301	101	56	0	0	1099
1985	0	1	18	78	112	247	365	280	163	34	3	0	1301

TABLE 6 SNOWFALL (inches)   ROANOKE, VIRGINIA

SEASON	JULY	AUG	SEP	OCT	NOV	DEC	JAN	FEB	MAR	APR	MAY	JUNE	TOTAL
1956-57	0.0	0.0	0.0	0.0	0.1	T	1.6	1.4	T	0.6	0.0	0.0	3.7
1957-58	0.0	0.0	0.0	1.0	2.2	T	4.1	6.2	14.8	T	0.0	0.0	28.3
1958-59	0.0	0.0	0.0	0.0	T	2.3	T	4.9	T	0.7	0.0	0.0	7.9
1959-60	0.0	0.0	0.0	0.0	T	T	4.8	27.6	30.3	T	0.0	0.0	62.7
1960-61	0.0	0.0	0.0	0.0	T	9.8	10.0	20.7	0.3	T	0.0	0.0	40.8
1961-62	0.0	0.0	0.0	0.0	2.9	4.5	8.3	3.3	11.3	T	0.0	0.0	30.3
1962-63	0.0	0.0	0.0	T	2.4	14.4	T	12.2	0.7	0.0	T	0.0	29.7
1963-64	0.0	0.0	0.0	0.0	10.1	15.7	20.2	4.3	0.0	0.0	0.0		50.3
1964-65	0.0	0.0	0.0	0.0	0.5	T	12.1	4.0	4.1	0.0	0.0	0.0	20.7
1965-66	0.0	0.0	0.0	0.0	T	0.3	41.2	8.4	T	0.0	0.0	0.0	49.9
1966-67	0.0	0.0	0.0	0.0	T	22.6	2.7	15.6	0.8	0.0	0.0	0.0	41.7
1967-68	0.0	0.0	0.0	0.0	T	14.8	15.4	3.8	T	0.0	0.0	0.0	34.0
1968-69	0.0	0.0	0.0	T	13.8	T	0.1	11.4	13.8	0.0	0.0	0.0	39.1
1969-70	0.0	0.0	0.0	0.0	T	16.8	4.4	5.4	T	0.2	0.0	0.0	26.8
1970-71	0.0	0.0	0.0	0.0	2.0	10.8	1.1	3.6	7.3	7.3	0.0	0.0	32.1
1971-72	0.0	0.0	0.0	0.0	10.2	0.2	T	12.8	T	T	0.0	0.0	23.2
1972-73	0.0	0.0	0.0	0.0	T	T	2.4	1.4	5.5	0.2	0.0	0.0	9.5
1973-74	0.0	0.0	0.0	T	T	7.9	T	9.7	T	T	0.0	0.0	17.6
1974-75	0.0	0.0	0.0	T	3.3	6.6	4.6	6.3	6.2	T	0.0	0.0	27.0
1975-76	0.0	0.0	0.0	0.0	T	0.1	T	T	2.2	0.0	0.0	0.0	2.3
1976-77	0.0	0.0	0.0	0.0	2.4	6.4	8.9	1.5	T	T	0.0	0.0	19.2
1977-78	0.0	0.0	0.0	T	2.4	0.4	14.5	9.6	10.4	0.0	0.0	0.0	37.3
1978-79	0.0	0.0	0.0	0.0	1.6	T	19.3	T	0.0	0.0	0.0		23.9
1979-80	0.0	0.0	0.0	0.3	T	0.2	15.1	4.2	12.0	T	0.0	0.0	31.8
1980-81	0.0	0.0	0.0	0.0	T	T	1.4	T	10.4	0.0	0.0	0.0	11.8
1981-82	0.0	0.0	0.0	0.0	4.0	3.9	8.4	12.2	0.5	1.9	0.0	0.0	30.9
1982-83	0.0	0.0	0.0	T	0.0	6.2	3.8	24.3	0.3	0.4	0.0	0.0	35.0
1983-84	0.0	0.0	0.0	0.0	T	0.6	7.2	1.5	0.3	0.2	0.0	0.0	9.8
1984-85	0.0	0.0	0.0	0.0	0.0	0.8	2.7	1.2	1.3	T	0.0	0.0	6.0
1985-86	0.0	0.0	0.0	0.0	0.0	1.2							
Record Mean	0.0	0.0	T	T	1.7	4.1	6.6	7.2	4.1	0.3	T	0.0	23.9

See Reference Notes, relative to all above tables, on preceding page.

The climate of Olympia and vicinity is characterized by warm, generally dry summers and wet, mild winters.

Fall rains usually begin about mid-October, borne inland to the Cascade Mountains by frequent maritime disturbances originating in the Pacific Ocean. These rains continue with few interruptions through spring. Daytime temperatures will be in the 40s and low 50s, with nighttime temperatures in the 30s. The progression of the wet and mild Pacific disturbances is usually broken once or twice each winter by the formation of large anticyclones, which originate in Alaska and northwestern Canada, and move southward over the state of Washington. These southerly migrations of polar-continental air from the interior will normally lower daytime temperatures to about freezing and nighttime temperatures to 10 - 20 degrees. In exceptional cases, temperatures slightly below zero have been recorded in this vicinity. Often the onset of the cold weather is accompanied by a little snow, but during these cold snaps the air is dry and skies are clear. Snow in an amount sufficient to seriously hinder highway travel occurs only once every two or three years.

During the spring months, the track of the Pacific storms moves gradually farther north, and the semi-permanent Pacific anticyclone also moves northward. The effects of the maritime disturbances lessen, and the periods of improving weather between storms lengthen. During the spring and early fall, clearing skies at night will usually be followed by fog or low stratus clouds in the early morning, which normally dissipate by noon.

Daily high temperatures average 70 to 80 degrees during July, August, and September. The temperature will equal or exceed 90 degrees about 6 days each summer, but as the warm weather is usually accompanied by lowering humidity, it is seldom uncomfortably hot. Rainfall is near 1 inch per month during July and August and about 2 inches per month during the transitional period of May, June, and September. About two-thirds of the days are sunny in July, August, and September, and about half are sunny during May and June.

Olympia and vicinity are quite well protected by the Coast Range from the strong south and southwest winds accompanying many of the Pacific storms during the fall and winter. Winds which reach hurricane force along the coast, only 45 miles away, will reach only 50 or 55 mph in gusts in this vicinity. Some damage to utility lines occurs every fall and winter from trees and limbs broken or felled by the wind, but damage rarely occurs to dwellings or buildings. The prevailing wind in Olympia is southerly during most of the year, but during the fair weather in the summers the wind is gentle and from the north to east.

The length of the growing season in the vicinity of Olympia varies with distance from the waterfront and elevation above sea level. The average length of the growing season is 166 days.

TABLE 1

NORMALS, MEANS AND EXTREMES

OLYMPIA WASHINGTON

LATITUDE: 46°58'N   LONGITUDE: 122°54'W   ELEVATION: FT. GRND 195 BARO 00196   TIME ZONE: PACIFIC   WBAN: 24227

	(a)	JAN	FEB	MAR	APR	MAY	JUNE	JULY	AUG	SEP	OCT	NOV	DEC	YEAR
TEMPERATURE °F:														
Normals														
−Daily Maximum		43.6	49.1	52.5	58.7	65.7	70.8	77.2	76.2	71.0	60.8	50.3	45.1	60.1
−Daily Minimum		30.8	32.5	32.8	35.8	40.5	46.0	48.7	48.8	45.1	39.4	34.8	32.8	39.0
−Monthly		37.2	40.8	42.7	47.3	53.1	58.4	63.0	62.5	58.0	50.1	42.5	39.0	49.5
Extremes														
−Record Highest	44	63	72	76	85	96	101	103	104	96	89	74	64	104
−Year		1942	1968	1969	1947	1983	1942	1941	1981	1981	1980	1949	1958	AUG 1981
−Record Lowest	44	−8	−1	13	23	25	30	35	33	25	20	−1	−7	−8
−Year		1979	1972	1955	1975	1954	1976	1962	1973	1972	1971	1955	1983	JAN 1979
NORMAL DEGREE DAYS:														
Heating (base 65°F)		862	678	691	531	369	208	101	115	214	462	672	806	5709
Cooling (base 65°F)		0	0	0	0	0	10	39	38	7	0	0	0	94
% OF POSSIBLE SUNSHINE														
MEAN SKY COVER (tenths)														
Sunrise − Sunset	41	8.7	8.4	8.1	7.7	7.2	6.9	5.3	5.9	6.2	7.7	8.5	8.8	7.5
MEAN NUMBER OF DAYS:														
Sunrise to Sunset														
−Clear	44	2.1	2.0	2.5	2.8	4.2	5.2	10.3	8.7	7.5	2.9	1.5	1.6	51.4
−Partly Cloudy	44	3.7	4.5	6.6	7.6	8.9	8.5	10.3	9.9	9.4	8.0	5.4	3.5	86.3
−Cloudy	44	25.2	21.8	21.9	19.5	17.9	16.4	10.5	12.4	13.0	20.1	23.1	25.9	227.6
Precipitation														
.01 inches or more	44	19.4	17.7	17.8	14.7	10.7	9.3	4.8	5.9	8.8	14.3	19.2	21.0	163.6
Snow, Ice pellets														
1.0 inches or more	44	2.6	0.9	0.7	0.*	0.0	0.0	0.0	0.0	0.0	0.0	0.4	1.2	5.7
Thunderstorms	42	0.*	0.2	0.2	0.5	0.7	0.6	0.4	0.9	0.7	0.5	0.3	0.1	5.2
Heavy Fog Visibility														
1/4 mile or less	42	9.5	8.2	7.6	5.2	3.4	2.4	3.4	6.0	9.8	14.3	11.2	9.8	90.9
Temperature °F														
−Maximum														
90° and above	26	0.0	0.0	0.0	0.0	0.1	0.6	2.6	2.5	0.6	0.0	0.0	0.0	6.3
32° and below	26	1.5	0.*	0.0	0.0	0.0	0.0	0.0	0.0	0.0	0.0	0.3	1.4	3.2
−Minimum														
32° and below	26	16.1	13.3	14.1	9.3	2.4	0.2	0.0	0.0	0.0	5.6	11.0	15.5	88.3
0° and below	26	0.2	0.*	0.0	0.0	0.0	0.0	0.0	0.0	0.0	0.0	0.1	0.2	0.5
AVG. STATION PRESS. (mb)	13	1011.0	1008.6	1007.9	1010.1	1010.3	1010.3	1010.4	1009.4	1009.5	1010.3	1008.7	1010.6	1009.8
RELATIVE HUMIDITY (%)														
Hour 04	26	91	91	90	90	89	90	90	90	92	93	92	91	91
Hour 10	26	88	86	80	71	66	67	64	67	74	85	88	89	77
Hour 16 (Local Time)	26	80	71	62	57	54	55	50	51	56	68	79	84	64
Hour 22	22	89	88	86	82	79	79	78	80	86	91	91	89	85
PRECIPITATION (inches):														
Water Equivalent														
−Normal		8.50	5.77	4.85	3.13	1.85	1.44	0.76	1.34	2.36	4.68	7.58	8.70	50.96
−Maximum Monthly	44	19.84	13.18	10.13	5.87	5.83	6.48	3.00	5.45	7.59	10.08	15.51	14.32	19.84
−Year		1953	1961	1950	1972	1948	1946	1983	1968	1978	1967	1962	1970	JAN 1953
−Minimum Monthly	44	0.29	1.71	0.48	0.37	0.15	0.04	T	0.00	T	0.78	1.37	2.28	0.00
−Year		1985	1973	1965	1956	1947	1945	1984	1946	1975	1975	1976	1944	AUG 1946
−Maximum in 24 hrs	44	3.32	4.93	3.92	2.31	1.54	1.91	1.50	1.39	2.48	3.60	4.33	3.83	4.93
−Year		1972	1951	1972	1965	1948	1985	1979	1977	1978	1981	1962	1956	FEB 1951
Snow, Ice pellets														
−Maximum Monthly	44	58.7	17.7	20.6	2.2	T	T			T	T	14.8	21.4	58.7
−Year		1969	1949	1951	1972	1985	1976			1972	1975	1978	1968	JAN 1969
−Maximum in 24 hrs	44	20.5	10.1	9.1	1.8	T	T			T	T	14.5	11.9	20.5
−Year		1972	1980	1966	1972	1985	1976			1972	1975	1978	1974	JAN 1972
WIND:														
Mean Speed (mph)	33	7.1	7.2	7.4	7.4	6.8	6.7	6.2	5.9	5.7	5.9	6.7	7.4	6.7
Prevailing Direction through 1963		SSW	SSW	SSW	SW	SW	SSW	SW	SW	SW	SSW	SW	SSW	SSW
Fastest Obs. 1 Min.														
−Direction (!!)	37	18	18	23	23	29	25	18	27	18	23	18	18	18
−Speed (MPH)	37	55	45	40	46	39	32	29	26	35	58	60	41	60
−Year		1956	1958	1956	1962	1960	1949	1957	1964	1967	1962	1958	1969	NOV 1958
Peak Gust														
−Direction (!!)	2	SW	S	SW	W	SW	W	W	S	S	S	SW	SW	SW
−Speed (mph)	2	32	43	43	36	32	32	25	26	35	38	47	39	47
−Date		1984	1984	1985	1985	1985	1984	1985	1985	1985	1985	1984	1984	NOV 1984

See Reference Notes to this table on the following page

OLYMPIA, WASHINGTON

TABLE 2   PRECIPITATION (inches)   OLYMPIA WASHINGTON

YEAR	JAN	FEB	MAR	APR	MAY	JUNE	JULY	AUG	SEP	OCT	NOV	DEC	ANNUAL
1956	10.75	3.93	8.27	0.37	0.30	2.57	0.38	0.88	2.30	9.52	2.81	9.36	51.44
1957	3.02	5.88	7.43	1.72	1.42	1.78	0.97	0.87	0.66	4.74	4.05	8.91	41.45
1958	7.87	6.40	2.29	4.39	1.47	1.65	T	0.62	1.52	5.41	12.35	8.43	52.40
1959	8.91	4.53	4.63	4.51	1.45	1.85	0.30	0.70	4.26	3.92	10.36	7.50	52.92
1960	6.35	5.93	6.12	4.53	3.50	0.59	T	1.16	1.21	5.84	11.33	3.89	50.45
1961	8.69	13.18	6.26	3.29	2.93	1.05	0.80	1.01	0.33	4.97	6.78	8.25	57.54
1962	3.22	3.72	3.82	4.50	1.81	0.89	0.14	3.17	2.45	6.00	15.51	5.81	51.04
1963	3.47	6.42	5.10	4.13	1.76	0.63	1.48	0.79	2.16	5.98	10.37	5.55	47.84
1964	15.13	2.54	4.47	1.58	0.98	2.35	1.07	1.47	2.26	1.79	9.18	9.11	51.93
1965	9.37	4.93	0.48	3.61	1.89	0.33	0.48	2.05	0.60	3.30	5.84	7.81	40.69
1966	7.89	3.38	7.28	1.71	1.30	1.28	1.34	0.68	1.95	4.83	8.16	11.53	51.33
1967	12.21	3.58	4.31	2.88	0.25	1.49	0.02	T	1.36	10.08	3.90	5.94	46.02
1968	9.04	7.83	6.53	3.02	2.57	2.43	0.89	5.45	2.51	6.07	7.96	9.95	64.25
1969	9.45	3.41	2.90	3.44	2.07	1.68	0.50	0.18	5.23	2.69	3.60	7.24	42.39
1970	12.48	4.30	3.07	4.76	1.21	0.14	0.16	0.15	3.20	2.71	7.40	14.32	53.90
1971	11.15	4.41	9.11	2.78	1.50	3.00	0.78	0.71	3.06	4.43	7.59	9.18	57.70
1972	12.43	11.06	10.01	5.87	0.83	1.07	1.72	0.70	5.04	0.85	4.17	10.66	64.41
1973	5.66	1.71	3.02	2.23	2.66	2.60	0.05	0.59	2.18	4.60	12.95	11.61	49.86
1974	10.57	5.68	6.45	4.77	2.65	1.57	2.29	0.07	0.50	1.38	7.44	8.86	52.43
1975	9.70	5.61	4.32	1.88	1.51	0.72	0.25	3.97	T	8.38	9.54	11.42	57.30
1976	9.40	7.25	4.24	2.79	2.66	1.07	1.26	2.71	1.22	2.64	1.37	3.00	39.61
1977	1.55	3.70	4.44	1.27	5.21	0.64	0.32	4.17	4.58	3.40	9.30	12.36	50.94
1978	6.61	4.39	3.45	3.97	2.90	1.54	1.49	1.47	7.59	0.78	6.81	2.95	43.95
1979	2.67	9.25	2.73	2.21	1.40	1.11	1.58	1.56	2.72	6.20	2.38	13.01	46.82
1980	6.27	6.18	3.98	4.14	0.93	2.46	0.41	0.36	2.21	1.84	9.77	10.92	49.47
1981	2.54	9.19	3.86	4.77	1.81	2.96	0.36	0.85	2.44	8.17	7.22	8.60	52.77
1982	7.79	9.91	4.67	3.89	0.51	1.15	0.54	0.55	1.89	6.12	5.74	10.64	53.40
1983	9.99	7.09	6.63	2.26	1.51	2.80	3.00	2.18	1.89	1.66	12.84	7.28	59.13
1984	6.97	5.49	6.42	3.67	5.48	3.74	T	0.21	1.76	5.13	12.19	4.97	56.03
1985	0.29	3.54	4.10	2.65	0.94	2.48	0.37	0.70	2.65	9.86	4.99	2.50	35.07
Record Mean	7.64	6.17	4.99	3.21	1.90	1.69	0.80	1.20	2.24	5.01	7.90	8.25	51.00

TABLE 3   AVERAGE TEMPERATURE (deg. F)   OLYMPIA WASHINGTON

YEAR	JAN	FEB	MAR	APR	MAY	JUNE	JULY	AUG	SEP	OCT	NOV	DEC	ANNUAL
1956	38.0	35.0	40.6	49.4	57.3	56.7	64.5	63.3	57.0	48.8	41.5	40.3	49.4
1957	32.2	39.0	43.9	50.6	57.6	60.0	61.5	62.0	62.6	50.7	43.0	42.4	50.5
1958	42.7	47.0	43.4	48.7	59.3	63.9	69.1	66.3	58.5	52.4	43.1	43.8	53.2
#1959	39.7	40.5	43.4	48.3	52.3	59.2	64.7	60.8	56.9	51.2	42.0	38.7	49.8
1960	36.5	40.2	43.6	49.1	52.1	58.4	64.7	61.4	56.5	51.1	41.7	37.8	49.4
1961	41.8	44.1	44.5	47.0	52.5	60.8	64.3	66.1	56.0	48.4	39.1	38.6	50.3
1962	36.5	40.8	40.7	48.9	49.6	57.8	61.9	62.0	58.3	51.0	45.6	41.2	49.5
1963	33.0	46.2	42.1	46.8	53.2	56.6	62.0	63.4	62.6	52.4	44.8	41.6	50.4
1964	39.8	39.0	42.1	44.6	49.9	56.1	61.1	60.1	55.9	50.3	41.0	34.4	47.8
1965	38.9	41.5	43.8	49.0	51.6	59.1	65.9	65.2	54.8	53.3	46.6	38.3	50.7
1966	39.4	40.6	43.8	49.1	52.6	56.3	60.9	61.8	59.4	49.1	44.0	42.1	49.9
1967	40.7	41.3	41.0	44.8	53.6	62.2	63.9	66.8	62.3	52.5	43.8	39.4	51.2
1968	38.3	43.9	46.2	47.0	54.0	58.3	64.8	61.7	57.7	48.6	44.5	35.6	50.1
1969	30.9	38.2	44.4	48.1	56.2	64.0	62.1	60.6	58.9	48.9	44.3	40.3	49.7
1970	38.1	42.7	43.4	45.1	51.8	60.8	64.1	62.1	55.1	47.6	42.4	36.8	49.2
1971	38.6	39.2	40.0	46.0	53.7	54.8	63.5	65.4	55.3	48.4	43.6	36.0	48.7
1972	34.0	38.8	45.3	43.8	54.9	58.0	62.8	64.2	55.1	46.5	43.0	35.5	48.5
1973	36.5	40.4	43.5	47.2	53.9	57.9	62.7	59.5	58.8	49.8	39.8	40.5	49.2
1974	37.4	41.9	44.7	46.6	51.8	59.6	62.4	64.7	63.2	49.5	43.6	40.1	50.5
1975	38.4	39.4	40.5	44.0	52.5	57.5	63.9	60.7	57.8	49.1	41.3	38.6	48.6
1976	40.1	39.3	40.3	46.0	51.5	54.9	61.7	60.7	58.9	49.6	43.2	39.1	48.8
1977	35.7	45.2	43.4	50.2	51.4	59.8	59.7	64.6	56.7	48.5	40.4	38.0	49.5
1978	41.8	43.6	45.9	48.1	52.9	63.1	65.2	63.5	56.8	50.6	37.9	34.1	50.3
1979	30.9	40.3	47.5	49.6	54.7	58.7	65.1	62.8	59.3	51.7	40.2	41.3	50.2
1980	31.6	39.8	42.0	48.9	52.8	56.3	62.9	62.3	59.6	51.9	46.5	43.7	49.9
1981	42.1	43.5	47.1	49.0	53.9	57.8	62.7	67.4	60.0	49.7	45.3	39.7	51.5
1982	38.0	40.6	42.6	45.8	54.1	62.9	63.1	64.4	59.5	50.8	40.2	38.6	50.1
1983	42.6	45.3	47.5	48.9	56.5	58.7	62.0	64.1	56.3	49.2	46.0	33.9	50.9
1984	41.7	43.0	46.7	46.7	51.6	57.8	63.3	63.8	58.2	48.4	43.3	35.4	50.0
1985	35.7	37.4	42.1	48.9	54.5	59.0	67.5	63.5	56.6	49.2	33.8	34.0	48.5
Record Mean	37.2	40.9	43.1	47.6	53.6	58.7	63.3	63.0	58.2	50.2	42.8	38.8	49.8
Max	43.7	49.0	52.9	58.9	66.0	70.8	77.3	76.7	71.1	60.5	50.3	44.9	60.2
Min	30.7	32.7	33.3	36.2	41.2	46.5	49.2	49.3	45.3	39.8	35.2	32.7	39.3

REFERENCE NOTES FOR TABLES 1, 2, 3 and 6   (OLYMPIA, WA)

GENERAL

T - TRACE AMOUNT
BLANK ENTRIES DENOTE MISSING/UNREPORTED DATA.
# INDICATES A STATION OR INSTRUMENT RELOCATION.

SPECIFIC

TABLE 1

(a) - LENGTH OF RECORD IN YEARS. ALTHOUGH
INDIVIDUAL MONTHS MAY BE MISSING.
* LESS THAN .05

NORMALS — BASED ON THE 1951-1980 RECORD PERIOD.
EXTREMES — DATES ARE THE MOST RECENT OCCURRENCE.
WIND DIR. — NUMERALS SHOW TENS OF DEGREES
CLOCKWISE FROM TRUE NORTH.
"00" INDICATES CALM.
RESULTANT WIND DIRECTIONS ARE GIVEN TO WHOLE DEGREES.

EXCEPTIONS

TABLE 1

1. MEAN WIND SPEED, THUNDERSTORMS, AND HEAVY FOG
ARE THROUGH 1969 AND 1972 TO DATE.

TABLES 2, 3, and 6

RECORD MEANS ARE THROUGH THE CURRENT YEAR,
BEGINNING IN   1942 FOR TEMPERATURE
1942 FOR PRECIPITATION
1942 FOR SNOWFALL

TABLE 4   HEATING DEGREE DAYS Base 65 deg. F   OLYMPIA WASHINGTON

SEASON	JULY	AUG	SEP	OCT	NOV	DEC	JAN	FEB	MAR	APR	MAY	JUNE	TOTAL
1956-57	80	75	236	493	695	756	1007	724	647	425	224	152	5514
1957-58	105	91	82	438	654	696	685	497	665	484	187	87	4671
1958-59	7	21	201	383	652	650	777	680	660	495	386	181	5093
#1959-60	75	133	236	422	681	809	879	711	652	475	392	192	5657
1960-61	59	140	251	424	690	835	708	579	629	534	380	137	5366
1961-62	62	19	266	508	771	816	878	674	748	476	469	212	5899
1962-63	145	98	194	428	579	731	984	523	705	536	368	247	5538
1963-64	92	55	87	383	603	721	772	747	705	604	461	263	5493
1964-65	130	150	266	450	715	945	801	652	650	470	408	177	5814
1965-66	52	52	298	356	546	820	788	676	649	471	377	258	5343
1966-67	129	99	161	485	624	700	745	653	739	599	350	96	5380
1967-68	51	13	94	379	631	786	820	604	575	534	334	201	5022
1968-69	64	113	214	503	610	903	1048	746	633	498	274	76	5682
1969-70	95	137	183	494	616	759	826	615	665	591	403	149	5533
1970-71	65	102	293	535	672	868	810	716	771	560	344	304	6040
1971-72	113	49	286	500	639	891	954	752	602	632	315	204	5937
1972-73	85	69	295	570	651	909	874	681	659	528	345	223	5889
1973-74	97	170	184	465	748	749	852	644	625	544	403	161	5642
1974-75	103	51	78	472	634	763	817	711	751	624	377	225	5606
1975-76	73	138	210	486	703	811	766	739	760	563	410	296	5955
1976-77	104	127	177	470	649	798	902	547	660	438	415	158	5445
1977-78	166	90	242	502	734	829	829	596	587	502	367	91	5417
1978-79	55	78	240	442	803	952	1048	683	536	454	314	188	5793
1979-80	59	71	165	404	737	726	1027	724	702	476	371	254	5716
1980-81	81	96	158	399	549	653	701	597	551	474	336	212	4807
·1981-82	100	50	161	465	585	777	831	678	686	570	330	103	5336
1982-83	78	63	170	435	738	813	687	545	535	476	275	182	4997
1983-84	98	38	255	482	566	958	712	631	560	545	408	211	5464
1984-85	80	64	205	507	644	911	903	767	704	476	320	181	5762
1985-86	15	68	245	484	932	954							

TABLE 5   COOLING DEGREE DAYS Base 65 deg. F   OLYMPIA WASHINGTON

YEAR	JAN	FEB	MAR	APR	MAY	JUNE	JULY	AUG	SEP	OCT	NOV	DEC	TOTAL
1969	0	0	0	0	7	51	13	6	8	0	0	0	85
1970	0	0	0	0	0	29	44	16	0	0	0	0	89
1971	0	0	0	0	0	3	71	67	0	0	0	0	141
1972	0	0	0	0	8	0	25	50	5	0	0	0	88
1973	0	0	0	0	6	14	35	7	5	0	0	0	67
1974	0	0	0	0	0	6	30	50	32	0	0	0	118
1975	0	0	0	0	0	4	45	9	1	0	0	0	59
1976	0	0	0	0	0	2	9	3	0	0	0	0	14
1977	0	0	0	0	0	14	7	87	1	0	0	0	109
1978	0	0	0	0	1	43	64	38	0	0	0	0	146
1979	0	0	0	0	0	6	73	12	2	0	0	0	93
1980	0	0	0	0	0	0	21	19	3	2	0	0	45
1981	0	0	0	0	0	2	35	131	18	0	0	0	186
1982	0	0	0	0	0	46	24	52	10	0	0	0	132
1983	0	0	0	0	18	1	10	18	0	0	0	0	47
1984	0	0	0	0	0	1	34	37	6	1	0	0	79
1985	0	0	0	0	3	7	97	30	0	0	0	0	137

TABLE 6   SNOWFALL (inches)   OLYMPIA WASHINGTON

SEASON	JULY	AUG	SEP	OCT	NOV	DEC	JAN	FEB	MAR	APR	MAY	JUNE	TOTAL
1956-57	0.0	0.0	0.0	T	0.0	1.1	12.4	5.5	T	0.0	0.0	0.0	19.0
1957-58	0.0	0.0	0.0	0.0	0.0	T	5.5	14.1	1.0	0.9	0.0	0.0	21.1
1958-59	0.0	0.0	0.0	0.0	0.5	0.0	7.1	8.4	T	0.0	0.0	0.0	16.2
1959-60	0.0	0.0	0.0	0.0	0.0	0.7	7.1	T	0.2	0.5	0.0	0.0	8.7
1960-61	0.0	0.0	0.0	0.0	8.0	0.0	0.0	0.0	0.0	0.0	0.0	0.0	8.0
1961-62	0.0	0.0	0.0	0.0	0.0	1.7	2.1	7.5	4.2	0.0	0.0	0.0	15.5
1962-63	0.0	0.0	0.0	0.0	T	T	2.0	T	T	0.0	0.0	0.0	2.0
1963-64	0.0	0.0	0.0	0.0	0.1	0.0	2.4	0.0	1.0	0.0	0.0	0.0	3.5
1964-65	0.0	0.0	0.0	0.0	1.3	16.5	14.8	T	0.2	0.0	0.0	0.0	32.8
1965-66	0.0	0.0	0.0	0.0	0.0	14.0	10.7	2.0	15.5	0.1	0.0	0.0	42.3
1966-67	0.0	0.0	0.0	0.0	0.0	1.7	3.3	T	1.1	T	0.0	0.0	6.1
1967-68	0.0	0.0	0.0	0.0	0.0	7.9	16.9	0.0	0.0	T	0.0	0.0	24.8
1968-69	0.0	0.0	0.0	0.0	0.0	21.4	58.7	1.4	0.0	0.0	0.0	0.0	81.5
1969-70	0.0	0.0	0.0	0.0	0.0	3.0	T	0.0	T	0.1	T	0.0	3.1
1970-71	0.0	0.0	0.0	T	0.4	9.0	15.4	5.1	6.1	0.1	T	T	36.1
1971-72	0.0	0.0	0.0	T	T	17.2	29.2	2.4	3.6	2.2	0.0	0.0	54.6
1972-73	0.0	0.0	T	0.0	0.0	5.8	2.9	0.0	T	0.0	T	0.0	8.7
1973-74	0.0	0.0	0.0	0.0	1.9	2.6	4.1	1.1	T	T	T	0.0	9.7
1974-75	0.0	0.0	0.0	0.0	T	13.5	7.2	1.5	0.3	T	T	T	22.5
1975-76	0.0	0.0	0.0	T	4.5	3.7	0.6	7.2	T	T	T	T	16.0
1976-77	0.0	0.0	0.0	0.0	0.0	0.0	1.8	T	0.8	T	T	0.0	2.6
1977-78	0.0	0.0	0.0	0.0	3.8	1.6	0.0	0.0	T	0.0	0.0	0.0	5.4
1978-79	0.0	0.0	0.0	0.0	14.8	2.5	0.1	1.5	0.0	0.0	0.0	0.0	18.9
1979-80	0.0	0.0	0.0	0.0	T	T	19.7	11.9	4.1	T	0.0	0.0	35.7
1980-81	0.0	0.0	0.0	0.0	T	3.0	0.0	0.9	0.0	0.4	0.0	0.0	4.3
1981-82	0.0	0.0	0.0	0.0	0.0	2.8	11.8	5.1	1.9	T	0.0	0.0	21.6
1982-83	0.0	0.0	0.0	0.0	T	T	0.0	0.0	0.0	T	0.0	0.0	T
1983-84	0.0	0.0	0.0	0.0	0.0	4.9	T	0.0	0.0	T	T	0.0	4.9
1984-85	0.0	0.0	0.0	0.0	0.0	4.7	T	13.8	T	T	T	0.0	18.5
1985-86	0.0	0.0	0.0	0.0	12.3	7.3							
Record Mean	0.0	0.0	T	T	1.7	3.8	8.1	2.8	1.9	0.1	T	T	18.4

See Reference Notes, relative to all above tables, on preceding page.

Quillayute Airport, located on the coastal plain between the Pacific Ocean and the Olympic Mountains, is 3 miles inland from the coast, and 10 miles west of the city of Forks. The terrain is slightly rolling with a gradual increase in elevation from sea level to 180 feet at the station, to 350 feet in the vicinity of Forks. Foothills of the Olympic Mountains begin near the eastern edge of Forks, and within 10 to 15 miles, the higher ridges reach elevations of 3,000 to 6,000 feet.

Timber is the primary economic product of this northwestern section of the Olympic Peninsula. Only small areas of the Quillayute plains and a few other localities are devoted to cattle raising and agriculture. Logging operations continue throughout the year in the lower elevations with little delay due to normal rain or snow. In the foothills and mountains, heavy snowfall, excessive precipitation, and high winds during the winter season result in shutdowns, but seldom for more than one or two days. Forests are closed to logging and recreation for short periods of time almost every summer when the relative humidity is low and the fire danger is high.

Maritime air from over the Pacific has an influence on the climate throughout the year. In the late fall and winter, the low pressure center in the Gulf of Alaska intensifies and is of major importance in controlling weather systems entering the Pacific Northwest. At this season of the year, storm systems crossing the Pacific follow a more southerly path striking the coast at frequent intervals. The prevailing flow of air is from the southwest and west. Air reaching this area is moist and near the temperature of the ocean water along the coast which ranges from 45 degrees in February to 57 degrees in August. The wet season begins in September or October. From October through January, rain may be expected on about 26 days per month, from February through March, on 20 days, from April to June, on 15 days, and from July to September, on 10 days. As the weather systems move inland, rainfall is usually of moderate intensity and continuous, rather than heavy downpours for brief periods. Gale force winds are not unusual. Most of the winter precipitation over the coastal plains falls as rain, however, snow can be expected each year. Snow seldom reaches depths in excess of 10 inches or remains on the ground longer than two weeks.

Annual precipitation increases from approximately 90 inches near the coast, to amounts in excess of 120 inches over the coastal plains, to 200 inches or more on the wettest slopes of the Olympic Mountains.

During the rainy season, temperatures show little diurnal or day-to-day change. Maximums are in the 40s and minimums in the mid-30s. A few brief outbreaks of cold air from the interior of Canada can be expected each winter. Clear, dry, cold weather generally prevails during periods of easterly winds.

In the late spring and summer, a clockwise circulation of air around the large high pressure center over the north Pacific brings a prevailing northwesterly and westerly flow of cool, comparatively dry, stable air into the northwest Olympic Peninsula. The dry season begins in May with the driest period between mid-July and mid-August. The total rainfall for July is less than .5 inch in one summer out of ten. It also exceeds 5 inches in one summer out of ten. During the warmest months, afternoon temperatures are in the upper 60s and lower 70s, reaching the upper 70s and the lower 80s on a few days. Occasionally, hot, dry air from the east of the Cascade Mountains reaches this area and temperatures are in the mid- or upper-90s for one to three days.

In summer and early fall, fog or low clouds form over the ocean and frequently move inland at night, but generally disappear by midday. In winter, under the influence of a surface high pressure system, centered off the coast, fog, low clouds, and drizzle are a daily occurrence as long as this type of pressure continues.

TABLE 1   NORMALS, MEANS AND EXTREMES

QUILLAYUTE AIRPORT WASHINGTON

LATITUDE: 47°57'N   LONGITUDE: 124°33'W   ELEVATION: FT. GRND   179 BARO   01816   TIME ZONE: PACIFIC   WBAN: 94240

	(a)	JAN	FEB	MAR	APR	MAY	JUNE	JULY	AUG	SEP	OCT	NOV	DEC	YEAR
TEMPERATURE °F:														
Normals														
-Daily Maximum		44.8	48.5	49.9	54.7	60.2	63.6	68.6	68.6	66.8	59.1	50.7	46.4	56.8
-Daily Minimum		33.2	34.8	34.4	37.2	41.7	46.5	49.4	49.7	46.7	41.6	36.9	35.1	40.6
-Monthly		39.0	41.7	42.2	46.0	51.0	55.0	59.0	59.2	56.8	50.4	43.8	40.8	48.7
Extremes														
-Record Highest	19	65	72	70	79	89	96	95	99	92	81	69	64	99
-Year		1981	1968	1979	1979	1973	1982	1979	1981	1972	1966	1976	1969	AUG 1981
-Record Lowest	19	7	15	19	24	29	33	38	36	28	24	5	7	5
-Year		1969	1972	1971	1975	1977	1976	1977	1985	1972	1984	1985	1972	NOV 1985
NORMAL DEGREE DAYS:														
Heating (base 65°F)		806	652	707	570	434	301	194	190	252	453	636	750	5945
Cooling (base 65°F)		0	0	0	0	0	0	8	10	6	0	0	0	24
% OF POSSIBLE SUNSHINE	19	22	27	32	36	36	35	43	43	44	33	21	18	33
MEAN SKY COVER (tenths)														
Sunrise - Sunset	19	8.4	8.3	7.9	8.0	7.8	7.7	6.6	6.7	6.5	7.3	8.1	8.2	7.6
MEAN NUMBER OF DAYS:														
Sunrise to Sunset														
-Clear	19	3.1	3.0	3.5	2.9	2.5	3.3	6.6	6.1	7.2	5.2	2.9	3.7	50.1
-Partly Cloudy	19	2.9	3.5	5.8	5.8	7.9	7.1	8.1	8.9	7.6	6.8	5.3	3.8	73.5
-Cloudy	19	25.0	21.8	21.6	21.3	20.6	19.6	16.3	16.0	15.2	19.1	21.7	23.4	241.6
Precipitation														
.01 inches or more	19	22.0	20.1	20.9	19.7	16.2	14.3	11.4	10.8	13.4	18.6	21.6	23.2	212.2
Snow,Ice pellets														
1.0 inches or more	19	2.1	1.0	0.6	0.1	0.0	0.0	0.0	0.0	0.0	0.0	0.4	1.2	5.4
Thunderstorms	18	0.7	0.2	0.5	0.5	0.2	0.1	0.6	0.3	0.4	1.2	1.7	1.1	7.4
Heavy Fog Visibility														
1/4 mile or less	18	4.3	2.9	2.7	2.1	2.9	2.5	4.4	7.2	6.3	5.6	3.7	3.8	48.4
Temperature °F														
-Maximum														
90° and above	19	0.0	0.0	0.0	0.0	0.0	0.0	0.2	0.5	0.1	0.0	0.0	0.0	1.3
32° and below	19	0.8	0.0	0.0	0.0	0.0	0.0	0.0	0.0	0.0	0.0	0.4	0.9	2.1
-Minimum														
32° and below	19	14.3	10.3	11.2	6.6	0.6	0.0	0.0	0.0	0.3	3.8	8.4	13.3	68.7
0° and below	19	0.0	0.0	0.0	0.0	0.0	0.0	0.0	0.0	0.0	0.0	0.0	0.0	0.0
AVG. STATION PRESS.(mb)	13	1009.7	1007.1	1007.1	1010.0	1010.6	1011.0	1011.3	1010.0	1009.6	1009.5	1007.2	1009.1	1009.4
RELATIVE HUMIDITY (%)														
Hour 04	19	91	92	92	92	93	93	94	95	94	93	92	91	93
Hour 10 (Local Time)	19	89	88	82	75	73	74	74	76	77	85	88	90	81
Hour 16	19	83	78	72	67	66	66	64	67	68	75	82	86	73
Hour 22	19	90	90	89	89	87	88	88	91	90	92	91	91	90
PRECIPITATION (inches):														
Water Equivalent														
-Normal		15.07	12.10	11.27	7.10	4.70	3.06	2.32	2.85	5.27	10.51	13.94	16.31	104.50
-Maximum Monthly	19	23.34	20.60	21.86	13.89	12.45	8.50	11.02	10.12	10.93	27.17	29.14	27.82	29.14
-Year		1971	1982	1974	1970	1974	1981	1983	1975	1969	1975	1983	1979	NOV 1983
-Minimum Monthly	19	1.22	5.09	7.37	2.94	1.05	0.40	0.36	0.45	0.36	2.30	4.41	3.63	0.36
-Year		1985	1970	1978	1972	1967	1985	1972	1972	1975	1972	1976	1985	JUL 1985
-Maximum in 24 hrs	19	8.32	5.07	4.23	2.77	3.54	2.01	6.45	3.12	4.13	5.54	5.36	6.76	8.32
-Year		1968	1982	1974	1968	1973	1978	1972	1975	1968	1975	1980	1972	JAN 1968
Snow,Ice pellets														
-Maximum Monthly	19	40.1	13.3	10.2	2.8	T	T			T	T	15.6	11.6	40.1
-Year		1969	1976	1971	1975	1985	1973			1972	1985	1985	1972	JAN 1969
-Maximum in 24 hrs	19	8.2	8.4	5.8	2.4	T	T			T	T	7.7	7.3	8.4
-Year		1969	1971	1976	1975	1985	1973			1972	1985	1985	1981	FEB 1971
WIND:														
Mean Speed (mph)	19	6.5	6.9	6.8	6.5	6.1	5.9	5.7	5.1	5.1	5.6	6.4	6.6	6.1
Prevailing Direction														
Fastest Mile														
-Direction (!!!)	19	S	SE	SE	SW	NE	S	NE	SE	SE	SE	SE	SW	SE
-Speed (MPH)	19	34	46	33	32	27	22	23	27	33	42	37	39	46
-Year		1968	1977	1974	1970	1972	1973	1972	1969	1972	1982	1973	1969	FEB 1977
Peak Gust														
-Direction (!!!)	2	SW	S	SW	S	S	S	S	S	NE	SE	SE	SE	S
-Speed (mph)	2	36	46	46	37	41	26	25	28	35	43	47	35	47
-Date		1984	1985	1985	1984	1985	1985	1985	1984	1985	1984	1984	1985	NOV 1984

See Reference Notes to this table on the following pages.

TABLE 2 PRECIPITATION (inches)   QUILLAYUTE AIRPORT WASHINGTON

YEAR	JAN	FEB	MAR	APR	MAY	JUNE	JULY	AUG	SEP	OCT	NOV	DEC	ANNUAL
1966								1.79	3.50	10.52	16.07	23.47	
1967	20.62	16.80	11.73	5.90	2.50	0.40	1.10	0.62	7.75	24.86	8.47	19.27	120.02
1968	22.59	12.89	15.17	8.15	4.33	4.58	1.74	4.61	8.46	17.46	13.54	15.90	129.42
1969	13.82	9.11	10.58	11.17	5.49	2.04	1.59	3.43	10.93	6.85	8.40	12.87	96.28
1970	13.67	5.09	7.43	13.89	0.55	1.20	2.45	0.55	7.21	9.09	11.32	16.55	91.27
1971	23.34	12.81	14.83	6.21	3.90	5.04	1.04	3.52	5.93	13.21	14.33	15.42	119.58
1972	12.92	15.51	15.38	11.46	1.05	1.49	9.33	0.45	6.91	2.30	10.88	22.20	109.88
1973	17.29	5.90	9.72	2.94	7.43	5.47	1.14	1.03	3.63	10.16	17.17	19.89	101.77
1974	19.91	17.45	21.86	9.26	12.45	3.14	5.42	0.51	2.71	3.13	19.02	127.76	
1975	14.25	11.89	9.61	5.14	5.68	3.02	0.43	10.12	0.36	27.17	24.28	19.69	131.64
1976	15.70	14.91	14.23	4.32	5.35	2.75	3.43	4.12	2.39	6.07	4.41	9.09	86.77
1977	8.26	11.22	10.82	3.88	8.87	2.10	1.92	3.15	6.28	9.21	18.10	15.03	98.84
1978	8.20	6.52	7.37	5.54	5.66	3.47	0.67	6.89	9.79	3.08	6.67	7.68	71.54
1979	3.64	19.13	8.39	5.41	4.00	2.58	2.18	0.84	8.64	8.86	7.53	27.82	99.0?
1980	6.30	13.10	10.52	8.11	3.15	1.85	4.69	2.00	7.74	4.41	22.17	18.31	102.35
1981	4.49	11.03	11.66	12.65	4.80	8.50	1.43	1.24	5.18	15.32	14.04	14.72	105.06
1982	19.34	20.60	9.10	9.72	1.55	2.39	1.53	2.54	3.81	13.61	12.71	12.25	109.15
1983	13.77	20.11	12.97	3.91	3.81	4.08	11.02	2.09	5.73	5.44	29.14	8.93	121.00
1984	18.27	13.77	9.79	8.89	10.56	2.83	0.55	1.61	3.69	13.91	14.65	13.89	112.41
1985	1.22	6.90	9.33	6.19	1.86	2.43	0.36	1.09	4.75	14.69	7.79	3.63	60.24
Record Mean	13.56	12.88	11.60	7.51	5.01	3.12	2.74	2.61	5.77	10.97	13.73	15.78	105.29

TABLE 3 AVERAGE TEMPERATURE (deg. F)   QUILLAYUTE AIRPORT WASHINGTON

YEAR	JAN	FEB	MAR	APR	MAY	JUNE	JULY	AUG	SEP	OCT	NOV	DEC	ANNUAL
1966								58.7	57.8	49.4	45.2	43.6	
1967	41.9	42.3	40.4	42.9	50.9	57.6	60.1	62.6	59.7	52.0	45.8	40.3	49.7
1968	40.2	44.7	45.6	45.1	51.5	54.8	60.6	58.6	56.2	49.0	45.6	36.7	49.0
1969	31.4	38.7	43.4	45.5	53.2	59.2	58.0	56.8	56.1	49.8	45.6	43.6	48.4
1970	40.3	45.6	43.7	43.8	49.7	57.0	58.3	58.4	54.8	49.4	43.8	39.0	48.4
1971	39.5	40.7	39.3	45.5	50.5	52.5	58.4	61.3	54.5	48.8	44.3	36.4	47.7
1972	35.7	40.1	43.5	42.7	53.7	55.8	60.5	61.3	52.5	48.4	44.5	37.5	48.0
1973	39.0	42.0	43.0	46.7	51.7	54.9	58.0	55.4	56.4	48.7	40.5	42.2	48.2
1974	38.1	40.4	42.0	46.5	48.2	55.0	57.9	59.4	60.4	50.6	45.0	43.6	48.9
1975	39.0	40.3	41.0	43.3	51.4	53.5	59.2	57.4	57.7	49.1	43.8	40.6	48.0
1976	41.9	40.0	39.9	46.7	49.9	52.9	58.4	58.1	57.9	51.0	45.8	43.8	48.8
1977	39.4	44.9	42.2	47.0	48.7	55.4	57.1	60.7	55.1	49.3	42.1	40.1	48.5
1978	42.9	43.9	45.4	47.5	50.7	54.7	59.9	60.4	55.9	51.7	40.4	35.9	49.4
1979	35.3	39.7	45.5	47.2	52.4	54.9	60.0	60.4	60.0	52.5	43.9	44.5	49.7
1980	35.7	44.4	43.4	48.1	50.5	54.2	59.7	56.7	56.6	52.3	46.3	43.9	49.3
1981	45.3	43.7	46.6	46.8	51.1	54.2	58.1	61.3	56.8	49.3	45.8	41.0	50.0
1982	37.8	40.6	41.5	43.1	49.8	56.3	57.1	59.5	57.7	50.3	41.2	39.4	47.9
1983	44.4	44.5	46.2	47.7	53.5	55.8	59.6	60.2	54.4	47.8	45.8	35.2	49.6
1984	42.0	40.6	46.4	45.7	49.4	53.2	58.6	58.9	55.1	47.9	43.5	36.7	48.5
1985	38.9	39.4	40.4	45.9	51.7	54.9	60.3	58.6	54.9	48.7	35.0	38.3	47.3
Record Mean	39.3	42.1	43.1	45.7	51.0	55.3	58.9	59.2	56.5	49.8	43.7	40.1	48.7
Max	45.6	49.2	51.2	54.6	60.1	64.0	68.5	68.9	66.7	58.8	50.6	46.1	57.0
Min	33.1	35.0	34.9	36.7	41.8	46.6	49.3	49.5	46.3	40.7	36.8	34.1	40.4

REFERENCE NOTES FOR TABLES 1, 2, 3 and 6   (QUILLAYUTE AIRPORT, WA)

GENERAL

T - TRACE AMOUNT
BLANK ENTRIES DENOTE MISSING/UNREPORTED DATA.
# INDICATES A STATION OR INSTRUMENT RELOCATION.

SPECIFIC

TABLE 1

(a) - LENGTH OF RECORD IN YEARS. ALTHOUGH
    INDIVIDUAL MONTHS MAY BE MISSING.
* LESS THAN .05

NORMALS — BASED ON THE 1951-1980 RECORD PERIOD.
EXTREMES — DATES ARE THE MOST RECENT OCCURRENCE.
WIND DIR. — NUMERALS SHOW TENS OF DEGREES
        CLOCKWISE FROM TRUE NORTH.
        "00" INDICATES CALM.
RESULTANT WIND DIRECTIONS ARE GIVEN TO WHOLE DEGREES.

EXCEPTIONS

TABLES 2, 3, and 6

RECORD MEANS ARE THROUGH THE CURRENT YEAR,
BEGINNING IN   1966 FOR TEMPERATURE
        1966 FOR PRECIPITATION
        1966 FOR SNOWFALL

TABLE 4 HEATING DEGREE DAYS Base 65 deg. F QUILLAYUTE AIRPORT WASHINGTON

SEASON	JULY	AUG	SEP	OCT	NOV	DEC	JAN	FEB	MAR	APR	MAY	JUNE	TOTAL
1966-67		191	208	479	586	656	709	628	755	655	431	214	5364
1967-68	153	83	163	396	570	760	761	580	596	589	412	301	6087
1968-69	147	195	261	487	576	871	1036	729	661	578	358	188	5690
1969-70	208	248	259	462	574	758	758	536	655	628	467	239	5690
1970-71	208	197	298	475	628	799	783	673	787	578	445	372	6243
1971-72	201	126	305	493	614	879	903	716	660	663	354	269	6183
1972-73	160	131	379	508	608	845	800	638	675	543	409	299	5995
1973-74	217	291	250	499	728	701	828	681	705	548	513	295	6256
1974-75	223	168	159	438	595	656	797	684	738	643	417	338	5856
1975-76	178	230	214	486	628	750	711	718	770	542	460	357	6044
1976-77	196	207	207	427	569	647	784	554	703	530	499	283	5606
1977-78	240	142	290	479	679	767	680	585	601	516	434	187	5600
1978-79	170	154	266	404	732	892	913	703	600	526	383	294	6037
1979-80	157	137	147	382	625	628	905	591	661	500	443	319	5495
1980-81	167	250	244	386	554	647	603	591	565	542	422	314	5285
1981-82	210	148	248	480	570	739	838	679	722	648	465	257	6004
1982-83	237	170	215	450	706	786	635	565	577	514	353	266	5474
1983-84	163	146	313	526	570	915	707	601	569	571	478	347	5906
1984-85	199	184	291	525	637	871	802	712	753	564	408	301	6247
1985-86	145	190	296	498	894	820							

TABLE 5 COOLING DEGREE DAYS Base 65 deg. F QUILLAYUTE AIRPORT WASHINGTON

YEAR	JAN	FEB	MAR	APR	MAY	JUNE	JULY	AUG	SEP	OCT	NOV	DEC	TOTAL
1969	0	0	0	0	2	18	0	1	0	0	0	0	21
1970	0	0	0	0	0	7	8	1	0	0	0	0	16
1971	0	0	0	0	0	3	5	18	0	0	0	0	26
1972	0	0	0	0	11	0	31	21	11	0	0	0	74
1973	0	0	0	0	2	2	7	0	1	0	0	0	12
1974	0	0	0	0	0	0	10	3	27	0	0	0	40
1975	0	0	0	0	0	0	5	0	1	0	0	0	6
1976	0	0	0	0	0	0	0	0	0	0	0	0	0
1977	0	0	0	0	0	1	1	15	0	0	0	0	17
1978	0	0	0	0	0	7	18	17	0	0	0	0	42
1979	0	0	0	0	0	0	11	1	3	0	0	0	15
1980	0	0	0	0	0	0	11	0	0	0	0	0	11
1981	0	0	0	0	0	0	1	38	7	0	0	0	46
1982	0	0	0	0	0	10	0	4	2	0	0	0	16
1983	0	0	0	0	4	0	0	6	0	0	0	0	10
1984	0	0	0	0	0	0	5	1	0	0	0	0	6
1985	0	0	0	0	1	6	6	1	0	0	0	0	14

TABLE 6 SNOWFALL (inches) QUILLAYUTE AIRPORT WASHINGTON

SEASON	JULY	AUG	SEP	OCT	NOV	DEC	JAN	FEB	MAR	APR	MAY	JUNE	TOTAL
1966-67	0.0	0.0	0.0	T	0.0	0.1	T	T	2.7	T	0.0	0.0	9.2
1967-68	0.0	0.0	0.0	0.0	T	1.8	6.8	0.0	T	0.6	0.0	0.0	50.7
1968-69	0.0	0.0	0.0	0.0	T	4.0	40.1	6.2	T	0.4	0.0	0.0	50.7
1969-70	0.0	0.0	0.0	0.0	0.1	0.0	T	0.0	0.4	1.1	0.0	0.0	1.6
1970-71	0.0	0.0	0.0	0.0	1.7	7.0	26.4	12.8	10.2	0.6	T	0.0	58.7
1971-72	0.0	0.0	0.0	T	0.0	8.3	13.7	1.0	0.2	2.5	0.0	0.0	25.7
1972-73	0.0	0.0	T	T	0.0	11.6	1.2	0.0	0.4	T	T	T	13.2
1973-74	0.0	0.0	0.0	0.0	4.4	T	2.4	6.7	8.6	0.0	T	0.0	22.1
1974-75	0.0	0.0	0.0	0.0	T	0.8	7.7	3.0	T	2.8	0.0	0.0	14.3
1975-76	0.0	0.0	0.0	0.0	3.8	2.3	0.3	13.3	8.7	0.1	0.0	0.0	28.5
1976-77	0.0	0.0	0.0	0.0	0.0	0.0	0.0	T	T	T	T	0.0	T
1977-78	0.0	0.0	0.0	0.0	T	4.5	1.4	T	T	T	0.0	0.0	5.9
1978-79	0.0	0.0	0.0	0.0	T	1.6	T	0.5	T	0.0	0.0	0.0	2.1
1979-80	0.0	0.0	0.0	0.0	0.0	0.0	11.0	T	2.1	T	T	0.0	13.1
1980-81	0.0	0.0	0.0	0.0	0.0	7.5	T	T	T	0.6	0.0	0.0	8.1
1981-82	0.0	0.0	0.0	0.0	T	8.3	13.1	7.2	0.6	0.8	T	0.0	30.0
1982-83	0.0	0.0	0.0	T	T	T	T	T	T	T	T	0.0	T
1983-84	0.0	0.0	0.0	0.0	T	0.9	0.0	T	4.7	0.4	T	0.0	0.9
1984-85	0.0	0.0	0.0	T	T	6.7	0.0	0.0	T	T	T	0.0	11.8
1985-86	0.0	0.0	0.0	T	15.6	T							
Record Mean	0.0	0.0	T	T	1.3	3.3	6.5	2.9	1.8	0.5	T	T	16.3

See Reference Notes, relative to all above tables, on preceding page.

SEATTLE-TACOMA, WASHINGTON

The Seattle-Tacoma International Airport is located 6 miles south of the Seattle city limits and 14 miles north of Tacoma. It is situated on a low ridge lying between Puget Sound on the west and the Green River valley on the east with terrain sloping moderately to the shores of Puget Sound some 2 miles to the west. The Olympic Mountains, rising sharply from Puget Sound, are about 50 miles to the northwest. Rather steep bluffs border the Green River Valley about 2.5 miles to the east and the foothills of the Cascade Range begin 10 to 15 miles to the east of the airport.

The mild climate of the Pacific Coast is modified by the Cascade Mountains and, to a lesser extent, by the Olympic Mountains. The climate is characterized by mild temperatures, a pronounced though not sharply defined rainy season, and considerable cloudiness, particularly during the winter months. The Cascades are very effective in shielding the Seattle-Tacoma area from the cold, dry continental air during the winter and the hot, dry continental air during the summer months. The extremes of temperature that occur in western Washington are the result of the occasional pressure distributions that force the continental air into the Puget Sound area. But the prevailing southwesterly circulation keeps the average winter daytime temperatures in the 40s and the nighttime readings in the 30s. During the summer, daytime temperatures are usually in the 70s with nighttime lows in the 50s. Extremes of temperatures, both in the winter and summer, are usually of short duration. The dry season is centered around July and early August with July being the driest month of the year. The rainy season extends from October to March with December normally the wettest month, however, precipitation is rather evenly distributed through the winter and early spring months with more than 75 percent of the yearly precipitation falling during the winter wet season. Most of the rainfall in the Seattle area comes from storms common to the middle latitudes. These disturbances are most vigorous during the winter as they move through western Washington. The storm track shifts to the north during the summer and those that reach the State are not the wind and rain producers of the winter months. Local summer afternoon showers and a few thunderstorms occur in the Seattle-Tacoma area but they do not contribute materially to the precipitation.

The occurrence of snow in the Seattle-Tacoma area is extremely variable and usually melts before accumulating measurable depths. There are winters on record with only a trace of snow, but at the other extreme, over 21 inches has fallen in a 24-hour period. Usually, winter storms do not produce snow unless the storm moves in such a way to bring cold air out of Canada directly or with only a short over water trajectory.

The highest winds recorded in the Seattle-Tacoma area were associated with strong storms crossing the state from the southwest. Prevailing winds are from the southwest but occasional severe winter storms will produce strong northerly winds. Winds during the summer months are relatively light with occasional land-sea breeze effects creating afternoon northerly winds of 8 to 15 miles an hour. Fog or low clouds that form over the southern Puget Sound area in the late summer, fall, and early winter months, often dominate the weather conditions during the late night and early morning hours with visibilities occasionally lower for a few hours near sunrise. Most of the summer clouds form along the coast and move into the Seattle area from the southwest.

Based on the 1951-1980 period, the average first occurrence of 32 degrees Fahrenheit in the fall is November 11 and the average last occurrence in the spring is March 24.

TABLE 1   NORMALS, MEANS AND EXTREMES

SEATTLE, WASHINGTON SEATTLE – TACOMA AIRPORT

LATITUDE: 47°27'N   LONGITUDE: 122°18'W   ELEVATION: FT. GRND   400 BARO   00451   TIME ZONE: PACIFIC   WBAN: 24233

	(a)	JAN	FEB	MAR	APR	MAY	JUNE	JULY	AUG	SEP	OCT	NOV	DEC	YEAR
TEMPERATURE °F:														
Normals														
–Daily Maximum		43.9	48.8	51.1	56.8	64.0	69.2	75.2	73.9	68.7	59.5	50.3	45.6	58.9
–Daily Minimum		34.3	36.8	37.2	40.5	46.0	51.1	54.3	54.3	51.2	45.3	39.3	36.3	43.9
–Monthly		39.1	42.8	44.2	48.7	55.0	60.2	64.8	64.1	60.0	52.4	44.8	41.0	51.4
Extremes														
–Record Highest	41	64	70	72	85	93	96	98	99	94	82	74	63	99
–Year		1981	1968	1947	1976	1963	1955	1979	1981	1981	1980	1949	1980	AUG 1981
–Record Lowest	41	0	1	11	29	28	38	43	44	35	28	6	6	0
–Year		1950	1950	1955	1975	1954	1952	1954	1955	1972	1949	1955	1968	JAN 1950
NORMAL DEGREE DAYS:														
Heating (base 65°F)		803	622	645	489	313	169	76	97	169	388	606	744	5121
Cooling (base 65°F)		0	0	0	0	0	25	70	70	19	0	0	0	184
% OF POSSIBLE SUNSHINE	19	25	37	49	52	56	54	65	64	59	43	29	21	46
MEAN SKY COVER (tenths)														
Sunrise - Sunset	41	8.5	8.3	7.9	7.7	7.1	7.0	5.3	5.8	6.2	7.5	8.3	8.7	7.4
MEAN NUMBER OF DAYS:														
Sunrise to Sunset														
–Clear	41	2.6	2.5	3.1	2.8	4.4	5.0	10.5	8.8	7.6	3.9	2.7	2.1	56.0
–Partly Cloudy	41	3.9	4.1	5.8	7.2	9.1	7.6	9.9	9.7	8.7	7.3	4.3	3.6	81.2
–Cloudy	41	24.4	21.6	22.0	19.9	17.5	17.3	10.3	12.6	13.7	19.7	23.0	25.4	227.6
Precipitation														
.01 inches or more	41	18.6	16.1	17.1	13.9	10.2	9.3	5.0	6.6	9.4	13.5	17.9	19.7	157.2
Snow,Ice pellets														
1.0 inches or more	41	1.8	0.5	0.5	0.*	0.0	0.0	0.0	0.0	0.0	0.*	0.4	1.0	4.2
Thunderstorms	41	0.2	0.3	0.6	0.9	0.9	0.7	0.7	0.8	0.8	0.3	0.6	0.4	7.2
Heavy Fog Visibility														
1/4 mile or less	41	5.8	3.4	2.3	1.1	0.9	0.8	1.8	2.9	5.6	7.8	6.0	6.4	44.8
Temperature °F														
–Maximum														
90° and above	26	0.0	0.0	0.0	0.0	0.2	0.3	1.2	1.2	0.2	0.0	0.0	0.0	3.0
32° and below	26	1.3	0.*	0.0	0.0	0.0	0.0	0.0	0.0	0.0	0.0	0.3	1.2	2.8
–Minimum														
32° and below	26	10.3	4.9	3.3	0.3	0.0	0.0	0.0	0.0	0.0	0.0	4.1	8.7	31.9
0° and below	26	0.0	0.0	0.0	0.0	0.0	0.0	0.0	0.0	0.0	0.0	0.0	0.0	0.0
AVG. STATION PRESS.(mb)	13	1001.6	999.3	998.5	1000.9	1001.3	1001.3	1001.5	1000.5	1000.6	1001.2	999.4	1001.2	1000.6
RELATIVE HUMIDITY (%)														
Hour 04	26	80	80	81	83	82	81	81	83	86	86	83	82	82
Hour 10 (Local Time)	26	78	76	74	71	67	66	65	69	74	78	79	80	73
Hour 16	26	74	67	62	57	54	53	49	51	58	67	74	77	62
Hour 22	26	77	76	75	73	71	69	67	71	76	81	80	80	75
PRECIPITATION (inches):														
Water Equivalent														
–Normal		6.04	4.22	3.59	2.40	1.58	1.38	0.74	1.27	2.02	3.43	5.60	6.33	38.60
–Maximum Monthly	41	12.92	9.11	8.40	4.19	4.76	3.90	2.39	4.59	5.95	8.95	9.69	11.85	12.92
–Year		1953	1961	1950	1978	1948	1946	1983	1975	1978	1947	1963	1979	JAN 1953
–Minimum Monthly	41	0.58	1.58	0.57	0.33	0.35	0.13	T	0.01	T	0.72	0.74	1.37	T
–Year		1985	1977	1965	1956	1947	1951	1960	1974	1975	1972	1976	1978	SEP 1975
–Maximum in 24 hrs	41	2.41	3.41	2.86	1.85	1.83	2.08	0.85	1.75	2.23	3.74	3.41	2.61	3.74
–Year		1967	1951	1972	1965	1969	1985	1981	1968	1978	1981	1959	1979	OCT 1981
Snow,Ice pellets														
–Maximum Monthly	41	57.2	13.1	18.2	2.3	T		T		T	2.0	17.5	22.1	57.2
–Year		1950	1949	1951	1972	1974		1980		1972	1971	1985	1968	JAN 1950
–Maximum in 24 hrs	41	21.4	7.2	5.6	2.3	T		T		T	2.0	9.4	13.0	21.4
–Year		1950	1962	1951	1972	1974		1980		1972	1971	1946	1968	JAN 1950
WIND:														
Mean Speed (mph)	37	9.8	9.6	9.9	9.6	9.0	8.7	8.3	7.9	8.1	8.6	9.2	9.7	9.0
Prevailing Direction														
through 1963		SSW	SW	SSW	SW	SW	SW	SW	SW	N	S	S	SSW	SW
Fastest Mile														
–Direction	18	SW	S	SW	SW	SW	SW	SW	SW	S	SW	S	S	S
–Speed (MPH)	18	45	51	44	38	32	29	26	29	33	38	66	49	66
–Year		1971	1981	1984	1972	1968	1974	1981	1977	1981	1982	1981	1982	NOV 1981
Peak Gust														
–Direction	2	S	S	SW	SW	SW	S	SW	N	SW	S	SW	N	SW
–Speed (mph)	2	39	37	43	38	36	31	26	30	38	36	45	39	45
–Date		1984	1985	1985	1985	1985	1984	1984	1985	1984	1984	1984	1984	NOV 1984

See Reference Notes to this table on the following page.

TABLE 2 PRECIPITATION (inches) SEATTLE, WASHINGTON SEATTLE – TACOMA AIRPORT

YEAR	JAN	FEB	MAR	APR	MAY	JUNE	JULY	AUG	SEP	OCT	NOV	DEC	ANNUAL
1956	8.67	2.17	4.95	0.33	0.83	2.47	0.33	0.76	2.42	6.71	1.59	5.62	36.85
1957	2.41	5.57	6.26	2.23	1.17	1.18	1.10	1.64	0.76	3.79	3.00	5.52	34.63
1958	8.72	5.36	2.26	3.51	0.94	0.90	T	0.31	1.42	3.99	8.07	7.15	42.63
1959	7.98	3.64	4.12	3.59	1.60	1.82	0.93	0.60	4.60	2.67	8.14	6.83	46.52
1960	5.48	4.01	4.08	2.88	3.04	0.70	T	1.92	1.17	4.22	8.03	3.75	39.28
1961	7.71	9.11	4.46	2.35	3.07	0.54	0.75	0.82	0.46	3.27	4.67	5.32	42.53
1962	2.43	2.29	2.86	2.03	1.82	0.68	0.69	1.96	2.31	4.16	9.34	5.22	35.79
1963	2.25	4.36	3.43	3.06	0.90	1.68	1.18	0.73	5.06	5.06	9.69	5.79	38.72
1964	9.76	1.66	2.96	1.56	0.91	3.82	0.99	1.23	2.27	1.00	9.65	5.53	41.34
1965	5.27	3.88	3.73	3.73	1.63	0.59	0.38	2.18	0.49	2.76	4.98	7.10	33.56
1966	5.43	2.31	4.38	1.99	1.35	1.15	1.35	0.42	1.77	2.92	6.85	8.31	38.23
1967	9.32	2.72	3.71	2.50	0.38	2.04	0.01	0.02	0.94	6.66	2.56	4.72	35.58
1968	6.90	6.08	5.08	1.33	1.67	3.02	0.83	4.58	1.93	4.32	5.86	8.55	50.15
1969	5.71	3.16	2.20	3.45	2.93	0.91	0.27	0.45	5.57	1.19	2.21	5.68	33.73
1970	8.22	2.26	3.16	3.31	1.17	0.43	0.48	0.32	2.52	2.52	5.03	8.28	37.41
1971	5.32	4.36	7.12	2.39	1.43	2.28	0.68	0.57	3.51	3.57	5.31	6.67	43.21
1972	7.24	8.11	6.74	4.12	0.69	1.81	1.34	1.13	4.10	0.72	8.98	48.36	48.36
1973	4.29	1.89	1.62	1.35	1.60	2.50	0.08	0.27	1.81	3.31	7.99	8.33	35.04
1974	7.78	4.01	5.84	2.39	1.37	1.25	1.51	0.01	0.21	1.99	5.06	6.45	37.87
1975	6.01	5.80	2.87	2.49	1.13	0.84	0.27	4.59	T	7.75	5.07	7.66	44.48
1976	5.55	4.74	2.71	1.67	1.61	0.63	1.17	2.71	1.25	2.06	0.74	1.86	26.70
1977	1.77	1.58	3.80	0.55	3.70	0.54	0.42	3.59	2.55	2.60	5.27	6.47	32.84
1978	4.30	3.59	2.43	4.19	1.79	0.75	1.40	1.19	5.95	0.98	6.05	1.37	33.99
1979	2.25	5.32	1.55	0.81	0.88	0.46	0.73	1.02	2.07	3.38	1.94	11.85	32.26
1980	4.09	5.04	2.10	3.23	0.97	1.77	0.46	0.64	1.43	1.32	7.16	7.39	35.60
1981	2.42	4.45	2.23	1.58	1.33	2.31	1.38	0.25	3.42	6.40	4.07	5.56	35.40
1982	5.35	7.57	3.73	2.07	0.63	1.03	0.59	0.62	1.49	4.07	5.31	6.86	39.32
1983	7.07	4.57	3.81	1.06	2.10	1.85	2.39	1.90	1.85	1.34	7.97	5.02	40.93
1984	3.62	3.91	3.91	2.87	3.38	2.81	0.17	0.13	1.01	2.14	8.09	4.95	36.99
1985	0.58	2.63	2.56	1.30	0.85	2.80	0.10	0.55	1.98	5.74	4.26	1.78	25.13
Record Mean	5.58	4.45	3.66	2.36	1.61	1.53	0.79	1.15	2.10	3.69	5.71	6.13	38.75

TABLE 3 AVERAGE TEMPERATURE (deg. F) SEATTLE, WASHINGTON SEATTLE – TACOMA AIRPORT

YEAR	JAN	FEB	MAR	APR	MAY	JUNE	JULY	AUG	SEP	OCT	NOV	DEC	ANNUAL
1956	39.0	35.7	40.8	49.6	56.8	56.9	64.6	63.0	57.9	49.5	42.6	40.2	49.7
1957	32.6	39.5	44.4	50.1	57.9	60.4	61.8	61.6	62.7	50.7	43.5	43.7	50.7
1958	43.6	47.5	44.0	49.2	60.4	64.0	68.8	66.6	59.8	53.9	43.4	44.9	53.8
#1959	40.8	41.0	44.2	49.6	53.5	59.6	66.1	62.3	57.4	52.0	43.9	40.3	50.9
1960	38.9	41.2	43.5	49.3	52.7	59.5	66.7	62.2	58.0	52.7	44.5	39.5	50.8
1961	43.6	44.4	45.3	47.0	53.8	63.5	67.1	68.4	58.7	50.8	41.9	39.3	52.0
1962	38.4	43.1	43.3	50.0	50.6	59.9	63.5	62.0	59.6	52.6	46.5	42.3	51.0
1963	33.9	48.2	43.8	48.3	57.7	59.9	62.4	64.6	63.5	54.5	44.4	40.8	51.8
1964	40.0	41.3	44.1	46.8	53.2	57.9	63.5	62.6	63.5	53.5	37.1	36.4	50.0
1965	40.2	43.0	47.0	49.5	51.9	60.8	67.8	65.7	58.4	56.4	49.4	40.4	52.6
1966	41.1	43.9	45.1	50.0	54.5	58.7	62.1	64.5	61.5	51.4	43.5	43.5	51.8
1967	42.4	42.8	42.2	46.6	55.4	62.7	66.5	71.1	65.7	54.8	47.3	41.6	53.2
1968	40.9	48.5	48.6	48.7	57.3	60.7	67.0	63.7	59.1	51.5	46.8	36.6	52.5
1969	33.1	42.3	46.9	48.9	58.0	64.3	64.7	64.0	61.0	52.4	46.6	45.2	52.3
1970	41.2	47.0	46.0	46.1	54.7	62.7	64.9	64.5	58.6	50.8	46.5	39.0	51.8
1971	39.7	42.3	41.3	48.9	54.5	55.9	65.5	67.7	57.6	51.0	45.7	37.5	50.6
1972	37.0	41.4	46.9	47.0	58.3	60.1	66.0	66.7	55.4	50.1	46.7	38.1	51.1
1973	38.7	43.9	44.1	48.6	56.5	59.3	64.7	61.6	61.9	52.2	43.7	44.4	51.6
1974	38.7	43.2	46.3	50.3	54.9	62.4	64.0	64.4	64.4	52.5	45.1	42.4	52.4
1975	38.8	40.8	42.9	45.8	54.6	60.7	67.5	63.2	63.0	51.4	44.9	41.5	51.3
1976	41.8	40.9	41.3	49.5	56.4	60.0	65.9	64.1	62.6	54.9	47.8	44.7	52.5
1977	39.4	48.7	45.7	53.6	54.5	63.0	65.1	68.5	58.9	52.2	43.9	42.2	53.0
1978	44.4	46.0	48.6	49.9	54.5	64.3	65.8	65.5	58.8	54.3	41.2	37.5	52.6
1979	37.8	42.3	49.3	50.8	57.2	62.5	67.4	64.0	62.6	54.2	43.9	44.1	53.0
1980	34.8	43.8	44.3	51.6	54.2	57.5	63.8	61.9	59.6	53.9	46.7	44.1	51.4
1981	44.4	44.2	48.8	49.6	54.7	57.5	63.3	68.1	61.1	50.9	47.2	41.7	52.7
1982	39.3	42.1	44.1	47.4	54.7	63.1	62.8	65.1	60.6	52.7	43.2	40.8	51.3
1983	45.0	46.9	49.4	50.7	57.7	59.9	63.3	65.6	58.3	51.7	47.8	36.1	52.7
1984	43.2	44.8	48.5	48.7	52.9	60.8	65.0	64.9	59.9	49.7	44.8	36.8	51.5
1985	37.1	39.0	43.3	49.2	54.8	60.0	68.6	65.2	58.1	51.4	35.8	36.2	49.9
Record Mean	38.8	42.5	44.5	48.6	55.0	60.0	64.6	64.2	59.7	51.9	44.5	40.5	51.2
Max	43.8	48.6	51.5	56.8	64.1	69.1	75.1	74.1	68.5	59.0	50.1	45.3	58.8
Min	33.8	36.5	37.4	40.4	46.0	50.9	54.1	54.2	50.8	44.8	38.9	35.7	43.6

REFERENCE NOTES FOR TABLES 1, 2, 3 and 6   (SEATTLE, WA)

GENERAL

T - TRACE AMOUNT
BLANK ENTRIES DENOTE MISSING/UNREPORTED DATA.
# INDICATES A STATION OR INSTRUMENT RELOCATION.

SPECIFIC

TABLE 1

(a) - LENGTH OF RECORD IN YEARS. ALTHOUGH
        INDIVIDUAL MONTHS MAY BE MISSING.
 * LESS THAN .05

NORMALS — BASED ON THE 1951-1980 RECORD PERIOD.
EXTREMES — DATES ARE THE MOST RECENT OCCURRENCE.
WIND DIR. — NUMERALS SHOW TENS OF DEGREES
              CLOCKWISE FROM TRUE NORTH.
              ''00'' INDICATES CALM.
RESULTANT WIND DIRECTIONS ARE GIVEN TO WHOLE DEGREES.

EXCEPTIONS

TABLES 2, 3, and 6

RECORD MEANS ARE THROUGH THE CURRENT YEAR,
BEGINNING IN   1945 FOR TEMPERATURE
                  1945 FOR PRECIPITATION
                  1945 FOR SNOWFALL

TABLE 4   HEATING DEGREE DAYS Base 65 deg. F   SEATTLE, WASHINGTON SEATTLE – TACOMA AIRPORT

SEASON	JULY	AUG	SEP	OCT	NOV	DEC	JAN	FEB	MAR	APR	MAY	JUNE	TOTAL
1956-57	70	89	208	474	663	760	995	706	629	441	213	135	5383
1957-58	97	95	72	434	638	656	656	482	645	468	163	81	4487
1958-59	15	19	165	338	640	617	742	667	637	456	355	166	4817
#1959-60	59	87	217	396	628	758	804	684	660	465	373	164	5295
1960-61	41	126	210	374	607	783	657	569	607	531	345	93	4943
1961-62	23	8	197	437	689	789	821	610	668	443	438	167	5290
1962-63	95	100	158	377	550	698	959	465	651	496	255	171	4975
1963-64	78	37	71	320	612	743	771	682	640	535	370	204	5063
1964-65	76	91	189	349	679	882	761	611	553	459	400	136	5186
1965-66	24	44	194	261	462	754	732	584	610	442	321	190	4618
1966-67	95	54	106	414	585	658	695	614	700	548	292	92	4853
1967-68	16	0	44	310	524	718	737	472	503	485	232	139	4180
1968-69	33	70	179	415	538	871	983	627	554	478	230	71	5049
1969-70	49	49	144	381	547	607	731	499	586	563	314	122	4592
1970-71	53	44	190	435	548	801	778	628	728	472	321	267	5265
1971-72	82	17	214	429	570	843	863	678	557	531	222	144	5150
1972-73	48	32	295	455	544	825	807	586	639	484	272	183	5170
1973-74	70	114	111	388	633	632	809	606	573	433	306	99	4774
1974-75	60	66	74	380	591	690	804	671	678	570	317	144	5045
1975-76	23	73	93	413	594	723	712	693	731	465	265	157	4942
1976-77	24	52	81	307	510	625	786	451	591	335	320	79	4161
1977-78	34	43	178	390	625	701	631	525	498	447	323	78	4473
1978-79	44	42	180	324	706	846	837	630	479	420	235	96	4839
1979-80	27	40	86	327	628	642	929	610	634	395	329	218	4865
1980-81	66	104	158	343	543	639	633	577	494	455	316	220	4548
1981-82	80	28	138	430	530	715	790	636	640	521	312	103	4923
1982-83	93	42	141	373	647	745	613	502	479	422	244	149	4450
1983-84	72	19	196	406	511	890	672	577	507	482	372	183	4887
1984-85	54	42	159	467	603	867	857	719	666	469	310	160	5373
1985-86	8	48	199	413	870	888							

TABLE 5   COOLING DEGREE DAYS Base 65 deg. F   SEATTLE, WASHINGTON SEATTLE – TACOMA AIRPORT

YEAR	JAN	FEB	MAR	APR	MAY	JUNE	JULY	AUG	SEP	OCT	NOV	DEC	TOTAL
1969	0	0	0	0	19	55	44	25	28	0	0	0	171
1970	0	0	0	0	1	60	58	36	6	0	0	0	161
1971	0	0	0	0	4	2	106	107	0	0	0	0	219
1972	0	0	0	0	22	3	85	91	11	0	0	0	212
1973	0	0	0	0	16	19	67	17	21	0	0	0	140
1974	0	0	0	0	0	36	38	62	60	0	0	0	196
1975	0	0	0	0	0	21	108	29	39	0	0	0	197
1976	0	0	0	8	4	14	59	29	15	0	0	0	129
1977	0	0	0	0	0	26	44	158	0	0	0	0	232
1978	0	0	0	0	4	66	76	64	0	0	0	0	210
1979	0	0	0	0	2	27	106	15	21	0	0	0	171
1980	0	0	0	0	0	0	34	15	3	2	0	0	54
1981	0	0	0	0	1	3	35	131	24	0	0	0	194
1982	0	0	0	0	0	53	31	55	15	0	0	0	154
1983	0	0	0	0	24	2	24	44	0	0	0	0	94
1984	0	0	0	0	1	5	62	45	11	0	0	0	124
1985	0	0	0	0	3	17	125	59	0	0	0	0	204

TABLE 6   SNOWFALL (inches)   SEATTLE, WASHINGTON SEATTLE – TACOMA AIRPORT

SEASON	JULY	AUG	SEP	OCT	NOV	DEC	JAN	FEB	MAR	APR	MAY	JUNE	TOTAL
1956-57	0.0	0.0	0.0	T	0.0	3.0	7.5	7.2	0.0	T	0.0	0.0	17.7
1957-58	0.0	0.0	0.0	0.0	0.0	T	0.0	0.0	T	0.0	0.0	0.0	T
1958-59	0.0	0.0	0.0	0.0	0.2	0.0	3.2	6.5	T	0.0	0.0	0.0	9.9
1959-60	0.0	0.0	0.0	0.0	T	T	2.2	0.0	5.1	0.0	0.0	0.0	7.3
1960-61	0.0	0.0	0.0	0.0	5.2	0.0	0.0	T	1.8	0.0	0.0	0.0	7.0
1961-62	0.0	0.0	0.0	0.0	T	0.9	1.0	7.0	1.7	0.0	0.0	0.0	10.6
1962-63	0.0	0.0	0.0	0.0	0.0	T	3.1	0.5	T	0.0	0.0	0.0	3.6
1963-64	0.0	0.0	0.0	0.0	1.0	T	0.5	T	T	T	0.0	0.0	1.5
1964-65	0.0	0.0	0.0	0.0	3.3	7.6	7.3	T	T	0.0	T	0.0	18.2
1965-66	0.0	0.0	0.0	0.0	0.0	15.3	2.1	T	5.5	T	0.0	0.0	22.9
1966-67	0.0	0.0	0.0	0.0	0.0	2.0	5.9	T	T	T	0.0	0.0	7.9
1967-68	0.0	0.0	0.0	0.0	0.0	3.6	7.5	0.0	0.0	0.5	0.0	0.0	11.6
1968-69	0.0	0.0	0.0	0.0	0.0	22.1	45.4	T	0.0	0.0	0.0	0.0	67.5
1969-70	0.0	0.0	0.0	0.0	T	0.0	T	T	T	T	0.0	0.0	T
1970-71	0.0	0.0	0.0	0.0	T	2.9	9.1	2.2	1.9	T	0.0	0.0	16.1
1971-72	0.0	0.0	0.0	2.0	T	10.6	14.0	0.3	T	2.3	0.0	0.0	29.2
1972-73	0.0	0.0	T	0.0	T	5.6	2.7	T	0.8	T	0.0	0.0	9.1
1973-74	0.0	0.0	0.0	0.0	0.2	0.3	3.7	T	T	0.0	T	0.0	4.2
1974-75	0.0	0.0	0.0	0.0	0.0	9.8	1.3	T	T	0.2	0.0	0.0	11.3
1975-76	0.0	0.0	0.0	0.0	1.6	2.6	T	0.5	0.2	T	0.0	0.0	4.9
1976-77	0.0	0.0	0.0	0.0	0.0	T	1.0	T	0.9	0.0	0.0	0.0	1.9
1977-78	0.0	0.0	0.0	0.0	3.5	T	T	0.0	T	T	0.0	0.0	3.5
1978-79	0.0	0.0	0.0	0.0	4.9	0.2	0.5	0.4	0.0	0.0	0.0	0.0	6.0
1979-80	0.0	0.0	0.0	0.0	0.0	1.2	8.8	2.5	0.1	T	0.0	0.0	12.6
1980-81	T	0.0	0.0	0.0	T	0.3	0.0	1.1	0.0	0.0	0.0	0.0	1.4
1981-82	0.0	0.0	0.0	0.0	0.0	T	7.0	T	2.0	T	0.0	0.0	9.0
1982-83	0.0	0.0	0.0	T	0.0	T	0.0	0.0	0.0	0.0	0.0	0.0	T
1983-84	0.0	0.0	0.0	0.0	T	0.3	T	0.0	0.0	T	0.0	0.0	0.3
1984-85	0.0	0.0	0.0	T	T	2.4	T	5.7	T	T	0.0	0.0	8.1
1985-86	0.0	0.0	0.0	T	17.5	1.7							
Record Mean	T	0.0	T	T	1.4	2.7	5.8	1.5	1.4	0.1	T	0.0	13.0

See Reference Notes, relative to all above tables, on preceding page.

SEATTLE (Urban Site) WASHINGTON

Seattle is located on a hill between the salt waters of Puget Sound to the west and the fresh waters of Lake Washington to the east. The lake shore roughly parallels the shore of Puget Sound at distances varying from about 2-1/2 to 6 miles. Hills rise abruptly from both shorelines and reach elevations of more than 300 feet in the central area to more than 500 feet in the northern and the southwestern sections. The north–south orientation of the city is matched on the east by the Cascade Mountains and the Olympic Mountains to the west and northwest. The Seattle Urban Climatological Station was relocated to the Portage Bay site in November 1972. Portage Bay is an arm of Lake Washington near the University of Washington. The observational site is on a grassy plot about ten yards from the edge of the Bay.

The climate of Seattle is mild and moist, the result of the prevailing westerly winds off the Pacific Ocean approximately 90 miles to the west, and the Cascade Mountains which tend to shield the city from the cold continental air from the east. Winters are comparatively warm and summers cool because of the steady influx of marine air. The daily temperature range is small and extremes of temperature, both hot and cold, are moderate and usually of short duration.

The warmest summer and the coldest winter days come with north winds from British Columbia or east winds from eastern Washington. An average year will have less than three days during the summer with a high temperature of 90 degrees or more with the maximum temperature rarely reaching 100 degrees. Nighttime temperatures during the warmest months seldom remain above 65 degrees. Daily highs during the winter fail to rise above 32 degrees on an average of about two days per year, while the number of days with minimum temperatures of 32 degrees or less averages only 15 days per year. Low temperatures may vary by several degrees throughout the city and depend upon the wind direction, distance from water, and site elevation.

The growing season often lasts from about the mid–March to late November and grass is usually green throughout the winter.

The city lies within the lee side dry area caused by the Olympic Mountains. As a result, the normal precipitation of less than 36 inches is relatively light when compared to the 50 inches or more that falls on the nearby Cascade foothills. The western slopes of these hills and mountains lift the moist marine air, causing very heavy precipitation on the seaward slopes and significantly less at the summits. The winter wet season, usually from October to March, is the result of the air flowing around the Aleutian low pressure system, but, in the summer the Eastern Pacific high pressure system moves north and forces the moist marine air to the north of Washington and brings relatively dry and cool air to the state. The warmest temperatures of the summer usually occur when the Pacific high extends into southwest Canada creating a hot and dry flow of continental air across the Cascade Mountains into the Puget Sound area. Less than 20 percent of the annual rainfall occurs during the summer dry season, April through September.

The average winter snowfall is about 9 inches but the snow seldom remains on the ground for more than two days at a time. Annual totals range from as little as a trace in several instances to over 36 inches in one season. Fog is a frequent occurrence during the late fall and winter months. Severe weather is relatively infrequent over Seattle with an average of just six thunderstorms per year and no tornado ever reported within the city.

TABLE 1 NORMALS, MEANS AND EXTREMES

SEATTLE, WASHINGTON URBAN CLIMATOLOGY STATION

LATITUDE: 47°39'N LONGITUDE: 122°18'W ELEVATION: FT. GRND 22 BARO 25 TIME ZONE: PACIFIC WBAN: 24281

	(a)	JAN	FEB	MAR	APR	MAY	JUNE	JULY	AUG	SEP	OCT	NOV	DEC	YEAR
TEMPERATURE °F:														
Normals														
-Daily Maximum		45.3	50.1	52.6	58.3	64.8	69.0	74.6	73.6	69.2	60.4	51.3	46.9	59.7
-Daily Minimum		35.9	38.2	38.8	42.4	47.7	53.0	56.0	56.3	52.9	47.1	41.1	38.1	45.6
-Monthly		40.6	44.2	45.7	50.4	56.3	61.0	65.3	65.0	61.0	53.8	46.2	42.5	52.7
Extremes														
-Record Highest	52	66	74	75	87	92	100	100	97	92	82	73	65	100
-Year		1935	1968	1941	1947	1963	1955	1941	1960	1967	1980	1970	1980	JUN 1955
-Record Lowest	52	11	12	22	31	35	42	47	48	40	30	13	10	10
-Year		1950	1950	1955	1936	1954	1976	1979	1980	1972	1935	1985	1968	DEC 1968
NORMAL DEGREE DAYS:														
Heating (base 65°F)		756	582	598	438	274	148	64	74	142	347	564	694	4681
Cooling (base 65°F)		0	0	0	0	0	28	73	74	25	0	0	0	200
% OF POSSIBLE SUNSHINE	31	28	34	42	47	52	49	63	56	53	37	28	23	43
MEAN SKY COVER (tenths)														
Sunrise - Sunset	24	8.0	7.7	7.4	6.9	6.4	6.4	4.9	5.3	5.6	7.2	8.0	8.1	6.8
MEAN NUMBER OF DAYS:														
Sunrise to Sunset														
-Clear	24	3.0	3.0	4.0	5.0	7.0	7.0	12.0	10.0	9.0	5.0	3.0	3.0	71.0
-Partly Cloudy	24	5.0	6.0	8.0	9.0	10.0	8.0	10.0	10.0	8.0	8.0	6.0	5.0	93.0
-Cloudy	24	23.0	19.0	19.0	16.0	14.0	15.0	9.0	11.0	13.0	18.0	21.0	23.0	201.0
Precipitation														
.01 inches or more	35	18.6	15.9	16.8	13.5	10.4	9.1	5.2	6.2	8.9	10.4	17.7	19.1	151.9
Snow, Ice pellets														
1.0 inches or more	35	1.3	0.3	0.1	0.*	0.0	0.0	0.0	0.0	0.0	0.0	0.2	0.6	2.6
Thunderstorms	25	0.3	0.3	0.3	0.4	0.6	0.6	0.7	0.8	0.7	0.6	0.4	0.3	6.0
Heavy Fog Visibility														
1/4 mile or less														
Temperature °F														
-Maximum														
90° and above	35	0.0	0.0	0.0	0.0	0.1	0.3	1.0	0.5	0.1	0.0	0.0	0.0	2.0
32° and below	35	1.0	0.1	0.0	0.0	0.0	0.0	0.0	0.0	0.0	0.0	0.3	0.7	2.1
-Minimum														
32° and below	35	6.5	2.8	1.5	0.0	0.0	0.0	0.0	0.0	0.0	0.*	2.4	4.3	17.5
0° and below	35	0.0	0.0	0.0	0.0	0.0	0.0	0.0	0.0	0.0	0.0	0.0	0.0	0.0
AVG. STATION PRESS.(mb)														
RELATIVE HUMIDITY (%)														
Hour 04														
Hour 10 (Local Time)														
Hour 16														
Hour 22														
PRECIPITATION (inches):														
Water Equivalent														
-Normal		5.94	4.20	3.70	2.46	1.66	1.53	0.89	1.38	2.03	3.40	5.36	6.29	38.84
-Maximum Monthly	52	10.93	7.75	7.23	4.56	4.67	3.68	2.16	5.49	5.62	8.04	11.20	10.41	11.20
-Year		1953	1961	1950	1937	1948	1964	1978	1977	1978	1975	1983	1939	NOV 1983
-Minimum Monthly	52	0.60	1.29	0.44	0.16	0.34	0.23	T	T	0.03	0.48	0.50	1.00	T
-Year		1985	1934	1965	1939	1972	1949	1960	1967	1975	1972	1976	1944	AUG 1967
-Maximum in 24 hrs	52	2.69	2.69	2.32	2.23	1.35	2.07	1.22	1.93	1.91	3.48	3.20	3.31	3.48
-Year		1983	1945	1950	1972	1948	1985	1954	1977	1953	1981	1937	1937	OCT 1981
Snow, Ice pellets														
-Maximum Monthly	52	31.0	10.4	7.5	1.0						T	9.6	13.5	31.0
-Year		1950	1949	1951	1972						1984	1985	1968	JAN 1950
-Maximum in 24 hrs	52	11.5	7.6	5.5	1.0						T	6.0	10.0	11.5
-Year		1943	1937	1960	1972						1984	1985	1968	JAN 1943
WIND:														
Mean Speed (mph)														
Prevailing Direction														
Fastest Mile														
-Direction (!!!)	11	S	SSW	WSW	WSW	WNW	SW	S	WNW	W	S	S	SSE	SSW
-Speed (MPH)	11	44	48	44	39	37	35	36	37	39	37	45	46	48
-Year		1978	1979	1980	1980	1983	1980	1976	1975	1981	1977	1981	1973	FEB 1979
Peak Gust														
-Direction (!!!)	2	SSW	S	WSW	SSW	WSW	WSW	WSW	SW	SSW	SW	SSW	N	WSW
-Speed (mph)	2	35	36	54	37	40	24	29	25	32	35	35	31	54
-Date		1984	1985	1985	1985	1985	1985	1984	1984	1985	1985	1985	1984	MAR 1985

See Reference Notes to this table on the following page.

SEATTLE (Urban Site) WASHINGTON

TABLE 2 PRECIPITATION (inches)   SEATTLE, WASHINGTON URBAN CLIMATOLOGY STATION

YEAR	JAN	FEB	MAR	APR	MAY	JUNE	JULY	AUG	SEP	OCT	NOV	DEC	ANNUAL
1956	7.82	2.16	5.69	0.59	0.68	2.66	0.15	0.61	2.22	4.41	1.53	3.59	32.11
1957	2.74	5.51	5.70	2.32	1.12	1.58	0.57	0.97	0.60	3.96	2.68	4.36	32.11
1958	7.62	5.27	1.91	2.82	0.92	0.72	T	0.32	1.47	3.12	6.53	5.51	36.21
1959	8.94	3.15	3.39	1.86	1.50	1.53	0.77	0.39	2.33	2.36	5.49	6.20	37.91
1960	5.59	3.45	3.93	2.68	2.66	0.57	T	1.43	0.99	4.10	8.01	3.21	36.62
1961	6.76	7.75	4.20	2.01	3.07	0.43	0.76	0.57	0.62	2.83	4.39	5.58	38.97
1962	2.50	1.93	3.43	1.82	1.50	0.69	1.11	1.41	2.02	3.41	6.87	3.83	30.52
1963	1.91	3.97	2.98	2.75	0.94	1.95	0.85	0.76	0.69	4.16	7.63	4.94	33.53
1964	8.16	1.55	3.20	1.29	1.07	3.68	0.84	1.46	2.01	0.83	8.11	4.86	37.06
1965	5.83	4.28	0.44	3.79	1.25	0.47	0.48	1.61	0.75	2.03	5.01	7.06	33.00
1966	6.40	2.36	4.76	2.02	1.34	0.75	1.39	0.17	1.57	2.21	7.18	7.73	37.88
1967	9.02	2.18	3.61	2.76	0.52	1.35	0.05	T	1.21	5.42	2.01	5.33	33.46
1968	7.39	4.87	5.11	2.06	1.34	2.52	0.48	4.28	1.63	3.52	4.87	10.07	48.14
1969	5.83	3.58	1.97	3.39	2.27	1.22	0.23	0.15	5.41	1.72	2.73	6.89	35.39
1970	8.11	2.01	3.07	2.73	1.00	0.78	0.53	0.74	2.01	3.42	4.30	8.34	37.04
1971	4.25	3.97	7.16	2.33	1.51	1.85	0.58	0.46	3.08	3.06	4.25	5.29	37.79
#1972	5.15	5.61	6.01	4.20	0.34	1.98	0.67	1.06	3.27	0.48	3.19	8.36	40.32
1973	4.61	2.00	1.55	0.93	1.11	1.73	0.25	0.57	1.56	2.74	8.40	9.56	35.01
1974	8.53	4.19	5.69	1.83	1.48	1.07	1.95	0.11	0.23	1.78	5.56	6.52	38.94
1975	5.03	4.91	4.82	2.51	1.51	1.26	0.11	2.81	0.03	8.04	6.40	7.57	45.00
1976	4.33	4.83	3.64	2.10	1.87	1.07	1.02	3.05	1.43	1.94	0.50	2.00	27.78
1977	2.13	2.11	5.24	1.35	4.24	0.59	0.80	5.49	2.82	2.72	5.03	5.81	38.33
1978	5.57	3.72	2.61	3.53	2.02	0.57	2.16	1.65	5.62	0.87	6.52	1.42	36.26
1979	2.20	6.32	1.32	2.69	1.07	0.45	0.97	0.61	1.86	4.81	2.28	10.29	34.87
1980	4.96	5.64	3.27	3.65	1.51	2.88	0.49	1.74	1.43	1.42	6.12	7.19	40.30
1981	2.58	4.07	2.54	2.09	2.25	2.07	1.52	0.34	3.25	6.45	5.21	7.40	39.77
1982	4.68	7.50	3.86	2.48	0.57	1.07	0.80	0.63	1.68	4.20	5.30	8.99	41.76
1983	6.84	5.29	4.37	0.91	2.18	1.99	2.07	2.08	2.55	11.20	1.11	5.52	46.11
1984	2.68	4.18	3.72	2.81	2.99	3.17	0.46	0.22	1.51	2.34	9.26	5.27	38.61
1985	0.60	2.88	2.92	2.59	1.99	3.05	0.22	0.72	1.77	6.81	3.77	1.22	28.54
Record Mean	5.45	4.00	3.54	2.27	1.56	1.50	0.74	1.13	1.88	3.19	5.35	5.84	36.44

TABLE 3 AVERAGE TEMPERATURE (deg. F)   SEATTLE, WASHINGTON URBAN CLIMATOLOGY STATION

YEAR	JAN	FEB	MAR	APR	MAY	JUNE	JULY	AUG	SEP	OCT	NOV	DEC	ANNUAL
1956	41.7	38.2	43.2	52.2	59.6	58.9	66.7	65.3	60.2	52.6	45.8	43.2	52.3
1957	36.1	42.2	46.7	52.3	60.4	62.2	63.6	64.6	65.6	53.8	46.3	46.0	53.3
1958	46.2	50.1	47.4	52.2	62.3	65.8	70.7	68.6	62.2	56.3	45.7	47.4	56.2
1959	42.7	43.5	46.3	52.1	56.2	61.9	67.8	64.5	59.3	54.2	46.7	43.5	53.3
1960	42.2	44.7	46.2	52.4	55.4	61.6	67.6	63.4	60.2	55.2	46.9	43.0	53.2
1961	46.5	46.9	47.9	50.3	56.6	64.7	67.8	69.3	60.9	53.2	45.0	42.4	54.3
1962	41.5	45.4	45.3	53.1	54.2	61.2	65.1	64.9	62.2	55.7	49.2	45.5	53.6
1963	37.4	50.3	46.8	50.7	58.6	59.9	63.6	65.3	64.5	56.1	47.6	44.6	53.8
1964	43.4	44.2	46.3	49.6	55.7	60.3	65.3	64.4	60.6	55.9	45.2	39.4	52.5
1965	42.8	44.6	48.9	52.5	54.6	62.5	67.1	66.0	59.7	57.3	51.5	43.4	54.2
1966	43.8	46.1	48.4	52.6	57.1	61.2	64.5	66.4	63.7	54.6	49.1	47.2	54.5
1967	45.0	45.9	45.1	49.0	57.7	65.3	67.3	70.6	66.4	55.9	50.0	43.7	54.7
1968	43.0	50.7	50.2	50.6	58.6	61.9	68.5	64.9	61.3	54.3	49.6	39.5	54.4
1969	34.8	43.2	48.4	50.7	59.9	65.0	64.9	63.5	61.1	54.0	47.7	46.0	53.3
1970	42.9	48.4	48.2	49.7	57.3	64.4	66.2	65.5	60.0	52.4	47.2	40.9	53.6
1971	41.6	42.6	43.5	50.9	56.1	58.1	66.1	66.9	60.6	51.9	47.0	39.1	52.1
#1972	38.2	43.2	48.3	48.1	58.9	60.1	65.9	66.3	57.6	51.6	46.8	39.1	52.1
1973	39.5	44.8	46.8	50.5	57.4	60.6	65.7	62.3	61.4	52.8	43.4	44.2	52.4
1974	39.3	43.5	45.4	49.9	53.8	61.1	64.3	66.4	64.8	54.1	47.9	44.4	52.9
1975	41.7	42.1	44.3	47.3	55.8	59.6	66.1	62.8	62.3	52.7	45.9	43.3	52.0
1976	44.0	42.9	44.0	51.2	56.0	59.1	65.0	63.1	61.7	54.0	47.9	44.4	52.8
1977	39.7	47.5	44.8	52.3	53.7	61.4	63.0	68.1	58.6	52.2	44.4	42.0	52.3
1978	43.6	46.2	49.1	50.6	54.9	63.8	65.4	65.0	59.0	54.8	42.0	38.3	52.7
1979	36.2	42.2	49.3	51.1	57.8	61.3	66.3	65.3	63.2	55.0	45.1	44.8	53.2
1980	36.6	45.2	46.9	53.4	56.1	59.2	64.5	63.2	61.5	55.0	48.4	45.7	53.0
1981	45.5	45.5	50.4	52.2	56.7	60.1	65.1	69.4	62.2	52.5	48.5	42.5	54.2
1982	40.9	43.3	45.6	48.7	56.1	64.2	64.0	65.1	61.3	53.6	43.7	42.1	52.4
1983	46.0	47.9	50.2	52.6	59.1	61.6	64.6	66.6	59.2	53.2	48.9	37.1	53.9
1984	44.6	46.3	49.9	50.6	60.3	60.3	64.7	65.5	60.3	51.5	46.1	38.4	52.9
1985	39.1	40.5	44.8	50.6	56.8	61.5	69.5	65.6	58.8	52.6	37.4	38.1	51.3
Record Mean	41.6	44.7	46.5	50.9	56.8	61.4	65.7	65.4	61.4	54.0	46.6	42.7	53.2
Max	45.9	50.1	52.9	58.3	64.8	69.3	74.8	73.9	68.9	60.1	51.4	46.9	59.8
Min	37.2	39.3	40.1	43.4	48.8	53.4	56.5	56.9	53.8	48.0	41.8	38.5	46.5

REFERENCE NOTES FOR TABLES 1, 2, 3 and 6 (SEATTLE, WA [URBAN CLIM. STAT.])

GENERAL

T - TRACE AMOUNT
BLANK ENTRIES DENOTE MISSING/UNREPORTED DATA.
# INDICATES A STATION OR INSTRUMENT RELOCATION.

SPECIFIC

TABLE 1

(a) - LENGTH OF RECORD IN YEARS. ALTHOUGH
INDIVIDUAL MONTHS MAY BE MISSING.
* LESS THAN .05

NORMALS — BASED ON THE 1951-1980 RECORD PERIOD.
EXTREMES — DATES ARE THE MOST RECENT OCCURRENCE.
WIND DIR. — NUMERALS SHOW TENS OF DEGREES
CLOCKWISE FROM TRUE NORTH.
"00" INDICATES CALM.
RESULTANT WIND DIRECTIONS ARE GIVEN TO WHOLE DEGREES.

EXCEPTIONS

TABLE 1

1. MEAN SKY COVER, AND DAYS CLEAR-PARTLY CLOUDY-
CLOUDY AND THUNDERSTORMS ARE 1934-1957.
2. PERCENT OF POSSIBLE SUNSHINE IS 1934-1964.

TABLES 2, 3, and 6

RECORD MEANS ARE THROUGH THE CURRENT YEAR,
BEGINNING IN 1951 FOR TEMPERATURE
1951 FOR PRECIPITATION
1951 FOR SNOWFALL

TABLE 4   HEATING DEGREE DAYS Base 65 deg. F   SEATTLE, WASHINGTON URBAN CLIMATOLOGY STATION

SEASON	JULY	AUG	SEP	OCT	NOV	DEC	JAN	FEB	MAR	APR	MAY	JUNE	TOTAL
1956-57	32	43	147	379	567	669	891	633	561	376	147	84	4529
1957-58	47	21	31	336	558	581	574	413	539	377	116	46	3639
1958-59	1	13	105	265	573	541	684	597	575	378	280	104	4116
1959-60	29	36	167	328	541	660	701	581	575	370	291	109	4388
1960-61	28	97	146	298	536	675	568	503	525	438	259	63	4136
1961-62	16	2	134	365	594	698	722	545	603	350	327	122	4478
1962-63	67	41	90	280	469	597	848	406	556	427	224	167	4172
1963-64	49	20	49	274	515	626	664	599	574	455	289	142	4256
1964-65	39	53	127	276	586	788	683	564	492	372	315	90	4385
1965-66	33	35	154	233	402	663	649	523	510	365	242	123	3932
1966-67	51	20	54	316	473	545	612	528	612	471	226	47	3955
1967-68	12	3	26	276	446	652	673	410	452	426	191	102	3669
1968-69	16	50	120	324	458	782	929	604	507	422	173	60	4445
1969-70	39	54	133	338	511	584	679	460	513	455	235	82	4083
1970-71	27	30	151	381	527	741	720	624	662	418	275	203	4759
1971-72	65	7	127	401	531	795	822	626	509	499	206	141	4729
#1972-73	49	30	225	407	538	797	783	558	560	429	235	148	4759
1973-74	42	100	116	371	642	637	790	595	600	448	338	124	4803
1974-75	58	26	55	331	504	634	714	635	638	522	279	165	4561
1975-76	28	81	93	374	566	666	644	633	645	411	271	176	4588
1976-77	32	66	96	336	507	635	780	483	617	374	346	105	4377
1977-78	78	41	185	387	609	704	657	522	487	423	307	70	4470
1978-79	45	45	170	310	685	822	886	630	479	411	215	118	4816
1979-80	33	12	68	304	591	620	872	569	554	341	270	169	4403
1980-81	47	69	107	307	491	588	597	540	443	376	255	148	3968
1981-82	39	7	103	383	489	691	743	604	593	483	271	77	4483
1982-83	55	31	119	347	633	699	583	474	452	367	194	95	4049
1983-84	38	6	167	360	477	859	628	537	459	427	297	145	4400
1984-85	32	29	145	413	558	816	797	678	618	427	253	114	4880
1985-86	3	27	180	378	822	827							

TABLE 5   COOLING DEGREE DAYS Base 65 deg. F   SEATTLE, WASHINGTON URBAN CLIMATOLOGY STATION

YEAR	JAN	FEB	MAR	APR	MAY	JUNE	JULY	AUG	SEP	OCT	NOV	DEC	TOTAL
1973	0	0	0	0	7	20	69	22	13	0	0	0	131
1974	0	0	0	0	0	12	42	76	54	0	0	0	184
1975	0	0	0	0	2	10	71	20	17	0	0	0	120
1976	0	0	0	4	1	6	41	14	7	0	0	0	73
1977	0	0	0	0	0	6	24	144	1	0	0	0	175
1978	0	0	0	0	2	44	66	49	0	0	0	0	161
1979	0	0	0	0	1	12	79	27	20	0	0	0	139
1980	0	0	0	0	0	0	38	22	8	5	0	0	73
1981	0	0	0	0	3	5	47	151	24	0	0	0	230
1982	0	0	0	0	1	61	33	43	16	0	0	0	154
1983	0	0	0	0	20	1	33	64	1	0	0	0	119
1984	0	0	0	0	2	13	61	50	10	2	0	0	138
1985	0	0	0	0	6	20	147	53	0	0	0	0	226

TABLE 6   SNOWFALL (inches)   SEATTLE, WASHINGTON URBAN CLIMATOLOGY STATION

SEASON	JULY	AUG	SEP	OCT	NOV	DEC	JAN	FEB	MAR	APR	MAY	JUNE	TOTAL
1956-57	0.0	0.0	0.0	0.0	0.0	4.3	10.1	6.8	0.0	0.0	0.0	0.0	21.2
1957-58	0.0	0.0	0.0	0.0	0.0	0.0	0.0	T	T	0.0	0.0	0.0	T
1958-59	0.0	0.0	0.0	0.0	T	T	5.0	1.5	T	0.0	0.0	0.0	6.5
1959-60	0.0	0.0	0.0	0.0	0.0	T	1.0	0.0	5.5	0.0	0.0	0.0	6.5
1960-61	0.0	0.0	0.0	0.0	4.5	0.0	0.0	T	T	0.0	0.0	0.0	4.5
1961-62	0.0	0.0	0.0	0.0	0.0	0.2	T	6.0	2.2	0.0	0.0	0.0	8.4
1962-63	0.0	0.0	0.0	0.0	T	0.0	1.8	T	T	0.0	0.0	0.0	1.8
1963-64	0.0	0.0	0.0	0.0	T	0.0	1.4	T	T	0.0	0.0	0.0	1.4
1964-65	0.0	0.0	0.0	0.0	1.5	6.4	4.3	0.0	0.0	0.0	0.0	0.0	12.2
1965-66	0.0	0.0	0.0	0.0	0.0	7.6	T	T	0.8	0.0	0.0	0.0	8.4
1966-67	0.0	0.0	0.0	0.0	0.0	1.0	5.0	0.0	0.0	0.0	0.0	0.0	6.0
1967-68	0.0	0.0	0.0	0.0	0.0	2.5	6.0	0.0	0.0	T	0.0	0.0	8.5
1968-69	0.0	0.0	0.0	0.0	0.0	13.5	22.7	0.0	0.0	0.0	0.0	0.0	36.2
1969-70	0.0	0.0	0.0	0.0	0.0	0.0	T	T	0.0	T	0.0	0.0	T
1970-71	0.0	0.0	0.0	0.0	0.3	9.1	5.2	1.0	0.0	0.0	0.0	15.6
1971-72	0.0	0.0	0.0	T	0.0	6.3	8.0	0.0	0.0	1.0	0.0	0.0	15.3
#1972-73	0.0	0.0	0.0	0.0	0.0	5.0	3.7	0.0	0.0	0.0	0.0	0.0	8.7
1973-74	0.0	0.0	0.0	0.0	T	2.0	0.0	1.0	T	0.0	0.0	3.0
1974-75	0.0	0.0	0.0	0.0	0.0	6.0	1.0	0.5	T	T	0.0	0.0	7.5
1975-76	0.0	0.0	0.0	0.0	0.5	0.5	0.0	0.0	T	0.0	0.0	0.0	1.0
1976-77	0.0	0.0	0.0	0.0	0.0	0.0	1.0	0.0	T	0.0	0.0	0.0	1.0
1977-78	0.0	0.0	0.0	0.0	0.3	T	T	0.0	T	0.0	0.0	0.0	0.3
1978-79	0.0	0.0	0.0	0.0	6.0	T	T	T	0.0	0.0	0.0	0.0	6.0
1979-80	0.0	0.0	0.0	0.0	0.0	T	11.9	2.5	0.0	0.0	0.0	0.0	14.4
1980-81	0.0	0.0	0.0	0.0	0.0	1.0	0.0	0.5	0.0	T	0.0	0.0	1.5
1981-82	0.0	0.0	0.0	0.0	T	T	5.2	T	0.0	0.0	0.0	0.0	5.2
1982-83	0.0	0.0	0.0	0.0	T	0.0	T	0.0	0.0	0.0	0.0	0.0	T
1983-84	0.0	0.0	0.0	0.0	T	1.4	T	0.0	0.0	0.0	0.0	0.0	1.4
1984-85	0.0	0.0	0.0	T	0.0	3.4	T	3.5	0.0	0.0	0.0	6.9
1985-86	0.0	0.0	0.0	0.0	9.6	1.0							
Record Mean	0.0	0.0	0.0	T	0.8	2.0	3.5	0.8	0.6	T	T	0.0	7.7

See Reference Notes, relative to all above tables, on preceding page.

Spokane lies on the eastern edge of the broad Columbia Basin area of Washington which is bounded by the Cascade Range on the west and the Rocky Mountains on the east. The elevations in eastern Washington vary from less than 400 feet above sea level near Pasco where the Columbia River flows out of Washington to over 5,000 feet in the mountain areas of the extreme eastern edge of the State. Spokane is located on the upper plateau area where the long gradual slope from the Columbia River meets the sharp rise of the Rocky Mountain Ranges.

Much of the urban area of Spokane lies along both sides of the Spokane River at an elevation of approximately 2,000 feet, but the residential areas have spread to the crests of the plateaus on either side of the river with elevations up to 2,500 feet above sea level. Spokane International Airport is situated on the plateau area 6 miles west-southwest and some 400 feet higher than the downtown business district.

The climate of Spokane combines some of the characteristics of damp coastal type weather and arid interior conditions. Most of the air masses which reach Spokane are brought in by the prevailing westerly and southwesterly circulations. Frequently, much of the moisture in the storms that move eastward and southeastward from the Gulf of Alaska and the eastern Pacific Ocean is precipitated out as the storms are lifted across the Coast and Cascade Ranges. Annual precipitation totals in the Spokane area are generally less than 20 inches and less than 50 percent of the amounts received west of the Cascades. However, the precipitation and total cloudiness in the Spokane vicinity is greater than that of the desert areas of south-central Washington. The lifting action of the air masses as they move up the east slope of the Columbia Basin frequently produces the cooling and condensation necessary for formation of clouds and precipitation.

Infrequently, the Spokane area comes under the influence of dry continental air masses from the north or east. On occasions when these air masses penetrate into eastern Washington the result is high temperatures and very low humidity in the summer and sub-zero temperatures in the winter. In the winter most of the severe arctic outbursts of cold air move southward on the east side of the Continental Divide and do not affect Spokane.

In general, Spokane weather has the characteristics of a mild, arid climate during the summer months and a cold, coastal type in the winter. Approximately 70 percent of the total annual precipitation falls between the first of October and the end of March and about half of that falls as snow. The growing season usually extends over nearly six months from mid-April to mid-October. Irrigation is required for all crops except dry-land type grains. The summer weather is ideal for full enjoyment of the many mountain and lake recreational areas in the immediate vicinity. Winter weather includes many cloudy or foggy days and below freezing temperatures with occasional snowfall of several inches in depth. Sub-zero temperatures and traffic-stopping snowfalls are infrequent.

Based on the 1951-1980 period, the average first occurrence of 32 degrees Fahrenheit in the fall is October 6 and the average last occurrence in the spring is May 4.

TABLE 1   NORMALS, MEANS AND EXTREMES

SPOKANE WASHINGTON

LATITUDE: 47°38'N   LONGITUDE: 117°32' W   ELEVATION: FT. GRND 2357 BARO   2360   TIME ZONE: PACIFIC   WBAN: 24157

	(a)	JAN	FEB	MAR	APR	MAY	JUNE	JULY	AUG	SEP	OCT	NOV	DEC	YEAR
TEMPERATURE °F:														
Normals														
-Daily Maximum		31.3	39.0	46.2	56.7	66.1	74.0	84.0	81.7	72.4	58.3	41.4	34.2	57.1
-Daily Minimum		20.0	25.7	29.0	34.9	42.5	49.3	55.3	54.3	46.5	36.7	28.5	23.7	37.2
-Monthly		25.6	32.4	37.6	45.8	54.3	61.7	69.7	68.0	59.5	47.5	35.0	29.0	47.2
Extremes														
-Record Highest	38	59	61	71	90	94	100	103	108	96	86	67	56	108
-Year		1971	1958	1960	1977	1983	1973	1967	1961	1950	1980	1975	1980	AUG 1961
-Record Lowest	38	-22	-17	-3	17	24	33	37	35	24	11	-21	-25	-25
-Year		1979	1979	1955	1966	1954	1984	1981	1965	1985	1984	1985	1968	DEC 1968
NORMAL DEGREE DAYS:														
Heating (base 65°F)		1218	913	849	576	339	140	17	63	209	539	903	1116	6882
Cooling (base 65°F)		0	0	0	0	8	41	162	159	41	0	0	0	411
% OF POSSIBLE SUNSHINE	37	26	38	53	60	63	66	80	77	70	52	28	21	53
MEAN SKY COVER (tenths)														
Sunrise - Sunset	38	8.2	8.0	7.4	7.1	6.6	6.1	3.7	4.2	4.9	6.4	8.0	8.5	6.6
MEAN NUMBER OF DAYS:														
Sunrise to Sunset														
-Clear	38	3.2	3.2	4.3	4.5	5.6	7.2	16.8	15.2	11.9	7.7	3.2	2.8	85.6
-Partly Cloudy	38	4.2	4.9	7.8	8.3	10.4	10.4	8.4	8.4	8.4	7.8	5.1	3.8	88.0
-Cloudy	38	23.5	20.1	18.9	17.2	15.0	12.4	5.8	7.4	9.8	15.6	21.6	24.4	191.7
Precipitation														
.01 inches or more	38	14.2	11.7	11.3	8.5	9.2	7.8	4.1	5.1	5.8	7.9	12.6	15.4	113.7
Snow, Ice pellets														
1.0 inches or more	38	5.4	2.9	1.6	0.2	0.*	0.0	0.0	0.0	0.0	0.2	2.1	5.1	17.5
Thunderstorms	38	0.*	0.*	0.2	0.6	1.4	2.9	2.1	2.3	0.7	0.3	0.1	0.0	10.6
Heavy Fog Visibility														
1/4 mile or less	38	9.4	7.3	2.8	1.3	0.9	0.5	0.3	0.3	0.9	4.4	8.8	12.2	48.8
Temperature °F														
-Maximum														
90° and above	26	0.0	0.0	0.0	0.*	0.2	2.0	9.3	7.3	0.8	0.0	0.0	0.0	19.6
32° and below	26	15.3	4.2	1.0	0.0	0.0	0.0	0.0	0.0	0.0	0.1	4.4	13.8	38.7
-Minimum														
32° and below	26	26.7	22.2	21.0	11.5	1.7	0.0	0.0	0.0	0.9	9.5	20.3	26.2	140.0
0° and below	26	2.9	0.5	0.0	0.0	0.0	0.0	0.0	0.0	0.0	0.0	0.3	1.9	5.6
AVG. STATION PRESS.(mb)	12	934.2	932.4	929.1	931.0	930.7	930.9	931.8	931.3	932.7	933.6	932.1	933.7	932.0
RELATIVE HUMIDITY (%)														
Hour 04	26	84	84	80	76	76	74	63	62	71	79	87	87	77
Hour 10	26	83	80	68	57	52	49	40	43	52	66	82	85	63
Hour 16 (Local Time)	26	78	69	55	44	41	36	26	28	35	49	75	82	52
Hour 22	26	83	81	73	65	63	58	44	46	56	71	84	86	68
PRECIPITATION (inches):														
Water Equivalent														
-Normal		2.47	1.61	1.36	1.08	1.38	1.23	0.50	0.74	0.71	1.08	2.06	2.49	16.71
-Maximum Monthly	38	4.96	3.94	3.75	3.08	5.71	3.06	1.85	1.83	2.05	4.05	5.10	5.13	5.71
-Year		1959	1961	1950	1948	1948	1964	1983	1976	1959	1950	1973	1964	MAY 1948
-Minimum Monthly	38	0.38	0.40	0.31	0.08	0.20	0.16	T	T	0.03	0.05	0.22	0.60	T
-Year		1985	1967	1965	1956	1982	1960	1973	1969	1975	1965	1976	1976	JUL 1973
-Maximum in 24 hrs	38	1.48	1.11	0.91	1.01	1.67	2.07	0.96	1.09	1.12	0.98	1.41	1.60	2.07
-Year		1954	1963	1950	1982	1948	1964	1983	1959	1973	1955	1960	1951	JUN 1964
Snow, Ice pellets														
-Maximum Monthly	38	56.9	28.5	15.3	6.6	3.5	T				6.1	.7	42.0	56.9
-Year		1950	1975	1962	1964	1967	1954				1957	i	1964	JAN 1950
-Maximum in 24 hrs	38	13.0	8.9	5.3	4.9	3.5	T				6.1	9.0	12.1	13.0
-Year		1950	1975	1970	1964	1967	1954				1957	1973	1951	JAN 1950
WIND:														
Mean Speed (mph)	38	8.6	9.1	9.5	9.9	9.0	9.0	8.3	8.1	8.2	8.1	8.4	8.6	8.7
Prevailing Direction														
through 1963		NE	SSW	SSW	SW	SSW	SSW	SW	SW	NE	SSW	NE	NE	SSW
Fastest Mile														
-Direction	38	SW	SW	SW	SW	W	W	SW	SW	SW	SW	SW	SW	SW
-Speed (MPH)	38	59	54	54	50	49	39	43	50	38	56	54	51	59
-Year		1972	1949	1971	1962	1957	1983	1970	1982	1961	1950	1949	1956	JAN 1972
Peak Gust														
-Direction	2	SW	SW	SW	SW	W	SW	SE	NW	SW	SE	SW	SW	SW
-Speed (mph)	2	43	41	41	54	40	46	41	47	45	49	51	43	54
-Date		1984	1985	1984	1985	1984	1985	1984	1984	1984	1985	1984	1984	APR 1985

See Reference Notes to this table on the following page.

SPOKANE, WASHINGTON

TABLE 2 PRECIPITATION (inches)   SPOKANE WASHINGTON

YEAR	JAN	FEB	MAR	APR	MAY	JUNE	JULY	AUG	SEP	OCT	NOV	DEC	ANNUAL
1956	2.87	1.44	1.29	0.08	0.59	1.18	0.50	1.41	0.09	1.87	0.34	1.22	12.88
1957	1.34	1.54	1.82	0.81	3.74	2.74	0.04	0.30	0.68	2.33	0.82	1.93	18.09
1958	3.55	3.27	0.84	1.72	0.71	1.63	1.15	0.13	0.47	0.79	3.72	2.93	20.91
1959	4.96	2.01	1.21	0.57	2.26	0.39	0.05	1.24	2.05	1.32	2.30	1.21	19.57
1960	1.05	1.64	2.36	1.51	2.73	0.16	T	0.56	0.72	0.95	4.64	1.37	17.69
1961	1.61	3.94	1.75	0.96	1.77	1.64	0.37	0.30	0.17	1.05	1.83	3.91	19.30
1962	1.39	1.72	2.56	1.02	1.65	0.78	0.29	0.63	0.90	1.62	3.02	1.44	17.02
1963	0.89	2.21	1.65	1.32	0.98	0.96	0.41	0.50	0.36	1.11	2.58	2.29	15.26
1964	3.15	0.98	1.53	0.98	0.45	3.06	0.39	1.46	1.03	0.46	2.89	5.13	21.51
1965	2.82	1.13	0.31	2.35	1.02	0.74	0.69	1.73	0.28	0.05	1.71	1.63	14.46
1966	1.94	0.50	2.43	0.13	0.49	0.70	0.95	0.15	0.51	0.36	3.01	2.96	14.13
1967	2.44	0.40	1.72	1.71	1.31	1.99	0.06	T	0.24	1.18	0.82	2.02	13.89
1968	1.57	2.12	0.71	0.10	1.16	0.87	0.23	1.35	0.63	2.24	2.35	2.93	16.26
1969	4.08	1.21	0.53	2.16	0.54	1.17	0.03	T	0.71	0.45	0.37	2.45	13.70
1970	4.15	1.83	1.30	0.93	0.94	1.60	0.59	0.10	0.48	2.13	2.04	1.43	17.52
1971	2.11	0.88	2.11	1.85	1.39	2.46	0.50	0.59	1.37	0.82	1.51	2.89	18.48
1972	1.74	1.13	1.05	1.09	1.99	1.56	0.25	0.87	0.86	0.19	0.88	1.92	13.53
1973	2.05	0.48	0.77	0.42	1.34	0.57	T	0.19	1.44	0.97	5.10	3.78	17.11
1974	3.79	1.79	2.22	0.80	1.03	0.23	0.71	0.04	0.18	0.12	2.59	2.54	16.04
1975	2.53	3.12	1.83	1.78	1.41	1.45	1.60	0.93	0.03	2.23	1.94	2.42	21.27
1976	1.28	2.04	0.83	0.97	1.24	0.78	0.79	1.83	0.05	0.59	0.22	0.60	11.22
1977	0.75	0.52	1.15	0.13	1.71	1.45	0.01	1.25	1.42	0.44	2.12	4.52	15.57
1978	2.53	1.64	0.77	2.62	2.81	1.22	1.76	1.71	0.93	0.13	2.02	1.05	19.19
1979	1.11	2.19	1.03	0.69	1.60	0.78	0.85	1.01	0.78	1.22	1.15	1.94	14.35
1980	1.96	1.90	0.91	1.06	2.34	0.99	0.21	0.79	0.84	0.64	1.67	3.72	17.03
1981	1.00	1.41	1.57	0.85	2.02	1.92	0.51	0.04	0.59	1.53	0.96	2.51	14.91
1982	1.61	1.67	1.49	2.23	0.20	0.85	1.05	0.25	1.77	1.48	1.86	2.79	17.25
1983	1.89	2.07	2.20	0.61	0.92	2.84	1.85	0.96	0.79	1.33	4.80	2.38	22.64
1984	0.99	1.37	1.80	1.75	2.01	1.89	0.07	0.27	0.56	0.76	4.26	2.28	18.01
1985	0.38	0.93	1.39	0.28	1.13	0.67	0.26	0.19	1.64	1.40	2.23	0.71	11.21
Record Mean	2.06	1.60	1.34	1.08	1.35	1.29	0.52	0.60	0.84	1.22	2.05	2.23	16.19

TABLE 3 AVERAGE TEMPERATURE (deg. F)   SPOKANE WASHINGTON

YEAR	JAN	FEB	MAR	APR	MAY	JUNE	JULY	AUG	SEP	OCT	NOV	DEC	ANNUAL
1956	28.9	22.6	36.1	49.9	58.1	58.9	71.6	67.6	60.7	46.3	32.7	30.7	47.0
1957	15.0	28.7	37.6	48.0	59.3	62.1	67.5	65.6	62.8	45.0	35.6	35.1	46.9
1958	32.0	39.8	38.9	44.9	62.7	65.6	73.0	73.7	58.7	50.2	35.7	32.4	50.7
#1959	29.0	30.5	39.2	47.4	50.6	61.7	70.6	64.4	55.1	46.1	31.5	28.8	46.3
1960	20.5	30.6	38.2	45.9	51.2	62.2	75.1	65.1	60.3	48.6	35.2	26.6	46.6
1961	30.3	37.0	39.9	45.1	53.0	66.6	71.9	74.0	55.9	45.2	30.7	26.6	48.0
1962	22.6	32.4	34.6	49.8	50.9	61.1	68.2	65.4	60.7	47.6	37.9	33.1	47.0
1963	19.3	37.4	40.6	45.0	54.8	61.7	66.7	69.1	66.0	51.1	38.1	26.5	48.0
1964	29.3	29.2	35.7	44.5	52.9	60.4	68.3	62.8	55.2	47.9	32.2	24.0	45.2
1965	28.6	32.1	34.4	47.2	52.4	61.3	70.1	67.9	53.8	52.7	38.0	30.0	47.4
1966	29.7	32.8	38.7	46.0	55.9	58.8	68.2	68.2	64.6	47.3	36.9	33.4	48.4
1967	33.9	36.2	37.1	42.3	52.8	63.5	70.6	74.5	65.3	48.4	35.4	27.8	49.0
1968	27.8	37.8	42.1	43.0	53.8	61.2	71.1	65.1	58.9	45.2	34.6	24.6	47.1
1969	16.3	26.1	35.6	46.2	57.4	65.2	67.4	67.1	59.8	43.7	36.3	29.4	45.9
1970	25.9	36.3	37.0	41.6	54.9	66.2	72.5	70.2	54.2	44.9	36.0	27.7	47.3
1971	31.8	33.6	35.2	45.3	56.3	58.2	69.7	74.1	55.2	44.2	35.4	25.8	47.0
1972	22.6	30.7	41.4	42.0	56.9	62.0	68.1	71.1	55.4	47.2	38.3	25.4	46.8
1973	27.0	34.9	41.1	46.2	56.5	62.0	71.2	69.1	59.7	47.2	33.7	33.3	48.5
1974	24.1	35.4	38.5	46.4	50.2	66.0	67.8	68.2	60.5	48.0	36.4	30.5	47.7
1975	23.6	24.7	34.0	41.7	52.7	59.2	72.4	64.1	61.0	46.9	33.8	30.9	45.4
1976	29.6	32.1	35.1	45.2	54.5	58.5	68.8	65.4	63.4	46.8	35.8	29.6	47.1
1977	22.0	35.1	38.2	50.9	51.6	65.0	67.0	71.2	55.1	46.5	34.0	26.2	46.9
1978	27.6	34.0	42.2	45.7	51.4	62.7	68.3	65.9	56.6	46.6	28.7	19.0	45.7
1979	10.5	28.8	40.4	45.5	54.7	62.7	70.4	70.0	63.1	51.1	30.5	35.2	46.9
1980	20.7	34.5	38.6	51.7	55.8	57.8	69.2	64.1	58.4	47.4	36.3	33.2	47.3
1981	32.8	33.9	40.9	45.7	52.0	57.0	65.1	71.5	59.7	45.9	39.9	29.7	47.8
1982	26.0	32.1	40.3	43.5	54.2	66.5	67.6	69.8	59.5	46.1	31.7	27.3	47.0
1983	35.8	38.1	43.0	46.3	57.1	61.9	65.5	72.3	57.1	49.7	39.3	16.2	48.5
1984	30.5	34.5	41.7	44.0	50.1	59.2	69.1	70.1	56.7	43.4	35.8	20.4	46.3
1985	21.4	24.9	35.9	48.0	56.2	61.8	75.0	64.9	53.3	44.7	19.5	19.3	43.7
Record Mean	26.6	31.8	39.4	47.5	55.6	62.4	70.1	68.7	59.3	48.5	36.6	29.9	48.0
Max	32.5	38.8	48.3	58.5	67.3	74.4	84.2	82.8	72.1	59.1	43.1	35.2	58.0
Min	20.8	24.7	30.5	36.6	43.9	50.3	55.9	54.5	46.5	37.8	30.1	24.6	38.0

REFERENCE NOTES FOR TABLES 1, 2, 3 and 6   (SPOKANE, WA)

GENERAL

T - TRACE AMOUNT
BLANK ENTRIES DENOTE MISSING/UNREPORTED DATA.
# INDICATES A STATION OR INSTRUMENT RELOCATION.

SPECIFIC

TABLE 1

(a) - LENGTH OF RECORD IN YEARS. ALTHOUGH
      INDIVIDUAL MONTHS MAY BE MISSING.
  * LESS THAN .05

NORMALS — BASED ON THE 1951-1980 RECORD PERIOD.
EXTREMES — DATES ARE THE MOST RECENT OCCURRENCE.
WIND DIR. — NUMERALS SHOW TENS OF DEGREES
            CLOCKWISE FROM TRUE NORTH.
            "00" INDICATES CALM.
RESULTANT WIND DIRECTIONS ARE GIVEN TO WHOLE DEGREES.

EXCEPTIONS

TABLES 2, 3, and 6

RECORD MEANS ARE THROUGH THE CURRENT YEAR,
BEGINNING IN   1882 FOR TEMPERATURE
                 1882 FOR PRECIPITATION
                 1948 FOR SNOWFALL

TABLE 4 HEATING DEGREE DAYS Base 65 deg. F   SPOKANE WASHINGTON

SEASON	JULY	AUG	SEP	OCT	NOV	DEC	JAN	FEB	MAR	APR	MAY	JUNE	TOTAL
.1956-57	17	66	149	572	963	1055	1547	1011	841	506	186	129	7042
1957-58	30	39	105	612	877	922	1015	697	804	596	143	76	5916
1958-59	5	10	226	452	868	1005	1111	960	794	517	441	131	6520
#1959-60	55	84	311	581	997	1113	1374	990	823	567	420	118	7433
1960-61	3	97	153	503	886	1186	1067	775	771	592	367	70	6470
1961-62	1	0	268	604	1025	1180	1309	905	932	450	430	149	7253
1962-63	60	65	159	531	804	981	1411	748	785	318	145	145	6583
1963-64	35	28	77	426	802	1186	1098	1030	897	609	369	149	6706
1964-65	31	118	290	524	976	1268	1121	915	942	528	387	129	7229
1965-66	31	62	330	377	804	1078	1088	896	808	561	291	190	6516
1966-67	30	42	67	544	838	975	956	799	859	677	370	96	6253
1967-68	8	2	71	508	882	1146	1149	783	702	654	343	138	6386
1968-69	19	89	199	607	897	1245	1504	1080	905	559	236	88	7428
1969-70	40	44	192	655	855	1097	1208	797	859	696	305	101	6849
1970-71	13	5	321	614	864	1146	1022	873	918	584	270	215	6845
1971-72	64	19	297	641	882	1208	1308	991	726	684	274	127	7221
1972-73	36	18	292	545	795	1219	1171	838	734	558	286	152	6644
1973-74	17	47	193	546	933	978	1265	824	814	554	455	97	6723
1974-75	41	22	134	519	852	1062	1276	1122	953	694	375	173	7223
1975-76	22	75	136	554	933	1048	1091	946	922	588	317	213	6845
1976-77	20	71	74	556	871	1089	1324	832	824	436	409	66	6572
1977-78	57	56	289	563	921	1197	1154	862	701	576	412	101	6889
1978-79	37	97	252	562	1083	1424	1684	1011	756	577	313	134	7930
1979-80	41	4	91	423	1029	918	1365	880	809	392	283	211	6446
1980-81	19	77	195	543	854	977	992	867	741	570	395	243	6473
1981-82	73	7	209	584	747	1088	1202	912	761	639	328	76	6626
1982-83	62	17	193	582	996	1163	890	747	672	558	285	113	6285
1983-84	55	2	230	468	765	1508	1065	880	715	621	460	194	6963
1984-85	21	18	264	662	870	1381	1345	1117	895	501	280	128	7482
1985-86	0	64	343	622	1363	1409							

TABLE 5 COOLING DEGREE DAYS Base 65 deg. F   SPOKANE WASHINGTON

YEAR	JAN	FEB	MAR	APR	MAY	JUNE	JULY	AUG	SEP	OCT	NOV	DEC	TOTAL
1969	0	0	0	0	7	99	121	112	40	0	0	0	379
1970	0	0	0	0	3	143	253	175	2	0	0	0	576
1971	0	0	0	0	10	17	216	306	9	0	0	0	558
1972	0	0	0	0	28	41	138	213	10	0	0	0	430
1973	0	0	0	0	31	67	216	177	39	0	0	0	530
1974	0	0	0	0	0	137	134	127	7	0	0	0	405
1975	0	0	0	0	0	7	256	57	20	0	0	0	340
1976	0	0	0	0	0	24	143	93	33	0	0	0	293
1977	0	0	0	18	2	72	126	254	0	0	0	0	472
1978	0	0	0	0	0	42	144	131	9	0	0	0	326
1979	0	0	0	0	1	73	217	166	39	0	0	0	496
1980	0	0	0	1	3	2	156	56	6	3	0	0	227
1981	0	0	0	0	0	9	82	213	60	0	0	0	364
1982	0	0	0	0	2	128	148	171	32	0	0	0	481
1983	0	0	0	0	46	26	77	235	1	0	0	0	385
1984	0	0	0	0	3	28	155	181	23	1	0	0	391
1985	0	0	0	0	15	36	317	68	0	0	0	0	436

TABLE 6 SNOWFALL (inches)   SPOKANE WASHINGTON

SEASON	JULY	AUG	SEP	OCT	NOV	DEC	JAN	FEB	MAR	APR	MAY	JUNE	TOTAL
1956-57	0.0	0.0	0.0	0.7	1.5	7.3	20.8	10.5	6.4	1.4	0.0	0.0	48.6
1957-58	0.0	0.0	0.0	6.1	2.0	10.8	7.4	0.7	1.9	T	0.0	0.0	28.9
1958-59	0.0	0.0	0.0	0.0	8.6	8.4	22.4	20.6	3.1	T	T	0.0	63.1
1959-60	0.0	0.0	0.0	0.0	10.7	6.5	8.3	3.5	9.0	0.6	T	0.0	38.6
1960-61	0.0	0.0	0.0	0.0	2.5	8.6	7.8	7.8	6.7	0.1	0.0	0.0	33.5
1961-62	0.0	0.0	0.0	T	8.9	26.2	12.2	4.5	15.3	T	0.0	0.0	67.1
1962-63	0.0	0.0	0.0	0.0	4.4	0.5	8.7	2.7	1.2	0.5	0.0	0.0	18.0
1963-64	0.0	0.0	0.0	0.0	0.4	19.6	26.3	5.2	5.2	6.6	0.0	0.0	63.3
1964-65	0.0	0.0	0.0	T	15.2	42.0	20.1	2.3	2.0	0.1	0.0	0.0	81.7
1965-66	0.0	0.0	0.0	T	6.3	15.4	13.9	1.6	7.2	T	T	0.0	44.4
1966-67	0.0	0.0	0.0	T	0.9	9.0	6.5	3.1	6.4	0.8	3.5	0.0	30.2
1967-68	0.0	0.0	0.0	T	4.8	12.7	11.8	0.4	T	T	T	0.0	29.7
1968-69	0.0	0.0	0.0	0.0	1.2	19.8	48.7	5.4	2.0	0.4	0.0	0.0	77.5
1969-70	0.0	0.0	0.0	0.0	T	10.4	19.4	2.8	6.9	0.3	0.1	0.0	39.9
1970-71	0.0	0.0	0.0	T	6.8	12.0	6.1	5.5	1.5	T	0.0	0.0	31.9
1971-72	0.0	0.0	0.0	3.1	4.0	34.2	17.2	5.9	2.5	0.2	0.0	0.0	67.1
1972-73	0.0	0.0	0.0	0.8	T	4.7	6.5	3.5	0.5	T	0.0	0.0	16.0
1973-74	0.0	0.0	0.0	0.8	23.6	9.1	15.0	4.4	2.5	0.4	0.4	0.0	56.2
1974-75	0.0	0.0	0.0	0.0	0.3	16.6	30.9	28.5	7.6	5.1	T	0.0	89.0
1975-76	0.0	0.0	0.0	3.9	11.4	15.3	6.3	4.6	0.4	0.0	0.0		48.8
1976-77	0.0	0.0	0.0	0.0	0.1	4.2	6.8	2.5	2.7	T	T	0.0	16.3
1977-78	0.0	0.0	0.0	0.0	11.2	30.3	19.1	6.6	2.2	T	T	0.0	69.4
1978-79	0.0	0.0	0.0	0.0	15.4	14.8	16.5	10.6	3.4	0.4	0.0	0.0	60.7
1979-80	0.0	0.0	0.0	0.0	3.9	10.4	16.6	5.9	1.1	0.4	0.0	0.0	38.3
1980-81	0.0	0.0	0.0	0.0	1.2	6.8	2.6	3.3	T	T	0.3	0.0	14.2
1981-82	0.0	0.0	0.0	T	0.8	13.0	23.3	2.2	2.1	6.0	T	0.0	47.4
1982-83	0.0	0.0	0.0	T	5.4	17.4	8.1	5.5	T	0.2	T	0.0	36.6
1983-84	0.0	0.0	0.0	0.0	5.7	24.8	5.3	8.0	1.9	1.3	0.8	0.0	47.8
1984-85	0.0	0.0	0.0	1.1	12.0	24.7	4.6	14.8	9.6	T	T	0.0	66.8
1985-86	0.0	0.0	0.0	0.4	23.7	8.3							
Record Mean	0.0	0.0	0.0	0.5	6.4	15.4	17.0	7.5	4.3	0.7	0.1	T	51.8

See Reference Notes, relative to all above tables, on preceding page.

The location of the station on top of the main divide of the Cascade Range greatly influences the climatic conditions of the area around the station, and at the same time makes the station climate vastly different from a place only 30 miles east or west of the station and at a lower elevation. To the west the land drops rapidly into the Green River Valley, and thence into the Puget Sound area. Eastward the land drops abruptly into the Yakima River Valley, and then opens out into the Columbia Basin to the southeast. The land around the station is covered with forests of fir, pine, and hemlock. Eastward the forests change to pine and then to open land covered with low brush. Westward the forests grow more dense with cedar replacing the hemlock. The station stands on the dividing line between a moist coastal and a semi-desert climate.

The lifting of air as it passes over the mountains both intensifies and lengthens any storm system. The mountains provide additional triggering action to unstable air with the increase of clouds, fog, and precipitation.

Temperatures seldom reach above the low 80s or drop below zero. Both high and low temperature extremes occur with a high pressure system centered in the Columbia Basin or southwest Canada that causes east winds at the station. The daily temperature range is small since the colder night air drains into the adjacent river valleys.

The precipitation regime is a simple curve, with the high in December and the low in July. From the middle of October until the middle of April, nearly all of the precipitation falls as snow. During the winter season the snow usually reaches a depth of at least 100 inches, while the maximum depth may be well over 220 inches.

As frost and snow has occurred in every month except August, a growing season in the normal sense does not exist.

Based on the 1951-1980 period, the average first occurrence of 32 degrees Fahrenheit in the fall is October 4 and the average last occurrence in the spring is June 2.

TABLE 1   NORMALS, MEANS AND EXTREMES

STAMPEDE PASS WASHINGTON

LATITUDE: 47°17'N   LONGITUDE: 121°20' W   ELEVATION: FT. GRND 3958 BARO 03967   TIME ZONE: PACIFIC   WBAN: 24237

	(a)	JAN	FEB	MAR	APR	MAY	JUNE	JULY	AUG	SEP	OCT	NOV	DEC	YEAR
TEMPERATURE °F:														
Normals														
-Daily Maximum		28.1	32.1	34.7	41.0	49.5	56.8	65.1	64.0	58.6	47.8	34.8	30.2	45.2
-Daily Minimum		20.1	24.1	25.3	29.5	35.5	41.4	47.1	47.2	43.6	36.3	27.1	22.7	33.3
-Monthly		24.1	28.1	30.0	35.3	42.5	49.1	56.1	55.6	51.1	42.0	31.0	26.5	39.3
Extremes														
-Record Highest	42	59	57	61	76	84	86	88	90	87	73	58	64	90
-Year		1968	1962	1960	1977	1983	1961	1985	1958	1944	1980	1969	1980	AUG 1958
-Record Lowest	42	-11	-9	1	16	20	28	30	34	26	15	-5	-21	-21
-Year		1950	1950	1960	1951	1954	1950	1955	1973	1972	1984	1955	1968	DEC 1968
NORMAL DEGREE DAYS:														
Heating (base 65°F)		1268	1033	1082	891	698	474	286	313	424	710	1023	1194	9396
Cooling (base 65°F)		0	0	0	0	0	0	10	22	7	0	0	0	39
% OF POSSIBLE SUNSHINE														
MEAN SKY COVER (tenths)														
Sunrise - Sunset	36	8.6	8.5	8.3	8.1	7.6	7.4	4.9	5.4	5.8	7.3	8.5	8.9	7.4
MEAN NUMBER OF DAYS:														
Sunrise to Sunset														
-Clear	36	2.8	2.4	3.1	3.1	4.4	5.4	13.0	11.4	10.4	5.9	2.8	1.8	66.5
-Partly Cloudy	36	2.6	3.0	4.0	4.7	5.8	5.4	7.4	7.6	5.8	4.7	3.1	2.5	56.6
-Cloudy	36	25.6	22.9	23.9	22.2	20.8	19.2	10.6	11.4	13.8	20.4	24.1	26.6	241.5
Precipitation														
.01 inches or more	42	21.8	19.8	21.0	18.9	16.1	15.1	8.7	10.9	12.5	16.0	20.3	22.8	203.8
Snow, Ice pellets														
1.0 inches or more	42	14.7	12.7	13.5	9.9	3.4	0.5	0.1	0.0	0.4	4.4	11.5	14.6	85.8
Thunderstorms	22	0.0	0.0	0.1	0.2	1.1	1.6	1.3	1.9	0.8	0.2	0.*	0.0	7.2
Heavy Fog Visibility														
1/4 mile or less	22	23.5	21.6	22.6	21.0	21.5	21.5	17.2	18.9	15.5	20.2	22.5	26.2	252.2
Temperature °F														
-Maximum														
90° and above	42	0.0	0.0	0.0	0.0	0.0	0.0	0.0	0.*	0.0	0.0	0.0	0.0	*
32° and below	42	22.1	15.7	13.0	4.6	0.6	0.0	0.0	0.0	0.0	1.1	12.5	21.0	90.7
-Minimum														
32° and below	42	29.7	26.5	28.2	22.9	11.3	1.4	0.1	0.0	1.2	10.9	25.3	29.5	186.7
0° and below	42	1.6	0.2	0.0	0.0	0.0	0.0	0.0	0.0	0.0	0.0	0.3	0.6	2.7
AVG. STATION PRESS.(mb)	11	877.4	876.3	875.1	878.2	879.3	880.1	881.5	880.9	881.2	880.8	877.9	877.9	878.9
RELATIVE HUMIDITY (%)														
Hour 04	41	92	91	90	88	89	90	86	86	81	86	92	94	89
Hour 10	42	91	90	85	80	75	74	66	69	70	79	91	93	80
Hour 16 (Local Time)	40	89	87	81	76	70	69	59	63	66	76	90	92	77
Hour 22														
PRECIPITATION (inches):														
Water Equivalent														
-Normal		14.59	10.19	8.88	6.28	3.97	3.84	1.56	2.85	4.65	7.74	12.14	15.88	92.57
-Maximum Monthly	42	30.42	20.78	19.54	12.50	9.12	7.98	5.96	7.17	15.24	23.55	25.43	29.06	30.42
-Year		1969	1961	1955	1970	1960	1947	1983	1975	1959	1967	1958	1953	JAN 1969
-Minimum Monthly	42	0.97	1.36	2.56	1.12	0.98	0.37	0.19	0.36	0.37	1.33	0.93	1.73	0.19
-Year		1985	1973	1965	1951	1958	1965	1960	1955	1975	1980	1952	1985	JUL 1960
-Maximum in 24 hrs	42	5.74	3.89	3.34	5.41	2.25	2.76	2.30	2.54	4.47	5.32	7.94	7.90	7.94
-Year		1945	1979	1954	1962	1971	1974	1983	1975	1959	1947	1962	1977	NOV 1962
Snow, Ice pellets														
-Maximum Monthly	42	192.9	181.7	154.8	107.2	45.9	9.4	6.4	T	26.7	65.4	138.5	163.3	192.9
-Year		1946	1949	1955	1970	1984	1954	1971	1964	1972	1984	1945	1949	JAN 1946
-Maximum in 24 hrs	42	37.7	36.2	25.3	25.8	18.8	6.5	6.4	T	11.9	17.2	28.6	32.8	37.7
-Year		1972	1949	1954	1970	1971	1955	1971	1964	1972	1977	1945	1951	JAN 1972
WIND:														
Mean Speed (mph)														
Prevailing Direction														
Fastest Obs. 1 Min.														
-Direction (!!)														
-Speed (MPH)														
-Year														
Peak Gust														
-Direction (!!)														
-Speed (mph)														
-Date														

See Reference Notes to this table on the following page.

TABLE 2   PRECIPITATION (inches)   STAMPEDE PASS WASHINGTON

YEAR	JAN	FEB	MAR	APR	MAY	JUNE	JULY	AUG	SEP	OCT	NOV	DEC	ANNUAL
1956	10.38	9.91	12.79	2.04	1.84	3.59	1.05	1.32	5.13	11.57	7.60	27.07	94.29
1957	6.68	9.44	9.08	4.28	2.51	3.71	1.23	1.38	0.92	4.86	8.41	19.13	71.63
1958	8.97	7.89	3.45	9.11	0.98	3.82	0.26	1.25	6.23	10.04	25.43	15.98	93.41
1959	14.99	6.13	13.69	6.27	5.53	2.93	2.20	3.79	15.24	13.98	19.34	13.66	117.75
1960	4.37	12.92	11.15	7.22	9.12	4.09	0.19	5.06	2.51	10.44	16.41	6.80	90.28
1961	10.93	20.78	9.24	9.87	3.70	2.19	1.36	1.44	5.55	9.78	12.05	18.43	105.32
1962	12.69	2.98	8.48	11.47	3.90	1.99	1.79	3.71	3.88	4.86	19.78	9.44	84.97
1963	5.52	9.06	8.03	6.39	2.35	3.38	2.95	1.56	1.64	6.03	11.16	7.54	65.61
1964	21.83	8.78	15.53	9.12	4.63	4.63	2.15	6.19	8.23	4.64	15.46	11.33	111.58
1965	18.64	11.07	2.56	4.00	4.04	0.37	0.31	3.93	2.24	5.01	8.23	6.20	66.60
1966	11.99	10.30	5.80	3.88	4.19	4.63	2.97	1.90	1.39	9.60	10.03	15.96	82.64
1967	19.81	15.51	8.49	4.28	2.05	1.17	0.33	0.47	2.40	23.55	8.62	16.06	102.74
1968	13.80	9.19	6.89	9.71	4.93	6.19	1.62	7.14	8.99	8.83	13.12	16.31	106.72
1969	30.42	8.23	4.35	7.09	3.07	5.21	1.59	1.19	7.51	3.75	5.29	8.74	86.44
1970	16.36	6.63	6.60	12.50	2.44	2.24	1.28	0.83	5.60	6.23	14.26	13.12	88.09
1971	23.29	10.89	11.08	2.43	4.62	3.38	2.01	1.42	3.95	9.92	11.01	18.68	102.68
1972	27.18	19.34	12.31	8.06	2.62	5.54	3.40	1.27	7.91	1.81	18.83	7.17	115.44
1973	7.97	1.36	5.08	3.69	3.63	5.33	0.31	0.87	4.02	9.65	13.56	16.00	71.47
1974	25.37	12.35	10.19	6.86	5.78	6.58	2.21	1.39	2.58	2.76	11.94	23.03	111.04
1975	23.88	9.19	6.24	3.32	5.50	4.56	1.28	7.17	0.37	12.93	21.10	28.56	124.10
1976	18.58	11.78	10.78	5.61	4.00	3.45	1.16	5.63	2.06	6.05	4.94	9.55	83.84
1977	5.38	8.15	11.87	2.67	5.81	1.79	2.36	5.76	7.99	5.96	20.84	25.52	104.10
1978	5.07	3.35	4.43	6.60	6.35	3.12	2.24	4.18	6.59	2.48	12.39	11.23	68.03
1979	4.72	13.85	6.25	6.04	2.85	2.61	2.13	3.24	2.39	4.96	2.19	18.97	70.20
1980	8.16	6.46	9.30	3.83	1.74	6.36	1.07	4.86	7.97	1.33	15.12	16.86	83.06
1981	2.61	13.73	5.03	11.38	4.66	7.64	1.72	1.49	3.93	5.92	5.59	12.02	75.72
1982	16.70	12.33	6.60	3.86	2.51	0.92	3.19	1.63	6.45	7.25	6.47	8.46	76.37
1983	13.65	5.12	5.98	2.23	2.98	3.49	5.96	1.87	4.69	2.90	14.92	7.24	71.03
1984	15.71	6.93	8.24	5.43	8.65	6.15	0.24	1.08	3.77	10.18	10.96	12.44	89.78
1985	0.97	7.14	6.18	7.58	3.55	4.39	0.20	0.89	7.22	16.39	14.00	1.73	70.24
Record Mean	13.48	10.16	8.72	6.28	4.02	3.99	1.69	2.55	4.76	8.16	12.16	14.66	90.62

TABLE 3   AVERAGE TEMPERATURE (deg. F)   STAMPEDE PASS WASHINGTON

YEAR	JAN	FEB	MAR	APR	MAY	JUNE	JULY	AUG	SEP	OCT	NOV	DEC	ANNUAL
1956	22.9	20.5	27.3	38.3	46.8	45.6	58.1	57.3	51.1	37.7	30.0	29.9	38.8
1957	16.3	26.7	29.7	37.5	48.4	49.4	52.1	52.9	57.3	39.6	33.0	29.7	39.4
1958	27.4	32.6	30.9	36.2	51.3	55.2	62.8	59.9	49.8	44.9	29.2	29.7	42.5
1959	24.6	26.0	29.9	35.8	40.2	49.2	57.9	52.3	46.4	41.9	30.8	26.8	38.5
1960	22.2	25.6	31.0	36.5	39.4	49.3	60.2	51.5	51.9	43.6	30.4	25.0	38.9
1961	27.8	29.7	31.1	33.9	41.8	55.5	58.4	61.8	46.5	39.4	27.6	24.8	39.9
1962	26.6	28.7	29.4	37.9	37.8	48.9	55.0	53.1	52.6	42.6	32.4	30.3	39.6
1963	21.8	36.1	32.0	34.3	44.5	47.0	50.9	56.7	56.5	42.9	31.0	26.6	40.0
1964	26.3	29.4	28.4	31.5	38.7	46.3	54.6	51.0	47.2	43.3	28.9	20.7	37.2
1965	26.8	28.6	32.5	37.1	40.2	49.9	58.8	56.0	49.2	46.3	34.3	25.5	40.4
1966	25.9	27.3	31.0	36.7	43.7	46.7	53.7	55.6	54.4	41.5	33.3	29.9	40.0
1967	28.4	30.4	28.2	32.2	42.8	54.6	57.4	64.8	56.6	40.8	32.9	26.4	41.3
1968	25.2	33.6	34.4	33.6	43.7	48.8	59.4	53.6	50.7	37.9	30.7	20.0	39.3
1969	15.6	24.6	32.8	35.3	47.5	56.1	54.2	53.3	50.0	39.3	33.9	26.6	39.1
1970	24.0	31.6	31.2	30.9	41.8	53.1	56.5	57.2	46.7	39.9	30.3	23.6	38.9
1971	27.1	27.9	26.1	33.9	42.9	46.0	59.6	60.5	46.9	39.9	31.3	23.9	38.9
1972	22.4	26.5	32.4	30.3	45.4	49.5	57.3	59.2	45.8	42.7	32.7	21.4	38.8
1973	21.8	29.8	31.4	36.0	44.0	47.3	56.6	52.5	50.7	39.7	26.8	28.5	38.8
1974	20.8	28.3	30.1	34.6	37.7	50.1	54.1	58.2	56.0	46.2	32.6	29.7	39.9
1975	26.1	25.2	28.0	31.6	40.6	45.2	57.5	50.1	54.1	37.8	29.5	28.4	37.8
1976	29.7	27.3	27.8	35.4	42.2	45.6	54.6	53.7	54.6	43.9	35.5	31.4	40.1
1977	27.2	32.3	29.8	41.5	39.9	52.5	53.0	61.5	46.9	40.4	27.9	24.7	39.8
1978	27.4	28.7	34.7	36.1	40.3	52.7	56.3	54.7	46.0	44.3	28.0	21.7	39.2
1979	16.6	26.2	34.9	36.7	44.6	49.4	56.3	55.8	53.7	45.3	28.6	29.6	39.8
1980	18.7	27.4	30.2	40.7	44.4	46.6	55.4	52.3	51.2	44.7	32.8	32.9	39.8
1981	29.3	29.6	35.2	36.3	42.0	45.2	54.1	61.3	51.6	42.0	34.1	26.2	40.6
1982	25.7	25.5	30.8	33.3	41.6	53.6	54.7	55.7	50.8	39.6	27.4	24.3	38.6
1983	30.6	30.6	35.1	38.4	46.0	47.4	51.8	56.6	46.5	41.6	31.2	17.5	39.4
1984	28.5	29.7	34.6	33.3	38.0	46.4	56.7	55.2	47.2	36.9	29.1	22.0	38.1
1985	22.0	24.9	30.3	37.5	43.7	49.3	62.5	54.9	46.5	38.0	19.3	22.9	37.7
Record Mean	24.0	27.9	30.4	35.3	42.8	49.0	55.9	55.7	50.8	41.5	30.6	29.9	39.2
Max	28.1	32.0	35.1	41.1	49.9	56.7	65.0	64.2	58.4	47.3	34.4	29.7	45.2
Min	19.9	23.7	25.6	29.5	35.7	41.2	46.8	47.2	43.2	35.8	26.7	21.9	33.1

REFERENCE NOTES FOR TABLES 1, 2, 3 and 6   (STAMPEDE PASS, WA)

GENERAL

T - TRACE AMOUNT
BLANK ENTRIES DENOTE MISSING/UNREPORTED DATA.
# INDICATES A STATION OR INSTRUMENT RELOCATION.

SPECIFIC

TABLE 1

(a) - LENGTH OF RECORD IN YEARS. ALTHOUGH INDIVIDUAL MONTHS MAY BE MISSING.
 # LESS THAN .05

NORMALS — BASED ON THE 1951-1980 RECORD PERIOD.
EXTREMES — DATES ARE THE MOST RECENT OCCURRENCE.
WIND DIR. — NUMERALS SHOW TENS OF DEGREES CLOCKWISE FROM TRUE NORTH. "00" INDICATES CALM.
RESULTANT WIND DIRECTIONS ARE GIVEN TO WHOLE DEGREES.

EXCEPTIONS

TABLE 1

1. THUNDERSTORMS AND HEAVY FOG ARE THROUGH 1964 AND MAY BE INCOMPLETE, DUE TO PART-TIME OPERATIONS.

TABLES 2, 3, and 6

RECORD MEANS ARE THROUGH THE CURRENT YEAR, BEGINNING IN   1944 FOR TEMPERATURE
     1944 FOR PRECIPITATION
     1944 FOR SNOWFALL

TABLE 4 HEATING DEGREE DAYS Base 65 deg. F   STAMPEDE PASS WASHINGTON

SEASON	JULY	AUG	SEP	OCT	NOV	DEC	JAN	FEB	MAR	APR	MAY	JUNE	TOTAL
1956-57	239	273	410	834	1043	1081	1506	1069	1085	818	509	464	9331
1957-58	393	368	233	778	951	1084	1156	900	1048	857	423	319	8510
1958-59	109	182	455	615	1069	1087	1244	1087	1083	866	761	470	9028
1959-60	260	388	553	709	1018	1176	1321	1135	1049	848	788	465	9710
1960-61	186	424	387	653	1030	1233	1148	982	1042	926	713	306	9030
1961-62	225	153	546	785	1117	1240	1182	1010	1095	809	834	477	9473
1962-63	330	362	368	684	972	1067	1333	803	1016	916	629	540	9020
1963-64	430	265	262	677	1015	1181	1192	1024	1129	997	809	553	9534
1964-65	322	431	525	665	1077	1365	1179	1015	1001	829	760	445	9614
1965-66	231	290	468	573	914	1219	1205	1051	1047	842	656	542	9038
1966-67	346	294	320	723	943	1081	1130	961	1134	979	683	318	8912
1967-68	244	100	260	742	955	1191	1227	904	942	931	653	487	8636
1968-69	212	359	424	833	1023	1389	1525	1126	990	885	536	286	9588
1969-70	328	360	446	791	927	1186	1263	928	1038	1015	715	360	9357
1970-71	272	254	541	770	1034	1274	1168	1033	1200	925	681	563	9715
1971-72	247	175	536	770	1004	1265	1313	1111	1002	1033	600	459	9515
1972-73	241	218	574	682	963	1345	1334	980	1033	862	644	529	9405
1973-74	265	390	426	777	1142	1124	1363	1021	1075	904	840	441	9768
1974-75	353	252	273	577	965	1087	1198	1109	1141	995	748	586	9284
1975-76	260	456	323	838	1060	1128	1088	1085	1147	880	700	577	9542
1976-77	318	355	303	646	878	1035	1166	910	1082	701	771	373	8538
1977-78	372	192	536	756	1105	1243	1159	1011	934	863	758	371	9300
1978-79	283	343	563	633	1101	1336	1494	1080	928	840	624	462	9687
1979-80	291	278	333	602	1086	1093	1425	1083	1072	721	633	546	9163
1980-81	298	387	407	622	960	990	1102	986	920	853	706	590	8821
'1981-82	341	194	413	704	918	1195	1213	1104	1055	942	717	351	9147
1982-83	339	298	422	782	1124	1254	1061	957	921	793	588	522	9061
1983-84	405	260	549	717	1005	1468	1124	1016	936	943	829	551	9803
1984-85	274	305	524	863	1068	1327	1322	1115	1067	818	651	469	9803
1985-86	120	314	547	828	1365	1299							

TABLE 5 COOLING DEGREE DAYS Base 65 deg. F   STAMPEDE PASS WASHINGTON

YEAR	JAN	FEB	MAR	APR	MAY	JUNE	JULY	AUG	SEP	OCT	NOV	DEC	TOTAL
1969	0	0	0	0	0	25	1	6	4	0	0	0	36
1970	0	0	0	0	0	9	17	21	0	0	0	0	47
1971	0	0	0	0	0	0	88	42	0	0	0	0	130
1972	0	0	0	0	0	0	9	47	3	0	0	0	59
1973	0	0	0	0	0	4	15	7	4	0	0	0	30
1974	0	0	0	0	0	0	22	48	7	0	0	0	77
1975	0	0	0	0	0	0	35	0	1	0	0	0	36
1976	0	0	0	0	0	0	2	12	0	0	0	0	14
1977	0	0	0	2	0	5	6	88	0	0	0	0	101
1978	0	0	0	0	0	12	19	29	0	0	0	0	60
1979	0	0	0	0	0	1	32	0	0	0	0	0	33
1980	0	0	0	0	0	0	6	3	0	0	0	0	9
1981	0	0	0	0	0	0	11	86	18	0	0	0	115
1982	0	0	0	0	0	16	29	14	2	0	0	0	61
1983	0	0	0	0	5	0	2	7	0	0	0	0	14
1984	0	0	0	0	0	0	25	7	0	0	0	0	32
1985	0	0	0	0	0	5	53	5	0	0	0	0	63

TABLE 6 SNOWFALL (inches)   STAMPEDE PASS WASHINGTON

SEASON	JULY	AUG	SEP	OCT	NOV	DEC	JAN	FEB	MAR	APR	MAY	JUNE	TOTAL
1956-57	0.0	0.0	0.0	27.2	40.1	48.0	57.1	72.7	72.2	16.2	T	0.0	333.5
1957-58	0.0	0.0	0.0	8.8	44.7	143.3	69.4	47.8	26.8	48.4	0.7	0.0	389.9
1958-59	0.0	0.0	1.3	8.1	90.7	37.5	84.5	41.5	97.3	21.9	25.4	T	408.2
1959-60	0.0	0.0	0.0	14.1	25.7	53.3	95.6	37.3	100.2	57.5	31.2	1.5	416.4
1960-61	0.0	0.0	0.0	31.4	110.6	57.8	64.5	154.7	60.8	51.0	8.0	T	538.8
1961-62	0.0	0.0	3.0	47.9	63.9	110.8	37.0	24.5	93.1	51.1	5.9	8.0	445.2
1962-63	2.1	0.0	0.9	11.3	84.0	44.4	26.1	27.5	65.2	29.6	14.0	T	305.1
1963-64	0.0	0.0	0.0	22.2	76.4	51.5	163.6	69.5	115.4	62.0	18.3	T	578.9
1964-65	0.0	T	1.6	6.9	52.8	83.2	99.6	73.6	23.7	11.7	17.3	0.0	370.4
1965-66	0.0	0.0	0.0	3.8	24.8	42.6	82.6	86.5	40.7	19.7	13.7	0.1	314.5
1966-67	0.0	0.0	0.0	26.4	50.3	75.2	114.6	102.0	67.5	38.2	8.8	0.0	483.0
1967-68	0.0	0.0	0.0	26.3	36.8	59.7	83.2	21.9	58.7	88.7	12.6	1.0	388.9
1968-69	0.0	0.0	1.8	38.2	82.6	6.7	98.9	34.6	31.3	59.1	3.0	0.0	356.2
1969-70	0.0	0.0	0.0	8.7	27.5	98.8	101.8	65.4	63.0	107.2	10.6	T	483.0
1970-71	0.0	0.0	1.5	24.6	109.3	143.1	59.6	88.7	21.0	29.7	T		579.2
1971-72	6.4	0.0	T	26.2	48.3	156.1	230.1	112.1	58.7	65.0	1.6	0.0	704.5
1972-73	0.0	0.0	26.7	12.8	30.6	96.6	43.7	9.1	43.9	26.1	19.0	2.1	310.6
1973-74	0.0	0.0	0.0	20.4	103.8	126.1	144.9	111.0	90.5	54.4	42.5	1.9	695.5
1974-75	0.0	0.0	0.0	1.4	87.5	131.3	109.5	74.4	42.3	27.2	12.5	T	486.1
1975-76	0.0	0.0	0.0	43.7	92.5	61.8	92.1	89.1	91.3	32.6	5.8	6.4	515.3
1976-77	0.0	0.0	T	0.3	6.0	38.1	10.9	39.3	107.0	18.5	26.2	1.7	248.0
1977-78	0.0	0.0	5.8	24.6	100.2	102.9	50.6	31.8	25.6	36.6	24.9	T	403.0
1978-79	0.0	0.0	5.4	8.6	56.9	66.1	42.4	101.5	27.8	54.2	10.0	1.6	374.5
1979-80	5.8	0.0	0.0	22.4	19.7	76.6	77.4	43.1	87.5	23.4	2.7	2.0	360.6
1980-81	0.0	0.0	T	3.8	28.8	34.2	21.6	38.1	44.8	79.1	19.6	1.7	271.7
1981-82	T	0.0	0.0	15.0	38.8	89.1	114.6	49.6	41.8	29.8	19.5	0.0	398.2
1982-83	T	0.0	T	33.6	60.3	53.1	77.2	42.0	37.1	19.0	9.1	0.0	331.4
1983-84	0.0	0.0	4.0	0.6	106.6	70.7	49.8	65.8	61.0	50.1	45.9	3.8	458.3
1984-85	0.0	0.0	5.1	65.4	104.4	114.7	7.4	53.8	51.1	51.3	16.3	T	469.5
1985-86	0.0	0.0	T	52.1	75.8	17.3							
Record Mean	0.4	T	1.5	19.4	62.7	79.1	77.3	68.3	65.2	43.1	13.9	1.4	432.4

See Reference Notes, relative to all above tables, on preceding page.

Walla Walla is located near the upper end of a wide valley which extends from west to east from the main valley of the Columbia River. The Blue Mountains rise to the south, southeast, and east of the city, with the foothills of those mountains 10 to 15 miles away. North and northeast of the valley there are rolling hills with elevations to about 2,000 feet. The Walla Walla River and a number of smaller tributary streams originating along the slopes of the Blue Mountains, flow westward through the valley and into the Columbia River.

The moderate climate of the Walla Walla area is due to the prevailing flow of air from over the Pacific Ocean. Cold spells are caused by outbreaks of frigid air from over Canada. They are broken by chinooks. Summer hot spells are caused by a northward drift of warm dry air. Most summers have about four days with temperatures of 100 degrees or higher. Hot spells are broken by a flow of air from over the ocean. Annual precipitation over the city usually ranges between 11 and 20 inches. Precipitation is quite light because the prevailing flow of air from over the ocean loses much of its moisture while crossing the Cascade Mountains. Terrain effects on most elements of local weather are frequently greater than those associated with migratory weather systems.

There is a wide variety of growing conditions because the elevation of farmland varies from about 360 feet to over 3,000 feet. Temperatures are lower and precipitation is heavier over the higher elevations. Some 15 to 20 miles west of Walla Walla, where the elevation is about 500 feet, the annual precipitation is less than 10 inches. Over the higher elevations to the south and east the annual precipitation is 35 to 40 inches and more. Wheat and peas, two of the main crops, are grown to an elevation of over 3,000 feet. Advantage is taken of this range in elevation in order to space the maturing and harvesting of green peas over a longer season by later plantings at higher levels. The local agricultural industry also includes a diversity of vegetables, fruits, grains, grasses, seeds and livestock. Use of irrigation over much of the area including the higher elevations is increasing.

Ice storms, also called silver thaws, are infrequent, but they occasionally cause hazardous conditions and some damage. Deposits of glaze are built up on exposed objects during falls of freezing precipitation. Hail is quite infrequent over the lower elevations. It is more likely to fall over the higher elevations. A few tornadoes, generally small-bore funnel clouds which touch the ground briefly have been observed over the Walla Walla area. The most likely season is in April, May, and June but they may occur in any month. Winds are generally quite light but occasional damaging windstorms and duststorms may be expected. Periods of air stagnation also occur.

Flash floods are rare but significant events. They are caused by cloudbursts from thunderstorm clouds mostly over the higher elevations in late spring and summer. Slow rising floods in small streams are generally preceded by snow pack on frozen ground followed by chinook winds and rains.

Humidity is moderate most of the time. Highest humidity is generally during periods of fog in late fall and winter. Lowest humidity usually occurs on hot summer afternoons.

Based on the 1951-1980 period, the average first occurrence of 32 degrees Fahrenheit in the fall is November 4 and the average last occurrence in the spring is March 27.

TABLE 1 **NORMALS, MEANS AND EXTREMES**

WALLA WALLA WASHINGTON

LATITUDE: 46°02'N   LONGITUDE: 118°20'W   ELEVATION: FT. GRND 949 BARO 00976   TIME ZONE: PACIFIC   WBAN: 94103

	(a)	JAN	FEB	MAR	APR	MAY	JUNE	JULY	AUG	SEP	OCT	NOV	DEC	YEAR
TEMPERATURE °F:														
Normals														
-Daily Maximum		40.1	47.3	54.3	62.2	70.9	79.3	88.8	86.0	77.3	64.0	49.0	43.0	63.5
-Daily Minimum		28.4	33.9	37.3	42.4	49.3	55.8	62.2	61.2	53.6	44.5	35.9	31.4	44.7
-Monthly		34.3	40.6	45.8	52.3	60.1	67.6	75.5	73.6	65.5	54.3	42.5	37.2	54.1
Extremes														
-Record Highest	71	71	71	71	79	93	99	106	112	113	103	90	80	73	113
-Year		1968	1963	1960	1977	1983	1983	1961	1928	1961	1955	1980	1975	1921	AUG 1961
-Record Lowest	71	-16	-14	13	19	28	41	46	45	26	15	-5	-14	-16
-Year		1957	1950	1955	1936	1954	1984	1971	1918	1926	1935	1985	1919	JAN 1957
NORMAL DEGREE DAYS:														
Heating (base 65°F)		952	680	595	381	175	54	0	16	85	332	675	862	4807
Cooling (base 65°F)		0	0	0	0	23	132	326	282	100	0	0	0	863
% OF POSSIBLE SUNSHINE	68	22	34	50	60	66	71	85	82	72	58	30	18	54
MEAN SKY COVER (tenths)														
Sunrise - Sunset	42	8.7	8.0	7.1	6.5	5.9	5.1	2.8	3.3	4.3	5.8	8.2	8.8	6.2
MEAN NUMBER OF DAYS:														
Sunrise to Sunset														
-Clear	69	2.2	3.3	5.9	7.4	9.4	12.0	20.9	20.0	14.8	10.8	3.8	2.1	112.7
-Partly Cloudy	69	4.8	5.7	8.3	9.8	10.7	9.8	7.0	6.8	7.8	7.2	5.8	4.2	88.0
-Cloudy	69	24.0	19.2	16.8	12.7	10.9	8.2	3.1	4.1	7.4	12.9	20.4	24.7	164.6
Precipitation														
.01 inches or more	71	13.4	11.5	11.5	9.4	8.5	6.9	2.6	3.5	5.4	8.4	11.9	14.1	107.0
Snow,Ice pellets														
1.0 inches or more	71	2.5	1.1	0.5	0.*	0.0	0.0	0.0	0.0	0.0	0.1	0.7	1.8	6.6
Thunderstorms	51	0.*	0.1	0.4	0.8	1.7	2.2	2.0	2.1	1.4	0.4	0.1	0.0	11.2
Heavy Fog Visibility														
1/4 mile or less	51	3.3	2.0	0.4	0.4	0.*	0.0	0.0	0.0	0.*	0.4	3.0	5.2	14.3
Temperature °F														
-Maximum														
90° and above	71	0.0	0.0	0.0	0.1	1.0	4.8	15.4	12.4	2.7	0.*	0.0	0.0	36.5
32° and below	71	9.9	2.5	0.2	0.0	0.0	0.0	0.0	0.0	0.0	0.*	2.0	7.3	21.9
-Minimum														
32° and below	71	19.4	12.0	5.0	0.5	0.*	0.0	0.0	0.0	0.*	0.9	9.3	17.5	64.6
0° and below	71	0.7	0.3	0.0	0.0	0.0	0.0	0.0	0.0	0.0	0.0	0.*	0.5	1.5
AVG. STATION PRESS.(mb)														
RELATIVE HUMIDITY (%)														
Hour 04														
Hour 10 (Local Time)	19	76	68	59	49	49	43	34	39	44	58	72	78	56
Hour 16														
Hour 22														
PRECIPITATION (inches):														
Water Equivalent														
-Normal		2.12	1.40	1.41	1.35	1.40	0.93	0.35	0.71	0.83	1.40	1.87	2.19	15.96
-Maximum Monthly	71	5.86	3.99	4.17	3.68	4.19	3.04	1.78	2.94	4.50	4.20	4.13	4.31	5.86
-Year		1970	1940	1983	1917	1957	1941	1966	1977	1927	1950	1973	1964	JAN 1970
-Minimum Monthly	71	0.34	0.10	0.34	0.11	T	0.04	0.00	T	T	0.01	0.01	0.29	0.00
-Year		1949	1920	1965	1956	1924	1919	1953	1974	1943	1917	1936	1965	JUL 1953
-Maximum in 24 hrs	70	1.42	1.38	1.35	1.58	1.91	2.02	1.15	1.57	1.39	3.08	1.41	1.41	3.08
-Year		1948	1970	1983	1915	1971	1923	1940	1977	1927	1980	1932	1971	OCT 1980
Snow,Ice pellets														
-Maximum Monthly	71	31.3	33.4	6.2	4.0	T					3.5	19.9	21.9	33.4
-Year		1969	1916	1944	1929	1908					1973	1921	1968	FEB 1916
-Maximum in 24 hrs	71	8.7	14.0	6.0	4.0	T					3.5	12.1	12.1	14.0
-Year		1966	1916	1944	1929	1968					1973	1921	1968	FEB 1916
WIND:														
Mean Speed (mph)	47	5.1	5.5	6.2	6.1	5.7	5.5	5.4	5.1	4.7	4.5	4.8	5.2	5.3
Prevailing Direction														
through 1962		S	S	S	S	S	S	S	S	S	S	S	S	S
Fastest Mile														
-Direction (!!)	68	W	W	W	NW	NW	S	W	NW	W	SE	SW	W	SW
-Speed (MPH)	68	49	47	62	41	37	37	36	35	51	54	67	47	67
-Year		1966	1961	1957	1957	1973	1921	1921	1966	1956	1950	1958	1961	NOV 1958
Peak Gust														
-Direction (!!)														
-Speed (mph)														
-Date														

See Reference Notes to this table on the following page.

TABLE 2 PRECIPITATION (inches) WALLA WALLA WASHINGTON

YEAR	JAN	FEB	MAR	APR	MAY	JUNE	JULY	AUG	SEP	OCT	NOV	DEC	ANNUAL
1956	3.29	1.63	0.97	0.11	2.03	0.74	0.08	1.52	0.19	2.08	0.53	1.82	14.99
1957	1.76	1.10	2.94	0.56	4.19	1.05	T	0.15	0.97	1.97	1.80	1.87	18.36
1958	2.06	1.60	1.49	3.17	2.36	0.80	0.08	0.01	0.49	0.44	1.39	3.35	17.24
1959	4.00	2.25	1.50	0.86	1.53	1.19	0.05	0.99	2.41	0.93	0.73	0.88	17.32
1960	1.15	1.28	1.69	1.31	2.27	0.93	T	0.92	0.79	1.51	2.29		15.02
1961	0.98	3.01	2.00	1.36	1.56	0.87	0.01	0.39	0.14	1.32	1.79	2.42	15.85
1962	0.63	0.92	2.36	1.20	3.78	0.45	T	0.41	1.44	2.99	1.76	2.40	18.34
1963	0.93	1.60	0.78	1.86	0.67	0.20	0.49	0.66	0.92	0.60	2.89	2.00	13.60
1964	0.97	0.17	0.60	0.90	0.23	1.22	1.47	0.30	0.94	0.66	3.51	4.31	15.28
1965	3.29	0.55	0.34	1.68	0.74	0.87	0.59	2.52	0.42	0.73	1.94	0.29	13.96
1966	2.22	0.83	1.85	0.19	0.45	1.09	1.78	0.29	0.15	0.87	2.72	2.60	15.04
1967	2.11	0.36	1.99	1.66	1.46	0.31	T	T	0.70	0.88	1.25	1.24	11.96
1968	0.79	2.42	0.88	0.43	0.69	0.91	0.48	0.98	1.45	1.62	2.64	3.77	17.06
1969	4.17	1.21	0.73	2.88	1.10	0.76	0.01	0.00	0.45	1.43	0.61	2.12	15.47
1970	5.86	2.56	1.54	1.84	0.48	1.27	0.15	0.01	1.92	1.56	2.74	0.67	20.60
1971	1.59	0.85	1.24	1.35	3.39	2.81	0.65	0.36	1.80	1.25	2.70	3.24	21.23
1972	1.20	1.61	2.63	1.14	1.89	1.12	0.57	0.34	0.66	0.60	0.83	2.66	15.25
1973	0.99	1.71	0.90	0.40	1.11	0.38	0.12	0.15	1.60	1.86	4.13	4.24	17.59
1974	1.68	1.62	1.64	2.25	0.40	0.32	0.91	T	0.01	0.39	1.85	2.01	13.08
1975	3.82	1.55	0.88	1.24	0.58	0.46	0.04	1.18	0.00	1.68	1.57	2.65	15.65
1976	1.35	1.70	1.57	1.39	1.17	0.61	0.30	1.52	0.11	1.61	0.31	0.75	12.39
1977	0.55	0.64	0.98	0.20	1.31	0.35	0.34	2.94	1.18	0.49	2.03	3.83	14.84
1978	2.42	1.42	1.07	3.26	0.68	0.61	0.81	2.29	0.99	0.03	2.38	2.29	18.25
1979	1.48	1.86	1.63	1.77	0.89	0.37	0.08	1.20	0.31	2.67	2.16	1.33	15.75
1980	3.11	1.62	1.92	0.61	1.95	0.89	0.31	0.42	1.39	3.67	1.74	2.89	20.52
1981	0.90	3.19	2.89	1.37	2.07	2.47	0.69	0.03	0.84	2.42	2.47	3.16	22.50
1982	2.35	2.87	2.69	1.70	0.58	0.32	0.82	0.59	2.10	3.06	1.59	2.90	21.57
1983	2.17	2.62	4.17	1.20	1.52	1.74	1.73	0.70	0.93	1.73	3.42	3.50	25.43
1984	0.83	2.62	3.68	1.48	1.37	3.01	0.00	0.23	1.26	1.71	3.75	2.21	22.15
1985	0.56	1.53	1.74	0.47	0.95	0.90	0.67	1.06	1.40	1.40	2.50	0.84	14.02
Record Mean	1.91	1.60	1.65	1.41	1.46	1.11	0.37	0.50	0.90	1.47	1.98	2.01	16.37

TABLE 3 AVERAGE TEMPERATURE (deg. F) WALLA WALLA WASHINGTON

YEAR	JAN	FEB	MAR	APR	MAY	JUNE	JULY	AUG	SEP	OCT	NOV	DEC	ANNUAL
1956	34.7	28.2	45.1	55.4	62.2	64.6	77.8	73.0	66.2	52.5	38.4	38.0	53.0
1957	20.6	38.0	44.7	54.3	62.9	68.5	72.8	71.1	67.9	50.4	42.4	42.4	53.0
1958	40.1	47.2	44.8	51.3	66.5	70.8	78.8	78.9	64.6	55.5	43.7	38.5	56.7
1959	37.2	38.8	47.2	53.6	56.8	67.8	76.9	70.9	62.3	54.5	39.2	35.3	53.4
1960	27.0	40.1	45.7	53.4	57.7	68.1	80.5	70.9	66.3	55.7	43.9	31.3	53.4
1961	38.7	47.1	48.3	52.0	59.3	73.2	78.4	79.1	61.8	52.5	37.3	37.3	55.4
1962	33.8	39.3	43.5	55.5	55.7	66.6	74.5	70.9	65.9	54.2	46.3	40.6	53.9
1963	28.8	41.5	47.8	50.7	60.9	67.3	70.8	74.6	69.8	58.0	48.3	33.5	54.1
1964	41.7	42.3	44.6	50.0	58.3	66.3	73.6	69.7	63.2	53.2	41.2	32.9	53.1
1965	36.6	43.1	43.6	55.2	59.0	66.9	75.8	73.4	61.6	58.8	45.5	38.5	54.9
1966	39.0	41.0	47.5	53.8	61.6	65.4	72.5	73.4	68.8	54.8	45.8	42.4	55.5
1967	44.8	45.1	45.9	48.0	59.9	70.6	77.8	80.9	70.9	56.7	43.2	38.1	56.8
1968	38.9	43.8	50.4	50.7	60.4	67.9	77.7	70.9	65.7	52.3	43.7	33.3	54.7
1969	24.3	35.7	46.6	52.2	63.4	71.3	73.8	71.6	66.1	50.7	44.5	36.2	53.0
1970	34.9	42.0	45.7	48.1	60.8	71.4	77.3	75.9	60.4	51.8	42.4	38.1	54.1
1971	42.2	41.5	43.8	51.0	60.5	63.2	76.4	78.9	60.2	52.0	45.0	37.5	54.4
1972	34.2	37.6	49.3	49.2	62.2	68.3	74.3	75.8	61.8	52.8	43.4	30.5	53.3
1973	33.4	39.6	48.6	53.2	63.0	68.0	76.3	73.8	64.8	54.4	42.2	41.8	55.0
1974	32.0	44.3	47.2	52.2	57.9	71.4	74.1	74.8	67.4	54.4	45.3	41.5	55.2
1975	36.1	37.1	43.3	48.3	59.5	64.8	78.0	70.3	66.6	54.6	44.5	39.4	53.5
1976	38.9	40.8	43.5	50.8	59.6	63.8	74.4	70.0	68.4	53.9	43.1	35.9	53.6
1977	26.0	43.2	45.5	57.3	56.5	71.8	72.8	77.5	62.0	53.4	42.3	37.6	53.9
1978	34.0	41.6	49.4	51.8	58.3	69.3	74.8	71.7	63.3	54.4	36.1	32.2	53.1
1979	18.5	40.0	48.1	53.3	62.3	68.9	76.6	74.3	69.1	58.0	37.1	41.1	54.0
1980	28.1	38.1	45.0	56.6	60.4	63.9	75.2	70.1	65.9	53.1	44.1	40.7	53.4
1981	37.9	40.7	48.6	53.9	59.0	64.4	72.7	78.0	66.9	53.4	45.5	38.2	54.9
1982	35.7	40.9	46.3	50.3	59.5	70.8	74.3	75.2	63.8	53.0	38.8	37.3	53.9
1983	42.7	45.2	49.4	52.6	62.9	66.0	71.8	76.6	61.5	54.3	47.9	25.5	54.7
1984	35.8	41.1	49.4	51.4	57.5	64.9	76.4	75.6	62.6	51.1	43.1	29.8	53.2
1985	27.1	36.0	45.8	56.5	62.3	68.6	80.8	71.1	60.0	53.2	28.7	21.5	51.0
Record Mean	33.2	38.8	46.2	53.1	60.2	67.0	75.3	73.9	66.4	54.5	42.7	36.2	53.8
Max	39.0	45.4	54.7	63.2	71.3	78.8	88.7	86.8	76.5	64.2	49.3	41.8	63.3
Min	27.4	32.1	37.6	42.9	49.1	55.2	62.0	61.0	53.1	44.8	36.0	30.6	44.3

REFERENCE NOTES FOR TABLES 1, 2, 3 and 6   (WALLA WALLA, WA)

GENERAL

T - TRACE AMOUNT
BLANK ENTRIES DENOTE MISSING/UNREPORTED DATA.
# INDICATES A STATION OR INSTRUMENT RELOCATION.

SPECIFIC

TABLE 1

(a) - LENGTH OF RECORD IN YEARS. ALTHOUGH
INDIVIDUAL MONTHS MAY BE MISSING.
* LESS THAN .05

NORMALS — BASED ON THE 1951-1980 RECORD PERIOD.
EXTREMES — DATES ARE THE MOST RECENT OCCURRENCE.
WIND DIR. — NUMERALS SHOW TENS OF DEGREES
CLOCKWISE FROM TRUE NORTH.
"00" INDICATES CALM.
RESULTANT WIND DIRECTIONS ARE GIVEN TO WHOLE DEGREES.

EXCEPTIONS

TABLE 1

1. MEAN WIND SPEED IS THROUGH 1962.
2. THUNDERSTORMS AND HEAVY FOG ARE THROUGH 1964
AND MAY BE INCOMPLETE, DUE TO PART-TIME OPERA-
TIONS.

TABLES 2, 3, and 6

RECORD MEANS ARE THROUGH THE CURRENT YEAR,
BEGINNING IN   1886 FOR TEMPERATURE
1886 FOR PRECIPITATION
1915 FOR SNOWFALL

TABLE 4 HEATING DEGREE DAYS Base 65 deg. F   WALLA WALLA WASHINGTON

SEASON	JULY	AUG	SEP	OCT	NOV	DEC	JAN	FEB	MAR	APR	MAY	JUNE	TOTAL
1956-57	0	15	52	382	794	830	1370	751	623	332	109	25	5283
1957-58	0	0	42	446	673	696	766	494	619	405	71	15	4227
1958-59	0	0	82	303	632	813	855	727	544	335	260	31	4582
1959-60	6	11	126	321	770	915	1171	719	592	347	236	34	5248
1960-61	0	26	41	292	624	1039	809	495	511	382	187	14	4420
1961-62	0	0	119	385	826	852	963	715	660	284	284	64	5152
1962-63	10	10	59	329	553	750	1115	653	526	425	163	47	4640
1963-64	0	0	36	231	569	971	714	654	625	442	215	54	4511
1964-65	3	22	81	356	710	988	872	607	657	292	206	36	4830
1965-66	5	13	119	203	577	818	799	664	535	328	160	63	4284
1966-67	18	4	22	312	568	692	619	551	587	505	182	11	4071
1967-68	0	0	19	252	647	827	801	607	447	430	156	40	4226
1968-69	0	12	67	384	633	975	1256	816	566	376	109	27	5221
1969-70	0	6	53	436	607	888	925	637	593	501	147	56	4849
1970-71	0	0	158	411	673	826	699	653	653	412	170	93	4748
1971-72	12	5	157	409	593	842	949	786	480	469	147	35	4884
1972-73	12	3	146	369	641	1062	972	705	500	346	138	61	4955
1973-74	0	9	80	321	676	714	1016	577	547	375	227	30	4572
1974-75	8	0	36	326	583	722	891	777	664	493	192	66	4758
1975-76	3	16	47	321	611	784	799	697	660	421	173	92	4624
1976-77	3	22	22	339	650	896	1202	602	598	250	261	15	4860
1977-78	10	18	133	353	677	846	955	650	478	388	207	26	4741
1978-79	5	17	83	322	863	1007	1436	693	517	345	119	41	5448
1979-80	7	0	9	223	829	737	1137	773	613	255	159	76	4818
1980-81	0	15	45	383	619	748	832	676	502	340	191	65	4416
1981-82	12	0	85	351	578	824	903	669	569	430	179	42	4642
1982-83	6	0	111	367	779	852	686	550	476	364	168	40	4399
1983-84	10	0	133	328	506	1219	896	685	474	408	240	87	4986
1984-85	0	0	145	432	649	1087	1168	806	589	256	147	35	5314
1985-86	0	4	159	358	1084	1343							

TABLE 5 COOLING DEGREE DAYS Base 65 deg. F   WALLA WALLA WASHINGTON

YEAR	JAN	FEB	MAR	APR	MAY	JUNE	JULY	AUG	SEP	OCT	NOV	DEC	TOTAL
1969	0	0	0	0	66	223	281	217	93	0	0	0	880
1970	0	0	0	0	23	254	388	343	25	8	0	0	1041
1971	0	0	0	0	35	48	373	443	22	10	0	0	931
1972	0	0	0	0	65	144	305	344	55	1	0	0	914
1973	0	0	0	0	81	154	375	290	79	0	0	0	979
1974	0	0	0	0	13	230	296	311	114	4	0	0	968
1975	0	0	0	0	28	67	412	185	103	2	3	0	800
1976	0	0	0	0	14	64	303	182	130	5	0	0	698
1977	0	0	0	26	4	222	258	415	52	0	0	0	977
1978	0	0	0	0	7	161	316	232	38	0	0	0	754
1979	0	0	0	3	44	165	375	299	135	11	0	0	1032
1980	0	0	0	12	27	48	324	181	77	22	0	2	693
1981	0	0	0	14	12	54	260	407	146	0	0	0	893
1982	0	0	0	0	17	221	303	323	83	0	0	0	947
1983	0	0	0	0	109	74	225	367	36	0	0	0	811
1984	0	0	0	5	15	91	361	334	78	5	0	0	889
1985	0	0	0	6	72	150	497	201	15	0	0	0	941

TABLE 6 SNOWFALL (inches)   WALLA WALLA WASHINGTON

SEASON	JULY	AUG	SEP	OCT	NOV	DEC	JAN	FEB	MAR	APR	MAY	JUNE	TOTAL
1956-57	0.0	0.0	0.0	0.0	1.0	0.6	20.8	5.9	0.1	0.0	0.0	0.0	28.4
1957-58	0.0	0.0	0.0	0.0	0.0	T	2.9	1.9	T	T	0.0	0.0	4.8
1958-59	0.0	0.0	0.0	0.0	1.2	2.6	9.4	6.9	T	T	0.0	0.0	20.1
1959-60	0.0	0.0	0.0	0.0	1.7	5.5	13.5	0.1	3.7	T	T	0.0	24.5
1960-61	0.0	0.0	0.0	0.0	0.0	2.3	0.5	0.0	0.7	T	0.0	0.0	3.5
1961-62	0.0	0.0	0.0	0.0	9.2	9.0	2.5	2.6	1.8	0.0	0.0	0.0	25.1
1962-63	0.0	0.0	0.0	0.0	0.0	T	5.5	T	0.7	0.0	0.0	0.0	5.5
1963-64	0.0	0.0	0.0	0.0	T	3.1	T	T	T	0.1	0.0	0.0	3.9
1964-65	0.0	0.0	0.0	0.0	0.7	10.8	7.3	T	1.0	0.0	0.0	0.0	20.5
1965-66	0.0	0.0	0.0	0.0	2.2	T	10.0	1.2	4.6	0.0	0.0	0.0	18.0
1966-67	0.0	0.0	0.0	0.0	T	0.4	0.7	0.6	4.3	T	0.0	0.0	6.0
1967-68	0.0	0.0	0.0	0.0	1.6	5.4	0.5	T	T	0.3	T	0.0	7.8
1968-69	0.0	0.0	0.0	0.0	0.8	21.9	31.3	3.7	0.0	0.0	0.0	0.0	57.7
1969-70	0.0	0.0	0.0	0.0	T	6.7	10.4	3.5	1.6	T	0.0	0.0	22.2
1970-71	0.0	0.0	0.0	0.0	1.8	3.3	5.7	T	4.0	T	0.0	0.0	15.4
1971-72	0.0	0.0	0.0	2.5	T	9.5	8.9	6.2	T	T	T	0.0	27.1
1972-73	0.0	0.0	0.0	0.0	T	7.2	5.1	9.7	T	T	T	0.0	22.0
1973-74	0.0	0.0	0.0	3.5	6.1	9.6	0.8	T	T	T	0.0	0.0	20.0
1974-75	0.0	0.0	0.0	0.0	T	0.0	11.1	8.9	T	0.5	T	0.0	20.0
1975-76	0.0	0.0	0.0	0.0	T	3.8	3.9	0.2	0.5	T	0.0	0.0	8.9
1976-77	0.0	0.0	0.0	0.0	0.0	1.8	7.5	T	0.9	0.0	0.0	0.0	10.2
1977-78	0.0	0.0	0.0	0.0	9.4	10.0	8.4	0.3	1.3	T	0.0	0.0	29.4
1978-79	0.0	0.0	0.0	0.0	9.7	9.4	15.6	3.0	0.4	T	0.0	0.0	38.1
1979-80	0.0	0.0	0.0	0.0	5.3	0.8	14.6	3.4	3.7	0.0	0.0	0.0	27.8
1980-81	0.0	0.0	0.0	0.0	2.2	1.2	T	2.0	0.0	0.0	0.0	0.0	5.4
1981-82	0.0	0.0	0.0	0.0	T	6.3	4.4	T	2.8	T	0.0	0.0	13.5
1982-83	0.0	0.0	0.0	0.0	0.0	3.1	0.2	0.6	T	0.0	0.0	0.0	3.9
1983-84	0.0	0.0	0.0	0.0	0.5	20.7	1.7	T	0.0	0.0	0.0	0.0	22.9
1984-85	0.0	0.0	0.0	0.0	0.0	5.4	3.5	12.9	T	0.0	0.0	0.0	21.8
1985-86	0.0	0.0	0.0	0.0	17.5	6.2							
Record Mean	0.0	0.0	0.0	0.1	2.1	5.0	7.6	3.9	1.2	0.2	T	0.0	20.1

See Reference Notes, relative to all above tables, on preceding page.

Yakima is located in a small east–west valley in the upper (northwestern) part of the Yakima Valley. Local topography is complex with a number of minor valleys and ridges giving a local relief of as much as 1,000 feet. This complex topography results in marked variations in air drainage, winds, and low temperatures within short distances.

The climate of the Yakima Valley is relatively mild and dry. It has characteristics of both maritime and continental climates, modified by the Cascade and the Rocky Mountains, respectively. Summers are dry and rather hot, and winters cool with only light snowfall. The maritime influence is strongest in winter when the prevailing westerlies are the strongest and most steady. The Selkirk and Rocky Mountains in British Columbia and Idaho shield the area from most of the very cold air masses that sweep down from Canada into the Great Plains and eastern United States. Sometimes a strong polar high pressure area over western Canada will occur at the same time that a low pressure area covers the southwestern United States. On these occasions, the cold arctic air will pour through the passes and down the river valleys of British Columbia, bringing very cold temperatures to Yakima. However, over one–half of the winters remain above zero.

The modifying influence of the Pacific Ocean is much less in summer. Afternoons are hot, but the dry air results in a rapid temperature fall after sunset, and nights are pleasantly cool with summertime low temperatures, usually in the 50s. Spells of 4 to 11 days of 100 degrees or more have occurred.

The length of the growing season varies depending on the immediate topography and the crop grown. Temperatures below 32 degrees are infrequent during the period from mid–May through September. Temperatures below 40 degrees during July and August have occurred in about half of the years.

Precipitation follows the pattern of a West Coast marine climate with the typical late fall and early winter high. However, since Yakima lies in the rain shadow of the Cascades, total amounts are small. The three months, November to January, total nearly half of the annual fall. Late June, July, and August are very dry.

Irrigation is necessary for nearly all crops. Ample water supplies are available from the snowmelt in the Cascade Mountains which is collected in storage reservoirs for summer use.

Snowfall in the Yakima area is light, averaging 20 to 25 inches.

Summers are sunny, with about 85 percent of the possible sunshine. Winters are generally cloudy, with only a third of the possible sunshine.

Winds are mostly light, averaging about 7 mph for the year, being somewhat stronger in late spring and weaker in winter. Speeds of 30 to 35 mph are reached at least once in about half the months and speeds over 40 mph occur in about 1 out of 5 months. The most common wind direction in downtown Yakima is northwest, while at the airport the wind is from the west in winter and the west–northwest in summer.

TABLE 1   NORMALS, MEANS AND EXTREMES

YAKIMA WASHINGTON

LATITUDE: 46°34'N   LONGITUDE: 120°32'W   ELEVATION: FT. GRND 1052 BARO 01068   TIME ZONE: PACIFIC   WBAN: 24243

	(a)	JAN	FEB	MAR	APR	MAY	JUNE	JULY	AUG	SEP	OCT	NOV	DEC	YEAR
TEMPERATURE °F:														
Normals														
-Daily Maximum		36.7	46.0	54.5	63.5	72.5	79.9	87.8	85.6	77.5	64.5	48.1	39.4	63.0
-Daily Minimum		19.7	26.1	29.2	34.7	42.1	49.1	53.0	51.5	44.3	35.1	28.2	23.6	36.4
-Monthly		28.2	36.0	41.9	49.1	57.3	64.5	70.4	68.6	60.9	49.8	38.2	31.5	49.7
Extremes														
-Record Highest	39	68	69	80	92	101	103	108	110	100	86	72	67	110
-Year		1977	1947	1960	1977	1983	1961	1971	1971	1949	1980	1975	1980	AUG 1971
-Record Lowest	39	-21	-25	-1	20	25	30	34	35	24	11	-13	-17	-25
-Year		1950	1950	1960	1985	1954	1984	1971	1960	1985	1971	1985	1964	FEB 1950
NORMAL DEGREE DAYS:														
Heating (base 65°F)		1141	809	716	474	254	101	18	46	161	468	804	1039	6031
Cooling (base 65°F)		0	0	0	0	16	86	186	158	38	0	0	0	484
% OF POSSIBLE SUNSHINE														
MEAN SKY COVER (tenths)														
Sunrise - Sunset	39	7.8	7.5	6.8	6.4	5.8	5.3	3.0	3.5	4.2	5.9	7.4	7.9	5.9
MEAN NUMBER OF DAYS:														
Sunrise to Sunset														
-Clear	39	4.5	4.3	6.3	6.3	8.5	10.5	19.2	17.6	14.5	9.1	4.9	4.0	109.7
-Partly Cloudy	39	5.1	5.9	8.1	9.5	10.5	9.6	7.7	7.8	8.1	8.2	6.0	5.4	91.9
-Cloudy	39	21.4	18.0	16.7	14.2	12.0	10.0	4.2	5.6	7.4	13.7	19.0	21.5	163.6
Precipitation														
.01 inches or more	39	9.5	7.3	6.3	4.4	5.0	4.7	2.1	3.0	3.2	5.3	8.6	9.8	69.1
Snow,Ice pellets														
1.0 inches or more	37	2.8	1.2	0.6	0.0	0.0	0.0	0.0	0.0	0.0	0.1	0.7	2.7	8.1
Thunderstorms	39	0.0	0.*	0.1	0.4	1.1	1.7	1.5	1.3	0.7	0.1	0.0	0.0	6.8
Heavy Fog Visibility														
1/4 mile or less	39	4.6	2.4	0.4	0.1	0.1	0.0	0.0	0.0	0.1	0.7	3.6	6.7	18.6
Temperature °F														
-Maximum														
90° and above	39	0.0	0.0	0.0	0.*	1.3	4.4	14.1	10.5	2.2	0.0	0.0	0.0	32.5
32° and below	39	10.8	2.4	0.1	0.0	0.0	0.0	0.0	0.0	0.0	0.0	1.8	7.9	23.1
-Minimum														
32° and below	39	27.9	23.6	20.6	12.3	2.8	0.1	0.0	0.0	1.1	10.6	21.1	27.7	147.8
0° and below	39	2.7	0.6	0.*	0.0	0.0	0.0	0.0	0.0	0.0	0.0	0.2	0.8	4.4
AVG. STATION PRESS.(mb)	13	982.4	979.1	976.0	977.5	976.8	976.2	976.3	975.7	977.7	979.4	979.5	981.6	978.2
RELATIVE HUMIDITY (%)														
Hour 04	39	82	82	76	71	69	69	68	70	76	80	84	84	76
Hour 10 (Local Time)	39	77	70	53	41	38	38	35	39	44	55	73	80	54
Hour 16	39	71	58	41	32	30	31	25	28	32	43	63	74	44
Hour 22	37	81	79	68	58	55	54	51	55	64	73	81	83	67
PRECIPITATION (inches):														
Water Equivalent														
-Normal		1.44	0.74	0.65	0.50	0.48	0.60	0.14	0.36	0.33	0.47	0.97	1.30	7.98
-Maximum Monthly	39	3.66	2.46	2.63	1.62	2.76	2.10	0.71	2.10	1.08	2.22	2.83	4.19	4.19
-Year		1970	1961	1957	1963	1948	1948	1966	1975	1982	1950	1973	1964	DEC 1964
-Minimum Monthly	39	0.09	T	0.01	T	0.03	0.01	T	0.00	T	0.00	0.07	0.00	0.00
-Year		1985	1971	1973	1985	1964	1970	1980	1955	1975	1978	1976	1976	OCT 1978
-Maximum in 24 hrs	39	1.37	0.87	0.62	1.25	0.77	1.56	0.66	1.47	0.83	1.05	1.08	1.58	1.58
-Year		1963	1961	1976	1974	1951	1982	1963	1975	1954	1982	1955	1977	DEC 1977
Snow,Ice pellets														
-Maximum Monthly	39	26.6	16.5	10.8	T	T					2.4	21.2	37.5	37.5
-Year		1950	1949	1971	1983	1984					1973	1955	1964	DEC 1964
-Maximum in 24 hrs	39	13.6	5.8	7.4	T	T					2.4	11.2	14.0	14.0
-Year		1963	1956	1951	1983	1984					1973	1984	1964	DEC 1964
WIND:														
Mean Speed (mph)	34	5.6	6.3	7.9	8.6	8.4	8.2	7.7	7.3	7.3	6.6	5.7	5.2	7.1
Prevailing Direction														
through 1963		W	W	W	WNW	WNW	NW	WNW	WNW	WNW	WNW	W	W	WNW
Fastest Obs. 1 Min.														
-Direction (!!!)	31	25	28	23	29	18	20	24	20	20	24	29	23	28
-Speed (MPH)	31	44	48	48	46	46	47	43	33	38	39	45	48	48
-Year		1962	1967	1956	1961	1961	1955	1968	1954	1959	1965	1955	1955	FEB 1967
Peak Gust														
-Direction (!!!)	2	NE	W	NW	W	NE	NW	NW	W	NW	W	NE	NE	NE
-Speed (mph)	2	40	56	49	51	69	44	43	41	43	48	44	40	69
-Date		1984	1985	1985	1985	1985	1985	1985	1984	1984	1984	1985	1985	MAY 1985

See reference Notes to this table on the following page.

TABLE 2 PRECIPITATION (inches) YAKIMA WASHINGTON

YEAR	JAN	FEB	MAR	APR	MAY	JUNE	JULY	AUG	SEP	OCT	NOV	DEC	ANNUAL
1956	3.25	0.81	0.21	T	0.48	1.81	0.02	0.10	0.54	0.35	0.09	0.55	8.21
1957	0.56	0.87	2.63	0.93	1.19	1.26	0.01	0.20	0.98	1.40	0.10	0.83	10.96
1958	1.37	1.84	0.81	0.81	0.59	0.21	0.22	0.16	0.05	0.29	1.14	0.82	8.31
1959	2.03	1.12	0.80	0.05	0.14	0.20	T	0.04	0.52	0.42	0.40	0.35	6.07
1960	0.89	1.43	0.65	0.54	0.87	0.19	0.05	0.25	0.11	0.16	1.55	0.90	7.59
1961	0.55	2.46	2.04	0.86	0.96	0.52	0.25	0.22	T	0.31	0.51	1.27	9.95
1962	0.16	1.48	0.65	0.62	1.09	0.07	0.01	0.33	0.30	1.49	0.79	0.47	7.46
1963	1.42	0.52	0.13	1.62	0.43	0.26	0.69	0.13	0.08	0.05	1.13	1.00	8.17
1964	0.60	T	0.14	0.25	0.03	1.18	0.08	0.20	0.03	0.15	0.70	4.19	7.55
1965	1.33	0.08	0.10	0.48	0.05	0.51	0.27	0.21	0.04	0.06	1.43	1.39	5.95
1966	1.73	0.11	0.81	T	0.10	0.17	0.71	T	0.87	0.41	2.14	0.95	8.00
1967	0.60	T	0.45	1.03	0.16	1.12	T	0.01	0.09	0.21	0.30	0.55	4.52
1968	1.76	0.88	0.11	T	0.47	0.02	0.02	1.71	0.32	0.94	1.32	1.91	9.46
1969	1.52	0.91	0.16	0.27	0.54	0.61	T	0.01	0.32	0.24	0.08	2.28	6.94
1970	3.66	0.49	0.22	0.16	0.06	0.01	0.13	T	0.07	0.54	1.25	1.41	8.00
1971	1.48	T	1.56	0.47	0.54	0.20	0.04	0.14	0.73	0.27	0.97	1.45	7.85
1972	0.88	0.31	1.05	0.09	0.60	1.50	0.04	0.65	0.06	0.12	0.72	1.31	7.33
1973	1.19	0.24	0.01	0.04	0.08	0.02	T	0.01	0.81	1.52	2.83	2.22	8.97
1974	1.67	0.85	1.21	1.46	0.80	0.12	0.18	T	0.02	0.45	0.30	1.14	8.20
1975	2.28	1.16	0.49	0.40	0.23	0.22	0.18	2.10	T	0.79	0.43	0.55	8.83
1976	0.56	0.78	0.70	0.33	0.09	0.69	0.26	0.50	0.13	0.07	T	0.07	4.18
1977	0.13	0.69	0.23	0.01	0.68	0.46	T	1.16	0.89	0.17	0.70	2.80	7.92
1978	2.30	1.30	0.52	0.91	0.28	0.32	0.29	0.38	0.64	0.00	0.14	0.42	8.02
1979	0.91	0.54	0.23	0.14	0.04	0.57	0.04	0.42	0.36	0.74	1.53	1.33	6.85
1980	2.23	1.30	0.29	0.80	0.84	1.12	T	0.29	0.48	0.23	1.00	2.69	11.27
1981	0.95	0.65	0.10	0.01	0.68	0.39	0.29	0.09	0.59	1.16	1.36	2.38	8.65
1982	0.58	1.48	0.34	0.30	0.37	1.70	0.12	0.39	1.08	1.46	0.90	2.15	10.87
1983	1.97	1.59	1.95	0.66	0.30	0.77	0.29	0.44	0.33	0.23	2.77	1.92	13.22
1984	0.13	0.92	1.04	1.05	0.51	1.45	0.13	0.04	0.46	0.16	2.62	0.51	9.02
1985	0.09	0.68	0.62	T	0.46	0.37	0.12	0.03	0.84	0.75	0.92	1.02	5.90
Record Mean	1.14	0.78	0.54	0.44	0.50	0.61	0.17	0.25	0.41	0.54	1.02	1.19	7.60

TABLE 3 AVERAGE TEMPERATURE (deg. F) YAKIMA WASHINGTON

YEAR	JAN	FEB	MAR	APR	MAY	JUNE	JULY	AUG	SEP	OCT	NOV	DEC	ANNUAL
1956	27.2	22.8	40.5	52.9	61.3	61.3	72.8	69.8	61.1	48.3	35.2	31.2	48.7
1957	16.0	32.5	41.5	51.4	62.4	65.6	66.9	65.0	63.1	47.9	36.7	35.3	48.7
1958	34.9	42.8	40.7	48.0	64.0	69.3	75.0	71.8	59.2	50.2	37.4	33.7	52.3
1959	31.0	34.9	42.1	50.6	53.1	63.5	71.2	65.8	58.3	50.4	35.1	32.2	49.0
1960	22.2	35.1	41.2	49.1	54.2	64.8	75.2	66.1	60.6	50.9	38.1	27.3	48.7
1961	32.7	40.7	42.9	49.7	55.9	68.4	73.1	73.6	58.2	47.9	33.7	31.4	50.7
1962	29.9	35.6	40.0	51.9	53.7	63.2	69.7	66.0	61.8	49.8	41.2	35.9	49.9
1963	25.2	37.8	43.6	47.1	57.8	64.5	67.1	69.2	65.5	51.7	40.7	29.6	50.0
1964	33.2	36.1	40.4	45.5	53.6	62.0	68.0	62.8	57.1	50.1	38.2	27.2	47.9
1965	28.9	37.5	41.2	50.5	56.4	64.5	70.0	68.4	57.7	52.5	42.4	31.0	50.1
1966	29.2	37.4	43.4	52.0	58.3	61.4	67.6	71.2	66.5	52.1	43.8	36.4	51.6
1967	38.6	41.3	41.7	44.9	57.6	67.8	71.4	73.3	65.3	51.8	40.3	31.5	52.2
1968	30.9	40.0	46.0	47.9	56.8	63.9	70.9	65.3	61.0	46.3	39.7	28.4	49.8
1969	18.2	28.5	41.8	48.3	59.9	68.9	68.4	66.2	61.2	47.3	39.3	33.5	48.5
1970	28.4	40.0	43.7	46.2	58.6	68.8	72.5	70.1	57.5	48.6	38.6	28.0	50.1
1971	34.8	37.8	38.7	47.8	60.7	60.7	71.3	72.3	55.5	47.1	38.5	27.8	49.4
1972	29.4	32.2	44.2	45.5	59.9	65.6	69.9	71.6	56.8	49.1	40.9	26.8	49.4
1973	27.6	37.7	45.0	51.7	58.5	64.4	72.2	69.3	61.9	49.2	37.6	35.9	50.9
1974	27.4	40.1	43.7	50.4	54.3	68.3	68.3	70.7	63.0	50.2	41.3	35.2	51.1
1975	30.1	33.0	40.9	46.7	57.7	63.8	75.0	66.3	62.1	49.4	37.8	34.7	49.8
1976	32.5	35.1	39.1	48.1	56.1	59.9	66.3	66.3	64.4	49.6	31.9	29.9	49.1
1977	24.5	38.7	42.8	54.1	52.7	68.8	68.8	74.3	57.7	48.9	35.9	33.5	50.1
1978	32.8	37.9	45.8	49.4	54.8	64.8	71.0	67.5	58.8	49.5	32.3	27.5	49.3
1979	14.8	32.6	44.0	49.6	59.5	65.4	72.1	70.2	64.0	52.9	34.1	34.8	49.5
1980	20.8	33.7	43.7	53.7	58.9	62.1	70.4	66.4	62.6	51.5	40.4	34.8	49.9
1981	39.9	39.3	47.7	49.7	56.8	61.8	69.0	73.1	61.6	48.8	41.2	30.2	51.6
1982	26.2	36.8	43.0	45.8	56.0	66.8	68.7	68.1	60.2	49.0	35.8	29.0	48.8
1983	37.6	40.2	47.8	49.8	60.9	62.9	67.6	70.9	58.1	50.6	42.4	22.2	50.9
1984	31.2	37.8	45.3	47.1	52.4	61.1	70.5	68.8	59.0	47.1	36.1	20.1	48.0
1985	26.4	28.9	41.9	50.8	58.8	64.6	74.9	66.0	55.3	46.7	23.0	19.1	46.4
Record Mean	28.4	35.5	43.7	51.0	58.8	65.6	72.1	70.3	61.7	50.9	38.5	31.0	50.7
Max	36.4	45.1	56.0	64.9	73.1	80.1	88.3	86.4	77.3	64.8	48.2	38.4	63.3
Min	20.4	26.0	31.3	37.1	44.4	51.0	55.8	54.1	46.1	36.9	28.8	23.6	38.0

REFERENCE NOTES FOR TABLES 1, 2, 3 and 6   (YAKIMA, WA)

GENERAL

T - TRACE AMOUNT
BLANK ENTRIES DENOTE MISSING/UNREPORTED DATA.
# INDICATES A STATION OR INSTRUMENT RELOCATION.

SPECIFIC

TABLE 1

(a) - LENGTH OF RECORD IN YEARS. ALTHOUGH INDIVIDUAL MONTHS MAY BE MISSING.

* LESS THAN .05

NORMALS — BASED ON THE 1951-1980 RECORD PERIOD.
EXTREMES — DATES ARE THE MOST RECENT OCCURRENCE.
WIND DIR. — NUMERALS SHOW TENS OF DEGREES CLOCKWISE FROM TRUE NORTH.
       "00" INDICATES CALM.
RESULTANT WIND DIRECTIONS ARE GIVEN TO WHOLE DEGREES.

EXCEPTIONS

TABLES 2, 3, and 6

RECORD MEANS ARE THROUGH THE CURRENT YEAR,
BEGINNING IN 1910 FOR TEMPERATURE
       1910 FOR PRECIPITATION
       1947 FOR SNOWFALL

TABLE 4 HEATING DEGREE DAYS Base 65 deg. F   YAKIMA WASHINGTON

SEASON	JULY	AUG	SEP	OCT	NOV	DEC	JAN	FEB	MAR	APR	MAY	JUNE	TOTAL
1956-57	2	22	131	511	890	1044	1515	904	722	402	104	64	6311
1957-58	28	43	82	522	844	914	924	615	744	503	112	34	5365
1958-59	3	12	192	452	821	964	1047	835	702	426	365	101	5920
1959-60	34	40	207	446	889	1011	1318	860	730	473	331	70	6409
1960-61	0	91	145	431	799	1164	996	673	676	452	280	55	5762
1961-62	6	3	202	524	931	1032	1081	818	768	388	344	117	6214
1962-63	52	38	120	463	709	894	1227	754	658	531	238	92	5776
1963-64	27	34	63	406	721	1093	982	830	757	578	358	110	5959
1964-65	32	94	229	456	798	1167	1114	763	731	430	264	49	6127
1965-66	29	38	207	378	671	1048	1104	767	662	385	218	129	5636
1966-67	30	11	28	394	629	880	814	659	716	593	228	30	5012
1967-68	4	0	49	401	732	1031	1049	583	508	253	79		5410
1968-69	20	64	132	572	751	1130	1447	1016	711	495	161	43	6542
1969-70	20	39	147	542	753	966	1127	693	655	556	200	54	5752
1970-71	13	5	221	504	785	1141	932	754	813	508	158	140	5974
1971-72	52	21	282	549	790	1146	1096	941	633	578	192	55	6335
1972-73	26	11	258	486	717	1179	1153	760	610	390	230	95	5915
1973-74	13	42	128	485	816	895	1157	692	652	433	329	72	5714
1974-75	44	8	79	452	706	917	1072	889	741	540	232	80	5760
1975-76	0	37	100	477	809	931	1004	859	795	500	267	173	5952
1976-77	26	53	61	472	769	1081	1248	730	684	323	371	34	5852
1977-78	32	24	217	491	862	968	992	753	588	464	311	73	5775
1978-79	6	53	184	470	975	1155	1549	901	641	456	178	72	6640
1979-80	29	4	59	371	919	929	1364	902	654	336	190	109	5866
1980-81	24	41	98	412	731	929	773	712	530	459	255	117	5081
1981-82	40	7	163	497	707	1072	1195	783	675	569	276	72	6056
1982-83	43	30	164	486	868	1111	840	690	527	452	191	90	5492
1983-84	32	2	206	440	672	1320	1043	781	602	531	384	142	6155
1984-85	13	23	199	546	860	1389	1189	1002	709	421	219	61	6631
1985-86	0	32	286	561	1255	1416							

TABLE 5 COOLING DEGREE DAYS Base 65 deg. F   YAKIMA WASHINGTON

YEAR	JAN	FEB	MAR	APR	MAY	JUNE	JULY	AUG	SEP	OCT	NOV	DEC	TOTAL
1969	0	0	0	0	8	166	132	81	37	0	0	0	424
1970	0	0	0	0	6	174	252	168	3	0	0	0	603
1971	0	0	0	0	34	16	257	256	1	0	0	0	564
1972	0	0	0	0	42	80	187	221	19	0	0	0	549
1973	0	0	0	0	34	84	245	181	41	0	0	0	585
1974	0	0	0	0	3	174	153	192	24	0	0	0	546
1975	0	0	0	0	10	52	314	86	22	0	0	0	484
1976	0	0	0	0	0	27	156	103	50	0	0	0	336
1977	0	0	0	5	0	153	158	319	4	0	0	0	639
1978	0	0	0	0	0	73	198	136	5	0	0	0	412
1979	0	0	0	0	12	91	255	174	37	2	0	0	571
1980	0	0	0	2	8	29	198	92	34	2	0	0	365
1981	0	0	0	6	9	28	169	263	66	0	0	0	541
1982	0	0	0	0	6	132	165	131	28	0	0	0	462
1983	0	0	0	0	72	35	119	191	6	0	0	0	423
1984	0	0	0	0	3	33	192	150	25	0	0	0	403
1985	0	0	0	0	32	57	312	71	2	0	0	0	474

TABLE 6 SNOWFALL (inches)   YAKIMA WASHINGTON

SEASON	JULY	AUG	SEP	OCT	NOV	DEC	JAN	FEB	MAR	APR	MAY	JUNE	TOTAL
1956-57	0.0	0.0	0.0	T	0.2	4.5	9.1	3.7	6.7	0.0	0.0	0.0	24.2
1957-58	0.0	0.0	0.0	0.4	0.0	1.3	0.4	T	T	0.0	0.0	0.0	2.1
1958-59	0.0	0.0	0.0	0.0	T	1.1	1.1	9.3	0.7	T	0.0	0.0	15.7
1959-60	0.0	0.0	0.0	0.0	0.7	1.9	7.6	1.2	6.1	T	T	0.0	17.5
1960-61	0.0	0.0	0.0	0.0	T	3.8	1.7	T	4.2	T	T	0.0	9.7
1961-62	0.0	0.0	0.0	0.0	1.0	10.8	2.2	3.6	2.8	0.0	0.0	0.0	20.4
1962-63	0.0	0.0	0.0	0.0	0.0	0.5	14.5	0.6	T	T	0.0	0.0	15.6
1963-64	0.0	0.0	0.0	0.0	0.0	5.9	4.3	T	T	T	0.0	0.0	10.2
1964-65	0.0	0.0	0.0	0.0	1.6	37.5	12.1	0.1	1.0	0.0	T	0.0	52.3
1965-66	0.0	0.0	0.0	0.0	1.4	14.0	14.6	0.8	1.1	0.0	0.0	0.0	31.9
1966-67	0.0	0.0	0.0	0.0	T	5.4	1.0	0.0	T	T	0.0	0.0	6.4
1967-68	0.0	0.0	0.0	0.0	2.0	4.7	11.1	0.4	0.0	T	0.0	0.0	18.2
1968-69	0.0	0.0	0.0	0.0	T	15.3	20.2	5.5	T	0.0	0.0	0.0	41.0
1969-70	0.0	0.0	0.0	0.0	T	9.0	15.0	0.5	T	0.0	0.0	0.0	24.5
1970-71	0.0	0.0	0.0	0.0	2.9	14.3	10.8	T	10.8	0.0	0.0	0.0	38.8
1971-72	0.0	0.0	0.0	1.1	0.0	14.7	10.1	3.6	5.5	T	0.0	0.0	35.0
1972-73	0.0	0.0	0.0	0.0	T	4.8	8.7	2.7	T	0.0	0.0	0.0	16.2
1973-74	0.0	0.0	0.0	2.4	7.5	18.6	6.1	0.0	0.6	0.0	T	0.0	35.2
1974-75	0.0	0.0	0.0	0.0	0.0	1.6	16.4	10.8	2.3	T	0.0	0.0	31.1
1975-76	0.0	0.0	0.0	0.0	2.4	4.7	5.4	5.2	T	0.0	0.0	0.0	17.7
1976-77	0.0	0.0	0.0	0.0	T	0.7	3.2	T	T	0.0	0.0	0.0	3.9
1977-78	0.0	0.0	0.0	0.0	3.6	6.6	3.0	2.8	0.9	T	T	0.0	16.9
1978-79	0.0	0.0	0.0	0.0	5.8	0.8	13.1	1.9	T	0.0	0.0	0.0	21.6
1979-80	0.0	0.0	0.0	0.0	9.3	11.4	17.6	7.6	1.7	T	0.0	0.0	47.6
1980-81	0.0	0.0	0.0	0.0	0.4	7.5	1.0	3.1	0.0	0.0	0.0	0.0	12.0
1981-82	0.0	0.0	0.0	0.0	0.0	20.6	5.6	0.1	1.9	T	0.0	0.0	28.2
1982-83	0.0	0.0	0.0	0.0	0.6	13.2	5.0	3.1	0.0	T	0.0	0.0	21.9
1983-84	0.0	0.0	0.0	0.0	3.7	16.3	1.2	T	0.0	0.0	T	0.0	21.2
1984-85	0.0	0.0	0.0	T	11.5	7.2	0.2	8.6	0.8	0.0	0.0	0.0	28.3
1985-86	0.0	0.0	0.0	T	10.0	11.5							
Record Mean	0.0	0.0	0.0	0.1	2.4	8.5	9.1	3.2	1.6	T	T	0.0	24.9

See Reference Notes, relative to all above tables, on preceding page.

San Juan, located on the north coast of the island of Puerto Rico, is surrounded by the waters of the Atlantic Ocean and San Juan Bay. Local custom assigns the name San Juan to the old city which lies right on the coast, but the modern metropolitan area extends inland about 12 miles. These inland sections have a temperature and rainfall regime significantly different from the coastal area. Isla Verde Airport, where weather observations are made, lies on the coast about 7 miles east of old San Juan. The surrounding terrain is level with a gradual upslope inland. Mountain ranges, with peak elevations of 4,000 feet, extend east and west through the central portion of Puerto Rico, and are located 15 to 20 miles east and south of San Juan. These mountain ranges have a decided influence on the rainfall of the San Juan metropolitan area, and on the entire island in general.

The climate is tropical maritime, characteristic of all tropical islands. The predominant easterly trade winds, modified by local effects such as the land and sea breeze and the particular island topography, are a primary feature of the climate of San Juan and have a significant influence on the temperature and rainfall. During daylight hours the wind blows almost constantly off the ocean. Usually, after sunset the wind shifts to the south or southeast, off land. This daily wind variation is a contributing factor to the delightful climate of the city. The annual temperature range is small with about a 5-6 degree difference between the temperatures of the warmest and coldest months. The inland sectors have warmer afternoons and cooler nights. In the interior mountain and valley regions even greater daily and annual ranges of temperature occur. The highest temperatures recorded in Puerto Rico have exceeded 105 degrees and the lowest have been near 40. Sea water temperatures range from 78 degrees in March to about 83 degrees in September.

Although rainfall in San Juan is nearly 60 inches, the geographical distribution of rainfall over the island shows the heaviest rainfall, of about 180 inches per year, in the Luquillo Range, only 23 miles distant from San Juan. The driest area, with annual rainfall of 30 to 35 inches, is located in the southwest corner of the island. Rain showers occur mostly in the afternoon and at night. The nocturnal showers, usually light, are a characteristic feature of the San Juan rainfall pattern. Rainfall is generally of the brief showery type except for the continuous rains occuring with the passage of tropical disturbances, or when the trailing edge of a cold front out of the United States reaches Puerto Rico. This normally occurs from about November to April.

Puerto Rico is in the tropical hurricane region of the eastern Caribbean. The hurricane season begins June 1 and ends November 30. Only a few hurricanes have passed close enough to San Juan to produce hurricane force winds or damage.

Mild temperatures, refreshing sea breezes in the daytime, plenty of sunshine, and adequate rainfall make the climate of San Juan most enjoyable for tourists and residents alike.

TABLE 1   NORMALS, MEANS AND EXTREMES

SAN JUAN, PUERTO RICO

LATITUDE: 18°26'N   LONGITUDE: 66°00'W   ELEVATION: FT. GRND   13 BARO 00069   TIME ZONE: ATLANTIC   WBAN: 11641

	(a)	JAN	FEB	MAR	APR	MAY	JUNE	JULY	AUG	SEP	OCT	NOV	DEC	YEAR
TEMPERATURE °F:														
Normals														
-Daily Maximum		82.7	83.2	84.2	85.2	86.7	88.0	87.9	88.2	88.2	87.9	85.7	83.6	86.0
-Daily Minimum		70.3	70.0	70.8	72.3	73.9	75.3	76.1	76.1	75.5	74.9	73.4	71.8	73.4
-Monthly		76.5	76.6	77.5	78.8	80.3	81.7	82.0	82.2	81.8	81.4	79.6	77.7	79.7
Extremes														
-Record Highest	31	92	96	96	97	96	97	95	97	97	98	96	91	98
-Year		1983	1983	1983	1983	1980	1983	1981	1980	1981	1981	1981	1981	OCT 1981
-Record Lowest	31	61	62	60	64	66	69	69	70	69	67	66	63	60
-Year		1962	1968	1957	1968	1962	1957	1959	1956	1960	1959	1969	1964	MAR 1957
NORMAL DEGREE DAYS:														
Heating (base 65°F)		0	0	0	0	0	0	0	0	0	0	0	0	0
Cooling (base 65°F)		357	325	388	414	474	501	527	533	507	508	438	394	5366
% OF POSSIBLE SUNSHINE	30	67	69	74	68	60	62	67	66	61	61	59	59	64
MEAN SKY COVER (tenths)														
Sunrise - Sunset	30	5.0	5.1	5.0	5.5	6.6	6.3	6.1	6.1	6.3	6.2	5.8	5.7	5.8
MEAN NUMBER OF DAYS:														
Sunrise to Sunset														
-Clear	30	8.1	6.6	8.7	6.7	3.1	3.4	4.1	4.2	3.3	3.8	4.8	5.3	62.0
-Partly Cloudy	30	18.8	17.2	18.1	17.0	16.0	16.3	18.0	17.8	16.8	17.1	18.1	19.2	210.4
-Cloudy	30	4.1	4.5	4.2	6.3	11.9	10.2	8.9	9.0	9.9	10.2	7.1	6.5	92.8
Precipitation														
.01 inches or more	30	16.4	12.7	11.9	13.0	16.8	15.7	18.9	18.5	17.2	17.3	17.9	19.5	195.8
Snow, Ice pellets														
1.0 inches or more	30	0.0	0.0	0.0	0.0	0.0	0.0	0.0	0.0	0.0	0.0	0.0	0.0	0.0
Thunderstorms	30	0.2	0.3	0.3	1.0	4.5	4.8	5.0	5.8	8.3	7.2	3.2	0.6	41.3
Heavy Fog Visibility														
1/4 mile or less	30	0.0	0.0	0.0	0.0	0.0	0.0	0.0	0.0	0.0	0.0	0.0	0.0	0.0
Temperature °F														
-Maximum														
90° and above	30	0.4	0.9	2.2	3.1	5.5	8.9	7.9	9.0	9.5	8.3	1.7	0.3	57.7
32° and below	30	0.0	0.0	0.0	0.0	0.0	0.0	0.0	0.0	0.0	0.0	0.0	0.0	0.0
-Minimum														
32° and below	30	0.0	0.0	0.0	0.0	0.0	0.0	0.0	0.0	0.0	0.0	0.0	0.0	0.0
0° and below	30	0.0	0.0	0.0	0.0	0.0	0.0	0.0	0.0	0.0	0.0	0.0	0.0	0.0
AVG. STATION PRESS.(mb)	13	1014.7	1014.6	1014.1	1013.4	1013.2	1014.5	1014.8	1013.8	1012.4	1011.5	1011.8	1013.8	1013.5
RELATIVE HUMIDITY (%)														
Hour 02	30	80	79	79	79	83	84	83	84	84	84	83	81	82
Hour 18	30	80	79	76	74	77	78	78	79	79	80	81	80	78
Hour 14 (Local Time)	30	63	62	60	61	65	66	66	66	67	66	66	66	65
Hour 20	30	75	73	72	73	76	77	77	77	78	79	77	76	76
PRECIPITATION (inches):														
Water Equivalent														
-Normal		3.01	2.02	2.31	3.62	5.64	4.66	4.87	5.93	5.99	5.89	5.59	4.46	53.99
-Maximum Monthly	31	7.60	6.69	5.41	8.54	14.99	10.96	9.35	11.76	11.44	15.06	15.96	16.81	16.81
-Year		1977	1982	1958	1983	1965	1965	1961	1955	1976	1970	1979	1981	DEC 1981
-Minimum Monthly	31	0.61	0.20	0.72	0.28	0.44	0.29	1.12	1.93	1.93	1.17	1.91	0.68	0.20
-Year		1978	1983	1970	1984	1972	1985	1974	1982	1961	1979	1980	1963	FEB 1983
-Maximum in 24 hrs	31	5.08	2.73	3.91	6.37	3.65	3.55	2.28	5.08	4.25	5.04	7.07	8.40	8.40
-Year		1969	1969	1969	1973	1979	1965	1969	1955	1976	1985	1979	1981	DEC 1981
Snow, Ice pellets														
-Maximum Monthly	31													
-Year														
-Maximum in 24 hrs	31													
-Year														
WIND:														
Mean Speed (mph)	30	8.6	9.0	9.4	9.2	8.4	8.9	9.7	9.0	7.5	6.8	7.5	8.5	8.5
Prevailing Direction														
through 1963		ENE	ENE	ENE	ENE	ENE	ENE	ENE	ENE	ENE	ENE	ENE	ENE	ENE
Fastest Mile														
-Direction	29	SE	E	S	E	E	NE	E	NE	SE	NE	NE	NE	NE
-Speed (MPH)	29	34	40	37	35	39	38	44	80	47	44	35	46	80
-Year		1974	1965	1983	1970	1976	1970	1975	1956	1973	1972	1961	1970	AUG 1956
Peak Gust														
-Direction	2	W	E	E	ENE	E	SW	ESE	ESE	E	SE	E	E	E
-Speed (mph)	2	32	36	30	30	28	38	31	33	39	35	39	41	41
-Date		1984	1984	1984	1984	1985	1984	1984	1985	1984	1985	1985	1984	DEC 1984

See Reference Notes to this table on the following page.

TABLE 2 PRECIPITATION (inches)   SAN JUAN, PUERTO RICO

YEAR	JAN	FEB	MAR	APR	MAY	JUNE	JULY	AUG	SEP	OCT	NOV	DEC	ANNUAL
1956	3.48	6.44	4.83	4.54	9.66	9.30	3.51	6.78	5.74	7.97	3.18	4.18	69.61
1957	3.78	1.29	1.51	1.14	2.25	6.51	7.16	7.68	4.53	6.40	4.13	6.63	53.01
1958	4.39	1.51	5.41	2.98	13.59	9.33	7.39	6.65	5.52	7.67	4.57	5.04	74.05
1959	5.79	1.61	1.34	6.18	7.01	4.67	2.80	4.29	4.29	6.59	4.85	3.50	51.65
1960	0.94	3.33	2.39	5.76	7.09	5.84	5.02	8.49	8.34	4.63	4.56	8.01	64.40
1961	3.51	1.31	2.64	2.82	1.77	5.26	9.35	5.19	1.93	8.47	9.26	10.00	61.51
1962	4.24	2.67	0.97	3.70	7.53	6.70	6.46	6.98	4.85	2.80	3.84	4.11	54.85
1963	3.13	1.39	4.68	5.21	6.85	2.74	5.02	3.43	10.85	1.63	3.00	0.68	48.61
1964	2.02	1.70	1.27	6.38	3.96	4.50	7.03	6.71	5.10	3.13	3.39	2.35	47.54
1965	2.62	0.79	0.86	2.19	14.99	10.96	5.88	8.66	4.80	5.08	3.93	5.05	65.81
1966	1.34	1.64	4.63	5.62	5.69	3.26	4.21	3.41	7.20	8.99	7.99	6.21	60.19
1967	3.07	2.93	1.46	0.85	4.15	3.38	4.79	4.20	5.12	4.47	5.00	3.13	42.55
1968	2.15	1.60	1.79	0.50	6.31	5.98	5.25	7.36	5.26	2.33	11.11	3.56	53.20
1969	7.49	3.97	2.89	2.42	5.79	4.04	7.49	6.89	4.86	6.99	6.70	2.28	61.81
1970	2.94	1.33	0.72	1.15	7.98	9.26	3.58	4.66	5.66	15.06	8.00	5.98	66.32
1971	2.18	3.67	1.78	2.93	3.87	1.24	1.69	5.18	2.19	4.61	2.31	3.93	35.58
1972	2.76	2.00	3.40	2.79	0.44	1.58	2.24	3.06	3.68	5.46	2.78	7.53	37.72
1973	2.27	0.92	4.66	8.48	0.48	4.71	2.44	7.00	3.13	3.29	3.01	4.16	44.55
1974	2.92	0.82	1.92	1.20	2.42	2.34	1.12	6.57	3.67	8.23	6.55	3.92	41.68
1975	2.69	0.71	1.13	1.01	1.04	2.64	3.35	4.08	9.29	6.60	10.90	7.82	51.26
1976	1.50	2.18	2.05	3.94	2.96	2.96	2.48	5.12	11.44	7.69	2.77	2.11	47.20
1977	7.60	1.02	1.73	0.96	4.04	1.49	4.64	4.42	4.71	5.94	12.44	3.82	52.81
1978	0.61	1.56	3.52	8.27	7.14	2.86	3.46	3.21	6.34	4.88	5.40	2.61	49.86
1979	1.29	1.80	2.25	4.28	12.13	5.76	6.61	9.38	10.11	1.17	15.96	3.81	74.55
1980	1.75	1.67	1.47	2.55	5.19	1.31	2.19	3.17	4.85	6.71	1.91	3.18	35.95
1981	2.55	2.72	4.39	2.89	11.02	5.48	7.04	3.32	2.98	9.32	4.94	16.81	73.46
1982	2.53	6.69	0.98	1.01	10.26	5.24	2.33	1.93	2.87	2.06	4.34	4.76	45.00
1983	0.69	0.20	1.47	8.54	3.85	1.91	6.53	5.15	2.75	4.06	3.25	3.50	41.90
1984	1.96	3.13	0.82	0.28	3.75	6.85	2.66	6.04	3.16	5.10	5.65	4.69	44.09
1985	2.80	2.40	1.84	1.02	5.95	0.29	2.85	4.33	5.44	11.10	4.54	2.80	45.36
Record Mean	2.93	2.29	2.22	3.45	6.17	4.70	4.84	5.82	5.71	5.89	5.72	4.96	54.69

TABLE 3 AVERAGE TEMPERATURE (deg. F)   SAN JUAN, PUERTO RICO

YEAR	JAN	FEB	MAR	APR	MAY	JUNE	JULY	AUG	SEP	OCT	NOV	DEC	ANNUAL
1956	73.9	74.3	75.7	77.1	78.6	79.1	80.3	80.6	80.3	79.4	77.7	75.6	77.7
1957	74.8	74.8	75.6	76.3	78.0	79.0	80.9	81.2	80.9	80.1	78.6	75.7	78.1
1958	75.7	75.8	77.6	79.2	80.2	81.4	80.5	81.6	81.4	80.0	78.0	76.1	79.0
1959	75.1	75.5	76.7	77.9	77.9	80.7	81.0	81.3	81.6	80.7	79.2	77.8	78.8
1960	75.9	76.3	77.3	78.4	80.0	81.2	81.1	81.1	80.8	80.6	79.4	76.5	79.0
1961	76.0	75.6	76.6	78.4	79.7	79.9	80.5	81.6	81.2	79.8	77.5	77.2	78.7
1962	75.2	74.2	75.7	77.2	78.5	80.5	81.1	81.3	81.0	80.4	78.9	77.3	78.5
1963	75.7	76.2	76.3	77.6	78.2	80.8	81.3	81.8	81.3	81.5	78.7	78.9	79.1
1964	76.6	77.5	78.8	78.5	80.9	81.6	81.8	82.2	82.2	80.6	79.7	75.7	79.6
1965	74.7	75.9	77.7	77.7	78.9	80.3	81.7	81.5	82.2	81.5	79.8	77.4	79.1
1966	77.4	76.7	78.1	78.5	79.5	81.0	82.6	81.1	82.6	80.0	77.5	77.1	79.3
1967	76.7	76.3	75.9	78.0	79.7	81.8	82.1	82.2	81.8	81.4	80.4	77.5	79.5
1968	76.4	76.5	77.0	77.0	79.9	80.4	81.2	82.0	81.8	81.6	79.5	77.5	79.2
1969	75.4	75.0	77.8	79.6	80.7	82.0	80.9	81.1	81.2	80.9	78.7	77.1	79.2
1970	76.5	76.2	77.9	80.0	80.6	81.6	82.4	82.3	82.2	81.9	80.2	78.2	80.0
1971	77.6	77.4	78.9	79.7	81.3	83.0	82.9	83.2	83.9	81.9	80.6	78.8	80.8
1972	77.7	77.8	78.7	80.1	81.7	83.8	83.9	83.4	82.9	82.6	81.1	79.2	81.1
1973	78.9	78.3	78.6	80.0	82.9	83.2	83.4	83.2	82.7	83.5	80.7	77.9	81.1
1974	77.3	78.0	78.5	79.6	81.4	83.4	83.7	83.2	82.9	82.5	80.0	78.3	80.7
1975	76.9	77.2	78.4	79.6	81.0	82.9	83.0	82.7	82.1	81.6	79.8	77.5	80.2
1976	76.3	76.6	76.8	78.5	80.5	81.7	83.1	83.5	83.3	82.5	80.9	78.3	80.2
1977	77.0	77.6	78.3	80.2	80.9	81.9	80.6	81.0	81.1	81.9	80.3	79.3	80.0
1978	78.3	79.2	79.8	80.1	80.8	82.3	83.1	83.2	83.9	82.7	81.8	79.6	81.2
1979	78.5	78.7	77.8	79.4	81.3	83.8	83.5	83.0	81.8	83.0	80.6	78.7	80.8
1980	78.2	78.6	78.2	80.8	83.8	85.1	83.2	85.1	84.5	84.4	82.2	80.7	82.3
1981	79.8	79.4	80.9	80.3	83.4	84.2	85.0	83.8	84.3	83.3	81.6	78.7	82.1
1982	78.5	77.8	78.0	80.4	81.0	82.9	83.3	84.6	84.6	83.8	80.2	78.0	81.1
1983	78.5	79.9	82.2	81.8	83.2	85.4	84.2	84.0	84.3	83.6	81.4	79.9	82.4
1984	78.1	77.8	80.2	81.8	81.1	82.4	82.6	82.6	82.2	81.3	78.7	77.3	80.5
1985	76.1	77.3	76.7	78.3	80.4	83.0	83.5	83.4	81.7	79.6	79.0	76.8	79.7
Record Mean	76.6	76.6	77.6	78.8	80.4	81.8	82.1	82.3	82.0	81.5	79.6	77.7	79.8
	82.8	83.1	84.2	85.3	86.8	88.1	88.0	88.4	88.4	88.0	85.7	83.4	86.0
	70.3	70.1	71.0	72.3	74.0	75.5	76.1	76.2	75.6	74.9	73.5	71.9	73.5

REFERENCE NOTES FOR TABLES 1, 2, 3 and 6   (SAN JUAN, PR)

GENERAL

T - TRACE AMOUNT
BLANK ENTRIES DENOTE MISSING/UNREPORTED DATA.
# INDICATES A STATION OR INSTRUMENT RELOCATION.

SPECIFIC

TABLE 1

(a) - LENGTH OF RECORD IN YEARS. ALTHOUGH
INDIVIDUAL MONTHS MAY BE MISSING.

* LESS THAN .05

NORMALS — BASED ON THE 1951-1980 RECORD PERIOD.
EXTREMES — DATES ARE THE MOST RECENT OCCURRENCE.
WIND DIR. — NUMERALS SHOW TENS OF DEGREES
CLOCKWISE FROM TRUE NORTH.
"00" INDICATES CALM.
RESULTANT WIND DIRECTIONS ARE GIVEN TO WHOLE DEGREES.

EXCEPTIONS

TABLE 1

1. FASTEST MILE WINDS ARE THROUGH MAY 1983

TABLES 2, 3, and 6

RECORD MEANS ARE THROUGH THE CURRENT YEAR,
BEGINNING IN   1951 FOR TEMPERATURE
1951 FOR PRECIPITATION

TABLE 4   HEATING DEGREE DAYS Base 65 deg. F   SAN JUAN, PUERTO RICO

SEASON	JULY	AUG	SEP	OCT	NOV	DEC	JAN	FEB	MAR	APR	MAY	JUNE	TOTAL
1983-84	0	0	0	0	0	0	0	0	0	0	0	0	0
1984-85	0	0	0	0	0	0	0	0	0	0	0	0	0
1985-86	0	0	0	0	0	0							

TABLE 5   COOLING DEGREE DAYS Base 65 deg. F   SAN JUAN, PUERTO RICO

YEAR	JAN	FEB	MAR	APR	MAY	JUNE	JULY	AUG	SEP	OCT	NOV	DEC	TOTAL
1969	328	287	401	447	496	517	501	506	493	500	417	382	5275
1970	362	321	409	460	490	506	548	540	523	528	461	416	5564
1971	398	353	438	445	511	545	562	573	577	531	473	438	5844
1972	400	376	431	458	525	569	592	578	543	554	490	446	5962
1973	438	379	428	458	562	553	578	570	539	579	475	406	5965
1974	390	368	424	442	511	558	586	575	544	549	455	418	5820
1975	375	346	420	446	501	543	565	556	521	519	450	395	5637
1976	357	342	375	413	485	506	566	579	557	548	485	419	5632
1977	382	360	419	463	499	514	491	502	487	532	469	452	5570
1978	420	404	464	462	499	526	566	570	575	555	513	459	6013
1979	426	393	404	436	512	571	582	564	511	569	477	432	5877
1980	414	402	446	479	590	610	633	628	591	608	522	494	6417
1981	467	407	499	468	578	584	626	587	588	574	505	429	6312
1982	423	364	412	470	504	543	573	615	593	590	466	405	5958
1983	426	424	541	509	573	621	604	593	583	582	502	468	6426
1984	415	377	479	508	503	528	553	553	522	517	417	389	5761
1985	352	349	368	403	483	547	579	577	508	459	429	371	5425

TABLE 6   SNOWFALL (inches)   SAN JUAN, PUERTO RICO

SEASON	JULY	AUG	SEP	OCT	NOV	DEC	JAN	FEB	MAR	APR	MAY	JUNE	TOTAL
1971-72	0.0	0.0	0.0	0.0	0.0	0.0	0.0	0.0	0.0	0.0	0.0	0.0	0.0
1972-73	0.0	0.0	0.0	0.0	0.0	0.0	0.0	0.0	0.0	0.0	0.0	0.0	0.0
1973-74	0.0	0.0	0.0	0.0	0.0	0.0	0.0	0.0	0.0	0.0	0.0	0.0	0.0
1974-75	0.0	0.0	0.0	0.0	0.0	0.0	0.0	0.0	0.0	0.0	0.0	0.0	0.0
1975-76	0.0	0.0	0.0	0.0	0.0	0.0	0.0	0.0	0.0	0.0	0.0	0.0	0.0
1976-77	0.0	0.0	0.0	0.0	0.0	0.0	0.0	0.0	0.0	0.0	0.0	0.0	0.0
1977-78	0.0	0.0	0.0	0.0	0.0	0.0	0.0	0.0	0.0	0.0	0.0	0.0	0.0
1978-79	0.0	0.0	0.0	0.0	0.0	0.0	0.0	0.0	0.0	0.0	0.0	0.0	0.0
1979-80	0.0	0.0	0.0	0.0	0.0	0.0	0.0	0.0	0.0	0.0	0.0	0.0	0.0
1980-81	0.0	0.0	0.0	0.0	0.0	0.0	0.0	0.0	0.0	0.0	0.0	0.0	0.0
1981-82	0.0	0.0	0.0	0.0	0.0	0.0	0.0	0.0	0.0	0.0	0.0	0.0	0.0
1982-83	0.0	0.0	0.0	0.0	0.0	0.0	0.0	0.0	0.0	0.0	0.0	0.0	0.0
1983-84	0.0	0.0	0.0	0.0	0.0	0.0	0.0	0.0	0.0	0.0	0.0	0.0	0.0
1984-85	0.0	0.0	0.0	0.0	0.0	0.0	0.0	0.0	0.0	0.0	0.0		
1985-86	0.0	0.0	0.0	0.0	0.0	0.0							
Record Mean	0.0	0.0	0.0	0.0	0.0	0.0	0.0	0.0	0.0	0.0	0.0	0.0	0.0

See Reference Notes, relative to all above tables, on preceding page.

BECKLEY, WEST VIRGINIA

The city of Beckley is located in the Appalachian Mountains about 30 miles northwest of the high ridges through eastern West Virginia. The Raleigh County Memorial Airport is on a plateau about 2.5 miles east of the city. Beckley is almost surrounded by distant peaks. The entire area is on a broad plateau composed of rough, hilly ground and lush valleys. The generalized 2,000-foot contour line is about 25 miles to the northwest and runs from southwest to northeast. The generalized 3,000-foot contour is about 10 miles to the southeast and follows the general configuration of the Appalachians.

One of the important results of this location is that only the air reaching Beckley from the northwest has had a trajectory which is primarily upslope. When a northwesterly circulation persists following the passage of a cold front, clearing often is delayed because of the upslope conditions causing clouds. The downslope effect is noticeable with an easterly circulation. When this occurs, the weather often remains clear over Beckley even though the low stratus to the east may obscure the ridges and be solid to the east of these ridges. To a lesser degree, the upslope effect from the northwest is largely responsible for the formation of heavy fogs at the Beckley Airport. Likewise, downslope motion resists formation of fog when the flow is from an easterly direction. The exception to this is a light southeasterly flow which frequently produces fog that is advected over the airport after forming over the large lake in New River Valley. Smoke is no problem due to the lack of any source region.

Beckley has a climate characterized by sharp temperature contrasts, both seasonal and day to day. The months of May through September are generally warm, those of November through March moderately cold, with April and October months of fairly rapid transition. Cold waves occur on an average of two or three times during the winter, but severe cold spells are seldom of more than two or three days duration. Below-zero temperatures, as well as temperatures in the 70s have been recorded during winter months. A low of −13 degrees may be expected once every 10 years and −16 degrees once every 25 years. Summer highs near 90 degrees have occurred, contrasting with lows in the 30s during the same months. Highs seldom reach above the mid-80s. Cool nights are common throughout the summer, with lowest temperatures usually ranging from the 50s to the low 60s.

Ample precipitation is well distributed throughout the year. July has the highest monthly average while October has the lowest average. Summer rainfall occurs mostly during thunderstorms or showery precipitation, while the heaviest winter precipitation usually is associated with storms originating to the southwest and moving northeastward over the Ohio Valley. The formation of storms in the eastern Gulf of Mexico which move up the east coast will sometimes bring heavy snow to the Beckley area. Snowfall occurs chiefly from November through March and occasionally in October and April. The average seasonal snowfall is greater than snowfall at stations to the west at lower elevations, and considerably less than the totals for stations at higher elevations to the east and northeast.

TABLE 1   NORMALS, MEANS AND EXTREMES

BECKLEY, WEST VIRGINIA

LATITUDE: 37°47'N   LONGITUDE: 81°07'W   ELEVATION: FT. GRND 2504 BARO 02512   TIME ZONE: EASTERN   WBAN: 03872

	(a)	JAN	FEB	MAR	APR	MAY	JUNE	JULY	AUG	SEP	OCT	NOV	DEC	YEAR
TEMPERATURE °F:														
Normals														
-Daily Maximum		38.6	41.4	50.5	62.2	70.5	76.5	79.6	78.7	73.3	62.9	51.3	42.1	60.6
-Daily Minimum		21.6	23.2	31.4	40.8	49.2	55.9	59.8	59.0	52.9	41.7	33.0	25.3	41.1
-Monthly		30.1	32.3	41.0	51.5	59.9	66.2	69.7	68.8	63.1	52.3	42.2	33.7	50.9
Extremes														
-Record Highest	22	69	74	80	86	85	90	90	90	89	81	75	73	90
-Year		1985	1977	1985	1976	1974	1964	1980	1965	1973	1969	1974	1971	JUL 1980
-Record Lowest	22	-22	-10	-5	11	23	32	41	36	30	18	4	-15	-22
-Year		1985	1970	1980	1985	1966	1972	1972	1965	1983	1976	1970	1983	JAN 1985
NORMAL DEGREE DAYS:														
Heating (base 65°F)		1082	916	744	405	186	57	11	14	110	398	684	970	5577
Cooling (base 65°F)		0	0	0	0	28	93	157	135	53	0	0	0	466
% OF POSSIBLE SUNSHINE														
MEAN SKY COVER (tenths)														
Sunrise - Sunset	22	7.6	7.5	7.6	7.1	7.1	7.2	7.2	6.9	6.6	6.2	7.2	7.6	7.1
MEAN NUMBER OF DAYS:														
Sunrise to Sunset														
-Clear	22	4.8	4.6	4.7	5.1	5.2	3.6	3.2	4.2	6.6	8.8	5.4	4.5	60.7
-Partly Cloudy	22	5.4	5.8	5.9	7.3	8.5	10.4	11.4	11.3	9.0	7.4	6.7	6.0	94.9
-Cloudy	22	20.9	17.9	20.5	17.6	17.3	16.0	16.4	15.5	14.4	14.8	18.0	20.5	209.7
Precipitation														
.01 inches or more	22	16.3	14.5	15.2	14.4	13.5	12.3	13.7	11.6	10.5	10.3	12.7	14.7	159.7
Snow,Ice pellets														
1.0 inches or more	22	6.1	5.4	3.0	0.4	0.0	0.0	0.0	0.0	0.0	0.2	1.3	3.3	19.6
Thunderstorms	22	0.3	0.5	2.1	3.8	6.3	8.4	10.5	7.6	3.1	0.8	0.5	0.2	44.2
Heavy Fog Visibility														
1/4 mile or less	22	4.7	3.4	3.6	2.0	3.0	3.7	5.1	6.2	6.0	4.0	3.0	3.4	48.2
Temperature °F														
-Maximum														
90° and above	22	0.0	0.0	0.0	0.0	0.0	0.*	0.1	0.1	0.0	0.0	0.0	0.0	0.2
32° and below	22	11.5	7.6	2.4	0.*	0.0	0.0	0.0	0.0	0.0	0.0	2.0	6.0	29.5
-Minimum														
32° and below	22	25.7	22.2	16.8	6.7	1.1	0.*	0.0	0.0	0.1	5.2	14.4	21.8	114.0
0° and below	22	2.7	1.4	0.1	0.0	0.0	0.0	0.0	0.0	0.0	0.0	0.0	0.5	4.6
AVG. STATION PRESS.(mb)	13	927.6	927.9	926.8	927.6	928.0	930.0	931.2	932.2	931.9	931.6	930.1	929.0	929.5
RELATIVE HUMIDITY (%)														
Hour 01	22	78	76	73	70	79	87	90	91	90	83	77	77	81
Hour 07	22	81	80	78	75	80	86	90	92	92	87	81	80	84
Hour 13 (Local Time)	22	68	64	59	51	54	59	63	64	63	58	61	67	61
Hour 19	22	73	68	63	55	60	68	73	77	79	71	69	73	69
PRECIPITATION (inches):														
Water Equivalent														
-Normal		3.44	3.19	4.13	3.59	3.86	3.82	4.46	3.68	3.38	2.54	2.81	3.23	42.13
-Maximum Monthly	22	6.36	6.01	9.18	6.27	7.13	7.05	9.61	5.93	8.27	5.88	6.33	6.14	9.61
-Year		1974	1972	1975	1977	1985	1969	1982	1977	1964	1976	1985	1969	JUL 1982
-Minimum Monthly	22	0.53	0.94	1.74	0.28	1.03	1.48	2.98	1.74	0.54	0.14	0.96	0.57	0.14
-Year		1983	1968	1966	1976	1977	1978	1978	1976	1985	1963	1965	1965	OCT 1963
-Maximum in 24 hrs	22	2.27	2.04	2.22	3.77	2.50	2.47	2.69	2.55	5.34	2.53	2.54	1.97	5.34
-Year		1974	1984	1967	1977	1985	1983	1984	1980	1964	1974	1985	1969	SEP 1964
Snow,Ice pellets														
-Maximum Monthly	22	35.5	30.8	20.4	8.2	T				T	3.2	13.1	22.1	35.5
-Year		1977	1964	1981	1971	1976				1967	1973	1974	1963	JAN 1977
-Maximum in 24 hrs	22	9.5	14.4	9.5	8.2	T				T	3.2	7.0	13.8	14.4
-Year		1971	1983	1980	1971	1976				1967	1973	1968	1967	FEB 1983
WIND:														
Mean Speed (mph)	22	10.8	10.9	11.5	10.8	9.1	7.8	7.0	6.6	7.3	8.7	9.9	10.8	9.3
Prevailing Direction														
Fastest Obs. 1 Min.														
-Direction (!!!)	22	24	26	27	27	24	27	36	32	30	30	16	28	27
-Speed (MPH)	22	46	40	58	44	41	40	46	40	46	30	44	41	58
-Year		1965	1965	1977	1970	1978	1973	1980	1965	1967	1981	1967	1964	MAR 1977
Peak Gust														
-Direction (!!!)	2	NW	SE	SW	W	W	W	W	N	W	SE	SE	25	SE
-Speed (mph)	2	48	62	47	46	39	53	43	29	37	38	41	41	62
-Date		1985	1984	1985	1985	1985	1985	1985	1984	1984	1985	1985	1985	FEB 1984

See Reference Notes to this table on the following page.

TABLE 2 PRECIPITATION (inches)   BECKLEY, WEST VIRGINIA

YEAR	JAN	FEB	MAR	APR	MAY	JUNE	JULY	AUG	SEP	OCT	NOV	DEC	ANNUAL
1956	1.88	6.00	5.73	4.73	3.76	5.77	5.91	3.04	6.24	1.40	2.80	3.35	50.61
1957	6.54	5.31	2.70	2.68	2.58	2.79	2.95	3.00	6.32	2.26	2.33	5.01	44.47
1958	2.34	3.48	4.80	5.08	5.61	3.40	8.57	3.44	2.22	1.45	2.88	0.85	44.12
1959	2.28	1.78	3.41	4.75	2.81	1.86	3.23	5.17	2.82	4.30	5.28	2.97	40.66
1960	1.93	3.16	2.93	1.75	3.28	3.24	3.99	3.80	3.88	1.97	2.19	2.38	34.50
1961	1.85	4.41	3.36	2.87	3.20	5.70	5.64	2.23	1.83	4.51	2.21	4.38	42.19
1962	3.60	5.68	2.51	2.32	2.24	5.72	8.66	2.26	2.51	3.25	4.58	3.85	47.18
#1963	2.20	2.56	7.59	1.21	3.51	4.78	4.45	2.58	0.14	4.68	1.68		40.91
1964	3.44	3.83	4.75	3.46	1.31	1.98	2.98	2.38	8.27	3.39	2.35	3.41	41.55
1965	5.31	2.11	4.70	3.98	1.71	3.13	3.03	2.26	2.19	2.67	0.96	0.57	32.62
1966	3.54	3.69	1.74	4.60	1.45	1.58	3.66	3.99	8.08	2.45	3.39	4.08	42.25
1967	1.40	2.85	6.50	2.84	6.25	2.25	7.62	3.81	2.83	2.42	3.91	4.96	47.64
1968	2.64	0.94	2.93	3.36	4.62	3.08	3.17	3.50	2.20	2.63	3.17	2.24	34.48
1969	1.91	1.61	2.48	2.55	2.81	7.05	5.14	4.66	2.19	1.18	2.45	6.14	40.17
1970	2.11	2.99	2.55	3.62	1.36	3.58	3.34	5.06	3.03	2.50	1.93	3.27	35.34
1971	3.74	3.67	2.87	2.97	6.24	4.49	4.23	1.74	4.51	3.01	2.03	1.91	41.41
1972	5.45	6.01	1.86	4.62	4.73	5.46	5.78	3.56	3.65	2.84	3.94	4.29	52.19
1973	1.75	2.66	4.49	5.12	4.63	2.11	5.21	2.79	2.33	3.22	4.52	4.62	43.45
1974	6.36	2.03	4.91	3.36	4.67	4.42	3.93	2.76	2.89	2.90	4.07	3.45	45.75
1975	4.54	4.34	9.18	3.92	5.65	4.61	7.26	4.21	4.62	2.75	2.51	3.51	57.10
1976	3.08	2.96	3.32	0.28	3.79	2.98	3.56	1.74	5.51	5.88	1.30	2.77	37.17
1977	2.02	1.58	2.48	6.27	1.03	5.65	3.28	5.93	4.06	4.09	2.94	2.11	41.44
1978	4.40	0.98	3.82	4.16	1.48	3.10	4.66	1.49	0.97	2.42	5.88		36.50
1979	5.31	2.54	2.60	3.11	4.99	6.40	4.89	3.40	4.59	2.59	3.68	1.48	45.58
1980	3.35	1.78	3.71	4.85	2.38	2.69	6.80	5.54	1.60	1.81	2.15	1.06	37.72
1981	0.57	2.11	1.79	3.40	5.16	5.65	3.29	2.50	5.00	3.15	1.25	3.42	37.29
1982	2.07	3.68	3.49	1.44	6.22	3.23	9.61	2.87	1.74	2.64	3.46	1.69	42.14
1983	0.53	1.43	1.83	3.43	3.98	7.00	3.72	2.52	1.60	4.55	1.84	2.27	34.70
1984	1.50	4.07	1.85	3.21	5.27	2.87	8.53	2.65	1.76	2.18	4.13	3.41	41.43
1985	2.25	2.02	2.14	0.87	7.13	4.11	3.14	4.32	0.54	1.57	6.33	1.71	36.13
Record Mean	3.14	3.09	3.80	3.40	3.92	4.08	4.77	3.72	3.39	2.62	2.96	3.10	41.98

TABLE 3 AVERAGE TEMPERATURE (deg. F)   BECKLEY, WEST VIRGINIA

YEAR	JAN	FEB	MAR	APR	MAY	JUNE	JULY	AUG	SEP	OCT	NOV	DEC	ANNUAL
1956	27.6	39.0	40.7	47.6	60.9	65.0	68.8	66.9	58.9		39.2	43.9	
1957	31.5	41.0	40.9	55.2	61.7	69.0	69.3	67.1	64.8	48.3	44.2	36.9	52.5
1958	28.4	25.0	36.0	50.8	58.8	64.7	72.0	68.0	61.7	50.1	43.3	27.7	48.9
1959	30.4	36.7	38.3	51.6	62.9	63.3	70.1	70.9	63.6	53.8	40.8	36.1	51.6
1960	34.2	31.1	27.2	55.1	57.5	65.2	67.8	70.0	64.1	53.8	42.4	27.2	49.7
1961	26.0	37.9	46.1	45.4	55.7	63.4	67.6	68.1	64.6	51.3	45.3	34.2	50.5
1962	31.1	38.7	39.3	48.1	65.4	65.3	67.2	67.2	60.0	53.9	41.1	27.7	50.4
#1963	26.9	26.4	46.1	52.6	58.1	64.6	66.1	66.4	59.7	54.2	43.0	23.4	49.0
1964	33.1	28.1	41.4	53.7	61.1	67.6	68.9	67.3	63.4	48.4	46.4	36.8	51.4
1965	31.2	32.8	36.0	52.2	64.6	65.0	68.7	68.0	65.9	49.7	44.0	37.1	51.3
1966	24.6	30.9	42.8	48.6	59.1	66.5	71.3	67.3	59.7	49.7	44.0	32.7	49.8
1967	35.5	28.6	44.8	53.8	55.2	66.6	66.5	65.4	57.3	51.9	38.1	36.9	50.0
1968	28.3	22.9	44.3	52.1	56.9	66.0	69.6	66.2	61.2	52.1	42.8	29.9	49.7
1969	28.6	30.7	33.8	53.2	61.3	67.7	71.5	67.8	60.8	52.2	38.0	28.1	49.5
1970	24.0	31.3	39.1	53.7	61.6	66.4	69.5	69.0	67.3	55.4	43.2	36.1	51.4
1971	27.7	33.3	36.0	49.4	56.2	69.1	68.3	67.5	66.0	58.3	41.2	42.5	51.3
1972	35.3	30.7	39.7	49.9	58.3	61.3	69.0	68.7	63.8	49.4	40.3	39.8	50.5
1973	31.3	31.4	48.5	49.1	56.8	68.8	70.2	69.9	65.6	55.3	44.3	34.1	52.1
1974	41.4	34.1	45.7	53.1	59.5	62.6	68.2	68.2	59.3	50.4	42.8	34.4	51.6
1975	34.3	36.5	37.3	47.7	62.6	66.6	69.5	71.9	61.1	55.5	47.1	35.3	52.1
1976	27.8	42.2	47.7	52.0	57.2	66.6	68.1	66.0	59.9	46.4	34.3	29.5	49.8
1977	16.5	29.9	46.7	54.8	63.0	63.5	71.9	69.8	65.1	49.3	46.1	32.6	50.7
1978	22.2	21.6	38.9	52.8	58.7	66.1	69.1	70.3	66.0	51.5	47.4	36.9	50.1
1979	25.6	26.6	45.5	51.8	60.1	64.0	67.7	68.6	62.2	50.5	44.4	36.2	50.3
1980	30.8	25.8	38.6	50.5	59.5	64.2	72.2	71.4	65.6	49.9	39.9	33.5	50.1
1981	24.2	34.4	36.2	55.6	56.6	68.6	70.6	67.3	60.7	50.1	42.4	29.9	49.7
1982	25.8	34.3	44.2	48.5	64.4	64.6	70.2	66.6	61.2	53.4	45.3	41.0	51.6
1983	30.2	34.2	42.8	48.6	57.8	66.0	70.3	70.9	63.0	51.8	41.5	28.5	50.5
1984	28.1	37.9	38.1	49.0	57.0	69.7	67.5	68.7	60.1	61.1	41.5	45.1	52.0
1985	24.2	31.6	43.2	54.8	60.5	65.1	69.5	67.5	63.0	57.2	52.2	28.8	51.5
Record Mean	29.7	32.8	40.9	51.5	59.6	65.8	69.4	68.4	62.5	52.1	42.3	33.9	50.8
Max	39.0	42.8	51.4	63.3	71.4	77.0	79.9	79.1	73.6	63.5	52.4	43.1	61.4
Min	20.4	22.8	30.4	39.6	47.8	54.6	58.9	57.8	51.3	40.7	32.2	24.7	40.1

REFERENCE NOTES FOR TABLES 1, 2, 3 and 6   (BECKLEY, WV)

GENERAL

T - TRACE AMOUNT
BLANK ENTRIES DENOTE MISSING/UNREPORTED DATA.
# INDICATES A STATION OR INSTRUMENT RELOCATION.

SPECIFIC

TABLE 1

(a) - LENGTH OF RECORD IN YEARS. ALTHOUGH INDIVIDUAL MONTHS MAY BE MISSING.

* LESS THAN .05

NORMALS — BASED ON THE 1951-1980 RECORD PERIOD.
EXTREMES — DATES ARE THE MOST RECENT OCCURRENCE.
WIND DIR. — NUMERALS SHOW TENS OF DEGREES CLOCKWISE FROM TRUE NORTH.
   "00" INDICATES CALM.
RESULTANT WIND DIRECTIONS ARE GIVEN TO WHOLE DEGREES.

EXCEPTIONS

TABLES 2, 3, and 6

RECORD MEANS ARE THROUGH THE CURRENT YEAR, BEGINNING IN   1951 FOR TEMPERATURE
   1951 FOR PRECIPITATION
   1964 FOR SNOWFALL

TABLE 4 HEATING DEGREE DAYS Base 65 deg. F   BECKLEY, WEST VIRGINIA

SEASON	JULY	AUG	SEP	OCT	NOV	DEC	JAN	FEB	MAR	APR	MAY	JUNE	TOTAL
1956-57	11	34	205		756	646	1032	666	739	311	133	8	
1957-58	6	38	91	512	614	867	1132	1113	889	419	193	51	5925
1958-59	0	24	121	457	647	1152	1064	787	819	395	104	94	5664
1959-60	2	2	96	354	720	887	949	976	1165	296	245	47	5739
1960-61	14	0	75	339	670	1165	1203	753	576	593	285	85	5758
1961-62	32	7	101	420	587	950	1046	728	791	503	51	27	5243
1962-63	21	17	182	346	705	1151	1177	1073	578	370	213	63	5896
#1963-64	43	31	161	249	653	1288	978	1063	726	340	137	56	5725
1964-65	8	48	93	509	550	868	1037	895	891	379	60	70	5408
1965-66	7	40	77	468	621	855	1243	946	680	490	199	53	5679
1966-67	5	27	172	469	625	993	905	1012	620	335	306	43	5512
1967-68	37	49	228	400	802	865	1133	1215	634	381	249	58	6051
1968-69	10	53	124	392	661	1078	1124	955	961	351	140	50	5899
1969-70	0	4	166	400	804	1139	1269	938	798	351	150	38	6057
1970-71	17	12	58	299	645	890	1149	883	890	461	280	4	5588
1971-72	6	10	46	212	714	692	916	987	775	446	204	139	5147
1972-73	35	10	74	475	737	776	1036	936	506	477	258	11	5331
1973-74	5	13	55	295	613	951	724	859	594	362	198	113	4782
1974-75	5	9	188	447	660	943	946	794	849	514	116	40	5511
1975-76	0	2	146	289	530	914	1148	655	533	421	241	37	4916
1976-77	24	49	151	566	914	1094	1496	978	565	310	113	97	6357
1977-78	8	11	67	479	564	997	1319	1211	801	366	216	56	6095
1978-79	14	3	58	411	521	864	1213	1070	597	389	175	69	5384
1979-80	28	37	116	448	611	890	1053	1130	809	435	186	81	5824
1980-81	0	1	69	463	746	970	1257	849	887	287	264	15	5808
1981-82	6	18	164	453	670	1084	1211	856	636	487	71	43	5699
1982-83	9	23	144	363	585	737	1071	856	680	490	230	46	5234
1983-84	14	12	138	400	699	1126	1138	781	827	483	266	18	5902
1984-85	24	14	189	137	698	608	1259	929	672	324	151	56	5061
1985-86	4	19	128	247	378	1116							

TABLE 5 COOLING DEGREE DAYS Base 65 deg. F   BECKLEY, WEST VIRGINIA

YEAR	JAN	FEB	MAR	APR	MAY	JUNE	JULY	AUG	SEP	OCT	NOV	DEC	TOTAL
1969	0	0	0	4	32	139	208	100	45	11	0	0	539
1970	0	0	0	16	53	88	165	142	131	11	0	0	606
1971	0	0	0	0	14	132	119	91	86	11	5	0	458
1972	0	0	0	2	4	34	165	132	42	0	0	0	379
1973	0	0	1	6	9	130	172	173	81	2	0	0	574
1974	0	0	1	12	35	51	112	112	22	0	0	0	345
1975	0	0	0	1	51	92	145	221	37	1	0	0	548
1976	0	0	2	38	7	90	127	86	6	0	0	0	356
1977	0	0	1	11	58	60	229	169	77	0	2	0	607
1978	0	0	0	5	33	96	147	175	96	0	0	0	552
1979	0	0	0	1	31	49	121	157	39	6	0	0	404
1980	0	0	0	4	24	66	235	207	94	2	0	0	632
1981	0	0	0	11	8	130	187	99	40	0	0	0	475
1982	0	0	0	0	61	37	176	83	37	12	0	0	406
1983	0	0	0	4	16	84	189	201	86	0	0	0	580
1984	0	0	0	8	26	166	107	134	52	24	0	0	517
1985	0	0	5	25	20	68	150	100	75	9	0	0	452

TABLE 6 SNOWFALL (inches)   BECKLEY, WEST VIRGINIA

SEASON	JULY	AUG	SEP	OCT	NOV	DEC	JAN	FEB	MAR	APR	MAY	JUNE	TOTAL
1956-57	0.0	0.0	0.0	0.0	1.0	7.0	15.5	8.0	1.5	3.0	0.0	0.0	36.0
1957-58	0.0	0.0	0.0	2.0	3.0	3.8	11.8	23.2	21.8	T	0.0	0.0	65.6
1958-59	0.0	0.0	0.0	0.0	1.0	5.0	9.0	T	3.5	0.5	0.0	0.0	19.0
1959-60	0.0	0.0	0.0	T	2.4	8.5	9.5	26.0	48.0	0.5	T	0.0	94.9
1960-61	0.0	0.0	0.0	0.0	1.0	8.0	21.0	14.1		4.0	0.0	0.0	
1961-62	0.0	0.0	0.0	7.0	2.0	11.0	12.4	11.1	9.2	2.5	0.0	0.0	46.1
1962-63	0.0	0.0	0.0	2.0	T	24.0	11.1	24.4	3.5	0.0	0.0	0.0	65.0
#1963-64	0.0	0.0	0.0	T	6.3	22.1	11.1	30.8	7.0	0.3	0.0	0.0	77.6
1964-65	0.0	0.0	0.0	0.0	3.3	1.4	21.3	11.0	10.5	T	0.0	0.0	47.5
1965-66	0.0	0.0	0.0	0.2	0.6	6.9	25.2	15.9	5.9	6.6	T	0.0	61.3
1966-67	0.0	0.0	0.0	T	4.0	18.1	9.2	26.6	6.3	T	T	0.0	64.2
1967-68	0.0	0.0	T	T	0.9	19.3	21.2	11.8	7.2	0.0	0.0	0.0	60.4
1968-69	0.0	0.0	0.0	T	11.0	12.0	4.0	10.9	15.4	T	T	0.0	53.3
1969-70	0.0	0.0	0.0	0.0	6.5	20.7	18.9	12.6	4.0	0.2	0.0	0.0	62.9
1970-71	0.0	0.0	0.0	0.0	4.5	10.3	14.0	13.4	19.8	8.2	T	0.0	70.2
1971-72	0.0	0.0	0.0	0.0	7.6	0.9	2.4	22.1	3.4	1.9	0.0	0.0	38.3
1972-73	0.0	0.0	0.0	0.2	3.6	2.3	5.5	10.9	5.0	1.4	0.0	0.0	28.9
1973-74	0.0	0.0	0.0	3.2	0.7	19.2	1.4	10.4	2.5	1.2	0.0	0.0	38.6
1974-75	0.0	0.0	0.0	1.4	13.1	16.5	11.7	7.8	17.3	1.0	T	0.0	68.8
1975-76	0.0	0.0	0.0	0.0	3.1	5.0	16.4	7.3	3.8	T	T	0.0	35.6
1976-77	0.0	0.0	0.0	T	12.7	19.3	35.5	15.4	3.5	2.2	0.0	0.0	88.6
1977-78	0.0	0.0	0.0	T	5.4	7.9	31.6	20.3	15.1	0.2	0.0	0.0	80.5
1978-79	0.0	0.0	0.0	0.0	2.4	3.8	27.7	29.5	5.4	T	0.0	0.0	68.8
1979-80	0.0	0.0	0.0	2.7	1.4	4.7	21.1	26.1	12.8	1.0	0.0	0.0	69.8
1980-81	0.0	0.0	0.0	T	1.4	4.4	21.4	8.4	20.4	T	0.0	0.0	56.0
1981-82	0.0	0.0	0.0	0.5	3.6	22.0	17.9	18.2	12.8	4.0	0.0	0.0	79.0
1982-83	0.0	0.0	0.0	0.0	0.3	14.8	10.9	25.6	6.8	0.7	0.0	0.0	59.1
1983-84	0.0	0.0	0.0	3.8	8.5	22.9	22.3	18.5	9.8	0.3	0.0	0.0	67.6
1984-85	0.0	0.0	0.0	0.0	0.5	4.8	34.1	18.5	2.2	6.9	0.0	0.0	67.0
1985-86	0.0	0.0	0.0	0.0	T	14.4							
Record Mean	0.0	0.0	T	0.4	4.1	10.8	17.5	17.1	9.0	1.6	T	0.0	60.5

See Reference Notes, relative to all above tables, on preceding page.

Charleston lies at the junction of the Kanawha and Elk Rivers in the western foothills of the Appalachian Mountains. The main urban and business areas have developed along the two river valleys, while some residential areas are in nearby valleys and on the surrounding hills. The hilltops are around 1,100 feet above sea level, about 500 feet higher than the valleys. The Kanawha Airport is just over 2 miles northeast of the center-city area, on an artificial plateau constructed from several hilltops.

Weather records are maintained at the Kanawha Airport by National Weather Service personnel. This site tends to be slightly cooler than the river valleys during the afternoons. Conversely, the valleys can become cooler than the hilltops during clear, calm nights. The weather at Charleston is highly changeable, especially from mid-autumn through the spring.

Winters can vary greatly from one season to the next. Snow does not favor any given winter month, heavy snowstorms are infrequent, and most snowfalls are in the 4-inch or less category. Snow and ice usually do not persist on valley roads, but can linger longer on nearby hills and outlying rural roads.

Afternoon temperatures in the 40s and morning readings in the 20s are common during the winter. Yet, every winter typically has two or three extended cold spells when temperatures stay below freezing for a few consecutive days. Northwesterly winds are associated with the cold weather. Air reaching Charleston from the northwest can cause cloudiness and flurries, even when there is no nearby organized storm system. Winter conditions are much more severe over the higher mountains less than 50 miles to the northeast through the southeast. Temperatures warm rapidly in the spring and are accompanied by low daytime humidities.

Summer and early autumn have more day-to-day consistency in the weather. Sunshine is more abundant than in winter. Summer precipitation falls mostly in brief, but sometimes heavy, showers. Flash flooding can occur along small streams, but flooding is rare on the dam-controlled Kanawha and Elk Rivers.

Afternoon summer temperatures are mostly in the 80s. Readings above 95 degrees are rare. However, during a hot spell, haze and humidity can add to the unpleasantness and indoor air conditioning is recommended. Cooler and less humid air often penetrates the area from the north to end a hot spell.

Early morning fog is common from late June into October. Industrial and vehicular pollutants can contribute to limited visibility any time of the year, especially when cooler air becomes trapped in the valleys. Autumn foliage is generally at its peak during the second and third weeks of October. By the end of October, the first 32 degree temperature has usually arrived.

Ample precipitation is well distributed throughout the year. July is quite often the wettest month of the year, while October averages the least rain. Droughts severe enough to limit water use are scarce. Any dry spells during the spring or autumn can cause conditions favorable for brush fires in outlying areas.

TABLE 1 — NORMALS, MEANS AND EXTREMES

CHARLESTON, WEST VIRGINIA

LATITUDE: 38°22'N   LONGITUDE: 81°36'W   ELEVATION: FT. GRND 1016 BARO 1019   TIME ZONE: EASTERN   WBAN: 13866

	(a)	JAN	FEB	MAR	APR	MAY	JUNE	JULY	AUG	SEP	OCT	NOV	DEC	YEAR
TEMPERATURE °F:														
Normals														
—Daily Maximum		41.8	45.4	55.4	67.3	76.0	82.5	85.2	84.2	78.7	67.7	55.6	45.9	65.5
—Daily Minimum		23.9	25.8	34.1	43.3	51.8	59.4	63.8	63.1	56.4	44.0	35.0	27.8	44.0
—Monthly		32.9	35.6	44.8	55.3	63.9	71.0	74.5	73.7	67.6	55.9	45.3	36.9	54.8
Extremes														
—Record Highest	38	79	78	87	92	93	98	102	100	102	92	85	80	102
—Year		1950	1977	1954	1985	1985	1953	1954	1953	1953	1951	1948	1982	JUL 1954
—Record Lowest	38	−15	−6	0	19	26	33	46	41	34	17	6	−10	−15
—Year		1985	1968	1980	1982	1966	1972	1963	1965	1983	1962	1950	1983	JAN 1985
NORMAL DEGREE DAYS:														
Heating (base 65°F)		995	823	626	298	125	16	0	0	51	301	591	871	4697
Cooling (base 65°F)		0	0	0	7	91	196	295	270	129	19	0	0	1007
% OF POSSIBLE SUNSHINE														
MEAN SKY COVER (tenths)														
Sunrise – Sunset	38	7.8	7.5	7.4	6.9	6.6	6.5	6.6	6.4	6.1	6.1	7.2	7.5	6.9
MEAN NUMBER OF DAYS:														
Sunrise to Sunset														
—Clear	38	3.5	4.3	4.4	5.8	5.8	4.7	4.2	4.8	6.7	8.7	5.2	4.6	62.7
—Partly Cloudy	38	6.8	6.1	7.8	7.7	10.2	13.0	13.3	14.5	11.3	9.2	7.1	6.6	113.6
—Cloudy	38	20.7	17.8	18.8	16.5	15.0	12.2	13.5	11.7	12.0	13.1	17.8	19.8	188.9
Precipitation														
.01 inches or more	38	15.6	13.6	15.2	13.9	13.2	11.3	12.9	11.0	9.2	9.7	12.0	14.0	151.5
Snow, Ice pellets														
1.0 inches or more	38	3.4	2.6	1.5	0.1	0.0	0.0	0.0	0.0	0.0	0.*	0.8	1.7	10.2
Thunderstorms	38	0.5	0.8	2.2	4.2	6.6	7.8	9.6	7.0	2.9	1.1	0.6	0.3	43.8
Heavy Fog Visibility 1/4 mile or less	38	4.4	3.3	2.8	2.8	7.8	12.4	15.9	19.4	16.8	11.2	4.7	4.0	105.6
Temperature °F														
—Maximum														
90° and above	38	0.0	0.0	0.0	0.3	1.1	4.8	7.6	4.9	2.4	0.1	0.0	0.0	21.3
32° and below	38	7.8	4.7	0.9	0.0	0.0	0.0	0.0	0.0	0.0	0.0	0.6	4.5	18.5
—Minimum														
32° and below	38	23.4	20.0	14.6	4.8	0.4	0.0	0.0	0.0	0.0	3.3	12.9	21.0	100.5
0° and below	38	1.1	0.3	0.*	0.0	0.0	0.0	0.0	0.0	0.0	0.0	0.0	0.2	1.7
AVG. STATION PRESS. (mb)														
RELATIVE HUMIDITY (%)														
Hour 01	38	74	72	68	67	79	86	90	91	89	83	75	75	79
Hour 07	38	77	77	74	74	82	86	90	92	91	87	80	77	82
Hour 13 (Local Time)	38	63	59	53	47	50	54	60	58	55	53	56	61	56
Hour 19	38	65	61	54	49	55	61	66	70	71	66	62	66	62
PRECIPITATION (inches):														
Water Equivalent														
—Normal		3.48	3.11	4.00	3.52	3.68	3.32	5.36	4.15	3.01	2.63	2.90	3.27	42.43
—Maximum Monthly	38	9.11	6.89	6.80	6.46	6.59	7.00	13.54	10.45	7.61	6.49	8.45	8.02	13.54
—Year		1950	1956	1967	1965	1968	1982	1961	1958	1971	1983	1985	1978	JUL 1961
—Minimum Monthly	38	1.09	0.64	1.43	0.50	0.84	0.70	2.16	0.66	0.65	0.09	0.64	0.45	0.09
—Year		1981	1968	1969	1976	1977	1966	1974	1957	1959	1963	1965	1965	OCT 1963
—Maximum in 24 hrs	38	1.91	2.45	2.86	2.72	3.31	2.24	5.60	4.17	2.40	2.48	2.45	2.47	5.60
—Year		1961	1951	1967	1948	1982	1962	1961	1958	1956	1961	1985	1978	JUL 1961
Snow, Ice pellets														
—Maximum Monthly	38	39.5	21.8	18.3	5.9	0.2					2.8	25.8	18.6	39.5
—Year		1978	1964	1960	1959	1963					1961	1950	1962	JAN 1978
—Maximum in 24 hrs	38	15.8	11.2	9.9	5.5	0.2					2.8	15.1	11.2	15.8
—Year		1978	1983	1954	1959	1963					1961	1950	1967	JAN 1978
WIND:														
Mean Speed (mph)	38	7.6	7.6	8.4	7.8	6.2	5.6	5.1	4.4	4.8	5.2	6.7	7.2	6.4
Prevailing Direction through 1963		WSW	WSW	WSW	SW	SW	SW	S	S	S	S	SW	SW	SW
Fastest Obs. 1 Min.														
—Direction	36	25	19	32	27	25	32	29	29	20	25	29	25	25
—Speed (MPH)	36	45	40	46	45	55	50	46	50	35	45	40	55	55
—Year		1951	1981	1955	1953	1953	1951	1957	1952	1956	1950	1954	1953	MAY 1953
Peak Gust														
—Direction	2	SW	W	NW	W	SW	W	SW	W	NW	NW	W	W	NW
—Speed (mph)	2	45	40	53	44	48	37	43	29	32	33	44	45	53
—Date		1985	1984	1985	1985	1984	1985	1984	1984	1985	1985	1984	1985	MAR 1985

See Reference Notes to this table on the following page.

TABLE 2 PRECIPITATION (inches) CHARLESTON, WEST VIRGINIA

YEAR	JAN	FEB	MAR	APR	MAY	JUNE	JULY	AUG	SEP	OCT	NOV	DEC	ANNUAL
1956	2.72	6.89	6.59	4.16	4.36	2.70	4.80	3.45	4.39	1.63	1.48	5.73	48.90
1957	5.78	4.69	2.52	2.60	2.83	1.69	4.27	0.66	3.51	3.14	1.99	3.80	37.48
1958	3.49	3.09	3.09	5.78	5.88	3.32	9.36	10.45	2.53	1.58	2.64	1.24	52.45
1959	3.27	2.90	2.28	4.77	2.40	1.90	6.74	2.02	0.65	3.64	3.81	2.02	36.40
1960	3.29	4.08	2.05	3.23	2.74	4.26	5.57	3.84	1.82	2.21	2.11	2.17	37.37
1961	3.79	3.63	4.86	3.22	4.23	5.22	13.54	1.19	1.50	6.11	2.89	4.74	54.92
1962	2.81	5.37	3.91	4.12	2.22	4.94	8.03	1.94	3.67	2.53	6.27	3.46	49.27
1963	1.85	2.70	6.37	1.21	3.48	2.67	3.06	2.85	1.34	0.09	3.19	1.44	30.25
1964	2.58	3.58	3.65	3.20	0.95	3.82	2.76	3.23	2.53	0.59	2.95	3.14	32.98
1965	4.65	2.02	4.52	6.46	1.90	2.18	2.46	4.38	3.21	2.09	0.64	0.45	34.96
1966	3.57	2.78	1.51	5.06	1.52	0.70	2.94	3.31	3.74	1.72	3.05	2.53	32.43
1967	1.21	2.95	6.80	3.21	6.45	1.83	4.59	1.85	1.68	2.09	4.30	4.72	41.68
1968	2.01	0.64	4.79	2.58	6.59	2.83	4.02	5.42	3.32	3.16	2.47	2.38	40.21
1969	1.50	1.27	1.43	2.35	1.95	2.43	6.13	8.20	3.27	1.52	2.22	4.85	37.12
1970	1.15	3.51	4.23	3.19	1.13	2.35	3.53	4.84	3.55	5.19	2.34	3.81	38.82
1971	2.35	3.40	1.97	1.19	5.17	2.58	6.59	2.12	7.61	1.30	2.83	1.71	38.82
1972	5.47	5.51	2.17	5.16	2.55	4.33	4.13	4.13	3.61	2.48	5.26	6.35	51.15
1973	1.52	2.41	3.40	5.44	5.36	4.48	6.88	2.07	3.91	4.75	5.42	3.68	49.32
1974	4.67	2.50	4.54	3.05	6.06	5.07	2.16	4.22	2.64	1.64	3.72	3.19	43.46
1975	4.84	3.10	6.01	4.03	6.44	4.25	2.71	5.14	4.99	3.08	2.66	3.74	50.99
1976	2.89	2.11	4.21	0.50	3.66	4.24	6.93	2.23	5.37	5.44	1.02	2.18	40.78
1977	1.90	1.08	3.16	4.06	0.84	5.93	4.92	6.58	1.14	4.16	3.78	2.07	39.62
1978	5.59	1.31	2.67	3.31	3.99	2.96	9.83	8.21	1.45	2.68	2.26	8.02	52.28
#1979	6.48	3.76	3.00	3.82	3.87	3.54	3.54	5.17	3.95	3.67	4.02	2.81	48.87
1980	2.85	2.25	5.32	4.49	2.67	2.17	8.47	10.32	2.37	2.03	3.02	1.85	47.81
1981	1.09	4.59	1.80	4.04	3.78	6.46	3.02	2.24	2.36	2.43	1.29	2.71	35.81
1982	3.74	3.23	4.96	1.14	6.19	7.00	2.68	2.65	2.58	1.65	4.65	2.71	43.18
1983	1.24	2.72	3.15	3.96	5.98	2.77	4.19	2.54	1.33	6.49	4.80	3.19	42.36
1984	1.67	2.56	2.72	4.00	3.71	2.56	4.37	4.57	2.95	3.28	4.73	3.78	40.90
1985	3.07	2.32	4.23	1.84	5.88	3.07	3.22	2.02	0.71	3.65	8.45	2.71	41.17
Record Mean	3.69	3.24	4.05	3.64	3.79	3.90	4.95	4.15	2.94	2.81	3.14	3.30	43.60

TABLE 3 AVERAGE TEMPERATURE (deg. F) CHARLESTON, WEST VIRGINIA

YEAR	JAN	FEB	MAR	APR	MAY	JUNE	JULY	AUG	SEP	OCT	NOV	DEC	ANNUAL
1956	30.9	41.8	44.2	52.0	64.4	70.0	73.5	73.1	64.1	60.0	45.3	46.9	55.5
1957	32.5	41.4	44.1	58.5	65.3	74.3	75.3	73.4	69.9	52.5	47.0	40.4	56.2
1958	31.2	27.5	39.3	54.3	63.4	68.5	76.0	72.3	66.8	55.6	48.2	31.5	52.9
1959	32.9	39.4	43.1	55.9	68.3	70.7	75.7	77.4	70.4	58.3	43.9	40.0	56.3
#1960	37.6	34.0	32.7	59.5	61.0	70.5	73.0	75.9	69.8	58.1	47.3	30.1	54.1
1961	27.5	40.0	48.0	49.7	59.4	67.9	72.4	73.7	71.1	56.1	46.3	37.2	54.1
1962	33.6	40.2	43.6	51.6	69.8	73.0	73.0	74.4	65.0	57.3	42.9	30.8	54.6
1963	27.9	29.0	49.5	56.1	61.3	69.1	72.0	70.2	64.5	59.2	46.2	26.1	52.6
1964	35.8	33.2	46.7	58.5	67.0	72.9	75.0	73.1	67.5	53.2	49.0	39.2	55.9
1965	33.3	35.6	39.6	55.1	69.2	70.0	73.2	72.5	69.3	52.6	46.1	39.8	54.7
1966	27.1	36.6	47.7	54.1	62.1	72.0	77.3	73.5	66.5	55.0	48.2	35.5	54.6
1967	38.5	31.6	48.5	56.3	58.3	71.6	71.1	70.2	62.1	55.9	41.3	38.9	53.7
1968	29.1	27.1	47.5	56.4	61.0	70.6	74.2	74.1	66.3	56.0	46.8	34.2	53.6
1969	32.7	35.2	38.7	56.2	64.1	73.7	76.6	72.1	64.1	53.9	41.3	32.3	53.4
1970	28.5	36.2	43.3	58.2	66.9	73.1	74.6	73.6	70.0	57.4	46.0	38.4	55.5
1971	29.4	35.2	39.7	51.9	59.9	72.4	71.9	70.7	63.8	45.9	47.3	35.2	55.2
1972	38.7	35.2	43.8	54.5	63.7	64.9	73.4	72.7	68.7	52.5	43.4	42.1	54.5
1973	34.3	35.1	52.2	53.8	60.8	73.3	74.5	74.6	70.1	58.9	46.8	38.0	56.0
1974	43.5	37.1	49.1	56.9	64.2	67.8	74.2	73.6	63.2	52.4	45.2	36.8	55.3
1975	35.7	38.0	39.8	50.2	66.1	71.7	74.2	76.7	63.9	57.6	50.2	38.6	55.2
1976	31.7	45.8	51.6	55.2	61.9	71.9	72.4	70.2	63.3	49.5	37.6	31.0	53.5
1977	18.6	33.2	49.8	58.4	67.0	68.6	76.9	73.6	70.2	53.4	49.1	35.1	54.5
1978	24.4	24.2	42.6	56.7	63.1	70.9	73.8	74.9	71.0	53.8	49.4	39.0	53.6
1979	28.1	27.9	50.5	55.1	63.0	68.8	72.9	73.4	67.0	54.6	47.5	38.3	54.0
1980	34.1	29.7	42.0	53.3	63.6	68.8	76.6	76.3	69.8	53.6	43.2	36.3	54.0
1981	28.0	37.2	41.4	59.1	60.4	73.2	75.7	72.8	66.5	54.2	45.3	34.6	54.0
1982	29.8	36.1	47.2	51.5	68.6	68.8	76.2	71.1	65.9	57.7	49.0	44.8	55.6
1983	34.0	37.7	47.0	52.1	61.1	71.6	77.0	78.0	68.4	58.1	47.4	32.0	55.4
1984	30.6	41.5	41.1	54.2	61.4	75.3	73.2	74.9	65.4	64.4	44.6	46.9	56.1
1985	27.2	34.0	49.4	60.8	66.3	71.0	75.8	74.0	69.6	62.3	55.5	33.8	56.6
Record Mean	35.5	37.6	46.0	56.0	64.7	72.1	75.8	74.7	68.1	57.7	46.7	38.1	56.1
Max	45.6	48.4	57.7	68.8	77.5	84.1	87.2	86.0	81.3	70.3	58.0	47.9	67.7
Min	25.5	26.7	34.2	43.1	51.8	60.0	64.5	63.4	56.9	45.1	35.4	28.2	44.6

REFERENCE NOTES FOR TABLES 1, 2, 3 and 6 (CHARLESTON, WV)

GENERAL

T - TRACE AMOUNT
BLANK ENTRIES DENOTE MISSING/UNREPORTED DATA.
# INDICATES A STATION OR INSTRUMENT RELOCATION.

SPECIFIC

TABLE 1

(a) - LENGTH OF RECORD IN YEARS. ALTHOUGH
      INDIVIDUAL MONTHS MAY BE MISSING.
  # LESS THAN .05

NORMALS — BASED ON THE 1951-1980 RECORD PERIOD.
EXTREMES — DATES ARE THE MOST RECENT OCCURRENCE.
WIND DIR. — NUMERALS SHOW TENS OF DEGREES
            CLOCKWISE FROM TRUE NORTH.
            "00" INDICATES CALM.
RESULTANT WIND DIRECTIONS ARE GIVEN TO WHOLE DEGREES.

EXCEPTIONS

TABLES 2, 3, and 6

RECORD MEANS ARE THROUGH THE CURRENT YEAR,
BEGINNING IN   1902 FOR TEMPERATURE
               1901 FOR PRECIPITATION
               1948 FOR SNOWFALL

TABLE 4 HEATING DEGREE DAYS Base 65 deg. F   CHARLESTON, WEST VIRGINIA

SEASON	JULY	AUG	SEP	OCT	NOV	DEC	JAN	FEB	MAR	APR	MAY	JUNE	TOTAL
1956-57	0	7	115	156	583	558	1002	657	645	268	89	3	4083
1957-58	0	0	43	385	534	756	1038	1042	784	328	93	15	5018
1958-59	0	1	54	289	511	1032	988	710	674	282	73	19	4633
1959-60	0	0	32	253	634	769	845	892	994	214	177	4	4814
#1960-61	0	0	16	215	523	1074	1154	696	522	471	201	33	4905
1961-62	3	0	57	275	559	853	965	687	656	424	37	0	4516
1962-63	3	0	114	273	661	1052	1144	1002	476	302	148	26	5201
1963-64	10	6	81	177	558	1200	899	917	563	226	54	20	4711
1964-65	0	13	40	364	474	791	974	820	780	299	24	15	4594
1965-66	0	15	58	382	561	776	1166	793	538	354	147	25	4815
1966-67	0	1	44	315	507	910	815	932	517	275	217	17	4550
1967-68	4	6	124	297	704	802	1104	1095	539	263	141	15	5094
1968-69	2	11	35	298	541	946	994	828	807	264	98	9	4833
1969-70	0	0	95	352	703	1007	1125	801	666	239	82	1	5071
1970-71	3	0	40	246	563	817	1097	828	778	387	184	2	4945
1971-72	0	1	13	78	578	543	809	856	649	333	90	81	4031
1972-73	16	2	14	378	642	701	945	832	394	351	157	4	4436
1973-74	0	1	19	202	541	833	659	775	500	277	115	25	3947
1974-75	0	0	110	388	590	869	899	749	772	445	59	4	4885
1975-76	0	0	106	227	441	813	1025	549	427	342	142	4	4076
1976-77	0	9	84	475	814	1047	1432	888	482	242	81	52	5606
1977-78	0	2	19	357	482	919	1249	1138	691	258	137	23	5275
1978-79	0	0	18	344	462	797	1137	1031	456	308	125	19	4697
1979-80	5	10	39	331	519	820	951	1017	707	349	106	27	4881
1980-81	0	0	33	356	650	882	1138	774	727	207	175	2	4944
1981-82	0	1	76	335	585	936	1086	801	545	405	36	2	4808
1982-83	1	2	69	268	480	626	955	757	554	388	153	16	4269
1983-84	4	0	66	227	521	1019	1059	674	734	346	171	5	4826
1984-85	1	0	98	74	613	563	1164	860	488	192	54	18	4125
1985-86	0	0	51	127	294	960							

TABLE 5 COOLING DEGREE DAYS Base 65 deg. F   CHARLESTON, WEST VIRGINIA

YEAR	JAN	FEB	MAR	APR	MAY	JUNE	JULY	AUG	SEP	OCT	NOV	DEC	TOTAL
1969	0	0	0	7	79	277	368	227	75	15	0	0	1048
1970	0	0	0	40	147	251	310	273	197	18	0	0	1236
1971	0	0	0	0	32	265	237	219	190	49	11	5	1008
1972	0	0	0	24	56	85	283	247	132	0	0	0	827
1973	0	0	4	22	34	256	304	305	181	22	2	0	1130
1974	0	0	14	43	99	118	292	275	62	6	1	0	910
1975	0	0	0	7	99	212	291	372	82	7	4	0	1074
1976	0	1	17	58	54	218	238	175	39	1	0	0	801
1977	0	0	18	50	148	165	373	277	180	4	12	0	1227
1978	0	0	0	16	88	207	279	314	205	4	1	0	1114
1979	0	0	13	18	69	138	257	277	105	17	0	0	894
1980	0	0	0	6	71	147	370	358	182	9	0	0	1143
1981	0	0	2	38	41	256	340	251	126	5	0	0	1059
1982	0	0	0	6	154	122	355	196	101	47	5	6	992
1983	0	0	2	6	39	222	385	407	177	18	0	8	1256
1984	0	0	0	27	64	318	261	312	116	65	7	8	1178
1985	0	0	9	72	105	204	339	285	194	52	14	0	1274

TABLE 6 SNOWFALL (inches)   CHARLESTON, WEST VIRGINIA

SEASON	JULY	AUG	SEP	OCT	NOV	DEC	JAN	FEB	MAR	APR	MAY	JUNE	TOTAL
1956-57	0.0	0.0	0.0	0.0	1.4	2.2	5.7	3.5	1.4	T	0.0	0.0	14.2
1957-58	0.0	0.0	0.0	0.8	0.8	2.7	6.8	16.7	9.4	T	0.0	0.0	37.2
1958-59	0.0	0.0	0.0	0.0	1.1	3.7	5.6	1.4	0.4	5.9	0.0	0.0	18.1
1959-60	0.0	0.0	0.0	0.0	6.5	3.4	3.7	21.5	18.3	T	0.0	0.0	53.4
1960-61	0.0	0.0	0.0	0.0	0.8	4.1	15.1	6.5	0.6	0.5	0.0	0.0	27.6
1961-62	0.0	0.0	0.0	2.8	0.6	5.6	3.9	4.8	8.5	T	0.0	0.0	26.2
1962-63	0.0	0.0	0.0	0.6	0.2	18.6	10.7	19.0	T	T	0.2	0.0	49.3
1963-64	0.0	0.0	0.0	T	3.6	12.6	11.3	21.8	2.2	T	0.0	0.0	51.5
1964-65	0.0	0.0	0.0	0.0	1.3	0.9	13.7	6.7	8.5	T	0.0	0.0	31.1
1965-66	0.0	0.0	0.0	T	T	2.7	19.8	8.0	1.7	T	0.0	0.0	32.2
1966-67	0.0	0.0	0.0	T	3.8	6.0	3.6	20.6	2.2	0.0	0.0	0.0	36.2
1967-68	0.0	0.0	0.0	0.0	0.2	16.6	13.3	9.2	2.0	0.0	0.0	0.0	41.3
1968-69	0.0	0.0	0.0	T	4.3	9.5	12.6	14.2	2.9	T	0.0	0.0	43.5
1969-70	0.0	0.0	0.0	0.0	0.2	11.6	10.5	11.7	10.9	T	0.0	0.0	44.9
1970-71	0.0	0.0	0.0	0.0	0.2	10.9	11.2	0.0	0.0	0.0	0.0	0.0	22.3
1971-72	0.0	0.0	0.0	0.0	4.2	0.8	4.0	9.8	3.5	T	0.0	0.0	22.3
1972-73	0.0	0.0	0.0	0.9	6.9	1.4	3.4	2.9	4.8	0.9	0.0	0.0	21.2
1973-74	0.0	0.0	0.0	0.0	T	6.5	0.3	13.0	0.9	1.2	0.0	0.0	21.9
1974-75	0.0	0.0	0.0	0.6	2.7	8.3	18.1	2.4	8.1	0.1	0.0	0.0	40.3
1975-76	0.0	0.0	0.0	0.0	T	4.0	14.3	2.6	5.4	0.0	0.0	0.0	26.3
1976-77	0.0	0.0	0.0	T	4.7	7.4	22.2	11.1	1.0	1.5	0.0	0.0	47.9
1977-78	0.0	0.0	0.0	0.0	4.4	5.4	39.5	15.6	11.7	T	0.0	0.0	76.6
#1978-79	0.0	0.0	0.0	0.0	0.9	0.4	27.5	20.1	5.5	T	0.0	0.0	54.6
1979-80	0.0	0.0	0.0	T	T	1.5	11.7	12.7	10.3	T	0.0	0.0	36.2
1980-81	0.0	0.0	0.0	T	0.5	2.6	9.1	6.8	7.5	T	0.0	0.0	26.5
1981-82	0.0	0.0	0.0	T	0.5	8.2	12.1	6.4	7.4	1.0	0.0	0.0	35.6
1982-83	0.0	0.0	0.0	0.0	T	2.9	5.8	15.0	5.2	0.1	0.0	0.0	29.0
1983-84	0.0	0.0	0.0	0.0	0.3	3.8	12.8	9.7	2.4	0.0	0.0	0.0	29.0
1984-85	0.0	0.0	0.0	0.0	T	3.7	17.6	20.1	0.9	1.7	0.0	0.0	44.0
1985-86	0.0	0.0	0.0	0.0	0.0	8.9							
Record Mean	0.0	0.0	0.0	0.2	2.4	5.0	10.6	8.9	4.8	0.4	T	0.0	32.1

See Reference Notes, relative to all above tables, on preceding page.

Elkins, West Virginia, is located near the principal storm tracks and is therefore subjected to frequent weather changes throughout the year. While changes may be rigorous, they bring relief from summer heat waves and winter cold waves. The airport and city are located near the middle of a valley with a narrow floor and ridges at or near 3,000 feet. The ridges are oriented north northeast to south southwest, 3 to 4 miles to the east and west. The valley is located on the general northwest slope of the Appalachian Mountains which crest about 20 miles to the southeast at about 4,500 feet, with some higher peaks.

The seasonal climates vary greatly from year to year. When the Atlantic High extends westward, warm weather with high humidities occur in both summer and winter. Conversely, if the Atlantic High is displaced eastward and the circulation is principally from the northwest, weather is colder than normal. Hottest weather is from warm westerlies which have been over land for a long time and have a path over the south central or southwestern areas of the United States.

Summers are characterized by warm, humid, showery weather, but the heat is moderated by elevation and orographically induced cloudiness. A daily high temperature of 90 or above may occasionally be expected during the summer months. Winters are moderately severe with rapid changes. Snowfall may be frequent, and at times, heavy. However, it seldom remains on the ground for extended periods. Snows often fall upon warm ground thereby causing preliminary melting, then freezing, resulting in slippery road conditions. Glaze formation upon the ground or upon wires and trees is rare. Cold spells alternate frequently with thaws, and snow is subject to frequent complete melting during the winter. Severe cold spells occur occasionally but they seldom last more than two or three days. A daily low of zero degrees or below can be expected several times annually.

Significant climatic characteristics are associated with air currents rising and descending over the mountains. Orographic lifting of air delays post-frontal clearing after the passage of a cold front, especially during the winter when low clouds and snow flurries sometimes persist for 24 hours or more after the front has passed. While this upslope effect prevails with winds from the north-west quadrant, a foehn effect prevails with easterly and southerly winds, tending to diminish existing low cloud layers and to keep ceilings higher than otherwise anticipated. Nocturnal radiation fog is common during the summer and the autumn but it usually dissipates rapidly after sunrise.

Tornadoes are rare in this area, and severe thunderstorms are very infrequent. However, occasionally intense local rainfall from warm-season thunderstorms causes flash flooding in the narrow valleys of the area. Due to the remote location of the city with respect to concentrated industry, the air is usually relatively unpolluted. There are no important smoke sources in the locality, and smoke or haze seldom reduces the visibility below four miles.

The average last occurrence in the spring of temperatures as low as 32 is early to mid-May, and the first occurrence in the autumn is early October. The length of the growing season averages about 148 days.

TABLE 1   NORMALS, MEANS AND EXTREMES

ELKINS, WEST VIRGINIA

LATITUDE: 38°53'N   LONGITUDE: 79°51'W   ELEVATION: FT. GRND 1948 BARO 01985   TIME ZONE: EASTERN   WBAN: 13729

	(a)	JAN	FEB	MAR	APR	MAY	JUNE	JULY	AUG	SEP	OCT	NOV	DEC	YEAR
TEMPERATURE °F:														
Normals														
-Daily Maximum		39.0	41.6	50.9	62.2	71.3	77.9	80.7	79.6	74.4	63.8	52.0	42.7	61.3
-Daily Minimum		17.6	19.1	27.3	36.0	44.8	52.4	57.0	56.3	49.7	37.1	29.1	21.6	37.3
-Monthly		28.3	30.4	39.1	49.1	58.0	65.2	68.8	68.0	62.0	50.5	40.5	32.2	49.3
Extremes														
-Record Highest	41	76	72	84	88	88	93	95	95	97	86	80	76	97
-Year		1950	1985	1954	1976	1979	1952	1954	1948	1953	1951	1958	1951	SEP 1953
-Record Lowest	41	-24	-22	-15	3	20	25	39	34	27	11	0	-17	-24
-Year		1984	1977	1978	1985	1978	1977	1963	1965	1963	1952	1958	1983	JAN 1984
NORMAL DEGREE DAYS:														
Heating (base 65°F)		1138	969	803	477	233	69	11	19	127	450	732	1017	6045
Cooling (base 65°F)		0	0	0	0	19	75	132	112	40	0	0	0	378
% OF POSSIBLE SUNSHINE	2	26	39	39	47	50	51	42	38	59	38	34	31	41
MEAN SKY COVER (tenths)														
Sunrise - Sunset	40	8.0	7.8	7.6	7.3	7.1	7.1	7.2	7.0	6.7	6.4	7.4	7.8	7.3
MEAN NUMBER OF DAYS:														
Sunrise to Sunset														
-Clear	40	3.2	3.5	4.0	4.3	4.4	3.2	2.5	2.5	4.6	7.6	4.7	4.1	48.6
-Partly Cloudy	40	5.8	5.8	6.7	7.7	9.5	11.3	12.6	13.8	11.8	7.8	6.4	5.5	104.6
-Cloudy	40	22.0	18.9	20.3	18.0	17.1	15.6	15.9	14.7	13.9	13.0	19.0	21.4	209.8
Precipitation														
.01 inches or more	41	18.1	15.5	17.1	15.1	14.5	13.2	13.7	11.9	10.4	10.7	13.6	16.5	170.2
Snow,Ice pellets														
1.0 inches or more	23	6.3	6.0	3.5	0.9	0.0	0.0	0.0	0.0	0.0	0.1	2.2	4.3	23.4
Thunderstorms	25	0.2	0.5	2.0	4.1	6.2	7.8	10.0	7.7	3.3	1.3	0.7	0.4	44.3
Heavy Fog Visibility														
1/4 mile or less	25	2.0	1.8	1.7	1.9	5.2	10.8	13.1	16.9	14.7	9.6	3.1	2.0	82.8
Temperature °F														
-Maximum														
90° and above	41	0.0	0.0	0.0	0.0	0.0	0.3	0.6	0.4	0.3	0.0	0.0	0.0	1.6
32° and below	41	10.2	7.1	2.8	0.2	0.0	0.0	0.0	0.0	0.0	0.*	2.1	6.7	29.2
-Minimum														
32° and below	41	26.8	23.8	21.9	11.9	2.8	0.1	0.0	0.0	0.8	9.8	19.8	25.8	143.4
0° and below	41	3.6	2.4	0.3	0.0	0.0	0.0	0.0	0.0	0.0	0.0	0.*	1.3	7.6
AVG. STATION PRESS.(mb)	10	945.7	945.9	944.7	945.1	945.4	947.3	948.3	949.6	949.2	948.9	947.9	947.0	947.1
RELATIVE HUMIDITY (%)														
Hour 01	6	81	78	82	81	85	95	97	97	96	91	84	85	88
Hour 07	23	80	80	82	83	87	91	94	96	95	89	84	82	87
Hour 13 (Local Time)	23	66	61	57	52	54	59	62	64	62	55	60	66	60
Hour 19	21	73	68	62	56	60	69	73	79	83	74	74	76	71
PRECIPITATION (inches):														
Water Equivalent														
-Normal		3.39	2.84	3.66	3.71	3.82	4.35	4.68	4.20	3.21	2.97	2.69	3.33	42.85
-Maximum Monthly	41	6.09	5.62	8.85	6.95	7.67	8.33	9.30	10.40	6.29	8.43	11.08	6.73	11.08
-Year		1949	1972	1963	1972	1967	1981	1958	1980	1971	1954	1985	1978	NOV 1985
-Minimum Monthly	41	1.05	0.79	1.39	1.02	1.45	1.66	1.93	1.09	0.32	0.31	1.19	0.90	0.31
-Year		1967	1978	1957	1971	1970	1960	1974	1976	1985	1963	1976	1965	OCT 1963
-Maximum in 24 hrs	23	1.85	1.73	2.94	2.02	2.07	2.44	2.70	3.21	2.68	3.62	5.10	2.22	5.10
-Year		1971	1984	1963	1966	1985	1974	1985	1969	1967	1985	1985	1970	NOV 1985
Snow,Ice pellets														
-Maximum Monthly	23	54.1	29.5	33.5	8.0	0.7					3.9	14.8	34.9	54.1
-Year		1985	1964	1971	1985	1963					1979	1976	1969	JAN 1985
-Maximum in 24 hrs	23	18.7	12.8	8.3	5.6	0.7					3.6	12.4	17.8	18.7
-Year		1971	1983	1971	1985	1963					1979	1970	1967	JAN 1971
WIND:														
Mean Speed (mph)	24	7.3	8.0	8.2	7.9	6.7	4.9	4.3	4.1	4.4	5.0	6.9	6.9	6.2
Prevailing Direction														
through 1963		W	NW	WNW	NW	SSE	NW	NW	NW	NW	NNW	WNW	W	NW
Fastest Obs. 1 Min.														
-Direction (!!!)	10	27	25	32	27	30	36	30	32	26	29	30	28	25
-Speed (MPH)	10	46	55	46	50	46	35	37	40	35	46	33	35	55
-Year		1965	1958	1955	1963	1967	1964	1965	1963	1967	1955	1965	1958	FEB 1958
Peak Gust														
-Direction (!!!)	2	W	SE	W	SW	W	N	NW	N	NW	SW		W	NW
-Speed (mph)	2	41	41	45	45	35	29	47	29	24	31	40	47	47
-Date		1985	1984	1985	1985	1985	1985	1984	1983	1984	1985	1984	1984	JUL 1984

See Reference Notes to this table on the following page.

TABLE 2 PRECIPITATION (inches)　　　　ELKINS, WEST VIRGINIA

YEAR	JAN	FEB	MAR	APR	MAY	JUNE	JULY	AUG	SEP	OCT	NOV	DEC	ANNUAL
1956	3.07	5.21	4.97	2.22	5.78	5.95	6.06	5.58	2.82	2.10	1.56	3.53	48.85
1957	4.57	5.43	1.39	3.86	1.93	5.81	3.32	1.67	2.38	4.59	1.71	4.40	41.06
1958	2.43	1.81	1.95	4.21	5.00	3.51	9.30	5.87	2.28	0.99	2.46	0.99	40.80
1959	3.33	2.31	2.45	3.41	4.55	3.16	6.50	3.31	0.89	4.59	3.94	2.95	41.39
1960	4.34	1.81	1.75	2.90	3.16	1.66	3.08	3.38	4.99	2.63	2.09	2.71	34.50
1961	2.34	4.25	4.50	5.13	3.81	5.31	5.82	3.89	3.67	3.94	2.51	4.30	49.47
1962	3.44	4.19	4.61	3.45	2.87	4.32	3.16	2.23	3.77	5.18	3.38	3.90	44.50
1963	1.66	2.72	8.85	2.40	2.33	4.45	3.77	2.39	3.34	0.31	4.78	1.65	38.65
1964	3.20	3.07	3.95	5.61	1.47	5.01	5.19	3.55	4.20	0.94	2.00	3.39	41.58
1965	5.31	1.94	4.51	6.90	1.45	2.21	2.27	1.74	2.89	2.51	1.43	0.90	34.06
1966	3.49	3.10	1.44	6.63	1.68	1.98	3.24	4.84	5.14	2.46	2.63	2.45	39.08
1967	1.05	3.03	7.69	3.24	7.67	2.83	4.55	5.10	3.33	3.44	3.55	4.47	49.95
1968	1.88	1.80	4.28	1.54	7.30	2.05	3.46	4.07	3.01	3.21	3.20	3.04	38.84
1969	3.01	2.13	2.26	3.34	2.76	3.83	6.82	5.60	5.94	1.60	1.81	5.84	44.94
1970	1.53	1.92	2.80	4.46	1.45	4.65	5.88	4.83	3.28	1.95	2.43	5.51	40.69
1971	4.33	3.16	2.90	1.02	4.23	2.34	4.10	3.87	6.29	1.85	2.76	1.65	38.50
1972	5.46	5.62	3.16	6.95	4.72	7.03	4.56	2.09	2.67	4.64	5.82	6.02	58.74
1973	2.05	2.52	2.13	5.87	3.75	3.53	3.82	3.47	3.66	4.22	3.28	4.53	42.83
1974	4.04	3.41	3.48	2.93	5.23	7.38	1.93	6.84	3.38	2.03	2.18	3.20	46.03
1975	4.17	2.84	4.64	4.38	7.28	4.87	2.56	3.55	3.71	2.49	2.04	3.76	46.29
1976	3.10	3.05	3.17	1.13	1.94	3.23	6.17	1.09	4.71	6.28	1.19	2.66	37.72
1977	1.73	1.13	3.40	2.83	1.96	5.08	4.49	5.60	2.14	3.74	3.30	2.59	37.99
1978	4.26	0.79	2.25	1.99	4.24	6.84	7.28	4.85	1.47	1.88	2.90	6.73	45.48
1979	5.88	3.63	2.03	3.40	5.71	4.94	6.22	5.06	4.61	5.16	3.54	2.33	52.51
1980	2.75	1.90	4.80	5.46	3.66	5.04	7.80	10.40	3.94	1.99	3.76	2.08	53.58
1981	1.15	4.22	2.76	3.72	5.79	8.33	3.43	1.72	6.26	3.48	1.54	3.17	45.57
1982	3.53	4.14	6.34	2.15	2.26	6.46	6.01	4.61	4.35	1.42	3.92	2.55	47.74
1983	1.59	1.50	3.69	5.26	5.49	4.05	2.03	3.59	1.98	4.13	3.65	3.57	40.53
1984	1.69	4.42	4.89	3.96	2.64	2.52	5.66	7.21	2.23	4.08	3.94	3.87	47.11
1985	3.26	1.99	4.68	2.44	6.95	3.56	5.58	2.78	0.32	6.00	11.08	2.59	51.23
Record Mean	3.65	3.08	4.05	3.75	4.24	5.02	5.18	4.12	3.36	3.04	2.92	3.41	45.82

TABLE 3 AVERAGE TEMPERATURE (deg. F)　　　　ELKINS, WEST VIRGINIA

YEAR	JAN	FEB	MAR	APR	MAY	JUNE	JULY	AUG	SEP	OCT	NOV	DEC	ANNUAL
1956	25.7	37.8	39.0	46.2	58.2	64.9	68.4	67.0	59.4	55.0	39.7	42.1	50.3
1957	28.8	37.5	40.2	53.0	60.3	68.7	69.0	66.4	63.9	46.9	42.0	35.8	51.1
1958	25.9	22.2	34.7	48.9	57.4	62.8	70.1	67.2	61.2	50.8	42.3	25.8	47.4
1959	28.1	33.3	37.1	50.1	62.1	64.3	69.7	71.0	63.3	52.7	39.1	35.0	50.5
1960	34.0	30.1	26.5	53.5	56.0	65.7	67.2	70.3	64.2	51.5	41.3	24.7	48.8
1961	24.6	36.8	43.4	44.2	54.3	63.9	69.1	68.9	65.3	51.1	43.1	33.0	49.8
1962	29.3	34.7	37.5	46.0	63.3	66.1	66.8	67.2	58.6	52.5	40.9	27.6	49.2
1963	25.9	23.6	42.0	49.3	56.4	64.4	66.2	65.2	58.8	52.9	41.7	22.6	47.4
1964	31.6	26.4	39.7	51.2	59.4	65.7	68.6	65.8	61.5	47.0	44.0	35.2	49.7
1965	28.1	31.6	35.3	49.2	62.3	63.8	67.9	67.3	64.9	48.7	41.9	36.0	49.8
1966	24.2	30.8	40.9	47.4	56.8	64.3	69.9	66.6	58.4	47.7	42.1	31.7	48.4
1967	33.7	27.6	42.5	50.5	52.7	66.5	66.7	65.4	57.5	50.4	37.2	35.8	48.9
1968	25.0	22.0	41.8	49.9	55.8	65.3	69.9	66.1	61.3	51.3	42.2	28.6	48.5
1969	27.8	30.4	32.6	50.4	59.0	66.9	70.7	67.2	61.0	50.1	37.1	25.7	48.2
1970	20.0	29.5	35.9	49.6	59.8	64.4	67.9	67.2	64.1	53.1	41.4	33.0	48.8
1971	24.2	31.5	34.5	45.4	55.4	68.8	68.1	67.0	66.2	57.7	39.7	41.1	50.0
1972	33.8	28.8	38.6	48.3	57.8	60.8	68.0	68.0	63.5	48.5	39.9	39.2	49.6
1973	30.3	30.2	48.0	47.9	55.4	67.2	69.1	68.5	62.5	52.8	41.5	32.6	50.5
1974	39.0	30.6	42.5	49.1	57.1	62.0	66.5	67.1	59.4	46.5	38.6	30.2	49.0
1975	31.3	33.7	35.7	43.7	60.1	65.6	68.4	69.8	60.3	53.8	44.9	34.5	50.2
1976	26.8	39.0	44.7	48.3	55.8	66.7	67.7	65.7	59.9	46.7	32.5	27.5	48.4
1977	15.0	28.3	45.5	52.5	60.8	62.0	69.2	68.8	65.9	48.6	44.6	28.9	49.2
1978	19.6	16.6	36.5	49.3	56.6	65.0	69.4	70.4	65.0	47.5	44.2	33.8	47.8
1979	25.7	22.4	42.9	49.2	58.0	63.5	68.0	68.1	61.6	49.0	43.0	34.0	48.8
1980	30.0	23.2	36.7	48.1	58.5	62.9	70.1	71.1	64.6	47.5	37.3	30.9	48.4
1981	20.4	31.7	35.0	52.2	53.7	66.0	68.1	65.1	59.6	47.0	37.4	27.6	47.0
1982	24.1	32.5	41.3	44.8	60.8	63.3	69.8	64.9	60.7	51.8	44.5	39.0	49.8
1983	29.3	30.8	38.3	46.0	53.9	63.5	67.8	69.0	59.8	51.9	40.4	27.5	48.2
1984	24.2	35.6	34.5	47.8	54.8	65.9	67.9	67.9	58.6	58.8	37.7	41.3	49.4
1985	21.6	28.4	41.0	50.2	59.1	63.2	68.9	66.7	61.6	54.8	50.6	27.6	49.5
Record Mean	30.3	31.1	39.7	49.0	58.2	65.8	69.5	68.4	62.6	51.8	40.9	32.5	50.0
Max	40.4	41.7	50.9	61.4	70.6	77.9	81.0	79.8	74.9	64.7	51.9	42.4	61.5
Min	20.2	20.5	28.4	36.5	45.7	53.8	58.0	57.0	50.3	39.0	29.8	22.6	38.5

REFERENCE NOTES FOR TABLES 1, 2, 3 and 6　　　　(ELKINS, WV)

GENERAL

T - TRACE AMOUNT
BLANK ENTRIES DENOTE MISSING/UNREPORTED DATA.
# INDICATES A STATION OR INSTRUMENT RELOCATION.

SPECIFIC

TABLE 1

(a) - LENGTH OF RECORD IN YEARS. ALTHOUGH
INDIVIDUAL MONTHS MAY BE MISSING.

* LESS THAN .05

NORMALS — BASED ON THE 1951-1980 RECORD PERIOD.
EXTREMES — DATES ARE THE MOST RECENT OCCURRENCE.
WIND DIR. — NUMERALS SHOW TENS OF DEGREES
CLOCKWISE FROM TRUE NORTH.
"00" INDICATES CALM.
RESULTANT WIND DIRECTIONS ARE GIVEN TO WHOLE DEGREES.

EXCEPTIONS

TABLE 1

1. FASTEST OBSERVED WINDS ARE 1955-1958 AND
1963-1968.
2. MEAN WIND SPEED, THUNDERSTORMS, AND HEAVY FOG
ARE THROUGH 1968.

TABLES 2, 3, and 6

RECORD MEANS ARE THROUGH THE CURRENT YEAR,
BEGINNING IN　　1899 FOR TEMPERATURE
　　　　　　　　1899 FOR PRECIPITATION
　　　　　　　　1963 FOR SNOWFALL

TABLE 4 HEATING DEGREE DAYS Base 65 deg. F   ELKINS, WEST VIRGINIA

SEASON	JULY	AUG	SEP	OCT	NOV	DEC	JAN	FEB	MAR	APR	MAY	JUNE	TOTAL
1956-57	11	35	189	302	751	700	1113	761	762	376	172	15	5187
1957-58	13	39	109	554	683	897	1206	1194	932	478	233	97	6435
1958-59	2	28	145	432	674	1209	1136	881	860	442	145	95	6049
1959-60	1	0	115	383	772	921	956	1006	1190	343	281	45	6013
1960-61	25	1	76	411	704	1234	1248	787	664	590	333	78	6151
1961-62	23	10	103	422	655	982	1100	841	843	565	106	15	5665
1962-63	21	19	215	392	714	1151	1208	1153	706	462	273	76	6390
1963-64	53	36	192	367	692	1306	1028	1113	779	411	174	70	6221
1964-65	23	55	124	553	624	914	1134	930	912	466	106	92	5933
1965-66	12	50	81	497	688	893	1258	951	741	525	254	81	6031
1966-67	15	29	207	531	679	1027	963	1039	692	430	378	47	6037
1967-68	33	37	226	443	826	900	1233	1239	711	444	279	67	6438
1968-69	16	46	122	425	673	1121	1147	964	996	430	194	57	6191
1969-70	0	8	160	462	828	1212	1390	986	892	456	182	65	6641
1970-71	26	14	97	362	702	987	1258	930	938	580	296	5	6195
1971-72	0	18	54	233	756	731	961	1047	815	494	217	156	5482
1972-73	30	13	76	504	745	793	1068	966	522	509	304	28	5558
1973-74	12	12	110	373	698	996	799	959	692	472	255	119	5497
1974-75	29	11	188	564	784	1074	1037	871	901	632	174	46	6311
1975-76	2	7	76	341	596	940	1175	746	625	492	292	44	5420
1976-77	24	44	152	564	968	1159	1548	1021	600	377	151	136	6744
1977-78	22	17	59	503	608	1111	1401	1349	877	465	265	70	6747
1978-79	13	6	75	534	617	961	1213	1187	680	467	231	83	6067
1979-80	39	46	130	491	652	954	1080	1203	872	500	210	113	6290
1980-81	4	5	72	538	823	1050	1375	928	924	384	339	42	6484
1981-82	21	54	187	553	822	1150	1265	904	728	600	166	76	6526
1982-83	13	57	149	403	609	798	1099	951	817	563	336	85	5880
1983-84	30	21	184	404	729	1153	1258	847	937	511	318	50	6442
1984-85	39	22	199	190	811	727	1338	1020	738	437	194	101	5816
1985-86	1	27	160	312	427	1153							

TABLE 5 COOLING DEGREE DAYS Base 65 deg. F   ELKINS, WEST VIRGINIA

YEAR	JAN	FEB	MAR	APR	MAY	JUNE	JULY	AUG	SEP	OCT	NOV	DEC	TOTAL
1969	0	0	0	0	18	122	184	83	45	7	0	0	459
1970	0	0	0	2	28	52	123	86	75	1	0	0	367
1971	0	0	0	0	6	128	105	88	99	11	3	0	440
1972	0	0	0	0	3	36	131	111	36	0	0	0	317
1973	0	0	0	3	11	98	147	125	41	3	0	0	428
1974	0	0	0	1	15	35	81	85	27	0	0	0	244
1975	0	0	0	0	0	25	70	114	161	27	0	0	397
1976	0	0	0	1	16	105	117	73	2	0	0	0	314
1977	0	0	1	6	26	55	160	142	91	0	4	0	485
1978	0	0	0	1	13	76	147	179	82	0	0	0	498
1979	0	0	0	2	21	45	142	149	33	0	0	0	392
1980	0	0	0	0	16	57	169	202	69	0	0	0	513
1981	0	0	0	6	0	76	126	65	32	0	0	0	305
1982	0	0	0	0	43	33	170	60	30	1	0	0	337
1983	0	0	0	0	0	46	124	153	33	4	0	0	360
1984	0	0	0	2	9	82	72	117	16	6	0	0	304
1985	0	0	0	0	17	54	129	87	63	5	0	0	355

TABLE 6 SNOWFALL (inches)   ELKINS WEST VIRGINIA

SEASON	JULY	AUG	SEP	OCT	NOV	DEC	JAN	FEB	MAR	APR	MAY	JUNE	TOTAL
1956-57	0.0	0.0	0.0	0.0	1.2	3.9	11.9	9.7	2.6	T	0.0	0.0	29.3
1957-58	0.0	0.0	0.0	2.0	T	7.7	7.8	22.5	14.0	0.1	0.0	0.0	54.1
1958-59	0.0	0.0	0.0	0.0	1.0	4.3	4.7	1.5	4.4	5.0	0.0	0.0	20.9
1959-60	0.0	0.0	0.0			17.9	19.1	14.5	5.3	8.6	T	0.0	
1960-61													
1961-62	0.0	0.0	0.0	0.0	0.7	16.2	7.6	11.9	21.8	13.3	0.0	0.0	71.5
1962-63	0.0	0.0	0.0	2.7	1.0	34.5	12.5	22.4	5.0	T	0.7	0.0	78.8
1963-64	0.0	0.0	0.0	T	4.9	14.5	13.0	29.5	7.7	0.7	0.0	0.0	70.3
1964-65	0.0	0.0	0.0	0.0	4.6	5.7	21.6	11.2	14.0	0.2	0.0	0.0	57.3
1965-66	0.0	0.0	0.0	0.4	1.6	4.6	21.1	21.0	8.1	5.0	T	0.0	61.8
1966-67	0.0	0.0	0.0	0.0	7.8	14.2	7.2	28.1	7.0	0.5	T	0.0	64.8
1967-68	0.0	0.0	0.0	T	9.0	23.2	14.1	24.1	10.3	0.1	0.0	0.0	80.8
1968-69	0.0	0.0	0.0	0.5	11.9	19.7	3.1	13.6	14.6	0.7	T	0.0	64.1
1969-70	0.0	0.0	0.0	T	8.9	34.9	20.0	13.8	8.3	0.7	0.0	0.0	86.6
1970-71	0.0	0.0	0.0	0.0	13.2	16.9	25.2	14.1	33.5	4.8	T	0.0	107.7
1971-72	0.0	0.0	0.0	0.0	11.0	1.6	4.1	26.9	7.2	4.6	0.0	0.0	55.4
1972-73	0.0	0.0	0.0	2.2	7.4	6.7	4.6	9.9	7.6	7.4	T	0.0	45.8
1973-74	0.0	0.0	0.0	2.5	3.2	25.0	0.8	15.1	1.9	2.0	0.0	0.0	50.5
1974-75	0.0	0.0	0.0	T	6.8	17.7	16.9	10.0	11.1	3.6	0.0	0.0	66.1
1975-76	0.0	0.0	0.0	0.0	3.9	6.5	21.0	8.1	8.6	1.8	T	0.0	49.9
1976-77	0.0	0.0	0.0	T	14.8	21.1	37.3	14.3	1.3	2.6	0.0	0.0	91.4
1977-78	0.0	0.0	0.0	T	5.3	12.6	32.8	13.2	13.8	T	0.0	0.0	77.7
1978-79	0.0	0.0	0.0	T	2.0	5.1	30.8	18.7	4.2	3.0	0.0	0.0	63.8
1979-80	0.0	0.0	0.0	3.9	3.8	10.2	23.0	19.7	14.1	3.6	0.0	0.0	78.3
1980-81	0.0	0.0	0.0	0.4	4.0	8.0	23.6	12.1	14.9	0.0	0.0	0.0	63.6
1981-82	0.0	0.0	0.0	0.4	8.4	20.6	18.2	10.1	13.4	7.1	0.0	0.0	78.2
1982-83	0.0	0.0	0.0	T	1.3	10.5	17.0	27.0	10.5	1.8	0.0	0.0	68.1
1983-84	0.0	0.0	0.0	T	11.1	7.0	26.4	23.7	21.6	2.9	0.0	0.0	92.7
1984-85	0.0	0.0	0.0	0.0	4.0	4.0	54.1	21.3	0.5	8.0	0.0	0.0	91.9
1985-86	0.0	0.0	0.0	0.0	0.0	22.6							
Record Mean	0.0	0.0	0.0	0.4	6.5	13.6	19.5	17.7	10.4	2.7	T	0.0	70.9

See Reference Notes, relative to all above tables, on preceding page.

The Tri-State Airport is near the confluence of the Ohio and Big Sandy Rivers, located on a man-made plateau constructed by cutting the tops off several hills and filling intervening valleys. The elevation of the ground at the National Weather Service Office is 260 feet higher than at the Federal Building in downtown Huntington.

The temperature record for the valley locations is not compatible with that for the airport, which is generally cooler throughout the year. The summer season is moderately warm and humid, with the valley locations considerably warmer and more humid than the Tri-State Airport site. The winter months are moderately cold, with an occasional severe cold wave lasting a few days. The four seasons are nearly equal in length and autumn is the most pleasant, with warm days and cool nights.

The heaviest rainfall occurs in July and August, mostly in thunderstorms, and flash floods are common in the area. The winter rainfall occurs mostly prior to and with a frontal passage and frequently lasts from two to four days, causing frequent general flooding on all streams.

Snow seldom remains on the ground more than two days in the valleys. However, at higher elevations surrounding the airport, roads are frequently blocked for several days during the winter months.

TABLE 1   # NORMALS, MEANS AND EXTREMES

HUNTINGTON, WEST VIRGINIA

LATITUDE: 38°22'N   LONGITUDE: 82°33'W   ELEVATION: FT. GRND 827 BARO 00837   TIME ZONE: EASTERN   WBAN: 03860

	(a)	JAN	FEB	MAR	APR	MAY	JUNE	JULY	AUG	SEP	OCT	NOV	DEC	YEAR
TEMPERATURE °F:														
Normals														
-Daily Maximum		41.1	45.0	55.2	67.2	75.7	82.6	85.6	84.4	78.7	67.6	55.2	45.2	65.3
-Daily Minimum		24.5	26.6	35.0	44.4	52.8	60.7	65.1	64.0	57.2	44.9	35.9	28.5	45.0
-Monthly		32.8	35.8	45.1	55.8	64.3	71.7	75.3	74.2	68.0	56.3	45.5	36.9	55.1
Extremes														
-Record Highest	25	74	79	85	92	92	95	100	100	97	86	82	80	100
-Year		1967	1977	1973	1985	1963	1980	1983	1983	1983	1962	1979	1982	JUL 1983
-Record Lowest	25	-16	-6	-2	20	27	40	46	43	31	16	8	-9	-16
-Year		1985	1970	1980	1985	1966	1977	1968	1965	1983	1962	1964	1983	JAN 1985
NORMAL DEGREE DAYS:														
Heating (base 65°F)		998	818	617	293	125	17	0	0	62	293	582	871	4676
Cooling (base 65°F)		0	0	0	17	103	218	322	285	152	24	0	0	1121
% OF POSSIBLE SUNSHINE														
MEAN SKY COVER (tenths)														
Sunrise - Sunset	24	7.6	7.6	7.5	7.1	6.9	6.8	6.8	6.8	6.5	6.2	7.4	7.7	7.1
MEAN NUMBER OF DAYS:														
Sunrise to Sunset														
-Clear	24	4.3	4.4	4.1	5.1	5.6	4.3	4.0	4.2	6.3	8.6	4.9	4.5	60.4
-Partly Cloudy	24	6.8	5.7	7.3	7.7	9.3	11.3	11.9	12.8	9.8	7.8	6.2	5.6	102.0
-Cloudy	24	19.9	18.2	19.6	17.2	16.2	14.4	15.0	14.0	13.9	14.6	18.9	20.9	202.8
Precipitation														
.01 inches or more	24	13.8	12.8	14.3	12.8	12.3	10.9	11.9	9.7	8.5	9.5	11.6	12.9	141.0
Snow,Ice pellets														
1.0 inches or more	24	3.1	2.4	1.2	0.0	0.0	0.0	0.0	0.0	0.0	0.0	0.4	1.0	8.1
Thunderstorms	23	0.3	0.6	2.7	4.3	6.3	7.0	9.2	7.1	2.2	1.1	0.9	0.3	42.0
Heavy Fog Visibility 1/4 mile or less	23	2.9	3.0	2.1	1.4	4.1	6.5	10.2	10.9	9.9	6.3	3.5	3.0	63.9
Temperature °F														
-Maximum														
90° and above	24	0.0	0.0	0.0	0.1	0.5	3.8	6.5	5.1	1.8	0.0	0.0	0.0	17.8
32° and below	24	10.2	6.1	1.0	0.0	0.0	0.0	0.0	0.0	0.0	0.0	0.4	5.0	22.7
-Minimum														
32° and below	24	24.5	21.0	13.5	3.8	0.3	0.0	0.0	0.0	0.*	3.2	11.3	20.2	97.9
0° and below	24	1.8	0.4	0.*	0.0	0.0	0.0	0.0	0.0	0.0	0.0	0.0	0.3	2.5
AVG. STATION PRESS.(mb)	13	988.7	988.4	985.8	985.6	984.9	986.1	987.1	988.1	988.6	989.4	988.9	989.0	987.6
RELATIVE HUMIDITY (%)														
Hour 01	23	73	71	67	66	78	86	88	89	89	81	74	74	78
Hour 07 (Local Time)	24	77	76	75	75	84	88	90	92	92	86	80	78	83
Hour 13	24	66	62	55	49	53	57	61	60	60	55	60	66	59
Hour 19	24	64	59	53	48	56	62	66	68	68	61	62	66	61
PRECIPITATION (inches):														
Water Equivalent														
-Normal		3.24	2.83	4.08	3.48	3.94	3.56	4.47	3.73	3.07	2.40	2.82	3.12	40.74
-Maximum Monthly	25	6.37	5.66	7.54	6.56	9.26	7.63	8.57	6.86	5.64	5.71	7.40	8.69	9.26
-Year		1978	1962	1963	1966	1974	1979	1962	1979	1966	1983	1985	1978	MAY 1974
-Minimum Monthly	25	0.64	0.53	1.12	0.74	0.93	0.41	1.37	0.68	0.35	T	0.73	0.31	T
-Year		1981	1968	1966	1976	1965	1966	1974	1962	1985	1963	1976	1965	OCT 1963
-Maximum in 24 hrs	25	2.63	2.43	3.43	2.26	2.60	3.42	4.27	2.90	2.74	2.95	2.28	3.36	4.27
-Year		1974	1966	1967	1978	1974	1979	1962	1964	1964	1985	1973	1978	JUL 1962
Snow,Ice pellets														
-Maximum Monthly	24	30.3	21.2	11.0	0.8	T					0.4	4.6	13.2	30.3
-Year		1978	1985	1971	1974	1963					1974	1969	1967	JAN 1978
-Maximum in 24 hrs	24	11.6	11.2	7.9	0.6	T					0.4	4.4	6.7	11.6
-Year		1978	1985	1971	1974	1963					1974	1969	1967	JAN 1978
WIND:														
Mean Speed (mph)	23	7.7	7.6	8.1	7.8	6.2	5.6	5.1	4.9	5.1	5.8	7.0	7.5	6.5
Prevailing Direction														
Fastest Obs. 1 Min.														
-Direction (!!!)	23	27	26	25	18	29	24	31	24	34	27	23	30	29
-Speed (MPH)	23	38	41	37	44	47	35	32	35	28	29	35	35	47
-Year		1971	1967	1971	1968	1967	1973	1976	1965	1963	1969	1966	1983	MAY 1967
Peak Gust														
-Direction (!!!)	2	NW	W	W	W	W	NW	NW	W	NW	NW	NW	SW	W
-Speed (mph)	2	44	39	47	51	38	51	49	37	30	25	35	39	51
-Date		1985	1984	1985	1985	1985	1985	1985	1984	1985	1985	1984	1984	APR 1985

See Reference Notes to this table on the following page.

TABLE 2 PRECIPITATION (inches)   HUNTINGTON, WEST VIRGINIA

YEAR	JAN	FEB	MAR	APR	MAY	JUNE	JULY	AUG	SEP	OCT	NOV	DEC	ANNUAL
1956	2.75	7.58	7.58	3.81	3.52	3.07	4.24	3.57	4.14	1.97	0.99	5.60	48.82
1957	4.53	3.60	2.15	2.79	2.76	2.14	1.61	1.83	5.04	1.97	3.69	4.44	36.55
1958	2.20	2.73	2.87	6.03	7.22	2.95	7.15	5.23	2.69	1.31	2.44	0.84	43.66
1959	3.75	2.31	1.91	4.90	3.46	2.01	7.78	2.68	3.96	3.29	3.61	2.06	41.72
1960	2.36	4.50	2.01	1.34	2.98	5.34	4.26	3.75	2.42	1.77	1.77	2.32	34.82
#1961	3.52	3.50	4.06	4.08	3.89	7.28	9.90	3.24	1.07	3.35	3.12	4.43	51.44
1962	2.62	5.66	4.57	4.06	2.58	3.81	8.57	0.68	2.51	2.80	4.66	2.97	45.49
1963	1.86	1.61	7.54	0.86	4.54	1.60	3.68	3.09	1.81	T	2.19	1.21	29.99
1964	2.11	3.07	4.54	3.09	1.25	1.97	2.18	6.21	4.53	0.80	3.37	3.90	37.02
1965	3.36	2.17	4.81	5.68	0.93	4.35	4.72	4.24	3.49	2.25	0.96	0.31	37.27
1966	3.26	3.97	1.12	6.56	1.58	0.41	6.37	4.15	5.64	2.28	3.06	2.81	41.21
1967	1.43	2.53	7.52	3.48	6.81	1.11	5.82	1.48	1.21	2.27	4.71	3.55	41.92
1968	1.72	0.53	6.05	2.81	6.10	1.67	2.59	6.03	1.14	2.89	2.41	2.20	36.14
1969	2.34	0.94	1.24	2.85	3.30	2.45	3.34	3.93	3.91	1.85	2.14	4.80	33.09
1970	1.12	3.25	3.33	3.68	3.03	3.28	3.75	4.28	2.80	5.23	2.11	4.11	39.97
1971	2.57	2.71	1.96	1.20	6.04	4.80	7.57	1.05	2.61	1.48	1.96	1.50	35.45
1972	4.79	4.90	2.76	6.07	3.82	4.39	2.69	2.06	3.32	2.02	4.38	5.52	46.72
1973	1.48	2.05	3.27	4.91	4.89	3.14	3.52	1.69	2.55	3.39	5.17	2.23	38.29
1974	5.57	1.85	4.45	2.19	9.26	4.86	1.37	4.94	3.63	1.64	4.01	2.87	46.64
1975	4.14	3.11	6.32	5.55	3.20	4.06	4.55	3.96	4.84	3.47	2.64	3.64	49.48
1976	2.84	2.38	3.90	0.74	2.41	4.25	5.48	5.72	4.27	5.54	0.73	2.17	40.43
1977	2.42	0.76	3.10	3.58	2.32	4.70	3.22	5.93	3.26	4.36	3.57	2.33	39.55
1978	6.37	1.06	3.11	4.06	4.30	1.66	4.49	4.22	1.37	3.11	2.74	8.69	45.18
1979	5.28	4.28	2.24	2.83	4.65	7.63	6.18	6.86	5.29	2.67	3.19	2.77	53.87
1980	2.46	1.71	5.04	3.11	2.64	1.95	7.94	5.91	1.89	1.57	2.96	1.82	39.00
1981	0.64	4.23	1.65	5.35	5.19	5.91	3.64	1.03	1.60	2.53	1.23	2.61	35.61
1982	4.49	2.36	4.40	1.49	6.28	4.53	3.76	4.45	2.00	1.80	4.33	3.34	43.23
1983	1.40	1.92	1.89	3.99	6.82	3.02	2.03	3.56	0.63	5.71	2.87	3.10	36.94
1984	1.82	2.09	2.45	4.46	4.45	1.75	3.37	4.66	2.67	4.87	3.85	4.36	40.80
1985	3.13	2.92	3.41	0.83	5.40	3.10	6.13	4.68	0.35	5.17	7.40	2.10	44.62
Record Mean	3.27	2.80	3.83	3.42	4.09	3.79	4.63	3.58	2.91	2.51	3.02	3.16	41.01

TABLE 3 AVERAGE TEMPERATURE (deg. F)   HUNTINGTON, WEST VIRGINIA

YEAR	JAN	FEB	MAR	APR	MAY	JUNE	JULY	AUG	SEP	OCT	NOV	DEC	ANNUAL
1956	32.0	42.4	45.2	52.7	65.7	72.3	75.3	74.9	66.2	61.3	46.3	47.1	56.8
1957	32.8	42.5	44.5	59.2	66.4	75.5	77.7	75.5	71.1	53.4	47.5	41.1	57.3
1958	32.7	28.6	40.4	55.9	64.4	70.0	77.4	74.8	68.3	56.8	48.8	31.8	54.1
1959	33.3	39.7	43.7	56.7	69.9	72.5	77.1	79.2	72.0	59.4	44.7	40.7	57.4
1960	38.8	34.7	33.8	60.0	62.4	71.7	74.9	77.6	70.5	58.7	47.4	31.2	55.2
#1961	29.8	42.0	49.4	50.4	60.4	69.8	75.0	75.3	72.7	58.3	47.2	38.1	55.7
1962	33.4	39.2	41.8	52.0	69.0	70.4	72.0	72.3	63.9	57.9	43.3	30.4	53.8
1963	27.0	28.1	50.8	59.0	63.6	69.8	71.4	70.4	65.3	60.6	46.5	25.3	53.1
1964	34.2	32.1	46.4	58.0	66.0	71.7	76.0	73.0	65.7	52.2	48.0	39.4	55.2
1965	33.9	35.9	40.9	56.2	69.3	71.9	74.5	73.2	69.2	54.2	47.8	41.6	55.7
1966	26.6	34.3	46.4	52.5	61.2	72.4	76.4	72.0	65.1	53.0	47.0	36.2	53.6
1967	39.3	32.0	48.1	56.3	59.2	73.1	71.4	70.4	62.9	55.2	40.6	38.9	54.0
1968	29.4	27.4	45.8	55.7	60.3	70.3	75.2	73.6	67.0	56.7	47.1	33.8	53.5
1969	31.7	35.6	38.6	56.1	64.1	72.6	77.9	75.0	66.3	55.3	42.1	30.7	53.9
1970	27.3	33.9	40.9	56.7	65.7	71.4	75.3	73.4	72.7	58.6	47.2	39.0	55.0
1971	29.6	37.3	41.9	54.0	60.6	73.3	71.9	71.8	71.0	63.9	44.8	45.6	55.5
1972	35.8	34.6	43.5	54.1	62.2	66.3	74.6	72.3	69.2	51.0	42.9	40.4	53.9
1973	34.2	33.8	53.2	52.5	61.1	74.3	76.4	75.9	71.2	61.0	48.1	37.6	56.6
1974	41.9	37.9	49.4	58.1	64.5	68.4	75.4	73.8	63.4	54.1	46.4	38.6	56.0
1975	37.3	39.6	42.4	52.7	68.1	73.2	75.0	78.4	65.1	58.0	50.1	38.6	56.6
1976	31.6	46.4	51.9	56.0	62.6	72.2	72.7	72.1	68.3	50.1	39.2	32.3	54.2
1977	19.7	35.1	50.4	60.3	68.2	69.1	78.6	75.4	68.8	54.2	50.2	35.1	55.4
1978	25.1	24.7	42.3	56.9	62.2	73.1	76.4	75.9	72.1	54.1	48.0	38.5	54.1
1979	26.3	28.6	49.5	55.6	63.7	70.1	74.0	74.4	67.4	55.4	48.4	39.1	54.4
1980	34.6	29.2	41.9	53.4	64.7	70.5	77.7	77.8	70.7	53.9	47.3	36.5	54.4
1981	28.2	37.1	42.7	60.5	60.8	73.6	75.6	74.1	67.2	55.2	45.4	34.2	54.6
1982	27.5	34.5	45.9	50.6	68.0	67.5	74.4	69.8	63.3	57.6	49.7	45.5	54.6
1983	34.7	37.9	48.0	52.3	60.8	71.7	78.6	79.8	69.1	57.3	46.9	30.3	55.6
1984	28.4	41.7	41.8	54.8	61.6	75.2	73.4	75.6	65.8	62.9	42.9	45.7	55.8
1985	25.9	32.1	48.5	59.6	64.9	70.2	74.3	72.1	67.9	61.0	53.3	32.6	55.2
Record Mean	33.3	36.2	45.2	56.0	64.4	72.3	75.8	74.7	68.3	57.5	46.1	37.1	55.6
Max	42.1	45.9	56.0	68.0	76.6	83.7	86.6	85.6	79.8	69.3	56.2	45.9	66.3
Min	24.5	26.4	34.3	43.9	52.3	60.8	64.9	63.7	56.7	45.7	35.9	28.3	44.8

REFERENCE NOTES FOR TABLES 1, 2, 3 and 6   (HUNTINGTON, WV)

GENERAL

T - TRACE AMOUNT
BLANK ENTRIES DENOTE MISSING/UNREPORTED DATA.
# INDICATES A STATION OR INSTRUMENT RELOCATION.

SPECIFIC

TABLE 1

(a) - LENGTH OF RECORD IN YEARS. ALTHOUGH
INDIVIDUAL MONTHS MAY BE MISSING.

* LESS THAN .05

NORMALS — BASED ON THE 1951-1980 RECORD PERIOD.
EXTREMES — DATES ARE THE MOST RECENT OCCURRENCE.
WIND DIR. — NUMERALS SHOW TENS OF DEGREES
CLOCKWISE FROM TRUE NORTH.
"00" INDICATES CALM.
RESULTANT WIND DIRECTIONS ARE GIVEN TO WHOLE DEGREES.

EXCEPTIONS

TABLES 2, 3, and 6

RECORD MEANS ARE THROUGH THE CURRENT YEAR,
BEGINNING IN   1941 FOR TEMPERATURE
1941 FOR PRECIPITATION
1962 FOR SNOWFALL

TABLE 4 HEATING DEGREE DAYS Base 65 deg. F   HUNTINGTON, WEST VIRGINIA

SEASON	JULY	AUG	SEP	OCT	NOV	DEC	JAN	FEB	MAR	APR	MAY	JUNE	TOTAL
1956-57	0	2	83	123	552	548	991	624	627	261	79	1	3891
1957-58	0	0	30	354	520	737	994	1014	756	283	84	7	4779
1958-59	0	0	34	257	490	1024	976	703	651	255	55	2	4453
1959-60	0	0	23	229	604	748	807	871	962	205	157	0	4606
1960-61	0	0	8	204	523	1041	1083	637	480	451	171	15	4613
#1961-62	0	0	35	213	535	828	972	715	709	409	36	10	4462
1962-63	7	0	123	266	643	1062	1170	1028	438	243	102	21	5103
1963-64	9	0	69	144	548	1222	950	947	570	237	56	27	4788
1964-65	0	17	69	396	504	786	956	806	741	264	17	7	4563
1965-66	0	10	59	337	510	719	1182	853	583	384	167	20	4824
1966-67	0	3	60	369	535	889	790	919	528	269	194	13	4569
1967-68	7	4	111	326	724	805	1095	1085	590	279	160	18	5204
1968-69	0	11	26	290	532	960	1028	819	810	268	95	18	4857
1969-70	0	0	72	320	681	1057	1163	866	737	264	103	3	5266
1970-71	4	0	28	215	529	800	1089	769	711	322	165	2	4634
1971-72	2	0	11	84	606	597	899	877	661	335	110	57	4239
1972-73	8	2	14	429	664	757	950	866	363	385	158	1	4597
1973-74	0	0	18	159	503	845	710	754	501	248	109	23	3870
1974-75	0	0	113	337	560	812	851	704	696	376	43	4	4496
1975-76	0	0	91	228	445	811	1032	535	422	314	125	3	4006
1976-77	0	1	76	460	765	1003	1397	829	475	200	67	45	5318
1977-78	0	0	35	332	455	919	1232	1122	698	254	164	12	5223
1978-79	0	0	15	334	504	815	1194	1014	484	292	110	12	4774
1979-80	2	5	36	312	494	798	936	1033	708	351	93	16	4784
1980-81	0	0	26	348	632	878	1132	777	690	183	167	0	4833
1981-82	0	0	67	307	577	947	1157	845	588	429	29	8	4954
1982-83	0	8	99	263	464	611	933	750	529	377	156	16	4207
1983-84	0	0	67	248	537	1070	1127	667	713	324	171	3	4927
1984-85	1	0	98	91	660	597	1205	918	507	206	72	18	4373
1985-86	0	0	63	164	348	1001							

TABLE 5 COOLING DEGREE DAYS Base 65 deg. F   HUNTINGTON, WEST VIRGINIA

YEAR	JAN	FEB	MAR	APR	MAY	JUNE	JULY	AUG	SEP	OCT	NOV	DEC	TOTAL
1969	0	0	0	8	73	255	405	316	115	26	0	0	1198
1970	0	0	0	21	130	201	275	268	265	25	0	0	1185
1971	0	0	0	0	35	258	225	215	197	56	7	2	995
1972	0	0	1	16	30	102	308	234	147	0	3	0	841
1973	0	0	4	14	40	284	357	344	212	41	0	0	1296
1974	0	0	21	45	100	130	332	279	69	8	7	0	991
1975	0	0	0	0	15	143	255	319	423	100	17	2	1274
1976	0	1	24	51	59	224	244	228	32	4	0	0	867
1977	0	0	25	66	173	173	427	330	156	4	19	0	1373
1978	0	0	0	18	85	261	361	344	234	5	0	0	1308
1979	0	0	13	17	76	173	286	301	113	22	3	0	1004
1980	0	0	0	10	89	190	403	401	206	11	0	0	1310
1981	0	0	3	55	45	262	334	289	140	10	0	0	1138
1982	0	0	0	7	131	88	299	162	72	42	10	11	822
1983	0	0	7	3	35	222	426	466	198	17	0	0	1374
1984	0	0	0	26	71	314	269	336	132	34	5	6	1193
1985	0	0	0	6	53	75	182	297	227	155	45	7	1047

TABLE 6 SNOWFALL (inches)   HUNTINGTON, WEST VIRGINIA

SEASON	JULY	AUG	SEP	OCT	NOV	DEC	JAN	FEB	MAR	APR	MAY	JUNE	TOTAL
1956-57	0.0	0.0	0.0	0.0	1.0	1.6	5.1	0.4	2.0	T	0.0	0.0	10.1
1957-58	0.0	0.0	0.0	T	T	0.5	2.5	9.3	4.4	0.0	0.0	0.0	16.7
1958-59	0.0	0.0	0.0	0.0	1.7	1.8	3.0	0.2	T	0.7	0.0	0.0	7.4
1959-60	0.0	0.0	0.0	0.0	4.0	0.4	2.1	17.1	14.6	T	0.0	0.0	38.2
1960-61	0.0	0.0	0.0	0.0	T	1.8	7.7	4.8	T	T	0.0	0.0	14.3
#1961-62	0.0	0.0	0.0	0.0	T	5.4	1.8	2.7	5.0	T	0.0	0.0	14.9
1962-63	0.0	0.0	0.0	T	T	7.2	7.9	7.9	0.4	T	T	0.0	23.4
1963-64	0.0	0.0	0.0	0.0	2.8	8.0	10.7	14.3	0.6	0.0	0.0	0.0	36.4
1964-65	0.0	0.0	0.0	0.0	1.0	1.1	10.0	2.7	7.3	0.0	0.0	0.0	22.1
1965-66	0.0	0.0	0.0	0.0	T	1.0	12.9	2.0	0.5	T	0.0	0.0	16.4
1966-67	0.0	0.0	0.0	T	4.1	5.0	3.4	11.7	3.0	0.0	0.0	0.0	27.2
1967-68	0.0	0.0	0.0	0.0	0.5	13.2	9.0	4.1	2.2	0.0	0.0	0.0	29.0
1968-69	0.0	0.0	0.0	0.0	1.5	2.5	2.1	2.8	4.1	0.0	0.0	0.0	13.0
1969-70	0.0	0.0	0.0	0.0	4.6	9.9	9.2	7.9	7.2	T	0.0	0.0	38.8
1970-71	0.0	0.0	0.0	0.0	0.1	4.7	7.1	9.8	11.0	0.0	0.0	0.0	32.7
1971-72	0.0	0.0	0.0	0.0	4.0	0.2	6.0	6.6	4.5	T	0.0	0.0	21.3
1972-73	0.0	0.0	0.0	0.2	3.6	0.8	2.6	1.5	3.4	0.7	0.0	0.0	12.8
1973-74	0.0	0.0	0.0	0.0	T	2.0	0.3	7.5	1.0	0.8	0.0	0.0	11.6
1974-75	0.0	0.0	0.0	0.4	T	5.9	14.7	1.9	6.3	T	0.0	0.0	29.2
1975-76	0.0	0.0	0.0	0.0	0.3	1.6	10.2	1.2	6.2	0.0	0.0	0.0	19.5
1976-77	0.0	0.0	0.0	T	2.2	4.8	23.7	5.3	0.8	T	0.0	0.0	36.8
1977-78	0.0	0.0	0.0	T	3.5	2.0	30.3	9.7	10.2	0.0	0.0	0.0	55.7
1978-79	0.0	0.0	0.0	0.0	T	0.6	14.7	18.1	0.3	T	0.0	0.0	33.7
1979-80	0.0	0.0	0.0	0.0	0.4	0.3	9.2	9.3	5.5	0.2	0.0	0.0	24.9
1980-81	0.0	0.0	0.0	T	0.7	1.1	4.6	2.4	3.3	0.0	0.0	0.0	12.1
1981-82	0.0	0.0	0.0	0.0	T	6.4	7.6	5.1	4.2	T	0.0	0.0	23.3
1982-83	0.0	0.0	0.0	0.0	T	0.7	3.3	11.1	3.9	0.0	0.0	0.0	19.0
1983-84	0.0	0.0	0.0	0.0	T	2.0	11.7	8.7	2.2	0.0	0.0	0.0	24.6
1984-85	0.0	0.0	0.0	0.0	0.1	2.5	19.5	21.2	0.2	0.4	0.0	0.0	43.9
1985-86	0.0	0.0	0.0	0.0	0.0	5.8							
Record Mean	0.0	0.0	0.0	T	1.2	3.7	9.7	7.3	3.9	0.1	T	0.0	25.9

See Reference Notes, relative to all above tables, on preceding page.

The Green Bay climate is modified by surrounding topography. The modification is caused by the Bay of Green Bay, Lakes Michigan, and Superior, and to a lesser extent, the slightly higher surrounding terrain terminating in the Fox River Valley. The city of Green Bay is located at the mouth of the Fox River, one of the largest rivers flowing northward in the United States. It empties into the south end of the Bay.

The modified continental climate of Green Bay is shown by the few occurrences of 90 degree temperatures in the summer season and the few occurrences of sub-zero temperatures in the winter season. The narrow temperature range stems from the lake effects and the limited hours of sunshine caused by cloudiness.

Precipitation normally falls in the five-month period May through September. Three-fifths of the annual total is in the growing season, most often falling during thunderstorms. During the winter months, snowfall is less than in nearby communities where the ground is slightly higher.

The comparatively low range in temperature along with the greater portion of the precipitation falling during the growing season is conducive to the development of the dairy industry. Cherry and apple orchards are important crops in nearby lake communities. The growing of potatoes and canning vegetables are predominant inland. Paper products are the major manufacturing industry.

High winds, excessive precipitation, and electrical storms cause occasional damage. Snowstorms are the principal winter hazard. While the winters are long in Green Bay, the extremes are never as severe as the northern latitude location would indicate.

Based on the 1951-1980 period, the average first occurrence of 32 degrees Fahrenheit in the fall is October 2 and the average last occurrence in the spring is May 12.

TABLE 1 NORMALS, MEANS AND EXTREMES

GREEN BAY, WISCONSIN

LATITUDE: 44°29'N   LONGITUDE: 88°08'W   ELEVATION: FT. GRND   682 BARO   00694   TIME ZONE: CENTRAL   WBAN: 14898

	(a)	JAN	FEB	MAR	APR	MAY	JUNE	JULY	AUG	SEP	OCT	NOV	DEC	YEAR
TEMPERATURE °F:														
Normals														
-Daily Maximum		22.5	26.9	37.0	53.7	66.6	76.2	80.9	78.7	69.8	58.5	42.0	28.5	53.4
-Daily Minimum		5.4	8.7	20.1	33.6	43.5	53.1	58.1	56.3	47.9	38.2	26.3	13.0	33.7
-Monthly		13.9	17.8	28.6	43.7	55.0	64.7	69.5	67.5	58.9	48.4	34.2	20.8	43.6
Extremes														
-Record Highest	36	50	55	73	89	91	97	99	99	95	88	72	62	99
-Year		1961	1981	1967	1980	1959	1971	1977	1955	1955	1963	1953	1970	JUL 1977
-Record Lowest	36	-31	-26	-29	7	21	32	40	38	24	15	-9	-27	-31
-Year		1951	1971	1962	1954	1966	1958	1965	1967	1949	1966	1976	1983	JAN 1951
NORMAL DEGREE DAYS:														
Heating (base 65°F)		1581	1322	1128	639	325	91	17	39	192	515	924	1370	8143
Cooling (base 65°F)		0	0	0	0	18	82	156	116	9	0	0	0	381
% OF POSSIBLE SUNSHINE	36	48	52	53	52	60	64	66	63	55	48	37	39	53
MEAN SKY COVER (tenths)														
Sunrise - Sunset	36	6.6	6.5	6.6	6.8	6.4	6.0	5.6	5.8	6.0	6.4	7.4	7.1	6.4
MEAN NUMBER OF DAYS:														
Sunrise to Sunset														
-Clear	36	8.0	7.2	7.2	6.3	6.9	7.4	8.5	8.8	8.1	7.3	5.0	6.4	87.2
-Partly Cloudy	36	6.3	6.5	7.6	7.7	9.6	11.1	12.1	10.5	9.7	8.4	6.4	6.0	101.9
-Cloudy	36	16.7	14.6	16.2	16.0	14.5	11.5	10.4	11.7	12.2	15.3	18.6	18.6	176.2
Precipitation														
.01 inches or more	36	10.3	8.3	10.9	11.0	11.2	10.6	9.7	10.4	10.2	8.8	9.3	10.9	121.7
Snow, Ice pellets														
1.0 inches or more	36	3.7	2.7	2.8	0.8	0.*	0.0	0.0	0.0	0.0	0.1	1.4	3.2	14.7
Thunderstorms	36	0.1	0.2	1.2	2.4	4.1	6.9	6.6	5.8	4.1	1.9	0.6	0.2	34.0
Heavy Fog Visibility														
1/4 mile or less	36	1.7	2.6	2.6	2.2	1.5	1.4	1.1	2.5	2.0	2.8	2.4	2.4	25.2
Temperature °F														
-Maximum														
90° and above	24	0.0	0.0	0.0	0.0	0.*	1.4	2.9	1.3	0.2	0.0	0.0	0.0	5.8
32° and below	24	24.5	19.2	8.4	0.5	0.0	0.0	0.0	0.0	0.0	0.0	4.7	20.3	77.5
-Minimum														
32° and below	24	30.7	27.5	26.8	14.2	2.8	0.0	0.0	0.0	0.6	7.8	21.9	29.4	161.5
0° and below	24	12.7	7.8	1.4	0.0	0.0	0.0	0.0	0.0	0.0	0.0	0.3	6.5	28.6
AVG. STATION PRESS.(mb)	13	991.3	992.0	989.3	989.3	988.4	988.2	989.8	991.0	991.4	991.5	990.7	991.0	990.3
RELATIVE HUMIDITY (%)														
Hour 00	24	75	76	77	75	75	80	82	86	86	81	81	79	79
Hour 06 (Local Time)	24	76	78	81	79	79	81	85	89	89	85	83	80	82
Hour 12	24	69	68	66	59	55	58	57	61	63	62	70	73	63
Hour 18	24	72	70	68	60	57	59	60	65	70	71	76	76	67
PRECIPITATION (inches):														
Water Equivalent														
-Normal		1.19	1.05	1.90	2.70	3.13	3.17	3.25	3.16	3.17	2.10	1.76	1.42	28.00
-Maximum Monthly	36	2.64	3.56	4.68	5.52	8.21	8.47	6.50	9.04	7.80	5.00	4.96	3.15	9.04
-Year		1950	1953	1977	1953	1973	1967	1950	1975	1965	1954	1985	1971	AUG 1975
-Minimum Monthly	36	0.12	0.04	0.31	0.98	0.56	0.31	0.83	0.90	0.28	0.16	0.10	0.10	T
-Year		1981	1969	1978	1963	1981	1976	1981	1955	1976	1952	1976	1960	OCT 1952
-Maximum in 24 hrs	36	1.14	1.78	1.44	2.00	3.28	2.65	2.95	4.60	2.99	3.68	2.30	1.55	4.60
-Year		1980	1966	1979	1981	1973	1969	1959	1975	1964	1954	1985	1959	AUG 1975
Snow, Ice pellets														
-Maximum Monthly	36	28.0	20.6	22.2	11.8	2.6				T	1.7	16.5	27.0	28.0
-Year		1982	1962	1972	1977	1960				1965	1959	1985	1977	JAN 1982
-Maximum in 24 hrs	36	8.8	9.2	10.1	10.2	2.2				T	1.6	8.2	11.1	11.1
-Year		1982	1959	1964	1977	1960				1965	1959	1977	1985	DEC 1985
WIND:														
Mean Speed (mph)	36	11.1	10.7	11.1	11.6	10.4	9.3	8.3	8.0	9.1	10.0	11.1	10.8	10.1
Prevailing Direction														
through 1963		SW	SW	NE	NE	NE	SW	SW	SW	SW	SW	SW	SW	SW
Fastest Mile														
-Direction (!!!)	35	W	W	W	SW	SW	SW	NE	SW	W	SW	W	W	SW
-Speed (MPH)	35	61	66	68	59	109	73	70	50	66	65	67	52	109
-Year		1950	1951	1951	1964	1950	1953	1957	1950	1951	1951	1955	1957	MAY 1950
Peak Gust														
-Direction (!!!)	2	NW	NW	E	NW	SW	NW	NW	W	SW	SW	W	SW	W
-Speed (mph)	2	38	46	46	49	52	49	52	53	47	44	41	52	53
-Date		1985	1985	1985	1984	1985	1984	1985	1985	1985	1984	1985	1984	AUG 1985

See Reference Notes to this table on the following page.

TABLE 2 PRECIPITATION (inches)   GREEN BAY, WISCONSIN

YEAR	JAN	FEB	MAR	APR	MAY	JUNE	JULY	AUG	SEP	OCT	NOV	DEC	ANNUAL
1956	0.56	0.60	1.86	1.45	4.66	3.90	5.85	3.09	1.65	0.65	2.13	0.87	27.27
1957	0.35	0.43	0.46	2.93	5.28	2.48	3.18	3.46	2.15	1.35	3.52	1.59	27.18
1958	0.42	0.16	0.50	2.56	1.27	1.82	2.07	2.68	3.20	2.25	1.34	0.16	18.43
1959	1.04	1.98	1.87	2.84	3.86	1.26	4.21	2.71	5.17	3.27	1.47	2.85	32.53
1960	1.04	0.48	1.21	3.13	7.75	3.07	1.87	3.52	3.09	2.32	0.69	0.10	28.27
1961	0.31	0.93	2.12	1.67	1.42	4.31	4.91	2.84	5.02	3.34	2.60	1.27	30.74
1962	1.27	2.02	1.13	2.55	2.86	4.35	2.70	2.86	3.87	1.94	0.84	1.03	27.42
1963	0.68	0.59	2.58	0.98	1.54	2.67	2.77	2.07	3.00	0.73	1.63	0.73	19.97
1964	1.14	0.26	1.76	2.55	4.14	1.05	4.55	2.72	6.74	0.44	2.07	0.70	28.12
1965	0.93	0.85	2.38	3.62	3.95	1.89	1.96	3.38	7.80	1.32	2.19	2.31	32.58
1966	1.18	2.25	2.46	1.38	1.28	1.09	4.19	2.65	1.21	0.72	1.58	1.65	21.64
1967	2.52	0.84	1.13	2.77	2.45	8.47	1.96	2.43	0.46	4.71	1.66	1.17	30.57
1968	0.94	0.45	0.97	4.84	3.10	6.97	2.00	2.66	3.31	1.01	1.01	2.69	29.95
1969	2.60	0.04	1.04	2.86	2.66	7.62	2.51	1.19	2.03	3.46	0.43	1.43	27.87
1970	0.73	0.23	1.07	1.61	5.76	1.11	4.02	1.25	6.11	2.98	2.68	1.24	28.79
1971	1.60	2.03	2.04	1.05	1.67	1.87	3.44	2.99	3.36	2.01	3.21	3.15	28.42
1972	0.65	0.96	2.19	1.45	0.82	2.25	1.85	5.86	5.76	1.84	1.15	2.49	27.27
1973	1.86	0.72	2.43	3.23	8.21	3.20	1.93	2.57	2.91	3.96	1.45	2.41	34.88
1974	1.71	1.17	1.07	2.62	4.46	4.91	4.25	1.61	1.05	1.72	2.09	1.67	28.33
1975	1.52	1.48	3.44	2.35	2.79	5.27	1.78	9.04	3.18	0.36	3.42	0.84	35.47
1976	1.72	1.33	3.65	2.44	2.42	0.31	2.96	1.15	0.28	0.82	0.16	0.61	17.85
1977	0.67	1.38	4.68	3.33	2.47	2.27	2.13	2.37	2.44	1.36	2.70	2.31	28.11
1978	1.33	0.35	0.31	3.44	3.38	2.72	6.03	4.36	4.82	2.33	2.93	1.30	33.30
1979	1.78	1.17	4.49	1.93	3.01	2.21	3.55	5.97	0.76	2.72	2.49	1.28	31.36
1980	1.92	0.35	1.00	2.73	1.77	3.82	1.87	7.31	3.42	1.79	1.25	1.35	28.58
1981	0.12	2.76	0.42	4.22	0.56	2.63	0.83	3.37	3.25	3.44	1.08	1.10	23.78
1982	1.34	0.14	1.95	2.66	2.74	2.67	5.10	2.91	1.43	1.20	4.51	2.50	29.15
1983	0.72	1.46	1.52	1.39	4.80	1.82	3.76	5.27	3.59	2.24	2.63	1.18	30.38
1984	0.59	1.59	1.64	3.33	1.65	5.60	3.17	3.78	5.66	4.92	2.55	1.72	36.20
1985	0.86	2.55	2.70	2.24	2.58	2.21	4.03	8.03	3.65	2.72	4.96	1.83	38.36
Record Mean	1.32	1.31	1.92	2.59	3.13	3.39	3.09	3.09	3.15	2.13	1.98	1.44	28.56

TABLE 3 AVERAGE TEMPERATURE (deg. F)   GREEN BAY, WISCONSIN

YEAR	JAN	FEB	MAR	APR	MAY	JUNE	JULY	AUG	SEP	OCT	NOV	DEC	ANNUAL
1956	17.8	17.6	24.0	39.3	52.1	66.0	65.0	66.8	55.2	53.1	33.6	22.5	42.8
1957	9.5	20.9	30.1	44.3	52.7	64.1	69.7	65.9	56.8	45.8	34.4	25.1	43.3
1958	19.5	15.0	31.5	43.9	53.0	58.8	67.2	67.9	59.1	50.6	36.2	15.0	43.1
1959	8.8	12.4	24.9	42.7	59.2	65.8	68.8	71.8	61.0	44.1	25.7	28.5	42.8
1960	21.0	20.4	19.5	44.3	53.1	61.3	66.9	67.8	60.7	47.2	37.1	19.4	43.2
#1961	15.8	25.2	32.8	41.4	52.9	64.6	69.2	68.1	61.5	50.1	34.8	16.9	44.5
1962	11.4	13.9	27.3	43.4	59.9	64.0	66.5	69.0	58.0	50.7	36.6	20.2	43.4
1963	7.0	11.1	30.1	47.5	54.4	68.4	70.0	65.6	58.5	56.5	37.9	12.1	43.3
1964	20.9	21.4	27.4	44.1	59.5	65.4	71.6	65.4	58.5	47.6	39.1	19.9	45.1
1965	14.4	15.3	24.4	40.4	60.0	63.6	68.8	67.0	57.5	49.3	35.3	28.7	43.7
1966	7.9	21.8	33.6	41.6	50.9	67.0	70.9	64.7	56.9	47.7	34.2	20.3	43.1
1967	20.1	13.2	29.5	44.2	49.8	66.2	68.3	63.9	58.0	46.7	30.9	22.5	42.8
1968	16.4	16.7	38.0	46.5	52.6	64.7	68.3	67.5	61.6	50.5	34.8	21.9	45.0
1969	14.0	20.7	24.4	43.5	54.3	57.2	68.0	70.4	58.7	45.0	31.8	20.9	42.4
1970	7.2	15.5	28.6	46.8	54.9	66.2	71.7	69.1	60.3	49.9	34.8	19.8	43.7
1971	6.9	15.2	25.9	43.7	53.0	69.8	67.8	65.3	62.3	54.6	34.7	23.5	43.6
1972	10.2	14.4	24.4	39.4	59.6	64.2	69.8	68.1	59.5	45.3	35.0	17.3	42.3
1973	21.2	22.9	39.5	43.8	53.0	68.2	71.7	71.8	59.5	54.5	34.3	20.9	46.8
1974	16.9	17.1	29.7	45.5	51.6	61.5	70.2	66.3	54.2	45.8	34.4	24.5	43.2
1975	20.3	18.4	24.1	38.9	60.0	65.3	71.0	68.0	54.6	49.9	39.1	21.2	44.2
1976	12.8	25.8	30.8	45.8	52.2	68.1	71.8	66.9	57.7	43.3	26.4	9.1	42.6
1977	3.1	20.0	37.0	48.2	63.4	63.8	73.3	65.5	59.5	47.3	33.3	18.0	44.4
1978	11.4	11.2	25.2	40.3	57.1	63.6	67.3	68.5	62.4	47.1	33.0	18.3	42.1
1979	5.9	9.0	28.8	41.7	52.4	64.6	70.4	66.5	60.0	45.8	33.8	27.7	42.2
1980	17.7	17.2	27.2	45.1	58.3	62.6	70.6	69.1	59.1	43.4	35.0	19.7	43.8
1981	15.3	22.9	34.7	45.7	53.6	65.6	69.3	68.2	56.8	45.0	36.9	22.9	44.7
1982	6.7	15.8	28.4	40.3	60.8	59.3	70.8	65.1	57.6	49.3	33.3	28.0	43.0
1983	21.4	26.3	31.1	40.5	48.8	64.5	72.9	70.8	59.8	48.5	36.5	10.6	44.3
1984	12.7	28.3	25.3	44.5	51.7	67.4	68.6	69.5	57.5	51.0	34.8	24.0	44.6
1985	12.2	17.6	34.2	47.6	58.5	62.3	69.1	66.4	60.8	48.3	30.9	9.4	43.1
Record Mean	15.7	18.0	29.2	43.3	55.0	65.1	70.3	67.9	60.1	48.9	34.5	21.8	44.1
Record Max	23.6	26.4	37.2	52.5	65.7	75.5	80.8	78.1	69.8	57.8	41.5	28.8	53.1
Record Min	7.7	9.7	21.2	34.1	44.2	54.7	59.7	57.8	50.3	39.9	27.5	14.8	35.1

REFERENCE NOTES FOR TABLES 1, 2, 3 and 6   (GREEN BAY, WI)

GENERAL

T - TRACE AMOUNT
BLANK ENTRIES DENOTE MISSING/UNREPORTED DATA.
# INDICATES A STATION OR INSTRUMENT RELOCATION.

SPECIFIC

TABLE 1

(a) - LENGTH OF RECORD IN YEARS. ALTHOUGH
      INDIVIDUAL MONTHS MAY BE MISSING.
  * LESS THAN .05

NORMALS — BASED ON THE 1951-1980 RECORD PERIOD.
EXTREMES — DATES ARE THE MOST RECENT OCCURRENCE.
WIND DIR. — NUMERALS SHOW TENS OF DEGREES
            CLOCKWISE FROM TRUE NORTH.
            "00" INDICATES CALM.
RESULTANT WIND DIRECTIONS ARE GIVEN TO WHOLE DEGREES.

EXCEPTIONS

TABLES 2, 3, and 6

RECORD MEANS ARE THROUGH THE CURRENT YEAR,
BEGINNING IN   1886 FOR TEMPERATURE
                1886 FOR PRECIPITATION
                1950 FOR SNOWFALL

TABLE 4 HEATING DEGREE DAYS Base 65 deg. F   GREEN BAY, WISCONSIN

SEASON	JULY	AUG	SEP	OCT	NOV	DEC	JAN	FEB	MAR	APR	MAY	JUNE	TOTAL
1956-57	46	46	300	367	935	1311	1718	1229	1078	616	385	91	8122
1957-58	14	50	263	586	912	1231	1402	1399	1034	629	366	208	8094
1958-59	27	52	194	437	856	1543	1741	1466	1236	663	213	72	8500
1959-60	20	10	187	642	1173	1128	1359	1286	1405	617	365	124	8316
1960-61	40	31	197	544	829	1408	1519	1109	991	703	373	86	7830
#1961-62	14	29	183	455	900	1484	1657	1426	1161	646	221	94	8270
1962-63	24	18	221	443	847	1384	1799	1505	1079	520	330	49	8219
1963-64	12	60	203	266	807	1635	1365	1258	1156	618	196	118	7694
1964-65	15	74	229	531	771	1392	1564	1388	1249	733	197	86	8229
1965-66	26	51	245	485	885	1118	1763	1207	967	695	437	79	7958
1966-67	10	74	253	531	917	1377	1386	1444	1093	620	467	44	8216
1967-68	40	100	235	563	1017	1313	1506	1395	832	551	378	97	8027
1968-69	39	82	130	467	899	1327	1580	1234	1252	641	346	247	8244
1969-70	38	11	223	619	991	1362	1788	1381	1121	550	333	73	8490
1970-71	17	17	185	466	898	1398	1797	1391	1206	636	365	43	8419
1971-72	32	57	155	328	902	1282	1694	1463	1253	762	215	92	8235
1972-73	37	54	180	606	895	1469	1350	1172	785	632	367	9	7556
1973-74	4	11	221	330	897	1363	1487	1334	1087	590	413	132	7869
1974-75	6	39	331	588	912	1248	1381	1297	1260	779	205	90	8136
1975-76	29	27	307	465	769	1351	1614	1130	1057	579	394	29	7751
1976-77	2	56	255	667	1152	1730	1918	1254	861	500	131	113	8639
1977-78	3	64	162	544	942	1454	1658	1500	1227	733	279	107	8673
1978-79	35	18	152	549	953	1442	1830	1564	1117	691	389	70	8810
1979-80	14	36	166	590	928	1151	1461	1380	1165	597	227	130	7845
1980-81	11	11	189	661	893	1398	1538	1175	932	573	350	45	7776
1981-82	26	21	248	614	839	1300	1805	1373	1127	733	152	178	8416
1982-83	3	75	250	483	946	1140	1344	1077	1046	727	495	100	7686
1983-84	17	2	210	507	847	1682	1617	1055	1223	611	406	13	8190
1984-85	18	20	237	430	899	1262	1632	1324	949	533	204	114	7622
1985-86	9	34	196	508	1016	1719							

TABLE 5 COOLING DEGREE DAYS Base 65 deg. F   GREEN BAY, WISCONSIN

YEAR	JAN	FEB	MAR	APR	MAY	JUNE	JULY	AUG	SEP	OCT	NOV	DEC	TOTAL
1969	0	0	0	0	21	17	138	184	40	3	0	0	403
1970	0	0	0	10	26	114	231	150	50	5	0	0	586
1971	0	0	0	0	1	195	125	71	77	15	0	0	484
1972	0	0	0	0	52	75	190	156	22	0	0	0	495
1973	0	0	0	2	0	112	220	228	63	13	0	0	638
1974	0	0	0	7	6	35	175	87	13	0	0	0	323
1975	0	0	0	0	56	106	221	126	3	2	0	0	514
1976	0	0	0	8	1	129	219	122	41	0	0	0	520
1977	0	0	0	5	87	81	269	87	5	0	0	0	534
1978	0	0	0	0	43	71	115	131	80	0	0	0	440
1979	0	0	0	0	6	68	191	91	23	1	0	0	380
1980	0	0	0	5	27	64	192	146	18	0	0	0	452
1981	0	0	0	0	5	71	168	127	9	0	0	0	380
1982	0	0	0	0	30	16	187	85	35	4	0	0	357
1983	0	0	0	0	0	95	270	188	60	4	0	0	617
1984	0	0	0	0	3	94	136	165	17	0	0	0	415
1985	0	0	0	16	11	41	141	85	79	0	0	0	373

TABLE 6 SNOWFALL (inches)   GREEN BAY, WISCONSIN

SEASON	JULY	AUG	SEP	OCT	NOV	DEC	JAN	FEB	MAR	APR	MAY	JUNE	TOTAL
1956-57	0.0	0.0	0.0	0.0	3.7	7.9	3.6	4.2	2.0	6.9	0.0	0.0	28.3
1957-58	0.0	0.0	0.0	T	2.9	7.2	4.7	1.6	5.6	1.4	0.0	0.0	23.4
1958-59	0.0	0.0	0.0	T	2.4	1.8	12.1	20.1	19.6	0.5	0.0	0.0	56.5
1959-60	0.0	0.0	0.0	1.7	11.6	13.4	4.4	6.6	2.7	2.5	2.6	0.0	45.5
1960-61	0.0	0.0	0.0	0.0	0.1	0.7	5.0	3.8	8.1	1.5	T	0.0	19.2
1961-62	0.0	0.0	0.0	0.0	3.4	16.8	15.8	20.6	6.7	T	T	0.0	63.3
1962-63	0.0	0.0	0.0	0.4	6.6	7.8	8.0	6.2	13.3	3.9	T	0.0	46.2
1963-64	0.0	0.0	0.0	0.0	T	4.1	2.9	3.2	14.7	0.5	0.0	0.0	25.4
1964-65	0.0	0.0	0.0	0.0	0.1	6.3	12.5	6.3	17.0	5.5	0.0	0.0	47.7
1965-66	0.0	0.0	T	0.0	0.3	2.8	14.6	1.1	4.5	0.9	T	0.0	24.2
1966-67	0.0	0.0	0.0	T	2.9	11.9	16.4	9.9	4.0	0.2	T	0.0	45.3
1967-68	0.0	0.0	0.0	0.5	2.9	3.7	9.2	3.0	0.6	0.7	0.0	0.0	20.6
1968-69	0.0	0.0	0.0	0.0	2.4	26.7	11.8	0.5	2.2	T	0.0	0.0	43.6
1969-70	0.0	0.0	0.0	0.1	1.7	16.5	9.8	2.9	8.8	T	0.0	0.0	39.8
1970-71	0.0	0.0	0.0	0.0	5.9	10.7	20.8	11.0	14.8	0.2	T	0.0	63.4
1971-72	0.0	0.0	0.0	0.0	12.6	13.2	8.2	12.1	22.2	2.3	0.0	0.0	70.6
1972-73	0.0	0.0	0.0	T	0.7	17.3	6.8	5.7	11.2	5.6	T	0.0	36.1
1973-74	0.0	0.0	0.0	0.0	2.4	15.5	4.4	18.5	13.4	2.0	0.0	0.0	54.0
1974-75	0.0	0.0	0.0	T	1.5	10.4	8.6	19.5	13.4	T	0.0	0.0	53.4
1975-76	0.0	0.0	0.0	0.0	4.4	9.3	12.4	11.5	8.1	0.3	0.1	0.0	62.9
1976-77	0.0	0.0	0.0	1.4	1.7	12.1	12.1	2.4	13.7	11.8	0.0	0.0	55.2
1977-78	0.0	0.0	0.0	0.0	14.1	27.0	18.0	7.1	1.8	T	0.0	0.0	68.0
1978-79	0.0	0.0	0.0	0.0	9.6	15.0	24.0	11.8	6.2	5.0	0.0	0.0	71.6
1979-80	0.0	0.0	0.0	T	6.2	1.4	7.3	4.4	11.4	7.4	0.0	0.0	38.1
1980-81	0.0	0.0	0.0	T	3.4	14.5	2.3	7.5	2.5	T	0.0	0.0	30.2
1981-82	0.0	0.0	0.0	1.0	2.7	11.4	28.0	2.0	6.0	2.9	0.0	0.0	54.0
1982-83	0.0	0.0	0.0	T	2.6	1.3	8.5	15.3	11.4	0.6	0.0	0.0	39.7
1983-84	0.0	0.0	0.0	0.0	7.2	12.7	10.5	1.8	6.5	T	0.0	0.0	38.7
1984-85	0.0	0.0	0.0	0.0	2.0	15.9	13.8	15.1	17.7	6.2	0.0	0.0	70.7
1985-86	0.0	0.0	0.0	0.0	16.5	22.7							
Record Mean	0.0	0.0	T	0.1	4.5	10.7	10.9	8.5	9.0	2.3	0.1	0.0	46.1

See Reference Notes, relative to all above tables, on preceding page.

The city of La Crosse is situated on the east bank of the Mississippi River at the confluence of the Mississippi, Black, and La Crosse Rivers. The official records are taken at the La Crosse Municipal Airport which is 6 1/2 miles north of the main Post Office, on the north end of French Island. This island is about 6 miles long from north to south and 2 to 4 miles wide with the Mississippi River to the west and the old channel of the Black River to the east. A rather level sandy plain exists on each side of the river extending between the Wisconsin and Minnesota bluffs which rise 450 to 500 feet above the valley floor. The distance from bluff to bluff averages about 5 miles. The Mississippi River bends to the northwest and continues directly southward from the city.

The prevailing winds in the area are from the northwest from January through April and southerly during the remainder of the year. The situation of the city and airport in a natural bowl between the hills results in somewhat colder temperatures at night due to the settling of cooler air. Valley fogs often persist to mid-forenoon. Steepsided hills with narrow valleys are characteristic of most of the surrounding area.

The flow of the Mississippi River is regulated by dams built for the purpose of navigation, but the reservoirs have limited storage capacity. La Crosse is in the area of Pool No. 8 with a mean sea level elevation of 631 feet. When the river reaches an elevation of 639 feet, with open gate operation, there is considerable flooding of land near the river and some industrial sections of the city.

The invigorating continental-type climate results in wide and frequent variations in temperature. General storms moving eastward or northeastward into our area bring warmer weather and supply most of our moisture. These are usually followed by cooler air from Canada. The winters are cold and humid. The summers are warm with moderate humidities, while periods of hot and humid weather occur occasionally, usually lasting from a few days to a week at a time.

Sixty percent of the precipitation falls during the main growing season, extending from May through September. Most of the summer rainfall occurs during scattered thunderstorms. Some damage from heavy rains, high winds, and hail occurs each year, but tornadoes are infrequent and cover very small areas. Snow is frequent and is the predominant form of precipitation in winter. Heavy snow sometimes falls with larger amounts over the ridges. Glaze storms are not numerous since La Crosse is north of the main path of freezing rain.

Farming is diversified with dairying the leading activity. The more important field crops are corn, oats, and hay. Some of the more specialized crops are soybeans, tobacco, small fruits, and cranberries. Commercial apple orchards are numerous across the Mississippi River in Minnesota.

Based on the 1951–1980 period, the average first occurrence of 32 degrees Fahrenheit in the fall is October 13 and the average last occurrence in the spring is April 29.

TABLE 1   NORMALS, MEANS AND EXTREMES

LA CROSSE, WISCONSIN

LATITUDE: 43°52'N   LONGITUDE: 91°15'W   ELEVATION: FT. GRND 651 BARO 00658   TIME ZONE: CENTRAL   WBAN: 14920

	(a)	JAN	FEB	MAR	APR	MAY	JUNE	JULY	AUG	SEP	OCT	NOV	DEC	YEAR
TEMPERATURE °F:														
Normals														
-Daily Maximum		23.0	29.4	40.0	57.5	70.2	79.1	83.5	81.4	72.0	60.8	43.1	29.2	55.8
-Daily Minimum		5.0	9.9	21.8	36.9	48.5	57.8	62.4	60.3	51.2	41.0	27.4	13.8	36.3
-Monthly		14.0	19.6	30.9	47.2	59.4	68.5	73.0	70.8	61.6	50.9	35.3	21.5	46.1
Extremes														
-Record Highest	35	57	64	80	93	94	98	104	103	100	93	75	64	104
-Year		1981	1981	1978	1980	1967	1985	1980	1955	1978	1963	1978	1982	JUL 1980
-Record Lowest	35	-37	-36	-28	7	26	37	33	40	28	15	-9	-30	-37
-Year		1951	1971	1962	1982	1971	1978	1982	1965	1967	1972	1977	1983	JAN 1951
NORMAL DEGREE DAYS:														
Heating (base 65°F)		1581	1268	1057	534	216	37	10	14	135	448	891	1349	7540
Cooling (base 65°F)		0	0	0	0	42	142	258	197	33	11	0	0	683
% OF POSSIBLE SUNSHINE														
MEAN SKY COVER (tenths)														
Sunrise - Sunset	17	6.7	6.2	6.7	6.7	6.5	6.2	5.6	5.7	5.8	5.7	7.1	7.1	6.3
MEAN NUMBER OF DAYS:														
Sunrise to Sunset														
-Clear	17	7.4	7.9	7.4	6.8	6.9	7.3	9.8	9.5	9.1	10.8	5.8	6.8	95.4
-Partly Cloudy	17	6.9	7.1	7.1	7.1	9.1	10.4	10.9	10.6	8.9	6.9	6.1	5.8	96.8
-Cloudy	17	16.7	13.2	16.5	16.1	15.0	12.3	10.4	10.9	11.9	13.4	18.1	18.5	173.0
Precipitation														
.01 inches or more	34	8.1	7.1	9.9	9.9	11.4	11.0	9.9	9.7	9.7	7.8	7.5	8.8	110.8
Snow, Ice pellets														
1.0 inches or more	34	3.0	2.4	2.8	0.5	0.0	0.0	0.0	0.0	0.0	0.*	1.1	2.9	12.8
Thunderstorms	34	0.1	0.3	1.3	2.9	5.5	7.9	7.4	7.0	4.7	2.2	0.6	0.3	40.2
Heavy Fog Visibility														
1/4 mile or less	34	1.1	1.5	1.6	0.6	0.6	1.2	1.7	3.7	3.4	2.0	1.4	1.2	20.0
Temperature °F														
-Maximum														
90° and above	34	0.0	0.0	0.0	0.1	0.7	3.2	6.5	4.4	1.3	0.1	0.0	0.0	16.2
32° and below	34	23.1	15.8	6.9	0.2	0.0	0.0	0.0	0.0	0.0	0.0	5.0	18.9	69.9
-Minimum														
32° and below	34	30.7	27.1	26.2	9.9	0.8	0.0	0.0	0.0	0.3	5.7	21.1	29.4	151.2
0° and below	34	12.2	7.3	1.3	0.0	0.0	0.0	0.0	0.0	0.0	0.0	0.4	5.9	27.2
AVG. STATION PRESS.(mb)	12	994.9	994.2	991.1	990.5	989.5	989.5	991.4	992.1	993.0	993.2	992.7	994.0	992.2
RELATIVE HUMIDITY (%)														
Hour 00	26	76	78	78	74	75	82	81	88	88	79	80	79	80
Hour 06 (Local Time)	34	77	79	81	79	80	84	88	91	91	84	83	82	83
Hour 12	34	67	65	62	53	52	56	57	60	62	59	67	72	61
Hour 18	34	70	68	64	54	53	57	57	61	66	65	72	75	64
PRECIPITATION (inches):														
Water Equivalent														
-Normal		0.94	0.89	1.96	3.05	3.61	4.15	3.83	3.70	3.47	2.08	1.50	1.07	30.25
-Maximum Monthly	34	2.86	2.58	3.82	7.31	8.83	9.53	9.17	9.84	10.52	5.09	3.72	2.55	10.52
-Year		1967	1959	1951	1973	1960	1968	1978	1980	1965	1984	1983	1971	SEP 1965
-Minimum Monthly	34	0.14	0.05	0.30	0.60	1.07	1.58	0.16	0.54	0.42	0.02	T	0.30	10.52
-Year		1981	1969	1978	1966	1958	1979	1967	1976	1952	1952	1976	1962	NOV 1976
-Maximum in 24 hrs	20	1.31	1.06	1.64	3.84	2.72	3.94	3.72	3.92	2.25	2.17	2.38	0.93	3.94
-Year		1967	1966	1966	1954	1960	1967	1952	1962	1968	1966	1958	1965	JUN 1967
Snow, Ice pellets														
-Maximum Monthly	34	29.7	31.0	33.5	17.0	0.8				T	1.4	13.0	26.6	33.5
-Year		1979	1959	1959	1973	1960				1985	1959	1957	1968	MAR 1959
-Maximum in 24 hrs	20	8.7	10.9	15.7	7.3	0.8				T	1.2	11.0	9.1	15.7
-Year		1963	1959	1959	1952	1960				1985	1959	1957	1966	MAR 1959
WIND:														
Mean Speed (mph)	34	8.6	8.5	9.2	10.4	9.5	8.4	7.6	7.4	8.2	9.2	9.6	8.7	8.8
Prevailing Direction														
through 1963		S	NW	NW	NW	S	S	S	S	S	S	S	S	S
Fastest Obs. 1 Min.														
-Direction (!!!)	18	32	34	34	25	09	34	27	32	27	34	18	34	34
-Speed (MPH)	18	45	37	40	53	58	63	52	63	40	39	46	43	63
-Year		1962	1963	1960	1964	1952	1963	1961	1963	1961	1964	1958	1957	JUN 1963
Peak Gust														
-Direction (!!!)														
-Speed (mph)														
-Date														

See Reference Notes to this table on the following page

TABLE 2 PRECIPITATION (inches) LA CROSSE, WISCONSIN

YEAR	JAN	FEB	MAR	APR	MAY	JUNE	JULY	AUG	SEP	OCT	NOV	DEC	ANNUAL
1956	0.46	0.43	3.64	2.09	5.68	2.75	2.47	3.47	1.56	1.52	1.66	0.53	26.26
1957	0.25	0.20	1.03	2.18	4.56	4.38	4.35	4.18	1.86	1.30	2.88	0.61	27.78
1958	0.33	0.07	0.30	2.53	1.07	2.29	2.98	2.06	3.28	0.80	2.56	0.30	18.57
1959	0.66	2.58	3.07	0.65	4.67	4.05	2.34	7.85	5.97	2.92	1.98	1.45	38.19
1960	0.78	0.38	0.80	3.25	8.83	4.88	1.38	4.86	4.47	3.00	0.74	0.41	33.78
1961	0.27	1.31	3.37	1.33	2.76	2.34	2.74	1.39	4.97	2.47	2.52	0.98	26.45
1962	0.19	1.82	1.92	1.69	3.78	2.23	3.22	7.16	2.41	2.24	0.09	0.30	27.05
1963	0.67	0.49	2.25	2.46	2.04	2.35	5.14	3.40	3.92	1.46	1.90	0.49	26.57
1964	0.34	0.09	1.30	2.37	3.72	3.13	3.39	3.34	5.48	0.25	1.09	0.86	25.36
1965	0.81	0.72	2.05	4.87	5.11	3.51	4.48	2.85	10.52	1.66	2.98	2.02	41.58
1966	0.77	1.51	3.31	0.60	2.28	3.47	4.60	1.87	0.96	3.32	0.30	1.20	24.19
1967	2.86	1.11	2.14	2.43	1.22	9.33	0.16	2.44	1.38	2.52	0.18	0.51	26.28
1968	0.87	0.06	0.98	4.46	3.20	9.53	4.06	1.96	5.10	2.59	0.61	2.24	35.66
1969	1.95	0.05	0.78	1.83	2.55	6.39	3.00	2.16	2.32	3.73	0.58	1.69	27.03
1970	0.56	0.33	2.17	2.58	5.82	2.46	4.48	1.02	5.30	3.43	2.60	0.97	31.72
1971	1.52	2.06	1.11	1.74	4.89	4.33	4.06	2.22	2.43	1.09	2.20	2.55	30.20
1972	0.62	0.50	1.68	2.38	1.80	2.81	6.03	5.97	6.72	4.01	1.45	2.46	36.43
1973	0.85	1.38	3.69	7.31	6.21	4.47	2.53	8.63	3.68	2.06	1.93	1.37	44.11
1974	0.40	1.65	2.13	1.83	3.50	5.30	1.76	4.20	1.40	1.91	1.11	1.39	26.58
1975	1.50	1.71	2.14	6.07	2.52	2.88	1.42	2.79	1.28	0.36	3.07	0.70	26.44
1976	0.69	0.64	2.82	4.37	2.95	1.82	2.15	0.54	0.54	0.28	T	0.67	17.47
1977	0.88	0.83	2.64	3.62	3.04	3.51	4.32	2.92	4.12	1.87	1.42	1.40	30.57
1978	0.76	0.68	0.30	4.01	3.97	7.02	9.17	1.20	5.12	0.80	2.35	0.93	36.31
1979	2.41	0.65	2.02	1.79	5.68	1.58	3.52	7.55	0.53	4.41	2.44	0.67	33.25
1980	1.61	0.35	0.65	1.74	2.66	3.57	2.26	2.26	8.51	2.36	0.14	0.61	34.30
1981	0.14	2.10	0.60	4.37	1.88	2.60	7.66	9.56	1.77	1.89	1.08	0.86	34.51
1982	1.34	0.17		1.65	4.71	1.35	1.89	2.91	2.67	2.83	3.64	2.03	
1983	0.89	2.27	1.60	2.37	4.50	1.67	3.16	3.06	4.94	3.35	3.72	0.68	32.21
1984	0.28	0.92	1.94	3.65	2.18	7.43	3.00	2.11	2.87	5.09	1.36	2.42	33.25
1985	0.88	1.27	2.66	2.85	1.08	2.82	2.39	3.21	5.63				22.79
Record Mean	1.12	1.07	1.83	2.63	3.62	4.24	3.77	3.62	3.82	2.27	1.69	1.25	30.93

TABLE 3 AVERAGE TEMPERATURE (deg. F) LA CROSSE, WISCONSIN

YEAR	JAN	FEB	MAR	APR	MAY	JUNE	JULY	AUG	SEP	OCT	NOV	DEC	ANNUAL
1956	15.6	17.5	26.1	42.7	58.6	73.3	70.3	71.6	59.5	57.5	34.3	24.6	46.0
1957	12.3	24.9	32.5	48.6	57.2	68.1	75.9	69.9	60.0	48.3	34.2	26.5	46.5
1958	21.2	16.3	33.4	48.7	61.2	64.1	70.5	71.7	62.7	53.9	37.5	15.7	46.4
1959	10.2	14.3	28.8	47.5	63.1	69.7	72.3	74.2	63.1	46.1	26.4	30.6	45.5
1960	21.3	21.6	22.1	48.8	58.8	65.0	72.0	72.3	64.1	50.9	38.2	21.5	46.4
1961	16.4	27.7	34.2	41.8	57.1	69.2	71.8	73.1	61.5	52.8	35.1	18.3	46.6
1962	12.8	16.8	29.2	45.4	63.5	68.0	68.6	70.5	58.8	53.1	37.3	21.5	45.4
1963	6.3	14.8	35.6	49.0	57.5	71.3	73.7	68.9	62.9	61.2	41.0	12.5	46.2
#1964	24.2	26.8	30.0	48.5	63.8	70.0	76.3	69.7	67.0	49.6	37.9	17.5	47.9
1965	11.0	14.0	23.0	43.5	60.7	66.3	70.4	67.5	56.6	50.3	35.3	30.5	44.1
1966	7.7	19.5	37.3	44.7	55.2	69.8	75.3	69.8	61.4	50.5	35.6	22.0	45.7
1967	19.4	12.3	33.3	48.4	54.2	68.5	71.3	67.9	61.1	49.3	32.9	25.3	45.3
1968	19.3	19.5	40.5	51.3	56.3	68.0	71.9	72.0	62.1	52.0	35.7	19.7	47.4
1969	12.3	20.7	26.0	49.9	60.6	61.4	72.5	73.9	62.8	47.8	35.2	22.3	45.4
1970	8.7	20.7	30.8	49.1	60.6	66.2	73.4	70.4	61.4	51.8	35.4	19.4	46.0
1971	7.4	17.0	28.1	47.1	55.3	72.1	68.5	68.1	64.3	55.7	36.4	24.5	45.4
1972	11.1	16.1	28.0	43.0	63.4	66.5	70.3	71.1	61.3	45.9	34.4	15.7	43.9
1973	20.1	24.8	41.3	44.5	55.2	70.3	73.9	72.2	62.6	57.1	36.7	18.9	48.1
1974	18.0	19.1	32.1	48.7	55.0	65.0	75.9	68.4	56.9	50.4	37.4	26.6	46.1
1975	18.1	18.7	24.5	41.4	62.3	68.9	74.0	71.7	57.4	53.0	40.0	23.7	46.2
1976	13.1	29.0	33.7	49.7	56.8	70.0	74.4	71.6	61.0	45.0	27.9	12.8	45.4
1977	2.9	22.0	39.9	54.6	67.4	68.0	75.2	66.8	61.7	49.0	33.6	17.0	46.5
1978	6.7	10.8	29.1	45.9	60.0	67.4	70.6	73.5	68.3	51.0	35.8	19.9	44.9
1979	6.0	12.6	32.8	45.7	58.0	67.3	72.1	68.8	62.8	47.7	34.9	28.3	44.7
1980	18.5	17.9	29.7	49.4	62.8	70.3	78.3	74.1	63.7	45.2	36.5	20.8	47.3
1981	19.5	25.3	37.7	51.2	58.4	69.1	73.6	71.4	60.8	47.6	38.0	19.6	47.7
1982	4.3	17.2		44.2	62.8	63.5	74.2	70.3	60.9	50.8	34.1	27.2	
1983	19.9	25.2	33.8	42.6	54.3	69.0	76.7	75.3	62.9	49.6	35.9	6.4	46.0
1984	14.1	30.3	27.3	48.2	55.9	72.9	71.1	72.9	59.0	52.4	35.0	21.5	46.4
1985	10.9	16.5	37.2	52.5	62.0	64.5	72.1	68.4	62.2				49.6
Record Mean	15.3	19.9	31.8	47.3	59.2	68.3	73.1	70.7	62.1	50.8	35.2	22.1	46.4
Max	24.3	29.0	40.6	57.1	69.4	78.3	83.4	80.8	72.2	60.3	42.8	29.5	55.7
Min	6.4	10.7	23.0	37.6	48.9	58.2	62.8	60.5	52.0	41.2	27.5	14.6	37.0

REFERENCE NOTES FOR TABLES 1, 2, 3 and 6 (LA CROSSE, WI)

GENERAL

T - TRACE AMOUNT
BLANK ENTRIES DENOTE MISSING/UNREPORTED DATA.
# INDICATES A STATION OR INSTRUMENT RELOCATION.

SPECIFIC

TABLE 1

(a) - LENGTH OF RECORD IN YEARS. ALTHOUGH
INDIVIDUAL MONTHS MAY BE MISSING.

* LESS THAN .05

NORMALS — BASED ON THE 1951-1980 RECORD PERIOD.
EXTREMES — DATES ARE THE MOST RECENT OCCURRENCE.
WIND DIR. — NUMERALS SHOW TENS OF DEGREES
CLOCKWISE FROM TRUE NORTH.
"00" INDICATES CALM.
RESULTANT WIND DIRECTIONS ARE GIVEN TO WHOLE DEGREES.

EXCEPTIONS

TABLE 1

1. MEAN SKY COVER, AND DAYS CLEAR-PARTLY CLOUDY-CLOUDY ARE THROUGH 1967.
2. MAXIMUM 24-HOUR PRECIPITATION AND SNOW, AND FASTEST OBSERVED WINDS ARE THROUGH SEPTEMBER 1968.

TABLES 2, 3, and 6

RECORD MEANS ARE THROUGH THE CURRENT YEAR,
BEGINNING IN   1873 FOR TEMPERATURE
1873 FOR PRECIPITATION
1951 FOR SNOWFALL

TABLE 4 HEATING DEGREE DAYS Base 65 deg. F   LA CROSSE, WISCONSIN

SEASON	JULY	AUG	SEP	OCT	NOV	DEC	JAN	FEB	MAR	APR	MAY	JUNE	TOTAL
1956-57	4	16	187	238	913	1247	1630	1116	1000	494	254	34	7133
1957-58	0	21	178	509	917	1189	1351	1360	973	484	151	72	7205
1958-59	4	30	116	350	818	1521	1694	1417	1114	519	142	33	7758
1959-60	5	0	156	577	1149	1057	1349	1250	1322	501	196	57	7619
1960-61	8	6	148	438	798	1342	1504	1040	951	688	272	28	7223
1961-62	1	0	188	375	891	1440	1613	1345	1105	598	158	38	7752
1962-63	18	4	208	379	823	1344	1819	1405	906	475	250	34	7665
1963-64	1	29	114	151	714	1623	1258	1104	1078	493	111	47	6723
#1964-65	0	37	179	469	806	1470	1672	1427	1299	640	182	25	8206
1965-66	8	61	256	456	884	1061	1774	1267	852	599	312	35	7565
1966-67	1	18	158	450	873	1325	1406	1472	978	494	355	17	7547
1967-68	21	42	153	491	957	1223	1411	1312	754	411	275	54	7104
1968-69	7	15	120	423	869	1398	1626	1236	1200	447	196	135	7672
1969-70	5	1	135	540	886	1315	1746	1236	1052	499	207	30	7652
1970-71	10	5	173	406	883	1406	1781	1340	1137	533	297	18	7989
1971-72	15	22	140	301	850	1244	1668	1414	1140	655	154	62	7665
1972-73	21	32	147	586	913	1523	1389	1119	727	608	299	8	7372
1973-74	1	7	123	267	844	1424	1453	1279	1011	490	316	69	7284
1974-75	0	24	261	445	822	1182	1446	1292	1248	702	162	56	7640
1975-76	20	2	236	384	741	1274	1600	1039	964	460	254	11	6985
1976-77	3	14	176	624	1104	1613	1923	1200	770	344	66	36	7873
1977-78	1	40	122	488	936	1483	1804	1515	1106	565	209	36	8305
1978-79	6	0	78	427	871	1388	1828	1466	992	569	239	29	7893
1979-80	1	37	179	535	894	1133	1438	1358	1085	480	155	15	7247
1980-81	0	3	112	607	849	1362	1402	1107	840	408	214	11	6915
1981-82	5	7	157	535	802	1402	1878	1333		619	112	77	
1982-83	0	41	177	433	919	1164	1391	1109	960	666	325	46	7231
1983-84	6	0	160	480	867	1814	1572	1002	1159	501	285	5	7851
1984-85	9	6	219	384	893	1344	1673	1355	854	409	129	92	7367
1985-86	0	24	203										

TABLE 5 COOLING DEGREE DAYS Base 65 deg. F   LA CROSSE, WISCONSIN

YEAR	JAN	FEB	MAR	APR	MAY	JUNE	JULY	AUG	SEP	OCT	NOV	DEC	TOTAL
1969	0	0	0	0	66	35	245	284	74	13	0	0	717
1970	0	0	0	29	75	208	279	178	71	5	0	0	845
1971	0	0	0	1	3	237	129	128	126	19	0	0	643
1972	0	0	0	0	113	112	191	230	39	0	0	0	685
1973	0	0	0	0	4	174	283	239	59	28	0	0	787
1974	0	0	0	10	15	74	343	137	23	0	0	0	602
1975	0	0	0	0	85	177	306	215	16	15	0	0	814
1976	0	0	0	5	7	167	304	228	60	10	0	0	781
1977	0	0	0	41	145	132	325	100	33	0	0	0	776
1978	0	0	0	0	61	116	187	272	183	1	0	0	820
1979	0	0	0	0	30	104	226	161	60	3	0	0	584
1980	0	0	0	20	96	180	416	293	79	0	0	0	1084
1981	0	0	0	0	19	139	277	211	37	0	0	0	683
1982	0	0	0	0	54	38	292	212	60	0	0	0	
1983	0	0	0	1	0	172	375	324	104	12	0	0	988
1984	0	0	0	2	8	149	206	257	48	0	0	0	670
1985	0	0	0	41	42	84	229	136	126				658

TABLE 6 SNOWFALL (inches)   LA CROSSE, WISCONSIN

SEASON	JULY	AUG	SEP	OCT	NOV	DEC	JAN	FEB	MAR	APR	MAY	JUNE	TOTAL
1956-57	0.0	0.0	0.0	0.0	6.7	4.2	3.4	3.0	7.0	0.6	0.0	0.0	24.9
1957-58	0.0	0.0	0.0	0.0	13.0	3.5	4.7	0.6	2.9	T	0.0	0.0	24.7
1958-59	0.0	0.0	0.0	0.0	0.9	4.3	8.6	31.0	33.5	T	0.0	0.0	78.3
1959-60	0.0	0.0	0.0	1.4	11.8	5.1	4.6	4.4	6.9	T	0.8	0.0	35.0
1960-61	0.0	0.0	0.0	0.0	1.0	1.4	5.1	4.5	16.3	4.2	0.0	0.0	32.5
1961-62	0.0	0.0	T	0.0	7.5	14.9	2.9	25.9	20.9	6.6	0.0	0.0	78.7
1962-63	0.0	0.0	0.0	T	0.2	4.7	13.3	5.3	14.3	0.7	0.0	0.0	38.5
1963-64	0.0	0.0	0.0	0.0	0.1	7.6	5.7	1.0	17.7	T	0.0	0.0	32.1
1964-65	0.0	0.0	0.0	T	4.7	6.7	12.7	6.6	15.0	1.6	0.0	0.0	47.3
1965-66	0.0	0.0	T	0.0	0.2	1.3	10.9	5.1	2.5	T	T	0.0	20.0
1966-67	0.0	0.0	0.0	T	0.2	11.7	18.3	15.2	6.1	T	0.0	0.0	51.5
1967-68	0.0	0.0	0.0	T	0.8	1.5	4.5	0.8	0.1	T	0.0	0.0	7.7
1968-69	0.0	0.0	0.0	0.0	1.0	26.6	17.0	0.8	3.7	0.0	0.0	0.0	49.1
1969-70	0.0	0.0	0.0	0.0	T	19.8	7.7	3.2	6.1	T	0.0	0.0	36.8
1970-71	0.0	0.0	0.0	T	1.1	13.5	26.5	20.4	5.6	0.5	0.0	0.0	67.6
1971-72	0.0	0.0	0.0	0.0	3.8	12.0	10.2	8.0	11.9	3.0	0.0	0.0	48.9
1972-73	0.0	0.0	0.0	T	0.8	18.6	7.3	9.0	T	17.0	0.0	0.0	52.7
1973-74	0.0	0.0	0.0	0.0	T	9.8	2.0	14.9	7.6	0.7	0.0	0.0	35.0
1974-75	0.0	0.0	T	0.0	2.2	10.1	18.8	20.1	18.1	3.9	0.0	0.0	73.2
1975-76	0.0	0.0	0.0	0.0	1.7	1.0	8.0	2.7	6.0	0.0	T	0.0	19.4
1976-77	0.0	0.0	0.0	0.2	T	6.8	10.1	2.9	6.6	0.5	0.0	0.0	27.1
1977-78	0.0	0.0	0.0	0.0	12.0	18.8	10.4	8.0	3.4	T	0.0	0.0	52.6
1978-79	0.0	0.0	0.0	0.0	6.6	12.0	29.7	7.3	3.0	1.0	0.0	0.0	59.6
1979-80	0.0	0.0	0.0	T	3.2	T	10.7	4.1	10.1	3.9	0.0	0.0	32.0
1980-81	0.0	0.0	0.0	T	0.4	4.6	1.9	13.9	1.0	T	0.0	0.0	21.8
1981-82	0.0	0.0	0.0	0.5	3.6	11.6	14.3	2.2			0.0	0.0	34.6
1982-83	0.0	0.0	0.0	0.0	T	0.0	9.8	18.5	4.2	2.1	0.0	0.0	
1983-84	0.0	0.0	0.0	4.0	9.2	3.5	1.7	11.0	T	0.0	0.0		29.4
1984-85	0.0	0.0	0.0	T	2.9	13.3	9.4	5.3	14.5	T	0.0	0.0	45.4
1985-86	0.0	0.0	T										
Record Mean	0.0	0.0	T	0.1	3.5	9.0	9.7	8.2	9.7	1.8	T	0.0	42.1

See Reference Notes, relative to all above tables, on preceding page.

Madison is set on a narrow isthmus of land between Lakes Mendota and Monona. Lake Mendota (15 square miles) lies northwest of Lake Monona (5 square miles) and the lakes are only two-thirds of a mile apart at one point. Drainage at Madison is southeast through two other lakes into the Rock River, which flows south into Illinois, and then west to the Mississippi. The westward flowing Wisconsin River is only 20 miles northwest of Madison. Madison lakes are normally frozen from mid-December to early April.

Madison has the typical continental climate of interior North America with a large annual temperature range and with frequent short period temperature changes. The range of extreme temperatures is from about 110 to -40 degrees. Winter temperatures (December-February) average near 20 degrees and the summer average (June-August) is in the upper 60s. Daily temperatures average below 32 degrees about 120 days and above 40 degrees for about 210 days of the year.

Madison lies in the path of the frequent cyclones and anticyclones which move eastward over this area during fall, winter and spring. In summer, the cyclones have diminished intensity and tend to pass farther north. The most frequent air masses are of polar origin. Occasional outbreaks of arctic air affect this area during the winter months. Although northward moving tropical air masses contribute considerable cloudiness and precipitation, the true Gulf air mass does not reach this area in winter, and only occasionally at other seasons. Summers are pleasant, with only occasional periods of extreme heat or high humidity.

There are no dry and wet seasons, but about 60 percent of the annual precipitation falls in the five months of May through September. Cold season precipitation is lighter, but lasts longer. Soil moisture is usually adequate in the first part of the growing season. During July, August, and September, the crops depend on current rainfall, which is mostly from thunderstorms and tends to be erratic and variable. Average occurrence of thunderstorms is just under 7 days per month during this period.

March and November are the windiest months. Tornadoes are infrequent. Dane County has about one tornado in every three to five years.

The ground is covered with 1 inch or more of snow about 60 percent of the time from about December 10 to near February 25 in an average winter. The soil is usually frozen from the first of December through most of March with an average frost penetration of 25 to 30 inches. The growing season averages 175 days.

Farming is diversified with the main emphasis on dairying. Field crops are mainly corn, oats, clover, and alfalfa, but barley, wheat, rye, and tobacco are also raised. Canning factories pack peas, sweet corn, and lima beans. Fruits are mainly apples, strawberries, and raspberries.

TABLE 1   NORMALS, MEANS AND EXTREMES

MADISON, WISCONSIN

LATITUDE: 43°08'N   LONGITUDE: 89°20'W   ELEVATION: FT. GRND 858 BARO 860   TIME ZONE: CENTRAL   WBAN: 14837

	(a)	JAN	FEB	MAR	APR	MAY	JUNE	JULY	AUG	SEP	OCT	NOV	DEC	YEAR
TEMPERATURE °F:														
Normals														
-Daily Maximum		24.5	30.0	40.8	57.5	69.8	78.8	82.8	80.6	72.3	61.1	44.1	30.6	56.1
-Daily Minimum		6.7	11.0	21.5	34.1	44.2	53.8	58.3	56.3	47.8	37.8	26.0	14.1	34.3
-Monthly		15.6	20.5	31.1	45.8	57.0	66.3	70.6	68.5	60.0	49.5	35.0	22.4	45.2
Extremes														
-Record Highest	46	55	61	82	94	93	97	104	101	99	90	76	62	104
-Year		1981	1981	1981	1980	1975	1953	1976	1947	1953	1976	1964	1984	JUL 1976
-Record Lowest	46	-37	-28	-29	0	19	31	36	35	25	14	-11	-25	-37
-Year		1951	1985	1962	1982	1978	1972	1965	1968	1974	1952	1947	1983	JAN 1951
NORMAL DEGREE DAYS:														
Heating (base 65°F)		1531	1246	1048	576	273	58	12	29	161	490	897	1321	7642
Cooling (base 65°F)		0	0	0	0	25	97	185	137	14	9	0	0	467
% OF POSSIBLE SUNSHINE	39	48	51	53	51	58	64	68	66	61	54	39	39	54
MEAN SKY COVER (tenths)														
Sunrise - Sunset	37	6.7	6.6	6.9	6.8	6.5	6.1	5.7	5.7	5.7	6.0	7.2	7.2	6.4
MEAN NUMBER OF DAYS:														
Sunrise to Sunset														
-Clear	39	7.6	7.2	6.4	6.4	6.7	7.2	9.3	9.4	9.6	9.5	5.7	6.2	91.0
-Partly Cloudy	39	6.3	5.8	7.7	7.6	9.6	10.2	11.2	10.6	8.5	7.4	6.1	6.1	97.0
-Cloudy	39	17.1	15.2	16.9	16.1	14.7	12.7	10.5	11.1	11.9	14.1	18.3	18.7	177.3
Precipitation														
.01 inches or more	37	10.2	7.9	11.1	11.4	11.4	10.6	9.5	9.5	9.2	8.7	9.1	9.9	118.5
Snow,Ice pellets														
1.0 inches or more	37	2.7	2.3	2.7	0.7	0.0	0.0	0.0	0.0	0.0	0.0	1.2	3.3	12.9
Thunderstorms	37	0.2	0.3	1.9	3.6	5.2	7.4	7.4	6.8	4.7	2.0	0.9	0.4	40.9
Heavy Fog Visibility														
1/4 mile or less	39	2.3	2.0	2.7	1.3	1.4	0.8	1.2	2.2	1.7	1.9	2.1	2.9	22.3
Temperature °F														
-Maximum														
90° and above	26	0.0	0.0	0.0	0.*	0.3	2.8	4.6	2.7	0.7	0.1	0.0	0.0	11.3
32° and below	26	22.3	15.7	6.5	0.4	0.0	0.0	0.0	0.0	0.0	0.*	3.8	17.8	66.4
-Minimum														
32° and below	26	30.2	27.1	26.1	14.3	3.5	0.*	0.0	0.0	1.3	9.6	21.6	29.1	162.8
0° and below	26	12.0	7.2	1.5	0.*	0.0	0.0	0.0	0.0	0.0	0.0	0.2	6.0	26.9
AVG. STATION PRESS.(mb)	13	986.0	986.0	983.2	983.0	982.3	982.7	984.4	985.5	985.9	985.9	985.0	985.3	984.6
RELATIVE HUMIDITY (%)														
Hour 00	26	77	77	78	76	77	81	84	87	88	82	83	82	81
Hour 06 (Local Time)	26	78	80	82	81	80	83	86	91	92	87	86	83	84
Hour 12	26	69	66	63	56	54	56	56	59	61	59	68	73	62
Hour 18	26	73	70	66	57	54	56	58	62	69	69	75	77	66
PRECIPITATION (inches):														
Water Equivalent														
-Normal		1.11	1.02	2.15	3.10	3.34	3.89	3.75	3.82	3.06	2.24	1.83	1.53	30.84
-Maximum Monthly	46	2.45	2.77	5.04	7.11	6.26	9.95	10.93	9.49	9.51	5.63	5.13	3.65	10.93
-Year		1974	1953	1973	1973	1960	1978	1950	1980	1941	1984	1985	1982	JUL 1950
-Minimum Monthly	46	0.14	0.08	0.28	0.96	0.64	0.81	1.38	0.70	0.11	0.06	0.11	0.25	0.06
-Year		1981	1958	1978	1946	1981	1973	1946	1948	1979	1952	1976	1960	OCT 1952
-Maximum in 24 hrs	37	1.27	1.58	2.52	2.83	3.64	3.67	5.25	2.90	3.57	2.78	2.36	1.76	5.25
-Year		1960	1981	1973	1975	1966	1963	1950	1965	1961	1984	1985	1985	JUL 1950
Snow,Ice pellets														
-Maximum Monthly	37	26.9	20.9	25.4	17.4	0.7				T	0.9	18.3	24.6	26.9
-Year		1979	1975	1959	1973	1966				1965	1967	1985	1977	JAN 1979
-Maximum in 24 hrs	37	11.6	10.3	13.6	12.9	0.7				T	0.9	9.0	16.0	16.0
-Year		1971	1950	1971	1973	1966				1965	1967	1985	1970	DEC 1970
WIND:														
Mean Speed (mph)	39	10.5	10.4	11.2	11.4	10.1	9.1	8.1	7.9	8.7	9.6	10.7	10.2	9.8
Prevailing Direction														
through 1963		WNW	WNW	NW	NW	S	S	S	S	S	S	S	W	S
Fastest Mile														
-Direction	38	E	W	SW	SW	SW	W	NW	W	SW	SW	SE	SW	SW
-Speed (MPH)	38	68	57	70	73	77	59	72	47	52	73	56	65	77
-Year		1947	1948	1954	1947	1950	1947	1951	1955	1948	1951	1947	1949	MAY 1950
Peak Gust														
-Direction	2	NW	W	SW	W	SW	W	NW	W	NW	SW	W	SW	W
-Speed (mph)	2	43	41	51	53	47	70	44	62	64	59	46	49	70
-Date		1985	1985	1985	1984	1985	1984	1984	1985	1985	1984	1985	1984	JUN 1984

See reference Notes to this table on the following page.

TABLE 2 PRECIPITATION (inches)   MADISON, WISCONSIN

YEAR	JAN	FEB	MAR	APR	MAY	JUNE	JULY	AUG	SEP	OCT	NOV	DEC	ANNUAL
1956	0.43	1.00	2.53	3.54	5.11	3.24	4.50	5.64	1.42	0.31	2.78	1.01	31.51
1957	0.41	0.38	1.19	2.40	5.80	6.41	4.00	0.95	2.14	2.91	1.41	32.86
1958	0.52	0.08	0.38	2.73	3.93	2.16	1.69	2.06	2.44	2.50	2.29	0.31	21.09
1959	1.40	1.58	2.90	4.01	3.06	3.86	4.12	5.68	3.44	5.55	2.29	2.45	40.34
1960	2.19	1.14	1.93	4.02	6.26	2.09	6.04	6.18	3.90	3.32	1.47	0.25	38.79
1961	0.19	1.01	3.42	1.33	1.17	1.84	3.67	1.78	7.92	3.75	3.94	1.02	31.04
1962	1.12	1.39	1.73	1.43	3.01	2.09	4.39	2.04	1.31	1.68	0.34	0.90	21.43
1963	0.76	0.39	2.33	1.67	1.82	8.15	2.29	3.23	2.30	0.64	1.96	0.65	26.19
1964	0.93	0.26	2.12	3.15	3.87	2.28	4.28	2.52	1.85	0.08	1.94	0.34	23.62
1965	1.80	0.74	2.51	2.94	1.86	2.31	3.30	6.77	9.22	1.69	1.96	2.50	37.60
1966	1.07	1.36	2.11	1.54	4.31	2.91	3.24	3.83	0.51	1.65	1.28	2.62	26.43
1967	1.63	1.17	1.49	2.57	3.53	6.46	2.51	2.71	2.68	5.52	1.83	1.89	33.99
1968	0.56	0.49	0.59	4.18	2.02	7.82	2.54	2.58	4.45	0.85	1.74	2.89	30.71
1969	2.26	0.18	1.47	2.72	3.45	7.96	4.28	0.96	1.35	2.65	0.70	1.66	29.64
1970	0.44	0.16	1.17	2.53	6.09	2.26	2.42	0.97	8.82	2.65	1.06	2.12	30.69
1971	1.48	2.59	1.52	2.42	0.98	2.27	1.65	3.96	1.87	1.30	3.48	3.64	27.16
1972	0.40	0.42	2.23	2.02	2.83	1.65	3.49	7.47	5.26	2.42	0.86	1.91	30.96
1973	1.54	1.20	5.04	7.11	5.27	0.81	2.68	2.53	3.59	2.30	1.48	1.98	35.53
1974	2.45	1.17	3.43	4.24	5.77	3.86	2.69	4.60	1.08	3.18	1.79	1.80	36.06
1975	0.98	1.54	3.09	4.19	4.57	4.30	6.05	5.25	0.84	0.64	2.79	0.29	34.53
1976	0.56	1.72	4.75	4.80	1.95	1.38	1.46	1.99	0.50	1.49	0.11	0.37	21.08
1977	0.53	1.44	3.03	2.59	2.52	2.63	6.63	5.19	2.84	1.41	2.12	1.60	32.53
1978	1.03	0.24	0.28	3.50	3.96	9.95	4.54	1.63	5.44	1.11	3.05	1.71	36.44
1979	1.69	0.90	2.67	2.46	2.70	2.53	2.80	4.96	0.11	3.10	2.27	1.93	28.12
1980	1.11	0.64	0.68	2.36	2.08	3.43	2.67	9.49	7.84	1.13	1.33	1.62	34.38
1981	0.14	2.47	0.33	3.42	0.64	4.99	4.81	7.06	3.10	2.68	1.71	0.75	32.10
1982	1.42	0.17	2.11	3.26	4.34	3.40	3.47	2.67	1.42	1.46	4.21	3.65	31.58
1983	0.53	2.26	2.70	2.23	4.21	1.17	1.92	5.05	2.85	2.59	3.18	2.30	31.67
1984	0.36	1.26	1.15	3.86	3.32	7.01	1.96	1.89	2.79	5.63	1.83	2.66	33.72
1985	1.43	1.89	3.13	1.52	3.35	3.06	4.48	2.98	5.00	4.58	5.13	2.39	38.94
Record Mean	1.18	1.07	2.06	2.87	3.33	4.09	3.66	3.58	3.18	2.22	2.04	1.61	30.90

TABLE 3 AVERAGE TEMPERATURE (deg. F)   MADISON, WISCONSIN

YEAR	JAN	FEB	MAR	APR	MAY	JUNE	JULY	AUG	SEP	OCT	NOV	DEC	ANNUAL
1956	21.0	22.4	30.0	43.6	57.0	71.8	70.0	70.7	59.8	55.8	35.7	25.7	47.0
1957	13.1	27.4	33.2	47.9	55.3	67.5	73.1	69.8	59.2	47.4	35.3	27.9	46.4
1958	20.9	17.0	34.0	46.8	58.4	62.7	69.4	71.1	61.5	57.6	37.6	16.4	45.7
#1959	10.6	16.3	29.3	46.3	62.9	69.0	71.0	74.5	63.7	46.5	27.1	31.1	45.7
1960	21.1	19.7	18.4	48.1	55.2	63.8	67.8	69.3	64.3	49.2	37.5	20.8	44.6
1961	16.9	28.3	33.8	40.6	54.0	66.4	69.9	69.4	61.7	50.5	35.6	18.8	45.5
1962	12.4	17.0	29.5	45.0	61.4	66.7	67.3	69.8	57.1	51.9	35.4	20.4	44.5
1963	5.4	14.1	34.0	47.8	55.4	69.6	73.1	67.6	60.2	58.1	39.2	11.3	44.6
1964	24.9	24.9	30.8	47.5	62.6	66.6	73.1	66.9	59.6	48.9	38.6	18.8	46.9
1965	15.9	18.7	25.0	43.7	59.6	64.2	68.3	66.7	58.3	49.2	35.6	29.8	44.6
1966	10.0	21.3	35.4	42.0	50.3	66.6	71.1	66.2	57.1	47.7	36.0	23.2	43.9
1967	22.2	14.7	33.3	45.6	50.2	66.2	67.1	62.0	57.6	48.0	33.0	25.1	43.7
1968	19.6	19.0	39.2	47.8	54.3	66.2	68.9	68.8	60.1	50.7	35.6	22.6	46.1
1969	14.8	23.6	27.9	47.0	56.4	59.5	69.6	70.3	59.5	46.4	33.1	21.3	44.1
1970	9.9	20.1	31.1	47.9	58.5	66.5	70.6	68.5	60.0	51.0	36.3	22.4	45.2
1971	9.6	19.9	28.6	45.4	55.1	71.7	68.5	68.3	65.1	55.9	35.2	26.8	45.9
1972	12.7	16.5	28.7	41.3	59.2	62.8	68.3	69.2	59.4	54.8	34.6	11.7	43.0
1973	23.4	24.0	41.6	44.9	54.4	67.9	71.6	70.6	60.7	54.1	36.6	21.5	47.6
1974	19.2	18.4	33.1	48.7	54.1	64.0	72.1	66.8	57.4	50.5	37.1	26.8	45.7
1975	21.9	21.3	26.1	41.0	62.5	69.2	72.4	70.6	57.5	52.2	41.9	25.5	46.8
1976	15.7	28.4	36.5	49.3	54.3	68.3	73.4	68.8	58.0	47.3	28.1	13.2	44.8
1977	3.7	22.4	39.8	51.9	65.2	64.9	73.6	64.6	59.7	47.6	34.0	19.6	45.6
1978	10.5	12.4	29.4	44.5	57.9	65.9	69.7	69.5	63.8	47.3	33.5	21.3	43.8
1979	6.9	11.7	32.1	42.4	56.7	66.0	69.8	66.6	61.1	47.5	35.1	28.8	43.7
1980	17.3	15.7	28.0	45.5	57.8	65.3	73.4	70.3	59.9	49.2	35.4	22.4	44.6
1981	20.5	25.3	36.9	48.7	55.3	67.4	70.6	68.7	59.1	46.6	36.7	22.0	46.5
1982	8.0	19.1	30.6	41.7	60.8	59.6	70.9	66.3	59.0	50.6	34.2	28.8	44.1
1983	21.4	26.3	33.1	41.6	51.9	67.5	75.0	72.2	60.1	48.2	37.3	10.8	45.5
1984	14.8	30.2	26.7	45.6	53.4	67.5	70.2	71.3	59.3	52.0	33.9	26.4	45.9
1985	12.2	19.0	37.7	52.2	60.7	63.8	70.0	66.4	61.6	49.4	31.0	11.3	44.6
Record Mean	16.5	21.1	31.9	46.2	57.1	66.7	71.4	69.4	60.5	50.3	35.3	22.3	45.7
Max	25.4	30.4	41.3	57.6	69.4	78.9	83.7	81.6	72.5	61.7	43.9	30.5	56.4
Min	7.6	11.8	22.4	34.7	44.7	54.5	59.1	57.3	48.6	38.8	26.7	14.1	35.0

REFERENCE NOTES FOR TABLES 1, 2, 3 and 6   (MADISON, WI)

GENERAL

T - TRACE AMOUNT
BLANK ENTRIES DENOTE MISSING/UNREPORTED DATA.
# INDICATES A STATION OR INSTRUMENT RELOCATION.

SPECIFIC

TABLE 1

(a) - LENGTH OF RECORD IN YEARS. ALTHOUGH
        INDIVIDUAL MONTHS MAY BE MISSING.
  * LESS THAN .05

NORMALS — BASED ON THE 1951-1980 RECORD PERIOD.
EXTREMES — DATES ARE THE MOST RECENT OCCURRENCE.
WIND DIR. — NUMERALS SHOW TENS OF DEGREES
            CLOCKWISE FROM TRUE NORTH.
            "00" INDICATES CALM.
RESULTANT WIND DIRECTIONS ARE GIVEN TO WHOLE DEGREES.

EXCEPTIONS

TABLES 2, 3, and 6

RECORD MEANS ARE THROUGH THE CURRENT YEAR,
BEGINNING IN   1940 FOR TEMPERATURE
                 1940 FOR PRECIPITATION
                 1949 FOR SNOWFALL

TABLE 4 HEATING DEGREE DAYS Base 65 deg. F   MADISON, WISCONSIN

SEASON	JULY	AUG	SEP	OCT	NOV	DEC	JAN	FEB	MAR	APR	MAY	JUNE	TOTAL
1956-57	10	20	169	285	870	1216	1605	1047	977	512	302	47	7060
1957-58	1	17	197	535	883	1142	1360	1339	955	545	221	104	7299
1958-59	12	26	137	371	814	1499	1684	1357	1098	557	152	32	7739
#1959-60	6	0	137	566	1129	1046	1355	1305	1440	513	296	77	7870
1960-61	23	23	134	484	816	1361	1484	1018	959	726	352	62	7442
1961-62	11	18	188	444	876	1426	1630	1337	1093	601	185	54	7863
1962-63	23	11	252	414	884	1376	1850	1420	957	508	299	48	8042
1963-64	2	47	174	228	769	1660	1237	1156	1052	519	133	76	7053
1964-65	13	68	216	556	789	1517	1427	1294	1234	631	208	79	8032
1965-66	27	69	231	486	878	1086	1702	1220	911	684	454	83	7831
1966-67	11	40	252	533	862	1290	1321	1407	978	574	462	46	7776
1967-68	61	120	239	535	955	1229	1401	1327	792	510	330	76	7575
1968-69	34	66	159	460	873	1305	1548	1152	1143	535	282	197	7754
1969-70	13	9	202	579	951	1346	1705	1252	1044	521	244	73	7939
1970-71	28	18	196	431	853	1310	1718	1258	1124	582	312	22	7852
1971-72	28	21	131	293	885	1179	1616	1401	1119	705	212	117	7707
1972-73	44	42	188	587	905	1475	1279	1143	720	596	325	15	7319
1973-74	4	25	180	349	847	1342	1416	1298	979	494	347	90	7371
1974-75	1	37	253	443	829	1179	1329	1220	1198	714	150	43	7396
1975-76	18	11	236	412	687	1217	1520	1056	877	477	333	32	6876
1976-77	4	40	236	656	1102	1602	1898	1188	772	409	110	95	8112
1977-78	6	95	161	533	925	1404	1688	1466	1096	608	269	59	8310
1978-79	19	22	130	543	940	1348	1800	1489	1013	671	283	52	8310
1979-80	14	62	144	546	890	1112	1471	1424	1138	586	255	84	7726
1980-81	2	11	178	651	881	1303	1373	1107	864	482	307	30	7189
1981-82	16	27	193	566	842	1327	1765	1281	1059	688	155	172	8091
1982-83	5	66	230	444	918	1117	1346	1078	978	693	400	57	7332
1983-84	11	6	193	519	823	1678	1550	1006	1181	575	358	20	7920
1984-85	9	21	215	397	927	1191	1632	1287	839	418	155	96	7187
1985-86	12	36	198	475	1012	1661							

TABLE 5 COOLING DEGREE DAYS Base 65 deg. F   MADISON, WISCONSIN

YEAR	JAN	FEB	MAR	APR	MAY	JUNE	JULY	AUG	SEP	OCT	NOV	DEC	TOTAL
1969	0	0	0	0	25	39	161	179	45	11	0	0	460
1970	0	0	0	12	47	125	210	133	53	4	0	0	584
1971	0	0	0	0	13	229	144	131	140	20	0	0	677
1972	0	0	0	0	41	61	156	180	27	0	0	0	465
1973	0	0	0	0	2	112	215	207	58	19	0	0	613
1974	0	0	0	9	17	68	228	102	31	2	0	0	457
1975	0	0	0	0	81	176	256	190	18	21	0	0	742
1976	0	0	0	14	6	136	270	165	34	2	0	0	627
1977	0	0	0	24	123	99	278	88	10	0	0	0	622
1978	0	0	0	0	56	92	171	168	102	0	0	0	589
1979	0	0	0	0	33	88	168	115	33	13	0	0	450
1980	0	0	0	8	39	100	268	183	31	0	0	0	629
1981	0	0	0	0	13	107	198	148	19	0	0	0	485
1982	0	0	0	0	29	16	194	114	53	3	0	0	409
1983	0	0	0	0	0	138	327	237	52	6	0	0	760
1984	0	0	0	1	5	102	177	224	50	0	0	0	559
1985	0	0	0	40	29	66	175	84	102	0	0	0	496

TABLE 6 SNOWFALL (inches)   MADISON, WISCONSIN

SEASON	JULY	AUG	SEP	OCT	NOV	DEC	JAN	FEB	MAR	APR	MAY	JUNE	TOTAL
1956-57	0.0	0.0	0.0	0.0	2.0	9.4	6.6	1.1	9.0	2.6	0.0	0.0	30.7
1957-58	0.0	0.0	0.0	T	2.3	6.8	6.1	1.1	5.2	0.4	0.0	0.0	21.9
1958-59	0.0	0.0	0.0	0.0	3.2	4.8	18.9	14.6	25.4	0.4	0.0	0.0	67.3
1959-60	0.0	0.0	0.0	T	8.7	12.4	9.0	12.3	8.0	1.8	T	0.0	52.2
1960-61	0.0	0.0	0.0	0.0	2.7	0.6	2.1	0.4	13.1	3.2	0.0	0.0	22.1
1961-62	0.0	0.0	0.0	0.1	1.4	12.0	10.0	16.1	7.0	2.7	0.0	0.0	49.3
1962-63	0.0	0.0	0.0	0.2	1.0	6.4	4.1	12.8	17.4	0.3	0.0	0.0	37.2
1963-64	0.0	0.0	0.0	0.0	0.1	8.2	1.4	3.7	17.4	0.3	0.0	0.0	31.1
1964-65	0.0	0.0	T	0.5	3.7	3.5	18.3	4.7	19.4	0.1	0.7	0.0	50.9
1965-66	0.0	0.0	T	0.0	T	5.7	9.6	3.8	5.5	0.1	0.7	0.0	25.4
1966-67	0.0	0.0	0.0	T	0.1	10.1	9.4	13.9	4.5	T	T	0.0	38.0
1967-68	0.0	0.0	0.0	0.9	0.7	2.4	3.9	3.9	0.5	0.4	T	0.0	12.7
1968-69	0.0	0.0	0.0	0.0	11.4	9.7	1.7	1.7	7.0	1.9	0.0	0.0	33.4
1969-70	0.0	0.0	0.0	0.0	1.0	19.5	6.4	1.7	7.0	1.9	0.0	0.0	37.5
1970-71	0.0	0.0	0.0	0.0	0.2	20.8	21.9	3.7	20.1	0.7	0.0	0.0	67.4
1971-72	0.0	0.0	0.0	T	8.8	6.0	3.6	1.9	18.8	11.1	T	0.0	50.3
1972-73	0.0	0.0	0.0	T	1.3	16.3	1.9	6.0	14.1	6.6	0.4	0.0	44.0
1973-74	0.0	0.0	0.0	0.0	0.4	10.9	10.5	5.2	10.0	5.9	0.0	0.0	42.9
1974-75	0.0	0.0	0.0	0.0	3.0	15.4	5.2	20.9	13.9	2.0	T	0.0	60.4
1975-76	0.0	0.0	0.0	0.0	5.5	2.8	10.1	10.4	2.0	T	0.0	0.0	30.8
1976-77	0.0	0.0	0.0	T	1.1	5.8	8.7	2.6	5.8	2.3	0.0	0.0	26.3
1977-78	0.0	0.0	0.0	0.0	10.4	24.6	13.5	4.7	3.0	0.5	0.0	0.0	56.7
1978-79	0.0	0.0	0.0	0.0	6.2	23.0	26.9	8.7	4.0	7.3	0.0	0.0	76.1
1979-80	0.0	0.0	0.0	0.2	4.4	1.3	4.9	7.5	5.6	7.1	0.0	0.0	31.0
1980-81	0.0	0.0	0.0	T	3.5	9.2	2.9	9.2	1.7	0.0	0.0	0.0	26.5
1981-82	0.0	0.0	0.0	0.1	2.0	7.2	19.4	2.4	8.6	10.3	0.0	0.0	50.0
1982-83	0.0	0.0	0.0	0.0	0.3	3.3	6.5	13.0	14.1	4.2	0.0	0.0	41.4
1983-84	0.0	0.0	0.0	0.0	2.1	22.6	6.0	0.8	6.8	3.9	0.0	0.0	42.2
1984-85	0.0	0.0	0.0	0.0	0.5	15.8	19.9	7.4	8.2	1.9	0.0	0.0	53.7
1985-86	0.0	0.0	0.0	0.0	18.3	24.0							
Record Mean	0.0	0.0	T	0.1	3.4	10.7	9.6	6.9	8.8	2.4	T	0.0	41.9

See Reference Notes, relative to all above tables, on preceding page.

Milwaukee possesses a continental climate characterized by a wide range of temperatures between summer and winter. Precipitation is moderate and occurs mostly in the spring, less in the autumn, and very little in the wintertime. Rainfall is well distributed for agricultural purposes, although spring planting is sometimes delayed by wet ground and cold weather.

Milwaukee is in a region of frequently changeable weather and its climate is influenced by general easterly−moving storms which traverse the nations midsection. The most severe winter storms, which produce in excess of 10 inches of snow, develop in the southern Great Plains and move northeast across Illinois and Indiana.

Occasionally during the cold season, frigid air masses from Canada push southeast across the Great Lakes region. These arctic air masses account for the coldest winter temperatures. Very low temperatures, zero degrees or lower, most often occur in air that flows southward to the west of Lake Superior before reaching the Milwaukee area. If northwesterly wind circulation persists, repeated incursions of arctic air will result in a period of bitterly cold weather lasting several days.

Summer temperatures, which reach into the 90s but rarely exceed 100 degrees, occur with brisk southwest winds that carry hot air from the plains and lower Mississippi River Valley across the city. A combination of high temperatures and humidity occasionally develops, usually building up over a period of several days when persistent southerly winds transport moisture from the Gulf of Mexico into the area.

The Gulf is a major source of moisture for Milwaukee in all seasons, but the type of precipitation which results is dependent upon the time of year. Cold−season precipitation (rain, snow, or a mixture) is usually of relatively long duration and low intensity, and occasionally persists for two days or more, whereas in the warm season, relatively short−duration and high−intensity showery rainfall, usually lasting a few hours or less, predominates.

The Great Lakes significantly influence the local climate. Temperature extremes are modified by Lake Michigan and, to a lesser extent, the other Great Lakes. In late autumn and winter, air masses that are initially very cold often reach the city only after being tempered by passage over one or more of the lakes. Similarly, air masses that approach from the northeast in the spring and summer are cooler because of movement over the Great Lakes.

The influence of Lake Michigan is variable and occasionally dramatic, especially when the temperature of the lake water differs strongly from the air temperature. During the spring and early summer, a wind shift from a westerly to an easterly direction frequently causes a sudden 10 to 20 degree temperature drop. When the breeze off the lake is light, this effect reaches inland only a mile or two. With stronger on−shore winds, the entire city is cooled. In the winter the relatively warm water of the lake moderates the temperature during easterly wind situations. Lake−induced snows usually occur a few times each winter, but snow accumulation is rarely heavy.

Topography does not significantly affect air flow, except that lesser frictional drag over Lake Michigan causes winds to be frequently stronger along the lake shore, and often permits air masses approaching from the north to reach shore areas one hour or more before affecting inland portions of the city.

TABLE 1 — NORMALS, MEANS AND EXTREMES

MILWAUKEE, WISCONSIN

LATITUDE: 42°57'N   LONGITUDE: 87°54'W   ELEVATION: FT. GRND   672 BARO   00691   TIME ZONE: CENTRAL   WBAN: 14839

	(a)	JAN	FEB	MAR	APR	MAY	JUNE	JULY	AUG	SEP	OCT	NOV	DEC	YEAR
TEMPERATURE °F:														
Normals														
-Daily Maximum		26.0	30.1	39.2	53.5	64.8	75.0	79.8	78.4	71.2	59.9	44.7	32.0	54.6
-Daily Minimum		11.3	15.8	24.9	35.6	44.7	54.7	61.1	60.2	52.5	41.9	29.9	18.2	37.6
-Monthly		18.6	23.0	32.0	44.5	54.8	64.8	70.5	69.3	61.9	50.9	37.3	25.1	46.1
Extremes														
-Record Highest	45	62	65	81	91	92	99	101	100	98	89	77	63	101
-Year		1944	1976	1945	1980	1975	1953	1955	1955	1953	1963	1950	1982	JUL 1955
-Record Lowest	45	-26	-19	-10	12	21	33	40	44	28	18	-5	-20	-26
-Year		1982	1951	1962	1982	1966	1945	1965	1982	1974	1981	1950	1983	JAN 1982
NORMAL DEGREE DAYS:														
Heating (base 65°F)		1435	1176	1020	612	334	84	11	25	117	444	831	1237	7326
Cooling (base 65°F)		0	0	0	0	18	81	182	158	24	7	0	0	470
% OF POSSIBLE SUNSHINE	45	44	47	50	53	59	64	70	66	59	54	40	38	54
MEAN SKY COVER (tenths)														
Sunrise - Sunset	45	6.8	6.8	6.9	6.7	6.3	6.0	5.3	5.4	5.6	5.8	7.2	7.2	6.3
MEAN NUMBER OF DAYS:														
Sunrise to Sunset														
-Clear	45	7.3	6.6	6.0	6.4	7.0	7.6	10.2	10.3	9.6	9.4	5.7	6.2	92.2
-Partly Cloudy	45	6.2	5.9	7.9	8.0	10.1	10.4	11.3	10.6	9.2	8.6	6.0	5.9	99.9
-Cloudy	45	17.6	15.7	17.1	15.6	14.0	12.0	9.5	10.1	11.2	13.0	18.3	18.9	173.2
Precipitation														
.01 inches or more	45	11.2	9.6	11.9	12.0	12.0	10.9	9.5	9.1	9.0	8.8	10.0	11.2	125.1
Snow, Ice pellets														
1.0 inches or more	45	3.7	2.6	2.7	0.5	0.0	0.0	0.0	0.0	0.0	0.*	1.0	3.3	13.8
Thunderstorms	45	0.3	0.4	1.4	3.5	4.5	6.6	6.4	5.7	3.9	1.6	1.0	0.4	35.8
Heavy Fog Visibility														
1/4 mile or less	45	2.1	2.0	3.1	3.0	3.2	2.5	1.1	1.7	1.2	2.2	2.3	2.2	26.7
Temperature °F														
-Maximum														
90° and above	25	0.0	0.0	0.0	0.*	0.1	1.6	3.7	2.1	0.6	0.0	0.0	0.0	8.1
32° and below	25	21.5	16.0	6.8	0.6	0.0	0.0	0.0	0.0	0.0	0.0	2.6	16.1	63.6
-Minimum														
32° and below	25	30.0	26.2	23.7	10.4	1.2	0.0	0.0	0.0	0.1	4.2	18.5	28.2	142.4
0° and below	25	8.4	3.8	0.2	0.0	0.0	0.0	0.0	0.0	0.0	0.0	0.1	3.4	15.9
AVG. STATION PRESS.(mb)	13	992.2	992.5	989.8	989.7	989.0	989.1	990.6	991.8	992.3	992.5	991.5	991.7	991.1
RELATIVE HUMIDITY (%)														
Hour 00	25	74	74	75	75	75	78	80	84	83	79	78	78	78
Hour 06	25	75	76	79	78	78	80	82	87	87	82	81	80	80
Hour 12 (Local Time)	25	68	67	65	61	60	61	60	63	63	63	67	72	64
Hour 18	25	71	70	68	64	62	62	63	68	71	70	74	75	68
PRECIPITATION (inches):														
Water Equivalent														
-Normal		1.64	1.33	2.58	3.37	2.66	3.59	3.54	3.09	2.88	2.25	1.98	2.03	30.94
-Maximum Monthly	45	4.04	3.10	6.93	7.31	5.83	8.28	7.66	7.07	9.87	6.42	7.11	4.34	9.87
-Year		1960	1974	1976	1973	1983	1954	1964	1960	1941	1959	1985	1971	SEP 1941
-Minimum Monthly	45	0.31	0.05	0.31	0.81	0.90	0.85	0.95	0.46	0.02	0.15	0.62	0.29	0.02
-Year		1981	1969	1968	1942	1977	1965	1946	1948	1979	1956	1949	1976	SEP 1979
-Maximum in 24 hrs	45	1.73	1.67	2.57	3.11	3.11	3.13	4.35	4.05	5.28	2.60	2.18	2.24	5.28
-Year		1985	1960	1960	1976	1978	1950	1959	1953	1941	1959	1943	1982	SEP 1941
Snow, Ice pellets														
-Maximum Monthly	45	33.6	42.0	26.7	15.8	0.4				T	4.0	16.1	27.9	42.0
-Year		1979	1974	1965	1973	1960				1960	1976	1977	1978	FEB 1974
-Maximum in 24 hrs	45	12.8	16.7	11.2	11.6	0.4				T	4.0	10.6	12.4	16.7
-Year		1962	1960	1961	1973	1960				1960	1976	1977	1978	FEB 1960
WIND:														
Mean Speed (mph)	45	12.8	12.5	13.1	13.0	11.7	10.5	9.6	9.4	10.6	11.4	12.5	12.4	11.6
Prevailing Direction														
through 1963		WNW	WNW	WNW	NNE	NNE	NNE	SW	SW	SSW	SSW	WNW	WNW	WNW
Fastest Mile														
-Direction	41	W	NE	SW	SW	SW	W	W	W	S	S	W	SW	SW
-Speed (MPH)	41	62	58	73	66	72	62	59	50	62	60	72	62	73
-Year		1950	1960	1954	1947	1950	1971	1952	1974	1941	1949	1955	1948	MAR 1954
Peak Gust														
-Direction	2	NE	SW	SW	SW	SW	SW	NW	N	SW	SW	W	SW	NW
-Speed (mph)	2	47	43	55	64	54	53	81	36	48	51	52	51	81
-Date		1985	1985	1985	1984	1985	1984	1984	1984	1985	1984	1985	1984	JUL 1984

See Reference Notes to this table on the following page.

TABLE 2 PRECIPITATION (inches) MILWAUKEE, WISCONSIN

YEAR	JAN	FEB	MAR	APR	MAY	JUNE	JULY	AUG	SEP	OCT	NOV	DEC	ANNUAL
1956	0.57	1.43	2.36	4.14	4.55	3.87	5.37	2.96	0.30	0.15	1.62	1.03	28.35
1957	0.88	0.96	1.59	2.70	3.82	4.01	1.50	2.03	0.88	1.34	2.88	2.36	24.95
1958	1.41	0.15	0.46	1.84	2.07	1.71	1.02	1.71	2.85	3.37	3.24	0.34	20.17
1959	2.48	1.98	3.03	3.29	1.28	1.67	6.82	3.47	2.31	6.42	2.08	2.85	37.68
1960	4.04	3.05	3.80	2.92	4.27	3.28	3.50	7.07	3.25	2.12	2.12	0.35	40.71
1961	0.31	1.22	3.80	3.89	1.25	1.53	2.91	2.35	9.41	2.75	2.37	1.02	32.81
1962	2.48	2.04	1.69	1.49	2.17	1.33	3.74	1.98	1.49	2.14	0.81	0.55	21.91
1963	0.66	0.42	2.20	2.54	1.95	1.50	2.36	2.48	1.78	0.34	2.17	0.70	19.10
1964	1.18	0.41	3.05	3.81	2.57	1.70	7.66	2.62	1.74	0.17	2.29	0.98	28.18
1965	3.33	1.04	3.61	3.47	2.12	0.85	2.64	6.15	6.85	2.68	2.02	3.73	38.49
1966	2.06	1.27	3.61	2.67	2.00	1.68	3.32	3.27	0.48	1.76	2.70	2.31	27.13
1967	1.49	1.81	1.35	2.70	1.80	7.38	1.35	1.23	1.69	2.70	1.52	1.33	25.85
1968	0.98	0.56	0.31	2.90	3.28	7.79	3.59	2.59	3.36	0.94	2.56	2.65	31.51
1969	1.83	0.05	1.05	3.42	3.05	7.53	6.61	0.53	2.18	4.48	1.14	1.18	33.05
1970	0.41	0.13	1.62	2.71	3.41	3.92	1.93	0.64	6.94	2.09	2.03	3.02	28.85
1971	1.37	2.50	2.83	1.31	0.90	2.67	2.60	2.28	1.30	1.90	2.45	4.34	26.45
1972	0.75	0.86	2.57	2.76	2.33	3.33	4.60	4.82	7.57	3.28	1.34	2.47	36.68
1973	1.12	1.51	2.86	7.31	3.39	1.96	1.55	0.95	4.50	2.97	1.83	3.80	33.75
1974	3.61	3.10	4.29	3.83	4.10	3.48	3.51	2.54	0.50	1.96	1.86	2.10	34.88
1975	2.25	2.53	3.01	4.08	2.01	3.99	1.14	3.89	1.00	0.72	2.83	1.70	29.15
1976	1.16	2.65	6.93	5.01	3.77	2.27	2.12	2.05	1.70	2.82	0.65	0.29	31.42
1977	0.90	0.59	4.56	2.09	0.90	5.78	5.99	3.82	4.11	2.02	2.56	3.27	36.59
1978	2.03	0.55	1.08	4.41	4.66	4.52	5.98	3.43	6.81	2.22	2.13	2.92	40.74
1979	3.00	0.97	4.17	5.43	1.82	2.84	1.06	4.85	0.02	1.77	2.67	2.27	30.87
1980	1.65	1.75	0.77	4.02	1.81	4.67	3.39	5.06	3.57	1.63	1.57	3.52	33.41
1981	0.31	2.88	0.51	4.87	3.05	2.39	4.35	4.26	5.47	2.71	2.05	1.03	33.88
1982	2.92	0.29	3.20	4.47	2.76	3.06	3.88	3.33	0.64	3.17	4.74	4.10	36.56
1983	0.75	2.23	4.12	4.66	5.83	1.41	1.34	4.70	2.79	2.65	4.10	2.89	37.47
1984	0.79	1.20	2.17	5.04	4.21	4.07	3.39	2.93	2.51	5.30	3.74	4.25	39.60
1985	1.94	2.34	4.11	1.93	2.73	1.27	2.18	2.23	3.44	5.39	7.11	2.62	37.29
Record Mean	1.81	1.57	2.51	2.90	3.18	3.50	2.94	2.87	3.13	2.33	2.09	1.85	30.67

TABLE 3 AVERAGE TEMPERATURE (deg. F) MILWAUKEE, WISCONSIN

YEAR	JAN	FEB	MAR	APR	MAY	JUNE	JULY	AUG	SEP	OCT	NOV	DEC	ANNUAL
#1956	24.9	25.0	31.4	43.0	53.7	68.7	67.9	70.1	59.3	55.8	36.7	28.1	47.0
1957	14.9	27.8	32.4	45.0	52.7	65.0	70.9	69.2	59.5	48.6	36.5	29.7	46.0
1958	21.9	16.8	33.3	45.7	55.9	60.2	69.2	70.8	62.2	53.6	39.2	18.7	45.6
1959	13.4	19.3	30.9	45.0	59.5	67.3	70.2	73.8	64.7	47.7	29.5	32.5	46.2
1960	24.1	23.4	21.5	48.0	51.8	61.4	67.1	67.8	63.4	48.8	39.2	22.5	44.9
#1961	19.4	29.8	35.2	41.2	50.8	65.3	69.9	70.3	64.8	51.2	37.6	22.8	46.5
1962	14.5	21.1	30.3	44.5	59.2	63.9	66.7	68.9	58.0	52.6	37.9	21.5	44.9
1963	8.7	15.9	35.2	45.8	52.3	65.6	70.8	67.1	61.2	58.6	41.7	13.2	44.7
1964	26.1	25.3	31.7	44.8	60.3	66.5	72.3	67.7	61.2	48.2	40.4	23.2	47.3
1965	19.5	21.2	25.3	42.7	58.6	63.1	68.9	66.8	61.7	50.8	38.5	32.9	45.8
1966	13.9	22.3	35.4	42.3	49.6	67.0	73.5	66.8	59.1	48.9	38.0	24.5	45.1
1967	24.5	17.5	33.2	44.9	50.2	66.5	68.6	66.0	60.8	50.4	35.2	28.9	45.6
1968	21.8	20.7	40.3	47.6	53.2	66.3	69.5	71.4	63.5	52.7	38.1	25.6	47.5
1969	18.6	27.3	29.8	44.9	55.3	67.8	67.8	70.9	62.0	47.5	34.3	25.1	45.2
1970	13.3	22.6	30.5	46.2	56.0	65.3	73.3	71.8	62.0	52.5	37.3	24.3	46.2
1971	12.7	21.6	28.6	40.9	51.1	67.0	68.0	67.0	64.9	55.5	37.3	29.5	45.4
1972	15.9	19.8	27.7	39.4	55.0	61.6	69.3	69.1	61.3	47.5	36.2	20.2	43.6
1973	24.7	25.4	39.9	42.7	51.0	69.2	71.4	72.5	63.9	54.6	38.5	25.5	48.3
1974	21.6	22.9	33.8	46.1	50.6	62.4	71.5	67.3	58.0	49.9	38.6	29.1	46.0
1975	24.0	23.5	28.6	39.7	57.1	65.7	71.7	71.2	58.5	54.1	44.5	27.9	47.0
1976	18.5	31.1	38.6	48.7	52.6	68.2	72.6	69.9	62.6	45.9	29.5	16.3	46.2
1977	8.3	23.7	39.2	48.9	61.3	62.5	73.1	67.2	61.9	49.0	37.1	22.8	46.2
1978	15.4	16.4	29.8	42.5	55.2	65.2	68.6	69.9	65.8	49.5	38.2	23.8	45.0
1979	11.6	15.1	33.2	42.1	54.9	64.7	70.9	68.8	64.7	51.1	38.2	31.4	45.6
1980	20.7	20.3	30.5	45.3	57.2	61.3	71.2	69.7	61.1	45.7	37.7	24.4	45.4
1981	18.9	25.3	35.6	46.5	51.5	65.2	67.3	67.8	59.1	45.8	37.4	24.2	45.4
1982	9.7	19.4	31.5	41.2	58.5	59.8	71.1	67.3	60.8	52.7	38.0	33.2	45.3
1983	26.4	29.4	35.0	41.7	50.2	66.3	76.2	74.4	62.9	52.0	39.9	14.4	47.4
1984	18.9	33.4	29.2	45.5	54.9	68.7	71.7	73.3	60.9	52.7	37.7	29.1	48.0
1985	15.2	21.3	37.9	50.8	58.8	63.8	72.4	68.4	64.3	50.7	36.7	15.7	46.3
Record Mean	20.5	23.3	32.7	44.3	54.4	64.5	70.8	69.7	62.5	51.3	37.6	25.8	46.5
Max	27.7	30.4	39.7	52.3	63.5	73.6	79.4	77.8	70.7	59.2	44.4	32.4	54.3
Min	13.2	16.1	25.7	36.3	45.3	55.4	62.2	61.5	54.3	43.4	30.7	19.2	38.6

REFERENCE NOTES FOR TABLES 1, 2, 3 and 6   (MILWAUKEE, WI)

GENERAL

T - TRACE AMOUNT
BLANK ENTRIES DENOTE MISSING/UNREPORTED DATA.
# INDICATES A STATION OR INSTRUMENT RELOCATION.

SPECIFIC

TABLE 1

(a) - LENGTH OF RECORD IN YEARS. ALTHOUGH
INDIVIDUAL MONTHS MAY BE MISSING.
* LESS THAN .05

NORMALS — BASED ON THE 1951-1980 RECORD PERIOD.
EXTREMES — DATES ARE THE MOST RECENT OCCURRENCE.
WIND DIR. — NUMERALS SHOW TENS OF DEGREES
CLOCKWISE FROM TRUE NORTH.
"00" INDICATES CALM.
RESULTANT WIND DIRECTIONS ARE GIVEN TO WHOLE DEGREES.

EXCEPTIONS

TABLE 1

1. FASTEST MILE WINDS ARE THROUGH JUNE 1982.

TABLES 2, 3, and 6

RECORD MEANS ARE THROUGH THE CURRENT YEAR,
BEGINNING IN   1875 FOR TEMPERATURE
1871 FOR PRECIPITATION
1941 FOR SNOWFALL

TABLE 4 HEATING DEGREE DAYS Base 65 deg. F   MILWAUKEE, WISCONSIN

SEASON	JULY	AUG	SEP	OCT	NOV	DEC	JAN	FEB	MAR	APR	MAY	JUNE	TOTAL
1956-57	27	18	188	287	840	1139	1544	1036	1006	598	381	88	7152
1957-58	8	8	193	503	848	1087	1330	1346	977	575	291	165	7331
1958-59	13	16	133	348	768	1429	1594	1273	1049	592	215	73	7503
1959-60	13	0	117	529	1059	999	1259	1200	1342	515	402	132	7567
#1960-61	40	20	142	494	766	1312	1407	978	918	708	434	86	7305
1961-62	18	9	134	422	815	1303	1561	1224	1067	623	254	119	7549
1962-63	24	22	227	395	804	1749	1749	915	1372	570	388	82	7890
1963-64	20	38	135	217	692	1601	1199	1141	1026	598	192	105	6964
1964-65	7	47	177	515	730	1290	1404	1222	1226	664	232	123	7637
1965-66	25	51	149	438	793	987	1579	1189	915	674	473	88	7361
1966-67	4	41	198	496	804	1249	1249	978	978	596	458	47	7445
1967-68	46	53	164	460	888	1112	1333	1277	758	521	363	74	7049
1968-69	31	23	82	403	799	1214	1434	1087	1053	596	317	215	7254
1969-70	34	5	143	539	913	1228	1600	1180	1062	561	301	104	7670
1970-71	7	7	145	383	823	1259	1615	1211	1122	716	421	65	7774
1971-72	20	37	119	308	824	1097	1518	1305	1149	758	305	139	7579
1972-73	40	32	133	534	859	1381	1242	1101	769	659	426	6	7182
1973-74	10	5	111	324	788	1218	1340	1173	959	560	448	106	7042
1974-75	0	20	237	461	786	1103	1260	1157	1122	814	267	69	7296
1975-76	17	4	203	353	610	1144	1438	978	813	507	382	43	6492
1976-77	2	21	124	589	1056	1504	1754	1152	790	490	173	151	7806
1977-78	8	47	106	485	827	1302	1531	1356	1086	667	335	83	7833
1978-79	21	5	76	473	796	1273	1654	1391	980	681	322	91	7763
1979-80	20	25	70	436	797	1036	1368	1063	1063	594	259	154	7112
1980-81	8	9	140	590	812	1250	1423	1106	905	548	417	69	7277
1981-82	44	21	187	590	820	1257	1712	1272	1032	707	215	172	8029
1982-83	3	44	170	381	802	983	1186	990	925	692	453	81	6710
1983-84	10	0	148	405	748	1565	1424	910	1103	579	318	35	7245
1984-85	3	7	179	373	812	1103	1542	1215	831	461	222	96	6844
1985-86	2	13	139	436	843	1523							

TABLE 5 COOLING DEGREE DAYS Base 65 deg. F   MILWAUKEE, WISCONSIN

YEAR	JAN	FEB	MAR	APR	MAY	JUNE	JULY	AUG	SEP	OCT	NOV	DEC	TOTAL
1969	0	0	0	0	24	38	126	197	59	5	0	0	449
1970	0	0	0	4	29	119	270	227	60	4	0	0	713
1971	0	0	0	0	0	148	120	105	123	20	0	0	516
1972	0	0	0	0	3	42	180	166	26	0	0	0	417
1973	0	0	0	0	0	140	216	247	84	6	0	0	693
1974	0	0	0	3	6	36	210	98	32	1	0	0	386
1975	0	0	0	0	30	98	230	203	16	21	0	0	598
1976	0	0	0	24	5	144	247	181	62	4	0	0	667
1977	0	0	0	12	65	81	264	122	20	0	0	0	564
1978	0	0	0	0	40	97	138	164	109	0	0	0	548
1979	0	0	0	0	16	87	209	147	68	11	0	0	538
1980	0	0	0	9	25	50	207	164	29	0	0	0	484
1981	0	0	0	2	3	84	121	112	16	0	0	0	338
1982	0	0	0	0	21	24	199	121	51	5	0	0	421
1983	0	0	0	0	0	127	364	299	92	9	0	0	891
1984	0	0	0	1	11	152	216	270	63	3	0	0	716
1985	0	0	0	42	35	68	240	127	127	0	0	0	639

TABLE 6 SNOWFALL (inches)   MILWAUKEE, WISCONSIN

SEASON	JULY	AUG	SEP	OCT	NOV	DEC	JAN	FEB	MAR	APR	MAY	JUNE	TOTAL
1956-57	0.0	0.0	0.0	0.0	0.3	10.6	11.2	3.0	9.0	0.3	0.0	0.0	34.4
1957-58	0.0	0.0	0.0	T	0.8	8.0	16.7	1.6	4.8	T	0.0	0.0	31.9
1958-59	0.0	0.0	0.0	0.0	2.1	6.9	27.5	10.8	14.5	1.5	0.0	0.0	63.3
1959-60	0.0	0.0	0.0	T	9.3	14.3	19.4	34.0	14.3	1.6	0.4	0.0	93.3
1960-61	0.0	0.0	T	0.2	2.8	2.3	3.9	2.4	14.3	7.0	0.0	0.0	32.9
1961-62	0.0	0.0	0.0	T	2.4	7.7	22.1	22.2	11.4	4.0	0.0	0.0	69.8
1962-63	0.0	0.0	0.0	0.1	0.9	6.5	8.1	6.4	7.1	T	T	0.0	29.1
1963-64	0.0	0.0	0.0	0.0	T	12.8	3.8	5.7	19.8	T	0.0	0.0	42.1
1964-65	0.0	0.0	0.0	T	1.4	8.1	23.6	10.1	26.7	4.1	0.0	0.0	74.0
1965-66	0.0	0.0	0.0	T	T	14.5	24.6	7.7	2.8	1.3	0.1	0.0	51.0
1966-67	0.0	0.0	0.0	0.0	2.0	9.9	13.1	27.1	7.4	T	T	0.0	59.5
1967-68	0.0	0.0	0.0	0.8	0.4	1.2	4.6	3.5	1.2	0.4	0.0	0.0	12.1
1968-69	0.0	0.0	0.0	0.3	11.6	0.7	11.1	0.7	6.2	0.0	T	0.0	29.9
1969-70	0.0	0.0	0.0	T	0.7	14.9	6.0	2.0	10.7	5.2	0.0	0.0	39.5
1970-71	0.0	0.0	0.0	0.0	0.6	19.6	15.8	2.5	18.1	0.7	T	0.0	57.3
1971-72	0.0	0.0	0.0	0.0	6.1	2.7	6.8	10.2	14.9	1.2	0.0	0.0	41.9
1972-73	0.0	0.0	0.0	T	3.4	13.7	0.2	9.9	1.8	15.8	T	0.0	44.8
1973-74	0.0	0.0	0.0	0.0	T	19.6	14.2	42.0	7.4	T	0.0	0.0	83.2
1974-75	0.0	0.0	0.0	T	2.0	9.1	3.5	12.2	15.1	10.4	0.0	0.0	52.3
1975-76	0.0	0.0	0.0	T	8.4	12.2	14.8	7.6	2.1	0.1	T	0.0	45.2
1976-77	0.0	0.0	0.0	4.0	3.6	5.3	15.6	5.6	12.4	2.1	0.0	0.0	48.6
1977-78	0.0	0.0	0.0	T	16.1	20.8	25.7	13.3	4.8	T	0.0	0.0	80.7
1978-79	0.0	0.0	0.0	0.0	5.3	27.9	33.6	9.1	6.2	0.8	0.0	0.0	82.9
1979-80	0.0	0.0	0.0	T	2.1	0.6	11.6	22.8	6.3	3.6	T	0.0	47.0
1980-81	0.0	0.0	0.0	T	2.3	17.5	4.9	15.7	1.5	T	0.0	0.0	41.9
1981-82	0.0	0.0	0.0	T	2.0	8.3	29.2	3.0	13.0	11.7	0.0	0.0	67.2
1982-83	0.0	0.0	0.0	T	0.4	3.1	6.3	13.8	1.0	1.0	0.0	0.0	38.1
1983-84	0.0	0.0	0.0	0.0	0.3	13.3	9.6	1.2	8.2	0.5	T	0.0	33.1
1984-85	0.0	0.0	0.0	0.0	T	20.8	15.3	9.0	2.5	0.0	0.0	0.0	66.6
1985-86	0.0	0.0	0.0	0.0	3.5	13.5							
Record Mean	0.0	0.0	T	0.1	2.9	10.6	13.2	9.4	9.2	1.9	T	0.0	47.4

See Reference Notes, relative to all above tables, on preceding page.

Casper is located in the central portion of Wyoming in the North Platte River Valley at an elevation of about 5,300 feet. The country immediately surrounding Casper is mostly rolling and hilly with considerable flat prairie land in each direction except toward the south where Casper Mountain rises some 3,500 feet above the valley floor. The prairie land is used mainly for grazing.

The National Weather Service Office is located at Natrona County International Airport, some 8 miles west-northwest of the Casper Post Office and about 200 feet higher in elevation.

The climate of Casper is semi-arid. Most of the air masses reaching this area move in from the Pacific and the mountains to the west are effective moisture barriers. About 70 percent of the annual precipitation occurs during the growing season of late spring and summer, mostly in thunderstorms. Monthly snowfall amounts are unusually uniform from November through February, and a bit heavier in March and April. Snow has occurred as early in the season as September and as late as early June.

Casper experiences large diurnal and annual temperature ranges. This is due to the advent of both warm and cold air masses and the relatively high elevation which permits rapid incoming and outgoing radiation. The mean daily temperature averages about 71 degrees in summer and 22 degrees in winter. Temperatures during winter months average a few degrees higher and summer temperatures average several degrees cooler than locations in the Missouri Valley to the east.

Windy days are quite frequent during winter and spring months. Usually the stronger winds are from the southwest and this tends to raise the temperature because the air is moving downslope.

Based on the 1951-1980 period, the average first occurrence of 32 degrees Fahrenheit in the fall is September 22 and the average last occurrence in the spring is May 22.

TABLE 1   NORMALS, MEANS AND EXTREMES

CASPER, WYOMING

LATITUDE: 42°55'N   LONGITUDE: 106°28'W   ELEVATION: FT. GRND 5338 BARO 05323   TIME ZONE: MOUNTAIN   WBAN: 24089

	(a)	JAN	FEB	MAR	APR	MAY	JUNE	JULY	AUG	SEP	OCT	NOV	DEC	YEAR
TEMPERATURE °F:														
Normals														
–Daily Maximum		32.5	37.4	43.4	54.9	66.2	78.1	87.1	84.8	74.2	61.0	43.9	35.6	58.3
–Daily Minimum		11.9	16.3	20.2	29.3	38.9	47.6	54.7	52.8	42.5	33.2	21.9	15.7	32.1
–Monthly		22.2	26.9	31.8	42.1	52.5	62.9	70.9	68.8	58.4	47.1	32.9	25.6	45.2
Extremes														
–Record Highest	35	60	68	73	81	92	101	104	102	96	85	71	61	104
–Year		1971	1982	1966	1980	1984	1970	1954	1979	1983	1957	1983	1980	JUL 1954
–Record Lowest	35	–40	–27	–21	–4	16	28	30	33	16	–3	–21	–39	–40
–Year		1972	1982	1965	1966	1953	1969	1972	1977	1983	1971	1985	1983	JAN 1972
NORMAL DEGREE DAYS:														
Heating (base 65°F)		1327	1067	1026	687	384	131	16	31	240	552	963	1218	7642
Cooling (base 65°F)		0	0	0	0	0	68	199	148	42	0	0	0	457
% OF POSSIBLE SUNSHINE														
MEAN SKY COVER (tenths)														
Sunrise – Sunset	35	6.7	6.6	6.8	6.8	6.7	5.4	4.3	4.5	4.6	5.3	6.3	6.4	5.9
MEAN NUMBER OF DAYS:														
Sunrise to Sunset														
–Clear	35	6.6	6.0	6.0	5.5	5.2	9.6	13.8	13.3	13.3	11.5	7.1	7.6	105.5
–Partly Cloudy	35	7.6	8.4	8.7	8.8	10.8	10.9	11.0	10.6	8.6	7.9	8.7	8.1	110.2
–Cloudy	35	16.8	13.9	16.3	15.7	15.0	9.5	6.1	7.1	8.1	11.6	14.2	15.3	149.5
Precipitation														
.01 inches or more	35	7.5	7.9	9.5	10.3	10.8	8.4	7.9	5.4	6.4	6.5	6.8	7.5	95.1
Snow, Ice pellets														
1.0 inches or more	35	3.6	3.4	4.5	3.7	1.0	0.1	0.0	0.0	0.4	1.8	3.1	3.1	24.7
Thunderstorms	35	0.0	0.*	0.2	1.2	6.1	8.1	9.0	6.5	2.8	0.5	0.1	0.0	34.5
Heavy Fog Visibility														
1/4 mile or less	35	0.7	0.7	1.0	1.5	1.0	0.5	0.4	0.3	0.7	0.9	0.9	0.6	9.1
Temperature °F														
–Maximum														
90° and above	21	0.0	0.0	0.0	0.0	0.1	3.2	12.6	10.8	1.4	0.0	0.0	0.0	28.0
32° and below	21	13.8	8.5	4.7	1.3	0.1	0.0	0.0	0.0	0.2	0.8	6.1	13.2	48.7
–Minimum														
32° and below	21	29.2	26.5	27.4	19.2	7.4	0.4	0.*	0.0	4.5	15.9	25.2	28.6	184.5
0° and below	21	7.5	3.3	1.6	0.2	0.0	0.0	0.0	0.0	0.0	0.1	2.1	5.6	20.4
AVG. STATION PRESS. (mb)	13	835.8	835.7	832.8	834.7	835.4	837.3	839.9	839.7	839.4	838.4	836.3	835.8	836.8
RELATIVE HUMIDITY (%)														
Hour 05	21	69	69	73	75	78	76	70	67	67	68	70	69	71
Hour 11	21	60	58	53	48	45	38	32	32	36	45	54	59	47
Hour 17 (Local Time)	21	62	56	48	42	41	33	26	25	30	41	56	62	44
Hour 23	21	70	69	70	69	69	63	55	53	58	64	67	68	65
PRECIPITATION (inches):														
Water Equivalent														
–Normal		0.50	0.56	0.99	1.51	2.13	1.24	1.06	0.63	0.76	0.88	0.66	0.51	11.43
–Maximum Monthly	35	1.19	1.01	2.43	3.92	6.46	4.15	3.05	2.66	3.40	2.49	2.72	3.71	6.46
–Year		1984	1955	1954	1974	1978	1982	1951	1979	1982	1962	1983	1982	MAY 1978
–Minimum Monthly	35	T	0.15	0.25	0.20	0.30	0.03	0.11	0.02	0.07	T	0.07	0.03	T
–Year		1952	1957	1953	1952	1966	1956	1971	1950	1956	1965	1965	1952	OCT 1965
–Maximum in 24 hrs	35	0.53	0.42	1.00	3.00	2.61	2.34	2.07	1.74	2.01	2.49	1.21	1.64	3.00
–Year		1972	1977	1958	1974	1978	1982	1983	1979	1973	1962	1983	1982	APR 1974
Snow, Ice pellets														
–Maximum Monthly	35	22.1	23.8	36.2	56.3	24.6	3.0		T	11.5	13.1	37.1	62.8	62.8
–Year		1980	1952	1975	1973	1978	1969		1964	1982	1971	1983	1982	DEC 1982
–Maximum in 24 hrs	35	9.7	10.4	14.6	16.5	14.1	3.0		T	6.8	8.2	14.3	31.1	31.1
–Year		1972	1952	1954	1973	1950	1969		1964	1982	1970	1983	1982	DEC 1982
WIND:														
Mean Speed (mph)	35	16.4	15.3	14.0	12.8	11.7	11.1	10.1	10.5	11.1	12.2	14.4	16.2	13.0
Prevailing Direction														
through 1963		SW	SW	SW	WSW	WSW	WSW	WSW	SW	WSW	SW	SW	SW	SW
Fastest Obs. 1 Min.														
–Direction (!!!)	32	20	23	25	25	32	36	25	25	32	25	25	20	25
–Speed (MPH)	32	58	58	81	54	58	52	52	50	53	55	49	63	81
–Year		1954	1957	1956	1967	1959	1959	1974	1954	1965	1954	1970	1955	MAR 1956
Peak Gust														
–Direction (!!!)	2	SW	SW	SW	NW	SW	SW	SW	27	SW	SW	SW	SW	SW
–Speed (mph)	2	53	62	53	60	64	59	52	51	63	62	60	66	66
–Date		1984	1985	1985	1985	1985	1985	1985	1984	1985	1985	1984	1984	DEC 1984

See Reference Notes to this table on the following pages.

TABLE 2 — PRECIPITATION (inches)   CASPER, WYOMING

YEAR	JAN	FEB	MAR	APR	MAY	JUNE	JULY	AUG	SEP	OCT	NOV	DEC	ANNUAL
1956	0.44	0.45	0.71	1.56	1.97	0.03	0.76	0.77	0.28	1.30		0.59	8.93
1957	0.42	0.15	0.91	2.45	3.85	1.93	1.00	1.13	1.08	2.00	0.77	0.43	16.12
#1958	0.11	0.38	2.27	1.95	0.78	1.32	0.96	1.00	0.34	0.02	1.06	0.23	10.42
1959	0.58	0.61	1.32	0.74	1.99	1.79	0.27	0.23	1.99	0.71	0.27	0.25	10.75
1960	0.71	0.85	0.42	1.06	0.56	0.40	0.59	0.57	0.41	0.27	1.03	0.47	7.34
1961	0.04	0.51	1.16	0.95	0.71	0.54	1.21	0.08	2.46	1.12	0.56	0.62	9.96
1962	0.85	0.44	0.55	0.93	3.80	1.75	1.24	0.25	1.14	2.49	0.34	0.23	14.01
1963	0.38	0.49	0.35	2.07	0.80	0.94	1.20	0.39	0.33	0.33	0.31	0.31	7.90
1964	0.75	0.72	0.66	3.09	2.17	0.59	0.73	0.47	0.34	0.23	0.66	0.47	10.88
1965	0.78	0.37	0.52	0.96	2.64	2.47	1.16	0.21	2.07	T	0.07	0.87	12.12
1966	0.30	0.43	0.49	1.29	0.30	1.12	0.46	1.03	0.39	1.33	0.43	0.57	8.14
1967	0.58	0.65	0.66	1.24	1.62	3.75	1.46	0.41	1.67	1.37	1.09	0.82	15.32
1968	0.33	0.46	0.92	1.18	3.35	1.86	1.02	0.61	0.30	0.58	0.56	0.71	11.88
1969	0.08	0.55	1.10	1.52	0.85	2.39	0.70	0.78	0.16	1.43	0.96	0.16	10.68
1970	0.59	0.27	1.53	1.57	2.03	2.03	0.73	0.16	0.21	1.01	1.01	0.89	12.03
1971	0.19	0.80	1.17	3.40	5.59	0.17	0.11	0.11	0.74	1.73	0.61	0.34	14.96
1972	0.99	0.41	1.41	1.44	0.76	0.83	1.04	1.52	0.37	2.04	0.66	0.41	11.88
1973	0.63	0.57	0.52	3.86	1.02	0.66	2.26	0.34	3.28	0.60	0.80	0.55	15.09
1974	0.69	0.42	0.37	3.92	0.62	0.60	0.75	1.00	0.83	1.15	0.46	0.22	11.03
1975	0.51	0.51	2.01	0.97	3.88	0.94	0.63	0.12	0.11	1.01	0.51	0.72	11.92
1976	0.44	0.89	0.50	1.52	2.69	1.66	1.07	0.31	1.15	1.21	0.79	0.38	11.91
1977	0.35	0.69	1.05	1.07	1.93	0.19	1.88	0.34	0.30	0.71	0.94	1.20	10.01
1978	0.70	0.75	0.93	1.05	6.46	1.40	2.62	0.93	0.26	0.40	0.75	0.36	17.64
1979	0.81	0.39	1.31	0.83	2.25	1.31	1.22	2.66	0.18	0.50	0.75	0.36	12.57
1980	0.81	0.63	1.19	0.35	2.82	0.10	0.85	0.65	0.10	0.64	0.74	0.37	9.25
1981	0.46	0.23	0.77	1.56	3.51	0.37	1.27	0.50	0.23	0.76	0.75	0.43	10.84
1982	0.41	0.33	0.60	1.25	2.10	4.15	1.92	0.88	3.40	1.18	0.55	3.71	20.48
1983	0.42	0.35	2.29	2.28	1.40	3.76	2.61	0.75	0.20	0.88	2.72	0.75	18.41
1984	1.19	0.48	1.59	2.23	1.33	1.34	2.26	0.25	0.50	0.71	0.70	0.78	13.36
1985	0.79	0.61	0.52	1.25	1.37	1.32	1.57	0.09	1.09	0.46	1.56	1.05	11.68
Record Mean	0.54	0.51	0.96	1.57	2.11	1.35	1.11	0.62	0.88	0.91	0.74	0.59	11.88

TABLE 3 — AVERAGE TEMPERATURE (deg. F)   CASPER, WYOMING

YEAR	JAN	FEB	MAR	APR	MAY	JUNE	JULY	AUG	SEP	OCT	NOV	DEC	ANNUAL
1956	27.4	24.1	33.2	40.7	55.9	68.3	70.8	67.0	61.0	49.8	29.8	28.4	46.4
1957	16.2	32.1	34.7	38.7	51.6	61.5	71.8	71.1	56.2	46.4	29.3	31.8	45.1
#1958	28.3	31.3	28.4	40.3	59.0	63.7	66.3	71.3	60.1	49.7	34.0	29.0	46.8
1959	23.7	23.0	32.6	42.0	50.6	66.8	70.0	70.3	56.2	43.7	30.6	30.4	45.0
1960	22.1	18.5	33.3	44.2	53.6	63.7	71.4	67.3	60.7	47.5	33.7	25.6	45.1
1961	28.7	31.4	37.0	41.0	54.4	66.1	70.3	72.1	51.4	44.8	30.3	21.6	45.7
1962	15.4	26.6	30.3	46.7	54.5	61.5	67.8	67.7	58.4	51.3	39.9	30.4	45.9
1963	13.7	32.4	34.8	42.4	55.2	64.1	72.0	70.4	64.1	54.6	38.3	24.9	47.3
#1964	23.0	21.2	25.1	39.7	54.1	61.1	74.1	66.1	56.1	47.1	31.3	23.4	43.5
1965	29.7	22.6	20.5	45.8	49.9	60.1	69.3	66.3	47.7	51.8	39.8	29.1	44.4
1966	21.8	24.3	36.2	38.9	55.8	61.8	74.9	66.6	62.2	45.2	36.1	25.3	45.8
1967	26.7	27.7	36.4	42.7	49.1	58.5	69.8	69.0	61.0	48.2	30.8	17.4	44.8
1968	23.2	30.3	36.7	38.4	48.1	60.5	69.1	64.9	56.5	47.9	32.0	18.8	43.8
1969	27.1	29.0	28.4	46.7	55.4	57.0	71.6	71.8	63.3	37.9	34.4	28.5	45.9
1970	23.3	32.7	27.6	38.2	53.3	62.6	71.2	72.3	54.7	41.2	34.9	25.2	44.8
1971	25.4	24.7	31.7	40.9	50.1	64.3	67.2	71.3	53.9	42.8	30.5	23.9	43.9
1972	17.7	28.5	38.2	43.0	52.1	65.0	68.0	65.9	56.7	44.1	30.6	17.7	44.0
1973	19.2	24.5	33.0	36.1	51.1	63.0	68.4	70.7	55.0	48.7	33.1	26.7	44.1
1974	18.7	27.6	36.2	43.1	51.4	64.9	72.6	64.4	54.4	48.5	34.3	24.3	45.0
1975	21.1	20.9	30.2	37.7	48.9	58.8	72.9	68.0	56.7	47.6	32.5	28.4	43.7
1976	23.0	30.2	32.0	44.0	53.6	60.7	72.8	68.4	59.2	43.8	34.3	28.3	45.9
1977	20.2	30.9	31.3	46.5	55.1	69.4	71.7	66.1	60.3	48.8	33.3	24.9	46.5
1978	17.1	22.3	37.2	44.6	49.6	63.6	69.6	66.8	60.2	47.9	26.9	13.2	43.3
1979	8.8	29.1	35.4	43.7	49.3	61.8	70.8	67.7	62.1	47.8	26.5	28.7	43.8
1980	16.4	27.5	31.0	44.2	51.6	64.4	72.2	66.4	59.7	46.5	33.8	34.3	45.7
1981	31.4	29.9	38.0	48.3	51.4	64.3	71.8	69.2	62.0	44.4	39.0	27.3	48.1
1982	20.1	25.7	36.6	41.1	51.1	60.3	70.7	73.7	56.9	45.2	32.2	23.3	44.7
1983	31.2	31.5	36.3	38.0	48.4	60.4	71.2	74.8	60.3	49.7	31.2	10.9	45.3
1984	21.9	27.5	34.4	39.1	54.5	61.5	70.4	71.2	52.5	40.1	34.8	22.9	44.2
1985	16.1	19.7	33.3	45.5	55.3	61.3	71.6	66.7	53.4	45.3	20.3	21.3	42.5
Record Mean	22.3	26.7	32.3	42.5	52.3	62.3	70.8	69.2	58.2	47.1	33.0	25.6	45.2
Max	32.6	37.2	43.7	55.2	65.8	77.2	87.0	85.2	73.8	60.7	43.9	35.5	58.2
Min	11.9	16.1	20.8	29.8	38.8	47.3	54.7	53.1	42.6	33.4	22.0	15.7	32.2

REFERENCE NOTES FOR TABLES 1, 2, 3 and 6   (CASPER, WY)

GENERAL

T - TRACE AMOUNT
BLANK ENTRIES DENOTE MISSING/UNREPORTED DATA.
# INDICATES A STATION OR INSTRUMENT RELOCATION.

SPECIFIC

TABLE 1

(a) - LENGTH OF RECORD IN YEARS. ALTHOUGH
INDIVIDUAL MONTHS MAY BE MISSING.
  * LESS THAN .05

NORMALS — BASED ON THE 1951-1980 RECORD PERIOD.
EXTREMES — DATES ARE THE MOST RECENT OCCURRENCE.
WIND DIR. — NUMERALS SHOW TENS OF DEGREES
  CLOCKWISE FROM TRUE NORTH.
  "00" INDICATES CALM.
RESULTANT WIND DIRECTIONS ARE GIVEN TO WHOLE DEGREES.

EXCEPTIONS

TABLES 2, 3, and 6

RECORD MEANS ARE THROUGH THE CURRENT YEAR,
BEGINNING IN   1940 FOR TEMPERATURE
            1940 FOR PRECIPITATION
            1951 FOR SNOWFALL

TABLE 4 HEATING DEGREE DAYS Base 65 deg. F CASPER, WYOMING

SEASON	JULY	AUG	SEP	OCT	NOV	DEC	JAN	FEB	MAR	APR	MAY	JUNE	TOTAL
1956-57	3	56	139	464	1049	1131	1506	916	934	782	408	135	7523
1957-58	5	14	267	570	1063	1021	1129	940	1129	735	207	88	7168
#1958-59	37	11	183	465	924	1108	1276	1169	999	684	441	63	7360
1959-60	23	12	312	654	1026	1066	1318	1342	978	615	349	80	7775
1960-61	11	61	185	535	931	1220	1118	936	863	714	331	71	6976
1961-62	24	0	409	622	1033	1341	1537	1071	1072	544	324	139	8116
1962-63	19	57	194	418	748	1066	1590	906	928	672	300	54	6952
1963-64	3	8	56	324	795	1236	1298	1264	1231	751	346	156	7468
#1964-65	0	101	272	545	1004	1283	1084	1181	1377	569	463	145	8024
1965-66	4	49	512	401	749	1108	1334	1132	887	775	289	141	7381
1966-67	0	47	122	609	862	1225	1180	1037	878	664	488	193	7305
1967-68	6	21	149	516	1017	1470	1289	999	868	794	518	166	7813
1968-69	30	81	251	520	984	1426	1164	998	1127	541	304	252	7678
1969-70	3	3	71	834	914	1125	1289	897	1149	797	355	143	7580
1970-71	1	0	322	733	894	1222	1222	1119	1025	714	456	71	7782
1971-72	36	0	352	681	1026	1268	1461	1052	824	653	393	48	7794
1972-73	67	42	251	641	1025	1461	1415	984	885	861	424	117	8412
1973-74	36	0	296	495	951	1181	1432	1044	885	650	416	105	7491
1974-75	3	61	319	504	912	1253	1352	1230	1066	813	493	193	8199
1975-76	1	18	247	532	969	1126	1295	1004	1017	625	346	149	7329
1976-77	3	14	199	654	913	1133	1385	949	1040	549	298	12	7149
1977-78	3	51	164	492	945	1240	1481	1191	854	605	471	107	7604
1978-79	17	60	212	522	1137	1602	1738	1153	910	631	479	151	8612
1979-80	0	41	118	523	1149	1116	1505	1081	1048	617	408	75	7681
1980-81	0	44	176	566	931	943	1036	979	829	494	418	86	6502
1981-82	15	17	129	633	775	1161	1385	1097	873	709	423	173	7390
1982-83	4	0	276	604	977	1287	1040	930	881	802	506	156	7463
1983-84	22	0	190	468	1009	1673	1331	1082	941	769	332	146	7963
1984-85	0	0	386	767	899	1298	1512	1264	973	578	297	151	8125
1985-86	0	51	355	606	1337	1348							

TABLE 5 COOLING DEGREE DAYS Base 65 deg. F CASPER, WYOMING

YEAR	JAN	FEB	MAR	APR	MAY	JUNE	JULY	AUG	SEP	OCT	NOV	DEC	TOTAL
1969	0	0	0	0	12	18	213	222	26	0	0	0	491
1970	0	0	0	0	0	77	200	234	24	0	0	0	535
1971	0	0	0	0	0	58	109	202	26	0	0	0	395
1972	0	0	0	0	0	54	101	142	7	0	0	0	304
1973	0	0	0	0	1	65	147	182	5	0	0	0	400
1974	0	0	0	0	0	109	243	47	7	0	0	0	406
1975	0	0	0	0	0	16	254	119	5	1	0	0	395
1976	0	0	0	0	0	29	250	127	33	0	0	0	439
1977	0	0	0	0	0	150	218	92	31	0	0	0	491
1978	0	0	0	0	2	72	170	124	74	0	0	0	442
1979	0	0	0	0	0	62	187	129	37	0	0	0	415
1980	0	0	0	0	1	64	232	93	20	0	0	0	410
1981	0	0	0	0	0	73	235	155	45	0	0	0	508
1982	0	0	0	0	0	41	188	275	37	0	0	0	541
1983	0	0	0	0	0	24	220	309	56	0	0	0	609
1984	0	0	0	0	14	48	175	197	15	0	0	0	449
1985	0	0	0	0	4	48	210	109	13	0	0	0	384

TABLE 6 SNOWFALL (inches) CASPER, WYOMING

SEASON	JULY	AUG	SEP	OCT	NOV	DEC	JAN	FEB	MAR	APR	MAY	JUNE	TOTAL
1956-57	0.0	0.0	0.0	1.4	19.9	7.9	10.3	1.5	13.1	21.2	T	0.0	75.3
1957-58	0.0	0.0	T	1.7	10.5	1.4	7.3	25.9	21.0	6.6	0.0	0.0	74.4
#1958-59	0.0	0.0	T	T	12.2	4.3	11.8	10.6	22.1	8.9	1.1	0.0	71.0
1959-60	0.0	0.0	1.2	4.1	4.4	4.8	10.8	11.9	5.1	9.7	T	0.0	52.0
1960-61	0.0	0.0	0.0	1.6	17.3	8.4	1.6	7.7	8.5	4.8	0.0	0.0	49.9
1961-62	0.0	0.0	6.3	5.3	7.0	11.7	16.2	9.7	8.0	1.1	0.0	T	65.3
1962-63	0.0	0.0	0.2	0.0	4.5	3.2	10.6	5.6	5.0	10.1	0.0	0.0	39.2
1963-64	0.0	0.0	0.0	1.4	1.2	5.1	13.2	12.7	12.8	22.8	T	0.0	69.2
1964-65	0.0	T	0.0	T	12.9	6.7	16.5	8.6	11.4	7.3	7.1	3.0	62.3
1965-66	0.0	0.0	8.8	0.0	0.5	10.0	5.5	6.2	9.2	17.0	2.1	3.0	62.3
1966-67	0.0	0.0	T	11.9	8.0	9.2	10.6	8.8	11.8	7.5	10.2	0.0	78.0
1967-68	0.0	0.0	0.0	4.3	13.1	17.8	5.2	5.0	13.3	14.5	3.4	T	76.6
1968-69	0.0	0.0	T	0.7	4.9	16.4	2.0	6.9	13.1	12.1		3.0	59.1
1969-70	0.0	0.0	0.0	12.5	12.0	2.7	9.0	4.8	22.3	10.9	1.7	T	75.9
1970-71	0.0	0.0	0.2	11.2	7.4	11.9	3.7	11.2	14.8	24.2	2.2	0.0	86.8
1971-72	0.0	0.0	2.8	13.1	9.7	4.7	19.2	7.2	14.5	13.9	0.0	0.0	85.1
1972-73	0.0	0.0	0.2	9.8	8.1	9.1	9.6	11.3	10.7	56.3	1.7	0.0	116.8
1973-74	0.0	0.0	T	5.9	13.2	9.6	14.0	8.1	7.5	19.4	2.0	0.0	79.7
1974-75	0.0	0.0	5.1	1.2	8.9	5.1	10.3	8.1	36.2	13.4	21.3	0.0	109.6
1975-76	0.0	0.0	T	11.8	6.8	16.1	6.9	16.1	7.3	4.1	1.4	T	70.5
1976-77	0.0	0.0	0.0	3.8	1.8	6.2	8.5	11.8	26.6	4.8	T	0.0	63.5
1977-78	0.0	0.0	0.0	4.3	18.2	16.8	14.2	14.6	8.0	5.4	24.6	0.0	106.1
1978-79	0.0	0.0	T	3.1	17.1	25.7	16.3	7.3	18.6	9.2	22.7	T	120.0
1979-80	0.0	0.0	0.0	16.5	8.8	5.1	22.1	13.5	17.3	3.7	15.7	0.0	101.2
1980-81	0.0	0.0	0.0	6.1	10.0	7.9	9.9	7.3	5.0	10.3	0.4	0.0	56.9
1981-82	0.0	0.0	0.0	5.8	7.3	6.1	10.0	6.1	9.3	17.8	6.3	T	68.7
1982-83	0.0	0.0	11.5	11.3	2.5	62.8	4.4	6.2	27.7	23.3	1.9	0.0	151.6
1983-84	0.0	0.0	0.3	1.3	37.1	12.0	19.4	7.6	26.1	33.6	5.7	0.0	143.1
1984-85	0.0	0.0	3.1	9.4	6.1	10.2	14.7	10.5	9.7	11.6	0.0	0.0	75.3
1985-86	0.0	0.0	7.0	4.8	28.3	14.4							
Record Mean	0.0	T	1.3	5.0	10.6	11.3	10.2	9.7	15.0	13.5	4.0	0.2	80.9

See Reference Notes, relative to all above tables, on preceding page.

CHEYENNE, WYOMING

The city of Cheyenne is located on a broad plateau between the North and South Platte Rivers in the extreme southeastern corner of Wyoming at an elevation of approximately 6,100 feet. The surrounding country is mostly rolling prairie which is used primarily for grazing. The ground level rises rapidly to a ridge approximately 9,000 feet in elevation about 30 miles west of the city. This ridge is known as the Laramie Mountains, one of the ranges of the Rockies, and extends in a north-south direction. Because of this ridge, winds from the northwest through west to southwest are downslope and produce a marked chinook effect in Cheyenne which is especially noticeable during the winter months. Also, winds from the north through east to south are upslope and may cause fog or low stratus clouds in the Cheyenne area throughout the year. Because of this terrain variation, the wind direction plays an important role in controlling the local temperature and weather.

Cheyenne experiences large diurnal and annual temperature ranges. This is due to the advent of both warm and cold air masses and the relatively high elevation of the city which permits rapid incoming and outgoing radiation. The daily temperature range averages about 30 degrees in the summer and 23 degrees in the winter. Many cold air masses from the north during the winter months miss Cheyenne. Because of the downslope of land to the east and the prevailing westerlies, some of the cold air masses do move over the city, but only about 13 percent of the days in an average January, the coldest month of the year, show temperatures dropping to zero or below. Temperatures during the winter months average a few degrees higher than over the Mississippi and Missouri Valleys at the same latitude.

Windy days are quire frequent during the winter and spring months. Since the wind is usually strongest during the daytime it is a very noticeable weather element. Usually the strong winds are from a westerly direction and this tends to raise the temperature because the air is moving downslope.

Most of the air masses reaching this area move in from the Pacific and since the mountains to the west are quite effective moisture barriers the climate is semi-arid. Fortunately, about 70 percent of normal annual precipitation occurs during the growing season. In the summer months, precipitation is mostly of the shower type and occurs mainly with thunderstorms. Hail is frequent and occasionally destructive in some thunderstorms. Most of the snow falls during the late winter and early spring months. It is not uncommon to have heavy snow in May.

The growing season in Cheyenne averages about 132 days a year and extends from around May 18th to September 27th. Freezing temperatures have occurred as late in the spring as mid-June, and as early in the fall as late August.

Relative humidity averages near 50 percent on an annual basis with large daily variations. Very seldom is the relative humidity above 30 percent when the temperature is above 80 degrees.

TABLE 1 — NORMALS, MEANS AND EXTREMES

CHEYENNE, WYOMING

LATITUDE: 41°09'N   LONGITUDE: 104°49'W   ELEVATION: FT. GRND 6126 BARO 06123   TIME ZONE: MOUNTAIN   WBAN: 24018

	(a)	JAN	FEB	MAR	APR	MAY	JUNE	JULY	AUG	SEP	OCT	NOV	DEC	YEAR
TEMPERATURE °F:														
Normals														
-Daily Maximum		37.3	40.7	43.6	54.0	64.6	75.4	83.1	80.8	72.1	61.0	46.5	40.4	58.3
-Daily Minimum		14.8	17.9	20.6	29.6	39.7	48.5	54.6	52.8	43.7	34.0	23.1	18.2	33.1
-Monthly		26.1	29.3	32.1	41.8	52.2	62.0	68.8	66.8	57.9	47.5	34.8	29.3	45.7
Extremes														
-Record Highest	50	66	71	73	82	90	100	100	96	93	83	73	69	100
-Year		1982	1962	1967	1981	1969	1954	1939	1979	1960	1967	1954	1939	JUN 1954
-Record Lowest	50	-29	-34	-21	-8	16	25	38	36	8	2	-14	-28	-34
-Year		1984	1936	1943	1975	1947	1951	1952	1975	1985	1935	1983	1983	FEB 1936
NORMAL DEGREE DAYS:														
Heating (base 65°F)		1206	1000	1020	696	397	139	24	37	235	543	906	1107	7310
Cooling (base 65°F)		0	0	0	0	0	49	145	93	22	0	0	0	309
% OF POSSIBLE SUNSHINE	46	62	66	66	61	60	65	68	67	70	69	60	60	65
MEAN SKY COVER (tenths)														
Sunrise - Sunset	50	5.8	6.2	6.3	6.5	6.7	5.5	5.0	5.1	4.6	4.7	5.6	5.8	5.6
MEAN NUMBER OF DAYS:														
Sunrise to Sunset														
-Clear	50	8.8	7.0	7.1	5.9	4.6	8.2	9.6	9.8	13.3	12.8	9.6	9.2	105.8
-Partly Cloudy	50	9.6	9.2	9.8	10.4	12.2	12.6	15.1	13.3	8.5	9.2	9.1	9.2	128.1
-Cloudy	50	13.0	12.0	13.9	13.7	14.2	9.2	6.3	7.8	8.3	9.0	11.3	12.8	131.8
Precipitation														
.01 inches or more	50	5.9	6.0	9.4	9.6	11.8	10.8	10.7	9.6	7.3	5.5	5.9	5.5	98.0
Snow, Ice pellets														
1.0 inches or more	50	2.1	1.9	3.6	2.5	0.9	0.1	0.0	0.0	0.3	1.2	2.4	1.8	16.8
Thunderstorms	50	0.0	0.1	0.2	2.1	7.6	11.1	13.3	10.5	4.3	0.8	0.*	0.0	50.1
Heavy Fog Visibility 1/4 mile or less	50	0.9	1.8	3.0	3.0	3.0	2.1	1.1	1.3	1.9	2.1	1.9	1.4	23.4
Temperature °F														
-Maximum														
90° and above	26	0.0	0.0	0.0	0.0	0.*	1.1	5.3	2.3	0.3	0.0	0.0	0.0	9.1
32° and below	26	9.6	7.1	5.5	1.5	0.*	0.0	0.0	0.0	0.2	0.7	4.2	8.6	37.3
-Minimum														
32° and below	26	29.0	26.4	27.8	18.1	3.4	0.0	0.0	0.0	2.3	12.0	24.7	28.4	172.1
0° and below	26	4.8	2.4	1.2	0.2	0.0	0.0	0.0	0.0	0.0	0.0	0.6	3.0	12.2
AVG. STATION PRESS.(mb)	13	808.7	808.8	806.4	808.6	810.0	812.6	815.5	815.3	814.4	812.9	810.0	809.0	811.0
RELATIVE HUMIDITY (%)														
Hour 05	26	57	60	64	66	70	70	69	68	65	60	60	57	64
Hour 11 (Local Time)	26	46	44	46	42	41	40	35	35	37	38	43	46	41
Hour 17	26	50	46	46	41	43	41	38	37	38	40	49	52	43
Hour 23	26	58	60	63	63	66	63	63	62	61	58	59	58	61
PRECIPITATION (inches):														
Water Equivalent														
-Normal		0.41	0.40	0.97	1.24	2.39	2.00	1.87	1.39	1.06	0.68	0.53	0.37	13.31
-Maximum Monthly	50	2.78	2.16	2.96	5.04	5.67	5.32	5.01	6.64	4.52	3.57	2.48	1.68	6.64
-Year		1949	1953	1983	1942	1981	1955	1973	1985	1973	1942	1979	1937	AUG 1985
-Minimum Monthly	50	T	T	0.12	0.35	0.11	0.07	0.58	0.03	0.10	0.03	T	0.03	T
-Year		1952	1983	1966	1946	1974	1980	1969	1944	1953	1964	1965	1959	FEB 1983
-Maximum in 24 hrs	50	1.41	1.60	1.88	1.94	2.00	2.68	3.42	6.06	2.75	1.70	1.66	1.19	6.06
-Year		1949	1953	1946	1984	1942	1955	1973	1985	1973	1947	1979	1979	AUG 1985
Snow, Ice pellets														
-Maximum Monthly	50	35.5	19.9	31.9	31.8	30.4	8.7			7.4	21.3	31.1	21.3	35.5
-Year		1980	1953	1983	1984	1943	1947			1985	1969	1979	1958	JAN 1980
-Maximum in 24 hrs	50	12.0	14.0	15.6	17.4	15.0	8.7			5.8	6.9	19.8	11.7	19.8
-Year		1980	1953	1973	1984	1942	1947			1985	1982	1979	1979	NOV 1979
WIND:														
Mean Speed (mph)	28	15.2	14.9	14.5	14.3	12.6	11.4	10.3	10.3	11.2	12.2	13.2	14.9	12.9
Prevailing Direction through 1963		WNW	W	WNW	WNW	WNW	WNW	WNW	W	W	W	WNW	WNW	WNW
Fastest Mile														
-Direction	45	NW	W	NW	NW	NW	NW	E	W	SW	W	W	W	NW
-Speed (MPH)	45	66	65	75	72	69	70	52	57	54	61	66	65	75
-Year		1975	1955	1972	1978	1967	1976	1955	1978	1979	1958	1967	1980	MAR 1972
Peak Gust														
-Direction	2	NW	NW	W	W	NW	W	N	SW	NW	SW	NW	W	SW
-Speed (mph)	2	52	56	69	62	62	51	55	56	58	71	59	71	71
-Date		1985	1984	1985	1984	1985	1985	1984	1985	1984	1985	1984	1984	OCT 1985

See Reference Notes to this table on the following page.

TABLE 2   PRECIPITATION (inches)   CHEYENNE, WYOMING

YEAR	JAN	FEB	MAR	APR	MAY	JUNE	JULY	AUG	SEP	OCT	NOV	DEC	ANNUAL
1956	0.70	0.46	0.24	1.31	2.56	0.99	1.43	1.73	0.14	0.16	1.27	0.41	11.40
1957	0.67	0.08	0.99	2.82	5.35	2.01	2.42	2.92	0.56	1.35	0.57	0.10	19.84
1958	0.06	0.50	2.08	1.13	2.98	1.79	3.85	1.35	0.91	0.20	0.45	1.27	16.57
#1959	0.52	0.30	1.09	1.54	4.16	0.71	0.80	0.77	1.61	1.38	0.16	0.03	13.07
1960	0.17	0.49	0.31	0.82	0.96	1.36	0.94	0.66	0.66	1.11	0.43	0.28	8.22
1961	0.06	0.37	2.08	0.83	2.92	2.91	1.53	3.12	2.17	0.56	0.35	0.09	16.99
1962	0.67	0.56	0.39	0.54	2.72	2.82	4.02	0.40	1.30	0.62	0.33	0.25	14.62
1963	0.51	0.17	0.74	1.66	1.14	2.84	0.77	2.28	3.28	1.06	0.02	0.42	14.89
1964	0.03	0.24	0.58	1.30	0.84	1.01	1.00	0.28	0.33	0.03	0.14	0.16	5.94
1965	0.57	0.26	0.76	0.57	3.11	4.03	0.89	1.54	1.32	0.75	T	0.22	14.02
1966	0.11	0.22	0.12	0.48	0.21	1.95	3.38	2.78	2.12	0.68	0.29	0.08	12.42
1967	0.45	0.54	0.76	2.15	4.04	2.63	1.71	0.99	0.88	0.46	0.32	0.46	15.39
1968	0.04	0.30	0.31	2.34	3.54	0.87	1.25	1.21	0.20	0.86	0.71	0.28	11.91
1969	0.23	0.25	0.27	0.82	1.77	2.70	0.58	0.99	0.84	2.04	0.23	0.21	10.93
1970	0.10	0.04	1.32	0.85	3.13	2.42	0.82	0.14	1.10	1.30	0.30	0.31	11.83
1971	0.51	0.62	1.08	2.81	2.38	0.97	1.08	1.41	1.76	1.16	0.05	0.07	13.90
1972	0.36	0.02	0.79	0.80	2.76	1.71	1.35	1.83	1.01	0.42	0.59	0.40	12.04
1973	0.23	0.07	1.85	1.75	0.31	1.20	5.01	0.27	4.52	0.06	1.25	1.06	17.58
1974	0.48	0.03	1.24	0.50	0.11	2.81	1.41	1.29	0.50	0.91	0.49	0.10	9.87
1975	0.40	0.17	1.17	0.47	2.27	1.49	2.62	0.39	0.52	0.49	0.20	0.52	10.71
1976	0.32	0.71	0.32	1.79	2.07	0.68	2.39	1.40	0.77	0.15	0.28	0.10	10.98
1977	0.14	0.08	1.21	1.86	2.50	2.44	3.49	1.07	0.19	0.08	0.35	0.24	13.65
1978	0.58	0.78	0.35	0.52	3.98	0.98	1.38	0.12	0.50	0.45	0.54		10.81
1979	0.27	0.14	1.34	0.77	2.90	3.32	1.83	1.86	0.32	0.46	2.48	1.50	17.19
1980	2.71	0.73	1.36	0.93	2.39	0.07	2.00	1.55	0.97	0.51	0.46	0.08	13.76
1981	0.30	0.20	0.70	0.73	5.67	1.66	2.85	2.90	0.31	0.85	0.09	0.45	16.71
1982	0.41	0.19	0.17	0.53	3.56	4.52	2.71	1.81	2.87	1.20	0.43	0.83	19.23
1983	0.02	T	2.96	4.45	2.31	2.81	2.12	1.95	0.78	0.49	2.34	0.46	20.69
1984	0.54	0.84	1.28	3.71	0.78	2.43	2.57	2.40	0.65	1.55	0.11	0.34	17.64
1985	0.66	0.19	0.36	1.10	1.05	1.59	3.99	6.64	1.78	0.94	0.84	0.80	19.94
Record Mean	0.43	0.53	1.03	1.76	2.42	1.82	2.01	1.57	1.16	0.88	0.55	0.46	14.62

TABLE 3   AVERAGE TEMPERATURE (deg. F)   CHEYENNE, WYOMING

YEAR	JAN	FEB	MAR	APR	MAY	JUNE	JULY	AUG	SEP	OCT	NOV	DEC	ANNUAL
1956	28.9	22.6	33.0	39.5	54.8	67.2	67.5	64.7	61.1	50.4	31.9	32.7	46.2
1957	20.7	34.9	33.2	36.9	48.1	60.4	69.0	67.4	55.4	46.9	29.8	35.0	44.8
1958	29.9	33.0	25.5	39.0	56.9	62.8	64.5	68.7	58.6	49.6	36.2	30.8	46.3
#1959	26.3	26.0	31.1	39.3	50.6	65.4	68.2	69.1	55.7	43.0	33.1	33.1	45.1
1960	25.6	19.9	33.4	46.8	55.7	64.4	70.8	68.7	61.2	48.8	35.6	26.8	46.5
1961	29.0	31.8	31.6	40.7	52.8	64.8	69.1	68.1	51.4	44.8	32.3	23.4	45.0
1962	19.0	27.2	29.0	45.6	55.1	59.9	65.6	67.3	58.9	51.0	39.8	31.5	45.8
1963	16.7	33.0	31.8	45.7	55.8	64.0	72.7	68.3	63.1	54.4	40.0	26.9	47.5
1964	26.1	22.3	27.4	40.8	54.6	60.5	70.9	64.3	56.9	48.7	35.4	29.5	44.8
1965	32.1	26.0	22.2	45.7	50.8	59.3	68.2	64.7	49.6	52.0	41.4	32.1	45.4
1966	25.8	26.8	37.7	39.6	55.6	61.5	74.5	66.2	59.9	47.8	38.8	31.0	47.1
1967	31.4	31.1	39.0	44.1	49.1	57.7	67.3	66.0	58.4	49.3	35.1	22.5	45.9
1968	28.7	32.6	38.0	38.9	49.2	62.6	68.2	64.5	57.6	49.2	34.0	28.0	46.0
1969	31.4	32.0	28.7	48.1	55.7	57.0	71.5	70.7	61.2	37.1	36.4	31.3	46.8
1970	28.7	34.6	28.9	37.8	54.3	60.9	69.5	70.7	55.0	41.2	35.7	29.1	45.5
1971	29.7	25.8	32.4	41.0	49.2	63.2	65.5	68.7	53.1	44.3	36.0	29.1	44.8
1972	25.3	33.2	39.3	42.5	51.2	63.2	64.3	64.6	56.6	45.5	29.5	21.0	44.7
1973	23.7	29.7	31.9	37.3	51.4	62.9	65.8	68.4	54.1	49.3	35.3	29.4	44.9
1974	24.1	31.9	37.1	43.3	55.0	64.1	70.1	65.2	55.3	48.3	35.6	26.6	46.4
1975	25.5	25.0	30.2	37.5	48.4	57.9	67.3	66.3	56.0	48.1	34.2	32.6	44.1
1976	26.9	32.9	30.9	42.7	51.1	59.7	69.2	65.0	57.7	43.4	33.5	30.6	45.3
1977	22.4	32.9	32.1	44.7	53.4	65.2	68.2	63.5	60.7	48.3	34.4	29.0	46.3
1978	22.1	25.0	37.7	43.9	49.0	61.0	68.7	64.3	59.6	47.9	33.5	21.1	44.5
1979	17.3	31.6	36.0	44.9	49.9	61.5	69.9	65.9	62.9	49.8	29.5	32.9	46.0
1980	22.2	29.4	33.0	42.3	51.2	65.5	71.4	66.5	60.4	46.7	36.8	38.5	47.0
1981	33.4	32.2	36.8	50.1	50.5	63.7	69.0	65.2	61.2	46.1	40.7	30.7	48.3
1982	25.5	28.2	36.1	41.8	50.5	57.5	67.7	69.1	56.7	45.1	31.4	28.3	44.8
1983	32.8	33.5	32.0	34.8	47.4	57.3	67.8	70.2	59.0	48.6	32.3	15.0	44.2
1984	24.3	28.4	32.4	35.7	53.6	59.4	68.4	66.6	53.0	40.0	35.0	27.0	43.7
1985	19.5	23.3	35.2	45.6	54.3	61.1	69.3	66.8	52.8	45.1	26.2	26.1	43.8
Record Mean	25.9	27.9	32.6	41.5	51.1	61.0	67.8	66.3	57.4	46.4	35.0	28.6	45.1
Max	37.1	39.1	43.9	53.4	63.4	74.5	81.8	80.2	71.4	59.4	46.7	39.4	57.5
Min	14.8	16.6	21.2	29.5	38.7	47.5	53.8	52.5	43.3	33.4	23.4	17.7	32.7

REFERENCE NOTES FOR TABLES 1, 2, 3 and 6   (CHEYENNE, WY)

GENERAL

T - TRACE AMOUNT
BLANK ENTRIES DENOTE MISSING/UNREPORTED DATA.
# INDICATES A STATION OR INSTRUMENT RELOCATION.

SPECIFIC

TABLE 1

(a) - LENGTH OF RECORD IN YEARS. ALTHOUGH INDIVIDUAL MONTHS MAY BE MISSING.

* LESS THAN .05

NORMALS — BASED ON THE 1951-1980 RECORD PERIOD.
EXTREMES — DATES ARE THE MOST RECENT OCCURRENCE.
WIND DIR. — NUMERALS SHOW TENS OF DEGREES CLOCKWISE FROM TRUE NORTH. "00" INDICATES CALM.
RESULTANT WIND DIRECTIONS ARE GIVEN TO WHOLE DEGREES.

EXCEPTIONS

TABLE 1

1.   FASTEST MILE WINDS ARE THROUGH MARCH 1981.

TABLES 2, 3, and 6

RECORD MEANS ARE THROUGH THE CURRENT YEAR, BEGINNING IN   1871 FOR TEMPERATURE
1871 FOR PRECIPITATION
1936 FOR SNOWFALL

TABLE 4 HEATING DEGREE DAYS Base 65 deg. F   CHEYENNE, WYOMING

SEASON	JULY	AUG	SEP	OCT	NOV	DEC	JAN	FEB	MAR	APR	MAY	JUNE	TOTAL
1956-57	19	69	136	445	989	994	1365	838	976	839	519	155	7344
1957-58	8	23	284	553	1049	923	1081	891	1218	774	255	100	7159
1958-59	61	23	203	467	854	1052	1195	1086	1045	767	441	71	7265
#1959-60	15	15	309	675	949	978	1214	1301	975	542	288	89	7350
1960-61	17	45	176	496	877	1176	1109	923	1028	722	377	90	7036
1961-62	17	8	405	622	977	1285	1419	1056	1106	576	303	167	7941
1962-63	40	57	187	425	751	1031	1492	890	1023	633	277	89	6895
1963-64	3	10	76	323	742	1174	1203	1229	1160	720	340	166	7146
1964-65	2	95	239	500	883	1095	1012	1088	1320	569	435	169	7407
1965-66	11	47	454	397	702	1016	1208	1061	841	758	291	143	6929
1966-67	0	51	166	526	781	1049	1036	945	800	622	490	221	6687
1967-68	12	45	202	481	891	1308	1117	932	833	777	487	107	7192
1968-69	40	82	221	481	924	1137	1034	919	1118	498	297	248	6999
1969-70	3	3	113	859	854	1038	1115	846	1113	811	326	168	7249
1970-71	3	5	302	732	874	1108	1087	1093	1007	713	481	99	7504
1971-72	56	6	364	633	863	1108	1226	916	791	668	419	61	7111
1972-73	85	75	248	599	1056	1358	1275	980	1017	824	414	122	8053
1973-74	80	4	323	482	883	1098	1264	922	862	643	304	110	6975
1974-75	4	55	302	509	873	1180	1215	1115	1070	819	506	212	7860
1975-76	11	39	274	515	920	998	1175	924	1048	661	425	158	7148
1976-77	11	44	224	664	937	1059	1314	894	1014	602	352	44	7159
1977-78	21	74	150	511	910	1108	1324	1115	840	627	491	157	7328
1978-79	28	73	200	523	937	1358	1471	933	893	597	459	139	7611
1979-80	2	62	105	468	1058	990	1321	1027	984	673	424	65	7179
1980-81	0	41	151	558	840	812	974	913	868	440	445	95	6137
1981-82	21	50	120	580	722	1058	1216	1025	892	687	446	227	7044
1982-83	29	7	264	608	1002	1131	992	875	1016	898	538	233	7593
1983-84	23	0	202	502	974	1547	1259	1058	1002	870	348	177	7962
1984-85	6	15	365	769	892	1171	1403	1163	917	576	323	162	7762
1985-86	11	37	364	612	1158	1199							

TABLE 5 COOLING DEGREE DAYS Base 65 deg. F   CHEYENNE, WYOMING

YEAR	JAN	FEB	MAR	APR	MAY	JUNE	JULY	AUG	SEP	OCT	NOV	DEC	TOTAL
1969	0	0	0	0	14	12	211	185	9	0	0	0	431
1970	0	0	0	0	2	52	149	189	11	0	0	0	403
1971	0	0	0	0	0	58	77	128	13	0	0	0	276
1972	0	0	0	0	0	11	69	69	2	0	0	0	151
1973	0	0	0	0	0	65	112	116	0	0	0	0	293
1974	0	0	0	0	4	88	173	67	17	0	0	0	349
1975	0	0	0	0	0	7	90	86	10	0	0	0	193
1976	0	0	0	0	0	7	145	50	15	0	0	0	217
1977	0	0	0	0	0	59	126	38	29	0	0	0	252
1978	0	0	0	0	1	43	150	59	44	0	0	0	297
1979	0	0	0	0	1	42	160	100	49	0	0	0	352
1980	0	0	0	0	0	88	205	94	21	0	0	0	408
1981	0	0	0	0	0	59	156	64	15	0	0	0	294
1982	0	0	0	0	0	8	120	140	21	0	0	0	289
1983	0	0	0	0	0	10	115	169	28	0	0	0	322
1984	0	0	0	0	1	14	118	72	8	0	0	0	213
1985	0	0	0	0	0	52	150	98	8	0	0	0	308

TABLE 6 SNOWFALL (inches)   CHEYENNE, WYOMING

SEASON	JULY	AUG	SEP	OCT	NOV	DEC	JAN	FEB	MAR	APR	MAY	JUNE	TOTAL
1956-57	0.0	0.0	0.0	T	13.9	4.0	7.6	1.0	10.6	21.8	7.7	T	66.6
1957-58	0.0	0.0	T	0.5	8.0	1.0	0.6	6.2	17.8	6.2	0.0	0.0	40.3
1958-59	0.0	0.0	T	T	5.3	21.3	7.3	4.3	14.3	14.0	1.3	0.0	67.8
1959-60	0.0	0.0	3.7	9.1	3.8	0.4	2.8	8.8	3.7	5.6	T	0.0	37.9
1960-61	0.0	0.0	T	T	5.8	5.2	1.2	3.8	23.2	7.8	11.5	0.0	58.5
1961-62	0.0	0.0	0.8	3.6	3.5	1.5	13.0	10.8	5.5	2.5	T	0.0	41.2
1962-63	0.0	0.0	T	T	3.8	3.5	8.9	3.1	11.8	3.2	T	0.0	34.3
1963-64	0.0	0.0	0.0	T	0.1	6.0	1.0	3.3	6.9	7.1	0.5	0.0	24.9
1964-65	0.0	0.0	0.0	0.0	2.0	1.6	8.5	5.5	10.0	2.5	1.4	0.0	31.5
1965-66	0.0	0.0	2.0	T	T	2.4	1.3	2.7	1.1	3.6	T	0.0	13.1
1966-67	0.0	0.0	T	4.5	2.3	0.8	5.4	8.0	3.4	15.9	11.0	0.0	51.3
1967-68	0.0	0.0	0.0	0.4	3.6	4.8	0.2	2.5	2.8	15.4	2.2	T	31.9
1968-69	0.0	0.0	0.0	0.5	6.0	5.5	3.0	4.7	2.9	3.0	0.0	T	25.6
1969-70	0.0	0.0	0.0	21.3	1.8	2.4	1.2	1.5	19.4	8.3	T	T	55.9
1970-71	0.0	0.0	T	11.0	4.7	4.9	8.6	9.3	15.2	13.0	1.2	0.0	67.9
1971-72	0.0	0.0	7.4	8.1	0.9	1.4	7.4	0.4	10.6	8.5	0.3	0.0	45.0
1972-73	0.0	0.0	T	5.0	8.8	7.6	4.6	0.6	27.0	12.8	1.2	0.0	67.6
1973-74	0.0	0.0	T	0.5	17.4	13.9	5.0	0.8	16.3	8.0	0.0	0.0	61.9
1974-75	0.0	0.0	1.9	1.4	2.4	1.5	5.8	4.3	14.0	5.4	0.8	0.0	37.5
1975-76	0.0	0.0	0.6	5.5	4.7	7.8	6.0	8.5	9.0	6.7	T	T	49.4
1976-77	0.0	0.0	0.0	1.5	3.5	1.2	2.5	0.8	12.9	7.0	T	0.0	29.4
1977-78	0.0	0.0	0.0	0.4	4.3	5.6	7.0	12.0	2.5	1.0	18.3	0.0	51.1
1978-79	0.0	0.0	0.1	2.0	9.3	17.6	6.7	2.2	21.1	4.0	14.1	T	77.1
1979-80	0.0	0.0	0.0	3.6	31.1	15.6	35.5	10.7	17.8	3.4	3.8	0.0	121.5
1980-81	0.0	0.0	0.0	1.1	6.3	2.1	3.4	2.9	9.0	2.0	0.8	0.0	27.6
1981-82	0.0	0.0	0.0	4.6	0.8	5.8	5.6	2.4	1.7	2.0	4.0	0.0	26.9
1982-83	0.0	0.0	0.0	12.2	7.9	13.1	0.1	T	31.9	25.7	10.1	0.0	101.0
1983-84	0.0	0.0	T	0.2	27.2	7.0	7.9	12.1	13.0	31.8	T	0.0	99.2
1984-85	0.0	0.0	1.6	3.8	1.5	5.6	9.8	1.9	3.6	2.6	0.4	0.0	30.8
1985-86	0.0	0.0	7.4	6.0	12.4	13.0							
Record Mean	0.0	0.0	0.8	3.5	7.0	5.9	6.4	5.5	12.0	9.3	3.7	0.3	54.3

See Reference Notes, relative to all above tables, on preceding page.

Lander, located in the central Wyoming valley of the Popo Agie River, lies at the foot and east of the Wind River Range. Situated on a flat-topped mesa, the airport station is 1 1/2 miles south-southeast and approximately 200 feet above the town.

The terrain to the north, east and south varies from rolling to broken with some grass covered hills 2 to 5 miles distant, rising approximately 400 feet above the station elevation. To the west and southwest the foothills of the Wind River Range begin about 3 miles from the station, sloping upward to over 12,000 feet above sea level along the Continental Divide, 20 miles distant.

Because Lander is in a pocket, winds from all directions except northeast are downslope and produce a Chinook effect, most noticable in winds from westerly quadrants. The airport, on its mesa, receives more wind than the town of Lander, the wind speed averaging 4.7 mph for the 56 years of record kept in the town. Because of light winds, steep temperature inversions are the rule during winter nights and early mornings. Temperatures in the valley will be as much as 15 degrees lower than at the airport on calm, clear nights when there is a snow cover. However, when the wind is calm and the humidities low, the chilling effect is much less than is usual in extreme cold. Winds are often so light that little or no mixing occurs between the cold surface air and the warmer layer 2,000 to 3,000 feet above the valley. For several days each winter, temperatures are 20 to 30 degrees lower than in the surrounding areas where higher wind speeds occur. The sheltered location, however, offers protection from most severe storms that sweep down from Canada.

Lander does not have a true spring season, and snow has been recorded in June.

Usually on 15 to 20 days a year the temperature reaches or exceeds 90 degrees. Even the warmest days are not oppressive, the humidity being low, and the nights being cool. The normal daily range of summer temperature is near 30 degrees.

Mountains block moisture from the Pacific, creating a semi-arid climate. The heaviest and most persistent precipitation comes when the wind in the lower levels is from easterly quadrants, through a combination of low pressure to the south, usually over Colorado, and high pressure to the north over Montana or the western Dakotas. Lander receives 45 percent more precipitation than the area 24 miles to the northeast and 83 percent more than areas 50 miles northeast. More than a third of the annual precipitation occurs in April and May, with another but lesser peak in September and October. Summer moisture comes from occasional showers but is very erratic and spotty. Since about one-third of the annual snowfall comes in March and April, when the temperature is comparatively high, the snow soon melts.

Hardier plants and vegetables do well in this area. Based on the 1951-1980 period, the average first occurrence of 32 degrees Fahrenheit in the fall is September 24 and the average last occurrence in the spring is May 22.

TABLE 1 — NORMALS, MEANS AND EXTREMES

LANDER, WYOMING

LATITUDE: 42°49'N LONGITUDE: 108°44'W ELEVATION: FT. GRND 5557 BARO 5562 TIME ZONE: MOUNTAIN WBAN: 24021

	(a)	JAN	FEB	MAR	APR	MAY	JUNE	JULY	AUG	SEP	OCT	NOV	DEC	YEAR
TEMPERATURE °F:														
Normals –Daily Maximum		31.3	37.7	44.2	54.7	65.5	76.5	86.0	83.7	73.1	60.3	42.6	35.0	57.6
–Daily Minimum		7.7	13.5	19.9	29.8	39.6	48.0	55.4	53.4	43.6	33.3	18.9	11.3	31.2
–Monthly		19.5	25.6	32.0	42.3	52.5	62.3	70.7	68.6	58.4	46.8	30.8	23.1	44.4
Extremes –Record Highest	39	63	68	76	82	91	100	101	101	94	85	70	64	101
–Year		1971	1951	1966	1962	1954	1954	1954	1979	1983	1963	1980	1980	AUG 1979
–Record Lowest	39	-37	-28	-16	-2	18	25	39	35	10	0	-18	-37	-37
–Year		1963	1949	1960	1973	1954	1951	1983	1962	1965	1971	1985	1983	DEC 1983
NORMAL DEGREE DAYS: Heating (base 65°F)		1407	1100	1020	681	384	142	12	32	241	564	1026	1296	7905
Cooling (base 65°F)		0	0	0	0	0	61	192	143	40	0	0	0	436
% OF POSSIBLE SUNSHINE	39	66	68	71	68	65	72	76	76	73	70	61	64	69
MEAN SKY COVER (tenths) Sunrise - Sunset	39	6.1	6.1	6.2	6.2	6.4	5.1	4.2	4.3	4.3	4.9	5.9	5.7	5.5
MEAN NUMBER OF DAYS: Sunrise to Sunset –Clear	39	7.6	6.7	7.2	5.9	6.1	10.1	13.6	13.1	13.8	11.9	7.9	9.3	113.3
–Partly Cloudy	39	10.3	10.1	9.9	10.5	11.1	11.0	11.8	12.1	8.9	9.3	9.9	9.9	124.8
–Cloudy	39	13.1	11.5	13.8	13.6	13.5	8.9	5.6	5.8	7.3	9.8	12.3	11.8	127.0
Precipitation .01 inches or more	39	4.6	5.2	7.5	8.4	9.2	6.4	5.8	4.5	5.3	4.9	5.1	4.6	71.5
Snow,Ice pellets 1.0 inches or more	39	2.6	2.9	4.4	4.3	1.5	0.2	0.0	0.0	0.8	2.2	3.0	2.7	24.6
Thunderstorms	39	0.0	0.0	0.2	0.8	4.5	7.3	9.5	6.9	2.9	0.4	0.*	0.0	32.4
Heavy Fog Visibility 1/4 mile or less	39	0.9	0.6	0.2	0.2	0.1	0.1	0.0	0.1	0.2	0.2	0.7	0.8	3.9
Temperature °F –Maximum 90° and above	39	0.0	0.0	0.0	0.0	0.1	2.1	10.1	6.6	0.9	0.0	0.0	0.0	19.8
32° and below	39	15.2	8.9	4.4	0.7	0.1	0.0	0.0	0.0	0.1	0.6	6.7	13.4	50.1
–Minimum 32° and below	39	30.5	27.8	28.6	18.7	5.2	0.3	0.0	0.0	3.1	14.1	27.7	30.6	186.6
0° and below	39	8.5	4.1	1.3	0.*	0.0	0.0	0.0	0.0	0.0	0.1	1.9	5.3	21.3
AVG. STATION PRESS.(mb)	13	827.6	827.4	824.4	826.4	827.1	829.1	831.8	831.5	831.1	830.2	828.1	827.7	828.5
RELATIVE HUMIDITY (%) Hour 05	39	67	68	66	66	67	62	55	54	59	64	69	68	64
Hour 11	39	60	57	51	46	44	41	34	33	39	45	57	60	47
Hour 17 (Local Time)	39	60	54	47	40	38	33	28	27	33	41	57	61	43
Hour 23	39	66	65	61	58	57	51	43	43	49	57	66	67	57
PRECIPITATION (inches): Water Equivalent –Normal		0.48	0.63	1.13	2.22	2.69	1.45	0.71	0.49	0.87	1.20	0.76	0.53	13.16
–Maximum Monthly	39	1.65	2.18	3.30	5.46	6.03	6.88	2.50	2.30	4.68	3.58	3.37	1.62	6.88
–Year		1949	1955	1977	1957	1957	1947	1977	1979	1973	1971	1983	1985	JUN 1947
–Minimum Monthly	39	T	T	0.35	0.76	0.27	T	0.05	T	0.01	0.10	0.01	0.03	T
–Year		1952	1970	1960	1982	1984	1971	1963	1970	1979	1958	1949	1954	JUN 1971
–Maximum in 24 hrs	39	0.81	0.88	1.28	2.16	2.75	3.56	2.13	1.08	2.21	1.71	1.38	1.25	3.56
–Year		1963	1955	1977	1971	1964	1947	1977	1979	1973	1966	1983	1985	JUN 1947
Snow,Ice pellets –Maximum Monthly	39	26.5	43.8	52.0	66.0	33.9	18.4			32.9	39.9	48.7	28.0	66.0
–Year		1962	1955	1977	1973	1975	1947			1982	1971	1983	1985	APR 1973
–Maximum in 24 hrs	39	13.8	19.8	20.3	21.9	20.8	18.4			16.9	19.4	23.1	20.5	23.1
–Year		1980	1955	1973	1967	1975	1947			1982	1966	1958	1985	NOV 1958
WIND: Mean Speed (mph)	39	6.0	6.1	7.1	7.9	7.9	7.9	7.7	7.5	7.0	6.1	5.6	5.8	6.9
Prevailing Direction through 1963		SW	SW	SW	SW	SW	SW	SW	SW	SW	SW	SW	SW	SW
Fastest Mile –Direction (!!)	39	SW	SW	SW	SW		SW	W	W	W	SW	N	SW	SW
–Speed (MPH)	39	73	77	80	72	66	61	57	56	56	70	75	73	80
–Year		1967	1957	1972	1955	1980	1960	1959	1962	1966	1950	1958	1964	MAR 1972
Peak Gust –Direction (!!)	2	W	SW	SW	NW	W	SW	W	SW	S	SW	SW	NW	W
–Speed (mph)	2	43	53	58	60	64	54	66	46	55	52	45	66	66
–Date		1985	1985	1985	1985	1985	1984	1985	1985	1984	1985	1985	1984	JUL 1985

See Reference Notes to this table on the following page.

TABLE 2 PRECIPITATION (inches) LANDER, WYOMING

YEAR	JAN	FEB	MAR	APR	MAY	JUNE	JULY	AUG	SEP	OCT	NOV	DEC	ANNUAL
1956	0.67	0.44	0.62	1.96	3.64	T	0.37	0.39	0.10	0.63	1.19	0.33	10.34
1957	0.15	0.04	1.08	5.46	6.03	2.49	0.69	1.02	1.76	0.95	2.08	0.14	21.89
1958	0.16	0.33	0.98	2.46	5.22	1.44	1.02	0.74	0.15	0.10	1.52	0.33	14.45
1959	0.76	0.85	2.17	1.24	2.85	0.44	0.15	0.07	1.21	1.32	0.80	0.68	12.54
1960	0.72	0.53	0.35	1.07	0.78	0.58	0.28	T	1.05	1.95	1.40	0.38	9.09
1961	0.13	1.19	0.66	0.85	2.21	1.72	0.38	0.42	1.59	2.98	0.90	0.20	13.23
1962	1.34	0.39	0.54	1.91	3.54	0.46	1.72	0.03	0.56	1.10	0.20	0.22	12.01
1963	1.13	0.31	1.69	3.38	0.59	3.84	0.05	0.54	0.19	0.36	0.13	0.61	12.82
1964	0.43	1.20	0.80	2.97	3.42	2.10	0.36	0.17	0.55	0.34	0.40	0.07	12.81
1965	0.19	0.83	0.43	1.60	3.15	2.07	1.67	0.18	2.62	0.34	0.44	0.44	13.96
1966	0.27	0.33	0.87	2.19	0.80	1.31	0.28	1.77	0.71	2.17	0.51	0.41	11.62
1967	0.25	1.18	0.73	2.42	3.97	5.06	0.91	0.11	1.36	0.21	1.66	1.12	18.98
1968	0.17	0.61	1.65	2.43	2.16	2.14	0.68	0.93	0.48	0.14	0.46	0.70	12.55
1969	0.12	0.13	0.92	2.32	1.02	5.29	0.07		0.01	1.89	0.21	0.38	12.36
1970	0.06	T	2.20	2.56	0.49	2.28	0.40	T	0.66	1.49	0.75	0.38	11.27
1971	0.36	0.45	0.85	5.22	4.95	T	0.27	0.27	2.27	3.58	0.41	0.36	18.99
1972	1.08	0.55	0.38	2.16	2.06	0.70	0.44	1.01	0.09	2.36	0.89	1.46	13.18
1973	0.89	0.28	3.02	4.02	0.91	0.22	2.10	0.33	4.68	1.47	0.74	0.64	19.30
1974	0.43	0.99	0.56	2.35	0.31	0.37	0.47	0.66	1.07	2.02	0.19	0.66	10.08
1975	0.74	0.43	1.18	1.68	4.44	1.88	0.29	0.10	0.38	1.37	0.73	0.79	14.01
1976	0.27	0.82	0.43	1.17	2.09	1.97	1.05	0.40	0.54	1.67	0.40	0.20	11.01
1977	0.53	0.11	3.30	1.47	1.11	0.66	2.50	0.47	0.18	0.91	0.65	0.33	12.22
1978	0.59	0.50	0.42	1.28	5.16	0.16	0.82	0.25	1.17	0.65	2.15	1.22	14.37
1979	0.75	0.05	0.56	1.69	3.00	0.83	0.18	2.30	0.01	0.48	0.57	1.09	11.51
1980	0.95	0.54	1.50	1.68	3.32	0.05	0.31	0.27	0.07	1.41	0.70	0.22	11.02
1981	0.67	0.27	1.98	1.11	3.20	0.04	0.88	0.92	0.61	0.61	0.05	0.05	10.39
1982	0.44	0.16	0.44	0.76	1.65	1.35	1.46	0.32	0.32	3.83	0.59	1.44	13.26
1983	0.08	0.49	2.11	3.34	3.02	1.21	0.20	0.32	0.50	0.57	3.37	0.60	15.81
1984	0.95	1.16	0.80	3.61	0.27	1.48	1.54	1.21	1.04	0.58	0.48	0.07	13.19
1985	0.45	0.26	0.48	0.83	0.83	1.70	1.00	0.07	1.74	0.16	1.55	1.62	10.69
Record Mean	0.50	0.67	1.21	2.26	2.47	1.30	0.78	0.54	1.08	1.35	0.79	0.62	13.57

TABLE 3 AVERAGE TEMPERATURE (deg. F) LANDER, WYOMING

YEAR	JAN	FEB	MAR	APR	MAY	JUNE	JULY	AUG	SEP	OCT	NOV	DEC	ANNUAL
1956	22.0	16.9	32.7	42.0	55.3	67.1	70.3	67.2	61.2	49.0	29.2	25.9	44.9
1957	14.8	32.5	35.0	38.0	51.4	60.1	70.2	69.9	56.2	46.2	24.9	28.9	44.0
1958	23.0	32.9	29.4	41.1	57.8	62.1	65.7	70.7	59.7	49.5	31.3	26.3	45.8
1959	20.5	21.0	32.9	43.4	49.0	66.0	70.9	70.9	56.6	43.9	27.5	25.9	44.0
1960	18.0	19.2	33.2	45.4	54.5	64.7	72.9	68.2	60.9	47.1	30.5	20.2	44.6
1961	22.1	29.2	36.2	41.8	53.7	66.9	71.6	71.6	50.6	42.5	23.1	17.7	43.9
1962	8.6	24.7	31.3	47.1	54.0	61.4	68.3	67.7	59.4	51.6	37.2	27.9	44.9
1963	6.9	34.2	33.5	41.1	55.5	62.1	70.8	69.8	63.5	53.6	36.3	19.2	45.5
1964	20.4	21.7	26.8	41.1	53.5	59.1	74.2	66.7	57.6	48.5	32.6	24.9	44.0
1965	29.7	24.0	23.7	45.4	49.6	60.4	70.0	67.1	48.0	51.9	38.6	25.8	44.5
1966	22.7	24.2	35.0	40.0	56.7	62.2	74.7	67.5	61.4	43.2	34.2	24.2	45.5
1967	26.6	27.8	36.7	41.4	50.0	57.5	69.0	69.2	60.6	48.1	28.1	15.1	44.2
1968	15.5	24.7	36.3	38.7	49.5	61.1	69.7	64.5	56.9	47.9	30.4	17.5	42.7
1969	25.3	28.2	29.5	47.0	57.0	56.8	71.4	73.1	63.8	37.8	33.9	25.6	45.8
1970	24.4	33.6	29.7	36.8	54.3	62.9	71.1	72.8	53.4	41.2	32.4	21.4	44.5
1971	26.5	26.1	32.6	41.0	50.9	63.9	68.3	72.6	53.4	41.0	27.0	21.5	43.7
1972	19.6	27.5	40.9	43.3	51.9	64.1	67.0	67.3	56.8	43.9	27.9	12.5	43.6
1973	9.4	13.6	26.4	34.1	52.5	63.7	68.5	69.9	53.8	48.0	31.9	25.6	41.5
1974	18.2	27.1	36.0	43.7	51.4	66.3	72.5	65.2	56.1	48.1	33.9	19.5	44.8
1975	20.9	22.9	31.9	36.9	48.1	58.1	71.5	66.8	58.0	46.7	28.8	23.6	42.8
1976	19.2	27.1	32.4	44.0	54.6	60.4	72.5	67.0	59.7	43.6	33.7	25.4	45.0
1977	15.9	31.3	30.6	46.7	53.2	68.3	70.7	65.3	59.7	48.0	32.0	26.3	45.7
1978	17.5	22.0	38.3	45.3	49.0	62.5	69.8	66.5	59.3	48.1	20.7	9.1	42.3
1979	1.3	20.3	33.8	44.1	51.4	62.9	70.7	66.5	63.8	49.8	26.0	24.9	43.0
1980	14.0	25.3	32.5	45.0	52.2	64.1	72.2	66.5	59.9	47.1	31.2	34.3	45.3
1981	28.5	27.2	39.3	48.4	52.4	64.8	71.2	69.7	62.3	43.7	37.7	27.1	47.7
1982	22.1	25.4	35.8	41.3	51.9	61.0	70.6	73.3	55.2	44.3	29.5	18.3	44.1
1983	26.3	28.1	36.2	37.9	48.6	61.5	70.0	74.0	62.3	49.0	28.2	6.1	43.9
1984	16.1	20.4	31.1	40.3	55.9	63.2	73.3	72.4	55.1	41.5	32.2	20.5	43.5
1985	16.2	24.5	36.0	48.3	59.1	65.4	73.7	67.1	53.2	45.4	17.8	16.1	43.6
Record Mean	19.1	23.7	32.3	42.5	52.1	61.5	69.3	67.4	57.2	45.5	30.9	21.1	43.5
Record Max	31.6	36.4	44.7	55.2	65.2	76.1	85.1	83.2	72.3	59.5	43.4	33.2	57.1
Record Min	6.5	11.0	19.9	29.9	39.0	46.9	53.6	51.6	42.1	31.6	18.3	8.9	30.0

REFERENCE NOTES FOR TABLES 1, 2, 3 and 6   (LANDER, WY)

GENERAL

T - TRACE AMOUNT
BLANK ENTRIES DENOTE MISSING/UNREPORTED DATA.
# INDICATES A STATION OR INSTRUMENT RELOCATION.

SPECIFIC

TABLE 1

(a) - LENGTH OF RECORD IN YEARS ALTHOUGH
INDIVIDUAL MONTHS MAY BE MISSING.

* LESS THAN .05

NORMALS — BASED ON THE 1951-1980 RECORD PERIOD.
EXTREMES — DATES ARE THE MOST RECENT OCCURRENCE.
WIND DIR. — NUMERALS SHOW TENS OF DEGREES
CLOCKWISE FROM TRUE NORTH.
"00" INDICATES CALM.
RESULTANT WIND DIRECTIONS ARE GIVEN TO WHOLE DEGREES.

EXCEPTIONS

TABLES 2, 3, and 6

RECORD MEANS ARE THROUGH THE CURRENT YEAR,
BEGINNING IN   1892 FOR TEMPERATURE
1892 FOR PRECIPITATION
1947 FOR SNOWFALL

TABLE 4 HEATING DEGREE DAYS Base 65 deg. F   LANDER, WYOMING

SEASON	JULY	AUG	SEP	OCT	NOV	DEC	JAN	FEB	MAR	APR	MAY	JUNE	TOTAL
1956-57	9	48	131	486	1068	1204	1553	903	924	802	414	166	7708
1957-58	6	24	267	576	1198	1111	1295	1098	990	641	230	108	7517
1958-59	51	9	193	472	1005	1192	1373	1228	990	641	487	74	7715
1959-60	15	8	297	648	1120	1206	1450	1324	982	580	322	67	8019
1960-61	13	49	164	549	1030	1381	1321	997	887	692	348	61	7492
1961-62	13	0	426	691	1250	1457	1746	1127	1038	531	337	149	8765
1962-63	19	58	179	417	826	1143	1800	856	970	708	286	103	7365
1963-64	4	10	76	351	853	1415	1376	1249	1175	713	358	195	7775
1964-65	0	84	230	505	965	1237	1088	1142	1275	582	469	143	7720
1965-66	1	39	507	395	786	1208	1304	1137	924	745	265	138	7449
1966-67	2	47	142	667	919	1261	1188	1036	869	703	466	225	7525
1967-68	0	9	157	515	1097	1541	1531	1165	881	780	474	153	8303
1968-69	18	89	251	522	1031	1467	1223	1025	1096	534	252	258	7766
1969-70	8	0	64	833	928	1212	1253	869	1087	842	323	136	7555
1970-71	3	0	357	733	970	1344	1187	1083	997	716	428	91	7909
1971-72	12	0	366	738	1133	1343	1404	1083	739	643	402	63	7926
1972-73	56	46	245	645	1105	1625	1720	1432	1186	918	386	122	9486
1973-74	35	3	332	521	988	1211	1451	1055	893	633	415	91	7628
1974-75	4	61	274	513	930	1401	1360	1172	1019	836	518	210	8298
1975-76	0	38	219	561	1079	1277	1411	1095	1006	624	316	170	7796
1976-77	1	16	182	657	933	1220	1514	940	1063	543	361	22	7452
1977-78	2	50	183	520	986	1193	1466	1192	823	586	490	127	7618
1978-79	26	54	226	519	1321	1730	1974	1248	962	620	415	137	9232
1979-80	0	61	84	463	1164	1234	1578	1145	1000	595	390	94	7808
1980-81	0	43	166	550	1005	943	1124	1051	789	487	383	83	6624
1981-82	18	15	104	655	812	1170	1324	1100	898	702	400	164	7362
1982-83	12	0	329	632	1058	1440	1194	1027	884	804	508	125	8013
1983-84	26	0	159	490	1096	1826	1510	1287	1046	736	290	135	8601
1984-85	0	0	316	721	977	1372	1507	1132	890	493	181	84	7673
1985-86	4	41	363	602	1413	1507							

TABLE 5 COOLING DEGREE DAYS Base 65 deg. F   LANDER, WYOMING

YEAR	JAN	FEB	MAR	APR	MAY	JUNE	JULY	AUG	SEP	OCT	NOV	DEC	TOTAL
1969	0	0	0	0	13	19	212	258	36	0	0	0	538
1970	0	0	0	0	1	81	196	247	14	0	0	0	539
1971	0	0	0	0	0	66	123	246	25	0	0	0	460
1972	0	0	0	0	1	43	123	127	3	0	0	0	297
1973	0	0	0	0	4	92	152	163	1	0	0	0	412
1974	0	0	0	0	1	138	243	73	12	0	0	0	467
1975	0	0	0	0	0	8	210	100	14	1	0	0	333
1976	0	0	0	0	1	41	238	84	30	0	0	0	394
1977	0	0	0	0	0	124	185	67	34	0	0	0	410
1978	0	0	0	0	0	56	179	109	64	0	0	0	408
1979	0	0	0	0	2	81	185	113	55	0	0	0	436
1980	0	0	0	0	3	74	231	97	19	0	0	0	424
1981	0	0	0	0	0	83	219	166	29	0	0	0	497
1982	0	0	0	0	1	51	188	266	41	0	0	0	547
1983	0	0	0	0	5	27	188	286	42	0	0	0	548
1984	0	0	0	0	13	87	264	237	26	0	0	0	627
1985	0	0	0	0	8	102	284	113	16	0	0	0	523

TABLE 6 SNOWFALL (inches)   LANDER, WYOMING

SEASON	JULY	AUG	SEP	OCT	NOV	DEC	JAN	FEB	MAR	APR	MAY	JUNE	TOTAL
1956-57	0.0	0.0	0.0	0.7	19.2	5.8	1.7	0.3	6.5	40.4	2.0	0.0	76.6
1957-58	0.0	0.0	3.6	0.0	32.5	3.2	3.4	4.9	18.4	29.8	0.3	0.0	96.1
1958-59	0.0	0.0	T	0.0	32.2	5.0	14.4	19.1	34.5	18.9	20.3	0.0	144.4
1959-60	0.0	0.0	2.3	12.8	17.1	11.0	12.7	9.7	7.6	13.0	T	0.0	86.2
1960-61	0.0	0.0	0.0	6.6	19.1	9.9	2.5	21.1	7.8	13.1	0.0	0.0	80.1
1961-62	0.0	0.0	4.0	38.6	21.4	3.3	26.5	6.3	29.8	28.7	0.0	0.0	111.6
1962-63	0.0	0.0	2.4	0.0	2.1	4.5	17.8	4.2	15.1	28.5	1.5	0.0	89.5
1963-64	0.0	0.0	0.0	2.4	1.8	12.9	8.3	24.6	10.2	14.2	16.0	0.0	95.1
1964-65	0.0	0.0	0.0	3.0	8.4	0.4	3.7	15.7	18.7	34.3	1.4	0.0	71.6
1965-66	0.0	0.0	23.6	0.0	5.2	12.0	9.3	7.0	18.7	37.0	24.5	0.0	111.5
1966-67	0.0	0.0	T	26.5	6.6	7.3	5.2	21.4	17.8	37.0	24.5	0.0	146.3
1967-68	0.0	0.0	0.0	1.4	21.0	26.3	2.4	9.9	26.5	39.5	5.2	0.0	132.2
1968-69	0.0	0.0	0.0	0.5	8.2	12.3	2.0	4.7	21.3	19.1	0.0	1.2	69.3
1969-70	0.0	0.0	0.0	32.4	4.2	9.3	1.9	T	40.3	37.2	0.0	0.0	125.3
1970-71	0.0	0.0	6.7	20.6	7.5	9.0	12.6	16.2	45.2	1.5	0.0	0.0	126.2
1971-72	0.0	0.0	10.1	39.9	9.3	6.6	18.7	10.7	4.6	28.5	0.0	0.0	128.4
1972-73	0.0	0.0	0.1	19.1	15.1	27.2	25.7	9.1	12.8	21.8	0.2	0.0	102.3
1973-74	0.0	0.0	T	16.1	13.5	12.4	7.2	8.1	22.9	34.7	33.9	0.0	133.8
1974-75	0.0	0.0	6.6	3.4	4.0	9.6	10.6	16.7	7.6	1.3	2.4	2.9	86.3
1975-76	0.0	0.0	0.0	15.9	14.9	19.7	4.9	16.7	17.4	16.7	4.4	0.0	110.6
1976-77	0.0	0.0	0.0	13.3	9.8	6.1	9.1	2.6	52.0	16.6	1.1	0.0	110.6
1977-78	0.0	0.0	0.0	3.7	12.8	7.7	9.3	9.1	1.1	T	30.3	2.6	80.4
1978-79	0.0	0.0	3.0	4.7	38.8	25.4	15.9	1.1	9.3	28.6	30.3	0.0	159.7
1979-80	0.0	0.0	0.0	5.3	10.0	20.4	26.2	10.6	25.4	12.6	13.9	0.0	124.4
1980-81	0.0	0.0	0.0	8.9	11.1	4.1	15.0	6.1	15.6	6.8	T	0.0	67.6
1981-82	0.0	0.0	0.0	0.2	1.8	1.7	8.1	2.8	7.4	7.7	12.1	0.0	41.8
1982-83	0.0	0.0	32.9	3.1	11.2	22.6	1.3	4.8	24.1	43.3	22.4	0.0	165.7
1983-84	0.0	0.0	1.9	1.2	48.7	11.9	18.0	17.4	16.3	45.4	1.1	0.0	161.9
1984-85	0.0	0.0	3.5	10.4	8.7	2.3	9.1	4.7	8.6	8.4	0.0	0.0	55.7
1985-86	0.0	0.0	7.6	2.2	32.3	28.0							
Record Mean	0.0	0.0	2.9	9.1	13.6	10.7	9.5	11.0	18.1	21.8	7.8	1.4	105.9

See Reference Notes, relative to all above tables, on preceding page.

Sheridan is located east of the Rocky Mountains at an elevation of a little less than 4,000 feet. To the northwest, east, and southeast are rolling hills, but to the southwest and west the Bighorn Mountains rise abruptly, oriented generally northwest-southeast. The foothills are only about 15 miles from Sheridan, and within 30 miles to the southwest the average elevation is near 10,000 feet, with Cloud Peak rising to 13,175 feet. This mountain range has a marked effect on the climate at Sheridan.

During the winter months, a few days after the outbreak of cold arctic air from Canada, the winds generally shift to the west or southwest and increase in velocity. These downslope winds produce a pronounced warming or Chinook. At other times, a gentle downslope flow will persist for several days and result in a prolonged period of mild weather. The Chinook is very effective in moderating the weather of the winter season, which otherwise would be more severe. On the other hand, winds from the east or northeast blowing toward the mountains are upslope and usually cause cooling, persistent low clouds, and often heavy precipitation. The upslope precipitation occurs at times all through the year, but most frequently during the winter and spring. Sheridan will often receive much heavier snow or rain with an easterly wind condition than the surrounding country farther away from the mountains. In the summer, the mountains act as a breeding ground for thunderstorms that frequently move away from the mountains toward the northeast and give afternoon or evening showers to Sheridan. Because of the close proximity to the Bighorn Mountains, the annual precipitation at Sheridan is greater, on the average, than in the neighboring area to the east and north.

Based on the 1951-1980 period, the average first occurrence of 32 degrees Fahrenheit in the fall is September 20 and the average last occurrence in the spring is May 20. Because of the short growing season and cold periods during winter, only the most hardy fruits can be grown successfully, but most varieties of vegetables will reach maturity.

The climate of Sheridan can be described generally as semi-arid with long cold winters and short hot summers. However, during all of the winter months, more than 50 percent of the possible sunshine is received, while the hot days in the summer are marked by very low humidity and nights are cool. There are few summer nights when the temperature remains above 60 degrees. During July, the warmest month, even though temperatures of 90 degrees or above occur frequently, the nights are cool. January is usually the coldest month. The cold weather comes from outbreaks of Canadian air moving southeastward down the east side of the Rockies, and the initial onslaught of arctic air is usually accompanied by strong northerly winds with drifting snow. The coldest nights, however, come after the skies have cleared and the wind becomes very light.

The yearly precipitation pattern for Sheridan is heavy in the spring and early summer. The three winter months constitute the period with the least moisture. Amounts of snowfall are quite generous during the winter, but the water content of the snow is usually low. This dry snow is ordinarily not injurious to livestock and does not result in serious inconvenience or discomfort to the public. During the spring months of March and April, however, precipitation often begins as rain, gradually turning to rain and snow mixed or to heavy wet snow. These snowstorms are frequently accompanied by strong winds and drifting. As a result, these two months are considered to have the most disagreeable weather of the year and are most likely to cause livestock loss. March has more snow than any other month.

TABLE 1 — NORMALS, MEANS AND EXTREMES

SHERIDAN, WYOMING

LATITUDE: 44°46'N   LONGITUDE: 106°58' W   ELEVATION: FT. GRND 3964 BARO 03946   TIME ZONE: MOUNTAIN   WBAN: 24029

	(a)	JAN	FEB	MAR	APR	MAY	JUNE	JULY	AUG	SEP	OCT	NOV	DEC	YEAR
TEMPERATURE °F:														
Normals														
–Daily Maximum		31.8	38.0	44.1	55.6	66.2	75.9	86.0	84.5	73.3	61.9	45.3	36.9	58.3
–Daily Minimum		7.3	14.1	19.7	29.4	39.7	47.6	53.8	52.1	42.0	32.1	19.7	12.3	30.8
–Monthly		19.6	26.1	31.9	42.5	53.0	61.8	69.9	68.3	57.7	47.0	32.5	24.6	44.6
Extremes														
–Record Highest	45	70	76	77	87	95	100	106	106	103	91	78	72	106
–Year		1974	1982	1978	1946	1960	1984	1983	1983	1983	1963	1975	1981	JUL 1983
–Record Lowest	45	-35	-31	-23	-2	13	27	35	34	6	1	-25	-37	-37
–Year		1963	1949	1965	1975	1954	1951	1971	1966	1984	1971	1959	1983	DEC 1983
NORMAL DEGREE DAYS:														
Heating (base 65°F)		1411	1089	1026	675	372	149	34	45	255	558	975	1252	7841
Cooling (base 65°F)		0	0	0	0	0	53	186	147	33	0	0	0	419
% OF POSSIBLE SUNSHINE	45	55	58	61	59	59	64	75	74	67	62	53	53	62
MEAN SKY COVER (tenths) Sunrise – Sunset	42	6.9	7.0	6.9	6.8	6.7	5.7	4.2	4.3	4.9	5.6	6.6	6.7	6.0
MEAN NUMBER OF DAYS:														
Sunrise to Sunset														
–Clear	45	5.7	4.8	4.9	5.3	5.6	7.8	13.8	13.9	11.8	10.2	6.1	6.2	96.2
–Partly Cloudy	45	8.0	8.1	9.4	9.0	10.6	11.7	12.2	10.9	9.1	8.7	8.2	7.8	113.8
–Cloudy	45	17.2	15.3	16.7	15.6	14.8	10.5	5.0	6.3	9.1	12.0	15.6	17.0	155.3
Precipitation .01 inches or more	45	9.0	8.6	10.8	10.8	11.7	11.2	7.2	6.5	7.6	7.4	7.7	8.9	107.3
Snow, Ice pellets 1.0 inches or more	42	3.7	3.7	4.4	2.9	0.5	0.*	0.0	0.0	0.5	1.2	2.8	3.8	23.6
Thunderstorms	45	0.0	0.*	0.0	0.7	4.8	9.4	9.5	7.3	2.6	0.3	0.*	0.0	34.6
Heavy Fog Visibility 1/4 mile or less	45	0.8	0.9	0.7	0.4	0.4	0.4	0.1	0.1	0.2	0.5	0.7	0.7	5.7
Temperature °F														
–Maximum 90° and above	21	0.0	0.0	0.0	0.0	0.2	2.1	10.5	11.0	2.3	0.0	0.0	0.0	26.1
32° and below	21	14.3	7.9	4.6	0.8	0.0	0.0	0.0	0.0	0.*	0.4	5.4	13.1	46.6
–Minimum 32° and below	21	30.3	27.4	28.0	19.2	5.6	0.3	0.0	0.0	3.8	16.9	28.0	30.0	189.5
0° and below	21	11.0	4.3	1.4	0.*	0.0	0.0	0.0	0.0	0.0	0.0	2.2	7.8	26.8
AVG. STATION PRESS.(mb)	7	877.8	877.5	874.8	876.2	876.7	877.9	880.1	879.6	880.5	880.0	879.0	877.2	878.1
RELATIVE HUMIDITY (%)														
Hour 05	21	69	70	72	72	75	77	72	68	70	69	72	70	71
Hour 11	21	61	59	53	48	48	46	37	34	41	46	57	61	49
Hour 17 (Local Time)	21	63	59	49	43	46	45	33	30	38	45	60	64	48
Hour 23	20	69	70	69	67	70	72	63	56	63	67	71	70	67
PRECIPITATION (inches):														
Water Equivalent –Normal		0.74	0.76	1.06	2.00	2.42	2.24	0.94	0.96	1.16	1.16	0.81	0.68	14.93
														9.54
–Maximum Monthly	45	1.79	2.68	3.26	4.80	6.80	9.54	3.78	3.02	3.08	3.16	2.23	2.03	9.54
–Year		1972	1955	1946	1963	1978	1944	1958	1968	1951	1971	1942	1955	JUN 1944
–Minimum Monthly	45	0.07	0.08	0.14	0.18	0.30	0.28	0.08	T	0.06	0.02	0.10	0.23	T
–Year		1983	1977	1978	1980	1958	1971	1959	1970	1964	1965	1981	1979	AUG 1970
–Maximum in 24 hrs	45	1.01	1.10	2.25	3.84	2.04	3.44	2.28	1.71	1.57	1.91	0.87	0.86	3.84
–Year		1972	1955	1946	1948	1956	1944	1948	1943	1982	1974	1978	1980	APR 1948
Snow, Ice pellets –Maximum Monthly	45	26.3	35.0	36.8	39.6	12.5	4.0			21.0	14.8	25.8	27.6	39.6
–Year		1977	1955	1954	1955	1979	1969			1984	1971	1964	1955	APR 1955
–Maximum in 24 hrs	45	13.5	11.0	13.3	26.7	10.9	4.0			12.9	8.4	12.0	11.7	26.7
–Year		1972	1955	1946	1955	1979	1969			1984	1977	1942	1980	APR 1955
WIND:														
Mean Speed (mph)	45	7.6	7.9	9.0	9.9	9.1	8.1	7.3	7.4	7.6	7.6	7.7	7.6	8.1
Prevailing Direction through 1963		NW	NW	NW	NW	NW	NW	NW	NW	NW	NW	NW	NW	NW
Fastest Mile –Direction (!!!)	44	NW	W	SW	NW	NW	SW	W	NW	NW	NW	SW	NW	SW
–Speed (MPH)	44	73	70	66	66	71	66	73	72	66	66	84	63	84
–Year		1946	1957	1963	1947	1946	1952	1947	1945	1945	1958	1949	1958	NOV 1949
Peak Gust –Direction (!!!)	2	NW	NW	SW	NW	NW	NW	NW	SE	NW	W	SW	NW	NW
–Speed (mph)	2	45	47	58	62	52	53	59	53	48	64	64	68	68
–Date		1985	1985	1985	1984	1985	1985	1985	1985	1985	1985	1984	1985	DEC 1985

See Reference Notes to this table on the following page.

TABLE 2 PRECIPITATION (inches) SHERIDAN, WYOMING

YEAR	JAN	FEB	MAR	APR	MAY	JUNE	JULY	AUG	SEP	OCT	NOV	DEC	ANNUAL
1956	0.55	0.41	1.99	1.91	4.94	1.07	0.35	1.40	0.72	0.88	0.97	0.23	15.42
1957	0.41	0.53	1.23	3.26	3.63	4.39	0.31	1.31	1.11	1.61	1.15	0.23	19.17
1958	0.34	0.84	0.70	4.22	0.30	2.01	3.78	0.26	0.17	0.35	0.76	0.70	14.43
1959	0.32	0.45	0.80	1.72	1.46	2.07	0.08	0.13	0.60	1.48	0.90	0.31	10.32
1960	0.16	0.54	0.24	0.60	0.83	1.00	0.24	2.32	0.27	0.87	0.69	0.57	8.23
1961	0.12	0.98	0.61	1.20	2.95	0.49	1.31	0.15	2.56	2.83	0.73	0.74	14.67
1962	0.72	0.42	0.66	1.04	2.28	3.03	1.57	1.10	1.43	0.37	0.63	0.53	13.78
1963	1.31	0.83	0.55	4.80	1.54	4.71	0.51	0.42	1.33	0.52	0.65	0.91	18.08
1964	0.35	0.81	0.51	2.74	1.76	5.11	0.11	1.18	0.06	0.34	1.99	0.66	15.62
1965	1.18	0.72	0.70	0.65	2.12	2.15	0.93	0.68	1.75	0.02	0.22	0.24	11.36
1966	0.36	0.40	1.04	2.25	1.08	1.29	0.29	0.86	0.98	1.02	0.62	0.82	11.01
1967	0.61	1.28	1.18	2.51	1.33	6.11	0.22	0.70	1.56	0.88	0.84	0.98	18.20
1968	1.05	0.75	0.92	0.71	2.29	3.89	0.33	3.02	2.12	0.63	0.91	0.97	17.59
1969	1.11	0.18	0.40	1.84	1.79	2.44	0.47	0.24	0.09	1.47	0.99	0.51	11.53
1970	0.72	0.71	2.31	1.67	5.20	2.10	1.06	T	1.87	0.94	1.43	0.62	18.63
1971	1.43	1.17	0.62	3.83	2.45	0.28	0.21	0.52	0.90	3.16	0.71	0.52	15.80
1972	1.79	0.66	0.97	1.02	1.35	1.96	1.62	0.96	0.94	1.01	0.32	0.90	13.50
1973	0.35	0.39	1.41	4.05	0.51	2.21	0.64	0.47	2.79	0.81	0.70	0.80	15.13
1974	0.67	0.46	0.96	1.58	1.44	0.48	0.65	0.70	1.48	2.96	0.64	0.24	12.26
1975	1.11	0.47	1.18	1.67	3.51	4.56	0.88	0.19	0.19	1.50	0.88	1.15	17.29
1976	0.51	0.66	0.44	2.24	1.11	2.32	0.73	0.80	1.62	1.05	0.93	0.26	12.67
1977	1.45	0.08	1.84	1.16	3.03	2.09	1.66	1.95	0.93	1.27	0.66	1.36	17.48
1978	1.29	0.84	0.14	2.26	6.80	0.46	1.73	1.50	1.67	0.27	1.45	0.95	19.36
1979	0.48	0.38	0.60	1.31	2.95	1.13	0.84	1.10	0.39	2.28	0.56	0.23	12.25
1980	0.60	0.86	1.25	0.18	3.65	1.11	0.33	1.12	1.31	0.88	1.27	0.72	13.28
1981	0.30	0.36	0.57	0.24	5.71	1.94	3.00	0.67	0.63	1.00	0.10	0.51	15.03
1982	0.70	0.25	1.33	0.93	1.46	3.87	0.51	0.58	2.90	0.65	0.26	1.12	14.56
1983	0.07	0.33	0.48	0.88	1.84	1.01	0.12	1.23	0.94	1.25	1.06	0.76	9.97
1984	0.69	0.46	2.11	2.22	1.55	2.34	0.40	0.96	2.28	0.28	0.57	0.35	14.21
1985	0.81	0.28	0.54	0.98	1.55	1.73	0.93	0.44	2.36	0.57	0.70	0.40	11.29
Record Mean	0.76	0.65	1.16	2.00	2.59	2.34	1.16	0.84	1.39	1.22	0.80	0.66	15.57

TABLE 3 AVERAGE TEMPERATURE (deg. F) SHERIDAN, WYOMING

YEAR	JAN	FEB	MAR	APR	MAY	JUNE	JULY	AUG	SEP	OCT	NOV	DEC	ANNUAL
1956	21.2	21.5	31.7	40.2	55.3	67.2	70.6	66.8	60.1	49.2	34.3	30.2	45.7
1957	9.9	25.7	34.7	40.0	53.5	60.3	72.6	70.9	56.8	44.5	32.3	35.0	44.7
1958	31.4	26.1	31.2	41.6	60.4	59.6	64.5	70.8	59.4	48.8	33.8	25.7	46.1
1959	19.8	19.4	34.7	41.8	49.0	65.4	69.7	70.5	57.0	44.2	29.2		43.9
1960	22.1	20.0	33.3	44.4	54.9	64.1	75.1	67.9	60.1	48.5	33.6	26.2	45.9
1961	27.0	34.1	38.0	40.8	53.8	68.5	71.7	73.1	50.7	43.7	27.7	20.9	45.8
1962	16.0	23.9	28.5	47.4	53.9	61.1	66.5	66.3	56.6	50.1	39.3	30.2	45.0
1963	12.1	33.0	38.1	42.9	62.0	62.0	70.8	70.8	65.3	54.7	37.4	23.0	46.9
#1964	26.4	26.7	28.8	43.4	54.3	60.1	73.4	65.9	55.2	48.0	28.0	20.0	44.2
1965	29.9	23.4	19.3	45.0	54.8	59.9	69.8	67.1	47.4	52.9	38.1	28.8	44.3
1966	18.4	23.4	36.8	38.5	53.8	60.2	72.5	65.3	61.6	45.9	29.9	24.6	44.3
1967	25.4	29.5	32.8	41.1	49.2	56.8	66.6	67.8	59.2	47.2	29.9	19.4	43.8
1968	16.9	29.7	38.5	40.2	49.4	58.8	67.6	64.5	55.9	46.8	33.1	15.8	43.1
1969	10.3	22.6	28.7	47.5	54.0	57.4	66.7	70.7	60.5	38.6	34.2	28.6	43.4
1970	19.2	31.1	28.1	38.4	53.6	62.1	68.8	70.4	53.9	42.8	33.2	21.4	43.6
1971	19.5	24.8	33.5	43.7	53.2	62.8	66.0	73.6	54.5	42.8	32.8	16.3	43.6
1972	14.0	27.4	40.2	44.5	54.1	64.8	64.1	68.5	55.1	44.4	33.4	16.5	43.9
1973	20.3	28.7	35.3	39.3	52.3	62.9	68.2	70.1	55.0	47.8	29.2	27.1	44.7
1974	19.7	33.1	35.7	45.8	49.9	65.5	72.8	63.0	54.6	48.2	34.3	26.7	45.8
1975	22.2	16.0	28.7	36.5	50.3	58.5	71.0	66.0	56.8	46.3	31.3	28.7	42.7
1976	23.5	32.0	31.3	45.2	54.3	60.3	71.5	67.9	59.6	42.3	33.2	27.6	45.7
1977	15.7	31.7	32.8	47.2	54.9	65.4	70.0	66.7	57.5	47.7	31.5	18.1	44.7
1978	11.4	17.5	34.2	46.7	49.9	60.6	66.7	65.4	59.2	47.8	23.6	12.5	41.3
1979	3.4	15.8	33.3	40.5	50.4	60.9	68.1	68.0	62.4	48.8	29.7	31.1	42.7
1980	15.8	24.5	30.5	47.9	55.0	62.1	71.3	65.6	58.8	46.8	36.9	28.0	45.3
1981	32.4	28.5	39.4	47.8	52.0	61.3	70.9	70.1	62.7	45.7	41.3	26.1	48.2
1982	16.4	26.0	34.8	42.0	50.7	59.6	69.8	73.5	58.1	46.9	32.1	22.8	44.4
1983	33.7	36.4	39.4	42.2	51.1	63.4	72.8	77.8	59.6	51.3	34.4	6.8	47.4
1984	25.0	32.5	37.1	43.2	52.8	61.3	71.0	72.8	53.5	41.2	33.1	14.1	44.8
1985	15.3	16.7	32.1	47.7	58.2	62.5	72.2	66.2	52.6	45.2	15.5	25.6	42.5
Record Mean	20.0	24.6	32.5	43.6	52.9	61.8	69.8	68.2	57.3	46.3	32.9	23.5	44.5
Max	32.6	37.0	44.8	56.5	66.1	75.8	85.9	84.6	72.9	61.0	45.7	35.8	58.2
Min	7.4	12.2	20.1	30.7	39.7	47.7	53.8	51.7	41.7	31.6	20.0	11.3	30.7

REFERENCE NOTES FOR TABLES 1, 2, 3 and 6   (SHERIDAN, WY)

GENERAL

T - TRACE AMOUNT
BLANK ENTRIES DENOTE MISSING/UNREPORTED DATA.
# INDICATES A STATION OR INSTRUMENT RELOCATION.

SPECIFIC

TABLE 1

(a) - LENGTH OF RECORD IN YEARS. ALTHOUGH INDIVIDUAL MONTHS MAY BE MISSING.

* LESS THAN .05

NORMALS — BASED ON THE 1951-1980 RECORD PERIOD.
EXTREMES — DATES ARE THE MOST RECENT OCCURRENCE.
WIND DIR. — NUMERALS SHOW TENS OF DEGREES
CLOCKWISE FROM TRUE NORTH.
"00" INDICATES CALM.
RESULTANT WIND DIRECTIONS ARE GIVEN TO WHOLE DEGREES.

EXCEPTIONS

TABLES 2, 3, and 6

RECORD MEANS ARE THROUGH THE CURRENT YEAR, BEGINNING IN 1908 FOR TEMPERATURE
1908 FOR PRECIPITATION
1941 FOR SNOWFALL

TABLE 4   HEATING DEGREE DAYS Base 65 deg. F   SHERIDAN, WYOMING

SEASON	JULY	AUG	SEP	OCT	NOV	DEC	JAN	FEB	MAR	APR	MAY	JUNE	TOTAL
1956-57	7	62	159	483	915	1073	1705	1093	933	742	352	160	7684
1957-58	3	14	248	637	975	921	1034	1079	1038	698	163	168	6978
1958-59	56	14	210	491	930	1212	1397	1270	934	690	488	83	7775
1959-60	30	7	296	638	1152	1101	1320	1301	977	612	315	73	7822
1960-61	6	50	205	506	·936	1194	1168	862	832	721	351	24	6855
1961-62	9	5	434	651	1111	1361	1515	1146	1127	521	339	138	8357
1962-63	26	81	245	454	765	1073	1639	891	827	656	360	115	7132
1963-64	8	7	52	328	820	1293	1189	1103	1116	643	339	169	7067
#1964-65	0	101	292	520	1103	1389	1078	1158	1412	594	464	153	8264
1965-66	7	44	523	372	800	1116	1438	1158	869	787	345	171	7630
1966-67	0	86	138	584	1023	1244	1221	988	991	707	482	238	7702
1967-68	17	21	183	546	1047	1404	1487	1016	813	740	477	199	7950
1968-69	35	75	267	557	947	1519	1692	1181	1121	522	345	236	8497
1969-70	20	6	151	811	918	1121	1414	944	1138	794	348	132	7797
1970-71	10	4	342	686	948	1347	1407	1122	968	631	358	107	7930
1971-72	59	5	328	683	963	1501	1577	1082	761	607	339	56	7961
1972-73	95	28	303	633	942	1502	1380	1010	915	766	387	130	8091
1973-74	23	4	299	523	1070	1169	1400	887	902	572	462	95	7406
1974-75	3	93	309	513	918	1178	1317	1370	1117	852	450	194	8314
1975-76	1	43	246	575	1003	1122	1282	949	1040	587	328	161	7337
1976-77	1	20	186	696	949	1154	1520	926	989	527	307	53	7328
1977-78	12	73	233	528	1001	1452	1654	1323	951	543	463	153	8386
1978-79	45	75	232	525	1235	1623	1911	1376	979	727	447	154	9329
1979-80	6	20	120	497	1051	1044	1523	1171	1061	507	312	111	7423
1980-81	0	50	195	557	836	1140	1004	1018	788	510	396	136	6630
1981-82	15	6	115	593	703	1200	1503	1087	932	685	436	186	7461
1982-83	26	2	257	556	983	1304	962	794	788	677	431	89	6869
1983-84	22	0	215	419	908	1802	1236	935	856	647	380	147	7567
1984-85	7	3	374	731	953	1574	1535	1351	1016	512	215	116	8387
1985-86	5	62	373	605	1483	1215							

TABLE 5   COOLING DEGREE DAYS Base 65 deg. F   SHERIDAN, WYOMING

YEAR	JAN	FEB	MAR	APR	MAY	JUNE	JULY	AUG	SEP	OCT	NOV	DEC	TOTAL
1969	0	0	0	0	9	16	114	191	25	0	0	0	355
1970	0	0	0	0	1	55	134	176	16	2	0	0	384
1971	0	0	0	0	0	49	96	276	21	0	0	0	442
1972	0	0	0	0	9	55	74	145	11	0	0	0	294
1973	0	0	0	0	1	73	129	168	5	0	0	0	376
1974	0	0	0	0	0	119	248	39	4	0	0	0	410
1975	0	0	0	0	0	6	194	83	10	0	0	0	293
1976	0	0	0	0	0	28	211	115	28	0	0	0	382
1977	0	0	0	0	3	73	176	48	16	0	0	0	316
1978	0	0	0	0	1	30	103	94	62	0	0	0	290
1979	0	0	0	0	1	42	108	121	46	0	0	0	318
1980	0	0	0	2	9	33	202	75	14	3	0	0	338
1981	0	0	0	0	0	31	207	173	56	0	0	0	467
1982	0	0	0	0	0	31	185	273	53	0	0	0	542
1983	0	0	0	0	6	47	272	402	57	0	0	0	784
1984	0	0	0	0	7	45	199	253	38	0	0	0	542
1985	0	0	0	0	13	47	238	104	10	0	0	0	412

TABLE 6   SNOWFALL (inches)   SHERIDAN, WYOMING

SEASON	JULY	AUG	SEP	OCT	NOV	DEC	JAN	FEB	MAR	APR	MAY	JUNE	TOTAL
1956-57	0.0	0.0	0.0	1.2	8.2	4.3	9.5	11.1	15.4	19.2	T	0.0	68.9
1957-58	0.0	0.0	T	4.8	10.6	2.1	4.4	14.3	6.4	28.3	0.0	0.0	70.9
1958-59	0.0	0.0	0.0	2.0	3.0	8.8	6.4	8.6	6.8	10.3	T	0.0	45.9
1959-60	0.0	0.0	T	4.1	10.3	4.2	3.1	10.2	5.4	5.4	T	0.0	42.7
1960-61	0.0	0.0	0.0	T	6.9	14.1	2.8	8.8	0.8	4.1	0.0	0.0	37.5
1961-62	0.0	0.0	T	12.6	9.1	5.9	9.6	4.6	7.1	1.1	0.0	0.0	50.0
1962-63	0.0	0.0	2.3	T	4.1	6.8	23.9	13.1	6.7	6.7	0.0	0.0	63.6
1963-64	0.0	0.0	0.0	1.3	2.2	18.4	4.9	13.3	7.8	7.1	0.0	0.0	55.7
1964-65	0.0	0.0	T	0.0	25.8	9.4	11.8	12.5	12.8	3.7	4.7	0.0	80.7
1965-66	0.0	0.0	6.0	0.0	4.5	5.3	8.6	5.3	12.9	12.9	3.4	0.0	58.9
1966-67	0.0	0.0	0.0	4.4	11.1	14.6	12.1	20.4	16.5	23.0	7.5	0.0	109.6
1967-68	0.0	0.0	0.0	0.2	11.5	20.6	13.5	7.3	14.0	7.1	0.6	0.0	74.8
1968-69	0.0	0.0	0.0	T	6.9	15.3	15.0	3.0	4.9	6.7	1.3	4.0	57.1
1969-70	0.0	0.0	0.0	4.9	7.8	5.7	11.5	10.1	26.1	9.4	2.0	0.0	77.5
1970-71	0.0	0.0	9.5	0.9	11.3	10.0	15.6	15.7	7.5	5.2	T	0.0	75.7
1971-72	0.0	0.0	T	14.8	9.3	9.2	25.1	9.4	9.1	7.8	T	0.0	84.7
1972-73	0.0	0.0	3.6	2.6	15.6	9.8	8.2	16.1	37.5	1.4	0.0	0.0	94.2
1973-74	0.0	0.0	3.2	4.9	11.7	10.7	10.0	6.8	9.8	7.7	0.4	0.0	65.2
1974-75	0.0	0.0	0.3	2.0	4.4	4.7	13.8	10.1	17.6	14.5	4.8	0.0	72.2
1975-76	0.0	0.0	0.0	8.5	14.1	18.0	8.5	12.9	8.1	8.7	0.0	0.0	78.8
1976-77	0.0	0.0	0.0	3.5	11.9	5.4	26.3	2.2	25.0	10.2	0.0	0.0	84.5
1977-78	0.0	0.0	0.0	8.4	9.1	24.0	23.2	15.5	1.5	0.7	5.2	0.0	87.6
1978-79	0.0	0.0	T	18.5	18.3	10.7	6.9	7.6	11.0	12.5	0.0	0.0	85.5
1979-80	0.0	0.0	0.0	10.1	8.9	4.2	12.1	18.2	19.3	3.0	0.0	0.0	75.8
1980-81	0.0	0.0	0.4	4.5	14.0	11.1	6.2	5.5	4.8	0.2	T	0.0	46.7
1981-82	0.0	0.0	0.0	5.6	1.0	8.2	12.8	5.2	18.8	7.3	T	0.0	58.9
1982-83	0.0	0.0	6.9	3.4	5.5	19.6	1.5	4.1	5.7	10.7	8.8	0.0	66.2
1983-84	0.0	0.0	3.3	0.0	12.3	17.6	12.4	7.7	20.5	25.6	5.4	0.0	104.8
1984-85	0.0	0.0	21.0	2.5	8.3	7.7	16.3	5.6	7.8	8.2	0.0	0.0	77.4
1985-86	0.0	0.0	5.2	1.8	13.7	6.3							
Record Mean	0.0	0.0	1.6	3.8	9.0	10.4	11.0	10.4	12.7	9.7	2.0	0.2	70.7

See Reference Notes, relative to all above tables, on preceding page.